STUDENT'S SOLUTIONS MANUAL

Cindy Trimble & Associates

Randy Gallaher
Lewis & Clark Community College

Kevin Bodden
Lewis & Clark Community College

ELEMENTARY & INTERMEDIATE ALGEBRA

Michael Sullivan, III
Joliet Junior College

Katherine Struve
Columbus State Community College

Janet Mazzarella
Southwestern Community College

Prentice Hall
is an imprint of

PEARSON

Reproduced by Pearson Prentice Hall from electronic files supplied by the author.

Copyright ©2010 Pearson Education, Inc.
Publishing as Pearson Prentice Hall, Upper Saddle River, NJ 07458.

ISBN-13: 978-0-321-59351-1   Standalone
ISBN-10: 0-321-59351-0   Standalone

ISBN-13: 978-0-321-59352-8   Component
ISBN-10: 0-321-59352-9   Component

1 2 3 4 5 6 BR 12 11 10 09

Prentice Hall
is an imprint of

www.pearsonhighered.com

Contents

Chapter 1

1.2 Quick Checks

1. In the statement $6 \cdot 8 = 48$, 6 and 8 are called <u>factors</u> and 48 is called the <u>product</u>.

2.

The prime factorization of 12 is $2 \cdot 2 \cdot 3$.

3.
```
      18
     /  \
    2    9
    |   / \
  2 · 3 · 3
```
The prime factorization of 18 is $2 \cdot 3 \cdot 3$.

4.
```
       75
      /  \
     3   25
     |   / \
   3 · 5 · 5
```
The prime factorization of 75 is $3 \cdot 5 \cdot 5$.

5.
```
           120
          /   \
         4     30
        / \   /  \
       2  2  2   15
       |  |  |  / \
     2 · 2 · 2 · 3 · 5
```
The prime factorization of 120 is $2 \cdot 2 \cdot 2 \cdot 3 \cdot 5$.

6. 131 is a prime number.

7.
```
        459
       /   \
      9     51
     / \   / \
    3 · 3 · 3 · 17
```
The prime factorization of 459 is $3 \cdot 3 \cdot 3 \cdot 17$.

8. $6 = 2 \cdot 3$
   $8 = 2 \cdot \ \ \cdot 2 \cdot 2$
   $\qquad 2 \cdot 3 \cdot 2 \cdot 2$
   The LCM is $2 \cdot 3 \cdot 2 \cdot 2 = 24$.

9. $5 = 5$
   $10 = 5 \cdot 2$
   $\qquad 5 \cdot 2$
   The LCM is $2 \cdot 5 = 10$.

10. $45 = 3 \cdot 3 \cdot 5$
    $72 = 3 \cdot 3 \cdot \ \ | \ \ 2 \cdot 2 \cdot 2$
    $\qquad 3 \cdot 3 \cdot 5 \cdot 2 \cdot 2 \cdot 2$
    The LCM is $2 \cdot 2 \cdot 2 \cdot 3 \cdot 3 \cdot 5 = 360$.

11. $7 = 7$
    $3 = \ | \ \ 3$
    $\qquad 7 \cdot 3$
    The LCM is $3 \cdot 7 = 21$.

12. $12 = 2 \cdot 2 \cdot 3$
    $18 = 2 \cdot \ | \ 3 \cdot 3$
    $30 = 2 \cdot \ | \ 3 \cdot \ | \ 5$
    $\qquad 2 \cdot 2 \cdot 3 \cdot 3 \cdot 5$
    The LCM is $2 \cdot 2 \cdot 3 \cdot 3 \cdot 5 = 180$.

13. Fractions which represent the same portion of a whole are called equivalent <u>fractions</u>.

14. Multiply the numerator and denominator of $\dfrac{1}{2}$ by 5.
    $$\frac{1}{2} = \frac{1 \cdot 5}{2 \cdot 5} = \frac{5}{10}$$

15. Multiply the numerator and denominator of $\dfrac{5}{8}$ by 6.
    $$\frac{5}{8} = \frac{5 \cdot 6}{8 \cdot 6} = \frac{30}{48}$$

16. The denominators are 4 and 6.
    $4 = 2 \cdot 2$
    $6 = 2 \cdot \ | \ 3$
    $\text{LCD} = 2 \cdot 2 \cdot 3 = 12$
    $$\frac{1}{4} = \frac{1 \cdot 3}{4 \cdot 3} = \frac{3}{12}$$
    $$\frac{5}{6} = \frac{5 \cdot 2}{6 \cdot 2} = \frac{10}{12}$$

17. The denominators are 12 and 15.
    $12 = 2 \cdot 2 \cdot 3$
    $15 = \ | \ \ | \ 3 \cdot 5$
    $\text{LCD} = 2 \cdot 2 \cdot 3 \cdot 5 = 60$

$$\frac{5}{12} = \frac{5 \cdot 5}{12 \cdot 5} = \frac{25}{60}$$

$$\frac{4}{15} = \frac{4 \cdot 4}{15 \cdot 4} = \frac{16}{60}$$

18. The denominators are 20 and 16.

$$20 = 2 \cdot 2 \cdot 5$$
$$16 = 2 \cdot 2 \cdot \mid 2 \cdot 2$$

$$LCD = 2 \cdot 2 \cdot 5 \cdot 2 \cdot 2 = 80$$

$$\frac{9}{20} = \frac{9 \cdot 4}{20 \cdot 4} = \frac{36}{80}$$

$$\frac{11}{16} = \frac{11 \cdot 5}{16 \cdot 5} = \frac{55}{80}$$

19. $\dfrac{45}{80} = \dfrac{3 \cdot 3 \cdot 5}{2 \cdot 2 \cdot 2 \cdot 2 \cdot 5} = \dfrac{3 \cdot 3 \cdot \cancel{5}}{2 \cdot 2 \cdot 2 \cdot 2 \cdot \cancel{5}} = \dfrac{9}{16}$

20. $\dfrac{4}{9} = \dfrac{2 \cdot 2}{3 \cdot 3} = \dfrac{4}{9}$

21. $\dfrac{20}{50} = \dfrac{2 \cdot 2 \cdot 5}{2 \cdot 5 \cdot 5} = \dfrac{\cancel{2} \cdot 2 \cdot \cancel{5}}{\cancel{2} \cdot 5 \cdot \cancel{5}} = \dfrac{2}{5}$

22. $\dfrac{30}{105} = \dfrac{2 \cdot 3 \cdot 5}{3 \cdot 5 \cdot 7} = \dfrac{2 \cdot \cancel{3} \cdot \cancel{5}}{\cancel{3} \cdot \cancel{5} \cdot 7} = \dfrac{2}{7}$

23. The 1 is two places to the right of the decimal; this is the hundredths place.

24. The 2 is one place to the right of the decimal; this the tenths place.

25. The 8 is four places to the left of the decimal; this is the thousands place.

26. The 9 is three places to the right of the decimal; this is the thousandths place.

27. The 3 is one place to the left of the decimal; this is the ones place.

28. The 2 is five places to the left of the decimal; this is the ten thousands place.

29. The number 1 is in the tenths place. The number to its right is 7. Since 7 is greater than 5, we round to 0.2.

30. The number 3 is in the hundredths place. The number to its right is 2. Since 2 is less than 5, we round to 0.93.

31. The number 9 is in the hundredths place. The number to its right is 6. Since 6 is greater than 5, we round to 1.40.

32. The number 8 is in the thousandths place. The number to its right is 3. Since 3 is less than 5, we round to 14.398.

33. The number 0 is in the hundredths place. The number to its right is 4. Since 4 is less than 5, we round to 690.00.

34. The number 9 is in the tenths place. The number to its right is 8. Since 8 is greater than 5, we round to 60.0.

35.
$$\begin{array}{r} 0.4 \\ 5\overline{)2.0} \\ \underline{2\,0} \\ 0 \end{array}$$

$$\frac{2}{5} = 0.4$$

36.
$$\begin{array}{r} 0.428571 \\ 7\overline{)3.000000} \\ \underline{2\,8} \\ 20 \\ \underline{14} \\ 60 \\ \underline{56} \\ 40 \\ \underline{35} \\ 50 \\ \underline{49} \\ 10 \\ \underline{7} \\ 3 \end{array}$$

$$\frac{3}{7} = 0.\overline{428571}$$

37.
$$\begin{array}{r} 1.375 \\ 8\overline{)11.000} \\ \underline{8} \\ 3\,0 \\ \underline{2\,4} \\ 60 \\ \underline{56} \\ 40 \\ \underline{40} \\ 0 \end{array}$$

$$\frac{11}{8} = 1.375$$

38.

$$6\overline{)5.000}$$ with quotient 0.833
$$\begin{array}{r} 0.833 \\ 6\overline{)5.000} \\ \underline{4\,8} \\ 20 \\ \underline{18} \\ 20 \\ \underline{18} \\ 2 \end{array}$$

$$\frac{5}{6} = 0.8\overline{3}$$

39.

$$\begin{array}{r} 0.55 \\ 9\overline{)5.00} \\ \underline{4\,5} \\ 50 \\ \underline{45} \\ 5 \end{array}$$

$$\frac{5}{9} = 0.\overline{5}$$

40. $0.65 = \dfrac{65}{100} = \dfrac{\cancel{5} \cdot 13}{\cancel{5} \cdot 20} = \dfrac{13}{20}$

41. $0.2 = \dfrac{2}{10} = \dfrac{\cancel{2}}{\cancel{2} \cdot 5} = \dfrac{1}{5}$

42. $0.625 = \dfrac{625}{1000} = \dfrac{5 \cdot \cancel{125}}{8 \cdot \cancel{125}} = \dfrac{5}{8}$

43. The word percent means parts per <u>hundred</u>, so 35% means <u>35</u> parts out of 100 parts or $\dfrac{35}{100}$.

44. $23\% = 23\% \cdot \dfrac{1}{100\%} = \dfrac{23}{100} = 0.23$

45. $1\% = 1\% \cdot \dfrac{1}{100\%} = \dfrac{1}{100} = 0.01$

46. $72.4\% = 72.4\% \cdot \dfrac{1}{100\%} = \dfrac{72.4}{100} = 0.724$

47. $127\% = 127\% \cdot \dfrac{1}{100\%} = \dfrac{127}{100} = 1.27$

48. $89.26\% = 89.26\% \cdot \dfrac{1}{100\%} = \dfrac{89.26}{100} = 0.8926$

49. To convert a decimal to a percent, multiply the decimal by <u>100%</u>.

50. $0.15 = 0.15 \cdot \dfrac{100\%}{1} = 15\%$

51. $0.8 = 0.8 \cdot \dfrac{100\%}{1} = 80\%$

52. $1.3 = 1.3 \cdot \dfrac{100\%}{1} = 130\%$

53. $0.398 = 0.398 \cdot \dfrac{100\%}{1} = 39.8\%$

54. $0.004 = 0.004 \cdot \dfrac{100\%}{1} = 0.4\%$

1.2 Exercises

55.

$$\begin{array}{c} 25 \\ \diagup \ \diagdown \\ 5 \ \cdot \ 5 \end{array}$$

The prime factorization of 25 is $5 \cdot 5$.

57.

$$\begin{array}{c} 28 \\ \diagup \ \diagdown \\ 4 \quad 7 \\ \diagup \diagdown \\ 2 \cdot 2 \cdot 7 \end{array}$$

The prime factorization of 28 is $2 \cdot 2 \cdot 7$.

59.

$$\begin{array}{c} 21 \\ \diagup \ \diagdown \\ 3 \ \cdot \ 7 \end{array}$$

The prime factorization of 21 is $3 \cdot 7$.

61.

$$\begin{array}{c} 36 \\ \diagup \quad \diagdown \\ 4 \qquad 9 \\ \diagup \diagdown \ \ \diagup \diagdown \\ 2 \cdot 2 \cdot 3 \cdot 3 \end{array}$$

The prime factorization of 36 is $2 \cdot 2 \cdot 3 \cdot 3$.

63.

$$\begin{array}{c} 20 \\ \diagup \ \diagdown \\ 4 \quad 5 \\ \diagup \diagdown \\ 2 \cdot 2 \cdot 5 \end{array}$$

The prime factorization of 20 is $2 \cdot 2 \cdot 5$.

65.

$$\begin{array}{c} 30 \\ \diagup \ \diagdown \\ 5 \quad 6 \\ \ \ \diagup \diagdown \\ 5 \cdot 2 \cdot 3 \end{array}$$

The prime factorization of 30 is $2 \cdot 3 \cdot 5$.

67.

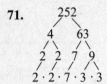

The prime factorization of 50 is $2 \cdot 5 \cdot 5$.

69. 53 is a prime number.

71.

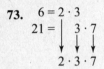

The prime factorization of 252 is
$2 \cdot 2 \cdot 3 \cdot 3 \cdot 7$.

73.
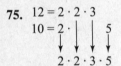
$$6 = 2 \cdot 3$$
$$21 = \quad 3 \cdot 7$$
$$\downarrow \quad \downarrow \quad \downarrow$$
$$2 \cdot 3 \cdot 7$$
The LCM is $2 \cdot 3 \cdot 7 = 42$.

75.
$$12 = 2 \cdot 2 \cdot 3$$
$$10 = 2 \cdot \quad \quad 5$$
$$\downarrow \quad \downarrow \quad \downarrow \quad \downarrow$$
$$2 \cdot 2 \cdot 3 \cdot 5$$
The LCM is $2 \cdot 2 \cdot 3 \cdot 5 = 60$.

77.
$$15 = 3 \cdot 5$$
$$14 = \quad \quad 2 \cdot 7$$
$$\downarrow \quad \downarrow \quad \downarrow \quad \downarrow$$
$$3 \cdot 5 \cdot 2 \cdot 7$$
The LCM is $2 \cdot 3 \cdot 5 \cdot 7 = 210$.

79.
$$30 = 2 \cdot 3 \cdot 5$$
$$45 = \quad 3 \cdot 5 \cdot 3$$
$$\downarrow \quad \downarrow \quad \downarrow \quad \downarrow$$
$$2 \cdot 3 \cdot 5 \cdot 3$$
The LCM is $2 \cdot 3 \cdot 3 \cdot 5 = 90$.

81.
$$5 = 5$$
$$6 = \quad 2 \cdot 3$$
$$12 = \quad 2 \cdot 3 \cdot 2$$
$$\downarrow \quad \downarrow \quad \downarrow \quad \downarrow$$
$$5 \cdot 2 \cdot 3 \cdot 2$$
The LCM is $2 \cdot 2 \cdot 3 \cdot 5 = 60$.

83.
$$3 = 3$$
$$8 = \quad 2 \cdot 2 \cdot 2$$
$$9 = 3 \cdot \quad \quad 3$$
$$\downarrow \quad \downarrow \quad \downarrow \quad \downarrow$$
$$3 \cdot 2 \cdot 2 \cdot 2 \cdot 3$$
The LCM is $2 \cdot 2 \cdot 2 \cdot 3 \cdot 3 = 72$.

85. $\dfrac{2}{3} = \dfrac{2 \cdot 4}{3 \cdot 4} = \dfrac{8}{12}$

87. $\dfrac{3}{4} = \dfrac{3 \cdot 6}{4 \cdot 6} = \dfrac{18}{24}$

89. $7 = \dfrac{7 \cdot 3}{1 \cdot 3} = \dfrac{21}{3}$

91.
$$2 = 2$$
$$8 = 2 \cdot 2 \cdot 2$$
$$\downarrow \quad \downarrow \quad \downarrow$$
$$\text{LCD} = 2 \cdot 2 \cdot 2 = 8$$
$$\frac{1}{2} = \frac{1 \cdot 4}{2 \cdot 4} = \frac{4}{8}$$
The equivalent fractions are $\dfrac{4}{8}$ and $\dfrac{3}{8}$.

93.
$$5 = \quad 5$$
$$3 = 3$$
$$\downarrow \quad \downarrow$$
$$\text{LCD} = 3 \cdot 5 = 15$$
$$\frac{3}{5} = \frac{3 \cdot 3}{5 \cdot 3} = \frac{9}{15}$$
$$\frac{2}{3} = \frac{2 \cdot 5}{3 \cdot 5} = \frac{10}{15}$$
The equivalent fractions are $\dfrac{9}{15}$ and $\dfrac{10}{15}$.

95.
$$6 = 2 \cdot \quad \quad 3$$
$$8 = 2 \cdot 2 \cdot 2$$
$$\downarrow \quad \downarrow \quad \downarrow \quad \downarrow$$
$$\text{LCD} = 2 \cdot 2 \cdot 2 \cdot 3 = 24$$
$$\frac{5}{6} = \frac{5 \cdot 4}{6 \cdot 4} = \frac{20}{24}$$
$$\frac{5}{8} = \frac{5 \cdot 3}{8 \cdot 3} = \frac{15}{24}$$
The equivalent fractions are $\dfrac{20}{24}$ and $\dfrac{15}{24}$.

97.
$$15 = \quad \quad 3 \cdot 5$$
$$20 = 2 \cdot 2 \cdot \quad 5$$
$$\downarrow \quad \downarrow \quad \downarrow \quad \downarrow$$
$$\text{LCD} = 2 \cdot 2 \cdot 3 \cdot 5 = 60$$
$$\frac{11}{15} = \frac{11 \cdot 4}{15 \cdot 4} = \frac{44}{60}$$
$$\frac{7}{20} = \frac{7 \cdot 3}{20 \cdot 3} = \frac{21}{60}$$
The equivalent fractions are $\dfrac{44}{60}$ and $\dfrac{21}{60}$.

99.
$$9 = 3 \cdot 3$$
$$18 = 2 \cdot 3 \cdot 3$$
$$30 = 2 \cdot 3 \cdot \quad 5$$

$$\text{LCD} = 2 \cdot 3 \cdot 3 \cdot 5 = 90$$

$$\frac{2}{9} = \frac{2 \cdot 10}{9 \cdot 10} = \frac{20}{90}$$

$$\frac{7}{18} = \frac{7 \cdot 5}{18 \cdot 5} = \frac{35}{90}$$

$$\frac{7}{30} = \frac{7 \cdot 3}{30 \cdot 3} = \frac{21}{90}$$

The equivalent fractions are $\dfrac{20}{90}$, $\dfrac{35}{90}$, and $\dfrac{21}{90}$.

101. $\dfrac{14}{21} = \dfrac{2 \cdot 7}{3 \cdot 7} = \dfrac{2 \cdot \cancel{7}}{3 \cdot \cancel{7}} = \dfrac{2}{3}$

103. $\dfrac{38}{18} = \dfrac{2 \cdot 19}{2 \cdot 9} = \dfrac{\cancel{2} \cdot 19}{\cancel{2} \cdot 9} = \dfrac{19}{9}$

105. $\dfrac{22}{44} = \dfrac{2 \cdot 11}{2 \cdot 2 \cdot 11} = \dfrac{\cancel{2} \cdot \cancel{11}}{\cancel{2} \cdot 2 \cdot \cancel{11}} = \dfrac{1}{2}$

107. $\dfrac{32}{40} = \dfrac{2 \cdot 2 \cdot 2 \cdot 2 \cdot 2}{2 \cdot 2 \cdot 2 \cdot 5} = \dfrac{\cancel{2} \cdot \cancel{2} \cdot \cancel{2} \cdot 2 \cdot 2}{\cancel{2} \cdot \cancel{2} \cdot \cancel{2} \cdot 5} = \dfrac{4}{5}$

109. The 0 is two places to the right of the decimal; this is the hundredths place.

111. The 5 is two places to the left of the decimal; this is the tens place.

113. The 6 is three places to the right of the decimal; this is the thousandths place.

115. The number 2 is in the tenths place. The number to the right of 2 is 0. Since 0 is less than 5, we round to 578.2.

117. The number 5 is in the tens place. The number to the right of 5 is 4. Since 4 is less than 5, we round to 350.

119. The number 9 is in the thousandths place. The number to the right of 9 is 8. Since 8 is greater than 5, we round to 3682.010.

121. The number 9 is in the ones place. The number to its right is also 9. Since 9 is greater than 5, we round to 30.

123.
$$\begin{array}{r} 0.625 \\ 8\overline{)5.000} \\ \underline{4\,8} \\ 20 \\ \underline{16} \\ 40 \\ \underline{40} \\ 0 \end{array}$$

$$\frac{5}{8} = 0.625$$

125.
$$\begin{array}{r} 0.285714 \\ 7\overline{)2.000000} \\ \underline{1\,4} \\ 60 \\ \underline{56} \\ 40 \\ \underline{35} \\ 50 \\ \underline{49} \\ 10 \\ \underline{7} \\ 30 \\ \underline{28} \\ 2 \end{array}$$

$$\frac{2}{7} = 0.\overline{285714}$$

127.
$$\begin{array}{r} 0.3125 \\ 16\overline{)5.0000} \\ \underline{4\,8} \\ 20 \\ \underline{16} \\ 40 \\ \underline{32} \\ 80 \\ \underline{80} \\ 0 \end{array}$$

$$\frac{5}{16} = 0.3125$$

129.

$$13 \overline{\smash{)}3.000000} \quad \begin{array}{r} 0.230769 \\ \end{array}$$

$$\begin{array}{r}
0.230769 \\
13\overline{\smash{)}3.000000} \\
\underline{2\,6} \\
40 \\
\underline{39} \\
100 \\
\underline{91} \\
90 \\
\underline{78} \\
120 \\
\underline{117} \\
3 \\
\end{array}$$

$$\frac{3}{13} = 0.\overline{230769}$$

131.

$$\begin{array}{r}
1.16 \\
25\overline{\smash{)}29.00} \\
\underline{25} \\
4\,0 \\
\underline{2\,5} \\
1\,50 \\
\underline{1\,50} \\
0 \\
\end{array}$$

$$\frac{29}{25} = 1.16$$

133.

$$\begin{array}{r}
2.16 \\
6\overline{\smash{)}13.0} \\
\underline{12} \\
1\,0 \\
\underline{6} \\
40 \\
\underline{36} \\
4 \\
\end{array}$$

$\dfrac{13}{6}$ rounded to the nearest tenth is 2.2.

135.

$$\begin{array}{r}
2.666 \\
3\overline{\smash{)}8.000} \\
\underline{6} \\
2\,0 \\
\underline{1\,8} \\
20 \\
\underline{18} \\
20 \\
\underline{18} \\
2 \\
\end{array}$$

$\dfrac{8}{3}$ rounded to the nearest hundredth is 2.67.

137.

$$\begin{array}{r}
0.5185 \\
27\overline{\smash{)}14.0000} \\
\underline{13\,5} \\
50 \\
\underline{27} \\
230 \\
\underline{216} \\
140 \\
\underline{135} \\
5 \\
\end{array}$$

$\dfrac{14}{27}$ rounded to the nearest thousandth is 0.519.

139. $0.75 = \dfrac{75}{100} = \dfrac{3 \cdot \cancel{25}}{4 \cdot \cancel{25}} = \dfrac{3}{4}$

141. $0.9 = \dfrac{9}{10}$

143. $0.982 = \dfrac{982}{1000} = \dfrac{\cancel{2} \cdot 491}{\cancel{2} \cdot 500} = \dfrac{491}{500}$

145. $0.2525 = \dfrac{2525}{10,000} = \dfrac{\cancel{25} \cdot 101}{\cancel{25} \cdot 400} = \dfrac{101}{400}$

147. $37\% = 37\% \cdot \dfrac{1}{100\%} = \dfrac{37}{100} = 0.37$

149. $6.02\% = 6.02\% \cdot \dfrac{1}{100\%} = \dfrac{6.02}{100} = 0.0602$

151. $0.1\% = 0.1\% \cdot \dfrac{1}{100\%} = \dfrac{0.1}{100} = 0.001$

153. $0.2 = 0.2 \cdot \dfrac{100\%}{1} = 20\%$

155. $0.275 = 0.275 \cdot \dfrac{100\%}{1} = 27.5\%$

157. $2 = 2 \cdot \dfrac{100\%}{1} = 200\%$

159. The least number of months is the LCM of 3, 7, and 12.

$$\begin{array}{l}
3 = 3 \\
7 = \quad\quad 7 \\
12 = 3 \cdot \quad 2 \cdot 2 \\
\end{array}$$

$$\text{LCM} = 3 \cdot 7 \cdot 2 \cdot 2$$

The least number of months is
$2 \cdot 2 \cdot 3 \cdot 7 = 84$.

161. The number of days is the LCM of 4 and 10.

$$4 = 2 \cdot 2$$
$$10 = 2 \cdot \quad 5$$

$$LCM = 2 \cdot 2 \cdot 5$$

Bob gives Sam both types of medication on the same day every $2 \cdot 2 \cdot 5 = 20$ days.

163. 325 of 500

$$\frac{325}{500} = \frac{5 \cdot 5 \cdot 13}{2 \cdot 2 \cdot 5 \cdot 5 \cdot 5} = \frac{\cancel{5} \cdot \cancel{5} \cdot 13}{2 \cdot 2 \cdot \cancel{5} \cdot \cancel{5} \cdot 5} = \frac{13}{20}$$

The fraction of students that work at least 25 hours per week is $\frac{13}{20}$.

165. 840 of 1200

$$\begin{array}{r} 0.7 \\ 1200\overline{)840.0} \\ 840\ 0 \\ \hline 0 \end{array}$$

$$\frac{840}{1200} = 0.7 = 0.7 \cdot \frac{100\%}{1} = 70\%$$

70% of adult Americans believe that they eat healthily.

167. 85 of 110

$$\begin{array}{r} 0.772 \\ 110\overline{)85.000} \\ 770 \\ \hline 800 \\ 770 \\ \hline 300 \\ 220 \\ \hline 80 \end{array}$$

$$\frac{85}{110} = 0.7\overline{72} \approx 0.7727 \cdot \frac{100\%}{1} = 77.27\%$$

The score is 77.27%.

169. (a) $\dfrac{8}{24} = 0.\overline{3}$

$$0.3333 \cdot \frac{100\%}{1} = 33.33\%$$

Jackson sleeps for 33.33% of the day.

(b) $\dfrac{4}{24} = 0.1\overline{6}$

$$0.1667 \cdot \frac{100\%}{1} = 16.67\%$$

Jackson works for 16.67% of the day.

(c) $\dfrac{6}{24} = 0.25$

$$0.25 \cdot \frac{100\%}{1} = 25\%$$

Jackson goes to school and studies 25% of the day.

171. 3 of 14

$$\begin{array}{r} 0.21428 \\ 14\overline{)3.00000} \\ 2\ 8 \\ \hline 20 \\ 14 \\ \hline 60 \\ 56 \\ \hline 40 \\ 28 \\ \hline 120 \\ 112 \\ \hline 8 \end{array}$$

$$\frac{3}{14} \approx 0.2143$$

$$0.2143 \cdot \frac{100\%}{1} = 21.43\%$$

21.43% is saturated fat.

173. The process results in the following 25 prime numbers: 2, 3, 5, 7, 11, 13, 17, 19, 23, 29, 31, 37, 41, 43, 47, 53, 59, 61, 67, 71, 73, 79, 83, 89, 97

Section 1.3

Preparing for the Number Systems and the Real Number Line

P1.
$$\begin{array}{r} 0.625 \\ 8\overline{)5.000} \\ 4\ 8 \\ \hline 20 \\ 16 \\ \hline 40 \\ 40 \\ \hline 0 \end{array}$$

$$\frac{5}{8} = 0.625$$

P2.

$$\begin{array}{r} 0.8181 \\ 11\overline{)9.0000} \\ \underline{8\ 8} \\ 20 \\ \underline{11} \\ 90 \\ \underline{88} \\ 20 \\ \underline{11} \\ 9 \end{array}$$

$\dfrac{9}{11} = 0.8181...$ or $0.\overline{81}$

1.3 Quick Checks

1. The first 4 positive odd numbers are 1, 3, 5, and 7. If we let O represent this set, then $O = \{1, 3, 5, 7\}$.

2. The states whose names begin with the letter A are Alabama, Alaska, Arizona, and Arkansas. If we let A represent this set, then $A = \{$Alabama, Alaska, Arizona, Arkansas$\}$.

3. There are no states whose names begin with the letter Z. If we let Z represent this set, then $Z = \{\ \}$ or $\varnothing$.

4. Every integer is a rational number. True

5. Real numbers that can be represented with a terminating decimal are called rational.

6. 12 is the only natural number.

7. 0 and 12 are the whole numbers.

8. −5, 12, and 0 are the integers.

9. $\dfrac{11}{5}$, −5, 12, $2.\overline{76}$, 0, and $\dfrac{18}{4}$ are the rational numbers.

10. π is the only irrational number.

11. All the numbers listed are real numbers.

12.

13. The symbols $<, >, \le, \ge$ are called inequality symbols.

14. $2 < 9$ because 2 lies to the left of 9 on the real number line.

15. $-5 < -3$ because −5 lies to the left of −3 on the real number line.

16. $\dfrac{4}{5} > \dfrac{1}{2}$ because $\dfrac{4}{5} = \dfrac{8}{10}$ and $\dfrac{1}{2} = \dfrac{5}{10}$. Having 8 parts out of 10 is more than having 5 parts out of 10. Also, $\dfrac{4}{5} = 0.8$ and $\dfrac{1}{2} = 0.5$ and 0.8 lies to the right of 0.5 on the real number line.

17. $\dfrac{4}{7} > 0.5$ because $\dfrac{4}{7} = 0.\overline{571428}$ and $0.\overline{571428}$ lies to the right of 0.5 on the real number line.

18. $\dfrac{4}{3} = \dfrac{20}{15}$

19. $-\dfrac{4}{3} < -\dfrac{5}{4}$

20. The distance from zero to a point on a real number line whose coordinate is a is called the absolute value of a.

21. $|-15| = 15$ because the distance from 0 to −15 on the real number line is 15.

22. $\left|-\dfrac{3}{4}\right| = \dfrac{3}{4}$ because the distance from 0 to $-\dfrac{3}{4}$ on the real number line is $\dfrac{3}{4}$.

1.3 Exercises

23. The set of whole numbers less than 5 is $A = \{0, 1, 2, 3, 4\}$.

25. The set of natural numbers less than 5 is $D = \{1, 2, 3, 4\}$.

27. The set of even natural numbers between 4 and 15 is $E = \{6, 8, 10, 12, 14\}$.

29. 3 is the only natural number.

31. −4, 3, and 0 are the integers.

33. 2.303003000... is the only irrational number.

35. All the numbers listed are real numbers.

37. π is the only irrational number.

39. $\dfrac{5}{5} = 1$, so $\dfrac{5}{5}$ is the only whole number.

41.

43. -2 lies to the right of -3 on the number line. The statement is true.

45. Since $-6 = -6$, the statement is true.

47. Since the decimal equivalent of $\dfrac{3}{2}$ is 1.5, the statement is true.

49. Since $\pi = 3.14159...$, $\pi > 3.14$. The statement is false.

51. Since -1 lies to the left of 0 on the number line, $-1 < 0$.

53. Since $\dfrac{5}{8} = 0.625$ and $\dfrac{6}{11} = 0.\overline{54}$, $\dfrac{5}{8} > \dfrac{6}{11}$.

55. Since $\dfrac{6}{13} = 0.\overline{461538}$, $\dfrac{6}{13} > 0.46$.

57. Since $\dfrac{42}{6}$ simplifies to 7, $\dfrac{42}{6} = 7$.

59. $|-12| = 12$ because the distance from 0 to -12 on the real number line is 12.

61. $|4| = 4$ because the distance from 0 to 4 on the real number line is 4.

63. $\left|-\dfrac{3}{8}\right| = \dfrac{3}{8}$ because the distance from 0 to $-\dfrac{3}{8}$ on the real number line is $\dfrac{3}{8}$.

65. $|-2.1| = 2.1$ because the distance from 0 to -2.1 on the real number line is 2.1.

67. **(a)**

(b) From left to right on the number line, the order of the numbers is -4.5, -1, $-\dfrac{1}{2}$, $\dfrac{3}{5}$, 1, 3.5, $|-7|$.

(c) **(i)** -1, 1, and $|-7| = 7$ are integers.

(ii) All numbers listed are rational numbers.

		Natural	Whole	Integers	Rational	Irrational	Real
69.	−100			√	√		√
71.	−10.5				√		√
73.	$\frac{75}{25}$	√	√	√	√		√
75.	7.56556555...					√	√

77. The whole numbers are a subset of the integers. The statement is true.

79. The set of rational numbers and the set of irrational numbers have no elements in common. The statement is false.

81. The natural numbers are a subset of the whole numbers. The statement is true.

83. Every terminating decimal can be expressed as the ratio of two integers. The statement is true.

85. 0 is an integer that is neither negative or positive. The statement is true.

87. Non-terminating and non-repeating decimals are irrational numbers.

89. The set of rational numbers combined with the irrational numbers comprises the set of real numbers.

91. The only number that is both nonnegative and non-positive is 0.

93. Both elements of Y are also elements of X, so $Y \subseteq X$. The statement is true.

95. Both elements of Y are also elements of Z, so $Y \subseteq Z$. The statement is true.

97. $A \cup B$ consists of all elements that are in either A or B, so $A \cup B = \{7, 8, 9, 10, 11, 12, 13, 14, 15\}$.

99. $B \cup C$ consists of all elements that are in either B or C, so $B \cup C = \{10, 11, 12, 13, 14, 15\}$.

101. $A \cap C$ is the set of all elements common to both A and C, so $A \cap C = \{11, 12\}$.

103. $A \cap B$ is the set of even whole numbers less than 11, so $A \cap B = \{2, 4, 6, 8, 10\}$.

105. **(a)** Answers may vary. One possibility: There are subsets with no elements, subsets with one element, subsets with two elements, subsets with three elements, and subsets with four elements.
0: { }
1: {1}, {2}, {3}, {4}
2: {1, 2}, {1, 3}, {1, 4}, {2, 3}, {2, 4}, {3, 4}
3: {1, 2, 3}, {1, 2, 4}, {1, 3, 4}, {2, 3 4}
4: {1, 2, 3, 4}

(b) There are a total of 16 subsets.

107. A rational number is any number that can be written as the quotient of two integers, denominator not equal to zero. Natural numbers, whole numbers and integers are rational numbers. Terminating and repeating decimals are also rational numbers.

Section 1.4

Preparing for Adding, Subtracting, Multiplying, and Dividing Integers

P1. $\dfrac{16}{36} = \dfrac{2 \cdot 2 \cdot 2 \cdot 2}{2 \cdot 2 \cdot 3 \cdot 3} = \dfrac{\cancel{2} \cdot \cancel{2} \cdot 2 \cdot 2}{\cancel{2} \cdot \cancel{2} \cdot 3 \cdot 3} = \dfrac{2 \cdot 2}{3 \cdot 3} = \dfrac{4}{9}$

1.4 Quick Checks

1. The answer to an addition problem is called the <u>sum</u>.

2.
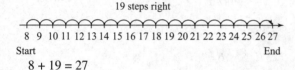
$8 + 19 = 27$

3.
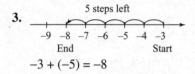
$-3 + (-5) = -8$

4.
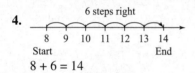
$8 + 6 = 14$

5.
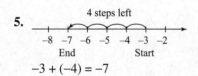
$-3 + (-4) = -7$

6.
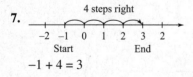
$-5 + (-12) = -17$

7.
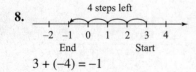
$-1 + 4 = 3$

8.
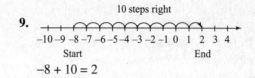
$3 + (-4) = -1$

9.
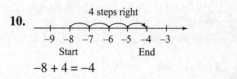
$-8 + 10 = 2$

10.
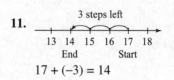
$-8 + 4 = -4$

11.
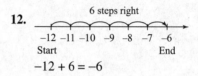
$17 + (-3) = 14$

12.
$-12 + 6 = -6$

13.
$15 + (-5) = 10$

14. The sum of two negative integers will be <u>negative</u>.

15. $|-11| = 11$
$|7| = 7$
The signs of -11 and 7 are different, so we subtract the absolute values: $11 - 7 = 4$.
The larger absolute value, 11, corresponds to a negative number in the original problem, so the sum is negative.
$-11 + 7 = -4$

16. $|5| = 5$
$|-8| = 8$
The signs of 5 and -8 are different, so we subtract the absolute values: $8 - 5 = 3$.
The larger absolute value, 8, corresponds to a negative number in the original problem, so the sum is negative.
$5 + (-8) = -3$

17. $|-8| = 8$
$|-16| = 16$
The signs of -8 and -16 are the same, so we add the absolute values: $8 + 16 = 24$.
Both numbers in the original problem are negative, so the sum is negative.
$-8 + (-16) = -24$

18. $|-94| = 94$

$|38| = 38$

The signs of -94 and 38 are different, so we subtract the absolute values: $94 - 38 = 56$. The larger absolute value, 94, corresponds to a negative number in the original problem, so the sum is negative.

$-94 + 38 = -56$

19. The additive inverse of 7 is -7 because $7 + (-7) = 0$.

20. The additive inverse of $\dfrac{3}{7}$ is $-\dfrac{3}{7}$ because

$\dfrac{3}{7} + \left(-\dfrac{3}{7}\right) = 0$.

21. The additive inverse of -21 is $-(-21) = 21$ because $-21 + 21 = 0$.

22. The additive inverse of $-\dfrac{8}{5}$ is $-\left(-\dfrac{8}{5}\right) = \dfrac{8}{5}$

because $-\dfrac{8}{5} + \dfrac{8}{5} = 0$.

23. The additive inverse of -5.75 is $-(-5.75) = 5.75$ because $-5.75 + 5.75 = 0$.

24. The answer to a subtraction problem is called the difference.

25. The subtraction problem $-3 - 10$ is equivalent to $-3 + (-10)$.

26. $59 - (-21) = 59 + 21 = 80$

27. $-32 - 146 = -32 + (-146) = -178$

28. 17 minus 35 is $17 - 35$.

$17 - 35 = 17 + (-35) = -18$

29. -382 subtracted from -2954 is

$-2954 - (-382)$.

$-2954 - (-382) = -2954 + 382 = -2572$

30. $8 - 13 + 5 - 21 = 8 + (-13) + 5 + (-21)$

$= -5 + 5 + (-21)$

$= 0 + (-21)$

$= -21$

31. $-27 - 49 + 18 = -27 + (-49) + 18$

$= -76 + 18$

$= -58$

32. $3 - (-14) - 8 + 3 = 3 + 14 + (-8) + 3$

$= 17 + (-8) + 3$

$= 9 + 3$

$= 12$

33. $-825 + 375 - (-735) + 265$

$= -825 + 375 + 735 + 265$

$= -450 + 735 + 265$

$= 285 + 265$

$= 550$

34. The product of two integers with the same sign is positive.

35. $-3(7) = -21$

36. $13(-4) = -52$

37. $5 \cdot 16 = 80$

38. $-9(-12) = 108$

39. $(-13)(-25) = 325$

40. The product of thirteen negative factors is negative. True

41. $-3 \cdot 9 \cdot (-4) = -27 \cdot (-4) = 108$

42. $(-3) \cdot (-4) \cdot (-5) \cdot (-6) = 12 \cdot (-5) \cdot (-6)$

$= -60 \cdot (-6)$

$= 360$

43. The reciprocal of 6 is $\dfrac{1}{6}$.

44. The reciprocal of -2 is $-\dfrac{1}{2}$.

45. The quotient of two negative numbers is positive. True

46. $\dfrac{20}{-4} = \dfrac{5 \cdot 4}{-1 \cdot 4} = \dfrac{5 \cdot \cancel{4}}{-1 \cdot \cancel{4}} = \dfrac{5}{-1} = -5$

47. $\dfrac{707}{-101} = \dfrac{7 \cdot 101}{-1 \cdot 101} = \dfrac{7 \cdot \cancel{101}}{-1 \cdot \cancel{101}} = \dfrac{7}{-1} = -7$

48. $-63 \div (-7) = \dfrac{-63}{-7} = \dfrac{9 \cdot (-7)}{1 \cdot (-7)} = \dfrac{9 \cdot \cancel{(-7)}}{1 \cdot \cancel{(-7)}} = 9$

1.4 Exercises

49. $8 + 7 = 15$

51. $-5 + 9 = 4$

53. $9 + (-5) = 4$

55. $-11 + (-8) = -19$

57. $-16 + 37 = 21$

59. $-119 + (-209) = -328$

61. $-14 + 21 + (-18) = 7 + (-18) = -11$

63. $74 + (-13) + (-23) + 5 = 61 + (-23) + 5$
$\qquad\qquad\qquad\qquad\quad = 38 + 5$
$\qquad\qquad\qquad\qquad\quad = 43$

65. The additive inverse of -325 is
$-(-325) = 325$ because $-325 + 325 = 0$.

67. The additive inverse of 125 is -125 because $125 + (-125) = 0$.

69. $23 - 12 = 23 + (-12) = 11$

71. $9 - 17 = 9 + (-17) = -8$

73. $-20 - 8 = -20 + (-8) = -28$

75. $13 - (-41) = 13 + 41 = 54$

77. $-36 - (-36) = -36 + 36 = 0$

79. $0 - 41 = 0 + (-41) = -41$

81. $-93 - (-62) = -93 + 62 = -31$

83. $86 - (-86) = 86 + 86 = 172$

85. $5 \cdot 8 = 40$

87. $8(-7) = -56$

89. $0 \cdot (-21) = 0$

91. $(-48)(-3) = 144$

93. $(-42)3 = -126$

95. $-5 \cdot 6 \cdot 3 = -30 \cdot 3 = -90$

97. $-10(3)(-7) = -30(-7) = 210$

99. $(-2)(4)(-1)(3)(5) = -8(-1)(3)(5)$
$\qquad\qquad\qquad\qquad\quad = 8(3)(5)$
$\qquad\qquad\qquad\qquad\quad = 24(5)$
$\qquad\qquad\qquad\qquad\quad = 120$

101. The reciprocal of 8 is $\dfrac{1}{8}$.

103. The reciprocal of -4 is $-\dfrac{1}{4}$.

105. The reciprocal of 1 is $\dfrac{1}{1}$ or 1.

107. $10 \div 2 = \dfrac{10}{2} = \dfrac{5 \cdot 2}{1 \cdot 2} = \dfrac{5 \cdot \cancel{2}}{1 \cdot \cancel{2}} = 5$

109. $\dfrac{-56}{-8} = \dfrac{7 \cdot (-8)}{1 \cdot (-8)} = \dfrac{7 \cdot \cancel{(-8)}}{1 \cdot \cancel{(-8)}} = 7$

111. $\dfrac{-45}{3} = \dfrac{-15 \cdot 3}{1 \cdot 3} = \dfrac{-15 \cdot \cancel{3}}{1 \cdot \cancel{3}} = -15$

113. $\dfrac{35}{10} = \dfrac{7 \cdot 5}{2 \cdot 5} = \dfrac{7 \cdot \cancel{5}}{2 \cdot \cancel{5}} = \dfrac{7}{2}$

115. $\dfrac{60}{-42} = \dfrac{10 \cdot 6}{-7 \cdot 6} = \dfrac{10 \cdot \cancel{6}}{-7 \cdot \cancel{6}} = \dfrac{10}{-7} = -\dfrac{10}{7}$

117. $\dfrac{-105}{-12} = \dfrac{35 \cdot (-3)}{4 \cdot (-3)} = \dfrac{35 \cdot \cancel{(-3)}}{4 \cdot \cancel{(-3)}} = \dfrac{35}{4}$

119. $-4 \cdot 18 = -72$

121. $-16 - (-76) = -16 + 76 = 60$

123. $-9 \cdot (-19) = 171$

125. $\dfrac{120}{-8} = \dfrac{15 \cdot 8}{-1 \cdot 8} = \dfrac{15 \cdot \cancel{8}}{-1 \cdot \cancel{8}} = -15$

127. $-98 + 56 = -42$

129. $\dfrac{75}{|-20|} = \dfrac{75}{20} = \dfrac{15 \cdot 5}{4 \cdot 5} = \dfrac{15 \cdot \cancel{5}}{4 \cdot \cancel{5}} = \dfrac{15}{4}$

131. $|-14| + |-26| = 14 + 26 = 40$

133. $|-389| - 627 = 389 - 627$
$$= 389 + (-627)$$
$$= -238$$

135. The sum of 28 and -21 is written as
$28 + (-21) = 7$.

137. -21 minus 47 is written as
$-21 - 47 = -21 + (-47) = -68$.

139. -12 multiplied by 18 is written as
$-12 \cdot 18 = -216$.

141. -36 divided by -108 is written as
$$-36 \div (-108) = \frac{-36}{-108}$$
$$= \frac{1 \cdot (-36)}{3 \cdot (-36)}$$
$$= \frac{1 \cdot \cancel{(-36)}}{3 \cdot \cancel{(-36)}}$$
$$= \frac{1}{3}.$$

143. -3.25 points; the number is negative because the price fell.

145. -6 yards; the number is negative because the team lost yards.

147. $-\$48$; the number is negative because the account is overdrawn.

149. $8 + 3 + 3 = 11 + 3 = 14$
Loren and Richard walked 14 miles.

151. $563 + (-46) + 233 + (-63) + (-32) = 655$
Martha's new balance is $655.

153. $725 + (-120) + (-590) + 310 = 325$
The company has 325 cases on hand. There is not enough stock to fill an order for 450 cases. The difference is
$325 + (-450) = -125$ cases.

155. $25,350 - (-375) = 25,350 + 375 = 25,725$
The distance between them is 25,725 feet.

157. $-3 + (-5) = -8$
$-3(-5) = 15$
The integers are -3 and -5.

159. $-12 + 2 = -10$
$-12(2) = -24$
The integers are -12 and 2.

161. (a) $\dfrac{1}{1} = 1,\ \dfrac{2}{1} = 2,\ \dfrac{3}{2} = 1.5,\ \dfrac{5}{3} \approx 1.667,\ \dfrac{8}{5} = 1.6,$

$\dfrac{13}{8} = 1.625,\ \dfrac{21}{13} \approx 1.615,$

$\dfrac{34}{21} \approx 1.619,\ \dfrac{55}{34} \approx 1.618, \ldots$

(b) The golden ratio is about 1.618.

(c) Answers may vary.

163. The problem $42 \div 4$ may be written equivalently as $42 \cdot \dfrac{1}{4}$.

Section 1.5

Preparing for Adding, Subtracting, Multiplying, and Dividing Rational Numbers Expressed as Fractions and Decimals

P1. $12 = 2 \cdot 2 \cdot 3$
$16 = 2 \cdot 2 \cdot 2 \cdot 2$
$\text{LCD} = 2 \cdot 2 \cdot 2 \cdot 2 \cdot 3 = 48$
The LCD of $\dfrac{5}{12}$ and $\dfrac{3}{16}$ is 48.

P2. $30 = 5 \cdot 6$
$$\frac{4}{5} = \frac{4}{5} \cdot \frac{6}{6} = \frac{4 \cdot 6}{5 \cdot 6} = \frac{24}{30}$$

1.5 Quick Checks

1. $\dfrac{-4}{14} = \dfrac{-2 \cdot 2}{7 \cdot 2} = \dfrac{-2 \cdot \cancel{2}}{7 \cdot \cancel{2}} = -\dfrac{2}{7}$

2. $-\dfrac{18}{30} = -\dfrac{3 \cdot 6}{5 \cdot 6} = -\dfrac{3 \cdot \cancel{6}}{5 \cdot \cancel{6}} = -\dfrac{3}{5}$

3. $\dfrac{24}{-4} = \dfrac{6 \cdot 4}{-1 \cdot 4} = \dfrac{6 \cdot \cancel{4}}{-1 \cdot \cancel{4}} = \dfrac{6}{-1} = -6$

4. $\dfrac{3}{4} \cdot \dfrac{9}{8} = \dfrac{3 \cdot 9}{4 \cdot 8} = \dfrac{27}{32}$

5. $\dfrac{-5}{7}\cdot\dfrac{56}{15}=\dfrac{-5\cdot56}{7\cdot15}$

$=\dfrac{-1\cdot5\cdot7\cdot8}{7\cdot3\cdot5}$

$=\dfrac{-1\cdot\cancel{5}\cdot\cancel{7}\cdot8}{\cancel{7}\cdot3\cdot\cancel{5}}$

$=-\dfrac{8}{3}$

6. $\dfrac{12}{45}\cdot\left(-\dfrac{18}{20}\right)=\dfrac{12}{45}\cdot\dfrac{-18}{20}$

$=\dfrac{12\cdot(-18)}{45\cdot20}$

$=\dfrac{2\cdot2\cdot3\cdot2\cdot3\cdot(-3)}{3\cdot3\cdot5\cdot2\cdot2\cdot5}$

$=\dfrac{\cancel{2}\cdot\cancel{2}\cdot\cancel{3}\cdot2\cdot\cancel{3}\cdot(-3)}{\cancel{3}\cdot\cancel{3}\cdot5\cdot\cancel{2}\cdot\cancel{2}\cdot5}$

$=\dfrac{2\cdot(-3)}{5\cdot5}$

$=-\dfrac{6}{25}$

7. $-\dfrac{25}{75}\cdot\left(-\dfrac{9}{4}\right)=\dfrac{-25}{75}\cdot\dfrac{-9}{4}$

$=\dfrac{-25\cdot(-9)}{75\cdot4}$

$=\dfrac{-1\cdot5\cdot5\cdot(-1)\cdot3\cdot3}{3\cdot5\cdot5\cdot2\cdot2}$

$=\dfrac{\cancel{5}\cdot\cancel{5}\cdot\cancel{3}\cdot3}{\cancel{3}\cdot\cancel{5}\cdot\cancel{5}\cdot2\cdot2}$

$=\dfrac{3}{2\cdot2}$

$=\dfrac{3}{4}$

8. $\dfrac{7}{3}\cdot\dfrac{1}{14}\cdot\left(-\dfrac{9}{11}\right)=\dfrac{7}{3}\cdot\dfrac{1}{14}\cdot\dfrac{-9}{11}$

$=\dfrac{7\cdot1\cdot(-9)}{3\cdot14\cdot11}$

$=\dfrac{7\cdot3\cdot(-3)}{3\cdot2\cdot7\cdot11}$

$=\dfrac{\cancel{7}\cdot\cancel{3}\cdot(-3)}{\cancel{3}\cdot2\cdot\cancel{7}\cdot11}$

$=\dfrac{-3}{2\cdot11}$

$=-\dfrac{3}{22}$

9. Two numbers are called multiplicative inverses, or reciprocals, if their product is equal to <u>one</u>.

10. The reciprocal of 12 is $\dfrac{1}{12}$ because $12\cdot\dfrac{1}{12}=1$.

11. The reciprocal of $\dfrac{7}{5}$ is $\dfrac{5}{7}$ because $\dfrac{7}{5}\cdot\dfrac{5}{7}=1$.

12. The reciprocal of $-\dfrac{1}{4}$ is -4 because

$-\dfrac{1}{4}\cdot(-4)=1$.

13. The reciprocal of $-\dfrac{31}{20}$ is $-\dfrac{20}{31}$ because

$-\dfrac{31}{20}\cdot\left(-\dfrac{20}{31}\right)=1$.

14. $\dfrac{5}{7}\div\dfrac{7}{10}=\dfrac{5}{7}\cdot\dfrac{10}{7}=\dfrac{5\cdot10}{7\cdot7}=\dfrac{50}{49}$

15. $-\dfrac{9}{12}\div\dfrac{14}{7}=-\dfrac{9}{12}\cdot\dfrac{7}{14}$

$=-\dfrac{3\cdot3\cdot7}{2\cdot2\cdot3\cdot2\cdot7}$

$=-\dfrac{\cancel{3}\cdot3\cdot\cancel{7}}{2\cdot2\cdot\cancel{3}\cdot2\cdot\cancel{7}}$

$=-\dfrac{3}{2\cdot2\cdot2}$

$=-\dfrac{3}{8}$

16. $\dfrac{8}{35}\div\left(\dfrac{-1}{10}\right)=\dfrac{8}{35}\cdot\left(\dfrac{10}{-1}\right)$

$=\dfrac{2\cdot2\cdot2\cdot5}{5\cdot7\cdot(-1)}$

$=\dfrac{2\cdot2\cdot2\cdot\cancel{5}}{\cancel{5}\cdot7\cdot(-1)}$

$=\dfrac{2\cdot2\cdot2}{7\cdot(-1)}$

$=-\dfrac{16}{7}$

17. $-\dfrac{18}{63} \div \left(-\dfrac{54}{35}\right) = \dfrac{-18}{63} \cdot \left(\dfrac{-35}{54}\right)$

$\qquad = \dfrac{-1 \cdot 2 \cdot 3 \cdot 3 \cdot (-1) \cdot 5 \cdot 7}{3 \cdot 3 \cdot 7 \cdot 2 \cdot 3 \cdot 3 \cdot 3}$

$\qquad = \dfrac{\cancel{2} \cdot \cancel{3} \cdot \cancel{3} \cdot 5 \cdot \cancel{7}}{\cancel{3} \cdot \cancel{3} \cdot \cancel{7} \cdot \cancel{2} \cdot 3 \cdot 3 \cdot 3}$

$\qquad = \dfrac{5}{3 \cdot 3 \cdot 3}$

$\qquad = \dfrac{5}{27}$

18. $\dfrac{-5}{7} + \dfrac{3}{7} = \dfrac{-5+3}{7}$

19. $-\dfrac{9}{10} - \dfrac{3}{10} = \dfrac{-9}{10} - \dfrac{3}{10}$

$\qquad = \dfrac{-9-3}{10}$

$\qquad = \dfrac{-12}{10}$

$\qquad = \dfrac{2 \cdot (-6)}{2 \cdot 5}$

$\qquad = \dfrac{\cancel{2} \cdot (-6)}{\cancel{2} \cdot 5}$

$\qquad = -\dfrac{6}{5}$

20. $\dfrac{8}{11} + \dfrac{2}{11} = \dfrac{8+2}{11} = \dfrac{10}{11}$

21. $-\dfrac{18}{35} + \dfrac{3}{35} = \dfrac{-18}{35} + \dfrac{3}{35}$

$\qquad = \dfrac{-18+3}{35}$

$\qquad = \dfrac{-15}{35}$

$\qquad = \dfrac{-3 \cdot 5}{5 \cdot 7}$

$\qquad = \dfrac{-3 \cdot \cancel{5}}{\cancel{5} \cdot 7}$

$\qquad = -\dfrac{3}{7}$

22. $\dfrac{19}{63} - \dfrac{10}{63} = \dfrac{19-10}{63} = \dfrac{9}{63} = \dfrac{1 \cdot 9}{7 \cdot 9} = \dfrac{1 \cdot \cancel{9}}{7 \cdot \cancel{9}} = \dfrac{1}{7}$

23. $12 = 2 \cdot 2 \cdot 3$

$18 = 2 \cdot 3 \cdot 3$

$\text{LCD} = 2 \cdot 2 \cdot 3 \cdot 3 = 36$

$\dfrac{5}{12} - \dfrac{5}{18} = \dfrac{5}{12} \cdot \dfrac{3}{3} - \dfrac{5}{18} \cdot \dfrac{2}{2}$

$\qquad = \dfrac{15}{36} - \dfrac{10}{36}$

$\qquad = \dfrac{15-10}{36}$

$\qquad = \dfrac{5}{36}$

24. $14 = 2 \cdot 7$

$21 = 3 \cdot 7$

$\text{LCD} = 2 \cdot 3 \cdot 7 = 42$

$\dfrac{3}{14} + \dfrac{10}{21} = \dfrac{3}{14} \cdot \dfrac{3}{3} + \dfrac{10}{21} \cdot \dfrac{2}{2}$

$\qquad = \dfrac{9}{42} + \dfrac{20}{42}$

$\qquad = \dfrac{9+20}{42}$

$\qquad = \dfrac{29}{42}$

25. $6 = 2 \cdot 3$

$12 = 2 \cdot 2 \cdot 3$

$\text{LCD} = 2 \cdot 2 \cdot 3 = 12$

$-\dfrac{23}{6} + \dfrac{7}{12} = \dfrac{-23}{6} \cdot \dfrac{2}{2} + \dfrac{7}{12}$

$\qquad = \dfrac{-46}{12} + \dfrac{7}{12}$

$\qquad = \dfrac{-46+7}{12}$

$\qquad = \dfrac{-39}{12}$

$\qquad = \dfrac{-1 \cdot 3 \cdot 13}{2 \cdot 2 \cdot 3}$

$\qquad = \dfrac{-1 \cdot \cancel{3} \cdot 13}{2 \cdot 2 \cdot \cancel{3}}$

$\qquad = \dfrac{-13}{2 \cdot 2}$

$\qquad = -\dfrac{13}{4}$

26. $5 = 5$

$11 = 11$

$\text{LCD} = 5 \cdot 11 = 55$

$$\frac{3}{5} + \left(-\frac{4}{11}\right) = \frac{3}{5} \cdot \frac{11}{11} + \frac{-4}{11} \cdot \frac{5}{5}$$

$$= \frac{33}{55} + \frac{-20}{55}$$

$$= \frac{33 + (-20)}{55}$$

$$= \frac{13}{55}$$

27. $-2 + \dfrac{7}{16} = \dfrac{-2}{1} + \dfrac{7}{16}$

$$= \frac{-2}{1} \cdot \frac{16}{16} + \frac{7}{16}$$

$$= \frac{-32}{16} + \frac{7}{16}$$

$$= \frac{-32 + 7}{16}$$

$$= \frac{-25}{16}$$

$$= -\frac{25}{16}$$

28. $6 + \left(\dfrac{-9}{4}\right) = \dfrac{6}{1} + \left(\dfrac{-9}{4}\right)$

$$= \frac{6}{1} \cdot \frac{4}{4} + \left(\frac{-9}{4}\right)$$

$$= \frac{24}{4} + \left(\frac{-9}{4}\right)$$

$$= \frac{24 + (-9)}{4}$$

$$= \frac{15}{4}$$

29.
$$
\begin{array}{r}
9.670 \\
+\ 11.344 \\
\hline
21.014
\end{array}
$$

So $9.67 + 11.344 = 21.014$.

30.
$$
\begin{array}{r}
81.96 \\
-\ 17.39 \\
\hline
64.57
\end{array}
$$

So $81.96 - 17.39 = 64.57$.

31.
$$
\begin{array}{r}
14.950 \\
7.118 \\
+\ 0.300 \\
\hline
22.368
\end{array}
$$

So $14.95 + 7.118 + 0.3 = 22.368$.

32.   345.6700
         $-$ 8.0912
         ‾‾‾‾‾‾‾‾
         337.5788
So $345.67 - 8.0912 = 337.5788$.

33.   180.782
         $-$ 100.300
         ‾‾‾‾‾‾‾‾
           80.482
$-180.782 + 100.3 + 9.07 = -80.482 + 9.07$
   80.482
 $-$ 9.070
 ‾‾‾‾‾‾‾
   71.412
So $-180.782 + 100.3 + 9.07 = -71.412$.

34.   74.280
         $+$ 14.832
         ‾‾‾‾‾‾‾
           89.112
So $-74.28 - 14.832 = -74.28 + (-14.832)$
$= -89.112$.

35.   23.9   one digit to the right of the decimal point
        $\times$ 0.2   one digit to the right of the decimal point
        ‾‾‾‾‾‾
         4.78   two digits to the right of the decimal point

36.   9.1   one digit to the right of the decimal point
        $\times$ 7.24   two digits to the right of the decimal point
        ‾‾‾‾‾‾
         364
         182
         637
        ‾‾‾‾‾‾
        65.884   three digits to the right of the decimal point

37. -3.45 two digits to the right of the decimal point
        $\times$ 0.03   two digits to the right of the decimal point
        ‾‾‾‾‾‾‾
        -0.1035 four digits to the right of the decimal point

38.   257   no digits to the right of the decimal point
        $\times -3.5$ one digit to the right of the decimal point
        ‾‾‾‾‾‾
        1285
         771
        ‾‾‾‾‾‾
        -899.5 one digit to the right of the decimal point

39. -0.03 two digits to the right of the decimal point
        $\times$ -0.45 two digits to the right of the decimal point
        ‾‾‾‾‾‾‾
         0.0135   four digits to the right of the decimal point

40.   9.9   one digit to the right of the decimal point
        $\times$ 0.002   three digits to the right of the decimal point
        ‾‾‾‾‾‾‾
         0.0198   four digits to the right of the decimal point

41.

$$73\overline{\smash{)}18.25}^{0.25}$$

$$\underline{146}$$
$$3\,65$$
$$\underline{3\,65}$$
$$0$$

So, $\dfrac{18.25}{73} = 0.25$.

42. $\dfrac{1.0032}{0.12} = \dfrac{1.0032}{0.12} \cdot \dfrac{100}{100} = \dfrac{100.32}{12}$

$$12\overline{\smash{)}100.32}^{8.36}$$
$$\underline{96}$$
$$4\,3$$
$$\underline{3\,6}$$
$$72$$
$$\underline{72}$$
$$0$$

So $\dfrac{1.0032}{0.12} = 8.36$.

43. $\dfrac{-4.2958}{45.7} = \dfrac{-4.2958}{45.7} \cdot \dfrac{10}{10} = \dfrac{-42.958}{457}$

$$457\overline{\smash{)}42.958}^{0.094}$$
$$\underline{41\,13}$$
$$1\,828$$
$$\underline{1\,828}$$
$$0$$

So $\dfrac{-4.2958}{45.7} = -0.094$.

44. $\dfrac{0.1515}{-5.05} = \dfrac{0.1515}{-5.05} \cdot \dfrac{100}{100} = \dfrac{15.15}{-505}$

$$505\overline{\smash{)}15.15}^{0.03}$$
$$\underline{15\,15}$$
$$0$$

So $\dfrac{0.1515}{-5.05} = -0.03$.

1.5 Exercises

45. $\dfrac{14}{21} = \dfrac{2 \cdot 7}{3 \cdot 7} = \dfrac{2 \cdot \cancel{7}}{3 \cdot \cancel{7}} = \dfrac{2}{3}$

47. $\dfrac{38}{-18} = \dfrac{2 \cdot 19}{2 \cdot (-9)} = \dfrac{\cancel{2} \cdot 19}{\cancel{2} \cdot (-9)} = \dfrac{19}{-9} = -\dfrac{19}{9}$

49. $-\dfrac{22}{44} = -\dfrac{1 \cdot 22}{2 \cdot 22} = -\dfrac{1 \cdot \cancel{22}}{2 \cdot \cancel{22}} = -\dfrac{1}{2}$

51. $\dfrac{32}{40} = \dfrac{4 \cdot 8}{5 \cdot 8} = \dfrac{4 \cdot \cancel{8}}{5 \cdot \cancel{8}} = \dfrac{4}{5}$

53. $\dfrac{6}{5} \cdot \dfrac{2}{5} = \dfrac{6 \cdot 2}{5 \cdot 5} = \dfrac{12}{25}$

55. $\dfrac{5}{-2} \cdot 10 = \dfrac{5}{-2} \cdot \dfrac{10}{1}$

$$= \dfrac{5 \cdot 10}{-2 \cdot 1}$$
$$= \dfrac{5 \cdot 2 \cdot 5}{-1 \cdot 2}$$
$$= \dfrac{5 \cdot \cancel{2} \cdot 5}{-1 \cdot \cancel{2}}$$
$$= \dfrac{5 \cdot 5}{-1}$$
$$= -25$$

57. $-\dfrac{3}{2} \cdot \dfrac{4}{9} = -\dfrac{3 \cdot 4}{2 \cdot 9}$

$$= -\dfrac{3 \cdot 2 \cdot 2}{2 \cdot 3 \cdot 3}$$
$$= -\dfrac{\cancel{3} \cdot \cancel{2} \cdot 2}{\cancel{2} \cdot \cancel{3} \cdot 3}$$
$$= -\dfrac{2}{3}$$

59. $-\dfrac{22}{3} \cdot \left(-\dfrac{12}{11}\right) = \dfrac{-22 \cdot (-12)}{3 \cdot 11}$

$$= \dfrac{-2 \cdot 11 \cdot 3 \cdot (-4)}{3 \cdot 11}$$
$$= \dfrac{-2 \cdot \cancel{11} \cdot \cancel{3} \cdot (-4)}{\cancel{3} \cdot \cancel{11}}$$
$$= \dfrac{-2 \cdot (-4)}{1}$$
$$= 8$$

61.
$$5 \cdot \frac{31}{15} = \frac{5}{1} \cdot \frac{31}{15}$$
$$= \frac{5 \cdot 31}{1 \cdot 15}$$
$$= \frac{5 \cdot 31}{1 \cdot 5 \cdot 3}$$
$$= \frac{\cancel{5} \cdot 31}{1 \cdot \cancel{5} \cdot 3}$$
$$= \frac{31}{1 \cdot 3}$$
$$= \frac{31}{3}$$

63.
$$\frac{3}{4} \cdot \frac{8}{11} = \frac{3 \cdot 8}{4 \cdot 11}$$
$$= \frac{3 \cdot 4 \cdot 2}{4 \cdot 11}$$
$$= \frac{3 \cdot \cancel{4} \cdot 2}{\cancel{4} \cdot 11}$$
$$= \frac{3 \cdot 2}{11}$$
$$= \frac{6}{11}$$

65. The reciprocal of $\frac{3}{5}$ is $\frac{5}{3}$.

67. The reciprocal of -5 or $-\frac{5}{1}$ is $-\frac{1}{5}$.

69.
$$\frac{4}{9} \div \frac{8}{15} = \frac{4}{9} \cdot \frac{15}{8}$$
$$= \frac{4 \cdot 15}{9 \cdot 8}$$
$$= \frac{4 \cdot 3 \cdot 5}{3 \cdot 3 \cdot 4 \cdot 2}$$
$$= \frac{\cancel{4} \cdot \cancel{3} \cdot 5}{\cancel{3} \cdot 3 \cdot \cancel{4} \cdot 2}$$
$$= \frac{5}{3 \cdot 2}$$
$$= \frac{5}{6}$$

71. $-\frac{1}{3} \div 3 = -\frac{1}{3} \div \frac{3}{1} = -\frac{1}{3} \cdot \frac{1}{3} = -\frac{1 \cdot 1}{3 \cdot 3} = -\frac{1}{9}$

73.
$$\frac{5}{6} \div \left(-\frac{5}{4}\right) = \frac{5}{6} \cdot \left(-\frac{4}{5}\right)$$
$$= \frac{5 \cdot (-4)}{6 \cdot 5}$$
$$= \frac{5 \cdot (-2) \cdot 2}{2 \cdot 3 \cdot 5}$$
$$= \frac{\cancel{5} \cdot (-2) \cdot \cancel{2}}{\cancel{2} \cdot 3 \cdot \cancel{5}}$$
$$= -\frac{2}{3}$$

75.
$$\frac{36}{28} \div \frac{22}{14} = \frac{36}{28} \cdot \frac{14}{22}$$
$$= \frac{36 \cdot 14}{28 \cdot 22}$$
$$= \frac{9 \cdot 4 \cdot 2 \cdot 7}{7 \cdot 4 \cdot 2 \cdot 11}$$
$$= \frac{9 \cdot \cancel{4} \cdot \cancel{2} \cdot \cancel{7}}{\cancel{7} \cdot \cancel{4} \cdot \cancel{2} \cdot 11}$$
$$= \frac{9}{11}$$

77.
$$-8 \div \frac{2}{3} = -\frac{8}{1} \cdot \frac{3}{2}$$
$$= \frac{-8 \cdot -3}{1 \cdot 2}$$
$$= \frac{-4 \cdot 2 \cdot 3}{1 \cdot 2}$$
$$= \frac{-4 \cdot \cancel{2} \cdot 3}{1 \cdot \cancel{2}}$$
$$= \frac{-4 \cdot 3}{1}$$
$$= -12$$

79. $-8 \div \left(-\frac{1}{4}\right) = -\frac{8}{1} \cdot \left(-\frac{4}{1}\right) = \frac{-8 \cdot (-4)}{1 \cdot 1} = 32$

81. $\frac{3}{4} + \frac{3}{4} = \frac{3+3}{4} = \frac{6}{4} = \frac{2 \cdot 3}{2 \cdot 2} = \frac{3}{2}$

83. $\frac{9}{8} - \frac{5}{8} = \frac{9-5}{8} = \frac{4}{8} = \frac{1 \cdot 4}{2 \cdot 4} = \frac{1}{2}$

85. $\frac{6}{7} - \left(-\frac{8}{7}\right) = \frac{6}{7} + \frac{8}{7} = \frac{6+8}{7} = \frac{14}{7} = \frac{2 \cdot 7}{7} = 2$

87. $-\frac{5}{3} + 2 = -\frac{5}{3} + \frac{2}{1} = -\frac{5}{3} + \frac{6}{3} = \frac{-5+6}{3} = \frac{1}{3}$

89. $6 - \dfrac{7}{2} = \dfrac{6}{1} - \dfrac{7}{2} = \dfrac{12}{2} - \dfrac{7}{2} = \dfrac{12-7}{2} = \dfrac{5}{2}$

91. $-\dfrac{4}{3} + \dfrac{1}{4} = -\dfrac{4}{3} \cdot \dfrac{4}{4} + \dfrac{1}{4} \cdot \dfrac{3}{3}$

$\qquad = -\dfrac{16}{12} + \dfrac{3}{12}$

$\qquad = \dfrac{-16+3}{12}$

$\qquad = -\dfrac{13}{12}$

93. $\dfrac{7}{5} + \left(-\dfrac{23}{20}\right) = \dfrac{7}{5} \cdot \dfrac{4}{4} + \left(-\dfrac{23}{20}\right)$

$\qquad = \dfrac{28}{20} + \left(-\dfrac{23}{20}\right)$

$\qquad = \dfrac{28 + (-23)}{20}$

$\qquad = \dfrac{5}{20}$

$\qquad = \dfrac{1 \cdot 5}{4 \cdot 5}$

$\qquad = \dfrac{1}{4}$

95. $\dfrac{7}{15} - \left(-\dfrac{4}{3}\right) = \dfrac{7}{15} + \dfrac{4}{3}$

$\qquad = \dfrac{7}{15} + \dfrac{4}{3} \cdot \dfrac{5}{5}$

$\qquad = \dfrac{7}{15} + \dfrac{20}{15}$

$\qquad = \dfrac{7+20}{15}$

$\qquad = \dfrac{27}{15}$

$\qquad = \dfrac{9 \cdot 3}{5 \cdot 3}$

$\qquad = \dfrac{9}{5}$

97. $15 = 3 \cdot 5$

$10 = 2 \cdot 5$

$\text{LCD} = 2 \cdot 3 \cdot 5 = 30$

$\dfrac{8}{15} - \dfrac{7}{10} = \dfrac{8}{15} \cdot \dfrac{2}{2} - \dfrac{7}{10} \cdot \dfrac{3}{3}$

$\qquad = \dfrac{16}{30} - \dfrac{21}{30}$

$\qquad = \dfrac{16-21}{30}$

$\qquad = \dfrac{-5}{30}$

$\qquad = \dfrac{-1 \cdot 5}{6 \cdot 5}$

$\qquad = -\dfrac{1}{6}$

99. $10 = 2 \cdot 5$

$8 = 2 \cdot 2 \cdot 2$

$\text{LCD} = 2 \cdot 2 \cdot 2 \cdot 5 = 40$

$-\dfrac{33}{10} - \left(-\dfrac{33}{8}\right) = -\dfrac{33}{10} + \dfrac{33}{8}$

$\qquad = -\dfrac{33}{10} \cdot \dfrac{4}{4} + \dfrac{33}{8} \cdot \dfrac{5}{5}$

$\qquad = -\dfrac{132}{40} + \dfrac{165}{40}$

$\qquad = \dfrac{-132 + 165}{40}$

$\qquad = \dfrac{33}{40}$

101. $12 = 2 \cdot 2 \cdot 3$

$18 = 2 \cdot 3 \cdot 3$

$\text{LCD} = 2 \cdot 2 \cdot 3 \cdot 3 = 36$

$\dfrac{19}{12} - \left(-\dfrac{41}{18}\right) = \dfrac{19}{12} + \dfrac{41}{18}$

$\qquad = \dfrac{19}{12} \cdot \dfrac{3}{3} + \dfrac{41}{18} \cdot \dfrac{2}{2}$

$\qquad = \dfrac{57}{36} + \dfrac{82}{36}$

$\qquad = \dfrac{57 + 82}{36}$

$\qquad = \dfrac{139}{36}$

103. $3 = 3$
$9 = 3 \cdot 3$
$6 = 2 \cdot 3$
$\text{LCD} = 2 \cdot 3 \cdot 3 = 18$

$$-\frac{2}{3} + \left(-\frac{5}{9}\right) + \frac{5}{6} = -\frac{2}{3} \cdot \frac{6}{6} + \left(-\frac{5}{9} \cdot \frac{2}{2}\right) + \frac{5}{6} \cdot \frac{3}{3}$$

$$= -\frac{12}{18} + \left(-\frac{10}{18}\right) + \frac{15}{18}$$

$$= \frac{-12 + (-10) + 15}{18}$$

$$= -\frac{7}{18}$$

105. $\quad 10.5$
$\underline{-\ 4.0}$
$\quad 6.5$

So, $-10.5 + 4 = -6.5$.

107. $\quad 3.5$
$\underline{+\ 4.9}$
$\quad 8.4$

So, $-(-3.5) + 4.9 = 3.5 + 4.9 = 8.4$.

109. $\quad 39.10$
$\underline{+\ 16.82}$
$\quad 55.92$

So, $39.1 - (-16.82) = 39.1 + 16.82 = 55.92$.

111. $\quad 6.70$
$\underline{-\ 5.21}$
$\quad 1.49$

So, $-5.21 - (-6.7) = -5.21 + 6.7 = 1.49$.

113. $\quad 45.00$
$\underline{-\ 2.45}$
$\quad 42.55$

So, $45 - 2.45 = 42.55$.

115. $\quad\ 4.3$
$\underline{\times\ 5.8}$
$\quad 344$
$\underline{215\quad}$
$\quad 24.94$

So, $4.3 \times 5.8 = 24.94$.

117. $\quad\quad 120$
$\underline{\times\ 0.075}$
$\quad\ 600$
$\underline{840\quad\ }$
$\quad 9.000$

So, $0.075 \times 120 = 9$.

119. $\dfrac{136.08}{5.6} = \dfrac{1360.8}{56} = 24.3$

$$
\begin{array}{r}
24.3 \\
56\overline{)1360.8} \\
\underline{112} \\
240 \\
\underline{224} \\
16\,8 \\
\underline{16\,8} \\
0
\end{array}
$$

121. $\dfrac{25.48}{0.052} = \dfrac{25,480}{52} = 490$

$$
\begin{array}{r}
490 \\
52\overline{)25480} \\
\underline{208} \\
468 \\
\underline{468} \\
0
\end{array}
$$

123. $-1.25 - (-0.6) + 1.6 = -1.25 + 0.6 + 1.6$
$\qquad\qquad\qquad\qquad = -0.65 + 1.6$
$\qquad\qquad\qquad\qquad = 0.95$

125. $6 = 2 \cdot 3$
$15 = 3 \cdot 5$
$\text{LCD} = 2 \cdot 3 \cdot 5 = 30$

$$-\frac{5}{6} + \frac{7}{15} = -\frac{5}{6} \cdot \frac{5}{5} + \frac{7}{15} \cdot \frac{2}{2}$$

$$= -\frac{25}{30} + \frac{14}{30}$$

$$= \frac{-25 + 14}{30}$$

$$= -\frac{11}{30}$$

127. $-\dfrac{10}{21} \cdot \dfrac{14}{5} = -\dfrac{10 \cdot 14}{21 \cdot 5}$

$$= -\frac{2 \cdot 5 \cdot 2 \cdot 7}{3 \cdot 7 \cdot 5}$$

$$= -\frac{2 \cdot \cancel{5} \cdot 2 \cdot \cancel{7}}{3 \cdot \cancel{7} \cdot \cancel{5}}$$

$$= -\frac{2 \cdot 2}{3}$$

$$= -\frac{4}{3}$$

129. $\dfrac{3}{8} \div \left(-\dfrac{9}{16}\right) = \dfrac{3}{8} \cdot \left(-\dfrac{16}{9}\right)$

$\qquad = \dfrac{3 \cdot (-16)}{8 \cdot 9}$

$\qquad = \dfrac{3 \cdot (-2) \cdot 8}{8 \cdot 3 \cdot 3}$

$\qquad = \dfrac{\cancel{3} \cdot (-2) \cdot \cancel{8}}{\cancel{8} \cdot \cancel{3} \cdot 3}$

$\qquad = -\dfrac{2}{3}$

131. $-\dfrac{5}{12} + \dfrac{2}{12} = \dfrac{-5+2}{12} = \dfrac{-3}{12} = \dfrac{-1 \cdot 3}{3 \cdot 4} = -\dfrac{1}{4}$

133. $-\dfrac{2}{7} - \dfrac{17}{5} = -\dfrac{2}{7} \cdot \dfrac{5}{5} - \dfrac{17}{5} \cdot \dfrac{7}{7}$

$\qquad = -\dfrac{10}{35} - \dfrac{119}{35}$

$\qquad = \dfrac{-10 - 119}{35}$

$\qquad = -\dfrac{129}{35}$

135.
$$
\begin{array}{r}
10.3 \\
-\ 8.7 \\
\hline
1.6
\end{array}
$$
So, $-8.7 - (-10.3) = -8.7 + 10.3 = 1.6$.

137. $12 = 2 \cdot 2 \cdot 3$
$28 = 2 \cdot 2 \cdot 7$
LCD $= 2 \cdot 2 \cdot 3 \cdot 7 = 84$

$\dfrac{1}{12} + \left(-\dfrac{5}{28}\right) = \dfrac{1}{12} \cdot \dfrac{7}{7} + \left(-\dfrac{5}{28}\right) \cdot \dfrac{3}{3}$

$\qquad = \dfrac{7}{84} + \left(-\dfrac{15}{84}\right)$

$\qquad = \dfrac{7 + (-15)}{84}$

$\qquad = \dfrac{-8}{84}$

$\qquad = \dfrac{-2 \cdot 4}{21 \cdot 4}$

$\qquad = -\dfrac{2}{21}$

139.
$$
\begin{array}{r}
-12.03 \\
\times \quad 4.2 \\
\hline
2406 \\
4812 \\
\hline
-50.526
\end{array}
$$
So, $-12.03 \times 4.2 = -50.526$.

141. $36 \cdot \left(-\dfrac{4}{9}\right) = \dfrac{36}{1} \cdot \left(-\dfrac{4}{9}\right)$

$\qquad = \dfrac{36 \cdot (-4)}{1 \cdot 9}$

$\qquad = \dfrac{4 \cdot 9 \cdot (-4)}{1 \cdot 9}$

$\qquad = \dfrac{4 \cdot \cancel{9} \cdot (-4)}{1 \cdot \cancel{9}}$

$\qquad = \dfrac{4 \cdot (-4)}{1}$

$\qquad = -16$

143. $-27 \div \dfrac{9}{5} = -\dfrac{27}{1} \div \dfrac{9}{5}$

$\qquad = -\dfrac{27}{1} \cdot \dfrac{5}{9}$

$\qquad = \dfrac{-27 \cdot 5}{1 \cdot 9}$

$\qquad = \dfrac{-3 \cdot 9 \cdot 5}{1 \cdot 9}$

$\qquad = \dfrac{-3 \cdot \cancel{9} \cdot 5}{1 \cdot \cancel{9}}$

$\qquad = \dfrac{-3 \cdot 5}{1}$

$\qquad = -15$

145.
$$
\begin{array}{r}
10.20 \\
-\ 3.62 \\
\hline
6.58
\end{array}
$$
So, $3.62 - 10.2 = -6.58$.

147. $\dfrac{-145.518}{18.42} = \dfrac{-14{,}551.8}{1842} = -7.9$

$$
\begin{array}{r}
7.9 \\
1842 \overline{)14551.8} \\
12894 \\
\hline
1657\ 8 \\
1657\ 8 \\
\hline
0
\end{array}
$$

149. $7 = 7$
$14 = 2 \cdot 7$
$21 = 3 \cdot 7$
LCD $= 2 \cdot 3 \cdot 7 = 42$

$$\frac{12}{7}-\frac{17}{14}-\frac{48}{21}=\frac{12}{7}\cdot\frac{6}{6}-\frac{17}{14}\cdot\frac{3}{3}-\frac{48}{21}\cdot\frac{2}{2}$$
$$=\frac{72}{42}-\frac{51}{42}-\frac{96}{42}$$
$$=\frac{72-51-96}{42}$$
$$=\frac{-75}{42}$$
$$=\frac{-3\cdot25}{3\cdot14}$$
$$=-\frac{25}{14}$$

151. $54.2-18.78-(-2.5)+20.47$
$$=54.2-18.78+2.5+20.47$$
$$=58.39$$

153. $400\times25.8\times0.003=10,320\times0.003$
$$=30.96$$

155. $12=2\cdot2\cdot3$
$6=2\cdot3$
$8=2\cdot2\cdot2$
$LCD=2\cdot2\cdot2\cdot3=24$
$$-\frac{11}{12}-\left(-\frac{1}{6}\right)+\frac{7}{8}=-\frac{11}{12}+\frac{1}{6}+\frac{7}{8}$$
$$=-\frac{11}{12}\cdot\frac{2}{2}+\frac{1}{6}\cdot\frac{4}{4}+\frac{7}{8}\cdot\frac{3}{3}$$
$$=-\frac{22}{24}+\frac{4}{24}+\frac{21}{24}$$
$$=\frac{-22+4+21}{24}$$
$$=\frac{3}{24}$$
$$=\frac{1\cdot3}{8\cdot3}$$
$$=\frac{1}{8}$$

157. $24\cdot7=168$ hours in one week
$$168\cdot\frac{1}{8}=\frac{168}{1}\cdot\frac{1}{8}$$
$$=\frac{168\cdot1}{1\cdot8}$$
$$=\frac{21\cdot8\cdot1}{1\cdot8}$$
$$=\frac{21\cdot\cancel{8}\cdot1}{1\cdot\cancel{8}}$$
$$=21$$
She spends 21 hours per week watching TV.

159. $36\cdot\frac{2}{3}\cdot\frac{3}{4}=\frac{36}{1}\cdot\frac{2}{3}\cdot\frac{3}{4}$
$$=\frac{36\cdot2\cdot3}{3\cdot4}$$
$$=\frac{9\cdot4\cdot2\cdot3}{3\cdot4}$$
$$=\frac{9\cdot\cancel{4}\cdot2\cdot\cancel{3}}{\cancel{3}\cdot\cancel{4}}$$
$$=\frac{9\cdot2}{1}$$
$$=18$$
This term, 18 students will pass the class.

161. $-43.29+(-25.50)=-68.79$
Maria has a balance of −$68.79.

163. $2.75+0.87+(-1.12)+0.52+(-0.62)$
$$=2.4$$
The net change was $2.40.

165. $d(P,Q)=|3.5-(-9.7)|$
$$=|3.5+9.7|$$
$$=|13.2|$$
$$=13.2$$

167. $d(P,Q)=\left|\frac{7}{5}-\left(-\frac{13}{3}\right)\right|$
$$=\left|\frac{7}{5}+\frac{13}{3}\right|$$
$$=\left|\frac{7}{5}\cdot\frac{3}{3}+\frac{13}{3}\cdot\frac{5}{5}\right|$$
$$=\left|\frac{21}{15}+\frac{65}{15}\right|$$
$$=\left|\frac{21+65}{15}\right|$$
$$=\left|\frac{86}{15}\right|$$
$$=\frac{86}{15}$$

Putting the Concepts Together (Sections 1.2–1.5)

1. $8=2\cdot2\cdot2$
$20=2\cdot2\cdot\quad5$
$LCD=2\cdot2\cdot2\cdot5=40$

$$\frac{7}{8} \cdot \frac{5}{5} = \frac{35}{40}$$

$$\frac{9}{20} \cdot \frac{2}{2} = \frac{18}{40}$$

2. $\dfrac{21}{63} = \dfrac{7 \cdot 3 \cdot 1}{7 \cdot 3 \cdot 3} = \dfrac{1}{3}$

3. $\dfrac{2}{7} = 7\overline{)2.000000}$ with quotient 0.285714

$$\begin{array}{r} 0.285714 \\ 7\overline{)2.000000} \\ 1\ 4 \\ \hline 60 \\ 56 \\ \hline 40 \\ 35 \\ \hline 50 \\ 49 \\ \hline 10 \\ 7 \\ \hline 30 \\ 28 \\ \hline \end{array}$$

$$\frac{2}{7} = 0.\overline{285714}$$

4. $0.375 = \dfrac{375}{1000} = \dfrac{3 \cdot 5 \cdot 5 \cdot 5}{2 \cdot 2 \cdot 2 \cdot 5 \cdot 5 \cdot 5} = \dfrac{3}{2 \cdot 2 \cdot 2} = \dfrac{3}{8}$

5. $12.3\% = 12.3\% \dfrac{1}{100\%} = \dfrac{12.3}{100} = 0.123$

6. $0.0625 = 0.0625 \cdot \dfrac{100\%}{1} = 6.25\%$

7. **(a)** -12, $-\dfrac{14}{7} = -2$, 0, and 3 are the integers.

   (b) -12, $-\dfrac{14}{7}$, -1.25, 0, 3, and 11.2 are the rational numbers.

   (c) $\sqrt{2}$ is the only irrational number.

   (d) All the numbers listed are real numbers.

8. $\dfrac{1}{8} < 0.5$ because $0.5 = \dfrac{1}{2} = \dfrac{4}{8}$ and $\dfrac{1}{8} < \dfrac{4}{8}$.

9. $17 + (-28) = -11$

10. $-23 + (-42) = -65$

11. $18 - 45 = 18 + (-45) = -27$

12. $3 - (-24) = 3 + 24 = 27$

13. $-18 - (-12.5) = -18 + 12.5 = -5.5$

14. $(-5)(2) = -10$

15. $25(-4) = -100$

16. $(-8)(-9) = 72$

17. $\dfrac{-35}{7} = \dfrac{-5 \cdot 7}{7} = \dfrac{-5 \cdot \cancel{7}}{\cancel{7}} = -5$

18. $\dfrac{-32}{-2} = \dfrac{-2 \cdot 16}{-2} = \dfrac{\cancel{-2} \cdot 16}{\cancel{-2}} = 16$

19. $27 \div -3 = \dfrac{27}{-3} = \dfrac{9 \cdot 3}{-1 \cdot 3} = \dfrac{9 \cdot \cancel{3}}{-1 \cdot \cancel{3}} = -9$

20. $-\dfrac{4}{5} - \dfrac{11}{5} = \dfrac{-4 - 11}{5}$

$$= \dfrac{-15}{5}$$

$$= \dfrac{-3 \cdot 5}{5}$$

$$= \dfrac{-3 \cdot \cancel{5}}{\cancel{5}}$$

$$= -3$$

21. $7 - \dfrac{4}{5} = \dfrac{7}{1} - \dfrac{4}{5}$

$$= \dfrac{7}{1} \cdot \dfrac{5}{5} - \dfrac{4}{5}$$

$$= \dfrac{35}{5} - \dfrac{4}{5}$$

$$= \dfrac{35 - 4}{5}$$

$$= \dfrac{31}{5}$$

22. $12 = 2 \cdot 2 \cdot 3$

$18 = 2 \cdot 3 \cdot 3$

$\text{LCD} = 2 \cdot 2 \cdot 3 \cdot 3 = 36$

$$\frac{7}{12}+\frac{5}{18}=\frac{7}{12}\cdot\frac{3}{3}+\frac{5}{18}\cdot\frac{2}{2}$$
$$=\frac{21}{36}+\frac{10}{36}$$
$$=\frac{21+10}{36}$$
$$=\frac{31}{36}$$

23. $12=2\cdot2\cdot3$
$18=2\cdot3\cdot3$
$LCD=2\cdot2\cdot3\cdot3=36$

$$-\frac{5}{12}-\frac{1}{18}=-\frac{5}{12}\cdot\frac{3}{3}-\frac{1}{18}\cdot\frac{2}{2}$$
$$=-\frac{15}{36}-\frac{2}{36}$$
$$=\frac{-15-2}{36}$$
$$=\frac{-17}{36}$$
$$=-\frac{17}{36}$$

24. $\frac{6}{25}\cdot15\cdot\frac{1}{2}=\frac{6}{25}\cdot\frac{15}{1}\cdot\frac{1}{2}$
$$=\frac{6\cdot15\cdot1}{25\cdot1\cdot2}$$
$$=\frac{2\cdot3\cdot3\cdot5\cdot1}{5\cdot5\cdot1\cdot2}$$
$$=\frac{\cancel{2}\cdot3\cdot3\cdot\cancel{5}}{\cancel{5}\cdot5\cdot\cancel{2}}$$
$$=\frac{3\cdot3}{5}$$
$$=\frac{9}{5}$$

25. $\frac{2}{7}\div(-8)=\frac{2}{7}\div\left(-\frac{8}{1}\right)$
$$=\frac{2}{7}\cdot\left(-\frac{1}{8}\right)$$
$$=\frac{2\cdot(-1)}{7\cdot8}$$
$$=\frac{2\cdot(-1)}{7\cdot2\cdot4}$$
$$=\frac{\cancel{2}\cdot(-1)}{7\cdot\cancel{2}\cdot4}$$
$$=\frac{-1}{7\cdot4}$$
$$=-\frac{1}{28}$$

26. $\frac{0}{-8}=0$

27. $\begin{array}{r}3.56\\+\ 7.20\\\hline 10.76\end{array}$

So $3.56-(-7.2)=3.56+7.2=10.76$.

28. $\begin{array}{r}18.946\\-\ 11.300\\\hline 7.646\end{array}$

So $18.946-11.3=7.646$.

29. $62.488\div42.8=\frac{62.488}{42.8}$
$$=\frac{62.488}{42.8}\cdot\frac{10}{10}$$
$$=\frac{624.88}{428}$$

$$\begin{array}{r}1.46\\428\overline{)624.88}\\\underline{428}\ \ \ \\196\ 8\ \\\underline{171\ 2}\ \\25\ 68\\\underline{25\ 68}\\0\end{array}$$

So $62.488\div42.8=1.46$.

30. $\begin{array}{r}7.94\\\times\ \ 2.8\\\hline 6\ 352\\15\ 88\ \ \\\hline 22.232\end{array}$

Section 1.6

Preparing for Properties of Real Numbers

P1. $12+3+(-12)=15+(-12)=3$

P2. $\frac{3}{4}\cdot11\cdot\frac{4}{3}=\frac{3}{4}\cdot\frac{11}{1}\cdot\frac{4}{3}$
$$=\frac{3\cdot11}{4\cdot1}\cdot\frac{4}{3}$$
$$=\frac{3\cdot11\cdot4}{4\cdot1\cdot3}$$
$$=\frac{\cancel{3}\cdot11\cdot\cancel{4}}{\cancel{4}\cdot1\cdot\cancel{3}}$$
$$=\frac{11}{1}$$
$$=11$$

1.6 Quick Checks

1. The product of any real number and the number <u>1</u> is that number.

2. $96 \text{ inches} = 96 \text{ inches} \cdot \dfrac{1 \text{ foot}}{12 \text{ inches}}$

   $= \dfrac{96}{12} \text{ feet}$

   $= \dfrac{2 \cdot 2 \cdot 2 \cdot 2 \cdot 2 \cdot 3}{2 \cdot 2 \cdot 3} \text{ feet}$

   $= \dfrac{\cancel{2} \cdot \cancel{2} \cdot 2 \cdot 2 \cdot 2 \cdot \cancel{3}}{\cancel{2} \cdot \cancel{2} \cdot \cancel{3}} \text{ feet}$

   $= 8 \text{ feet}$

3. $500 \text{ minutes} = 500 \text{ minutes} \cdot \dfrac{1 \text{ hour}}{60 \text{ minutes}}$

   $= \dfrac{500}{60} \text{ hours}$

   $= \dfrac{2 \cdot 2 \cdot 5 \cdot 5 \cdot 5}{2 \cdot 2 \cdot 3 \cdot 5} \text{ hours}$

   $= \dfrac{\cancel{2} \cdot \cancel{2} \cdot \cancel{5} \cdot 5 \cdot 5}{\cancel{2} \cdot \cancel{2} \cdot 3 \cdot \cancel{5}} \text{ hours}$

   $= \dfrac{25}{3} \text{ hours}$

   $= 8\dfrac{1}{3} \text{ hours or 8 hours, 20 minutes}$

4. $88 \text{ ounces} = 88 \text{ ounces} \cdot \dfrac{1 \text{ pound}}{16 \text{ ounces}}$

   $= \dfrac{88}{16} \text{ pounds}$

   $= \dfrac{2 \cdot 2 \cdot 2 \cdot 11}{2 \cdot 2 \cdot 2 \cdot 2} \text{ pounds}$

   $= \dfrac{\cancel{2} \cdot \cancel{2} \cdot \cancel{2} \cdot 11}{\cancel{2} \cdot \cancel{2} \cdot \cancel{2} \cdot 2} \text{ pounds}$

   $= \dfrac{11}{2} \text{ pounds}$

   $= 5\dfrac{1}{2} \text{ pounds or 5 pounds, 8 ounces}$

5. The Commutative Property of Addition states that for any real numbers a and b, $a + b = \underline{b + a}$.

6. The sum of any real number and its opposite is equal to <u>0</u>.

7. $(-8) + 22 + 8 = (-8) + 8 + 22 = 0 + 22 = 22$

8. $\dfrac{8}{15} + \dfrac{3}{20} + \left(-\dfrac{8}{15}\right) = \dfrac{8}{15} + \left(-\dfrac{8}{15}\right) + \dfrac{3}{20}$

   $= 0 + \dfrac{3}{20}$

   $= \dfrac{3}{20}$

9. $2.1 + 11.98 + (-2.1) = 2.1 + (-2.1) + 11.98$

   $= 0 + 11.98$

   $= 11.98$

10. $-8 \cdot (-13) \cdot \left(-\dfrac{3}{4}\right) = -8 \cdot \left(-\dfrac{3}{4}\right) \cdot (-13)$

    $= -\overset{2}{\cancel{8}} \cdot \left(-\dfrac{3}{\underset{1}{\cancel{4}}}\right) \cdot (-13)$

    $= -2 \cdot (-3) \cdot (-13)$

    $= 6 \cdot (-13)$

    $= -78$

11. $\dfrac{5}{22} \cdot \dfrac{18}{331} \cdot \left(-\dfrac{44}{5}\right) = \dfrac{5}{22} \cdot \left(-\dfrac{44}{5}\right) \cdot \dfrac{18}{331}$

    $= \dfrac{\overset{1}{\cancel{5}}}{\underset{1}{\cancel{22}}} \cdot \left(-\dfrac{\overset{2}{\cancel{44}}}{\underset{1}{\cancel{5}}}\right) \cdot \dfrac{18}{331}$

    $= \dfrac{1}{1} \cdot \left(-\dfrac{2}{1}\right) \cdot \dfrac{18}{331}$

    $= -\dfrac{2}{1} \cdot \dfrac{18}{331}$

    $= -\dfrac{36}{331}$

12. $100,000 \cdot 349 \cdot 0.00001$

    $= 100,000 \cdot 0.00001 \cdot 349$

    $= 1 \cdot 349$

    $= 349$

13. $14 + 101 + (-101) = 14 + (101 + (-101))$

    $= 14 + 0$

    $= 14$

14. $14 \cdot \dfrac{1}{5} \cdot 5 = 14 \cdot \left(\dfrac{1}{5} \cdot 5\right) = 14 \cdot 1 = 14$

15. $-34.2 + 12.6 + (-2.6)$

    $= -34.2 + (12.6 + (-2.6))$

    $= -34.2 + 10$

    $= -24.2$

16. $\dfrac{19}{2} \cdot \dfrac{4}{38} \cdot \dfrac{50}{13} = \left(\dfrac{19}{2} \cdot \dfrac{4}{38} \right) \cdot \dfrac{50}{13}$

$$= \left(\dfrac{\overset{1}{\cancel{19}}}{\underset{1}{\cancel{2}}} \cdot \dfrac{\overset{2}{\cancel{4}}}{\underset{2}{\cancel{38}}} \right) \cdot \dfrac{50}{13}$$

$$= \dfrac{2}{2} \cdot \dfrac{50}{13}$$

$$= 1 \cdot \dfrac{50}{13}$$

$$= \dfrac{50}{13}$$

17. $\dfrac{0}{22} = 0$ because 0 is the dividend.

18. $\dfrac{-11}{0}$ is undefined because 0 is the divisor.

19. $-\dfrac{0}{5} = 0$ because 0 is the dividend.

20. $\dfrac{5678}{0}$ is undefined because 0 is the divisor.

1.6 Exercises

21. $13 \text{ feet} = 13 \text{ feet} \cdot \dfrac{12 \text{ inches}}{1 \text{ foot}}$

$\qquad = 13 \cdot 12 \text{ inches}$

$\qquad = 156 \text{ inches}$

23. 4500 centimeters

$= 4500 \text{ centimeters} \cdot \dfrac{1 \text{ meter}}{100 \text{ centimeters}}$

$= \dfrac{4500}{100} \text{ meters}$

$= 45 \text{ meters}$

25. $42 \text{ quarts} = 42 \text{ quarts} \cdot \dfrac{1 \text{ gallon}}{4 \text{ quarts}}$

$\qquad = \dfrac{42}{4} \text{ gallons}$

$\qquad = 10\dfrac{1}{2} \text{ gallons}$

$\qquad = 10 \text{ gallons, 2 quarts}$

27. $180 \text{ ounces} = 180 \text{ ounces} \cdot \dfrac{1 \text{ pound}}{16 \text{ ounces}}$

$\qquad = \dfrac{180}{16} \text{ pounds}$

$\qquad = 11\dfrac{1}{4} \text{ pounds}$

$\qquad = 11 \text{ pounds, 4 ounces}$

29. 16,200 seconds

$= 16,200 \text{ seconds} \cdot \dfrac{1 \text{ hour}}{3600 \text{ seconds}}$

$= \dfrac{16,200}{3600} \text{ hours}$

$= 4\dfrac{1}{2} \text{ hours}$

$= 4 \text{ hours, 30 minutes}$

31. $16 + (-16) = 0$ illustrates the Additive Inverse Property.

33. $\dfrac{3}{4} \cdot \dfrac{5}{5} = \dfrac{3}{4} \cdot 1 = \dfrac{3}{4}$ illustrates the Multiplicative Identity Property.

35. $12 \cdot \dfrac{1}{12} = 1$ illustrates the Multiplicative Inverse Property.

37. $34.2 + (-34.2) = 0$ illustrates the Additive Inverse Property.

39. $\dfrac{0}{a} = 0$

41. $\dfrac{2}{3} \cdot \left(-\dfrac{12}{43} \right) \cdot \dfrac{3}{2} = \dfrac{2}{3} \cdot \dfrac{3}{2} \cdot \left(-\dfrac{12}{43} \right)$ illustrates the Commutative Property of Multiplication since the order in which the numbers are multiplied changes.

43. $5.23 + 4.98 + (-5.23)$
$= 5.23 + (-5.23) + 4.98$ illustrates the Commutative Property of Addition since the order in which the numbers are added changes.

45. $\dfrac{a}{0}$ is undefined.

47. $54 + 29 + (-54) = 54 + (-54) + 29$
$\qquad\qquad\qquad\qquad\quad = 0 + 29$
$\qquad\qquad\qquad\qquad\quad = 29$

49. $\dfrac{9}{5} \cdot \dfrac{5}{9} \cdot 18 = 1 \cdot 18 = 18$

51. $-25 \cdot 13 \cdot \dfrac{1}{5} = -25 \cdot \dfrac{1}{5} \cdot 13 = -5 \cdot 13 = -65$

53. $\begin{aligned} 347 + 456 + (-456) &= 347 + (456 + (-456)) \\ &= 347 + 0 \\ &= 347 \end{aligned}$

55. $\dfrac{9}{2} \cdot \left(-\dfrac{10}{3}\right) \cdot 6 = \dfrac{\overset{3}{\cancel{9}}}{\cancel{2}} \cdot \left(-\dfrac{\overset{5}{\cancel{10}}}{\cancel{3}}\right) \cdot 6 = 3 \cdot (-5) \cdot 6 = -15 \cdot 6 = -90$

57. $\dfrac{7}{0}$ is undefined because 0 is the divisor.

59. $100(-34)(0.01) = 100(0.01)(-34) = 1(-34) = -34$

61. $569.003 \cdot 0 = 0$

63. $\dfrac{45}{3902} + \left(-\dfrac{45}{3902}\right) = 0$

65. $\begin{aligned} -\dfrac{5}{44} \cdot \dfrac{80}{3} \cdot \dfrac{11}{5} &= -\dfrac{5}{44} \cdot \dfrac{11}{5} \cdot \dfrac{80}{3} \\[2mm] &= -\dfrac{\overset{1}{\cancel{5}}}{\underset{4}{\cancel{44}}} \cdot \dfrac{\overset{1}{\cancel{11}}}{\underset{1}{\cancel{5}}} \cdot \dfrac{80}{3} \\[2mm] &= -\dfrac{1}{4} \cdot \dfrac{80}{3} \\[2mm] &= -\dfrac{1}{\underset{1}{\cancel{4}}} \cdot \dfrac{\overset{20}{\cancel{80}}}{3} \\[2mm] &= -\dfrac{20}{3} \end{aligned}$

67. $\begin{aligned} 321.03 + (-32.84) + (-85.03) + (-120.56) + 120.56 &= 321.03 + (-32.84) + (-85.03) + [(-120.56) + 120.56] \\ &= 321.03 + (-32.84) + (-85.03) + 0 \\ &= 288.19 + (-85.03) \\ &= 203.16 \end{aligned}$

Alberto's balance is $203.16.

69. $\begin{aligned} -3 - (4 - 10) &= -3 - [4 + (-10)] \\ &= -3 - (-6) \\ &= -3 + 6 \\ &= 3 \end{aligned}$

71.
$$\begin{aligned}
-15+10-(4-8) &= -15+10-[4+(-8)] \\
&= -15+10-(-4) \\
&= -15+10+4 \\
&= -5+4 \\
&= -1
\end{aligned}$$

73.
$$\begin{aligned}
&\frac{30 \text{ miles}}{1 \text{ hour}} \\
&= \frac{30 \text{ miles}}{1 \text{ hour}} \cdot \frac{1 \text{ hour}}{3600 \text{ seconds}} \cdot \frac{5280 \text{ feet}}{1 \text{ mile}} \\
&= \frac{30 \cdot 5280 \text{ feet}}{3600 \text{ seconds}} \\
&= 44 \text{ feet per second}
\end{aligned}$$

75. Zero does not have a multiplicative inverse because two numbers are multiplicative inverses if their product is one. There is no real number such that the product of that number and 0 is equal to 1.

77. The quotient $\dfrac{0}{4} = 0$ because $4 \cdot 0 = 0$. The quotient $\dfrac{4}{0}$ is undefined because we should be able to determine a real number $\square$ such that $0 \cdot \square = 4$. But because the product of 0 and every real number is 0, there is no replacement value for $\square$.

79. The product of a nonzero real number and its multiplicative inverse (reciprocal) equals 1, the multiplicative identity.

Section 1.7

Preparing for Exponents and the Order of Operations

P1. $9 + (-19) = -10$

P2. $28 - (-7) = 28 + 7 = 35$

P3.
$$\begin{aligned}
-7 \cdot \frac{8}{3} \cdot 36 &= -7 \cdot \left(\frac{8}{3} \cdot \frac{36}{1} \right) \\
&= -7 \cdot \left(\frac{8 \cdot 36}{3 \cdot 1} \right) \\
&= -7 \cdot \left(\frac{8 \cdot 3 \cdot 12}{3 \cdot 1} \right) \\
&= -7 \cdot \left(\frac{8 \cdot \cancel{3} \cdot 12}{\cancel{3} \cdot 1} \right) \\
&= -7 \cdot \left(\frac{96}{1} \right) \\
&= -672
\end{aligned}$$

P4. $\dfrac{100}{-15} = \dfrac{5 \cdot 20}{-3 \cdot 5} = \dfrac{\cancel{5} \cdot 20}{-3 \cdot \cancel{5}} = \dfrac{20}{-3} = -\dfrac{20}{3}$

1.7 Quick Checks

1. The expression $11 \cdot 11 \cdot 11 \cdot 11 \cdot 11$ contains five factors of 11, so $11 \cdot 11 \cdot 11 \cdot 11 \cdot 11 = 11^5$.

2. The expression $(-7)(-7)(-7)(-7)$ contains four factors of -7, so $(-7)^4$.

3. The expression $(-2) \cdot (-2) \cdot (-2)$ contains three factors of -2, so $(-2) \cdot (-2) \cdot (-2) = (-2)^3$.

4. $2^4 = 2 \cdot 2 \cdot 2 \cdot 2 = 16$

5. $(-7)^2 = (-7) \cdot (-7) = 49$

6. $\left(-\dfrac{1}{6}\right)^3 = \left(-\dfrac{1}{6}\right)\left(-\dfrac{1}{6}\right)\left(-\dfrac{1}{6}\right) = -\dfrac{1}{216}$

7. $(0.9)^2 = (0.9) \cdot (0.9) = 0.81$

8. $-2^4 = -(2 \cdot 2 \cdot 2 \cdot 2) = -16$

9. $(-2)^4 = (-2) \cdot (-2) \cdot (-2) \cdot (-2) = 16$

10. $1 + 7 \cdot 2 = 1 + 14 = 15$

11. $-11 \cdot 3 + 2 = -33 + 2 = -31$

12.
$$\begin{aligned}
18 + 3 \div \left(-\frac{1}{2}\right) &= 18 + 3 \cdot \left(-\frac{2}{1}\right) \\
&= 18 + (-6) \\
&= 12
\end{aligned}$$

13. $9 \cdot 4 - 5 = 36 - 5 = 31$

14.
$$\frac{15}{2} \div (-5) - \frac{3}{2} = \frac{15}{2} \cdot \left(-\frac{1}{5}\right) - \frac{3}{2}$$
$$= -\frac{3}{2} - \frac{3}{2}$$
$$= -\frac{6}{2}$$
$$= -3$$

15. $8(2+3) = 8(5) = 40$

16. $(2-9) \cdot (5+4) = (-7) \cdot (9) = -63$

17. $\left(\frac{6}{7} + \frac{8}{7}\right) \cdot \left(\frac{11}{8} + \frac{5}{8}\right) = \left(\frac{14}{7}\right) \cdot \left(\frac{16}{8}\right) = 2 \cdot 2 = 4$

18. $\dfrac{2+5 \cdot 6}{-3 \cdot 8 - 4} = \dfrac{2+30}{-24-4} = \dfrac{32}{-28} = \dfrac{8 \cdot \cancel{4}}{-7 \cdot \cancel{4}} = -\dfrac{8}{7}$

19.
$$\frac{(12+14) \cdot 2}{13 \cdot 2 + 13 \cdot 5} = \frac{26 \cdot 2}{13 \cdot 2 + 13 \cdot 5}$$
$$= \frac{52}{26+65}$$
$$= \frac{52}{91}$$
$$= \frac{4 \cdot \cancel{13}}{7 \cdot \cancel{13}}$$
$$= \frac{4}{7}$$

20.
$$\frac{4+3 \div \frac{1}{7}}{2 \cdot 9 - 3} = \frac{4+3 \cdot 7}{2 \cdot 9 - 3}$$
$$= \frac{4+21}{18-3}$$
$$= \frac{25}{15}$$
$$= \frac{5 \cdot \cancel{5}}{3 \cdot \cancel{5}}$$
$$= \frac{5}{3}$$

21.
$$4 \cdot [2 \cdot (3+7) - 15] = 4 \cdot [2 \cdot 10 - 15]$$
$$= 4 \cdot [20 - 15]$$
$$= 4 \cdot [5]$$
$$= 20$$

22.
$$2 \cdot \{4 \cdot [26 - (9+7)] - 15\} - 10$$
$$= 2 \cdot \{4[26-16] - 15\} - 10$$
$$= 2 \cdot \{4[10] - 15\} - 10$$
$$= 2 \cdot \{40 - 15\} - 10$$
$$= 2 \cdot \{25\} - 10$$
$$= 50 - 10$$
$$= 40$$

23. $\dfrac{7-5^2}{2} = \dfrac{7-25}{2} = \dfrac{-18}{2} = \dfrac{-9 \cdot \cancel{2}}{\cancel{2}} = -9$

24. $3(7-3)^2 = 3(4)^2 = 3 \cdot 16 = 48$

25.
$$\frac{(-3)^2 + 7(1-3)}{3 \cdot 2 + 5} = \frac{9 + 7(1-3)}{6+5}$$
$$= \frac{9 + 7(-2)}{11}$$
$$= \frac{9 + (-14)}{11}$$
$$= \frac{-5}{11}$$
$$= -\frac{5}{11}$$

26.
$$2 + 5 \cdot 3^2 - \frac{3}{2} \cdot 2^2 = 2 + 5 \cdot 9 - \frac{3}{2} \cdot 4$$
$$= 2 + 45 - 6$$
$$= 47 - 6$$
$$= 41$$

27. $\dfrac{(4-10)^2}{2^3 - 5} = \dfrac{(-6)^2}{2^3 - 5} = \dfrac{36}{8-5} = \dfrac{36}{3} = \dfrac{12 \cdot \cancel{3}}{\cancel{3}} = 12$

28.
$$-3[(-4)^2 - 5(8-6)]^2 = -3[(-4)^2 - 5(2)]^2$$
$$= -3[16 - 5(2)]^2$$
$$= -3[16 - 10]^2$$
$$= -3[6]^2$$
$$= -3[36]$$
$$= -108$$

29. $\dfrac{(2.9+7.1)^2}{5^2 - 15} = \dfrac{(10)^2}{25-15} = \dfrac{100}{10} = \dfrac{10 \cdot \cancel{10}}{\cancel{10}} = 10$

30. $\left(\dfrac{4^2-4(-3)(1)}{7\cdot 2}\right)^2 = \left(\dfrac{16-4(-3)}{14}\right)^2$

$= \left(\dfrac{16+12}{14}\right)^2$

$= \left(\dfrac{28}{14}\right)^2$

$= (2)^2$

$= 2^2$

$= 4$

1.7 Exercises

31. The expression $5 \cdot 5$ contains two factors of 5, so $5 \cdot 5 = 5^2$.

33. The expression $\dfrac{3}{5} \cdot \dfrac{3}{5} \cdot \dfrac{3}{5}$ contains three factors of $\dfrac{3}{5}$, so $\dfrac{3}{5} \cdot \dfrac{3}{5} \cdot \dfrac{3}{5} = \left(\dfrac{3}{5}\right)^3$.

35. $8^2 = 8 \cdot 8 = 64$

37. $(-8)^2 = (-8)(-8) = 64$

39. $10^3 = 10 \cdot 10 \cdot 10 = 1000$

41. $\left(\dfrac{3}{4}\right)^3 = \left(\dfrac{3}{4}\right)\left(\dfrac{3}{4}\right)\left(\dfrac{3}{4}\right) = \dfrac{27}{64}$

43. $(1.5)^2 = (1.5)(1.5) = 2.25$

45. $-3^2 = -(3 \cdot 3) = -9$

47. $-1^{20} = -\underbrace{(1 \cdot 1 \cdot 1 \cdot \cdots \cdot 1)}_{20 \text{ factors}} = -1$

49. $0^4 = 0 \cdot 0 \cdot 0 \cdot 0 = 0$

51. $\left(-\dfrac{1}{2}\right)^6$

$= \left(-\dfrac{1}{2}\right)\left(-\dfrac{1}{2}\right)\left(-\dfrac{1}{2}\right)\left(-\dfrac{1}{2}\right)\left(-\dfrac{1}{2}\right)\left(-\dfrac{1}{2}\right)$

$= \dfrac{1}{64}$

53. $\left(-\dfrac{1}{3}\right)^3 = \left(-\dfrac{1}{3}\right)\left(-\dfrac{1}{3}\right)\left(-\dfrac{1}{3}\right) = -\dfrac{1}{27}$

55. $2 + 3 \cdot 4 = 2 + 12 = 14$

57. $-5 \cdot 3 + 12 = -15 + 12 = -3$

59. $100 \div 2 \cdot 50 = 50 \cdot 50 = 2500$

61. $156 - 3 \cdot 2 + 10 = 156 - 6 + 10$

$= 150 + 10$

$= 160$

63. $(2+3) \cdot 4 = 5 \cdot 4 = 20$

65. $8 \div 4 \cdot 2 = 2 \cdot 2 = 4$

67. $\dfrac{4+2}{2+8} = \dfrac{6}{10} = \dfrac{2 \cdot 3}{2 \cdot 5} = \dfrac{3}{5}$

69. $\dfrac{14-6}{6-14} = \dfrac{8}{-8} = \dfrac{1 \cdot 8}{-1 \cdot 8} = -1$

71. $13 - [3+(-8)4] = 13 - [3+(-32)]$

$= 13 - [-29]$

$= 13 + 29$

$= 42$

73. $(-8.75 - 1.25) \div (-2) = -10 \div (-2) = 5$

75. $4 - 2^3 = 4 - 8 = -4$

77. $15 + 4 \cdot 5^2 = 15 + 4 \cdot 25 = 15 + 100 = 115$

79. $-2^3 + 3^2 \div (2^2 - 1) = -8 + 9 \div (4-1)$

$= -8 + 9 \div 3$

$= -8 + 3$

$= -5$

81. $\left(\dfrac{4^2-3}{12-2\cdot 5}\right)^2 = \left(\dfrac{16-3}{12-10}\right)^2 = \left(\dfrac{13}{2}\right)^2 = \dfrac{169}{4}$

83. $-2 \cdot [5 \cdot (9-3) - 3 \cdot 6] = -2 \cdot [5 \cdot 6 - 3 \cdot 6]$

$= -2 \cdot [30 - 18]$

$= -2 \cdot 12$

$= -24$

85. $\left(\dfrac{4}{3}+\dfrac{5}{6}\right)\left(\dfrac{2}{5}-\dfrac{9}{10}\right)=\left(\dfrac{8}{6}+\dfrac{5}{6}\right)\left(\dfrac{4}{10}-\dfrac{9}{10}\right)$

$=\left(\dfrac{13}{6}\right)\left(\dfrac{-5}{10}\right)$

$=\left(\dfrac{13}{6}\right)\left(\dfrac{-1}{2}\right)$

$=\dfrac{13\cdot(-1)}{6\cdot 2}$

$=-\dfrac{13}{12}$

87. $-2.5+4.5\div 1.5=-2.5+3=0.5$

89. $4+2\cdot(6-2)=4+2\cdot 4=4+8=12$

91. $\dfrac{12-16\div 4+(-24)}{16\cdot 2-4\cdot 0}=\dfrac{12-4+(-24)}{32-0}$

$=\dfrac{8+(-24)}{32}$

$=\dfrac{-16}{32}$

$=\dfrac{-1\cdot 16}{2\cdot 16}$

$=-\dfrac{1}{2}$

93. $\left(\dfrac{2-(-4)^3}{5^2-7\cdot 2}\right)^2=\left(\dfrac{2-(-64)}{25-14}\right)^2$

$=\left(\dfrac{2+64}{11}\right)^2$

$=\left(\dfrac{66}{11}\right)^2$

$=6^2$

$=36$

95. $\dfrac{5^2-10}{3^2+6}=\dfrac{25-10}{9+6}=\dfrac{15}{15}=1$

97. $\left|6\cdot(5-3^2)\right|=\left|6\cdot(5-9)\right|$

$=\left|6\cdot(-4)\right|$

$=\left|-24\right|$

$=24$

99. $\dfrac{81}{8}+\dfrac{13}{4}\div\dfrac{1}{2}=\dfrac{81}{8}+\dfrac{13}{4}\cdot\dfrac{2}{1}$

$=\dfrac{81}{8}+\dfrac{13}{2}$

$=\dfrac{81}{8}+\dfrac{52}{8}$

$=\dfrac{133}{8}$

101. $\dfrac{-7}{20}+\dfrac{3}{8}\div\dfrac{1}{2}=\dfrac{-7}{20}+\dfrac{3}{8}\cdot\dfrac{2}{1}$

$=\dfrac{-7}{20}+\dfrac{6}{8}$

$=\dfrac{-7}{20}+\dfrac{3}{4}$

$=\dfrac{-7}{20}+\dfrac{15}{20}$

$=\dfrac{8}{20}$

$=\dfrac{2\cdot 4}{5\cdot 4}$

$=\dfrac{2}{5}$

103. $\dfrac{21-3^2}{1+3}=\dfrac{21-9}{1+3}=\dfrac{12}{4}=\dfrac{3\cdot 4}{1\cdot 4}=3$

105. $\dfrac{3}{4}\cdot\left[\dfrac{5}{4}\div\left(\dfrac{3}{8}-\dfrac{1}{8}\right)-3\right]=\dfrac{3}{4}\cdot\left[\dfrac{5}{4}\div\dfrac{2}{8}-3\right]$

$=\dfrac{3}{4}\cdot\left[\dfrac{5}{4}\div\dfrac{1}{4}-3\right]$

$=\dfrac{3}{4}\cdot\left[\dfrac{5}{4}\cdot\dfrac{4}{1}-3\right]$

$=\dfrac{3}{4}\cdot[5-3]$

$=\dfrac{3}{4}\cdot 2$

$=\dfrac{6}{4}$

$=\dfrac{3\cdot 2}{2\cdot 2}$

$=\dfrac{3}{2}$

107. $\left(\dfrac{4}{3}\right)^3 - \left(\dfrac{1}{2}\right)^2 \cdot \left(\dfrac{8}{3}\right) + 2 \div 3$

$= \dfrac{64}{27} - \dfrac{1}{4} \cdot \left(\dfrac{8}{3}\right) + 2 \div 3$

$= \dfrac{64}{27} - \dfrac{8}{12} + 2 \div 3$

$= \dfrac{64}{27} - \dfrac{2}{3} + \dfrac{2}{3}$

$= \dfrac{64}{27} + \left(-\dfrac{2}{3} + \dfrac{2}{3}\right)$

$= \dfrac{64}{27} + 0$

$= \dfrac{64}{27}$

109. $\dfrac{5^2 - 3^3}{\left|4 - 4^2\right|} = \dfrac{25 - 27}{\left|4 - 16\right|}$

$= \dfrac{-2}{\left|-12\right|}$

$= \dfrac{-2}{12}$

$= \dfrac{-1 \cdot 2}{6 \cdot 2}$

$= -\dfrac{1}{6}$

111. $72 = 8 \cdot 9 = (2 \cdot 2 \cdot 2) \cdot (3 \cdot 3) = 2^3 \cdot 3^2$

113. $48 = 16 \cdot 3 = (2 \cdot 2 \cdot 2 \cdot 2) \cdot 3 = 2^4 \cdot 3$

115. $(4 \cdot 3 + 6) \cdot 2 = (12 + 6) \cdot 2 = 18 \cdot 2 = 36$

117. $(4 + 3) \cdot (4 + 2) = 7 \cdot 6 = 42$

119. $(6 - 4) + (3 - 1) = 2 + 2 = 4$

121. $479 + 0.075(479) = 479 + 35.925 = 514.925$
The total amount is $514.93.

123. $2 \cdot 3.1416 \cdot 6^2 + 2 \cdot 3.1416 \cdot 6 \cdot 10$
$= 2 \cdot 3.1416 \cdot 36 + 2 \cdot 3.1416 \cdot 6 \cdot 10$
$= 226.1952 + 376.992$
$= 603.1872$
The surface area is about 603.19 square inches.

125. $1000(1 + 0.03)^2 = 1000(1.03)^2$
$= 1000(1.0609)$
$= 1060.9$
The amount of money is $1060.90.

127. $\angle XYZ = \angle XYQ + \angle QYZ$
$= 46.5° + 69.25°$
$= 115.75°$

129. The sum $2x^2 + 4x^2$ is not equal to $6x^4$ because when we combine like terms, we add the coefficients of the like terms and keep the variables and exponents the same. Put another way, $2x^2 + 4x^2 = (2 + 4)x^2 = 6x^2$.

Section 1.8

Preparing for Simplifying Algebraic Expressions

P1. $-3 + 8 = 5$

P2. $-7 - 8 = -7 + (-8) = -15$

P3. $-\dfrac{4}{3}(27) = -\dfrac{4}{3}\left(\dfrac{27}{1}\right)$

$= -\dfrac{4 \cdot 27}{3 \cdot 1}$

$= -\dfrac{4 \cdot 3 \cdot 9}{3 \cdot 1}$

$= -\dfrac{4 \cdot \cancel{3} \cdot 9}{\cancel{3} \cdot 1}$

$= -\dfrac{4 \cdot 9}{1}$

$= -36$

1.8 Quick Checks

1. To evaluate an algebraic expression means to substitute the numerical value for each variable into the expression and simplify.

2. Substitute 4 for k.
$-3k + 5 = -3(4) + 5 = -12 + 5 = -7$

3. Substitute 12 for t.
$\dfrac{5}{4}t - 6 = \dfrac{5}{4}(12) - 6 = \dfrac{60}{4} - 6 = 15 - 6 = 9$

4. Substitute -2 for y.

$$-2y^2 - y + 8 = -2(-2)^2 - (-2) + 8$$
$$= -2(4) - (-2) + 8$$
$$= -8 - (-2) + 8$$
$$= -8 + 2 + 8$$
$$= -8 + 8 + 2$$
$$= 0 + 2$$
$$= 2$$

5. Substitute 8 for x and 16 for y.

$$7.50x + 10y = 7.50(8) + 10(16)$$
$$= 60 + 160$$
$$= 220$$

The value is $220.

6. The algebraic expression $5x^2 + 3xy$ has two terms: $5x^2$ and $3xy$.

7. The algebraic expression $9ab - 3bc + 5ac - ac^2$ has four terms: $9ab$, $-3bc$, $5ac$, and $-ac^2$.

8. The algebraic expression $\dfrac{2mn}{5} - \dfrac{3n}{7}$ has two terms: $\dfrac{2mn}{5}$ and $-\dfrac{3n}{7}$.

9. The algebraic expression $\dfrac{m^2}{3} - 8$ has two terms:

$\dfrac{m^2}{3}$ and -8.

10. The coefficient of $2z^2$ is 2.

11. The coefficient of $xy = 1 \cdot xy$ is 1.

12. The coefficient of $-b = -1 \cdot b$ is -1.

13. The coefficient of 5 is 5.

14. The coefficient of $-\dfrac{2}{3}z$ is $-\dfrac{2}{3}$.

15. The coefficient of $\dfrac{x}{6} = \dfrac{1}{6} \cdot x$ is $\dfrac{1}{6}$.

16. $-\dfrac{2}{3}p^2$ and $\dfrac{4}{5}p^2$ are like terms. They have the same variable raised to the same power.

17. $\dfrac{m}{6} = \dfrac{1}{6}m$ and $4m$ are like terms. They have the same variable raised to the same power.

18. $3a^2b$ and $-2ab^2$ are unlike terms. The variable a is raised to the second power in $3a^2b$ and to the first power in $-2ab^2$.

19. $8a$ and 11 are unlike terms. $8a$ has a variable and 11 does not.

20. $6(x + 2) = 6 \cdot x + 6 \cdot 2 = 6x + 12$

21. $-5(x + 2) = -5 \cdot x + (-5) \cdot 2$
$$= -5x + (-10)$$
$$= -5x - 10$$

22. $-2(k - 7) = -2 \cdot k - (-2) \cdot 7$
$$= -2k - (-14)$$
$$= -2k + 14$$

23. $(8x + 12)\dfrac{3}{4} = 8x \cdot \dfrac{3}{4} + 12 \cdot \dfrac{3}{4}$

$$= \overset{2}{\cancel{8}}x \cdot \dfrac{3}{\underset{1}{\cancel{4}}} + \overset{3}{\cancel{12}} \cdot \dfrac{3}{\underset{1}{\cancel{4}}}$$

$$= 2x \cdot 3 + 3 \cdot 3$$
$$= 6x + 9$$

24. $3x - 8x = (3 - 8)x = -5x$

25. $-5x^2 + x^2 = -5x^2 + 1x^2 = (-5 + 1)x^2 = -4x^2$

26. $-7x - x + 6 - 3 = -7x - 1x + 6 - 3$
$$= (-7 - 1)x + (6 - 3)$$
$$= -8x + 3$$

27. $4x - 12x - 3 + 17 = (4 - 12)x + (-3 + 17)$
$$= -8x + 14$$

28. $3a + 2b - 5a + 7b - 4 = 3a - 5a + 2b + 7b - 4$
$$= (3 - 5)a + (2 + 7)b - 4$$
$$= -2a + 9b - 4$$

29. $(5ac + 2b) + (7ac - 5a) + (-b)$
$$= 5ac + 2b + 7ac - 5a + (-1)b$$
$$= 5ac + 7ac + 2b + (-1)b - 5a$$
$$= (5 + 7)ac + (2 + (-1))b - 5a$$
$$= 12ac + 1b - 5a$$
$$= 12ac + b - 5a$$

30. $5ab^2 + 7a^2b + 3ab^2 - 8a^2b$
$= 5ab^2 + 3ab^2 + 7a^2b - 8a^2b$
$= (5+3)ab^2 + (7-8)a^2b$
$= 8ab^2 - 1a^2b$
$= 8ab^2 - a^2b$

31. $\dfrac{4}{3}rs - \dfrac{3}{2}r^2 + \dfrac{2}{3}rs - 5$
$= \dfrac{4}{3}rs + \dfrac{2}{3}rs - \dfrac{3}{2}r^2 - 5$
$= \left(\dfrac{4}{3} + \dfrac{2}{3}\right)rs - \dfrac{3}{2}r^2 - 5$
$= \dfrac{6}{3}rs - \dfrac{3}{2}r^2 - 5$
$= 2rs - \dfrac{3}{2}r^2 - 5$

32. To simplify an algebraic expression means to remove all parentheses and combine like terms.

33. $3x + 2(x-1) - 7x + 1 = 3x + 2x - 2 - 7x + 1$
$= 3x + 2x - 7x - 2 + 1$
$= -2x - 1$

34. $m + 2n - 3(m+2n) - (7-3n)$
$= m + 2n - 3m - 6n - 7 + 3n$
$= m - 3m + 2n - 6n + 3n - 7$
$= -2m - 1n - 7$
$= -2m - n - 7$

35. $2(a-4b) - (a+4b) + b$
$= 2a - 8b - a - 4b + b$
$= 2a - a - 8b - 4b + b$
$= 2a - 1a - 8b - 4b + 1b$
$= (2-1)a + (-8-4+1)b$
$= 1a + (-11)b$
$= a - 11b$

36. $\dfrac{1}{2}(6x+4) - \dfrac{1}{3}(12-9x) = 3x + 2 - 4 + 3x$
$= 3x + 3x + 2 - 4$
$= 6x - 2$

1.8 Exercises

37. Substitute 4 for x.
$2x + 5 = 2(4) + 5 = 8 + 5 = 13$

39. Substitute 3 for x.
$x^2 + 3x - 1 = 3^2 + 3(3) - 1$
$= 9 + 3(3) - 1$
$= 9 + 9 - 1$
$= 18 - 1$
$= 17$

41. Substitute -5 for k.
$4 - k^2 = 4 - (-5)^2$
$= 4 - 25$
$= -21$

43. Substitute 8 for x and 10 for y.
$\dfrac{5x}{y} + y^2 = \dfrac{5(8)}{10} + 10^2$
$= \dfrac{5(8)}{10} + 100$
$= \dfrac{40}{10} + 100$
$= 4 + 100$
$= 104$

45. Substitute 3 for x and 2 for y.
$\dfrac{9x - 5y}{x + y} = \dfrac{9(3) - 5(2)}{3 + 2} = \dfrac{27 - 10}{5} = \dfrac{17}{5}$

47. Substitute 3 for x and 4 for y.
$(x+3y)^2 = (3 + 3 \cdot 4)^2$
$= (3 + 12)^2$
$= 15^2$
$= 225$

49. Substitute 3 for x and 4 for y.
$x^2 + 9y^2 = 3^2 + 9(4^2)$
$= 9 + 9(16)$
$= 9 + 144$
$= 153$

51. $2x^3 + 3x^2 - x + 6$ can be written as $2x^3 + 3x^2 + (-1 \cdot x) + 6$. The terms are $2x^3$, $3x^2$, $-x$, and 6. The coefficient of $2x^3$ is 2. The coefficient of $3x^2$ is 3. The coefficient of $-x$ is -1. The coefficient of 6 is 6.

53. $z^2 + \dfrac{2y}{3}$ can be written as $1 \cdot z^2 + \dfrac{2}{3}y$. The terms are z^2 and $\dfrac{2y}{3}$. The coefficient of z^2 is 1. The coefficient of $\dfrac{2y}{3}$ is $\dfrac{2}{3}$.

55. $8x$ and 8 are unlike terms. $8x$ has a variable and 8 does not.

57. 54 and -21 are like terms. They are both constants.

59. $12b$ and $-b$ are like terms. They have the same variable raised to the same power.

61. $r^2 s$ and rs^2 are unlike terms. The variable r is raised to the second power in $r^2 s$ and the first power in rs^2.

63. $3(m + 2) = 3 \cdot m + 3 \cdot 2 = 3m + 6$

65. $(3n^2 + 2n - 1)6 = 3n^2 \cdot 6 + 2n \cdot 6 - 1 \cdot 6$
$$= 18n^2 + 12n - 6$$

67. $-(x - y) = -1 \cdot (x - y)$
$$= -1 \cdot x - (-1) \cdot y$$
$$= -x - (-y)$$
$$= -x + y$$

69. $(8x - 6y)(-0.5) = 8x \cdot (-0.5) - 6y(-0.5)$
$$= -4x - (-3y)$$
$$= -4x + 3y$$

71. $5x - 2x = (5 - 2)x = 3x$

73. $4z - 6z + 8z = (4 - 6 + 8)z = 6z$

75. $2m + 3n + 8m + 7n = 2m + 8m + 3n + 7n$
$$= (2 + 8)m + (3 + 7)n$$
$$= 10m + 10n$$

77. $0.3x^7 + x^7 + 0.9x^7 = (0.3 + 1 + 0.9)x^7$
$$= 2.2x^7$$

79. $-3y^6 + 13y^6 = (-3 + 13)y^6 = 10y^6$

81. $-(6w - 12y - 13z)$
$$= -1 \cdot 6w - (-1) \cdot 12y - (-1) \cdot 13z$$
$$= -6w - (-12y) - (-13z)$$
$$= -6w + 12y + 13z$$

83. $5(k + 3) - 8k = 5k + 15 - 8k$
$$= 5k - 8k + 15$$
$$= -3k + 15$$

85. $7n - (3n + 8) = 7n - 3n - 8 = 4n - 8$

87. $(7 - 2x) - (x + 4) = 7 - 2x - x - 4$
$$= 7 - 3x - 4$$
$$= -3x + 3$$

89. $(7n - 8) - (3n - 6) = 7n - 8 - 3n + 6$
$$= 4n - 8 + 6$$
$$= 4n - 2$$

91. $-6(n - 3) + 2(n + 1) = -6n + 18 + 2n + 2$
$$= -6n + 2n + 18 + 2$$
$$= -4n + 20$$

93. $\dfrac{2}{3}x + \dfrac{1}{6}x = \dfrac{4}{6}x + \dfrac{1}{6}x = \dfrac{5}{6}x$

95. $\dfrac{1}{2}(8x + 5) - \dfrac{2}{3}(6x + 12) = 4x + \dfrac{5}{2} - 4x - 8$
$$= 4x - 4x + \dfrac{5}{2} - 8$$
$$= \dfrac{5}{2} - \dfrac{16}{2}$$
$$= -\dfrac{11}{2}$$

97. $2(0.5x + 9) - 3(1.5x + 8) = x + 18 - 4.5x - 24$
$$= x - 4.5x + 18 - 24$$
$$= -3.5x - 6$$

99. $3.2(x + 1.6) + 1.4(2x - 3.7)$
$$= 3.2x + 5.12 + 2.8x - 5.18$$
$$= 3.2x + 2.8x + 5.12 - 5.18$$
$$= 6x - 0.06$$

101. **(a)** $5x + 3x = 5(4) + 3(4) = 20 + 12 = 32$

   (b) $5x + 3x = 8x = 8(4) = 32$

103. (a)
$$-2a^2 + 5a^2 = -2(-3)^2 + 5(-3)^2$$
$$= -2(9) + 5(9)$$
$$= -18 + 45$$
$$= 27$$

(b) $-2a^2 + 5a^2 = 3a^2 = 3(-3)^2 = 3(9) = 27$

105. (a)
$$4z - 3(z + 2) = 4(6) - 3(6 + 2)$$
$$= 4(6) - 3(8)$$
$$= 24 - 24$$
$$= 0$$

(b)
$$4z - 3(z + 2) = 4z - 3z - 6$$
$$= z - 6$$
$$= 6 - 6$$
$$= 0$$

107. (a)
$$5y^2 + 6y - 2y^2 + 5y - 3$$
$$= 5(-2)^2 + 6(-2) - 2(-2)^2 + 5(-2) - 3$$
$$= 5(4) + 6(-2) - 2(4) + 5(-2) - 3$$
$$= 20 - 12 - 8 - 10 - 3$$
$$= -13$$

(b)
$$5y^2 + 6y - 2y^2 + 5y - 3$$
$$= 5y^2 - 2y^2 + 6y + 5y - 3$$
$$= 3y^2 + 11y - 3$$
$$= 3(-2)^2 + 11(-2) - 3$$
$$= 3(4) + 11(-2) - 3$$
$$= 12 - 22 - 3$$
$$= -13$$

109. (a)
$$\frac{1}{2}(4x - 2) - \frac{2}{3}(3x + 9)$$
$$= \frac{1}{2}(4 \cdot 3 - 2) - \frac{2}{3}(3 \cdot 3 + 9)$$
$$= \frac{1}{2}(12 - 2) - \frac{2}{3}(9 + 9)$$
$$= \frac{1}{2}(10) - \frac{2}{3}(18)$$
$$= 5 - 12$$
$$= -7$$

(b)
$$\frac{1}{2}(4x - 2) - \frac{2}{3}(3x + 9) = 2x - 1 - 2x - 6$$
$$= 2x - 2x - 1 - 6$$
$$= -7$$

111. (a)
$$3a + 4b - 7a + 3(a - 2b)$$
$$= 3(2) + 4(5) - 7(2) + 3(2 - 2 \cdot 5)$$
$$= 3(2) + 4(5) - 7(2) + 3(2 - 10)$$
$$= 3(2) + 4(5) - 7(2) + 3(-8)$$
$$= 6 + 20 - 14 - 24$$
$$= -12$$

(b)
$$3a + 4b - 7a + 3(a - 2b)$$
$$= 3a + 4b - 7a + 3a - 6b$$
$$= 3a - 7a + 3a + 4b - 6b$$
$$= -a - 2b$$
$$= -2 - 2(5)$$
$$= -2 - 10$$
$$= -12$$

113. Let $h = 4$, $b = 5$, $B = 17$.
$$\frac{1}{2}h(b + B) = \frac{1}{2}(4)(5 + 17)$$
$$= \frac{1}{2}(4)(22)$$
$$= 2(22)$$
$$= 44$$

115. Let $a = 6$, $b = 3$, $c = -4$, $d = -2$.
$$\frac{a - b}{c - d} = \frac{6 - 3}{-4 - (-2)} = \frac{6 - 3}{-4 + 2} = \frac{3}{-2} = -\frac{3}{2}$$

117. Let $a = 7$, $b = 8$, $c = 1$.
$$b^2 - 4ac = 8^2 - 4(7)(1) = 64 - 28 = 36$$

119. Let $m = 125$.
$$59.95 + 0.15m = 59.95 + 0.15(125)$$
$$= 59.95 + 18.75$$
$$= 78.70$$
The cost of renting the truck is $78.70.

121. Let $a = 156$ and $c = 421$.
$$12a + 7c = 12(156) + 7(421)$$
$$= 1872 + 2947$$
$$= 4819$$
The revenue for 156 adult tickets and 421 children's tickets is $4819.

123. (a)
$$2w + 2(3w - 4) = 2w + 6w - 8$$
$$= 8w - 8$$

(b) Let $w = 5$.
$$8w - 8 = 8(5) - 8 = 40 - 8 = 32$$
The perimeter is 32 yards.

125. Let $s = 2950$ and $b = 2050$.

$0.055s + 0.0325b$

$= 0.055(2950) + 0.0325(2050)$

$= 162.25 + 66.625$

$= 228.875$

The annual interest is about \$228.88.

127. Answers may vary. One possibility:

$2.75(-3x^2 + 7x - 3) - 1.75(-3x^2 + 7x - 3)$

$= (2.75 - 1.75)(-3x^2 + 7x - 3)$

$= (1)(-3x^2 + 7x - 3)$

$= -3x^2 + 7x - 3$

129. The sum $2x^2 + 4x^2$ is not equal to $6x^4$ because when we combine like terms, we add the coefficients of the like terms and keep the variables and exponents the same. Put another way, $2x^2 + 4x^2 = (2 + 4)x^2 = 6x^2$.

Chapter 1 Review

1.
```
      24
     /  \
    4    6
   /\   / \
  2·2·2·3
```
$24 = 2 \cdot 2 \cdot 2 \cdot 3$

2.
```
  87
  / \
 3·29
```
$87 = 3 \cdot 29$

3.
```
     81
    /  \
   9    9
  / \  / \
 3·3·3·3
```
$81 = 3 \cdot 3 \cdot 3 \cdot 3$

4.
```
   124
   / \
  4   31
 /\    \
2·2·31
```
$124 = 2 \cdot 2 \cdot 31$

5. 17 is prime.

6.
$18 = 3 \cdot 3 \cdot 2$
$24 = 3 \quad | \quad \cdot 2 \cdot 2 \cdot 2$
$3 \cdot 3 \cdot 2 \cdot 2 \cdot 2 = 72$

7.
$4 = 2 \cdot 2$
$8 = 2 \cdot 2 \cdot 2$
$18 = 2 \cdot \quad | \quad | \quad 3 \cdot 3$
$2 \cdot 2 \cdot 2 \cdot 3 \cdot 3 = 72$

8. $\dfrac{7}{15} = \dfrac{7}{15} \cdot \dfrac{2}{2} = \dfrac{14}{30}$

9. $3 = \dfrac{3}{1} = \dfrac{3}{1} \cdot \dfrac{4}{4} = \dfrac{12}{4}$

10.
$6 = 2 \cdot 3$
$8 = 2 \quad | \quad \cdot 2 \cdot 2$
$\text{LCD} = 2 \cdot 3 \cdot 2 \cdot 2 = 24$
$\dfrac{1}{6} = \dfrac{1}{6} \cdot \dfrac{4}{4} = \dfrac{4}{24}$
$\dfrac{3}{8} = \dfrac{3}{8} \cdot \dfrac{3}{3} = \dfrac{9}{24}$

11.
$16 = 2 \cdot 2 \cdot 2 \quad \cdot 2$
$24 = 2 \cdot 2 \cdot 2 \cdot 3 \quad |$
$\text{LCD} = 2 \cdot 2 \cdot 2 \cdot 3 \cdot 2 = 48$
$\dfrac{9}{16} = \dfrac{9}{16} \cdot \dfrac{3}{3} = \dfrac{27}{48}$
$\dfrac{7}{24} = \dfrac{7}{24} \cdot \dfrac{2}{2} = \dfrac{14}{48}$

12. $\dfrac{25}{60} = \dfrac{5 \cdot 5}{5 \cdot 3 \cdot 2 \cdot 2} = \dfrac{5}{3 \cdot 2 \cdot 2} = \dfrac{5}{12}$

13. $\dfrac{125}{250} = \dfrac{5 \cdot 5 \cdot 5}{5 \cdot 5 \cdot 5 \cdot 2} = \dfrac{1}{2}$

14. $\dfrac{96}{120} = \dfrac{3 \cdot 2 \cdot 2 \cdot 2 \cdot 2 \cdot 2}{3 \cdot 2 \cdot 2 \cdot 2 \cdot 5} = \dfrac{2 \cdot 2}{5} = \dfrac{4}{5}$

15. 21.76

16. 15

17. $\dfrac{8}{9} = 9\overline{)\begin{array}{l} 0.88 \\ 8.00 \end{array}}$

$\phantom{\dfrac{8}{9} = 9)}\;\underline{7\,2}$

$\phantom{\dfrac{8}{9} = 9)8.00\,}\;80$

$\dfrac{8}{9} = 0.88... = 0.\overline{8}$

18. $\dfrac{9}{32} = 32\overline{)\begin{array}{l} 0.28125 \\ 9.00000 \end{array}}$

$\phantom{\dfrac{9}{32} = 32)}\;\underline{64}$

$\phantom{\dfrac{9}{32} = 32)}\;2\,60$

$\phantom{\dfrac{9}{32} = 32)}\;\underline{2\,56}$

$\phantom{\dfrac{9}{32} = 32)9}\;40$

$\phantom{\dfrac{9}{32} = 32)9}\;\underline{32}$

$\phantom{\dfrac{9}{32} = 32)99}\;80$

$\phantom{\dfrac{9}{32} = 32)99}\;\underline{64}$

$\phantom{\dfrac{9}{32} = 32)99}\;160$

$\phantom{\dfrac{9}{32} = 32)99}\;\underline{160}$

$\phantom{\dfrac{9}{32} = 32)999}\;0$

$\dfrac{9}{32} = 0.28125$

19. $6\overline{)\begin{array}{l} 1.833 \\ 11.000 \end{array}}$

$\;\underline{6}$

$\;5\,0$

$\;\underline{4\,8}$

$\;20$

$\;\underline{18}$

$\;20$

$\dfrac{11}{6}$ rounded to the nearest hundredth is 1.83.

20. $8\overline{)\begin{array}{l} 2.37 \\ 19.00 \end{array}}$

$\;\underline{16}$

$\;3\,0$

$\;\underline{2\,4}$

$\;60$

$\;\underline{56}$

$\;40$

$\dfrac{19}{8}$ rounded to the nearest tenth is 2.4.

21. $0.6 = \dfrac{6}{10} = \dfrac{3 \cdot 2}{5 \cdot 2} = \dfrac{3}{5}$

22. $0.375 = \dfrac{375}{1000} = \dfrac{5 \cdot 5 \cdot 5 \cdot 3}{5 \cdot 5 \cdot 5 \cdot 2 \cdot 2 \cdot 2} = \dfrac{3}{2 \cdot 2 \cdot 2} = \dfrac{3}{8}$

23. $0.864 = \dfrac{864}{1000}$

$= \dfrac{3 \cdot 3 \cdot 3 \cdot 2 \cdot 2 \cdot 2 \cdot 2 \cdot 2}{5 \cdot 5 \cdot 5 \cdot 2 \cdot 2 \cdot 2}$

$= \dfrac{3 \cdot 3 \cdot 3 \cdot 2 \cdot 2}{5 \cdot 5 \cdot 5}$

$= \dfrac{108}{125}$

24. $41\% = 41\% \cdot \dfrac{1}{100\%} = \dfrac{41}{100} = 0.41$

25. $760\% = 760\% \cdot \dfrac{1}{100\%} = \dfrac{760}{100} = 7.60$

26. $9.03\% = 9.03\% \cdot \dfrac{1}{100\%} = \dfrac{9.03}{100} = 0.0903$

27. $0.35\% = 0.35\% \cdot \dfrac{1}{100\%} = \dfrac{0.35}{100} = 0.0035$

28. $0.23 = 0.23 \cdot \dfrac{100\%}{1} = 23\%$

29. $1.17 = 1.17 \cdot \dfrac{100\%}{1} = 117\%$

30. $0.045 = 0.045 \cdot \dfrac{100\%}{1} = 4.5\%$

31. $3 = 3 \cdot \dfrac{100\%}{1} = 300\%$

32. (a) $\dfrac{12}{20} = \dfrac{3 \cdot 2 \cdot 2}{5 \cdot 2 \cdot 2} = \dfrac{3}{5}$

The student earned $\dfrac{3}{5}$ of the point.

(b) $\dfrac{3}{5} = 0.6$

$0.6 \cdot \dfrac{100\%}{1} = 60\%$

The student earned 60% of the points.

33. The set of whole numbers less than 7 is $A = \{0, 1, 2, 3, 4, 5, 6\}$.

34. The set of natural numbers less than or equal to 3 is $B = \{1, 2, 3\}$.

35. The set of integers greater than −3 and less than or equal to 5 is
$C = \{-2, -1, 0, 1, 2, 3, 4, 5\}$.

36. The set of integers greater than or equal to −2 and less than 4 is
$D = \{-2, -1, 0, 1, 2, 3\}$.

37. $\frac{9}{3} = 3$ and 11 are the natural numbers.

38. $0, \frac{9}{3} = 3$, and 11 are the whole numbers.

39. $-6, 0, \frac{9}{3} = 3$, and 11 are the integers.

40. $-6, -3.25, 0, \frac{9}{3}, 11,$ and $\frac{5}{7}$ are the rational numbers.

41. 5.030030003... is the only irrational number.

42. All the numbers listed are real numbers.

43.

44.

45. Since −3 lies to the left of −1 on the real number line, $-3 < -1$. The statement is false.

46. Since $5 = 5$, the statement $5 \le 5$ is true.

47. Since −5 lies to the left of −3 on the real number line, $-5 \le -3$ is a true statement.

48. Since $\frac{1}{2} = 0.5$, the statement is true.

49. $-\left|\frac{1}{2}\right| = -\frac{1}{2}$

50. $|-7| = 7$

51. $-|-6| = -6$

52. $-|-8.2| = -8.2$

53. $\frac{1}{4} = 0.25$

54. Since −6 lies to the left of 0 on the real number line, $-6 < 0$.

55. Since $\frac{3}{4} = 0.75$ and $0.83 > 0.75$, then $0.83 > \frac{3}{4}$.

56. Since −2 lies to the right of −10 on the real number line, $-2 > -10$.

57. $|-4| = 4$
$|-3| = 3$
Since 4 lies to the right of 3 on the real number line $4 > 3$ and $|-4| > |-3|$.

58. $\frac{4}{5} = \frac{4 \cdot 6}{5 \cdot 6} = \frac{24}{30}$
$\left|-\frac{5}{6}\right| = \frac{5}{6} = \frac{5 \cdot 5}{6 \cdot 5} = \frac{25}{30}$
Since $\frac{24}{30} < \frac{25}{30}$, then $\frac{4}{5} < \left|-\frac{5}{6}\right|$.

59. A rational number is any number that may be written as the quotient of two integers where the denominator does not equal zero. Both terminating decimals and repeating decimals are rational numbers. An irrational number is a non-repeating, non-terminating decimal.

60. The set of positive integers is called the natural numbers.

61. $-2 + 9 = 7$

62. $6 + (-10) = -4$

63. $-23 + (-11) = -34$

64. $-120 + 25 = -95$

65. $-|-2 + 6| = -|4| = -4$

66. $-|-15| + |-62| = -15 + 62 = 47$

67. $-110 + 50 + (-18) + 25 = -60 + (-18) + 25$
$\qquad\qquad\qquad\qquad = -78 + 25$
$\qquad\qquad\qquad\qquad = -53$

68. $-28 + (-35) + (-52) = -63 + (-52) = -115$

69. $-10 - 12 = -10 + (-12) = -22$

70. $18 - 25 = 18 + (-25) = -7$

71. $-11 - (-32) = -11 + 32 = 21$

72. $0 - (-67) = 0 + 67 = 67$

73. $34 - 18 + 10 = 34 + (-18) + 10 = 16 + 10 = 26$

74. $\begin{aligned} -49 - 8 + 21 &= -49 + (-8) + 21 \\ &= -57 + 21 \\ &= -36 \end{aligned}$

75. $-6(-2) = 12$

76. $4(-10) = -40$

77. $13(-86) = -1118$

78. $-19 \times 423 = -8037$

79. $(11)(13)(-5) = 143(-5) = -715$

80. $(-53)(-21)(-10) = 1113(-10) = -11,130$

81. $\dfrac{-20}{-4} = \dfrac{-4 \cdot 5}{-4 \cdot 1} = \dfrac{\cancel{-4} \cdot 5}{\cancel{-4} \cdot 1} = \dfrac{5}{1} = 5$

82. $\dfrac{60}{-5} = \dfrac{5 \cdot 12}{5 \cdot (-1)} = \dfrac{\cancel{5} \cdot 12}{\cancel{5} \cdot (-1)} = \dfrac{12}{-1} = -12$

83. $\dfrac{|-55|}{11} = \dfrac{55}{11} = \dfrac{11 \cdot 5}{11 \cdot 1} = \dfrac{\cancel{11} \cdot 5}{\cancel{11} \cdot 1} = \dfrac{5}{1} = 5$

84. $\begin{aligned} -\left|\dfrac{-100}{4}\right| &= -\left|\dfrac{4 \cdot (-25)}{4}\right| \\ &= -\left|\dfrac{\cancel{4} \cdot (-25)}{\cancel{4}}\right| \\ &= -|-25| \\ &= -25 \end{aligned}$

85. $\dfrac{120}{-15} = \dfrac{15 \cdot 8}{15 \cdot (-1)} = \dfrac{\cancel{15} \cdot 8}{\cancel{15} \cdot (-1)} = \dfrac{8}{-1} = -8$

86. $\dfrac{64}{-20} = \dfrac{4 \cdot 16}{4 \cdot (-5)} = \dfrac{\cancel{4} \cdot 16}{\cancel{4} \cdot (-5)} = -\dfrac{16}{5}$

87. $\dfrac{-180}{54} = \dfrac{-10 \cdot 18}{3 \cdot 18} = \dfrac{-10 \cdot \cancel{18}}{3 \cdot \cancel{18}} = -\dfrac{10}{3}$

88. $\dfrac{-450}{105} = \dfrac{-30 \cdot 15}{7 \cdot 15} = \dfrac{-30 \cdot \cancel{15}}{7 \cdot \cancel{15}} = -\dfrac{30}{7}$

89. The additive inverse of 13 is -13 since $13 + (-13) = 0$.

90. The additive inverse of -45 is 45 since $-45 + 45 = 0$.

91. -43 plus 101 is written as $-43 + 101 = 58$.

92. 45 plus -28 is written as $45 + (-28) = 17$.

93. -10 minus -116 is written as $-10 - (-116) = -10 + 116 = 106$.

94. 74 minus 56 is written as $74 - 56 = 74 + (-56) = 18$.

95. The sum of 13 and -8 is written as $13 + (-8) = 5$.

96. The difference between -60 and -10 is written as $-60 - (-10) = -60 + 10 = -50$.

97. -21 multiplied by -3 is written as $-21 \cdot (-3) = 63$.

98. 54 multiplied by -18 is written as $54 \cdot (-18) = -972$.

99. -34 divided by -2 is written as $-34 \div (-2) = \dfrac{-34}{-2} = \dfrac{-2 \cdot 17}{-2 \cdot 1} = \dfrac{17}{1} = 17$.

100. -49 divided by 14 is written as $-49 \div 14 = \dfrac{-49}{14} = \dfrac{-7 \cdot 7}{2 \cdot 7} = -\dfrac{7}{2}$.

101. $20 + (-6) + 12 = 14 + 12 = 26$
His total yardage was a gain of 26 yards.

102. $10 + 12 + (-25) = 22 + (-25) = -3$
The temperature at midnight was $-3°F$.

103. $6 - (-18) = 6 + 18 = 24$
The difference in temperature was $24°F$.

104. $11 \cdot 5 = 55$
$8 \cdot 4 = 32$
$55 + 32 = 87$
Sarah's test score was 87 points.

105. $\dfrac{32}{64} = \dfrac{1 \cdot 32}{2 \cdot 32} = \dfrac{1 \cdot \cancel{32}}{2 \cdot \cancel{32}} = \dfrac{1}{2}$

106. $-\dfrac{27}{81} = -\dfrac{1 \cdot 27}{3 \cdot 27} = -\dfrac{1 \cdot \cancel{27}}{3 \cdot \cancel{27}} = -\dfrac{1}{3}$

107. $\dfrac{-100}{150} = \dfrac{-2 \cdot 50}{3 \cdot 50} = \dfrac{-2 \cdot \cancel{50}}{3 \cdot \cancel{50}} = \dfrac{-2}{3} = -\dfrac{2}{3}$

108. $\dfrac{35}{-25} = \dfrac{5 \cdot 7}{5 \cdot (-5)} = \dfrac{\cancel{5} \cdot 7}{\cancel{5} \cdot (-5)} = \dfrac{7}{-5} = -\dfrac{7}{5}$

109. $\dfrac{2}{3} \cdot \dfrac{15}{8} = \dfrac{2 \cdot 15}{3 \cdot 8} = \dfrac{2 \cdot 3 \cdot 5}{3 \cdot 2 \cdot 4} = \dfrac{\cancel{2} \cdot \cancel{3} \cdot 5}{\cancel{3} \cdot \cancel{2} \cdot 4} = \dfrac{5}{4}$

110. $-\dfrac{3}{8} \cdot \dfrac{10}{21} = -\dfrac{3 \cdot 10}{8 \cdot 21}$

$= -\dfrac{3 \cdot 2 \cdot 5}{2 \cdot 4 \cdot 3 \cdot 7}$

$= -\dfrac{\cancel{3} \cdot \cancel{2} \cdot 5}{\cancel{2} \cdot 4 \cdot \cancel{3} \cdot 7}$

$= -\dfrac{5}{4 \cdot 7}$

$= -\dfrac{5}{28}$

111. $\dfrac{5}{8} \cdot \left(-\dfrac{2}{25}\right) = \dfrac{5 \cdot (-2)}{8 \cdot 25}$

$= \dfrac{5 \cdot 2 \cdot (-1)}{2 \cdot 4 \cdot 5 \cdot 5}$

$= \dfrac{\cancel{5} \cdot \cancel{2} \cdot (-1)}{\cancel{2} \cdot 4 \cdot \cancel{5} \cdot 5}$

$= \dfrac{-1}{4 \cdot 5}$

$= -\dfrac{1}{20}$

112. $5 \cdot \left(-\dfrac{3}{10}\right) = \dfrac{5}{1} \cdot \left(-\dfrac{3}{10}\right)$

$= \dfrac{5 \cdot (-3)}{1 \cdot 10}$

$= \dfrac{5 \cdot (-3)}{1 \cdot 2 \cdot 5}$

$= \dfrac{\cancel{5} \cdot (-3)}{1 \cdot 2 \cdot \cancel{5}}$

$= -\dfrac{3}{2}$

113. $\dfrac{24}{17} \div \dfrac{18}{3} = \dfrac{24}{17} \cdot \dfrac{3}{18}$

$= \dfrac{24 \cdot 3}{17 \cdot 18}$

$= \dfrac{6 \cdot 4 \cdot 3}{17 \cdot 6 \cdot 3}$

$= \dfrac{\cancel{6} \cdot 4 \cdot \cancel{3}}{17 \cdot \cancel{6} \cdot \cancel{3}}$

$= \dfrac{4}{17}$

114. $-\dfrac{5}{12} \div \dfrac{10}{16} = -\dfrac{5}{12} \cdot \dfrac{16}{10}$

$= -\dfrac{5 \cdot 16}{12 \cdot 10}$

$= -\dfrac{5 \cdot 4 \cdot 2 \cdot 2}{4 \cdot 3 \cdot 5 \cdot 2}$

$= -\dfrac{\cancel{5} \cdot \cancel{4} \cdot 2 \cdot \cancel{2}}{\cancel{4} \cdot 3 \cdot \cancel{5} \cdot \cancel{2}}$

$= -\dfrac{2}{3}$

115. $-\dfrac{27}{10} \div 9 = -\dfrac{27}{10} \div \dfrac{9}{1}$

$= -\dfrac{27}{10} \cdot \dfrac{1}{9}$

$= -\dfrac{27 \cdot 1}{10 \cdot 9}$

$= -\dfrac{3 \cdot 9 \cdot 1}{10 \cdot 9}$

$= -\dfrac{3 \cdot \cancel{9} \cdot 1}{10 \cdot \cancel{9}}$

$= -\dfrac{3}{10}$

116. $20 \div \left(-\dfrac{5}{8}\right) = \dfrac{20}{1} \div \left(-\dfrac{5}{8}\right)$

$= \dfrac{20}{1} \cdot \left(-\dfrac{8}{5}\right)$

$= \dfrac{20 \cdot (-8)}{1 \cdot 5}$

$= \dfrac{4 \cdot 5 \cdot (-8)}{1 \cdot 5}$

$= \dfrac{4 \cdot \cancel{5} \cdot (-8)}{1 \cdot \cancel{5}}$

$= \dfrac{4 \cdot (-8)}{1}$

$= -32$

117. $\dfrac{2}{9}+\dfrac{1}{9}=\dfrac{2+1}{9}=\dfrac{3}{9}=\dfrac{3\cdot 1}{3\cdot 3}=\dfrac{\cancel{3}\cdot 1}{\cancel{3}\cdot 3}=\dfrac{1}{3}$

118. $-\dfrac{6}{5}+\dfrac{4}{5}=\dfrac{-6+4}{5}=\dfrac{-2}{5}=-\dfrac{2}{5}$

119. $\dfrac{5}{7}-\dfrac{2}{7}=\dfrac{5-2}{7}=\dfrac{3}{7}$

120. $\dfrac{7}{5}-\left(-\dfrac{8}{5}\right)=\dfrac{7}{5}+\dfrac{8}{5}$

$=\dfrac{7+8}{5}$

$=\dfrac{15}{5}$

$=\dfrac{5\cdot 3}{5}$

$=\dfrac{\cancel{5}\cdot 3}{\cancel{5}}$

$=3$

121. $\dfrac{3}{10}+\dfrac{1}{20}=\dfrac{3}{10}\cdot\dfrac{2}{2}+\dfrac{1}{20}$

$=\dfrac{6}{20}+\dfrac{1}{20}$

$=\dfrac{6+1}{20}$

$=\dfrac{7}{20}$

122. $12=4\cdot 3$
$9=3\cdot 3$
$\text{LCD}=4\cdot 3\cdot 3=36$
$\dfrac{5}{12}+\dfrac{4}{9}=\dfrac{5}{12}\cdot\dfrac{3}{3}+\dfrac{4}{9}\cdot\dfrac{4}{4}$

$=\dfrac{15}{36}+\dfrac{16}{36}$

$=\dfrac{15+16}{36}$

$=\dfrac{31}{36}$

123. $35=5\cdot 7$
$49=7\cdot 7$
$\text{LCD}=5\cdot 7\cdot 7=245$
$-\dfrac{7}{35}-\dfrac{2}{49}=-\dfrac{7}{35}\cdot\dfrac{7}{7}-\dfrac{2}{49}\cdot\dfrac{5}{5}$

$=-\dfrac{49}{245}-\dfrac{10}{245}$

$=\dfrac{-49-10}{245}$

$=-\dfrac{59}{245}$

124. $6=2\cdot 3$
$4=2\cdot 2$
$\text{LCD}=2\cdot 2\cdot 3=12$
$\dfrac{5}{6}-\left(-\dfrac{1}{4}\right)=\dfrac{5}{6}+\dfrac{1}{4}$

$=\dfrac{5}{6}\cdot\dfrac{2}{2}+\dfrac{1}{4}\cdot\dfrac{3}{3}$

$=\dfrac{10}{12}+\dfrac{3}{12}$

$=\dfrac{10+3}{12}$

$=\dfrac{13}{12}$

125. $-2-\left(-\dfrac{5}{12}\right)=-2+\dfrac{5}{12}$

$=-\dfrac{2}{1}\cdot\dfrac{12}{12}+\dfrac{5}{12}$

$=-\dfrac{24}{12}+\dfrac{5}{12}$

$=\dfrac{-24+5}{12}$

$=-\dfrac{19}{12}$

126. $-5+\dfrac{9}{4}=-\dfrac{5}{1}\cdot\dfrac{4}{4}+\dfrac{9}{4}$

$=-\dfrac{20}{4}+\dfrac{9}{4}$

$=\dfrac{-20+9}{4}$

$=-\dfrac{11}{4}$

127. $10 = 2 \cdot 5$
$5 = 5$
$2 = 2$
$LCD = 2 \cdot 5 = 10$

$$-\frac{1}{10} + \left(-\frac{2}{5}\right) + \frac{1}{2} = -\frac{1}{10} + \left(-\frac{2}{5} \cdot \frac{2}{2}\right) + \frac{1}{2} \cdot \frac{5}{5}$$
$$= -\frac{1}{10} + \left(-\frac{4}{10}\right) + \frac{5}{10}$$
$$= \frac{-1 + (-4) + 5}{10}$$
$$= \frac{0}{10}$$
$$= 0$$

128. $6 = 2 \cdot 3$
$4 = 2 \cdot 2$
$24 = 2 \cdot 2 \cdot 2 \cdot 3$
$LCD = 2 \cdot 2 \cdot 2 \cdot 3 = 24$

$$-\frac{5}{6} - \frac{1}{4} + \frac{3}{24} = -\frac{5}{6} \cdot \frac{4}{4} - \frac{1}{4} \cdot \frac{6}{6} + \frac{3}{24}$$
$$= -\frac{20}{24} - \frac{6}{24} + \frac{3}{24}$$
$$= \frac{-20 - 6 + 3}{24}$$
$$= -\frac{23}{24}$$

129. $\begin{array}{r} 30.3 \\ + 18.2 \\ \hline 48.5 \end{array}$

So, $30.3 + 18.2 = 48.5$.

130. $\begin{array}{r} 43.02 \\ - 18.36 \\ \hline 24.66 \end{array}$

So, $-43.02 + 18.36 = -24.66$.

131. $\begin{array}{r} 201.37 \\ - 118.39 \\ \hline 82.98 \end{array}$

So, $201.37 - 118.39 = 82.98$.

132. $\begin{array}{r} 35.10 \\ + 18.64 \\ \hline 53.74 \end{array}$

So, $-35.1 - 18.64 = -35.1 + (-18.64)$
$= -53.74$.

133. $\begin{array}{r} 2.01 \\ \times 0.04 \\ \hline 0.0804 \end{array}$

So, $(-0.04)(-2.01) = 0.0804$.

134. $\begin{array}{r} 87.3 \\ \times 2.98 \\ \hline 6984 \\ 7857 \\ 1746 \\ \hline 260.154 \end{array}$

So, $(87.3)(-2.98) = -260.154$.

135. $\dfrac{69.92}{3.8} = \dfrac{699.2}{38} = 18.4$

$$\begin{array}{r} 18.4 \\ 38\overline{)699.2} \\ \underline{38} \\ 319 \\ \underline{304} \\ 152 \\ \underline{152} \\ 0 \end{array}$$

136. $-\dfrac{1.08318}{0.042} = -\dfrac{1083.18}{42} = -25.79$

$$\begin{array}{r} 25.79 \\ 42\overline{)1083.18} \\ \underline{84} \\ 243 \\ \underline{210} \\ 33\;1 \\ \underline{29\;4} \\ 3\;78 \\ \underline{3\;78} \\ 0 \end{array}$$

137. $12.5 - 18.6 + 8.4 = 12.5 + (-18.6) + 8.4$
$= -6.1 + 8.4$
$= 2.3$

138. $-13.5 + 10.8 - 20.2 = -13.5 + 10.8 + (-20.2)$
$= -2.7 + (-20.2)$
$= -22.9$

139. $12.9 \times 1.4 \times (-0.3) = 18.06 \times (-0.3) = -5.418$

140. $2.4 \times 6.1 \times (-0.05) = 14.64 \times (-0.05)$
$= -0.732$

141. $256.75 + (-175.68) + (-180.00)$
$= 81.07 + (-180.00)$
$= -98.93$

Lee's checking account balance is −$98.93. The account is overdrawn.

142. $36 \cdot \dfrac{2}{3} = \dfrac{36}{1} \cdot \dfrac{2}{3}$

$= \dfrac{36 \cdot 2}{1 \cdot 3}$

$= \dfrac{3 \cdot 12 \cdot 2}{1 \cdot 3}$

$= \dfrac{\cancel{3} \cdot 12 \cdot 2}{1 \cdot \cancel{3}}$

$= \dfrac{12 \cdot 2}{1}$

$= 24$

Jarred had 24 friends that wanted the Panthers to win.

143. $15 - 3\dfrac{1}{2} = 15 - \dfrac{7}{2}$

$= \dfrac{15}{1} - \dfrac{7}{2}$

$= \dfrac{15}{1} \cdot \dfrac{2}{2} - \dfrac{7}{2}$

$= \dfrac{30}{2} - \dfrac{7}{2}$

$= \dfrac{30 - 7}{2}$

$= \dfrac{23}{2}$ or $11\dfrac{1}{2}$

The length of the remaining piece is $11\dfrac{1}{2}$ inches.

144. net price $=$ price $\times$ quantity

$= \$35 \times 5$

$= \$175$

sales tax $= \$175 \times 6.75\%$

$= \$175 \times 0.0675$

$= \$11.81$

Sierra spent $\$175 + \$11.81 = \$186.81$ on the clothes.

145. $(5 \cdot 12) \cdot 10 = 5 \cdot (12 \cdot 10)$ illustrates the associative property of multiplication since the grouping of multiplication changes.

146. $20 \cdot \dfrac{1}{20} = 1$ illustrates the multiplicative inverse property.

147. $\dfrac{8}{3} \cdot \dfrac{3}{8} = 1$ illustrates the multiplicative inverse property.

148. $\dfrac{5}{3} \cdot \left(-\dfrac{18}{61}\right) \cdot \dfrac{3}{5} = \dfrac{5}{3} \cdot \dfrac{3}{5} \cdot \left(-\dfrac{18}{61}\right)$ illustrates the commutative property of multiplication since the order in which the numbers are multiplied changes.

149. $9 \cdot 73 \cdot \dfrac{1}{9} = 9 \cdot \dfrac{1}{9} \cdot 73$ illustrates the commutative property of multiplication since the order in which the numbers are multiplied changes.

150. $23.9 + (-23.9) = 0$ illustrates the additive inverse property.

151. $36 + 0 = 36$ illustrates the identity property of addition.

152. $-49 + 0 = -49$ illustrates the identity property of addition.

153. $23 + 5 + (-23) = 23 + (-23) + 5$ illustrates the commutative property of addition since the order of the addition changes.

154. $\dfrac{7}{8} = \dfrac{7}{8} \cdot \dfrac{3}{3}$ illustrates the multiplicative identity property since $\dfrac{3}{3} = 1$.

155. $14 \cdot 0 = 0$ illustrates the multiplication property of zero.

156. $-5.3 + (5.3 + 2.8) = (-5.3 + 5.3) + 2.8$ illustrates the associative property of addition since the grouping of the addition changes.

157. $144 + 29 + (-144) = 144 + (-144) + 29$

$= 0 + 29$

$= 29$

158. $76 + 99 + (-76) = 76 + (-76) + 99$

$= 0 + 99$

$= 99$

159. $\dfrac{19}{3} \cdot 18 \cdot \dfrac{3}{19} = \dfrac{19}{3} \cdot \dfrac{3}{19} \cdot 18 = 1 \cdot 18 = 18$

160. $\dfrac{14}{9} \cdot 121 \cdot \dfrac{9}{14} = \dfrac{14}{9} \cdot \dfrac{9}{14} \cdot 121 = 1 \cdot 121 = 121$

161. $3.4 + 42.56 + (-42.56)$
$= 3.4 + [42.56 + (-42.56)]$
$= 3.4 + 0$
$= 3.4$

162. $5.3 + 3.6 + (-3.6) = 5.3 + [3.6 + (-3.6)]$
$= 5.3 + 0$
$= 5.3$

163. $\dfrac{9}{7} \cdot \left(-\dfrac{11}{3}\right) \cdot 7 = \dfrac{9}{7} \cdot 7 \cdot \left(-\dfrac{11}{3}\right)$

$= 9 \cdot \left(-\dfrac{11}{3}\right)$

$= \overset{3}{\cancel{9}} \cdot \left(-\dfrac{11}{\cancel{3}}\right)$

$= 3 \cdot (-11)$

$= -33$

164. $\dfrac{13}{5} \cdot \dfrac{18}{39} \cdot 5 = \dfrac{13}{5} \cdot 5 \cdot \dfrac{18}{39}$

$= 13 \cdot \dfrac{18}{39}$

$= \cancel{13} \cdot \dfrac{18}{\underset{3}{\cancel{39}}}$

$= \dfrac{18}{3}$

$= \dfrac{6 \cdot 3}{1 \cdot 3}$

$= 6$

165. $\dfrac{7}{0}$ is undefined because 0 is the divisor.

166. $\dfrac{0}{100} = 0$ because 0 is the dividend.

167. $1000(-334)(0.001) = 1000(0.001)(-334)$
$= 1(-334)$
$= -334$

168. $400(0.5)(0.01) = 400(0.01)(0.5) = 4(0.5) = 2$

169. $43{,}569{,}003 \cdot 0 = 0$

170. $154 \cdot \dfrac{1}{154} = 1$

171. $\dfrac{3445}{302} + \left(-\dfrac{3445}{302}\right) = 0$

172. $130 \cdot \dfrac{42}{42} = 130 \cdot 1 = 130$

173. $-\dfrac{7}{48} \cdot \dfrac{20}{3} \cdot \dfrac{12}{7} = -\dfrac{7}{48} \cdot \dfrac{12}{7} \cdot \dfrac{20}{3}$

$= -\dfrac{\overset{1}{\cancel{7}}}{\underset{4}{\cancel{48}}} \cdot \dfrac{\overset{1}{\cancel{12}}}{\cancel{7}} \cdot \dfrac{20}{3}$

$= -\dfrac{1}{4} \cdot \dfrac{20}{3}$

$= -\dfrac{1}{\underset{1}{\cancel{4}}} \cdot \dfrac{\overset{5}{\cancel{20}}}{3}$

$= -\dfrac{5}{3}$

174. $\dfrac{9}{8} \cdot \left(-\dfrac{25}{13}\right) \cdot \dfrac{48}{9} = \dfrac{9}{8} \cdot \dfrac{48}{9} \cdot \left(-\dfrac{25}{13}\right)$

$= \dfrac{\overset{1}{\cancel{9}}}{\underset{1}{\cancel{8}}} \cdot \dfrac{\overset{6}{\cancel{48}}}{\underset{1}{\cancel{9}}} \cdot \left(-\dfrac{25}{13}\right)$

$= 6 \cdot \left(-\dfrac{25}{13}\right)$

$= -\dfrac{150}{13}$

175. The expression $3 \cdot 3 \cdot 3 \cdot 3$ contains four factors of 3, so $3 \cdot 3 \cdot 3 \cdot 3 = 3^4$.

176. The expression $\dfrac{2}{3} \cdot \dfrac{2}{3} \cdot \dfrac{2}{3}$ contains three factors of $\dfrac{2}{3}$, so $\dfrac{2}{3} \cdot \dfrac{2}{3} \cdot \dfrac{2}{3} = \left(\dfrac{2}{3}\right)^3$.

177. The expression $(-4)(-4)$ contains two factors of -4, so $(-4)(-4) = (-4)^2$.

178. The expression $(-3)(-3)(-3)$ contains three factors of -3, so $(-3)(-3)(-3) = (-3)^3$.

179. $5^3 = 5 \cdot 5 \cdot 5 = 125$

180. $2^5 = 2 \cdot 2 \cdot 2 \cdot 2 \cdot 2 = 32$

181. $(-3)^4 = (-3) \cdot (-3) \cdot (-3) \cdot (-3) = 81$

182. $(-4)^3 = (-4) \cdot (-4) \cdot (-4) = -64$

183. $-3^4 = -(3 \cdot 3 \cdot 3 \cdot 3) = -81$

184. $\left(\dfrac{1}{2}\right)^6 = \dfrac{1}{2} \cdot \dfrac{1}{2} \cdot \dfrac{1}{2} \cdot \dfrac{1}{2} \cdot \dfrac{1}{2} \cdot \dfrac{1}{2} = \dfrac{1}{64}$

185.
$$\begin{aligned}
-2 + 16 \div 4 \cdot 2 - 10 &= -2 + 4 \cdot 2 - 10 \\
&= -2 + 8 - 10 \\
&= 6 - 10 \\
&= -4
\end{aligned}$$

186.
$$\begin{aligned}
-4 + 3[2^3 + 4(2 - 10)] &= -4 + 3[2^3 + 4(-8)] \\
&= -4 + 3[8 + 4(-8)] \\
&= -4 + 3[8 - 32] \\
&= -4 + 3[-24] \\
&= -4 - 72 \\
&= -76
\end{aligned}$$

187.
$$\begin{aligned}
(12 - 7)^3 + (19 - 10)^2 &= 5^3 + 9^2 \\
&= 125 + 81 \\
&= 206
\end{aligned}$$

188.
$$\begin{aligned}
5 - (-12 \div 2 \cdot 3) + (-3)^2 &= 5 - (-6 \cdot 3) + (-3)^2 \\
&= 5 - (-18) + (-3)^2 \\
&= 5 - (-18) + 9 \\
&= 5 + 18 + 9 \\
&= 32
\end{aligned}$$

189. $\dfrac{2 \cdot (4 + 8)}{3 + 3^2} = \dfrac{2 \cdot (12)}{3 + 9} = \dfrac{24}{12} = \dfrac{2 \cdot 12}{1 \cdot 12} = 2$

190.
$$\begin{aligned}
\frac{3 \cdot (5 + 2^2)}{2 \cdot 3^3} &= \frac{3 \cdot (5 + 4)}{2 \cdot 27} \\
&= \frac{3 \cdot 9}{2 \cdot 27} \\
&= \frac{27}{2 \cdot 27} \\
&= \frac{1}{2}
\end{aligned}$$

191.
$$\begin{aligned}
\frac{6 \cdot [12 - 3 \cdot (5 - 2)]}{5 \cdot [21 - 2 \cdot (4 + 5)]} &= \frac{6 \cdot [12 - 3 \cdot 3]}{5 \cdot [21 - 2 \cdot 9]} \\
&= \frac{6 \cdot [12 - 9]}{5 \cdot [21 - 18]} \\
&= \frac{6 \cdot 3}{5 \cdot 3} \\
&= \frac{6}{5}
\end{aligned}$$

192.
$$\begin{aligned}
\frac{4 \cdot [3 + 2 \cdot (8 - 6)]}{5 \cdot [14 - 2 \cdot (2 + 3)]} &= \frac{4 \cdot [3 + 2 \cdot 2]}{5 \cdot [14 - 2 \cdot 5]} \\
&= \frac{4 \cdot [3 + 4]}{5 \cdot [14 - 10]} \\
&= \frac{4 \cdot 7}{5 \cdot 4} \\
&= \frac{7}{5}
\end{aligned}$$

193. Let $x = 5$ and $y = -2$.
$$x^2 - y^2 = 5^2 - (-2)^2 = 25 - 4 = 21$$

194. Let $x = 3$ and $y = -3$.
$$\begin{aligned}
x^2 - 3y^2 &= 3^2 - 3(-3)^2 \\
&= 9 - 3(9) \\
&= 9 - 27 \\
&= -18
\end{aligned}$$

195. Let $x = -1$ and $y = -4$.
$$\begin{aligned}
(x + 2y)^3 &= (-1 + 2(-4))^3 \\
&= (-1 - 8)^3 \\
&= (-9)^3 \\
&= -729
\end{aligned}$$

196. Let $a = 5$, $b = -10$, $x = -3$, $y = 2$.
$$\begin{aligned}
\frac{a - b}{x - y} &= \frac{5 - (-10)}{-3 - 2} \\
&= \frac{5 + 10}{-3 - 2} \\
&= \frac{15}{-5} \\
&= \frac{3 \cdot 5}{-1 \cdot 5} \\
&= -3
\end{aligned}$$

197. $3x^2 - x + 6$ can be written as $3x^2 + (-1) \cdot x + 6$.

The terms are $3x^2$, $-x$, and 6. The coefficient of $3x^2$ is 3. The coefficient of $-x$ is -1. The coefficient of 6 is 6.

198. $2x^2y^3 - \dfrac{y}{5}$ can be written as $2x^2y^3 + \left(-\dfrac{1}{5}\right) \cdot y$.

The terms are $2x^2y^3$ and $-\dfrac{y}{5}$. The coefficient

of $2x^2y^3$ is 2. The coefficient of $-\dfrac{y}{5}$ is $-\dfrac{1}{5}$.

199. $4xy^2$ and $-6xy^2$ are like terms. They have the same variables raised to the same power.

200. $-3x$ and $4x^2$ are unlike terms. They have the same variable, but it is raised to different powers.

201. $-6y$ and -6 are unlike terms. $-6y$ has a variable and -6 does not.

202. -10 and 4 are like terms. They are both constants.

203. $4x - 6x - x = (4 - 6 - 1)x = -3x$

204. $6x - 10 - 10x - 5 = 6x - 10x - 10 - 5$
$$= (6-10)x + (-10-5)$$
$$= -4x - 15$$

205. $0.2x^4 + 0.3x^3 - 4.3x^4$
$$= 0.2x^4 - 4.3x^4 + 0.3x^3$$
$$= (0.2 - 4.3)x^4 + 0.3x^3$$
$$= -4.1x^4 + 0.3x^3$$

206. $-3(x^4 - 2x^2 - 4)$
$$= -3 \cdot x^4 - (-3) \cdot 2x^2 - (-3) \cdot 4$$
$$= -3x^4 - (-6x^2) - (-12)$$
$$= -3x^4 + 6x^2 + 12$$

207. $20 - (x + 2) = 20 - x - 2 = 20 - 2 - x = 18 - x$

208. $-6(2x + 5) + 4(4x + 3)$
$$= -6 \cdot 2x + (-6) \cdot 5 + 4 \cdot 4x + 4 \cdot 3$$
$$= -12x - 30 + 16x + 12$$
$$= -12x + 16x - 30 + 12$$
$$= 4x - 18$$

209. $5 - (3x - 1) + 2(6x - 5)$
$$= 5 - 3x + 1 + 2 \cdot 6x - 2 \cdot 5$$
$$= 5 - 3x + 1 + 12x - 10$$
$$= -3x + 12x + 5 + 1 - 10$$
$$= 9x - 4$$

210. $\dfrac{1}{6}(12x + 18) - \dfrac{2}{5}(5x + 10)$
$$= \dfrac{1}{6} \cdot 12x + \dfrac{1}{6} \cdot 18 + \left(-\dfrac{2}{5}\right) \cdot 5x + \left(-\dfrac{2}{5}\right) \cdot 10$$
$$= 2x + 3 - 2x - 4$$
$$= 2x - 2x + 3 - 4$$
$$= -1$$

211. Let $m = 315$.
$$19.95 + 0.25m = 19.95 + 0.25(315)$$
$$= 19.95 + 78.75$$
$$= 98.7$$
The total daily cost is $98.70.

Chapter 1 Test

1. $2 = 2$
$6 = 2 \cdot 3$
$14 = 2 \quad\ \cdot 7$

$2 \cdot 3 \cdot 7$

The LCM is $2 \cdot 3 \cdot 7 = 42$.

2. $\dfrac{21}{66} = \dfrac{3 \cdot 7}{2 \cdot 3 \cdot 11} = \dfrac{\cancel{3} \cdot 7}{2 \cdot \cancel{3} \cdot 11} = \dfrac{7}{22}$

3.
$$
\begin{array}{r}
1.444 \\
9\overline{)13.000} \\
\end{array}
$$
$\underline{9}$
$4\ 0$
$\underline{3\ 6}$
40
$\underline{36}$
40

$\dfrac{13}{9}$ rounded to the nearest hundredth is 1.44.

4. $0.425 = \dfrac{425}{1000} = \dfrac{17 \cdot \cancel{25}}{40 \cdot \cancel{25}} = \dfrac{17}{40}$

5. $0.6\% = 0.6\% \cdot \dfrac{1}{100\%} = \dfrac{0.6}{100} = 0.006$

6. $0.183 = 0.183 \cdot \dfrac{100\%}{1} = 18.3\%$

7. $\dfrac{4}{15} - \left(-\dfrac{2}{30}\right) = \dfrac{4}{15} \cdot \dfrac{2}{2} - \left(-\dfrac{2}{30}\right)$

$\quad = \dfrac{8}{30} - \left(-\dfrac{2}{30}\right)$

$\quad = \dfrac{8}{30} + \dfrac{2}{30}$

$\quad = \dfrac{8+2}{30}$

$\quad = \dfrac{10}{30}$

$\quad = \dfrac{1 \cdot 10}{3 \cdot 10}$

$\quad = \dfrac{1}{3}$

8. $\dfrac{21}{4} \cdot \dfrac{3}{7} = \dfrac{21 \cdot 3}{4 \cdot 7}$

$\quad = \dfrac{3 \cdot 7 \cdot 3}{4 \cdot 7}$

$\quad = \dfrac{3 \cdot \cancel{7} \cdot 3}{4 \cdot \cancel{7}}$

$\quad = \dfrac{3 \cdot 3}{4}$

$\quad = \dfrac{9}{4}$

9. $-16 \div \dfrac{3}{20} = -16 \cdot \dfrac{20}{3}$

$\quad = \dfrac{-16}{1} \cdot \dfrac{20}{3}$

$\quad = \dfrac{-16 \cdot 20}{1 \cdot 3}$

$\quad = -\dfrac{320}{3}$

10. $14 - 110 - (-15) + (-21)$
$= 14 + (-110) + 15 + (-21)$
$= -96 + 15 + (-21)$
$= -81 + (-21)$
$= -102$

11. $\begin{array}{r} 14.50 \\ -\ 2.34 \\ \hline 12.16 \end{array}$

So, $-14.5 + 2.34 = -12.16$.

12. $(-4)(-1)(-5) = 4(-5) = -20$

13. $16 \div 0 = \dfrac{16}{0}$ is undefined because 0 is the divisor.

14. $-20 - (-6) = -20 + 6 = -14$

15. $-110 \div (-2) = \dfrac{-110}{-2} = \dfrac{-2 \cdot 55}{-2 \cdot 1} = 55$

16. **(a)** 6 is the only natural number.

(b) 0 and 6 are the whole numbers

(c) -2, 0, and 6 are the integers.

(d) All those listed, -2, $-\dfrac{1}{2}$, 0, 2.5, and 6, are the rational numbers.

(e) There are no irrational numbers.

(f) All those listed are real numbers.

17. $-|-14| = -14$
Since $-14 < -12$, $-|-14| < -12$.

18. $\left|-\dfrac{2}{5}\right| = |-0.4| = 0.4$

So, $\left|-\dfrac{2}{5}\right| = 0.4$.

19. $-16 \div 2^2 \cdot 4 + (-3)^2 = -16 \div 4 \cdot 4 + 9$
$= -4 \cdot 4 + 9$
$= -16 + 9$
$= -7$

20. $\dfrac{4(-9) - 3^2}{25 + 4(-6-1)} = \dfrac{4(-9) - 9}{25 + 4(-7)}$

$\quad = \dfrac{-36 - 9}{25 - 28}$

$\quad = \dfrac{-45}{-3}$

$\quad = \dfrac{-3 \cdot 15}{-3 \cdot 1}$

$\quad = 15$

21. $8 - 10[6^2 - 5(2+3)] = 8 - 10[6^2 - 5(5)]$
$= 8 - 10[36 - 25]$
$= 8 - 10[11]$
$= 8 - 110$
$= -102$

22. Let $x = -1$ and $y = 3$.

$$(x - 2y)^3 = (-1 - 2 \cdot 3)^3$$
$$= (-1 - 6)^3$$
$$= (-7)^3$$
$$= (-7)(-7)(-7)$$
$$= -343$$

23. $-6(2x + 5) - (4x - 2)$

$$= -6 \cdot 2x - 6 \cdot 5 - 4x - (-2)$$
$$= -12x - 30 - 4x + 2$$
$$= -12x - 4x - 30 + 2$$
$$= -16x - 28$$

24. $\dfrac{1}{2}(4x^2 + 8) - 6x^2 + 5x$

$$= \frac{1}{2} \cdot 4x^2 + \frac{1}{2} \cdot 8 - 6x^2 + 5x$$
$$= 2x^2 + 4 - 6x^2 + 5x$$
$$= 2x^2 - 6x^2 + 5x + 4$$
$$= -4x^2 + 5x + 4$$

25. $675.15 + (-175.50) + (-78) + 110.20$

$$= 499.65 + (-78) + 110.20$$
$$= 421.65 + 110.20$$
$$= 531.85$$

Latoya has $531.85 in her bank account.

26. $2(x + 5) + 2x = 2 \cdot x + 2 \cdot 5 + 2x$

$$= 2x + 10 + 2x$$
$$= 2x + 2x + 10$$
$$= 4x + 10$$

Chapter 2

Section 2.1

Preparing for Linear Equations: The Addition and Multiplication Properties of Equality

P1. The additive inverse of 3 is -3 because $3 + (-3) = 0$.

P2. The multiplicative inverse of $-\dfrac{4}{3}$ is $-\dfrac{3}{4}$

because $-\dfrac{4}{3}\left(-\dfrac{3}{4}\right) = 1$.

P3. $\dfrac{2}{3}\left(\dfrac{3}{2}\right) = \dfrac{2 \cdot 3}{3 \cdot 2} = \dfrac{\overset{1}{\cancel{2}} \cdot \overset{1}{\cancel{3}}}{\underset{1}{\cancel{3}} \cdot \underset{1}{\cancel{2}}} = \dfrac{1 \cdot 1}{1 \cdot 1} = 1$

P4. $-4(2x+3) = -4 \cdot 2x + (-4) \cdot 3$
$= -8x + (-12)$
$= -8x - 12$

P5. $11 - (x+6) = 11 - x - 6 = 5 - x$ or $-x + 5$

2.1 Quick Checks

1. The values of the variable that result in a true statement are called <u>solutions</u>.

2. $a - 4 = -7$
$-3 - 4 \overset{?}{=} -7$
$-7 = -7$ True
Since the left side equals the right side when we replace a by -3, $a = -3$ is a solution of the equation.

3. $\dfrac{1}{2} + x = 10$

$\dfrac{1}{2} + \dfrac{21}{2} \overset{?}{=} 10$

$\dfrac{22}{2} \overset{?}{=} 10$

$11 = 10$ False
Since the left side does not equal the right side

when we replace x by $\dfrac{21}{2}$, $x = \dfrac{21}{2}$ is *not* a

solution of the equation.

4. $3x - (x+4) = 8$
$3(6) - (6+4) \overset{?}{=} 8$
$3(6) - 10 \overset{?}{=} 8$
$18 - 10 \overset{?}{=} 8$
$8 = 8$ True
Since the left side equals the right side when we replace x by 6, $x = 6$ is a solution of the equation.

5. $-9b + 3 + 7b = -3b + 8$
$-9(-3) + 3 + 7(-3) \overset{?}{=} -3(-3) + 8$
$27 + 3 + (-21) \overset{?}{=} 9 + 8$
$9 = 17$ False
Since the left side does not equal the right side when we replace b by -3, $b = -3$ is *not* a solution of the equation.

6. $x - 11 = 21$
$x - 11 + 11 = 21 + 11$
$x = 32$
Check: $x - 11 = 21$
$32 - 11 \overset{?}{=} 21$
$21 = 21$ True
Because $x = 32$ satisfies the original equation, the solution is 32, or the solution set is $\{32\}$.

7. $y + 7 = 21$
$y + 7 - 7 = 21 - 7$
$y = 14$
Check: $y + 7 = 21$
$14 + 7 \overset{?}{=} 21$
$21 = 21$ True
Because $y = 14$ satisfies the original equation, the solution is 14, or the solution set is $\{14\}$.

8. $-8 + a = 4$
$-8 + a + 8 = 4 + 8$
$a = 12$
Check: $-8 + a = 4$
$-8 + 12 \overset{?}{=} 4$
$4 = 4$ True
Because $a = 12$ satisfies the original equation, the solution is 12, or the solution set is $\{12\}$.

9. $-3 = 12 + c$
$-3 - 12 = 12 + c - 12$
$-15 = c$
Check: $-3 = 12 + c$
$-3 \overset{?}{=} 12 + (-15)$
$-3 = -3$ True
Because $c = -15$ satisfies the original equation, the solution is -15, or the solution set is $\{-15\}$.

10.
$$z - \frac{2}{3} = \frac{5}{3}$$
$$z - \frac{2}{3} + \frac{2}{3} = \frac{5}{3} + \frac{2}{3}$$
$$z = \frac{7}{3}$$

Check: $z - \frac{2}{3} = \frac{5}{3}$
$$\frac{7}{3} - \frac{2}{3} \stackrel{?}{=} \frac{5}{3}$$
$$\frac{5}{3} = \frac{5}{3} \quad \text{True}$$

Because $z = \frac{7}{3}$ satisfies the original equation,

the solution is $\frac{7}{3}$, or the solution set is $\left\{\frac{7}{3}\right\}$.

11.
$$p + \frac{5}{4} = \frac{1}{4}$$
$$p + \frac{5}{4} - \frac{5}{4} = \frac{1}{4} - \frac{5}{4}$$
$$p = \frac{-4}{4}$$
$$p = -1$$

Check: $p + \frac{5}{4} = \frac{1}{4}$
$$-1 + \frac{5}{4} \stackrel{?}{=} \frac{1}{4}$$
$$-\frac{4}{4} + \frac{5}{4} \stackrel{?}{=} \frac{1}{4}$$
$$\frac{1}{4} = \frac{1}{4} \quad \text{True}$$

Because $p = -1$ satisfies the original equation, the solution is -1, or the solution set is $\{-1\}$.

12.
$$\frac{3}{8} = w - \frac{1}{4}$$
$$\frac{3}{8} + \frac{1}{4} = w - \frac{1}{4} + \frac{1}{4}$$
$$\frac{3}{8} + \frac{2}{8} = w$$
$$\frac{5}{8} = w$$

Check: $\frac{3}{8} = w - \frac{1}{4}$
$$\frac{3}{8} \stackrel{?}{=} \frac{5}{8} - \frac{1}{4}$$
$$\frac{3}{8} \stackrel{?}{=} \frac{5}{8} - \frac{2}{8}$$
$$\frac{3}{8} = \frac{3}{8} \quad \text{True}$$

Because $w = \frac{5}{8}$ satisfies the original equation,

the solution is $\frac{5}{8}$, or the solution set is $\left\{\frac{5}{8}\right\}$.

13.
$$\frac{5}{4} + x = \frac{1}{6}$$
$$\frac{5}{4} + x - \frac{5}{4} = \frac{1}{6} - \frac{5}{4}$$
$$x = \frac{2}{12} - \frac{15}{12}$$
$$x = -\frac{13}{12}$$

Check: $\frac{5}{4} + x = \frac{1}{6}$
$$\frac{5}{4} + \left(-\frac{13}{12}\right) \stackrel{?}{=} \frac{1}{6}$$
$$\frac{15}{12} + \left(-\frac{13}{12}\right) \stackrel{?}{=} \frac{1}{6}$$
$$\frac{2}{12} \stackrel{?}{=} \frac{1}{6}$$
$$\frac{1}{6} = \frac{1}{6} \quad \text{True}$$

Because $x = -\frac{13}{12}$ satisfies the original equation,

the solution is $-\frac{13}{12}$, or the solution set is

$$\left\{-\frac{13}{12}\right\}.$$

14.
$$p + 1472.25 = 13,927.25$$
$$p + 1472.25 - 1472.25 = 13,927.25 - 1472.25$$
$$p = 12,455$$

Since $p = 12,455$, the price of the car was $12,455 before the extra charges.

15. Dividing both sides of an equation by 3 is the

same as multiplying both sides by $\frac{1}{3}$.

16.
$$8p = 16$$
$$\frac{1}{8}(8p) = \frac{1}{8}(16)$$
$$\left(\frac{1}{8}\cdot 8\right)p = \frac{1}{8}(16)$$
$$p = 2$$

Check: $8p = 16$
$$8(2) \stackrel{?}{=} 16$$
$$16 = 16 \quad \text{True}$$

Because $p = 2$ satisfies the original equation, the solution is 2, or the solution set is $\{2\}$.

17.
$$-7n = 14$$
$$-\frac{1}{7}(-7n) = -\frac{1}{7}(14)$$
$$\left(-\frac{1}{7}\cdot -7\right)n = -\frac{1}{7}(14)$$
$$n = -2$$

Check: $-7n = 14$
$$-7(-2) \stackrel{?}{=} 14$$
$$14 = 14 \quad \text{True}$$

Because $n = -2$ satisfies the original equation, the solution is -2, or the solution set is $\{-2\}$.

18.
$$6z = 15$$
$$\frac{1}{6}(6z) = \frac{1}{6}(15)$$
$$\left(\frac{1}{6}\cdot 6\right)z = \frac{15}{6}$$
$$z = \frac{5}{2}$$

Check: $6z = 15$
$$6\left(\frac{5}{2}\right) \stackrel{?}{=} 15$$
$$\overset{3}{\cancel{6}}\left(\frac{5}{\underset{1}{\cancel{2}}}\right) \stackrel{?}{=} 15$$
$$15 = 15 \quad \text{True}$$

Because $z = \frac{5}{2}$ satisfies the original equation, the solution is $\frac{5}{2}$, or the solution set is $\left\{\frac{5}{2}\right\}$.

19.
$$-12b = 28$$
$$-\frac{1}{12}(-12b) = -\frac{1}{12}(28)$$
$$\left(-\frac{1}{12}\cdot -12\right)b = -\frac{28}{12}$$
$$b = -\frac{7}{3}$$

Check: $-12b = 28$
$$-12\left(-\frac{7}{3}\right) \stackrel{?}{=} 28$$
$$-\overset{4}{\cancel{12}}\left(-\frac{7}{\underset{1}{\cancel{3}}}\right) \stackrel{?}{=} 28$$
$$28 = 28 \quad \text{True}$$

Because $b = -\frac{7}{3}$ satisfies the original equation, the solution is $-\frac{7}{3}$, or the solution set is $\left\{-\frac{7}{3}\right\}$.

20. False: To solve $-\frac{4}{3}z = 16,$ multiply each side of the equation by $-\frac{3}{4}.$

21.
$$\frac{4}{3}n = 12$$
$$\frac{3}{4}\left(\frac{4}{3}n\right) = \frac{3}{4}(12)$$
$$\left(\frac{3}{4}\cdot\frac{4}{3}\right)n = \frac{3}{4}(12)$$
$$n = 9$$

Check: $\frac{4}{3}n = 12$
$$\frac{4}{3}(9) \stackrel{?}{=} 12$$
$$12 = 12 \quad \text{True}$$

Because $n = 9$ satisfies the original equation, the solution is 9, or the solution set is $\{9\}$.

22.
$$-21 = \frac{7}{3}k$$
$$\frac{3}{7}(-21) = \frac{3}{7}\left(\frac{7}{3}k\right)$$
$$\frac{3}{7}(-21) = \left(\frac{3}{7}\cdot\frac{7}{3}\right)k$$
$$-9 = k$$

Check: $-21 = \dfrac{7}{3}k$

$$-21 \overset{?}{=} \dfrac{7}{3}(-9)$$

$$-21 = -21 \quad \text{True}$$

Because $k = -9$ satisfies the original equation, the solution is -9, or the solution set is $\{-9\}$.

23. $\qquad 15 = -\dfrac{z}{2}$

$$-2(15) = -2\left(-\dfrac{z}{2}\right)$$

$$-30 = z$$

Check: $15 = -\dfrac{z}{2}$

$$15 \overset{?}{=} -\dfrac{-30}{2}$$

$$15 = 15 \quad \text{True}$$

Because $z = -30$ satisfies the original equation, the solution is -30 or the solution set is $\{-30\}$.

24. $\qquad \dfrac{3}{8}b = \dfrac{9}{4}$

$$\dfrac{8}{3}\left(\dfrac{3}{8}b\right) = \dfrac{8}{3} \cdot \dfrac{9}{4}$$

$$\left(\dfrac{8}{3} \cdot \dfrac{3}{8}\right)b = \dfrac{\overset{2}{\cancel{8}}}{\cancel{3}} \cdot \dfrac{\overset{3}{\cancel{9}}}{\cancel{4}}$$

$$b = 6$$

Check: $\qquad \dfrac{3}{8}b = \dfrac{9}{4}$

$$\dfrac{3}{8}(6) \overset{?}{=} \dfrac{9}{4}$$

$$\dfrac{3}{\underset{4}{\cancel{8}}}(\overset{3}{\cancel{6}}) \overset{?}{=} \dfrac{9}{4}$$

$$\dfrac{9}{4} = \dfrac{9}{4} \quad \text{True}$$

The solution is 6, or the solution set is $\{6\}$.

25. $\qquad -\dfrac{4}{9} = \dfrac{-t}{6}$

$$-\dfrac{4}{9} = -\dfrac{1}{6}t$$

$$-6\left(-\dfrac{4}{9}\right) = -6\left(-\dfrac{1}{6}t\right)$$

$$-\overset{2}{\cancel{6}}\left(-\dfrac{4}{\underset{3}{\cancel{9}}}\right) = \left(-6 \cdot -\dfrac{1}{6}\right)t$$

$$\dfrac{8}{3} = t$$

Check: $-\dfrac{4}{9} = \dfrac{-t}{6}$

$$-\dfrac{4}{9} \overset{?}{=} -\dfrac{1}{6}\left(\dfrac{8}{3}\right)$$

$$-\dfrac{4}{9} \overset{?}{=} -\dfrac{1}{\cancel{6}}\left(\dfrac{\overset{4}{\cancel{8}}}{3}\right)$$

$$-\dfrac{4}{9} = -\dfrac{4}{9} \quad \text{True}$$

The solution is $\dfrac{8}{3}$, or the solution set is $\left\{\dfrac{8}{3}\right\}$.

26. $\qquad \dfrac{1}{4} = -\dfrac{7}{10}m$

$$-\dfrac{10}{7}\left(\dfrac{1}{4}\right) = -\dfrac{10}{7}\left(-\dfrac{7}{10}m\right)$$

$$-\dfrac{\overset{5}{\cancel{10}}}{7}\left(\dfrac{1}{\underset{2}{\cancel{4}}}\right) = \left(-\dfrac{10}{7} \cdot -\dfrac{7}{10}\right)m$$

$$-\dfrac{5}{14} = m$$

Check: $\dfrac{1}{4} = -\dfrac{7}{10}m$

$$\dfrac{1}{4} \overset{?}{=} -\dfrac{7}{10}\left(-\dfrac{5}{14}\right)$$

$$\dfrac{1}{4} \overset{?}{=} -\dfrac{\overset{1}{\cancel{7}}}{\underset{2}{\cancel{10}}}\left(-\dfrac{\overset{1}{\cancel{5}}}{\underset{2}{\cancel{14}}}\right)$$

$$\dfrac{1}{4} = \dfrac{1}{4} \quad \text{True}$$

The solution is $-\dfrac{5}{14}$, or the solution set is

$\left\{-\dfrac{5}{14}\right\}.$

2.1 Exercises

27. $3x - 1 = 5;\ x = 2$

$3(2) - 1 \stackrel{?}{=} 5$

$6 - 1 \stackrel{?}{=} 5$

$5 = 5$ True

Yes, $x = 2$ is a solution of the equation.

29. $4 - (m + 2) = 3(2m - 1);\ m = 1$

$4 - (1 + 2) \stackrel{?}{=} 3(2(1) - 1)$

$4 - 3 \stackrel{?}{=} 3(2 - 1)$

$1 \stackrel{?}{=} 3(1)$

$1 = 3$ False

No, $m = 1$ is not a solution of the equation.

31. $8k - 2 = 4;\ k = \dfrac{3}{4}$

$8\left(\dfrac{3}{4}\right) - 2 \stackrel{?}{=} 4$

$6 - 2 \stackrel{?}{=} 4$

$4 = 4$ True

Yes, $k = \dfrac{3}{4}$ is a solution of the equation.

33. $r + 1.6 = 2r + 1;\ r = 0.6$

$0.6 + 1.6 \stackrel{?}{=} 2(0.6) + 1$

$2.2 \stackrel{?}{=} 1.2 + 1$

$2.2 = 2.2$ True

Yes, $r = 0.6$ is a solution of the equation.

35. $x - 9 = 11$

$x - 9 + 9 = 11 + 9$

$x = 20$

Check: $x - 9 = 11$

$20 - 9 \stackrel{?}{=} 11$

$11 = 11$ True

The solution is 20, or the solution set is $\{20\}$.

37. $x + 4 = -8$

$x + 4 - 4 = -8 - 4$

$x = -12$

Check: $x + 4 = -8$

$-12 + 4 \stackrel{?}{=} -8$

$-8 = -8$ True

The solution is -12, or the solution set is $\{-12\}$.

39. $12 = n - 7$

$12 + 7 = n - 7 + 7$

$19 = n$

Check: $12 = n - 7$

$12 \stackrel{?}{=} 19 - 7$

$12 = 12$ True

The solution is 19 or the solution set is $\{19\}$.

41. $-8 = x + 5$

$-8 - 5 = x + 5 - 5$

$-13 = x$

Check: $-8 = x + 5$

$-8 \stackrel{?}{=} -13 + 5$

$-8 = -8$ True

The solution is -13 or the solution set is $\{-13\}$.

43. $x - \dfrac{2}{3} = \dfrac{4}{3}$

$x - \dfrac{2}{3} + \dfrac{2}{3} = \dfrac{4}{3} + \dfrac{2}{3}$

$x = \dfrac{6}{3}$

$x = 2$

Check: $x - \dfrac{2}{3} = \dfrac{4}{3}$

$2 - \dfrac{2}{3} \stackrel{?}{=} \dfrac{4}{3}$

$\dfrac{6}{3} - \dfrac{2}{3} \stackrel{?}{=} \dfrac{4}{3}$

$\dfrac{4}{3} = \dfrac{4}{3}$ True

The solution is 2 or the solution set is $\{2\}$.

45. $z + \dfrac{1}{2} = \dfrac{3}{4}$

$z + \dfrac{1}{2} - \dfrac{1}{2} = \dfrac{3}{4} - \dfrac{1}{2}$

$z = \dfrac{3}{4} - \dfrac{2}{4}$

$z = \dfrac{1}{4}$

Check: $z + \dfrac{1}{2} = \dfrac{3}{4}$

$\dfrac{1}{4} + \dfrac{1}{2} \stackrel{?}{=} \dfrac{3}{4}$

$\dfrac{1}{4} + \dfrac{2}{4} \stackrel{?}{=} \dfrac{3}{4}$

$\dfrac{3}{4} = \dfrac{3}{4}$ True

The solution is $\dfrac{1}{4}$ or the solution set is $\left\{\dfrac{1}{4}\right\}$.

47. $\dfrac{5}{12} = x - \dfrac{3}{8}$

$\dfrac{5}{12} + \dfrac{3}{8} = x - \dfrac{3}{8} + \dfrac{3}{8}$

$\dfrac{10}{24} + \dfrac{9}{24} = x$

$\dfrac{19}{24} = x$

Check: $\dfrac{5}{12} = x - \dfrac{3}{8}$

$\dfrac{5}{12} \stackrel{?}{=} \dfrac{19}{24} - \dfrac{3}{8}$

$\dfrac{5}{12} \stackrel{?}{=} \dfrac{19}{24} - \dfrac{9}{24}$

$\dfrac{5}{12} \stackrel{?}{=} \dfrac{10}{24}$

$\dfrac{5}{12} = \dfrac{5}{12}$ True

The solution is $\dfrac{19}{24}$ or the solution set is $\left\{\dfrac{19}{24}\right\}$.

49. $w + 3.5 = -2.6$

$w + 3.5 - 3.5 = -2.6 - 3.5$

$w = -6.1$

Check: $w + 3.5 = -2.6$

$-6.1 + 3.5 \stackrel{?}{=} -2.6$

$-2.6 = -2.6$ True

The solution is -6.1 or the solution set is $\{-6.1\}$.

51. $5c = 25$

$\dfrac{5c}{5} = \dfrac{25}{5}$

$c = 5$

Check: $5c = 25$

$5(5) \stackrel{?}{=} 25$

$25 = 25$ True

The solution is 5 or the solution set is $\{5\}$.

53. $-7n = 28$

$\dfrac{-7n}{-7} = \dfrac{28}{-7}$

$n = -4$

Check: $-7n = 28$

$-7(-4) \stackrel{?}{=} 28$

$28 = 28$ True

The solution is -4 or the solution set is $\{-4\}$.

55. $4k = 14$

$\dfrac{4k}{4} = \dfrac{14}{4}$

$k = \dfrac{7}{2}$

Check: $4k = 14$

$4 \cdot \dfrac{7}{2} \stackrel{?}{=} 14$

$2 \cdot 7 \stackrel{?}{=} 14$

$14 = 14$ True

The solution is $\dfrac{7}{2}$ or the solution set is $\left\{\dfrac{7}{2}\right\}$.

57. $-6w = 15$

$\dfrac{-6w}{-6} = \dfrac{15}{-6}$

$w = -\dfrac{5}{2}$

Check: $-6w = 15$

$-6 \cdot \left(-\dfrac{5}{2}\right) \stackrel{?}{=} 15$

$3 \cdot 5 \stackrel{?}{=} 15$

$15 = 15$ True

The solution is $-\dfrac{5}{2}$ or the solution set is $\left\{-\dfrac{5}{2}\right\}$.

59. $\dfrac{5}{3}a = 35$

$\dfrac{3}{5}\left(\dfrac{5}{3}a\right) = \dfrac{3}{5}(35)$

$a = 3(7)$

$a = 21$

Check: $\dfrac{5}{3}a = 35$

$\dfrac{5}{3}(21) \stackrel{?}{=} 35$

$5(7) \stackrel{?}{=} 35$

$35 = 35$ True

The solution is 21 or the solution set is $\{21\}$.

61.
$$-\frac{3}{11}p = -33$$

$$-\frac{11}{3}\left(-\frac{3}{11}p\right) = -\frac{11}{3}(-33)$$

$$p = -11(-11)$$

$$p = 121$$

Check: $-\dfrac{3}{11}p = -33$

$$-\frac{3}{11}(121) \stackrel{?}{=} -33$$

$$-3(11) \stackrel{?}{=} -33$$

$$-33 = -33 \quad \text{True}$$

The solution is 121 or the solution set is {121}.

63.
$$\frac{n}{5} = 8$$

$$\frac{5}{1} \cdot \frac{n}{5} = \frac{5}{1} \cdot 8$$

$$n = 40$$

Check: $\dfrac{n}{5} = 8$

$$\frac{40}{5} \stackrel{?}{=} 8$$

$$8 = 8 \quad \text{True}$$

The solution is 40, or solution set is {40}.

65.
$$\frac{6}{5} = 2x$$

$$\frac{1}{2} \cdot \frac{6}{5} = \frac{1}{2} \cdot 2x$$

$$\frac{3}{5} = x$$

Check: $\dfrac{6}{5} = 2x$

$$\frac{6}{5} \stackrel{?}{=} 2 \cdot \frac{3}{5}$$

$$\frac{6}{5} = \frac{6}{5} \quad \text{True}$$

The solution is $\dfrac{3}{5}$ or the solution set is $\left\{\dfrac{3}{5}\right\}$.

67.
$$5y = -\frac{5}{3}$$

$$\frac{1}{5} \cdot 5y = \frac{1}{5} \cdot \left(-\frac{5}{3}\right)$$

$$y = -\frac{1}{3}$$

Check: $5y = -\dfrac{5}{3}$

$$5 \cdot \left(-\frac{1}{3}\right) \stackrel{?}{=} -\frac{5}{3}$$

$$-\frac{5}{3} = -\frac{5}{3} \quad \text{True}$$

The solution is $-\dfrac{1}{3}$ or the solution set is $\left\{-\dfrac{1}{3}\right\}$.

69.
$$\frac{1}{2}m = \frac{9}{2}$$

$$2 \cdot \frac{1}{2}m = 2 \cdot \frac{9}{2}$$

$$m = 9$$

Check: $\dfrac{1}{2}m = \dfrac{9}{2}$

$$\frac{1}{2} \cdot 9 \stackrel{?}{=} \frac{9}{2}$$

$$\frac{9}{2} = \frac{9}{2} \quad \text{True}$$

The solution is 9 or the solution set is {9}.

71.
$$-\frac{3}{8}t = \frac{1}{6}$$

$$-\frac{8}{3} \cdot \left(-\frac{3}{8}t\right) = -\frac{8}{3} \cdot \frac{1}{6}$$

$$t = -\frac{8}{18}$$

$$t = -\frac{4}{9}$$

Check: $-\dfrac{3}{8}t = \dfrac{1}{6}$

$$-\frac{3}{8} \cdot \left(-\frac{4}{9}\right) \stackrel{?}{=} \frac{1}{6}$$

$$-\frac{1}{2} \cdot \left(-\frac{1}{3}\right) \stackrel{?}{=} \frac{1}{6}$$

$$\frac{1}{6} = \frac{1}{6} \quad \text{True}$$

The solution is $-\dfrac{4}{9}$ or the solution set is $\left\{-\dfrac{4}{9}\right\}$.

73.
$$\frac{5}{24} = \frac{-y}{8}$$

$$-8 \cdot \frac{5}{24} = -8 \cdot \frac{-y}{8}$$

$$-\frac{5}{3} = y$$

Check: $\dfrac{5}{24} = \dfrac{-y}{8}$

$$\dfrac{5}{24} \overset{?}{=} \dfrac{-\left(-\dfrac{5}{3}\right)}{8}$$

$$\dfrac{5}{24} \overset{?}{=} \dfrac{5}{3} \cdot \dfrac{1}{8}$$

$$\dfrac{5}{24} = \dfrac{5}{24} \quad \text{True}$$

The solution is $-\dfrac{5}{3}$ or the solution set is $\left\{-\dfrac{5}{3}\right\}$.

75. $\quad n - 4 = -2$

$n - 4 + 4 = -2 + 4$

$n = 2$

Check: $n - 4 = -2$

$2 - 4 \overset{?}{=} -4$

$-2 = -2 \quad \text{True}$

The solution is 2 or the solution set is $\{2\}$.

77. $\quad b + 12 = 9$

$b + 12 - 12 = 9 - 12$

$b = -3$

Check: $\quad b + 12 = 9$

$-3 + 12 \overset{?}{=} 9$

$9 = 9 \quad \text{True}$

The solution is -3 or the solution set is $\{-3\}$.

79. $\quad 2 = 3x$

$$\dfrac{2}{3} = \dfrac{3x}{3}$$

$$\dfrac{2}{3} = x$$

Check: $2 = 3x$

$$2 \overset{?}{=} 3 \cdot \dfrac{2}{3}$$

$2 = 2 \quad \text{True}$

The solution is $\dfrac{2}{3}$ or the solution set is $\left\{\dfrac{2}{3}\right\}$.

81. $\quad -4q = 24$

$$\dfrac{-4q}{-4} = \dfrac{24}{-4}$$

$q = -6$

Check: $\quad -4q = 24$

$-4(-6) \overset{?}{=} 24$

$24 = 24 \quad \text{True}$

The solution is -6 or the solution set is $\{-6\}$.

83. $\quad -39 = x - 58$

$-39 + 58 = x - 58 + 58$

$19 = x$

Check: $-39 = x - 58$

$-39 \overset{?}{=} 19 - 58$

$-39 = -39 \quad \text{True}$

The solution is 19 or the solution set is $\{19\}$.

85. $\quad -18 = -301 + x$

$301 - 18 = 301 - 301 + x$

$283 = x$

Check: $-18 = -301 + x$

$-18 \overset{?}{=} -301 + 283$

$-18 = -18 \quad \text{True}$

The solution is 283 or the solution set is $\{283\}$.

87. $\quad \dfrac{x}{5} = -10$

$$5 \cdot \dfrac{x}{5} = 5 \cdot (-10)$$

$x = -50$

Check: $\quad \dfrac{x}{5} = -10$

$$\dfrac{-50}{5} \overset{?}{=} -10$$

$-10 = -10 \quad \text{True}$

The solution is -50 or the solution set is $\{-50\}$.

89. $\quad m - 56.3 = -15.2$

$m - 56.3 + 56.3 = -15.2 + 56.3$

$m = 41.1$

Check: $\quad m - 56.3 = -15.2$

$41.1 - 56.3 \overset{?}{=} -15.2$

$-15.2 = -15.2 \quad \text{True}$

The solution is 41.1 or the solution set is $\{41.1\}$.

91. $\quad -40 = -6c$

$$\dfrac{-40}{-6} = \dfrac{-6c}{-6}$$

$$\dfrac{20}{3} = c$$

Check: $-40 = -6c$

$$-40 \overset{?}{=} -6 \cdot \dfrac{20}{3}$$

$-40 \overset{?}{=} -2 \cdot 20$

$-40 = -40 \quad \text{True}$

The solution is $\dfrac{20}{3}$ or the solution set is $\left\{\dfrac{20}{3}\right\}$.

93.
$$14 = -\frac{7}{2}c$$
$$-\frac{2}{7} \cdot 14 = -\frac{2}{7} \cdot \left(-\frac{7}{2}c\right)$$
$$-2 \cdot 2 = c$$
$$-4 = c$$

Check: $14 = -\dfrac{7}{2}c$

$$14 \overset{?}{=} -\frac{7}{2} \cdot (-4)$$
$$14 \overset{?}{=} -7 \cdot (-2)$$
$$14 = 14 \quad \text{True}$$

The solution is -4 or the solution set is $\{-4\}$.

95.
$$\frac{3}{4} = -\frac{x}{16}$$
$$-16 \cdot \frac{3}{4} = -16 \cdot \left(-\frac{x}{16}\right)$$
$$-4 \cdot 3 = x$$
$$-12 = x$$

Check: $\dfrac{3}{4} = -\dfrac{x}{16}$

$$\frac{3}{4} \overset{?}{=} -\frac{-12}{16}$$
$$\frac{3}{4} = \frac{3}{4} \quad \text{True}$$

The solution is -12 or the solution set is $\{-12\}$.

97.
$$x - \frac{5}{16} = \frac{3}{16}$$
$$x - \frac{5}{16} + \frac{5}{16} = \frac{3}{16} + \frac{5}{16}$$
$$x = \frac{8}{16}$$
$$x = \frac{1}{2}$$

Check: $x - \dfrac{5}{16} = \dfrac{3}{16}$

$$\frac{1}{2} - \frac{5}{16} \overset{?}{=} \frac{3}{16}$$
$$\frac{8}{16} - \frac{5}{16} \overset{?}{=} \frac{3}{16}$$
$$\frac{3}{16} = \frac{3}{16} \quad \text{True}$$

The solution is $\dfrac{1}{2}$ or the solution set is $\left\{\dfrac{1}{2}\right\}$.

99.
$$-\frac{3}{16} = -\frac{3}{8} + z$$
$$\frac{3}{8} - \frac{3}{16} = \frac{3}{8} - \frac{3}{8} + z$$
$$\frac{6}{16} - \frac{3}{16} = z$$
$$\frac{3}{16} = z$$

Check: $-\dfrac{3}{16} = -\dfrac{3}{8} + z$

$$-\frac{3}{16} \overset{?}{=} -\frac{3}{8} + \frac{3}{16}$$
$$-\frac{3}{16} \overset{?}{=} -\frac{6}{16} + \frac{3}{16}$$
$$-\frac{3}{16} = -\frac{3}{16} \quad \text{True}$$

The solution is $\dfrac{3}{16}$ or the solution set is $\left\{\dfrac{3}{16}\right\}$.

101.
$$\frac{5}{6} = -\frac{2}{3}z$$
$$-\frac{3}{2} \cdot \frac{5}{6} = -\frac{3}{2} \cdot \left(-\frac{2}{3}z\right)$$
$$-\frac{1}{2} \cdot \frac{5}{2} = z$$
$$-\frac{5}{4} = z$$

Check: $\dfrac{5}{6} = -\dfrac{2}{3}z$

$$\frac{5}{6} \overset{?}{=} -\frac{2}{3}\left(-\frac{5}{4}\right)$$
$$\frac{5}{6} \overset{?}{=} \frac{10}{12}$$
$$\frac{5}{6} = \frac{5}{6} \quad \text{True}$$

The solution is $-\dfrac{5}{4}$ or the solution set is $\left\{-\dfrac{5}{4}\right\}$.

103.
$$y + 1562.35 = 20,062.15$$
$$y + 1562.35 - 1562.35 = 20,062.15 - 1562.35$$
$$y = 18,499.80$$

The price of the car before the extra charges was $18,499.80.

105.
$$p - 17 = 51$$
$$p - 17 + 17 = 51 + 17$$
$$p = 68$$

The original price of the sleeping bag was $68.

107. $4h = 48$

$$\frac{4h}{4} = \frac{48}{4}$$

$$h = 12$$

Rebecca purchased 12 Happy Meals.

109. $45 = \dfrac{3000}{12} \cdot r$

$$\frac{12}{3000} \cdot 45 = \frac{12}{3000} \cdot \frac{3000}{12} \cdot r$$

$$\frac{540}{3000} = r$$

$$0.18 = r$$

The annual interest rate is 0.18 or 18%.

111. $x + \lambda = 48$

$$x + \lambda - \lambda = 48 - \lambda$$

$$x = 48 - \lambda$$

113. $14 = \theta x$

$$\frac{14}{\theta} = \frac{\theta x}{\theta}$$

$$\frac{14}{\theta} = x$$

115. Let $x = -\dfrac{2}{9}$.

$$x + \lambda = \frac{16}{3}$$

$$-\frac{2}{9} + \lambda = \frac{16}{3}$$

$$-\frac{2}{9} + \lambda + \frac{2}{9} = \frac{16}{3} + \frac{2}{9}$$

$$\lambda = \frac{48}{9} + \frac{2}{9}$$

$$\lambda = \frac{50}{9}$$

117. Let $x = \dfrac{7}{8}$.

$$-\frac{3}{4} = \theta x$$

$$-\frac{3}{4} = \theta \left(\frac{7}{8} \right)$$

$$\frac{8}{7} \left(-\frac{3}{4} \right) = \frac{8}{7} \left(\frac{7}{8} \theta \right)$$

$$-\frac{2(3)}{7} = \theta$$

$$-\frac{6}{7} = \theta$$

119. The solution of an equation is the value of the variable that satisfies the equation.

121. An algebraic expression differs from an equation in that the expression does not contain an equals sign and an equation does. An equation using the expression $x - 10$ is $x - 10 = 22$ solving for x, $x = 32$.

Section 2.2

Preparing for Linear Equations: Using the Properties Together

P1. $6 - (4 + 3x) + 8 = 6 - 4 - 3x + 8$

$$= 6 - 4 + 8 - 3x$$

$$= 2 + 8 - 3x$$

$$= 10 - 3x$$

P2. $2(3x + 4) - 5$ for $x = -1$:

$$2[3(-1) + 4] - 5 = 2[-3 + 4] - 5$$

$$= 2[1] - 5$$

$$= 2 - 5$$

$$= -3$$

2.2 Quick Checks

1. To solve the equation $2x - 11 = 40$, the first step is to <u>add 11 to each side of the equation</u>.

2. $5x - 4 = 11$

$$5x - 4 + 4 = 11 + 4$$

$$5x = 15$$

$$\frac{5x}{5} = \frac{15}{5}$$

$$x = 3$$

Check: $5x - 4 = 11$

$$5(3) - 4 \overset{?}{=} 11$$

$$15 - 4 \overset{?}{=} 11$$

$$11 = 11 \quad \text{True}$$

Because $x = 3$ satisfies the equation, the solution is $x = 3$, or the solution set is $\{3\}$.

3. $8 - 5r = -2$

$$-8 + 8 - 5r = -8 + (-2)$$

$$-5r = -10$$

$$\frac{-5r}{-5} = \frac{-10}{-5}$$

$$r = 2$$

Check: $8 - 5r = -2$

$$8 - 5(2) \overset{?}{=} -2$$

$$8 - 10 \overset{?}{=} -2$$

$$-2 = -2 \quad \text{True}$$

Because $r = 2$ satisfies the equation, the solution is $r = 2$, or the solution set is $\{2\}$.

4.
$$8 = \frac{2}{3}k - 4$$
$$8 + 4 = \frac{2}{3}k - 4 + 4$$
$$12 = \frac{2}{3}k$$
$$\frac{3}{2}(12) = \frac{3}{2}\left(\frac{2}{3}k\right)$$
$$18 = k$$

Check: $8 = \frac{2}{3}k - 4$
$$8 \overset{?}{=} \frac{2}{3}(18) - 4$$
$$8 \overset{?}{=} 12 - 4$$
$$8 = 8 \quad \text{True}$$

Because $k = 18$ satisfies the equation, the solution is $k = 18$, or the solution set is $\{18\}$.

5.
$$-\frac{3}{2}n + 2 = -\frac{1}{4}$$
$$-\frac{3}{2}n + 2 - 2 = -\frac{1}{4} - 2$$
$$-\frac{3}{2}n = -\frac{1}{4} - \frac{8}{4}$$
$$-\frac{3}{2}n = -\frac{9}{4}$$
$$-\frac{2}{3}\left(-\frac{3}{2}n\right) = -\frac{2}{3}\left(-\frac{9}{4}\right)$$
$$n = \frac{3}{2}$$

Check: $-\frac{3}{2}n + 2 = -\frac{1}{4}$
$$-\frac{3}{2}\left(\frac{3}{2}\right) + 2 \overset{?}{=} -\frac{1}{4}$$
$$-\frac{9}{4} + 2 \overset{?}{=} -\frac{1}{4}$$
$$-\frac{9}{4} + \frac{8}{4} \overset{?}{=} -\frac{1}{4}$$
$$-\frac{1}{4} = -\frac{1}{4} \quad \text{True}$$

Because $n = \frac{3}{2}$ satisfies the equation, the solution is $n = \frac{3}{2}$, or the solution set is $\left\{\frac{3}{2}\right\}$.

6.
$$7b - 3b + 3 = 11$$
$$4b + 3 = 11$$
$$4b + 3 - 3 = 11 - 3$$
$$4b = 8$$
$$\frac{4b}{4} = \frac{8}{4}$$
$$b = 2$$

Check: $7b - 3b + 3 = 11$
$$7(2) - 3(2) + 3 \overset{?}{=} 11$$
$$14 - 6 + 3 \overset{?}{=} 11$$
$$11 = 11 \quad \text{True}$$

Since $b = 2$ results in a true statement, the solution of the equation is 2, or the solution set is $\{2\}$.

7.
$$-3a + 4 + 4a = 13 - 27$$
$$4 + a = -14$$
$$-4 + 4 + a = -4 + (-14)$$
$$a = -18$$

Check: $-3a + 4 + 4a = 13 - 27$
$$-3(-18) + 4 + 4(-18) \overset{?}{=} 13 - 27$$
$$54 + 4 - 72 \overset{?}{=} 13 - 27$$
$$-14 = -14 \quad \text{True}$$

Since $a = -18$ results in a true statement, the solution of the equation is -18, or the solution set is $\{-18\}$.

8.
$$6c - 2 + 2c = 18$$
$$8c - 2 = 18$$
$$8c - 2 + 2 = 18 + 2$$
$$8c = 20$$
$$\frac{8c}{8} = \frac{20}{8}$$
$$c = \frac{5}{2}$$

Check: $6c - 2 + 2c = 18$
$$6\left(\frac{5}{2}\right) - 2 + 2\left(\frac{5}{2}\right) \overset{?}{=} 18$$
$$15 - 2 + 5 \overset{?}{=} 18$$
$$18 = 18 \quad \text{True}$$

Since $c = \frac{5}{2}$ results in a true statement, the solution of the equation is $\frac{5}{2}$, or the solution set is $\left\{\frac{5}{2}\right\}$.

9.
$$-12 = 5x - 3x + 4$$
$$-12 = 2x + 4$$
$$-12 - 4 = 2x + 4 - 4$$
$$-16 = 2x$$
$$\frac{-16}{2} = \frac{2x}{2}$$
$$-8 = x$$
Check: $-12 = 5x - 3x + 4$
$$-12 \stackrel{?}{=} 5(-8) - 3(-8) + 4$$
$$-12 \stackrel{?}{=} -40 + 24 + 4$$
$$-12 = -12 \quad \text{True}$$
Since $x = -8$ results in a true statement, the solution of the equation is -8, or the solution set is $\{-8\}$.

10.
$$2(y+5) - 3 = 11$$
$$2y + 10 - 3 = 11$$
$$2y + 7 = 11$$
$$2y + 7 - 7 = 11 - 7$$
$$2y = 4$$
$$\frac{2y}{2} = \frac{4}{2}$$
$$y = 2$$

Check: $2(y+5) - 3 = 11$
$$2(2+5) - 3 \stackrel{?}{=} 11$$
$$2(7) - 3 \stackrel{?}{=} 11$$
$$14 - 3 \stackrel{?}{=} 11$$
$$11 = 11 \quad \text{True}$$
Since $y = 2$ results in a true statement, the solution of the equation is 2, or the solution set is $\{2\}$.

11. $\frac{1}{2}(4 - 6x) + 5 = 3$
$$2 - 3x + 5 = 3$$
$$-3x + 7 = 3$$
$$-3x + 7 - 7 = 3 - 7$$
$$-3x = -4$$
$$\frac{-3x}{-3} = \frac{-4}{-3}$$
$$x = \frac{4}{3}$$

Check: $\frac{1}{2}(4 - 6x) + 5 = 3$
$$\frac{1}{2}\left(4 - 6 \cdot \frac{4}{3}\right) + 5 \stackrel{?}{=} 3$$
$$\frac{1}{2}(4 - 8) + 5 \stackrel{?}{=} 3$$
$$\frac{1}{2}(-4) + 5 \stackrel{?}{=} 3$$
$$-2 + 5 \stackrel{?}{=} 3$$
$$3 = 3 \quad \text{True}$$
Since $x = \frac{4}{3}$ results in a true statement, the solution of the equation is $\frac{4}{3}$, or the solution set is $\left\{\frac{4}{3}\right\}$.

12. $4 - (6 - x) = 11$
$$4 - 6 + x = 11$$
$$-2 + x = 11$$
$$2 - 2 + x = 2 + 11$$
$$x = 13$$
Check: $4 - (6 - x) = 11$
$$4 - (6 - 13) \stackrel{?}{=} 11$$
$$4 - (-7) \stackrel{?}{=} 11$$
$$4 + 7 \stackrel{?}{=} 11$$
$$11 = 11 \quad \text{True}$$
Since $x = 13$ results in a true statement, the solution of the equation is 13, or the solution set is $\{13\}$.

13. $8 + \frac{2}{3}(2n - 9) = 10$
$$8 + \frac{4}{3}n - 6 = 10$$
$$\frac{4}{3}n + 2 = 10$$
$$\frac{4}{3}n + 2 - 2 = 10 - 2$$
$$\frac{4}{3}n = 8$$
$$\frac{3}{4}\left(\frac{4}{3}n\right) = \frac{3}{4}(8)$$
$$n = 6$$

Check: $8 + \dfrac{2}{3}(2n-9) = 10$

$8 + \dfrac{2}{3}(2 \cdot 6 - 9) \stackrel{?}{=} 10$

$8 + \dfrac{2}{3}(12-9) \stackrel{?}{=} 10$

$8 + \dfrac{2}{3}(3) \stackrel{?}{=} 10$

$8 + 2 \stackrel{?}{=} 10$

$10 = 10$ True

Since $n = 6$ results in a true statement, the solution of the equation is 6, or the solution set is $\{6\}$.

14. $\dfrac{1}{3}(2x+9) + \dfrac{x}{3} = 5$

$\dfrac{2x}{3} + \dfrac{9}{3} + \dfrac{x}{3} = 5$

$3 + \dfrac{3x}{3} = 5$

$3 + \dfrac{x}{1} = 5$

$3 + x = 5$

$3 - 3 + x = 5 - 3$

$x = 2$

Check: $\dfrac{1}{3}(2x+9) + \dfrac{x}{3} = 5$

$\dfrac{1}{3}[2(2)+9] + \dfrac{2}{3} \stackrel{?}{=} 5$

$\dfrac{1}{3}(4+9) + \dfrac{2}{3} \stackrel{?}{=} 5$

$\dfrac{1}{3}(13) + \dfrac{2}{3} \stackrel{?}{=} 5$

$\dfrac{13}{3} + \dfrac{2}{3} \stackrel{?}{=} 5$

$\dfrac{15}{3} \stackrel{?}{=} 5$

$5 = 5$ True

Since $x = 2$ results in a true statement, the solution of the equation is 2, or the solution set is $\{2\}$.

15. $3x + 4 = 5x - 8$

$3x + 4 - 3x = 5x - 8 - 3x$

$4 = 2x - 8$

$4 + 8 = 2x - 8 + 8$

$12 = 2x$

$\dfrac{12}{2} = \dfrac{2x}{2}$

$6 = x$

Check: $3x + 4 = 5x - 8$

$3(6) + 4 \stackrel{?}{=} 5(6) - 8$

$18 + 4 \stackrel{?}{=} 30 - 8$

$22 = 22$ True

Since $x = 6$ results in a true statement, the solution of the equation is 6, or the solution set is $\{6\}$.

16. $10m + 3 = 6m - 11$

$10m + 3 - 6m = 6m - 11 - 6m$

$4m + 3 = -11$

$4m + 3 - 3 = -11 - 3$

$4m = -14$

$\dfrac{4m}{4} = \dfrac{-14}{4}$

$m = -\dfrac{7}{2}$

Check: $10m + 3 = 6m - 11$

$10\left(-\dfrac{7}{2}\right) + 3 \stackrel{?}{=} 6\left(-\dfrac{7}{2}\right) - 11$

$-35 + 3 \stackrel{?}{=} -21 - 11$

$-32 = -32$ True

Since $m = -\dfrac{7}{2}$ results in a true statement, the

solution of the equation is $-\dfrac{7}{2}$, or the solution

set is $\left\{-\dfrac{7}{2}\right\}$.

17. False: To solve the equation $13 - 2(7x + 1) + 8x = 12$, the first step is to remove the parentheses using the Distributive Property.

18. $-9x + 3(2x - 3) = -10 - 2x$

$-9x + 6x - 9 = -10 - 2x$

$-3x - 9 = -10 - 2x$

$3x - 3x - 9 = 3x - 10 - 2x$

$-9 = x - 10$

$-9 + 10 = x - 10 + 10$

$1 = x$

The solution to the equation is $x = 1$, or the solution set is $\{1\}$.

19. $3 - 4(p+5) = 5(p+2) - 12$
$3 - 4p - 20 = 5p + 10 - 12$
$-4p - 17 = 5p - 2$
$4p - 4p - 17 = 4p + 5p - 2$
$-17 = 9p - 2$
$-17 + 2 = 9p - 2 + 2$
$-15 = 9p$
$\dfrac{-15}{9} = \dfrac{9p}{9}$
$-\dfrac{5}{3} = p$

The solution to the equation is $p = -\dfrac{5}{3}$, or the

solution set is $\left\{ -\dfrac{5}{3} \right\}$.

20. $400 + 20(h - 40) = 640$
$400 + 20h - 800 = 640$
$20h - 400 = 640$
$20h - 400 + 400 = 640 + 400$
$20h = 1040$
$\dfrac{20h}{20} = \dfrac{1040}{20}$
$h = 52$

Marcella worked 52 hours that week.

2.2 Exercises

21. $3x + 4 = 7$
$3x + 4 - 4 = 7 - 4$
$3x = 3$
$\dfrac{3x}{3} = \dfrac{3}{3}$
$x = 1$
Check: $3x + 4 = 7$
$3(1) + 4 \overset{?}{=} 7$
$3 + 4) \overset{?}{=} 7$
$7 = 7$ True
The solution is 1 or the solution set is $\{1\}$.

23. $2y - 1 = -5$
$2y - 1 + 1 = -5 + 1$
$2y = -4$
$\dfrac{2y}{2} = \dfrac{-4}{2}$
$y = -2$

Check: $2y - 1 = -5$
$2(-2) - 1 \overset{?}{=} -5$
$-4 - 1 \overset{?}{=} -5$
$-5 = -5$ True
The solution is -2 or the solution set is $\{-2\}$.

25. $-3p + 1 = 10$
$-3p + 1 - 1 = 10 - 1$
$-3p = 9$
$\dfrac{-3p}{-3} = \dfrac{9}{-3}$
$p = -3$
Check: $-3p + 1 = 10$
$-3(-3) + 1 \overset{?}{=} 10$
$9 + 1 \overset{?}{=} 10$
$10 = 10$ True
The solution is -3, or the solution set is $\{-3\}$.

27. $8y + 3 = 15$
$8y + 3 - 3 = 15 - 3$
$8y = 12$
$\dfrac{8y}{8} = \dfrac{12}{8}$
$y = \dfrac{3}{2}$
Check: $8y + 3 = 15$
$8\left(\dfrac{3}{2}\right) + 3 \overset{?}{=} 15$
$4 \cdot 3 + 3 \overset{?}{=} 15$
$12 + 3 \overset{?}{=} 15$
$15 = 15$ True
The solution is $\dfrac{3}{2}$, or the solution set is $\left\{ \dfrac{3}{2} \right\}$.

29. $5 - 2z = 11$
$-5 + 5 - 2z = -5 + 11$
$-2z = 6$
$\dfrac{-2z}{-2} = \dfrac{6}{-2}$
$z = -3$
Check: $5 - 2z = 11$
$5 - 2(-3) \overset{?}{=} 11$
$5 + 6 \overset{?}{=} 11$
$11 = 11$ True
The solution is -3 or the solution set is $\{-3\}$.

31. $\dfrac{2}{3}x + 1 = 9$

$\dfrac{2}{3}x + 1 - 1 = 9 - 1$

$\dfrac{2}{3}x = 8$

$\dfrac{3}{2} \cdot \dfrac{2}{3}x = \dfrac{3}{2} \cdot 8$

$x = 12$

Check: $\dfrac{2}{3}x + 1 = 9$

$\dfrac{2}{3} \cdot 12 + 1 \overset{?}{=} 9$

$8 + 1 \overset{?}{=} 9$

$9 = 9$ True

The solution is 12 or the solution set is $\{12\}$.

33. $\dfrac{7}{2}y - 1 = 13$

$\dfrac{7}{2}y - 1 + 1 = 13 + 1$

$\dfrac{7}{2}y = 14$

$\dfrac{2}{7} \cdot \dfrac{7}{2}y = \dfrac{2}{7} \cdot 14$

$y = 4$

Check: $\dfrac{7}{2}y - 1 = 13$

$\dfrac{7}{2}(4) - 1 \overset{?}{=} 13$

$14 - 1 \overset{?}{=} 13$

$13 = 13$ True

The solution is 4, or the solution set is $\{4\}$.

35. $3x - 7 + 2x = -17$

$5x - 7 = -17$

$5x - 7 + 7 = -17 + 7$

$5x = -10$

$\dfrac{5x}{5} = \dfrac{-10}{5}$

$x = -2$

Check: $3x - 7 + 2x = -17$

$3(-2) - 7 + 2(-2) \overset{?}{=} -17$

$-6 - 7 + (-4) \overset{?}{=} -17$

$-13 - 4 \overset{?}{=} -17$

$-17 = -17$ True

The solution is -2, or the solution set is $\{-2\}$.

37. $2k - 7k - 8 = 17$

$-5k - 8 = 17$

$-5k - 8 + 8 = 17 + 8$

$-5k = 25$

$\dfrac{-5k}{-5} = \dfrac{25}{-5}$

$k = -5$

Check: $2k - 7k - 8 = 17$

$2(-5) - 7(-5) - 8 \overset{?}{=} 17$

$-10 + 35 - 8 \overset{?}{=} 17$

$17 = 17$ True

The solution is -5, or the solution set is $\{-5\}$.

39. $2(x + 1) = -14$

$2x + 2 = -14$

$2x + 2 - 2 = -14 - 2$

$2x = -16$

$\dfrac{2x}{2} = \dfrac{-16}{2}$

$x = -8$

Check: $2(x + 1) = -14$

$2(-8 + 1) \overset{?}{=} -14$

$2(-7) \overset{?}{=} -14$

$-14 = -14$ True

The solution is -8 or the solution set is $\{-8\}$.

41. $-3(2 + r) = 9$

$-6 - 3r = 9$

$-6 - 3r + 6 = 9 + 6$

$-3r = 15$

$\dfrac{-3r}{-3} = \dfrac{15}{-3}$

$r = -5$

Check: $-3(2 + r) = 9$

$-3[2 + (-5)] \overset{?}{=} 9$

$-3(-3) \overset{?}{=} 9$

$9 = 9$ True

The solution is -5, or the solution set is $\{-5\}$.

43. $17 = 2 - (n + 6)$

$17 = 2 - n - 6$

$17 = -n - 4$

$4 + 17 = -n - 4 + 4$

$21 = -n$

$-1 \cdot 21 = -n \cdot -1$

$-21 = n$

Check: $17 = 2 - (n + 6)$
$17 \overset{?}{=} 2 - (-21 + 6)$
$17 \overset{?}{=} 2 - (-15)$
$17 \overset{?}{=} 2 + 15$
$17 = 17$ True

The solution is -21, or the solution set is $\{-21\}$.

45. $-8 = 5 - (7 - z)$
$-8 = 5 - 7 + z$
$-8 = -2 + z$
$2 - 8 = -2 + 2 + z$
$-6 = z$

Check: $-8 = 5 - (7 - z)$
$-8 \overset{?}{=} 5 - [7 - (-6)]$
$-8 \overset{?}{=} 5 - (13)$
$-8 = -8$ True

The solution is -6, or the solution set is $\{-6\}$.

47. $2x + 9 = x + 1$
$2x + 9 - x = x + 1 - x$
$x + 9 = 1$
$x + 9 - 9 = 1 - 9$
$x = -8$

Check: $2x + 9 = x + 1$
$2(-8) + 9 \overset{?}{=} -8 + 1$
$-16 + 9 \overset{?}{=} -7$
$-7 = -7$ True

The solution is -8, or the solution set is $\{-8\}$.

49. $2t - 6 = 3 - t$
$2t - 6 + t = 3 - t + t$
$3t - 6 = 3$
$3t - 6 + 6 = 3 + 6$
$3t = 9$
$\dfrac{3t}{3} = \dfrac{9}{3}$
$t = 3$

Check: $2t - 6 = 3 - t$
$2(3) - 6 \overset{?}{=} 3 - 3$
$6 - 6 \overset{?}{=} 0$
$0 = 0$ True

The solution is 3, or the solution set is $\{3\}$.

51. $14 - 2n = -4n + 7$
$14 - 2n + 4n = -4n + 7 + 4n$
$14 + 2n = 7$
$14 + 2n - 14 = 7 - 14$
$2n = -7$
$\dfrac{2n}{2} = -\dfrac{7}{2}$
$n = -\dfrac{7}{2}$

Check: $14 - 2n = -4n + 7$
$14 - 2\left(-\dfrac{7}{2}\right) \overset{?}{=} -4\left(-\dfrac{7}{2}\right) + 7$
$14 + 7 \overset{?}{=} 2(7) + 7$
$21 \overset{?}{=} 14 + 7$
$21 = 21$ True

The solution is $-\dfrac{7}{2}$, or the solution set is

$\left\{-\dfrac{7}{2}\right\}$.

53. $-3(5 - 3k) = 6k + 6$
$-15 + 9k = 6k + 6$
$-15 + 9k - 6k = 6k + 6 - 6k$
$-15 + 3k = 6$
$15 - 15 + 3k = 15 + 6$
$3k = 21$
$\dfrac{3k}{3} = \dfrac{21}{3}$
$k = 7$

Check: $-3(5 - 3k) = 6k + 6$
$-3(5 - 3 \cdot 7) \overset{?}{=} 6 \cdot 7 + 6$
$-3(5 - 21) \overset{?}{=} 42 + 6$
$-3(-16) \overset{?}{=} 48$
$48 = 48$ True

The solution is 7, or the solution set is $\{7\}$.

55. $2(2x + 3) = 3(x - 4)$
$4x + 6 = 3x - 12$
$4x + 6 - 3x = 3x - 12 - 3x$
$x + 6 = -12$
$x + 6 - 6 = -12 - 6$
$x = -18$

Check: $2(2x + 3) = 3(x - 4)$
$2[2(-18) + 3] \overset{?}{=} 3(-18 - 4)$
$2(-36 + 3) \overset{?}{=} 3(-22)$
$2(-33) \overset{?}{=} -66$
$-66 = -66$ True

The solution is -18, or the solution set is $\{-18\}$.

57.

$$3+2(x-1)=5x$$
$$3+2x-2=5x$$
$$2x+1=5x$$
$$-2x+2x+1=5x-2x$$
$$1=3x$$
$$\frac{1}{3}=\frac{3x}{3}$$
$$\frac{1}{3}=x$$

Check: $3+2(x-1)=5x$
$$3+2x-2=5x$$
$$2x+1=5x$$
$$2\left(\frac{1}{3}\right)+1 \stackrel{?}{=} 5\left(\frac{1}{3}\right)$$
$$\frac{2}{3}+1 \stackrel{?}{=} \frac{5}{3}$$
$$\frac{2}{3}+\frac{3}{3} \stackrel{?}{=} \frac{5}{3}$$
$$\frac{5}{3}=\frac{5}{3} \quad \text{True}$$

The solution is $\frac{1}{3}$, or the solution set is $\left\{\frac{1}{3}\right\}$.

59.

$$9(6+a)+33a=10a$$
$$54+9a+33a=10a$$
$$54+42a=10a$$
$$-42a+54+42a=-42a+10a$$
$$54=-32a$$
$$\frac{54}{-32}=\frac{-32a}{-32}$$
$$-\frac{27}{16}=a$$

Check: $9(6+a)+33a=10a$
$$9\left[6\left(-\frac{27}{16}\right)\right]+33\left(-\frac{27}{16}\right) \stackrel{?}{=} 10\left(-\frac{27}{16}\right)$$
$$9\left(\frac{69}{16}\right)+33\left(-\frac{27}{16}\right) \stackrel{?}{=} 10\left(-\frac{27}{16}\right)$$
$$\frac{621}{16}-\frac{891}{16} \stackrel{?}{=} \frac{-270}{16}$$
$$\frac{-270}{16}=\frac{-270}{16} \quad \text{True}$$

The solution is $-\frac{27}{16}$, or the solution set is $\left\{-\frac{27}{16}\right\}$.

61.

$$-5x+11=1$$
$$-5x+11-11=1-11$$
$$-5x=-10$$
$$\frac{-5x}{-5}=\frac{-10}{-5}$$
$$x=2$$

Check: $-5x+11=1$
$$-5(2)+11 \stackrel{?}{=} 1$$
$$-10+11 \stackrel{?}{=} 1$$
$$1=1 \quad \text{True}$$

The solution is 2, or the solution set is $\{2\}$.

63.

$$4m+5=2$$
$$4m+5-5=2-5$$
$$4m=-3$$
$$\frac{4m}{4}=\frac{-3}{4}$$
$$m=-\frac{3}{4}$$

Check: $4m+5=2$
$$4\cdot\left(-\frac{3}{4}\right)+5 \stackrel{?}{=} 2$$
$$-3+5 \stackrel{?}{=} 2$$
$$2=2 \quad \text{True}$$

The solution is $-\frac{3}{4}$, or the solution set is $\left\{-\frac{3}{4}\right\}$.

65.

$$-2(3n-2)=2$$
$$-6n+4=2$$
$$-6n+4-4=2-4$$
$$-6n=-2$$
$$\frac{-6n}{-6}=\frac{-2}{-6}$$
$$n=\frac{1}{3}$$

Check: $-2(3n-2)=2$
$$-2\left(3\cdot\frac{1}{3}-2\right) \stackrel{?}{=} 2$$
$$-2(1-2) \stackrel{?}{=} 2$$
$$-2(-1) \stackrel{?}{=} 2$$
$$2=2 \quad \text{True}$$

The solution is $\frac{1}{3}$, or the solution set is $\left\{\frac{1}{3}\right\}$.

67. $4k - (3+k) = -(2k+3)$
$$4k - 3 - k = -2k - 3$$
$$-3 + 3k = -2k - 3$$
$$3 - 3 + 3k = -2k - 3 + 3$$
$$3k = -2k$$
$$2k + 3k = 0$$
$$5k = 0$$
$$\frac{5k}{5} = \frac{0}{5}$$
$$k = 0$$
Check: $4k - (3+k) = -(2k+3)$
$$4(0) - (3+0) \overset{?}{=} -(2 \cdot 0 + 3)$$
$$0 - 3 \overset{?}{=} -3$$
$$-3 = -3 \quad \text{True}$$
The solution is 0, or the solution set is $\{0\}$.

69.
$$2y + 36 = 6 + 6y$$
$$2y + 36 - 2y = 6 + 6y - 2y$$
$$36 = 6 + 4y$$
$$36 - 6 = 6 + 4y - 6$$
$$30 = 4y$$
$$\frac{30}{4} = \frac{4y}{4}$$
$$\frac{15}{2} = y$$
Check: $2y + 36 = 6 + 6y$
$$2 \cdot \frac{15}{2} + 36 \overset{?}{=} 6 + 6 \cdot \frac{15}{2}$$
$$15 + 36 \overset{?}{=} 6 + 45$$
$$51 = 51 \quad \text{True}$$
The solution is $\frac{15}{2}$, or the solution set is $\left\{\frac{15}{2}\right\}$.

71. $\frac{1}{2}(-4k + 28) = 6 + 14k$
$$-2k + 14 = 6 + 14k$$
$$-2k + 14 + 2k = 6 + 14k + 2k$$
$$14 = 6 + 16k$$
$$14 - 6 = 6 + 16k - 6$$
$$8 = 16k$$
$$\frac{8}{16} = \frac{16k}{16}$$
$$\frac{1}{2} = k$$

Check: $\frac{1}{2}(-4k + 28) = 6 + 14k$
$$\frac{1}{2}\left[-4\left(\frac{1}{2}\right) + 28\right] \overset{?}{=} 6 + 14\left(\frac{1}{2}\right)$$
$$\frac{1}{2}(-2 + 28) \overset{?}{=} 6 + 7$$
$$\frac{1}{2}(26) \overset{?}{=} 13$$
$$13 = 13 \quad \text{True}$$
The solution is $\frac{1}{2}$, or the solution set is $\left\{\frac{1}{2}\right\}$.

73. $-\frac{5}{2}(x+6) + \frac{3}{2}x = -8$
$$-\frac{5}{2}x - \frac{30}{2} + \frac{3}{2}x = -8$$
$$-15 - x = -8$$
$$15 - 15 - x = 15 - 8$$
$$-x = 7$$
$$\frac{-x}{-1} = \frac{7}{-1}$$
$$x = -7$$
Check: $-\frac{5}{2}(x+6) + \frac{3}{2}x = -8$
$$-\frac{5}{2}(-7+6) + \frac{3}{2}(-7) \overset{?}{=} -8$$
$$-\frac{5}{2}(-1) - \frac{21}{2} \overset{?}{=} -8$$
$$\frac{5}{2} - \frac{21}{2} \overset{?}{=} -8$$
$$-\frac{16}{2} \overset{?}{=} -8$$
$$-8 = -8 \quad \text{True}$$
The solution is -7, or the solution set is $\{-7\}$.

75. $-3(2y+3) - 1 = -4(y+6) + 2y$
$$-6y - 9 - 1 = -4y - 24 + 2y$$
$$-6y - 10 = -2y - 24$$
$$2y - 6y - 10 = 2y - 2y - 24$$
$$-4y - 10 = -24$$
$$10 - 4y - 10 = 10 - 24$$
$$-4y = -14$$
$$\frac{-4y}{-4} = \frac{-14}{-4}$$
$$y = \frac{7}{2}$$

Check: $-3(2y+3)-1=-4(y+6)+2y$

$$-3\left(2\cdot\frac{7}{2}+3\right)-1 \stackrel{?}{=} -4\left(\frac{7}{2}+6\right)+2\cdot\frac{7}{2}$$

$$-3(7+3)-1 \stackrel{?}{=} -4\left(\frac{19}{2}\right)+7$$

$$-3(10)-1 \stackrel{?}{=} -38+7$$

$$-30-1 \stackrel{?}{=} -31$$

$$-31=-31 \quad \text{True}$$

The solution is $\dfrac{7}{2}$, or the solution set is $\left\{\dfrac{7}{2}\right\}$.

77. $x+(x+6)=38$

$x+x+6=38$

$2x+6=38$

$-6+2x+6=38-6$

$2x=32$

$\dfrac{2x}{2}=\dfrac{32}{2}$

$x=16$

$x+6=16+6=22$

There are 16 grams of fat in McDonald's Southwestern salad and 22 grams of fat in a Burger King Tendercrisp Chicken garden salad.

79. $2w+2(2w+2)=30$

$2w+4w+4=30$

$6w+4=30$

$6w+4-4=30-4$

$6w=26$

$\dfrac{6w}{6}=\dfrac{26}{6}$

$w=\dfrac{13}{3}$ or $4\dfrac{1}{3}$

$2w+2=2\left(\dfrac{13}{3}\right)+2=\dfrac{26}{3}+\dfrac{6}{3}=\dfrac{32}{3}$ or $10\dfrac{2}{3}$

The width is $\dfrac{13}{3}$ or $4\dfrac{1}{3}$ feet and the length is $\dfrac{32}{3}$ or $10\dfrac{2}{3}$ feet.

81. $40x+4(1.5x)=368$

$40x+6x=368$

$46x=368$

$\dfrac{46x}{46}=\dfrac{368}{46}$

$x=8$

Jennifer's regular pay rate is $8 per hour.

83. $2x + 2(x + 5) = 42$

$2x + 2x + 10 = 42$

$4x + 10 = 42$

$4x + 10 - 10 = 42 - 10$

$4x = 32$

$\dfrac{4x}{4} = \dfrac{32}{4}$

$x = 8$

$x + 5 = 8 + 5 = 13$

Yes, Becky has enough wallpaper.

85. $8[4 - 6(x - 1)] + 5[(2x + 3) - 5] = 18x - 338$

$8[4 - 6x + 6] + 5[2x + 3 - 5] = 18x - 338$

$8(10 - 6x) + 5(2x - 2) = 18x - 338$

$80 - 48x + 10x - 10 = 18x - 338$

$70 - 38x = 18x - 338$

$70 - 38x + 38x = 18x - 338 + 38x$

$70 = 56x - 338$

$70 + 338 = 56x - 338 + 338$

$408 = 56x$

$\dfrac{408}{56} = \dfrac{56x}{56}$

$\dfrac{51}{7} = x$

Check: $\qquad 8[4 - 6(x - 1)] + 5[(2x + 3) - 5] = 18x - 338$

$8\left[4 - 6\left(\dfrac{51}{7} - 1\right)\right] + 5\left[\left(2 \cdot \dfrac{51}{7} + 3\right) - 5\right] \overset{?}{=} 18 \cdot \dfrac{51}{7} - 338$

$8\left[4 - 6\left(\dfrac{44}{7}\right)\right] + 5\left[\left(\dfrac{102}{7} + \dfrac{21}{7}\right) - 5\right] \overset{?}{=} \dfrac{918}{7} - 338$

$8\left[\dfrac{28}{7} - \dfrac{264}{7}\right] + 5\left[\dfrac{123}{7} - \dfrac{35}{7}\right] \overset{?}{=} \dfrac{918}{7} - 338$

$8\left(-\dfrac{236}{7}\right) + 5\left(\dfrac{88}{7}\right) \overset{?}{=} \dfrac{918}{7} - \dfrac{2366}{7}$

$-\dfrac{1888}{7} + \dfrac{440}{7} \overset{?}{=} -\dfrac{1448}{7}$

$-\dfrac{1448}{7} = -\dfrac{1448}{7}$

The solution is $\dfrac{51}{7}$, or the solution set is $\left\{\dfrac{51}{7}\right\}$.

87. $\qquad 3(36.7 - 4.3x) - 10 = 4(10 - 2.5x) - 8(3.5 - 4.1x)$

$110.1 - 12.9x - 10 = 40 - 10x - 28 + 32.8x$

$-12.9x + 100.1 = 22.8x + 12$

$-22.8x - 12.9x + 100.1 = -22.8x + 22.8x + 12$

$-35.7x + 100.1 = 12$

$-35.7x + 100.1 - 100.1 = 12 - 100.1$

$-35.7x = -88.1$

$\dfrac{-35.7x}{-35.7} = \dfrac{-88.1}{-35.7}$

$x \approx 2.47$

89. $3.5\{4-[6-(2x+3)]+5\} = -18.4$
$3.5\{4-[6-2x-3]+5\} = -18.4$
$3.5\{4-[3-2x]+5\} = -18.4$
$3.5\{4-3+2x+5\} = -18.4$
$3.5\{6+2x\} = -18.4$
$21+7x = -18.4$
$7x = -39.4$
$x = \dfrac{-39.4}{7}$
$x \approx -5.6$

91. $3d+2x=12;\ x=-4$
$3d+2(-4)=12$
$3d-8=12$
$3d-8+8=12+8$
$3d=20$
$\dfrac{3d}{3}=\dfrac{20}{3}$
$d=\dfrac{20}{3}$

93. $\dfrac{2}{3}x-d=1;\ x=-\dfrac{3}{8}$
$\dfrac{2}{3}\left(-\dfrac{3}{8}\right)-d=1$
$-\dfrac{1}{4}-d=1$
$-\dfrac{1}{4}-d+\dfrac{1}{4}=1+\dfrac{1}{4}$
$-d=\dfrac{5}{4}$
$d=-\dfrac{5}{4}$

95. Answers may vary. Possible answer:
$6x - 2(x + 1)$ is an expression, while
$6x - 2(x + 1) = 6$ is an equation. An algebraic equation involves an equals sign, while an algebraic expression does not.

97. Answers may vary. Possible answer: This can lead to a correct answer if the next step is to add $-7x$ to both sides of the equation. However, the first step should be to combine like terms.

Section 2.3

Preparing for Solving Linear Equations Involving Fractions and Decimals; Classifying Equations

P1. $5 = 5$
$4 = 2 \cdot 2$
$\text{LCD} = 2 \cdot 2 \cdot 5 = 20$
The LCD of $\dfrac{3}{5}$ and $\dfrac{3}{4}$ is 20.

P2. $8 = 2 \cdot 2 \cdot 2$
$12 = 2 \cdot 2 \cdot 3$
$\text{LCD} = 2 \cdot 2 \cdot 2 \cdot 3 = 24$
The LCD of $\dfrac{3}{8}$ and $-\dfrac{7}{12}$ is 24.

2.3 Quick Checks

1. The <u>least</u> <u>common</u> <u>denominator</u> is the smallest number that each denominator has as a common multiple.

2. The LCD of 5, 4, and 2 is 20.
$\dfrac{2x}{5}-\dfrac{x}{4}=\dfrac{3}{2}$
$20\left(\dfrac{2x}{5}-\dfrac{x}{4}\right)=20\left(\dfrac{3}{2}\right)$
$20\left(\dfrac{2x}{5}\right)-20\left(\dfrac{x}{4}\right)=20\left(\dfrac{3}{2}\right)$
$8x-5x=30$
$3x=30$
$\dfrac{3x}{3}=\dfrac{30}{3}$
$x=10$

Check: $\dfrac{2x}{5}-\dfrac{x}{4}=\dfrac{3}{2}$
$\dfrac{2(10)}{5}-\dfrac{10}{4}\overset{?}{=}\dfrac{3}{2}$
$4-\dfrac{5}{2}\overset{?}{=}\dfrac{3}{2}$
$\dfrac{8}{2}-\dfrac{5}{2}\overset{?}{=}\dfrac{3}{2}$
$\dfrac{3}{2}=\dfrac{3}{2}$ True

The solution of the equation is 10, or the solution set is $\{10\}$.

3. The LCD of 6 and 9 is 18.

$$\frac{5}{6}x + \frac{1}{9} = -\frac{1}{6}x - \frac{1}{6}$$

$$18\left(\frac{5}{6}x + \frac{1}{9}\right) = 18\left(-\frac{1}{6}x - \frac{1}{6}\right)$$

$$18\left(\frac{5}{6}x\right) + 18\left(\frac{1}{9}\right) = 18\left(-\frac{1}{6}x\right) - 18\left(\frac{1}{6}\right)$$

$$15x + 2 = -3x - 3$$

$$3x + 15x + 2 = 3x - 3x - 3$$

$$18x + 2 = -3$$

$$18x + 2 - 2 = -3 - 2$$

$$18x = -5$$

$$\frac{18x}{18} = \frac{-5}{18}$$

$$x = -\frac{5}{18}$$

Check: $\dfrac{5}{6}x + \dfrac{1}{9} = -\dfrac{1}{6}x - \dfrac{1}{6}$

$$\frac{5}{6}\left(-\frac{5}{18}\right) + \frac{1}{9} \stackrel{?}{=} -\frac{1}{6}\left(-\frac{5}{18}\right) - \frac{1}{6}$$

$$-\frac{25}{108} + \frac{1}{9} \stackrel{?}{=} \frac{5}{108} - \frac{1}{6}$$

$$-\frac{25}{108} + \frac{12}{108} \stackrel{?}{=} \frac{5}{108} - \frac{18}{108}$$

$$-\frac{13}{108} = -\frac{13}{108} \quad \text{True}$$

The solution of the equation is $-\dfrac{5}{18}$, or the

solution set is $\left\{-\dfrac{5}{18}\right\}$.

4. The result of multiplying the equation
$\dfrac{1}{5}x + 7 = \dfrac{3}{10}$ by 10 is $2x + \underline{70} = 3$.

5.
$$\frac{a}{3} - \frac{1}{3} = -5$$

$$3\left(\frac{a}{3} - \frac{1}{3}\right) = 3(-5)$$

$$3\left(\frac{a}{3}\right) - 3\left(\frac{1}{3}\right) = 3(-5)$$

$$a - 1 = -15$$

$$a - 1 + 1 = -15 + 1$$

$$a = -14$$

Check: $\dfrac{a}{3} - \dfrac{1}{3} = -5$

$$\frac{-14}{3} - \frac{1}{3} \stackrel{?}{=} -5$$

$$\frac{-15}{3} \stackrel{?}{=} -5$$

$$-5 = -5 \quad \text{True}$$

The solution is -14, or the solution set is $\{-14\}$.

6. The LCD of 4 and 5 is 20.

$$\frac{3x-3}{4} - 1 = \frac{3}{5}x$$

$$20\left(\frac{3x-3}{4} - 1\right) = 20\left(\frac{3}{5}x\right)$$

$$20\left(\frac{3x-3}{4}\right) - 20(1) = 20\left(\frac{3}{5}x\right)$$

$$5(3x - 3) - 20 = 4(3x)$$

$$15x - 15 - 20 = 12x$$

$$15x - 35 = 12x$$

$$15x - 35 - 15x = 12x - 15x$$

$$-35 = -3x$$

$$\frac{-35}{-3} = \frac{-3x}{-3}$$

$$\frac{35}{3} = x$$

Check: $\dfrac{3x-3}{4} - 1 = \dfrac{3}{5}x$

$$\frac{3\left(\frac{35}{3}\right) - 3}{4} - 1 \stackrel{?}{=} \frac{3}{5}\left(\frac{35}{3}\right)$$

$$\frac{35 - 3}{4} - 1 \stackrel{?}{=} 7$$

$$\frac{32}{4} - 1 \stackrel{?}{=} 7$$

$$8 - 1 \stackrel{?}{=} 7$$

$$7 = 7 \quad \text{True}$$

The solution is $\dfrac{35}{3}$, or the solution set is $\left\{\dfrac{35}{3}\right\}$.

7. To clear the decimals in the equation
$0.25x + 5 = 7 - 0.3x$, multiply both sides of the
equation by $\underline{100}$.

8.
$$0.2z = 20$$

$$10 \cdot 0.2z = 10 \cdot 20$$

$$2z = 200$$

$$\frac{2z}{2} = \frac{200}{2}$$

$$z = 100$$

Check: $0.2z = 20$

$0.2(100) \overset{?}{=} 20$

$20 = 20$ True

The solution is 100, or the solution set is $\{100\}$.

9.

$$0.15p - 2.5 = 5$$
$$100(0.15p - 2.5) = 100(5)$$
$$100 \cdot 0.15p - 100 \cdot 2.5 = 100 \cdot 5$$
$$15p - 250 = 500$$
$$15p - 250 + 250 = 500 + 250$$
$$15p = 750$$
$$\frac{15p}{15} = \frac{750}{15}$$
$$p = 50$$

Check: $0.15p - 2.5 = 5$

$0.15(50) - 2.5 \overset{?}{=} 5$

$7.5 - 2.5 \overset{?}{=} 5$

$5 = 5$ True

The solution is 50, or the solution set is $\{50\}$.

10. The coefficient of the first term of the equation $n + 0.25n = 50$ is $\underline{1}$.

11. $p + 0.05p = 52.5$

$$1p + 0.05p = 52.5$$
$$1.05p = 52.5$$
$$100 \cdot 1.05p = 100 \cdot 52.5$$
$$105p = 5250$$
$$\frac{105p}{105} = \frac{5250}{105}$$
$$p = 50$$

The solution of the equation is 50, or the solution set is $\{50\}$.

12. $c - 0.25c = 120$

$$1c - 0.25c = 120$$
$$0.75c = 120$$
$$100 \cdot 0.75c = 100 \cdot 120$$
$$75c = 12,000$$
$$\frac{75c}{75} = \frac{12,000}{75}$$
$$c = 160$$

The solution of the equation is 160, or the solution set is $\{160\}$.

13. $0.36y - 0.5 = 0.16y + 0.3$

$$0.20y - 0.5 = 0.3$$
$$0.2y = 0.8$$
$$10(0.2y) = 10(0.8)$$
$$2y = 8$$
$$y = 4$$

The solution to the equation is 4, or the solution set is $\{4\}$.

14. $0.12x + 0.05(5000 - x) = 460$

$$0.12x + 250 - 0.05x = 460$$
$$0.07x + 250 = 460$$
$$0.07x = 210$$
$$100(0.07x) = 100(210)$$
$$7x = 21,000$$
$$x = 3000$$

The solution to the equation is 3000, or the solution set is $\{3000\}$.

15. True

16.

$$3(x + 4) = 4 + 3x + 18$$
$$3x + 12 = 3x + 22$$
$$3x + 12 - 3x = 3x + 22 - 3x$$
$$12 = 22$$

The statement $12 = 22$ is false, so the equation is a contradiction. The solution set is $\varnothing$ or $\{\ \}$.

17. $\frac{1}{3}(6x - 9) - 1 = 6x - [4x - (-4)]$

$$2x - 3 - 1 = 6x - [4x + 4]$$
$$2x - 4 = 6x - 4x - 4$$
$$2x - 4 = 2x - 4$$
$$2x - 4 - 2x = 2x - 4 - 2x$$
$$-4 = -4$$

The statement $-4 = -4$ is true for all real numbers x. The solution set is the set of all real numbers.

18. $-5 - (9x + 8) + 23 = 7 + x - (10x - 3)$

$$-5 - 9x - 8 + 23 = 7 + x - 10x + 3$$
$$-9x + 10 = -9x + 10$$
$$-9x + 10 + 9x = -9x + 10 + 9x$$
$$10 = 10$$

The statement $10 = 10$ is true for all real numbers x. The solution set is the set of all real numbers.

19.

$$\frac{3}{2}x - 8 = x + 7 + \frac{1}{2}x$$

$$\frac{3}{2}x - 8 = \frac{2}{2}x + 7 + \frac{1}{2}x$$

$$\frac{3}{2}x - 8 = \frac{3}{2}x + 7$$

$$\frac{3}{2}x - 8 - \frac{3}{2}x = \frac{3}{2}x + 7 - \frac{3}{2}x$$

$$-8 = 7$$

The statement $-8 = 7$ is false, so the equation is a contradiction. The solution set is $\varnothing$ or { }.

20. When the variable is eliminated from a linear equation and a true statement results, the solution set is <u>all real numbers</u>.

21. When the variable is eliminated from a linear equation and a false statement results, the solution set is <u>the empty set ($\varnothing$)</u>.

22. $2(x - 7) + 8 = 6x - (4x + 2) - 4$

$$2x - 14 + 8 = 6x - 4x - 2 - 4$$

$$2x - 6 = 2x - 6$$

$$2x - 6 - 2x = 2x - 6 - 2x$$

$$-6 = -6$$

The statement $-6 = -6$ is true for all values of x, so the equation is an identity. The solution set is the set of all real numbers.

23.

$$\frac{4(7 - x)}{3} = x$$

$$3\left(\frac{4(7 - x)}{3}\right) = 3x$$

$$4(7 - x) = 3x$$

$$28 - 4x = 3x$$

$$28 - 4x + 4x = 3x + 4x$$

$$28 = 7x$$

$$4 = x$$

The equation has solution $x = 4$, so it is a conditional equation. The solution set is {4}.

24. $\frac{1}{2}(4x - 6) = 6\left(\frac{1}{3}x - \frac{1}{2}\right) + 4$

$$2x - 3 = 2x - 3 + 4$$

$$2x - 3 = 2x + 1$$

$$2x - 3 - 2x = 2x + 1 - 2x$$

$$-3 = 1$$

The statement $-3 = 1$ is false, so the equation is a contradiction. The solution set is $\varnothing$ or { }.

25. $4(5x - 4) + 1 = -2 + 20x$

$$20x - 16 + 1 = -2 + 20x$$

$$20x - 15 = -2 + 20x$$

$$20x - 15 - 20x = -2 + 20x - 20x$$

$$-15 = -2$$

The statement $-15 = -2$ is false, so the equation is a contradiction. The solution set is $\varnothing$ or { }.

26. $0.04x + 0.06(x + 250) = 65$

$$0.04x + 0.06x + 15 = 65$$

$$0.10x + 15 = 65$$

$$0.1x = 50$$

$$10(0.1x) = 10(50)$$

$$x = 500$$

She invested \$500 in the savings account.

2.3 Exercises

27.

$$\frac{2k - 1}{4} = 2$$

$$4\left(\frac{2k - 1}{4}\right) = 4(2)$$

$$2k - 1 = 8$$

$$2k - 1 + 1 = 8 + 1$$

$$2k = 9$$

$$\frac{2k}{2} = \frac{9}{2}$$

$$k = \frac{9}{2}$$

Check: $\frac{2k - 1}{4} = 2$

$$\frac{2\left(\frac{9}{2}\right) - 1}{4} \stackrel{?}{=} 2$$

$$\frac{9 - 1}{4} \stackrel{?}{=} 2$$

$$\frac{8}{4} \stackrel{?}{=} 2$$

$$2 = 2 \quad \text{True}$$

The solution is $\frac{9}{2}$, or the solution set is $\left\{\frac{9}{2}\right\}$.

29.

$$\frac{3x + 2}{4} = \frac{x}{2}$$

$$4\left(\frac{3x + 2}{4}\right) = 4\left(\frac{x}{2}\right)$$

$$3x + 2 = 2x$$

$$3x + 2 - 3x = 2x - 3x$$

$$2 = -x$$

$$-1(2) = -1(-x)$$

$$-2 = x$$

Check: $\dfrac{3x+2}{4} = \dfrac{x}{2}$

$\dfrac{3(-2)+2}{4} = \dfrac{-2}{2}$

$\dfrac{-6+2}{4} = -1$

$\dfrac{-4}{4} = -1$

$-1 = -1$ True

The solution is -2, or the solution set is $\{-2\}$.

31. $\dfrac{1}{5}x + \dfrac{3}{2} = \dfrac{3}{10}$

$10\left(\dfrac{1}{5}x + \dfrac{3}{2}\right) = 10\left(\dfrac{3}{10}\right)$

$10\left(\dfrac{1}{5}x\right) + 10\left(\dfrac{3}{2}\right) = 3$

$2x + 15 = 3$

$2x + 15 - 15 = 3 - 15$

$2x = -12$

$\dfrac{2x}{2} = \dfrac{-12}{2}$

$x = -6$

Check: $\dfrac{1}{5}x + \dfrac{3}{2} = \dfrac{3}{10}$

$\dfrac{1}{5}(-6) + \dfrac{3}{2} = \dfrac{3}{10}$

$-\dfrac{12}{10} + \dfrac{15}{10} = \dfrac{3}{10}$

$\dfrac{3}{10} = \dfrac{3}{10}$ True

The solution is -6, or the solution set is $\{-6\}$.

33. $\dfrac{-2x}{3} + 1 = \dfrac{5}{9}$

$9\left(\dfrac{-2x}{3} + 1\right) = 9\left(\dfrac{5}{9}\right)$

$9\left(\dfrac{-2x}{3}\right) + 9(1) = 5$

$-6x + 9 = 5$

$-6x + 9 - 9 = 5 - 9$

$-6x = -4$

$\dfrac{-6x}{-6} = \dfrac{-4}{-6}$

$x = \dfrac{2}{3}$

Check: $\dfrac{-2x}{3} + 1 = \dfrac{5}{9}$

$\dfrac{-2 \cdot \frac{2}{3}}{3} + 1 \stackrel{?}{=} \dfrac{5}{9}$

$-\dfrac{4}{9} + 1 \stackrel{?}{=} \dfrac{5}{9}$

$-\dfrac{4}{9} + \dfrac{9}{9} \stackrel{?}{=} \dfrac{5}{9}$

$\dfrac{5}{9} = \dfrac{5}{9}$ True

The solution is $\dfrac{2}{3}$, or the solution set is $\left\{\dfrac{2}{3}\right\}$.

35. $\dfrac{a}{4} - \dfrac{a}{3} = -\dfrac{1}{2}$

$12\left(\dfrac{a}{4} - \dfrac{a}{3}\right) = 12\left(-\dfrac{1}{2}\right)$

$12\left(\dfrac{a}{4}\right) - 12\left(\dfrac{a}{3}\right) = -6$

$3a - 4a = -6$

$-a = -6$

$\dfrac{-a}{-1} = \dfrac{-6}{-1}$

$a = 6$

Check: $\dfrac{a}{4} - \dfrac{a}{3} = -\dfrac{1}{2}$

$\dfrac{6}{4} - \dfrac{6}{3} \stackrel{?}{=} -\dfrac{1}{2}$

$\dfrac{18}{12} - \dfrac{24}{12} \stackrel{?}{=} -\dfrac{6}{12}$

$-\dfrac{6}{12} = -\dfrac{6}{12}$ True

The solution is 6, or the solution set is $\{6\}$.

37. $\dfrac{5}{4}(2a - 10) = -\dfrac{3}{2}a$

$4 \cdot \dfrac{5}{4}(2a - 10) = 4 \cdot \left(-\dfrac{3}{2}a\right)$

$5(2a - 10) = 2(-3a)$

$10a - 50 = -6a$

$10a - 50 - 10a = -6a - 10a$

$-50 = -16a$

$\dfrac{-50}{-16} = \dfrac{-16a}{-16}$

$\dfrac{25}{8} = a$

Check: $\dfrac{5}{4}(2a-10)=-\dfrac{3}{2}a$

$\dfrac{5}{4}\left[2\left(\dfrac{25}{8}\right)-10\right]\overset{?}{=}-\dfrac{3}{2}\cdot\dfrac{25}{8}$

$\dfrac{5}{4}\left(\dfrac{25}{4}-10\right)\overset{?}{=}-\dfrac{75}{16}$

$\dfrac{125}{16}-\dfrac{200}{16}\overset{?}{=}-\dfrac{75}{16}$

$-\dfrac{75}{16}=-\dfrac{75}{16}$ True

The solution is $\dfrac{25}{8}$, or the solution set is $\left\{\dfrac{25}{8}\right\}$.

39. $\dfrac{y}{10}+3=\dfrac{y}{4}+6$

$20\left(\dfrac{y}{10}+3\right)=20\left(\dfrac{y}{4}+6\right)$

$20\left(\dfrac{y}{10}\right)+20(3)=20\left(\dfrac{y}{4}\right)+20(6)$

$2y+60=5y+120$

$-2y+2y+60=-2y+5y+120$

$60=3y+120$

$60-120=3y+120-120$

$-60=3y$

$\dfrac{-60}{3}=\dfrac{3y}{3}$

$-20=y$

Check: $\dfrac{y}{10}+3=\dfrac{y}{4}+6$

$\dfrac{-20}{10}+3\overset{?}{=}\dfrac{-20}{4}+6$

$-2+3\overset{?}{=}-5+6$

$1=1$ True

The solution is -20, or the solution set is $\{-20\}$.

41. $\dfrac{4x-9}{3}+\dfrac{x}{6}=\dfrac{x}{2}-2$

$6\left(\dfrac{4x-9}{3}+\dfrac{x}{6}\right)=6\left(\dfrac{x}{2}-2\right)$

$6\left(\dfrac{4x-9}{3}\right)+6\left(\dfrac{x}{6}\right)=6\left(\dfrac{x}{2}\right)-6(2)$

$2(4x-9)+x=3x-12$

$8x-18+x=3x-12$

$9x-18=3x-12$

$9x-18-3x=3x-12-3x$

$6x-18=-12$

$6x-18+18=-12+18$

$6x=6$

$\dfrac{6x}{6}=\dfrac{6}{6}$

$x=1$

Check: $\dfrac{4x-9}{3}+\dfrac{x}{6}=\dfrac{x}{2}-2$

$\dfrac{4\cdot1-9}{3}+\dfrac{1}{6}\overset{?}{=}\dfrac{1}{2}-2$

$-\dfrac{10}{6}+\dfrac{1}{6}\overset{?}{=}\dfrac{3}{6}-\dfrac{12}{6}$

$-\dfrac{9}{6}=-\dfrac{9}{6}$ True

The solution is 1, or the solution set is $\{1\}$.

43. $0.4w=12$

$10(0.4w)=10(12)$

$4w=120$

$\dfrac{4w}{4}=\dfrac{120}{4}$

$w=30$

Check: $0.4w=12$

$0.4(30)\overset{?}{=}12$

$12=12$ True

The solution is 30, or the solution set is $\{30\}$.

45. $-1.3c=5.2$

$10(-1.3c)=10(5.2)$

$-13c=52$

$\dfrac{-13c}{-13}=\dfrac{52}{-13}$

$c=-4$

Check: $-1.3c=5.2$

$-1.3(-4)\overset{?}{=}5.2$

$5.2=5.2$ True

The solution is -4, or the solution set is $\{-4\}$.

47.
$$1.05p = 52.5$$
$$100(1.05p) = 100(52.5)$$
$$105p = 5250$$
$$\frac{105p}{105} = \frac{5250}{105}$$
$$p = 50$$
Check: $1.05p = 52.5$
$$1.05(50) \stackrel{?}{=} 52.5$$
$$52.5 = 52.5 \quad \text{True}$$
The solution is 50, or the solution set is $\{50\}$.

49. $p + 1.5p = 12$
$$2.5p = 12$$
$$10(2.5p) = 10(12)$$
$$25p = 120$$
$$\frac{25p}{25} = \frac{120}{25}$$
$$p = 4.8$$
Check: $p + 1.5p = 12$
$$4.8 + 1.5(4.8) \stackrel{?}{=} 12$$
$$4.8 + 7.2 \stackrel{?}{=} 12$$
$$12 = 12 \quad \text{True}$$
The solution is 4.8, or the solution set is $\{4.8\}$.

51. $p + 0.05p = 157.5$
$$1.05p = 157.5$$
$$100(1.05p) = 100(157.5)$$
$$105p = 15,750$$
$$\frac{105p}{105} = \frac{15,750}{105}$$
$$p = 150$$
Check: $p + 0.05p = 157.5$
$$150 + 0.05(150) \stackrel{?}{=} 157.5$$
$$150 + 7.5 \stackrel{?}{=} 157.5$$
$$157.5 = 157.5 \quad \text{True}$$
The solution is 150, or the solution set is $\{150\}$.

53.
$$0.3x + 2.3 = 0.2x + 1.1$$
$$10(0.3x + 2.3) = 10(0.2x + 1.1)$$
$$10(0.3x) + 10(2.3) = 10(0.2x) + 10(1.1)$$
$$3x + 23 = 2x + 11$$
$$-2x + 3x + 23 = -2x + 2x + 11$$
$$x + 23 = 11$$
$$x + 23 - 23 = 11 - 23$$
$$x = -12$$
Check: $0.3x + 2.3 = 0.2x + 1.1$
$$0.3(-12) + 2.3 \stackrel{?}{=} 0.2(-12) + 1.1$$
$$-3.6 + 2.3 \stackrel{?}{=} -2.4 + 1.1$$
$$-1.3 = -1.3 \quad \text{True}$$
The solution is -12, or the solution set is $\{-12\}$.

55.
$$0.65x + 0.3x = x - 3$$
$$0.95x = x - 3$$
$$100(0.95x) = 100(x - 3)$$
$$95x = 100x - 300$$
$$95x - 100x = 100x - 300 - 100x$$
$$-5x = -300$$
$$\frac{-5x}{-5} = \frac{-300}{-5}$$
$$x = 60$$
Check: $0.65x + 0.3x = x - 3$
$$0.65(60) + 0.3(60) \stackrel{?}{=} 60 - 3$$
$$39 + 18 \stackrel{?}{=} 57$$
$$57 = 57 \quad \text{True}$$
The solution is 60, or the solution set is $\{60\}$.

57.
$$3 + 1.5(z + 2) = 3.5z - 4$$
$$3 + 1.5z + 3 = 3.5z - 4$$
$$1.5z + 6 = 3.5z - 4$$
$$-1.5z + 1.5z + 6 = -1.5z + 3.5z - 4$$
$$6 = 2z - 4$$
$$6 + 4 = 2z - 4 + 4$$
$$10 = 2z$$
$$\frac{10}{2} = \frac{2z}{2}$$
$$5 = z$$
Check: $3 + 1.5(z + 2) = 3.5z - 4$
$$3 + 1.5(5 + 2) \stackrel{?}{=} 3.5(5) - 4$$
$$3 + 1.5(7) \stackrel{?}{=} 17.5 - 4$$
$$3 + 10.5 \stackrel{?}{=} 13.5$$
$$13.5 = 13.5 \quad \text{True}$$
The solution is 5, or the solution set is $\{5\}$.

59.
$$0.02(2c - 24) = -0.4(c - 1)$$
$$100[0.02(2c - 24)] = 100[-0.4(c - 1)]$$
$$2(2c - 24) = -40(c - 1)$$
$$4c - 48 = -40c + 40$$
$$4c - 48 + 40c = -40c + 40 + 40c$$
$$44c - 48 = 40$$
$$44c - 48 + 48 = 40 + 48$$
$$44c = 88$$
$$\frac{44c}{44} = \frac{88}{44}$$
$$c = 2$$
Check: $0.02(2c - 24) = -0.4(c - 1)$
$$0.02[2(2) - 24] \stackrel{?}{=} -0.4(2 - 1)$$
$$0.02(4 - 24) \stackrel{?}{=} -0.4(1)$$
$$0.02(-20) \stackrel{?}{=} -0.4$$
$$-0.4 = -0.4 \quad \text{True}$$
The solution is 2, or the solution set is $\{2\}$.

61.
$$0.15x + 0.10(250 - x) = 28.75$$
$$0.15x + 25 - 0.1x = 28.75$$
$$25 + 0.05x = 28.75$$
$$-25 + 25 + 0.05x = -25 + 28.75$$
$$0.05x = 3.75$$
$$100(0.05x) = 100(3.75)$$
$$5x = 375$$
$$\frac{5x}{5} = \frac{375}{5}$$
$$x = 75$$

Check:
$$0.15x + 0.10(250 - x) = 28.75$$
$$0.15(75) + 0.10(250 - 75) \overset{?}{=} 28.75$$
$$11.25 + 0.10(175) \overset{?}{=} 28.75$$
$$11.25 + 17.5 \overset{?}{=} 28.75$$
$$28.75 = 28.75 \quad \text{True}$$

The solution is 75, or the solution set is {75}.

63.
$$4z - 3(z + 1) = 2(z - 3) - z$$
$$4z - 3z - 3 = 2z - 6 - z$$
$$z - 3 = z - 6$$
$$-z + z - 3 = -z + z - 6$$
$$-3 = -6$$

This is a false statement, so the equation is a contraction. The solution set is $\varnothing$ or { }.

65.
$$6q - (q - 3) = 2q + 3(q + 1)$$
$$6q - q + 3 = 2q + 3q + 3$$
$$5q + 3 = 5q + 3$$
$$-5q + 5q + 3 = -5q + 5q + 3$$
$$3 = 3$$

This is a true statement. The equation is an identity. The solution set is the set of all real numbers.

67.
$$9a - 5(a + 1) = 2(a - 3)$$
$$9a - 5a - 5 = 2a - 6$$
$$4a - 5 = 2a - 6$$
$$-2a + 4a - 5 = -2a + 2a - 6$$
$$2a - 5 = -6$$
$$2a - 5 + 5 = -6 + 5$$
$$2a = -1$$
$$\frac{2a}{2} = \frac{-1}{2}$$
$$a = -\frac{1}{2}$$

This is a conditional equation. The solution set is $\left\{-\dfrac{1}{2}\right\}$.

69.
$$\frac{4x - 9}{6} - \frac{x}{2} = \frac{x}{6} + 3$$
$$6\left(\frac{4x - 9}{6} - \frac{x}{2}\right) = 6\left(\frac{x}{6} + 3\right)$$
$$6\left(\frac{4x - 9}{6}\right) - 6\left(\frac{x}{2}\right) = 6\left(\frac{x}{6}\right) + 6(3)$$
$$4x - 9 - 3x = x + 18$$
$$x - 9 = x + 18$$
$$-x + x - 9 = -x + x + 18$$
$$-9 = 18$$

This is a false statement, so the equation is a contradiction. The solution set is $\varnothing$ or { }.

71.
$$\frac{5z + 1}{5} = \frac{2z - 3}{2}$$
$$10\left(\frac{5z + 1}{5}\right) = 10\left(\frac{2z - 3}{2}\right)$$
$$2(5z + 1) = 5(2z - 3)$$
$$10z + 2 = 10z - 15$$
$$-10z + 10z + 2 = -10z + 10z - 15$$
$$2 = -15$$

This is a false statement, so the equation is a contradiction. The solution set is $\varnothing$ or { }.

73.
$$\frac{q}{3} + \frac{4}{5} = \frac{5q + 12}{15}$$
$$15\left(\frac{q}{3} + \frac{4}{5}\right) = 15\left(\frac{5q + 12}{15}\right)$$
$$15\left(\frac{q}{3}\right) + 15\left(\frac{4}{5}\right) = 5q + 12$$
$$5q + 12 = 5q + 12$$
$$-5q + 5q + 12 = -5q + 5q + 12$$
$$12 = 12$$

This is a true statement. The equation is an identity. The solution set is the set of all real numbers.

75.
$$-3(2n + 4) = 10n$$
$$-6n - 12 = 10n$$
$$6n - 6n - 12 = 6n + 10n$$
$$-12 = 16n$$
$$\frac{-12}{16} = \frac{16n}{16}$$
$$-\frac{3}{4} = n$$

The solution is $-\dfrac{3}{4}$ or the solution set is $\left\{-\dfrac{3}{4}\right\}$.

77. $-2x+5x = 4(x+2)-(x+8)$
$3x = 4x+8-x-8$
$3x = 3x$
$-3x+3x = -3x+3x$
$0 = 0$

This is a true statement. The equation is an identity. The solution set is the set of all real numbers.

79. $-6(x-2)+8x = -x+10-3x$
$-6x+12+8x = -x+10-3x$
$12+2x = 10-4x$
$12+2x+4x = 10-4x+4x$
$12+6x = 10$
$-12+12+6x = -12+10$
$6x = -2$
$\dfrac{6x}{6} = \dfrac{-2}{6}$
$x = -\dfrac{1}{3}$

The solution is $-\dfrac{1}{3}$ or the solution set is $\left\{-\dfrac{1}{3}\right\}$.

81. $\dfrac{3}{4}x = \dfrac{1}{2}x-5$
$4\left(\dfrac{3}{4}x\right) = 4\left(\dfrac{1}{2}x-5\right)$
$3x = 2x-20$
$-2x+3x = -2x+2x-20$
$x = -20$

The solution is -20, or the solution set is $\{-20\}$.

83. $\dfrac{1}{2}x+2 = \dfrac{4x+1}{4}$
$4\left(\dfrac{1}{2}x+2\right) = 4\left(\dfrac{4x+1}{4}\right)$
$2x+8 = 4x+1$
$-2x+2x+8 = -2x+4x+1$
$8 = 2x+1$
$8-1 = 2x+1-1$
$7 = 2x$
$\dfrac{7}{2} = \dfrac{2x}{2}$
$\dfrac{7}{2} = x$

The solution is $\dfrac{7}{2}$, or the solution set is $\left\{\dfrac{7}{2}\right\}$.

85. $0.3p+2 = 0.1(p+5)+0.2(p+1)$
$0.3p+2 = 0.1p+0.5+0.2p+0.2$
$0.3p+2 = 0.3p+0.7$
$-0.3p+0.3p+2 = -0.3p+0.3p+0.7$
$2 = 0.7$

This is a false statement, so the equation is a contradiction. The solution set is $\varnothing$ or { }.

87. $-0.7x = 1.4$
$10(-0.7x) = 10(1.4)$
$-7x = 14$
$\dfrac{-7x}{-7} = \dfrac{14}{-7}$
$x = -2$

The solution is -2, or the solution set is $\{-2\}$.

89. $\dfrac{3(2y-1)}{5} = 2y-3$
$5\left[\dfrac{3(2y-1)}{5}\right] = 5(2y-3)$
$3(2y-1) = 5(2y-3)$
$6y-3 = 10y-15$
$-6y+6y-3 = -6y+10y-15$
$-3 = 4y-15$
$-3+15 = 4y-15+15$
$12 = 4y$
$\dfrac{12}{4} = \dfrac{4y}{4}$
$3 = y$

The solution is 3, or the solution set is $\{3\}$.

91. $0.6x-0.2(x-4) = 0.4(x-2)$
$0.6x-0.2x+0.8 = 0.4x-0.8$
$0.4x+0.8 = 0.4x-0.8$
$-0.4x+0.4x+0.8 = -0.4x+0.4x-0.8$
$0.8 = -0.8$

This is a false statement, so the equation is a contradiction. The solution set is $\varnothing$ or { }.

93. $\dfrac{3x-2}{4} = \dfrac{5x-1}{6}$
$12\left(\dfrac{3x-2}{4}\right) = 12\left(\dfrac{5x-1}{6}\right)$
$3(3x-2) = 2(5x-1)$
$9x-6 = 10x-2$
$-9x+9x-6 = -9x+10x-2$
$-6 = x-2$
$-6+2 = x-2+2$
$-4 = x$

The solution is -4, or the solution set is $\{-4\}$.

95. $0.3x + 2.6x = 5.7 - 1.8 + 2.8x$

$2.9x = 3.9 + 2.8x$

$2.9x - 2.8x = 3.9 + 2.8x - 2.8x$

$0.1x = 3.9$

$10(0.1x) = 10(3.9)$

$x = 39$

The solution is 39, or the solution set is $\{39\}$.

97. $\dfrac{3}{2}x - 6 = \dfrac{2(x-9)}{3} + \dfrac{1}{6}x$

$6\left(\dfrac{3}{2}x - 6\right) = 6\left[\dfrac{2(x-9)}{3} + \dfrac{1}{6}x\right]$

$9x - 36 = 4(x-9) + x$

$9x - 36 = 4x - 36 + x$

$9x - 36 = 5x - 36$

$-5x + 9x - 36 = -5x + 5x - 36$

$4x - 36 = -36$

$4x - 36 + 36 = -36 + 36$

$4x = 0$

$\dfrac{4x}{4} = \dfrac{0}{4}$

$x = 0$

The solution is 0, or the solution set is $\{0\}$.

99. $\dfrac{2}{3}\left[4 - \left(\dfrac{x}{2} + 6\right) - 2x\right] + 3 = \dfrac{5x}{6}$

$6\left\{\dfrac{2}{3}\left[4 - \left(\dfrac{x}{2} + 6\right) - 2x\right] + 3\right\} = 6\left(\dfrac{5x}{6}\right)$

$4\left[4 - \left(\dfrac{x}{2} + 6\right) - 2x\right] + 18 = 5x$

$16 - 4\left(\dfrac{x}{2} + 6\right) - 8x + 18 = 5x$

$16 - 2x - 24 - 8x + 18 = 5x$

$10 - 10x = 5x$

$10 = 15x$

$\dfrac{2}{3} = x$

The solution is $\dfrac{2}{3}$, or the solution set is $\left\{\dfrac{2}{3}\right\}$.

101. $2.8x + 13.754 = 4 - 2.95x$

$5.75x + 13.754 = 4$

$5.75x = -9.754$

$x = \dfrac{-9.754}{5.75}$

$x \approx -1.70$

103. $x - \{1.5x - 2[x - 3.1(x+10)]\} = 0$

$x - \{1.5x - 2[x - 3.1x - 31]\} = 0$

$x - \{1.5x - 2[-2.1x - 31]\} = 0$

$x - \{1.5x + 4.2x + 62\} = 0$

$x - \{5.7x + 62\} = 0$

$x - 5.7x - 62 = 0$

$-4.7x - 62 = 0$

$-4.7x = 62$

$x = \dfrac{62}{-4.7}$

$x \approx -13.2$

105. $1.06p = 53$

$100(1.06p) = 100(53)$

$106p = 5300$

$\dfrac{106p}{106} = \dfrac{5300}{106}$

$p = 50$

The price of the jeans is $50.

107. $x + 0.06x = 19{,}080$

$1.06x = 19{,}080$

$100(1.06x) = 100(19{,}080)$

$106x = 1{,}908{,}000$

$\dfrac{106x}{106} = \dfrac{1{,}908{,}000}{106}$

$x = 18{,}000$

The cost of the car before taxes was $18,000.

109. $w + 0.04w = 8.84$

$1.04w = 8.84$

$w = \dfrac{8.84}{1.04}$

$w = 8.5$

Bob's hourly wage was $8.50 before the raise.

111. $p - 0.25p = 60$

$0.75p = 60$

$p = \dfrac{60}{0.75}$

$p = 80$

The original price of the MP3 player was $80.

113. $0.25q + 0.10(2q + 3) = 7.05$

$100[0.25q + 0.10(2q + 3)] = 100(7.05)$

$25q + 10(2q + 3) = 705$

$25q + 20q + 30 = 705$

$45q + 30 = 705$

$45q = 675$

$q = 15$

There were 15 quarters in the piggy bank.

115. $2x + 2(x+3) = 2\left(\dfrac{1}{2}x\right) + 2(x+6)$

$\qquad 2x + 2x + 6 = x + 2x + 12$

$\qquad\quad 4x + 6 = 3x + 12$

$\qquad\qquad x + 6 = 12$

$\qquad\qquad\quad x = 6$

The first rectangle has width 6 units.

117. $\qquad 1442.50 = 0.15(x - 7300) + 730$

$100(1442.50) = 100[0.15(x - 7300) + 730]$

$\qquad 144{,}250 = 15(x - 7300) + 73{,}000$

$\qquad 144{,}250 = 15x - 109{,}500 + 73{,}000$

$\qquad 144{,}250 = 15x - 36{,}500$

$\qquad 180{,}750 = 15x$

$\qquad\quad 12{,}050 = x$

Your adjusted gross income was \$12,050.

119. A linear equation with one solution is $2x + 5 = 11$. A linear equation with no solution is $2x + 5 = 6 + 2x - 9$. A linear equation that is an identity is $2(x + 5) - 3 = 4x - (2x - 7)$. To form an identity or a contradiction, the variable expressions must be eliminated, leaving either a true (identity) or a false (contradiction) statement.

121. The student didn't multiply each term of the equation by 6. Solving using that (incorrect) method gives the second step $4x - 5 = 3x$, and solving for x gives $x = 5$. The correct method to solve the equation is to multiply ALL terms by 6. The correct second step is $4x - 30 = 3x$, producing the correct solution $x = 30$.

Section 2.4

Preparing for Evaluating Formulas and Solving Formulas for a Variable

P1. $2L + 2W$ for $L = 7$ and $W = 5$:
$2(7) + 2(5) = 14 + 10 = 24$

P2. In 0.5873, the number 8 is in the hundredths place. The number to the right of 8 is 7. Since 7 is greater than or equal to 5, we round 0.5873 to 0.59.

2.4 Quick Checks

1. A formula is an equation that describes how two or more variables are related.

2. $F = \dfrac{9}{5}C + 32$

$F = \dfrac{9}{5}(15) + 32$

$F = 27 + 32$

$F = 59$

The temperature is 59° Fahrenheit.

3. $c = a + 30$

$c = 10 + 30$

$c = 40$

A size 10 dress in the United States is a Continental dress size 40.

4. $E = 250 + 0.05S$

$E = 250 + 0.05(1250)$

$E = 250 + 62.5$

$E = 312.5$

The earnings of the salesman were \$312.50.

5. $N = p + 0.06p$

$N = 5600 + 0.06(5600)$

$N = 5600 + 336$

$N = 5936$

The new population is 5936 persons.

6. The total amount borrowed in a loan is called principal. Interest is the money paid for the use of the money.

7. The amount Bill invested, P, is \$2500. The interest rate, r, is 3% = 0.03. Because 8 months is $\dfrac{2}{3}$ of a year, $t = \dfrac{2}{3}$.

$I = Prt$

$I = 2500 \cdot 0.03 \cdot \dfrac{2}{3}$

$I = 50$

Bill earned \$50 on his investment. At the end of 8 months he had \$2500 + \$50 = \$2550.

8. $A = \dfrac{1}{2}h(B + b)$

$A = \dfrac{1}{2} \cdot 4.5(9 + 7)$

$A = \dfrac{1}{2} \cdot 4.5 \cdot 16$

$A = 36$

The area of the trapezoid is 36 square inches.

9. The area of a circle is found using the formula $\underline{A = \pi r^2}$.

10. (a) The radius of the circle is

$\frac{1}{2}(4 \text{ feet}) = 2 \text{ feet}.$

Area of remaining garden
= Area of rectangle − Area of circle

$= lw - \pi r^2$

$= (20 \text{ feet})(10 \text{ feet}) - \pi(2 \text{ feet})^2$

$= 200 \text{ feet}^2 - 4\pi \text{ feet}^2$

$\approx 187 \text{ feet}^2$

Approximately 187 square feet of garden
will receive grass.

(b) Cost for sod

$= 187 \text{ square feet} \cdot \frac{\$0.25}{1 \text{ square foot}}$

$= \$46.75$

The sod will cost $46.75.

11. The radius of the pad is $\frac{1}{2}(6 \text{ feet}) = 3 \text{ feet}.$

$\text{Area} = \pi r^2$

$\quad = \pi(3 \text{ feet})^2$

$\quad = 9\pi \text{ feet}^2$

$\quad \approx 28.27 \text{ feet}^2$

The area of the pad is about 28.27 square feet.

12. Area of 18" pizza $= \pi r^2$

$\quad\quad\quad\quad = \pi(9 \text{ inches})^2$

$\quad\quad\quad\quad = 81\pi \text{ inches}^2$

$\quad\quad\quad\quad \approx 254.47 \text{ inches}^2$

Area of 9" pizza $= \pi r^2$

$\quad\quad\quad\quad = \pi(4.5 \text{ inches})^2$

$\quad\quad\quad\quad = 20.25\pi \text{ inches}^2$

$\quad\quad\quad\quad \approx 63.62 \text{ inches}^2$

Cost per square inch of 18" pizza:

$\frac{\$16.99}{254.47} \approx \$0.07$

Cost per square inch of 9" pizza:

$\frac{\$8.99}{63.62} \approx \$0.14$

The 18" pizza is the better buy.

13.

$$F = \frac{9}{5}C + 32$$

$$F - 32 = \frac{9}{5}C + 32 - 32$$

$$F - 32 = \frac{9}{5}C$$

$$\frac{5}{9}(F - 32) = \frac{5}{9}\left(\frac{9}{5}C\right)$$

$$\frac{5}{9}(F - 32) = C \text{ or } C = \frac{5F - 160}{9}$$

14.

$$S = 2\pi rh + 2\pi r^2$$

$$S - 2\pi r^2 = 2\pi rh + 2\pi r^2 - 2\pi r^2$$

$$S - 2\pi r^2 = 2\pi rh$$

$$\frac{S - 2\pi r^2}{2\pi r} = \frac{2\pi rh}{2\pi r}$$

$$\frac{S - 2\pi r^2}{2\pi r} = h$$

15. False: Solving $x - y = 6$ for y results in $y = x - 6$.

16.

$$x + 2y = 7$$

$$x + 2y - x = 7 - x$$

$$2y = 7 - x$$

$$\frac{2y}{2} = \frac{7 - x}{2}$$

$$y = \frac{7 - x}{2}$$

$$y = \frac{7}{2} - \frac{x}{2}$$

17.

$$5x - 3y = 15$$

$$5x - 3y - 5x = 15 - 5x$$

$$-3y = 15 - 5x$$

$$\frac{-3y}{-3} = \frac{15 - 5x}{-3}$$

$$y = \frac{15 - 5x}{-3}$$

$$y = -5 + \frac{5}{3}x$$

18. $\dfrac{3}{4}a + 2b = 7$

$$\dfrac{3}{4}a + 2b - \dfrac{3}{4}a = 7 - \dfrac{3}{4}a$$

$$2b = 7 - \dfrac{3}{4}a$$

$$\dfrac{1}{2}(2b) = \dfrac{1}{2}\left(7 - \dfrac{3}{4}a\right)$$

$$b = \dfrac{7}{2} - \dfrac{3}{8}a \text{ or } b = \dfrac{28 - 3a}{8}$$

19. $3rs + \dfrac{1}{2}t = 12$

$$3rs + \dfrac{1}{2}t - 3rs = 12 - 3rs$$

$$\dfrac{1}{2}t = 12 - 3rs$$

$$2\left(\dfrac{1}{2}t\right) = 2(12 - 3rs)$$

$$t = 24 - 6rs$$

20. (a) $d = rt$

$$\dfrac{d}{r} = \dfrac{rt}{r}$$

$$\dfrac{d}{r} = t$$

   (b) Use $t = \dfrac{d}{r}$ with $d = 550$ and $r = 60$.

$$t = \dfrac{d}{r}$$

$$t = \dfrac{550}{60}$$

$$t = \dfrac{55}{6} = 9\dfrac{1}{6}$$

   It will take them $9\dfrac{1}{6}$ hours, or 9 hours and

   10 minutes.

21. (a) $I = Prt$

$$\dfrac{I}{Pr} = \dfrac{Prt}{Pr}$$

$$\dfrac{I}{Pr} = t$$

   (b) Use $t = \dfrac{I}{Pr}$ with $I = 35$, $P = 1000$, and

   $r = 7\% = 0.07$.

$$t = \dfrac{I}{Pr}$$

$$t = \dfrac{35}{1000(0.07)}$$

$$t = \dfrac{35}{70}$$

$$t = \dfrac{1}{2}$$

The $1000 must be invested for

$\dfrac{1}{2}$ year, or 6 months, at 7% interest to earn

$35 in interest.

22. (a) $R = qp$

$$\dfrac{R}{q} = \dfrac{qp}{q}$$

$$\dfrac{R}{q} = p$$

   (b) Use $p = \dfrac{R}{q}$ with $R = 5000$ and $q = 125$.

$$p = \dfrac{R}{q}$$

$$p = \dfrac{5000}{125}$$

$$p = 40$$

The price is $40.

2.4 Exercises

23. $f = 3.281m$
   $m = 335$:
   $f = 3.281(335) \approx 1099.14$
   $m = 300$:
   $f = 3.281(300) = 984.3$
   The Rogun dam is about 1099.14 feet high, the Nurek dam is 984.3 feet high.

25. $S = P - 0.20P$; $P = 130$
   $S = 130 - 0.20(130) = 130 - 26 = 104$
   The sale price is $104.00.

27. $E = 500 + 0.15S$; $S = 1000$
   $E = 500 + 0.15(1000) = 500 + 150 = 650$
   The earnings are $650.

29. $C = \dfrac{5}{9}(F - 32)$; $F = 68$

$$C = \dfrac{5}{9}(68 - 32) = \dfrac{5}{9}(36) = 20$$

   The temperature is 20°C.

31. $I = Prt$; $P = 200$, $r = 3\% = 0.03$, $t = \dfrac{6}{12} = 0.5$

$I = 200(0.03)(0.5)$
$I = 3$
Therese's investment will earn \$3.

33. (a) $P = 2l + 2w$; $l = 16$, $w = 9$
$P = 2(16) + 2(9) = 32 + 18 = 50$
The perimeter is 50 units.

    (b) $A = lw = 16 \cdot 9 = 144$
The area is 144 square units.

35. (a) $P = 2l + 2w$; $l = 12.5$, $w = 5.6$
$P = 2(12.5) + 2(5.6) = 25 + 11.2 = 36.2$
The perimeter is 36.2 meters.

    (b) $A = lw = 12.5(5.6) = 70$
The area is 70 square meters.

37. (a) $P = 4s$; $s = 9$
$P = 4 \cdot 9 = 36$
The perimeter is 36 units.

    (b) $A = s^2 = 9^2 = 81$
The area is 81 square units.

39. (a) $C = 2\pi r$; $r = 5$
$C = 2\pi(5) = 10\pi = 10(3.14) = 31.4$
The circumference is 31.4 centimeters.

    (b) $A = \pi r^2 = 3.14(5)^2 = 3.14(25) = 78.5$
The area is 78.5 square centimeters.

41. $A = \pi r^2$; $r = \dfrac{14}{3}$

$A = \pi\left(\dfrac{14}{3}\right)^2 = \dfrac{22}{7}\left(\dfrac{196}{9}\right) = \dfrac{616}{9} \approx 68.44$

The area is 68.44 square inches.

43. $d = rt$

$\dfrac{d}{t} = \dfrac{rt}{t}$

$\dfrac{d}{t} = r$

45. $C = \pi d$

$\dfrac{C}{\pi} = \dfrac{\pi d}{\pi}$

$\dfrac{C}{\pi} = d$

47. $I = Prt$

$\dfrac{I}{Pr} = \dfrac{Prt}{Pr}$

$\dfrac{I}{Pr} = t$

49. $A = \dfrac{1}{2}bh$

$\dfrac{2}{h}(A) = \dfrac{2}{h}\left(\dfrac{1}{2}bh\right)$

$\dfrac{2A}{h} = b$

51. $P = a + b + c$
$P - b - c = a + b + c - b - c$
$P - b - c = a$

53. $A = P + Prt$
$A - P = Prt$

$\dfrac{A - P}{Pt} = \dfrac{Prt}{Pt}$

$\dfrac{A - P}{Pt} = r$

55. $A = \dfrac{1}{2}h(B + b)$

$2A = h(B + b)$

$\dfrac{2A}{h} = B + b$

$\dfrac{2A}{h} - B = b$

57. $3x + y = 12$
$-3x + 3x + y = -3x + 12$
$y = -3x + 12$

59. $10x - 5y = 25$
$-10x + 10x - 5y = -10x + 25$
$-5y = -10x + 25$

$\dfrac{-5y}{-5} = \dfrac{-10x + 25}{-5}$

$y = 2x - 5$

61. $4x + 3y = 13$
$-4x + 4x + 3y = -4x + 13$
$3y = -4x + 13$

$\dfrac{3y}{3} = \dfrac{-4x + 13}{3}$

$y = \dfrac{-4x + 13}{3}$

63.

$$\frac{1}{2}x - \frac{1}{6}y = 2$$

$$6\left(\frac{1}{2}x - \frac{1}{6}y\right) = 6 \cdot 2$$

$$6\left(\frac{1}{2}x\right) - 6\left(\frac{1}{6}y\right) = 12$$

$$3x - y = 12$$

$$-3x + 3x - y = -3x + 12$$

$$-y = -3x + 12$$

$$-1 \cdot (-y) = -1(-3x + 12)$$

$$y = 3x - 12$$

65. (a)

$$P = R - C$$

$$-R + P = -R + R - C$$

$$-R + P = -C$$

$$\frac{-R + P}{-1} = \frac{-C}{-1}$$

$$R - P = C$$

(b) $C = R - P$; $P = 1200$, $R = 1650$

$C = 1650 - 1200 = 450$

The cost is \$450.

67. (a)

$$I = Prt$$

$$\frac{I}{Pt} = \frac{Prt}{Pt}$$

$$\frac{I}{Pt} = r$$

(b) $r = \dfrac{I}{Pt}$; $I = 225$, $P = 5000$, $t = 1.5$

$$r = \frac{225}{5000(1.5)} = \frac{225}{7500} = 0.03$$

The rate is 0.03 or 3%.

69. (a)

$$Z = \frac{x - \mu}{\sigma}$$

$$Z \cdot \sigma = \left(\frac{x - \mu}{\sigma}\right) \cdot \sigma$$

$$Z\sigma = x - \mu$$

$$Z\sigma + \mu = x - \mu + \mu$$

$$Z\sigma + \mu = x$$

(b) $x = Z\sigma + \mu$; $Z = 2$, $\mu = 100$, $\sigma = 15$

$x = 2(15) + 100 = 30 + 100 = 130$

71. (a)

$$y = mx + 5$$

$$y - 5 = mx + 5 - 5$$

$$y - 5 = mx$$

$$\frac{y - 5}{x} = \frac{mx}{x}$$

$$\frac{y - 5}{x} = m$$

(b) $m = \dfrac{y - 5}{x}$; $x = 3$, $y = -1$

$$m = \frac{-1 - 5}{3} = \frac{-6}{3} = -2$$

73. (a)

$$A = P + Prt$$

$$A - P = P + Prt - P$$

$$A - P = Prt$$

$$\frac{A - P}{Pt} = \frac{Prt}{Pt}$$

$$\frac{A - P}{Pt} = r$$

(b) $r = \dfrac{A - P}{Pt}$; $A = 540$, $P = 500$, $t = 2$

$$r = \frac{540 - 500}{500(2)} = \frac{40}{1000} = 0.04$$

The rate is 0.04 or 4%.

75. (a)

$$V = \pi r^2 h$$

$$\frac{V}{\pi r^2} = \frac{\pi r^2 h}{\pi r^2}$$

$$\frac{V}{\pi r^2} = h$$

(b) $h = \dfrac{V}{\pi r^2}$; $V = 320\pi$, $r = 8$

$$h = \frac{320\pi}{\pi(8)^2} = \frac{320\pi}{64\pi} = 5$$

The height is 5 mm.

77. (a)

$$A = \frac{1}{2}bh$$

$$\frac{2}{h} \cdot A = \frac{2}{h} \cdot \frac{1}{2}bh$$

$$\frac{2A}{h} = b$$

(b) $b = \dfrac{2A}{h}$; $A = 45, h = 5$

$$b = \frac{2(45)}{5} = \frac{90}{5} = 18$$

The base is 18 feet.

79. $A = 37, H = 178, W = 82$
$E = 66.67 + 13.75W + 5H - 6.76A$
$\quad = 66.67 + 13.75(82) + 5(178) - 6.76(37)$
$\quad = 66.67 + 1127.5 + 890 - 250.12$
$\quad = 1834.05$

The Basal Energy Expenditure is 1834.05.

81. (a) $V = \pi r^2 h$

$$\frac{V}{\pi r^2} = \frac{\pi r^2 h}{\pi r^2}$$

$$\frac{V}{\pi r^2} = h$$

(b) $h = \dfrac{V}{\pi r^2}$; $V = 90\pi, r = 3$

$$h = \frac{90\pi}{\pi(3)^2} = \frac{90}{9} = 10$$

The height is 10 inches.

83.

Size	Area: $A = \pi r^2$	Price per square inch
12"	$A = \pi \cdot 6^2 \approx 113.097$	$\frac{\$9.99}{113.097} = \$0.088$
8"	$A = \pi \cdot 4^2 \approx 50.265$	$\frac{\$4.49}{50.265} = \$0.089$

Since $\$0.088 < \$0.089$, the medium pizza is a better deal.

85. (a) $d = rt$

$$\frac{d}{r} = \frac{rt}{r}$$

$$\frac{d}{r} = t$$

Let $d = 600, r = 50$.

$$t = \frac{600}{50} = 12$$

Jason expects the trip to take 12 hours.

(b) $12(28) = 336$
Jason can expect \$336.

87. Region = Rectangle + Trapezoid

$$\text{Area} = lw + \frac{1}{2}h(B+b)$$
$$= 8(5) + \frac{1}{2}(3)(5+2)$$
$$= 40 + \frac{21}{2}$$
$$= 50.5$$

The area is 50.5 square inches.

89. figure $= \text{cone} + \frac{1}{2} \cdot \text{sphere}$

$$\text{Volume} = \frac{1}{3}\pi r^2 h + \frac{1}{2} \cdot \frac{4}{3}\pi r^3$$
$$= \frac{1}{3}\pi(4)^2(10) + \frac{2}{3}\pi(4)^3$$
$$= \frac{160}{3}\pi + \frac{128}{3}\pi$$
$$= \frac{288}{3}\pi$$
$$= 96\pi$$

The amount of ice cream is
$96\pi \approx 301.59$ cubic cm.

91. (a)
$$P = D - 0.02(I - 234{,}600)$$
$$-D + P = -D + D - 0.02(I - 234{,}600)$$
$$-D + P = -0.02(I - 234{,}600)$$
$$\frac{-D+P}{-0.02} = \frac{-0.02(I-234{,}600)}{-0.02}$$
$$\frac{D-P}{0.02} = I - 234{,}600$$
$$\frac{D-P}{0.02} + 234{,}600 = I - 234{,}600 + 234{,}600$$
$$\frac{D-P}{0.02} + 234{,}600 = I$$

(b) $I = \dfrac{D-P}{0.02} + 234{,}600$
$$= \frac{15{,}821 - 15{,}500}{0.02} + 234{,}600$$
$$= \frac{321}{0.02} + 234{,}600$$
$$= 16{,}050 + 234{,}600$$
$$= 250{,}650$$

The adjusted gross income is $250,650.

93. (a)   7 feet 6 inches $= 7\dfrac{1}{2}$ feet

8 feet 2 inches $= 8\dfrac{1}{6}$ feet

$$A = lw = \left(7\dfrac{1}{2}\right)\left(8\dfrac{1}{6}\right) = \left(\dfrac{15}{2}\right)\left(\dfrac{49}{6}\right) = \dfrac{245}{4} = 61\dfrac{1}{4}$$

You need 62 tiles to cover the floor.

(b) $\$6(62) = \$372$
It will cost $\$372$.

(c)   Yes; $\$372 > \$350$

95. (a)   Region = Rectangle − Circle

Area $= lw - \pi r^2 \approx (90)(60) - 3.14159(12)^2 = 5400 - 452.38896 = 4947.61104$

The grass area is 4948 square feet.

(b) $4948(0.25) = 1237$
It will cost $\$1237$ to sod the lawn.

97. (a) $l = 5$ ft $= 5 \times 12$ in. $= 60$ in.; $w = 18$ in.
$A = lw = 60(18) = 1080$
The area is 1080 square inches.

(b) $l = 5$ ft; $w = 18$ in. $= \dfrac{18}{12}$ ft $= 1.5$ ft

$A = lw = 5(1.5) = 7.5$
The area is 7.5 square feet.

99.   To convert from square inches to square feet, multiply square inches by $\dfrac{1 \text{ ft}^2}{144 \text{ in.}^2}$.

101.   Answers may vary. One possibility: Both are correct. The first answer can be expanded into the second answer.

Putting the Concepts Together (Sections 2.1–2.4)

1. (a) $4 - (6 - x) = 5x - 8$; $x = \dfrac{3}{2}$

$$4 - \left(6 - \dfrac{3}{2}\right) \stackrel{?}{=} 5\left(\dfrac{3}{2}\right) - 8$$

$$4 - \left(\dfrac{12}{2} - \dfrac{3}{2}\right) \stackrel{?}{=} \dfrac{15}{2} - 8$$

$$\dfrac{8}{2} - \dfrac{9}{2} \stackrel{?}{=} \dfrac{15}{2} - \dfrac{16}{2}$$

$$-\dfrac{1}{2} = -\dfrac{1}{2} \quad \text{True}$$

Yes, $x = \dfrac{3}{2}$ is a solution of the equation.

(b) $4-(6-x)=5x-8; \ x=-\dfrac{5}{2}$

$$4-\left[6-\left(-\dfrac{5}{2}\right)\right] \stackrel{?}{=} 5\left(-\dfrac{5}{2}\right)-8$$

$$4-\left(\dfrac{12}{2}+\dfrac{5}{2}\right) \stackrel{?}{=} \dfrac{-25}{2}-8$$

$$\dfrac{8}{2}-\dfrac{17}{2} \stackrel{?}{=} -\dfrac{25}{2}-\dfrac{16}{2}$$

$$-\dfrac{9}{2}=-\dfrac{41}{2} \quad \text{False}$$

No, $x=-\dfrac{5}{2}$ is *not* a solution of the equation.

2. (a) $\dfrac{1}{2}(x-4)+3x=x+\dfrac{1}{2}; \ x=-4$

$$\dfrac{1}{2}(-4-4)+3(-4) \stackrel{?}{=} -4+\dfrac{1}{2}$$

$$\dfrac{1}{2}(-8)+(-12) \stackrel{?}{=} -\dfrac{8}{2}+\dfrac{1}{2}$$

$$-4+(-12) \stackrel{?}{=} -\dfrac{7}{2}$$

$$-16=-\dfrac{7}{2} \quad \text{False}$$

No, $x=-4$ is *not* a solution of the equation.

(b) $\dfrac{1}{2}(x-4)+3x=x+\dfrac{1}{2}; \ x=1$

$$\dfrac{1}{2}(1-4)+3(1) \stackrel{?}{=} 1+\dfrac{1}{2}$$

$$\dfrac{1}{2}(-3)+3 \stackrel{?}{=} \dfrac{2}{2}+\dfrac{1}{2}$$

$$-\dfrac{3}{2}+\dfrac{6}{2} \stackrel{?}{=} \dfrac{3}{2}$$

$$\dfrac{3}{2}=\dfrac{3}{2} \quad \text{True}$$

Yes, $x=1$ is a solution of the equation.

3. $x+\dfrac{1}{2}=-\dfrac{1}{6}$

$$6\left(x+\dfrac{1}{2}\right)=6\left(-\dfrac{1}{6}\right)$$

$$6x+3=-1$$

$$6x+3-3=-1-3$$

$$6x=-4$$

$$\dfrac{6x}{6}=\dfrac{-4}{6}$$

$$x=-\dfrac{2}{3}$$

Check: $x+\dfrac{1}{2}=-\dfrac{1}{6}$

$$-\dfrac{2}{3}+\dfrac{1}{2} \stackrel{?}{=} -\dfrac{1}{6}$$

$$-\dfrac{4}{6}+\dfrac{3}{6} \stackrel{?}{=} -\dfrac{1}{6}$$

$$-\dfrac{1}{6}=-\dfrac{1}{6} \quad \text{True}$$

The solution is $-\dfrac{2}{3}$ or the solution set is $\left\{-\dfrac{2}{3}\right\}$.

4. $-0.4m=16$

$$10(-0.4m)=10(16)$$

$$-4m=160$$

$$\dfrac{-4m}{-4}=\dfrac{160}{-4}$$

$$m=-40$$

Check: $-0.4m=16$

$$-0.4(-40) \stackrel{?}{=} 16$$

$$16=16 \quad \text{True}$$

The solution is -40, or the solution set is $\{-40\}$.

5. $14=-\dfrac{7}{3}p$

$$-\dfrac{3}{7}\cdot 14=-\dfrac{3}{7}\cdot\left(-\dfrac{7}{3}p\right)$$

$$-3\cdot 2=p$$

$$-6=p$$

Check: $14=-\dfrac{7}{3}p$

$$14 \stackrel{?}{=} -\dfrac{7}{3}(-6)$$

$$14=14 \quad \text{True}$$

The solution is -6, or the solution set is $\{-6\}$.

6. $8n-11=13$

$$8n-11+11=13+11$$

$$8n=24$$

$$\dfrac{8n}{8}=\dfrac{24}{8}$$

$$n=3$$

Check: $8n-11=13$

$$8(3)-11 \stackrel{?}{=} 13$$

$$24-11 \stackrel{?}{=} 13$$

$$13=13 \quad \text{True}$$

The solution is 3, or the solution set is $\{3\}$.

7. $\dfrac{5}{2}n - 4 = -19$

$\dfrac{5}{2}n - 4 + 4 = -19 + 4$

$\dfrac{5}{2}n = -15$

$\dfrac{2}{5} \cdot \dfrac{5}{2}n = \dfrac{2}{5} \cdot (-15)$

$n = -6$

Check: $\dfrac{5}{2}n - 4 = -19$

$\dfrac{5}{2} \cdot (-6) - 4 \stackrel{?}{=} -19$

$-15 - 4 \stackrel{?}{=} -19$

$-19 = -19$ True

The solution is -6, or the solution set is $\{-6\}$.

8. $-(5 - x) = 2(5x + 8)$

$-5 + x = 10x + 16$

$-16 - 5 + x = 10x + 16 - 16$

$-21 + x = 10x$

$-21 + x - x = 10x - x$

$-21 = 9x$

$\dfrac{-21}{9} = \dfrac{9x}{9}$

$-\dfrac{7}{3} = x$

Check: $-(5 - x) = 2(5x + 8)$

$-\left[5 - \left(-\dfrac{7}{3}\right)\right] \stackrel{?}{=} 2\left[5\left(-\dfrac{7}{3}\right) + 8\right]$

$-\left(\dfrac{15}{3} + \dfrac{7}{3}\right) \stackrel{?}{=} 2\left(-\dfrac{35}{3} + \dfrac{24}{3}\right)$

$-\dfrac{22}{3} \stackrel{?}{=} 2\left(-\dfrac{11}{3}\right)$

$-\dfrac{22}{3} = -\dfrac{22}{3}$ True

The solution is $-\dfrac{7}{3}$, or the solution set is

$\left\{-\dfrac{7}{3}\right\}$.

9. $7(x + 6) = 2x + 3x - 15$

$7x + 42 = 5x - 15$

$7x + 42 - 42 = 5x - 15 - 42$

$7x = 5x - 57$

$-5x + 7x = -5x + 5x - 57$

$2x = -57$

$\dfrac{2x}{2} = \dfrac{-57}{2}$

$x = -\dfrac{57}{2}$

Check: $7(x + 6) = 2x + 3x - 15$

$7\left(-\dfrac{57}{2} + 6\right) \stackrel{?}{=} 2\left(-\dfrac{57}{2}\right) + 3\left(-\dfrac{57}{2}\right) - 15$

$7\left(-\dfrac{45}{2}\right) \stackrel{?}{=} -\dfrac{114}{2} - \dfrac{171}{2} - \dfrac{30}{2}$

$-\dfrac{315}{2} = -\dfrac{315}{2}$ True

The solution is $-\dfrac{57}{2}$, or the solution set is

$\left\{-\dfrac{57}{2}\right\}$.

10. $-7a + 5 + 8a = 2a + 8 - 28$

$5 + a = 2a - 20$

$5 + a + 20 = 2a - 20 + 20$

$25 + a = 2a$

$25 + a - a = 2a - a$

$25 = a$

Check: $-7a + 5 + 8a = 2a + 8 - 28$

$-7(25) + 5 + 8(25) \stackrel{?}{=} 2(25) + 8 - 28$

$-175 + 5 + 200 \stackrel{?}{=} 50 + 8 - 28$

$30 = 30$ True

The solution is 25, or the solution set is $\{25\}$.

11. $-\dfrac{1}{2}(x - 6) + \dfrac{1}{6}(x + 6) = 2$

$6\left[-\dfrac{1}{2}(x - 6) + \dfrac{1}{6}(x + 6)\right] = 6 \cdot 2$

$-3(x - 6) + 1(x + 6) = 12$

$-3x + 18 + x + 6 = 12$

$-2x + 24 = 12$

$-2x + 24 - 24 = 12 - 24$

$-2x = -12$

$\dfrac{-2x}{-2} = \dfrac{-12}{-2}$

$x = 6$

Check: $-\dfrac{1}{2}(x-6)+\dfrac{1}{6}(x+6)=2$

$-\dfrac{1}{2}(6-6)+\dfrac{1}{6}(6+6)\overset{?}{=}2$

$-\dfrac{1}{2}(0)+\dfrac{1}{6}(12)\overset{?}{=}2$

$0+2\overset{?}{=}2$

$2=2$ True

The solution is 6, or the solution set is $\{6\}$.

12.
$$0.3x-1.4=-0.2x+6$$
$$10(0.3x-1.4)=10(-0.2x+6)$$
$$3x-14=-2x+60$$
$$3x-14+14=-2x+60+14$$
$$3x=-2x+74$$
$$2x+3x=2x-2x+74$$
$$5x=74$$
$$\dfrac{5x}{5}=\dfrac{74}{5}$$
$$x=14.8$$

Check: $0.3x-1.4=-0.2x+6$

$0.3(14.8)-1.4\overset{?}{=}-0.2(14.8)+6$

$4.44-1.4\overset{?}{=}-2.96+6$

$3.04=3.04$ True

The solution is $\dfrac{74}{5}=14.8,$ or the solution set is $\{14.8\}$.

13. $5+3(2x+1)=5x+x-10$
$$5+6x+3=5x+x-10$$
$$8+6x=6x-10$$
$$8+6x-6x=6x-10-6x$$
$$8=-10$$

This is a false statement, so the equation is a contradiction. The solution set is $\varnothing$ or $\{\ \}$.

14. $3-2(x+5)=-2(x+2)-3$
$$3-2x-10=-2x-4-3$$
$$-2x-7=-2x-7$$
$$2x-2x-7=2x-2x-7$$
$$-7=-7$$

This is a true statement. The equation is an identity. The solution set is the set of all real numbers.

15. $0.024x+0.04(7500-x)=220$
$$0.024x+300-0.04x=220$$
$$300-0.016x=220$$
$$-300+300-0.016x=-300+220$$
$$-0.016x=-80$$
$$1000(-0.016x)=1000(-80)$$
$$-16x=-80,000$$
$$x=5000$$

You should invest $5000 in the CD.

16. (a)
$$A=\dfrac{1}{2}h(B+b)$$
$$\dfrac{2}{h}\cdot A=\dfrac{2}{h}\cdot\dfrac{1}{2}h(B+b)$$
$$\dfrac{2A}{h}=B+b$$
$$\dfrac{2A}{h}-B=B+b-B$$
$$\dfrac{2A}{h}-B=b$$

(b) $b=\dfrac{2A}{h}-B;\ A=76,\ h=8,\ B=13$

$b=\dfrac{2(76)}{8}-13=\dfrac{152}{8}-13=19-13=6$

The length of base b is 6 inches.

17. (a) $V=\pi r^2 h$
$$\dfrac{V}{\pi r^2}=\dfrac{\pi r^2 h}{\pi r^2}$$
$$\dfrac{V}{\pi r^2}=h$$

(b) $h=\dfrac{V}{\pi r^2};\ V=117\pi,\ r=3$

$h=\dfrac{117\pi}{\pi(3)^2}=\dfrac{117\pi}{9\pi}=13$

The height is 13 inches.

18.
$$3x+2y=14$$
$$-3x+3x+2y=-3x+14$$
$$2y=-3x+14$$
$$\dfrac{2y}{2}=\dfrac{-3x+14}{2}$$
$$y=-\dfrac{3}{2}x+7$$

Section 2.5

Preparing for Introduction to Problem Solving: Direct Translation Problems

P1.
$$x + 34.95 = 60.03$$
$$x + 34.95 - 34.95 = 60.03 - 34.95$$
$$x = 25.08$$
The solution set is $\{25.08\}$.

P2.
$$x + 0.25x = 60$$
$$1x + 0.25x = 60$$
$$1.25x = 60$$
$$100 \cdot 1.25x = 100 \cdot 60$$
$$125x = 6000$$
$$\frac{125x}{125} = \frac{6000}{125}$$
$$x = 48$$
The solution set is $\{48\}$.

2.5 Quick Checks

1. The sum of 5 and 17 is represented mathematically as $5 + 17$.

2. The product of -2 and 6 is represented mathematically as $-2 \cdot 6$.

3. The quotient of 25 and 3 is represented mathematically as $\frac{25}{3}$.

4. The difference of 7 and 4 is represented mathematically as $7 - 4$.

5. Twice a less 2 is represented mathematically as $2a - 2$.

6. Three plus the quotient of z and 4 is represented mathematically as $3 + \frac{z}{4}$.

7. Anne earned $z + 50$ dollars.

8. Melissa paid $x - 15$ dollars for her sociology book.

9. Tim has $75 - d$ quarters.

10. The width of the platform is $3l - 2$ feet.

11. The number of dimes is $2q + 3$.

12. The number of red M&Ms in the bowl is $3b - 5$.

13. To translate an English sentence into a mathematical statement we use <u>equations</u>.

14. "The product of 3 and y is equal to 21" is represented mathematically as $3y = 21$.

15. "The sum of 3 and x is equivalent to the product of 5 and x" is represented mathematically as $3 + x = 5x$.

16. "The difference of x and 10 equals the quotient of x and 2" is represented mathematically as
$$x - 10 = \frac{x}{2}.$$

17. "Three less than a number y is five times y" is represented mathematically as $y - 3 = 5y$.

18. Letting variables represent unknown quantities and then expressing relationships among the variables in the form of equations is called <u>mathematical modeling</u>.

19. We want to know how much each person paid for the pizza. Let s be the amount that Sean pays. Then Connor pays $\frac{2}{3}s$. The total amount they pay is \$15, so $s + \frac{2}{3}s = 15$.

$$s + \frac{2}{3}s = 15$$
$$\frac{5}{3}s = 15$$
$$s = \frac{3}{5}(15)$$
$$s = 9$$
$$\frac{2}{3}s = \frac{2}{3}(9) = 6$$

Sean pays \$9 for the pizza, and Connor pays \$6.

20. False: If n represents the first of three consecutive odd integers, then $n + 2$ and $n + 4$ represent the next two odd integers.

21. We want to find three consecutive even integers that sum to 270. Let n be the first even integer. Then the next even integer is $n + 2$, and the even integer after that is $n + 4$.

$$n+(n+2)+(n+4)=270$$
$$n+n+2+n+4=270$$
$$3n+6=270$$
$$3n=264$$
$$n=88$$
$$n+2=88+2=90$$
$$n+4=88+4=92$$
The integers are 88, 90, and 92.

22. We want to find four consecutive odd integers that sum to 72. Let n be the first odd integers. Then the next three odd integers are $n + 2$, $n + 4$, and $n + 6$.

$$n+(n+2)+(n+4)+(n+6)=72$$
$$4n+12=72$$
$$4n+12-12=72-12$$
$$4n=60$$
$$\frac{4n}{4}=\frac{60}{4}$$
$$n=15$$

The integers are 15, 17, 19, and 21.

23. We want to find the lengths of the three pieces of ribbon. Let x be the length of the shortest piece. Then the length of the longest piece is $x + 24$, and the length of the third piece is $\frac{1}{2}(x+24)$.

The sum of the lengths of the pieces is the length of the original ribbon, 76 inches.

$$x+(x+24)+\frac{1}{2}(x+24)=76$$
$$x+x+24+\frac{1}{2}x+12=76$$
$$2\left(x+x+24+\frac{1}{2}x+12\right)=2(76)$$
$$2x+2x+48+x+24=152$$
$$5x+72=152$$
$$5x=80$$
$$x=16$$
$$x+24=16+24=40$$
$$\frac{1}{2}(x+24)=\frac{1}{2}(16+24)=\frac{1}{2}(40)=20$$

The lengths of the pieces are 16 inches, 20 inches, and 40 inches.

24. We want to know the amount invested in each type of investment. Let s be the amount invested in stocks. Then the amount invested in bonds is $2s$. The total amount invested is $18,000.
$$s+2s=18,000$$
$$3s=18,000$$
$$s=6000$$

$$2s=2(6000)=12,000$$
The amount invested in stocks is $6000, and the amount invested in bonds is $12,000.

25. We are looking for the number of miles for which the rental costs are the same. Let m be the number of miles driven. Renting from E-Z Rental would cost $30 + 0.15m$. Renting from Do It Yourself Rental would cost $15 + 0.25m$.
$$30+0.15m=15+0.25m$$
$$30=15+0.10m$$
$$15=0.1m$$
$$10\cdot15=10\cdot0.1m$$
$$150=m$$
The costs are the same if 150 miles are driven.

26. We are looking for the number of minutes for which the monthly costs of the plans will be the same. Let m be the number of minutes. The monthly cost of plan A is $15 + 0.05m$, and the monthly cost of plan B is $0.20m$.
$$15+0.05m=0.20m$$
$$15=0.15m$$
$$100=m$$
The monthly costs are the same for 100 minutes.

2.5 Exercises

27. The sum of -5 and a number: $-5 + x$

29. The product of a number and $\frac{2}{3}$: $x\left(\frac{2}{3}\right)$ or $\frac{2}{3}x$

31. Half of a number: $\frac{1}{2}x$

33. A number less -25: $x - (-25)$

35. The quotient of a number and 3: $\frac{x}{3}$

37. $\frac{1}{2}$ more than a number: $x+\frac{1}{2}$

39. 9 more than 6 times a number: $6x + 9$

41. Twice the sum of 13.7 and a number: $2(13.7 + x)$

43. The sum of twice a number and 31: $2x + 31$

45. Let r be the number of runs scored by the Richmond Braves. Then $r + 5$ is the number of runs scored by the Columbus Clippers.

47. Let b be the amount Bill has in his bank. Then $b + 0.55$ is the amount Jan has in her bank.

49. Let j be the amount Janet will get. Then $200 - j$ is the amount Kathy will get.

51. Let a be the number of adults who visited the show. Then $1433 - a$ is the number of children who visited the show.

53. $x + 15 = -34$

55. $35 = 3x - 7$

57. $\dfrac{x}{-4} + 5 = 36$

59. $2(x + 6) = x + 3$

61. We want to find the number. Let n be the number.
$$n + (-12) = 71$$
$$n - 12 = 71$$
$$n = 83$$
Is the sum of 83 and -12 equal to 71? Yes. The number is 83.

63. We want to find the number. Let n be the number.
$$2n - 25 = -53$$
$$2n = -28$$
$$n = -14$$
Is 25 less than twice -14 equal to -53? Yes. The number is -14.

65. We want to find three consecutive integers that sum to 165. Let n be the first integer. Then the other integers are $n + 1$ and $n + 2$, respectively.
$$n + (n + 1) + (n + 2) = 165$$
$$n + n + 1 + n + 2 = 165$$
$$3n + 3 = 165$$
$$3n = 162$$
$$n = 54$$
If $n = 54$, then $n + 1 = 55$, and $n + 2 = 56$. Are 54, 55, and 56 consecutive integers? Yes. Do they sum to 165? Yes. The numbers are 54, 55 and 56.

67. We want to find the lengths of the bridges. If x is the length of the Verrazano-Narrows Bridge, then the length of the Golden Gate Bridge is

$x - 60$.
$$x + (x - 60) = 8460$$
$$2x - 60 = 8460$$
$$2x = 8520$$
$$x = 4260$$
If $x = 4260$, then $x - 60 = 4200$. Is 4200 equal to 60 less than 4260? Yes. Do 4260 and 4200 sum to 8460? Yes.
The Verrazano-Narrows Bridge is 4260 feet long and the Golden Gate Bridge is 4200 feet long.

69. We want to find the price of the motorcycle before the extra charges. Let m be the price of the motorcycle before the extra charges. Then the total price is $m + 679.79$, which is also 11,894.79.
$$m + 679.79 = 11,894.79$$
$$m = 11,215$$
Is 11,215 plus 679.79 equal to 11,894.79? Yes. The price of the motorcycle before the extra charges is $11,215.

71. We want to find the amount invested in each type of investment. Let x be the amount invested in CDs. Then $x + 3000$ is the amount invested in bonds. The total invested is 20,000.
$$x + (x + 3000) = 20,000$$
$$2x + 3000 = 20,000$$
$$2x = 17,000$$
$$x = 8500$$
If $x = 8500$, $x + 3000 = 11,500$. Is 11,500 3000 greater than 8500? Yes. Is the sum of 8500 and 11,500 equal to 20,000? Yes. The amount invested in CDs is $8500, the amount invested in bonds is $11,500.

73. We want to find the amount invested in each type of investment. Let x be the amount in stocks. Then $\dfrac{3}{5}x$ is the amount invested in bonds. The total invested in 32,000.
$$x + \frac{3}{5}x = 32,000$$
$$\frac{5}{5}x + \frac{3}{5}x = 32,000$$
$$\frac{8}{5}x = 32,000$$
$$\frac{5}{8} \cdot \frac{8}{5}x = \frac{5}{8} \cdot 32,000$$
$$x = 20,000$$
$$\frac{3}{5}x = 12,000$$

Is 12,000 $\frac{3}{5}$ of 20,000. Yes. Is the sum of

20,000 and 12,000 equal to 32,000? Yes. You should invest $20,000 in stocks and $12,000 in bonds.

75. We want to find the amount of dietary fiber in each cereal. Let x be the amount of dietary fiber in Kellogg's Smart Start Cereal. Then $4x$ is the amount of dietary fiber in Kashi Go Lean Crunch.

$$x + 4x = 10$$
$$5x = 10$$
$$x = 2$$

If $x = 2$, then $4x = 8$. Is 8 four times 2? Yes. Is the sum of 8 and 2 equal to 10? Yes.
There are 2 grams of dietary fiber in Kellogg's Smart Start Cereal, and 8 grams in Kashi Go Lean Crunch.

77. We want to find Elizabeth Morrell's adjusted gross income (AGI). Let x be her AGI. Then her husband's AGI was
$x - 2549$.

$$x + (x - 2549) = 55,731$$
$$2x - 2549 = 55,731$$
$$2x = 58,280$$
$$x = 29,140$$

If $x = 29,140$, then $x - 2549 = 26,591$. Do 29,140 and 26,591 sum to 55,731? Yes. Elizabeth Morrell's adjusted gross income was $29,140.

79. We want to find the number of miles for which the companies charge the same amount. Let x be the number of miles. Then EZ-Rental charges $35 + 0.15x$ and Do It Yourself Rental charges $20 + 0.25x$.

$$35 + 0.15x = 20 + 0.25x$$
$$35 = 20 + 0.10x$$
$$15 = 0.10x$$
$$150 = x$$

If 150 miles are driven, EZ-Rental charges
$35 + 0.15(150) = 35 + 22.5 = \$57.50$ and Do It Yourself Rental charges
$20 + 0.25(150) = 20 + 37.5 = \$57.50$, so the charges are the same. The cost of renting is the same when 150 miles are driven.

81. We want to find the number of pages for which the cost is the same. Let x be the number of pages. Then Hewlett-Packard costs $200 + 0.03x$ and the Brother costs $240 + 0.01x$.

$$200 + 0.03x = 240 + 0.01x$$
$$200 + 0.02x = 240$$
$$0.02x = 40$$
$$x = 2000$$

If 2000 pages are printed, the Hewlett-Packard costs $200 + 0.03(2000) = 200 + 60 = \$260$ and the Brother costs
$240 + 0.01(2000) = 240 + 20 = \$260$, so the costs are the same. The cost is the same for 2000 pages.

83. We want to find the adjusted gross incomes (AGIs). Let j be Jenson Beck's AGI. Then $j - 249$ is Maureen Beck's AGI, since Jensen Beck's AGI is 249 more than Maureen's.

$$j + (j - 249) = 72,193$$
$$2j - 249 = 72,193$$
$$2j = 72,442$$
$$j = 36,221$$

If $j = 36,221$, then $j - 249 = 35,972$. Is 36,221 less 249 equal to 35,972? Yes. Do 36,221 and 35,972 sum to 72,193? Yes. Jensen Beck's adjusted gross income was $36,221, Maureen Beck's was $35,972.

85. We want to find the number of runs scored in each game. There are 4 numbers since there is a losing number of runs and a winning number of runs for each game. Let n be the losing number of runs in the first game. Since the scores were consecutive integers, the numbers of runs scored were n,
$n + 1$, $n + 2$, and $n + 3$.

$$n + (n + 1) + (n + 2) + (n + 3) = 26$$
$$4n + 6 = 26$$
$$4n = 20$$
$$n = 5$$

If $n = 5$, then $n + 1 = 6$, $n + 2 = 7$, and $n + 3 = 8$.
Are 5, 6, 7, and 8 consecutive integers that sum to 26? Yes. The number of runs scored were 5, 6, 7, and 8.

87. Answers may vary.

89. Answers may vary.

91. We want to find the measure of each angle. Let x be the measure of the second angle. Then the

smallest angle has measure $\frac{1}{2}x$ and the third

angle has measure $4\left(\frac{1}{2}x\right) + 40$. Their measures

sum to 180.

$$x + \left(\frac{1}{2}x\right) + \left[4\left(\frac{1}{2}x\right) + 40\right] = 180$$

$$x + \frac{1}{2}x + 2x + 40 = 180$$

$$\frac{2}{2}x + \frac{1}{2}x + \frac{4}{2}x = 140$$

$$\frac{7}{2}x = 140$$

$$x = 40$$

If $x = 40$, then $\frac{1}{2}x = 20$ and $4\left(\frac{1}{2}x\right) + 40 = 120$.

Is 20 $\frac{1}{2}$ of 40? Yes. Is 120 40 more than 4 times 20? Yes. Do 40, 20, and 120 sum to 180? Yes. The angles measure 20°, 40°, and 120°.

93. The process of taking a verbal description of a problem and developing a mathematical equation that can be used to solve the problem is mathematical modeling. We often make assumptions to make the mathematics more manageable in the mathematical models.

95. Both students are correct, and the value for n will be the same for both students. Answers may vary.

Section 2.6

Preparing for Problem Solving: Direct Translation Problems Involving Percent

P1. $45\% = 45\% \cdot \frac{1}{100\%} = \frac{45}{100} = 0.45$

P2. $0.2875 = 0.2875 \cdot \frac{100\%}{1} = 28.75\%$

2.6 Quick Checks

1. Percent means "divided by 100."

2. The word "of" translates into multiplication in mathematics, so 40% of 120 means 0.40 times 120.

3. We want to know the unknown number. Let n represent the number.
   $n = 0.89 \cdot 900$
   $n = 801$
   801 is 89% of 900.

4. We want to know the unknown number. Let n represent the number.
   $n = 0.035 \cdot 72$
   $n = 2.52$
   2.52 is 3.5% of 72.

5. We want to know the unknown number. Let n represent the number.
   $n = 1.50 \cdot 24$
   $n = 36$
   36 is 150% of 24.

6. We want to know the unknown number. Let n represent the number.
   $8\frac{3}{4}\% = 8.75\% = 0.0875$
   $n = 0.0875 \cdot 40$
   $n = 3.5$
   3.5 is $8\frac{3}{4}\%$ of 40.

7. We want to know the percentage. Let x represent the percent.
   $8 = x \cdot 20$
   $\frac{8}{20} = \frac{20x}{20}$
   $0.4 = x$
   $40\% = x$
   The number 8 is 40% of 20.

8. We want to know the percentage. Let x represent the percent.
   $15 = x \cdot 40$
   $\frac{15}{40} = \frac{40x}{40}$
   $0.375 = x$
   $37.5\% = x$
   The number 15 is 37.5% of 40.

9. We want to know the percentage. Let x represent the percent.
   $12.3 = x \cdot 60$
   $\frac{12.3}{60} = \frac{60x}{60}$
   $0.205 = x$
   $20.5\% = x$
   The number 12.3 is 20.5% of 60.

10. We want to know the percentage. Let x represent the percent.

$$44 = x \cdot 40$$

$$\frac{44}{40} = \frac{40x}{40}$$

$$1.10 = x$$

$$110\% = x$$

The number 44 is 110% of 40.

11. We want to know a number. Let x represent the number.

$$14 = 0.28 \cdot x$$

$$\frac{14}{0.28} = \frac{0.28x}{0.28}$$

$$50 = x$$

14 is 28% of 50.

12. We want to know a number. Let x represent the number.

$$111 = 0.74 \cdot x$$

$$\frac{111}{0.74} = \frac{0.74x}{0.74}$$

$$150 = x$$

111 is 74% of 150.

13. We want to know a number. Let x represent the number.

$$14.8 = 0.185 \cdot x$$

$$\frac{14.8}{0.185} = \frac{0.185x}{0.185x}$$

$$80 = x$$

14.8 is 18.5% of 80.

14. We want to know a number. Let x represent the number.

$$102 = 1.36 \cdot x$$

$$\frac{102}{1.36} = \frac{1.36x}{1.36}$$

$$75 = x$$

102 is 136% of 75.

15. We want to know the number of U.S. residents 25 years of age or older in 2007 who have bachelor's degrees. Let x represent the number of people 25 years of age or older in 2007 who have bachelor's degrees.

$$0.17 \cdot 195,000,000 = x$$

$$33,150,000 = x$$

The number of residents 25 years of age or older in 2007 who have bachelor's degrees is 33,150,000.

16. We want to know Janet's new salary. Let n represent her new salary, which is her old salary plus the 2.5% raise.

$$n = 39,000 + 0.025 \cdot 39,000$$

$$n = 39,000 + 975$$

$$n = 39,975$$

Janet's new salary is $39,975.

17. We want to know the price of the car before the sales tax. Let p represent the price of the car. The sales tax is 7% of the price of the car before sales tax.

$$p + 0.7p = 7811$$

$$1.07p = 7811$$

$$\frac{1.07p}{1.07} = \frac{7811}{1.07}$$

$$p = 7300$$

The price of the car before sales tax was $7300.

18. We want to know the wholesale price of the gasoline. Let p represent the wholesale price of the gasoline. The selling price is the sum of the wholesale price and the markup.

$$p + 0.80p = 4.50$$

$$1.80p = 4.50$$

$$\frac{1.80p}{1.80} = \frac{4.50}{1.80}$$

$$p = 2.50$$

The gas station pays $2.50 per gallon for the gasoline.

19. We want to know the original price of the recliners. Let p represent the original price. The sale price is the original price less the 25% discount.

$$p - 0.25p = 494.25$$

$$0.75p = 494.25$$

$$\frac{0.75p}{0.75} = \frac{494.25}{0.75}$$

$$p = 659$$

The original price of the recliners was $659.

20. We want to know the value of Albert's house one year ago. Let x represent the value one year ago. The value now is the value a year ago minus the 2% loss.

$$x - 0.02x = 148,000$$

$$0.98x = 148,000$$

$$\frac{0.98x}{0.98} = \frac{148,000}{0.98}$$

$$x \approx 151,020$$

Albert's house was worth $151,020 one year ago.

2.6 Exercises

21. $n = 0.50(160)$
$n = 80$
80 is 50% of 160.

23. $0.07(200) = n$
$14 = n$
7% of 200 is 14.

25. $n = 0.16(30)$
$n = 4.8$
4.8 is 16% of 30.

27. $31.5 = 0.15x$
$\dfrac{31.5}{0.15} = \dfrac{0.15x}{0.15}$
$210 = x$
31.5 is 15% of 210.

29. $0.60(120) = x$
$72 = x$
60% of 120 is 72.

31. $24 = 1.20x$
$\dfrac{24}{1.20} = \dfrac{1.20x}{1.20}$
$20 = x$
24 is 120% of 20.

33. $p \cdot 60 = 24$
$\dfrac{60p}{60} = \dfrac{24}{60}$
$p = 0.4$
$p = 40\%$
24 is 40% of 60.

35. $1.5 = p \cdot 20$
$\dfrac{1.5}{20} = \dfrac{20p}{20}$
$0.075 = p$
$7.5\% = p$
1.5 is 7.5% of 20.

37. $p \cdot 300 = 600$
$\dfrac{300p}{300} = \dfrac{600}{300}$
$p = 2$
$p = 200\%$
600 is 200% of 300.

39. Let x be the price of the tennis racket before tax. Then the sales tax is $0.06x$. The total cost is $57.24.
$x + 0.06x = 57.24$
$1.06x = 57.24$
$x = 54$
The tennis racket cost $54 before tax.

41. Let x be Todd's salary after the pay cut.
$x = 120,000 - 0.15(120,000)$
$x = 120,000 - 18,000$
$x = 102,000$
Todd's salary is $102,000 after the pay cut.

43. Let x be Mrs. Fisher's original investment. Then the amount she lost is $0.09x$.
$x - 0.09x = 22,750$
$0.91x = 22,750$
$x = 25,000$
Mrs. Fisher's original investment was $25,000.

45. Let x be the price before the discount. The amount of the discount is $0.25x$.
$x - 0.25x = 51$
$0.75x = 51$
$x = 68$
The price of the merchandise was $68 before the discount.

47. Let x be the original price of the suit. Then the discount is $0.30x$.
$x - 0.30x = 399$
$0.70x = 399$
$x = 570$
The original price of the suit was $570.

49. Let x be the original price of the table. Then the discount is $0.4x$.
$x = 0.4x - 240$
$0.6x = 240$
$x = 400$
The original price of the table was $400.

51. Let v be the number of votes that the winner received. Then the loser received $0.60v$ votes.
$v + 0.60v = 848$
$1.60v = 848$
$v = 530$
If $v = 530$, then $0.60v = 318$.
The winner received 530 votes and the loser received 318 votes.

53. Let h be the value of the house. Then $0.03h$ is the amount of Melanie's commission.
$$0.03h = 8571$$
$$h = 285,700$$
The value of the house was $285,700.

55. Let $x =$ number of votes that separated the two candidates.
$220,369 =$ total number of votes cast.
0.48%, or 0.0048 is the fraction of votes separating the two candidates.
$$x = (0.0048)220,369$$
$$x = 1057.7 \text{ rounded up}$$
$$x = 1058 \text{ votes}$$

57. Let x be the number of males that have never married.
$$x = 0.29(106)$$
$$x = 30.74$$
30.74 million males aged 18 years or older have never married.

59. Let $x =$ percent of population that held an Associate's degree in 2004.
$$x = \frac{13,243,927}{186,534,177} \cdot 100\% = 7.1\%$$

61. Let $x =$ percent of population that held a Bachelor's degree in 2004.
$$x = \frac{32,083,878}{186,534,177} \cdot 100\% = 17.2\%$$

63. Let x be the initial selling price. Then the sale price is $x - 0.25x$. This needs to be $6 more than the purchase price of $12.
$$x - 0.25x = 12 + 6$$
$$0.75x = 18$$
$$x = 24$$
The shirts should be priced at $24.

65. The amount of change is $21 - 15 = 6$.
$$\frac{6}{21} \approx 0.286$$
The gas mileage decreases by 28.6% due to the extra weight.

67. (a) value after 1st year $= 55,670 - 0.25(55,670)$
$\qquad = 55,670 - 13,917.50$
$\qquad = 41,752.50$
value after 2nd year
$\qquad = 41,752.50 - 0.25(41,752.50)$
$\qquad = 41,752.50 - 10,438.125$
$\qquad = 31,314.375$
The car will be worth $31,314.38 after two years.

(b) The amount of change is
$55,670 - 31,314.38 = 24,355.62$.
$$\frac{24,355.62}{55,670} \approx 0.437$$
The overall decrease is 43.7%.

69. The amount of change is $4.39 - 3.89 = 0.50$.
$$\frac{0.50}{3.89} \approx 0.1285$$
The percent increase is 12.9%.

71. The equation should be $x + 0.05x = 12.81$. Jack's new hourly wage is a percentage of his current hourly wage, so multiplying 5% to his original wage gives his hourly raise.

Section 2.7

Preparing for Problem Solving: Geometry and Uniform Motion

P1. $q + 2q - 30 = 180$
$\qquad 3q - 30 = 180$
$\qquad\quad 3q = 210$
$\qquad\quad\ \ q = 70$
The solution set is $\{70\}$.

P2. $30w + 20(w + 5) = 300$
$\quad 30w + 20w + 100 = 300$
$\qquad\quad 50w + 100 = 300$
$\qquad\qquad\ \ 50w = 200$
$\qquad\qquad\quad\ w = 4$
The solution set is $\{4\}$.

2.7 Quick Checks

1. Complementary angles are angles whose measures sum to <u>90</u> degrees.

2. This is a complementary angle problem. We are looking for the measures of two angles whose sum is $90°$. Let x represent the measure of the smaller angle. Then $x + 12$ represents the measure of the larger angle.
$$x + (x + 12) = 90$$
$$2x + 12 = 90$$
$$2x = 78$$
$$x = 39$$
$$x + 12 = 39 + 12 = 51$$
The two complementary angles measure $39°$ and $51°$.

3. This is a supplementary angle problem. We are looking for the measures of two angles whose sum is 180°. Let x represent the measure of the smaller angle. Then $2x - 30$ represents the measure of the larger angle.
$$x + (2x - 30) = 180$$
$$3x - 30 = 180$$
$$3x = 210$$
$$x = 70$$
$2x - 30 = 2(70) - 30 = 140 - 30 = 110$
The two supplementary angles measure 70° and 110°.

4. The sum of the measures of the angles of a triangle is 180 angles.

5. This is an "angles of a triangle" problem. We know that the sum of the measures of the interior angles of a triangle is 180°. Let x represent the measure of the largest angle. Then $\frac{1}{3}x$ represents the measure of the smallest angle, and $x - 65$ represents the measure of the middle angle.
$$x + \left(\frac{1}{3}x\right) + (x - 65) = 180$$
$$2\frac{1}{3}x - 65 = 180$$
$$\frac{7}{3}x = 245$$
$$x = 105$$
$\frac{1}{3}x = \frac{1}{3}(105) = 35$
$x - 65 = 105 - 65 = 40$
The measures of the angles of the triangle are 35°, 40°, and 105°.

6. True

7. False; the perimeter of a rectangle can be found by adding twice the length of the rectangle to twice the width of the rectangle.

8. This is a perimeter problem. We want to find the width and length of a garden. Let w represent the width of the garden. Then $2w$ represents the length of the garden. We know that the perimeter is 9 feet, and that the formula for the perimeter of a rectangle is $P = 2l + 2w$.

$$2(2w) + 2w = 9$$
$$4w + 2w = 9$$
$$6w = 9$$
$$w = \frac{9}{6}$$
$$w = \frac{3}{2}$$
$$2w = 2\left(\frac{3}{2}\right) = 3$$

The width of the garden is $\frac{3}{2} = 1.5$ feet and the length is 3 feet.

9. This problem is about the surface area of a rectangular box. The formula for the surface area of a rectangular box is $SA = 2lw + 2lh + 2hw$, where l is the length of the box, w is the width of the box, and h is the height of the box. We are given the surface area, the length, and the width of the rectangular box.
$$SA = 2lw + 2lh + 2hw$$
$$62 = 2 \cdot 3 \cdot 2 + 2 \cdot 3 \cdot h + 2 \cdot h \cdot 2$$
$$62 = 12 + 6h + 4h$$
$$62 = 12 + 10h$$
$$50 = 10h$$
$$5 = h$$
The height of the box is 5 feet.

10. False; when using $d = rt$ to calculate the distance traveled, it is necessary to travel at a constant speed.

11. This is a uniform motion problem. We want to know the average speed of each biker. Let r represent the average speed of Luis. Then $r + 5$ represents José's average speed. Each biker rides for 3 hours.

	Rate	·	Time	= Distance
Luis	r		3	$3r$
José	$r + 5$		3	$3(r + 5)$

Since the bikers are 63 miles apart after 3 hours, the sum of the distances they biked is 63.
$$3r + 3(r + 5) = 63$$
$$3r + 3r + 15 = 63$$
$$6r + 15 = 63$$
$$6r = 48$$
$$r = 8$$

$r + 5 = 8 + 5 = 13$
Luis' average speed was 8 miles per hour and
José's average speed was 13 miles per hour.

12. This is a uniform motion problem. We want to
know how long it takes to catch up with Tanya,
and how far you are from your house when you
catch her. Let x represent the amount of time you
drive before catching Tanya. Then $x + 2$
represents the number of hours that Tanya runs
before you catch her.

	Rate	· Time	= Distance
Tanya	8	$x + 2$	$8(x + 2)$
You	40	x	$40x$

When you catch up with Tanya, the distances the
two of you have traveled is the same.
$$8(x + 2) = 40x$$
$$8x + 16 = 40x$$
$$16 = 32x$$
$$\frac{1}{2} = x$$

You catch up to Tanya after driving for
$\frac{1}{2}$ hour.

$$40x = 40\left(\frac{1}{2}\right) = 20$$

You and Tanya are 20 miles from your house
when you catch up to her.

2.7 Exercises

13. This is a supplementary angle problem. We are
looking for the measures of two angles whose
sum is 180°. Let x represent the measure of the
second angle. Then $3x - 10$ represents the
measure of the first angle.
$$(3x - 10) + x = 180$$
$$4x - 10 = 180$$
$$4x = 190$$
$$x = 47.5$$
$3x - 10 = 3(47.5) - 10 = 142.5 - 10 = 132.5$
The angles measure 47.5° and 132.5°.

15. We want to find complementary angles. The
measures of complementary angles sum to 90°.
Let x be the measure of the smaller angle. Then
the measure of the larger angle is $x + 2$.

$$x + (x + 2) = 90$$
$$2x + 2 = 90$$
$$2x = 88$$
$$x = 44$$
If $x = 44$, then $x + 2 = 46$.
Are 44 and 46 consecutive even integers that
sum to 90? Yes. The measures of the angles are
44° and 46°.

17. $x + (x + 2) + (2x + 10) = 180$
$$4x + 12 = 180$$
$$4x = 168$$
$$x = 42$$
The angles measure 42°, (42 + 2)° = 44°, and (2 ·
42 + 10)° = (84 + 10)° = 94°.

19. We want to find the measures of the angles of a
triangle. The measures of the angels of a triangle
sum to 180°. Let x be the measure of the first
angle. Then $x + 2$ and $x + 4$ are the measures of
the next two angles, respectively.
$$x + (x + 2) + (x + 4) = 180$$
$$3x + 6 = 180$$
$$3x = 174$$
$$x = 58$$
If $x = 58$, then $x + 2 = 60$, and $x + 4 = 62$. Do 58,
60, and 62 sum to 180? Yes. The measures of the
angles are 58°, 60°, and 62°.

21. We want the dimensions of a rectangle. The
perimeter of a rectangle is twice the length plus
twice the width. Let w be the width of the
rectangle. Then the length is $2w + 8$. The
perimeter is 88 feet.
$$2w + 2(2w + 8) = 88$$
$$2w + 4w + 16 = 88$$
$$6w + 16 = 88$$
$$6w = 72$$
$$w = 12$$
If $w = 12$, then $2w + 8 = 32$. Is the sum of twice
12 and twice 32 equal to 88? Yes. The length of
the rectangle is 32 feet, and the width is 12 feet.

23. We want to find the dimensions of the
rectangular field. The perimeter of a rectangle is
twice the length plus twice the width.
Let w be the width (shorter dimension) of the
original field. then the length of the field is $2w$
and the sides of the smaller squares are all w.
There are 7 sides that need fencing and 294 feet
of fencing were used.

$7w = 294$

$w = 42$

If $w = 42$, then $2w = 84$. Would it require 294 feet of fencing to enclose a field that is 42 feet by 84 feet and divide it into two equal squares? Yes. The field is 42 feet by 84 feet.

25. We want to find the sides of a triangle. The perimeter of a triangle is the sum of the lengths of the legs. Let x be the length of each of the congruent legs. The base is 45 inches and the perimeter is 98 inches.

    $x + x + 45 = 98$

    $2x = 53$

    $x = 26.5$

    Is the sum of 26.5, 26.5, and 45 equal to 98? Yes. Each leg is 26.5 inches.

27. **(a)** Let t be the time since the cars left Chicago. The distance traveled by the car going east is $62t$.

    (b) The distance traveled by the car going west is $68t$.

    (c) The total distance is $62t + 68t$.

    (d) The equation is $62t + 68t = 585$.

29.

	Rate	· Time	= Distance
Martha	528	$t + 10$	$528(t + 10)$
Mom	880	t	$880t$

$528(t + 10) = 880t$

31. Let l be the length of the rectangle. Then the width is $\frac{1}{2}l - 3$. The perimeter is 36 inches.

    $2l + 2\left(\frac{1}{2}l - 3\right) = 36$

    $2l + l - 6 = 36$

    $3l - 6 = 36$

    $3l = 42$

    $l = 14$

    If $l = 14$, then $\frac{1}{2}l - 3 = 4$.

    The length is 14 inches, the width is 4 inches.

33. Let l be the length of the billboard. The height is 15 feet and the perimeter is 110 feet.

$2l + 2(15) = 110$

$2l + 30 = 110$

$2l = 80$

$l = 40$

The length of the billboard is 40 feet.

35. The area of a trapezoid is $A = \frac{1}{2}h(B + b)$, where h is the height and the bases are B and b. Let B be the longer base. Then the shorter base is $B - 8$. The height is 60 feet and the area is 2160 square feet.

    $\frac{1}{2}(60)[B + (B - 8)] = 2160$

    $30[2B - 8] = 2160$

    $60B - 240 = 2160$

    $60B = 2400$

    $B = 40$

    If $B = 40$, then $B - 8 = 32$. The bases are 40 feet and 32 feet.

37. **(a)** Let l be the length of the garden. Then the width is $2l - 3$. The perimeter is 60 yards.

    $2l + 2(2l - 3) = 60$

    $2l + 4l - 6 = 60$

    $6l - 6 = 60$

    $6l = 66$

    $l = 11$

    If $l = 11$, then $2l - 3 = 19$. The length of the garden is 11 feet and the width is 19 feet.

    (b) $A = lw = 11(19) = 209$

    The area of the garden is 209 square feet.

39. Since the southbound boat is traveling at 47 mph, the northbound boat is traveling at $47 + 16 = 63$ mph.

    After t hours, the northbound boat will have gone $63t$ miles and the southbound boat will have gone $47t$ miles.

    $63t + 47t = 1430$

    $110t = 1430$

    $t = 13$

    The boats will be 1430 miles apart after 13 hours.

41. Let x be the speed of the slower car. Then $x + 12$ is the speed of the faster car. The slower car stops after 4 hours and 30 minutes (4.5 hours) and has traveled $4.5x$ miles. The faster car stops after 4 hours and has traveled $4(x + 12)$ miles. The difference between their stopping points is

24 miles.

$$4(x+12)-4.5x=24$$
$$4x+48-4.5x=24$$
$$-0.5x=-24$$
$$x=48$$

If $x = 48$, then $x + 12 = 60$. The slower car is traveling at 48 mph, and the faster car is traveling at 60 mph.

43. Let t be the amount of time spent traveling at 62 mph. Then $6 - t$ is the amount of time spent traveling at 54 mph. The distance traveled at 62 mph is $62t$, and the distance at 54 mph is $54(6 - t)$.

$$62t+54(6-t)=360$$
$$62t+324-54t=360$$
$$8t+324=360$$
$$8t=36$$
$$t=4.5$$

If $t = 4.5$, then $6 - t = 1.5$.
The trip consisted of 4.5 hours on the freeway and 1.5 hours on the 2-lane highway.

45. Let r be the rate at which she jogs.

	Rate	· Time	= Distance
jog	r	$\frac{10}{60}=\frac{1}{6}$	$\frac{1}{6}r$
walk	$r-4$	$\frac{30}{60}=\frac{1}{2}$	$\frac{1}{2}(r-4)$

The distances are equal.

$$\frac{1}{6}r=\frac{1}{2}(r-4)$$
$$6\left(\frac{1}{6}r\right)=6\left[\frac{1}{2}(r-4)\right]$$
$$r=3(r-4)$$
$$r=3r-12$$
$$12=2r$$
$$6=r$$

Carol jogs at 6 mph.

47. Let x = measure of base angles.
Let $2x - 16$ = measure of vertex angle.
180 = sum of all angles

$$x+x+2x-16=180$$
$$4x-16=180$$
$$16+4x-16=180+16$$
$$4x=196$$
$$\frac{4x}{4}=\frac{196}{4}$$
$$x=49°$$

Substitute x into $2x - 16$ for vertex angle of 82°.
$49°, 49°, 82°$

49. The marked angles are equal in measure because they are alternate interior angles.

$$20-2x=3x+5$$
$$20=5x+5$$
$$15=5x$$
$$3=x$$

51. The marked angles are supplementary because they are interior angles on the same side of the transversal.

$$(8x+12)+(3x+3)=180$$
$$11x+15=180$$
$$11x=165$$
$$x=15$$

53. The marked angles are equal in measure because they are corresponding angles.

$$4x+2=\frac{3}{2}x+32$$
$$2(4x+2)=2\left(\frac{3}{2}x+32\right)$$
$$8x+4=3x+64$$
$$5x=60$$
$$x=12$$

55. Complementary angles are those whose sum 90° and supplementary angles sum to 180°.

57. Answers may vary. The equation $65t + 40t = 115$ is describing the sum of distances whereas the equation $65t - 40t = 115$ represents the difference of distances.

Section 2.8

Preparing for Solving Linear Inequalities in One Variable

P1. Since 4 is to the left of 19 on the number line, $4 < 19$.

P2. Since -11 is to the right of -24 on the number line, $-11 > -24$.

P3. $\frac{1}{4}=0.25$

P4. Since $\dfrac{5}{6} = \dfrac{25}{30}$, $\dfrac{4}{5} = \dfrac{24}{30}$, and $25 > 24$, then

$\dfrac{5}{6} > \dfrac{4}{5}$.

2.8 Quick Checks

1. True: When graphing an inequality that contains a > or a < symbol, we use parentheses.

2. $n \geq 8$

3. $a < -6$

4. $x > -1$

5. $p \leq 0$

6. True

7. False; the inequality $x < -4$ is written in interval notation as $(-\infty, -4)$.

8. $[-3, \infty)$

9. $(-\infty, 12)$

10. $(-\infty, 2.5]$

11. $(125, \infty)$

12. To <u>solve</u> an inequality means to find the set of all replacement values of the variable for which the statement is true.

13. The <u>Addition</u> <u>Property</u> of <u>Inequality</u> states that the direction, or sense, of each inequality remains the same when the same quantity is added to each side of the inequality.

14. $\quad 5n - 4 > 11$

$\quad 5n - 4 + 4 > 11 + 4$

$\quad\quad\quad 5n > 15$

$\quad\quad\quad \dfrac{5n}{5} > \dfrac{15}{5}$

$\quad\quad\quad\quad n > 3$

The solution set is $\{n \mid n > 3\}$ or $(3, \infty)$.

15. $\quad -2x + 3 < 7 - 3x$

$\quad -2x + 3 + 3x < 7 - 3x + 3x$

$\quad\quad\quad x + 3 < 7$

$\quad\quad\quad x + 3 - 3 < 7 - 3$

$\quad\quad\quad\quad x < 4$

The solution set is $\{x \mid x < 4\}$ or $(-\infty, 4)$.

16. $\quad 5n + 8 \leq 4n + 4$

$\quad 5n + 8 - 4n \leq 4n + 4 - 4n$

$\quad\quad\quad n + 8 \leq 4$

$\quad\quad\quad n + 8 - 8 \leq 4 - 8$

$\quad\quad\quad\quad n \leq -4$

The solution set is $\{n \mid n \leq -4\}$ or $(-\infty, -4]$.

17. $\quad 3(4x - 8) + 12 > 11x - 13$

$\quad 12x - 24 + 12 > 11x - 13$

$\quad\quad 12x - 12 > 11x - 13$

$\quad 12x - 12 - 11x > 11x - 13 - 11x$

$\quad\quad\quad x - 12 > -13$

$\quad\quad x - 12 + 12 > -13 + 12$

$\quad\quad\quad\quad x > -1$

The solution set is $\{x \mid x > -1\}$ or $(-1, \infty)$.

18. When solving an inequality, we reverse the direction of the inequality symbol when we multiply or divide by a <u>negative</u> number.

19. False; the solution to the inequality $-\dfrac{1}{2}x > 9$ is $\{x \mid x < -18\}$. This solution is written in interval notation as $(-\infty, -18)$.

20. $\quad 6k < -36$

$\quad \dfrac{6k}{6} < \dfrac{-36}{6}$

$\quad\quad k < -6$

The solution set is $\{k \mid k < -6\}$ or $(-\infty, -6)$.

21. $2n \geq -5$

$$\frac{2n}{2} \geq \frac{-5}{2}$$

$$n \geq -\frac{5}{2}$$

The solution set is $\left\{ n \middle| n \geq -\frac{5}{2} \right\}$ or $\left[-\frac{5}{2}, \infty \right)$.

22. $-\frac{3}{2}k > 12$

$$-\frac{2}{3}\left(-\frac{3}{2}k \right) < -\frac{2}{3}(12)$$

$$k < -8$$

The solution set is $\{k | k < -8\}$ or $(-\infty, -8)$.

23. $-\frac{4}{3}p \leq -\frac{4}{5}$

$$-\frac{3}{4}\left(-\frac{4}{3}p \right) \geq -\frac{3}{4}\left(-\frac{4}{5} \right)$$

$$p \geq \frac{3}{5}$$

The solution set is $\left\{ p \middle| p \geq \frac{3}{5} \right\}$ or $\left[\frac{3}{5}, \infty \right)$.

24. $3x - 7 > 14$

$$3x - 7 + 7 > 14 + 7$$

$$3x > 21$$

$$\frac{3x}{3} > \frac{21}{3}$$

$$x > 7$$

The solution set is $\{x | x > 7\}$ or $(7, \infty)$.

25. $-4n - 3 < 9$

$$-4n - 3 + 3 < 9 + 3$$

$$-4n < 12$$

$$\frac{-4n}{-4} > \frac{12}{-4}$$

$$n > -3$$

The solution set is $\{n | n > -3\}$ or $(-3, \infty)$.

26. $2x - 6 < 3(x + 1) - 5$

$$2x - 6 < 3x + 3 - 5$$

$$2x - 6 < 3x - 2$$

$$2x - 6 - 2x < 3x - 2 - 2x$$

$$-6 < x - 2$$

$$-6 + 2 < x - 2 + 2$$

$$-4 < x$$

The solution set is $\{x | x > -4\}$ or $(-4, \infty)$.

27. $-4(x + 6) + 18 \geq -2x + 6$

$$-4x - 24 + 18 \geq -2x + 6$$

$$-4x - 6 \geq -2x + 6$$

$$-4x - 6 + 2x \geq -2x + 6 + 2x$$

$$-2x - 6 \geq 6$$

$$-2x - 6 + 6 \geq 6 + 6$$

$$-2x \geq 12$$

$$\frac{-2x}{-2} \leq \frac{12}{-2}$$

$$x \leq -6$$

The solution set is $\{x | x \leq -6\}$ or $(-\infty, -6]$.

28. $\frac{1}{2}(x + 2) > \frac{1}{5}(x + 17)$

$$10\left[\frac{1}{2}(x + 2) \right] > 10\left[\frac{1}{5}(x + 17) \right]$$

$$5(x + 2) > 2(x + 17)$$

$$5x + 10 > 2x + 34$$

$$5x + 10 - 2x > 2x + 34 - 2x$$

$$3x + 10 > 34$$

$$3x + 10 - 10 > 34 - 10$$

$$3x > 24$$

$$\frac{3x}{3} > \frac{24}{3}$$

$$x > 8$$

The solution set is $\{x | x > 8\}$ or $(8, \infty)$.

29.
$$\frac{4}{3}x - \frac{2}{3} \leq \frac{4}{5}x + \frac{3}{5}$$
$$15\left(\frac{4}{3}x - \frac{2}{3}\right) \leq 15\left(\frac{4}{5}x + \frac{3}{5}\right)$$
$$20x - 10 \leq 12x + 9$$
$$20x - 10 - 12x \leq 12x + 9 - 12x$$
$$8x - 10 \leq 9$$
$$8x - 10 + 10 \leq 9 + 10$$
$$8x \leq 19$$
$$\frac{8x}{8} \leq \frac{19}{8}$$
$$x \leq \frac{19}{8}$$

The solution set is $\left\{x \middle| x \leq \frac{19}{8}\right\}$ or $\left(-\infty, \frac{19}{8}\right]$.

30. When solving an inequality, if the variable is eliminated and the result is a true statement, the solution is <u>all real numbers or $(-\infty, \infty)$.</u>

31. When solving an inequality, if the variable is eliminated and the result is a false statement, the solution is <u>the empty set or $\varnothing$.</u>

32.
$$-2x + 7(x - 5) \leq 6x + 32$$
$$-2x + 7x - 35 \leq 6x + 32$$
$$5x - 35 \leq 6x + 32$$
$$5x - 35 - 5x \leq 6x + 32 - 5x$$
$$-35 \leq x + 32$$
$$-35 - 32 \leq x + 32 - 32$$
$$-67 \leq x$$

The solution set is $\{x | x \geq -67\}$ or $[-67, \infty)$.

33.
$$-x + 7 - 8x \geq 2(8 - 5x) + x$$
$$-9x + 7 \geq 16 - 10x + x$$
$$-9x + 7 \geq 16 - 9x$$
$$-9x + 7 + 9x \geq 16 - 9x + 9x$$
$$7 \geq 16$$

The statement $7 \geq 16$ is a false statement, so this inequality has no solution. The solution set is $\varnothing$ or $\{\ \}$.

34.
$$\frac{3}{2}x + 5 - \frac{5}{2}x < 4x - 3(x + 1)$$
$$-\frac{2}{2}x + 5 < 4x - 3x - 3$$
$$-x + 5 < x - 3$$
$$-x + 5 + x < x - 3 + x$$
$$5 < 2x - 3$$
$$5 + 3 < 2x - 3 + 3$$
$$8 < 2x$$
$$\frac{8}{2} < \frac{2x}{2}$$
$$4 < x$$

The solution set is $\{x | x > 4\}$ or $(4, \infty)$.

35.
$$0.8x + 3.2(x + 4) \geq 2x + 12.8 + 3x - x$$
$$0.8x + 3.2x + 12.8 \geq 4x + 12.8$$
$$4x + 12.8 \geq 4x + 12.8$$
$$4x + 12.8 - 4x \geq 4x + 12.8 - 4x$$
$$12.8 \geq 12.8$$

Since 12.8 is always greater than or equal to 12.8, the solution set for this inequality is $\{x | x$ is any real number$\}$ or $(-\infty, \infty)$.

36. We want to know the maximum number of boxes of supplies that the worker can move on the elevator. Let b represent the number of boxes. Then the weight of the boxes is $91b$, and the weight of the worker and the boxes is $180 + 91b$. This weight cannot be more than 2000 pounds.
$$180 + 91b \leq 2000$$
$$91b \leq 1820$$
$$\frac{91b}{91} \leq \frac{1820}{91}$$
$$b \leq 20$$

The number of boxes must be less than or equal to 20, so the maximum number of boxes the worker can move in the elevator is 20.

2.8 Exercises

37. $x > 2$

$(2, \infty)$

39. $x \leq -1$

$(-\infty, -1]$

41. $z \geq -3$

$[-3, \infty)$

43. $x < 4$

$(-\infty, 4)$

45. $(-\infty, 2)$

47. $\varnothing$ or $\{\ \}$

49. $(-\infty, \infty)$

51. Adding 7 to each side does not change the direction of the inequality symbol. The symbol remains <. We used the Addition Principle of Inequality.

53. Multiplying each side by 3, a positive number, does not change the direction of the inequality symbol. The symbol remains >. We used the Multiplication Principle of Inequality.

55. Subtracting 2 or adding −2 to each side does not change the direction of the inequality symbol. The symbol remains ≤. We used the Addition Principle of Inequality.

57. Dividing each side by −3 or multiplying each side by $-\dfrac{1}{3}$, a negative number, reverses the direction of the inequality symbol. The symbol becomes ≤. We used the Multiplication Principle of Inequality.

59. $x + 1 < 5$
  $x < 4$
  $\{x | x < 4\}$
  $(-\infty, 4)$

61. $x - 6 \geq -4$
  $x \geq 2$
  $\{x | x \geq 2\}$
  $[2, \infty)$

63. $3x \leq 15$
  $x \leq 5$
  $\{x | x \leq 5\}$
  $(-\infty, 5]$

65. $-5x < 35$
  $x > -7$
  $\{x | x > -7\}$
  $(-7, \infty)$

67. $3x - 7 > 2$
  $3x > 9$
  $x > 3$
  $\{x | x > 3\}$
  $(3, \infty)$

69. $3x - 1 \geq 3 + x$
  $2x - 1 \geq 3$
  $2x \geq 4$
  $x \geq 2$
  $\{x | x \geq 2\}$
  $[2, \infty)$

71. $1 - 2x \leq 3$
  $-2x \leq 2$
  $x \geq -1$
  $\{x | x \geq -1\}$
  $[-1, \infty)$

73. $-2(x + 3) < 8$
  $-2x - 6 < 8$
  $-2x < 14$
  $x > -7$
  $\{x | x > -7\}$
  $(-7, \infty)$

75.
$$4 - 3(1 - x) \le 3$$
$$4 - 3 + 3x \le 3$$
$$1 + 3x \le 3$$
$$3x \le 2$$
$$x \le \frac{2}{3}$$
$$\left\{ x \mid x \le \frac{2}{3} \right\}$$
$$\left(-\infty, \frac{2}{3} \right]$$

77.
$$\frac{1}{2}(x - 4) > x + 8$$
$$\frac{1}{2}x - 2 > x + 8$$
$$-2 > \frac{1}{2}x + 8$$
$$-10 > \frac{1}{2}x$$
$$-20 > x$$
$$\{ x \mid x < -20 \}$$
$$(-\infty, -20)$$

79.
$$4(x - 1) > 3(x - 1) + x$$
$$4x - 4 > 3x - 3 + x$$
$$4x - 4 > 4x - 3$$
$$-4 > -3$$
This is a false statement. Therefore, there is no solution. The solution set is $\varnothing$ or $\{\ \}$.

81.
$$5(n + 2) - 2n \le 3(n + 4)$$
$$5n + 10 - 2n \le 3n + 12$$
$$3n + 10 \le 3n + 12$$
$$10 \le 12$$
The solution to this inequality is all real numbers, since 10 is always less than 12.
$\{ n \mid n$ is any real number $\}$
$(-\infty, \infty)$

83.
$$2n - 3(n - 2) < n - 4$$
$$2n - 3n + 6 < n - 4$$
$$-n + 6 < n - 4$$
$$-2n + 6 < -4$$
$$-2n < -10$$
$$n > 5$$
$\{ n \mid n > 5 \}$
$(5, \infty)$

85.
$$4(2w - 1) \ge 3(w + 2) + 5(w - 2)$$
$$8w - 4 \ge 3w + 6 + 5w - 10$$
$$8w - 4 \ge 8w - 4$$
$$-4 \ge -4$$
The solution to this inequality is all real numbers, since -4 is always equal to -4.
$\{ w \mid w$ is any real number $\}$
$(-\infty, \infty)$

87.
$$3y - (5y + 2) > 4(y + 1) - 2y$$
$$3y - 5y - 2 > 4y + 4 - 2y$$
$$-2y - 2 > 2y + 4$$
$$-4y - 2 > 4$$
$$-4y > 6$$
$$y < -\frac{3}{2}$$
$$\left\{ y \mid y < -\frac{3}{2} \right\}$$
$$\left(-\infty, -\frac{3}{2} \right)$$

89. $x \ge 16{,}000$

91. $x \le 20{,}000$

93. $x > 12{,}000$

95. $x > 0$

97. $x \le 0$

99.
$$-1 < x - 5$$
$$4 < x$$
$\{ x \mid x > 4 \}$
$(4, \infty)$

101.
$$-\frac{3}{4}x > -\frac{9}{16}$$
$$-\frac{4}{3}\cdot\left(-\frac{3}{4}\right)x < -\frac{4}{3}\cdot\left(-\frac{9}{16}\right)$$
$$x < \frac{3}{4}$$
$$\left\{x \,\middle|\, x < \frac{3}{4}\right\}$$
$$\left(-\infty, \frac{3}{4}\right)$$

103. $3(x+1) > 2(x+1)+x$
$$3x+3 > 2x+2+x$$
$$3x+3 > 3x+2$$
$$3 > 2$$
The solution to this inequality is all real numbers, since 3 is always greater than 2.
$\{x|x \text{ is any real number}\}$
$(-\infty, \infty)$

105. $-4a+1 > 9+3(2a+1)+a$
$$-4a+1 > 9+6a+3+a$$
$$-4a+1 > 7a+12$$
$$-11a+1 > 12$$
$$-11a > 11$$
$$a < -1$$
$\{a|a < -1\}$
$(-\infty, -1)$

107. $n+3(2n+3) > 7n-3$
$$n+6n+9 > 7n-3$$
$$7n+9 > 7n-3$$
$$9 > -3$$
The solution to this inequality is all real numbers, since 9 is always greater than -3.
$\{n|n \text{ is any real number}\}$
$(-\infty, \infty)$

109.
$$\frac{x}{2} \ge 1 - \frac{x}{4}$$
$$4\left(\frac{x}{2}\right) \ge 4\left(1-\frac{x}{4}\right)$$
$$2x \ge 4-x$$
$$3x \ge 4$$
$$x \ge \frac{4}{3}$$
$$\left\{x \,\middle|\, x \ge \frac{4}{3}\right\}$$
$$\left[\frac{4}{3}, \infty\right)$$

111.
$$\frac{x+5}{2}+4 > \frac{2x+1}{3}+2$$
$$6\left(\frac{x+5}{2}+4\right) > 6\left(\frac{2x+1}{3}+2\right)$$
$$3(x+5)+24 > 2(2x+1)+12$$
$$3x+15+24 > 4x+2+12$$
$$3x+39 > 4x+14$$
$$39 > x+14$$
$$25 > x$$
$\{x|x < 25\}$
$(-\infty, 25)$

113. $-5z-(3+2z) > 3-7z$
$$-5z-3-2z > 3-7z$$
$$-3-7z > 3-7z$$
$$-3 > 3$$
This inequality is false. Therefore, there is no solution. The solution set is $\varnothing$ or { }.

115. $1.3x+3.1 < 4.5x-15.9$
$$3.1 < 3.2x-15.9$$
$$19 < 3.2x$$
$$5.9375 < x$$
$\{x|x > 5.9375\}$
$(5.9375, \infty)$

117. Let x be the number of miles. Then the charge for 1 week is $(55 + 0.18x)$ dollars.
$$55 + 0.18x \leq 280$$
$$0.18x \leq 225$$
$$x \leq 1250$$
You can drive at most 1250 miles.

119. Let f be Yvette's score on the final exam. Then she has six scores total $(72, 78, 66, 81, f,$ and $f)$, since the final exam counts as two tests.
$$72 + 78 + 66 + 81 + f + f \geq 360$$
$$297 + 2f \geq 360$$
$$2f \geq 63$$
$$f \geq 31.5$$
Assuming that only whole-number scores are possible, the minimum score on the final is 32.

121. Let m be the number of minutes. Then Imperial Telephone charges $(10 + 0.03m)$ dollars and Mayflower Communications charges $(6 + 0.04m)$ dollars.
$$10 + 0.03m < 6 + 0.04m$$
$$4 < 0.01m$$
$$400 < m$$
Imperial Telephone is cheaper for more than 400 minutes.

123.
$$150,000 \leq 2.98I - 76.11$$
$$150,076.11 \leq 2.98I$$
$$50,361.11 \leq I$$
The bank will lend at least $150,000 for an annual income of $50,361.11 or greater.

125. Let x be the score on the fifth test. The average is then $\dfrac{68 + 82 + 87 + 89 + x}{5}$.
$$80 \leq \frac{68 + 82 + 87 + 89 + x}{5}$$
$$80 \leq \frac{326 + x}{5}$$
$$400 \leq 326 + x$$
$$74 \leq x$$
You need a score of at least 74 to get a B.

127.
$$-3 < x + 30 < 16$$
$$-3 - 30 < x + 30 - 30 < 16 - 30$$
$$-33 < x < -14$$
The solution set is $\{x|-33 < x < -14\}$.

129.
$$-6 \leq \frac{3x}{2} \leq 9$$
$$2(-6) \leq 2\left(\frac{3x}{2}\right) \leq 2(9)$$
$$-12 \leq 3x \leq 18$$
$$\frac{-12}{3} \leq \frac{3x}{3} \leq \frac{18}{3}$$
$$-4 \leq x \leq 6$$
The solution set is $\{x|-4 \leq x \leq 6\}$.

131.
$$-7 \leq 2x - 3 < 15$$
$$-7 + 3 \leq 2x - 3 + 3 < 15 + 3$$
$$-4 \leq 2x < 18$$
$$-\frac{4}{2} \leq \frac{2x}{2} < \frac{18}{2}$$
$$-2 \leq x < 9$$
The solution set is $\{x|-2 \leq x < 9\}$.

133.
$$4 < 6 - \frac{x}{2} \leq 10$$
$$4 - 6 < 6 - \frac{x}{2} - 6 \leq 10 - 6$$
$$-2 < -\frac{x}{2} \leq 4$$
$$-2(-2) > -2\left(-\frac{x}{2}\right) \geq -2(4)$$
$$4 > x \geq -8$$
This is the same as $-8 \leq x < 4$. The solution set is $\{x|-8 \leq x < 4\}$.

135. A left parenthesis is used to indicate that the solution is greater than a number. A left bracket is used to show that the solution is greater than or equal to a given number.

137. When solving an inequality and the variables are eliminated and a true statement results, the solution is all real numbers. When solving an inequality and the variables are eliminated and a false statement results, the solution is the empty set.

Chapter 2 Review

1. $3x + 2 = 7; x = 5$
$$3(5) + 2 \overset{?}{=} 7$$
$$15 + 2 \overset{?}{=} 7$$
$$17 = 7 \quad \text{False}$$
No, $x = 5$ is *not* a solution to the equation.

2. $5m - 1 = 17$; $m = 4$

$5(4) - 1 \overset{?}{=} 17$

$20 - 1 \overset{?}{=} 17$

$19 = 17$ False

No, $m = 4$ is *not* a solution to the equation.

3. $6x + 6 = 12$; $x = \dfrac{1}{2}$

$6\left(\dfrac{1}{2}\right) + 6 \overset{?}{=} 12$

$3 + 6 \overset{?}{=} 12$

$9 = 12$ False

No, $x = \dfrac{1}{2}$ is *not* a solution to the equation.

4. $9k + 3 = 9$; $k = \dfrac{2}{3}$

$9\left(\dfrac{2}{3}\right) + 3 \overset{?}{=} 9$

$6 + 3 \overset{?}{=} 9$

$9 = 9$ True

Yes, $k = \dfrac{2}{3}$ is a solution to the equation.

5. $n - 6 = 10$

$n - 6 + 6 = 10 + 6$

$n = 16$

Check: $n - 6 = 10$

$16 - 16 \overset{?}{=} 10$

$10 = 10$ True

The solution is 16, or the solution set is $\{16\}$.

6. $n - 8 = 12$

$n - 8 + 8 = 12 + 8$

$n = 20$

Check: $n - 8 = 12$

$20 - 8 \overset{?}{=} 12$

$12 = 12$ True

The solution is 20, or the solution set is $\{20\}$.

7. $x + 6 = -10$

$x + 6 - 6 = -10 - 6$

$x = -16$

Check: $x + 6 = -10$

$-16 + 6 \overset{?}{=} -10$

$-10 = -10$ True

The solution is −16, or the solution set is $\{-16\}$.

8. $x + 2 = -5$

$x + 2 - 2 = -5 - 2$

$x = -7$

Check: $x + 2 = -5$

$-7 + 2 \overset{?}{=} -5$

$-5 = -5$ True

The solution is −7, or the solution set is $\{-7\}$.

9. $-100 = m - 5$

$-100 + 5 = m - 5 + 5$

$-95 = m$

Check: $-100 = m - 5$

$-100 \overset{?}{=} -95 - 5$

$-100 = -100$ True

The solution is −95, or the solution set is $\{-95\}$.

10. $-26 = m - 76$

$-26 + 76 = m - 76 + 76$

$50 = m$

Check: $-26 = m - 76$

$-26 \overset{?}{=} 50 - 76$

$-26 = -26$ True

The solution is 50, or the solution set is $\{50\}$.

11. $\dfrac{2}{3}y = 16$

$\dfrac{3}{2} \cdot \dfrac{2}{3}y = \dfrac{3}{2} \cdot 16$

$y = 3 \cdot 8$

$y = 24$

Check: $\dfrac{2}{3}y = 16$

$\dfrac{2}{3} \cdot 24 \overset{?}{=} 16$

$2 \cdot 8 \overset{?}{=} 16$

$16 = 16$ True

The solution is 24, or the solution set is $\{24\}$.

12. $\dfrac{x}{4} = 20$

$4 \cdot \dfrac{x}{4} = 4 \cdot 20$

$x = 80$

Check: $\dfrac{x}{4} = 20$

$\dfrac{80}{4} \overset{?}{=} 20$

$20 = 20$ True

The solution is 80, or the solution set is $\{80\}$.

13. $-6x = 36$

$$\frac{-6x}{-6} = \frac{36}{-6}$$

$$x = -6$$

Check: $\quad -6x = 36$

$$-6(-6) \stackrel{?}{=} 36$$

$$36 = 36 \quad \text{True}$$

The solution is -6 or the solution set is $\{-6\}$.

14. $\quad -4x = -20$

$$\frac{-4x}{-4} = \frac{-20}{-4}$$

$$x = 5$$

Check: $\quad -4x = -20$

$$-4 \cdot 5 \stackrel{?}{=} -20$$

$$-20 = -20 \quad \text{True}$$

The solution is 5, or the solution set is $\{5\}$.

15. $\qquad z + \dfrac{5}{6} = \dfrac{1}{2}$

$$z + \frac{5}{6} - \frac{5}{6} = \frac{1}{2} - \frac{5}{6}$$

$$z = \frac{3}{6} - \frac{5}{6}$$

$$z = -\frac{2}{6}$$

$$z = -\frac{1}{3}$$

Check: $\qquad z + \dfrac{5}{6} = \dfrac{1}{2}$

$$-\frac{1}{3} + \frac{5}{6} \stackrel{?}{=} \frac{1}{2}$$

$$-\frac{2}{6} + \frac{5}{6} \stackrel{?}{=} \frac{1}{2}$$

$$\frac{3}{6} \stackrel{?}{=} \frac{1}{2}$$

$$\frac{1}{2} = \frac{1}{2} \quad \text{True}$$

The solution is $-\dfrac{1}{3}$ or the solution set is $\left\{ -\dfrac{1}{3} \right\}$.

16. $\qquad m - \dfrac{1}{8} = \dfrac{1}{4}$

$$m - \frac{1}{8} + \frac{1}{8} = \frac{1}{4} + \frac{1}{8}$$

$$m = \frac{2}{8} + \frac{1}{8}$$

$$m = \frac{3}{8}$$

Check: $\quad m - \dfrac{1}{8} = \dfrac{1}{4}$

$$\frac{3}{8} - \frac{1}{8} \stackrel{?}{=} \frac{1}{4}$$

$$\frac{2}{8} \stackrel{?}{=} \frac{1}{4}$$

$$\frac{1}{4} = \frac{1}{4} \quad \text{True}$$

The solution is $\dfrac{3}{8}$ or the solution set is $\left\{ \dfrac{3}{8} \right\}$.

17. $\quad 1.6x = 6.4$

$$\frac{1.6x}{1.6} = \frac{6.4}{1.6}$$

$$x = 4$$

Check: $\quad 1.6x = 6.4$

$$1.6(4) \stackrel{?}{=} 6.4$$

$$6.4 = 6.4 \quad \text{True}$$

The solution is 4, or the solution set is $\{4\}$.

18. $\quad 1.8m = 9$

$$\frac{1.8m}{1.8} = \frac{9}{1.8}$$

$$m = 5$$

Check: $\quad 1.8m = 9$

$$1.8(5) \stackrel{?}{=} 9$$

$$9 = 9 \quad \text{True}$$

The solution is 5, or the solution set is $\{5\}$.

19. $\qquad p - 1200 = 18,900$

$$p - 1200 + 1200 = 18,900 + 1200$$

$$p = 20,100$$

The original price was \$20,100.

20. $\quad 3c = 7.65$

$$\frac{3c}{3} = \frac{7.65}{3}$$

$$c = 2.55$$

Each cup of coffee cost \$2.55.

21. $\quad 5x - 1 = -21$

$$5x - 1 + 1 = -21 + 1$$

$$5x = -20$$

$$\frac{5x}{5} = \frac{-20}{5}$$

$$x = -4$$

Check: $\quad 5x - 1 = -21$

$$5(-4) - 1 \stackrel{?}{=} -21$$

$$-20 - 1 \stackrel{?}{=} -21$$

$$-21 = -21 \quad \text{True}$$

The solution is -4, or the solution set is $\{-4\}$.

22.
$$-3x + 7 = -5$$
$$-3x + 7 - 7 = -5 - 7$$
$$-3x = -12$$
$$\frac{-3x}{-3} = \frac{-12}{-3}$$
$$x = 4$$
Check: $-3x + 7 = -5$
$$-3(4) + 7 \stackrel{?}{=} -5$$
$$-12 + 7 \stackrel{?}{=} -5$$
$$-5 = -5 \quad \text{True}$$
The solution is 4, or the solution set is {4}.

23.
$$\frac{2}{3}x + 5 = 11$$
$$\frac{2}{3}x + 5 - 5 = 11 - 5$$
$$\frac{2}{3}x = 6$$
$$\frac{3}{2} \cdot \frac{2}{3}x = \frac{3}{2} \cdot 6$$
$$x = 9$$

Check: $\frac{2}{3}x + 5 = 11$
$$\frac{2}{3} \cdot 9 + 5 \stackrel{?}{=} 11$$
$$6 + 5 \stackrel{?}{=} 11$$
$$11 = 11 \quad \text{True}$$
The solution is 9, or the solution set is {9}.

24.
$$\frac{5}{7}x - 2 = -17$$
$$\frac{5}{7}x - 2 + 2 = -17 + 2$$
$$\frac{5}{7}x = -15$$
$$\frac{7}{5} \cdot \frac{5}{7}x = \frac{7}{5} \cdot (-15)$$
$$x = -21$$
Check: $\frac{5}{7}x - 2 = -17$
$$\frac{5}{7}(-21) - 2 \stackrel{?}{=} -17$$
$$-15 - 2 \stackrel{?}{=} -17$$
$$-17 = -17 \quad \text{True}$$
The solution is −21, or the solution set is {−21}.

25.
$$-2x + 5 + 6x = -11$$
$$4x + 5 = -11$$
$$4x + 5 - 5 = -11 - 5$$
$$4x = -16$$
$$\frac{4x}{4} = \frac{-16}{4}$$
$$x = -4$$
Check: $-2x + 5 + 6x = -11$
$$-2(-4) + 5 + 6(-4) \stackrel{?}{=} -11$$
$$8 + 5 - 24 \stackrel{?}{=} -11$$
$$-11 = -11 \quad \text{True}$$
The solution is −4, or the solution set is {−4}.

26.
$$3x - 5x + 6 = 18$$
$$-2x + 6 = 18$$
$$-2x + 6 - 6 = 18 - 6$$
$$-2x = 12$$
$$\frac{-2x}{-2} = \frac{12}{-2}$$
$$x = -6$$
Check: $3x + 5x + 6 = 18$
$$3(-6) - 5(-6) + 6 \stackrel{?}{=} 18$$
$$-18 + 30 + 6 \stackrel{?}{=} 18$$
$$18 = 18 \quad \text{True}$$
The solution is −6, or the solution set is {−6}.

27.
$$2m + 0.5m = 10$$
$$2.5m = 10$$
$$10 \cdot 2.5m = 10 \cdot 10$$
$$25m = 100$$
$$\frac{25m}{25} = \frac{100}{25}$$
$$m = 4$$
Check: $2m + 0.5m = 10$
$$2(4) + 0.5(4) \stackrel{?}{=} 10$$
$$8 + 2 \stackrel{?}{=} 10$$
$$10 = 10 \quad \text{True}$$
The solution is 4, or the solution set is {4}.

28.
$$1.4m + m = -12$$
$$2.4m = -12$$
$$\frac{2.4m}{2.4} = \frac{-12}{2.4}$$
$$m = -5$$
Check: $1.4m + m = -12$
$$1.4(-5) + (-5) \stackrel{?}{=} -12$$
$$-7 + (-5) \stackrel{?}{=} -12$$
$$-12 = -12 \quad \text{True}$$
The solution is −5, or the solution set is {−5}.

29.
$$-2(x+5)=-22$$
$$-2x-10=-22$$
$$-2x-10+10=-22+10$$
$$-2x=-12$$
$$\frac{-2x}{-2}=\frac{-12}{-2}$$
$$x=6$$
Check: $-2(x+5)=-22$
$$-2(6+5)\overset{?}{=}-22$$
$$-2(11)\overset{?}{=}-22$$
$$-22=-22 \quad \text{True}$$
The solution is 6, or the solution set is $\{6\}$.

30.
$$3(2x+5)=-21$$
$$6x+15=-21$$
$$6x+15-15=-21-15$$
$$6x=-36$$
$$\frac{6x}{6}=\frac{-36}{6}$$
$$x=-6$$
Check: $3(2x+5)=-21$
$$3[2(-6)+5]\overset{?}{=}-21$$
$$3(-12+5)\overset{?}{=}-21$$
$$3(-7)\overset{?}{=}-21$$
$$-21=-21 \quad \text{True}$$
The solution is -6, or the solution set is $\{-6\}$.

31.
$$5x+4=-7x+20$$
$$5x+4-4=-7x+20-4$$
$$5x=-7x+16$$
$$7x+5x=7x-7x+16$$
$$12x=16$$
$$\frac{12x}{12}=\frac{16}{12}$$
$$x=\frac{4}{3}$$
Check: $5x+4=-7x+20$
$$5\left(\frac{4}{3}\right)+4\overset{?}{=}-7\left(\frac{4}{3}\right)+20$$
$$\frac{20}{3}+\frac{12}{3}\overset{?}{=}\frac{-28}{3}+\frac{60}{3}$$
$$\frac{32}{3}=\frac{32}{3} \quad \text{True}$$

The solution is $\dfrac{4}{3}$, or the solution set is $\left\{\dfrac{4}{3}\right\}$.

32.
$$-3x+5=x-15$$
$$-3x+5-5=x-15-5$$
$$-3x=x-20$$
$$-x-3x=-x+x-20$$
$$-4x=-20$$
$$\frac{-4x}{-4}=\frac{-20}{-4}$$
$$x=5$$
Check: $-3x+5=x-15$
$$-3(5)+5\overset{?}{=}5-15$$
$$-15+5\overset{?}{=}-10$$
$$-10=-10 \quad \text{True}$$
The solution is 5, or the solution set is $\{5\}$.

33.
$$4(x-5)=-3x+5x-16$$
$$4x-20=2x-16$$
$$-2x+4x-20=-2x+2x-16$$
$$2x-20=-16$$
$$2x-20+20=-16+20$$
$$2x=4$$
$$\frac{2x}{2}=\frac{4}{2}$$
$$x=2$$
Check: $4(x-5)=-3x+5x-16$
$$4(2-5)\overset{?}{=}-3(2)+5(2)-16$$
$$4(-3)\overset{?}{=}-6+10-16$$
$$-12=-12 \quad \text{True}$$
The solution is 2, or the solution set is $\{2\}$.

34.
$$4(m+1)=m+5m-10$$
$$4m+4=6m-10$$
$$-4m+4m+4=-4m+6m-10$$
$$4=2m-10$$
$$4+10=2m-10+10$$
$$14=2m$$
$$\frac{14}{2}=\frac{2m}{2}$$
$$7=m$$
Check: $4(m+1)=m+5m-10$
$$4(7+1)\overset{?}{=}7+5(7)-10$$
$$4(8)\overset{?}{=}7+35-10$$
$$32=32 \quad \text{True}$$
The solution is 7, or the solution sct is $\{7\}$.

35.
$$x + x + 4 = 24$$
$$2x + 4 = 24$$
$$2x + 4 - 4 = 24 - 4$$
$$2x = 20$$
$$\frac{2x}{2} = \frac{20}{2}$$
$$x = 10$$
$$x + 4 = 14$$
Skye is 14 years old.

36.
$$2w + 2(w + 10) = 96$$
$$2w + 2w + 20 = 96$$
$$4w + 20 = 96$$
$$4w + 20 - 20 = 96 - 20$$
$$4w = 76$$
$$\frac{4w}{4} = \frac{76}{4}$$
$$w = 19$$
$$w + 10 = 29$$
The width is 19 yards and the length is 29 yards.

37.
$$\frac{6}{7}x + 3 = \frac{1}{2}$$
$$14\left(\frac{6}{7}x + 3\right) = 14\left(\frac{1}{2}\right)$$
$$14\left(\frac{6}{7}x\right) + 14(3) = 7$$
$$12x + 42 = 7$$
$$12x + 42 - 42 = 7 - 42$$
$$12x = -35$$
$$\frac{12x}{12} = \frac{-35}{12}$$
$$x = -\frac{35}{12}$$

Check:
$$\frac{6}{7}x + 3 = \frac{1}{2}$$
$$\frac{6}{7}\left(-\frac{35}{12}\right) + 3 \stackrel{?}{=} \frac{1}{2}$$
$$-\frac{5}{2} + \frac{6}{2} \stackrel{?}{=} \frac{1}{2}$$
$$\frac{1}{2} = \frac{1}{2} \quad \text{True}$$

The solution is $-\dfrac{35}{12}$, or the solution set is
$$\left\{-\frac{35}{12}\right\}.$$

38.
$$\frac{1}{4}x + 6 = \frac{5}{6}$$
$$12\left(\frac{1}{4}x + 6\right) = 12\left(\frac{5}{6}\right)$$
$$12\left(\frac{1}{4}x\right) + 12(6) = 2 \cdot 5$$
$$3x + 72 = 10$$
$$3x + 72 - 72 = 10 - 72$$
$$3x = -62$$
$$\frac{3x}{3} = \frac{-62}{3}$$
$$x = -\frac{62}{3}$$

Check:
$$\frac{1}{4}x + 6 = \frac{5}{6}$$
$$\frac{1}{4}\left(-\frac{62}{3}\right) + 6 \stackrel{?}{=} \frac{5}{6}$$
$$-\frac{31}{6} + \frac{36}{6} \stackrel{?}{=} \frac{5}{6}$$
$$\frac{5}{6} = \frac{5}{6} \quad \text{True}$$

The solution is $-\dfrac{62}{3}$, or the solution set is
$$\left\{-\frac{62}{3}\right\}.$$

39.
$$\frac{n}{2} + \frac{2}{3} = \frac{n}{6}$$
$$6\left(\frac{n}{2} + \frac{2}{3}\right) = 6\left(\frac{n}{6}\right)$$
$$6\left(\frac{n}{2}\right) + 6\left(\frac{2}{3}\right) = n$$
$$3n + 4 = n$$
$$-3n + 3n + 4 = -3n + n$$
$$4 = -2n$$
$$\frac{4}{-2} = \frac{-2n}{-2}$$
$$-2 = n$$

Check:
$$\frac{n}{2} + \frac{2}{3} = \frac{n}{6}$$
$$\frac{-2}{2} + \frac{2}{3} \stackrel{?}{=} \frac{-2}{6}$$
$$-\frac{3}{3} + \frac{2}{3} \stackrel{?}{=} -\frac{1}{3}$$
$$-\frac{1}{3} = -\frac{1}{3} \quad \text{True}$$

The solution is -2, or the solution set is $\{-2\}$.

40.
$$\frac{m}{8}+\frac{m}{2}=\frac{3}{4}$$
$$8\left(\frac{m}{8}+\frac{m}{2}\right)=8\left(\frac{3}{4}\right)$$
$$8\left(\frac{m}{8}\right)+8\left(\frac{m}{2}\right)=2\cdot3$$
$$m+4m=6$$
$$5m=6$$
$$\frac{5m}{5}=\frac{6}{5}$$
$$m=\frac{6}{5}$$

Check:
$$\frac{m}{8}+\frac{m}{2}=\frac{3}{4}$$
$$\frac{\frac{6}{5}}{8}+\frac{\frac{6}{5}}{2}\overset{?}{=}\frac{3}{4}$$
$$\frac{3}{20}+\frac{6}{10}\overset{?}{=}\frac{3}{4}$$
$$\frac{15}{20}=\frac{15}{20}\quad\text{True}$$

The solution is $\frac{6}{5}$, or the solution set is $\left\{\frac{6}{5}\right\}$.

41.
$$1.2r=-1+2.8$$
$$1.2r=1.8$$
$$10(1.2r)=10(1.8)$$
$$12r=18$$
$$\frac{12r}{12}=\frac{18}{12}$$
$$r=\frac{3}{2}$$

Check:
$$1.2r=-1+2.8$$
$$1.2\left(\frac{3}{2}\right)\overset{?}{=}1.8$$
$$1.8=1.8\quad\text{True}$$

The solution is $\frac{3}{2}$, or the solution set is $\left\{\frac{3}{2}\right\}$.

42.
$$0.2x+0.5x=2.1$$
$$0.7x=2.1$$
$$10(0.7x)=10(2.1)$$
$$7x=21$$
$$\frac{7x}{7}=\frac{21}{7}$$
$$x=3$$

Check:
$$0.2x+0.5x=2.1$$
$$0.2(3)+0.5(3)\overset{?}{=}2.1$$
$$0.6+1.5\overset{?}{=}2.1$$
$$2.1=2.1\quad\text{True}$$

The solution is 3, or the solution set is {3}.

43.
$$1.2m-3.2=0.8m-1.6$$
$$10(1.2m-3.2)=10(0.8m-1.6)$$
$$12m-32=8m-16$$
$$-8m+12m-32=-8m+8m-16$$
$$4m-32=-16$$
$$4m-32+32=-16+32$$
$$4m=16$$
$$\frac{4m}{4}=\frac{16}{4}$$
$$m=4$$

Check:
$$1.2m-3.2=0.8m-1.6$$
$$1.2(4)-3.2\overset{?}{=}0.8(4)-1.6$$
$$4.8-3.2\overset{?}{=}3.2-1.6$$
$$1.6=1.6\quad\text{True}$$

The solution is 4, or the solution set is {4}.

44.
$$0.3m+0.8=0.5m+1$$
$$10(0.3m+0.8)=10(0.5m+1)$$
$$3m+8=5m+10$$
$$-3m+3m+8=-3m+5m+10$$
$$8=2m+10$$
$$8-10=2m+10-10$$
$$-2=2m$$
$$\frac{-2}{2}=\frac{2m}{2}$$
$$-1=m$$

Check:
$$0.3m+0.8=0.5m+1$$
$$0.3(-1)+0.8\overset{?}{=}0.5(-1)+1$$
$$-0.3+0.8\overset{?}{=}-0.5+1$$
$$0.5=0.5\quad\text{True}$$

The solution is -1, or the solution set is $\{-1\}$

45.
$$\frac{1}{2}(x+5)=\frac{3}{4}$$
$$4\cdot\frac{1}{2}(x+5)=4\cdot\frac{3}{4}$$
$$2(x+5)=3$$
$$2x+10=3$$
$$2x+10-10=3-10$$
$$2x=-7$$
$$\frac{2x}{2}=\frac{-7}{2}$$
$$x=-\frac{7}{2}$$

Check: $\dfrac{1}{2}(x+5) = \dfrac{3}{4}$

$\dfrac{1}{2}\left(-\dfrac{7}{2}+5\right) \overset{?}{=} \dfrac{3}{4}$

$\dfrac{1}{2}\left(-\dfrac{7}{2}+\dfrac{10}{2}\right) \overset{?}{=} \dfrac{3}{4}$

$\dfrac{1}{2}\left(\dfrac{3}{2}\right) \overset{?}{=} \dfrac{3}{4}$

$\dfrac{3}{4} = \dfrac{3}{4}$ True

The solution is $-\dfrac{7}{2}$, or the solution set is

$\left\{-\dfrac{7}{2}\right\}$.

46. $-\dfrac{1}{6}(x-1) = \dfrac{2}{3}$

$-6\cdot\left[-\dfrac{1}{6}(x-1)\right] = -6\cdot\dfrac{2}{3}$

$x-1 = -4$

$x-1+1 = -4+1$

$x = -3$

Check: $-\dfrac{1}{6}(x-1) = \dfrac{2}{3}$

$-\dfrac{1}{6}(-3-1) \overset{?}{=} \dfrac{2}{3}$

$-\dfrac{1}{6}(-4) \overset{?}{=} \dfrac{2}{3}$

$\dfrac{2}{3} = \dfrac{2}{3}$

The solution is -3, or the solution set is $\{-3\}$.

47. $0.1(x+80) = -0.2+14$

$10[0.1(x+80)] = 10(-0.2+14)$

$1(x+80) = -2+140$

$x+80 = 138$

$x+80-80 = 138-80$

$x = 58$

Check: $0.1(x+80) = -0.2+14$

$0.1(58+80) \overset{?}{=} -0.2+14$

$0.1(138) \overset{?}{=} 13.8$

$13.8 = 13.8$ True

The solution is 58, or the solution set is $\{58\}$.

48. $0.35(x+6) = 0.45(x+7)$

$100[0.35(x+6)] = 100[0.45(x+7)]$

$35(x+6) = 45(x+7)$

$35x+210 = 45x+315$

$-35x+35x+210 = -35x+45x+315$

$210 = 10x+315$

$210-315 = 10x+315-315$

$-105 = 10x$

$\dfrac{-105}{10} = \dfrac{10x}{10}$

$-10.5 = x$

Check: $0.35(x+6) = 0.45(x+7)$

$0.35(-10.5+6) \overset{?}{=} 0.45(-10.5+7)$

$0.35(-4.5) \overset{?}{=} 0.45(-3.5)$

$-1.575 = -1.575$ True

The solution is -10.5, or the solution set is $\{-10.5\}$.

49. $4x+2x-10 = 6x+5$

$6x-10 = 6x+5$

$-6x+6x-10 = -6x+6x+5$

$-10 = 5$

This is a false statement, so the equation is a contradiction. The solution set is $\varnothing$ or $\{\ \}$.

50. $-2(x+5) = -5x+3x+2$

$-2x-10 = -2x+2$

$2x-2x-10 = 2x-2x+2$

$-10 = 2$

This is a false statement, so the equation is a contradiction. The solution set is $\varnothing$ or $\{\ \}$.

51. $-5(2n+10) = 6n-50$

$-10n-50 = 6n-50$

$10n-10n-50 = 10n+6n-50$

$-50 = 16n-50$

$-50+50 = 16n-50+50$

$0 = 16n$

$\dfrac{0}{16} = \dfrac{16n}{16}$

$0 = n$

This is a conditional equation. The solution set is $\{0\}$.

52.
$$8m + 10 = -2(7m - 5)$$
$$8m + 10 = -14m + 10$$
$$14m + 8m + 10 = 14m - 14m + 10$$
$$22m + 10 = 10$$
$$22m + 10 - 10 = 10 - 10$$
$$22m = 0$$
$$\frac{22m}{22} = \frac{0}{22}$$
$$m = 0$$

This is a conditional equation. The solution set is {0}.

53.
$$10x - 2x + 18 = 2(4x + 9)$$
$$8x + 18 = 8x + 18$$
$$-8x + 8x + 18 = -8x + 8x + 18$$
$$18 = 18$$

This is a true statement. The equation is an identity. The solution set is the set of all real numbers.

54.
$$-3(2x - 8) = -3x - 3x + 24$$
$$-6x + 24 = -6x + 24$$
$$6x - 6x + 24 = 6x - 6x + 24$$
$$24 = 24$$

This is a true statement. The equation is an identity. The solution set is the set of all real numbers.

55.
$$p - 0.20p = 12.60$$
$$0.8p = 12.60$$
$$10(0.8p) = 10(12.60)$$
$$8p = 126$$
$$\frac{8p}{8} = \frac{126}{8}$$
$$p = 15.75$$

The shirt's original price was $15.75.

56.
$$0.10x + 0.05(2x - 1) = 0.55$$
$$0.10x + 0.10x - 0.05 = 0.55$$
$$0.2x - 0.05 = 0.55$$
$$0.2x = 0.6$$
$$10 \cdot 0.2x = 10 \cdot 0.6$$
$$2x = 6$$
$$\frac{2x}{2} = \frac{6}{2}$$
$$x = 3$$

Juanita found 3 dimes.

57. $A = lw; l = 8, w = 6$
$$A = 8(6) = 48$$
The area is 48 square inches.

58. $P = 4s; s = 16$
$$P = 4(16) = 64$$
The perimeter is 64 centimeters.

59. $P = 2l + 2w; P = 16, l = \dfrac{13}{2}$
$$16 = 2\left(\frac{13}{2}\right) + 2w$$
$$16 = 13 + 2w$$
$$-13 + 16 = -13 + 13 + 2w$$
$$3 = 2w$$
$$\frac{3}{2} = \frac{2w}{2}$$
$$\frac{3}{2} = w$$

The width is $\dfrac{3}{2}$ yards.

60. $C = \pi d; d = \dfrac{15}{\pi}$
$$C = \pi\left(\frac{15}{\pi}\right) = 15$$
The circumference is 15 millimeters.

61.
$$V = LWH$$
$$\frac{V}{LW} = \frac{LWH}{LW}$$
$$\frac{V}{LW} = H$$

62.
$$I = Prt$$
$$\frac{I}{rt} = \frac{Prt}{rt}$$
$$\frac{I}{rt} = P$$

63.
$$S = 2LW + 2LH + 2WH$$
$$S - 2LH = 2LW + 2LH + 2WH - 2LH$$
$$S - 2LH = 2LW + 2WH$$
$$S - 2LH = W(2L + 2H)$$
$$\frac{S - 2LH}{2L + 2H} = \frac{W(2L + 2H)}{2L + 2H}$$
$$\frac{S - 2LH}{2L + 2H} = W$$

64.
$$\rho = mv + MV$$
$$\rho - mv = mv + MV - mv$$
$$\rho - mv = MV$$
$$\frac{\rho - mv}{V} = \frac{MV}{V}$$
$$\frac{\rho - mv}{V} = M$$

65.
$$2x + 3y = 10$$
$$-2x + 2x + 3y = -2x + 10$$
$$3y = -2x + 10$$
$$\frac{3y}{3} = \frac{-2x + 10}{3}$$
$$y = \frac{-2x + 10}{3}$$

66.
$$6x - 7y = 14$$
$$6x - 7y + 7y = 14 + 7y$$
$$6x = 14 + 7y$$
$$\frac{6x}{6} = \frac{14 + 7y}{6}$$
$$x = \frac{14 + 7y}{6}$$

67. (a)
$$A = P(1 + r)^t$$
$$\frac{A}{(1 + r)^t} = \frac{P(1 + r)^t}{(1 + r)^t}$$
$$\frac{A}{(1 + r)^t} = P$$

(b) $P = \dfrac{A}{(1 + r)^t}$; $A = 3000$, $t = 6$,

$r = 5\% = 0.05$

$$P = \frac{3000}{(1 + 0.05)^6} = \frac{3000}{1.05^6} \approx 2238.65$$

You would have to deposit \$2238.65.

68. (a)
$$A = 2\pi rh + 2\pi r^2$$
$$A - 2\pi r^2 = 2\pi rh + 2\pi r^2 - 2\pi r^2$$
$$A - 2\pi r^2 = 2\pi rh$$
$$\frac{A - 2\pi r^2}{2\pi r} = \frac{2\pi rh}{2\pi r}$$
$$\frac{A - 2\pi r^2}{2\pi r} = h$$

(b) $h = \dfrac{A - 2\pi r^2}{2\pi r}$; $A = 72\pi$, $r = 4$

$$h = \frac{72\pi - 2\pi(4^2)}{2\pi(4)}$$
$$h = \frac{\pi(72 - 32)}{\pi(8)}$$
$$h = \frac{40}{8}$$
$$h = 5$$

The height is 5 centimeters.

69. $I = Prt$; $P = 500$, $r = 3\% = 0.03$, $t = \dfrac{9}{12} = \dfrac{3}{4}$

$$I = 500(0.03)\left(\frac{3}{4}\right) = 11.25$$

Samuel's investment will earn \$11.25.

70. $d = 3$, $r = \dfrac{3}{2}$

$$A = \pi r^2 = \pi\left(\frac{3}{2}\right)^2 = \frac{9}{4}\pi \approx 7.069$$

The area is about 7.1 square feet.

71. the difference between a number and 6:
$x - 6$

72. eight subtracted from a number: $x - 8$

73. the product of -8 and a number: $-8x$

74. the quotient of a number and 10: $\dfrac{x}{10}$

75. twice the sum of 6 and a number: $2(6 + x)$

76. four times the difference of 5 and a number:
$4(5 - x)$

77. $6 + x = 2x + 5$

78. $6x - 10 = 2x + 1$

79. $x - 8 = \dfrac{1}{2}x$

80. $\dfrac{6}{x} = 10 + x$

81. $4(2x + 8) = 16$

82. $5(2x - 8) = -24$

83. Let s be Sarah's age. Then $s + 7$ is Jacob's age.

84. Let c be Consuelo's speed. Then $2c$ is José's speed.

85. Let m be Max's amount. Then $m - 6$ is Irene's amount.

86. Let v be Victor's amount. Then $350 - v$ is Larry's amount.

87. We want to find Lee Lai's weight one year ago. Let n be the weight.
$$n - 28 = 125$$
$$n = 153$$
Is the difference between 153 and 125 28? Yes. Lee Lai's weight was 153 pounds one year ago.

88. We want to find three consecutive integers. Let n be the first integer. Then $n + 1$ and $n + 2$ are the next two integers.
$$n + (n+1) + (n+2) = 39$$
$$3n + 3 = 39$$
$$3n = 36$$
$$n = 12$$
If $n = 12$, then $n + 1 = 13$ and $n + 2 = 14$. Are the numbers 12, 13, 14 consecutive integers? Yes. Do they sum to 39? Yes. The integers are 12, 13, and 14.

89. We want to find how much each will receive. Let j be the amount received by Juan. Then $j - 2000$ is the amount received by Roberto.
$$j + (j - 2000) = 20,000$$
$$2j - 2000 = 20,000$$
$$2j = 22,000$$
$$j = 11,000$$
If $j = 11,000$, then $j - 2000 = 9000$. Do 11,000 and 9000 differ by 2000? Yes. Do 11,000 and 9000 sum to 20,000? Yes. Juan will receive $11,000 and Roberto will receive $9000.

90. We want to find the number of miles for which the cost will be the same. Let x be the number of miles driven. ABC-Rental charges $30 + 0.15x$ and U-Do-It Rental charges $15 + 0.3x$.
$$30 + 0.15x = 15 + 0.3x$$
$$30 = 15 + 0.15x$$
$$15 = 0.15x$$
$$100 = x$$
ABC-Rental's cost will be $30 + 0.15(100) = 30 + 15 = \$45$ and U-Do-It-Rental's cost will be $15 + 0.3(100) = 15 + 30 = \$45$, and they are the same. The cost will be the same for 100 miles.

91. $x = 0.065(80)$
$x = 5.2$
5.2 is 6.5% of 80.

92. $18 = 0.3x$
$$\frac{18}{0.3} = \frac{0.3x}{0.3}$$
$$60 = x$$
18 is 30% of 60.

93. $15.6 = p \cdot 120$
$$\frac{15.6}{120} = \frac{120p}{120}$$
$$0.13 = p$$
$$13\% = p$$
15.6 is 13% of 120.

94. $1.1 \cdot x = 55$
$$\frac{1.1x}{1.1} = \frac{55}{1.1}$$
$$x = 50$$
110% of 50 is 55.

95. Let x be the cost before tax. Then $0.06x$ is the tax amount.
$$x + 0.06x = 19.61$$
$$1.06x = 19.61$$
$$x = 18.5$$
The leotard cost $18.50 before sales tax.

96. Let x be the previous hourly fee. Then $0.085x$ is the amount of the increase.
$$x + 0.085x = 32.55$$
$$1.085x = 32.55$$
$$x = 30$$
Mei Ling's previous hourly fee was $30.

97. Let x be the sweater's original price. Then $0.70x$ is the discounted amount.
$$x - 0.70x = 12$$
$$0.3x = 12$$
$$x = 40$$
The sweater's original price was $40.

98. Let x be the store's price. Then $0.80x$ is the markup amount.
$$x + 0.80x = 360$$
$$1.8x = 360$$
$$x = 200$$
The store paid $200 for the suit.

99. Let x be the total value of the computers. Then Tanya earns \$500 plus $0.02x$.

$$500 + 0.02x = 3000$$
$$0.02x = 2500$$
$$x = 125,000$$

Tanya must sell computers worth a total of \$125,000.

100. Let x be the winner's amount. then $0.80x$ is the loser's amount.

$$x + 0.80x = 900$$
$$1.8x = 900$$
$$x = 500$$

If $x = 500$, then $0.8(500) = 400$. The winner received 500 votes, whereas the loser received 400 votes.

101. We want to find complementary angles. The measures of complementary angles sum to 90°. Let x be the measure of the second angle. Then $6x + 20$ is the measure of the first angle.

$$x + (6x + 20) = 90$$
$$7x + 20 = 90$$
$$7x = 70$$
$$x = 10$$

If $x = 10$, then
$6x + 20 = 6(10) + 20 = 60 + 20 = 80$. The measures of the angles are 10° and 80°.

102. We want to find supplementary angles. The measures of supplementary angles sum to 180°. Let x be the measure of the second angle. Then $2x - 60$ is the measure of the first angle.

$$x + (2x - 60) = 180$$
$$3x - 60 = 180$$
$$3x = 240$$
$$x = 80$$

If $x = 80$, then
$2x - 60 = 2(80) - 60 = 160 - 60 = 100$. The measures of the angles are 80° and 100°.

103. We want to find the measures of the angles of the triangle. The measures of the angles of a triangle sum to 180°. Let x be the measure of the first angle. Then $2x$ is the measure of the second and $2x + 30$ is the measure of the third.

$$x + (2x) + (2x + 30) = 180$$
$$5x + 30 = 180$$
$$5x = 150$$
$$x = 30$$

If $x = 30$, then $2x = 60$, and
$2x + 30 = 60 + 30 = 90$. The measures of the angles are 30°, 60°, and 90°.

104. We want to find the measures of the angles of the triangle. The measures of the angles of a triangle sum to 180°. Let x be the measure of the first angle. Then $x - 5$ is the measure of the second angle and $2(x - 5) - 5$ is the measure of the third angle.

$$x + (x - 5) + 2(x - 5) - 5 = 180$$
$$x + x - 5 + 2x - 10 - 5 = 180$$
$$4x - 20 = 180$$
$$4x = 200$$
$$x = 50$$

If $x = 50$, then $x - 5 = 45$ and
$2(x - 5) - 5 = 2(45) - 5 = 85$. The measures of the angles are 50°, 45°, and 85°.

105. We want to find the dimensions of the rectangle. Let w be the width of the rectangle. Then $2w + 15$ is the length of the rectangle. The perimeter of a rectangle is the sum of twice the length and twice the width. The perimeter is 78 inches.

$$2(2w + 15) + 2w = 78$$
$$4w + 30 + 2w = 78$$
$$6w + 30 = 78$$
$$6w = 48$$
$$w = 8$$

If $w = 8$, then $2w + 15 = 2(8) + 15 = 31$. The length is 31 inches and the width is 8 inches.

106. We want to find the dimensions of the rectangle. Let w be the width of the rectangle. then $4w$ is the length of the rectangle. The perimeter of a rectangle is the sum of twice the length and twice the width. The perimeter is 70 cm.

$$2(4w) + 2w = 70$$
$$8w + 2w = 70$$
$$10w = 70$$
$$w = 7$$

If $w = 7$, then $4w = 28$. The width of the rectangle is 7 cm and the length is 28 cm.

107. (a) We want to find the dimensions of the rectangular garden. Let l be the length. Then the width is $2l$. The perimeter is twice length plus twice the width and is 120 feet.

$$2l + 2(2l) = 120$$
$$2l + 4l = 120$$
$$6l = 120$$
$$l = 20$$

If $l = 20$, then $2l = 40$. The garden's length is 20 feet and the width is 40 feet.

(b) $A = lw = 20(40) = 800$
The area of the garden is 800 square feet.

108. The area of a trapezoid is $A = \frac{1}{2}h(B+b)$, where h is the height and the bases are B and b. Let B be the longer base. Then $B - 10$ is the shorter base. The height is 80 feet and the area is 3600 square feet.

$$\frac{1}{2}(80)[B+(B-10)] = 3600$$
$$40(2B-10) = 3600$$
$$80B - 400 = 3600$$
$$80B = 4000$$
$$B = 50$$

If $B = 50$, then $B - 10 = 40$. The bases are 50 feet and 40 feet.

109. Let t be the time at which they are 35 miles apart.

	Rate	· Time	= Distance
slow	18	t	$18t$
fast	25	t	$25t$

The difference of the distances is 35, since they are traveling in the same direction.
$$25t - 18t = 35$$
$$7t = 35$$
$$t = 5$$
They will be 35 miles apart after 5 hours.

110. Let r be the speed of the faster train.

	Rate	· Time	= Distance
East	r	6	$6r$
West	$r-10$	6	$6(r-10)$

The sum of their distances is 720, since they are traveling in opposite directions.
$$6r + 6(r-10) = 720$$
$$6r + 6r - 60 = 720$$
$$12r = 780$$
$$r = 65$$
The faster train is traveling at 65 mph.

111. $x \le -3$

112. $x > 4$

113. $m < 2$

114. $m \ge -5$

115. $0 < n$

116. $-3 \le n$

117. $x < -4$
$(-\infty, -4)$

118. $x \ge 7$
$[7, \infty)$

119. $[2, \infty)$

120. $(-\infty, 3)$

121. $4x + 3 < 2x - 10$
$$2x + 3 < -10$$
$$2x < -13$$
$$x < -\frac{13}{2}$$
$$\left\{ x \,\middle|\, x < -\frac{13}{2} \right\}$$
$$\left(-\infty, -\frac{13}{2} \right)$$

122. $3x - 5 \ge -12$
$$3x \ge -7$$
$$x \ge -\frac{7}{3}$$
$$\left\{ x \,\middle|\, x \ge -\frac{7}{3} \right\}$$
$$\left[-\frac{7}{3}, \infty \right)$$

123. $-4(x-1) \le x+8$
$-4x+4 \le x+8$
$-5x+4 \le 8$
$-5x \le 4$
$x \ge -\dfrac{4}{5}$

$\left\{ x \middle| x \ge -\dfrac{4}{5} \right\}$

$\left[-\dfrac{4}{5}, \infty \right)$

124. $6x-10 < 7x+2$
$-10 < x+2$
$-12 < x$
$\{x | x > -12\}$
$(-12, \infty)$

125. $-3(x+7) > -x-2x$
$-3x-21 > -3x$
$-21 > 0$
This is a false statement. Therefore, there is no solution. The solution set is $\varnothing$ or $\{\ \}$.

126. $4x+10 \le 2(2x+7)$
$4x+10 \le 4x+14$
$10 \le 14$
The solution to this inequality is all real numbers, since 10 is always less than or equal to 14.
$\{x | x \text{ is any real number}\}$

127. $\dfrac{1}{2}(3x-1) > \dfrac{2}{3}(x+3)$
$6 \cdot \dfrac{1}{2}(3x-1) > 6 \cdot \dfrac{2}{3}(x+3)$
$3(3x-1) > 4(x+3)$
$9x-3 > 4x+12$
$5x-3 > 12$
$5x > 15$
$x > 3$
$\{x | x > 3\}$
$(3, \infty)$

128. $\dfrac{5}{4}x+2 < \dfrac{5}{6}x - \dfrac{7}{6}$
$12\left(\dfrac{5}{4}x+2\right) < 12\left(\dfrac{5}{6}x-\dfrac{7}{6}\right)$
$15x+24 < 10x-14$
$5x+24 < -14$
$5x < -38$
$x < -\dfrac{38}{5}$

$\left\{ x \middle| x < -\dfrac{38}{5} \right\}$

$\left(-\infty, -\dfrac{38}{5} \right)$

129. Let m be the number of miles driven.
$19.95 + 0.2m \le 32.95$
$0.2m \le 13$
$m \le 65$
A customer can drive at most 65 miles.

130. Let s be the score on his third game.
$\dfrac{148+155+s}{3} > 151$
$\dfrac{303+s}{3} > 151$
$303+s > 453$
$s > 150$
Travis must score more than 150.

Chapter 2 Test

1. $x+3 = -14$
$x+3-3 = -14-3$
$x = -17$
Check:
$x+3 = -14$
$-17+3 \overset{?}{=} -14$
$-14 = -14$ True

The solution is -17, or the solution set is $\{-17\}$.

2. $-\dfrac{2}{3}m = \dfrac{8}{27}$
$-\dfrac{3}{2} \cdot \left(-\dfrac{2}{3}m\right) = -\dfrac{3}{2} \cdot \dfrac{8}{27}$
$m = -\dfrac{4}{9}$

Check: $-\dfrac{2}{3}m = \dfrac{8}{27}$

$$-\dfrac{2}{3}\cdot\left(-\dfrac{4}{9}\right)\stackrel{?}{=}\dfrac{8}{27}$$

$$\dfrac{8}{27}=\dfrac{8}{27}\quad\text{True}$$

The solution is $-\dfrac{4}{9}$, or the solution set is

$\left\{-\dfrac{4}{9}\right\}$.

3. $\qquad 5(2x-4)=5x$

$$10x-20=5x$$
$$-10x+10x-20=-10x+5x$$
$$-20=-5x$$
$$\dfrac{-20}{-5}=\dfrac{-5x}{-5}$$
$$4=x$$

Check: $\quad 5(2x-4)=5x$

$$5(2\cdot4-4)\stackrel{?}{=}5\cdot4$$
$$5(8-4)\stackrel{?}{=}20$$
$$5(4)\stackrel{?}{=}20$$
$$20=20\quad\text{True}$$

The solution is 4, or the solution set is $\{4\}$.

4. $\qquad -2(x-5)=5(-3x+4)$

$$-2x+10=-15x+20$$
$$2x-2x+10=2x-15x+20$$
$$10=-13x+20$$
$$10-20=-13x+20-20$$
$$-10=-13x$$
$$\dfrac{-10}{-13}=\dfrac{-13x}{-13}$$
$$\dfrac{10}{13}=x$$

Check: $\qquad -2(x-5)=5(-3x+4)$

$$-2\left(\dfrac{10}{13}-5\right)\stackrel{?}{=}5\left(-3\cdot\dfrac{10}{13}+4\right)$$
$$-2\left(\dfrac{10}{13}-\dfrac{65}{13}\right)\stackrel{?}{=}5\left(-\dfrac{30}{13}+\dfrac{52}{13}\right)$$
$$-2\left(-\dfrac{55}{13}\right)\stackrel{?}{=}5\left(\dfrac{22}{13}\right)$$
$$\dfrac{110}{13}=\dfrac{110}{13}\quad\text{True}$$

The solution is $\dfrac{10}{13}$ or the solution set is $\left\{\dfrac{10}{13}\right\}$.

5. $\qquad -\dfrac{2}{3}x+\dfrac{3}{4}=\dfrac{1}{3}$

$$12\left(-\dfrac{2}{3}x+\dfrac{3}{4}\right)=12\left(\dfrac{1}{3}\right)$$
$$12\left(-\dfrac{2}{3}x\right)+12\left(\dfrac{3}{4}\right)=12\left(\dfrac{1}{3}\right)$$
$$-8x+9=4$$
$$-8x+9-9=4-9$$
$$-8x=-5$$
$$\dfrac{-8x}{8}=\dfrac{-5}{-8}$$
$$x=\dfrac{5}{8}$$

Check: $\qquad -\dfrac{2}{3}x+\dfrac{3}{4}=\dfrac{1}{3}$

$$-\dfrac{2}{3}\left(\dfrac{5}{8}\right)+\dfrac{3}{4}\stackrel{?}{=}\dfrac{1}{3}$$
$$-\dfrac{5}{12}+\dfrac{9}{12}\stackrel{?}{=}\dfrac{1}{3}$$
$$\dfrac{4}{12}\stackrel{?}{=}\dfrac{1}{3}$$
$$\dfrac{1}{3}=\dfrac{1}{3}\quad\text{True}$$

The solution is $\dfrac{5}{8}$ or the solution set is $\left\{\dfrac{5}{8}\right\}$.

6. $\qquad -0.6+0.4y=1.4$

$$10(-0.6+0.4y)=10(1.4)$$
$$-6+4y=14$$
$$6-6+4y=6+14$$
$$4y=20$$
$$\dfrac{4y}{4}=\dfrac{20}{4}$$
$$y=5$$

Check: $\quad -0.6+0.4y=1.4$

$$-0.6+0.4(5)\stackrel{?}{=}1.4$$
$$-0.6+2\stackrel{?}{=}1.4$$
$$1.4=1.4\quad\text{True}$$

The solution is 5, or the solution set is $\{5\}$.

7. $8x+3(2-x)=5(x+2)$

$$8x+6-3x=5x+10$$
$$6+5x=5x+10$$
$$6+5x-5x=5x+10-5x$$
$$6=10$$

This is a false statement. The equation is a contradiction. The solution set is $\varnothing$ or $\{\ \}$.

8.
$$2(x+7) = 2x - 2 + 16$$
$$2x + 14 = 2x + 14$$
$$-2x + 2x + 14 = -2x + 2x + 14$$
$$14 = 14$$
This is a true statement. The equation is an identity. The solution set is the set of all real numbers.

9. (a)
$$V = lwh$$
$$\frac{V}{wh} = \frac{lwh}{wh}$$
$$\frac{V}{wh} = l$$

(b) $l = \dfrac{V}{wh}$; $V = 540$, $w = 6$, $h = 10$
$$l = \frac{540}{6(10)} = \frac{540}{60} = 9$$
The length is 9 inches.

10. (a)
$$2x + 3y = 12$$
$$-2x + 2x + 3y = -2x + 12$$
$$3y = -2x + 12$$
$$\frac{3y}{3} = \frac{-2x + 12}{3}$$
$$y = -\frac{2}{3}x + 4$$

(b) $y = -\dfrac{2}{3}x + 4$; $x = 8$
$$y = -\frac{2}{3}(8) + 4$$
$$y = -\frac{16}{3} + \frac{12}{3}$$
$$y = -\frac{4}{3}$$

11. Let x be the number.
$$6(x - 8) = 2x - 5$$

12.
$$18 = 0.30x$$
$$\frac{18}{0.30} = \frac{0.30x}{0.30}$$
$$60 = x$$
18 is 30% of 60.

13. We want to find three consecutive integers. Let n be the first integer. Then $n + 1$ and $n + 2$ are the next two integers, respectively. They sum to 48.
$$n + (n+1) + (n+2) = 48$$
$$3n + 3 = 48$$
$$3n = 45$$
$$n = 15$$
If $n = 15$, then $n + 1 = 16$ and $n + 2 = 17$. Do 15, 16, and 17 sum to 48? Yes. Are 15, 16, and 17 consecutive integers? Yes. The integers are 15, 16, and 17.

14. We need to find the lengths of the three sides. Let m be the length of the middle side. Then the length of the longest side is $m + 2$, and the length of the shortest side is $m - 14$. The perimeter, or the sum of the three sides, is 60.
$$m + (m+2) + (m-14) = 60$$
$$3m - 12 = 60$$
$$3m = 72$$
$$m = 24$$
If $m = 24$, then $m + 2 = 26$, and $m - 14 = 10$. Do 24, 26, and 10 sum to 60? Yes. The lengths of the sides are 10 inches, 24 inches, and 26 inches.

15. Let t be the time at which they are 350 miles apart.

	Rate	· Time	= Distance
Kimberly	40	t	$40t$
Clay	60	t	$60t$

The sum of their distances is 350 since they are traveling in opposite directions.
$$40t + 60t = 350$$
$$100t = 350$$
$$t = 3.5$$
They will be 350 miles apart in 3.5 hours.

16. Let x be the length of the shorter piece. Then $3x + 1$ is the length of the longer piece. The sum of the lengths of the pieces is 21.
$$x + (3x + 1) = 21$$
$$4x + 1 = 21$$
$$4x = 20$$
$$x = 5$$
$3x + 1 = 3(5) + 1 = 15 + 1 = 16$
The shorter piece is 5 feet and the longer piece is 16 feet.

17. Let x be the original price of the backpack. Then the discount amount was $0.20x$

$$x - 0.20x = 28.80$$
$$0.80x = 28.80$$
$$10 \cdot 0.8x = 10 \cdot 28.8$$
$$8x = 288$$
$$x = 36$$

The original price of the backpack was $36.

18. $3(2x-5) \le x+15$

$$6x - 15 \le x + 15$$
$$5x - 15 \le 15$$
$$5x \le 30$$
$$x \le 6$$

$\{x \mid x \le 6\}$

$(-\infty, 6]$

19. $-6x - 4 < 2(x-7)$

$$-6x - 4 < 2x - 14$$
$$-4 < 8x - 14$$
$$10 < 8x$$
$$\frac{5}{4} < x$$

$\left\{ x \mid x > \dfrac{5}{4} \right\}$

$\left(\dfrac{5}{4}, \infty \right)$

20. Let m be the number of minutes Danielle can use.

$$30 + 0.35m \le 100$$
$$0.35m \le 70$$
$$m \le 200$$

She can use her cell phone at most 200 minutes.

Chapter 3

Section 3.1

Preparing for the Rectangular Coordinate System and Equations in Two Variables

P1.

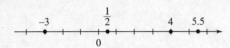

P2. **(a)** $3x + 5$ for $x = 4$:

$3(4) + 5 = 12 + 5 = 17$

(b) $3x + 5$ for $x = -1$:

$3(-1) + 5 = -3 + 5 = 2$

P3. **(a)** $2x - 5y$ for $x = 3$, $y = 2$:

$2(3) - 5(2) = 6 - 10 = -4$

(b) $2x - 5y$ for $x = 1$, $y = -4$:

$2(1) - 5(-4) = 2 + 20 = 22$

P4. $\begin{aligned} 3x + 5 &= 14 \\ 3x + 5 - 5 &= 14 - 5 \\ 3x &= 9 \\ \frac{3x}{3} &= \frac{9}{3} \\ x &= 3 \end{aligned}$

The solution set is $\{3\}$.

P5. $\begin{aligned} 5(x - 3) - 2x &= 3x + 12 \\ 5x - 15 - 2x &= 3x + 12 \\ 3x - 15 &= 3x + 12 \\ 3x - 15 - 3x &= 3x + 12 - 3x \\ -15 &= 12 \end{aligned}$

The statement $-15 = 12$ is false, so the equation is a contradiction. The solution set is $\varnothing$.

3.1 Quick Checks

1. In the rectangular coordinate system, we call the horizontal number line the _x-axis_ and we call the vertical real number line the _y-axis_. The point where these two axes intersect is called the _origin_.

2. If (x, y) are the coordinates of a point P, then x is called the _x-coordinate_ of P and y is called the _y-coordinate_ of P.

3. False; the point whose ordered pair is $(-2, 4)$ is located in Quadrant II.

4. False; the ordered pairs $(3, 2)$ and $(2, 3)$ do not represent the same point in the Cartesian plane.

5.

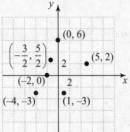

(a) $(5, 2)$ lies in quadrant I.

(b) $(-4, -3)$ lies in quadrant III.

(c) $(1, -3)$ lies in quadrant IV.

(d) $(-2, 0)$ lies on the _x_-axis.

(e) $(0, 6)$ lies on the _y_-axis.

(f) $\left(-\frac{3}{2}, \frac{5}{2}\right) = (-1.5, 2.5)$ lies in quadrant II.

6.

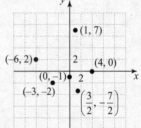

(a) $(-6, 2)$ lies in quadrant II.

(b) $(1, 7)$ lies in quadrant I.

(c) $(-3, -2)$ lies in quadrant III.

(d) $(4, 0)$ lies on the _x_-axis.

(e) $(0, -1)$ lies on the _y_-axis.

(f) $\left(\frac{3}{2}, -\frac{7}{2}\right) = (1.5, -3.5)$ lies in quadrant IV.

7.

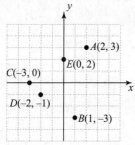

(a) $A(2, 3)$

(b) $B(1, -3)$

(c) $C(-3, 0)$

(d) $D(-2, -1)$

(e) $E(0, 2)$

8. True

9. (a) $x + 4y = 12$
$$4 + 4(2) \stackrel{?}{=} 12$$
$$4 + 8 \stackrel{?}{=} 12$$
$$12 = 12 \quad \text{True}$$
The statement is true, so (4, 2) satisfies the equation $x + 4y = 12$.

(b) $x + 4y = 12$
$$-2 + 4(4) \stackrel{?}{=} 12$$
$$-2 + 16 \stackrel{?}{=} 12$$
$$14 = 12 \quad \text{False}$$
The statement $14 = 12$ is false, so $(-2, 4)$ does not satisfy the equation $x + 4y = 12$.

(c) $x + 4y = 12$
$$1 + 4(8) \stackrel{?}{=} 12$$
$$1 + 32 \stackrel{?}{=} 12$$
$$33 = 12 \quad \text{False}$$
The statement $33 = 12$ is false, so (1, 8) does not satisfy the equation $x + 4y = 12$.

10. (a) $y = 4x + 3$
$$3 \stackrel{?}{=} 4(1) + 3$$
$$3 \stackrel{?}{=} 4 + 3$$
$$3 = 7 \quad \text{False}$$
The statement $3 = 7$ is false, so (1, 3) does not satisfy the equation $y = 4x + 3$.

(b) $y = 4x + 3$
$$-5 \stackrel{?}{=} 4(-2) + 3$$
$$-5 \stackrel{?}{=} -8 + 3$$
$$-5 = -5 \quad \text{True}$$
The statement is true, so $(-2, -5)$ satisfies the equation $y = 4x + 3$.

(c) $y = 4x + 3$
$$-3 \stackrel{?}{=} 4\left(-\frac{3}{2}\right) + 3$$
$$-3 \stackrel{?}{=} -6 + 3$$
$$-3 = -3 \quad \text{True}$$
The statement is true, so $\left(-\dfrac{3}{2}, -3\right)$ satisfies the equation $y = 4x + 3$.

11. Substitute 3 for x and solve for y.
$$2x + y = 10$$
$$2(3) + y = 10$$
$$6 + y = 10$$
$$6 - 6 + y = 10 - 6$$
$$y = 4$$
The ordered pair that satisfies the equation is (3, 4).

12. Substitute 1 for y and solve for x.
$$-3x + 2y = 11$$
$$-3x + 2(1) = 11$$
$$-3x + 2 = 11$$
$$-3x + 2 - 2 = 11 - 2$$
$$-3x = 9$$
$$\frac{-3x}{-3} = \frac{9}{-3}$$
$$x = -3$$
The ordered pair that satisfies the equation is $(-3, 1)$.

13. Substitute $\dfrac{1}{2}$ for x and solve for y.
$$4x + 3y = 0$$
$$4\left(\frac{1}{2}\right) + 3y = 0$$
$$2 + 3y = 0$$
$$-2 + 2 + 3y = -2$$
$$3y = -2$$
$$\frac{3y}{3} = \frac{-2}{3}$$
$$y = -\frac{2}{3}$$
The ordered pair that satisfies the equation is $\left(\dfrac{1}{2}, -\dfrac{2}{3}\right)$.

14. $y = 5x - 2$

$x = -2$: $y = 5(-2) - 2$
$\qquad\quad y = -10 - 2$
$\qquad\quad y = -12$

$x = 0$: $y = 5(0) - 2$
$\qquad\quad y = 0 - 2$
$\qquad\quad y = -2$

$x = 1$: $y = 5(1) - 2$
$\qquad\quad y = 5 - 2$
$\qquad\quad y = 3$

x	y	(x, y)
-2	-12	$(-2, -12)$
0	-2	$(0, -2)$
1	3	$(1, 3)$

15. $y = -3x + 4$

$x = -1$: $y = -3(-1) + 4$
$\qquad\quad\ y = 3 + 4$
$\qquad\quad\ y = 7$

$x = 2$: $y = -3(2) + 4$
$\qquad\quad y = -6 + 4$
$\qquad\quad y = -2$

$x = 5$: $y = -3(5) + 4$
$\qquad\quad y = -15 + 4$
$\qquad\quad y = -11$

x	y	(x, y)
-1	7	$(-1, 7)$
2	-2	$(2, -2)$
5	-11	$(5, -11)$

16. $2x + y = -8$

$x = -5$: $2(-5) + y = -8$
$\qquad\qquad -10 + y = -8$
$\qquad\qquad\qquad\ y = 2$

$y = -4$: $2x + (-4) = -8$
$\qquad\qquad 2x - 4 = -8$
$\qquad\qquad\quad 2x = -4$
$\qquad\qquad\qquad x = -2$

$x = 2$: $2(2) + y = -8$
$\qquad\quad\ 4 + y = -8$
$\qquad\qquad\ \ y = -12$

x	y	(x, y)
-5	2	$(-5, 2)$
-2	-4	$(-2, -4)$
2	-12	$(2, -12)$

17. $2x - 5y = 18$

$x = -6$: $2(-6) - 5y = 18$
$\qquad\qquad -12 - 5y = 18$
$\qquad\qquad\qquad -5y = 30$
$\qquad\qquad\qquad\quad\ y = -6$

$y = -4$: $2x - 5(-4) = 18$
$\qquad\qquad 2x + 20 = 18$
$\qquad\qquad\qquad 2x = -2$
$\qquad\qquad\qquad\ \ x = -1$

$x = 2$: $2(2) - 5y = 18$
$\qquad\quad\ 4 - 5y = 18$
$\qquad\qquad -5y = 14$
$\qquad\qquad\quad\ y = -\dfrac{14}{5}$

x	y	(x, y)
-6	-6	$(-6, -6)$
-1	-4	$(-1, -4)$
2	$-\dfrac{14}{5}$	$\left(2, -\dfrac{14}{5}\right)$

18. **(a)** $C = 10 + 1.32607x$

$x = 50$: $C = 10 + 1.32607(50)$
$\qquad\qquad C = 76.30$

$x = 100$: $C = 10 + 1.32607(100)$
$\qquad\qquad\ C = 142.61$

$x = 150$: $C = 10 + 1.32607(150)$
$\qquad\qquad\ C = 208.91$

x therms	50 therms	100 therms	150 therms
C ($)	$76.30	$142.61	$208.91

(b)

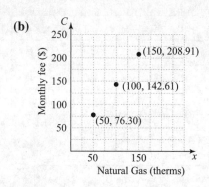

3.1 Exercises

19. Quadrant I: B
Quadrant II: A, E
Quadrant III: C
Quadrant IV: D, F

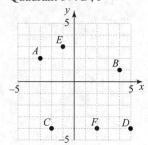

21. Quadrant I: C, E
Quadrant III: F
Quadrant IV: B
x-axis: A, G
y-axis: D, G

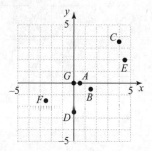

23. Positive x-axis: A
Negative x-axis: D
Positive y-axis: C
Negative y-axis: B

25. $A(4, 0)$; $B(-3, 2)$; $C(1, -4)$; $D(-2, -4)$;
$E(3, 5)$; $F(0, -3)$;
Quadrant I: E; Quadrant II: B;
Quadrant III: D; Quadrant IV: C;
Positive x-axis: A; Negative y-axis: F

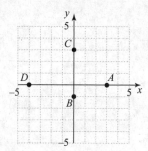

27. $y = -3x + 5$
$A(-2, -1)$
$-1 \stackrel{?}{=} -3(-2) + 5$
$-1 \stackrel{?}{=} 6 + 5$
$-1 \neq 11$
No
$B(2, -1)$
$-1 \stackrel{?}{=} -3(2) + 5$
$-1 \stackrel{?}{=} -6 + 5$
$-1 = -1$
Yes
$C\left(\dfrac{1}{3}, 4\right)$
$4 \stackrel{?}{=} -3\left(\dfrac{1}{3}\right) + 5$
$4 \stackrel{?}{=} -1 + 5$
$4 = 4$
Yes

29. $3x + 2y = 4$
$A(0, 2)$
$3(0) + 2(2) \stackrel{?}{=} 4$
$0 + 4 \stackrel{?}{=} 4$
$4 = 4$
Yes
$B(1, 0)$
$3(1) + 2(0) \stackrel{?}{=} 4$
$3 + 0 \stackrel{?}{=} 4$
$3 \neq 4$
No
$C(4, -4)$
$3(4) + 2(-4) \stackrel{?}{=} 4$
$12 - 8 \stackrel{?}{=} 4$
$4 = 4$
Yes

31. $\frac{4}{3}x + y - 1 = 0$

$A(3, -3)$

$\frac{4}{3}(3) + (-3) - 1 \stackrel{?}{=} 0$

$\qquad 4 - 3 - 1 \stackrel{?}{=} 0$

$\qquad\qquad 0 = 0$

Yes

$B(-6, -9)$

$\frac{4}{3}(-6) + (-9) - 1 \stackrel{?}{=} 0$

$\qquad -8 - 9 - 1 \stackrel{?}{=} 0$

$\qquad\qquad -18 \neq 0$

No

$C\left(\frac{3}{4}, 0\right)$

$\frac{4}{3}\left(\frac{3}{4}\right) + 0 - 1 \stackrel{?}{=} 0$

$\qquad 1 + 0 - 1 \stackrel{?}{=} 0$

$\qquad\qquad 0 = 0$

Yes

33. Let $x = 4$.

$x + y = 5$

$4 + y = 5$

$\quad y = 1$

The ordered pair is $(4, 1)$.

35. Let $y = -1$.

$2x + y = 9$

$2x + (-1) = 9$

$\quad 2x = 10$

$\quad x = 5$

The ordered pair is $(5, -1)$.

37. Let $x = -3$.

$-3x + 2y = 15$

$-3(-3) + 2y = 15$

$\quad 9 + 2y = 15$

$\quad 2y = 6$

$\quad y = 3$

The ordered pair is $(-3, 3)$.

39.

x	$y = -x$	(x, y)
-3	$-(-3) = 3$	$(-3, 3)$
0	$-0 = 0$	$(0, 0)$
1	-1	$(1, -1)$

41.

x	$y = -3x + 1$	(x, y)
-2	$-3(-2) + 1 = 7$	$(-2, 7)$
-1	$-3(-1) + 1 = 4$	$(-1, 4)$
4	$-3(4) + 1 = -11$	$(4, -11)$

43. $2x + y = 6$

$\quad y = -2x + 6$

x	$y = -2x + 6$	(x, y)
-1	$-2(-1) + 6 = 8$	$(-1, 8)$
2	$-2(2) + 6 = 2$	$(2, 2)$
3	$-2(3) + 6 = 0$	$(3, 0)$

45. $y = 6$

x	$y = 6$	(x, y)
-4	6	$(-4, 6)$
1	6	$(1, 6)$
12	6	$(12, 6)$

47. $x - 2y + 6 = 0$

$\quad x = 2y - 6$

or

$x - 2y + 6 = 0$

$\quad -2y = -x - 6$

$\quad y = \frac{1}{2}x + 3$

$x = 2y - 6$	$y = \frac{1}{2}x + 3$	(x, y)
1	$\frac{1}{2}(1) + 3 = \frac{7}{2}$	$\left(1, \frac{7}{2}\right)$
$2(1) - 6 = -4$	1	$(-4, 1)$
-2	$\frac{1}{2}(-2) + 3 = 2$	$(-2, 2)$

49. $y = 5 + \frac{1}{2}x$

$y - 5 = \frac{1}{2}x$

$2(y - 5) = x$

$x = 2(y-5)$	$y = 5 + \frac{1}{2}x$	(x, y)
$2(7-5)=4$	7	$(4, 7)$
-4	$5 + \frac{1}{2}(-4)=3$	$(-4, 3)$
$2(2-5)=-6$	2	$(-6, 2)$

51. $\dfrac{x}{2} + \dfrac{y}{3} = -1$

$\qquad \dfrac{x}{2} = -\dfrac{y}{3} - 1$

$\qquad x = -\dfrac{2y}{3} - 2$

or

$\dfrac{x}{2} + \dfrac{y}{3} = -1$

$\qquad \dfrac{y}{3} = -\dfrac{x}{2} - 1$

$\qquad y = -\dfrac{3x}{2} - 3$

$x = -\frac{2}{3}y - 2$	$y = -\frac{3}{2}x - 3$	(x, y)
0	$-\frac{3}{2}(0)-3=-3$	$(0, -3)$
$-\frac{2}{3}(0)-2=-2$	0	$(-2, 0)$
$-\frac{2}{3}(-6)-2=2$	-6	$(2, -6)$

53. $y = -3x - 10$

In $A(\underline{\ \ }, -16)$, $y = -16$.

$-16 = -3x - 10$

$\quad -6 = -3x$

$\qquad 2 = x$

$A(2, -16)$

In $B(\ 0, \underline{\ \ })$, $x = -3$.

$y = -3(-3) - 10 = 9 - 10 = -1$

$B(-3, -1)$

In $C(\underline{\ \ }, -9)$, $y = -9$.

$\quad -9 = -3x - 10$

$\qquad 1 = -3x$

$\quad -\dfrac{1}{3} = x$

$C\left(-\dfrac{1}{3}, -9\right)$

55. $x = -\dfrac{1}{3}y$

In $A(2, \underline{\ \ })$, $x = 2$

$\quad 2 = -\dfrac{1}{3}y$

$-6 = y$

$A(2, -6)$

In $B(\underline{\ \ }, 0)$, $y = 0$

$x = -\dfrac{1}{3}(0) = 0$

$B(0, 0)$

In $C\left(\underline{\ \ }, -\dfrac{1}{2}\right)$, $y = -\dfrac{1}{2}$

$x = -\dfrac{1}{3}\left(-\dfrac{1}{2}\right) = \dfrac{1}{6}$

$C\left(\dfrac{1}{6}, -\dfrac{1}{2}\right)$

57. $x = 4$

Here the x-coordinate will be 4, regardless of the value of the y-coordinate.

$A(4, -8)$

$B(4, -19)$

$C(4, 5)$

59. $y = \dfrac{2}{3}x + 2$

In $A(\underline{\ \ }, 4)$, $y = 4$.

$4 = \dfrac{2}{3}x + 2$

$2 = \dfrac{2}{3}x$

$3 = x$

$A(3, 4)$

In $B(-6, \underline{\ \ })$, $x = -6$.

$y = \dfrac{2}{3}(-6) + 2 = -4 + 2 = -2$

$B(-6, -2)$

In $C\left(\dfrac{1}{2}, \underline{\ \ }\right)$, $x = \dfrac{1}{2}$

$y = \dfrac{2}{3}\left(\dfrac{1}{2}\right) + 2 = \dfrac{1}{3} + 2 = \dfrac{7}{3}$

$C\left(\dfrac{1}{2}, \dfrac{7}{3}\right)$

61. $\dfrac{1}{2}x - 3y = 2$

In $A(-4, \_), x = -4$

$\dfrac{1}{2}(-4) - 3y = 2$

$-2 - 3y = 2$

$-3y = 4$

$y = -\dfrac{4}{3}$

$A\left(-4, -\dfrac{4}{3}\right)$

In $B(\_, -1), y = -1$

$\dfrac{1}{2}x - 3(-1) = 2$

$\dfrac{1}{2}x + 3 = 2$

$\dfrac{1}{2}x = -1$

$x = -2$

$B = (-2, -1)$

In $C\left(-\dfrac{2}{3}, \_\right), x = -\dfrac{2}{3}$

$\dfrac{1}{2}\left(-\dfrac{2}{3}\right) - 3y = 2$

$-\dfrac{1}{3} - 3y = 2$

$-3y = \dfrac{7}{3}$

$y = -\dfrac{7}{9}$

$C\left(-\dfrac{2}{3}, -\dfrac{7}{9}\right)$

63. $0.5x - 0.3y = 3.1$

In $A(20, \_), x = 20$

$0.5(20) - 0.3y = 3.1$

$10 - 0.3y = 3.1$

$-0.3y = -6.9$

$y = 23$

$A(20, 23)$

In $B(\_, -17), y = -17$

$0.5x - 0.3(-17) = 3.1$

$0.5x + 5.1 = 3.1$

$0.5x = -2$

$x = -4$

$B(-4, -17)$

In $C(2.6, \_), x = 2.6$

$0.5(2.6) - 0.3y = 3.1$

$1.3 - 0.3y = 3.1$

$-0.3y = 1.8$

$y = -6$

$C(2.6, -6)$

65. $C = 9.95n + 4.95$

(a) $n = 2$

$C = 9.95(2) + 4.95$

$= 19.9 + 4.95$

$= 24.85$

It will cost \$24.85 to order 2 CDs.

(b) $n = 5$

$C = 9.95(5) + 4.95$

$= 49.75 + 4.95$

$= 54.7$

It will cost \$54.70 to order 5 CDs.

(c) Let $C = 64.65$

$64.65 = 9.95n + 4.95$

$59.70 = 9.95n$

$6 = n$

If you have \$64.65, you can order 6 CDs.

(d) It costs \$34.80 to order 3 CDs.

67. $P = 0.444n + 21.14$

(a) $n = 0$

$P = 0.444(0) + 21.14 = 0.21.14 = 21.14$

21.14% of U.S. population 25 years or older had a bachelor's degree in 1990.

(b) $n = 10$

$P = 0.444(10) + 21.14$

$= 4.44 + 21.14$

$= 25.58$

25.58% of U.S. population 25 years or older had a bachelor's degree in 2000.

(c) $n = 30$

$P = 0.444(30) + 21.14$

$= 13.32 + 21.14$

$= 34.46$

34.46% of U.S. population 25 years or older will have a bachelor's degree in 2020.

(d) Let $P = 50$.

$$50 = 0.444n + 21.14$$
$$28.86 = 0.444n$$
$$65 = n$$

65 years after 1990 = 2055.

(e) Answers will vary.

69. $4a + 2b = -8$

$$4a = -2b - 8$$
$$a = -\frac{1}{2}b - 2$$

or

$$4a + 2b = -8$$
$$2b = -4a - 8$$
$$b = -2a - 4$$

$a = -\frac{1}{2}b - 2$	$b = -2a - 4$	(a, b)
2	$-2(2) - 4 = -8$	$(2, -8)$
$-\frac{1}{2}(-4) - 2 = 0$	-4	$(0, -4)$
$-\frac{1}{2}(6) - 2 = -5$	6	$(-5, 6)$

71. $\dfrac{2p}{5} + \dfrac{3q}{10} = 1$

$$\frac{2p}{5} = -\frac{3q}{10} + 1$$
$$p = -\frac{3}{4}q + \frac{5}{2}$$

or

$$\frac{2p}{5} + \frac{3q}{10} = 1$$
$$\frac{3q}{10} = -\frac{2p}{5} + 1$$
$$q = -\frac{4}{3}p + \frac{10}{3}$$

$p = -\frac{3}{4}q + \frac{5}{2}$	$q = -\frac{4}{3}p + \frac{10}{3}$	(p, q)
0	$-\frac{4}{3}(0) + \frac{10}{3} = \frac{10}{3}$	$\left(0, \frac{10}{3}\right)$
$-\frac{3}{4}(0) + \frac{5}{2} = \frac{5}{2}$	0	$\left(\frac{5}{2}, 0\right)$
-10	$-\frac{4}{3}(-10) + \frac{10}{3} = \frac{50}{3}$	$\left(-10, \frac{50}{3}\right)$

73. $y = -2x + k$; (1, 2) is a solution.
$$2 = -2(1) + k$$
$$2 = -2 + k$$
$$4 = k$$

75. $7x - ky = -4$; (2, 9) is a solution.
$$7(2) - k(9) = -4$$
$$14 - 9k = -4$$
$$-9k = -18$$
$$k = 2$$

77. $kx - 4y = 6$; $\left(-8, -\dfrac{5}{2}\right)$ is a solution.

$$k(-8) - 4\left(-\dfrac{5}{2}\right) = 6$$
$$-8k + 10 = 6$$
$$-8k = -4$$
$$k = \dfrac{1}{2}$$

79. $3x - 2y = -6$
Points may vary.
$x = 0$: $3(0) - 2y = -6$
$$-2y = -6$$
$$y = 3$$
$$(0, 3)$$
$y = 0$: $3x - 2(0) = -6$
$$3x = -6$$
$$x = -2$$
$$(-2, 0)$$
$x = -4$: $3(-4) - 2y = -6$
$$-12 - 2y = -6$$
$$-2y = 6$$
$$y = -3$$
$$(-4, -3)$$

x	y	(x, y)
0	3	(0, 3)
-2	0	(-2, 0)
-4	-3	(-4, -3)

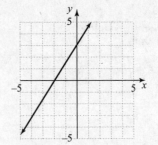

The figure is a line.

81.

x	$y = x^2 - 4$	(x, y)
-2	$(-2)^2 - 4 = 0$	(-2, 0)
-1	$(-1)^2 - 4 = -3$	(-1, -3)
0	$0^2 - 4 = -4$	(0, -4)
1	$1^2 - 4 = -3$	(1, -3)
2	$2^2 - 4 = 0$	(2, 0)

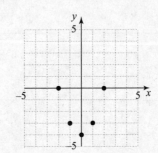

83.

x	$y = -x^3 + 2$	(x, y)
-2	$-(-2)^3 + 2 = 10$	(-2, 10)
-1	$-(-1)^3 + 2 = 3$	(-1, 3)
0	$-(0)^3 + 2 = 2$	(0, 2)
1	$-(1)^3 + 2 = 1$	(1, 1)
2	$-(2)^3 + 2 = -6$	(2, -6)

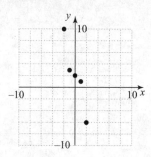

85. Answers may vary. One possibility: The quadrants are numbered I, II, III, and IV, counterclockwise starting from the upper right. The signs of the coordinates of a point determine the quadrant.
Quadrant I: $x > 0$, $y > 0$
Quadrant II: $x < 0$, $y > 0$
Quadrant III: $x < 0$, $y < 0$
Quadrant IV: $x > 0$, $y < 0$
A point with one coordinate of 0 lies on either the x-axis (second coordinate 0) or the y-axis (first coordinate 0).

87.

X	Y1
-3	-15
-2	-13
-1	-11
0	-9
1	-7
2	-5
3	-3

Y1∎2X-9

89.

X	Y1
-3	11
-2	10
-1	9
0	8
1	7
2	6
3	5

Y1∎-X+8

91. $y + 2x = 13$
$y = 13 - 2x$

X	Y1
-3	19
-2	17
-1	15
0	13
1	11
2	9
3	7

Y1∎13-2X

93.

X	Y1
-3	-53
-2	-23
-1	-5
0	1
1	-5
2	-23
3	-53

Y1∎-6X²+1

Section 3.2

Preparing for Graphing Equations in Two Variables

P1. $4x = 24$
$\dfrac{4x}{4} = \dfrac{24}{4}$
$x = 6$
The solution set is $\{6\}$.

P2. $-3y = 18$
$\dfrac{-3y}{-3} = \dfrac{18}{-3}$
$y = -6$
The solution set is $\{-6\}$.

P3. $2x + 5 = 13$
$2x + 5 - 5 = 13 - 5$
$2x = 8$
$\dfrac{2x}{2} = \dfrac{8}{2}$
$x = 4$
The solution set is $\{4\}$.

3.2 Quick Checks

1. Points may vary.

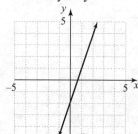

2. Points may vary.

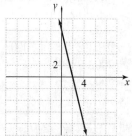

3. A <u>linear</u> equation is an equation of the form $Ax + By = C$, where A, B, and C are real numbers, and A and B are not both zero. Equations written in this form are said to be in <u>standard form</u>.

4. The equation $4x - y = 12$ is a linear equation in two variables because it is written in the form $Ax + By = C$ with $A = 4$, $B = -1$, and $C = 12$.

5. The equation $5x - y^2 = 10$ is not a linear equation because y is squared.

6. The equation $5x = 20$ is a linear equation in two variables because it is written in the form $Ax + By = C$ with $A = 5$, $B = 0$, and $C = 20$.

7. Points may vary.

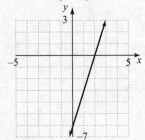

8. Points may vary.

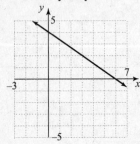

9. (a) $S = 0.08x + 3000$

$x = 0$: $S = 0.08(0) + 3000$
$= 0 + 3000$
$= 3000$

$x = 10,000$: $S = 0.08(10,000) + 3000$
$= 800 + 3000$
$= 3800$

$x = 25,000$: $S = 0.08(25,000) + 3000$
$= 2000 + 3000$
$= 5000$

x	S	(x, S)
0	3000	(0, 3000)
10,000	3800	(10,000, 3800)
25,000	5000	(25,000, 5000)

(b)

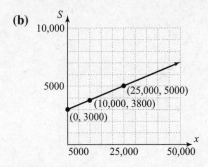

10. The <u>intercepts</u> are the points, if any, where a graph crosses or touches the coordinate axes.

11. The intercepts of the graph are the points $(0, 3)$ and $(4, 0)$. The x-intercept is 4. The y-intercept is 3.

12. The only intercept of the graph is the point $(0, -2)$. There is no x-intercept. The y-intercept is -2.

13. False; to find the y-intercept(s), if any, of the graph of an equation, let $x = 0$ in the equation and solve for y.

For 14–18, additional points may vary.

14. $x + y = 3$
x-intercept, let $y = 0$:
$x + 0 = 3$
$x = 3$
y-intercept, let $x = 0$:
$0 + y = 3$
$y = 3$

Plot the points $(3, 0)$, $(0, 3)$, and an additional point.

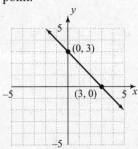

15. $2x - 5y = 20$
x-intercept, let $y = 0$:
$2x - 5(0) = 20$
$2x = 20$
$x = 10$
y-intercept, let $x = 0$:

$2(0) - 5y = 20$

$-5y = 20$

$y = -4$

Plot the points (10, 0), (0, −4), and an additional point.

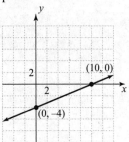

16. $\frac{3}{2}x - 2y = 9$

x-intercept, let $y = 0$:

$\frac{3}{2}x - 2(0) = 9$

$\frac{3}{2}x = 9$

$x = \frac{2}{3} \cdot 9$

$x = 6$

y-intercept, let $x = 0$:

$\frac{3}{2}(0) - 2y = 9$

$-2y = 9$

$y = -\frac{9}{2}$

Plot the points (6, 0), $\left(0, -\frac{9}{2}\right)$, and an additional point.

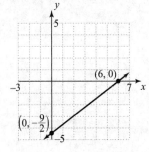

17. $y = \frac{1}{2}x$

x-intercept, let $y = 0$:

$0 = \frac{1}{2}x$

$0 = x$

y-intercept, let $x = 0$:

$y = \frac{1}{2}(0)$

$y = 0$

The intercepts are the same point, (0, 0), so let $x = 2$ to get a second point:

$y = \frac{1}{2}(2)$

$y = 1$

Plot the points (0, 0), (2, 1), and an additional point.

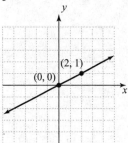

18. $4x + y = 0$

x-intercept, let $y = 0$:

$4x + 0 = 0$

$4x = 0$

$x = 0$

y-intercept, let $x = 0$:

$4(0) + y = 0$

$y = 0$

The intercepts are the same point, (0, 0), so let $x = 1$ to get a second point:

$4(1) + y = 0$

$4 + y = 0$

$y = -4$

Plot the points (0, 0), (1, −4), and an additional point.

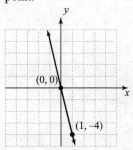

19. The graph of $x = -5$ is a vertical line whose x-intercept is -5.

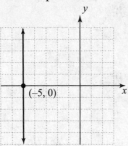

20. The graph of $y = -4$ is a horizontal line whose y-intercept is -4.

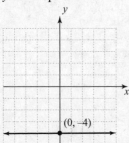

3.2 Exercises

21. Yes; the equation is written in the form $Ax + By = C$ with $A = 2$, $B = -5$, and $C = 10$.

23. No; variables cannot be squared in a linear equation.

25. No; variables cannot be in the denominator of a fraction in a linear equation.

27. Yes; the equation can be rewritten as $y = 1$, which is in the form $Ax + By = C$ with $A = 0$, $B = 1$, and $C = 1$.

In Problems 29–45, points may vary.

29. $y = 2x$

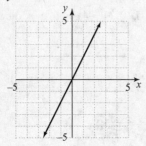

31. $y = 4x - 2$

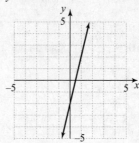

33. $y = -2x + 5$

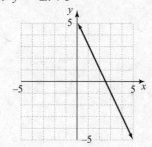

35. $x + y = 5$

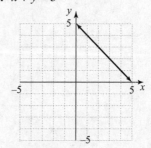

37. $-2x + y = 6$

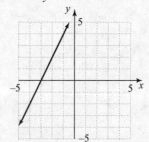

39. $4x - 2y = -8$

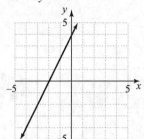

41. $x = -4y$

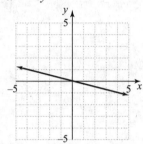

43. $y + 7 = 0$

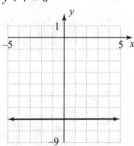

45. $y - 2 = 3(x + 1)$

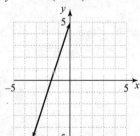

47. The graph crosses the y-axis at $(0, -5)$ and the x-axis at $(5, 0)$. The intercepts are $(0, -5)$ and $(5, 0)$.

49. The graph crosses the y-axis at $(0, 4)$ and the x-axis at $(2, 0)$. The intercepts are $(0, 4)$ and $(2, 0)$.

51. The graph crosses the y-axis at $(0, -3)$ and does not cross the x-axis (no x-intercept). The only intercept is $(0, -3)$.

53. The graph does not cross the y-axis (no y-intercept) and crosses the x-axis at $(-5, 0)$. The only intercept is $(-5, 0)$.

55. $2x + 3y = -12$
y-intercept, let $x = 0$:
$$2(0) + 3y = -12$$
$$3y = -12$$
$$y = -4$$
x-intercept, let $y = 0$:
$$2x + 3(0) = -12$$
$$2x = -12$$
$$x = -6$$
The intercepts are $(0, -4)$ and $(-6, 0)$.

57. $x = -6y$
y-intercept, let $x = 0$:
$$0 = -6y$$
$$0 = y$$
x-intercept, let $y = 0$:
$$x = -6(0)$$
$$x = 0$$
The only intercept is $(0, 0)$.

59. $y = x - 5$
y-intercept, let $x = 0$:
$$y = 0 - 5$$
$$y = -5$$
x-intercept, let $y = 0$:
$$0 = x - 5$$
$$5 = x$$
The intercepts are $(0, -5)$ and $(5, 0)$.

61. $\dfrac{x}{6} + \dfrac{y}{8} = 1$
y-intercept, let $x = 0$:
$$\frac{0}{6} + \frac{y}{8} = 1$$
$$\frac{y}{8} = 1$$
$$y = 8$$
x-intercept, let $y = 0$:
$$\frac{x}{6} + \frac{0}{8} = 1$$
$$\frac{x}{6} = 1$$
$$x = 6$$
The intercepts are $(0, 8)$ and $(6, 0)$.

63. $x = 4$ is the equation of a vertical line which has *x*-intercept (4, 0) and no *y*-intercept.

65. $y + 2 = 0$, or $y = -2$, is the equation of a horizontal line which has *y*-intercept (0, −2) and no *x*-intercept.

In Problems 67–81, additional points may vary.

67. $3x + 6y = 18$
y-intercept, $x = 0$:
$$3(0) + 6y = 18$$
$$6y = 18$$
$$y = 3$$
(0, 3)
x-intercept, $y = 0$:
$$3x + 6(0) = 18$$
$$3x = 18$$
$$x = 6$$
(6, 0)

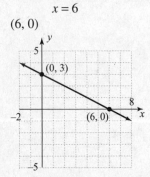

69. $-x + 5y = 15$
y-intercept, $x = 0$:
$$-0 + 5y = 15$$
$$5y = 15$$
$$y = 3$$
(0, 3)
x-intercept, $y = 0$:
$$-x + 5(0) = 15$$
$$-x = 15$$
$$x = -15$$
(−15, 0)

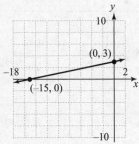

71. $\frac{1}{2}x = y + 3$
y-intercept, $x = 0$:
$$\frac{1}{2}(0) = y + 3$$
$$0 = y + 3$$
$$-3 = y$$
(0, −3)
x-intercept, $y = 0$:
$$\frac{1}{2}x = 0 + 3$$
$$\frac{1}{2}x = 3$$
$$x = 6$$
(6, 0)

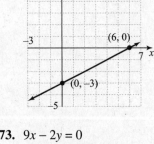

73. $9x - 2y = 0$
y-intercept, $x = 0$:
$$9(0) - 2y = 0$$
$$-2y = 0$$
$$y = 0$$
(0, 0)
x-intercept, $y = 0$:
$$9x - 2(0) = 0$$
$$9x = 0$$
$$x = 0$$
(0, 0)

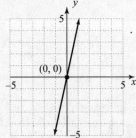

75. $y = -\dfrac{1}{2}x + 3$

y-intercept, $x = 0$

$y = -\dfrac{1}{2}(0) + 3$

$y = 3$

$(0, 3)$

x-intercept, $y = 0$

$0 = -\dfrac{1}{2}x + 3$

$-3 = -\dfrac{1}{2}x$

$6 = x$

$(6, 0)$

77. $\dfrac{1}{3}y + 2 = 2x$

y-intercept, $x = 0$:

$\dfrac{1}{3}y + 2 = 2(0)$

$\dfrac{1}{3}y = -2$

$y = -6$

$(0, -6)$

x-intercept, $y = 0$:

$\dfrac{1}{3}(0) + 2 = 2x$

$2 = 2x$

$1 = x$

$(1, 0)$

79. $\dfrac{x}{2} + \dfrac{y}{3} = 1$

y-intercept, $x = 0$:

$\dfrac{0}{2} + \dfrac{y}{3} = 1$

$\dfrac{y}{3} = 1$

$y = 3$

$(0, 3)$

x-intercept, $y = 0$

$\dfrac{x}{2} + \dfrac{0}{3} = 1$

$\dfrac{x}{2} = 1$

$x = 2$

$(2, 0)$

81. $4y - 2x + 1 = 0$

y-intercept, $x = 0$:

$4y - 2(0) + 1 = 0$

$4y + 1 = 0$

$4y = -1$

$y = -\dfrac{1}{4}$

$\left(0, -\dfrac{1}{4}\right)$

x-intercept, $y = 0$

$4(0) - 2x + 1 = 0$

$-2x + 1 = 0$

$-2x = -1$

$x = \dfrac{1}{2}$

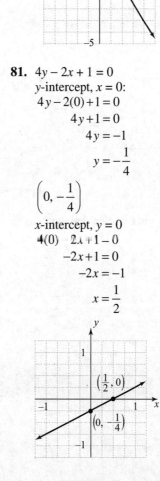

83. $x = 5$ is a vertical line with an x-intercept of $(5, 0)$.

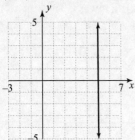

85. $y = -6$ is a horizontal line with a y-intercept of $(0, -6)$.

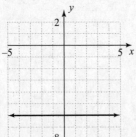

87. $y - 12 = 0$ or $y = 12$ is a horizontal line with a y-intercept of $(0, 12)$.

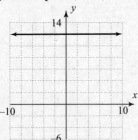

89. $3x - 5 = 0$ or $x = \dfrac{5}{3}$ is a vertical line with an x-intercept of $\left(\dfrac{5}{3}, 0\right)$.

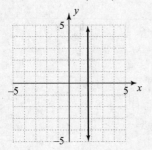

91. $y = 2x - 5$

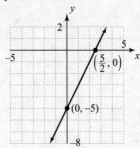

93. $y = -5$

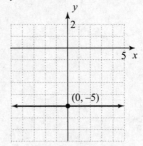

95. $2x + 5y = -20$

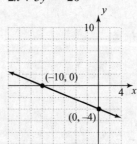

97. $2x = -6y + 4$

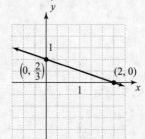

99. $x - 3 = 0$

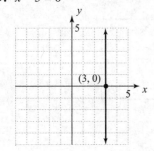

101. $3y - 12 = 0$

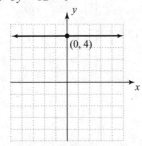

103. $4x + 3y = 18$; $(3, y)$
$4(3) + 3y = 18$
$12 + 3y = 18$
$3y = 6$
$y = 2$

105. $3x + 5y = 11$; $(x, -2)$
$3x + 5(-2) = 11$
$3x - 10 = 11$
$3x = 21$
$x = 7$

107.

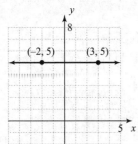

The line is horizontal. The equation is $y = 5$.

109.

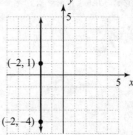

The line is vertical. The equation is $x = -2$.

111. The line is horizontal and has y-intercept $(0, 4)$. The equation is $y = 4$.

113. The line is vertical and has x-intercept $(-9, 0)$. The equation is $x = -9$.

115. If the x-coordinate is twice the y-coordinate, then $x = 2y$.

117. If the y-coordinate is two more than the x-coordinate, then $y = x + 2$.

119. (a) $E = 100n + 500$
$n = 0$: $E = 100(0) + 500$
$= 0 + 500$
$= 500$
$n = 4$: $E = 100(4) + 500$
$= 400 + 500$
$= 900$
$n = 10$: $E = 100(10) + 500$
$= 1000 + 500$
$= 1500$
The ordered pairs are $(0, 500)$, $(4, 900)$, and $(10, 1500)$.

(b)

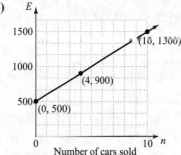

(c) If she sells 0 cars, her earnings are $500.

121. The "steepness" of the lines is the same.

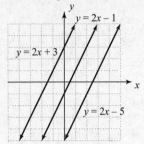

123. The lines get steeper as the coefficient of x gets larger.

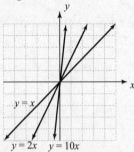

125. The graph crosses the y-axis at $(0, -6)$ and the x-axis at $(-2, 0)$ and $(3, 0)$. The intercepts are $(0, -6)$, $(-2, 0)$, and $(3, 0)$.

127. The graph crosses the y-axis at $(0, 14)$ and the x-axis at $(-3, 0)$, $(2, 0)$, and $(5, 0)$. The intercepts are $(0, 14)$, $(-3, 0)$, $(2, 0)$, and $(5, 0)$.

129. The graph of an equation is the set of all ordered pairs (x, y) that make the equation a true statement.

131. Two points are required, but it is good to plot a third point to check the line.

133. $y = 2x - 9$

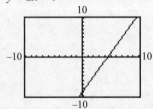

135. $y + 2x = 13$ or $y = 13 - 2x$

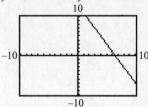

137. $y = -6x^2 + 1$

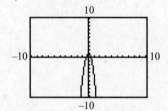

Section 3.3

Preparing for Slope

P1. $\dfrac{5-2}{8-7} = \dfrac{3}{1} = 3$

P2. $\dfrac{3-7}{9-3} = \dfrac{-4}{6} = \dfrac{-2 \cdot 2}{2 \cdot 3} = -\dfrac{2}{3}$

P3. $\dfrac{-3-4}{6-(-1)} = \dfrac{-7}{6+1} = \dfrac{-7}{7} = -1$

3.3 Quick Checks

1. Let run = 10, rise = 6

   $\text{slope} = \dfrac{\text{rise}}{\text{run}} = \dfrac{6}{10} = \dfrac{3}{5}$

2. False; if $P = (x_1, y_1)$ and $Q = (x_2, y_2)$, then the slope of the line that contains P and Q is

   $\dfrac{y_2 - y_1}{x_2 - x_1}$, not $\dfrac{x_2 - x_1}{y_2 - y_1}$.

3. True

4. If the graph of a line goes up as you move to the right, then the slope of this line must be <u>positive</u>.

5.

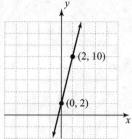

Let $(0, 2) = (x_1, y_1)$ and $(2, 10) = (x_2, y_2)$.

$$m = \frac{y_2 - y_1}{x_2 - x_1} = \frac{10 - 2}{2 - 0} = \frac{8}{2} = \frac{4}{1} = 4$$

The value of y increases by 4 when x increases by 1.

6.

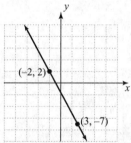

Let $(-2, 2) = (x_1, y_1)$ and $(3, -7) = (x_2, y_2)$.

$$m = \frac{y_2 - y_1}{x_2 - x_1} = \frac{-7 - 2}{3 - (-2)} = \frac{-9}{5} = -\frac{9}{5}$$

The value of y decreases by 9 when x increases by 5.

7. The slope of a horizontal line is <u>0</u>, while the slope of a vertical line is <u>undefined</u>.

8.

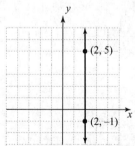

Let $(2, 5) = (x_1, y_1)$ and $(2, -1) = (x_2, y_2)$.

$$m = \frac{y_2 - y_1}{x_2 - x_1} = \frac{-1 - 5}{2 - 2} = \frac{-6}{0}$$

The slope of the line is undefined. When y increases by 1, there is no change in x.

9.

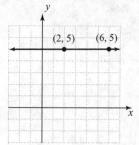

Let $(2, 5) = (x_1, y_1)$ and $(6, 5) = (x_2, y_2)$.

$$m = \frac{y_2 - y_1}{x_2 - x_1} = \frac{5 - 5}{6 - 2} = \frac{0}{4} = 0$$

There is no change in y when x increases by 1.

10. (a) $m = \dfrac{\text{rise}}{\text{run}} = \dfrac{1}{2}$

y will increase by 1 unit when x increases by 2 units. Starting at $(1, 2)$ and moving 1 unit up and 2 units to the right, we end up at $(3, 3)$.

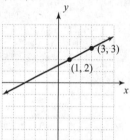

(b) $m = \dfrac{\text{rise}}{\text{run}} = -3 = \dfrac{-3}{1}$

y will decrease by 3 units when x increases by 1 unit. Starting at $(1, 2)$ and moving 3 units down and 1 unit to the right, we end up at $(2, -1)$.

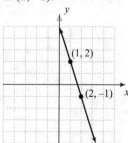

(c) $m = \dfrac{\text{rise}}{\text{run}} = 0 = \dfrac{0}{1}$

There is no change in y when x increases or decreases; the line is horizontal and passes through (1, 2).

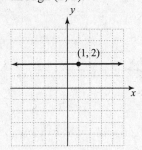

11. $\text{grade} = \dfrac{\text{rise}}{\text{run}} = \dfrac{4 \text{ feet}}{50 \text{ feet}} = 0.08 = 8\%$

The grade of the road is 8%.

12.   Let $(x_1, y_1) = (10,000, 1370)$ and $(x_2, y_2) = (14,000, 1850)$.

$$m = \frac{y_2 - y_1}{x_2 - x_1}$$

$$= \frac{1850 - 1370}{14,000 - 10,000}$$

$$= \frac{480}{4000}$$

$$= 0.12$$

The unit of measure of y is dollars while the unit of measure of x is miles driven. Between 10,000 and 14,000 miles driven, the annual cost of operating a Cobalt is $0.12 per mile, on average.

3.3 Exercises

13.   The line passes through $(0, 3) = (x_1, y_1)$ and $(2, 0) = (x_2, y_2)$.

$$m = \frac{y_2 - y_1}{x_2 - x_1} = \frac{0 - 3}{2 - 0} = \frac{-3}{2} = -\frac{3}{2}$$

15.   The line passes through $(-4, -4) = (x_1, y_1)$ and $(8, 2) = (x_2, y_2)$.

$$m = \frac{y_2 - y_1}{x_2 - x_1} = \frac{2 - (-4)}{8 - (-4)} = \frac{6}{12} = \frac{1}{2}$$

17.   The line passes through $(-3, 3) = (x_1, y_1)$ and $(3, -1) = (x_2, y_2)$.

$$m = \frac{y_2 - y_1}{x_2 - x_1} = \frac{-1 - 3}{3 - (-3)} = \frac{-4}{6} = -\frac{2}{3}$$

19.   **(a), (b)**

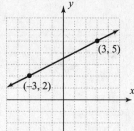

(c)   Let $(-3, 2) = (x_1, y_1)$ and $(3, 5) = (x_2, y_2)$.

$$m = \frac{y_2 - y_1}{x_2 - x_1} = \frac{5 - 2}{3 - (-3)} = \frac{3}{6} = \frac{1}{2}$$

The value of y increases by 1 unit when x increases by 2 units.

21.   **(a), (b)**

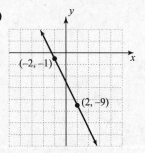

(c)   Let $(2, -9) = (x_1, y_1)$ and $(-2, -1) = (x_2, y_2)$.

$$m = \frac{-1 - (-9)}{-2 - 2} = \frac{8}{-4} = -2$$

The value of y decreases by 2 units when x increases by 1 unit.

23. $(10, 4) = (x_1, y_1); (6, 12) = (x_2, y_2)$

$$m = \frac{y_2 - y_1}{x_2 - x_1} = \frac{12 - 4}{6 - 10} = \frac{8}{-4} = -2 = \frac{-2}{1}$$

The value of y decreases by 2 units when x increases by 1 unit.

25. $(4, -4) = (x_1, y_1); (12, -12) = (x_2, y_2)$

$$m = \frac{y_2 - y_1}{x_2 - x_1} = \frac{-12 - (-4)}{12 - 4} = \frac{-8}{8} = -1 = \frac{-1}{1}$$

The value of y decreases by 1 unit when x increases by 1 unit.

27. $(7, -2) = (x_1, y_1); (4, 3) = (x_2, y_2)$

$$m = \frac{y_2 - y_1}{x_2 - x_1} = \frac{3 - (-2)}{4 - 7} = \frac{5}{-3} = \frac{-5}{3}$$

The value of y decreases by 5 units when x increases by 3 units.

29. $(0, 6) = (x_1, y_1); (-4, 0) = (x_2, y_2)$

$$m = \frac{y_2 - y_1}{x_2 - x_1} = \frac{0 - 6}{-4 - 0} = \frac{-6}{-4} = \frac{3}{2}$$

The value of y increases by 3 units when x increases by 2 units.

31. $(-4, -1) = (x_1, y_1); (2, 3) = (x_2, y_2)$

$$m = \frac{y_2 - y_1}{x_2 - x_1} = \frac{3 - (-1)}{2 - (-4)} = \frac{4}{6} = \frac{2}{3}$$

The value of y increases by 2 units when x increases by 3 units.

33. $\left(\frac{1}{2}, \frac{3}{4}\right) = (x_1, y_1); \left(-\frac{5}{2}, -\frac{1}{4}\right) = (x_2, y_2)$

$$m = \frac{y_2 - y_1}{x_2 - x_1} = \frac{-\frac{1}{4} - \frac{3}{4}}{-\frac{5}{2} - \frac{1}{2}} = \frac{-1}{-3} = \frac{1}{3}$$

The value of y increases by 1 unit when x increases by 3 units.

35. $\left(\frac{1}{2}, \frac{1}{3}\right) = (x_1, y_1); \left(\frac{3}{4}, \frac{5}{6}\right) = (x_2, y_2)$

$$m = \frac{y_2 - y_1}{x_2 - x_1}$$

$$= \frac{\frac{5}{6} - \frac{1}{3}}{\frac{3}{4} - \frac{1}{2}}$$

$$= \frac{\frac{5}{6} - \frac{2}{6}}{\frac{3}{4} - \frac{2}{4}}$$

$$= \frac{\frac{3}{6}}{\frac{1}{4}}$$

$$= \frac{\frac{1}{2}}{\frac{1}{4}}$$

$$= \frac{1}{2} \div \frac{1}{4}$$

$$= \frac{1}{2} \cdot \frac{4}{1}$$

$$= 2$$

$$= \frac{2}{1}$$

The value of y increases by 2 units when x increases by 1 unit.

37. The line is vertical, so the slope is undefined.

39. The line is horizontal, so the slope is 0.

41. $(4, -6) = (x_1, y_1); (-1, -6) = (x_2, y_2)$

$$m = \frac{y_2 - y_1}{x_2 - x_1} = \frac{-6 - (-6)}{-1 - 4} = \frac{0}{-5} = 0$$

There is no change in the y values, the line is horizontal.

43. $(3, 9) = (x_1, y_1); (3, -2) = (x_2, y_2)$

$$m = \frac{y_2 - y_1}{x_2 - x_1} = \frac{-2 - 9}{3 - 3} = \frac{-11}{0}$$

The slope is undefined. The line is vertical.

45. $(4, 2); m = 1$

Plot $(4, 2)$. Since $m = 1 = \frac{1}{1}$, there is a 1-unit increase in y for every 1-unit increase in x.

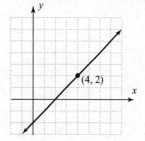

47. $(0, 6); m = -2$

Plot $(0, 6)$. Since $m = -2 = \frac{-2}{1}$, there is a 2-unit decrease in y for every 1-unit increase in x.

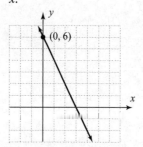

49. $(-1, 0); m = \frac{1}{4}$

Plot $(-1, 0)$. Since $m = \frac{1}{4}$, there is a 1-unit increase in y for every 4-unit increase in x.

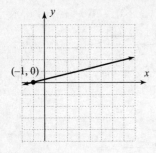

51. $(2, -3); m = 0$

Plot $(2, -3)$. Since $m = 0$, the line is horizontal.

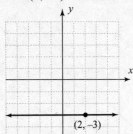

53. $(2, 1); m = \dfrac{2}{3}$

Plot $(2, 1)$. Since $m = \dfrac{2}{3}$, there is a 2-unit

increase in y for every 3-unit increase in x.

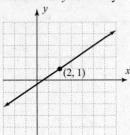

55. $(-1, 4); m = -\dfrac{5}{3}$

Plot $(-1, 4)$. Since $m = -\dfrac{5}{3} = -\dfrac{5}{3}$, there is a

5-unit decrease in y for every 3-unit increase in
x.

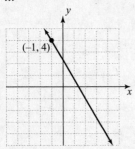

57. $(0, 0); m$ is undefined.
Plot $(0, 0)$, which is the origin. Since m is
undefined, the line is vertical—it is the
y-axis.

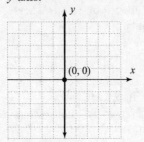

59. $(0, 2); m = -4$

Plot $(0, 2)$. Since $m = -4 = \dfrac{-4}{1}$, there is a

4-unit decrease in y for every 1-unit increase in
x.

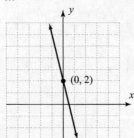

61. $(2, -3); m = \dfrac{3}{4}$

Plot $(2, -3)$. Since $m = \dfrac{3}{4}$, there is a 3-unit

increase in y for every 4-unit increase in x.

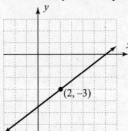

63. $(2, -1)$

$$m_1 = 2 = \dfrac{2}{1}$$

$$m_2 = -\dfrac{1}{2} = \dfrac{-1}{2}$$

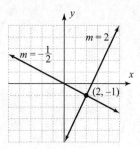

65. $m = \dfrac{3}{4}$

$(-1, -2)$

$(2, 1)$

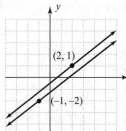

67. For every 1 foot = 12 inches horizontally (run), there is a 4-inch elevation (rise). The pitch (slope) is $\dfrac{4}{12} = \dfrac{1}{3}$.

69. The pitch (slope) is $\dfrac{2}{5}$ with a horizontal distance (run) of 30 in. Let x be the height.

$$\dfrac{x}{30} = \dfrac{2}{5}$$

$$x = \dfrac{2}{5}(30) = 12$$

He should add 12 in. = 1 ft to the height.

71. $\dfrac{\text{rise}}{\text{run}} = \dfrac{200}{1250} = 0.16 = 16\%$

The grade is 16%.

73. $(0, 123 \text{ million})$; $(70, 281 \text{ million})$

$$m = \dfrac{(281 \text{ million}) - (123 \text{ million})}{70 - 0}$$

$$= 158 \text{ million}$$

$$\approx 2.26 \text{ million}$$

The population is increasing at an average rate of about 2.26 million people per year.

In 75–77 points may vary.

75. $(-2, 1)$, $(0, -5)$ lie on the line.

$$m = \dfrac{-5 - 1}{0 - (-2)} = \dfrac{-6}{2} = -3$$

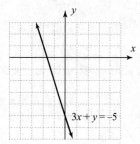

77. $(-2, -2)$ and $(0, 4)$ lie on the line.

$$m = \dfrac{4 - (-2)}{0 - (-2)} = \dfrac{6}{2} = 3$$

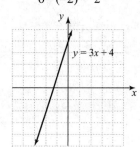

79. $(2a, a) = (x_1, y_1)$; $(3a, -a) = (x_2, y_2)$

$$m = \dfrac{y_2 - y_1}{x_2 - x_1} = \dfrac{-a - a}{3a - 2a} = \dfrac{-2a}{a} = -2$$

81. $(2p + 1, q - 4) = (x_1, y_1)$;

$(3p + 1, 2q - 4) = (x_2, y_2)$

$$m = \dfrac{y_2 - y_1}{x_2 - x_1} = \dfrac{2q - 4 - (q - 4)}{3p + 1 - (2p + 1)}$$

$$= \dfrac{2q - 4 - q + 4}{3p + 1 - 2p - 1}$$

$$= \dfrac{q}{p}$$

83. $(a + 1, b - 1) = (x_1, y_1)$;

$(2a - 5, b + 5) = (x_2, y_2)$

$$m = \dfrac{y_2 - y_1}{x_2 - x_1} = \dfrac{b + 5 - (b - 1)}{2a - 5 - (a + 1)}$$

$$= \dfrac{b + 5 - b + 1}{2a - 5 - a - 1}$$

$$= \dfrac{6}{a - 6}$$

85. Let $R_1 = 1000$ and $Q_1 = 400$, so $R_2 = 1200$ and $Q_2 = 500$.

$$MR = \frac{R_2 - R_1}{Q_2 - Q_1} = \frac{1200 - 1000}{500 - 400}$$
$$= \frac{200}{100}$$
$$= 2$$

The marginal revenue, or rate of change, is 2. For every hot dog sold, revenue increases by \$2.

87. Answers may vary. One possibility: A line with one x-intercept, say $(x_1, 0)$, but no y-intercept is a vertical line. $(x_1, 1)$ and $(x_1, 3)$ could lie on the line. Since the line is vertical, the slope is undefined.

Section 3.4

Preparing for Slope-Intercept Form of a Line

P1. $4x + 2y = 10$
$$-4x + 4x + 2y = -4x + 10$$
$$2y = -4x + 10$$
$$\frac{2y}{2} = \frac{-4x + 10}{2}$$
$$y = -2x + 5$$

P2. $10 = 2x - 8$
$$10 + 8 = 2x - 8 + 8$$
$$18 = 2x$$
$$\frac{18}{2} = \frac{2x}{2}$$
$$9 = x$$
The solution set is $\{9\}$.

3.4 Quick Checks

1. Comparing $y = 4x - 3$ to $y = mx + b$, we see that $m = 4$ and $b = -3$. The slope is 4 and the y-intercept is -3.

2. $3x + y = 7$
$$y = -3x + 7$$

Comparing $y = -3x + 7$ to $y = mx + b$, we see that $m = -3$ and $b = 7$. The slope is -3 and the y-intercept is 7.

3. $2x + 5y = 15$
$$5y = -2x + 15$$
$$y = -\frac{2}{5}x + 3$$

Comparing $y = -\frac{2}{5}x + 3$ to $y = mx + b$, we see that $m = -\frac{2}{5}$ and $b = 3$. The slope is $-\frac{2}{5}$ and the y-intercept is 3.

4. Comparing $y = 8$ to $y = mx + b$, we see that $m = 0$ and $b = 8$. The slope is 0 and the y-intercept is 8.

5. $x = 3$ cannot be compared to $y = mx + b$. It has an undefined slope, and no y-intercept.

6. $y = 2x - 5$
The slope is $m = 2$ and the y-intercept is $b = -5$. Plot $(0, -5)$, then use $m = 2 = \frac{2}{1} = \frac{\text{rise}}{\text{run}}$ to find a second point on the graph.

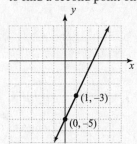

7. $y = \frac{1}{2}x - 5$

The slope is $m = \frac{1}{2}$ and the y-intercept is $b = -5$. Plot $(0, -5)$, then use $m = \frac{1}{2} = \frac{\text{rise}}{\text{run}}$ to find a second point on the graph.

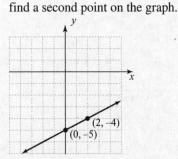

8. $y = -3x + 1$

The slope is $m = -3$ and the y-intercept is $b = 1$.

Plot $(0, 1)$, then use $m = -3 = \dfrac{-3}{1} = \dfrac{\text{rise}}{\text{run}}$ to find a

second point on the graph.

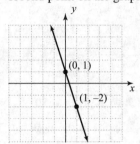

9. $y = -\dfrac{3}{2}x + 4$

The slope is $m = -\dfrac{3}{2}$ and the y-intercept is

$b = 4$. Plot $(0, 4)$, then use $m = -\dfrac{3}{2} = \dfrac{-3}{2} = \dfrac{\text{rise}}{\text{run}}$

to find a second point on the graph.

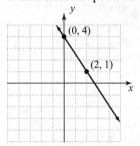

10. $-2x + y = -3$

$\qquad y = 2x - 3$

The slope is $m = 2$ and the y-intercept is

$b = -3$. Plot $(0, -3)$, then use $m = 2 = \dfrac{2}{1} = \dfrac{\text{rise}}{\text{run}}$

to find a second point on the graph

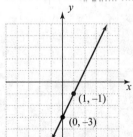

11. $6x - 2y = 2$

$\qquad -2y = -6x + 2$

$\qquad y = 3x - 1$

The slope is $m = 3$ and the y-intercept is

$b = -1$. Plot $(0, -1)$, then use $m = 3 = \dfrac{3}{1} = \dfrac{\text{rise}}{\text{run}}$ to

find a second point on the graph.

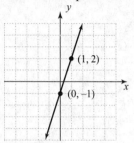

12. $3x + 5y = 0$

$\qquad 5y = -3x$

$\qquad y = -\dfrac{3}{5}x$

The slope is $m = -\dfrac{3}{5}$ and the y-intercept is

$b = 0$. Plot $(0, 0)$, then use $m = -\dfrac{3}{5} = \dfrac{-3}{5} = \dfrac{\text{rise}}{\text{run}}$

to find a second point on the graph.

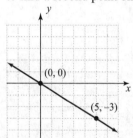

13. List three techniques that can be used to graph a line. <u>point plotting, using intercepts, using slope and a point</u>.

14. Substitute 3 for m and -2 for b in

$y = mx + b$.

$y = 3x - 2$

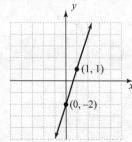

15. Substitute $-\frac{1}{4}$ for m and 3 for b in

$y = mx + b$.

$y = -\frac{1}{4}x + 3$

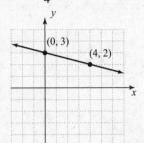

16. Substitute 0 for m and -1 for b in
$y = mx + b$.
$y = 0x - 1$
$y = -1$

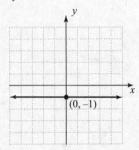

17. **(a)** $y = 143x - 2215$
$x = 30$: $y = 143(30) - 2215 = 2075$
After a gestation period of 30 weeks, the birth weight is predicted to be 2075 grams.

(b) $y = 143x - 2215$
$x = 36$: $y = 143(36) - 2215 = 2933$
After a gestation period of 36 weeks, the birth weight is predicted to be 2933 grams.

(c) The slope is $m = 143 = \dfrac{\text{rise}}{\text{run}} = \dfrac{143 \text{ grams}}{1 \text{ week}}$.
The birth weight increases by 143 grams as the length of the gestation period increases by 1 week.

(d) The y-intercept corresponds to a gestation period of 0 weeks, which makes no sense.

(e)

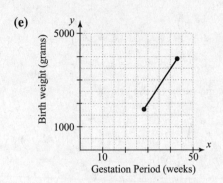

18. **(a)** The rate of change is given as $0.38 per mile or $\dfrac{\$0.38}{1 \text{ mile}}$, which is the slope m. The cost of $50 does not change with the number of miles driven, so this is the y-intercept, b.
$y = 0.38x + 50$

(b) $x = 75$: $y = 0.38(75) + 50 = 78.5$
If the truck is driven for 75 miles, the cost of the rental is $78.50.

(c) $y = 84.20$: $84.20 = 0.38x + 50$
$34.20 = 0.38x$
$90 = x$
If the cost of renting the truck was $84.20, then 90 miles were driven.

(d)

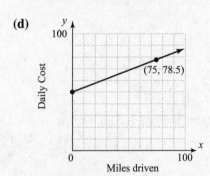

3.4 Exercises

19. $y = 5x + 2$
Slope: $m = 5$
y-intercept: $b = 2$

21. $y = x - 9$
Slope: $m = 1$
y-intercept: $b = -9$

23. $y = -10x + 7$
Slope: $m = -10$
y-intercept: $b = 7$

25. $y = -x - 9$
Slope: $m = -1$
y-intercept: $b = -9$

27. $2x + y = 4$
$\qquad y = -2x + 4$
Slope: $m = -2$
y-intercept: $b = 4$

29. $2x + 3y = 24$
$\qquad 3y = -2x + 24$
$\qquad y = -\dfrac{2}{3}x + 8$

Slope: $m = -\dfrac{2}{3}$
y-intercept: $b = 8$

31. $5x - 3y = 9$
$\qquad -3y = -5x + 9$
$\qquad y = \dfrac{5}{3}x - 3$

Slope: $m = \dfrac{5}{3}$
y-intercept: $b = -3$

33. $x - 2y = 5$
$\qquad -2y = -x + 5$
$\qquad y = \dfrac{1}{2}x - \dfrac{5}{2}$

Slope: $m = \dfrac{1}{2}$

y-intercept: $b = -\dfrac{5}{2}$

35. $y = -5$
This is a horizontal line.
slope: $m = 0$
y-intercept: -5

37. $x = 6$
This is a vertical line.
Slope: $m =$ undefined
y-intercept: none

39. $y = x + 3$
$m = 1$
$b = 3$

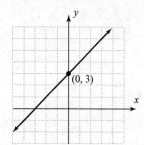

41. $y = -2x - 3$
$m = -2$
$b = -3$

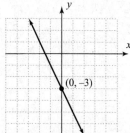

43. $y = -\dfrac{2}{3}x + 2$

$m = -\dfrac{2}{3}$

$b = 2$

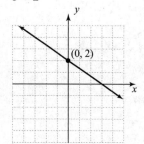

45. $y = -\dfrac{5}{2}x - 2$

$m = -\dfrac{5}{2}$

$b = -2$

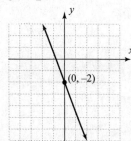

47. $4x + y = 5$
$\qquad y = -4x + 5$

$m = -4 = \dfrac{-4}{1}$

$b = 5$

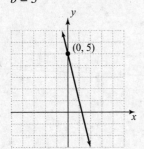

49. $x + 2y = -6$
$\qquad 2y = -x - 6$

$\qquad y = -\dfrac{1}{2}x - 3$

$m = -\dfrac{1}{2}$

$b = -3$

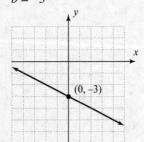

51. $3x - 2y = 10$
$\qquad -2y = -3x + 10$

$\qquad y = \dfrac{3}{2}x - 5$

$m = \dfrac{3}{2}$

$b = -5$

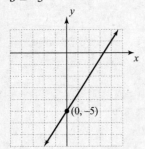

53. $6x + 3y = -15$
$\qquad 3y = -6x - 15$
$\qquad y = -2x - 5$

$m = -2 = \dfrac{-2}{1}$

$b = -5$

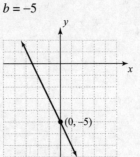

55. $m = -1; b = 8$
$y = -1x + 8$
$y = -x + 8$

57. $m = \dfrac{6}{7}; \ b = -6$

$y = \dfrac{6}{7}x - 6$

59. $m = -\dfrac{1}{3}; b = \dfrac{2}{3}$

$y = -\dfrac{1}{3}x + \dfrac{2}{3}$

61. m is undefined; x-intercept -5
This is a vertical line.
$x = -5$

63. $m = 0; b = 3$
$y = 0x + 3$
$y = 3$

65. $m = 5; b = 0$
$y = 5x + 0$
$y = 5x$

67. $y = 2x - 7$

$m = 2, b = -7$

x-intercept $\left(\dfrac{7}{2}, 0\right)$

y-intercept $(0, -7)$

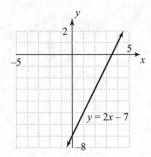

69. $3x - 2y = 24$

$\quad -2y = -3x + 24$

$\quad\quad y = \dfrac{3}{2}x - 12$

$m = \dfrac{3}{2}, \ b = -12$

x-intercept $(8, 0)$

y-intercept $(0, -12)$

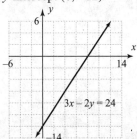

71. $y = -5$ or $y = 0x - 5$

$m = 0, b = -5$

Horizontal line

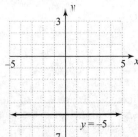

73. $x = -6$

Vertical line

Undefined slope

no y-intercept

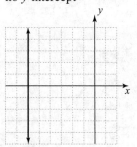

75. $6x - 4y = 0$

$\quad -4y = -6x + 0$

$\quad\quad y = \dfrac{6}{4}x + 0$

$\quad\quad y = \dfrac{3}{2}x + 0$

$m = \dfrac{3}{2}, \ b = 0$

x- and y-intercept $(0, 0)$

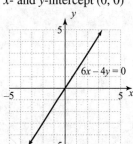

77. $y = -\dfrac{5}{3}x + 6$

$m = -\dfrac{5}{3}, b = 6$

x-intercept $\left(\dfrac{18}{5}, 0\right)$

y-intercept $(0, 6)$

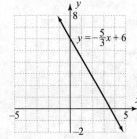

157

79. $2y = x + 4$ or $y = \frac{1}{2}x + 2$

$m = \frac{1}{2}, \ b = 2$

x-intercept $(-4, 0)$
y-intercept $(0, 2)$

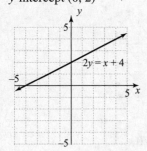

81. $y = \frac{x}{3} = \frac{1}{3}x + 0$

$m = \frac{1}{3}$

$b = 0$

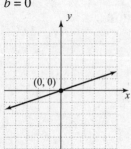

83. $2x = -8y$

$8y = -2x$

$y = -\frac{2}{8}x = -\frac{1}{4}x + 0$

$m = -\frac{1}{4}$

$b = 0$

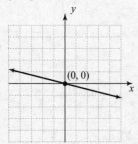

85. $y = -\frac{2x}{3} + 1 = -\frac{2}{3}x + 1$

$m = -\frac{2}{3}$

$b = 1$

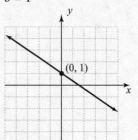

87. $x + 2 = -7$

$x = -9$

Vertical line
Undefined slope
No y-intercept

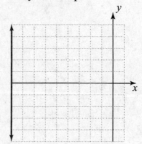

89. $5x + y + 1 = 0$

$y = -5x - 1$

$m = -5$

$b = -1$

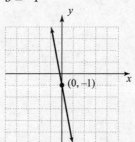

91. (a) The rate of change is 8% or 0.08 of the amount of sales. With no sales, the salary is $400. Thus, $m = 0.08$ and $b = 400$.

$y = 0.08x + 400$

(b) For sales of $1200, $x = 1200$.
$y = 0.08(1200) + 400 = 96 + 400 = 496$
Dien's income is $496 if he sells $1200 worth of merchandise.

(c)

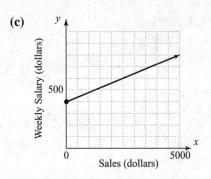

93. $y = -3.583x + 56$

    (a) In 1999, $x = 4$.
        $y = -3.583(4) + 56 = -14.332 + 56 = 41.668$
        In 1999, the cost per minute was 47¢.

    (b) Find x when $y = 20.17$.
        $20.17 = -3.583x + 56$
        $-35.83 = -3.583x$
           $10 = x$
        $x = 10$ corresponds to 10 years after 1995,
        so the cost was 20.17¢ per minute in 2005.

    (c) The cost per minute is decreasing by 3.583¢
        per year.

    (d) No, the cost will never be 0 or negative.

    (e)

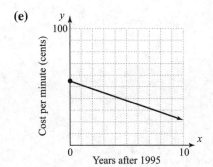

95. $2x + By = 12$
       $By = -2x + 12$
       $y = \dfrac{-2}{B}x + \dfrac{12}{B}$

    $m = -\dfrac{2}{B}$

    $\dfrac{1}{2} = -\dfrac{2}{B}$

    $\dfrac{B}{2} = -2$

    $B = -4$

97. $Ax - 2y = 10$
       $-2y = -Ax + 10$
          $y = \dfrac{A}{2}x - 5$

    $m = \dfrac{A}{2}$

    $-2 = \dfrac{A}{2}$

    $-4 = A$

99. $x + By = \dfrac{1}{2}$

    $By = -x + \dfrac{1}{2}$

    $y = -\dfrac{1}{B}x + \dfrac{1}{2B}$

    $b = \dfrac{1}{2B}$

    $-\dfrac{1}{6} = \dfrac{1}{2B}$

    $-\dfrac{B}{6} = \dfrac{1}{2}$

    $B = -3$

101. (a) Variable cost $= m = 40$
        Fixed cost $= b = 4000$
        $y = 40x + 4000$

    (b) $x = 500$
        $y = 40(500) + 4000$
          $= 20,000 + 4000$
          $= 24,000$
        The daily cost of manufacturing
        500 calculators is $24,000.

    (c) $y = 19,000$
        $19,000 = 40x + 4000$
        $15,000 = 40x$
          $375 = x$
        375 calculators were manufactured.

    (d)

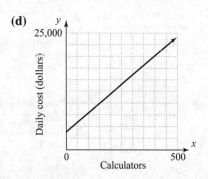

103. The slope of the line shown is positive and the y-intercept is negative.

(a) $y = 3x - 2$

(b) $y = -2x + 5$

(c) $y = 3$

(d) $2x + 3y = 6$ or $3y = -2x + 6$ or $y = -\frac{2}{3}x + 2$

(e) $3x - 2y = 8$ or $-2y = -3x + 8$ or $y = \frac{3}{2}x - 4$

(f) $4x - y = -4$ or $y = 4x + 4$

(g) $-5x + 2y = 12$ or $2y = 5x + 12$ or
$$y = \frac{5}{2}x + 6$$

(h) $x - y = -3$ or $y = x + 3$

Using the slope-intercept forms of the lines, the equations (a) or (e) could have the graph.

Section 3.5

Preparing for Point-Slope Form of a Line

P1.
$$y - 3 = 2(x + 1)$$
$$y - 3 = 2x + 2$$
$$y - 3 + 3 = 2x + 2 + 3$$
$$y = 2x + 5$$

P2. $\dfrac{7 - 3}{4 - 2} = \dfrac{4}{2} = 2$

3.5 Quick Checks

1. The point-slope form of a nonvertical line whose slope is m that contains the point (x_1, y_1) is
$$\underline{y - y_1 = m(x - x_1)}.$$

2. True

3. $m = 3$; $(x_1, y_1) = (2, 1)$
$$y - y_1 = m(x - x_1)$$
$$y - 1 = 3(x - 2)$$
$$y - 1 = 3x - 6$$
$$y = 3x - 5$$

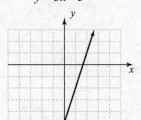

(0, −5)

4. $m = \dfrac{1}{3}$; $(x_1, y_1) = (3, -4)$
$$y - y_1 = m(x - x_1)$$
$$y - (-4) = \frac{1}{3}(x - 3)$$
$$y + 4 = \frac{1}{3}x - 1$$
$$y = \frac{1}{3}x - 5$$

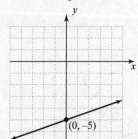

(0, −5)

5. $m = -4$; $(x_1, y_1) = (-2, 5)$
$$y - y_1 = m(x - x_1)$$
$$y - 5 = -4(x - (-2))$$
$$y - 5 = -4(x + 2)$$
$$y - 5 = -4x - 8$$
$$y = -4x - 3$$

(0, −3)

6. $m = -\dfrac{5}{2}$; $(x_1, y_1) = (-4, 5)$

$$y - y_1 = m(x - x_1)$$

$$y - 5 = -\frac{5}{2}(x - (-4))$$

$$y - 5 = -\frac{5}{2}(x + 4)$$

$$y - 5 = -\frac{5}{2}x - 10$$

$$y = -\frac{5}{2}x - 5$$

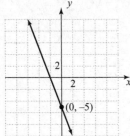

7. Since the line is horizontal, the slope, m, is 0.

$m = 0$; $(x_1, y_1) = (-2, 3)$

$$y - y_1 = m(x - x_1)$$

$$y - 3 = 0(x - (-2))$$

$$y - 3 = 0$$

$$y = 3$$

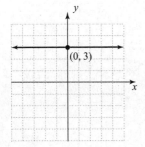

8. Let $(x_1, y_1) = (0, 2)$ and $(x_2, y_2) = (3, 5)$

$$m = \frac{y_2 - y_1}{x_2 - x_1} = \frac{5 - 2}{3 - 0} = \frac{3}{3} = 1$$

$$y - y_1 = m(x - x_1)$$

$$y - 2 = 1(x - 0)$$

$$y - 2 = x$$

$$y = x + 2$$

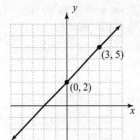

9. Let $(x_1, y_1) = (-1, 4)$ and $(x_2, y_2) = (1, -2)$.

$$m = \frac{y_2 - y_1}{x_2 - x_1} = \frac{-2 - 4}{1 - (-1)} = \frac{-6}{1 + 1} = \frac{-6}{2} = -3$$

$$y - y_1 = m(x - x_1)$$

$$y - 4 = -3(x - (-1))$$

$$y - 4 = -3(x + 1)$$

$$y - 4 = -3x - 3$$

$$y = -3x + 1$$

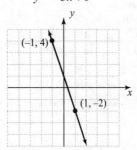

10. Let $(x_1, y_1) = (3, 2)$ and $(x_2, y_2) = (3, -4)$.

$$m = \frac{y_2 - y_1}{x_2 - x_1} = \frac{-4 - 2}{3 - 3} = \frac{-6}{0}$$

The slope is undefined, so the line is vertical. The equation of the line is $x = 3$.

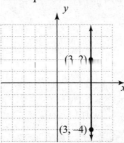

11. List the five forms that are used when writing the equation of a line: <u>Horizontal line, $y = b$</u>; <u>Vertical line, $x = a$</u>; <u>Point-slope, $y - y_1 = m(x - x_1)$</u>; <u>Slope-intercept, $y = mx + b$</u>; <u>Standard form, $Ax + By = C$</u>.

12. (a) When $x = 3.90$, then $y = 400$, so
$(x_1, y_1) = (3.9, 400)$. When $x = 4.10$, then
$y = 380$, so $(x_2, y_2) = (4.1, 380)$.

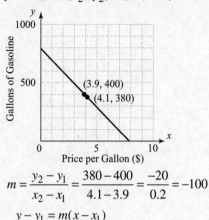

$$m = \frac{y_2 - y_1}{x_2 - x_1} = \frac{380 - 400}{4.1 - 3.9} = \frac{-20}{0.2} = -100$$

$$y - y_1 = m(x - x_1)$$
$$y - 400 = -100(x - 3.9)$$
$$y - 400 = -100x + 390$$
$$y = -100x + 790$$

(b) Let $x = 4.00$.
$y = -100x + 790$
$y = -100(4) + 790 = 390$
390 gallons will be sold if the price is \$4.00
per gallon.

(c) The slope is -100. The number of gallons of
gasoline sold will decrease by 100 if the
price per gallon increases by \$1.

3.5 Exercises

13. $(2, 5)$; slope $= 3$
$$y - y_1 = m(x - x_1)$$
$$y - 5 = 3(x - 2)$$
$$y - 5 = 3x - 6$$
$$y = 3x - 1$$

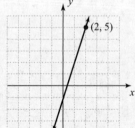

15. $(-1, 2)$; slope $= -2$
$$y - y_1 = m(x - x_1)$$
$$y - 2 = -2[x - (-1)]$$
$$y - 2 = -2(x + 1)$$
$$y - 2 = -2x - 2$$
$$y = -2x$$

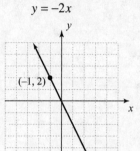

17. $(8, -1)$; slope $= \dfrac{1}{4}$
$$y - y_1 = m(x - x_1)$$
$$y - (-1) = \frac{1}{4}(x - 8)$$
$$y + 1 = \frac{1}{4}x - 2$$
$$y = \frac{1}{4}x - 3$$

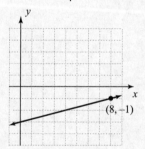

19. $(0, 13)$; slope $= -6$
$$y - y_1 = m(x - x_1)$$
$$y - 13 = -6(x - 0)$$
$$y - 13 = -6x$$
$$y = -6x + 13$$

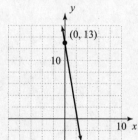

21. $(5, -7)$; slope $= 0$

$$y - y_1 = m(x - x_1)$$
$$y - (-7) = 0(x - 5)$$
$$y + 7 = 0$$
$$y = -7$$

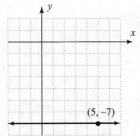

23. $(-4, 5)$; undefined slope
Since the slope is undefined, the line is vertical.
The equation is $x = -4$.

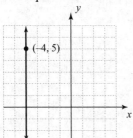

25. $(-3, 0)$; slope $= \dfrac{2}{3}$

$$y - y_1 = m(x - x_1)$$
$$y - 0 = \frac{2}{3}[x - (-3)]$$
$$y = \frac{2}{3}(x + 3)$$
$$y = \frac{2}{3}x + 2$$

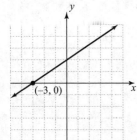

27. $(-8, 6)$; slope $= -\dfrac{3}{4}$

$$y - y_1 = m(x - x_1)$$
$$y - 6 = -\frac{3}{4}[x - (-8)]$$
$$y - 6 = -\frac{3}{4}(x + 8)$$
$$y - 6 = -\frac{3}{4}x - 6$$
$$y = -\frac{3}{4}x$$

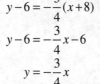
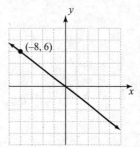

29. $(-3, 10)$; vertical
$x = -3$

31. $(-1, -5)$; horizontal
$y = -5$

33. $(0.2, -4.3)$; horizontal
$y = -4.3$

35. $\left(\dfrac{1}{2}, \dfrac{7}{4}\right)$; vertical

$x = \dfrac{1}{2}$

37. $(0, 4)$, $(-2, 0)$

$$m = \frac{y_2 - y_1}{x_2 - x_1} = \frac{0 - 4}{-2 - 0} = \frac{-4}{-2} = 2$$
$$y - y_1 = m(x - x_1)$$
$$y - 4 = 2(x - 0)$$
$$y - 4 = 2x$$
$$y = 2x + 4$$

39. $(1, 2)$, $(0, 6)$

$$m = \frac{y_2 - y_1}{x_2 - x_1} = \frac{6 - 2}{0 - 1} = \frac{4}{-1} = -4$$
$$y - y_1 = m(x - x_1)$$
$$y - 2 = -4(x - 1)$$
$$y - 2 = -4x + 4$$
$$y = -4x + 6$$

41. $(-3, 2), (1, -4)$

$$m = \frac{y_2 - y_1}{x_2 - x_1} = \frac{-4 - 2}{1 - (-3)} = \frac{-6}{4} = -\frac{3}{2}$$

$$y - y_1 = m(x - x_1)$$

$$y - 2 = -\frac{3}{2}[x - (-3)]$$

$$y - 2 = -\frac{3}{2}(x + 3)$$

$$y - 2 = -\frac{3}{2}x - \frac{9}{2}$$

$$y = -\frac{3}{2}x - \frac{9}{2} + \frac{4}{2}$$

$$y = -\frac{3}{2}x - \frac{5}{2}$$

43. $(-3, -11), (2, -1)$

$$m = \frac{y_2 - y_1}{x_2 - x_1} = \frac{-1 - (-11)}{2 - (-3)} = \frac{-1 + 11}{2 + 3} = \frac{10}{5} = 2$$

$$y - y_1 = m(x - x_1)$$

$$y - (-11) = 2[x - (-3)]$$

$$y + 11 = 2(x + 3)$$

$$y + 11 = 2x + 6$$

$$y = 2x - 5$$

45. $(4, -3), (-3, -3)$

$$m = \frac{y_2 - y_1}{x_2 - x_1} = \frac{-3 - (-3)}{-3 - 4} = \frac{-3 + 3}{-7} = \frac{0}{-7} = 0$$

$$y - y_1 = m(x - x_1)$$

$$y - (-3) = 0(x - 4)$$

$$y + 3 = 0$$

$$y = -3$$

47. $(2, -1), (2, -9)$

$$m = \frac{y_2 - y_1}{x_2 - x_1} = \frac{-9 - (-1)}{2 - 2} = \frac{-9 + 1}{0}$$

Since the slope is undefined, the line is vertical.
$x = 2$

49. $(0.1, 0.6), (0.5, 0.7)$

$$m = \frac{y_2 - y_1}{x_2 - x_1} = \frac{0.7 - 0.6}{0.5 - 0.1} = \frac{0.1}{0.4} = 0.25$$

$$y - y_1 = m(x - x_1)$$

$$y - 0.6 = 0.25(x - 0.1)$$

$$y - 0.6 = 0.25x - 0.025$$

$$y = 0.25x + 0.575$$

51. $\left(\frac{1}{2}, -\frac{9}{4}\right), \left(\frac{5}{2}, -\frac{1}{4}\right)$

$$m = \frac{y_2 - y_1}{x_2 - x_1} = \frac{-\frac{1}{4} - \left(-\frac{9}{4}\right)}{\frac{5}{2} - \frac{1}{2}} = \frac{-\frac{1}{4} + \frac{9}{4}}{\frac{4}{2}} = \frac{2}{2} = 1$$

$$y - y_1 = m(x - x_1)$$

$$y - \left(-\frac{9}{4}\right) = 1\left(x - \frac{1}{2}\right)$$

$$y + \frac{9}{4} = x - \frac{1}{2}$$

$$y = x - \frac{2}{4} - \frac{9}{4}$$

$$y = x - \frac{11}{4}$$

53. $(4, -2), m = 5$

$$y - y_1 = m(x - x_1)$$

$$y - (-2) = 5(x - 4)$$

$$y + 2 = 5x - 20$$

$$y = 5x - 22$$

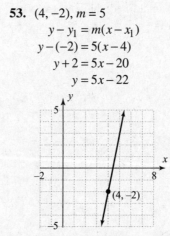

55. $(-3, 5)$, horizontal
$y = 5$

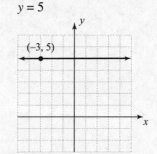

57. $(1, 3), (-4, -2)$

$$m = \frac{y_2 - y_1}{x_2 - x_1} = \frac{-2 - 3}{-4 - 1} = \frac{-5}{-5} = 1$$

$$y - y_1 = m(x - x_1)$$

$$y - 3 = 1(x - 1)$$

$$y - 3 = x - 1$$

$$y = x + 2$$

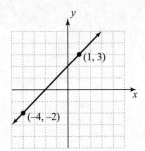

59. $(-2, 3)$, $m = \dfrac{1}{2}$

$$y - y_1 = m(x - x_1)$$
$$y - 3 = \frac{1}{2}[x - (-2)]$$
$$y - 3 = \frac{1}{2}(x + 2)$$
$$y - 3 = \frac{1}{2}x + 1$$
$$y = \frac{1}{2}x + 4$$

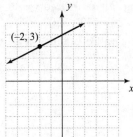

61. $(5, 2)$, vertical line
$x = 5$

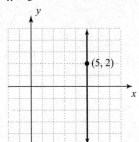

63. $(3, -19)$, $(-1, 9)$

$$m = \frac{y_2 - y_1}{x_2 - x_1} = \frac{9 - (-19)}{-1 - 3} = \frac{28}{-4} = -7$$
$$y - y_1 = m(x - x_1)$$
$$y - (-19) = -7(x - 3)$$
$$y + 19 = -7x + 21$$
$$y = -7x + 2$$

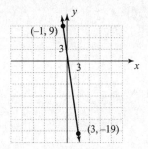

65. $(6, 3)$, $m = -\dfrac{2}{3}$

$$y - y_1 = m(x - x_1)$$
$$y - 3 = -\frac{2}{3}(x - 6)$$
$$y - 3 = -\frac{2}{3}x + 4$$
$$y = -\frac{2}{3}x + 7$$

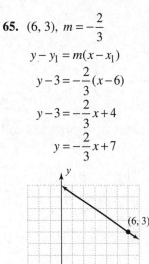

67. $(-2, 3)$, $(4, -6)$

$$m = \frac{y_2 - y_1}{x_2 - x_1} = \frac{-6 - 3}{4 - (-2)} = \frac{-9}{6} = -\frac{3}{2}$$
$$y - y_1 = m(x - x_1)$$
$$y - 3 = -\frac{3}{2}[x - (-2)]$$
$$y - 3 = -\frac{3}{2}(x + 2)$$
$$y - 3 = -\frac{3}{2}x - 3$$
$$y = -\frac{3}{2}x$$

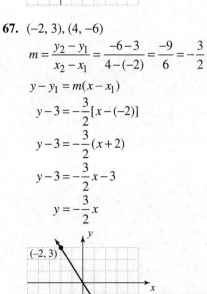

69. *x*-intercept: 5 (5, 0)

y-intercept: −2 (0, −2)

$$m = \frac{y_2 - y_1}{x_2 - x_1} = \frac{-2 - 0}{0 - 5} = \frac{-2}{-5} = \frac{2}{5}$$

$$y = \frac{2}{5}x - 2$$

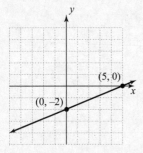

71. (a) (60, 1635) indicates that when 60 packages are shipped, the expenses for the department are $1635.

(b)

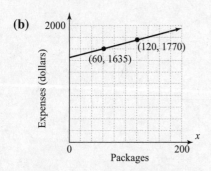

(c) Use (60, 1635) and (120, 1770).

$$m = \frac{y_2 - y_1}{x_2 - x_1} = \frac{1770 - 1635}{120 - 60} = \frac{135}{60} = \frac{9}{4}$$

$$y - y_1 = m(x - x_1)$$

$$y - 1635 = \frac{9}{4}(x - 60)$$

$$y - 1635 = \frac{9}{4}x - 135$$

$$y = \frac{9}{4}x + 1500$$

(d) For 200 packages, *x* = 200.

$$y = \frac{9}{4}(200) + 1500 = 450 + 1500 = 1950$$

The total expenses are $1950 when 200 packages are sent.

(e) Expenses increase by $\$\frac{9}{4} = 2.25$ for each additional package sent.

73. (a) 1980 is 0 years after 1980.
2005 is 25 years after 1980.
The ordered pairs are (0, 820) and (25, 985).

(b)

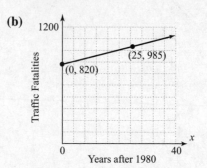

(c) Use (0, 820) and (25, 985).

$$m = \frac{y_2 - y_1}{x_2 - x_1} = \frac{985 - 820}{25 - 0} = \frac{165}{25} = 6.6$$

$$y = mx + b;\ b = 820$$

$$y = 6.6x + 820$$

(d) 2000 is 20 years after 1980, so *x* = 20.

$$y = 6.6(20) + 820 = 132 + 820 = 952$$

There were 952 traffic fatalities in 2000.

(e) The number of traffic fatalities in Kentucky increases by 6.6 each year.

75. (−4, 2); slope = 3

$$y = mx + b$$
$$y = 3x + b$$
$$2 = 3(-4) + b$$
$$2 = -12 + b$$
$$14 = b$$
$$y = 3x + 14$$

77. (3, −8); slope = −2

$$y = mx + b$$
$$y = -2x + b$$
$$-8 = -2(3) + b$$
$$-8 = -6 + b$$
$$-2 = b$$
$$y = -2x - 2$$

79. $\left(\dfrac{2}{3}, \dfrac{1}{2}\right)$; slope $= 6$

$$y = mx + b$$
$$y = 6x + b$$
$$\frac{1}{2} = 6\left(\frac{2}{3}\right) + b$$
$$\frac{1}{2} = 4 + b$$
$$-\frac{7}{2} = b$$

$$y = 6x - \frac{7}{2}$$

81. $(6, -13)$ and $(-2, -5)$

$$m = \frac{y_2 - y_1}{x_2 - x_1}$$
$$= \frac{-5 - (-13)}{-2 - 6}$$
$$= \frac{-5 + 13}{-8}$$
$$= \frac{8}{-8}$$
$$= -1$$
$$y = mx + b$$
$$y = -x + b$$
$$-13 = -(6) + b$$
$$-13 = -6 + b$$
$$-7 = b$$
$$y = -x - 7$$

83. $(5, -1)$ and $(-10, -4)$

$$m = \frac{y_2 - y_1}{x_2 - x_1} = \frac{-4 - (-1)}{-10 - 5} = \frac{-4 + 1}{-15} = \frac{-3}{-15} = \frac{1}{5}$$
$$y = mx + b$$
$$y = \frac{1}{5}x + b$$
$$-1 = \frac{1}{5}(5) + b$$
$$-1 = 1 + b$$
$$-2 = b$$
$$y = \frac{1}{5}x - 2$$

85. $(-4, 8)$ and $(2, -1)$

$$m = \frac{y_2 - y_1}{x_2 - x_1} = \frac{-1 - 8}{2 - (-4)} = \frac{-9}{2 + 4} = \frac{-9}{6} = -\frac{3}{2}$$

$$y = mx + b$$
$$y = -\frac{3}{2}x + b$$
$$8 = -\frac{3}{2}(-4) + b$$
$$8 = 6 + b$$
$$2 = b$$
$$y = -\frac{3}{2}x + 2$$

87. Answers may vary. One possibility: Yes, although the equations (in point-slope form) will be different, they will simplify to the same slope-intercept form.

Section 3.6

Preparing for Parallel and Perpendicular Lines

P1. The reciprocal of 3 is $\dfrac{1}{3}$ since $3\left(\dfrac{1}{3}\right) = 1$.

P2. The reciprocal of $-\dfrac{3}{5}$ is $-\dfrac{5}{3}$ since

$$-\frac{3}{5}\left(-\frac{5}{3}\right) = 1.$$

3.6 Quick Checks

1. Two nonvertical lines are parallel if and only if their <u>slopes</u> are equal and they have different <u>y-intercepts</u>. Vertical lines are parallel if they have different <u>x-intercepts</u>.

2. The slope of $y = 2x + 1$ is 2 and the slope of $y = -2x - 3$ is -2. Since the slopes are different, the lines are not parallel.

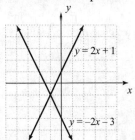

3. $6x + 3y = 3$
$$3y = -6x + 3$$
$$y = -2x + 1$$

The slope of the line is -2 and the y-intercept is 1.
$$10x + 5y = 10$$
$$5y = -10x + 10$$
$$y = -2x + 2$$

The slope of the line is -2 and the y-intercept is 2. Because the lines have the same slope, but different y-intercepts, the lines are parallel.

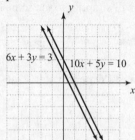

4. $4x + 5y = 10$
$$5y = -4x + 10$$
$$y = -\frac{4}{5}x + 2$$

The slope of the line is $-\frac{4}{5}$ and the y-intercept is 2.
$$8x + 10y = 20$$
$$10y = -8x + 20$$
$$y = -\frac{4}{5}x + 2$$

The slope of the line is $-\frac{4}{5}$ and the y-intercept is 2.
Because the lines have the same slope and y-intercept, they are the same line, so they are not parallel.

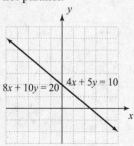

5. The slope of $y = 2x + 1$ is 2.
$m = 2$; (2, 3)
$$y - y_1 = m(x - x_1)$$
$$y - 3 = 2(x - 2)$$
$$y - 3 = 2x - 4$$
$$y = 2x - 1$$

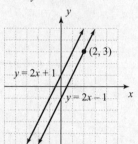

6. $3x + 2y = 4$
$$2y = -3x + 4$$
$$y = -\frac{3}{2}x + 2$$

The slope is $-\frac{3}{2}$.

$m = -\frac{3}{2}$; (-2, 3)
$$y - y_1 = m(x - x_1)$$
$$y - 3 = -\frac{3}{2}(x - (-2))$$
$$y - 3 = -\frac{3}{2}(x + 2)$$
$$y - 3 = -\frac{3}{2}x - 3$$
$$y = -\frac{3}{2}x$$

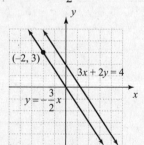

7. $x = -2$ is a vertical line, so any line parallel to $x = -2$ will also be vertical. The vertical line through (3, 1) has equation $x = 3$.

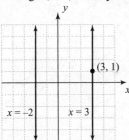

8. $y + 3 = 0$
$$y = -3$$

$y = -3$ is a horizontal line, so any line parallel to $y = -3$ will also be horizontal. The horizontal line through (−2, 5) has equation $y = 5$.

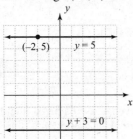

9. Given any two nonvertical lines, if the product of their slopes is −1, then the lines are <u>perpendicular</u>.

10. False; L_1 and L_2 are not perpendicular because their slopes are not negative reciprocals.

11. The negative reciprocal of −4 is $\dfrac{-1}{-4} = \dfrac{1}{4}$. Any line whose slope is $\dfrac{1}{4}$ will be perpendicular to a line whose slope is −4.

12. The negative reciprocal of $\dfrac{5}{4}$ is $\dfrac{-1}{\frac{5}{4}} = -\dfrac{4}{5}$. Any line whose slope is $-\dfrac{4}{5}$ will be perpendicular to a line whose slope is $\dfrac{5}{4}$.

13. The negative reciprocal of $-\dfrac{1}{5}$ is $\dfrac{-1}{-\frac{1}{5}} = \dfrac{5}{1} = 5$. Any line whose slope is 5 will be perpendicular to a line whose slope is $-\dfrac{1}{5}$.

14. The slope of $y = 4x - 3$ is 4. The slope of $y = -\dfrac{1}{4}x - 4$ is $-\dfrac{1}{4}$. Since $4 \cdot \left(-\dfrac{1}{4}\right) = -1$, the lines are perpendicular.

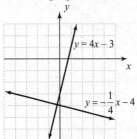

15. $2x - y = 3$
$$-y = -2x + 3$$
$$y = 2x - 3$$
The slope of the line is 2.
$$x - 2y = 2$$
$$-2y = -x + 2$$
$$y = \dfrac{1}{2}x - 1$$

The slope of the line is $\dfrac{1}{2}$. Since $2 \cdot \dfrac{1}{2} = 1 \neq -1$, the lines are not perpendicular.

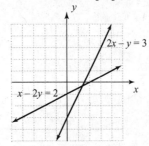

16. $5x + 2y = 8$
$$2y = -5x + 8$$
$$y = -\dfrac{5}{2}x + 4$$

The slope of the line is $-\dfrac{5}{2}$.

$$2x - 5y = 10$$
$$-5y = -2x + 10$$
$$y = \frac{2}{5}x - 2$$

The slope of the line is $\frac{2}{5}$. Since $-\frac{5}{2} \cdot \frac{2}{5} = -1$, the lines are perpendicular.

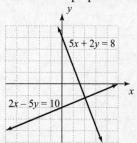

17. The slope of $y = 2x + 1$ is 2. A line perpendicular to $y = 2x + 1$ will have slope $\frac{-1}{2} = -\frac{1}{2}$.

$$m = -\frac{1}{2}; \ (-4,\ 2)$$
$$y - y_1 = m(x - x_1)$$
$$y - 2 = -\frac{1}{2}(x - (-4))$$
$$y - 2 = -\frac{1}{2}(x + 4)$$
$$y - 2 = -\frac{1}{2}x - 2$$
$$y = -\frac{1}{2}x$$

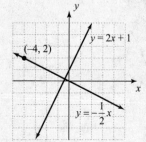

18. $2x + 3y = 3$
$$3y = -2x + 3$$
$$y = -\frac{2}{3}x + 1$$

The slope of $y = -\frac{2}{3}x + 1$ is $-\frac{2}{3}$. A line perpendicular to $y = -\frac{2}{3}x + 1$ will have slope

$$\frac{-1}{-\frac{2}{3}} = \frac{3}{2}.$$

$$m = \frac{3}{2}; \ (-2,\ -1)$$
$$y - y_1 = m(x - x_1)$$
$$y - (-1) = \frac{3}{2}(x - (-2))$$
$$y + 1 = \frac{3}{2}(x + 2)$$
$$y + 1 = \frac{3}{2}x + 3$$
$$y = \frac{3}{2}x + 2$$

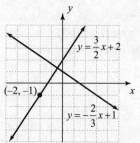

19. The line $x = -4$ is vertical, so a perpendicular line will be horizontal. The horizontal line through $(-1, -5)$ has equation $y = -5$.

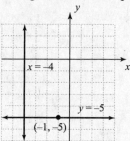

20. $y + 2 = 0$
$$y = -2$$

The line $y = -2$ is horizontal, so a perpendicular line will be vertical. The vertical line through $(3, -2)$ has equation $x = 3$.

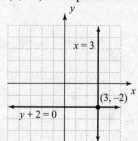

3.6 Exercises

	Slope of the Given Line	Slope of a Line Parallel to the Given Line	Slope of a Line Perpendicular to the Given Line
21.	$m = -3$	$m_1 = m = -3$	$m_2 = \frac{-1}{m} = \frac{-1}{-3} = \frac{1}{3}$
23.	$m = \frac{1}{2}$	$m_1 = m = \frac{1}{2}$	$m_2 = \frac{-1}{m} = \frac{-1}{\frac{1}{2}} = -1\left(\frac{2}{1}\right) = -2$
25.	$m = -\frac{4}{9}$	$m_1 = m = -\frac{4}{9}$	$m_2 = \frac{-1}{m} = \frac{-1}{-\frac{4}{9}} = -1\left(-\frac{9}{4}\right) = \frac{9}{4}$
27.	$m = 0$	$m_1 = m = 0$	$m_2 = \frac{-1}{m} = \frac{-1}{0} = \text{undefined}$

29. $L_1 : y = x - 3; \; m_1 = 1$

$L_2 : y = 1 - x = -x + 1; \; m_2 = -1$

Since $m_1 m_2 = (1)(-1) = -1$, the lines are perpendicular.

31. $L_1 : \; y = \frac{3}{4}x + 2; \; m_1 = \frac{3}{4}, \; b_1 = 2$

$L_2 : y = 0.75x - 1 = \frac{3}{4}x - 1; \; m_2 = \frac{3}{4}, \; b_2 = 1$

Since $m_1 = m_2$ and $b_1 \neq b_2$, the lines are parallel.

33. $L_1 : y = -\frac{5}{3}x - 6; \; m_1 = -\frac{5}{3}$

$L_2 : y = \frac{3}{5}x - 1; \; m_2 = \frac{3}{5}$

Since $m_1 m_2 = \left(-\frac{5}{3}\right)\left(\frac{3}{5}\right) = -1$, the lines are perpendicular.

35. $L_1 : x + y = -3 \text{ or } y = -x - 3; \; m_1 = -1$

$L_2 : y - x = 1 \text{ or } y = x + 1; \; m_2 = 1$

Since $m_1 m_2 = (-1)(1) = -1$, the lines are perpendicular.

37. $L_1 : 2x - 5y = 5 \text{ or } y = \frac{2}{5}x - 1; \; m_1 = \frac{2}{5}$

$L_2 : 5x + 2y = 4 \text{ or } y = -\frac{5}{2}x + 2; \; m_2 = -\frac{5}{2}$

Since $m_1 m_2 = \left(\frac{2}{5}\right)\left(-\frac{5}{2}\right) = -1$, the lines are perpendicular.

39. $L_1 : 4x - 5y - 15 = 0$ or $y = \dfrac{4}{5}x - 3$; $m_1 = \dfrac{4}{5}$,

$b_1 = -3$

$L_2 : 8x - 10y + 5 = 0$ or $y = \dfrac{4}{5}x + \dfrac{1}{2}$;

$m_2 = \dfrac{4}{5}$, $b_2 = \dfrac{1}{2}$

Since $m_1 = m_2$ and $b_1 \neq b_2$, the lines are parallel.

41. $L_1 : 4x = 3y + 3$ or $y = \dfrac{4}{3}x - 1$; $m_1 = \dfrac{4}{3}$, $b_1 = -1$

$L_2 : 6y = 8x + 36$ or $y = \dfrac{4}{3}x + 6$; $m_2 = \dfrac{4}{3}$,

$b_2 = 6$

Since $m_1 = m_2$ and $b_1 \neq b_2$, the lines are parallel.

43. $(4, -2); y = 3x - 1$

$y = 3x - 1$ has slope 3. A parallel line also has slope 3.

$y - y_1 = m(x - x_1)$

$y - (-2) = 3(x - 4)$

$y + 2 = 3x - 12$

$y = 3x - 14$

45. $(-3, 8); y = -4x + 5$

$y = -4x + 5$ has slope -4. A parallel line also has slope -4.

$y - y_1 = m(x - x_1)$

$y - 8 = -4[x - (-3)]$

$y - 8 = -4(x + 3)$

$y - 8 = -4x - 12$

$y = -4x - 4$

47. $(3, -7); y = 4$

$y = 4$ is a horizontal line and has slope 0. A parallel line has equation $y = b$.

$y = -7$

49. $(-1, 10); x = 10$

$x = 10$ is a vertical line and the slope is undefined. A parallel line has equation $x = c$.

$x = -1$

51. $(10, 2); 3x - 2y = 5$

$3x - 2y = 5$

$-2y = -3x + 5$

$y = \dfrac{3}{2}x - \dfrac{5}{2}$

The line has slope $\dfrac{3}{2}$. A parallel line also has

slope $\dfrac{3}{2}$.

$y - y_1 = m(x - x_1)$

$y - 2 = \dfrac{3}{2}(x - 10)$

$y - 2 = \dfrac{3}{2}x - 15$

$y = \dfrac{3}{2}x - 13$

53. $(-1, -10); x + 2y = 4$

$x + 2y = 4$

$2y = -x + 4$

$y = -\dfrac{1}{2}x + 2$

The line has slope $-\dfrac{1}{2}$. A parallel line also has

slope $-\dfrac{1}{2}$.

$y - y_1 = m(x - x_1)$

$y - (-10) = -\dfrac{1}{2}[x - (-1)]$

$y + 10 = -\dfrac{1}{2}(x + 1)$

$y + 10 = -\dfrac{1}{2}x - \dfrac{1}{2}$

$y = -\dfrac{1}{2}x - \dfrac{21}{2}$

55. $(3, 5); y = \dfrac{1}{2}x - 2$

$y = \dfrac{1}{2}x - 2$ has slope $\dfrac{1}{2}$. A perpendicular line

has slope $\dfrac{-1}{\frac{1}{2}} = -2$.

$y - y_1 = m(x - x_1)$

$y - 5 = -2(x - 3)$

$y - 5 = -2x + 6$

$y = -2x + 11$

57. $(-4, -1)$; $y = -4x + 1$

$y = -4x + 1$ has slope -4. A perpendicular line

has slope $\dfrac{-1}{-4} = \dfrac{1}{4}$.

$$y - y_1 = m(x - x_1)$$
$$y - (-1) = \frac{1}{4}[x - (-4)]$$
$$y + 1 = \frac{1}{4}(x + 4)$$
$$y + 1 = \frac{1}{4}x + 1$$
$$y = \frac{1}{4}x$$

59. $(-2, 1)$; x-axis

The x-axis is horizontal. A perpendicular line is vertical. The vertical line through $(-2, 1)$ is $x = -2$.

61. $(7, 5)$; y-axis

The y-axis is vertical. A perpendicular line is horizontal. The horizontal line through $(7, 5)$ is $y = 5$.

63. $(0, 0)$; $2x + 5y = 7$

$$2x + 5y = 7$$
$$5y = -2x + 7$$
$$y = -\frac{2}{5}x + \frac{7}{5}$$

The line has slope $-\dfrac{2}{5}$. A perpendicular line has

slope $\dfrac{-1}{-\frac{2}{5}} = (-1)\left(-\dfrac{5}{2}\right) = \dfrac{5}{2}$.

$$y - y_1 = m(x - x_1)$$
$$y - 0 = \frac{5}{2}(x - 0)$$
$$y = \frac{5}{2}x$$

65. $(-10, -3)$; $5x - 3y = 4$

$$5x - 3y = 4$$
$$-3y = -5x + 4$$
$$y = \frac{5}{3}x - \frac{4}{3}$$

The line has slope $\dfrac{5}{3}$. A perpendicular line has

slope $\dfrac{-1}{\frac{5}{3}} = -\dfrac{3}{5}$.

$$y - y_1 = m(x - x_1)$$
$$y - (-3) = -\frac{3}{5}[x - (-10)]$$
$$y + 3 = -\frac{3}{5}(x + 10)$$
$$y + 3 = -\frac{3}{5}x - 6$$
$$y = -\frac{3}{5}x - 9$$

67. $(3, -5) = (x_1, y_1)$; $m = 7$

$$y - y_1 = m(x - x_1)$$
$$y - (-5) = 7(x - 3)$$
$$y + 5 = 7x - 21$$
$$y = 7x - 26$$

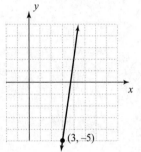

69. The slope of $y = -5x + 3$ is -5. A perpendicular

line has slope $\dfrac{-1}{-5} = \dfrac{1}{5}$.

$$(2, 9) = (x_1, y_1)$$
$$y - y_1 = m(x - x_1)$$
$$y - 9 = \frac{1}{5}(x - 2)$$
$$y - 9 = \frac{1}{5}x - \frac{2}{5}$$
$$y = \frac{1}{5}x + \frac{43}{5}$$

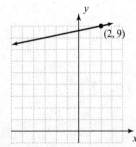

71. The slope of $y = -7x + 2$ is -7. A parallel line will also have slope -7.
$$(6, -1) = (x_1, y_1)$$
$$y - y_1 = m(x - x_1)$$
$$y - (-1) = -7(x - 6)$$
$$y + 1 = -7x + 42$$
$$y = -7x + 41$$

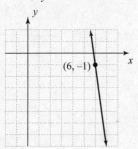

73. $(-6, 2)$ and $(-1, -8)$
$$m = \frac{-8 - 2}{-1 - (-6)} = \frac{-10}{-1 + 6} = \frac{-10}{5} = -2$$
$$y - y_1 = m(x - x_1)$$
$$y - 2 = -2[x - (-6)]$$
$$y - 2 = -2(x + 6)$$
$$y - 2 = -2x - 12$$
$$y = -2x - 10$$

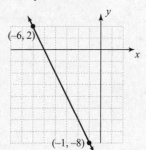

75. $m = 3, b = -2$
$$y = mx + b$$
$$y = 3x - 2$$

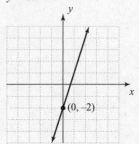

77. The line $x = -6$ is vertical. A parallel line will also be vertical. The vertical line through $(5, 1)$ is $x = 5$.

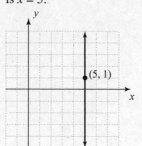

79. $4x + 3y = 9$
$$3y = -4x + 9$$
$$y = -\frac{4}{3}x + 3$$

The slope of the line is $-\frac{4}{3}$. A parallel line will also have slope $-\frac{4}{3}$.
$$(3, -2) = (x_1, y_1)$$
$$y - y_1 = m(x - x_1)$$
$$y - (-2) = -\frac{4}{3}(x - 3)$$
$$y + 2 = -\frac{4}{3}x + 4$$
$$y = -\frac{4}{3}x + 2$$

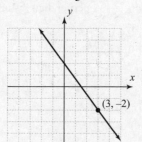

81. $x - 2y = -10$
$$-2y = -x - 10$$
$$y = \frac{1}{2}x + 5$$

The line has slope $\frac{1}{2}$. A perpendicular line has slope $\frac{-1}{\frac{1}{2}} = -1\left(\frac{2}{1}\right) = -2$.

$(-1, -3) = (x_1, y_1)$

$y - y_1 = m(x - x_1)$

$y - (-3) = -2[x - (-1)]$

$y + 3 = -2(x + 1)$

$y + 3 = -2x - 2$

$y = -2x - 5$

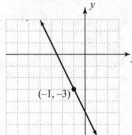

83. (a) L_1 : (0, −1) and (−2, −7)

$$m_1 = \frac{-7 - (-1)}{-2 - 0} = \frac{-7 + 1}{-2} = \frac{-6}{-2} = 3$$

L_2 : (−1, 5) and (2, −4)

$$m_2 = \frac{-4 - 5}{2 - (-1)} = \frac{-9}{2 + 1} = \frac{-9}{3} = -3$$

(b) Since $m_1 \neq m_2$ and $m_1 \neq \dfrac{-1}{m_2}$, the lines are

neither parallel nor perpendicular.

85. (a) L_1 : (2, 8) and (7, 18)

$$m_1 = \frac{18 - 8}{7 - 2} = \frac{10}{5} = 2$$

L_2 : (−2, −3) and (6, 13)

$$m_2 = \frac{13 - (-3)}{6 - (-2)} = \frac{13 + 3}{6 + 2} = \frac{16}{8} = 2$$

(b) Since $m_1 = m_2$, the lines are parallel.

87. (a) L_1 : (−2, −5) and (4, −2)

$$m = \frac{-2 - (-5)}{4 - (-2)} = \frac{-2 + 5}{4 + 2} = \frac{3}{6} = \frac{1}{2}$$

L_2 : (−8, −5) and (0, −1)

$$m_2 = \frac{-1 - (-5)}{0 - (-8)} = \frac{-1 + 5}{0 + 8} = \frac{4}{8} = \frac{1}{2}$$

(b) Since $m_1 = m_2$, the lines are parallel.

89. (a) L_1 : (−6, −9) and (3, 6)

$$m_1 = \frac{6 - (-9)}{3 - (-6)} = \frac{6 + 9}{3 + 6} = \frac{15}{9} = \frac{5}{3}$$

L_2 : (10, −8) and (−5, 1)

$$m_2 = \frac{1 - (-8)}{-5 - 10} = \frac{1 + 8}{-15} = \frac{9}{-15} = -\frac{3}{5}$$

(b) Since $m_1 m_2 = \left(\dfrac{5}{3}\right)\left(-\dfrac{3}{5}\right) = -1$, the lines are

perpendicular.

91.

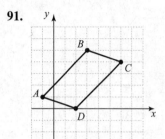

Slope $AB = m_1 = \dfrac{5 - 1}{3 - (-1)} = \dfrac{4}{3 + 1} = \dfrac{4}{4} = 1$

Slope $BC = m_2 = \dfrac{4 - 5}{6 - 3} = \dfrac{-1}{3} = -\dfrac{1}{3}$

Slope $CD = m_3 = \dfrac{0 - 4}{2 - 6} = \dfrac{-4}{-4} = 1 = m_1$

Slope $DA = m_4 = \dfrac{1 - 0}{-1 - 2} = \dfrac{1}{-3} = -\dfrac{1}{3} = m_2$

Since the slopes of the opposite sides are the same, the figure is a parallelogram.

93.

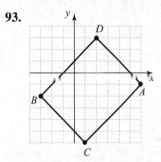

Slope $AC = m_1 = \dfrac{-6 - (-1)}{1 - 6}$

$= \dfrac{-6 + 1}{-5}$

$= \dfrac{-5}{-5}$

$= 1$

Slope $CB = m_2 = \dfrac{-2-(-6)}{-3-1}$

$= \dfrac{-2+6}{-4}$

$= \dfrac{4}{-4}$

$= -1$

$= \dfrac{-1}{m_1}$

Slope $BD = m_3 = \dfrac{3-(-2)}{2-(-3)}$

$= \dfrac{3+2}{2+3}$

$= \dfrac{5}{5}$

$= 1$

$= m_1$

Slope $DA = m_4 = \dfrac{-1-3}{6-2} = \dfrac{-4}{4} = -1 = m_2$

Since the slopes of the opposite sides are the same, the figure is a parallelogram.
Since the slopes of the adjacent sides AC and CB are negative reciprocals, the sides are perpendicular so the figure is a rectangle.

95.

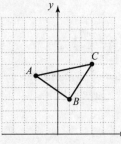

Slope $AB = m_1 = \dfrac{3-5}{1-(-2)} = \dfrac{-2}{1+2} = -\dfrac{2}{3}$

Slope $BC = m_2 = \dfrac{6-3}{3-1} = \dfrac{3}{2} = \dfrac{-1}{m}$,

Slope $CA = m_3 = \dfrac{5-6}{-2-3} = \dfrac{-1}{-5} = \dfrac{1}{5}$

Since the slopes of AC and BC are negative reciprocals, the triangle is a right triangle.

97.

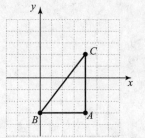

Slope $AB = m_1 = \dfrac{-3-(-3)}{0-4}$

$= \dfrac{-3+3}{-4}$

$= \dfrac{0}{-4}$

$= 0$

Slope $BC = m_2 = \dfrac{2-(-3)}{4-0} = \dfrac{2+3}{4} = \dfrac{5}{4}$

Slope $CA = m_3 = \dfrac{-3-2}{4-4} = \dfrac{-5}{0} = $ undefined

A vertical line (with undefined slope) is perpendicular to a line with 0 slope, thus the triangle is a right triangle.

99. $-3y = 6x-12$

$y = -2x+4$

Slope $= -2$

$4x+By = -2$

$By = -4x-2$

$y = -\dfrac{4}{B}x - \dfrac{2}{B}$

Slope $= -\dfrac{4}{B}$

Parallel lines have the same slope.

$-2 = -\dfrac{4}{B}$

$-2B = -4$

$B = 2$

101. $Ax+6y = -6$

$6y = -Ax-6$

$y = -\dfrac{A}{6}x - 1$

Slope $= -\dfrac{A}{6}$

$12-6y = -9x$

$-6y = -9x-12$

$y = \dfrac{3}{2}x+2$

$\text{Slope} = \dfrac{3}{2}$

The slopes of perpendicular lines are negative reciprocals.

$-\dfrac{A}{6} = \dfrac{-1}{\frac{3}{2}}$

$-\dfrac{A}{6} = -\dfrac{2}{3}$

$3A = 12$

$A = 4$

103.

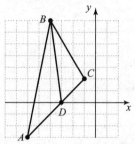

$\text{Slope } AC = m_1 = \dfrac{2-(-3)}{-1-(-6)} = \dfrac{2+3}{-1+6} = \dfrac{5}{5} = 1$

$\text{Slope } BD = m_2 = \dfrac{0-7}{-3-(-4)}$

$= \dfrac{-7}{-3+4}$

$= \dfrac{-7}{1}$

$= -7$

Since $m_1 m_2 = 1(-7) = -7 \neq 1$, $\overline{BD}$ is not a perpendicular to $\overline{AC}$, so it is not an altitude of triangle ABC.

105. Answers may vary. One possibility:

If $m_1 = m_2$ and $b_1 = b_2$, the lines are identical.

If $m_1 = m_2$ and $b_1 \neq b_2$, the lines are parallel.

If $m_1 \neq m_2$ and $m_1 \neq \dfrac{-1}{m_2}$ the lines intersect in one point, but they are not perpendicular.

If $m_1 = \dfrac{-1}{m_2}$ the lines intersect in one point and are perpendicular.

Putting the Concepts Together (Sections 3.1–3.6)

1. $(1, -2); 4x - 3y = 10$

$4x - 3y = 10$

$4(1) - 3(-2) \stackrel{?}{=} 10$

$4 + 6 \stackrel{?}{=} 10$

$10 = 10$ True

$(1, -2)$ is a solution to the equation

$4x - 3y = 10$.

In Problems 2 and 3, points may vary.

2. $y = \dfrac{2}{3}x - 1$

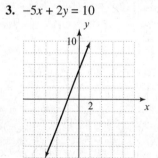

3. $-5x + 2y = 10$

4. $-8x + 2y = 6$

(a) For the x-intercept, let $y = 0$.

$-8x + 2(0) = 6$

$-8x = 6$

$x = -\dfrac{6}{8}$

$x = -\dfrac{3}{4}$

The x-intercept is $-\dfrac{3}{4}$.

(b) For the y-intercept, let $x = 0$.

$-8(0) + 2y = 6$

$2y = 6$

$y = 3$

The y-intercept is 3.

5. $4x + 3y = 6$

$y = 0$: $4x + 3(0) = 6$

$\qquad\qquad 4x = 6$

$\qquad\qquad x = \dfrac{6}{4}$

$\qquad\qquad x = \dfrac{3}{2}$

The x-intercept is $\dfrac{3}{2}$.

$x = 0$: $4(0) + 3y = 6$

$\qquad\qquad 3y = 6$

$\qquad\qquad y = 2$

The y-intercept is 2.

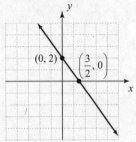

6. $6x + 9y = -12$

$9y = -6x - 12$

$y = -\dfrac{6}{9}x - \dfrac{12}{9}$

$y = -\dfrac{2}{3}x - \dfrac{4}{3}$

(a) The slope is $m = -\dfrac{2}{3}$.

(b) The y-intercept is $b = -\dfrac{4}{3}$.

7. $(3, -5) = (x_1, y_1)$; $(-6, -2) = (x_2, y_2)$

$m = \dfrac{y_2 - y_1}{x_2 - x_1}$

$\quad = \dfrac{-2 - (-5)}{-6 - 3}$

$\quad = \dfrac{-2 + 5}{-9}$

$\quad = \dfrac{3}{-9}$

$\quad = -\dfrac{1}{3}$

The slope is $-\dfrac{1}{3}$.

8. $2y = -5x - 4$

$y = -\dfrac{5}{2}x - 2$

(a) The given line has slope $-\dfrac{5}{2}$. A perpendicular line will have slope $\dfrac{-1}{-\frac{5}{2}} = \dfrac{2}{5}$.

(b) The given line has slope $-\dfrac{5}{2}$. A parallel line will also have slope $-\dfrac{5}{2}$.

9. L_1: $10x + 5y = 2$

$\qquad\qquad 5y = -10x + 2$

$\qquad\qquad y = -2x + \dfrac{2}{5}$

L_1 has slope -2 and y-intercept $\dfrac{2}{5}$.

L_2: $y = -2x + 3$

L_2 has slope -2 and y-intercept 3.

Since the lines have the same slope but different y-intercepts, they are parallel.

10. $m = 3$; $b = 1$

$y = mx + b$

$y = 3x + 1$

11. $m = -6$; $(x_1, y_1) = (-1, 4)$

$y - y_1 = m(x - x_1)$

$y - 4 = -6[x - (-1)]$

$y - 4 = -6(x + 1)$

$y - 4 = -6x - 6$

$\qquad y = -6x - 2$

12. $(x_1, y_1) = (4, -1)$; $(x_2, y_2) = (-2, 11)$

$m = \dfrac{y_2 - y_1}{x_2 - x_1} = \dfrac{11 - (-1)}{-2 - 4} = \dfrac{11 + 1}{-6} = \dfrac{12}{-6} = -2$

$y - y_1 = m(x - x_1)$

$y - (-1) = -2(x - 4)$

$\qquad y + 1 = -2x + 8$

$\qquad\quad y = -2x + 7$

13. The slope of $y = \dfrac{2}{5}x - 5$ is $\dfrac{2}{5}$. A perpendicular

line will have slope $\dfrac{-1}{\frac{2}{5}} = -\dfrac{5}{2}$.

$m = -\dfrac{5}{2};\ (x_1,\ y_1) = (-8,\ 0)$

$y - y_1 = m(x - x_1)$

$y - 0 = -\dfrac{5}{2}[x - (-8)]$

$y = -\dfrac{5}{2}(x + 8)$

$y = -\dfrac{5}{2}x - 20$

14. $-8y + 2x = -1$

$\quad -8y = -2x - 1$

$\quad\ y = \dfrac{1}{4}x + \dfrac{1}{8}$

The slope of the line is $\dfrac{1}{4}$. A parallel line will

also have slope $\dfrac{1}{4}$.

$m = \dfrac{1}{4};\ (x_1,\ y_1) = (-8,\ 3)$

$y - y_1 = m(x - x_1)$

$y - 3 = \dfrac{1}{4}[x - (-8)]$

$y - 3 = \dfrac{1}{4}(x + 8)$

$y - 3 = \dfrac{1}{4}x + 2$

$\quad\ y = \dfrac{1}{4}x + 5$

15. The horizontal line through $(-6, -8)$ has
equation $y = -8$.

16. A line with undefined slope is a vertical line.
The vertical line through $(2, 6)$ has equation
$x = 2$.

17. Let $(80, 1180) = (x_1,\ y_1)$ and

$(50, 850) = (x_2,\ y_2)$.

$m = \dfrac{y_2 - y_1}{x_2 - x_1} = \dfrac{850 - 1180}{50 - 80} = \dfrac{-330}{-30} = 11$

The average rate to ship an additional package is
$11 per package.

18. **(a)** Let $(0.7, 3543) = (x_1,\ y_1)$ and

$(0.8, 4378) = (x_2,\ y_2)$.

$m = \dfrac{y_2 - y_1}{x_2 - x_1}$

$\ = \dfrac{4378 - 3543}{0.8 - 0.7}$

$\ = \dfrac{835}{0.1}$

$\ = 8350$

$y - y_1 = m(x - x_1)$

$y - 3543 = 8350(x - 0.7)$

$y - 3543 = 8350x - 5845$

$\quad\quad\ y = 8350x - 2302$

The equation is $y = 8350x - 2302$ where x is
the weight, in carats, of the diamond and y is
the price.

(b) For every 1-carat increase in the weight of a
diamond, the cost increases by $8350.

(c) $y = 8350x - 2302;\ x = 0.76$
$y = 8350(0.76) - 2302 = 4044$
A 0.76-carat diamond costs $4044.

Section 3.7

Preparing for Linear Inequalities in Two Variables

P1. $\quad x - 4 > 5$

$\quad x - 4 + 4 > 5 + 4$

$\quad\quad\quad\ x > 9$

The solution set is $\{x | x > 9\}$.

P2. $\quad 3x + 1 \le 10$

$\quad 3x + 1 - 1 \le 10 - 1$

$\quad\quad\quad 3x \le 9$

$\quad\quad\ \dfrac{3x}{3} \le \dfrac{9}{3}$

$\quad\quad\quad\ x \le 3$

The solution set is $\{x | x \le 3\}$.

P3. $\quad 2(x + 1) - 6x > 18$

$\quad\ 2x + 2 - 6x > 18$

$\quad\quad\ -4x + 2 > 18$

$\quad -4x + 2 - 2 > 18 - 2$

$\quad\quad\quad\ -4x > 16$

$\quad\quad\ \dfrac{-4x}{-4} < \dfrac{16}{-4}$

$\quad\quad\quad\quad\ x < -4$

The solution set is $\{x | x < -4\}$.

3.7 Quick Checks

1. (a) Let $x = 2$ and $y = 1$ in the inequality.
$$2x + y > 7$$
$$2(2) + 1 > 7?$$
$$4 + 1 > 7?$$
$$5 > 7 \quad \text{False}$$
The statement $5 > 7$ is false, so $(2, 1)$ is not a solution to the inequality.

(b) Let $x = 3$ and $y = 4$ in the inequality.
$$2x + y > 7$$
$$2(3) + 4 > 7?$$
$$6 + 4 > 7?$$
$$10 > 7 \quad \text{True}$$
The statement $10 > 7$ is true, so $(3, 4)$ is a solution to the inequality.

(c) Let $x = -1$ and $y = 10$ in the inequality.
$$2x + y > 7$$
$$2(-1) + 10 > 7?$$
$$-2 + 10 > 7?$$
$$8 > 7 \quad \text{True}$$
The statement $8 > 7$ is true, so $(-1, 10)$ is a solution to the inequality.

2. (a) Let $x = 2$ and $y = 1$ in the inequality.
$$-3x + 2y \le 8$$
$$-3(2) + 2(1) \le 8?$$
$$-6 + 2 \le 8?$$
$$-4 \le 8 \quad \text{True}$$
The statement $-4 \le 8$ is true, so $(2, 1)$ is a solution to the inequality.

(b) Let $x = 3$ and $y = 4$ in the inequality.
$$-3x + 2y \le 8$$
$$-3(3) + 2(4) \le 8?$$
$$-9 + 8 \le 8?$$
$$-1 \le 8 \quad \text{True}$$
The statement $-1 \le 8$ is true, so $(3, 4)$ is a solution to the inequality.

(c) Let $x = -1$ and $y = 10$ in the inequality.
$$-3x + 2y \le 8$$
$$-3(-1) + 2(10) \le 8?$$
$$3 + 20 \le 8?$$
$$23 \le 8 \quad \text{False}$$
The statement $23 \le 8$ is false, so $(-1, 10)$ is not a solution to the inequality.

3. When drawing the boundary line for the graph of $Ax + By \ge C$, we use a <u>solid</u> line. When drawing the boundary line for the graph of $Ax + By < C$, we use a <u>dashed</u> line.

4. The boundary line separates the xy-plane into two regions, called <u>half-planes</u>.

5. Graph the line $y = -2x + 1$ with a dashed line since the inequality is strict.
Use $(0, 0)$ as a test point.
$$y < -2x + 1$$
$$0 < -2(0) + 1?$$
$$0 < 1 \quad \text{True}$$
Since $0 < 1$ is a true statement, shade the half-plane containing $(0, 0)$.

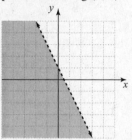

6. Graph the line $y = 3x + 2$ with a solid line since the inequality is nonstrict.
Use $(0, 0)$ as a test point.
$$y \ge 3x + 2$$
$$0 \ge 3(0) + 2?$$
$$0 \ge 2 \quad \text{False}$$
Since $0 \ge 2$ is a false statement, shade the half-plane that does not contain $(0, 0)$.

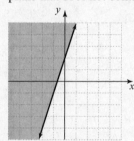

7. Graph the line $2x + 3y = 6$ with a solid line since the inequality is nonstrict.
x-intercept, $y = 0$:
$$2x + 3(0) = 6$$
$$2x = 6$$
$$x = 3$$
y-intercept, $x = 0$:
$$2(0) + 3y = 6$$
$$3y = 6$$
$$y = 2$$
Use $(0, 0)$ as a test point.
$$2x + 3y \le 6$$
$$2(0) + 3(0) \le 6?$$
$$0 \le 6 \quad \text{True}$$
Since $0 \le 6$ is a true statement, shade the half-

plane containing (0, 0).

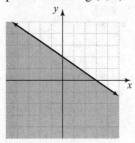

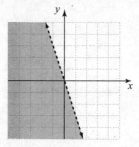

8. Graph the line $4x - 6y = 12$ with a dashed line since the inequality is strict.
x-intercept, $y = 0$:
$$4x - 6(0) = 12$$
$$4x = 12$$
$$x = 3$$
y-intercept, $x = 0$:
$$4(0) - 6y = 12$$
$$-6y = 12$$
$$y = -2$$
Use (0, 0) as a test point.
$$4x - 6y > 12$$
$$4(0) - 6(0) > 12?$$
$$0 > 12 \quad \text{False}$$
Since $0 > 12$ is a false statement, shade the half-plane that does not contain (0, 0).

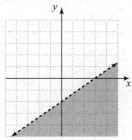

9. Graph the line $3x + y = 0$ with a dashed line since the inequality is strict.
$$3x + y = 0$$
$$y = -3x$$
Slope -3, y-intercept 0.
Use (1, 1) as a test point.
$$3x + y < 0$$
$$3(1) + 1 < 0?$$
$$4 < 0 \quad \text{False}$$
Since $4 < 0$ is a false statement, shade the half-plane that does not contain (1, 1).

10. Graph the line $2x - 5y = 0$ with a solid line since the inequality is nonstrict.
$$2x - 5y = 0$$
$$-5y = -2x$$
$$y = \frac{2}{5}x$$
Slope $\frac{2}{5}$, y-intercept 0.
Use (1, 1) as a test point.
$$2x - 5y \leq 0$$
$$2(1) - 5(1) \leq 0?$$
$$2 - 5 \leq 0?$$
$$-3 \leq 0 \quad \text{True}$$
Since $-3 \leq 0$ is a true statement, shade the half-plane containing (1, 1).

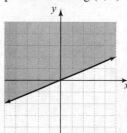

11. Graph the horizontal line $y = -1$ with a dashed line, since the inequality is strict. Use (0, 0) as a test point.
$$y < -1$$
$$0 < -1 \quad \text{False}$$
Since $0 < -1$ is a false statement, shade the half-plane that does not contain (0, 0).

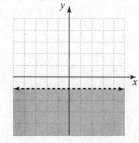

12. $3y - 9 = 0$

$3y = 9$

$y = 3$

Graph the horizontal line $y = 3$ with a solid line since the inequality is nonstrict.
Use $(0, 0)$ as a test point.

$3y - 9 \geq 0$

$3(0) - 9 \geq 0$?

$-9 \geq 0$ False

Since $-9 \geq 0$ is a false statement, shade the half-plane that does not contain $(0, 0)$.

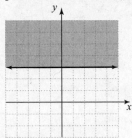

13. Graph the vertical line $x = 6$ with a dashed line since the inequality is strict.
Use $(0, 0)$ as a test point.

$x > 0$

$0 > 6$ False

Since $0 > 6$ is a false statement, shade the half-plane that does not contain $(0, 0)$.

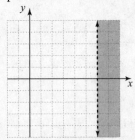

14. (a) We want to determine the number of suckers and taffy sticks Kevin can buy. Since each sucker costs $0.20, s suckers will cost $0.20s$ dollars. Since each taffy stick costs $0.25, t taffy sticks will cost $0.25t$ dollars.
Since Kevin can spend no more than $2 we use a nonstrict inequality.

$0.2s + 0.25t \leq 2$

(b) Let $s = 6$ and $t = 3$.

$0.2(6) + 0.25(3) \leq 2$?

$1.2 + 0.75 \leq 2$?

$1.95 \leq 2$ True

Since $1.95 \leq 2$ is a true statement, Kevin can buy 6 suckers and 3 taffy sticks.

(c) Let $s = 5$ and $t = 5$.

$0.2(5) + 0.25(5) \leq 2$?

$1 + 1.25 \leq 2$?

$2.25 \leq 2$ False

Since $2.25 \leq 2$ is a false statement, Kevin cannot buy 5 suckers and 5 taffy sticks.

3.7 Exercises

15. $y > -x + 2$

$A(2, 4)$: $4 > -2 + 2$?

$4 > 0$ True

$B(3, -6)$: $-6 > 3 + 2$?

$-6 > 5$ False

$C(0, 0)$: $0 > -0 + 2$?

$0 > 2$ False

$A(2, 4)$ is a solution.

17. $y \leq 3x - 1$

$A(-6, -15)$: $-15 \leq 3(-6) - 1$?

$-15 \leq -18 - 1$?

$-15 \leq -19$ False

$B(0, 0)$: $0 \leq 3(0) - 1$?

$0 \leq 0 - 1$?

$0 \leq -1$ False

$C(-1, -6)$: $-6 \leq 3(-1) - 1$?

$-6 \leq -3 - 1$?

$-6 \leq -4$ True

$C(-1, -6)$ is a solution.

19. $3x \geq 2y$

$A(-8, -12)$: $3(-8) \geq 2(-12)$?

$-24 \geq -24$ True

$B(3, 5)$: $3(3) \geq 2(5)$?

$9 \geq 10$ False

$C(-5, -8)$: $3(-5) \geq 2(-8)$?

$-15 \geq -16$ True

$A(-8, -12)$ and $C(-5, -8)$ are solutions.

21. $2x - 3y < -6$

$A(2, -1)$: $2(2) - 3(-1) < -6$?

$4 + 3 < -6$?

$7 < -6$ False

$B(4, 8)$: $2(4) - 3(8) < -6$?

$8 - 24 < -6$?

$-16 < -6$ True

$C(-3, 0)$: $2(-3) - 3(0) < -6$?

$-6 - 0 < -6$?

$-6 < -6$ False

$B(4, 8)$ is a solution.

23. $x \le 2$

A(7, 2): $7 \le 2$ False
B(2, 5): $2 \le 2$ True
C(4, 2): $4 \le 2$ False
B(2, 5) is a solution.

25. $y > -1$

A(−1, 1): $1 > -1$ True
B(3, −1): $-1 > -1$ False
C(4, −2): $-2 > -1$ False
A(−1, 1) is a solution.

27. $y > 3x - 2$

Graph $y = 3x - 2$ with a dashed line.
Test (0, 0): $0 > 3(0) - 2$?
             $0 > -2$ True
Shade the half-plane containing (0, 0).

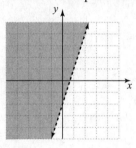

29. $y \le -x + 1$

Graph $y = -x + 1$ with a solid line.
Test (0, 0): $0 \le -0 + 1$?
          $0 \le 1$ True
Shade the half-plane containing (0, 0).

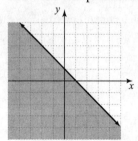

31. $y < \dfrac{x}{2}$

Graph $y = \dfrac{x}{2}$ with a dashed line.

Test (1, 1): $1 < \dfrac{1}{2}$ False
Shade the half-plane *not* containing (1, 1).

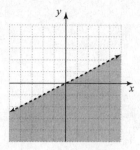

33. $y > 5$

Graph $y = 5$ with a dashed line.
Test (0, 0): $0 > 5$ False
Shade the half-plane *not* containing (0, 0).

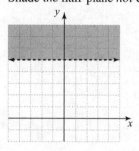

35. $y \le \dfrac{2}{5}x + 3$

Graph $y = \dfrac{2}{5}x + 3$ with a solid line.

Test (0, 0): $0 \le \dfrac{2}{5}(0) + 3$?
          $0 \le 3$ True
Shade the half-plane containing (0, 0).

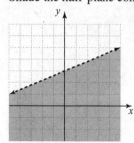

37. $y \ge -\dfrac{4}{3}x + 2$

Graph $y = -\dfrac{4}{3}x + 2$ with a solid line.

Test (0, 0): $0 \ge -\dfrac{4}{3}(0) + 2$?
          $0 \ge 2$ False
Shade the half-plane *not* containing (0, 0).

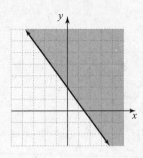

39. $x < 2$

Graph $x = 2$ with a dashed line.
Test $(0, 0)$: $0 < 2$ True
Shade the half-plane containing $(0, 0)$.

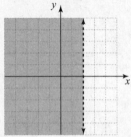

41. $3x - 4y < 12$

Graph $3x - 4y = 12$ with a dashed line.
Test $(0, 0)$: $3(0) - 4(0) < 12$?
$\qquad\qquad\quad 0 < 12$ True
Shade the half-plane containing $(0, 0)$.

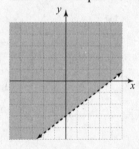

43. $2x + y \geq -4$

Graph $2x + y = 4$ with a solid line.
Test $(0, 0)$: $2(0) + 0 \geq -4$?
$\qquad\qquad\quad 0 \geq -4$ True
Shade the half-plane containing $(0, 0)$.

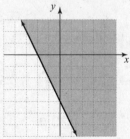

45. $x + y > 0$

Graph $x + y = 0$ with a dashed line.
Test $(1, 1)$: $1 + 1 > 0$?
$\qquad\qquad\quad 2 > 0$ True
Shade the half-plane containing $(1, 1)$.

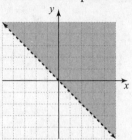

47. $5x - 2y < -8$

Graph $5x - 2y = -8$ with a dashed line.
Test $(0, 0)$: $5(0) - 2(0) < -8$?
$\qquad\qquad\qquad 0 < -8$ False
Shade the half-plane *not* containing $(0, 0)$.

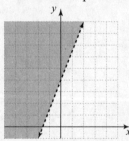

49. $x > -1$

Graph $x = -1$ with a dashed line.
Test $(0, 0)$: $0 > -1$ True
Shade the half-plane containing $(0, 0)$.

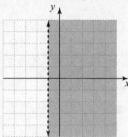

51. $y \le 4$

Graph $y = 4$ with a solid line.
Test (0, 0): $0 \le 4$ True
Shade the half-plane containing (0, 0).

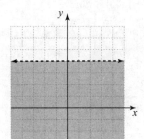

53. $\dfrac{x}{3} - \dfrac{y}{5} \ge 1$

Graph $\dfrac{x}{3} - \dfrac{y}{5} = 1$ with a solid line.

Test (0, 0): $\dfrac{0}{3} - \dfrac{0}{5} \ge 1$?

$\qquad 0 \ge 1$ False
Shade the half-plane *not* containing (0, 0).

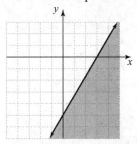

55. $-3 \ge x - y$

Graph $-3 = x - y$ with a solid line.
Test (0, 0): $-3 \ge 0 - 0$?

$\qquad -3 \ge 0$ False
Shade the half-plane *not* containing (0, 0).

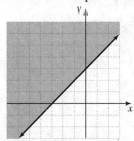

57. $x + y \ge 26$

Graph $x + y = 26$ with a solid line.
Test (0, 0): $0 + 0 \ge 26$?

$\qquad 0 \ge 26$ False
Shade the half-plane *not* containing (0, 0).

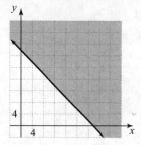

59. $\dfrac{y}{-2} \le 4$ or $y \ge -8$

Graph $y = -8$ with a solid line.
Test (0, 0): $0 \ge -8$ True
Shade the half-plane containing (0, 0).

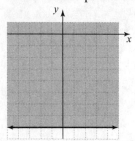

61. $x \le y - 3$

Graph $x = y - 3$ with a solid line.
Test (0, 0): $0 \le 0 - 3$?

$\qquad 0 \le -3$ False
Shade the half-plane *not* containing (0, 0).

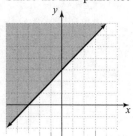

63. $x + 3y < 0$

Graph $x + 3y = 0$ with a dashed line.
Test (1, 1): $1 + 3(1) < 0$?

$\qquad 4 < 0$ False
Shade the half-plane *not* containing (1, 1).

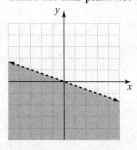

65. $2x - \dfrac{1}{2}y \geq 5$

Graph $2x - \dfrac{1}{2}y = 5$ with a solid line.

Test $(0, 0)$: $2(0) - \dfrac{1}{2}(0) \geq 5$?

$\qquad\qquad\qquad 0 \geq 5$ False

Shade the half-plane *not* containing $(0, 0)$.

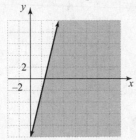

67. $x(-2) > -1$

$\qquad -2x > -1$

$\qquad\quad x < \dfrac{1}{2}$

Graph $x = \dfrac{1}{2}$ with a dashed line.

Test $(0, 0)$: $0 < \dfrac{1}{2}$ True

Shade the half-plane containing $(0, 0)$.

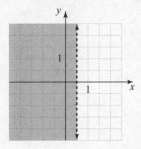

69. **(a)** Let s be the number of students and a the number of adults. The cost of s students is $3s$; the cost for a adults is $5a$.
$3s + 5a \leq 120$

(b) Test $s = 32$, $a = 6$:
$3(32) + 5(6) \leq 120$?
$\qquad 96 + 30 \leq 120$?
$\qquad\quad 126 \leq 120$ False
No, there is not enough money.

(c) Test $s = 29$, $a = 4$:
$3(29) + 5(4) \leq 120$?
$\qquad 87 + 20 \leq 120$?
$\qquad\quad 107 \leq 120$ True
Yes, there is enough money.

71. **(a)** Let s be the number of camping stoves and b be the number of sleeping bags he carries. The weight of s stoves is $3.1s$; the weight of b bags is $5.7b$.
$3.1s + 5.7b \leq 22$

(b) Test $s = 3$, $b = 2$:
$3.1(3) + 5.7(2) \leq 22$?
$\qquad 9.3 + 11.4 \leq 22$?
$\qquad\quad 20.7 \leq 22$ True
No, it will not be too heavy.

(c) Test $s = 2$, $b = 3$:
$3.1(2) + 5.7(3) \leq 22$?
$\qquad 6.2 + 17.1 \leq 22$?
$\qquad\quad 23.3 \leq 22$ False
Yes, it will be too heavy.

73. $3x - 2y > 6$ and $x + y < 2$

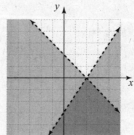

75. $y > \dfrac{3}{4}x - 1$ and $x \geq 0$

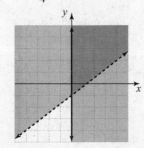

77. $x < -3$ and $y \le 4$

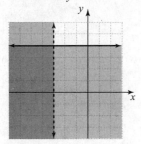

79. Answers may vary.

81. $y > 3$

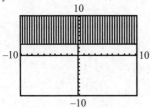

83. $y < 5x$

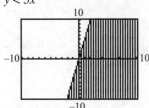

85. $y > 2x + 3$

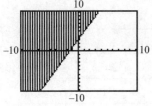

87. $y \le \dfrac{1}{2}x - 5$

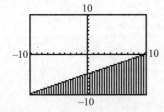

89. $3x + y \le 4$

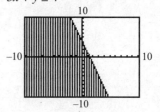

91. $2x + 5y \le -10$

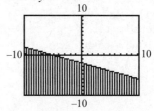

Chapter 3 Review

1. Plot $A(3, -2)$, which lies in quadrant IV.

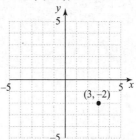

2. Plot $B(-1, -3)$, which lies in quadrant III.

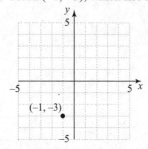

3. Plot $C(-4, 0)$, which lies on the x-axis.

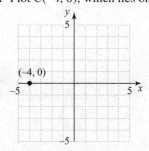

4. Plot $D(0, 2)$ which lies on the y-axis.

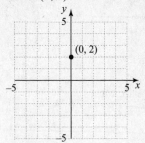

5. $A(1, 4)$ lies in quadrant I.
$B(-3, 0)$ lies on the x-axis.

6. $A(0, 1)$ lies on the y-axis.
$B(-2, 2)$ lies in quadrant II.

7. $y = 3x - 7$
$A(-1, -10)$
$-10 \overset{?}{=} 3(-1) - 7$
$-10 \overset{?}{=} -3 - 7$
$-10 = -10$
Yes
$B(-7, 0)$
$0 \overset{?}{=} 3(-7) - 7$
$0 \overset{?}{=} -21 - 7$
$0 \neq -28$
No

8. $4x - 3y = 2$
$A(2, 2)$
$4(2) - 3(2) \overset{?}{=} 2$
$8 - 6 \overset{?}{=} 2$
$2 = 2$
Yes
$B(6, 5)$
$4(6) - 3(5) \overset{?}{=} 2$
$24 - 15 \overset{?}{=} 2$
$9 \neq 2$
No

9. $-x = 3y$

(a) Let $x = 4$.
$-4 = 3y$
$-\dfrac{4}{3} = y$

Therefore, $\left(4, -\dfrac{4}{3}\right)$ satisfies the equation.

(b) Let $y = 2$.
$-x = 3(2)$
$-x = 6$
$x = -6$
Therefore, $(-6, 2)$ satisfies the equation.

10. $x - y = 0$

(a) Let $x = 4$.
$4 - y = 0$
$-y = -4$
$y = 4$
Therefore, $(4, 4)$ satisfies the equation.

(b) Let $y = -2$.
$x - (-2) = 0$
$x + 2 = 0$
$x = -2$
Therefore, $(-2, -2)$ satisfies the equation.

11. $3x - 2y = 10$ or $-2y = -3x + 10$ or $y = \dfrac{3}{2}x - 5$

x	$y = \frac{3}{2}x - 5$	(x, y)
-2	$\frac{3}{2}(-2) - 5 = -8$	$(-2, -8)$
0	$\frac{3}{2}(0) - 5 = -5$	$(0, -5)$
4	$\frac{3}{2}(4) - 5 = 1$	$(4, 1)$

12.

x	$y = -x + 2$	(x, y)
-3	$-(-3) + 2 = 5$	$(-3, 5)$
2	$-2 + 2 = 0$	$(2, 0)$
4	$-4 + 2 = -2$	$(4, -2)$

13. $y = -\dfrac{1}{3}x - 4$ or $\dfrac{1}{3}x = -y - 4$ or

$x = -3y - 12$

$x = -3y - 12$	$y = -\dfrac{1}{3}x - 4$	(x, y)
$-3(2) - 12 = -18$	2	$(-18, 2)$
-6	$-\dfrac{1}{3}(-6) - 4 = -2$	$(-6, -2)$
$-3(-12) - 12 = 24$	-12	$(24, -12)$

14. $3x - 2y = 7$ $3x - 2y = 7$
        $-2y = -3x + 7$ $3x = 2y + 7$
        $y = \dfrac{3}{2}x - \dfrac{7}{2}$ $x = \dfrac{2}{3}y + \dfrac{7}{3}$

$x = \dfrac{2}{3}y + \dfrac{7}{3}$	$y = \dfrac{3}{2}x - \dfrac{7}{2}$	(x, y)
$\dfrac{2}{3}(-8) + \dfrac{7}{3} = -3$	-8	$(-3, -8)$
3	$\dfrac{3}{2}(3) - \dfrac{7}{2} = 1$	$(3, 1)$
$\dfrac{2}{3}(4) + \dfrac{7}{3} = 5$	4	$(5, 4)$

15.

x	$C = 5 + 2x$	(x, C)
1	$5 + 2(1) = 7$	$(1, 7)$
2	$5 + 2(2) = 9$	$(2, 9)$
3	$5 + 2(3) = 11$	$(3, 11)$

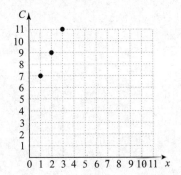

16.

x	$E = 1000 + 0.10$	(x, E)
500	$1000 + 0.10(500) = 1050$	$(500, 1050)$
1000	$1000 + 0.10(1000) = 1100$	$(1000, 1100)$
2000	$1000 + 0.10(2000) = 1200$	$(2000, 1200)$

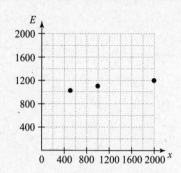

In Problems 17–20, points may vary.

17. $y = -2x$

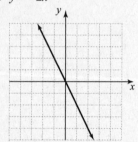

18. $y = x$

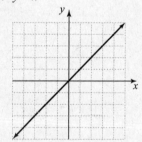

19. $4x + y = -2$

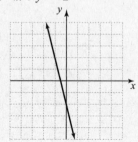

20. $3x - y = -1$

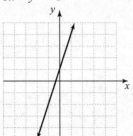

21.

p	$C = 40 + 2p$	(p, C)
20	$40 + 2(20) = 80$	(20, 80)
50	$40 + 2(50) = 140$	(50, 140)
80	$40 + 2(80) = 200$	(80, 200)

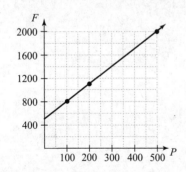

22.

p	$F = 500 + 3p$	(p, F)
100	$500 + 3(100) = 800$	(100, 800)
200	$500 + 3(200) = 1100$	(200, 1100)
500	$500 + 3(500) = 2000$	(500, 2000)

23. The graph crosses the *x*-axis at $(-2, 0)$ and the *y*-axis at $(0, -4)$. The intercepts are $(-2, 0)$ and $(0, -4)$.

24. The graph crosses the *y*-axis at $(0, 1)$ and does not cross the *x*-axis. The only intercept is $(0, 1)$.

25. $-3x + y = 9$

y-intercept: let $x = 0$.

$-3(0) + y = 9$

$\qquad y = 9$

x-intercept: let $y = 0$.

$-3x + 0 = 9$

$\qquad -3x = 9$

$\qquad\quad x = -3$

The intercepts are $(0, 9)$ and $(-3, 0)$.

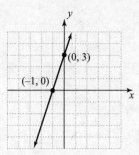

26. $y = 2x - 6$

y-intercept: let $x = 0$.

$y = 2(0) - 6$

$y = -6$

x-intercept: let $y = 0$.

$0 = 2x - 6$

$6 = 2x$

$3 = x$

The intercepts are $(0, -6)$ and $(3, 0)$.

27. $x = 3$ is the equation of a vertical line which has x-intercept $(3, 0)$ and no y-intercept.

28. $2x - 5y = 2$

y-intercept: let $x = 0$.

$2(0) - 5y = 2$

$\qquad -5y = 2$

$\qquad\quad y = -\dfrac{2}{5}$

x-intercept: let $y = 0$.

$2x - 5(0) = 2$

$\qquad 2x = 2$

$\qquad\ x = 1$

The intercepts are $\left(0, -\dfrac{2}{5}\right)$ and $(1, 0)$.

29. $y - 3x = 3$

y-intercept: lct $x = 0$.

$y - 3(0) = 3$

$\qquad y = 3$

$(0, 3)$

x-intercept: let $y = 0$.

$0 - 3x = 3$

$\quad -3x = 3$

$\qquad x = -1$

$(-1, 0)$

30. $2x + 5y = 0$

y-intercept: let $x = 0$.

$2(0) + 5y = 0$

$\qquad 5y = 0$

$\qquad\ y = 0$

$(0, 0)$

x-intercept: let $y = 0$.

$2x + 5(0) = 0$

$\qquad 2x = 0$

$\qquad\ x = 0$

$(0, 0)$

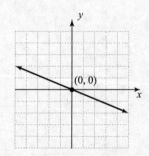

31. $\dfrac{x}{3} + \dfrac{y}{2} = 1$

y-intercept: let $x = 0$.

$\dfrac{0}{3} + \dfrac{y}{2} = 1$

$\qquad \dfrac{y}{2} = 1$

$\qquad\ y = 2$

$(0, 2)$

x-intercept: let $y = 0$.

$\dfrac{x}{3} + \dfrac{0}{2} = 1$

$\qquad \dfrac{x}{3} = 1$

$\qquad\ x = 3$

$(3, 0)$

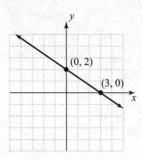

32. $y = -\dfrac{3}{4}x + 3$

y-intercept: let $x = 0$.

$y = -\dfrac{3}{4}(0) + 3$

$y = 3$

(0, 3)

x-intercept: let $y = 0$.

$0 = -\dfrac{3}{4}x + 3$

$-3 = -\dfrac{3}{4}x$

$4 = x$

(4, 0)

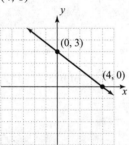

33. $x = -2$ is a vertical line with an *x*-intercept of (−2, 0).

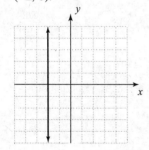

34. $y = 3$ is a horizontal line with a *y*-intercept of (0, 3).

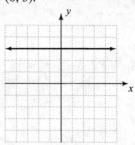

35. $y = -4$ is a horizontal line with a *y*-intercept of (0, −4).

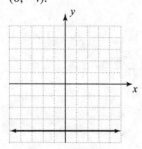

36. $x = 1$ is a vertical line with an *x*-intercept of (1, 0).

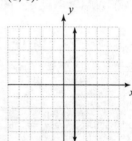

37. The line passes through the points (−3, 0) and (0, 4). Let $(-3, 0) = (x_1, y_1)$ and $(0, 4) = (x_2, y_2)$.

$$m = \frac{y_2 - y_1}{x_2 - x_1} = \frac{4 - 0}{0 - (-3)} = \frac{4}{3}$$

38. The line passes through the points (−4, 4) and (−1, −2). Let $(-4, 4) = (x_1, y_1)$ and $(-1, -2) = (x_2, y_2)$.

$$m = \frac{y_2 - y_1}{x_2 - x_1} = \frac{-2 - 4}{-1 - (-4)} = \frac{-6}{3} = -2$$

39. Let $(-4, 6) = (x_1, y_1)$ and $(-3, -2) = (x_2, y_2)$.

$$m = \frac{y_2 - y_1}{x_2 - x_1} = \frac{-2 - 6}{-3 - (-4)} = \frac{-8}{1} = -8$$

40. Let $(4, 1) = (x_1, y_1)$ and $(0, -7) = (x_2, y_2)$.

$$m = \frac{y_2 - y_1}{x_2 - x_1} = \frac{-7 - 1}{0 - 4} = \frac{-8}{-4} = 2$$

41. Let $\left(\frac{1}{2}, -\frac{3}{4}\right) = (x_1, y_1)$ and

$\left(\frac{5}{2}, -\frac{1}{4}\right) = (x_2, y_2)$.

$$m = \frac{y_2 - y_1}{x_2 - x_1} = \frac{-\frac{1}{4} - \left(-\frac{3}{4}\right)}{\frac{5}{2} - \frac{1}{2}} = \frac{-\frac{1}{4} + \frac{3}{4}}{\frac{4}{2}} = \frac{\frac{1}{2}}{2} = \frac{1}{4}$$

42. Let $\left(-\frac{1}{2}, \frac{2}{3}\right) = (x_1, y_1)$ and $\left(\frac{3}{2}, \frac{1}{3}\right) = (x_2, y_2)$.

$$m = \frac{y_2 - y_1}{x_2 - x_1} = \frac{\frac{1}{3} - \frac{2}{3}}{\frac{3}{2} - \left(-\frac{1}{2}\right)} = \frac{-\frac{1}{3}}{\frac{4}{2}} = \frac{-\frac{1}{3}}{2} = -\frac{1}{6}$$

43. Let $(-3, -6) = (x_1, y_1)$ and
$(-3, -10) = (x_2, y_2)$.

$$m = \frac{y_2 - y_1}{x_2 - x_1} = \frac{-10 - (-6)}{-3 - (-3)} = \frac{-10 + 6}{-3 + 3} = \frac{-4}{0} \text{ is}$$
undefined.

44. Let $(-5, -1) = (x_1, y_1)$ and $(-1, -1) = (x_2, y_2)$.

$$m = \frac{y_2 - y_1}{x_2 - x_1} = \frac{-1 - (-1)}{-1 - (-5)} = \frac{-1 + 1}{-1 + 5} = \frac{0}{4} = 0$$

45. Let $\left(\frac{3}{4}, \frac{1}{2}\right) = (x_1, y_1)$ and $\left(-\frac{1}{4}, \frac{1}{2}\right) = (x_2, y_2)$.

$$m = \frac{y_2 - y_1}{x_2 - x_1} = \frac{\frac{1}{2} - \frac{1}{2}}{-\frac{1}{4} - \frac{3}{4}} = \frac{0}{-1} = 0$$

46. Let $\left(\frac{1}{3}, -\frac{3}{5}\right) = (x_1, y_1)$ and

$\left(\frac{3}{9}, -\frac{1}{5}\right) = (x_2, y_2)$.

$$m = \frac{y_2 - y_1}{x_2 - x_1} = \frac{-\frac{1}{5} - \left(-\frac{3}{5}\right)}{\frac{3}{9} - \frac{1}{3}} = \frac{-\frac{1}{5} + \frac{3}{5}}{\frac{1}{3} - \frac{1}{3}} = \frac{\frac{2}{5}}{0} \text{ is}$$
undefined.

47. $(-2, -3)$; $m = 4$

Plot $(-2, -3)$. Since $m = 4 = \frac{4}{1}$, there is a

4-unit increase in y for every 1-unit increase in x.

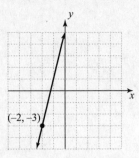

48. $(1, -3)$; $m = -2$

Plot $(1, -3)$. Since $m = -2 = \frac{-2}{1}$, there is a 2-

unit decrease in y for every 1-unit increase in x.

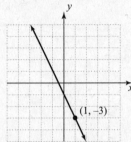

49. $(0, 1)$; $m = -\frac{2}{3}$

Plot $(0, 1)$. Since $m = -\frac{2}{3} = \frac{-2}{3}$, there is a 2-unit

decrease in y for every 3-unit increase in x.

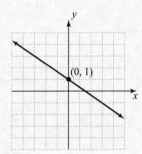

50. $(2, 3)$; $m = 0$

Plot $(2, 3)$. Since $m = 0$, the line is horizontal.

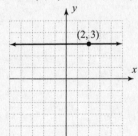

51. Let $(x_1, y_1) = (20, 1400)$ and
$(x_2, y_2) = (50, 2750)$.

$$m = \frac{y_2 - y_1}{x_2 - x_1} = \frac{2750 - 1400}{50 - 20} = \frac{1350}{30} = 45$$

The cost to produce 1 additional bicycle is $45.

52. $\text{grade} = \text{slope} = \dfrac{\text{rise}}{\text{run}} = \dfrac{-5}{100} = -\dfrac{1}{20} = -0.05$

or 5%

53. $y = -x + \dfrac{1}{2}$

slope: $m = -1$

y-intercept: $b = \dfrac{1}{2}$

54. $y = x - \dfrac{3}{2}$

slope: $m = 1$

y-intercept: $b = -\dfrac{3}{2}$

55. $3x - 4y = -4$

$\qquad -4y = -3x - 4$

$\qquad\quad y = \dfrac{3}{4}x + 1$

slope: $m = \dfrac{3}{4}$

y-intercept: $b = 1$

56. $2x + 5y = 8$

$\qquad 5y = -2x + 8$

$\qquad\ y = -\dfrac{2}{5}x + \dfrac{8}{5}$

slope: $m = -\dfrac{2}{5}$

y-intercept: $b = \dfrac{8}{5}$

57. $y = \dfrac{1}{3}x + 1$

slope: $m = \dfrac{1}{3}$

y-intercept: $b = 1$

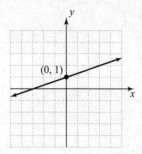

58. $y = -\dfrac{x}{2} - 1$

slope: $m = -\dfrac{1}{2}$

y-intercept: $b = -1$

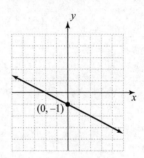

59. $y = -\dfrac{2x}{3} - 2$

slope: $m = -\dfrac{2}{3}$

y-intercept: $b = -2$

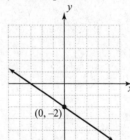

60. $y = \dfrac{3x}{4} + 3$

slope: $m = \dfrac{3}{4}$

y-intercept: $b = 3$

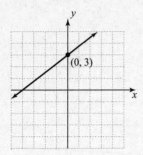

61. $y = x$
slope: $m = 1$
y-intercept: $b = 0$

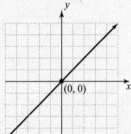

62. $y = -2x$
slope: $m = -2$
y-intercept: $b = 0$

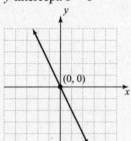

63. $2x - y = -4$
$-y = -2x - 4$
$y = 2x + 4$
slope: $m = 2$
y-intercept: $b = 4$

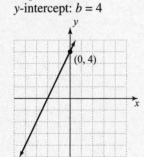

64. $-4x + 2y = 2$
$2y = 4x + 2$
$y = 2x + 1$
slope: $m = 2$
y-intercept: $b = 1$

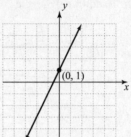

65. $m = -\dfrac{3}{4}$; $b = \dfrac{2}{3}$
$y = mx + b$
$y = -\dfrac{3}{4}x + \dfrac{2}{3}$

66. $m = \dfrac{1}{5}$; $b = 10$
$y = mx + b$
$y = \dfrac{1}{5}x + 10$

67. m undefined; x-intercept $= -12$
This is a vertical line.
$x = -12$

68. $m = 0$; $b = -4$
$y = mx + b$
$y = 0x - 4$
$y = -4$

69. $m = 1$; $b = -20$
$y = mx + b$
$y = 1x - 20$
$y = x - 20$

70. $m = -1$; $b = -8$
$y = mx + b$
$y = -1x - 8$
$y = -x - 8$

71. $C = 120 + 80d$

(a) Let $d = 3$.
$C = 120 + 80(3) = 120 + 240 = 360$
It will cost \$360 to rent for 3 days.

(b) Let $C = 680$.
$$680 = 120 + 80d$$
$$560 = 80d$$
$$7 = d$$
The car was rented for 7 days.

(c)

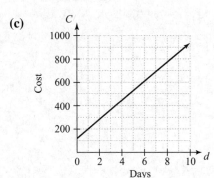

72. (a) $d = 22$ days, $C = \$418$: $(22, 418)$
$d = 35$ days, $C = \$665$: $(35, 665)$
$$m = \frac{665 - 418}{35 - 22} = \frac{247}{13} = \frac{19}{1} \text{ or } 19$$

(b) It costs \$19 more for each additional day.

(c) $C = 19d$

(d) Let $d = 8$.
$$C = 19(8) = 152$$
It will cost \$152 to rent for 8 days.

73. $(0, -3)$; slope $= 6$
$$y - y_1 = m(x - x_1)$$
$$y - (-3) = 6(x - 0)$$
$$y + 3 = 6x$$
$$y = 6x - 3$$

74. $(4, 0)$; slope $= -2$
$$y - y_1 = m(x - x_1)$$
$$y - 0 = -2(x - 4)$$
$$y = -2x + 8$$

75. $(3, -1)$; slope $= -\dfrac{1}{2}$
$$y - y_1 = m(x - x_1)$$
$$y - (-1) = -\frac{1}{2}(x - 3)$$
$$y + 1 = -\frac{1}{2}x + \frac{3}{2}$$
$$y = -\frac{1}{2}x + \frac{1}{2}$$

76. $(-1, -3)$; slope $= \dfrac{2}{3}$
$$y - y_1 = m(x - x_1)$$
$$y - (-3) = \frac{2}{3}[x - (-1)]$$
$$y + 3 = \frac{2}{3}(x + 1)$$
$$y + 3 = \frac{2}{3}x + \frac{2}{3}$$
$$y = \frac{2}{3}x - \frac{7}{3}$$

77. $\left(-\dfrac{4}{3}, -\dfrac{1}{2}\right)$; horizontal

A horizontal line has slope 0. The horizontal line through $\left(-\dfrac{4}{3}, -\dfrac{1}{2}\right)$ has equation $y = -\dfrac{1}{2}$.

78. $\left(-\dfrac{4}{7}, \dfrac{8}{5}\right)$; vertical

A vertical line has undefined slope. The vertical line through $\left(-\dfrac{4}{7}, \dfrac{8}{5}\right)$ has equation $x = -\dfrac{4}{7}$.

79. $(-5, 2)$; slope is undefined.
A line with undefined slope is vertical. The vertical line through $(-5, 2)$ has equation $x = -5$.

80. $(6, 0)$; slope $= 0$
A line with 0 slope is horizontal. The horizontal line through $(6, 0)$ has equation $y = 0$.

81. $(x_1, y_1) = (-7, 0)$ and $(x_2, y_2) = (0, 8)$
$$m = \frac{y_2 - y_1}{x_2 - x_1} = \frac{8 - 0}{0 - (-7)} = \frac{0}{7}$$
$$y - y_1 = m(x - x_1)$$
$$y - 0 = \frac{8}{7}[x - (-7)]$$
$$y = \frac{8}{7}(x + 7)$$
$$y = \frac{8}{7}x + 8$$

82. $(x_1, y_1) = (0, -6)$ and $(x_2, y_2) = (4, 0)$

$$m = \frac{y_2 - y_1}{x_2 - x_1} = \frac{0 - (-6)}{4 - 0} = \frac{6}{4} = \frac{3}{2}$$

$$y - y_1 = m(x - x_1)$$

$$y - (-6) = \frac{3}{2}(x - 0)$$

$$y + 6 = \frac{3}{2}x$$

$$y = \frac{3}{2}x - 6$$

83. $(x_1, y_1) = (3, 5)$ and $(x_2, y_2) = (-2, -10)$

$$m = \frac{y_2 - y_1}{x_2 - x_1} = \frac{-10 - 5}{-2 - 3} = \frac{-15}{-5} = 3$$

$$y - y_1 = m(x - x_1)$$

$$y - 5 = 3(x - 3)$$

$$y - 5 = 3x - 9$$

$$y = 3x - 4$$

84. $(x_1, y_1) = (-15, 1)$ and $(x_2, y_2) = (-5, -3)$

$$m = \frac{y_2 - y_1}{x_2 - x_1} = \frac{-3 - 1}{-5 - (-15)}$$

$$= \frac{-4}{-5 + 15}$$

$$= \frac{-4}{10}$$

$$= -\frac{2}{5}$$

$$y - y_1 = m(x - x_1)$$

$$y - 1 = -\frac{2}{5}[x - (-15)]$$

$$y - 1 = -\frac{2}{5}(x + 15)$$

$$y - 1 = -\frac{2}{5}x - 6$$

$$y = -\frac{2}{5}x - 5$$

85. $(x_1, y_1) = (3, 12)$ and $(x_2, y_2) = (5, 4)$

$$m = \frac{y_2 - y_1}{x_2 - x_1} = \frac{4 - 12}{5 - 3} = \frac{-8}{2} = -4$$

$$y - y_1 = m(x - x_1)$$

$$y - 12 = -4(x - 3)$$

$$y - 12 = -4x + 12$$

$$y = -4x + 24$$

$$A = -4d + 24$$

86. $(x_1, y_1) = (0, 15)$ and $(x_2, y_2) = (6, 13)$

$$m = \frac{y_2 - y_1}{x_2 - x_1} = \frac{13 - 15}{6 - 0} = \frac{-2}{6} = -\frac{1}{3}$$

$$y - y_1 = m(x - x_1)$$

$$y - 15 = -\frac{1}{3}(x - 0)$$

$$y - 15 = -\frac{1}{3}x$$

$$y = -\frac{1}{3}x + 15$$

$$F = -\frac{1}{3}m + 15$$

87. $y = -\frac{1}{3}x + 2$

$$m_1 = -\frac{1}{3}$$

$$x - 3y = 3 \text{ or } y = \frac{1}{3}x - 1$$

$$m_2 = \frac{1}{3}$$

Since $m_1 \neq m_2$, the lines are not parallel.

88. $y = \frac{1}{2}x - 4$

$$m_1 = \frac{1}{2}$$

$$x - 2y = 6 \text{ or } y = \frac{1}{2}x - 3$$

$$m_2 = \frac{1}{2}$$

Since $m_1 = m_2$, the lines are parallel.

89. $(3, -1); y = -x + 5$

The slope of the line is -1. A parallel line also has slope -1.

$$y - y_1 = m(x - x_1)$$

$$y - (-1) = -1(x - 3)$$

$$y + 1 = -x + 3$$

$$y = -x + 2$$

90. $(-2, 4); y = 2x - 1$

The slope of the line is 2. A parallel line also has slope 2.

$$y - y_1 = m(x - x_1)$$
$$y - 4 = 2[x - (-2)]$$
$$y - 4 = 2(x + 2)$$
$$y - 4 = 2x + 4$$
$$y = 2x + 8$$

91. $(-1, 10); 3x + y = -7$ or $y = -3x - 7$

The slope of the line is -3. A parallel line also has slope -3.

$$y - y_1 = m(x - x_1)$$
$$y - 10 = -3[x - (-1)]$$
$$y - 10 = -3(x + 1)$$
$$y - 10 = -3x - 3$$
$$y = -3x + 7$$

92. $(4, -5); 6x + 2y = 5$ or $y = -3x + \dfrac{5}{2}$

The slope of the line is -3. A parallel line also has slope -3.

$$y - y_1 = m(x - x_1)$$
$$y - (-5) = -3(x - 4)$$
$$y + 5 = -3x + 12$$
$$y = -3x + 7$$

93. $(5, 19); y$-axis

The y-axis is vertical. A parallel line will also be vertical. The vertical line through $(5, 19)$ has equation $x = 5$.

94. $(-1, -12); x$-axis

The x-axis is horizontal. A parallel line will also be horizontal. The horizontal line through $(-1, -12)$ has equation $y = -12$.

95. $3x - 2y = 5$

$$-2y = -3x + 5$$
$$y = \frac{3}{2}x - \frac{5}{2}$$

The slope of the line is $\dfrac{3}{2}$. The slope of a perpendicular line is $\dfrac{-1}{\frac{3}{2}} = -\dfrac{2}{3}$.

96. $4x - 9y = 1$

$$-9y = -4x + 1$$
$$y = \frac{4}{9}x - \frac{1}{9}$$

The slope of the line is $\dfrac{4}{9}$. The slope of a perpendicular line is $\dfrac{-1}{\frac{4}{9}} = -\dfrac{9}{4}$.

97. $x + 3y = 3$

$$3y = -x + 3$$
$$y = -\frac{1}{3}x + 1$$

$$m_1 = -\frac{1}{3}$$

$$y = 3x + 1$$
$$m_2 = 3$$

Since $m_1 m_2 = \left(-\dfrac{1}{3}\right)(3) = -1,$ the lines are perpendicular.

98. $5x - 2y = 2$

$$-2y = -5x + 2$$
$$y = \frac{5}{2}x - 1$$

$$m_1 = \frac{5}{2}$$

$$y = \frac{2}{5}x + 12$$

$$m_2 = \frac{2}{5}$$

Since $m_1 m_2 = \left(\dfrac{5}{2}\right)\left(\dfrac{2}{5}\right) = 1 \neq -1,$ the lines are not perpendicular.

99. $(-3, 4); y = -3x + 1$

The slope of the line is -3. The slope of a perpendicular line is $\dfrac{-1}{-3} = \dfrac{1}{3}$.

$$y - y_1 = m(x - x_1)$$
$$y - 4 = \frac{1}{3}[x - (-3)]$$
$$y - 4 = \frac{1}{3}(x + 3)$$
$$y - 4 = \frac{1}{3}x + 1$$
$$y = \frac{1}{3}x + 5$$

100. $(4, -1); \; y = 2x - 1$
The slope of the line is 2. The slope of a
perpendicular line is $\dfrac{-1}{2}$.

$$y - y_1 = m(x - x_1)$$
$$y - (-1) = -\frac{1}{2}(x - 4)$$
$$y + 1 = -\frac{1}{2}x + 2$$
$$y = -\frac{1}{2}x + 1$$

101. $(1, -3); \; 2x - 3y = 6$
$$-3y = -2x + 6$$
$$y = \frac{2}{3}x - 2$$

The slope of the line is $\dfrac{2}{3}$. The slope of a

perpendicular line is $\dfrac{-1}{\frac{2}{3}} = -\dfrac{3}{2}$.

$$y - y_1 = m(x - x_1)$$
$$y - (-3) = -\frac{3}{2}(x - 1)$$
$$y + 3 = -\frac{3}{2}x + \frac{3}{2}$$
$$y = -\frac{3}{2}x - \frac{3}{2}$$

102. $\left(-\dfrac{3}{5}, \dfrac{2}{5}\right); \; x + y = -7$
$$y = -x - 7$$
The slope of the line is -1. The slope of a
perpendicular line is $\dfrac{-1}{-1} = 1$.

$$y - y_1 = m(x - x_1)$$
$$y - \frac{2}{5} = 1\left[x - \left(-\frac{3}{5}\right)\right]$$
$$y - \frac{2}{5} = x + \frac{3}{5}$$
$$y = x + 1$$

103. $y \le 3x + 4$
$A(2, 0): \; 0 \le 3(2) + 4?$
$0 \le 6 + 4?$
$0 \le 10$ True
$B(-4, -8): \; -8 \le 3(-4) + 4$
$-8 \le -12 + 4$
$-8 \le -8$ True

$C(7, 26): \; 26 \le 3(7) + 4?$
$26 \le 21 + 4?$
$26 \le 25$ False
$A(2, 0)$ and $B(-4, -8)$ are solutions.

104. $y > \dfrac{1}{3}x + 4$

$A(6, -2): \; -2 > \dfrac{1}{3}(6) + 4?$
$-2 > 2 + 4?$
$-2 > 6$ False

$B(0, 4): \; 4 > \dfrac{1}{3}(0) + 4?$
$4 > 4$ False

$C(-18, -1): \; -1 > \dfrac{1}{3}(-18) + 4?$
$-1 > -6 + 4?$
$-1 > -2$ True
$C(-18, -1)$ is a solution.

105. $y < -\dfrac{1}{4}x + 2$

Graph $y = -\dfrac{1}{4}x + 2$ with a dashed line.

Test $(0, 0): \; 0 < -\dfrac{1}{4}(0) + 2?$
$0 < 2$ True
Shade the half-plane containing $(0, 0)$.

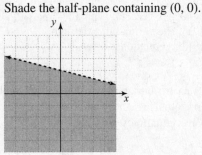

106. $y > 2x - 1$
Graph the line $y = 2x - 1$ with a dashed line.
Test $(0, 0): \; 0 > 2(0) - 1?$
$0 > -1$ True
Shade the half-plane containing $(0, 0)$.

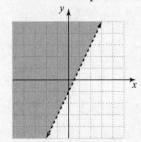

107. $3x + 2y \geq -6$

Graph the line $3x + 2y = -6$ with a solid line.

Test $(0, 0)$: $3(0) + 2(0) \geq -6$

$\qquad\qquad 0 \geq -6$ True

Shade the half-plane containing $(0, 0)$.

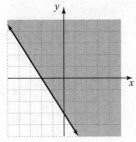

108. $-2x + y \geq 4$

Graph the line $-2x + y = 4$ with a solid line.

Test $(0, 0)$: $-2(0) + 0 \geq 4$?

$\qquad\qquad 0 \geq 4$ False

Shade the half-plane *not* containing $(0, 0)$.

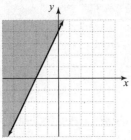

109. $x - 3y \leq 0$

Graph the line $x - 3y = 0$ with a solid line.

Test $(0, 1)$: $0 - 3(1) \leq 0$?

$\qquad\qquad -3 \leq 0$ True

Shade the half-plane containing $(0, 1)$.

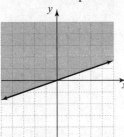

110. $x - 4y \geq 4$

Graph $x - 4y = 4$ with a solid line.

Test $(0, 0)$: $0 - 4(0) \geq 4$?

$\qquad\qquad 0 \geq 4$ False

Shade the half-plane *not* containing $(0, 0)$.

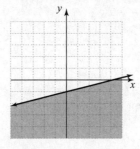

111. $x < -3$

Graph $x = -3$ with a dashed line.

Test $(0, 0)$: $0 < -3$ False

Shade the half-plane *not* containing $(0, 0)$.

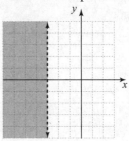

112. $y > 2$

Graph $y = 2$ with a dashed line.

Test $(0, 0)$: $0 > 2$ False

Shade the half-plane *not* containing $(0, 0)$.

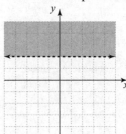

113. Each quarter is worth \$0.25, so x quarters are worth $0.25x$. Each dime is worth \$0.10, so y dimes are worth $0.1y$.

$0.25x + 0.1y \geq 12$

114. $2x - \dfrac{1}{2}y \leq 10$

Chapter 3 Test

1. $3x - 4y = -17$; $(-3, -2)$

$3(-3) - 4(-2) \stackrel{?}{=} -17$

$-9 + 8 \stackrel{?}{=} -17$

$-1 = -17$ False

$(-3, -2)$ is not a solution to the equation.

2. $3x - 9y = 12$

(a) Let $y = 0$.
$$3x - 9(0) = 12$$
$$3x = 12$$
$$x = 4$$
The x-intercept is $(4, 0)$.

(b) Let $x = 0$.
$$3(0) - 9y = 12$$
$$-9y = 12$$
$$y = -\frac{4}{3}$$
The y-intercept is $\left(0, -\frac{4}{3}\right)$.

3. $4x - 3y = -24$
$$-3y = -4x - 24$$
$$y = \frac{4}{3}x + 8$$
$$y = mx + b$$

(a) The slope is $m = \frac{4}{3}$.

(b) The y-intercept is $b = 8$, or $(0, 8)$.

4. $y = -\frac{3}{4}x + 2$

The slope is $m = -\frac{3}{4}$ and the y-intercept is $b = 2$.

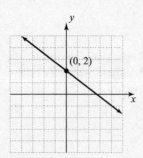

5. $3x - 6y = -12$
$$-6y = -3x - 12$$
$$y = \frac{1}{2}x + 2$$

The slope is $m = \frac{1}{2}$ and the y-intercept is $b = 2$.

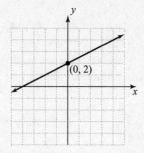

6. Let $(2, -2) = (x_1, y_1)$ and $(-4, -1) = (x_2, y_2)$.
$$m = \frac{y_2 - y_1}{x_2 - x_1} = \frac{-1 - (-2)}{-4 - 2} = \frac{-1 + 2}{-6} = -\frac{1}{6}$$

7. $3y = 2x - 1$
$$y = \frac{2}{3}x - \frac{1}{3}$$
$$y = mx + b$$

The slope of the line is $m = \frac{2}{3}$.

(a) A line perpendicular to the given line has slope $\dfrac{-1}{\frac{2}{3}} = -\dfrac{3}{2}$.

(b) A line parallel to the given line has the same slope, $\dfrac{2}{3}$.

8. L_1: $3x - 7y = 2$
$$-7y = -3x + 2$$
$$y = \frac{3}{7}x - \frac{2}{7}$$

The line has slope $m_1 = \frac{3}{7}$.

L_2: $y = \frac{7}{3}x + 4$

The line has slope $m_2 = \frac{7}{3}$.

Since $m_1 \neq m_2$ and $m_1 m_2 \neq -1$, the lines are neither parallel nor perpendicular.

9. slope $= -4$ and y-intercept is -15
$m = -4$ and $b = -15$
$y = mx + b$
$y = -4x - 15$

10. slope = 2 and contains $(-3, 8)$

$$y - y_1 = m(x - x_1)$$
$$y - 8 = 2[x - (-3)]$$
$$y - 8 = 2(x + 3)$$
$$y - 8 = 2x + 6$$
$$y = 2x + 14$$

11. $(x_1, y_1) = (-3, -2)$ and $(x_2, y_2) = (-4, 1)$

$$m = \frac{y_2 - y_1}{x_2 - x_1} = \frac{1 - (-2)}{-4 - (-3)}$$
$$= \frac{1 + 2}{-4 + 3}$$
$$= \frac{3}{-1}$$
$$= -3$$

$$y - y_1 = m(x - x_1)$$
$$y - (-2) = -3[x - (-3)]$$
$$y + 2 = -3(x + 3)$$
$$y + 2 = -3x - 9$$
$$y = -3x - 11$$

12. $y = \frac{1}{2}x + 2$ has slope $m = \frac{1}{2}$.

A parallel line also has slope $m = \frac{1}{2}$.

Let $(x_1, y_1) = (4, 0)$.

$$y - y_1 = m(x - x_1)$$
$$y - 0 = \frac{1}{2}(x - 4)$$
$$y = \frac{1}{2}x - 2$$

13. $4x - 6y = 5$

$$-6y = -4x + 5$$
$$y = \frac{2}{3}x - \frac{5}{6}$$

This line has slope $m = \frac{2}{3}$. A perpendicular line

has slope $m = \frac{-1}{\frac{2}{3}} = -\frac{3}{2}$.

Let $(x_1, y_1) = (4, 2)$.

$$y - y_1 = m(x - x_1)$$
$$y - 2 = -\frac{3}{2}(x - 4)$$
$$y - 2 = -\frac{3}{2}x + 6$$
$$y = -\frac{3}{2}x + 8$$

14. A horizontal line has slope $m = 0$.
The horizontal line through $(3, 5)$ has equation $y = 5$.

15. A line with undefined slope is vertical. The vertical line through $(-2, -1)$ has equation $x = -2$.

16. Let $(x_1, y_1) = (20, 560)$ and
$(x_2, y_2) = (30, 640)$.

$$m = \frac{y_2 - y_1}{x_2 - x_1} = \frac{640 - 560}{30 - 20} = \frac{80}{10} = \frac{8}{1} \text{ or } 8$$

The average rate to ship a package is \$8.

17. $y \geq x - 3$
Graph $y = x - 3$ with a solid line.
Test $(0, 0)$: $0 \geq 0 - 3$
$\qquad\qquad\quad 0 \geq -3$ True
Shade the half-plane containing $(0, 0)$.

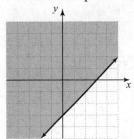

18. $-2x - 4y < 8$
Graph $-2x - 4y = 8$ with a dashed line.
Test $(0, 0)$: $-2(0) - 4(0) < 8$?
$\qquad\qquad\qquad\qquad 0 < 8$ True
Shade the half-plane containing $(0, 0)$.

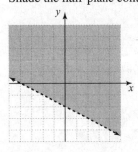

19. $x \leq -4$
Graph $x = -4$ with a solid line.
Test $(0, 0)$: $0 \leq -4$ False
Shade the half-plane *not* containing $(0, 0)$.

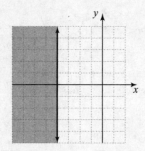

Cumulative Review Chapters 1–3

1. $200 \div 25 \cdot (-2) = 8 \cdot (-2) = -16$

2. $\dfrac{3}{4} + \dfrac{1}{6} - \dfrac{2}{3} = \dfrac{9}{12} + \dfrac{2}{12} - \dfrac{8}{12} = \dfrac{9+2-8}{12} = \dfrac{3}{12} = \dfrac{1}{4}$

3. $\dfrac{8-3\left(5-3^2\right)}{7-2\cdot 6} = \dfrac{8-3(5-9)}{7-12}$

   $= \dfrac{8-3(-4)}{-5}$

   $= \dfrac{8+12}{-5}$

   $= \dfrac{20}{-5}$

   $= -4$

4. $(-3)^3 + 3(-3)^2 - 5(-3) - 7$

   $= -27 + 3(9) - 5(-3) - 7$

   $= -27 + 27 + 15 - 7$

   $= 0 + 15 - 7$

   $= 15 - 7$

   $= 8$

5. $8m - 5m^2 - 3 + 9m^2 - 3m - 6$

   $= -5m^2 + 9m^2 + 8m - 3m - 3 - 6$

   $= 4m^2 + 5m - 9$

6. $8(n+2) - 7 = 6n - 5$

   $8n + 16 - 7 = 6n - 5$

   $8n + 9 = 6n - 5$

   $8n = 6n - 14$

   $2n = -14$

   $n = -7$

   The solution set is $\{-7\}$.

7. $\dfrac{2}{5}x + \dfrac{1}{6} = -\dfrac{2}{3}$

   $30\left(\dfrac{2}{5}x + \dfrac{1}{6}\right) = 30\left(-\dfrac{2}{3}\right)$

   $12x + 5 = -20$

   $12x = -25$

   $x = \dfrac{-25}{12}$

   The solution set is $\left\{-\dfrac{25}{12}\right\}$.

8. $A = \dfrac{1}{2}h(b+B)$

   $2(A) = 2\left(\dfrac{1}{2}h(b+B)\right)$

   $2A = h(b+B)$

   $2A = hb + hB$

   $2A - hb = hB$

   $\dfrac{2A - hb}{h} = B$

   $B = \dfrac{2A-hb}{h}$ or $B = \dfrac{2A}{h} - b$

9. $6x - 7 > -31$

   $6x > -24$

   $x > -4$

   $\{x \mid x > -4\}$ or $(-4, \infty)$

   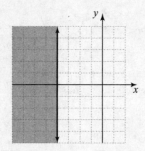

10. $5(x-3) \geq 7(x-4) + 3$

    $5x - 15 \geq 7x - 28 + 3$

    $5x - 15 \geq 7x - 25$

    $5x \geq 7x - 10$

    $-2x \geq -10$

    $x \leq 5$

    $\{x \mid x \leq 5\}$ or $(-\infty, 5]$

11.

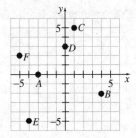

12. For $y = -\dfrac{1}{2}x + 4$, the slope is $-\dfrac{1}{2}$ and the
y-intercept is 4. Begin at $(0, 4)$ and move to the
right 2 units and down 1 units to find the point
$(2, 3)$. We can also move 2 units to the left and 1
units up to find the point $(-2, 5)$.

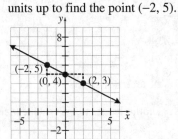

13. $4x - 5y = 15$

$$-5y = -4x + 15$$

$$y = \frac{-4x + 15}{-5}$$

$$y = \frac{4}{5}x - 3$$

The slope is $\dfrac{4}{5}$ and the y-intercept is -3. Begin

at the point $(0, -3)$ and move to the right 5 units
and up 4 units to find the point $(5, 1)$.

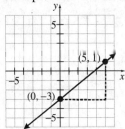

14. $m = \dfrac{y_2 - y_1}{x_2 - x_1} = \dfrac{10 - (-2)}{-6 - 3} = \dfrac{12}{-9} = -\dfrac{4}{3}$

$$y - y_1 = m(x - x_1)$$

$$y - 10 = -\frac{4}{3}(x - (-6))$$

$$y - 10 = -\frac{4}{3}(x + 6)$$

$$y - 10 = -\frac{4}{3}x - 8$$

$$y = -\frac{4}{3}x + 2 \quad \text{or} \quad 4x + 3y = 6$$

15. The slope of the line we seek is $m = -3$, the
same as the slope of the line $y = -3x + 10$.
Thus, the equation of the line we seek is:

$$y - 7 = -3(x - (-5))$$

$$y - 7 = -3(x + 5)$$

$$y - 7 = -3x - 15$$

$$y = -3x - 8 \quad \text{or} \quad 3x + y = -8$$

16. Replace the inequality symbol with an equal sign
to obtain $x - 3y = 12$. Because the inequality is

strict, graph $x - 3y = 12$ $\left(y = \dfrac{1}{3}x - 4 \right)$ using a

dashed line.

Test Point: $(0, 0)$: $0 - 3(0) \overset{?}{>} 12$

$$0 \overset{?}{>} 12 \quad \text{False}$$

Therefore, $(0, 0)$ is a not a solution to
$x - 3y > 12$. Shade the half-plane that does not
contain $(0, 0)$.

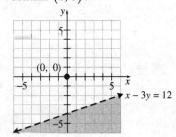

17. Let x = score on final exam.

$$93 \le \frac{94+95+90+97+2x}{6} \le 100$$

$$93 \le \frac{376+2x}{6} \le 100$$

$$6(93) \le 6\left(\frac{376+2x}{6}\right) \le 6(100)$$

$$558 \le 376+2x \le 600$$

$$558-376 \le 376+2x-376 \le 600-376$$

$$182 \le 2x \le 224$$

$$\frac{182}{2} \le \frac{2x}{2} \le \frac{224}{2}$$

$$91 \le x \le 112$$

Shawn needs to score at least 91 on the final exam to earn an A (assuming the maximum score on the exam is 100).

18. Let x = weight in pounds.

$$0.2x-2 \ge 30$$

$$0.2x-2+2 \ge 30+2$$

$$0.2x \ge 32$$

$$\frac{0.2x}{0.2} \ge \frac{32}{0.2}$$

$$x \ge 160$$

A person 62 inches tall would be considered obese if they weighed 160 pounds or more.

19. Let x = measure of the smaller angle.

Larger angle: $15+2x$

$$x+(15+2x)=180$$

$$3x+15=180$$

$$3x+15-15=180-15$$

$$3x=165$$

$$\frac{3x}{3}=\frac{165}{3}$$

$$x=55$$

The angles measure $55°$ and $125°$.

20. Let h = height of cylinder in inches.

$$S=2\pi r^2+2\pi rh$$

$$100=2\pi(2)^2+2\pi(2)h$$

$$100=8\pi+4\pi h$$

$$100-8\pi=4\pi h$$

$$h=\frac{100-8\pi}{4\pi}\approx 5.96$$

The cylinder should be about 5.96 inches tall.

21. Let x = the first even integer.

$x+2$ = the second even integer:

$x+4$ = the third even integer:

$$x+(x+2)=22+(x+4)$$

$$2x+2=x+26$$

$$2x+2-x=x+26-x$$

$$x+2=26$$

$$x+2-2=26-2$$

$$x=24$$

The three consecutive even integers are 24, 26, and 28.

Chapter 4

Section 4.1

Preparing for Solving Systems of Linear Equations by Graphing

P1. $y = 2x - 3$

The slope is $m = 2$ and the y-intercept is $b = -3$. Plot the y-intercept $(0, -3)$. Use the slope $m = \dfrac{2}{1} = \dfrac{\text{rise}}{\text{run}}$ to find a second point on the graph.

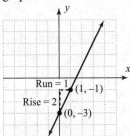

P2. $3x + 4y = 12$

x-intercept, let $y = 0$:

$3x + 4(0) = 12$

$3x = 12$

$x = 4$

y-intercept, let $x = 0$:

$3(0) + 4y = 12$

$4y = 12$

$y = 3$

Additional point: Let $x = 2$.

$3(2) + 4y = 12$

$6 + 4y = 12$

$4y = 6$

$y = \dfrac{6}{4} = \dfrac{3}{2}$

Plot the points $(4, 0)$, $(0, 3)$, and $\left(2, \dfrac{3}{2}\right)$.

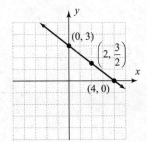

P3. $2x + 6y = 12$

$6y = -2x + 12$

$y = -\dfrac{1}{3}x + 2$

The slope is $-\dfrac{1}{3}$ and the y-intercept is 2.

$-3x - 9y = 18$

$-9y = 3x + 18$

$y = -\dfrac{1}{3}x - 2$

The slope is $-\dfrac{1}{3}$ and the y-intercept is -2.

Because the lines have the same slope, $-\dfrac{1}{3}$, but different y-intercepts, they are parallel.

4.1 Quick Checks

1. A <u>system of linear equations</u> is a grouping of two or more linear equations, each of which contains one or more variables.

2. A <u>solution</u> of a system of equations consists of values of the variables that satisfy each equation of the system.

3. $\begin{cases} 2x + 3y = 7 & (1) \\ 3x + y = -7 & (2) \end{cases}$

   (a) $(3, 1)$

   (1): $2(3) + 3(1) \overset{?}{=} 7$

   $6 + 3 \overset{?}{=} 7$

   $9 = 7$ False

   $(3, 1)$ is not a solution.

   (b) $(-4, 5)$

   (1): $2(-4) + 3(5) \overset{?}{=} 7$

   $-8 + 15 \overset{?}{=} 7$

   $7 = 7$ True

   (2): $3(-4) + 5 \overset{?}{=} -7$

   $-12 + 5 \overset{?}{=} -7$

   $-7 = -7$ True

   $(-4, 5)$ is a solution.

(c) $(-2, -1)$

(1): $2(-2) + 3(-1) \overset{?}{=} 7$

$-4 - 3 \overset{?}{=} 7$

$-7 = 7$ False

$(-2, -1)$ is not a solution.

4. $\begin{cases} 3x - 6y = 6 & (1) \\ -2x + 4y = -4 & (2) \end{cases}$

(a) $(2, 0)$

(1): $3(2) - 6(0) \overset{?}{=} 6$

$6 - 0 \overset{?}{=} 6$

$6 = 6$ True

(2): $-2(2) + 4(0) \overset{?}{=} -4$

$-4 + 0 \overset{?}{=} -4$

$-4 = -4$ True

$(2, 0)$ is a solution.

(b) $(0, -1)$

(1): $3(0) - 6(-1) \overset{?}{=} 6$

$0 + 6 \overset{?}{=} 6$

$6 = 6$ True

(2): $-2(0) + 4(-1) \overset{?}{=} -4$

$0 - 4 \overset{?}{=} -4$

$-4 = -4$ True

$(0, -1)$ is a solution.

(c) $(4, 1)$

(1): $3(4) - 6(1) \overset{?}{=} 6$

$12 - 6 \overset{?}{=} 6$

$6 = 6$ True

(2): $-2(4) + 4(1) \overset{?}{=} -4$

$-8 + 4 \overset{?}{=} -4$

$-4 = -4$ True

$(4, 1)$ is a solution.

5. $\begin{cases} y = -2x + 9 & (1) \\ y = 3x - 11 & (2) \end{cases}$

We graph $y = -2x + 9$ using slope -2 and y-intercept 9. We graph $y = 3x - 11$ using slope 3 and y-intercept -11.

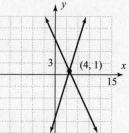

The point of intersection is $(4, 1)$.

Check:

(1): $1 \overset{?}{=} -2(4) + 9$

$1 \overset{?}{=} -8 + 9$

$1 = 1$ True

(2): $1 \overset{?}{=} 3(4) - 11$

$1 \overset{?}{=} 12 - 11$

$1 = 1$ True

The solution is $(4, 1)$.

6. $\begin{cases} 4x + y = -3 & (1) \\ 3x - y = -11 & (2) \end{cases}$

(1): $4x + y = -3$

$y = -4x - 3$

Graph using slope -4 and y-intercept -3.

(2): $3x - y = -11$

$-y = -3x - 11$

$y = 3x + 11$

Graph using slope 3 and y-intercept 11.

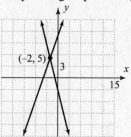

The intersection point is $(-2, 5)$.

Check:

(1): $4(-2) + 5 \overset{?}{=} -3$

$-8 + 5 \overset{?}{=} -3$

$-3 = -3$ True

(2): $3(-2) - 5 \overset{?}{=} -11$

$-6 - 11 \overset{?}{=} -11$

$-11 \overset{?}{=} -11$ True

The solution is $(-2, 5)$.

7. $\begin{cases} y = 2x + 4 & (1) \\ 2x - y = 1 & (2) \end{cases}$

Graph (1) using slope 2 and y-intercept 4. Graph

(2) using intercepts $\left(\dfrac{1}{2}, 0 \right)$ and $(0, -1)$.

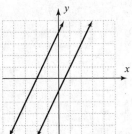

The lines are parallel and do not intersect. The system of equations has no solution or the solution set is $\varnothing$.

8. $\begin{cases} y = 3x + 2 & (1) \\ -6x + 2y = 4 & (2) \end{cases}$

Graph (1) using slope 3 and y-intercept 2.

(2): $-6x + 2y = 4$
$$2y = 6x + 4$$
$$y = 3x + 2$$

Graph using slope 3 and y-intercept 2.

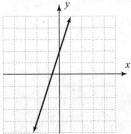

The equations are the same, so the lines coincide. The system of equations has infinitely many solutions.

9. If a system of linear equations in two variables has at least one solution, the system is <u>consistent</u>.

10. If a system of linear equations in two variables has exactly one solution, the lines are <u>independent</u>.

11. A system of linear equations in two variables that has no solutions is <u>inconsistent</u>.

12. True

13. $\begin{cases} 7x - 2y = 4 & (1) \\ 2x + 7y = 7 & (2) \end{cases}$

(1): $7x - 2y = 4$
$$-2y = -7x + 4$$
$$y = \frac{7}{2}x - 2$$

(2): $2x + 7y = 7$
$$7y = -2x + 7$$
$$y = -\frac{2}{7}x + 1$$

The equations have different slopes, so the lines intersect in exactly one point. The system is consistent and the equations are independent.

14. $\begin{cases} 6x + 4y = 4 & (1) \\ -12x - 8y = -8 & (2) \end{cases}$

(1): $6x + 4y = 4$
$$4y = -6x + 4$$
$$y = -\frac{3}{2}x + 1$$

(2): $-12x - 8y = -8$
$$-8y = 12x - 8$$
$$y = -\frac{3}{2}x + 1$$

The equations have the same slope, $-\dfrac{3}{2}$, and the same y-intercept, 1, so the lines coincide. There are infinitely many solutions. The system is consistent and the equations are dependent.

15. $\begin{cases} 3x - 4y = 8 & (1) \\ -6x + 8y = 8 & (2) \end{cases}$

(1): $3x - 4y = 8$
$$-4y = -3x + 8$$
$$y = \frac{3}{4}x - 2$$

(2): $-6x + 8y = 8$
$$8y = 6x + 8$$
$$y = \frac{3}{4}x + 1$$

The equations have the same slope, $\dfrac{3}{4}$, but different y-intercepts, so the lines are parallel. Therefore, the system has no solution and is inconsistent.

16. We want to know the number of miles for which the cost of both trucks is the same. Let m represent the number of miles and let C represent the cost.

EZ-Rental: $C = 0.40m + 20$

U-Move-It: $C = 0.25m + 35$

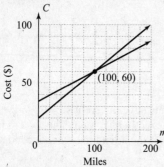

The charge will be the same for 100 miles.

EZ-Rental: $C = 0.40(100) + 20$

$C = 40 + 20$

$C = 60$

U-Move-It: $C = 0.25(100) + 35$

$C = 25 + 35$

$C = 60$

For 100 miles, both costs are the same, $60.

4.1 Exercises

17. $\begin{cases} x - y = -4 & (1) \\ 3x + y = -4 & (2) \end{cases}$

(a) $(2, 6)$

(1): $2 - 6 \stackrel{?}{=} -4$

$-4 = -4$ True

(2): $3(2) + 6 \stackrel{?}{=} -4$

$6 + 6 \stackrel{?}{=} -4$

$12 = -4$ False

$(2, 6)$ is not a solution.

(b) $(-2, 2)$

(1): $-2 - 2 \stackrel{?}{=} -4$

$-4 = -4$ True

(2): $3(-2) + 2 \stackrel{?}{=} -4$

$-6 + 2 \stackrel{?}{=} -4$

$-4 = -4$ True

$(-2, 2)$ is a solution.

(c) $(2, -2)$

(1): $2 - (-2) \stackrel{?}{=} -4$

$2 + 2 \stackrel{?}{=} -4$

$4 = -4$ False

$(2, -2)$ is not a solution.

19. $\begin{cases} 3x - y = 2 & (1) \\ -15x + 5y = -10 & (2) \end{cases}$

(a) $(1, -1)$

(1): $3(1) - (-1) \stackrel{?}{=} 2$

$3 + 1 \stackrel{?}{=} 2$

$4 = 2$ False

$(1, -1)$ is not a solution.

(b) $(-2, -8)$

(1): $3(-2) - (-8) \stackrel{?}{=} 2$

$-6 + 8 \stackrel{?}{=} 2$

$2 = 2$ True

(2): $-15(-2) + 5(-8) \stackrel{?}{=} -10$

$30 - 40 \stackrel{?}{=} -10$

$-10 = -10$ True

$(-2, -8)$ is a solution.

(c) $(0, -2)$

(1): $3(0) - (-2) \stackrel{?}{=} 2$

$0 + 2 \stackrel{?}{=} 2$

$2 = 2$ True

(2): $-15(0) + 5(-2) \stackrel{?}{=} -10$

$0 - 10 \stackrel{?}{=} -10$

$-10 = -10$ True

$(0, -2)$ is a solution.

21. $\begin{cases} 6x - 2y = 1 & (1) \\ y = 3x + 2 & (2) \end{cases}$

(a) $\left(0, -\dfrac{1}{2}\right)$

(1): $6(0) - 2\left(-\dfrac{1}{2}\right) \stackrel{?}{=} 1$

$0 + 1 \stackrel{?}{=} 1$

$1 = 1$ True

(2): $-\dfrac{1}{2} \stackrel{?}{=} 3(0) + 2$

$-\dfrac{1}{2} = 2$ False

$\left(0, -\dfrac{1}{2}\right)$ is not a solution.

(b) $(-2, -4)$

    (1): $6(-2) - 2(-4) \stackrel{?}{=} 1$

             $-12 + 8 \stackrel{?}{=} 1$

                  $-4 = 1$ False

    $(-2, -4)$ is not a solution.

(c) $(0, 2)$

    (1): $6(0) - 2(2) \stackrel{?}{=} 1$

             $0 - 4 \stackrel{?}{=} 1$

               $-4 = 1$ False

    $(0, 2)$ is not a solution.

23. $\begin{cases} 2x - y = -1 & (1) \\ 3x + 2y = -5 & (2) \end{cases}$

(1): $2x - y = -1$

    $-y = -2x - 1$

     $y = 2x + 1$

(2): $3x + 2y = -5$

    $2y = -3x - 5$

     $y = -\dfrac{3}{2}x - \dfrac{5}{2}$

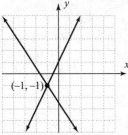

Check $(-1, -1)$:

(1): $2(-1) - (-1) \stackrel{?}{=} -1$

       $-2 + 1 \stackrel{?}{=} -1$

            $-1 = -1$ True

(2): $3(-1) + 2(-1) \stackrel{?}{=} -5$

         $-3 - 2 \stackrel{?}{=} -5$

             $-5 = -5$ True

The solution is $(-1, -1)$.

25. $\begin{cases} y = x + 5 & (1) \\ y = -\dfrac{1}{5}x - 1 & (2) \end{cases}$

Both equations are in slope-intercept form.

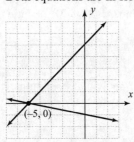

Check $(-5, 0)$:

(1): $0 \stackrel{?}{=} -5 + 5$

     $0 = 0$ True

(2): $0 \stackrel{?}{=} -\dfrac{1}{5}(-5) - 1$

     $0 \stackrel{?}{=} 1 - 1$

     $0 = 0$ True

The solution is $(-5, 0)$.

27. $\begin{cases} y = \dfrac{3}{4}x - 4 & (1) \\ y = -\dfrac{1}{2}x + 1 & (2) \end{cases}$

Both equations are in slope-intercept form.

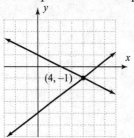

Check $(4, -1)$:

(1): $-1 \stackrel{?}{=} \dfrac{3}{4}(4) - 4$

     $-1 \stackrel{?}{=} 3 - 4$

        $-1 = -1$ True

(2): $-1 \stackrel{?}{=} -\dfrac{1}{2}(4) + 1$

     $-1 \stackrel{?}{=} -2 + 1$

        $-1 = -1$ True

The solution is $(4, -1)$.

29. $\begin{cases} x - 4 = 0 & (1) \\ 3x - 5y = 22 & (2) \end{cases}$

(1): $x - 4 = 0$

      $x = 4$

(2): $3x - 5y = 22$

    $-5y = -3x + 22$

      $y = \dfrac{3}{5}x - \dfrac{22}{5}$

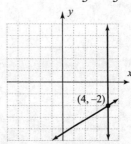

Check $(4, -2)$:

(1): $4 - 4 \stackrel{?}{=} 0$

$\qquad 0 = 0$ True

(2): $3(4) - 5(-2) \stackrel{?}{=} 22$

$\qquad 12 + 10 \stackrel{?}{=} 22$

$\qquad\qquad 22 = 22$ True

The solution is $(4, -2)$.

31. $\begin{cases} 3x - y = -1 & (1) \\ -6x + 2y = -4 & (2) \end{cases}$

(1): $3x - y = -1$

$\qquad -y = -3x - 1$

$\qquad y = 3x + 1$

(2): $-6x + 2y = -4$

$\qquad 2y = 6x - 4$

$\qquad y = 3x - 2$

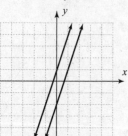

The lines are parallel, so there is no solution.

33. $\begin{cases} x + y = -2 & (1) \\ 3x - 4y = 8 & (2) \end{cases}$

(1): $x + y = -2$

$\qquad y = -x - 2$

(2): $3x - 4y = 8$

$\qquad -4y = -3x + 8$

$\qquad y = \dfrac{3}{4}x - 2$

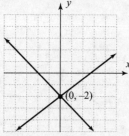

Check $(0, -2)$:

(1): $0 + (-2) \stackrel{?}{=} -2$

$\qquad -2 = -2$ True

(2): $3(0) - 4(-2) \stackrel{?}{=} 8$

$\qquad 0 + 8 \stackrel{?}{=} 8$

$\qquad\qquad 8 = 8$ True

The solution is $(0, -2)$.

35. $\begin{cases} 2y = 6 - 4x & (1) \\ 6x = 9 - 3y & (2) \end{cases}$

(1): $2y = 6 - 4x$

$\qquad 2y = -4x + 6$

$\qquad y = -2x + 3$

(2): $6x = 9 - 3y$

$\qquad 3y = -6x + 9$

$\qquad y = -2x + 3$

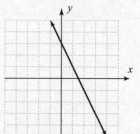

The lines coincide, so there are infinitely many solutions.

37. $\begin{cases} y = -x + 3 & (1) \\ 3y = 2x + 9 & (2) \end{cases}$

(1): $y = -x + 3$

(2): $3y = 2x + 9$

$\qquad y = \dfrac{2}{3}x + 3$

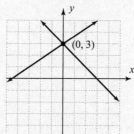

Check $(0, 3)$:

(1): $3 \stackrel{?}{=} -0 + 3$

$\qquad 3 = 3$ True

(2): $3(3) \stackrel{?}{=} 2(0) + 9$

$\qquad 9 \stackrel{?}{=} 0 + 9$

$\qquad 9 = 9$ True

The solution is $(0, 3)$.

39. The system has one solution. It is consistent and the equations are independent. The solution is $(3, 2)$.

41. The system has no solution. It is inconsistent.

43. The system has infinitely many solutions. It is consistent and the equations are dependent.

45. The system has one solution. It is consistent and the equations are independent. The solution is $(-1, -2)$.

47. Since the lines have different slopes, they will intersect. The system has one solution. It is consistent and the equations are independent.

49. $\begin{cases} 6x + 2y = 12 & (1) \\ 3x + y = 12 & (2) \end{cases}$

(1): $6x + 2y = 12$
$$2y = -6x + 12$$
$$y = -3x + 6$$

(2): $3x + y = 12$
$$y = -3x + 12$$

The lines have the same slope and different y-intercepts, so the lines are parallel. The system has no solution. It is inconsistent.

51. $\begin{cases} x + 2y = 2 & (1) \\ 2x + 4y = 4 & (2) \end{cases}$

(1): $x + 2y = 2$
$$2y = -x + 2$$
$$y = -\frac{1}{2}x + 1$$

(2): $2x + 4y = 4$
$$4y = -2x + 4$$
$$y = -\frac{1}{2}x + 1$$

The lines have the same slope and y-intercept, so the lines coincide. The system has infinitely many solutions. It is consistent and the equations are dependent.

53. $\begin{cases} y = 4 & (1) \\ x = 4 & (2) \end{cases}$

Equation (1) is a horizontal line and equation (2) is a vertical line. They intersect. There is one solution. The system is consistent and the equations are independent.

55. $\begin{cases} x - 2y = -4 & (1) \\ -x + 2y = -4 & (2) \end{cases}$

(1): $x - 2y = -4$
$$-2y = -x - 4$$
$$y = \frac{1}{2}x + 2$$

(2): $-x + 2y = -4$
$$2y = x - 4$$
$$y = \frac{1}{2}x - 2$$

The lines have the same slope but different y-intercepts, so the lines are parallel. The system has no solution. It is inconsistent.

57. $\begin{cases} x + 2y = 4 & (1) \\ x + 1 = 5 - 2y & (2) \end{cases}$

(1): $x + 2y = 4$
$$2y = -x + 4$$
$$y = -\frac{1}{2}x + 2$$

(2): $x + 1 = 5 - 2y$
$$2y = -x + 4$$
$$y = -\frac{1}{2}x + 2$$

The lines have the same slope and y-intercept, so the lines coincide. The system has infinitely many solutions. It is consistent and the equations are dependent.

59. $\begin{cases} y = 2x + 9 \\ y = -3x - 6 \end{cases}$

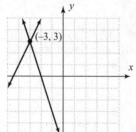

The system is consistent. The equations are independent. The solution is $(-3, 3)$.

61. $\begin{cases} x - 2y = 6 \\ 2x - 4y = 0 \end{cases}$

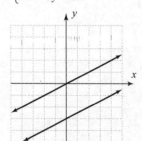

The system is inconsistent. There is no solution.

63. $\begin{cases} 3x - 2y = -2 \\ 2x + y = 8 \end{cases}$

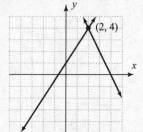

The system is consistent. The equations are independent. The solution is (2, 4).

65. $\begin{cases} y = x + 2 \\ 3x - 3y = -6 \end{cases}$

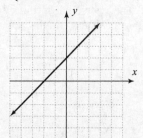

The system is consistent. The equations are dependent. There are infinitely many solutions.

67. $\begin{cases} -5x - 2y = -2 \\ 10x + 4y = 4 \end{cases}$

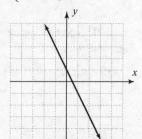

The system is consistent. The equations are dependent. There are infinitely many solutions.

69. $\begin{cases} 3x = 4y - 12 \\ 2x = -4y - 8 \end{cases}$

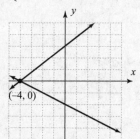

The system is consistent. The equations are independent. The solution is (−4, 0).

71. $\begin{cases} y = 15x + 1000 \\ y = 25x \end{cases}$

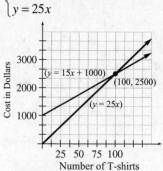

The break-even point is (100, 2500). The company needs to sell 100 T-shirts to break even at a cost/revenue of $2500.

73. $\begin{cases} y = 0.8x + 3.5 \\ y = 1.5x \end{cases}$

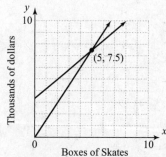

The break-even point is (5, 7.5). The company needs to sell 5 boxes of skates to break even at a cost/revenue of $7500.

75. Let *x* be the number of miles Liza drives, and let *y* be the cost. The equation for Wheels-to-Go is $y = 0.20x + 50$. The equation for Acme is $y = 0.12x + 62$. The system is:

$\begin{cases} y = 0.2x + 50 \\ y = 0.12x + 62 \end{cases}$

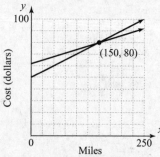

The solution is (150, 80), so the cost for driving 150 miles is the same ($80) for both companies. The per-mile charge for Acme is less, so she should use Acme if she will drive more than 150 miles.

77. Let x be the number of minutes and y be the cost per month. The equation for Plan A is $y = 0.05x + 8.95$. The equation for Plan B is $y = 0.07x + 5.95$.

$$\begin{cases} y = 0.05x + 8.95 \\ y = 0.07x + 5.95 \end{cases}$$

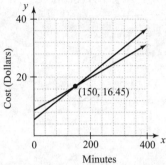

The solution is (150, 16.45). The monthly cost is the same ($16.45) when 150 minutes are used. The monthly fee for Plan B is less when fewer than 150 minutes are used, so choose Plan B if you typically use 100 minutes per month.

79. The system is dependent when the equations represent the same line.

$3x - y = -4$
$-y = -3x - 4$
$y = 3x + 4$

The slope is 3, the y-intercept is 4.
$y = cx + 4$
The slope is c, the y-intercept is 4.
The system is dependent when $c = 3$.

81. The system is dependent when the equations represent the same line.

$x + 3 = 3(x - y)$
$x + 3 = 3x - 3y$
$3y = 2x - 3$
$y = \dfrac{2}{3}x - 1$

The slope is $\dfrac{2}{3}$, the y-intercept is -1.

$2y + 3 = 2cx - y$
$3y = 2cx - 3$
$y = \dfrac{2c}{3}x - 1$

The slope is $\dfrac{2c}{3}$, the y-intercept is -1. The

system is dependent when $\dfrac{2c}{3} = \dfrac{2}{3}$ or when

$c = 1$.

83. $\begin{cases} y = 3x - 1 & (1) \\ y = -2x + 5 & (2) \end{cases}$

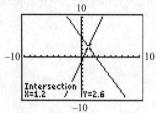

The solution is (1.2, 2.6).

85. $\begin{cases} 3x - y = -1 & (1) \\ -4x + y = -3 & (2) \end{cases}$

(1) $3x - y = -1$
$\quad\quad y = 3x + 1$
(2) $-4x + y = -3$
$\quad\quad y = 4x - 3$

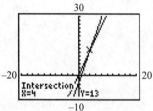

The solution is (4, 13).

87. $\begin{cases} 4x - 3y = 1 & (1) \\ -8x + 6y = -2 & (2) \end{cases}$

(1) $4x - 3y = 1$
$\quad -3y = -4x + 1$
$\quad\quad y = \dfrac{4}{3}x - \dfrac{1}{3}$
(2) $-8x + 6y = -2$
$\quad\quad 6y = 8x - 2$
$\quad\quad y = \dfrac{4}{3}x - \dfrac{1}{3}$

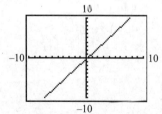

The lines coincide. There are infinitely many solutions.

89. The three possibilities for a solution of a system of two linear equation containing two variables are (1) lines intersecting at a single point (one solution); (2) the lines are coincident (a single line in the plane) so an infinite number of solutions; (3) parallel lines (no solution).

Section 4.2

Preparing for Solving Systems of Linear Equations Using Substitution

P1. $3x - y = 2$

$-y = -3x + 2$

$y = 3x - 2$

P2. $2x + 5y = 8$

$2x = -5y + 8$

$x = -\dfrac{5}{2}y + 4$

P3. $3x - 2(5x + 1) = 12$

$3x - 10x - 2 = 12$

$-7x - 2 = 12$

$-7x = 14$

$x = -2$

The solution set is $\{-2\}$.

4.2 Quick Checks

1. When solving a system of equations by substitution, whenever possible solve one of the equations for the variable that has the coefficient of $\underline{1}$ or $\underline{-1}$.

2. $\begin{cases} y = 3x - 2 & (1) \\ 2x - 3y = -8 & (2) \end{cases}$

Equation (1) is already solved for y. Substitute $3x - 2$ for y in equation (2).

$2x - 3(3x - 2) = -8$

$2x - 9x + 6 = -8$

$-7x + 6 = -8$

$-7x = -14$

$x = 2$

Let $x = 2$ in equation (1).

$y = 3(2) - 2$

$y = 6 - 2$

$y = 4$

Check (2, 4):

(1): $4 \stackrel{?}{=} 3(2) - 2$

$4 \stackrel{?}{=} 6 - 2$

$4 = 4$ True

(2): $2(2) - 3(4) \stackrel{?}{=} -8$

$4 - 12 \stackrel{?}{=} -8$

$-8 = -8$ True

The solution is (2, 4).

3. $\begin{cases} 2x + y = -1 & (1) \\ 4x + 3y = 3 & (2) \end{cases}$

Solve equation (1) for y.

$2x + y = -1$

$y = -2x - 1$

Substitute $-2x - 1$ for y in equation (2).

$4x + 3(-2x - 1) = 3$

$4x - 6x - 3 = 3$

$-2x - 3 = 3$

$-2x = 6$

$x = -3$

Let $x = -3$ in equation (1).

$2(-3) + y = -1$

$-6 + y = -1$

$y = 5$

Check (−3, 5):

(1): $2(-3) + 5 \stackrel{?}{=} -1$

$-6 + 5 \stackrel{?}{=} -1$

$-1 = -1$ True

(2): $4(-3) + 3(5) \stackrel{?}{=} 3$

$-12 + 15 \stackrel{?}{=} 3$

$3 = 3$ True

The solution is (−3, 5).

4. $\begin{cases} -4x + y = 1 & (1) \\ 8x - y = 5 & (2) \end{cases}$

Solve equation (1) for y.

$-4x + y = 1$

$y = 4x + 1$

Substitute $4x + 1$ for y in equation (2).

$8x - (4x + 1) = 5$

$8x - 4x - 1 = 5$

$4x - 1 = 5$

$4x = 6$

$x = \dfrac{3}{2}$

Let $x = \dfrac{3}{2}$ in equation (1).

$-4\left(\dfrac{3}{2}\right) + y = 1$

$-6 + y = 1$

$y = 7$

Check $\left(\dfrac{3}{2}, 7\right)$:

(1): $-4\left(\dfrac{3}{2}\right)+7 \overset{?}{=} 1$

$-6+7 \overset{?}{=} 1$

$1 = 1$ True

(2): $8\left(\dfrac{3}{2}\right)-7 \overset{?}{=} 5$

$12-7 \overset{?}{=} 5$

$5 = 5$ True

The solution is $\left(\dfrac{3}{2}, 7\right)$.

5. $\begin{cases} 3x+2y=-3 & (1) \\ -x+y=\dfrac{11}{6} & (2) \end{cases}$

Solve equation (2) for y.

$-x+y=\dfrac{11}{6}$

$y=x+\dfrac{11}{6}$

Substitute $x+\dfrac{11}{6}$ for y in equation (1).

$3x+2\left(x+\dfrac{11}{6}\right)=-3$

$3x+2x+\dfrac{11}{3}=-3$

$5x+\dfrac{11}{3}=-\dfrac{9}{3}$

$5x=-\dfrac{20}{3}$

$x=-\dfrac{4}{3}$

Let $x=-\dfrac{4}{3}$ in equation (2).

$-\left(-\dfrac{4}{3}\right)+y=\dfrac{11}{6}$

$\dfrac{4}{3}+y=\dfrac{11}{6}$

$y=\dfrac{11}{6}-\dfrac{4}{3}$

$y=\dfrac{11}{6}-\dfrac{8}{6}=\dfrac{3}{6}$ or $\dfrac{1}{2}$

Check $\left(-\dfrac{4}{3}, \dfrac{1}{2}\right)$:

(1): $3\left(-\dfrac{4}{3}\right)+2\left(\dfrac{1}{2}\right) \overset{?}{=} -3$

$-4+1 \overset{?}{=} -3$

$-3 = -3$ True

(2): $-\left(-\dfrac{4}{3}\right)+\dfrac{1}{2} \overset{?}{=} \dfrac{11}{6}$

$\dfrac{8}{6}+\dfrac{3}{6} \overset{?}{=} \dfrac{11}{6}$

$\dfrac{11}{6}=\dfrac{11}{6}$ True

The solution is $\left(-\dfrac{4}{3}, \dfrac{1}{2}\right)$.

6. While solving a system of equations by substitution, the variable expression has been eliminated and a *false* statement such as $-3 = 0$ results. This means that the solution of the system is $\varnothing$ or $\{\ \}$.

7. While solving a system of equations by substitution, the variable expression has been eliminated and a *true* statement such as $0 = 0$ results. This means that the system has infinitely many solutions.

8. False; the value of x found from equation (1) must be substituted into equation (2), not back into equation (1).

9. $\begin{cases} y=5x+2 & (1) \\ -10x+2y=4 & (2) \end{cases}$

Let $y = 5x + 2$ in equation (2).

$-10x+2(5x+2)=4$

$-10x+10x+4=4$

$4=4$

Since $4 = 4$ is a true statement, the system is consistent and the equations are dependent. The system has infinitely many solutions.

10. $\begin{cases} 3x-2y=0 & (1) \\ -9x+6y=5 & (2) \end{cases}$

Solve equation (1) for x.

$3x-2y=0$

$3x=2y$

$x=\dfrac{2}{3}y$

Let $x=\dfrac{2}{3}y$ in equation (2).

$-9\left(\dfrac{2}{3}y\right)+6y=5$

$-6y+6y=5$

$0=5$

The statement $0 = 5$ is false. The system is inconsistent and has no solution.

11. $\begin{cases} 2x - 6y = 2 & (1) \\ -3x + 9y = 4 & (2) \end{cases}$

Solve equation (1) for x.

$2x - 6y = 2$

$\quad 2x = 6y + 2$

$\quad\quad x = 3y + 1$

Let $x = 3y + 1$ in equation (2).

$-3(3y + 1) + 9y = -4$

$\quad -9y - 3 + 9y = -4$

$\quad\quad\quad\quad -3 = -4$

The statement $-3 = -4$ is false. The system is inconsistent and has no solution.

12. We want the year when we can expect the participation rate of women in the workforce to be equal to that of men.

The variables are P, which represents the participation rate, and t, which represents the year. The system is

$\begin{cases} P = -0.08t + 243.96 & (1) \\ P = 0.24t - 415.10 & (2) \end{cases}$

Substitute $-0.08t + 243.96$ for P in equation (2).

$-0.08t + 243.96 = 0.24t - 415.10$

$\quad\quad -0.08t = 0.24t - 659.06$

$\quad\quad -0.32t = -659.06$

$\quad\quad\quad\quad t \approx 2059.6$

We round this result to the nearest year, 2060.

Check:

(1): $P = -0.08(2060) + 243.96$

$\quad\quad P = 79.16$

(2): $P = 0.24(2060) - 415.10$

$\quad\quad P = 79.3$

The answers differ slightly due to rounding, but the answer checks. The participation rate of men and women in the workforce will be the same in 2060.

4.2 Exercises

13. $\begin{cases} x + 2y = 2 \\ y = 2x - 9 \end{cases}$

Let $y = 2x - 9$ in the first equation.

$x + 2(2x - 9) = 2$

$\quad x + 4x - 18 = 2$

$\quad\quad\quad 5x = 20$

$\quad\quad\quad\quad x = 4$

$y = 2(4) - 9 = 8 - 9 = -1$

The solution is $(4, -1)$.

15. $\begin{cases} -2x + 5y = 7 \\ x = 3y - 4 \end{cases}$

Let $x = 3y - 4$ in the first equation.

$-2(3y - 4) + 5y = 7$

$\quad -6y + 8 + 5y = 7$

$\quad\quad\quad\quad -y = -1$

$\quad\quad\quad\quad\quad y = 1$

$x = 3(1) - 4 = 3 - 4 = -1$

The solution is $(-1, 1)$.

17. $\begin{cases} x + y = -7 \\ 2x - y = -2 \end{cases}$

Solve the first equation for y.

$y = -7 - x$

Let $y = -7 - x$ in the second equation.

$2x - (-7 - x) = -2$

$\quad 2x + 7 + x = -2$

$\quad\quad\quad\quad 3x = -9$

$\quad\quad\quad\quad\quad x = -3$

$-3 + y = -7$

$\quad\quad\quad y = -4$

The solution is $(-3, -4)$.

19. $\begin{cases} y = \dfrac{1}{2}x - 5 \\ y = -\dfrac{3}{4}x - 10 \end{cases}$

Let $y = \dfrac{1}{2}x - 5$ in the second equation.

$\dfrac{1}{2}x - 5 = -\dfrac{3}{4}x - 10$

$\quad\quad\quad 5 = -\dfrac{5}{4}x$

$\left(-\dfrac{4}{5}\right)5 = x$

$\quad\quad -4 = x$

$y = \dfrac{1}{2}(-4) - 5 = -2 - 5 = -7$

The solution is $(-4, -7)$.

21. $\begin{cases} y = 3x + 4 \\ y = -\dfrac{1}{2}x + \dfrac{5}{3} \end{cases}$

Let $y = 3x + 4$ in the second equation.

$$3x + 4 = -\frac{1}{2}x + \frac{5}{3}$$
$$18x + 24 = -3x + 10$$
$$21x = -14$$
$$x = -\frac{14}{21} = -\frac{2}{3}$$
$$y = -\frac{1}{2}\left(-\frac{2}{3}\right) + \frac{5}{3} = \frac{1}{3} + \frac{5}{3} = \frac{6}{3} = 2$$

The solution is $\left(-\frac{2}{3}, 2\right)$.

23. $\begin{cases} x = -6y \\ x - 3y = 3 \end{cases}$

Let $x = -6y$ in the second equation.
$$-6y - 3y = 3$$
$$-9y = 3$$
$$y = \frac{3}{-9} = -\frac{1}{3}$$
$$x = -6\left(-\frac{1}{3}\right) = 2$$

The solution is $\left(2, -\frac{1}{3}\right)$.

25. $\begin{cases} 2x - 3y = 0 \\ 8x + 6y = 3 \end{cases}$

Solve the first equation for x.
$$2x = 3y$$
$$x = \frac{3}{2}y$$

Let $x = \frac{3}{2}y$ in the second equation.
$$8\left(\frac{3}{2}y\right) + 6y = 3$$
$$12y + 6y = 3$$
$$18y = 3$$
$$y = \frac{3}{18} = \frac{1}{6}$$
$$2x - 3\left(\frac{1}{6}\right) = 0$$
$$2x - \frac{1}{2} = 0$$
$$2x = \frac{1}{2}$$
$$x = \frac{1}{4}$$

The solution is $\left(\frac{1}{4}, \frac{1}{6}\right)$.

27. $\begin{cases} x + 3y = -12 \\ x + 3y = 6 \end{cases}$

Solve the first equation for x.
$$x = -3y - 12$$
Let $x = -3y - 12$ in the second equation.
$$-3y - 12 + 3y = 6$$
$$-12 = 6 \quad \text{False}$$
The system is inconsistent and has no solution.

29. $\begin{cases} y = 4x - 1 \\ 8x - 2y = 2 \end{cases}$

Let $y = 4x - 1$ in the second equation.
$$8x - 2(4x - 1) = 2$$
$$8x - 8x + 2 = 2$$
$$2 = 2 \quad \text{True}$$
The system is dependent and has infinitely many solutions.

31. $\begin{cases} x + 2y = 6 \\ x = 3 - 2y \end{cases}$

Let $x = 3 - 2y$ in the first equation.
$$3 - 2y + 2y = 6$$
$$3 = 6 \quad \text{False}$$
The system is inconsistent and has no solution.

33. $\begin{cases} 5x + 6 = 2 - y \\ y = -5x - 4 \end{cases}$

Let $y = -5x - 4$ in the first equation.
$$5x + 6 = 2 - (-5x - 4)$$
$$5x + 6 = 2 + 5x + 4$$
$$6 = 6 \quad \text{True}$$
The system is dependent and has infinitely many solutions.

35. $\begin{cases} x = 2 \\ x - 6y = 4 \end{cases}$

Let $x = 2$ in the second equation.
$$2 - 6y = 4$$
$$-6y = 2$$
$$y = \frac{2}{-6} = -\frac{1}{3}$$

The solution is $\left(2, -\frac{1}{3}\right)$.

37. $\begin{cases} y = \frac{1}{2}x \\ x = 2(y + 1) \end{cases}$

Let $y = \frac{1}{2}x$ in the second equation.

$$x = 2\left(\frac{1}{2}x + 1\right)$$
$$x = x + 2$$
$$0 = 2 \quad \text{False}$$
The system is inconsistent and has no solution.

39. $\begin{cases} y = 2x - 8 \\ x - \dfrac{1}{2}y = 4 \end{cases}$

Let $y = 2x - 8$ in the second equation.
$$x - \frac{1}{2}(2x - 8) = 4$$
$$x - x + 4 = 4$$
$$4 = 4 \quad \text{True}$$
The system is dependent and has infinitely many solutions.

41. $\begin{cases} 3x + 2y = 4 \\ 3x + y = \dfrac{9}{2} \end{cases}$

Solve the second equation for y.
$$y = -3x + \frac{9}{2}$$

Let $y = -3x + \dfrac{9}{2}$ in the first equation.
$$3x + 2\left(-3x + \frac{9}{2}\right) = 4$$
$$3x - 6x + 9 = 4$$
$$-3x + 9 = 4$$
$$-3x = -5$$
$$x = \frac{5}{3}$$

$$3\left(\frac{5}{3}\right) + y = \frac{9}{2}$$
$$5 + y = \frac{9}{2}$$
$$y = \frac{9}{2} - \frac{10}{2}$$
$$y = -\frac{1}{2}$$

The solution is $\left(\dfrac{5}{3}, -\dfrac{1}{2}\right)$.

43. $\begin{cases} x - 5y = 3 \\ -2x + 10y = 8 \end{cases}$

Solve the first equation for x.
$$x = 5y + 3$$
Let $x = 5y + 3$ in the second equation.

$$-2(5y + 3) + 10y = 8$$
$$-10y - 6 + 10y = 8$$
$$-6 = 8 \quad \text{False}$$
The system is inconsistent and has no solution.

45. $\begin{cases} 3x - y = 1 \\ -6x + 2y = -2 \end{cases}$

Solve the first equation for y.
$$-y = -3x + 1$$
$$y = 3x - 1$$

Let $y = 3x - 1$ in the second equation.
$$-6x + 2(3x - 1) = -2$$
$$-6x + 6x - 2 = -2$$
$$-2 = -2 \quad \text{True}$$
The system is dependent and has infinitely many solutions.

47. $\begin{cases} 4x + 8y = -9 \\ 2x + y = \dfrac{3}{4} \end{cases}$

Solve the second equation for y.
$$y = -2x + \frac{3}{4}$$

Let $y = -2x + \dfrac{3}{4}$ in the first equation.
$$4x + 8\left(-2x + \frac{3}{4}\right) = -9$$
$$4x - 16x + 6 = -9$$
$$-12x + 6 = -9$$
$$-12x = -15$$
$$x = \frac{-15}{-12} = \frac{5}{4}$$

$$2\left(\frac{5}{4}\right) + y = \frac{3}{4}$$
$$\frac{10}{4} + y = \frac{3}{4}$$
$$y = -\frac{7}{4}$$

The solution is $\left(\dfrac{5}{4}, -\dfrac{7}{4}\right)$.

49. $\begin{cases} \dfrac{3}{2}x - y = 1 \\ 3x - 2y = 2 \end{cases}$

Solve the first equation for y.
$$y = \frac{3}{2}x - 1$$

Let $y = \dfrac{3}{2}x - 1$ in the second equation.

$$3x - 2\left(\frac{3}{2}x - 1\right) = 2$$
$$3x - 3x + 2 = 2$$
$$2 = 2 \quad \text{True}$$

The system is dependent and has infinitely many solutions.

51. $\begin{cases} \dfrac{x}{2} + \dfrac{y}{3} = \dfrac{1}{12} \\ \dfrac{2x}{3} + \dfrac{y}{3} = -\dfrac{1}{3} \end{cases}$

Solve the first equation for y.

$$\frac{y}{3} = -\frac{x}{2} + \frac{1}{12}$$
$$y = 3\left(-\frac{x}{2} + \frac{1}{12}\right)$$

Let $y = 3\left(-\dfrac{x}{2} + \dfrac{1}{12}\right)$ in the second equation.

$$\frac{2}{3}x + \frac{1}{3}\left[3\left(-\frac{x}{2} + \frac{1}{12}\right)\right] = -\frac{1}{3}$$
$$\frac{2}{3}x + \left(-\frac{x}{2} + \frac{1}{12}\right) = -\frac{1}{3}$$
$$\frac{4}{6}x - \frac{3}{6}x + \frac{1}{12} = -\frac{1}{3}$$
$$\frac{1}{6}x + \frac{1}{12} = -\frac{4}{12}$$
$$\frac{1}{6}x = -\frac{5}{12}$$
$$x = 6\left(-\frac{5}{12}\right)$$
$$x = -\frac{5}{2}$$

$$\frac{1}{2}\left(-\frac{5}{2}\right) + \frac{1}{3}y = \frac{1}{12}$$
$$-\frac{5}{4} + \frac{1}{3}y = \frac{1}{12}$$
$$-15 + 4y = 1$$
$$4y = 16$$
$$y = 4$$

The solution is $\left(-\dfrac{5}{2}, 4\right)$.

53. $\begin{cases} 2l + 2w = 34 & (1) \\ l = w + 3 & (2) \end{cases}$

Let $l = w + 3$ in (1).

$$2(w + 3) + 2w = 34$$
$$2w + 6 + 2w = 34$$
$$4w = 28$$
$$w = 7$$
$$l = 7 + 3 = 10$$

The length is 10 feet and the width is 7 feet.

55. $\begin{cases} x = 2y & (1) \\ 0.025x + 0.09y = 560 & (2) \end{cases}$

Let $x = 2y$ in (2).

$$0.025(2y) + 0.09y = 560$$
$$0.05y + 0.09y = 560$$
$$0.14y = 560$$
$$y = \frac{560}{0.14} = 4000$$
$$x = 2(4000) = 8000$$

He should invest \$8000 in the money market fund and \$4000 in the international fund.

57. $y = 15{,}000 + 0.02x$
$y = 25{,}000 + 0.01x$

To find the sales (x) at which both salaries are equal, set (y) from option A equal to (y) from option B.

$$15{,}000 + 0.02x = 25{,}000 + 0.01x$$
$$0.01x = 10{,}000$$
$$x = \frac{10{,}000}{0.01}$$
$$x = 1{,}000{,}000$$

The options result in the same salary when annual sales are \$1,000,000.

59. $\begin{cases} x + y = 17 & (1) \\ x - y = 7 & (2) \end{cases}$

Solve (2) for x.
$x = y + 7$
Let $x = y + 7$ in (1).
$$y + 7 + y = 17$$
$$2y = 10$$
$$y = 5$$
$$x - 5 = 7$$
$$x = 12$$

The numbers are 5 and 12.

61. $\begin{cases} Ax + 3By = 2 \\ -3Ax + By = -11 \end{cases}$

Let $x = 3$ and $y = 1$ in the system.
$\begin{cases} 3A + 3B = 2 \\ -9A + B = -11 \end{cases}$

Solve the second equation for B.
$B = 9A - 11$
Let $B = 9A - 11$ in the first equation.

$$3A + 3(9A - 11) = 2$$
$$3A + 27A - 33 = 2$$
$$30A - 33 = 2$$
$$30A = 35$$
$$A = \frac{7}{6}$$

Let $A = \frac{7}{6}$ in $-9A + B = -11$.

$$-9\left(\frac{7}{6}\right) + B = -11$$
$$-\frac{21}{2} + B = -11$$
$$B = -\frac{22}{2} + \frac{21}{2}$$
$$B = -\frac{1}{2}$$

Therefore, $A = \frac{7}{6}$ and $B = -\frac{1}{2}$.

63. Answers may vary. One possibility:
$$\begin{cases} 2x - y = 1 \\ -x + 2y = 7 \end{cases}$$

65. Answers may vary. One possibility:
$$\begin{cases} 2x - y = 4 \\ 6x - 3y = 12 \end{cases}$$

67. A reasonable first step is to multiply the first equation by 6 to clear fractions and multiply the second equation by 4 to clear fractions. Then the system is $\begin{cases} 2x - y = 4 \\ 6x + 2y = -5 \end{cases}$.

Section 4.3

Preparing for Solving Systems of Linear Equations Using Elimination

P1. The additive inverse of 5 is -5 since $5 + (-5) = 0$.

P2. The additive inverse of -8 is $-(-8) = 8$ since $-8 + 8 = 0$.

P3. $\frac{2}{3}(3x - 9y) = \frac{2}{3}(3x) - \frac{2}{3}(9y) = 2x - 6y$

P4. $\quad 2y - 5y = 12$
$$-3y = 12$$
$$y = -4$$
The solution set is $\{-4\}$.

P5. $\quad 4 = 2 \cdot 2$
$$\underline{5 = \qquad\quad 5}$$
$$2 \cdot 2 \cdot 5$$
The LCM is $2 \cdot 2 \cdot 5 = 20$.

4.3 Quick Checks

1. The basic idea in using the elimination method is to get the coefficients of one of the variables to be <u>additive inverses</u>, such as 3 and -3.

2. $\begin{cases} x - 3y = 2 & (1) \\ 2x + 3y = -14 & (2) \end{cases}$

Add the equations.
$$\begin{cases} x - 3y = 2 \\ 2x + 3y = -14 \end{cases}$$
$$\overline{\quad 3x \qquad = -12}$$
$$x = -4$$
Let $x = -4$ in equation (1).
$$-4 - 3y = 2$$
$$-3y = 6$$
$$y = -2$$
Check $(-4, -2)$:
(1): $-4 - 3(-2) \overset{?}{=} 2$
$$-4 + 6 \overset{?}{=} 2$$
$$2 = 2 \quad \text{True}$$

(2): $2(-4) + 3(-2) \overset{?}{=} -14$
$$-8 - 6 \overset{?}{=} -14$$
$$-14 = -14 \quad \text{True}$$
The solution is $(-4, -2)$.

3. $\begin{cases} x - 2y = 2 & (1) \\ -2x + 5y = -1 & (2) \end{cases}$

Multiply (1) by 2. Then add.
$$\begin{cases} 2x - 4y = 4 \\ -2x + 5y = -1 \end{cases}$$
$$\overline{\qquad\quad y = 3}$$
Let $y = 3$ in equation (1).
$$x - 2(3) = 2$$
$$x - 6 = 2$$
$$x = 8$$

Check (8, 3):

(1): $8 - 2(3) \stackrel{?}{=} 2$

$\qquad 8 - 6 \stackrel{?}{=} 2$

$\qquad\qquad 2 = 2$ True

(2): $-2(8) + 5(3) \stackrel{?}{=} -1$

$\qquad -16 + 15 \stackrel{?}{=} -1$

$\qquad\qquad -1 = -1$ True

The solution is (8, 3).

4. $\begin{cases} 5x + 4y = 10 & (1) \\ -2x + 3y = -27 & (2) \end{cases}$

Multiply (1) by 2 and (2) by 5. Then add.

$\begin{cases} 10x + 8y = 20 \\ -10x + 15y = -135 \end{cases}$

$\qquad\qquad 23y = -115$

$\qquad\qquad\quad y = -5$

Let $y = -5$ in equation (1).

$5x + 4(-5) = 10$

$\quad 5x - 20 = 10$

$\qquad\; 5x = 30$

$\qquad\;\; x = 6$

Check (6, –5):

(1): $5(6) + 4(-5) \stackrel{?}{=} 10$

$\qquad 30 - 20 \stackrel{?}{=} 10$

$\qquad\qquad 10 = 10$ True

(2): $-2(6) + 3(-5) \stackrel{?}{=} -27$

$\qquad -12 - 15 \stackrel{?}{=} -27$

$\qquad\qquad -27 = -27$ True

The solution is (6, –5).

5. $\begin{cases} 4y = -6x + 30 & (1) \\ 7x + 10y = 35 & (2) \end{cases}$

Start by writing equation (1) in standard form.

$\begin{cases} 6x + 4y = 30 & (1) \\ 7x + 10y = 35 & (2) \end{cases}$

Multiply (1) by 10 and (2) by –4. Then add.

$\begin{cases} 60x + 40y = 300 \\ -28x - 40y = -140 \end{cases}$

$\quad 32x \qquad\;\; = 160$

$\qquad\;\; x = 5$

Let $x = 5$ in equation (1).

$4y = -6(5) + 30$

$4y = -30 + 30$

$4y = 0$

$\;\; y = 0$

Check (5, 0):

(1): $4(0) \stackrel{?}{=} -6(5) + 30$

$\qquad 0 \stackrel{?}{=} -30 + 30$

$\qquad 0 = 0$ True

(2): $7(5) + 10(0) \stackrel{?}{=} 35$

$\qquad 35 + 0 \stackrel{?}{=} 35$

$\qquad\qquad 35 = 35$ True

The solution is (5, 0).

6. When using the elimination method to solve a system of equations, you add equation (1) and equation (2) resulting in the statement $-50 = -50$. This means that the equations are dependent have infinitely many solutions.

7. $\begin{cases} 2x - 6y = 10 & (1) \\ 5x - 15y = 4 & (2) \end{cases}$

Multiply (1) by 5 and (2) by –2. Then add.

$\begin{cases} 10x - 30y = 50 \\ -10x + 30y = -8 \end{cases}$

$\qquad\qquad 0 = 42$

The statement $0 = 42$ is false. The system is inconsistent and has no solution.

8. $\begin{cases} -x + 3y = 2 & (1) \\ 3x - 9y = -6 & (2) \end{cases}$

Multiply (1) by 3 and then add.

$\begin{cases} -3x + 9y = 6 \\ 3x - 9y = -6 \end{cases}$

$\qquad\qquad 0 = 0$

The statement $0 = 0$ is true. The system is consistent but the equations are dependent. The system has infinitely many solutions.

9. $\begin{cases} -4x + 8y = 4 & (1) \\ 3x - 6y = -3 & (2) \end{cases}$

Multiply (1) by 3 and (2) by 4. Then add.

$\begin{cases} -12x + 24y = 12 \\ 12x - 24y = -12 \end{cases}$

$\qquad\qquad 0 = 0$

The statement $0 = 0$ is true. The system is consistent but the equations are dependent. The system has infinitely many solutions.

10. $\begin{cases} 5c + 3s = 15.50 & (1) \\ 3c + 2s = 9.75 & (2) \end{cases}$

Multiply (1) by 2 and (2) by –3. Then add.

$\begin{cases} 10c + 6s = 31.00 \\ -9c - 6s = -29.25 \end{cases}$

$\quad c \qquad\quad = 1.75$

Let $c = 1.75$ in (2).
$$3(1.75) + 2s = 9.75$$
$$5.25 + 2s = 9.75$$
$$2s = 4.50$$
$$s = 2.25$$
A cheeseburger costs \$1.75, and a medium shake costs \$2.25.

4.3 Exercises

11. $\begin{cases} 2x + y = 3 & (1) \\ 5x - y = 11 & (2) \end{cases}$

Add the equations, then solve for x.
$$\begin{cases} 2x + y = 3 \\ 5x - y = 11 \end{cases}$$
$$\overline{7x = 14}$$
$$x = 2$$
Let $x = 2$ in (1).
$$2(2) + y = 3$$
$$4 + y = 3$$
$$y = -1$$
The solution is $(2, -1)$.

13. $\begin{cases} 3x - 2y = 10 & (1) \\ -3x + 12y = 30 & (2) \end{cases}$

Add the equations, then solve for y.
$$\begin{cases} 3x - 2y = 10 \\ -3x + 12y = 30 \end{cases}$$
$$\overline{10y = 40}$$
$$y = 4$$
Let $y = 4$ in (1).
$$3x - 2(4) = 10$$
$$3x - 8 = 10$$
$$3x = 18$$
$$x = 6$$
The solution is $(6, 4)$.

15. $\begin{cases} 2x + 3y = -4 & (1) \\ -2x + y = 6 & (2) \end{cases}$

Add the equations, then solve for y.
$$\begin{cases} 2x + 3y = -4 \\ -2x + y = 6 \end{cases}$$
$$\overline{4y = 2}$$
$$y = \frac{1}{2}$$
Let $y = \frac{1}{2}$ in (2).

$$-2x + \frac{1}{2} = 6$$
$$-2x = \frac{11}{2}$$
$$x = -\frac{11}{4}$$
The solution is $\left(-\frac{11}{4}, \frac{1}{2}\right)$.

17. $\begin{cases} 6x - 2y = 0 & (1) \\ -9x - 4y = 21 & (2) \end{cases}$

Multiply (1) by -2, then add the equations and solve for x.
$$\begin{cases} -12x + 4y = 0 \\ -9x - 4y = 21 \end{cases}$$
$$\overline{-21x = 21}$$
$$x = -1$$
Let $x = -1$ in (1).
$$6(-1) - 2y = 0$$
$$-6 - 2y = 0$$
$$-2y = 6$$
$$y = -3$$
The solution is $(-1, -3)$.

19. $\begin{cases} 2x + 2y = 1 & (1) \\ -2x - 2y = 1 & (2) \end{cases}$

Add the equations.
$$\begin{cases} 2x + 2y = 1 \\ -2x - 2y = 1 \end{cases}$$
$$\overline{0 = 2 \quad \text{False}}$$
The system is inconsistent and has no solution.

21. $\begin{cases} 3x + y = -1 & (1) \\ 6x + 2y = -2 & (2) \end{cases}$

Multiply (1) by -2, then add.
$$\begin{cases} -6x - 2y = 2 \\ 6x + 2y = -2 \end{cases}$$
$$\overline{0 = 0 \quad \text{True}}$$
The system is dependent and has infinitely many solutions.

23. $\begin{cases} 2x - 3y = 10 & (1) \\ -4x + 6y = -20 & (2) \end{cases}$

Multiply (1) by 2 and add the equations.
$$\begin{cases} 4x - 6y = 20 \\ -4x + 6y = -20 \end{cases}$$
$$\overline{0 = 0 \quad \text{True}}$$
The system is dependent and has infinitely many solutions.

25. $\begin{cases} -4x+8y=1 & (1) \\ 3x-6y=1 & (2) \end{cases}$

Multiply (1) by 3 and (2) by 4 and add the equations.

$\begin{cases} -12x+24y=3 \\ 12x-24y=4 \end{cases}$

$\qquad\qquad 0=7$　False

The system is inconsistent and has no solution.

27. $\begin{cases} 2x+3y=14 & (1) \\ -3x+y=23 & (2) \end{cases}$

Multiply (2) by -3, then add.

$\begin{cases} 2x+3y=14 \\ 9x-3y=-69 \end{cases}$

$\overline{11x \qquad =-55}$

$\qquad x=-5$

Let $x=-5$ in (2).

$-3(-5)+y=23$

$\qquad 15+y=23$

$\qquad\qquad y=8$

The solution is $(-5, 8)$.

29. $\begin{cases} 2x+4y=0 & (1) \\ 5x+2y=6 & (2) \end{cases}$

Multiply (2) by -2, then add.

$\begin{cases} 2x+4y=0 \\ -10x-4y=-12 \end{cases}$

$\overline{-8x \qquad =-12}$

$\qquad x=\dfrac{-12}{-8}=\dfrac{3}{2}$

Let $x=\dfrac{3}{2}$ in (1).

$2\left(\dfrac{3}{2}\right)+4y=0$

$\qquad 3+4y=0$

$\qquad\quad 4y=-3$

$\qquad\quad y=-\dfrac{3}{4}$

The solution is $\left(\dfrac{3}{2}, -\dfrac{3}{4}\right)$.

31. $\begin{cases} x-3y=4 & (1) \\ -2x+6y=3 & (2) \end{cases}$

Multiply (1) by 2, then add.

$\begin{cases} 2x-6y=8 \\ -2x+6y=3 \end{cases}$

$\qquad\qquad 0=11$　False

The system is inconsistent and has no solution.

33. $\begin{cases} 2x+3y=-3 & (1) \\ 3x+5y=-9 & (2) \end{cases}$

Multiply (1) by -3 and (2) by 2, then add.

$\begin{cases} -6x-9y=9 \\ 6x+10y=-18 \end{cases}$

$\qquad\qquad y=-9$

Let $y=-9$ in (1).

$2x+3(-9)=-3$

$\qquad 2x-27=-3$

$\qquad\quad 2x=24$

$\qquad\quad x=12$

The solution is $(12, -9)$.

35. $\begin{cases} 10y=4x-2 & (1) \\ 2x-5y=1 & (2) \end{cases}$

Rearrange (1).

$\begin{cases} -4x+10y=-2 & (1) \\ 2x-5y=1 & (2) \end{cases}$

Multiply (1) by $\dfrac{1}{2}$ and add.

$\begin{cases} -2x+5y=-1 \\ 2x-5y=1 \end{cases}$

$\qquad\qquad 0=0$　True

The system is dependent and has infinitely many solutions.

37. $\begin{cases} 4x+3y=0 & (1) \\ 3x-5y=2 & (2) \end{cases}$

Multiply (1) by 5 and (2) by 3, then add.

$\begin{cases} 20x+15y=0 \\ 9x-15y=6 \end{cases}$

$\overline{29x \qquad =6}$

$\qquad\quad x=\dfrac{6}{29}$

Let $x=\dfrac{6}{29}$ in (1).

$$4\left(\frac{6}{29}\right)+3y=0$$

$$\frac{24}{29}+3y=0$$

$$3y=-\frac{24}{29}$$

$$y=-\frac{8}{29}$$

The solution is $\left(\frac{6}{29}, -\frac{8}{29}\right)$.

39. $\begin{cases} 4x-3y=-10 & (1) \\ -\dfrac{2}{3}x+y=\dfrac{11}{3} & (2) \end{cases}$

Multiply (2) by 3, then add.

$$\begin{cases} 4x-3y=-10 \\ -2x+3y=11 \end{cases}$$
$$2x=1$$
$$x=\frac{1}{2}$$

Let $x=\dfrac{1}{2}$ in (1).

$$4\left(\frac{1}{2}\right)-3y=-10$$
$$2-3y=-10$$
$$-3y=-12$$
$$y=4$$

The solution is $\left(\dfrac{1}{2}, 4\right)$.

41. $\begin{cases} 1.5x+0.5y=-0.45 & (1) \\ -0.3x-0.4y=-0.54 & (2) \end{cases}$

Multiply (2) by 5, then add.

$$\begin{cases} 1.5x+0.5y=-0.45 \\ -1.5x-2y=-2.7 \end{cases}$$
$$-1.5y=-3.15$$
$$y=\frac{-3.15}{-1.5}=2.1$$

Let $y=2.1$ in (1).
$$1.5x+0.5(2.1)=-0.45$$
$$1.5x+1.05=-0.45$$
$$1.5x=-1.5$$
$$x=-1$$
The solution is $(-1, 2.1)$.

43. $\begin{cases} \dfrac{1}{2}x+\dfrac{2}{3}y=-5 & (1) \\ \dfrac{5}{2}x+\dfrac{5}{6}y=-10 & (2) \end{cases}$

Multiply (1) by -5, then add.

$$\begin{cases} -\dfrac{5x}{2}-\dfrac{10y}{3}=25 \\ \dfrac{5x}{2}+\dfrac{5y}{6}=-10 \end{cases}$$
$$-\frac{10y}{3}+\frac{5y}{6}=15$$
$$-\frac{20y}{6}+\frac{5y}{6}=15$$
$$-\frac{15y}{6}=15$$
$$y=\left(-\frac{6}{15}\right)15=-6$$

Let $y=-6$ in (2).

$$\frac{5x}{2}+\frac{5(-6)}{6}=-10$$
$$\frac{5x}{2}-5=-10$$
$$\frac{5}{2}x=-5$$
$$x=\left(\frac{2}{5}\right)(-5)=-2$$

The solution is $(-2, -6)$.

45. $\begin{cases} 0.05x+0.10y=5.50 & (1) \\ x+y=80 & (2) \end{cases}$

Multiply (1) by 100 and (2) by -5, then add.
$$\begin{cases} 5x+10y=550 \\ -5x-5y=-400 \end{cases}$$
$$5y=150$$
$$y=30$$

Let $y=30$ in (2).
$$x+30=80$$
$$x=50$$
The solution is $(50, 30)$.

47. $\begin{cases} x-y=-4 & (1) \\ 3x+y=8 & (2) \end{cases}$

$$4x=4$$
$$x=1$$

Let $x=1$ in (1).
$$1-y=-4$$
$$-y=-5$$
$$y=5$$

The solution is $(1, 5)$.

49. $\begin{cases} 3x - 10y = -5 & (1) \\ 6x - 8y = 14 & (2) \end{cases}$

Multiply (1) by –2, then add.

$\begin{cases} -6x + 20y = 10 \\ 6x - 8y = 14 \end{cases}$

$12y = 24$

$y = 2$

Let $y = 2$ in (1).

$3x - 10(2) = -5$

$3x - 20 = -5$

$3x = 15$

$x = 5$

The solution is $(5, 2)$.

51. $\begin{cases} -x + 3y = 6 & (1) \\ 4x + 5y = 7 & (2) \end{cases}$

Solve (1) for x.

$-x = -3y + 6$

$x = 3y - 6$

Let $x = 3y - 6$ in (2).

$4(3y - 6) + 5y = 7$

$12y - 24 + 5y = 7$

$17y = 31$

$y = \dfrac{31}{17}$

Let $y = \dfrac{31}{17}$ in (1).

$-x + 3\left(\dfrac{31}{17}\right) = 6$

$-x + \dfrac{93}{17} = 6$

$-x = \dfrac{102}{17} - \dfrac{93}{17}$

$-x = \dfrac{9}{17}$

$x = -\dfrac{9}{17}$

The solution is $\left(-\dfrac{9}{17}, \dfrac{31}{17}\right)$.

53. $\begin{cases} 0.3x - 0.7y = 1.2 & (1) \\ 1.2x + 2.1y = 2 & (2) \end{cases}$

Multiply (1) by –4, then add.

$\begin{cases} -1.2x + 2.8y = -4.8 \\ 1.2x + 2.1y = 2 \end{cases}$

$4.9y = -2.8$

$y = -\dfrac{2.8}{4.9} = -\dfrac{4}{7}$

Let $y = -\dfrac{4}{7}$ in (1).

$0.3x - 0.7\left(-\dfrac{4}{7}\right) = 1.2$

$0.3x + 0.4 = 1.2$

$0.3x = 0.8$

$x = \dfrac{8}{3}$

The solution is $\left(\dfrac{8}{3}, -\dfrac{4}{7}\right)$.

55. $\begin{cases} 3x - 2y = 6 & (1) \\ \dfrac{3}{2}x - y = 3 & (2) \end{cases}$

Multiply (2) by –2 and then add.

$\begin{cases} 3x - 2y = 6 \\ -3x + 2y = -6 \end{cases}$

$0 = 0 \quad \text{True}$

The system is dependent. There are infinitely many solutions.

57. $\begin{cases} y = -\dfrac{2}{3}x - \dfrac{7}{3} & (1) \\ y = \dfrac{3}{4}x - \dfrac{15}{4} & (2) \end{cases}$

Let $y = -\dfrac{2}{3}x - \dfrac{7}{3}$ in the second equation.

$-\dfrac{2}{3}x - \dfrac{7}{3} = \dfrac{3}{4}x - \dfrac{15}{4}$

$12\left(-\dfrac{2}{3}x - \dfrac{7}{3}\right) = 12\left(\dfrac{3}{4}x - \dfrac{15}{4}\right)$

$-8x - 28 = 9x - 45$

$-17x = -17$

$x = 1$

Let $x = 1$ in (1).

$y = -\dfrac{2}{3}(1) - \dfrac{7}{3} = -\dfrac{2}{3} - \dfrac{7}{3} = -\dfrac{9}{3} = -3$

The solution is $(1, -3)$.

59. $\begin{cases} \dfrac{x}{2} + \dfrac{y}{4} = -2 & (1) \\ \dfrac{3x}{2} + \dfrac{y}{5} = -6 & (2) \end{cases}$

Multiply (1) by –8 and (2) by 10, then add.

$\begin{cases} -4x - 2y = 16 \\ 15x + 2y = -60 \end{cases}$

$11x = -44$

$x = -4$

Let $x = -4$ in (1).

$$-\frac{4}{2} + \frac{y}{4} = -2$$

$$-2 + \frac{y}{4} = -2$$

$$\frac{y}{4} = 0$$

$$y = 0$$

The solution is $(-4, 0)$.

61. $\begin{cases} x - 2y = -7 & (1) \\ 3x + 4y = 6 & (2) \end{cases}$

Multiply (1) by 2, then add.

$\begin{cases} 2x - 4y = -14 \\ 3x + 4y = 6 \end{cases}$

$$5x \quad\quad = -8$$

$$x = -\frac{8}{5}$$

Multiply (1) by -3, then add.

$\begin{cases} -3x + 6y = 21 \\ 3x + 4y = 6 \end{cases}$

$$10y = 27$$

$$y = \frac{27}{10}$$

The solution is $\left(-\frac{8}{5}, \frac{27}{10}\right)$.

63. $\begin{cases} 6x - 5y = 1 & (1) \\ 8x - 2y = -22 & (2) \end{cases}$

Multiply (1) by -2 and (2) by 5, then add.

$\begin{cases} -12x + 10y = -2 \\ 40x - 10y = -110 \end{cases}$

$$28x \quad\quad = -112$$

$$x = -4$$

Let $x = -4$ in (1).

$$6(-4) - 5y = 1$$

$$-24 - 5y = 1$$

$$-5y = 25$$

$$y = -5$$

The solution is $(-4, -5)$.

65. $\begin{cases} y = 2x - 4y \\ 4x + 1 = 10y + 3 \end{cases}$

$\begin{cases} -2x + 5y = 0 & (1) \\ 4x - 10y = 2 & (2) \end{cases}$

Multiply (1) by 2, then add.

$\begin{cases} -4x + 10y = 0 \\ 4x - 10y = 2 \end{cases}$

$$0 = 2$$

The system is inconsistent and has no solution.

67. $\begin{cases} \dfrac{x}{2} - y = 1 \\ \dfrac{x}{5} + \dfrac{5y}{6} = \dfrac{14}{15} \end{cases}$

Multiply the first equation by 2 and the second equation by 30 to clear the fractions.

$\begin{cases} x - 2y = 2 & (1) \\ 6x + 25y = 28 & (2) \end{cases}$

Multiply (1) by -6, then add.

$\begin{cases} -6x + 12y = -12 \\ 6x + 25y = 28 \end{cases}$

$$37y = 16$$

$$y = \frac{16}{37}$$

Multiply (1) by 25 and (2) by 2, then add.

$\begin{cases} 25x - 50y = 50 \\ 12x + 50y = 56 \end{cases}$

$$37x \quad\quad = 106$$

$$x = \frac{106}{37}$$

The solution is $\left(\dfrac{106}{37}, \dfrac{16}{37}\right)$.

69. $\begin{cases} 2h + c = 770 & (1) \\ 3h + 2c = 1260 & (2) \end{cases}$

Multiply (1) by -2, then add.

$\begin{cases} -4h - 2c = -1540 \\ 3h + 2c = 1260 \end{cases}$

$$-h \quad\quad = -280$$

$$h = 280$$

Let $h = 280$ in (1).

$$2(280) + c = 770$$

$$560 + c = 770$$

$$c = 210$$

A hamburger contains 280 calories and a medium Coke contains 210 calories.

71. $\begin{cases} 2t + 3z = 65 & (1) \\ 3t + 4z = 90 & (2) \end{cases}$

Multiply (1) by -3 and (2) by 2, then add.

$\begin{cases} -6t - 9z = -195 \\ 6t + 8z = 180 \end{cases}$

$\begin{aligned} -z &= -15 \\ z &= 15 \end{aligned}$

Let $z = 15$ in (1).

$\begin{aligned} 2t + 3(15) &= 65 \\ 2t + 45 &= 65 \\ 2t &= 20 \\ t &= 10 \end{aligned}$

It takes the farmer 10 hours to plant an acre of tomatoes and 15 hours to plant an acre of zucchini.

To plant 5 acres of tomatoes and 2 acres of zucchini would take

$5(10) + 2(15) = 50 + 30 = 80$ hours, so he won't get the crops in before the rain.

73. $\begin{cases} a + r = 100 & (1) \\ 9a + 11.50r = 1000 & (2) \end{cases}$

Multiply (1) by -9, then add.

$\begin{cases} -9a - 9r = -900 \\ 9a + 11.5r = 1000 \end{cases}$

$\begin{aligned} 2.5r &= 100 \\ r &= 40 \end{aligned}$

Let $r = 40$ in (1).

$\begin{aligned} a + 40 &= 100 \\ a &= 60 \end{aligned}$

The blend will contain 60 pounds of Arabica beans and 40 pounds of Robusta beans.

75. $\begin{cases} A + B = 90 \\ A = 10 + 3B \end{cases}$

$\begin{cases} A + B = 90 & (1) \\ A - 3B = 10 & (2) \end{cases}$

Multiply (1) by 3, then add.

$\begin{cases} 3A + 3B = 270 \\ A - 3B = 10 \end{cases}$

$\begin{aligned} 4A &= 280 \\ A &= 70 \end{aligned}$

Let $A = 70$ in (1).

$\begin{aligned} 70 + B &= 90 \\ B &= 20 \end{aligned}$

The angles measure $70°$ and $20°$.

77. $\begin{cases} ax + 4y = 1 & (1) \\ -2ax - y = 3 & (2) \end{cases}$

Multiply (1) by 2, then add.

$\begin{cases} 2ax + 8y = 2 \\ -2ax - 3y = 3 \end{cases}$

$\begin{aligned} 5y &= 5 \\ y &= 1 \end{aligned}$

Let $y = 1$ in (1).

$\begin{aligned} ax + 4(1) &= 1 \\ ax + 4 &= 1 \\ ax &= -3 \\ x &= -\frac{3}{a} \end{aligned}$

The solution is $\left(-\frac{3}{a}, 1 \right)$.

79. $\begin{cases} -3x + 2y = 6a & (1) \\ x - 2y = 2b & (2) \end{cases}$

$\begin{aligned} -2x &= 6a + 2b \\ x &= \frac{6a + 2b}{-2} = -3a - b \end{aligned}$

Multiply (2) by 3, then add.

$\begin{cases} -3x + 2y = 6a \\ 3x - 6y = 6b \end{cases}$

$\begin{aligned} -4y &= 6a + 6b \\ y &= \frac{6a + 6b}{-4} = -\frac{3}{2}a - \frac{3}{2}b \end{aligned}$

The solution is $\left(-3a - b, -\frac{3}{2}a - \frac{3}{2}b \right)$.

81. It is easier to eliminate y. Multiply the second equation by 3, to form $6x + 3y = 15$. Then add the equations, obtaining $7x = 21$. So $x = 3$. Substitute $x = 3$ into either the first equation or the second equation and solve for y. $y = -1$.

83. Both of you will obtain the correct solution. Other strategies are: multiply equation (1) by one-third and adding; solving equation (2) for y and substituting this expression into equation (1); and graphing the equation and finding the point of intersection.

Putting the Concepts Together (Sections 4.1–4.3)

1. $\begin{cases} 4x + y = -20 & (1) \\ y = -\dfrac{1}{6}x + 3 & (2) \end{cases}$

    (a) $\left(3, \dfrac{5}{2}\right)$

        (1): $4(3) + \dfrac{5}{2} \stackrel{?}{=} -20$

               $12 + \dfrac{5}{2} \stackrel{?}{=} -20$

             $\dfrac{24}{2} + \dfrac{5}{2} \stackrel{?}{=} -20$

                 $\dfrac{29}{2} = -20$ False

         $\left(3, \dfrac{5}{2}\right)$ is not a solution.

    (b) $(-6, 4)$
        (1): $4(-6) + 4 \stackrel{?}{=} -20$
              $-24 + 4 \stackrel{?}{=} -20$
                 $-20 = -20$ True
        (2): $4 \stackrel{?}{=} -\dfrac{1}{6}(-6) + 3$
             $4 \stackrel{?}{=} 1 + 3$
             $4 = 4$ True
        $(-6, 4)$ is a solution.

    (c) $(-4, -4)$
        (1): $4(-4) + (-4) \stackrel{?}{=} -20$
               $-16 - 4 \stackrel{?}{=} -20$
                 $-20 = -20$ True
        (2): $-4 \stackrel{?}{=} -\dfrac{1}{6}(-4) + 3$
            $-4 \stackrel{?}{=} \dfrac{2}{3} + 3$
            $-4 = \dfrac{11}{3}$ False

        $(-4, -4)$ is not a solution.

2. (a) The lines coincide. Therefore, there are infinitely many solutions.

    (b) The system is consistent.

    (c) The system is dependent.

3. (a) Since the slopes are not equal, the lines intersect at one point. There is one solution.

(b) The system is consistent.

(c) The system is independent.

4. $\begin{cases} 4x + y = 5 & (1) \\ -x + y = 0 & (2) \end{cases}$
Solve (2) for y.
$y = x$
Let $y = x$ in (1).
$4x + x = 5$
    $5x = 5$
      $x = 1$
Let $x = 1$ in (2).
$-1 + y = 0$
     $y = 1$
The solution is $(1, 1)$.

5. $\begin{cases} y = -\dfrac{2}{5}x + 1 & (1) \\ y = -x + 4 & (2) \end{cases}$
Let $y = -x + 4$ in (1).
    $-x + 4 = -\dfrac{2}{5}x + 1$

    $-x + \dfrac{2}{5}x = 1 - 4$

        $-\dfrac{3}{5}x = -3$

          $x = 5$
Let $x = 5$ in (2).
$y = -5 + 4$
$y = -1$
The solution is $(5, -1)$.

6. $\begin{cases} x = 2y + 11 & (1) \\ 3x - y = 8 & (2) \end{cases}$
Let $x = 2y + 11$ in (2).
$3(2y + 11) - y = 8$
    $6y + 33 - y = 8$
          $5y = -25$
           $y = -5$
Let $y = -5$ in (1).
$x = 2(-5) + 11$
$x = -10 + 11$
$x = 1$
The solution is $(1, -5)$.

7. $\begin{cases} 4x + 3y = -4 & (1) \\ x + 5y = -1 & (2) \end{cases}$
Solve (2) for x.
$x = -5y - 1$
Let $x = -5y - 1$ in (1).

$$4(-5y-1)+3y=-4$$
$$-20y-4+3y=-4$$
$$-17y=0$$
$$y=0$$

Let $y = 0$ in (2).
$$x+5(0)=-1$$
$$x=-1$$
The solution is $(-1, 0)$.

8. $\begin{cases} y=-2x-3 & (1) \\ y=\dfrac{1}{2}x+7 & (2) \end{cases}$

Let $y = -2x - 3$ in (2).
$$-2x-3=\frac{1}{2}x+7$$
$$-2x-\frac{1}{2}x=7+3$$
$$-\frac{5}{2}x=10$$
$$x=-4$$

Let $x = -4$ in (1).
$$y=-2(-4)-3$$
$$y=8-3$$
$$y=5$$
The solution is $(-4, 5)$.

9. $\begin{cases} -2x+4y=2 & (1) \\ 3x+5y=-14 & (2) \end{cases}$

Multiply (1) by 3 and (2) by 2. Then add.
$$\begin{cases} -6x+12y=6 \\ 6x+10y=-28 \end{cases}$$
$$\overline{\qquad 22y=-22}$$
$$y=-1$$

Let $y = -1$ in (1).
$$-2x+4(-1)=2$$
$$-2x-4=2$$
$$-2x=6$$
$$x=-3$$
The solution is $(-3, -1)$.

10. $\begin{cases} -\dfrac{3}{4}x+\dfrac{2}{3}y=\dfrac{9}{4} & (1) \\ 3x-\dfrac{1}{2}y=-\dfrac{5}{2} & (2) \end{cases}$

Multiply (1) by 4 and add.

$$\begin{cases} -3x+\dfrac{8}{3}y=9 \\ 3x-\dfrac{1}{2}y=-\dfrac{5}{2} \end{cases}$$
$$\overline{\quad \dfrac{8}{3}y-\dfrac{1}{2}y=9-\dfrac{5}{2}}$$
$$\frac{16}{6}y-\frac{3}{6}y=\frac{18}{2}-\frac{5}{2}$$
$$\frac{13}{6}y=\frac{13}{2}$$
$$y=3$$

Let $y = 3$ in (2).
$$3x-\frac{1}{2}(3)=-\frac{5}{2}$$
$$3x-\frac{3}{2}=-\frac{5}{2}$$
$$3x=-\frac{5}{2}+\frac{3}{2}$$
$$3x=-1$$
$$x=-\frac{1}{3}$$
The solution is $\left(-\dfrac{1}{3}, 3\right)$.

11. $\begin{cases} 3(y+3)=1+4(3x-1) & (1) \\ x=\dfrac{y}{4}+1 & (2) \end{cases}$

Let $x=\dfrac{y}{4}+1$ in (1).

$$3(y+3)=1+4\left[3\left(\frac{y}{4}+1\right)-1\right]$$
$$3y+9=1+4\left(\frac{3y}{4}+3-1\right)$$
$$3y+9=1+4\left(\frac{3y}{4}+2\right)$$
$$3y+9=1+3y+8$$
$$3y+9=3y+9$$
$$9=9 \quad \text{True}$$
The system is dependent. There are infinitely many solutions.

12. $\begin{cases} 0.4x-2.5y=-6.5 & (1) \\ x+y=5.5 & (2) \end{cases}$

Multiply (1) by 10 and (2) by -4. Then add.
$$\begin{cases} 4x-25y=-65 \\ -4x-4y=-22 \end{cases}$$
$$\overline{\quad -29y=-87}$$
$$y=3$$

Let $y = 3$ in (2).
$$x + 3 = 5.5$$
$$x = 2.5$$
The solution is (2.5, 3).

13. $\begin{cases} 2(y+1) = 3x + 4 & (1) \\ x = \dfrac{2}{3}y - 3 & (2) \end{cases}$

Let $x = \dfrac{2}{3}y - 3$ in (1).

$$2(y+1) = 3\left(\dfrac{2}{3}y - 3\right) + 4$$
$$2y + 2 = 2y - 9 + 4$$
$$2y + 2 = 2y - 5$$
$$2 = -5 \quad \text{False}$$

The system is inconsistent. There is no solution.

Section 4.4

Preparing for Solving Direct Translation, Geometry, and Uniform Motion Problems Using Systems of Linear Equations

P1. $r = 45, t = 3$
$$d = rt$$
$$d = 45 \cdot 3$$
$$d = 135$$
You will travel 135 miles.

P2. $P = 3600, r = 0.015, t = \dfrac{1}{12}$

$$I = Prt$$
$$I = 3600 \cdot 0.015 \cdot \dfrac{1}{12}$$
$$I = 4.5$$
The interest paid after 1 month is $4.50.

4.4 Quick Checks

1. We are looking for two unknown numbers. Let x represent the first number and y represent the second number. The sum of the numbers is 104, so $x + y = 104$. The second number is 25 less than twice the first number, so $y = 2x - 25$. The system is
$$\begin{cases} x + y = 104 & (1) \\ y = 2x - 25 & (2) \end{cases}$$
Let $y = 2x - 25$ in equation (1).
$$x + (2x - 25) = 104$$
$$3x - 25 = 104$$
$$3x = 129$$
$$x = 43$$

Let $x = 43$ in equation (2).
$$y = 2(43) - 25$$
$$y = 86 - 25$$
$$y = 61$$
The sum of 43 and 61 is 104. Twice 43 minus 25 is 61. The numbers are 43 and 61.

2. We are looking for the length and the width of the yard. Let l represent the length and w represent the width of the yard. The perimeter is 400 yards, so $2l + 2w = 400$. The length is three times the width, so $l = 3w$. The system is
$$\begin{cases} 2l + 2w = 400 & (1) \\ l = 3w & (2) \end{cases}$$
Let $l = 3w$ in equation (1).
$$2(3w) + 2w = 400$$
$$6w + 2w = 400$$
$$8w = 400$$
$$w = 50$$
Let $w = 50$ in equation (2).
$$l = 3(50) = 150$$
With $l = 150$ and $w = 50$, the perimeter is $2(150) + 2(50) = 400$ yards. The length (150 yards) is three times the width (50 yards). The length of the yard is 150 yards and the width is 50 yards.

3. True

4. Supplementary angles are angles whose measures sum to 180°.

5. This is a complementary angle problem. We are looking for the measures of two angles whose sum is 90°. Let x represent the measure of the smaller angle and y represent the measure of the larger angle. Since the angles are complementary, we have $x + y = 90$. Since the larger angle measures 18° more than the smaller angle, we have
$y = x + 18$. The system is
$$\begin{cases} x + y = 90 & (1) \\ y = x + 18 & (2) \end{cases}$$
Let $y = x + 18$ in equation (1).
$$x + (x + 18) = 90$$
$$2x + 18 = 90$$
$$2x = 72$$
$$x = 36$$
Let $x = 36$ in equation (2).
$$y = 36 + 18 = 54$$
The sum of 36 and 54 is 90, and 54 is 18 more than 36. The angles measure 36° and 54°.

6. This is a supplementary angle problem. We are looking for the measures of two angles whose sum is 180°. Let x represent the measure of the smaller angle and y represent the measure of the larger angle. Since the angles are supplementary, $x + y = 180$. Since the larger angle measures 16° less than three times the measure of the smaller angle,
$y = 3x - 16$.
The system is
$$\begin{cases} x + y = 180 & (1) \\ y = 3x - 16 & (2) \end{cases}$$
Let $y = 3x - 16$ in equation (1).
$$x + (3x - 16) = 180$$
$$4x - 16 = 180$$
$$4x = 196$$
$$x = 49$$
Let $x = 49$ in equation (2).
$y = 3(49) - 16 = 147 - 16 = 131$
The sum of 49 and 131 is 180, and 131 is 16 less than three times 49. The angles measure 49° and 131°.

7. True

8. This is a uniform motion problem. We want to determine the air speed of the plane and the effect of wind resistance. Let a represent the airspeed of the plane and w represent the impact of wind resistance.

	Distance	Rate	Time
East	1200	$a + w$	3
West	1200	$a - w$	4

The system is
$$\begin{cases} 3(a + w) = 1200 \\ 4(a - w) = 1200 \end{cases}$$
or
$$\begin{cases} 3a + 3w = 1200 & (1) \\ 4a - 4w = 1200 & (2) \end{cases}$$
Multiply (1) by 4 and (2) by 3. Then add.
$$\begin{cases} 12a + 12w = 4800 \\ 12a - 12w = 3600 \end{cases}$$
$$\overline{ 24a = 8400}$$
$$a = 350$$
Let $a = 350$ in equation (1).

$$3(350) + 3w = 1200$$
$$1050 + 3w = 1200$$
$$3w = 150$$
$$w = 50$$
Flying west the groundspeed of the plane is $350 - 50 = 300$ miles per hour. Flying west, the plane flies 1200 miles in 4 hours for an average speed of 300 miles per hour. Flying east the groundspeed of the plane is $350 + 50 = 400$ miles per hour. Flying east, the plane flies 1200 miles in 3 hours for an average speed of 400 miles per hour. The airspeed of the plane is 350 miles per hour. The impact of wind resistance on the plane is 50 miles per hour.

4.4 Exercises

9. Two times the smaller number is written as $2x$. Twelve more than one-half the larger number is written as $12 + \dfrac{1}{2}y$. The equation is $12 + \dfrac{1}{2}y$.

11. Since the perimeter is 59 inches and the formula for perimeter is $P = 2l + 2w$, the equation is $2l + 2w = 59$.

13. The rate of the boat going upstream is $r - c$ and the time going upstream is 8 hours. The distance is 16 miles. Multiply the rate by the time and set equal to the distance. The equation is $8(r - c) = 16$.

15. Let x represent the first number and y represent the second number. The sum of x and y, $x + y$, is 82, while the difference of x and y, $x - y$, is 16.
$$\begin{cases} x + y = 82 \\ x - y = 16 \end{cases}$$
$$\overline{2x = 98}$$
$$x = 49$$
$$49 + y = 82$$
$$y = 33$$
The numbers are 33 and 49.

17. Let x represent the first number and y represent the second number. The sum of x and y, $x + y$, is 51, while twice the first subtracted from the second, $y - 2x$, is 9.
$$\begin{cases} x + y = 51 \\ y - 2x = 9 \end{cases}$$
$$\begin{cases} x + y = 51 & (1) \\ -2x + y = 9 & (2) \end{cases}$$
Multiply (1) by 2, then add.

$$\begin{cases} 2x + 2y = 102 \\ -2x + y = 9 \end{cases}$$

$$\begin{aligned} 3y &= 111 \\ y &= 37 \end{aligned}$$

$$\begin{aligned} x + 37 &= 51 \\ x &= 14 \end{aligned}$$

The numbers are 14 and 37.

19. Let t represent Thursday night's attendance, and let f represent Friday night's attendance. The total attendance for the two nights, $t + f$, was 77,000, while 7000 more than two-thirds of Friday's attendance, $7000 + \frac{2}{3}f$, was Thursday's attendance, t.

$$\begin{cases} t + f = 77,000 & (1) \\ 7000 + \frac{2}{3}f = t & (2) \end{cases}$$

Let $t = 7000 + \frac{2}{3}f$ in (1).

$$7000 + \frac{2}{3}f + f = 77,000$$

$$\frac{5}{3}f = 70,000$$

$$f = \frac{3}{5}(70,000) = 42,000$$

$$t + 42,000 = 77,000$$

$$t = 35,000$$

Thursday night's attendance was 35,000; Friday night's attendance was 42,000.

21. Let s be the amount in stocks and b be the amount in bonds. The total amount of money, $s + b$, is \$21,000. Four times the amount in bonds equals three times the amount in stocks, or $4b = 3s$.

$$\begin{cases} s + b = 21,000 & (1) \\ 4b = 3s & (2) \end{cases}$$

Let $b = \frac{3s}{4}$ in (1).

$$s + \frac{3s}{4} = 21,000$$

$$\frac{7s}{4} = 21,000$$

$$s = 12,000$$

$$12,000 + b = 21,000$$

$$b = 9000$$

Therefore, \$12,000 should be invested in stocks and \$9000 in bonds.

23. Let l represent the length of the garden along the garage, and w represent the width. Since the length is 3 feet more than the width, $l = w + 3$. The sides that need to be fenced are w feet, l feet, and w feet, which total $2w + l$, or 30 feet.

$$\begin{cases} l = w + 3 & (1) \\ 2w + l = 30 & (2) \end{cases}$$

Let $l = w + 3$ in (2).

$$2w + w + 3 = 30$$

$$3w = 27$$

$$w = 9$$

$$l = 9 + 3 = 12$$

The length is 12 feet and the width is 9 feet.

25. Let l and w represent the length and width of the rectangle, respectively. The perimeter of 70 meters is $2l + 2w$. Since the width is 40% of the length, $w = 0.40l$.

$$\begin{cases} 2l + 2w = 70 & (1) \\ w = 0.40l & (2) \end{cases}$$

Let $w = 0.40l$ in (1).

$$2l + 2(0.40l) = 70$$

$$2l + 0.8l = 70$$

$$2.8l = 70$$

$$l = 25$$

$$w = 0.40(25) = 10$$

The length is 25 meters and the width is 10 meters.

27. Let x and y represent the measures of the angles. Since the angles are complementary, the sum of their measures, $x + y$, is 90. Since one angle, say x, is $15°$ more than half its complement y,

$$x = \frac{1}{2}y + 15.$$

$$\begin{cases} x + y = 90 & (1) \\ x = \frac{1}{2}y + 15 & (2) \end{cases}$$

Let $x = \frac{1}{2}y + 15$ in (1).

$$\frac{1}{2}y + 15 + y = 90$$

$$\frac{3}{2}y = 75$$

$$y = \frac{2}{3}(75) = 2(25) = 50$$

$$x + 50 = 90$$

$$x = 40$$

The angles measure $40°$ and $50°$.

29. Let x and y represent the measures of the angles. Since the angles are supplementary, the sum of their measures, $x + y$, is $180°$. Since one angle, say x, is $30°$ less than one-third of its supplement y, $x = \dfrac{1}{3}y - 30$.

$$\begin{cases} x + y = 180 & (1) \\ x = \dfrac{1}{3}y - 30 & (2) \end{cases}$$

Let $x = \dfrac{1}{3}y - 30$ in (1).

$$\dfrac{1}{3}y - 30 + y = 180$$

$$\dfrac{4}{3}y = 210$$

$$y = \dfrac{3}{4}(210) = 157.5$$

$$x + 157.5 = 180$$

$$x = 22.5$$

The angles measure $22.5°$ and $157.5°$.

31. Let s be his paddling speed and c the speed of the current. Paddling with the current, his speed is $s + c$, or 4.3 mph. Paddling against the current, his speed is $s - c$, or 3.5 mph.

$$\begin{cases} s + c = 4.3 \\ s - c = 3.5 \end{cases}$$

$$\begin{aligned} 2s &= 7.8 \\ s &= 3.9 \end{aligned}$$

$$3.9 - c = 3.5$$

$$-c = -0.4$$

$$c = 0.4$$

The current is 0.4 mph and his still-water speed is 3.9 mph.

33. Let b be his biking speed and w the wind speed. Biking with the wind, his speed is $b + w$; biking against the wind, his speed is $b - w$. Biking for 6 hours against the wind is a distance of $6(b - w)$, while biking for 5 hours with the wind is a distance of $5(b + w)$. The distance is 60 miles.

$$\begin{cases} 6(b - w) = 60 & (1) \\ 5(b + w) = 60 & (2) \end{cases}$$

Divide (1) by 6 and (2) by 5, then add.

$$\begin{cases} b - w = 10 \\ b + w = 12 \end{cases}$$

$$\begin{aligned} 2b &= 22 \\ b &= 11 \end{aligned}$$

$$11 - w = 10$$

$$w = 1$$

The biking speed is 11 mph and the wind is 1 mph.

35. Let n be the speed of the northbound train and s the speed of the southbound train. Since the northbound train is going 12 mph slower, $n = s - 12$. After 4 hours, the northbound train will have gone $4n$ miles and the southbound train will have gone $4s$ miles. The total distance, $4n + 4s$, is 528 miles.

$$\begin{cases} n = s - 12 & (1) \\ 4n + 4s = 528 & (2) \end{cases}$$

Let $n = s - 12$ in (2).

$$4(s - 12) + 4s = 528$$

$$4s - 48 + 4s = 528$$

$$8s = 576$$

$$s = 72$$

$$n = 72 - 12 = 60$$

The northbound train is going 60 mph; the southbound train is going 72 mph.

37. Let v be the length of time that Vanessa rides, and r the length of time that Richie rides. Since Vanessa leaves the camp 30 minutes (0.5 hour) later, $r = v + 0.5$. In v hours, Vanessa goes $10v$ miles, while Richie goes $7r$ miles in r hours. The distance between them is 7 miles when $10v - 7r = 7$.

$$\begin{cases} r = v + 0.5 & (1) \\ 10v - 7r = 7 & (2) \end{cases}$$

Let $r = v + 0.5$ in (2).

$$10v - 7(v + 0.5) = 7$$

$$10v - 7v - 3.5 = 7$$

$$3v = 10.5$$

$$v = 3.5$$

$$r = 3.5 + 0.5 = 4.0$$

Richie has been riding for 4 hours when Vanessa is 7 miles ahead.

39. Let p be the speed of the Piper aircraft, and w the speed of the wind. Flying with the wind, the rate is $p + w$, and the distance traveled in 3 hours is $3(p + w)$, which is 600. Flying against the wind, the rate is $p - w$, and the distance traveled in 4 hours is $4(p - w)$, which is 600.

$$\begin{cases} 3(p + w) = 600 & (1) \\ 4(p - w) = 600 & (2) \end{cases}$$

Divide (1) by 3 and (2) by 4, then add.

$$\begin{cases} p+w=200 \\ p-w=150 \end{cases}$$
$$\overline{2p \quad\;\; =350}$$
$$p=175$$
$$3(175+w)=600$$
$$175+w=200$$
$$w=25$$

The average wind speed was 25 mph and the speed of the Piper was 175 mph.

41. Let x be the digit in the tens place and y the digit in the ones place. Since the sum of the digits is 6, we have $x+y=6$.
The original number equals $10x+y$. With the digits reversed, the value is $10y+x$. The difference between the new number and the original number is 18, so we have $(10y+x)-(10x+y)=18$. Simplify this equation.
$$(10y+x)-(10x+y)=18$$
$$10y+x-10x-y=18$$
$$-9x+9y=18$$
$$-9(x-y)=18$$
$$x-y=-2$$

We have the following system. Add the equations.
$$\begin{cases} x+y=6 & (1) \\ x-y=-2 & (2) \end{cases}$$
$$\overline{2x \quad\;\; =4}$$
$$x=2$$

Let $x=2$ in equation (1).
$$2+y=6$$
$$y=4$$

The tens digit is 2 and the ones digit is 4. The number is 24.

Section 4.5

Preparing for Solving Mixture Problems Using Systems of Linear Equations

P1. $P=1200$, $r=0.14$, $t=\dfrac{1}{12}$

$I=Prt$

$I=1200 \cdot 0.14 \cdot \dfrac{1}{12}$

$I=14$

Roberta will be charged $14 interest for one month. After one month, her balance will be $1200 + $14 = $1214.

P2. $0.25x=80$
$$\dfrac{0.25x}{0.25}=\dfrac{80}{0.25}$$
$$x=320$$
The solution set is $\{320\}$.

4.5 Quick Checks

1. The general equation we use to solve mixture problems is
   <u>number of units of the same kind</u> · <u>rate</u>
   = <u>amount</u>.

2. Let a represent the number of adult tickets and c represent the number of children's tickets.

	Number	Cost per person	
Adult	a	42.95	$42.95a$
Children	c	15.95	$15.95c$
Total	9		$332.55

3. We want to know the number of dimes in the piggy bank. Let q represent the number of quarters and d represent the number of dimes.

	Number	Value	Total Value
Quarters	q	0.25	$0.25q$
Dimes	d	0.10	$0.10d$
Total	85		14.50

The system is
$$\begin{cases} q+d=85 & (1) \\ 0.25q+0.10d=14.50 & (2) \end{cases}$$

Multiply (1) by -0.10 and then add.
$$\begin{cases} -0.10q-0.10d=-8.50 \\ 0.25q+0.10d=14.50 \end{cases}$$
$$\overline{0.15q \quad\quad\;\; =6.00}$$
$$q=40$$

Let $q=40$ in equation (1).
$$40+d=85$$
$$d=45$$

The total number of coins is $40 + 45 = 85$. The value of the coins is
$0.25(40) + 0.10(45) = 14.50$. There are 45 dimes.

4. The simple interest formula states that interest = <u>principal</u> · <u>rate</u> · <u>time</u>.

5. *True or False:* When we solve a simple interest problem using the mixture model, we assume that $t = 1$. <u>True</u>

6. We need to determine how much should be placed in each investment. Let a represent the amount invested in Aa-bonds and b represent the amount invested in B-rated bonds.

	Principal	Rate	Interest
Aa-bonds	a	0.05	$0.05a$
B-rated	b	0.07	$0.07b$
Total	90,000		5500

The system is
$$\begin{cases} a + b = 90,000 & (1) \\ 0.05a + 0.07b = 5500 & (2) \end{cases}$$
Multiply (1) by -0.05 and then add.
$$\begin{cases} -0.05a - 0.05b = -4500 \\ 0.05a + 0.07b = 5500 \end{cases}$$
$$\begin{aligned} 0.02b &= 1000 \\ b &= 50,000 \end{aligned}$$
Let $b = 50,000$ in equation (1).
$$a + 50,000 = 90,000$$
$$a = 40,000$$
The sum of \$40,000 and \$50,000 is \$90,000, and the total interest is $0.05(40,000) + 0.07(50,000) = \$5500$. Faye should invest \$40,000 in Aa-bonds and \$50,000 in B-rated bonds.

7. We need to know the number of pounds of each type of coffee that are required in the mix. Let b represent the pounds of Brazilian coffee and c represent the pounds of Colombian coffee.

	Price per pound	Number of pounds	Revenue
Brazilian	6	b	$6b$
Colombian	10	c	$10c$
Blend	9	20	$9(20) = 180$

The system is
$$\begin{cases} b + c = 20 & (1) \\ 6b + 10c = 180 & (2) \end{cases}$$
Multiply (1) by -6 and then add.
$$\begin{cases} -6b - 6c = -120 \\ 6b + 10c = 180 \end{cases}$$
$$\begin{aligned} 4c &= 60 \\ c &= 15 \end{aligned}$$
Let $c = 15$ in equation (1).
$$b + 15 = 20$$
$$b = 5$$
The total weight is $5 + 15 = 20$ pounds. The total revenue is $6(5) + 10(15) = \$180$. Mix 5 pounds of Brazilian coffee with 15 pounds of Colombian coffee.

8. We need to know how many gallons of each type of ice cream are required for the mixture. Let x represent the gallons of 5% butterfat ice cream and y represent the gallons of 15% butterfat ice cream.

	Gallons	Concentration	Amount of Alcohol
5% Butterfat	x	0.05	$0.05x$
15% Butterfat	y	0.15	$0.15y$
Total	200	0.09	$0.09(200) = 18$

The system is
$$\begin{cases} x + y = 200 & (1) \\ 0.05x + 0.15y = 18 & (2) \end{cases}$$

Multiply (1) by -0.05 and then add.
$$\begin{cases} -0.05x - 0.05y = -10 \\ 0.05x + 0.15y = 18 \end{cases}$$
$$\overline{0.10y = 8}$$
$$y = 80$$

Let $y = 80$ in equation (1).
$$x + 80 = 200$$
$$x = 120$$

The total number of gallons is $120 + 80 = 200$. The total butterfat is $0.05(120) + 0.15(80) = 18$ gallons. Mix 120 gallons of 5% butterfat ice cream with 80 gallons of 15% butterfat ice cream.

4.5 Exercises

9. Let a be the number of adult tickets and s be the number of student tickets.

	number	cost per person	total value
adults' tickets	a	4	$4a$
students' tickets	s	1.5	$1.5s$
total	215		580

$$\begin{cases} a + s = 215 \\ 4a + 1.5s = 580 \end{cases}$$

11. Let s be the amount in savings and m be the amount in a money market.

	Principal	Rate	Interest
Savings account	s	0.05	$0.05s$
money market	m	0.03	$0.03m$
total	1600		50

$$\begin{cases} s + m = 1600 \\ 0.05s + 0.03m = 50 \end{cases}$$

13. Let m be the pounds of mild coffee and r be the pounds of robust coffee.

	number of pounds	price per pound	total value
mild coffee	m	7.50	$7.5m$
robust coffee	r	10	$10r$
total	12	8.75	8.75(12)

$$\begin{cases} m + r = 12 \\ 7.5m + 10r = 8.75(12) \end{cases}$$

15.

	number	cost	total
adults	a	32	$32a$
children	c	24	$24c$
total	11		296

$$\begin{cases} a + c = 11 \\ 32a + 24c = 296 \end{cases}$$

17.

	P	R	I
A account	A	0.10	$0.10A$
B account	B	0.07	$0.07B$
total	2250		195

$$\begin{cases} A + B = 2250 \\ 0.01A + 0.07B = 195 \end{cases}$$

19.

	number bouquets	price per pound	total
Red	r	5.85	$5.85r$
Yellow	y	4.20	$4.20y$
total	$r + y$		128.85

$$\begin{cases} r = 3 + 2y \\ 5.85r + 4.20y = 128.85 \end{cases}$$

21. Let s be the number of student tickets and n be the number of nonstudents.

	number	price	amount
students	s	8	$8s$
nonstudents	n	10	$10n$
total	390		3270

$$\begin{cases} s + n = 390 & (1) \\ 8s + 10n = 3270 & (2) \end{cases}$$

Solve (1) for n.
$n = 390 - s$
Let $n = 390 - s$ in (2).
$8s + 10(390 - s) = 3270$
$8s + 3900 - 10s = 3270$
$\qquad\qquad -2s = -630$
$\qquad\qquad\quad s = 315$
There were 315 student tickets sold.

23. Let b be the flats of bedding plants and h be the number of hanging baskets.

	number	price	amount
bedding	b	13	$13b$
hanging	h	18	$18h$
total			8800

$$\begin{cases} b = 2h & (1) \\ 13b + 18h = 8800 & (2) \end{cases}$$

Solve (1) for h.

$$h = \frac{b}{2}$$

Let $h = \frac{b}{2}$ in (2).

$$13b + 18\left(\frac{b}{2}\right) = 8800$$
$$13b + 9b = 8800$$
$$22b = 8800$$
$$b = 400$$

They hope to sell 400 flats of bedding plants.

25. Let n be the number of nickels and d be the number of dimes.

	number	value	total amount
nickels	n	0.05	$0.05n$
dimes	d	0.10	$0.10d$
total	150		12

$$\begin{cases} n+d=150 & (1) \\ 0.05n+0.10d=12 & (2) \end{cases}$$

Solve (1) for n.

$n = 150 - d$

Let $n = 150 - d$ in (2).

$0.05(150-d)+0.10d=12$

$7.5-0.05d+0.10d=12$

$0.05d = 4.5$

$d = 90$

Let $d = 90$ in (1).

$n+90=150$

$n = 60$

There are 60 nickels and 90 dimes in the jar.

27. Let f be the cost of each first-class stamp and p be the cost of each postcard stamp.

	number	price	amount
first-class	20	f	$20f$
postcard	10	p	$10p$
total			11.10

	number	price	amount
first-class	80	f	$80f$
postcard	50	p	$50p$
total			47.10

$$\begin{cases} 20f+10p=11.10 & (1) \\ 80f+50p=47.10 & (2) \end{cases}$$

Multiply (1) by -5, then add.

$$\begin{cases} -100f-50p=-55.50 \\ 80f+50p=47.10 \end{cases}$$

$-20f =-8.4$

$f = 0.42$

Let $f = 0.42$ in (1).

$20(0.37)+10p=9.30$

$7.4+10p=9.30$

$10p=1.9$

$p=0.19$

A first-class stamp was $0.42 and a postcard stamp was $0.19.

29. Let x be the amount invested at 5% and y be the amount invested at 8%.

	P	R	I
5% account	x	0.05	$0.05x$
8% account	y	0.08	$0.08y$
total	10,000		575

$$\begin{cases} x+y=10,000 & (1) \\ 0.05x+0.08y=575 & (2) \end{cases}$$

Solve (1) for x.

$x = 10,000 - y$

Let $x = 10,000 - y$ in (2).

$0.05(10,000-y)+0.08y=575$

$500-0.05y+0.08y=575$

$0.03y=75$

$y=2500$

Let $y = 2500$ in (1).

$x+2500=10,000$

$x=7500$

He should invest $7500 in the 5% account and $2500 in the 8% account.

31. Let r be the amount invested in the risky plan and s be the amount invested in the safer plan.

	P	R	I
risky	r	0.12	$0.12r$
safe	s	0.08	$0.08s$
total	5000		528

$$\begin{cases} r+s=5000 & (1) \\ 0.12r+0.08s=528 & (2) \end{cases}$$

Solve (1) for r.

$r = 5000 - s$

Let $r = 5000 - s$ in (2).

$0.12(5000-s)+0.08s=528$

$600-0.12s+0.08s=528$

$-0.04s=-72$

$s=1800$

Let $s = 1800$ in (1).

$$r + 1800 = 5000$$
$$r = 3200$$

She invested $3200 in the risky plan and $1800 in the safer plan.

33. Let a be the pounds of arbequina olives and g be the pounds of green olives.

	number pounds	cost	total amount
arbequina	a	9	$9a$
green	g	4	$4g$
total	5	6	$5(6)$

$$\begin{cases} a + g = 5 & (1) \\ 9a + 4g = 30 & (2) \end{cases}$$

Multiply (1) by –4, then add.

$$\begin{cases} -4a - 4g = -20 \\ 9a + 4g = 30 \end{cases}$$
$$\overline{\quad 5a \qquad = 10}$$
$$a = 2$$
$$2 + g = 5$$
$$g = 3$$

2 pounds of arbequina and 3 pounds of green olives should be mixed.

35. Let x be the pounds of $2.75 per pound coffee and y be the pounds of $5 per pound coffee.

	number of pounds	price	total amount
$2.75/lb	x	2.75	$2.75x$
$5/lb	y	5	$5y$
total	100	3.90	$3.90(100)$

$$\begin{cases} x + y = 100 & (1) \\ 2.75x + 5y = 390 & (2) \end{cases}$$

Solve (1) for x.
$$x = 100 - y$$
Let $x = 100 - y$ in (2).
$$2.75(100 - y) + 5y = 390$$
$$275 - 2.75y + 5y = 390$$
$$2.25y = 115$$
$$y = 51.1$$
Let $y = 51.1$ in (1).
$$x + 51.1 = 100$$
$$x = 48.9$$

About 48.9 pounds of the $2.75 per pound coffee should be blended with about 51.1 pounds of the $5 per pound coffee.

37. Let r be the pounds of rye seed and b be the pounds of blue-grass seed.

	number of pounds	price	total amount
rye	r	4.20	$4.20r$
blue-grass	b	3.75	$3.75b$
total	180	3.95	$3.95(180)$

$$\begin{cases} r + b = 180 & (1) \\ 4.2r + 3.75b = 711 & (2) \end{cases}$$

Solve (1) for r.
$$r = 180 - b$$
Let $r = 180 - b$ in (2).
$$4.2(180 - b) + 3.75b = 711$$
$$756 - 4.2b + 3.75b = 711$$
$$-0.45b = -45$$
$$b = 100$$
Let $b = 100$ in (1).
$$r + 100 = 180$$
$$r = 80$$

The mixture has 80 pounds of rye seed and 100 pounds of blue-grass seed.

39. Let x be the ml of 30% saline solution and y be the ml of 60% saline solution.

	ml	concentration	amount
30%	x	0.3	$0.3x$
60%	y	0.6	$0.6y$
total	60	0.5	$0.5(60)$

$$\begin{cases} x + y = 60 & (1) \\ 0.3x + 0.6y = 30 & (2) \end{cases}$$

Solve (1) for x.
$$x = 60 - y$$
Let $x = 60 - y$ in (2).
$$0.3(60 - y) + 0.6y = 30$$
$$18 - 0.3y + 0.6y = 30$$
$$0.3y = 12$$
$$y = 40$$
Let $y = 40$ in (1).
$$x + 40 = 60$$
$$x = 20$$

She should add 20 ml of 30% saline solution to 40 ml of 60% saline solution.

41. Let x be the liters of 10% silver and y be the total liters of the 30% silver.

	liters	concentration	amount
10%	x	0.1	$0.1x$
50%	70	0.5	70(0.5)
total (30%)	y	0.3	$0.3y$

$$\begin{cases} x+70=y & (1) \\ 0.1x+35=0.3y & (2) \end{cases}$$

Let $y = x + 70$ in (2).
$0.1x+35=0.3(x+70)$
$0.1x+35=0.3x+21$
$\qquad 14 = 0.2x$
$\qquad 70 = x$

So 70 liters of 10% silver should be added.

43. Let x be the gallons of 25% antifreeze that must remain. Let y be the gallons of water (or the amount of antifreeze drained).

	gallons	concentration	amount
25%	x	0.25	$0.25x$
water	y	0	0
total (15%)	3	0.15	0.15(3)

$$\begin{cases} x+y=3 & (1) \\ 0.25x=0.45 & (2) \end{cases}$$

Solve (2) for x.
$$x = \frac{0.45}{0.25} = 1.8$$

Let $x = 1.8$ in (1).
$1.8 + y = 3$
$\qquad y = 1.2$

Therefore, 1.8 gallons of 25% antifreeze should remain and 1.2 gallons should be drained and replaced with water.

45. Let x be the amount invested at 5% and y be the amount invested at a loss of 7.5%.

	P	R	I
5%	x	0.05	$0.05x$
7.5%	y	0.075	$0.075y$
total	10,000		25

$$\begin{cases} x+y=10,000 & (1) \\ 0.05x-0.075y=25 & (2) \end{cases}$$

Solve (1) for x.
$x = 10,000 - y$
Let $x = 10,000 - y$ in (2).
$0.05(10,000-y)-0.075y=25$
$\quad 500-0.05y-0.075y=25$
$\qquad\qquad -0.125y=-475$
$\qquad\qquad\qquad y=3800$

Let $y = 3800$ in (1).
$x+3800=10,000$
$\qquad x = 6200$

He invested \$6200 at 5% and \$3800 at $7\frac{1}{2}$% loss.

47. The percentage ethanol in the final solution is greater than the percentage of ethanol in either of the two original solutions.

Section 4.6

Preparing for Systems of Linear Inequalities

P1. $3x-2 \ge 7$
$\quad 3x \ge 9$
$\quad\; x \ge 3$
The solution set is $\{x|x \ge 3\}$ or $[3, \infty)$.

P2. $4(x-1) < 6x+4$
$\quad 4x-4 < 6x+4$
$\quad -2x-4 < 4$
$\qquad -2x < 8$
$\qquad\; x > -4$
The solution set is $\{x|x > -4\}$ or $(-4, \infty)$.

P3. $y > 2x-5$
Graph the line $y = 2x - 5$ using a dashed line. Because (0, 0) satisfies the inequality, we shade the half-plane containing (0, 0).

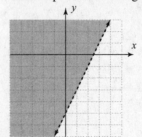

P4. $2x + 3y \leq 9$

$3y \leq -2x + 9$

$y \leq -\dfrac{2}{3}x + 3$

Graph $y = -\dfrac{2}{3}x + 3$ using a solid line.

Because (0, 0) satisfies the inequality, we shade the half-plane containing (0, 0).

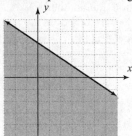

4.6 Quick Checks

1. An ordered pair is a <u>solution</u> of a system of linear inequalities if it makes each inequality in the system a true statement.

2. $\begin{cases} 4x + y \leq 6 & (1) \\ 2x - 5y < 10 & (2) \end{cases}$

   (a) (1, 2)
   (1): $4(1) + 2 \leq 6$?
   $4 + 2 \leq 6$?
   $6 \leq 6$ True

   (2): $2(1) - 5(2) < 10$?
   $2 - 10 < 10$?
   $-8 < 10$ True
   (1, 2) is a solution.

   (b) (−1, −3)
   (1): $4(-1) + (-3) \leq 6$?
   $-4 - 3 \leq 6$?
   $-7 \leq 6$ True

   (2): $2(-1) - 5(-3) < 10$?
   $-2 + 15 < 10$?
   $13 < 10$ False
   (−1, −3) is not a solution.

3. When graphing linear inequalities, we use a <u>dashed</u> line when graphing strict inequalities (> or <) and a <u>solid</u> line when graphing nonstrict inequalities (≥ or ≤).

4. $\begin{cases} y \geq -3x + 8 \\ y \geq 2x - 7 \end{cases}$

   Graph $y \geq -3x + 8$ by graphing the line
   $y = -3x + 8$ as a solid line because the inequality is nonstrict (≥). We choose the test point (0, 0). Since (0, 0) does not make the inequality true shade the half-plane not containing (0, 0). Graph $y \geq 2x - 7$ by graphing the line
   $y = 2x - 7$ as a solid line because the inequality is nonstrict (≥). We choose the test point (0, 0). Since (0, 0) makes the inequality true shade the half-plane containing (0, 0).
   The overlapping shaded region is the solution of the system.

   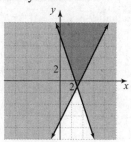

5. $\begin{cases} 4x + 2y < -9 \\ x + 3y < -1 \end{cases}$

   Graph $4x + 2y < -9 \left(y < -2x - \dfrac{9}{2} \right)$. Use a
   dashed line because the inequality is strict
   (<).Graph $x + 3y < -1 \left(y < -\dfrac{1}{3}x - \dfrac{1}{3} \right)$. Use a
   dashed line because the inequality is strict (<).
   The overlapping shaded region is the solution of the system.

   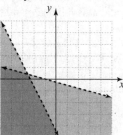

6. $\begin{cases} x + y \leq 4 \\ -x + y > -4 \end{cases}$

   Graph $x + y \leq 4 \,(y \leq -x + 4)$ Use a solid line
   because the inequality is nonstrict (≤). Graph
   $-x + y > -4 \,(y > x - 4)$ Use a dashed line
   because the inequality is strict (>). The

overlapping shaded region is the solution of the system.

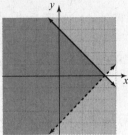

7. $\begin{cases} 3x + y > -5 \\ x + 2y \le 0 \end{cases}$

Graph $3x + y > -5 \, (y > -3x - 5)$ Use a dashed line because the inequality is strict (>). Graph $x + 2y \le 0 \left(y \le -\dfrac{1}{2}x \right)$. Use a solid line because the inequality is nonstrict ($\le$). The overlapping shaded region is the solution of the system.

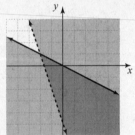

8. $\begin{cases} c + t \le 75,000 \\ c \le 50,000 \\ t \ge 25,000 \end{cases}$

(a) We draw a rectangular coordinate system with the horizontal axis labeled c and the vertical axis labeled t. Each axis is in thousands, so we draw the line $c + t = 75$ and shade below. We draw the line $c = 50$ and shade to the left. We draw the line $t = 25$ and shade above.

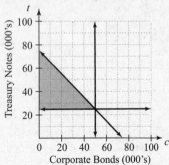

(b) Yes, Jack and Mary can invest \$30,000 in corporate bonds and \$35,000 in Treasury notes because these values lie within the shaded region. Put another way, $c = 30,000$ and $t = 35,000$ satisfies all three inequalities.

(c) No, Jack and Mary cannot invest \$60,000 in corporate bonds and \$15,000 in Treasury notes because these values do not lie within the shaded region. Put another way, $c = 60,000$ and $t = 15,000$ does not satisfy the inequality $c \le 50,000$.

4.6 Exercises

9. $\begin{cases} x \ge 5 & (1) \\ y < -\dfrac{1}{2}x + 3 & (2) \end{cases}$

(a) $(5, -2)$

(1): $5 \ge 5$ True

(2): $-2 < -\dfrac{1}{2}(5) + 3$?

$-2 < -\dfrac{5}{2} + 3$?

$-2 < \dfrac{1}{2}$ True

$(5, -2)$ is a solution.

(b) $(10, -4)$

(1): $10 \ge 5$ True

(2): $-4 < -\dfrac{1}{2}(10) + 3$?

$-4 < -5 + 3$?

$-4 < -2$ True

$(10, -4)$ is a solution.

(c) $(8, -3)$

(1): $8 \ge 5$ True

(2): $-3 < -\dfrac{1}{2}(8) + 3$?

$-3 < -4 + 3$?

$-3 < -1$ True

$(8, -3)$ is a solution.

11. $\begin{cases} 2x + y > -4 & (1) \\ x - y \le 1 & (2) \end{cases}$

(a) $(-2, 1)$

(1): $2(-2) + 1 > -4$?

$-4 + 1 > -4$?

$-3 > -4$ True

(2): $-2-1 \le 1$?

$\qquad -3 \le 1$ True

$(-2, 1)$ is a solution.

(b) $(-1, -2)$

(1): $2(-1)+(-2) > -4$?

$\qquad -2-2 > -4$?

$\qquad\qquad -4 > -4$ False

$(-1, -2)$ is not a solution.

(c) $(2, -3)$

(1): $2(2)+(-3) > -4$?

$\qquad 4-3 > -4$?

$\qquad\qquad 1 > -4$ True

(2): $2-(-3) \le 1$?

$\qquad 2+3 \le 1$?

$\qquad\qquad 5 \le 1$ False

$(2, -3)$ is not a solution.

13. $\begin{cases} x > 2 \\ y \le -1 \end{cases}$

Graph $x > 2$ with a dashed line and $y \le -1$ with a solid line. Graph on the same rectangular coordinate system. The overlapping shaded region represents the solution.

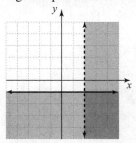

15. $\begin{cases} y > -2 \\ x > -3 \end{cases}$

Graph $y > -2$ with a dashed line and $x > -3$ with a dashed line. Graph on the same rectangular coordinate system. The overlapping shaded region represents the solution.

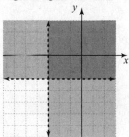

17. $\begin{cases} x+y < 3 \\ x-y > 5 \end{cases}$

Graph $x + y < 3$ $(y < -x + 3)$ with a dashed line and $x - y > 5$ $(y < x - 5)$ with a dashed line. Graph on the same rectangular coordinate system. The overlapping shaded region represents the solution.

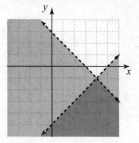

19. $\begin{cases} x+y > 3 \\ 2x-y > 4 \end{cases}$

Graph $x + y > 3$ $(y > -x + 3)$ with a dashed line and $2x - y > 4$ $(y < 2x - 4)$ with a dashed line. Graph on the same rectangular coordinate system. The overlapping shaded region represents the solution.

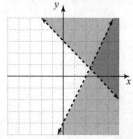

21. $\begin{cases} x < 2 \\ y < \dfrac{1}{2}x+3 \end{cases}$

Graph $x < 2$ with a dashed line and $y < \dfrac{1}{2}x+3$ with a dashed line. Graph on the same rectangular coordinate system. The overlapping shaded region represents the solution.

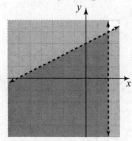

23. $\begin{cases} x \geq -2 \\ y < 2x+3 \end{cases}$

Graph $x \geq -2$ with a solid line and $y < 2x + 3$ with a dashed line. Graph on the same rectangular coordinate system. The overlapping shaded region represents the solution.

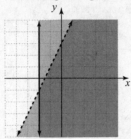

25. $\begin{cases} x > 0 \\ y \leq \dfrac{2}{5}x - 1 \end{cases}$

Graph $x > 0$ with a dashed line and $y \leq \dfrac{2}{5}x - 1$

with a solid line. Graph on the same rectangular coordinate system. The overlapping shaded region represents the solution.

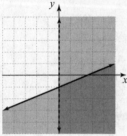

27. $\begin{cases} -y \leq x \\ 3x - y \geq -5 \end{cases}$

Graph $-y \leq x$ $(y \geq -x)$ with a solid line and $3x - y \geq -5$ $(y \leq 3x + 5)$ with a solid line. Graph on the same rectangular coordinate system. The overlapping shaded region represents the solution.

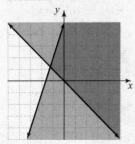

29. $\begin{cases} x + y \leq -2 \\ y \geq x+3 \end{cases}$

Graph $x + y \leq -2$ $(y \leq -x - 2)$ with a solid line and $y \geq x + 3$ with a solid line. Graph on the same rectangular coordinate system. The overlapping shaded region represents the solution.

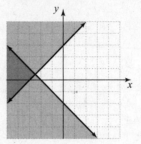

31. $\begin{cases} x + 3y \geq 0 \\ 2y < x+1 \end{cases}$

Graph $x + 3y \geq 0$ $\left(y \geq -\dfrac{1}{3}x \right)$ with a solid line

and $2y < x + 1$ $\left(y < \dfrac{1}{2}x + \dfrac{1}{2} \right)$ with a dashed line.

Graph on the same rectangular coordinate system. The overlapping shaded region represents the solution.

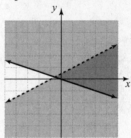

33. $\begin{cases} x + y \geq 0 \\ x < 2y+4 \end{cases}$

Graph $x + y \geq 0$ $(y \geq -x)$ with a solid line and

$x < 2y + 4$ $\left(y > \dfrac{1}{2}x - 2 \right)$ with a dashed line.

Graph on the same rectangular coordinate system. The overlapping shaded region represents the solution.

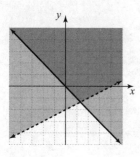

35. $\begin{cases} x + 3y > 6 \\ 2x - y \le 4 \end{cases}$

Graph $x + 3y > 6$ $\left(y > -\dfrac{1}{3}x + 2 \right)$ with a dashed

line and $2x - y \le 4$ $(y \ge 2x - 4)$ with a solid line. Graph on the same rectangular coordinate system. The overlapping shaded region represents the solution.

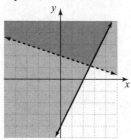

37. $\begin{cases} -y \le 3x - 4 \\ 2x + 3y \ge -3 \end{cases}$

Graph $-y \le 3x - 4$ $(y \ge -3x + 4)$ with a solid line

and $2x + 3y \ge -3$ $\left(y \ge -\dfrac{2}{3}x - 1 \right)$ with a solid

line. Graph on the same rectangular coordinate system. The overlapping shaded region represents the solution.

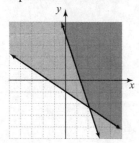

39. (a) $\begin{cases} f + h \le 30 \\ f \ge 2h \\ f \ge 0 \\ h \ge 0 \end{cases}$

Graph $f + h \le 30$ with a solid line and $f \ge 2h$ with a solid line.

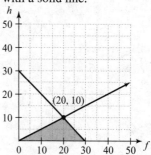

(b) $(f, h) = (18, 11)$
No, these values do not lie within the shaded region.

(c) $(f, h) = (8, 4)$
Yes, these values lie within the shaded region.

41. (a) $\begin{cases} 12c + 18t \le 360 \\ t < 2c - 20 \\ c \le 24 \end{cases}$

Graph $12c + 18t \le 360$ and $c \le 24$ with solid lines and $t < 2c - 20$ with a dashed line.

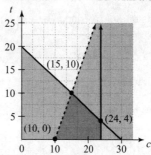

(b) $(c, t) = (17, 9)$
No, these values do not lie in the shaded region.

(c) $(c, t) = (12, 11)$
No, these values do not lie in the shaded region.

43. $\begin{cases} \dfrac{y}{2}-\dfrac{x}{6}\ge 1 & (1) \\[2mm] \dfrac{x}{3}-\dfrac{y}{1}\ge 1 & (2) \end{cases}$

(1): $\dfrac{y}{2}-\dfrac{x}{6}\ge 1$

$\qquad 6\left(\dfrac{y}{2}-\dfrac{x}{6}\right)\ge 6(1)$

$\qquad\quad 3y-x\ge 6$

$\qquad\qquad 3y\ge x+6$

$\qquad\qquad\quad y\ge \dfrac{1}{3}x+2$

(2): $\dfrac{x}{3}-\dfrac{y}{1}\ge 1$

$\qquad \dfrac{x}{3}\ge \dfrac{y}{1}+1$

$\qquad \dfrac{x}{3}-1\ge y$

$\qquad\qquad y\le \dfrac{1}{3}x-1$

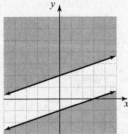

The regions do not overlap. There is no solution.

45. $\begin{cases} x<\dfrac{3}{2}y+\dfrac{9}{2} & (1) \\[2mm] -2x<-3(y+2) & (2) \end{cases}$

(1): $x<\dfrac{3}{2}y+\dfrac{9}{2}$

$\qquad 2x<2\left(\dfrac{3}{2}y+\dfrac{9}{2}\right)$

$\qquad 2x<3y+9$

$\quad 2x-9<3y$

$\qquad \dfrac{2}{3}x-3<y$ or $y>\dfrac{2}{3}x-3$

(2): $\quad -2x<-3(y+2)$

$\qquad -2x<-3y-6$

$\qquad -2x+6<-3y$

$\qquad \dfrac{2}{3}x-2>y$ or $y<\dfrac{2}{3}x+2$

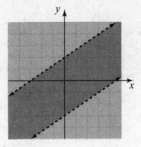

47. $\begin{cases} y\ge 0 \\ y\ge x \\ y\le \dfrac{1}{2}(x+6) \end{cases}$

49. Answers will vary. Graph each inequality in the system. The overlapping shaded region represents the solution of the system.

51. **a.** $\begin{cases} y\ge x \\ y\ge -2x \end{cases}$

      b. $\begin{cases} y\le x \\ y\ge -2x \end{cases}$

      c. $\begin{cases} y\ge x \\ y\le -2x \end{cases}$

      d. $\begin{cases} y\le x \\ y\le -2x \end{cases}$

Chapter 4 Review

1. $\begin{cases} x+2y=6 & (1) \\ 3x-y=-10 & (2) \end{cases}$

   (a) $(3,-1)$

$\qquad$ (1): $3+2(-1)\overset{?}{=}6$

$\qquad\qquad\quad 3-2\overset{?}{=}6$

$\qquad\qquad\qquad 1=6$ False

$\qquad (3,-1)$ is not a solution.

   (b) $(-2,4)$

$\qquad$ (1): $-2+2(4)\overset{?}{=}6$

$\qquad\qquad -2+8\overset{?}{=}6$

$\qquad\qquad\qquad 6=6$ True

$\qquad$ (2): $3(-2)-4\overset{?}{=}-10$

$\qquad\qquad -6-4\overset{?}{=}-10$

$\qquad\qquad -10=-10$ True

$\qquad (-2,4)$ is a solution.

(c) $(4, 1)$

(1): $4 + 2(1) \overset{?}{=} 6$

$\qquad 4 + 2 \overset{?}{=} 6$

$\qquad\quad 6 = 6$ True

(2): $3(4) - 1 \overset{?}{=} -10$

$\qquad 12 - 1 \overset{?}{=} -10$

$\qquad\quad 11 = -10$ False

$(4, 1)$ is not a solution.

2. $\begin{cases} y = 3x - 5 & (1) \\ 3y = 6x - 5 & (2) \end{cases}$

(a) $(2, 1)$

(1): $1 \overset{?}{=} 3(2) - 5$

$\quad 1 \overset{?}{=} 6 - 5$

$\quad 1 = 1$ True

(2): $3(1) \overset{?}{=} 6(2) - 5$

$\quad\ 3 \overset{?}{=} 12 - 5$

$\quad\ 3 = 7$ False

$(2, 1)$ is not a solution.

(b) $\left(0, -\dfrac{5}{3}\right)$

(1): $-\dfrac{5}{3} \overset{?}{=} 3(0) - 5$

$\quad -\dfrac{5}{3} = -5$ False

$\left(0, -\dfrac{5}{3}\right)$ is not a solution.

(c) $\left(\dfrac{10}{3}, 5\right)$

(1): $5 \overset{?}{=} 3\left(\dfrac{10}{3}\right) - 5$

$\quad 5 \overset{?}{=} 10 - 5$

$\quad 5 = 5$ True

(2): $3(5) \overset{?}{=} 6\left(\dfrac{10}{3}\right) - 5$

$\quad 15 \overset{?}{=} 20 - 5$

$\quad 15 = 15$ True

$\left(\dfrac{10}{3}, 5\right)$ is a solution.

3. $\begin{cases} 3x - 4y = 2 & (1) \\ 20y = 15x - 10 & (2) \end{cases}$

(a) $\left(\dfrac{1}{2}, -\dfrac{1}{8}\right)$

(1): $3\left(\dfrac{1}{2}\right) - 4\left(-\dfrac{1}{8}\right) \overset{?}{=} 2$

$\qquad\quad \dfrac{3}{2} + \dfrac{1}{2} \overset{?}{=} 2$

$\qquad\qquad\qquad 2 = 2$ True

(2): $20\left(-\dfrac{1}{8}\right) \overset{?}{=} 15\left(\dfrac{1}{2}\right) - 10$

$\qquad -\dfrac{5}{2} \overset{?}{=} \dfrac{15}{2} - \dfrac{20}{2}$

$\qquad -\dfrac{5}{2} = -\dfrac{5}{2}$ True

$\left(\dfrac{1}{2}, -\dfrac{1}{8}\right)$ is a solution.

(b) $(6, 4)$

(1): $3(6) - 4(4) \overset{?}{=} 2$

$\qquad 18 - 16 \overset{?}{=} 2$

$\qquad\qquad\ 2 = 2$ True

(2): $20(4) \overset{?}{=} 15(6) - 10$

$\qquad 80 \overset{?}{=} 90 - 10$

$\qquad 80 = 80$ True

$(6, 4)$ is a solution.

(c) $(0.4, -0.2)$

(1): $3(0.4) - 4(-0.2) \overset{?}{=} 2$

$\qquad\quad 1.2 + 0.8 \overset{?}{=} 2$

$\qquad\qquad\qquad 2 = 2$ True

(2): $20(-0.2) \overset{?}{=} 15(0.4) - 10$

$\qquad\quad -4 \overset{?}{=} 6 - 10$

$\qquad\quad -4 = -4$ True

$(0.4, -0.2)$ is a solution.

4. $\begin{cases} x = -4y + 2 & (1) \\ 2x + 8y = 12 & (2) \end{cases}$

(a) $(10, -1)$

(1): $10 \overset{?}{=} -4(-1) + 2$

$\quad 10 \overset{?}{=} 4 + 2$

$\quad 10 = 8$ False

$(10, -1)$ is not a solution.

(b) $\left(-\dfrac{1}{2}, \dfrac{1}{2}\right)$

(1): $-\dfrac{1}{2} \overset{?}{=} -4\left(\dfrac{1}{2}\right) + 2$

$-\dfrac{1}{2} \overset{?}{=} -2 + 2$

$-\dfrac{1}{2} = 0$ False

$\left(-\dfrac{1}{2}, \dfrac{1}{2}\right)$ is not a solution.

(c)   (2, 1)

(1): $2 \overset{?}{=} -4(1) + 2$

$2 \overset{?}{=} -4 + 2$

$2 = -2$ False

(2, 1) is not a solution.

5. $\begin{cases} 2x - 4y = 8 & (1) \\ x + y = 7 & (2) \end{cases}$

(1): $2x - 4y = 8$

$-4y = -2x + 8$

$y = \dfrac{1}{2}x - 2$

(2): $x + y = 7$

$y = -x + 7$

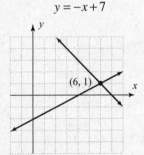

Check (6, 1):

(1): $2(6) - 4(1) \overset{?}{=} 8$

$12 - 4 \overset{?}{=} 8$

$8 = 8$ True

(2): $6 + 1 \overset{?}{=} 7$

$7 = 7$ True

The solution is (6, 1).

6. $\begin{cases} x - y = -3 & (1) \\ 3x + 2y = 6 & (2) \end{cases}$

(1): $x - y = -3$

$-y = -x - 3$

$y = x + 3$

(2): $3x + 2y = 6$

$2y = -3x + 6$

$y = -\dfrac{3}{2}x + 3$

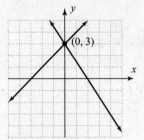

Check (0, 3):

(1): $0 - 3 \overset{?}{=} -3$

$-3 = -3$ True

(2): $3(0) + 2(3) \overset{?}{=} 6$

$6 = 6$ True

The solution is (0, 3).

7. $\begin{cases} y = -\dfrac{x}{2} + 2 & (1) \\ y = x + 8 & (2) \end{cases}$

The equations are already in slope-intercept form.

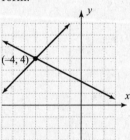

Check (−4, 4):

(1): $4 \overset{?}{=} -\dfrac{-4}{2} + 2$

$4 \overset{?}{=} 2 + 2$

$4 = 4$ True

(2): $4 \overset{?}{=} -4 + 8$

$4 = 4$ True

The solution is (−4, 4).

8. $\begin{cases} y = -x - 5 & (1) \\ y = \dfrac{3x}{4} + 2 & (2) \end{cases}$

The equations are already in slope-intercept form.

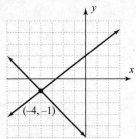

Check $(-4, -1)$:

(1): $-1 \overset{?}{=} -(-4) - 5$

$-1 \overset{?}{=} 4 - 5$

$-1 = -1$ True

(2): $-1 \overset{?}{=} \dfrac{3(-4)}{4} + 2$

$-1 \overset{?}{=} -3 + 2$

$-1 = -1$ True

The solution is $(-4, -1)$.

9. $\begin{cases} 4x - 8 = 0 & (1) \\ 3y + 9 = 0 & (2) \end{cases}$

(1): $4x - 8 = 0$

$4x = 8$

$x = 2$

(2): $3y + 9 = 0$

$3y = -9$

$y = -3$

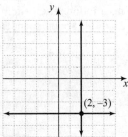

Check $(2, -3)$:

(1): $4(2) - 8 \overset{?}{=} 0$

$8 - 8 \overset{?}{=} 0$

$0 = 0$ True

(2): $3(-3) + 9 \overset{?}{=} 0$

$-9 + 9 \overset{?}{=} 0$

$0 = 0$ True

The solution is $(2, -3)$.

10. $\begin{cases} x = y & (1) \\ x + y = 0 & (2) \end{cases}$

(1): $y = x$

(2): $x + y = 0$

$y = -x$

Check $(0, 0)$:

(1): $0 = 0$ True

(2): $0 + 0 \overset{?}{=} 0$

$0 = 0$ True

The solution is $(0, 0)$.

11. $\begin{cases} 0.6x + 0.5y = 2 & (1) \\ 10y = -12x + 20 & (2) \end{cases}$

(1): $0.6x + 0.5y = 2$

$0.5y = -0.6x + 2$

$y = -1.2x + 4$

(2): $10y = -12x + 20$

$y = -\dfrac{6}{5}x + 2$

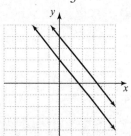

The lines are parallel, so there is no solution.

12. $\begin{cases} \dfrac{1}{4}x - \dfrac{1}{2}y = 1 & (1) \\ 3x - 6y = 12 & (2) \end{cases}$

(1): $\dfrac{1}{4}x - \dfrac{1}{2}y = 1$

$-\dfrac{1}{2}y = -\dfrac{1}{4}x + 1$

$y = \dfrac{1}{2}x - 2$

(2): $3x - 6y = 12$

$-6y = -3x + 12$

$y = \dfrac{1}{2}x - 2$

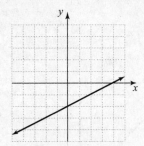

The lines coincide. There are infinitely many solutions.

13. $\begin{cases} 3x = y + 4 & (1) \\ 3x - y = -4 & (2) \end{cases}$

(1): $3x = y + 4$

$\quad\quad y = 3x - 4$

(2): $3x - y = -4$

$\quad\quad y = 3x + 4$

The lines have the same slope, but different y-intercepts, so the lines are parallel. The system has no solution. It is inconsistent.

14. $\begin{cases} 4y = 2x - 8 & (1) \\ x - 2y = -4 & (2) \end{cases}$

(1): $4y = 2x - 8$

$\quad\quad y = \dfrac{1}{2}x - 2$

(2): $x - 2y = -4$

$\quad\quad -2y = -x - 4$

$\quad\quad y = \dfrac{1}{2}x + 2$

The lines have the same slope, but different y-intercepts, so the lines are parallel. The system has no solution. It is inconsistent.

15. $\begin{cases} -3x + 3y = -3 & (1) \\ \dfrac{1}{2}x - \dfrac{1}{2}y = 0.5 & (2) \end{cases}$

(1): $-3x + 3y = -3$

$\quad\quad 3y = 3x - 3$

$\quad\quad y = x - 1$

(1): $\dfrac{1}{2}x - \dfrac{1}{2}y = 0.5$

$\quad\quad -\dfrac{1}{2}y = -\dfrac{1}{2}x + 0.5$

$\quad\quad y = x - 1$

The lines have the same slope and y-intercept, so the lines coincide. The system has infinitely many solutions. It is consistent and the equations are dependent.

16. $\begin{cases} x - 2 = -\dfrac{2}{3}y - \dfrac{2}{3} & (1) \\ 3x = 4 - 2y \end{cases}$

(1): $x - 2 = -\dfrac{2}{3}y - \dfrac{2}{3}$

$\quad\quad x - \dfrac{4}{3} = -\dfrac{2}{3}y$

$\quad\quad -\dfrac{3}{2}x + 2 = y$ or $y = -\dfrac{3}{2}x + 2$

(2): $\quad 3x = 4 - 2y$

$\quad\quad 3x - 4 = -2y$

$\quad\quad -\dfrac{3}{2}x + 2 = y$ or $y = -\dfrac{3}{2}x + 2$

The lines have the same slope and y-intercept, so the lines coincide. The system has infinitely many solutions. It is consistent and the equations are dependent.

17. $\begin{cases} 3 - 2x = y & (1) \\ \dfrac{y}{2} = x + 1.5 & (2) \end{cases}$

(1): $3 - 2x = y$ or $y = -2x + 3$

(2): $\dfrac{y}{2} = x + 1.5$

$\quad\quad y = 2x + 3$

The lines have different slopes, so the lines intersect. There is one solution. The system is consistent and the equations are independent.

18. $\begin{cases} \dfrac{y}{2} = \dfrac{x}{4} + 2 & (1) \\ \dfrac{x}{8} + \dfrac{y}{4} = -1 & (2) \end{cases}$

(1): $\dfrac{y}{2} = \dfrac{x}{4} + 2$

$\quad\quad y = \dfrac{1}{2}x + 4$

(2): $\dfrac{x}{8} + \dfrac{y}{4} = -1$

$\quad\quad \dfrac{y}{4} = -\dfrac{x}{8} - 1$

$\quad\quad y = -\dfrac{1}{2}x - 4$

The lines have different slopes, so the lines intersect. There is one solution. The system is consistent and the equations are independent.

19. (a) Let x be the number of fliers and y be the cost.
Printer A: $y = 0.10x + 70$
Printer B: $y = 0.04x + 100$

(b)

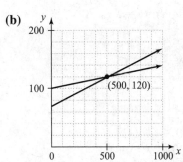

The cost is the same for 500 fliers.

(c) Since the graph for printer A is below that of printer B for fliers less than 500, she should choose Printer A.

20. (a) Let x be the square yards and y be the cost.
carpet: $y = 40x + 50$
tile: $y = 10x + 350$

(b)

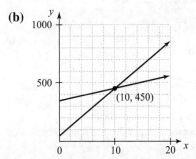

The cost is the same for 10 square yards.

(c) Since the graph for tile is below that for carpet for square yards greater than 10, he should choose tile.

21. $\begin{cases} x + 4y = 6 & (1) \\ y = 2x - 3 & (2) \end{cases}$

Let $y = 2x - 3$ in (1).
$x + 4(2x - 3) = 6$
$x + 8x - 12 = 6$
$9x - 12 = 6$
$9x = 18$
$x = 2$
Let $x = 2$ in (2).
$y = 2(2) - 3$
$y = 4 - 3$
$y = 1$
The solution is $(2, 1)$.

22. $\begin{cases} 7x - 3y = 10 & (1) \\ y = 3x - 4 & (2) \end{cases}$

Let $y = 3x - 4$ in (1).
$7x - 3(3x - 4) = 10$
$7x - 9x + 12 = 10$
$-2x + 12 = 10$
$-2x = -2$
$x = 1$
Let $x = 1$ in (2).
$y = 3(1) - 4$
$y = 3 - 4$
$y = -1$
The solution is $(1, -1)$.

23. $\begin{cases} 2x + 5y = 4 & (1) \\ x = 3 - 2y & (2) \end{cases}$

Let $x = 3 - 2y$ in (1).
$2(3 - 2y) + 5y = 4$
$6 - 4y + 5y = 4$
$6 + y = 4$
$y = -2$
Let $y = -2$ in (2).
$x = 3 - 2(-2)$
$x = 3 + 4$
$x = 7$
The solution is $(7, -2)$.

24. $\begin{cases} 3x + y = 10 & (1) \\ x = 8 + 2y & (2) \end{cases}$

Let $x = 8 + 2y$ in (1).
$3(8 + 2y) + y = 10$
$24 + 6y + y = 10$
$24 + 7y = 10$
$7y = -14$
$y = -2$
Let $y = -2$ in (2).
$x = 8 + 2(-2)$
$x = 8 - 4$
$x = 4$
The solution is $(4, -2)$.

25. $\begin{cases} y = \dfrac{2}{3}x - 1 & (1) \\ y = \dfrac{1}{2}x + 2 & (2) \end{cases}$

Let $y = \dfrac{2}{3}x - 1$ in (2).

$$\frac{2}{3}x - 1 = \frac{1}{2}x + 2$$

$$\frac{2}{3}x - \frac{1}{2}x = 2 + 1$$

$$\frac{4}{6}x - \frac{3}{6}x = 3$$

$$\frac{1}{6}x = 3$$

$$x = 18$$

Let $x = 18$ in (2).

$$y = \frac{1}{2}(18) + 2$$

$$y = 9 + 2$$

$$y = 11$$

The solution is (18, 11).

26. $\begin{cases} y = -\dfrac{5}{6}x + 3 & (1) \\ y = -\dfrac{4}{3}x & (2) \end{cases}$

Let $y = -\dfrac{4}{3}x$ in (1).

$$-\frac{4}{3}x = -\frac{5}{6}x + 3$$

$$-\frac{4}{3}x + \frac{5}{6}x = 3$$

$$-\frac{8}{6}x + \frac{5}{6}x = 3$$

$$-\frac{1}{2}x = 3$$

$$x = -6$$

Let $x = -6$ in (2).

$$y = -\frac{4}{3}(-6)$$

$$y = 8$$

The solution is (−6, 8).

27. $\begin{cases} 2x - y = 6 & (1) \\ 4x + 3y = 2 & (2) \end{cases}$

Solve (1) for y.

$$2x - y = 6$$

$$-y = -2x + 6$$

$$y = 2x - 6$$

Let $y = 2x - 6$ in (2).

$$4x + 3(2x - 6) = 2$$

$$4x + 6x - 18 = 2$$

$$10x - 18 = 2$$

$$10x = 20$$

$$x = 2$$

Let $x = 2$ in (1).

$$2(2) - y = 6$$

$$4 - y = 6$$

$$-y = 2$$

$$y = -2$$

The solution is (2, −2).

28. $\begin{cases} 5x + 2y = 13 & (1) \\ x + 4y = -1 & (2) \end{cases}$

Solve (2) for x.

$$x + 4y = -1$$

$$x = -1 - 4y$$

Let $x = -1 - 4y$ in (1).

$$5(-1 - 4y) + 2y = 13$$

$$-5 - 20y + 2y = 13$$

$$-5 - 18y = 13$$

$$-18y = 18$$

$$y = -1$$

Let $y = -1$ in (2).

$$x + 4(-1) = -1$$

$$x - 4 = -1$$

$$x = 3$$

The solution is (3, −1).

29. $\begin{cases} 6x + 3y = 12 & (1) \\ y = -2x + 4 & (2) \end{cases}$

Let $y = -2x + 4$ in (1).

$$6x + 3(-2x + 4) = 12$$

$$6x - 6x + 12 = 12$$

$$12 = 12 \quad \text{True}$$

The system is dependent. There are infinitely many solutions.

30. $\begin{cases} x = 4y - 2 & (1) \\ 8y - 2x = 4 & (2) \end{cases}$

Let $x = 4y - 2$ in (2).

$$8y - 2(4y - 2) = 4$$

$$8y - 8y + 4 = 4$$

$$4 = 4 \quad \text{True}$$

The system is dependent. There are infinitely many solutions.

31. $\begin{cases} -6 - 2(3x - 6y) = 0 & (1) \\ 6 - 12(x - 2y) = 0 & (2) \end{cases}$

(1): $-6 - 6x + 12y = 0$

$$-6x + 12y = 6$$

$$x - 2y = -1$$

$$x = 2y - 1$$

Let $x = 2y - 1$ in (2).

$$6 - 12[(2y-1) - 2y] = 0$$
$$6 - 12(2y - 1 - 2y) = 0$$
$$6 - 12(-1) = 0$$
$$6 + 12 = 0$$
$$18 = 0 \quad \text{False}$$

The system is inconsistent. There is no solution.

32. $\begin{cases} 6 - 2(3y + 4x) = 0 & (1) \\ 9(y-1) + 12x = 0 & (2) \end{cases}$

(1): $6 - 2(3y + 4x) = 0$
$$6 - 6y - 8x = 0$$
$$-6y - 8x = -6$$
$$-6y = 8x - 6$$
$$y = -\frac{4}{3}x + 1$$

Let $y = -\frac{4}{3}x + 1$ in (2).

$$9\left(-\frac{4}{3}x + 1 - 1\right) + 12x = 0$$
$$9\left(-\frac{4}{3}x\right) + 12x = 0$$
$$-12x + 12x = 0$$
$$0 = 0$$

The system is dependent. There are infinitely many solutions.

33. $\begin{cases} \dfrac{1}{2}x - \dfrac{1}{4}y = \dfrac{1}{2} & (1) \\ \dfrac{1}{3}x - \dfrac{3}{4}y = -\dfrac{1}{4} & (2) \end{cases}$

Solve (1) for y.
$$\frac{1}{2}x - \frac{1}{4}y = \frac{1}{2}$$
$$-\frac{1}{4}y = -\frac{1}{2}x + \frac{1}{2}$$
$$y = 2x - 2$$

Let $y = 2x - 2$ in (2).
$$\frac{1}{3}x - \frac{3}{4}(2x - 2) = -\frac{1}{4}$$
$$\frac{1}{3}x - \frac{3}{2}x + \frac{3}{2} = -\frac{1}{4}$$
$$\frac{2}{6}x - \frac{9}{6}x = -\frac{6}{4} - \frac{1}{4}$$
$$-\frac{7}{6}x = -\frac{7}{4}$$
$$x = \frac{3}{2}$$

Let $x = \frac{3}{2}$ in (1).

$$\frac{1}{2}\left(\frac{3}{2}\right) - \frac{1}{4}y = \frac{1}{2}$$
$$\frac{3}{4} - \frac{1}{4}y = \frac{1}{2}$$
$$-\frac{1}{4}y = -\frac{1}{4}$$
$$y = 1$$

The solution is $\left(\dfrac{3}{2}, 1\right)$.

34. $\begin{cases} -\dfrac{5x}{4} + \dfrac{y}{6} = -\dfrac{7}{12} \\ \dfrac{3x}{2} - \dfrac{y}{10} = \dfrac{3}{5} \end{cases}$

Multiply the first equation by 12 and the second equation by 10 to clear fractions.
$$\begin{cases} -15x + 2y = -7 & (1) \\ 15x - y = 6 & (2) \end{cases}$$

Solve (2) for y.
$$15x - y = 6$$
$$y = 15x - 6$$

Let $y = 15x - 6$ in (1).
$$-15x + 2(15x - 6) = -7$$
$$-15x + 30x - 12 = -7$$
$$15x = 5$$
$$x = \frac{1}{3}$$

Let $x = \frac{1}{3}$ in (2).

$$15\left(\frac{1}{3}\right) - y = 6$$
$$5 - y = 6$$
$$-y = 1$$
$$y = -1$$

The solution is $\left(\dfrac{1}{3}, -1\right)$.

35. Let w be the width and l be the length.
$$\begin{cases} 2w + 2l = 650 & (1) \\ l = w + 75 & (2) \end{cases}$$

Let $l = w + 75$ in (1).
$$2w + 2(w + 75) = 650$$
$$2w + 2w + 150 = 650$$
$$4w = 500$$
$$w = 125$$

Let $w = 125$ in (2).
$$l = 125 + 75$$
$$l = 200$$

The width is 125 meters and the length is 200 meters.

36. $\begin{cases} x+y=12 & (1) \\ x-2y=21 & (2) \end{cases}$

Solve (1) for y.

$y = 12 - x$

Let $y = 12 - x$ in (2).

$x - 2(12 - x) = 21$

$x - 24 + 2x = 21$

$-24 + 3x = 21$

$3x = 45$

$x = 15$

Let $x = 15$ in (1).

$15 + y = 12$

$y = -3$

The smaller number is -3.

37. $\begin{cases} 4x-y=12 & (1) \\ 2x+y=-12 & (2) \end{cases}$

Add the equations.

$\begin{cases} 4x-y=12 \\ 2x+y=-12 \end{cases}$

$\overline{6x = 0}$

$x = 0$

Let $x = 0$ in (2).

$2(0) + y = -12$

$y = -12$

The solution is $(0, -12)$.

38. $\begin{cases} -2x+3y=27 & (1) \\ 2x-5y=-41 & (2) \end{cases}$

Add the equations.

$\begin{cases} -2x+3y=27 \\ 2x-5y=-41 \end{cases}$

$\overline{-2y=-14}$

$y = 7$

Let $y = 7$ in (2).

$2x - 5(7) = -41$

$2x - 35 = -41$

$2x = -6$

$x = -3$

The solution is $(-3, 7)$.

39. $\begin{cases} -3x+4y=25 & (1) \\ x-5y=-23 & (2) \end{cases}$

Multiply (2) by 3 and add.

$\begin{cases} -3x+4y=25 \\ 3x-15y=-69 \end{cases}$

$\overline{-11y=-44}$

$y = 4$

Let $y = 4$ in (2).

$x - 5(4) = -23$

$x - 20 = -23$

$x = -3$

The solution is $(-3, 4)$.

40. $\begin{cases} 5x+8y=-15 & (1) \\ -2x+y=6 & (2) \end{cases}$

Multiply (2) by -8 and add.

$\begin{cases} 5x+8y=-15 \\ 16x-8y=-48 \end{cases}$

$\overline{21x=-63}$

$x = -3$

Let $x = -3$ in (2).

$-2(-3) + y = 6$

$6 + y = 6$

$y = 0$

The solution is $(-3, 0)$.

41. $\begin{cases} 4x-3y=-1 & (1) \\ 2x-\dfrac{3}{2}y=3 & (2) \end{cases}$

Multiply (2) by -2 and add.

$\begin{cases} 4x-3y=-1 \\ -4x+3y=-6 \end{cases}$

$\overline{0=-7}$

The system is inconsistent. There is no solution.

42. $\begin{cases} -2x+5y=3 & (1) \\ 4x-10y=-6 & (2) \end{cases}$

Multiply (1) by 2. Then add.

$\begin{cases} -4x+10y=6 \\ 4x-10y=-6 \end{cases}$

$\overline{0=0 \quad \text{True}}$

The system is consistent and dependent. There are infinitely many solutions.

43. $\begin{cases} 1.3x-0.2y=-3 & (1) \\ -0.1x+0.5y=1.2 & (2) \end{cases}$

Multiply (1) by 5 and (2) by 2, then add.

$$\begin{cases} 6.5x - y = -15 \\ -0.2x + y = 2.4 \end{cases}$$

$$\begin{array}{l} \overline{6.3x \quad\;\; = -12.6} \\ \quad\; x = -2 \end{array}$$

Let $x = -2$ in (2).
$$-0.1(-2) + 0.5y = 1.2$$
$$0.2 + 0.5y = 1.2$$
$$0.5y = 1$$
$$y = 2$$

The solution is $(-2, 2)$.

44. $\begin{cases} 2.5x + 0.5y = 6.25 & (1) \\ -0.5x - 1.2y = 1.5 & (2) \end{cases}$

Multiply (1) by 2 and (2) by 10, then add.
$$\begin{cases} 5x + y = 12.5 \\ -5x - 12y = 15 \end{cases}$$
$$\overline{\quad -11y = 27.5 \quad}$$
$$y = -\frac{27.5}{11} = -2.5$$

Let $y = -2.5$ in equation (1).
$$2.5x + 0.5(-2.5) = 6.25$$
$$2.5x - 1.25 = 6.25$$
$$2.5x = 7.5$$
$$x = 3$$

The solution is $(3, -2.5)$.

45. $\begin{cases} 2x + y = -1 & (1) \\ -6x - 8y = 13 & (2) \end{cases}$

Solve (1) for y.
$$2x + y = -1$$
$$y = -2x - 1$$

Let $y = -2x - 1$ in (2).
$$-6x - 8(-2x - 1) = 13$$
$$-6x + 16x + 8 = 13$$
$$10x + 8 = 13$$
$$10x = 5$$
$$x = \frac{1}{2}$$

Let $x = \frac{1}{2}$ in (1).

$$2\left(\frac{1}{2}\right) + y = -1$$
$$1 + y = -1$$
$$y = -2$$

The solution is $\left(\frac{1}{2}, -2\right)$.

46. $\begin{cases} \dfrac{1}{6}x + y = \dfrac{3}{4} \\ \dfrac{2y - 8}{3} = 4x + 4 \end{cases}$

Multiply the first equation by 12 and rearrange.

$$12\left(\frac{1}{6}x + y\right) = 12\left(\frac{3}{4}\right)$$
$$2x + 12y = 9$$

Multiply the second equation by 3 and rearrange.

$$3\left(\frac{2y - 8}{3}\right) = 3(4x + 4)$$
$$2y - 8 = 12x + 12$$
$$-12x + 2y = 20$$
$$-6x + y = 10$$

$$\begin{cases} 2x + 12y = 9 & (1) \\ -6x + y = 10 & (2) \end{cases}$$

Multiply (1) by 3, then add.
$$\begin{cases} 6x + 36y = 27 \\ -6x + y = 10 \end{cases}$$
$$\overline{\quad 37y = 37 \quad}$$
$$y = 1$$

Let $y = 1$ in (1).
$$2x + 12(1) = 9$$
$$2x = -3$$
$$x = -\frac{3}{2}$$

The solution is $\left(-\frac{3}{2}, 1\right)$.

47. $\begin{cases} y + 5 = \dfrac{2}{3}x + 3 \\ \dfrac{1}{3}x - \dfrac{1}{2}y = 1 \end{cases}$

Multiply the first equation by 3 and the second equation by 6.
$$\begin{cases} 3y + 15 = 2x + 9 \\ 2x - 3y = 6 \end{cases}$$

Rearrange the first equation, then add.
$$\begin{cases} -2x + 3y = -6 \\ 2x - 3y = 6 \end{cases}$$
$$\overline{\quad 0 = 0 \quad \text{True}}$$

The system is dependent. There are infinitely many solutions.

48. $\begin{cases} \dfrac{1}{14} - \dfrac{x}{2} = -\dfrac{y}{7} \\ y = \dfrac{1}{2} + \dfrac{7x}{3} \end{cases}$

Multiply the first equation by 14 and the second equation by 6.

$\begin{cases} 1 - 7x = -2y & (1) \\ 6y = 3 + 14x & (2) \end{cases}$

Multiply (1) by 2 and rearrange both, then add.

$\begin{cases} -14x + 4y = -2 \\ \underline{14x - 6y = -3} \end{cases}$
$\qquad\qquad -2y = -5$
$\qquad\qquad\quad y = \dfrac{5}{2}$

Let $y = \dfrac{5}{2}$ in (2) and solve for x.

$6\left(\dfrac{5}{2}\right) = 3 + 14x$

$15 = 3 + 14x$

$12 = 14x$

$\dfrac{6}{7} = x$

The solution is $\left(\dfrac{6}{7}, \dfrac{5}{2}\right)$.

49. $\begin{cases} -x + y = 7 & (1) \\ -3x + 4y = 8 & (2) \end{cases}$

Solve (1) for y.

$y = x + 7$

Let $y = x + 7$ in (2).

$-3x + 4(x + 7) = 8$

$-3x + 4x + 28 = 8$

$x + 28 = 8$

$x = -20$

Let $x = -20$ in (1).

$-(-20) + y = 7$

$20 + y = 7$

$y = -13$

The solution is $(-20, -13)$.

50. $\begin{cases} 4x + y = 2 & (1) \\ 9y - 3x = 5 & (2) \end{cases}$

Solve (1) for y.

$y = 2 - 4x$

Let $y = 2 - 4x$ in (2).

$9(2 - 4x) - 3x = 5$

$18 - 36x - 3x = 5$

$18 - 39x = 5$

$-39x = -13$

$x = \dfrac{1}{3}$

Let $x = \dfrac{1}{3}$ in (2).

$9y - 3\left(\dfrac{1}{3}\right) = 5$

$9y - 1 = 5$

$9y = 6$

$y = \dfrac{2}{3}$

The solution is $\left(\dfrac{1}{3}, \dfrac{2}{3}\right)$.

51. $\begin{cases} 4x + 6 = 3y + 5 \\ 4(-2x - 4) = 6(-y - 3) \end{cases}$

Simplify and rearrange.

$\begin{cases} 4x - 3y = -1 & (1) \\ -8x + 6y = -2 & (2) \end{cases}$

Multiply (1) by 2 and add.

$\begin{cases} 8x - 6y = -2 \\ \underline{-8x + 6y = -2} \end{cases}$
$\qquad\qquad 0 = -4 \quad \text{False}$

The system is inconsistent. There is no solution.

52. $\begin{cases} 3y + 2x = 16x + 2 & (1) \\ x = -\dfrac{3}{14}y - \dfrac{1}{7} & (2) \end{cases}$

Rearrange and multiply the second equation by 14.

$\begin{cases} -14x + 3y = 2 \\ \underline{14x + 3y = -2} \end{cases}$
$\qquad\qquad 6y = 0$
$\qquad\qquad\ y = 0$

Let $y = 0$ in (2).

$x = -\dfrac{3}{14}(0) - \dfrac{1}{7}$

$x = -\dfrac{1}{7}$

The solution is $\left(-\dfrac{1}{7}, 0\right)$.

53. $\begin{cases} h+c=4 & (1) \\ 1.50h+7.75c=16 & (2) \end{cases}$

Solve (1) for c.

$c = 4 - h$

Let $c = 4 - h$ in (2).

$1.5h + 7.75(4-h) = 4(4)$

$1.5h + 31 - 7.75h = 16$

$-6.25h = -15$

$h = 2.4$

Let $h = 2.4$ in (1).

$2.4 + c = 4$

$c = 1.6$

Therefore, 2.4 pounds of cookies and 1.6 pounds of chocolates should be included.

54. $\begin{cases} t+b=725 & (1) \\ t+5b=2025 & (2) \end{cases}$

Multiply (1) by -1, then add.

$\begin{cases} -t-b=-725 \\ t+5b=2025 \end{cases}$

$\overline{}$

$\quad 4b = 1300$

$\quad\ b = 325$

Let $b = 325$ in (1).

$t + 325 = 725$

$\quad\ t = 400$

Therefore, 400 individual tickets and 325 block tickets were sold.

55. Let x be the first number and y be the second number. Their sum, $x + y$, is $\dfrac{17}{24}$. Their difference, $x - y$, is $\dfrac{1}{24}$.

$\begin{cases} x+y = \dfrac{17}{24} & (1) \\[2mm] x-y = \dfrac{1}{24} & (2) \end{cases}$

Add the equations.

$\begin{cases} x+y = \dfrac{17}{24} \\[2mm] x-y = \dfrac{1}{24} \end{cases}$

$\overline{}$

$2x \quad = \dfrac{18}{24}$

$\quad x = \dfrac{3}{8}$

Let $x = \dfrac{3}{8}$ in (1).

$\dfrac{3}{8} + y = \dfrac{17}{24}$

$\quad y = \dfrac{1}{3}$

The numbers are $\dfrac{3}{8}$ and $\dfrac{1}{3}$.

56. Let x be the smaller number. Then y is the larger number. Their sum, $x + y$, is 58. If twice the smaller is subtracted from the larger, $y - 2x$, the difference is -20.

$\begin{cases} x+y = 58 & (1) \\ y-2x = -20 & (2) \end{cases}$

Solve (2) for y.

$y = 2x - 20$

Let $y = 2x - 20$ in (1).

$x + 2x - 20 = 58$

$\quad 3x = 78$

$\quad\ x = 26$

The smaller number is 26.

57. Let s be the amount invested in stocks and b be the amount invested in bonds. The sum, $s + b$, is 50,000. The amount in stocks equals \$4,000 less than twice the amount in bonds.

$\begin{cases} s+b = 50,000 & (1) \\ s = 2b-4000 & (2) \end{cases}$

Let $s = 2b - 4000$ in (1).

$2b - 4000 + b = 50,000$

$\quad 3b = 54,000$

$\quad\ b = 18,000$

Let $b = 18,000$ in (2).

$s = 2(18,000) - 4000$

$s = 36,000 - 4000$

$s = 32,000$

He should invest \$32,000 in stocks and \$18,000 in bonds.

58. Let n be the number of notebooks sold and c be the number of scientific calculators sold. If a calculator sells for \$7.50 and a notebook costs $\dfrac{1}{3}$ of that, notebooks sell for \$2.50. The total number of items sold, $n + c$, is 24. The total amount, $2.5n + 7.5c$, is \$110.

$\begin{cases} n+c = 24 & (1) \\ 2.5n+7.5c = 110 & (2) \end{cases}$

Solve (1) for n.

$n = 24 - c$

Let $n = 24 - c$ in (2).

$$2.5(24-c)+7.5c = 110$$
$$60-2.5c+7.5c = 110$$
$$5c = 50$$
$$c = 10$$

Let $c = 10$ in (1).
$$n+10 = 24$$
$$n = 14$$
They sold 14 notebooks and 10 calculators.

59. Let x be the measure of the angle and y be the measure of the other angle. Supplementary angles sum to 180, $x + y = 180$. One angle is 25° less than its supplement.
$$\begin{cases} x+y=180 & (1) \\ x=y-25 & (2) \end{cases}$$
Let $x = y - 25$ in (1).
$$y-25+y = 180$$
$$2y = 205$$
$$y = 102.5$$
Let $y = 102.5$ in (2).
$$x = 102.5 - 25 = 77.5$$
The angles measure 77.5° and 102.5°.

60. Let x be the measure of one angle and y be the measure of the other angle. Complementary angles sum to 90°, $x + y = 90$. One angle is 15° more than $\dfrac{1}{2}$ of its complement.
$$\begin{cases} x+y=90 & (1) \\ x=\dfrac{1}{2}y+15 & (2) \end{cases}$$
Let $x = \dfrac{1}{2}y+15$ in (1).
$$\frac{1}{2}y+15+y = 90$$
$$\frac{3}{2}y = 75$$
$$y = 50$$
Let $y = 50$ in (2).
$$x = \frac{1}{2}(50)+15$$
$$x = 25+15$$
$$x = 40$$
The angles measure 40° and 50°.

61. Let w be the width and l be the length. The three sides, $2w + l$, require 52 meters of fencing. The longest side is 8 meters less than twice the short side.
$$\begin{cases} 2w+l=52 & (1) \\ l=2w-8 & (2) \end{cases}$$

Let $l = 2w - 8$ in (1).
$$2w+2w-8 = 52$$
$$4w = 60$$
$$w = 15$$
Let $w = 15$ in (2).
$$l = 2(15)-8$$
$$l = 30-8$$
$$l = 22$$
The corral is 15 meters by 22 meters.

62. Let x be the measure of one angle and y be the measure of the other angle. The three angles sum to 180, $x + y + 30 = 180$. The first angle is 10° less than three times the second.
$$\begin{cases} x+y+30=180 & (1) \\ x=3y-10 & (2) \end{cases}$$
Let $x = 3y - 10$ in (1).
$$3y-10+y+30 = 180$$
$$4y = 160$$
$$y = 40$$
Let $y = 40$ in (2).
$$x = 3(40)-10$$
$$x = 120-10$$
$$x = 110$$
The angles measure 110° and 40°.

63. Let s be the speed of the plane in still air and w be the speed of the wind.

	d	r	t
with wind	2000	$s+w$	4
against wind	2000	$s-w$	5

$$\begin{cases} 4(s+w)=2000 & (1) \\ 5(s-w)=2000 & (2) \end{cases}$$
Divide (1) by 4 and (2) by 5, then add.
$$\begin{cases} s+w=500 \\ s-w=400 \end{cases}$$
$$\overline{2s \qquad = 900}$$
$$s = 450$$
Let $s = 450$ in (1).
$$4(450+w) = 2000$$
$$1800+4w = 2000$$
$$4w = 200$$
$$w = 50$$
The plane's speed is 450 mph and the wind's speed is 50 mph.

64. Let c be the speed of the current and p be the paddling speed.

	d	r	t
upstream	10	$p - c$	2.5
downstream	10	$p + c$	1.25

$$\begin{cases} 10 = 2.5(p-c) & (1) \\ 10 = 1.25(p+c) & (2) \end{cases}$$
Divide (1) by 2.5 and (2) by 1.25, then add.
$$\begin{cases} 4 = p - c \\ 8 = p + c \end{cases}$$
$$12 = 2p$$
$$\ \ 6 = p$$
Let $p = 6$ in (2).
$$10 = 1.25(6 + c)$$
$$\ \ 8 = 6 + c$$
$$\ \ 2 = c$$
The current is 2 mph and the paddling speed is 6 mph.

65. Let c be the speed of the cyclist and w be the speed of the wind.

	d	r	t
with wind	36	$c + w$	3
against wind	32	$c - w$	4

$$\begin{cases} 3(c+w) = 36 & (1) \\ 4(c-w) = 32 & (2) \end{cases}$$
Divide (1) by 3 and (2) by 4, then add.
$$\begin{cases} c + w = 12 \\ c - w = 8 \end{cases}$$
$$2c\ \ \ \ \ = 20$$
$$\ \ c = 10$$
Let $c = 10$ in (1).
$$3(10 + w) = 36$$
$$\ \ 10 + w = 12$$
$$\ \ \ \ \ \ \ w = 2$$
The cyclist travels at 10 mph and the wind is 2 mph.

66. Let f be the speed of the faster jogger, and s be the speed of the slower jogger.

	d	r	$t = \dfrac{d}{r}$
faster	12	f	$\dfrac{12}{f}$
slower	9	s	$\dfrac{9}{s}$

The rate of the faster jogger is 2 mph more than that of the slower jogger. The times are the same.
$$\begin{cases} f = s + 2 & (1) \\ \dfrac{12}{f} = \dfrac{9}{s} & (2) \end{cases}$$
Let $f = s + 2$ in (2).
$$\frac{12}{s+2} = \frac{9}{s}$$
$$12s = 9(s + 2)$$
$$12s = 9s + 18$$
$$\ \ 3s = 18$$
$$\ \ \ s = 6$$
Let $s = 6$ in (1).
$$f = 6 + 2 = 8$$
The faster jogger's speed is 8 mph and the slower jogger's speed is 6 mph.

67. Let d be the number of dimes and n be the number of nickels.

	number of coins	· value of each coin	= total value
dimes	d	0.10	$0.10d$
nickels	n	0.05	$0.05n$
total	35		2.25

68. Let s be the amount invested in savings and m be the amount invested in a mutual fund.

	Principal	Rate	Interest
Savings account	s	0.065	$0.065s$
mutual fund	m	0.08	$0.08m$
total	15,000		2200

69. Let r be the number of regular tickets and m the number of matinee tickets.

	Number	Cost	Amount
Regular	r	$8	$8r$
Matinee	m	$5.50	$5.5m$
Total	1498		$9929

$$\begin{cases} r+m=1498 & (1) \\ 8r+5.5m=9929 & (2) \end{cases}$$

Solve (1) for m.

$m = 1498 - r$

Let $m = 1498 - r$ = (2).

$8r + 5.5(1498 - r) = 9929$

$8r + 8239 - 5.5r = 9929$

$2.5r + 8239 = 9929$

$2.5r = 1690$

$r = 676$

676 regular-priced tickets were sold.

70. Let d be the number of dimes and q be the number of quarters.

	number	value	amount
dimes	d	0.10	$0.10d$
quarters	q	0.25	$0.25q$
			1.70

$$\begin{cases} d=q+3 & (1) \\ 0.10d+0.25q=1.70 & (2) \end{cases}$$

Let $d = q + 3$ in (2).

$0.10(q+3) + 0.25q = 1.70$

$0.10q + 0.3 + 0.25q = 1.70$

$0.35q = 1.40$

$q = 4$

There are 4 quarters.

71. Let x be the amount invested at 5% and y be the amount invested at 9%.

	P	R	I
5%	x	0.05	$0.05x$
9%	y	0.09	$0.09y$
total			1430

$$\begin{cases} x=y-5000 & (1) \\ 0.05x+0.09y=1430 & (2) \end{cases}$$

Let $x = y - 5000$ in (2).

$0.05(y-5000) + 0.09y = 1430$

$0.05y - 250 + 0.09y = 1430$

$0.14y = 1680$

$y = 12,000$

Let $y = 12,000$ in (1).

$x = 12,000 - 5000 = 7000$

Carlos invested $7000 at 5% and $12,000 at 9%.

72. Let b be the amount invested in bonds and s be the amount invested in stocks.

	P	R	I
bonds	b	0.07	$0.07b$
stocks	s	0.08	$0.08s$
total	25,000		1900

$$\begin{cases} b+s=25,000 & (1) \\ 0.07b+0.08s=1900 & (2) \end{cases}$$

Solve (1) for b.

$b = 25,000 - s$

Let $b = 25,000 - s$ in (2).

$0.07(25,000 - s) + 0.08s = 1900$

$1750 - 0.07s + 0.08s = 1900$

$0.01s = 150$

$s = 15,000$

Let $s = 15,000$ in (1).

$b + 15,000 = 25,000$

$b = 10,000$

Hilda invested $10,000 in bonds and $15,000 in stocks.

73. Let x be the quarts of 60% sugar solution and y be the quarts of 30% sugar solution.

	quarts	concentration	amount
60%	x	0.60	$0.60x$
30%	y	0.30	$0.30y$
total (51%)	10	0.51	0.51(10)

$$\begin{cases} x+y=10 & (1) \\ 0.60x+0.30y=5.1 & (2) \end{cases}$$

Solve (1) for x.

$x = 10 - y$

Let $x = 10 - y$ in (2).

$$0.60(10 - y) + 0.30y = 5.1$$
$$6 - 0.6y + 0.3y = 5.1$$
$$-0.3y = -0.9$$
$$y = 3$$

The baker should use 3 quarts of 30% sugar solution.

74. Let x be the pints of 25% peroxide solution and y be the pints of total solution (30% peroxide solution).

	pints	concentration	amount
25%	x	0.25	0.25x
60%	10	0.60	0.6(10)
total (30%)	y	0.30	0.3y

$$\begin{cases} x + 10 = y & (1) \\ 0.25x + 6 = 0.3y & (2) \end{cases}$$

Let $y = x + 10$ in (2).
$$0.25x + 6 = 0.3(x + 10)$$
$$0.25x + 6 = 0.3x + 3$$
$$3 = 0.05x$$
$$60 = x$$

Add 60 pints of 25% peroxide solution.

75. Let p be the pounds of peanuts and a be the pounds of almonds.

	pounds	value	amount
peanuts	p	4	4p
almonds	a	6.5	6.5a
total	5	5	5(5)

$$\begin{cases} p + a = 5 & (1) \\ 4p + 6.5a = 25 & (2) \end{cases}$$

Solve (1) for p.
$$p = 5 - a$$
Let $p = 5 - a$ in (2).
$$4(5 - a) + 6.5a = 25$$
$$20 - 4a + 6.5a = 25$$
$$2.5a = 5$$
$$a = 2$$
Let $a = 2$ in (1).
$$p + 2 = 5$$
$$p = 3$$

She should use 3 pounds of peanuts and 2 pounds of almonds.

76. Let x be the liters of 35% acid and y be the liters of 60% acid.

	liters	concentration	amount
35%	x	0.35	0.35x
60%	y	0.60	0.60y
total (55%)	20	0.55	0.55(20)

$$\begin{cases} x + y = 20 & (1) \\ 0.35x + 0.60y = 11 & (2) \end{cases}$$

Solve (1) for x.
$$x = 20 - y$$
Let $x = 20 - y$ in (2).
$$0.35(20 - y) + 0.60y = 11$$
$$7 - 0.35y + 0.60y = 11$$
$$0.25y = 4$$
$$y = 16$$

Let $y = 16$ in (1).
$$x + 16 = 20$$
$$x = 4$$

He should use 4 liters of 35% acid and 16 liters of 60% acid.

77. $\begin{cases} x + y \le 2 & (1) \\ 3x - 2y > 6 & (2) \end{cases}$

(a) $(-1, -5)$
(1): $-1 + (-5) \le 2$?
$$-6 \le 2 \quad \text{True}$$
(2): $3(-1) - 2(-5) > 6$?
$$-3 + 10 > 6$?$$
$$7 > 6 \quad \text{True}$$
$(-1, -5)$ is a solution.

(b) $(3, -1)$
(1): $3 + (-1) \le 2$?
$$2 \le 2 \quad \text{True}$$
(2): $3(3) - 2(-1) > 6$?
$$9 + 2 > 6$?$$
$$11 > 6 \quad \text{True}$$
$(3, -1)$ is a solution.

(c) $(4, 1)$
(1): $4 + 1 \le 2$?
$$5 \le 2 \quad \text{False}$$
$(4, 1)$ is not a solution.

78. $\begin{cases} y \geq 3x+5 & (1) \\ y \geq -2x & (2) \end{cases}$

    (a) $(-1, 0)$

        (1): $0 \geq 3(-1)+5$?

           $0 \geq -3+5$?

           $0 \geq 2$ False

        $(-1, 0)$ is not a solution.

    (b) $(-2, 4)$

        (1): $4 \geq 3(-2)+5$?

           $4 \geq -6+5$?

           $4 \geq -1$ True

        (2): $4 \geq -2(-2)$?

           $4 \geq 4$ True

        $(-2, 4)$ is a solution.

    (c) $(1, 8)$

        (1): $8 \geq 3(1)+5$?

           $8 \geq 3+5$?

           $8 \geq 8$ True

        (2): $8 \geq -2(1)$?

           $8 \geq -2$ True

        $(1, 8)$ is a solution.

79. $\begin{cases} x > 5 & (1) \\ y < -2 & (2) \end{cases}$

    (a) $(10, -10)$

        (1): $10 > 5$ True

        (2): $-10 < -2$ True

        $(10, -10)$ is a solution.

    (b) $(5, -3)$

        (1): $5 > 5$ False

        $(5, -3)$ is not a solution.

    (c) $(7, -2)$

        (1): $7 > 5$ True

        (2): $-2 < -2$ False

        $(7, -2)$ is not a solution.

80. $\begin{cases} y > x & (1) \\ 2x - y \leq 3 & (2) \end{cases}$

    (a) $(4, 3)$

        (1): $3 > 4$ False

        $(4, 3)$ is not a solution.

    (b) $(-3, -2)$

        (1): $-2 > -3$ True

        (2): $2(-3)-(-2) \leq 3$?

           $-6+2 \leq 3$?

             $-4 \leq 3$ True

        $(-3, -2)$ is a solution.

    (c) $(-1, 4)$

        (1): $4 > -1$ True

        (2): $2(-1)-4 \leq 3$?

           $-2-4 \leq 3$?

             $-6 \leq 3$ True

        $(-1, 4)$ is a solution.

81. $\begin{cases} x + 2y < 6 & (1) \\ 4y - 2x > 16 & (2) \end{cases}$

    (a) $(-4, 2)$

        (1): $-4 + 2(2) < 6$?

           $-4 + 4 < 6$?

             $0 < 6$ True

        (2): $4(2) - 2(-4) > 16$?

             $8 + 8 > 16$?

               $16 > 16$ False

        $(-4, 2)$ is not a solution.

    (b) $(-8, 6)$

        (1): $-8 + 2(6) < 6$?

             $-8 + 12 < 6$?

               $4 < 6$ True

        (2): $4(6) - 2(-8) > 16$?

              $24 + 16 > 16$?

               $40 > 16$ True

        $(-8, 6)$ is a solution.

    (c) $(-2, -3)$

        (1): $-2 + 2(-3) < 6$?

             $-2 - 6 < 6$?

               $-8 < 6$ True

        (2): $4(-3) - 2(-2) > 16$?

             $-12 + 4 > 16$?

               $-8 > 16$ False

        $(-2, -3)$ is not a solution.

82. $\begin{cases} 2x - y \geq 3 & (1) \\ y \leq 2x + 1 & (2) \end{cases}$

(a) (1, 2)

(1): $2(1) - 2 \geq 3$?

$2 - 2 \geq 3$?

$0 \geq 3$ False

(1, 2) is not a solution.

(b) (3, −2)

(1): $2(3) - (-2) \geq 3$?

$6 + 2 \geq 3$?

$8 \geq 3$ True

(2): $-2 \leq 2(3) + 1$?

$-2 \leq 6 + 1$?

$-2 \leq 7$ True

(3, −2) is a solution.

(c) (−1, 4)

(1): $2(-1) - 4 \geq 3$?

$-2 - 43 \geq 3$?

$-6 \geq 3$ False

(−1, 4) is not a solution.

83. $\begin{cases} x > -2 \\ y > 1 \end{cases}$

Graph $x > -2$ and $y > 1$ with dashed lines. Graph on the same rectangular coordinate system. The overlapping shaded region represents the solution.

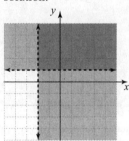

84. $\begin{cases} x \leq 3 \\ y > -1 \end{cases}$

Graph $x \leq 3$ with a solid line and $y > -1$ with a dashed line. Graph on the same rectangular coordinate system. The overlapping shaded region represents the solution.

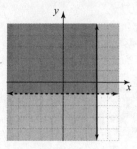

85. $\begin{cases} x + y \geq -2 \\ 2x - y \leq -4 \end{cases}$

Graph $x + y \geq -2$ ($y \geq -x - 2$) and $2x - y \leq -4$ ($y \geq 2x + 4$) with solid lines. Graph on the same rectangular coordinate system. The overlapping shaded region represents the solution.

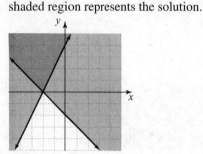

86. $\begin{cases} 3x + 2y < -6 \\ x - y < 2 \end{cases}$

Graph $3x + 2y < -6$ $\left(y < -\dfrac{3}{2}x - 3 \right)$ and

$x - y < 2$ ($y > x - 2$) with dashed lines. Graph on the same rectangular coordinate system. The overlapping shaded region represents the solution.

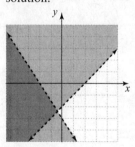

87. $\begin{cases} x > 0 \\ y \leq \dfrac{3}{4}x + 1 \end{cases}$

Graph $x > 0$ with a dashed line and $y \leq \dfrac{3}{4}x + 1$

with a solid line. Graph on the same rectangular coordinate system. The overlapping shaded region represents the solution.

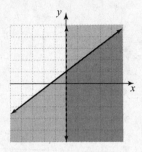

88. $\begin{cases} y \le 0 \\ y \le -\dfrac{1}{2}x - 3 \end{cases}$

Graph each with a solid line. Graph on the same rectangular coordinate system. The overlapping shaded region represents the solution.

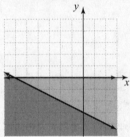

89. $\begin{cases} -y \ge x \\ 4x - 3y > -12 \end{cases}$

Graph $-y \ge x \ (y \le -x)$ with a solid line and

$4x - 3y > -12 \left(y < \dfrac{4}{3}x + 4 \right)$ with a dashed line.

Graph on the same rectangular coordinate system. The overlapping shaded region represents the solution.

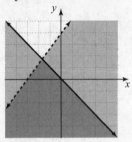

90. $\begin{cases} y \ge -x + 2 \\ x - y \ge -4 \end{cases}$

Graph $y \ge -x + 2$ with a solid line and $x - y \ge -4 \ (y \le x + 4)$ with a solid line. Graph on the same rectangular coordinate system. The overlapping shaded region represents the solution.

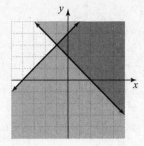

91. $\begin{cases} 2x + 3y \ge -3 \\ y > \dfrac{3}{2}x + 1 \end{cases}$

Graph $2x + 3y \ge -3 \left(y \ge -\dfrac{2}{3}x - 1 \right)$ with a solid

line and $y > \dfrac{3}{2}x + 1$ with a dashed line. Graph on

the same rectangular coordinate system. The overlapping shaded region represents the solution.

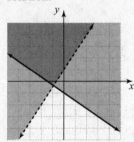

92. $\begin{cases} x + 4y \le -4 \\ y \ge \dfrac{1}{4}x + 3 \end{cases}$

Graph $x + 4y \le -4 \left(y \le -\dfrac{1}{4}x - 1 \right)$ and

$y \ge \dfrac{1}{4}x + 3$ with solid lines. Graph on the same

rectangular coordinate system. The overlapping shaded region represents the solution.

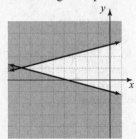

93. $\begin{cases} y > 2x - 5 \\ y - 2x \leq 0 \end{cases}$

Graph $y > 2x - 5$ with a dashed line and $y - 2x \leq 0$ ($y \leq 2x$) with a solid line. Graph on the same rectangular coordinate system. The overlapping shaded region represents the solution.

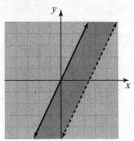

94. $\begin{cases} y < x + 2 \\ y > -\dfrac{5}{2}x + 2 \end{cases}$

Graph both with dashed lines. Graph on the same rectangular coordinate system. The overlapping shaded region represents the solution.

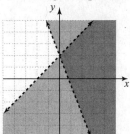

95. Let x be the pounds of fish and y be the pounds of carne asada.

$\begin{cases} 8x + 5y \leq 40 \\ x > 2y \\ x \geq 0 \\ y \geq 0 \end{cases}$

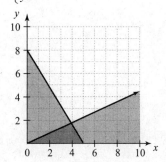

Chapter 4 Test

1. $\begin{cases} 3x - y = -5 & (1) \\ y = \dfrac{2}{3}x - 2 & (2) \end{cases}$

   (a) $(-1, 2)$

      (1): $3(-1) - 2 \stackrel{?}{=} -5$

           $-3 - 2 \stackrel{?}{=} -5$

               $-5 = -5$ True

      (2): $2 \stackrel{?}{=} \dfrac{2}{3}(-1) - 2$

           $2 \stackrel{?}{=} -\dfrac{2}{3} - 2$

           $2 = -\dfrac{8}{3}$ False

      $(-1, 2)$ is not a solution.

   (b) $(-9, -8)$

      (1): $3(-9) - (-8) \stackrel{?}{=} -5$

           $-27 + 8 \stackrel{?}{=} -5$

             $-19 = -5$ False

      $(-9, -8)$ is not a solution.

   (c) $(-3, -4)$

      (1): $3(-3) - (-4) \stackrel{?}{=} -5$

           $-9 + 4 \stackrel{?}{=} -5$

              $-5 = -5$ True

      (2): $-4 \stackrel{?}{=} \dfrac{2}{3}(-3) - 2$

          $-4 \stackrel{?}{=} -2 - 2$

          $-4 = -4$ True

      $(-3, -4)$ is a solution.

2. $\begin{cases} 2x + y \geq 10 & (1) \\ y < \dfrac{x}{2} + 1 & (2) \end{cases}$

   (a) $(5, 3)$

      (1): $2(5) + 3 \geq 10$?

           $10 + 3 \geq 10$?

             $13 \geq 10$ True

      (2): $3 < \dfrac{5}{2} + 1$?

          $3 < \dfrac{7}{2}$ True

      $(5, 3)$ is a solution.

(b) $(-2, 14)$

(1): $2(-2) + 14 \geq 10$?

$-4 + 14 \geq 10$?

$10 \geq 10$ True

(2): $14 < \dfrac{-2}{2} + 1$?

$14 < -1 + 1$?

$14 < 0$ False

$(-2, 14)$ is not a solution.

(c) $(-3, -1)$

(1): $2(-3) + (-1) \geq 10$?

$-6 - 1 \geq 10$?

$-7 \geq 10$ False

$(-3, -1)$ is not a solution.

3. (a) Since the slopes are negative reciprocals. The lines are perpendicular to each other. They intersect at one point, so there is one solution.

(b) The system is consistent.

(c) The equations are independent.

4. (a) Since the slopes are the same, but the y-intercepts are different, the lines are parallel. There is no solution.

(b) The system is inconsistent.

(c) The system is inconsistent.

5. $\begin{cases} 2x + 3y = 0 & (1) \\ x + 4y = 5 & (2) \end{cases}$

(1): $2x + 3y = 0$

$3y = -2x$

$y = -\dfrac{2}{3}x$

(2): $x + 4y = 5$

$4y = -x + 5$

$y = -\dfrac{1}{4}x + \dfrac{5}{4}$

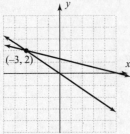

The solution is $(-3, 2)$.

6. $\begin{cases} y = 2x - 6 \\ y = -\dfrac{1}{4}x + 3 \end{cases}$

The equations are already written in slope-intercept form.

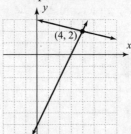

The solution is $(4, 2)$.

7. $\begin{cases} 3x - y = 3 & (1) \\ 4x + 5y = -15 & (2) \end{cases}$

Solve (1) for y.

$y = 3x - 3$

Let $y = 3x - 3$ in (2).

$4x + 5(3x - 3) = -15$

$4x + 15x - 15 = -15$

$4x + 15x = 0$

$19x = 0$

$x = 0$

Let $x = 0$ in (1).

$3(0) - y = 3$

$-y = 3$

$y = -3$

The solution is $(0, -3)$.

8. $\begin{cases} y = \dfrac{2}{3}x + 7 & (1) \\ y = \dfrac{1}{4}x + 2 & (2) \end{cases}$

Let $y = \dfrac{2}{3}x + 7$ in (2).

$\dfrac{2}{3}x + 7 = \dfrac{1}{4}x + 2$

$\dfrac{2}{3}x - \dfrac{1}{4}x = 2 - 7$

$\dfrac{5}{12}x = -5$

$x = -12$

Let $x = -12$ in (1).

$y = \frac{2}{3}(-12) + 7$

$y = -8 + 7$

$y = -1$

The solution is $(-12, -1)$.

9. $\begin{cases} 3x + 2y = -3 & (1) \\ 5x - y = -18 & (2) \end{cases}$

Multiply (2) by 2, then add.

$\begin{cases} 3x + 2y = -3 \\ \underline{10x - 2y = -36} \\ 13x \quad\quad = -39 \end{cases}$

$x = -3$

Let $x = -3$ in (2).

$5(-3) - y = -18$

$-15 - y = -18$

$-y = -3$

$y = 3$

The solution is $(-3, 3)$.

10. $\begin{cases} 4x - 5y = 12 & (1) \\ 3x + 4y = -22 & (2) \end{cases}$

Multiply (1) by 4 and (2) by 5.

$\begin{cases} 16x - 20y = 48 \\ \underline{15x + 20y = -110} \\ 31x \quad\quad = -62 \end{cases}$

$x = -2$

Let $x = -2$ in (2).

$3(-2) + 4y = -22$

$-6 + 4y = -22$

$4y = -16$

$y = -4$

The solution is $(-2, -4)$.

11. $\begin{cases} \frac{2}{3}x - \frac{1}{6}y = \frac{25}{12} & (1) \\ -\frac{1}{4}x = \frac{3}{2}y & (2) \end{cases}$

Solve (2) for x.

$-\frac{1}{4}x = \frac{3}{2}y$

$-4\left(-\frac{1}{4}x\right) = -4\left(\frac{3}{2}y\right)$

$x = -6y$

Let $x = -6y$ in (1).

$\frac{2}{3}(-6y) - \frac{1}{6}y = \frac{25}{12}$

$-4y - \frac{1}{6}y = \frac{25}{12}$

$-\frac{25}{6}y = \frac{25}{12}$

$y = -\frac{1}{2}$

Let $y = -\frac{1}{2}$ in (2).

$-\frac{1}{4}x = \frac{3}{2}\left(-\frac{1}{2}\right)$

$-\frac{1}{4}x = -\frac{3}{4}$

$x = 3$

The solution is $\left(3, -\frac{1}{2}\right)$.

12. $\begin{cases} 0.4x - 2.5y = -6.5 & (1) \\ x + y = 5.5 & (2) \end{cases}$

Solve (2) for x.

$x = 5.5 - y$

Let $x = 5.5 - y$ in (1).

$0.4(5.5 - y) - 2.5y = -6.5$

$2.2 - 0.4y - 2.5y = -6.5$

$-2.9y = -8.7$

$y = 3$

Let $y = 3$ in (2).

$x + 3 = 5.5$

$x = 2.5$

The solution is $(2.5, 3)$.

13. $\begin{cases} 4 + 3(y - 3) = -2(x + 1) & (1) \\ x + \frac{3y}{2} = \frac{3}{2} & (2) \end{cases}$

Simplify (1).

$4 + 3(y - 3) = -2(x + 1)$

$4 + 3y - 9 = -2x - 2$

$2x + 3y = -2 + 9 - 4$

$2x + 3y = 3$

Multiply (2) by -2, then add.

$\begin{cases} 2x + 3y = 3 \\ \underline{-2x - 3y = -3} \\ 0 = 0 \quad \text{True} \end{cases}$

The system is dependent. There are infinitely many solutions.

14. Let p be the speed of the plane in still air and w be the speed of the wind.

	d	r	t
with wind	1575	$p + w$	7
against wind	1575	$p - w$	9

$$\begin{cases} 7(p+w) = 1575 & (1) \\ 9(p-w) = 1575 & (2) \end{cases}$$

Divide (1) by 7 and (2) by 9, then add.

$$\begin{cases} p + w = 225 \\ p - w = 175 \end{cases}$$
$$2p = 400$$
$$p = 200$$

Let $p = 200$ in (1).
$$7(200 + w) = 1575$$
$$200 + w = 225$$
$$w = 25$$

The plane's speed is 200 mph and the wind's speed is 25 mph.

15. Let v be the number of containers of vanilla ice cream and p be the number of containers of peach ice cream.

	containers	price	amount
vanilla	v	6	$6v$
peach	p	11.50	$11.5p$
total	11	8	8(11)

$$\begin{cases} v + p = 11 & (1) \\ 6v + 11.5p = 88 & (2) \end{cases}$$

Solve (1) for v.
$$v = 11 - p$$

Let $v = 11 - p$ in (2).
$$6(11 - p) + 11.5p = 88$$
$$66 - 6p + 11.5p = 88$$
$$5.5p = 22$$
$$p = 4$$

Let $p = 4$ in (1).
$$v + 4 = 11$$
$$v = 7$$

He should use 7 containers of vanilla and 4 containers of peach ice cream.

16. Let x be the measure of one angle and y be the measure of the other angle. Supplementary angles sum to $180°$, $x + y = 180$. One angle is $30°$ less than three times its supplement, $x = 3y - 30$.

$$\begin{cases} x + y = 180 & (1) \\ x = 3y - 30 & (2) \end{cases}$$

Let $x = 3y - 30$ in (1).
$$3y - 30 + y = 180$$
$$4y = 210$$
$$y = 52.5$$

Let $y = 52.5$ in (1).
$$x + 52.5 = 180$$
$$x = 127.5$$

The angles are $52.5°$ and $127.5°$.

17. Let b be the cost of a basketball and v be the cost of a volleyball. The total number of balls, $b + v$, is 40. The total cost, $25b + 15v$, is 750.

$$\begin{cases} b + v = 40 & (1) \\ 25b + 15v = 750 & (2) \end{cases}$$

Solve (1) for b.
$$b = 40 - v$$

Let $b = 40 - v$ in (2).
$$25(40 - v) + 15v = 750$$
$$1000 - 25v + 15v = 750$$
$$-10v = -250$$
$$v = 25$$

Let $v = 25$ in (1).
$$b + 25 = 40$$
$$b = 15$$

The school will receive 15 basketballs and 25 volleyballs.

18. $\begin{cases} 4x - 2y < 8 \\ x + 3y < 6 \end{cases}$

Graph $4x - 2y < 8$ ($y > 2x - 4$) and

$x + 3y < 6 \left(y < -\dfrac{1}{3}x + 2 \right)$ with dashed lines.

Graph on the same rectangular coordinate system. The overlapping shaded region represents the solution.

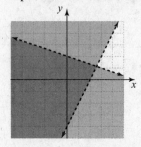

19. $\begin{cases} x \le -2 \\ -2x - 4y \le 8 \end{cases}$

Graph $x \le -2$ and $-2x - 4y \le 8$ $\left(y \ge -\dfrac{1}{2}x - 2 \right)$

with solid lines. Graph on the same rectangular coordinate system. The overlapping shaded region represents the solution.

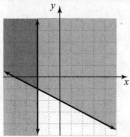

20. $\begin{cases} y \le \dfrac{2}{3}x + 4 \\ x + 4y < 0 \end{cases}$

Graph $y \le \dfrac{2}{3}x + 4$ with a solid line and

$x + 4y < 0$ $\left(y < -\dfrac{1}{4}x \right)$ with a dashed line. Graph

on the same rectangular coordinate system. The overlapping shaded region represents the solution.

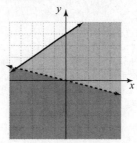

Chapter 5

Preparing for Adding and Subtracting Polynomials

P1. The coefficient of $-4x^5$ is -4.

P2. $-3x + 2 - 2x - 6x - 7 = -3x - 2x - 6x + 2 - 7$
$$= -5x - 6x - 5$$
$$= -11x - 5$$

P3. $-4(x - 3) = -4 \cdot x - (-4) \cdot 3$
$$= -4x - (-12)$$
$$= -4x + 12$$

P4. $5 - 3x$ for $x = -2$:
$$5 - 3(-2) = 5 - (-6) = 5 + 6 = 11$$

5.1 Quick Checks

1. A <u>monomial</u> in one variable is the product of a number and a variable raised to a whole number power.

2. The coefficient of a monomial such as x^2 or z is <u>1</u>.

3. $12x^6$ is a monomial. The coefficient is 12 and the degree is 6.

4. $3x^{-3}$ is not a monomial because the exponent of the variable x is -3 and -3 is not a whole number.

5. 10 is a monomial. The coefficient is 10 and the degree is 0.

6. $n^{1/3}$ is not a monomial because the exponent of the variable n is $\dfrac{1}{3}$ and $\dfrac{1}{3}$ is not a whole number.

7. The degree of a monomial in the form $ax^m y^n$ is <u>$m + n$</u>.

8. $3x^5 y^2$ is a monomial in x and y of degree $5 + 2 = 7$. The coefficient is 3.

9. $4ab^{1/2}$ is not a monomial because the exponent on the variable b is $\dfrac{1}{2}$, and $\dfrac{1}{2}$ is not a whole number.

10. $-x^2 y$ is a monomial in x and y of degree $2 + 1 = 3$. The coefficient is -1.

11. True

12. A polynomial is said to be written in <u>standard form</u> if it is written with the terms in descending order according to degree.

13. $-4x^3 + 2x^2 - 5x + 3$ is a polynomial of degree 3.

14. $2m^{-1} + 7$ is not a polynomial because the exponent on the first term, -1, is not a whole number.

15. $\dfrac{-1}{x^2 + 1}$ is not a polynomial because there is a variable expression in the denominator of the fraction.

16. $5p^3 q - 8pq^2 + pq$ is a polynomial of degree $3 + 1 = 4$.

17. True or False: $4x^3 + 7x^3 = 11x^6$ <u>False</u>

18. $(9x^2 - x + 5) + (3x^2 + 4x - 2)$
$$= 9x^2 - x + 5 + 3x^2 + 4x - 2$$
$$= 9x^2 + 3x^2 - x + 4x + 5 - 2$$
$$= (9 + 3)x^2 + (-1 + 4)x + (5 - 2)$$
$$= 12x^2 + 3x + 3$$

19.
$$
\begin{array}{r}
4z^4 - 2z^3 + z - 5 \\
+ (-2z^4 + 6z^3 - z^2 + 4) \\
\hline
2z^4 + 4z^3 - z^2 + z - 1
\end{array}
$$

20.
$$
\begin{aligned}
(7x^2y + x^2y^2 - 5xy^2) + (-2x^2y + 5x^2y^2 + 4xy^2) &= 7x^2y + x^2y^2 - 5xy^2 - 2x^2y + 5x^2y^2 + 4xy^2 \\
&= 7x^2y - 2x^2y + x^2y^2 + 5x^2y^2 - 5xy^2 + 4xy^2 \\
&= (7-2)x^2y + (1+5)x^2y^2 + (-5+4)xy^2 \\
&= 5x^2y + 6x^2y^2 - xy^2
\end{aligned}
$$

21.
$$
\begin{array}{r}
7x^3 - 3x^2 + 2x + 8 \\
- (2x^3 + 12x^2 - x + 1) \\
\hline
\end{array}
$$
Rewrite the subtraction as addition.
$$
\begin{array}{r}
7x^3 - 3x^2 + 2x + 8 \\
+ (-2x^3 - 12x^2 + x - 1) \\
\hline
5x^3 - 15x^2 + 3x + 7
\end{array}
$$

22.
$$
\begin{aligned}
(6y^3 - 3y^2 + 2y + 4) - (-2y^3 + 5y + 10) &= (6y^3 - 3y^2 + 2y + 4) + (2y^3 - 5y - 10) \\
&= 6y^3 - 3y^2 + 2y + 4 + 2y^3 - 5y - 10 \\
&= 6y^3 + 2y^3 - 3y^2 + 2y - 5y + 4 - 10 \\
&= 8y^3 - 3y^2 - 3y - 6
\end{aligned}
$$

23.
$$
\begin{aligned}
(9x^2y + 6x^2y^2 - 3xy^2) - (-2x^2y + 4x^2y^2 + 6xy^2) &= (9x^2y + 6x^2y^2 - 3xy^2) + (2x^2y - 4x^2y^2 - 6xy^2) \\
&= 9x^2y + 6x^2y^2 - 3xy^2 + 2x^2y - 4x^2y^2 - 6xy^2 \\
&= 9x^2y + 2x^2y + 6x^2y^2 - 4x^2y^2 - 3xy^2 - 6xy^2 \\
&= 11x^2y + 2x^2y^2 - 9xy^2
\end{aligned}
$$

24.
$$
\begin{aligned}
(4a^2b - 3a^2b^3 + 2ab^2) - (3a^2b - 4a^2b^3 + 5ab) &= (4a^2b - 3a^2b^3 + 2ab^2) + (-3a^2b + 4a^2b^3 - 5ab) \\
&= 4a^2b - 3a^2b^3 + 2ab^2 - 3a^2b + 4a^2b^3 - 5ab \\
&= a^2b + a^2b^3 + 2ab^2 - 5ab
\end{aligned}
$$

25.
$$
\begin{aligned}
(3a^2 - 5ab + 3b^2) + (4a^2 - 7b^2) - (8b^2 - ab) &= 3a^2 - 5ab + 3b^2 + 4a^2 - 7b^2 - 8b^2 + ab \\
&= 3a^2 + 4a^2 - 5ab + ab + 3b^2 - 7b^2 - 8b^2 \\
&= 7a^2 - 4ab - 12b^2
\end{aligned}
$$

26. (a) Let $x = 0$.
$$
\begin{aligned}
-2x^3 + 7x + 1 &= -2(0)^3 + 7(0) + 1 \\
&= 0 + 0 + 1 \\
&= 1
\end{aligned}
$$

(b) Let $x = 5$.
$$
\begin{aligned}
-2x^3 + 7x + 1 &= -2(5)^3 + 7(5) + 1 \\
&= -250 + 35 + 1 \\
&= -214
\end{aligned}
$$

(c) Let $x = -4$.
$$
\begin{aligned}
-2x^3 + 7x + 1 &= -2(-4)^3 + 7(-4) + 1 \\
&= 128 - 28 + 1 \\
&= 101
\end{aligned}
$$

27. (a) Let $m = -2$ and $n = -4$.

$-2m^2 n + 3mn - n^2$
$= -2(-2)^2(-4) + 3(-2)(-4) - (-4)^2$
$= -2(4)(-4) + 3(-2)(-4) - (16)$
$= 32 + 24 - 16$
$= 40$

(b) Let $m = 1$ and $n = 5$.

$-2m^2 n + 3mn - n^2$
$= -2(1)^2(5) + 3(1)(5) - (5)^2$
$= -2(1)(5) + 3(1)(5) - (25)$
$= -10 + 15 - 25$
$= -20$

28. $40x - 0.2x^2 = 40(75) - 0.2(75)^2$
$= 40(75) - 0.2(5625)$
$= 3000 - 1125$
$= 1875$

The revenue is $1875.

5.1 Exercises

29. Yes, $\frac{1}{2}y^3$ is a monomial of degree 3 with coefficient $\frac{1}{2}$.

31. Yes, $\frac{x^2}{7} = \frac{1}{7}x^2$ is a monomial of degree 2 with coefficient $\frac{1}{7}$.

33. No, z^{-6} is not a monomial because the exponent of z is not a whole number.

35. Yes, $12mn^4$ is a monomial in m and n of degree $1 + 4 = 5$. The coefficient is 12.

37. No, $\frac{3}{n^2}$ is not a monomial because there is a variable in the denominator of the fraction.

39. Yes, 4 is a monomial of degree 0 with coefficient 4.

41. Yes, $6x^2 - 10$ is a binomial of degree 2.

43. No, $\frac{-20}{n}$ is not a polynomial because there is a variable in the denominator of a fraction.

45. No, $3y^{1/3} + 2$ is not a polynomial because a variable has an exponent that is not a whole number.

47. Yes, $\frac{1}{8}$ is a monomial of degree 0.

49. Yes, $3t^2 - \frac{1}{2}t^4 + 6t = -\frac{1}{2}t^4 + 3t^2 + 6t$ is a trinomial of degree 4.

51. No, $7x^{-1} + 4$ is not a polynomial because a variable has an exponent that is not a whole number.

53. Yes, $5z^3 - 10z^2 + z + 12$ is a polynomial of degree 3.

55. Yes, $3x^2 y^2 + 2xy^4 + 4 = 2xy^4 + 3x^2 y^2 + 4$ is a trinomial of degree 5 (1 + 4).

57. $(4x - 3) + (3x - 7) = 4x - 3 + 3x - 7$
$= 4x + 3x - 3 - 7$
$= (4 + 3)x + (-3 - 7)$
$= 7x - 10$

59. $(-4m^2 + 2m - 1) + (2m^2 - 2m + 6)$
$= -4m^2 + 2m - 1 + 2m^2 - 2m + 6$
$= -4m^2 + 2m^2 + 2m - 2m - 1 + 6$
$= (-4 + 2)m^2 + (2 - 2)m + (-1 + 6)$
$= -2m^2 + 0m + 5$
$= -2m^2 + 5$

61. $(p - p^3 + 2) + (6 - 2p^2 + p^3)$
$= p - p^3 + 2 + 6 - 2p^2 + p^3$
$= -p^3 + p^3 - 2p^2 + p + 2 + 6$
$= (-1 + 1)p^3 - 2p^2 + p + (2 + 6)$
$= 0p^3 - 2p^2 + p + 8$
$= -2p^2 + p + 8$

63. $(2y - 10) + (-3y^2 - 4y + 6)$
$= 2y - 10 - 3y^2 - 4y + 6$
$= -3y^2 + 2y - 4y - 10 + 6$
$= -3y^2 + (2 - 4)y + (-10 + 6)$
$= -3y^2 - 2y - 4$

65. $\left(\dfrac{1}{2}p^2 - \dfrac{2}{3}p + 2\right) + \left(\dfrac{3}{4}p^2 + \dfrac{5}{6}p - 5\right)$

$= \dfrac{1}{2}p^2 - \dfrac{2}{3}p + 2 + \dfrac{3}{4}p^2 + \dfrac{5}{6}p - 5$

$= \dfrac{1}{2}p^2 + \dfrac{3}{4}p^2 - \dfrac{2}{3}p + \dfrac{5}{6}p + 2 - 5$

$= \left(\dfrac{1}{2} + \dfrac{3}{4}\right)p^2 + \left(-\dfrac{2}{3} + \dfrac{5}{6}\right)p + (2 - 5)$

$= \dfrac{5}{4}p^2 + \dfrac{1}{6}p - 3$

67. $(5m^2 - 6mn + 2n^2) + (m^2 + 2mn - 3n^2)$

$= 5m^2 - 6mn + 2n^2 + m^2 + 2mn - 3n^2$

$= 5m^2 + m^2 - 6mn + 2mn + 2n^2 - 3n^2$

$= (5 + 1)m^2 + (-6 + 2)mn + (2 - 3)n^2$

$= 6m^2 - 4mn - n^2$

69. $\quad \begin{array}{r} 4n^2 - 2n + 1 \\ + \quad (-6n + 4) \\ \hline 4n^2 - 8n + 5 \end{array}$

71. $(3x - 10) - (4x + 6) = (3x - 10) + (-4x - 6)$

$\qquad\qquad\qquad\quad = 3x - 10 - 4x - 6$

$\qquad\qquad\qquad\quad = 3x - 4x - 10 - 6$

$\qquad\qquad\qquad\quad = -x - 16$

73. $(12x^2 - 2x - 4) - (-2x^2 + x + 1)$

$= (12x^2 - 2x - 4) + (2x^2 - x - 1)$

$= 12x^2 - 2x - 4 + 2x^2 - x - 1$

$= 12x^2 + 2x^2 - 2x - x - 4 - 1$

$= 14x^2 - 3x - 5$

75. $(y^3 - 2y + 1) - (-3y^3 + y + 5)$

$= (y^3 - 2y + 1) + (3y^3 - y - 5)$

$= y^3 - 2y + 1 + 3y^3 - y - 5$

$= y^3 + 3y^3 - 2y - y + 1 - 5$

$= 4y^3 - 3y - 4$

77. $(3y^3 - 2y) - (2y + y^2 + y^3)$

$= (3y^3 - 2y) + (-2y - y^2 - y^3)$

$= 3y^3 - 2y - 2y - y^2 - y^3$

$= 3y^3 - y^3 - y^2 - 2y - 2y$

$= 2y^3 - y^2 - 4y$

79. $\left(\dfrac{5}{3}q^2 - \dfrac{5}{2}q + 4\right) - \left(\dfrac{1}{9}q^2 + \dfrac{3}{8}q + 2\right)$

$= \left(\dfrac{5}{3}q^2 - \dfrac{5}{2}q + 4\right) + \left(-\dfrac{1}{9}q^2 - \dfrac{3}{8}q - 2\right)$

$= \dfrac{5}{3}q^2 - \dfrac{5}{2}q + 4 - \dfrac{1}{9}q^2 - \dfrac{3}{8}q - 2$

$= \dfrac{5}{3}q^2 - \dfrac{1}{9}q^2 - \dfrac{5}{2}q - \dfrac{3}{8}q + 4 - 2$

$= \dfrac{14}{9}q^2 - \dfrac{23}{8}q + 2$

81. $(-4m^2n^2 - 2mn + 3) - (4m^2n^2 + 2mn + 10)$

$= (-4m^2n^2 - 2mn + 3) + (-4m^2n^2 - 2mn - 10)$

$= -4m^2n^2 - 2mn + 3 - 4m^2n^2 - 2mn - 10$

$= -4m^2n^2 - 4m^2n^2 - 2mn - 2mn + 3 - 10$

$= -8m^2n^2 - 4mn - 7$

83. $\begin{array}{r} 6x - 3 \\ - \,(10x + 2) \\ \hline \end{array} \quad \rightarrow \quad \begin{array}{r} 6x - 3 \\ + \,(-10x - 2) \\ \hline -4x - 5 \end{array}$

85. $(2x^2 - 3x + 1) + (x^2 + 9) - (4x^2 - 2x - 5)$

$= (2x^2 - 3x + 1) + (x^2 + 9) + (-4x^2 + 2x + 5)$

$= 2x^2 - 3x + 1 + x^2 + 9 - 4x^2 + 2x + 5$

$= 2x^2 + x^2 - 4x^2 - 3x + 2x + 1 + 9 + 5$

$= -x^2 - x + 15$

87. $(p^2 + 25) - (3p^2 - p + 4) - (-4p^2 - 9p + 5)$

$= (p^2 + 25) + (-3p^2 + p - 4) + (4p^2 + 9p - 5)$

$= p^2 + 25 - 3p^2 + p - 4 + 4p^2 + 9p - 5$

$= p^2 - 3p^2 + 4p^2 + p + 9p + 25 - 4 - 5$

$= 2p^2 + 10p + 16$

89. $(4y - 3) + (y^3 - 8) - (2y^2 - 7y + 3)$

$= (4y - 3) + (y^3 - 8) + (-2y^2 + 7y - 3)$

$= 4y - 3 + y^3 - 8 - 2y^2 + 7y - 3$

$= y^3 - 2y^2 + 4y + 7y - 3 - 8 - 3$

$= y^3 - 2y^2 + 11y - 14$

91. $2x^2 - x + 3$

     (a) $x = 0:\ 2(0)^2 - 0 + 3 = 2(0) - 0 + 3 = 3$

(b) $x = 5$: $2(5)^2 - 5 + 3 = 2(25) - 5 + 3$
$$= 50 - 5 + 3$$
$$= 48$$

(c) $x = -2$: $2(-2)^2 - (-2) + 3 = 2(4) + 2 + 3$
$$= 8 + 2 + 3$$
$$= 13$$

93. $7 - x^2$

(a) $x = 3$: $7 - (3)^2 = 7 - 9 = -2$

(b) $x = -\dfrac{5}{2}$: $7 - \left(-\dfrac{5}{2}\right)^2 = 7 - \dfrac{25}{4} = \dfrac{3}{4}$

(c) $x = -1.5$: $7 - (-1.5)^2 = 7 - 2.25 = 4.75$

95. $-x^2 y + 2xy^2 - 3$; $x = 2, y = -3$:

$-(2)^2(-3) + 2(2)(-3)^2 - 3$
$$= -(4)(-3) + 2(2)(9) - 3$$
$$= 12 + 36 - 3$$
$$= 45$$

97. $st + 2s^2 t + 3st^2 - t^4$; $s = -2, t = 4$:

$(-2)(4) + 2(-2)^2(4) + 3(-2)(4)^2 - (4)^4$
$$= (-2)(4) + 2(4)(4) + 3(-2)(16) - 256$$
$$= -8 + 32 - 96 - 256$$
$$= -328$$

99. $(7t - 3) - 4t = 7t - 3 - 4t = 7t - 4t - 3 = 3t - 3$

101. $(5x^2 + x - 4) + (-2x^2 - 4x + 1)$
$$= 5x^2 + x - 4 - 2x^2 - 4x + 1$$
$$= 5x^2 - 2x^2 + x - 4x - 4 + 1$$
$$= 3x^2 - 3x - 3$$

103. $(2xy^2 - 3) + (7xy^2 + 4) = 2xy^2 - 3 + 7xy^2 + 4$
$$= 2xy^2 + 7xy^2 - 3 + 4$$
$$= 9xy^2 + 1$$

105. $(4 + 8y - 2y^2) - (3 - 7y - y^2)$
$$= (4 + 8y - 2y^2) + (-3 + 7y + y^2)$$
$$= 4 + 8y - 2y^2 - 3 + 7y + y^2$$
$$= -2y^2 + y^2 + 8y + 7y + 4 - 3$$
$$= -y^2 + 15y + 1$$

107. $\left(\dfrac{5}{6}q^2 - \dfrac{1}{3}\right) + \left(\dfrac{3}{2}q^2 + 2\right) = \dfrac{5}{6}q^2 - \dfrac{1}{3} + \dfrac{3}{2}q^2 + 2$

$$= \dfrac{5}{6}q^2 + \dfrac{3}{2}q^2 - \dfrac{1}{3} + 2$$
$$= \dfrac{7}{3}q^2 + \dfrac{5}{3}$$

109. $14d^2 - (2d - 10) - (d^2 - 3d)$
$$= 14d^2 + (-2d + 10) + (-d^2 + 3d)$$
$$= 14d^2 - 2d + 10 - d^2 + 3d$$
$$= 14d^2 - d^2 - 2d + 3d + 10$$
$$= 13d^2 + d + 10$$

111. $(4a^2 - 1) + (a^2 + 5a + 2) - (-a^2 + 4)$
$$= (4a^2 - 1) + (a^2 + 5a + 2) + (a^2 - 4)$$
$$= 4a^2 - 1 + a^2 + 5a + 2 + a^2 - 4$$
$$= 4a^2 + a^2 + a^2 + 5a - 1 + 2 - 4$$
$$= 6a^2 + 5a - 3$$

113. $(x^2 - 2xy - y^2) - (3x^2 + xy - y^2) + (xy + y^2)$
$$= (x^2 - 2xy - y^2) + (-3x^2 - xy + y^2) + (xy + y^2)$$
$$= x^2 - 2xy - y^2 - 3x^2 - xy + y^2 + xy + y^2$$
$$= x^2 - 3x^2 - 2xy - xy + xy - y^2 + y^2 + y^2$$
$$= -2x^2 - 2xy + y^2$$

115. $(3x + 10) + (-8x + 2) = 3x + 10 - 8x + 2$
$$= 3x - 8x + 10 + 2$$
$$= -5x + 12$$

117. $(-x^2 + 2x + 3) - (-4x^2 - 2x + 6)$
$$= (-x^2 + 2x + 3) + (4x^2 + 2x - 6)$$
$$= -x^2 + 2x + 3 + 4x^2 + 2x - 6$$
$$= -x^2 + 4x^2 + 2x + 2x + 3 - 6$$
$$= 3x^2 + 4x - 3$$

119. $(2x - 10) - (14x^2 - 2x + 3)$
$$= (2x - 10) + (-14x^2 + 2x - 3)$$
$$= 2x - 10 - 14x^2 + 2x - 3$$
$$= -14x^2 + 2x + 2x - 10 - 3$$
$$= -14x^2 + 4x - 13$$

121. $-(3x - 5) = -3x + 5$:
$3x - 5 + (-3x + 5) = 3x - 5 - 3x + 5$
$$= 3x - 3x - 5 + 5$$
$$= 0$$

123. $-16t^2 + 30; \; t = \dfrac{1}{2}$

$-16\left(\dfrac{1}{2}\right)^2 + 30 = -16\left(\dfrac{1}{4}\right) + 30 = -4 + 30 = 26$

The height of the ball is 26 feet.

125. **(a)** Let $x = 0$.

$-32.4x^2 + 905.6x + 438$

$= -32.4(0)^2 + 905.6(0) + 438$

$= 438$

Worldwide online social network advertising spending for 2006 is estimated to be \$438 million.

(b) Let $x = 2$.

$-32.4x^2 + 905.6x + 438$

$= -32.4(2)^2 + 905.6(2) + 438$

$= -32.4(4) + 905.6(2) + 438$

$= -129.6 + 1811.2 + 438$

$= 2119.6$

Worldwide online social network advertising spending for 2008 is estimated to be \$2119.6 million.

(c) Let $x = 5$.

$-32.4x^2 + 905.6x + 438$

$= -32.4(5)^2 + 905.6(5) + 438$

$= -32.4(25) + 905.6(5) + 438$

$= -810 + 4528 + 438$

$= 4156$

Worldwide online social network advertising spending for 2011 is estimated to be \$4156 million.

127. **(a)** profit $=$ (revenue) $-$ (cost)

$= (-2x^2 + 120x) - (0.125x^2 + 15x)$

$= (-2x^2 + 120x) + (-0.125x^2 - 15x)$

$= -2x^2 + 120x - 0.125x^2 - 15x$

$= -2x^2 - 0.125x^2 + 120x - 15x$

$= -2.125x^2 + 105x$

(b) $-2.125x^2 + 105x; \; x = 20$

$-2.125(20)^2 + 105(20)$

$= -2.125(400) + 105(20)$

$= -850 + 2100$

$= 1250$

The profit is \$1250 if 20 calculators are produced and sold each day.

129. **(a)** For x lawns, Marissa's cost is $5x + 10x = 15x$, and her revenue is $25x$.

profit $=$ (revenue) $-$ (cost)

$\quad\quad = (25x) - (15x)$

$\quad\quad = 10x$

(b) $10x; \; x = 50$

$10(50) = 500$

Her profit is \$500 per week if she fertilizes 50 lawns.

(c) Since her profit for one week is \$500, Marissa's profit for 30 weeks is $30(\$500) = \$15,000$.

131. $(4x - 3) + (4x - 3) + (4x - 3) + (4x - 3)$

$= 4x + 4x + 4x + 4x - 3 - 3 - 3 - 3$

$= 16x - 12$

133. $(3x - 4) + (2x + 1) + (10) + (10) + (2x + 1)$

$= 3x + 2x + 2x - 4 + 1 + 10 + 10 + 1$

$= 7x + 18$

135. $(3x - 10) - (x - 5) = (3x - 10) + (-x + 5)$

$\quad\quad = 3x - 10 - x + 5$

$\quad\quad = 3x - x - 10 + 5$

$\quad\quad = 2x - 5$

137. Answers may vary.

139. To find the degree of a polynomial in one variable, use the largest exponent on any term in the polynomial. To find the degree of a polynomial in more than one variable, find the degree of each term in the polynomial. The degree of the term $ax^m y^n$ is the sum of the exponents, $m + n$. The highest degree of all of the terms is the degree of the polynomial.

Section 5.2

Preparing for Multiplying Monomials: The Product and Power Rules

P1. $4^3 = 4 \cdot 4 \cdot 4 = 16 \cdot 4 = 64$

P2. $3(4 - 5x) = 3 \cdot 4 - 3 \cdot 5x = 12 - 15x$

5.2 Quick Checks

1. The expression 12^3 is written in <u>exponential</u> form.

2. $a^m \cdot a^m = a^{\underline{m+m}}$

3. False; $3^4 \cdot 3^2 = 3^{4+2} = 3^6$.

4. $3^2 \cdot 3 = 3^{2+1} = 3^3 = 27$

5. $(-5)^2 (-5)^3 = (-5)^{2+3} = (-5)^5 = -3125$

6. $c^6 \cdot c^2 = c^{6+2} = c^8$

7. $y^3 \cdot y \cdot y^5 = y^3 \cdot y^1 \cdot y^5 = y^{3+1+5} = y^9$

8. $a^4 \cdot a^5 \cdot b^6 = a^{4+5} \cdot b^6 = a^9 b^6$

9. $(2^2)^4 = 2^{2 \cdot 4} = 2^8$

10. $[(-3)^3]^2 = (-3)^{3 \cdot 2} = (-3)^6$

11. $(b^2)^5 = b^{2 \cdot 5} = b^{10}$

12. False;
$$(ab^3)^2 \overset{?}{=} ab^6$$
$$(a)^2 (b^3)^2 \overset{?}{=} ab^6$$
$$a^2 b^6 \neq ab^6$$

13. $(2n)^3 = (2)^3 (n)^3 = 8n^3$

14. $(-5x^4)^3 = (-5)^3 (x^4)^3$
$$= -125 x^{4 \cdot 3}$$
$$= -125 x^{12}$$

15. $(-7a^3 b)^2 = (-7)^2 (a^3)^2 (b)^2$
$$= 49 a^{3 \cdot 2} b^2$$
$$= 49 a^6 b^2$$

16. $(2a^6)(4a^5) = 2 \cdot 4 \cdot a^6 \cdot a^5 = 8a^{6+5} = 8a^{11}$

17. $(3m^2 n^4)(-6mn^5) = 3 \cdot (-6) \cdot m^2 \cdot m \cdot n^4 \cdot n^5$
$$= -18 \cdot m^{2+1} \cdot n^{4 \cdot 5}$$
$$= -18 m^3 n^9$$

18. $\left(\frac{8}{3} xy^2\right)\left(\frac{1}{2} x^2 y\right)(-12xy)$
$$= \left(\frac{8}{3}\right)\left(\frac{1}{2}\right)\left(-\frac{12}{1}\right) x \cdot x^2 \cdot x \cdot y^2 \cdot y \cdot y$$
$$= \left(\frac{4}{3}\right)\left(-\frac{12}{1}\right) x^{1+2+1} y^{2+1+1}$$
$$= -16 x^4 y^4$$

5.2 Exercises

19. $4^2 \cdot 4^3 = 4^{2+3} = 4^5 = 1024$

21. $(-2)^3 (-2)^4 = (-2)^{3+4} = (-2)^7 = -128$

23. $m^4 \cdot m^5 = m^{4+5} = m^9$

25. $b^9 \cdot b^{11} = b^{9+11} = b^{20}$

27. $x^7 \cdot x = x^7 \cdot x^1 = x^{7+1} = x^8$

29. $p \cdot p^2 \cdot p^6 = p^1 \cdot p^2 \cdot p^6 = p^{1+2+6} = p^9$

31. $(-n)^3 (-n)^4 = (-n)^{3+4} = (-n)^7$

33. $(2^3)^2 = 2^{3 \cdot 2} = 2^6 = 64$

35. $[(-2)^2]^3 = (-2)^{2 \cdot 3} = (-2)^6 = 64$

37. $[(-m)^9]^2 = (-m)^{9 \cdot 2} = (-m)^{18}$

39. $(m^2)^7 = m^{2 \cdot 7} = m^{14}$

41. $[(-b)^4]^5 = (-b)^{4 \cdot 5} = (-b)^{20}$

43. $(3x^2)^3 = 3^3 \cdot (x^2)^3$
$$= 27 \cdot x^{2 \cdot 3}$$
$$= 27 x^6$$

45. $(-5z^2)^2 = (-5)^2 (z^2)^2 = (25)(z^{2 \cdot 2}) = 25 z^4$

47. $\left(\frac{1}{2} a\right)^2 = \left(\frac{1}{2}\right)^2 (a)^2 = \left(\frac{1}{4}\right) a^2 = \frac{1}{4} a^2$

49. $(-3p^7q^2)^4 = (-3)^4(p^7)^4(q^2)^4$
$\quad\quad = 81 \cdot p^{7\cdot4} \cdot q^{2\cdot4}$
$\quad\quad = 81p^{28}q^8$

51. $(-2m^2n)^3 = (-2)^3(m^2)^3 \cdot n^3$
$\quad\quad = -8 \cdot m^{2\cdot3} \cdot n^3$
$\quad\quad = -8m^6n^3$

53. $(-5xy^2z^3)^2 = (-5)^2 \cdot x^2 \cdot (y^2)^2 \cdot (z^3)^2$
$\quad\quad = 25 \cdot x^2 \cdot y^{2\cdot2} \cdot z^{3\cdot2}$
$\quad\quad = 25x^2y^4z^6$

55. $(4x^2)(3x^3) = 4 \cdot 3 \cdot x^2 \cdot x^3$
$\quad\quad = 12 \cdot x^{2+3}$
$\quad\quad = 12x^5$

57. $(10a^3)(-4a^7) = 10 \cdot (-4) \cdot a^3 \cdot a^7$
$\quad\quad = -40 \cdot a^{3+7}$
$\quad\quad = -40a^{10}$

59. $(-m^3)(7m) = (-1 \cdot m^3) \cdot (7m)$
$\quad\quad = -1 \cdot 7 \cdot m^3 \cdot m$
$\quad\quad = -7 \cdot m^{3+1}$
$\quad\quad = -7m^4$

61. $\left(\frac{4}{5}x^4\right)\left(\frac{15}{2}x^3\right) = \frac{4}{5} \cdot \frac{15}{2} \cdot x^4 \cdot x^3$
$\quad\quad = 6 \cdot x^{4+3}$
$\quad\quad = 6x^7$

63. $(2x^2y^3)(3x^4y) = 2 \cdot 3 \cdot x^2 \cdot x^4 \cdot y^3 \cdot y$
$\quad\quad = 6 \cdot x^{2+4} \cdot y^{3+1}$
$\quad\quad = 6x^6y^4$

65. $\left(\frac{1}{4}mn^3\right)(-20mn) = \frac{1}{4} \cdot (-20) \cdot m \cdot m \cdot n^3 \cdot n$
$\quad\quad = -5 \cdot m^{1+1} \cdot n^{3+1}$
$\quad\quad = -5m^2n^4$

67. $(4x^2y)(-5xy^3)(2x^2y^2)$
$\quad = 4 \cdot (-5) \cdot 2 \cdot x^2 \cdot x \cdot x^2 \cdot y \cdot y^3 \cdot y^2$
$\quad = (-20) \cdot 2 \cdot x^{2+1+2} \cdot y^{1+3+2}$
$\quad = -40x^5y^6$

69. $x^2 \cdot 4x = 4 \cdot x^2 \cdot x = 4 \cdot x^{2+1} = 4x^3$

71. $(-x)(5x^2) = (-1) \cdot (5) \cdot x \cdot x^2$
$\quad\quad = -5 \cdot x^{1+2}$
$\quad\quad = -5x^3$

73. $(b^3)^2 = b^{3\cdot2} = b^6$

75. $(-3b)^4 = (-3)^4 \cdot b^4 = 81b^4$

77. $(-6p)^2\left(\frac{1}{4}p^3\right) = (-6)^2 \cdot p^2 \cdot \left(\frac{1}{4}p^3\right)$
$\quad\quad = 36p^2 \cdot \left(\frac{1}{4}p^3\right)$
$\quad\quad = 36 \cdot \frac{1}{4} \cdot p^2 \cdot p^3$
$\quad\quad = 9 \cdot p^{2+3}$
$\quad\quad = 9p^5$

79. $(-xy)(x^2y)(-3x)^3$
$\quad = (-xy)(x^2y)(-3)^3(x)^3$
$\quad = (-xy)(x^2y)(-27)(x^3)$
$\quad = (-27) \cdot (-1) \cdot x \cdot x^2 \cdot x^3 \cdot y \cdot y$
$\quad = (27) \cdot x^{1+2+3} \cdot y^{1+1}$
$\quad = 27x^6y^2$

81. $\left(\frac{3}{5}\right)(5c)(-10d^2) = \left(\frac{3}{5}\right) \cdot (5) \cdot (-10) \cdot c \cdot d^2$
$\quad\quad = (3) \cdot (-10) \cdot c \cdot d^2$
$\quad\quad = -30cd^2$

83. $(5x)^2(x^2)^3 = 5^2 \cdot x^2 \cdot x^{2\cdot3}$
$\quad\quad = 25 \cdot x^2 \cdot x^6$
$\quad\quad = 25 \cdot x^{2+6}$
$\quad\quad = 25x^8$

85. $(x^2y)^3(-2xy^3) = x^{2\cdot3} \cdot y^3 \cdot (-2) \cdot x \cdot y^3$
$\quad\quad = x^6 \cdot y^3 \cdot (-2) \cdot x \cdot y^3$
$\quad\quad = -2 \cdot x^6 \cdot x \cdot y^3 \cdot y^3$
$\quad\quad = -2 \cdot x^{6+1}y^{3+3}$
$\quad\quad = -2x^7y^6$

87. $(-3x)^2(2x^4)^3 = (-3)^2 \cdot x^2 \cdot 2^3 \cdot (x^4)^3$

$\qquad\qquad\qquad = 9 \cdot x^2 \cdot 8 \cdot x^{4 \cdot 3}$

$\qquad\qquad\qquad = 9 \cdot x^2 \cdot 8 \cdot x^{12}$

$\qquad\qquad\qquad = 9 \cdot 8 \cdot x^2 \cdot x^{12}$

$\qquad\qquad\qquad = 72 \cdot x^{2+12}$

$\qquad\qquad\qquad = 72x^{14}$

89. $\left(\dfrac{4}{5}q\right)^2 (-5q)^2 \left(\dfrac{1}{4}q^2\right)$

$\qquad = \left(\dfrac{4}{5}\right)^2 \cdot q^2 \cdot (-5)^2 \cdot q^2 \cdot \dfrac{1}{4} \cdot q^2$

$\qquad = \dfrac{16}{25} \cdot q^2 \cdot 25 \cdot q^2 \cdot \dfrac{1}{4} \cdot q^2$

$\qquad = \dfrac{16}{25} \cdot 25 \cdot \dfrac{1}{4} \cdot q^2 \cdot q^2 \cdot q^2$

$\qquad = 4 \cdot q^{2+2+2}$

$\qquad = 4q^6$

91. $(3x)^2(-2x)^4 \left(\dfrac{1}{3}x^4\right)^2$

$\qquad = 3^2 \cdot x^2 \cdot (-2)^4 \cdot x^4 \cdot \left(\dfrac{1}{3}\right)^2 \cdot (x^4)^2$

$\qquad = 9 \cdot x^2 \cdot 16 \cdot x^4 \cdot \dfrac{1}{9} \cdot x^{4 \cdot 2}$

$\qquad = 9 \cdot x^2 \cdot 16 \cdot x^4 \cdot \dfrac{1}{9} \cdot x^8$

$\qquad = 9 \cdot 16 \cdot \dfrac{1}{9} \cdot x^2 \cdot x^4 \cdot x^8$

$\qquad = 16 \cdot x^{2+4+8}$

$\qquad = 16x^{14}$

93. $-3(-5mn^3)^2 \left(\dfrac{2}{5}m^5 n\right)^3$

$\qquad = -3 \cdot (-5)^2 \cdot m^2 \cdot (n^3)^2 \cdot \left(\dfrac{2}{5}\right)^3 \cdot (m^5)^3 \cdot n^3$

$\qquad = -3 \cdot 25 \cdot m^2 \cdot n^{3 \cdot 2} \cdot \dfrac{8}{125} \cdot m^{5 \cdot 3} \cdot n^3$

$\qquad = -3 \cdot 25 \cdot m^2 \cdot n^6 \cdot \dfrac{8}{125} \cdot m^{15} \cdot n^3$

$\qquad = -3 \cdot 25 \cdot \dfrac{8}{125} \cdot m^2 \cdot m^{15} \cdot n^6 \cdot n^3$

$\qquad = -\dfrac{24}{5} \cdot m^{2+15} \cdot n^{6+3}$

$\qquad = -\dfrac{24}{5} m^{17} n^9$

95. volume $= (x^2)^3 = x^{2 \cdot 3} = x^6$

97. area $= (4x) \cdot (12x)$

$\qquad\quad = 4 \cdot 12 \cdot x \cdot x$

$\qquad\quad = 48 \cdot x^{1+1}$

$\qquad\quad = 48x^2$

99. The product rule for exponential expressions is a "shortcut" for writing out each exponential expression in expanded form. For example, $x^2 \cdot x^3 = (x \cdot x) \cdot (x \cdot x \cdot x) = x^5$.

101. The product to a power rule generalizes the result of using a product as a factor several times. That is, $(a \cdot b)^n = (a \cdot b) \cdot (a \cdot b) \cdot (a \cdot b) \ldots$

$\qquad\qquad\qquad\quad = a \cdot a \cdot a \cdot \ldots \cdot b \cdot b \cdot b \cdot \ldots$

$\qquad\qquad\qquad\quad = a^n \cdot b^n$.

103. The expression $4x^2 + 3x^2$ is a sum of terms: add the coefficients and retain the common base.

The expression $(4x^2)(3x^2)$ is a product: multiply the constant terms and add the exponents on the variable bases.

Section 5.3

Preparing for Multiplying Polynomials

P1. $2(4x - 3) = 2 \cdot 4x - 2 \cdot 3 = 8x - 6$

P2. $(3x^2)(-5x^3) = 3 \cdot (-5) \cdot x^2 \cdot x^3$

$\qquad\qquad\qquad = -15x^{2+3}$

$\qquad\qquad\qquad = -15x^5$

P3. $(9a)^2 = 9a \cdot 9a = 9 \cdot 9 \cdot a \cdot a = 81a^2$

5.3 Quick Checks

1. $3x(x^2 - 2x + 4) = 3x(x^2) - 3x(2x) + 3x(4)$

$\qquad\qquad\qquad\qquad = 3x^3 - 6x^2 + 12x$

2. $-a^2 b(2a^2 b^2 - 4ab^2 + 3ab)$

$\qquad = -a^2 b(2a^2 b^2) - (-a^2 b)(4ab^2) + (-a^2 b)(3ab)$

$\qquad = -2a^4 b^3 + 4a^3 b^3 - 3a^3 b^2$

3. $\left(5n^3 - \dfrac{15}{8}n^2 - \dfrac{10}{7}n\right)\dfrac{3}{5}n^2$

$= 5n^3\left(\dfrac{3}{5}n^2\right) - \dfrac{15}{8}n^2\left(\dfrac{3}{5}n^2\right) - \dfrac{10}{7}n\left(\dfrac{3}{5}n^2\right)$

$= 3n^5 - \dfrac{9}{8}n^4 - \dfrac{6}{7}n^3$

4. $(x+7)(x+2) = (x+7)x + (x+7)(2)$
$= x^2 + 7x + 2x + 14$
$= x^2 + 9x + 14$

5. $(2n+1)(n-3) = (2n+1)n + (2n+1)(-3)$
$= 2n^2 + n - 6n - 3$
$= 2n^2 - 5n - 3$

6. $(3p-2)(2p-1) = (3p-2)(2p) + (3p-2)(-1)$
$= 6p^2 - 4p - 3p + 2$
$= 6p^2 - 7p + 2$

7. $(x+3)(x+4) = (x)(x) + (x)(4) + (3)(x) + (3)(4)$
$= x^2 + 4x + 3x + 12$
$= x^2 + 7x + 12$

8. $(y+5)(y-3)$
$= (y)(y) + (y)(-3) + (5)(y) + (5)(-3)$
$= y^2 - 3 + 5y - 15$
$= y^2 + 2y - 15$

9. $(a-1)(a-5)$
$= (a)(a) + (a)(-5) + (-1)(a) + (-1)(-5)$
$= a^2 - 5a - a + 5$
$= a^2 - 6a + 5$

10. $(2x+3)(x+4)$
$= (2x)(x) + (2x)(4) + (3)(x) + (3)(4)$
$= 2x^2 + 8x + 3x + 12$
$= 2x^2 + 11x + 12$

11. $(3y+5)(2y-3)$
$= (3y)(2y) + (3y)(-3) + (5)(2y) + (5)(-3)$
$= 6y^2 - 9y + 10y - 15$
$= 6y^2 + y - 15$

12. $(2a-1)(3a-4) = 6a^2 - 8a - 3a + 4$
$= 6a^2 - 11a + 4$

13. $(5x+2y)(3x-4y)$
$= (5x)(3x) + (5x)(-4y) + (2y)(3x) + (2y)(-4y)$
$= 15x^2 - 20xy + 6xy - 8y^2$
$= 15x^2 - 14xy - 8y^2$

14. $(A+B)(A-B) = \underline{A^2 - B^2}$.

15. $(a-4)(a+4) = a^2 - 4^2 = a^2 - 16$

16. $(3w+7)(3w-7) = (3w)^2 - 7^2 = 9w^2 - 49$

17. $(x-2y^3)(x+2y^3) = x^2 - (2y^3)^2$
$= x^2 - 2^2(y^3)^2$
$= x^2 - 4y^6$

18. Let $x = 0$ and $y = 1$.
$(x-y)^2 \overset{?}{=} x^2 - y^2$
$(0-1)^2 \overset{?}{=} 0^2 - 1^2$
$(-1)^2 \overset{?}{=} -1^2$
$1 \neq -1$
The statement is false.

19. $x^2 + 2xy + y^2$ is referred to as a <u>perfect square trinomial</u>.

20. False; consider $(a+b)(a-b) = a^2 - b^2$.

21. $(z-9)^2 = z^2 - 2(z)(9) + 9^2$
$= z^2 - 18z + 81$

22. $(p+1)^2 = p^2 + 2(p)(1) + 1^2$
$= p^2 + 2p + 1$

23. $(4-a)^2 = 4^2 - 2(4)(a) + a^2 = 16 - 8a + a^2$

24. $(3z-4)^2 = (3z)^2 - 2(3z)(4) + 4^2$
$= 9z^2 - 24z + 16$

25. $(5p+1)^2 = (5p)^2 + 2(5p)(1) + (1)^2$
$= 25p^2 + 10p + 1$

26. $(2w+7y)^2 = (2w)^2 + 2(2w)(7y) + (7y)^2$
$= 4w^2 + 28wy + 49y^2$

27. False; the product $(x-y)(x^2+2xy+y^2)$ cannot be found using the FOIL method. The FOIL method can only be used with two binomials.

28.
$$(x-2)(x^2+2x+4) = (x-2)x^2+(x-2)\cdot 2x+(x-2)\cdot 4$$
$$= x(x^2)-2(x^2)+x(2x)-2(2x)+x(4)-2(4)$$
$$= x^3-2x^2+2x^2-4x+4x-8$$
$$= x^3-8$$

29.
$$(3y-2)(y^2+2y+4) = (3y-2)y^2+(3y-2)\cdot 2y+(3y-2)\cdot 4$$
$$= 3y(y^2)-2(y^2)+3y(2y)-2(2y)+3y(4)-2(4)$$
$$= 3y^3-2y^2+6y^2-4y+12y-8$$
$$= 3y^3+4y^2+8y-8$$

30.

$$
\begin{array}{r}
x^2+2x+4 \\
\times \quad\quad x-2 \\
\hline
-2x^2-4x-8 \\
x^3+2x^2+4x \quad\quad \\
\hline
x^3+0x^2+0x-8
\end{array}
$$

The product is x^3-8.

31.

$$
\begin{array}{r}
y^2+2y-5 \\
\times \quad\quad 3y-1 \\
\hline
-y^2-\ 2y+5 \\
3y^3+6y^2-15y \quad\quad \\
\hline
3y^3+5y^2-17y+5
\end{array}
$$

32.
$$-2a(4a-1)(3a+5) = (-8a^2+2a)(3a+5)$$
$$= -8a^2\cdot 3a+(-8a^2)\cdot 5+2a\cdot 3a+2a\cdot 5$$
$$= -24a^3-40a^2+6a^2+10a$$
$$= -24a^3-34a^2+10a$$

33.
$$\frac{3}{2}(5x-2)(2x+8) = \frac{3}{2}[5x\cdot 2x+5x\cdot 8+(-2)\cdot 2x+(-2)\cdot 8]$$
$$= \frac{3}{2}(10x^2+40x-4x-16)$$
$$= \frac{3}{2}(10x^2+36x-16)$$
$$= \frac{3}{2}\cdot 10x^2+\frac{3}{2}\cdot 36x-\frac{3}{2}\cdot 16$$
$$= 15x^2+54x-24$$

34.
$$(x+2)(x-1)(x+3) = (x^2-x+2x-2)(x+3)$$
$$= (x^2+x-2)(x+3)$$
$$= x^3+x^2-2x+3x^2+3x-6$$
$$= x^3+4x^2+x-6$$

5.3 Exercises

35. $2x(3x-5) = 2x(3x) + 2x(-5) = 6x^2 - 10x$

37. $\frac{1}{2}n(4n-6) = \frac{1}{2}n(4n) + \frac{1}{2}n(-6) = 2n^2 - 3n$

39. $3n^2(4n^2 + 2n - 5) = 3n^2(4n^2) + 3n^2(2n) + 3n^2(-5)$
$$= 12n^4 + 6n^3 - 15n^2$$

41. $(4x^2y - 3xy^2)x^2y = 4x^2y(x^2y) + (-3xy^2)(x^2y)$
$$= 4x^4y^2 - 3x^3y^3$$

43. $(x+5)(x+7) = (x+5)(x) + (x+5)(7)$
$$= x \cdot x + 5 \cdot x + 7 \cdot x + 5 \cdot 7$$
$$= x^2 + 5x + 7x + 35$$
$$= x^2 + 12x + 35$$

45. $(y-5)(y+7) = (y-5)y + (y-5)7$
$$= y \cdot y - 5 \cdot y + 7 \cdot y - 5 \cdot 7$$
$$= y^2 - 5y + 7y - 35$$
$$= y^2 + 2y - 35$$

47. $(3m-2y)(2m+5y) = (3m-2y) \cdot (2m) + (3m-2y) \cdot (5y)$
$$= (3m) \cdot (2m) - (2y) \cdot (2m) + (3m) \cdot (5y) - (2y) \cdot (5y)$$
$$= 6m^2 - 4my + 15my - 10y^2$$
$$= 6m^2 + 11my - 10y^2$$

49. $(x+2)(x+3) = x^2 + 3x + 2x + 6$
$$= x^2 + 5x + 6$$

51. $(q-6)(q-7) = q^2 - 7q - 6q + 42$
$$= q^2 - 13q + 42$$

53. $(2x+3)(3x-1) = 6x^2 - 2x + 9x - 3$
$$= 6x^2 + 7x - 3$$

55. $(x^2+3)(x^2+1) = x^4 + x^2 + 3x^2 + 3$
$$= x^4 + 4x^2 + 3$$

57. $(7-x)(6-x) = 42 - 7x - 6x + x^2$
$$= 42 - 13x + x^2$$

59. $(5u+6v)(2u+v) = 10u^2 + 5uv + 12uv + 6v^2$
$$= 10u^2 + 17uv + 6v^2$$

61. $(2a - b)(5a + 2b) = 10a^2 + 4ab - 5ab - 2b^2$
$$= 10a^2 - ab - 2b^2$$

63. $(x - 3)(x + 3) = (x)^2 - (3)^2 = x^2 - 9$

65. $(2z + 5)(2z - 5) = (2z)^2 - (5)^2 = 4z^2 - 25$

67. $(4x^2 + 1)(4x^2 - 1) = (4x^2)^2 - (1)^2 = 16x^4 - 1$

69. $(2x - 3y)(2x + 3y) = (2x)^2 - (3y)^2 = 4x^2 - 9y^2$

71. $\left(x - \dfrac{1}{3}\right)\left(x + \dfrac{1}{3}\right) = (x)^2 - \left(\dfrac{1}{3}\right)^2 = x^2 - \dfrac{1}{9}$

73. $(x - 2)^2 = (x)^2 - 2(x)(2) + (2)^2 = x^2 - 4x + 4$

75. $(5k - 3)^2 = (5k)^2 - (2)(5k)(3) + (3)^2$
$$= 25k^2 - 30k + 9$$

77. $(x + 2y)^2 = (x)^2 + 2(x)(2y) + (2y)^2$
$$= x^2 + 4xy + 4y^2$$

79. $(2a - 3b)^2 = (2a)^2 - 2(2a)(3b) + (3b)^2$
$$= 4a^2 - 12ab + 9b^2$$

81. $\left(x + \dfrac{1}{2}\right)^2 = (x)^2 + 2(x)\left(\dfrac{1}{2}\right) + \left(\dfrac{1}{2}\right)^2 = x^2 + x + \dfrac{1}{4}$

83. $(x - 2)(x^2 + 3x + 1) = (x - 2)x^2 + (x - 2)3x + (x - 2)1$
$$= x \cdot x^2 + (-2) \cdot x^2 + x \cdot 3x + (-2) \cdot 3x + x \cdot 1 + (-2) \cdot 1$$
$$= x^3 - 2x^2 + 3x^2 - 6x + x - 2$$
$$= x^3 + x^2 - 5x - 2$$

85. $(2y^2 - 6y + 1)(y - 3) = (2y^2 - 6y + 1)y + (2y^2 - 6y + 1)(-3)$
$$= 2y^2 \cdot y + (-6y) \cdot y + 1 \cdot y + 2y^2 \cdot (-3) + (-6y) \cdot (-3) + 1 \cdot (-3)$$
$$= 2y^3 - 6y^2 + y - 6y^2 + 18y - 3$$
$$= 2y^3 - 12y^2 + 19y - 3$$

87. $(2x - 3)(x^2 - 2x - 1) = (2x - 3)x^2 + (2x - 3)(-2x) + (2x - 3)(-1)$
$$= 2x \cdot x^2 + (-3) \cdot x^2 + 2x \cdot (-2x) + (-3) \cdot (-2x) + 2x \cdot (-1) + (-3) \cdot (-1)$$
$$= 2x^3 - 3x^2 - 4x^2 + 6x - 2x + 3$$
$$= 2x^3 - 7x^2 + 4x + 3$$

89. $2b(b-3)(b+4) = (2b^2 - 6b)(b+4)$
$$= (2b^2 - 6b)b + (2b^2 - 6b)4$$
$$= 2b^2 \cdot b + (-6b) \cdot b + 2b^2 \cdot 4 + (-6b) \cdot 4$$
$$= 2b^3 - 6b^2 + 8b^2 - 24b$$
$$= 2b^3 + 2b^2 - 24b$$

91. $-\dfrac{1}{2}x(2x+6)(x-3) = (-x^2 - 3x)(x-3)$
$$= (-x^2 - 3x)x + (-x^2 - 3x)(-3)$$
$$= (-x^2) \cdot x + (-3x) \cdot x + (-x^2) \cdot (-3) + (-3x) \cdot (-3)$$
$$= -x^3 - 3x^2 + 3x^2 + 9x$$
$$= -x^3 + 9x$$

93. $(5y^3 - y^2 + 2)(2y^2 + y + 1) = (5y^3 - y^2 + 2)2y^2 + (5y^3 - y^2 + 2)y + (5y^3 - y^2 + 2)1$
$$= 5y^3 \cdot 2y^2 + (-y^2) \cdot 2y^2 + 2 \cdot 2y^2 + 5y^3 \cdot y + (-y^2) \cdot y + 2 \cdot y + 5y^3 \cdot 1 + (-y^2) \cdot 1 + 2 \cdot 1$$
$$= 10y^5 - 2y^4 + 4y^2 + 5y^4 - y^3 + 2y + 5y^3 - y^2 + 2$$
$$= 10y^5 + 3y^4 + 4y^3 + 3y^2 + 2y + 2$$

95. $(b+1)(b-2)(b+3) = (b^2 - 2b + b - 2)(b+3)$
$$= (b^2 - b - 2) \cdot (b+3)$$
$$= b^2 \cdot (b+3) + (-b) \cdot (b+3) + (-2) \cdot (b+3)$$
$$= b^3 + 3b^2 - b^2 - 3b - 2b - 6$$
$$= b^3 + 2b^2 - 5b - 6$$

97. $4n(3 - 2n - n^2) = 4n(3) + 4n(-2n) + 4n(-n^2)$
$$= 12n - 8n^2 - 4n^3$$

99. $(2a - 3b)(4a + b) = 8a^2 + 2ab - 12ab - 3b^2$
$$= 8a^2 - 10ab - 3b^2$$

101. $\left(\dfrac{1}{2}x + 3\right)\left(\dfrac{1}{2}x - 3\right) = \left(\dfrac{1}{2}x\right)^2 - (3)^2$
$$= \dfrac{1}{4}x^2 - 9$$

103. $(0.5b + 3)^2 = (0.5b)^2 + (2) \cdot (0.5b) \cdot (3) + (3)^2$
$$= 0.25b^2 + 3b + 9$$

105. $(x^2 + 1)(x^4 - 3) = x^6 - 3x^2 + x^4 - 3$
$$= x^6 + x^4 - 3x^2 - 3$$

107. $(x+2)(2x^2 - 3x - 1) = x \cdot (2x^2 - 3x - 1) + 2(2x^2 - 3x - 1)$
$$= 2x^3 - 3x^2 - x + 4x^2 - 6x - 2$$
$$= 2x^3 + x^2 - 7x - 2$$

109.
$$-\frac{1}{2}x^2(10x^5 - 6x^4 + 12x^3) + (x^3)^2 = -\frac{1}{2}x^2(10x^5) + \left(-\frac{1}{2}x^2\right)(-6x^4) + \left(-\frac{1}{2}x^2\right)(12x^3) + x^6$$
$$= -5x^7 + 3x^6 - 6x^5 + x^6$$
$$= -5x^7 + 4x^6 - 6x^5$$

111.
$$7x^2(x+3) - 2x(x^2-1) = 7x^2(x) + 7x^2(3) + (-2x)(x^2) + (-2x)(-1)$$
$$= 7x^3 + 21x^2 - 2x^3 + 2x$$
$$= 5x^3 + 21x^2 + 2x$$

113.
$$-3w(w-4)(w+3) = (-3w^2 + 12w)(w+3)$$
$$= (-3w^2)(w) + (-3w^2)(3) + (12w)(w) + (12w)(3)$$
$$= -3w^3 - 9w^2 + 12w^2 + 36w$$
$$= -3w^3 + 3w^2 + 36w$$

115. $3a(a+4)^2 = 3a[(a)^2 + 2(a)(4) + (4)^2] = 3a(a^2 + 8a + 16) = 3a(a^2) + 3a(8a) + 3a(16) = 3a^3 + 24a^2 + 48a$

117. $(n+3)(n-3) + (n+3)^2 = (n)^2 - (3)^2 + (n)^2 + 2(n)(3) + (3)^2 = n^2 - 9 + n^2 + 6n + 9 = 2n^2 + 6n$

119.
$$(a+6b)^2 - (a-6b)^2 = (a)^2 + 2(a)(6b) + (6b)^2 - [(a)^2 - 2(a)(6b) + (6b)^2]$$
$$= a^2 + 12ab + 36b^2 - (a^2 - 12ab + 36b^2)$$
$$= a^2 + 12ab + 36b^2 - a^2 + 12ab - 36b^2$$
$$= 24ab$$

121.
$$(x+1)^2 - (2x+1)(x-1) = (x)^2 + 2(x)(1) + (1)^2 - [(2x)(x) + (2x)(-1) + (1)(x) + (1)(-1)]$$
$$= x^2 + 2x + 1 - (2x^2 - 2x + x - 1)$$
$$= x^2 + 2x + 1 - (2x^2 - x - 1)$$
$$= x^2 + 2x + 1 - 2x^2 + x + 1$$
$$= -x^2 + 3x + 2$$

123. $(2x+1)^2 = (2x)^2 + 2(2x)(1) + (1)^2 = 4x^2 + 4x + 1$

125.
$$(x-1)^3 = (x-1)^2(x-1)$$
$$= [(x)^2 - 2(x)(1) + (1)^2](x-1)$$
$$= (x^2 - 2x + 1)(x-1)$$
$$= x^2(x-1) + (-2x)(x-1) + 1(x-1)$$
$$= x^2(x) + x^2(-1) + (-2x)(x) + (-2x)(-1) + 1(x) + 1(-1)$$
$$= x^3 - x^2 - 2x^2 + 2x + x - 1$$
$$= x^3 - 3x^2 + 3x - 1$$

127. $(x+3)(2x-5) - (x-6) = (x)(2x) + (x)(-5) + (3)(2x) + 3(-5) - x + 6 = 2x^2 - 5x + 6x - 15 - x + 6 = 2x^2 - 9$

129. area $= (2x-3)(x+5) = (2x)(x) + (2x)(5) + (-3)(x) + (-3)(5) = 2x^2 + 10x - 3x - 15 = 2x^2 + 7x - 15$

131. $\text{area} = x(3x-1)-(x-1)^2$

$\qquad = x(3x)+x(-1)-[(x)^2-2(x)(1)+(1)^2]$

$\qquad = 3x^2-x-(x^2-2x+1)$

$\qquad = 3x^2-x-x^2+2x-1$

$\qquad = 2x^2+x-1$

133. $\text{Volume} = x(4x-3)(x+1)$

$\qquad = (4x^2-3x)(x+1)$

$\qquad = 4x^3+4x^2-3x^2-3x$

$\qquad = 4x^3+x^2-3x$

135. Find the area of the rectangle by adding the areas of the interior rectangles.

$\text{area} = (x)(x)+(x)(10)+(2)(x)+(2)(10)$

$\qquad = x^2+10x+2x+20$

$\qquad = x^2+12x+20$

Find the area of the rectangle by multiplying width and height.

$\text{area} = (x+10)(x+2)$

$\qquad = (x)(x)+(x)(2)+(10)(x)+(10)(2)$

$\qquad = x^2+2x+10x+20$

$\qquad = x^2+12x+20$

137. The second consecutive integer is $x+1$; the third is $(x+1)+1 = x+2$.

$(x+1)(x+2) = x(x)+x(2)+1(x)+1(2)$

$\qquad = x^2+2x+x+2$

$\qquad = x^2+3x+2$

139. The area of a triangle is one-half the product of the base and the altitude. In this exercise, the altitude is x and the base is $2x-2$.

$\dfrac{1}{2}(2x-2)x \text{ ft}^2 = (x-1)x \text{ ft}^2 = (x^2-x) \text{ ft}^2$

141. The area of a circle with radius r is πr^2.

$\pi(x+2)^2 \text{ square feet} = \pi[(x)^2+2(x)(2)+(2)^2] \text{ square feet}$

$\qquad = \pi[x^2+4x+4] \text{ square feet}$

$\qquad = \pi x^2+4\pi x+4\pi \text{ square feet}$

143. (a) Area of upper left quadrilateral: $a \cdot a = a^2$

Area of upper right quadrilateral: $a \cdot b = ab$

Area of lower left quadrilateral: $b \cdot a = ab$

Area of lower right quadrilateral: $b \cdot b = b^2$

(b) Area of entire region: $a^2+ab+ab+b^2 = a^2+2ab+b^2$

    (c) Length of entire region: $a + b$

        Width of entire region: $a + b$

        Area of entire region: $(a+b)(a+b) = (a+b)^2$

        Therefore, $(a+b)^2 = a^2 + 2ab + b^2$

145. $\begin{aligned}(x-2)(x+2)^2 &= (x-2)[(x)^2 + 2(x)(2) + (2)^2] \\ &= (x-2)[x^2 + 4x + 4] \\ &= x(x^2 + 4x + 4) + (-2)(x^2 + 4x + 4) \\ &= x^3 + 4x^2 + 4x - 2x^2 - 8x - 8 \\ &= x^3 + 2x^2 - 4x - 8\end{aligned}$

147. $\begin{aligned}[3-(x+y)][3+(x+y)] &= (3)^2 - (x+y)^2 \\ &= 9 - [(x)^2 + 2(x)(y) + (y)^2] \\ &= 9 - (x^2 + 2xy + y^2) \\ &= 9 - x^2 - 2xy - y^2\end{aligned}$

149. $\begin{aligned}(x+2)^3 &= (x+2)^2(x+2) \\ &= [(x)^2 + 2(x)(2) + (2)^2](x+2) \\ &= (x^2 + 4x + 4)(x+2) \\ &= x^2(x+2) + (4x)(x+2) + 4(x+2) \\ &= x^2(x) + x^2(2) + 4x(x) + 4x(2) + 4(x) + 4(2) \\ &= x^3 + 2x^2 + 4x^2 + 8x + 4x + 8 \\ &= x^3 + 6x^2 + 12x + 8\end{aligned}$

151. $\begin{aligned}(z+3)^4 &= (z+3)^2(z+3)^2 \\ &= [(z)^2 + 2(z)(3) + (3)^2][(z)^2 + 2(z)(3) + (3)^2] \\ &= (z^2 + 6z + 9)(z^2 + 6z + 9) \\ &= z^2(z^2 + 6z + 9) + (6z)(z^2 + 6z + 9) + 9(z^2 + 6z + 9) \\ &= z^2(z^2) + z^2(6z) + z^2(9) + 6z(z^2) + 6z(6z) + 6z(9) + 9(z^2) + 9(6z) + 9(9) \\ &= z^4 + 6z^3 + 9z^2 + 6z^3 + 36z^2 + 54z + 9z^2 + 54z + 81 \\ &= z^4 + 12z^3 + 54z^2 + 108z + 81\end{aligned}$

153. To square the binomial $(3a-5)^2$, use the pattern $A^2 - 2AB + B^2$. This means, square the first term, take the product of -2 times the first term times the second term, and square the second term. So
$(3a-5)^2 = (3a)^2 - 2(3a)(5) + 5^2 = 9a^2 - 30a + 25.$

155. The FOIL method can be used to multiply only binomials.

Section 5.4

Preparing for Dividing Monomials: The Quotient Rule and Integer Exponents

  P1. $(3a^2)^3 = 3^3(a^2)^3 = 27a^{2\cdot3} = 27a^6$

P2. $\left(\dfrac{2}{3}\right)^2 = \dfrac{2}{3}\cdot\dfrac{2}{3} = \dfrac{2\cdot 2}{3\cdot 3} = \dfrac{4}{9}$

P3. **(a)** The reciprocal of $5 = \dfrac{5}{1}$ is $\dfrac{1}{5}$.

     (b) The reciprocal of $-\dfrac{6}{7}$ is $-\dfrac{7}{6}$.

5.4 Quick Checks

1. $\dfrac{a^m}{a^n} = a^{\underline{m-n}}$ provided that $a \neq 0$.

2. False; $\dfrac{6^{10}}{6^4} \overset{?}{=} 1^6$

$\qquad\qquad 6^{10-4} \overset{?}{=} 1^6$

$\qquad\qquad\quad 6^6 \overset{?}{=} 1^6$

$\qquad\quad 46{,}656 \neq 1$

3. $\dfrac{3^7}{3^5} = 3^{7-5} = 3^2 = 9$

4. $\dfrac{14c^6}{10c^5} = \dfrac{14}{10}\cdot\dfrac{c^6}{c^5} = \dfrac{7}{5}\cdot c^{6-5} = \dfrac{7}{5}c^1 = \dfrac{7}{5}c$ or $\dfrac{7c}{5}$

5. $\dfrac{-21w^4z^8}{14w^3z} = \dfrac{-21}{14}\cdot w^{4-3}\cdot z^{8-1}$

$\qquad\qquad = -\dfrac{3}{2}w^1z^7$

$\qquad\qquad = -\dfrac{3}{2}wz^7$ or $-\dfrac{3wz^7}{2}$

6. $\left(\dfrac{a}{b}\right)^n = \dfrac{a^n}{\underline{b^n}}$ provided that $\underline{b \neq 0}$.

7. $\left(\dfrac{p}{2}\right)^4 = \dfrac{p^4}{2^4} = \dfrac{p^4}{16}$

8. $\left(-\dfrac{2a^2}{b^4}\right)^3 = \left(\dfrac{-2a^2}{b^4}\right)^3$

$\qquad\qquad = \dfrac{(-2a^2)^3}{(b^4)^3}$

$\qquad\qquad = \dfrac{(-2)^3(a^2)^3}{b^{4\cdot 3}}$

$\qquad\qquad = \dfrac{-8a^{2\cdot 3}}{b^{12}}$

$\qquad\qquad = -\dfrac{8a^6}{b^{12}}$

9. **(a)** $10^0 = 1$

     (b) $-10^0 = -1\cdot 10^0 = -1$

     (c) $(-10)^0 = 1$

10. **(a)** $(2b)^0 = 1$

     (b) $2b^0 = 2\cdot 1 = 2$

     (c) $(-2b)^0 = 1$

11. **(a)** $2^{-4} = \dfrac{1}{2^4} = \dfrac{1}{16}$

     (b) $(-2)^{-4} = \dfrac{1}{(-2)^4} = \dfrac{1}{16}$

     (c) $-2^{-4} = -1\cdot 2^{-4} = \dfrac{-1}{2^4} = \dfrac{-1}{16} = -\dfrac{1}{16}$

12. $4^{-1} - 2^{-3} = \dfrac{1}{4^1} - \dfrac{1}{2^3} = \dfrac{1}{4} - \dfrac{1}{8} = \dfrac{2}{8} - \dfrac{1}{8} = \dfrac{1}{8}$

13. **(a)** $y^{-8} = \dfrac{1}{y^8}$

     (b) $(-y)^{-8} = \dfrac{1}{(-y)^8} = \dfrac{1}{(-1\cdot y)^8} = \dfrac{1}{(-1)^8 y^8} = \dfrac{1}{y^8}$

     (c) $-y^{-8} = -1\cdot y^{-8} = -1\cdot\dfrac{1}{y^8} = -\dfrac{1}{y^8}$

14. (a) $2m^{-5} = 2 \cdot \dfrac{1}{m^5} = \dfrac{2}{m^5}$

(b) $(2m)^{-5} = \dfrac{1}{(2m)^5} = \dfrac{1}{2^5 \cdot m^5} = \dfrac{1}{32m^5}$

(c) $(-2m)^{-5} = \dfrac{1}{(-2m)^5}$

$= \dfrac{1}{(-2)^5 \cdot m^5}$

$= \dfrac{1}{-32m^5}$

$= -\dfrac{1}{32m^5}$

15. $\dfrac{1}{3^{-2}} = 3^2 = 9$

16. $\dfrac{1}{-10^{-2}} = \dfrac{1}{-1 \cdot 10^{-2}} = \dfrac{10^2}{-1} = \dfrac{100}{-1} = -100$

17. $\dfrac{5}{2z^{-2}} = \dfrac{5z^2}{2}$

18. $\left(\dfrac{7}{8}\right)^{-1} = \left(\dfrac{8}{7}\right)^1 = \dfrac{8}{7}$

19. $\left(\dfrac{3a}{5}\right)^{-2} = \left(\dfrac{5}{3a}\right)^2 = \dfrac{5^2}{(3a)^2} = \dfrac{25}{3^2 a^2} = \dfrac{25}{9a^2}$

20. $\left(-\dfrac{2}{3n^4}\right)^{-3} = \left(-\dfrac{3n^4}{2}\right)^3$

$= \dfrac{(-3n^4)^3}{2^3}$

$= \dfrac{(-3)^3 (n^4)^3}{8}$

$= \dfrac{-27n^{4\cdot3}}{8}$

$= -\dfrac{27n^{12}}{8}$

21. $(-4a^{-3})(5a) = (-4 \cdot 5)(a^{-3} \cdot a)$

$= -20a^{-3+1}$

$= -20a^{-2}$

$= -20 \cdot \dfrac{1}{a^2}$

$= -\dfrac{20}{a^2}$

22. $\left(-\dfrac{2}{5}m^{-2}n^{-1}\right)\left(-\dfrac{15}{2}mn^0\right)$

$= \left[-\dfrac{2}{5} \cdot \left(-\dfrac{15}{2}\right)\right](m^{-2} \cdot m)(n^{-1} \cdot n^0)$

$= \dfrac{30}{10}m^{-2+1}n^{-1+0}$

$= 3m^{-1}n^{-1}$

$= 3 \cdot \dfrac{1}{m} \cdot \dfrac{1}{n}$

$= \dfrac{3}{mn}$

23. $-\dfrac{16a^4b^{-1}}{12ab^{-4}} = -\dfrac{16}{12} \cdot \dfrac{a^4}{a} \cdot \dfrac{b^{-1}}{b^{-4}}$

$= -\dfrac{4}{3} \cdot a^{4-1} \cdot b^{-1-(-4)}$

$= -\dfrac{4}{3} \cdot a^3 \cdot b^3$

$= -\dfrac{4a^3b^3}{3}$

24. $\dfrac{45x^{-2}y^{-2}}{35x^{-4}y} = \dfrac{45}{35} \cdot \dfrac{x^{-2}}{x^{-1}} \cdot \dfrac{y^{-2}}{y}$

$= \dfrac{9}{7} \cdot x^{-2-(-4)} \cdot y^{-2-1}$

$= \dfrac{9}{7} \cdot x^2 \cdot y^{-3}$

$= \dfrac{9x^2}{7y^3}$

25. $(3y^2z^{-3})^{-2} = \left(\dfrac{3y^2}{z^3}\right)^{-2}$

$\qquad = \left(\dfrac{z^3}{3y^2}\right)^2$

$\qquad = \dfrac{(z^3)^2}{(3y^2)^2}$

$\qquad = \dfrac{z^6}{9y^4}$

26. $\left(\dfrac{2wz^{-3}}{7w^{-1}}\right)^2 = \left(\dfrac{2\cdot w\cdot w}{7\cdot z^3}\right)^2$

$\qquad = \left(\dfrac{2w^2}{7z^3}\right)^2$

$\qquad = \dfrac{(2w^2)^2}{(7z^3)^2}$

$\qquad = \dfrac{4w^4}{49z^6}$

27. $\left(\dfrac{-6p^{-2}}{p}\right)(3p^8)^{-1} = \left(\dfrac{-6}{p\cdot p^2}\right)\left(\dfrac{1}{3p^8}\right)$

$\qquad = \left(\dfrac{-6}{p^3}\right)\left(\dfrac{1}{3p^8}\right)$

$\qquad = \dfrac{-6}{3}\cdot\dfrac{1}{p^3\cdot p^8}$

$\qquad = -2\cdot\dfrac{1}{p^{11}}$

$\qquad = -\dfrac{2}{p^{11}}$

28. $(-25k^5r^{-2})\left(\dfrac{2}{5}k^{-3}r\right)^2 = \left(-\dfrac{25k^5}{r^2}\right)\left(\dfrac{2r}{5k^3}\right)^2$

$\qquad = \left(-\dfrac{25k^5}{r^2}\right)\dfrac{(2r)^2}{(5k^3)^2}$

$\qquad = -\dfrac{25k^5}{r^2}\cdot\dfrac{4r^2}{25k^6}$

$\qquad = -4\cdot\dfrac{k^5}{k^6}\cdot\dfrac{r^2}{r^2}$

$\qquad = -4\cdot k^{5-6}\cdot 1$

$\qquad = -4\cdot k^{-1}$

$\qquad = -4\cdot\dfrac{1}{k}$

$\qquad = -\dfrac{4}{k}$

5.4 Exercises

29. $\dfrac{2^{23}}{2^{19}} = 2^{23-19} = 2^4 = 16$

31. $\dfrac{x^{15}}{x^6} = x^{15-6} = x^9$

33. $\dfrac{16y^4}{4y} = \dfrac{16}{4}\cdot\dfrac{y^4}{y^1} = 4\cdot y^{4-1} = 4y^3$

35. $\dfrac{-16m^{10}}{24m^3} = \dfrac{-16}{24}\cdot\dfrac{m^{10}}{m^3} = -\dfrac{2}{3}\cdot m^{10-3} = -\dfrac{2}{3}m^7$

37. $\dfrac{-12m^9n^3}{-6mn} = \dfrac{-12}{-6}\cdot\dfrac{m^9}{m^1}\cdot\dfrac{n^3}{n^1}$

$\qquad = 2m^{9-1}n^{3-1}$

$\qquad = 2m^8n^2$

39. $\left(\dfrac{3}{2}\right)^3 = \dfrac{3^3}{2^3} = \dfrac{27}{8}$

41. $\left(\dfrac{x^5}{3}\right)^3 = \dfrac{(x^5)^3}{3^3} = \dfrac{x^{5\cdot3}}{27} = \dfrac{x^{15}}{27}$

43. $\left(-\dfrac{x^5}{y^7}\right)^4 = (-1)^4\dfrac{(x^5)^4}{(y^7)^4} = 1\cdot\dfrac{x^{5\cdot4}}{y^{7\cdot4}} = \dfrac{x^{20}}{y^{28}}$

45. $\left(\dfrac{7a^2b}{c^3}\right)^2 = \dfrac{(7a^2b)^2}{(c^3)^2}$

$\qquad = \dfrac{7^2(a^2)^2b^2}{(c^3)^2}$

$\qquad = \dfrac{49a^{2\cdot 2}b^2}{c^{3\cdot 2}}$

$\qquad = \dfrac{49a^4b^2}{c^6}$

47. $3^0 = 1$

49. $-\left(\dfrac{1}{2}\right)^0 = -1\left(\dfrac{1}{2}\right)^0 = -1(1) = -1$

51. $18 \cdot 2^0 = 18(1) = 18$

53. $(-10)^0 = 1$

55. $(24ab)^0 = 1$

57. $24ab^0 = (24a)\cdot b^0 = 24a\cdot 1 = 24a$

59. $10^{-3} = \dfrac{1}{10^3} = \dfrac{1}{1000}$

61. $m^{-2} = \dfrac{1}{m^2}$

63. $-a^{-2} = -1\cdot a^{-2} = -1\cdot\dfrac{1}{a^2} = -\dfrac{1}{a^2}$

65. $-4y^{-3} = -4\cdot\dfrac{1}{y^3} = -\dfrac{4}{y^3}$

67. $2^{-1} + 3^{-2} = \dfrac{1}{2^1} + \dfrac{1}{3^2} = \dfrac{1}{2} + \dfrac{1}{9} = \dfrac{9}{18} + \dfrac{2}{18} = \dfrac{11}{18}$

69. $\left(\dfrac{2}{5}\right)^{-2} = \left(\dfrac{5}{2}\right)^2 = \dfrac{5^2}{2^2} = \dfrac{25}{4}$

71. $\left(\dfrac{3}{z^2}\right)^{-1} = \left(\dfrac{z^2}{3}\right)^1 = \dfrac{z^{2\cdot 1}}{3^1} = \dfrac{z^2}{3}$

73. $\left(-\dfrac{2n}{m^2}\right)^{-3} = \left(-\dfrac{m^2}{2n}\right)^3$

$\qquad = (-1)^3\dfrac{(m^2)^3}{(2n)^3}$

$\qquad = -1\cdot\dfrac{m^{2\cdot 3}}{2^3 n^3}$

$\qquad = -\dfrac{m^6}{8n^3}$

75. $\dfrac{1}{4^{-2}} = 4^2 = 16$

77. $\dfrac{6}{x^{-4}} = 6\cdot x^4 = 6x^4$

79. $\dfrac{5}{2m^{-3}} = \dfrac{5m^3}{2}$

81. $\dfrac{5}{(2m)^{-3}} = 5\cdot(2m)^3$

$\qquad = 5\cdot 2^3\cdot m^3$

$\qquad = 5\cdot 8\cdot m^3$

$\qquad = 40m^3$

83. $\left(\dfrac{4}{3}y^{-2}z\right)\left(\dfrac{5}{8}y^{-2}z^4\right) = \left(\dfrac{4}{3}\right)\cdot\left(\dfrac{5}{8}\right)y^{-2-2}z^{1+4}$

$\qquad = \dfrac{20}{24}y^{-4}z^5$

$\qquad = \dfrac{5}{6}\cdot\dfrac{z^5}{y^4}$

$\qquad = \dfrac{5z^5}{6y^4}$

85. $\dfrac{-7m^7n^6}{3m^{-3}n^0} = -\dfrac{7}{3}m^{7+3}n^{6-0}$

$\qquad = -\dfrac{7}{3}m^{10}n^6$

$\qquad = -\dfrac{7m^{10}n^6}{3}$

87. $\dfrac{21y^2z^{-3}}{3y^{-2}z^{-1}} = \dfrac{21}{3}y^{2+2}z^{-3+1} = 7y^4z^{-2} = \dfrac{7y^4}{z^2}$

89. $(4x^2 y^{-2})^{-2} = \left(\dfrac{4x^2}{y^2}\right)^{-2}$

$= \left(\dfrac{y^2}{4x^2}\right)^{2}$

$= \dfrac{(y^2)^2}{(4x^2)^2}$

$= \dfrac{y^4}{4^2(x^2)^2}$

$= \dfrac{y^4}{16x^4}$

91. $(3a^2 b^{-1})\left(\dfrac{4a^{-1}b^2}{b^3}\right)^2 = (3a^2 b^{-1})(4a^{-1}b^{2-3})^2$

$= (3a^2 b^{-1})(4a^{-1}b^{-1})^2$

$= \left(3a^2 \cdot \dfrac{1}{b}\right)\left(4 \cdot \dfrac{1}{a} \cdot \dfrac{1}{b}\right)^2$

$= \left(\dfrac{3a^2}{6}\right)\left(\dfrac{4}{ab}\right)^2$

$= \left(\dfrac{3a^2}{b}\right)\left(\dfrac{4^2}{(ab)^2}\right)$

$= \left(\dfrac{3a^2}{b}\right)\left(\dfrac{16}{a^2 b^2}\right)$

$= \dfrac{48a^2}{a^2 b^3}$

$= \dfrac{48}{b^3}$

93. $2^5 \cdot 2^{-3} = 2^{5-3} = 2^2 = 4$

95. $2^{-7} \cdot 2^4 = 2^{-7+4} = 2^{-3} = \dfrac{1}{2^3} = \dfrac{1}{8}$

97. $\dfrac{3}{3^{-3}} = \dfrac{3^1}{3^{-3}} = 3^{1-(-3)} = 3^{1+3} = 3^4 = 81$

99. $\dfrac{x^6}{x^{15}} = x^{6-15} = x^{-9} = \dfrac{1}{x^9}$

101. $\dfrac{x^{10}}{x^{-3}} = x^{10-(-3)} = x^{10+3} = x^{13}$

103. $\dfrac{8x^2}{2x^5} = \dfrac{8}{2} \cdot \dfrac{x^2}{x^5} = 4 \cdot x^{2-5} = 4x^{-3} = \dfrac{4}{x^3}$

105. $\dfrac{-27xy^3 z^4}{18x^4 y^3 z} = \dfrac{-27}{18} \cdot \dfrac{x}{x^4} \cdot \dfrac{y^3}{y^3} \cdot \dfrac{z^4}{z}$

$= -\dfrac{3}{2} \cdot x^{1-4} \cdot y^{3-3} \cdot z^{4-1}$

$= -\dfrac{3}{2} \cdot x^{-3} \cdot y^0 \cdot z^3$

$= -\dfrac{3}{2} \cdot \dfrac{1}{x^3} \cdot 1 \cdot z^3$

$= -\dfrac{3z^3}{2x^3}$

107. $(3x^2 y^{-3})(12^{-1} x^{-5} y^{-6})$

$= 3 \cdot \dfrac{1}{12} \cdot x^2 \cdot x^{-5} \cdot y^{-3} \cdot y^{-6}$

$= \dfrac{1}{4} \cdot x^{2-5} \cdot y^{-3-6}$

$= \dfrac{1}{4} \cdot x^{-3} \cdot y^{-9}$

$= \dfrac{1}{4x^3 y^9}$

109. $(-a^4)^{-3} = (-1)^{-3}(a^4)^{-3}$

$= -1 \cdot a^{4(-3)}$

$= -1 \cdot a^{-12}$

$= -\dfrac{1}{a^{12}}$

111. $(3m^{-2})^3 = 3^3 (m^{-2})^3$

$= 27m^{-2 \cdot 3}$

$= 27m^{-6}$

$= \dfrac{27}{m^6}$

113. $(-3x^{-2} y^{-3})^{-2} = (-3)^{-2}(x^{-2})^{-2}(y^{-3})^{-2}$

$= \dfrac{1}{(-3)^2} \cdot x^{(-2)(-2)} \cdot y^{(-3)(-2)}$

$= \dfrac{1}{9} \cdot x^4 \cdot y^6$

$= \dfrac{x^4 y^6}{9}$

115. $2p^{-4} \cdot p^{-3} \cdot p^0 = 2p^{-4-3+0} = 2p^{-7} = \dfrac{2}{p^7}$

117. $(-16a^3)(-3a^4)\left(\dfrac{1}{4}a^{-7}\right)$

$= (-16)(-3)\left(\dfrac{1}{4}\right) \cdot a^3 \cdot a^4 \cdot a^{-7}$

$= 12a^{3+4-7}$

$= 12a^{7-7}$

$= 12a^0$

$= 12(1)$

$= 12$

119. $\dfrac{8x^2 \cdot x^7}{12x^{-3} \cdot x^4} = \dfrac{8}{12} \cdot \dfrac{x^{2+7}}{x^{-3+4}}$

$= \dfrac{2}{3} \cdot \dfrac{x^9}{x^1}$

$= \dfrac{2}{3} \cdot x^{9-1}$

$= \dfrac{2x^8}{3}$

121. $(2x^{-3}y^{-2})^4(3x^2y^{-3})^{-3}$

$= 2^4(x^{-3})^4(y^{-2})^4(3)^{-3}(x^2)^{-3}(y^{-3})^{-3}$

$= 16 \cdot \dfrac{1}{3^3} \cdot x^{-3 \cdot 4} \cdot x^{2(-3)} \cdot y^{-2(4)} \cdot y^{-3(-3)}$

$= \dfrac{16}{27} \cdot x^{-12} \cdot x^{-6} \cdot y^{-8} \cdot y^9$

$= \dfrac{16}{27} \cdot x^{-12-6} \cdot y^{-8+9}$

$= \dfrac{16}{27} x^{-18} y^1$

$= \dfrac{16y}{27x^{18}}$

123. $\left(\dfrac{y}{2z}\right)^{-3} \cdot \left(\dfrac{3y^2}{4z^3}\right)^2 = \left(\dfrac{2z}{y}\right)^3 \cdot \left(\dfrac{3y^2}{4z^3}\right)^2$

$= \dfrac{(2z)^3}{y^3} \cdot \dfrac{(3y^2)^2}{(4z^3)^2}$

$= \dfrac{8z^3}{y^3} \cdot \dfrac{9y^4}{16z^6}$

$= 8 \cdot \dfrac{9}{16} \cdot z^{3-6} y^{4-3}$

$= \dfrac{9}{2} z^{-3} y^1$

$= \dfrac{9}{2} \cdot \dfrac{1}{z^3} \cdot y$

$= \dfrac{9y}{2z^3}$

125. $(4x^2y)^3 \cdot \left(\dfrac{2x}{3y}\right)^{-3} = (4x^2y)^3 \cdot \left(\dfrac{3y}{2x}\right)^3$

$= 4^3(x^2)^3 y^3 \cdot \dfrac{3^3 y^3}{2^3 x^3}$

$= 64x^{2 \cdot 3} y^3 \cdot \dfrac{27y^3}{8x^3}$

$= 64x^6 y^3 \cdot \dfrac{27y^3}{8x^3}$

$= 64 \cdot \dfrac{27}{8} \cdot x^{6-3} \cdot y^{3+3}$

$= 216x^3 y^6$

127. $\dfrac{(5a^{-3}b^2)^2}{a^{-4}b^{-4}}(15a^{-3}b)^{-1}$

$= \dfrac{5^2(a^{-3})^2(b^2)^2}{a^{-4}b^{-4}} \cdot 15^{-1}(a^{-3})^{-1}b^{-1}$

$= \dfrac{25a^{-3 \cdot 2}b^{2 \cdot 2}}{a^{-4}b^{-4}} \cdot 15^{-1} a^{-3(-1)}b^{-1}$

$= \dfrac{25a^{-6}b^4}{a^{-4}b^{-4}} \cdot \dfrac{1}{15} a^3 b^{-1}$

$= \dfrac{25}{15} a^{-6-(-4)+3} b^{4-(-4)-1}$

$= \dfrac{5}{3} a^1 b^7$

$= \dfrac{5ab^7}{3}$

129. $\dfrac{4x^{-3}(2x^3)}{20(x^{-3})^2} = \dfrac{4 \cdot 2 \cdot x^{-3} \cdot x^3}{20x^{-3 \cdot 2}}$

$= \dfrac{8x^{-3+3}}{20x^{-6}}$

$= \dfrac{2x^0}{5x^{-6}}$

$= \dfrac{2x^{0-(-6)}}{5}$

$= \dfrac{2x^6}{5}$

131. $x \cdot x \cdot 3x = 3 \cdot x^1 \cdot x^1 \cdot x^1 = 3x^{1+1+1} = 3x^3$

The volume is $3x^3$ cubic meters.

133. The radius is one-half of the diameter, so $r = \dfrac{d}{2}$.

$$V = \pi r^2 h = \pi \left(\dfrac{d}{2}\right)^2 h = \pi \cdot \dfrac{d^2}{2^2} \cdot h = \dfrac{\pi d^2 h}{4}$$

135. The area of a circle is πr^2, so each button has

area $\pi(3x)^2$. Since there are x buttons,

$x \cdot \pi(3x)^2$ square units of material are needed.

$x \cdot \pi(3x)^2 = x \cdot \pi \cdot 3^2 \cdot x^2 = 9\pi x^{1+2} = 9\pi x^3$

$9\pi x^3$ square units of material are needed.

137. $\dfrac{x^{2n}}{x^{3n}} = x^{2n-3n} = x^{-n} = \dfrac{1}{x^n}$

139. $\left(\dfrac{x^n y^m}{x^{4n-1} y^{m+1}}\right)^{-2} = \left(\dfrac{x^{4n-1} y^{m+1}}{x^n y^m}\right)^2$

$= (x^{4n-1-n} y^{m+1-m})^2$

$= (x^{3n-1} y^1)^2$

$= (x^{3n-1})^2 y^2$

$= x^{(3n-1)(2)} y^2$

$= x^{6n-2} y^2$

141. $(x^{2a} y^b z^{-c})^{3a} = (x^{2a})^{3a} (y^b)^{3a} (z^{-c})^{3a}$

$= x^{2a \cdot 3a} y^{b \cdot 3a} z^{-c \cdot 3a}$

$= x^{6a^2} y^{3ab} z^{-3ac}$

$= \dfrac{x^{6a^2} y^{3ab}}{z^{3ac}}$

143. $\dfrac{(3a^n)^2}{(2a^{4n})^3} = \dfrac{3^2 (a^n)^2}{2^3 (a^{4n})^3}$

$= \dfrac{9a^{n \cdot 2}}{8a^{4n \cdot 3}}$

$= \dfrac{9a^{2n}}{8a^{12n}}$

$= \dfrac{9}{8} a^{2n-12n}$

$= \dfrac{9}{8} a^{-10n}$

$= \dfrac{9}{8a^{10n}}$

145. The quotient $\dfrac{11^6}{11^4} \neq 1^2$ because

$\dfrac{11^6}{11^4} = \dfrac{11 \cdot 11 \cdot 11 \cdot 11 \cdot 11 \cdot 11}{11 \cdot 11 \cdot 11 \cdot 11}$

$= \dfrac{\cancel{11} \cdot \cancel{11} \cdot \cancel{11} \cdot \cancel{11} \cdot 11 \cdot 11}{\cancel{11} \cdot \cancel{11} \cdot \cancel{11} \cdot \cancel{11}}$

$= 11^2$.

147. When simplifying the expression $-12x^0$, the exponent 0 applies to the base x, so $-12x^0 = -12 \cdot 1 = -12$. However, $(-12x)^0 = 1$ because the entire expression is raised to the 0 power.

149. His answer is incorrect. To raise a factor to a power, multiply exponents. Your friend added exponents.

Putting the Concepts Together (Sections 5.1–5.4)

1. Yes, $6x^2 y^4 - 8x^5 + 3$ is trinomial of degree 6 $(2 + 4)$.

2. $-x^2 + 3x$

  (a) $x = 0$: $-(0)^2 + 3(0) = 0 + 0 = 0$

  (b) $x = -1$: $-(-1)^2 + 3(-1) = -1 - 3 = -4$

  (c) $x = 2$: $-(2)^2 + 3(2) = -4 + 6 = 2$

3. $(6x^4 - 2x^2 + 7) - (-2x^4 - 7 + 2x^2)$

$= (6x^4 - 2x^2 + 7) + (2x^4 + 7 - 2x^2)$

$= 6x^4 - 2x^2 + 7 + 2x^4 + 7 - 2x^2$

$= 8x^4 - 4x^2 + 14$

4. $(2x^2 y - xy + 3y^2) + (4xy - y^2 + 3x^2 y)$

$= 2x^2 y - xy + 3y^2 + 4xy - y^2 + 3x^2 y$

$= 5x^2 y + 3xy + 2y^2$

5. $-2mn(3m^2 n - mn^3)$

$= -2mn(3m^2 n) + (-2mn)(-mn^3)$

$= -6m^3 n^2 + 2m^2 n^4$

6. $(5x+3)(x-4) = (5x)(x) + (5x)(-4) + (3)(x) + (3)(-4)$
$$= 5x^2 - 20x + 3x - 12$$
$$= 5x^2 - 17x - 12$$

7. $(2x-3y)(4x-7y) = (2x)(4x) + (2x)(-7y) + (-3y)(4x) + (-3y)(-7y)$
$$= 8x^2 - 14xy - 12xy + 21y^2$$
$$= 8x^2 - 26xy + 21y^2$$

8. $(5x+8)(5x-8) = (5x)^2 - (8)^2 = 25x^2 - 64$

9. $(2x+3y)^2 = (2x)^2 + 2(2x)(3y) + (3y)^2 = 4x^2 + 12xy + 9y^2$

10. $2a(3a-4)(a+5) = 2a[(3a)(a) + (3a)(5) + (-4)(a) + (-4)(5)]$
$$= 2a(3a^2 + 15a - 4a - 20)$$
$$= 2a(3a^2 + 11a - 20)$$
$$= 2a(3a^2) + 2a(11a) + 2a(-20)$$
$$= 6a^3 + 22a^2 - 40a$$

11. $(4m+3)(4m^3 - 2m^2 + 4m - 8) = 4m(4m^3 - 2m^2 + 4m - 8) + 3(4m^3 - 2m^2 + 4m - 8)$
$$= 4m(4m^3) + 4m(-2m^2) + 4m(4m) + 4m(-8) + 3(4m^3) + 3(-2m^2) + 3(4m) + 3(-8)$$
$$= 16m^4 - 8m^3 + 16m^2 - 32m + 12m^3 - 6m^2 + 12m - 24$$
$$= 16m^4 - 8m^3 + 12m^3 + 16m^2 - 6m^2 - 32m + 12m - 24$$
$$= 16m^4 + 4m^3 + 10m^2 - 20m - 24$$

12. $(-2x^7 y^0 z)(4xz^8) = -2 \cdot 4 \cdot x^7 \cdot x^1 \cdot y^0 \cdot z^1 \cdot z^8 = -8 \cdot x^{7+1} \cdot y^0 \cdot z^{1+8} = -8 \cdot x^8 \cdot 1 \cdot z^9 = -8x^8 z^9$

13. $(5m^3 n^{-2})(-3m^{-4}n) = 5 \cdot (-3) \cdot m^3 \cdot m^{-4} \cdot n^{-2} \cdot n^1 = -15 \cdot m^{3-4} \cdot n^{-2+1} = -15m^{-1}n^{-1} = -\dfrac{15}{mn}$

14. $\dfrac{-18a^8 b^3}{6a^5 b} = \dfrac{-18}{6} \cdot \dfrac{a^8}{a^5} \cdot \dfrac{b^3}{b} = -3 \cdot a^{8-5} \cdot b^{3-1} = -3 \cdot a^3 \cdot b^2 = -3a^3 b^2$

15. $\dfrac{16x^7 y}{32x^9 y^3} = \dfrac{16}{32} \cdot \dfrac{x^7}{x^9} \cdot \dfrac{y}{y^3} = \dfrac{1}{2} \cdot x^{7-9} \cdot y^{1-3} = \dfrac{1}{2} \cdot x^{-2} \cdot y^{-2} = \dfrac{1}{2x^2 y^2}$

16. $\dfrac{7}{2ab^{-2}} = \dfrac{7b^2}{2a}$

17. $\dfrac{q^{-6} r t^5}{qr^{-4} t^7} = q^{-6-1} r^{1+4} t^{5-7} = q^{-7} r^5 t^{-2} = \dfrac{r^5}{q^7 t^2}$

18. $\left(\dfrac{3}{2}r^2\right)^{-3} = \left(\dfrac{3}{2}\right)^{-3} \cdot (r^2)^{-3} = \left(\dfrac{2}{3}\right)^3 \cdot r^{2(-3)} = \dfrac{8}{27} \cdot r^{-6} = \dfrac{8}{27r^6}$

19. $(4y^{-2}z^3)^{-2} = 4^{-2}(y^{-2})^{-2}(z^3)^{-2}$

$$= \frac{1}{4^2} \cdot y^{-2(-2)} \cdot z^{3(-2)}$$

$$= \frac{1}{16} \cdot y^4 \cdot z^{-6}$$

$$= \frac{y^4}{16z^6}$$

20. $\left(\dfrac{3x^3 y^{-3}}{2x^{-1} y^0}\right)^{-4} \cdot (2x^{-4} y^{-2})^2$

$$= \left(\frac{3}{2} x^{3+1} y^{-3-0}\right)^{-4} \cdot (2x^{-4} y^{-2})^2$$

$$= \left(\frac{3}{2} x^4 y^{-3}\right)^{-4} \cdot (2x^{-4} y^{-2})^2$$

$$= \left[\left(\frac{3}{2}\right)^{-4} x^{-4(4)} y^{-4(-3)}\right] \cdot (4x^{-8} y^{-4})$$

$$= \left[\left(\frac{81}{16}\right)^{-1} x^{-16} y^{12}\right] \cdot (4x^{-8} y^{-4})$$

$$= \left(\frac{16}{81} x^{-16} y^{12}\right) \cdot (4x^{-8} y^{-4})$$

$$= (4)\left(\frac{16}{81}\right) x^{-16-8} y^{12-4}$$

$$= \frac{64}{81} x^{-24} y^8$$

$$= \frac{64y^8}{81x^{24}}$$

Section 5.5

Preparing for Dividing Polynomials

P1. $\dfrac{24a^3}{6a} = \dfrac{24}{6} \cdot \dfrac{a^3}{a} = 4 \cdot a^{3-1} = 4a^2$

P2. $\dfrac{-49x^3}{21x^4} = \dfrac{-49}{21} \cdot \dfrac{x^3}{x^4}$

$$= \frac{-7}{3} \cdot x^{3-4}$$

$$= -\frac{7}{3} \cdot x^{-1}$$

$$= -\frac{7}{3} \cdot \frac{1}{x}$$

$$= -\frac{7}{3x}$$

P3. $3x(7x-2) = 3x \cdot 7x - 3x \cdot 2 = 21x^2 - 6x$

P4. The degree of $4x^2 - 2x + 1$ is 2.

5.5 Quick Checks

1. The first step to simplify $\dfrac{4x^4 + 8x^2}{2x}$ would be to rewrite $\dfrac{4x^4 + 8x^2}{2x}$ as $\dfrac{4x^4}{2x} + \dfrac{8x^2}{2x}$.

2. $\dfrac{10n^4 - 20n^3 + 5n^2}{5n^2}$

$$= \frac{10n^4}{5n^2} - \frac{20n^3}{5n^2} + \frac{5n^2}{5n^2}$$

$$= \frac{10}{5} n^{4-2} - \frac{20}{5} n^{3-2} + \frac{5}{5} n^{2-2}$$

$$= 2n^2 - 4n^1 + 1n^0$$

$$= 2n^2 - 4n + 1$$

3. $\dfrac{12k^4 - 18k^2 + 5}{2k^2}$

$$= \frac{12k^4}{2k^2} - \frac{18k^2}{2k^2} + \frac{5}{2k^2}$$

$$= \frac{12}{2} k^{4-2} - \frac{18}{2} k^{2-2} + \frac{5}{2} k^{-2}$$

$$= 6k^2 - 9k^0 + \frac{5}{2} k^{-2}$$

$$= 6k^2 - 9 + \frac{5}{2k^2}$$

4. $\dfrac{x^4y^4 + 8x^2y^2 - 4xy}{4x^3y}$

$= \dfrac{x^4y^4}{4x^3y} + \dfrac{8x^2y^2}{4x^3y} - \dfrac{4xy}{4x^3y}$

$= \dfrac{1}{4}x^{4-3}y^{4-1} + \dfrac{8}{4}x^{2-3}y^{2-1} - \dfrac{4}{4}x^{1-3}y^{1-1}$

$= \dfrac{1}{4}x^1y^3 + 2x^{-1}y^1 - 1x^{-2}y^0$

$= \dfrac{xy^3}{4} + \dfrac{2y}{x} - \dfrac{1}{x^2}$

5. To begin a polynomial division problem, write the divisor and the dividend in <u>standard</u> form.

6. To check the result of long division, multiply the <u>quotient</u> and the divisor and add this result to the <u>remainder</u>. If correct, this result will be equal to the <u>dividend</u>.

7.
$$
\begin{array}{r}
x-8 \\
x+5\overline{)\;x^2-3x-40} \\
\underline{-(x^2+5x)} \\
-8x-40 \\
\underline{-(-8x-40)} \\
0
\end{array}
$$

$\dfrac{x^2-3x-40}{x+5} = x-8$

8.
$$
\begin{array}{r}
x-4 \\
2x+3\overline{)\;2x^2-5x-12} \\
\underline{-(2x^2+3x)} \\
-8x-12 \\
\underline{-(-8x-12)} \\
0
\end{array}
$$

$\dfrac{2x^2-5x-12}{2x+3} = x-4$

9.
$$
\begin{array}{r}
4x+5 \\
x+3\overline{)\;4x^2+17x+21} \\
\underline{-(4x^2+12x)} \\
5x+21 \\
\underline{-(5x+15)} \\
6
\end{array}
$$

$\dfrac{4x^2+17x+21}{x+3} = 4x+5+\dfrac{6}{x+3}$

10. $\dfrac{x+1-3x^2+4x^3}{x+2} = \dfrac{4x^3-3x^2+x+1}{x+2}$

$$
\begin{array}{r}
4x^2-11x+23 \\
x+2\overline{)\;4x^3-3x^2+\;\;x+\;1} \\
\underline{-(4x^3+8x^2)} \\
-11x^2+\;\;x+\;1 \\
\underline{-(-11x^2-22x)} \\
23x+\;1 \\
\underline{-(23x+46)} \\
-45
\end{array}
$$

$\dfrac{x+1-3x^2+4x^3}{x+2} = 4x^2-11x+23+\dfrac{-45}{x+2}$

11. $\dfrac{2x^3+3x^2+10}{2x-5} = \dfrac{2x^3+3x^2+0x+10}{2x-5}$

$$
\begin{array}{r}
x^2+4x+10 \\
2x-5\overline{)\;2x^3+3x^2+\;0x+10} \\
\underline{-(2x^3-5x^2)} \\
8x^2+\;0x+10 \\
\underline{-(8x^2-20x)} \\
20x+10 \\
\underline{-(20x-50)} \\
60
\end{array}
$$

$\dfrac{2x^3+3x^2+10}{2x-5} = x^2+4x+10+\dfrac{60}{2x-5}$

12.
$$
\begin{array}{r}
4x-3 \\
x^2+2\overline{)\;4x^3-3x^2+\;x+1} \\
\underline{-(4x^3\;\;\;\;\;\;+8x)} \\
-3x^2-7x+1 \\
\underline{-(-3x^2\;\;\;\;\;-6)} \\
-7x+7
\end{array}
$$

$\dfrac{4x^3-3x^2+x+1}{x^2+2} = 4x-3+\dfrac{-7x+7}{x^2+2}$

5.5 Exercises

13. $\dfrac{4x^2-2x}{2x} = \dfrac{4x^2}{2x} - \dfrac{2x}{2x}$

$= \dfrac{4}{2}x^{2-1} - \dfrac{2}{2}x^{1-1}$

$= 2x^1 - 1\cdot x^0$

$= 2x-1$

15. $\dfrac{9a^3 + 27a^2 - 3}{3a^2} = \dfrac{9a^3}{3a^2} + \dfrac{27a^2}{3a^2} - \dfrac{3}{3a^2}$

$\qquad = \dfrac{9}{3}a^{3-2} + \dfrac{27}{3}a^{2-2} - \dfrac{3}{3}a^{-2}$

$\qquad = 3a^1 + 9a^0 - 1 \cdot a^{-2}$

$\qquad = 3a + 9 - \dfrac{1}{a^2}$

17. $\dfrac{5n^5 - 10n^3 - 25n}{25n}$

$= \dfrac{5n^5}{25n} - \dfrac{10n^3}{25n} - \dfrac{25n}{25n}$

$= \dfrac{5}{25}n^{5-1} - \dfrac{10}{25}n^{3-1} - \dfrac{25}{25}n^{1-1}$

$= \dfrac{1}{5}n^4 - \dfrac{2}{5}n^2 - 1 \cdot n^0$

$= \dfrac{n^4}{5} - \dfrac{2n^2}{5} - 1$

19. $\dfrac{15r^5 - 27r^3}{9r^3} = \dfrac{15r^5}{9r^3} - \dfrac{27r^3}{9r^3}$

$\qquad = \dfrac{15}{9}r^{5-3} - \dfrac{27}{9}r^{3-3}$

$\qquad = \dfrac{5}{3}r^2 - 3r^0$

$\qquad = \dfrac{5r^2}{3} - 3$

21. $\dfrac{3x^7 - 9x^6 + 27x^3}{-3x^5}$

$= \dfrac{3x^7}{-3x^5} - \dfrac{9x^6}{-3x^5} + \dfrac{27x^3}{-3x^5}$

$= \dfrac{3}{0}x^{7-5} - \dfrac{9}{0}x^{6-5} + \dfrac{27}{-3}x^{3-5}$

$= -1 \cdot x^2 - (-3)x^1 + (-9)x^{-2}$

$= -x^2 + 3x - \dfrac{9}{x^2}$

23. $\dfrac{3z + 4z^3 - 2z^2}{8z} = \dfrac{3z}{8z} + \dfrac{4z^3}{8z} - \dfrac{2z^2}{8z}$

$\qquad = \dfrac{3}{8}z^{1-1} + \dfrac{4}{8}z^{3-1} - \dfrac{2}{8}z^{2-1}$

$\qquad = \dfrac{3}{8}z^0 + \dfrac{1}{2}z^2 - \dfrac{1}{4}z^1$

$\qquad = \dfrac{3}{8} + \dfrac{z^2}{2} - \dfrac{z}{4}$

25. $\dfrac{14x - 10y}{-2} = \dfrac{14x}{-2} - \dfrac{10y}{-2}$

$\qquad = \dfrac{14}{-2}x - \dfrac{10}{-2}y$

$\qquad = -7x - (-5)y$

$\qquad = -7x + 5y$

27. $\dfrac{12y - 30x}{-2x} = \dfrac{12y}{-2x} - \dfrac{30x}{-2x}$

$\qquad = \dfrac{12}{-2} \cdot \dfrac{y}{x} - \dfrac{30}{-2}x^{1-1}$

$\qquad = -6 \cdot \dfrac{y}{x} - (-15)x^0$

$\qquad = -\dfrac{6y}{x} + 15$

29. $\dfrac{25a^3b^2c + 10a^2bc^3}{-5a^4b^2c}$

$= \dfrac{25a^3b^2c}{-5a^4b^2c} + \dfrac{10a^2bc^3}{-5a^4b^2c}$

$= \dfrac{25}{-5}a^{3-4}b^{2-2}c^{1-1} + \dfrac{10}{-5}a^{2-4}b^{1-2}c^{3-1}$

$= -5a^{-1}b^0c^0 + (-2)a^{-2}b^{-1}c^2$

$= -\dfrac{5}{a} - \dfrac{2c^2}{a^2b}$

31.

$$\begin{array}{r} x - 7 \\ x+3\overline{)\,x^2 - 4x - 21} \\ \underline{-(x^2 + 3x)} \\ -7x - 21 \\ \underline{-(-7x - 21)} \\ 0 \end{array}$$

$\dfrac{x^2 - 4x - 21}{x + 3} = x - 7$

33.

$$\begin{array}{r} x - 5 \\ x-4\overline{)\,x^2 - 9x + 20} \\ \underline{-(x^2 - 4x)} \\ -5x + 20 \\ \underline{-(-5x + 20)} \\ 0 \end{array}$$

$\dfrac{x^2 - 9x + 20}{x - 4} = x - 5$

35.

$$x-2\overline{)\begin{array}{l} x^2+6x-3 \\ x^3+4x^2-15x+6 \end{array}}$$
$$\underline{-(x^3-2x^2)}$$
$$6x^2-15x$$
$$\underline{-(6x^2-12x)}$$
$$-3x+6$$
$$\underline{-(-3x+6)}$$
$$0$$

$$\frac{x^3+4x^2-15x+6}{x-2}=x^2+6x-3$$

37.

$$x+2\overline{)\begin{array}{l} x^3-3x^2+6x-2 \\ x^4-\ x^3+0x^2+10x-4 \end{array}}$$
$$\underline{-(x^4+2x^3)}$$
$$-3x^3+0x^2$$
$$\underline{-(-3x^3-6x^2)}$$
$$6x^2+10x$$
$$\underline{-(6x^2+12x)}$$
$$-2x-4$$
$$\underline{-(-2x-4)}$$
$$0$$

$$\frac{x^4-x^3+10x-4}{x+2}=x^3-3x^2+6x-2$$

39.

$$x+1\overline{)\begin{array}{l} x^2-2x+3 \\ x^3-x^2\ +x+8 \end{array}}$$
$$\underline{-(x^3+x^2)}$$
$$-2x^2\ +x$$
$$\underline{-(-2x^2-2x)}$$
$$3x+8$$
$$\underline{-(3x+3)}$$
$$5$$

$$\frac{x^3-x^2+x+8}{x+1}=x^2-2x+3+\frac{5}{x+1}$$

41.

$$x-5\overline{)\begin{array}{l} 2x+3 \\ 2x^2\ -7x-15 \end{array}}$$
$$\underline{-(2x^2-10x)}$$
$$3x-15$$
$$\underline{-(3x-15)}$$
$$0$$

$$\frac{2x^2-7x-15}{x-5}=2x+3$$

43.

$$x-2\overline{)\begin{array}{l} x^2+6x+7 \\ x^3+4x^2-\ 5x+\ 2 \end{array}}$$
$$\underline{-(x^3-2x^2)}$$
$$6x^2-\ 5x$$
$$\underline{-(6x^2-12x)}$$
$$7x+\ 2$$
$$\underline{-(7x-14)}$$
$$16$$

$$\frac{x^3+4x^2-5x+2}{x-2}=x^2+6x+7+\frac{16}{x-2}$$

45.

$$x-4\overline{)\begin{array}{l} 2x^3\ +5x^2\ +9x-4 \\ 2x^4-3x^3-11x^2-40x-1 \end{array}}$$
$$\underline{-(2x^4-8x^3)}$$
$$5x^3-11x^2$$
$$\underline{-(5x^3-20x^2)}$$
$$9x^2-40x$$
$$\underline{-(9x^2-36x)}$$
$$-4x\ -1$$
$$\underline{-(-4x+16)}$$
$$-17$$

$$\frac{2x^4-3x^3-11x^2-40x-1}{x-4}$$
$$=2x^3+5x^2+9x-4+\frac{-17}{x-4}$$

47.

$$2x-1\overline{)\begin{array}{l} x^2+4x-3 \\ 2x^3+7x^2-10x+5 \end{array}}$$
$$\underline{-(2x^3\ -x^2)}$$
$$8x^2-10x$$
$$\underline{-(8x^2\ -4x)}$$
$$-6x+5$$
$$\underline{-(-6x+3)}$$
$$2$$

$$\frac{2x^3+7x^2-10x+5}{2x-1}=x^2+4x-3+\frac{2}{2x-1}$$

49. $\dfrac{-24+x^2+x}{5+x} = \dfrac{x^2+x-24}{x+5}$

$$
\begin{array}{r}
x-4 \\
x+5\overline{\smash{\big)}\ x^2+x-24} \\
\underline{-(x^2+5x)} \\
-4x-24 \\
\underline{-(-4x-20)} \\
-4
\end{array}
$$

$\dfrac{-24+x^2+x}{5+x} = x-4+\dfrac{-4}{5+x}$

51. $\dfrac{4x^2+5}{1+2x} = \dfrac{4x^2+0x+5}{2x+1}$

$$
\begin{array}{r}
2x-1 \\
2x+1\overline{\smash{\big)}\ 4x^2+0x+5} \\
\underline{-(4x^2+2x)} \\
-2x+5 \\
\underline{-(-2x-1)} \\
6
\end{array}
$$

$\dfrac{4x^2+5}{1+2x} = 2x-1+\dfrac{6}{1+2x}$

53. $\dfrac{x^3+2x^2-8}{x^2-2} = \dfrac{x^3+2x^2+0x-8}{x^2-2}$

$$
\begin{array}{r}
x+2 \\
x^2-2\overline{\smash{\big)}\ x^3+2x^2+0x-8} \\
\underline{-(x^3-2x)} \\
2x^2+2x-8 \\
\underline{-(2x^2-4)} \\
2x-4
\end{array}
$$

$\dfrac{x^3+2x^2-8}{x^2-2} = x+2+\dfrac{2x-4}{x^2-2}$

55. $(a-5)(a+6)$
$= a(a)+a(6)+(-5)(a)+(-5)(6)$
$= a^2+6a-5a-30$
$= a^2+a-30$

57. $(2x-8)-(3x+x^2-2)$
$= (2x-8)+(-3x-x^2+2)$
$= 2x-8-3x-x^2+2$
$= -x^2+2x-3x-8+2$
$= -x^2-x-6$

59. $(2ab+b^2-a^2)+(b^2-4ab+a^2)$
$= 2ab+b^2-a^2+b^2-4ab+a^2$
$= -a^2+a^2+2ab-4ab+b^2+b^2$
$= -2ab+2b^2$

61. $\dfrac{4+7x^2-3x^4+6x^3}{2x^2}$

$= \dfrac{-3x^4+6x^3+7x^2+4}{2x^2}$

$= \dfrac{-3x^4}{2x^2}+\dfrac{6x^3}{2x^2}+\dfrac{7x^2}{2x^2}+\dfrac{4}{2x^2}$

$= -\dfrac{3}{2}x^{4-2}+\dfrac{6}{2}x^{3-2}+\dfrac{7}{2}x^{2-2}+\dfrac{4}{2}x^{-2}$

$= -\dfrac{3}{2}x^2+\dfrac{6}{2}x^1+\dfrac{7}{2}x^0+\dfrac{4}{2}x^{-2}$

$= \dfrac{-3x^2}{2}+3x+\dfrac{7}{2}+\dfrac{2}{x^2}$

63. $\dfrac{18x^3y^{-4}z^6}{27x^{-4}y^{-12}z^{-6}}$

$= \dfrac{18}{27}x^{3-(-4)}y^{-4-(-12)}z^{6-(-6)}$

$= \dfrac{2}{3}x^{3+4}y^{-4+12}z^{6+6}$

$= \dfrac{2}{3}x^7y^8z^{12}$

65.
$$
\begin{array}{r}
2x-6 \\
3x-5\overline{\smash{\big)}\ 6x^2-28x+30} \\
\underline{-(6x^2-10x)} \\
-18x+30 \\
\underline{-(-18x+30)} \\
0
\end{array}
$$

$\dfrac{6x^2-28x+30}{3x-5} = 2x-6$

67. $(n-3)^2 = (n)^2-2(n)(3)+(3)^2 = n^2-6n+9$

69. $(x^3+x-4x^4)(-10x^2)$
$= x^3(-10x^2)+x(-10x^2)+(-4x^4)(-10x^2)$
$= -10x^5-10x^3+40x^6$

71. $(2pq - q^2) + (4pq - p^2 - q^2)$

$= 2pq - q^2 + 4pq - p^2 - q^2$

$= 2pq + 4pq - q^2 - q^2 - p^2$

$= 6pq - 2q^2 - p^2$

73. $(x^2 + x - 1)(x + 5)$

$= x^2(x + 5) + x(x + 5) + (-1)(x + 5)$

$= x^3 + 5x^2 + x^2 + 5x - x - 5$

$= x^3 + 6x^2 + 4x - 5$

75. $(7rs^2 - 2r^2s) - (2r^2s - 8rs^2)$

$= (7rs^2 - 2r^2s) + (-2r^2s + 8rs^2)$

$= 7rs^2 - 2r^2s - 2r^2s + 8rs^2$

$= 7rs^2 + 8rs^2 - 2r^2s - 2r^2s$

$= 15rs^2 - 4r^2s$

77. $(x^4 - 2x^2 + x)(-3x)$

$= (x^4)(-3x) + (-2x^2)(-3x) + (x)(-3x)$

$= -3x^5 + 6x^3 - 3x^2$

79. First perform the division.

$$
\begin{array}{r}
3x - 1 \\
2x - 3 \overline{\smash{\big)}\ 6x^2 - 11x + 3} \\
\underline{-(6x^2 - 9x)} \\
-2x + 3 \\
\underline{-(-2x + 3)} \\
0
\end{array}
$$

Replace $\dfrac{3 + 6x^2 - 11x}{2x - 3}$ with $3x - 1$.

$\dfrac{3 + 6x^2 - 11x}{2x - 3} + (x + 1) = (3x - 1) + (x + 1)$

$\qquad\qquad\qquad\qquad = 3x - 1 + x + 1$

$\qquad\qquad\qquad\qquad = 4x$

81. $\dfrac{3x^4 - 6x + 12x^2}{-3x^3}$

$= \dfrac{3x^4}{-3x^3} - \dfrac{6x}{-3x^3} + \dfrac{12x^2}{-3x^3}$

$= \dfrac{3}{-3}x^{4-3} - \dfrac{6}{-3}x^{1-3} + \dfrac{12}{-3}x^{2-3}$

$= -1 \cdot x^1 - (-2)x^{-2} + (-4)x^{-1}$

$= -x + \dfrac{2}{x^2} - \dfrac{4}{x}$

83. $\dfrac{(x^2 + 3x - 1) + (x - 1)}{-2x^3}$

$= \dfrac{x^2 + 3x - 1 + x - 1}{-2x^3}$

$= \dfrac{x^2 + 4x - 2}{-2x^3}$

$= \dfrac{x^2}{-2x^3} + \dfrac{4x}{-2x^3} - \dfrac{2}{-2x^3}$

$= \dfrac{1}{-2}x^{2-3} + \dfrac{4}{-2}x^{1-3} - \dfrac{2}{-2} \cdot \dfrac{1}{x^3}$

$= -\dfrac{1}{2}x^{-1} + (-2)x^{-2} - (-1)\dfrac{1}{x^3}$

$= -\dfrac{1}{2x} - \dfrac{2}{x^2} + \dfrac{1}{x^3}$

85. The volume is the product of the lengths of the sides. Thus the length of the third side is the quotient of $x^3 - 5x^2 + 6x$ and the product of $x - 2$ and $x - 3$.

$(x - 2)(x - 3)$

$= x(x) + x(-3) + (-2)(x) + (-2)(-3)$

$= x^2 - 3x - 2x + 6$

$= x^2 - 5x + 6$

$$
\begin{array}{r}
x \\
x^2 - 5x + 6 \overline{\smash{\big)}\ x^3 - 5x^2 + 6x + 0} \\
\underline{-(x^3 - 5x^2 + 6x)} \\
0
\end{array}
$$

Since $\dfrac{x^3 - 5x^2 + 6x}{(x - 2)(x - 3)} = \dfrac{x^3 - 5x^2 + 6x}{x^2 - 5x + 6} = x$, the

length of the third side is x feet.

87. The area is the product of the length and the width. Thus the width is the quotient of $z^2 + 6z + 9$ and $z + 3$.

$$
\begin{array}{r}
z + 3 \\
z + 3 \overline{\smash{\big)}\ z^2 + 6z + 9} \\
\underline{-(z^2 + 3z)} \\
3z + 9 \\
\underline{-(3z + 9)} \\
0
\end{array}
$$

Since $\dfrac{z^2 + 6z + 9}{z + 3} = z + 3$, the width of the

rectangle is $z + 3$ inches.

89. The area is one-half of the product of the base and the height $\left(A = \dfrac{1}{2}bh \right),$ so twice the area is the product of the base and the height. Thus the base is the quotient of $2(6x^3 - 2x^2 - 8x)$ and $3x^2 - 4x.$

$$2(6x^3 - 2x^2 - 8x)$$
$$= 2(6x^3) + 2(-2x^2) + 2(-8x)$$
$$= 12x^3 - 4x^2 - 16x$$

$$
\begin{array}{r}
4x+4 \\
3x^2-4x\overline{\smash{)}\;12x^3\;-4x^2-16x+0} \\
\underline{-(12x^3-16x^2)} \\
12x^2-16x \\
\underline{-(12x^2-16x)} \\
0
\end{array}
$$

Since $\dfrac{2(6x^3 - 2x^2 - 8x)}{3x^2 - 4x} = 4x + 4,$ the length of the base is $4x + 4$ feet.

91. (a) $\dfrac{0.004x^3 - 0.8x^2 + 180x + 5000}{x}$

$$= \frac{0.004x^3}{x} - \frac{0.8x^2}{x} + \frac{180x}{x} + \frac{5000}{x}$$

$$= 0.004x^{3-1} - 0.8x^{2-1} + 180x^{1-1} + \frac{5000}{x}$$

$$= 0.004x^2 - 0.8x + 180 + \frac{5000}{x}$$

(b) When $x = 140,$ the value of the expression is
$$0.004(140)^2 - 0.8(140) + 180 + \frac{5000}{140}$$

$$= 78.4 - 112 + 180 + \frac{250}{7}$$

$$\approx 182.11$$

The average cost of manufacturing $x = 140$ computers in a day is $182.11 per computer.

93.
$$
\begin{array}{r}
3x-2 \\
2x-3\overline{\smash{)}\;6x^2-13x+\,?} \\
\underline{-(6x^2\;-9x)} \\
-4x+\,? \\
\underline{-(-4x+6)} \\
?-6
\end{array}
$$

If the remainder is zero, then $? - 6 = 0$ or $? = 6.$

95. When the divisor is a monomial, divide each term in the numerator by the denominator. When dividing by a binomial, use long division. Answers may vary.

Section 5.6

Preparing for Applying Exponent Rules: Scientific Notation

P1. $(3a^6)(4.5a^4) = 3 \cdot 4.5 \cdot a^6 \cdot a^4$
$$= 13.5 \cdot a^{6+4}$$
$$= 13.5a^{10}$$

P2. $(7n^3)(2n^{-2}) = 7 \cdot 2 \cdot n^3 \cdot n^{-2}$
$$= 14 \cdot n^{3+(-2)}$$
$$= 14n^1$$
$$= 14n$$

P3. $\dfrac{3.6b^9}{0.9b^{-2}} = \dfrac{3.6}{0.9} \cdot \dfrac{b^9}{b^{-2}} = 4 \cdot b^{9-(-2)} = 4b^{11}$

5.6 Quick Checks

1. A number written as 3.2×10^{-6} is said to be written in <u>scientific notation</u>.

2. When writing 47,000,000 in scientific notation, the power of 10 will be <u>positive</u> (positive or negative).

3. False; when a number is expressed in scientific notation, it is expressed as the product of a number $x,$ $1 \le x < 10,$ and a power of 10.

4. $432 = 4.32 \times 100 = 4.32 \times 10^2$

5. $10,302 = 1.0302 \times 10,000 = 1.0302 \times 10^4$

6. $5,432,000 = 5.432 \times 1,000,000 = 5.432 \times 10^6$

7. $0.093 = 9.3 \times 0.01 = 9.3 \times 10^{-2}$

8. $0.0000459 = 4.59 \times 0.00001 = 4.59 \times 10^{-5}$

9. $0.00000008 = 8.0 \times 0.00000001$
$$= 8.0 \times 10^{-8}$$

10. To write 3.2×10^{-6} in decimal notation, move the decimal point in 3.2 six places to the <u>left</u>.

11. True

12. $3.1 \times 10^2 = 3.1 \times 100 = 310$

13. $9.01 \times 10^{-1} = 9.01 \times 0.1 = 0.901$

14. $1.7 \times 10^5 = 1.7 \times 100,000 = 170,000$

15. $7 \times 10^0 = 7 \times 1 = 7$

16. $8.9 \times 10^{-4} = 8.9 \times 0.0001 = 0.00089$

17. $(3 \times 10^4) \cdot (2 \times 10^3) = (3 \cdot 2) \times (10^4 \cdot 10^3)$
$$= 6 \times 10^7$$

18. $(2 \times 10^{-2}) \cdot (4 \times 10^{-1}) = (2 \cdot 4) \times (10^{-2} \cdot 10^{-1})$
$$= 8 \times 10^{-3}$$

19. $(5 \times 10^{-4}) \cdot (3 \times 10^7) = (5 \cdot 3) \times (10^{-4} \cdot 10^7)$
$$= 15 \times 10^3$$
$$= (1.5 \times 10^1) \times 10^3$$
$$= 1.5 \times 10^4$$

20. $(8 \times 10^{-4}) \cdot (3.5 \times 10^{-2})$
$$= (8 \cdot 3.5) \times (10^{-4} \cdot 10^{-2})$$
$$= 28 \times 10^{-6}$$
$$= (2.8 \times 10^1) \times 10^{-6}$$
$$= 2.8 \times 10^{-5}$$

21. $\dfrac{8 \times 10^6}{2 \times 10^1} = \dfrac{8}{2} \times \dfrac{10^6}{10^1} = 4 \times 10^5$

22. $\dfrac{2.8 \times 10^{-7}}{1.4 \times 10^{-3}} = \dfrac{2.8}{1.4} \times \dfrac{10^{-7}}{10^{-3}}$
$$= 2 \times 10^{-7-(-3)}$$
$$= 2 \times 10^{-4}$$

23. $\dfrac{3.6 \times 10^3}{7.2 \times 10^{-1}} = \dfrac{3.6}{7.2} \times \dfrac{10^3}{10^{-1}}$
$$= 0.5 \times 10^{3-(-1)}$$
$$= (5 \times 10^{-1}) \times 10^4$$
$$= 5 \times 10^3$$

24. $\dfrac{5 \times 10^{-2}}{8 \times 10^2} = \dfrac{5}{8} \times \dfrac{10^{-2}}{10^2}$
$$= 0.625 \times 10^{-2-2}$$
$$= (6.25 \times 10^{-1}) \times 10^{-4}$$
$$= 6.25 \times 10^{-5}$$

25. There are 31 days in August.
$$31 = 3.1 \times 10^1$$
$$(3.1 \times 10^1) \cdot (8.72 \times 10^6)$$
$$= (3.1 \cdot 8.72) \times (10^1 \cdot 10^6)$$
$$= 27.032 \times 10^7$$
$$= (2.7032 \times 10^1) \times 10^7$$
$$= 2.7032 \times 10^8$$
$$= 270,320,000$$
Saudi Arabia produced
$2.7032 \times 10^8 = 270,320,000$ barrels of oil in August 2007.

26. There are 31 days in August: $31 = 3.1 \times 10^1$. In scientific notation, 12 million is written as 1.2×10^7.
$$(3.1 \times 10^1)(1.2 \times 10^7) = (3.1 \times 1.2) \times 10^{1+7}$$
$$= 3.72 \times 10^8$$
$$= 372,000,000$$
Saudi Arabia is expected to produce 3.72×10^8 barrels or 372,000,000 barrels of oil in August 2009.

5.6 Exercises

27. $300,000 = 3.0 \times 100,000 = 3 \times 10^5$

29. $64,000,000 = 6.4 \times 10,000,000$
$$= 6.4 \times 10^7$$

31. $0.00051 = 5.1 \times 0.0001 = 5.1 \times 10^{-4}$

33. $0.000000001 = 1.0 \times 0.000000001$
$$= 1 \times 10^{-9}$$

35. $8,007,000,000 = 8.007 \times 1,000,000,000$
$$= 8.007 \times 10^9$$

37. $0.0000309 = 3.09 \times 0.00001 = 3.09 \times 10^{-5}$

39. $620 = 6.2 \times 100 = 6.2 \times 10^2$

41. $4 = 4.0 \times 1 = 4 \times 10^0$

43. $6,656,000,000 = 6.656 \times 10^9$

45. $70,510,000,000 = 7.051 \times 10^{10}$

47. $170,200,000 \text{ km} = 1.702 \times 10^8 \text{ km}$

49. $0.00003 = 3 \times 0.00001 = 3 \times 10^{-5}$

51. $0.00000025 \text{ m} = 2.5 \times 10^{-7} \text{ m}$

53. $0.00311 \text{ kg} = 3.11 \times 10^{-3} \text{ kg}$

55. $4.2 \times 10^5 = 4.2 \times 100,000 = 420,000$

57. $1 \times 10^8 = 1 \times 100,000,000$
$= 100,000,000$

59. $3.9 \times 10^{-3} = 3.9 \times 0.001 = 0.0039$

61. $4 \times 10^{-1} = 4 \times 0.1 = 0.4$

63. $3.76 \times 10^3 = 3.76 \times 1000 = 3760$

65. $8.2 \times 10^{-3} = 8.2 \times 0.001 = 0.0082$

67. $6 \times 10^{-5} = 6 \times 0.00001 = 0.00006$

69. $7.05 \times 10^6 = 7.05 \times 1,000,000 = 7,050,000$

71. $1 \times 10^{-15} = 1 \times 0.000000000000001$
$= 0.000000000000001$

73. $2.25 \times 10^{-3} = 2.25 \times 0.001 = 0.00225$

75. $5 \times 10^5 = 5 \times 100,000 = 500,000$

77. $(2 \times 10^6)(1.5 \times 10^3) = (2 \cdot 1.5) \times (10^6 \cdot 10^3)$
$= 3 \times 10^9$

79. $(1.2 \times 10^0)(7 \times 10^{-3}) = (1.2 \cdot 7) \times (10^0 \cdot 10^{-3})$
$= 8.4 \times 10^{-3}$

81. $\dfrac{9 \times 10^4}{3 \times 10^{-4}} = \dfrac{9}{3} \times \dfrac{10^4}{10^{-4}} = 3 \times 10^{4-(-4)} = 3 \times 10^8$

83. $\dfrac{2 \times 10^{-3}}{8 \times 10^{-5}} = \dfrac{2}{8} \times \dfrac{10^{-3}}{10^{-5}}$
$= \dfrac{1}{4} \times 10^{-3-(-5)}$
$= 0.25 \times 10^2$
$= (2.5 \times 10^{-1}) \times 10^2$
$= 2.5 \times 10^1$

85. $\dfrac{56,000}{0.00007} = \dfrac{5.6 \times 10^4}{7 \times 10^{-5}}$
$= \dfrac{5.6}{7} \times 10^{4-(-5)}$
$= 0.8 \times 10^9$
$= (8 \times 10^{-1}) \times 10^9$
$= 8 \times 10^8$

87. $\dfrac{300,000 \times 15,000,000}{0.0005}$
$= \dfrac{(3 \times 10^5) \cdot (1.5 \times 10^7)}{5 \times 10^{-4}}$
$= \dfrac{(3 \cdot 1.5) \times (10^5 \cdot 10^7)}{5 \times 10^{-4}}$
$= \dfrac{4.5 \times 10^{12}}{5 \times 10^{-4}}$
$= \dfrac{4.5}{5} \times 10^{12-(-4)}$
$= 0.9 \times 10^{16}$
$= (9 \times 10^{-1}) \times 10^{16}$
$= 9 \times 10^{15}$

89. $(1.86 \times 10^5)(6.0 \times 10^1)$
$= (1.86 \cdot 6.0) \times (10^5 \cdot 10^1)$
$= 11.16 \times 10^6$
$= (1.116 \times 10^1) \times 10^6$
$= 1.116 \times 10^7$
Light travels 1.116×10^7 miles in one minute.

91. $\dfrac{5.026 \times 10^{11} \text{ lb}}{3.0 \times 10^8 \text{ people}} = \dfrac{5.026}{3} \times 10^{11-8} \left(\dfrac{\text{lb}}{\text{person}} \right)$
$= \dfrac{5.026}{3} \times 10^3 \cdot \left(\dfrac{\text{lb}}{\text{person}} \right)$
$\approx 1.68 \times 10^3$ lb per person

93. (a) $1.55 \text{ billion} = 1.55 \times 10^9$

(b) $300,000,000 = 3 \times 10^8$

(c) $\dfrac{1.55 \times 10^9}{3 \times 10^8} \cdot \left(\dfrac{\text{gallons}}{\text{person}}\right)$

$= \dfrac{1.55}{3} \times 10^{9-8} \cdot \left(\dfrac{\text{gallons}}{\text{person}}\right)$

$\approx 0.52 \times 10 \text{ gallons per person}$

$= 5.2 \text{ gallons per person}$

95. (a) $179.1 \text{ million} = 179.1 \times 10^6$

$= 1.791 \times 10^8$

(b) $44.8 \text{ million} = 44.8 \times 10^6 = 4.48 \times 10^7$

(c) $\dfrac{4.48 \times 10^7}{1.791 \times 10^8} \cdot 100\%$

$= \left(\dfrac{4.48}{1.791} \times 10^{7-8}\right) \cdot 100\%$

$\approx (2.5 \times 10^{-1}) \cdot 100\%$

$= 0.25 \cdot 100\%$

$= 25\%$

97. $250 \ \mu\text{m} = 250 \cdot (1 \times 10^{-6}) \text{ m}$

$= (2.5 \times 10^2) \cdot (1 \times 10^{-6}) \text{ m}$

$= (2.5 \cdot 1) \times (10^2 \cdot 10^{-6}) \text{ m}$

$= 2.5 \times 10^{-4} \text{ m}$

99. $800 \ \text{pm} = 800 \cdot (1 \times 10^{-12}) \text{ m}$

$= (8 \times 10^2) \cdot (1 \times 10^{-12}) \text{ m}$

$= (8 \cdot 1) \times (10^2 \cdot 10^{-12}) \text{ m}$

$= 8 \times 10^{-10} \text{ m}$

101. $71.5 \ \text{nm} = 71.5 \cdot (1 \times 10^{-9}) \text{ m}$

$= (7.15 \times 10^1) \cdot (1 \times 10^{-9}) \text{ m}$

$= (7.15 \cdot 1) \times (10^1 \cdot 10^{-9}) \text{ m}$

$= 7.15 \times 10^{-8} \text{ m}$

103. $21 \ \mu\text{m} = 21 \cdot (1 \times 10^{-6}) \text{ m}$

$= (2.1 \times 10^1) \cdot (1 \times 10^{-6}) \text{ m}$

$= (2.1 \cdot 1) \times (10^1 \cdot 10^{-6}) \text{ m}$

$= 2.1 \times 10^{-5} \text{ m}$

$V = \dfrac{4}{3}\pi (21 \ \mu\text{m})^3$

$= \dfrac{4}{3}\pi (2.1 \times 10^{-5} \text{ m})^3$

$= \dfrac{4}{3}\pi [2.1^3 \times (10^{-5})^3 \text{ m}^3]$

$= \dfrac{4}{3}\pi (9.261 \times 10^{-15}) \text{ m}^3$

$= \pi (12.348 \times 10^{-15}) \text{ m}^3$

$= \pi [(1.2348 \times 10^1) \times 10^{-15}] \text{ m}^3$

$= \pi (1.2348 \times 10^{-14}) \text{ m}^3$

$= 1.2348\pi \times 10^{-14} \text{ m}^3$

105. $6 \ \text{nm} = 6 \cdot (1 \times 10^{-9}) \text{ m}$

$= (6 \times 10^0) \cdot (1 \times 10^{-9}) \text{ m}$

$= (6 \cdot 1) \times (10^0 \cdot 10^{-9}) \text{ m}$

$= 6 \times 10^{-9} \text{ m}$

$V = \dfrac{4}{3}\pi (6 \ \text{nm})^3$

$= \dfrac{4}{3}\pi (6 \times 10^{-9} \text{ m})^3$

$= \dfrac{4}{3}\pi [6^3 \times (10^{-9})^3 \text{ m}^3]$

$= \dfrac{4}{3}\pi (216 \times 10^{-27}) \text{ m}^3$

$= \pi (288 \times 10^{-27}) \text{ m}^3$

$= \pi [(2.88 \times 10^2) \times 10^{-27}] \text{ m}^3$

$= \pi (2.88 \times 10^{-25}) \text{ m}^3$

$= 2.88\pi \times 10^{-25} \text{ m}^3$

107. To convert a number written in decimal notation to scientific notation, count the number of decimal places, N, that the decimal point must be moved to arrive at a number x such that $1 \le x < 10$. If the number is greater than or equal to 1, move the decimal point to the left that many places and write the number in the form $x \times 10^N$. If the number is between 0 and 1, move the decimal point to the right that many places and write the number in the form $x \times 10^{-N}$.

109. The number 34.5×10^4 is incorrect because the number 34.5 is not a number between 1 (inclusive) and 10. The correct answer is 3.45×10^5.

Chapter 5 Review

1. Yes, $4x^3$ is a monomial of degree 3 with coefficient 4.

2. No, $6x^{-3}$ is not a monomial because the exponent of x is not a whole number.

3. No, $m^{1/2}$ is not a monomial because the exponent of m is not a whole number.

4. Yes, mn^2 is monomial of degree 3 (1 + 2) with coefficient 1.

5. No, $4x^6 - 4x^{1/2}$ is not a polynomial because a variable has an exponent that is not a whole number.

6. No, $\dfrac{3}{x} - \dfrac{1}{x^2}$ is not a polynomial because there is a variable in the denominator of a fraction.

7. Yes, 6 is monomial of degree 0.

8. Yes, $3x^3 - 4xy^4$ is a binomial of degree 5 (1 + 4).

9. Yes, $-2x^5 y - 7x^4 y + 7$ is a trinomial of degree 6 (5 + 1).

10. Yes, $\dfrac{1}{2}x^3 + 2x^{10} - 5$ is trinomial of degree 10.

11. $(6x^2 - 2x + 1) + (3x^2 + 10x - 3)$
$= 6x^2 - 2x + 1 + 3x^2 + 10x - 3$
$= 6x^2 + 3x^2 - 2x + 10x + 1 - 3$
$= 9x^2 + 8x - 2$

12. $(-7m^3 - 2mn) + (8m^3 - 5m + 3mn)$
$= -7m^3 - 2mn + 8m^3 - 5m + 3mn$
$= -7m^3 + 8m^3 - 2mn + 3mn - 5m$
$= m^3 + mn - 5m$

13. $(4x^2 y + 10x) - (5x^2 y - 2x)$
$= (4x^2 y + 10x) + (-5x^2 y + 2x)$
$= 4x^2 y + 10x - 5x^2 y + 2x$
$= 4x^2 y - 5x^2 y + 10x + 2x$
$= -x^2 y + 12x$

14. $(3y^2 - yz + 3z^2) - (10y^2 + 5yz - 6z^2)$
$= (3y^2 - yz + 3z^2) + (-10y^2 - 5yz + 6z^2)$
$= 3y^2 - yz + 3z^2 - 10y^2 - 5yz + 6z^2$
$= 3y^2 - 10y^2 - yz - 5yz + 3z^2 + 6z^2$
$= -7y^2 - 6yz + 9z^2$

15. $(-6x^2 + 5) + (4x^2 - 7) = -6x^2 + 5 + 4x^2 - 7$
$\qquad\qquad\qquad\qquad = -6x^2 + 4x^2 + 5 - 7$
$\qquad\qquad\qquad\qquad = -2x^2 - 2$

16. $(-18y + 10) - (20y^2 - 10y + 5)$
$= (-18y + 10) + (-20y^2 + 10y - 5)$
$= -18y + 10 - 20y^2 + 10y - 5$
$= -20y^2 - 18y + 10y + 10 - 5$
$= -20y^2 - 8y + 5$

17. $3x^2 - 5x$

   (a) $x = 0$: $3(0)^2 - 5(0) = 0 - 0 = 0$

   (b) $x = -1$: $3(-1)^2 - 5(-1) = 3 + 5 = 8$

   (c) $x = 2$: $3(2)^2 - 5(2) = 12 - 10 = 2$

18. $-x^2 + 3$

   (a) $x = 0$: $-(0)^2 + 3 = 0 + 3 = 3$

   (b) $x = -1$: $-(-1)^2 + 3 = -1 + 3 = 2$

   (c) $x = \dfrac{1}{2}$: $-\left(\dfrac{1}{2}\right)^2 + 3 = -\dfrac{1}{4} + 3 = -\dfrac{1}{4} + \dfrac{12}{4} = \dfrac{11}{4}$

19. $x^2 y + 2xy^2$ for $x = -2$ and $y = 1$:
$(-2)^2(1) + 2(-2)(1)^2 = 4(1) + 2(-2)(1)$
$\qquad\qquad\qquad\qquad\quad = 4 - 4$
$\qquad\qquad\qquad\qquad\quad = 0$

20. $4a^2 b^2 - 3ab + 2$ for $a = -1$ and $b = -3$:
$4(-1)^2(-3)^2 - 3(-1)(-3) + 2$
$= 4(1)(9) - 3(-1)(-3) + 2$
$= 36 - 9 + 2$
$= 29$

21. $6^2 \cdot 6^5 = 6^{2+5} = 6^7 = 279,936$

22. $\left(-\dfrac{1}{3}\right)^2 \left(-\dfrac{1}{3}\right)^3 = \left(-\dfrac{1}{3}\right)^{2+3} = \left(-\dfrac{1}{3}\right)^5 = -\dfrac{1}{243}$

23. $(4^2)^6 = 4^{2 \cdot 6} = 4^{12} = 16,777,216$

24. $[(-1)^4]^3 = (-1)^{4 \cdot 3} = (-1)^{12} = 1$

25. $x^4 \cdot x^8 \cdot x = x^4 \cdot x^8 \cdot x^1 = x^{4+8+1} = x^{13}$

26. $m^4 \cdot m^2 = m^{4+2} = m^6$

27. $(r^3)^4 = r^{3 \cdot 4} = r^{12}$

28. $(m^8)^3 = m^{8 \cdot 3} = m^{24}$

29. $(4x)^3 (4x)^2 = (4x)^{3+2}$
$= (4x)^5$
$= 4^5 \cdot x^5$
$= 1024x^5$

30. $(-2n)^3 (-2n)^3 = (-2n)^{3+3}$
$= (-2n)^6$
$= (-2)^6 \cdot n^6$
$= 64n^6$

31. $(-3x^2 y)^4 = (-3)^4 \cdot (x^2)^4 \cdot y^4$
$= 81 \cdot x^{2 \cdot 4} \cdot y^4$
$= 81x^8 y^4$

32. $(2x^3 y^4)^2 = 2^2 \cdot (x^3)^2 \cdot (y^4)^2$
$= 4 \cdot x^{3 \cdot 2} \cdot y^{4 \cdot 2}$
$= 4x^6 y^8$

33. $3x^2 \cdot 5x^4 = 3 \cdot 5 \cdot x^2 \cdot x^4 = 15x^{2+4} = 15x^6$

34. $-4a \cdot 9a^3 = -4 \cdot 9 \cdot a^1 \cdot a^3 = -36a^{1+3} = -36a^4$

35. $-8y^4 \cdot (-2y) = -8 \cdot (-2) \cdot y^4 \cdot y^1$
$= 16y^{4+1}$
$= 16y^5$

36. $12p \cdot (-p^5) = 12p^1 \cdot (-1 \cdot p^5)$
$= 12 \cdot (-1) \cdot p^1 \cdot p^5$
$= -12p^{1+5}$
$= -12p^6$

37. $\dfrac{8}{3} w^3 \cdot \dfrac{9}{2} w = \dfrac{8}{3} \cdot \dfrac{9}{2} \cdot w^3 \cdot w^1 = 12w^{3+1} = 12w^4$

38. $\dfrac{1}{3} z^2 \cdot \left(-\dfrac{9}{4} z\right) = \dfrac{1}{3} \cdot \left(-\dfrac{9}{4}\right) \cdot z^2 \cdot z^1$
$= -\dfrac{3}{4} z^{2+1}$
$= -\dfrac{3}{4} z^3$

39. $(3x^2)^3 \cdot (2x)^2 = 3^3 \cdot (x^2)^3 \cdot 2^2 \cdot x^2$
$= 27 \cdot x^{2 \cdot 3} \cdot 4 \cdot x^2$
$= 27 \cdot 4 \cdot x^6 \cdot x^2$
$= 108x^{6+2}$
$= 108x^8$

40. $(-4a)^2 \cdot (5a^4) = (-4)^2 \cdot a^2 \cdot 5 \cdot a^4$
$= 16 \cdot 5 \cdot a^2 \cdot a^4$
$= 80a^{2+4}$
$= 80a^6$

41. $-2x^3 (4x^2 - 3x + 1)$
$= -2x^3 (4x^2) + (-2x^3)(-3x) + (-2x^3)(1)$
$= -8x^5 + 6x^4 - 2x^3$

42. $\dfrac{1}{2} x^4 (4x^3 + 8x^2 - 2)$
$= \dfrac{1}{2} x^4 (4x^3) + \dfrac{1}{2} x^4 (8x^2) + \dfrac{1}{2} x^4 (-2)$
$= 2x^7 + 4x^6 - x^4$

43. $(3x - 5)(2x + 1)$
$= (3x - 5)(2x) + (3x - 5)(1)$
$= 3x \cdot 2x + (-5) \cdot 2x + 3x \cdot 1 + (-5) \cdot 1$
$= 6x^2 - 10x + 3x - 5$
$= 6x^2 - 7x - 5$

44. $(4x + 3)(x - 2) = (4x + 3)(x) + (4x + 3)(-2)$
$= 4x \cdot x + 3 \cdot x + 4x \cdot (-2) + 3 \cdot (-2)$
$= 4x^2 + 3x - 8x - 6$
$= 4x^2 - 5x - 6$

45. $(x+5)(x-8) = (x+5)x + (x+5)(-8)$
$$= x \cdot x + 5 \cdot x + x \cdot (-8) + 5 \cdot (-8)$$
$$= x^2 + 5x - 8x - 40$$
$$= x^2 - 3x - 40$$

46. $(w-1)(w+10) = (w-1)(w) + (w-1)(10)$
$$= w \cdot w + (-1) \cdot w + w \cdot 10 + (-1) \cdot 10$$
$$= w^2 - w + 10w - 10$$
$$= w^2 + 9w - 10$$

47. $(4m-3)(6m^2 - m + 1) = (4m-3)(6m^2) + (4m-3)(-m) + (4m-3)(1)$
$$= 4m \cdot 6m^2 + (-3) \cdot 6m^2 + 4m \cdot (-m) + (-3) \cdot (-m) + 4m \cdot 1 + (-3) \cdot 1$$
$$= 24m^3 - 18m^2 - 4m^2 + 3m + 4m - 3$$
$$= 24m^3 - 22m^2 + 7m - 3$$

48. $(2y+3)(4y^4 + 2y^2 - 3) = (2y+3)(4y^4) + (2y+3)(2y^2) + (2y+3)(-3)$
$$= 2y \cdot 4y^4 + 3 \cdot 4y^4 + 2y \cdot 2y^2 + 3 \cdot 2y^2 + 2y \cdot (-3) + 3 \cdot (-3)$$
$$= 8y^5 + 12y^4 + 4y^3 + 6y^2 - 6y - 9$$

49. $(x+5)(x+3) = (x)(x) + (x)(3) + (5)(x) + (5)(3) = x^2 + 3x + 5x + 15 = x^2 + 8x + 15$

50. $(2x-1)(x-8) = (2x)(x) + (2x)(-8) + (-1)(x) + (-1)(-8) = 2x^2 - 16x - x + 8 = 2x^2 - 17x + 8$

51. $(2m+7)(3m-2) = (2m)(3m) + (2m)(-2) + (7)(3m) + (7)(-2) = 6m^2 - 4m + 21m - 14 = 6m^2 + 17m - 14$

52. $(6m-4)(8m+1) = (6m)(8m) + (6m)(1) + (-4)(8m) + (-4)(1) = 48m^2 + 6m - 32m - 4 = 48m^2 - 26m - 4$

53. $(3x+2y)(7x-3y) = (3x)(7x) + (3x)(-3y) + (2y)(7x) + (2y)(-3y)$
$$= 21x^2 - 9xy + 14xy - 6y^2$$
$$= 21x^2 + 5xy - 6y^2$$

54. $(4x-y)(5x+3y) = (4x)(5x) + (4x)(3y) + (-y)(5x) + (-y)(3y)$
$$= 20x^2 + 12xy - 5xy - 3y^2$$
$$= 20x^2 + 7xy - 3y^2$$

55. $(x-4)(x+4) = (x)^2 - (4)^2 = x^2 - 16$

56. $(2x+5)(2x-5) = (2x)^2 - (5)^2 = 4x^2 - 25$

57. $(2x+3)^2 = (2x)^2 + 2(2x)(3) + (3)^2$
$$= 4x^2 + 12x + 9$$

58. $(7x-2)^2 = (7x)^2 - 2(7x)(2) + (2)^2$
$$= 49x^2 - 28x + 4$$

59. $(3x+4y)(3x-4y) = (3x)^2 - (4y)^2$
$$= 9x^2 - 16y^2$$

60. $(8m-6n)(8m+6n) = (8m)^2 - (6n)^2$
$$= 64m^2 - 36n^2$$

61. $(5x-2y)^2 = (5x)^2 - 2(5x)(2y) + (2y)^2$
$$= 25x^2 - 20xy + 4y^2$$

62. $(2a+3b)^2 = (2a)^2 + 2(2a)(3b) + (3b)^2$
$$= 4a^2 + 12ab + 9b^2$$

63. $(x-0.5)(x+0.5) = (x)^2 - (0.5)^2 = x^2 - 0.25$

64. $(r+0.25)(r-0.25) = (r)^2 - (0.25)^2$
$$= r^2 - 0.0625$$

65. $\left(y+\dfrac{2}{3}\right)^2 = (y)^2 + 2(y)\left(\dfrac{2}{3}\right) + \left(\dfrac{2}{3}\right)^2$
$$= y^2 + \dfrac{4}{3}y + \dfrac{4}{9}$$

66. $\left(y-\dfrac{1}{2}\right)^2 = (y)^2 - 2(y)\left(\dfrac{1}{2}\right) + \left(\dfrac{1}{2}\right)^2$
$$= y^2 - y + \dfrac{1}{4}$$

67. $\dfrac{6^5}{6^3} = 6^{5-3} = 6^2 = 36$

68. $\dfrac{7}{7^4} = \dfrac{7^1}{7^4} = 7^{1-4} = 7^{-3} = \dfrac{1}{7^3} = \dfrac{1}{343}$

69. $\dfrac{x^{16}}{x^{12}} = x^{16-12} = x^4$

70. $\dfrac{x^3}{x^{11}} = x^{3-11} = x^{-8} = \dfrac{1}{x^8}$

71. $5^0 = 1$

72. $-5^0 = -1 \cdot 5^0 = -1$

73. $m^0 = 1,\ m \neq 0$

74. $-m^0 = -1 \cdot m^0 = -1,\ m \neq 0$

75. $\dfrac{25x^3 y^7}{10xy^{10}} = \dfrac{25}{10} \cdot x^{3-1} \cdot y^{7-10}$
$$= \dfrac{5}{2} \cdot x^2 \cdot y^{-3}$$
$$= \dfrac{5x^2}{2y^3}$$

76. $\dfrac{3x^4 y^2}{9x^2 y^{10}} = \dfrac{3}{9} \cdot x^{4-2} \cdot y^{2-10}$
$$= \dfrac{1}{3} \cdot x^2 \cdot y^{-8}$$
$$= \dfrac{x^2}{3y^8}$$

77. $\left(\dfrac{x^3}{y^2}\right)^5 = \dfrac{(x^3)^5}{(y^2)^5} = \dfrac{x^{3\cdot5}}{y^{2\cdot5}} = \dfrac{x^{15}}{y^{10}}$

78. $\left(\dfrac{7}{x^2}\right)^3 = \dfrac{7^3}{(x^2)^3} = \dfrac{343}{x^{2\cdot3}} = \dfrac{343}{x^6}$

79. $\left(\dfrac{2m^2 n^2}{p^4}\right)^3 = \dfrac{2^3 (m^2)^3 (n^2)^3}{(p^4)^3}$
$$= \dfrac{8 \cdot m^{2\cdot3} \cdot n^{2\cdot3}}{p^{4\cdot3}}$$
$$= \dfrac{8m^6 n^6}{p^{12}}$$

80. $\left(\dfrac{3mn^2}{p^5}\right)^4 = \dfrac{3^4 m^4 (n^2)^4}{(p^5)^4}$
$$= \dfrac{81 \cdot m^4 \cdot n^{2\cdot4}}{p^{5\cdot4}}$$
$$= \dfrac{81 m^4 n^8}{p^{20}}$$

81. $-5^{-2} = -1 \cdot 5^{-2} = -1 \cdot \dfrac{1}{5^2} = -\dfrac{1}{25}$

82. $\dfrac{1}{4^{-3}} = 4^3 = 64$

83. $\left(\dfrac{2}{3}\right)^{-4} = \left(\dfrac{3}{2}\right)^4 = \dfrac{3^4}{2^4} = \dfrac{81}{16}$

84. $\left(\dfrac{1}{3}\right)^{-3} = \left(\dfrac{3}{1}\right)^3 = 3^3 = 27$

85. $2^{-2} + 3^{-1} = \dfrac{1}{2^2} + \dfrac{1}{3^1} = \dfrac{1}{4} + \dfrac{1}{3} = \dfrac{3}{12} + \dfrac{4}{12} = \dfrac{7}{12}$

86. $4^{-1} - 2^{-3} = \dfrac{1}{4^1} - \dfrac{1}{2^3} = \dfrac{1}{4} - \dfrac{1}{8} = \dfrac{2}{8} - \dfrac{1}{8} = \dfrac{1}{8}$

87. $\dfrac{16x^{-3}y^4}{24x^{-6}y^{-1}} = \dfrac{16}{24} \cdot x^{-3-(-6)} \cdot y^{4-(-1)}$

$\qquad = \dfrac{2}{3} \cdot x^3 \cdot y^5$

$\qquad = \dfrac{2x^3 y^5}{3}$

88. $\dfrac{15x^0 y^{-6}}{35xy^4} = \dfrac{15}{35} \cdot x^{0-1} \cdot y^{-6-4}$

$\qquad = \dfrac{3}{7} x^{-1} y^{-10}$

$\qquad = \dfrac{3}{7xy^{10}}$

89. $(2m^{-3}n)^{-4}(3m^{-4}n^2)^2$

$\qquad = 2^{-4} \cdot (m^{-3})^{-4} \cdot n^{-4} \cdot 3^2 \cdot (m^{-4})^2 \cdot (n^2)^2$

$\qquad = \dfrac{1}{2^4} \cdot m^{-3 \cdot (-4)} \cdot n^{-4} \cdot 9 \cdot m^{-4 \cdot 2} \cdot n^{2 \cdot 2}$

$\qquad = \dfrac{1}{16} \cdot m^{12} \cdot n^{-4} \cdot 9 \cdot m^{-8} \cdot n^4$

$\qquad = \dfrac{9}{16} \cdot m^{12} \cdot m^{-8} \cdot n^{-4} \cdot n^4$

$\qquad = \dfrac{9}{16} m^4 n^0$

$\qquad = \dfrac{9m^4}{16}$

90. $(4m^{-6}n^0)^3 (3m^{-6}n^3)^{-2}$

$\qquad = 4^3 \cdot (m^{-6})^3 \cdot 1^3 \cdot 3^{-2} \cdot (m^{-6})^{-2} \cdot (n^3)^{-2}$

$\qquad = 64 \cdot m^{-6 \cdot 3} \cdot 1 \cdot \dfrac{1}{3^2} \cdot m^{-6 \cdot (-2)} \cdot n^{3 \cdot (-2)}$

$\qquad = 64 \cdot m^{-18} \cdot 1 \cdot \dfrac{1}{9} \cdot m^{12} \cdot n^{-6}$

$\qquad = \dfrac{64}{9} \cdot m^{-18} \cdot m^{12} \cdot n^{-6}$

$\qquad = \dfrac{64}{9} m^{-6} n^{-6}$

$\qquad = \dfrac{64}{9m^6 n^6}$

91. $\left(\dfrac{3rs^{-1}}{4s^2}\right)^{-2} \cdot (2r^{-6}t^0)^{-1}$

$\qquad = \left(\dfrac{3}{4} r \cdot s^{-1-2}\right)^{-2} \cdot (2r^{-6} \cdot 1)^{-1}$

$\qquad = \left(\dfrac{3}{4} r \cdot s^{-3}\right)^{-2} \cdot (2r^{-6})^{-1}$

$\qquad = \left(\dfrac{3}{4}\right)^{-2} \cdot r^{-2} \cdot (s^{-3})^{-2} \cdot 2^{-1} \cdot (r^{-6})^{-1}$

$\qquad = \left(\dfrac{4}{3}\right)^2 \cdot r^{-2} \cdot s^{-3 \cdot (-2)} \cdot \dfrac{1}{2} \cdot r^{-6 \cdot (-1)}$

$\qquad = \dfrac{16}{9} \cdot r^{-2} \cdot s^6 \cdot \dfrac{1}{2} \cdot r^6$

$\qquad = \dfrac{16}{9} \cdot \dfrac{1}{2} \cdot r^{-2} \cdot r^6 \cdot s^6$

$\qquad = \dfrac{8}{9} \cdot r^4 \cdot s^6$

$\qquad = \dfrac{8r^4 s^6}{9}$

92. $(6r^4s^{-3})^2 \cdot \left(\dfrac{3r^4s}{2r^{-2}s^{-2}} \right)^{-3}$

$= (6r^4s^{-3})^2 \cdot \left(\dfrac{3}{2} r^{4-(-2)} s^{1-(-2)} \right)^{-3}$

$= (6r^4s^{-3})^2 \cdot \left(\dfrac{3}{2} r^6 s^3 \right)^{-3}$

$= 6^2 \cdot (r^4)^2 \cdot (s^{-3})^2 \cdot \left(\dfrac{3}{2} \right)^{-3} \cdot (r^6)^{-3} \cdot (s^3)^{-3}$

$= 36 \cdot r^{4 \cdot 2} \cdot s^{-3 \cdot 2} \cdot \left(\dfrac{2}{3} \right)^3 \cdot r^{6 \cdot (-3)} \cdot s^{3 \cdot (-3)}$

$= 36 \cdot r^8 \cdot s^{-6} \cdot \dfrac{8}{27} \cdot r^{-18} \cdot s^{-9}$

$= 36 \cdot \dfrac{8}{27} \cdot r^8 \cdot r^{-18} \cdot s^{-6} \cdot s^{-9}$

$= \dfrac{32}{3} \cdot r^{-10} \cdot s^{-15}$

$= \dfrac{32}{3r^{10}s^{15}}$

93. $\dfrac{36x^7 - 24x^6 + 30x^2}{6x^2}$

$= \dfrac{36x^7}{6x^2} - \dfrac{24x^6}{6x^2} + \dfrac{30x^2}{6x^2}$

$= \dfrac{36}{6} x^{7-2} - \dfrac{24}{6} x^{6-2} + \dfrac{30}{6} x^{2-2}$

$= 6x^5 - 4x^4 + 5x^0$

$= 6x^5 - 4x^4 + 5$

94. $\dfrac{15x^5 + 25x^3 - 30x^2}{5x}$

$= \dfrac{15x^5}{5x} + \dfrac{25x^3}{5x} - \dfrac{30x^2}{5x}$

$= \dfrac{15}{5} x^{5-1} + \dfrac{25}{5} x^{3-1} - \dfrac{30}{5} x^{2-1}$

$= 3x^4 + 5x^2 - 6x^1$

$= 3x^4 + 5x^2 - 6x$

95. $\dfrac{16n^8 + 4n^5 - 10n}{4n^5}$

$= \dfrac{16n^8}{4n^5} + \dfrac{4n^5}{4n^5} - \dfrac{10n}{4n^5}$

$= \dfrac{16}{4} n^{8-5} + \dfrac{4}{4} n^{5-5} - \dfrac{10}{4} n^{1-5}$

$= 4n^3 + n^0 - \dfrac{5}{2} n^{-4}$

$= 4n^3 + 1 - \dfrac{5}{2n^4}$

96. $\dfrac{30n^6 - 20n^5 - 16n^3}{5n^5}$

$= \dfrac{30n^6}{5n^5} - \dfrac{20n^5}{5n^5} - \dfrac{16n^3}{5n^5}$

$= \dfrac{30}{5} n^{6-5} - \dfrac{20}{5} n^{5-5} - \dfrac{16}{5} n^{3-5}$

$= 6n^1 - 4n^0 - \dfrac{16}{5} n^{-2}$

$= 6n - 4 - \dfrac{16}{5n^2}$

97. $\dfrac{2p^8 + 4p^5 - 8p^3}{-16p^5}$

$= \dfrac{2p^8}{-16p^5} + \dfrac{4p^5}{-16p^5} - \dfrac{8p^3}{-16p^5}$

$= -\dfrac{2}{16} p^{8-5} - \dfrac{4}{16} p^{5-5} + \dfrac{8}{16} p^{3-5}$

$= -\dfrac{1}{8} p^3 - \dfrac{1}{4} p^0 + \dfrac{1}{2} p^{-2}$

$= -\dfrac{p^3}{8} - \dfrac{1}{4} + \dfrac{1}{2p^2}$

98. $\dfrac{3p^4 - 6p^2 + 9}{-6p^2}$

$= \dfrac{3p^4}{-6p^2} - \dfrac{6p^2}{-6p^2} + \dfrac{9}{-6p^2}$

$= -\dfrac{3}{6} p^{4-2} + \dfrac{6}{6} p^{2-2} - \dfrac{9}{6} p^{-2}$

$= -\dfrac{1}{2} p^2 + p^0 - \dfrac{3}{2} p^{-2}$

$= -\dfrac{p^2}{2} + 1 - \dfrac{3}{2p^2}$

99.

$$2x+3 \overline{\smash{\big)}\ 8x^2 - 2x - 21} \quad \begin{array}{c} 4x-7 \end{array}$$

$$\begin{array}{r} -(8x^2+12x) \\ \hline -14x-21 \\ -(-14x-21) \\ \hline 0 \end{array}$$

$$\frac{8x^2-2x-21}{2x+3} = 4x-7$$

100.

$$3x-1 \overline{\smash{\big)}\ 3x^2+17x-6} \quad \begin{array}{c} x+6 \end{array}$$

$$\begin{array}{r} -(3x^2 - x) \\ \hline 18x-6 \\ -(18x-6) \\ \hline 0 \end{array}$$

$$\frac{3x^2+17x-6}{3x-1} = x+6$$

101. $\dfrac{6x^2+x^3-2x+1}{x-1} = \dfrac{x^3+6x^2-2x+1}{x-1}$

$$x-1 \overline{\smash{\big)}\ x^3+6x^2-2x+1} \quad \begin{array}{c} x^2+7x+5 \end{array}$$

$$\begin{array}{r} -(x^3 - x^2) \\ \hline 7x^2-2x \\ -(7x^2-7x) \\ \hline 5x+1 \\ -(5x-5) \\ \hline 6 \end{array}$$

$$\frac{6x^2+x^3-2x+1}{x-1} = x^2+7x+5+\frac{6}{x-1}$$

102. $\dfrac{-6x+2x^3-7x^2+8}{x-2} = \dfrac{2x^3-7x^2-6x+8}{x-2}$

$$x-2 \overline{\smash{\big)}\ 2x^3-7x^2-6x+8} \quad \begin{array}{c} 2x^2-3x-12 \end{array}$$

$$\begin{array}{r} -(2x^3-4x^2) \\ \hline -3x^2-6x \\ -(-3x^2+6x) \\ \hline -12x+8 \\ -(-12x+24) \\ \hline -16 \end{array}$$

$$\frac{-6x+2x^3-7x^2+8}{x-2} = 2x^2-3x-12+\frac{-16}{x-2}$$

103. $\dfrac{x^3+8}{x+2} = \dfrac{x^3+0x^2+0x+8}{x+2}$

$$x+2 \overline{\smash{\big)}\ x^3+0x^2+0x+8} \quad \begin{array}{c} x^2-2x+4 \end{array}$$

$$\begin{array}{r} -(x^3+2x^2) \\ \hline -2x^2+0x \\ -(-2x^2-4x) \\ \hline 4x+8 \\ -(4x+8) \\ \hline 0 \end{array}$$

$$\frac{x^3+8}{x+2} = x^2-2x+4$$

104. $\dfrac{3x^3+2x-7}{x-5} = \dfrac{3x^3+0x^2+2x-7}{x-5}$

$$x-5 \overline{\smash{\big)}\ 3x^3+0x^2+2x-7} \quad \begin{array}{c} 3x^2+15x+77 \end{array}$$

$$\begin{array}{r} -(3x^3-15x^2) \\ \hline 15x^2+2x \\ -(15x^2-75x) \\ \hline 77x-7 \\ -(77x-385) \\ \hline 378 \end{array}$$

$$\frac{3x^3+2x-7}{x-5} = 3x^2+15x+77+\frac{378}{x-5}$$

105. $27,000,000 = 2.7 \times 10,000,000 = 2.7 \times 10^7$

106. $1,230,000,000 = 1.23 \times 1,000,000,000$
$$= 1.23 \times 10^9$$

107. $0.00006 = 6 \times 0.00001 = 6 \times 10^{-5}$

108. $0.00000305 = 3.05 \times 0.000001 = 3.05 \times 10^{-6}$

109. $3 = 3 \times 1 = 3 \times 10^0$

110. $8 = 8 \times 1 = 8 \times 10^0$

111. $6 \times 10^{-4} = 6 \times 0.0001 = 0.0006$

112. $1.25 \times 10^{-3} = 1.25 \times 0.001 = 0.00125$

113. $6.13 \times 10^5 = 6.13 \times 100,000 = 613,000$

114. $8 \times 10^4 = 8 \times 10,000 = 80,000$

115. $3.7 \times 10^{-1} = 3.7 \times 0.1 = 0.37$

116. $5.4 \times 10^7 = 5.4 \times 10,000,000$
$= 54,000,000$

117. $(1.2 \times 10^{-5})(5 \times 10^8) = (1.2 \cdot 5) \times (10^{-5} \cdot 10^8)$
$= 6 \times 10^3$

118. $(1.4 \times 10^{-10})(3 \times 10^2)$
$= (1.4 \cdot 3) \times (10^{-10} \cdot 10^2)$
$= 4.2 \times 10^{-8}$

119. $\dfrac{2.4 \times 10^{-6}}{1.2 \times 10^{-8}} = \dfrac{2.4}{1.2} \times 10^{-6-(-8)} = 2 \times 10^2$

120. $\dfrac{5 \times 10^6}{25 \times 10^{-3}} = \dfrac{5}{25} \times 10^{6-(-3)}$
$= 0.2 \times 10^9$
$= (2 \times 10^{-1}) \times 10^9$
$= 2 \times 10^8$

121. $\dfrac{200,000 \times 4,000,000}{0.0002} = \dfrac{(2 \times 10^5) \cdot (4 \times 10^6)}{2 \times 10^{-4}}$
$= \dfrac{(2 \cdot 4) \times 10^{5+6}}{2 \times 10^{-4}}$
$= \dfrac{8 \times 10^{11}}{2 \times 10^{-4}}$
$= \dfrac{8}{2} \times 10^{11-(-4)}$
$= 4 \times 10^{15}$

122. $\dfrac{1,200,000}{0.003 \times 2,000,000} = \dfrac{1.2 \times 10^6}{(3 \times 10^{-3}) \cdot (2 \times 10^6)}$
$= \dfrac{1.2 \times 10^6}{(3 \cdot 2) \times 10^{-3+6}}$
$= \dfrac{1.2 \times 10^6}{6 \times 10^3}$
$= \dfrac{1.2}{6} \times 10^{6-3}$
$= 0.2 \times 10^3$
$= (2 \times 10^{-1}) \times 10^3$
$= 2 \times 10^2$

Chapter 5 Test

1. Yes, $6x^5 - 2x^4$ is a binomial of degree 5.

2. $3x^2 - 2x + 5$

   (a) $x = 0$: $3(0)^2 - 2(0) + 5 = 0 - 0 + 5 = 5$

   (b) $x = -2$: $3(-2)^2 - 2(-2) + 5 = 12 + 4 + 5 = 21$

   (c) $x = 3$: $3(3)^2 - 2(3) + 5 = 27 - 6 + 5 = 26$

3. $(3x^2 y^2 - 2x + 3y) + (-4x - 6y + 4x^2 y^2)$
$= 3x^2 y^2 - 2x + 3y - 4x - 6y + 4x^2 y^2$
$= 3x^2 y^2 + 4x^2 y^2 - 2x - 4x + 3y - 6y$
$= 7x^2 y^2 - 6x - 3y$

4. $(8m^3 + 6m^2 - 4) - (5m^2 - 2m^3 + 2)$
$= (8m^3 + 6m^2 - 4) + (-5m^2 + 2m^3 - 2)$
$= 8m^3 + 6m^2 - 4 - 5m^2 + 2m^3 - 2$
$= 8m^3 + 2m^3 + 6m^2 - 5m^2 - 4 - 2$
$= 10m^3 + m^2 - 6$

5. $-3x^3(2x^2 - 6x + 5)$
$= -3x^3(2x^2) + (-3x^3)(-6x) + (-3x^3)(5)$
$= -6x^5 + 18x^4 - 15x^3$

6. $(x - 5)(2x + 7)$
$= (x)(2x) + (x)(7) + (-5)(2x) + (-5)(7)$
$= 2x^2 + 7x - 10x - 35$
$= 2x^2 - 3x - 35$

7. $(2x - 7)^2 = (2x)^2 - 2(2x)(7) + (7)^2$
$= 4x^2 - 28x + 49$

8. $(4x - 3y)(4x + 3y) = (4x)^2 - (3y)^2$
$= 16x^2 - 9y^2$

9. $(3x-1)(2x^2+x-8) = (3x-1)(2x^2)+(3x-1)(x)+(3x-1)(-8)$

$\qquad\qquad\qquad\qquad = 3x\cdot 2x^2+(-1)\cdot 2x^2+3x\cdot x+(-1)\cdot x+3x\cdot(-8)+(-1)\cdot(-8)$

$\qquad\qquad\qquad\qquad = 6x^3-2x^2+3x^2-x-24x+8$

$\qquad\qquad\qquad\qquad = 6x^3+x^2-25x+8$

10. $\dfrac{6x^4-8x^3+9}{3x^3} = \dfrac{6x^4}{3x^3}-\dfrac{8x^3}{3x^3}+\dfrac{9}{3x^3} = \dfrac{6}{3}x^{4-3}-\dfrac{8}{3}x^{3-3}+\dfrac{9}{3}x^{-3} = 2x^1-\dfrac{8}{3}x^0+3x^{-3} = 2x-\dfrac{8}{3}+\dfrac{3}{x^3}$

11. $\dfrac{3x^3-2x^2+5}{x+3} = \dfrac{3x^3-2x^2+0x+5}{x+3}$

$$
\begin{array}{r}
3x^2-11x+33 \\
x+3\overline{\smash{)}3x^3-2x^2+\ 0x+\ \ 5} \\
\underline{-(3x^3+9x^2)} \\
-11x^2+\ 0x \\
\underline{-(-11x^2-33x)} \\
33x+\ 5 \\
\underline{-(33x+99)} \\
-94
\end{array}
$$

$\dfrac{3x^3-2x^2+5}{x+3} = 3x^2-11x+33+\dfrac{-94}{x+3}$

12. $(4x^3y^2)(-3xy^4) = 4\cdot x^3\cdot y^2\cdot(-3)\cdot x\cdot y^4 = 4\cdot(-3)\cdot x^3\cdot x\cdot y^2\cdot y^4 = -12x^{3+1}y^{2+4} = -12x^4y^6$

13. $\dfrac{18m^5n}{27m^2n^6} = \dfrac{18}{27}\cdot m^{5-2}\cdot n^{1-6} = \dfrac{2}{3}m^3n^{-5} = \dfrac{2m^3}{n^5}$

14. $\left(\dfrac{m^{-2}n^0}{m^{-7}n^4}\right)^{-6} = (m^{-2-(-7)}n^{0-4})^{-6} = (m^5n^{-4})^{-6} = m^{5\cdot(-6)}n^{-4\cdot(-6)} = m^{-30}n^{24} = \dfrac{n^{24}}{m^{30}}$

15. $(4x^{-3}y)^{-2}(2x^4y^{-3})^4 = 4^{-2}\cdot(x^{-3})^{-2}\cdot y^{-2}\cdot 2^4\cdot(x^4)^4\cdot(y^{-3})^4$

$\qquad\qquad\qquad\qquad = \dfrac{1}{4^2}\cdot x^{-3\cdot(-2)}\cdot y^{-2}\cdot 16\cdot x^{4\cdot 4}\cdot y^{-3\cdot 4}$

$\qquad\qquad\qquad\qquad = \dfrac{1}{16}\cdot x^6\cdot y^{-2}\cdot 16\cdot x^{16}\cdot y^{-12}$

$\qquad\qquad\qquad\qquad = \dfrac{1}{16}\cdot 16\cdot x^6\cdot x^{16}\cdot y^{-2}\cdot y^{-12}$

$\qquad\qquad\qquad\qquad = x^{22}y^{-14}$

$\qquad\qquad\qquad\qquad = \dfrac{x^{22}}{y^{14}}$

16. $(2m^{-4}n^2)^{-1} \cdot \left(\dfrac{16m^0 n^{-3}}{m^{-3} n^2} \right)$

$= (2m^{-4}n^2)^{-1} \cdot (16m^{0-(-3)} n^{-3-2})$

$= (2m^{-4}n^2)^{-1} \cdot (16m^3 n^{-5})$

$= 2^{-1} \cdot (m^{-4})^{-1} \cdot (n^2)^{-1} \cdot 16 \cdot m^3 \cdot n^{-5}$

$= \dfrac{1}{2} \cdot m^{-4 \cdot (-1)} \cdot n^{2 \cdot (-1)} \cdot 16 \cdot m^3 \cdot n^{-5}$

$= \dfrac{1}{2} \cdot m^4 \cdot n^{-2} \cdot 16 \cdot m^3 \cdot n^{-5}$

$= \dfrac{1}{2} \cdot 16 \cdot m^4 \cdot m^3 \cdot n^{-2} \cdot n^{-5}$

$= 8m^7 n^{-7}$

$= \dfrac{8m^7}{n^7}$

17. $0.000012 = 1.2 \times 0.00001 = 1.2 \times 10^{-5}$

18. $2.101 \times 10^5 = 2.101 \times 100,000 = 210,100$

19. $(2.1 \times 10^{-6}) \cdot (1.7 \times 10^{10})$

$= (2.1 \cdot 1.7) \times (10^{-6} \cdot 10^{10})$

$= 3.57 \times 10^4$

20. $\dfrac{3 \times 10^{-4}}{15 \times 10^2} = \dfrac{3}{15} \times 10^{-4-2}$

$= 0.2 \times 10^{-6}$

$= (2 \times 10^{-1}) \times 10^{-6}$

$= 2 \times 10^{-7}$

Cumulative Review Chapters 1–5

1. $-\dfrac{4}{2} = 2, \sqrt{25} = 5$

(a) $\left\{ \sqrt{25} \right\}$

(b) $\left\{ 0, \sqrt{25} \right\}$

(c) $\left\{ -6, -\dfrac{4}{2}, 0, \sqrt{25} \right\}$

(d) $\left\{ -6, -\dfrac{4}{2}, 0, 1.4, \sqrt{25} \right\}$

(e) $\left\{ \sqrt{7} \right\}$

(f) $\left\{ -6, -\dfrac{4}{2}, 0, 1.4, \sqrt{7}, \sqrt{25} \right\}$

2. $-\dfrac{1}{2} + \dfrac{2}{3} \div 4 \cdot \dfrac{1}{3} = -\dfrac{1}{2} + \dfrac{2}{3} \cdot \dfrac{1}{4} \cdot \dfrac{1}{3}$

$= -\dfrac{1}{2} + \dfrac{1}{6} \cdot \dfrac{1}{3}$

$= -\dfrac{1}{2} + \dfrac{1}{18}$

$= -\dfrac{9}{18} + \dfrac{1}{18}$

$= -\dfrac{8}{18}$

$= -\dfrac{4}{9}$

3. $2 + 3[3 + 10(-1)] = 2 + 3[3 + (-10)]$

$= 2 + 3[-7]$

$= 2 + (-21)$

$= -19$

4. $6x^3 - (-2x^2 + 3x) + 3x^2$

$= 6x^3 + (2x^2 - 3x) + 3x^2$

$= 6x^3 + 2x^2 - 3x + 3x^2$

$= 6x^3 + 2x^2 + 3x^2 - 3x$

$= 6x^3 + 5x^2 - 3x$

5. $-4(6x - 1) + 2(3x + 2)$

$= -4(6x) + (-4)(-1) + 2(3x) + 2(2)$

$= -24x + 4 + 6x + 4$

$= -24x + 6x + 4 + 4$

$= -18x + 8$

6. $\qquad -2(3x - 4) + 6 = 4x - 6x + 10$

$-2(3x) + (-2)(-4) + 6 = -2x + 10$

$\qquad -6x + 8 + 6 = -2x + 10$

$\qquad -6x + 14 = -2x + 10$

$-6x + 2x + 14 = -2x + 2x + 10$

$\qquad -4x + 14 = 10$

$\qquad -4x + 14 - 14 = 10 - 14$

$\qquad -4x = -4$

$\qquad \dfrac{-4x}{-4} = \dfrac{-4}{-4}$

$\qquad x = 1$

The solution set is $\{1\}$.

7. $4(x - 5) = 10 + 2x$

8. Let x be Kathy's earnings before the 3% raise.
Then the amount of the raise is $0.03x$.
$$x + 0.3x = 659.20$$
$$1.03x = 659.20$$
$$\frac{1.03x}{1.03} = \frac{659.20}{1.03}$$
$$x = 640$$
Her earnings before the raise were $640 per month.

9. Let x be Amber's speed. Then $x + 12$ is
Cheyenne's speed. In 3 hours, Amber drives $3x$
miles and Cheyenne drives $3(x + 12)$ miles.
$$3x + 3(x+12) = 306$$
$$3x + 3x + 36 = 306$$
$$6x + 36 = 306$$
$$6x + 36 - 36 = 306 - 36$$
$$6x = 270$$
$$\frac{6x}{6} = \frac{270}{6}$$
$$x = 45$$
Amber's speed is 45 miles per hour.

10.
$$-5x + 2 > 17$$
$$-5x + 2 - 2 > 17 - 2$$
$$-5x > 15$$
$$\frac{-5x}{-5} < \frac{15}{-5}$$
$$x < -3$$
$$\{x | x < -3\}$$
$$(-\infty, -3)$$

11. $\dfrac{x^2 - y^2}{z}$ when $x = 3$, $y = -2$, and $z = -10$:
$$\frac{(3)^2 - (-2)^2}{-10} = \frac{9 - 4}{-10} = \frac{5}{-10} = -\frac{1}{2}$$

12. Let $(x_1, y_1) = (-3, 8)$ and $(x_2, y_2) = (1, 2)$.
$$m = \frac{y_2 - y_1}{x_2 - x_1} = \frac{2 - 8}{1 - (-3)} = \frac{-6}{4} = -\frac{3}{2}$$

13. $2x + 3y = 24$
Let $x = 0$.
$$2(0) + 3y = 24$$
$$3y = 24$$
$$y = 8$$
The y-intercept is 8 or $(0, 8)$.
Let $y = 0$.
$$2x + 3(0) = 24$$
$$2x = 24$$
$$x = 12$$
The x-intercept is 12 or $(12, 0)$.

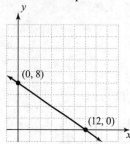

14. $y = -3x + 8$
The y-intercept is 8. Plot $(0, 8)$. The slope is
$-3 = \dfrac{-3}{1}$. Move 3 units down and 1 unit to the
right to plot another point.

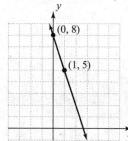

15. $\begin{cases} 2x - 3y = 27 & (1) \\ -4x + 2y = -27 & (2) \end{cases}$
Multiply (1) by 2. Then add.
$$\begin{cases} 4x - 6y = 54 \\ -4x + 2y = -27 \end{cases}$$
$$-4y = 27$$
$$y = \frac{27}{-4} = -\frac{27}{4}$$
Let $y = -\dfrac{27}{4}$ in (1).

$$2x - 3\left(-\frac{27}{4}\right) = 27$$

$$2x + \frac{81}{4} = 27$$

$$2x = \frac{108}{4} - \frac{81}{4}$$

$$2x = \frac{27}{4}$$

$$x = \frac{27}{8}$$

The solution is $\left(\dfrac{27}{8}, -\dfrac{27}{4}\right)$.

16. $\begin{cases} 3x - 2y = 8 & (1) \\ -6x + 4y = 8 & (2) \end{cases}$

Multiply (1) by 2. Then add.

$\begin{cases} 6x - 4y = 16 \\ -6x + 4y = 8 \end{cases}$

$ 0 = 24 \quad$ False

The system is inconsistent. There is no solution.

17. $(4x^2 + 6x) - (-x + 5x^2) + (6x^3 - 2x^2) = (4x^2 + 6x) + (x - 5x^2) + (6x^3 - 2x^2)$

$$= 4x^2 + 6x + x - 5x^2 + 6x^3 - 2x^2$$

$$= 6x^3 + 4x^2 - 5x^2 - 2x^2 + 6x + x$$

$$= 6x^3 - 3x^2 + 7x$$

18. $(4m - 3)(7m + 2) = (4m)(7m) + (4m)(2) + (-3)(7m) + (-3)(2) = 28m^2 + 8m - 21m - 6 = 28m^2 - 13m - 6$

19. $(3m - 2n)(3m + 2n) = (3m)^2 - (2n)^2 = 9m^2 - 4n^2$

20. $(7x + y)^2 = (7x)^2 + 2(7x)(y) + (y)^2 = 49x^2 + 14xy + y^2$

21. $(2m + 5)(2m^2 - 5m + 3) = (2m + 5)(2m^2) + (2m + 5)(-5m) + (2m + 5)(3)$

$$= 2m(2m^2) + 5(2m^2) + 2m(-5m) + 5(-5m) + 2m(3) + 5(3)$$

$$= 4m^3 + 10m^2 - 10m^2 - 25m + 6m + 15$$

$$= 4m^3 - 19m + 15$$

22. $\dfrac{14xy^2 + 7x^2 y}{7x^2 y^2} = \dfrac{14xy^2}{7x^2 y^2} + \dfrac{7x^2 y}{7x^2 y^2} = \dfrac{14}{7} x^{1-2} y^{2-2} + \dfrac{7}{7} x^{2-2} y^{1-2} = 2x^{-1} y^0 + 1x^0 y^{-1} = \dfrac{2}{x} + \dfrac{1}{y}$

23. $\dfrac{x^3+27}{x+3}=\dfrac{x^3+0x^2+0x+27}{x+3}$

$$\begin{array}{r} x^2-3x+9 \\ x+3\overline{\smash{\big)}\,x^3+0x^2+0x+27} \end{array}$$

$$\begin{array}{r} -(x^3+3x^2) \\ \hline -3x^2+0x \\ -(-3x^2-9x) \\ \hline 9x+27 \\ -(9x+27) \\ \hline 0 \end{array}$$

$\dfrac{x^3+27}{x+3}=x^2-3x+9$

24. $(4m^0n^3)(-6n)=4\cdot m^0\cdot n^3\cdot(-6)\cdot n$

$\qquad\qquad = 4\cdot(-6)\cdot m^0\cdot n^3\cdot n$

$\qquad\qquad = -24\cdot 1\cdot n^{3+1}$

$\qquad\qquad = -24n^4$

25. $\dfrac{25m^{-6}n^{-2}}{-10m^{-4}n^{-10}}=\dfrac{25}{-10}\cdot m^{-6-(-4)}\cdot n^{-2-(-10)}$

$\qquad\qquad = -\dfrac{5}{2}\cdot m^{-2}\cdot n^8$

$\qquad\qquad = -\dfrac{5n^8}{2m^2}$

26. $\left(\dfrac{2xy^4}{z^{-2}}\right)^{-6}=\left(\dfrac{z^{-2}}{2xy^4}\right)^6$

$\qquad\qquad = \dfrac{(z^{-2})^6}{2^6x^6(y^4)^6}$

$\qquad\qquad = \dfrac{z^{-2\cdot 6}}{64x^6y^{4\cdot 6}}$

$\qquad\qquad = \dfrac{z^{-12}}{64x^6y^{24}}$

$\qquad\qquad = \dfrac{1}{64x^6y^{24}z^{12}}$

27. $(x^4y^{-2})^{-4}\cdot\left(\dfrac{6x^{-4}y^3}{3y^{-8}}\right)^{-1}$

$\quad = (x^4y^{-2})^{-4}\cdot\left(\dfrac{3y^{-8}}{6x^{-4}y^3}\right)$

$\quad = (x^4)^{-4}(y^{-2})^{-4}\cdot\left(\dfrac{3}{6}\cdot x^4\cdot y^{-8-3}\right)$

$\quad = x^{4\cdot(-4)}\cdot y^{-2\cdot(-4)}\cdot\dfrac{1}{2}\cdot x^4\cdot y^{-11}$

$\quad = x^{-16}\cdot y^8\cdot\dfrac{1}{2}\cdot x^4\cdot y^{-11}$

$\quad = \dfrac{1}{2}\cdot x^{-16}\cdot x^4\cdot y^8\cdot y^{-11}$

$\quad = \dfrac{1}{2}x^{-12}y^{-3}$

$\quad = \dfrac{1}{2x^{12}y^3}$

28. $0.0000605=6.05\times 0.00001$

$\qquad\qquad = 6.05\times 10^{-5}$

29. $2.175\times 10^6=2.175\times 1,000,000$

$\qquad\qquad = 2,175,000$

30. $(3.4\times 10^8)(2.1\times 10^{-3})$

$\quad = (3.4\cdot 2.1)\times(10^8\cdot 10^{-3})$

$\quad = 7.14\times 10^5$

Chapter 6

Section 6.1

Preparing for Greatest Common Factor and Factoring by Grouping

P1. $48 = 2 \cdot 24$
$= 2 \cdot 2 \cdot 12$
$= 2 \cdot 2 \cdot 2 \cdot 6$
$= 2 \cdot 2 \cdot 2 \cdot 2 \cdot 3$
$= 2^4 \cdot 3$

P2. $2(5x - 3) = 2 \cdot 5x - 2 \cdot 3 = 10x - 6$

P3. $(2x + 5)(x - 3)$
$= 2x \cdot x + 2x \cdot (-3) + 5 \cdot x + 5 \cdot (-3)$
$= 2x^2 - 6x + 5x - 15$
$= 2x^2 - x - 15$

6.1 Quick Checks

1. The largest expression that divides evenly into a set of numbers is called the <u>greatest common factor</u>.

2. In the product $(3x - 2)(x + 4) = 3x^2 + 10x - 8$, the polynomials $(3x - 2)$ and $(x + 4)$ are called <u>factors</u> of the polynomial $3x^2 + 10x - 8$.

3. $32 = 8 \cdot 4 = 2 \cdot 2 \cdot 2 \cdot 2 \cdot 2$
$40 = 8 \cdot 5 = 2 \cdot 2 \cdot 2 \qquad \cdot 5$
$\text{GCF} = 2 \cdot 2 \cdot 2 = 2^3 = 8$

4. $12 = 4 \cdot 3 = 2 \cdot 2 \cdot 3$
$45 = 3 \cdot 15 = \qquad 3 \cdot 3 \cdot 5$
$\text{GCF} = 3$

5. $21 = 3 \cdot 7$
$35 = 5 \cdot 7$
$84 = 21 \cdot 4 = 3 \cdot 7 \cdot 2 \cdot 2$
$\text{GCF} = 7$

6. $14 = 2 \cdot 7$
$35 = 5 \cdot 7$
The GCF of the coefficients is 7.
The GCF of y^3 and y^2 is y^2.
The GCF of $14y^3$ and $35y^2$ is $7y^2$.

7. $6 = 2 \cdot 3$
$8 = 2 \cdot 2 \cdot 2$
$12 = 2 \cdot 2 \cdot 3$
The GCF of the coefficients is 2.
The GCF of z^3, z^2, and z is z.
The GCF of $6z^3$, $8z^2$, and $12z$ is $2z$.

8. $4 = 2 \cdot 2$
$8 = 4 \cdot 2 = 2 \cdot 2 \cdot 2$
$24 = 4 \cdot 6 = 2 \cdot 2 \cdot 2 \cdot 3$
The GCF of the coefficients is $2 \cdot 2 = 2^2 = 4$.

The GCF of x^3, x^2, and x is x.
The GCF of y^5, y^3, and y^4 is y^3.
The GCF of $4x^3y^5$, $8x^2y^3$, and $24xy^4$ is $4xy^3$.

9. There is no common factor between the coefficients. Since each expression has $2x + 3$ as a factor, the GCF of $7(2x + 3)$ and $-4(2x + 3)$ is $2x + 3$.

10. The GCF between 9 and 12 is 3. The GCF between $(k + 8)(3k - 2)$ and $(k - 1)(k + 8)^2$ is $k + 8$. The GCF of the expressions $9(k + 8)(3k - 2)$ and $12(k - 1)(k + 8)^2$ is $3(k + 8)$.

11. To <u>factor</u> a polynomial means to write the polynomial as a product of two or more polynomials.

12. When we factor a polynomial using the GCF, we use the <u>Distributive</u> Property in reverse.

13. The GCF of 5 and 30 is 5.
The GCF of z^2 and z is z.
The GCF of $5z^2 + 30z$ is $5z$.
$5z^2 + 30z = 5z(z) + 5z(6) = 5z(z + 6)$

14. The GCF of 12 and 12 is 12.
The GCF of p^2 and p is p.
The GCF of $12p^2 - 12p$ is $12p$.
$12p^2 - 12p = 12p(p) - 12p(1) = 12p(p - 1)$

15. The GCF of 16, 12, and 4 is 4.

The GCF of y^3, y^2, and y is y.

The GCF of $16y^3 - 12y^2 + 4y$ is $4y$.

$16y^3 - 12y^2 + 4y$
$= 4y(4y^2) - 4y(3y) + 4y(1)$
$= 4y(4y^2 - 3y + 1)$

16. The GCF of 6, 18, and 22 is 2.

The GCF of m^4, m^3, and m^2 is m^2.

The GCF of n^2, n^4, and n^5 is n^2.

The GCF of $6m^4n^2 + 18m^3n^4 - 22m^2n^5$ is $2m^2n^2$.

$6m^4n^2 + 18m^3n^4 - 22m^2n^5$
$= 2m^2n^2(3m^2) + 2m^2n^2(9mn^2) - 2m^2n^2(11n^3)$
$= 2m^2n^2(3m^2 + 9mn^2 - 11n^3)$

17. Use $-4y$ as the GCF.

$-4y^2 + 8y = -4y(y) + (-4y)(-2) = -4y(y - 2)$

18. Use $-3a$ as the GCF.

$-6a^3 + 12a^2 - 3a$
$= -3a(2a^2) + (-3a)(-4a) + (-3a)(1)$
$= -3a(2a^2 - 4a + 1)$

19. The GCF is $(2x + 1)$.
$(2x+1)(x-3) + (2x+1)(2x+7)$
$= (2x+1)(x-3+2x+7)$
$= (2x+1)(3x+4)$

$(2x+1)(3x+4) \neq (2x+1)^2(3x+4)$
False

20. The GCF is the binomial $a - 5$.
$2a(a-5) + 3(a-5) = (a-5)(2a+3)$

21. The GCF is the binomial $z + 5$.
$7z(z+5) - 4(z+5) = (z+5)(7z-4)$

22. $4x + 4y + bx + by = (4x+4y) + (bx+by)$
$= 4(x+y) + b(x+y)$
$= (x+y)(4+b)$

23. $6az - 2a - 9bz + 3b = (6az - 2a) + (-9bz + 3b)$
$= 2a(3z-1) + (-3b)(3z-1)$
$= (3z-1)(2a-3b)$

24. $8b + 4 - 10ab - 5a = (8b+4) + (-10ab - 5a)$
$= 4(2b+1) + (-5a)(2b+1)$
$= (2b+1)(4-5a)$

25. The GCF of $3z^3 + 12z^2 + 6z + 24$ is 3.
$3z^3 + 12z^2 + 6z + 24 = 3(z^3 + 4z^2 + 2z + 8)$
$= 3[(z^3 + 4z^2) + (2z + 8)]$
$= 3[z^2(z+4) + 2(z+4)]$
$= 3(z+4)(z^2+2)$

26. The GCF of $2n^4 + 2n^3 - 4n^2 - 4n$ is $2n$.
$2n^4 + 2n^3 - 4n^2 - 4n$
$= 2n(n^3 + n^2 - 2n - 2)$
$= 2n[(n^3 + n^2) + (-2n - 2)]$
$= 2n[n^2(n+1) + (-2)(n+1)]$
$= 2n(n+1)(n^2-2)$

6.1 Exercises

27. $8 = 2 \cdot 2 \cdot 2$
$6 = 2 \cdot \quad\quad 3$
The only common factor is 2. The GCF is 2.

29. $15 = 3 \cdot 5$
$14 = \quad\quad 2 \cdot 7$
There are no common prime factors. The GCF is 1.

31. $12 = 2 \cdot 2 \cdot 3$
$28 = 2 \cdot 2 \cdot \quad\quad 7$
$48 = 2 \cdot 2 \cdot 3 \cdot \quad 2 \cdot 2$
Because all three numbers contain two factors of 2, the GCF is $2 \cdot 2 = 4$.

33. $x^{10} = x \cdot x \cdot x \cdot x \cdot x \cdot x \cdot x \cdot x \cdot x \cdot x$
$x^2 = x \cdot x$
$x^8 = x \cdot x \cdot x \cdot x \cdot x \cdot x \cdot x \cdot x$
Each expression contains two factors of x, so the GCF is x^2.

35. The GCF of 7 and 14 is 7.
The GCF of x and x^3 is x.
The GCF is $7x$.

37. The GCF of 45 and 75 is 15.

The GCF of a^2 and a is a.

The GCF of b^3 and b^2 is b^2.

The GCF is $15ab^2$.

39. The GCF of 4, 6, and 8 is 2.

The GCF of a^2, a, and a^2 is a.

The GCF of b, b^2, and b^2 is b.

The GCF of c^3, c^2, and c^4 is c^2.

The GCF is $2abc^2$.

41. $3(x-1) = \quad 3 \cdot (x-1)$
$6(x+1) = 2 \cdot 3 \cdot (x+1)$
The GCF is 3.

43. $2(x-4)^2 = \quad 2 \cdot (x-4)^2$
$4(x-4)^3 = 2 \cdot 2 \cdot (x-4)^3$

The GCF of the variable expression is $(x-4)^2$. The GCF is $2(x-4)^2$.

45. $12(x+2)(x-3)^2 = 2 \cdot 2 \cdot 3 \cdot \quad (x+2)(x-3)^2$
$18(x-3)^2(x-2) = 2 \cdot \quad 3 \cdot 3 \cdot \quad \quad (x-3)^2(x-2)$

The GCF of the variable expression is $(x-3)^2$. The GCF is $2 \cdot 3 \cdot (x-3)^2 = 6(x-3)^2$.

47. The GCF is 6.
$12x - 18 = 6(2x) - 6(3) = 6(2x-3)$

49. The GCF is $-x$.
$-3x^2 + 12x = -3x(x) + (-3x)(-4) = -3x(x-4)$

51. The GCF is $5x^2y$.
$5x^2y - 15x^3y^2 = 5x^2y(1) - 5x^2y(3xy) = 5x^2y(1-3xy)$

53. The GCF is $3x$.
$3x^3 + 6x^2 - 3x = 3x(x^2) + 3x(2x) - 3x(1) = 3x(x^2 + 2x - 1)$

55. The GCF is $-5x$.
$-5x^3 + 10x^2 - 15x$
$= -5x(x^2) + (-5x)(-2x) + (-5x)(3)$
$= -5x(x^2 - 2x + 3)$

57. The GCF is 3.
$9m^5 - 18m^3 - 12m^2 + 81 = 3(3m^5) - 3(6m^3) - 3(4m^2) + 3(27) = 3(3m^5 - 6m^3 - 4m^2 + 27)$

59. The GCF is $-4z$.

$-12z^3 + 16z^2 - 8z$
$= -4z(3z^2) + (-4z)(-4z) + (-4z)(2)$
$= -4z(3z^2 - 4z + 2)$

61. The GCF is -5.

$10 - 5b - 15b^3 = -15b^3 - 5b + 10$
$\qquad = -5(3b^3) + (-5)b + (-5)(-2)$
$\qquad = -5(3b^3 + b - 2)$

63. The GCF is $15ab^2$.

$15a^2b^4 - 60ab^3 + 45a^3b^2$
$= 15ab^2(ab^2) - 15ab^2(4b) + 15ab^2(3a^2)$
$= 15ab^2(ab^2 - 4b + 3a^2)$

65. The GCF is $x - 3$.
$(x-3)x - (x-3)5 = (x-3)(x-5)$

67. The GCF is $x - 1$.
$x^2(x-1) + y^2(x-1) = (x-1)(x^2 + y^2)$

69. The GCF is $4x + 1$.
$x^2(4x+1) + 2x(4x+1) + 5(4x+1)$
$= (4x+1)(x^2 + 2x + 5)$

71. $xy + 3y + 4x + 12 = (xy + 3y) + (4x + 12)$
$\qquad = y(x+3) + 4(x+3)$
$\qquad = (x+3)(y+4)$

73. $yz + z - y - 1 = (yz + z) + (-y - 1)$
$\qquad = z(y+1) + (-1)(y+1)$
$\qquad = (y+1)(z-1)$

75. $x^3 - x^2 + 2x - 2 = (x^3 - x^2) + (2x - 2)$
$\qquad = x^2(x-1) + 2(x-1)$
$\qquad = (x-1)(x^2 + 2)$

77. $2t^3 - t^2 - 4t + 2 = (2t^3 - t^2) + (-4t + 2)$
$\qquad = t^2(2t-1) + (-2)(2t-1)$
$\qquad = (2t-1)(t^2 - 2)$

79. $2t^4 - t^3 - 6t + 3 = (2t^4 - t^3) + (-6t + 3)$
$\qquad = t^3(2t-1) + (-3)(2t-1)$
$\qquad = (2t-1)(t^3 - 3)$

81. $4y - 20 = 4(y) - 4(5) = 4(y - 5)$

83. $28m^3 + 7m^2 + 63m$
$= 7m(4m^2) + 7m(m) + 7m(9)$
$= 7m(4m^2 + m + 9)$

85. $12m^3n^2p - 18m^2n$
$= 6m^2n(2mnp) - 6m^2n(3)$
$= 6m^2n(2mnp - 3)$

87. $(2p-1)(p+3) + (7p+4)(p+3)$
$= (p+3)[(2p-1) + (7p+4)]$
$= (p+3)(9p+3)$
$= (p+3)[3(3p+1)]$
$= 3(p+3)(3p+1)$

89. $18ax - 9ay - 12bx + 6by$
$= 3(6ax - 3ay - 4bx + 2by)$
$= 3[(6ax - 3ay) + (-4bx + 2by)]$
$= 3[3a(2x - y) + (-2b)(2x - y)]$
$= 3(2x - y)(3a - 2b)$

91. $(x-2)(x-3) + (x-2)$
$= (x-2)[(x-3) + 1]$
$= (x-2)(x-2)$ or $(x-2)^2$

93. $15x^4 - 6x^3 + 30x^2 - 12x$
$= 3x(5x^3 - 2x^2 + 10x - 4)$
$= 3x[(5x^3 - 2x^2) + (10x - 4)]$
$= 3x[x^2(5x - 2) + 2(5x - 2)]$
$= 3x(5x - 2)(x^2 + 2)$

95. $-3x^3 + 6x^2 - 9x$
$= -3x(x^2) + (-3x)(-2x) + (-3x)(3)$
$= -3x(x^2 - 2x + 3)$

97. $-12b + 16b^2 = 16b^2 - 12b$
$\qquad = 4b(4b) - 4b(3)$
$\qquad = 4b(4b - 3)$

99. $12xy + 9x - 8y - 6$
$= (12xy + 9x) + (-8y - 6)$
$= 3x(4y + 3) + (-2)(4y + 3)$
$= (4y + 3)(3x - 2)$

101. $\dfrac{1}{3}x^3 - \dfrac{2}{9}x^2 = \dfrac{1}{3}x^2(x) + \dfrac{1}{3}x^2\left(-\dfrac{2}{3}\right)$

$$= \dfrac{1}{3}x^2\left(x - \dfrac{2}{3}\right)$$

103. $-16t^2 + 150t = -2t(8t) + (-2t)(-75)$

$$= -2t(8t - 75)$$

105. $21,000 - 150p = 150(140) - 150(p)$

$$= 150(140 - p)$$

107. $8x^5 - 28x^3 = 4x^3(2x^2) - 4x^3(7)$

$$= 4x^3(2x^2 - 7)$$

109. $S = 2\pi r^2 + 8\pi r$

$$= 2\pi r(r) + 2\pi r(4)$$

$$= 2\pi r(r + 4)$$

111. $4x^2 - 8x = 4x(x) - 4x(2) = 4x(x - 2)$

The missing length is $x - 2$.

113. $8x^{3n} + 10x^n = 2x^n(4x^{2n}) + 2x^n(5)$

$$= 2x^n(4x^{2n} + 5)$$

The missing factor is $4x^{2n} + 5$.

115. $3 - 4x^{-1} + 2x^{-3}$

$$= x^{-3}(3x^3) - x^{-3}(4x^2) + x^{-3}(2)$$

$$= x^{-3}(3x^3 - 4x^2 + 2)$$

The missing factor is $3x^3 - 4x^2 + 2$.

117. $x^2 - 3x^{-1} + 2x^{-2}$

$$= x^{-2}(x^4) - x^{-2}(3x) + x^{-2}(2)$$

$$= x^{-2}(x^4 - 3x + 2)$$

The missing factor is $x^4 - 3x + 2$.

119. $\dfrac{6}{35}x^4 - \dfrac{1}{7}x^2 + \dfrac{2}{7}x$

$$= \dfrac{2}{7}x\left(\dfrac{3}{5}x^3\right) - \dfrac{2}{7}x\left(\dfrac{1}{2}x\right) + \dfrac{2}{7}x(1)$$

$$= \dfrac{2}{7}x\left(\dfrac{3}{5}x^3 - \dfrac{1}{2}x + 1\right)$$

The missing factor is $\dfrac{3}{5}x^3 - \dfrac{1}{2}x + 1$.

121. To find the greatest common factor, (1) determine the GCF of the coefficients of the variable expressions (the largest number that divides evenly into a set of numbers). (2) For each variable expression common to all the terms, determine the smallest exponent that the variable expression is raised to. (3) Find the product of these common factors. This is the GCF.

To factor the GCF from a polynomial, (1) identify the GCF of the terms that make up the polynomial. (2) Rewrite each term as the product of the GCF and the remaining factor. (3) Use the Distributive Property "in reverse" to factor out the GCF. (4) Use the Distributive Property to verify that the factorization is correct.

123. $3a(x + y) - 4b(x + y) \neq (3a - 4b)(x + y)^2$ because when checking, the product $(3a - 4b)(x + y)^2$ is equal to

$(3a - 4b)(x + y)^2 = (3a - 4b)(x^2 + 2xy + y^2)$, not

$3a(x + y) - 4b(x + y)$

$= 3ax + 3ay - 4bx - 4by$.

Section 6.2

Preparing for Factoring Trinomials of the Form $x^2 + bx + c$

P1. $2 \cdot 9 = 18$ and $2 + 9 = 11$, so the factors are 2 and 9.

P2. $4 \cdot (-6) = -24$ and $4 + (-6) = -2$, so the factors are 4 and −6.

P3. $4 \cdot (-3) = -12$ and $4 + (-3) = 1$, so the factors are 4 and −3.

P4. $(-7) \cdot (-5) = 35$ and $(-7) + (-5) = -12$, so the factors are −7 and −5.

P5. The coefficients of

$3x^2 - x - 4 = 3x^2 + (-1)x + (-4)$ are 3, −1, and −4.

P6. $-5p(p + 4) = -5p(p) + (-5p)(4)$

$$= -5p^2 + (-20p)$$

$$= -5p^2 - 20p$$

P7. $(z-1)(z+4) = z(z) + z(4) + (-1)(z) + (-1)(4)$
$$= z^2 + 4z - z - 4$$
$$= z^2 + 3z - 4$$

6.2 Quick Checks

1. A <u>quadratic trinomial</u> is a polynomial of the form $ax^2 + bx^2 + c,\ a \neq 0$.

2. When factoring $x^2 - 10x + 24$, the signs of the factors of 24 must both be <u>negative</u>.

3. False; $4 + 4x + x^2 = x^2 + 4x + 4$
   The leading coefficient is 1 not 4.

4. In $y^2 + 9y + 20$, $b = 9$ and $c = 20$.

Integers Whose Product is 20	1, 20	2, 10	4, 5
Sum	21	12	9

   $4 \cdot 5 = 20 = c$ and $4 + 5 = 9 = b$ so $y^2 + 9y + 20 = (y+4)(y+5)$.

   Check: $(y+4)(y+5) = y^2 + 5y + 4y + 4(5)$
   $$= y^2 + 9y + 20$$

5. In $z^2 - 9z + 14$, $b = -9$ and $c = 14$.

Integers Whose Product is 14	−1, −14	−2, −7
Sum	−15	−9

   $-2 \cdot (-7) = 14 = c$ and $-2 + (-7) = -9 = b$ so
   $z^2 - 9z + 14 = (z + (-2))(z + (-7))$
   $$= (z-2)(z-7)$$

   Check: $(z-2)(z-7) = z^2 - 7z - 2z + (-2)(-7)$
   $$= z^2 - 9z + 14$$

6. True

7. In $y^2 - 2y - 15$, $b = -2$ and $c = -15$.

Integers Whose Product is −15	−1, 15	−3, 5	1, −15	3, −5
Sum	14	2	−14	−2

   $3 \cdot (-5) = -15 = c$ and $3 + (-5) = -2 = b$, so

$$y^2 - 2y - 15 = (y+3)(y+(-5))$$
$$= (y+3)(y-5)$$

Check: $(y+3)(y-5) = y^2 - 5y + 3y + 3(-5)$
$$= y^2 - 2y - 15$$

8. In $w^2 + w - 12$, $b = 1$ and $c = -12$.

Integers Whose Product is –12	–1, 12	–2, 6	–3, 4	1, –12	2, –6	3, –4
Sum	11	4	1	–11	–4	–1

$-3 \cdot 4 = -12 = c$ and $-3 + 4 = 1 = b$, so
$$w^2 + w - 12 = (w+(-3))(w+4) = (w-3)(w+4)$$

Check: $(w-3)(w+4) = w^2 + 4w - 3w + (-3)(4) = w^2 + w - 12$

9. In $z^2 - 5z + 8$, $b = -5$ and $c = 8$.

Integers Whose Product is 8	–1, –8	–2, –4
Sum	–9	–6

There are no factors of 8 whose sum is –5. Therefore $z^2 - 5z + 8$ is prime.

10. In $q^2 + 4q - 45$, $b = 4$ and $c = -45$.

Integers Whose Product is –45	–1, 45	–3, 15	–5, 9
Sum	44	12	4

$-5 \cdot 9 = -45 = c$ and $-5 + 9 = 4 = b$, so
$$q^2 + 4q - 45 = (q+(-5))(q+9)$$
$$= (q-5)(q+9)$$

Check: $(q-5)(q+9) = q^2 + 9q - 5q + (-5)(9)$
$$= q^2 + 4q - 45$$

11. In $x^2 + 9xy + 20y^2$, $b = 9$ and $c = 20$.

Integers Whose Product is 20	1, 20	2, 10	4, 5
Sum	21	12	9

$4 \cdot 5 = 20 = c$ and $4 + 5 = 9 = b$, so $x^2 + 9xy + 20y^2 = (x+4y)(x+5y)$

Check: $(x+4y)(x+5y) = x^2 + 5xy + 4xy + 20y^2$
$$= x^2 + 9xy + 20y^2$$

12. In $m^2 + mn - 42n^2$, $b = 1$ and $c = -42$.

Integers Whose Product is −42	−1, 42	−2, 21	−3, 14	−6, 7
Sum	41	19	11	1

$-6 \cdot 7 = -42 = c$ and $-6 + 7 = 1 = b$, so

$m^2 + mn - 42n^2 = (m + (-6)n)(m + 7n)$
$= (m - 6n)(m + 7n)$

Check:

$(m - 6n)(m + 7n) = m^2 + 7mn - 6mn - 42n^2$
$= m^2 + mn - 42n^2$

13. (a) $-12 - x + x^2 = x^2 - x - 12$

(b) $9n + n^2 - 10 = n^2 + 9n - 10$

14. In $-56 + n^2 + n = n^2 + n - 56$, $b = 1$ and $c = -56$.

Integers Whose Product is −56	−1, 56	−2, 28	−4, 14	−7, 8
Sum	55	26	10	1

$-7 \cdot 8 = -56 = c$ and $-7 + 8 = 1 = b$, so

$n^2 + n - 56 = (n - 7)(n + 8)$

Check: $(n - 7)(n + 8) = n^2 + 8n - 7n + (-7)(8)$
$= n^2 + n - 56$

15. In $y^2 + 35 - 12y = y^2 - 12y + 35$, $b = -12$ and $c = 35$.

Integers Whose Product is 35	−1, −35	−5, −7
Sum	−36	−12

$-5 \cdot (-7) = 35 = c$ and $-5 + (-7) = -12 = b$, so

$y^2 - 12y + 35 = (y - 5)(y - 7)$

Check: $(y - 5)(y - 7) = y^2 - 7y - 5y + (-5)(-7)$
$= y^2 - 12y + 35$

16. $(x - 3)(3x + 6) = (x - 3)[3(x + 2)]$
$= 3(x - 3)(x + 2)$

The polynomial was not factored completely.
False

17. The GCF of $4m^2 - 16m - 84$ is 4.

$4m^2 - 16m - 84 = 4(m^2 - 4m - 21)$

In $m^2 - 4m - 21$, $b = -4$ and $c = -21$.

Integers Whose Product is −21	1, −21	3, −7
Sum	−20	−4

$3 \cdot (-7) = -21 = c$ and $3 + (-7) = -4 = b$, so

$m^2 - 4m - 21 = (m + 3)(m - 7)$

$4m^2 - 16m - 84 = 4(m^2 - 4m - 21)$
$= 4(m + 3)(m - 7)$

Check: $4(m + 3)(m - 7) = 4(m^2 - 7m + 3m - 21)$
$= 4(m^2 - 4m - 21)$
$= 4m^2 - 16m - 84$

18. The GCF of $3z^3 + 12z^2 - 15z$ is $3z$.

$3z^3 + 12z^2 - 15z = 3z(z^2 + 4z - 5)$

In $z^2 + 4z - 5$, $b = 4$ and $c = -5$.

Integers Whose Product is −5	−1, 5	1, −5
Sum	4	−4

$-1 \cdot 5 = -5 = c$ and $-1 + 5 = 4 = b$, so

$z^2 + 4z - 5 = (z - 1)(z + 5)$

$3z^3 + 12z^2 - 15z = 3z(z^2 + 4z - 5)$
$= 3z(z - 1)(z + 5)$

Check: $3z(z - 1)(z + 5) = 3z(z^2 + 5z - z - 5)$
$= 3z(z^2 + 4z - 5)$
$= 3z^3 + 12z^2 - 15z$

19. $-w^2 - 3w + 10 = -1(w^2 + 3w - 10)$

In $w^2 + 3w - 10$, $b = 3$ and $c = -10$.

Integers Whose Product is -10	$-1, 10$	$-2, 5$
Sum	9	3

$-2 \cdot 5 = -10 = c$ and $-2 + 5 = 3 = b$, so

$w^2 + 3w - 10 = (w - 2)(w + 5)$.

$-w^2 - 3w + 10 = -1(w^2 + 3w - 10)$
$= -1(w - 2)(w + 5)$
$= -(w - 2)(w + 5)$

20. The GCF of $-2a^2 - 8a + 24$ is -2.

$-2a^2 - 8a + 24 = -2(a^2 + 4a - 12)$

In $a^2 + 4a - 24$, $b = 4$ and $c = -12$.

Integers Whose Product is -12	$-1, 12$	$-2, 6$	$-3, 4$
Sum	11	4	1

$-2 \cdot 6 = -12 = c$ and $-2 + 6 = 4 = b$, so

$a^2 + 4a - 12 = (a - 2)(a + 6)$.

$-2a^2 - 8a + 24 = -2(a^2 + 4a - 12)$
$= -2(a - 2)(a + 6)$

6.2 Exercises

21. $2 \cdot 3 = 6$ and $2 + 3 = 5$
$x^2 + 5x + 6 = (x + 2)(x + 3)$

23. $3 \cdot 6 = 18$ and $3 + 6 = 9$
$m^2 + 9m + 18 = (m + 3)(m + 6)$

25. $-3 \cdot (-12) = 36$ and $-3 + (-12) = -15$
$x^2 - 15x + 36 = (x - 3)(x - 12)$

27. $-2 \cdot (-6) = 12$ and $-2 + (-6) = -8$
$p^2 - 8p + 12 = (p - 2)(p - 6)$

29. $-4 \cdot 3 = -12$ and $-4 + 3 = -1$
$x^2 - x - 12 = (x - 4)(x + 3)$

31. $-4 \cdot 5 = -20$ and $-4 + 5 = 1$
$x^2 + x - 20 = (x - 4)(x + 5)$

33. $-3 \cdot 15 = -45$ and $-3 + 15 = 12$
$z^2 + 12z - 45 = (z - 3)(z + 15)$

35. $-2 \cdot (-3) = 6$ and $-2 + (-3) = -5$
$x^2 - 5xy + 6y^2 = (x - 2y)(x - 3y)$

37. $-2 \cdot 3 = -6$ and $-2 + 3 = 1$
$r^2 + rs - 6s^2 = (r - 2s)(r + 3s)$

39. $-4 \cdot 1 = -4$ and $-4 + 1 = -3$
$x^2 - 3xy - 4y^2 = (x - 4y)(x + y)$

41. There are no factors of 1 whose sum is 7.
$z^2 + 7zy + y^2$ is prime.

43. $3x^2 + 3x - 6 = 3(x^2 + x - 2) = 3(x + 2)(x - 1)$

45. $3n^3 - 24n^2 + 45n = 3n(n^2 - 8n + 15)$
$= 3n(n - 3)(n - 5)$

47. $5x^2z - 20xz - 160z = 5z(x^2 - 4x - 32)$
$= 5z(x - 8)(x + 4)$

49. $-3x^2 + x^3 - 18x = x^3 - 3x^2 - 18x$
$= x(x^2 - 3x - 18)$
$= x(x - 6)(x + 3)$

51. $-2y^2 + 8y - 8$
$= -2(y^2 - 4y + 4)$
$= -2(y - 2)(y - 2)$ or $-2(y - 2)^2$

53. $4x^3 - 32x^2 + x^4 = x^4 + 4x^3 - 32x^2$
$= x^2(x^2 + 4x - 32)$
$= x^2(x + 8)(x - 4)$

55. $2x^2 + x^3 - 15x = x^3 + 2x^2 - 15x$
$= x(x^2 + 2x - 15)$
$= x(x + 5)(x - 3)$

57. $-7 \cdot 4 = -28$ and $-7 + 4 = -3$
$x^2 - 3xy - 28y^2 = (x + 4y)(x - 7y)$

59. There are no factors of 6 whose product is 1.
$x^2 + x + 6$ is prime.

61. $-5 \cdot 4 = -20$ and $-5 + 4 = -1$
$k^2 - k - 20 = (k-5)(k+4)$

63. $-5 \cdot 6 = -30$ and $-5 + 6 = 1$
$x^2 + xy - 30y^2 = (x-5y)(x+6y)$

65. $-3 \cdot (-5) = 15$ and $-3 + (-5) = -8$
$s^2 t^2 - 8st + 15 = (st-3)(st-5)$

67. $-3p^3 + 3p^2 + 6p = -3p(p^2 - p - 2)$
$\qquad\qquad = -3p(p-2)(p+1)$

69. $-x^2 - x + 6 = -(x^2 + x - 6) = -(x+3)(x-2)$

71. $g^2 - 4g + 21$ cannot be factored; it is prime.

73. $n^4 - 30n^2 - n^3 = n^4 - n^3 - 30n^2$
$\qquad\qquad = n^2(n^2 - n - 30)$
$\qquad\qquad = n^2(n-6)(n+5)$

75. $m^2 + 2mn - 15n^2 = (m+5n)(m-3n)$

77. $35 + 12s + s^2 = s^2 + 12s + 35 = (s+5)(s+7)$

79. $n^2 - 9n - 45$ cannot be factored; it is prime.

81. $m^2 n^2 - 8mn + 12 = (mn-6)(mn-2)$

83. $-x^3 + 12x^2 + 28x = -x(x^2 - 12x - 28)$
$\qquad\qquad = -x(x-14)(x+2)$

85. $y^2 - 12y + 36 = (y-6)(y-6)$ or $(y-6)^2$

87. $-36x + 20x^2 - 2x^3 = -2x^3 + 20x^2 - 36x$
$\qquad\qquad = -2x(x^2 - 10x + 18)$

89. $-21x^3 y - 14xy^2 = -7xy(3x^2 + 2y)$

91. $-16t^2 + 64t + 80 = -16(t^2 - 4t - 5)$
$\qquad\qquad = -16(t+1)(t-5)$

93. $x^2 + 9x + 18 = (x+6)(x+3)$
The binomials $x + 6$ and $x + 3$ represent the length and the width.

95. $\dfrac{1}{2}x^2 + x - \dfrac{15}{2} = \dfrac{1}{2}(x^2 + 2x - 15)$
$\qquad\qquad = \dfrac{1}{2}(x+5)(x-3)$

The binomials $x + 5$ and $x - 3$ represent the base and height.

97. If $b > 0$, then the constant 6 factors as (1)(6) or (2)(3). Thus, the coefficient of x could be $1 + 6 = 7$ or $2 + 3 = 5$. If $b < 0$, then the constant 6 factors as (−1)(−6) or (−2)(−3). Thus, the coefficient of x could be $-1 + (-6) = -7$ or $-2 + (-3) = -5$. So b could be −7, −5, 5, or 7.

99. If $b > 0$, then the constant −21 factors as (−1)(21) or (−3)(7). Thus, the coefficient of x could be $(-1) + 21 = 20$ or $(-3) + 7 = 4$. If $b < 0$, then the constant 21 factors as (1)(−21) or (3)(−7). Thus the coefficient of x could be $1 + (-21) = -20$ or $3 + (-7) = -4$. So, b could be −20, −4, 4, or 20.

101. Since $c > 0$, then c must have two negative factors that sum to −2.
$-1, -1: c = (-1)(-1) = 1$
Thus, c must be 1.

103. Since $c > 0$, then c must have two negative factors that sum to −7.
$-1, -6: c = (-1)(-6) = 6$
$-2, -5: c = (-2)(-5) = 10$
$-3, -4: c = (-3)(-4) = 12$
Thus, c could be 6, 10, or 12.

105. Your answer is marked correctly.
The product
$(1-x)(2-x) = 2 - x - 2x + x^2 = 2 - 3x + x^2$.
The product
$(x-1)(x-2) = x^2 - 2x - x + 2 = x^2 - 3x + 2$

Section 6.3

Preparing for Factoring Trinomials of the Form $ax^2 + bx + c, a \neq 1$

P1. $24 = 2 \cdot 12 = 2 \cdot 2 \cdot 6 = 2 \cdot 2 \cdot 2 \cdot 3 = 2^3 \cdot 3$

P2. The coefficients of
$5x^2 - 3x + 7 = 5x^2 + (-3)x + 7$ are 5, −3, and 7.

P3. $(2x+7)(3x-1)$
$= 2x \cdot 3x + 2x \cdot (-1) + 7 \cdot 3x + 7 \cdot (-1)$
$= 6x^2 - 2x + 21x - 7$
$= 6x^2 + 19x - 7$

6.3 Quick Checks

1. True

2. $3x^2 = 3x \cdot x$
$2 = 2 \cdot 1 \text{ or } -2 \cdot (-1)$
Since the coefficient of x, 5, is positive, we do not use the factorization $-2 \cdot (-1)$.
$(3x+2)(x+1) = 3x^2 + 3x + 2x + 2$
$\qquad\qquad\quad = 3x^2 + 5x + 2$
$(3x+1)(x+2) = 3x^2 + 6x + x + 2$
$\qquad\qquad\quad = 3x^2 + 7x + 2$
$3x^2 + 5x + 2 = (3x+2)(x+1)$

3. $7y^2 = 7y \cdot y$
$3 = 3 \cdot 1 \text{ or } -3 \cdot (-1)$
Since the coefficient of y, 22, is positive, we do not use the factorization $-3 \cdot (-1)$.
$(7y+3)(y+1) = 7y^2 + 7y + 3y + 3$
$\qquad\qquad\quad = 7y^2 + 10y + 3$
$(7y+1)(y+3) = 7y^2 + 21y + y + 3$
$\qquad\qquad\quad = 7y^2 + 22y + 3$
$7y^2 + 22y + 3 = (7y+1)(y+3)$

4. $3x^2 = 3x \cdot x$
$12 = 1 \cdot 12, 2 \cdot 6, 3 \cdot 4, -1 \cdot (-12), -2 \cdot (-6),$ or $-3 \cdot (-4)$
Since the coefficient of x, −13, is negative, we do not use the factorizations of 12 into positive factors.
$(3x-1)(x-12) = 3x^2 - 36x - x + 12$
$\qquad\qquad\qquad = 3x^2 - 37x + 12$
$(3x-12)(x-1) = 3x^2 - 3x - 12x + 12$
$\qquad\qquad\qquad = 3x^2 - 15x + 12$

$(3x-2)(x-6) = 3x^2 - 18x - 2x + 12$
$\qquad\qquad\qquad = 3x^2 - 20x + 12$
$(3x-6)(x-2) = 3x^2 - 6x - 6x + 12$
$\qquad\qquad\qquad = 3x^2 - 12x + 12$
$(3x-3)(x-4) = 3x^2 - 12x - 3x + 12$
$\qquad\qquad\qquad = 3x^2 - 15x + 12$
$(3x-4)(x-3) = 3x^2 - 9x - 4x + 12$
$\qquad\qquad\qquad = 3x^2 - 13x + 12$
$3x^2 - 13x + 12 = (3x-4)(x-3)$

5. $5p^2 = 5p \cdot p$
$4 = 1 \cdot 4, 2 \cdot 2, -1 \cdot (-4), \text{ or } -2 \cdot (-2)$
Since the coefficient of p, −21, is negative, we do not use the factorizations of 4 into positive factors.
$(5p-1)(p-4) = 5p^2 - 20p - p + 4$
$\qquad\qquad\qquad = 5p^2 - 21p + 4$
$(5p-4)(p-1) = 5p^2 - 5p - 4p + 4$
$\qquad\qquad\qquad = 5p^2 - 9p + 4$
$(5p-2)(p-2) = 5p^2 - 10p - 2p + 4$
$\qquad\qquad\qquad = 5p^2 - 12p + 4$
$5p^2 - 21p + 4 = (5p-1)(p-4)$

6. $2n^2 = 2n \cdot n$
$-9 = -1 \cdot 9, -3 \cdot 3, \text{ or } 1 \cdot (-9)$
$(2n-1)(n+9) = 2n^2 + 18n - n - 9$
$\qquad\qquad\qquad = 2n^2 + 17n - 9$
$(2n+9)(n-1) = 2n^2 - 2n + 9n - 9$
$\qquad\qquad\qquad = 2n^2 + 7n - 9$
$(2n-3)(n+3) = 2n^2 + 6n - 3n - 9$
$\qquad\qquad\qquad = 2n^2 + 3n - 9$
$(2n+3)(n-3) = 2n^2 - 6n + 3n - 9$
$\qquad\qquad\qquad = 2n^2 - 3n - 9$
$(2n+1)(n-9) = 2n^2 - 18n + n - 9$
$\qquad\qquad\qquad = 2n^2 - 17n - 9$
$(2n-9)(n+1) = 2n^2 + 2n - 9n - 9$
$\qquad\qquad\qquad = 2n^2 - 7n - 9$
$2n^2 - 17n - 9 = (2n+1)(n-9)$

7. $4w^2 = 4w \cdot w$ or $2w \cdot 2w$

$-6 = -1 \cdot 6, -2 \cdot 3, 1 \cdot (-6),$ or $2 \cdot (-3)$

$(4w-1)(w+6) = 4w^2 + 24w - w - 6$
$$= 4w^2 + 23w - 6$$

$(4w+6)(w-1) = 4w^2 - 4w + 6w - 6$
$$= 4w^2 + 2w - 6$$

$(4w-2)(w+3) = 4w^2 + 12w - 2w - 6$
$$= 4w^2 + 10w - 6$$

$(4w+3)(w-2) = 4w^2 - 8w + 3w - 6$
$$= 4w^2 - 5w - 6$$

$(4w+1)(w-6) = 4w^2 - 24w + w - 6$
$$= 4w^2 - 23w - 6$$

$(4w-6)(w+1) = 4w^2 + 4w - 6w - 6$
$$= 4w^2 - 2w - 6$$

$(4w+2)(w-3) = 4w^2 - 12w + 2w - 6$
$$= 4w^2 - 10w - 6$$

$(4w-3)(w+2) = 4w^2 + 8w - 3w - 6$
$$= 4w^2 + 5w - 6$$

$(2w-1)(2w+6) = 4w^2 + 12w - 2w - 6$
$$= 4w^2 + 10w - 6$$

$(2w-2)(2w+3) = 4w^2 + 6w - 4w - 6$
$$= 4w^2 + 2w - 6$$

$(2w+1)(2w-6) = 4w^2 - 12w + 2w - 6$
$$= 4w^2 - 10w - 6$$

$(2w+2)(2w-3) = 4w^2 - 6w + 4w - 6$
$$= 4w^2 - 2w - 6$$

$4w^2 - 5w - 6 = (4w+3)(w-2)$

8. When factoring $ax^2 + bx + c$ and b is a small number, choose factors of a and factors of c that are <u>close</u> to each other.

9. $12x^2 = 12x \cdot x,\ 6x \cdot 2x,$ or $4x \cdot 3x$

$6 = 6 \cdot 1, 3 \cdot 2, -6 \cdot (-1),$ or $-3 \cdot (-2)$

Since the coefficient of x, 17, is positive, we do not use the factorizations of 6 into negative factors. Since 17 is neither very large nor very small, try factors of 12 and 6 that are similar in size: $12x^2 = 4x \cdot 3x$ and $6 = 3 \cdot 2$.

$(4x+3)(3x+2) = 12x^2 + 8x + 9x + 6$
$$= 12x^2 + 17x + 6$$

$12x^2 + 17x + 6 = (4x+3)(3x+2)$

10. $12y^2 = 12y \cdot y,\ 6y \cdot 2y,$ or $4y \cdot 3y$

$-35 = -1 \cdot 35, -5 \cdot 7, 1 \cdot (-35),$ or $5 \cdot (-7)$

Since the coefficient of y, 32, is closer to

$6 \cdot 5 = 30$ and $6 \cdot 7 = 42$ than to

$12 \cdot 35 = 420$ or $12 \cdot 1 = 12$, try $12y^2 = 6y \cdot 2y$

and $-35 = -5 \cdot 7$.

$(6y-5)(2y+7) = 12y^2 + 42y - 10y - 35$
$$= 12y^2 + 32y - 35$$

$12y^2 + 32y - 35 = (6y-5)(2y+7)$

11. When factoring any polynomial, the first step is to look for a common factor.

12. The GCF of $8x^2 - 28x - 60$ is 4.

$8x^2 - 28x - 60 = 4(2x^2 - 7x - 15)$

$2x^2 = 2x \cdot x$

$-15 = -15 \cdot 1, 15 \cdot (-1), -5 \cdot 3,$ or $5 \cdot (-3)$

Try $2x^2 = 2x \cdot x$

$-15 = -5 \cdot 3$

$(2x+3)(x-5) = 2x^2 - 10x + 3x - 15$
$$= 2x^2 - 7x - 15$$

$8x^2 - 28x - 60 = 4(2x+3)(x-5)$

13. The GCF of $90x^2 + 21xy - 6y^2$ is 3.

$90x^2 + 21xy - 6y^2 = 3(30x^2 + 7xy - 2y^2)$

$30x^2 = 30x \cdot x,\ 15x \cdot 2x,\ 10x \cdot 3x,$ or $6x \cdot 5x$

$-2y^2 = -2y \cdot y$ or $2y \cdot (-y)$

Since the coefficient of xy, 7, is fairly small, try factors of $30x^2$ that are close together:

$30x^2 = 6x \cdot 5x$.

$(6x-2y)(5x+y) = 30x^2 + 6xy - 10xy - 2y^2$
$$= 30x^2 - 4xy - 2y^2$$

$(6x+y)(5x-2y) = 30x^2 - 12xy + 5xy - 2y^2$
$$= 30x^2 - 7xy - 2y^2$$

To get the coefficient of xy to be 7, reverse the signs on the y-terms.

$(6x-y)(5x+2y) = 30x^2 + 12xy - 5xy - 2y^2$
$$= 30x^2 + 7xy - 2y^2$$

$90x^2 + 21xy - 6y^2 = 3(30x^2 + 7xy - 2y^2)$
$$= 3(6x-y)(5x+2y)$$

14. The GCF of $12x^2 + 22x + 6$ is 2.

$12x^2 + 22x + 6 = 2(6x^2 + 11x + 3)$

$6x^2 = 6x \cdot x, \ 3x \cdot 2x$

$3 = 3 \cdot 1$

$12x^2 + 22x + 6 = 2(3x+1)(2x+3)$

Check

$2(3x+1)(2x+3) = 2(6x^2 + 9x + 2x + 3)$

$\qquad\qquad\qquad\quad = 2(6x^2 + 11x + 3)$

$12x^2 + 2x + 6$ was not completely factored as $(4x+6)(3x+1)$. False

15. $-6y^2 + 23y + 4 = -1(6y^2 - 23y - 4)$

$6y^2 = 6y \cdot y$ or $3y \cdot 2y$

$-4 = -1 \cdot 4, -2 \cdot 2,$ or $1 \cdot (-4)$

Since the coefficient of y, -23, is close to

$6 \cdot (-4)$, try $6y^2 = 6y \cdot y$ and $-4 = 1 \cdot (-4)$.

$(6y+1)(y-4) = 6y^2 - 24y + y - 4$

$\qquad\qquad\qquad = 6y^2 - 23y - 4$

$-6y^2 + 23y + 4 = -1(6y^2 - 23y - 4)$

$\qquad\qquad\qquad\quad = -1(6y+1)(y-4)$

$\qquad\qquad\qquad\quad = -(6y+1)(y-4)$

16. The GCF of $-6x^2 - 3x + 45$ is -3.

$-6x^2 - 3x + 45 = -3(2x^2 + x - 15)$

$2x^2 = 2x \cdot x$

$-15 = -1 \cdot 15, -3 \cdot 5, 1 \cdot (-15), 3 \cdot (-5)$

Since the coefficient of x, 1, is small, try factors of -15 whose difference is small: $-15 = -3 \cdot 5$.

$(2x-3)(x+5) = 2x^2 + 10x - 3x - 15$

$\qquad\qquad\qquad = 2x^2 + 7x - 15$

$(2x+5)(x-3) = 2x^2 - 6x + 5x - 15$

$\qquad\qquad\qquad = 2x^2 - x - 15$

To get the coefficient of x to be 1, reverse the signs on the constants.

$(2x-5)(x+3) = 2x^2 + 6x - 5x - 15$

$\qquad\qquad\qquad = 2x^2 + x - 15$

$-6x^2 - 3x + 45 = -3(2x^2 + x - 15)$

$\qquad\qquad\qquad\quad = -3(2x-5)(x+3)$

17. When factoring $6x^2 + x - 1$ using grouping,

$ac = \underline{-6}$ and $b = \underline{1}$.

18. False; since $ac = 12$, we want to determine the integers whose product is 12 and whose sum is -13.

19. In $3x^2 - 2x - 8$, $a = 3$, $b = -2$, and $c = -8$.

$ac = 3 \cdot (-8) = -24$

Integers Whose Product is −24	1, −24	2, −12	3, −8	4, −6
Sum	−23	−10	−5	−2

$4 \cdot (-6) = -24 = ac$ and $4 + (-6) = -2 = b$, so we write $-2x$ as $4x - 6x$.

$$3x^2 - 2x - 8 = 3x^2 + 4x - 6x - 8$$
$$= (3x^2 + 4x) + (-6x - 8)$$
$$= x(3x + 4) + (-2)(3x + 4)$$
$$= (3x + 4)(x - 2)$$

20. In $10z^2 + 21z + 9$, $a = 10$, $b = 21$, and $c = 9$.

$ac = 10 \cdot 9 = 90$

Integers Whose Product is 90	1, 90	2, 45	3, 30	5, 18	6, 15	9, 10
Sum	91	47	33	23	21	19

$6 \cdot 15 = 90 = ac$ and $6 + 15 = 21 = b$, so we write $21z$ as $6z + 15z$.

$$10z^2 + 21z + 9 = 10z^2 + 6z + 15z + 9$$
$$= (10z^2 + 6z) + (15z + 9)$$
$$= 2z(5z + 3) + 3(5z + 3)$$
$$= (5z + 3)(2z + 3)$$

21. The GCF of $24x^2 + 6x - 9$ is 3.

$$24x^2 + 6x - 9 = 3(8x^2 + 2x - 3)$$

In $8x^2 + 2x - 3$, $a = 8$, $b = 2$, and $c = -3$.

$ac = 8 \cdot (-3) = -24$

Integers Whose Product is −24	−1, 24	−2, 12	−3, 8	−4, 6
Sum	23	10	5	2

$-4 \cdot 6 = -24 = ac$ and $-4 + 6 = 2 = b$, so we rewrite $2x$ as $-4x + 6x$.

$$8x^2 + 2x - 3 = 8x^2 - 4x + 6x - 3$$
$$= (8x^2 - 4x) + (6x - 3)$$
$$= 4x(2x - 1) + 3(2x - 1)$$
$$= (2x - 1)(4x + 3)$$

$$24x^2 + 6x - 9 = 3(8x^2 + 2x - 3)$$
$$= 3(2x - 1)(4x + 3)$$

22. $-10n^2 + 17n - 3 = -1(10n^2 - 17n + 3)$

In $10n^2 - 17n + 3$, $a = 10$, $b = -17$, and $c = 3$.

$ac = 10 \cdot 3 = 30$

Integers Whose Product is 30	−1, −30	−2, −15	−3, −10	−5, −6
Sum	−31	−17	−13	−11

$-2 \cdot (-15) = 30 = ac$ and $-2 + (-15) = -17 = b$, so we rewrite $-17n$ as $-2n - 15n$.

$$10n^2 - 17n + 3 = 10n^2 - 2n - 15n + 3$$
$$= (10n^2 - 2n) + (-15n + 3)$$
$$= 2n(5n - 1) + (-3)(5n - 1)$$
$$= (5n - 1)(2n - 3)$$

$$-10n^2 + 17n - 3 = -1(10n^2 - 17n + 3)$$
$$= -1(5n - 1)(2n - 3)$$
$$= -(5n - 1)(2n - 3)$$

6.3 Exercises

23. $2x^2 + 5x + 3$

$2x^2 = 2x \cdot x$

$3 = 3 \cdot 1$

Since the coefficient of x is positive, only the positive factors of 3 are listed.

$(2x + 3)(x + 1) = 2x^2 + 2x + 3x + 3 = 2x^2 + 5x + 3$

So, $2x^2 + 5x + 3 = (2x + 3)(x + 1)$.

25. $5n^2 + 7n + 2$

$5n^2 = 5n \cdot n$

$2 = 2 \cdot 1$

Since the coefficient of n is positive, only the positive factors of 2 are listed.

$(5n + 2)(n + 1) = 5n^2 + 5n + 2n + 2 = 5n^2 + 7n + 2$

So, $5n^2 + 7n + 2 = (5n + 2)(n + 1)$.

27. $5y^2 + 2y - 3$

$5y^2 = 5y \cdot y$

$-3 = -3 \cdot 1, -1 \cdot 3$

$(5y - 3)(y + 1) = 5y^2 + 5y - 3y - 3 = 5y^2 + 2y - 3$

So, $5y^2 + 2y - 3 = (5y - 3)(y + 1)$.

29. $-4p^2 + 11p + 3 = -(4p^2 - 11p - 3)$

$4p^2 = 4p \cdot p, 2p \cdot 2p$

$-3 = 3 \cdot (-1), (-3) \cdot 1$

$$(4p+1)(p-3) = 4p^2 - 12p + p - 3$$
$$= 4p^2 - 11p - 3$$

So, $-4p^2 + 11p + 3 = -(4p+1)(p-3)$.

31. $5w^2 + 13w - 6$

$5w^2 = 5w \cdot w$

$-6 = -6 \cdot 1, 6 \cdot (-1), -3 \cdot 2, 3 \cdot (-2)$

$$(5w-2)(w+3) = 5w^2 + 15w - 2w - 6$$
$$= 5w^2 + 13w - 6$$

So, $5w^2 + 13w - 6 = (5w-2)(w+3)$.

33. $7t^2 + 37t + 10$

$7t^2 = 7t \cdot t$

$10 = 10 \cdot 1, 5 \cdot 2$

Since the coefficient of t is positive, only the positive factors of 10 are listed.

$$(7t+2)(t+5) = 7t^2 + 35t + 2t + 10$$
$$= 7t^2 + 37t + 10$$

So, $7t^2 + 37t + 10 = (7t+2)(t+5)$.

35. $6n^2 - 17n + 10$

$6n^2 = 6n \cdot n, \ 3n \cdot 2n$

$10 = -10 \cdot (-1), -5 \cdot (-2)$

Since the coefficient of n is negative, only the negative factors of 10 are listed.

$$(6n-5)(n-2) = 6n^2 - 12n - 5n + 10$$
$$= 6n^2 - 17n + 10$$

So, $6n^2 - 17n + 10 = (6n-5)(n-2)$.

37. $2 - 11x + 5x^2 = 5x^2 - 11x + 2$

$5x^2 = 5x \cdot x$

$2 = -2 \cdot (-1)$

Since the coefficient of x is negative, only the negative factors of 2 are listed.

$$(5x-1)(x-2) = 5x^2 - 10x - x + 2$$
$$= 5x^2 - 11x + 2$$

So, $2 - 11x + 5x^2 = (5x-1)(x-2)$.

39. $2x^2 + 3xy + y^2$

$2x^2 = 2x \cdot x$

$y^2 = y \cdot y$

$$(2x+y)(x+y) = 2x^2 + 2xy + xy + y^2$$
$$= 2x^2 + 3xy + y^2$$

So, $2x^2 + 3xy + y^2 = (2x+y)(x+y)$.

41. $2m^2 - 3mn - 2n^2$

$2m^2 = 2m \cdot m$

$-2n^2 = 2n \cdot (-n), \ (-2n) \cdot n$

$(2m + n)(m - 2n) = 2m^2 - 4mn + mn - 2n^2$
$$= 2m^2 - 3mn - 2n^2$$

So, $2m^2 - 3mn - 2n^2 = (2m + n)(m - 2n)$.

43. $6x^2 + 2xy - 4y^2 = 2(3x^2 + xy - 2y^2)$

$3x^2 = 3x \cdot x$

$-2y^2 = -2y \cdot y, \ 2y \cdot (-y)$

$(3x - 2y)(x + y) = 3x^2 + 3xy - 2xy - 2y^2$
$$= 3x^2 + xy - 2y^2$$

So, $6x^2 + 2xy - 4y^2 = 2(3x - 2y)(x + y)$.

45. $-2x^2 - 7x + 15 = -(2x^2 + 7x - 15)$

$2x^2 = 2x \cdot x$

$-15 = 15 \cdot (-1), \ (-15) \cdot 1, \ 5 \cdot (-3), \ (-5) \cdot 3$

$(2x - 3)(x + 5) = 2x^2 + 10x - 3x - 15$
$$= 2x^2 + 7x - 15$$

So, $-2x^2 - 7x + 15 = -(2x - 3)(x + 5)$.

47. $2x^2 + 13x + 15, \ a = 2, b = 13, c = 15$

$ac = (2)(15) = 30$

Integers whose product is 30	30, 1	−30, −1	15, 2	−15, −2	10, 3	−10, −3	6, 5	−6, −5
Sum	31	−31	17	−17	13	−13	11	−11

The integers are 10 and 3.

$2x^2 + 13x + 15 = 2x^2 + 10x + 3x + 15$
$$= (2x^2 + 10x) + (3x + 15)$$
$$= 2x(x + 5) + 3(x + 5)$$
$$= (2x + 3)(x + 5)$$

49. $5w^2 + 13w - 6, \ a = 5, b = 13, c = -6$

$ac = (5)(-6) = -30$

Integers whose product is −30	−30, 1	30, −1	−15, 2	15, −2	−10, 3	10, −3	−6, 5	6, −5
Sum	−29	29	−13	13	−7	7	−1	1

The integers are 15 and −2.

$$5w^2 + 13w - 6 = 5w^2 + 15w - 2w - 6$$
$$= (5w^2 + 15w) + (-2w - 6)$$
$$= 5w(w + 3) + (-2)(w + 3)$$
$$= 5w(w + 3) - 2(w + 3)$$
$$= (5w - 2)(w + 3)$$

51. $4w^2 - 8w - 5$, $a = 4$, $b = -8$, $c = -5$

$ac = (4)(-5) = -20$

Integers whose product is −20	−20, 1	20, −1	−10, 2	10, −2	−5, 4	5, −4
Sum	−19	19	−8	8	−1	1

The integers are −10 and 2.

$$4w^2 - 8w - 5 = 4w^2 + 2w - 10w - 5$$
$$= (4w^2 + 2w) + (-10w - 5)$$
$$= 2w(2w + 1) - 5(2w + 1)$$
$$= (2w - 5)(2w + 1)$$

53. $27z^2 + 3z - 2$, $a = 27$, $b = 3$, $c = -2$

$ac = (27)(-2) = -54$

Integers whose product is −54	−54, 1	54, −1	−27, 2	27, −2	−18, 3	18, −3	−9, 6	9, −6
Sum	−53	53	−25	25	−15	15	−3	3

The integers are 9 and −6.

$$27z^2 + 3z - 2 = 27z^2 + 9z - 6z - 2$$
$$= (27z^2 + 9z) + (-6z - 2)$$
$$= 9z(3z + 1) - 2(3z + 1)$$
$$= (9z - 2)(3z + 1)$$

55. $6y^2 - 5y - 6$, $a = 6$, $b = -5$, $c = -6$

$ac = (6)(-6) = -36$

Integers whose product is −36	−36, 1	36, −1	−18, 2	18, −2	−12, 3	12, −3	−9, 4	9, −4	6, −6
Sum	−35	35	−16	16	−9	9	−5	5	0

The integers are −9 and 4.

$$6y^2 - 5y - 6 = 6y^2 - 9y + 4y - 6$$
$$= (6y^2 - 9y) + (4y - 6)$$
$$= 3y(2y - 3) + 2(2y - 3)$$
$$= (3y + 2)(2y - 3)$$

57. $4m^2 + 8m - 5$, $a = 4$, $b = 8$, $c = -5$

$ac = (4)(-5) = -20$

Integers whose product is −20	−20, 1	20, −1	−10, 2	10, −2	−5, 4	5, −4
Sum	−19	19	−8	8	−1	1

The integers are 10 and −2.

$$4m^2 + 8m - 5 = 4m^2 - 2m + 10m - 5$$
$$= (4m^2 - 2m) + (10m - 5)$$
$$= 2m(2m - 1) + 5(2m - 1)$$
$$= (2m + 5)(2m - 1)$$

59. $12n^2 + 19n + 5$, $a = 12$, $b = 19$, $c = 5$

$ac = (12)(5) = 60$

Integers whose product is 60	60, 1	30, 2	20, 3	15, 4	12, 5	10, 6
Sum	61	32	23	19	17	16

The integers are 15 and 4.

$$12n^2 + 19n + 5 = 12n^2 + 4n + 15n + 5$$
$$= (12n^2 + 4n) + (15n + 5)$$
$$= 4n(3n + 1) + 5(3n + 1)$$
$$= (4n + 5)(3n + 1)$$

61. $-5 - 9x + 18x^2 = 18x^2 - 9x - 5$

$a = 18$, $b = -9$, $c = -5$

$ac = (18)(-5) = -90$

Integers whose product is −90	−90 1	90 −1	−45 2	45 −2	−30 3	30 −3	−18 5	18 −5	−15 6	15 −6	−10 9	10 −9
Sum	−89	89	−43	43	−27	27	−13	13	−9	9	−1	1

The integers are −15 and 6.

$$18x^2 - 9x - 5 = 18x^2 + 6x - 15x - 5$$
$$= (18x^2 + 6x) + (-15x - 5)$$
$$= 6x(3x + 1) - 5(3x + 1)$$
$$= (6x - 5)(3x + 1)$$

63. $12x^2 + 2xy - 4y^2 = 2(6x^2 + xy - 2y^2)$

$a = 6, b = 1, c = -2$

$ac = (6)(-2) = -12$

Integers whose product is −12	−12, 1	12, −1	−6, 2	6, −2	−4, 3	4, −3
Sum	−11	11	−4	4	−1	1

The integers are 4 and −3.

$$6x^2 + xy - 2y^2 = 6x^2 - 3xy + 4xy - 2y^2$$
$$= 3x(2x - y) + 2y(2x - y)$$
$$= (3x + 2y)(2x - y)$$

So, $12x^2 + 2xy - 4y^2 = 2(3x + 2y)(2x - y)$.

65. $8x^2 + 28x + 12 = 4(2x^2 + 7x + 3)$

$a = 2, b = 7, c = 3$

$ac = (2)(3) = 6$

Integers whose product is 6	6, 1	3, 2
Sum	7	5

The integers are 6 and 1.

$$2x^2 + 7x + 3 = 2x^2 + 6x + x + 3$$
$$= 2x(x + 3) + (x + 3)$$
$$= (2x + 1)(x + 3)$$

So, $8x^2 + 28x + 12 = 4(2x + 1)(x + 3)$.

67. $7x - 5 + 6x^2 = 6x^2 + 7x - 5$

$a = 6, b = 7, c = -5$

$ac = (6)(-5) = -30$

Integers whose product is −30	−30, 1	30, −1	−15, 2	15, −2	−10, 3	10, −3	−6, 5	6, −5
Sum	−29	29	−13	13	−7	7	−1	1

The integers are 10 and -3.

$$
\begin{aligned}
6x^2 + 7x - 5 &= 6x^2 - 3x + 10x - 5 \\
&= 3x(2x-1) + 5(2x-1) \\
&= (3x+5)(2x-1)
\end{aligned}
$$

69. $-8p^2 + 6p + 9 = -(8p^2 - 6p - 9)$

$a = 8, b = -6, c = -9$

$ac = (8)(-9) = -72$

Integers whose product is -72	-72 1	72 -1	-36 2	36 -2	-24 3	24 -3	-18 4	18 -4	-12 6	12 -6	-9 8	9 -8
Sum	-71	71	-34	34	-21	21	-14	14	-6	6	-1	1

The integers are -12 and 6.

$$
\begin{aligned}
8p^2 - 6p - 9 &= 8p^2 - 12p + 6p - 9 \\
&= 4p(2p-3) + 3(2p-3) \\
&= (4p+3)(2p-3)
\end{aligned}
$$

So, $-8p^2 + 6p + 9 = -(4p+3)(2p-3)$.

71. $15 \cdot 4 = 60 = -20 \cdot (-3)$ and

$-20 + (-3) = -23$

$$
\begin{aligned}
15x^2 - 23x + 4 &= 15x^2 - 20x - 3x + 4 \\
&= 5x(3x-4) - 1(3x-4) \\
&= (3x-4)(5x-1)
\end{aligned}
$$

73. $-13y + 12 - 4y^2 = -4y^2 - 13y + 12$

$$
\begin{aligned}
&= -(4y^2 + 13y - 12) \\
&= -(4y^2 + 16y - 3y - 12) \\
&= -[4y(y+4) - 3(y+4)] \\
&= -(y+4)(4y-3)
\end{aligned}
$$

75. $10x^2 - 8xy - 24y^2 = 2(5x^2 - 4xy - 12y^2)$

$$
\begin{aligned}
&= 2(5x^2 - 10xy + 6xy - 12y^2) \\
&= 2[5x(x-2y) + 6y(x-2y)] \\
&= 2(x-2y)(5x+6y)
\end{aligned}
$$

77. $8 - 18x + 9x^2 = 9x^2 - 18x + 8$

$$
\begin{aligned}
&= 9x^2 - 12x - 6x + 8 \\
&= 3x(3x-4) - 2(3x-4) \\
&= (3x-4)(3x-2)
\end{aligned}
$$

79. $-12x + 9 - 24x^2 = -24x^2 - 12x + 9$

$$
= -3(8x^2 + 4x - 3)
$$

81. $4x^3y^2 - 8x^2y^3 - 4x^2y^2 = 4x^2y^2(x - 2y - 1)$

83. $4m^2 + 13mn + 3n^2 = 4m^2 + 12mn + mn + 3n^2$
$$= 4m(m + 3n) + n(m + 3n)$$
$$= (m + 3n)(4m + n)$$

85. $6x^2 - 17x - 12$ cannot be factored; it is prime.

87. $48xy + 24x^2 - 30y^2$
$$= 24x^2 + 48xy - 30y^2$$
$$= 6(4x^2 + 8xy - 5y^2)$$
$$= 6(4x^2 + 10xy - 2xy - 5y^2)$$
$$= 6[2x(2x + 5y) - y(2x + 5y)]$$
$$= 6(2x + 5y)(2x - y)$$

89. $18m^2 + 39mn - 24n^2$
$$= 3(6m^2 + 13mn - 8n^2)$$
$$= 3(6m^2 + 16mn - 3mn - 8n^2)$$
$$= 3[2m(3m + 8n) - n(3m + 8n)]$$
$$= 3(3m + 8n)(2m - n)$$

91. $-6x^3 + 10x^2 - 4x^4 = -4x^4 - 6x^3 + 10x^2$
$$= -2x^2(2x^2 + 3x - 5)$$
$$= -2x^2(2x + 5)(x - 1)$$

93. $30x + 22x^2 - 24x^3 = -24x^3 + 22x^2 + 30x$
$$= -2x(12x^2 - 11x - 15)$$
$$= -2x(4x + 3)(3x - 5)$$

95. $6x^2(x^2 + 1) - 25x(x^2 + 1) + 14(x^2 + 1)$
$$= (x^2 + 1)(6x^2 - 25x + 14)$$
$$= (x^2 + 1)(6x^2 - 21x - 4x + 14)$$
$$= (x^2 + 1)[3x(2x - 7) - 2(2x - 7)]$$
$$= (x^2 + 1)(2x - 7)(3x - 2)$$

97. $3x^2 + \dfrac{13}{2}x - 14$
$$= \frac{1}{2}(6x^2 + 13x - 28)$$
$$= \frac{1}{2}(6x^2 + 21x - 8x - 28)$$
$$= \frac{1}{2}[(6x^2 + 21x) + (-8x - 28)]$$
$$= \frac{1}{2}[3x(2x + 7) - 4(2x + 7)]$$
$$= \frac{1}{2}(2x + 7)(3x - 4)$$

The base and height are represented by $(2x + 7)$ meters and $(3x - 4)$ meters.

99. $6 = 3 \cdot 2$ and $-10 = 2 \cdot (-5)$, so the other factor is $2x - 5$.

101. $27z^4 + 42z^2 + 16 = (9z^2 + 8)(3z^2 + 2)$

103. $3x^{2n} + 19x^n + 6 = 3x^{2n} + x^n + 18x^n + 6$
$$= x^n(3x^n + 1) + 6(3x^n + 1)$$
$$= (3x^n + 1)(x^n + 6)$$

105. $3x^2 = 3x \cdot x$ and $-5 = -1 \cdot 5$ or $1 \cdot (-5)$.
$$(3x + 5)(x - 1) = 3x^2 + 2x - 5$$
$$(3x - 5)(x + 1) = 3x^2 - 2x - 5$$
$$(3x + 1)(x - 5) = 3x^2 - 14x - 5$$
$$(3x - 1)(x + 5) = 3x^2 + 14x - 5$$
Thus, b could be ± 2 or ± 14.

107. The trial and error method may be used when the value of a and/or the value of c are prime numbers, or numbers with few factors. The grouping method can be used when the product of a and c is not a very large number. Examples may vary.

Section 6.4

Preparing for Factoring Special Products

P1. $5^2 = 5 \cdot 5 = 25$

P2. $(-2)^3 = (-2) \cdot (-2) \cdot (-2) = -8$

P3. $(5p^2)^3 = 5^3(p^2)^3 = 125p^{2 \cdot 3} = 125p^6$

P4. $(3z + 2)^2 = (3z)^2 + 2(3z)(2) + 2^2 = 9z^2 + 12z + 4$

P5. $(4m+5)(4m-5) = (4m)^2 - 5^2 = 16m^2 - 25$

6.4 Quick Checks

1. The expression $A^2 + 2AB + B^2$ is called a <u>perfect square trinomial</u>.

2. $A^2 - 2AB + B^2 = \underline{(A-B)^2}$

3. $x^2 - 12x + 36 = x^2 - 2 \cdot x \cdot 6 + 6^2$
   $= (x-6)^2$

4. $16x^2 + 40x + 25 = (4x)^2 + 2 \cdot 4x \cdot 5 + 5^2$
   $= (4x+5)^2$

5. $9a^2 - 60ab + 100b^2$
   $= (3a)^2 - 2 \cdot 3a \cdot 10b + (10b)^2$
   $= (3a - 10b)^2$

6. $z^2 - 8z + 16 = z^2 - 2 \cdot z \cdot 4 + 4^2$
   $= (z-4)^2$

7. $4n^2 + 12n + 9 = (2n)^2 + 2 \cdot 2n \cdot 3 + 3^2$
   $= (2n+3)^2$

8. $35 = 1 \cdot 35$ or $5 \cdot 7$
   $1 + 35 = 36$ and $5 + 7 = 12$, so none of the factors of 35 sum to 13. The polynomial is prime.

9. The first thing we look for when factoring a trinomial is the <u>greatest common factor</u>.

10. $4z^2 + 24z + 36 = 4(z^2 + 6z + 9)$
    $= 4(z^2 + 2 \cdot z \cdot 3 + 3^2)$
    $= 4(z+3)^2$

11. $50a^3 + 80a^2 + 32a$
    $= 2a(25a^2 + 40a + 16)$
    $= 2a[(5a)^2 + 2 \cdot 5a \cdot 4 + 4^2]$
    $= 2a(5a+4)^2$

12. $4m^2 - 81n^2$ is called the <u>difference</u> of <u>two squares</u> and factors into two binomials.

13. $P^2 - Q^2 = \underline{(P-Q)(P+Q)}$

14. $z^2 - 25 = z^2 - 5^2 = (z-5)(z+5)$

15. $81m^2 - 16n^2 = (9m)^2 - (4n)^2$
    $= (9m-4n)(9m+4n)$

16. $16a^2 - \dfrac{4}{9}b^2 = (4a)^2 - \left(\dfrac{2}{3}b\right)^2$
    $= \left(4a - \dfrac{2}{3}b\right)\left(4a + \dfrac{2}{3}b\right)$

17. True: $x^2 + 9$ is prime.

18. $100k^4 - 81w^2 = (10k^2)^2 - (9w)^2$
    $= (10k^2 - 9w)(10k^2 + 9w)$

19. $x^4 - 16 = (x^2)^2 - 4^2$
    $= (x^2 - 4)(x^2 + 4)$
    $= (x^2 - 2^2)(x^2 + 4)$
    $= (x-2)(x+2)(x^2 + 4)$

20. False;
    $4x^2 - 16y^2 = 4(x^2 - 4y^2) = 4(x-2y)(x+2y)$

21. $147x^2 - 48 = 3(49x^2 - 16)$
    $= 3[(7x)^2 - 4^2]$
    $= 3(7x-4)(7x+4)$

22. $-27a^3b + 75ab^3 = -3ab(9a^2 - 25b^2)$
    $= -3ab[(3a)^2 - (5b)^2]$
    $= -3ab(3a-5b)(3a+5b)$

23. The binomial $27x^3 + 64y^3$ is called the <u>sum</u> of <u>two cubes</u>.

24. False; $(x-1)(x^2 - x + 1)$ cannot be factored further.

25. $z^3 + 125 = z^3 + 5^3$
    $= (z+5)(z^2 - z \cdot 5 + 5^2)$
    $= (z+5)(z^2 - 5z + 25)$

26. $8p^3 - 27q^6$
    $= (2p)^3 - (3q^2)^3$
    $= (2p - 3q^2)[(2p)^2 + 2p \cdot 3q^2 + (3q^2)^2]$
    $= (2p - 3q^2)(4p^2 + 6pq^2 + 9q^4)$

27. $54a - 16a^4 = 2a(27 - 8a^3)$
$$= 2a[3^3 - (2a)^3]$$
$$= 2a(3 - 2a)[3^2 + 3 \cdot 2a + (2a)^2]$$
$$= 2a(3 - 2a)(9 + 6a + 4a^2)$$

28. $-375b^3 + 3 = -3(125b^3 - 1)$
$$= -3[(5b)^3 - 1^3]$$
$$= -3(5b - 1)[(5b)^2 + 5b \cdot 1 + 1^2]$$
$$= -3(5b - 1)(25b^2 + 5b + 1)$$

6.4 Exercises

29. $x^2 + 10x + 25 = x^2 + 2 \cdot x \cdot 5 + 5^2 = (x + 5)^2$

31. $4p^2 - 4p + 1 = (2p)^2 - 2 \cdot 2p \cdot 1 + 1^2$
$$= (2p - 1)^2$$

33. $16x^2 + 24x + 9 = (4x)^2 + 2 \cdot 4x \cdot 3 + 3^2$
$$= (4x + 3)^2$$

35. $x^2 - 4xy + 4y^2 = x^2 - 2 \cdot x \cdot 2y + (2y)^2$
$$= (x - 2y)^2$$

37. $4z^2 - 12z + 9 = (2z)^2 - 2 \cdot 2z \cdot 3 + 3^2$
$$= (2z - 3)^2$$

39. $16x^2 - 49y^2 = (4x)^2 - (7y)^2$
$$= (4x - 7y)(4x + 7y)$$

41. $4x^2 - 25 = (2x)^2 - 5^2 = (2x - 5)(2x + 5)$

43. $100n^8 - 81p^4 = (10n^4)^2 - (9p^2)^2$
$$= (10n^4 - 9p^2)(10n^4 + 9p^2)$$

45. $k^8 - 256 = (k^4)^2 - 16^2$
$$= (k^4 - 16)(k^4 + 16)$$
$$= [(k^2)^2 - 4^2](k^4 + 16)$$
$$= (k^2 - 4)(k^2 + 4)(k^4 + 16)$$
$$= (k^2 - 2^2)(k^2 + 4)(k^4 + 16)$$
$$= (k - 2)(k + 2)(k^2 + 4)(k^4 + 16)$$

47. $25p^4 - 49q^2 = (5p^2)^2 - (7q)^2$
$$= (5p^2 - 7q)(5p^2 + 7q)$$

49. $27 + x^3 = 3^3 + x^3$
$$= (3 + x)(3^2 - 3 \cdot x + x^2)$$
$$= (3 + x)(9 - 3x + x^2)$$

51. $8x^3 - 27y^3$
$$= (2x)^3 - (3y)^3$$
$$= (2x - 3y)[(2x)^2 + (2x)(3y) + (3y)^2]$$
$$= (2x - 3y)(4x^2 + 6xy + 9y^2)$$

53. $x^6 - 8y^3$
$$= (x^2)^3 - (2y)^3$$
$$= (x^2 - 2y)[(x^2)^2 + (x^2)(2y) + (2y)^2]$$
$$= (x^2 - 2y)(x^4 + 2x^2 y + 4y^2)$$

55. $27c^3 + 64d^9$
$$= (3c)^3 + (4d^3)^3$$
$$= (3c + 4d^3)[(3c)^2 - (3c)(4d^3) + (4d^3)^2]$$
$$= (3c + 4d^3)(9c^2 - 12cd^3 + 16d^6)$$

57. $16x^3 + 250y^3$
$$= 2(8x^3 + 125y^3)$$
$$= 2[(2x)^3 + (5y)^3]$$
$$= 2(2x + 5y)[(2x)^2 - (2x)(5y) + (5y)^2]$$
$$= 2(2x + 5y)(4x^2 - 10xy + 25y^2)$$

59. $16m^2 + 40mn + 25n^2$
$$= (4m)^2 + 2 \cdot 4m \cdot 5n + (5n)^2$$
$$= (4m + 5n)^2$$

61. $18 - 12x + 2x^2 = 2(9 - 6x + x^2)$
$$= 2(3^2 - 2 \cdot 3 \cdot x + x^2)$$
$$= 2(3 - x)^2 \text{ or } 2(x - 3)^2$$

63. $48a^3 + 72a^2 + 27a$
$$= 3a(16a^2 + 24a + 9)$$
$$= 3a[(4a)^2 + 2 \cdot 4a \cdot 3 + 3^2]$$
$$= 3a(4a + 3)^2$$

65. $2x^3 + 10x^2 + 16x = 2x(x^2 + 5x + 8)$

67. $x^4 y^2 - x^2 y^4 = x^2 y^2(x^2 - y^2)$
$$= x^2 y^2(x - y)(x + y)$$

69. $x^8 - 25y^{10} = (x^4)^2 - (5y^5)^2$
$$= (x^4 - 5y^5)(x^4 + 5y^5)$$

71. $2t^4 - 54t = 2t(t^3 - 27)$
$$= 2t[(t)^3 - (3)^3]$$
$$= 2t(t-3)(t^2 + 3t + 9)$$

73. $3s^7 + 24s = 3s(s^6 + 8)$
$$= 3s[(s^2)^3 + (2)^3]$$
$$= 3s(s^2 + 2)[(s^2)^2 - 2s^2 + (2)^2]$$
$$= 3s(s^2 + 2)(s^4 - 2s^2 + 4)$$

75. $2x^5 - 162x = 2x(x^4 - 81)$
$$= 2x[(x^2)^2 - (9)^2]$$
$$= 2x(x^2 - 9)(x^2 + 9)$$
$$= 2x[(x)^2 - (3)^2](x^2 + 9)$$
$$= 2x(x-3)(x+3)(x^2 + 9)$$

77. $x^3 y - xy^3 = xy(x^2 - y^2) = xy(x-y)(x+y)$

79. $x^3 y^3 + 1 = (xy)^3 + (1)^3$
$$= (xy+1)[(xy)^2 - xy(1) + (1)^2]$$
$$= (xy+1)(x^2 y^2 - xy + 1)$$

81. $3n^2 + 14n + 36$ cannot be factored; it is prime.

83. $2x^2 - 8x + 8 = 2(x^2 - 4x + 4)$
$$= 2(x-2)^2$$

85. $9x^2 + y^2$ cannot be factored; it is prime.

87. $x^4 y^3 + 216xy^3 = xy^3(x^3 + 216)$
$$= xy^3[(x)^3 + (6)^3]$$
$$= xy^3(x+6)[(x)^2 - 6x + (6)^2]$$
$$= xy^3(x+y)(x^2 - 6x + 36)$$

89. $48n^4 - 24n^3 + 3n^2 = 3n^2(16n^2 - 8n + 1)$
$$= 3n^2(4n-1)^2$$

91. $2x(x^2 - 4) + 5(x^2 - 4)$
$$= (x^2 - 4)(2x + 5)$$
$$= (x^2 - 2^2)(2x + 5)$$
$$= (x-2)(x+2)(2x+5)$$

93. $2y^3 + 5y^2 - 32y - 80$
$$= (2y^3 + 5y^2) + (-32y - 80)$$
$$= y^2(2y+5) - 16(2y+5)$$
$$= (2y+5)(y^2 - 16)$$
$$= (2y+5)(y^2 - 4^2)$$
$$= (2y+5)(y-4)(y+4)$$

95. $4x^2 + 20x + 25 = (2x)^2 + 2 \cdot 2x \cdot 5 + 5^2$
$$= (2x+5)^2$$
The length of one side is $2x + 5$.

97. $(x-2)^2 - (x+1)^2$
$$= [(x-2) - (x+1)][(x-2) + (x+1)]$$
$$= -3(2x-1)$$

99. $(x-y)^3 + y^3$
$$= [(x-y) + y][(x-y)^2 - (x-y)y + y^2]$$
$$= x(x^2 - 2xy + y^2 - xy + y^2 + y^2)$$
$$= x(x^2 - 3xy + 3y^2)$$

101. $(x+1)^2 - 9 = [(x+1) - 3][(x+1) + 3]$
$$= (x-2)(x+4)$$

103. $2a^2(x+1) - 17a(x+1) + 30(x+1)$
$$= (x+1)(2a^2 - 17a + 30)$$
$$= (x+1)(2a-5)(a-6)$$

105. $5(x+2)^2 - 7(x+2) - 6$
$$= [5(x+2) + 3][(x+2) - 2]$$
$$= (5x+13)(x)$$
$$= x(5x+13)$$

107. First identify the problem as a sum or a difference of cubes. Identify A and B and rewrite the problem as $A^3 + B^3$ or $A^3 - B^3$. The sum of cubes can be factored into the product of a binomial and a trinomial where the binomial factor is $A + B$ and the trinomial factor is A^2 minus the product AB plus B^2. The difference of cubes also can be factored into the product of a binomial and a trinomial, but here the binomial factor is $A - B$ and the trinomial factor is the sum of A^2, the product AB, and B^2.

109. The correct factorization of $x^3 + y^3$ is

$(x+y)(x^2 - xy + y^2)$. The trinomial factor

$(x^2 - xy + y^2)$ is not a perfect square trinomial.

A perfect square trinomial is

$(x-y)^2 = (x^2 - 2xy + y^2)$.

Section 6.5

6.5 Quick Checks

1. The first step in any factoring problem is to look for the <u>greatest common factor</u>.

2. $2p^2 + 8p - 90 = 2(p^2 + 4p - 45)$

$= 2(p-5)(p+9)$

Check: $2(p-5)(p+9) = 2(p^2 + 9p - 5p - 45)$

$= 2(p^2 + 4p - 45)$

$= 2p^2 + 8p - 90$

3. $-45x^2 + 3xy + 6y^2$

$= -3(15x^2 - xy - 2y^2)$

$= -3(15x^2 + 5xy - 6xy - 2y^2)$

$= -3[5x(3x+y) - 2y(3x+y)]$

$= -3(3x+y)(5x-2y)$

Check: $-3(3x+y)(5x-2y)$

$= -3(15x^2 - 6xy + 5xy - 2y^2)$

$= -3(15x^2 - xy - 2y^2)$

$= -45x^2 + 3xy + 6y^2$

4. $100x^2 - 81y^2 = (10x)^2 - (9y)^2$

$= (10x - 9y)(10x + 9y)$

Check: $(10x - 9y)(10x + 9y)$

$= 100x^2 + 90xy - 90xy - 81y^2$

$= 100x^2 - 81y^2$

5. $2ab^2 - 242a = 2a(b^2 - 121)$

$= 2a(b^2 - 11^2)$

$= 2a(b-11)(b+11)$

Check:

$2a(b-11)(b+11) = 2a(b^2 + 11b - 11b - 121)$

$= 2a(b^2 - 121)$

$= 2ab^2 - 242a$

6. $p^2 - 12pq + 36q^2 = p^2 - 2 \cdot p \cdot 6q + (6q)^2$

$= (p-6q)^2$

Check: $(p-6q)^2 = (p-6q)(p-6q)$

$= p^2 - 6pq - 6pq + 36q^2$

$= p^2 - 12pq + 36q^2$

7. $75x^2 + 90x + 27 = 3(25x^2 + 30x + 9)$

$= 3[(5x)^2 + 2 \cdot 5x \cdot 3 + 3^2]$

$= 3(5x+3)^2$

Check: $3(5x+3)^2 = 3(5x+3)(5x+3)$

$= 3(25x^2 + 15x + 15x + 9)$

$= 3(25x^2 + 30x + 9)$

$= 75x^2 + 90x + 27$

8. $125y^3 - 64 = (5y)^3 - 4^3$

$= (5y-4)[(5y)^2 + 5y \cdot 4 + 4^2]$

$= (5y-4)(25y^2 + 20y + 16)$

Check:

$(5y-4)(25y^2 + 20y + 16)$

$= 125y^3 + 100y^2 + 80y - 100y^2 - 80y - 64$

$= 125y^3 - 64$

9. $-24a^3 + 3b^3$

$= -3(8a^3 - b^3)$

$= -3[(2a)^3 - b^3]$

$= -3(2a-b)[(2a)^2 + 2a \cdot b + b^2]$

$= -3(2a-b)(4a^2 + 2ab + b^2)$

Check:

$-3(2a-b)(4a^2 + 2ab + b^2)$

$= -3(8a^3 + 4a^2b + 2ab^2 - 4a^2b - 2ab^2 - b^3)$

$= -3(8a^3 - b^3)$

$= -24a^3 + 3b^3$

10. False; $x^4 - 81 = (x^2 - 9)(x^2 + 9)$

$= (x+3)(x-3)(x^2 + 9)$

11. $x^4 - 1 = (x^2)^2 - 1^2$

$= (x^2 - 1)(x^2 + 1)$

$= (x^2 - 1^2)(x^2 + 1)$

$= (x-1)(x+1)(x^2 + 1)$

Check: $(x-1)(x+1)(x^2+1)$
$$= (x^2+x-x-1)(x^2+1)$$
$$= (x^2-1)(x^2+1)$$
$$= x^4+x^2-x^2-1$$
$$= x^4-1$$

12. $-36x^2y+16y = -4y(9x^2-4)$
$$= -4y[(3x)^2-2^2]$$
$$= -4y(3x-2)(3x+2)$$

Check: $-4y(3x-2)(3x+2)$
$$= -4y(9x^2+6x-6x-4)$$
$$= -4y(9x^2-4)$$
$$= -36x^2y+16y$$

13. When factoring a polynomial with four terms, try factoring by <u>grouping</u>.

14. True

15. $2x^3+3x^2+4x+6 = (2x^3+3x^2)+(4x+6)$
$$= x^2(2x+3)+2(2x+3)$$
$$= (2x+3)(x^2+2)$$

Check: $(2x+3)(x^2+2) = 2x^3+4x+3x^2+6$
$$= 2x^3+3x^2+4x+6$$

16. $6x^3-9x^2-6x-9$
$$= 3(2x^3+3x^2-2x-3)$$
$$= 3[(2x^3+3x^2)+(-2x-3)]$$
$$= 3[x^2(2x+3)-1(2x+3)]$$
$$= 3(2x+3)(x^2-1)$$
$$= 3(2x+3)(x^2-1^2)$$
$$= 3(2x+3)(x-1)(x+1)$$

Check: $3(2x+3)(x-1)(x+1)$
$$= 3(2x+3)(x^2+x-x-1)$$
$$= 3(2x+3)(x^2-1)$$
$$= 3(2x^3-2x+3x^2-3)$$
$$= 3(2x^3+3x^2-2x-3)$$
$$= 6x^3+9x^2-6x-9$$

17. $-3z^2+9z-21 = -3(z^2-3z+7)$
No further factorization is possible since z^2-3z+7 is prime.

18. $6xy^2+15x^3 = 3x(2y^2+5x^2)$
No further factorization is possible.

6.5 Exercises

19. $x^2-100 = x^2-10^2 = (x-10)(x+10)$

21. $t^2+t-6 = (t+3)(t-2)$

23. $x+y+2ax+2ay = (x+y)+2a(x+y)$
$$= (x+y)(1+2a)$$

25. $a^3-8 = a^3-2^3 = (a-2)(a^2+2a+4)$

27. $a^2-ab-6b^2 = (a-3b)(a+2b)$

29. $2x^2-5x-7 = (2x-7)(x+1)$

31. $2x^2-6xy-20y^2 = 2(x^2-3xy-10y^2)$
$$= 2(x-5y)(x+2y)$$

33. $9-a^2 = 3^2-a^2 = (3-a)(3+a)$

35. $u^2-14u+33 = (u-11)(u-3)$

37. $xy-ay-bx+ab = y(x-a)-b(x-a)$
$$= (x-a)(y-b)$$

39. $w^2+6w+8 = (w+2)(w+4)$

41. $36a^2-49b^4 = (6a)^2-(7b^2)^2$
$$= (6a-7b^2)(6a+7b^2)$$

43. $x^2+2xm-8m^2 = (x+4m)(x-2m)$

45. $6x^2y^2-13xy+6 = (3xy-2)(2xy-3)$

47. $x^3+x^2+x+1 = x^2(x+1)+(x+1)$
$$= (x+1)(x^2+1)$$

49. $12z^2+12z+18 = 6(2z^2+2z+3)$

51. $14c^2+19c-3 = (7c-1)(2c+3)$

53. $27m^3 + 64n^6$
$= (3m)^3 + (4n^2)^3$
$= (3m + 4n^2)(9m^2 - 12mn^2 + 16n^4)$

55. $2j^6 - 2j^2 = 2j^2(j^4 - 1)$
$= 2j^2[(j^2)^2 - 1^2]$
$= 2j^2(j^2 - 1)(j^2 + 1)$
$= 2j^2(j^2 - 1^2)(j^2 + 1)$
$= 2j^2(j - 1)(j + 1)(j^2 + 1)$

57. $8a^2 + 18ab - 5b^2 = (4a - b)(2a + 5b)$

59. $2a^3 + 6a = 2a(a^2 + 3)$

61. $12z^2 - 3 = 3(4z^2 - 1)$
$= 3[(2z)^2 - 1^2]$
$= 3(2z - 1)(2z + 1)$

63. $x^2 - x + 6$ cannot be factored; it is prime.

65. $16a^4 + 2ab^3 = 2a(8a^3 + b^3)$
$= 2a[(2a)^3 + b^3]$
$= 2a(2a + b)(4a^2 - 2ab + b^2)$

67. $p^2q^2 + 6pq - 7 = (pq + 7)(pq - 1)$

69. $s^2(s + 2) - 4(s + 2)$
$= (s + 2)(s^2 - 4)$
$= (s + 2)(s^2 - 2^2)$
$= (s + 2)(s - 2)(s + 2)$ or $(s + 2)^2(s - 2)$

71. $-12x^3 + 2x^2 + 2x = -2x(6x^2 - x - 1)$
$= -2x(3x + 1)(2x - 1)$

73. $10v^2 - 2 - v = 10v^2 - v - 2 = (5v + 2)(2v - 1)$

75. $4n^2 - n^4 + 3n^3 = -n^4 + 3n^3 + 4n^2$
$= -n^2(n^2 - 3n - 4)$
$= -n^2(n - 4)(n + 1)$

77. $-4a^3b + 2a^2b - 2ab = -2ab(2a^2 - a + 1)$

79. $12p - p^3 + p^2 = -p^3 + p^2 + 12p$
$= -p(p^2 - p - 12)$
$= -p(p - 4)(p + 3)$

81. $-32x^3 + 72xy^2 = -8x(4x^2 - 9y^2)$
$= -8x[(2x)^2 - (3y)^2]$
$= -8x(2x - 3y)(2x + 3y)$

83. $2n^3 - 10n^2 - 6n + 30$
$= 2(n^3 - 5n^2 - 3n + 15)$
$= 2[n^2(n - 5) - 3(n - 5)]$
$= 2(n - 5)(n^2 - 3)$

85. $16x^2 + 4x - 12 = 4(4x^2 + x - 3)$
$= 4(4x - 3)(x + 1)$

87. $14x^2 + 3x^4 + 8 = 3x^4 + 14x^2 + 8$
$= (3x^2 + 2)(x^2 + 4)$

89. $-2x^3y + x^2y^2 + 3xy^3 = -xy(2x^2 - xy - 3y^2)$
$= -xy(2x - 3y)(x + y)$

91. $(x^3 + 8x) - (8x^2 - 7x) = x^3 + 8x - 8x^2 + 7x$
$= x^3 - 8x^2 + 15x$
$= x(x^2 - 8x + 15)$
$= x(x - 5)(x - 3)$

93. $4x^3 - 10x^2 - 6x = 2x(2x^2 - 5x - 3)$
$= 2x(2x + 1)(x - 3)$
Possible dimensions for the box are $2x$, $2x + 1$, and $x - 3$.

95. $4m^2 - 4mn + n^2 - p^2$
$= (4m^2 - 4mn + n^2) - p^2$
$= (2m - n)^2 - p^2$
$= (2m - n + p)(2m - n - p)$

97. $x^2 - y^2 + 2yz - z^2$
$= x^2 - (y^2 - 2yz + z^2)$
$= x^2 - (y - z)^2$
$= [x + (y - z)][x - (y - z)]$
$= (x + y - z)(x - y + z)$

99. $x^2 - 2xy + y^2 - 6x + 6y + 8$
$= (x^2 - 2xy + y^2) + (-6x + 6y) + 8$
$= (x-y)^2 - 6(x-y) + 8$
$= [(x-y) - 4][(x-y) - 2]$
$= (x-y-4)(x-y-2)$

101. $x^2 + 2xy + y^2 - a^2 + 2ab - b^2$
$= (x^2 + 2xy + y^2) - (a^2 - 2ab + b^2)$
$= (x+y)^2 - (a-b)^2$
$= [(x+y) - (a-b)][(x+y) + (a-b)]$
$= (x+y-a+b)(x+y+a-b)$

103. The student initially wrote the sum of the terms as a product of terms by enclosing the binomial $-3x - 12$ in parentheses. Then in the second step, the student writes the factor $-3(x-4)$ and concludes that the factorization of

$x^2 + 4x - 3x - 12$ is $(x-3)(x+4)$, even though both terms do not have the common factor of $(x+4)$. The student obtained the correct answer, but the first two steps are incorrect, the correct factorization is

$x^2 + 4x - 3x - 12 = x(x+4) - 3(x+4)$
$= (x+4)(x-3)$

Putting the Concepts Together (Sections 6.1–6.5)

1. The GCF of 10, 15, and 25 is 5.
The GCF of x^3, x^5, and x^2 is x^2.
The GCF of y^4, y, and y^7 is y.
The GCF is $5x^2 y$.

2. $x^2 - 3x - 4 = (x-4)(x+1)$

3. $x^6 - 27 = (x^2)^3 - 3^3$
$= (x^2 - 3)[(x^2)^2 + x^2 \cdot 3 + 3^2]$
$= (x^2 - 3)(x^4 + 3x^2 + 9)$

4. $6x(2x + 1) + 5z(2x + 1) = (2x + 1)(6x + 5z)$

5. $x^2 + 5xy - 6y^2 = (x+6y)(x-y)$

6. $x^3 + 64 = x^3 + 4^3 = (x+4)(x^2 - 4x + 16)$

7. $4x^2 + 49y^2$ cannot be factored; it is prime.

8. $3x^2 + 12xy - 36y^2 = 3(x^2 + 4xy - 12y^2)$
$= 3(x+6y)(x-2y)$

9. $12z^5 - 44z^3 - 24z^2 = 4z^2(3z^3 - 11z - 6)$

10. $x^2 + 6x - 5$ cannot be factored; it is prime.

11. $4m^4 + 5m^3 - 6m^2$
$= m^2(4m^2 + 5m - 6)$
$= m^2(4m^2 + 8m - 3m - 6)$
$= m^2[(4m^2 + 8m) + (-3m - 6)]$
$= m^2[4m(m+2) - 3(m+2)]$
$= m^2(m+2)(4m-3)$

12. $5p^2 - 17p + 6 = (5p - 2)(p - 3)$

13. $10m^2 + 25m - 6m - 15$
$= (10m^2 + 25m) + (-6m - 15)$
$= 5m(2m+5) - 3(2m+5)$
$= (2m+5)(5m-3)$

14. $36m^2 + 6m - 6 = 6(6m^2 + m - 1)$
$= 6(3m-1)(2m+1)$

15. $4m^2 - 20m + 25 = (2m)^2 - 2 \cdot 2m \cdot 5 + 5^2$
$= (2m-5)^2$

16. $5x^2 - xy - 4y^2 = (5x + 4y)(x - y)$

17. $S = 2\pi rh + 2\pi r^2$
$= 2\pi r \cdot h + 2\pi r \cdot r$
$= 2\pi r(h + r)$

18. $h = 48t - 16t^2$
$= 16t \cdot 3 + 16t \cdot (-t)$
$= 16t(3 - t)$

Section 6.6

Preparing for Solving Polynomial Equations by Factoring

P1. $x + 5 = 0$
$x + 5 - 5 = 0 - 5$
$x = -5$
The solution set is $\{-5\}$.

P2. $2(x-4)-10=0$
$$2x-8-10=0$$
$$2x-18=0$$
$$2x-18+18=0+18$$
$$2x=18$$
$$\frac{2x}{2}=\frac{18}{2}$$
$$x=9$$
The solution set is $\{9\}$.

P3. $2x^2+3x-4$

(a) $x=2$: $2(2)^2+3\cdot2-4=8+6-4=10$

(b) $x=-1$: $2(-1)^2+3(-1)-4=2-3-4=-5$

6.6 Quick Checks

1. The Zero-Product Property states that if $ab=0$, then either $\underline{a=0}$ or $\underline{b=0}$.

2. $x(x+3)=0$
$$x=0 \quad\text{or}\quad x+3=0$$
$$x=-3$$
Check $x=0$: $x(x+3)=0$
$$0(0+3)\overset{?}{=}0$$
$$0(3)\overset{?}{=}0$$
$$0=0 \quad\text{True}$$
Check $x=-3$: $x(x+3)=0$
$$-3(-3+3)\overset{?}{=}0$$
$$-3(0)\overset{?}{=}0$$
$$0=0 \quad\text{True}$$
The solution set is $\{0,-3\}$.

3. $(x-2)(4x+5)=0$
$$x-2=0 \quad\text{or}\quad 4x+5=0$$
$$x-2=0 \quad\text{or}\quad 4x=-5$$
$$x=-\frac{5}{4}$$
Check $x=2$: $(x-2)(4x+5)=0$
$$(2-2)(4\cdot2+5)\overset{?}{=}0$$
$$0(13)\overset{?}{=}0$$
$$0=0 \quad\text{True}$$

Check $x=-\dfrac{5}{4}$:
$$(x-2)(4x+5)=0$$
$$\left(-\frac{5}{4}-2\right)\left(4\left(-\frac{5}{4}\right)+5\right)\overset{?}{=}0$$
$$\left(-\frac{13}{4}\right)(-5+5)\overset{?}{=}0$$
$$-\frac{13}{4}\cdot0\overset{?}{=}0$$
$$0=0 \quad\text{True}$$
The solution set is $\left\{2,-\dfrac{5}{4}\right\}$.

4. A <u>quadratic</u> equation is an equation that can be written in the form $ax^2+bx+c=0$, where a, b, and c are real numbers and $a\neq0$.

5. Quadratic equations are also known as <u>second-degree</u> equations.

6. False; the standard form of the equation $3x+x^2=6$ is $x^2+3x-6=0$.

7. $p^2-6p+8=0$
$$(p-2)(p-4)=0$$
$$p-2=0 \quad\text{or}\quad p-4=0$$
$$p=2 \quad\text{or}\quad p=4$$
Check $p=2$: $p^2-6p+8=0$
$$2^2-6\cdot2+8\overset{?}{=}0$$
$$4-12+8\overset{?}{=}0$$
$$0=0 \quad\text{True}$$
Check $p=4$: $p^2-6p+8=0$
$$4^2-6\cdot4+8\overset{?}{=}0$$
$$16-24+8\overset{?}{=}0$$
$$0=0 \quad\text{True}$$
The solution set is $\{2,4\}$.

8. $2t^2-5t=3$
$$2t^2-5t-3=0$$
$$(2t+1)(t-3)=0$$
$$2t+1=0 \quad\text{or}\quad t-3=0$$
$$2t=-1 \quad\text{or}\quad t=3$$
$$t=-\frac{1}{2}$$

Check $t = -\dfrac{1}{2}$:　　　　　$2t^2 - 5t = 3$

$$2\left(-\frac{1}{2}\right)^2 - 5\left(-\frac{1}{2}\right) \stackrel{?}{=} 3$$

$$2\left(\frac{1}{4}\right) + \frac{5}{2} \stackrel{?}{=} 3$$

$$\frac{1}{2} + \frac{5}{2} \stackrel{?}{=} 3$$

$$3 = 3 \quad \text{True}$$

Check $t = 3$:　　　$2t^2 - 5t = 3$

$$2(3)^2 - 5 \cdot 3 \stackrel{?}{=} 3$$

$$18 - 15 \stackrel{?}{=} 3$$

$$3 = 3 \quad \text{True}$$

The solution set is $\left\{-\dfrac{1}{2}, 3\right\}$.

9.　　　$2x^2 + 3x = 5$

$$2x^2 + 3x - 5 = 0$$

$$(2x + 5)(x - 1) = 0$$

$$2x + 5 = 0 \quad \text{or} \quad x - 1 = 0$$

$$2x = -5 \quad \text{or} \quad x = 1$$

$$x = -\frac{5}{2}$$

Check $x = -\dfrac{5}{2}$:　　　$2x^2 + 3x = 5$

$$2\left(-\frac{5}{2}\right)^2 + 3\left(-\frac{5}{2}\right) \stackrel{?}{=} 5$$

$$2\left(\frac{25}{4}\right) - \frac{15}{2} \stackrel{?}{=} 5$$

$$\frac{25}{2} - \frac{15}{2} \stackrel{?}{=} 5$$

$$5 = 5 \quad \text{True}$$

Check $x = 1$:　$2x^2 + 3x = 5$

$$2(1)^2 + 3 \cdot 1 \stackrel{?}{=} 5$$

$$2 + 3 \stackrel{?}{=} 5$$

$$5 = 5 \quad \text{True}$$

The solution set is $\left\{-\dfrac{5}{2}, 1\right\}$.

10.　　　$z^2 + 20 = -9z$

$$z^2 + 9z + 20 = 0$$

$$(z + 5)(z + 4) = 0$$

$$z + 5 = 0 \quad \text{or} \quad z + 4 = 0$$

$$z = -5 \quad \text{or} \quad z = -4$$

Check $z = -5$:　　$z^2 + 20 = -9z$

$$(-5)^2 + 20 \stackrel{?}{=} -9(-5)$$

$$25 + 20 \stackrel{?}{=} 45$$

$$45 = 45 \quad \text{True}$$

Check $z = -4$:　　$z^2 + 20 = -9z$

$$(-4)^2 + 20 \stackrel{?}{=} -9(-4)$$

$$16 + 20 \stackrel{?}{=} 36$$

$$36 = 36 \quad \text{True}$$

The solution set is $\{-5, -4\}$.

11.　　　　$5k^2 + 3k - 1 = 3 - 5k$

$$5k^2 + 3k - 1 + 5k - 3 = 3 - 5k + 5k - 3$$

$$5k^2 + 8k - 4 = 0$$

$$(5k - 2)(k + 2) = 0$$

$$5k - 2 = 0 \quad \text{or} \quad k + 2 = 0$$

$$5k = 2 \quad \text{or} \quad k = -2$$

$$k = \frac{2}{5}$$

Substitute $k = \dfrac{2}{5}$ and $k = -2$ into the original equation to check. The solution set is $\left\{-2, \dfrac{2}{5}\right\}$.

12.　　　　$3x^2 + 9x = 4 - 2x$

$$3x^2 + 9x + 2x - 4 = 4 - 2x + 2x - 4$$

$$3x^2 + 11x - 4 = 0$$

$$(3x - 1)(x + 4) = 0$$

$$3x - 1 = 0 \quad \text{or} \quad x + 4 = 0$$

$$3x = 1 \quad \text{or} \quad x = -4$$

$$x = \frac{1}{3}$$

Substitute $x = \dfrac{1}{3}$ and $x = -4$ into the original equation to check. The solution set is $\left\{\dfrac{1}{3}, -4\right\}$.

13. False;　　$x(x - 3) = 4$

$$x^2 - 3x = 4$$

$$x^2 - 3x - 4 = 4 - 4$$

$$x^2 - 3x - 4 = 0$$

$$(x - 4)(x + 1) = 0$$

$$x - 4 = 0 \quad \text{or} \quad x + 1 = 0$$

$$x = 4 \quad \text{or} \quad x = -1$$

The solution set is $\{-1, 4\}$.

14. $(x-3)(x+5)=9$

$x^2+2x-15=9$

$x^2+2x-24=0$

$(x+6)(x-4)=0$

$x+6=0$ or $x-4=0$

$x=-6$ or $x=4$

Substitute $x=-6$ and $x=4$ into the original equation to check. The solution set is $\{-6,4\}$.

15. $(x+3)(2x-1)=7x-3x^2$

$2x^2+5x-3=7x-3x^2$

$5x^2-2x-3=0$

$(5x+3)(x-1)=0$

$5x+3=0$ or $x-1=0$

$5x=-3$ or $x=1$

$x=-\dfrac{3}{5}$

Substitute $x=-\dfrac{3}{5}$ and $x=1$ into the original

equation to check. The solution set is $\left\{-\dfrac{3}{5},1\right\}$.

16. $9p^2+16=24p$

$9p^2-24p+16=0$

$(3p-4)^2=0$

$3p-4=0$ or $3p-4=0$

$3p=4$ or $3p=4$

$p=\dfrac{4}{3}$ or $p=\dfrac{4}{3}$

Substitute $p=\dfrac{4}{3}$ into the original equation to

check. The solution set is $\left\{\dfrac{4}{3}\right\}$.

17. $4x^2+12x-72=0$

$4(x^2+3x-18)=0$

$4(x+6)(x-3)=0$

$4=0$ or $x+6=0$ or $x-3=0$

$x=-6$ or $x=3$

The statement $4=0$ is false. Substitute $x=-6$ and $x=3$ into the original equation to check. The solution set is $\{-6,3\}$.

18. $-2x^2+2x=-12$

$-2x^2+2x+12=0$

$-2(x^2-x-6)=0$

$-2(x+2)(x-3)=0$

$-2=0$ or $x+2=0$ or $x-3=0$

$x=-2$ or $x=3$

The statement $-2=0$ is false. Substitute $x=-2$ and $x=3$ into the original equation to check. The solution set is $\{-2,3\}$.

19. $-16t^2+80t=64$

$-16t^2+80t-64=0$

$-16(t^2-5t+4)=0$

$-16(t-1)(t-4)=0$

$-16=0$ or $t-1=0$ or $t-4=0$

$t=1$ or $t=4$

The statement $-16=0$ is false. The toy rocket is 64 feet from the ground after 1 second and 4 seconds.

20. $x^3-4x^2-12x=0$

$x(x^2-4x-12)=0$

$x(x+2)(x-6)=0$

$x=0$ or $x+2=0$ or $x-6=0$

$x=2$ $x=6$

The solution set is $\{0,2,6\}$.

True, x^3-4x^2-12x can be solved using the Zero-Product Property.

21. $(4x-5)(x^2-9)=0$

$(4x-5)(x-3)(x+3)=0$

$4x-5=0$ or $x-3=0$ or $x+3=0$

$4x=5$ or $x=3$ or $x=-3$

$x=\dfrac{5}{4}$

Substitute $x=\dfrac{5}{4}$, $x=3$, and $x=-3$ into the

original equation to check. The solution set is $\left\{\dfrac{5}{4},3,-3\right\}$.

22. $3x^3+9x^2+6x=0$

$3x(x^2+3x+2)=0$

$3x(x+1)(x+2)=0$

$3x=0$ or $x+1=0$ or $x+2=0$

$x=0$ or $x=-1$ or $x=-2$

Substitute $x=0$, $x=-1$, and $x=-2$ into the original equation to check. The solution set is $\{0,-1,-2\}$.

6.6 Exercises

23. $2x(x + 4) = 0$
$2x = 0$ or $x + 4 = 0$
$x = 0$ $\qquad$ $x = -4$
The solution set is $\{-4, 0\}$.

25. $(n + 3)(n - 9) = 0$
$n + 3 = 0$ or $n - 9 = 0$
$n = -3$ $\qquad$ $n = 9$
The solution set is $\{-3, 9\}$.

27. $(3p + 1)(p - 5) = 0$
$3p + 1 = 0$ or $p - 5 = 0$
$3p = -1$ $\qquad$ $p = 5$
$p = -\dfrac{1}{3}$
The solution set is $\left\{-\dfrac{1}{3}, 5\right\}$.

29. $(5y + 3)(2 - 7y) = 0$
$5y + 3 = 0$ or $2 - 7y = 0$
$5y = -3$ $\qquad$ $-7y = -2$
$y = -\dfrac{3}{5}$ $\qquad$ $y = \dfrac{2}{7}$
The solution set is $\left\{-\dfrac{3}{5}, \dfrac{2}{7}\right\}$.

31. The highest exponent is 1; the equation is linear.

33. The highest exponent is 2; the equation is quadratic.

35. $x^2 - 3x - 4 = 0$
$(x - 4)(x + 1) = 0$
$x - 4 = 0$ or $x + 1 = 0$
$x = 4$ $\qquad$ $x = -1$
The solution set is $\{-1, 4\}$.

37. $n^2 + 9n + 14 = 0$
$(n + 7)(n + 2) = 0$
$n + 7 = 0$ or $n + 2 = 0$
$n = -7$ $\qquad$ $n = -2$
The solution set is $\{-7, -2\}$.

39. $4x^2 + 2x = 0$
$2x(2x + 1) = 0$
$2x = 0$ or $2x + 1 = 0$
$x = 0$ $\qquad$ $2x = -1$
$x = -\dfrac{1}{2}$
The solution set is $\left\{-\dfrac{1}{2}, 0\right\}$.

41. $2x^2 - 3x - 2 = 0$
$(2x + 1)(x - 2) = 0$
$2x + 1 = 0$ or $x - 2 = 0$
$2x = -1$ $\qquad$ $x = 2$
$x = -\dfrac{1}{2}$
The solution set is $\left\{-\dfrac{1}{2}, 2\right\}$.

43. $a^2 - 6a + 9 = 0$
$(a - 3)^2 = 0$
$a - 3 = 0$ or $a - 3 = 0$
$a = 3$ $\qquad$ $a = 3$
The solution set is $\{3\}$.

45. $6x^2 = 36x$
$6x^2 - 36x = 0$
$6x(x - 6) = 0$
$6x = 0$ or $x - 6 = 0$
$x = 0$ $\qquad$ $x = 6$
The solution set is $\{0, 6\}$.

47. $n^2 - n = 6$
$n^2 - n - 6 = 0$
$(n - 3)(n + 2) = 0$
$n - 3 = 0$ or $n + 2 = 0$
$n = 3$ $\qquad$ $n = -2$
The solution set is $\{-2, 3\}$.

49. $b^2 + 18 = 11b$
$b^2 - 11b + 18 = 0$
$(b - 9)(b - 2) = 0$
$b - 9 = 0$ or $b - 2 = 0$
$b = 9$ $\qquad$ $b = 2$
The solution set is $\{2, 9\}$.

51.
$$1 - 5m = -4m^2$$
$$4m^2 - 5m + 1 = 0$$
$$(4m - 1)(m - 1) = 0$$
$$4m - 1 = 0 \quad \text{or} \quad m - 1 = 0$$
$$4m = 1 \qquad\qquad m = 1$$
$$m = \frac{1}{4}$$

The solution set is $\left\{\dfrac{1}{4}, 1\right\}$.

53.
$$n(n - 2) = 24$$
$$n^2 - 2n = 24$$
$$n^2 - 2n - 24 = 0$$
$$(n - 6)(n + 4) = 0$$
$$n - 6 = 0 \quad \text{or} \quad n + 4 = 0$$
$$n = 6 \qquad\qquad n = -4$$
The solution set is $\{-4, 6\}$.

55.
$$(x - 2)(x - 3) = 56$$
$$x^2 - 5x + 6 = 56$$
$$x^2 - 5x - 50 = 0$$
$$(x - 10)(x + 5) = 0$$
$$x - 10 = 0 \quad \text{or} \quad x + 5 = 0$$
$$x = 10 \qquad\qquad x = -5$$
The solution set is $\{-5, 10\}$.

57.
$$(c + 2)^2 = 9$$
$$c^2 + 4c + 4 = 9$$
$$c^2 + 4c - 5 = 0$$
$$(c + 5)(c - 1) = 0$$
$$c + 5 = 0 \quad \text{or} \quad c - 1 = 0$$
$$c = -5 \qquad\qquad c = 1$$
The solution set is $\{-5, 1\}$.

59. $2x^3 + 2x^2 - 12x = 0$
$$2x(x^2 + x - 6) = 0$$
$$2x(x + 3)(x - 2) = 0$$
$$2x = 0 \quad \text{or} \quad x + 3 = 0 \quad \text{or} \quad x - 2 = 0$$
$$x = 0 \qquad\quad x = -3 \qquad\qquad x = 2$$
The solution set is $\{-3, 0, 2\}$.

61. $y^3 + 3y^2 - 4y - 12 = 0$
$$y^2(y + 3) - 4(y + 3) = 0$$
$$(y + 3)(y^2 - 4) = 0$$
$$(y + 3)(y - 2)(y + 2) = 0$$
$$y + 3 = 0 \quad \text{or} \quad y - 2 = 0 \quad \text{or} \quad y + 2 = 0$$
$$y = -3 \qquad\quad y = 2 \qquad\qquad y = -2$$
The solution set is $\{-3, -2, 2\}$.

63.
$$2x^3 + 3x^2 = 8x + 12$$
$$2x^3 + 3x^2 - 8x - 12 = 0$$
$$x^2(2x + 3) - 4(2x + 3) = 0$$
$$(2x + 3)(x^2 - 4) = 0$$
$$(2x + 3)(x - 2)(x + 2) = 0$$
$$2x + 3 = 0 \quad \text{or} \quad x - 2 = 0 \quad \text{or} \quad x + 2 = 0$$
$$2x = -3 \qquad\qquad x = 2 \qquad\qquad x = -2$$
$$x = -\frac{3}{2}$$
The solution set is $\left\{-\dfrac{3}{2}, 2, -2\right\}$.

65. $(5x + 3)(x - 4) = 0$
$$5x + 3 = 0 \quad \text{or} \quad x - 4 = 0$$
$$5x = -3 \qquad\qquad x = 4$$
$$x = -\frac{3}{5}$$
The solution set is $\left\{-\dfrac{3}{5}, 4\right\}$.

67. $p^2 - p - 20 = 0$
$$(p - 5)(p + 4) = 0$$
$$p - 5 = 0 \quad \text{or} \quad p + 4 = 0$$
$$p = 5 \qquad\qquad p = -4$$
The solution set is $\{-4, 5\}$.

69. $4w + 3 = 2w - 7$
$$2w + 3 = -7$$
$$2w = -10$$
$$w = -5$$
The solution set is $\{-5\}$.

71. $4a^2 - 25a = 21$

$4a^2 - 25a - 21 = 0$

$(4a + 3)(a - 7) = 0$

$4a + 3 = 0$ or $a - 7 = 0$

    $4a = -3$ $a = 7$

      $a = -\dfrac{3}{4}$

The solution set is $\left\{ -\dfrac{3}{4}, 7 \right\}$.

73. $2a(a + 1) = a^2 + 8$

   $2a^2 + 2a = a^2 + 8$

   $a^2 + 2a - 8 = 0$

   $(a + 4)(a - 2) = 0$

   $a + 4 = 0$ or $a - 2 = 0$

     $a = -4$ $a = 2$

The solution set is $\{-4, 2\}$.

75. $2x^3 + x^2 = 32x + 16$

   $2x^3 + x^2 - 32x - 16 = 0$

  $x^2(2x + 1) - 16(2x + 1) = 0$

      $(2x + 1)(x^2 - 16) = 0$

   $(2x + 1)(x - 4)(x + 4) = 0$

  $2x + 1 = 0$ or $x - 4 = 0$ or $x + 4 = 0$

    $2x = -1$ $x = 4$ $x = -4$

      $x = -\dfrac{1}{2}$

The solution set is $\left\{ -4, -\dfrac{1}{2}, 4 \right\}$.

77. $4(b - 3) - 3b = 8$

   $4b - 12 - 3b = 8$

      $b - 12 = 8$

          $b = 20$

The solution set is $\{20\}$.

79. $y^2 + 5y = 5(y + 20)$

  $y^2 + 5y - 5(y + 20) = 0$

  $y^2 + 5y - 5y - 100 = 0$

        $y^2 - 100 = 0$

    $(y + 10)(y - 10) = 10$

  $y + 10 = 0$ or $y - 10 = 0$

    $y = -10$ $y = 10$

The solution set is $\{-10, 10\}$.

81. $(a + 3)(a - 5)(3a + 2) = 0$

   $a + 3 = 0$ or $a - 5 = 0$ or $3a + 2 = 0$

     $a = -3$ $a = 5$ $3a = -2$

                                   $a = -\dfrac{2}{3}$

The solution set is $\left\{ -3, -\dfrac{2}{3}, 5 \right\}$.

83. $(2k - 3)(2k^2 - 9k - 5) = 0$

   $(2k - 3)(2k + 1)(k - 5) = 0$

   $2k - 3 = 0$ or $2k + 1 = 0$ or $k - 5 = 0$

     $2k = 3$ $2k = -1$ $k = 5$

      $k = \dfrac{3}{2}$ $k = -\dfrac{1}{2}$

The solution set is $\left\{ -\dfrac{1}{2}, \dfrac{3}{2}, 5 \right\}$.

85. $(w - 3)^2 = 9 + 2w$

  $w^2 - 6w + 9 = 9 + 2w$

     $w^2 - 8w = 0$

     $w(w - 8) = 0$

  $w = 0$ or $w - 8 = 0$

               $w = 8$

The solution set is $\{0, 8\}$.

87. $\dfrac{1}{2}x^2 + \dfrac{5}{4}x = 3$

  $4\left(\dfrac{1}{2}x^2 + \dfrac{5}{4}x \right) = 4(3)$

       $2x^2 + 5x = 12$

   $2x^2 + 5x - 12 = 0$

   $(2x - 3)(x + 4) = 0$

   $2x - 3 = 0$ or $x + 4 = 0$

     $2x = 3$ $x = -4$

      $x = \dfrac{3}{2}$

The solution set is $\left\{ -4, \dfrac{3}{2} \right\}$.

89.
$$8x^2 + 44x = 24$$
$$8x^2 + 44x - 24 = 0$$
$$4(2x^2 + 11x - 6) = 0$$
$$4(2x-1)(x+6) = 0$$
$$4 = 0 \quad \text{or} \quad 2x-1=0 \quad \text{or} \quad x+6=0$$
$$\text{False} \qquad\qquad 2x=1 \qquad\qquad x=-6$$
$$x = \frac{1}{2}$$

The solution set is $\left\{-6, \dfrac{1}{2}\right\}$.

91.
$$-16t^2 + 64t + 80 = 128$$
$$-16t^2 + 64t - 48 = 0$$
$$-16(t^2 - 4t + 3) = 0$$
$$-16(t-3)(t-1) = 0$$
$$-16 = 0 \quad \text{or} \quad t-3=0 \quad \text{or} \quad t-1=0$$
$$\text{false} \qquad\qquad t=3 \qquad\qquad t=1$$

The ball is 128 feet from the ground after 1 second and after 3 seconds.

93.
$$-16t^2 + 40 = 24$$
$$-16t^2 + 16 = 0$$
$$-16(t^2 - 1) = 0$$
$$-16(t-1)(t+1) = 0$$
$$-16 = 0 \quad \text{or} \quad t-1=0 \quad \text{or} \quad t+1=0$$
$$\text{false} \qquad\qquad t=1 \qquad\qquad t=-1$$

The negative value has no meaning. The balloon is 24 feet from the ground after 1 second.

95. Let n and $n + 1$ be the integers.
$$n(n+1) = 12$$
$$n^2 + n - 12 = 0$$
$$(n+4)(n-3) = 0$$
$$n+4=0 \quad \text{or} \quad n-3=0$$
$$n=-4 \qquad\qquad n=3$$
$$n+1=-3 \qquad\quad n+1=4$$

The numbers are either -4 and -3 or 3 and 4.

97. Let n, $n + 2$, and $n + 4$ be the integers.
$$n(n+4) = 96$$
$$n^2 + 4n - 96 = 0$$
$$(n+12)(n-8) = 0$$
$$n+12=0 \quad \text{or} \quad n-8=0$$
$$n=-12 \qquad\qquad n=8$$
$$n+2=-10 \qquad\quad n+2=10$$
$$n+4=-8 \qquad\quad n+4=12$$

The numbers are either -12, -10, and 8 or 8, 10, and 12.

99. Let n and $n + 2$ be the width and length of the rectangle.
$$n(n+2) = 255$$
$$n^2 + 2n - 255 = 0$$
$$(n+17)(n-15) = 0$$
$$n+17=0 \quad \text{or} \quad n-15=0$$
$$n=-17 \qquad\qquad n=15$$
$$\qquad\qquad\qquad n+2=17$$

The negative value has no meaning. The rectangle is 15 by 17.

101.
$$\frac{t^2 - t}{2} = 28$$
$$t^2 - t = 56$$
$$t^2 - t - 56 = 0$$
$$(t-8)(t+7) = 0$$
$$t-8=0 \quad \text{or} \quad t+7=0$$
$$t=8 \qquad\qquad t=-7$$

The negative value has no meaning. The league has 8 teams.

103. Answers may vary. One solution is $(x-3)(x+5) = 0$; 2nd degree

105. Answers may vary. One solution is $(z-6)^2 = 0$; 2nd degree

107. Answers may vary. One solution is $(x+3)(x-1)(x-5) = 0$; 3rd degree

109.
$$x^2 - ax + bx - ab = 0$$
$$x(x-a) + b(x-a) = 0$$
$$(x+b)(x-a) = 0$$
$$x = -b,\ a$$

111.
$$2x^3 - 4ax^2 = 0$$
$$2x^2(x-2a) = 0$$
$$x = 0,\ 2a$$

113. The student divided by the variable x instead of writing the quadratic equation in standard form, factoring, and using the Zero-Product Property. The correct solution is

$$15x^2 = 5x$$
$$15x^2 - 5x = 0$$
$$5x(3x - 1) = 0$$
$$5x = 0 \quad \text{or} \quad 3x - 1 = 0$$
$$x = 0 \quad \text{or} \quad 3x = 1$$
$$x = \frac{1}{3}$$

115. We write a quadratic equation in standard form, $ax^2 + bx + c = 0$, so that when the quadratic polynomial on the left side of the equation is factored, we can apply the Zero-Product Property, which says that if $a \cdot b = 0$, then $a = 0$ or $b = 0$.

Section 6.7

Preparing for Modeling and Solving Problems with Quadratic Equations

P1. $15^2 = 15 \cdot 15 = 225$

P2. $x^2 - 5x - 14 = 0$
$(x + 2)(x - 7) = 0$
$x + 2 = 0 \quad \text{or} \quad x - 7 = 0$
$x = -2 \qquad x = 7$
The solution set is $\{-2, 7\}$.

6.7 Quick Checks

1. (a) Let $h = 384$ and solve for t.
$$-16t^2 + 160t = 384$$
$$-16t^2 + 160t - 384 = 0$$
$$-16(t^2 - 10t + 24) = 0$$
$$-16(t - 4)(t - 6) = 0$$
$$-16 = 0 \quad \text{or} \quad t - 4 = 0 \quad \text{or} \quad t - 6 = 0$$
$$t = 4 \quad \text{or} \qquad t = 6$$

The equation $-16 = 0$ is false, so the solutions are $t = 4$ and $t = 6$. The height of the rocket will be 384 feet after 4 seconds and after 6 seconds.

(b) The rocket will strike the ground when $h = 0$, so we let $h = 0$ and solve for t.
$$-16t^2 + 160t = 0$$
$$-16t(t - 10) = 0$$
$$-16t = 0 \quad \text{or} \quad t - 10 = 0$$
$$t = 0 \quad \text{or} \qquad t = 10$$

$t = 0$ indicates that the rocket was fired from the height of 0 feet (ground level). The rocket strikes the ground after 10 seconds.

2. Let w represent the width of the plot. Then the length is represented by $2w - 3$. We are given that the area is 104 square kilometers.

(length)(width) = area
$$(2w - 3)(w) = 104$$
$$2w^2 - 3w = 104$$
$$2w^2 - 3w - 104 = 0$$
$$(2w + 13)(w - 8) = 0$$
$$2w + 13 = 0 \quad \text{or} \quad w - 8 = 0$$
$$2w = -13 \quad \text{or} \qquad w = 8$$
$$w = -\frac{13}{2}$$

Since w represents the width of the plot, we discard the negative solution.
$w = 8$: $2w - 3 = 2(8) - 3 = 16 - 3 = 13$
The width of the plot is 8 kilometers and the length is 13 kilometers.

3. Let h represent the height of the triangle. Then the base is represented by $h + 4$. We are given that the area is 48 square yards.

$\frac{1}{2}$(base)(height) = area
$$\frac{1}{2}(h + 4)(h) = 48$$
$$h^2 + 4h = 96$$
$$h^2 + 4h - 96 = 0$$
$$(h + 12)(h - 8) = 0$$
$$h + 12 = 0 \quad \text{or} \quad h - 8 = 0$$
$$h = -12 \quad \text{or} \qquad h = 8$$

Since h represents the height of the triangle, we discard the negative solution.
$h = 8$: $h + 4 = 8 + 4 = 12$
The base is 12 yards and the height is 8 yards.

4. In the triangle pictured, the legs have lengths x and $x - 3$, while the hypotenuse has length 15.
$$a^2 + b^2 = c^2$$
$$x^2 + (x - 3)^2 = 15^2$$
$$x^2 + x^2 - 6x + 9 = 225$$
$$2x^2 - 6x + 9 = 225$$
$$2x^2 - 6x - 216 = 0$$
$$2(x^2 - 3x - 108) = 0$$
$$2(x + 9)(x - 12) = 0$$
$$2 = 0 \quad \text{or} \quad x + 9 = 0 \quad \text{or} \quad x - 12 = 0$$
$$x = -9 \quad \text{or} \qquad x = 12$$

The equation $2 = 0$ is false and the solution $x = -9$ makes no sense, so $x = 12$.
$x = 12$: $x - 3 = 12 - 3 = 9$
The legs have lengths $x = 12$ and $x - 3 = 9$.

5. Let x = length of first leg in inches
$x + 17$ = length of second leg in inches
hypotenuse = 25 inches

$$x^2 + (x+17)^2 = 25^2$$
$$x^2 + x^2 + 34x + 289 = 625$$
$$2x^2 + 34x - 336 = 0$$
$$2(x^2 + 17x - 168) = 0$$
$$2(x+24)(x-7) = 0$$

x cannot be negative, so the only solution for
$x = 7$ inches
$x + 17 = 24$ inches

6. Let x represent the height of the screen. Then $x + 2$ represents the width of the screen. We are given that the length of the diagonal is 10 inches.

$$a^2 + b^2 = c^2$$
$$x^2 + (x+2)^2 = 10^2$$
$$x^2 + x^2 + 4x + 4 = 100$$
$$2x^2 + 4x + 4 = 100$$
$$2x^2 + 4x - 96 = 0$$
$$2(x^2 + 2x - 48) = 0$$
$$2(x+8)(x-6) = 0$$

$2 = 0$ or $x + 8 = 0$ or $x - 6 = 0$
              $x = -8$ or $x = 6$

The equation $2 = 0$ is false and the solution $x = -8$ makes no sense, so $x = 6$.
$x = 6$: $x + 2 = 6 + 2 = 8$
The height of the television screen is 6 inches and the width is 8 inches.

6.7 Exercises

7. $\qquad x(2x+6) = 56$
$$2x^2 + 6x - 56 = 0$$
$$2(x^2 + 3x - 28) = 0$$
$$2(x+7)(x-4) = 0$$
$2 = 0$ or $x + 7 = 0$ or $x - 4 = 0$
false $x = -7$ $x = 4$
                               $2x + 6 = 14$
Discard $x = -7$. The sides are 4 and 14.

9. $\qquad x(x-7) = 18$
$$x^2 - 7x - 18 = 0$$
$$(x-9)(x+2) = 0$$
$\qquad x = 9$ or $x = -2$
$x - 7 = 2$
Discard $x = -2$. The sides are 2 and 9.

11. $\qquad \dfrac{1}{2}x(3x+2) = 104$
$$x(3x+2) = 208$$
$$3x^2 + 2x - 208 = 0$$
$$(3x+26)(x-8) = 0$$
$3x + 26 = 0$ or $x = 8$
$3x = -26$ $3x + 2 = 26$
$x = -\dfrac{26}{3}$

Discard $x = -\dfrac{26}{3}$. The base is 26, and the height is 8.

13. $\qquad \dfrac{1}{2}(3x-6)(2x) = 144$
$$x(3x-6) = 144$$
$$3x^2 - 6x - 144 = 0$$
$$3(x^2 - 2x - 48) = 0$$
$$3(x-8)(x+6) = 0$$
$3 = 0$ or $x - 8 = 0$ or $x + 6 = 0$
false $x = 8$ $x = -6$
Discard $x = -6$. The base is $3x - 6 = 18$, and the height is $2x = 16$.

15. $(2x+1)(2x-1) = 143$
$$4x^2 - 1 = 143$$
$$4x^2 - 144 = 0$$
$$4(x^2 - 36) = 0$$
$$4(x+6)(x-6) = 0$$
$4 = 0$ or $x + 6 = 0$ or $x - 6 = 0$
false $x = -6$ $x = 6$
Discard $x = -6$. The base is $2x + 1 = 13$ and the height is $2x - 1 = 11$.

17. $\qquad \dfrac{1}{2}(4)[(2x-10) + 2x] = 192$
$$2(4x-10) = 192$$
$$8x - 20 = 192$$
$$8x = 212$$
$$x = \dfrac{53}{2}$$
The base $B = 2x = 53$ and the base $b = 2x - 10 = 43$.

19. $x^2 + (x+3)^2 = 15^2$

$x^2 + x^2 + 6x + 9 = 225$

$2x^2 + 6x - 216 = 0$

$2(x^2 + 3x - 108) = 0$

$2(x+12)(x-9) = 0$

$2 = 0$ or $x+12 = 0$ or $x-9 = 0$

false $x = -12$ $x = 9$

Discard $x = -12$. The legs are 9 and

$x + 3 = 12$.

21. $(2x)^2 + (2x-7)^2 = 13^2$

$4x^2 + 4x^2 - 28x + 49 = 169$

$8x^2 - 28x - 120 = 0$

$4(2x^2 - 7x - 30) = 0$

$4(2x+5)(x-6) = 0$

$4 = 0$ or $2x+5 = 0$ or $x-6 = 0$

false $2x = -5$ $x = 6$

$$x = -\frac{5}{2}$$

Discard $x = -\dfrac{5}{2}$. The legs are $2x = 12$ and

$2x - 7 = 5$.

23.

Time, in Sec.	0	0.5	1	1.5	2	2.5	3	3.5	4
Height, in Feet	256	252	240	220	192	156	112	60	0

25. $0 = -16t^2 + 96t$

$0 = -16t(t-6)$

$-16t = 0$ or $t-6 = 0$

$t = 0$ $t = 6$

Since $t = 0$ corresponds to the time that the rocket is fired, the rocket will hit the ground after 6 seconds.

27. Let w = width of the room. Then $w + 8$ = length of the room.

$w(w+8) = 48$

$w^2 + 8w - 48 = 0$

$(w+12)(w-4) = 0$

$w+12 = 0$ or $w-4 = 0$

$w = -12$ $w = 4$

Discard $w = -12$. The width is 4 meters, and the length is $w + 8 = 12$ meters.

29. Let b = base of the triangle. Then $3b$ = height of the sail.

$$\frac{1}{2}b(3b) = 54$$

$$3b^2 = 108$$

$$b^2 = 36$$

$$b^2 - 36 = 0$$

$$(b-6)(b+6) = 0$$

$$b-6 = 0 \quad \text{or} \quad b+6 = 0$$

$$b = 6 \qquad b = -6$$

Discard $b = -6$. The base of the sail is 6 feet, and the height is $3b$ = 18 feet.

31. Let h = height of the TV.

$$h^2 + 40^2 = 50^2$$

$$h^2 + 1600 = 2500$$

$$h^2 - 900 = 0$$

$$(h+30)(h-30) = 0$$

$$h+30 = 0 \quad \text{or} \quad h-30 = 0$$

$$h = -30 \qquad h = 30$$

Discard $h = -30$. The height of the TV is 30 inches.

33. Let w = width of the rectangle. Then $2w + 1$ = length of the rectangle.

$$w(2w+1) = 300$$

$$2w^2 + w - 300 = 0$$

$$(2w+25)(w-12) = 0$$

$$2w+25 = 0 \quad \text{or} \quad w-12 = 0$$

$$2w = -25 \qquad w = 12$$

$$w = -\frac{25}{2}$$

Discard $w = -\frac{25}{2}$. The width is 12 mm, and the length is $2w + 1 = 25$ mm.

35. Let s = a side of the square. Then $s - 6$ = width of the rectangle, and $2s + 5$ = length of the rectangle.

$$(2s+5)(s-6) = s^2$$

$$2s^2 - 7s - 30 = s^2$$

$$s^2 - 7s - 30 = 0$$

$$(s-10)(s+3) = 0$$

$$s-10 = 0 \quad \text{or} \quad s+3 = 0$$

$$s = 10 \qquad s = -3$$

Discard $s = -3$. The width of the rectangle is $s - 6 = 4$ cm, and the length is $2s + 5 = 25$ cm.

37. (a) Let w = width of the TV screen. Then $w + 17$ = length of the TV screen. Including the casing, the width is $w + 3$ and the length is $w + 17 + 3 = w + 20$.

$$(w+3)^2 + (w+20)^2 = 53^2$$

$$w^2 + 6w + 9 + w^2 + 40w + 400 = 2809$$

$$2w^2 + 46w - 2400 = 0$$

$$2(w^2 + 23w - 1200) = 0$$

$$2(w+48)(w-25) = 0$$

$$2 = 0 \quad \text{or} \quad w+48 = 0 \quad \text{or} \quad w-25 = 0$$

$$\text{false} \qquad w = -48 \qquad w = 25$$

Discard $w = -48$. The TV screen is 25 inches wide and $w + 17 = 42$ inches long.

(b) $25 + 3 = 28$; $42 + 3 = 45$
The opening should be 28 inches by 45 inches.

39. Let h = height of the pole. Then $h + 2$ is the distance along the ground.

$$h^2 + (h+2)^2 = 10^2$$

$$h^2 + h^2 + 4h + 4 = 100$$

$$2h^2 + 4h - 96 = 0$$

$$2(h^2 + 2h - 48) = 0$$

$$2(h-6)(h+8) = 0$$

$$2 = 0 \quad \text{or} \quad h-6 = 0 \quad \text{or} \quad h+8 = 0$$

$$\text{false} \qquad h = 6 \qquad h = -8$$

Discard $h = -8$. The pole is 6 feet.

41. Let x = width of the garden. Then $28 - 2x$ = length of the garden.

$$x(28 - 2x) = 98$$

$$28x - 2x^2 - 98 = 0$$

$$-2(x^2 - 14x + 49) = 0$$

$$-2(x-7)(x-7) = 0$$

$$-2 = 0 \quad \text{or} \quad x-7 = 0$$

$$\text{false} \qquad x = 7$$

The width is 7 feet and the length is $28 - 2x = 14$ feet.

43. If you have two positive numbers that satisfy the projectile motion word problem, then one answer is the height of the object one its upward path, and the other represents the height of the object on its downward path.

Chapter 6 Review

1. $24 = 2 \cdot 2 \cdot 2 \cdot 3$
    $36 = 2 \cdot 2 \cdot 3 \cdot 3$
    The common factors are 2, 2, and 3. The GCF is
    $2 \cdot 2 \cdot 3 = 12$.

2. $27 = 3 \cdot 3 \cdot 3$
    $54 = 2 \cdot 3 \cdot 3 \cdot 3$
    The common factors are 3, 3, and 3. The GCF is
    $3 \cdot 3 \cdot 3 = 27$.

3. $10 = 2 \cdot 5$
    $20 = 2 \cdot 2 \cdot 5$
    $30 = 2 \cdot 3 \cdot 5$
    The common factors are 2 and 5. The GCF is
    $2 \cdot 5 = 10$.

4. $8 = 2 \cdot 2 \cdot 2$
    $16 = 2 \cdot 2 \cdot 2 \cdot 2$
    $28 = 2 \cdot 2 \cdot 7$
    The common factors are 2 and 2. The GCF is
    $2 \cdot 2 = 4$.

5. $x^4 = x \cdot x \cdot x \cdot x$
    $x^2 = x \cdot x$
    $x^8 = x \cdot x \cdot x \cdot x \cdot x \cdot x \cdot x \cdot x$
    Each expression contains two factors of x. The
    GCF is x^2.

6. $m^3 = m \cdot m \cdot m$
    $m = m$
    $m^5 = m \cdot m \cdot m \cdot m \cdot m$
    The only common factor is m. The GCF is m.

7. The GCF of 30 and 45 is 15.
    The GCF of a^2 and a is a.
    The GCF of b^4 and b^2 is b^2.
    The GCF is $15ab^2$.

8. The GCF of 18 and 24 is 6.
    The GCF of x^4 and x^3 is x^3.
    The GCF of y^2 and y^5 is y^2.
    The GCF of z^3 and z is z.
    The GCF is $6x^3y^2z$.

9. The GCF of 4 and 6 is 2.
    The GCF of $(2a+1)^2$ and $(2a+1)^3$ is $(2a+1)^2$.
    The GCF is $2(2a+1)^2$.

10. The GCF of 9 and 18 is 9.
    The GCF of $(x-y)$ and $(x-y)$ is $(x-y)$.
    The GCF is $9(x-y)$.

11. $-18a^3 - 24a^2 = -6a^2(3a) - 6a^2(4)$
    $ = -6a^2(3a+4)$

12. $-9x^2 + 12x = -3x(3x) - 3x(-4)$
    $ = -3x(3x-4)$

13. $15y^2z + 5y^7z + 20y^3z$
    $= 5y^2z(3) + 5y^2z(y^5) + 5y^2z(4y)$
    $= 5y^2z(3 + y^5 + 4y)$

14. $7x^3y - 21x^2y^2 + 14xy^3$
    $= 7xy(x^2) + 7xy(-3xy) + 7xy(2y^2)$
    $= 7xy(x^2 - 3xy + 2y^2)$

15. $x(5-y) + 2(5-y) = (5-y)x + (5-y)2$
    $ = (5-y)(x+2)$

16. $z(a+b) + y(a+b) = (a+b)z + (a+b)y$
    $ = (a+b)(z+y)$

17. $5m^2 + 2mn + 15mn + 6n^2$
    $= (5m^2 + 2mn) + (15mn + 6n^2)$
    $= m(5m + 2n) + 3n(5m + 2n)$
    $= (5m + 2n)(m + 3n)$

18. $2xy + y^2 + 2x^2 + xy$
    $= (2xy + y^2) + (2x^2 + xy)$
    $= y(2x + y) + x(2x + y)$
    $= (2x + y)(y + x)$

19. $8x + 16 - xy - 2y = (8x + 16) + (-xy - 2y)$
    $ = 8(x+2) - y(x+2)$
    $ = (x+2)(8-y)$

20. $xy^2 + x - 3y^2 - 3 = (xy^2 + x) + (-3y^2 - 3)$
    $ = x(y^2 + 1) - 3(y^2 + 1)$
    $ = (y^2 + 1)(x - 3)$

21. $3(2) = 6$ and $3 + 2 = 5$

$x^2 + 5x + 6 = (x+3)(x+2)$

22. $2(4) = 8$ and $2 + 4 = 6$

$x^2 + 6x + 8 = (x+2)(x+4)$

23. $x^2 - 21 - 4x = x^2 - 4x - 21 = (x-7)(x+3)$

24. $3x + x^2 - 10 = x^2 + 3x - 10 = (x+5)(x-2)$

25. $m^2 + m + 20$ cannot be factored; it is prime.

26. $m^2 - 6m - 5$ cannot be factored; it is prime.

27. $x^2 - 8xy + 15y^2 = (x-3y)(x-5y)$

28. $m^2 + 4mn - 5n^2 = (m+5n)(m-n)$

29. $-p^2 - 11p - 30 = -(p^2 + 11p + 30)$

$= -(p+6)(p+5)$

30. $-y^2 + 2y + 15 = -(y^2 - 2y - 15)$

$= -(y-5)(y+3)$

31. $3x^3 + 33x^2 + 36x = 3x(x^2 + 11x + 12)$

32. $4x^2 + 36x + 32 = 4(x^2 + 9x + 8)$

$= 4(x+1)(x+8)$

33. $2x^2 - 2xy - 84y^2 = 2(x^2 - xy - 42y^2)$

$= 2(x-7y)(x+6y)$

34. $4y^3 + 12y^2 - 40y = 4y(y^2 + 3y - 10)$

$= 4y(y+5)(y-2)$

35. $20(-6) = -120 = 5(-24)$ and $20 - 6 = 14$

$5y^2 + 14y - 24 = 5y^2 + 20y - 6y - 24$

$= 5y(y+4) - 6(y+4)$

$= (y+4)(5y-6)$

36. $-42(1) = -42 = 6(-7)$ and $-42 + 1 = -41$

$6y^2 - 41y - 7 = 6y^2 - 42y + y - 7$

$= 6y(y-7) + 1(y-7)$

$= (y-7)(6y+1)$

37. $-5x + 2x^2 + 3 = 2x^2 - 5x + 3$

$= (2x-3)(x-1)$

38. $23x + 6x^2 + 7 = 6x^2 + 23x + 7$

$21 \cdot 2 = 42 = 6 \cdot 7$ and $21 + 2 = 23$

$6x^2 + 23x + 7 = 6x^2 + 21x + 2x + 7$

$= 3x(2x+7) + 1(2x+7)$

$= (2x+7)(3x+1)$

39. $2x^2 - 7x - 6$ cannot be factored; it is prime.

40. $6 \cdot 12 = 72 = 8 \cdot 9$ and $6 + 12 = 18$

$8m^2 + 18m + 9 = 8m^2 + 6m + 12m + 9$

$= 2m(4m+3) + 3(4m+3)$

$= (4m+3)(2m+3)$

41. $9m^3 + 30m^2n + 21mn^2$

$= 3m(3m^2 + 10mn + 7n^2)$

$= 3m(3m^2 + 3mn + 7mn + 7n^2)$

$= 3m[3m(m+n) + 7n(m+n)]$

$= 3m(m+n)(3m+7n)$

42. $14m^2 + 16mn + 2n^2$

$= 2(7m^2 + 8mn + n^2)$

$= 2(7m^2 + mn + 7mn + n^2)$

$= 2[m(7m+n) + n(7m+n)]$

$= 2(7m+n)(m+n)$

43. $15x^3 + x^2 - 2x = x(15x^2 + x - 2)$

$= x(15x^2 + 6x - 5x - 2)$

$= x[3x(5x+2) - 1(5x+2)]$

$= x(5x+2)(3x-1)$

44. $6p^4 + p^3 - p^2 = p^2(6p^2 + p - 1)$

$= p^2(3p-1)(2p+1)$

45. $4x^2 - 12x + 9 = (2x)^2 - 2 \cdot 2x \cdot 3 + 3^2$

$= (2x-3)^2$

46. $x^2 - 10x + 25 = x^2 - 2 \cdot x \cdot 5 + 5^2 = (x-5)^2$

47. $x^2 + 6xy + 9y^2 = x^2 + 2 \cdot x \cdot 3y + (3y)^2$

$= (x+3y)^2$

48. $9x^2 + 24xy + 4y^2$ cannot be factored; it is prime.

49. $8m^2 + 8m + 2 = 2(4m^2 + 4m + 1)$
$$= 2[(2m)^2 + 2 \cdot 2m \cdot 1 + 1^2]$$
$$= 2(2m+1)^2$$

50. $2m^2 - 24m + 72 = 2(m^2 - 12m + 36)$
$$= 2(m^2 - 2 \cdot m \cdot 6 + 6^2)$$
$$= 2(m-6)^2$$

51. $4x^2 - 25y^2 = (2x)^2 - (5y)^2$
$$= (2x - 5y)(2x + 5y)$$

52. $49x^2 - 36y^2 = (7x)^2 - (6y)^2$
$$= (7x - 6y)(7x + 6y)$$

53. $x^2 + 25$ cannot be factored; it is prime.

54. $x^2 + 100$ cannot be factored; it is prime.

55. $x^4 - 81 = (x^2)^2 - 9^2$
$$= (x^2 - 9)(x^2 + 9)$$
$$= (x^2 - 3^2)(x^2 + 9)$$
$$= (x-3)(x+3)(x^2 + 9)$$

56. $x^4 - 625 = (x^2)^2 - (25)^2$
$$= (x^2 - 25)(x^2 + 25)$$
$$= (x^2 - 5^2)(x^2 + 25)$$
$$= (x-5)(x+5)(x^2 + 25)$$

57. $m^3 + 27 = m^3 + 3^3$
$$= (m+3)(m^2 - m \cdot 3 + 3^2)$$
$$= (m+3)(m^2 - 3m + 9)$$

58. $m^3 + 125 = m^3 + 5^3$
$$= (m+5)(m^2 - m \cdot 5 + 5^2)$$
$$= (m+5)(m^2 - 5m + 25)$$

59. $27p^3 - 8 = (3p)^3 - 2^3$
$$= (3p-2)[(3p)^2 + 3p \cdot 2 + 2^2]$$
$$= (3p-2)(9p^2 + 6p + 4)$$

60. $64p^3 - 1 = (4p)^3 - 1^3$
$$= (4p-1)[(4p)^2 + 4p \cdot 1 + 1^2]$$
$$= (4p-1)(16p^2 + 4p + 1)$$

61. $y^9 + 64z^6$
$$= (y^3)^3 + (4z^2)^3$$
$$= (y^3 + 4z^2)[(y^3)^2 - y^3 \cdot 4z^2 + (4z^2)^2]$$
$$= (y^3 + 4z^2)(y^6 - 4y^3z^2 + 16z^4)$$

62. $8y^3 + 27z^6$
$$= (2y)^3 + (3z^2)^3$$
$$= (2y + 3z^2)[(2y)^2 - 2y \cdot 3z^2 + (3z^2)^2]$$
$$= (2y + 3z^2)(4y^2 - 6yz^2 + 9z^4)$$

63. $15a^3 - 6a^2b - 25ab^2 + 10b^3$
$$= 3a^2(5a - 2b) - 5b^2(5a - 2b)$$
$$= (5a - 2b)(3a^2 - 5b^2)$$

64. $12a^2 - 9ab + 4ab - 3b^2$
$$= 3a(4a - 3b) + b(4a - 3b)$$
$$= (4a - 3b)(3a + b)$$

65. $x^2 - xy - 48y^2$ cannot be factored; it is prime.

66. $x^2 - 10xy - 24y^2 = (x - 12y)(x + 2y)$

67. $x^3 - x^2 - 42x = x(x^2 - x - 42)$
$$= x(x - 7)(x + 6)$$

68. $3x^6 - 30x^5 + 63x^4 = 3x^4(x^2 - 10x + 21)$
$$= 3x^4(x - 7)(x - 3)$$

69. $6x^2 + 11x + 3 = 6x^2 + 2x + 9x + 3$
$$= 2x(3x + 1) + 3(3x + 1)$$
$$= (3x + 1)(2x + 3)$$

70. $10z^2 + 9z - 9 = 10z^2 + 15z - 6z - 9$
$$= 5z(2z + 3) - 3(2z + 3)$$
$$= (2z + 3)(5z - 3)$$

71. $27x^3 + 8 = (3x)^3 + 2^3$
$$= (3x + 2)[(3x)^2 - (3x) \cdot 2 + 2^2]$$
$$= (3x + 2)(9x^2 - 6x + 4)$$

72. $8z^3 - 1 = (2z)^3 - 1^3$
$$= (2z - 1)[(2z)^2 + 2z \cdot 1 + 1^2]$$
$$= (2z - 1)(4z^2 + 2z + 1)$$

73. $4z^2 + 18y - 10 = 2(2y^2 + 9y - 5)$
$\qquad\qquad\qquad = 2(2y-1)(y+5)$

74. $5x^3y^2 - 8x^2y^2 + 3xy^2 = xy^2(5x^2 - 8x + 3)$
$\qquad\qquad\qquad\qquad = xy^2(5x-3)(x-1)$

75. $25k^2 - 81m^2 = (5k)^2 - (9m)^2$
$\qquad\qquad\qquad = (5k - 9m)(5k + 9m)$

76. $x^4 - 9 = (x^2)^2 - 3^2 = (x^2 - 3)(x^2 + 3)$

77. $16m^2 + 1$ cannot be factored; it is prime.

78. $m^4 + 25$ cannot be factored; it is prime.

79. $(x-4)(2x-3) = 0$
$x - 4 = 0 \quad\text{or}\quad 2x - 3 = 0$
$\qquad x = 4 \qquad\qquad 2x = 3$
$\qquad\qquad\qquad\qquad\quad x = \dfrac{3}{2}$

The solution set is $\left\{\dfrac{3}{2}, 4\right\}$.

80. $(2x + 1)(x + 7) = 0$
$2x + 1 = 0 \quad\text{or}\quad x + 7 = 0$
$\quad 2x = -1 \qquad\qquad x = -7$
$\qquad x = -\dfrac{1}{2}$

The solutions et is $\left\{-7, -\dfrac{1}{2}\right\}$.

81. $x^2 - 12x - 45 = 0$
$(x - 15)(x + 3) = 0$
$x - 15 = 0 \quad\text{or}\quad x + 3 = 0$
$\quad x = 15 \qquad\qquad x = -3$
The solution set is $\{-3, 15\}$.

82. $x^2 - 7x + 10 = 0$
$(x - 2)(x - 5) = 0$
$x - 2 = 0 \quad\text{or}\quad x - 5 = 0$
$\quad x = 2 \qquad\qquad x = 5$
The solution set is $\{2, 5\}$.

83. $3x^2 + 6x = 0$
$3x(x + 2) = 0$
$3x = 0 \quad\text{or}\quad x + 2 = 0$
$\quad x = 0 \qquad\qquad x = -2$
The solution set is $\{0, -2\}$.

84. $4x^2 + 18x = 0$
$2x(2x + 9) = 0$
$2x = 0 \quad\text{or}\quad 2x + 9 = 0$
$\quad x = 0 \qquad\qquad 2x = -9$
$\qquad\qquad\qquad\qquad x = -\dfrac{9}{2}$

The solution set is $\left\{-\dfrac{9}{2}, 0\right\}$.

85. $\quad 3x(x + 1) = 2x^2 + 5x + 3$
$\quad 3x^2 + 3x = 2x^2 + 5x + 3$
$\quad x^2 - 2x - 3 = 0$
$\quad (x - 3)(x + 1) = 0$
$\quad x - 3 = 0 \quad\text{or}\quad x + 1 = 0$
$\qquad x = 3 \qquad\qquad x = -1$
The solution set is $\{-1, 3\}$.

86. $\quad 2x^2 + 6x = (3x + 1)(x + 3)$
$\qquad 2x^2 + 6x = 3x^2 + 10x + 3$
$\quad -x^2 - 4x - 3 = 0$
$\quad -1(x^2 + 4x + 3) = 0$
$\quad -1(x + 3)(x + 1) = 0$
$\quad -1 = 0 \quad\text{or}\quad x + 3 = 0 \quad\text{or}\quad x + 1 = 0$
$\quad\text{false} \qquad\qquad x = -3 \qquad\qquad x = -1$
The solution set is $\{-3, -1\}$.

87. $5x^2 + 7x + 16 = 3x^2 - 5x$
$2x^2 + 12x + 16 = 0$
$2(x^2 + 6x + 8) = 0$
$2(x + 4)(x + 2) = 0$
$x + 4 = 0 \quad\text{or}\quad x + 2 = 0$
$\quad x = -4 \quad\text{or}\qquad x = -2$
The solution set is $\{-4, -2\}$.

88. $\qquad 8x^2 - 10x = -2x + 6$
$\qquad 8x^2 - 8x - 6 = 0$
$\quad 2(4x^2 - 4x - 3) = 0$
$\quad 2(2x + 1)(2x - 3) = 0$
$\quad 2 = 0 \quad\text{or}\quad 2x + 1 = 0 \quad\text{or}\quad 2x - 3 = 0$
$\quad\text{false} \qquad\qquad 2x = -1 \qquad\qquad 2x = 3$
$\qquad\qquad\qquad\qquad x = -\dfrac{1}{2} \qquad\qquad x = \dfrac{3}{2}$

The solution set is $\left\{-\dfrac{1}{2}, \dfrac{3}{2}\right\}$.

89.
$$x^3 = -11x^2 + 42x$$
$$x^3 + 11x^2 - 42x = 0$$
$$x(x^2 + 11x - 42) = 0$$
$$x(x + 14)(x - 3) = 0$$
$x = 0$ or $x + 14 = 0$ or $x - 3 = 0$
$$x = -14 \qquad x = 3$$
The solution set is $\{-14, 0, 3\}$.

90.
$$-3x^2 = -x^3 + 18x$$
$$x^3 - 3x^2 - 18x = 0$$
$$x(x^2 - 3x - 18) = 0$$
$$x(x - 6)(x + 3) = 0$$
$x = 0$ or $x - 6 = 0$ or $x + 3 = 0$
$$x = 6 \qquad x = -3$$
The solution set is $\{-3, 0, 6\}$.

91.
$$(3x - 4)(x^2 - 9) = 0$$
$$(3x - 4)(x + 3)(x - 3) = 0$$
$3x - 4 = 0$ or $x + 3 = 0$ or $x - 3 = 0$
$$3x = 4 \qquad x = -3 \qquad x = 3$$
$$x = \frac{4}{3}$$
The solution set is $\left\{-3, \dfrac{4}{3}, 3\right\}$.

92. $(2x + 5)(x^2 + 4x + 4) = 0$
$$(2x + 5)(x + 2)(x + 2) = 0$$
$2x + 5 = 0$ or $x + 2 = 0$
$$2x = -5 \qquad x = -2$$
$$x = -\frac{5}{2}$$
The solution set is $\left\{-\dfrac{5}{2}, -2\right\}$.

93. $h = -16t^2 + 80t$
$$0 = -16t^2 + 80t$$
$$0 = -16t(t - 5)$$
$-16t = 0$ or $t - 5 = 0$
$$t = 0 \qquad t = 5$$
After 5 seconds, the water hits the ground.

94.
$$h = -16t^2 + 80t$$
$$96 = -16t^2 + 80t$$
$$16t^2 - 80t + 96 = 0$$
$$16(t^2 - 5t + 6) = 0$$
$$16(t - 3)(t - 2) = 0$$
$16 = 0$ or $t - 3 = 0$ or $t - 2 = 0$
false $\qquad t = 3 \qquad t = 2$
The water is 96 feet high after 2 seconds and 3 seconds.

95. Let w = width. Then $2w - 3$ = length.
$$w(2w - 3) = 54$$
$$2w^2 - 3w - 54 = 0$$
$$(2w + 9)(w - 6) = 0$$
$2w + 9 = 0$ or $w - 6 = 0$
$$2w = -9 \qquad w = 6$$
$$w = -\frac{9}{2}$$
Discard $w = -\dfrac{9}{2}$. The width is 6 feet and the length is $2w - 3 = 9$ feet.

96. Let w = width. Then $2w - 1$ = length.
$$w(2w - 1) = 15$$
$$2w^2 - w - 15 = 0$$
$$(2w + 5)(w - 3) = 0$$
$2w + 5 = 0$ or $w - 3 = 0$
$$2w = -5 \qquad w = 3$$
$$w = -\frac{5}{2}$$
Discard $w = -\dfrac{5}{2}$. The width is 3 yards and the length is $2w - 1 = 5$ yards.

97. Let l = length of longer leg. Then $l - 2$ = length of shorter leg.
$$l^2 + (l - 2)^2 = 10^2$$
$$l^2 + l^2 - 4l + 4 = 100$$
$$2l^2 - 4l - 96 = 0$$
$$2(l^2 - 2l - 48) = 0$$
$$2(l - 8)(l + 6) = 0$$
$2 = 0$ or $l - 8 = 0$ or $l + 6 = 0$
false $\qquad l = 8 \qquad l = -6$
Discard $l = -6$. The longer leg is 8 feet and the shorter leg is $l - 2 = 6$ feet.

98. Let l = length of longer leg. Then
$l - 14$ = length of shorter leg.

$$l^2 + (l-14)^2 = 26^2$$
$$l^2 + l^2 - 28l + 196 = 676$$
$$2l^2 - 28l - 480 = 0$$
$$2(l^2 - 14l - 240) = 0$$
$$2(l-24)(l+10) = 0$$

$2 = 0$ or $l - 24 = 0$ or $l + 10 = 0$
false $l = 24$ $l = -10$

Discard $l = -10$. The longer leg is 24 feet and the shorter leg is $l - 14 = 10$ feet.

Chapter 6 Test

1. The GCF of 16, 20, and 24 is 4.
The GCF of x^5, x^4, and x^6 is x^4.
The GCF of y^2, y^6, and y^8 is y^2.
The GCF is $4x^4y^2$.

2. $x^4 - 256 = (x^2)^2 - 16^2$
$ = (x^2 - 16)(x^2 + 16)$
$ = (x^2 - 4^2)(x^2 + 16)$
$ = (x+4)(x-4)(x^2 + 16)$

3. $18x^3 - 9x^2 - 27x = 9x(2x^2 - x - 3)$
$ = 9x(2x-3)(x+1)$

4. $xy - 7y - 4x + 28 = (xy - 7y) + (-4x + 28)$
$ = y(x-7) - 4(x-7)$
$ = (x-7)(y-4)$

5. $27x^3 + 125 = (3x)^3 + 5^3$
$ = (3x+5)[(3x)^2 - (3x)5 + 5^2]$
$ = (3x+5)(9x^2 - 15x + 25)$

6. $y^2 - 8y - 48 = (y - 12)(y + 4)$

7. $6m^2 - m - 5 = (6m+5)(m-1)$

8. $4x^2 + 25 = (2x)^2 + 5^2$

$4x^2 + 25$ cannot be factored; it is prime.

9. $4(x-5) + y(x-5) = (x-5)(4+y)$

10. $3x^2y - 15xy - 42y = 3y(x^2 - 5x - 14)$
$ = 3y(x-7)(x+2)$

11. $x^2 + 4x + 12$ cannot be factored; it is prime.

12. $2x^6 - 54y^3 = 2(x^6 - 27y^3)$
$ = 2[(x^2)^3 - (3y)^3]$
$ = 2(x^3 - 3y)(x^4 + 3x^2y + 9y^2)$

13. $9x^3 + 39x^2 + 12x = 3x(3x^2 + 13x + 4)$
$ = 3x(3x+1)(x+4)$

14. $6m^2 + 7m + 2 = (3m+2)(2m+1)$

15. $4m^2 - 6mn + 4 = 2(2m^2 - 3mn + 2)$

16. $25x^2 + 70xy + 49y^2$
$ = (5x)^2 + 2(5x)(7y) + (7y)^2$
$ = (5x + 7y)^2$

17. $ 5x^2 = -16x - 3$
$ 5x^2 + 16x + 3 = 0$
$ (5x+1)(x+3) = 0$
$ 5x + 1 = 0$ or $x + 3 = 0$
$ 5x = -1$ $ x = -3$
$ x = -\dfrac{1}{5}$

The solution set is $\left\{ -\dfrac{1}{5}, -3 \right\}$.

18. $5x^3 - 20x^2 + 20x = 0$
$ 5x(x^2 - 4x + 4) = 0$
$ 5x(x-2)^2 = 0$
$ 5x = 0$ or $x - 2 = 0$
$ x = 0$ $ x = 2$
The solution set is $\{0, 2\}$.

19. Let w = width. Then $3w - 8$ = length.
$ w(3w - 8) = 35$
$ 3w^2 - 8w - 35 = 0$
$ (3w + 7)(w - 5) = 0$
$ 3w + 7 = 0$ or $w - 5 = 0$
$ 3w = -7$ $ w = 5$
$ w = -\dfrac{7}{3}$

Discard $w = -\dfrac{7}{3}$. The width is 5 inches and the
length is $3w - 8 = 7$ inches.

20. Let l = length of longer leg. Then
$l + 1$ = length of hypotenuse and
$l - 7$ = length of shorter leg.

$$l^2 + (l-7)^2 = (l+1)^2$$
$$l^2 + l^2 - 14l + 49 = l^2 + 2l + 1$$
$$l^2 - 16l + 48 = 0$$
$$(l-12)(l-4) = 0$$
$$l - 12 = 0 \quad \text{or} \quad l - 4 = 0$$
$$l = 12 \qquad\qquad l = 4$$

Discard $l = 4$ since $l - 7 = -3$. The legs are
12 inches and $l - 7 = 5$ inches and the
hypotenuse is $l + 1 = 13$ inches.

Getting Ready for Intermediate Algebra:
A Review of Chapters 1 – 6

1. **(a)** 1

(b) $-6, 0, 1$

(c) $-6, -0.83, 0, \dfrac{2}{5}, 0.5454..., 1$

(d) $1.010010001...$

(e) $-6, -0.83, 0, \dfrac{2}{5}, 0.5454..., 1, 1.010010001...$

2. $4(-5) = -20$

3. $\dfrac{-72}{60} = \dfrac{-1 \cdot \cancel{2} \cdot \cancel{2} \cdot 2 \cdot \cancel{3} \cdot 3}{\cancel{2} \cdot \cancel{2} \cdot \cancel{3} \cdot 5} = \dfrac{-1 \cdot 2 \cdot 3}{5} = -\dfrac{6}{5}$

4. $\dfrac{12}{5} \cdot \left(-\dfrac{35}{8}\right) = \dfrac{\cancel{2} \cdot \cancel{2} \cdot 3}{\cancel{8}} \cdot \dfrac{-1 \cdot \cancel{5} \cdot 7}{\cancel{2} \cdot \cancel{2} \cdot 2} = -\dfrac{21}{2}$

5. $\dfrac{5}{12} + \dfrac{11}{12} = \dfrac{5+11}{12} = \dfrac{16}{12} = \dfrac{\cancel{2} \cdot \cancel{2} \cdot 2 \cdot 2}{\cancel{2} \cdot \cancel{2} \cdot 3} = \dfrac{4}{3}$

6. $6 = 2 \cdot 3$
$30 = 2 \cdot 3 \cdot 5$
The LCD $= 2 \cdot 3 \cdot 5 = 30$

$\dfrac{5}{6} - \dfrac{7}{30} = \dfrac{5}{6} \cdot \dfrac{5}{5} - \dfrac{7}{30}$

$\qquad = \dfrac{25}{30} - \dfrac{7}{30}$

$\qquad = \dfrac{18}{30} = \dfrac{3 \cdot \cancel{6}}{5 \cdot \cancel{6}} = \dfrac{3}{5}$

7. $5 - (1-4)^3 + 5 \cdot 3 = 5 - 2(-3)^3 + 5 \cdot 3$
$\qquad = 5 - 2(-27) + 5 \cdot 3$
$\qquad = 5 + 54 + 15$
$\qquad = 74$

8. $\dfrac{|3 - 3 \cdot 5|}{-2^3} = \dfrac{|3 - 15|}{-8} = \dfrac{|-12|}{-8} = \dfrac{12}{-8} = \dfrac{3 \cdot \cancel{4}}{-2 \cdot \cancel{4}} = -\dfrac{3}{2}$

9. $5(x+1) - (2x+1) = x+10$
$5x + 5 - 2x - 1 = x + 10$
$3x + 4 = x + 10$
$2x + 4 = 10$
$2x = 6$
$x = 3$
The solution set is $\{3\}$.

10. $\dfrac{1}{6}(x+1) = 3 - \dfrac{1}{2}(x+4)$

$\dfrac{1}{6}x + \dfrac{1}{6} = 3 - \dfrac{1}{2}x - 2$

$\dfrac{1}{6}x + \dfrac{1}{6} = -\dfrac{1}{2}x + 1$

$6\left(\dfrac{1}{6}x + \dfrac{1}{6}\right) = 6\left(-\dfrac{1}{2}x + 1\right)$

$x + 1 = -3x + 6$
$4x + 1 = 6$
$4x = 5$
$x = \dfrac{5}{4}$

The solution set is $\left\{\dfrac{5}{4}\right\}$.

11. Let x = the measure of the smaller angle. Then $2x + 15$ = the measure of the larger angle.
$x + (2x + 15) = 180$
$3x + 15 = 180$
$3x = 165$
$x = 55$
Thus, $2x + 15 = 2(55) + 15 = 110 + 15 = 125$.
The two angles measure $55°$ and $125°$.

12. $2(x+3) \le 3x - 1$
$2x + 6 \le 3x - 1$
$-x + 6 \le -1$
$-x \le -7$
$x \ge 7$
The solution set is $\{x \mid x \ge 7\}$ or $[7, \infty)$.

367

13. $0.2x - 2 \geq 30$

$\quad\;\; 0.2x \geq 32$

$\quad\qquad x \geq 160$

A person 62 inches tall would be considered obese if he or she weighed 160 pounds or more.

14. The degree of $3x^5 - 4x^3 + 2x^2 - 6x - 1$ is 5.

15. $3(-2)^2 + 2(-2) - 7 = 3(4) + 2(-2) - 7$

$\qquad\qquad\qquad\qquad = 12 - 4 - 7$

$\qquad\qquad\qquad\qquad = 1$

16. $\left(4y^3 + 8y^2 - y + 3\right) + \left(y^3 - 3y^2 - 9\right)$

$\quad = 4y^3 + y^3 + 8y^2 - 3y^2 - y + 3 - 9$

$\quad = 5y^3 + 5y^2 - y - 6$

17. $\left(-3x^4\right)^2 = (-3)^2\left(x^4\right)^2 = 9x^8$

18. $-4x\left(x^2 - 5x + 3\right) = -4x\left(x^2\right) - 4x(-5x) - 4x(3)$

$\qquad\qquad\qquad\qquad = -4x^3 + 20x^2 - 12x$

19. $(4x + 1)(3x - 5)$

$\quad = 4x(3x) + 4x(-5) + 1(3x) + 1(-5)$

$\quad = 12x^2 - 20x + 3x - 5$

$\quad = 12x^2 - 17x - 5$

20. $(2x - 7)^2 = (2x)^2 - 2(2x)(7) + (7)^2$

$\qquad\qquad\; = 4x^2 - 28x + 49$

21. $(5x + 3y)(5x - 3y) = (5x)^2 - (3y)^2$

$\qquad\qquad\qquad\quad = 25x^2 - 9y^2$

22. $\dfrac{\left(4y^2z\right)^{-1}}{y^{-3}z^2} = \dfrac{y^3}{\left(4y^2z\right)^1 z^2}$

$\qquad\quad = \dfrac{y^3}{4y^2z \cdot z^2} = \dfrac{y^{3-2}}{4z^{1+2}} = \dfrac{y^1}{4z^3} = \dfrac{y}{4z^3}$

23. $\dfrac{4a^3 - 12a^2 + 2a}{2a} = \dfrac{4a^3}{2a} - \dfrac{12a^2}{2a} + \dfrac{2a}{2a}$

$\qquad\qquad\qquad\quad = 2a^2 - 6a + 1$

24.

$$\begin{array}{r} 3x^2 - 5x + 2 \\ x+5\overline{\smash)3x^3 + 10x^2 - 23x + 1} \end{array}$$

$\qquad \dfrac{-\left(3x^3 + 15x^2\right)}{}$

$\qquad\qquad -5x^2 - 23x$

$\qquad\qquad \dfrac{-\left(-5x^2 - 25x\right)}{}$

$\qquad\qquad\qquad 2x + 1$

$\qquad\qquad\qquad \dfrac{-(2x + 10)}{}$

$\qquad\qquad\qquad\qquad -9$

$\dfrac{3x^3 + 10x^2 - 23x + 1}{x + 5} = 3x^2 - 5x + 2 + \dfrac{-9}{x + 5}$

$\qquad\qquad\; \text{or } 3x^2 - 5x + 2 - \dfrac{9}{x + 5}$

25. $-4x^4 + 16x^2 - 20x = -4x\left(x^3 - 4x + 5\right)$

26. $8z^3 - 4z^2 + 6z - 3 = 4z^2(2z - 1) + 3(2z - 1)$

$\qquad\qquad\qquad\qquad\quad = (2z - 1)\left(4z^2 + 3\right)$

27. $5(-7) = -35$ and $5 + (-7) = -2$

$w^2 - 2w - 35 = (w + 5)(w - 7)$

28. $-3c^3 + 15c^2 + 72c = -3c\left(c^2 - 5c - 24\right)$

$\qquad\qquad\qquad\qquad\; = -3c(c - 8)(c + 3)$

29. There are no factors of -28 whose sum is -4.

$m^2 - 4m - 28$ is prime.

30. $3y^2 - 8y - 16 = 3y^2 - 4y + 12y - 16$

$\qquad\qquad\qquad = y(3y - 4) + 4(3y - 4)$

$\qquad\qquad\qquad = (3y - 4)(y + 4)$

31. $p^4 - 16 = \left(p^2\right)^2 - 4^2$

$\qquad\quad = \left(p^2 - 4\right)\left(p^2 + 4\right)$

$\qquad\quad = \left(p^2 - 2^2\right)\left(p^2 + 4\right)$

$\qquad\quad = (p - 2)(p + 2)\left(p^2 + 4\right)$

32. $4a^2 + 20ab + 25b^2 = (2a)^2 + 2(2a)(5b) + (5b)^2$

$\qquad\qquad\qquad\qquad = (2a + 5b)^2$

33. $(b-3)(2b+5) = 0$

$b - 3 = 0$ or $2b + 5 = 0$

$b = 3$ $\qquad 2b = -5$

$\qquad\qquad b = -\dfrac{5}{2}$

The solution set is $\left\{-\dfrac{5}{2}, 3\right\}$.

34. $\qquad x^2 + 20 = 9x$

$x^2 - 9x + 20 = 0$

$(x-5)(x-4) = 0$

$x - 5 = 0$ or $x - 4 = 0$

$x = 5$ $\qquad x = 4$

The solution set is $\{4, 5\}$.

35. $\quad 6k^2 + 17k - 14 = 0$

$(3k-2)(2k+7) = 0$

$3k - 2 = 0$ or $2k + 7 = 0$

$3k = 2$ $\qquad 2k = -7$

$k = \dfrac{2}{3}$ $\qquad k = -\dfrac{7}{2}$

The solution set is $\left\{-\dfrac{7}{2}, \dfrac{2}{3}\right\}$.

Chapter 7

Preparing for Simplifying Rational Expressions

P1. Substitute -1 for x and 7 for y.

$$\frac{3x+y}{2} = \frac{3(-1)+7}{2} = \frac{-3+7}{2} = \frac{4}{2} = 2$$

P2. $2x^2 + x - 3 = (2x+3)(x-1)$

P3.
$$3x^2 - 5x - 2 = 0$$
$$(3x+1)(x-2) = 0$$
$$3x+1 = 0 \quad \text{or} \quad x-2 = 0$$
$$3x = -1 \quad \text{or} \quad x = 2$$
$$x = -\frac{1}{3} \quad \text{or} \quad x = 2$$

The solution set is $\left\{-\dfrac{1}{3}, 2\right\}$.

P4. $\dfrac{21}{70} = \dfrac{3 \cdot 7}{10 \cdot 7} = \dfrac{3}{10}$

P5. $\dfrac{x^3 y^4}{xy^2} = x^{3-1} y^{4-2} = x^2 y^2$

7.1 Quick Checks

1. The quotient of two polynomials is called a <u>rational</u> <u>expression</u>.

2. (a)
$$\frac{3}{5x+1} = \frac{3}{5(-5)+1}$$
$$= \frac{3}{-25+1}$$
$$= \frac{3}{-24}$$
$$= \frac{1 \cdot 3}{-8 \cdot 3}$$
$$= -\frac{1}{8}$$

(b) $\dfrac{3}{5x+1} = \dfrac{3}{5(3)+1} = \dfrac{3}{15+1} = \dfrac{3}{16}$

3. (a)
$$\frac{x^2+6x+9}{x+1} = \frac{(-5)^2 + 6(-5)+9}{-5+1}$$
$$= \frac{25-30+9}{-4}$$
$$= \frac{4}{-4}$$
$$= -1$$

(b)
$$\frac{x^2+6x+9}{x+1} = \frac{(3)^2 + 6(3)+9}{3+1}$$
$$= \frac{9+18+9}{4}$$
$$= \frac{36}{4}$$
$$= \frac{4 \cdot 9}{4}$$
$$= 9$$

4. Substitute 2 for m, -1 for n, and 6 for p.
$$\frac{3m-5n}{p-4} = \frac{3(2)-5(-1)}{6-4} = \frac{6+5}{2} = \frac{11}{2}$$

5. Substitute 2 for y and 1 for z.
$$\frac{5y+2}{2y-z} = \frac{5(2)+2}{2(2)-1} = \frac{10+2}{4-1} = \frac{12}{3} = 4$$

6. A rational expression is <u>undefined</u> for those values of the variable(s) that make the denominator zero.

7. We want to find all values of x in the rational expression $\dfrac{3}{x+7}$ that cause $x+7$ to equal 0.
$$x+7 = 0$$
$$x = -7$$
So -7 causes the denominator, $x+7$, to equal 0.

Therefore, the expression $\dfrac{3}{x+7}$ is undefined for $x = -7$.

8. We want to find all values of n in the rational expression $\dfrac{-4}{3n+5}$ that cause $3n+5$ to equal 0.
$$3n+5 = 0$$
$$3n = -5$$
$$n = -\frac{5}{3}$$

So $-\dfrac{5}{3}$ causes the denominator, $3n + 5$, to equal

0. Therefore, the expression $\dfrac{-4}{3n+5}$ is undefined

for $n = -\dfrac{5}{3}$.

9. We want to find all values of x in the rational

expression $\dfrac{8x}{x^2+2x-3}$ that cause x^2+2x-3 to

equal 0.

$x^2 + 2x - 3 = 0$
$(x+3)(x-1) = 0$
$x + 3 = 0$ or $x - 1 = 0$
     $x = -3$ or $x = 1$

Since -3 or 1 cause the denominator to equal 0,

the rational expression $\dfrac{8x}{x^2+2x-3}$ is undefined

for $x = -3$ or $x = 1$.

10. To <u>simplify</u> a rational expression means to write

the rational expression in the form $\dfrac{p}{q}$ where p

and q are polynomials that have no common
factors.

11. False; $\dfrac{2n+3}{2n+6} = \dfrac{2n+3}{2(n+3)} \neq \dfrac{3}{6}$ or $\dfrac{1}{2}$.

12. $\dfrac{3n^2+12n}{6n} = \dfrac{3n(n+4)}{3n(2)} = \dfrac{n+4}{2}$

13. $\dfrac{2z^2+6z+4}{-4z-8} = \dfrac{2(z^2+3z+2)}{-4(z+2)}$

$= \dfrac{2(z+1)(z+2)}{-2\cdot2(z+2)}$

$= -\dfrac{z+1}{2}$

14. $\dfrac{a^2+3a-28}{2a^2-a-28} = \dfrac{(a+7)(a-4)}{(2a+7)(a-4)} = \dfrac{a+7}{2a+7}$

15. $\dfrac{z^3+z^2-3z-3}{z^2-z-2} = \dfrac{z^2(z+1)-3(z+1)}{(z+1)(z-2)}$

$= \dfrac{(z+1)(z^2-3)}{(z+1)(z-2)}$

$= \dfrac{z^2-3}{z-2}$

16. $\dfrac{4k^2+4k+1}{4k^2-1} = \dfrac{(2k+1)^2}{(2k+1)(2k-1)} = \dfrac{2k+1}{2k-1}$

17. False; $\dfrac{a+b}{a-b} = \dfrac{a+b}{-1(-a+b)} \neq -1$.

18. $\dfrac{7a-7b}{b-a} = \dfrac{7(a-b)}{b-a} = \dfrac{7(-1)(b-a)}{b-a} = -7$

19. $\dfrac{12-4x}{4x^2-13x+3} = \dfrac{-4(x-3)}{(x-3)(4x-1)} = \dfrac{-4}{4x-1}$

20. $\dfrac{25z^2-1}{3-15z} = \dfrac{(5z+1)(5z-1)}{3(1-5z)}$

$= \dfrac{(5z+1)(5z-1)}{3(-1)(5z-1)}$

$= -\dfrac{5z+1}{3}$

7.1 Exercises

21. $\dfrac{x}{x-5}$

     (a) $x = 10$: $\dfrac{10}{10-5} = \dfrac{10}{5} = 2$

     (b) $x = -5$: $\dfrac{-5}{-5-5} = \dfrac{-5}{-10} = \dfrac{1}{2}$

     (c) $x = 0$: $\dfrac{0}{0-5} = \dfrac{0}{-5} = 0$

23. $\dfrac{2a-3}{a}$

     (a) $a = 3$: $\dfrac{2(3)-3}{3} = \dfrac{6-3}{3} = \dfrac{3}{3} = 1$

     (b) $a = -3$: $\dfrac{2(-3)-3}{-3} = \dfrac{-6-3}{-3} = \dfrac{-9}{-3} = 3$

(c) $a = 9$: $\dfrac{2(9) - 3}{9} = \dfrac{18 - 3}{9} = \dfrac{15}{9} = \dfrac{5}{3}$

25. $\dfrac{x^2 - 2x}{x - 4}$

(a) $x = 3$: $\dfrac{3^2 - 2(3)}{3 - 4} = \dfrac{9 - 6}{-1} = \dfrac{3}{-1} = -3$

(b) $x = 2$: $\dfrac{2^2 - 2(2)}{2 - 4} = \dfrac{4 - 4}{-2} = \dfrac{0}{-2} = 0$

(c) $x = -3$: $\dfrac{(-3)^2 - 2(-3)}{-3 - 4} = \dfrac{9 + 6}{-7} = -\dfrac{15}{7}$

27. $\dfrac{x^2 - y^2}{2x - y}$

(a) $x = 2, y = 2$: $\dfrac{2^2 - 2^2}{2(2) - 2} = \dfrac{4 - 4}{4 - 2} = \dfrac{0}{2} = 0$

(b) $x = 3, y = 4$: $\dfrac{3^2 - 4^2}{2(3) - 4} = \dfrac{9 - 16}{6 - 4} = \dfrac{-7}{2} = -\dfrac{7}{2}$

(c) $x = 1, y = -2$: $\dfrac{1^2 - (-2)^2}{2(1) - (-2)} = \dfrac{1 - 4}{2 + 2} = \dfrac{-3}{4} = -\dfrac{3}{4}$

29. $\dfrac{2 - 4x}{3x}$ is undefined when $3x = 0$.

$3x = 0$

$x = 0$

31. $\dfrac{3p}{p - 5}$ is undefined when $p - 5 = 0$.

$p - 5 = 0$

$p = 5$

33. $\dfrac{8}{3 - 2x}$ is undefined when $3 - 2x = 0$.

$3 - 2x = 0$

$-2x = -3$

$x = \dfrac{3}{2}$

35. $\dfrac{6z}{z^2 - 36}$ is undefined when $z^2 - 36 = 0$.

$z^2 - 36 = 0$

$(z + 6)(z - 6) = 0$

$z + 6 = 0 \quad$ or $\quad z - 6 = 0$

$z = -6 \quad$ or $\quad z = 6$

37. $\dfrac{x}{x^2 - 7x + 10}$ is undefined when

$x^2 - 7x + 10 = 0$.

$x^2 - 7x + 10 = 0$

$(x - 2)(x - 5) = 0$

$x - 2 = 0 \quad$ or $\quad x - 5 = 0$

$x = 2 \quad$ or $\quad x = 5$

39. $\dfrac{12x + 5}{x^3 - x^2 - 6x}$ is undefined when

$x^3 - x^2 - 6x = 0$.

$x^3 - x^2 - 6x = 0$

$x(x + 2)(x - 3) = 0$

$x = 0 \quad$ or $\quad x + 2 = 0 \quad$ or $\quad x - 3 = 0$

$x = 0 \quad$ or $\quad\quad x = -2 \quad$ or $\quad\quad x = 3$

41. $\dfrac{5x - 10}{15} = \dfrac{5(x - 2)}{5 \cdot 3} = \dfrac{x - 2}{3}$

43. $\dfrac{3z^2 + 6z}{3z} = \dfrac{3z(z + 2)}{3z} = z + 2$

45. $\dfrac{p - 3}{p^2 - p - 6} = \dfrac{(p - 3) \cdot 1}{(p - 3)(p + 2)} = \dfrac{1}{p + 2}$

47. $\dfrac{2 - x}{x - 2} = \dfrac{-1(x - 2)}{x - 2} = -1$

49. $\dfrac{2k^2 - 14k}{7 - k} = \dfrac{2k(k - 7)}{-1 \cdot (k - 7)} = -2k$

51. $\dfrac{x^2 - 1}{x^2 + 5x + 4} = \dfrac{(x - 1)(x + 1)}{(x + 4)(x + 1)} = \dfrac{x - 1}{x + 4}$

53. $\dfrac{6x + 30}{x^2 - 25} = \dfrac{6(x + 5)}{(x + 5)(x - 5)} = \dfrac{6}{x - 5}$

55. $\dfrac{-3x-3y}{x+y} = \dfrac{-3(x+y)}{x+y} = -3$

57. $\dfrac{b^2-25}{4b+20} = \dfrac{(b-5)(b+5)}{4(b+5)} = \dfrac{b-5}{4}$

59. $\dfrac{x^2-2x-15}{x^2-8x+15} = \dfrac{(x+3)(x-5)}{(x-3)(x-5)} = \dfrac{x+3}{x-3}$

61. $\dfrac{x^3+x^2-12x}{x^3-x^2-20x} = \dfrac{x(x+4)(x-3)}{x(x+4)(x-5)} = \dfrac{x-3}{x-5}$

63. $\dfrac{x^2-y^2}{y-x} = \dfrac{(x+y)(x-y)}{y-x}$

$\qquad = \dfrac{-(x+y)(y-x)}{y-x}$

$\qquad = -(x+y)$

65. $\dfrac{x^3+4x^2+x+4}{x^2+5x+4} = \dfrac{x^2(x+4)+1\cdot(x+4)}{(x+1)(x+4)}$

$\qquad = \dfrac{(x+4)(x^2+1)}{(x+1)(x+4)}$

$\qquad = \dfrac{x^2+1}{x+1}$

67. $\dfrac{16-c^2}{(c-4)^2} = \dfrac{(4+c)(4-c)}{(c-4)(c-4)}$

$\qquad = \dfrac{-(4+c)(c-4)}{(c-4)(c-4)}$

$\qquad = \dfrac{-(4+c)}{c-4}$

69. $\dfrac{4x^2-20x+24}{6x^2-48x+90} = \dfrac{4(x-2)(x-3)}{6(x-3)(x-5)}$

$\qquad = \dfrac{2(x-3)\cdot 2(x-2)}{2(x-3)\cdot 3(x-5)}$

$\qquad = \dfrac{2(x-2)}{3(x-5)}$

71. $\dfrac{6+x-x^2}{x^2-4} = \dfrac{-1(x^2-x-6)}{x^2-4}$

$\qquad = \dfrac{-1(x-3)(x+2)}{(x-2)(x+2)}$

$\qquad = \dfrac{-(x-3)}{x-2}$

73. $\dfrac{2t^2-18}{t^4-81} = \dfrac{2(t^2-9)}{(t^2-9)(t^2+9)} = \dfrac{2}{t^2+9}$

75. $\dfrac{12w-3w^2}{w^3-5w^2+4w} = \dfrac{-3w(w-4)}{w(w-1)(w-4)} = \dfrac{-3}{w-1}$

77. (a) $C = \dfrac{50t}{t^2+25}$; find C when $t=5$.

$\dfrac{50(5)}{5^2+25} = \dfrac{250}{25+25} = \dfrac{250}{50} = 5$

The concentration is 5 mg/mL after 5 minutes.

(b) Find C when $t=10$.

$C = \dfrac{50(10)}{10^2+25} = \dfrac{500}{100+25} = \dfrac{500}{125} = 4$

The concentration is 4 mg/mL after 10 minutes.

79. $\text{BMI} = \dfrac{k}{m^2}$; find BMI when $k=110$ and $m=2$.

$\dfrac{110}{2^2} = \dfrac{110}{4} = 27.5$

Yes, the person should consider looking for a weight-loss program.

81. $\bar{C} = \dfrac{0.2x^3-2.3x^2+14.3x+10.2}{x}$; find $\bar{C}$ when $x=2$.

$\dfrac{0.2(2)^3-2.3(2)^2+14.3(2)+10.2}{2}$

$= \dfrac{1.6-9.2+28.6+10.2}{2}$

$= \dfrac{31.2}{2}$

$= 15.6$

The average cost of producing 2 Cobalts is \$15,600 per car.

83. $\dfrac{c^8-1}{(c^4-1)(c^6+1)} = \dfrac{(c^4+1)(c^4-1)}{(c^4-1)(c^6+1)} = \dfrac{c^4+1}{c^6+1}$

85. $\dfrac{x^4 - x^2 - 12}{x^4 + 2x^3 - 9x - 18} = \dfrac{(x^2 - 4)(x^2 + 3)}{x^3(x + 2) - 9(x + 2)}$

$\qquad\qquad = \dfrac{(x + 2)(x - 2)(x^2 + 3)}{(x + 2)(x^3 - 9)}$

$\qquad\qquad = \dfrac{(x - 2)(x^2 + 3)}{x^3 - 9}$

87. $\dfrac{(t + 2)^3(t^4 - 16)}{(t^3 + 8)(t + 2)(t^2 - 4)}$

$\qquad = \dfrac{(t + 2)^3(t^2 + 4)(t^2 - 4)}{(t + 2)(t^2 - 2t + 4)(t + 2)(t + 2)(t - 2)}$

$\qquad = \dfrac{(t + 2)^3(t^2 + 4)(t + 2)(t - 2)}{(t + 2)(t^2 - 2t + 4)(t + 2)(t + 2)(t - 2)}$

$\qquad = \dfrac{(t + 2)^3(t - 2) \cdot (t^2 + 4)(t + 2)}{(t + 2)^3(t - 2) \cdot (t^2 - 2t + 4)}$

$\qquad = \dfrac{(t^2 + 4)(t + 2)}{t^2 - 2t + 4}$

89. (a) The x^2s cannot be divided out because they are terms, not factors.

(b) When the factors of $x - 2$ are divided out, the result is $\dfrac{1}{3}$, not 3.

91. A rational expression is undefined if the denominator is equal to zero. Division by zero is not allowed in the real number system. The expression $\dfrac{x^2 - 9}{3x - 6}$ is undefined for $x = 2$ because when 2 is substituted for x, we get $\dfrac{2^2 - 9}{3(2) - 6} = \dfrac{4 - 9}{6 - 6} = \dfrac{-5}{0}$, which is undefined.

93. $\dfrac{x - 7}{7 - x} = -1$ because $x - 7$ and $7 - x$ are opposites:

$\dfrac{x - 7}{7 - x} = \dfrac{x - 7}{-1(-7 + x)}$

$\qquad = \dfrac{x - 7}{-1(x - 7)}$

$\qquad = \dfrac{\cancel{x - 7}}{-1\cancel{(x - 7)}}$

$\qquad = \dfrac{1}{-1}$

$\qquad = 1.$

The expression $\dfrac{x - 7}{x + 7}$ is in simplest form because $x - 7$ and $x + 7$ are not opposites and the terms cannot be divided out.

Section 7.2

Preparing for Multiplying and Dividing Rational Expressions

P1. $\dfrac{3}{14} \cdot \dfrac{28}{9} = \dfrac{3 \cdot 28}{14 \cdot 9} = \dfrac{3 \cdot 2 \cdot 14}{14 \cdot 3 \cdot 3} = \dfrac{2}{3}$

P2. The reciprocal of $\dfrac{5}{8}$ is $\dfrac{8}{5}$.

P3. $\dfrac{12}{25} \div \dfrac{12}{5} = \dfrac{12}{25} \cdot \dfrac{5}{12} = \dfrac{12 \cdot 5}{25 \cdot 12} = \dfrac{12 \cdot 5}{5 \cdot 5 \cdot 12} = \dfrac{1}{5}$

P4. $3x^3 - 27x = 3x(x^2 - 9)$
$\qquad\qquad\quad = 3x(x + 3)(x - 3)$

P5. $\dfrac{3x + 12}{5x^2 + 20x} = \dfrac{3(x + 4)}{5x(x + 4)} = \dfrac{3}{5x}$

7.2 Quick Checks

1. To multiply two rational expressions, <u>factor</u> each numerator and denominator and then divide out any <u>common</u> <u>factors</u>.

2. False; $\dfrac{x+4}{24} \cdot \dfrac{24}{(x+4)(x-4)} = \dfrac{(x+4)\cdot 24 \cdot 1}{24(x+4)(x-4)}$

$\qquad = \dfrac{\cancel{(x+4)} \cdot \cancel{24} \cdot 1}{\cancel{24}\,\cancel{(x+4)}(x-4)}$

$\qquad = \dfrac{1}{(x-4)}$

$\qquad \neq x-4$

3. $\dfrac{p^2-9}{5} \cdot \dfrac{25p}{2p-6} = \dfrac{(p+3)(p-3)}{5} \cdot \dfrac{5^2 p}{2(p-3)}$

$\qquad = \dfrac{(p+3)(p-3)\cdot 5^2 p}{5 \cdot 2(p-3)}$

$\qquad = \dfrac{5p(p+3)}{2}$

4. $\dfrac{2x+8}{x^2+3x-4} \cdot \dfrac{7x-7}{6x+30}$

$\qquad = \dfrac{2(x+4)}{(x+4)(x-1)} \cdot \dfrac{7(x-1)}{2\cdot 3(x+5)}$

$\qquad = \dfrac{2(x+4)\cdot 7(x-1)}{(x+4)(x-1)\cdot 2\cdot 3(x+5)}$

$\qquad = \dfrac{7}{3(x+5)}$

5. False; $\dfrac{8}{x-7} \cdot \dfrac{7-x}{16x} = \dfrac{\overset{1}{\cancel{8}}}{-1(7-x)} \cdot \dfrac{7-6}{\underset{2}{\cancel{16}x}}$

$\qquad = -\dfrac{1}{7-x} \cdot \dfrac{7-x}{2x}$

$\qquad = -\dfrac{1}{\underset{1}{\cancel{(7-x)}}} \cdot \dfrac{\overset{1}{\cancel{(7-x)}}}{2x}$

$\qquad = -\dfrac{1}{2x} \neq \dfrac{1}{2x}$

6. $\dfrac{15a-3a^2}{7} \cdot \dfrac{3+2a}{2a^2-7a-15}$

$\qquad = \dfrac{3a(5-a)}{7} \cdot \dfrac{3+2a}{(2a+3)(a-5)}$

$\qquad = \dfrac{3a(-1)(a-5)(2a+3)}{7(2a+3)(a-5)}$

$\qquad = -\dfrac{3a}{7}$

7. $\dfrac{a^2+2ab+b^2}{3a+3b} \cdot \dfrac{a-b}{a^2-b^2}$

$\qquad = \dfrac{(a+b)^2}{3(a+b)} \cdot \dfrac{a-b}{(a+b)(a-b)}$

$\qquad = \dfrac{(a+b)^2(a-b)}{3(a+b)^2(a-b)}$

$\qquad = \dfrac{1}{3}$

8. $\dfrac{8}{9} \div \dfrac{2}{3} = \dfrac{8}{9} \cdot \dfrac{3}{2} = \dfrac{4\cdot 2\cdot 3}{3\cdot 3\cdot 2} = \dfrac{4\cdot \cancel{2}\cdot \cancel{3}}{\cancel{3}\cdot 3\cdot \cancel{2}} = \dfrac{4}{3}$

9. $\dfrac{12}{x^2-x} \div \dfrac{4x-2}{x^2-1} = \dfrac{12}{x^2-x} \cdot \dfrac{x^2-1}{4x-2}$

$\qquad = \dfrac{12}{x(x-1)} \cdot \dfrac{(x+1)(x-1)}{2(2x-1)}$

$\qquad = \dfrac{12(x+1)(x-1)}{2x(x-1)(2x-1)}$

$\qquad = \dfrac{6(x+1)}{x(2x-1)}$

10. $\dfrac{x^2-9}{x^2-16} \div \dfrac{x^2-x-12}{x^2+x-12}$

$\qquad = \dfrac{x^2-9}{x^2-16} \cdot \dfrac{x^2+x-12}{x^2-x-12}$

$\qquad = \dfrac{(x+3)(x-3)}{(x+4)(x-4)} \cdot \dfrac{(x+4)(x-3)}{(x+3)(x-4)}$

$\qquad = \dfrac{(x+3)(x+4)(x-3)^2}{(x+3)(x+4)(x-4)^2}$

$\qquad = \dfrac{(x-3)^2}{(x-4)^2}$

11. $\dfrac{q^2-6q-7}{q^2-25} \div (q-7) = \dfrac{q^2-6q-7}{q^2-25} \div \dfrac{q-7}{1}$

$\qquad = \dfrac{q^2-6q-7}{q^2-25} \cdot \dfrac{1}{q-7}$

$\qquad = \dfrac{(q+1)(q-7)}{(q+5)(q-5)} \cdot \dfrac{1}{q-7}$

$\qquad = \dfrac{(q+1)(q-7)}{(q+5)(q-5)(q-7)}$

$\qquad = \dfrac{q+1}{(q+5)(q-5)}$

12.
$$\frac{\frac{2}{x+1}}{\frac{8}{x^2-1}} = \frac{2}{x+1} \cdot \frac{x^2-1}{8}$$
$$= \frac{2}{x+1} \cdot \frac{(x+1)(x-1)}{2 \cdot 4}$$
$$= \frac{2 \cdot (x+1)(x-1)}{2 \cdot 4 \cdot (x+1)}$$
$$= \frac{x+1}{4}$$

13.
$$\frac{\frac{x+3}{x^2-4}}{\frac{4x+12}{7x^2+14x}} = \frac{x+3}{x^2-4} \cdot \frac{7x^2+14x}{4x+12}$$
$$= \frac{x+3}{(x+2)(x-2)} \cdot \frac{7x(x+2)}{4(x+3)}$$
$$= \frac{(x+3) \cdot 7x(x+2)}{(x+2)(x-2) \cdot 4(x+3)}$$
$$= \frac{7x}{4(x-2)}$$

14.
$$\frac{3m-6n}{5n} \div \frac{m^2-4n^2}{10mn}$$
$$= \frac{3m-6n}{5n} \cdot \frac{10mn}{m^2-4n^2}$$
$$= \frac{3(m-2n)}{5 \cdot n} \cdot \frac{5 \cdot 2 \cdot m \cdot n}{(m+2n)(m-2n)}$$
$$= \frac{3(m-2n) \cdot 5 \cdot 2 \cdot m \cdot n}{5 \cdot n \cdot (m+2n)(m-2n)}$$
$$= \frac{3 \cdot 2 \cdot m}{m+2n}$$
$$= \frac{6m}{m+2n}$$

7.2 Exercises

15.
$$\frac{x+5}{7} \cdot \frac{14x}{x^2-25} = \frac{x+5}{7} \cdot \frac{7 \cdot 2x}{(x+5)(x-5)}$$
$$= \frac{7(x+5) \cdot 2x}{7(x+5) \cdot (x-5)}$$
$$= \frac{2x}{x-5}$$

17.
$$\frac{x^2-x}{x^2-x-2} \cdot \frac{x-2}{x^2-1}$$
$$= \frac{x(x-1)}{(x+1)(x-2)} \cdot \frac{x-2}{(x+1)(x-1)}$$
$$= \frac{(x-1)(x-2) \cdot x}{(x-1)(x-2) \cdot (x+1)(x+1)}$$
$$= \frac{x}{(x+1)^2}$$

19.
$$\frac{p^2-1}{2p-3} \cdot \frac{2p^2+p-6}{p^2+3p+2}$$
$$= \frac{(p+1)(p-1)}{2p-3} \cdot \frac{(2p-3)(p+2)}{(p+2)(p+1)}$$
$$= \frac{(p+1)(2p-3)(p+2) \cdot (p-1)}{(p+1)(2p-3)(p+2)}$$
$$= p-1$$

21.
$$(x+1) \cdot \frac{x-6}{x^2-5x-6} = \frac{x+1}{1} \cdot \frac{x-6}{(x+1)(x-6)}$$
$$= \frac{(x+1)(x-6)}{(x+1)(x-6)}$$
$$= 1$$

23.
$$\frac{x-4}{2x-8} \div \frac{3x}{2} = \frac{x-4}{2x-8} \cdot \frac{2}{3x}$$
$$= \frac{x-4}{2(x-4)} \cdot \frac{2}{3x}$$
$$= \frac{2(x-4) \cdot 1}{2(x-4) \cdot 3x}$$
$$= \frac{1}{3x}$$

25.
$$\frac{m^2-16}{6m} \div \frac{m^2+8m+16}{12m}$$
$$= \frac{m^2-16}{6m} \cdot \frac{12m}{m^2+8m+16}$$
$$= \frac{(m+4)(m-4)}{6m} \cdot \frac{2 \cdot 6m}{(m+4)(m+4)}$$
$$= \frac{6m(m+4) \cdot 2(m-4)}{6m(m+4) \cdot (m+4)}$$
$$= \frac{2(m-4)}{m+4}$$

27. $\dfrac{x^2-x}{x^2-1} \div \dfrac{x+2}{x^2+3x+2}$

$= \dfrac{x^2-x}{x^2-1} \cdot \dfrac{x^2+3x+2}{x+2}$

$= \dfrac{x(x-1)}{(x+1)(x-1)} \cdot \dfrac{(x+2)(x+1)}{x+2}$

$= \dfrac{(x-1)(x+2)(x+1)\cdot x}{(x-1)(x+2)(x+1)\cdot 1}$

$= \dfrac{x}{1}$

$= x$

29. $\dfrac{(x+2)^2}{x^2-4} \div \dfrac{x^2-x-6}{x^2-5x+6}$

$= \dfrac{(x+2)^2}{x^2-4} \cdot \dfrac{x^2-5x+6}{x^2-x-6}$

$= \dfrac{(x+2)(x+2)}{(x-2)(x+2)} \cdot \dfrac{(x-3)(x-2)}{(x-3)(x+2)}$

$= \dfrac{(x+2)(x+2)(x-3)(x-2)}{(x+2)(x+2)(x-3)(x-2)}$

$= 1$

31. $\dfrac{14}{9} \cdot \dfrac{15}{7} = \dfrac{2\cdot 7 \cdot 3 \cdot 5}{3\cdot 3 \cdot 7} = \dfrac{3\cdot 7 \cdot 2 \cdot 5}{3\cdot 7 \cdot 3} = \dfrac{10}{3}$

33. $-\dfrac{20}{16} \div \left(-\dfrac{30}{24}\right) = -\dfrac{20}{16} \cdot \left(-\dfrac{24}{30}\right)$

$= \dfrac{-2\cdot 2 \cdot 5}{2\cdot 2 \cdot 2 \cdot 2} \cdot \left(\dfrac{2\cdot 2 \cdot 2 \cdot 3}{-2\cdot 3 \cdot 5}\right)$

$= \dfrac{-2\cdot 2 \cdot 2 \cdot 2 \cdot 2 \cdot 3 \cdot 5}{-2\cdot 2 \cdot 2 \cdot 2 \cdot 2 \cdot 3 \cdot 5}$

$= 1$

35. $\dfrac{8}{11} \div (-2) = \dfrac{8}{11} \cdot \left(-\dfrac{1}{2}\right)$

$= \dfrac{2\cdot 4}{11} \cdot \left(-\dfrac{1}{2}\right)$

$= -\dfrac{2\cdot 4}{2\cdot 11}$

$= -\dfrac{4}{11}$

37. $\dfrac{3y}{y^2-y-6} \cdot \dfrac{4y+8}{9y^2} = \dfrac{3y}{(y+2)(y-3)} \cdot \dfrac{4(y+2)}{3y\cdot 3y}$

$= \dfrac{4}{3y(y-3)}$

39. $\dfrac{4a+8b}{a^2+2ab} \cdot \dfrac{a^2}{12} = \dfrac{4(a+2b)}{a(a+2b)} \cdot \dfrac{a\cdot a}{3\cdot 4} = \dfrac{a}{3}$

41. $\dfrac{3x^2-6x}{x^2-2x-8} \div \dfrac{x-2}{x+2} = \dfrac{3x^2-6x}{x^2-2x-8} \cdot \dfrac{x+2}{x-2}$

$= \dfrac{3x(x-2)}{(x-4)(x+2)} \cdot \dfrac{x+2}{x-2}$

$= \dfrac{(x-2)(x+2)\cdot 3x}{(x-2)(x+2)\cdot (x-4)}$

$= \dfrac{3x}{x-4}$

43. $\dfrac{(w-4)^2}{4-w^2} \div \dfrac{w^2-16}{w-2}$

$= \dfrac{(w-4)^2}{4-w^2} \cdot \dfrac{w-2}{w^2-16}$

$= \dfrac{(w-4)^2}{(2+w)(2-w)} \cdot \dfrac{-(2-w)}{(w+4)(w-4)}$

$= \dfrac{-(2-w)(w-4)^2}{(2-w)(2+w)(w+4)(w-4)}$

$= \dfrac{4-w}{(w+2)(w+4)}$

45. $\dfrac{3xy-2y^2-x^2}{x+y} \cdot \dfrac{x^2-y^2}{x^2-2xy}$

$= \dfrac{-(x-y)(x-2y)}{x+y} \cdot \dfrac{(x+y)(x-y)}{x(x-2y)}$

$= -\dfrac{(x-y)^2(x+y)(x-2y)}{x(x+y)(x-2y)}$

$= -\dfrac{(x-y)^2}{x}$

47. $\dfrac{\frac{2c-4}{8}}{\frac{2-c}{2}} = \dfrac{2c-4}{8} \cdot \dfrac{2}{2-c}$

$= \dfrac{2(c-2)}{2\cdot 2 \cdot 2} \cdot \dfrac{2}{-(c-2)}$

$= -\dfrac{2\cdot 2(c-2)}{2\cdot 2(c-2)\cdot 2}$

$= -\dfrac{1}{2}$

49. $\dfrac{4n^2-9}{6n+18} \cdot \dfrac{9n^2-81}{2n^2+5n-12}$

$= \dfrac{(2n+3)(2n-3)}{2\cdot 3\cdot (n+3)} \cdot \dfrac{3\cdot 3\cdot (n+3)(n-3)}{(n+4)(2n-3)}$

$= \dfrac{(2n+3)(2n-3)\cdot 3\cdot 3\cdot (n+3)(n-3)}{2\cdot 3\cdot (n+3)(n+4)(2n-3)}$

$= \dfrac{3(2n+3)(n-3)}{2(n+4)}$

51. $\dfrac{x^2 y}{2x^2-5xy+2y^2} \div \dfrac{(2xy^2)^2}{2x^2 y-xy^2}$

$= \dfrac{x^2 y}{2x^2-5xy+2y^2} \cdot \dfrac{2x^2 y-xy^2}{(2xy^2)^2}$

$= \dfrac{x^2 y}{(2x-y)(x-2y)} \cdot \dfrac{xy(2x-y)}{4x^2 y^4}$

$= \dfrac{x^2 y\cdot xy(2x-y)}{4x^2 y^4(2x-y)(x-2y)}$

$= \dfrac{x}{4y^2(x-2y)}$

53. $\dfrac{2a^2+3ab-2b^2}{a^2-b^2} \cdot \dfrac{a^2-ab}{2a^3+4a^2 b}$

$= \dfrac{(a+2b)(2a-b)}{(a+b)(a-b)} \cdot \dfrac{a(a-b)}{2a^2(a+2b)}$

$= \dfrac{a(a+2b)(a-b)\cdot (2a-b)}{a(a+2b)(a-b)\cdot 2a(a+b)}$

$= \dfrac{2a-b}{2a(a+b)}$

55. $\dfrac{3t^2-27}{t+2} \cdot \dfrac{t^2-4}{9t-27}$

$= \dfrac{3(t+3)(t-3)}{t+2} \cdot \dfrac{(t+2)(t-2)}{3\cdot 3(t-3)}$

$= \dfrac{3(t+2)(t-3)\cdot (t+3)(t-2)}{3(t+2)(t-3)\cdot 3}$

$= \dfrac{(t+3)(t-2)}{3}$

57. $\dfrac{(x+2)^2}{x^2-4} \div \dfrac{-x^2+x+6}{x^2-5x+6}$

$= \dfrac{(x+2)^2}{x^2-4} \cdot \dfrac{x^2-5x+6}{-x^2+x+6}$

$= \dfrac{(x+2)^2}{(x+2)(x-2)} \cdot \dfrac{(x-2)(x-3)}{-(x+2)(x-3)}$

$= -\dfrac{(x+2)^2(x-2)(x-3)}{(x+2)^2(x-2)(x-3)}$

$= -1$

59. $\dfrac{9-x^2}{x^2+5x+4} \div \dfrac{x^2-2x-3}{x^2+4x}$

$= \dfrac{9-x^2}{x^2+5x+4} \cdot \dfrac{x^2+4x}{x^2-2x-3}$

$= \dfrac{-(x+3)(x-3)}{(x+4)(x+1)} \cdot \dfrac{x(x+4)}{(x+1)(x-3)}$

$= -\dfrac{(x-3)(x+4)\cdot x(x+3)}{(x-3)(x+4)\cdot (x+1)(x+1)}$

$= -\dfrac{x(x+3)}{(x+1)^2}$

61. $\dfrac{\frac{a^2-b^2}{a^2+b^2}}{\frac{4a-4b}{2a^2+2b^2}} = \dfrac{a^2-b^2}{a^2+b^2} \cdot \dfrac{2a^2+2b^2}{4a-4b}$

$= \dfrac{(a+b)(a-b)}{a^2+b^2} \cdot \dfrac{2(a^2+b^2)}{2\cdot 2(a-b)}$

$= \dfrac{2(a-b)(a^2+b^2)\cdot (a+b)}{2(a-b)(a^2+b^2)\cdot 2}$

$= \dfrac{a+b}{2}$

63. $\dfrac{x^3-1}{x^4-1} \div \dfrac{3x^2+3x+3}{x^3+x^2+x+1}$

$= \dfrac{x^3-1}{x^4-1} \cdot \dfrac{x^3+x^2+x+1}{3x^2+3x+3}$

$= \dfrac{(x-1)(x^2+x+1)}{(x^2+1)(x+1)(x-1)} \cdot \dfrac{x^2(x+1)+1(x+1)}{3(x^2+x+1)}$

$= \dfrac{(x+1)(x-1)(x^2+1)(x^2+x+1)}{3(x+1)(x-1)(x^2+1)(x^2+x+1)}$

$= \dfrac{1}{3}$

65.
$$\frac{xy-ay+xb-ab}{xy+ay-xb-ab}\cdot\frac{2xy-2ay-2xb+2ab}{4b+4y}$$

$$=\frac{y(x-a)+b(x-a)}{y(x+a)-b(x+a)}\cdot\frac{2y(x-a)-2b(x-a)}{4(y+b)}$$

$$=\frac{(x-a)(y+b)}{(x+a)(y-b)}\cdot\frac{2(x-a)(y-b)}{2\cdot 2(y+b)}$$

$$=\frac{2(y+b)(y-b)\cdot(x-a)^2}{2(y+b)(y-b)\cdot 2(x+a)}$$

$$=\frac{(x-a)^2}{2(x+a)}$$

67.
$$\frac{x^2+3xy+2y^2}{x^2-y^2}\cdot\frac{3x-3y}{9x^2+9xy-18y^2}$$

$$=\frac{(x+y)(x+2y)}{(x+y)(x-y)}\cdot\frac{3(x-y)}{3\cdot 3\cdot(x+2y)(x-y)}$$

$$=\frac{3(x-y)(x+y)(x+2y)}{3(x-y)(x+y)(x+2y)\cdot 3(x-y)}$$

$$=\frac{1}{3(x-y)}$$

69.
$$\frac{\frac{x}{2y}}{\frac{(2xy)^2}{9xy^3}}=\frac{x}{2y}\cdot\frac{9xy^3}{(2xy)^2}$$

$$=\frac{x}{2y}\cdot\frac{9xy^3}{4x^2y^2}$$

$$=\frac{x^2y^3\cdot 9}{x^2y^3\cdot 8}$$

$$=\frac{9}{8}$$

71.
$$\frac{\left(\frac{x-y}{x+y}\right)^2}{x^2-y^2}=\frac{(x-y)^2}{(x+y)^2}\cdot\frac{1}{x^2-y^2}$$

$$=\frac{(x-y)^2}{(x+y)^2}\cdot\frac{1}{(x+y)(x-y)}$$

$$=\frac{(x-y)\cdot(x-y)}{(x-y)\cdot(x+y)^2}$$

$$=\frac{x-y}{(x+y)^3}$$

73.
$$\frac{\frac{6x}{x^2-4}}{\frac{3x-9}{2x+4}}=\frac{6x}{x^2-4}\cdot\frac{2x+4}{3x-9}$$

$$=\frac{2\cdot 3\cdot x}{(x+2)(x-2)}\cdot\frac{2(x+2)}{3(x-3)}$$

$$=\frac{3(x+2)\cdot 4x}{3(x+2)\cdot(x-2)(x-3)}$$

$$=\frac{4x}{(x-2)(x-3)}$$

75. In a rectangle,

$$\text{Area}=(\text{length})(\text{width})=\frac{3x+9}{27x^2}\cdot\frac{9x}{x+3}$$

$$=\frac{3(x+3)}{27x^2}\cdot\frac{9x}{x+3}$$

$$=\frac{27x(x+3)}{27x(x+3)\cdot x}$$

$$=\frac{1}{x}$$

The area is $\dfrac{1}{x}$ square feet.

77. In a triangle,

$$\text{Area}=\frac{1}{2}(\text{base})(\text{height})$$

$$=\frac{1}{2}\cdot\frac{1}{x^2-9}\cdot\frac{4x^2+20x+24}{1}$$

$$=\frac{1}{2}\cdot\frac{1}{(x+3)(x-3)}\cdot\frac{2\cdot 2(x+3)(x+2)}{1}$$

$$=\frac{2(x+3)\cdot 2(x+2)}{2(x+3)\cdot(x-3)}$$

$$=\frac{2(x+2)}{x-3}$$

The area is $\dfrac{2(x+2)}{x-3}$ square inches.

79. $\dfrac{x^2+x-12}{x^2-2x-35} \div \dfrac{x+4}{x^2+4x-5} \div \dfrac{12-4x}{x-7} = \dfrac{x^2+x-12}{x^2-2x-35} \cdot \dfrac{x^2+4x-5}{x+4} \cdot \dfrac{x-7}{12-4x}$

$$= \dfrac{(x+4)(x-3)}{(x+5)(x-7)} \cdot \dfrac{(x+5)(x-1)}{x+4} \cdot \dfrac{x-7}{-4(x-3)}$$

$$= -\dfrac{(x+4)(x-3)(x+5)(x-7) \cdot (x-1)}{(x+4)(x-3)(x+5)(x-7) \cdot 4}$$

$$= \dfrac{1-x}{4}$$

81. $\dfrac{a^2-2ab}{2b-3a} \div \dfrac{3a^2-4ab-4b^2}{16a^2b^2-36a^4} \div (6a) = \dfrac{a^2-2ab}{2b-3a} \cdot \dfrac{16a^2b^2-36a^4}{3a^2-4ab-4b^2} \cdot \dfrac{1}{6a}$

$$= \dfrac{a(a-2b)}{2b-3a} \cdot \dfrac{4a^2(4b^2-9a^2)}{(3a+2b)(a-2b)} \cdot \dfrac{1}{6a}$$

$$= \dfrac{a(a-2b)}{2b-3a} \cdot \dfrac{2 \cdot 2a^2 \cdot (2b+3a)(2b-3a)}{(3a+2b)(a-2b)} \cdot \dfrac{1}{3 \cdot 2 \cdot a}$$

$$= \dfrac{2a(a-2b)(2b-3a)(2b+3a) \cdot 2a^2}{2a(a-2b)(2b-3a)(2b+3a) \cdot 3}$$

$$= \dfrac{2a^2}{3}$$

83. $\dfrac{x^2+xy-3x-3y}{x^3+y^3} \cdot \dfrac{x^2+2xy+y^2}{x^2-x-6} = \dfrac{x(x+y)-3(x+y)}{x^3+y^3} \cdot \dfrac{x^2+2xy+y^2}{x^2-x-6}$

$$= \dfrac{(x+y)(x-3)}{(x+y)(x^2-xy+y^2)} \cdot \dfrac{(x+y)^2}{(x+2)(x-3)}$$

$$= \dfrac{(x+y)(x-3) \cdot (x+y)^2}{(x+y)(x-3) \cdot (x^2-xy+y^2)(x+2)}$$

$$= \dfrac{(x+y)^2}{(x^2-xy+y^2)(x+2)}$$

85. $\dfrac{p^3-27q^3}{9pq} \cdot \dfrac{(3p^2q)^3}{p^2-9q^2} \div \dfrac{1}{p^2+2pq-3q^2} = \dfrac{p^3-27q^3}{9pq} \cdot \dfrac{27p^6q^3}{p^2-9q^2} \cdot \dfrac{p^2+2pq-3q^2}{1}$

$$= \dfrac{(p-3q)(p^2+3pq+9q^2)}{9pq} \cdot \dfrac{9pq \cdot 3p^5q^2}{(p+3q)(p-3q)} \cdot \dfrac{(p+3q)(p-q)}{1}$$

$$= \dfrac{9pq(p-3q)(p+3q) \cdot 3p^5q^2(p^2+3pq+9q^2)(p-q)}{9pq(p-3q)(p+3q)}$$

$$= 3p^5q^2(p^2+3pq+9q^2)(p-q)$$

87.
$$\frac{2x}{x^3-3x^2} \cdot \frac{x^2-x-6}{?} = \frac{1}{3x}$$
$$\frac{2x}{x \cdot x(x-3)} \cdot \frac{(x+2)(x-3)}{?} = \frac{1}{3x}$$
$$\frac{x(x-3) \cdot 2(x+2)}{x(x-3) \cdot x \cdot ?} = \frac{1}{3x}$$
$$\frac{2(x+2)}{x \cdot ?} = \frac{1}{3x}$$
$$? = 3 \cdot 2(x+2)$$
$$= 6(x+2)$$
$$= 6x+12$$

89. To multiply two rational expressions, (1) factor the polynomials in the numerators and denominators; (2) multiply the numerators and denominators using $\dfrac{a}{b} \cdot \dfrac{c}{d} = \dfrac{a \cdot c}{b \cdot d}$; (3) divide out common factors in the numerators and denominators. Leave the result in factored form.

91. The error is in incorrectly dividing out the factors $(n-6)$ and $(6-n)$.
$$\frac{n^2-2n-24}{6n-n^2} = \frac{(n-6)(n+4)}{n(6-n)}$$
$$= \frac{(n-6)(n+4)}{n(-1)(n-6)}$$
$$= \frac{n+4}{-n}$$
$$= -\frac{n+4}{n}$$

Section 7.3

Preparing for Adding and Subtracting Rational Expressions with a Common Denominator

P1. $\dfrac{12}{15} = \dfrac{2 \cdot 2 \cdot 3}{3 \cdot 5} = \dfrac{2 \cdot 2}{5} = \dfrac{4}{5}$

P2. (a) $\dfrac{7}{5} + \dfrac{2}{5} = \dfrac{7+2}{5} = \dfrac{9}{5}$

(b) $\dfrac{5}{6} + \dfrac{11}{6} = \dfrac{5+11}{6}$
$$= \frac{16}{6}$$
$$= \frac{2 \cdot 2 \cdot 2 \cdot 2}{2 \cdot 3}$$
$$= \frac{2 \cdot 2 \cdot 2}{3}$$
$$= \frac{8}{3}$$

P3. (a) $\dfrac{7}{9} - \dfrac{5}{9} = \dfrac{7-5}{9} = \dfrac{2}{9}$

(b) $\dfrac{7}{8} - \dfrac{5}{8} = \dfrac{7-5}{8} = \dfrac{2}{8} = \dfrac{2}{2 \cdot 2 \cdot 2} = \dfrac{1}{2 \cdot 2} = \dfrac{1}{4}$

P4. The additive inverse of 5 is -5 because $5 + (-5) = 0$.

P5. $-(x-2) = -1 \cdot x + (-1)(-2) = -x + 2 = 2 - x$

7.3 Quick Checks

1. If $\dfrac{a}{c}$ and $\dfrac{b}{c}$, $c \neq 0$, are two rational expressions, then $\dfrac{a}{c} + \dfrac{b}{c} = \dfrac{a+b}{c}$.

2. False; $\dfrac{2x}{a} + \dfrac{4x}{a} = \dfrac{2x+4x}{a} = \dfrac{6x}{a} \neq \dfrac{6x}{2a}$.

3. $\dfrac{1}{x-2} + \dfrac{3}{x-2} = \dfrac{1+3}{x-2} = \dfrac{4}{x-2}$

4. $\dfrac{2x+1}{x+1} + \dfrac{x^2}{x+1} = \dfrac{2x+1+x^2}{x+1} = \dfrac{(x+1)^2}{x+1} = x+1$

5. $\dfrac{9x}{6x-5} + \dfrac{2x-3}{6x-5} = \dfrac{9x+2x-3}{6x-5} = \dfrac{11x-3}{6x-5}$

6. $\dfrac{2x-2}{2x^2-7x-15} + \dfrac{5}{2x^2-7x-15}$
$$= \frac{2x-2+5}{2x^2-7x-15}$$
$$= \frac{2x+3}{(2x+3)(x-5)}$$
$$= \frac{1}{x-5}$$

7. $\dfrac{8y}{2y-5} - \dfrac{6}{2y-5} = \dfrac{8y-6}{2y-5} = \dfrac{2(4y-3)}{2y-5}$

8. $\dfrac{10+3z}{6z} - \dfrac{7}{6z} = \dfrac{10+3z-7}{6z}$
$= \dfrac{3z+3}{6z}$
$= \dfrac{3(z+1)}{6z}$
$= \dfrac{z+1}{2z}$

9. $\dfrac{3x+1}{x-4} - \dfrac{x+3}{x-4} = \dfrac{3x+1-(x+3)}{x-4}$

10. $\dfrac{2x^2-5x}{3x} - \dfrac{x^2-13x}{3x} = \dfrac{2x^2-5x-(x^2-13x)}{3x}$
$= \dfrac{2x^2-5x-x^2+13x}{3x}$
$= \dfrac{x^2+8x}{3x}$
$= \dfrac{x(x+8)}{3x}$
$= \dfrac{x+8}{3}$

11. $\dfrac{3x^2+8x-1}{x^2-3x-28} - \dfrac{2x^2+2x-9}{x^2-3x-28}$
$= \dfrac{3x^2+8x-1-(2x^2+2x-9)}{x^2-3x-28}$
$= \dfrac{3x^2+8x-1-2x^2-2x+9}{x^2-3x-28}$
$= \dfrac{x^2+6x+8}{x^2-3x-28}$
$= \dfrac{(x+2)(x+4)}{(x-7)(x+4)}$
$= \dfrac{x+2}{x-7}$

12. True

13. $\dfrac{3x}{x-5} + \dfrac{1}{5-x} = \dfrac{3x}{x-5} + \dfrac{1}{-1(x-5)}$
$= \dfrac{3x}{x-5} + \dfrac{-1}{x-5}$
$= \dfrac{3x-1}{x-5}$

14. $\dfrac{a^2+2a}{a-7} + \dfrac{a^2+14}{7-a} = \dfrac{a^2+2a}{a-7} + \dfrac{a^2+14}{-1(a-7)}$
$= \dfrac{a^2+2a}{a-7} + \dfrac{-a^2-14}{a-7}$
$= \dfrac{a^2+2a-a^2-14}{a-7}$
$= \dfrac{2a-14}{a-7}$
$= \dfrac{2(a-7)}{a-7}$
$= 2$

15. $\dfrac{2n}{n^2-9} - \dfrac{6}{9-n^2} = \dfrac{2n}{n^2-9} - \dfrac{6}{-1(n^2-9)}$
$= \dfrac{2n}{n^2-9} - \dfrac{-6}{n^2-9}$
$= \dfrac{2n-(-6)}{n^2-9}$
$= \dfrac{2n+6}{n^2-9}$
$= \dfrac{2(n+3)}{(n+3)(n-3)}$
$= \dfrac{2}{n-3}$

16. $\dfrac{2k}{6k-6} - \dfrac{9+4k}{6-6k} = \dfrac{2k}{6k-6} - \dfrac{9+4k}{-1(6k-6)}$
$= \dfrac{2k}{6k-6} - \dfrac{-9-4k}{6k-6}$
$= \dfrac{2k-(-9-4k)}{6k-6}$
$= \dfrac{2k+9+4k}{6k-6}$
$= \dfrac{6k+9}{6k-6}$
$= \dfrac{3(2k+3)}{2 \cdot 3(k-1)}$
$= \dfrac{2k+3}{2(k-1)}$

7.3 Exercises

17. $\dfrac{3p}{8} + \dfrac{11p}{8} = \dfrac{3p+11p}{8} = \dfrac{14p}{8} = \dfrac{7p}{4}$

19. $\dfrac{n}{2} + \dfrac{3n}{2} = \dfrac{n+3n}{2} = \dfrac{4n}{2} = 2n$

21. $\dfrac{4a-1}{3a} + \dfrac{2a-2}{3a} = \dfrac{4a-1+2a-2}{3a}$

$\qquad = \dfrac{6a-3}{3a}$

$\qquad = \dfrac{3(2a-1)}{3a}$

$\qquad = \dfrac{2a-1}{a}$

23. $\dfrac{8c-3}{c-1} + \dfrac{2c-1}{c-1} = \dfrac{8c-3+2c-1}{c-1}$

$\qquad = \dfrac{10c-4}{c-1}$

$\qquad = \dfrac{2(5c-2)}{c-1}$

25. $\dfrac{2x}{x+y} + \dfrac{2y}{x+y} = \dfrac{2x+2y}{x+y} = \dfrac{2(x+y)}{x+y} = 2$

27. $\dfrac{14x}{x+2} + \dfrac{7x^2}{x+2} = \dfrac{14x+7x^2}{x+2} = \dfrac{7x(x+2)}{x+2} = 7x$

29. $\dfrac{12a-1}{2a+6} + \dfrac{13-8a}{2a+6} = \dfrac{12a-1+13-8a}{2a+6}$

$\qquad = \dfrac{4a+12}{2a+6}$

$\qquad = \dfrac{4(a+3)}{2(a+3)}$

$\qquad = \dfrac{4}{2}$

$\qquad = 2$

31. $\dfrac{a}{a^2-3a-10} + \dfrac{2}{a^2-3a-10} = \dfrac{a+2}{a^2-3a-10}$

$\qquad = \dfrac{a+2}{(a+2)(a-5)}$

$\qquad = \dfrac{1}{a-5}$

33. $\dfrac{x^2}{x^2-9} - \dfrac{3x}{x^2-9} = \dfrac{x^2-3x}{x^2-9}$

$\qquad = \dfrac{x(x-3)}{(x-3)(x+3)}$

$\qquad = \dfrac{x}{x+3}$

35. $\dfrac{x^2-x}{2x} - \dfrac{2x^2-x}{2x} = \dfrac{x^2-x-(2x^2-x)}{2x}$

$\qquad = \dfrac{x^2-x-2x^2+x}{2x}$

$\qquad = \dfrac{-x^2}{2x}$

$\qquad = -\dfrac{x}{2}$

37. $\dfrac{2c-3}{4c} - \dfrac{6c+9}{4c} = \dfrac{2c-3-(6c+9)}{4c}$

$\qquad = \dfrac{2c-3-6c-9}{4c}$

$\qquad = \dfrac{-4c-12}{4c}$

$\qquad = \dfrac{-4(c+3)}{4c}$

$\qquad = -\dfrac{c+3}{c}$

39. $\dfrac{x^2-6}{x-2} - \dfrac{x^2-3x}{x-2} = \dfrac{x^2-6-(x^2-3x)}{x-2}$

$\qquad = \dfrac{x^2-6-x^2+3x}{x-2}$

$\qquad = \dfrac{-6+3x}{x-2}$

$\qquad = \dfrac{3(x-2)}{x-2}$

$\qquad = 3$

41. $\dfrac{2}{c^2-4} - \dfrac{c^2-2}{c^2-4} = \dfrac{2-(c^2-2)}{c^2-4}$

$\qquad = \dfrac{2-c^2+2}{c^2-4}$

$\qquad = \dfrac{4-c^2}{c^2-4}$

$\qquad = \dfrac{-(c^2-4)}{c^2-4}$

$\qquad = -1$

43. $\dfrac{2x^2+5x}{2x+3}-\dfrac{4x+3}{2x+3}=\dfrac{2x^2+5x-(4x+3)}{2x+3}$

$\qquad\qquad=\dfrac{2x^2+5x-4x-3}{2x+3}$

$\qquad\qquad=\dfrac{2x^2+x-3}{2x+3}$

$\qquad\qquad=\dfrac{(2x+3)(x-1)}{2x+3}$

$\qquad\qquad=x-1$

45. $\dfrac{n}{n-3}+\dfrac{3}{3-n}=\dfrac{n}{n-3}+\dfrac{3}{-1(n-3)}$

$\qquad\qquad=\dfrac{n}{n-3}+\dfrac{-3}{n-3}$

$\qquad\qquad=\dfrac{n-3}{n-3}$

$\qquad\qquad=1$

47. $\dfrac{12q}{2p-2q}-\dfrac{8p}{2q-2p}=\dfrac{12q}{2p-2q}-\dfrac{8p}{-1(2p-2q)}$

$\qquad\qquad=\dfrac{12q}{2p-2q}-\dfrac{-8p}{2p-2q}$

$\qquad\qquad=\dfrac{12q-(-8p)}{2p-2q}$

$\qquad\qquad=\dfrac{12q+8p}{2p-2q}$

$\qquad\qquad=\dfrac{4(2p+3q)}{2(p-q)}$

$\qquad\qquad=\dfrac{2(2p+3q)}{p-q}$

49. $\dfrac{2p^2-1}{p-1}-\dfrac{2p+2}{1-p}=\dfrac{2p^2-1}{p-1}-\dfrac{2p+2}{-1(p-1)}$

$\qquad\qquad=\dfrac{2p^2-1}{p-1}-\dfrac{-(2p+2)}{p-1}$

$\qquad\qquad=\dfrac{2p^2-1+(2p+2)}{p-1}$

$\qquad\qquad=\dfrac{2p^2+2p+1}{p-1}$

51. $\dfrac{2a}{a-b}+\dfrac{2a-4b}{b-a}=\dfrac{2a}{a-b}+\dfrac{2a-4b}{-1(a-b)}$

$\qquad\qquad=\dfrac{2a}{a-b}+\dfrac{-(2a-4b)}{a-b}$

$\qquad\qquad=\dfrac{2a-(2a-4b)}{a-b}$

$\qquad\qquad=\dfrac{2a-2a+4b}{a-b}$

$\qquad\qquad=\dfrac{4b}{a-b}$

53. $\dfrac{3}{p^2-3}+\dfrac{p^2}{3-p^2}=\dfrac{3}{p^2-3}+\dfrac{p^2}{-1(p^2-3)}$

$\qquad\qquad=\dfrac{3}{p^2-3}+\dfrac{-p^2}{p^2-3}$

$\qquad\qquad=\dfrac{3-p^2}{p^2-3}$

$\qquad\qquad=\dfrac{-(p^2-3)}{p^2-3}$

$\qquad\qquad=-1$

55. $\dfrac{2x}{x^2-y^2}-\dfrac{2y}{y^2-x^2}=\dfrac{2x}{x^2-y^2}-\dfrac{2y}{-1(x^2-y^2)}$

$\qquad\qquad=\dfrac{2x}{x^2-y^2}-\dfrac{-2y}{x^2-y^2}$

$\qquad\qquad=\dfrac{2x-(-2y)}{x^2-y^2}$

$\qquad\qquad=\dfrac{2x+2y}{x^2-y^2}$

$\qquad\qquad=\dfrac{2(x+y)}{(x+y)(x-y)}$

$\qquad\qquad=\dfrac{2}{x-y}$

57. $\dfrac{4}{7x}-\dfrac{11}{7x}=\dfrac{4-11}{7x}=\dfrac{-7}{7x}=-\dfrac{1}{x}$

59. $\dfrac{5}{2x-5}-\dfrac{2x}{2x-5}=\dfrac{5-2x}{2x-5}=\dfrac{-(2x-5)}{2x-5}=-1$

61. $\dfrac{2n^3}{n-1}-\dfrac{2n^3}{n-1}=\dfrac{2n^3-2n^3}{n-1}=\dfrac{0}{n-1}=0$

63.
$$\frac{n^2-3}{2n+3}+\frac{n^2+n}{2n+3}=\frac{n^2-3+n^2+n}{2n+3}$$
$$=\frac{2n^2+n-3}{2n+3}$$
$$=\frac{(2n+3)(n-1)}{2n+3}$$
$$=n-1$$

65.
$$\frac{3v-1}{v^2-9}-\frac{8}{v^2-9}=\frac{3v-1-8}{v^2-9}$$
$$=\frac{3v-9}{v^2-9}$$
$$=\frac{3(v-3)}{(v+3)(v-3)}$$
$$=\frac{3}{v+3}$$

67.
$$\frac{x^2}{x^2-1}+\frac{2x+1}{x^2-1}=\frac{x^2+2x+1}{x^2-1}$$
$$=\frac{(x+1)^2}{(x+1)(x-1)}$$
$$=\frac{x+1}{x-1}$$

69.
$$\frac{2x^2+3}{x^2-3x}-\frac{x^2}{x^2-3x}=\frac{2x^2+3-x^2}{x^2-3x}$$
$$=\frac{x^2+3}{x^2-3x}$$
$$=\frac{x^2+3}{x(x-3)}$$

71.
$$\frac{3x^2}{x+1}-\frac{x^2+2}{x+1}=\frac{3x^2-(x^2+2)}{x+1}$$
$$=\frac{3x^2-x^2-2}{x+1}$$
$$=\frac{2x^2-2}{x+1}$$
$$=\frac{2(x+1)(x-1)}{x+1}$$
$$=2(x-1)$$

73.
$$\frac{3x-1}{x-y}+\frac{1-3y}{y-x}=\frac{3x-1}{x-y}+\frac{1-3y}{-1(x-y)}$$
$$=\frac{3x-1}{x-y}+\frac{-(1-3y)}{x-y}$$
$$=\frac{3x-1-(1-3y)}{x-y}$$
$$=\frac{3x-1-1+3y}{x-y}$$
$$=\frac{3x+3y-2}{x-y}$$

75.
$$\frac{2p-3q}{3p^2-18q^2}-\frac{9q+p}{18q^2-3p^2}$$
$$=\frac{2p-3q}{3p^2-18q^2}-\frac{9q+p}{-1(3p^2-18q^2)}$$
$$=\frac{2p-3q}{3p^2-18q^2}-\frac{-(9q+p)}{3p^2-18q^2}$$
$$=\frac{2p-3q+9q+p}{3p^2-18q^2}$$
$$=\frac{3p+6q}{3p^2-18q^2}$$
$$=\frac{3(p+2q)}{3(p^2-6q^2)}$$
$$=\frac{p+2q}{p^2-6q^2}$$

77.
$$\frac{7a^2}{a}+\frac{5a^2}{a}=\frac{7a^2+5a^2}{a}=\frac{12a^2}{a}=12a$$

79.
$$\frac{2n}{n+1}-\frac{n+3}{n+1}=\frac{2n-(n+3)}{n+1}$$
$$=\frac{2n-n-3}{n+1}$$
$$=\frac{n-3}{n+1}$$

81.
$$\frac{11}{2n}-\frac{3}{2n}=\frac{11-3}{2n}=\frac{8}{2n}=\frac{4}{n}$$

83. Let $\dfrac{?}{x^2-9}$ be the unknown rational expression.

$$\frac{x^2}{x^2-9}-\frac{?}{x^2-9}=\frac{3x}{x^2-9}$$

$$\frac{x^2-?}{x^2-9}=\frac{3x}{x^2-9}$$

$$x^2-?=3x$$

$$x^2-?+?=3x+?$$

$$x^2=3x+?$$

$$x^2-3x=3x+?-3x$$

$$x^2-3x=?$$

The rational expression is

$$\frac{x^2-3x}{x^2-9}=\frac{x(x-3)}{(x+3)(x-3)}=\frac{x}{x+3}.$$

85. Let $\dfrac{?}{x+2}$ be the unknown rational expression.

Note that $1=\dfrac{x+2}{x+2}$.

$$\frac{-3x+4}{x+2}+\frac{?}{x+2}=1$$

$$\frac{-3x+4+?}{x+2}=\frac{x+2}{x+2}$$

$$-3x+4+?=x+2$$

$$-3x+4+?+3x=x+2+3x$$

$$4+?=4x+2$$

$$4+?-4=4x+2-4$$

$$?=4x-2$$

The rational expression is $\dfrac{4x-2}{x+2}$.

87. In a rectangle,
Perimeter = 2[(Length) + (Width)].

$$2\left(\frac{2x-3}{x}+\frac{4x+2}{x}\right)=2\left(\frac{2x-3+4x+2}{x}\right)$$

$$=2\left(\frac{6x-1}{x}\right)$$

$$=\frac{2}{1}\cdot\frac{6x-1}{x}$$

$$=\frac{2(6x-1)}{1\cdot x}$$

$$=\frac{2(6x-1)}{x}\ \text{cm}$$

89.
$$\frac{7x-3y}{x^2-y^2}-\left[\frac{2x-3y}{x^2-y^2}-\frac{x+7y}{x^2-y^2}\right]$$

$$=\frac{7x-3y-(2x-3y)+(x+7y)}{x^2-y^2}$$

$$=\frac{7x-3y-2x+3y+x+7y}{x^2-y^2}$$

$$=\frac{6x+7y}{x^2-y^2}$$

91.
$$\frac{x^2}{3x^2+5x-2}-\frac{x}{3x-1}\cdot\frac{2x-1}{x+2}$$

$$=\frac{x^2}{3x^2+5x-2}-\frac{x(2x-1)}{(3x-1)(x+2)}$$

$$=\frac{x^2}{3x^2+5x-2}-\frac{2x^2-x}{3x^2+5x-2}$$

$$=\frac{x^2-(2x^2-x)}{3x^2+5x-2}$$

$$=\frac{x^2-2x^2+x}{3x^2+5x-2}$$

$$=\frac{-x^2+x}{3x^2+5x-2}$$

$$=\frac{-x(x-1)}{(3x-1)(x+2)}$$

93.
$$\frac{5x}{x-2}+\frac{2(x+3)}{x-2}-\frac{6(2x-1)}{x-2}$$

$$=\frac{5x+2(x+3)-6(2x-1)}{x-2}$$

$$=\frac{5x+2x+6-12x+6}{x-2}$$

$$=\frac{-5x+12}{x-2}$$

95.
$$\frac{2n+1}{n-3}-\frac{?}{n-3}=\frac{6n+7}{n-3}$$

$$\frac{2n+1-?}{n-3}=\frac{6n+7}{n-3}$$

$$2n+1-?=6n+7$$

$$2n+1-?-2n=6n+7-2n$$

$$1-?=4n+7$$

$$1-?-1=4n+7-1$$

$$-?=4n+6$$

$$-1(-?)=-1(4n+6)$$

$$?=-4n-6$$

97. The expression is incorrect because the subtraction symbol was not applied to both terms of the second numerator.

$$\frac{x-2}{x} - \frac{x+4}{x} = \frac{x-2-(x+4)}{x}$$

$$= \frac{x-2-x-4}{x}$$

$$= \frac{-6}{x}$$

99. The correct answer is (c). The answer

(a) $\dfrac{7}{2x}$ is incorrect because the denominators

were added; and the answer (b) $\dfrac{7}{x^2}$ is incorrect

because the denominators were multiplied.

Section 7.4

Preparing for Finding the Least Common Denominator and Forming Equivalent Rational Expressions

P1. $\dfrac{5}{12} = \dfrac{5}{12} \cdot \dfrac{2}{2} = \dfrac{10}{24}$

P2. $15 = 3 \cdot 5$ and $25 = 5 \cdot 5$, so
LCD $= 3 \cdot 5 \cdot 5 = 75$.

7.4 Quick Checks

1. The <u>least</u> <u>common</u> <u>denominator</u> of two or more rational expressions is the smallest polynomial that is a multiple of each denominator in the rational expressions to be added or subtracted.

2. False; $4a^2b = 2 \cdot 2 \cdot a^2 \cdot b$
$4ab = 2 \cdot 2 \cdot a \cdot b$
LCD $= 2 \cdot 2 \cdot a^2 \cdot b = 4a^2b \neq 16a^2b^2$

3. $8x^2y = 2^3 \quad \cdot x^2 \cdot y$
$12xy^3 = 2^2 \cdot 3 \cdot x \ \cdot y^3$
LCD $= 2^3 \cdot 3 \cdot x^2 \cdot y^3 = 24x^2y^3$

4. $6a^3b^2 = 2 \cdot 3 \cdot \quad a^3 \cdot b^2$
$21ab^3 = \quad 3 \cdot 7 \cdot a \ \cdot b^3$
LCD $= 2 \cdot 3 \cdot 7 \cdot a^3 \cdot b^3 = 42a^3b^3$

5. True

6. $15z = 3 \cdot 5 \cdot z$
$5z^2 + 5z = 5 \cdot z \cdot (z+1)$
LCD $= 3 \cdot 5 \cdot z \cdot (z+1) = 15z(z+1)$

7. $x^2 + 4x - 5 = (x+5)(x-1)$
$x^2 + 10x + 25 = (x+5)^2$
LCD $= (x+5)^2(x-1)$

8. $21 - 3x = -3(x-7)$
$x^2 - 49 = \quad (x-7)(x+7)$
LCD $= -3(x-7)(x+7)$

9. If we want to rewrite the rational expression

$\dfrac{2x+1}{x-1}$ with a denominator of $(x-1)(x+3)$, we

multiply the numerator and denominator of

$\dfrac{2x+1}{x-1}$ by <u>$x+3$</u>.

10. $4p^2 - 8p = 4p(p-2)$
$16p^3(p-2) = 4^2 p^3(p-2)$

The "missing factor" is $4p^2$.

$$\frac{3}{4p^2 - 8p} = \frac{3}{4p(p-2)} \cdot \frac{4p^2}{4p^2}$$

$$= \frac{12p^2}{16p^3(p-2)}$$

11. $6a^2b = 2 \quad \cdot 3 \quad \cdot a \cdot a \cdot b$
$20ab^3 = 2 \cdot 2 \quad \cdot 5 \cdot a \quad \cdot b \cdot b \cdot b$
LCD $= 2 \cdot 2 \cdot 3 \cdot 5 \cdot a \cdot a \cdot b \cdot b \cdot b$
$\qquad = 60a^2b^3$

$$\frac{3}{6a^2b} = \frac{3}{6a^2b} \cdot \frac{10b^2}{10b^2} = \frac{30b^2}{60a^2b^3}$$

$$\frac{-5}{20ab^3} = \frac{-5}{20ab^3} \cdot \frac{3a}{3a} = \frac{-15a}{60a^2b^3}$$

12. $x^2 - 4x - 5 = (x+1)(x-5)$
$x^2 - 7x + 10 = \quad (x-5)(x-2)$
LCD $= (x+1)(x-5)(x-2)$

$$\frac{5}{x^2-4x-5} = \frac{5}{(x+1)(x-5)} \cdot \frac{x-2}{x-2}$$
$$= \frac{5(x-2)}{(x+1)(x-5)(x-2)}$$

$$\frac{-3}{x^2-7x+10} = \frac{-3}{(x-5)(x-2)} \cdot \frac{x+1}{x+1}$$
$$= \frac{-3(x+1)}{(x+1)(x-5)(x-2)}$$

7.4 Exercises

13. $5x^2 = 5 \cdot x^2$
$\quad 5x = 5 \cdot x$
$\quad \text{LCD} = 5x^2$

15. $12xy^2 = 2^2 \cdot 3 \cdot \quad x \cdot \ y^2$
$\quad 15x^3y = \quad\ \ 3 \cdot 5 \cdot x^3 \cdot y$
$\quad \text{LCD} = 2^2 \cdot 3 \cdot 5 \cdot x^3 \cdot y^2 = 60x^3y^2$

17. $x = x$
$\quad x+1 = x+1$
$\quad \text{LCD} = x(x+1)$

19. Note that $2 = \dfrac{2}{1}$.
$\quad 2x+1 = 2x+1$
$\quad 1 = 1$
$\quad \text{LCD} = 1(2x+1) = 2x+1$

21. $2b-6 = 2(b-3)$
$\quad 4b-12 = 4(b-3) = 2^2(b-3)$
$\quad \text{LCD} = 2^2(b-3) = 4(b-3)$

23. $p^2+p = p(p+1)$
$\quad p^2-p-2 = (p+1)(p-2)$
$\quad \text{LCD} = p(p+1)(p-2)$

25. $r^2+4r+4 = (r+2)^2$
$\quad r^2-r-2 = (r+1)(r-2)$
$\quad \text{LCD} = (r+2)^2(r+1)(r-2)$

27. $x-4 = x-4$
$\quad 4-x = -1(x-4)$
$\quad \text{LCD} = -1(x-4) = -(x-4)$

29. $x^2-9 = (x+3)(x-3)$
$\quad 6-2x = -2(x-3)$
$\quad \text{LCD} = -2(x+3)(x-3)$

31. $(x-1)(x+2) = (x-1)(x+2)$
$\quad (1-x)(2+x) = -1(x-1)(x+2)$
$\quad \text{LCD} = -1(x-1)(x+2) = -(x-1)(x+2)$

33. $x = x$
$\quad 3x^3 = 3 \cdot x^3 = 3x^2 \cdot x$
The "missing factor" is $3x^2$.
$$\frac{4}{x} = \frac{4}{x} \cdot \frac{3x^2}{3x^2} = \frac{4 \cdot 3x^2}{x \cdot 3x^2} = \frac{12x^2}{3x^3}$$

35. $a^2b^2c = a^2 \cdot b^2 \cdot c$
$\quad a^2b^2c^2 = a^2 \cdot b^2 \cdot c^2 = c \cdot a^2b^2c$
The "missing factor" is c.
$$\frac{3+c}{a^2b^2c} = \frac{3+c}{a^2b^2c} \cdot \frac{c}{c} = \frac{(3+c)c}{a^2b^2c \cdot c} = \frac{3c+c^2}{a^2b^2c^2}$$

37. $x+4 = x+4$
$\quad x^2-16 = (x+4)(x-4)$
The "missing factor" is $x-4$.
$$\frac{x-4}{x+4} = \frac{x-4}{x+4} \cdot \frac{x-4}{x-4}$$
$$= \frac{(x-4)(x-4)}{(x+4)(x-4)}$$
$$= \frac{(x-4)^2}{x^2-16}$$

39. $2n+2 = 2(n+1)$
$\quad 6n^2-6 = 2 \cdot 3(n+1)(n-1) = 3(n-1) \cdot 2(n+1)$
The "missing factor" is $3(n-1)$.
$$\frac{3n}{2n+2} = \frac{3n}{2(n+1)} \cdot \frac{3(n-1)}{3(n-1)}$$
$$= \frac{3n \cdot 3(n-1)}{2(n+1) \cdot 3(n-1)}$$
$$= \frac{9n^2-9n}{6n^2-6}$$

41. $4t = \dfrac{4t}{1} = \dfrac{4t}{1} \cdot \dfrac{t-1}{t-1} = \dfrac{4t(t-1)}{t-1} = \dfrac{4t^2-4t}{t-1}$

43. $3y = 3 \cdot y$

$9y^2 = 3^2 \cdot y^2$

$\text{LCD} = 3^2 \cdot y^2 = 9y^2$

$\dfrac{2x}{3y} = \dfrac{2x}{3y} \cdot \dfrac{3y}{3y} = \dfrac{2x \cdot 3y}{3y \cdot 3y} = \dfrac{6xy}{9y^2}$

$\dfrac{4}{9y^2} = \dfrac{4}{9y^2} \cdot \dfrac{1}{1} = \dfrac{4}{9y^2}$

45. $2a^3 = 2 \cdot a^3$

$4a = 2^2 \cdot a$

$\text{LCD} = 2^2 \cdot a^3 = 4a^3$

$\dfrac{3a+1}{2a^3} = \dfrac{3a+1}{2a^3} \cdot \dfrac{2}{2} = \dfrac{(3a+1) \cdot 2}{2a^3 \cdot 2} = \dfrac{6a+2}{4a^3}$

$\dfrac{4a-1}{4a} = \dfrac{4a-1}{4a} \cdot \dfrac{a^2}{a^2} = \dfrac{(4a-1)a^2}{4a \cdot a^2} = \dfrac{4a^3 - a^2}{4a^3}$

47. $m = m$

$m + 1 = m + 1$

$\text{LCD} = m(m+1)$

$\dfrac{2}{m} = \dfrac{2}{m} \cdot \dfrac{m+1}{m+1} = \dfrac{2(m+1)}{m(m+1)} = \dfrac{2m+2}{m(m+1)}$

$\dfrac{3}{m+1} = \dfrac{3}{m+1} \cdot \dfrac{m}{m} = \dfrac{3 \cdot m}{(m+1)m} = \dfrac{3m}{m(m+1)}$

49. $4y = 2^2 \cdot y$

$8y - 4 = 2^2 \cdot (2y-1)$

$\text{LCD} = 2^2 \cdot y \cdot (2y-1) = 4y(2y-1)$

$\dfrac{y-2}{4y} = \dfrac{y-2}{4y} \cdot \dfrac{2y-1}{2y-1}$

$\qquad = \dfrac{(y-2)(2y-1)}{4y(2y-1)}$

$\qquad = \dfrac{2y^2 - 5y + 2}{4y(2y-1)}$

$\dfrac{y}{8y-4} = \dfrac{y}{4(2y-1)} \cdot \dfrac{y}{y}$

$\qquad = \dfrac{y \cdot y}{4(2y-1) \cdot y}$

$\qquad = \dfrac{y^2}{4y(2y-1)}$

51. $x - 1 = x - 1$

$1 - x = -(x-1)$

$\text{LCD} = -(x-1)$

$\dfrac{1}{x-1} = \dfrac{1}{x-1} \cdot \dfrac{-1}{-1} = \dfrac{1(-1)}{(x-1)(-1)} = \dfrac{-1}{-(x-1)}$

$\dfrac{2x}{1-x} = \dfrac{2x}{-(x-1)}$

53. $x - 7 = x - 7$

$7 - x = -(x-7)$

$\text{LCD} = -(x-7)$

$\dfrac{3x}{x-7} = \dfrac{3x}{x-7} \cdot \dfrac{-1}{-1} = \dfrac{3x(-1)}{(x-7)(-1)} = \dfrac{-3x}{-(x-7)}$

$\dfrac{-5}{7-x} = \dfrac{-5}{-(x-7)}$

55. $x^2 - 4 = (x+2)(x-2)$

$x + 2 = x + 2$

$\text{LCD} = (x+2)(x-2)$

$\dfrac{4x}{x^2-4} = \dfrac{4x}{(x+2)(x-2)}$

$\dfrac{2}{x+2} = \dfrac{2}{x+2} \cdot \dfrac{x-2}{x-2}$

$\qquad = \dfrac{2(x-2)}{(x+2)(x-2)}$

$\qquad = \dfrac{2x-4}{(x+2)(x-2)}$

57. $x^2 - 9 = (x+3)(x-3)$

$x^2 - 2x - 3 = (x+1)(x-3)$

$\text{LCD} = (x+3)(x-3)(x+1)$

$\dfrac{x+1}{x^2-9} = \dfrac{x+1}{(x+3)(x-3)} \cdot \dfrac{x+1}{x+1}$

$\qquad = \dfrac{(x+1)(x+1)}{(x+3)(x-3)(x+1)}$

$\qquad = \dfrac{x^2 + 2x + 1}{(x+3)(x-3)(x+1)}$

$\dfrac{x+2}{x^2-2x-3} = \dfrac{x+2}{(x+1)(x-3)} \cdot \dfrac{x+3}{x+3}$

$\qquad = \dfrac{(x+2)(x+3)}{(x+1)(x-3)(x+3)}$

$\qquad = \dfrac{x^2 + 5x + 6}{(x+3)(x-3)(x+1)}$

59. $x + 4 = x + 4$

$2x^2 + 7x - 4 = (x+4)(2x-1)$

$\text{LCD} = (x+4)(2x-1)$

$$\frac{3}{x+4} = \frac{3}{x+4} \cdot \frac{2x-1}{2x-1}$$

$$= \frac{3(2x-1)}{(x+4)(2x-1)}$$

$$= \frac{6x-3}{(x+4)(2x-1)}$$

$$\frac{2x-1}{2x^2+7x-4} = \frac{2x-1}{(x+4)(2x-1)}$$

61. $\dfrac{1}{x}$ and $\dfrac{1}{3x}$

$x = x$

$3x = 3 \cdot x$

$\text{LCD} = 3x$

$$\frac{1}{x} = \frac{1}{x} \cdot \frac{3}{3} = \frac{3}{3x}$$

$$\frac{1}{3x} = \frac{1}{3x} \cdot \frac{1}{1} = \frac{1}{3x}$$

63. $\dfrac{12}{r-4}$ and $\dfrac{12}{r+4}$

$r - 4 = r - 4$

$r + 4 = r + 4$

$\text{LCD} = (r-4)(r+4)$

$$\frac{12}{r-4} = \frac{12}{r-4} \cdot \frac{r+4}{r+4}$$

$$= \frac{12(r+4)}{(r-4)(r+4)}$$

$$= \frac{12r+48}{(r-4)(r+4)}$$

$$\frac{12}{r+4} = \frac{12}{r+4} \cdot \frac{r-4}{r-4}$$

$$= \frac{12(r-4)}{(r+4)(r-4)}$$

$$= \frac{12r-48}{(r-4)(r+4)}$$

65. $x^3 + 8 = (x+2)(x^2 - 2x + 4)$

$x^2 - 4 = (x+2)(x-2)$

$x^3 - 8 = (x-2)(x^2 + 2x + 4)$

$\text{LCD} = (x+2)(x-2)(x^2 - 2x + 4)(x^2 + 2x + 4)$

67. $p^2 - p = p(p-1)$

$p^3 - p^2 = p^2(p-1)$

$p^2 - 4p + 3 = (p-1)(p-3)$

$\text{LCD} = p^2(p-1)(p-3)$

69. $2a^4 - 2a^2 b^2 = 2a^2(a^2 - b^2)$

$$= 2a^2(a+b)(a-b)$$

$4ab^2 + 4b^3 = 2^2 b^2(a+b)$

$a^3 b - b^4 = b(a^3 - b^3)$

$$= b(a-b)(a^2 + ab + b^2)$$

$\text{LCD} = 2^2 a^2 b^2 (a+b)(a-b)(a^2 + ab + b^2)$

$$= 4a^2 b^2 (a+b)(a-b)(a^2 + ab + b^2)$$

71. To write equivalent rational expressions with a common denominator, (1) write each denominator in factored form; (2) determine the "missing factors" (3) Multiply the original expression by $1 = \dfrac{\text{missing factors}}{\text{missing factors}}$; (3) find the product leaving the denominator in factored form.

Section 7.5

Preparing for Adding and Subtracting Rational Expressions with Unlike Denominators

P1. $15 = \qquad 3 \cdot 5$

$24 = 2^3 \cdot 3$

$\text{LCD} = 2^3 \cdot 3 \cdot 5 = 120$

P2. $2x^2 + 3x + 1 = (2x+1)(x+1)$

P3. $\dfrac{3x+9}{6x} = \dfrac{3(x+3)}{3 \cdot 2x} = \dfrac{x+3}{2x}$

P4. $\dfrac{1-x^2}{2x-2} = \dfrac{(1-x)(1+x)}{2(x-1)}$

$$= \frac{-1(x-1)(1+x)}{2(x-1)}$$

$$= \frac{-(1+x)}{2}$$

P5. $\dfrac{5}{2x} + \dfrac{7}{2x} = \dfrac{5+7}{2x} = \dfrac{12}{2x} = \dfrac{6}{x}$

7.5 Quick Checks

1. The least common denominator of $\dfrac{1}{8}$ and $\dfrac{5}{18}$ is

   <u>72</u>.

2. $12 = 2^2 \cdot 3$

   $18 = 2 \cdot 3^2$

   $\text{LCD} = 2^2 \cdot 3^2 = 36$

   $\dfrac{5}{12} + \dfrac{5}{18} = \dfrac{5}{12} \cdot \dfrac{3}{3} + \dfrac{5}{18} \cdot \dfrac{2}{2} = \dfrac{15}{36} + \dfrac{10}{36} = \dfrac{25}{36}$

3. $15 = 3 \cdot 5$

   $10 = 2 \cdot 5$

   $\text{LCD} = 2 \cdot 3 \cdot 5 = 30$

   $\dfrac{8}{15} - \dfrac{3}{10} = \dfrac{8}{15} \cdot \dfrac{2}{2} - \dfrac{3}{10} \cdot \dfrac{3}{3} = \dfrac{16}{30} - \dfrac{9}{30} = \dfrac{7}{30}$

4. The first step in adding rational expressions with unlike denominators is to determine the <u>least</u> <u>common</u> <u>denominator</u>.

5. True

6. $8a^3b = 2^3 \cdot a^3 \cdot b$

   $12ab^2 = 2^2 \cdot 3 \cdot a \cdot b^2$

   $\text{LCD} = 2^3 \cdot 3 \cdot a^3 \cdot b^2 = 24a^3b^2$

   $\dfrac{1}{8a^3b} + \dfrac{5}{12ab^2} = \dfrac{1}{8a^3b} \cdot \dfrac{3b}{3b} + \dfrac{5}{12ab^2} \cdot \dfrac{2a^2}{2a^2}$

   $\phantom{\dfrac{1}{8a^3b} + \dfrac{5}{12ab^2}} = \dfrac{3b}{24a^3b^2} + \dfrac{10a^2}{24a^3b^2}$

   $\phantom{\dfrac{1}{8a^3b} + \dfrac{5}{12ab^2}} = \dfrac{10a^2 + 3b}{24a^3b^2}$

7. $15xy^2 = 3 \cdot 5 \cdot x \cdot y^2$

   $18x^3y = 2 \cdot 3^2 \cdot x^3 \cdot y$

   $\text{LCD} = 2 \cdot 3^2 \cdot 5 \cdot x^3 \cdot y^2 = 90x^3y^2$

   $\dfrac{1}{15xy^2} + \dfrac{7}{18x^3y} = \dfrac{1}{15xy^2} \cdot \dfrac{6x^2}{6x^2} + \dfrac{7}{18x^3y} \cdot \dfrac{5y}{5y}$

   $\phantom{\dfrac{1}{15xy^2} + \dfrac{7}{18x^3y}} = \dfrac{6x^2}{90x^3y^2} + \dfrac{35y}{90x^3y^2}$

   $\phantom{\dfrac{1}{15xy^2} + \dfrac{7}{18x^3y}} = \dfrac{6x^2 + 35y}{90x^3y^2}$

8. $\dfrac{5}{x-4} + \dfrac{3}{x+2}$

   $= \dfrac{5}{x-4} \cdot \dfrac{x+2}{x+2} + \dfrac{3}{x+2} \cdot \dfrac{x-4}{x-4}$

   $= \dfrac{5x+10}{(x+2)(x-4)} + \dfrac{3x-12}{(x+2)(x-4)}$

   $= \dfrac{5x+10+3x-12}{(x+2)(x-4)}$

   $= \dfrac{8x-2}{(x+2)(x-4)}$

   $= \dfrac{2(4x-1)}{(x+2)(x-4)}$

9. $\dfrac{-1}{n-3} + \dfrac{4}{n+1} = \dfrac{-1}{n-3} \cdot \dfrac{n+1}{n+1} + \dfrac{4}{n+1} \cdot \dfrac{n-3}{n-3}$

   $= \dfrac{-n-1}{(n+1)(n-3)} + \dfrac{4n-12}{(n+1)(n-3)}$

   $= \dfrac{-n-1+4n-12}{(n+1)(n-3)}$

   $= \dfrac{3n-13}{(n+1)(n-3)}$

10. $z^2 + 7z + 10 = (z+5)(z+2)$

    $z+5 = z+5$

    $\text{LCD} = (z+5)(z+2)$

    $\dfrac{-z+1}{z^2+7z+10} + \dfrac{2}{z+5}$

    $= \dfrac{-z+1}{(z+5)(z+2)} + \dfrac{2}{z+5}$

    $= \dfrac{-z+1}{(z+5)(z+2)} + \dfrac{2}{z+5} \cdot \dfrac{z+2}{z+2}$

    $= \dfrac{-z+1}{(z+5)(z+2)} + \dfrac{2z+4}{(z+5)(z+2)}$

    $= \dfrac{-z+1+2z+4}{(z+5)(z+2)}$

    $= \dfrac{z+5}{(z+5)(z+2)}$

    $= \dfrac{1}{z+2}$

11. $x^2 + 5x = x(x+5)$

$x^2 - 5x = x \quad\quad \cdot (x-5)$

$\text{LCD} = x(x+5)(x-5)$

$\dfrac{1}{x^2+5x} + \dfrac{1}{x^2-5x}$

$= \dfrac{1}{x(x+5)} + \dfrac{1}{x(x-5)}$

$= \dfrac{1}{x(x+5)} \cdot \dfrac{x-5}{x-5} + \dfrac{1}{x(x-5)} \cdot \dfrac{x+5}{x+5}$

$= \dfrac{x-5}{x(x+5)(x-5)} + \dfrac{x+5}{x(x+5)(x-5)}$

$= \dfrac{x-5+x+5}{x(x+5)(x-5)}$

$= \dfrac{2x}{x(x+5)(x-5)}$

$= \dfrac{2}{(x+5)(x-5)}$

12. $5ab^2 = \quad 5 \cdot a \cdot b^2$

$4a^2b^3 = 2^2 \quad \cdot a^2 \cdot b^3$

$\text{LCD} = 2^2 \cdot 5 \cdot a^2 \cdot b^3 = 20a^2b^3$

$\dfrac{-4}{5ab^2} - \dfrac{3}{4a^2b^3} = \dfrac{-4}{5ab^2} \cdot \dfrac{4ab}{4ab} - \dfrac{3}{4a^2b^3} \cdot \dfrac{5}{5}$

$\quad\quad = \dfrac{-16ab}{20a^2b^3} - \dfrac{15}{20a^2b^3}$

$\quad\quad = \dfrac{-16ab-15}{20a^2b^3}$

13. $\text{LCD} = x \cdot (x-4) = x(x-4)$

$\dfrac{5}{x} - \dfrac{3}{x-4} = \dfrac{5}{x} \cdot \dfrac{x-4}{x-4} - \dfrac{3}{x-4} \cdot \dfrac{x}{x}$

$\quad\quad = \dfrac{5x-20}{x(x-4)} - \dfrac{3x}{x(x-4)}$

$\quad\quad = \dfrac{5x-20-3x}{x(x-4)}$

$\quad\quad = \dfrac{2x-20}{x(x-4)}$

$\quad\quad = \dfrac{2(x-10)}{x(x-4)}$

14. $\text{LCD} = (x-5)(x+4)(x-1)$

$\dfrac{3}{(x-5)(x+4)} - \dfrac{2}{(x-5)(x-1)}$

$= \dfrac{3}{(x-5)(x+4)} \cdot \dfrac{x-1}{x-1} - \dfrac{2}{(x-5)(x-1)} \cdot \dfrac{x+4}{x+4}$

$= \dfrac{3x-3}{(x-5)(x+4)(x-1)} - \dfrac{2x+8}{(x-5)(x+4)(x-1)}$

$= \dfrac{3x-3-2x-8}{(x-5)(x+4)(x-1)}$

$= \dfrac{x-11}{(x-5)(x+4)(x-1)}$

15. $x^2 - 3x = x(x-3)$

$4x - 12 = 4(x-3)$

$\text{LCD} = 4x(x-3)$

$\dfrac{x-2}{x^2-3x} + \dfrac{x+3}{4x-12}$

$= \dfrac{x-2}{x(x-3)} \cdot \dfrac{4}{4} + \dfrac{x+3}{4(x-3)} \cdot \dfrac{x}{x}$

$= \dfrac{4x-8}{4x(x-3)} + \dfrac{x^2+3x}{4x(x-3)}$

$= \dfrac{4x-8+x^2+3x}{4x(x-3)}$

$= \dfrac{x^2+7x-8}{4x(x-3)}$

$= \dfrac{(x+8)(x-1)}{4x(x-3)}$

16. True

17. $\dfrac{7}{3p^2-3p} + \dfrac{5}{6-6p} = \dfrac{7}{3p^2-3p} + \dfrac{5}{-1(6p-6)}$

$\quad\quad\quad\quad\quad = \dfrac{7}{3p^2-3p} + \dfrac{-5}{6p-6}$

$3p^2 - 3p = 3p(p-1)$

$6p - 6 = 6(p-1) = 2 \cdot 3(p-1)$

$\text{LCD} = 2 \cdot 3p(p-1) = 6p(p-1)$

$\dfrac{7}{3p^2-3p} + \dfrac{-5}{6p-6}$

$= \dfrac{7}{3p^2-3p} \cdot \dfrac{2}{2} + \dfrac{-5}{6p-6} \cdot \dfrac{p}{p}$

$= \dfrac{14}{6p(p-1)} + \dfrac{-5p}{6p(p-1)}$

$= \dfrac{14-5p}{6p(p-1)}$

18. $2 - \dfrac{3}{x-1} = \dfrac{2}{1} - \dfrac{3}{x-1}$

LCD $= x - 1$

$$\dfrac{2}{1} \cdot \dfrac{x-1}{x-1} - \dfrac{3}{x-1} = \dfrac{2(x-1)}{x-1} - \dfrac{3}{x-1}$$

$$= \dfrac{2x-2}{x-1} - \dfrac{3}{x-1}$$

$$= \dfrac{2x-2-3}{x-1}$$

$$= \dfrac{2x-5}{x-1}$$

19. $x + 1 = x + 1$

$x^2 - 1 = (x+1)(x-1)$

$x - 1 = x - 1$

LCD $= (x+1)(x-1)$

$$\dfrac{1}{x+1} - \dfrac{2}{x^2-1} + \dfrac{3}{x-1}$$

$$= \dfrac{1}{x+1} \cdot \dfrac{x-1}{x-1} - \dfrac{2}{(x+1)(x-1)} + \dfrac{3}{x-1} \cdot \dfrac{x+1}{x+1}$$

$$= \dfrac{1(x-1)}{(x+1)(x-1)} - \dfrac{2}{(x+1)(x-1)} + \dfrac{3(x+1)}{(x+1)(x-1)}$$

$$= \dfrac{x-1-2+3x+3}{(x+1)(x-1)}$$

$$= \dfrac{4x}{(x+1)(x-1)}$$

20. $ x = x$

$ x - 2 = x - 2$

$x^2 - 2x = x(x-2)$

 LCD $= x(x-2)$

$$\dfrac{3}{x} - \left(\dfrac{1}{x-2} - \dfrac{6}{x^2-2x} \right)$$

$$= \dfrac{3}{x} - \dfrac{1}{x-2} + \dfrac{6}{x(x-2)}$$

$$= \dfrac{3}{x} \cdot \dfrac{x-2}{x-2} - \dfrac{1}{x-2} \cdot \dfrac{x}{x} + \dfrac{6}{x(x-2)}$$

$$= \dfrac{3(x-2)}{x(x-2)} - \dfrac{x}{x(x-2)} + \dfrac{6}{x(x-2)}$$

$$= \dfrac{3x-6-x+6}{x(x-2)}$$

$$= \dfrac{2x}{x(x-2)}$$

$$= \dfrac{2}{x-2}$$

7.5 Exercises

21. LCD $= 3 \cdot 2 = 6$

$$\dfrac{-4}{3} + \dfrac{1}{2} = \dfrac{-4}{3} \cdot \dfrac{2}{2} + \dfrac{1}{2} \cdot \dfrac{3}{3}$$

$$= \dfrac{-8}{6} + \dfrac{3}{6}$$

$$= \dfrac{-8+3}{6}$$

$$= \dfrac{-5}{6}$$

23. LCD $= 3 \cdot x = 3x$

$$\dfrac{2}{3x} + \dfrac{1}{x} = \dfrac{2}{3x} + \dfrac{1}{x} \cdot \dfrac{3}{3} = \dfrac{2}{3x} + \dfrac{3}{3x} = \dfrac{2+3}{3x} = \dfrac{5}{3x}$$

25. LCD $= (2a-1)(2a+1)$

$$\dfrac{a}{2a-1} + \dfrac{3}{2a+1}$$

$$= \dfrac{a}{2a-1} \cdot \dfrac{2a+1}{2a+1} + \dfrac{3}{2a+1} \cdot \dfrac{2a-1}{2a-1}$$

$$= \dfrac{a(2a+1)}{(2a-1)(2a+1)} + \dfrac{3(2a-1)}{(2a-1)(2a+1)}$$

$$= \dfrac{2a^2+a+6a-3}{(2a-1)(2a+1)}$$

$$= \dfrac{2a^2+7a-3}{(2a-1)(2a+1)}$$

27. $x - 4 = x - 4; \; 4 - x = -(x-4)$

LCD $= -(x-4)$

$$\dfrac{3}{x-4} + \dfrac{5}{4-x} = \dfrac{3}{x-4} \cdot \dfrac{-1}{-1} + \dfrac{5}{-(x-4)}$$

$$= \dfrac{-3}{-(x-4)} + \dfrac{5}{-(x-4)}$$

$$= \dfrac{-3+5}{-(x-4)}$$

$$= \dfrac{2}{-(x-4)}$$

$$= \dfrac{-2}{x-4}$$

29. $5a - a^2 = -a(a-5);\ 4a - 20 = 4(a-5)$

$\quad$ LCD $= -4a(a-5)$

$\quad \dfrac{a+3}{5a-a^2} + \dfrac{2a+1}{4a-20}$

$\quad = \dfrac{a+3}{-a(a-5)} \cdot \dfrac{4}{4} + \dfrac{2a+1}{4(a-5)} \cdot \dfrac{-a}{-a}$

$\quad = \dfrac{4a+12}{-4a(a-5)} + \dfrac{-2a^2-a}{-4a(a-5)}$

$\quad = \dfrac{4a+12-2a^2-a}{-4a(a-5)}$

$\quad = \dfrac{-2a^2+3a+12}{-4a(a-5)}$

$\quad = \dfrac{2a^2-3a-12}{4a(a-5)}$

31. $x^2 + 2x = x(x+2);\ x = x$

$\quad$ LCD $= x(x+2)$

$\quad \dfrac{2x+4}{x^2+2x} + \dfrac{3}{x} = \dfrac{2x+4}{x(x+2)} + \dfrac{3}{x} \cdot \dfrac{x+2}{x+2}$

$\quad = \dfrac{2x+4}{x(x+2)} + \dfrac{3x+6}{x(x+2)}$

$\quad = \dfrac{2x+4+3x+6}{x(x+2)}$

$\quad = \dfrac{5x+10}{x(x+2)}$

$\quad = \dfrac{5(x+2)}{x(x+2)}$

$\quad = \dfrac{5}{x}$

33. $15 = 3 \cdot 5;\ 25 = 5 \cdot 5$

$\quad$ LCD $= 3 \cdot 5 \cdot 5 = 75$

$\quad \dfrac{7}{15} - \dfrac{9}{25} = \dfrac{7}{15} \cdot \dfrac{5}{5} - \dfrac{9}{25} \cdot \dfrac{3}{3}$

$\quad = \dfrac{35}{75} - \dfrac{27}{75}$

$\quad = \dfrac{35-27}{75}$

$\quad = \dfrac{8}{75}$

35. LCD $= m$

$\quad m - \dfrac{16}{m} = \dfrac{m}{1} \cdot \dfrac{m}{m} - \dfrac{16}{m}$

$\quad = \dfrac{m^2}{m} - \dfrac{16}{m}$

$\quad = \dfrac{m^2-16}{m}$

$\quad = \dfrac{(m-4)(m+4)}{m}$

37. LCD $= (x-3)(x+3)$

$\quad \dfrac{x}{x-3} - \dfrac{x-2}{x+3} = \dfrac{x}{x-3} \cdot \dfrac{x+3}{x+3} - \dfrac{x-2}{x+3} \cdot \dfrac{x-3}{x-3}$

$\quad = \dfrac{x(x+3)}{(x-3)(x+3)} - \dfrac{(x-2)(x-3)}{(x-3)(x+3)}$

$\quad = \dfrac{x^2+3x-(x^2-5x+6)}{(x-3)(x+3)}$

$\quad = \dfrac{x^2+3x-x^2+5x-6}{(x-3)(x+3)}$

$\quad = \dfrac{8x-6}{(x-3)(x+3)}$

$\quad = \dfrac{2(4x-3)}{(x-3)(x+3)}$

39. $x + 3 = x + 3;\ x^2 + 6x + 9 = (x+3)^2$

$\quad$ LCD $= (x+3)^2$

$\quad \dfrac{x+2}{x+3} - \dfrac{x^2-x}{x^2+6x+9} = \dfrac{x+2}{x+3} \cdot \dfrac{x+3}{x+3} - \dfrac{x^2-x}{(x+3)^2}$

$\quad = \dfrac{x^2+5x+6}{(x+3)^2} - \dfrac{x^2-x}{(x+3)^2}$

$\quad = \dfrac{x^2+5x+6-(x^2-x)}{(x+3)^2}$

$\quad = \dfrac{x^2+5x+6-x^2+x}{(x+3)^2}$

$\quad = \dfrac{6x+6}{(x+3)^2}$

$\quad = \dfrac{6(x+1)}{(x+3)^2}$

41. $2a + 6 = 2(a + 3); \; a + 3 = a + 3$
LCD $= 2(a + 3)$

$$\frac{6}{2a+6} - \frac{4a-1}{a+3} = \frac{6}{2(a+3)} - \frac{4a-1}{a+3} \cdot \frac{2}{2}$$

$$= \frac{6}{2(a+3)} - \frac{8a-2}{2(a+3)}$$

$$= \frac{6-(8a-2)}{2(a+3)}$$

$$= \frac{6-8a+2}{2(a+3)}$$

$$= \frac{-8a+8}{2(a+3)}$$

$$= \frac{-4 \cdot 2(a-1)}{2(a+3)}$$

$$= \frac{-4(a-1)}{a+3}$$

43. $x^2 + x - 6 = (x+3)(x-2); \; 2 - x = -(x - 2)$
LCD $= -(x + 3)(x - 2)$

$$\frac{-3x-9}{x^2+x-6} - \frac{x+3}{2-x}$$

$$= \frac{-3x-9}{(x+3)(x-2)} \cdot \frac{-1}{-1} - \frac{x+3}{-(x-2)} \cdot \frac{x+3}{x+3}$$

$$= \frac{3x+9}{-(x+3)(x-2)} - \frac{x^2+6x+9}{-(x+3)(x-2)}$$

$$= \frac{3x+9-(x^2+6x+9)}{-(x+3)(x-2)}$$

$$= \frac{3x+9-x^2-6x-9}{-(x+3)(x-2)}$$

$$= \frac{-x^2-3x}{-(x+3)(x-2)}$$

$$= \frac{-x(x+3)}{-(x+3)(x-2)}$$

$$= \frac{x}{x-2}$$

45. LCD $= 3 \cdot 4 \cdot y^2 = 12y^2$

$$\frac{5}{3y^2} - \frac{3}{4y} = \frac{5}{3y^2} \cdot \frac{4}{4} - \frac{3}{4y} \cdot \frac{3y}{3y}$$

$$= \frac{20}{12y^2} - \frac{9y}{12y^2}$$

$$= \frac{20-9y}{12y^2}$$

47. LCD $= 2 \cdot 5x = 10x$

$$\frac{9}{5x} - \frac{6}{10x} = \frac{9}{5x} \cdot \frac{2}{2} - \frac{6}{10x}$$

$$= \frac{18}{10x} - \frac{6}{10x}$$

$$= \frac{18-6}{10x}$$

$$= \frac{12}{10x}$$

$$= \frac{2 \cdot 6}{2 \cdot 5x}$$

$$= \frac{6}{5x}$$

49. LCD $= 1(2x + 3) = 2x + 3$

$$\frac{2x}{2x+3} - 1 = \frac{2x}{2x+3} - \frac{1}{1} \cdot \frac{2x+3}{2x+3}$$

$$= \frac{2x}{2x+3} - \frac{2x+3}{2x+3}$$

$$= \frac{2x-(2x+3)}{2x+3}$$

$$= \frac{2x-2x-3}{2x+3}$$

$$= \frac{-3}{2x+3}$$

51. LCD $= (n - 2) \cdot n = n(n - 2)$

$$\frac{n}{n-2} + \frac{n+2}{n} = \frac{n}{n-2} \cdot \frac{n}{n} + \frac{n+2}{n} \cdot \frac{n-2}{n-2}$$

$$= \frac{n^2}{n(n-2)} + \frac{n^2-4}{n(n-2)}$$

$$= \frac{n^2+n^2-4}{n(n-2)}$$

$$= \frac{2n^2-4}{n(n-2)}$$

$$= \frac{2(n^2-2)}{n(n-2)}$$

53. LCD $= (x - 3) \cdot x = x(x - 3)$

$$\frac{2x}{x-3} - \frac{5}{x} = \frac{2x}{x-3} \cdot \frac{x}{x} - \frac{5}{x} \cdot \frac{x-3}{x-3}$$

$$= \frac{2x^2}{x(x-3)} - \frac{5x-15}{x(x-3)}$$

$$= \frac{2x^2-(5x-15)}{x(x-3)}$$

$$= \frac{2x^2-5x+15}{x(x-3)}$$

55. $\text{LCD} = a^2(a-1)$

$$\frac{a}{a^2} - \frac{1}{a-1} = \frac{a}{a^2} \cdot \frac{a-1}{a-1} - \frac{1}{a-1} \cdot \frac{a^2}{a^2}$$

$$= \frac{a^2-a}{a^2(a-1)} - \frac{a^2}{a^2(a-1)}$$

$$= \frac{a^2-a-a^2}{a^2(a-1)}$$

$$= \frac{-a}{a^2(a-1)}$$

$$= \frac{-1}{a(a-1)}$$

57. $\text{LCD} = 1 \cdot (2n+1) = 2n+1$

$$\frac{4}{2n+1} + 1 = \frac{4}{2n+1} + \frac{1}{1} \cdot \frac{2n+1}{2n+1}$$

$$= \frac{4}{2n+1} + \frac{2n+1}{2n+1}$$

$$= \frac{4+2n+1}{2n+1}$$

$$= \frac{2n+5}{2n+1}$$

59. $3x-6 = 3(x-2);\ x-2 = x-2$

$\text{LCD} = 3(x-2)$

$$\frac{-12}{3x-6} + \frac{4x-1}{x-2} = \frac{-12}{3(x-2)} + \frac{4x-1}{x-2} \cdot \frac{3}{3}$$

$$= \frac{-12}{3(x-2)} + \frac{12x-3}{3(x-2)}$$

$$= \frac{-12+12x-3}{3(x-2)}$$

$$= \frac{12x-15}{3(x-2)}$$

$$= \frac{3(4x-5)}{3(x-2)}$$

$$= \frac{4x-5}{x-2}$$

61. $x = x;\ x^2 - 9x = x(x-9)$

$\text{LCD} = x(x-9)$

$$\frac{3x-1}{x} - \frac{9}{x^2-9x} = \frac{3x-1}{x} \cdot \frac{x-9}{x-9} - \frac{9}{x(x-9)}$$

$$= \frac{3x^2-28x+9}{x(x-9)} - \frac{9}{x(x-9)}$$

$$= \frac{3x^2-28x+9-9}{x(x-9)}$$

$$= \frac{3x^2-28x}{x(x-9)}$$

$$= \frac{x(3x-28)}{x(x-9)}$$

$$= \frac{3x-28}{x-9}$$

63. $n^2 - 3n = n(n-3);\ n^3 - n^2 = n^2(n-1)$

$\text{LCD} = n^2(n-3)(n-1)$

$$\frac{4}{n^2-3n} - \frac{3}{n^3-n^2}$$

$$= \frac{4}{n(n-3)} \cdot \frac{n(n-1)}{n(n-1)} - \frac{3}{n^2(n-1)} \cdot \frac{n-3}{n-3}$$

$$= \frac{4n^2-4n}{n^2(n-3)(n-1)} - \frac{3n-9}{n^2(n-1)(n-3)}$$

$$= \frac{4n^2-4n-(3n-9)}{n^2(n-3)(n-1)}$$

$$= \frac{4n^2-4n-3n+9}{n^2(n-3)(n-1)}$$

$$= \frac{4n^2-7n+9}{n^2(n-3)(n-1)}$$

65. $2a-a^2 = -a(a-2);\ 2a^2-4a = 2a(a-2)$

$\text{LCD} = -2a(a-2)$

$$\frac{5}{2a-a^2} - \frac{3}{2a^2-4a}$$

$$= \frac{5}{-a(a-2)} \cdot \frac{2}{2} - \frac{3}{2a(a-2)} \cdot \frac{-1}{-1}$$

$$= \frac{10}{-2a(a-2)} - \frac{-3}{-2a(a-2)}$$

$$= \frac{10-(-3)}{-2a(a-2)}$$

$$= \frac{13}{-2a(a-2)}$$

$$= \frac{-13}{2a(a-2)}$$

67. $n^2 - 4 = (n+2)(n-2); \quad 6 - n^2 - n = -n^2 - n + 6 = -(n+3)(n-2)$

$\text{LCD} = -(n+2)(n-2)(n+3)$

$$\frac{2n+1}{n^2-4} + \frac{3n}{6-n^2-n} = \frac{2n+1}{(n+2)(n-2)} \cdot \frac{-(n+3)}{-(n+3)} + \frac{3n}{-(n+3)(n-2)} \cdot \frac{n+2}{n+2}$$

$$= \frac{-2n^2 - 7n - 3}{-(n+2)(n-2)(n+3)} + \frac{3n^2 + 6n}{-(n+2)(n-2)(n+3)}$$

$$= \frac{-2n^2 - 7n - 3 + 3n^2 + 6n}{-(n+2)(n-2)(n+3)}$$

$$= \frac{n^2 - n - 3}{-(n+2)(n-2)(n+3)}$$

69. $x^2 - 1 = (x+1)(x-1); \quad x^2 - 2x + 1 = (x-1)^2$

$\text{LCD} = (x+1)(x-1)^2$

$$\frac{x}{x^2-1} + \frac{x+1}{x^2-2x+1} = \frac{x}{(x+1)(x-1)} \cdot \frac{x-1}{x-1} + \frac{x+1}{(x-1)^2} \cdot \frac{x+1}{x+1}$$

$$= \frac{x^2 - x}{(x+1)(x-1)^2} + \frac{x^2 + 2x + 1}{(x+1)(x-1)^2}$$

$$= \frac{x^2 - x + x^2 + 2x + 1}{(x+1)(x-1)^2}$$

$$= \frac{2x^2 + x + 1}{(x+1)(x-1)^2}$$

71. $n + 3 = n + 3; \quad n^2 + n - 6 = (n+3)(n-2)$

$\text{LCD} = (n+3)(n-2)$

$$\frac{2n+1}{n+3} + \frac{7-2n^2}{n^2+n-6} = \frac{2n+1}{n+3} \cdot \frac{n-2}{n-2} + \frac{7-2n^2}{(n+3)(n-2)}$$

$$= \frac{2n^2 - 3n - 2}{(n+3)(n-2)} + \frac{7 - 2n^2}{(n+3)(n-2)}$$

$$= \frac{2n^2 - 3n - 2 + 7 - 2n^2}{(n+3)(n-2)}$$

$$= \frac{-3n + 5}{(n+3)(n-2)}$$

73. $x^2 + x - 6 = (x+3)(x-2); \ 2-x = -(x-2)$

LCD $= -(x+3)(x-2)$

$$\frac{-3x-9}{x^2+x-6} - \frac{x+3}{2-x} = \frac{-3x-9}{(x+3)(x-2)} \cdot \frac{-1}{-1} - \frac{x+3}{-(x-2)} \cdot \frac{x+3}{x+3}$$

$$= \frac{3x+9}{-(x+3)(x-2)} - \frac{x^2+6x+9}{-(x+3)(x-2)}$$

$$= \frac{3x+9-(x^2+6x+9)}{-(x+3)(x-2)}$$

$$= \frac{3x+9-x^2-6x-9}{-(x+3)(x-2)}$$

$$= \frac{-x^2-3x}{-(x+3)(x-2)}$$

$$= \frac{-x(x+3)}{-(x+3)(x-2)}$$

$$= \frac{x}{x-2}$$

75. $x+2 = x+2; \ x^2+2x = x(x+2); \ x = x$

LCD $= x(x+2)$

$$\frac{4}{x+2} + \frac{-5x-2}{x^2+2x} - \frac{3-x}{x} = \frac{4}{x+2} \cdot \frac{x}{x} + \frac{-5x-2}{x(x+2)} - \frac{3-x}{x} \cdot \frac{x+2}{x+2}$$

$$= \frac{4x}{x(x+2)} + \frac{-5x-2}{x(x+2)} - \frac{-x^2+x+6}{x(x+2)}$$

$$= \frac{4x-5x-2-(-x^2+x+6)}{x(x+2)}$$

$$= \frac{x^2-2x-8}{x(x+2)}$$

$$= \frac{(x+2)(x-4)}{x(x+2)}$$

$$= \frac{x-4}{x}$$

77. $m + 2 = m + 2;\ m = m;\ m^2 - 4 = (m + 2)(m - 2)$

LCD $= m(m + 2)(m - 2)$

$$\frac{2}{m+2} - \frac{3}{m} + \frac{m+10}{m^2-4} = \frac{2}{m+2} \cdot \frac{m(m-2)}{m(m-2)} - \frac{3}{m} \cdot \frac{(m+2)(m-2)}{(m+2)(m-2)} + \frac{m+10}{(m+2)(m-2)} \cdot \frac{m}{m}$$

$$= \frac{2m^2 - 4m}{m(m+2)(m-2)} - \frac{3m^2 - 12}{m(m+2)(m-2)} + \frac{m^2 + 10m}{m(m+2)(m-2)}$$

$$= \frac{2m^2 - 4m - (3m^2 - 12) + m^2 + 10m}{m(m+2)(m-2)}$$

$$= \frac{2m^2 - 4m - 3m^2 + 12 + m^2 + 10m}{m(m+2)(m-2)}$$

$$= \frac{6m + 12}{m(m+2)(m-2)}$$

$$= \frac{6(m+2)}{m(m+2)(m-2)}$$

$$= \frac{6}{m(m-2)}$$

79. $\dfrac{\frac{x-3}{x+2}}{\frac{x^2-9}{x^2+4}} = \dfrac{x-3}{x+2} \div \dfrac{x^2-9}{x^2+4} = \dfrac{x-3}{x+2} \cdot \dfrac{x^2+4}{x^2-9} = \dfrac{x-3}{x+2} \cdot \dfrac{x^2+4}{(x+3)(x-3)} = \dfrac{x^2+4}{(x+2)(x+3)}$

81. $2x^2 + x - 3 = (2x+3)(x-1);\ x^2 - 1 = (x+1)(x-1)$

LCD $= (2x+3)(x+1)(x-1)$

$$\frac{4x+6}{2x^2+x-3} - \frac{x-1}{x^2-1} = \frac{4x+6}{(2x+3)(x-1)} \cdot \frac{x+1}{x+1} - \frac{x-1}{(x+1)(x-1)} \cdot \frac{2x+3}{2x+3}$$

$$= \frac{4x^2 + 10x + 6}{(2x+3)(x+1)(x-1)} - \frac{2x^2 + x - 3}{(2x+3)(x+1)(x-1)}$$

$$= \frac{4x^2 + 10x + 6 - (2x^2 + x - 3)}{(2x+3)(x+1)(x-1)}$$

$$= \frac{4x^2 + 10x + 6 - 2x^2 - x + 3}{(2x+3)(x+1)(x-1)}$$

$$= \frac{2x^2 + 9x + 9}{(2x+3)(x+1)(x-1)}$$

$$= \frac{(2x+3)(x+3)}{(2x+3)(x+1)(x-1)}$$

$$= \frac{x+3}{(x+1)(x-1)}$$

83. $x^2 + x - 6 = (x+3)(x-2); \; x^2 + 5x + 6 = (x+3)(x+2)$

LCD $= (x+3)(x-2)(x+2)$

$$\frac{x+2}{x^2+x-6} + \frac{x-3}{x^2+5x+6} = \frac{x+2}{(x+3)(x-2)} \cdot \frac{x+2}{x+2} + \frac{x-3}{(x+3)(x+2)} \cdot \frac{x-2}{x-2}$$

$$= \frac{(x+2)^2}{(x+3)(x-2)(x+2)} + \frac{(x-3)(x-2)}{(x+3)(x-2)(x+2)}$$

$$= \frac{x^2+4x+4}{(x+3)(x-2)(x+2)} + \frac{x^2-5x+6}{(x+3)(x-2)(x+2)}$$

$$= \frac{x^2+4x+4+x^2-5x+6}{(x+3)(x-2)(x+2)}$$

$$= \frac{2x^2-x+10}{(x+3)(x-2)(x+2)}$$

85. $\dfrac{2x-3}{x+6} \cdot \dfrac{x-1}{x-7} = \dfrac{(2x-3)(x-1)}{(x+6)(x-7)} = \dfrac{2x^2-5x+3}{(x+6)(x-7)}$

87. In a rectangle, Perimeter = [(Length) + (Width)]

$$2\left(\frac{2x+1}{4} + \frac{3x-7}{8}\right) = 2\left(\frac{2x+1}{4} \cdot \frac{2}{2} + \frac{3x-7}{8}\right)$$

$$= 2\left(\frac{4x+2}{8} + \frac{3x-7}{8}\right)$$

$$= 2\left(\frac{4x+2+3x-7}{8}\right)$$

$$= 2\left(\frac{7x-5}{8}\right)$$

$$= \frac{7x-5}{4} \text{ units}$$

89. $a = a; \; a - 1 = a - 1; \; (a-1)^2 = (a-1)^2$

LCD $= a(a-1)^2$

$$\frac{2}{a} - \left(\frac{2}{a-1} - \frac{3}{(a-1)^2}\right) = \frac{2}{a} \cdot \frac{(a-1)^2}{(a-1)^2} - \left(\frac{2}{a-1} \cdot \frac{a(a-1)}{a(a-1)} - \frac{3}{(a-1)^2} \cdot \frac{a}{a}\right)$$

$$= \frac{2a^2-4a+2}{a(a-1)^2} - \left(\frac{2a^2-2a}{a(a-1)^2} - \frac{3a}{a(a-1)^2}\right)$$

$$= \frac{2a^2-4a+2}{a(a-1)^2} - \left(\frac{2a^2-2a-3a}{a(a-1)^2}\right)$$

$$= \frac{2a^2-4a+2}{a(a-1)^2} - \frac{2a^2-5a}{a(a-1)^2}$$

$$= \frac{2a^2-4a+2-2a^2+5a}{a(a-1)^2}$$

$$= \frac{a+2}{a(a-1)^2}$$

91. $x = x;\ x^2 + x = x(x+1);\ x^3 - x^2 = x^2(x-1)$

LCD $= x^2(x+1)(x-1)$

$$\frac{1}{x} - \frac{2}{x^2+x} + \frac{3}{x^3-x^2} = \frac{1}{x} \cdot \frac{x(x+1)(x-1)}{x(x+1)(x-1)} - \frac{2}{x(x+1)} \cdot \frac{x(x-1)}{x(x-1)} + \frac{3}{x^2(x-1)} \cdot \frac{x+1}{x+1}$$

$$= \frac{x^3-x}{x^2(x+1)(x-1)} - \frac{2x^2-2x}{x^2(x+1)(x-1)} + \frac{3x+3}{x^2(x+1)(x-1)}$$

$$= \frac{x^3-x-(2x^2-2x)+3x+3}{x^2(x+1)(x-1)}$$

$$= \frac{x^3-x-2x^2+2x+3x+3}{x^2(x+1)(x-1)}$$

$$= \frac{x^3-2x^2+4x+3}{x^2(x+1)(x-1)}$$

93. $\dfrac{2a+b}{a-b} \cdot \dfrac{2}{a+b} - \dfrac{3a+3b}{a^2-b^2} = \dfrac{4a+2b}{(a-b)(a+b)} - \dfrac{3a+3b}{(a-b)(a+b)} = \dfrac{4a+2b-(3a+3b)}{(a-b)(a+b)} = \dfrac{a-b}{(a-b)(a+b)} = \dfrac{1}{a+b}$

95. LCD $= (x-4)(x+1)$

$$\frac{x-3}{x-4} + \frac{x+2}{x-4} \cdot \frac{4}{x+1} = \frac{x-3}{x-4} \cdot \frac{x+1}{x+1} + \frac{x+2}{x-4} \cdot \frac{4}{x+1}$$

$$= \frac{x^2-2x-3}{(x-4)(x+1)} + \frac{4x+8}{(x-4)(x+1)}$$

$$= \frac{x^2-2x-3+4x+8}{(x-4)(x+1)}$$

$$= \frac{x^2+2x+5}{(x-4)(x+1)}$$

97. LCD $= x-2$

$$\frac{x+2}{x-2} - \frac{x-2}{x+2} \div \frac{1}{x^2-4} = \frac{x+2}{x-2} - \frac{x-2}{x+2} \cdot \frac{(x+2)(x-2)}{1}$$

$$= \frac{x+2}{x-2} - \frac{(x-2)(x-2)}{1}$$

$$= \frac{x+2}{x-2} - \frac{(x^2-4x+4)}{1} \cdot \frac{x-2}{x-2}$$

$$= \frac{x+2}{x-2} - \frac{x^3-6x^2+12x-8}{x-2}$$

$$= \frac{x+2-x^3+6x^2-12x+8}{x-2}$$

$$= \frac{-x^3+6x^2-11x+10}{x-2}$$

99. The steps to add or subtract rational expressions with unlike denominators are:
(1) find the least common denominator; (2) rewrite each expression with the common denominator; (3) add or subtract the rational expressions from step 2; (4) simplify the result.

101. Since the fractions have a common denominator, subtract the second numerator from the first and write the result over the common denominator. Simplify the numerator and look for factors common to the numerator and denominator.

$$\frac{3x+1}{(2x+5)(x-1)} - \frac{x-4}{(2x+5)(x-1)}$$

$$= \frac{3x+1-(x-4)}{(2x+5)(x-1)}$$

$$= \frac{2x+5}{(2x+5)(x-1)}$$

$$= \frac{1}{x-1}$$

Section 7.6

Preparing for Complex Rational Expressions

P1. $6y^2 - 5y - 6 = (3y+2)(2y-3)$

P2. $\dfrac{x+3}{12} \div \dfrac{x^2-9}{15} = \dfrac{x+3}{12} \cdot \dfrac{15}{x^2-9}$

$$= \frac{15(x+3)}{12(x+3)(x-3)}$$

$$= \frac{15}{12(x-3)}$$

$$= \frac{3 \cdot 5}{3 \cdot 4(x-3)}$$

$$= \frac{5}{4(x-3)}$$

7.6 Quick Checks

1. An expression such as $\dfrac{\frac{x}{2}+\frac{5}{x}}{\frac{2x-1}{3}}$ is called a

complex rational expression.

2. To simplify a complex rational expression means

to write the rational expression in the form $\dfrac{p}{q}$,

where p and q are polynomials that have no common factors.

3. $\dfrac{\frac{2}{k+3}}{\frac{4}{k^2+4k+3}} = \dfrac{2}{k+3} \cdot \dfrac{k^2+4k+3}{4}$

$$= \frac{2}{k+3} \cdot \frac{(k+3)(k+1)}{2 \cdot 2}$$

$$= \frac{2(k+3)(k+1)}{2(k+3) \cdot 2}$$

$$= \frac{k+1}{2}$$

4. $\dfrac{\frac{1}{n+3}}{\frac{8n}{2n+6}} = \dfrac{1}{n+3} \cdot \dfrac{2n+6}{8n}$

$$= \frac{1}{n+3} \cdot \frac{2(n+3)}{2 \cdot 4n}$$

$$= \frac{2(n+3)}{2(n+3) \cdot 4n}$$

$$= \frac{1}{4n}$$

5. False; $\dfrac{x-y}{\frac{1}{x}+\frac{2}{y}} = \dfrac{x-y}{\frac{1}{x} \cdot \frac{y}{y} + \frac{2}{y} \cdot \frac{x}{x}}$

$$= \frac{x-y}{\frac{y}{xy}+\frac{2x}{xy}}$$

$$= \frac{x-y}{\frac{y+2x}{xy}}$$

$$= \frac{x-y}{1} \cdot \frac{xy}{y+2x}$$

$$\neq \frac{x-y}{1} \cdot \left(\frac{x}{1}+\frac{y}{2}\right)$$

6. $\dfrac{\frac{1}{2}+\frac{2}{5}}{1-\frac{2}{5}} = \dfrac{\frac{1}{2} \cdot \frac{5}{5}+\frac{2}{5} \cdot \frac{2}{2}}{\frac{5}{5}-\frac{2}{5}}$

$$= \frac{\frac{5}{10}+\frac{4}{10}}{\frac{5-2}{5}}$$

$$= \frac{\frac{9}{10}}{\frac{3}{5}}$$

$$= \frac{9}{10} \cdot \frac{5}{3}$$

$$= \frac{3 \cdot 3}{2 \cdot 5} \cdot \frac{5}{3}$$

$$= \frac{3 \cdot 5 \cdot 3}{3 \cdot 5 \cdot 2}$$

$$= \frac{3}{2}$$

7.

$$\frac{\frac{3}{y}-1}{\frac{9}{y}-y} = \frac{\frac{3}{y}-1\cdot\frac{y}{y}}{\frac{9}{y}-y\cdot\frac{y}{y}}$$

$$= \frac{\frac{3}{y}-\frac{y}{y}}{\frac{9}{y}-\frac{y^2}{y}}$$

$$= \frac{\frac{3-y}{y}}{\frac{9-y^2}{y}}$$

$$= \frac{3-y}{y}\cdot\frac{y}{9-y^2}$$

$$= \frac{3-y}{y}\cdot\frac{y}{(3-y)(3+y)}$$

$$= \frac{y(3-y)}{y(3-y)(3+y)}$$

$$= \frac{1}{3+y}$$

8.

$$\frac{\frac{1}{3}+\frac{1}{x+5}}{\frac{x+8}{9}} = \frac{\frac{1}{3}\cdot\frac{x+5}{x+5}+\frac{1}{x+5}\cdot\frac{3}{3}}{\frac{x+8}{9}}$$

$$= \frac{\frac{x+5}{3(x+5)}+\frac{3}{3(x+5)}}{\frac{x+8}{9}}$$

$$= \frac{\frac{x+5+3}{3(x+5)}}{\frac{x+8}{3\cdot3}}$$

$$= \frac{x+8}{3(x+5)}\cdot\frac{3\cdot3}{x+8}$$

$$= \frac{3(x+8)\cdot3}{3(x+8)(x+5)}$$

$$= \frac{3}{x+5}$$

9.

$$\frac{\frac{3}{2}+\frac{2}{3}}{\frac{1}{2}-\frac{1}{3}} = \frac{\frac{3}{2}+\frac{2}{3}}{\frac{1}{2}-\frac{1}{3}}\cdot\frac{6}{6} \qquad \text{LCD}=2\cdot3=6$$

$$= \frac{\frac{3}{2}\cdot6+\frac{2}{3}\cdot6}{\frac{1}{2}\cdot6-\frac{1}{3}\cdot6}$$

$$= \frac{9+4}{3-2}$$

$$= \frac{13}{1}$$

$$= 13$$

10. $\text{LCD}=x\cdot y = xy$

$$\frac{\frac{1}{x}+\frac{2}{y}}{\frac{2}{x}-\frac{1}{y}} = \frac{\frac{1}{x}+\frac{2}{y}}{\frac{2}{x}-\frac{1}{y}}\cdot\frac{xy}{xy}$$

$$= \frac{\frac{1}{x}\cdot xy+\frac{2}{y}\cdot xy}{\frac{2}{x}\cdot xy-\frac{1}{y}\cdot xy}$$

$$= \frac{y+2x}{2y-x}$$

Exercises 7.6

11. $\dfrac{1-\frac{3}{4}}{\frac{1}{8}+2} = \dfrac{\frac{4}{4}-\frac{3}{4}}{\frac{1}{8}+\frac{16}{8}} = \dfrac{\frac{1}{4}}{\frac{17}{8}} = \dfrac{1}{4}\cdot\dfrac{8}{17} = \dfrac{2}{17}$

13.

$$\frac{\frac{x^2}{12}-\frac{1}{3}}{\frac{x+2}{18}} = \frac{\frac{x^2}{12}-\frac{1}{3}\cdot\frac{4}{4}}{\frac{x+2}{18}}$$

$$= \frac{\frac{x^2}{12}-\frac{4}{12}}{\frac{x+2}{18}}$$

$$= \frac{\frac{x^2-4}{12}}{\frac{x+2}{18}}$$

$$= \frac{x^2-4}{12}\cdot\frac{18}{x+2}$$

$$= \frac{(x+2)(x-2)}{2\cdot6}\cdot\frac{3\cdot6}{x+2}$$

$$= \frac{6(x+2)\cdot3(x-2)}{6(x+2)\cdot2}$$

$$= \frac{3(x-2)}{2}$$

15.

$$\frac{x+3}{\frac{x^2}{9}-1} = \frac{x+3}{\frac{x^2}{9}-1\cdot\frac{9}{9}}$$

$$= \frac{x+3}{\frac{x^2-9}{9}}$$

$$= \frac{x+3}{1}\cdot\frac{9}{x^2-9}$$

$$= \frac{x+3}{1}\cdot\frac{9}{(x+3)(x-3)}$$

$$= \frac{9(x+3)}{(x+3)(x-3)}$$

$$= \frac{9}{x-3}$$

17.
$$\frac{\frac{m}{2}+n}{\frac{m}{n}}=\frac{\frac{m}{n}+n\cdot\frac{2}{2}}{\frac{m}{n}}$$

$$=\frac{\frac{m}{2}+\frac{2n}{2}}{\frac{m}{n}}$$

$$=\frac{\frac{m+2n}{2}}{\frac{m}{n}}$$

$$=\frac{m+2n}{2}\cdot\frac{n}{m}$$

$$=\frac{n(m+2n)}{2m}$$

19.
$$\frac{\frac{5}{a}+\frac{4}{b^2}}{\frac{5b+4}{b^2}}=\frac{\frac{5}{a}\cdot\frac{b^2}{b^2}+\frac{4}{b^2}\cdot\frac{a}{a}}{\frac{5b+4}{b^2}}$$

$$=\frac{\frac{5b^2}{ab^2}+\frac{4a}{ab^2}}{\frac{5b+4}{b^2}}$$

$$=\frac{\frac{5b^2+4a}{ab^2}}{\frac{5b+4}{b^2}}$$

$$=\frac{5b^2+4a}{ab^2}\cdot\frac{b^2}{5b+4}$$

$$=\frac{b^2(5b^2+4a)}{b^2\cdot a(5b+4)}$$

$$=\frac{5b^2+4a}{a(5b+4)}$$

21.
$$\frac{\frac{8}{y+3}-2}{y-\frac{4}{y+3}}=\frac{\frac{8}{y+3}-2\cdot\frac{y+3}{y+3}}{y\cdot\frac{y+3}{y+3}-\frac{4}{y+3}}$$

$$=\frac{\frac{8-2y-6}{y+3}}{\frac{y^2+3y-4}{y+3}}$$

$$=\frac{8-2y-6}{y+3}\cdot\frac{y+3}{y^2+3y-4}$$

$$=\frac{-2y+2}{y+3}\cdot\frac{y+3}{y^2+3y-4}$$

$$=\frac{-2(y-1)}{y+3}\cdot\frac{y+3}{(y+4)(y-1)}$$

$$=\frac{(y-1)(y+3)\cdot(-2)}{(y-1)(y+3)\cdot(y+4)}$$

$$=-\frac{2}{y+4}$$

23.
$$\frac{\frac{1}{4}-\frac{6}{y}}{\frac{5}{6y}-y}=\frac{\frac{1}{4}\cdot\frac{y}{y}-\frac{6}{y}\cdot\frac{4}{4}}{\frac{5}{6y}-y\cdot\frac{6y}{6y}}$$

$$=\frac{\frac{y}{4y}-\frac{24}{4y}}{\frac{5}{6y}-\frac{6y^2}{y}}$$

$$=\frac{\frac{y-24}{4y}}{\frac{5-6y^2}{6y}}$$

$$=\frac{y-24}{4y}\cdot\frac{6y}{5-6y^2}$$

$$=\frac{y-24}{2\cdot2y}\cdot\frac{3\cdot2y}{5-6y^2}$$

$$=\frac{2y\cdot3(y-24)}{2y\cdot2(5-6y^2)}$$

$$=\frac{3(y-24)}{2(5-6y^2)}$$

25. LCD $= 12$
$$\frac{\frac{3}{2}-\frac{1}{4}}{\frac{5}{6}+\frac{1}{2}}=\frac{\frac{3}{2}-\frac{1}{4}}{\frac{5}{6}+\frac{1}{2}}\cdot\frac{12}{12}=\frac{\frac{3}{2}\cdot12-\frac{1}{4}\cdot12}{\frac{5}{6}\cdot12+\frac{1}{2}\cdot12}=\frac{18-3}{10+6}=\frac{15}{16}$$

27. LCD $= m^2$
$$\frac{\frac{3}{m}+\frac{2}{m^2}}{\frac{6}{m}+\frac{4}{m^2}}=\frac{\frac{3}{m}+\frac{2}{m^2}}{\frac{6}{m}+\frac{4}{m^2}}\cdot\frac{m^2}{m^2}$$

$$=\frac{3m+2}{6m+4}$$

$$=\frac{3m+2}{2(3m+2)}$$

$$=\frac{1}{2}$$

29. LCD $= b^2$
$$\frac{1-\frac{49}{b^2}}{1+\frac{7}{b}}=\frac{1-\frac{49}{b^2}}{1+\frac{7}{b}}\cdot\frac{b^2}{b^2}$$

$$=\frac{b^2-49}{b^2+7b}$$

$$=\frac{(b+7)(b-7)}{b(b+7)}$$

$$=\frac{b-7}{b}$$

31. $\text{LCD} = x(x+4)$

$$\frac{1+\frac{5}{x}}{1+\frac{1}{x+4}} = \frac{1+\frac{5}{x}}{1+\frac{1}{x+4}} \cdot \frac{x(x+4)}{x(x+4)}$$

$$= \frac{x(x+4)+5(x+4)}{x(x+4)+x}$$

$$= \frac{x^2+4x+5x+20}{x^2+4x+x}$$

$$= \frac{x^2+9x+20}{x^2+5x}$$

$$= \frac{(x+4)(x+5)}{x(x+5)}$$

$$= \frac{x+4}{x}$$

33. $\text{LCD} = x^2 y^2$

$$\frac{\frac{1}{x^2}-\frac{1}{y^2}}{x-y} = \frac{\frac{1}{x^2}-\frac{1}{y^2}}{x-y} \cdot \frac{x^2 y^2}{x^2 y^2}$$

$$= \frac{y^2-x^2}{x^2 y^2(x-y)}$$

$$= \frac{-(x+y)(x-y)}{x^2 y^2(x-y)}$$

$$= -\frac{x+y}{x^2 y^2}$$

35. $\text{LCD} = b$

$$\frac{a+\frac{1}{b}}{a+\frac{2}{b}} = \frac{a+\frac{1}{b}}{a+\frac{2}{b}} \cdot \frac{b}{b} = \frac{ab+1}{ab+2}$$

37. $\text{LCD} = 3n$

$$\frac{12}{\frac{4}{n}-\frac{2}{3n}} = \frac{12}{\frac{4}{n}-\frac{2}{3n}} \cdot \frac{3n}{3n} = \frac{36n}{12-2} = \frac{36n}{10} = \frac{18n}{5}$$

39.

$$\frac{\frac{x}{x-y}+\frac{y}{x+y}}{\frac{xy}{x^2-y^2}} = \frac{(x+y)(x-y)\left(\frac{x}{x-y}+\frac{y}{x+y}\right)}{(x+y)(x-y)\left(\frac{xy}{(x+y)(x-y)}\right)}$$

$$= \frac{x(x+y)+y(x-y)}{xy}$$

$$= \frac{x^2+xy+xy-y^2}{xy}$$

$$= \frac{x^2+2xy-y^2}{xy}$$

41.

$$\frac{\frac{2}{x+4}}{\frac{2}{x+4}-4} = \frac{(x+4)\left(\frac{2}{x+4}\right)}{(x+4)\left(\frac{2}{x+4}-4\right)}$$

$$= \frac{2}{2-4(x+4)}$$

$$= \frac{2}{2-4x-16}$$

$$= \frac{2}{-4x-14}$$

$$= \frac{2}{-2(2x+7)}$$

$$= -\frac{1}{2x+7}$$

43.

$$\frac{\frac{b^2}{b^2-16}-\frac{b}{b+4}}{\frac{b}{b^2-16}-\frac{1}{b-4}}$$

$$= \frac{\frac{b^2}{(b+4)(b-4)}-\frac{b}{b+4}}{\frac{b}{(b+4)(b-4)}-\frac{1}{b-4}} \cdot \frac{(b+4)(b-4)}{(b+4)(b-4)}$$

$$= \frac{b^2-b(b-4)}{b-(b+4)}$$

$$= \frac{b^2-b^2+4b}{b-b-4}$$

$$= \frac{4b}{-4}$$

$$= -b$$

45.

$$\frac{1-\frac{2}{x}-\frac{3}{x^2}}{1-\frac{9}{x^2}} = \frac{x^2\left(1-\frac{2}{x}-\frac{3}{x^2}\right)}{x^2\left(1-\frac{9}{x^2}\right)}$$

$$= \frac{x^2-2x-3}{x^2-9}$$

$$= \frac{(x+1)(x-3)}{(x+3)(x-3)}$$

$$= \frac{x+1}{x+3}$$

47.

$$\frac{1-\frac{a^2}{4b^2}}{1+\frac{a}{2b}} = \frac{4b^2\left(1-\frac{a^2}{4b^2}\right)}{4b^2\left(1+\frac{a}{2b}\right)}$$

$$= \frac{4b^2-a^2}{4b^2+2ab}$$

$$= \frac{(2b+a)(2b-a)}{2b(2b+a)}$$

$$= \frac{2b-a}{2b}$$

49. $\dfrac{\frac{1}{y+z}+\frac{1}{y-z}}{\frac{1}{y^2-z^2}} = \dfrac{\frac{1}{y+z}+\frac{1}{y-z}}{\frac{1}{(y+z)(y-z)}}$

$\qquad = \dfrac{(y+z)(y-z)\left(\frac{1}{y+z}+\frac{1}{y-z}\right)}{(y+z)(y-z)\frac{1}{(y+z)(y-z)}}$

$\qquad = \dfrac{y-z+y+z}{1}$

$\qquad = \dfrac{2y}{1}$

$\qquad = 2y$

51. $\dfrac{\frac{3}{n-2}+1}{5+\frac{1}{n-2}} = \dfrac{(n-2)\left(\frac{3}{n-2}+1\right)}{(n-2)\left(5+\frac{1}{n-2}\right)}$

$\qquad = \dfrac{3+n-2}{5(n-2)+1}$

$\qquad = \dfrac{1+n}{5n-10+1}$

$\qquad = \dfrac{n+1}{5n-9}$

53. $\dfrac{\frac{n}{6}+\frac{n+3}{2}+\frac{2n-1}{8}}{3} = \dfrac{24\left(\frac{n}{6}+\frac{n+3}{2}+\frac{2n-1}{8}\right)}{24(3)}$

$\qquad = \dfrac{4n+12(n+3)+3(2n-1)}{72}$

$\qquad = \dfrac{4n+12n+36+6n-3}{72}$

$\qquad = \dfrac{22n+33}{72}$

$\qquad = \dfrac{11(2n+3)}{72}$

55. $\dfrac{\frac{2x+3}{x^2}-\frac{3}{x}}{\frac{x^2-9}{x^5}} = \dfrac{x^5\left(\frac{2x+3}{x^2}-\frac{3}{x}\right)}{x^5\left(\frac{x^2-9}{x^5}\right)}$

$\qquad = \dfrac{x^3(2x+3)-3x^4}{x^2-9}$

$\qquad = \dfrac{2x^4+3x^3-3x^4}{x^2-9}$

$\qquad = \dfrac{-x^4+3x^3}{x^2-9}$

$\qquad = \dfrac{-x^3(x-3)}{(x+3)(x-3)}$

$\qquad = -\dfrac{x^3}{x+3}$

The length is $-\dfrac{x^3}{x+3}$ feet.

57. **(a)** $R = \dfrac{1}{\frac{1}{R_1}+\frac{1}{R_2}}$

$\qquad = \dfrac{R_1 R_2(1)}{R_1 R_2\left(\frac{1}{R_1}+\frac{1}{R_2}\right)}$

$\qquad = \dfrac{R_1 R_2}{R_2+R_1}$

(b) $R_1 = 6$ ohms; $R_2 = 10$ ohms

$\qquad R = \dfrac{6(10)}{10+6} = \dfrac{60}{16} = \dfrac{15}{4}$ ohms

59. $\dfrac{x^{-1}-3}{x^{-2}-9} = \dfrac{\frac{1}{x}-3}{\frac{1}{x^2}-9}$

$\qquad = \dfrac{x^2\left(\frac{1}{x}-3\right)}{x^2\left(\frac{1}{x^2}-9\right)}$

$\qquad = \dfrac{x-3x^2}{1-9x^2}$

$\qquad = \dfrac{x(1-3x)}{(1+3x)(1-3x)}$

$\qquad = \dfrac{x}{1+3x}$

61. $\dfrac{2x^{-1}+5}{4x^{-2}-25}=\dfrac{\frac{2}{x}+5}{\frac{4}{x^2}-25}$

$\phantom{\dfrac{2x^{-1}+5}{4x^{-2}-25}}=\dfrac{x^2\left(\frac{2}{x}+5\right)}{x^2\left(\frac{4}{x^2}-25\right)}$

$\phantom{\dfrac{2x^{-1}+5}{4x^{-2}-25}}=\dfrac{2x+5x^2}{4-25x^2}$

$\phantom{\dfrac{2x^{-1}+5}{4x^{-2}-25}}=\dfrac{x(2+5x)}{(2+5x)(2-5x)}$

$\phantom{\dfrac{2x^{-1}+5}{4x^{-2}-25}}=\dfrac{x}{2-5x}$

63. $1+\dfrac{1}{1+\frac{1}{x}}=1+\dfrac{1}{\frac{x+1}{x}}$

$\phantom{1+\dfrac{1}{1+\frac{1}{x}}}=1+\dfrac{x}{x+1}$

$\phantom{1+\dfrac{1}{1+\frac{1}{x}}}=\dfrac{x+1+x}{x+1}$

$\phantom{1+\dfrac{1}{1+\frac{1}{x}}}=\dfrac{2x+1}{x+1}$

65. $\dfrac{1}{1-\dfrac{1}{2-\frac{1}{3-x}}}=\dfrac{1}{1-\dfrac{1}{\frac{2(3-x)-1}{3-x}}}$

$=\dfrac{1}{1-\dfrac{1}{\frac{5-2x}{3-x}}}$

$=\dfrac{1}{1-\dfrac{3-x}{5-2x}}$

$=\dfrac{1}{\frac{5-2x-(3-x)}{5-2x}}$

$=\dfrac{1}{\frac{2-x}{5-2x}}$

$=\dfrac{5-2x}{2-x}$

67. The expression $\dfrac{\frac{2}{x+1}}{\frac{1}{x-5}}$ is not in simplified form,

since the numerator and denominator are rational expressions, not polynomials. Simplified form is

the form $\dfrac{p}{q}$ where p and q are polynomials that

have no common factors.

69. Answers may vary.

Putting the Concepts Together (Sections 7.1–7.6)

1. $\dfrac{a^3-b^3}{a+b}=\dfrac{2^3-(-3)^3}{2+(-3)}$

$\phantom{\dfrac{a^3-b^3}{a+b}}=\dfrac{8-(-27)}{2-3}$

$\phantom{\dfrac{a^3-b^3}{a+b}}=\dfrac{8+27}{-1}$

$\phantom{\dfrac{a^3-b^3}{a+b}}=-35$

2. (a) $\dfrac{-3a}{a-6}$ is undefined when $a-6=0$, so the

expression is undefined for $a=6$.

(b) $\dfrac{y+2}{y^2+4y}$ is undefined when $y^2+4y=0$.

$y^2+4y=0$

$y(y+4)=0$

$y=0\quad\text{or}\quad y+4=0$

$y=0\quad\text{or}\qquad y=-4$

The expression is undefined for $y=0$ or $y=-4$.

3. (a) $\dfrac{ax+ay-4bx-4by}{2x+2y}$

$=\dfrac{a(x+y)-4b(x+y)}{2(x+y)}$

$=\dfrac{(x+y)(a-4b)}{2(x+y)}$

$=\dfrac{a-4b}{2}$

(b) $\dfrac{x^2+x-2}{1-x^2}=\dfrac{(x+2)(x-1)}{(1+x)(1-x)}$

$=\dfrac{(x+2)(x-1)}{(x+1)(-1)(x-1)}$

$=\dfrac{-(x+2)}{x+1}$

4. $2a+4b=2(a+2b)$

$4a+8b=4(a+2b)=2^2(a+2b)$

$8a-32b=8(a-4b)=2^3(a-4b)$

$\text{LCD}=2^3(a+2b)(a-4b)$

$\phantom{\text{LCD}}=8(a+2b)(a-4b)$

5. $\dfrac{7}{3x^2-x}=\dfrac{7}{x(3x-1)}\cdot\dfrac{5x}{5x}=\dfrac{35x}{5x^2(3x-1)}$

6. $\dfrac{y^2-y}{3y}\cdot\dfrac{6y^2}{1-y^2}=\dfrac{y(y-1)}{3y}\cdot\dfrac{3y(2y)}{-1(y+1)(y-1)}$

$\qquad=\dfrac{3y(y-1)\cdot2y^2}{3y(y-1)\cdot(-1)(y+1)}$

$\qquad=\dfrac{-2y^2}{y+1}$

7. $\dfrac{m^2+m-2}{m^3-6m^2}\cdot\dfrac{2m^2-14m+12}{m+2}$

$\qquad=\dfrac{(m+2)(m-1)}{m^2(m-6)}\cdot\dfrac{2(m-6)(m-1)}{m+2}$

$\qquad=\dfrac{(m+2)(m-6)\cdot2(m-1)^2}{(m+2)(m-6)\cdot m^2}$

$\qquad=\dfrac{2(m-1)^2}{m^2}$

8. $\dfrac{4y+12}{5y-5}\div\dfrac{2y^2-18}{y^2-2y+1}$

$\qquad=\dfrac{4y+12}{5y-5}\cdot\dfrac{y^2-2y+1}{2y^2-18}$

$\qquad=\dfrac{2^2(y+3)}{5(y-1)}\cdot\dfrac{(y-1)^2}{2(y+3)(y-3)}$

$\qquad=\dfrac{2(y-1)(y+3)\cdot2(y-1)}{2(y-1)(y+3)\cdot5(y-3)}$

$\qquad=\dfrac{2(y-1)}{5(y-3)}$

9. $\dfrac{4x^2+x}{x+3}+\dfrac{12x+3}{x+3}=\dfrac{4x^2+x+12x+3}{x+3}$

$\qquad=\dfrac{4x^2+13x+3}{x+3}$

$\qquad=\dfrac{(4x+1)(x+3)}{x+3}$

$\qquad=4x+1$

10. $\dfrac{x^2}{x^3+1}-\dfrac{x-1}{x^3+1}=\dfrac{x^2-(x-1)}{(x+1)(x^2-x+1)}$

$\qquad=\dfrac{x^2-x+1}{(x+1)(x^2-x+1)}$

$\qquad=\dfrac{1}{x+1}$

11. $\dfrac{-8}{2x-1}-\dfrac{9}{1-2x}=\dfrac{-8}{2x-1}-\dfrac{9}{-1(2x-1)}$

$\qquad=\dfrac{-8}{2x-1}-\dfrac{-9}{2x-1}$

$\qquad=\dfrac{-8-(-9)}{2x-1}$

$\qquad=\dfrac{1}{2x-1}$

12. $\dfrac{4}{m+3}+\dfrac{3}{3m+2}$

$\qquad=\dfrac{4}{m+3}\cdot\dfrac{3m+2}{3m+2}+\dfrac{3}{3m+2}\cdot\dfrac{m+3}{m+3}$

$\qquad=\dfrac{4(3m+2)}{(m+3)(3m+2)}+\dfrac{3(m+3)}{(m+3)(3m+2)}$

$\qquad=\dfrac{12m+8+3m+9}{(m+3)(3m+2)}$

$\qquad=\dfrac{15m+17}{(m+3)(3m+2)}$

13. $\dfrac{3m}{m^2+7m+10}-\dfrac{2m}{m^2+6m+8}$

$\qquad=\dfrac{3m}{(m+5)(m+2)}-\dfrac{2m}{(m+4)(m+2)}$

$\qquad=\dfrac{3m}{(m+5)(m+2)}\cdot\dfrac{m+4}{m+4}$

$\qquad\qquad-\dfrac{2m}{(m+4)(m+2)}\cdot\dfrac{m+5}{m+5}$

$\qquad=\dfrac{3m(m+4)-2m(m+5)}{(m+5)(m+4)(m+2)}$

$\qquad=\dfrac{3m^2+12m-2m^2-10m}{(m+5)(m+4)(m+2)}$

$\qquad=\dfrac{m^2+2m}{(m+5)(m+4)(m+2)}$

$\qquad=\dfrac{m(m+2)}{(m+5)(m+4)(m+2)}$

$\qquad=\dfrac{m}{(m+5)(m+4)}$

14.
$$\frac{\frac{-1}{m+1}-1}{m-\frac{2}{m+1}} = \frac{\frac{-1}{m+1}-1}{m-\frac{2}{m+1}} \cdot \frac{m+1}{m+1}$$
$$= \frac{-1-(m+1)}{m(m+1)-2}$$
$$= \frac{-1-m-1}{m^2+m-2}$$
$$= \frac{-m-2}{(m+2)(m-1)}$$
$$= \frac{-(m+2)}{(m+2)(m-1)}$$
$$= -\frac{1}{m-1}$$

15.
$$\frac{\frac{4}{a}+\frac{1}{6}}{\frac{3}{a^2}+\frac{1}{2}} = \frac{\frac{4}{a}+\frac{1}{6}}{\frac{3}{a^2}+\frac{1}{2}} \cdot \frac{6a^2}{6a^2}$$
$$= \frac{24a+a^2}{18+3a^2}$$
$$= \frac{a(24+a)}{3(6+a^2)}$$

Section 7.7

Preparing for Rational Equations

P1. $3k-2(k+1)=6$
$3k-2k-2=6$
$k-2=6$
$k-2+2=6+2$
$k=8$
The solution set is $\{8\}$.

P2. $3p^2-7p-6=(3p+2)(p-3)$

P3. $8z^2-10z-3=0$
$(4z+1)(2z-3)=0$
$4z+1=0$ or $2z-3=0$
$4z=-1$ or $2z=3$
$z=-\frac{1}{4}$ or $z=\frac{3}{2}$
The solution set is $\left\{-\frac{1}{4},\frac{3}{2}\right\}$.

P4. $\dfrac{x+4}{x^2-2x-24}$ is undefined when
$x^2-2x-24=0$.
$x^2-2x-24=0$
$(x+4)(x-6)=0$
$x+4=0$ or $x-6=0$
$x=-4$ or $x=6$
The expression is undefined for $x=-4$ or $x=6$.

P5. $4x-2y=10$
$-4x+4x-2y=-4x+10$
$-2y=-4x+10$
$\dfrac{-2y}{-2}=\dfrac{-4x+10}{-2}$
$y=2x-5$

Quick Checks 7.7

1. A <u>rational equation</u> is an equation that contains a rational expression.

2. Undefined value: $z=0$
LCD $=2z$
$$\frac{5}{2}+\frac{1}{z}=4$$
$$2z\left(\frac{5}{2}+\frac{1}{z}\right)=2z(4)$$
$$5z+2=8z$$
$$2=3z$$
$$\frac{2}{3}=z$$

Check: $\dfrac{5}{2}+\dfrac{1}{\frac{2}{3}} \overset{?}{=} 4$
$\dfrac{5}{2}+\dfrac{3}{2} \overset{?}{=} 4$
$\dfrac{5+3}{2} \overset{?}{=} 4$
$\dfrac{8}{2} \overset{?}{=} 4$
$4=4$ True
The solution set is $\left\{\dfrac{2}{3}\right\}$.

3. Undefined values: $x = -4$ and $x = 3$

LCD $= (x+4)(x-3)$

$$\frac{8}{x+4} = \frac{12}{x-3}$$

$$(x+4)(x-3)\left(\frac{8}{x+4}\right) = (x+4)(x-3)\left(\frac{12}{x-3}\right)$$

$$8(x-3) = 12(x+4)$$

$$8x - 24 = 12x + 48$$

$$-72 = 4x$$

$$-18 = x$$

Check: $\dfrac{8}{-18+4} \overset{?}{=} \dfrac{12}{-18-3}$

$\dfrac{8}{-14} \overset{?}{=} \dfrac{12}{-21}$

$-\dfrac{4}{7} = -\dfrac{4}{7}$ True

The solution set is $\{-18\}$.

4. Undefined value: $b = 0$

LCD $= 6b$

$$\frac{4}{3b} + \frac{1}{6b} = \frac{7}{2b} + \frac{1}{3}$$

$$6b\left(\frac{4}{3b} + \frac{1}{6b}\right) = 6b\left(\frac{7}{2b} + \frac{1}{3}\right)$$

$$8 + 1 = 21 + 2b$$

$$9 = 21 + 2b$$

$$-12 = 2b$$

$$-6 = b$$

Check: $\dfrac{4}{3(-6)} + \dfrac{1}{6(-6)} \overset{?}{=} \dfrac{7}{2(-6)} + \dfrac{1}{3}$

$-\dfrac{4}{18} - \dfrac{1}{36} \overset{?}{=} -\dfrac{7}{12} + \dfrac{1}{3}$

$-\dfrac{8}{36} - \dfrac{1}{36} \overset{?}{=} -\dfrac{7}{12} + \dfrac{4}{12}$

$-\dfrac{9}{36} \overset{?}{=} -\dfrac{3}{12}$

$-\dfrac{1}{4} = -\dfrac{1}{4}$ True

The solution set is $\{-6\}$.

5. $x - 3 = 0$ when $x = 3$. $2x - 6 = 2(x-3) = 0$ when $x = 3$.

The equation is undefined for $x = 3$.

LCD $= 2 \cdot (x-3) = 2x - 6 = 2(x-3)$

$$\frac{3}{2} + \frac{5}{x-3} = \frac{x+9}{2x-6}$$

$$2(x-3)\left(\frac{3}{2} + \frac{5}{x-3}\right) = 2(x-3)\left(\frac{x+9}{2x-6}\right)$$

$$3(x-3) + 2(5) = (2x-6)\left(\frac{x+9}{2x-6}\right)$$

$$3x - 9 + 10 = x + 9$$

$$3x + 1 = x + 9$$

$$2x = 8$$

$$x = 4$$

Check: $\dfrac{3}{2} + \dfrac{5}{4-3} \overset{?}{=} \dfrac{4+9}{2(4)-6}$

$\dfrac{3}{2} + \dfrac{5}{1} \overset{?}{=} \dfrac{13}{8-6}$

$\dfrac{3}{2} + \dfrac{10}{2} \overset{?}{=} \dfrac{13}{2}$

$\dfrac{13}{2} = \dfrac{13}{2}$ True

The solution set is $\{4\}$.

6. Undefined values: $x = 3$ and $x = -3$

LCD $= (x+3)(x-3)$

$$\frac{4}{x-3} - \frac{3}{x+3} = 1$$

$$(x+3)(x-3)\left(\frac{4}{x-3} - \frac{3}{x+3}\right) = (x+3)(x-3)(1)$$

$$4(x+3) - 3(x-3) = x^2 - 9$$

$$4x + 12 - 3x + 9 = x^2 - 9$$

$$x + 21 = x^2 - 9$$

$$0 = x^2 - x - 30$$

$$0 = (x+5)(x-6)$$

$x + 5 = 0$ or $x - 6 = 0$

$x = -5$ or $x = 6$

Check: $\dfrac{4}{-5-3} - \dfrac{3}{-5+3} \overset{?}{=} 1$

$\dfrac{4}{-8} - \dfrac{3}{-2} \overset{?}{=} 1$

$-\dfrac{1}{2} + \dfrac{3}{2} \overset{?}{=} 1$

$\dfrac{2}{2} \overset{?}{=} 1$

$1 = 1$ True

$$\frac{4}{6-3}-\frac{3}{6+3}\stackrel{?}{=}1$$

$$\frac{4}{3}-\frac{3}{9}\stackrel{?}{=}1$$

$$\frac{12}{9}-\frac{3}{9}\stackrel{?}{=}1$$

$$\frac{9}{9}\stackrel{?}{=}1$$

$$1=1 \quad \text{True}$$

The solution set is $\{-5, 6\}$.

7. Solutions obtained through the solving process that do not satisfy the original equation are called <u>extraneous solutions</u>.

8. $z-4=0$ when $z=4$. $z-2=0$ when $z=2$.

$z^2-6z+8=(z-4)(z-2)=0$ when $z=4$ or $z=2$.

The equation is undefined for $z=4$ and $z=2$.

LCD $=(z-4)(z-2)$

$$\frac{5}{z-4}+\frac{3}{z-2}=\frac{z^2-z-2}{z^2-6z+8}$$

$$(z-4)(z-2)\left(\frac{5}{z-4}+\frac{3}{z-2}\right)=(z-4)(z-2)\left(\frac{z^2-z-2}{(z-4)(z-2)}\right)$$

$$5(z-2)+3(z-4)=z^2-z-2$$

$$5z-10+3z-12=z^2-z-2$$

$$8z-22=z^2-z-2$$

$$0=z^2-9z+20$$

$$0=(z-5)(z-4)$$

$z-5=0 \quad \text{or} \quad z-4=0$

$z=5 \quad \text{or} \qquad z=4$

$z=4$ is extraneous, so the solution set is $\{5\}$.

9. True

10. $y+1=0$ when $y=-1$. $y-1=0$ when $y=1$. $y^2-1=(y+1)(y-1)=0$ when $y=1$ or $y=-1$.

The equation is undefined for $y=1$ and $y=-1$.

LCD $=(y+1)(y-1)$

$$\frac{4}{y+1}=\frac{7}{y-1}-\frac{8}{y^2-1}$$

$$(y+1)(y-1)\left(\frac{4}{y+1}\right)=(y+1)(y-1)\left(\frac{7}{y-1}-\frac{8}{(y+1)(y-1)}\right)$$

$$4(y-1)=7(y+1)-8$$

$$4y-4=7y+7-8$$

$$4y-4=7y-1$$

$$-3=3y$$

$$-1=y$$

Since y cannot equal -1, we reject $y=-1$ as a solution. Therefore, the equation has no solution. The solution set is $\{\ \}$ or $\varnothing$.

11.
$$C = \frac{50t}{t^2 + 25}$$
$$4 = \frac{50t}{t^2 + 25}$$
$$(t^2 + 25)(4) = (t^2 + 25)\left(\frac{50t}{t^2 + 25}\right)$$
$$4t^2 + 100 = 50t$$
$$4t^2 - 50t + 100 = 0$$
$$2t^2 - 25t + 50 = 0$$
$$(2t - 5)(t - 10) = 0$$
$$2t - 5 = 0 \quad \text{or} \quad t - 10 = 0$$
$$2t = 5 \quad \text{or} \quad t = 10$$
$$t = \frac{5}{2} \quad \text{or} \quad t = 10$$

The concentration of the drug will be

4 milligrams per liter at $\frac{5}{2}$ hours and 10 hours.

12. $x \neq 0$
LCD $= x \cdot 1 = x$
$$R = \frac{4g}{x}$$
$$x(R) = x\left(\frac{4g}{x}\right)$$
$$xR = 4g$$
$$x = \frac{4g}{R}$$

13. $r \neq 1$
LCD $= 1 \cdot (1 - r) = 1 - r$
$$S = \frac{a}{1 - r}$$
$$(1 - r)(S) = (1 - r)\left(\frac{a}{1 - r}\right)$$
$$S(1 - r) = a$$
$$1 - r = \frac{a}{S}$$
$$-r = -1 + \frac{a}{S}$$
$$r = 1 - \frac{a}{S}$$

14. $f \neq 0, p \neq 0, q \neq 0$
LCD $= fpq$
$$\frac{1}{f} = \frac{1}{p} + \frac{1}{q}$$
$$fpq\left(\frac{1}{f}\right) = fpq\left(\frac{1}{p} + \frac{1}{q}\right)$$
$$pq = fq + fp$$
$$pq - fp = fq$$
$$p(q - f) = fq$$
$$p = \frac{fq}{q - f}$$

Exercises 7.7

15. Undefined value: $y = 0$.
$$\frac{5}{3y} - \frac{1}{2} = \frac{5}{6y} - \frac{1}{12}$$
$$12y\left(\frac{5}{3y} - \frac{1}{2}\right) = 12y\left(\frac{5}{6y} - \frac{1}{12}\right)$$
$$4(5) - 6y = 2(5) - y$$
$$20 - 6y = 10 - y$$
$$20 = 10 + 5y$$
$$10 = 5y$$
$$2 = y$$

The solution set is $\{2\}$.

17. Undefined value: $x = 0$
$$\frac{6}{x} + \frac{2}{3} = \frac{4}{2x} - \frac{14}{3}$$
$$6x\left(\frac{6}{x} + \frac{2}{3}\right) = 6x\left(\frac{4}{2x} - \frac{14}{3}\right)$$
$$6(6) + 2x(2) = 3(4) - 2x(14)$$
$$36 + 4x = 12 - 28x$$
$$36 + 32x = 12$$
$$32x = -24$$
$$x = -\frac{24}{32} = -\frac{3}{4}$$

The solution set is $\left\{-\frac{3}{4}\right\}$.

19. Undefined values: $x = 1$ and $x = -1$.
$$\frac{4}{x - 1} = \frac{3}{x + 1}$$
$$(x - 1)(x + 1)\left(\frac{4}{x - 1}\right) = (x - 1)(x + 1)\left(\frac{3}{x + 1}\right)$$
$$4(x + 1) = 3(x - 1)$$
$$4x + 4 = 3x - 3$$
$$x + 4 = -3$$
$$x = -7$$

The solution set is $\{-7\}$.

21. Undefined value: $x = -2$.

$$\frac{2}{x+2} + 2 = \frac{7}{x+2}$$

$$(x+2)\left(\frac{2}{x+2} + 2\right) = (x+2)\left(\frac{7}{x+2}\right)$$

$$2 + 2(x+2) = 7$$

$$2 + 2x + 4 = 7$$

$$2x + 6 = 7$$

$$2x = 1$$

$$x = \frac{1}{2}$$

The solution set is $\left\{\frac{1}{2}\right\}$.

23. Undefined value: $r = 0$.

$$\frac{r-4}{3r} + \frac{2}{5r} = \frac{1}{5}$$

$$15r\left(\frac{r-4}{3r} + \frac{2}{5r}\right) = 15r\left(\frac{1}{5}\right)$$

$$5(r-4) + 3(2) = 3r$$

$$5r - 20 + 6 = 3r$$

$$5r - 14 = 3r$$

$$-14 = -2r$$

$$7 = r$$

The solution set is $\{7\}$.

25. Undefined values: $a = 1$ and $a = -1$.

$$\frac{2}{a-1} + \frac{3}{a+1} = \frac{-6}{a^2 - 1}$$

$$(a-1)(a+1)\left(\frac{2}{a-1} + \frac{3}{a+1}\right) = (a-1)(a+1)\left(\frac{-6}{(a-1)(a+1)}\right)$$

$$2(a+1) + 3(a-1) = -6$$

$$2a + 2 + 3a - 3 = -6$$

$$5a - 1 = -6$$

$$5a = -5$$

$$a = -1$$

$a = -1$ is an extraneous solution. The solution set is $\{\ \}$ or $\varnothing$.

27. Undefined values: $x = 4$ and $x = -4$.

$$\frac{1}{4-x} + \frac{2}{x^2 - 16} = \frac{1}{x-4}$$

$$(x-4)(x+4)\left(\frac{1}{-(x-4)} + \frac{2}{(x-4)(x+4)}\right) = (x-4)(x+4)\left(\frac{1}{x-4}\right)$$

$$-(x+4)+2 = x+4$$
$$-x-4+2 = x+4$$
$$-x-2 = x+4$$
$$-2 = 2x+4$$
$$-6 = 2x$$
$$-3 = x$$

The solution set is $\{-3\}$.

29. $2t - 2 = 2(t - 1)$; $3t - 3 = 3(t - 1)$
Undefined value: $t = 1$.

$$\frac{3}{2t-2} - \frac{2t}{3t-3} = -4$$

$$2 \cdot 3(t-1)\left(\frac{3}{2(t-1)} - \frac{2t}{3(t-1)}\right) = 2 \cdot 3(t-1)(-4)$$

$$3(3) - 2(2t) = -24(t-1)$$
$$9 - 4t = -24t + 24$$
$$9 + 20t = 24$$
$$20t = 15$$
$$t = \frac{3}{4}$$

The solution set is $\left\{\dfrac{3}{4}\right\}$.

31. $10a - 20 = 10(a - 2)$
Undefined value: $a = 2$.

$$\frac{2}{5} + \frac{3-2a}{10a-20} = \frac{2a+1}{a-2}$$

$$10(a-2)\left(\frac{2}{5} + \frac{3-2a}{10(a-2)}\right) = 10(a-2)\left(\frac{2a+1}{a-2}\right)$$

$$2 \cdot 2(a-2) + 3 - 2a = 10(2a+1)$$
$$4a - 8 + 3 - 2a = 20a + 10$$
$$2a - 5 = 20a + 10$$
$$-5 = 18a + 10$$
$$-15 = 18a$$
$$-\frac{15}{18} = a$$
$$-\frac{5}{6} = a$$

The solution set is $\left\{-\dfrac{5}{6}\right\}$.

33. $j^2 - 1 = (j+1)(j-1); \quad j^2 - 5j + 4 = (j-1)(j-4)$

Undefined values: $j = -1, j = 1,$ and $j = 4.$

$$\frac{6}{j^2 - 1} - \frac{4j}{j^2 - 5j + 4} = -\frac{4}{j-1}$$

$$(j+1)(j-1)(j-4)\left(\frac{6}{(j+1)(j-1)} - \frac{4j}{(j-1)(j-4)}\right) = (j+1)(j-1)(j-4)\left(-\frac{4}{j-1}\right)$$

$$6(j-4) - 4j(j+1) = (-4j-4)(j-4)$$

$$6j - 24 - 4j^2 - 4j = -4j^2 + 16j - 4j + 16$$

$$-4j^2 + 2j - 24 = -4j^2 + 12j + 16$$

$$-10j = 40$$

$$j = -4$$

The solution set is $\{-4\}$.

35.

$$\frac{x}{x-2} = \frac{3}{x+8}$$

$$(x-2)(x+8)\left(\frac{x}{x-2}\right) = (x-2)(x+8)\left(\frac{3}{x+8}\right)$$

$$x(x+8) = 3(x-2)$$

$$x^2 + 8x = 3x - 6$$

$$x^2 + 5x + 6 = 0$$

$$(x+2)(x+3) = 0$$

$$x + 2 = 0 \quad \text{or} \quad x + 3 = 0$$

$$x = -2 \quad \text{or} \quad x = -3$$

Neither value is extraneous. The solution set is $\{-3, -2\}$.

37.

$$\frac{x}{x+3} = \frac{6}{x-3} + 1$$

$$(x+3)(x-3)\left(\frac{x}{x+3}\right) = (x+3)(x-3)\left(\frac{6}{x-3} + 1\right)$$

$$x(x-3) = 6(x+3) + (x+3)(x-3)$$

$$x^2 - 3x = 6x + 18 + x^2 - 9$$

$$x^2 - 3x = x^2 + 6x + 9$$

$$-3x = 6x + 9$$

$$0 = 9x + 9$$

$$-9 = 9x$$

$$-1 = x$$

The value is not extraneous. The solution set is $\{-1\}$.

39.
$$x = \frac{2-x}{6x}$$
$$6x(x) = 6x\left(\frac{2-x}{6x}\right)$$
$$6x^2 = 2-x$$
$$6x^2 + x - 2 = 0$$
$$(3x+2)(2x-1) = 0$$
$$3x+2 = 0 \quad \text{or} \quad 2x-1 = 0$$
$$x = -\frac{2}{3} \quad \text{or} \quad x = \frac{1}{2}$$

Neither value is extraneous. The solution set is $\left\{-\frac{2}{3}, \frac{1}{2}\right\}$.

41.
$$\frac{2x+3}{x-1} = \frac{x-2}{x+1} + \frac{6x}{x^2-1}$$
$$(x-1)(x+1)\left(\frac{2x+3}{x-1}\right) = (x-1)(x+1)\left(\frac{x-2}{x+1} + \frac{6x}{(x-1)(x+1)}\right)$$
$$(x+1)(2x+3) = (x-1)(x-2) + 6x$$
$$2x^2 + 5x + 3 = x^2 - 3x + 2 + 6x$$
$$2x^2 + 5x + 3 = x^2 + 3x + 2$$
$$x^2 + 2x + 1 = 0$$
$$(x+1)^2 = 0$$
$$x+1 = 0$$
$$x = -1$$

Since $x = -1$ causes a denominator to be zero, it is an extraneous solution. The solution set is { } or $\varnothing$.

43.
$$\frac{2x}{x+3} - \frac{2x^2+2}{x^2-9} = \frac{-6}{x-3} + 1$$
$$(x+3)(x-3)\left(\frac{2x}{x+3} - \frac{2x^2+2}{x^2-9}\right) = (x+3)(x-3)\left(\frac{-6}{x-3} + 1\right)$$
$$2x(x-3) - (2x^2+2) = -6(x+3) + (x+3)(x-3)$$
$$2x^2 - 6x - 2x^2 - 2 = -6x - 18 + x^2 - 9$$
$$-6x - 2 = x^2 - 6x - 27$$
$$0 = x^2 - 25$$
$$0 = (x+5)(x-5)$$
$$x+5 = 0 \quad \text{or} \quad x-5 = 0$$
$$x = -5 \quad \text{or} \quad x = 5$$

Neither value is extraneous. The solution set is $\{-5, 5\}$.

45.

$$\frac{1}{n-3} = \frac{3n-1}{9-n^2}$$

$$-(n-3)(n+3)\left(\frac{1}{n-3}\right) = -(n-3)(n+3)\left(\frac{3n-1}{-(n-3)(n+3)}\right)$$

$$-1(n+3) = 3n-1$$

$$-n-3 = 3n-1$$

$$-2 = 4n$$

$$-\frac{1}{2} = n$$

The value is not extraneous. The solution set is $\left\{-\frac{1}{2}\right\}$.

47.

$$x = \frac{2}{y}$$

$$y(x) = y\left(\frac{2}{y}\right)$$

$$xy = 2$$

$$\frac{xy}{x} = \frac{2}{x}$$

$$y = \frac{2}{x}$$

49.

$$I = \frac{E}{R}$$

$$R(I) = R\left(\frac{E}{R}\right)$$

$$IR = E$$

$$\frac{IR}{I} = \frac{E}{I}$$

$$R = \frac{E}{I}$$

51.

$$h = \frac{2A}{B+b}$$

$$(B+b)(h) = (B+b)\left(\frac{2A}{B+b}\right)$$

$$Bh + bh = 2A$$

$$bh = 2A - Bh$$

$$\frac{bh}{h} = \frac{2A - Bh}{h}$$

$$b = \frac{2A - Bh}{h} \text{ or } b = \frac{2A}{h} - B$$

53.

$$\frac{x}{3+y} = z$$

$$(3+y)\left(\frac{x}{3+y}\right) = (3+y)z$$

$$x = 3z + yz$$

$$x - 3z = yz$$

$$\frac{x-3z}{z} = \frac{yz}{z}$$

$$y = \frac{x-3z}{z} \text{ or } y = \frac{x}{z} - 3$$

55.

$$\frac{1}{R} = \frac{1}{S} + \frac{1}{T}$$

$$RST\left(\frac{1}{R}\right) = RST\left(\frac{1}{S} + \frac{1}{T}\right)$$

$$ST = RT + RS$$

$$ST - RS = RT$$

$$S(T-R) = RT$$

$$\frac{S(T-R)}{T-R} = \frac{RT}{T-R}$$

$$S = \frac{RT}{T-R}$$

57.

$$m = \frac{n}{y} - \frac{p}{ay}$$

$$ay(m) = ay\left(\frac{n}{y} - \frac{p}{ay}\right)$$

$$amy = an - p$$

$$\frac{amy}{am} = \frac{an-p}{am}$$

$$y = \frac{an-p}{am}$$

59.

$$A = \frac{xy}{x+y}$$

$$(x+y)A = (x+y)\left(\frac{xy}{x+y}\right)$$

$$Ax + Ay = xy$$

$$Ay = xy - Ax$$

$$Ay = x(y-A)$$

$$\frac{Ay}{y-A} = \frac{x(y-A)}{y-A}$$

$$x = \frac{Ay}{y-A}$$

61.

$$\frac{2}{x} - \frac{1}{y} = \frac{6}{z}$$

$$xyz\left(\frac{2}{x} - \frac{1}{y}\right) = xyz\left(\frac{6}{z}\right)$$

$$2yz - xz = 6xy$$

$$2yz - 6xy = xz$$

$$y(2z - 6x) = xz$$

$$\frac{y(2z-6x)}{2z-6x} = \frac{xz}{2z-6x}$$

$$y = \frac{xz}{2z-6x}$$

63.

$$y = \frac{x}{x-c}$$

$$(x-c)y = (x-c)\left(\frac{x}{x-c}\right)$$

$$xy - cy = x$$

$$-cy = x - xy$$

$$-cy = x(1-y)$$

$$\frac{-cy}{1-y} = \frac{x(1-y)}{1-y}$$

$$\frac{-cy}{1-y} = x$$

$$x = \frac{cy}{y-1}$$

65.

$$\frac{1}{x} + \frac{3}{x+5} = \frac{1}{x} \cdot \frac{x+5}{x+5} + \frac{3}{x+5} \cdot \frac{x}{x}$$

$$= \frac{x+5}{x(x+5)} + \frac{3x}{x(x+5)}$$

$$= \frac{x+5+3x}{x(x+5)}$$

$$= \frac{4x+5}{x(x+5)}$$

67.

$$x - \frac{6}{x} = 1$$

$$x\left(x - \frac{6}{x}\right) = x(1)$$

$$x^2 - 6 = x$$

$$x^2 - x - 6 = 0$$

$$(x+2)(x-3) = 0$$

$$x+2 = 0 \quad \text{or} \quad x-3 = 0$$

$$x = -2 \quad \text{or} \quad x = 3$$

Neither value is extraneous. The solution set is $\{-2, 3\}$.

69.
$$\frac{3}{x-1} \cdot \frac{x^2-1}{6} + 3 = \frac{3}{x-1} \cdot \frac{(x+1)(x-1)}{2 \cdot 3} + 3$$
$$= \frac{x+1}{2} + 3$$
$$= \frac{x+1}{2} + \frac{6}{2}$$
$$= \frac{x+1+6}{2}$$
$$= \frac{x+7}{2}$$

71.
$$2b - \frac{5}{3} = \frac{1}{3b}$$
$$3b\left(2b - \frac{5}{3}\right) = 3b\left(\frac{1}{3b}\right)$$
$$6b^2 - 5b = 1$$
$$6b^2 - 5b - 1 = 0$$
$$(6b+1)(b-1) = 0$$
$$6b + 1 = 0 \quad \text{or} \quad b - 1 = 0$$
$$b = -\frac{1}{6} \quad \text{or} \quad b = 1$$

Neither value is extraneous. The solution set is $\left\{-\frac{1}{6}, 1\right\}$.

73.
$$\frac{5x}{2x-3} = \frac{3x}{x-1} - \frac{5}{2x^2 - 5x + 3}$$
$$\frac{5x}{2x-3} = \frac{3x}{x-1} - \frac{5}{(2x-3)(x-1)}$$
$$(2x-3)(x-1)\left(\frac{5x}{2x-3}\right) = (2x-3)(x-1)\left(\frac{3x}{x-1} - \frac{5}{(2x-3)(x-1)}\right)$$
$$5x(x-1) = 3x(2x-3) - 5$$
$$5x^2 - 5x = 6x^2 - 9x - 5$$
$$0 = x^2 - 4x - 5$$
$$0 = (x-5)(x+1)$$
$$x - 5 = 0 \quad \text{or} \quad x + 1 = 0$$
$$x = 5 \qquad\qquad x = -1$$

Neither value is extraneous. The solution set is $\{-1, 5\}$.

75.
$$\frac{x^2}{x^2-4} - \frac{1}{x} = \frac{x^2}{x^2-4} \cdot \frac{x}{x} - \frac{1}{x} \cdot \frac{x^2-4}{x^2-4}$$
$$= \frac{x^3}{x(x^2-4)} - \frac{x^2-4}{x(x^2-4)}$$
$$= \frac{x^3 - (x^2-4)}{x(x+2)(x-2)}$$
$$= \frac{x^3 - x^2 + 4}{x(x+2)(x-2)}$$

77. $\dfrac{x^2-9}{x+1} \div \dfrac{3x^2+9x}{x^2-1} = \dfrac{x^2-9}{x+1} \cdot \dfrac{x^2-1}{3x^2+9x}$

$\qquad\qquad = \dfrac{(x+3)(x-3)}{x+1} \cdot \dfrac{(x+1)(x-1)}{3x(x+3)}$

$\qquad\qquad = \dfrac{(x+3)(x+1)\cdot(x-3)(x-1)}{(x+3)(x+1)\cdot 3x}$

$\qquad\qquad = \dfrac{(x-3)(x-1)}{3x}$

79. $\dfrac{\frac{1}{a}+3}{\frac{3a+1}{5}} = \dfrac{\frac{1}{a}+\frac{3a}{a}}{\frac{3a+1}{5}}$

$\qquad = \dfrac{\frac{1+3a}{a}}{\frac{3a+1}{5}}$

$\qquad = \dfrac{1+3a}{a} \cdot \dfrac{5}{3a+1}$

$\qquad = \dfrac{(3a+1)\cdot 5}{(3a+1)\cdot a}$

81. $2a-2 = 2(a-1);\ 3a+3 = 3(a+1);\ 12a^2-12 = 2\cdot 2\cdot 3(a+1)(a-1)$

$$\dfrac{a}{2a-2} - \dfrac{2}{3a+3} = \dfrac{5a^2-2a+9}{12a^2-12}$$

$$12(a+1)(a-1)\left(\dfrac{a}{2(a-1)} - \dfrac{2}{3(a+1)}\right) = 12(a+1)(a-1)\left(\dfrac{5a^2-2a+9}{12(a+1)(a-1)}\right)$$

$$6a(a+1) - 8(a-1) = 5a^2-2a+9$$

$$6a^2+6a-8a+8 = 5a^2-2a+9$$

$$6a^2-2a+8 = 5a^2-2a+9$$

$$a^2-1 = 0$$

$$(a+1)(a-1) = 0$$

$a+1=0 \quad$ or $\quad a-1=0$

$\quad a=-1 \quad$ or $\qquad a=1$

Since both $a = -1$ and $a = 1$ cause denominators to be zero, they are extraneous solutions. The solution set is { } or $\varnothing$.

83. $\left(\dfrac{3}{b+c}+\dfrac{5}{b-c}\right)\div\left(\dfrac{4b+c}{b^2-c^2}\right)=\left(\dfrac{3}{b+c}\cdot\dfrac{b-c}{b-c}+\dfrac{5}{b-c}\cdot\dfrac{b+c}{b+c}\right)\div\left(\dfrac{4b+c}{(b+c)(b-c)}\right)$

$$=\left(\dfrac{3b-3c}{(b+c)(b-c)}+\dfrac{5b+5c}{(b+c)(b-c)}\right)\div\left(\dfrac{4b+c}{(b+c)(b-c)}\right)$$

$$=\left(\dfrac{3b-3c+5b+5c}{(b+c)(b-c)}\right)\div\left(\dfrac{4b+c}{(b+c)(b-c)}\right)$$

$$=\left(\dfrac{8b+2c}{(b+c)(b-c)}\right)\div\left(\dfrac{4b+c}{(b+c)(b-c)}\right)$$

$$=\dfrac{8b+2c}{(b+c)(b-c)}\cdot\dfrac{(b+c)(b-c)}{4b+c}$$

$$=\dfrac{2(4b+c)}{(b+c)(b-c)}\cdot\dfrac{(b+c)(b-c)}{4b+c}$$

$$=\dfrac{(b+c)(b-c)(4b+c)\cdot 2}{(b+c)(b-c)(4b+c)}$$

$$=2$$

85. $C=\dfrac{40t}{t^2+9}$; find t when $C=4$.

$$4=\dfrac{40t}{t^2+9}$$

$$(t^2+9)(4)=(t^2+9)\left(\dfrac{40t}{t^2+9}\right)$$

$$4t^2+36=40t$$

$$4t^2-40t+36=0$$

$$t^2-10t+9=0$$

$$(t-9)(t-1)=0$$

$$t-9=0 \quad\text{or}\quad t-1=0$$

$$t=9 \quad\text{or}\quad\quad t=1$$

The concentration is 4 milligrams per liter after 1 hour and 9 hours.

87. $C=\dfrac{25x}{100-x}$; find x when $C=100$.

$$100=\dfrac{25x}{100-x}$$

$$(100-x)\cdot 100=(100-x)\cdot\dfrac{25x}{100-x}$$

$$10,000-100x=25x$$

$$10,000=125x$$

$$80=x$$

If the government budgets $100 million, then 80% of the pollutants can be removed.

89. If the solution set is {2}, the value of k can be found by letting $x = 2$ and solving for k.

$$\frac{4x+3}{k} = \frac{x-1}{3}$$

$$\frac{4(2)+3}{k} = \frac{2-1}{3}$$

$$\frac{11}{k} = \frac{1}{3}$$

$$3k\left(\frac{11}{k}\right) = 3k\left(\frac{1}{3}\right)$$

$$33 = k$$

91. The error occurred when the student incorrectly multiplied $-2[(x-1)(x+1)]$ on the right side of the equation. The correct solution is

$$\frac{2}{x-1} - \frac{4}{x^2-1} = -2$$

$$\frac{2}{x-1} - \frac{4}{(x-1)(x+1)} = -2$$

$$[(x-1)(x+1)]\left(\frac{2}{x-1} - \frac{4}{(x-1)(x+1)}\right) = -2[(x-1)(x+1)]$$

$$2(x+1) - 4 = -2(x-1)(x+1)$$

$$2x+2-4 = -2(x^2-1)$$

$$2x-2 = -2x^2+2$$

$$2x^2+2x-4 = 0$$

$$x^2+x-2 = 0$$

$$(x+2)(x-1) = 0$$

$$x+2 = 0 \quad \text{or} \quad x-1 = 0$$
$$x = -2 \quad\quad\quad x = 1$$

The value of $x = 1$ is extraneous. The solution set is $\{-2\}$.

93. To *simplify* a rational expression means to add, subtract, multiply, or divide the rational expressions and express the result in a form in which there are no common factors in the numerator or denominator. To *solve* an equation means to find the value(s) of the variable that satisfy the equation.

Section 7.8

Preparing for Models Involving Rational Equations

P1.
$$\frac{150}{r} = \frac{250}{r+20}$$

$$r(r+20)\left(\frac{150}{r}\right) = r(r+20)\left(\frac{250}{r+20}\right)$$

$$150(r+20) = 250r$$

$$150r+3000 = 250r$$

$$3000 = 100r$$

$$r = 30$$

The solution set is {30}.

7.8 Quick Checks

1. A <u>proportion</u> is an equation of the form $\dfrac{a}{b} = \dfrac{c}{d}$, where $b \neq 0$ and $d \neq 0$.

2. $\dfrac{2p+1}{4} = \dfrac{p}{8}$

$8(2p+1) = 4p$

$16p + 8 = 4p$

$12p = -8$

$p = -\dfrac{8}{12}$

$p = -\dfrac{2}{3}$

The solution set is $\left\{-\dfrac{2}{3}\right\}$.

3. $\dfrac{6}{x^2} = \dfrac{2}{x}$

$6x = 2x^2$

$0 = 2x^2 - 6x$

$0 = 2x(x-3)$

$2x = 0 \quad \text{or} \quad x - 3 = 0$

$x = 0 \quad \text{or} \qquad x = 3$

$x = 0$ is extraneous since it makes the denominators zero. The solution set is $\{3\}$.

4. Let x be the monthly payment.

$\dfrac{1000}{16.67} = \dfrac{14,000}{x}$

$1000x = 16.67(14,000)$

$\dfrac{1000x}{1000} = \dfrac{16.67(14,000)}{1000}$

$x = 16.67(14)$

$x = 233.38$

Clem's monthly payment is $233.38.

5. Let x be the number of miles between the two cities.

$\dfrac{\frac{1}{4}}{15} = \dfrac{\frac{7}{2}}{x}$

$\dfrac{1}{4}x = \dfrac{7}{2}(15)$

$x = 4\left(\dfrac{7}{2}\right)(15)$

$x = 14(15)$

$x = 210$

The cities are 210 miles apart.

6. In geometry, two figures are <u>similar</u> if their corresponding angle measures are equal and their corresponding sides are proportional.

7. $\dfrac{XY}{XZ} = \dfrac{MN}{MP}$

$\dfrac{a}{10} = \dfrac{12}{15}$

$15a = 12(10)$

$15a = 120$

$\dfrac{15a}{15} = \dfrac{120}{15}$

$a = 8$

The length of side *XY* is 8 units.

8. Let h be the height of the tree.

$\dfrac{6}{2.5} = \dfrac{h}{25}$

$2.5h = 25(6)$

$2.5h = 150$

$\dfrac{2.5h}{2.5} = \dfrac{150}{2.5}$

$h = 60$

The tree is 60 feet tall.

9. Let t be the number of hours working together. Molly shovels $\dfrac{1}{3}$ of the driveway in 1 hour. The neighbor shovels $\dfrac{1}{6}$ of the driveway in 1 hour. Together they shovel $\dfrac{1}{t}$ of the driveway in 1 hour.

$\dfrac{1}{3} + \dfrac{1}{6} = \dfrac{1}{t}$

$6t\left(\dfrac{1}{3} + \dfrac{1}{6}\right) = 6t\left(\dfrac{1}{t}\right)$

$2t + t = 6$

$3t = 6$

$t = 2$

Molly and her neighbor would take 2 hours to shovel the driveway together.

10. Let t be the number of hours it would take Michael to seal the driveway alone. Leon seals $\dfrac{1}{5}$ of the driveway in 1 hour. Michael seals $\dfrac{1}{t}$ of the driveway in 1 hour. Together they seal $\dfrac{1}{2}$ of the driveway in 1 hour.

$$\frac{1}{5} + \frac{1}{t} = \frac{1}{2}$$

$$10t\left(\frac{1}{5} + \frac{1}{t}\right) = 10t\left(\frac{1}{2}\right)$$

$$2t + 10 = 5t$$

$$10 = 3t$$

$$t = \frac{10}{3}$$

It would take Michael $3\frac{1}{3}$ hours or 3 hours, 20 minutes to seal the driveway alone.

11. Let s be the speed of the wind. The speed into the wind is $120 - s$. The speed with the wind is $120 + s$. Since $d = rt$, then $t = \frac{d}{r}$. The times are the same for each trip. The plane flies 700 miles at $120 + s$ miles per hour and 500 miles at $120 - s$ miles per hour.

$$\frac{500}{120-s} = \frac{700}{120+s}$$

$$500(120+s) = 700(120-s)$$

$$5(120+s) = 7(120-s)$$

$$600 + 5s = 840 - 7s$$

$$12s = 240$$

$$s = 20$$

The speed of the wind is 20 mph.

12. Let Sue's running speed be s. Sue's walking speed is $\frac{s}{12}$. Since $d = rt$, then $t = \frac{d}{r}$.

time running + time walking = total time

$$\frac{18}{s} + \frac{1}{\frac{s}{12}} = 5$$

$$\frac{18}{s} + \frac{12}{s} = 5$$

$$\frac{18+12}{s} = 5$$

$$\frac{30}{s} = 5$$

$$30 = 5s$$

$$s = 6$$

Sue's running speed was 6 mph.

7.8 Exercises

13. $$\frac{9}{x} = \frac{3}{4}$$
$$9(4) = 3x$$
$$36 = 3x$$
$$12 = x$$
The solution set is $\{12\}$.

15. $$\frac{4}{7} = \frac{2x}{9}$$
$$4(9) = 7(2x)$$
$$36 = 14x$$
$$\frac{36}{14} = x$$
$$\frac{18}{7} = x$$
The solution set is $\left\{\frac{18}{7}\right\}$.

17. $$\frac{6}{5} = \frac{x+2}{15}$$
$$15\left(\frac{6}{5}\right) = 15\left(\frac{x+2}{15}\right)$$
$$3(6) = x + 2$$
$$18 = x + 2$$
$$16 = x$$
The solution set is $\{16\}$.

19. $$\frac{b}{b+6} = \frac{4}{9}$$
$$9(b) = 4(b+6)$$
$$9b = 4b + 24$$
$$5b = 24$$
$$b = \frac{24}{5}$$
The solution set is $\left\{\frac{24}{5}\right\}$.

21. $\dfrac{y}{y-10} = \dfrac{2}{27}$

$27(y) = 2(y-10)$

$27y = 2y - 20$

$25y = -20$

$y = \dfrac{-20}{25}$

$y = -\dfrac{4}{5}$

The solution set is $\left\{-\dfrac{4}{5}\right\}$.

23. $\dfrac{p+2}{4} = \dfrac{2p+4}{5}$

$5(p+2) = 4(2p+4)$

$5p + 10 = 8p + 16$

$10 = 3p + 16$

$-6 = 3p$

$-2 = p$

The solution set is $\{-2\}$.

25. $\dfrac{2z-1}{z} = \dfrac{3}{5}$

$5(2z-1) = 3(z)$

$10z - 5 = 3z$

$10z = 3z + 5$

$7z = 5$

$z = \dfrac{5}{7}$

The solution set is $\left\{\dfrac{5}{7}\right\}$.

27. $\dfrac{2}{v^2 - v} = \dfrac{1}{3-v}$

$2(3-v) = 1(v^2 - v)$

$6 - 2v = v^2 - v$

$0 = v^2 + v - 6$

$0 = (v+3)(v-2)$

$v + 3 = 0$ or $v - 2 = 0$

$v = -3$ or $v = 2$

The solution set is $\{-3, 2\}$.

29. $\dfrac{10-x}{4x} = \dfrac{1}{x-1}$

$(10-x)(x-1) = 4x(1)$

$-x^2 + 11x - 10 = 4x$

$0 = x^2 - 7x + 10$

$0 = (x-2)(x-5)$

$x - 2 = 0$ or $x - 5 = 0$

$x = 2$ or $x = 5$

The solution set is $\{2, 5\}$.

31. $\dfrac{2p-3}{p^2 + 12p + 6} = \dfrac{1}{p+4}$

$(2p-3)(p+4) = (p^2 + 12p + 6)(1)$

$2p^2 + 5p - 12 = p^2 + 12p + 6$

$p^2 - 7p - 18 = 0$

$(p+2)(p-9) = 0$

$p + 2 = 0$ or $p - 9 = 0$

$p = -2$ or $p = 9$

The solution set is $\{-2, 9\}$.

33. $\dfrac{AB}{AC} = \dfrac{XY}{XZ}$

$\dfrac{6}{8} = \dfrac{9}{XZ}$

$6(XZ) = 9(8)$

$6(XZ) = 72$

$XZ = 12$

35. $\dfrac{XY}{ZY} = \dfrac{AB}{BC}$

$\dfrac{n}{2n-1} = \dfrac{3}{5}$

$5(n) = 3(2n-1)$

$5n = 6n - 3$

$-n = -3$

$n = 3$

$ZY = 2n - 1 = 2(3) - 1 = 5$

37. $\dfrac{AB}{BC} = \dfrac{EF}{FG}$

$\dfrac{x-2}{2x+3} = \dfrac{4}{9}$

$9(x-2) = 4(2x+3)$

$9x - 18 = 8x + 12$

$x - 18 = 12$

$x = 30$

$AB = x - 2 = 30 - 2 = 28$

39. hours Natalie worked $= 2x$

41. days David painted $= t - 3$

43. Let x be the number of hours working together. Christina paints $\frac{1}{5}$ chair in one hour, Victoria paints $\frac{1}{3}$ chair in one hour, and together they paint $\frac{1}{t}$ chair in one hour.

$$\frac{1}{5} + \frac{1}{3} = \frac{1}{t}$$

45. Let b be the number of hours for Pipe B alone to fill the tank. Pipe A takes $(b + 4)$ hours. Pipe A fills $\frac{1}{b+4}$ tank in one hour, Pipe B fills $\frac{1}{b}$ tank in one hour, and together they fill $\frac{1}{7}$ tank in one hour.

$$\frac{1}{b+4} + \frac{1}{b} = \frac{1}{7}$$

47. Joe's rate upstream $= r - 2$

49. rate of the plan against the wind $= r - 48$

51. Let c be the speed of the current. Bob's speed upstream is $14 - c$, and his speed downstream is $14 + c$. Since $d = rt$, then $t = \frac{d}{r}$. The times are the same for 4 miles upstream and 7 miles downstream.

$$\frac{4}{14-c} = \frac{7}{14+c}$$

53. Let r be the speed of the boat in still water. Assen's speed downstream is $r + 3$, and his speed upstream is $r - 3$. Since $d = rt$, then $t = \frac{d}{r}$. The times are the same for 12 miles downstream and 8 miles upstream.

$$\frac{12}{r+3} = \frac{8}{r-3}$$

55. Let x be the distance between the two towns.

$$\frac{\frac{1}{4}}{10} = \frac{3\frac{5}{8}}{x}$$
$$\frac{1}{4}x = 10\left(3\frac{5}{8}\right)$$
$$\frac{x}{4} = 10\left(\frac{29}{8}\right)$$
$$\frac{x}{4} = \frac{145}{4}$$
$$x = 145$$

The distance between the towns is 145 miles.

57. Let f be the flour for 5 loaves.

$$\frac{5}{3} = \frac{f}{5}$$
$$5(5) = 3f$$
$$25 = 3f$$
$$\frac{25}{3} = f$$

It takes $\frac{25}{3} = 8\frac{1}{3}$ lb of flour for 5 loaves.

59. Let r be the number of rubles.

$$\frac{5}{143.25} = \frac{1300}{r}$$
$$5r = 186,225$$
$$r = 37,245$$

She can purchase 37,245 rubles.

61. Let t be the height of the tree.

$$\frac{6}{2.5} = \frac{t}{10}$$
$$60 = 2.5t$$
$$24 = t$$

The tree is 24 feet tall.

63. Let t be the number of hours working together. Josh cleans $\frac{1}{3}$ building in one hour, Ken cleans $\frac{1}{5}$ building one hour, and they clean $\frac{1}{t}$ building in one hour working together.

$$\frac{1}{3}+\frac{1}{5}=\frac{1}{t}$$

$$15t\left(\frac{1}{3}+\frac{1}{5}\right)=15t\left(\frac{1}{t}\right)$$

$$5t+3t=15$$

$$8t=15$$

$$t=\frac{15}{8}$$

It takes them $\frac{15}{8}=1.875$ hours working together.

65. Let t be the time working together. Dyanne collects $\frac{1}{8}$ of the balls in one minute, Makini collects $\frac{1}{6}$ of the balls in one minute, and they collect $\frac{1}{t}$ of the balls in one minute working together.

$$\frac{1}{8}+\frac{1}{6}=\frac{1}{t}$$

$$24t\left(\frac{1}{8}+\frac{1}{6}\right)=24t\left(\frac{1}{t}\right)$$

$$3t+4t=24$$

$$7t=24$$

$$t=\frac{24}{7}$$

It takes them $\frac{24}{7}\approx 3.4$ minutes together.

67. Since the part that is left is equal to what each of them has already completed, José can paint $\frac{1}{9}$ of the remaining part in one hour, and Joaquín can paint $\frac{1}{12}$ in one hour. Let t be the time it takes them working together. Then they paint $\frac{1}{t}$ of the remaining part in one hour.

$$\frac{1}{9}+\frac{1}{12}=\frac{1}{t}$$

$$36t\left(\frac{1}{9}+\frac{1}{12}\right)=36t\left(\frac{1}{t}\right)$$

$$4t+3t=36$$

$$7t=36$$

$$t=\frac{36}{7}$$

It takes them $\frac{36}{7}\approx 5.1$ hours to finish the job.

69. Let p be the hours it takes the experienced plumber. Then it takes the apprentice $2p$ hours. They complete $\frac{1}{5}$ of the job in one hour working together, the plumber completes $\frac{1}{p}$ of the job in one hour, and the apprentice completes $\frac{1}{2p}$ of the job in one hour.

$$\frac{1}{p}+\frac{1}{2p}=\frac{1}{5}$$

$$10p\left(\frac{1}{p}+\frac{1}{2p}\right)=10p\left(\frac{1}{5}\right)$$

$$10+5=2p$$

$$15=2p$$

It would take the apprentice $2p = 15$ hours to do the job alone.

71. Let t be the time to fill the pool. Then $\frac{1}{6}$ of the pool will be filled in one hour, while $\frac{1}{8}$ of the pool will be drained in one hour.

$$\frac{1}{6}-\frac{1}{8}=\frac{1}{t}$$

$$24t\left(\frac{1}{6}-\frac{1}{8}\right)=24t\left(\frac{1}{t}\right)$$

$$4t-3t=24$$

$$t=24$$

It will take 24 hours to fill the pool.

73. Let b be the speed of the boat in still water. The boat's speed upstream is $b - 2$ and its speed downstream is $b + 2$. Since $d = rt$, then $t = \frac{d}{r}$. The times are the same for 12 km downstream and 4 km upstream.

$$\frac{12}{b+2} = \frac{4}{b-2}$$
$$12(b-2) = 4(b+2)$$
$$12b - 24 = 4b + 8$$
$$8b - 24 = 8$$
$$8b = 32$$
$$b = 4$$

The speed of the boat in still water is 4 km per hr.

75. Let b be his running speed, so $8b$ is his biking speed. Since $d = rt$, then $t = \frac{d}{r}$. Tony's time running is $\frac{15}{b}$ and his time biking is $\frac{60}{8b} = \frac{15}{2b}$.

The total time is 5 hours.

$$\frac{15}{b} + \frac{15}{2b} = 5$$
$$2b\left(\frac{15}{b} + \frac{15}{2b}\right) = 2b(5)$$
$$30 + 15 = 10b$$
$$45 = 10b$$
$$4.5 = b$$

The time he spent biking is

$$\frac{15}{2b} = \frac{15}{2(4.5)} = \frac{15}{9} = 1\frac{2}{3} \text{ hours}$$

or 1 hr, 40 min.

77. Let r be the average rate.

	d	r	$t = \frac{d}{r}$
slow	60	$r-20$	$\frac{60}{r-20}$
fast	90	r	$\frac{90}{r}$

The trip took a total of 3 hours.

$$\frac{60}{r-20} + \frac{90}{r} = 3$$
$$r(r-20)\left(\frac{60}{r-20} + \frac{90}{r}\right) = 3r(r-20)$$
$$60r + 90(r-20) = 3r^2 - 60r$$
$$60r + 90r - 1800 = 3r^2 - 60r$$
$$0 = 3r^2 - 210r + 1800$$
$$0 = r^2 - 70r + 600$$
$$0 = (r-60)(r-10)$$
$$r - 60 = 0 \quad \text{or} \quad r - 10 = 0$$
$$r = 60 \quad \text{or} \quad r = 10$$

$r = 10$ is extraneous since the slow speed would be negative ($r - 20 = 10 - 20 = -10$). The average rate during the last 60 miles was $r - 20 = 60 - 20 = 40$ mph.

79. Let r be the average speed going to work. Then $r - 7$ is the average speed going home. Since $d = rt$, then $t = \frac{d}{r}$. The time for 16 miles at $(r - 7)$ mph is the same as the time for 20 miles at r mph.

$$\frac{20}{r} = \frac{16}{r-7}$$
$$20(r-7) = 16r$$
$$20r - 140 = 16r$$
$$4r = 140$$
$$r = 35$$

The average speed going to work is 35 miles per hour.

81.
$$\frac{XA}{YA} = \frac{BA}{CA}$$
$$\frac{4}{x} = \frac{12}{18}$$
$$18(4) = 12x$$
$$72 = 12x$$
$$6 = x$$

83. Note that $AB = AX + XB$.
$$\frac{XA}{XY} = \frac{AB}{BC}$$
$$\frac{XA}{XY} = \frac{AX + XB}{BC}$$
$$\frac{5}{7} = \frac{5+9}{x}$$
$$\frac{5}{7} = \frac{14}{x}$$
$$5x = 7(14)$$
$$5x = 98$$
$$x = \frac{98}{5}$$

85. Note that $AB = AX + XB$ and
$AC = AY + YC$.

$$\frac{AX}{AY} = \frac{AB}{AC}$$

$$\frac{AX}{AY} = \frac{AX + XB}{AY + YC}$$

$$\frac{6}{5} = \frac{6+x}{5+12}$$

$$\frac{6}{5} = \frac{6+x}{17}$$

$$17(6) = 5(6+x)$$

$$102 = 30 + 5x$$

$$72 = 5x$$

$$\frac{72}{5} = x$$

87. The time component is not correctly placed. The times in the table should be $\frac{10}{r+1}$ and $\frac{10}{r-1}$. The equation will incorporate the 5 hours: time traveled downstream + time traveled upstream = 5 hours.

89. One equation to solve the problem is $\frac{1}{8} + \frac{1}{6} = \frac{1}{t}$ where t is the number of hours required to complete the job together. Answers may vary.

Chapter 7 Review

1. $\dfrac{2x}{x+3}$

   (a) $x = 1$: $\dfrac{2(1)}{1+3} = \dfrac{2}{4} = \dfrac{1}{2}$

   (b) $x = -2$: $\dfrac{2(-2)}{-2+3} = \dfrac{-4}{1} = -4$

   (c) $x = 3$: $\dfrac{2(3)}{3+3} = \dfrac{6}{6} = 1$

2. $\dfrac{3x}{x-4}$

   (a) $x = 0$: $\dfrac{3(0)}{0-4} = \dfrac{0}{-4} = 0$

   (b) $x = 1$: $\dfrac{3(1)}{1-4} = \dfrac{3}{-3} = -1$

   (c) $x = -2$: $\dfrac{3(-2)}{-2-4} = \dfrac{-6}{-6} = 1$

3. $\dfrac{x^2 + 2xy + y^2}{x-y}$

   (a) $x = 1, y = -1$:
$$\frac{1^2 + 2(1)(-1) + (-1)^2}{1-(-1)} = \frac{1-2+1}{2} = \frac{0}{2} = 0$$

   (b) $x = 2, y = 1$: $\dfrac{2^2 + 2(2)(1) + 1^2}{2-1} = \dfrac{4+4+1}{1} = 9$

   (c) $x = -3, y = -4$:
$$\frac{(-3)^2 + 2(-3)(-4) + (-4)^2}{-3-(-4)} = \frac{9+24+16}{-3+4}$$
$$= \frac{49}{1}$$
$$= 49$$

4. $\dfrac{2x^2 - x - 3}{x+z}$

   (a) $x = 1, z = 1$:
$$\frac{2(1)^2 - 1 - 3}{1+1} = \frac{2-1-3}{2} = \frac{-2}{2} = -1$$

   (b) $x = 1, z = 2$:
$$\frac{2(1)^2 - 1 - 3}{1+2} = \frac{2-1-3}{3} = \frac{-2}{3} = -\frac{2}{3}$$

   (c) $x = 5, z = -4$: $\dfrac{2(5)^2 - 5 - 3}{5+(-4)} = \dfrac{50-5-3}{1} = 42$

5. $\dfrac{3x}{3x-7}$ is undefined when $3x - 7 = 0$.
$$3x - 7 = 0$$
$$3x = 7$$
$$x = \frac{7}{3}$$

6. $\dfrac{x+1}{4x-2}$ is undefined when $4x - 2 = 0$.
$$4x - 2 = 0$$
$$4x = 2$$
$$x = \frac{2}{4} = \frac{1}{2}$$

7. $\dfrac{5}{x^2+25}$ is undefined when $x^2+25=0$.

$x^2+25=0$

$x^2=-25$

x^2 is always nonnegative. There are no values

for which $\dfrac{5}{x^2+25}$ is undefined.

8. $\dfrac{17}{4x^2+49}$ is undefined when $4x^2+49=0$.

$4x^2+49=0$

$4x^2=-49$

$x^2=-\dfrac{49}{4}$

x^2 is always nonnegative. There are no values

for which $\dfrac{17}{4x^2+49}$ is undefined.

9. $\dfrac{5x+2}{x^2+12x+20}$ is undefined when

$x^2+12x+20=0$.

$x^2+12x+20=0$

$(x+10)(x+2)=0$

$x+10=0 \quad \text{or} \quad x+2=0$

$x=-10 \quad \text{or} \qquad x=-2$

10. $\dfrac{3x-1}{x^2-3x-4}$ is undefined when $x^2-3x-4=0$.

$x^2-3x-4=0$

$(x+1)(x-4)=0$

$x+1=0 \quad \text{or} \quad x-4=0$

$x=-1 \quad \text{or} \qquad x=4$

11. $\dfrac{5y^2+10y}{25y}=\dfrac{5y(y+2)}{5y\cdot5}=\dfrac{y+2}{5}$

12. $\dfrac{2x^3-8x^2}{10x}=\dfrac{2x\cdot x(x-4)}{2x\cdot5}=\dfrac{x(x-4)}{5}$

13. $\dfrac{3k-21}{k^2-5k-14}=\dfrac{3(k-7)}{(k+2)(k-7)}=\dfrac{3}{k+2}$

14. $\dfrac{-2x-2}{x^2-2x-3}=\dfrac{-2(x+1)}{(x-3)(x+1)}=-\dfrac{2}{x-3}$

15. $\dfrac{x^2+8x+15}{2x^2+5x-3}=\dfrac{(x+5)(x+3)}{(2x-1)(x+3)}=\dfrac{x+5}{2x-1}$

16. $\dfrac{3x^2+5x-2}{2x^3+16}=\dfrac{(3x-1)(x+2)}{2(x+2)(x^2-2x+4)}$

$=\dfrac{3x-1}{2(x^2-2x+4)}$

17. $\dfrac{12m^4n^3}{7}\cdot\dfrac{21}{18m^2n^5}$

$=\dfrac{2\cdot6\cdot m^2\cdot m^2\cdot n^3}{7}\cdot\dfrac{7\cdot3}{3\cdot6\cdot m^2\cdot n^3\cdot n^2}$

$=\dfrac{3\cdot6\cdot7\cdot m^2\cdot n^3\cdot2\cdot m^2}{3\cdot6\cdot7\cdot m^2\cdot n^3\cdot n^2}$

$=\dfrac{2m^2}{n^2}$

18. $\dfrac{3x^2y^4}{4}\cdot\dfrac{2}{9x^3y}=\dfrac{3\cdot x^2\cdot y\cdot y^3}{2\cdot2}\cdot\dfrac{2}{3\cdot3\cdot x^2\cdot x\cdot y}$

$=\dfrac{2\cdot3\cdot x^2\cdot y\cdot y^3}{2\cdot3\cdot x^2\cdot y\cdot2\cdot3\cdot x}$

$=\dfrac{y^3}{6x}$

19. $\dfrac{10m^2n^4}{9m^3n}\div\dfrac{15mn^6}{21m^2n}$

$=\dfrac{10m^2n^4}{9m^3n}\cdot\dfrac{21m^2n}{15mn^6}$

$=\dfrac{2\cdot5\cdot m^2\cdot n^4}{3\cdot3\cdot m^2\cdot m\cdot n}\cdot\dfrac{3\cdot7\cdot m\cdot m\cdot n}{3\cdot5\cdot m\cdot n^4\cdot n^2}$

$=\dfrac{3\cdot5\cdot m^2\cdot m\cdot m\cdot n^4\cdot n\cdot2\cdot7}{3\cdot5\cdot m^2\cdot m\cdot m\cdot n^4\cdot n\cdot3\cdot3\cdot n^2}$

$=\dfrac{14}{9n^2}$

20. $\dfrac{5ab^3}{3b^4} \div \dfrac{10a^2b^8}{6b^2}$

$= \dfrac{5ab^3}{3b^4} \cdot \dfrac{6b^2}{10a^2b^8}$

$= \dfrac{5 \cdot a \cdot b^3}{3 \cdot b^2 \cdot b^2} \cdot \dfrac{2 \cdot 3 \cdot b^2}{2 \cdot 5 \cdot a \cdot a \cdot b^3 \cdot b^5}$

$= \dfrac{3 \cdot 2 \cdot 5 \cdot a \cdot b^3 \cdot b^2 \cdot 1}{3 \cdot 2 \cdot 5 \cdot a \cdot b^3 \cdot b^2 \cdot a \cdot b^2 \cdot b^5}$

$= \dfrac{1}{ab^7}$

21. $\dfrac{5x-15}{x^2-x-12} \cdot \dfrac{x^2-6x+8}{3-x}$

$= \dfrac{5(x-3)}{(x-4)(x+3)} \cdot \dfrac{(x-4)(x-2)}{-1(x-3)}$

$= \dfrac{(x-3)(x-4) \cdot 5(x-2)}{(x-3)(x-4) \cdot (-1)(x+3)}$

$= -\dfrac{5(x-2)}{x+3}$

22. $\dfrac{4x-24}{x^2-18x+81} \cdot \dfrac{x^2-9}{6-x}$

$= \dfrac{4(x-6)}{(x-9)(x-9)} \cdot \dfrac{(x-3)(x+3)}{-1(x-6)}$

$= \dfrac{(x-6) \cdot 4(x-3)(x+3)}{(x-6) \cdot (-1)(x-9)(x-9)}$

$= -\dfrac{4(x-3)(x+3)}{(x-9)(x-9)}$

23. $\dfrac{\frac{x^2-4}{x^2-8x+15}}{\frac{12x+24}{3x-15}} = \dfrac{x^2-4}{x^2-8x+15} \cdot \dfrac{3x-15}{12x+24}$

$= \dfrac{(x-2)(x+2)}{(x-3)(x-5)} \cdot \dfrac{3(x-5)}{12(x+2)}$

$= \dfrac{3(x+2)(x-5) \cdot (x-2)}{3(x+2)(x-5) \cdot 4(x-3)}$

$= \dfrac{x-2}{4(x-3)}$

24. $\dfrac{\frac{5x^3+10x^2}{3x}}{\frac{2x+4}{18x^2}} = \dfrac{5x^3+10x^2}{3x} \cdot \dfrac{18x^2}{2x+4}$

$= \dfrac{5x^2(x+2)}{3x} \cdot \dfrac{2 \cdot 3 \cdot 3 \cdot x \cdot x}{2(x+2)}$

$= \dfrac{2 \cdot 3 \cdot x \cdot (x+2) \cdot 3 \cdot 5 \cdot x \cdot x^2}{2 \cdot 3 \cdot x \cdot (x+2) \cdot 1}$

$= 15x^3$

25. $\dfrac{3x^2+14x-5}{x^2+x-30} \cdot \dfrac{x^2-2x-15}{3x^2+8x-3}$

$= \dfrac{(3x-1)(x+5)}{(x+6)(x-5)} \cdot \dfrac{(x-5)(x+3)}{(3x-1)(x+3)}$

$= \dfrac{(3x-1)(x-5)(x+3) \cdot (x+5)}{(3x-1)(x-5)(x+3) \cdot (x+6)}$

$= \dfrac{x+5}{x+6}$

26. $\dfrac{y^2-5y-14}{y^2-2y-35} \cdot \dfrac{y^2+6y+5}{y^2-y-6}$

$= \dfrac{(y-7)(y+2)}{(y-7)(y+5)} \cdot \dfrac{(y+5)(y+1)}{(y-3)(y+2)}$

$= \dfrac{(y-7)(y+2)(y+5) \cdot (y+1)}{(y-7)(y+2)(y+5) \cdot (y-3)}$

$= \dfrac{y+1}{y-3}$

27. $\dfrac{x^2-9x}{x^2+3x+2} \div \dfrac{x^2-81}{x^2+2x}$

$= \dfrac{x^2-9x}{x^2+3x+2} \cdot \dfrac{x^2+2x}{x^2-81}$

$= \dfrac{x(x-9)}{(x+2)(x+1)} \cdot \dfrac{x(x+2)}{(x-9)(x+9)}$

$= \dfrac{x(x-9)(x+2) \cdot x}{(x-9)(x+2) \cdot (x+1)(x+9)}$

$= \dfrac{x^2}{(x+1)(x+9)}$

28.

$$\frac{y^2-9}{2y^2-y-15} \div \frac{3y^2+10y+3}{2y^2+y-10}$$

$$=\frac{y^2-9}{2y^2-y-15} \cdot \frac{2y^2+y-10}{3y^2+10y+3}$$

$$=\frac{(y-3)(y+3)}{(2y+5)(y-3)} \cdot \frac{(2y+5)(y-2)}{(3y+1)(y+3)}$$

$$=\frac{(y-3)(y+3)(2y+5)\cdot(y-2)}{(y-3)(y+3)(2y+5)\cdot(3y+1)}$$

$$=\frac{y-2}{3y+1}$$

29. $\dfrac{4}{x-3}+\dfrac{5}{x-3}=\dfrac{4+5}{x-3}=\dfrac{9}{x-3}$

30. $\dfrac{3}{x+4}+\dfrac{7}{x+4}=\dfrac{3+7}{x+4}=\dfrac{10}{x+4}$

31. $\dfrac{m^2}{m+3}+\dfrac{3m}{m+3}=\dfrac{m^2+3m}{m+3}=\dfrac{m(m+3)}{m+3}=m$

32. $\dfrac{1}{6m}+\dfrac{5}{6m}=\dfrac{1+5}{6m}=\dfrac{6}{6m}=\dfrac{1}{m}$

33.

$$\frac{-m+1}{m^2-4}+\frac{2m+1}{m^2-4}=\frac{-m+1+2m+1}{m^2-4}$$

$$=\frac{m+2}{(m+2)(m-2)}$$

34.

$$\frac{2m^2}{m+1}+\frac{5m+3}{m+1}=\frac{2m^2+5m+3}{m+1}$$

$$=\frac{(2m+3)(m+1)}{m+1}$$

$$=2m+3$$

35. $\dfrac{11}{15m}-\dfrac{6}{15m}=\dfrac{11-6}{15m}=\dfrac{5}{15m}=\dfrac{1}{3m}$

36. $\dfrac{15b}{2b^2}-\dfrac{11b}{2b^2}=\dfrac{15b-11b}{2b^2}=\dfrac{4b}{2b^2}=\dfrac{2b\cdot2}{2b\cdot b}=\dfrac{2}{b}$

37.

$$\frac{2y^2}{y-7}-\frac{y^2+49}{y-7}=\frac{2y^2-(y^2+49)}{y-7}$$

$$=\frac{2y^2-y^2-49}{y-7}$$

$$=\frac{y^2-49}{y-7}$$

$$=\frac{(y-7)(y+7)}{y-7}$$

$$=y+7$$

38.

$$\frac{6m+5}{m^2-36}-\frac{5m-1}{m^2-36}=\frac{6m+5-(5m-1)}{m^2-36}$$

$$=\frac{6m+5-5m+1}{m^2-36}$$

$$=\frac{m+6}{(m+6)(m-6)}$$

$$=\frac{1}{m-6}$$

39.

$$\frac{3}{x-y}+\frac{10}{y-x}=\frac{3}{x-y}+\frac{10}{-1(x-y)}$$

$$=\frac{3}{x-y}+\frac{-10}{x-y}$$

$$=\frac{3-10}{x-y}$$

$$=-\frac{7}{x-y}$$

40.

$$\frac{7}{a-b}+\frac{3}{b-a}=\frac{7}{a-b}+\frac{3}{-1(a-b)}$$

$$=\frac{7}{a-b}+\frac{-3}{a-b}$$

$$=\frac{7-3}{a-b}$$

$$=\frac{4}{a-b}$$

41.

$$\frac{2x}{x^2-25}-\frac{x-5}{25-x^2}=\frac{2x}{x^2-25}-\frac{x-5}{-1(x^2-25)}$$

$$=\frac{2x}{x^2-25}-\frac{-(x-5)}{x^2-25}$$

$$=\frac{2x+x-5}{x^2-25}$$

$$=\frac{3x-5}{x^2-25}$$

42.
$$\frac{x+5}{2x-6}-\frac{x+3}{6-2x}=\frac{x+5}{2x-6}-\frac{x+3}{-(2x-6)}$$
$$=\frac{x+5}{2x-6}-\frac{-(x+3)}{2x-6}$$
$$=\frac{x+5+x+3}{2x-6}$$
$$=\frac{2x+8}{2x-6}$$
$$=\frac{2(x+4)}{2(x-3)}$$
$$=\frac{x+4}{x-3}$$

43. $4x^2y^7=2^2\cdot x^2\cdot y^7$
$6x^4y=2\cdot3\cdot x^4\cdot y$
$\text{LCD}=2^2\cdot3\cdot x^4\cdot y^7=12x^4y^7$

44. $20a^3bc^4=2^2\cdot5\cdot a^3\cdot b\cdot c^4$
$30ab^4c^7=2\cdot3\cdot5\cdot a\cdot b^4\cdot c^7$
$\text{LCD}=2^2\cdot3\cdot5\cdot a^3\cdot b^4\cdot c^7=60a^3b^4c^7$

45. $4a=2^2\cdot a$
$8a+16=2^3\cdot\ \ (a+2)$
$\text{LCD}=2^3\cdot a\cdot(a+2)=8a(a+2)$

46. $5a=5\cdot a$
$a^2+6a=\ \ a\cdot(a+6)$
$\text{LCD}=5a\cdot(a+6)=5a(a+6)$

47. $4x-12=2^2\cdot(x-3)$
$x^2-2x-3=\ \ (x-3)\cdot(x+1)$
$\text{LCD}=2^2\cdot(x-3)\cdot(x+1)=4(x-3)(x+1)$

48. $x^2-7x=x\cdot\ (x-7)$
$x^2-49=\ \ (x-7)\cdot\ (x+7)$
$\text{LCD}=x(x-7)(x+7)$

49. $x^4y^7=xy^6\cdot x^3y$
The "missing factor" is xy^6.
$$\frac{6}{x^3y}=\frac{6}{x^3y}\cdot\frac{xy^6}{xy^6}=\frac{6xy^6}{x^4y^7}$$

50. $a^7b^5=a^4b^3\cdot a^3b^2$
The "missing factor" is a^4b^3.
$$\frac{11}{a^3b^2}=\frac{11}{a^3b^2}\cdot\frac{a^4b^3}{a^4b^3}=\frac{11a^4b^3}{a^7b^5}$$

51. $x^2-4=(x+2)(x-2)$
The "missing factor" is $x+2$.
$$\frac{x-1}{x-2}=\frac{x-1}{x-2}\cdot\frac{x+2}{x+2}=\frac{(x-1)(x+2)}{(x-2)(x+2)}$$

52. $m^2+5m-14=(m-2)(m+7)$
The "missing factor" is $m-2$.
$$\frac{m+2}{m+7}=\frac{m+2}{m+7}\cdot\frac{m-2}{m-2}=\frac{(m+2)(m-2)}{(m+7)(m-2)}$$

53. $6x^3=2\cdot3\cdot x^3$
$8x^5=2^3\cdot\ x^5$
$\text{LCD}=2^3\cdot3\cdot x^5=24x^5$
$$\frac{5y}{6x^3}=\frac{5y}{6x^3}\cdot\frac{4x^2}{4x^2}=\frac{5y\cdot4x^2}{6x^3\cdot4x^2}=\frac{20x^2y}{24x^5}$$
$$\frac{7}{8x^5}=\frac{7}{8x^5}\cdot\frac{3}{3}=\frac{7\cdot3}{8x^5\cdot3}=\frac{21}{24x^5}$$

54. $5a^3=\ \ 5\cdot a^3$
$10a=2\cdot5\cdot a$
$\text{LCD}=2\cdot5\cdot a^3=10a^3$
$$\frac{6}{5a^3}=\frac{6}{5a^3}\cdot\frac{2}{2}=\frac{6\cdot2}{5a^3\cdot2}=\frac{12}{10a^3}$$
$$\frac{11b}{10a}=\frac{11b}{10a}\cdot\frac{a^2}{a^2}=\frac{11b\cdot a^2}{10a\cdot a^2}=\frac{11a^2b}{10a^3}$$

55. $x-2=x-2$
$2-x=-(x-2)$
$\text{LCD}=-(x-2)$
$$\frac{4x}{x-2}=\frac{4x}{x-2}\cdot\frac{-1}{-1}=\frac{-4x}{-(x-2)}$$
$$\frac{6}{2-x}=\frac{6}{-(x-2)}$$

56. $m - 5 = m - 5$

$5 - m = -(m - 5)$

$LCD = -(m - 5)$

$$\frac{3}{m-5} = \frac{3}{m-5} \cdot \frac{-1}{-1} = \frac{-3}{-(m-5)}$$

$$\frac{-2m}{5-m} = \frac{-2m}{-(m-5)}$$

57. $m^2 + 5m - 14 = (m+7)(m-2)$

$m^2 + 9m + 14 = (m+7)(m+2)$

$LCD = (m+7)(m-2)(m+2)$

$$\frac{2}{m^2+5m-14} = \frac{2}{(m+7)(m-2)} \cdot \frac{m+2}{m+2}$$

$$= \frac{2m+4}{(m+7)(m-2)(m+2)}$$

$$\frac{m+1}{m^2+9m+14} = \frac{m+1}{(m+7)(m+2)} \cdot \frac{m-2}{m-2}$$

$$= \frac{m^2-m-2}{(m+7)(m-2)(m+2)}$$

58. $n^2 - 5n = n(n-5)$

$n^2 - 25 = (n-5)(n+5)$

$LCD = n(n-5)(n+5)$

$$\frac{n-2}{n^2-5n} = \frac{n-2}{n(n-5)} \cdot \frac{n+5}{n+5} = \frac{n^2+3n-10}{n(n-5)(n+5)}$$

$$\frac{n}{n^2-25} = \frac{n}{(n-5)(n+5)} \cdot \frac{n}{n} = \frac{n^2}{n(n-5)(n+5)}$$

59. $xy^3 = x \cdot y^3$; $x^2 z = x^2 \cdot z$

$LCD = x^2 y^3 z$

$$\frac{4}{xy^3} + \frac{8y}{x^2 z} = \frac{4}{xy^3} \cdot \frac{xz}{xz} + \frac{8y}{x^2 z} \cdot \frac{y^3}{y^3}$$

$$= \frac{4xz}{x^2 y^3 z} + \frac{8y^4}{x^2 y^3 z}$$

$$= \frac{4xz + 8y^4}{x^2 y^3 z}$$

60. $2x^3 y = 2 \cdot x^3 \cdot y$; $10xy^3 = 2 \cdot 5 \cdot x \cdot y^3$

$LCD = 2 \cdot 5 \cdot x^3 \cdot y^3 = 10x^3 y^3$

$$\frac{x}{2x^3 y} + \frac{y}{10xy^3} = \frac{x}{2x^3 y} \cdot \frac{5y^2}{5y^2} + \frac{y}{10xy^3} \cdot \frac{x^2}{x^2}$$

$$= \frac{5xy^2}{10x^3 y^3} + \frac{x^2 y}{10x^3 y^3}$$

$$= \frac{5xy^2 + x^2 y}{10x^3 y^3}$$

61. $LCD = (x+7)(x-7)$

$$\frac{x}{x+7} + \frac{2}{x-7}$$

$$= \frac{x}{x+7} \cdot \frac{x-7}{x-7} + \frac{2}{x-7} \cdot \frac{x+7}{x+7}$$

$$= \frac{x^2-7x}{(x+7)(x-7)} + \frac{2x+14}{(x+7)(x-7)}$$

$$= \frac{x^2-7x+2x+14}{(x+7)(x-7)}$$

$$= \frac{x^2-5x+14}{(x+7)(x-7)}$$

62. $LCD = (2x+3)(2x-3)$

$$\frac{4}{2x+3} + \frac{x+1}{2x-3}$$

$$= \frac{4}{2x+3} \cdot \frac{2x-3}{2x-3} + \frac{x+1}{2x-3} \cdot \frac{2x+3}{2x+3}$$

$$= \frac{8x-12}{(2x+3)(2x-3)} + \frac{2x^2+5x+3}{(2x+3)(2x-3)}$$

$$= \frac{8x-12+2x^2+5x+3}{(2x+3)(2x-3)}$$

$$= \frac{2x^2+13x-9}{(2x+3)(2x-3)}$$

63. $LCD = x(x-2)$

$$\frac{x+5}{x} - \frac{x+7}{x-2} = \frac{x+5}{x} \cdot \frac{x-2}{x-2} - \frac{x+7}{x-2} \cdot \frac{x}{x}$$

$$= \frac{x^2+3x-10}{x(x-2)} - \frac{x^2+7x}{x(x-2)}$$

$$= \frac{x^2+3x-10-x^2-7x}{x(x-2)}$$

$$= \frac{-4x-10}{x(x-2)}$$

64. $\text{LCD} = x(x+5)$

$$\frac{3x}{x+5} - \frac{x+1}{x} = \frac{3x}{x+5} \cdot \frac{x}{x} - \frac{x+1}{x} \cdot \frac{x+5}{x+5}$$

$$= \frac{3x^2}{x(x+5)} - \frac{x^2+6x+5}{x(x+5)}$$

$$= \frac{3x^2 - (x^2+6x+5)}{x(x+5)}$$

$$= \frac{3x^2 - x^2 - 6x - 5}{x(x+5)}$$

$$= \frac{2x^2 - 6x - 5}{x(x+5)}$$

65. $4x + 1 = 4x + 1; \ 4x^2 + 9x + 2 = (4x+1)(x+2)$

$\text{LCD} = (4x + 1)(x + 2)$

$$\frac{3x-4}{4x+1} + \frac{3x+6}{4x^2+9x+2}$$

$$= \frac{3x-4}{4x+1} \cdot \frac{x+2}{x+2} + \frac{3x+6}{(4x+1)(x+2)}$$

$$= \frac{3x^2+2x-8}{(4x+1)(x+2)} + \frac{3x+6}{(4x+1)(x+2)}$$

$$= \frac{3x^2+2x-8+3x+6}{(4x+1)(x+2)}$$

$$= \frac{3x^2+5x-2}{(4x+1)(x+2)}$$

$$= \frac{(3x-1)(x+2)}{(4x+1)(x+2)}$$

$$= \frac{3x-1}{4x+1}$$

66. $3x^2 + x - 4 = (3x+4)(x-1);$

$3x^2 - 2x - 8 = (3x+4)(x-2)$

$\text{LCD} = (3x + 4)(x - 1)(x - 2)$

$$\frac{7}{3x^2+x-4} + \frac{9x+2}{3x^2-2x-8}$$

$$= \frac{7}{(3x+4)(x-1)} \cdot \frac{x-2}{x-2} + \frac{9x+2}{(3x+4)(x-2)} \cdot \frac{x-1}{x-1}$$

$$= \frac{7x-14}{(3x+4)(x-1)(x-2)} + \frac{9x^2-7x-2}{(3x+4)(x-1)(x-2)}$$

$$= \frac{7x-14+9x^2-7x-2}{(3x+4)(x-1)(x-2)}$$

$$= \frac{9x^2-16}{(3x+4)(x-1)(x-2)}$$

$$= \frac{(3x+4)(3x-4)}{(3x+4)(x-1)(x-2)}$$

$$= \frac{3x-4}{(x-1)(x-2)}$$

67. $m^2 - 9 = (m+3)(m-3); \ m + 3 = m + 3$

$\text{LCD} = (m + 3)(m - 3)$

$$\frac{m}{m^2-9} - \frac{4m-12}{m+3}$$

$$= \frac{m}{(m+3)(m-3)} - \frac{4m-12}{m+3} \cdot \frac{m-3}{m-3}$$

$$= \frac{m}{(m+3)(m-3)} - \frac{4m^2-24m+36}{(m+3)(m-3)}$$

$$= \frac{m-4m^2+24m-36}{(m+3)(m-3)}$$

$$= \frac{-4m^2+25m-36}{(m+3)(m-3)}$$

$$= \frac{-(4m-9)(m-4)}{(m+3)(m-3)}$$

68. $m - 5 = m - 5; \ m^2 - 3m - 10 = (m-5)(m+2)$

$\text{LCD} = (m - 5)(m + 2)$

$$\frac{2m+1}{m-5} - \frac{4}{m^2-3m-10}$$

$$= \frac{2m+1}{m-5} \cdot \frac{m+2}{m+2} - \frac{4}{(m-5)(m+2)}$$

$$= \frac{2m^2+5m+2}{(m-5)(m+2)} - \frac{4}{(m-5)(m+2)}$$

$$= \frac{2m^2+5m+2-4}{(m-5)(m+2)}$$

$$= \frac{2m^2+5m-2}{(m-5)(m+2)}$$

69. $m - 2 = m - 2; \ 2 - m = -(m - 2)$

$\text{LCD} = -(m - 2)$

$$\frac{3}{m-2} - \frac{1}{2-m} = \frac{3}{m-2} \cdot \frac{-1}{-1} - \frac{1}{-(m-2)}$$

$$= \frac{-3}{-(m-2)} - \frac{1}{-(m-2)}$$

$$= \frac{-3-1}{-(m-2)}$$

$$= \frac{-4}{-(m-2)}$$

$$= \frac{4}{m-2}$$

70. $m - n = m - n; \ n - m = -(m - n)$

LCD $= -(m - n)$

$$\frac{m}{m-n} - \frac{n}{n-m} = \frac{m}{m-n} \cdot \frac{-1}{-1} - \frac{n}{-(m-n)}$$

$$= \frac{-m}{-(m-n)} - \frac{n}{-(m-n)}$$

$$= \frac{-m-n}{-(m-n)}$$

$$= \frac{-(m+n)}{-(m-n)}$$

$$= \frac{m+n}{m-n}$$

71. LCD $= x + 3$

$$4 + \frac{x}{x+3} = \frac{4}{1} \cdot \frac{x+3}{x+3} + \frac{x}{x+3}$$

$$= \frac{4x+12}{x+3} + \frac{x}{x+3}$$

$$= \frac{4x+12+x}{x+3}$$

$$= \frac{5x+12}{x+3}$$

72. LCD $= x + 2$

$$7 - \frac{1}{x+2} = \frac{7}{1} \cdot \frac{x+2}{x+2} - \frac{1}{x+2}$$

$$= \frac{7x+14}{x+2} - \frac{1}{x+2}$$

$$= \frac{7x+14-1}{x+2}$$

$$= \frac{7x+13}{x+2}$$

73. $\dfrac{\frac{1}{2} - \frac{2}{3}}{\frac{4}{9} + \frac{5}{6}} = \dfrac{\frac{3}{6} - \frac{4}{6}}{\frac{8}{18} + \frac{15}{18}} = \dfrac{-\frac{1}{6}}{\frac{23}{18}} = -\frac{1}{6} \cdot \frac{18}{23} = -\frac{3}{23}$

74. $\dfrac{\frac{1}{4} + \frac{1}{2}}{\frac{5}{8} - \frac{1}{6}} = \dfrac{\frac{1}{4} + \frac{2}{4}}{\frac{15}{24} - \frac{4}{24}} = \dfrac{\frac{3}{4}}{\frac{11}{24}} = \frac{3}{4} \cdot \frac{24}{11} = \frac{18}{11}$

75. $\dfrac{\frac{1}{5} - \frac{1}{m}}{\frac{1}{10} + \frac{1}{m^2}} = \dfrac{10m^2\left(\frac{1}{5} - \frac{1}{m}\right)}{10m^2\left(\frac{1}{10} + \frac{1}{m^2}\right)} = \dfrac{2m^2 - 10m}{m^2 + 10}$

76. $\dfrac{\frac{1}{m^2} + \frac{2}{3}}{\frac{1}{m} - \frac{5}{6}} = \dfrac{6m^2\left(\frac{1}{m^2} + \frac{2}{3}\right)}{6m^2\left(\frac{1}{m} - \frac{5}{6}\right)} = \dfrac{6 + 4m^2}{6m - 5m^2}$

77. $\dfrac{\frac{x}{4} - \frac{1}{2}}{\frac{3x}{2} - 3} = \dfrac{\frac{x}{4} - \frac{1}{2}}{\frac{3x}{2} - 3} \cdot \frac{4}{4} = \dfrac{x-2}{6x-12} = \dfrac{x-2}{6(x-2)} = \dfrac{1}{6}$

78. $\dfrac{\frac{7x}{3} + 7}{\frac{3x+9}{8}} = \dfrac{\frac{7x+21}{3}}{\frac{3x+9}{8}}$

$$= \frac{7x+21}{3} \cdot \frac{8}{3x+9}$$

$$= \frac{7(x+3) \cdot 8}{3 \cdot 3(x+3)}$$

$$= \frac{56}{9}$$

79. $\dfrac{\frac{8}{y+4} + 2}{\frac{12}{y+4} - 2} = \dfrac{(y+4)\left(\frac{8}{y+4} + 2\right)}{(y+4)\left(\frac{12}{y+4} - 2\right)}$

$$= \frac{8 + 2y + 8}{12 - 2y - 8}$$

$$= \frac{2y + 16}{-2y + 4}$$

$$= \frac{2(y+8)}{2(-y+2)}$$

$$= \frac{y+8}{-y+2}$$

80. $\dfrac{\frac{25}{y+5} + 5}{\frac{3}{y+5} - 5} = \dfrac{(y+5)\left(\frac{25}{y+5} + 5\right)}{(y+5)\left(\frac{3}{y+5} - 5\right)}$

$$= \frac{25 + 5y + 25}{3 - 5y - 25}$$

$$= \frac{5y + 50}{-5y - 22}$$

$$= \frac{5(y+10)}{-5y - 22}$$

81. The rational equation is undefined when $x + 5 = 0$ or $x = -5$ and when $x = 0$.

82. The rational equation is undefined when

$$x^2 - 36 = 0 \quad \text{or} \quad x + 6 = 0$$
$$(x+6)(x-6) = 0 \qquad\qquad x = -6$$
$$x + 6 = 0 \quad \text{or} \quad x - 6 = 0$$
$$x = -6 \quad \text{or} \qquad x = 6$$

83.
$$\frac{2}{x} - \frac{3}{4} = \frac{5}{x}$$
$$4x\left(\frac{2}{x} - \frac{3}{4}\right) = 4x\left(\frac{5}{x}\right)$$
$$8 - 3x = 20$$
$$-3x = 12$$
$$x = -4$$
The value is not extraneous. The solution set is $\{-4\}$.

84.
$$\frac{4}{x} + \frac{3}{4} = \frac{2}{3x} + \frac{23}{4}$$
$$12x\left(\frac{4}{x} + \frac{3}{4}\right) = 12x\left(\frac{2}{3x} + \frac{23}{4}\right)$$
$$48 + 9x = 8 + 69x$$
$$40 = 60x$$
$$x = \frac{40}{60}$$
$$x = \frac{2}{3}$$
The value is not extraneous. The solution set is $\left\{\frac{2}{3}\right\}$.

85.
$$\frac{4}{m} - \frac{3}{2m} = \frac{1}{2}$$
$$2m\left(\frac{4}{m} - \frac{3}{2m}\right) = 2m\left(\frac{1}{2}\right)$$
$$8 - 3 = m$$
$$5 = m$$
The value is not extraneous. The solution set is $\{5\}$.

86.
$$\frac{3}{m} + \frac{5}{3m} = 1$$
$$3m\left(\frac{3}{m} + \frac{5}{3m}\right) = 3m(1)$$
$$9 + 5 = 3m$$
$$14 = 3m$$
$$\frac{14}{3} = m$$
The value is not extraneous. The solution set is $\left\{\frac{14}{3}\right\}$.

87.
$$\frac{m+4}{m-3} = \frac{m+10}{m+2}$$
$$(m-3)(m+2)\left(\frac{m+4}{m-3}\right) = (m-3)(m+2)\left(\frac{m+10}{m+2}\right)$$
$$(m+2)(m+4) = (m-3)(m+10)$$
$$m^2 + 6m + 8 = m^2 + 7m - 30$$
$$6m + 8 = 7m - 30$$
$$8 = m - 30$$
$$38 = m$$
The value is not extraneous. The solution set is $\{38\}$.

88.
$$\frac{m-6}{m+5} = \frac{m-3}{m+1}$$
$$(m+5)(m+1)\left(\frac{m-6}{m+5}\right) = (m+5)(m+1)\left(\frac{m-3}{m+1}\right)$$
$$(m+1)(m-6) = (m+5)(m-3)$$
$$m^2 - 5m - 6 = m^2 + 2m - 15$$
$$-5m - 6 = 2m - 15$$
$$-7m = -9$$
$$m = \frac{9}{7}$$
The value is not extraneous. The solution set is $\left\{\frac{9}{7}\right\}$.

89.
$$\frac{2x}{x-1} - 5 = \frac{2}{x-1}$$
$$(x-1)\left(\frac{2x}{x-1} - 5\right) = (x-1)\left(\frac{2}{x-1}\right)$$
$$2x - 5x + 5 = 2$$
$$-3x = -3$$
$$x = 1$$
The value is extraneous. The solution set is $\{\ \}$ or $\varnothing$.

90.
$$\frac{2x}{x-2} - 3 = \frac{4}{x-2}$$
$$(x-2)\left(\frac{2x}{x-2} - 3\right) = (x-2)\left(\frac{4}{x-2}\right)$$
$$2x - 3x + 6 = 4$$
$$-x = -2$$
$$x = 2$$
The value is extraneous. The solution set is $\{\ \}$ or $\varnothing$.

91.
$$\frac{1}{x+3}+\frac{1}{x-3}=\frac{-5}{x^2-9}$$

$$(x+3)(x-3)\left(\frac{1}{x+3}+\frac{1}{x-3}\right)=(x+3)(x-3)\left(\frac{-5}{(x+3)(x-3)}\right)$$

$$x-3+x+3=-5$$
$$2x=-5$$
$$x=-\frac{5}{2}$$

The value is not extraneous. The solution set is $\left\{-\frac{5}{2}\right\}$.

92.
$$\frac{3}{x-5}-\frac{11}{x^2-25}=\frac{4}{x+5}$$

$$(x-5)(x+5)\left(\frac{3}{x-5}-\frac{11}{x^2-25}\right)=(x-5)(x+5)\left(\frac{4}{x+5}\right)$$

$$3(x+5)-11=4(x-5)$$
$$3x+15-11=4x-20$$
$$3x+4=4x-20$$
$$4=x-20$$
$$24=x$$

The value is not extraneous. The solution set is $\{24\}$.

93. $$y=\frac{4}{k}$$
$$yk=\frac{4}{k}\cdot k$$
$$yk=4$$
$$\frac{yk}{y}=\frac{4}{y}$$
$$k=\frac{4}{y}$$

94. $$6=\frac{x}{y}$$
$$6y=\frac{x}{y}\cdot y$$
$$6y=x$$
$$\frac{6y}{6}=\frac{x}{6}$$
$$y=\frac{x}{6}$$

95.

$$\frac{1}{x}+\frac{1}{y}=\frac{1}{z}$$

$$xyz\left(\frac{1}{x}+\frac{1}{y}\right)=xyz\left(\frac{1}{z}\right)$$

$$yz+xz=xy$$

$$yz-xy=-xz$$

$$y(z-x)=-xz$$

$$\frac{y(z-x)}{z-x}=\frac{-xz}{z-x}$$

$$y=\frac{-xz}{z-x}$$

$$y=\frac{xz}{x-z}$$

96.

$$\frac{1}{x}+\frac{1}{y}=\frac{1}{z}$$

$$xyz\left(\frac{1}{x}+\frac{1}{y}\right)=xyz\left(\frac{1}{z}\right)$$

$$yz+xz=xy$$

$$z(y+x)=xy$$

$$\frac{z(y+x)}{y+x}=\frac{xy}{y+x}$$

$$z=\frac{xy}{x+y}$$

97.

$$\frac{6}{4y+5}=\frac{2}{7}$$

$$6(7)=2(4y+5)$$

$$42=8y+10$$

$$32=8y$$

$$4=y$$

The solution set is $\{4\}$.

98.

$$\frac{2}{y-3}=\frac{5}{y}$$

$$2y=5(y-3)$$

$$2y=5y-15$$

$$-3y=-15$$

$$y=5$$

The solution set is $\{5\}$.

99.

$$\frac{y+1}{8}=\frac{1}{4}$$

$$4(y+1)=8\cdot1$$

$$4y+4=8$$

$$4y=4$$

$$y=1$$

The solution set is $\{1\}$.

100.

$$\frac{6y+7}{10}=\frac{2y+9}{6}$$

$$6(6y+7)=10(2y+9)$$

$$36y+42=20y+90$$

$$16y+42=90$$

$$16y=48$$

$$y=3$$

The solution set is $\{3\}$.

101.

$$\frac{2}{3}=\frac{10}{x}$$

$$2x=3\cdot10$$

$$2x=30$$

$$x=15$$

102.

$$\frac{14}{7}=\frac{7}{x}$$

$$14x=7\cdot7$$

$$14x=49$$

$$x=\frac{49}{14}$$

$$x=3.5$$

103. Let x be the number of tanks needed for 30 bags.

$$\frac{4}{15}=\frac{x}{30}$$

$$4\cdot30=15x$$

$$120=15x$$

$$8=x$$

Therefore, 8 tanks of water are needed for 30 bags of cement.

104. Let x be the cost of 7 small pizzas.

$$\frac{4}{15}=\frac{7}{x}$$

$$4x=15\cdot7$$

$$4x=105$$

$$x=26.25$$

The cost of 7 small pizzas is $26.25.

105. Let t be the time working together. Lucille can complete $\frac{1}{3}$ of the job in one hour. Teresa can complete $\frac{1}{2}$ of the job in one hour.

$$\frac{1}{3}+\frac{1}{2}=\frac{1}{t}$$
$$6t\left(\frac{1}{3}+\frac{1}{2}\right)=6t\left(\frac{1}{t}\right)$$
$$2t+3t=6$$
$$5t=6$$
$$t=\frac{6}{5}$$

It will take them $\frac{6}{5}=1.2$ hours working together.

106. Let t be the time working together. Fred can carpet $\frac{1}{3}$ of the room in an hour and Barney can carpet $\frac{1}{5}$ of the room in one hour.

$$\frac{1}{3}+\frac{1}{5}=\frac{1}{t}$$
$$15t\left(\frac{1}{3}+\frac{1}{5}\right)=15t\left(\frac{1}{t}\right)$$
$$5t+3t=15$$
$$8t=15$$
$$t=\frac{15}{8}$$

It will take them $\frac{15}{8}\approx1.9$ hours working together.

107. Let t be the time it takes Adrienne to wash the dishes. Then $t+5$ is the time it takes Jake to wash the dishes. Adrienne can complete $\frac{1}{t}$ of the job in an hour, while Jake can complete $\frac{1}{t+5}$ of the job in an hour.

$$\frac{1}{t}+\frac{1}{t+5}=\frac{1}{6}$$
$$6t(t+5)\left(\frac{1}{t}+\frac{1}{t+5}\right)=6t(t+5)\cdot\frac{1}{6}$$
$$6(t+5)+6t=t(t+5)$$
$$6t+30+6t=t^2+5t$$
$$12t+30=t^2+5t$$
$$0=t^2-7t-30$$
$$0=(t-10)(t+3)$$

$$t-10=0 \quad\text{or}\quad t+3=0$$
$$t=10 \quad\text{or}\quad t=-3$$

Disregard the negative value. Jake can wash the dishes is $t+5=10+5=15$ minutes.

108. Let t be the time it took Ben to paint the room alone. Donovan can complete $\frac{1}{7}$ of the room in one hour whereas Ben can complete $\frac{1}{t}$ of the room in one hour. It takes 4 hours working together.

$$\frac{1}{7}+\frac{1}{t}=\frac{1}{4}$$
$$28t\left(\frac{1}{7}+\frac{1}{t}\right)=28t\left(\frac{1}{4}\right)$$
$$4t+28=7t$$
$$28=3t$$
$$\frac{28}{3}=t$$

It will take Ben $\frac{28}{3}\approx9.3$ hours working alone.

109. Let p be Paul's speed.

	d	r	$t=\frac{d}{r}$
upstream	15	$p-2$	$\frac{15}{p-2}$
downstream	27	$p+2$	$\frac{27}{p+2}$

The times are the same.
$$\frac{15}{p-2}=\frac{27}{p+2}$$
$$15(p+2)=27(p-2)$$
$$15p+30=27p-54$$
$$30=12p-54$$
$$84=12p$$
$$7=p$$

Paul travels $p+2=7+2=9$ mph when traveling downstream.

110. Let r be the rate of the ship.

	d	r	$t=\frac{d}{r}$
with current	275	$r+10$	$\frac{275}{r+10}$
against current	175	$r-10$	$\frac{175}{r-10}$

The times are the same.

$$\frac{275}{r+10} = \frac{175}{r-10}$$
$$275(r-10) = 175(r+10)$$
$$275r - 2750 = 175r + 1750$$
$$100r = 4500$$
$$r = 45$$

The ship cruises at $r + 10 = 45 + 10 = 55$ mph.

111. Let r be the speed of the train.

	d	r	$t = \frac{d}{r}$
train	135	r	$t = \frac{135}{r}$
plane	855	$3r$	$t = \frac{855}{3}$

The total time was 6 hours.
$$\frac{135}{r} + \frac{855}{3r} = 6$$
$$3r\left(\frac{135}{r} + \frac{855}{3r}\right) = 3r(6)$$
$$3(135) + 855 = 18r$$
$$405 + 855 = 18r$$
$$1260 = 18r$$
$$70 = r$$

The train's speed was 70 mph.

112. Let r be Tamika's speed in the city.

	d	r	$t = \frac{d}{r}$
city	90	r	$\frac{90}{r}$
highway	130	$r + 20$	$\frac{130}{r+20}$

Her total time was 4 hours.
$$\frac{90}{r} + \frac{130}{r+20} = 4$$
$$r(r+20)\left(\frac{90}{r} + \frac{130}{r+20}\right) = r(r+20) \cdot 4$$
$$90(r+20) + 130r = 4r(r+20)$$
$$90r + 1800 + 130r = 4r^2 + 80r$$
$$0 = 4r^2 - 140r - 1800$$
$$0 = 4(r-45)(r+10)$$
$$r - 45 = 0 \quad \text{or} \quad r + 10 = 0$$
$$r = 45 \quad \text{or} \quad r = -10$$

Disregard the negative rate. Tamika's speed in the city was 45 mph.

Chapter 7 Test

1. $\dfrac{3x - 2y^2}{6z}$; $x = 2, y = -3, z = -1$

$$\frac{3(2) - 2(-3)^2}{6(-1)} = \frac{3(2) - 2(9)}{6(-1)}$$
$$= \frac{6 - 18}{-6}$$
$$= \frac{-12}{-6}$$
$$= 2$$

2. $\dfrac{x+5}{x^2 - 3x - 10}$ is undefined when $x^2 - 3x - 10 = 0$.

$$x^2 - 3x - 10 = 0$$
$$(x-5)(x+2) = 0$$
$$x - 5 = 0 \quad \text{or} \quad x + 2 = 0$$
$$x = 5 \quad \text{or} \quad x = -2$$

3. $\dfrac{x^2 - 4x - 21}{14 - 2x} = \dfrac{(x-7)(x+3)}{-2(x-7)}$

$$= \frac{x+3}{-2}$$
$$= -\frac{x+3}{2}$$

4. $\dfrac{\frac{35x^6}{9x^4}}{\frac{25x^5}{18x}} = \dfrac{35x^6}{9x^4} \cdot \dfrac{18x}{25x^5}$

$$= \frac{35x^6 \cdot 18x}{9x^4 \cdot 25x^5}$$
$$= \frac{5 \cdot 9 \cdot x^5 \cdot x \cdot x \cdot 2 \cdot 7}{5 \cdot 9 \cdot x^5 \cdot x \cdot x \cdot 5 \cdot x^2}$$
$$= \frac{14}{5x^2}$$

5. $\dfrac{5x - 15}{3x + 9} \cdot \dfrac{5x + 15}{3x - 9} = \dfrac{5(x-3)}{3(x+3)} \cdot \dfrac{5(x+3)}{3(x-3)}$

$$= \frac{5(x-3) \cdot 5(x+3)}{3 \cdot (x+3) \cdot 3(x-3)}$$
$$= \frac{(x+3) \cdot 5 \cdot 5 \cdot (x-3)}{(x+3) \cdot 3 \cdot 3 \cdot (x-3)}$$
$$= \frac{25}{9}$$

6. $\dfrac{\dfrac{2x^2-5xy-12y^2}{x^2+xy-20y^2}}{\dfrac{4x^2-9y^2}{x^2+4xy-5y^2}}$

$= \dfrac{2x^2-5xy-12y^2}{x^2+xy-20y^2}\cdot\dfrac{x^2+4xy-5y^2}{4x^2-9y^2}$

$= \dfrac{(2x+3y)(x-4y)\cdot(x+5y)(x-y)}{(x+5y)(x-4y)\cdot(2x-3y)(2x+3y)}$

$= \dfrac{x-y}{2x-3y}$

7. $\dfrac{y^2}{y+3}+\dfrac{3y}{y+3}=\dfrac{y^2+3y}{y+3}=\dfrac{y(y+3)}{y+3}=y$

8. $\dfrac{x^2}{x^2-9}-\dfrac{8x-15}{x^2-9}=\dfrac{x^2-(8x-15)}{x^2-9}$

$= \dfrac{x^2-8x+15}{x^2-9}$

$= \dfrac{(x-5)(x-3)}{(x+3)(x-3)}$

$= \dfrac{x-5}{x+3}$

9. $\dfrac{6}{y-z}+\dfrac{7}{z-y}=\dfrac{6}{y-z}+\dfrac{7}{-1\cdot(y-z)}$

$= \dfrac{6}{y-z}+\dfrac{-7}{y-z}$

$= \dfrac{6-7}{y-z}$

$= \dfrac{-1}{y-z}$ or $\dfrac{1}{z-y}$

10. LCD $=(x-2)(2x+1)$

$\dfrac{x}{x-2}+\dfrac{3}{2x+1}$

$= \dfrac{x}{x-2}\cdot\dfrac{2x+1}{2x+1}+\dfrac{3}{2x+1}\cdot\dfrac{x-2}{x-2}$

$= \dfrac{2x^2+x}{(x-2)(2x+1)}+\dfrac{3x-6}{(2x+1)(x-2)}$

$= \dfrac{2x^2+x+3x-6}{(x-2)(2x+1)}$

$= \dfrac{2x^2+4x-6}{(x-2)(2x+1)}$

$= \dfrac{2(x+3)(x-1)}{(x-2)(2x+1)}$

11. $x^2+5x+6=(x+2)(x+3)$

$x^2+2x-3=(x-1)(x+3)$

LCD $=(x+2)(x+3)(x-1)$

$\dfrac{2x}{x^2+5x+6}-\dfrac{x+1}{x^2+2x-3}$

$= \dfrac{2x}{(x+2)(x+3)}\cdot\dfrac{x-1}{x-1}-\dfrac{x+1}{(x-1)(x+3)}\cdot\dfrac{x+2}{x+2}$

$= \dfrac{2x^2-2x}{(x+2)(x+3)(x-1)}-\dfrac{x^2+3x+2}{(x+2)(x+3)(x-1)}$

$= \dfrac{2x^2-2x-(x^2+3x+2)}{(x+2)(x+3)(x-1)}$

$= \dfrac{2x^2-2x-x^2-3x-2}{(x+2)(x+3)(x-1)}$

$= \dfrac{x^2-5x-2}{(x+2)(x+3)(x-1)}$

12. $\dfrac{\dfrac{1}{9}-\dfrac{1}{y^2}}{\dfrac{1}{3}+\dfrac{1}{y}}=\dfrac{9y^2\left(\dfrac{1}{9}-\dfrac{1}{y^2}\right)}{9y^2\left(\dfrac{1}{3}+\dfrac{1}{y}\right)}$

$= \dfrac{y^2-9}{3y^2+9y}$

$= \dfrac{(y+3)(y-3)}{3y(y+3)}$

$= \dfrac{y-3}{3y}$

13. Undefined value: $m=0$.

$\dfrac{m}{5}+\dfrac{5}{m}=\dfrac{m+3}{4}$

$\dfrac{m}{5}\cdot\dfrac{m}{m}+\dfrac{5}{m}\cdot\dfrac{5}{5}=\dfrac{m+3}{4}$

$\dfrac{m^2}{5m}+\dfrac{25}{5m}=\dfrac{m+3}{4}$

$\dfrac{m^2+25}{5m}=\dfrac{m+3}{4}$

$4(m^2+25)=5m(m+3)$

$4m^2+100=5m^2+15m$

$0=m^2+15m-100$

$0=(m+20)(m-5)$

$m+20=0 \quad$ or $\quad m-5=0$

$m=-20 \quad$ or $\qquad m=5$

Neither value is extraneous. The solution set is $\{-20, 5\}$.

14. Undefined values: $x = -3$ and $x = 6$.

$$\frac{4}{x+3}+\frac{5}{x-6}=\frac{4x+1}{x^2-3x-18}$$

$$\frac{4}{x+3}\cdot\frac{x-6}{x-6}+\frac{5}{x-6}\cdot\frac{x+3}{x+3}=\frac{4x+1}{(x-6)(x+3)}$$

$$\frac{4x-24}{(x+3)(x-6)}+\frac{5x+15}{(x-6)(x+3)}=\frac{4x+1}{(x-6)(x+3)}$$

$$\frac{4x-24+5x+15}{(x-6)(x+3)}=\frac{4x+1}{(x-6)(x+3)}$$

$$\frac{9x-9}{(x-6)(x+3)}=\frac{4x+1}{(x-6)(x+3)}$$

$$9x-9=4x+1$$

$$5x=10$$

$$x=2$$

The value is not extraneous. The solution set is $\{2\}$.

15.

$$\frac{1}{x}+\frac{1}{y}=\frac{1}{z}$$

$$xyz\left(\frac{1}{x}+\frac{1}{y}\right)=xyz\left(\frac{1}{z}\right)$$

$$yz+xz=xy$$

$$yz-xy=-xz$$

$$y(z-x)=-xz$$

$$y=\frac{-xz}{z-x}$$

$$y=\frac{xz}{x-z}$$

16. Undefined values: $y = -1$ and $y = 2$.

$$\frac{2}{y+1}=\frac{1}{y-2}$$

$$2(y-2)=1(y+1)$$

$$2y-4=y+1$$

$$2y-y=1+4$$

$$y=5$$

The solution set is $\{5\}$.

17.

$$\frac{5}{12}=\frac{2}{x}$$

$$5x=2(12)$$

$$5x=24$$

$$x=\frac{24}{5}$$

18. Let x be the cost of 12 hair barrettes.

$$\frac{8}{22}=\frac{12}{x}$$

$$8x=22(12)$$

$$8x=264$$

$$x=33$$

The cost is $33.

19. Let j be the time for Juan to wash the car alone. Then $j + 18$ is the time for Frank to wash the car alone. Juan can wash $\frac{1}{j}$ of the car in one minute, whereas Frank can wash $\frac{1}{j+18}$ of the car in one minute. Together, it takes them 12 minutes.

$$\frac{1}{j}+\frac{1}{j+18}=\frac{1}{12}$$

$$12j(j+18)\left(\frac{1}{j}+\frac{1}{j+18}\right)=12j(j+18)\frac{1}{12}$$

$$12(j+18)+12j=j(j+18)$$

$$12j+216+12j=j^2+18j$$

$$0=j^2-6j-216$$

$$0=(j-18)(j+12)$$

$$j-18=0 \quad\text{or}\quad j+12=0$$

$$j=18 \quad\text{or}\quad j=-12$$

Disregard the negative time. It takes Frank $j + 18 = 18 + 18 = 36$ minutes to wash the car alone.

20. Let r be the rate at which they hiked.

	d	r	$t=\frac{d}{r}$
nature path	7	$r+3$	$t=\frac{7}{r+3}$
mountainside	12	r	$t=\frac{12}{r}$

The total time was 4 hours.

$$\frac{7}{r+3}+\frac{12}{r}=4$$

$$r(r+3)\left(\frac{7}{r+3}+\frac{12}{r}\right)=r(r+3)\cdot 4$$

$$7r+12(r+3)=4r(r+3)$$

$$7r+12r+36=4r^2+12r$$

$$19r+36=4r^2+12r$$

$$0=4r^2-7r-36$$

$$0=(4r+9)(r-4)$$

$$4r+9=0 \quad \text{or} \quad r-4=0$$

$$r=-\frac{9}{4} \quad \text{or} \quad r=4$$

Disregard the negative rate. The tourists hiked at 4 mph.

Cumulative Review Chapters 1–7

1. $-6^2+4(-5+2)^3=-36+4(-3)^3$
$$=-36+4(-27)$$
$$=-36+(-108)$$
$$=-144$$

2. $3(4x-2)-(3x+5)=3\cdot 4x-3\cdot 2-3x-5$
$$=12x-6-3x-5$$
$$=12x-3x-6-5$$
$$=9x-11$$

3. $-3(x-5)+2x=5x-4$
$$-3x+15+2x=5x-4$$
$$-x+15=5x-4$$
$$x-x+15=x+5x-4$$
$$15=6x-4$$
$$15+4=6x-4+4$$
$$19=6x$$
$$\frac{19}{6}=\frac{6x}{6}$$
$$\frac{19}{6}=x$$

The solution set is $\left\{\frac{19}{6}\right\}$.

4. $3(2x-1)+5=6x+2$
$$6x-3+5=6x+2$$
$$6x+2=6x+2$$
$$-6x+6x+2=-6x+6x+2$$
$$2=2$$

This is a true statement. It is an identity. The solution set is the set of all real numbers.

5. $0.25x+0.10(x-3)=0.05(22)$
$$100[0.25x+0.10(x-3)]=100[0.05(22)]$$
$$25x+10(x-3)=5(22)$$
$$25x+10x-30=110$$
$$35x-30=110$$
$$35x-30+30=110+30$$
$$35x=140$$
$$\frac{35x}{35}=\frac{140}{35}$$
$$x=4$$

The solution set is {4}.

6. Let w be the width of the poster.
Then $2w-3$ is the length of the poster.
Perimeter = 2(length + width)
$$24=2(2w-3+w)$$
$$24=2(3w-3)$$
$$24=6w-6$$
$$30=6w$$
$$5=w$$

The length of the poster is
$2w-3=2(5)-3=10-3=7$ feet.

7. Let n be the first even integer. Then $n+2$ and $n+4$ are the second and third consecutive even integers, respectively. Their sum is 138.
$$n+(n+2)+(n+4)=138$$
$$n+n+2+n+4=138$$
$$3n+6=138$$
$$3n=132$$
$$n=44$$
$$n+2=46$$
$$n+4=48$$

The three integers are 44, 46, and 48.

8. $2(x-3)-5\le 3(x+2)-18$
$$2x-6-5\le 3x+6-18$$
$$2x-11\le 3x-12$$
$$-3x+2x-11\le -3x+3x-12$$
$$-x-11\le -12$$
$$-x-11+11\le -12+11$$
$$-x\le -1$$
$$x\ge 1$$

9. $(3x-2y)^2=(3x)^2-2(3x)(2y)+(2y)^2$
$$=9x^2-12xy+4y^2$$

10. $\left(\dfrac{2a^5 b}{4ab^{-2}}\right)^{-4} = \dfrac{2^{-4} a^{-20} b^{-4}}{4^{-4} a^{-4} b^8}$

$\qquad = \dfrac{4^4}{2^4} \cdot a^{-20-(-4)} b^{-4-8}$

$\qquad = \dfrac{256}{16} a^{-16} b^{-12}$

$\qquad = \dfrac{16}{a^{16} b^{12}}$

11. $(3x^0 y^{-4} z^3)^3 = (3 \cdot 1 \cdot y^{-4} \cdot z^3)^3$

$\qquad = \left(\dfrac{3z^3}{y^4}\right)^3$

$\qquad = \dfrac{3^3 z^{3 \cdot 3}}{y^{4 \cdot 3}}$

$\qquad = \dfrac{27 z^9}{y^{12}}$

12. $8a^2 b + 34ab - 84b = 2b(4a^2 + 17a - 42)$

$\qquad\qquad\qquad\qquad = 2b(4a - 7)(a + 6)$

13. $\qquad 6x^3 - 31x^2 = -5x$

$6x^3 - 31x^2 + 5x = 0$

$x(6x^2 - 31x + 5) = 0$

$x(6x - 1)(x - 5) = 0$

$x = 0 \quad \text{or} \quad 6x - 1 = 0 \quad \text{or} \quad x - 5 = 0$

$\qquad\qquad x = \dfrac{1}{6} \quad \text{or} \qquad x = 5$

The solution set is $\left\{0, \dfrac{1}{6}, 5\right\}$.

14. Let w be the width of the rectangle. Then $2w + 2$ is the length. The area is 40 square centimeters.

Area = (length)(width)

$\qquad 40 = (2w + 2)(w)$

$\qquad 40 = 2w^2 + 2w$

$\qquad 0 = 2w^2 + 2w - 40$

$\qquad 0 = 2(w^2 + w - 20)$

$\qquad 0 = 2(w + 5)(w - 4)$

$2 = 0 \quad \text{or} \quad w + 5 = 0 \quad \text{or} \quad w - 4 = 0$

false   or $w = -5$ or $w = 4$

Disregard a negative width. The length is $2w + 2 = 2(4) + 2 = 8 + 2 = 10$ centimeters.

15. $\dfrac{x+5}{x^2 + 25}; \; x = -5$

$\dfrac{-5+5}{(-5)^2 + 25} = \dfrac{0}{25 + 25} = \dfrac{0}{50} = 0$

16. $\dfrac{3x+3}{5x - 5x^2} \cdot \dfrac{2x^2 + x - 3}{4x^2 - 9}$

$= \dfrac{3(x+1)}{-5x(x-1)} \cdot \dfrac{(2x+3)(x-1)}{(2x-3)(2x+3)}$

$= -\dfrac{3(x+1)}{5x(2x-3)}$

17. $\dfrac{x^2 - x - 2}{10} \div \dfrac{2x+4}{5}$

$= \dfrac{x^2 - x - 2}{10} \cdot \dfrac{5}{2x+4}$

$= \dfrac{(x-2)(x+1)}{5 \cdot 2} \cdot \dfrac{5}{2(x+2)}$

$= \dfrac{5 \cdot (x-2)(x+1)}{5 \cdot 4(x+2)}$

$= \dfrac{(x-2)(x+1)}{4(x+2)}$

18. $\dfrac{2x+3}{x^2 - x - 30} - \dfrac{x-2}{x^2 - x - 30} = \dfrac{2x+3-(x-2)}{x^2 - x - 30}$

$= \dfrac{2x+3-x+2}{x^2 - x - 30}$

$= \dfrac{x+5}{(x-6)(x+5)}$

$= \dfrac{1}{x-6}$

19. $\dfrac{15}{2x-4} + \dfrac{x}{x^2 - 4}$

$= \dfrac{15}{2(x-2)} + \dfrac{x}{(x+2)(x-2)}$

$= \dfrac{15}{2(x-2)} \cdot \dfrac{x+2}{x+2} + \dfrac{x}{(x+2)(x-2)} \cdot \dfrac{2}{2}$

$= \dfrac{15x+30}{2(x-2)(x+2)} + \dfrac{2x}{2(x-2)(x+2)}$

$= \dfrac{15x+30+2x}{2(x-2)(x+2)}$

$= \dfrac{17x+30}{2(x-2)(x+2)}$

20. $\dfrac{\frac{2}{x^2}-\frac{3}{5x}}{\frac{4}{x}+\frac{1}{4x}}=\dfrac{20x^2\left(\frac{2}{x^2}-\frac{3}{5x}\right)}{20x^2\left(\frac{4}{x}+\frac{1}{4x}\right)}$

$=\dfrac{40-12x}{80x+5x}$

$=\dfrac{4(10-3x)}{85x}$

21. LCD $=(x-4)(x+4)$

Undefined values: $x=4$ and $x=-4$.

$\dfrac{3}{x-4}=\dfrac{5x+4}{x^2-16}-\dfrac{4}{x+4}$

$\dfrac{3}{x-4}=\dfrac{5x+4}{(x+4)(x-4)}-\dfrac{4}{x+4}\cdot\dfrac{x-4}{x-4}$

$\dfrac{3}{x-4}=\dfrac{5x+4}{(x+4)(x-4)}-\dfrac{4x-16}{(x+4)(x-4)}$

$\dfrac{3}{x-4}=\dfrac{5x+4-4x+16}{(x+4)(x-4)}$

$\dfrac{3}{x-4}=\dfrac{x+20}{(x+4)(x-4)}$

$3(x+4)(x-4)=(x-4)(x+20)$

$3x^2-48=x^2+16x-80$

$2x^2-16x+32=0$

$2(x-4)(x-4)=0$

$2=0$ or $x-4=0$

false or $x=4$

Since $x=4$ is undefined, there is no solution.
The solution set is { } or $\varnothing$.

22. $\dfrac{x-5}{3}=\dfrac{x+2}{2}$

$2(x-5)=3(x+2)$

$2x-10=3x+6$

$-10=x+6$

$-16=x$

The solution set is $\{-16\}$.

23. Let x represent the number of inches for 125 miles.

$\dfrac{4}{50}=\dfrac{x}{125}$

$4(125)=50x$

$500=50x$

$10=x$

There are 10 inches for 125 miles.

24. Let s be the time for Sharona working alone. Then $s+9$ is the time for Trent working alone.

Sharona can do $\dfrac{1}{s}$ of the job in an hour, whereas

Trent can do $\dfrac{1}{s+9}$ of the job in an hour. It takes

then 6 hours working together.

$\dfrac{1}{s}+\dfrac{1}{s+9}=\dfrac{1}{6}$

$6s(s+9)\left(\dfrac{1}{s}+\dfrac{1}{s+9}\right)=6s(s+9)\left(\dfrac{1}{6}\right)$

$6(s+9)+6s=s(s+9)$

$6s+54+6s=s^2+9s$

$0=s^2-3s-54$

$0=(s+6)(s-9)$

$s+6=0$ or $s-9=0$

$s=-6$ or $s=9$

Disregard the negative time. It takes Sharona 9 hours working alone.

25. Let r be the rate he walks.

	d	r	$t=\frac{d}{r}$
jogged	35	$r+4$	$\frac{35}{r+4}$
walked	6	r	$\frac{6}{r}$

It took him 7 hours total.

$\dfrac{35}{r+4}+\dfrac{6}{r}=7$

$r(r+4)\left(\dfrac{35}{r+4}+\dfrac{6}{r}\right)=r(r+4)\cdot 7$

$35r+6(r+4)=7r(r+4)$

$35r+6r+24=7r^2+28r$

$0=7r^2-13r-24$

$0=(7r+8)(r-3)$

$7r+8=0$ or $r-3=0$

$r=-\dfrac{8}{7}$ or $r=3$

Disregard the negative rate. Francisco jogs at $r+4=3+4=7$ mph.

26. $(x_1,\ y_1)=(-1,\ 3);\ (x_2,\ y_2)=(5,\ 11)$

$m=\dfrac{y_2-y_1}{x_2-x_1}=\dfrac{11-3}{5-(-1)}=\dfrac{8}{6}=\dfrac{4}{3}$

27. $m = -\dfrac{3}{2}; \; (x_1, y_1) = (4, -1)$

$$y - y_1 = m(x - x_1)$$

$$y - (-1) = -\dfrac{3}{2}(x - 4)$$

$$y + 1 = -\dfrac{3}{2}x + 6$$

$$y = -\dfrac{3}{2}x + 5 \text{ or } 3x + 2y = 10$$

28. $2x - 3y = -12$

y-intercept: Let $x = 0$.

$$2(0) - 3y = -12$$

$$-3y = -12$$

$$y = 4$$

$(0, 4)$

x-intercept: Let $y = 0$.

$$2x - 3(0) = -12$$

$$2x = -12$$

$$x = -6$$

$(-6, 0)$

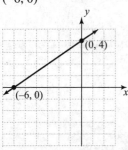

29. $3x + 5y = 15$

$$5y = -3x + 15$$

$$y = -\dfrac{3}{5}x + 3$$

$$m_{\text{parallel}} = -\dfrac{3}{5}$$

$$m_{\text{perpendicular}} = \dfrac{5}{3}$$

30. $\begin{cases} 2x + 3y = 1 & (1) \\ -3x + 2y = 18 & (2) \end{cases}$

Multiply equation (1) by 3 and equation (2) by 2.

$$\begin{cases} 6x + 9y = 3 \\ -6x + 4y = 36 \end{cases}$$

$$13y = 39$$

$$y = 3$$

Let $y = 3$ in equation (1).

$$2x + 3(3) = 1$$

$$2x + 9 = 1$$

$$2x = -8$$

$$x = -4$$

The solution is $(-4, 3)$.

Getting Ready for Intermediate Algebra:
A Review of Chapters 1 – 7

1. $\dfrac{4-(-6)}{8-2}=\dfrac{4+6}{8-2}=\dfrac{10}{6}=\dfrac{\cancel{2}\cdot 5}{\cancel{2}\cdot 3}=\dfrac{5}{3}$

2. Substitute 6 for x in the expression $-\dfrac{2}{3}x+5$:

$-\dfrac{2}{3}(6)+5=-4+5=1$

3.
$$4x+3=17$$
$$4x+3-3=17-3$$
$$4x=14$$
$$\dfrac{4x}{4}=\dfrac{14}{4}$$
$$x=\dfrac{7}{2}$$

The solution set is $\left\{\dfrac{7}{2}\right\}$. The equation is conditional.

4.
$$5(x-2)-2x=3x+4$$
$$5x-10-2x=3x+4$$
$$3x-10=3x+4$$
$$3x-3x-10=3x-3x+4$$
$$-10=4$$

The last statement is false, so the equation is a contradiction. Thus, the equation has no solution. The solution set is { } or $\varnothing$.

5.
$$\dfrac{1}{2}(x-4)+\dfrac{2}{3}x=\dfrac{1}{6}(x-4)$$
$$\dfrac{1}{2}x-2+\dfrac{2}{3}x=\dfrac{1}{6}x-\dfrac{2}{3}$$
$$6\left(\dfrac{1}{2}x-2+\dfrac{2}{3}x\right)=6\left(\dfrac{1}{6}x-\dfrac{2}{3}\right)$$
$$3x-12+4x=x-4$$
$$7x-12=x-4$$
$$6x-12=-4$$
$$6x=8$$
$$x=\dfrac{8}{6}=\dfrac{4}{3}$$

The solution set is $\left\{\dfrac{4}{3}\right\}$. The equation is conditional.

6. $0.3(x+1)-0.1(x-7)=0.4x-0.2(x-5)$
$$0.3x+0.3-0.1x+0.7=0.4x-0.2x+1$$
$$0.2x+1=0.2x+1$$
$$0.2x-0.2x+1=0.2x-0.2x+1$$
$$1=1$$

The last statement is true, indicating that the equation is true for all real numbers x. The solution set is the set of all real numbers. The equation is an identity.

7.
$$6x-7>-31$$
$$6x-7+7>-31+7$$
$$6x>-24$$
$$\dfrac{6x}{6}>\dfrac{-24}{6}$$
$$x>-4$$

The solution set is $\{x\,|\,x>-4\}$ or $(-4,\infty)$.

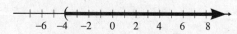

8.
$$5(x-3)\ge 7(x-4)+3$$
$$5x-15\ge 7x-28+3$$
$$5x-15\ge 7x-25$$
$$5x-7x-15\ge 7x-7x-25$$
$$-2x-15\ge -25$$
$$-2x-15+15\ge -25+15$$
$$-2x\ge -10$$
$$\dfrac{-2x}{-2}\le \dfrac{-10}{-2}$$
$$x\le 5$$

The solution set is $\{x\,|\,x\le 5\}$ or $(-\infty,5]$.

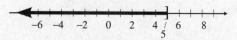

9.

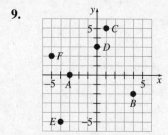

448

10. To find the *y*-intercept, let $x = 0$:

$$-4(0) + 3y = 24$$
$$0 + 3y = 24$$
$$3y = 24$$
$$y = 8$$

To find the *x*-intercept, let $y = 0$:

$$-4x + 3(0) = 24$$
$$-4x + 0 = 24$$
$$-4x = 24$$
$$x = -6$$

The *y*-intercept is 8 and the *x*-intercept is -6, so the points $(0, 8)$ and $(-6, 0)$ are on the graph.

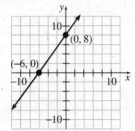

11. The graph of $x = 5$ is a vertical line that passes through the point $(5, 0)$.

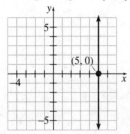

12. Let $(x_1, y_1) = (3, 6)$ and $(x_2, y_2) = (-1, -4)$.

$$m = \frac{y_2 - y_1}{x_2 - x_1} = \frac{-4 - 6}{-1 - 3} = \frac{-10}{-4} = \frac{5}{2}$$

13. $y = 3x + 1$

The slope is $m = 3$ and the *y*-intercept is $b = 1$. Plot the point $(0, 1)$. Use the slope

$$m = \frac{3}{1} = \frac{\text{rise}}{\text{run}}$$ to find a second point on the line.

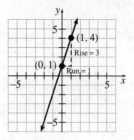

14. We use point-slope form of a line with $m = -\dfrac{4}{3}$ and $(-3, 1)$.

$$y - y_1 = m(x - x_1)$$
$$y - 1 = -\frac{4}{3}(x - (-3))$$
$$y - 1 = -\frac{4}{3}(x + 3)$$
$$y - 1 = -\frac{4}{3}x - 4$$
$$y = -\frac{4}{3}x - 3$$

15. Let $(x_1, y_1) = (-2, 5)$ and $(x_2, y_2) = (2, 3)$.

$$m = \frac{y_2 - y_1}{x_2 - x_1} = \frac{3 - 5}{2 - (-2)} = \frac{-2}{4} = -\frac{1}{2}$$
$$y - y_1 = m(x - x_1)$$
$$y - 5 = -\frac{1}{2}(x - (-2))$$
$$y - 5 = -\frac{1}{2}(x + 2)$$
$$y - 5 = -\frac{1}{2}x - 1$$
$$y = -\frac{1}{2}x + 4$$

16. Let $(x_1, y_1) = (-5, 7)$. The slope of the line $y = -3x + 10$ is -3, so the slope of the parallel line is also -3. Thus, the equation of the parallel line is:

$$y - y_1 = m(x - x_1)$$
$$y - 7 = -3(x - (-5))$$
$$y - 7 = -3(x + 5)$$
$$y - 7 = -3x - 15$$
$$y = -3x - 8$$

17. $x - 3y > 12$

Replace the inequality symbol with an equal sign to obtain $x - 3y = 12$. Because the inequality is strict, graph $x - 3y = 12$ $\left(y = \frac{1}{3}x - 4 \right)$ using a dashed line.

Test Point: $(0,0)$: $(0) - 3(0) \overset{?}{>} 12$
$$0 - 0 \overset{?}{>} 12$$
$$0 > 12 \quad \text{False}$$

Therefore, the half-plane not containing $(0,0)$ is the solution set of $x - 3y > 12$.

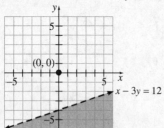

18. $\begin{cases} y = 4x - 3 & (1) \\ 4x - 3y = 5 & (2) \end{cases}$

Substituting $4x - 3$ for y in equation (2), we get
$$4x - 3(4x - 3) = 5$$
$$4x - 12x + 9 = 5$$
$$-8x + 9 = 5$$
$$-8x = -4$$
$$x = \frac{-4}{-8} = \frac{1}{2}$$

Substituting $\frac{1}{2}$ for x in equation (1), we obtain
$$y = 4\left(\frac{1}{2} \right) - 3 = 2 - 3 = -1.$$

The solution is the ordered pair $\left(\frac{1}{2}, -1 \right)$.

19. $\begin{cases} x + y = 3 & (1) \\ 3x + 2y = 2 & (2) \end{cases}$

Multiply both sides of equation (1) by -2, and add the result to equation (2).
$$\begin{cases} -2x - 2y = -6 \\ 3x + 2y = 2 \end{cases}$$
$$\overline{ x = -4}$$

Substituting -4 for x in equation (1), we obtain
$$-4 + y = 3$$
$$y = 7$$

The solution is the ordered pair $(-4,\ 7)$.

20. $\begin{cases} x + y \ge 2 \\ -3x + y \le 10 \end{cases}$

First, graph the inequality $x + y \ge 2$. To do so, replace the inequality symbol with an equal sign to obtain $x + y = 2$. Because the inequality is non-strict, graph $x + y = 2$ $(y = -x + 2)$ using a solid line.

Test point: $(0,0)$: $0 + 0 \overset{?}{\ge} 2$
$$0 \ge 2 \quad \text{False}$$

Therefore, the half-plane not containing $(0,0)$ is the solution set of $x + y \ge 2$.

Second, graph the inequality $-3x + y \le 10$. To do so, replace the inequality symbol with an equal sign to obtain $-3x + y = 10$. Because the inequality is non-strict, graph $-3x + y = 10$ $(y = 3x + 10)$ using a solid line.

Test point: $(0,0)$: $-3(0) + 0 \overset{?}{\le} 10$
$$0 + 0 \overset{?}{\le} 10$$
$$0 \le 10 \quad \text{True}$$

Therefore, the half-plane containing $(0, 0)$ is the solution set of $-3x + y \le 10$.

The overlapping shaded region (that is, the shaded region in the graph below) is the solution to the system of linear inequalities.

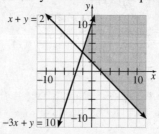

21. $\left(12x^3 + 5x^2 - 3x + 1 \right) - \left(2x^3 - 4x + 8 \right)$
$$= 12x^3 + 5x^2 - 3x + 1 - 2x^3 + 4x - 8$$
$$= 12x^3 - 2x^3 + 5x^2 - 3x + 4x + 1 - 8$$
$$= 10x^3 + 5x^2 + x - 7$$

22. $(2x-5)(x+3)$

$= 2x(x) + 2x(3) - 5(x) - 5(3)$

$= 2x^2 + 6x - 5x - 15$

$= 2x^2 + x - 15$

23.
$$x-4 \overline{)\begin{array}{r} x^2+2x+3 \\ x^3-2x^2-5x-3 \end{array}}$$

$\quad\quad\ \underline{-\left(x^3-4x^2\right)}$

$\quad\quad\quad\quad\ 2x^2-5x$

$\quad\quad\quad\quad\ \underline{-\left(2x^2-8x\right)}$

$\quad\quad\quad\quad\quad\quad 3x-3$

$\quad\quad\quad\quad\quad\quad \underline{-(3x-12)}$

$\quad\quad\quad\quad\quad\quad\quad\quad 9$

$\dfrac{x^3-2x^2-5x-3}{x-4} = x^2 + 2x + 3 + \dfrac{9}{x-4}$

24. $-2x^3 + 6x^2 - 8x + 24 = -2\left(x^3 - 3x^2 + 4x - 12\right)$

$\quad\quad\quad\quad\quad\quad\quad\quad = -2\left[x^2(x-3) + 4(x-3)\right]$

$\quad\quad\quad\quad\quad\quad\quad\quad = -2\left[(x-3)\left(x^2+4\right)\right]$

$\quad\quad\quad\quad\quad\quad\quad\quad = -2(x-3)\left(x^2+4\right)$

25. $5p^3 + 50p^2 + 80p = 5p\left(p^2 + 10p + 16\right)$

$\quad\quad\quad\quad\quad\quad\quad = 5p(p+8)(p+2)$

26. $8y^2 - 2y - 3 = 8y^2 + 4y - 6y - 3$

$\quad\quad\quad\quad\quad\ = 4y(2y+1) - 3(2y+1)$

$\quad\quad\quad\quad\quad\ = (2y+1)(4y-3)$

27. $9a^2 + 24a + 16 = (3a)^2 + 2(3a)(4) + (4)^2$

$\quad\quad\quad\quad\quad\quad\ = (3a+4)^2$

28. $3k^4 - 27k^2 = 3k^2\left(k^2 - 9\right)$

$\quad\quad\quad\quad\ = 3k^2\left(k^2 - 3^2\right)$

$\quad\quad\quad\quad\ = 3k^2(k+3)(k-3)$

29. $\quad\quad 8b^2 + 10b = 3$

$\quad\quad 8b^2 + 10b - 3 = 0$

$\quad (4b-1)(2b+3) = 0$

$\quad 4b - 1 = 0 \quad\text{or}\quad 2b + 3 = 0$

$\quad\quad 4b = 1 \quad\quad\quad\quad 2b = -3$

$\quad\quad b = \dfrac{1}{4} \quad\quad\quad\quad b = -\dfrac{3}{2}$

The solution set is $\left\{-\dfrac{3}{2}, \dfrac{1}{4}\right\}$.

30. $\quad (2x+3)(x-1) = 6x$

$\quad 2x^2 - 2x + 3x - 3 = 6x$

$\quad\quad 2x^2 - 5x - 3 = 0$

$\quad (2x+1)(x-3) = 0$

$\quad 2x + 1 = 0 \quad\text{or}\quad x - 3 = 0$

$\quad\quad 2x = -1 \quad\quad\quad\quad x = 3$

$\quad\quad x = -\dfrac{1}{2}$

The solution set is $\left\{-\dfrac{1}{2}, 3\right\}$.

31. $\dfrac{2x^2 + x - 21}{x^2 + 6x - 27} = \dfrac{(2x+7)(x-3)}{(x+9)(x-3)}$

$\quad\quad\quad\quad\quad = \dfrac{(2x+7)\,\cancel{(x-3)}}{(x+9)\,\cancel{(x-3)}}$

$\quad\quad\quad\quad\quad = \dfrac{2x+7}{x+9}$

32. $\dfrac{3x^2 + 14x - 5}{x^2 + x - 30} \cdot \dfrac{x^2 - 2x - 15}{3x^2 + 8x - 3}$

$\quad = \dfrac{(3x-1)(x+5)}{(x+6)(x-5)} \cdot \dfrac{(x-5)(x+3)}{(3x-1)(x+3)}$

$\quad = \dfrac{\cancel{(3x-1)}(x+5)}{(x+6)\cancel{(x-5)}} \cdot \dfrac{\cancel{(x-5)}\,\cancel{(x+3)}}{\cancel{(3x-1)}\,\cancel{(x+3)}}$

$\quad = \dfrac{x+5}{x+6}$

33. $\dfrac{\dfrac{y^2-9}{2y^2-y-15}}{\dfrac{3y^2+10y+3}{2y^2+y-10}} = \dfrac{y^2-9}{2y^2-y-15} \cdot \dfrac{2y^2+y-10}{3y^2+10y+3}$

$$= \frac{(y-3)(y+3)}{(2y+5)(y-3)} \cdot \frac{(2y+5)(y-2)}{(3y+1)(y+3)}$$

$$= \frac{(y-3)(y+3)}{(2y+5)(y-3)} \cdot \frac{(2y+5)(y-2)}{(3y+1)(y+3)}$$

$$= \frac{y-2}{3y+1}$$

34. $x-3$
$x+2$
$\text{LCD} = (x-3)(x+2)$

$$\frac{2x}{x-3} - \frac{x+1}{x+2} = \frac{2x}{x-3} \cdot \frac{x+2}{x+2} - \frac{x+1}{x+2} \cdot \frac{x-3}{x-3}$$

$$= \frac{2x^2+4x}{(x-3)(x+2)} - \frac{x^2-2x-3}{(x-3)(x+2)}$$

$$= \frac{2x^2+4x-(x^2-2x-3)}{(x-3)(x+2)}$$

$$= \frac{2x^2+4x-x^2+2x+3}{(x-3)(x+2)}$$

$$= \frac{x^2+6x+3}{(x-3)(x+2)}$$

35. $\dfrac{9}{k-2} = \dfrac{6}{k}+3$

$$k(k-2)\left(\frac{9}{k-2}\right) = k(k-2)\left(\frac{6}{k}+3\right)$$

$$9k = 6(k-2)+3k(k-2)$$

$$9k = 6k-12+3k^2-6k$$

$$0 = 3k^2-9k-12$$

$$0 = k^2-3k-4$$

$$0 = (k-4)(k+1)$$

$$k-4=0 \quad \text{or} \quad k+1=0$$

$$k=4 \qquad\qquad k=-1$$

Check: $k=-1$ 　　　　 Check: $k=4$

$\dfrac{9}{-1-2} \overset{?}{=} \dfrac{6}{-1}+3$ 　　　 $\dfrac{9}{4-2} \overset{?}{=} \dfrac{6}{4}+3$

$\dfrac{9}{-3} \overset{?}{=} -6+3$ 　　　　 $\dfrac{9}{2} \overset{?}{=} \dfrac{3}{2}+\dfrac{6}{2}$

　　　　　　　　　　　　 $\dfrac{9}{2} = \dfrac{9}{2}$ True

$-3=-3$ True

Both check. The solution set is $\{-1, 4\}$.

36. $\dfrac{7}{y^2+y-12} - \dfrac{4y}{y^2+7y+12} = \dfrac{6}{y^2-9}$

$$\frac{7}{(y-3)(y+4)} - \frac{4y}{(y+3)(y+4)} = \frac{6}{(y-3)(y+3)}$$

$$(y-3)(y+3)(y+4)\left(\frac{7}{(y-3)(y+4)} - \frac{4y}{(y+3)(y+4)}\right) =$$

$$(y-3)(y+3)(y+4)\left(\frac{6}{(y-3)(y+3)}\right)$$

$$7(y+3)-4y(y-3) = 6(y+4)$$

$$7y+21-4y^2+12y = 6y+24$$

$$-4y^2+19y+21 = 6y+24$$

$$0 = 4y^2-13y+3$$

$$0 = (4y-1)(y-3)$$

$$4y-1=0 \quad \text{or} \quad y-3=0$$

$$y=\frac{1}{4} \quad \text{or} \qquad y=3$$

Since $y=3$ is not in the domain of the variable, it is an extraneous solution.

Check $y=\dfrac{1}{4}$:

$$\frac{7}{\left(\frac{1}{4}\right)^2+\left(\frac{1}{4}\right)-12} - \frac{4\left(\frac{1}{4}\right)}{\left(\frac{1}{4}\right)^2+7\left(\frac{1}{4}\right)+12} \overset{?}{=} \frac{6}{\left(\frac{1}{4}\right)^2-9}$$

$$\frac{7}{\frac{1}{16}+\frac{1}{4}-12} - \frac{1}{\frac{1}{16}+\frac{7}{4}+12} \overset{?}{=} \frac{6}{\frac{1}{16}-9}$$

$$\frac{7}{\frac{1}{16}+\frac{1}{4}-12}\cdot\frac{16}{16} - \frac{1}{\frac{1}{16}+\frac{7}{4}+12}\cdot\frac{16}{16} \overset{?}{=} \frac{6}{\frac{1}{16}-9}\cdot\frac{16}{16}$$

$$\frac{112}{1+4-192} - \frac{16}{1+28+192} \overset{?}{=} \frac{96}{1-144}$$

$$\frac{112}{-187} - \frac{16}{221} \overset{?}{=} \frac{96}{-143}$$

$$-\frac{1456}{2431} - \frac{176}{2431} \overset{?}{=} \frac{96}{143}$$

$$-\frac{1632}{2431} \overset{?}{=} \frac{96}{143}$$

$$-\frac{96}{143} = -\frac{96}{143} \quad \text{True}$$

The solution set is $\left\{\dfrac{1}{4}\right\}$.

Chapter 8

Getting Ready for Chapter 8

Getting Ready for Chapter 8 Quick Checks

1. closed interval

2. left endpoint; right endpoint

3. $-3 \le x \le 2$
Interval: $[-3, 2]$
Graph:

4. $3 \le x < 6$
Interval: $[3, 6)$
Graph:

5. $x \le 3$
Interval: $(-\infty, 3]$
Graph:

6. $\dfrac{1}{2} < x < \dfrac{7}{2}$
Interval: $\left(\dfrac{1}{2}, \dfrac{7}{2}\right)$
Graph:

7. $(0, 5]$
Inequality: $0 < x \le 5$
Graph:

8. $(-6, 0)$
Inequality: $-6 < x < 0$
Graph:

9. $(5, \infty)$
Inequality: $x > 5$
Graph:

10. $\left(-\infty, \dfrac{8}{3}\right]$
Inequality: $x \le \dfrac{8}{3}$
Graph:

Getting Ready for Chapter 8 Exercises

11. $[2, 10]$

13. $[-4, 0)$

15. $[6, \infty)$

17. $\left(-\infty, \dfrac{3}{2}\right)$

19. $1 < x < 8$

21. $-5 < x \le 1$

23. $x < 5$

25. $x \ge 3$

453

Section 8.1

Preparing for Graphs of Equations

P1.

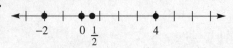

P2. $3x - 5(x + 2) = 4$

(a) Let $x = 0$ in the equation.
$$3(0) - 5(0 + 2) \stackrel{?}{=} 4$$
$$0 - 5(2) \stackrel{?}{=} 4$$
$$-10 = 4 \text{ False}$$
$x = 0$ is not a solution to the equation.

(b) Let $x = -3$ in the equation.
$$3(-3) - 5(-3 + 2) \stackrel{?}{=} 4$$
$$-9 - 5(-1) \stackrel{?}{=} 4$$
$$-9 + 5 \stackrel{?}{=} 4$$
$$-4 = 4 \text{ False}$$
$x = -3$ is not a solution to the equation.

(c) Let $x = -7$ in the equation.
$$3(-7) - 5(-7 + 2) \stackrel{?}{=} 4$$
$$-21 - 5(-5) \stackrel{?}{=} 4$$
$$-21 + 25 \stackrel{?}{=} 4$$
$$4 = 4 \text{ True}$$
$x = -7$ is a solution to the equation.

P3. **(a)** Let $x = 0$:
$$2x^2 - 3x + 1 = 2(0)^2 - 3(0) + 1$$
$$= 0 - 0 + 1$$
$$= 1$$

(b) Let $x = 2$:
$$2x^2 - 3x + 1 = 2(2)^2 - 3(2) + 1$$
$$= 2(4) - 6 + 1$$
$$= 8 - 6 + 1$$
$$= 3$$

(c) Let $x = -3$:
$$2x^2 - 3x + 1 = 2(-3)^2 - 3(-3) + 1$$
$$= 2(9) + 9 + 1$$
$$= 18 + 9 + 1$$
$$= 28$$

P4.
$$3x + 2y = 8$$
$$3x + 2y - 3x = 8 - 3x$$
$$2y = 8 - 3x$$
$$\frac{2y}{2} = \frac{8 - 3x}{2}$$
$$y = \frac{8}{2} - \frac{3x}{2}$$
$$y = 4 - \frac{3}{2}x \quad \text{or} \quad y = -\frac{3}{2}x + 4$$

P5. $|-4| = 4$ because -4 is 4 units away from 0 on a real number line.

P6. (3, 5): Quadrant I
(−2, 3): Quadrant II
(−1, −2): Quadrant III
(5, −3): Quadrant IV
(0, 4): y-axis

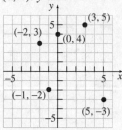

P7. $3x - 2y = 7$

(a) (1, −2):
$$3(1) - 2(-2) \stackrel{?}{=} 7$$
$$3 + 4 \stackrel{?}{=} 7$$
$$7 = 7 \quad \text{True}$$
(1, −2) is a solution.

(b) (3, 1):
$$3(3) - 2(1) \stackrel{?}{=} 7$$
$$9 - 2 \stackrel{?}{=} 7$$
$$7 = 7 \quad \text{True}$$
(3, 1) is a solution.

(c) (−2, −5):
$$3(-2) - 2(-5) \stackrel{?}{=} 7$$
$$-6 + 10 \stackrel{?}{=} 7$$
$$4 = 7 \quad \text{False}$$
(−2, −5) is not a solution.

P8. $y = -4x + 5$
$m = -4$
$b = 5$

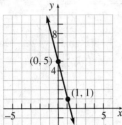

P9. $3x - 5y = 15$
Let $x = 0$:
$3(0) - 5y = 15$
$0 - 5y = 15$
$-5y = 15$
$y = -3$
y-intercept is -3.
Let $y = 0$:
$3x - 5(0) = 15$
$3x - 0 = 15$
$3x = 15$
$x = 5$

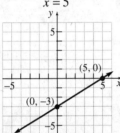

Section 8.1 Quick Checks

1. origin

2. True

3. A: quadrant I
   B: quadrant IV
   C: y-axis
   D: quadrant III

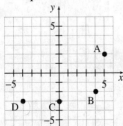

4. A: quadrant II
   B: x-axis
   C: quadrant IV
   D: quadrant I

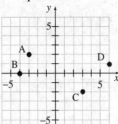

5. True

6. **(a)** Let $x = 2$ and $y = -3$.
$$2(2) - 4(-3) \overset{?}{=} 12$$
$$4 + 12 \overset{?}{=} 12$$
$$16 = 12 \text{ False}$$
The point $(2, -3)$ is not on the graph.

   (b) Let $x = 2$ and $y = -2$.
$$2(2) - 4(-2) \overset{?}{=} 12$$
$$4 + 8 \overset{?}{=} 12$$
$$12 = 12 \text{ T}$$
The point $(2, -2)$ is on the graph.

   (c) Let $x = \dfrac{3}{2}$ and $y = -\dfrac{9}{4}$.
$$2\left(\frac{3}{2}\right) - 4\left(-\frac{9}{4}\right) \overset{?}{=} 12$$
$$3 + 9 \overset{?}{=} 12$$
$$12 = 12 \text{ T}$$
The point $\left(\dfrac{3}{2}, -\dfrac{9}{4}\right)$ is on the graph.

7. **(a)** Let $x = 1$ and $y = 4$.
$$(4) \overset{?}{=} (1)^2 + 3$$
$$4 \overset{?}{=} 1 + 3$$
$$4 = 4 \text{ T}$$
The point $(1, 4)$ is on the graph.

   (b) Let $x = -2$ and $y = -1$.
$$(-1) \overset{?}{=} (-2)^2 + 3$$
$$-1 \overset{?}{=} 4 + 3$$
$$-1 = 7 \text{ False}$$
The point $(-2, -1)$ is not on the graph.

(c) Let $x = -3$ and $y = 12$.

$$(12) \overset{?}{=} (-3)^2 + 3$$
$$12 \overset{?}{=} 9 + 3$$
$$12 = 12 \text{ T}$$

The point $(-3, 12)$ is on the graph.

8. $y = 3x + 1$

x	$y = 3x + 1$	(x, y)
-2	$y = 3(-2) + 1 = -5$	$(-2, -5)$
-1	$y = 3(-1) + 1 = -2$	$(-1, -2)$
0	$y = 3(0) + 1 = 1$	$(0, 1)$
1	$y = 3(1) + 1 = 4$	$(1, 4)$
2	$y = 3(2) + 1 = 7$	$(2, 7)$

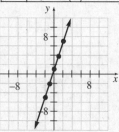

9. $2x + 3y = 8$
$$3y = -2x + 8$$
$$y = -\frac{2}{3}x + \frac{8}{3}$$

x	$y = -\frac{2}{3}x + \frac{8}{3}$	(x, y)
-5	$y = -\frac{2}{3}(-5) + \frac{8}{3} = 6$	$(-5, 6)$
-2	$y = -\frac{2}{3}(-2) + \frac{8}{3} = 4$	$(-2, 4)$
1	$y = -\frac{2}{3}(1) + \frac{8}{3} = 2$	$(1, 2)$
4	$y = -\frac{2}{3}(4) + \frac{8}{3} = 0$	$(4, 0)$
7	$y = -\frac{2}{3}(7) + \frac{8}{3} = -2$	$(7, -2)$

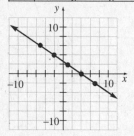

10. $y = x^2 + 3$

x	$y = x^2 + 3$	(x, y)
-2	$y = (-2)^2 + 3 = 7$	$(-2, 7)$
-1	$y = (-1)^2 + 3 = 4$	$(-1, 4)$
0	$y = (0)^2 + 3 = 3$	$(0, 3)$
1	$y = (1)^2 + 3 = 4$	$(1, 4)$
2	$y = (2)^2 + 3 = 7$	$(2, 7)$

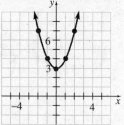

11. $x = y^2 + 2$

y	$x = y^2 + 2$	(x, y)
-2	$x = (-2)^2 + 2 = 6$	$(6, -2)$
-1	$x = (-1)^2 + 2 = 3$	$(3, -1)$
0	$x = (0)^2 + 2 = 2$	$(2, 0)$
1	$x = (1)^2 + 2 = 3$	$(3, 1)$
2	$x = (2)^2 + 2 = 6$	$(6, 2)$

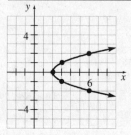

12. $x = (y - 1)^2$

y	$x = (y - 1)^2$	(x, y)
-1	$x = (-1 - 1)^2 = 4$	$(4, -1)$
0	$x = (0 - 1)^2 = 1$	$(1, 0)$
1	$x = (1 - 1)^2 = 0$	$(0, 1)$
2	$x = (2 - 1)^2 = 1$	$(1, 2)$
3	$x = (3 - 1)^2 = 4$	$(4, 3)$

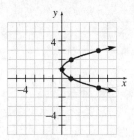

13. intercepts

14. False

15. To find the intercepts, we look for the points where the graph crosses or touches either coordinate axis. From the graph, we see that the intercepts are $(-5, 0)$, $(0, -0.9)$, $(1, 0)$, and $(6.7, 0)$.

The x-intercepts are the x-coordinates of the points where the graph crosses or touches the x-axis. From the graph we see that the x-intercepts are -5, 1, and 6.7.

The y-intercept is the y-coordinate of the point where the graph crosses the y-axis. From the graph we see that the y-intercept is -0.9.

16. **(a)** Locate the value 250 along the x-axis, go up to the graph, and then read the corresponding value on the y-axis. According to the graph, the cost of refining 250 thousand gallons of gasoline per hour is $200 thousand.

(b) Locate the value 400 along the x-axis, go up to the graph, and then read the corresponding value on the y-axis. According to the graph, the cost of refining 400 thousand gallons of gasoline per hour is $350 thousand.

(c) Since the horizontal axis represents the number of gallons per hour that can be refined, the graph ending at 700 thousand gallons per hour represents the capacity of the refinery per hour.

(d) The intercept is $(0, 100)$. This represents the fixed costs of operating the refinery. That is, refining 0 gallons per hour costs $100 thousand.

8.1 Exercises

17. $A : (2, 3)$; in quadrant I
$B : (-5, 2)$; in quadrant II
$C : (0, -2)$; on the y-axis
$D : (-4, -3)$; in quadrant III
$E : (3, -4)$; in quadrant IV
$F : (4, 0)$; on the x-axis

19. A: quadrant I; B: quadrant III; C: x-axis; D: quadrant IV; E: y-axis; F: quadrant II

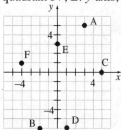

21. **(a)** Let $x = 1$ and $y = 2$.
$$2(1) + 5(2) \stackrel{?}{=} 12$$
$$2 + 10 \stackrel{?}{=} 12$$
$$12 = 12 \text{ True}$$
The point $(1, 2)$ is on the graph.

(b) Let $x = -2$ and $y = 3$.
$$2(-2) + 5(3) \stackrel{?}{=} 12$$
$$-4 + 15 \stackrel{?}{=} 12$$
$$11 = 12 \text{ False}$$
The point $(-2, 3)$ is not on the graph.

(c) Let $x = -4$ and $y = 4$.
$$2(-4) + 5(4) \stackrel{?}{=} 12$$
$$-8 + 20 \stackrel{?}{=} 12$$
$$12 = 12 \text{ True}$$
The point $(-4, 4)$ is on the graph.

(d) Let $x = -\dfrac{3}{2}$ and $y = 3$.
$$2\left(-\dfrac{3}{2}\right) + 5(3) \stackrel{?}{=} 12$$
$$-3 + 15 \stackrel{?}{=} 12$$
$$12 = 12 \text{ True}$$
The point $\left(-\dfrac{3}{2}, 3\right)$ is on the graph.

23. (a) Let $x = -2$ and $y = -15$.

$$(-15) \stackrel{?}{=} -2(-2)^2 + 3(-2) - 1$$
$$-15 \stackrel{?}{=} -2(4) - 6 - 1$$
$$-15 \stackrel{?}{=} -8 - 6 - 1$$
$$-15 = -15 \text{ True}$$

The point $(-2, -15)$ is on the graph.

(b) Let $x = 3$ and $y = 10$.

$$(10) \stackrel{?}{=} -2(3)^2 + 3(3) - 1$$
$$10 \stackrel{?}{=} -2(9) + 9 - 1$$
$$10 \stackrel{?}{=} -18 + 9 - 1$$
$$10 = -10 \text{ False}$$

The point $(3, 10)$ is not on the graph.

(c) Let $x = 0$ and $y = 1$.

$$(1) \stackrel{?}{=} -2(0)^2 + 3(0) - 1$$
$$1 \stackrel{?}{=} 0 + 0 - 1$$
$$1 = -1 \text{ False}$$

The point $(0, 1)$ is not on the graph.

(d) Let $x = 2$ and $y = -3$.

$$(-3) \stackrel{?}{=} -2(2)^2 + 3(2) - 1$$
$$-3 \stackrel{?}{=} -2(4) + 6 - 1$$
$$-3 \stackrel{?}{=} -8 + 6 - 1$$
$$-3 = -3 \text{ True}$$

The point $(2, -3)$ is on the graph.

25. (a) Let $x = 1$ and $y = 4$.

$$(4) \stackrel{?}{=} |1 - 3|$$
$$4 \stackrel{?}{=} |-2|$$
$$4 = 2 \text{ False}$$

The point $(1, 4)$ is not on the graph.

(b) Let $x = 4$ and $y = 1$.

$$(1) \stackrel{?}{=} |4 - 3|$$
$$1 \stackrel{?}{=} |1|$$
$$1 = 1 \text{ True}$$

The point $(4, 1)$ is on the graph.

(c) Let $x = -6$ and $y = 9$.

$$(9) \stackrel{?}{=} |-6 - 3|$$
$$9 \stackrel{?}{=} |-9|$$
$$9 = 9 \text{ True}$$

The point $(-6, 9)$ is on the graph.

(d) Let $x = 0$ and $y = 3$.

$$(3) \stackrel{?}{=} |0 - 3|$$
$$3 \stackrel{?}{=} |-3|$$
$$3 = 3 \text{ True}$$

The point $(0, 3)$ is on the graph.

27. $y = 4x$

x	$y = 4x$	(x, y)
-2	$y = 4(-2) = -8$	$(-2, -8)$
-1	$y = 4(-1) = -4$	$(-1, -4)$
0	$y = 4(0) = 0$	$(0, 0)$
1	$y = 4(1) = 4$	$(1, 4)$
2	$y = 4(2) = 8$	$(2, 8)$

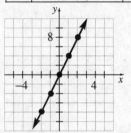

29. $y = -\dfrac{1}{2}x$

x	$y = -\dfrac{1}{2}x$	(x, y)
-4	$y = -\dfrac{1}{2}(-4) = 2$	$(-4, 2)$
-2	$y = -\dfrac{1}{2}(-2) = 1$	$(-2, 1)$
0	$y = -\dfrac{1}{2}(0) = 0$	$(0, 0)$
2	$y = -\dfrac{1}{2}(2) = -1$	$(2, -1)$
4	$y = -\dfrac{1}{2}(4) = -2$	$(4, -2)$

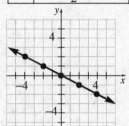

31. $y = x + 3$

x	$y = x + 3$	(x, y)
–4	$y = (-4) + 3 = -1$	$(-4, -1)$
–2	$y = (-2) + 3 = 1$	$(-2, 1)$
0	$y = (0) + 3 = 3$	$(0, 3)$
2	$y = (2) + 3 = 5$	$(2, 5)$
4	$y = (4) + 3 = 7$	$(4, 7)$

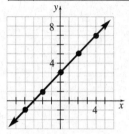

33. $y = -3x + 1$

x	$y = -3x + 1$	(x, y)
–2	$y = -3(-2) + 1 = 7$	$(-2, 7)$
–1	$y = -3(-1) + 1 = 4$	$(-1, 4)$
0	$y = -3(0) + 1 = 1$	$(0, 1)$
1	$y = -3(1) + 1 = -2$	$(1, -2)$
2	$y = -3(2) + 1 = -5$	$(2, -5)$

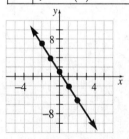

35. $y = \dfrac{1}{2}x - 4$

x	$y = \dfrac{1}{2}x - 4$	(x, y)
–4	$y = \dfrac{1}{2}(-4) - 4 = -6$	$(-4, -6)$
–2	$y = \dfrac{1}{2}(-2) - 4 = -5$	$(-2, -5)$
0	$y = \dfrac{1}{2}(0) - 4 = -4$	$(0, -4)$
2	$y = \dfrac{1}{2}(2) - 4 = -3$	$(2, -3)$
4	$y = \dfrac{1}{2}(4) - 4 = -2$	$(4, -2)$

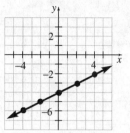

37. $2x + y = 7$
$y = -2x + 7$

x	$y = -2x + 7$	(x, y)
–5	$y = -2(-5) + 7 = 17$	$(-5, 17)$
–3	$y = -2(-3) + 7 = 13$	$(-3, 13)$
0	$y = -2(0) + 7 = 7$	$(0, 7)$
3	$y = -2(3) + 7 = 1$	$(3, 1)$
5	$y = -2(5) + 7 = -3$	$(5, -3)$

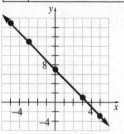

39. $y = -x^2$

x	$y = -x^2$	(x, y)
–3	$y = -(-3)^2 = -9$	$(-3, -9)$
–2	$y = -(-2)^2 = -4$	$(-2, -4)$
0	$y = -(0)^2 = 0$	$(0, 0)$
2	$y = -(2)^2 = -4$	$(2, -4)$
3	$y = -(3)^2 = -9$	$(3, -9)$

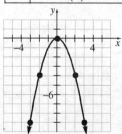

41. $y = 2x^2 - 8$

x	$y = 2x^2 - 8$	(x, y)
-3	$y = 2(-3)^2 - 8 = 10$	$(-3, 10)$
-2	$y = 2(-2)^2 - 8 = 0$	$(-2, 0)$
-1	$y = 2(-1)^2 - 8 = -6$	$(-1, -6)$
0	$y = 2(0)^2 - 8 = -8$	$(0, -8)$
1	$y = 2(1)^2 - 8 = -6$	$(1, -6)$
2	$y = 2(2)^2 - 8 = 0$	$(2, 0)$
3	$y = 2(3)^2 - 8 = 10$	$(3, 10)$

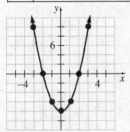

43. $y = |x|$

x	$y = |x|$	(x, y)
-4	$y = |-4| = 4$	$(-4, 4)$
-2	$y = |-2| = 2$	$(-2, 2)$
0	$y = |0| = 0$	$(0, 0)$
2	$y = |2| = 2$	$(2, 2)$
4	$y = |4| = 4$	$(4, 4)$

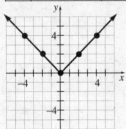

45. $y = |x - 1|$

x	$y = |x - 1|$	(x, y)
-4	$y = |-4 - 1| = 5$	$(-4, 5)$
-1	$y = |-1 - 1| = 2$	$(-1, 2)$
0	$y = |0 - 1| = 1$	$(0, 1)$
1	$y = |1 - 1| = 0$	$(1, 0)$
3	$y = |3 - 1| = 2$	$(3, 2)$
5	$y = |5 - 1| = 4$	$(5, 4)$

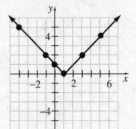

47. $y = x^3$

x	$y = x^3$	(x, y)
-3	$y = (-3)^3 = -27$	$(-3, -27)$
-2	$y = (-2)^3 = -8$	$(-2, -8)$
-1	$y = (-1)^3 = -1$	$(-1, -1)$
0	$y = (0)^3 = 0$	$(0, 0)$
1	$y = (1)^3 = 1$	$(1, 1)$
2	$y = (2)^3 = 8$	$(2, 8)$
3	$y = (3)^3 = 27$	$(3, 27)$

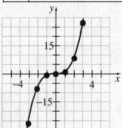

49. $y = x^3 + 1$

x	$y = x^3 + 1$	(x, y)
-2	$y = (-2)^3 + 1 = -7$	$(-2, -7)$
-1	$y = (-1)^3 + 1 = 0$	$(-1, 0)$
0	$y = (0)^3 + 1 = 1$	$(0, 1)$
1	$y = (1)^3 + 1 = 2$	$(1, 2)$
2	$y = (2)^3 + 1 = 9$	$(2, 9)$

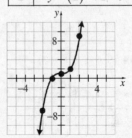

51. $x^2 - y = 4$

$x^2 = y + 4$

$x^2 - 4 = y$

$y = x^2 - 4$

x	$y = x^2 - 4$	(x, y)
-3	$y = (-3)^2 - 4 = 5$	$(-3, 5)$
-2	$y = (-2)^2 - 4 = 0$	$(-2, 0)$
0	$y = (0)^2 - 4 = -4$	$(0, -4)$
2	$y = (2)^2 - 4 = 0$	$(2, 0)$
3	$y = (3)^2 - 4 = 5$	$(3, 5)$

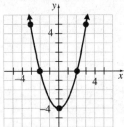

53. $x = y^2 - 1$

y	$x = y^2 - 1$	(x, y)
-2	$x = (-2)^2 - 1 = 3$	$(3, -2)$
-1	$x = (-1)^2 - 1 = 0$	$(0, -1)$
0	$x = (0)^2 - 1 = -1$	$(-1, 0)$
1	$x = (1)^2 - 1 = 0$	$(0, 1)$
2	$x = (2)^2 - 1 = 3$	$(3, 2)$

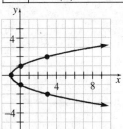

55. The intercepts are $(-2, 0)$ and $(0, 3)$.

57. The intercepts are $(-2, 0)$, $(1, 0)$, and $(0, -4)$.

59. If $(a, 4)$ is a point on the graph, then we have

$4 = 4(a) - 3$

$4 = 4a - 3$

$7 = 4a$

$\dfrac{7}{4} = a$

We need $a = \dfrac{7}{4}$.

61. If $(3, b)$ is a point on the graph, then we have

$b = (3)^2 - 2(3) + 1 = 9 - 6 + 1 = 4$

We need $b = 4$.

63. **(a)** According to the graph, the area of the opening will be 400 ft^2 when the width is 10 feet.

(b) According to the graph, the area of the opening will have a maximum value of 625 ft^2 when the width is 25 feet.

(c) The *x*-intercepts are $x = 0$ and $x = 50$. These values indicate that the window opening will have 0 area when the width is 0 feet or 50 feet. These values form the bounds for the width of the opening. The *y*-intercept is $y = 0$. This indicates that the area of the opening will be 0 ft^2 when the width is 0 feet.

65. **(a)** According to the graph, the Sprint PCS 2000-minute plan would cost $100 for 200 minutes of talking in a month and $100 for 500 minutes of talking in a month.

(b) According to the graph the plan would cost $1600 for 8000 minutes of talking in a month.

(c) The only intercept for the graph is the *y*-intercept, $y = 100$. This indicates that the monthly cost will be $100 if no minutes are used. In other words, the base charge for the plan is $100 per month.

67. The set of all points of the form $(4, y)$ form a vertical line with an x-intercept of 4.

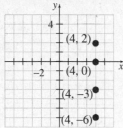

69. Answers will vary. One possible graph is below.

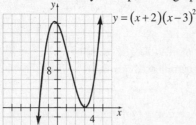

71. Answers will vary. One possibility:
All three points have a y-coordinate of 0. Therefore, they all lie on the horizontal line with equation $y = 0$.

73. A graph is considered to be complete when enough is shown to indicate main features and allow the viewer to 'see' the rest of the graph as an obvious continuation of what is shown.

75. The point-plotting method is when we choose values for one variable and find the corresponding value of the other variable. The resulting ordered pairs form points that lie on the graph of the equation.

77. $y = 3x - 9$

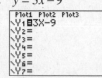

79. $y = -x^2 + 8$

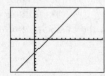

81. $y + 2x^2 = 13$

$y = -2x^2 + 13$

 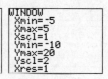

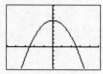

83. $y = x^3 - 6x + 1$

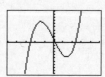

Section 8.2

Preparing for Relations

P1. Inequality: $-4 \le x \le 4$
Interval: $[-4, 4]$
We use square brackets in interval notation because the inequalities are not strict.

P2. Interval: $[2, \infty)$
Inequality: $x \ge 2$
The square bracket indicates a non-strict inequality.

Section 8.2 Quick Checks

1. corresponds; depends

2. The first element of the ordered pair comes from the set 'Friend' and the second element is the corresponding element from the set 'Birthday'.
{(Max, November 8), (Alesia, January 20), (Trent, March 3), (Yolanda, November 8), (Wanda, July 6), (Elvis, January 8)}

3. The first elements of the ordered pairs make up the first step and the second elements make up the second set.

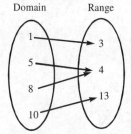

Domain   Range

4. domain; range

5. The domain is the set of all inputs and the range is the set of all outputs. The inputs are the elements in the set 'Friend' and the outputs are the elements in the set 'Birthday'.
Domain:
{Max, Alesia, Trent, Yolanda, Wanda, Elvis}

Range:
{January 20, March 3, July 6, November 8, January 8}

6. The domain is the set of all inputs and the range is the set of all outputs. The inputs are the first elements in the ordered pairs and the outputs are the second elements in the ordered pairs.
Domain: Range:
{1, 5, 8, 10}   {3, 4, 13}

7. First we notice that the ordered pairs on the graph are (–2, 0), (–1, 2), (–1, –2), (2, 3), (3, 0), and (4, –3).
The domain is the set of all *x*-coordinates and the range is the set of all *y*-coordinates.
Domain: Range:
{–2, –1, 2, 3, 4}   {–3, –2, 0, 2, 3}

8. True

9. False

10. To find the domain, we first determine the *x*-values for which the graph exists. The graph exists for all *x*-values between –2 and 4, inclusive. Thus, the domain is $\{x \mid -2 \le x \le 4\}$, or [–2, 4] if we use interval notation.

To find the range, we first determine the *y*-values for which the graph exists. The graph exists for all *y*-values between –2 and 2, inclusive. Thus, the range is $\{y \mid -2 \le y \le 2\}$, or [–2, 2] if we use interval notation.

11. To find the domain, we first determine the *x*-values for which the graph exists. The graph exists for all *x*-values on a real number line. Thus, the domain is $\{x \mid x \text{ is any real number}\}$, or (–∞, ∞) if we use interval notation.
To find the range, we first determine the *y*-values for which the graph exists. The graph exists for all *y*-values on a real number line. Thus, the range is $\{y \mid y \text{ is any real number}\}$, or (–∞, ∞) if we use interval notation.

12. $y = 3x - 8$

x	$y = 3x - 8$	(x, y)
–1	$y = 3(-1) - 8 = -11$	$(-1, -11)$
0	$y = 3(0) - 8 = -8$	$(0, -8)$
1	$y = 3(1) - 8 = -5$	$(1, -5)$
2	$y = 3(2) - 8 = -2$	$(2, -2)$
3	$y = 3(3) - 8 = 1$	$(3, 1)$

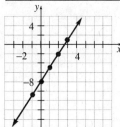

Domain: $\{x \mid x \text{ is any real number}\}$ or $(-\infty, \infty)$

Range: $\{y \mid y \text{ is any real number}\}$ or $(-\infty, \infty)$

13. $y = x^2 - 8$

x	$y = x^2 - 8$	(x, y)
–3	$y = (-3)^2 - 8 = 1$	$(-3, 1)$
–2	$y = (-2)^2 - 8 = -4$	$(-2, -4)$
0	$y = (0)^2 - 8 = -8$	$(0, -8)$
2	$y = (2)^2 - 8 = -4$	$(2, -4)$
3	$y = (3)^2 - 8 = 1$	$(3, 1)$

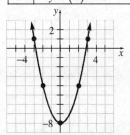

Domain: $\{x \mid x \text{ is any real number}\}$ or $(-\infty, \infty)$

Range: $\{y \mid y \ge -8\}$ or $[-8, \infty)$

14. $x = y^2 + 1$

y	$x = y^2 + 1$	(x, y)
-2	$x = (-2)^2 + 1 = 5$	$(5, -2)$
-1	$x = (-1)^2 + 1 = 2$	$(2, -1)$
0	$x = (0)^2 + 1 = 1$	$(1, 0)$
1	$x = (1)^2 + 1 = 2$	$(2, 1)$
2	$x = (2)^2 + 1 = 5$	$(5, 2)$

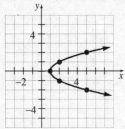

Domain: $\{x \mid x \geq 1\}$ or $[1, \infty)$

Range: $\{y \mid y \text{ is any real number}\}$ or $(-\infty, \infty)$

8.2 Exercises

15. {(USA Today, 2.5), (Wall Street Journal, 2.1),
(New York Times, 1.7),
(Los Angeles Times, 1.2),
(Washington Post, 1.0)}

Domain: {USA Today, Wall Street Journal,
New York Times, Los Angeles Times,
Washington Post}

Range: {1.0, 1.2, 1.7, 2.1, 2.5}

17. {(Less than 9th Grade, \$16,321),
(9th-12th Grade – No Diploma, \$20,934),
(High School Graduate, \$30,134),
(Associate's Degree, \$41,934),
(Bachelor's Degree or Higher, \$58,114)}

Domain: {Less than 9th Grade,
9th-12th Grade – No Diploma,
High School Graduate, Associate's Degree,
Bachelor's Degree or Higher}

Range:
{\$16,321, \$20,934, \$30,134, \$41,934, \$58,114}

19.

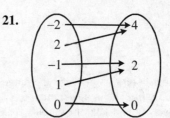

Domain: $\{-3, -2, -1, 0, 1\}$
Range: $\{4, 6, 8, 10, 12\}$

21.

Domain: $\{-2, -1, 0, 1, 2\}$
Range: $\{0, 2, 4\}$

23.

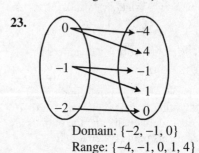

Domain: $\{-2, -1, 0\}$
Range: $\{-4, -1, 0, 1, 4\}$

25. Domain: $\{-3, -2, 0, 2, 3\}$
Range: $\{-3, -1, 2, 3\}$

27. Domain: $\{x \mid -4 \leq x \leq 4\}$ or $[-4, 4]$
Range: $\{y \mid -2 \leq y \leq 2\}$ or $[-2, 2]$

29. Domain: $\{x \mid -1 \leq x \leq 3\}$ or $[-1, 3]$
Range: $\{y \mid 0 \leq y \leq 4\}$ or $[0, 4]$

31. Domain: $\{x \mid x \text{ is a real number}\}$ or $(-\infty, \infty)$
Range: $\{y \mid y \geq -3\}$ or $[-3, \infty)$

33. $y = -3x + 1$

x	$y = -3x + 1$	(x, y)
-2	$y = -3(-2) + 1 = 7$	$(-2, 7)$
-1	$y = -3(-1) + 1 = 4$	$(-1, 4)$
0	$y = -3(0) + 1 = 1$	$(0, 1)$
1	$y = -3(1) + 1 = -2$	$(1, -2)$
2	$y = -3(2) + 1 = -5$	$(2, -5)$

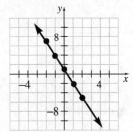

Domain: $\{x \mid x$ is a real number$\}$ or $(-\infty, \infty)$

Range: $\{y \mid y$ is a real number$\}$ or $(-\infty, \infty)$

35. $y = \dfrac{1}{2}x - 4$

x	$y = \dfrac{1}{2}x - 4$	(x, y)
-4	$y = \dfrac{1}{2}(-4) - 4 = -6$	$(-4, -6)$
-2	$y = \dfrac{1}{2}(-2) - 4 = -5$	$(-2, -5)$
0	$y = \dfrac{1}{2}(0) - 4 = -4$	$(0, -4)$
2	$y = \dfrac{1}{2}(2) - 4 = -3$	$(2, -3)$
4	$y = \dfrac{1}{2}(4) - 4 = -2$	$(4, -2)$

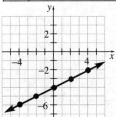

Domain: $\{x \mid x$ is a real number$\}$ or $(-\infty, \infty)$

Range: $\{y \mid y$ is a real number$\}$ or $(-\infty, \infty)$

37. $2x + y = 7$

$y = -2x + 7$

x	$y = -2x + 7$	(x, y)
-5	$y = -2(-5) + 7 = 17$	$(-5, 17)$
-3	$y = -2(-3) + 7 = 13$	$(-3, 13)$
0	$y = -2(0) + 7 = 7$	$(0, 7)$
3	$y = -2(3) + 7 = 1$	$(3, 1)$
5	$y = -2(5) + 7 = -3$	$(5, -3)$

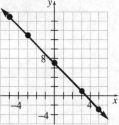

Domain: $\{x \mid x$ is a real number$\}$ or $(-\infty, \infty)$

Range: $\{y \mid y$ is a real number$\}$ or $(-\infty, \infty)$

39. $y = -x^2$

x	$y = -x^2$	(x, y)
-3	$y = -(-3)^2 = -9$	$(-3, -9)$
-2	$y = -(-2)^2 = -4$	$(-2, -4)$
0	$y = -(0)^2 = 0$	$(0, 0)$
2	$y = -(2)^2 = -4$	$(2, -4)$
3	$y = -(3)^2 = -9$	$(3, -9)$

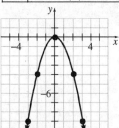

Domain: $\{x \mid x$ is a real number$\}$ or $(-\infty, \infty)$

Range: $\{y \mid y \le 0\}$ or $(-\infty, 0]$

41. $y = 2x^2 - 8$

x	$y = 2x^2 - 8$	(x, y)
-3	$y = 2(-3)^2 - 8 = 10$	$(-3, 10)$
-2	$y = 2(-2)^2 - 8 = 0$	$(-2, 0)$
-1	$y = 2(-1)^2 - 8 = -6$	$(-1, -6)$
0	$y = 2(0)^2 - 8 = -8$	$(0, -8)$
1	$y = 2(1)^2 - 8 = -6$	$(1, -6)$
2	$y = 2(2)^2 - 8 = 0$	$(2, 0)$
3	$y = 2(3)^2 - 8 = 10$	$(3, 10)$

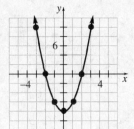

Domain: $\{x \mid x$ is a real number$\}$ or $(-\infty, \infty)$

Range: $\{y \mid y \geq -8\}$ or $[-8, \infty)$

43. $y = |x|$

x	$y = |x|$	(x, y)
-4	$y = |-4| = 4$	$(-4, 4)$
-2	$y = |-2| = 2$	$(-2, 2)$
0	$y = |0| = 0$	$(0, 0)$
2	$y = |2| = 2$	$(2, 2)$
4	$y = |4| = 4$	$(4, 4)$

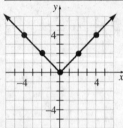

Domain: $\{x \mid x$ is a real number$\}$ or $(-\infty, \infty)$

Range: $\{y \mid y \geq 0\}$ or $[0, \infty)$

45. $y = |x - 1|$

x	$y = |x-1|$	(x, y)
-4	$y = |-4-1| = 5$	$(-4, 5)$
-1	$y = |-1-1| = 2$	$(-1, 2)$
0	$y = |0-1| = 1$	$(0, 1)$
1	$y = |1-1| = 0$	$(1, 0)$
3	$y = |3-1| = 2$	$(3, 2)$
5	$y = |5-1| = 4$	$(5, 4)$

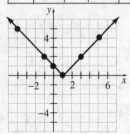

Domain: $\{x \mid x$ is a real number$\}$ or $(-\infty, \infty)$

Range: $\{y \mid y \geq 0\}$ or $[0, \infty)$

47. $y = x^3$

x	$y = x^3$	(x, y)
-3	$y = (-3)^3 = -27$	$(-3, -27)$
-2	$y = (-2)^3 = -8$	$(-2, -8)$
-1	$y = (-1)^3 = -1$	$(-1, -1)$
0	$y = (0)^3 = 0$	$(0, 0)$
1	$y = (1)^3 = 1$	$(1, 1)$
2	$y = (2)^3 = 8$	$(2, 8)$
3	$y = (3)^3 = 27$	$(3, 27)$

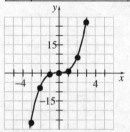

Domain: $\{x \mid x$ is a real number$\}$ or $(-\infty, \infty)$

Range: $\{y \mid y$ is a real number$\}$ or $(-\infty, \infty)$

49. $y = x^3 + 1$

x	$y = x^3 + 1$	(x, y)
-2	$y = (-2)^3 + 1 = -7$	$(-2, -7)$
-1	$y = (-1)^3 + 1 = 0$	$(-1, 0)$
0	$y = (0)^3 + 1 = 1$	$(0, 1)$
1	$y = (1)^3 + 1 = 2$	$(1, 2)$
2	$y = (2)^3 + 1 = 9$	$(2, 9)$

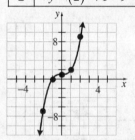

Domain: $\{x \mid x$ is a real number$\}$ or $(-\infty, \infty)$

Range: $\{y \mid y$ is a real number$\}$ or $(-\infty, \infty)$

51. $x^2 - y = 4$

$$x^2 = y + 4$$

$$x^2 - 4 = y$$

$$y = x^2 - 4$$

x	$y = x^2 - 4$	(x, y)
-3	$y = (-3)^2 - 4 = 5$	$(-3, 5)$
-2	$y = (-2)^2 - 4 = 0$	$(-2, 0)$
0	$y = (0)^2 - 4 = -4$	$(0, -4)$
2	$y = (2)^2 - 4 = 0$	$(2, 0)$
3	$y = (3)^2 - 4 = 5$	$(3, 5)$

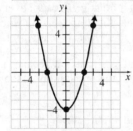

Domain: $\{x \mid x \text{ is a real number}\}$ or $(-\infty, \infty)$

Range: $\{y \mid y \geq -4\}$ or $[-4, \infty)$

53. $x = y^2 - 1$

y	$x = y^2 - 1$	(x, y)
-2	$x = (-2)^2 - 1 = 3$	$(3, -2)$
-1	$x = (-1)^2 - 1 = 0$	$(0, -1)$
0	$x = (0)^2 - 1 = -1$	$(-1, 0)$
1	$x = (1)^2 - 1 = 0$	$(0, 1)$
2	$x = (2)^2 - 1 = 3$	$(3, 2)$

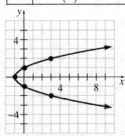

Domain: $\{x \mid x \geq -1\}$ or $[-1, \infty)$

Range: $\{y \mid y \text{ is a real number}\}$ or $(-\infty, \infty)$

55. (a) Domain: $\{x \mid 0 \leq x \leq 50\}$ or $[0, 50]$

Range: $\{y \mid 0 \leq y \leq 625\}$ or $[0, 625]$

The window can be between 0 and 50 feet wide (inclusive) and will have an area of between 0 and 625 square feet (inclusive).

(b) Answers may vary;
Assuming the window is rectangular, the perimeter is given by the equation

$$P = 2l + 2w$$

where l is the length and w is the width. Since the window cannot have a negative width, it is reasonable for the domain to begin with 0. If the width were greater than 50, this would require a negative value for the length in order to have a perimeter of 100 in the above equation. Since length cannot be negative, it is reasonable for the domain to stop at 50.

57. (a) Domain: $\{m \mid 0 \leq m \leq 15{,}120\}$ or $[0, 15{,}120]$
Range: $\{c \mid 100 \leq c \leq 3380\}$ or $[100, 3380]$
The monthly cost will be at least $100, but no more than $3380.

(b) Answers may vary. It is not possible to talk a negative number of minutes so the lower bound of 0 makes sense. Assuming there are 21 non-weekend days in a month, we get:

$$\frac{21 \text{ days}}{1} \cdot \frac{12 \text{ hrs}}{1 \text{ day}} \cdot \frac{60 \text{ min}}{1 \text{ hr}} = 15{,}120$$

That is, in 21 non-weekend days there are 15,120 'anytime' minutes so this is the most someone could use in a month.

59. Actual graphs will vary but each graph should be a horizontal line.

61. Answers will vary. A relation could be thought of as a rule that tells us what to do to a value in the domain (input value) to obtain its corresponding value in the range (output value).

Section 8.3

Preparing for an Introduction to Functions

P1. (a) Let $x = 1$:

$$2x^2 - 5x = 2(1)^2 - 5(1) = 2 - 5 = -3$$

(b) Let $x = 4$:
$$2x^2 - 5x = 2(4)^2 - 5(4)$$
$$= 2(16) - 20$$
$$= 32 - 20$$
$$= 12$$

(c) Let $x = -3$:
$$2x^2 - 5x = 2(-3)^2 - 5(-3)$$
$$= 2(9) + 15$$
$$= 18 + 15$$
$$= 33$$

P2. Inequality: $x \le 5$
Interval: $(-\infty, 5]$

P3. Interval: $(2, \infty)$
Set notation: $\{x \mid x > 2\}$

The inequality is strict since the parenthesis was used instead of a square bracket.

Section 8.3 Quick Checks

1. function

2. False

3. The relation is a function because each element in the domain (Friend) corresponds to exactly one element in the range (Birthday).
Domain: {Max, Alesia, Trent, Yolanda, Wanda, Elvis}
Range: {January 20, March 3, July 6, November 8, January 8}

4. The relation is not a function because there is an element in the domain, 210, that corresponds to more than one element in the range. If 210 is selected from the domain, a single sugar content cannot be determined.

5. The relation is a function because there are no ordered pairs with the same first coordinate but different second coordinates.
Domain: {−3, −2, −1, 0, 1}
Range: {0, 1, 2, 3}

6. The relation is not a function because there are two ordered pairs, (−3, 2) and (−3, 6), with the same first coordinate but different second coordinates.

7. $y = -2x + 5$
The relation is a function since there is only one output than can result for each input.

8. $y = \pm 3x$
The relation is not a function since a single input for x will yield two output values for y. For example, if $x = 1$, then $y = \pm 3$.

9. $y = x^2 + 5x$
The relation is a function since there is only one output than can result for each input.

10. True

11. The graph is that of a function because every vertical line will cross the graph in at most one point.

12. The graph is not that of a function because a vertical line can cross the graph in more than one point.

13. dependent; independent; argument

14. $f(x) = 3x + 2$
$$f(4) = 3(4) + 2$$
$$= 12 + 2$$
$$= 14$$

15. $g(x) = -2x^2 + x - 3$
$$g(-2) = -2(-2)^2 + (-2) - 3$$
$$= -2(4) - 5$$
$$= -8 - 5$$
$$= -13$$

16. $f(x) = 3x + 2$
$$f(x-2) = 3(x-2) + 2$$
$$= 3x - 6 + 2$$
$$= 3x - 4$$

17. $f(x) - f(2) = [3x+2] - [3(2)+2]$
$$= 3x + 2 - 8$$
$$= 3x - 6$$

18. **(a)** Independent variable: t (number of days)
Dependent variable: A (square miles)

(b) $A(t) = 0.25\pi t^2$
$$A(30) = 0.25\pi(30)^2 \approx 706.86 \text{ sq. miles}$$
After the tanker has been leaking for 30 days, the circular oil slick will cover about 706.86 square miles.

8.3 Exercises

19. Function. Each state corresponds to exactly one number of Representatives.
Domain: {Virginia, Nevada, New Mexico, Tennessee, Texas}
Range: {3, 9, 11, 32}

21. Not a function. The domain element 174 for horsepower corresponds to two different top speeds in the range.
Domain: {150, 174, 180|
Range: {118, 130, 140}

23. Function. There are no ordered pairs that have the same first coordinate but different second coordinates.
Domain: {0, 1, 2, 3]
Range: {3, 4, 5, 6}

25. Function. There are no ordered pairs that have the same first coordinate but different second coordinates.
Domain: {−3, 1, 4, 7}
Range: {5}

27. Not a function. There are two ordered pairs that have the same first coordinate but different second coordinates.
Domain: {−10, −5, 0}
Range: {1, 2, 3, 4}

29. $y = 2x + 9$

Since there is only one output y that can result from any given input x, this relation is a function.

31. $2x + y = 10$
$y = -2x + 10$

Since there is only one output y that can result from any given input x, this relation is a function.

33. $y = \pm 5x$

Since a given input x can result in more than one output y, this relation is not a function.

35. $y = x^2 + 2$

Since there is only one output y that can result from any given input x, this relation is a function.

37. $x + y^2 = 10$
$y^2 = 10 - x$

Since a given input x can result in more than one output y, this relation is not a function. For example, if $x = 1$ then $y = 3$ or $y = -3$.

39. Function. The graph passes the vertical line test so it is the graph of a function.

41. Not a function. The graph fails the vertical line test so it is not the graph of a function.

43. Function. The graph passes the vertical line test so it is the graph of a function.

45. Function. The graph passes the vertical line test so it is the graph of a function.

47. (a) $f(0) = 2(0) + 3 = 0 + 3 = 3$

(b) $f(3) = 2(3) + 3 = 6 + 3 = 9$

(c) $f(-2) = 2(-2) + 3 = -4 + 3 = -1$

(d) $f(-x) = 2(-x) + 3 = -2x + 3$

(e) $-f(x) = -(2x + 3) = -2x - 3$

(f) $f(x + 2) = 2(x + 2) + 3 = 2x + 4 + 3 = 2x + 7$

(g) $f(2x) = 2(2x) + 3 = 4x + 3$

(h) $f(x + h) = 2(x + h) + 3 = 2x + 2h + 3$

49. (a) $f(0) = -5(0) + 2 = 0 + 2 = 2$

(b) $f(3) = -5(3) + 2 = -15 + 2 = -13$

(c) $f(-2) = -5(-2) + 2 = 10 + 2 = 12$

(d) $f(-x) = -5(-x) + 2 = 5x + 2$

(e) $-f(x) = -(-5x + 2) = 5x - 2$

(f) $f(x + 2) = -5(x + 2) + 2$
$= -5x - 10 + 2$
$= -5x - 8$

(g) $f(2x) = -5(2x) + 2 = -10x + 2$

(h) $f(x+h) = -5(x+h) + 2$
$$= -5x - 5h + 2$$

51. $f(x) = x^2 + 3$

$f(2) = (2)^2 + 3 = 4 + 3 = 7$

53. $s(t) = -t^3 - 4t$

$s(-2) = -(-2)^3 - 4(-2)$
$$= -(-8) + 8$$
$$= 8 + 8$$
$$= 16$$

55. $F(x) = |x - 2|$

$F(-3) = |(-3) - 2| = |-5| = 5$

57. $F(z) = \dfrac{z+2}{z-5}$

$F(4) = \dfrac{(4)+2}{(4)-5} = \dfrac{6}{-1} = -6$

59. $f(x) = 3x^2 - x + C; f(3) = 18$

$18 = 3(3)^2 - (3) + C$
$18 = 3(9) - 3 + C$
$18 = 27 - 3 + C$
$18 = 24 + C$
$-6 = C$

61. $f(x) = \dfrac{2x+5}{x-A}; f(0) = -1$

$-1 = \dfrac{2(0)+5}{0-A}$

$-1 = \dfrac{0+5}{0-A}$

$-1 = \dfrac{5}{-A}$

$A = 5$

63. $A(r) = \pi r^2$

$A(4) = \pi(4)^2 = 16\pi \approx 50.27$

The area is roughly 50.27 square inches.

65. $h =$ number of hours worked
$G =$ gross salary

$G(h) = 15h$

$G(25) = 15(25) = 375$

Jackie's gross salary for 25 hours is $375.

67. (a) The dependent variable is the population, P, and the independent variable is the age, a.

(b)
$$P(20) = 18.75(20)^2 - 5309.62(20) + 321,783.32$$
$$= 18.75(400) - 106,192.4 + 321,783.32$$
$$= 7500 - 106,192.4 + 321,783.32$$
$$= 223,090.92$$

The population of Americans that were 20 years of age or older in 2007 was roughly 223,091 thousand (223,091,000).

(c)
$$P(0) = 18.75(0)^2 - 5309.62(0) + 321,783.32$$
$$= 321,783.32$$

$P(0)$ represents the entire population of the U.S. since every member of the population is at least 0 years of age. The population of the U.S. in 2007 was roughly 321,783 thousand (321,783,000).

69. (a) The dependent variable is revenue, R, and the independent variable is price, p.

(b)
$$R(50) = -(50)^2 + 200(50)$$
$$= -2500 + 10,000$$
$$= 7500$$
Selling PDAs for $50 will yield a daily revenue of $7500 for the company.

(c)
$$R(120) = -(120)^2 + 200(120)$$
$$= -14,400 + 24,000$$
$$= 9600$$
Selling PDAs for $120 will yield a daily revenue of $9600 for the company.

71. (a) i. $3 \geq 0$ so use $f(x) = -2x + 1$

$$f(3) = -2(3) + 1 = -6 + 1 = -5$$
$$f(3) = -5$$

ii. $-2 < 0$ so use $f(x) = x + 3$

$$f(-2) = -2 + 3 = 1$$
$$f(-2) = 1$$

iii. $0 \geq 0$ so use $f(x) = -2x + 1$

$$f(0) = -2(0) + 1 = 0 + 1 = 1$$
$$f(0) = 1$$

(b) i. $-4 < -2$ so use $f(x) = -3x + 1$

$$f(-4) = -3(-4) + 1 = 12 + 1 = 13$$
$$f(-4) = 13$$

ii. $2 \geq -2$ so use $f(x) = x^2$

$$f(2) = (2)^2 = 4$$
$$f(2) = 4$$

iii $-2 \geq -2$ so use $f(x) = x^2$

$$f(-2) = (-2)^2 = 4$$
$$f(-2) = 4$$

73. Answers will vary.

75. Answers will vary. A function could be thought of as a relation that tells us what to do to a value in the domain (input value) to obtain exactly one corresponding value in the range (output value).

77. The four forms of functions presented in this section are: maps, ordered pairs, equations, and graphs.

79. $f(x) = x^2 + 3$

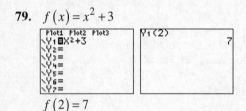

$f(2) = 7$

81. $F(x) = |x - 2|$

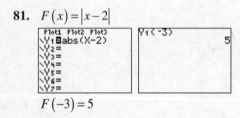

$F(-3) = 5$

83. $H(x) = \sqrt{4x - 3}$

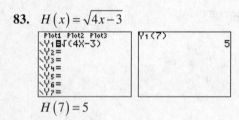

$H(7) = 5$

85. $F(z) = \dfrac{z+2}{z-5}$

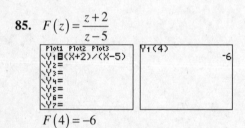

$F(4) = -6$

Section 8.4

Preparing for Functions and Their Graphs

P1.
$$3x - 12 = 0$$
$$3x - 12 + 12 = 0 + 12$$
$$3x = 12$$
$$\frac{3x}{3} = \frac{12}{3}$$
$$x = 4$$

The solution set is $\{4\}$.

P2. $y = x^2$

x	$y = x^2$	(x, y)
-2	$y = (-2)^2 = 4$	$(-2, 4)$
-1	$y = (-1)^2 = 1$	$(-1, 1)$
0	$y = (0)^2 = 0$	$(0, 0)$
1	$y = (1)^2 = 1$	$(1, 1)$
2	$y = (2)^2 = 4$	$(2, 4)$

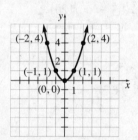

Section 8.4 Quick Checks

1. domain

2. $f(x) = 3x^2 + 2$

The function tells us to square a number x, multiply by 3, and then add 2. Since these operations can be performed on any real number, the domain of f is the set of all real numbers. The domain can be written as $\{x \mid x \text{ is any real number}\}$, or $(-\infty, \infty)$ in interval notation.

3. $h(x) = \dfrac{x+1}{x-3}$

The function tells us to divide $x + 1$ by $x - 3$. Since division by 0 is not defined, the denominator $x - 3$ can never be 0. Therefore, x can never equal 3. The domain can be written as $\{x \mid x \neq 3\}$.

4. $A(r) = \pi r^2$

Since r represents the radius of the circle, it must take on positive values. Therefore, the domain is $\{r \mid r > 0\}$, or $(0, \infty)$ in interval notation.

5. $f(x) = -2x + 9$

x	$y = f(x) = -2x + 9$	(x, y)
-2	$f(-2) = -2(-2) + 9 = 13$	$(-2, 13)$
0	$f(0) = -2(0) + 9 = 9$	$(0, 9)$
2	$f(2) = -2(2) + 9 = 5$	$(2, 5)$
4	$f(4) = -2(4) + 9 = 1$	$(4, 1)$
6	$f(6) = -2(6) + 9 = -3$	$(6, -3)$

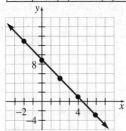

6. $f(x) = x^2 + 2$

x	$y = f(x) = x^2 + 2$	(x, y)
-3	$f(-3) = (-3)^2 + 2 = 11$	$(-3, 11)$
-1	$f(-1) = (-1)^2 + 2 = 3$	$(-1, 3)$
0	$f(0) = (0)^2 + 2 = 2$	$(0, 2)$
1	$f(1) = (1)^2 + 2 = 3$	$(1, 3)$
3	$f(3) = (3)^2 + 2 = 11$	$(3, 11)$

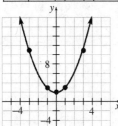

7. $f(x) = |x - 2|$

x	$y = f(x) = |x - 2|$	(x, y)
-2	$f(-2) = |-2 - 2| = 4$	$(-2, 4)$
0	$f(0) = |0 - 2| = 2$	$(0, 2)$
2	$f(2) = |2 - 2| = 0$	$(2, 0)$
4	$f(4) = |4 - 2| = 2$	$(4, 2)$
6	$f(6) = |6 - 2| = 4$	$(6, 4)$

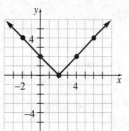

8. **(a)** The arrows on the ends of the graph indicate that the graph continues indefinitely. Therefore, the domain is $\{x \mid x \text{ is any real number}\}$, or $(-\infty, \infty)$ in interval notation.
The function reaches a maximum value of 2, but has no minimum value. Therefore, the range is $\{y \mid y \leq 2\}$, or $(-\infty, 2]$ in interval notation.

(b) The intercepts are $(-2, 0)$, $(0, 2)$, and $(2, 0)$. The x-intercepts are -2 and 2, and the y-intercept is 2.

9. $f(3) = 8$; $(-2, 4)$

10. **(a)** Since $(-3, -15)$ and $(1, -3)$ are on the graph of f, then $f(-3) = -15$ and $f(1) = -3$.

(b) To determine the domain, notice that the graph exists for all real numbers. Thus, the domain is $\{x \mid x \text{ is any real number}\}$, or $(-\infty, \infty)$ in interval notation.

(c) To determine the range, notice that the function can assume any real number. Thus, the range is $\{y \mid y \text{ is any real number}\}$, or $(-\infty, \infty)$ in interval notation.

(d) The intercepts are $(-2, 0)$, $(0, 0)$, and $(2, 0)$. The x-intercepts are -2, 0, and 2. The y-intercept is 0.

(e) Since $(3, 15)$ is the only point on the graph where $y = f(x) = 15$, the solution set to $f(x) = 15$ is $\{3\}$.

11. **(a)** When $x = -2$, then
$$f(x) = -3x + 7$$
$$f(-2) = -3(-2) + 7$$
$$= 6 + 7$$
$$= 13$$
Since $f(-2) = 13$, the point $(-2, 13)$ is on the graph. This means the point $(-2, 1)$ is **not** on the graph.

(b) If $x = 3$, then
$$f(x) = -3x + 7$$
$$f(3) = -3(3) + 7$$
$$= -9 + 7$$
$$= -2$$
The point $(3, -2)$ is on the graph.

(c) If $f(x) = -8$, then
$$f(x) = -8$$
$$-3x + 7 = -8$$
$$-3x = -15$$
$$x = 5$$
If $f(x) = -8$, then $x = 5$. The point $(5, -8)$ is on the graph.

12. $f(x) = 2x + 6$
$$f(-3) = 2(-3) + 6 = -6 + 6 = 0$$
Yes, -3 is a zero of f.

13. $g(x) = x^2 - 2x - 3$
$$g(1) = (1)^2 - 2(1) - 3 = 1 - 2 - 3 = -4$$
No, 1 is not a zero of g.

14. $h(z) = -z^3 + 4z$
$$h(2) = -(2)^3 + 4(2) = -8 + 8 = 0$$
Yes, 2 is a zero of h.

15. The zeros of the function are the x-intercepts: -2 and 2.

16. Maria's distance from home is a function of time so we put time (in minutes) on the horizontal axis and distance (in blocks) on the vertical axis. Starting at the origin $(0, 0)$, we draw a straight line to the point $(5, 5)$. The ordered pair $(5, 5)$ represents Maria being 5 blocks from home after 5 minutes. From the point $(5, 5)$, we draw a straight line to the point $(7, 0)$ that represents her trip back home. The ordered pair $(7, 0)$ represents Maria being back at home after 7 minutes. Draw a line segment from $(7, 0)$ to $(8, 0)$ to represent the time it takes Maria to find her keys and lock the door. Next, draw a line segment from $(8, 0)$ to $(13, 8)$ that represents her 8 block run in 5 minutes. Then draw a line segment from $(13, 8)$ to $(14, 11)$ that represents her 3 block run in 1 minute. Now draw a horizontal line from $(14, 11)$ to $(16, 11)$ that represents Maria's resting period. Finally, draw a line segment from $(16, 11)$ to $(26, 0)$ that represents her walk home.

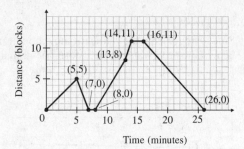

8.4 Exercises

17. $f(x) = 4x + 7$

Since each operation in the function can be performed for any real number, the domain of the function is all real numbers.

Domain: $\{x \mid x \text{ is a real number}\}$ or $(-\infty, \infty)$

19. $F(z) = \dfrac{2z + 1}{z - 5}$

The function tells us to divide $2z + 1$ by $z - 5$. Since division by 0 is not defined, the denominator can never equal 0. Thus, z can never equal 5.

Domain: $\{z \mid z \neq 5\}$

21. $f(x) = 3x^4 - 2x^2$

Since each operation in the function can be performed for any real number, the domain of the function is all real numbers.

Domain: $\{x \mid x \text{ is a real number}\}$ or $(-\infty, \infty)$

23. $G(x) = \dfrac{3x - 5}{3x + 1}$

The function tells us to divide $3x - 5$ by $3x + 1$. Since division by 0 is not defined, the denominator can never equal 0.
$$3x + 1 = 0$$
$$3x = -1$$
$$x = -\frac{1}{3}$$

Thus, x can never equal $-\dfrac{1}{3}$.

Domain: $\left\{ x \mid x \neq -\dfrac{1}{3} \right\}$

25. $f(x) = 4x - 6$

x	$y = f(x) = 4x - 6$	(x, y)
-2	$f(-2) = 4(-2) - 6 = -14$	$(-2, -14)$
-1	$f(-1) = 4(-1) - 6 = -10$	$(-1, -10)$
0	$f(0) = 4(0) - 6 = -6$	$(0, -6)$
1	$f(1) = 4(1) - 6 = -2$	$(1, -2)$
2	$f(2) = 4(2) - 6 = 2$	$(2, 2)$

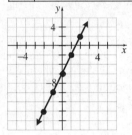

27. $h(x) = x^2 - 2$

x	$y = h(x) = x^2 - 2$	(x, y)
-3	$h(-3) = (-3)^2 - 2 = 7$	$(-3, 7)$
-2	$h(-2) = (-2)^2 - 2 = 2$	$(-2, 2)$
-1	$h(-1) = (-1)^2 - 2 = -1$	$(-1, -1)$
0	$h(0) = (0)^2 - 2 = -2$	$(0, -2)$
1	$h(1) = (1)^2 - 2 = -1$	$(1, -1)$
2	$h(2) = (2)^2 - 2 = 2$	$(2, 2)$
3	$h(3) = (3)^2 - 2 = 7$	$(3, 7)$

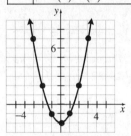

29. $G(x) = |x - 1|$

x	$y = G(x) = |x - 1|$	(x, y)
-3	$G(-3) = |-3 - 1| = 4$	$(-3, 4)$
-1	$G(-1) = |-1 - 1| = 2$	$(-1, 2)$
1	$G(1) = |1 - 1| = 0$	$(1, 0)$
3	$G(3) = |3 - 1| = 2$	$(3, 2)$
5	$G(5) = |5 - 1| = 4$	$(5, 4)$

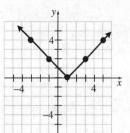

31. $g(x) = x^3$

x	$y = g(x) = x^3$	(x, y)
-3	$y = (-3)^3 = -27$	$(-3, -27)$
-2	$y = (-2)^3 = -8$	$(-2, -8)$
-1	$y = (-1)^3 = -1$	$(-1, -1)$
0	$y = (0)^3 = 0$	$(0, 0)$
1	$y = (1)^3 = 1$	$(1, 1)$
2	$y = (2)^3 = 8$	$(2, 8)$
3	$y = (3)^3 = 27$	$(3, 27)$

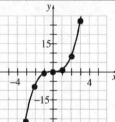

33. (a) Domain: $\{x \mid x \text{ is a real number}\}$ or $(-\infty, \infty)$

Range: $\{y \mid y \text{ is a real number}\}$ or $(-\infty, \infty)$

(b) The intercepts are $(0, 2)$ and $(1, 0)$. The x-intercept is 1 and the y-intercept is 2.

(c) Zero: 1

35. (a) Domain: $\{x \mid x \text{ is a real number}\}$ or $(-\infty, \infty)$

Range: $\{y \mid y \geq -2.25\}$ or $[-2.25, \infty)$

(b) The intercepts are $(-2, 0)$, $(4, 0)$, and $(0, -2)$. The x-intercepts are -2 and 4, and the y-intercept is -2.

(c) Zeros: -2, 4

37. (a) Domain: $\{x \mid x \text{ is a real number}\}$ or $(-\infty, \infty)$

Range: $\{y \mid y \text{ is a real number}\}$ or $(-\infty, \infty)$

(b) The intercepts are $(-3, 0)$, $(-1, 0)$, $(2, 0)$, and $(0, -3)$. The x-intercepts are -3, -1, and 2, and the y-intercept is -3.

(c) Zeros: -3, -1, 2

39. (a) Domain: $\{x \mid x \text{ is a real number}\}$ or $(-\infty, \infty)$

Range: $\{y \mid y \geq 0\}$ or $[0, \infty)$

(b) The intercepts are $(-3, 0)$, $(3, 0)$, and $(0, 9)$. The x-intercepts are -3 and 3, and the y-intercept is 9.

(c) Zeros: -3, 3

41. (a) Domain: $\{x \mid x \leq 4\}$ or $(-\infty, 4]$

Range: $\{y \mid y \leq 3\}$ or $(-\infty, 3]$

(b) The intercepts are $(-2, 0)$ and $(0, 2)$. The x-intercept is -2 and the y-intercept is 2.

(c) Zero: -2

43. (a) $f(-7) = -2$

(b) $f(-3) = 3$

(c) $f(6) = 2$

(d) negative

(e) $f(x) = 0$ for $\{-6, -1, 4\}$

(f) Domain: $\{x \mid -7 \leq x \leq 6\}$ or $[-7, 6]$

(g) Range: $\{y \mid -2 \leq y \leq 3\}$ or $[-2, 3]$

(h) The x-intercepts are -6, -1, and 4.

(i) The y-intercept is -1.

(j) $f(x) = -2$ for $\{-7, 2\}$.

(k) $f(x) = 3$ for $\{-3\}$.

(l) The zeros are -6, -1, and 4.

45. (a) From the table, when $x = -2$ the value of the function is 3. Therefore,
$F(-2) = 3$

(b) From the table, when $x = 3$ the value of the function is -6. Therefore,
$F(3) = -6$

(c) From the table, $F(x) = 5$ when $x = -1$.

(d) The x-intercept is the value of x that makes the function equal 0. From the table,
$F(x) = 0$ when $x = -4$. Therefore, the x-intercept is -4.

(e) The y-intercept is the value of the function when $x = 0$. From the table, when $x = 0$ the value of the function is 2. Therefore, the y-intercept is 2.

47. (a) $f(2) = 4(2) - 9 = 8 - 9 = -1$

Since $f(2) = -1$, the point $(2, 1)$ is not on the graph of the function.

(b) $f(3) = 4(3) - 9 = 12 - 9 = 3$
The point $(3, 3)$ is on the graph.

(c) $4x - 9 = 7$
$4x = 16$
$x = 4$
The point $(4, 7)$ is on the graph.

(d) $f(2) = 4(2) - 9 = 8 - 9 = -1$
2 is not a zero of f.

49. (a) $g(4) = -\dfrac{1}{2}(4) + 4 = -2 + 4 = 2$

Since $g(4) = 2$, the point $(4, 2)$ is on the graph of the function.

(b) $g(6) = -\dfrac{1}{2}(6) + 4 = -3 + 4 = 1$
The point $(6, 1)$ is on the graph.

(c) $-\dfrac{1}{2}x + 4 = 10$
$-\dfrac{1}{2}x = 6$
$x = -12$
The point $(-12, 10)$ is on the graph.

(d) $g(8) = -\dfrac{1}{2}(8) + 4 = -4 + 4 = 0$
8 is a zero of g.

51. Square function, (c)

53. Square root function, (e)

55. Linear function, (b)

57. Reciprocal function, (f)

59. $f(x) = x^2$

x	$y = f(x) = x^2$	(x, y)
-3	$y = (-3)^2 = 9$	$(-3, 9)$
-2	$y = (-2)^2 = 4$	$(-2, 4)$
0	$y = (0)^2 = 0$	$(0, 0)$
2	$y = (2)^2 = 4$	$(2, 4)$
3	$y = (3)^2 = 9$	$(3, 9)$

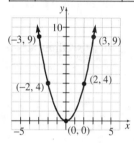

61. $f(x) = \sqrt{x}$

x	$y = f(x) = \sqrt{x}$	(x, y)
0	$y = \sqrt{0} = 0$	$(0, 0)$
1	$y = \sqrt{1} = 1$	$(1, 1)$
4	$y = \sqrt{4} = 2$	$(4, 2)$
9	$y = \sqrt{9} = 3$	$(9, 3)$

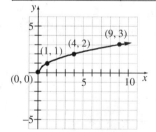

63. $f(x) = \dfrac{1}{x}$

x	$y = f(x) = \frac{1}{x}$	(x, y)
-5	$y = \frac{1}{-5} = -\frac{1}{5}$	$\left(-5, -\frac{1}{5}\right)$
-1	$y = \frac{1}{-1} = -1$	$(-1, -1)$
$-\frac{1}{2}$	$y = \frac{1}{-\frac{1}{2}} = -2$	$\left(-\frac{1}{2}, -2\right)$
$\frac{1}{2}$	$y = \frac{1}{\frac{1}{2}} = 2$	$\left(\frac{1}{2}, 2\right)$
1	$y = \frac{1}{1} = 1$	$(1, 1)$
5	$y = \frac{1}{5}$	$\left(5, \frac{1}{5}\right)$

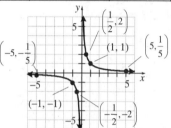

65. $V(r) = \dfrac{4}{3}\pi r^3$

Since the radius must have a positive length, the domain is all positive real numbers.
Domain: $\{r \mid r > 0\}$ or $(0, \infty)$

67. $G(h) = 22.5h$

Since Jackie cannot work a negative number of hours and she can work a maximum of 60 hours in a week, the domain is all real numbers between 0 and 60 (inclusive).
Domain: $\{h \mid 0 \le h \le 60\}$ or $[0, 60]$

69. $D(p) = 1200 - 10p$

A graph of the function would indicate that the domain is all real numbers. However, the context of the problem needs to be considered. It is not feasible that the price would be negative, nor that the demand for hot dogs would be negative. Thus, the domain is all real numbers between 0 and 120, inclusive.
Domain: $\{p \mid 0 \le p \le 120\}$ or $[0, 120]$

71. (a) Graph (III). The height should fluctuate up and down as the person jumps. Thus, we expect to see a graph that oscillates.

(b) Graph (I). Billed phone calls are generally charged per minute so the cost will continue to increase the longer you talk. Thus, we expect to see a graph that continually increases.

(c) Graph (IV). As a person grows, his or her height increases until they reach adulthood at which time the height tends to stabilize. Thus, we expect to see a graph that starts increasing but then levels off.

(d) Graph (V). As price increases, we expect that revenues will increase to a point. If prices get too high, few people will be able to afford the cars and little or no revenue will be generated. We would look for a graph that starts increasing but then begins to decrease.

(e) Graph (II). Starting at some initial value, we expect the value to decrease over time until it reaches 0. We expect to see a graph that is constantly decreasing but stops when it hits 0.

73.

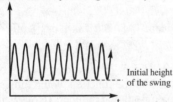

75. Answers will vary. One possibility:

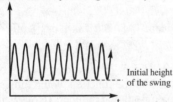

Initial height of the swing

77. Answers will vary. One possibility: Starting from the person's birth weight, the individual gained weight as they grew to adulthood. As an adult, the individual's weight fluctuated a little and then started to decline slightly after middle-age.

79. Answers will vary. One possibility:

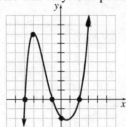

81. If a graph has more than one y-intercept, then a single input, $x = 0$, would yield more than one output (the y-intercepts). Thus, the graph could not be of a function.

83. Answers will vary. The range is the set of all possible output values.

Putting the Concepts Together (Sections 8.1–8.4)

1. $A(7, 0)$: x-axis
$B(-2, 6)$: Quadrant II
$C(8, 4)$: Quadrant I
$D(0, -9)$: y-axis
$E(-5, -10)$: Quadrant III
$F(6, -3)$: Quadrant IV

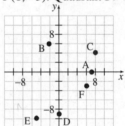

2. $y = 4x - \dfrac{3}{2}$

(a) $\left(1, \dfrac{5}{2}\right)$:

$\dfrac{5}{2} \stackrel{?}{=} 4(1) - \dfrac{3}{2}$

$\dfrac{5}{2} \stackrel{?}{=} \dfrac{8}{2} - \dfrac{3}{2}$

$\dfrac{5}{2} = \dfrac{5}{2}$ True

$\left(1, \dfrac{5}{2}\right)$ is a point on the graph.

(b) $\left(\dfrac{1}{2}, \dfrac{1}{2}\right)$:

$$\dfrac{1}{2} \stackrel{?}{=} 4\left(\dfrac{1}{2}\right) - \dfrac{3}{2}$$

$$\dfrac{1}{2} \stackrel{?}{=} \dfrac{4}{2} - \dfrac{3}{2}$$

$$\dfrac{1}{2} = \dfrac{1}{2} \quad \text{True}$$

$\left(\dfrac{1}{2}, \dfrac{1}{2}\right)$ is a point on the graph.

(c) $\left(\dfrac{1}{4}, \dfrac{1}{4}\right)$:

$$\dfrac{1}{4} \stackrel{?}{=} 4\left(\dfrac{1}{2}\right) - \dfrac{3}{2}$$

$$\dfrac{1}{4} \stackrel{?}{=} 1 - \dfrac{3}{2}$$

$$\dfrac{1}{4} \stackrel{?}{=} \dfrac{2}{2} - \dfrac{3}{2}$$

$$\dfrac{1}{4} = -\dfrac{1}{2} \quad \text{False}$$

$\left(\dfrac{1}{4}, \dfrac{1}{4}\right)$ is not a point on the graph.

3. $y = |x| + 3$

x	$y = |x| + 3$	(x, y)
-2	$y = |-2| + 3 = 5$	$(-2, 5)$
-1	$y = |-1| + 3 = 4$	$(-1, 4)$
0	$y = |0| + 3 = 3$	$(0, 3)$
1	$y = |1| + 3 = 4$	$(1, 4)$
2	$y = |2| + 3 = 5$	$(2, 5)$

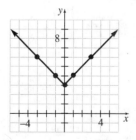

4. $y = \dfrac{1}{2}x^2 - 1$

x	$y = \frac{1}{2}x^2 - 1$	(x, y)
-4	$y = \frac{1}{2}(-4)^2 - 1 = 7$	$(-4, 7)$
-2	$y = \frac{1}{2}(-2)^2 - 1 = 1$	$(-2, 1)$
0	$y = \frac{1}{2}(0)^2 - 1 = -1$	$(0, -1)$
2	$y = \frac{1}{2}(2)^2 - 1 = 1$	$(2, 1)$
4	$y = \frac{1}{2}(4)^2 - 1 = 7$	$(4, 7)$

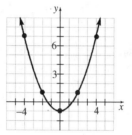

5. The intercepts are $(-3, 0)$, $(-1, 0)$, $(0, -3)$ and $(2, 0)$. The graph has x-intercepts -3, -1, and 2. The graph has y-intercept -3.

6. The relation is a function because each element in the domain corresponds to exactly one element in the range.
$\{(-2, 1), (-1, 0), (0, 1), (1, 2), (2, 3)\}$

7. (a) $y = x^3 - 4x$ is a function because any specific value of x (input) yields exactly one value of y (output).

(b) $y = \pm 4x + 3$ is not a function because with the exception of 0, any value of x can yield two values of y. For instance, if $x = 1$, then $y = 7$ or $y = -1$.

8. Yes, the graph represents a function.
Domain: $\{-4, -1, 0, 3, 6\}$
Range: $\{-3, -2, 2, 6\}$

9. This relation is a function because it passes the vertical line test.
$f(5) = -6$

10. The zero is 4.

11. (a) $f(4) = -5(4) + 3 = -20 + 3 = -17$

(b) $g(-3) = -2(-3)^2 + 5(-3) - 1$
$$= -2(9) - 15 - 1$$
$$= -18 - 15 - 1$$
$$= -34$$

(c) $f(x) - f(4) = [-5x + 3] - [-17]$
$$= -5x + 3 + 17$$
$$= -5x + 20$$

(d) $f(x - 4) = -5(x - 4) + 3$
$$= (-5)x - (-5)4 + 3$$
$$= -5x + 20 + 3$$
$$= -5x + 23$$

12. (a) Domain: $\{h \mid h \text{ is a real number}\}$ or $(-\infty, \infty)$

(b) Since we cannot divide by zero, we must find the values of w which make the denominator equal to zero.
$$3w + 1 = 0$$
$$3w + 1 - 1 = 0 - 1$$
$$3w = -1$$
$$\frac{3w}{3} = \frac{-1}{3}$$
$$w = -\frac{1}{3}$$
Domain: $\left\{w \mid w \neq -\frac{1}{3}\right\}$

13. $y = |x| - 2$

x	$y = |x| - 2$	(x, y)
-4	$y = |-4| - 2 = 2$	$(-4, 2)$
-2	$y = |-2| - 2 = 0$	$(-2, 0)$
0	$y = |0| - 2 = -2$	$(0, -2)$
2	$y = |2| - 2 = 0$	$(2, 0)$
4	$y = |4| - 2 = 2$	$(4, 2)$

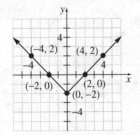

14. (a) $h(2.5) = 80$
The ball is 80 feet high after 2.5 seconds.

(b) $[0, 3.8]$

(c) $[0, 105]$

(d) 1.25 seconds

15. (a) $f(3) = 5(3) - 2 = 15 - 2 = 13$
Since the point $(3, 13)$ is on the graph, the point $(3, 12)$ is not on the graph of the function.

(b) $f(-2) = 5(-2) - 2 = -10 - 2 = -12$
The point $(-2, -12)$ is on the graph of the function.

(c) $f(x) = -22$
$$5x - 2 = -22$$
$$5x - 2 + 2 = -22 + 2$$
$$5x = -20$$
$$\frac{5x}{5} = \frac{-20}{5}$$
$$x = -4$$
The point $(-4, -22)$ is on the graph of f.

(d) $f\left(\dfrac{2}{5}\right) = \cancel{5}\left(\dfrac{2}{\cancel{5}}\right) - 2 = 2 - 2 = 0$

$\dfrac{2}{5}$ is a zero of f.

Section 8.5

Preparing for Linear Functions and Models

P1. $y = 2x - 3$
Let $x = -1, 0, 1,$ and 2.

$x = -1:$ $y = 2(-1) - 3$
$$y = -2 - 3$$
$$y = -5$$

$x = 0:$ $y = 2(0) - 3$
$$y = 0 - 3$$
$$y = -3$$

$x = 1:$ $y = 2(1) - 3$
$$y = 2 - 3$$
$$y = -1$$

$x = 2$: $\quad y = 2(2) - 3$
$\qquad\qquad y = 4 - 3$
$\qquad\qquad y = 1$

Thus, the points $(-1, -5)$, $(0, -3)$, $(1, -1)$, and $(2, 1)$ are on the graph.

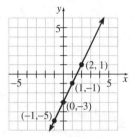

P2. $\dfrac{1}{2}x + y = 2$

Let $x = -2, 0, 2$, and 4.

$x = -2$: $\quad \dfrac{1}{2}(-2) + y = 2$
$\qquad\qquad\qquad -1 + y = 2$
$\qquad\qquad\qquad\qquad y = 3$

$x = 0$: $\quad \dfrac{1}{2}(0) + y = 2$
$\qquad\qquad\qquad 0 + y = 2$
$\qquad\qquad\qquad\quad y = 2$

$x = 2$: $\quad \dfrac{1}{2}(2) + y = 2$
$\qquad\qquad\qquad 1 + y = 2$
$\qquad\qquad\qquad\quad y = 1$

$x = 4$: $\quad \dfrac{1}{2}(4) + y = 2$
$\qquad\qquad\qquad 2 + y = 2$
$\qquad\qquad\qquad\quad y = 0$

Thus, the points $(-2, 3)$, $(0, 2)$, $(2, 1)$, and $(4, 0)$ are on the graph.

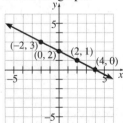

P3. The graph of $y = -4$ is a horizontal line with y-intercept -4.

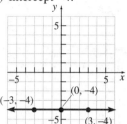

P4. The graph of $x = 5$ is a vertical line with x-intercept 5. It consists of all ordered pairs whose x-coordinate is 5.

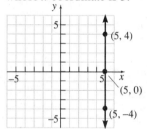

P5. $m = \dfrac{-4 - 3}{3 - (-1)} = \dfrac{-7}{4} = -\dfrac{7}{4}$

Using $m = \dfrac{-7}{4}$ we would interpret the slope as saying that y will decrease 7 units if we increase x by 4 units. We could also say $m = \dfrac{7}{-4}$ in which case we would interpret the slope as saying that y will increase by 7 units if we decrease x by 4 units. In either case, the slope is the average rate of change of y with respect to x.

P6. We start by finding the slope of the line using the two given points.

$$m = \frac{9 - 3}{4 - 1} = \frac{6}{3} = 2$$

Now we use the point-slope form of the equation of a line:

$y - y_1 = m(x - x_1)$
$\quad y - 3 = 2(x - 1)$
$\quad y - 3 = 2x - 2$
$\qquad\quad y = 2x + 1$

The equation of the line is $y = 2x + 1$.

P7.
$$0.5(x-40)+100=84$$
$$(0.5)x-(0.5)40+100=84$$
$$0.5x-20+100=84$$
$$0.5x+80=84$$
$$0.5x+80-80=84-80$$
$$0.5x=4$$
$$\frac{0.5x}{0.5}=\frac{4}{0.5}$$
$$x=8$$

P8.
$$4x+20\geq32$$
$$4x+20-20\geq32-20$$
$$4x\geq12$$
$$\frac{4x}{4}\geq\frac{12}{4}$$
$$x\geq3$$
$$\{x\mid x\geq3\}\text{ or }[3,\infty)$$

Section 8.5 Quick Checks

1. slope; *y*-intercept

2. line

3. False

4. −2; 3

5. Comparing $f(x)=2x-3$ to $f(x)=mx+b$, we see that the slope *m* is 2 and the *y*-intercept *b* is −3. We begin by plotting the point (0, −3). Because $m=2=\frac{2}{1}=\frac{\Delta y}{\Delta x}=\frac{\text{Rise}}{\text{Run}}$, from the point (0, −3) we go up 2 units and to the right 1 unit and end up at (1, −1). We draw a line through these points and obtain the graph of $f(x)=2x-3$.

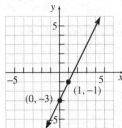

6. Comparing $G(x)=-5x+4$ to $G(x)=mx+b$, we see that the slope *m* is −5 and the *y*-intercept *b* is 4. We begin by plotting the point (0, 4). Because $m=-5=\frac{-5}{1}=\frac{\Delta y}{\Delta x}=\frac{\text{Rise}}{\text{Run}}$, from the

point (0, 4) we go down 5 units and to the right 1 unit and end up at (1, −1). We draw a line through these points and obtain the graph of $G(x)=-5x+4$.

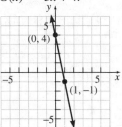

7. Comparing $h(x)=\frac{3}{2}x+1$ to $h(x)=mx+b$, we see that the slope *m* is $\frac{3}{2}$ and the *y*-intercept *b* is 1. We begin by plotting the point (0, 1). Because $m=\frac{3}{2}=\frac{\Delta y}{\Delta x}=\frac{\text{Rise}}{\text{Run}}$, from the point (0, 1) we go up 3 units and to the right 2 units and end up at (2, 4). We draw a line through these points and obtain the graph of $h(x)=\frac{3}{2}x+1$.

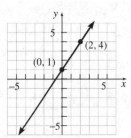

8. Comparing $f(x)=4$ to $f(x)=mx+b$, we see that the slope *m* is 0 and the *y*-intercept *b* is 4. Since the slope is 0, we have a horizontal line. We draw a horizontal line through the point (0, 4) to obtain the graph of $f(x)=4$.

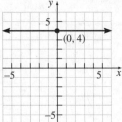

9.
$$f(x)=0$$
$$3x-15=0$$
$$3x=15$$
$$x=5$$
5 is the zero.

10. $G(x) = 0$

$\dfrac{1}{2}x + 4 = 0$

$\dfrac{1}{2}x = -4$

$x = -8$

−8 is the zero.

11. $F(p) = 0$

$-\dfrac{2}{3}p + 8 = 0$

$-\dfrac{2}{3}p = -8$

$-2p = -24$

$p = 12$

12 is the zero.

12. (a) The independent variable is the number of miles driven, x. It does not make sense to drive a negative number of miles, we have that the domain of the function is $\{x \mid x \geq 0\}$ or, using interval notation, $[0, \infty)$.

(b) To determine the C-intercept, we find $C(0) = 0.35(0) + 40 = 40$. The C-intercept is 40, so the point $(0, 40)$ is on the graph.

(c) $C(80) = 0.35(80) + 40 = 28 + 40 = 68$. If the truck is driven 80 miles, the rental cost will be $68.

(d) We solve $C(x) = 85.50$:

$0.35x + 40 = 85.50$

$0.35x = 45.50$

$x = 130$

If the rental cost is $85.50, then the truck was driven 130 miles.

(e) We plot the independent variable, *number of miles driven*, on the horizontal axis and the dependent variable, *rental cost*, on the vertical axis. From parts (b) and (c), we have that the points $(0, 40)$ and $(80, 68)$ are on the graph. We find one more point by evaluating the function for $x = 200$: $C(200) = 0.35(200) + 40 = 70 + 40 = 110$. The point $(200, 110)$ is also on the graph.

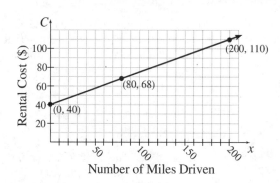

(f) We solve $C(x) \leq 127.50$:

$0.35x + 40 \leq 127.50$

$0.35x \leq 87.50$

$x \leq 250$

You can drive up to 250 miles if you can spend up to $127.50.

13. (a) From Example 4, the daily fixed costs were $2000 with a variable cost of $80 per bicycle. The tax of $1 per bicycle changes the variable cost to $81 per bicycle. Thus, the cost function is $C(x) = 81x + 2000$.

(b) $C(5) = 81(5) + 2000 = 2405$

So, the cost of manufacturing 5 bicycles in a day is $2405.

(c) $C(x) = 2810$

$81x + 2000 = 2810$

$81x = 810$

$x = 10$

So, 10 bicycles can be manufactured for a cost of $2810.

(d) Label the horizontal axis x and the vertical axis C.

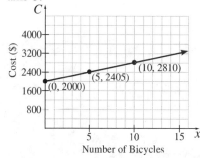

14. (a) We let $C(x)$ represent the monthly cost of operating the car after driving x miles, so $C(x) = mx + b$. The monthly cost before the car is driven is $250, so $C(0) = 250$. The C-intercept of the linear function is 250. Because the maintenance and gas cost is

\$0.18 per mile, the slope of the linear function is 0.18. The linear function that relates the monthly cost of operating the car as a function of miles driven is $C(x) = 0.18x + 250$.

(b) The car cannot be driven a negative distance, the number of miles driven, x, must be greater than or equal to zero. In addition, there is no definite maximum number of miles that the car can be driven. Therefore, the implied domain of the function is $\{x \mid x \geq 0\}$, or using interval notation $[0, \infty)$.

(c) $C(320) = 0.18(320) + 250 = 307.6$
So, the monthly cost of driving 320 miles is \$307.60.

(d)
$$C(x) = 282.40$$
$$0.18x + 250 = 282.40$$
$$0.18x = 32.40$$
$$x = 180$$
So, Roberta can drive 180 miles each month for the monthly cost of \$282.40.

(e) Label the horizontal axis x and the vertical axis C. From part (a) we know $C(0) = 250$, and from part (c) we know $C(320) = 307.6$, so (0, 250) and (320, 307.60) are on the graph.

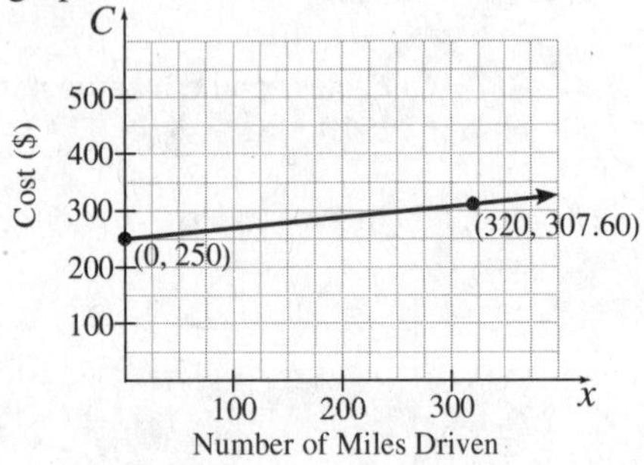

15. (a)

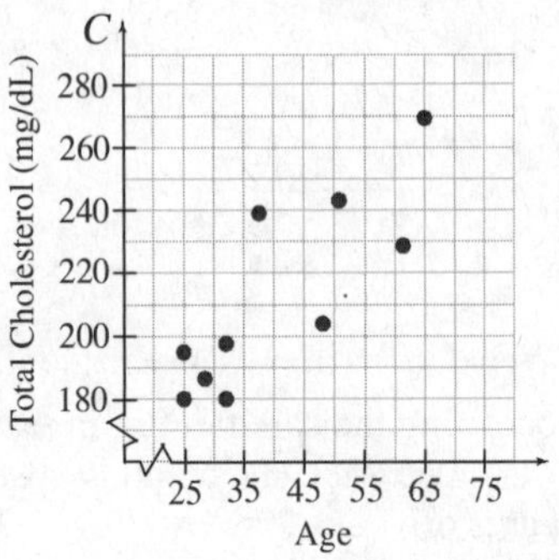

(b) From the scatter diagram, we can see that as the age increases, the total cholesterol also increases.

16. Non-linear

17. Linear with a positive slope.

18. (a) Answers will vary. We will use the points (25, 180) and (65, 269).
$$m = \frac{269-180}{65-25} = \frac{89}{40} = 2.225$$
$$y - 180 = 2.225(x - 25)$$
$$y - 180 = 2.225x - 55.625$$
$$y = f(x) = 2.225x + 124.375$$

(b)

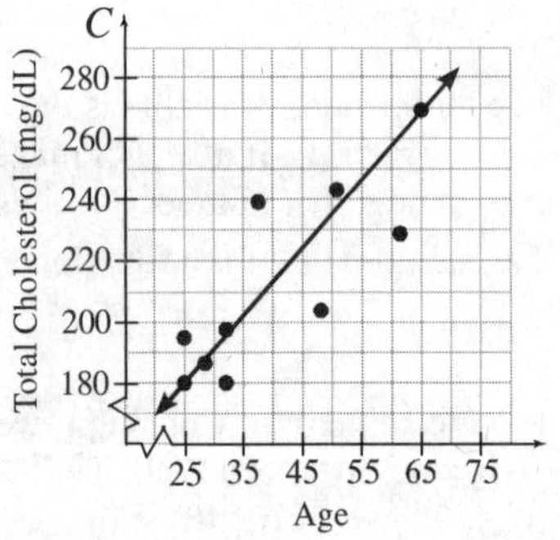

(c) $f(39) = 2.225(39) + 124.375 = 211.15$
We predict that the total cholesterol of a 39-year-old male will be approximately 211 mg/dL.

(d) The slope of the linear function is 2.225. This means that, for males, the total cholesterol increases by 2.225 mg/dL for each one-year increase in age. The y-intercept, 124.375, would represent the total cholesterol of a male who is 0 years old. Thus, it does not make sense to interpret this y-intercept.

8.5 Exercises

19. Comparing $F(x) = 5x - 2$ to $F(x) = mx + b$, we see that the slope m is 5 and the y-intercept b is −2. We begin by plotting the point (0, −2). Because $m = 5 = \frac{5}{1} = \frac{\Delta y}{\Delta x} = \frac{\text{Rise}}{\text{Run}}$, from the point (0, −2) we go up 5 units and to the right 1 unit and end up at (1, 3). We draw a line through these points and obtain the graph of $F(x) = 5x - 2$.

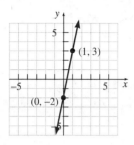

21. Comparing $G(x) = -3x + 7$ to $G(x) = mx + b$,

we see that the slope m is -3 and the y-intercept b is 7. We begin by plotting the point (0, 7).

Because $m = -3 = \dfrac{-3}{1} = \dfrac{\Delta y}{\Delta x} = \dfrac{\text{Rise}}{\text{Run}}$, from the

point (0, 7) we go down 3 units and to the right 1 unit and end up at (1, 4). We draw a line through these points and obtain the graph of $G(x) = -3x + 7$.

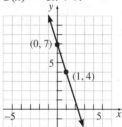

23. Comparing $H(x) = -2$ to $H(x) = mx + b$, we see

that the slope m is 0 and the y-intercept b is -2. The graph is a horizontal line through the point (0, -2). We draw a horizontal line through this point and obtain the graph of $H(x) = -2$.

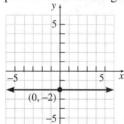

25. Comparing $f(x) - \dfrac{1}{2}x - 4$ to $f(x) = mx + b$, we

see that the slope m is $\dfrac{1}{2}$ and the y-intercept b is

-4. We begin by plotting the point (0, -4).

Because $m = \dfrac{1}{2} = \dfrac{\Delta y}{\Delta x} = \dfrac{\text{Rise}}{\text{Run}}$, from the point

(0, -4) we go up 1 unit and to the right 2 units and end up at (2, -3). We draw a line through these points and obtain the graph of

$$f(x) = \frac{1}{2}x - 4.$$

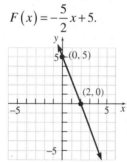

27. Comparing $F(x) = -\dfrac{5}{2}x + 5$ to $F(x) = mx + b$,

we see that the slope m is $-\dfrac{5}{2}$ and the

y-intercept b is 5. We begin by plotting the point

(0, 5). Because $m = -\dfrac{5}{2} = \dfrac{-5}{2} = \dfrac{\Delta y}{\Delta x} = \dfrac{\text{Rise}}{\text{Run}}$, from

the point (0, 5) we go down 5 units and to the right 2 units and end up at (2, 0). We draw a line through these points and obtain the graph of

$$F(x) = -\frac{5}{2}x + 5.$$

29. Comparing $G(x) = -\dfrac{3}{2}x$ to $G(x) = mx + b$, we

see that the slope m is $-\dfrac{3}{2}$ and the y-intercept b

is 0. We begin by plotting the point (0, 0).

Because $m = -\dfrac{3}{2} = \dfrac{-3}{2} = \dfrac{\Delta y}{\Delta x} = \dfrac{\text{Rise}}{\text{Run}}$, from the

point (0, 0) we go down 3 units and to the right 2 units and end up at (2, -3). We draw a line through these points and obtain the graph of

$$G(x) = -\frac{3}{2}x.$$

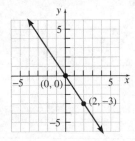

31. $f(x) = 0$
$2x + 10 = 0$
$2x = -10$
$x = -5$
−5 is the zero.

33. $G(x) = 0$
$-5x + 40 = 0$
$-5x = -40$
$x = 8$
8 is the zero.

35. $s(t) = 0$
$\frac{1}{2}t - 3 = 0$
$\frac{1}{2}t = 3$
$t = 6$
6 is the zero.

37. $P(z) = 0$
$-\frac{4}{3}z + 12 = 0$
$-\frac{4}{3}z = -12$
$-4z = -36$
$z = 9$
9 is the zero.

39. Nonlinear

41. Linear with positive slope

43. (a)

(b) Answers will vary. We will use the points
(4, 1.8) and (9, 2.6).
$m = \dfrac{2.6-1.8}{9-4} = \dfrac{0.8}{5} = 0.16$
$y - 1.8 = 0.16(x-4)$
$y - 1.8 = 0.16x - 0.64$
$y = 0.16x + 1.16$

(c)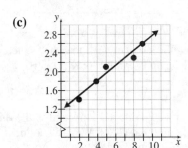

45. (a)

(b) Answers will vary. We will use the points
(1.2, 8.4) and (4.1, 2.4).
$m = \dfrac{2.4-8.4}{4.1-1.2} = \dfrac{-6.0}{2.9} \approx -2.1$
$y - 8.4 = -2.1(x-1.2)$
$y - 8.4 = -2.1x + 2.52$
$y = -2.1x + 10.92$

(c)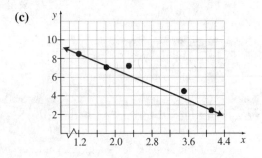

47. (a) 3

(b) 2

(c) $f(x) = 0$

$3x + 2 = 0$

$3x = -2$

$x = -\dfrac{2}{3}$

$-\dfrac{2}{3}$ is the zero.

(d) $f(x) = 5$

$3x + 2 = 5$

$3x = 3$

$x = 1$

The point $(1, 5)$ is on the graph of f.

(e) $f(x) < -1$

$3x + 2 < -1$

$3x < -3$

$x < -1$

$\{x \mid x < -1\}$ or $(-\infty, -1)$

(f)

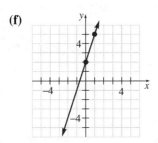

49. (a) $f(x) = g(x)$

$x - 5 = -3x + 7$

$x + 3x = 7 + 5$

$4x = 12$

$x = 3$

$f(3) = (3) - 5 = -2$

$(3, -2)$ is on the graph of $f(x)$ and $g(x)$.

(b) $f(x) > g(x)$

$x - 5 > -3x + 7$

$x + 3x > 7 + 5$

$4x > 12$

$x > 3$

$\{x \mid x > 3\}$ or $(3, \infty)$

(c)

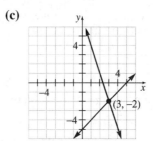

51. Since $f(2) = 6$ and $f(5) = 12$, the points $(2, 6)$ and $(5, 12)$ are on the graph of f. Thus,

$m = \dfrac{y_2 - y_1}{x_2 - x_1} = \dfrac{12 - 6}{5 - 2} = \dfrac{6}{3} = 2$

$y - y_1 = m(x - x_1)$

$y - 6 = 2(x - 2)$

$y - 6 = 2x - 4$

$y = 2x + 2$ or $f(x) = 2x + 2$

Finally, $f(-2) = 2(-2) + 2 = -4 + 2 = -2$.

53. Since $h(3) = 7$ and $h(-1) = 14$, the points $(3, 7)$ and $(-1, 14)$ are on the graph of h. Thus,

$m = \dfrac{y_2 - y_1}{x_2 - x_1} = \dfrac{14 - 7}{-1 - 3} = \dfrac{7}{-4} = -\dfrac{7}{4}$

$y - y_1 = m(x - x_1)$

$y - 7 = -\dfrac{7}{4}(x - 3)$

$y - 7 = -\dfrac{7}{4}x + \dfrac{21}{4}$

$y = -\dfrac{7}{4}x + \dfrac{49}{4}$ or $h(x) = -\dfrac{7}{4}x + \dfrac{49}{4}$

Finally, $h\left(\dfrac{1}{2}\right) = -\dfrac{7}{4}\left(\dfrac{1}{2}\right) + \dfrac{49}{4} = -\dfrac{7}{8} + \dfrac{49}{4} = \dfrac{91}{8}$.

55. (a) The point $(3, 1)$ is on the graph of f, so $f(3) = 1$. Thus, the solution of $f(x) = 1$ is $x = 3$.

(b) The point $(-1, -3)$ is on the graph of f, so $f(-1) = -3$. Thus, the solution of $f(x) = -3$ is $x = -1$.

(c) The point $(4, 2)$ is on the graph of f, so $f(4) = 2$.

(d) The intercepts of $y = f(x)$ are $(0, -2)$ and $(2, 0)$. The y-intercept is -2 and the x-intercept is 2.

(e) Use any two points to determine the slope. Here we use (3, 1) and (4, 2):

$$m = \frac{2-1}{4-3} = \frac{1}{1} = 1$$

From part (d), we know the *y*-intercept is −2, so the equation of the function is $f(x) = x - 2$.

57. (a) We are told that the tax function *T* is for adjusted gross incomes *x* between $8025 and $32,550, inclusive. Thus the domain is $\{x \mid 8025 \le x \le 32{,}550\}$ or, using interval notation, [8025, 32,550].

(b) Evaluate *T* at *x* = 20,000.
$$T(20{,}000) = 0.15(20{,}000) - 8025) + 802.50$$
$$= 2598.75$$
A single filer will pay $2598.75 in taxes if his or her adjusted gross income is $20,000.

(c) The independent variable is adjusted gross income, *x*. The dependent variable is the tax bill, *T*.

(d) Evaluate *T* at *x* = 8025 and 32,550.
$$T(8025) = 0.15(8025 - 8025) + 802.50$$
$$= 802.50$$
$$T(32{,}550) = 0.15(32{,}550 - 8025) + 802.50$$
$$= 4481.25$$
Thus the points (8025, 802.50), (32,550, 4481.25), and (20,000, 2598.75) (from part b) are on the graph.

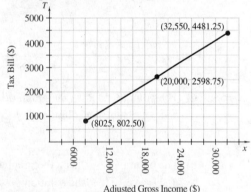

Adjusted Gross Income ($)

(e) We must solve *T*(*x*) = 3352.50.
$$0.15(x - 8025) + 802.50 = 3352.50$$
$$0.15x - 1203.75 + 802.50 = 3352.50$$
$$0.15x - 401.25 = 3352.50$$
$$0.15x = 3753.75$$
$$x = 25{,}025$$
A single filer with an adjusted gross income of $25,025 will have a tax bill of $3352.50.

59. (a) The independent variable is the number of miles traveled, *m*. It would not make sense to travel a negative number of miles. Thus, the domain of *C* is $\{m \mid m \ge 0\}$ or, using interval notation, [0, ∞).

(b) $C(0) = 1.5(0) + 2 = 2$
The base fare is $2.00 before any distance is driven.

(c) Evaluate *C* at *m* = 5.
$$C(5) = 1.5(5) + 2 = 9.5$$
The cab fare for a 5-mile ride is $9.50.

(d) Evaluate *C* at *m* = 0, 10, and 15.
$$C(0) = 1.5(0) + 2 = 2$$
$$C(10) = 1.5(10) + 2 = 17$$
$$C(15) = 1.5(15) + 2 = 24.5$$
Thus, the points (0, 2), (10, 17), and (15, 24.5) are on the graph.

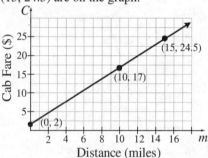

Distance (miles)

(e) We must solve *C*(*m*) = 13.25.
$$1.50m + 2.00 = 13.25$$
$$1.50m = 11.25$$
$$m = 7.5$$
A person can travel 7.5 miles in a cab for $13.25.

(f)
$$C(m) \le 39.50$$
$$1.5m + 2 \le 39.50$$
$$1.5m \le 37.50$$
$$m \le 25$$
You can ride from 0 to 25 miles, inclusive, if you can spend no more than $39.50, which is [0, 25] in interval notation.

61. (a) The independent variable is age, *a*. The dependent variable is the annual cost of health insurance, *H*.

(b) We are told in the problem that *a* is restricted from 15 to 90, inclusive. Thus, the domain of *H* is $\{a \mid 15 \le a \le 90\}$ or, using interval notation, [15, 90].

(c) Evaluate *H* at *a* = 30.
$$H(30) = 22.8(30) - 117.5 = 566.5$$
The health insurance premium of a 30 year old is $566.50.

(d) Evaluate *H* at *a* = 15, 50, and 90.
$$H(15) = 22.8(15) - 117.5 = 224.5$$
$$H(50) = 22.8(50) - 117.5 = 1022.5$$
$$H(90) = 22.8(90) - 117.5 = 1934.5$$
Thus, the points (15, 224.5), (50, 1022.5), and (90, 1934.5) are on the graph.

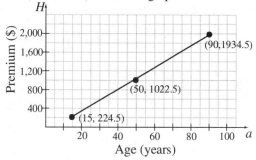

(e) We must solve *H*(*a*) = 976.9.
$$22.8a - 117.5 = 976.9$$
$$22.8a = 1094.4$$
$$a = 48$$
A 48-year-old individual will have a health insurance premium of $976.90.

63. (a) $B(m) = 0.05m + 5.95$

(b) The number of minutes, *m*, is the independent variable. The amount of the bill, *B*, is the dependent variable.

(c) Because the number of minutes can not be negative, *m* must be greater than or equal to zero. The implied domain is $\{m \mid m \ge 0\}$, or using interval notation $[0, \infty)$.

(d) $B(300) = 0.05(300) + 5.95 = 20.95$
If 300 minutes are used, the monthly bill will be $20.95.

(e) $0.05m + 5.95 = 17.95$
$$0.05m = 12$$
$$m = 240$$
If the monthly bill is $17.95, then 240 minutes were used.

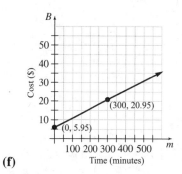

(f)

(g) $0.05m + 5.95 \le 18.45$
$$0.05m \le 12.50$$
$$m \le 250$$
You can speak from 0 to 250 minutes, included, if you don't want to spend more than $18.45. Or using interval notation, [0, 250].

65. (a) The computer will depreciate by $\dfrac{\$2700}{3} = \$900$ per year. Thus, the slope is -900. The *y*-intercept will be $2700, the initial value of the computer. The linear function that represents book value, *V*, of the computer after *x* years is $V(x) = -900x + 2700$.

(b) Because the computer cannot have a negative age, the age, *x*, must be greater than or equal to 0. After 3 years, the book value will be $V(3) = -900(3) + 2700 = 0$, and the book value cannot be negative. Therefore the implied domain of function is $\{x \mid 0 \le x < 3\}$, or using interval notation [0, 3].

(c) $V(1) = -900(1) + 2700 = 1800$
After one year, the book value of the computer will be $1800.

(d) The intercepts are (0, 2700) and (3, 0). The *V*-intercept is 2700 and the *x*-intercept is 3.

(e) $-900x + 2700 = 900$
$$-900x = -1800$$
$$x = 2$$
After two years, the book value of the computer will be $900.

(f)

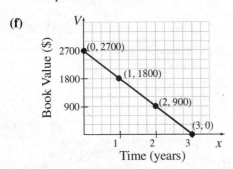

67. (a) Let x represent the weight of the diamond and C represent the cost.
$$m = \frac{4378 - 3543}{0.8 - 0.7} = \$8350 \text{ per carat}$$
$$C - 3543 = 8350(x - 0.7)$$
$$C - 3543 = 8350x - 5845$$
$$C = 8350x - 2302$$
Using function notation,
$C(x) = 8350x - 2302$.

(b) $C(0.77) = 8350(0.77) - 2302 = 4127.5$

The price of a 0.77 carat diamond would be $4127.50.

(c) The slope indicates that the cost of diamonds increases at a rate of $8350 per carat.

(d) $8350x - 2302 = 5300$
$$8350x = 7602$$
$$x = \frac{7602}{8350} \approx 0.91$$
A diamond weighing approximately 0.91 carat should cost $5300.

69. (a) Let x represent disposable income (in billions of dollars) and C represent personal consumption expenditures (in billions of dollars).
$$m = \frac{9269 - 6739}{9523 - 7194} \approx 1.086$$

$$C - 6739 = \left(\frac{9269 - 6739}{9523 - 7194}\right)(x - 7194)$$
$$C - 6739 \approx 1.086x - 7814.865$$
$$C \approx 1.086x - 1075.865$$
In function notation, and replacing the approximately equals sign, we have
$C(x) = 1.086x - 1075.865$.

(b) $C(9742) = 1.086(9742) - 1075.865$
$$\approx 9503.9$$
In 2007, personal consumption expenditures would have been $9503.9 billion.

(c) The slope indicates that the personal consumption expenditures are increasing at a rate of approximately $1.09 for every $1.00 increase in personal disposable income.

(d) $1.086x - 1075.865 = 9520$
$$1.086x = 10,595.865$$
$$x \approx 9757$$
Personal disposable income would be approximately $9757 billion if personal consumption expenditures were $9520 billion.

71. (a)

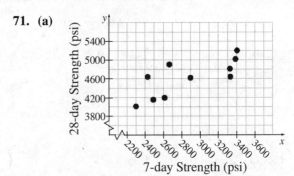

(b) Linear

(c) Answers will vary. We will use the points (2300, 4070) and (3390, 5220).
$$m = \frac{5220 - 4070}{3390 - 2300} = \frac{1150}{1090} \approx 1.055$$

$$y - 4070 = 1.06(x - 2300)$$
$$y - 4070 = 1.06x - 2438$$
$$y = 1.06x + 1632$$

(d)

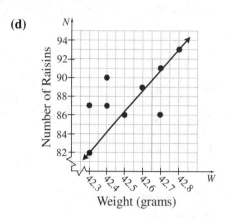

7-day Strength (psi)

(e) $x = 3000$: $y = 1.06(3000) + 1632 = 4812$
We predict that the 28-day strength will be
4812 psi if the 7-day strength is 3000 psi.

(f) The slope of the line found is 1.06. This
means that if the 7-day strength is increased
by 1 psi, then the 28-day strength will
increase by 1.06 psi.

73. (a) No, the relation does not represent a
function. Several w-coordinates are paired
with multiple N-coordinates. For example,
the w-coordinate 42.3 is paired with the two
different N-coordinates 87 and 82.

(b)

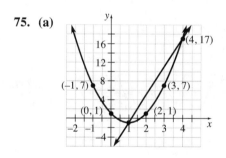

Weight (grams)

(c) Answers will vary. We will use the points
(42.3, 82) and (42.8, 93).
$$m = \frac{93 - 82}{42.8 - 42.3} = \frac{11}{0.5} = 22$$
$$N - 82 = 22(w - 42.3)$$
$$N - 82 = 22w - 930.6$$
$$N = 22w - 848.6$$

(d)

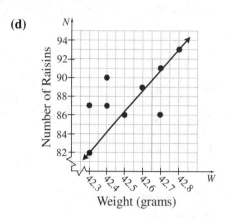

Weight (grams)

(e) Let N represent the number of raisins in the
box, and let w represent the weight (in
grams) of the box of raisins.
$$N(w) = 22w - 848.6$$

(f) $N(42.5) = 22(42.5) - 848.6 = 86.4$

We predict that approximately 86 raisins
will be in a box weighing 42.5 grams.

(g) The slope of the line found is 22 raisins per
gram. This means that if the weight is to be
increased by one gram, then the number of
raisins must be increased by 22 raisins.

75. (a)

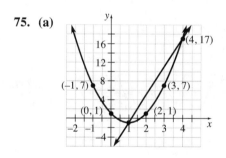

(b) $f(1) = -1$
$f(4) = 17$
$$m = \frac{17 - (-1)}{4 - 1} = \frac{18}{3} = 6$$
The slope is 6.

(c) $y - 17 = 6(x - 4)$
$y - 17 = 6x - 24$
$ y = 6x - 7$

(d)

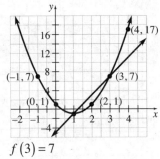

$$f(3) = 7$$
$$m = \frac{7 - (-1)}{3 - 1} = \frac{8}{2} = 4$$
$$y = 4x - 5$$

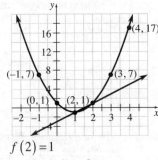

$$f(2) = 1$$
$$m = \frac{1 - (-1)}{2 - 1} = \frac{2}{1} = 2$$
$$y = 2x - 3$$

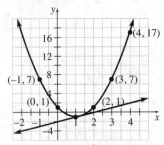

$$f(1.5) = 2(1.5)^2 - 4(1.5) + 1 = 4.5 - 6 + 1 = -0.5$$
$$m = \frac{-0.5 - (-1)}{1.5 - 1} = \frac{0.5}{0.5} = 1$$
$$y = x - 2$$

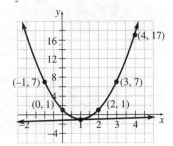

$$f(1.1) = 2(1.1)^2 - 4(1.1) + 1 = -0.98$$
$$m = \frac{-0.98 - (-1)}{1.1 - 1} = \frac{0.02}{0.1} = 0.2$$
$$y = 0.2x - 1.2$$

(e) The slope is getting closer to 0 as x gets closer to 1.

77. (a) The scatter diagram and window settings are shown below.

 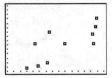

(b) As shown below the line of best fit is approximately $y = 0.676x + 2675.562$.

79. (a) The scatter diagram and window settings are shown below.

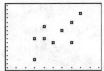

(b) As shown below the line of best fit is approximately $y = 11.449x - 399.123$.

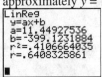

Section 8.6

Preparing for Compound Inequalities

P1. Set-builder: $\{x \mid -2 \le x \le 5\}$

Interval: [2, 5]

P2. $x \ge 4$

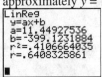

P3. The parenthesis indicates that we do *not* include -1 in our interval, while the square bracket indicates that we *do* include 3. Our interval is $(-1, 3]$.

P4. $2(x+3)-5x=15$
$2x+6-5x=15$
$-3x+6=15$
$-3x=9$
$x=-3$
The solution set is $\{-3\}$.

P5. $2x+3>11$
$2x+3-3>11-3$
$2x>8$
$\dfrac{2x}{2}>\dfrac{8}{2}$
$x>4$

Set-builder: $\{x\mid x>4\}$

Interval: $(4,\infty)$
Graph:

P6. $x+8\geq 4(x-1)-x$
$x+8\geq 4x-4-x$
$x+8\geq 3x-4$
$x+8-x\geq 3x-4-x$
$8\geq 2x-4$
$8+4\geq 2x-4+4$
$12\geq 2x$
$\dfrac{12}{2}\geq\dfrac{2x}{2}$
$6\geq x$ or $x\leq 6$

Set-builder: $\{x\mid x\leq 6\}$

Interval: $(-\infty,6]$
Graph:

Section 8.6 Quick Checks

1. intersection

2. and; or

3. True. If the two sets have no elements in common, the intersection will be the empty set.

4. False. The symbol for the union of two sets is $\cup$ while the symbol for intersection is $\cap$.

5. $A\cap B=\{1,3,5\}$

6. $A\cap C=\{2,4,6\}$

7. $A\cup B=\{1,2,3,4,5,6,7\}$

8. $A\cup C=\{1,2,3,4,5,6,8\}$

9. $B\cap C=\{\ \ \}$ or $\varnothing$

10. $B\cup C=\{1,2,3,4,5,6,7,8\}$

11. $A\cap B$ is the set of all real numbers that are greater than 2 and less than 7.

Set-builder: $\{x\mid 2<x<7\}$

Interval: $(2,7)$

12. $A\cup C$ is the set of real numbers that are greater than 2 or less than or equal to -3.

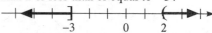

Set-builder: $\{x\mid x\leq -3 \text{ or } x>2\}$

Interval: $(-\infty,-3]\cup(2,\infty)$

13. $2x+1\geq 5$ and $-3x+2<5$
$\qquad 2x\geq 4 \qquad\qquad -3x<3$
$\qquad\ x\geq 2 \qquad\qquad\ \ x>-1$
We need $x\geq 2$ and $x>-1$.
Set-builder: $\{x\mid x\geq 2\}$

Interval: $[2,\infty)$

14. $4x-5<7$ and $3x-1>-10$
$\quad 4x<12 \qquad\qquad 3x>-9$
$\quad\ x<3 \qquad\qquad\ \ x>-3$
We need $x<3$ and $x>-3$.
Set-builder: $\{x\mid -3<x<3\}$

Interval: $(-3,3)$

15. $-8x+3<-5$ and $\dfrac{2}{3}x+1<3$
$\quad -8x<-8$
$\qquad x>1 \qquad\qquad\ \dfrac{2}{3}x<2$
$\qquad\qquad\qquad\qquad\quad x<3$
We need $x>1$ and $x<3$.
Set-builder: $\{x\mid 1<x<3\}$

Interval: $(1,3)$

16. $3x - 5 < -8$　and　$2x + 1 > 5$
　　$3x < -3$　　　　$2x > 4$
　　$x < -1$　　　　　$x > 2$
We need $x < -1$ and $x > 2$. Looking at the graphs of the inequalities separately we see that there are no such numbers that satisfy both inequalities. Therefore, the solution set is empty.
Solution set: { } or $\varnothing$

17. $5x + 1 \le 6$　and　$3x + 2 \ge 5$
　　$5x \le 5$　　　　　$3x \ge 3$
　　$x \le 1$　　　　　$x \ge 1$
We need $x \le 1$ and $x \ge 1$. Looking at the graphs of the inequalities separately we see that the only number that is both less than or equal to 1 and greater than or equal to 1 is the number 1.
Solution set: $\{1\}$

18.　　$-2 < 3x + 1 < 10$
　　$-2 - 1 < 3x + 1 - 1 < 10 - 1$
　　　$-3 < 3x < 9$
　　　$\dfrac{-3}{3} < \dfrac{3x}{3} < \dfrac{9}{3}$
　　　$-1 < x < 3$

Set-builder: $\{x \mid -1 < x < 3\}$

Interval: $(-1, 3)$

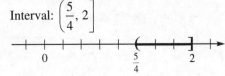

19.　　$0 < 4x - 5 \le 3$
　　$0 + 5 < 4x - 5 + 5 \le 3 + 5$
　　　$5 < 4x \le 8$
　　　$\dfrac{5}{4} < \dfrac{4x}{4} \le \dfrac{8}{4}$
　　　$\dfrac{5}{4} < x \le 2$

Set-builder: $\left\{x \mid \dfrac{5}{4} < x \le 2\right\}$

Interval: $\left(\dfrac{5}{4}, 2\right]$

20.　　$3 \le -2x - 1 \le 11$
　　$3 + 1 \le -2x - 1 + 1 \le 11 + 1$
　　　$4 \le -2x \le 12$
　　　$\dfrac{4}{-2} \ge \dfrac{-2x}{-2} \ge \dfrac{12}{-2}$
　　　$-2 \ge x \ge -6$　or　$-6 \le x \le -2$
Set-builder: $\{x \mid -6 \le x \le -2\}$

Interval: $[-6, -2]$

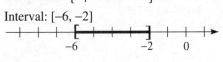

21. $x + 3 < 1$　　or　$x - 2 > 3$
　　$x < -2$　　　　$x > 5$
We need $x < -2$ or $x > 5$.
Set-builder: $\{x \mid x < -2 \text{ or } x > 5\}$

Interval: $(-\infty, -2) \cup (5, \infty)$

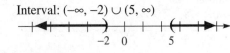

22. $3x + 1 \le 7$　or　$2x - 3 > 9$
　　$3x \le 6$　　　　$2x > 12$
　　$x \le 2$　　　　　$x > 6$
We need $x \le 2$ or $x > 6$.
Set-builder: $\{x \mid x \le 2 \text{ or } x > 6\}$

Interval: $(-\infty, 2] \cup (6, \infty)$

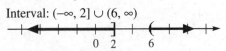

23. $2x - 3 \ge 1$　or　$6x - 5 \ge 1$
　　$2x \ge 4$　　　　$6x \ge 6$
　　$x \ge 2$　　　　　$x \ge 1$
We need $x \ge 2$ or $x \ge 1$.
Set-builder: $\{x \mid x \ge 1\}$

Interval: $[1, \infty)$

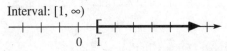

24. $\dfrac{3}{4}(x + 4) < 6$　or　$\dfrac{3}{2}(x + 1) > 15$
　　$x + 4 < 8$　　　　$x + 1 > 10$
　　$x < 4$　　　　　　$x > 9$
We need $x < 4$ or $x > 9$.
Set-builder: $\{x \mid x < 4 \text{ or } x > 9\}$

Interval: $(-\infty, 4) \cup (9, \infty)$

25. $3x - 2 > -5$　or　$2x - 5 \le 1$
　　$3x > -3$　　　　$2x \le 6$
　　$x > -1$　　　　　$x \le 3$
If we look at the graphs of the inequalities

separately, we see that the union of their solution sets is the set of real numbers.

Set-builder: $\{x \mid x \text{ is any real number}\}$

Interval: $(-\infty, \infty)$

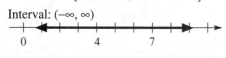

26. $-5x - 2 \leq 3$ or $7x - 9 > 5$
$-5x \leq 5$ $7x > 14$
$x \geq -1$ $x > 2$

We need $x \geq -1$ or $x > 2$. Since the solution set of the inequality $x > 2$ is a subset of the solution set for the inequality $x \geq -1$, we only need to consider $x \geq -1$.

Set-builder: $\{x \mid x \geq -1\}$

Interval: $[-1, \infty)$

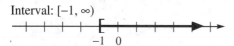

27. Let x = taxable income (in dollars). The federal income tax in the 25% bracket is $4481.25 plus 25% of the amount over $32,550. In general, the income tax for the 25% bracket is given by
$4481.25 + 0.25(x - 32,550)$

Because the federal income tax is between $4481.25 and $16,056.25, we have
$4481.25 \leq 4481.25 + 0.25(x - 32,550) \leq 16,056.25$
$4481.25 \leq 4481.25 + 0.25x - 8137.50 \leq 16,056.25$
$4481.25 \leq 0.25x - 3656.25 \leq 16,056.25$
$8137.50 \leq 0.25x \leq 19,712.50$
$32,550 \leq x \leq 78,850$

To be in the 25% tax bracket, an individual would have an income between $32,550 and $78,850.

28. Let x = number of minutes. The long distance plan charges $2.00 per month and $0.10 per minute. In general, the charge is given by $2.00 + 0.10x$. Sophia's charges ranged from $6.50 to $26.50, so we get the inequality
$6.50 \leq 2.00 + 0.10x \leq 26.50$
$4.50 \leq 0.10x \leq 24.50$
$45 \leq x \leq 245$

Over the course of the year, Sophia's monthly minutes were between 45 minutes and 245 minutes.

8.6 Exercises

29. $A \cup B = \{1, 4, 5, 6, 7, 8, 9\}$

31. $A \cap B = \{5, 7, 9\}$

33. $B \cap C = \varnothing$ or $\{ \ \}$

35. **(a)** $A \cap B = \{x \mid -2 < x \leq 5\}$

Interval: $(-2, 5]$

(b) $A \cup B = \{x \mid x \text{ is a real number}\}$

Interval: $(-\infty, \infty)$

37. **(a)** $E \cap F = \varnothing$ or $\{ \ \}$

(b) $E \cup F = \{x \mid x < -1 \text{ or } x > 3\}$

Interval: $(-\infty, -1) \cup (3, \infty)$

39. We need the set of all values for x such that $2x - 1 \leq 3$ and $2x - 1 \geq -5$. On the graph, we look for where the graph of $g(x) = 2x - 1$ is between the horizontal lines $f(x) = -5$ and $h(x) = 3$ (inclusive).

Set-builder: $\{x \mid -2 \leq x \leq 2\}$

Interval: $[-2, 2]$

41. We need the set of all values for x such that $-\frac{5}{3}x + 1 < 6$ and $-\frac{5}{3}x + 1 > -4$. On the graph, we look for where the graph of $g(x) = -\frac{5}{3}x + 1$ is between the horizontal lines $f(x) = -4$ and $h(x) = 6$, not inclusive.

Set-builder: $\{x \mid -3 < x < 3\}$

Interval: $(-3, 3)$

43. $x < 3$ and $x \geq -2$

Set-builder: $\{x \mid -2 \leq x < 3\}$

Interval: $[-2, 3)$

45. $4x-4<0$ and $-5x+1\le-9$

$\qquad 4x<4 \qquad\qquad -5x\le-10$

$\qquad\ x<1 \qquad\qquad\ \ x\ge2$

We need $x<1$ and $x\ge2$.

Solution set: $\varnothing$ or $\{\ \}$

47. $4x-3<5$ and $-5x+3>13$

$\qquad 4x<8 \qquad\qquad -5x>10$

$\qquad\ x<2 \qquad\qquad\ \ x<-2$

We need $x<2$ and $x<-2$.

Set-builder: $\{x\,|\,x<-2\}$

Interval: $(-\infty,-2)$

49. $7x+2\ge9$ and $4x+3\le7$

$\qquad 7x\ge7 \qquad\qquad 4x\le4$

$\qquad\ x\ge1 \qquad\qquad\ x\le1$

We need $x\ge1$ and $x\le1$.

Solution set: $\{1\}$

51. $\qquad -3\le5x+2<17$

$\quad -3-2\le5x+2-2<17-2$

$\qquad\quad -5\le5x<15$

$\qquad\quad \dfrac{-5}{5}\le\dfrac{5x}{5}<\dfrac{15}{5}$

$\qquad\quad -1\le x<3$

Set-builder: $\{x\,|\,-1\le x<3\}$

Interval: $[-1,3)$

53. $\qquad -3\le6x+1\le10$

$\quad -3-1\le6x+1-1\le10-1$

$\qquad\quad -4\le6x\le9$

$\qquad\quad \dfrac{-4}{6}\le\dfrac{6x}{6}\le\dfrac{9}{6}$

$\qquad\quad -\dfrac{2}{3}\le x\le\dfrac{3}{2}$

Set-builder: $\left\{x\,\Big|\,-\dfrac{2}{3}\le x\le\dfrac{3}{2}\right\}$

Interval: $\left[-\dfrac{2}{3},\dfrac{3}{2}\right]$

55. $\qquad 3\le-5x+7<12$

$\quad 3-7\le-5x+7-7<12-7$

$\qquad\quad -4\le-5x<5$

$\qquad\quad \dfrac{-4}{-5}\ge\dfrac{-5x}{-5}>\dfrac{5}{-5}$

$\qquad\quad \dfrac{4}{5}\ge x>-1$ or $-1<x\le\dfrac{4}{5}$

Set-builder: $\left\{x\,\Big|\,-1<x\le\dfrac{4}{5}\right\}$

Interval: $\left(-1,\dfrac{4}{5}\right]$

57. $\qquad -1\le\dfrac{1}{2}x-1\le3$

$\quad -1+1\le\dfrac{1}{2}x-1+1\le3+1$

$\qquad\quad 0\le\dfrac{1}{2}x\le4$

$\quad 2(0)\le2\cdot\dfrac{1}{2}x\le2\cdot4$

$\qquad\quad 0\le x\le8$

Set-builder: $\{x\,|\,0\le x\le8\}$

Interval: $[0,8]$

59. $\qquad 3\le-2x-1\le11$

$\quad 3+1\le-2x-1+1\le11+1$

$\qquad\quad 4\le-2x\le12$

$\qquad\quad \dfrac{4}{-2}\ge\dfrac{-2x}{-2}\ge\dfrac{12}{-2}$

$\qquad\quad -2\ge x\ge-6$ or $-6\le x\le-2$

Set-builder: $\{x\,|\,-6\le x\le-2\}$

Interval: $[-6,-2]$

61.　　$\dfrac{2}{3}x+\dfrac{1}{2}<\dfrac{5}{6}$　and　$-\dfrac{1}{5}x+1<\dfrac{3}{10}$

$6\left(\dfrac{2}{3}x+\dfrac{1}{2}\right)<6\left(\dfrac{5}{6}\right)$　　$10\left(-\dfrac{1}{5}x+1\right)<10\left(\dfrac{3}{10}\right)$

　　$4x+3<5$　　　　　　$-2x+10<3$

　　　$4x<2$　　　　　　　$-2x<-7$

　　　　$x<\dfrac{1}{2}$　　　　　　　$x>\dfrac{7}{2}$

We need $x<\dfrac{1}{2}$ and $x>\dfrac{7}{2}$.

Solution set: $\varnothing$ or $\{\ \}$

63.　　$-2<\dfrac{3x+1}{2}\le 8$

$2(-2)<2\left(\dfrac{3x+1}{2}\right)\le 2(8)$

　　$-4<3x+1\le 16$

　$-4-1<3x+1-1\le 16-1$

　　$-5<3x\le 15$

　　$\dfrac{-5}{3}<\dfrac{3x}{3}\le\dfrac{15}{3}$

　　$-\dfrac{5}{3}<x\le 5$

Set-builder: $\left\{x\,\Big|\,-\dfrac{5}{3}<x\le 5\right\}$

Interval: $\left(-\dfrac{5}{3},\,5\right]$

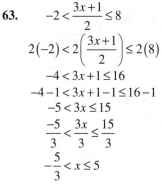

65.　　$-8\le -2(x+1)<6$

　　$-8\le -2x-2<6$

$-8+2\le -2x-2+2<6+2$

　　$-6\le -2x<8$

　　$\dfrac{-6}{-2}\ge\dfrac{-2x}{-2}>\dfrac{8}{-2}$

　　$3\ge x>-4$　or　$-4<x\le 3$

Set-builder: $\{x\,|\,-4<x\le 3\}$

Interval: $(-4,\,3]$

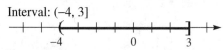

67.　$x<-2$ or $x>3$

Solution set: $\{x\,|\,x<-2\text{ or }x>3\}$

Interval: $(-\infty,\,-2)\cup(3,\,\infty)$

69.　$x-2<-4$　or　$x+3>8$

　　$x<-2$　　　　$x>5$

We need $x<-2$ or $x>5$.

Set-builder: $\{x\,|\,x<-2\text{ or }x>5\}$

Interval: $(-\infty,\,-2)\cup(5,\,\infty)$

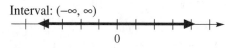

71.　$6(x-2)<12$　or　$4(x+3)>12$

　　$6x-12<12$　　　$4x+12>12$

　　　$6x<24$　　　　　$4x>0$

　　　　$x<4$　　　　　　$x>0$

We need $x<4$ or $x>0$.

Set-builder: $\{x\,|\,x\text{ is a real number}\}$

Interval: $(-\infty,\,\infty)$

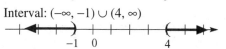

73.　$-8x+6x-2>0$　or　$5x>3x+8$

　　$-2x-2>0$　　　　$2x>8$

　　$-2x>2$　　　　　$x>4$

　　　$x<-1$

We need $x<-1$ or $x>4$.

Set-builder: $\{x\,|\,x<-1\text{ or }x>4\}$

Interval: $(-\infty,\,-1)\cup(4,\,\infty)$

75.　$2x+5\le -1$　or　$\dfrac{4}{3}x-3>5$

　　$2x\le -6$

　　$x\le -3$　　　　　$\dfrac{4}{3}x>8$

　　　　　　　　　　$4x>24$

　　　　　　　　　　$x>6$

We need $x\le -3$ or $x>6$.

Set-builder: $\{x\,|\,x\le -3\text{ or }x>6\}$

Interval: $(-\infty,\,-3]\cup(6,\,\infty)$

77.　$\dfrac{1}{2}x<3$　or　$\dfrac{3x-1}{2}>4$

　　$x<6$　　　$3x-1>8$

　　　　　　　$3x>9$

　　　　　　　$x>3$

We need $x<6$ or $x>3$.

Set-builder: $\{x\,|\,x\text{ is a real number}\}$

Interval: $(-\infty,\,\infty)$

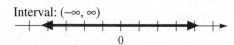

79. $3(x-1)+5<2$ or $-2(x-3)<1$

$\quad\; 3x-3+5<2 \qquad\quad -2x+6<1$

$\qquad\; 3x+2<2 \qquad\qquad -2x<-5$

$\qquad\quad\; 3x<0 \qquad\qquad\quad x>\dfrac{5}{2}$

$\qquad\qquad x<0$

We need $x<0$ or $x>\dfrac{5}{2}$.

Set-builder: $\left\{x\,|\,x<0 \text{ or } x>\dfrac{5}{2}\right\}$

Interval: $(-\infty, 0)\cup\left(\dfrac{5}{2}, \infty\right)$

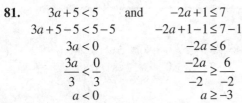

81. $\quad 3a+5<5$ and $\quad -2a+1\le 7$

$\;\; 3a+5-5<5-5 \qquad -2a+1-1\le 7-1$

$\qquad\quad 3a<0 \qquad\qquad\quad -2a\le 6$

$\qquad\; \dfrac{3a}{3}<\dfrac{0}{3} \qquad\qquad \dfrac{-2a}{-2}\ge\dfrac{6}{-2}$

$\qquad\quad\; a<0 \qquad\qquad\quad a\ge -3$

We need $a<0$ and $a\ge -3$.

Set-builder: $\{a\,|\,-3\le a<0\}$

Interval: $[-3, 0)$

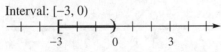

83. $5(x+2)<20$ or $4(x-4)>-20$

$\;\; \dfrac{5(x+2)}{5}<\dfrac{20}{5} \qquad \dfrac{4(x-4)}{4}>\dfrac{-20}{4}$

$\qquad x+2<4 \qquad\qquad x-4>-5$

$\;\; x+2-2<4-2 \qquad x-4+4>-5+4$

$\qquad\;\; x<2 \qquad\qquad\quad x>-1$

We need $x<2$ or $x>-1$.

Set-builder: $\{x\,|\,x \text{ is a real number}\}$

Interval: $(-\infty, \infty)$

85. $\qquad -4\le 3x+2\le 10$

$\;\; -4-2\le 3x+2-2\le 10-2$

$\qquad\quad -6\le 3x\le 8$

$\qquad\;\; \dfrac{-6}{3}\le\dfrac{3x}{3}\le\dfrac{8}{3}$

$\qquad\quad -2\le x\le\dfrac{8}{3}$

Set-builder: $\left\{x\,|\,-2\le x\le\dfrac{8}{3}\right\}$

Interval: $\left[-2, \dfrac{8}{3}\right]$

87. $\quad 2x+7<-13 \qquad$ or $\qquad 5x-3>7$

$\;\; 2x+7-7<-13-7 \qquad 5x-3+3>7+3$

$\qquad 2x<-20 \qquad\qquad\quad 5x>10$

$\qquad \dfrac{2x}{2}<\dfrac{-20}{2} \qquad\qquad \dfrac{5x}{5}>\dfrac{10}{5}$

$\qquad\;\; x<-10 \qquad\qquad\quad x>2$

We need $x<-10$ or $x>2$.

Set-builder: $\{x\,|\,x<-10 \text{ or } x>2\}$

Interval: $(-\infty, -10)\cup(2, \infty)$

89. $\quad 5<3x-1<14$

$\;\; 5+1<3x-1+1<14+1$

$\qquad\; 6<3x<15$

$\qquad\; \dfrac{6}{3}<\dfrac{3x}{3}<\dfrac{15}{3}$

$\qquad\; 2<x<5$

Set-builder: $\{x\,|\,2<x<5\}$

Interval: $(2, 5)$

91. $\quad \dfrac{x}{3}\le -1 \qquad$ or $\qquad \dfrac{4x-1}{2}>7$

$\;\; 3\cdot\dfrac{x}{3}\le 3\cdot(-1) \qquad 2\cdot\dfrac{4x-1}{2}>2\cdot 7$

$\qquad\;\; x\le -3 \qquad\qquad\quad 4x-1>14$

$\qquad\qquad\qquad\qquad\quad 4x-1+1>14+1$

$\qquad\qquad\qquad\qquad\qquad\;\; 4x>15$

$\qquad\qquad\qquad\qquad\qquad\;\; \dfrac{4x}{4}>\dfrac{15}{4}$

$\qquad\qquad\qquad\qquad\qquad\quad\;\; x>\dfrac{15}{4}$

We need $x\le -3$ or $x>\dfrac{15}{4}$.

Set-builder: $\left\{x\,|\,x\le -3 \text{ or } x>\dfrac{15}{4}\right\}$

Interval: $(-\infty, -3]\cup\left(\dfrac{15}{4}, \infty\right)$

93.
$$-3 \le -2(x+1) < 8$$
$$-3 \le -2x - 2 < 8$$
$$-3 + 2 \le -2x - 2 + 2 < 8 + 2$$
$$-1 \le -2x < 10$$
$$\frac{-1}{-2} \ge \frac{-2x}{-2} > \frac{10}{-2}$$
$$\frac{1}{2} \ge x > -5 \quad \text{or} \quad -5 < x \le \frac{1}{2}$$

Set-builder: $\left\{ x \mid -5 < x \le \frac{1}{2} \right\}$

Interval: $\left(-5, \frac{1}{2} \right]$

95.
$$-3 < x < 4$$
$$-3 + 4 < x + 4 < 4 + 4$$
$$1 < x + 4 < 8$$
We need $a = 1$ and $b = 8$.

97.
$$4 < x < 10$$
$$3 \cdot 4 < 3 \cdot x < 3 \cdot 10$$
$$12 < 3x < 30$$
We need $a = 12$ and $b = 30$.

99.
$$-2 < x < 6$$
$$3 \cdot (-2) < 3 \cdot x < 3 \cdot 6$$
$$-6 < 3x < 18$$
$$-6 + 5 < 3x + 5 < 18 + 5$$
$$-1 < 3x + 5 < 23$$
We need $a = -1$ and $b = 23$.

101. Let x = systolic blood pressure.
$$90 < x < 140$$

103. Let x = final exam score.
$$80 \le \frac{74 + 86 + 77 + 89 + 2x}{6} \le 89$$
$$80 \le \frac{326 + 2x}{6} \le 89$$
$$6 \cdot 80 \le 6 \cdot \frac{326 + 2x}{6} \le 6 \cdot 89$$
$$480 \le 326 + 2x \le 534$$
$$480 - 326 \le 326 + 2x - 326 \le 534 - 326$$
$$154 \le 2x \le 208$$
$$\frac{154}{2} \le \frac{2x}{2} \le \frac{208}{2}$$
$$77 \le x \le 104$$

Joanna needs to score at least a 77 on the final. That is, $77 \le x \le 100$ (assuming 100 is the max score, otherwise $77 \le x \le 104$).

105. Let x = weekly wages.
$$700 \le x \le 800$$
$$700 - 645 \le x - 645 \le 800 - 645$$
$$55 \le x - 645 \le 155$$
$$0.25(55) \le 0.25(x - 645) \le 0.25(155)$$
$$13.75 \le 0.25(x - 645) \le 38.75$$
$$13.75 + 81.90 \le 0.25(x - 645) + 81.90 \le 38.75 + 81.90$$
$$95.65 \le 0.25(x - 645) + 81.90 \le 120.65$$

The amount withheld ranges between \$95.65 and \$120.65, inclusive.

107. Let x = number of therms.
$$157.73 \le 1.15855(x - 70) + 65.05 \le 175.11$$
$$92.68 \le 1.15855(x - 70) \le 110.06$$
$$\frac{92.68}{1.15855} \le x - 70 \le \frac{110.06}{1.15855}$$
$$\frac{92.68}{1.15855} + 70 \le x \le \frac{110.06}{1.15855} + 70$$
$$150 \le x \le 165 \quad \text{(approx.)}$$

The gas usage ranged from 150 to 165 therms.

109. Step 1:
$$a < b$$
$$a + a < a + b$$
$$2a < a + b$$
$$\frac{2a}{2} < \frac{a + b}{2}$$
$$a < \frac{a + b}{2}$$

Step 2:
$$a < b$$
$$a + b < b + b$$
$$a + b < 2b$$
$$\frac{a + b}{2} < \frac{2b}{2}$$
$$\frac{a + b}{2} < b$$

Step 3:
Since $a < \dfrac{a + b}{2}$ and $\dfrac{a + b}{2} < b$, it follows that
$$a < \frac{a + b}{2} < b.$$

111. $2x+1 \le 5x+7 \le x-5$

We can rewrite the inequality as

$2x+1 \le 5x+7$ and $5x+7 \le x-5$

$\quad 2x+1 \le 5x+7 \qquad\qquad 5x+7 \le x-5$

$2x+1-5x \le 5x+7-5x \quad 5x+7-x \le x-5-x$

$\quad\quad -3x+1 \le 7 \qquad\qquad\quad 4x+7 \le -5$

$-3x+1-1 \le 7-1 \qquad\quad 4x+7-7 \le -5-7$

$\quad\quad\quad -3x \le 6 \qquad\qquad\qquad 4x \le -12$

$\quad\quad\quad \dfrac{-3x}{-3} \ge \dfrac{6}{-3} \qquad\qquad\quad \dfrac{4x}{4} \le \dfrac{-12}{4}$

$\quad\quad\quad\quad x \ge -2 \qquad\qquad\qquad\quad x \le -3$

Therefore, we need $x \ge -2$ AND $x \le -3$. Since $-2 > -3$, there are no values for x that can satisfy both inequalities. Thus, the solution set is $\{\ \}$ or $\varnothing$.

113. $\quad\quad 4x+1 > 2(2x+1)$

$\quad\quad\quad\quad 4x+1 > 4x+2$

$\quad 4x+1-4x > 4x+2-4x$

$\quad\quad\quad\quad\quad 1 > 2$

This is a contradiction. There is no solution. If, during simplification, the variable terms all cancel out and a contradiction results, then there is no solution to the inequality.

115. If $x < 2$ then $x-2 < 2-2 \implies x-2 < 0$.
When multiplying both sides of the inequality by $x-2$ in the second step, the direction of the inequality must switch.

Section 8.7

Preparing for Absolute Value Equations and Inequalities

P1. $|3| = 3$ because the distance from 0 to 3 on a real number line is 3 units.

P2. $|-4| = 4$ because the distance from 0 to -4 on a real number line is 4 units.

P3. $|-1.6| = 1.6$ because the distance from 0 to -1.6 on a real number line is 1.6 units.

P4. $|0| = 0$ because the distance from 0 to 0 on a real number line is 0 units.

P5. The distance between 0 and 5 on a real number line can be expressed as $|5|$.

P6. The distance between 0 and -8 on a real number line can be expressed as $|-8|$.

P7. $\quad\quad 4x+5 = -9$

$\quad 4x+5-5 = -9-5$

$\quad\quad\quad\quad 4x = -14$

$\quad\quad\quad \dfrac{4x}{4} = \dfrac{-14}{4}$

$\quad\quad\quad\quad x = -\dfrac{7}{2}$

The solution set is $\left\{ -\dfrac{7}{2} \right\}$.

P8. $\quad\quad -2x+1 > 5$

$\quad -2x+1-1 > 5-1$

$\quad\quad\quad\quad -2x > 4$

$\quad\quad\quad \dfrac{-2x}{-2} < \dfrac{4}{-2}$

$\quad\quad\quad\quad x < -2$

Set-builder: $\{x \mid x < -2\}$

Interval: $(-\infty, -2)$

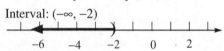

Section 8.7 Quick Checks

1. $|x| = 7$

$x = 7$ or $x = -7$ because both numbers are 7 units away from 0 on a real number line.

Solution set: $\{-7, 7\}$

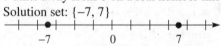

2. $|z| = 1$

$z = 1$ or $z = -1$ because both numbers are 1 unit away from 0 on a real number line.

Solution set: $\{-1, 1\}$

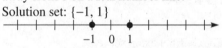

3. $u = a$ or $u = -a$

4. $|2x-3| = 7$

$\quad 2x-3 = 7 \quad$ or $\quad 2x-3 = -7$

$\quad\quad 2x = 10 \qquad\qquad 2x = -4$

$\quad\quad\quad x = 5 \qquad\qquad\quad x = -2$

Check:

Let $x = 5$:   Let $x = -2$:

$|2(5)-3| \overset{?}{=} 7 \qquad\qquad |2(-2)-3| \overset{?}{=} 7$

$\quad |10-3| \overset{?}{=} 7 \qquad\qquad\quad |-4-3| \overset{?}{=} 7$

$\quad\quad\quad 7 = 7 \text{ T} \qquad\qquad\qquad 7 = 7 \text{ T}$

Solution set: $\{-2, 5\}$

5. $|3x-2|+3=10$

$\qquad|3x-2|=7$

$3x-2=7 \quad$ or $\quad 3x-2=-7$

$\qquad 3x=9 \qquad\qquad 3x=-5$

$\qquad x=3 \qquad\qquad\quad x=-\dfrac{5}{3}$

Check:

Let $x=3$: $\qquad\qquad$ Let $x=-\dfrac{5}{3}$:

$|3(3)-2|+3 \stackrel{?}{=} 10 \qquad \left|3\left(-\dfrac{5}{3}\right)-2\right|+3 \stackrel{?}{=} 10$

$\quad |9-2|+3 \stackrel{?}{=} 10 \qquad\qquad |-5-2|+3 \stackrel{?}{=} 10$

$\qquad\quad 7+3 \stackrel{?}{=} 10 \qquad\qquad\qquad 7+3 \stackrel{?}{=} 10$

$\qquad\qquad 10=10 \;\; T \qquad\qquad\qquad\quad 10=10 \;\; T$

Solution set: $\left\{-\dfrac{5}{3},\, 3\right\}$

6. $|-5x+2|-2=5$

$\qquad |-5x+2|=7$

$-5x+2=7 \quad$ or $\quad -5x+2=-7$

$\quad -5x=5 \qquad\qquad -5x=-9$

$\qquad x=-1 \qquad\qquad\; x=\dfrac{9}{5}$

Check:

Let $x=-1$: $\qquad\qquad$ Let $x=\dfrac{9}{5}$

$|-5(-1)+2|-2 \stackrel{?}{=} 5 \qquad \left|-5\left(\dfrac{9}{5}\right)+2\right|-2 \stackrel{?}{=} 5$

$\quad |5+2|-2 \stackrel{?}{=} 5 \qquad\qquad |-9+2|-2 \stackrel{?}{=} 5$

$\qquad\; 7-2 \stackrel{?}{=} 5 \qquad\qquad\qquad 7-2 \stackrel{?}{=} 5$

$\qquad\qquad 5=5 \;\; T \qquad\qquad\qquad\;\; 5=5 \;\; T$

Solution set: $\left\{-1,\, \dfrac{9}{5}\right\}$

7. $3|x+2|-4=5$

$\qquad 3|x+2|=9$

$\qquad\quad |x+2|=3$

$x+2=3 \quad$ or $\quad x+2=-3$

$\quad x=1 \qquad\qquad x=-5$

Check:

Let $x=1$: $\qquad\qquad$ Let $x=-5$:

$3|1+2|-4 \stackrel{?}{=} 5 \qquad 3|-5+2|-4 \stackrel{?}{=} 5$

$\;\; 3(3)-4 \stackrel{?}{=} 5 \qquad\quad 3(3)-4 \stackrel{?}{=} 5$

$\quad\; 9-4 \stackrel{?}{=} 5 \qquad\qquad 9-4 \stackrel{?}{=} 5$

$\qquad 5=5 \;\; T \qquad\qquad\quad 5=5 \;\; T$

Solution set: $\{-5,\, 1\}$

8. True. The absolute value of a number represents the distance of the number from 0 on a real number line. Since distance is never negative, absolute value is never negative.

9. $|5x+3|=-2$

Since absolute values are never negative, this equation has no solution.

Solution set: $\{\ \}$ or $\varnothing$

10. $|2x+5|+7=3$

$\qquad |2x+5|=-4$

Since absolute values are never negative, this equation has no solution.

Solution set: $\{\ \}$ or $\varnothing$

11. $|x+1|+3=3$

$\qquad |x+1|=0$

$\qquad x+1=0$

$\qquad\quad x=-1$

Check:

Let $x=-1$:

$|-1+1|+3 \stackrel{?}{=} 3$

$\quad 0+3 \stackrel{?}{=} 3$

$\qquad 3 \stackrel{?}{=} 3 \;\; T$

Solution set: $\{-1\}$

12. $|x-3|=|2x+5|$

$x-3=2x+5 \quad$ or $\quad x-3=-(2x+5)$

$\quad x=2x+8 \qquad\qquad x-3=-2x-5$

$\;\; -x=8 \qquad\qquad\qquad x=-2x-2$

$\quad\; x=-8 \qquad\qquad\qquad 3x=-2$

$\qquad\qquad\qquad\qquad\qquad\qquad\; x=-\dfrac{2}{3}$

Check:

Let $x=-8$: $\qquad\qquad$ Let $x=-\dfrac{2}{3}$:

$|-8-3| \stackrel{?}{=} |2(-8)+5| \qquad \left|-\dfrac{2}{3}-3\right| \stackrel{?}{=} \left|2\left(-\dfrac{2}{3}\right)+5\right|$

$\quad |-11| \stackrel{?}{=} |-16+5| \qquad\qquad \left|-\dfrac{11}{3}\right| \stackrel{?}{=} \left|-\dfrac{4}{3}+5\right|$

$\qquad 11=11 \;\; T \qquad\qquad\qquad\qquad \dfrac{11}{3}=\dfrac{11}{3} \;\; T$

Solution set: $\left\{-8,\, -\dfrac{2}{3}\right\}$

13. $|8z+11|=|6z+17|$

$8z+11=6z+17$ or $8z+11=-(6z+17)$
$8z=6z+6$ $8z+11=-6z-17$
$2z=6$ $8z=-6z-28$
$z=3$ $14z=-28$
                         $z=-2$

Check:
Let $z=-2$:
$|8(-2)+11| \overset{?}{=} |6(-2)+17|$
$|-16+11| \overset{?}{=} |-12+17|$
$5=5$ T
Let $z=3$:
$|8(3)+11| \overset{?}{=} |6(3)+17|$
$|24+11| \overset{?}{=} |18+17|$
$35=35$ T
Solution set: $\{-2, 3\}$

14. $|3-2y|=|4y+3|$

$3-2y=4y+3$ or $3-2y=-(4y+3)$
$-2y=4y$ $3-2y=-4y-3$
$-6y=0$ $-2y=-4y-6$
$y=0$ $2y=-6$
                       $y=-3$

Check:
Let $y=0$:   Let $y=-3$:
$|3-2(0)| \overset{?}{=} |4(0)+3|$ $|3-2(-3)| \overset{?}{=} |4(-3)+3|$
$|3| \overset{?}{=} |3|$ $|3+6| \overset{?}{=} |-12+3|$
$3=3$ T $9=9$ T
Solution set: $\{-3, 0\}$

15. $|2x-3|=|5-2x|$

$2x-3=5-2x$ or $2x-3=-(5-2x)$
$2x=8-2x$ $2x-3=-5+2x$
$4x=8$ $2x=-2+2x$
$x=2$ $0=-2$ false

The second equation leads to a contradiction.
Therefore, the only solution is $x=2$.
Check:
Let $x=2$:
$|2(2)-3| \overset{?}{=} |5-2(2)|$
$|4-3| \overset{?}{=} |5-4|$
$1=1$ T
Solution set: $\{2\}$

16. $-a < u < a$

17. $a < 0$

18. $|x| \le 5$

$-5 \le x \le 5$
Set-builder: $\{x \mid -5 \le x \le 5\}$
Interval: $[-5, 5]$

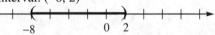

19. $|x| < \dfrac{3}{2}$

$-\dfrac{3}{2} < x < \dfrac{3}{2}$
Set-builder: $\left\{x \mid -\dfrac{3}{2} < x < \dfrac{3}{2}\right\}$
Interval: $\left(-\dfrac{3}{2}, \dfrac{3}{2}\right)$

20. $|x+3| < 5$

$-5 < x+3 < 5$
$-5-3 < x+3-3 < 5-3$
$-8 < x < 2$
Set-builder: $\{x \mid -8 < x < 2\}$
Interval: $(-8, 2)$

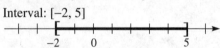

21. $|2x-3| \le 7$

$-7 \le 2x-3 \le 7$
$-7+3 \le 2x-3+3 \le 7+3$
$-4 \le 2x \le 10$
$\dfrac{-4}{2} \le \dfrac{2x}{2} \le \dfrac{10}{2}$
$-2 \le x \le 5$
Set-builder: $\{x \mid -2 \le x \le 5\}$
Interval: $[-2, 5]$

22. $|7x+2| < -3$

Since absolute values are never negative, this
inequality has no solution.
Solution set: $\{\ \}$ or $\varnothing$

23. $|x| + 4 < 6$

$\quad |x| < 2$

$-2 < x < 2$

Set-builder: $\{x \mid -2 < x < 2\}$; Interval: $(-2, 2)$

24. $|x - 3| + 4 \leq 8$

$\quad |x - 3| \leq 4$

$\quad -4 \leq x - 3 \leq 4$

$\quad -4 + 3 \leq x - 3 + 3 \leq 4 + 3$

$\quad -1 \leq x \leq 7$

Set-builder: $\{x \mid -1 \leq x \leq 7\}$

Interval: $[-1, 7]$

25. $3|2x + 1| \leq 9$

$\quad |2x + 1| \leq 3$

$\quad -3 \leq 2x + 1 \leq 3$

$\quad -3 - 1 \leq 2x + 1 - 1 \leq 3 - 1$

$\quad -4 \leq 2x \leq 2$

$\quad \dfrac{-4}{2} \leq \dfrac{2x}{2} \leq \dfrac{2}{2}$

$\quad -2 \leq x \leq 1$

Set-builder: $\{x \mid -2 \leq x \leq 1\}$

Interval: $[-2, 1]$

26. $|-3x + 1| - 5 < 3$

$\quad |-3x + 1| < 8$

$\quad -8 < -3x + 1 < 8$

$\quad -8 - 1 < -3x + 1 - 1 < 8 - 1$

$\quad -9 < -3x < 7$

$\quad \dfrac{-9}{-3} > \dfrac{-3x}{-3} > \dfrac{7}{-3}$

$\quad 3 > x > -\dfrac{7}{3}$ or $-\dfrac{7}{3} < x < 3$

Set-builder: $\left\{ x \mid -\dfrac{7}{3} < x < 3 \right\}$

Interval: $\left(-\dfrac{7}{3}, 3 \right)$

27. $u < -a$ or $u > a$

28. False. $|x| > -2$ has the entire real number line as solutions.

29. $|x| \geq 6$

$\quad x \leq -6$ or $x \geq 6$

Set-builder: $\{x \mid x \leq -6 \text{ or } x \geq 6\}$

Interval: $(-\infty, -6] \cup [6, \infty)$

30. $|x| > \dfrac{5}{2}$

$\quad x < -\dfrac{5}{2}$ or $x > \dfrac{5}{2}$

Set-builder: $\left\{ x \mid x < -\dfrac{5}{2} \text{ or } x > \dfrac{5}{2} \right\}$

Interval: $\left(-\infty, -\dfrac{5}{2} \right) \cup \left(\dfrac{5}{2}, \infty \right)$

31. $|x + 3| > 4$

$\quad x + 3 < -4$ or $x + 3 > 4$

$\quad\quad x < -7 \quad\quad\quad x > 1$

Set-builder: $\{x \mid x < -7 \text{ or } x > 1\}$

Interval: $(-\infty, -7) \cup (1, \infty)$

32. $|4x - 3| \geq 5$

$\quad 4x - 3 \leq -5$ or $4x - 3 \geq 5$

$\quad\quad 4x \leq -2 \quad\quad\quad 4x \geq 8$

$\quad\quad x \leq -\dfrac{1}{2} \quad\quad\quad x \geq 2$

Set-builder: $\left\{ x \mid x \leq -\dfrac{1}{2} \text{ or } x \geq 2 \right\}$

Interval: $\left(-\infty, -\dfrac{1}{2} \right] \cup [2, \infty)$

33. $|-3x + 2| > 7$

$\quad -3x + 2 < -7$ or $-3x + 2 > 7$

$\quad\quad -3x < -9 \quad\quad\quad -3x > 5$

$\quad\quad\quad x > 3 \quad\quad\quad\quad x < -\dfrac{5}{3}$

Set-builder: $\left\{ x \mid x < -\dfrac{5}{3} \text{ or } x > 3 \right\}$

Interval: $\left(-\infty, -\dfrac{5}{3} \right) \cup (3, \infty)$

34. $|2x+5| - 2 > -2$

$|2x+5| > 0$

$2x+5 < 0 \quad$ or $\quad 2x+5 > 0$

$\quad 2x < -5 \qquad\qquad 2x > -5$

$\quad x < -\dfrac{5}{2} \qquad\qquad x > -\dfrac{5}{2}$

Set-builder: $\left\{ x \mid x \neq -\dfrac{5}{2} \right\}$

Interval: $\left(-\infty, -\dfrac{5}{2} \right) \cup \left(-\dfrac{5}{2}, \infty \right)$

35. $|6x-5| \geq 0$

Since absolute values are always nonnegative, all real numbers are solutions to this inequality.

Set-builder: $\left\{ x \mid x \text{ is any real number} \right\}$

Interval: $(-\infty, \infty)$

36. $|2x+1| > -3$

Since absolute values are always nonnegative, all real numbers are solutions to this inequality.

Set-builder: $\left\{ x \mid x \text{ is any real number} \right\}$

Interval: $(-\infty, \infty)$

37. $|x-4| \leq \dfrac{1}{32}$

$-\dfrac{1}{32} \leq x-4 \leq \dfrac{1}{32}$

$-\dfrac{1}{32} + 4 \leq x-4+4 \leq \dfrac{1}{32} + 4$

$-\dfrac{1}{32} + \dfrac{128}{32} \leq x \leq \dfrac{1}{32} + \dfrac{128}{32}$

$\dfrac{127}{32} \leq x \leq \dfrac{129}{32}$

The acceptable belt widths are between

$\dfrac{127}{32}$ inches and $\dfrac{129}{32}$ inches.

38. $|p-9| \leq 1.7$

$-1.7 \leq p-9 \leq 1.7$

$-1.7+9 \leq p-9+9 \leq 1.7+9$

$7.3 \leq p \leq 10.7$

The percentage of people that have been shot at is between 7.3 percent and 10.7 percent.

8.7 Exercises

39. $|x| = 10$

$x = 10 \text{ or } x = -10$

Solution set: $\{-10, 10\}$

41. $|y-3| = 4$

$y-3 = 4 \quad$ or $\quad y-3 = -4$

$\quad y = 7 \quad$ or $\quad y = -1$

Solution set: $\{-1, 7\}$

43. $|-3x+5| = 8$

$-3x+5 = 8 \quad$ or $\quad -3x+5 = -8$

$\quad -3x = 3 \quad$ or $\qquad -3x = -13$

$\qquad x = -1 \quad$ or $\qquad x = \dfrac{13}{3}$

Solution set: $\left\{ -1, \dfrac{13}{3} \right\}$

45. $|y| - 7 = -2$

$|y| = 5$

$y = 5 \text{ or } y = -5$

Solution set: $\{-5, 5\}$

47. $|2x+3| - 5 = 3$

$|2x+3| = 8$

$2x+3 = 8 \quad$ or $\quad 2x+3 = -8$

$\quad 2x = 5 \quad$ or $\qquad 2x = -11$

$\qquad x = \dfrac{5}{2} \quad$ or $\qquad x = -\dfrac{11}{2}$

Solution set: $\left\{ -\dfrac{11}{2}, \dfrac{5}{2} \right\}$

49. $-2|x-3| + 10 = -4$

$-2|x-3| = -14$

$|x-3| = 7$

$x-3 = 7 \quad$ or $\quad x-3 = -7$

$\quad x = 10 \quad$ or $\qquad x = -4$

Solution set: $\{-4, 10\}$

51. $|-3x|-5=-5$

   $|-3x|=0$

    $-3x=0$

     $x=0$

  Solution set: $\{0\}$

53. $\left|\dfrac{3x-1}{4}\right|=2$

  $\dfrac{3x-1}{4}=2$ or $\dfrac{3x-1}{4}=-2$

  $3x-1=8$ or $3x-1=-8$

   $3x=9$ or $3x=-7$

    $x=3$ or $x=-\dfrac{7}{3}$

  Solution set: $\left\{-\dfrac{7}{3},3\right\}$

55. $|3x+2|=|2x-5|$

  $3x+2=2x-5$ or $3x+2=-(2x-5)$

   $3x=2x-7$ or $3x+2=-2x+5$

    $x=-7$ or $3x=-2x+3$

    $x=-7$ or $5x=3$

    $x=-7$ or $x=\dfrac{3}{5}$

  Solution set: $\left\{-7,\dfrac{3}{5}\right\}$

57. $|8-3x|=|2x-7|$

  $8-3x=2x-7$ or $8-3x=-(2x-7)$

  $-3x=2x-15$ or $8-3x=-2x+7$

  $-5x=-15$ or $-3x=-2x-1$

   $x=3$ or $-x=-1$

   $x=3$ or $x=1$

  Solution set: $\{1,3\}$

59. $|4y-7|=|9-4y|$

  $4y-7=9-4y$ or $4y-7=-(9-4y)$

  $4y=16-4y$ or $4y-7=-9+4y$

  $8y=16$ or $4y=-2+4y$

   $y=2$ or $0=-2$

  Solution set: $\{2\}$

61. $|x|<9$

  $-9<x<9$

  Set-builder: $\{x\,|-9<x<9\}$

  Interval: $(-9,9)$

63. $|x-4|\le7$

  $-7\le x-4\le7$

  $-3\le x\le11$

  Set-builder: $\{x\,|-3\le x\le11\}$

  Interval: $[-3,11]$

65. $|3x+1|<8$

  $-8<3x+1<8$

  $-9<3x<7$

  $-3<x<\dfrac{7}{3}$

  Set-builder: $\left\{x\,|-3<x<\dfrac{7}{3}\right\}$

  Interval: $\left(-3,\dfrac{7}{3}\right)$

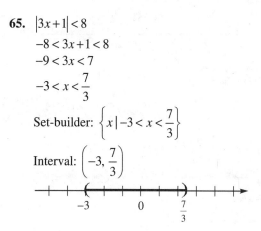

67. $|6x+5|<-1$

  No solution. Absolute value is never negative.

  Solution set: $\varnothing$ or $\{\ \}$

69. $2|x-3|+3<9$

   $2|x-3|<6$

    $|x-3|<3$

    $-3<x-3<3$

     $0<x<6$

  Set-builder: $\{x\,|0<x<6\}$

  Interval: $(0,6)$

71. $|2-5x|+3<10$

   $|2-5x|<7$

  $-7<2-5x<7$

  $-9<-5x<5$

  $\dfrac{9}{5}>x>-1$ or $-1<x<\dfrac{9}{5}$

  Set-builder: $\left\{x\,|-1<x<\dfrac{9}{5}\right\}$

  Interval: $\left(-1,\dfrac{9}{5}\right)$

73. $\left|(2x-3)-1\right|<0.01$

$\left|2x-3-1\right|<0.01$

$\left|2x-4\right|<0.01$

$-0.01<2x-4<0.01$

$3.99<2x<4.01$

$1.995<x<2.005$

Set-builder: $\left\{x\,|\,1.995<x<2.005\right\}$

Interval: $(1.995,\ 2.005)$

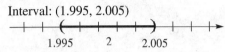

$1.995 \qquad 2 \qquad 2.005$

75. $\left|y-5\right|>2$

$y-5<-2\ $ or $\ y-5>2$

$y<3\quad$ or $\quad y>7$

Set-builder: $\left\{y\,|\,y<3\text{ or }y>7\right\}$

Interval: $(-\infty,\ 3)\cup(7,\ \infty)$

$0 \qquad 3 \qquad\qquad 7$

77. $\left|-4x-3\right|\ge 5$

$-4x-3\le-5\ $ or $\ -4x-3\ge 5$

$-4x\le-2\ $ or $\quad -4x\ge 8$

$x\ge\dfrac{1}{2}\quad$ or $\quad x\le-2$

Set-builder: $\left\{x\,\Big|\,x\le-2\text{ or }x\ge\dfrac{1}{2}\right\}$

Interval: $\left(-\infty,\ -2\right]\cup\left[\dfrac{1}{2},\ \infty\right)$

$-2 \qquad 0\ \ \dfrac{1}{2}$

79. $2\left|y\right|+3>1$

$2\left|y\right|>-2$

$\left|y\right|>-1$

Since $\left|y\right|\ge 0>-1$ for all y, all real numbers are solutions.

Set-builder: $\left\{y\,|\,y\text{ is a real number}\right\}$

Interval: $(-\infty,\ \infty)$

0

81. $\left|-5x-3\right|-7>0$

$\left|-5x-3\right|>7$

$-5x-3<-7\ $ or $\ -5x-3>7$

$-5x<-4\ $ or $\quad -5x>10$

$x>\dfrac{4}{5}\quad$ or $\qquad x<-2$

Set-builder: $\left\{x\,\Big|\,x<-2\text{ or }x>\dfrac{4}{5}\right\}$

Interval: $(-\infty,\ -2)\cup\left(\dfrac{4}{5},\ \infty\right)$

$-2 \qquad\quad 0 \qquad \dfrac{4}{5}$

83. $4\left|-2x+1\right|>4$

$\left|-2x+1\right|>1$

$-2x+1<-1\ $ or $\ -2x+1>1$

$-2x<-2\ $ or $\quad -2x>0$

$x>1\quad$ or $\qquad x<0$

Set-builder: $\left\{x\,|\,x<0\text{ or }x>1\right\}$

Interval: $(-\infty,\ 0)\cup(1,\ \infty)$

$0 \qquad 1$

85. $\left|1-2x\right|\ge\left|-5\right|$

$\left|1-2x\right|\ge 5$

$1-2x\le-5\ $ or $\ 1-2x\ge 5$

$-2x\le-6\ $ or $\ -2x\ge 4$

$x\ge 3\quad$ or $\qquad x\le-2$

Set-builder: $\left\{x\,|\,x\le-2\text{ or }x\ge 3\right\}$

Interval: $(-\infty,\ -2]\cup[3,\ \infty)$

$-2 \qquad 0 \qquad\quad 3$

87. (a) $f(x)=g(x)$ when $x=-5$ and $x=5$.

The solution set is $\{-5,5\}$.

(b) $f(x)\le g(x)$ when $-5\le x\le 5$.

Set-builder: $\left\{x\,|\,-5\le x\le 5\right\}$

Interval: $[-5,\ 5]$

(c) $f(x)>g(x)$ for $x<-5$ or $x>5$.

Set-builder: $\left\{x\,|\,x<-5\text{ or }x>5\right\}$

Interval: $(-\infty,\ -5)\cup(5,\ \infty)$

89. (a) $f(x) = g(x)$ when $x = -5$ and $x = 1$.

The solution set is $\{-5, 1\}$.

(b) $f(x) < g(x)$ when $-5 < x < 1$.

Set-builder: $\{x \mid -5 < x < 1\}$

Interval: $(-5, 1)$

(c) $f(x) \geq g(x)$ for $x \leq -5$ or $x \geq 1$.

Set-builder: $\{x \mid x \leq -5 \text{ or } x \geq 1\}$

Interval: $(-\infty, -5] \cup [1, \infty)$

91. $|x| > 5$

$x < -5$ or $x > 5$

Set-builder: $\{x \mid x < -5 \text{ or } x > 5\}$

Interval: $(-\infty, -5) \cup (5, \infty)$

93. $|2x + 5| = 3$

$2x + 5 = 3$ or $2x + 5 = -3$

$2x = -2$ or $2x = -8$

$x = -1$ or $x = -4$

Solution set: $\{-4, -1\}$

95. $7|x| = 35$

$|x| = 5$

$x = 5$ or $x = -5$

Solution set: $\{-5, 5\}$

97. $|5x + 2| \leq 8$

$-8 \leq 5x + 2 \leq 8$

$-10 \leq 5x \leq 6$

$-2 \leq x \leq \dfrac{6}{5}$

Set-builder: $\left\{x \mid -2 \leq x \leq \dfrac{6}{5}\right\}$

Interval: $\left[-2, \dfrac{6}{5}\right]$

99. $|-2x + 3| = -4$

No solution. Absolute value is never negative.

Solution set: $\varnothing$ or $\{\ \}$

101. $|3x + 2| \geq 5$

$3x + 2 \leq -5$ or $3x + 2 \geq 5$

$3x \leq -7$ or $3x \geq 3$

$x \leq -\dfrac{7}{3}$ or $x \geq 1$

Set-builder: $\left\{x \mid x \leq -\dfrac{7}{3} \text{ or } x \geq 1\right\}$

Interval: $\left(-\infty, -\dfrac{7}{3}\right] \cup [1, \infty)$

103. $|3x - 2| + 7 > 9$

$|3x - 2| > 2$

$3x - 2 < -2$ or $3x - 2 > 2$

$3x < 0$ or $3x > 4$

$x < 0$ or $x > \dfrac{4}{3}$

Set-builder: $\left\{x \mid x < 0 \text{ or } x > \dfrac{4}{3}\right\}$

Interval: $\left(-\infty, 0\right) \cup \left(\dfrac{4}{3}, \infty\right)$

105. $|5x + 3| = |3x + 5|$

$5x + 3 = 3x + 5$ or $5x + 3 = -(3x + 5)$

$5x = 3x + 2$ or $5x + 3 = -3x - 5$

$2x = 2$ or $8x = -8$

$x = 1$ or $x = -1$

Solution set: $\{-1, 1\}$

107. $|4x + 7| + 6 < 5$

$|4x + 7| < -1$

Since absolute value is always nonnegative, that is ≥ 0, this inequality has no solution.

Solution set: $\varnothing$ or $\{\ \}$

109. $\left|\dfrac{x-2}{4}\right| = \left|\dfrac{2x+1}{6}\right|$

$\dfrac{x-2}{4} = \dfrac{2x+1}{6}$ or $\dfrac{x-2}{4} = -\dfrac{2x+1}{6}$

$12 \cdot \dfrac{x-2}{4} = 12 \cdot \dfrac{2x+1}{6}$ or $12 \cdot \dfrac{x-2}{4} = 12 \cdot \dfrac{-2x-1}{6}$

$3x - 6 = 4x + 2$ or $3x - 6 = -4x - 2$

$3x = 4x + 8$ or $3x = -4x + 4$

$-x = 8$ or $7x = 4$

$x = -8$ or $x = \dfrac{4}{7}$

Solution set: $\left\{-8, \dfrac{4}{7}\right\}$

111. $|5 - x| < 3$

$-3 < 5 - x < 3$

$-8 < -x < -2$

$8 > x > 2$ or $2 < x < 8$

Set-builder: $\{x \mid 2 < x < 8\}$

Interval: $(2, 8)$

113. $|2x - (-6)| > 3$

$|2x + 6| > 3$

$2x + 6 < -3$ or $2x + 6 > 3$

$2x < -9$ or $2x > -3$

$x < -\dfrac{9}{2}$ or $x > -\dfrac{3}{2}$

Set-builder: $\left\{x \mid x < -\dfrac{9}{2} \text{ or } x > -\dfrac{3}{2}\right\}$

Interval: $\left(-\infty, -\dfrac{9}{2}\right) \cup \left(-\dfrac{3}{2}, \infty\right)$

115. $|x - 5.7| \le 0.0005$

$-0.0005 \le x - 5.7 \le 0.0005$

$5.6995 \le x \le 5.7005$

The acceptable rod lengths are between 5.6995 inches and 5.7005 inches, inclusive.

117. $\left|\dfrac{x - 100}{15}\right| > 1.96$

$\dfrac{x - 100}{15} < -1.96$ or $\dfrac{x - 100}{15} > 1.96$

$x - 100 < -29.4$ or $x - 100 > 29.4$

$x < 70.6$ or $x > 129.4$

An unusual IQ score would be less than 70.6 or greater than 129.4.

119. $|x| - x = 5$

$|x| = x + 5$

$x = x + 5$ or $x = -(x + 5)$

$0 \ne 5$ or $x = -x - 5$

$2x = -5$

$x = -\dfrac{5}{2}$

Check: $-\dfrac{5}{2}$:

$\left|-\dfrac{5}{2}\right| - \left(-\dfrac{5}{2}\right) \stackrel{?}{=} 5$

$\dfrac{5}{2} + \dfrac{5}{2} \stackrel{?}{=} 5$

$5 = 5$ T

Solution set: $\left\{-\dfrac{5}{2}\right\}$

121. $z + |-z| = 4$

$|-z| = 4 - z$

$-z = 4 - z$ or $-z = -(4 - z)$

$0 \ne 4$ or $-z = -4 + z$

$-2z = -4$

$z = 2$

Check 2: $2 + |-(2)| \stackrel{?}{=} 4$

$2 + 2 \stackrel{?}{=} 4$

$4 = 4$ T

Solution set: $\{2\}$

123. $|4x + 1| = x - 2$

$4x + 1 = x - 2$ or $4x + 1 = -(x - 2)$

$4x = x - 3$ or $4x + 1 = -x + 2$

$3x = -3$ or $4x = -x + 1$

$x = -1$ or $5x = 1$

$x = -1$ or $x = \dfrac{1}{5}$

Check:

$|4(-1) + 1| \stackrel{?}{=} (-1) - 2$ $\left|4\left(\dfrac{1}{5}\right) + 1\right| \stackrel{?}{=} \dfrac{1}{5} - 2$

$|-4 + 1| \stackrel{?}{=} -3$ $\left|\dfrac{4}{5} + 1\right| \stackrel{?}{=} -\dfrac{9}{5}$

$|-3| \stackrel{?}{=} -3$ $\left|\dfrac{9}{5}\right| \stackrel{?}{=} -\dfrac{9}{5}$

$3 \ne -3$ $\dfrac{9}{5} \ne -\dfrac{9}{5}$

Solution set: $\varnothing$ or $\{\ \}$

125. $|x+5| = -(x+5)$

Since we have $|u| = -u$, we need $u \le 0$ so the absolute value will not be negative. Thus, we can say that $x + 5 \le 0$ or $x \le -5$.

Set-builder: $\{x \mid x \le -5\}$

Interval: $(-\infty, -5]$

127. $|2x-3| + 1 = 0$ has no solution because the absolute value, when isolated, is equal to a negative number which is not possible.

129. $|4x+3| + 3 < 0$ has the empty set as the solution set because the absolute value, when isolated, is less than a negative number (-3). Since absolute values are always nonnegative, this is not possible.

Section 8.8

Preparing for Variation

P1. $30 = 5x$

$$\frac{30}{5} = \frac{5x}{5}$$

$$x = 6$$

The solution set is $\{6\}$.

P2. $4 = \dfrac{k}{3}$

$$4 \cdot 3 = \frac{k}{\cancel{3}} \cdot \cancel{3}$$

$$k = 12$$

The solution set is $\{12\}$.

P3. $y = 3x$

$m = 3$, $b = 0$

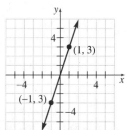

Section 8.8 Quick Checks

1. Variation

2. $y = kx$

3. (a) Because C varies directly with g, we know that $C = kg$ for some constant k. Because $C = 25.20$ when $g = 8$ gallons, we obtain

$$C = kg$$

$$25.20 = k \cdot 8$$

$$k = \frac{25.20}{8} = 3.15$$

So, we have that $C = 3.15g$, or writing this as a linear function, $C(g) = 3.15g$.

(b) $C(5.6) = 3.15(5.6) = 17.64$

So, the cost of 5.6 gallons would be $17.64.

(c) Label the horizontal axis g and the vertical axis C. From the problem we know the point $(8, 25.20)$ is on the graph, and from part (b), we know the point $(5.6, 17.64)$ is on the graph.

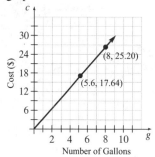

4. $y = \dfrac{k}{x}$

5. (a) $V = \dfrac{k}{l}$

$$500 = \frac{k}{30}$$

$$k = 500(30) = 15,000$$

Thus, the function that relates the rate of vibration to the length of the string is

$$V(l) = \frac{15,000}{l}.$$

(b) $V(50) = \dfrac{15,000}{50} = 300$

The rate of vibration is 300 oscillations per second.

6. combined variation

7. Let K represent the kinetic energy, m represent the mass, and v represent the velocity. Then,

$$K = kmv^2$$
$$4455 = k \cdot 110 \cdot 9^2$$
$$4455 = 8910k$$
$$\frac{1}{2} = k$$

Thus, $K = \frac{1}{2}mv^2$.

For $m = 140$ and $v = 5$,

$$K = \frac{1}{2}(140)(5)^2 = 1750.$$

The kinetic energy of a 140-kg lineman running at 5 meters per second is 1750 joules.

8. $R = \dfrac{kl}{d^2}$

$$1.24 = \frac{k \cdot 432}{4^2}$$
$$1.24 = \frac{432k}{16}$$
$$1.24 = 27k$$
$$k = \frac{1.24}{27} \approx 0.04593$$

Thus, $R = \dfrac{0.04593l}{d^2}$.

For $l = 282$ and $d = 3$, $R = \dfrac{0.04593(282)}{3^2} \approx 1.44$.

The resistance is approximately 1.44 ohms when the length of the wire is 282 feet and the diameter is 3 millimeters.

8.8 Exercises

9. (a) $y = kx$
$$30 = k \cdot 5$$
$$k = \frac{30}{5} = 6$$

(b) $y = 6x$

(c) $y = 6(7) = 42$

11. (a) $y = kx$
$$3 = k \cdot 7$$
$$k = \frac{3}{7}$$

(b) $y = \dfrac{3}{7}x$

(c) $y = \dfrac{3}{7}(28) = 12$

13. (a) $y = kx$
$$4 = k \cdot 8$$
$$k = \frac{4}{8} = \frac{1}{2}$$

(b) $y = \dfrac{1}{2}x$

(c) $y = \dfrac{1}{2}(30) = 15$

15. (a) $y = \dfrac{k}{x}$
$$2 = \frac{k}{10}$$
$$10 \cdot 2 = 10\left(\frac{k}{10}\right)$$
$$20 = k$$

(b) $y = \dfrac{20}{x}$

(c) When $x = 5$, $y = \dfrac{20}{5} = 4$.

17. (a) $y = \dfrac{k}{x}$
$$3 = \frac{k}{7}$$
$$7 \cdot 3 = 7\left(\frac{k}{7}\right)$$
$$21 = k$$

(b) $y = \dfrac{21}{x}$

(c) When $x = 28$, $y = \dfrac{21}{28} = \dfrac{3}{4}$.

19. (a) $y = kxz$
$$10 = k \cdot 8 \cdot 5$$
$$10 = 40k$$
$$k = \frac{10}{40} = \frac{1}{4}$$

(b) $y = \dfrac{1}{4}xz$

(c) When $x = 12$ and $z = 9$, $y = \dfrac{1}{4} \cdot 12 \cdot 9 = 27$.

21. (a)

$$Q = \frac{kx}{y}$$

$$\frac{13}{12} = \frac{k \cdot 5}{6}$$

$$12\left(\frac{13}{12}\right) = 12\left(\frac{k \cdot 5}{6}\right)$$

$$13 = 10k$$

$$\frac{13}{10} = k$$

(b) $Q = \dfrac{(13/10)x}{y} = \dfrac{13x}{10y}$

(c) When $x = 9$ and $y = 4$, $Q = \dfrac{13 \cdot 9}{10 \cdot 4} = \dfrac{117}{40}$.

23. (a) Because p varies directly with b, we know that $p = kb$ for some constant k. We know that $p = \$700.29$ when $b = \$120,000$, so,

$$700.29 = k \cdot 120,000$$
$$k \approx 0.0058358$$

Therefore, we have $p = 0.0058358b$ or, using function notation,
$p(b) = 0.00538358b$.

(b) $p(140,000) = 0.0058358(140,000)$
$$\approx 817.01$$

The monthly payment would be approximately $817.01 if $140,000 is borrowed.

(c)

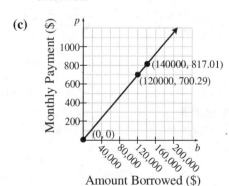

Amount Borrowed ($)

25. (a) Because C varies directly with w, we know that $C = kw$ for some constant k. We know that $C = \$28$ when $w = 5$ lb, so,

$$28 = k \cdot 5$$
$$k = 5.6$$

Therefore, we have $C = 5.6w$ or, using function notation, $C(w) = 5.6w$.

(b) $C(3.5) = 5.6(3.5) = \$19.60$

The cost of 3.5 pounds of chocolate covered almonds would be $19.60.

(c)

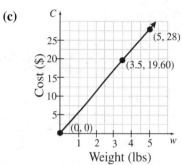

Weight (lbs)

27. Because v varies directly with t, we know that $v = kt$ for some constant k. We know that $v = 64$ feet per second when $t = 2$ seconds, so

$$64 = k \cdot 2$$
$$k = 32$$

Thus, $v = 32t$ or, using function notation, $v(t) = 32t$. Finally, $v(3) = 32(3) = 96$.

Therefore, the velocity of a falling object will be 96 feet per second after 3 seconds.

29. (a)

$$D = \frac{k}{p}$$

$$150 = \frac{k}{2.50}$$

$$2.50 \cdot 150 = 2.50\left(\frac{k}{2.50}\right)$$

$$375 = k$$

Thus, the demand of the candy as a function of its price is $D(p) = \dfrac{375}{p}$.

(b) $D(3) = \dfrac{375}{3} = 125$

If the price of the candy is $3.00, then 125 bags of candy will be sold.

31.
$$V = \frac{k}{P}$$
$$600 = \frac{k}{150}$$
$$150 \cdot 600 = 150\left(\frac{k}{150}\right)$$
$$90,000 = k$$
$$V(P) = \frac{90,000}{P}$$
$$V(200) = \frac{90,000}{200} = 450$$

When the pressure is 200 mm Hg, the volume of the gas will be 450 cc.

33. Let w represent the weight of an object that is a distance d from the center of the earth.
$$w = \frac{k}{d^2}$$
$$120 = \frac{k}{3960^2}$$
$$3960^2 \cdot 120 = 3960^2\left(\frac{k}{3960^2}\right)$$
$$1,881,792,000 = k$$
$$w(d) = \frac{1,881,792,000}{d^2}$$

Now, Maria is $3960 + 3.8 = 3963.8$ miles from the center of Earth.
$$w(3963.8) = \frac{1,881,792,000}{3963.8^2} \approx 119.8$$

Maria's weight at the top of Mount McKinley is 119.8 pounds.

35. Let D represent drag force, A represent surface area, and v represent velocity.
$$D = kAv^2$$
$$1152 = k \cdot 2 \cdot 40^2$$
$$1152 = 3200k$$
$$0.36 = k$$
$$D = 0.36Av^2$$

For $A = 2.5$ and $v = 50$,
$$D = 0.36 \cdot 2.5 \cdot 50^2 = 2250.$$

The drag force on a parachutist whose surface area is 2.5 square meters falling at 50 meters per second is 2250 newtons.

37.
$$F = \frac{km_1 m_2}{r^2}$$
$$2.24112 \times 10^{-8} = \frac{k \cdot 105 \cdot 80}{5^2}$$
$$2.24112 \times 10^{-8} = 336k$$
$$\frac{2.24112 \times 10^{-8}}{336} = k$$
$$6.67 \times 10^{-11} = k$$
$$F = \frac{\left(6.67 \times 10^{-11}\right)m_1 m_2}{r^2}$$

For $m_1 = 105$, $m_2 = 80$, and $r = 2$,
$$F = \frac{\left(6.67 \times 10^{-11}\right)m_1 m_2}{r^2}$$
$$= \frac{\left(6.67 \times 10^{-11}\right)(105)(80)}{2^2}$$
$$= 1.4007 \times 10^{-7}.$$

When they are 2 meters apart, the force of gravity between the couple is 1.4007×10^{-7} newtons.

39. Let s represent stress, p represent internal pressure, d represent internal diameter, and x represent thickness.
$$s = \frac{kpd}{x}$$
$$100 = \frac{k \cdot 25 \cdot 5}{0.75}$$
$$0.75(100) = 0.75\left(\frac{k \cdot 25 \cdot 5}{0.75}\right)$$
$$75 = 125k$$
$$0.6 = k$$
$$s = \frac{0.6pd}{x}$$

For $p = 50$, $d = 6$, and $x = 0.5$,
$$s = \frac{0.6 \cdot 50 \cdot 6}{0.5} = 360.$$

The stress is 360 pounds per square inch when the internal pressure is 50 pounds per square inch, the diameter is 6 inches, and the thickness is 0.5 inch.

41. (a) $\omega = 2\pi \cdot 50 = 100\pi \approx 314.16$

(b) $v = \omega r = 314.16 \cdot 3 = 942.48$ meters per minute

$v = \dfrac{942.48 \text{ m}}{1 \text{ min}} \cdot \dfrac{1 \text{ min}}{60 \text{ sec}} = 15.708$ meters per second

(c)
$$F = \frac{kmv^2}{r}$$
$$2.3 = \frac{k \cdot 0.5 \cdot 15.708^2}{3}$$
$$2.3 = k \cdot 41.123544$$
$$\frac{2.3}{4.1123544} = k$$
$$0.056 \approx k$$

(d) For spinning rate of 80 revolutions per minute, $\omega = 2\pi \cdot 80 = 160\pi \approx 502.65$.
$v = \omega r = 502.65 \cdot 4 = 2010.60$ meters per minute

$v = \dfrac{2010.60 \text{ m}}{1 \text{ min}} \cdot \dfrac{1 \text{ min}}{60 \text{ sec}} = 33.51$ meters per second

$F = \dfrac{0.056mv^2}{r} = \dfrac{0.056 \cdot 0.5 \cdot 33.51^2}{4} \approx 7.86$

The force required to keep the stone in motion is approximately 7.86 newtons.

Chapter 8 Review

1. A: quadrant IV; B: quadrant III; C: y-axis; D: quadrant II; E: x-axis; F: quadrant I

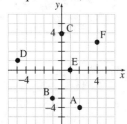

2. A: x-axis; B: quadrant I; C: quadrant III; D: quadrant II; E: quadrant IV; F: y-axis

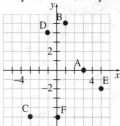

3. (a) Let $x = 3$ and $y = 1$.
$$3(3) - 2(1) \overset{?}{=} 7$$
$$9 - 2 \overset{?}{=} 7$$
$$7 = 7 \text{ True}$$
The point $(3, 1)$ is on the graph.

(b) Let $x = 2$ and $y = -1$.
$$3(2) - 2(-1) \overset{?}{=} 7$$
$$6 + 2 \overset{?}{=} 7$$
$$8 = 7 \text{ False}$$
The point $(2, -1)$ is not on the graph.

(c) Let $x = 4$ and $y = 0$.
$$3(4) - 2(0) \overset{?}{=} 7$$
$$12 - 0 \overset{?}{=} 7$$
$$12 = 7 \text{ False}$$
The point $(4, 0)$ is not on the graph.

(d) Let $x = \dfrac{1}{3}$ and $y = -3$.
$$3\left(\frac{1}{3}\right) - 2(-3) \overset{?}{=} 7$$
$$1 + 6 \overset{?}{=} 7$$
$$7 = 7 \text{ True}$$
The point $\left(\dfrac{1}{3}, -3\right)$ is on the graph.

4. (a) Let $x = -1$ and $y = 3$.
$$3 \overset{?}{=} 2(-1)^2 - 3(-1) + 2$$
$$3 \overset{?}{=} 2(1) + 3 + 2$$
$$3 \overset{?}{=} 2 + 3 + 2$$
$$3 = 7 \text{ False}$$
The point $(-1, 3)$ is not on the graph.

(b) Let $x = 1$ and $y = 1$.
$$1 \overset{?}{=} 2(1)^2 - 3(1) + 2$$
$$1 \overset{?}{=} 2(1) - 3 + 2$$
$$1 \overset{?}{=} 2 - 3 + 2$$
$$1 = 1 \text{ True}$$
The point $(1, 1)$ is on the graph.

(c) Let $x = -2$ and $y = 16$.
$$16 \overset{?}{=} 2(-2)^2 - 3(-2) + 2$$
$$16 \overset{?}{=} 2(4) + 6 + 2$$
$$16 \overset{?}{=} 8 + 6 + 2$$
$$16 = 16 \text{ True}$$
The point $(-2, 16)$ is on the graph.

(d) Let $x = \dfrac{1}{2}$ and $y = \dfrac{3}{2}$.

$$\frac{3}{2} \overset{?}{=} 2\left(\frac{1}{2}\right)^2 - 3\left(\frac{1}{2}\right) + 2$$

$$\frac{3}{2} \overset{?}{=} 2\left(\frac{1}{4}\right) - \frac{3}{2} + 2$$

$$\frac{3}{2} \overset{?}{=} \frac{1}{2} - \frac{3}{2} + 2$$

$$\frac{3}{2} = 1 \text{ False}$$

The point $\left(\dfrac{1}{2}, \dfrac{3}{2}\right)$ is not on the graph.

5. $y = x + 2$

x	$y = x + 2$	(x, y)
-3	$y = (-3) + 2 = -1$	$(-3, -1)$
-1	$y = (-1) + 2 = 1$	$(-1, 1)$
0	$y = (0) + 2 = 2$	$(0, 2)$
1	$y = (1) + 2 = 3$	$(1, 3)$
3	$y = (3) + 2 = 5$	$(3, 5)$

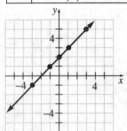

6. $2x + y = 3$
$y = -2x + 3$

x	$y = -2x + 3$	(x, y)
-3	$y = -2(-3) + 3 = 9$	$(-3, 9)$
-1	$y = -2(-1) + 3 = 5$	$(-1, 5)$
0	$y = -2(0) + 3 = 3$	$(0, 3)$
1	$y = -2(1) + 3 = 1$	$(1, 1)$
3	$y = -2(3) + 3 = -3$	$(3, -3)$

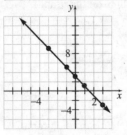

7. $y = -x^2 + 4$

x	$y = -x^2 + 4$	(x, y)
-3	$y = -(-3)^2 + 4 = -5$	$(-3, -5)$
-1	$y = -(-1)^2 + 4 = 3$	$(-1, 3)$
0	$y = -(0)^2 + 4 = 4$	$(0, 4)$
1	$y = -(1)^2 + 4 = 3$	$(1, 3)$
3	$y = -(3)^2 + 4 = -5$	$(3, -5)$

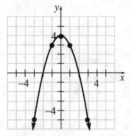

8. $y = |x + 2| - 1$

x	$y = |x + 2| - 1$	(x, y)
-5	$y = |-5 + 2| - 1 = 2$	$(-5, 2)$
-3	$y = |-3 + 2| - 1 = 0$	$(-3, 0)$
-1	$y = |-1 + 2| - 1 = 0$	$(-1, 0)$
0	$y = |0 + 2| - 1 = 1$	$(0, 1)$
1	$y = |1 + 2| - 1 = 2$	$(1, 2)$

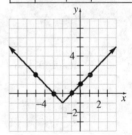

9. $y = x^3 + 2$

x	$y = x^3 + 2$	(x, y)
-2	$y = (-2)^3 + 2 = -6$	$(-2, -6)$
-1	$y = (-1)^3 + 2 = 1$	$(-1, 1)$
0	$y = (0)^3 + 2 = 2$	$(0, 2)$
1	$y = (1)^3 + 2 = 3$	$(1, 3)$
2	$y = (2)^3 + 2 = 10$	$(2, 10)$

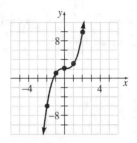

10. $x = y^2 + 1$

y	$x = y^2 + 1$	(x, y)
-2	$x = (-2)^2 + 1 = 5$	$(5, -2)$
-1	$x = (-1)^2 + 1 = 2$	$(2, -1)$
0	$x = (0)^2 + 1 = 1$	$(1, 0)$
1	$x = (1)^2 + 1 = 2$	$(2, 1)$
2	$x = (2)^2 + 1 = 5$	$(5, 2)$

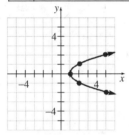

11. The intercepts are $(-3, 0)$, $(0, -1)$, and $(0, 3)$.

12. (a) Since 2250 is less than 3000, the monthly bill will still be $40. Using the graph, we find that when $x = 2.25$ (2250 minutes) the y-value is 40, or $40.

(b) For 12,000 minutes, we have $x = 12$. According to the graph, the monthly bill for 12,000 minutes will be about $500.

13. {(Cent, 2.500), (Nickel, 5.000), (Dime, 2.268), (Quarter, 5.670), (Half Dollar, 11.340), (Dollar, 8.100)}
Domain:
{Cent, Nickel, Dime, Quarter, Half Dollar, Dollar}
Range:
{2.268, 2.500, 5.000, 5.670, 8.100, 11.340}

14. {(70, $6.99), (90, $9.99), (120, $9.99), (128, $12.99), (446, $49.99)}
Domain:
{70, 90, 120, 128, 446}
Range:
{$6.99, $9.99, $12.99, $49.99}

15. Domain: $\{-4, -2, 2, 3, 6\}$
Range: $\{-9, -1, 5, 7, 8\}$

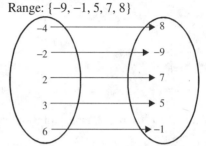

16. Domain: $\{-2, 1, 3, 5\}$
Range: $\{1, 4, 7, 8\}$

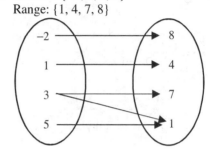

17. Domain: $\{x \mid x \text{ is a real number}\}$ or $(-\infty, \infty)$
Range: $\{y \mid y \text{ is a real number}\}$ or $(-\infty, \infty)$

18. Domain: $\{x \mid -6 \leq x \leq 4\}$ or $[-6, 4]$
Range: $\{y \mid -4 \leq y \leq 6\}$ or $[-4, 6]$

19. Domain: $\{2\}$
Range: $\{y \mid y \text{ is a real number}\}$ or $(-\infty, \infty)$

20. Domain: $\{x \mid x \geq -1\}$ or $[-1, \infty)$
Range: $\{y \mid y \geq -2\}$ or $[-2, \infty)$

21. $y = x + 2$

x	$y = x + 2$	(x, y)
-3	$y = (-3) + 2 = -1$	$(-3, -1)$
-1	$y = (-1) + 2 = 1$	$(-1, 1)$
0	$y = (0) + 2 = 2$	$(0, 2)$
1	$y = (1) + 2 = 3$	$(1, 3)$
3	$y = (3) + 2 = 5$	$(3, 5)$

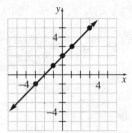

Domain: $\{x \mid x$ is a real number$\}$ or $(-\infty, \infty)$

Range: $\{y \mid y$ is a real number$\}$ or $(-\infty, \infty)$

22. $2x + y = 3$
$y = -2x + 3$

x	$y = -2x + 3$	(x, y)
-3	$y = -2(-3) + 3 = 9$	$(-3, 9)$
-1	$y = -2(-1) + 3 = 5$	$(-1, 5)$
0	$y = -2(0) + 3 = 3$	$(0, 3)$
1	$y = -2(1) + 3 = 1$	$(1, 1)$
3	$y = -2(3) + 3 = -3$	$(3, -3)$

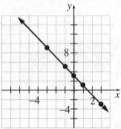

Domain: $\{x \mid x$ is a real number$\}$ or $(-\infty, \infty)$

Range: $\{y \mid y$ is a real number$\}$ or $(-\infty, \infty)$

23. $y = -x^2 + 4$

x	$y = -x^2 + 4$	(x, y)
-3	$y = -(-3)^2 + 4 = -5$	$(-3, -5)$
-1	$y = -(-1)^2 + 4 = 3$	$(-1, 3)$
0	$y = -(0)^2 + 4 = 4$	$(0, 4)$
1	$y = -(1)^2 + 4 = 3$	$(1, 3)$
3	$y = -(3)^2 + 4 = -5$	$(3, -5)$

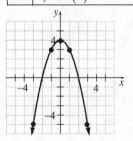

Domain: $\{x \mid x$ is a real number$\}$ or $(-\infty, \infty)$

Range: $\{y \mid y \le 4\}$ or $(-\infty, 4]$

24. $y = |x + 2| - 1$

x	$y = |x + 2| - 1$	(x, y)
-5	$y = |-5 + 2| - 1 = 2$	$(-5, 2)$
-3	$y = |-3 + 2| - 1 = 0$	$(-3, 0)$
-1	$y = |-1 + 2| - 1 = 0$	$(-1, 0)$
0	$y = |0 + 2| - 1 = 1$	$(0, 1)$
1	$y = |1 + 2| - 1 = 2$	$(1, 2)$

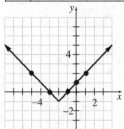

Domain: $\{x \mid x$ is a real number$\}$ or $(-\infty, \infty)$

Range: $\{y \mid y \ge -1\}$ or $[-1, \infty)$

25. $y = x^3 + 2$

x	$y = x^3 + 2$	(x, y)
-2	$y = (-2)^3 + 2 = -6$	$(-2, -6)$
-1	$y = (-1)^3 + 2 = 1$	$(-1, 1)$
0	$y = (0)^3 + 2 = 2$	$(0, 2)$
1	$y = (1)^3 + 2 = 3$	$(1, 3)$
2	$y = (2)^3 + 2 = 10$	$(2, 10)$

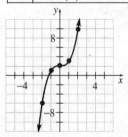

Domain: $\{x \mid x$ is a real number$\}$ or $(-\infty, \infty)$

Range: $\{y \mid y$ is a real number$\}$ or $(-\infty, \infty)$

26. $x = y^2 + 1$

y	$x = y^2 + 1$	(x, y)
-2	$x = (-2)^2 + 1 = 5$	$(5, -2)$
-1	$x = (-1)^2 + 1 = 2$	$(2, -1)$
0	$x = (0)^2 + 1 = 1$	$(1, 0)$
1	$x = (1)^2 + 1 = 2$	$(2, 1)$
2	$x = (2)^2 + 1 = 5$	$(5, 2)$

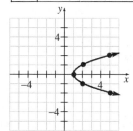

Domain: $\{x \mid x \geq 1\}$ or $[1, \infty)$

Range: $\{y \mid y$ is a real number$\}$ or $(-\infty, \infty)$

27. (a) Domain: $\{x \mid 0 \leq x \leq 44.64\}$ or $[0, 44.64]$

Range: $\{y \mid 40 \leq y \leq 2122\}$ or $[40, 2122]$

The monthly cost will be at least $40 but no more than $2122. The number of minutes used must be between 0 and 44,640.

(b) Answers may vary. There are at most 31 days in a month. Since 31 days is equivalent to 44,640 minutes, this must be the largest value in the domain. It is not possible to talk for a negative number of minutes, so the domain should begin at 0.

28. Domain: $\{t \mid 0 \leq t \leq 4\}$ or $[0, 4]$

Range: $\{y \mid 0 \leq y \leq 121\}$ or $[0, 121]$

The ball will be in the air from 0 to 4 seconds and will reach heights from 0 feet up to a maximum of 121 feet.

29. (a) Not a function. The domain element -1 corresponds to two different values in the range.
Domain: $\{-1, 5, 7, 9\}$
Range: $\{-2, 0, 2, 3, 4\}$

(b) Function. Each animal corresponds to exactly one typical lifespan.
Domain: {Camel, Macaw, Deer, Fox, Tiger, Crocodile}
Range: {14, 22, 35, 45, 50}

30. (a) Function. There are no ordered pairs that have the same first coordinate but different second coordinates.
Domain: $\{-3, -2, 2, 4, 5\}$
Range: $\{-1, 3, 4, 7\}$

(b) Not a function. The domain element 'Blue' corresponds to two different types of cars in the range.
Domain: {Red, Blue, Green, Black}
Range: {Camry, Taurus, Windstar, Durango}

31. $3x - 5y = 18$
$$-5y = -3x + 18$$
$$y = \frac{-3x + 18}{-5}$$
$$y = \frac{3}{5}x - \frac{18}{5}$$

Since there is only one output y that can result from any given input x, this relation is a function.

32. $x^2 + y^2 = 81$
$$y^2 = 81 - x^2$$

Since a given input x can result in more than one output y, this relation is not a function. For example, if $x = 0$ then $y = 9$ or $y = -9$.

33. $y = \pm 10x$

Since a given input x can result in more than one output y, this relation is not a function.

34. $y = x^2 - 14$

Since there is only one output y that can result from any given input x, this relation is a function.

35. Not a function. The graph fails the vertical line test so it is not the graph of a function.

36. Function. The graph passes the vertical line test so it is the graph of a function.

37. Function. The graph passes the vertical line test so it is the graph of a function.

38. Not a function. The graph fails the vertical line test so it is not the graph of a function.

39. (a) $f(-2) = (-2)^2 + 2(-2) - 5$
$= 4 - 4 - 5$
$= -5$

(b) $f(3) = (3)^2 + 2(3) - 5$
$= 9 + 6 - 5$
$= 10$

40. (a) $g(0) = \dfrac{2(0) + 1}{(0) - 3}$
$= \dfrac{0 + 1}{-3}$
$= -\dfrac{1}{3}$

(b) $g(2) = \dfrac{2(2) + 1}{(2) - 3}$
$= \dfrac{4 + 1}{-1}$
$= -5$

41. (a) $F(5) = -2(5) + 7$
$= -10 + 7$
$= -3$

(b) $F(-x) = -2(-x) + 7$
$= 2x + 7$

42. (a) $G(7) = 2(7) + 1$
$= 14 + 1$
$= 15$

(b) $G(x + h) = 2(x + h) + 1$
$= 2x + 2h + 1$

43. (a) The dependent variable is the population, P, and the independent variable is the number of years after 1900, t.

(b) $P(110)$
$= 0.213(110)^2 - 18.474(110) + 597.372$
$= 1142.532$
According to the model, the population of Orange County will be roughly 1,142,532 in 2010.

(c) $P(-70)$
$= 0.213(-70)^2 - 18.474(-70) + 597.372$
$= 2934.252$
According to the model, the population of Orange County was roughly 2,934,252 in 1830. This is not reasonable. (The population of the entire Florida territory was roughly 35,000 in 1830.)

44. (a) The dependent variable is the percent of the population with an advanced degree, P, and the independent variable is the age of the population, a.

(b) $P(30) = -0.0064(30)^2 + 0.6826(30) - 6.82$
$= 7.898$
Approximately 7.9% of 30 year olds have an advanced degree.

45. $f(x) = -\dfrac{3}{2}x + 5$

Since each operation in the function can be performed for any real number, the domain of the function is all real numbers.
Domain: $\{x \,|\, x \text{ is a real number}\}$ or $(-\infty, \infty)$

46. $g(w) = \dfrac{w - 9}{2w + 5}$

The function tells us to divide $w - 9$ by $2w + 5$. Since division by 0 is not defined, the denominator can never equal 0.
$2w + 5 = 0$
$2w = -5$
$w = -\dfrac{5}{2}$

Thus, the domain is all real numbers except $-\dfrac{5}{2}$.

Domain: $\left\{ w \,\middle|\, w \neq -\dfrac{5}{2} \right\}$

47. $h(t) = \dfrac{t + 2}{t - 5}$

The function tells us to divide $t + 2$ by $t - 5$. Since division by 0 is not defined, the denominator can never equal 0.
$t - 5 = 0$
$t = 5$
Thus, the domain of the function is all real numbers except 5.
Domain: $\{t \,|\, t \neq 5\}$

48. $G(t) = 3t^2 + 4t - 9$

Since each operation in the function can be performed for any real number, the domain of the function is all real numbers.

Domain: $\{t \mid t \text{ is a real number}\}$ or $(-\infty, \infty)$

49. $f(x) = 2x - 5$

x	$y = f(x) = 2x - 5$	(x, y)
-1	$f(-1) = 2(-1) - 5 = -7$	$(-1, -7)$
0	$f(0) = 2(0) - 5 = -5$	$(0, -5)$
1	$f(1) = 2(1) - 5 = -3$	$(1, -3)$
2	$f(2) = 2(2) - 5 = -1$	$(2, -1)$
3	$f(3) = 2(3) - 5 = 1$	$(3, 1)$

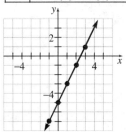

50. $g(x) = x^2 - 3x + 2$

x	$y = g(x) = x^2 - 3x + 2$	(x, y)
-1	$g(-1) = (-1)^2 - 3(-1) + 2 = 6$	$(-1, 6)$
0	$g(0) = (0)^2 - 3(0) + 2 = 2$	$(0, 2)$
1	$g(1) = (1)^2 - 3(1) + 2 = 0$	$(1, 0)$
2	$g(2) = (2)^2 - 3(2) + 2 = 0$	$(2, 0)$
3	$g(3) = (3)^2 - 3(3) + 2 = 2$	$(3, 2)$
4	$g(4) = (4)^2 - 3(4) + 2 = 6$	$(4, 6)$

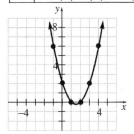

51. $h(x) = (x - 1)^3 - 3$

x	$y = h(x) = (x-1)^3 - 3$	(x, y)
-1	$h(-1) = (-1-1)^3 - 3 = -11$	$(-1, -11)$
0	$h(0) = (0-1)^3 - 3 = -4$	$(0, -4)$
1	$h(1) = (1-1)^3 - 3 = -3$	$(1, -3)$
2	$h(2) = (2-1)^3 - 3 = -2$	$(2, -2)$
3	$h(3) = (3-1)^3 - 3 = 5$	$(3, 5)$

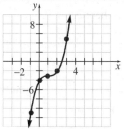

52. $f(x) = |x + 1| - 4$

x	$y = f(x) = |x+1| - 4$	(x, y)
-5	$f(-5) = |-5+1| - 4 = 0$	$(-5, 0)$
-3	$f(-3) = |-3+1| - 4 = -2$	$(-3, -2)$
-1	$f(-1) = |-1+1| - 4 = -4$	$(-1, -4)$
1	$f(1) = |1+1| - 4 = -2$	$(1, -2)$
3	$f(3) = |3+1| - 4 = 0$	$(3, 0)$

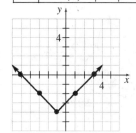

53. **(a)** Domain: $\{x \mid x \text{ is a real number}\}$ or $(-\infty, \infty)$

Range: $\{y \mid y \text{ is a real number}\}$ or $(-\infty, \infty)$

(b) The intercepts are $(0, 2)$ and $(4, 0)$. The x-intercept is 4 and the y-intercept is 2.

54. **(a)** Domain: $\{x \mid x \text{ is a real number}\}$ or $(-\infty, \infty)$

Range: $\{y \mid y \geq -3\}$ or $[-3, \infty)$

(b) The intercepts are $(-2, 0)$, $(2, 0)$, and $(0, -3)$. The x-intercepts are -2 and 2, and the y-intercept is -3.

55. (a) Domain: $\{x \mid x \text{ is a real number}\}$ or $(-\infty, \infty)$

Range: $\{y \mid y \text{ is a real number}\}$ or $(-\infty, \infty)$

(b) The intercepts are $(0, 0)$ and $(2, 0)$. The x-intercepts are 0 and 2; the y-intercept is 0.

56. (a) Domain: $\{x \mid x \geq -3\}$ or $[-3, \infty)$

Range: $\{y \mid y \geq 1\}$ or $[1, \infty)$

(b) The only intercept is $(0, 3)$. There are no x-intercepts, but there is a y-intercept of 3.

57. (a) Since the point $(-3, 4)$ is on the graph, $f(-3) = 4$.

(b) Since the point $(1, -4)$ is on the graph, when $x = 1$, $f(x) = 4$.

(c) Since the x-intercepts are -1 and 3, the zeroes of f are -1 and 3.

58. (a) $y = x^2$

x	$y = x^2$	(x, y)
-2	$y = (-2)^2 = 4$	$(-2, 4)$
-1	$y = (-1)^2 = 1$	$(-1, 1)$
0	$y = (0)^2 = 0$	$(0, 0)$
1	$y = (1)^2 = 1$	$(1, 1)$
2	$y = (2)^2 = 4$	$(2, 4)$

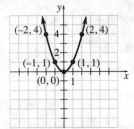

(b) $y = \sqrt{x}$

x	$y = \sqrt{x}$	(x, y)
0	$y = \sqrt{0} = 0$	$(0, 0)$
1	$y = \sqrt{1} = 1$	$(1, 1)$
4	$y = \sqrt{4} = 2$	$(4, 2)$

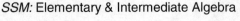
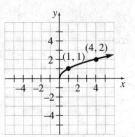

59. (a) $h(3) = 2(3) - 7 = 6 - 7 = -1$

Since $h(3) = -1$, the point $(3, -1)$ is on the graph of the function.

(b) $h(-2) = 2(-2) - 7 = -4 - 7 = -11$

The point $(-2, -11)$ is on the graph of the function.

(c) $h(x) = 4$

$2x - 7 = 4$

$2x = 11$

$x = \dfrac{11}{2}$

The point $\left(\dfrac{11}{2}, 4\right)$ is on the graph of h.

60. (a) $g(-5) = \dfrac{3}{5}(-5) + 4 = -3 + 4 = 1$

Since $g(-5) = 1$, the point $(-5, 2)$ is not on the graph of the function.

(b) $g(3) = \dfrac{3}{5}(3) + 4 = \dfrac{9}{5} + 4 = \dfrac{29}{5}$

The point $\left(3, \dfrac{29}{5}\right)$ is on the graph of the function.

(c) $g(x) = -2$

$\dfrac{3}{5}x + 4 = -2$

$\dfrac{3}{5}x = -6$

$x = -10$

The point $(-10, -2)$ is on the graph of g.

61.

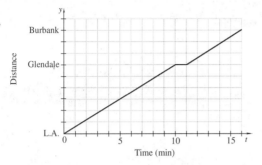

62.

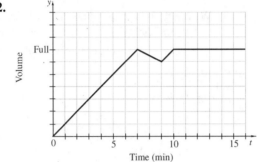

63. Comparing $g(x) = 2x - 6$ to $g(x) = mx + b$, we see that the slope m is 2 and the y-intercept b is −6. We begin by plotting the point $(0, -6)$.

Because $m = 2 = \dfrac{2}{1} = \dfrac{\Delta y}{\Delta x} = \dfrac{\text{Rise}}{\text{Run}}$, from the point $(0, -6)$ we go up 2 units and to the right 1 unit and end up at $(1, -4)$. We draw a line through these points and obtain the graph of $g(x) = 2x - 6$.

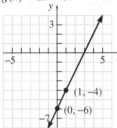

$$f(x) = 0$$
$$2x - 6 = 0$$
$$2x = 6$$
$$x = 3$$

The zero of f is 3.

64. Comparing $H(x) = -\dfrac{4}{3}x + 5$ to $H(x) = mx + b$,

we see that the slope m is $-\dfrac{4}{3}$ and the

y-intercept b is 5. We begin by plotting the point

$(0, 5)$. Because $m = -\dfrac{4}{3} = \dfrac{-4}{3} = \dfrac{\Delta y}{\Delta x} = \dfrac{\text{Rise}}{\text{Run}}$,

from the point $(0, 5)$ we go down 4 units and to the right 3 units and end up at $(3, 1)$. We draw a line through these points and obtain the graph of

$H(x) = -\dfrac{4}{3}x + 5.$

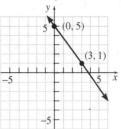

$$H(x) = 0$$
$$-\frac{4}{3}x + 5 = 0$$
$$-\frac{4}{3}x = -5$$
$$x = \frac{15}{4}$$

The zero of H is $\dfrac{15}{4}$.

65. Comparing $F(x) = -x - 3$ to $F(x) = mx + b$, we see that the slope m is −1 and the y-intercept b is −3. We begin by plotting the point $(0, -3)$.

Because $m = -1 = \dfrac{-1}{1} = \dfrac{\Delta y}{\Delta x} = \dfrac{\text{Rise}}{\text{Run}}$, from the

point $(0, -3)$ we go down 1 unit and to the right 1 unit and end up at $(1, -4)$. We draw a line through these points and obtain the graph of $F(x) = -x - 3.$

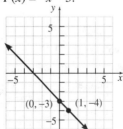

$$F(x) = 0$$
$$-x - 3 = 0$$
$$-x = 3$$
$$x = -3$$

The zero of F is −3.

66. Comparing $f(x) = \frac{3}{4}x - 3$ to $f(x) = mx + b$, we

see that the slope m is $\frac{3}{4}$ and the y-intercept b is

-3. We begin by plotting the point $(0, -3)$.

Because $m = \frac{3}{4} = \frac{\Delta y}{\Delta x} = \frac{\text{Rise}}{\text{Run}}$, from the point

$(0, -3)$ we go up 3 units and to the right 4 units and end up at $(4, 0)$. We draw a line through these points and obtain the graph of

$f(x) = \frac{3}{4}x - 3.$

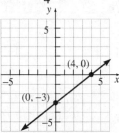

From the graph, we see that the x-intercept is 4, so the zero of f is 4.

67. (a) The independent variable is the number of long-distance minutes used, x. It would not make sense to talk for a negative number of minutes. Thus, the domain of C is

$\{x \mid x \geq 0\}$ or, using interval notation, $[0, \infty)$.

(b) $C(235) = 0.07(235) + 5 = 21.45$

The cost for 235 minutes of long-distance calls during one month is $21.45.

(c) Evaluate C at $x = 0$, 100, and 500.

$C(0) = 0.07(0) + 5 = 5$

$C(100) = 0.07(100) + 5 = 12$

$C(500) = 0.07(500) + 5 = 40$

Thus, the points $(0, 5)$, $(100, 12)$, and $(500, 40)$ are on the graph.

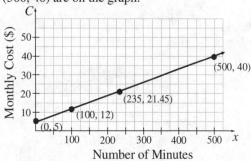

(d) We must solve $C(x) = 75$.

$0.07x + 5 = 75$

$0.07x = 70$

$x = 1000$

In one month, a person can purchase 1000 minutes of long distance for $75.

68. (a) The independent variable is the number of years after purchase, x. The dependent variable is the value of the computer, V.

(b) We are told that the value function V is for 0 to 5 years, inclusive. Thus, the domain is $\{x \mid 0 \leq x \leq 5\}$ or, using interval notation, $[0, 5]$.

(c) The initial value of the computer will be the value at $x = 0$ years.

$V(0) = 1800 - 360(0) = 1800$

The initial value of the computer is $1800.

(d) $V(2) = 1800 - 360(2) = 1080$

After 2 years, the value of the computer is $1080.

(e) Evaluate V at $x = 3$, 4, and 5.

$V(3) = 1800 - 360(3) = 720$

$V(4) = 1800 - 360(4) = 360$

$V(5) = 1800 - 360(5) = 0$

Thus, the points $(3, 720)$, $(4, 360)$, and $(5, 0)$ are on the graph.

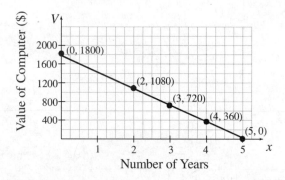

(f) We must solve $V(x) = 0$.

$1800 - 360x = 0$

$-360x = -1800$

$x = 5$

After 5 years, the computer's value will be $0.

69. (a) Let x represent the number of years since 1996. Then $x = 0$ represents 1996 and $x = 9$ represents 2005. Let E represent the percentage of electronically-filed returns.

$$m = \frac{30.1 - 12.6}{9 - 0} \approx 1.94 \text{ percent}$$

$$E - 12.6 = 1.94(x - 0)$$
$$E - 12.6 = 1.94x$$
$$E = 1.94x + 12.6$$

In function notation, $E(x) = 1.94x + 12.6$.

(b) The year 2008 corresponds to $x = 12$.
$$E(12) = 1.94(12) + 12.6 = 35.88$$

If the linear trend continues, 35.88% of U.S. Federal Tax Returns will be filed electronically in 2008.

(c) The slope (1.94) indicates that electronically-filed tax returns are increasing at a rate of 1.94% per year.

(d)
$$1.94x + 12.6 = 47.52$$
$$1.94x = 34.92$$
$$x = 18$$

Now, $x = 18$ corresponds to the year 2014. Thus, 47.52% of U.S. Federal Tax Returns will be filed electronically in the year 2014.

70. (a) Let x represent the age of men and H represent the maximum recommended heart rate for men under stress.

$$m = \frac{160 - 200}{60 - 20}$$
$$= -1 \text{ beat per minute per year}$$

$$H - 200 = -1(x - 20)$$
$$H - 200 = -x + 20$$
$$H = -x + 220$$

In function notation, $H(x) = -x + 220$.

(b) $H(45) = -(45) + 220 = 175$

The maximum recommended heart rate for a 45 year old man under stress is 175 beats per minute.

(c) The slope (-1) indicates that the maximum recommended heart rate for men under stress decreases at a rate of 1 beat per minute per year.

(d)
$$-x + 220 = 168$$
$$-x = -52$$
$$x = 52$$

The maximum recommended heart rate under stress is 168 beats per minute for 52-year-old men.

71. (a) $C(m) = 0.12m + 35$

(b) The number of miles driven, m, is the independent variable. The rental cost, C, is the dependent variable.

(c) Because the number of miles cannot be negative, the it must be greater than or equal to zero. That is, the implied domain is $\{m \mid m \geq 0\}$ or, using interval notation, $[0, \infty)$.

(d) $C(124) = 0.12(124) + 35 = 49.88$

If 124 miles are driven during a one-day rental, the charge will be $49.88.

(e)
$$0.12m + 35 = 67.16$$
$$0.12m = 32.16$$
$$m = 268$$

If the charge for a one-day rental is $67.16, then 268 miles were driven.

(f)

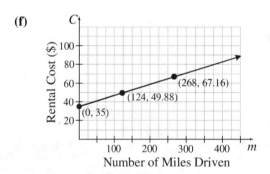

72. (a) $B(x) = 3.50x + 33.99$

(b) The number of pay-per-view movies watched, x, is the independent variable. The monthly bill, B, is the dependent variable.

(c) Because the number pay-per-view movies watched cannot be negative, it must be greater than or equal to zero. That is, the implied domain is $\{x \mid x \geq 0\}$ or, using interval notation, $[0, \infty)$.

(d) $B(5) = 3.50(5) + 33.99 = 51.49$

If 5 pay-per-view movies are watched one month, the bill will be $51.49.

(e) $3.50x + 33.99 = 58.49$

$3.50x = 24.50$

$x = 7$

If the bill one month is $58.49, then 7 pay-per-view movies were watched.

(f)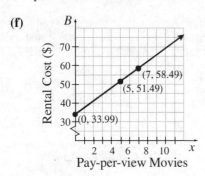

73. (a)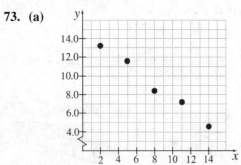

(b) Answers will vary. We will use the points (2, 13.3) and (14, 4.6).

$m = \dfrac{4.6 - 13.3}{14 - 2} = \dfrac{-8.7}{12} = -0.725$

$y - 13.3 = -0.725(x - 2)$

$y - 13.3 = -0.725x + 1.45$

$y = -0.725x + 14.75$

(c)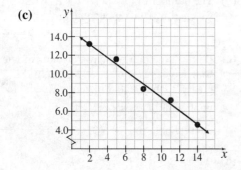

74. (a)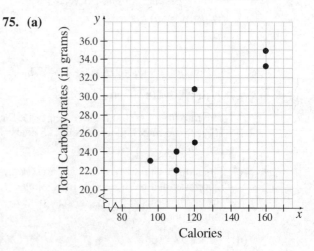

(b) Answers will vary. We will use the points (0, 0.6) and (4.2, 3.0).

$m = \dfrac{3.0 - 0.6}{4.2 - 0} = \dfrac{2.4}{4.2} = \dfrac{4}{7}$

$y - 0.6 = \dfrac{4}{7}(x - 0)$

$y - 0.6 = \dfrac{4}{7}x$

$y = \dfrac{4}{7}x + 0.6$

(c)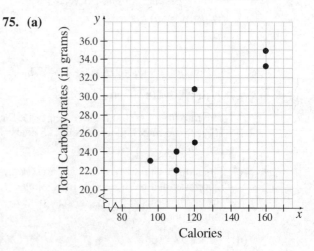

75. (a)

(b) Approximately linear

(c) Answers will vary. We will use the points (96, 23.2) and (160, 33.3).

$$m = \frac{33.3 - 23.2}{160 - 96} = \frac{10.1}{64} \approx 0.158$$

$$y - 23.2 = 0.158(x - 96)$$
$$y - 23.2 = 0.158x - 15.168$$
$$y = 0.158x + 8.032$$

(d)

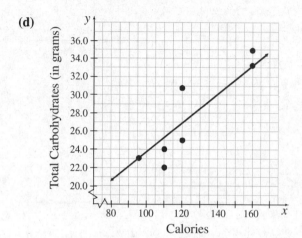

(e) $x = 140$: $\quad y = 0.158(140) + 8.032$
$$= 30.152$$

We predict that a one-cup serving of cereal having 140 calories will have approximately 30.2 grams of total carbohydrates.

(f) The slope of the line found is 0.158. This means that, in a one-cup serving of cereal, total carbohydrates will increase by 0.158 gram for each one-calorie increase.

76. (a)

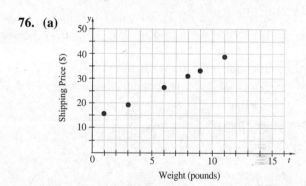

(b) Approximately linear

(c) Answers will vary. We will use the point (1, 15.88) and (11, 38.10).

$$m = \frac{38.10 - 15.88}{11 - 1} = \frac{22.22}{10} = 2.222$$

$$y - 15.88 = 2.222(x - 1)$$
$$y - 15.88 = 2.222x - 2.222$$
$$y = 2.222x + 13.658$$

(d)

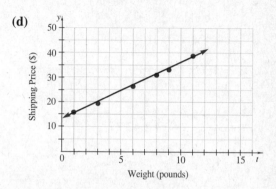

(e) $x = 5$; $\;y = 2.222(5) + 13.658$
$$= 24.768$$

We predict that the FedEx 2Day ® price for shipping a 5-pound package will be \$24.77.

(f) The slope of the line found is 2.222. This means that the FedEx 2Day ® shipping price increases by \$2.222 for each one-pound increase in the weight of a package.

77. $A \cup B = \{-1, 0, 1, 2, 3, 4, 6, 8\}$

78. $A \cap C = \{2, 4\}$

79. $B \cap C = \{1, 2, 3, 4\}$

80. $A \cup C = \{1, 2, 3, 4, 6, 8\}$

81. (a) $A \cap B = \{x \mid 2 < x \le 4\}$

Interval: (2, 4]

(b) $A \cup B = \{x \mid x \text{ is a real number}\}$

Interval: $(-\infty, \infty)$

82. (a) $E \cap F = \{\ \}$ or $\varnothing$

(b) $E \cup F = \{x \mid x < -2 \text{ or } x \geq 3\}$

Interval: $(-\infty, -2) \cup [3, \infty)$

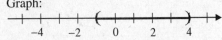

83. $x < 4$ and $x + 3 > 2$

$\qquad\qquad\qquad x > -1$

We need $x < 4$ and $x > -1$.

Set-builder: $\{x \mid -1 < x < 4\}$

Interval: $(-1, 4)$

Graph:

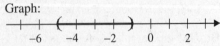

84. $\qquad 3 < 2 - x < 7$

$3 - 2 < 2 - x - 2 < 7 - 2$

$\qquad 1 < -x < 5$

$-1(1) > -1(-x) > -1(5)$

$\qquad -1 > x > -5$

$\qquad -5 < x < -1$

Set-builder: $\{x \mid -5 < x < -1\}$

Interval: $(-5, -1)$

Graph:

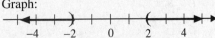

85. $x + 3 < 1$ or $x > 2$

$\quad x < -2$

We need $x < -2$ or $x > 2$.

Set-builder: $\{x \mid x < -2 \text{ or } x > 2\}$

Interval: $(-\infty, -2) \cup (2, \infty)$

Graph:

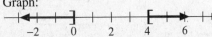

86. $x + 6 \geq 10$ or $x \leq 0$

$\quad x \geq 4$

We need $x \geq 4$ or $x \leq 0$.

Set-builder: $\{x \mid x \leq 0 \text{ or } x \geq 4\}$

Interval: $(-\infty, 0] \cup [4, \infty)$

Graph:

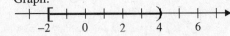

87. $3x + 2 \leq 5$ and $-4x + 2 \leq -10$

$\quad 3x \leq 3 \qquad\qquad -4x \leq -12$

$\quad x \leq 1 \qquad\qquad\quad x \geq 3$

We need $x \leq 1$ and $x \geq 3$.

Solution set: $\{\ \}$ or $\varnothing$

88. $\qquad 1 \leq 2x + 5 < 13$

$1 - 5 \leq 2x + 5 - 5 < 13 - 5$

$\qquad -4 \leq 2x < 8$

$\qquad \dfrac{-4}{2} \leq \dfrac{2x}{2} < \dfrac{8}{2}$

$\qquad -2 \leq x < 4$

Set-builder: $\{x \mid -2 \leq x < 4\}$

Interval: $[-2, 4)$

Graph:

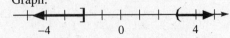

89. $x - 3 \leq -5$ or $2x + 1 > 7$

$\quad x \leq -2 \qquad\qquad 2x > 6$

$\qquad\qquad\qquad\qquad x > 3$

We need $x \leq -2$ or $x > 3$.

Set-builder: $\{x \mid x \leq -2 \text{ or } x > 3\}$

Interval: $(-\infty, -2] \cup (3, \infty)$

Graph:

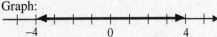

90. $3x + 4 > -2$ or $4 - 2x \geq -6$

$\quad 3x > -6 \qquad\qquad -2x \geq -10$

$\quad x > -2 \qquad\qquad\quad x \leq 5$

We need $x > -2$ or $x \leq 5$.

Set-builder: $\{x \mid x \text{ is any real number}\}$

Interval: $(-\infty, \infty)$

Graph:

91. $\dfrac{1}{3}x > 2$ or $\dfrac{2}{5}x < -4$

$\quad x > 6 \qquad\quad x < -10$

We need $x < -10$ or $x > 6$.

Set-builder: $\{x \mid x < -10 \text{ or } x > 6\}$

Interval: $(-\infty, -10) \cup (6, \infty)$

Graph:

92. $x + \dfrac{3}{2} \geq 0$ and $-2x + \dfrac{3}{2} > \dfrac{1}{4}$

$\quad x \geq -\dfrac{3}{2} \qquad\qquad -2x > -\dfrac{5}{4}$

$\qquad\qquad\qquad\qquad\qquad x < \dfrac{5}{8}$

We need $x \geq -\dfrac{3}{2}$ and $x < \dfrac{5}{8}$.

Set-builder: $\left\{ x \mid -\dfrac{3}{2} \le x < \dfrac{5}{8} \right\}$

Interval: $\left[-\dfrac{3}{2}, \dfrac{5}{8} \right)$

Graph:

93. $70 \le x \le 75$

94. Let x = number of kilowatt-hours. Then, the number of kilowatt-hours *above* 300 is given by the expression $x - 300$. Therefore, we need to solve the following inequality:

$50.28 \le 23.12 + 0.05947(x - 300) \le 121.43$

$27.16 \le 0.05947(x - 300) \le 98.31$

$456.7 \le x - 300 \le 1653.1$

$756.7 \le x \le 1953.1$

The electric usage varied from roughly 756.7 kilowatt-hours up to roughly 1953.1 kilowatt-hours.

95. $|x| = 4$

$x = 4 \ $ or $\ x = -4$

Solution set: $\{-4, 4\}$

96. $|3x - 5| = 4$

$3x - 5 = 4 \ $ or $\ 3x - 5 = -4$

$3x = 9 \qquad\qquad 3x = 1$

$x = 3 \qquad\qquad x = \dfrac{1}{3}$

Solution set: $\left\{ \dfrac{1}{3}, 3 \right\}$

97. $|-y + 4| = 9$

$-y + 4 = 9 \ $ or $\ -y + 4 = -9$

$-y = 5 \qquad\qquad -y = -13$

$y = -5 \qquad\qquad y = 13$

Solution set: $\{-5, 13\}$

98. $-3|x + 2| - 5 = -8$

$-3|x + 2| = -3$

$|x + 2| = 1$

$x + 2 = 1 \ $ or $\ x + 2 = -1$

$x = -1 \qquad\qquad x = -3$

Solution set: $\{-3, -1\}$

99. $|2w - 7| = -3$

This equation has no solution since an absolute value can never yield a negative result.

Solution set: $\{ \ \}$ or $\varnothing$

100. $|x + 3| = |3x - 1|$

$x + 3 = 3x - 1 \ $ or $\ x + 3 = -(3x - 1)$

$-2x = -4 \qquad\qquad x + 3 = -3x + 1$

$x = 2 \qquad\qquad 4x = -2$

$x = -\dfrac{1}{2}$

Solution set: $\left\{ -\dfrac{1}{2}, 2 \right\}$

101. $|x| < 2$

$-2 < x < 2$

Set-builder: $\{ x \mid -2 < x < 2 \}$

Interval: $(-2, 2)$

102. $|x| \ge \dfrac{7}{2}$

$x \le -\dfrac{7}{2} \ $ or $\ x \ge \dfrac{7}{2}$

Set-builder: $\left\{ x \mid x \le -\dfrac{7}{2} \ \text{or} \ x \ge \dfrac{7}{2} \right\}$

Interval: $\left(-\infty, -\dfrac{7}{2} \right] \cup \left[\dfrac{7}{2}, \infty \right)$

103. $|x + 2| \le 3$

$-3 \le x + 2 \le 3$

$-3 - 2 \le x + 2 - 2 \le 3 - 2$

$-5 \le x \le 1$

Set-builder: $\{ x \mid -5 \le x \le 1 \}$

Interval: $[-5, 1]$

104. $|4x - 3| \ge 1$

$4x - 3 \le -1 \ $ or $\ 4x - 3 \ge 1$

$4x \le 2 \qquad\qquad 4x \ge 4$

$x \le \dfrac{1}{2} \qquad\qquad x \ge 1$

Set-builder: $\left\{ x \mid x \le \dfrac{1}{2} \ \text{or} \ x \ge 1 \right\}$

Interval: $\left(-\infty, \dfrac{1}{2} \right] \cup [1, \infty)$

105. $3|x| + 6 \geq 1$

$$3|x| \geq -5$$

$$|x| \geq -\frac{5}{3}$$

Since the result of an absolute value is always nonnegative, any real number is a solution to this inequality.

Set-builder: $\{x \mid x \text{ is a real number}\}$

Interval: $(-\infty, \infty)$

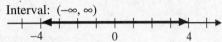

106. $|7x + 5| + 4 < 3$

$$|7x + 5| < -1$$

Since the result of an absolute value is never negative, this inequality has no solutions.

Solution set: $\{\ \}$ or $\varnothing$

107. $|(x - 3) - 2| \leq 0.01$

$$|x - 3 - 2| \leq 0.01$$

$$|x - 5| \leq 0.01$$

$$-0.01 \leq x - 5 \leq 0.01$$

$$-0.01 + 5 \leq x - 5 + 5 \leq 0.01 + 5$$

$$4.99 \leq x \leq 5.01$$

Set-builder: $\{x \mid 4.99 \leq x \leq 5.01\}$

Interval: $[4.99, 5.01]$

108. $\left|\dfrac{2x - 3}{4}\right| > 1$

$$\frac{2x - 3}{4} < -1 \quad \text{or} \quad \frac{2x - 3}{4} > 1$$

$$2x - 3 < -4 \qquad\qquad 2x - 3 > 4$$

$$2x < -1 \qquad\qquad 2x > 7$$

$$x < -\frac{1}{2} \qquad\qquad x > \frac{7}{2}$$

Solution set: $\left\{x \mid x < -\dfrac{1}{2} \text{ or } x > \dfrac{7}{2}\right\}$

Interval: $\left(-\infty, -\dfrac{1}{2}\right) \cup \left(\dfrac{7}{2}, \infty\right)$

109. $|x - 0.503| \leq 0.001$

$$-0.001 \leq x - 0.503 \leq 0.001$$

$$0.502 \leq x \leq 0.504$$

The acceptable diameters of the bearing are between 0.502 inch and 0.504 inch, inclusive.

110. $\left|\dfrac{x - 40}{2}\right| > 1.96$

$$\frac{x - 40}{2} < -1.96 \quad \text{or} \quad \frac{x - 40}{2} > 1.96$$

$$x - 40 < -3.92 \qquad\qquad x - 40 > 3.92$$

$$x < 36.08 \qquad\qquad x > 43.92$$

Tensile strengths below 36.08 lb/in.^{2} or above 43.92 lb/in.^{2} would be considered unusual.

111. (a) $y = kx$

$$30 = k \cdot 6$$

$$k = 5$$

(b) $y = 5x$

(c) $y = 5(10)$

$$y = 50$$

112. (a) $y = kx$

$$18 = k \cdot (-3)$$

$$k = -6$$

(b) $y = -6x$

(c) $y = -6(8)$

$$y = -48$$

113. (a) $y = \dfrac{k}{x}$

$$15 = \frac{k}{4}$$

$$60 = k$$

(b) $y = \dfrac{60}{x}$

(c) If $x = 5$, then $y = \dfrac{60}{5} = 12$.

114. (a) $y = kxz$

$$45 = k \cdot 6 \cdot 10$$

$$45 = 60k$$

$$k = \frac{45}{60} = \frac{3}{4}$$

(b) $y = \dfrac{3}{4}xz$

(c) If $x = 8$ and $z = 7$, then $y = \frac{3}{4} \cdot 8 \cdot 7 = 42$.

115. (a) $s = \dfrac{k}{t^2}$

$18 = \dfrac{k}{2^2}$

$18 = \dfrac{k}{4}$

$72 = k$

(b) $s = \dfrac{72}{t^2}$

(c) If $t = 3$, then $s = \dfrac{72}{3^2} = \dfrac{72}{9} = 8$.

116. (a) $w = k \cdot \dfrac{x}{z}$

$\dfrac{4}{3} = k \cdot \dfrac{10}{12}$

$k = \dfrac{4}{3} \cdot \dfrac{12}{10} = \dfrac{8}{5}$

(b) $w = \dfrac{8}{5} \cdot \dfrac{x}{z}$ or $w = \dfrac{8x}{5z}$

(c) If $x = 9$ and $z = 16$, then $w = \dfrac{8 \cdot 9}{5 \cdot 16} = \dfrac{9}{10}$.

117. Let w represent the amount of water in d inches of snow.

$w = kd$

$4.8 = k \cdot 40$

$k = 0.12$

$w = 0.12d$

$w = 0.12(50)$

$w = 6$

There are 6 inches of water contained in 50 inches of snow.

118. $p = kb$

$293.49 = k \cdot 15,000$

$k = 0.019566$

$p = 0.019566b$

$p = 0.019566(18,000)$

$p = 352.188$

If Roberta borrows \$18,000, her payment would be \$352.19.

119. Let f represent the frequency of a radio signal with wavelength l.

$f = \dfrac{k}{l}$

$800 = \dfrac{k}{375}$

$375 \cdot 800 = 375\left(\dfrac{k}{375}\right)$

$300,000 = k$

$f(l) = \dfrac{300,000}{l}$

$f(250) = \dfrac{300,000}{250} = 1200$

When the wavelength is 250 meters, the frequency is 1200 kilohertz.

120. Let C represent the electrical current when the resistance is R.

$C = \dfrac{k}{R}$

$8 = \dfrac{k}{15}$

$15 \cdot 8 = 15\left(\dfrac{k}{15}\right)$

$120 = k$

$C(R) = \dfrac{120}{R}$

$C(R) = 10$

$\dfrac{120}{R} = 10$

$120 = 10R$

$12 = R$

If the electrical current is 10 amperes, then the resistance is 12 ohms.

121. $V = khd^2$

$231 = k \cdot 6 \cdot 7^2$

$231 = 294k$

$k = \dfrac{231}{294} = \dfrac{11}{14}$

$V = \dfrac{11}{14}hd^2$

For $h = 14$ and $d = 8$, $V = \dfrac{11}{14}(14)(8)^2 = 704$.

If the diameter is 8 centimeters and the height is 14 centimeters, then the volume of the cylinder is 704 cubic centimeters.

122. $V = kBh$

$270 = k(81)(10)$

$270 = 810k$

$k = \dfrac{270}{810} = \dfrac{1}{3}$

$V = \dfrac{1}{3}Bh$

For $B = 125$ and $h = 9$, $V = \dfrac{1}{3}(125)(9) = 375$.

When the area of the base is 125 square inches and the height is 9 inches, the volume of the pyramid is 375 cubic inches.

Chapter 8 Test

1. *A*: quadrant IV; *B*: *y*-axis; *C*: *x*-axis; *D*: quadrant I; *E*: quadrant III; *F*: quadrant II

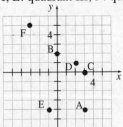

2. (a) Let $x = -2$ and $y = 4$.

$4 \stackrel{?}{=} 3(-2)^2 + (-2) - 5$

$4 \stackrel{?}{=} 3(4) - 7$

$4 \stackrel{?}{=} 12 - 7$

$4 = 5$ False

The point $(-2, 4)$ is not on the graph.

(b) Let $x = -1$ and $y = -3$.

$-3 \stackrel{?}{=} 3(-1)^2 + (-1) - 5$

$-3 \stackrel{?}{=} 3(1) - 6$

$-3 \stackrel{?}{=} 3 - 6$

$-3 = -3$ True

The point $(-1, -3)$ is on the graph.

(c) Let $x = 2$ and $y = 9$.

$9 \stackrel{?}{=} 3(2)^2 + (2) - 5$

$9 \stackrel{?}{=} 3(4) - 3$

$9 \stackrel{?}{=} 12 - 3$

$9 = 9$ True

The point $(2, 9)$ is on the graph.

3. The intercepts are $(-3, 0)$, $(0, 1)$, and $(0, 3)$. The *x*-intercept is -3 and the *y*-intercepts are 1 and 3.

4. (a) 30 miles per hour

(b) The graph passes through the origin and the point $(32, 0)$. This means the car was stopped (its speed was 0 mph) at 0 seconds and at 32 seconds.

5. Domain: $\{-4, 2, 5, 7\}$

Range: $\{-7, -2, -1, 3, 8, 12\}$

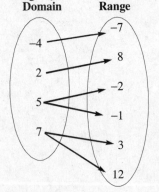

6. Domain: $\left\{x \mid -\dfrac{5\pi}{2} \le x \le \dfrac{5\pi}{2}\right\}$ or $\left[-\dfrac{5\pi}{2}, \dfrac{5\pi}{2}\right]$

Range: $\{y \mid 1 \le y \le 5\}$ or $[1, 5]$

7. $y = x^2 - 3$

x	$y = x^2 - 3$	(x, y)
-2	$y = (-2)^2 - 3 = 1$	$(-2, 1)$
-1	$y = (-1)^2 - 3 = -2$	$(-1, -2)$
0	$y = (0)^2 - 3 = -3$	$(0, -3)$
1	$y = (1)^2 - 3 = -2$	$(1, -2)$
2	$y = (2)^2 - 3 = 1$	$(2, 1)$

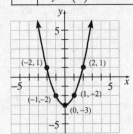

Domain: $\{x \mid x$ is a real number$\}$ or $(-\infty, \infty)$

Range: $\{y \mid y \ge -3\}$ or $[-3, \infty)$

8. Function. Each element in the domain corresponds to exactly one element in the range.
Domain: $\{-5, -3, 0, 2\}$
Range: $\{3, 7\}$

9. Not a function. The graph fails the vertical line test so it is not the graph of a function.
Domain: $\{x \mid x \le 3\}$ or $(-\infty, 3]$

Range: $\{y \mid y \text{ is a real number}\}$ or $(-\infty, \infty)$

10. No, $y = \pm 5x$ is not a function because a single input, x, can yield two different outputs. For example, if $x = 1$ then $y = -5$ or $y = 5$.

11. $f(x+h) = -3(x+h) + 11 = -3x - 3h + 11$

12. (a) $g(-2) = 2(-2)^2 + (-2) - 1 = 2(4) - 3$
$= 8 - 3 = 5$

(b) $g(0) = 2(0)^2 + (0) - 1 = 0 + 0 - 1 = -1$

(c) $g(3) = 2(3)^2 + (3) - 1$
$= 2(9) + 2$
$= 18 + 2$
$= 20$

13. $f(x) = x^2 + 3$

x	$y = f(x) = x^2 + 3$	(x, y)
-2	$f(-2) = (-2)^2 + 3 = 7$	$(-2, 7)$
-1	$f(-1) = (-1)^2 + 3 = 4$	$(-1, 4)$
0	$f(0) = (0)^2 + 3 = 3$	$(0, 3)$
1	$f(1) = (1)^2 + 3 = 4$	$(1, 4)$
2	$f(2) = (2)^2 + 3 = 7$	$(2, 7)$

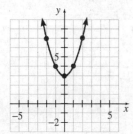

14. (a) The dependent variable is the ticket price, P, and the independent variable is the number of years after 1989, x.

(b) $P(20) = 0.16(20) + 3.63 = 6.83$
According to the model, the average ticket price in 2009 ($x = 20$) was $6.83.

(c) $7.63 = 0.16x + 3.63$
$4.00 = 0.16x$
$x = 25$
According to the model, the average movie ticket price will be $7.63 in 2014 ($x = 25$).

15. The function tells us to divide -15 by $x + 2$. Since we can't divide by zero, we need $x \ne -2$.
Domain: $\{x \mid x \ne -2\}$

16. (a) $h(2) = -5(2) + 12$
$= -10 + 12$
$= 2$
Since $h(2) = 2$, the point $(2, 2)$ is on the graph of the function.

(b) $h(3) = -5(3) + 12$
$= -15 + 12$
$= -3$
Since $h(3) = -3$, the point $(3, -3)$ is on the graph of the function.

(c) $h(x) = 27$
$-5x + 12 = 27$
$-5x = 15$
$x = -3$
The point $(-3, 27)$ is on the graph of h.

(d) $h(x) = 0$
$-5x + 12 = 0$
$-5x = -12$
$x = \dfrac{12}{5}$

$\dfrac{12}{5}$ is the zero of h.

17. (a) The profit is $30 - $12 = $18 times the number of shelves sold x, minus the $100 for renting the display. Thus, the function is $P(x) = 18x - 100$.

(b) The independent variable is the number of shelves sold, x. Henry could not sell a negative number of shelves. Thus, the domain of P is $\{x \mid x \ge 0\}$ or, using interval notation, $[0, \infty)$.

(c) $P(34) = 18(34) - 100$
$$= 512$$
If Henry sells 34 shelves, his profit will be $512.

(d) Evaluate P at $x = 0, 20,$ and 50.
$$P(0) = 18(0) - 100$$
$$= -100$$
$$P(20) = 18(20) - 100$$
$$= 260$$
$$P(50) = 18(50) - 100$$
$$= 800$$
Thus, the points $(0, -100)$, $(20, 260)$, and $(50, 800)$ are on the graph.

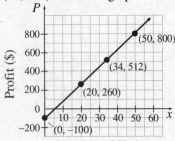

Number of Shelves

(e) We must solve $P(x) = 764$.
$$18x - 100 = 764$$
$$18x = 864$$
$$x = 48$$
If Henry sells 48 shelves, his profit will be $764.

18. (a)

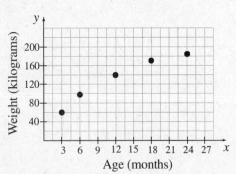

Age (months)

(b) Approximately linear

(c) Answers will vary. We will use the points $(6, 95)$ and $(18, 170)$.
$$m = \frac{170 - 95}{18 - 6} = \frac{75}{12} = 6.25$$
$$y - 95 = 6.25(x - 6)$$
$$y - 95 = 6.25x - 37.5$$
$$y = 6.25x + 57.5$$

(d)

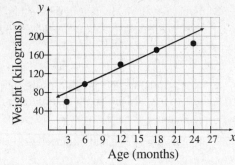

Age (months)

(e) $x = 9:$ $\quad y = 6.25(9) + 57.5$
$$= 113.75$$
We predict that a 9-month-old Shetland pony will weigh 113.75 kilograms.

(f) The slope of the line found is 6.25. This means that a Shetland pony's weight will increase by 6.25 kilograms for each one-month increase in age.

19. $|2x + 5| - 3 = 0$
$$|2x + 5| = 3$$
$$2x + 5 = -3 \quad \text{or} \quad 2x + 5 = 3$$
$$2x = -8 \qquad\qquad 2x = -2$$
$$x = -4 \qquad\qquad x = -1$$
Solution set: $\{-4, -1\}$

20. $x + 2 < 8 \quad \text{and} \quad 2x + 5 \geq 1$
$$x < 6 \qquad\qquad 2x \geq -4$$
$$x \geq -2$$
We need $x \geq -2$ and $x < 6$.
Set-builder: $\{x \mid -2 \leq x < 6\}$
Interval: $[-2, 6)$

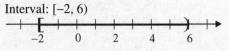

21. $x > 4 \quad \text{or} \quad 2(x - 1) + 3 < -2$
$$2x - 2 + 3 < -2$$
$$2x + 1 < -2$$
$$2x + 1 - 1 < -2 - 1$$
$$2x < -3$$
$$\frac{2x}{2} < \frac{-3}{2}$$
$$x < -\frac{3}{2}$$
We need $x > 4$ or $x < -\frac{3}{2}$.

Set-builder: $\left\{x \mid x < -\frac{3}{2} \text{ or } x > 4\right\}$

Interval: $\left(-\infty, -\frac{3}{2}\right) \cup (4, \infty)$

22. $2|x-5|+1 < 7$

$\qquad 2|x-5| < 6$

$\qquad |x-5| < 3$

$\quad -3 < x-5 < 3$

$\qquad 2 < x < 8$

Set-builder: $\{x \mid 2 < x < 8\}$

Interval: (2, 8)

23. $|-2x+1| \geq 5$

$-2x+1 \leq -5$ or $-2x+1 \geq 5$

$\quad -2x \leq -6 \qquad\quad -2x \geq 4$

$\qquad x \geq 3 \qquad\qquad\quad x \leq -2$

Set-builder: $\{x \mid x \leq -2 \text{ or } x \geq 3\}$

Interval: $(-\infty, -2] \cup [3, \infty)$

24. $\quad F = \dfrac{k}{l}$

$\quad 50 = \dfrac{k}{4}$

$200 = k$

$\quad F = \dfrac{200}{l}$

If $l = 10$, then $F = \dfrac{200}{10} = 20$.

If the length of the length of the force arm of the lever is 10 feet, the force required to lift the boulder will be 20 pounds.

25. $\qquad L = krh$

$\quad 528 = k(7)(12)$

$\quad 528 = 84k$

$\qquad k = \dfrac{528}{84} = \dfrac{44}{7}$

$\qquad L = \dfrac{44}{7}rh$

For $r = 9$ and $h = 14$, $L = \dfrac{44}{7}(9)(14) = 792$.

When the radius is 9 centimeters and the height is 14 centimeters, the lateral surface area of the right circular cylinder is 792 square centimeters.

Chapter 9

Preparing for Square Roots

P1. $-4, 0, 13$

P2. $-4, 0, \dfrac{5}{3}, 6.95, 13$

P3. $\sqrt{2}, \pi$

P4. **(a)** $\left(\dfrac{3}{2}\right)^2 = \dfrac{3}{2} \cdot \dfrac{3}{2} = \dfrac{9}{4}$

(b) $(0.4)^2 = (0.4)(0.4) = 0.16$

Section 9.1 Quick Checks

1. radical sign

2. principal square root

3. $-4; 4$

4. $\sqrt{81} = 9$ because $9^2 = 81$.

5. $\sqrt{900} = 30$ because $30^2 = 900$.

6. $\sqrt{\dfrac{9}{4}} = \dfrac{3}{2}$ because $\left(\dfrac{3}{2}\right)^2 = \dfrac{9}{4}$.

7. $\sqrt{0.16} = 0.4$ because $(0.4)^2 = 0.16$

8. $\left(\sqrt{13}\right)^2 = 13$ because $\left(\sqrt{c}\right)^2 = c$ if $c \geq 0$.

9. $5\sqrt{9} = 5\sqrt{3^2} = 5 \cdot 3 = 15$

10. $\sqrt{36 + 64} = \sqrt{100} = \sqrt{10^2} = 10$

11. $\sqrt{36} + \sqrt{64} = \sqrt{6^2} + \sqrt{8^2} = 6 + 8 = 14$

12. $\sqrt{25 - 4 \cdot 3 \cdot (-2)} = \sqrt{25 + 24} = \sqrt{49} = \sqrt{7^2} = 7$

13. True

14. $\sqrt{400}$ is rational because $20^2 = 400$. Thus, $\sqrt{400} = 20$.

15. $\sqrt{40}$ is irrational because 40 is not a perfect square. There is no rational number whose square is 40. Using a calculator, we find $\sqrt{40} \approx 6.32$.

16. $\sqrt{-25}$ is not a real number. There is no real number whose square is -25.

17. $-\sqrt{196}$ is a rational number because $14^2 = 196$. Thus, $-\sqrt{196} = -14$.

18. $|a|$

19. $\sqrt{(-14)^2} = |-14| = 14$

20. $\sqrt{z^2} = |z|$

21. $\sqrt{(2x+3)^2} = |2x+3|$

22. $\sqrt{p^2 - 12p + 36} = \sqrt{(p-6)^2} = |p-6|$

Section 9.1 Exercises

23. $\sqrt{1} = \sqrt{1^2} = 1$

25. $-\sqrt{100} = -\sqrt{10^2} = -10$

27. $\sqrt{\dfrac{1}{4}} = \sqrt{\left(\dfrac{1}{2}\right)^2} = \dfrac{1}{2}$

29. $\sqrt{0.36} = \sqrt{(0.6)^2} = 0.6$

31. $\left(\sqrt{1.6}\right)^2 = 1.6$

33. $\sqrt{-14}$ is not a real number because the radicand of the square root is negative.

35. $\sqrt{64}$ is rational because 64 is a perfect square.

$\sqrt{64} = \sqrt{(8)^2} = 8$

37. $\sqrt{\dfrac{1}{16}}$ is rational because $\dfrac{1}{16}$ is a perfect square.

$\sqrt{\dfrac{1}{16}} = \sqrt{\left(\dfrac{1}{4}\right)^2} = \dfrac{1}{4}$

39. $\sqrt{44}$ is irrational because 44 is not a perfect square.

$\sqrt{44} \approx 6.63$

41. $\sqrt{50}$ is irrational because 50 is not a perfect square.

$\sqrt{50} \approx 7.07$

43. $\sqrt{-16}$ is not a real number because the radicand of the square root is negative.

45. $\sqrt{8^2} = 8$

47. $\sqrt{(-19)^2} = |-19| = 19$

49. $\sqrt{r^2} = |r|$

51. $\sqrt{(x+4)^2} = |x+4|$

53. $\sqrt{(4x-3)^2} = |4x-3|$

55. $\sqrt{4y^2 + 12y + 9} = \sqrt{(2y+3)^2} = |2y+3|$

57. $\sqrt{25+144} = \sqrt{169} = \sqrt{(13)^2} = 13$

59. $\sqrt{25} + \sqrt{144} = \sqrt{5^2} + \sqrt{12^2} = 5 + 12 = 17$

61. $\sqrt{-144}$ is not a real number.

63. $3\sqrt{25} = 3\sqrt{(5)^2} = 3 \cdot 5 = 15$

65. $5\sqrt{\dfrac{16}{25}} - \sqrt{144} = 5\sqrt{\left(\dfrac{4}{5}\right)^2} - \sqrt{12^2} = 5 \cdot \dfrac{4}{5} - 12$

$= 4 - 12 = -8$

67. $\sqrt{8^2 - 4 \cdot 1 \cdot 7} = \sqrt{64-28} = \sqrt{36} = \sqrt{6^2} = 6$

69. $\sqrt{(-5)^2 - 4 \cdot 2 \cdot 5} = \sqrt{25-40} = \sqrt{-15}$ is not a real number.

71. $\dfrac{-(-1) + \sqrt{(-1)^2 - 4 \cdot 6 \cdot (-2)}}{2 \cdot (-1)} = \dfrac{1 + \sqrt{1+48}}{-2}$

$= \dfrac{1 + \sqrt{49}}{-2} = \dfrac{1+7}{-2} = \dfrac{8}{-2} = -4$

73. $\sqrt{(6-1)^2 + (15-3)^2} = \sqrt{5^2 + 12^2}$

$= \sqrt{25+144} = \sqrt{169} = \sqrt{13^2} = 13$

75. The square roots of 36 are 6 and -6 because

$(6)^2 = 36$ and $(-6)^2 = 36$.

$\sqrt{36} = 6$ (the principal square root).

77. $Z = \dfrac{X-\mu}{\dfrac{\sigma}{\sqrt{n}}} = \dfrac{120-100}{\dfrac{15}{\sqrt{13}}} = \dfrac{20}{\dfrac{15}{\sqrt{13}}}$

$= 20 \cdot \dfrac{\sqrt{13}}{15} = \dfrac{4\sqrt{13}}{3} \approx 4.81$

79. Answers may vary. When taking a square root, we are looking for a number that we can multiply by itself to obtain the radicand. Since $a \cdot a = a^2$, it should follow that $\sqrt{a^2} = a$. However, since the product of two negative numbers is positive, it is impossible to determine from a^2 whether a was positive or negative. We define $\sqrt{a^2} = |a|$ to ensure that the principle square root is positive even if $a < 0$.

For example:

$2^2 = 4$ and $\sqrt{4} = 2$

$(-2)^2 = 4$ and $\sqrt{4} = 2$ not -2 since

$\sqrt{(-2)^2} = |-2| = 2$.

Section 9.2

Preparing for *n*th Roots and Rational Exponents

P1. $\left(\dfrac{x^2 y}{xy^{-2}}\right)^{-3} = \left(x^{2-1} y^{1-(-2)}\right)^{-3}$

$\qquad\qquad = \left(xy^3\right)^{-3}$

$\qquad\qquad = \left(\dfrac{1}{xy^3}\right)^{3}$

$\qquad\qquad = \dfrac{1}{x^3 y^9}$

P2. $\left(\sqrt{7}\right)^2 = 7$

P3. $\sqrt{64} = 8$ since $8^2 = 64$.

P4. $\sqrt{(x+1)^2} = |x+1|$

P5. **(a)** $3^{-2} = \dfrac{1}{3^2} = \dfrac{1}{9}$

$\qquad$ **(b)** $x^{-4} = \dfrac{1}{x^4}$

Section 9.2 Quick Checks

1. index

2. $\sqrt[3]{64} = 4$ since $4^3 = 64$

3. $\sqrt[4]{81} = 3$ since $3^4 = 81$.

4. $\sqrt[3]{-216} = -6$ since $(-6)^3 = -216$.

5. $\sqrt[4]{-32}$ is not a real number.

6. $\sqrt[5]{\dfrac{1}{32}} = \dfrac{1}{2}$ since $\left(\dfrac{1}{2}\right)^5 = \dfrac{1}{32}$.

7. $\sqrt[3]{50} \approx 3.68$

```
³√(50)
       3.684031499
```

8. $\sqrt[4]{80} \approx 2.99$

```
4*√(80)
       2.990697562
```

9. Because the index is even, we have
$\sqrt[4]{5^4} = |5| = 5$.

10. Because the index is even, we have $\sqrt[6]{z^6} = |z|$.

11. Because the index is odd, we have
$\sqrt[7]{(3x-2)^7} = 3x-2$.

12. Because the index is even, we have
$\sqrt[8]{(-2)^8} = |-2| = 2$.

13. Because the index is odd, we have
$\sqrt[5]{\left(-\dfrac{2}{3}\right)^5} = -\dfrac{2}{3}$

14. $\sqrt[n]{a}$

15. $25^{1/2} = \sqrt{25} = \sqrt{5^2} = 5$

16. $(-27)^{1/3} = \sqrt[3]{-27} = \sqrt[3]{(-3)^3} = -3$

17. $-64^{1/2} = -\sqrt{64} = -\sqrt{8^2} = -8$

18. $(-64)^{1/2} = \sqrt{-64}$ is not a real number.

19. $b^{1/2} = \sqrt{b}$

20. $\sqrt[5]{8b} = (8b)^{1/5}$

21. $\sqrt[8]{\dfrac{mn^5}{3}} = \left(\dfrac{mn^5}{3}\right)^{1/8}$

22. $\sqrt[n]{a^m}$ or $\left(\sqrt[n]{a}\right)^m$

23. $16^{3/2} = \left(\sqrt{16}\right)^3 = 4^3 = 64$

24. $27^{2/3} = \left(\sqrt[3]{27}\right)^2 = 3^2 = 9$

25. $-16^{3/4} = -\left(\sqrt[4]{16}\right)^3 = -2^3 = -8$

26. $(-64)^{2/3} = \left(\sqrt[3]{-64}\right)^2 = (-4)^2 = 16$

27. $(-25)^{5/2} = \left(\sqrt{-25}\right)^5$ is not a real number.

28. $50^{2/3} \approx 13.57$

```
50^(2/3)
        13.57208808
```

29. $40^{0.15} \approx 1.74$

```
40^0.15
        1.739037707
```

30. $\sqrt[8]{a^3} = \left(a^3\right)^{1/8} = a^{3/8}$

31. $\left(\sqrt[4]{12ab^3}\right)^9 = \left(\left(12ab^3\right)^{1/4}\right)^9 = \left(12ab^3\right)^{9/4}$

32. $81^{-1/2} = \dfrac{1}{81^{1/2}} = \dfrac{1}{\sqrt{81}} = \dfrac{1}{9}$

33. $\dfrac{1}{8^{-2/3}} = 8^{2/3} = \left(\sqrt[3]{8}\right)^2 = 2^2 = 4$

34. $(13x)^{-3/2} = \dfrac{1}{(13x)^{3/2}}$

9.2 Exercises

35. $\sqrt[3]{125} = \sqrt[3]{5^3} = 5$

37. $\sqrt[3]{-27} = \sqrt[3]{(-3)^3} = -3$

39. $-\sqrt[4]{625} = -\sqrt[4]{5^4} = -5$

41. $\sqrt[3]{-\dfrac{1}{8}} = \sqrt[3]{\left(-\dfrac{1}{2}\right)^3} = -\dfrac{1}{2}$

43. $-\sqrt[5]{-243} = -\sqrt[5]{(-3)^5} = -(-3) = 3$

45. $\sqrt[3]{25} = 25^{\wedge}(1/3) \approx 2.92$

47. $\sqrt[4]{12} = 12^{\wedge}(1/4) \approx 1.86$

49. $\sqrt[3]{5^3} = 5$

51. $\sqrt[4]{m^4} = |m|$

53. $\sqrt[9]{(x-3)^9} = x-3$

55. $-\sqrt[4]{(3p+1)^4} = -|3p+1|$

57. $4^{1/2} = \sqrt{4} = \sqrt{2^2} = 2$

59. $-36^{1/2} = -\sqrt{36} = -\sqrt{6^2} = -6$

61. $8^{1/3} = \sqrt[3]{8} = \sqrt[3]{2^3} = 2$

63. $-16^{1/4} = -\sqrt[4]{16} = -\sqrt[4]{2^4} = -2$

65. $\left(\dfrac{4}{25}\right)^{1/2} = \sqrt{\dfrac{4}{25}} = \dfrac{\sqrt{4}}{\sqrt{25}} = \dfrac{\sqrt{2^2}}{\sqrt{5^2}} = \dfrac{2}{5}$

67. $(-125)^{1/3} = \sqrt[3]{-125} = \sqrt[3]{(-5)^3} = -5$

69. $(-4)^{1/2} = \sqrt{-4}$ is not a real number.

71. $\sqrt[3]{3x} = (3x)^{1/3}$

73. $\sqrt[4]{\dfrac{x}{3}} = \left(\dfrac{x}{3}\right)^{1/4}$

75. $4^{5/2} = \left(\sqrt{4}\right)^5 = 2^5 = 32$

77. $-16^{3/2} = -\left(\sqrt{16}\right)^3 = -4^3 = -64$

79. $8^{4/3} = \left(\sqrt[3]{8}\right)^4 = 2^4 = 16$

81. $(-64)^{2/3} = \left(\sqrt[3]{-64}\right)^2 = (-4)^2 = 16$

83. $-(-32)^{3/5} = -\left(\sqrt[5]{-32}\right)^3 = -(-2)^3 = -(-8) = 8$

85. $144^{-1/2} = \left(\sqrt{144}\right)^{-1} = (12)^{-1} = \dfrac{1}{12}$

87. $\dfrac{1}{25^{-3/2}} = 25^{3/2} = \left(\sqrt{25}\right)^3 = 5^3 = 125$

89. $\dfrac{1}{8^{-5/3}} = 8^{5/3} = \left(\sqrt[3]{8}\right)^5 = 2^5 = 32$

91. $\sqrt[4]{x^3} = x^{3/4}$

93. $\left(\sqrt[5]{3x}\right)^2 = (3x)^{2/5}$

95. $\sqrt{\left(\dfrac{5x}{y}\right)^3} = \left(\dfrac{5x}{y}\right)^{3/2}$

97. $\sqrt[3]{(9ab)^4} = (9ab)^{4/3}$

99. $20^{1/2} = \sqrt{20} \approx 4.47$

101. $4^{5/3} = 4 \wedge (5/3) \approx 10.08$

103. $10^{0.1} = 10 \wedge (0.1) \approx 1.26$

105. $\sqrt[3]{x^3} + 4\sqrt[6]{x^6} = x + 4|x|$

If $x \geq 0$, $\sqrt[3]{x^3} + 4\sqrt[6]{x^6} = x + 4x = 5x$.

If $x < 0$, $\sqrt[3]{x^3} + 4\sqrt[6]{x^6} = x + 4(-x) = -3x$.

107. $-16^{3/4} - \sqrt[5]{32} = -\left(\sqrt[4]{16}\right)^3 - 2$

$\qquad = -2^3 - 2$

$\qquad = -8 - 2$

$\qquad = -10$

109. $\sqrt[3]{512} = \sqrt[3]{8^3} = 8$

111. $9^{5/2} = \left(\sqrt{9}\right)^5 = 3^5 = 243$

113. $\sqrt[4]{-16}$ is not a real number because the index is even and the radicand is negative.

115. $144^{-1/2} \cdot 3^2 = \dfrac{1}{144^{1/2}} \cdot 9 = \dfrac{1}{\sqrt{144}} \cdot 9 = \dfrac{1}{12} \cdot 9 = \dfrac{3}{4}$

117. $\sqrt[3]{0.008} = \sqrt[3]{(0.2)^3} = 0.2$

119. $4^{1/2} + 25^{3/2} = \sqrt{4} + \left(\sqrt{25}\right)^3 = 2 + 5^3$

$\qquad = 2 + 125 = 127$

121. $(-25)^{5/2} = \left(\sqrt{-25}\right)^5$ is not a real number.

123. $\sqrt[3]{(3p-5)^3} = 3p - 5$

125. $-9^{\frac{3}{2}} + \dfrac{1}{27^{-\frac{2}{3}}} = -\left(9^{1/2}\right)^3 + \dfrac{1}{\left(27^{1/3}\right)^{-2}}$

$\qquad = -(3)^3 + \dfrac{1}{(3)^{-2}}$

$\qquad = -27 + 3^2$

$\qquad = -27 + 9$

$\qquad = -18$

127. $\sqrt[6]{(-2)^6} = \sqrt[6]{64} = \sqrt[6]{2^6} = 2$

129. $f(x) = x^{3/2}$

$f(4) = 4^{3/2} = \left(\sqrt{4}\right)^3 = 2^3 = 8$

131. $F(z) = z^{4/3}$

$F(-8) = (-8)^{4/3} = \left(\sqrt[3]{-8}\right)^4 = (-2)^4 = 16$

133. 10 is the only cube root of 1000.

$\sqrt[3]{1000} = \sqrt[3]{(10)^3} = 10$

135. **(a)** $W = 35.74 + 0.6215(30) - 35.75(10)^{0.16}$

$\qquad\qquad + 0.4275(30)(10)^{0.16}$

$\qquad \approx 21.25$

The windchill would be about $21.25°F$.

(b) $W = 35.74 + 0.6215(30) - 35.75(20)^{0.16}$
$$+ 0.4275(30)(20)^{0.16}$$
$$\approx 17.36$$
The windchill would be about $17.36°F$.

(c) $W = 35.74 + 0.6215(0) - 35.75(10)^{0.16}$
$$+ 0.4275(0)(10)^{0.16}$$
$$\approx -15.93$$
The windchill would be about $-15.93°F$.

137. (a) $v_t = \sqrt{\dfrac{2mg}{C\rho A}}$

$$= \sqrt{\dfrac{2 \cdot \frac{4}{3}\pi r^3 \rho_w \cdot g}{C\rho \cdot \pi r^2}}$$

$$= \sqrt{\dfrac{8r\rho_w g}{3C\rho}} \text{ m/s}$$

(b) The radius is $1.5 \text{ mm} = 0.0015 \text{ m}$. We also have $\rho_w = 1000 \text{ kg/m}^3$, $g \approx 9.81 \text{ m/s}^2$, $C = 0.6$, and $\rho \approx 1.2 \text{ kg/m}^3$.

$$v_t = \sqrt{\dfrac{8(0.0015)(1000)(9.81)}{3(0.6)(1.2)}} \approx 7.38 \text{ m/s}$$

The terminal velocity is about 7.38 meters per second.

139. Answers will vary.

$(-9)^{1/2}$ and $-9^{1/2}$ are different because of order of operations. This is easier seen in a different form:

$(-9)^{1/2} = \sqrt{-9}$ (radicand is negative, so the result is not a real number)

$-9^{1/2} = -\left(9^{1/2}\right) = -\sqrt{9} = -3$ (radicand is positive, so the result is a real number)

141. If $\dfrac{m}{n}$, in lowest terms, is positive, then $a^{\frac{m}{n}}$ is a real number provided $\sqrt[n]{a}$ is a real number.

If $\dfrac{m}{n}$, in lowest terms, is negative, then $a^{\frac{m}{n}}$ is a real number provided $a \neq 0$ and $\sqrt[n]{a}$ is a real number.

143. $\dfrac{(x+2)^2 (x-1)^4}{(x+2)(x-1)} = (x+2)^{2-1}(x-1)^{4-1}$
$$= (x+2)(x-1)^3$$

145. $\dfrac{\left(4z^2 - 7z + 3\right) + \left(-3z^2 - z + 9\right)}{\left(4z^2 - 2z - 7\right) + \left(-3z^2 - z + 9\right)}$

$$= \dfrac{4z^2 - 7z + 3 - 3z^2 - z + 9}{4z^2 - 2z - 7 - 3z^2 - z + 9}$$

$$= \dfrac{z^2 - 8z + 12}{z^2 - 3z + 2} = \dfrac{(z-6)(z-2)}{(z-1)(z-2)}$$

$$= \dfrac{z-6}{z-1}$$

Section 9.3

Preparing for Simplifying Expressions Using the Laws of Exponents

P1. $z^{-3} = \dfrac{1}{z^3}$

P2. $x^{-2} \cdot x^5 = x^{-2+5} = x^3$

P3. $\left(\dfrac{2a^2}{b^{-1}}\right)^3 = \left(2a^2b\right)^3 = 2^3 \left(a^2\right)^3 b^3 = 8a^6b^3$

P4. $\sqrt{64} = 8$ because $8^2 = 64$.

Section 9.3 Quick Checks

1. $a^r b^r$

2. a^{r+s}

3. $5^{3/4} \cdot 5^{1/6} = 5^{\frac{3}{4}+\frac{1}{6}} = 5^{\frac{9}{12}+\frac{2}{12}} = 5^{\frac{11}{12}}$

4. $\dfrac{32^{6/5}}{32^{3/5}} = 32^{\frac{6}{5}-\frac{3}{5}} = 32^{3/5} = \left(\sqrt[5]{32}\right)^3 = 2^3 = 8$

5. $\left(100^{3/8}\right)^{4/3} = 100^{\frac{3}{8}\cdot\frac{4}{3}} = 100^{1/2} = \sqrt{100} = 10$

6. $\left(a^{3/2} \cdot b^{5/4}\right)^{2/3} = a^{\frac{3}{2}\cdot\frac{2}{3}} \cdot b^{\frac{5}{4}\cdot\frac{2}{3}} = ab^{5/6}$

7. $\dfrac{x^{1/2} \cdot x^{1/3}}{\left(x^{1/12}\right)^2} = \dfrac{x^{\frac{1}{2}+\frac{1}{3}}}{x^{\frac{1}{12} \cdot 2}}$

$= \dfrac{x^{\frac{3}{6}+\frac{2}{6}}}{x^{\frac{1}{6}}}$

$= \dfrac{x^{\frac{5}{6}}}{x^{\frac{1}{6}}}$

$= x^{\frac{5}{6}-\frac{1}{6}}$

$= x^{4/6}$

$= x^{2/3}$

8. $\left(8x^{3/4}y^{-1}\right)^{2/3} = 8^{2/3}\left(x^{3/4}\right)^{2/3}\left(y^{-1}\right)^{2/3}$

$= 4x^{\frac{3}{4} \cdot \frac{2}{3}}y^{-1 \cdot \frac{2}{3}}$

$= 4x^{1/2}y^{-2/3}$

$= \dfrac{4x^{1/2}}{y^{2/3}}$

9. $\left(\dfrac{25x^{1/2}y^{3/4}}{x^{-3/4}y}\right)^{1/2} = \left(25x^{\frac{1}{2}-\left(-\frac{3}{4}\right)}y^{\frac{3}{4}-1}\right)^{1/2}$

$= \left(25x^{5/4}y^{-1/4}\right)^{1/2}$

$= 25^{1/2}\left(x^{5/4}\right)^{1/2}\left(y^{-1/4}\right)^{1/2}$

$= 5x^{\frac{5}{4} \cdot \frac{1}{2}}y^{-\frac{1}{4} \cdot \frac{1}{2}}$

$= 5x^{5/8}y^{-1/8}$

$= \dfrac{5x^{5/8}}{y^{1/8}}$

10. $8\left(125a^{3/4}b^{-1}\right)^{2/3} = 8 \cdot 125^{2/3}\left(a^{3/4}\right)^{2/3}\left(b^{-1}\right)^{2/3}$

$= 8 \cdot 25a^{\frac{3}{4} \cdot \frac{2}{3}}b^{-1 \cdot \frac{2}{3}}$

$= 200a^{1/2}b^{-2/3}$

$= \dfrac{200a^{1/2}}{b^{2/3}}$

11. $\sqrt[10]{36^5} = \left(36^5\right)^{1/10} = 36^{5 \cdot \frac{1}{10}} = 36^{1/2} = \sqrt{36} = 6$

12. $\sqrt[4]{16a^8b^{12}} = \left(16a^8b^{12}\right)^{1/4}$

$= 16^{1/4}\left(a^8\right)^{1/4}\left(b^{12}\right)^{1/4}$

$= 2a^{8 \cdot \frac{1}{4}}b^{12 \cdot \frac{1}{4}}$

$= 2a^2b^3$

13. $\dfrac{\sqrt[3]{x^2}}{\sqrt[4]{x}} = \dfrac{x^{2/3}}{x^{1/4}} = x^{\frac{2}{3}-\frac{1}{4}} = x^{5/12} = \sqrt[12]{x^5}$

14. $\sqrt[4]{\sqrt[3]{a^2}} = \sqrt[4]{a^{2/3}} = \left(a^{2/3}\right)^{1/4} = a^{\frac{2}{3} \cdot \frac{1}{4}} = a^{1/6} = \sqrt[6]{a}$

15. $8x^{3/2} + 3x^{1/2}\left(4x+3\right) = 8x^{\frac{2}{2}+\frac{1}{2}} + 3x^{1/2}\left(4x+3\right)$

$= 8x \cdot x^{1/2} + 3x^{1/2}\left(4x+3\right)$

$= x^{1/2}\left(8x+3\left(4x+3\right)\right)$

$= x^{1/2}\left(8x+12x+9\right)$

$= x^{1/2}\left(20x+9\right)$

16. $9x^{1/3} + x^{-2/3}\left(3x+1\right) = 9x^{\frac{3}{3}-\frac{2}{3}} + x^{-2/3}\left(3x+1\right)$

$= 9x \cdot x^{-2/3} + x^{-2/3}\left(3x+1\right)$

$= x^{-2/3}\left(9x+\left(3x+1\right)\right)$

$= x^{-2/3}\left(12x+1\right)$

$= \dfrac{12x+1}{x^{2/3}}$

9.3 Exercises

17. $5^{1/2} \cdot 5^{3/2} = 5^{\frac{1}{2}+\frac{3}{2}} = 5^{\frac{4}{2}} = 5^2 = 25$

19. $\dfrac{8^{5/4}}{8^{1/4}} = 8^{\frac{5}{4}-\frac{1}{4}} = 8^{4/4} = 8^1 = 8$

21. $2^{1/3} \cdot 2^{-3/2} = 2^{\frac{1}{3}+\left(-\frac{3}{2}\right)} = 2^{\frac{2}{6}-\frac{9}{6}} = 2^{-7/6} = \dfrac{1}{2^{7/6}}$

23. $\dfrac{x^{1/4}}{x^{5/6}} = x^{\frac{1}{4}-\frac{5}{6}} = x^{\frac{3}{12}-\frac{10}{12}} = x^{-7/12} = \dfrac{1}{x^{7/12}}$

25. $\left(4^{4/3}\right)^{3/8} = 4^{\frac{4}{3} \cdot \frac{3}{8}} = 4^{1/2} = \sqrt{4} = 2$

27. $\left(25^{3/4} \cdot 4^{-3/4}\right)^2 = \left[\left(25 \cdot 4^{-1}\right)^{3/4}\right]^2$

$$= \left(\frac{25}{4}\right)^{\frac{3}{4} \cdot 2} = \left(\frac{25}{4}\right)^{3/2}$$

$$= \left(\sqrt{\frac{25}{4}}\right)^3 = \left(\frac{5}{2}\right)^3 = \frac{125}{8}$$

29. $\left(x^{3/4} \cdot y^{1/3}\right)^{2/3} = \left(x^{3/4}\right)^{2/3} \cdot \left(y^{1/3}\right)^{2/3}$

$$= x^{\frac{3}{4} \cdot \frac{2}{3}} \cdot y^{\frac{1}{3} \cdot \frac{2}{3}}$$

$$= x^{1/2} y^{2/9}$$

31. $\left(x^{-1/3} \cdot y\right)\left(x^{1/2} \cdot y^{-4/3}\right) = x^{-\frac{1}{3}+\frac{1}{2}} \cdot y^{1+\left(-\frac{4}{3}\right)}$

$$= x^{-\frac{2}{6}+\frac{3}{6}} \cdot y^{\frac{3}{3}-\frac{4}{3}}$$

$$= x^{1/6} \cdot y^{-1/3}$$

$$= \frac{x^{1/6}}{y^{1/3}}$$

33. $\left(4a^2 b^{-3/2}\right)^{1/2} = 4^{1/2} \cdot \left(a^2\right)^{1/2} \cdot \left(b^{-3/2}\right)^{1/2}$

$$= 2 \cdot a^{2 \cdot \frac{1}{2}} \cdot b^{-\frac{3}{2} \cdot \frac{1}{2}}$$

$$= 2a^1 b^{-3/4}$$

$$= \frac{2a}{b^{3/4}}$$

35. $\left(\frac{x^{2/3} y^{-1/3}}{8x^{1/2} y}\right)^{1/3} = \left(\frac{1}{8} x^{(2/3)-(1/2)} y^{(-1/3)-1}\right)^{1/3}$

$$= \left(\frac{1}{8} x^{1/6} y^{-4/3}\right)^{1/3}$$

$$= \left(\frac{1}{8}\right)^{1/3} \left(x^{1/6}\right)^{1/3} \left(y^{-4/3}\right)^{1/3}$$

$$= \frac{1}{2} x^{1/18} y^{-4/9}$$

$$= \frac{x^{1/18}}{2y^{4/9}}$$

37. $\left(\frac{50x^{3/4} y}{2x^{1/2}}\right)^{1/2} + \left(\frac{x^{1/2} y^{1/2}}{9x^{3/4} y^{3/2}}\right)^{-1/2}$

$$= \left(25x^{\frac{3}{4}-\frac{1}{2}} y\right)^{1/2} + \left(\frac{1}{9} x^{\frac{1}{2}-\frac{3}{4}} y^{\frac{1}{2}-\frac{3}{2}}\right)^{-1/2}$$

$$= \left(25x^{1/4} y\right)^{1/2} + \left(\frac{1}{9} x^{-1/4} y^{-1}\right)^{-1/2}$$

$$= \left(25x^{1/4} y\right)^{1/2} + \left(\frac{1}{9x^{1/4} y^1}\right)^{-1/2}$$

$$= \left(25x^{1/4} y\right)^{1/2} + \left(9x^{1/4} y\right)^{1/2}$$

$$= 25^{1/2} \left(x^{1/4}\right)^{1/2} y^{1/2} + 9^{1/2} \left(x^{1/4}\right)^{1/2} y^{1/2}$$

$$= 5x^{1/8} y^{1/2} + 3x^{1/8} y^{1/2}$$

$$= 8x^{1/8} y^{1/2}$$

39. $\sqrt{x^8} = \left(x^8\right)^{1/2} = x^{8 \cdot \frac{1}{2}} = x^4$

41. $\sqrt[12]{8^4} = \left(8^4\right)^{1/12} = 8^{4 \cdot \frac{1}{12}} = 8^{1/3} = \sqrt[3]{8} = 2$

43. $\sqrt[3]{8a^3 b^{12}} = \left(8a^3 b^{12}\right)^{1/3}$

$$= 8^{1/3} \left(a^3\right)^{1/3} \left(b^{12}\right)^{1/3}$$

$$= 2ab^4$$

45. $\dfrac{\sqrt{x}}{\sqrt[4]{x}} = \dfrac{x^{1/2}}{x^{1/4}} = x^{\frac{1}{2}-\frac{1}{4}} = x^{\frac{2}{4}-\frac{1}{4}} = x^{1/4} = \sqrt[4]{x}$

47. $\sqrt{x} \cdot \sqrt[3]{x} = x^{1/2} \cdot x^{1/3} = x^{\frac{1}{2}+\frac{1}{3}}$

$$= x^{\frac{3}{6}+\frac{2}{6}} = x^{5/6}$$

$$= \sqrt[6]{x^5}$$

49. $\sqrt{\sqrt[4]{x^3}} = \left(\left(x^3\right)^{1/4}\right)^{1/2} = \left(x^{3/4}\right)^{1/2} = x^{3/8} = \sqrt[8]{x^3}$

51. $\sqrt{3} \cdot \sqrt[3]{9} = \sqrt{3} \cdot \sqrt[3]{3^2}$

$\quad\quad = 3^{1/2} \cdot 3^{2/3}$

$\quad\quad = 3^{\frac{1}{2}+\frac{2}{3}}$

$\quad\quad = 3^{7/6}$

$\quad\quad = \sqrt[6]{3^7}$

53. $\dfrac{\sqrt{6}}{\sqrt[4]{36}} = \dfrac{\sqrt{6}}{\sqrt[4]{6^2}} = \dfrac{6^{1/2}}{6^{2/4}} = 6^{\frac{1}{2}-\frac{2}{4}} = 6^{\frac{1}{2}-\frac{1}{2}} = 6^0 = 1$

55. $2x^{3/2} + 3x^{1/2}(x+5) = 2x^{1/2} \cdot x + 3x^{1/2}(x+5)$

$\quad\quad = x^{1/2}(2x + 3(x+5))$

$\quad\quad = x^{1/2}(2x + 3x + 15)$

$\quad\quad = x^{1/2}(5x + 15)$

$\quad\quad = 5x^{1/2}(x+3)$

57. $5(x+2)^{2/3}(3x-2) + 9(x+2)^{5/3}$

$\quad = 5(x+2)^{2/3}(3x-2) + 9(x+2)^{2/3} \cdot (x+2)^{3/3}$

$\quad = (x+2)^{2/3}(5(3x-2) + 9(x+2))$

$\quad = (x+2)^{2/3}(15x - 10 + 9x + 18)$

$\quad = (x+2)^{2/3}(24x + 8)$

$\quad = 8(x+2)^{2/3}(3x+1)$

59. $x^{-1/2}(2x+5) + 4x^{1/2}$

$\quad = x^{-1/2}(2x+5) + 4x^{-1/2} \cdot x^{2/2}$

$\quad = x^{-1/2}(2x + 5 + 4x)$

$\quad = x^{-1/2}(6x+5)$

$\quad = \dfrac{6x+5}{x^{1/2}}$

61. $2(x-4)^{-1/3}(4x-3) + 12(x-4)^{2/3}$

$\quad = 2(x-4)^{-1/3}(4x-3) + 12(x-4)^{-1/3}(x-4)^{3/3}$

$\quad = 2(x-4)^{-1/3}((4x-3) + 6(x-4))$

$\quad = 2(x-4)^{-1/3}(4x-3+6x-24)$

$\quad = 2(x-4)^{-1/3}(10x-27)$

$\quad = \dfrac{2(10x-27)}{(x-4)^{1/3}}$

63. $15x(x^2+4)^{1/2} + 5(x^2+4)^{3/2}$

$\quad = 15x(x^2+4)^{1/2} + 5(x^2+4)^{1/2}(x^2+4)^{2/2}$

$\quad = 5(x^2+4)^{1/2}(3x + x^2 + 4)$

$\quad = 5(x^2+4)^{1/2}(x^2 + 3x + 4)$

65. $\sqrt[8]{4^4} = 4^{4/8} = 4^{1/2} = (2^2)^{1/2} = 2$

67. $2^{1/2} \cdot 2^{3/2} = 2^{4/2} = 2^2 = 4$

69. $(100^{1/3})^{3/2} = 100^{\frac{1}{3} \cdot \frac{3}{2}} = 100^{1/2} = \sqrt{100} = 10$

71. $(\sqrt[4]{25})^2 = 25^{2/4} = 25^{1/2} = \sqrt{25} = 5$

73. $\sqrt[4]{x^2} - \dfrac{\sqrt[4]{x^6}}{x} = x^{2/4} - \dfrac{x^{6/4}}{x}$

$\quad\quad = x^{1/2} - x^{\frac{3}{2}-1}$

$\quad\quad = x^{1/2} - x^{1/2}$

$\quad\quad = 0$

75. $(4 \cdot 9^{1/4})^{-2} = 4^{-2} \cdot (9^{1/4})^{-2}$

$\quad\quad = \dfrac{1}{4^2} \cdot 9^{-1/2} = \dfrac{1}{16 \cdot 9^{1/2}}$

$\quad\quad = \dfrac{1}{16\sqrt{9}} = \dfrac{1}{16(3)} = \dfrac{1}{48}$

77. $x^{1/2}(x^{3/2}-2) = x^{1/2} \cdot x^{3/2} - x^{1/2} \cdot 2$

$\quad\quad = x^{4/2} - 2x^{1/2}$

$\quad\quad = x^2 - 2x^{1/2}$

79. $2y^{-1/3}(1+3y) = 2y^{-1/3} \cdot 1 + 2y^{-1/3} \cdot 3y$

$\quad\quad = 2y^{-1/3} + 6y^{-\frac{1}{3}+1}$

$\quad\quad = \dfrac{2}{y^{1/3}} + 6y^{2/3}$

81. $4z^{3/2}\left(z^{3/2}-8z^{-3/2}\right)$

$= 4z^{3/2} \cdot z^{3/2} - 4z^{3/2} \cdot 8z^{-3/2}$

$= 4z^{\frac{3}{2}+\frac{3}{2}} - 32z^{\frac{3}{2}+\left(-\frac{3}{2}\right)}$

$= 4z^{6/2} - 32z^{0}$

$= 4z^{3} - 32$

$= 4\left(z^{3}-8\right)$

$= 4(z-2)\left(z^{2}+2z+4\right)$

83. $3^{x} = 25$

$3^{x/2} = 3^{x \cdot \frac{1}{2}} = \left(3^{x}\right)^{1/2} = (25)^{1/2} = 5$

Thus, $3^{x/2} = 5$.

85. $7^{x} = 9$

$\sqrt{7^{x}} = \sqrt{9} = 3$

87. $\sqrt[4]{\sqrt[3]{\sqrt{x}}} = \left(\sqrt[3]{\sqrt{x}}\right)^{1/4}$

$= \left(\left(\sqrt{x}\right)^{1/3}\right)^{1/4}$

$= \left(\sqrt{x}\right)^{1/12}$

$= \left(x^{1/2}\right)^{1/12}$

$= x^{1/24}$

$= \sqrt[24]{x}$

89. $\left(6^{\sqrt{2}}\right)^{\sqrt{2}} = 6^{\sqrt{2}\cdot\sqrt{2}} = 6^{2} = 36$

91. $f(x) = (x+3)^{1/2}(x+1)^{-1/2} = \dfrac{\sqrt{x+3}}{\sqrt{x+1}}$

We need $x+3 \geq 0$ and $x+1 > 0$.

$x+3 \geq 0$ and $x+1 > 0$

    $x \geq -3$ $x > -1$

The domain is the set of values that satisfy both inequalities. Thus, the domain is $x > -1$.

Domain: $\{x \mid x > -1\}$ or $(-1, \infty)$

We can check this graphically:

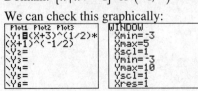

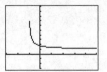

93. $3a(a-3)+(a+3)(a-2)$

$= 3a^{2} - 9a + a^{2} - 2a + 3a - 6$

$= 4a^{2} - 8a - 6$

$= 2\left(2a^{2} - 4a - 3\right)$

95. $\dfrac{x^{2}-4}{x+2} \cdot (x+5) - (x+4)(x-1)$

$= \dfrac{(x+2)(x-2)(x+5)}{(x+2)} - (x+4)(x-1)$

$= (x-2)(x+5) - (x+4)(x-1)$

$= \left(x^{2}+3x-10\right) - \left(x^{2}+3x-4\right)$

$= x^{2}+3x-10-x^{2}-3x+4$

$= -6$

Section 9.4

Preparing for Simplifying Radical Expressions

P1. $1^{2}=1$, $2^{2}=4$, $3^{2}=9$, $4^{2}=16$, $5^{2}=25$, $6^{2}=36$, $7^{2}=49$, $8^{2}=64$, $9^{2}=81$, $10^{2}=100$, $11^{2}=121$, $12^{2}=144$, $13^{2}=169$, and $14^{2}=196$.

P2. $1^{3}=1$, $2^{3}=8$, $3^{3}=27$, $4^{3}=64$, and $5^{3}=125$.

P3. (a) $\sqrt{16} = \sqrt{4^{2}} = 4$

     (b) $\sqrt{p^{2}} = |p|$

Section 9.4 Quick Checks

1. $\sqrt[n]{ab}$

2. $\sqrt{11} \cdot \sqrt{7} = \sqrt{11 \cdot 7} = \sqrt{77}$

3. $\sqrt[4]{6} \cdot \sqrt[4]{7} = \sqrt[4]{6 \cdot 7} = \sqrt[4]{42}$

4. $\sqrt{x-5} \cdot \sqrt{x+5} = \sqrt{(x-5)(x+5)} = \sqrt{x^{2}-25}$

5. $\sqrt[7]{5p} \cdot \sqrt[7]{4p^3} = \sqrt[7]{5p \cdot 4p^3} = \sqrt[7]{20p^4}$

6. $0 = 0^2$, $1 = 1^2$, $4 = 2^2$, $9 = 3^2$, $16 = 4^2$, $25 = 5^2$

7. $1 = 1^3$, $8 = 2^3$, $27 = 3^3$, $64 = 4^3$, $125 = 5^3$, $216 = 6^3$

8. $\sqrt{48} = \sqrt{16 \cdot 3} = \sqrt{16} \cdot \sqrt{3} = 4\sqrt{3}$

9. $4\sqrt[3]{54} = 4\sqrt[3]{27 \cdot 2} = 4\sqrt[3]{27} \cdot \sqrt[3]{2} = 4 \cdot 3\sqrt[3]{2} = 12\sqrt[3]{2}$

10. $\sqrt{200a^2} = \sqrt{200} \cdot \sqrt{a^2}$
$= \sqrt{100 \cdot 2} \cdot \sqrt{a^2}$
$= \sqrt{100} \cdot \sqrt{2} \cdot \sqrt{a^2}$
$= 10\sqrt{2}|a|$ or $10|a|\sqrt{2}$

11. $\sqrt[4]{40}$ cannot be simplified further.

12. $\dfrac{6 + \sqrt{45}}{3} = \dfrac{6 + \sqrt{9 \cdot 5}}{3} = \dfrac{6 + \sqrt{9} \cdot \sqrt{5}}{3}$
$= \dfrac{6 + 3\sqrt{5}}{3} = \dfrac{3\left(2 + \sqrt{5}\right)}{3}$
$= 2 + \sqrt{5}$

13. $\dfrac{-2 + \sqrt{32}}{4} = \dfrac{-2 + \sqrt{16 \cdot 2}}{4} = \dfrac{-2 + \sqrt{16} \cdot \sqrt{2}}{4}$
$= \dfrac{-2 + 4\sqrt{2}}{4} = \dfrac{2\left(-1 + 2\sqrt{2}\right)}{4}$
$= \dfrac{-1 + 2\sqrt{2}}{2}$

14. $\sqrt{75a^6} = \sqrt{25 \cdot 3 \cdot a^6}$
$= \sqrt{25} \cdot \sqrt{3} \cdot \sqrt{a^6}$
$= \sqrt{25} \cdot \sqrt{3} \cdot \sqrt{\left(a^3\right)^2}$
$= 5\sqrt{3} \cdot a^3$
$= 5a^3\sqrt{3}$

15. $\sqrt{18a^5} = \sqrt{9 \cdot 2 \cdot a^4 \cdot a}$
$= \sqrt{9} \cdot \sqrt{2} \cdot \sqrt{a^4} \cdot \sqrt{a}$
$= \sqrt{9} \cdot \sqrt{2} \cdot \sqrt{\left(a^2\right)^2} \cdot \sqrt{a}$
$= 3\sqrt{2} \cdot a^2 \sqrt{a}$
$= 3a^2\sqrt{2a}$

16. $\sqrt[3]{128x^6y^{10}} = \sqrt[3]{64 \cdot 2 \cdot x^6 \cdot y^9 \cdot y}$
$= \sqrt[3]{64} \cdot \sqrt[3]{2} \cdot \sqrt[3]{x^6} \cdot \sqrt[3]{y^9} \cdot \sqrt[3]{y}$
$= \sqrt[3]{64} \cdot \sqrt[3]{2} \cdot \sqrt[3]{\left(x^2\right)^3} \cdot \sqrt[3]{\left(y^3\right)^3} \cdot \sqrt[3]{y}$
$= 4x^2y^3\sqrt[3]{2y}$

17. $\sqrt[4]{16a^5b^{11}} = \sqrt[4]{16 \cdot a^4 \cdot a \cdot b^8 \cdot b^3}$
$= \sqrt[4]{16} \cdot \sqrt[4]{a^4} \cdot \sqrt[4]{a} \cdot \sqrt[4]{b^8} \cdot \sqrt[4]{b^3}$
$= \sqrt[4]{16} \cdot \sqrt[4]{a^4} \cdot \sqrt[4]{a} \cdot \sqrt[4]{\left(b^2\right)^4} \cdot \sqrt[4]{b^3}$
$= 2ab^2\sqrt[4]{ab^3}$

18. $\sqrt{6} \cdot \sqrt{8} = \sqrt{6 \cdot 8}$
$= \sqrt{48}$
$= \sqrt{16 \cdot 3}$
$= \sqrt{16} \cdot \sqrt{3}$
$= 4\sqrt{3}$

19. $\sqrt[3]{12a^2} \cdot \sqrt[3]{10a^4} = \sqrt[3]{12a^2 \cdot 10a^4}$
$= \sqrt[3]{120a^6}$
$= \sqrt[3]{8 \cdot 15 \cdot a^6}$
$= \sqrt[3]{8} \cdot \sqrt[3]{a^6} \cdot \sqrt[3]{15}$
$= 2a^2\sqrt[3]{15}$

20. $4\sqrt[3]{8a^2b^5} \cdot \sqrt[3]{6a^2b^4} = 4\sqrt[3]{8a^2b^5 \cdot 6a^2b^4}$
$= 4\sqrt[3]{48a^4b^9}$
$= 4\sqrt[3]{8 \cdot 6 \cdot a^3 \cdot a \cdot b^9}$
$= 4\sqrt[3]{8} \cdot \sqrt[3]{6} \cdot \sqrt[3]{a^3} \cdot \sqrt[3]{a} \cdot \sqrt[3]{b^9}$
$= 4 \cdot 2ab^3 \cdot \sqrt[3]{6a}$
$= 8ab^3\sqrt[3]{6a}$

21. $\sqrt{\dfrac{13}{49}} = \dfrac{\sqrt{13}}{\sqrt{49}} = \dfrac{\sqrt{13}}{7}$

22. $\sqrt[3]{\dfrac{27p^3}{8}} = \dfrac{\sqrt[3]{27p^3}}{\sqrt[3]{8}} = \dfrac{3p}{2}$

23. $\sqrt[4]{\dfrac{3q^4}{16}} = \dfrac{\sqrt[4]{3q^4}}{\sqrt[4]{16}} = \dfrac{q\sqrt[4]{3}}{2}$

24. $\dfrac{\sqrt{12a^5}}{\sqrt{3a}} = \sqrt{\dfrac{12a^5}{3a}} = \sqrt{4a^4} = 2a^2$

25. $\dfrac{\sqrt[3]{-24x^2}}{\sqrt[3]{3x^{-1}}} = \sqrt[3]{\dfrac{-24x^2}{3x^{-1}}} = \sqrt[3]{-8x^3} = -2x$

26. $\dfrac{\sqrt[3]{250a^5b^{-2}}}{\sqrt[3]{2ab}} = \sqrt[3]{\dfrac{250a^5b^{-2}}{2ab}}$

$= \sqrt[3]{125a^4b^{-3}}$

$= \sqrt[3]{\dfrac{125a^4}{b^3}}$

$= \dfrac{5a\sqrt[3]{a}}{b}$

27. $\sqrt[4]{5} \cdot \sqrt[3]{3} = 5^{1/4} \cdot 3^{1/3}$

$= 5^{3/12} \cdot 3^{4/12}$

$= \left(5^3\right)^{1/12} \cdot \left(3^4\right)^{1/12}$

$= \left[\left(5^3\right)\left(3^4\right)\right]^{1/12}$

$= (10,125)^{1/12}$

$= \sqrt[12]{10,125}$

28. $\sqrt{10} \cdot \sqrt[3]{12} = 10^{1/2} \cdot 12^{1/3}$

$= 10^{3/6} \cdot 12^{2/6}$

$= \left(10^3\right)^{1/6} \cdot \left(12^2\right)^{1/6}$

$= \left[\left(10^3\right)\left(12^2\right)\right]^{1/6}$

$= \left(2^3 \cdot 5^3 \cdot 4^2 \cdot 3^2\right)^{1/6}$

$= \left(2^3 \cdot 5^3 \cdot 2^4 \cdot 3^2\right)^{1/6}$

$= \left(2^7 \cdot 5^3 \cdot 3^2\right)^{1/6}$

$= 2\left(2 \cdot 5^3 \cdot 3^2\right)^{1/6}$

$= 2(2250)^{1/6}$

$= 2\sqrt[6]{2250}$

9.4 Exercises

29. $\sqrt[3]{6} \cdot \sqrt[3]{10} = \sqrt[3]{6 \cdot 10} = \sqrt[3]{60}$

31. $\sqrt{3a} \cdot \sqrt{5b} = \sqrt{3a \cdot 5b} = \sqrt{15ab}$ if $a, b \geq 0$.

(If $a, b < 0$, then the result is $-\sqrt{15ab}$. If a and b have opposite signs, the result is not a real number.)

33. $\sqrt{x-7} \cdot \sqrt{x+7} = \sqrt{(x-7)(x+7)} = \sqrt{x^2-49}$

if $|x| \geq 7$, otherwise the result is not a real number.

35. $\sqrt{\dfrac{5x}{3}} \cdot \sqrt{\dfrac{3}{x}} = \sqrt{\dfrac{5x}{3} \cdot \dfrac{3}{x}} = \sqrt{5}$ if $x > 0$.

37. $\sqrt{50} = \sqrt{25 \cdot 2} = \sqrt{25} \cdot \sqrt{2} = 5\sqrt{2}$

39. $\sqrt[3]{54} = \sqrt[3]{27 \cdot 2} = \sqrt[3]{27} \cdot \sqrt[3]{2} = 3\sqrt[3]{2}$

41. $\sqrt{48x^2} = \sqrt{16x^2 \cdot 3} = \sqrt{16x^2} \cdot \sqrt{3} = 4|x|\sqrt{3}$

43. $\sqrt[3]{-27x^3} = -3x$

45. $\sqrt[4]{32m^4} = \sqrt[4]{16m^4 \cdot 2} = \sqrt[4]{16m^4} \cdot \sqrt[4]{2} = 2|m|\sqrt[4]{2}$

47. $\sqrt{12p^2q} = \sqrt{4p^2 \cdot 3q} = \sqrt{4p^2} \cdot \sqrt{3q} = 2|p|\sqrt{3q}$

49. $\sqrt{162m^4} = \sqrt{81m^4 \cdot 2} = \sqrt{81m^4} \cdot \sqrt{2} = 9m^2\sqrt{2}$

51. $\sqrt{y^{13}} = \sqrt{y^{12} \cdot y} = \sqrt{y^{12}} \cdot \sqrt{y} = y^6 \sqrt{y}$

53. $\sqrt[3]{c^8} = \sqrt[3]{c^6 \cdot c^2} = \sqrt[3]{c^6} \cdot \sqrt[3]{c^2} = c^2 \sqrt[3]{c^2}$

55. $\sqrt{125 p^3 q^4} = \sqrt{25 p^2 q^4 \cdot 5p} = \sqrt{25 p^2 q^4} \cdot \sqrt{5p}$
$= 5|p|q^2 \sqrt{5p}$

57. $\sqrt[3]{-16 x^9} = \sqrt[3]{-8x^9 \cdot 2} = \sqrt[3]{-8x^9} \cdot \sqrt[3]{2} = -2x^3 \sqrt[3]{2}$

59. $\sqrt[5]{-16 m^8 n^2} = \sqrt[5]{-m^5 \cdot 16 m^3 n^2}$
$= \sqrt[5]{-m^5} \cdot \sqrt[5]{16 m^3 n^2}$
$= -m \sqrt[5]{16 m^3 n^2}$

61. $\sqrt[4]{(x-y)^5} = \sqrt[4]{(x-y)^4 \cdot (x-y)}$
$= \sqrt[4]{(x-y)^4} \cdot \sqrt[4]{x-y}$
$= (x-y) \sqrt[4]{x-y}$

63. $\sqrt[3]{8x^3 - 8y^3} = \sqrt[3]{8(x^3 - y^3)}$
$= \sqrt[3]{8} \cdot \sqrt[3]{x^3 - y^3}$
$= 2 \sqrt[3]{x^3 - y^3}$

65. $\dfrac{4 + \sqrt{36}}{2} = \dfrac{4+6}{2} = \dfrac{10}{2} = 5$

67. $\dfrac{9 + \sqrt{18}}{3} = \dfrac{9 + \sqrt{9 \cdot 2}}{3} = \dfrac{9 + \sqrt{9} \cdot \sqrt{2}}{3}$
$= \dfrac{9 + 3\sqrt{2}}{3} = \dfrac{3(3 + \sqrt{2})}{3}$
$= 3 + \sqrt{2}$

69. $\dfrac{7 - \sqrt{98}}{14} = \dfrac{7 - \sqrt{49 \cdot 2}}{14} = \dfrac{7 - \sqrt{49} \cdot \sqrt{2}}{14}$
$= \dfrac{7 - 7\sqrt{2}}{14} = \dfrac{7(1 - \sqrt{2})}{7 \cdot 2}$
$= \dfrac{1 - \sqrt{2}}{2}$

71. $\sqrt{5} \cdot \sqrt{5} = \sqrt{5 \cdot 5} = \sqrt{25} = 5$

73. $\sqrt{2} \cdot \sqrt{8} = \sqrt{2 \cdot 8} = \sqrt{16} = 4$

75. $\sqrt[3]{4} \cdot \sqrt[3]{2} = \sqrt[3]{4 \cdot 2} = \sqrt[3]{8} = 2$

77. $\sqrt{5x} \cdot \sqrt{15x} = \sqrt{5x \cdot 15x}$
$= \sqrt{75 x^2}$
$= \sqrt{25 x^2 \cdot 3}$
$= \sqrt{25 x^2} \cdot \sqrt{3}$
$= 5x\sqrt{3}$

79. $\sqrt[3]{4b^2} \cdot \sqrt[3]{6b^2} = \sqrt[3]{4b^2 \cdot 6b^2}$
$= \sqrt[3]{24 b^4}$
$= \sqrt[3]{8b^3 \cdot 3b}$
$= \sqrt[3]{8b^3} \cdot \sqrt[3]{3b}$
$= 2b \sqrt[3]{3b}$

81. $2\sqrt{6ab} \cdot 3\sqrt{15 ab^3} = 6\sqrt{6ab \cdot 15 ab^3}$
$= 6\sqrt{90 a^2 b^4}$
$= 6\sqrt{9 a^2 b^4 \cdot 10}$
$= 6\sqrt{9 a^2 b^4} \cdot \sqrt{10}$
$= 18 ab^2 \sqrt{10}$

83. $\sqrt[4]{27 p^3 q^2} \cdot \sqrt[4]{12 p^2 q^2} = \sqrt[4]{27 p^3 q^2 \cdot 12 p^2 q^2}$
$= \sqrt[4]{324 p^5 q^4}$
$= \sqrt[4]{81 p^4 q^4 \cdot 4p}$
$= \sqrt[4]{81 p^4 q^4} \cdot \sqrt[4]{4p}$
$= 3pq \sqrt[4]{4p}$

85. $\sqrt[5]{-8 a^3 b^4} \cdot \sqrt[5]{12 a^3 b} = \sqrt[5]{-8 a^3 b^4 \cdot 12 a^3 b}$
$= \sqrt[5]{-96 a^6 b^5}$
$= \sqrt[5]{-32 a^5 b^5 \cdot 3a}$
$= \sqrt[5]{-32 a^5 b^5} \cdot \sqrt[5]{3a}$
$= -2ab \sqrt[5]{3a}$

87. $\sqrt[4]{8(x-y)^2} \cdot \sqrt[4]{6(x-y)^3}$
$= \sqrt[4]{8(x-y)^2 \cdot 6(x-y)^3}$
$= \sqrt[4]{48(x-y)^5}$
$= \sqrt[4]{16(x-y)^4 \cdot 3(x-y)}$
$= \sqrt[4]{16(x-y)^4} \cdot \sqrt[4]{3(x-y)}$
$= 2(x-y) \sqrt[4]{3(x-y)}$

89. $\sqrt{\dfrac{3}{16}} = \dfrac{\sqrt{3}}{\sqrt{16}} = \dfrac{\sqrt{3}}{4}$

91. $\sqrt[4]{\dfrac{5x^4}{16}} = \dfrac{\sqrt[4]{5x^4}}{\sqrt[4]{16}} = \dfrac{x\sqrt[4]{5}}{2}$

93. $\sqrt{\dfrac{9y^2}{25x^2}} = \dfrac{\sqrt{9y^2}}{\sqrt{25x^2}} = \dfrac{3y}{5x}$

95. $\sqrt[3]{\dfrac{-27x^9}{64y^{12}}} = \dfrac{\sqrt[3]{-27x^9}}{\sqrt[3]{64y^{12}}} = \dfrac{-3x^3}{4y^4} = -\dfrac{3x^3}{4y^4}$

97. $\dfrac{\sqrt{8}}{\sqrt{2}} = \sqrt{\dfrac{8}{2}} = \sqrt{4} = 2$

99. $\dfrac{\sqrt[3]{128}}{\sqrt[3]{2}} = \sqrt[3]{\dfrac{128}{2}} = \sqrt[3]{64} = 4$

101. $\dfrac{\sqrt{48a^3}}{\sqrt{6a}} = \sqrt{\dfrac{48a^3}{6a}} = \sqrt{8a^2} = 2a\sqrt{2}$

103. $\dfrac{\sqrt{24a^5b}}{\sqrt{3ab^3}} = \sqrt{\dfrac{24a^5b}{3ab^3}} = \sqrt{\dfrac{8a^4}{b^2}} = \dfrac{\sqrt{8a^4}}{\sqrt{b^2}} = \dfrac{2a^2\sqrt{2}}{b}$

105. $\dfrac{\sqrt{512a^7b}}{3\sqrt{2ab^3}} = \dfrac{1}{3}\sqrt{\dfrac{512a^7b}{2ab^3}} = \dfrac{1}{3}\sqrt{\dfrac{256a^6}{b^2}} = \dfrac{16a^3}{3b}$

107. $\dfrac{\sqrt[3]{104a^5}}{\sqrt[3]{4a^{-1}}} = \sqrt[3]{\dfrac{104a^5}{4a^{-1}}} = \sqrt[3]{26a^6} = a^2\sqrt[3]{26}$

109. $\dfrac{\sqrt{90x^3y^{-1}}}{\sqrt{2x^{-3}y}} = \sqrt{\dfrac{90x^3y^{-1}}{2x^{-3}y}}$

$= \sqrt{\dfrac{45x^6}{y^2}}$

$= \dfrac{\sqrt{45x^6}}{\sqrt{y^2}}$

$= \dfrac{3x^3\sqrt{5}}{y}$

111. $\sqrt{3} \cdot \sqrt[3]{4} = 3^{1/2} \cdot 4^{1/3}$

$= 3^{3/6} \cdot 4^{2/6}$

$= \left(3^3\right)^{1/6} \cdot \left(4^2\right)^{1/6}$

$= \left(3^3 \cdot 4^2\right)^{1/6}$

$= (432)^{1/6}$

$= \sqrt[6]{432}$

113. $\sqrt[3]{2} \cdot \sqrt[6]{3} = 2^{1/3} \cdot 3^{1/6}$

$= 2^{2/6} \cdot 3^{1/6}$

$= \left(2^2\right)^{1/6} \cdot (3)^{1/6}$

$= \left(2^2 \cdot 3\right)^{1/6}$

$= (12)^{1/6}$

$= \sqrt[6]{12}$

115. $\sqrt{3} \cdot \sqrt[3]{18} = 3^{1/2} \cdot 18^{1/3}$

$= 3^{3/6} \cdot 18^{2/6}$

$= \left(3^3\right)^{1/6} \cdot \left(18^2\right)^{1/6}$

$= \left(3^3 \cdot 18^2\right)^{1/6}$

$= \left(3^3 \cdot (2 \cdot 3 \cdot 3)^2\right)^{1/6}$

$= \left(3^3 \cdot 2^2 \cdot 3^2 \cdot 3^2\right)^{1/6}$

$= \left(3^7 \cdot 2^2\right)^{1/6}$

$= 3\left(3 \cdot 2^2\right)^{1/6}$

$= 3(12)^{1/6}$

$= 3\sqrt[6]{12}$

117. $\sqrt[4]{9} \cdot \sqrt[6]{12} = 9^{1/4} \cdot 12^{1/6}$

$$= \left(3^2\right)^{1/4} \cdot 12^{1/6}$$

$$= 3^{1/2} \cdot 12^{1/6}$$

$$= 3^{3/6} \cdot 12^{1/6}$$

$$= \left(3^3\right)^{1/6} \cdot \left(12\right)^{1/6}$$

$$= \left(3^3 \cdot 12\right)^{1/6}$$

$$= \left(3^3 \cdot 3 \cdot 2^2\right)^{1/6}$$

$$= \left(3^4 \cdot 2^2\right)^{1/6}$$

$$= \left(\left(3^2 \cdot 2\right)^2\right)^{1/6}$$

$$= \left(3^2 \cdot 2\right)^{1/3}$$

$$= \sqrt[3]{18}$$

119. $\sqrt[3]{\dfrac{5x}{8}} = \dfrac{\sqrt[3]{5x}}{\sqrt[3]{8}} = \dfrac{\sqrt[3]{5x}}{2}$

121. $\sqrt[3]{5a} \cdot \sqrt[3]{9a} = \sqrt[3]{5a \cdot 9a} = \sqrt[3]{45a^2}$

123. $\sqrt{72a^4} = \sqrt{36a^4 \cdot 2} = \sqrt{36a^4} \cdot \sqrt{2} = 6a^2 \sqrt{2}$

125. $\sqrt[3]{6a^2 b} \cdot \sqrt[3]{9ab} = \sqrt[3]{6a^2 b \cdot 9ab}$

$$= \sqrt[3]{54a^3 b^2}$$

$$= \sqrt[3]{27a^3 \cdot 2b^2}$$

$$= \sqrt[3]{27a^3} \cdot \sqrt[3]{2b^2}$$

$$= 3a\sqrt[3]{2b^2}$$

127. $\dfrac{\sqrt[3]{-32a}}{\sqrt[3]{2a^4}} = \sqrt[3]{\dfrac{-32a}{2a^4}} = \sqrt[3]{\dfrac{-16}{a^3}} = \dfrac{\sqrt[3]{-16}}{\sqrt[3]{a^3}} = \dfrac{-2\sqrt[3]{2}}{a}$

129. $-5\sqrt[3]{32m^3} = -5\sqrt[3]{8m^3 \cdot 4}$

$$= -5 \cdot \sqrt[3]{8m^3} \cdot \sqrt[3]{4}$$

$$= -5 \cdot 2m \cdot \sqrt[3]{4}$$

$$= -10m\sqrt[3]{4}$$

131. $\sqrt[3]{81a^4 b^7} = \sqrt[3]{27a^3 b^6 \cdot 3ab}$

$$= \sqrt[3]{27a^3 b^6} \cdot \sqrt[3]{3ab}$$

$$= 3ab^2 \sqrt[3]{3ab}$$

133. $\sqrt[3]{12} \cdot \sqrt[3]{18} = \sqrt[3]{12 \cdot 18}$

$$= \sqrt[3]{216}$$

$$= 6$$

135. (a)

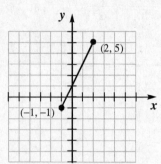

(b) $\sqrt{\left(5-(-1)\right)^2 + \left(2-(-1)\right)^2}$

$$= \sqrt{\left(5+1\right)^2 + \left(2+1\right)^2}$$

$$= \sqrt{6^2 + 3^2}$$

$$= \sqrt{36+9}$$

$$= \sqrt{45}$$

$$= 3\sqrt{5}$$

The line segment has a length of $3\sqrt{5}$ units.

137. (a) $R(8) = \sqrt[3]{\dfrac{8}{2}} = \sqrt[3]{4} \approx 1.587$

The company's annual revenue will be roughly \$1,587,000 after 8 years.

(b) $R(27) = \sqrt[3]{\dfrac{27}{2}} = \dfrac{\sqrt[3]{27}}{\sqrt[3]{2}} = \dfrac{3}{\sqrt[3]{2}} \approx 2.381$

The company's annual revenue will be roughly \$2,381,000 after 27 years.

139. (a) $(f \cdot g)(x) = f(x) \cdot g(x)$

$$= \sqrt{2x} \cdot \sqrt{8x^3}$$

$$= \sqrt{2x \cdot 8x^3}$$

$$= \sqrt{16x^4}$$

$$= 4x^2$$

(b) Using the result from part (a), we get

$$(f \cdot g)(3) = 4(3)^2 = 36$$

141. $a = 1, b = 6, c = 3$

$$x = \frac{-6 \pm \sqrt{6^2 - 4(1)(3)}}{2(1)}$$

$$= \frac{-6 \pm \sqrt{36 - 12}}{2}$$

$$= \frac{-6 \pm \sqrt{24}}{2}$$

$$= \frac{-6 \pm 2\sqrt{6}}{2}$$

$$= -3 \pm \sqrt{6}$$

$$x = -3 - \sqrt{6} \quad \text{or} \quad x = -3 + \sqrt{6}$$

143. $a = 3, b = 4, c = -1$

$$x = \frac{-4 \pm \sqrt{4^2 - 4(3)(-1)}}{2(3)}$$

$$= \frac{-4 \pm \sqrt{16 + 12}}{6}$$

$$= \frac{-4 \pm \sqrt{28}}{6}$$

$$= \frac{-4 \pm 2\sqrt{7}}{6}$$

$$= \frac{-2 \pm \sqrt{7}}{3}$$

$$x = \frac{-2 - \sqrt{7}}{3} \quad \text{or} \quad x = \frac{-2 + \sqrt{7}}{3}$$

145. The indexes must be the same.

147. $2|7x - 1| + 4 = 16$

$$2|7x - 1| = 12$$

$$|7x - 1| = 6$$

$$7x - 1 = -6 \quad \text{or} \quad 7x - 1 = 6$$

$$7x = -5 \qquad\qquad 7x = 7$$

$$x = -\frac{5}{7} \qquad\qquad x = 1$$

The solution set is $\left\{ -\frac{5}{7}, 1 \right\}$.

149. $\frac{5}{2}|x + 1| + 1 \le 11$

$$\frac{5}{2}|x + 1| \le 10$$

$$|x + 1| \le 4$$

$$-4 \le x + 1 \le 4$$

$$-4 - 1 \le x + 1 - 1 \le 4 - 1$$

$$-5 \le x \le 3$$

The solution set is $\{ x \mid -5 \le x \le 3 \}$, or in interval notation we would write $[-5, 3]$.

151. (a)

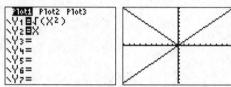

The graphs of $y = \sqrt{x^2}$ and $y = x$ are not the same so the expressions cannot be equal.

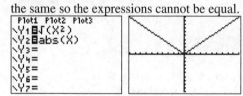

The graphs of $y = \sqrt{x^2}$ and $y = |x|$ appear to be the same. $\sqrt{x^2} = |x|$

(b)

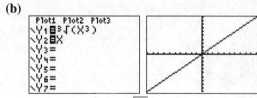

The graphs of $y = \sqrt[3]{x^3}$ and $y = x$ appear to be the same. $\sqrt[3]{x^3} = x$

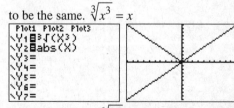

The graph of $y = \sqrt[3]{x^3}$ and $y = |x|$ are not the same so the expressions cannot be equal.

(c)

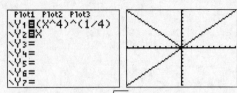

The graphs of $y = \sqrt[4]{x^4}$ and $y = x$ are not the same so the expressions cannot be equal.

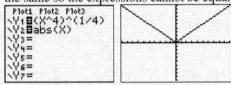

The graphs of $y = \sqrt[4]{x^4}$ and $y = |x|$ appear to be the same. $\sqrt[4]{x^4} = |x|$

(d) Answers will vary.

$\sqrt[n]{x^n} = x$ if the index is odd and $\sqrt[n]{x^n} = |x|$ if the index is even.

Section 9.5

Preparing for Adding, Subtracting, and Multiplying Radical Expressions

P1. $4y^3 - 2y^2 + 8y - 1 + \left(-2y^3 + 7y^2 - 3y + 9\right)$

$= 4y^3 - 2y^3 - 2y^2 + 7y^2 + 8y - 3y - 1 + 9$

$= 2y^3 + 5y^2 + 5y + 8$

P2. $5z^2 + 6 - \left(3z^2 - 8z - 3\right)$

$= 5z^2 + 6 - 3z^2 + 8z + 3$

$= 5z^2 - 3z^2 + 8z + 6 + 3$

$= 2z^2 + 8z + 9$

P3. $\left(4x + 3\right)\left(x - 5\right) = 4x \cdot x - 4x \cdot 5 + 3 \cdot x - 3 \cdot 5$

$= 4x^2 - 20x + 3x - 15$

$= 4x^2 - 17x - 15$

P4. $\left(2y - 3\right)\left(2y + 3\right) = \left(2y\right)^2 - 3^2$

$= 4y^2 - 9$

Section 9.5 Quick Checks

1. like radicals

2. $9\sqrt{13y} + 4\sqrt{13y} = \left(9 + 4\right)\sqrt{13y} = 13\sqrt{13y}$

3. $\sqrt[4]{5} + 9\sqrt[4]{5} - 3\sqrt[4]{5} = \left(1 + 9 - 3\right)\sqrt[4]{5} = 7\sqrt[4]{5}$

4. $4\sqrt{18} - 3\sqrt{8} = 4 \cdot \sqrt{9} \cdot \sqrt{2} - 3 \cdot \sqrt{4} \cdot \sqrt{2}$

$= 4 \cdot 3\sqrt{2} - 3 \cdot 2\sqrt{2}$

$= 12\sqrt{2} - 6\sqrt{2}$

$= \left(12 - 6\right)\sqrt{2}$

$= 6\sqrt{2}$

5. $-5x\sqrt[3]{54x} + 7\sqrt[3]{2x^4}$

$= -5x \cdot \sqrt[3]{27} \cdot \sqrt[3]{2x} + 7 \cdot \sqrt[3]{x^3} \cdot \sqrt[3]{2x}$

$= -5x \cdot 3\sqrt[3]{2x} + 7x \cdot \sqrt[3]{2x}$

$= -15x\sqrt[3]{2x} + 7x\sqrt[3]{2x}$

$= \left(-15 + 7\right)x\sqrt[3]{2x}$

$= -8x\sqrt[3]{2x}$

6. $7\sqrt{10} - 6\sqrt{3}$ cannot be simplified further.

7. $\sqrt[3]{8z^4} - 2z\sqrt[3]{-27z} + \sqrt[3]{125z}$

$= \sqrt[3]{8z^3} \cdot \sqrt[3]{z} - 2z \cdot \sqrt[3]{-27} \cdot \sqrt[3]{z} + \sqrt[3]{125} \cdot \sqrt[3]{z}$

$= 2z\sqrt[3]{z} - 2z\left(-3\right)\sqrt[3]{z} + 5\sqrt[3]{z}$

$= 2z\sqrt[3]{z} + 6z\sqrt[3]{z} + 5\sqrt[3]{z}$

$= \left(2z + 6z + 5\right)\sqrt[3]{z}$

$= \left(8z + 5\right)\sqrt[3]{z}$

8. $\sqrt{25m} - 3\sqrt[4]{m^2} = \sqrt{25m} - 3m^{2/4}$

$= \sqrt{25m} - 3m^{1/2}$

$= \sqrt{25m} - 3\sqrt{m}$

$= \sqrt{25} \cdot \sqrt{m} - 3\sqrt{m}$

$= 5\sqrt{m} - 3\sqrt{m}$

$= \left(5 - 3\right)\sqrt{m}$

$= 2\sqrt{m}$

9. $\sqrt{6}\left(3 - 5\sqrt{6}\right) = \sqrt{6} \cdot 3 - 5\sqrt{6} \cdot \sqrt{6}$

$= 3\sqrt{6} - 5\sqrt{36}$

$= 3\sqrt{6} - 5 \cdot 6$

$= 3\sqrt{6} - 30$

$= 3\left(\sqrt{6} - 10\right)$

10. $\sqrt[3]{12}\left(3 - \sqrt[3]{2}\right) = \sqrt[3]{12} \cdot 3 - \sqrt[3]{12} \cdot \sqrt[3]{2}$

$= 3\sqrt[3]{12} - \sqrt[3]{24}$

$= 3\sqrt[3]{12} - 2\sqrt[3]{3}$

11. $\left(2-7\sqrt{3}\right)\left(5+4\sqrt{3}\right)$

$=2\cdot5+2\cdot4\sqrt{3}-7\sqrt{3}\cdot5-7\sqrt{3}\cdot4\sqrt{3}$

$=10+8\sqrt{3}-35\sqrt{3}-28\sqrt{9}$

$=10-27\sqrt{3}-28\cdot3$

$=10-27\sqrt{3}-84$

$=-74-27\sqrt{3}$

12. conjugates

13. False. The conjugate of $-5+\sqrt{2}$ is $-5-\sqrt{2}$.

14. $\left(5\sqrt{2}+\sqrt{3}\right)^2=\left(5\sqrt{2}+\sqrt{3}\right)\left(5\sqrt{2}+\sqrt{3}\right)$

$=\left(5\sqrt{2}\right)^2+2\left(5\sqrt{2}\right)\left(\sqrt{3}\right)+\left(\sqrt{3}\right)^2$

$=25\sqrt{4}+10\sqrt{6}+\sqrt{9}$

$=25\cdot2+10\sqrt{6}+3$

$=50+10\sqrt{6}+3$

$=53+10\sqrt{6}$

15. $\left(\sqrt{7}-3\sqrt{2}\right)^2=\left(\sqrt{7}-3\sqrt{2}\right)\left(\sqrt{7}-3\sqrt{2}\right)$

$=\left(\sqrt{7}\right)^2-2\left(\sqrt{7}\right)\left(3\sqrt{2}\right)+\left(3\sqrt{2}\right)^2$

$=\sqrt{49}-6\sqrt{14}+9\sqrt{4}$

$=7-6\sqrt{14}+9\cdot2$

$=7-6\sqrt{14}+18$

$=25-6\sqrt{14}$

16. $\left(\sqrt{3}+\sqrt{2}\right)\left(\sqrt{3}-\sqrt{2}\right)=\left(\sqrt{3}\right)^2-\left(\sqrt{2}\right)^2$

$=\sqrt{9}-\sqrt{4}$

$=3-2$

$=1$

9.5 Exercises

17. $3\sqrt{2}+7\sqrt{2}=\left(3+7\right)\sqrt{2}=10\sqrt{2}$

19. $5\sqrt[3]{x}-3\sqrt[3]{x}=\left(5-3\right)\sqrt[3]{x}=2\sqrt[3]{x}$

21. $8\sqrt{5x}-3\sqrt{5x}+9\sqrt{5x}=\left(8-3+9\right)\sqrt{5x}$

$=14\sqrt{5x}$

23. $4\sqrt[3]{5}-3\sqrt{5}+7\sqrt[3]{5}-8\sqrt{5}$

$=\left(4+7\right)\sqrt[3]{5}+\left(-3-8\right)\sqrt{5}$

$=11\sqrt[3]{5}-11\sqrt{5}$

25. $\sqrt{8}+6\sqrt{2}=\sqrt{4}\cdot\sqrt{2}+6\sqrt{2}$

$=2\sqrt{2}+6\sqrt{2}$

$=\left(2+6\right)\sqrt{2}$

$=8\sqrt{2}$

27. $\sqrt[3]{24}-4\sqrt[3]{3}=\sqrt[3]{8}\cdot\sqrt[3]{3}-4\sqrt[3]{3}$

$=2\sqrt[3]{3}-4\sqrt[3]{3}$

$=\left(2-4\right)\sqrt[3]{3}$

$=-2\sqrt[3]{3}$

29. $\sqrt[3]{54}-7\sqrt[3]{128}=\sqrt[3]{27}\cdot\sqrt[3]{2}-7\cdot\sqrt[3]{64}\cdot\sqrt[3]{2}$

$=3\sqrt[3]{2}-7\cdot4\cdot\sqrt[3]{2}$

$=3\sqrt[3]{2}-28\sqrt[3]{2}$

$=\left(3-28\right)\sqrt[3]{2}$

$=-25\sqrt[3]{2}$

31. $5\sqrt{54x}-3\sqrt{24x}$

$=5\cdot\sqrt{9}\cdot\sqrt{6x}-3\cdot\sqrt{4}\cdot\sqrt{6x}$

$=5\cdot3\cdot\sqrt{6x}-3\cdot2\cdot\sqrt{6x}$

$=15\sqrt{6x}-6\sqrt{6x}$

$=\left(15-6\right)\sqrt{6x}$

$=9\sqrt{6x}$

33. $2\sqrt{8}+3\sqrt{10}=2\sqrt{4}\cdot\sqrt{2}+3\sqrt{10}$

$=2\cdot2\sqrt{2}+3\sqrt{10}$

$=4\sqrt{2}+3\sqrt{10}$

Cannot be simplified further.

35. $\sqrt{12x^3}+5x\sqrt{108x}$

$=\sqrt{4x^2}\cdot\sqrt{3x}+5x\cdot\sqrt{36}\cdot\sqrt{3x}$

$=2x\sqrt{3x}+5x\cdot6\cdot\sqrt{3x}$

$=2x\sqrt{3x}+30x\sqrt{3x}$

$=\left(2x+30x\right)\sqrt{3x}$

$=32x\sqrt{3x}$

37. $\sqrt{12x^2}+3x\sqrt{2}-2\sqrt{98x^2}$

$=\sqrt{4x^2}\cdot\sqrt{3}+3x\sqrt{2}-2\sqrt{49x^2}\cdot\sqrt{2}$

$=2x\sqrt{3}+3x\sqrt{2}-2\cdot7x\sqrt{2}$

$=2x\sqrt{3}+3x\sqrt{2}-14x\sqrt{2}$

$=2x\sqrt{3}+\left(3-14\right)x\sqrt{2}$

$=2x\sqrt{3}-11x\sqrt{2}$

39. $\sqrt[3]{-54x^3} + 3x\sqrt[3]{16} - 2\sqrt[3]{128}$

$\quad = \sqrt[3]{-27x^3} \cdot \sqrt[3]{2} + 3x\sqrt[3]{8} \cdot \sqrt[3]{2} - 2\sqrt[3]{64} \cdot \sqrt[3]{2}$

$\quad = -3x\sqrt[3]{2} + 3x \cdot 2\sqrt[3]{2} - 2 \cdot 4\sqrt[3]{2}$

$\quad = -3x\sqrt[3]{2} + 6x\sqrt[3]{2} - 8\sqrt[3]{2}$

$\quad = (-3x + 6x - 8)\sqrt[3]{2}$

$\quad = (3x - 8)\sqrt[3]{2}$

41. $\sqrt{9x - 9} + \sqrt{4x - 4}$

$\quad = \sqrt{9(x-1)} + \sqrt{4(x-1)}$

$\quad = \sqrt{9} \cdot \sqrt{x-1} + \sqrt{4} \cdot \sqrt{x-1}$

$\quad = 3\sqrt{x-1} + 2\sqrt{x-1}$

$\quad = (3 + 2)\sqrt{x-1}$

$\quad = 5\sqrt{x-1}$

43. $\sqrt{16x} - \sqrt[6]{x^3} = \sqrt{16x} - x^{3/6}$

$\quad = \sqrt{16} \cdot \sqrt{x} - x^{1/2}$

$\quad = 4\sqrt{x} - \sqrt{x}$

$\quad = (4 - 1)\sqrt{x}$

$\quad = 3\sqrt{x}$

45. $\sqrt[3]{27x} + 2\sqrt[9]{x^3} = \sqrt[3]{27} \cdot \sqrt[3]{x} + 2 \cdot x^{3/9}$

$\quad = 3\sqrt[3]{x} + 2x^{1/3}$

$\quad = 3\sqrt[3]{x} + 2\sqrt[3]{x}$

$\quad = (3 + 2)\sqrt[3]{x}$

$\quad = 5\sqrt[3]{x}$

47. $\sqrt{3}(2 - 3\sqrt{2}) = \sqrt{3} \cdot 2 - \sqrt{3} \cdot 3\sqrt{2}$

$\quad = 2\sqrt{3} - 3\sqrt{2 \cdot 3}$

$\quad = 2\sqrt{3} - 3\sqrt{6}$

49. $\sqrt{3}(\sqrt{2} + \sqrt{6}) = \sqrt{3} \cdot \sqrt{2} + \sqrt{3} \cdot \sqrt{6}$

$\quad = \sqrt{3 \cdot 2} + \sqrt{3 \cdot 6}$

$\quad = \sqrt{6} + \sqrt{18}$

$\quad = \sqrt{6} + \sqrt{9} \cdot \sqrt{2}$

$\quad = \sqrt{6} + 3\sqrt{2}$

51. $\sqrt[3]{4}(\sqrt[3]{3} - \sqrt[3]{6}) = \sqrt[3]{4} \cdot \sqrt[3]{3} - \sqrt[3]{4} \cdot \sqrt[3]{6}$

$\quad = \sqrt[3]{4 \cdot 3} - \sqrt[3]{4 \cdot 6}$

$\quad = \sqrt[3]{12} - \sqrt[3]{24}$

$\quad = \sqrt[3]{12} - \sqrt[3]{8} \cdot \sqrt[3]{3}$

$\quad = \sqrt[3]{12} - 2\sqrt[3]{3}$

53. $\sqrt{2x}(3 - \sqrt{10x}) = \sqrt{2x} \cdot 3 - \sqrt{2x} \cdot \sqrt{10x}$

$\quad = 3\sqrt{2x} - \sqrt{20x^2}$

$\quad = 3\sqrt{2x} - \sqrt{4x^2} \cdot \sqrt{5}$

$\quad = 3\sqrt{2x} - 2x\sqrt{5}$

55. $(3 + \sqrt{2})(4 + \sqrt{3})$

$\quad = 3 \cdot 4 + 3 \cdot \sqrt{3} + \sqrt{2} \cdot 4 + \sqrt{2} \cdot \sqrt{3}$

$\quad = 12 + 3\sqrt{3} + 4\sqrt{2} + \sqrt{6}$

57. $(6 + \sqrt{3})(2 - \sqrt{7})$

$\quad = 6 \cdot 2 - 6 \cdot \sqrt{7} + \sqrt{3} \cdot 2 - \sqrt{3} \cdot \sqrt{7}$

$\quad = 12 - 6\sqrt{7} + 2\sqrt{3} - \sqrt{21}$

59. $(4 - 2\sqrt{7})(3 + 3\sqrt{7})$

$\quad = 4 \cdot 3 + 4 \cdot 3\sqrt{7} - 2\sqrt{7} \cdot 3 - 2\sqrt{7} \cdot 3\sqrt{7}$

$\quad = 12 + 12\sqrt{7} - 6\sqrt{7} - 6\sqrt{49}$

$\quad = 12 + (12 - 6)\sqrt{7} - 6 \cdot 7$

$\quad = 12 + 6\sqrt{7} - 42$

$\quad = 6\sqrt{7} - 30$

61. $(\sqrt{2} + 3\sqrt{6})(\sqrt{3} - 2\sqrt{2})$

$\quad = \sqrt{2} \cdot \sqrt{3} - \sqrt{2} \cdot 2\sqrt{2} + 3\sqrt{6} \cdot \sqrt{3} - 3\sqrt{6} \cdot 2\sqrt{2}$

$\quad = \sqrt{6} - 2\sqrt{4} + 3\sqrt{18} - 6\sqrt{12}$

$\quad = \sqrt{6} - 2 \cdot 2 + 3 \cdot 3\sqrt{2} - 6 \cdot 2\sqrt{3}$

$\quad = \sqrt{6} - 4 + 9\sqrt{2} - 12\sqrt{3}$

$\quad = \sqrt{6} - 12\sqrt{3} + 9\sqrt{2} - 4$

63. $(2\sqrt{5} + \sqrt{3})(4\sqrt{5} - 3\sqrt{3})$

$\quad = 2\sqrt{5} \cdot 4\sqrt{5} - 2\sqrt{5} \cdot 3\sqrt{3} + \sqrt{3} \cdot 4\sqrt{5} - \sqrt{3} \cdot 3\sqrt{3}$

$\quad = 8\sqrt{25} - 6\sqrt{15} + 4\sqrt{15} - 3\sqrt{9}$

$\quad = 8 \cdot 5 + (-6 + 4)\sqrt{15} - 3 \cdot 3$

$\quad = 40 - 2\sqrt{15} - 9$

$\quad = 31 - 2\sqrt{15}$

65. $(1 + \sqrt{3})^2 = (1)^2 + 2(1)(\sqrt{3}) + (\sqrt{3})^2$

$\quad = 1 + 2\sqrt{3} + \sqrt{9}$

$\quad = 1 + 2\sqrt{3} + 3$

$\quad = 4 + 2\sqrt{3}$

67. $\left(\sqrt{2}-\sqrt{5}\right)^2 = \left(\sqrt{2}\right)^2 - 2\left(\sqrt{2}\right)\left(\sqrt{5}\right) + \left(\sqrt{5}\right)^2$

$\quad\quad = \sqrt{4} - 2\sqrt{10} + \sqrt{25}$

$\quad\quad = 2 - 2\sqrt{10} + 5$

$\quad\quad = 7 - 2\sqrt{10}$

69. $\left(\sqrt{x}-\sqrt{2}\right)^2 = \left(\sqrt{x}\right)^2 - 2\left(\sqrt{x}\right)\left(\sqrt{2}\right) + \left(\sqrt{2}\right)^2$

$\quad\quad = \sqrt{x^2} - 2\sqrt{2x} + \sqrt{4}$

$\quad\quad = x - 2\sqrt{2x} + 2$

71. $\left(\sqrt{2}-1\right)\left(\sqrt{2}+1\right) = \left(\sqrt{2}\right)^2 - (1)^2$

$\quad\quad = \sqrt{4} - 1$

$\quad\quad = 2 - 1$

$\quad\quad = 1$

73. $\left(3-2\sqrt{5}\right)\left(3+2\sqrt{5}\right) = (3)^2 - \left(2\sqrt{5}\right)^2$

$\quad\quad = 9 - 4\sqrt{25}$

$\quad\quad = 9 - 4 \cdot 5$

$\quad\quad = 9 - 20$

$\quad\quad = -11$

75. $\left(\sqrt{2x}+\sqrt{3y}\right)\left(\sqrt{2x}-\sqrt{3y}\right)$

$\quad = \left(\sqrt{2x}\right)^2 - \left(\sqrt{3y}\right)^2$

$\quad = \sqrt{4x^2} - \sqrt{9y^2}$

$\quad = 2x - 3y$

77. $\left(\sqrt[3]{x}+4\right)\left(\sqrt[3]{x}-3\right)$

$\quad = \sqrt[3]{x} \cdot \sqrt[3]{x} - \sqrt[3]{x} \cdot 3 + 4 \cdot \sqrt[3]{x} - 4 \cdot 3$

$\quad = \sqrt[3]{x^2} - 3\sqrt[3]{x} + 4\sqrt[3]{x} - 12$

$\quad = \sqrt[3]{x^2} + (-3+4)\sqrt[3]{x} - 12$

$\quad = \sqrt[3]{x^2} + \sqrt[3]{x} - 12$

79. $\left(\sqrt[3]{2a}-5\right)\left(\sqrt[3]{2a}+5\right) = \left(\sqrt[3]{2a}\right)^2 - (5)^2$

$\quad\quad = \sqrt[3]{4a^2} - 25$

81. $\sqrt{5}\left(\sqrt{3}+\sqrt{10}\right) = \sqrt{5} \cdot \sqrt{3} + \sqrt{5} \cdot \sqrt{10}$

$\quad\quad = \sqrt{15} + \sqrt{50}$

$\quad\quad = \sqrt{15} + \sqrt{25} \cdot \sqrt{2}$

$\quad\quad = \sqrt{15} + 5\sqrt{2}$

83. $\sqrt{28x^5} - x\sqrt{7x^3} + 5\sqrt{175x^5}$

$\quad = \sqrt{4x^4} \cdot \sqrt{7x} - x\sqrt{x^2} \cdot \sqrt{7x} + 5\sqrt{25x^4} \cdot \sqrt{7x}$

$\quad = 2x^2\sqrt{7x} - x^2\sqrt{7x} + 25x^2\sqrt{7x}$

$\quad = \left(2x^2 - x^2 + 25x^2\right)\sqrt{7x}$

$\quad = 26x^2\sqrt{7x}$

85. $\left(2\sqrt{3}+5\right)\left(2\sqrt{3}-5\right) = \left(2\sqrt{3}\right)^2 - 5^2$

$\quad\quad\quad = 4\sqrt{9} - 25$

$\quad\quad\quad = 4 \cdot 3 - 25$

$\quad\quad\quad = 12 - 25$

$\quad\quad\quad = -13$

87. $\sqrt[3]{7}\left(2+\sqrt[3]{4}\right) = \sqrt[3]{7} \cdot 2 + \sqrt[3]{7} \cdot \sqrt[3]{4}$

$\quad\quad = 2\sqrt[3]{7} + \sqrt[3]{7 \cdot 4}$

$\quad\quad = 2\sqrt[3]{7} + \sqrt[3]{28}$

89. $\left(2\sqrt{2}+5\right)\left(4\sqrt{2}-4\right)$

$\quad = 2\sqrt{2} \cdot 4\sqrt{2} - 2\sqrt{2} \cdot 4 + 5 \cdot 4\sqrt{2} - 5 \cdot 4$

$\quad = 8\sqrt{4} - 8\sqrt{2} + 20\sqrt{2} - 20$

$\quad = 8 \cdot 2 + (-8 + 20)\sqrt{2} - 20$

$\quad = 16 + 12\sqrt{2} - 20$

$\quad = -4 + 12\sqrt{2}$

91. $4\sqrt{18} + 2\sqrt{32} = 4 \cdot \sqrt{9} \cdot \sqrt{2} + 2 \cdot \sqrt{16} \cdot \sqrt{2}$

$\quad\quad = 4 \cdot 3 \cdot \sqrt{2} + 2 \cdot 4 \cdot \sqrt{2}$

$\quad\quad = 12\sqrt{2} + 8\sqrt{2}$

$\quad\quad = (12+8)\sqrt{2}$

$\quad\quad = 20\sqrt{2}$

93. $\left(\sqrt{5}-\sqrt{3}\right)^2 = \left(\sqrt{5}\right)^2 - 2\left(\sqrt{5}\right)\left(\sqrt{3}\right) + \left(\sqrt{3}\right)^2$

$\quad\quad = \sqrt{25} - 2\sqrt{15} + \sqrt{9}$

$\quad\quad = 5 - 2\sqrt{15} + 3$

$\quad\quad = 8 - 2\sqrt{15}$

95. $3\sqrt[3]{5x^3y} + \sqrt[3]{40y}$

$\quad = 3 \cdot \sqrt[3]{x^3} \cdot \sqrt[3]{5y} + \sqrt[3]{8} \cdot \sqrt[3]{5y}$

$\quad = 3x \cdot \sqrt[3]{5y} + 2 \cdot \sqrt[3]{5y}$

$\quad = (3x+2)\sqrt[3]{5y}$

97. $\left(\sqrt{2x}-\sqrt{7y}\right)\left(\sqrt{2x}+\sqrt{7y}\right)=\left(\sqrt{2x}\right)^2-\left(\sqrt{7y}\right)^2$
$$=\sqrt{4x^2}-\sqrt{49y^2}$$
$$=2x-7y$$

99. $-\dfrac{3}{5}\cdot\left(-\dfrac{\sqrt{5}}{5}\right)-\dfrac{4}{5}\cdot\left(-\dfrac{2\sqrt{5}}{5}\right)$
$$=\dfrac{3\sqrt{5}}{25}+\dfrac{8\sqrt{5}}{25}$$
$$=\dfrac{3\sqrt{5}+8\sqrt{5}}{25}$$
$$=\dfrac{(3+8)\sqrt{5}}{25}$$
$$=\dfrac{11\sqrt{5}}{25}$$

101. (a) $(f+g)(x)=\sqrt{3x}+\sqrt{12x}$
$$=\sqrt{3x}+\sqrt{4}\cdot\sqrt{3x}$$
$$=\sqrt{3x}+2\sqrt{3x}$$
$$=(1+2)\sqrt{3x}$$
$$=3\sqrt{3x}$$

(b) Use the result from part (a).
$$(f+g)(4)=3\sqrt{3(4)}$$
$$=3\sqrt{12}$$
$$=3\sqrt{4}\cdot\sqrt{3}$$
$$=3\cdot2\sqrt{3}$$
$$=6\sqrt{3}$$

(c) $(f\cdot g)(x)=\sqrt{3x}\cdot\sqrt{12x}$
$$=\sqrt{36x^2}$$
$$=6x$$

103. Check $x=-2+\sqrt{5}$:
$$0\overset{?}{=}x^2+4x-1$$
$$0\overset{?}{=}\left(-2+\sqrt{5}\right)^2+4\left(-2+\sqrt{5}\right)-1$$
$$0\overset{?}{=}(-2)^2+2(-2)\left(\sqrt{5}\right)+\left(\sqrt{5}\right)^2-8+4\sqrt{5}-1$$
$$0\overset{?}{=}4-4\sqrt{5}+\sqrt{25}-8+4\sqrt{5}-1$$
$$0\overset{?}{=}9-9$$
$$0=0\ \text{true}$$
Substituting $-2+\sqrt{5}$ for x yields a true statement, therefore the value is a solution.
Check $x=-2-\sqrt{5}$:

$$0\overset{?}{=}x^2+4x-1$$
$$0\overset{?}{=}\left(-2-\sqrt{5}\right)^2+4\left(-2-\sqrt{5}\right)-1$$
$$0\overset{?}{=}(-2)^2-2(-2)\left(\sqrt{5}\right)+\left(\sqrt{5}\right)^2-8-4\sqrt{5}-1$$
$$0\overset{?}{=}4+4\sqrt{5}+\sqrt{25}-8-4\sqrt{5}-1$$
$$0\overset{?}{=}9-9$$
$$0=0\ \text{true}$$
Substituting $-2-\sqrt{5}$ for x yields a true statement, therefore the value is a solution.

105. $A=l\cdot w$
$$=\sqrt{162}\cdot\sqrt{72}$$
$$=\sqrt{81}\cdot\sqrt{2}\cdot\sqrt{36}\cdot\sqrt{2}$$
$$=9\sqrt{2}\cdot6\sqrt{2}$$
$$=54\sqrt{4}$$
$$=54\cdot2$$
$$=108$$
The area is 108 square units.
$$P=2l+2w$$
$$=2\left(\sqrt{162}\right)+2\left(\sqrt{72}\right)$$
$$=2\sqrt{81}\cdot\sqrt{2}+2\sqrt{36}\cdot\sqrt{2}$$
$$=18\sqrt{2}+12\sqrt{2}$$
$$=(18+12)\sqrt{2}$$
$$=30\sqrt{2}$$
The perimeter is $30\sqrt{2}$ units.

107. Area of larger triangle:
$$s=\dfrac{1}{2}(14+10+8)=\dfrac{1}{2}(32)=16$$
$$A=\sqrt{16(16-14)(16-10)(16-8)}$$
$$=\sqrt{16(2)(6)(8)}=\sqrt{16\cdot16\cdot6}$$
$$=\sqrt{16}\cdot\sqrt{16}\cdot\sqrt{6}=4\cdot4\cdot\sqrt{6}$$
$$=16\sqrt{6}$$
Area of smaller triangle:
$$s=\dfrac{1}{2}(7+5+4)=\dfrac{1}{2}(16)=8$$
$$A=\sqrt{8(8-7)(8-5)(8-4)}$$
$$=\sqrt{8(1)(3)(4)}=\sqrt{96}=\sqrt{16}\cdot\sqrt{6}$$
$$=4\sqrt{6}$$
Area of shaded region:
area of larger $-$ area of smaller
$$=16\sqrt{6}-4\sqrt{6}=(16-4)\sqrt{6}=12\sqrt{6}$$

The area of the shaded region is $12\sqrt{6}$ square units.

109. Answers may vary.

111. $\left(3a^3b\right)\left(4a^2b^4\right) = 3 \cdot 4a^{3+2}b^{1+4} = 12a^5b^5$

113. $(3y+2)(2y-1) = 3y(2y-1) + 2(2y-1)$
$= 6y^2 - 3y + 4y - 2$
$= 6y^2 + y - 2$

115. $(5w+2)(5w-2) = (5w)^2 - 2^2 = 25w^2 - 4$

Section 9.6

Preparing for Rationalizing Radical Expressions

P1. Start by finding the prime factors of 12.
$12 = 3 \cdot 4 = 3 \cdot 2 \cdot 2$
To make a perfect square, we need to have each unique prime factor occur an even number of times. To make the smallest perfect square, we want to use the smallest even number possible for each factor. In this case, the factor 2 occurs twice so no additional factors of 2 are needed. However, the factor 3 only occurs once. Therefore, we need to multiply by one more factor of 3.
The smallest perfect square that is a multiple of 12 is $12 \cdot 3 = 36$.

P2. $\sqrt{25x^2} = \sqrt{(5x)^2} = 5x$

(since $x > 0$, we do not need $|x|$)

Section 9.6 Quick Checks

1. rationalizing the denominator

2. $\sqrt{11}$

3. $\dfrac{1}{\sqrt{3}} = \dfrac{1}{\sqrt{3}} \cdot \dfrac{\sqrt{3}}{\sqrt{3}} = \dfrac{\sqrt{3}}{\sqrt{9}} = \dfrac{\sqrt{3}}{3}$

4. $\dfrac{\sqrt{5}}{\sqrt{8}} = \dfrac{\sqrt{5}}{2\sqrt{2}} = \dfrac{\sqrt{5}}{2\sqrt{2}} \cdot \dfrac{\sqrt{2}}{\sqrt{2}} = \dfrac{\sqrt{10}}{2\sqrt{4}} = \dfrac{\sqrt{10}}{2 \cdot 2} = \dfrac{\sqrt{10}}{4}$

5. $\dfrac{5}{\sqrt{10x}} = \dfrac{5}{\sqrt{10x}} \cdot \dfrac{\sqrt{10x}}{\sqrt{10x}}$
$= \dfrac{5\sqrt{10x}}{\sqrt{100x^2}}$
$= \dfrac{5\sqrt{10x}}{10x}$
$= \dfrac{\sqrt{10x}}{2x}$

6. $\dfrac{4}{\sqrt[3]{3}} = \dfrac{4}{\sqrt[3]{3}} \cdot \dfrac{\sqrt[3]{3^2}}{\sqrt[3]{3^2}} = \dfrac{4\sqrt[3]{9}}{\sqrt[3]{27}} = \dfrac{4\sqrt[3]{9}}{3}$

7. $\sqrt[3]{\dfrac{3}{20}} = \dfrac{\sqrt[3]{3}}{\sqrt[3]{20}} = \dfrac{\sqrt[3]{3}}{\sqrt[3]{2^2 \cdot 5}} \cdot \dfrac{\sqrt[3]{2 \cdot 5^2}}{\sqrt[3]{2 \cdot 5^2}}$
$= \dfrac{\sqrt[3]{150}}{\sqrt[3]{1000}}$
$= \dfrac{\sqrt[3]{150}}{10}$

8. $\dfrac{3}{\sqrt[4]{p}} = \dfrac{3}{\sqrt[4]{p}} \cdot \dfrac{\sqrt[4]{p^3}}{\sqrt[4]{p^3}} = \dfrac{3\sqrt[4]{p^3}}{p}$

9. $-2 - \sqrt{7}$

10. $\dfrac{4}{\sqrt{3}+1} = \dfrac{4}{\sqrt{3}+1} \cdot \dfrac{\sqrt{3}-1}{\sqrt{3}-1}$
$= \dfrac{4\left(\sqrt{3}-1\right)}{\left(\sqrt{3}+1\right)\left(\sqrt{3}-1\right)}$
$= \dfrac{4\left(\sqrt{3}-1\right)}{\left(\sqrt{3}\right)^2 - 1^2}$
$= \dfrac{4\left(\sqrt{3}-1\right)}{3-1}$
$= \dfrac{4\left(\sqrt{3}-1\right)}{2}$
$= 2\left(\sqrt{3}-1\right)$

11. $\dfrac{\sqrt{2}}{\sqrt{6}-\sqrt{2}} = \dfrac{\sqrt{2}}{\sqrt{6}-\sqrt{2}} \cdot \dfrac{\sqrt{6}+\sqrt{2}}{\sqrt{6}+\sqrt{2}}$

$= \dfrac{\sqrt{2}\left(\sqrt{6}+\sqrt{2}\right)}{\left(\sqrt{6}-\sqrt{2}\right)\left(\sqrt{6}+\sqrt{2}\right)}$

$= \dfrac{\sqrt{2}\left(\sqrt{6}+\sqrt{2}\right)}{\left(\sqrt{6}\right)^2 - \left(\sqrt{2}\right)^2}$

$= \dfrac{\sqrt{12}+\sqrt{4}}{6-2}$

$= \dfrac{2\sqrt{3}+2}{4}$

$= \dfrac{2\left(\sqrt{3}+1\right)}{4}$

$= \dfrac{\sqrt{3}+1}{2}$

12. $\dfrac{\sqrt{5}+4}{\sqrt{5}-\sqrt{2}} = \dfrac{\sqrt{5}+4}{\sqrt{5}-\sqrt{2}} \cdot \dfrac{\sqrt{5}+\sqrt{2}}{\sqrt{5}+\sqrt{2}}$

$= \dfrac{\left(\sqrt{5}+4\right)\left(\sqrt{5}+\sqrt{2}\right)}{\left(\sqrt{5}-\sqrt{2}\right)\left(\sqrt{5}+\sqrt{2}\right)}$

$= \dfrac{\sqrt{25}+\sqrt{10}+4\sqrt{5}+4\sqrt{2}}{\left(\sqrt{5}\right)^2 - \left(\sqrt{2}\right)^2}$

$= \dfrac{5+\sqrt{10}+4\sqrt{5}+4\sqrt{2}}{5-2}$

$= \dfrac{5+\sqrt{10}+4\sqrt{5}+4\sqrt{2}}{3}$

9.6 Exercises

13. $\dfrac{1}{\sqrt{2}} = \dfrac{1}{\sqrt{2}} \cdot \dfrac{\sqrt{2}}{\sqrt{2}} = \dfrac{\sqrt{2}}{\sqrt{4}} = \dfrac{\sqrt{2}}{2}$

15. $-\dfrac{6}{5\sqrt{3}} = -\dfrac{6}{5\sqrt{3}} \cdot \dfrac{\sqrt{3}}{\sqrt{3}}$

$= -\dfrac{6\sqrt{3}}{5\sqrt{9}}$

$= -\dfrac{6\sqrt{3}}{5\cdot 3}$

$= -\dfrac{2\sqrt{3}}{5}$

17. $\dfrac{3}{\sqrt{12}} = \dfrac{3}{2\sqrt{3}} = \dfrac{3}{2\sqrt{3}} \cdot \dfrac{\sqrt{3}}{\sqrt{3}}$

$= \dfrac{3\sqrt{3}}{2\sqrt{9}}$

$= \dfrac{3\sqrt{3}}{2\cdot 3}$

$= \dfrac{\sqrt{3}}{2}$

19. $\dfrac{\sqrt{2}}{\sqrt{6}} = \dfrac{\sqrt{2}}{\sqrt{6}} \cdot \dfrac{\sqrt{6}}{\sqrt{6}} = \dfrac{\sqrt{12}}{\sqrt{36}} = \dfrac{2\sqrt{3}}{6} = \dfrac{\sqrt{3}}{3}$

21. $\sqrt{\dfrac{2}{p}} = \dfrac{\sqrt{2}}{\sqrt{p}} = \dfrac{\sqrt{2}}{\sqrt{p}} \cdot \dfrac{\sqrt{p}}{\sqrt{p}} = \dfrac{\sqrt{2p}}{\sqrt{p^2}} = \dfrac{\sqrt{2p}}{p}$

23. $\dfrac{\sqrt{8}}{\sqrt{y^3}} = \dfrac{2\sqrt{2}}{y\sqrt{y}} = \dfrac{2\sqrt{2}}{y\sqrt{y}} \cdot \dfrac{\sqrt{y}}{\sqrt{y}}$

$= \dfrac{2\sqrt{2y}}{y\sqrt{y^2}} = \dfrac{2\sqrt{2y}}{y\cdot y}$

$= \dfrac{2\sqrt{2y}}{y^2}$

25. $\dfrac{2}{\sqrt[3]{2}} = \dfrac{2}{\sqrt[3]{2}} \cdot \dfrac{\sqrt[3]{4}}{\sqrt[3]{4}} = \dfrac{2\sqrt[3]{4}}{\sqrt[3]{8}} = \dfrac{2\sqrt[3]{4}}{2} = \sqrt[3]{4}$

27. $\sqrt[3]{\dfrac{7}{q}} = \dfrac{\sqrt[3]{7}}{\sqrt[3]{q}} = \dfrac{\sqrt[3]{7}}{\sqrt[3]{q}} \cdot \dfrac{\sqrt[3]{q^2}}{\sqrt[3]{q^2}}$

$= \dfrac{\sqrt[3]{7q^2}}{\sqrt[3]{q^3}}$

$= \dfrac{\sqrt[3]{7q^2}}{q}$

29. $\sqrt[3]{\dfrac{-3}{50}} = \dfrac{\sqrt[3]{-3}}{\sqrt[3]{50}} = \dfrac{\sqrt[3]{-3}}{\sqrt[3]{50}} \cdot \dfrac{\sqrt[3]{20}}{\sqrt[3]{20}}$

$= \dfrac{\sqrt[3]{-60}}{\sqrt[3]{1000}}$

$= -\dfrac{\sqrt[3]{60}}{10}$

31. $\dfrac{2}{\sqrt[3]{20y}} = \dfrac{2}{\sqrt[3]{20y}} \cdot \dfrac{\sqrt[3]{50y^2}}{\sqrt[3]{50y^2}}$

$\qquad = \dfrac{2\sqrt[3]{50y^2}}{\sqrt[3]{1000y^3}} = \dfrac{2\sqrt[3]{50y^2}}{10y}$

$\qquad = \dfrac{\sqrt[3]{50y^2}}{5y}$

33. $\dfrac{-4}{\sqrt[4]{3x^3}} = \dfrac{-4}{\sqrt[4]{3x^3}} \cdot \dfrac{\sqrt[4]{27x}}{\sqrt[4]{27x}}$

$\qquad = \dfrac{-4\sqrt[4]{27x}}{\sqrt[4]{81x^4}}$

$\qquad = -\dfrac{4\sqrt[4]{27x}}{3x}$

35. $\dfrac{12}{\sqrt[5]{m^3n^2}} = \dfrac{12}{\sqrt[5]{m^3n^2}} \cdot \dfrac{\sqrt[5]{m^2n^3}}{\sqrt[5]{m^2n^3}}$

$\qquad = \dfrac{12\sqrt[5]{m^2n^3}}{\sqrt[5]{m^5n^5}}$

$\qquad = \dfrac{12\sqrt[5]{m^2n^3}}{mn}$

37. $\dfrac{4}{\sqrt{6}-2} = \dfrac{4}{\sqrt{6}-2} \cdot \dfrac{\sqrt{6}+2}{\sqrt{6}+2} = \dfrac{4\left(\sqrt{6}+2\right)}{\left(\sqrt{6}\right)^2 - 2^2}$

$\qquad = \dfrac{4\left(\sqrt{6}+2\right)}{6-4} = \dfrac{4\left(\sqrt{6}+2\right)}{2}$

$\qquad = 2\left(\sqrt{6}+2\right)$

39. $\dfrac{5}{\sqrt{5}+2} = \dfrac{5}{\sqrt{5}+2} \cdot \dfrac{\sqrt{5}-2}{\sqrt{5}-2}$

$\qquad = \dfrac{5\left(\sqrt{5}-2\right)}{\left(\sqrt{5}\right)^2 - 2^2} = \dfrac{5\left(\sqrt{5}-2\right)}{5-4}$

$\qquad = 5\left(\sqrt{5}-2\right)$

41. $\dfrac{8}{\sqrt{7}-\sqrt{3}} = \dfrac{8}{\sqrt{7}-\sqrt{3}} \cdot \dfrac{\sqrt{7}+\sqrt{3}}{\sqrt{7}+\sqrt{3}}$

$\qquad = \dfrac{8\left(\sqrt{7}+\sqrt{3}\right)}{\left(\sqrt{7}\right)^2 - \left(\sqrt{3}\right)^2}$

$\qquad = \dfrac{8\left(\sqrt{7}+\sqrt{3}\right)}{7-3}$

$\qquad = \dfrac{8\left(\sqrt{7}+\sqrt{3}\right)}{4}$

$\qquad = 2\left(\sqrt{7}+\sqrt{3}\right)$

43. $\dfrac{\sqrt{2}}{\sqrt{10}-\sqrt{6}} = \dfrac{\sqrt{2}}{\sqrt{2}\sqrt{5}-\sqrt{2}\sqrt{3}}$

$\qquad = \dfrac{1}{\sqrt{5}-\sqrt{3}}$

$\qquad = \dfrac{1}{\sqrt{5}-\sqrt{3}} \cdot \dfrac{\sqrt{5}+\sqrt{3}}{\sqrt{5}+\sqrt{3}}$

$\qquad = \dfrac{\sqrt{5}+\sqrt{3}}{\left(\sqrt{5}\right)^2 - \left(\sqrt{3}\right)^2}$

$\qquad = \dfrac{\sqrt{5}+\sqrt{3}}{5-3}$

$\qquad = \dfrac{\sqrt{5}+\sqrt{3}}{2}$

45. $\dfrac{\sqrt{p}}{\sqrt{p}+\sqrt{q}} = \dfrac{\sqrt{p}}{\sqrt{p}+\sqrt{q}} \cdot \dfrac{\sqrt{p}-\sqrt{q}}{\sqrt{p}-\sqrt{q}}$

$\qquad = \dfrac{\sqrt{p}\left(\sqrt{p}-\sqrt{q}\right)}{\left(\sqrt{p}\right)^2 - \left(\sqrt{q}\right)^2}$

$\qquad = \dfrac{\sqrt{p^2}-\sqrt{pq}}{p-q}$

$\qquad = \dfrac{p-\sqrt{pq}}{p-q}$

47.
$$\frac{18}{2\sqrt{3}+3\sqrt{2}} = \frac{18}{2\sqrt{3}+3\sqrt{2}} \cdot \frac{2\sqrt{3}-3\sqrt{2}}{2\sqrt{3}-3\sqrt{2}}$$
$$= \frac{18\left(2\sqrt{3}-3\sqrt{2}\right)}{\left(2\sqrt{3}\right)^2 - \left(3\sqrt{2}\right)^2}$$
$$= \frac{18\left(2\sqrt{3}-3\sqrt{2}\right)}{12-18}$$
$$= \frac{18\left(2\sqrt{3}-3\sqrt{2}\right)}{-6}$$
$$= -3\left(2\sqrt{3}-3\sqrt{2}\right) \text{ or } 3\left(3\sqrt{2}-2\sqrt{3}\right)$$

49.
$$\frac{\sqrt{7}+3}{\sqrt{7}-3} = \frac{\sqrt{7}+3}{\sqrt{7}-3} \cdot \frac{\sqrt{7}+3}{\sqrt{7}+3}$$
$$= \frac{\left(\sqrt{7}+3\right)\left(\sqrt{7}+3\right)}{\left(\sqrt{7}-3\right)\left(\sqrt{7}+3\right)}$$
$$= \frac{\left(\sqrt{7}\right)^2 + 2\left(\sqrt{7}\right)(3) + (3)^2}{\left(\sqrt{7}\right)^2 - (3)^2}$$
$$= \frac{7+6\sqrt{7}+9}{7-9}$$
$$= \frac{16+6\sqrt{7}}{-2}$$
$$= -8-3\sqrt{7} \text{ or } -3\sqrt{7}-8$$

51.
$$\frac{\sqrt{3}-4\sqrt{2}}{2\sqrt{3}+5\sqrt{2}} = \frac{\sqrt{3}-4\sqrt{2}}{2\sqrt{3}+5\sqrt{2}} \cdot \frac{2\sqrt{3}-5\sqrt{2}}{2\sqrt{3}-5\sqrt{2}}$$
$$= \frac{\left(\sqrt{3}-4\sqrt{2}\right)\left(2\sqrt{3}-5\sqrt{2}\right)}{\left(2\sqrt{3}+5\sqrt{2}\right)\left(2\sqrt{3}-5\sqrt{2}\right)}$$
$$= \frac{2\sqrt{9}-5\sqrt{6}-8\sqrt{6}+20\sqrt{4}}{\left(2\sqrt{3}\right)^2 - \left(5\sqrt{2}\right)^2}$$
$$= \frac{6-13\sqrt{6}+40}{12-50}$$
$$= \frac{46-13\sqrt{6}}{-38}$$
$$= \frac{13\sqrt{6}-46}{38}$$

53.
$$\frac{\sqrt{p}+2}{\sqrt{p}-2} = \frac{\sqrt{p}+2}{\sqrt{p}-2} \cdot \frac{\sqrt{p}+2}{\sqrt{p}+2}$$
$$= \frac{\left(\sqrt{p}+2\right)\left(\sqrt{p}+2\right)}{\left(\sqrt{p}-2\right)\left(\sqrt{p}+2\right)}$$
$$= \frac{\left(\sqrt{p}\right)^2 + 2\left(\sqrt{p}\right)(2) + 2^2}{\left(\sqrt{p}\right)^2 - 2^2}$$
$$= \frac{p+4\sqrt{p}+4}{p-4}$$

55.
$$\frac{\sqrt{2}-3}{\sqrt{8}-\sqrt{2}} = \frac{\sqrt{2}-3}{2\sqrt{2}-\sqrt{2}}$$
$$= \frac{\sqrt{2}-3}{\sqrt{2}}$$
$$= \frac{\sqrt{2}-3}{\sqrt{2}} \cdot \frac{\sqrt{2}}{\sqrt{2}}$$
$$= \frac{\sqrt{4}-3\sqrt{2}}{\sqrt{4}}$$
$$= \frac{2-3\sqrt{2}}{2}$$

57.
$$\sqrt{3}+\frac{1}{\sqrt{3}} = \sqrt{3}+\frac{1}{\sqrt{3}} \cdot \frac{\sqrt{3}}{\sqrt{3}} = \sqrt{3}+\frac{\sqrt{3}}{3}$$
$$= \frac{3\sqrt{3}}{3}+\frac{\sqrt{3}}{3} = \frac{3\sqrt{3}+\sqrt{3}}{3}$$
$$= \frac{4\sqrt{3}}{3}$$

59.
$$\frac{\sqrt{10}}{2}-\frac{1}{\sqrt{2}} = \frac{\sqrt{10}}{2}-\frac{1}{\sqrt{2}} \cdot \frac{\sqrt{2}}{\sqrt{2}}$$
$$= \frac{\sqrt{10}}{2}-\frac{\sqrt{2}}{2}$$
$$= \frac{\sqrt{10}-\sqrt{2}}{2}$$

61. $\sqrt{\dfrac{1}{3}}+\sqrt{12}+\sqrt{75}=\dfrac{\sqrt{1}}{\sqrt{3}}+2\sqrt{3}+5\sqrt{3}$

$\qquad\qquad\qquad = \dfrac{1}{\sqrt{3}}+7\sqrt{3}$

$\qquad\qquad\qquad = \dfrac{1}{\sqrt{3}}\cdot\dfrac{\sqrt{3}}{\sqrt{3}}+7\sqrt{3}$

$\qquad\qquad\qquad = \dfrac{\sqrt{3}}{3}+7\sqrt{3}$

$\qquad\qquad\qquad = \dfrac{\sqrt{3}}{3}+\dfrac{21\sqrt{3}}{3}$

$\qquad\qquad\qquad = \dfrac{\sqrt{3}+21\sqrt{3}}{3}$

$\qquad\qquad\qquad = \dfrac{22\sqrt{3}}{3}$

63. $\dfrac{3}{\sqrt{18}}-\sqrt{\dfrac{1}{2}}=\dfrac{3}{3\sqrt{2}}-\dfrac{\sqrt{1}}{\sqrt{2}}$

$\qquad\qquad\quad = \dfrac{1}{\sqrt{2}}-\dfrac{1}{\sqrt{2}}$

$\qquad\qquad\quad = 0$

65. $\dfrac{\sqrt{3}}{\sqrt{12}}=\dfrac{\sqrt{3}}{2\sqrt{3}}=\dfrac{1}{2}$

67. $\dfrac{3}{\sqrt{72}}=\dfrac{3}{6\sqrt{2}}=\dfrac{1}{2\sqrt{2}}$

$\qquad\quad = \dfrac{1}{2\sqrt{2}}\cdot\dfrac{\sqrt{2}}{\sqrt{2}}=\dfrac{\sqrt{2}}{2\cdot 2}$

$\qquad\quad = \dfrac{\sqrt{2}}{4}$

69. $\sqrt{\dfrac{4}{3}}=\dfrac{\sqrt{4}}{\sqrt{3}}=\dfrac{2}{\sqrt{3}}=\dfrac{2}{\sqrt{3}}\cdot\dfrac{\sqrt{3}}{\sqrt{3}}=\dfrac{2\sqrt{3}}{3}$

71. $\dfrac{\sqrt{3}-3}{\sqrt{3}+3}=\dfrac{\sqrt{3}-3}{\sqrt{3}+3}\cdot\dfrac{\sqrt{3}-3}{\sqrt{3}-3}$

$\qquad\qquad = \dfrac{\left(\sqrt{3}\right)^2-2\left(\sqrt{3}\right)(3)+3^2}{\left(\sqrt{3}\right)^2-3^2}$

$\qquad\qquad = \dfrac{3-6\sqrt{3}+9}{3-9}=\dfrac{12-6\sqrt{3}}{-6}$

$\qquad\qquad = \sqrt{3}-2$

73. $\dfrac{2}{\sqrt{5}+2}=\dfrac{2}{\sqrt{5}+2}\cdot\dfrac{\sqrt{5}-2}{\sqrt{5}-2}$

$\qquad\qquad = \dfrac{2\left(\sqrt{5}-2\right)}{\left(\sqrt{5}\right)^2-2^2}=\dfrac{2\left(\sqrt{5}-2\right)}{5-4}$

$\qquad\qquad = 2\left(\sqrt{5}-2\right)$

75. $\dfrac{\sqrt{8}}{\sqrt{2}}=\dfrac{2\sqrt{2}}{\sqrt{2}}=2$

77. $\dfrac{1}{\sqrt{3}}=\dfrac{1}{\sqrt{3}}\cdot\dfrac{\sqrt{3}}{\sqrt{3}}=\dfrac{\sqrt{3}}{3}$

79. $\dfrac{1}{\sqrt[3]{12}}=\dfrac{1}{\sqrt[3]{12}}\cdot\dfrac{\sqrt[3]{18}}{\sqrt[3]{18}}=\dfrac{\sqrt[3]{18}}{\sqrt[3]{216}}=\dfrac{\sqrt[3]{18}}{6}$

(Since $12=2^2\cdot 3$, we need to multiply by $2\cdot 3^2$ to get powers of 3 on the factors.)

81. $\dfrac{1}{\sqrt{3}+5}=\dfrac{1}{\sqrt{3}+5}\cdot\dfrac{\sqrt{3}-5}{\sqrt{3}-5}$

$\qquad\qquad = \dfrac{\sqrt{3}-5}{\left(\sqrt{3}+5\right)\left(\sqrt{3}-5\right)}$

$\qquad\qquad = \dfrac{\sqrt{3}-5}{\left(\sqrt{3}\right)^2-5^2}$

$\qquad\qquad = \dfrac{\sqrt{3}-5}{3-25}$

$\qquad\qquad = \dfrac{\sqrt{3}-5}{-22}$

$\qquad\qquad = \dfrac{5-\sqrt{3}}{22}$

83. $\dfrac{1}{\sqrt{2}}\cdot\dfrac{\sqrt{3}}{2}-\dfrac{1}{\sqrt{2}}\cdot\dfrac{1}{2}=\dfrac{\sqrt{3}}{2\sqrt{2}}-\dfrac{1}{2\sqrt{2}}$

$\qquad\qquad\qquad = \dfrac{\sqrt{3}-1}{2\sqrt{2}}$

$\qquad\qquad\qquad = \dfrac{\sqrt{3}-1}{2\sqrt{2}}\cdot\dfrac{\sqrt{2}}{\sqrt{2}}$

$\qquad\qquad\qquad = \dfrac{\sqrt{6}-\sqrt{2}}{4}$

85. $\dfrac{\sqrt{2}+1}{3} = \dfrac{\sqrt{2}+1}{3}\cdot\dfrac{\sqrt{2}-1}{\sqrt{2}-1}$

$= \dfrac{\left(\sqrt{2}+1\right)\left(\sqrt{2}-1\right)}{3\left(\sqrt{2}-1\right)}$

$= \dfrac{\left(\sqrt{2}\right)^2 - 1^2}{3\sqrt{2}-3}$

$= \dfrac{2-1}{3\sqrt{2}-3}$

$= \dfrac{1}{3\sqrt{2}-3} = \dfrac{1}{3\left(\sqrt{2}-1\right)}$

87. $\dfrac{\sqrt{x}-\sqrt{h}}{\sqrt{x}} = \dfrac{\sqrt{x}-\sqrt{h}}{\sqrt{x}}\cdot\dfrac{\sqrt{x}+\sqrt{h}}{\sqrt{x}+\sqrt{h}}$

$= \dfrac{\left(\sqrt{x}-\sqrt{h}\right)\left(\sqrt{x}+\sqrt{h}\right)}{\sqrt{x}\left(\sqrt{x}+\sqrt{h}\right)}$

$= \dfrac{\left(\sqrt{x}\right)^2 - \left(\sqrt{h}\right)^2}{\sqrt{x^2}+\sqrt{xh}}$

$= \dfrac{x-h}{x+\sqrt{xh}}$

89. $\left(\dfrac{\sqrt{6}+\sqrt{2}}{4}\right)^2 \stackrel{?}{=} \left(\dfrac{\sqrt{2+\sqrt{3}}}{2}\right)^2$

$\dfrac{\left(\sqrt{6}\right)^2 + 2\cdot\sqrt{6}\cdot\sqrt{2}+\left(\sqrt{2}\right)^2}{4^2} \stackrel{?}{=} \dfrac{2+\sqrt{3}}{4}$

$\dfrac{6+2\sqrt{12}+2}{16} \stackrel{?}{=} \dfrac{2+\sqrt{3}}{4}$

$\dfrac{8+2\cdot2\sqrt{3}}{16} \stackrel{?}{=} \dfrac{2+\sqrt{3}}{4}$

$\dfrac{8+4\sqrt{3}}{16} \stackrel{?}{=} \dfrac{2+\sqrt{3}}{4}$

$\dfrac{2+\sqrt{3}}{4} = \dfrac{2+\sqrt{3}}{4}$

91. (a) $\dfrac{\sqrt{x+h}-\sqrt{x}}{h}$

$= \dfrac{\sqrt{x+h}-\sqrt{x}}{h}\cdot\dfrac{\sqrt{x+h}+\sqrt{x}}{\sqrt{x+h}+\sqrt{x}}$

$= \dfrac{\left(\sqrt{x+h}-\sqrt{x}\right)\left(\sqrt{x+h}+\sqrt{x}\right)}{h\left(\sqrt{x+h}+\sqrt{x}\right)}$

$= \dfrac{(x+h)+\sqrt{x}\sqrt{x+h}-\sqrt{x}\sqrt{x+h}-(x)}{h\left(\sqrt{x+h}+\sqrt{x}\right)}$

$= \dfrac{x+h-x}{h\left(\sqrt{x+h}+\sqrt{x}\right)}$

$= \dfrac{\cancel{h}}{\cancel{h}\left(\sqrt{x+h}+\sqrt{x}\right)}$

$= \dfrac{1}{\sqrt{x+h}+\sqrt{x}}$

(b) $\dfrac{1}{\sqrt{x+0}+\sqrt{x}} = \dfrac{1}{\sqrt{x}+\sqrt{x}} = \dfrac{1}{2\sqrt{x}}$

(c) $m = \dfrac{1}{2\sqrt{4}} = \dfrac{1}{2\cdot2} = \dfrac{1}{4}$

(d) $f(x)=\sqrt{x}$

$f(4)=\sqrt{4}=2$

The point $(4, 2)$ is on the graph.

(e) $y-y_1 = m\left(x-x_1\right)$

$y-2 = \dfrac{1}{4}(x-4)$

$y-2 = \dfrac{1}{4}x-1$

$y = \dfrac{1}{4}x+1$

(f)

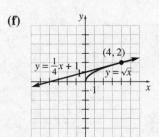

93. Answers will vary. The goal is to express a fraction containing irrational numbers in the denominator as a fraction containing only rational numbers in the denominator.

95. $g(x) = -3x + 9$

x	$y = -3x + 9$	(x, y)
-1	$y = -3(-1) + 9 = 12$	$(-1, 12)$
0	$y = -3(0) + 9 = 9$	$(0, 9)$
1	$y = -3(1) + 9 = 6$	$(1, 6)$
2	$y = -3(2) + 9 = 3$	$(2, 3)$
3	$y = -3(3) + 9 = 0$	$(3, 0)$

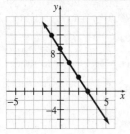

97. $F(x) = x^3$

x	$y = x^3$	(x, y)
-2	$y = (-2)^3 = -8$	$(-2, -8)$
-1	$y = (-1)^3 = -1$	$(-1, -1)$
0	$y = 0^3 = 0$	$(0, 0)$
1	$y = 1^3 = 1$	$(1, 1)$
2	$y = 2^3 = 8$	$(2, 8)$

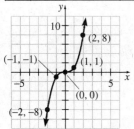

Putting the Concepts Together (Sections 9.1–9.6)

1. $-25^{1/2} = -\sqrt{25} = -5$

2. $(-64)^{-2/3} = \dfrac{1}{(-64)^{2/3}} = \dfrac{1}{\left(\sqrt[3]{-64}\right)^2} = \dfrac{1}{(-4)^2} = \dfrac{1}{16}$

3. $\sqrt[4]{3x^3} = \left(3x^3\right)^{1/4}$

4. $7z^{4/5} = 7\left(z^4\right)^{1/5} = 7\sqrt[5]{z^4}$

5. $\sqrt[3]{\sqrt{64x^3}} = \left(\left(64x^3\right)^{1/2}\right)^{1/3}$

$\qquad = \left(64x^3\right)^{1/6}$

$\qquad = \left(2^6 x^3\right)^{1/6}$

$\qquad = 2^{6/6} x^{3/6}$

$\qquad = 2x^{1/2}$ or $2\sqrt{x}$

6. $c^{1/2}\left(c^{3/2} + c^{5/2}\right) = c^{1/2} \cdot c^{3/2} + c^{1/2} \cdot c^{5/2}$

$\qquad = c^{\frac{1}{2} + \frac{3}{2}} + c^{\frac{1}{2} + \frac{5}{2}}$

$\qquad = c^{\frac{4}{2}} + c^{\frac{6}{2}}$

$\qquad = c^2 + c^3$

7. $\left(a^{2/3} b^{-1/3}\right)\left(a^{4/3} b^{-5/3}\right) = a^{\frac{2}{3} + \frac{4}{3}} b^{-\frac{1}{3} + \left(-\frac{5}{3}\right)}$

$\qquad = a^{6/3} b^{-6/3}$

$\qquad = a^2 b^{-2}$

$\qquad = \dfrac{a^2}{b^2}$

8. $\dfrac{x^{3/4}}{x^{1/8}} = x^{\frac{3}{4} - \frac{1}{8}} = x^{\frac{6}{8} - \frac{1}{8}} = x^{5/8}$ or $\sqrt[8]{x^5}$

9. $\left(x^{3/4} y^{-1/8}\right)^8 = \left(x^{3/4}\right)^8 \left(y^{-1/8}\right)^8$

$\qquad = x^{\frac{3}{4} \cdot 8} y^{-\frac{1}{8} \cdot 8}$

$\qquad = x^6 y^{-1}$

$\qquad = \dfrac{x^6}{y}$

10. $\sqrt{15a} \cdot \sqrt{2b} = \sqrt{15a \cdot 2b} = \sqrt{30ab}$

11. $\sqrt{10m^3 n^2} \cdot \sqrt{20mn} = \sqrt{10m^3 n^2 \cdot 20mn}$

$\qquad = \sqrt{200m^4 n^3}$

$\qquad = \sqrt{100m^4 n^2 \cdot 2n}$

$\qquad = \sqrt{100m^4 n^2} \cdot \sqrt{2n}$

$\qquad = 10m^2 n \sqrt{2n}$

12. $\sqrt[3]{\dfrac{-32xy^4}{4x^{-2}y}} = \sqrt[3]{-8x^3 y^3} = \sqrt[3]{(-2)^3 x^3 y^3} = -2xy$

13. $2\sqrt{108} - 3\sqrt{75} + \sqrt{48}$

$= 2 \cdot \sqrt{36 \cdot 3} - 3\sqrt{25 \cdot 3} + \sqrt{16 \cdot 3}$

$= 2\sqrt{36} \cdot \sqrt{3} - 3\sqrt{25} \cdot \sqrt{3} + \sqrt{16} \cdot \sqrt{3}$

$= 2 \cdot 6\sqrt{3} - 3 \cdot 5\sqrt{3} + 4\sqrt{3}$

$= 12\sqrt{3} - 15\sqrt{3} + 4\sqrt{3}$

$= (12 - 15 + 4)\sqrt{3}$

$= \sqrt{3}$

14. $-5b\sqrt{8b} + 7\sqrt{18b^3} = -5b\sqrt{4 \cdot 2b} + 7\sqrt{9b^2 \cdot 2b}$

$= -5b\sqrt{4} \cdot \sqrt{2b} + 7\sqrt{9b^2} \cdot \sqrt{2b}$

$= -5b \cdot 2\sqrt{2b} + 7 \cdot 3b\sqrt{2b}$

$= -10b\sqrt{2b} + 21b\sqrt{2b}$

$= (-10 + 21)b\sqrt{2b}$

$= 11b\sqrt{2b}$

15. $\sqrt[3]{16y^4} - y\sqrt[3]{2y} = \sqrt[3]{8y^3 \cdot 2y} - y\sqrt[3]{2y}$

$= \sqrt[3]{8y^3} \cdot \sqrt[3]{2y} - y\sqrt[3]{2y}$

$= 2y\sqrt[3]{2y} - y\sqrt[3]{2y}$

$= (2 - 1)y\sqrt[3]{2y}$

$= y\sqrt[3]{2y}$

16. $\left(3\sqrt{x}\right)\left(4\sqrt{x}\right) = 3 \cdot 4 \cdot \sqrt{x \cdot x}$

$= 12\sqrt{x^2}$

$= 12x$

17. $3\sqrt{x} + 4\sqrt{x} = (3 + 4)\sqrt{x} = 7\sqrt{x}$

18. $\left(2 - 3\sqrt{2}\right)\left(10 + \sqrt{2}\right)$

$= 2 \cdot 10 + 2 \cdot \sqrt{2} - 3\sqrt{2} \cdot 10 - 3\sqrt{2} \cdot \sqrt{2}$

$= 20 + 2\sqrt{2} - 30\sqrt{2} - 3 \cdot 2$

$= 20 - 28\sqrt{2} - 6$

$= 14 - 28\sqrt{2} = 14\left(1 - 2\sqrt{2}\right)$

19. $\left(4\sqrt{2} - 3\right)^2 = \left(4\sqrt{2}\right)^2 - 2\left(4\sqrt{2}\right)(3) + (3)^2$

$= 16 \cdot 2 - 24\sqrt{2} + 9$

$= 32 - 24\sqrt{2} + 9$

$= 41 - 24\sqrt{2}$

20. $\dfrac{3}{2\sqrt{32}} = \dfrac{3}{2 \cdot 4\sqrt{2}} = \dfrac{3}{8\sqrt{2}}$

$= \dfrac{3}{8\sqrt{2}} \cdot \dfrac{\sqrt{2}}{\sqrt{2}} = \dfrac{3\sqrt{2}}{8 \cdot 2}$

$= \dfrac{3\sqrt{2}}{16}$

21. $\dfrac{4}{\sqrt{3} - 8} = \dfrac{4}{\sqrt{3} - 8} \cdot \dfrac{\sqrt{3} + 8}{\sqrt{3} + 8}$

$= \dfrac{4\left(\sqrt{3} + 8\right)}{\left(\sqrt{3} - 8\right)\left(\sqrt{3} + 8\right)}$

$= \dfrac{4\left(\sqrt{3} + 8\right)}{\left(\sqrt{3}\right)^2 - 8^2}$

$= \dfrac{4\sqrt{3} + 32}{3 - 64}$

$= \dfrac{4\sqrt{3} + 32}{-61}$

$= -\dfrac{4\sqrt{3} + 32}{61} = -\dfrac{4\left(\sqrt{3} + 8\right)}{61}$

Section 9.7

Preparing for Functions Involving Radicals

P1. $\sqrt{121} = \sqrt{11^2} = 11$

P2. $\sqrt{p^2} = |p|$

P3. $f(x) = x^2 - 4$

$f(3) = (3)^2 - 4 = 9 - 4 = 5$

P4.
$-2x + 3 \geq 0$

$-2x + 3 - 3 \geq 0 - 3$

$-2x \geq -3$

$\dfrac{-2x}{-2} \leq \dfrac{-3}{-2}$

$x \leq \dfrac{3}{2}$

The solution set is $\left\{ x \mid x \leq \dfrac{3}{2} \right\}$, or $\left(-\infty, \dfrac{3}{2} \right]$ in interval notation.

P5. $f(x) = x^2 + 1$

x	$y = x^2 + 1$	(x, y)
-2	$y = (-2)^2 + 1 = 5$	$(-2, 5)$
-1	$y = (-1)^2 + 1 = 2$	$(-1, 2)$
0	$y = (0)^2 + 1 = 1$	$(0, 1)$
1	$y = (1)^2 + 1 = 2$	$(1, 2)$
2	$y = (2)^2 + 1 = 5$	$(2, 5)$

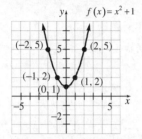

Section 9.7 Quick Checks

1. $f(x) = \sqrt{3x + 7}$

   (a) $f(3) = \sqrt{3(3) + 7} = \sqrt{9 + 7} = \sqrt{16} = 4$

   (b) $f(7) = \sqrt{3(7) + 7} = \sqrt{21 + 7} = \sqrt{28} = 2\sqrt{7}$

2. $g(x) = \sqrt[3]{2x + 7}$

   (a) $g(-4) = \sqrt[3]{2(-4) + 7} = \sqrt[3]{-8 + 7} = \sqrt[3]{-1} = -1$

   (b) $g(10) = \sqrt[3]{2(10) + 7} = \sqrt[3]{20 + 7} = \sqrt[3]{27} = 3$

3. even; odd

4. $H(x) = \sqrt{x + 6}$

The function tells us to take the square root of $x + 6$. We can only take the square root of numbers that are greater than or equal to 0. Thus, we need
$$x + 6 \geq 0$$
$$x \geq -6$$
The domain of H is $\{x \mid x \geq -6\}$ or the interval $[-6, \infty)$.

5. $g(t) = \sqrt[5]{3t - 1}$

The function tells us to take the fifth root of $3t - 1$. Since we can take the fifth root of any real number, the domain of g is all real numbers $\{t \mid t$ is any real number$\}$ or the interval $(-\infty, \infty)$.

6. $F(m) = \sqrt[4]{6 - 3m}$

The function tells us to take the fourth root of $6 - 3m$. We can only take the fourth root of numbers that are greater than or equal to 0. Thus, we need
$$6 - 3m \geq 0$$
$$-3m \geq -6$$
$$m \leq 2$$
The domain of F is $\{m \mid m \leq 2\}$ or the interval $(-\infty, 2]$.

7. $f(x) = \sqrt{x + 3}$

   (a) The function tells us to take the square root of $x + 3$. We can only take the square root of numbers that are greater than or equal to 0. Thus, we need
$$x + 3 \geq 0$$
$$x \geq -3$$
The domain of f is $\{x \mid x \geq -3\}$ or the interval $[-3, \infty)$.

   (b)

x	$f(x) = \sqrt{x + 3}$	(x, y)
-3	$f(-3)\sqrt{-3 + 3} = 0$	$(-3, 0)$
-2	$f(-2) = \sqrt{-2 + 3} = 1$	$(-2, 1)$
1	$f(1) = \sqrt{1 + 3} = 2$	$(1, 2)$
6	$f(6) = \sqrt{6 + 3} = 3$	$(6, 3)$

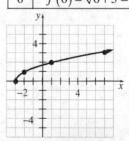

   (c) Based on the graph, the range is $\{y \mid y \geq 0\}$ or the interval $[0, \infty)$.

8. $G(x) = \sqrt[3]{x} - 1$

(a) The function tells us to take the cube root of x and then subtract 1. We can take the cube root of any real number so the domain is all real numbers, $\{x \mid x \text{ is any real number}\}$ or the interval $(-\infty, \infty)$.

(b)

x	$G(x) = \sqrt[3]{x} - 1$	(x, y)
-8	$G(-8) = \sqrt[3]{-8} - 1 = -3$	$(-8, -3)$
-1	$G(-1) = \sqrt[3]{-1} - 1 = -2$	$(-1, -2)$
0	$G(0) = \sqrt[3]{0} - 1 = -1$	$(0, -1)$
1	$G(1) = \sqrt[3]{1} - 1 = 0$	$(1, 0)$
8	$G(8) = \sqrt[3]{8} - 1 = 1$	$(8, 1)$

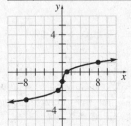

(c) Based on the graph, the range is all real numbers, $\{y \mid y \text{ is any real number}\}$ or the interval $(-\infty, \infty)$.

9.7 Exercises

9. $f(x) = \sqrt{x + 6}$

(a) $f(3) = \sqrt{3 + 6} = \sqrt{9} = 3$

(b) $f(8) = \sqrt{8 + 6} = \sqrt{14}$

(c) $f(-2) = \sqrt{-2 + 6} = \sqrt{4} = 2$

11. $g(x) = -\sqrt{2x + 3}$

(a) $g(11) = -\sqrt{2(11) + 3}$
$= -\sqrt{25}$
$= -5$

(b) $g(-1) = -\sqrt{2(-1) + 3}$
$= -\sqrt{1}$
$= -1$

(c) $g\left(\dfrac{1}{8}\right) = -\sqrt{2\left(\dfrac{1}{8}\right) + 3}$
$= -\sqrt{\dfrac{13}{4}}$
$= -\dfrac{\sqrt{13}}{\sqrt{4}}$
$= -\dfrac{\sqrt{13}}{2}$

13. $G(m) = 2\sqrt{5m - 1}$

(a) $G(1) = 2\sqrt{5(1) - 1}$
$= 2\sqrt{4}$
$= 2 \cdot 2$
$= 4$

(b) $G(5) = 2\sqrt{5(5) - 1}$
$= 2\sqrt{24}$
$= 4\sqrt{6}$

(c) $G\left(\dfrac{1}{2}\right) = 2\sqrt{5\left(\dfrac{1}{2}\right) - 1}$
$= 2\sqrt{\dfrac{3}{2}}$
$= \dfrac{2\sqrt{3}}{\sqrt{2}} \cdot \dfrac{\sqrt{2}}{\sqrt{2}}$
$= \dfrac{2\sqrt{6}}{2}$
$= \sqrt{6}$

15. $H(z) = \sqrt[3]{z + 4}$

(a) $H(4) = \sqrt[3]{4 + 4} = \sqrt[3]{8} = 2$

(b) $H(-12) = \sqrt[3]{-12 + 4} = \sqrt[3]{-8} = -2$

(c) $H(-20) = \sqrt[3]{-20 + 4}$
$= \sqrt[3]{-16}$
$= \sqrt[3]{-8} \cdot \sqrt[3]{2}$
$= -2\sqrt[3]{2}$

17. $f(x) = \sqrt{\dfrac{x-2}{x+2}}$

(a) $f(7) = \sqrt{\dfrac{7-2}{7+2}} = \sqrt{\dfrac{5}{9}} = \dfrac{\sqrt{5}}{\sqrt{9}} = \dfrac{\sqrt{5}}{3}$

(b) $f(6) = \sqrt{\dfrac{6-2}{6+2}}$

$= \sqrt{\dfrac{1}{2}}$

$= \dfrac{1}{\sqrt{2}} \cdot \dfrac{\sqrt{2}}{\sqrt{2}}$

$= \dfrac{\sqrt{2}}{2}$

(c) $f(10) = \sqrt{\dfrac{10-2}{10+2}} = \sqrt{\dfrac{2}{3}}$

$= \dfrac{\sqrt{2}}{\sqrt{3}} \cdot \dfrac{\sqrt{3}}{\sqrt{3}}$

$= \dfrac{\sqrt{6}}{3}$

19. $g(z) = \sqrt[3]{\dfrac{2z}{z-4}}$

(a) $g(-4) = \sqrt[3]{\dfrac{2(-4)}{-4-4}} = \sqrt[3]{\dfrac{-8}{-8}} = \sqrt[3]{1} = 1$

(b) $g(8) = \sqrt[3]{\dfrac{2(8)}{8-4}} = \sqrt[3]{\dfrac{16}{4}} = \sqrt[3]{4}$

(c) $g(12) = \sqrt[3]{\dfrac{2(12)}{12-4}} = \sqrt[3]{\dfrac{24}{8}} = \sqrt[3]{3}$

21. $f(x) = \sqrt{x-7}$

$x - 7 \geq 0$

$x \geq 7$

The domain of the function is $\{x \mid x \geq 7\}$ or the interval $[7, \infty)$.

23. $g(x) = \sqrt{2x+7}$

$2x + 7 \geq 0$

$2x \geq -7$

$x \geq -\dfrac{7}{2}$

The domain of the function is $\left\{x \mid x \geq -\dfrac{7}{2}\right\}$ or the interval $\left[-\dfrac{7}{2}, \infty\right)$.

25. $F(x) = \sqrt{4-3x}$

$4 - 3x \geq 0$

$-3x \geq -4$

$x \leq \dfrac{4}{3}$

The domain of the function is $\left\{x \mid x \leq \dfrac{4}{3}\right\}$ or the interval $\left(-\infty, \dfrac{4}{3}\right]$.

27. $H(z) = \sqrt[3]{2z+1}$

Since the index is odd, the domain of the function is all real numbers.

$\{z \mid z \text{ is any real number}\}$ or $(-\infty, \infty)$.

29. $W(p) = \sqrt[4]{7p-2}$

$7p - 2 \geq 0$

$7p \geq 2$

$p \geq \dfrac{2}{7}$

The domain of the function is $\left\{p \mid p \geq \dfrac{2}{7}\right\}$ or the interval $\left[\dfrac{2}{7}, \infty\right)$.

31. $g(x) = \sqrt[5]{x-3}$

Since the index is odd, the domain of the function is all real numbers.

$\{x \mid x \text{ is any real number}\}$ or $(-\infty, \infty)$.

33. $f(x) = \sqrt{\dfrac{3}{x+5}}$

The index is even so we need the radicand to be nonnegative. The numerator is positive which means we need the denominator to be positive as well.

$x + 5 > 0$

$x > -5$

The domain of the function is $\{x \mid x > -5\}$ or the interval $(-5, \infty)$.

35. $H(x) = \sqrt{\dfrac{x+3}{x-3}}$

The index is even so we need the radicand to be nonnegative. That is, we need to solve

$\dfrac{x+3}{x-3} \geq 0$

The radicand equals 0 when $x = -3$ and it is undefined when $x = 3$. We can use these two values to split up the real number line into subintervals.

			-3		3		

Interval	$(-\infty, -3)$	$(-3, 3)$	$(3, \infty)$
Num. chosen	-4	0	4
Value of radicand	$\frac{1}{7}$	-1	7
Conclusion	positive	negative	positive

Since we need the radicand to be positive or 0, the domain is $\{x \mid x \leq -3 \text{ or } x > 3\}$ or the interval $(-\infty, -3] \cup (3, \infty)$.

37. $f(x) = \sqrt{x-4}$

(a) $x - 4 \geq 0$
     $x \geq 4$

The domain is $\{x \mid x \geq 4\}$ or the interval $[4, \infty)$.

(b)

x	$f(x) = \sqrt{x-4}$	(x, y)
4	$f(4) = \sqrt{4-4} = 0$	$(4, 0)$
5	$f(5) = 1$	$(5, 1)$
8	$f(8) = 2$	$(8, 2)$
13	$f(13) = 3$	$(13, 3)$
20	$f(20) = 4$	$(20, 4)$

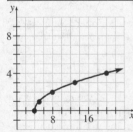

(c) Based on the graph, the range is $[0, \infty)$.

39. $g(x) = \sqrt{x+2}$

(a) $x + 2 \geq 0$
     $x \geq -2$

The domain is $\{x \mid x \geq -2\}$ or the interval $[-2, \infty)$.

(b)

x	$g(x) = \sqrt{x+2}$	(x, y)
-2	$g(-2) = \sqrt{-2+2} = 0$	$(-2, 0)$
-1	$g(-1) = 1$	$(-1, 1)$
2	$g(2) = 2$	$(2, 2)$
7	$g(7) = 3$	$(7, 3)$
14	$g(14) = 4$	$(14, 4)$

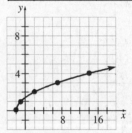

(c) Based on the graph, the range is $[0, \infty)$.

41. $G(x) = \sqrt{2-x}$

(a) $2 - x \geq 0$
     $-x \geq -2$
     $x \leq 2$

The domain is $\{x \mid x \leq 2\}$ or the interval $(-\infty, 2]$.

(b)

x	$G(x) = \sqrt{2-x}$	(x, y)
-14	$G(-14) = \sqrt{2+14} = 4$	$(-14, 4)$
-7	$G(-7) = 3$	$(-7, 3)$
-2	$G(-2) = 2$	$(-2, 2)$
1	$G(1) = 1$	$(1, 1)$
2	$G(2) = 0$	$(2, 0)$

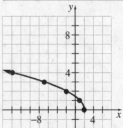

(c) Based on the graph, the range is $[0,\infty)$.

43. $f(x) = \sqrt{x} + 3$

(a) The domain is $\{x \mid x \geq 0\}$ or the interval $[0,\infty)$.

(b)

x	$f(x) = \sqrt{x} + 3$	(x,y)
0	$f(0) = \sqrt{0} + 3 = 3$	$(0,3)$
1	$f(1) = 4$	$(1,4)$
4	$f(4) = 5$	$(4,5)$
9	$f(9) = 6$	$(9,6)$
16	$f(16) = 7$	$(16,7)$

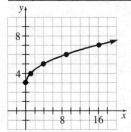

(c) Based on the graph, the range is $[3,\infty)$.

45. $g(x) = \sqrt{x} - 4$

(a) The domain is $\{x \mid x \geq 0\}$ or the interval $[0,\infty)$.

(b)

x	$g(x) = \sqrt{x} - 4$	(x,y)
0	$g(0) = \sqrt{0} - 4 = -4$	$(0,-4)$
1	$g(1) = -3$	$(1,-3)$
4	$g(4) = -2$	$(4,-2)$
9	$g(9) = -1$	$(9,-1)$
16	$g(16) = 0$	$(16,0)$

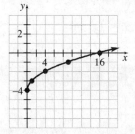

(c) Based on the graph, the range is $[-4,\infty)$.

47. $H(x) = 2\sqrt{x}$

(a) The domain is $\{x \mid x \geq 0\}$ or the interval $[0,\infty)$.

(b)

x	$H(x) = 2\sqrt{x}$	(x,y)
0	$H(0) = 2\sqrt{0} = 0$	$(0,0)$
1	$H(1) = 2$	$(1,2)$
4	$H(4) = 4$	$(4,4)$
9	$H(9) = 6$	$(9,6)$
16	$H(16) = 8$	$(16,8)$

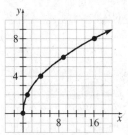

(c) Based on the graph, the range is $[0,\infty)$.

49. $f(x) = \dfrac{1}{2}\sqrt{x}$

(a) The domain is $\{x \mid x \geq 0\}$ or the interval $[0,\infty)$.

(b)

x	$f(x) = \dfrac{1}{2}\sqrt{x}$	(x,y)
0	$f(0) = \dfrac{1}{2}\sqrt{0} = 0$	$(0,0)$
1	$f(1) = \dfrac{1}{2}$	$\left(1,\dfrac{1}{2}\right)$
4	$f(4) = 1$	$(4,1)$
9	$f(9) = \dfrac{3}{2}$	$\left(9,\dfrac{3}{2}\right)$
16	$f(16) = 2$	$(16,2)$

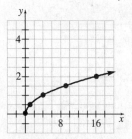

(c) Based on the graph, the range is $[0,\infty)$.

51. $G(x) = -\sqrt{x}$

(a) The domain is $\{x \mid x \geq 0\}$ or the interval $[0,\infty)$.

(b)

x	$G(x) = -\sqrt{x}$	(x,y)
0	$G(0) = -\sqrt{0} = 0$	$(0,0)$
1	$G(1) = -1$	$(1,-1)$
4	$G(4) = -2$	$(4,-2)$
9	$G(9) = -3$	$(9,-3)$
16	$G(16) = -4$	$(16,-4)$

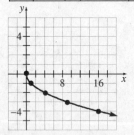

(c) Based on the graph, the range is $(-\infty,0]$.

53. $h(x) = \sqrt[3]{x+2}$

(a) The index is odd so the domain is all real numbers, $\{x \mid x \text{ is any real number}\}$ or $(-\infty,\infty)$.

(b)

x	$h(x) = \sqrt[3]{x+2}$	(x,y)
-10	$h(-10) = \sqrt[3]{-10+2} = -2$	$(-10,-2)$
-3	$h(-3) = -1$	$(-3,-1)$
-2	$h(-2) = 0$	$(-2,0)$
-1	$h(-1) = 1$	$(-1,1)$
6	$h(6) = 2$	$(6,2)$

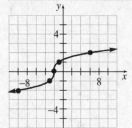

(c) Based on the graph, the range is $(-\infty,\infty)$.

55. $f(x) = \sqrt[3]{x} - 3$

(a) The index is odd so the domain is all real numbers, $\{x \mid x \text{ is any real number}\}$ or $(-\infty,\infty)$.

(b)

x	$f(x) = \sqrt[3]{x} - 3$	(x,y)
-8	$f(-8) = \sqrt[3]{-8} - 3 = -5$	$(-8,-5)$
-1	$f(-1) = -4$	$(-1,-4)$
0	$f(0) = -3$	$(0,-3)$
1	$f(1) = -2$	$(1,-2)$
8	$f(8) = -1$	$(8,-1)$

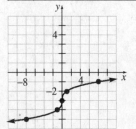

(c) Based on the graph, the range is $(-\infty,\infty)$.

57. $G(x) = 2\sqrt[3]{x}$

(a) The index is odd so the domain is all real numbers, $\{x \mid x \text{ is any real number}\}$ or $(-\infty,\infty)$.

(b)

x	$G(x) = 2\sqrt[3]{x}$	(x,y)
-8	$G(-8) = 2\sqrt[3]{-8} = -4$	$(-8,-4)$
-1	$G(-1) = -2$	$(-1,-2)$
0	$G(0) = 0$	$(0,0)$
1	$G(1) = 2$	$(1,2)$
8	$G(8) = 4$	$(8,4)$

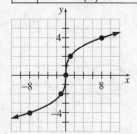

(c) Based on the graph, the range is $(-\infty,\infty)$.

59. (a) $d(0) = \sqrt{0^4 - 9(0)^2 + 25} = \sqrt{25} = 5$

The distance is 5 units.

(b) $d(1) = \sqrt{1^4 - 9(1)^2 + 25} = \sqrt{17}$

The distance is $\sqrt{17} \approx 4.123$ units.

(c) $d(5) = \sqrt{5^4 - 9(5)^2 + 25} = \sqrt{425} = 5\sqrt{17}$

The distance is $5\sqrt{17} \approx 20.616$ units.

61. (a) $x = 1$

$A(1) = 2(1) \cdot \sqrt{9 - 1^2} = 2\sqrt{8} = 4\sqrt{2}$

The area is $4\sqrt{2} \approx 5.657$ square units.

(b) $x = 2$

$A(2) = 2(2) \cdot \sqrt{9 - 2^2} = 4\sqrt{5}$

The area is $4\sqrt{5} \approx 8.944$ square units.

(c) $x = \sqrt{2}$

$A(\sqrt{2}) = 2(\sqrt{2}) \cdot \sqrt{9 - (\sqrt{2})^2}$

$= 2\sqrt{2} \cdot \sqrt{7} = 2\sqrt{14}$

The area is $2\sqrt{14} \approx 7.483$ square units.

63. The graph of $g(x) = \sqrt{x + c}$ can be obtained from the graph of $f(x) = \sqrt{x}$ by shifting the graph of $f(x)$ c units to the right (if $c < 0$) or left (if $c > 0$).

65. $\dfrac{1}{3} + \dfrac{1}{2} = \dfrac{2}{6} + \dfrac{3}{6} = \dfrac{2+3}{6} = \dfrac{5}{6}$

67. $\dfrac{1}{x} + \dfrac{3}{x+1} = \dfrac{1}{x} \cdot \dfrac{(x+1)}{(x+1)} + \dfrac{3}{x+1} \cdot \dfrac{x}{x}$

$= \dfrac{x+1}{x(x+1)} + \dfrac{3x}{x(x+1)}$

$= \dfrac{x+1+3x}{x(x+1)}$

$= \dfrac{4x+1}{x(x+1)}$

69. $\dfrac{4}{x-1} + \dfrac{3}{x+1} = \dfrac{4}{x-1} \cdot \dfrac{x+1}{x+1} + \dfrac{3}{x+1} \cdot \dfrac{x-1}{x-1}$

$= \dfrac{4(x+1) + 3(x-1)}{(x-1)(x+1)}$

$= \dfrac{4x+4+3x-3}{(x-1)(x+1)}$

$= \dfrac{7x+1}{(x-1)(x+1)}$ or $\dfrac{7x+1}{x^2-1}$

71. $f(x) = \sqrt{x-4}$

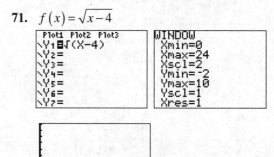

73. $g(x) = \sqrt{x+2}$

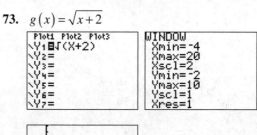

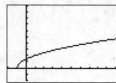

75. $G(x) = \sqrt{2-x}$

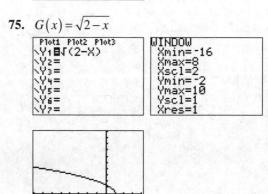

77. $f(x) = \sqrt{x} + 3$

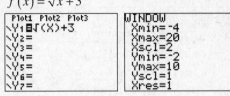

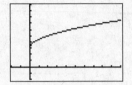

79. $g(x) = \sqrt{x} - 4$

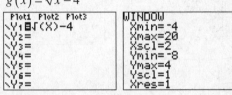

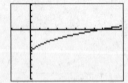

81. $H(x) = 2\sqrt{x}$

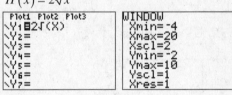

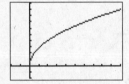

83. $f(x) = \frac{1}{2}\sqrt{x}$

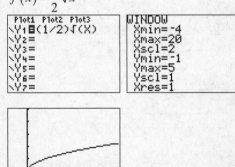

85. $G(x) = -\sqrt{x}$

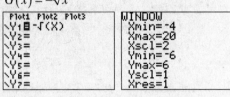

87. $h(x) = \sqrt[3]{x+2}$

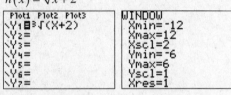

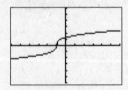

89. $f(x) = \sqrt[3]{x} - 3$

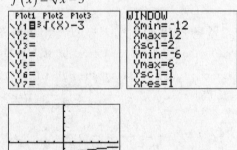

91. $G(x) = 2\sqrt[3]{x}$

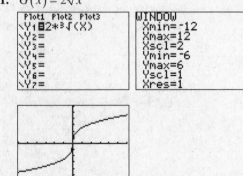

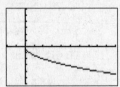

Section 9.8

Preparing for Radical Equations and Their Applications

P1. $3x - 5 = 0$
$$3x = 5$$
$$x = \frac{5}{3}$$

The solution set is $\left\{ \frac{5}{3} \right\}$.

P2. $2p^2 + 4p - 6 = 0$
$$p^2 + 2p - 3 = 0$$
$$(p+3)(p-1) = 0$$
$$p + 3 = 0 \quad \text{or} \quad p - 1 = 0$$
$$p = -3 \qquad\qquad p = 1$$

The solution set is $\{-3, 1\}$.

P3. $\left(\sqrt[3]{x-5} \right)^3 = x - 5$

Section 9.8 Quick Checks

1. radical equation

2. extraneous

3. False. The first step would be to isolate the radical by subtracting x from both sides.

4. $\sqrt{3x+1} - 4 = 0$
$$\sqrt{3x+1} = 4$$
$$\left(\sqrt{3x+1} \right)^2 = 4^2$$
$$3x + 1 = 16$$
$$3x + 1 - 1 = 16 - 1$$
$$3x = 15$$
$$\frac{3x}{3} = \frac{15}{3}$$
$$x = 5$$

Check:
$$\sqrt{3(5)+1} \overset{?}{=} 4$$
$$\sqrt{15+1} \overset{?}{=} 4$$
$$\sqrt{16} \overset{?}{=} 4$$
$$4 = 4 \, \text{T}$$

The solution set is $\{5\}$.

5. $\sqrt{2x+35} - 2 = 3$
$$\sqrt{2x+35} - 2 + 2 = 3 + 2$$
$$\sqrt{2x+35} = 5$$
$$\left(\sqrt{2x+35} \right)^2 = 5^2$$
$$2x + 35 = 25$$
$$2x + 35 - 35 = 25 - 35$$
$$2x = -10$$
$$\frac{2x}{2} = \frac{-10}{2}$$
$$x = -5$$

Check:
$$\sqrt{2(-5)+35} - 2 \overset{?}{=} 3$$
$$\sqrt{25} - 2 \overset{?}{=} 3$$
$$5 - 2 \overset{?}{=} 3$$
$$3 = 3 \, \text{T}$$

The solution set is $\{-5\}$.

6. $\sqrt{2x+3} + 8 = 6$
$$\sqrt{2x+3} + 8 - 8 = 6 - 8$$
$$\sqrt{2x+3} = -2$$
$$\left(\sqrt{2x+3} \right)^2 = (-2)^2$$
$$2x + 3 = 4$$
$$2x + 3 - 3 = 4 - 3$$
$$2x = 1$$
$$\frac{2x}{2} = \frac{1}{2}$$
$$x = \frac{1}{2}$$

Check:
$$\sqrt{2\left(\frac{1}{2}\right)+3} + 8 \overset{?}{=} 6$$
$$\sqrt{4} + 8 \overset{?}{=} 6$$
$$2 + 8 \overset{?}{=} 6$$
$$10 \neq 6$$

The solution does not check. Since there are no other possible solutions, the equation has no real solution.

7. $\sqrt{2x+1} = x-1$

$\left(\sqrt{2x+1}\right)^2 = (x-1)^2$

$2x+1 = x^2 - 2x + 1$

$x^2 - 4x = 0$

$x(x-4) = 0$

$x = 0$ or $x - 4 = 0$

$x = 4$

Check:

$\sqrt{2(0)+1} \stackrel{?}{=} 0-1$ $\sqrt{2(4)+1} \stackrel{?}{=} 4-1$

$\sqrt{1} \stackrel{?}{=} -1$ $\sqrt{9} \stackrel{?}{=} 3$

$1 \neq -1$ $3 = 3$ T

The solution set is $\{4\}$.

8. $\sqrt[3]{3x+1} - 4 = -6$

$\sqrt[3]{3x+1} - 4 + 4 = -6 + 4$

$\sqrt[3]{3x+1} = -2$

$\left(\sqrt[3]{3x+1}\right)^3 = (-2)^3$

$3x + 1 = -8$

$3x + 1 - 1 = -8 - 1$

$3x = -9$

$\dfrac{3x}{3} = \dfrac{-9}{3}$

$x = -3$

Check:

$\sqrt[3]{3(-3)+1} - 4 \stackrel{?}{=} -6$

$\sqrt[3]{-8} - 4 \stackrel{?}{=} -6$

$-2 - 4 \stackrel{?}{=} -6$

$-6 = -6$ T

The solution set is $\{-3\}$.

9. $(2x-3)^{1/3} - 7 = -4$

$(2x-3)^{1/3} - 7 + 7 = -4 + 7$

$(2x-3)^{1/3} = 3$

$\left[(2x-3)^{1/3}\right]^3 = 3^3$

$2x - 3 = 27$

$2x - 3 + 3 = 27 + 3$

$2x = 30$

$\dfrac{2x}{2} = \dfrac{30}{2}$

$x = 15$

Check:

$\left(2(15)-3\right)^{1/3} - 7 \stackrel{?}{=} -4$

$(27)^{1/3} - 7 \stackrel{?}{=} -4$

$3 - 7 \stackrel{?}{=} -4$

$-4 = -4$ T

The solution set is $\{15\}$.

10. $\sqrt[3]{m^2 + 4m + 4} = \sqrt[3]{2m+7}$

$\left(\sqrt[3]{m^2 + 4m + 4}\right)^3 = \left(\sqrt[3]{2m+7}\right)^3$

$m^2 + 4m + 4 = 2m + 7$

$m^2 + 4m + 4 - 2m - 7 = 2m + 7 - 2m - 7$

$m^2 + 2m - 3 = 0$

$(m+3)(m-1) = 0$

$m + 3 = 0$ or $m - 1 = 0$

$m = -3$ $m = 1$

Check:

$\sqrt[3]{(-3)^2 + 4(-3) + 4} \stackrel{?}{=} \sqrt[3]{2(-3)+7}$

$\sqrt[3]{9 - 12 + 4} \stackrel{?}{=} \sqrt[3]{-6+7}$

$\sqrt[3]{1} \stackrel{?}{=} \sqrt[3]{1}$

$1 = 1$ T

$\sqrt[3]{(1)^2 + 4(1) + 4} \stackrel{?}{=} \sqrt[3]{2(1)+7}$

$\sqrt[3]{1 + 4 + 4} \stackrel{?}{=} \sqrt[3]{2+7}$

$\sqrt[3]{9} = \sqrt[3]{9}$ T

The solution set is $\{-3, 1\}$.

11. $\sqrt{2x+1} - \sqrt{x+4} = 1$

$\sqrt{2x+1} = 1 + \sqrt{x+4}$

$\left(\sqrt{2x+1}\right)^2 = \left(1 + \sqrt{x+4}\right)^2$

$2x + 1 = 1 + 2\sqrt{x+4} + \left(\sqrt{x+4}\right)^2$

$2x + 1 = 1 + 2\sqrt{x+4} + x + 4$

$2x + 1 = 5 + x + 2\sqrt{x+4}$

$x - 4 = 2\sqrt{x+4}$

$(x-4)^2 = \left(2\sqrt{x+4}\right)^2$

$x^2 - 8x + 16 = 4(x+4)$

$x^2 - 8x + 16 = 4x + 16$

$x^2 - 12x = 0$

$x(x-12) = 0$

$x = 0$ or $x - 12 = 0$

$x = 12$

Check:
$$\sqrt{2(0)+1}-\sqrt{0+4} \overset{?}{=} 1$$
$$\sqrt{1}-\sqrt{4} \overset{?}{=} 1$$
$$1-2 \overset{?}{=} 1$$
$$-1 \neq 1$$
$$\sqrt{2(12)+1}-\sqrt{(12)+4} \overset{?}{=} 1$$
$$\sqrt{25}-\sqrt{16} \overset{?}{=} 1$$
$$5-4 \overset{?}{=} 1$$
$$1 = 1 \text{ T}$$

The solution set is $\{12\}$.

12. (a)
$$T = 2\pi\sqrt{\frac{L}{32}}$$
$$\frac{T}{2\pi} = \sqrt{\frac{L}{32}}$$
$$\left(\frac{T}{2\pi}\right)^2 = \left(\sqrt{\frac{L}{32}}\right)^2$$
$$\frac{T^2}{4\pi^2} = \frac{L}{32}$$
$$L = \frac{32T^2}{4\pi^2}$$
$$L = \frac{8T^2}{\pi^2}$$

(b) $L = \dfrac{8T^2}{\pi^2} = \dfrac{8(2\pi)^2}{\pi^2} = \dfrac{8 \cdot 4\pi^2}{\pi^2} = 32 \text{ feet}$

9.8 Exercises

13.
$$\sqrt{x} = 4$$
$$\left(\sqrt{x}\right)^2 = 4^2$$
$$x = 16$$
Check: $\sqrt{16} = 4$?
$$4 = 4 \text{ T}$$

The solution set is $\{16\}$.

15.
$$\sqrt{x-3} = 2$$
$$\left(\sqrt{x-3}\right)^2 = 2^2$$
$$x-3 = 4$$
$$x = 7$$
Check: $\sqrt{7-3} = 2$?
$$\sqrt{4} = 2$$?
$$2 = 2 \text{ T}$$

The solution set is $\{7\}$.

17.
$$\sqrt{2t+3} = 5$$
$$\left(\sqrt{2t+3}\right)^2 = 5^2$$
$$2t+3 = 25$$
$$2t = 22$$
$$t = 11$$

Check: $\sqrt{2(11)+3} = 5$?
$$\sqrt{25} = 5$$?
$$5 = 5 \text{ T}$$

The solution set is $\{11\}$.

19. $\sqrt{4x+3} = -2$

Since the principal square root is never negative, this equation has no real solution. Solving this equation by squaring both sides will yield an extraneous solution.
$$\sqrt{4x+3} = -2$$
$$\left(\sqrt{4x+3}\right)^2 = (-2)^2$$
$$4x+3 = 4$$
$$4x = 1$$
$$x = \frac{1}{4}$$

Check: $\sqrt{4\left(\dfrac{1}{4}\right)+3} = -2$?
$$\sqrt{4} = -2$$?
$$2 = -2 \text{ False}$$

The solution does not check so the problem has no real solution.

21.
$$\sqrt[3]{4t} = 2$$
$$\left(\sqrt[3]{4t}\right)^3 = 2^3$$
$$4t = 8$$
$$t = 2$$

Check: $\sqrt[3]{4(2)} = 2$?
$$\sqrt[3]{8} = 2$$?
$$2 = 2 \text{ T}$$

The solution set is $\{2\}$.

23. $\sqrt[3]{5q+4} = 4$

$\left(\sqrt[3]{5q+4}\right)^3 = 4^3$

$5q + 4 = 64$

$5q = 60$

$q = 12$

Check: $\sqrt[3]{5(12)+4} = 4$?

$\sqrt[3]{64} = 4$?

$4 = 4$ T

The solution set is $\{12\}$.

25. $\sqrt{y} + 3 = 8$

$\sqrt{y} = 5$

$\left(\sqrt{y}\right)^2 = 5^2$

$y = 25$

Check: $\sqrt{25} + 3 = 8$?

$5 + 3 = 8$?

$8 = 8$ T

The solution set is $\{25\}$.

27. $\sqrt{x+5} - 3 = 1$

$\sqrt{x+5} = 4$

$\left(\sqrt{x+5}\right)^2 = 4^2$

$x + 5 = 16$

$x = 11$

Check: $\sqrt{11+5} - 3 = 1$?

$\sqrt{16} - 3 = 1$?

$4 - 3 = 1$?

$1 = 1$ T

The solution set is $\{11\}$.

29. $\sqrt{2x+9} + 5 = 6$

$\sqrt{2x+9} = 1$

$\left(\sqrt{2x+9}\right)^2 = (1)^2$

$2x + 9 = 1$

$2x = -8$

$x = -4$

Check: $\sqrt{2(-4)+9} + 5 = 6$?

$\sqrt{1} + 5 = 6$?

$1 + 5 = 6$?

$6 = 6$ ✓

The solution set is $\{-4\}$.

31. $3\sqrt{x} + 5 = 8$

$3\sqrt{x} = 3$

$\sqrt{x} = 1$

$\left(\sqrt{x}\right)^2 = 1^2$

$x = 1$

Check: $3\sqrt{1} + 5 = 8$?

$3 + 5 = 8$?

$8 = 8$ T

The solution set is $\{1\}$.

33. $\sqrt{4-x} - 3 = 0$

$\sqrt{4-x} = 3$

$\left(\sqrt{4-x}\right)^2 = 3^2$

$4 - x = 9$

$-x = 5$

$x = -5$

Check: $\sqrt{4-(-5)} - 3 = 0$?

$\sqrt{9} - 3 = 0$?

$3 - 3 = 0$?

$0 = 0$ T

The solution set is $\{-5\}$.

35. $\sqrt{p} = 2p$

$\left(\sqrt{p}\right)^2 = (2p)^2$

$p = 4p^2$

$0 = 4p^2 - p$

$0 = p(4p-1)$

$p = 0$ or $4p - 1 = 0$

$p = 0$ or $p = \dfrac{1}{4}$

Check:

$\sqrt{0} = 2(0)$?

$0 = 0$ T

$$\sqrt{\frac{1}{4}} = 2\left(\frac{1}{4}\right) \ ?$$

$$\frac{1}{2} = \frac{1}{2} \ \text{T}$$

The solution set is $\left\{0, \frac{1}{4}\right\}$.

37. $\sqrt{x+6} = x$

$$\left(\sqrt{x+6}\right)^2 = x^2$$

$$x+6 = x^2$$

$$0 = x^2 - x - 6$$

$$0 = (x-3)(x+2)$$

$$x-3 = 0 \ \text{ or } \ x+2 = 0$$

$$x = 3 \ \text{ or } \ x = -2$$

Check:

$$\sqrt{3+6} = 3 \ ?$$

$$\sqrt{9} = 3 \ ?$$

$$3 = 3 \ \text{T}$$

$$\sqrt{-2+6} = -2 \ ?$$

$$\sqrt{4} = -2 \ ?$$

$$2 = -2 \ \text{False}$$

The second solution does not check. It is extraneous. The solution set is $\{3\}$.

39. $\sqrt{w} = 6 - w$

$$\left(\sqrt{w}\right)^2 = (6-w)^2$$

$$w = 36 - 12w + w^2$$

$$0 = w^2 - 13w + 36$$

$$0 = (w-4)(w-9)$$

$$w - 4 = 0 \ \text{ or } \ w - 9 = 0$$

$$w = 4 \ \text{ or } \quad w = 9$$

Check:

$$\sqrt{4} = 6 - 4 \ ?$$

$$2 = 2 \ \checkmark$$

$$\sqrt{9} = 6 - 9 \ ?$$

$$3 \neq -3 \ \text{False}$$

The second solution does not check. It is extraneous. The solution set is $\{4\}$.

41. $\sqrt{17 - 2x} + 1 = x$

$$\sqrt{17 - 2x} = x - 1$$

$$\left(\sqrt{17 - 2x}\right)^2 = (x-1)^2$$

$$17 - 2x = x^2 - 2x + 1$$

$$0 = x^2 - 16$$

$$0 = (x-4)(x+4)$$

$$x - 4 = 0 \ \text{ or } \ x + 4 = 0$$

$$x = 4 \ \text{ or } \ x = -4$$

Check:

$$\sqrt{17 - 2(4)} + 1 = 4 \ ?$$

$$\sqrt{9} + 1 = 4 \ ?$$

$$3 + 1 = 4 \ ?$$

$$4 = 4 \ \text{T}$$

$$\sqrt{17 - 2(-4)} + 1 = -4 \ ?$$

$$\sqrt{17 + 8} + 1 = -4 \ ?$$

$$\sqrt{25} + 1 = -4 \ ?$$

$$5 + 1 = -4 \ ?$$

$$6 = -4 \ \text{False}$$

The second solution is extraneous, so the solution set is $\{4\}$.

43. $\sqrt{w^2 - 11} + 5 = w + 4$

$$\sqrt{w^2 - 11} = w - 1$$

$$\left(\sqrt{w^2 - 11}\right)^2 = (w-1)^2$$

$$w^2 - 11 = w^2 - 2w + 1$$

$$0 = -2w + 12$$

$$2w = 12$$

$$w = 6$$

Check: $\sqrt{6^2 - 11} + 5 = 6 + 4 \ ?$

$$\sqrt{36 - 11} + 5 = 10 \ ?$$

$$\sqrt{25} + 5 = 10 \ ?$$

$$5 + 5 = 10 \ ?$$

$$10 = 10 \ \text{T}$$

The solution set is $\{6\}$.

45. $\sqrt{x+9} = \sqrt{2x+5}$

$$\left(\sqrt{x+9}\right)^2 = \left(\sqrt{2x+5}\right)^2$$

$$x + 9 = 2x + 5$$

$$-x = -4$$

$$x = 4$$

Check: $\sqrt{4+9} = \sqrt{2(4)+5}$?

$\sqrt{13} = \sqrt{8+5}$?

$\sqrt{13} = \sqrt{13}$ T

The solution set is $\{4\}$.

47. $\sqrt[3]{4x-3} = \sqrt[3]{2x-9}$

$\left(\sqrt[3]{4x-3}\right)^3 = \left(\sqrt[3]{2x-9}\right)^3$

$4x-3 = 2x-9$

$2x = -6$

$x = -3$

Check: $\sqrt[3]{4(-3)-3} = \sqrt[3]{2(-3)-9}$?

$\sqrt[3]{-12-3} = \sqrt[3]{-6-9}$?

$\sqrt[3]{-15} = \sqrt[3]{-15}$ T

The solution set is $\{-3\}$.

49. $\sqrt{2w^2-3w-4} = \sqrt{w^2+6w+6}$

$\left(\sqrt{2w^2-3w-4}\right)^2 = \left(\sqrt{w^2+6w+6}\right)^2$

$2w^2-3w-4 = w^2+6w+6$

$w^2-9w-10 = 0$

$(w-10)(w+1) = 0$

$w-10 = 0$ or $w+1 = 0$

$w = 10$ or $w = -1$

Check:

$\sqrt{2(10)^2-3(10)-4} = \sqrt{10^2+6(10)+6}$?

$\sqrt{200-30-4} = \sqrt{100+60+6}$?

$\sqrt{166} = \sqrt{166}$ T

$\sqrt{2(-1)^2-3(-1)-4} = \sqrt{(-1)^2+6(-1)+6}$?

$\sqrt{2+3-4} = \sqrt{1-6+6}$?

$\sqrt{1} = \sqrt{1}$

$1 = 1$ T

The solution set is $\{-1,10\}$.

51. $\sqrt{3w+4} = 2+\sqrt{w}$

$\left(\sqrt{3w+4}\right)^2 = \left(2+\sqrt{w}\right)^2$

$3w+4 = 2^2 + 2\cdot2\sqrt{w} + \left(\sqrt{w}\right)^2$

$3w+4 = 4+4\sqrt{w}+w$

$2w = 4\sqrt{w}$

$w = 2\sqrt{w}$

$w^2 = \left(2\sqrt{w}\right)^2$

$w^2 = 4w$

$w^2-4w = 0$

$w(w-4) = 0$

$w = 0$ or $w-4 = 0$

$w = 0$ or $w = 4$

Check:

$\sqrt{3(0)+4} = 2+\sqrt{0}$?

$\sqrt{4} = 2$?

$2 = 2$ T

$\sqrt{3(4)+4} = 2+\sqrt{4}$?

$\sqrt{16} = 2+2$?

$4 = 4$ T

The solution set is $\{0,4\}$.

53. $\sqrt{x+1}-\sqrt{x-2} = 1$

$\sqrt{x+1} = \sqrt{x-2}+1$

$\left(\sqrt{x+1}\right)^2 = \left(\sqrt{x-2}+1\right)^2$

$x+1 = (x-2)+2\sqrt{x-2}+1$

$x+1 = x-1+2\sqrt{x-2}$

$2 = 2\sqrt{x-2}$

$1 = \sqrt{x-2}$

$1^2 = \left(\sqrt{x-2}\right)^2$

$1 = x-2$

$3 = x$

Check: $\sqrt{3+1}-\sqrt{3-2} = 1$?

$\sqrt{4}-\sqrt{1} = 1$?

$2-1 = 1$?

$1 = 1$ T

The solution set is $\{3\}$.

55. $\sqrt{2x+6} - \sqrt{x-1} = 2$

$\sqrt{2x+6} = \sqrt{x-1} + 2$

$\left(\sqrt{2x+6}\right)^2 = \left(\sqrt{x-1}+2\right)^2$

$2x+6 = (x-1) + 2 \cdot 2\sqrt{x-1} + 2^2$

$2x+6 = x-1 + 4\sqrt{x-1} + 4$

$x+3 = 4\sqrt{x-1}$

$(x+3)^2 = \left(4\sqrt{x-1}\right)^2$

$x^2 + 6x + 9 = 16(x-1)$

$x^2 + 6x + 9 = 16x - 16$

$x^2 - 10x + 25 = 0$

$(x-5)^2 = 0$

$x - 5 = 0$

$x = 5$

Check: $\sqrt{2(5)+6} - \sqrt{5-1} = 2$?

$\sqrt{16} - \sqrt{4} = 2$?

$4 - 2 = 2$?

$2 = 2$ T

The solution set is $\{5\}$.

57. $\sqrt{2x+5} - \sqrt{x-1} = 2$

$\sqrt{2x+5} = \sqrt{x-1} + 2$

$\left(\sqrt{2x+5}\right)^2 = \left(\sqrt{x-1}+2\right)^2$

$2x+5 = (x-1) + 2 \cdot 2\sqrt{x-1} + 2^2$

$2x+5 = x-1 + 4\sqrt{x-1} + 4$

$x+2 = 4\sqrt{x-1}$

$(x+2)^2 = \left(4\sqrt{x-1}\right)^2$

$x^2 + 4x + 4 = 16(x-1)$

$x^2 + 4x + 4 = 16x - 16$

$x^2 - 12x + 20 = 0$

$(x-10)(x-2) = 0$

$x - 10 = 0$ or $x - 2 = 0$

$x = 10$ or $x = 2$

Check:

$\sqrt{2(10)+5} - \sqrt{10-1} = 2$?

$\sqrt{25} - \sqrt{9} = 2$?

$5 - 3 = 2$?

$2 = 2$ T

$\sqrt{2(2)+5} - \sqrt{2-1} = 2$?

$\sqrt{9} - \sqrt{1} = 2$?

$3 - 1 = 2$?

$2 = 2$ T

The solution set is $\{2, 10\}$.

59. $(2x+3)^{1/2} = 3$

$\left((2x+3)^{1/2}\right)^2 = 3^2$

$2x + 3 = 9$

$2x = 6$

$x = 3$

Check: $(2(3)+3)^{1/2} = 3$?

$(9)^{1/2} = 3$?

$3 = 3$ T

The solution set is $\{3\}$.

61. $(6x-1)^{1/4} = (2x+15)^{1/4}$

$\left((6x-1)^{1/4}\right)^4 = \left((2x+15)^{1/4}\right)^4$

$6x - 1 = 2x + 15$

$4x = 16$

$x = 4$

Check: $(6(4)-1)^{1/4} = (2(4)+15)^{1/4}$?

$(23)^{1/4} = (23)^{1/4}$ T

The solution set is $\{4\}$.

63. $(x+3)^{1/2} - (x-5)^{1/2} = 2$

$(x+3)^{1/2} = (x-5)^{1/2} + 2$

$\left((x+3)^{1/2}\right)^2 = \left((x-5)^{1/2}+2\right)^2$

$x+3 = (x-5) + 2 \cdot 2(x-5)^{1/2} + 2^2$

$x+3 = x-5 + 4(x-5)^{1/2} + 4$

$4 = 4(x-5)^{1/2}$

$1 = (x-5)^{1/2}$

$1^2 = \left((x-5)^{1/2}\right)^2$

$1 = x - 5$

$6 = x$

Check: $(6+3)^{1/2} - (6-5)^{1/2} = 2$?

$$9^{1/2} - 1^{1/2} = 2 \text{ ?}$$
$$3 - 1 = 2 \text{ ?}$$
$$2 = 2 \text{ T}$$

The solution set is $\{6\}$.

65. $\quad A = P\sqrt{1+r}$

$$\frac{A}{P} = \sqrt{1+r}$$
$$\left(\frac{A}{P}\right)^2 = \left(\sqrt{1+r}\right)^2$$
$$\frac{A^2}{P^2} = 1 + r$$
$$\frac{A^2}{P^2} - 1 = r \quad \text{or} \quad r = \frac{A^2}{P^2} - 1 = \frac{A^2 - P^2}{P^2}$$

67. $\quad r = \sqrt[3]{\frac{3V}{4\pi}}$

$$r^3 = \left(\sqrt[3]{\frac{3V}{4\pi}}\right)^3$$
$$r^3 = \frac{3V}{4\pi}$$
$$4\pi r^3 = 3V$$
$$\frac{4\pi r^3}{3} = V \quad \text{or} \quad V = \frac{4}{3}\pi r^3$$

69. $\quad r = \sqrt{\frac{4F\pi\varepsilon_0}{q_1 q_2}}$

$$r^2 = \left(\sqrt{\frac{4F\pi\varepsilon_0}{q_1 q_2}}\right)^2$$
$$r^2 = \frac{4F\pi\varepsilon_0}{q_1 q_2}$$
$$q_1 q_2 r^2 = 4F\pi\varepsilon_0$$
$$\frac{q_1 q_2 r^2}{4\pi\varepsilon_0} = F \quad \text{or} \quad F = \frac{q_1 q_2 r^2}{4\pi\varepsilon_0}$$

71. $\sqrt{5p-3} + 7 = 3$

$$\sqrt{5p-3} = -4$$

At this point we see that there is no real solution. Since the principal square root is never negative, $\sqrt{5p-3} = -4$ has no real solution. Solving the usual way yields the same result.

$$\left(\sqrt{5p-3}\right)^2 = (-4)^2$$
$$5p - 3 = 16$$
$$5p = 19$$
$$p = \frac{19}{5}$$

Check: $\sqrt{5\left(\dfrac{19}{5}\right) - 3} + 7 = 3$?

$$\sqrt{19-3} + 7 = 3 \text{ ?}$$
$$\sqrt{16} + 7 = 3 \text{ ?}$$
$$4 + 7 = 3 \text{ ?}$$
$$11 = 3 \text{ False}$$

The solution does not check, so the equation has no real solutions.

73. $\quad \sqrt{x+12} = x$

$$\left(\sqrt{x+12}\right)^2 = x^2$$
$$x + 12 = x^2$$
$$0 = x^2 - x - 12$$
$$0 = (x-4)(x+3)$$
$$x - 4 = 0 \quad \text{or} \quad x + 3 = 0$$
$$x = 4 \qquad\qquad x = -3$$

Check:

$$\sqrt{4+12} = 4 \text{ ?} \qquad \sqrt{-3+12} = -3 \text{ ?}$$
$$\sqrt{16} = 4 \text{ ?} \qquad\quad \sqrt{9} = -3 \text{ ?}$$
$$4 = 4 \text{ T} \qquad\qquad 3 = -3 \text{ False}$$

The second solution does not check. The solution set is $\{4\}$.

75. $\quad \sqrt{2p+12} = 4$

$$\left(\sqrt{2p+12}\right)^2 = 4^2$$
$$2p + 12 = 16$$
$$2p = 4$$
$$p = 2$$

Check: $\sqrt{2(2)+12} = 4$?

$$\sqrt{16} = 4 \text{ ?}$$
$$4 = 4 \text{ T}$$

The solution set is $\{2\}$.

77. $\sqrt[4]{x+7} = 2$

$\left(\sqrt[4]{x+7}\right)^4 = 2^4$

$x+7 = 16$

$x = 9$

Check: $\sqrt[4]{9+7} = 2$?

$\sqrt[4]{16} = 2$?

$2 = 2 \text{ T}$

The solution set is $\{9\}$.

79. $(3x+1)^{1/3} + 2 = 0$

$(3x+1)^{1/3} = -2$

$\left((3x+1)^{1/3}\right)^3 = (-2)^3$

$3x+1 = -8$

$3x = -9$

$x = -3$

Check: $(3(-3)+1)^{1/3} + 2 = 0$?

$(-8)^{1/3} + 2 = 0$?

$-2 + 2 = 0$?

$0 = 0 \text{ T}$

The solution set is $\{-3\}$.

81. $\sqrt{10-x} - x = 10$

$\sqrt{10-x} = x+10$

$\left(\sqrt{10-x}\right)^2 = (x+10)^2$

$10 - x = x^2 + 20x + 100$

$0 = x^2 + 21x + 90$

$0 = (x+6)(x+15)$

$x+6 = 0 \quad \text{or} \quad x+15 = 0$

$x = -6 \quad \text{or} \quad x = -15$

Check:

$\sqrt{10-(-6)} - (-6) = 10$?

$\sqrt{16} + 6 = 10$?

$4 + 6 = 10$?

$10 = 10 \checkmark$

$\sqrt{10-(-15)} - (-15) = 10$?

$\sqrt{25} + 15 = 10$?

$5 + 15 = 10$?

$20 \neq 10 \text{ False}$

The second solution does not check. It is extraneous. The solution set is $\{-6\}$.

83. $\sqrt{2x+5} = \sqrt{3x-4}$

$\left(\sqrt{2x+5}\right)^2 = \left(\sqrt{3x-4}\right)^2$

$2x+5 = 3x-4$

$-x+5 = -4$

$-x = -9$

$x = 9$

Check: $\sqrt{2(9)+5} = \sqrt{3(9)-4}$?

$\sqrt{23} = \sqrt{23} \text{ T}$

The solution set is $\{9\}$.

85. $\sqrt{x-1} + \sqrt{x+4} = 5$

$\sqrt{x-1} = 5 - \sqrt{x+4}$

$\left(\sqrt{x-1}\right)^2 = \left(5 - \sqrt{x+4}\right)^2$

$x-1 = 5^2 - 2 \cdot 5\sqrt{x+4} + (x+4)$

$x-1 = 25 - 10\sqrt{x+4} + x + 4$

$-30 = -10\sqrt{x+4}$

$3 = \sqrt{x+4}$

$3^2 = \left(\sqrt{x+4}\right)^2$

$9 = x+4$

$5 = x$

Check: $\sqrt{5-1} + \sqrt{5+4} = 5$?

$\sqrt{4} + \sqrt{9} = 5$?

$2 + 3 = 5$?

$5 = 5 \text{ T}$

The solution set is $\{5\}$.

87. $\sqrt{x^2} = x+4$

$\left(\sqrt{x^2}\right)^2 = (x+4)^2$

$x^2 = x^2 + 8x + 16$

$0 = 8x + 16$

$8x = -16$

$x = -2$

Check:

$\sqrt{(-2)^2} = -2+4$?

$\sqrt{4} = 2$?

$2 = 2 \checkmark$

The solution set is $\{-2\}$.

$$\sqrt{x^2} = x+4$$
$$|x| = x+4$$
$$x = -(x+4) \quad \text{or} \quad x = x+4$$
$$x = -x-4 \quad \text{or} \quad 0 = 4 \quad \text{False}$$
$$2x = -4$$
$$x = -2$$
Preference will vary.

89. $f(x) = \sqrt{x-2}$

(a)
$$f(x) = 0$$
$$\sqrt{x-2} = 0$$
$$\left(\sqrt{x-2}\right)^2 = 0^2$$
$$x-2 = 0$$
$$x = 2$$
The point $(2,0)$ is on the graph of f.

(b)
$$f(x) = 1$$
$$\sqrt{x-2} = 1$$
$$\left(\sqrt{x-2}\right)^2 = 1^2$$
$$x-2 = 1$$
$$x = 3$$
The point $(3,1)$ is on the graph of f.

(c)
$$f(x) = 2$$
$$\sqrt{x-2} = 2$$
$$\left(\sqrt{x-2}\right)^2 = 2^2$$
$$x-2 = 4$$
$$x = 6$$
The point $(6,2)$ is on the graph of f.

(d) The points $(2,0)$, $(3,1)$, and $(6,2)$ are on the graph.

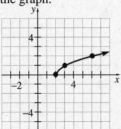

(e) The equation $f(x) = -1$ has no solution because the graph of the function does not go below the x-axis. Therefore, the value of the function will never be negative.

91. (a)
$$\sqrt{4^2 + (y-2)^2} = 5$$
$$\left(\sqrt{4^2 + (y-2)^2}\right)^2 = 5^2$$
$$4^2 + (y-2)^2 = 5^2$$
$$16 + (y-2)^2 = 25$$
$$(y-2)^2 = 9$$
$$\sqrt{(y-2)^2} = \sqrt{9}$$
$$|y-2| = 3$$
$$y-2 = 3 \quad \text{or} \quad y-2 = -3$$
$$y = 5 \quad \text{or} \quad y = -1$$

(b) The points are $(3,2)$, $(-1,-1)$, and $(-1,5)$.

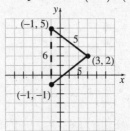

The figure is an isosceles triangle whose common side has a length of 5 units. The base has a length of 6 units. If the triangle were bisected from the vertex to the base, two 3-4-5 right triangles would be formed.

93. (a)
$$\sqrt[3]{\frac{t}{2}} = 1$$
$$\left(\sqrt[3]{\frac{t}{2}}\right)^3 = 1^3$$
$$\frac{t}{2} = 1$$
$$t = 2$$
Annual revenue will be \$1 million after 2 years.

(b)
$$\sqrt[3]{\frac{t}{2}} = 2$$
$$\left(\sqrt[3]{\frac{t}{2}}\right)^3 = 2^3$$
$$\frac{t}{2} = 8$$
$$t = 16$$
Annual revenue will be \$2 million after 16 years.

95. $R(t) = 26 \cdot \sqrt[10]{t}$

(a) $26 \cdot \sqrt[10]{t} = 39$

$\sqrt[10]{t} = \dfrac{39}{26}$

$\left(\sqrt[10]{t}\right)^{10} = \left(\dfrac{39}{26}\right)^{10}$

$t \approx 57.67$

The plural birth rate is predicted to be 39 in the year 2053.

(b) $26 \cdot \sqrt[10]{t} = 36$

$\sqrt[10]{t} = \dfrac{18}{13}$

$\left(\sqrt[10]{t}\right)^{10} = \left(\dfrac{18}{13}\right)^{10}$

$t \approx 25.9$

The plural birth rate is predicted to be 36 in the year 2021.

97. $\sqrt{3\sqrt{x+1}} = \sqrt{2x+3}$

$\left(\sqrt{3\sqrt{x+1}}\right)^2 = \left(\sqrt{2x+3}\right)^2$

$3\sqrt{x+1} = 2x+3$

$\left(3\sqrt{x+1}\right)^2 = (2x+3)^2$

$9(x+1) = (2x)^2 + 2 \cdot 3 \cdot 2x + 3^2$

$9x+9 = 4x^2 + 12x + 9$

$0 = 4x^2 + 3x$

$0 = x(4x+3)$

$x = 0$ or $4x+3 = 0$

$x = 0$ or $4x = -3$

$x = 0$ or $x = -\dfrac{3}{4}$

Check:

$\sqrt{3\sqrt{0+1}} = \sqrt{2(0)+3}$?

$\sqrt{3\sqrt{1}} = \sqrt{3}$?

$\sqrt{3} = \sqrt{3}$ T

$\sqrt{3\sqrt{-\dfrac{3}{4}+1}} = \sqrt{2\left(-\dfrac{3}{4}\right)+3}$?

$\sqrt{3 \cdot \dfrac{1}{2}} = \sqrt{-\dfrac{3}{2}+3}$?

$\sqrt{\dfrac{3}{2}} = \sqrt{\dfrac{3}{2}}$ T

The solution set is $\left\{-\dfrac{3}{4}, 0\right\}$.

99. Answers will vary. It is necessary to check solutions to radical equations because manipulations of the equation may have introduced extraneous solutions.

101. Answers will vary. Radical equations with an even index may have extraneous solutions because often the domain is not all real numbers. When we clear the radical, we lose this restriction. Radical equations with an odd equation do not have this problem, because they do not have domain restrictions.

103. 0, -4, and 12 are the integers.

105. $\sqrt{2^3}$, π, and $\sqrt[3]{-4}$ are irrational numbers.

107. Answers will vary.
A rational number is a real number that can be expressed as the ratio of two integers. An irrational number is a real number that cannot be written as the ratio of two integers. In decimal form, a rational number will terminate or repeat (e.g. 3.7 or $1.\overline{3}$) while an irrational number will be non-terminating and non-repeating (e.g. $\sqrt{2} = 1.414213562...$).

109.

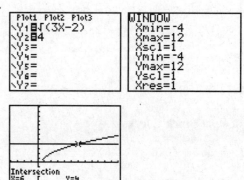

111.

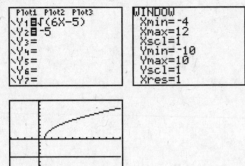

Section 9.9

Preparing for The Complex Number System

P1. **(a)** 8 is the only natural number in the set.

(b) 8 and 0 are the whole numbers in the set.

(c) -23, $-\dfrac{12}{3}$, 0, and 8 are the integers in the set.

(d) -23, $-\dfrac{12}{3}$, $-\dfrac{1}{3}$, 0, $1.\overline{26}$, and 8 are the rational numbers in the set.

(e) $\sqrt{2}$ is the only irrational number in the set.

(f) -23, $-\dfrac{12}{3}$, $-\dfrac{1}{3}$, 0, $1.\overline{26}$, $\sqrt{2}$, and 8 are the real numbers in the set.

P2. $3x(4x-3) = 3x \cdot 4x - 3x \cdot 3 = 12x^2 - 9x$

P3. $(z+4)(3z-2) = z \cdot 3z + 4 \cdot 3z - 2 \cdot z - 4 \cdot 2$
$$= 3z^2 + 12z - 2z - 8$$
$$= 3z^2 + 10z - 8$$

P4. $(2y+5)(2y-5) = (2y)^2 - 5^2 = 4y^2 - 25$

Section 9.9 Quick Checks

1. imaginary unit

2. pure imaginary number

3. $\sqrt{N}\,i$

4. True

5. $\sqrt{-36} = \sqrt{36 \cdot (-1)} = \sqrt{36} \cdot \sqrt{-1} = 6i$

6. $\sqrt{-5} = \sqrt{5 \cdot (-1)} = \sqrt{5} \cdot \sqrt{-1} = \sqrt{5}\,i$

7. $\sqrt{-12} = \sqrt{12 \cdot (-1)} = \sqrt{12} \cdot \sqrt{-1} = 2\sqrt{3}\,i$

8. $4 + \sqrt{-100} = 4 + \sqrt{100 \cdot (-1)}$
$$= 4 + \sqrt{100} \cdot \sqrt{-1}$$
$$= 4 + 10i$$

9. $-2 - \sqrt{-8} = -2 - \sqrt{8 \cdot (-1)}$
$$= -2 - \sqrt{8} \cdot \sqrt{-1}$$
$$= -2 - 2\sqrt{2}\,i$$

10. $\dfrac{6 - \sqrt{-72}}{3} = \dfrac{6 - \sqrt{72 \cdot (-1)}}{3}$
$$= \dfrac{6 - \sqrt{72} \cdot \sqrt{-1}}{3}$$
$$= \dfrac{6 - 6\sqrt{2}\,i}{3}$$
$$= \dfrac{3\left(2 - 2\sqrt{2}\,i\right)}{3}$$
$$= 2 - 2\sqrt{2}\,i$$

11. $(4+6i)+(-3+5i) = 4+6i-3+5i$
$$= (4-3)+(6+5)i$$
$$= 1+11i$$

12. $(4-2i)-(-2+7i) = 4-2i+2-7i$
$$= (4+2)+(-2-7)i$$
$$= 6-9i$$

13. $\left(4-\sqrt{-4}\right)+\left(-7+\sqrt{-9}\right) = (4-2i)+(-7+3i)$
$$= 4-2i-7+3i$$
$$= (4-7)+(-2+3)i$$
$$= -3+i$$

14. $3i(5-4i) = 3i \cdot 5 - 3i \cdot 4i$
$$= 15i - 12i^2$$
$$= 15i - 12(-1)$$
$$= 12 + 15i$$

15. $(-2+5i)(4-2i) = -2 \cdot 4 - 2 \cdot (-2i) + 5i \cdot 4 - 5i \cdot 2i$
$$= -8 + 4i + 20i - 10i^2$$
$$= -8 + 24i - 10(-1)$$
$$= 2 + 24i$$

16. $\sqrt{-9} \cdot \sqrt{-36} = 3i \cdot 6i = 18i^2 = -18$

17. $\left(2 + \sqrt{-36}\right)\left(4 - \sqrt{-25}\right)$
$$= (2 + 6i)(4 - 5i)$$
$$= 2 \cdot 4 - 2 \cdot 5i + 6i \cdot 4 - 6i \cdot 5i$$
$$= 8 - 10i + 24i - 30i^2$$
$$= 8 + 14i - 30(-1)$$
$$= 38 + 14i$$

18. $-3 - 5i$

19. $(3 - 8i)(3 + 8i) = 3^2 - (8i)^2$
$$= 9 - 64i^2$$
$$= 9 - 64(-1)$$
$$= 73$$

20. $(-2 + 5i)(-2 - 5i) = (-2)^2 - (5i)^2$
$$= 4 - 25i^2$$
$$= 4 - 25(-1)$$
$$= 29$$

21. $\dfrac{-4 + i}{3i} = \dfrac{-4 + i}{3i} \cdot \dfrac{3i}{3i} = \dfrac{-12i + 3i^2}{9i^2}$
$$= \dfrac{-12i - 3}{-9} = \dfrac{-3}{-9} + \dfrac{-12}{-9}i$$
$$= \dfrac{1}{3} + \dfrac{4}{3}i$$

22. $\dfrac{4 + 3i}{1 - 3i} = \dfrac{4 + 3i}{1 - 3i} \cdot \dfrac{1 + 3i}{1 + 3i}$
$$= \dfrac{(4 + 3i)(1 + 3i)}{(1 - 3i)(1 + 3i)}$$
$$= \dfrac{4 + 12i + 3i + 9i^2}{1 - 9i^2}$$
$$= \dfrac{4 + 15i - 9}{1 + 9}$$
$$= \dfrac{-5 + 15i}{10}$$
$$= -\dfrac{5}{10} + \dfrac{15}{10}i$$
$$= -\dfrac{1}{2} + \dfrac{3}{2}i$$

23. $i^{43} = i^{40} \cdot i^3$
$$= \left(i^4\right)^{10} \cdot i^3$$
$$= (1)^{10} \cdot (-i)$$
$$= -i$$

24. $i^{98} = i^{96} \cdot i^2$
$$= \left(i^4\right)^{24} \cdot i^2$$
$$= (1)^{24} \cdot (-1)$$
$$= -1$$

9.9 Exercises

25. $\sqrt{-4} = \sqrt{-1} \cdot \sqrt{4} = i \cdot 2 = 2i$

27. $-\sqrt{-81} = -\sqrt{-1} \cdot \sqrt{81} = -i \cdot 9 = -9i$

29. $\sqrt{-45} = \sqrt{-1} \cdot \sqrt{9} \cdot \sqrt{5} = i \cdot 3 \cdot \sqrt{5} = 3\sqrt{5}\,i$

31. $\sqrt{-300} = \sqrt{-1} \cdot \sqrt{3} \cdot \sqrt{100} = i\sqrt{3} \cdot 10 = 10\sqrt{3}\,i$

33. $\sqrt{-7} = \sqrt{-1} \cdot \sqrt{7} = i\sqrt{7} = \sqrt{7}\,i$

35. $5 + \sqrt{-49} = 5 + \sqrt{-1} \cdot \sqrt{49}$
$$= 5 + i7$$
$$= 5 + 7i$$

37. $-2 - \sqrt{-28} = -2 - \sqrt{-1} \cdot \sqrt{4} \cdot \sqrt{7}$
$$= -2 - i \cdot 2\sqrt{7}$$
$$= -2 - 2\sqrt{7}\,i$$

39. $\dfrac{4+\sqrt{-4}}{2} = \dfrac{4+\sqrt{-1}\cdot\sqrt{4}}{2}$

$\qquad\qquad = \dfrac{4+i\cdot 2}{2}$

$\qquad\qquad = 2+i$

41. $\dfrac{4+\sqrt{-8}}{12} = \dfrac{4+\sqrt{-1}\cdot\sqrt{4}\cdot\sqrt{2}}{12}$

$\qquad\qquad = \dfrac{4+i\cdot 2\sqrt{2}}{12}$

$\qquad\qquad = \dfrac{2+i\sqrt{2}}{6}$

$\qquad\qquad = \dfrac{1}{3}+\dfrac{\sqrt{2}}{6}i$

43. $(4+5i)+(2-7i) = 4+5i+2-7i$

$\qquad\qquad = (4+2)+(5-7)i$

$\qquad\qquad = 6-2i$

45. $(4+i)-(8-5i) = 4+i-8+5i$

$\qquad\qquad = (4-8)+(1+5)i$

$\qquad\qquad = -4+6i$

47. $\left(4-\sqrt{-4}\right)-\left(2+\sqrt{-9}\right) = (4-2i)-(2+3i)$

$\qquad\qquad\qquad = 4-2i-2-3i$

$\qquad\qquad\qquad = (4-2)+(-2-3)i$

$\qquad\qquad\qquad = 2-5i$

49. $\left(-2+\sqrt{-18}\right)+\left(5-\sqrt{-50}\right)$

$\qquad = \left(-2+3\sqrt{2}\,i\right)+\left(5-5\sqrt{2}\,i\right)$

$\qquad = -2+3\sqrt{2}\,i+5-5\sqrt{2}\,i$

$\qquad = (-2+5)+\left(3\sqrt{2}-5\sqrt{2}\right)i$

$\qquad = 3-2\sqrt{2}\,i$

51. $6i(2-4i) = 6i\cdot 2-6i\cdot 4i$

$\qquad\qquad = 12i-24i^2$

$\qquad\qquad = 12i-24(-1)$

$\qquad\qquad = 24+12i$

53. $-\dfrac{1}{2}i(4-10i) = -\dfrac{1}{2}i\cdot 4+\dfrac{1}{2}i\cdot 10i$

$\qquad\qquad\qquad = -2i+5i^2$

$\qquad\qquad\qquad = -2i+5(-1)$

$\qquad\qquad\qquad = -5-2i$

55. $(2+i)(4+3i) = 2\cdot 4+2\cdot 3i+i\cdot 4+i\cdot 3i$

$\qquad\qquad = 8+6i+4i+3i^2$

$\qquad\qquad = 8+10i-3$

$\qquad\qquad = 5+10i$

57. $(-3-5i)(2+4i) = -3\cdot 2-3\cdot 4i-5i\cdot 2-5i\cdot 4i$

$\qquad\qquad = -6-12i-10i-20i^2$

$\qquad\qquad = -6-22i-20(-1)$

$\qquad\qquad = -6-22i+20$

$\qquad\qquad = 14-22i$

59. $(2-3i)(4+6i) = 2\cdot 4+2\cdot 6i-3i\cdot 4-3i\cdot 6i$

$\qquad\qquad = 8+12i-12i-18i^2$

$\qquad\qquad = 8-18(-1)$

$\qquad\qquad = 8+18$

$\qquad\qquad = 26$

61. $\left(3-\sqrt{2}i\right)\left(-2+\sqrt{2}i\right)$

$\qquad = 3(-2)+3\left(\sqrt{2}\,i\right)-\sqrt{2}\,i(-2)-\left(\sqrt{2}\,i\right)^2$

$\qquad = -6+3\sqrt{2}\,i+2\sqrt{2}\,i-2i^2$

$\qquad = -6+5\sqrt{2}\,i-2(-1)$

$\qquad = -4+5\sqrt{2}\,i$

63. $\left(\dfrac{1}{2}-\dfrac{1}{4}i\right)\left(\dfrac{2}{3}+\dfrac{3}{4}i\right)$

$\qquad = \dfrac{1}{2}\cdot\dfrac{2}{3}+\dfrac{1}{2}\cdot\dfrac{3}{4}i-\dfrac{1}{4}i\cdot\dfrac{2}{3}-\dfrac{1}{4}i\cdot\dfrac{3}{4}i$

$\qquad = \dfrac{1}{3}+\dfrac{3}{8}i-\dfrac{1}{6}i-\dfrac{3}{16}i^2$

$\qquad = \dfrac{1}{3}+\dfrac{5}{24}i-\dfrac{3}{16}(-1)$

$\qquad = \dfrac{25}{48}+\dfrac{5}{24}i$

65. $(3+2i)^2 = 3^2+2(3)(2i)+(2i)^2$

$\qquad\qquad = 9+12i+4i^2$

$\qquad\qquad = 9+12i+4(-1)$

$\qquad\qquad = 5+12i$

67. $(-4-5i)^2 = (-4)^2-2(-4)(5i)+(5i)^2$

$\qquad\qquad = 16+40i+25i^2$

$\qquad\qquad = 16+40i+25(-1)$

$\qquad\qquad = -9+40i$

69. $\sqrt{-9}\cdot\sqrt{-4} = 3i\cdot 2i = 6i^2 = -6$

71. $\sqrt{-8} \cdot \sqrt{-10} = 2\sqrt{2}\,i \cdot \sqrt{10}\,i$
$$= 2\sqrt{20}\,i^2$$
$$= 2 \cdot 2\sqrt{5} \cdot (-1)$$
$$= -4\sqrt{5}$$

73. $\left(2 + \sqrt{-81}\right)\left(-3 - \sqrt{-100}\right)$
$$= (2 + 9i)(-3 - 10i)$$
$$= 2(-3) - 2(10i) - 9i(3) - 9i(10i)$$
$$= -6 - 20i - 27i - 90i^2$$
$$= -6 - 47i - 90(-1)$$
$$= 84 - 47i$$

75. **(a)** The conjugate of $3 + 5i$ is $3 - 5i$.

    (b) $(3 + 5i)(3 - 5i) = 3^2 - (5i)^2$
$$= 9 - 25i^2$$
$$= 9 - 25(-1)$$
$$= 9 + 25$$
$$= 34$$

77. **(a)** The conjugate of $2 - 7i$ is $2 + 7i$.

    (b) $(2 - 7i)(2 + 7i) = 2^2 - (7i)^2$
$$= 4 - 49i^2$$
$$= 4 - 49(-1)$$
$$= 4 + 49$$
$$= 53$$

79. **(a)** The conjugate of $-7 + 2i$ is $-7 - 2i$.

    (b) $(-7 + 2i)(-7 - 2i) = (-7)^2 - (2i)^2$
$$= 49 - 4i^2$$
$$= 49 - 4(-1)$$
$$= 49 + 4$$
$$= 53$$

81. $\dfrac{1+i}{3i} = \dfrac{(1+i)}{3i} \cdot \dfrac{i}{i}$
$$= \frac{i + i^2}{3i^2}$$
$$= \frac{i + (-1)}{3(-1)}$$
$$= \frac{i - 1}{-3}$$
$$= \frac{-1}{-3} + \frac{1}{-3}i$$
$$= \frac{1}{3} - \frac{1}{3}i$$

83. $\dfrac{-5 + 2i}{5i} = \dfrac{(-5 + 2i)}{5i} \cdot \dfrac{i}{i}$
$$= \frac{-5i + 2i^2}{5i^2}$$
$$= \frac{-5i + 2(-1)}{5(-1)}$$
$$= \frac{-5i - 2}{-5}$$
$$= \frac{-2}{-5} + \frac{-5}{-5}i$$
$$= \frac{2}{5} + i$$

85. $\dfrac{3}{2+i} = \dfrac{3}{2+i} \cdot \dfrac{2-i}{2-i}$
$$= \frac{6 - 3i}{2^2 - i^2}$$
$$= \frac{6 - 3i}{4 - (-1)}$$
$$= \frac{6 - 3i}{5}$$
$$= \frac{6}{5} - \frac{3}{5}i$$

87.
$$\frac{-2}{-3-7i} = \frac{-2}{-3-7i} \cdot \frac{-3+7i}{-3+7i}$$
$$= \frac{6-14i}{(-3)^2 - (7i)^2}$$
$$= \frac{6-14i}{9-49i^2}$$
$$= \frac{6-14i}{9-49(-1)}$$
$$= \frac{6-14i}{58}$$
$$= \frac{6}{58} - \frac{14}{58}i$$
$$= \frac{3}{29} - \frac{7}{29}i$$

89.
$$\frac{2+3i}{3-2i} = \frac{2+3i}{3-2i} \cdot \frac{3+2i}{3+2i}$$
$$= \frac{2\cdot3 + 2\cdot2i + 3i\cdot3 + 3i\cdot2i}{3^2 - (2i)^2}$$
$$= \frac{6+4i+9i+6i^2}{9-4i^2}$$
$$= \frac{6+13i-6}{9+4}$$
$$= \frac{13i}{13}$$
$$= i$$

91.
$$\frac{4+2i}{1-i} = \frac{4+2i}{1-i} \cdot \frac{1+i}{1+i}$$
$$= \frac{4\cdot1 + 4\cdot i + 2i\cdot1 + 2i\cdot i}{1^2 - i^2}$$
$$= \frac{4+4i+2i+2i^2}{1+1}$$
$$= \frac{4+6i-2}{2}$$
$$= \frac{2+6i}{2}$$
$$= 1+3i$$

93.
$$\frac{4-2i}{1+3i} = \frac{4-2i}{1+3i} \cdot \frac{1-3i}{1-3i}$$
$$= \frac{4\cdot1 - 4\cdot3i - 2i\cdot1 + 2i\cdot3i}{1^2 - (3i)^2}$$
$$= \frac{4-12i-2i+6i^2}{1-9i^2}$$
$$= \frac{4-14i-6}{1+9}$$
$$= \frac{-2-14i}{10}$$
$$= \frac{-2}{10} + \frac{-14i}{10}$$
$$= -\frac{1}{5} - \frac{7}{5}i$$

95. $i^{53} = i^{52} \cdot i = \left(i^4\right)^{13} \cdot i = 1 \cdot i = i$

97. $i^{43} = i^{40} \cdot i^3 = \left(i^4\right)^{10} \cdot i^3 = 1 \cdot (-i) = -i$

99. $i^{153} = i^{152} \cdot i = \left(i^4\right)^{38} \cdot i = 1 \cdot i = i$

101. $i^{-45} = i^{-48} \cdot i^3 = \left(i^4\right)^{-12} \cdot i^3 = 1 \cdot i^3 = i^3 = -i$

103.
$$(-4-i)(4+i) = -4\cdot4 - 4\cdot i - i\cdot4 - i\cdot i$$
$$= -16 - 4i - 4i - i^2$$
$$= -16 - 8i - (-1)$$
$$= -15 - 8i$$

105.
$$(3+2i)^2 = 3^2 + 2(3)(2i) + (2i)^2$$
$$= 9 + 12i + 4i^2$$
$$= 9 + 12i - 4$$
$$= 5 + 12i$$

107.
$$\frac{-3+2i}{3i} = \frac{-3+2i}{3i} \cdot \frac{i}{i}$$
$$= \frac{-3i + 2i^2}{3i^2}$$
$$= \frac{-3i - 2}{-3}$$
$$= \frac{2}{3} + i$$

109. $\dfrac{-4+i}{-5-3i} = \dfrac{-4+i}{-5-3i} \cdot \dfrac{-5+3i}{-5+3i}$

$= \dfrac{-4(-5)-4\cdot 3i - i\cdot 5 + i\cdot 3i}{(-5)^2 - (3i)^2}$

$= \dfrac{20 - 12i - 5i + 3i^2}{25 - 9i^2}$

$= \dfrac{20 - 17i - 3}{25 + 9}$

$= \dfrac{17 - 17i}{34}$

$= \dfrac{17}{34} - \dfrac{17}{34}i$

$= \dfrac{1}{2} - \dfrac{1}{2}i$

111. $(10-3i)+(2+3i) = 10-3i+2+3i$

$= (10+2)+(-3+3)i$

$= 12 + 0i$

$= 12$

113. $5i^{37}(-4+3i) = 5\cdot i^{36}\cdot i(-4+3i)$

$= 5\left(i^4\right)^9 \cdot i(-4+3i)$

$= 5\cdot 1\cdot i(-4+3i)$

$= 5i(-4+3i)$

$= 5i\cdot(-4)+5i\cdot 3i$

$= -20i + 15i^2$

$= -20i - 15$

$= -15 - 20i$

115. $\sqrt{-10}\cdot\sqrt{-15} = \sqrt{-1}\cdot\sqrt{10}\cdot\sqrt{-1}\cdot\sqrt{15}$

$= i\cdot\sqrt{10}\cdot i\cdot\sqrt{15}$

$= i^2\cdot\sqrt{150}$

$= -1\cdot\sqrt{25}\cdot\sqrt{6}$

$= -5\sqrt{6}$

117. $\dfrac{1}{5i} = \dfrac{1}{5i}\cdot\dfrac{i}{i} = \dfrac{i}{5i^2} = \dfrac{i}{-5} = -\dfrac{1}{5}i$

119. $\dfrac{1}{2-i} = \dfrac{1}{2-i}\cdot\dfrac{2+i}{2+i}$

$= \dfrac{2+i}{2^2 - i^2}$

$= \dfrac{2+i}{4-(-1)}$

$= \dfrac{2+i}{5}$

$= \dfrac{2}{5} + \dfrac{1}{5}i$

121. $\dfrac{1}{-4+5i} = \dfrac{1}{-4+5i}\cdot\dfrac{-4-5i}{-4-5i}$

$= \dfrac{-4-5i}{(-4)^2 - (5i)^2}$

$= \dfrac{-4-5i}{16 - 25i^2}$

$= \dfrac{-4-5i}{16 + 25}$

$= \dfrac{-4-5i}{41}$

$= -\dfrac{4}{41} - \dfrac{5}{41}i$

123. $f(x) = x^2$

(a) $f(i) = i^2 = -1$

(b) $f(1+i) = (1+i)^2$

$= 1^2 + 2i + i^2$

$= 1 + 2i - 1$

$= 2i$

125. $f(x) = x^2 + 2x + 2$

(a) $f(3i) = (3i)^2 + 2(3i) + 2$

$= 9i^2 + 6i + 2$

$= -9 + 6i + 2$

$= -7 + 6i$

(b) $f(1-i) = (1-i)^2 + 2(1-i) + 2$

$= 1^2 - 2i + i^2 + 2 - 2i + 2$

$= 1 - 2i - 1 + 2 - 2i + 2$

$= 4 - 4i$

127. (a) To find the total impedance, we add the individual impedances.

$$(7+3i)+(3-4i) = 7+3i+3-4i$$
$$= (7+3)+(3-4)i$$
$$= 10-i$$

The total impedance is $10-i$ ohms.

(b) The total resistance is the real part of the impedance. Thus, the total resistance is 10 ohms.

(c) The total reactance is the imaginary part of the impedance. Thus, the total reactance is -1 ohm. (Since the reactance is negative, this would be called *capacitive reactance*.)

129. $f(x) = x^2 + 4x + 5$

(a) $f(-2+i) = (-2+i)^2 + 4(-2+i) + 5$
$$= (-2)^2 + 2(-2)i + i^2 - 8 + 4i + 5$$
$$= 4 - 4i - 1 - 8 + 4i + 5$$
$$= 0$$

(b) $f(-2-i) = (-2-i)^2 + 4(-2-i) + 5$
$$= (-2)^2 - 2(-2)i + i^2 - 8 - 4i + 5$$
$$= 4 + 4i - 1 - 8 - 4i + 5$$
$$= 0$$

131. $f(x) = x^3 + 1$

(a) $f(-1) = (-1)^3 + 1 = -1 + 1 = 0$

(b) $f\left(\dfrac{1}{2}+\dfrac{\sqrt{3}}{2}i\right) = \left(\dfrac{1}{2}+\dfrac{\sqrt{3}}{2}i\right)^3 + 1$

$$= \left(\dfrac{1}{2}+\dfrac{\sqrt{3}}{2}i\right)\left(\dfrac{1}{2}+\dfrac{\sqrt{3}}{2}i\right)^2 + 1$$

$$= \left(\dfrac{1}{2}+\dfrac{\sqrt{3}}{2}i\right)\left(\dfrac{1}{4}+\dfrac{\sqrt{3}}{2}i+\dfrac{3}{4}i^2\right) + 1$$

$$= \left(\dfrac{1}{2}+\dfrac{\sqrt{3}}{2}i\right)\left(\dfrac{1}{4}+\dfrac{\sqrt{3}}{2}i-\dfrac{3}{4}\right) + 1$$

$$= \left(\dfrac{1}{2}+\dfrac{\sqrt{3}}{2}i\right)\left(-\dfrac{1}{2}+\dfrac{\sqrt{3}}{2}i\right) + 1$$

$$= \left(\dfrac{\sqrt{3}}{2}i\right)^2 - \left(\dfrac{1}{2}\right)^2 + 1$$

$$= \dfrac{3}{4}i^2 - \dfrac{1}{4} + 1$$

$$= -\dfrac{3}{4} + \dfrac{3}{4}$$

$$= 0$$

(c) $f\left(\dfrac{1}{2}-\dfrac{\sqrt{3}}{2}i\right) = \left(\dfrac{1}{2}-\dfrac{\sqrt{3}}{2}i\right)^3 + 1$

$$= \left(\dfrac{1}{2}-\dfrac{\sqrt{3}}{2}i\right)\left(\dfrac{1}{2}-\dfrac{\sqrt{3}}{2}i\right)^2 + 1$$

$$= \left(\dfrac{1}{2}-\dfrac{\sqrt{3}}{2}i\right)\left(\dfrac{1}{4}-\dfrac{\sqrt{3}}{2}i+\dfrac{3}{4}i^2\right) + 1$$

$$= \left(\dfrac{1}{2}-\dfrac{\sqrt{3}}{2}i\right)\left(\dfrac{1}{4}-\dfrac{\sqrt{3}}{2}i-\dfrac{3}{4}\right) + 1$$

$$= \left(\dfrac{1}{2}-\dfrac{\sqrt{3}}{2}i\right)\left(-\dfrac{1}{2}-\dfrac{\sqrt{3}}{2}i\right) + 1$$

$$= \left(-\dfrac{\sqrt{3}}{2}i\right)^2 - \left(\dfrac{1}{2}\right)^2 + 1$$

$$= \dfrac{3}{4}i^2 - \dfrac{1}{4} + 1$$

$$= -\dfrac{3}{4} + \dfrac{3}{4}$$

$$= 0$$

133. For a polynomial with real coefficients, the zeros will be real numbers or will occur in conjugate pairs. If the complex number $a+bi$ is a complex zero of the polynomial, then its conjugate $a-bi$ is also a complex conjugate.

135. Answers will vary.
The counting numbers are a subset of the whole numbers; the whole numbers are a subset of the integers; the integers are a subset of the rational numbers; rational numbers are a subset of the real numbers; and the real numbers are a subset of the complex numbers.

137. Answers will vary.
Complex numbers, in standard form, resemble binomials in that there are two terms separated by a $+$ or a $-$.

139. $(x+2)^3 = x^3 + 3x^2 \cdot 2 + 3x \cdot 2^2 + 2^3$
$$= x^3 + 6x^2 + 12x + 8$$

141. $(3+i)^3 = 3^3 + 3 \cdot 3^2 \cdot i + 3 \cdot 3 \cdot i^2 + i^3$
$$= 27 + 27i + 9i^2 + i^3$$
$$= 27 + 27i - 9 - i$$
$$= 18 + 26i$$

143. Answers will vary.

145.

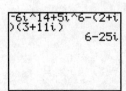

```
(-3.4+1.9i)-(6.5
-5.3i)
        -9.9+7.2i
```

147.

```
(-4.3+0.2i)(7.2-
0.5i)
       -30.86+3.59i
```

149.

```
(1-7i)/(-4+i)▶Fr
ac
      -11/17+27/17i
```

151.

```
-6i^14+5i^6-(2+i
)(3+11i)
            6-25i
```

1. The square roots of 4 are 2 and -2 because $(2)^2 = 4$ and $(-2)^2 = 4$.

2. The square roots of 81 are 9 and -9 because $(9)^2 = 81$ and $(-9)^2 = 81$.

3. $-\sqrt{1} = -\sqrt{1^2} = -1$

4. $-\sqrt{25} = -\sqrt{5^2} = -5$

5. $\sqrt{0.16} = \sqrt{(0.4)^2} = 0.4$

6. $\sqrt{0.04} = \sqrt{(0.2)^2} = 0.2$

7. $\dfrac{3}{2}\sqrt{\dfrac{25}{36}} = \dfrac{3}{2} \cdot \dfrac{5}{6} = \dfrac{15}{12} = \dfrac{5}{4}$

8. $\dfrac{4}{3}\sqrt{\dfrac{81}{4}} = \dfrac{4}{3} \cdot \dfrac{9}{2} = \dfrac{36}{6} = 6$

9. $\sqrt{25-9} = \sqrt{16} = 4$

10. $\sqrt{169-25} = \sqrt{144} = 12$

11. $\sqrt{9^2 - (4)(5)(-18)} = \sqrt{81 + 360} = \sqrt{441} = 21$

12. $\sqrt{13^2 - (4)(-3)(-4)} = \sqrt{169 - 48} = \sqrt{121} = 11$

13. $-\sqrt{9}$ is rational because 9 is a perfect square.
$-\sqrt{9} = -\sqrt{(3)^2} = -3$

14. $-\dfrac{1}{2}\sqrt{48}$ is irrational because 48 is not a perfect square. $-\dfrac{1}{2}\sqrt{48} \approx -3.46$

15. $\sqrt{14}$ is irrational because 14 is not a perfect square. $\sqrt{14} \approx 3.74$

16. $\sqrt{-2}$ is not a real number because the radicand of the square root is negative.

17. $\sqrt{(4x-9)^2} = |4x-9|$

18. $\sqrt{(16m-25)^2} = |16m-25|$

19. $\sqrt[3]{343} = \sqrt[3]{7^3} = 7$

20. $\sqrt[3]{-125} = \sqrt[3]{(-5)^3} = -5$

21. $\sqrt[3]{\dfrac{8}{27}} = \sqrt[3]{\left(\dfrac{2}{3}\right)^3} = \dfrac{2}{3}$

22. $\sqrt[4]{81} = \sqrt[4]{3^4} = 3$

23. $-\sqrt[5]{-243} = -\sqrt[5]{(-3)^5} = -(-3) = 3$

24. $\sqrt[3]{10^3} = 10$

25. $\sqrt[5]{z^5} = z$

26. $\sqrt[4]{(5p-3)^4} = |5p-3|$

27. $81^{1/2} = \sqrt{81} = \sqrt{9^2} = 9$

28. $(-256)^{1/4} = \sqrt[4]{-256}$ is not a real number.

29. $-4^{1/2} = -\sqrt{4} = -\sqrt{2^2} = -2$

30. $729^{1/3} = \sqrt[3]{729} = \sqrt[3]{9^3} = 9$

31. $16^{7/4} = \left(\sqrt[4]{16}\right)^7 = \left(\sqrt[4]{2^4}\right)^7 = 2^7 = 128$

32. $-(-27)^{2/3} = -\left(\sqrt[3]{-27}\right)^2$
$\qquad = -\left(\sqrt[3]{(-3)^3}\right)^2$
$\qquad = -(-3)^2$
$\qquad = -9$

33. $-121^{3/2} = -\left(\sqrt{121}\right)^3$
$\qquad = -\left(\sqrt{11^2}\right)^3$
$\qquad = -(11)^3$
$\qquad = -1331$

34. $\dfrac{1}{36^{-1/2}} = 36^{1/2} = \sqrt{36} = \sqrt{6^2} = 6$

35. $(-65)^{1/3} \approx -4.02$

36. $4^{3/5} \approx 2.30$

37. $\sqrt[3]{100} \approx 4.64$

38. $\sqrt[4]{10} \approx 1.78$

39. $\sqrt[3]{5a} = (5a)^{1/3}$

40. $\sqrt[5]{p^7} = p^{7/5}$

41. $\left(\sqrt[4]{10z}\right)^3 = (10z)^{3/4}$

42. $\sqrt[6]{(2ab)^5} = (2ab)^{5/6}$

43. $4^{2/3} \cdot 4^{7/3} = 4^{\frac{2}{3}+\frac{7}{3}} = 4^{\frac{9}{3}} = 4^3 = 64$

44. $\dfrac{k^{1/2}}{k^{3/4}} = k^{\frac{1}{2}-\frac{3}{4}} = k^{\frac{2}{4}-\frac{3}{4}} = k^{-1/4} = \dfrac{1}{k^{1/4}}$

45. $\left(p^{4/3} \cdot q^4\right)^{3/2} = \left(p^{4/3}\right)^{3/2} \cdot \left(q^4\right)^{3/2}$
$\qquad = p^{\frac{4}{3}\cdot\frac{3}{2}} \cdot q^{4\cdot\frac{3}{2}}$
$\qquad = p^2 \cdot q^6 \quad \text{or} \quad \left(p \cdot q^3\right)^2$

46. $\left(32a^{-3/2} \cdot b^{1/4}\right)^{1/5} = (32)^{1/5} \cdot \left(a^{-3/2}\right)^{1/5} \cdot \left(b^{1/4}\right)^{1/5}$
$\qquad = 2 \cdot a^{-3/10} \cdot b^{1/20}$
$\qquad = \dfrac{2b^{1/20}}{a^{3/10}} \quad \text{or} \quad 2\left(\dfrac{b}{a^6}\right)^{1/20}$

47. $5m^{-2/3}\left(2m + m^{-1/3}\right)$
$\qquad = 5m^{-2/3} \cdot 2m + 5m^{-2/3} \cdot m^{-1/3}$
$\qquad = 10m^{-2/3} \cdot m^{3/3} + 5m^{-3/3}$
$\qquad = 10m^{1/3} + 5m^{-1}$
$\qquad = 10m^{1/3} + \dfrac{5}{m}$

48. $\left(\dfrac{16x^{1/3}}{x^{-1/3}}\right)^{-1/2} + \left(\dfrac{x^{-3/2}}{64x^{-1/2}}\right)^{1/3}$

$= \left(16x^{\frac{1}{3}-\left(-\frac{1}{3}\right)}\right)^{-1/2} + \left(64^{-1}x^{-\frac{3}{2}-\left(-\frac{1}{2}\right)}\right)^{1/3}$

$= \left(16x^{2/3}\right)^{-1/2} + \left(\dfrac{1}{64}x^{-1}\right)^{1/3}$

$= \left(\dfrac{1}{16}x^{-2/3}\right)^{1/2} + \left(\dfrac{1}{64}x^{-1}\right)^{1/3}$

$= \left(\dfrac{1}{16}\right)^{1/2}\left(x^{-2/3}\right)^{1/2} + \left(\dfrac{1}{64}\right)^{1/3}\left(x^{-1}\right)^{1/3}$

$= \dfrac{1}{4}x^{-1/3} + \dfrac{1}{4}x^{-1/3}$

$= \dfrac{1}{2}x^{-1/3}$

$= \dfrac{1}{2x^{1/3}}$

49. $\sqrt[8]{x^6} = x^{6/8} = x^{3/4} = \sqrt[4]{x^3}$

50. $\sqrt{121x^4y^{10}} = \left(121x^4y^{10}\right)^{1/2}$

$= 121^{1/2}\left(x^4\right)^{1/2}\left(y^{10}\right)^{1/2}$

$= 11x^{4/2}y^{10/2}$

$= 11x^2y^5$

51. $\sqrt[3]{m^2} \cdot \sqrt{m^3} = m^{2/3} \cdot m^{3/2}$

$= m^{\frac{2}{3}+\frac{3}{2}}$

$= m^{\frac{4}{6}+\frac{9}{6}}$

$= m^{13/6}$

$= \sqrt[6]{m^{13}}$

$= m^2\sqrt[6]{m}$

52. $\dfrac{\sqrt[3]{c}}{\sqrt[6]{c^4}} = \dfrac{c^{1/3}}{c^{4/6}} = \dfrac{c^{1/3}}{c^{2/3}} = c^{\frac{1}{3}-\frac{2}{3}} = c^{-1/3} = \dfrac{1}{c^{1/3}} = \dfrac{1}{\sqrt[3]{c}}$

53. $2(3m-1)^{1/4} + (m-7)(3m-1)^{5/4}$

$= 2\cdot(3m-1)^{1/4} + (m-7)(3m-1)^{1/4}\cdot(3m-1)$

$= (3m-1)^{1/4}\left(2+(m-7)(3m-1)\right)$

$= (3m-1)^{1/4}\left(2+3m^2-21m-m+7\right)$

$= (3m-1)^{1/4}\left(3m^2-22m+9\right)$

54. $3\left(x^2-5\right)^{1/3} - 4x\left(x^2-5\right)^{-2/3}$

$= 3\left(x^2-5\right)^{-2/3}\cdot\left(x^2-5\right) - 4x\left(x^2-5\right)^{-2/3}$

$= \left(x^2-5\right)^{-2/3}\left(3\left(x^2-5\right)-4x\right)$

$= \left(x^2-5\right)^{-2/3}\left(3x^2-15-4x\right)$

$= \left(x^2-5\right)^{-2/3}\left(3x^2-4x-15\right)$

$= \dfrac{3x^2-4x-15}{\left(x^2-5\right)^{2/3}} = \dfrac{(3x+5)(x-3)}{\left(x^2-5\right)^{2/3}}$

55. $\sqrt{15}\cdot\sqrt{7} = \sqrt{15\cdot 7} = \sqrt{105}$

56. $\sqrt[4]{2ab^2}\cdot\sqrt[4]{6a^2b} = \sqrt[4]{2ab^2\cdot 6a^2b} = \sqrt[4]{12a^3b^3}$

57. $\sqrt{80} = \sqrt{16\cdot 5} = \sqrt{16}\cdot\sqrt{5} = 4\sqrt{5}$

58. $\sqrt[3]{-500} = \sqrt[3]{-125\cdot 4} = \sqrt[3]{-125}\cdot\sqrt[3]{4} = -5\sqrt[3]{4}$

59. $\sqrt[3]{162m^6n^4} = \sqrt[3]{27m^6n^3\cdot 6n}$

$= \sqrt[3]{27m^6n^3}\cdot\sqrt[3]{6n}$

$= 3m^2n\sqrt[3]{6n}$

60. $\sqrt[4]{50p^8q^4} = \sqrt[4]{p^8q^4\cdot 50}$

$= \sqrt[4]{p^8q^4}\cdot\sqrt[4]{50}$

$= p^2|q|\sqrt[4]{50}$

61. $2\sqrt{16x^6y} = 2\sqrt{16x^6\cdot y}$

$= 2\sqrt{16x^6}\cdot\sqrt{y}$

$= 2\cdot 4\left|x^3\right|\sqrt{y}$

$= 8\left|x^3\right|\sqrt{y}$

as long as $y \geq 0$. Otherwise the result is not a real number.

62. $\sqrt{(2x+1)^3} = \sqrt{(2x+1)^2 \cdot (2x+1)}$
$$= \sqrt{(2x+1)^2} \cdot \sqrt{2x+1}$$
$$= (2x+1)\sqrt{2x+1}$$

as long as $2x+1 \geq 0$. Otherwise the result is not a real number.

63. $\sqrt{w^3 z^2} = \sqrt{w^2 z^2 \cdot w} = \sqrt{w^2 z^2} \cdot \sqrt{w} = wz\sqrt{w}$

64. $\sqrt{45x^4 yz^3} = \sqrt{9x^4 z^2 \cdot 5yz}$
$$= \sqrt{9x^4 z^2} \cdot \sqrt{5yz}$$
$$= 3x^2 z \sqrt{5yz}$$

65. $\sqrt[3]{16a^{12}b^5} = \sqrt[3]{8a^{12}b^3 \cdot 2b^2}$
$$= \sqrt[3]{8a^{12}b^3} \cdot \sqrt[3]{2b^2}$$
$$= 2a^4 b \sqrt[3]{2b^2}$$

66. $\sqrt{4x^2 + 8x + 4} = \sqrt{4\left(x^2 + 2x + 1\right)}$
$$= \sqrt{4(x+1)^2}$$
$$= 2(x+1)$$

67. $\sqrt{15} \cdot \sqrt{18} = \sqrt{15 \cdot 18}$
$$= \sqrt{270}$$
$$= \sqrt{9 \cdot 30}$$
$$= \sqrt{9} \cdot \sqrt{30}$$
$$= 3\sqrt{30}$$

68. $\sqrt[3]{20} \cdot \sqrt[3]{30} = \sqrt[3]{20 \cdot 30}$
$$= \sqrt[3]{600}$$
$$= \sqrt[3]{8 \cdot 75}$$
$$= \sqrt[3]{8} \cdot \sqrt[3]{75}$$
$$= 2\sqrt[3]{75}$$

69. $\sqrt[3]{-3x^4 y^7} \cdot \sqrt[3]{24x^3 y^2} = \sqrt[3]{-3x^4 y^7 \cdot 24x^3 y^2}$
$$= \sqrt[3]{-72x^7 y^9}$$
$$= \sqrt[3]{-8x^6 y^9 \cdot 9x}$$
$$= \sqrt[3]{-8x^6 y^9} \cdot \sqrt[3]{9x}$$
$$= -2x^2 y^3 \sqrt[3]{9x}$$

70. $3\sqrt{4xy^2} \cdot 5\sqrt{3x^2 y} = 15\sqrt{4xy^2 \cdot 3x^2 y}$
$$= 15\sqrt{12x^3 y^3}$$
$$= 15\sqrt{4x^2 y^2 \cdot 3xy}$$
$$= 15\sqrt{4x^2 y^2} \cdot \sqrt{3xy}$$
$$= 15 \cdot 2xy\sqrt{3xy}$$
$$= 30xy\sqrt{3xy}$$

71. $\sqrt{\dfrac{121}{25}} = \dfrac{\sqrt{121}}{\sqrt{25}} = \dfrac{11}{5}$

72. $\sqrt{\dfrac{5a^4}{64b^2}} = \dfrac{\sqrt{5a^4}}{\sqrt{64b^2}} = \dfrac{a^2\sqrt{5}}{8b}$

73. $\sqrt[3]{\dfrac{54k^2}{9k^5}} = \sqrt[3]{\dfrac{6}{k^3}} = \dfrac{\sqrt[3]{6}}{\sqrt[3]{k^3}} = \dfrac{\sqrt[3]{6}}{k}$

74. $\sqrt[3]{\dfrac{-160w^{11}}{343w^{-4}}} = \sqrt[3]{\dfrac{-160w^{15}}{343}}$
$$= \dfrac{\sqrt[3]{-160w^{15}}}{\sqrt[3]{343}}$$
$$= \dfrac{\sqrt[3]{-8w^{15} \cdot 20}}{\sqrt[3]{7^3}}$$
$$= \dfrac{-2w^5 \sqrt[3]{20}}{7}$$

75. $\dfrac{\sqrt{12h^3}}{\sqrt{3h}} = \sqrt{\dfrac{12h^3}{3h}} = \sqrt{4h^2} = 2h$

76. $\dfrac{\sqrt{50a^3 b^3}}{\sqrt{8a^5 b^{-3}}} = \sqrt{\dfrac{50a^3 b^3}{8a^5 b^{-3}}} = \sqrt{\dfrac{25b^6}{4a^2}} = \dfrac{\sqrt{25b^6}}{\sqrt{4a^2}} = \dfrac{5b^3}{2a}$

77. $\dfrac{\sqrt[3]{-8x^7 y}}{\sqrt[3]{27xy^4}} = \sqrt[3]{\dfrac{-8x^7 y}{27xy^4}}$
$$= \sqrt[3]{\dfrac{-8x^6}{27y^3}}$$
$$= \dfrac{\sqrt[3]{-8x^6}}{\sqrt[3]{27y^3}}$$
$$= \dfrac{-2x^2}{3y}$$

78. $\dfrac{\sqrt[4]{48m^2n^7}}{\sqrt[4]{3m^6n}} = \sqrt[4]{\dfrac{48m^2n^7}{3m^6n}}$

$= \sqrt[4]{\dfrac{16n^6}{m^4}}$

$= \dfrac{\sqrt[4]{16n^6}}{\sqrt[4]{m^4}}$

$= \dfrac{2n\sqrt[4]{n^2}}{m}$

$= \dfrac{2n\sqrt{n}}{m}$

79. $\sqrt{5} \cdot \sqrt[3]{2} = 5^{1/2} \cdot 2^{1/3}$

$= 5^{3/6} \cdot 2^{2/6}$

$= \left(5^3\right)^{1/6} \cdot \left(2^2\right)^{1/6}$

$= \left(5^3 \cdot 2^2\right)^{1/6}$

$= (125 \cdot 4)^{1/6}$

$= 500^{1/6}$

$= \sqrt[6]{500}$

80. $\sqrt[4]{8} \cdot \sqrt[6]{4} = 8^{1/4} \cdot 4^{1/6}$

$= \left(2^3\right)^{1/4} \cdot \left(2^2\right)^{1/6}$

$= 2^{3/4} \cdot 2^{1/3} = 2^{\frac{3}{4}+\frac{1}{3}}$

$= 2^{13/12} = 2^{\frac{12}{12}} \cdot 2^{\frac{1}{12}}$

$= 2\sqrt[12]{2}$

81. $2\sqrt[4]{x} + 6\sqrt[4]{x} = (2+6)\sqrt[4]{x} = 8\sqrt[4]{x}$

82. $7\sqrt[3]{4y} + 2\sqrt[3]{4y} - 3\sqrt[3]{4y} = (7+2-3)\sqrt[3]{4y}$

$= 6\sqrt[3]{4y}$

83. $5\sqrt{2} - 2\sqrt{12} = 5\sqrt{2} - 2 \cdot \sqrt{4} \cdot \sqrt{3}$

$= 5\sqrt{2} - 2 \cdot 2 \cdot \sqrt{3}$

$= 5\sqrt{2} - 4\sqrt{3}$

Cannot be simplified further.

84. $\sqrt{18} + 2\sqrt{50} = \sqrt{9} \cdot \sqrt{2} + 2\sqrt{25} \cdot \sqrt{2}$

$= 3 \cdot \sqrt{2} + 2 \cdot 5 \cdot \sqrt{2}$

$= 3\sqrt{2} + 10\sqrt{2}$

$= (3+10)\sqrt{2}$

$= 13\sqrt{2}$

85. $\sqrt[3]{-16z} + \sqrt[3]{54z} = \sqrt[3]{-8} \cdot \sqrt[3]{2z} + \sqrt[3]{27} \cdot \sqrt[3]{2z}$

$= -2\sqrt[3]{2z} + 3\sqrt[3]{2z}$

$= (-2+3)\sqrt[3]{2z}$

$= \sqrt[3]{2z}$

86. $7\sqrt[3]{8x^2} - \sqrt[3]{-27x^2} = 7 \cdot \sqrt[3]{8} \cdot \sqrt[3]{x^2} - \sqrt[3]{-27} \cdot \sqrt[3]{x^2}$

$= 7 \cdot 2 \cdot \sqrt[3]{x^2} - (-3) \cdot \sqrt[3]{x^2}$

$= 14\sqrt[3]{x^2} + 3\sqrt[3]{x^2}$

$= (14+3)\sqrt[3]{x^2}$

$= 17\sqrt[3]{x^2}$

87. $\sqrt{16a} + \sqrt[6]{729a^3} = \sqrt{16} \cdot \sqrt{a} + \sqrt[6]{729} \cdot \sqrt[6]{a^3}$

$= \sqrt{16} \cdot \sqrt{a} + \sqrt[6]{729} \cdot a^{3/6}$

$= 4\sqrt{a} + 3 \cdot a^{1/2}$

$= 4\sqrt{a} + 3\sqrt{a}$

$= (4+3)\sqrt{a}$

$= 7\sqrt{a}$

88. $\sqrt{27x^2} - x\sqrt{48} + 2\sqrt{75x^2}$

$= \sqrt{9x^2} \cdot \sqrt{3} - x \cdot \sqrt{16} \cdot \sqrt{3} + 2 \cdot \sqrt{25x^2} \cdot \sqrt{3}$

$= 3x\sqrt{3} - 4x\sqrt{3} + 10x\sqrt{3}$

$= (3+10-4)x\sqrt{3}$

$= 9x\sqrt{3}$

89. $5\sqrt[3]{4m^5y^2} - \sqrt[6]{16m^{10}y^4}$

$= 5 \cdot \sqrt[3]{m^3} \cdot \sqrt[3]{4m^2y^2} - \sqrt[6]{m^6} \cdot \sqrt[6]{16m^4y^4}$

$= 5m\sqrt[3]{4m^2y^2} - m \cdot \sqrt[6]{\left(4m^2y^2\right)^2}$

$= 5m\sqrt[3]{4m^2y^2} - m \cdot \left(4m^2y^2\right)^{2/6}$

$= 5m\sqrt[3]{4m^2y^2} - m \cdot \left(4m^2y^2\right)^{1/3}$

$= 5m\sqrt[3]{4m^2y^2} - m\sqrt[3]{4m^2y^2}$

$= (5-1)m\sqrt[3]{4m^2y^2}$

$= 4m\sqrt[3]{4m^2y^2}$

90. $\sqrt{y^3 - 4y^2} - 2\sqrt{y-4} + \sqrt[4]{y^2 - 8y + 16}$

$= \sqrt{y^2} \cdot \sqrt{(y-4)} - 2\sqrt{y-4} + \sqrt[4]{(y-4)^2}$

$= y\sqrt{y-4} - 2\sqrt{y-4} + (y-4)^{2/4}$

$= y\sqrt{y-4} - 2\sqrt{y-4} + (y-4)^{1/2}$

$= y\sqrt{y-4} - 2\sqrt{y-4} + \sqrt{y-4}$

$= (y - 2 + 1)\sqrt{y-4}$

$= (y-1)\sqrt{y-4}$

91. $\sqrt{3}\left(\sqrt{5} - \sqrt{15}\right) = \sqrt{3} \cdot \sqrt{5} - \sqrt{3} \cdot \sqrt{15}$

$= \sqrt{3} \cdot \sqrt{5} - \sqrt{3 \cdot 15}$

$= \sqrt{3} \cdot \sqrt{5} - \sqrt{45}$

$= \sqrt{3} \cdot \sqrt{5} - \sqrt{9} \cdot \sqrt{5}$

$= \sqrt{3} \cdot \sqrt{5} - 3 \cdot \sqrt{5}$

$= \sqrt{15} - 3\sqrt{5}$

92. $\sqrt[3]{5}\left(3 + \sqrt[3]{4}\right) = \sqrt[3]{5} \cdot 3 + \sqrt[3]{5} \cdot \sqrt[3]{4} = 3\sqrt[3]{5} + \sqrt[3]{20}$

93. $\left(3 + \sqrt{5}\right)\left(4 - \sqrt{5}\right) = 3 \cdot 4 - 3 \cdot \sqrt{5} + 4 \cdot \sqrt{5} - \sqrt{5} \cdot \sqrt{5}$

$= 12 - 3\sqrt{5} + 4\sqrt{5} - \sqrt{25}$

$= 12 + \sqrt{5} - 5$

$= 7 + \sqrt{5}$

94. $\left(7 + \sqrt{3}\right)\left(6 + \sqrt{2}\right)$

$= 7 \cdot 6 + 7 \cdot \sqrt{2} + \sqrt{3} \cdot 6 + \sqrt{3} \cdot \sqrt{2}$

$= 42 + 7\sqrt{2} + 6\sqrt{3} + \sqrt{6}$

95. $\left(1 - 3\sqrt{5}\right)\left(1 + 3\sqrt{5}\right) = (1)^2 - \left(3\sqrt{5}\right)^2$

$= 1 - 9\sqrt{25}$

$= 1 - 9 \cdot 5$

$= 1 - 45$

$= -44$

96. $\left(\sqrt[3]{x} + 1\right)\left(9\sqrt[3]{x} - 4\right)$

$= \sqrt[3]{x} \cdot 9\sqrt[3]{x} + \sqrt[3]{x} \cdot (-4) + 1 \cdot 9\sqrt[3]{x} + 1 \cdot (-4)$

$= 9\sqrt[3]{x^2} - 4\sqrt[3]{x} + 9\sqrt[3]{x} - 4$

$= 9\sqrt[3]{x^2} + 5\sqrt[3]{x} - 4$

97. $\left(\sqrt{x} - \sqrt{5}\right)^2 = \left(\sqrt{x}\right)^2 - 2\left(\sqrt{x}\right)\left(\sqrt{5}\right) + \left(\sqrt{5}\right)^2$

$= \sqrt{x^2} - 2\sqrt{5x} + \sqrt{25}$

$= x - 2\sqrt{5x} + 5$

98. $\left(11\sqrt{2} + \sqrt{5}\right)^2$

$= \left(11\sqrt{2}\right)^2 + 2\left(11\sqrt{2}\right)\left(\sqrt{5}\right) + \left(\sqrt{5}\right)^2$

$= 121\sqrt{4} + 22\sqrt{10} + \sqrt{25}$

$= 121 \cdot 2 + 22\sqrt{10} + 5$

$= 242 + 22\sqrt{10} + 5$

$= 247 + 22\sqrt{10}$

99. $\left(\sqrt{2a} - b\right)\left(\sqrt{2a} + b\right) = \left(\sqrt{2a}\right)^2 - (b)^2$

$= \sqrt{4a^2} - b^2$

$= 2a - b^2$

100. $\left(\sqrt[3]{6s} + 2\right)\left(\sqrt[3]{6s} - 7\right)$

$= \sqrt[3]{6s} \cdot \sqrt[3]{6s} + \sqrt[3]{6s} \cdot (-7) + 2 \cdot \sqrt[3]{6s} + 2 \cdot (-7)$

$= \sqrt[3]{36s^2} - 7\sqrt[3]{6s} + 2\sqrt[3]{6s} - 14$

$= \sqrt[3]{36s^2} - 5\sqrt[3]{6s} - 14$

101. $\dfrac{2}{\sqrt{6}} = \dfrac{2}{\sqrt{6}} \cdot \dfrac{\sqrt{6}}{\sqrt{6}} = \dfrac{2\sqrt{6}}{\sqrt{36}} = \dfrac{2\sqrt{6}}{6} = \dfrac{\sqrt{6}}{3}$

102. $\dfrac{6}{\sqrt{3}} = \dfrac{6}{\sqrt{3}} \cdot \dfrac{\sqrt{3}}{\sqrt{3}} = \dfrac{6\sqrt{3}}{\sqrt{9}} = \dfrac{6\sqrt{3}}{3} = 2\sqrt{3}$

103. $\dfrac{\sqrt{48}}{\sqrt{p^3}} = \dfrac{4\sqrt{3}}{p\sqrt{p}} = \dfrac{4\sqrt{3}}{p\sqrt{p}} \cdot \dfrac{\sqrt{p}}{\sqrt{p}}$

$= \dfrac{4\sqrt{3p}}{p\sqrt{p^2}} = \dfrac{4\sqrt{3p}}{p \cdot p}$

$= \dfrac{4\sqrt{3p}}{p^2}$

104. $\dfrac{5}{\sqrt{2a}} = \dfrac{5}{\sqrt{2a}} \cdot \dfrac{\sqrt{2a}}{\sqrt{2a}} = \dfrac{5\sqrt{2a}}{\sqrt{4a^2}} = \dfrac{5\sqrt{2a}}{2a}$

105.
$$\frac{-2}{\sqrt{6y^3}} = \frac{-2}{y\sqrt{6y}} = \frac{-2}{y\sqrt{6y}} \cdot \frac{\sqrt{6y}}{\sqrt{6y}}$$
$$= \frac{-2\sqrt{6y}}{y\sqrt{36y^2}} = \frac{-2\sqrt{6y}}{y \cdot 6y}$$
$$= \frac{-2\sqrt{6y}}{6y^2}$$
$$= -\frac{\sqrt{6y}}{3y^2}$$

106.
$$\frac{3}{\sqrt[3]{5}} = \frac{3}{\sqrt[3]{5}} \cdot \frac{\sqrt[3]{25}}{\sqrt[3]{25}} = \frac{3\sqrt[3]{25}}{\sqrt[3]{125}} = \frac{3\sqrt[3]{25}}{5}$$

107.
$$\sqrt[3]{\frac{-4}{45}} = \frac{\sqrt[3]{-4}}{\sqrt[3]{45}} = \frac{\sqrt[3]{-4}}{\sqrt[3]{45}} \cdot \frac{\sqrt[3]{75}}{\sqrt[3]{75}} = \frac{\sqrt[3]{-300}}{\sqrt[3]{3375}} = -\frac{\sqrt[3]{300}}{15}$$

108.
$$\frac{27}{\sqrt[5]{8p^3q^4}} = \frac{27}{\sqrt[5]{8p^3q^4}} \cdot \frac{\sqrt[5]{4p^2q}}{\sqrt[5]{4p^2q}}$$
$$= \frac{27\sqrt[5]{4p^2q}}{\sqrt[5]{32p^5q^5}}$$
$$= \frac{27\sqrt[5]{4p^2q}}{2pq}$$

109.
$$\frac{6}{7-\sqrt{6}} = \frac{6}{7-\sqrt{6}} \cdot \frac{7+\sqrt{6}}{7+\sqrt{6}}$$
$$= \frac{6\left(7+\sqrt{6}\right)}{7^2-\left(\sqrt{6}\right)^2}$$
$$= \frac{42+6\sqrt{6}}{49-6}$$
$$= \frac{42+6\sqrt{6}}{43}$$

110.
$$\frac{3}{\sqrt{3}-9} = \frac{3}{\sqrt{3}-9} \cdot \frac{\sqrt{3}+9}{\sqrt{3}+9}$$
$$= \frac{3\left(\sqrt{3}+9\right)}{\left(\sqrt{3}\right)^2-9^2}$$
$$= \frac{3\sqrt{3}+27}{3-81}$$
$$= \frac{3\sqrt{3}+27}{-78}$$
$$= -\frac{3\sqrt{3}+27}{78}$$
$$= -\frac{\sqrt{3}+9}{26}$$

111.
$$\frac{\sqrt{3}}{3+\sqrt{2}} = \frac{\sqrt{3}}{3+\sqrt{2}} \cdot \frac{3-\sqrt{2}}{3-\sqrt{2}}$$
$$= \frac{\sqrt{3}\left(3-\sqrt{2}\right)}{3^2-\left(\sqrt{2}\right)^2}$$
$$= \frac{\sqrt{3}\left(3-\sqrt{2}\right)}{9-2}$$
$$= \frac{\sqrt{3}\left(3-\sqrt{2}\right)}{7} \text{ or } \frac{3\sqrt{3}-\sqrt{6}}{7}$$

112.
$$\frac{\sqrt{k}}{\sqrt{k}-\sqrt{m}} = \frac{\sqrt{k}}{\sqrt{k}-\sqrt{m}} \cdot \frac{\sqrt{k}+\sqrt{m}}{\sqrt{k}+\sqrt{m}}$$
$$= \frac{\sqrt{k}\left(\sqrt{k}+\sqrt{m}\right)}{\left(\sqrt{k}\right)^2-\left(\sqrt{m}\right)^2}$$
$$= \frac{k+\sqrt{km}}{k-m}$$

113.
$$\frac{\sqrt{10}+2}{\sqrt{10}-2} = \frac{\sqrt{10}+2}{\sqrt{10}-2} \cdot \frac{\sqrt{10}+2}{\sqrt{10}+2}$$
$$= \frac{\left(\sqrt{10}\right)^2+2 \cdot \sqrt{10} \cdot 2+2^2}{\left(\sqrt{10}\right)^2-2^2}$$
$$= \frac{10+4\sqrt{10}+4}{10-4} = \frac{14+4\sqrt{10}}{6}$$
$$= \frac{2\left(7+2\sqrt{10}\right)}{2(3)} = \frac{7+2\sqrt{10}}{3}$$

114. $\dfrac{3-\sqrt{y}}{3+\sqrt{y}} = \dfrac{3-\sqrt{y}}{3+\sqrt{y}} \cdot \dfrac{3-\sqrt{y}}{3-\sqrt{y}}$

$\qquad = \dfrac{3^2 - 2\cdot 3 \cdot \sqrt{y} + \left(\sqrt{y}\right)^2}{3^2 - \left(\sqrt{y}\right)^2}$

$\qquad = \dfrac{9 - 6\sqrt{y} + y}{9-y} \quad \text{or} \quad \dfrac{y - 6\sqrt{y} + 9}{9-y}$

115. $\dfrac{4}{2\sqrt{3}+5\sqrt{2}} = \dfrac{4}{2\sqrt{3}+5\sqrt{2}} \cdot \dfrac{2\sqrt{3}-5\sqrt{2}}{2\sqrt{3}-5\sqrt{2}}$

$\qquad = \dfrac{4\left(2\sqrt{3}-5\sqrt{2}\right)}{\left(2\sqrt{3}\right)^2 - \left(5\sqrt{2}\right)^2}$

$\qquad = \dfrac{8\sqrt{3}-20\sqrt{2}}{4\cdot 3 - 25\cdot 2}$

$\qquad = \dfrac{8\sqrt{3}-20\sqrt{2}}{12-50}$

$\qquad = \dfrac{8\sqrt{3}-20\sqrt{2}}{-38}$

$\qquad = \dfrac{-2\left(-4\sqrt{3}+10\sqrt{2}\right)}{-2(19)}$

$\qquad = \dfrac{10\sqrt{2}-4\sqrt{3}}{19}$

116. $\dfrac{\sqrt{5}-\sqrt{6}}{\sqrt{10}+\sqrt{3}} = \dfrac{\sqrt{5}-\sqrt{6}}{\sqrt{10}+\sqrt{3}} \cdot \dfrac{\sqrt{10}-\sqrt{3}}{\sqrt{10}-\sqrt{3}}$

$\qquad = \dfrac{\left(\sqrt{5}-\sqrt{6}\right)\left(\sqrt{10}-\sqrt{3}\right)}{\left(\sqrt{10}+\sqrt{3}\right)\left(\sqrt{10}-\sqrt{3}\right)}$

$\qquad = \dfrac{\sqrt{50}-\sqrt{15}-\sqrt{60}+\sqrt{18}}{\left(\sqrt{10}\right)^2 - \left(\sqrt{3}\right)^2}$

$\qquad = \dfrac{5\sqrt{2}-\sqrt{15}-2\sqrt{15}+3\sqrt{2}}{10-3}$

$\qquad = \dfrac{8\sqrt{2}-3\sqrt{15}}{7}$

117. $\dfrac{\sqrt{7}}{3} + \dfrac{6}{\sqrt{7}} = \dfrac{\sqrt{7}}{3} \cdot \dfrac{\sqrt{7}}{\sqrt{7}} + \dfrac{6}{\sqrt{7}} \cdot \dfrac{3}{3}$

$\qquad = \dfrac{7}{3\sqrt{7}} + \dfrac{18}{3\sqrt{7}}$

$\qquad = \dfrac{25}{3\sqrt{7}}$

$\qquad = \dfrac{25}{3\sqrt{7}} \cdot \dfrac{\sqrt{7}}{\sqrt{7}}$

$\qquad = \dfrac{25\sqrt{7}}{3\cdot 7}$

$\qquad = \dfrac{25\sqrt{7}}{21}$

118. $\left(4-\sqrt{7}\right)^{-1} = \dfrac{1}{4-\sqrt{7}}$

$\qquad = \dfrac{1}{4-\sqrt{7}} \cdot \dfrac{4+\sqrt{7}}{4+\sqrt{7}}$

$\qquad = \dfrac{4+\sqrt{7}}{4^2 - \left(\sqrt{7}\right)^2}$

$\qquad = \dfrac{4+\sqrt{7}}{16-7}$

$\qquad = \dfrac{4+\sqrt{7}}{9}$

119. $f(x) = \sqrt{x+4}$

(a) $f(-3) = \sqrt{-3+4} = \sqrt{1} = 1$

(b) $f(0) = \sqrt{0+4} = \sqrt{4} = 2$

(c) $f(5) = \sqrt{5+4} = \sqrt{9} = 3$

120. $g(x) = \sqrt{3x-2}$

(a) $g\left(\dfrac{2}{3}\right) = \sqrt{3\left(\dfrac{2}{3}\right)-2} = \sqrt{2-2} = \sqrt{0} = 0$

(b) $g(2) = \sqrt{3(2)-2} = \sqrt{6-2} = \sqrt{4} = 2$

(c) $g(6) = \sqrt{3(6)-2} = \sqrt{18-2} = \sqrt{16} = 4$

121. $H(t) = \sqrt[3]{t+3}$

(a) $H(-2) = \sqrt[3]{-2+3} = \sqrt[3]{1} = 1$

(b) $H(-4) = \sqrt[3]{-4+3} = \sqrt[3]{-1} = -1$

(c) $H(5) = \sqrt[3]{5+3} = \sqrt[3]{8} = 2$

122. $G(z) = \sqrt{\dfrac{z-1}{z+2}}$

(a) $G(1) = \sqrt{\dfrac{1-1}{1+2}} = \sqrt{\dfrac{0}{3}} = \sqrt{0} = 0$

(b) $G(-3) = \sqrt{\dfrac{-3-1}{-3+2}} = \sqrt{\dfrac{-4}{-1}} = \sqrt{4} = 2$

(c) $G(2) = \sqrt{\dfrac{2-1}{2+2}} = \sqrt{\dfrac{1}{4}} = \dfrac{1}{2}$

123. $f(x) = \sqrt{3x-5}$

$3x - 5 \geq 0$

$\quad 3x \geq 5$

$\quad\quad x \geq \dfrac{5}{3}$

The domain of the function is $\left\{ x \mid x \geq \dfrac{5}{3} \right\}$ or the

interval $\left[\dfrac{5}{3}, \infty \right)$.

124. $g(x) = \sqrt[3]{2x-7}$

Since the index is odd, the domain of the function is all real numbers, $\{ x \mid x \text{ is any real number} \}$ or the interval $(-\infty, \infty)$.

125. $h(x) = \sqrt[4]{6x+1}$

$6x + 1 \geq 0$

$\quad 6x \geq -1$

$\quad\quad x \geq -\dfrac{1}{6}$

The domain of the function is $\left\{ x \mid x \geq -\dfrac{1}{6} \right\}$ or

the interval $\left[-\dfrac{1}{6}, \infty \right)$.

126. $F(x) = \sqrt[5]{2x-9}$

Since the index is odd, the domain of the function is all real numbers, $\{ x \mid x \text{ is any real number} \}$, or the interval $(-\infty, \infty)$.

127. $G(x) = \sqrt{\dfrac{4}{x-2}}$

The index is even so we need the radicand to be nonnegative. The numerator is positive, so we need the denominator to be positive as well.

$x - 2 > 0$

$\quad x > 2$

The domain of the function is $\{ x \mid x > 2 \}$ or the

interval $(2, \infty)$.

128. $H(x) = \sqrt{\dfrac{x-3}{x}}$

The index is even so we need the radicand to be nonnegative. The radicand will equal zero when the numerator equals zero. The radicand will be positive when the numerator and denominator have the same sign.

$x - 3 = 0$

$\quad x = 3$

$x - 3 > 0 \quad \text{and} \quad x > 0$

$\quad x > 3 \quad \text{and} \quad x > 0 \quad \rightarrow \quad x > 3$

$x - 3 < 0 \quad \text{and} \quad x < 0$

$\quad x < 3 \quad \text{and} \quad x < 0 \quad \rightarrow \quad x < 0$

The domain of the function is $\{ x \mid x < 0 \text{ or } x \geq 3 \}$ or the interval $(-\infty, 0) \cup [3, \infty)$.

129. $f(x) = \dfrac{1}{2}\sqrt{1-x}$

(a) $1 - x \geq 0$

$\quad -x \leq 1$

The domain of the function is $\{ x \mid x \leq 1 \}$ or the interval $(-\infty, 1]$.

(b)

x	$f(x) = \frac{1}{2}\sqrt{1-x}$	(x,y)
-15	$f(-15) = \frac{1}{2}\sqrt{1-(-15)} = 2$	$(-15, 2)$
-8	$f(-8) = \frac{3}{2}$	$\left(-8, \frac{3}{2}\right)$
-3	$f(-3) = 1$	$(-3, 1)$
0	$f(0) = \frac{1}{2}$	$\left(0, \frac{1}{2}\right)$
1	$f(1) = 0$	$(1, 0)$

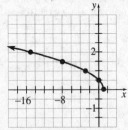

(c) Based on the graph, the range is $[0, \infty)$.

130. $g(x) = \sqrt{x+1} - 2$

(a) $x + 1 \ge 0$ or $x \ge -1$
The domain of the function is $\{x \mid x \ge -1\}$
or the interval $[-1, \infty)$.

(b)

x	$g(x) = \sqrt{x+1} - 2$	(x,y)
-1	$g(-1) = \sqrt{-1+1} - 2 = -2$	$(-1, -2)$
0	$g(0) = -1$	$(0, -1)$
3	$g(3) = 0$	$(3, 0)$
8	$g(8) = 1$	$(8, 1)$
15	$g(15) = 2$	$(15, 2)$

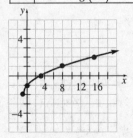

(c) Based on the graph, the range is $[-2, \infty)$.

131. $h(x) = -\sqrt{x+3}$

(a) $x + 3 \ge 0$
$x \ge -3$
The domain of the function is $\{x \mid x \ge -3\}$
or the interval $[-3, \infty)$.

(b)

x	$h(x) = -\sqrt{x+3}$	(x,y)
-3	$h(-3) = -\sqrt{-3+3} = 0$	$(-3, 0)$
-2	$h(-2) = -1$	$(-2, -1)$
1	$h(1) = -2$	$(1, -2)$
6	$h(6) = -3$	$(6, -3)$
13	$h(13) = -4$	$(13, -4)$

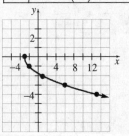

(c) Based on the graph, the range is $(-\infty, 0]$.

132. $F(x) = \sqrt[3]{x+1}$

(a) Since the index is odd, the domain of the
function is all real numbers,
$\{x \mid x \text{ is any real number}\}$, or the interval
$(-\infty, \infty)$.

(b)

x	$F(x) = \sqrt[3]{x+1}$	(x,y)
-9	$F(-9) = \sqrt[3]{-9+1} = -2$	$(-9, -2)$
-2	$F(-2) = -1$	$(-2, -1)$
-1	$F(-1) = 0$	$(-1, 0)$
0	$F(0) = 1$	$(0, 1)$
7	$F(7) = 2$	$(7, 2)$

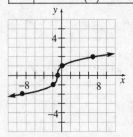

(c) Based on the graph, the range is $(-\infty, \infty)$.

133. $\sqrt{m} = 13$

$\left(\sqrt{m}\right)^2 = 13^2$

$m = 169$

Check: $\sqrt{169} = 13$?

$13 = 13$ T

The solution set is $\{169\}$.

134. $\sqrt[3]{3t+1} = -2$

$\left(\sqrt[3]{3t+1}\right)^3 = (-2)^3$

$3t + 1 = -8$

$3t = -9$

$t = -3$

Check: $\sqrt[3]{3(-3)+1} = -2$?

$\sqrt[3]{-9+1} = -2$?

$\sqrt[3]{-8} = -2$?

$-2 = -2$ T

The solution set is $\{-3\}$.

135. $\sqrt[4]{3x-8} = 3$

$\left(\sqrt[4]{3x-8}\right)^4 = 3^4$

$3x - 8 = 81$

$3x = 89$

$x = \dfrac{89}{3}$

Check: $\sqrt[4]{3\left(\dfrac{89}{3}\right)-8} = 3$?

$\sqrt[4]{89-8} = 3$?

$\sqrt[4]{81} = 3$?

$3 = 3$ Y

The solution set is $\left\{\dfrac{89}{3}\right\}$.

136. $\sqrt{2x+5} + 4 = 2$

$\sqrt{2x+5} = -2$

$\left(\sqrt{2x+5}\right)^2 = (-2)^2$

$2x + 5 = 4$

$2x = -1$

$x = -\dfrac{1}{2}$

Check: $\sqrt{2\left(-\dfrac{1}{2}\right)+5} + 4 = 2$?

$\sqrt{-1+5} + 4 = 2$?

$\sqrt{4} + 4 = 2$?

$2 + 4 = 2$?

$6 = 2$ False

The solution does not check so the problem has no real solution.

137. $\sqrt{4-k} - 3 = 0$

$\sqrt{4-k} = 3$

$\left(\sqrt{4-k}\right)^2 = 3^2$

$4 - k = 9$

$-5 = k$

Check: $\sqrt{4-(-5)} - 3 = 0$?

$\sqrt{9} - 3 = 0$?

$3 - 3 = 0$?

$0 = 0$ ✓

The solution set is $\{-5\}$.

138. $3\sqrt{t} - 4 = 11$

$3\sqrt{t} = 15$

$\sqrt{t} = 5$

$\left(\sqrt{t}\right)^2 = 5^2$

$t = 25$

Check: $3\sqrt{25} - 4 = 11$?

$3 \cdot 5 - 4 = 11$?

$15 - 4 = 11$?

$11 = 11$ T

The solution set is $\{25\}$.

139. $2\sqrt[3]{m}+5=-11$

$2\sqrt[3]{m}=-16$

$\sqrt[3]{m}=-8$

$\left(\sqrt[3]{m}\right)^3=(-8)^3$

$m=-512$

Check: $2\sqrt[3]{-512}+5=-11$?

$2(-8)+5=-11$?

$-16+5=-11$?

$-11=-11$ T

The solution set is $\{-512\}$.

140. $\sqrt{q+2}=q$

$\left(\sqrt{q+2}\right)^2=q^2$

$q+2=q^2$

$q^2-q-2=0$

$(q-2)(q+1)=0$

$q-2=0$ or $q+1=0$

$q=2$ or $q=-1$

Check: $\sqrt{2+2}=2$?

$\sqrt{4}=2$?

$2=2$ T

$\sqrt{-1+2}=-1$?

$\sqrt{1}=-1$?

$1=-1$ False

The second solution is extraneous, so the solution set is $\{2\}$.

141. $\sqrt{w+11}+3=w+2$

$\sqrt{w+11}=w-1$

$\left(\sqrt{w+11}\right)^2=(w-1)^2$

$w+11=w^2-2w+1$

$w^2-3w-10=0$

$(w-5)(w+2)=0$

$w-5=0$ or $w+2=0$

$w=5$ or $w=-2$

Check: $\sqrt{5+11}+3=5+2$?

$\sqrt{16}+3=7$?

$4+3=7$?

$7=7$ T

$\sqrt{-2+11}+3=-2+2$?

$\sqrt{9}+3=0$?

$3+3=0$?

$6=0$ False

The second solution is extraneous, so the solution set is $\{5\}$.

142. $\sqrt{p^2-2p+9}=p+1$

$\left(\sqrt{p^2-2p+9}\right)^2=(p+1)^2$

$p^2-2p+9=p^2+2p+1$

$-4p+8=0$

$-4p=-8$

$p=2$

Check: $\sqrt{2^2-2(2)+9}=2+1$?

$\sqrt{4-4+9}=3$?

$\sqrt{9}=3$?

$3=3$ T

The solution set is $\{2\}$.

143. $\sqrt{a+10}=\sqrt{2a-1}$

$\left(\sqrt{a+10}\right)^2=\left(\sqrt{2a-1}\right)^2$

$a+10=2a-1$

$11=a$

Check: $\sqrt{11+10}=\sqrt{2(11)-1}$?

$\sqrt{21}=\sqrt{22-1}$?

$\sqrt{21}=\sqrt{21}$ T

The solution set is $\{11\}$.

144. $\sqrt{5x+9}=\sqrt{7x-3}$

$\left(\sqrt{5x+9}\right)^2=\left(\sqrt{7x-3}\right)^2$

$5x+9=7x-3$

$12=2x$

$6=x$

Check: $\sqrt{5(6)+9}=\sqrt{7(6)-3}$?

$\sqrt{30+9}=\sqrt{42-3}$?

$\sqrt{39}=\sqrt{39}$ T

The solution set is $\{6\}$.

145. $\sqrt{c-8}+\sqrt{c}=4$

$$\sqrt{c-8}=4-\sqrt{c}$$

$$\left(\sqrt{c-8}\right)^2=\left(4-\sqrt{c}\right)^2$$

$$c-8=4^2-2(4)\sqrt{c}+\left(\sqrt{c}\right)^2$$

$$c-8=16-8\sqrt{c}+c$$

$$8\sqrt{c}=24$$

$$\sqrt{c}=3$$

$$\left(\sqrt{c}\right)^2=3^2$$

$$c=9$$

Check: $\sqrt{9-8}+\sqrt{9}=4$?

$$\sqrt{1}+\sqrt{9}=4 \ ?$$

$$1+3=4 \ ?$$

$$4=4 \ \text{T}$$

The solution set is $\{9\}$.

146. $\sqrt{x+2}-\sqrt{x+9}=7$

$$\sqrt{x+2}=\sqrt{x+9}+7$$

$$\left(\sqrt{x+2}\right)^2=\left(\sqrt{x+9}+7\right)^2$$

$$x+2=\left(\sqrt{x+9}\right)^2+2(7)\sqrt{x+9}+7^2$$

$$x+2=x+9+14\sqrt{x+9}+49$$

$$x+2=x+14\sqrt{x+9}+58$$

$$-56=14\sqrt{x+9}$$

$$-4=\sqrt{x+9}$$

$$(-4)^2=\left(\sqrt{x+9}\right)^2$$

$$16=x+9$$

$$7=x$$

Check: $\sqrt{7+2}-\sqrt{7+9}=7$?

$$\sqrt{9}-\sqrt{16}=7 \ ?$$

$$3-4=7 \ ?$$

$$-1=7 \ \text{False}$$

The solution does not check so the equation has no real solution.

147. $(4x-3)^{1/3}-3=0$

$$(4x-3)^{1/3}=3$$

$$\left((4x-3)^{1/3}\right)^3=3^3$$

$$4x-3=27$$

$$4x=30$$

$$x=\frac{30}{4}=\frac{15}{2}$$

Check: $\left(4\left(\frac{15}{2}\right)-3\right)^{1/3}-3=0$?

$$(30-3)^{1/3}-3=0 \ ?$$

$$27^{1/3}-3=0 \ ?$$

$$3-3=0 \ ?$$

$$0=0 \ \text{T}$$

The solution set is $\left\{\frac{15}{2}\right\}$.

148. $\left(x^2-9\right)^{1/4}=2$

$$\left(\left(x^2-9\right)^{1/4}\right)^4=2^4$$

$$x^2-9=16$$

$$x^2=25$$

$$x=\pm5$$

Check:

$$\left(5^2-9\right)^{1/4}=2 \ ?$$

$$(25-9)^{1/4}=2 \ ?$$

$$16^{1/4}=2 \ ?$$

$$2=2 \ \text{T}$$

$$\left((-5)^2-9\right)^{1/4}=2 \ ?$$

$$(25-9)^{1/4}=2 \ ?$$

$$16^{1/4}=2 \ ?$$

$$2=2 \ \text{T}$$

The solution set is $\{-5,5\}$.

149.
$$r = \sqrt{\frac{3V}{\pi h}}$$
$$(r)^2 = \left(\sqrt{\frac{3V}{\pi h}}\right)^2$$
$$r^2 = \frac{3V}{\pi h}$$
$$h \cdot r^2 = \frac{3V}{\pi}$$
$$h = \frac{3V}{\pi r^2}$$

150.
$$f_s = \sqrt[3]{\frac{30}{v}}$$
$$(f_s)^3 = \left(\sqrt[3]{\frac{30}{v}}\right)^3$$
$$f_s^{\;3} = \frac{30}{v}$$
$$v \cdot f_s^{\;3} = 30$$
$$v = \frac{30}{f_s^{\;3}}$$

151. $\sqrt{-29} = \sqrt{-1} \cdot \sqrt{29} = i \cdot \sqrt{29} = \sqrt{29}\, i$

152. $\sqrt{-54} = \sqrt{-1} \cdot \sqrt{9} \cdot \sqrt{6} = i \cdot 3 \cdot \sqrt{6} = 3\sqrt{6}\, i$

153. $14 - \sqrt{-162} = 14 - \sqrt{-1} \cdot \sqrt{81} \cdot \sqrt{2}$
$$= 14 - i \cdot 9 \cdot \sqrt{2}$$
$$= 14 - 9\sqrt{2}\, i$$

154.
$$\frac{6 + \sqrt{-45}}{3} = \frac{6 + \sqrt{-1} \cdot \sqrt{9} \cdot \sqrt{5}}{3}$$
$$= \frac{6 + i \cdot 3 \cdot \sqrt{5}}{3}$$
$$= \frac{6 + 3\sqrt{5}\, i}{3}$$
$$= \frac{3\left(2 + \sqrt{5}\, i\right)}{3}$$
$$= 2 + \sqrt{5}\, i$$

155. $(3 - 7i) + (-2 + 5i) = 3 - 2 - 7i + 5i$
$$= 1 - 2i$$

156. $(4 + 2i) - (9 - 8i) = 4 + 2i - 9 + 8i$
$$= 4 - 9 + 2i + 8i$$
$$= -5 + 10i$$

157. $\left(8 - \sqrt{-45}\right) - \left(3 + \sqrt{-80}\right)$
$$= \left(8 - 3\sqrt{5}\, i\right) - \left(3 + 4\sqrt{5}\, i\right)$$
$$= 8 - 3\sqrt{5}\, i - 3 - 4\sqrt{5}\, i$$
$$= 8 - 3 - 3\sqrt{5}\, i - 4\sqrt{5}\, i$$
$$= 5 - 7\sqrt{5}\, i$$

158. $\left(1 + \sqrt{-9}\right) + \left(-6 + \sqrt{-16}\right) = (1 + 3i) + (-6 + 4i)$
$$= 1 + 3i - 6 + 4i$$
$$= 1 - 6 + 3i + 4i$$
$$= -5 + 7i$$

159. $(4 - 5i)(3 + 7i) = 4 \cdot 3 + 4 \cdot 7i - 5i \cdot 3 - 5i \cdot 7i$
$$= 12 + 28i - 15i - 35i^2$$
$$= 12 + 13i - 35(-1)$$
$$= 12 + 13i + 35$$
$$= 47 + 13i$$

160. $\left(\frac{1}{2} + \frac{2}{3}i\right)(4 - 9i)$
$$= \frac{1}{2} \cdot 4 + \frac{1}{2}(-9i) + \frac{2}{3}i \cdot 4 + \frac{2}{3}i(-9i)$$
$$= 2 - \frac{9}{2}i + \frac{8}{3}i - 6i^2$$
$$= 2 - \frac{11}{6}i - 6(-1)$$
$$= 2 - \frac{11}{6}i + 6$$
$$= 8 - \frac{11}{6}i$$

161. $\sqrt{-3} \cdot \sqrt{-27} = \sqrt{3}\, i \cdot 3\sqrt{3}\, i$
$$= 3\left(\sqrt{3}\right)^2 \cdot i^2$$
$$= 3 \cdot 3 \cdot (-1)$$
$$= -9$$

Note: It is necessary to write each radical in terms of i prior to performing the operation. Otherwise the sign of the answer will be wrong.
$$\sqrt{(-3)(-27)} = \sqrt{81} = 9 \neq -9 = \sqrt{-3} \cdot \sqrt{-27}$$

162. $\left(1+\sqrt{-36}\right)\left(-5-\sqrt{-144}\right)$

$= \left(1+6i\right)\left(-5-12i\right)$

$= 1\cdot(-5)+1\cdot(-12i)+6i\cdot(-5)+6i\cdot(-12i)$

$= -5-12i-30i-72i^2$

$= -5-42i-72(-1)$

$= -5-42i+72$

$= 67-42i$

163. $\left(1+12i\right)\left(1-12i\right) = 1^2-\left(12i\right)^2$

$= 1-144i^2$

$= 1-144(-1)$

$= 1+144$

$= 145$

164. $\left(7+2i\right)\left(5+4i\right) = 7\cdot5+7\cdot4i+2i\cdot5+2i\cdot4i$

$= 35+28i+10i+8i^2$

$= 35+38i+8(-1)$

$= 35+38i-8$

$= 27+38i$

165. $\dfrac{4}{3+5i} = \dfrac{4}{3+5i}\cdot\dfrac{3-5i}{3-5i}$

$= \dfrac{4(3-5i)}{3^2-\left(5i\right)^2}$

$= \dfrac{4(3-5i)}{9-25i^2}$

$= \dfrac{4(3-5i)}{9+25}$

$= \dfrac{4(3-5i)}{34}$

$= \dfrac{2\cdot2(3-5i)}{2\cdot17}$

$= \dfrac{2(3-5i)}{17}$

$= \dfrac{6-10i}{17}$

$= \dfrac{6}{17}-\dfrac{10}{17}i$

166. $\dfrac{-3}{7-2i} = \dfrac{-3}{7-2i}\cdot\dfrac{7+2i}{7+2i}$

$= \dfrac{-3(7+2i)}{7^2-\left(2i\right)^2}$

$= \dfrac{-21-6i}{49-4i^2}$

$= \dfrac{-21-6i}{49-4(-1)}$

$= \dfrac{-21-6i}{49+4}$

$= \dfrac{-21-6i}{53}$

$= -\dfrac{21}{53}-\dfrac{6}{53}i$

167. $\dfrac{2-3i}{5+2i} = \dfrac{2-3i}{5+2i}\cdot\dfrac{5-2i}{5-2i}$

$= \dfrac{(2-3i)(5-2i)}{(5+2i)(5-2i)}$

$= \dfrac{10-4i-15i+6i^2}{25-4i^2}$

$= \dfrac{10-19i+6(-1)}{25-4(-1)}$

$= \dfrac{10-19i-6}{25+4}$

$= \dfrac{4-19i}{29}$

$= \dfrac{4}{29}-\dfrac{19}{29}i$

168. $\dfrac{4+3i}{1-i} = \dfrac{4+3i}{1-i}\cdot\dfrac{1+i}{1+i}$

$= \dfrac{(4+3i)(1+i)}{(1-i)(1+i)}$

$= \dfrac{4+4i+3i+3i^2}{1-i^2}$

$= \dfrac{4+7i+3(-1)}{1-(-1)}$

$= \dfrac{4+7i-3}{1+1}$

$= \dfrac{1+7i}{2}$

$= \dfrac{1}{2}+\dfrac{7}{2}i$

169. $i^{59} = i^{56}\cdot i^3 = \left(i^4\right)^{14}\cdot i^3 = 1\cdot i^3 = i^2\cdot i = -1\cdot i = -i$

170. $i^{173} = i^{172} \cdot i = \left(i^4\right)^{43} \cdot i = 1 \cdot i = i$

Chapter 9 Test

1. $49^{-1/2} = \dfrac{1}{49^{1/2}} = \dfrac{1}{\sqrt{49}} = \dfrac{1}{7}$

2. $\sqrt[3]{8x^{1/2}y^3} \cdot \sqrt{9xy^{1/2}}$

$= \left(8x^{1/2}y^3\right)^{1/3} \cdot \left(9xy^{1/2}\right)^{1/2}$

$= 8^{1/3}\left(x^{1/2}\right)^{1/3}\left(y^3\right)^{1/3} \cdot 9^{1/2}x^{1/2}\left(y^{1/2}\right)^{1/2}$

$= 2x^{1/6}y \cdot 3x^{1/2}y^{1/4}$

$= 6x^{\frac{1}{6}+\frac{1}{2}}y^{1+\frac{1}{4}}$

$= 6x^{2/3}y^{5/4}$

$= 6x^{8/12}y^{15/12}$

$= 6y\sqrt[12]{x^8 y^3}$

3. $\sqrt[5]{\left(2a^4b^3\right)^7} = \left(2a^4b^3\right)^{7/5} = 2^{7/5}a^{28/5}b^{21/5}$

$= 2a^5b^4 \cdot 2^{2/5}a^{3/5}b^{1/5}$

$= 2a^5b^4\sqrt[5]{4a^3b}$

4. $\sqrt{3m} \cdot \sqrt{13n} = \sqrt{3m \cdot 13n} = \sqrt{39mn}$

5. $\sqrt{32x^7y^4} = \sqrt{16x^6y^4 \cdot 2x}$

$= \sqrt{16x^6y^4} \cdot \sqrt{2x}$

$= 4x^3y^2\sqrt{2x}$

6. $\dfrac{\sqrt{9a^3b^{-3}}}{\sqrt{4ab}} = \sqrt{\dfrac{9a^3b^{-3}}{4ab}}$

$= \sqrt{\dfrac{9a^2}{4b^4}}$

$= \dfrac{\sqrt{9a^2}}{\sqrt{4b^4}}$

$= \dfrac{3a}{2b^2}$

7. $\sqrt{5x^3} + 2\sqrt{45x} = \sqrt{x^2 \cdot 5x} + 2\sqrt{9 \cdot 5x}$

$= \sqrt{x^2} \cdot \sqrt{5x} + 2\sqrt{9} \cdot \sqrt{5x}$

$= x\sqrt{5x} + 2 \cdot 3 \cdot \sqrt{5x}$

$= x\sqrt{5x} + 6\sqrt{5x}$

$= (x+6)\sqrt{5x}$

8. $\sqrt{9a^2b} - \sqrt[4]{16a^4b^2} = \sqrt{9a^2} \cdot \sqrt{b} - \sqrt[4]{16a^4} \cdot \sqrt[4]{b^2}$

$= 3a\sqrt{b} - 2a \cdot b^{2/4}$

$= 3a\sqrt{b} - 2a \cdot b^{1/2}$

$= 3a\sqrt{b} - 2a\sqrt{b}$

$= (3-2)a\sqrt{b}$

$= a\sqrt{b}$

9. $\left(11+2\sqrt{x}\right)\left(3-\sqrt{x}\right)$

$= 11 \cdot 3 + 11\left(-\sqrt{x}\right) + 2\sqrt{x} \cdot 3 + 2\sqrt{x}\left(-\sqrt{x}\right)$

$= 33 - 11\sqrt{x} + 6\sqrt{x} - 2\left(\sqrt{x}\right)^2$

$= 33 + (-11+6)\sqrt{x} - 2x$

$= 33 - 5\sqrt{x} - 2x$

10. $\dfrac{-2}{3\sqrt{72}} = \dfrac{-2}{3 \cdot 6\sqrt{2}} = \dfrac{-2}{18\sqrt{2}}$

$= \dfrac{-1}{9\sqrt{2}} = \dfrac{-1}{9\sqrt{2}} \cdot \dfrac{\sqrt{2}}{\sqrt{2}}$

$= \dfrac{-\sqrt{2}}{9 \cdot 2}$

$= \dfrac{-\sqrt{2}}{18}$

11. $\dfrac{\sqrt{5}}{\sqrt{5}+2} = \dfrac{\sqrt{5}}{\sqrt{5}+2} \cdot \dfrac{\sqrt{5}-2}{\sqrt{5}-2}$

$= \dfrac{\sqrt{5}\left(\sqrt{5}-2\right)}{\left(\sqrt{5}\right)^2 - 2^2}$

$= \dfrac{5 - 2\sqrt{5}}{5-4}$

$= 5 - 2\sqrt{5}$

12. $f(x) = \sqrt{-2x+3}$

(a) $f(1) = \sqrt{-2(1)+3} = \sqrt{-2+3} = \sqrt{1} = 1$

(b) $f(-3) = \sqrt{-2(-3)+3} = \sqrt{6+3} = \sqrt{9} = 3$

13. $g(x) = \sqrt{-3x+5}$

$-3x+5 \geq 0$

$-3x \geq -5$

$x \leq \dfrac{5}{3}$

The domain of the function is $\left\{ x \mid x \leq \dfrac{5}{3} \right\}$ or the

interval $\left(-\infty, \dfrac{5}{3} \right]$.

14. $f(x) = \sqrt{x} - 3$

(a) $x \geq 0$

The domain of the function is $\{ x \mid x \geq 0 \}$ or

the interval $[0, \infty)$.

(b)

x	$f(x) = \sqrt{x} - 3$	(x, y)
0	$f(0) = \sqrt{0} - 3 = -3$	$(0, -3)$
1	$f(1) = -2$	$(1, -2)$
4	$f(4) = -1$	$(4, -1)$
9	$f(9) = 0$	$(9, 0)$

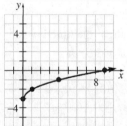

(c) Based on the graph, the range is $[-3, \infty)$.

15. $\sqrt{x+3} = 4$

$\left(\sqrt{x+3} \right)^2 = 4^2$

$x + 3 = 16$

$x = 13$

Check: $\sqrt{13+3} = 4$?

$\sqrt{16} = 4$?

$4 = 4$ T

The solution set is $\{ 13 \}$.

16. $\sqrt{x+13} - 4 = x - 3$

$\sqrt{x+13} = x + 1$

$\left(\sqrt{x+13} \right)^2 = (x+1)^2$

$x + 13 = x^2 + 2x + 1$

$x^2 + x - 12 = 0$

$(x+4)(x-3) = 0$

$x + 4 = 0 \quad \text{or} \quad x - 3 = 0$

$x = -4 \quad \text{or} \quad x = 3$

Check: $\sqrt{-4+13} - 4 = -4 - 3$?

$\sqrt{9} - 4 = -7$?

$3 - 4 = -7$?

$-1 = -7$ False

$\sqrt{3+13} - 4 = 3 - 3$?

$\sqrt{16} - 4 = 0$?

$4 - 4 = 0$?

$0 = 0$ T

The first solution is extraneous, so the solution set is $\{ 3 \}$.

17. $\sqrt{x-1} + \sqrt{x+2} = 3$

$\sqrt{x-1} = 3 - \sqrt{x+2}$

$\left(\sqrt{x-1} \right)^2 = \left(3 - \sqrt{x+2} \right)^2$

$x - 1 = 3^2 - 2(3)\sqrt{x+2} + \left(\sqrt{x+2} \right)^2$

$x - 1 = 9 - 6\sqrt{x+2} + x + 2$

$-12 = -6\sqrt{x+2}$

$2 = \sqrt{x+2}$

$2^2 = \left(\sqrt{x+2} \right)^2$

$4 = x + 2$

$2 = x$

Check: $\sqrt{2-1} + \sqrt{2+2} = 3$?

$\sqrt{1} + \sqrt{4} = 3$?

$1 + 2 = 3$?

$3 = 3$ T

The solution set is $\{ 2 \}$.

18. $(13 + 2i) + (4 - 15i) = 13 + 2i + 4 - 15i$

$= 13 + 4 + 2i - 15i$

$= 17 - 13i$

19. $(4-7i)(2+3i) = 4\cdot2+4\cdot3i-7i\cdot2-7i\cdot3i$
$$= 8+12i-14i-21i^2$$
$$= 8-2i-21(-1)$$
$$= 8-2i+21$$
$$= 29-2i$$

20. $\dfrac{7-i}{12+11i} = \dfrac{7-i}{12+11i}\cdot\dfrac{12-11i}{12-11i}$
$$= \frac{(7-i)(12-11i)}{(12+11i)(12-11i)}$$
$$= \frac{84-77i-12i+11i^2}{144-121i^2}$$
$$= \frac{84-89i+11(-1)}{144-121(-1)}$$
$$= \frac{84-89i-11}{144+121}$$
$$= \frac{73-89i}{265}$$
$$= \frac{73}{265}-\frac{89}{265}i$$

Cumulative Review Chapters 1–9

1. $6-3^2\div(9-3) = 6-3^2\div6$
$$= 6-9\div6$$
$$= 6-\frac{3}{2}$$
$$= \frac{9}{2}$$

2. $(3x+2y)-(2x-5y+3)+9$
$$= 3x+2y-2x+5y-3+9$$
$$= 3x-2x+2y+5y-3+9$$
$$= x+7y+6$$

3. $(3x+5)-2 = 7x-13$
$$3x+5-2 = 7x-13$$
$$3x+3 = 7x-13$$
$$3x+16 = 7x$$
$$16 = 4x$$
$$\frac{16}{4} = \frac{4x}{4}$$
$$4 = x$$
The solution set is $\{4\}$.

4. $6x+\dfrac{1}{2}(4x-2) \le 3x+9$
$$6x+2x-1 \le 3x+9$$
$$8x-1 \le 3x+9$$
$$5x-1 \le 9$$
$$5x \le 10$$
$$x \le 2$$
Interval: $(-\infty,2]$

5. $f(x) = 3x^2-x+5$

  (a) $f(-2) = 3(-2)^2-(-2)+5$
$$= 3(4)+2+5 = 12+7 = 19$$

  (b) $f(3) = 3(3)^2-(3)+5$
$$= 3(9)-3+5 = 27+2 = 29$$

6. The function g tells us to divide x^2-9 by x^2-2x-8. Since division by 0 is not defined, the denominator can never be 0.
$$x^2-2x-8 = 0$$
$$(x-4)(x+2) = 0$$
$$x-4 = 0 \text{ or } x+2 = 0$$
$$x = 4 \text{ or } x = -2$$
Thus, the domain of the function is $\{x\,|\,x\neq4,-2\}$.

7. **(a)** $n(50) = -50(50)+6,000$
$$= -2,500+6,000$$
$$= 3,500$$
If the price of the game were \$50, you would sell 3500 games per year.

  (b) We want to solve $n(p) = 0$.
$$-50p+6,000 = 0$$
$$6,000 = 50p$$
$$\frac{6,000}{50} = \frac{50p}{50}$$
$$120 = p$$
When the price of the game reaches \$120, no games will be sold.

8. First we use the two points to find the slope of the line.
$$m = \frac{-2-6}{3-(-1)} = \frac{-8}{4} = -2$$

Next, we use the slope and one of the points to determine the y-intercept.

$y = mx + b$
$6 = -2(-1) + b$
$6 = 2 + b$
$4 = b$

The slope is $m = -2$ and the y-intercept is $b = 4$. Therefore, the equation of the line is $y = -2x + 4$.

9. Start by solving the inequality for y.
$6x + 3y > 24$
$3y > -6x + 24$
$\dfrac{3y}{3} > \dfrac{-6x + 24}{3}$
$y > -2x + 8$

Since the inequality is strict, graph the line $y = -2x + 8$ with a dashed line. We will use the point $(0, 0)$ as our test point.

$6(0) + 3(0) > 24$?
$0 > 24$ false

Since we obtained a contradiction, we shade the region that does not contain the point $(0, 0)$.

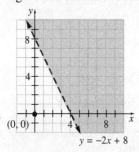

$y = -2x + 8$

10. $4x - y = 17$
$5x + 6y = 14$

Solve the first equation for y.
$4x - y = 17$
$-y = -4x + 17$
$y = 4x - 17$

Substitute this result for y in the second equation and solve for x.
$5x + 6(4x - 17) = 14$
$5x + 24x - 102 = 14$
$29x - 102 = 14$
$29x = 116$
$x = 4$

Use this result to solve for y.
$y = 4x - 17$
$y = 4(4) - 17$
$= 16 - 17$
$= -1$

The ordered pair solution is $(4, -1)$.

Check: $4(4) - (-1) = 17$?
$16 + 1 = 17$?
$17 = 17$ True

$5(4) + 6(-1) = 14$?
$20 - 6 = 14$?
$14 = 14$ True

11. Let x = pounds of dried fruit and y = pounds of nuts. Since the total pounds will be 10, we have the equation
$x + y = 10$
We want the total revenue to be the same. Thus, our second equation will be
$3.45x + 2.10y = 2.64(10)$
$3.45x + 2.10y = 26.40$

Putting the two equations together gives us the following system:
$x + y = 10$
$3.45x + 2.10y = 26.40$

We can solve this system by using substitution. Start by solving the first equation for y.
$x + y = 10$
$y = -x + 10$

Substitute this result for y in the second equation and solve for x.
$3.45x + 2.10(-x + 10) = 26.40$
$3.45x - 2.10x + 21.00 = 26.40$
$1.35x + 21.00 = 26.40$
$1.35x = 5.40$
$x = 4$

Use this result to solve for y.
$y = -x + 10$
$= -4 + 10$
$= 6$

The trail mix should contain 4 pounds of dried fruit and 6 pounds of nuts.

12. $g(-4) = (-4)^3 + 4(-4) - 16$
$= -64 - 16 - 16$
$= -96$

13. $(8x^3 - 4x^2 + 5x + 3) + (2x^2 - 8x + 7)$
$= 8x^3 - 4x^2 + 5x + 3 + 2x^2 - 8x + 7$
$= 8x^3 - 4x^2 + 2x^2 + 5x - 8x + 3 + 7$
$= 8x^3 - 2x^2 - 3x + 10$

14. $(2x-1)\left(4x^2+2x-9\right)$

$=8x^3+4x^2-18x-4x^2-2x+9$

$=8x^3+4x^2-4x^2-18x-2x+9$

$=8x^3-20x+9$

15.

$$2x^2+3x-5\overline{\smash{\big)}\,6x^4+13x^3-21x^2-28x+37}$$

with quotient $3x^2+2x-6$

$-\left(6x^4+9x^3-15x^2\right)$

$4x^3-6x^2-28x$

$-\left(4x^3+6x^2-10x\right)$

$-12x^2-18x+37$

$-\left(-12x^2-18x+30\right)$

7

$$\frac{6x^4+13x^3-21x^2-28x+37}{2x^2+3x-5}$$

$$=3x^2+2x-6+\frac{7}{2x^2+3x-5}$$

16. $8x^2-44x-84$

Begin by factoring out the greatest common factor.

$4\left(2x^2-11x-21\right)$

For the reduced polynomial, we can use the AC method.

$AC=2(-21)=-42$

We need two factors of -42 whose sum is -11. Since the product is negative, the factors will have opposite signs. Since the sum is also negative, the factor with the larger absolute value will be negative.

factor 1	factor 2	sum
2	−21	−19 too small
6	−7	−1 too large
3	−14	−11 okay

$8x^2-44x-84=4\left(2x^2-11x-21\right)$

$=4\left(2x^2+3x-14x-21\right)$

$=4\left(x(2x+3)+(-7)(2x+3)\right)$

$=4(2x+3)(x-7)$

17. $\dfrac{2x}{x-3}-\dfrac{x+1}{x+2}$

To subtract, we need a common denominator. Since both denominators are prime, the LCD will be the product of the denominators.

$LCD=(x-3)(x+2)$

Next we write equivalent fractions using the LCD.

$$\frac{2x(x+2)}{(x-3)(x+2)}-\frac{(x+1)(x-3)}{(x-3)(x+2)}$$

Now we can combine the numerators and simplify.

$$\frac{2x(x+2)-(x+1)(x-3)}{(x-3)(x+2)}$$

$$=\frac{2x^2+4x-\left(x^2-2x-3\right)}{(x-3)(x+2)}$$

$$=\frac{2x^2+4x-x^2+2x+3}{(x-3)(x+2)}$$

$$=\frac{x^2+6x+3}{(x-3)(x+2)}$$

18. $\dfrac{9}{k-2}=\dfrac{6}{k}+3$

$LCD=k(k-2)$

The solution set cannot contain the restricted values 0 or 2 since these values will make one of the terms have a 0 in its denominator. Multiply each term by the LCD.

$$k(k-2)\left(\frac{9}{k-2}\right)=k(k-2)\frac{6}{k}+k(k-2)(3)$$

$$k\cancel{(k-2)}\left(\frac{9}{\cancel{k-2}}\right)=\cancel{k}(k-2)\frac{6}{\cancel{k}}+k(k-2)(3)$$

$9k=6(k-2)+3k(k-2)$

$9k=6k-12+3k^2-6k$

$9k=3k^2-12$

$0=3k^2-9k-12$

$0=k^2-3k-4$

$0=(k-4)(k+1)$

$k-4=0$ or $k+1=0$

$k=4$ or $k=-1$

Neither of these are restricted values, so the solution set is $\{-1,4\}$.

19.

$$3x+1<13 \quad \text{or} \quad -2x+3 \le 23$$
$$3x+1-1<13-1 \quad \text{or} \quad -2x+3-3 \le 23-3$$
$$3x<12 \quad \text{or} \quad -2x \le 20$$
$$\frac{3x}{3}<\frac{12}{3} \quad \text{or} \quad \frac{-2x}{-2} \ge \frac{20}{-2}$$
$$x<4 \quad \text{or} \quad x \ge -10$$

The solution is $\{x \mid x<4 \text{ or } x \ge 10\}$.

Interval: $(-\infty, 4) \cup [10, \infty)$

Graph:

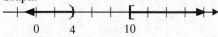

20. Let $x =$ hours needed to paint the room together.

$$\frac{1 \text{ room}}{4 \text{ hours}} + \frac{1 \text{ room}}{6 \text{ hours}} = \frac{1 \text{ room}}{x \text{ hours}}$$
$$12x\left(\frac{1}{4}\right) + 12x\left(\frac{1}{6}\right) = 12x\left(\frac{1}{x}\right)$$
$$3x+2x = 12$$
$$5x = 12$$
$$x = \frac{12}{5} = 2.4$$

It will take Shawn and Payton 2.4 hours to paint the room together.

21. $\dfrac{\sqrt{50a^3b}}{\sqrt{2a^{-1}b^3}} = \sqrt{\dfrac{50a^3b}{2a^{-1}b^3}} = \sqrt{\dfrac{25a^4}{b^2}} = \dfrac{\sqrt{25a^4}}{\sqrt{b^2}} = \dfrac{5a^2}{b}$

22. $f(x) = \sqrt[4]{8-3x}$

Since the index is even, we need the radicand to be greater than or equal to zero.
$$8-3x \ge 0$$
$$-3x \ge -8$$
$$\frac{-3x}{-3} \le \frac{-8}{-3}$$
$$x \le \frac{8}{3}$$

The domain of the function is $\left\{x \mid x \le \dfrac{8}{3}\right\}$ or the

interval $\left(-\infty, \dfrac{8}{3}\right]$.

23. $\sqrt{x+7}-8 = x-7$
$$\sqrt{x+7} = x+1$$
$$\left(\sqrt{x+7}\right)^2 = (x+1)^2$$
$$x+7 = x^2+2x+1$$
$$0 = x^2+x-6$$
$$0 = (x+3)(x-2)$$
$$x+3=0 \text{ or } x-2=0$$
$$x=-3 \text{ or } x=2$$

Check: $\sqrt{-3+7}-8 = -3-7$?
$$\sqrt{4}-8 = -10 \text{ ?}$$
$$2-8 = -10 \text{ ?}$$
$$-6 = -10 \text{ False}$$

$$\sqrt{2+7}-8 = 2-7 \text{ ?}$$
$$\sqrt{9}-8 = -5 \text{ ?}$$
$$3-8 = -5 \text{ ?}$$
$$-5 = -5 \text{ True}$$

The first solution is extraneous, so the solution set is $\{2\}$.

24. $\dfrac{3i}{1-7i} = \dfrac{3i}{1-7i} \cdot \dfrac{1+7i}{1+7i}$

$$= \frac{3i+21i^2}{1-49i^2}$$
$$= \frac{3i+21(-1)}{1-49(-1)}$$
$$= \frac{-21+3i}{1+49}$$
$$= \frac{-21+3i}{50}$$
$$= -\frac{21}{50} + \frac{3}{50}i$$

25. (a) $\dfrac{2}{4-\sqrt{11}} = \dfrac{2}{4-\sqrt{11}} \cdot \dfrac{4+\sqrt{11}}{4+\sqrt{11}}$

$$= \frac{2\left(4+\sqrt{11}\right)}{4^2-\left(\sqrt{11}\right)^2}$$
$$= \frac{8+2\sqrt{11}}{16-11}$$
$$= \frac{8+2\sqrt{11}}{5}$$

Chapter 10

Preparing for Solving Quadratic Equations by Completing the Square

P1. $(2p+3)^2 = (2p)^2 + 2(2p)(3) + 3^2$
$$= 4p^2 + 12p + 9$$

P2. $y^2 - 8y + 16 = y^2 - 2 \cdot y \cdot 4 + 4^2 = (y-4)^2$

P3. $x^2 + 5x - 14 = 0$
$(x+7)(x-2) = 0$
$x + 7 = 0$ or $x - 2 = 0$
$x = -7$ or $x = 2$
The solution set is $\{-7, 2\}$.

P4. $x^2 - 16 = 0$
$(x+4)(x-4) = 0$
$x + 4 = 0$ or $x - 4 = 0$
$x = -4$ or $x = 4$
The solution set is $\{-4, 4\}$.

P5. (a) $\sqrt{36} = 6$ because $6^2 = 36$.

(b) $\sqrt{45} = \sqrt{9 \cdot 5} = 3\sqrt{5}$

(c) $\sqrt{-12} = \sqrt{-1 \cdot 4 \cdot 3} = 2\sqrt{3}\, i$

P6. The complex conjugate is $-3 - 2i$.

P7. $\sqrt{x^2} = |x|$ by definition.

P8. The absolute value of a number x, written $|x|$ is the distance from 0 to x on the real number line.
Alternatively, $|x| = \begin{cases} -x & \text{if } x < 0 \\ x & \text{if } x \geq 0 \end{cases}$.

10.1 Quick Checks

1. $\sqrt{p}$ or $-\sqrt{p}$

2. $p^2 = 48$
$p = \pm\sqrt{48}$
$p = \pm 4\sqrt{3}$
The solution set is $\left\{-4\sqrt{3},\ 4\sqrt{3}\right\}$.

3. $3b^2 = 75$
$b^2 = 25$
$b = \pm\sqrt{25}$
$b = \pm 5$
The solution set is $\{-5, 5\}$.

4. $s^2 - 81 = 0$
$s^2 = 81$
$s = \pm\sqrt{81}$
$s = \pm 9$
The solution set is $\{-9, 9\}$.

5. $d^2 = -72$
$d = \pm\sqrt{-72}$
$d = \pm 6\sqrt{2}\, i$
The solution set is $\left\{-6\sqrt{2}\,i,\ 6\sqrt{2}\,i\right\}$.

6. $3q^2 + 27 = 0$
$3q^2 = -27$
$q^2 = -9$
$q = \pm\sqrt{-9}$
$q = \pm 3i$
The solution set is $\{-3i,\ 3i\}$.

7. $(y+3)^2 = 100$
$y + 3 = \pm\sqrt{100}$
$y + 3 = \pm 10$
$y = -3 \pm 10$
$y = -3 - 10$ or $y = -3 + 10$
$y = -13$ or $y = 7$
The solution set is $\{-13, 7\}$.

8. $(q-5)^2 + 20 = 4$

$\qquad (q-5)^2 = -16$

$\qquad q - 5 = \pm\sqrt{-16}$

$\qquad q - 5 = \pm 4i$

$\qquad q = 5 \pm 4i$

The solution set is $\{5-4i,\ 5+4i\}$.

9. Start: $p^2 + 14p$

Add: $\left(\dfrac{1}{2} \cdot 14\right)^2 = 49$

Result: $p^2 + 14p + 49$

Factored Form: $(p+7)^2$

10. Start: $w^2 + 3w$

Add: $\left(\dfrac{1}{2} \cdot 3\right)^2 = \dfrac{9}{4}$

Result: $w^2 + 3w + \dfrac{9}{4}$

Factored Form: $\left(w + \dfrac{3}{2}\right)^2$

11. $\qquad b^2 + 2b - 8 = 0$

$\qquad b^2 + 2b = 8$

$\qquad b^2 + 2b + \left(\dfrac{1}{2} \cdot 2\right)^2 = 8 + \left(\dfrac{1}{2} \cdot 2\right)^2$

$\qquad b^2 + 2b + 1 = 8 + 1$

$\qquad (b+1)^2 = 9$

$\qquad b + 1 = \pm\sqrt{9}$

$\qquad b + 1 = \pm 3$

$\qquad b = -1 \pm 3$

$\qquad b = -4 \ \text{or} \ b = 2$

The solution set is $\{-4,\ 2\}$.

12. $\qquad z^2 - 8z + 9 = 0$

$\qquad z^2 - 8z = -9$

$\qquad z^2 - 8z + \left(\dfrac{1}{2} \cdot (-8)\right)^2 = -9 + \left(\dfrac{1}{2} \cdot (-8)\right)^2$

$\qquad z^2 - 8z + 16 = -9 + 16$

$\qquad (z-4)^2 = 7$

$\qquad z - 4 = \pm\sqrt{7}$

$\qquad z = 4 \pm \sqrt{7}$

The solution set is $\{4-\sqrt{7}, 4+\sqrt{7}\}$.

13. $\qquad 2q^2 + 6q - 1 = 0$

$\qquad \dfrac{2q^2 + 6q - 1}{2} = \dfrac{0}{2}$

$\qquad q^2 + 3q - \dfrac{1}{2} = 0$

$\qquad q^2 + 3q = \dfrac{1}{2}$

$\qquad q^2 + 3q + \left[\dfrac{1}{2} \cdot 3\right]^2 = \dfrac{1}{2} + \left[\dfrac{1}{2} \cdot 3\right]^2$

$\qquad q^2 + 3q + \dfrac{9}{4} = \dfrac{1}{2} + \dfrac{9}{4}$

$\qquad \left(q + \dfrac{3}{2}\right)^2 = \dfrac{11}{4}$

$\qquad q + \dfrac{3}{2} = \pm\sqrt{\dfrac{11}{4}}$

$\qquad q + \dfrac{3}{2} = \pm\dfrac{\sqrt{11}}{2}$

$\qquad q = -\dfrac{3}{2} \pm \dfrac{\sqrt{11}}{2}$

The solution set is $\left\{\dfrac{-3}{2} - \dfrac{\sqrt{11}}{2},\ \dfrac{-3}{2} + \dfrac{\sqrt{11}}{2}\right\}$, or

$\left\{\dfrac{-3-\sqrt{11}}{2},\ \dfrac{-3+\sqrt{11}}{2}\right\}$.

14.
$$3m^2 + 2m + 7 = 0$$
$$\frac{3m^2 + 2m + 7}{3} = \frac{0}{3}$$
$$m^2 + \frac{2}{3}m + \frac{7}{3} = 0$$
$$m^2 + \frac{2}{3}m \qquad = -\frac{7}{3}$$
$$m^2 + \frac{2}{3}m + \left[\frac{1}{2} \cdot \frac{2}{3}\right]^2 = -\frac{7}{3} + \left[\frac{1}{2} \cdot \frac{2}{3}\right]^2$$
$$m^2 + \frac{2}{3}m + \frac{1}{9} = -\frac{7}{3} + \frac{1}{9}$$
$$\left(m + \frac{1}{3}\right)^2 = -\frac{20}{9}$$
$$m + \frac{1}{3} = \pm\sqrt{-\frac{20}{9}}$$
$$m + \frac{1}{3} = \pm\frac{2\sqrt{5}}{3}i$$
$$m = -\frac{1}{3} \pm \frac{2\sqrt{5}}{3}i$$

The solution set is $\left\{-\frac{1}{3} - \frac{2\sqrt{5}}{3}i, \ -\frac{1}{3} + \frac{2\sqrt{5}}{3}i\right\}$.

15. hypotenuse; legs

16. False. The Pythagorean Theorem states that, in a *right* triangle, the *square* of the length of the hypotenuse is equal to the sum of the squares of the length of the legs.

17.
$$c^2 = a^2 + b^2$$
$$c^2 = 3^2 + 4^2$$
$$ = 9 + 16$$
$$ = 25$$
$$c = \sqrt{25}$$
$$ = 5$$
The hypotenuse is 5 units long.

18. Let d represent the unknown distance.

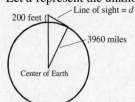

Since the line of sight and the two lines drawn from the center of Earth form a right triangle, we know $\text{Hypotenuse}^2 = \text{Leg}^2 + \text{Leg}^2$. We also

know $200 \text{ feet} = \frac{200}{5280}$ mile, so the hypotenuse is $\left(3960 + \frac{200}{5280}\right)$ miles. Since one leg is the line of sight, d, and the other leg is 3960 miles, we have that

$$d^2 + 3960^2 = \left(3960 + \frac{200}{5280}\right)^2$$
$$d^2 = \left(3960 + \frac{200}{5280}\right)^2 - 3960^2$$
$$d^2 \approx 300.001435$$
$$d \approx \sqrt{300.001435}$$
$$ \approx 17.32$$

The sailor can see approximately 17.32 miles.

10.1 Exercises

19. $y^2 = 100$
$$y = \pm\sqrt{100}$$
$$y = \pm 10$$
The solution set is $\{-10, \ 10\}$.

21. $p^2 = 50$
$$p = \pm\sqrt{50}$$
$$p = \pm 5\sqrt{2}$$
The solution set is $\left\{-5\sqrt{2}, \ 5\sqrt{2}\right\}$.

23. $m^2 = -25$
$$m = \pm\sqrt{-25}$$
$$m = \pm 5i$$
The solution set is $\{-5i, \ 5i\}$.

25. $w^2 = \frac{5}{4}$
$$w = \pm\sqrt{\frac{5}{4}}$$
$$w = \pm\frac{\sqrt{5}}{2}$$

The solution set is $\left\{-\frac{\sqrt{5}}{2}, \ \frac{\sqrt{5}}{2}\right\}$.

27. $x^2 + 5 = 13$

$x^2 = 8$

$x = \pm\sqrt{8}$

$x = \pm 2\sqrt{2}$

The solution set is $\left\{-2\sqrt{2},\ 2\sqrt{2}\right\}$.

29. $3z^2 = 48$

$z^2 = 16$

$z = \pm\sqrt{16}$

$z = \pm 4$

The solution set is $\{-4,\ 4\}$.

31. $3x^2 = 8$

$x^2 = \dfrac{8}{3}$

$x = \pm\sqrt{\dfrac{8}{3}}$

$x = \pm\dfrac{\sqrt{8}}{\sqrt{3}}$

$x = \pm\dfrac{2\sqrt{2}}{\sqrt{3}} \cdot \dfrac{\sqrt{3}}{\sqrt{3}}$

$x = \pm\dfrac{2\sqrt{6}}{3}$

The solution set is $\left\{-\dfrac{2\sqrt{6}}{3},\ \dfrac{2\sqrt{6}}{3}\right\}$.

33. $2p^2 + 23 = 15$

$2p^2 = -8$

$p^2 = -4$

$p = \pm\sqrt{-4}$

$p = \pm 2i$

The solution set is $\{-2i,\ 2i\}$.

35. $(d-1)^2 = -18$

$d - 1 = \pm\sqrt{-18}$

$d - 1 = \pm 3\sqrt{2}i$

$d = 1 \pm 3\sqrt{2}i$

The solution set is $\left\{1 - 3\sqrt{2}i,\ 1 + 3\sqrt{2}i\right\}$.

37. $3(q+5)^2 - 1 = 8$

$3(q+5)^2 = 9$

$(q+5)^2 = 3$

$q + 5 = \pm\sqrt{3}$

$q = -5 \pm\sqrt{3}$

The solution set is $\left\{-5 - \sqrt{3},\ -5 + \sqrt{3}\right\}$.

39. $(3q+1)^2 = 9$

$3q + 1 = \pm\sqrt{9}$

$3q + 1 = \pm 3$

$3q = -1 \pm 3$

$3q = -1 - 3$ or $3q = -1 + 3$

$3q = -4$ or $3q = 2$

$q = -\dfrac{4}{3}$ or $q = \dfrac{2}{3}$

The solution set is $\left\{-\dfrac{4}{3},\ \dfrac{2}{3}\right\}$.

41. $\left(x - \dfrac{2}{3}\right)^2 = \dfrac{5}{9}$

$x - \dfrac{2}{3} = \pm\sqrt{\dfrac{5}{9}}$

$x - \dfrac{2}{3} = \pm\dfrac{\sqrt{5}}{3}$

$x = \dfrac{2}{3} \pm \dfrac{\sqrt{5}}{3}$

The solution set is $\left\{\dfrac{2}{3} - \dfrac{\sqrt{5}}{3},\ \dfrac{2}{3} + \dfrac{\sqrt{5}}{3}\right\}$.

43. $x^2 + 8x + 16 = 81$

$(x+4)^2 = 81$

$x + 4 = \pm\sqrt{81}$

$x + 4 = \pm 9$

$x = -4 \pm 9$

$x = -4 - 9$ or $x = -4 + 9$

$x = -13$ or $x = 5$

The solution set is $\{-13,\ 5\}$.

45. Start: $x^2 + 10x$

Add: $\left(\dfrac{1}{2} \cdot 10\right)^2 = 25$

Result: $x^2 + 10x + 25$

Factored Form: $(x+5)^2$

47. Start: $z^2 - 18z$

Add: $\left[\dfrac{1}{2} \cdot (-18)\right]^2 = 81$

Result: $z^2 - 18z + 81$

Factored Form: $(z - 9)^2$

49. Start: $y^2 + 7y$

Add: $\left(\dfrac{1}{2} \cdot 7\right)^2 = \dfrac{49}{4}$

Result: $y^2 + 7y + \dfrac{49}{4}$

Factored Form: $\left(y + \dfrac{7}{2}\right)^2$

51. Start: $w^2 + \dfrac{1}{2}w$

Add: $\left(\dfrac{1}{2} \cdot \dfrac{1}{2}\right)^2 = \dfrac{1}{16}$

Result: $w^2 + \dfrac{1}{2}w + \dfrac{1}{16}$

Factored Form: $\left(w + \dfrac{1}{4}\right)^2$

53.
$$x^2 + 4x = 12$$
$$x^2 + 4x + \left(\dfrac{1}{2} \cdot 4\right)^2 = 12 + \left(\dfrac{1}{2} \cdot 4\right)^2$$
$$x^2 + 4x + 4 = 12 + 4$$
$$(x + 2)^2 = 16$$
$$x + 2 = \pm\sqrt{16}$$
$$x + 2 = \pm 4$$
$$x = -2 \pm 4$$
$$x = -6 \text{ or } x = 2$$

The solution set is $\{-6, \ 2\}$.

55.
$$x^2 - 4x + 1 = 0$$
$$x^2 - 4x = -1$$
$$x^2 - 4x + \left(\dfrac{1}{2} \cdot (-4)\right)^2 = -1 + \left(\dfrac{1}{2} \cdot (-4)\right)^2$$
$$x^2 - 4x + 4 = -1 + 4$$
$$(x - 2)^2 = 3$$
$$x - 2 = \pm\sqrt{3}$$
$$x = 2 \pm \sqrt{3}$$

The solution set is $\left\{2 - \sqrt{3}, \ 2 + \sqrt{3}\right\}$.

57.
$$a^2 - 4a + 5 = 0$$
$$a^2 - 4a = -5$$
$$a^2 - 4a + \left(\dfrac{1}{2} \cdot (-4)\right)^2 = -5 + \left(\dfrac{1}{2} \cdot (-4)\right)^2$$
$$a^2 - 4a + 4 = -5 + 4$$
$$(a - 2)^2 = -1$$
$$a - 2 = \pm\sqrt{-1}$$
$$a - 2 = \pm i$$
$$a = 2 \pm i$$

The solution set is $\{2 - i, \ 2 + i\}$.

59.
$$b^2 + 5b - 2 = 0$$
$$b^2 + 5b = 2$$
$$b^2 + 5b + \left(\dfrac{1}{2} \cdot 5\right)^2 = 2 + \left(\dfrac{1}{2} \cdot 5\right)^2$$
$$b^2 + 5b + \dfrac{25}{4} = 2 + \dfrac{25}{4}$$
$$\left(b + \dfrac{5}{2}\right)^2 = \dfrac{33}{4}$$
$$b + \dfrac{5}{2} = \pm\sqrt{\dfrac{33}{4}}$$
$$b + \dfrac{5}{2} = \pm\dfrac{\sqrt{33}}{2}$$
$$b = -\dfrac{5}{2} \pm \dfrac{\sqrt{33}}{2}$$

The solution set is $\left\{-\dfrac{5}{2} - \dfrac{\sqrt{33}}{2}, \ -\dfrac{5}{2} + \dfrac{\sqrt{33}}{2}\right\}$.

61.
$$m^2 = 8m + 3$$
$$m^2 - 8m = 3$$
$$m^2 - 8m + \left(\frac{1}{2} \cdot (-8)\right)^2 = 3 + \left(\frac{1}{2} \cdot (-8)\right)^2$$
$$m^2 - 8m + 16 = 3 + 16$$
$$(m - 4)^2 = 19$$
$$m - 4 = \pm\sqrt{19}$$
$$m = 4 \pm \sqrt{19}$$

The solution set is $\left\{4 - \sqrt{19},\ 4 + \sqrt{19}\right\}$.

63.
$$p^2 - p + 3 = 0$$
$$p^2 - p = -3$$
$$p^2 - p + \left(\frac{1}{2} \cdot (-1)\right)^2 = -3 + \left(\frac{1}{2} \cdot (-1)\right)^2$$
$$p^2 - p + \frac{1}{4} = -3 + \frac{1}{4}$$
$$\left(p - \frac{1}{2}\right)^2 = -\frac{11}{4}$$
$$p - \frac{1}{2} = \pm\sqrt{-\frac{11}{4}}$$
$$p - \frac{1}{2} = \pm\frac{\sqrt{11}}{2}i$$
$$p = \frac{1}{2} \pm \frac{\sqrt{11}}{2}i$$

The solution set is $\left\{\frac{1}{2} - \frac{\sqrt{11}}{2}i,\ \frac{1}{2} + \frac{\sqrt{11}}{2}i\right\}$.

65.
$$2y^2 - 5y - 12 = 0$$
$$\frac{2y^2 - 5y - 12}{2} = \frac{0}{2}$$
$$y^2 - \frac{5}{2}y - 6 = 0$$
$$y^2 - \frac{5}{2}y = 6$$
$$y^2 - \frac{5}{2}y + \left[\frac{1}{2} \cdot \left(-\frac{5}{2}\right)\right]^2 = 6 + \left[\frac{1}{2} \cdot \left(-\frac{5}{2}\right)\right]^2$$
$$y^2 - \frac{5}{2}y + \frac{25}{16} = 6 + \frac{25}{16}$$
$$\left(y - \frac{5}{4}\right)^2 = \frac{121}{16}$$
$$y - \frac{5}{4} = \pm\sqrt{\frac{121}{16}}$$
$$y - \frac{5}{4} = \pm\frac{11}{4}$$
$$y = \frac{5}{4} \pm \frac{11}{4}$$
$$y = -\frac{3}{2} \text{ or } y = 4$$

The solution set is $\left\{-\frac{3}{2},\ 4\right\}$.

67.
$$3y^2 - 6y + 2 = 0$$
$$\frac{3y^2 - 6y + 2}{3} = \frac{0}{3}$$
$$y^2 - 2y + \frac{2}{3} = 0$$
$$y^2 - 2y \quad = -\frac{2}{3}$$
$$y^2 - 2y + \left(\frac{1}{2} \cdot (-2)\right)^2 = -\frac{2}{3} + \left(\frac{1}{2} \cdot (-2)\right)^2$$
$$y^2 - 2y + 1 = -\frac{2}{3} + 1$$
$$(y-1)^2 = \frac{1}{3}$$
$$y - 1 = \pm\sqrt{\frac{1}{3}}$$
$$y - 1 = \pm\frac{1}{\sqrt{3}} \cdot \frac{\sqrt{3}}{\sqrt{3}}$$
$$y - 1 = \pm\frac{\sqrt{3}}{3}$$
$$y = 1 \pm \frac{\sqrt{3}}{3}$$

The solution set is $\left\{1 - \frac{\sqrt{3}}{3},\ 1 + \frac{\sqrt{3}}{3}\right\}$.

69.
$$2z^2 - 5z + 1 = 0$$
$$\frac{2z^2 - 5z + 1}{2} = \frac{0}{2}$$
$$z^2 - \frac{5}{2}z + \frac{1}{2} = 0$$
$$y^2 - \frac{5}{2}y \quad = -\frac{1}{2}$$
$$z^2 - \frac{5}{2}z + \left[\frac{1}{2}\cdot\left(-\frac{5}{2}\right)\right]^2 = -\frac{1}{2} + \left[\frac{1}{2}\cdot\left(-\frac{5}{2}\right)\right]^2$$
$$z^2 - \frac{5}{2}z + \frac{25}{16} = -\frac{1}{2} + \frac{25}{16}$$
$$\left(z - \frac{5}{4}\right)^2 = \frac{17}{16}$$
$$z - \frac{5}{4} = \pm\sqrt{\frac{17}{16}}$$
$$z - \frac{5}{4} = \pm\frac{\sqrt{17}}{4}$$
$$z = \frac{5}{4} \pm \frac{\sqrt{17}}{4}$$

The solution set is $\left\{\frac{5}{4} - \frac{\sqrt{17}}{4},\ \frac{5}{4} + \frac{\sqrt{17}}{4}\right\}$.

71.
$$2x^2 + 4x + 5 = 0$$
$$\frac{2x^2 + 4x + 5}{2} = \frac{0}{2}$$
$$x^2 + 2x + \frac{5}{2} = 0$$
$$x^2 + 2x \quad = -\frac{5}{2}$$
$$x^2 + 2x + \left(\frac{1}{2}\cdot 2\right)^2 = -\frac{5}{2} + \left(\frac{1}{2}\cdot 2\right)^2$$
$$x^2 + 2x + 1 = -\frac{5}{2} + 1$$
$$(x+1)^2 = -\frac{3}{2}$$
$$x + 1 = \pm\sqrt{-\frac{3}{2}}$$
$$x + 1 = \pm\frac{\sqrt{3}}{\sqrt{2}}i$$
$$x + 1 = \pm\frac{\sqrt{6}}{2}i$$
$$x = -1 \pm \frac{\sqrt{6}}{2}i$$

The solution set is $\left\{-1 - \frac{\sqrt{6}}{2}i,\ -1 + \frac{\sqrt{6}}{2}i\right\}$.

73.
$$\begin{aligned}c^2 &= 6^2 + 8^2 \\ &= 36 + 64 \\ &= 100 \\ c &= \sqrt{100} = 10\end{aligned}$$

75.
$$\begin{aligned}c^2 &= 12^2 + 16^2 \\ &= 144 + 256 \\ &= 400 \\ c &= \sqrt{400} = 20\end{aligned}$$

77.
$$\begin{aligned}c^2 &= 5^2 + 5^2 \\ &= 25 + 25 \\ &= 50 \\ c &= \sqrt{50} = 5\sqrt{2} \approx 7.07\end{aligned}$$

79.
$$\begin{aligned}c^2 &= 1^2 + \left(\sqrt{3}\right)^2 \\ &= 1 + 3 \\ &= 4 \\ c &= \sqrt{4} = 2\end{aligned}$$

81. $c^2 = 6^2 + 10^2$
$\quad = 36 + 100$
$\quad = 136$
$\quad c = \sqrt{136} = 2\sqrt{34} \approx 11.66$

83. $c^2 = a^2 + b^2$
$\quad 8^2 = 4^2 + b^2$
$\quad 64 = 16 + b^2$
$\quad 48 = b^2$
$\quad\quad b = \sqrt{48}$
$\quad\quad b = 4\sqrt{3} \approx 6.93$

85. $c^2 = a^2 + b^2$
$\quad 12^2 = a^2 + 8^2$
$\quad 144 = a^2 + 64$
$\quad 80 = a^2$
$\quad\quad a = \sqrt{80}$
$\quad\quad a = 4\sqrt{5} \approx 8.94$

87. $\quad f(x) = 36$
$\quad (x-3)^2 = 36$
$\quad\quad x - 3 = \pm\sqrt{36}$
$\quad\quad x - 3 = \pm 6$
$\quad\quad\quad x = 3 \pm 6$
$\quad\quad\quad x = 3 - 6 \ \text{ or } \ x = 3 + 6$
$\quad\quad\quad x = -3 \quad \text{ or } \ x = 9$

The solution set is $\{-3, \ 9\}$.

The points $(-3, 36)$ and $(9, 36)$ are on the graph of f.

89. $\quad g(x) = 18$
$\quad (x+2)^2 = 18$
$\quad\quad x + 2 = \pm\sqrt{18}$
$\quad\quad x + 2 = \pm 3\sqrt{2}$
$\quad\quad\quad x = -2 \pm 3\sqrt{2}$

The solution set is $\left\{-2 - 3\sqrt{2}, \ -2 + 3\sqrt{2}\right\}$.

The points $\left(-2 - 3\sqrt{2}, 18\right)$ and $\left(-2 + 3\sqrt{2}, 18\right)$ are on the graph of g.

91. Let x represent the diagonal.
$\quad x^2 = 8^2 + 4^2$
$\quad x^2 = 64 + 16$
$\quad x^2 = 80$
$\quad\quad x = \sqrt{80} = 4\sqrt{5} \approx 8.944$

93. Let x represent the distance from the ball to the center of the green.
$\quad x^2 = 30^2 + 100^2$
$\quad x^2 = 900 + 10{,}000$
$\quad x^2 = 10{,}900$
$\quad\quad x = \sqrt{10{,}900} \approx 104.403$

The ball is approximately 104.403 yards from the center of the green.

95. Let x represent the length of the guy wire.
$\quad x^2 = 10^2 + 30^2$
$\quad x^2 = 100 + 900$
$\quad x^2 = 1000$
$\quad\quad x = \sqrt{1000} \approx 31.623$

The guy wire is approximately 31.623 feet long.

97. (a) Let x represent the height up the wall to which the ladder can reach.

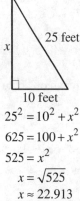

$\quad 25^2 = 10^2 + x^2$
$\quad 625 = 100 + x^2$
$\quad 525 = x^2$
$\quad\quad x = \sqrt{525}$
$\quad\quad x \approx 22.913$

The ladder can reach approximately 22.913 feet up the wall.

(b) Let x represent the distance that the base of the ladder can be from the wall.

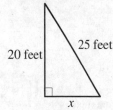

$$25^2 = x^2 + 20^2$$
$$625 = x^2 + 400$$
$$225 = x^2$$
$$x = \sqrt{225}$$
$$x = 15$$

The base of the ladder can be at most 15 feet from the wall.

99. (a)
$$s = 16t^2$$
$$16 = 16t^2$$
$$1 = t^2$$
$$t = \sqrt{1}$$
$$t = 1$$

It takes 1 second for an object to fall 16 feet.

(b)
$$s = 16t^2$$
$$48 = 16t^2$$
$$3 = t^2$$
$$t = \sqrt{3}$$
$$t \approx 1.732$$

It takes approximately 1.732 seconds for an object to fall 48 feet.

(c)
$$s = 16t^2$$
$$64 = 16t^2$$
$$4 = t^2$$
$$t = \sqrt{4}$$
$$t = 2$$

It takes 2 seconds for an object to fall 64 feet.

101.
$$1200 = 1000(1+r)^2$$
$$1.2 = (1+r)^2$$
$$1+r = \sqrt{1.2}$$
$$r = -1 + \sqrt{1.2}$$
$$r \approx 0.0954$$

The required rate of interest is approximately 9.54%.

103.
$$17^2 \overset{?}{=} 8^2 + 15^2$$
$$289 \overset{?}{=} 64 + 225$$
$$289 = 289 \quad \leftarrow \text{True}$$

Because $c^2 = a^2 + b^2$, the triangle is a right triangle. The hypotenuse is 17.

105.
$$20^2 \overset{?}{=} 14^2 + 18^2$$
$$400 \overset{?}{=} 196 + 324$$
$$400 = 520 \quad \leftarrow \text{False}$$

Because $c^2 \neq a^2 + b^2$, the triangle is not a right triangle.

107.
$$c^2 = \left(m^2 + n^2\right)^2 = m^4 + 2m^2n^2 + n^4$$
$$a^2 + b^2 = \left(m^2 - n^2\right)^2 + \left(2mn\right)^2$$
$$= m^4 - 2m^2n^2 + n^4 + 4m^2n^2$$
$$= m^4 + 2m^2n^2 + n^4$$

Because c^2 and $a^2 + b^2$ result in the same expression, a, b, and c are the lengths of the sides of a right triangle.

109.
$$a^2 - 5a - 36 = 0$$
$$(a-9)(a+4) = 0$$
$$a - 9 = 0 \quad \text{or} \quad a + 4 = 0$$
$$a = 9 \quad \text{or} \quad a = -4$$

The solution set is $\{-4,\ 9\}$.

111.
$$|4q+1| = 3$$
$$4q + 1 = 3 \quad \text{or} \quad 4q + 1 = -3$$
$$4q = 2 \quad \text{or} \quad 4q = -4$$
$$q = \frac{1}{2} \quad \text{or} \quad q = -1$$

The solution set is $\left\{-1,\ \dfrac{1}{2}\right\}$.

113. In both cases, the simpler equations are linear.

Section 10.2

Preparing for Solving Quadratic Equations by the Quadratic Formula

P1. (a) $\sqrt{54} = \sqrt{9 \cdot 6} = 3\sqrt{6}$

(b) $\sqrt{121} = 11$ because $11^2 = 121$.

P2. (a) $\sqrt{-9} = \sqrt{-1 \cdot 9} = 3i$

(b) $\sqrt{-72} = \sqrt{-1 \cdot 36 \cdot 2} = 6\sqrt{2}\,i$

P3. $\dfrac{3+\sqrt{18}}{6} = \dfrac{3+3\sqrt{2}}{6}$

$\qquad = \dfrac{3\left(1+\sqrt{2}\right)}{6}$

$\qquad = \dfrac{1+\sqrt{2}}{2}$ or $\dfrac{1}{2}+\dfrac{\sqrt{2}}{2}$

Section 10.2 Quick Checks

1. $\dfrac{-b \pm \sqrt{b^2 - 4ac}}{2a}$

2. $2x^2 - 3x - 9 = 0$

For this equation, $a = 2$, $b = -3$, and $c = -9$.

$x = \dfrac{-(-3) \pm \sqrt{(-3)^2 - 4(2)(-9)}}{2(2)}$

$\quad = \dfrac{3 \pm \sqrt{9 + 72}}{4}$

$\quad = \dfrac{3 \pm \sqrt{81}}{4}$

$\quad = \dfrac{3 \pm 9}{4}$

$x = \dfrac{3 - 9}{4}$ or $x = \dfrac{3 + 9}{4}$

$\quad = \dfrac{-6}{4}$ or $\quad = \dfrac{12}{4}$

$\quad = -\dfrac{3}{2}$ or $\quad = 3$

The solution set is $\left\{-\dfrac{3}{2},\ 3\right\}$.

3. $\qquad 2x^2 + 7x = 4$

$2x^2 + 7x - 4 = 0$

For this equation, $a = 2$, $b = 7$, and $c = -4$.

$x = \dfrac{-7 \pm \sqrt{7^2 - 4(2)(-4)}}{2(2)}$

$\quad = \dfrac{-7 \pm \sqrt{49 + 32}}{4}$

$\quad = \dfrac{-7 \pm \sqrt{81}}{4}$

$\quad = \dfrac{-7 \pm 9}{4}$

$x = \dfrac{-7 - 9}{4}$ or $x = \dfrac{-7 + 9}{4}$

$\quad = \dfrac{-16}{4}$ or $\quad = \dfrac{2}{4}$

$\quad = -4$ or $\quad = \dfrac{1}{2}$

The solution set is $\left\{-4,\ \dfrac{1}{2}\right\}$.

4. $\qquad 4z^2 + 1 = 8z$

$4z^2 - 8z + 1 = 0$

For this equation, $a = 4$, $b = -8$, and $c = 1$.

$z = \dfrac{-(-8) \pm \sqrt{(-8)^2 - 4(4)(1)}}{2(4)}$

$\quad = \dfrac{8 \pm \sqrt{64 - 16}}{8}$

$\quad = \dfrac{8 \pm \sqrt{48}}{8}$

$\quad = \dfrac{8 \pm 4\sqrt{3}}{8}$

$\quad = \dfrac{8}{8} \pm \dfrac{4\sqrt{3}}{8}$

$\quad = 1 \pm \dfrac{\sqrt{3}}{2}$

The solution set is $\left\{1 - \dfrac{\sqrt{3}}{2},\ 1 + \dfrac{\sqrt{3}}{2}\right\}$.

5.
$$4w + \frac{25}{w} = 20$$
$$w\left(4w + \frac{25}{w}\right) = w(20)$$
$$4w^2 + 25 = 20w$$
$$4w^2 - 20w + 25 = 0$$

For this equation, $a = 4$, $b = -20$, and $c = 25$.
$$w = \frac{-(-20) \pm \sqrt{(-20)^2 - 4(4)(25)}}{2(4)}$$
$$= \frac{20 \pm \sqrt{400 - 400}}{8}$$
$$= \frac{20 \pm \sqrt{0}}{8}$$
$$= \frac{20}{8}$$
$$= \frac{5}{2}$$

The solution set is $\left\{\dfrac{5}{2}\right\}$.

6.
$$2x = 8 - \frac{3}{x}$$
$$x(2x) = x\left(8 - \frac{3}{x}\right)$$
$$2x^2 = 8x - 3$$
$$2x^2 - 8x + 3 = 0$$
For this equation, $a = 2$, $b = -8$, and $c = 3$.
$$x = \frac{-(-8) \pm \sqrt{(-8)^2 - 4(2)(3)}}{2(2)}$$
$$= \frac{8 \pm \sqrt{64 - 24}}{4}$$
$$= \frac{8 \pm \sqrt{40}}{4}$$
$$= \frac{8 \pm 2\sqrt{10}}{4}$$
$$= \frac{8}{4} \pm \frac{2\sqrt{10}}{4}$$
$$= 2 \pm \frac{\sqrt{10}}{2}$$

The solution set is $\left\{2 - \dfrac{\sqrt{10}}{2}, \; 2 + \dfrac{\sqrt{10}}{2}\right\}$.

7. $z^2 + 2z + 26 = 0$
For this equation, $a = 1$, $b = 2$, and $c = 26$.
$$z = \frac{-2 \pm \sqrt{2^2 - 4(1)(26)}}{2(1)}$$
$$= \frac{-2 \pm \sqrt{4 - 104}}{2}$$
$$= \frac{-2 \pm \sqrt{-100}}{2}$$
$$= \frac{-2 \pm 10i}{2}$$
$$= \frac{-2}{2} \pm \frac{10i}{2}$$
$$= -1 \pm 5i$$

The solution set is $\{-1 - 5i, \; -1 + 5i\}$.

8. discriminant

9. negative

10. False. The equation will have one repeated real solution.

11. True.

12. $2z^2 + 5z + 4 = 0$
For this equation, $a = 2$, $b = 5$, and $c = 4$.
$$b^2 - 4ac = 5^2 - 4(2)(4) = 25 - 32 = -7$$

Because $b^2 - 4ac = -7$ is negative, the quadratic equation will have two complex solutions that are not real. The solutions will be complex conjugates of each other.

13.
$$4y^2 + 12y = -9$$
$$4y^2 + 12y + 9 = 0$$
For this equation, $a = 4$, $b = 12$, and $c = 9$.
$$b^2 - 4ac = 12^2 - 4(4)(9) = 144 - 144 = 0$$

Because $b^2 - 4ac = 0$, the quadratic equation will have one repeated real solution.

14. $2x^2 - 4x + 1 = 0$
For this equation, $a = 2$, $b = -4$, and $c = 1$.
$$b^2 - 4ac = (-4)^2 - 4(2)(1) = 16 - 8 = 8$$

Because $b^2 - 4ac = 8$ is positive, but not a perfect square, the quadratic equation will have two irrational solutions.

15. True.

16. $5n^2 - 45 = 0$

Because this equation has no linear term, solve by using the square root method.

$$5n^2 = 45$$
$$n^2 = 9$$
$$n = \pm\sqrt{9}$$
$$= \pm 3$$

The solution set is $\{-3, 3\}$.

17. $-2y^2 + 5y - 6 = 0$

$$2y^2 - 5y + 6 = 0$$

Because this equation does not easily factor, solve by using the quadratic formula. For this equation, $a = 2$, $b = -5$, and $c = 6$.

$$y = \frac{-(-5) \pm \sqrt{(-5)^2 - 4(2)(6)}}{2(2)}$$
$$= \frac{5 \pm \sqrt{25 - 48}}{4}$$
$$= \frac{5 \pm \sqrt{-23}}{4}$$
$$= \frac{5 \pm \sqrt{23}\,i}{4}$$

The solution set is $\left\{ \dfrac{5 - \sqrt{23}\,i}{4}, \dfrac{5 + \sqrt{23}\,i}{4} \right\}$ or $\left\{ \dfrac{5}{4} - \dfrac{\sqrt{23}}{4}i, \dfrac{5}{4} + \dfrac{\sqrt{23}}{4}i \right\}$.

18. $3w^2 + 2w = 5$

$$3w^2 + 2w - 5 = 0$$

Because this equation factors easily, solve by factoring.

$$(3w + 5)(w - 1) = 0$$
$$3w + 5 = 0 \quad \text{or} \quad w - 1 = 0$$
$$3w = -5 \quad \text{or} \quad w = 1$$
$$w = -\frac{5}{3}$$

The solution set is $\left\{ -\dfrac{5}{3}, 1 \right\}$.

19. (a) $R(x) = 600$

$$-0.005x^2 + 4x = 600$$
$$-0.005x^2 + 4x - 600 = 0$$

For this equation, $a = -0.005$, $b = 4$, and $c = -600$.

$$x = \frac{-4 \pm \sqrt{4^2 - 4(-0.005)(-600)}}{2(-0.005)}$$
$$= \frac{-4 \pm \sqrt{4}}{-0.01}$$
$$= \frac{-4 \pm 2}{-0.01}$$
$$x = \frac{-6}{-0.01} \quad \text{or} \quad x = \frac{-2}{-0.01}$$
$$= 600 \quad \text{or} \quad = 200$$

If revenue is to be \$600 per day, then either 200 or 600 DVDs must be rented.

(b) $R(x) = 800$

$$-0.005x^2 + 4x = 800$$
$$-0.005x^2 + 4x - 800 = 0$$

For this equation, $a = -0.005$, $b = 4$, and $c = -800$.

$$x = \frac{-4 \pm \sqrt{4^2 - 4(-0.005)(-800)}}{2(-0.005)}$$
$$= \frac{-4 \pm \sqrt{0}}{-0.01}$$
$$= \frac{-4 \pm 0}{-0.01}$$
$$x = \frac{-4}{-0.01} = 400$$

If revenue is to be \$800 per day, then 400 DVDs must be rented.

20. Let w represent the width of the rectangle. Then $w + 14$ will represent the length.

$$w^2 + (w + 14)^2 = 34^2$$
$$w^2 + w^2 + 28w + 196 = 1156$$
$$2w^2 + 28w - 960 = 0$$
$$w^2 + 14w - 480 = 0$$
$$(w - 16)(w + 30) = 0$$
$$w - 16 = 0 \quad \text{or} \quad w + 30 = 0$$
$$w = 16 \quad \text{or} \quad w = -30$$

We disregard $w = -30$ because w represents the width of the rectangle, which must be positive. Thus, $w = 16$ is the only viable answer. Now, $w + 14 = 16 + 14 = 30$. Thus, the dimensions of the rectangle are 16 meters by 30 meters.

10.2 Exercises

21. $x^2 - 4x - 12 = 0$

For this equation, $a = 1$, $b = -4$, and $c = -12$.

$$x = \frac{-(-4) \pm \sqrt{(-4)^2 - 4(1)(-12)}}{2(1)}$$

$$= \frac{4 \pm \sqrt{16 + 48}}{2}$$

$$= \frac{4 \pm \sqrt{64}}{2}$$

$$= \frac{4 \pm 8}{2}$$

$$x = \frac{4 - 8}{2} \quad \text{or} \quad x = \frac{4 + 8}{2}$$

$$= \frac{-4}{2} \quad \text{or} \quad = \frac{12}{2}$$

$$= -2 \quad \text{or} \quad = 6$$

The solution set is $\{-2, 6\}$.

23. $6y^2 - y - 15 = 0$

For this equation, $a = 6$, $b = -1$, and $c = -15$.

$$y = \frac{-(-1) \pm \sqrt{(-1)^2 - 4(6)(-15)}}{2(6)}$$

$$= \frac{1 \pm \sqrt{1 + 360}}{12}$$

$$= \frac{1 \pm \sqrt{361}}{12}$$

$$= \frac{1 \pm 19}{12}$$

$$y = \frac{1 - 19}{12} \quad \text{or} \quad y = \frac{1 + 19}{12}$$

$$= \frac{-18}{12} \quad \text{or} \quad = \frac{20}{12}$$

$$= -\frac{3}{2} \quad \text{or} \quad = \frac{5}{3}$$

The solution set is $\left\{-\frac{3}{2}, \frac{5}{3}\right\}$.

25. $4m^2 - 8m + 1 = 0$

For this equation, $a = 4$, $b = -8$, and $c = 1$.

$$m = \frac{-(-8) \pm \sqrt{(-8)^2 - 4(4)(1)}}{2(4)}$$

$$= \frac{8 \pm \sqrt{64 - 16}}{8}$$

$$= \frac{8 \pm \sqrt{48}}{8} = \frac{8}{8} \pm \frac{4\sqrt{3}}{8} = 1 \pm \frac{\sqrt{3}}{2}$$

The solution set is $\left\{1 - \frac{\sqrt{3}}{2}, 1 + \frac{\sqrt{3}}{2}\right\}$.

27.

$$3w - 6 = \frac{1}{w}$$

$$(3w - 6)w = \left(\frac{1}{w}\right)w$$

$$3w^2 - 6w = 1$$

$$3w^2 - 6w - 1 = 0$$

For this equation, $a = 3$, $b = -6$, and $c = -1$.

$$w = \frac{-(-6) \pm \sqrt{(-6)^2 - 4(3)(-1)}}{2(3)}$$

$$= \frac{6 \pm \sqrt{36 + 12}}{6}$$

$$= \frac{6 \pm \sqrt{48}}{6}$$

$$= \frac{6 \pm 4\sqrt{3}}{6} = \frac{6}{6} \pm \frac{4\sqrt{3}}{6} = 1 \pm \frac{2\sqrt{3}}{3}$$

The solution set is $\left\{1 - \frac{2\sqrt{3}}{3}, 1 + \frac{2\sqrt{3}}{3}\right\}$.

29.

$$3p^2 = -2p + 4$$

$$3p^2 + 2p - 4 = 0$$

For this equation, $a = 3$, $b = 2$, and $c = -4$.

$$p = \frac{-2 \pm \sqrt{2^2 - 4(3)(-4)}}{2(3)}$$

$$= \frac{-2 \pm \sqrt{4 + 48}}{6}$$

$$= \frac{-2 \pm \sqrt{52}}{6}$$

$$= \frac{-2 \pm 2\sqrt{13}}{6} = -\frac{2}{6} \pm \frac{2\sqrt{13}}{6} = -\frac{1}{3} \pm \frac{\sqrt{13}}{3}$$

The solution set is $\left\{-\frac{1}{3} - \frac{\sqrt{13}}{3}, -\frac{1}{3} + \frac{\sqrt{13}}{3}\right\}$.

31. $x^2 - 2x + 7 = 0$

For this equation, $a = 1$, $b = -2$, and $c = 7$.

$$x = \frac{-(-2) \pm \sqrt{(-2)^2 - 4(1)(7)}}{2(1)}$$

$$= \frac{2 \pm \sqrt{4 - 28}}{2}$$

$$= \frac{2 \pm \sqrt{-24}}{2}$$

$$= \frac{2 \pm 2\sqrt{6}\, i}{2}$$

$$= \frac{2}{2} \pm \frac{2\sqrt{6}\, i}{2} = 1 \pm \sqrt{6}\, i$$

The solution set is $\left\{ 1 - \sqrt{6}\, i, \, 1 + \sqrt{6}\, i \right\}$.

33. $2z^2 + 7 = 2z$

$2z^2 - 2z + 7 = 0$

For this equation, $a = 2$, $b = -2$, and $c = 7$.

$$z = \frac{-(-2) \pm \sqrt{(-2)^2 - 4(2)(7)}}{2(2)}$$

$$= \frac{2 \pm \sqrt{4 - 56}}{4}$$

$$= \frac{2 \pm \sqrt{-52}}{4}$$

$$= \frac{2 \pm 2\sqrt{13}\, i}{4}$$

$$= \frac{2}{4} \pm \frac{2\sqrt{13}\, i}{4} = \frac{1}{2} \pm \frac{\sqrt{13}}{2} i$$

The solution set is $\left\{ \frac{1}{2} - \frac{\sqrt{13}}{2} i, \, \frac{1}{2} + \frac{\sqrt{13}}{2} i \right\}$.

35. $4x^2 = 2x + 1$

$4x^2 - 2x - 1 = 0$

For this equation, $a = 4$, $b = -2$, and $c = -1$.

$$x = \frac{-(-2) \pm \sqrt{(-2)^2 - 4(4)(-1)}}{2(4)}$$

$$= \frac{2 \pm \sqrt{4 + 16}}{8}$$

$$= \frac{2 \pm \sqrt{20}}{8}$$

$$= \frac{2 \pm 2\sqrt{5}}{8} = \frac{2}{8} \pm \frac{2\sqrt{5}}{8} = \frac{1}{4} \pm \frac{\sqrt{5}}{4}$$

The solution set is $\left\{ \frac{1}{4} - \frac{\sqrt{5}}{4}, \, \frac{1}{4} + \frac{\sqrt{5}}{4} \right\}$.

37. $1 = 3q^2 + 4q$

$0 = 3q^2 + 4q - 1$

For this equation, $a = 3$, $b = 4$, and $c = -1$.

$$q = \frac{-4 \pm \sqrt{4^2 - 4(3)(-1)}}{2(3)}$$

$$= \frac{-4 \pm \sqrt{16 + 12}}{6}$$

$$= \frac{-4 \pm \sqrt{28}}{6}$$

$$= \frac{-4 \pm 2\sqrt{7}}{6} = \frac{-4}{6} \pm \frac{2\sqrt{7}}{6} = -\frac{2}{3} \pm \frac{\sqrt{7}}{3}$$

The solution set is $\left\{ -\frac{2}{3} - \frac{\sqrt{7}}{3}, \, -\frac{2}{3} + \frac{\sqrt{7}}{3} \right\}$.

39. $x^2 - 5x + 1 = 0$

For this equation, $a = 1$, $b = -5$, and $c = 1$.

$b^2 - 4ac = (-5)^2 - 4(1)(1) = 25 - 4 = 21$

Because $b^2 - 4ac = 21$ is positive, but not a perfect square, the quadratic equation will have two irrational solutions.

41. $3z^2 + 2z + 5 = 0$

For this equation, $a = 3$, $b = 2$, and $c = 5$.

$b^2 - 4ac = 2^2 - 4(3)(5) = 4 - 60 = -56$

Because $b^2 - 4ac = -56$ is negative, the quadratic equation will have two complex solutions that are not real. The solutions will be complex conjugates of each other.

43. $9q^2 - 6q + 1 = 0$

For this equation, $a = 9$, $b = -6$, and $c = 1$.

$b^2 - 4ac = (-6)^2 - 4(9)(1) = 36 - 36 = 0$

Because $b^2 - 4ac = 0$, the quadratic equation will have one repeated real solution.

45. $3w^2 = 4w - 2$

$3w^2 - 4w + 2 = 0$

For this equation, $a = 3$, $b = -4$, and $c = 2$.

$b^2 - 4ac = (-4)^2 - 4(3)(2) = 16 - 24 = -8$

Because $b^2 - 4ac = -8$ is negative, the quadratic equation will have two complex solutions that are not real. The solutions will be complex conjugates of each other.

47. $6x = 2x^2 - 1$

$0 = 2x^2 - 6x - 1$

For this equation, $a = 2$, $b = -6$, and $c = -1$.

$b^2 - 4ac = (-6)^2 - 4(2)(-1) = 36 + 8 = 44$

Because $b^2 - 4ac = 44$ is positive, but not a perfect square, the quadratic equation will have two irrational solutions.

49. $w^2 - 5w + 5 = 0$

Because this equation does not easily factor, solve by using the quadratic formula. For this equation, $a = 1$, $b = -5$, and $c = 5$.

$w = \dfrac{-(-5) \pm \sqrt{(-5)^2 - 4(1)(5)}}{2(1)}$

$= \dfrac{5 \pm \sqrt{25 - 20}}{2} = \dfrac{5 \pm \sqrt{5}}{2} = \dfrac{5}{2} \pm \dfrac{\sqrt{5}}{2}$

The solution set is $\left\{ \dfrac{5}{2} - \dfrac{\sqrt{5}}{2}, \dfrac{5}{2} + \dfrac{\sqrt{5}}{2} \right\}$.

51. $3x^2 + 5x = 8$

$3x^2 + 5x - 8 = 0$

Because this equation factors easily, solve by factoring.

$(3x + 8)(x - 1) = 0$

$3x + 8 = 0 \quad$ or $\quad x - 1 = 0$

$3x = -8 \quad$ or $\quad x = 1$

$x = -\dfrac{8}{3}$

The solution set is $\left\{ -\dfrac{8}{3}, 1 \right\}$.

53. $2x^2 = 3x + 35$

$2x^2 - 3x - 35 = 0$

Because this equation factors easily, solve by factoring.

$(2x + 7)(x - 5) = 0$

$2x + 7 = 0 \quad$ or $\quad x - 5 = 0$

$2x = -7 \quad$ or $\quad x = 5$

$x = -\dfrac{7}{2}$

The solution set is $\left\{ -\dfrac{7}{2}, 5 \right\}$.

55. $q^2 + 2q + 8 = 0$

Because this equation does not easily factor, solve by using the quadratic formula. For this equation, $a = 1$, $b = 2$, and $c = 8$.

$q = \dfrac{-2 \pm \sqrt{2^2 - 4(1)(8)}}{2(1)}$

$= \dfrac{-2 \pm \sqrt{4 - 32}}{2}$

$= \dfrac{-2 \pm \sqrt{-28}}{2}$

$= \dfrac{-2 \pm 2\sqrt{7}i}{2} = \dfrac{-2}{2} \pm \dfrac{2\sqrt{7}i}{2} = -1 \pm \sqrt{7}i$

The solution set is $\left\{ -1 - \sqrt{7}i, -1 + \sqrt{7}i \right\}$.

57. $2z^2 = 2(z + 3)^2$

$2z^2 = 2\left(z^2 + 6z + 9 \right)$

$2z^2 = 2z^2 + 12z + 18$

$0 = 12z + 18$

$12z = -18$

$z = -\dfrac{18}{12} = -\dfrac{3}{2}$

The solution set is $\left\{ -\dfrac{3}{2} \right\}$.

59. $7q - 2 = \dfrac{4}{q}$

$(7q - 2)q = \left(\dfrac{4}{q} \right) q$

$7q^2 - 2q = 4$

$7q^2 - 2q - 4 = 0$

Because this equation does not easily factor, solve by using the quadratic formula. For this equation, $a = 7$, $b = -2$, and $c = -4$.

$q = \dfrac{-(-2) \pm \sqrt{(-2)^2 - 4(7)(-4)}}{2(7)}$

$= \dfrac{2 \pm \sqrt{4 + 112}}{14} = \dfrac{2 \pm \sqrt{116}}{14}$

$= \dfrac{2 \pm 2\sqrt{29}}{14} = \dfrac{2}{14} \pm \dfrac{2\sqrt{29}}{14} = \dfrac{1}{7} \pm \dfrac{\sqrt{29}}{7}$

The solution set is $\left\{ \dfrac{1}{7} - \dfrac{\sqrt{29}}{7}, \dfrac{1}{7} + \dfrac{\sqrt{29}}{7} \right\}$.

61. $5a^2 - 80 = 0$

Because this equation has no linear term, solve by using the square root method.

$$5a^2 = 80$$
$$a^2 = 16$$
$$a = \pm\sqrt{16} = \pm 4$$

The solution set is $\{-4, 4\}$.

63. $8n^2 + 1 = 4n$

$$8n^2 - 4n + 1 = 0$$

Because this equation does not easily factor, solve by using the quadratic formula. For this equation, $a = 8$, $b = -4$, and $c = 1$.

$$n = \frac{-(-4) \pm \sqrt{(-4)^2 - 4(8)(1)}}{2(8)}$$
$$= \frac{4 \pm \sqrt{16 - 32}}{16}$$
$$= \frac{4 \pm \sqrt{-16}}{16}$$
$$= \frac{4 \pm 4i}{16} = \frac{4}{16} \pm \frac{4i}{16} = \frac{1}{4} \pm \frac{1}{4}i$$

The solution set is $\left\{\frac{1}{4} - \frac{1}{4}i, \ \frac{1}{4} + \frac{1}{4}i\right\}$.

65. $27x^2 + 36x + 12 = 0$

$$\frac{27x^2 + 36x + 12}{3} = \frac{0}{3}$$
$$9x^2 + 12x + 4 = 0$$

Because this equation factors easily, solve by factoring.

$$(3x + 2)(3x + 2) = 0$$
$$3x + 2 = 0 \quad \text{or} \quad 3x + 2 = 0$$
$$3x = -2 \quad \text{or} \qquad 3x = -2$$
$$x = -\frac{2}{3} \quad \text{or} \qquad x = -\frac{2}{3}$$

The solution set is $\left\{-\frac{2}{3}\right\}$.

67. $\frac{1}{3}x^2 + \frac{2}{9}x - 1 = 0$

$$9\left(\frac{1}{3}x^2 + \frac{2}{9}x - 1\right) = 9(0)$$
$$3x^2 + 2x - 9 = 0$$

Because this equation does not easily factor, solve by using the quadratic formula. For this equation, $a = 3$, $b = 2$, and $c = -9$.

$$x = \frac{-2 \pm \sqrt{2^2 - 4(3)(-9)}}{2(3)}$$
$$= \frac{-2 \pm \sqrt{4 + 108}}{6}$$
$$= \frac{-2 \pm \sqrt{112}}{6}$$
$$= \frac{-2 \pm 4\sqrt{7}}{6}$$
$$= \frac{-2}{6} \pm \frac{4\sqrt{7}}{6} = -\frac{1}{3} \pm \frac{2\sqrt{7}}{3}$$

The solution set is $\left\{-\frac{1}{3} - \frac{2\sqrt{7}}{3}, \ -\frac{1}{3} + \frac{2\sqrt{7}}{3}\right\}$.

69. $(x - 5)(x + 1) = 4$

$$x^2 - 4x - 5 = 4$$
$$x^2 - 4x - 9 = 0$$

Because this equation does not easily factor, solve by using the quadratic formula. For this equation, $a = 1$, $b = -4$, and $c = -9$.

$$x = \frac{-(-4) \pm \sqrt{(-4)^2 - 4(1)(-9)}}{2(1)}$$
$$= \frac{4 \pm \sqrt{16 + 36}}{2}$$
$$= \frac{4 \pm \sqrt{52}}{2}$$
$$= \frac{4 \pm 2\sqrt{13}}{2}$$
$$= \frac{4}{2} \pm \frac{2\sqrt{13}}{2} = 2 \pm \sqrt{13}$$

The solution set is $\left\{2 - \sqrt{13}, \ 2 + \sqrt{13}\right\}$.

71. Note: $x \neq -2$.

$$\frac{x - 2}{x + 2} = x - 3$$
$$(x + 2)\left(\frac{x - 2}{x + 2}\right) = (x + 2)(x - 3)$$
$$x - 2 = x^2 - x - 6$$
$$0 = x^2 - 2x - 4$$

Because this equation does not easily factor, solve by using the quadratic formula. For this equation, $a = 1$, $b = -2$, and $c = -4$.

$$x = \frac{-(-2) \pm \sqrt{(-2)^2 - 4(1)(-4)}}{2(1)}$$

$$= \frac{2 \pm \sqrt{4 + 16}}{2}$$

$$= \frac{2 \pm \sqrt{20}}{2}$$

$$= \frac{2 \pm 2\sqrt{5}}{2}$$

$$= \frac{2}{2} \pm \frac{2\sqrt{5}}{2} = 1 \pm \sqrt{5}$$

The solution set is $\{1 - \sqrt{5}, 1 + \sqrt{5}\}$.

73.
$$\frac{x - 4}{x^2 + 2} = 2$$

$$(x^2 + 2)\left(\frac{x - 4}{x^2 + 2}\right) = (x^2 + 2)(2)$$

$$x - 4 = 2x^2 + 4$$

$$0 = 2x^2 - x + 8$$

Because this equation does not easily factor, solve by using the quadratic formula. For this equation, $a = 2$, $b = -1$, and $c = 8$.

$$x = \frac{-(-1) \pm \sqrt{(-1)^2 - 4(2)(8)}}{2(2)}$$

$$= \frac{1 \pm \sqrt{1 - 64}}{4}$$

$$= \frac{1 \pm \sqrt{-63}}{4}$$

$$= \frac{1 \pm 3\sqrt{7}\, i}{4} = \frac{1}{4} \pm \frac{3\sqrt{7}}{4} i$$

The solution set is $\left\{\frac{1}{4} - \frac{3\sqrt{7}}{4} i, \frac{1}{4} + \frac{3\sqrt{7}}{4} i\right\}$.

75. (a)
$$f(x) = 0$$

$$x^2 + 4x - 21 = 0$$

$$(x + 7)(x - 3) = 0$$

$$x + 7 = 0 \quad \text{or} \quad x - 3 = 0$$

$$x = -7 \quad \text{or} \quad x = 3$$

The solution set is $\{-7, 3\}$.

(b)
$$f(x) = -21$$

$$x^2 + 4x - 21 = -21$$

$$x^2 + 4x = 0$$

$$x(x + 4) = 0$$

$$x = 0 \quad \text{or} \quad x + 4 = 0$$

$$x = -4$$

The solution set is $\{-4, 0\}$.

The points $(-7, 0)$, $(3, 0)$, $(-4, 21)$ and $(0, -21)$ are on the graph of f.

77. (a)
$$H(x) = 0$$

$$-2x^2 - 4x + 1 = 0$$

$$2x^2 + 4x - 1 = 0$$

For this equation, $a = 2$, $b = 4$, and $c = -1$.

$$x = \frac{-4 \pm \sqrt{4^2 - 4(2)(-1)}}{2(2)}$$

$$= \frac{-4 \pm \sqrt{16 + 8}}{4}$$

$$= \frac{-4 \pm \sqrt{24}}{4}$$

$$= \frac{-4 \pm 2\sqrt{6}}{4} = \frac{-4}{4} \pm \frac{2\sqrt{6}}{4} = -1 \pm \frac{\sqrt{6}}{2}$$

The solution set is $\left\{-1 - \frac{\sqrt{6}}{2}, -1 + \frac{\sqrt{6}}{2}\right\}$.

(b)
$$H(x) = 2$$

$$-2x^2 - 4x + 1 = 2$$

$$0 = 2x^2 + 4x + 1$$

For this equation, $a = 2$, $b = 4$, and $c = 1$.

$$x = \frac{-4 \pm \sqrt{4^2 - 4(2)(1)}}{2(2)}$$

$$= \frac{-4 \pm \sqrt{16 - 8}}{4}$$

$$= \frac{-4 \pm \sqrt{8}}{4}$$

$$= \frac{-4 \pm 2\sqrt{2}}{4} = \frac{-4}{4} \pm \frac{2\sqrt{2}}{4} = -1 \pm \frac{\sqrt{2}}{2}$$

The solution set is $\left\{-1 - \frac{\sqrt{2}}{2}, -1 + \frac{\sqrt{2}}{2}\right\}$.

79. $G(x) = 3x^2 + 2x - 2$

$$3x^2 + 2x - 2 = 0$$

For this equation, $a = 3$, $b = 2$, and $c = -2$.

$$x = \frac{-2 \pm \sqrt{(2)^2 - 4(3)(-2)}}{2(3)}$$

$$= \frac{-2 \pm \sqrt{4 + 24}}{6}$$

$$= \frac{-2 \pm \sqrt{28}}{6}$$

$$= \frac{-2 \pm 2\sqrt{7}}{6}$$

$$= \frac{-2}{6} \pm \frac{2\sqrt{7}}{6}$$

$$= -\frac{1}{3} \pm \frac{\sqrt{7}}{3}$$

The solution set is $\left\{ -\frac{1}{3} - \frac{\sqrt{7}}{3}, -\frac{1}{3} + \frac{\sqrt{7}}{3} \right\}$.

The zeros of $G(x)$ are $x = \frac{-1 \pm \sqrt{7}}{3}$.

81.
$$x^2 + (x+1)^2 = (2x-1)^2$$
$$x^2 + x^2 + 2x + 1 = 4x^2 - 4x + 1$$
$$2x^2 + 2x + 1 = 4x^2 - 4x + 1$$
$$0 = 2x^2 - 6x$$
$$0 = 2x(x-3)$$
$$2x = 0 \quad \text{or} \quad x - 3 = 0$$
$$x = 0 \quad \text{or} \qquad x = 3$$

Disregard $x = 0$ because x represents the length of one leg of the triangle. Thus, $x = 3$ is the only viable answer. Now, $x + 1 = 3 + 1 = 4$ and $2x - 1 = 2(3) - 1 = 5$. The three measurements are 3, 4, and 5.

83.
$$(5x-1)^2 + (x+2)^2 = (5x)^2$$
$$25x^2 - 10x + 1 + x^2 + 4x + 4 = 25x^2$$
$$26x^2 - 6x + 5 = 25x^2$$
$$x^2 - 6x + 5 = 0$$
$$(x-1)(x-5) = 0$$
$$x - 1 = 0 \quad \text{or} \quad x - 5 = 0$$
$$x = 1 \quad \text{or} \qquad x = 5$$

For $x = 1$, $x + 2 = 1 + 2 = 3$, $5x - 1 = 5(1) - 1 = 4$, and $5x = 5(1) = 5$.

For $x = 5$, $x + 2 = 5 + 2 = 7$, $5x - 1 = 5(5) - 1 = 24$, and $5x = 5(5) = 25$.

The three measurements can be either 3, 4, and 5, or 7, 24, and 25.

85. Let x represent the length of the rectangle. Then $x + 4$ will represent the width.
$$x(x+4) = 40$$
$$x^2 + 4x = 40$$
$$x^2 + 4x - 40 = 0$$

For this equation, $a = 1$, $b = 4$, and $c = -40$.

$$x = \frac{-4 \pm \sqrt{4^2 - 4(1)(-40)}}{2(1)}$$

$$= \frac{-4 \pm \sqrt{16 + 160}}{2} = \frac{-4 \pm \sqrt{176}}{2}$$

$$= \frac{-4 \pm 4\sqrt{11}}{2} = -\frac{4}{2} \pm \frac{4\sqrt{11}}{2} = -2 \pm 2\sqrt{11}$$

Disregard $x = -2 - 2\sqrt{11} \approx -8.633$ because x represents the length of the rectangle, which must be positive. Thus, $x = -2 + 2\sqrt{11} \approx 4.633$ is the only viable answer. Now,

$x + 4 = -2 + 2\sqrt{11} + 4 = 2 + 2\sqrt{11} \approx 8.633$. Thus, the dimensions of the rectangle are $-2 + 2\sqrt{11}$ inches by $2 + 2\sqrt{11}$ inches, which is approximately 4.633 inches by 8.633 inches.

87. Let x represent the base of the triangle. Then $x - 3$ will represent the height.

$$\frac{1}{2}x(x-3) = 25$$

$$\frac{1}{2}x^2 - \frac{3}{2}x - 25 = 0$$

$$2\left(\frac{1}{2}x^2 - \frac{3}{2}x - 25 \right) = 2(0)$$

$$x^2 - 3x - 50 = 0$$

For this equation, $a = 1$, $b = -3$, and $c = -50$.

$$x = \frac{-(-3) \pm \sqrt{(-3)^2 - 4(1)(-50)}}{2(1)}$$

$$= \frac{3 \pm \sqrt{9 + 200}}{2} = \frac{3 \pm \sqrt{209}}{2} = \frac{3}{2} \pm \frac{\sqrt{209}}{2}$$

Disregard $x = \frac{3}{2} - \frac{\sqrt{209}}{2} \approx -5.728$ because x represents the base of the triangle, which must be positive. Thus, $x = \frac{3}{2} + \frac{\sqrt{209}}{2} \approx 8.728$ is the only viable answer. Now,

$$x - 3 = \frac{3}{2} + \frac{\sqrt{209}}{2} - 3 = -\frac{3}{2} + \frac{\sqrt{209}}{2} \approx 5.728.$$

Thus, the base of the triangle is $\frac{3}{2} + \frac{\sqrt{209}}{2}$

inches, which is approximately 8.728 inches. The height of the triangle is $-\dfrac{3}{2}+\dfrac{\sqrt{209}}{2}$ inches, which is approximately 5.728 inches.

89. (a) $R(17)=-0.1(17)^2+70(17)=1161.1$

If 17 pairs of sunglasses are sold per week, then the company's revenue will be $1161.10.

$R(25)=-0.1(25)^2+70(25)=1687.5$

If 25 pairs of sunglasses are sold per week, then the company's revenue will be $1687.50.

(b) $R(x)=10,000$

$-0.1x^2+70x=10,000$

$0=0.1x^2-70x+10,000$

For this equation, $a=0.1$, $b=-70$, and $c=10,000$.

$x=\dfrac{-(-70)\pm\sqrt{(-70)^2-4(0.1)(10,000)}}{2(0.1)}$

$=\dfrac{70\pm\sqrt{4900-4000}}{0.2}$

$=\dfrac{70\pm\sqrt{900}}{0.2}=\dfrac{70\pm30}{0.2}$

$x=\dfrac{70-30}{0.2}$ or $x=\dfrac{70+30}{0.2}$

$=\dfrac{40}{0.2}=200$ or $=\dfrac{100}{0.2}=500$

The revenue will be $10,000 per week if either 200 or 500 pairs of sunglasses are sold per week.

(c) $R(x)=12,250$

$-0.1x^2+70x=12,250$

$0=0.1x^2-70x+12,250$

For this equation, $a=0.1$, $b=-70$, and $c=12,250$.

$x=\dfrac{-(-70)\pm\sqrt{(-70)^2-4(0.1)(12,250)}}{2(0.1)}$

$=\dfrac{70\pm\sqrt{4900-4900}}{0.2}$

$=\dfrac{70\pm\sqrt{0}}{0.2}$

$=\dfrac{70\pm0}{0.2}=\dfrac{70}{0.2}=350$

The revenue will be $12,250 per week if 350 pairs of sunglasses are sold per week.

91. (a) $s(t)=40$

$40=-16t^2+70t+5$

$16t^2-70t+35=0$

For this equation, $a=16$, $b=-70$, and $c=35$.

$x=\dfrac{-(-70)\pm\sqrt{(-70)^2-4(16)(35)}}{2(16)}$

$=\dfrac{70\pm\sqrt{4900-2240}}{32}$

$=\dfrac{70\pm\sqrt{2660}}{32}$

$=\dfrac{70\pm2\sqrt{665}}{32}$

$x=\dfrac{35-\sqrt{665}}{16}$ or $x=\dfrac{35+\sqrt{665}}{16}$

≈0.576 or ≈3.799

Rounding to the nearest tenth, the height of the ball will be 40 feet after approximately 0.6 second and after approximately 3.8 seconds.

(b) $s(t)=70$

$70=-16t^2+70t+5$

$16t^2-70t+65=0$

For this equation, $a=16$, $b=-70$, and $c=65$.

$$x = \frac{-(-70) \pm \sqrt{(-70)^2 - 4(16)(65)}}{2(16)}$$

$$= \frac{70 \pm \sqrt{4900 - 4160}}{32}$$

$$= \frac{70 \pm \sqrt{740}}{32}$$

$$= \frac{70 \pm 2\sqrt{185}}{32}$$

$$= \frac{35 \pm \sqrt{185}}{16}$$

$$x = \frac{35 - \sqrt{185}}{16} \quad \text{or} \quad x = \frac{35 + \sqrt{185}}{16}$$

$$\approx 1.337 \qquad \text{or} \qquad \approx 3.038$$

Rounding to the nearest tenth, the height of the ball will be 70 feet after approximately 1.3 seconds and after approximately 3.0 seconds.

(c)
$$s(t) = 150$$
$$150 = -16t^2 + 70t + 5$$
$$16t^2 - 70t + 145 = 0$$

For this equation, $a = 16$, $b = -70$, and $c = 145$.

$$x = \frac{-(-70) \pm \sqrt{(-70)^2 - 4(16)(145)}}{2(16)}$$

$$= \frac{70 \pm \sqrt{4900 - 9280}}{32}$$

$$= \frac{70 \pm \sqrt{-4380}}{32}$$

$$= \frac{70 \pm 2\sqrt{1095}\, i}{32}$$

$$= \frac{35 \pm \sqrt{1095}\, i}{16}$$

$$x = \frac{35}{16} - \frac{\sqrt{185}}{16}i \quad \text{or} \quad x = \frac{35}{16} + \frac{\sqrt{185}}{16}i$$

The ball will never reach a height of 150 feet. This is clear because the solutions to the equation above are complex solutions that are not real.

93. Because $\triangle ABC \sim \triangle DEC$, we know $\dfrac{\overline{AB}}{\overline{BC}} = \dfrac{\overline{DE}}{\overline{EC}}$.

$\overline{BC} = 24$, $\overline{DE} = 6$, $\overline{AB} = x$, and $\overline{EC} = x$,

$$\frac{x}{24} = \frac{6}{x}$$
$$x \cdot x = 24 \cdot 6$$
$$x^2 = 144$$
$$x = \pm\sqrt{144} = \pm 12$$

Disregard $x = -12$ because x represents a length, which must be positive. Thus, $x = 12$ is the only viable answer.

95. (a) $I(a) = 40,000$

$$40,000 = -55a^2 + 5119a - 54,448$$
$$55a^2 - 5119a + 94,448 = 0$$

For this equation, $a = 55$, $b = -5119$, and $c = 94,448$.

$$x = \frac{-(-5119) \pm \sqrt{(-5119)^2 - 4(55)(94,448)}}{2(55)}$$

$$= \frac{5119 \pm \sqrt{26,204,161 - 20,778,560}}{110}$$

$$= \frac{5119 \pm \sqrt{5,425,601}}{110}$$

$$x = \frac{5119 - \sqrt{5,425,601}}{110} \approx 25.361$$

or

$$x = \frac{5119 + \sqrt{5,425,601}}{110} \approx 67.712$$

Rounding to the nearest year, the average income equals $40,000 at ages 25 and 68.

(b) $I(a) = 50,000$

$$50,000 = -55a^2 + 5119a - 54,448$$
$$55a^2 - 5119a + 104,448 = 0$$

For this equation, $a = 55$, $b = -5119$, and $c = 104,448$.

$$x = \frac{-(-5119) \pm \sqrt{(-5119)^2 - 4(55)(104,448)}}{2(55)}$$

$$= \frac{5119 \pm \sqrt{26,204,161 - 22,978,560}}{110}$$

$$= \frac{5119 \pm \sqrt{3,225,601}}{110}$$

$$x = \frac{5119 - \sqrt{3,225,601}}{110} \approx 30.209$$

or

$$x = \frac{5119 + \sqrt{3,225,601}}{110} \approx 62.864$$

Rounding to the nearest year, the average income equals $50,000 at ages 30 and 63.

97. Let x represent speed of the current.

	Distance	Rate	Time
Up Stream	4	$5 - x$	$\dfrac{4}{5-x}$
Down Stream	4	$5 + x$	$\dfrac{4}{5+x}$

$$\frac{4}{5-x} + \frac{4}{5+x} = 6$$

$$(5-x)(5+x)\left(\frac{4}{5-x} + \frac{4}{5+x}\right) = (5-x)(5+x)(6)$$

$$4(5+x) + 4(5-x) = (25 - x^2)(6)$$

$$20 + 4x + 20 - 4x = 150 - 6x^2$$

$$40 = 150 - 6x^2$$

$$6x^2 = 110$$

$$x^2 = \frac{110}{6}$$

$$x = \pm\sqrt{\frac{110}{6}}$$

$$x \approx \pm 4.282$$

The speed of the current is approximately 4.3 miles per hour.

99. Let t represent the time required for Susan to finish the route alone. Then $t - 1$ will represent the time required for Robert to finish the route alone.

$$\begin{pmatrix} \text{Part done} \\ \text{by Susan} \\ \text{in 1 hour} \end{pmatrix} + \begin{pmatrix} \text{Part done} \\ \text{by Robert} \\ \text{in 1 hour} \end{pmatrix} = \begin{pmatrix} \text{Part done} \\ \text{together} \\ \text{in 1 hour} \end{pmatrix}$$

$$\frac{1}{t} + \frac{1}{t-1} = \frac{1}{2}$$

$$2t(t-1)\left(\frac{1}{t} + \frac{1}{t-1}\right) = 2t(t-1)\left(\frac{1}{2}\right)$$

$$2(t-1) + 2t = t(t-1)$$

$$2t - 2 + 2t = t^2 - t$$

$$4t - 2 = t^2 - t$$

$$0 = t^2 - 5t + 2$$

For this equation, $a = 1$, $b = -5$, and $c = 2$.

$$t = \frac{-(-5) \pm \sqrt{(-5)^2 - 4(1)(2)}}{2(1)}$$

$$= \frac{5 \pm \sqrt{25 - 8}}{2} = \frac{5 \pm \sqrt{17}}{2}$$

$$t = \frac{5}{2} - \frac{\sqrt{17}}{2} \approx 0.438 \quad \text{or} \quad t = \frac{5}{2} + \frac{\sqrt{17}}{2} \approx 4.562$$

Disregard $t \approx 0.438$ because this value makes Robert's time negative: $t - 1 = 0.438 - 1 = -0.562$. The only viable answer is $t \approx 4.562$ hours. Working alone, it will take Susan approximately 4.6 hours to finish the route.

101. By the quadratic formula, the solutions of the equation $ax^2 + bx + c = 0$ are

$$x = \frac{-b - \sqrt{b^2 - 4ac}}{2a} \quad \text{and} \quad x = \frac{-b + \sqrt{b^2 - 4ac}}{2a}.$$

The sum of these two solutions is:

$$\frac{-b - \sqrt{b^2 - 4ac}}{2a} + \frac{-b + \sqrt{b^2 - 4ac}}{2a} = \frac{-2b}{2a} = -\frac{b}{a}.$$

103. Assume $b^2 - 4ac \geq 0$. The solutions of

$$ax^2 + bx + c = 0 \quad \text{are} \quad x = \frac{-b \pm \sqrt{b^2 - 4ac}}{2a}.$$

The solutions of $ax^2 - bx + c = 0$ are

$$x = \frac{-(-b) \pm \sqrt{(-b)^2 - 4ac}}{2a} = \frac{b \pm \sqrt{b^2 - 4ac}}{2a}.$$

Now, the negatives of the solutions to $ax^2 - bx + c = 0$ are

$$-\left(\frac{b \pm \sqrt{b^2 - 4ac}}{2a}\right) = \frac{-b \mp \sqrt{b^2 - 4ac}}{2a}.$$

$$= \frac{-b \pm \sqrt{b^2 - 4ac}}{2a}$$

which are the solutions to $ax^2 + bx + c = 0$.

Thus, the real solutions of $ax^2 + bx + c = 0$ are the negatives of the real solutions of $ax^2 - bx + c = 0$.

105. Answers may vary. One possibility follows: If the quadratic equation is in the form

$ax^2 + bx + c = 0$ (or if it can easily be put in that form) and if the left side is easy to factor, then use factoring to solve the equation. Note that the discriminant must be a perfect square.

107. (a) $f(x) = x^2 + 3x + 2$

Let $x = -3, -2, -1.5, -1,$ and 0.

$$f(-3) = (-3)^2 + 3(-3) + 2 = 9 - 9 + 2 = 2$$

$$f(-2) = (-2)^2 + 3(-2) + 2 = 4 - 6 + 2 = 0$$

$$f(-1.5) = (-1.5)^2 + 3(-1.5) + 2$$
$$= 2.25 - 4.5 + 2$$
$$= -0.25$$

$$f(-1) = (-1)^2 + 3(-1) + 2 = 1 - 3 + 2 = 0$$

$$f(0) = 0^2 + 3(0) + 2 = 0 + 0 + 2 = 2$$

Thus, the points $(-3, 2)$, $(-2, 0)$, $(-1.5, -0.25)$, $(-1, 0)$, and $(0, 2)$ are on the graph of f. We plot the points and connect them with a smooth curve.

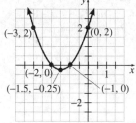

(b) $x^2 + 3x + 2 = 0$
$$(x+1)(x+2) = 0$$
$$x + 1 = 0 \quad \text{or} \quad x + 2 = 0$$
$$x = -1 \quad \text{or} \qquad x = -2$$

The solution set is $\{-2, -1\}$.

(c) From the graph in part (a), the x-intercepts of the function $f(x) = x^2 + 3x + 2$ are -2 and -1, which are the same as the solutions of the equation $x^2 + 3x + 2 = 0$. To find the x-intercepts of a function, set the function equal to zero and solve for x.

109. (a) $g(x) = x^2 - 2x + 1$

Let $x = -1, 0, 1, 2,$ and 3.

$$f(-1) = (-1)^2 - 2(-1) + 1 = 1 + 2 + 1 = 4$$

$$f(0) = 0^2 - 2(0) + 1 = 0 - 0 + 1 = 1$$

$$f(1) = 1^2 - 2(1) + 1 = 1 - 2 + 1 = 0$$

$$f(2) = 2^2 - 2(2) + 1 = 4 - 4 + 1 = 1$$

$$f(3) = 3^2 - 2(3) + 1 = 9 - 6 + 1 = 4$$

Thus, the points $(-1, 4)$, $(0, 1)$, $(1, 0)$, $(2, 1)$, and $(3, 4)$ are on the graph of g.

We plot the points and connect them with a smooth curve.

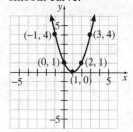

(b) $x^2 - 2x + 1 = 0$
$$(x-1)^2 = 0$$
$$x - 1 = 0$$
$$x = 1$$

The solution set is $\{1\}$.

(c) From the graph in part (a), the x-intercept of the function $g(x) = x^2 - 2x + 1$ is 1, which is the same as the solution of the equation $x^2 - 2x + 1 = 0$. To find the x-intercepts of a function, set the function equal to zero and solve for x.

111. $f(x) = 0$

$$x^2 - 7x + 3 = 0$$

For this equation, $a = 1$, $b = -7$, and $c = 3$.

$$b^2 - 4ac = (-7)^2 - 4(1)(3) = 49 - 12 = 37$$

Because $b^2 - 4ac = 37$ is positive, but not a perfect square, the equation has two irrational solutions. This conclusion based on the discriminant is apparent in the graph because the graph crosses the x-axis in two places. That is, the graph has two x-intercepts.

113. $f(x) = 0$

$$-x^2 - 3x - 4 = 0$$

For this equation, $a = -1$, $b = -3$, and $c = -4$.

$$b^2 - 4ac = (-3)^2 - 4(-1)(-4) = 9 - 16 = -7$$

Because $b^2 - 4ac = -7$ is negative, the equation has two complex solutions that are not real. This conclusion based on the discriminant is apparent in the graph because the graph does not cross the x-axis. That is, the graph has no x-intercept.

115. (a) $x^2 - 5x - 24 = 0$

$(x+3)(x-8) = 0$

$x+3 = 0$ or $x-8 = 0$

$x = -3$ or $x = 8$

The solution set is $\{-3, 8\}$.

(b) The *x*-intercepts are –3 and 8 (see graph).

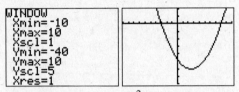

The *x*-intercepts of $y = x^2 - 5x - 24$ are the same as the solutions of $x^2 - 5x - 24 = 0$.

117. (a) $x^2 - 6x + 9 = 0$

$(x-3)(x-3) = 0$

$x-3 = 0$ or $x-3 = 0$

$x = 3$ or $x = 3$

The solution set is $\{3\}$.

(b) The *x*-intercept is 3 (see graph).

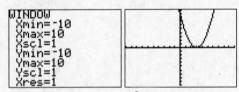

The *x*-intercept of $y = x^2 - 6x + 9$ is the same as the solutions of $x^2 - 6x + 9 = 0$.

119. (a) $x^2 + 5x + 8 = 0$

For this equation, $a = 1$, $b = 5$, and $c = 8$.

$$x = \frac{-5 \pm \sqrt{5^2 - 4(1)(8)}}{2(1)}$$

$$= \frac{-5 \pm \sqrt{25 - 32}}{2}$$

$$= \frac{-5 \pm \sqrt{-7}}{2}$$

$$= \frac{-5 \pm \sqrt{7}\,i}{2} = -\frac{5}{2} \pm \frac{\sqrt{7}}{2}i$$

The solution set is

$$\left\{ -\frac{5}{2} - \frac{\sqrt{7}}{2}i,\ -\frac{5}{2} + \frac{\sqrt{7}}{2}i \right\}.$$

(b) The graph has no *x*-intercepts (see graph).

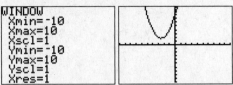

$y = x^2 + 5x + 8$ has no *x*-intercepts, and the solutions of $x^2 + 5x + 8 = 0$ are not real.

Section 10.3

Preparing for Solving Equations Quadratic in Form

P1. $x^4 - 5x^2 - 6 = \left(x^2 - 6\right)\left(x^2 + 1\right)$

P2. $2u^2 + 3u - 5 = 2u^2 - 2u + 5u - 5$

$\qquad = 2u(u-1) + 5(u-1)$

$\qquad = (u-1)(2u+5)$

P3. (a) $\left(x^2\right)^2 = x^{2 \cdot 2} = x^4$

(b) $\left(p^{-1}\right)^2 = p^{-1 \cdot 2} = p^{-2} = \dfrac{1}{p^2}$

Section 10.3 Quick Checks

1. quadratic in form

2. $3x + 1$

3. True. Letting $u = \dfrac{x}{x-2}$ yields the quadratic equation $3u^2 - 5u + 3 = 0$.

4. $u = \dfrac{1}{x}$

5. $\qquad x^4 - 13x^2 + 36 = 0$

$\left(x^2\right)^2 - 13\left(x^2\right) + 36 = 0$

Let $u = x^2$.

$u^2 - 13u + 36 = 0$

$(u-4)(u-9) = 0$

$u - 4 = 0$ or $u - 9 = 0$

$u = 4$ or $u = 9$

$x^2 = 4$ or $x^2 = 9$

$x = \pm\sqrt{4}$ or $x = \pm\sqrt{9}$

$x = \pm 2$ or $x = \pm 3$

Check:

$x = -2:\ (-2)^4 - 13(-2)^2 + 36 \stackrel{?}{=} 0$

$16 - 13 \cdot 4 + 36 \stackrel{?}{=} 0$

$16 - 52 + 36 \stackrel{?}{=} 0$

$0 = 0\ \checkmark$

$x = 2:\ 2^4 - 13(2)^2 + 36 \stackrel{?}{=} 0$

$16 - 13 \cdot 4 + 36 \stackrel{?}{=} 0$

$16 - 52 + 36 \stackrel{?}{=} 0$

$0 = 0\ \checkmark$

$x = -3:\ (-3)^4 - 13(-3)^2 + 36 \stackrel{?}{=} 0$

$81 - 13 \cdot 9 + 36 \stackrel{?}{=} 0$

$81 - 117 + 36 \stackrel{?}{=} 0$

$0 = 0\ \checkmark$

$x = 3:\ 3^4 - 13(3)^2 + 36 \stackrel{?}{=} 0$

$81 - 13 \cdot 9 + 36 \stackrel{?}{=} 0$

$81 - 117 + 36 \stackrel{?}{=} 0$

$0 = 0\ \checkmark$

All check; the solution set is $\{-3, -2,\ 2,\ 3\}$.

6. $p^4 - 7p^2 = 18$

$p^4 - 7p^2 - 18 = 0$

$\left(p^2\right)^2 - 7\left(p^2\right) - 18 = 0$

Let $u = p^2$.

$u^2 - 7u - 18 = 0$

$(u + 2)(u - 9) = 0$

$u + 2 = 0$ or $u - 9 = 0$

$u = -2$ or $u = 9$

$p^2 = -2$ or $p^2 = 9$

$p = \pm\sqrt{-2}$ or $p = \pm\sqrt{9}$

$p = \pm\sqrt{2}\,i$ or $p = \pm 3$

Check:

$p = -\sqrt{2}\,i:\ \left(\sqrt{2}\,i\right)^4 - 7\left(\sqrt{2}\,i\right)^2 \stackrel{?}{=} 18$

$4i^4 - 7 \cdot 2i^2 \stackrel{?}{=} 18$

$4(1) - 7 \cdot 2(-1) \stackrel{?}{=} 18$

$4 - 7 \cdot 2(-1) \stackrel{?}{=} 18$

$4 + 14 \stackrel{?}{=} 18$

$18 = 18\ \checkmark$

$p = \sqrt{2}\,i:\ \left(-\sqrt{2}\,i\right)^4 - 7\left(-\sqrt{2}\,i\right)^2 \stackrel{?}{=} 18$

$4i^4 - 7 \cdot 2i^2 \stackrel{?}{=} 18$

$4(1) - 7 \cdot 2(-1) \stackrel{?}{=} 18$

$4 - 7 \cdot 2(-1) \stackrel{?}{=} 18$

$4 + 14 \stackrel{?}{=} 18$

$18 = 18\ \checkmark$

$p = -3:\ (-3)^4 - 7(-3)^2 \stackrel{?}{=} 18$

$81 - 7 \cdot 9 \stackrel{?}{=} 18$

$81 - 63 \stackrel{?}{=} 18$

$18 = 18\ \checkmark$

$p = 3:\ 3^4 - 7(3)^2 \stackrel{?}{=} 18$

$81 - 7 \cdot 9 \stackrel{?}{=} 18$

$81 - 63 \stackrel{?}{=} 18$

$18 = 18\ \checkmark$

All check; the solution set is

$\left\{-\sqrt{2}\,i,\ \sqrt{2}\,i,\ -3,\ 3\right\}$.

7. $\left(p^2 - 2\right)^2 - 9\left(p^2 - 2\right) + 14 = 0$

Let $u = p^2 - 2$.

$u^2 - 9u + 14 = 0$

$(u - 2)(u - 7) = 0$

$u - 2 = 0$ or $u - 7 = 0$

$u = 2$ or $u = 7$

$p^2 - 2 = 2$ or $p^2 - 2 = 7$

$p^2 = 4$ or $p^2 = 9$

$p = \pm\sqrt{4}$ or $p = \pm\sqrt{9}$

$p = \pm 2$ or $p = \pm 3$

Check:

$$p = -2: \left((-2)^2 - 2\right)^2 - 9\left((-2)^2 - 2\right) + 14 \overset{?}{=} 0$$

$$(4-2)^2 - 9(4-2) + 14 \overset{?}{=} 0$$

$$2^2 - 9 \cdot 2 + 14 \overset{?}{=} 0$$

$$4 - 18 + 14 \overset{?}{=} 0$$

$$0 = 0 \checkmark$$

$$p = 2: \left(2^2 - 2\right)^2 - 9\left(2^2 - 2\right) + 14 \overset{?}{=} 0$$

$$(4-2)^2 - 9(4-2) + 14 \overset{?}{=} 0$$

$$2^2 - 9 \cdot 2 + 14 \overset{?}{=} 0$$

$$4 - 18 + 14 \overset{?}{=} 0$$

$$0 = 0 \checkmark$$

$$p = -3: \left((-3)^2 - 2\right)^2 - 9\left((-3)^2 - 2\right) + 14 \overset{?}{=} 0$$

$$(9-2)^2 - 9(9-2) + 14 \overset{?}{=} 0$$

$$7^2 - 9 \cdot 7 + 14 \overset{?}{=} 0$$

$$49 - 63 + 14 \overset{?}{=} 0$$

$$0 = 0 \checkmark$$

$$p = 3: \left(3^2 - 2\right)^2 - 9\left(3^2 - 2\right) + 14 \overset{?}{=} 0$$

$$(9-2)^2 - 9(9-2) + 14 \overset{?}{=} 0$$

$$7^2 - 9 \cdot 7 + 14 \overset{?}{=} 0$$

$$49 - 63 + 14 \overset{?}{=} 0$$

$$0 = 0 \checkmark$$

All check; the solution set is $\{-3, -2, 2, 3\}$.

8. $2\left(2z^2 - 1\right)^2 + 5\left(2z^2 - 1\right) - 3 = 0$

Let $u = 2z^2 - 1$.

$$2u^2 + 5u - 3 = 0$$

$$(2u - 1)(u + 3) = 0$$

$$2u - 1 = 0 \quad \text{or} \quad u + 3 = 0$$

$$u = \frac{1}{2} \quad \text{or} \quad u = -3$$

$$2z^2 - 1 = \frac{1}{2} \quad \text{or} \quad 2z^2 - 1 = -3$$

$$2z^2 = \frac{3}{2} \quad \text{or} \quad 2z^2 = -2$$

$$z^2 = \frac{3}{4} \quad \text{or} \quad z^2 = -1$$

$$z = \pm\sqrt{\frac{3}{4}} \quad \text{or} \quad z = \pm\sqrt{-1}$$

$$z = \pm\frac{\sqrt{3}}{2} \quad \text{or} \quad z = \pm i$$

Check:

$$z = -\frac{\sqrt{3}}{2}:$$

$$2\left(2\left(-\frac{\sqrt{3}}{2}\right)^2 - 1\right)^2 + 5\left(2\left(-\frac{\sqrt{3}}{2}\right)^2 - 1\right) - 3 \overset{?}{=} 0$$

$$2\left(2\left(\frac{3}{4}\right) - 1\right)^2 + 5\left(2\left(\frac{3}{4}\right) - 1\right) - 3 \overset{?}{=} 0$$

$$2\left(\frac{3}{2} - 1\right)^2 + 5\left(\frac{3}{2} - 1\right) - 3 \overset{?}{=} 0$$

$$2\left(\frac{1}{2}\right)^2 + 5\left(\frac{1}{2}\right) - 3 \overset{?}{=} 0$$

$$2\left(\frac{1}{4}\right) + 5\left(\frac{1}{2}\right) - 3 \overset{?}{=} 0$$

$$\frac{1}{2} + \frac{5}{2} - 3 \overset{?}{=} 0$$

$$0 = 0 \checkmark$$

$z = \dfrac{\sqrt{3}}{2}$:

$$2\left(2\left(\dfrac{\sqrt{3}}{2}\right)^2 - 1\right)^2 + 5\left(2\left(\dfrac{\sqrt{3}}{2}\right)^2 - 1\right) - 3 \stackrel{?}{=} 0$$

$$2\left(2\left(\dfrac{3}{4}\right) - 1\right)^2 + 5\left(2\left(\dfrac{3}{4}\right) - 1\right) - 3 \stackrel{?}{=} 0$$

$$2\left(\dfrac{3}{2} - 1\right)^2 + 5\left(\dfrac{3}{2} - 1\right) - 3 \stackrel{?}{=} 0$$

$$2\left(\dfrac{1}{2}\right)^2 + 5\left(\dfrac{1}{2}\right) - 3 \stackrel{?}{=} 0$$

$$2\left(\dfrac{1}{4}\right) + 5\left(\dfrac{1}{2}\right) - 3 \stackrel{?}{=} 0$$

$$\dfrac{1}{2} + \dfrac{5}{2} - 3 \stackrel{?}{=} 0$$

$$0 = 0 \checkmark$$

$z = -i$:

$$2\left(2(-i)^2 - 1\right)^2 + 5\left(2(-i)^2 - 1\right) - 3 \stackrel{?}{=} 0$$

$$2\left(2i^2 - 1\right)^2 + 5\left(2i^2 - 1\right) - 3 \stackrel{?}{=} 0$$

$$2\left(2(-1) - 1\right)^2 + 5\left(2(-1) - 1\right) - 3 \stackrel{?}{=} 0$$

$$2(-2 - 1)^2 + 5(-2 - 1) - 3 \stackrel{?}{=} 0$$

$$2(-3)^2 + 5(-3) - 3 \stackrel{?}{=} 0$$

$$2(9) + 5(-3) - 3 \stackrel{?}{=} 0$$

$$18 - 15 - 3 \stackrel{?}{=} 0$$

$$0 = 0 \checkmark$$

$z = i$:

$$2\left(2i^2 - 1\right)^2 + 5\left(2i^2 - 1\right) - 3 \stackrel{?}{=} 0$$

$$2\left(2(-1) - 1\right)^2 + 5\left(2(-1) - 1\right) - 3 \stackrel{?}{=} 0$$

$$2(-2 - 1)^2 + 5(-2 - 1) - 3 \stackrel{?}{=} 0$$

$$2(-3)^2 + 5(-3) - 3 \stackrel{?}{=} 0$$

$$2(9) + 5(-3) - 3 \stackrel{?}{=} 0$$

$$18 - 15 - 3 \stackrel{?}{=} 0$$

$$0 = 0 \checkmark$$

All check; the solution set is

$$\left\{ -\dfrac{\sqrt{3}}{2},\ \dfrac{\sqrt{3}}{2}, -i,\ i \right\}.$$

9. $\quad\quad\quad 3w - 14\sqrt{w} + 8 = 0$

$$3\left(\sqrt{w}\right)^2 - 14\left(\sqrt{w}\right) + 8 = 0$$

Let $u = \sqrt{w}$.

$$3u^2 - 14u + 8 = 0$$

$$(3u - 2)(u - 4) = 0$$

$$3u - 2 = 0 \quad\quad \text{or} \quad u - 4 = 0$$

$$u = \dfrac{2}{3} \quad\quad \text{or} \quad\quad u = 4$$

$$\sqrt{w} = \dfrac{2}{3} \quad\quad \text{or} \quad \sqrt{w} = 4$$

$$w = \left(\dfrac{2}{3}\right)^2 \quad \text{or} \quad\quad w = 4^2$$

$$w = \dfrac{4}{9} \quad\quad \text{or} \quad\quad w = 16$$

<u>Check:</u>

$$w = \dfrac{4}{9}: \quad 3\left(\dfrac{4}{9}\right) - 14\sqrt{\dfrac{4}{9}} + 8 \stackrel{?}{=} 0$$

$$\dfrac{4}{3} - 14 \cdot \dfrac{2}{3} + 8 \stackrel{?}{=} 0$$

$$\dfrac{4}{3} - \dfrac{28}{3} + 8 \stackrel{?}{=} 0$$

$$0 = 0 \checkmark$$

$$w = 16: \quad 3(16) - 14\sqrt{16} + 8 \stackrel{?}{=} 0$$

$$48 - 14 \cdot 4 + 8 \stackrel{?}{=} 0$$

$$48 - 56 + 8 \stackrel{?}{=} 0$$

$$0 = 0 \checkmark$$

Both check; the solution set is $\left\{ \dfrac{4}{9},\ 16 \right\}.$

10. $\quad\quad\quad 2q - 9\sqrt{q} - 5 = 0$

$$2\left(\sqrt{q}\right)^2 - 9\left(\sqrt{q}\right) - 5 = 0$$

Let $u = \sqrt{q}$.

$$2u^2 - 9u - 5 = 0$$

$$(2u + 1)(u - 5) = 0$$

$$2u + 1 = 0 \quad \text{or} \quad u - 5 = 0$$

$$u = -\dfrac{1}{2} \quad \text{or} \quad\quad u = 5$$

$$\sqrt{q} = -\frac{1}{2} \quad \text{or} \quad \sqrt{q} = 5$$

$$q = \left(-\frac{1}{2}\right)^2 \quad \text{or} \quad q = 5^2$$

$$q = \frac{1}{4} \quad \text{or} \quad q = 25$$

Check:

$$q = \frac{1}{4}: \quad 2\left(\frac{1}{4}\right) - 9\sqrt{\frac{1}{4}} - 5 \stackrel{?}{=} 0$$

$$\frac{1}{2} - 9 \cdot \frac{1}{2} - 5 \stackrel{?}{=} 0$$

$$\frac{1}{2} - \frac{9}{2} - 5 \stackrel{?}{=} 0$$

$$-9 \neq 0 \quad \text{✗}$$

$$q = 25: \quad 2(25) - 9\sqrt{25} - 5 \stackrel{?}{=} 0$$

$$50 - 9 \cdot 5 - 5 \stackrel{?}{=} 0$$

$$50 - 45 - 5 \stackrel{?}{=} 0$$

$$0 = 0 \quad \checkmark$$

$q = \frac{1}{4}$ does not check; the solution set is $\{25\}$.

11.
$$5x^{-2} + 12x^{-1} + 4 = 0$$

$$5\left(x^{-1}\right)^2 + 12\left(x^{-1}\right) + 4 = 0$$

Let $u = x^{-1}$.

$$5u^2 + 12u + 4 = 0$$

$$(5u + 2)(u + 2) = 0$$

$$5u + 2 = 0 \quad \text{or} \quad u + 2 = 0$$

$$u = -\frac{2}{5} \quad \text{or} \quad u = -2$$

$$x^{-1} = -\frac{2}{5} \quad \text{or} \quad x^{-1} = -2$$

$$\frac{1}{x} = -\frac{2}{5} \quad \text{or} \quad \frac{1}{x} = -2$$

$$x = -\frac{5}{2} \quad \text{or} \quad x = -\frac{1}{2}$$

Check:

$$x = -\frac{5}{2}: \quad 5\left(-\frac{5}{2}\right)^{-2} + 12\left(-\frac{5}{2}\right)^{-1} + 4 \stackrel{?}{=} 0$$

$$5\left(-\frac{2}{5}\right)^2 + 12\left(-\frac{2}{5}\right) + 4 \stackrel{?}{=} 0$$

$$5\left(\frac{4}{25}\right) + 12\left(-\frac{2}{5}\right) + 4 \stackrel{?}{=} 0$$

$$\frac{4}{5} - \frac{24}{5} + 4 \stackrel{?}{=} 0$$

$$0 = 0 \quad \checkmark$$

$$x = -\frac{1}{2}: \quad 5\left(-\frac{1}{2}\right)^{-2} + 12\left(-\frac{1}{2}\right)^{-1} + 4 \stackrel{?}{=} 0$$

$$5(-2)^2 + 12(-2) + 4 \stackrel{?}{=} 0$$

$$5(4) + 12(-2) + 4 \stackrel{?}{=} 0$$

$$20 - 24 + 4 \stackrel{?}{=} 0$$

$$0 = 0 \quad \checkmark$$

Both check; the solution set is $\left\{-\frac{5}{2}, -\frac{1}{2}\right\}$.

12.
$$p^{2/3} - 4p^{1/3} - 5 = 0$$

$$\left(p^{1/3}\right)^2 - 4\left(p^{1/3}\right) - 5 = 0$$

Let $u = p^{1/3}$.

$$u^2 - 4u - 5 = 0$$

$$(u + 1)(u - 5) = 0$$

$$u + 1 = 0 \quad \text{or} \quad u - 5 = 0$$

$$u = -1 \quad \text{or} \quad u = 5$$

$$p^{1/3} = -1 \quad \text{or} \quad p^{1/3} = 5$$

$$\left(p^{1/3}\right)^3 = (-1)^3 \quad \text{or} \quad \left(p^{1/3}\right)^3 = 5^3$$

$$p = -1 \quad \text{or} \quad p = 125$$

Check:

$$p = -1: \quad (-1)^{2/3} - 4 \cdot (-1)^{1/3} - 5 \stackrel{?}{=} 0$$

$$\left(\sqrt[3]{-1}\right)^2 - 4\left(\sqrt[3]{-1}\right) - 5 \stackrel{?}{=} 0$$

$$(-1)^2 - 4(-1) - 5 \stackrel{?}{=} 0$$

$$1 + 4 - 5 \stackrel{?}{=} 0$$

$$0 = 0 \quad \checkmark$$

$p = 125$: $(125)^{2/3} - 4 \cdot (125)^{1/3} - 5 \stackrel{?}{=} 0$

$$\left(\sqrt[3]{125}\right)^2 - 4\left(\sqrt[3]{125}\right) - 5 \stackrel{?}{=} 0$$

$$(5)^2 - 4(5) - 5 \stackrel{?}{=} 0$$

$$25 - 20 - 5 \stackrel{?}{=} 0$$

$$0 = 0 \checkmark$$

Both check; the solution set is $\{-1,\ 125\}$.

10.3 Exercises

13. $x^4 - 5x^2 + 4 = 0$

$$\left(x^2\right)^2 - 5\left(x^2\right) + 4 = 0$$

Let $u = x^2$.

$u^2 - 5u + 4 = 0$

$(u-1)(u-4) = 0$

$u - 1 = 0$ or $u - 4 = 0$

$u = 1$ or $u = 4$

$x^2 = 1$ or $x^2 = 4$

$x = \pm\sqrt{1}$ or $x = \pm\sqrt{4}$

$x = \pm 1$ or $x = \pm 2$

Check:

$x = -1$: $(-1)^4 - 5(-1)^2 + 4 \stackrel{?}{=} 0$

$$1 - 5 \cdot 1 + 4 \stackrel{?}{=} 0$$

$$1 - 5 + 4 \stackrel{?}{=} 0$$

$$0 = 0 \checkmark$$

$x = 1$: $1^4 - 5(1)^2 + 4 \stackrel{?}{=} 0$

$$1 - 5 \cdot 1 + 4 \stackrel{?}{=} 0$$

$$1 - 5 + 4 \stackrel{?}{=} 0$$

$$0 = 0 \checkmark$$

$x = -2$: $(-2)^4 - 5(-2)^2 + 4 \stackrel{?}{=} 0$

$$16 - 5 \cdot 4 + 4 \stackrel{?}{=} 0$$

$$16 - 20 + 4 \stackrel{?}{=} 0$$

$$0 = 0 \checkmark$$

$x = 2$: $2^4 - 5(2)^2 + 4 \stackrel{?}{=} 0$

$$16 - 5 \cdot 4 + 4 \stackrel{?}{=} 0$$

$$16 - 20 + 4 \stackrel{?}{=} 0$$

$$0 = 0 \checkmark$$

All check; the solution set is $\{-2,\ -1,\ 1,\ 2\}$.

15. $q^4 + 13q^2 + 36 = 0$

$$\left(q^2\right)^2 + 13\left(q^2\right) + 36 = 0$$

Let $u = q^2$.

$u^2 + 13u + 36 = 0$

$(u+4)(u+9) = 0$

$q^2 = -4$ or $q^2 = -9$

$q = \pm\sqrt{-4}$ or $q = \pm\sqrt{-9}$

$q = \pm 2i$ or $q = \pm 3i$

Check:

$q = -3i$: $(-3i)^4 + 13(-3i)^2 + 36 \stackrel{?}{=} 0$

$$81i^4 + 13 \cdot 9i^2 + 36 \stackrel{?}{=} 0$$

$$81(1) + 13 \cdot 9 \cdot (-1) + 36 \stackrel{?}{=} 0$$

$$81 - 117 + 36 \stackrel{?}{=} 0$$

$$0 = 0 \checkmark$$

$q = 3i$: $(3i)^4 + 13(3i)^2 + 36 \stackrel{?}{=} 0$

$$81i^4 + 13 \cdot 9i^2 + 36 \stackrel{?}{=} 0$$

$$81(1) + 13 \cdot 9 \cdot (-1) + 36 \stackrel{?}{=} 0$$

$$81 - 117 + 36 \stackrel{?}{=} 0$$

$$0 = 0 \checkmark$$

$q = -2i$: $(-2i)^4 + 13(-2i)^2 + 36 \stackrel{?}{=} 0$

$$16i^4 + 13 \cdot 4i^2 + 36 \stackrel{?}{=} 0$$

$$16(1) + 13 \cdot 4 \cdot (-1) + 36 \stackrel{?}{=} 0$$

$$16 - 52 + 36 \stackrel{?}{=} 0$$

$$0 = 0 \checkmark$$

$q = 2i$: $(2i)^4 + 13(2i)^2 + 36 \stackrel{?}{=} 0$

$$16i^4 + 13 \cdot 4i^2 + 36 \stackrel{?}{=} 0$$

$$16(1) + 13 \cdot 4 \cdot (-1) + 36 \stackrel{?}{=} 0$$

$$16 - 52 + 36 \stackrel{?}{=} 0$$

$$0 = 0 \checkmark$$

All check; the solution set is $\{-3i,\ 3i, -2i,\ 2i\}$.

17. $4a^4 - 17a^2 + 4 = 0$

$$4\left(a^2\right)^2 - 17\left(a^2\right) + 4 = 0$$

Let $u = a^2$.

$$4u^2 - 17u + 4 = 0$$
$$(4u-1)(u-4) = 0$$
$$4u - 1 = 0 \quad \text{or} \quad u - 4 = 0$$
$$4u = 1 \quad \text{or} \quad u = 4$$
$$u = \frac{1}{4}$$

$$a^2 = \frac{1}{4} \quad \text{or} \quad a^2 = 4$$
$$a = \pm\sqrt{\frac{1}{4}} \quad \text{or} \quad a = \pm\sqrt{4}$$
$$a = \pm\frac{1}{2} \quad \text{or} \quad a = \pm 2$$

Check:

$$a = -\frac{1}{2}: \quad 4\left(-\frac{1}{2}\right)^4 - 17\left(-\frac{1}{2}\right)^2 + 4 \stackrel{?}{=} 0$$
$$4\left(\frac{1}{16}\right) - 17\left(\frac{1}{4}\right) + 4 \stackrel{?}{=} 0$$
$$\frac{1}{4} - \frac{17}{4} + 4 \stackrel{?}{=} 0$$
$$0 = 0 \checkmark$$

$$a = \frac{1}{2}: \quad 4\left(\frac{1}{2}\right)^4 - 17\left(\frac{1}{2}\right)^2 + 4 \stackrel{?}{=} 0$$
$$4\left(\frac{1}{16}\right) - 17\left(\frac{1}{4}\right) + 4 \stackrel{?}{=} 0$$
$$\frac{1}{4} - \frac{17}{4} + 4 \stackrel{?}{=} 0$$
$$0 = 0 \checkmark$$

$$a = -2: \quad 4(-2)^4 - 17(-2)^2 + 4 \stackrel{?}{=} 0$$
$$4 \cdot 16 - 17 \cdot 4 + 4 \stackrel{?}{=} 0$$
$$64 - 68 + 4 \stackrel{?}{=} 0$$
$$0 = 0 \checkmark$$

$$a = 2: \quad 4(2)^4 - 17(2)^2 + 4 \stackrel{?}{=} 0$$
$$4 \cdot 16 - 17 \cdot 4 + 4 \stackrel{?}{=} 0$$
$$64 - 68 + 4 \stackrel{?}{=} 0$$
$$0 = 0 \checkmark$$

All check; the solution set is $\left\{-2, -\frac{1}{2}, \frac{1}{2}, 2\right\}$.

19.
$$p^4 + 6 = 5p^2$$
$$p^4 - 5p^2 + 6 = 0$$
$$\left(p^2\right)^2 - 5\left(p^2\right) + 6 = 0$$

Let $u = p^2$.
$$u^2 - 5u + 6 = 0$$
$$(u-2)(u-3) = 0$$
$$u - 2 = 0 \quad \text{or} \quad u - 3 = 0$$
$$u = 2 \quad \text{or} \quad u = 3$$

$$p^2 = 2 \quad \text{or} \quad p^2 = 3$$
$$p = \pm\sqrt{2} \quad \text{or} \quad p = \pm\sqrt{3}$$

Check:

$$p = -\sqrt{2}: \quad \left(-\sqrt{2}\right)^4 + 6 \stackrel{?}{=} 5\left(-\sqrt{2}\right)^2$$
$$4 + 6 \stackrel{?}{=} 5 \cdot 2$$
$$10 = 10 \checkmark$$

$$p = \sqrt{2}: \quad \left(\sqrt{2}\right)^4 + 6 \stackrel{?}{=} 5\left(\sqrt{2}\right)^2$$
$$4 + 6 \stackrel{?}{=} 5 \cdot 2$$
$$10 = 10 \checkmark$$

$$p = -\sqrt{3}: \quad \left(-\sqrt{3}\right)^4 + 6 \stackrel{?}{=} 5\left(-\sqrt{3}\right)^2$$
$$9 + 6 \stackrel{?}{=} 5 \cdot 3$$
$$15 = 15 \checkmark$$
$$15 = 15 \checkmark$$

$$p = \sqrt{3}: \quad \left(\sqrt{3}\right)^4 + 6 \stackrel{?}{=} 5\left(\sqrt{3}\right)^2$$
$$9 + 6 \stackrel{?}{=} 5 \cdot 3$$
$$15 = 15 \checkmark$$

All check; the solution set is
$\left\{-\sqrt{3}, -\sqrt{2}, \sqrt{2}, \sqrt{3}\right\}$.

21. $(x-3)^2 - 6(x-3) - 7 = 0$
Let $u = x - 3$.
$$u^2 - 6u - 7 = 0$$
$$(u-7)(u+1) = 0$$
$$u - 7 = 0 \quad \text{or} \quad u + 1 = 0$$
$$u = 7 \quad \text{or} \quad u = -1$$
$$x - 3 = 7 \quad \text{or} \quad x - 3 = -1$$
$$x = 10 \quad \text{or} \quad x = 2$$

Check:

$$x = 10: \quad (10-3)^2 - 6(10-3) - 7 \overset{?}{=} 0$$
$$7^2 - 6(7) - 7 \overset{?}{=} 0$$
$$49 - 42 - 7 \overset{?}{=} 0$$
$$0 = 0 \ \checkmark$$

$$x = 2: \quad (2-3)^2 - 6(2-3) - 7 \overset{?}{=} 0$$
$$(-1)^2 - 6(-1) - 7 \overset{?}{=} 0$$
$$1 + 6 - 7 \overset{?}{=} 0$$
$$0 = 0 \ \checkmark$$

Both check; the solution set is $\{2, 10\}$.

23. $\left(x^2 - 1\right)^2 - 11\left(x^2 - 1\right) + 24 = 0$

Let $u = x^2 - 1$.

$$u^2 - 11u + 24 = 0$$
$$(u - 3)(u - 8) = 0$$

$$u - 3 = 0 \quad \text{or} \quad u - 8 = 0$$
$$u = 3 \quad \text{or} \quad u = 8$$
$$x^2 - 1 = 3 \quad \text{or} \quad x^2 - 1 = 8$$
$$x^2 = 4 \quad \text{or} \quad x^2 = 9$$
$$x = \pm\sqrt{4} \ \text{or} \ x = \pm\sqrt{9}$$
$$x = \pm 2 \quad \text{or} \ x = \pm 3$$

Check:

$$x = -2: \quad \left((-2)^2 - 1\right)^2 - 11\left((-2)^2 - 1\right) + 24 \overset{?}{=} 0$$
$$(4-1)^2 - 11(4-1) + 24 \overset{?}{=} 0$$
$$3^2 - 11 \cdot 3 + 24 \overset{?}{=} 0$$
$$9 - 33 + 24 \overset{?}{=} 0$$
$$0 = 0 \ \checkmark$$

$$x = 2: \quad \left(2^2 - 1\right)^2 - 11\left(2^2 - 1\right) + 24 \overset{?}{=} 0$$
$$(4-1)^2 - 11(4-1) + 24 \overset{?}{=} 0$$
$$3^2 - 11 \cdot 3 + 24 \overset{?}{=} 0$$
$$9 - 33 + 24 \overset{?}{=} 0$$
$$0 = 0 \ \checkmark$$

$$x = -3: \quad \left((-3)^2 - 1\right)^2 - 11\left((-3)^2 - 1\right) + 24 \overset{?}{=} 0$$
$$(9-1)^2 - 11(9-1) + 24 \overset{?}{=} 0$$
$$8^2 - 11 \cdot 8 + 24 \overset{?}{=} 0$$
$$64 - 88 + 24 \overset{?}{=} 0$$
$$0 = 0 \ \checkmark$$

$$x = 3: \quad \left(3^2 - 1\right)^2 - 11\left(3^2 - 1\right) + 24 \overset{?}{=} 0$$
$$(9-1)^2 - 11(9-1) + 24 \overset{?}{=} 0$$
$$8^2 - 11 \cdot 8 + 24 \overset{?}{=} 0$$
$$64 - 88 + 24 \overset{?}{=} 0$$
$$0 = 0 \ \checkmark$$

All check; the solution set is $\{-3, -2, \ 2, \ 3\}$.

25. $\left(y^2 + 2\right)^2 + 7\left(y^2 + 2\right) + 10 = 0$

Let $u = y^2 + 2$.

$$u^2 + 7u + 10 = 0$$
$$(u + 2)(u + 5) = 0$$

$$u + 2 = 0 \quad \text{or} \quad u + 5 = 0$$
$$u = -2 \quad \text{or} \quad u = -5$$
$$y^2 + 2 = -2 \quad \text{or} \ y^2 + 2 = -5$$
$$y^2 = -4 \quad \text{or} \quad y^2 = -7$$
$$y = \pm\sqrt{-4} \ \text{or} \ y = \pm\sqrt{-7}$$
$$y = \pm 2i \quad \text{or} \ y = \pm\sqrt{7} \ i$$

Check:
$$y = -2i:$$

$$\left((-2i)^2 + 2\right)^2 + 7\left((-2i)^2 + 2\right) + 10 \overset{?}{=} 0$$
$$\left(4i^2 + 2\right)^2 + 7\left(4i^2 + 2\right) + 10 \overset{?}{=} 0$$
$$(-4 + 2)^2 + 7(-4 + 2) + 10 \overset{?}{=} 0$$
$$(-2)^2 + 7(-2) + 10 \overset{?}{=} 0$$
$$4 - 14 + 10 \overset{?}{=} 0$$
$$0 = 0 \ \checkmark$$

$y = 2i:$ $\left((2i)^2 + 2\right)^2 + 7\left((2i)^2 + 2\right) + 10 \overset{?}{=} 0$

$\left(4i^2 + 2\right)^2 + 7\left(4i^2 + 2\right) + 10 \overset{?}{=} 0$

$(-4 + 2)^2 + 7(-4 + 2) + 10 \overset{?}{=} 0$

$(-2)^2 + 7(-2) + 10 \overset{?}{=} 0$

$4 - 14 + 10 \overset{?}{=} 0$

$0 = 0 \checkmark$

$y = -\sqrt{7}i:$

$\left((-\sqrt{7}i)^2 + 2\right)^2 + 7\left((-\sqrt{7}i)^2 + 2\right) + 10 \overset{?}{=} 0$

$\left(7i^2 + 2\right)^2 + 7\left(7i^2 + 2\right) + 10 \overset{?}{=} 0$

$(-7 + 2)^2 + 7(-7 + 2) + 10 \overset{?}{=} 0$

$(-5)^2 + 7(-5) + 10 \overset{?}{=} 0$

$25 - 35 + 10 \overset{?}{=} 0$

$0 = 0 \checkmark$

$y = \sqrt{7}i:$

$\left((\sqrt{7}i)^2 + 2\right)^2 + 7\left((\sqrt{7}i)^2 + 2\right) + 10 \overset{?}{=} 0$

$\left(7i^2 + 2\right)^2 + 7\left(7i^2 + 2\right) + 10 \overset{?}{=} 0$

$(-7 + 2)^2 + 7(-7 + 2) + 10 \overset{?}{=} 0$

$(-5)^2 + 7(-5) + 10 \overset{?}{=} 0$

$25 - 35 + 10 \overset{?}{=} 0$

$0 = 0 \checkmark$

All check; the solution set is
$\left\{-\sqrt{7}\, i,\ \sqrt{7}\, i, -2i,\ 2i\right\}$.

27. $x - 3\sqrt{x} - 4 = 0$

$\left(\sqrt{x}\right)^2 - 3\left(\sqrt{x}\right) - 4 = 0$

Let $u = \sqrt{x}$.

$u^2 - 3u - 4 = 0$

$(u - 4)(u + 1) = 0$

$u - 4 = 0 \quad \text{or} \quad u + 1 = 0$

$u = 4 \quad \text{or} \quad u = -1$

$\sqrt{x} = 4 \quad \text{or} \quad \sqrt{x} = -1$

$x = 4^2 \quad \text{or} \quad x = (-1)^2$

$x = 16 \quad \text{or} \quad x = 1$

Check:

$x = 16:$ $16 - 3\sqrt{16} - 4 \overset{?}{=} 0$

$16 - 3 \cdot 4 - 4 \overset{?}{=} 0$

$16 - 12 - 4 \overset{?}{=} 0$

$0 = 0 \checkmark$

$x = 1:$ $1 - 3\sqrt{1} - 4 \overset{?}{=} 0$

$1 - 3 \cdot 1 - 4 \overset{?}{=} 0$

$1 - 3 - 4 \overset{?}{=} 0$

$-6 \neq 0 \ ✗$

$x = 1$ does not check; the solution set is $\{16\}$.

29. $w + 5\sqrt{w} + 6 = 0$

$\left(\sqrt{w}\right)^2 + 5\left(\sqrt{w}\right) + 6 = 0$

Let $u = \sqrt{w}$.

$u^2 + 5u + 6 = 0$

$(u + 2)(u + 3) = 0$

$u + 2 = 0 \quad \text{or} \quad u + 3 = 0$

$u = -2 \quad \text{or} \quad u = -3$

$\sqrt{w} = -2 \quad \text{or} \quad \sqrt{w} = -3$

$w = (-2)^2 \quad \text{or} \quad w = (-3)^2$

$w = 4 \quad \text{or} \quad w = 9$

Check:

$w = 4:$ $4 + 5\sqrt{4} + 6 \overset{?}{=} 0$

$4 + 5 \cdot 2 + 6 \overset{?}{=} 0$

$4 + 10 + 6 \overset{?}{=} 0$

$20 \neq 0 \ ✗$

$w = 9:$ $9 + 5\sqrt{9} + 6 \overset{?}{=} 0$

$9 + 5 \cdot 3 + 6 \overset{?}{=} 0$

$9 + 15 + 6 \overset{?}{=} 0$

$30 \neq 0 \ ✗$

Neither possibility checks; the equation has no solution. The solution set is $\{\ \}$ or $\varnothing$.

31. $2x + 5\sqrt{x} = 3$

$2x + 5\sqrt{x} - 3 = 0$

$2\left(\sqrt{x}\right)^2 + 5\left(\sqrt{x}\right) - 3 = 0$

Let $u = \sqrt{x}$.

$$2u^2 + 5u - 3 = 0$$
$$(2u - 1)(u + 3) = 0$$
$$2u - 1 = 0 \quad \text{or} \quad u + 3 = 0$$
$$2u = 1 \quad \text{or} \quad u = -3$$
$$u = \frac{1}{2}$$

$$\sqrt{x} = \frac{1}{2} \quad \text{or} \quad \sqrt{x} = -3$$
$$x = \left(\frac{1}{2}\right)^2 \quad \text{or} \quad x = (-3)^2$$
$$x = \frac{1}{4} \quad \text{or} \quad x = 9$$

Check:

$$x = \frac{1}{4}: \quad 2 \cdot \frac{1}{4} + 5\sqrt{\frac{1}{4}} \overset{?}{=} 3$$
$$\frac{1}{2} + 5 \cdot \frac{1}{2} \overset{?}{=} 3$$
$$\frac{1}{2} + \frac{5}{2} \overset{?}{=} 3$$
$$3 = 3 \checkmark$$

$$x = 9: \quad 2 \cdot 9 + 5\sqrt{9} \overset{?}{=} 3$$
$$2 \cdot 9 + 5 \cdot 3 \overset{?}{=} 3$$
$$18 + 15 \overset{?}{=} 3$$
$$33 \neq 3 \; \times$$

$x = 9$ does not check; the solution set is $\left\{\dfrac{1}{4}\right\}$.

33.
$$x^{-2} + 3x^{-1} = 28$$
$$x^{-2} + 3x^{-1} - 28 = 0$$
$$\left(x^{-1}\right)^2 + 3\left(x^{-1}\right) - 28 = 0$$

Let $u = x^{-1}$.
$$u^2 + 3u - 28 = 0$$
$$(u - 4)(u + 7) = 0$$
$$u - 4 = 0 \quad \text{or} \quad u + 7 = 0$$
$$u = 4 \quad \text{or} \quad u = -7$$
$$x^{-1} = 4 \quad \text{or} \quad x^{-1} = -7$$
$$\frac{1}{x} = 4 \quad \text{or} \quad \frac{1}{x} = -7$$
$$x = \frac{1}{4} \quad \text{or} \quad x = -\frac{1}{7}$$

Check:

$$x = \frac{1}{4}: \quad \left(\frac{1}{4}\right)^{-2} + 3\left(\frac{1}{4}\right)^{-1} \overset{?}{=} 28$$
$$(4)^2 + 3(4) \overset{?}{=} 28$$
$$16 + 12 \overset{?}{=} 28$$
$$28 = 28 \checkmark$$

$$x = -\frac{1}{7}: \quad \left(-\frac{1}{7}\right)^{-2} + 3\left(-\frac{1}{7}\right)^{-1} \overset{?}{=} 28$$
$$(-7)^2 + 3(-7) \overset{?}{=} 28$$
$$49 - 21 \overset{?}{=} 28$$
$$28 = 28 \checkmark$$

Both check; the solution set is $\left\{-\dfrac{1}{7}, \dfrac{1}{4}\right\}$.

35.
$$10z^{-2} + 11z^{-1} = 6$$
$$10z^{-2} + 11z^{-1} - 6 = 0$$
$$10\left(z^{-1}\right)^2 + 11\left(z^{-1}\right) - 6 = 0$$

Let $u = z^{-1}$.
$$10u^2 + 11u - 6 = 0$$
$$(5u - 2)(2u + 3) = 0$$
$$5u - 2 = 0 \quad \text{or} \quad 2u + 3 = 0$$
$$5u = 2 \quad \text{or} \quad 2u = -3$$
$$u = \frac{2}{5} \quad \text{or} \quad u = -\frac{3}{2}$$
$$z^{-1} = \frac{2}{5} \quad \text{or} \quad z^{-1} = -\frac{3}{2}$$
$$\frac{1}{z} = \frac{2}{5} \quad \text{or} \quad \frac{1}{z} = -\frac{3}{2}$$
$$z = \frac{5}{2} \quad \text{or} \quad z = -\frac{2}{3}$$

Check:

$z = \dfrac{5}{2}$:

$$10\left(\dfrac{5}{2}\right)^{-2} + 11\left(\dfrac{5}{2}\right)^{-1} \stackrel{?}{=} 6$$

$$10\left(\dfrac{2}{5}\right)^{2} + 11\left(\dfrac{2}{5}\right) \stackrel{?}{=} 6$$

$$10\left(\dfrac{4}{25}\right) + 11\left(\dfrac{2}{5}\right) \stackrel{?}{=} 6$$

$$\dfrac{8}{5} + \dfrac{22}{5} \stackrel{?}{=} 6$$

$$\dfrac{30}{5} \stackrel{?}{=} 6$$

$$6 = 6 \checkmark$$

$z = -\dfrac{2}{3}$:

$$10\left(-\dfrac{2}{3}\right)^{-2} + 11\left(-\dfrac{2}{3}\right)^{-1} \stackrel{?}{=} 6$$

$$10\left(-\dfrac{3}{2}\right)^{2} + 11\left(-\dfrac{3}{2}\right) \stackrel{?}{=} 6$$

$$10\left(\dfrac{9}{4}\right) + 11\left(-\dfrac{3}{2}\right) \stackrel{?}{=} 6$$

$$\dfrac{45}{2} - \dfrac{33}{2} \stackrel{?}{=} 6$$

$$\dfrac{12}{2} \stackrel{?}{=} 6$$

$$6 = 6 \checkmark$$

Both check; the solution set is $\left\{ -\dfrac{2}{3},\ \dfrac{5}{2} \right\}$.

37. $x^{2/3} + 3x^{1/3} - 4 = 0$

$$\left(x^{1/3}\right)^{2} + 3\left(x^{1/3}\right) - 4 = 0$$

Let $u = x^{1/3}$.

$$u^{2} + 3u - 4 = 0$$
$$(u - 1)(u + 4) = 0$$

$u - 1 = 0$ or $u + 4 = 0$

$u = 1$ or $u = -4$

$x^{1/3} = 1$ or $x^{1/3} = -4$

$\left(x^{1/3}\right)^{3} = 1^{3}$ or $\left(x^{1/3}\right)^{3} = (-4)^{3}$

$x = 1$ or $x = -64$

Check:

$x = 1:\ (1)^{2/3} + 3 \cdot (1)^{1/3} - 4 \stackrel{?}{=} 0$

$$\left(\sqrt[3]{1}\right)^{2} + 3\left(\sqrt[3]{1}\right) - 4 \stackrel{?}{=} 0$$

$$(1)^{2} + 3(1) - 4 \stackrel{?}{=} 0$$

$$1 + 3 - 4 \stackrel{?}{=} 0$$

$$0 = 0 \checkmark$$

$x = -64:\ (-64)^{2/3} + 3 \cdot (-64)^{1/3} - 4 \stackrel{?}{=} 0$

$$\left(\sqrt[3]{-64}\right)^{2} + 3\left(\sqrt[3]{-64}\right) - 4 \stackrel{?}{=} 0$$

$$(-4)^{2} + 3(-4) - 4 \stackrel{?}{=} 0$$

$$16 - 12 - 4 \stackrel{?}{=} 0$$

$$0 = 0 \checkmark$$

Both check; the solution set is $\{-64,\ 1\}$.

39. $z^{2/3} - z^{1/3} = 2$

$$\left(z^{1/3}\right)^{2} - \left(z^{1/3}\right) - 2 = 0$$

Let $u = z^{1/3}$.

$$u^{2} - u - 2 = 0$$
$$(u - 2)(u + 1) = 0$$

$u - 2 = 0$ or $u + 1 = 0$

$u = 2$ or $u = -1$

$z^{1/3} = 2$ or $z^{1/3} = -1$

$z = 2^{3} = 8$ or $z = (-1)^{3} = -1$

Check:

$z = 8:\ (8)^{2/3} - (8)^{1/3} \stackrel{?}{=} 2$

$$\left(\sqrt[3]{8}\right)^{2} - \left(\sqrt[3]{8}\right) \stackrel{?}{=} 2$$

$$(2)^{2} - 2 \stackrel{?}{=} 2$$

$$4 - 2 \stackrel{?}{=} 2$$

$$2 = 2 \checkmark$$

$z = -1:\ (-1)^{2/3} - (-1)^{1/3} \stackrel{?}{=} 2$

$$\left(\sqrt[3]{-1}\right)^{2} - \left(\sqrt[3]{-1}\right) \stackrel{?}{=} 2$$

$$(-1)^{2} - (-1) \stackrel{?}{=} 2$$

$$1 + 1 \stackrel{?}{=} 2$$

$$2 = 2 \checkmark$$

Both check; the solution set is $\{-1,\ 8\}$.

41.
$$a + a^{1/2} = 30$$
$$\left(a^{1/2}\right)^2 + \left(a^{1/2}\right) - 30 = 0$$

Let $u = a^{1/2}$.
$$u^2 + u - 30 = 0$$
$$(u - 5)(u + 6) = 0$$

$$u - 5 = 0 \quad \text{or} \quad u + 6 = 0$$
$$u = 5 \quad \text{or} \quad u = -6$$
$$a^{1/2} = 5 \quad \text{or} \quad a^{1/2} = -6$$
$$\left(a^{1/2}\right)^2 = 5^2 \quad \text{or} \quad \left(a^{1/2}\right)^2 = (-6)^2$$
$$a = 25 \quad \text{or} \quad a = 36$$

Check:
$$a = 25: \quad 25 + (25)^{1/2} \stackrel{?}{=} 30$$
$$25 + \sqrt{25} \stackrel{?}{=} 30$$
$$25 + 5 \stackrel{?}{=} 30$$
$$30 = 30 \checkmark$$

$$a = 36: \quad 36 + (36)^{1/2} \stackrel{?}{=} 30$$
$$36 + \sqrt{36} \stackrel{?}{=} 30$$
$$36 + 6 \stackrel{?}{=} 30$$
$$42 \neq 30 \text{ ✗}$$

$a = 36$ does not check; the solution set is $\{25\}$.

43.
$$\frac{1}{x^2} - \frac{5}{x} + 6 = 0$$
$$\left(\frac{1}{x}\right)^2 - 5\left(\frac{1}{x}\right) + 6 = 0$$

Let $u = \frac{1}{x}$.
$$u^2 - 5u + 6 = 0$$
$$(u - 2)(u - 3) = 0$$
$$u - 2 = 0 \quad \text{or} \quad u - 3 = 0$$
$$u = 2 \quad \text{or} \quad u = 3$$
$$\frac{1}{x} = 2 \quad \text{or} \quad \frac{1}{x} = 3$$
$$x = \frac{1}{2} \quad \text{or} \quad x = \frac{1}{3}$$

Check:
$$x = \frac{1}{2}: \quad \frac{1}{\left(\frac{1}{2}\right)^2} - \frac{5}{\frac{1}{2}} + 6 \stackrel{?}{=} 0$$
$$\frac{1}{\frac{1}{4}} - \frac{5}{\frac{1}{2}} + 6 \stackrel{?}{=} 0$$
$$4 - 10 + 6 \stackrel{?}{=} 0$$
$$0 = 0 \checkmark$$

$$x = \frac{1}{3}: \quad \frac{1}{\left(\frac{1}{3}\right)^2} - \frac{5}{\frac{1}{3}} + 6 \stackrel{?}{=} 0$$
$$\frac{1}{\frac{1}{9}} - \frac{5}{\frac{1}{3}} + 6 \stackrel{?}{=} 0$$
$$9 - 15 + 6 \stackrel{?}{=} 0$$
$$0 = 0 \checkmark$$

Both check; the solution set is $\left\{\frac{1}{3}, \frac{1}{2}\right\}$.

45.
$$\left(\frac{1}{x+2}\right)^2 + \frac{4}{x+2} = 5$$
$$\left(\frac{1}{x+2}\right)^2 + 4\left(\frac{1}{x+2}\right) - 5 = 0$$

Let $u = \frac{1}{x+2}$.
$$u^2 + 4u - 5 = 0$$
$$(u - 1)(u + 5) = 0$$
$$u - 1 = 0 \quad \text{or} \quad u + 5 = 0$$
$$u = 1 \quad \text{or} \quad u = -5$$
$$\frac{1}{x+2} = 1 \quad \text{or} \quad \frac{1}{x+2} = -5$$
$$x + 2 = 1 \quad \text{or} \quad x + 2 = -\frac{1}{5}$$
$$x = -1 \quad \text{or} \quad x = -\frac{11}{5}$$

Check:
$$x = -1: \quad \left(\frac{1}{-1+2}\right)^2 + \frac{4}{-1+2} \stackrel{?}{=} 5$$
$$\left(\frac{1}{1}\right)^2 + \frac{4}{1} \stackrel{?}{=} 5$$
$$(1)^2 + 4 \stackrel{?}{=} 5$$
$$1 + 4 \stackrel{?}{=} 5$$
$$5 = 5 \checkmark$$

$$x = -\frac{11}{5}: \quad \left(\frac{1}{-\frac{11}{5}+2}\right)^2 + \frac{4}{-\frac{11}{5}+2} \overset{?}{=} 5$$

$$\left(\frac{1}{-\frac{1}{5}}\right)^2 + \frac{4}{-\frac{1}{5}} \overset{?}{=} 5$$

$$(-5)^2 + 4(-5) \overset{?}{=} 5$$

$$25 - 20 \overset{?}{=} 5$$

$$5 = 5 \checkmark$$

Both check; the solution set is $\left\{-\dfrac{11}{5}, -1\right\}$.

47.
$$p^6 - 28p^3 + 27 = 0$$
$$\left(p^3\right)^2 - 28\left(p^3\right) + 27 = 0$$

Let $u = p^3$.
$$u^2 - 28u + 27 = 0$$
$$(u-1)(u-27) = 0$$
$$u - 1 = 0 \quad \text{or} \quad u - 27 = 0$$
$$p^3 - 1 = 0 \quad \text{or} \quad p^3 - 27 = 0$$

First consider $p^3 - 1 = 0$. Note that the expression on the left is a difference of cubes.
$$p^3 - 1 = 0$$
$$(p-1)\left(p^2 + p + 1\right) = 0$$

$$p - 1 = 0 \quad \text{or} \quad p^2 + p + 1 = 0$$

$$p = 1 \quad \text{or} \quad p = \frac{-1 \pm \sqrt{1^2 - 4(1)(1)}}{2(1)}$$

$$= \frac{-1 \pm \sqrt{-3}}{2}$$

$$= \frac{-1 \pm \sqrt{3}\,i}{2}$$

$$= -\frac{1}{2} \pm \frac{\sqrt{3}}{2}i$$

Now consider $p^3 - 27 = 0$. Note that the expression on the left is a difference of cubes.
$$p^3 - 27 = 0$$
$$(p-3)\left(p^2 + 3p + 9\right) = 0$$

$$p - 3 = 0 \quad \text{or} \quad p^2 + 3p + 9 = 0$$

$$p = 3 \quad \text{or} \quad p = \frac{-3 \pm \sqrt{3^2 - 4(1)(9)}}{2(1)}$$

$$= \frac{-3 \pm \sqrt{-27}}{2}$$

$$= \frac{-3 \pm 3\sqrt{3}\,i}{2}$$

$$= -\frac{3}{2} \pm \frac{3\sqrt{3}}{2}i$$

All will check; the solution set is
$$\left\{1, 3, -\frac{1}{2} - \frac{\sqrt{3}}{2}i, -\frac{1}{2} + \frac{\sqrt{3}}{2}i, -\frac{3}{2} - \frac{3\sqrt{3}}{2}i,\right.$$
$$\left.-\frac{3}{2} + \frac{3\sqrt{3}}{2}i\right\}.$$

49.
$$8a^{-2} + 2a^{-1} = 1$$
$$8a^{-2} + 2a^{-1} - 1 = 0$$
$$8\left(a^{-1}\right)^2 + 2\left(a^{-1}\right) - 1 = 0$$

Let $u = a^{-1}$.
$$8u^2 + 2u - 1 = 0$$
$$(4u - 1)(2u + 1) = 0$$
$$4u - 1 = 0 \quad \text{or} \quad 2u + 1 = 0$$
$$4u = 1 \quad \text{or} \quad 2u = -1$$
$$u = \frac{1}{4} \quad \text{or} \quad u = -\frac{1}{2}$$
$$a^{-1} = \frac{1}{4} \quad \text{or} \quad a^{-1} = -\frac{1}{2}$$
$$\frac{1}{a} = \frac{1}{4} \quad \text{or} \quad \frac{1}{a} = -\frac{1}{2}$$
$$a = 4 \quad \text{or} \quad a = -2$$

Check:
$$a = 4: \quad 8(4)^{-2} + 2(4)^{-1} \overset{?}{=} 1$$
$$8\left(\frac{1}{4}\right)^2 + 2\left(\frac{1}{4}\right) \overset{?}{=} 1$$
$$8\left(\frac{1}{16}\right) + 2\left(\frac{1}{4}\right) \overset{?}{=} 1$$
$$\frac{1}{2} + \frac{1}{2} \overset{?}{=} 1$$
$$1 = 1 \checkmark$$

$a = -2: \ 8(-2)^{-2} + 2(-2)^{-1} \stackrel{?}{=} 1$

$8\left(-\dfrac{1}{2}\right)^{2} + 2\left(-\dfrac{1}{2}\right) \stackrel{?}{=} 1$

$8\left(\dfrac{1}{4}\right) + 2\left(-\dfrac{1}{2}\right) \stackrel{?}{=} 1$

$2 - 1 \stackrel{?}{=} 1$

$1 = 1 \ \checkmark$

Both check; the solution set is $\{-2, \ 4\}$.

51.
$$z^4 = 4z^2 + 32$$
$$z^4 - 4z^2 - 32 = 0$$
$$\left(z^2\right)^2 - 4\left(z^2\right) - 32 = 0$$

Let $u = z^2$.

$u^2 - 4u - 32 = 0$

$(u - 8)(u + 4) = 0$

$u - 8 = 0 \qquad \text{or} \quad u + 4 = 0$

$u = 8 \qquad \text{or} \qquad u = -4$

$z^2 = 8 \qquad \text{or} \qquad z^2 = -4$

$z = \pm\sqrt{8} \quad \text{or} \qquad z = \pm\sqrt{-4}$

$z = \pm 2\sqrt{2} \ \text{ or} \qquad z = \pm 2i$

Check:

$z = -2\sqrt{2}: \ \left(-2\sqrt{2}\right)^4 \stackrel{?}{=} 4\left(-2\sqrt{2}\right)^2 + 32$

$64 \stackrel{?}{=} 4 \cdot 8 + 32$

$64 \stackrel{?}{=} 32 + 32$

$64 = 64 \ \checkmark$

$z = 2\sqrt{2}: \ \left(2\sqrt{2}\right)^4 \stackrel{?}{=} 4\left(2\sqrt{2}\right)^2 + 32$

$64 \stackrel{?}{=} 4 \cdot 8 + 32$

$64 \stackrel{?}{=} 32 + 32$

$64 = 64 \ \checkmark$

$z = -2i: \ (-2i)^4 \stackrel{?}{=} 4(-2i)^2 + 32$

$16i^4 \stackrel{?}{=} 4 \cdot 4i^2 + 32$

$16(1) \stackrel{?}{=} 4 \cdot (-4) + 32$

$16 \stackrel{?}{=} -16 + 32$

$16 = 16 \ \checkmark$

$z = 2i: \ (2i)^4 \stackrel{?}{=} 4(2i)^2 + 32$

$16i^4 \stackrel{?}{=} 4 \cdot 4i^2 + 32$

$16(1) \stackrel{?}{=} 4 \cdot (-4) + 32$

$16 \stackrel{?}{=} -16 + 32$

$16 = 16 \ \checkmark$

All check; the solution set is
$\left\{-2\sqrt{2}, \ 2\sqrt{2}, -2i, \ 2i\right\}$.

53.
$$x^{1/2} + x^{1/4} - 6 = 0$$
$$\left(x^{1/4}\right)^2 + \left(x^{1/4}\right) - 6 = 0$$

Let $u = x^{1/4}$.

$u^2 + u - 6 = 0$

$(u - 2)(u + 3) = 0$

$u - 2 = 0 \qquad \text{or} \qquad u + 3 = 0$

$u = 2 \qquad \text{or} \qquad u = -3$

$x^{1/4} = 2 \qquad \text{or} \qquad x^{1/4} = -3$

$\left(x^{1/4}\right)^4 = 2^4 \quad \text{or} \ \left(x^{1/4}\right)^4 = (-3)^4$

$x = 16 \qquad \text{or} \qquad x = 81$

Check:

$x = 16: \ (16)^{1/2} + (16)^{1/4} - 6 \stackrel{?}{=} 0$

$\sqrt{16} + \sqrt[4]{16} - 6 \stackrel{?}{=} 0$

$4 + 2 - 6 \stackrel{?}{=} 0$

$0 = 0 \ \checkmark$

$x = 81: \ (81)^{1/2} + (81)^{1/4} - 6 \stackrel{?}{=} 0$

$\sqrt{81} + \sqrt[4]{81} - 6 \stackrel{?}{=} 0$

$9 + 3 - 6 \stackrel{?}{=} 0$

$6 \neq 0 \ \times$

$x = 81$ does not check; the solution set is $\{16\}$.

55.
$$w^4 - 5w^2 - 36 = 0$$
$$\left(w^2\right)^2 - 5\left(w^2\right) - 36 = 0$$

Let $u = w^2$.

$u^2 - 5u - 36 = 0$

$(u - 9)(u + 4) = 0$

$u-9=0 \quad$ or $\quad u+4=0$

$\quad u=9 \quad$ or $\quad u=-4$

$\quad w^2=9 \quad$ or $\quad w^2=-4$

$\quad w=\pm\sqrt{9} \quad$ or $\quad w=\pm\sqrt{-4}$

$\quad w=\pm 3 \quad$ or $\quad w=\pm 2i$

Check:

$w=-3: \ (-3)^4-5(-3)^2-36\overset{?}{=}0$

$\qquad\qquad\ 81-5\cdot9-36\overset{?}{=}0$

$\qquad\qquad\ 81-45-36\overset{?}{=}0$

$\qquad\qquad\qquad\qquad\quad 0=0 \ \checkmark$

$w=3: \ 3^4-5(3)^2-36\overset{?}{=}0$

$\qquad\quad\ 81-5\cdot9-36\overset{?}{=}0$

$\qquad\quad\ 81-45-36\overset{?}{=}0$

$\qquad\qquad\qquad\qquad 0=0 \ \checkmark$

$w=-2i: \ (-2i)^4-5(-2i)^2-36\overset{?}{=}0$

$\qquad\qquad\ 16i^4-5\cdot4i^2-36\overset{?}{=}0$

$\qquad\qquad\ 16(1)-5\cdot4(-1)-36\overset{?}{=}0$

$\qquad\qquad\qquad\ 16+20-36\overset{?}{=}0$

$\qquad\qquad\qquad\qquad\qquad 0=0 \ \checkmark$

$w=2i: \ (2i)^4-5(2i)^2-36\overset{?}{=}0$

$\qquad\qquad 16i^4-5\cdot4i^2-36\overset{?}{=}0$

$\qquad\ 16(1)-5\cdot4(-1)-36\overset{?}{=}0$

$\qquad\qquad\quad 16+20-36\overset{?}{=}0$

$\qquad\qquad\qquad\qquad\quad 0=0 \ \checkmark$

All check; the solution set is $\{-3,\ 3,-2i,\ 2i\}$.

57. $\qquad \left(\dfrac{1}{x+3}\right)^2+\dfrac{2}{x+3}=3$

$\left(\dfrac{1}{x+3}\right)^2+2\left(\dfrac{1}{x+3}\right)-3=0$

Let $u=\dfrac{1}{x+3}$.

$\quad u^2+2u-3=0$

$\quad (u-1)(u+3)=0$

$u-1=0 \quad$ or $\quad u+3=0$

$\quad u=1 \quad$ or $\quad u=-3$

$\dfrac{1}{x+3}=1 \quad$ or $\quad \dfrac{1}{x+3}=-3$

$x+3=1 \quad$ or $\quad x+3=-\dfrac{1}{3}$

$\qquad x=-2 \quad$ or $\qquad x=-\dfrac{10}{3}$

Check:

$x=-2: \ \left(\dfrac{1}{-2+3}\right)^2+\dfrac{2}{-2+3}\overset{?}{=}3$

$\qquad\qquad \left(\dfrac{1}{1}\right)^2+\dfrac{2}{1}\overset{?}{=}3$

$\qquad\qquad\qquad 1+2\overset{?}{=}3$

$\qquad\qquad\qquad\qquad 3=3 \ \checkmark$

$x=-\dfrac{10}{3}: \ \left(\dfrac{1}{-\frac{10}{3}+3}\right)^2+\dfrac{2}{-\frac{10}{3}+3}\overset{?}{=}3$

$\qquad\qquad \left(\dfrac{1}{-\frac{1}{3}}\right)^2+\dfrac{2}{-\frac{1}{3}}\overset{?}{=}3$

$\qquad\qquad (-3)^2+2(-3)\overset{?}{=}3$

$\qquad\qquad\qquad 9-6\overset{?}{=}3$

$\qquad\qquad\qquad\qquad 3=3 \ \checkmark$

Both check; the solution set is $\left\{-\dfrac{10}{3},-2\right\}$.

59. $\qquad x-7\sqrt{x}+12=0$

$\left(\sqrt{x}\right)^2-7\left(\sqrt{x}\right)+12=0$

Let $u=\sqrt{x}$.

$\quad u^2-7u+12=0$

$\quad (u-4)(u-3)=0$

$u-4=0 \quad$ or $\quad u-3=0$

$\quad u=4 \quad$ or $\quad u=3$

$\sqrt{x}=4 \quad$ or $\quad \sqrt{x}=3$

$x=4^2 \quad$ or $\qquad x=3^2$

$x=16 \quad$ or $\qquad x=9$

Check:

$x = 16:$ $16 - 7\sqrt{16} + 12 \overset{?}{=} 0$

$16 - 7 \cdot 4 + 12 \overset{?}{=} 0$

$16 - 28 + 12 \overset{?}{=} 0$

$0 = 0 \checkmark$

$x = 9:$ $9 - 7\sqrt{9} + 12 \overset{?}{=} 0$

$9 - 7 \cdot 3 + 12 \overset{?}{=} 0$

$9 - 21 + 12 \overset{?}{=} 0$

$0 = 0 \checkmark$

Both check; the solution set is $\{9, 16\}$.

61. $2(x-1)^2 - 7(x-1) = 4$

$2(x-1)^2 - 7(x-1) - 4 = 0$

Let $u = x - 1$.

$2u^2 - 7u - 4 = 0$

$(2u+1)(u-4) = 0$

$2u + 1 = 0$ or $u - 4 = 0$

$u = -\dfrac{1}{2}$ or $u = 4$

$x - 1 = -\dfrac{1}{2}$ or $x - 1 = 4$

$x = \dfrac{1}{2}$ or $x = 5$

Check:

$x = \dfrac{1}{2}:$ $2\left(\dfrac{1}{2} - 1\right)^2 - 7\left(\dfrac{1}{2} - 1\right) \overset{?}{=} 4$

$2\left(-\dfrac{1}{2}\right)^2 - 7\left(-\dfrac{1}{2}\right) \overset{?}{=} 4$

$2\left(\dfrac{1}{4}\right) - 7\left(-\dfrac{1}{2}\right) \overset{?}{=} 4$

$\dfrac{1}{2} + \dfrac{7}{2} \overset{?}{=} 4$

$4 = 4 \checkmark$

$x = 5:$ $2(5-1)^2 - 7(5-1) \overset{?}{=} 4$

$2(4)^2 - 7(4) \overset{?}{=} 4$

$2 \cdot 16 - 7 \cdot 4 \overset{?}{=} 4$

$32 - 28 \overset{?}{=} 4$

$4 = 4 \checkmark$

Both check; the solution set is $\left\{\dfrac{1}{2}, 5\right\}$.

63. (a) $f(x) = 12$

$x^4 + 7x^2 + 12 = 12$

$\left(x^2\right)^2 + 7\left(x^2\right) = 0$

Let $u = x^2$.

$u^2 + 7u = 0$

$u(u+7) = 0$

$u + 7 = 0$ or $u = 0$

$u = -7$

$x^2 = -7$ or $x^2 = 0$

$x = \pm\sqrt{-7}$ or $x = \pm\sqrt{0}$

$x = \pm\sqrt{7}\, i$ or $x = 0$

Check:

$f\left(-\sqrt{7}\, i\right) = \left(-\sqrt{7}\, i\right)^4 + 7\left(-\sqrt{7}\, i\right)^2 + 12$

$= 49i^4 + 7 \cdot 7i^2 + 12$

$= 49(1) + 7 \cdot 7(-1) + 12$

$= 49 - 49 + 12$

$= 12 \checkmark$

$f\left(\sqrt{7}\, i\right) = \left(\sqrt{7}\, i\right)^4 + 7\left(\sqrt{7}\, i\right)^2 + 12$

$= 49i^4 + 7 \cdot 7i^2 + 12$

$= 49(1) + 7 \cdot 7(-1) + 12$

$= 49 - 49 + 12 = 12 \checkmark$

$f(0) = 0^4 + 7 \cdot 0^2 + 12$

$= 12 \checkmark$

All check; the values that make $f(x) = 12$

are $\left\{-\sqrt{7}\, i, \sqrt{7}\, i, 0\right\}$.

(b) $f(x) = 6$

$x^4 + 7x^2 + 12 = 6$

$\left(x^2\right)^2 + 7\left(x^2\right) + 6 = 0$

Let $u = x^2$.

$u^2 + 7u + 6 = 0$

$(u+6)(u+1) = 0$

$u + 6 = 0$ or $u + 1 = 0$

$u = -6$ or $u = -1$

$x^2 = -6$ or $x^2 = -1$

$x = \pm\sqrt{-6}$ or $x = \pm\sqrt{-1}$

$x = \pm\sqrt{6}\, i$ or $x = \pm i$

Check:

$$f\left(-\sqrt{6}\ i\right) = \left(-\sqrt{6}\ i\right)^4 + 7\left(-\sqrt{6}\ i\right)^2 + 12$$
$$= 36i^4 + 7\cdot 6i^2 + 12$$
$$= 36(1) + 7\cdot 6(-1) + 12$$
$$= 36 - 42 + 12$$
$$= 6\ \checkmark$$

$$f\left(\sqrt{6}\ i\right) = \left(\sqrt{6}\ i\right)^4 + 7\left(\sqrt{6}\ i\right)^2 + 12$$
$$= 36i^4 + 7\cdot 6i^2 + 12$$
$$= 36(1) + 7\cdot 6(-1) + 12$$
$$= 36 - 42 + 12$$
$$= 6\ \checkmark$$

$$f(-i) = (-i)^4 + 7(-i)^2 + 12$$
$$= i^4 + 7i^2 + 12$$
$$= 1 + 7(-1) + 12$$
$$= 1 - 7 + 12$$
$$= 6\ \checkmark$$

$$f(i) = i^4 + 7i^2 + 12$$
$$= 1 + 7(-1) + 12$$
$$= 1 - 7 + 12$$
$$= 6\ \checkmark$$

All check; the values that make $f(x) = 6$
are $\left\{-\sqrt{6}\ i,\ \sqrt{6}\ i,\ -i,\ i\right\}$.

65. (a)
$$g(x) = -5$$
$$2x^4 - 6x^2 - 5 = -5$$
$$2\left(x^2\right)^2 - 6\left(x^2\right) = 0$$

Let $u = x^2$.
$$2u^2 - 6u = 0$$
$$2u(u - 3) = 0$$
$$2u = 0 \quad \text{or} \quad u - 3 = 0$$
$$u = 0 \quad \text{or} \quad u = 3$$
$$x^2 = 0 \quad \text{or} \quad x^2 = 3$$
$$x = \pm\sqrt{0} \quad \text{or} \quad x = \pm\sqrt{3}$$
$$x = 0$$

Check:
$$g(0) = 2\cdot 0^4 - 6\cdot 0^2 - 5$$
$$= -5\ \checkmark$$

$$g\left(-\sqrt{3}\right) = 2\left(-\sqrt{3}\right)^4 - 6\left(-\sqrt{3}\right)^2 - 5$$
$$= 2\cdot 9 - 6\cdot 3 - 5$$
$$= 18 - 18 - 5$$
$$= -5\ \checkmark$$

$$g\left(\sqrt{3}\right) = 2\left(\sqrt{3}\right)^4 - 6\left(\sqrt{3}\right)^2 - 5$$
$$= 2\cdot 9 - 6\cdot 3 - 5$$
$$= 18 - 18 - 5$$
$$= -5\ \checkmark$$

All check; the values that make $g(x) = -5$
are $\left\{-\sqrt{3},\ 0,\ \sqrt{3}\right\}$.

(b)
$$g(x) = 15$$
$$2x^4 - 6x^2 - 5 = 15$$
$$2x^4 - 6x^2 - 20 = 0$$
$$\frac{2x^4 - 6x^2 - 20}{2} = \frac{0}{2}$$
$$x^4 - 3x^2 - 10 = 0$$
$$\left(x^2\right)^2 - 3\left(x^2\right) - 10 = 0$$

Let $u = x^2$.
$$u^2 - 3u - 10 = 0$$
$$(u + 2)(u - 5) = 0$$

$$u + 2 = 0 \quad \text{or} \quad u - 5 = 0$$
$$u = -2 \quad \text{or} \quad u = 5$$
$$x^2 = -2 \quad \text{or} \quad x^2 = 5$$
$$x = \pm\sqrt{-2} \quad \text{or} \quad x = \pm\sqrt{5}$$
$$x = \pm\sqrt{2}\ i$$

Check:
$$g\left(-\sqrt{2}\ i\right) = 2\left(-\sqrt{2}\ i\right)^4 - 6\left(-\sqrt{2}\ i\right)^2 - 5$$
$$= 2\cdot 4i^4 - 6\cdot 2i^2 - 5$$
$$= 2\cdot 4(1) - 6\cdot 2(-1) - 5$$
$$= 8 + 12 - 5$$
$$= 15\ \checkmark$$

$$g\left(\sqrt{2}\ i\right) = 2\left(\sqrt{2}\ i\right)^4 - 6\left(\sqrt{2}\ i\right)^2 - 5$$
$$= 2\cdot 4i^4 - 6\cdot 2i^2 - 5$$
$$= 2\cdot 4(1) - 6\cdot 2(-1) - 5$$
$$= 8 + 12 - 5$$
$$= 15\ \checkmark$$

$$g\left(-\sqrt{5}\right) = 2\left(-\sqrt{5}\right)^4 - 6\left(-\sqrt{5}\right)^2 - 5$$
$$= 2 \cdot 25 - 6 \cdot 5 - 5$$
$$= 50 - 30 - 5$$
$$= 15 \ \checkmark$$

$$g\left(\sqrt{5}\right) = 2\left(\sqrt{5}\right)^4 - 6\left(\sqrt{5}\right)^2 - 5$$
$$= 2 \cdot 25 - 6 \cdot 5 - 5$$
$$= 50 - 30 - 5$$
$$= 15 \ \checkmark$$

All check; the values that make $g(x) = 15$ are $\left\{-\sqrt{2}\,i,\ \sqrt{2}\,i,\ -\sqrt{5},\ \sqrt{5}\right\}$.

67. (a)
$$F(x) = 6$$
$$x^{-2} - 5x^{-1} = 6$$
$$\left(x^{-1}\right)^2 - 5\left(x^{-1}\right) - 6 = 0$$

Let $u = x^{-1}$.
$$u^2 - 5u - 6 = 0$$
$$(u - 6)(u + 1) = 0$$

$$u - 6 = 0 \quad \text{or} \quad u + 1 = 0$$
$$u = 6 \quad \text{or} \quad u = -1$$
$$x^{-1} = 6 \quad \text{or} \quad x^{-1} = -1$$
$$\frac{1}{x} = 6 \quad \text{or} \quad \frac{1}{x} = -1$$
$$x = \frac{1}{6} \quad \text{or} \quad x = -1$$

Check:
$$F\left(\frac{1}{6}\right) = \left(\frac{1}{6}\right)^{-2} - 5\left(\frac{1}{6}\right)^{-1}$$
$$= 6^2 - 5 \cdot 6^1$$
$$= 36 - 30$$
$$= 6 \ \checkmark$$

$$F(-1) = (-1)^{-2} - 5(-1)^{-1}$$
$$= (-1)^2 - 5(-1)^1$$
$$= 1 + 5$$
$$= 6 \ \checkmark$$

Both check; the values that make $F(x) = 6$ are $\left\{-1,\ \dfrac{1}{6}\right\}$.

(b)
$$F(x) = 14$$
$$x^{-2} - 5x^{-1} = 14$$
$$\left(x^{-1}\right)^2 - 5\left(x^{-1}\right) - 14 = 0$$

Let $u = x^{-1}$.
$$u^2 - 5u - 14 = 0$$
$$(u - 7)(u + 2) = 0$$

$$u - 7 = 0 \quad \text{or} \quad u + 2 = 0$$
$$u = 7 \quad \text{or} \quad u = -2$$
$$x^{-1} = 7 \quad \text{or} \quad x^{-1} = -2$$
$$\frac{1}{x} = 7 \quad \text{or} \quad \frac{1}{x} = -2$$
$$x = \frac{1}{7} \quad \text{or} \quad x = -\frac{1}{2}$$

Check:
$$F\left(\frac{1}{7}\right) = \left(\frac{1}{7}\right)^{-2} - 5\left(\frac{1}{7}\right)^{-1}$$
$$= 7^2 - 5 \cdot 7^1$$
$$= 49 - 35$$
$$= 14 \ \checkmark$$

$$F\left(-\frac{1}{2}\right) = \left(-\frac{1}{2}\right)^{-2} - 5\left(-\frac{1}{2}\right)^{-1}$$
$$= (-2)^2 - 5 \cdot (-2)^1$$
$$= 4 + 10$$
$$= 14 \ \checkmark$$

Both check; the values that make $F(x) = 14$ are $\left\{-\dfrac{1}{2},\ \dfrac{1}{7}\right\}$.

69.
$$x^4 + 9x^2 + 14 = 0$$
$$\left(x^2\right)^2 + 9\left(x^2\right) + 14 = 0$$

Let $u = x^2$.
$$u^2 + 9u + 14 = 0$$
$$(u + 7)(u + 2) = 0$$
$$u + 7 = 0 \quad \text{or} \quad u + 2 = 0$$
$$u = -7 \quad \text{or} \quad u = -2$$
$$x^2 = -7 \quad \text{or} \quad x^2 = -2$$
$$x = \pm\sqrt{-7} \quad \text{or} \quad x = \pm\sqrt{-2}$$
$$x = \pm\sqrt{7}\,i \quad \text{or} \quad x = \pm\sqrt{2}\,i$$

Check:

$$f\left(-\sqrt{7}\ i\right) = \left(-\sqrt{7}\ i\right)^4 + 9\left(-\sqrt{7}\ i\right)^2 + 14$$
$$= 49i^4 + 9 \cdot 7i^2 + 14$$
$$= 49(1) + 9 \cdot 7(-1) + 14$$
$$= 49 - 63 + 14$$
$$= 0 \ \checkmark$$

$$f\left(\sqrt{7}\ i\right) = \left(\sqrt{7}\ i\right)^4 + 9\left(\sqrt{7}\ i\right)^2 + 14$$
$$= 49i^4 + 9 \cdot 7i^2 + 14$$
$$= 49(1) + 9 \cdot 7(-1) + 14$$
$$= 49 - 63 + 14$$
$$= 0 \ \checkmark$$

$$f\left(-\sqrt{2}\ i\right) = \left(-\sqrt{2}\ i\right)^4 + 9\left(-\sqrt{2}\ i\right)^2 + 14$$
$$= 4i^4 + 9 \cdot 2i^2 + 14$$
$$= 4(1) + 9 \cdot 2(-1) + 14$$
$$= 4 - 18 + 14$$
$$= 0 \ \checkmark$$

$$f\left(\sqrt{2}\ i\right) = \left(\sqrt{2}\ i\right)^4 + 9\left(\sqrt{2}\ i\right)^2 + 14$$
$$= 4i^4 + 9 \cdot 2i^2 + 14$$
$$= 4(1) + 9 \cdot 2(-1) + 14$$
$$= 4 - 18 + 14$$
$$= 0 \ \checkmark$$

All check; the zeros of *f* are
$$\left\{-\sqrt{7}\ i,\ \sqrt{7}\ i, -\sqrt{2}\ i,\ \sqrt{2}\ i\right\}.$$

71.
$$6t - 25\sqrt{t} - 9 = 0$$
$$6\left(\sqrt{t}\right)^2 - 25\left(\sqrt{t}\right) - 9 = 0$$

Let $u = \sqrt{t}$.
$$6u^2 - 25u - 9 = 0$$
$$(2u - 9)(3u + 1) = 0$$

$$2u - 9 = 0 \quad \text{or} \quad 3u + 1 = 0$$
$$2u = 9 \quad \text{or} \quad 3u = -1$$
$$u = \frac{9}{2} \quad \text{or} \quad u = -\frac{1}{3}$$

$$\sqrt{t} = \frac{9}{2} \quad \text{or} \quad \sqrt{t} = -\frac{1}{3}$$
$$t = \left(\frac{9}{2}\right)^2 \quad \text{or} \quad t = \left(-\frac{1}{3}\right)^2$$
$$t = \frac{81}{4} \quad \text{or} \quad t = \frac{1}{9}$$

Check:

$$g\left(\frac{81}{4}\right) = 6 \cdot \frac{81}{4} - 25\sqrt{\frac{81}{4}} - 9$$
$$= 6 \cdot \frac{81}{4} - 25 \cdot \frac{9}{2} - 9$$
$$= \frac{243}{2} - \frac{225}{2} - 9$$
$$= 0 \ \checkmark$$

$$g\left(\frac{1}{9}\right) = 6 \cdot \frac{1}{9} - 25\sqrt{\frac{1}{9}} - 9$$
$$= 6 \cdot \frac{1}{9} - 25 \cdot \frac{1}{3} - 9$$
$$= \frac{2}{3} - \frac{25}{3} - 9$$
$$= -\frac{50}{3}$$
$$\neq 0 \ \boldsymbol{\times}$$

$t = \frac{1}{9}$ does not check; the zero of *g* is $\left\{\frac{81}{4}\right\}$.

73.
$$\frac{1}{(d+3)^2} - \frac{4}{d+3} + 3 = 0$$
$$\left(\frac{1}{d+3}\right)^2 - 4\left(\frac{1}{d+3}\right) + 3 = 0$$

Let $u = \frac{1}{d+3}$.
$$u^2 - 4u + 3 = 0$$
$$(u - 1)(u - 3) = 0$$
$$u - 1 = 0 \quad \text{or} \quad u - 3 = 0$$
$$u = 1 \quad \text{or} \quad u = 3$$
$$\frac{1}{d+3} = 1 \quad \text{or} \quad \frac{1}{d+3} = 3$$
$$d + 3 = 1 \quad \text{or} \quad d + 3 = \frac{1}{3}$$
$$d = -2 \quad \text{or} \quad d = -\frac{8}{3}$$

Check:

$$s(-2) = \frac{1}{(-2+3)^2} - \frac{4}{-2+3} + 3$$

$$= \frac{1}{1^2} - \frac{4}{1} + 3$$

$$= 1 - 4 + 3$$

$$= 0 \checkmark$$

$$s\left(-\frac{8}{3}\right) = \frac{1}{\left(-\frac{8}{3}+3\right)^2} - \frac{4}{-\frac{8}{3}+3} + 3$$

$$= \frac{1}{\left(\frac{1}{3}\right)^2} - \frac{4}{\frac{1}{3}} + 3$$

$$= \frac{1}{\frac{1}{9}} - \frac{4}{\frac{1}{3}} + 3$$

$$= 9 - 12 + 3$$

$$= 0 \checkmark$$

Both check; the zeros of s are $\left\{-\frac{8}{3}, -2\right\}$.

75. (a) $x^2 - 5x + 6 = 0$
$$(x-2)(x-3) = 0$$
$$x - 2 = 0 \quad \text{or} \quad x - 3 = 0$$
$$x = 2 \quad \text{or} \quad x = 3$$
Both check; the solution set is $\{2, 3\}$.

(b) $(x-3)^2 - 5(x-3) + 6 = 0$
Let $u = x - 3$.
$$u^2 - 5u + 6 = 0$$
$$(u-2)(u-3) = 0$$
$$u - 2 = 0 \quad \text{or} \quad u - 3 = 0$$
$$u = 2 \quad \text{or} \quad u = 3$$
$$x - 3 = 2 \quad \text{or} \quad x - 3 = 3$$
$$x = 5 \quad \text{or} \quad x = 6$$
Both check; the solution set is $\{5, 6\}$.

Comparing these solutions to those in part (a), we note that $5 = 2+3$ and $6 = 3+3$.

(c) $(x+2)^2 - 5(x+2) + 6 = 0$
Let $u = x + 2$.
$$u^2 - 5u + 6 = 0$$
$$(u-2)(u-3) = 0$$
$$u - 2 = 0 \quad \text{or} \quad u - 3 = 0$$
$$u = 2 \quad \text{or} \quad u = 3$$
$$x + 2 = 2 \quad \text{or} \quad x + 2 = 3$$
$$x = 0 \quad \text{or} \quad x = 1$$
Both check; the solution set is $\{0, 1\}$.

Comparing these solutions to those in part (a), we note that $0 = 2-2$ and $1 = 3-2$.

(d) $(x-5)^2 - 5(x-5) + 6 = 0$
Let $u = x - 5$.
$$u^2 - 5u + 6 = 0$$
$$(u-2)(u-3) = 0$$
$$u - 2 = 0 \quad \text{or} \quad u - 3 = 0$$
$$u = 2 \quad \text{or} \quad u = 3$$
$$x - 5 = 2 \quad \text{or} \quad x - 5 = 3$$
$$x = 7 \quad \text{or} \quad x = 8$$
Both check; the solution set is $\{7, 8\}$.

Comparing these solutions to those in part (a), we note that $7 = 2+5$ and $8 = 3+5$.

(e) Conjecture: The solution set of the equation $(x-a)^2 - 5(x-a) + 6 = 0$ is $\{2+a, 3+a\}$.

NOTE: This conjecture can be shown to be true by using the techniques from parts (b) through (d).

77. (a) $f(x) = 2x^2 - 3x + 1$
$$2x^2 - 3x + 1 = 0$$
$$(2x-1)(x-1) = 0$$
$$2x - 1 = 0 \quad \text{or} \quad x - 1 = 0$$
$$2x = 1 \quad \text{or} \quad x = 1$$
$$x = \frac{1}{2}$$

Both check; the zeros of $f(x)$ are $\left\{\frac{1}{2}, 1\right\}$.

(b) $f(x-2) = 2(x-2)^2 - 3(x-2)+1$

Let $u = x-2$.

$$2u^2 - 3u + 1 = 0$$
$$(2u-1)(u-1) = 0$$
$$2u-1 = 0 \quad \text{or} \quad u-1 = 0$$
$$2u = 1 \quad \text{or} \quad u = 1$$
$$u = \frac{1}{2}$$
$$x-2 = \frac{1}{2} \quad \text{or} \quad x-2 = 1$$
$$x = \frac{5}{2} \quad \text{or} \quad x = 3$$

Both check; the zeros of $f(x-2)$ are

$\left\{\frac{5}{2}, 3\right\}$. Comparing these solutions to those

in part (a), we note that $\frac{5}{2} = \frac{1}{2} + 2$ and

$3 = 1 + 2$.

(c) $f(x-5) = 2(x-5)^2 - 3(x-5)+1$

Let $u = x-5$.

$$2u^2 - 3u + 1 = 0$$
$$(2u-1)(u-1) = 0$$
$$2u-1 = 0 \quad \text{or} \quad u-1 = 0$$
$$2u = 1 \quad \text{or} \quad u = 1$$
$$u = \frac{1}{2}$$
$$x-5 = \frac{1}{2} \quad \text{or} \quad x-5 = 1$$
$$x = \frac{11}{2} \quad \text{or} \quad x = 6$$

Both check; the zeros of $f(x-5)$ are

$\left\{\frac{11}{2}, 6\right\}$. Comparing these solutions to those

in part (a), we note that $\frac{11}{2} = \frac{1}{2} + 5$ and

$6 = 1 + 5$.

(d) Conjecture: For $f(x) = 2x^2 - 3x + 1$, the

zeros of the $f(x-a)$ are $\left\{\frac{1}{2} + a, 1 + a\right\}$.

NOTE: This conjecture can be shown to be true by using the techniques from parts (b) and (c).

79. (a) $R(1990)$

$$= \frac{(1990-1990)^2}{2} + \frac{3(1990-1990)}{2} + 3000$$
$$= \frac{0^2}{2} + \frac{3(0)}{2} + 3000$$
$$= 3000$$

Interpretation: The revenue in 1990 was $3000 thousand (or $3,000,000).

(b) $\dfrac{(x-1990)^2}{2} + \dfrac{3(x-1990)}{2} + 3000 = 3065$

$$\frac{1}{2}(x-1990)^2 + \frac{3}{2}(x-1990) - 65 = 0$$

Let $u = x - 1990$.

$$\frac{1}{2}u^2 + \frac{3}{2}u - 65 = 0$$
$$2\left(\frac{1}{2}u^2 + \frac{3}{2}u - 65\right) = 2(0)$$
$$u^2 + 3u - 130 = 0$$
$$(u-10)(u+13) = 0$$

$$u-10 = 0 \quad \text{or} \quad u+13 = 0$$
$$u = 10 \quad \text{or} \quad u = -13$$
$$x-1990 = 10 \quad \text{or} \quad x-1990 = -13$$
$$x = 2000 \quad \text{or} \quad x = 1977$$

Since 1977 is before 1990, disregard it. The solution set is $\{2000\}$.

Interpretation: In the year 2000, revenue was $3065 thousand (or $3,065,000).

(c) $\dfrac{(x-1990)^2}{2} + \dfrac{3(x-1990)}{2} + 3000 = 3350$

$$\frac{1}{2}(x-1990)^2 + \frac{3}{2}(x-1990) - 350 = 0$$

Let $u = x - 1990$.

$$\frac{1}{2}u^2 + \frac{3}{2}u - 350 = 0$$
$$2\left(\frac{1}{2}u^2 + \frac{3}{2}u - 350\right) = 2(0)$$
$$u^2 + 3u - 700 = 0$$
$$(u-25)(u+28) = 0$$

$$u-25 = 0 \quad \text{or} \quad u+28 = 0$$
$$u = 25 \quad \text{or} \quad u = -28$$
$$x-1990 = 25 \quad \text{or} \quad x-1990 = -28$$
$$x = 2015 \quad \text{or} \quad x = 1962$$

Since 1962 is before 1990, disregard it. The solution set is $\{2015\}$. The model predicts that revenue will be \$3350 thousand (or \$3,350,000) in the year 2015.

81.
$$x^4 + 5x^2 + 2 = 0$$
$$\left(x^2\right)^2 + 5\left(x^2\right) + 2 = 0$$
Let $u = x^2$.
$$u^2 + 5u + 2 = 0$$
For this equation, $a = 1$, $b = 5$, and $c = 2$.
$$u = \frac{-5 \pm \sqrt{5^2 - 4(1)(2)}}{2(1)}$$
$$= \frac{-5 \pm \sqrt{25 - 8}}{2}$$
$$= \frac{-5 \pm \sqrt{17}}{2}$$

$$u = \frac{-5 - \sqrt{17}}{2} \quad \text{or} \quad u = \frac{-5 + \sqrt{17}}{2}$$
$$x^2 = \frac{-5 - \sqrt{17}}{2} \quad \text{or} \quad x^2 = \frac{-5 + \sqrt{17}}{2}$$
$$x = \pm\sqrt{\frac{-5 - \sqrt{17}}{2}} \quad \text{or} \quad x = \pm\sqrt{\frac{-5 + \sqrt{17}}{2}}$$
$$x = \pm\sqrt{\frac{-1\left(5 + \sqrt{17}\right)}{2}} \qquad x = \pm\sqrt{\frac{-1\left(5 - \sqrt{17}\right)}{2}}$$
$$x = \pm i\sqrt{\frac{5 + \sqrt{17}}{2}} \qquad x = \pm i\sqrt{\frac{5 - \sqrt{17}}{2}}$$
$$x = \pm i\sqrt{\frac{5 + \sqrt{17}}{2} \cdot \frac{2}{2}} \qquad x = \pm i\sqrt{\frac{5 - \sqrt{17}}{2} \cdot \frac{2}{2}}$$
$$x = \pm\frac{\sqrt{10 + 2\sqrt{17}}}{2}i \qquad x = \pm\frac{\sqrt{10 - 2\sqrt{17}}}{2}i$$

All check, the solution set is $\left\{-\dfrac{\sqrt{10 + 2\sqrt{17}}}{2}i, \right.$

$\left. \dfrac{\sqrt{10 + 2\sqrt{17}}}{2}i, -\dfrac{\sqrt{10 - 2\sqrt{17}}}{2}i, \dfrac{\sqrt{10 - 2\sqrt{17}}}{2}i \right\}$.

83. $2(x-2)^2 + 8(x-2) - 1 = 0$
Let $u = x - 2$.
$$2u^2 + 8u - 1 = 0$$
For this equation, $a = 2$, $b = 8$, and $c = -1$.

$$u = \frac{-8 \pm \sqrt{8^2 - 4(2)(-1)}}{2(2)}$$
$$= \frac{-8 \pm \sqrt{64 + 8}}{4}$$
$$= \frac{-8 \pm \sqrt{72}}{4}$$
$$= \frac{-8 \pm 6\sqrt{2}}{4} = -2 \pm \frac{3\sqrt{2}}{2}$$

$$u = -2 - \frac{3\sqrt{2}}{2} \quad \text{or} \quad u = -2 + \frac{3\sqrt{2}}{2}$$
$$x - 2 = -2 - \frac{3\sqrt{2}}{2} \quad \text{or} \quad x - 2 = -2 + \frac{3\sqrt{2}}{2}$$
$$x = -\frac{3\sqrt{2}}{2} \quad \text{or} \quad x = \frac{3\sqrt{2}}{2}$$

Both check, the solution set is $\left\{-\dfrac{3\sqrt{2}}{2}, \dfrac{3\sqrt{2}}{2}\right\}$.

85. Method 1:
$$x - 5\sqrt{x} - 6 = 0$$
$$\left(\sqrt{x}\right)^2 - 5\sqrt{x} - 6 = 0$$
Let $u = \sqrt{x}$
$$u^2 - 5u - 6 = 0$$
$$(u - 6)(u + 1) = 0$$
$$u - 6 = 0 \quad \text{or} \quad u + 1 = 0$$
$$u = 6 \quad \text{or} \quad u = -1$$
$$\sqrt{x} = 6 \quad \text{or} \quad \sqrt{x} = -1$$
$$x = 36 \quad \text{or} \quad x = 1$$
Check:
$$x = 36: \ 36 - 5\sqrt{36} - 6 \stackrel{?}{=} 0$$
$$36 - 5 \cdot 6 - 6 \stackrel{?}{=} 0$$
$$36 - 30 - 6 \stackrel{?}{=} 0$$
$$0 = 0 \ \checkmark$$
$$x = 1: \ 1 - 5\sqrt{1} - 6 \stackrel{?}{=} 0$$
$$1 - 5 \cdot 1 - 6 \stackrel{?}{=} 0$$
$$1 - 5 - 6 \stackrel{?}{=} 0$$
$$-10 \neq 0 \ \times$$
The solution set is $\{36\}$.

Method 2:
$$x - 5\sqrt{x} - 6 = 0$$
$$x - 6 = 5\sqrt{x}$$
$$(x-6)^2 = \left(5\sqrt{x}\right)^2$$
$$x^2 - 12x + 36 = 25x$$
$$x^2 - 37x + 36 = 0$$
$$(x-36)(x-1) = 0$$
$$x - 36 = 0 \quad \text{or} \quad x - 1 = 0$$
$$x = 36 \quad \text{or} \quad x = 1$$

Check:
$$x = 36: \quad 36 - 5\sqrt{36} - 6 \stackrel{?}{=} 0$$
$$36 - 5 \cdot 6 - 6 \stackrel{?}{=} 0$$
$$36 - 30 - 6 \stackrel{?}{=} 0$$
$$0 = 0 \checkmark$$

$$x = 1: \quad 1 - 5\sqrt{1} - 6 \stackrel{?}{=} 0$$
$$1 - 5 \cdot 1 - 6 \stackrel{?}{=} 0$$
$$1 - 5 - 6 \stackrel{?}{=} 0$$
$$-10 \neq 0 \quad \times$$

The solution set is $\{36\}$.
Preferences will vary.

87. Answers will vary. One possibility follows: Extraneous solutions may be introduced when we raise both sides of the equation to an even power. They may also occur when the equation involves rational expressions.

89. $\left(3p^{-2} - 4p^{-1} + 8\right) - \left(2p^{-2} - 8p^{-1} - 1\right)$
$$= 3p^{-2} - 4p^{-1} + 8 - 2p^{-2} + 8p^{-1} + 1$$
$$= p^{-2} + 4p^{-1} + 9$$

91. $\sqrt[3]{16a} + \sqrt[3]{54a} - \sqrt[3]{128a^4}$
$$= \sqrt[3]{8 \cdot 2a} + \sqrt[3]{27 \cdot 2a} - \sqrt[3]{64a^3 \cdot 2a}$$
$$= 2\sqrt[3]{2a} + 3\sqrt[3]{2a} - 4a\sqrt[3]{2a}$$
$$= 5\sqrt[3]{2a} - 4a\sqrt[3]{2a}$$

93. Let $Y_1 = x^4 + 5x^2 - 14$.

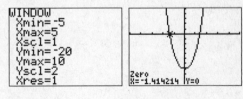

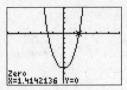

The solution set is approximately $\{-1.41, 1.41\}$.

95. Let $Y_1 = 2(x-2)^2$ and $Y_2 = 5(x-2)+1$.

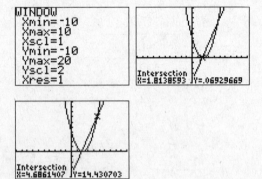

The solution set is approximately $\{1.81, 4.69\}$.

97. Let $Y_1 = x - 5\sqrt{x}$ and $Y_2 = -3$.

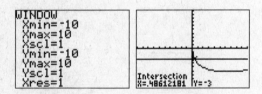

The solution set is approximately $\{0.49\}$.

99. For the graphs shown in parts (a) through (c) the WINDOW setting is the following:

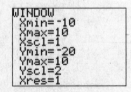

(a) $Y_1 = x^2 - 5x - 6$

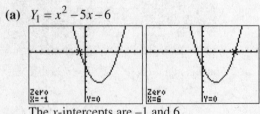

The x-intercepts are -1 and 6.

(b) $Y_1 = (x+2)^2 - 5(x+2) - 6$

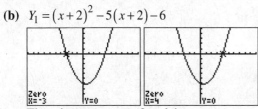

The x-intercepts are -3 and 4.

(c) $Y_1 = (x+5)^2 - 5(x+5) - 6$

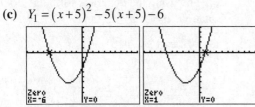

The x-intercepts are -6 and 1.

(d) The x-intercepts of the graph of
$y = f(x) = x^2 - 5x - 6$ are -1 and 6.
The x-intercepts of the graph of
$y = f(x+a) = (x+a)^2 - 5(x+a) - 6$ are
$-1-a$ and $6-a$.

Putting the Concepts Together (Sections 10.1–10.3)

1. Start: $z^2 + 10z$

Add: $\left[\dfrac{1}{2} \cdot 10\right]^2 = 25$

Result: $z^2 + 10z + 25$

Factored Form: $(z+5)^2$

2. Start: $x^2 + 7x$

Add: $\left[\dfrac{1}{2} \cdot 7\right]^2 = \dfrac{49}{4}$

Result: $x^2 + 7x + \dfrac{49}{4}$

Factored Form: $\left(x + \dfrac{7}{2}\right)^2$

3. Start: $n^2 - \dfrac{1}{4}n$

Add: $\left[\dfrac{1}{2}\cdot\left(-\dfrac{1}{4}\right)\right]^2 = \dfrac{1}{64}$

Result: $n^2 - \dfrac{1}{4}n + \dfrac{1}{64}$

Factored Form: $\left(n - \dfrac{1}{8}\right)^2$

4. $(2x-3)^2 - 5 = -1$

$(2x-3)^2 = 4$

$2x - 3 = \pm\sqrt{4}$

$2x - 3 = \pm 2$

$2x = 3 \pm 2$

$2x = 1$ or $2x = 5$

$x = \dfrac{1}{2}$ or $x = \dfrac{5}{2}$

The solution set is $\left\{\dfrac{1}{2}, \dfrac{5}{2}\right\}$.

5. $\quad x^2 + 8x + 4 = 0$

$x^2 + 8x \quad = -4$

$x^2 + 8x + \left(\dfrac{1}{2}\cdot 8\right)^2 = -4 + \left(\dfrac{1}{2}\cdot 8\right)^2$

$x^2 + 8x + 16 = -4 + 16$

$(x+4)^2 = 12$

$x + 4 = \pm\sqrt{12}$

$x + 4 = \pm 2\sqrt{3}$

$x = -4 \pm 2\sqrt{3}$

The solution set is $\left\{-4 - 2\sqrt{3}, \ -4 + 2\sqrt{3}\right\}$.

6. $\quad x(x-6) = -7$

$x^2 - 6x + 7 = 0$

For this equation, $a = 1$, $b = -6$, and $c = 7$.

$x = \dfrac{-(-6) \pm \sqrt{(-6)^2 - 4(1)(7)}}{2(1)}$

$= \dfrac{6 \pm \sqrt{36 - 28}}{2}$

$= \dfrac{6 \pm \sqrt{8}}{2}$

$= \dfrac{6 \pm 2\sqrt{2}}{2} = \dfrac{6}{2} \pm \dfrac{2\sqrt{2}}{2} = 3 \pm \sqrt{2}$

The solution set is $\left\{3 - \sqrt{2}, \ 3 + \sqrt{2}\right\}$.

7. $49x^2 - 80 = 0$

$$49x^2 = 80$$

$$x^2 = \frac{80}{49}$$

$$x = \pm\sqrt{\frac{80}{49}} = \pm\frac{\sqrt{80}}{\sqrt{49}} = \pm\frac{4\sqrt{5}}{7}$$

The solution set is $\left\{ -\frac{4\sqrt{5}}{7}, \ \frac{4\sqrt{5}}{7} \right\}$.

8. $p^2 - 8p + 6 = 0$

Because this equation does not easily factor, solve by using the quadratic formula. For this equation, $a = 1$, $b = -8$, and $c = 6$.

$$p = \frac{-(-8) \pm \sqrt{(-8)^2 - 4(1)(6)}}{2(1)}$$

$$= \frac{8 \pm \sqrt{64 - 24}}{2}$$

$$= \frac{8 \pm \sqrt{40}}{2}$$

$$= \frac{8 \pm 2\sqrt{10}}{2}$$

$$= \frac{8}{2} \pm \frac{2\sqrt{10}}{2} = 4 \pm \sqrt{10}$$

The solution set is $\left\{ 4 - \sqrt{10}, \ 4 + \sqrt{10} \right\}$.

9. $3y^2 + 6y + 4 = 0$

Because this equation does not easily factor, solve by using the quadratic formula. For this equation, $a = 3$, $b = 6$, and $c = 4$.

$$y = \frac{-6 \pm \sqrt{6^2 - 4(3)(4)}}{2(3)}$$

$$= \frac{-6 \pm \sqrt{36 - 48}}{6}$$

$$= \frac{-6 \pm \sqrt{-12}}{6}$$

$$= \frac{-6 \pm 2\sqrt{3}i}{6} = \frac{-6}{6} \pm \frac{2\sqrt{3}}{6}i = -1 \pm \frac{\sqrt{3}}{3}i$$

The solution set is $\left\{ -1 - \frac{\sqrt{3}}{3}i, \ -1 + \frac{\sqrt{3}}{3}i \right\}$.

10. $\frac{1}{4}n^2 + n = \frac{1}{6}$

$$12\left(\frac{1}{4}n^2 + n \right) = 12\left(\frac{1}{6} \right)$$

$$3n^2 + 12n = 2$$

$$3n^2 + 12n - 2 = 0$$

Because this equation does not easily factor, solve by using the quadratic formula. For this equation, $a = 3$, $b = 12$, and $c = -2$.

$$n = \frac{-12 \pm \sqrt{12^2 - 4(3)(-2)}}{2(3)}$$

$$= \frac{-12 \pm \sqrt{144 + 24}}{6}$$

$$= \frac{-12 \pm \sqrt{168}}{6}$$

$$= \frac{-12 \pm 2\sqrt{42}}{6}$$

$$= -\frac{12}{6} \pm \frac{2\sqrt{42}}{6} = -2 \pm \frac{\sqrt{42}}{3}$$

The solution set is $\left\{ -2 - \frac{\sqrt{42}}{3}, \ -2 + \frac{\sqrt{42}}{3} \right\}$.

11. $9x^2 + 12x + 4 = 0$

For this equation, $a = 9$, $b = 12$, and $c = 4$.

$$b^2 - 4ac = 12^2 - 4(9)(4) = 144 - 144 = 0$$

Because $b^2 - 4ac = 0$, the quadratic equation will have one repeated real solution.

12. $3x^2 + 6x - 2 = 0$

For this equation, $a = 3$, $b = 6$, and $c = -2$.

$$b^2 - 4ac = 6^2 - 4(3)(-2) = 36 + 24 = 60$$

Because $b^2 - 4ac = 60$ is positive, but not a perfect square, the quadratic equation will have two irrational solutions.

13. $2x^2 + 6x + 5 = 0$

For this equation, $a = 2$, $b = 6$, and $c = 5$.

$$b^2 - 4ac = 6^2 - 4(2)(5) = 36 - 40 = -4$$

Because $b^2 - 4ac = -4$ is negative, the quadratic equation will have two complex solutions that are not real. The solutions will be complex conjugates of each other.

14. $c^2 = 4^2 + 10^2 = 16 + 100 = 116$

$$c = \sqrt{116} = 2\sqrt{29}$$

15.
$$2m + 7\sqrt{m} - 15 = 0$$
$$2\left(\sqrt{m}\right)^2 + 7\left(\sqrt{m}\right) - 15 = 0$$
Let $u = \sqrt{m}$.
$$2u^2 + 7u - 15 = 0$$
$$(2u - 3)(u + 5) = 0$$
$$2u - 3 = 0 \quad \text{or} \quad u + 5 = 0$$
$$2u = 3 \quad \text{or} \quad u = -5$$
$$u = \frac{3}{2}$$
$$\sqrt{m} = \frac{3}{2} \quad \text{or} \quad \sqrt{m} = -5$$
$$m = \left(\frac{3}{2}\right)^2 \quad \text{or} \quad m = (-5)^2$$
$$m = \frac{9}{4} \quad \text{or} \quad m = 25$$
Check:
$$m = \frac{9}{4}: \quad 2 \cdot \frac{9}{4} + 7\sqrt{\frac{9}{4}} - 15 \stackrel{?}{=} 0$$
$$2 \cdot \frac{9}{4} + 7 \cdot \frac{3}{2} - 15 \stackrel{?}{=} 0$$
$$\frac{9}{2} + \frac{21}{2} - 15 \stackrel{?}{=} 0$$
$$0 = 0 \checkmark$$

$$m = 25: \quad 2 \cdot 25 + 7\sqrt{25} - 15 \stackrel{?}{=} 0$$
$$2 \cdot 25 + 7 \cdot 5 - 15 \stackrel{?}{=} 0$$
$$50 + 35 - 15 \stackrel{?}{=} 0$$
$$70 \neq 0 \;\; \textbf{✗}$$

$m = 25$ does not check; the solution set is $\left\{\dfrac{9}{4}\right\}$.

16.
$$p^{-2} - 3p^{-1} - 18 = 0$$
$$\left(p^{-1}\right)^2 - 3\left(p^{-1}\right) - 18 = 0$$
Let $u = p^{-1}$.
$$u^2 - 3u - 18 = 0$$
$$(u + 3)(u - 6) = 0$$
$$u + 3 = 0 \quad \text{or} \quad u - 6 = 0$$
$$u = -3 \quad \text{or} \quad u = 6$$
$$p^{-1} = -3 \quad \text{or} \quad p^{-1} = 6$$
$$\frac{1}{p} = -3 \quad \text{or} \quad \frac{1}{p} = 6$$
$$p = -\frac{1}{3} \quad \text{or} \quad p = \frac{1}{6}$$

Check:
$$p = -\frac{1}{3}: \quad \left(-\frac{1}{3}\right)^{-2} - 3\left(-\frac{1}{3}\right)^{-1} - 18 \stackrel{?}{=} 0$$
$$(-3)^2 - 3(-3) - 18 \stackrel{?}{=} 0$$
$$9 + 9 - 18 \stackrel{?}{=} 0$$
$$0 = 0 \checkmark$$

$$p = \frac{1}{6}: \quad \left(\frac{1}{6}\right)^{-2} - 3\left(\frac{1}{6}\right)^{-1} - 18 \stackrel{?}{=} 0$$
$$(6)^2 - 3(6) - 18 \stackrel{?}{=} 0$$
$$36 - 18 - 18 \stackrel{?}{=} 0$$
$$0 = 0 \checkmark$$

Both check; the solution set is $\left\{\dfrac{1}{6}, \ -\dfrac{1}{3}\right\}$.

17.
$$12{,}000 = -0.4x^2 + 140x$$
$$0.4x^2 - 140x + 12{,}000 = 0$$
For this equation, $a = 0.4$, $b = -140$, and $c = 12{,}000$.
$$x = \frac{-(-140) \pm \sqrt{(-140)^2 - 4(0.4)(12{,}000)}}{2(0.4)}$$
$$= \frac{140 \pm \sqrt{400}}{0.8}$$
$$= \frac{140 \pm 20}{0.8}$$
$$x = \frac{140 + 20}{0.8} \quad \text{or} \quad x = \frac{140 - 20}{0.8}$$
$$= 200 \quad \text{or} \quad = 150$$
Revenue will be $12,000 when either 150 microwaves or 200 microwaves are sold.

18. Let x represent the speed of the wind.

	Distance	Rate	Time
Against Wind	300	$140 - x$	$\dfrac{300}{140 - x}$
With Wind	300	$140 + x$	$\dfrac{300}{140 + x}$

$$\frac{300}{140-x}+\frac{300}{140+x}=5$$

$$(140-x)(140+x)\left(\frac{300}{140-x}+\frac{300}{140+x}\right)$$
$$=(140-x)(140+x)(5)$$

$$300(140+x)+300(140-x)=\left(19{,}600-x^2\right)(5)$$

$$42{,}000+300x+42{,}000-300x=98{,}000-5x^2$$

$$84{,}000=98{,}000-5x^2$$

$$-14{,}000=-5x^2$$

$$2800=x^2$$

$$x=\pm\sqrt{2800}$$

$$=\pm20\sqrt{7}$$

$$\approx\pm52.915$$

Because the speed should be positive, we disregard $x\approx-52.915$. Thus, the only viable answer is $x\approx52.915$. Rounding to the nearest tenth, the speed of the wind was approximately 52.9 miles per hour.

Section 10.4

Preparing for Graphing Quadratic Functions Using Transformations

P1. Locate some points on the graph of $f(x)=x^2$.

x	$f(x)=x^2$	$(x, f(x))$
-3	$f(-3)=(-3)^2=9$	$(-3, 9)$
-2	$f(-2)=(-2)^2=4$	$(-2, 4)$
-1	$f(-1)=(-1)^2=1$	$(-1, 1)$
0	$f(0)=0^2=0$	$(0, 0)$
1	$f(1)=1^2=1$	$(1, 1)$
2	$f(2)=2^2=4$	$(2, 4)$
3	$f(3)=3^2=9$	$(3, 9)$

Plot the points and connect them with a smooth curve.

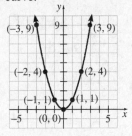

P2. Locate some points on the graph of $f(x)=x^2-3$.

x	$f(x)=x^2-3$	$(x, f(x))$
-3	$f(-3)=(-3)^2-3=6$	$(-3, 6)$
-2	$f(-2)=(-2)^2-3=1$	$(-2, 1)$
-1	$f(-1)=(-1)^2-3=-2$	$(-1,-2)$
0	$f(0)=0^2-3=-3$	$(0,-3)$
1	$f(1)=1^2-3=-2$	$(1,-2)$
2	$f(2)=2^2-3=1$	$(2, 1)$
3	$f(3)=3^2-3=6$	$(3, 6)$

Plot the points and connect them with a smooth curve.

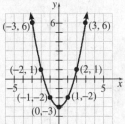

P3. For $f(x)=2x^2+5x+1$, the independent variable x can take on any value on the real number line. That is, f is defined for all real x values. Thus, the domain is the set of all real number, or $(-\infty,\infty)$.

Section 10.4 Quick Checks

1. quadratic function

2. up; down

3. Begin with the graph of $y=x^2$, then shift the graph up 5 unit to obtain the graph of $f(x)=x^2+5$.

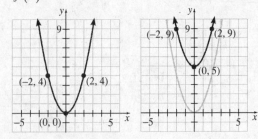

4. Begin with the graph of $y = x^2$, then shift the graph down 2 units to obtain the graph of $f(x) = x^2 - 2$.

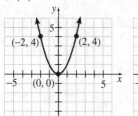 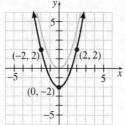

5. False. To obtain the graph of $f(x) = (x + 12)^2$ we shift the graph of $y = x^2$ horizontally to the *left* 12 units.

6. Begin with the graph of $y = x^2$, then shift the graph to the left 5 units to obtain the graph of $f(x) = (x + 5)^2$.

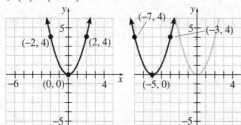

7. Begin with the graph of $y = x^2$, then shift the graph to the right 1 unit to obtain the graph of $f(x) = (x - 1)^2$.

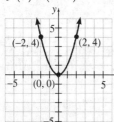 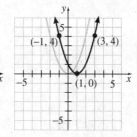

8. Begin with the graph of $y = x^2$, then shift the graph 3 units to the right to obtain the graph of $y = (x - 3)^2$. Shift this graph up 2 units to obtain the graph of $f(x) = (x - 3)^2 + 2$.

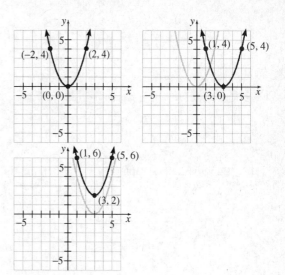

9. Begin with the graph of $y = x^2$, then shift the graph 1 unit to the left to obtain the graph of $y = (x + 1)^2$. Shift this graph down 4 units to obtain the graph of $f(x) = (x + 1)^2 - 4$.

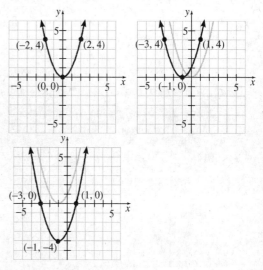

10. y-coordinate; a; vertically stretched; vertically compressed

11. Begin with the graph of $y = x^2$, then vertically stretch the graph by a factor of 3 (multiply each y-coordinate by 3) to obtain the graph of $f(x) = 3x^2$.

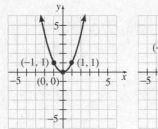 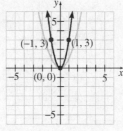

12. Begin with the graph of $y = x^2$, then multiply the y-coordinates by $-\dfrac{1}{4}$ to obtain the graph of $f(x) = -\dfrac{1}{4}x^2$.

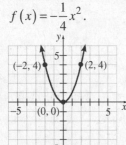 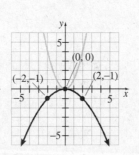

13. True. In this case, $a = -3$ which is less than zero, so the graph opens down.

14. Begin with the graph of $y = x^2$, then shift the graph 2 units to the left to obtain the graph of $y = (x+2)^2$. Multiply the y-coordinates by -3 to obtain the graph of $y = -3(x+2)^2$. Lastly, shift the graph up 1 unit to obtain the graph of $f(x) = -3(x+2)^2 + 1$.

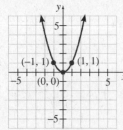 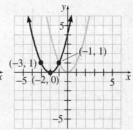

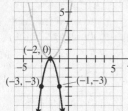 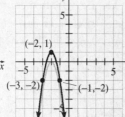

The domain is $\{x \mid x$ is and real number$\}$ or, using interval notation, $(-\infty, \infty)$. The range is $\{y \mid y \le 1\}$ or using interval notation $(-\infty, 1]$.

15. Use completing the square to write the function in the form $y = a(x-h)^2 + k$.

$$f(x) = 2x^2 - 8x + 5$$
$$= \left(2x^2 - 8x\right) + 5$$
$$= 2\left(x^2 - 4x\right) + 5$$
$$= 2\left(x^2 - 4x + 4\right) + 5 - 8$$
$$= 2(x-2)^2 - 3$$

Begin with the graph of $y = x^2$, then shift the graph right 2 units to obtain the graph of $y = (x-2)^2$. Vertically stretch this graph by a factor of 2 (multiply the y-coordinates by 2) to obtain the graph of $y = 2(x-2)^2$. Lastly, shift the graph down 3 units to obtain the graph of $f(x) = 2(x-2)^2 - 3$.

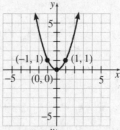 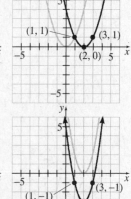

The domain is $\{x \mid x$ is and real number$\}$ or, using interval notation, $(-\infty, \infty)$. The range is $\{y \mid y \ge -3\}$ or, using interval notation, $[-3, \infty)$.

16. Consider the form $f(x) = a(x-h)^2 + k$. From the graph, we know that the vertex is $(-1, 2)$. So $h = -1$ and $k = 2$, and we have that
$$f(x) = a(x-(-1))^2 + 2$$
$$f(x) = a(x+1)^2 + 2$$

The graph also passes through the point $(0, 1)$ which means that $f(0) = 1$. Substituting these values into the function, we can solve for a:

$$f(x) = a(x+1)^2 + 2$$
$$1 = a(0+1)^2 + 2$$
$$1 = a(1)^2 + 2$$
$$1 = a + 2$$
$$-1 = a$$

The quadratic function is $f(x) = -(x+1)^2 + 2$.

10.4 Exercises

17. (I)The graph of $f(x) = x^2 + 3$ is the graph of $y = x^2$ shifted 3 units up. Thus, the graph of the function is graph (D).

(II)The graph of $f(x) = (x+3)^2$ is the graph of $y = x^2$ shifted 3 units to the left. Thus, the graph of the function is graph (A).

(III)The graph of $f(x) = x^2 - 3$ is the graph of $y = x^2$ shifted 3 units down. Thus, the graph of the function is graph (C).

(IV)The graph of $f(x) = (x-3)^2$ is the graph of $y = x^2$ shifted 3 units to the right. Thus, the graph of the function is graph (B).

19. To obtain the graph of $f(x) = (x+10)^2$, begin with the graph of $y = x^2$ and shift it 10 units to the left.

21. To obtain the graph of $F(x) = x^2 + 12$, begin with the graph of $y = x^2$ and shift it 12 units up.

23. To obtain the graph of $H(x) = 2(x-5)^2$, begin with the graph of $y = x^2$, shift it 5 units to the right, and vertically stretch it by a factor of 2 (multiply the y-coordinates by 2).

25. To obtain the graph of $f(x) = -3(x+5)^2 + 8$, begin with the graph of $y = x^2$, shift it 5 units to the left, multiply the y-coordinates by -3 (which means it opens down and is stretched vertically by a factor of 3), and shift the graph up 8 units.

27. Begin with the graph of $y = x^2$, then shift the graph up 1 unit to obtain the graph of $f(x) = x^2 + 1$.

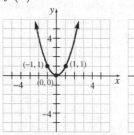 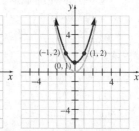

29. Begin with the graph of $y = x^2$, then shift the graph down 1 unit to obtain the graph of $f(x) = x^2 - 1$.

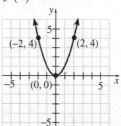 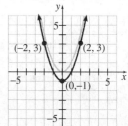

31. Begin with the graph of $y = x^2$, then shift the graph to the right 3 units to obtain the graph of $F(x) = (x-3)^2$.

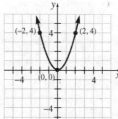 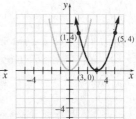

33. Begin with the graph of $y = x^2$, then shift the graph 2 units to the left to obtain the graph of $h(x) = (x+2)^2$.

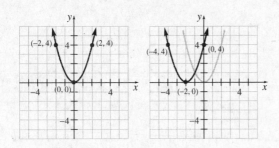

35. Begin with the graph of $y = x^2$, then vertically stretch the graph by a factor of 4 (multiply each y-coordinate by 4) to obtain the graph of $g(x) = 4x^2$.

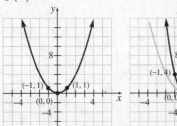

37. Begin with the graph of $y = x^2$, then vertically compress the graph by a factor of $\frac{1}{3}$ (multiply each y-coordinate by $\frac{1}{3}$) to obtain the graph of $H(x) = \frac{1}{3}x^2$.

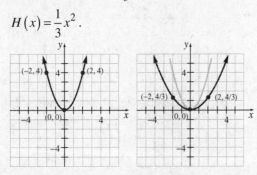

39. Begin with the graph of $y = x^2$, then multiply each y-coordinate by -1 to obtain the graph of $p(x) = -x^2$.

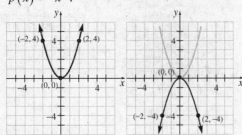

41. Begin with the graph of $y = x^2$, then shift the graph 1 unit to the right to obtain the graph of $y = (x-1)^2$. Shift this graph down 3 units to obtain the graph of $f(x) = (x-1)^2 - 3$.

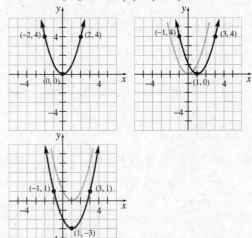

43. Begin with the graph of $y = x^2$, then shift the graph 3 units to the left to obtain the graph of $y = (x+3)^2$. Shift this graph up 1 unit to obtain the graph of $F(x) = (x+3)^2 + 1$.

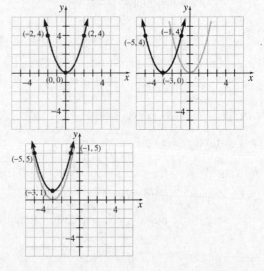

45. Begin with the graph of $y = x^2$, then shift the graph 3 units to the left to obtain the graph of $y = (x+3)^2$. Multiply the y-coordinates by -1 to obtain the graph of $y = -(x+3)^2$. Lastly, shift the graph up 2 units to obtain the graph of

$h(x) = -(x+3)^2 + 2$.

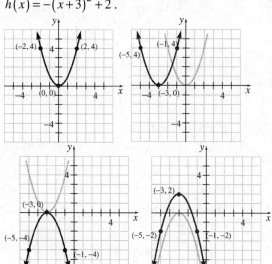

$H(x) = -\dfrac{1}{2}(x+5)^2 + 3$.

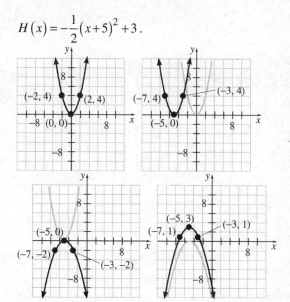

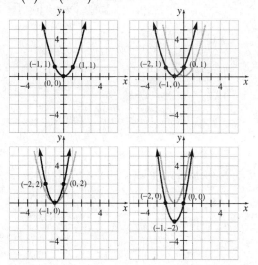

47. Begin with the graph of $y = x^2$, then shift the graph 1 unit to the left to obtain the graph of $y = (x+1)^2$. Vertically stretch this graph by a factor of 2 (multiply the y-coordinates by 2) to obtain the graph of $y = 2(x+1)^2$. Lastly, shift the graph down 2 units to obtain the graph of $G(x) = 2(x+1)^2 - 2$.

51. Use completing the square to write the function in the form $y = a(x-h)^2 + k$.

$$f(x) = x^2 + 2x - 4$$
$$= \left(x^2 + 2x + 1\right) - 4 - 1$$
$$= (x+1)^2 - 5$$

Begin with the graph of $y = x^2$, the shift the graph 1 unit to the left to obtain the graph of $y = (x+1)^2$. Shift this graph down 5 units to obtain the graph of $f(x) = (x+1)^2 - 5$.

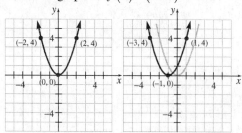

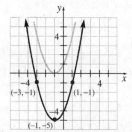

49. Begin with the graph of $y = x^2$, then shift the graph 5 units to the left to obtain the graph of $y = (x+5)^2$. Multiply the y-coordinates by $-\dfrac{1}{2}$ to obtain the graph of $y = -\dfrac{1}{2}(x+5)^2$. Lastly, shift the graph up 3 units to obtain the graph of

The vertex is $(h,k)=(-1,-5)$ and the axis of symmetry is $x=-1$. The domain is the set of all real numbers or, using interval notation, $(-\infty,\infty)$. The range is $\{y \mid y \geq -5\}$ or, using interval notation, $[-5,\infty)$.

53. Use completing the square to write the function in the form $y=a(x-h)^2+k$.

$$g(x)=x^2-4x+8=\left(x^2-4x\right)+8$$
$$=\left(x^2-4x+4\right)+8-4$$
$$=(x-2)^2+4$$

Begin with the graph of $y=x^2$, then shift the graph right 2 units to obtain the graph of $y=(x-2)^2$. Shift this graph up 4 units to obtain the graph of $g(x)=(x-2)^2+4$.

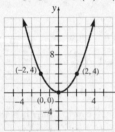 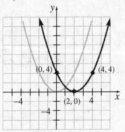

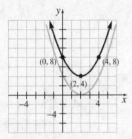

The vertex is $(h,k)=(2,4)$ and the axis of symmetry is $x=2$. The domain is the set of all real numbers or, using interval notation, $(-\infty,\infty)$. The range is $\{y \mid y \geq 4\}$ or, using interval notation, $[4,\infty)$.

55. Consider the form $y=a(x-h)^2+k$. From the graph we know that the vertex is $(-1,-3)$ so we have $h=-1$ and $k=-3$. The graph also passes through the point $(x,y)=(0,-2)$. Substituting these values for x, y, h, and k, we can solve for a:

$$-2=a\left(0-(-1)\right)^2+(-3)$$
$$-2=a(1)^2-3$$
$$-2=a-3$$
$$1=a$$

The quadratic function is $f(x)=(x+1)^2-3$.

57. Consider the form $y=a(x-h)^2+k$. From the graph we know that the vertex is $(3,7)$ so we have $h=3$ and $k=7$. The graph also passes through the point $(x,y)=(0,-11)$. Substituting these values for x, y, h, and k, we can solve for a:

$$-11=a(0-3)^2+7$$
$$-11=a(-3)^2+7$$
$$-18=9a$$
$$-2=a$$

The quadratic function is $f(x)=-2(x-3)^2+7$.

59. Consider the form $y=a(x-h)^2+k$. From the graph we know that the vertex is $(-4,0)$ so we have $h=-4$ and $k=0$. The graph also passes through the point $(x,y)=(-3,1)$. Substituting these values for x, y, h, and k, we can solve for a:

$$1=a\left(-3-(-4)\right)^2+0$$
$$1=a(-3+4)^2$$
$$1=a(1)^2$$
$$1=a$$

The quadratic function is $f(x)=(x+4)^2$.

61. Use completing the square to write the function in the form $y=a(x-h)^2+k$.

$$f(x)=x^2+6x-16$$
$$=\left(x^2+6x\right)-16$$
$$=\left(x^2+6x+9\right)-16-9$$
$$=(x+3)^2-25$$

Begin with the graph of $y=x^2$, then shift the graph 3 units left to obtain the graph of $y=(x+3)^2$. Shift this result down 25 units to obtain the graph of $f(x)=(x+3)^2-25$.

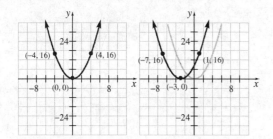

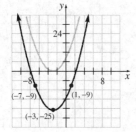

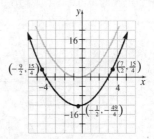

The vertex is $(h,k) = \left(-3, -25\right)$ and the axis of symmetry is $x = -3$. The domain is the set of all real numbers or, using interval notation, $(-\infty, \infty)$. The range is $\{y \mid y \geq -25\}$ or, using interval notation, $[-25, \infty)$.

The vertex is $(h,k) = \left(-\dfrac{1}{2}, -\dfrac{49}{4}\right)$ and the axis of symmetry is $x = -\dfrac{1}{2}$. The domain is the set of all real numbers or, using interval notation, $(-\infty, \infty)$. The range is $\left\{ y \mid y \geq -\dfrac{49}{4} \right\}$ or, using interval notation, $\left[-\dfrac{49}{4}, \infty\right)$.

63. Use completing the square to write the function in the form $y = a(x-h)^2 + k$.

$$F(x) = x^2 + x - 12$$
$$= \left(x^2 + x\right) - 12$$
$$= \left(x^2 + x + \frac{1}{4}\right) - 12 - \frac{1}{4}$$
$$= \left(x + \frac{1}{2}\right)^2 - \frac{49}{4}$$

Begin with the graph of $y = x^2$, then shift the graph left $\dfrac{1}{2}$ unit to obtain the graph of $y = \left(x + \dfrac{1}{2}\right)^2$. Shift this graph down $\dfrac{49}{4}$ units to obtain the graph of $F(x) = \left(x + \dfrac{1}{2}\right)^2 - \dfrac{49}{4}$.

65. Use completing the square to write the function in the form $y = a(x-h)^2 + k$.

$$H(x) = 2x^2 - 4x - 1$$
$$= \left(2x^2 - 4x\right) - 1$$
$$= 2\left(x^2 - 2x\right) - 1$$
$$= 2\left(x^2 - 2x + 1\right) - 1 - 2$$
$$= 2(x-1)^2 - 3$$

Begin with the graph of $y = x^2$, then shift the graph right 1 unit to obtain the graph of $y = (x-1)^2$. Vertically stretch this graph by a factor of 2 (multiply the y-coordinates by 2) to obtain the graph of $y = 2(x-1)^2$. Lastly, shift the graph down 3 units to obtain the graph of $H(x) = 2(x-1)^2 - 3$.

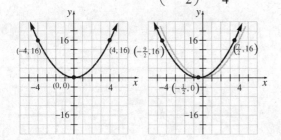

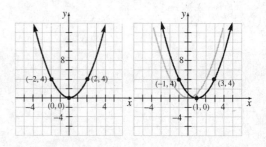

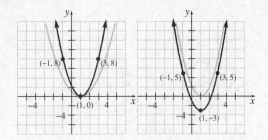   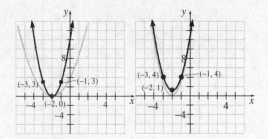

The vertex is $(h,k)=(1,-3)$ and the axis of symmetry is $x=1$. The domain is the set of all real numbers or, using interval notation, $(-\infty,\infty)$. The range is $\{y\,|\,y\ge-3\}$ or, using interval notation, $[-3,\infty)$.

The vertex is $(h,k)=(-2,1)$ and the axis of symmetry is $x=-2$. The domain is the set of all real numbers or, using interval notation, $(-\infty,\infty)$. The range is $\{y\,|\,y\ge1\}$ or, using interval notation, $[1,\infty)$.

67. Use completing the square to write the function in the form $y=a(x-h)^2+k$.

$$P(x)=3x^2+12x+13$$
$$=\left(3x^2+12x\right)+13$$
$$=3\left(x^2+4x\right)+13$$
$$=3\left(x^2+4x+4\right)+13-12$$
$$=3(x+2)^2+1$$

Begin with the graph of $y=x^2$, then shift the graph left 2 units to obtain the graph of $y=(x+2)^2$. Vertically stretch this graph by a factor of 3 (multiply the y-coordinates by 3) to obtain the graph of $y=3(x+2)^2$. Lastly, shift the graph up 1 unit to obtain the graph of $P(x)=3(x+2)^2+1$.

69. Use completing the square to write the function in the form $y=a(x-h)^2+k$.

$$F(x)=-x^2-10x-21$$
$$=\left(-x^2-10x\right)-21$$
$$=-\left(x^2+10x\right)-21$$
$$=-\left(x^2+10x+25\right)-21+25$$
$$=-(x+5)^2+4$$

Begin with the graph of $y=x^2$, then shift the graph left 5 units to obtain the graph of $y=(x+5)^2$. Multiply the y-coordinates by -1 to obtain the graph of $y=-(x+5)^2$. Lastly, shift the graph up 4 units to obtain the graph of $F(x)=-(x+5)^2+4$.

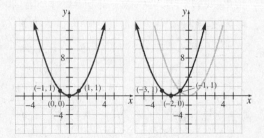   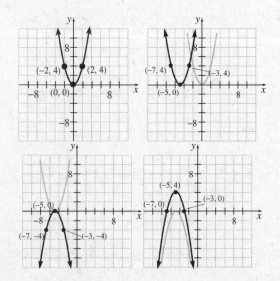

The vertex is $(h,k) = (-5,4)$ and the axis of symmetry is $x = -5$. The domain is the set of all real numbers or, using interval notation, $(-\infty, \infty)$. The range is $\{y \mid y \le 4\}$ or, using interval notation, $(-\infty, 4]$.

71. Use completing the square to write the function in the form $y = a(x-h)^2 + k$.

$$\begin{aligned} g(x) &= -x^2 + 6x - 1 \\ &= \left(-x^2 + 6x\right) - 1 \\ &= -\left(x^2 - 6x\right) - 1 \\ &= -\left(x^2 - 6x + 9\right) - 1 + 9 \\ &= -(x-3)^2 + 8 \end{aligned}$$

Begin with the graph of $y = x^2$, then shift the graph right 3 units to obtain the graph of $y = (x-3)^2$. Multiply the y-coordinates by -1 to obtain the graph of $y = -(x-3)^2$. Lastly, shift the graph up 8 units to obtain the graph of $g(x) = -(x-3)^2 + 8$.

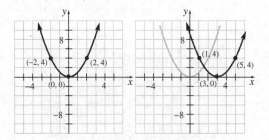

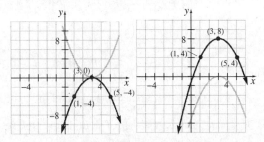

The vertex is $(h,k) = (3,8)$ and the axis of symmetry is $x = 3$. The domain is the set of all real numbers or, using interval notation, $(-\infty, \infty)$. The range is $\{y \mid y \le 8\}$ or, using interval notation, $(-\infty, 8]$.

73. Use completing the square to write the function in the form $y = a(x-h)^2 + k$.

$$\begin{aligned} H(x) &= -2x^2 + 8x - 4 \\ &= \left(-2x^2 + 8x\right) - 4 \\ &= -2\left(x^2 - 4x\right) - 4 \\ &= -2\left(x^2 - 4x + 4\right) - 4 + 8 \\ &= -2(x-2)^2 + 4 \end{aligned}$$

Begin with the graph of $y = x^2$, then shift the graph right 2 units to obtain the graph of $y = (x-2)^2$. Multiply the y-coordinates by -2 to obtain the graph of $y = -2(x-2)^2$. Lastly, shift the graph up 4 units to obtain the graph of $H(x) = -2(x-2)^2 + 4$.

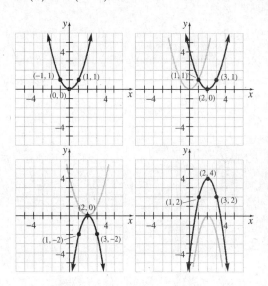

The vertex is $(h,k) = (2,4)$ and the axis of symmetry is $x = 2$. The domain is the set of all real numbers or, using interval notation, $(-\infty, \infty)$. The range is $\{y \mid y \le 4\}$ or, using interval notation, $(-\infty, 4]$.

75. Use completing the square to write the function in the form $y = a(x-h)^2 + k$.

$$f(x) = \frac{1}{3}x^2 - 2x + 4$$
$$= \left(\frac{1}{3}x^2 - 2x\right) + 4$$
$$= \frac{1}{3}\left(x^2 - 6x\right) + 4$$
$$= \frac{1}{3}\left(x^2 - 6x + 9\right) + 4 - 3$$
$$= \frac{1}{3}(x-3)^2 + 1$$

Begin with the graph of $y = x^2$, then shift the graph right 3 units to obtain the graph of $y = (x-3)^2$. Vertically compress the graph by a factor of $\frac{1}{3}$ (multiply the y-coordinates by $\frac{1}{3}$) to obtain the graph of $y = \frac{1}{3}(x-3)^2$. Lastly, shift the graph up 1 unit to obtain the graph of $f(x) = \frac{1}{3}(x-3)^2 + 1$.

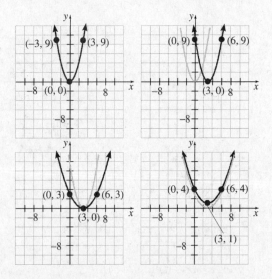

The vertex is $(h,k) = (3,1)$ and the axis of symmetry is $x = 1$. The domain is the set of all real numbers or, using interval notation, $(-\infty,\infty)$. The range is $\{y \mid y \geq 1\}$ or, using interval notation, $[1,\infty)$.

77. Use completing the square to write the function in the form $y = a(x-h)^2 + k$.

$$G(x) = -12x^2 - 12x + 1$$
$$= \left(-12x^2 - 12x\right) + 1$$
$$= -12\left(x^2 + x\right) + 1$$
$$= -12\left(x^2 + x + \frac{1}{4}\right) + 1 + 3$$
$$= -12\left(x + \frac{1}{2}\right)^2 + 4$$

Begin with the graph of $y = x^2$, then shift left $\frac{1}{2}$ unit to obtain the graph of $y = \left(x + \frac{1}{2}\right)^2$.

Multiply the y-coordinates by -12 to obtain the graph of $y = -12\left(x + \frac{1}{2}\right)^2$. Lastly, shift the graph up 4 units to obtain the graph of $G(x) = -12\left(x + \frac{1}{2}\right)^2 + 4$.

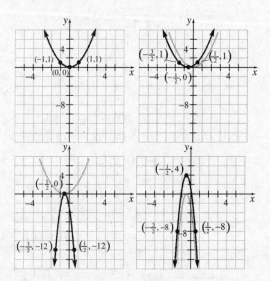

The vertex is $(h,k) = \left(-\frac{1}{2}, 4\right)$ and the axis of symmetry is $x = -\frac{1}{2}$. The domain is the set of all real numbers or, using interval notation, $(-\infty,\infty)$. The range is $\{y \mid y \leq 4\}$ or, using interval notation, $(-\infty, 4]$.

79. Answers may vary. Since the graph opens up, one possibility is to let $a = 1$. The vertex is $(3,0)$ so we have $h = 3$ and $k = 0$. Substituting these values into the form $y = a(x-h)^2 + k$ gives $y = f(x) = (x-3)^2$.

81. Answers may vary. Since the graph opens up, one possibility is to let $a = 1$. The vertex is $(-3,1)$ so we have $h = -3$ and $k = 1$. Substituting these values into the form $y = a(x-h)^2 + k$ give $y = f(x) = (x+3)^2 + 1$.

83. Answers may vary. Since the graph opens down, one possibility is to let $a = -1$. The vertex is $(5,-1)$ so we have $h = 5$ and $k = -1$. Substituting these values into the form $y = a(x-h)^2 + k$ gives $y = f(x) = -(x-5)^2 - 1$.

85. Consider the form $y = a(x-h)^2 + k$. Since the graph opens up, we know that $a > 0$. The graph is vertically stretched by a factor of 4 so we know that $a = 4$. The vertex is $(9,-6)$ so we have $h = 9$ and $k = -6$. Substituting these values gives $y = f(x) = 4(x-9)^2 - 6$.

87. Consider the form $y = a(x-h)^2 + k$. Since the graph opens down, we know that $a < 0$. The graph is vertically compressed by a factor of $\frac{1}{3}$ so we know that $a = -\frac{1}{3}$. The vertex is $(0,6)$ so we have $h = 0$ and $k = 6$. Substituting these values gives $y = f(x) = -\frac{1}{3}x^2 + 6$.

89. The lowest or highest point on a parabola (the graph of $y = ax^2 + bx + c$) is called the *vertex*. If $a > 0$, the graph opens up and the vertex is the low point. If $a < 0$, the graph opens down and the vertex is the high point.

91. A quadratic function can never have a range of $(-\infty, \infty)$. The graph of a quadratic always has a turning point so it always has either a lowest point or a highest point. In either case, the range is limited by k, the y-coordinate of the vertex. The range will either be $(-\infty, k]$ if the graph opens down or $[k, \infty)$ if the graph opens up.

93.
$$\begin{array}{r} 29 \\ 12\overline{)349} \\ \underline{24} \\ 109 \\ \underline{108} \\ 1 \end{array}$$
So, $\frac{349}{12} = 29 + \frac{1}{12} = 29\frac{1}{12}$.

95.
$$\begin{array}{r} x^3 - 5x^2 + 4x - 2 \\ 2x-1\overline{)2x^4 - 11x^3 + 13x^2 - 8x} \\ \underline{2x^4 - \ \ x^3} \\ -10x^3 + 13x^2 - 8x \\ \underline{-10x^3 + 5x^2} \\ 8x^2 - 8x \\ \underline{8x^2 - 4x} \\ -4x \\ \underline{-4x + 2} \\ -2 \end{array}$$

So,
$$\frac{2x^4 - 11x^3 + 13x^2 - 8x}{2x-1} = x^3 - 5x^2 + 4x - 2 + \frac{-2}{2x-1}$$

97. $f(x) = x^2 + 1.3$

```
WINDOW
Xmin=-10
Xmax=10
Xscl=1
Ymin=-10
Ymax=10
Yscl=1
Xres=1
```

Vertex: $(0, 1.3)$

Axis of symmetry: $x = 0$

Range: $\{y \mid y \geq 1.3\} = [1.3, \infty)$

99. $g(x) = (x-2.5)^2$

 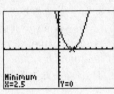

Vertex: $(2.5, 0)$

Axis of symmetry: $x = 2.5$

Range: $\{y \mid y \geq 0\} = [0, \infty)$

101. $h(x) = 2.3(x-1.4)^2 + 0.5$

 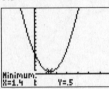

Vertex: $(1.4, 0.5)$

Axis of symmetry: $x = 1.4$

Range: $\{y \mid y \geq 0.5\} = [0.5, \infty)$

103. $F(x) = -3.4(x-2.8)^2 + 5.9$

 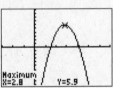

Vertex: $(2.8, 5.9)$

Axis of symmetry: $x = 2.8$

Range: $\{y \mid y \leq 5.9\} = (-\infty, 5.9]$

Section 10.5

Preparing for Graphing Quadratic Functions Using Properties

P1. To find the y-intercept, let $x = 0$ and solve for y:
$$2(0) + 5y = 20$$
$$5y = 20$$
$$y = 4$$
To find the x-intercept, let $y = 0$ and solve for x:
$$2x + 5(0) = 20$$
$$2x = 20$$
$$x = 10$$
The intercepts are $(0, 4)$ and $(10, 0)$.

P2. $2x^2 - 3x - 20 = 0$
$$(2x + 5)(x - 4) = 0$$
$$2x + 5 = 0 \quad \text{or} \quad x - 4 = 0$$
$$x = -\frac{5}{2} \quad \text{or} \quad x = 4$$
The solution set is $\left\{ -\dfrac{5}{2}, 4 \right\}$.

P3. $f(x) = 0$
$$x^2 - 3x - 4 = 0$$
$$(x + 1)(x - 4) = 0$$
$$x + 1 = 0 \quad \text{or} \quad x - 4 = 0$$
$$x = -1 \quad \text{or} \quad x = 4$$
The zeros are -1 and 4.

Section 10.5 Quick Checks

1. $x = -\dfrac{b}{2a}$

2. $>$

3. $f(x) = -2x^2 - 3x + 6$
$$b^2 - 4ac = (-3)^2 - 4(-2)(6) = 9 + 48 = 57 > 0$$
Since the discriminant is greater than zero, the graph will have two x-intercepts.

4. $f(x) = x^2 + 4x - 3$
$$x = -\frac{b}{2a} = -\frac{4}{2} = -2$$
$$f(-2) = (-2)^2 + 4(-2) - 3$$
$$= 4 - 8 - 3 = -7$$
The vertex is $(-2, -7)$.

5. For the function $f(x) = x^2 - 4x - 12$, we see that $a = 1$, $b = -4$, and $c = -12$. The parabola opens up because $a = 1 > 0$. The x-coordinate of the vertex is $x = -\dfrac{b}{2a} = -\dfrac{(-4)}{2(1)} = 2$. The y-coordinate of the vertex is
$$f\left(-\frac{b}{2a}\right) = f(2)$$
$$= (2)^2 - 4(2) - 12$$
$$= 4 - 8 - 12$$
$$= -16$$
Thus, the vertex is $(2, -16)$ and the axis of

symmetry is the line $x = 2$.
The y-intercept is

$f(0) = (0)^2 - 4(0) - 12 = -12$.

Now, $b^2 - 4ac = (-4)^2 - 4(1)(-12) = 64 > 0$.
The parabola will have two distinct x-intercepts.
We find these by solving

$$f(x) = 0$$
$$x^2 - 4x - 12 = 0$$
$$(x - 6)(x + 2) = 0$$
$$x - 6 = 0 \text{ or } x + 2 = 0$$
$$x = 6 \text{ or } \quad x = -2$$

Finally, the y-intercept point, $(0, -12)$, is two
units to the left of the axis of symmetry.
Therefore, if we move two units to the right of
the axis of symmetry, we obtain the point
$(4, -12)$ which must also be on the graph.

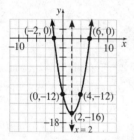

6. For the function $f(x) = -3x^2 + 12x - 7$, we see
    that $a = -3$, $b = 12$, and $c = -7$. The parabola
    opens down because $a = -3 < 0$. The
    x-coordinate of the vertex is
    $x = -\dfrac{b}{2a} = -\dfrac{12}{2(-3)} = 2$. The y-coordinate of
    the vertex is

    $$f\left(-\frac{b}{2a}\right) = f(2)$$
    $$= -3(2)^2 + 12(2) - 7$$
    $$= -12 + 24 - 7$$
    $$= 5$$

    Thus, the vertex is $(2, 5)$ and the axis of
    symmetry is the line $x = 2$. The y-intercept is
    $f(0) = -3(0)^2 + 12(0) - 7 = -7$.

    Now, $b^2 - 4ac = 12^2 - 4(-3)(-7) = 60 > 0$. The
    parabola will have two distinct x-intercepts. We
    find these by solving

    $$f(x) = 0$$
    $$-3x^2 + 12x - 7 = 0$$

$$x = \frac{-12 \pm \sqrt{60}}{2(-3)}$$
$$= \frac{-12 \pm 2\sqrt{15}}{-6}$$
$$= \frac{6 \pm \sqrt{15}}{3}$$
$$x \approx 0.71 \text{ or } x \approx 3.29$$

Finally, the y-intercept point, $(0, -7)$, is two
units to the left of the axis of symmetry.
Therefore, if we move two units to the right of
the axis of symmetry, we obtain the point
$(4, -7)$ which must also be on the graph.

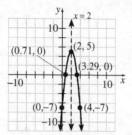

7. For the function $f(x) = x^2 + 6x + 9$, we see that
    $a = 1$, $b = 6$, and $c = 9$. The parabola opens up
    because $a = 1 > 0$. The x-coordinate of the
    vertex is $x = -\dfrac{b}{2a} = -\dfrac{6}{2(1)} = -3$. The
    y-coordinate of the vertex is

    $$f\left(-\frac{b}{2a}\right) = f(-3)$$
    $$= (-3)^2 + 6(-3) + 9$$
    $$= 9 - 18 + 9$$
    $$= 0$$

    Thus, the vertex is $(-3, 0)$ and the axis of
    symmetry is the line $x = -3$. The y-intercept is
    $f(0) = 0^2 + 6(0) + 9 = 9$. Now,
    $b^2 - 4ac = 6^2 - 4(1)(9) = 36 - 36 = 0$. Since the
    discriminant is 0, the x-coordinate of the vertex
    is the only x-intercept, $x = -3$. Finally, the y-
    intercept point, $(0, 9)$, is three units to the right
    of the axis of symmetry. Therefore, if we move
    three units to the left of the axis of symmetry, we
    obtain the point $(-6, 9)$ which must also be on
    the graph.

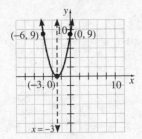

8. For the function $G(x) = -3x^2 + 9x - 8$, we see that $a = -3$, $b = 9$, and $c = -8$. The parabola opens down because $a = -3 < 0$. The x-coordinate of the vertex is

$$x = -\frac{b}{2a} = -\frac{9}{2(-3)} = \frac{9}{6} = \frac{3}{2}.$$

The y-coordinate of the vertex is

$$\begin{aligned}
G\left(-\frac{b}{2a}\right) = G\left(\frac{3}{2}\right) \\
= -3\left(\frac{3}{2}\right)^2 + 9\left(\frac{3}{2}\right) - 8 \\
= -\frac{27}{4} + \frac{27}{2} - 8 \\
= -\frac{5}{4}
\end{aligned}$$

Thus, the vertex is $\left(\frac{3}{2}, -\frac{5}{4}\right)$ and the axis of symmetry is the line $x = \frac{3}{2}$. The y-intercept is $G(0) = -3(0)^2 + 9(0) - 8 = -8$. Now,

$b^2 - 4ac = 9^2 - 4(-3)(-8) = 81 - 96 = -15$.

Since the discriminant is negative, there are no x-intercepts. Finally, the y-intercept point, $(0, -8)$, is three-halves units to the left of the axis of symmetry. Therefore, if we move three-halves units to the right of the axis of symmetry, we obtain the point $(3, -8)$ which must also be on the graph.

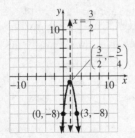

9. True. Since $a < 0$, the graph opens down. The vertex is the highest point on the graph, and therefore $f\left(-\frac{b}{2a}\right)$ is the maximum value of f.

10. If we compare $f(x) = 2x^2 - 8x + 1$ to $f(x) = ax^2 + bx + c$, we find that $a = 2$, $b = -8$, and $c = 1$. Because $a > 0$, we know the graph will open up, so the function will have a minimum value. The minimum value occurs at

$$x = -\frac{b}{2a} = -\frac{(-8)}{2(2)} = 2.$$

The minimum value of the function is

$$f\left(-\frac{b}{2a}\right) = f(2) = 2(2)^2 - 8(2) + 1 = -7.$$ So, the minimum value is -7 and it occurs when $x = 2$.

11. If we compare $G(x) = -x^2 + 10x + 8$ to $G(x) = ax^2 + bx + c$, we find that $a = -1$, $b = 10$, and $c = 8$. Because $a < 0$, we know the graph will open down, so the function will have a maximum value. The maximum value occurs at $x = -\frac{b}{2a} = -\frac{10}{2(-1)} = 5$.

The maximum value of the function is

$$G\left(-\frac{b}{2a}\right) = G(5) = -5^2 + 10(5) + 8 = 33.$$ So, the maximum value is 33 and it occurs when $x = 5$.

12. (a) We first recognize that the revenue function is a quadratic function whose graph opens down since $a = -0.5 < 0$. This means that the function indeed has a maximum value. The maximum value occurs when

$$p = -\frac{b}{2a} = -\frac{75}{2(-0.5)} = 75.$$

The revenue will be maximized when the calculators are sold at a price of $75.

(b) The maximum revenue is obtained by evaluating the revenue function at the price found in part (a).

$$R(75) = -0.5(75)^2 + 75(75) = 2812.5$$

The maximum daily revenue is $2812.50.

13. Let l = length and w = width.
The area of a rectangle is given by $A = l \cdot w$.
Before we can work on maximizing the area, we need to get the function in terms of one independent variable.
The 1000 yards of fence will form the perimeter of the rectangle. That is, we have
$2l + 2w = 1000$.
We can solve this equation for l and substitute the result in the area equation.
$2l + 2w = 1000$
$l + w = 500$
$l = 500 - w$
Thus, the area equation becomes
$$A = l \cdot w$$
$$= (500 - w) \cdot w$$
$$= -w^2 + 500w$$

The area function is a quadratic function whose graph opens down since $a = -1 < 0$. This means that the function indeed has a maximum. This

occurs when $w = -\dfrac{b}{2a} = -\dfrac{500}{2(-1)} = 250$.

The maximum area can be found by substituting this value for w in the area function.
$A = -250^2 + 500(250) = 62,500$ square yards

Since $l = 500 - w$ and $w = 250$, the length will be $l = 500 - 250 = 250$ yards.

The rectangular field will have a maximum area of 62,500 square yards when the field measures 250 yards by 250 yards.

14. Let x represent the number of boxes in excess of 30. Revenue is price times quantity. If 30 boxes of CDs are sold, the revenue will be $\$100(30)$.

If 31 boxes of CDs are sold, the revenue will be $\$99(31)$. If 32 boxes of CDs are sold, the

revenue will be $\$98(32)$. In general, if x boxes

in excess of 30 are sold, then the number of boxes will be $30 + x$, and the price per box will be $100 - x$. Thus, revenue will be
$$R(x) = (100 - x)(30 + x)$$
$$= 3000 + 100x - 30x - x^2$$
$$= -x^2 + 70x + 3000$$
Now, this revenue function is a quadratic function whose graph opens down since $a = -1 < 0$. This means that the function indeed has a maximum value. The maximum value

occurs when $x = -\dfrac{b}{2a} = -\dfrac{70}{2(-1)} = 35$.

The maximum revenue will be
$$R(35) = -35^2 + 70(35) + 3000 = 4225.$$

Now recall that x is the number of boxes in excess of 30. Therefore, $30 + 30 = 65$ boxes of CDs should be sold in order to maximize revenue. The maximum revenue will be $\$4,225$.

10.5 Exercises

15. (a) $f(x) = x^2 - 6x - 16$

$$x = -\frac{b}{2a} = -\frac{-6}{2} = 3$$

$$f(3) = (3)^2 - 6(3) - 16$$
$$= 9 - 18 - 16$$
$$= -25$$
The vertex is $(3, -25)$.

(b) $f(x) = x^2 - 6x - 16$;

$a = 1, b = -6, c = -16;$

$$b^2 - 4ac = (-6)^2 - 4(1)(-16)$$
$$= 36 + 64$$
$$= 100.$$
Because the discriminant is positive, the parabola will have two distinct x-intercepts.

The x-intercepts are:
$$f(x) = 0$$
$$x^2 - 6x - 16 = 0$$
$$(x - 8)(x + 2) = 0$$
$$x - 8 = 0 \quad \text{or} \quad x + 2 = 0$$
$$x = 8 \qquad\qquad x = -2$$

17. (a) $G(x) = -2x^2 + 4x - 5$

$$x = -\frac{b}{2a} = -\frac{4}{-4} = 1$$

$$G(1) = -2(1)^2 + 4(1) - 5$$
$$= -2 + 4 - 5$$
$$= -3$$
The vertex is $(1, -3)$.

(b) $G(x) = -2x^2 + 4x - 5$

$a = -2, b = 4, c = -5;$

$b^2 - 4ac = (4)^2 - 4(-2)(-5)$
$= 16 - 40$
$= -24$

Because the discriminant is negative, the parabola will have no x-intercepts.

19. (a) $h(x) = 4x^2 + 4x + 1$

$x = -\dfrac{b}{2a} = -\dfrac{4}{8} = -\dfrac{1}{2}$

$h\left(-\dfrac{1}{2}\right) = 4\left(-\dfrac{1}{2}\right)^2 + 4\left(-\dfrac{1}{2}\right) + 1$
$= 1 - 2 + 1$
$= 0$

The vertex is $\left(-\dfrac{1}{2}, 0\right)$.

(b) $h(x) = 4x^2 + 4x + 1$;

$a = 4, b = 4, c = 1;$

$b^2 - 4ac = 4^2 - 4(4)(1) = 16 - 16 = 0$.

Because the discriminant is zero, the parabola will have one x-intercept.

The x-intercept is:

$h(x) = 0$

$4x^2 + 4x + 1 = 0$

$(2x + 1)^2 = 0$

$2x + 1 = 0$

$2x = -1$

$x = -\dfrac{1}{2}$

21. (a) $F(x) = 4x^2 - x - 1$

$x = -\dfrac{b}{2a} = -\dfrac{-1}{8} = \dfrac{1}{8}$

$F\left(\dfrac{1}{8}\right) = 4\left(\dfrac{1}{8}\right)^2 - \left(\dfrac{1}{8}\right) - 1$

$= \dfrac{1}{16} - \dfrac{2}{16} - \dfrac{16}{16} = -\dfrac{17}{16}$

The vertex is $\left(\dfrac{1}{8}, -\dfrac{17}{16}\right)$.

(b) $F(x) = 4x^2 - x - 1$;

$a = 4, b = -1, c = -1;$

$b^2 - 4ac = (-1)^2 - 4(4)(-1) = 1 + 16 = 17$.

Because the discriminant is positive, the parabola will have two distinct x-intercepts.

The x-intercepts are:

$F(x) = 0$

$4x^2 - x - 1 = 0$

$x = \dfrac{-(-1) \pm \sqrt{(-1)^2 - 4(4)(-1)}}{2(4)}$

$= \dfrac{1 \pm \sqrt{17}}{8}$

$x \approx -0.39 \quad \text{or} \quad x \approx 0.64$

23. $f(x) = x^2 - 4x - 5$

$a = 1, b = -4, c = -5$

The graph opens up because $a > 0$.

vertex:

$x = -\dfrac{b}{2a} = -\dfrac{(-4)}{2(1)} = 2$

$f(2) = (2)^2 - 4(2) - 5 = -9$

The vertex is $(2, -9)$ and the axis of symmetry is $x = 2$.

y-intercept:

$f(0) = (0)^2 - 4(0) - 5 = -5$

x-intercepts:

$b^2 - 4ac = (-4)^2 - 4(1)(-5) = 36 > 0$

There are two distinct x-intercepts. We find these by solving

$f(x) = 0$

$x^2 - 4x - 5 = 0$

$(x - 5)(x + 1) = 0$

$x - 5 = 0 \quad \text{or} \quad x + 1 = 0$

$x = 5 \quad \text{or} \quad x = -1$

Graph:

The y-intercept point, $(0, -5)$, is two units to the left of the axis of symmetry. Therefore, if we move two units to the right of the axis of symmetry, we obtain the point $(4, -5)$ which must also be on the graph.

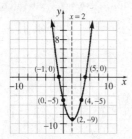

The domain is the set of all real numbers or, using interval notation, $(-\infty, \infty)$. The range is $\{y \mid y \geq -9\}$ or, using interval notation, $[-9, \infty)$.

25. $G(x) = x^2 + 12x + 32$

$a = 1, b = 12, c = 32$
The graph opens up because $a > 0$.

<u>vertex:</u>

$$x = -\frac{b}{2a} = -\frac{(12)}{2(1)} = -6$$

$$G(-6) = (-6)^2 + 12(-6) + 32 = -4$$

The vertex is $(-6, -4)$ and the axis of symmetry is $x = -6$.

<u>y-intercept:</u>

$$G(0) = (0)^2 + 12(0) + 32 = 32$$

<u>x-intercepts:</u>

$$b^2 - 4ac = (12)^2 - 4(1)(32) = 16 > 0$$

There are two distinct x-intercepts. We find these by solving

$$G(x) = 0$$
$$x^2 + 12x + 32 = 0$$
$$(x+8)(x+4) = 0$$
$$x+8 = 0 \text{ or } x+4 = 0$$
$$x = -8 \text{ or } x = -4$$

<u>Graph:</u>
The y-intercept point, $(0, 32)$, is six units to the right of the axis of symmetry. Therefore, if we move six units to the left of the axis of symmetry, we obtain the point $(-12, 32)$ which must also be on the graph.

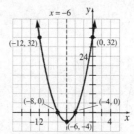

The domain is the set of all real numbers or, using interval notation, $(-\infty, \infty)$. The range is $\{y \mid y \geq -4\}$ or, using interval notation, $[-4, \infty)$.

27. $F(x) = -x^2 + 2x + 8$

$a = -1, b = 2, c = 8$
The graph opens down because $a < 0$.

<u>vertex:</u>

$$x = -\frac{b}{2a} = -\frac{(2)}{2(-1)} = 1$$

$$F(1) = -(1)^2 + 2(1) + 8 = 9$$

The vertex is $(1, 9)$ and the axis of symmetry is $x = 1$.

<u>y-intercept:</u>

$$F(0) = -(0)^2 + 2(0) + 8 = 8$$

<u>x-intercepts:</u>

$$b^2 - 4ac = (2)^2 - 4(-1)(8) = 36 > 0$$

There are two distinct x-intercepts. We find these by solving

$$F(x) = 0$$
$$-x^2 + 2x + 8 = 0$$
$$x^2 - 2x - 8 = 0$$
$$(x+2)(x-4) = 0$$
$$x+2 = 0 \text{ or } x-4 = 0$$
$$x = -2 \text{ or } x = 4$$

<u>Graph:</u>
The y-intercept point, $(0, 8)$, is one unit to the left of the axis of symmetry. Therefore, if we move one unit to the right of the axis of symmetry, we obtain the point $(2, 8)$ which must also be on the graph.

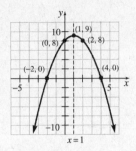

The domain is the set of all real numbers or, using interval notation, $(-\infty,\infty)$. The range is $\{y\,|\,y\le 9\}$ or, using interval notation, $(-\infty,9]$.

29. $H(x)=x^2-4x+4$

$a=1,b=-4,c=4$

The graph opens up because $a>0$.

vertex:

$$x=-\frac{b}{2a}=-\frac{(-4)}{2(1)}=2$$

$$H(2)=(2)^2-4(2)+4=0$$

The vertex is $(2,0)$ and the axis of symmetry is $x=2$.

y-intercept:

$$H(0)=(0)^2-4(0)+4=4$$

x-intercepts:
Since the discriminant is 0, the x-coordinate of the vertex is the only x-intercept, $x=2$.

Graph:
The y-intercept point, $(0,4)$, is two units to the left of the axis of symmetry. Therefore, if we move two units to the right of the axis of symmetry, we obtain the point $(4,4)$ which must also be on the graph.

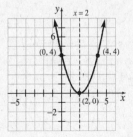

The domain is the set of all real numbers or, using interval notation, $(-\infty,\infty)$. The range is $\{y\,|\,y\ge 0\}$ or, using interval notation, $[0,\infty)$.

31. $g(x)=x^2+2x+5$

$a=1,b=2,c=5$

The graph opens up because $a>0$.

vertex:

$$x=-\frac{b}{2a}=-\frac{(2)}{2(1)}=-1$$

$$g(-1)=(-1)^2+2(-1)+5=4$$

The vertex is $(-1,4)$ and the axis of symmetry is $x=-1$.

y-intercept:

$$g(0)=(0)^2+2(0)+5=5$$

x-intercepts:

$$b^2-4ac=(2)^2-4(1)(5)=-16<0$$

There are no x-intercepts since the discriminant is negative.

Graph:
The y-intercept point, $(0,5)$, is one unit to the right of the axis of symmetry. Therefore, if we move one unit to the left of the axis of symmetry, we obtain the point $(-2,5)$ which must also be on the graph.

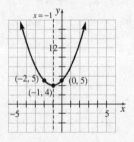

The domain is the set of all real numbers or, using interval notation, $(-\infty,\infty)$. The range is $\{y\,|\,y\ge 4\}$ or, using interval notation, $[4,\infty)$.

33. $h(x) = -x^2 - 10x - 25$

$a = -1, b = -10, c = -25$

The graph opens down because $a < 0$.

vertex:

$$x = -\frac{b}{2a} = -\frac{(-10)}{2(-1)} = -5$$

$$h(-5) = -(-5)^2 - 10(-5) - 25 = 0$$

The vertex is $(-5, 0)$ and the axis of symmetry is $x = -5$.

y-intercept:

$$h(0) = -(0)^2 - 10(0) - 25 = -25$$

x-intercepts:

$$b^2 - 4ac = (-10)^2 - 4(-1)(-25) = 100 - 100 = 0$$

Since the discriminant is 0, the x-coordinate of the vertex is the only x-intercept. $x = -5$.

Graph:

The y-intercept point, $(0, -25)$, is five units to the right of the axis of symmetry. Therefore, if we move five units to the left of the axis of symmetry, we obtain the point $(-10, -25)$ which must also be on the graph.

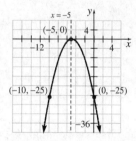

The domain is the set of all real numbers or, using interval notation, $(-\infty, \infty)$. The range is $\{y \mid y \le 0\}$ or, using interval notation, $(-\infty, 0]$.

35. $p(x) = -x^2 + 2x - 5$

$a = -1, b = 2, c = -5$

The graph opens down because $a < 0$.

vertex:

$$x = -\frac{b}{2a} = -\frac{(2)}{2(-1)} = 1$$

$$p(1) = -(1)^2 + 2(1) - 5 = -4$$

The vertex is $(1, -4)$ and the axis of symmetry is $x = 1$.

y-intercept:

$$p(0) = -(0)^2 + 2(0) - 5 = -5$$

x-intercepts:

$$b^2 - 4ac = (2)^2 - 4(-1)(-5) = -16 < 0$$

There are no x-intercepts since the discriminant is negative.

Graph:

The y-intercept point, $(0, -5)$, is one unit to the left of the axis of symmetry. Therefore, if we move one unit to the right of the axis of symmetry, we obtain the point $(2, -5)$ which must also be on the graph.

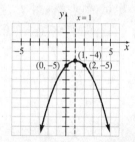

The domain is the set of all real numbers or, using interval notation, $(-\infty, \infty)$. The range is $\{y \mid y \le -4\}$ or, using interval notation, $(-\infty, -4]$.

37. $F(x) = 4x^2 - 4x - 3$

$a = 4, b = -4, c = -3$

The graph opens up because $a > 0$.

vertex:

$$x = -\frac{b}{2a} = -\frac{(-4)}{2(4)} = \frac{1}{2}$$

$$F\left(\tfrac{1}{2}\right) = 4\left(\tfrac{1}{2}\right)^2 - 4\left(\tfrac{1}{2}\right) - 3 = -4$$

The vertex is $\left(\tfrac{1}{2}, -4\right)$ and the axis of symmetry is $x = 2$.

y-intercept:

$$F(0) = 4(0)^2 - 4(0) - 3 = -3$$

x-intercepts:

$$b^2 - 4ac = (-4)^2 - 4(4)(-3) = 64 > 0$$

There are two distinct x-intercepts. We find these by solving

$$F(x) = 0$$
$$4x^2 - 4x - 3 = 0$$
$$(2x+1)(2x-3) = 0$$
$$2x+1 = 0 \quad \text{or} \quad 2x-3 = 0$$
$$2x = -1 \quad \text{or} \quad 2x = 3$$
$$x = -\frac{1}{2} \quad \text{or} \quad x = \frac{3}{2}$$

Graph:

The y-intercept point, $(0,-3)$, is $\frac{1}{2}$ unit to the left of the axis of symmetry. Therefore, if we move $\frac{1}{2}$ unit to the right of the axis of symmetry, we obtain the point $(1,-3)$ which must also be on the graph.

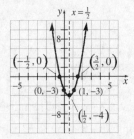

The domain is the set of all real numbers or, using interval notation, $(-\infty, \infty)$. The range is $\{y \mid y \geq -4\}$ or, using interval notation, $[-4, \infty)$.

39. $G(x) = -9x^2 + 18x + 7$

$a = -9, b = 18, c = 7$

The graph opens down because $a < 0$.

vertex:

$$x = -\frac{b}{2a} = -\frac{(18)}{2(-9)} = 1$$

$$G(1) = -9(1)^2 + 18(1) + 7 = 16$$

The vertex is $(1,16)$ and the axis of symmetry is $x = 1$.

y-intercept:

$$G(0) = -9(0)^2 + 18(0) + 7 = 7$$

x-intercepts:

$$b^2 - 4ac = (18)^2 - 4(-9)(7) = 576 > 0$$

There are two distinct x-intercepts. We find these by solving

$$G(x) = 0$$
$$-9x^2 + 18x + 7 = 0$$
$$9x^2 - 18x - 7 = 0$$
$$(3x+1)(3x-7) = 0$$
$$3x+1 = 0 \quad \text{or} \quad 3x-7 = 0$$
$$3x = -1 \quad \text{or} \quad 3x = 7$$
$$x = -\frac{1}{3} \quad \text{or} \quad x = \frac{7}{3}$$

Graph:

The y-intercept point, $(0,7)$, is one unit to the left of the axis of symmetry. Therefore, if we move one unit to the right of the axis of symmetry, we obtain the point $(2,7)$ which must also be on the graph.

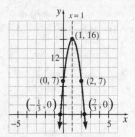

The domain is the set of all real numbers or, using interval notation, $(-\infty, \infty)$. The range is $\{y \mid y \leq 16\}$ or, using interval notation, $(-\infty, 16]$.

41. $H(x) = 4x^2 - 4x + 1$

$a = 4, b = -4, c = 1$

The graph opens up because $a > 0$.

vertex:

$$x = -\frac{b}{2a} = -\frac{(-4)}{2(4)} = \frac{1}{2}$$

$$H\left(\frac{1}{2}\right) = 4\left(\frac{1}{2}\right)^2 - 4\left(\frac{1}{2}\right) + 1 = 0$$

The vertex is $\left(\frac{1}{2},0\right)$ and the axis of symmetry is $x=\frac{1}{2}$.

y-intercept:

$H(0)=4(0)^2-4(0)+1=1$

x-intercepts:

$b^2-4ac=(-4)^2-4(4)(1)=0$

Since the discriminant is 0, the x-coordinate of the vertex is the only x-intercept, $x=\frac{1}{2}$.

Graph:

The y-intercept point, $(0,1)$, is $\frac{1}{2}$ unit to the left of the axis of symmetry. If we move $\frac{1}{2}$ unit to the right of the axis of symmetry, we obtain the point $(1,1)$ which must also be on the graph.

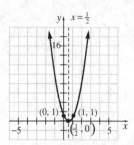

The domain is the set of all real numbers or, using interval notation, $(-\infty,\infty)$. The range is $\{y\,|\,y\ge 0\}$ or, using interval notation, $[0,\infty)$.

43. $f(x)=-16x^2-24x-9$

$a=-16, b=-24, c=-9$

The graph opens down because $a<0$.

vertex:

$x=-\dfrac{b}{2a}=-\dfrac{(-24)}{2(-16)}=-\dfrac{3}{4}$

$f\left(-\frac{3}{4}\right)=-16\left(-\frac{3}{4}\right)^2-24\left(-\frac{3}{4}\right)-9=0$

The vertex is $\left(-\frac{3}{4},0\right)$ and the axis of symmetry is $x=-\frac{3}{4}$.

y-intercept:

$f(0)=-16(0)^2-24(0)-9=-9$

x-intercepts:

$b^2-4ac=(-24)^2-4(-16)(-9)=0$

Since the discriminant is 0, the only x-intercept is the x-coordinate of the vertex, $x=-\frac{3}{4}$.

Graph:

The y-intercept point, $(0,-9)$, is $\frac{3}{4}$ unit to the right of the axis of symmetry. Therefore, if we move $\frac{3}{4}$ unit to the left of the axis of symmetry, we obtain the point $\left(-\frac{3}{2},-9\right)$ which must also be on the graph.

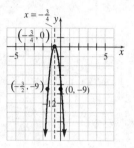

The domain is the set of all real numbers or, using interval notation, $(-\infty,\infty)$. The range is $\{y\,|\,y\le 0\}$ or, using interval notation, $(-\infty,0]$.

45. $f(x)=2x^2+8x+11$

$a=2, b=8, c=11$

The graph opens up because $a>0$.

vertex:

$x=-\dfrac{b}{2a}=-\dfrac{(8)}{2(2)}=-2$

$f(-2)=2(-2)^2+8(-2)+11=3$

The vertex is $(-2,3)$ and the axis of symmetry is $x=-2$.

y-intercept:

$f(0)=2(0)^2+8(0)+11=11$

x-intercepts:

$b^2-4ac=(8)^2-4(2)(11)=-24<0$

There are no x-intercepts since the discriminant is negative.

Graph:

The y-intercept point, $(0,11)$, is two units to the right of the axis of symmetry. Therefore, if we move two unit to the left of the axis of symmetry, we obtain the point $(-4,11)$ which must also be on the graph.

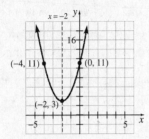

The domain is the set of all real numbers or, using interval notation, $(-\infty,\infty)$. The range is $\{y \mid y \geq 3\}$ or, using interval notation, $[3,\infty)$.

47. $P(x) = -4x^2 + 6x - 3$

$a = -4, b = 6, c = -3$

The graph opens down because $a < 0$.

vertex:

$$x = -\frac{b}{2a} = -\frac{(6)}{2(-4)} = \frac{3}{4}$$

$$P\left(\tfrac{3}{4}\right) = -4\left(\tfrac{3}{4}\right)^2 + 6\left(\tfrac{3}{4}\right) - 3 = -\frac{3}{4}$$

The vertex is $\left(\tfrac{3}{4}, -\tfrac{3}{4}\right)$ and the axis of symmetry is $x = \tfrac{3}{4}$.

y-intercept:

$$P(0) = -4(0)^2 + 6(0) - 3 = -3$$

x-intercepts:

$$b^2 - 4ac = (6)^2 - 4(-4)(-3) = -12 < 0$$

There are no x-intercepts since the discriminant is negative.

Graph:

The y-intercept point, $(0,-3)$, is $\tfrac{3}{4}$ unit to the left of the axis of symmetry. Therefore, if we move $\tfrac{3}{4}$ unit to the right of the axis of symmetry, we obtain the point $\left(\tfrac{3}{2}, -3\right)$ which must also be on the graph.

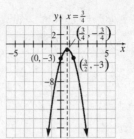

The domain is the set of all real numbers or, using interval notation, $(-\infty,\infty)$. The range is $\left\{y \mid y \leq -\dfrac{3}{4}\right\}$ or, using interval notation, $\left(-\infty, -\dfrac{3}{4}\right]$.

49. $h(x) = x^2 + 5x + 3$

$a = 1, b = 5, c = 3$

The graph opens $a > 0$.

vertex:

$$x = -\frac{b}{2a} = -\frac{(5)}{2(1)} = -\frac{5}{2}$$

$$h\left(-\tfrac{5}{2}\right) = \left(-\tfrac{5}{2}\right)^2 + 5\left(-\tfrac{5}{2}\right) + 3 = -\frac{13}{4}$$

The vertex is $\left(-\tfrac{5}{2}, -\tfrac{13}{4}\right)$ and the axis of symmetry is $x = -\tfrac{5}{2}$.

y-intercept:

$$h(0) = (0)^2 + 5(0) + 3 = 3$$

x-intercepts:

$$b^2 - 4ac = (5)^2 - 4(1)(3) = 13 > 0$$

There are two distinct x-intercepts. We find these by solving

$$h(x) = 0$$

$$x^2 + 5x + 3 = 0$$

$$x = \frac{-5 \pm \sqrt{13}}{2}$$

$$x \approx -0.70 \quad \text{or} \quad x \approx -4.30$$

Graph:

The y-intercept point, $(0,3)$, is $\tfrac{5}{2}$ units to the right of the axis of symmetry. Therefore, if we move $\tfrac{5}{2}$ units to the left of the axis of symmetry, we obtain the point $(-5, 3)$ which must also be on the graph.

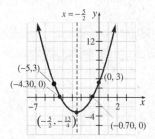

The domain is the set of all real numbers or, using interval notation, $(-\infty, \infty)$. The range is

$$\left\{ y \mid y \geq -\frac{13}{4} \right\} \text{ or, using interval notation,}$$

$$\left[-\frac{13}{4}, \infty \right).$$

51. $G(x) = -3x^2 + 8x + 2$

$a = -3, b = 8, c = 2$

The graph opens down because $a < 0$.

vertex:

$$x = -\frac{b}{2a} = -\frac{(8)}{2(-3)} = \frac{4}{3}$$

$$G\left(\frac{4}{3}\right) = -3\left(\frac{4}{3}\right)^2 + 8\left(\frac{4}{3}\right) + 2 = \frac{22}{3}$$

The vertex is $\left(\frac{4}{3}, \frac{22}{3}\right)$ and the axis of symmetry is $x = \frac{4}{3}$.

y-intercept:

$$G(0) = -3(0)^2 + 8(0) + 2 = 2$$

x-intercepts:

$$b^2 - 4ac = (8)^2 - 4(-3)(2) = 88 > 0$$

There are two distinct x-intercepts. We find these by solving

$$G(x) = 0$$

$$-3x^2 + 8x + 2 = 0$$

$$x = \frac{-8 \pm \sqrt{88}}{2(-3)} = \frac{-8 \pm 2\sqrt{22}}{-6} = \frac{4 \pm \sqrt{22}}{3}$$

$$x \approx -0.23 \text{ or } x \approx 2.90$$

Graph:

The y-intercept point, $(0, 2)$, is $\frac{4}{3}$ units to the left of the axis of symmetry. Therefore, if we move $\frac{4}{3}$ units to the right of the axis of symmetry, we obtain the point $\left(\frac{8}{3}, 2\right)$ which must also be on the graph.

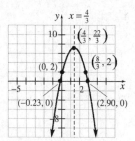

The domain is the set of all real numbers or, using interval notation, $(-\infty, \infty)$. The range is

$$\left\{ y \mid y \leq \frac{22}{3} \right\} \text{ or, using interval notation,}$$

$$\left(-\infty, \frac{22}{3} \right].$$

53. $f(x) = 5x^2 - 5x + 2$

$a = 5, b = -5, c = 2$

The graph opens up because $a > 0$.

vertex:

$$x = -\frac{b}{2a} = -\frac{(-5)}{2(5)} = \frac{1}{2}$$

$$f\left(\frac{1}{2}\right) = 5\left(\frac{1}{2}\right)^2 - 5\left(\frac{1}{2}\right) + 2 = \frac{3}{4}$$

The vertex is $\left(\frac{1}{2}, \frac{3}{4}\right)$ and the axis of symmetry is $x = \frac{1}{2}$.

y-intercept:

$$f(0) = 5(0)^2 - 5(0) + 2 = 2$$

x-intercepts:

$$b^2 - 4ac = (-5)^2 - 4(5)(2) = -15 < 0$$

There are no x-intercepts since the discriminant is negative.

Graph:

The y-intercept point, $(0,2)$, is $\frac{1}{2}$ unit to the left of the axis of symmetry. Therefore, if we move one unit to the right of the axis of symmetry, we obtain the point $(1,2)$ which must also be on the graph.

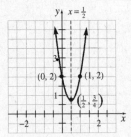

The domain is the set of all real numbers or, using interval notation, $(-\infty, \infty)$. The range is $\left\{ y \mid y \geq \frac{3}{4} \right\}$ or, using interval notation, $\left[\frac{3}{4}, \infty \right)$.

55. $H(x) = -3x^2 + 6x$

$a = -3, b = 6, c = 0$

The graph opens down because $a < 0$.

vertex:

$$x = -\frac{b}{2a} = -\frac{(6)}{2(-3)} = 1$$

$$H(1) = -3(1)^2 + 6(1) = 3$$

The vertex is $(1,3)$ and the axis of symmetry is $x = 1$.

y-intercept:

$$H(0) = -3(0)^2 + 6(0) = 0$$

x-intercepts:

$$b^2 - 4ac = (6)^2 - 4(-3)(0) = 36 > 0$$

There are two distinct x-intercepts. We find these by solving

$$H(x) = 0$$
$$-3x^2 + 6x = 0$$
$$-3x(x - 2) = 0$$
$$-3x = 0 \text{ or } x - 2 = 0$$
$$x = 0 \text{ or } x = 2$$

Graph:

The y-intercept point, $(0,0)$, is one unit to the left of the axis of symmetry. Therefore, if we move one unit to the right of the axis of symmetry, we obtain the point $(2,0)$ which must also be on the graph (these points are actually the x-intercept points).

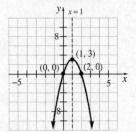

The domain is the set of all real numbers or, using interval notation, $(-\infty, \infty)$. The range is $\{ y \mid y \leq 3 \}$ or, using interval notation, $(-\infty, 3]$.

57. $f(x) = x^2 - \frac{5}{2}x - \frac{3}{2}$

$a = 1, b = -\frac{5}{2}, c = -\frac{3}{2}$

The graph opens up because $a > 0$.

vertex:

$$x = -\frac{b}{2a} = -\frac{\left(-\frac{5}{2}\right)}{2(1)} = \frac{5}{4}$$

$$f\left(\frac{5}{4}\right) = \left(\frac{5}{4}\right)^2 - \frac{5}{2}\left(\frac{5}{4}\right) - \frac{3}{2} = -\frac{49}{16}$$

The vertex is $\left(\frac{5}{4}, -\frac{49}{16}\right)$ and the axis of symmetry is $x = \frac{5}{4}$.

y-intercept:

$$f(0) = (0)^2 - \frac{5}{2}(0) - \frac{3}{2} = -\frac{3}{2}$$

x-intercepts:

$$b^2 - 4ac = \left(-\frac{5}{2}\right)^2 - 4(1)\left(-\frac{3}{2}\right) = \frac{49}{4} > 0$$

There are two distinct x-intercepts. We find these by solving

$$f(x) = 0$$

$$x^2 - \frac{5}{2}x - \frac{3}{2} = 0$$

$$2x^2 - 5x - 3 = 0$$

$$(2x+1)(x-3) = 0$$

$$2x+1 = 0 \text{ or } x-3 = 0$$

$$2x = -1 \text{ or } x = 3$$

$$x = -\frac{1}{2} \text{ or } x = 3$$

Graph:

The y-intercept point, $\left(0, -\frac{3}{2}\right)$, is $\frac{5}{4}$ units to the left of the axis of symmetry. Therefore, if we move $\frac{5}{4}$ units to the right of the axis of symmetry, we obtain the point $\left(\frac{5}{2}, -\frac{3}{2}\right)$ which must also be on the graph.

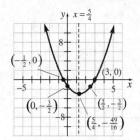

The domain is the set of all real numbers or, using interval notation, $(-\infty, \infty)$. The range is $\left\{y \mid y \geq -\frac{49}{16}\right\}$ or, using interval notation, $\left[-\frac{49}{16}, \infty\right)$.

59. $G(x) = \frac{1}{2}x^2 + 2x - 6$

$$a = \frac{1}{2}, b = 2, c = -6$$

The graph opens up because $a > 0$.

vertex:

$$x = -\frac{b}{2a} = -\frac{(2)}{2\left(\frac{1}{2}\right)} = -2$$

$$G(-2) = \frac{1}{2}(-2)^2 + 2(-2) - 6 = -8$$

The vertex is $(-2, -8)$ and the axis of symmetry is $x = -2$.

y-intercept:

$$G(0) = \frac{1}{2}(0)^2 + 2(0) - 6 = -6$$

x-intercepts:

$$b^2 - 4ac = (2)^2 - 4\left(\frac{1}{2}\right)(-6) = 16 > 0$$

There are two distinct x-intercepts. We find these by solving

$$G(x) = 0$$

$$\frac{1}{2}x + 2x - 6 = 0$$

$$x^2 + 4x - 12 = 0$$

$$(x+6)(x-2) = 0$$

$$x+6 = 0 \text{ or } x-2 = 0$$

$$x = -6 \text{ or } x = 2$$

Graph:

The y-intercept point, $(0, -6)$, is two units to the right of the axis of symmetry. Therefore, if we move two units to the left of the axis of symmetry, we obtain the point $(-4, -6)$ which must also be on the graph.

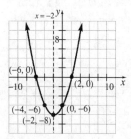

The domain is the set of all real numbers or, using interval notation, $(-\infty, \infty)$. The range is $\{y \mid y \geq -8\}$ or, using interval notation, $[-8, \infty)$.

61. $F(x) = -\frac{1}{4}x^2 + x + 15$

$$a = -\frac{1}{4}, b = 1, c = 15$$

The graph opens down because $a < 0$.

vertex:

$$x = -\frac{b}{2a} = -\frac{(1)}{2\left(-\frac{1}{4}\right)} = 2$$

$$F(2) = -\frac{1}{4}(2)^2 + (2) + 15 = 16$$

The vertex is $(2,16)$ and the axis of symmetry is $x = 2$.

y-intercept:
$$F(0) = -\frac{1}{4}(0)^2 + (0) + 15 = 15$$

x-intercepts:
$$b^2 - 4ac = (1)^2 - 4\left(-\frac{1}{4}\right)(15) = 16 > 0$$

There are two distinct *x*-intercepts. We find these by solving
$$F(x) = 0$$
$$-\frac{1}{4}x^2 + x + 15 = 0$$
$$x^2 - 4x - 60 = 0$$
$$(x-10)(x+6) = 0$$
$$x - 10 = 0 \ \text{ or } \ x + 6 = 0$$
$$x = 10 \ \text{ or } \ x = -6$$

Graph:
The *y*-intercept point, $(0,15)$, is two units to the left of the axis of symmetry. Therefore, if we move two units to the right of the axis of symmetry, we obtain the point $(4,15)$ which must also be on the graph.

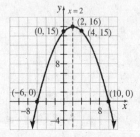

The domain is the set of all real numbers or, using interval notation, $(-\infty, \infty)$. The range is $\{y \mid y \le 16\}$ or, using interval notation, $(-\infty, 16]$.

63. If we compare $f(x) = x^2 + 8x + 13$ to

$f(x) = ax^2 + bx + c$, we find that $a = 1$, $b = 8$, and $c = 13$. Because $a > 0$, we know the graph will open up, so the function will have a minimum value.
The minimum value occurs at
$$x = -\frac{b}{2a} = -\frac{8}{2(1)} = -4.$$

The minimum value is
$$f(-4) = (-4)^2 + 8(-4) + 13 = -3.$$
So, the minimum value is -3 and it occurs when $x = -4$.

65. If we compare $G(x) = -x^2 - 10x + 3$ to

$G(x) = ax^2 + bx + c$, we find that $a = -1$, $b = -10$, and $c = 3$. Because $a < 0$, we know the graph will open down, so the function will have a maximum value.
The maximum value occurs at
$$x = -\frac{b}{2a} = -\frac{(-10)}{2(-1)} = -5.$$
The maximum value is
$$G(-5) = -(-5)^2 - 10(-5) + 3 = 28.$$
So, the maximum value is 28 and it occurs when $x = -5$.

67. If we compare $F(x) = -2x^2 + 12x + 5$ to

$F(x) = ax^2 + bx + c$, we find that $a = -2$, $b = 12$, and $c = 5$. Because $a < 0$, we know the graph will open down, so the function will have a maximum value.
The maximum value occurs at
$$x = -\frac{b}{2a} = -\frac{12}{2(-2)} = 3.$$
The maximum value is
$$F(3) = -2(3)^2 + 12(3) + 5 = 23.$$
So, the maximum value is 23 and it occurs when $x = 3$.

69. If we compare $h(x) = 4x^2 + 16x - 3$ to

$h(x) = ax^2 + bx + c$, we find that $a = 4$, $b = 16$, and $c = -3$. Because $a > 0$, we know the graph will open up, so the function will have a minimum value.
The minimum value occurs at
$$x = -\frac{b}{2a} = -\frac{16}{2(4)} = -2.$$
The minimum value is
$$h(-2) = 4(-2)^2 + 16(-2) - 3 = -19.$$
So, the minimum value is -19 and it occurs when $x = -2$.

71. If we compare $f(x) = 2x^2 - 5x + 1$ to

$f(x) = ax^2 + bx + c$, we find that $a = 2$,

$b = -5$, and $c = 1$. Because $a > 0$, we know the graph will open up, so the function will have a minimum value.
The minimum value occurs at

$$x = -\frac{b}{2a} = -\frac{(-5)}{2(2)} = \frac{5}{4} \text{ or } 1.25.$$

The minimum value is

$$f\left(\frac{5}{4}\right) = 2\left(\frac{5}{4}\right)^2 - 5\left(\frac{5}{4}\right) + 1 = -\frac{17}{8} \text{ or } -2.125.$$

So, the minimum value is $-\frac{17}{8}$ and it occurs

when $x = \frac{5}{4}$.

73. If we compare $H(x) = -3x^2 + 4x + 1$ to

$H(x) = ax^2 + bx + c$, we find that $a = -3$,

$b = 4$, and $c = 1$. Because $a < 0$, we know the graph will open down, so the function will have a maximum value.
The maximum value occurs at

$$x = -\frac{b}{2a} = -\frac{4}{2(-3)} = \frac{2}{3}.$$

The maximum value is

$$H\left(\frac{2}{3}\right) = -3\left(\frac{2}{3}\right)^2 + 4\left(\frac{2}{3}\right) + 1 = \frac{7}{3}.$$

So, the maximum value is $\frac{7}{3}$ and it occurs

when $x = \frac{2}{3}$.

75. (a) We first recognize that the quadratic function has the leading coefficient $a = -2.5 < 0$. This means that the graph will open down and the function indeed has a maximum value. The maximum value

occurs when $p = -\frac{b}{2a} = -\frac{600}{2(-2.5)} = 120$.

The revenue will be maximized when the DVD players are sold at a price of $120.

(b) The maximum revenue is obtained by evaluating the revenue function at the price found in part (a).

$$R(120) = -2.5(120)^2 + 600(120) = 36,000$$

The maximum revenue is $36,000.

77. First we recognize that the quadratic function has the leading coefficient $a = 0.05 > 0$. This means that the graph will open up and the function will indeed have a minimum value. The minimum

value occurs when $x = -\frac{b}{2a} = -\frac{(-6)}{2(0.05)} = 60$.

The marginal cost will be minimized when 60 digital cameras are produced.
To find the minimum marginal cost, we evaluate the marginal cost function for $x = 60$.

$$C(60) = 0.05(60)^2 - 6(60) + 215 = 35$$

The minimum marginal cost is $35.

79. (a) First we recognize that the quadratic function has the negative leading coefficient $a = -16 < 0$. This means the graph will open down and the function will indeed have a maximum.
The maximum height will occur when

$$t = -\frac{b}{2a} = -\frac{240}{2(-16)} = 7.5.$$

The pumpkin will reach a maximum height after 7.5 seconds.

(b) The maximum height can be found by evaluating $s(t)$ for the value of t found in part (a).

$$s(7.5) = -16(7.5)^2 + 240(7.5) + 10 = 910$$

The pumpkin will reach a maximum height of 910 feet.

(c) When the pumpkin is on the ground, it will have a height of 0 feet. Thus, we need to solve $s(t) = 0$.

$$-16t^2 + 240t + 10 = 0$$
$$8t^2 - 120t - 5 = 0$$

$$a = 8, b = -120, c = -5$$

$$t = \frac{-(-120) \pm \sqrt{(-120)^2 - 4(8)(-5)}}{2(8)}$$

$$= \frac{120 \pm \sqrt{14560}}{16}$$

$t \approx -0.042 \text{ or } t \approx 15.042$

Since the time of flight cannot be negative, we discard the negative solution. The pumpkin will hit the ground in about 15.042 seconds.

81. (a) We first recognize that the quadratic function has the leading coefficient

$a = \dfrac{-32}{335^2} < 0$. This means the graph will open down and the function will indeed have a maximum.

The maximum height will occur when

$x = -\dfrac{b}{2a} = -\dfrac{1}{2\left(-32/335^2\right)} \approx 1753.52$.

The pumpkin will reach a maximum height when it is about 1753.52 feet from the cannon.

(b) The maximum height is obtained by evaluating $h(x)$ for the value of x found in part (a).

$h(1753.52) = \dfrac{-32}{335^2}(1753.52)^2 + (1753.52) + 10$
≈ 886.76

The pumpkin will reach a maximum height of about 886.76 feet.

(c) When the pumpkin is on the ground, it will have a height of 0 feet. Thus, we need to solve $h(x) = 0$.

$\dfrac{-32}{335^2}x^2 + x + 10 = 0$

$a = \dfrac{-32}{335^2}$, $b = 1$, $c = 10$

$x = \dfrac{-1 \pm \sqrt{1^2 - 4\left(\dfrac{-32}{335^2}\right)(10)}}{2\left(\dfrac{-32}{335^2}\right)}$

$x \approx -9.97$ or $x \approx 3517.00$

Since the horizontal distance traveled cannot be negative, we discard the negative result. The pumpkin will hit the ground at a distance of about 3517 feet from the cannon.

(d) The two answers are close. The difference is because the initial velocity component used in the formula for problem 73 was an approximation while the value used in the formula for this problem was exact. The exact formula for problem 73 would be

$s(t) = -16t^2 + \dfrac{335}{\sqrt{2}}t + 10$.

If we had used this formula, the two results would be the same.

Note: The answer in part (b) of this problem is also an approximation because we approximated the time in part (a).

83. (a) We first recognize that the quadratic function has the leading coefficient $a = -55 < 0$. This means the graph will open down and the function will indeed have a maximum.

The maximum value occurs when

$x = -\dfrac{5119}{2(-55)} = \dfrac{5119}{110} \approx 46.5$.

Average income will be maximized at an age of about 46.5 years.

(b) To determine the maximum average income, we evaluate $I(a)$ for the value of a found in part (a).

$I(46.5)$
$= -55(46.5)^2 + 5119(46.5) - 54,448$
$= 64,661.75$

The maximum average income is about $64,661.75.

85. Let x represent the first number. Then the second number must be $36 - x$. We can express the product of the two numbers as the function

$p(x) = x(36 - x) = -x^2 + 36x$

This is a quadratic function with $a = -1$, $b = 36$, and $c = 0$. The function is maximized when $x = -\dfrac{b}{2a} = -\dfrac{36}{2(-1)} = 18$.

The maximum product can be obtained by evaluating $p(x)$ when $x = 18$.

$p(x) = 18(36 - 18) = 18(18) = 324$

Two numbers that sum to 36 have a maximum product of 324 when both numbers are 18.

87. Let x represent the smaller number. Then the larger number must be $x + 18$. We can express the product of the two numbers as the function

$p(x) = x(x + 18) = x^2 + 18x$.

This is a quadratic function with $a = 1$, $b = 18$, and $c = 0$. The product will be a minimum when

$x = -\dfrac{b}{2a} = -\dfrac{18}{2(1)} = -9$.

The minimum product can be found by evaluating $p(x)$ when $x = -9$.

$p(-9) = -9(-9+18) = -9(9) = -81$

Two numbers whose difference is 18 have a minimum product of -81 when the smaller number is -9 and the larger number is 9.

89. Let l = length and w = width.
The area of a rectangle is given by $A = l \cdot w$. Before we can work on maximizing area, we need to get the equation in terms of one independent variable.
The 500 yards of fencing will form the perimeter of the rectangle. That is, we have
$2l + 2w = 500$.
We can solve this equation for l and substitute the result in the area equation.
$2l + 2w = 500$
$l + w = 250$
$l = 250 - w$
Thus, the area equation becomes
$A = l \cdot w$
$ = (250 - w) \cdot w$
$ = -w^2 + 250w$

Since $a = -1 < 0$, we know the graph opens down, so there will be a maximum area. This

occurs when $w = -\dfrac{b}{2a} = -\dfrac{250}{2(-1)} = 125$.

The maximum area can be found by substituting this value for w in the area equation.
$A = 125(250 - 125) = 125(125) = 15,625$

The rectangular field will have a maximum area of 15,625 square yards when the field measures 125 yards $\times$ 125 yards.

91. The area of a rectangular region is the product of the length and width. From the figure, we can see that the area would be
$A = (2000 - 2x) \cdot x = -2x^2 + 2000x$

Since $a = -2 < 0$, we know the graph opens down and there will be a maximum area. This value occurs when

$x = -\dfrac{b}{2a} = -\dfrac{2000}{2(-2)} = 500$.

The maximum area can be found by substituting this value for x into the area equation.
$A = 500(2000 - 2(500)) = 500,000$

The rectangular field will have a maximum area of 500,000 square meters when the field measures 500 m $\times$ 1000 m and the long side is parallel to the river.

93. From the diagram, we can see that the cross-sectional area will be rectangular with a width of x inches and a length of $20 - 2x$ inches. The cross-sectional area is then
$A = (20 - 2x) \cdot x = -2x^2 + 20x$

Since $a = -2 < 0$, the graph will open down and the function will have a maximum value. This value occurs when

$x = -\dfrac{b}{2a} = -\dfrac{20}{2(-2)} = 5$

The maximum cross-sectional area can be found by substituting this value for x in the area equation.
$A = (20 - 2(5)) \cdot 5 = 10 \cdot 5 = 50$

The gutter will have a maximum cross-sectional area of 50 square inches if the gutter has a depth of 5 inches.

95. (a) Since $R = x \cdot p$, we have
$R = (-p + 110) \cdot p = -p^2 + 110p$

(b) The revenue function is quadratic with $a = -1 < 0$. This means the graph will open down and the function will have a maximum value. This value occurs when

$p = -\dfrac{b}{2a} = -\dfrac{110}{2(-1)} = 55$.

The maximum revenue can be found by substituting this value for p in the revenue equation.

$R = (-(55) + 110) \cdot (55) = 55 \cdot 55 = 3025$

There will be a maximum revenue of \$3025 if the price is set at \$55 for each pair of jeans.

(c) To determine how many pairs of jeans will be sold, we substitute the maximizing price into the demand equation.
$x = -p + 110$
$ = -(55) + 110 = 55$

When the price per pair is \$55, the department store will sell 55 pairs.

97. (a) $a = 1$:
$f(x) = 1(x-2)(x-6) = x^2 - 8x + 12$

$a = 2$:
$f(x) = 2(x-2)(x-6) = 2x^2 - 16x + 24$

$a = -2$:

$$f(x) = -2(x-2)(x-6) = -2x^2 + 16x - 24$$

(b) The value of a has no effect on the x-intercepts. These depend only on the factors and are given in the problem.
The value of a does have an effect on the y-intercept which can be expressed as
$c = a \cdot r_1 \cdot r_2 = 12a$.

(c) The value of a has no effect on the axis of symmetry. The axis of symmetry lies halfway between the two x-intercepts which are fixed. Note that in this case we have

$$f(x) = ax^2 - 8ax + 12a$$

The axis of symmetry would be

$$x = -\frac{(-8a)}{2(a)} = \frac{8a}{2a} = 4$$

which does not depend on a.

(d) Consider the general function in this case written in the form $f(x) = a(x-h)^2 + k$.

$$\begin{aligned} f(x) &= ax^2 - 8ax + 12a \\ &= a\left(x^2 - 8x\right) + 12a \\ &= a\left(x^2 - 8x + 16\right) + 12a - 16a \\ &= a(x-4)^2 - 4a \end{aligned}$$

The x-coordinate of the vertex is 4, which does not depend on a. However, the y-coordinate is $-4a$ which does depend on a.

99. Answers may vary. One possibility follows:

If the discriminant $b^2 - 4ac > 0$, the graph of the quadratic function will have two different x-intercepts. If the discriminant $b^2 - 4ac = 0$, the graph will have one x-intercept. If the discriminant $b^2 - 4ac < 0$, the graph will not have any x-intercepts.

101. A revenue of \$0 when the price charged is some positive number is in an indication that the price was too high to keep any consumer demand. Regardless of the price charged, if no items are sold, then no revenue can be generated.

103. $G(x) = \frac{1}{4}x - 2$

Let $x = -4,\ 0,$ and 4.

$$G(-4) = \frac{1}{4}(-4) - 2 = -1 - 2 = -3$$

$$G(0) = \frac{1}{4}(0) - 2 = 0 - 2 = -2$$

$$G(4) = \frac{1}{4}(4) - 2 = 1 - 2 = -1$$

Thus, the points $(-4, -3)$, $(0, -2)$, and $(4, -1)$ are on the graph.

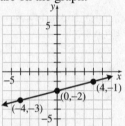

105. $f(x) = (x+2)^2 + 4$

Let $x = -4,\ -3,\ -2,\ -1,$ and 0.

$$f(-4) = (-4+2)^2 + 4 = (-2)^2 + 4 = 4 + 4 = 8$$

$$f(-3) = (-3+2)^2 + 4 = (-1)^2 + 4 = 1 + 4 = 5$$

$$f(-2) = (-2+2)^2 + 4 = (0)^2 + 4 = 0 + 4 = 4$$

$$f(-1) = (-1+2)^2 + 4 = (1)^2 + 4 = 1 + 4 = 5$$

$$f(0) = (0+2)^2 + 4 = (2)^2 + 4 = 4 + 4 = 8$$

Thus, the points $(-4, 8)$, $(-3, 5)$, $(-2, 4)$, $(-1, 5)$, and $(0, 8)$ are on the graph.

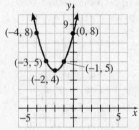

107. Vertex: $(3.5, -9.25)$

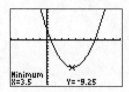

109. Vertex: $(3.5, 37.5)$

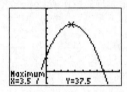

111. Vertex: $(-0.3, -20.45)$

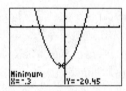

113. Vertex: $(0.67, 4.78)$

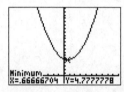

115. c is the y-intercept. In terms of transformations, c is also 1 larger than the vertical shift and thus, 1 larger than the y-coordinate of the vertex. This is because we can write:

$$f(x) = x^2 + 2x + c$$
$$= (x^2 + 2x) + c$$
$$= (x^2 + 2x + 1) + c - 1$$
$$= (x+1)^2 + (c-1)$$

 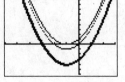

Section 10.6

Preparing for Quadratic Inequalities

P1. The inequality $-4 \le x < 5$ in interval notation is $[-4, 5)$.

P2.
$$3x + 5 > 5x - 3$$
$$3x + 5 - 5x > 5x - 3 - 5x$$
$$-2x + 5 > -3$$
$$-2x + 5 - 5 > -3 - 5$$
$$-2x > -8$$
$$\frac{-2x}{-2} < \frac{-8}{-2}$$
$$x < 4$$

The solution set is $\{x \mid x < 4\}$ or, using interval notation, $(-\infty, 4)$.

Section 10.6 Quick Checks

1. We graph $f(x) = x^2 + 3x - 10$. We see that $a = 1$, $b = 3$, and $c = -10$. The parabola opens up because $a = 1 > 0$. The x-coordinate of the vertex is $x = -\dfrac{b}{2a} = -\dfrac{3}{2(1)} = -\dfrac{3}{2}$. The y-coordinate of the vertex is

$$f\left(-\frac{b}{2a}\right) = f\left(-\frac{3}{2}\right)$$
$$= \left(-\frac{3}{2}\right)^2 + 3\left(-\frac{3}{2}\right) - 10$$
$$= \frac{9}{4} - \frac{9}{2} - 10$$
$$= -\frac{49}{4}$$

Thus, the vertex is $\left(-\dfrac{3}{2}, -\dfrac{49}{4}\right)$ and the axis of symmetry is the line $x = -\dfrac{3}{2}$.

The y-intercept is $f(0) = 0^2 + 3(0) - 10 = -10$.

Now, $b^2 - 4ac = 3^2 - 4(1)(-10) = 49 > 0$. The parabola will have two distinct x-intercepts. We find these by solving

$$f(x) = 0$$
$$x^2 + 3x - 10 = 0$$
$$(x+5)(x-2) = 0$$
$$x + 5 = 0 \quad \text{or} \quad x - 2 = 0$$
$$x = -5 \quad \text{or} \quad x = 2$$

Finally, the y-intercept point, $(0,-10)$, is three-haves units to the right of the axis of symmetry. Therefore, if we move three-halves units to the left of the axis of symmetry, we obtain the point $(-3,-10)$ which must also be on the graph.

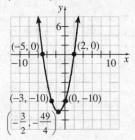

From the graph, we can see that

$f(x) = x^2 + 3x - 10$ is greater than 0 for $x < -5$

or $x > 2$. Because the inequality is non-strict, we include the x-intercepts in the solution. So, the solution is $\{x \mid x \leq -5 \text{ or } x \geq 2\}$ using set-builder notation; the solution is

$(-\infty, -5] \cup [2, \infty)$ using interval notation.

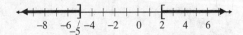

2. $x^2 + 3x - 10 \geq 0$

Solve: $x^2 + 3x - 10 = 0$

$(x-2)(x+5) = 0$

$x - 2 = 0 \quad \text{or} \quad x + 5 = 0$

$x = 2 \quad \text{or} \quad x = -5$

Determine where each factor is positive and negative and where the product of these factors is positive and negative.

Interval	$(-\infty, -5)$	-5	$(-5, 2)$	2	$(2, \infty)$
$x - 2$	Neg	Neg	Neg	0	Pos
$x + 5$	Neg	0	Pos	Pos	Pos
$(x-2)(x+5)$	Pos	0	Neg	0	Pos

The inequality is non-strict, so -5 and 2 are part of the solution. Now, $(x-2)(x+5)$ is greater than zero where the product is positive. The solution is $\{x \mid x \leq -5 \text{ or } x \geq 2\}$ in set-builder notation; the solution is $(-\infty, -5] \cup [2, \infty)$ in interval notation.

3.
$$-x^2 > 2x - 24$$
$$0 > x^2 + 2x - 24$$

$x^2 + 2x - 24 < 0$

Solve: $x^2 + 2x - 24 = 0$

$(x-4)(x+6) = 0$

$x - 4 = 0 \quad \text{or} \quad x + 6 = 0$

$x = 4 \quad \text{or} \quad x = -6$

Determine where each factor is positive and negative and where the product of these factors is positive and negative.

Interval	$(-\infty, -6)$	-6	$(-6, 4)$	4	$(4, \infty)$
$x - 4$	Neg	Neg	Neg	0	Pos
$x + 6$	Neg	0	Pos	Pos	Pos
$(x-4)(x+6)$	Pos	0	Neg	0	Pos

The inequality is strict, so -6 and 4 are not part of the solution. Now, $(x+6)(x-4)$ is less than zero where the product is negative. The solution is $\{x \mid -6 < x < 4\}$ in set-builder notation; the solution is $(-6, 4)$ in interval notation.

4.
$$3x^2 > -x + 5$$

$3x^2 + x - 5 > 0$

Graphical Method:

To graph $f(x) = 3x^2 + x - 5$, we notice that $a = 3$, $b = 1$, and $c = -5$. The parabola opens up because $a = 3 > 0$. The x-coordinate of the vertex is $x = -\dfrac{b}{2a} = -\dfrac{1}{2(3)} = -\dfrac{1}{6}$. The y-coordinate of the vertex is

$$f\left(-\frac{b}{2a}\right) = f\left(-\frac{1}{6}\right)$$
$$= 3\left(-\frac{1}{6}\right)^2 + \left(-\frac{1}{6}\right) - 5$$
$$= -\frac{61}{12}$$

Thus, the vertex is $\left(-\dfrac{1}{6}, -\dfrac{61}{12}\right)$ and the axis of symmetry is the line $x = -\dfrac{1}{6}$. The y-intercept is

$f(0) = 3(0)^2 + 0 - 5 = -5$.

Now, $b^2 - 4ac = 1^2 - 4(3)(-5) = 61 > 0$. The parabola will have two distinct x-intercepts. We find these by solving

$$f(x) = 0$$

$$3x^2 + x - 5 = 0$$

$$x = \frac{-1 \pm \sqrt{61}}{2(3)}$$

$$= \frac{-1 \pm \sqrt{61}}{6}$$

$$x \approx -1.47 \quad \text{or} \quad x \approx 1.14$$

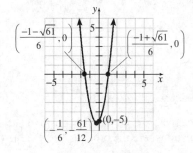

From the graph, we can see that

$f(x) = 3x^2 + x - 5$ is greater than 0 for

$x < \dfrac{-1 - \sqrt{61}}{6}$ or $x > \dfrac{-1 + \sqrt{61}}{6}$. Because the inequality is strict, we do not include the x-intercepts in the solution. So, the solution is

$\left\{ x \,\middle|\, x < \dfrac{-1 - \sqrt{61}}{6} \text{ or } x > \dfrac{-1 + \sqrt{61}}{6} \right\}$ using set-builder notation; the solution is

$\left(-\infty, \dfrac{-1 - \sqrt{61}}{6} \right) \cup \left(\dfrac{-1 + \sqrt{61}}{6}, \infty \right)$ using interval notation.

Algebraic Method:

Solve: $3x^2 + x - 5 = 0$

$$x = \frac{-1 \pm \sqrt{61}}{2(3)}$$

$$= \frac{-1 + \sqrt{61}}{6}$$

$$x \approx -1.47 \quad \text{or} \quad x \approx 1.14$$

Determine where each factor is positive and negative and where the product of these factors is positive and negative.

Interval	$(-\infty, -1.47)$		$(-1.47, 1.14)$		$(1.14, \infty)$
		$\frac{-1-\sqrt{61}}{6}$		$\frac{-1+\sqrt{61}}{6}$	
$x - \left(\frac{-1-\sqrt{61}}{6}\right)$	Neg	0	Pos	Pos	Pos
$x - \left(\frac{-1+\sqrt{61}}{6}\right)$	Neg	Neg	Neg	0	Pos
$\left[x - \left(\frac{-1-\sqrt{61}}{6}\right)\right]\left[x - \left(\frac{-1+\sqrt{61}}{6}\right)\right]$	Pos	0	Neg	0	Pos

The inequality is strict, so $\dfrac{-1 - \sqrt{61}}{6}$ and

$\dfrac{-1 + \sqrt{61}}{6}$ are not part of the solution. Now,

$\left[x - \left(\dfrac{-1 - \sqrt{61}}{6} \right) \right]\left[x - \left(\dfrac{-1 + \sqrt{61}}{6} \right) \right]$ is greater

than zero where the product is positive. So, the

solution is $\left\{ x \,\middle|\, x < \dfrac{-1 - \sqrt{61}}{6} \text{ or } x > \dfrac{-1 + \sqrt{61}}{6} \right\}$

using set-builder notation; the solution is

$\left(-\infty, \dfrac{-1 - \sqrt{61}}{6} \right) \cup \left(\dfrac{-1 + \sqrt{61}}{6}, \infty \right)$ using

interval notation.

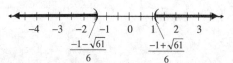

10.6 Exercises

5. (a) The graph is greater than 0 for $x < -6$ or $x > 5$. The solution is $\{ x \,|\, x < -6 \text{ or } x > 5 \}$ using set-builder notation; the solution is $(-\infty, -6) \cup (5, \infty)$ using interval notation.

(b) The graph is 0 or less for $-6 \le x \le 5$. The solution is $\{ x \,|\, -6 \le x \le 5 \}$ using set-builder notation; the solution is $[-6, 5]$ using interval notation.

7. (a) The graph is 0 or greater for $-6 \le x \le \dfrac{5}{2}$.

The solution is $\left\{ x \Big| -6 \le x \le \dfrac{5}{2} \right\}$ using set-builder notation; the solution is $\left[-6, \dfrac{5}{2} \right]$ using interval notation.

(b) The graph is less than 0 for $x < -6$ or $x > \dfrac{5}{2}$. The solution is $\left\{ x \Big| x < -6 \text{ or } x > \dfrac{5}{2} \right\}$ using set-builder notation; the solution is $\left(-\infty, -6 \right) \cup \left(\dfrac{5}{2}, \infty \right)$ using interval notation.

9. $(x-5)(x+2) \ge 0$

Solve: $(x-5)(x+2)=0$

$x-5=0$ or $x+2=0$
$x=5$ or $x=-2$

Determine where each factor is positive and negative and where the product of these factors is positive and negative.

Interval	$(-\infty, -2)$	-2	$(-2, 5)$	5	$(5, \infty)$
$x-5$	$----$	$-$	$----$	0	$++++$
$x+2$	$----$	0	$++++$	$+$	$++++$
$(x-5)(x+2)$	$++++$	0	$----$	0	$++++$

The inequality is non-strict, so -2 and 5 are part of the solution. Now, $(x-5)(x+2)$ is greater than zero where the product is positive. The solution is $\{ x \,|\, x \le -2 \text{ or } x \ge 5 \}$ in set-builder notation; the solution is $(-\infty, -2] \cup [5, \infty)$ in interval notation.

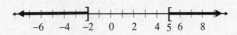

11. $(x+3)(x+7)<0$

Solve: $(x+3)(x+7)=0$

$x+3=0$ or $x+7=0$
$x=-3$ or $x=-7$

Determine where each factor is positive and negative and where the product of these factors is positive and negative.

Interval	$(-\infty, -7)$	-7	$(-7, -3)$	-3	$(-3, \infty)$
$x+7$	$----$	0	$++++$	$+$	$++++$
$x+3$	$----$	$-$	$----$	0	$++++$
$(x+7)(x+3)$	$++++$	0	$----$	0	$++++$

The inequality is strict, so -7 and -3 are not part of the solution. Now, $(x+3)(x+7)$ is less than zero where the product is negative. The solution is $\{ x \,|\, -7 < x < -3 \}$ in set-builder notation; the solution is $(-7, -3)$ in interval notation.

13. $x^2 - 2x - 35 > 0$

Solve: $x^2 - 2x - 35 = 0$
$(x+5)(x-7)=0$

$x+5=0$ or $x-7=0$
$x=-5$ or $x=7$

Determine where each factor is positive and negative and where the product of these factors is positive and negative.

Interval	$(-\infty, -5)$	-5	$(-5, 7)$	7	$(7, \infty)$
$x+5$	$----$	0	$++++$	$+$	$++++$
$x-7$	$----$	$-$	$----$	0	$++++$
$(x+5)(x-7)$	$++++$	0	$----$	0	$++++$

The inequality is strict, so -5 and 7 are not part of the solution. Now, $(x+5)(x-7)$ is greater than zero where the product is positive. The solution is $\{ x \,|\, x < -5 \text{ or } x > 7 \}$ in set-builder notation; the solution is $(-\infty, -5) \cup (7, \infty)$ in interval notation.

15. $n^2 - 6n - 8 \le 0$

Solve: $n^2 - 6n - 8 = 0$

$n = \dfrac{-(-6) \pm \sqrt{(-6)^2 - 4(1)(-8)}}{2(1)}$

$= \dfrac{6 \pm \sqrt{68}}{2}$

$= \dfrac{6 \pm 2\sqrt{17}}{2}$

$= 3 \pm \sqrt{17}$

$n = 3 - \sqrt{17}$ or $n = 3 + \sqrt{17}$

≈ -1.12 or ≈ 7.12

Determine where each factor is positive and negative and where the product of these factors is positive and negative.

Interval	$(-\infty, -1.12)$		$(-1.12, 7.12)$		$(7.12, \infty)$
		$3 - \sqrt{17}$ ≈ -1.12		$3 + \sqrt{17}$ ≈ 7.12	
$n - (3 - \sqrt{17})$	$- - - -$	0	$+ + + +$	$+$	$+ + + +$
$n - (3 + \sqrt{17})$	$- - - -$	$-$	$- - - -$	0	$+ + + +$
$[n - (3 - \sqrt{17})][n - (3 + \sqrt{17})]$	$+ + + +$	0	$- - - -$	0	$+ + + +$

The inequality is non-strict, so $3 - \sqrt{17}$ and $3 + \sqrt{17}$ are part of the solution. Now, $\left[n - (3 - \sqrt{17}) \right]\left[n - (3 + \sqrt{17}) \right]$ is less than zero where the product is negative. The solution is $\{ n \mid 3 - \sqrt{17} \le n \le 3 + \sqrt{17} \}$ in set-builder notation; the solution is $\left[3 - \sqrt{17}, \; 3 + \sqrt{17} \right]$ in interval notation.

17. $m^2 + 5m \ge 14$

$m^2 + 5m - 14 \ge 0$

Solve: $m^2 + 5m - 14 = 0$

$(m + 7)(m - 2) = 0$

$m + 7 = 0$ or $m - 2 = 0$

$m = -7$ or $m = 2$

Determine where each factor is positive and negative and where the product of these factors is positive and negative.

Interval	$(-\infty, -7)$	-7	$(-7, 2)$	2	$(2, \infty)$
$m + 7$	$- - - -$	0	$+ + + +$	$+$	$+ + + +$
$m - 2$	$- - - -$	$-$	$- - - -$	0	$+ + + +$
$(m + 7)(m - 2)$	$+ + + +$	0	$- - - -$	0	$+ + + +$

The inequality is non-strict, so -7 and 2 are part of the solution. Now, $(m + 7)(m - 2)$ is greater than zero where the product is positive. The solution is $\{ m \mid m \le -7 \text{ or } m \ge 2 \}$ in set-builder notation; the solution is $(-\infty, -7] \cup [2, \infty)$ in

interval notation.

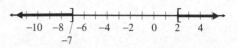

19. $2q^2 \ge q + 15$

$2q^2 - q - 15 \ge 0$

Solve: $2q^2 - q - 15 = 0$

$(2q + 5)(q - 3) = 0$

$2q + 5 = 0$ or $q - 3 = 0$

$q = -\dfrac{5}{2}$ or $q = 3$

Determine where each factor is positive and negative and where the product of these factors is positive and negative.

Interval	$\left(-\infty, -\frac{5}{2}\right)$	$-\frac{5}{2}$	$\left(-\frac{5}{2}, 3\right)$	3	$(3, \infty)$
$2q + 5$	$- - - -$	0	$+ + + +$	$+$	$+ + + +$
$q - 3$	$- - - -$	$-$	$- - - -$	0	$+ + + +$
$(2q + 5)(q - 3)$	$+ + + +$	0	$- - - -$	0	$+ + + +$

The inequality is non-strict, so $-\dfrac{5}{2}$ and 3 are part of the solution. Now, $(2q + 5)(q - 3)$ is greater than zero where the product is positive. The solution is $\left\{ q \mid q \le -\dfrac{5}{2} \text{ or } q \ge 3 \right\}$ in set-builder notation; the solution is $\left(-\infty, \; -\dfrac{5}{2} \right] \cup [3, \infty)$ in interval notation.

21. $3x + 4 \ge x^2$

$0 \ge x^2 - 3x - 4$

$x^2 - 3x - 4 \le 0$

Solve: $x^2 - 3x - 4 = 0$

$(x + 1)(x - 4) = 0$

$x + 1 = 0$ or $x - 4 = 0$

$x = -1$ or $x = 4$

Determine where each factor is positive and negative and where the product of these factors is positive and negative.

Interval

	$(-\infty, -1)$	-1		4	$(4, \infty)$
$x+1$	$----$	0	$++++$	$+$	$++++$
$x-4$	$----$	$-$	$----$	0	$++++$
$(x+1)(x-4)$	$++++$	0	$----$	0	$++++$

The inequality is non-strict, so -1 and 4 are part of the solution. Now, $(x+1)(x-4)$ is less than zero where the product is negative. The solution is $\{x \mid -1 \le x \le 4\}$ in set-builder notation; the solution is $[-1,\ 4]$ in interval notation.

23. $-x^2 + 3x < -10$

$0 < x^2 - 3x - 10$

$x^2 - 3x - 10 > 0$

Solve: $x^2 - 3x - 10 = 0$

$(x+2)(x-5) = 0$

$x+2 = 0$ or $x-5 = 0$

$x = -2$ or $x = 5$

Determine where each factor is positive and negative and where the product of these factors is positive and negative.

Interval

	$(-\infty, -2)$	-2	$(-2, 5)$	5	$(5, \infty)$
$x+2$	$----$	0	$++++$	$+$	$++++$
$x-5$	$----$	$-$	$----$	0	$++++$
$(x+2)(x-5)$	$++++$	0	$----$	0	$++++$

The inequality is strict, so -2 and 5 are not part of the solution. Now, $(x+2)(x-5)$ is greater than zero where the product is positive. The solution is $\{x \mid x < -2 \text{ or } x > 5\}$ in set-builder notation; the solution is $(-\infty, -2) \cup (5, \infty)$ in interval notation.

25. $-3x^2 \le -10x - 8$

$0 \le 3x^2 - 10x - 8$

$3x^2 - 10x - 8 \ge 0$

Solve: $3x^2 - 10x - 8 = 0$

$(3x+2)(x-4) = 0$

$3x + 2 = 0$ or $x - 4 = 0$

$x = -\dfrac{2}{3}$ or $x = 4$

Determine where each factor is positive and negative and where the product of these factors is positive and negative.

Interval

	$\left(-\infty, -\frac{2}{3}\right)$	$-\frac{2}{3}$	$\left(-\frac{2}{3}, 4\right)$	4	$(4, \infty)$
$3x+2$	$----$	0	$++++$	$+$	$++++$
$x-4$	$----$	$-$	$----$	0	$++++$
$(3x+2)(x-4)$	$++++$	0	$----$	0	$++++$

The inequality is non-strict, so $-\dfrac{2}{3}$ and 4 are part of the solution. Now, $(3x+2)(x-4)$ is greater than zero where the product is positive. The solution is $\left\{x \mid x \le -\dfrac{2}{3} \text{ or } x \ge 4\right\}$ in set-builder notation; the solution is $\left(-\infty,\ -\dfrac{2}{3}\right] \cup [4,\ \infty)$ in interval notation.

27. $x^2 + 4x + 1 < 0$

Solve: $x^2 + 4x + 1 = 0$

$x = \dfrac{-4 \pm \sqrt{4^2 - 4(1)(1)}}{2(1)}$

$= \dfrac{-4 \pm \sqrt{12}}{2}$

$= \dfrac{-4 \pm 2\sqrt{3}}{2}$

$= -2 \pm \sqrt{3}$

$x = -2 - \sqrt{3}$ or $x = -2 + \sqrt{3}$

≈ -3.73 or ≈ -0.27

Determine where each factor is positive and negative and where the product of these factors is positive and negative.

Interval

	$(-\infty, -3.73)$	$(-3.73, -0.27)$	$(-0.27, \infty)$
	$-2-\sqrt{3}$	$-2+\sqrt{3}$	
	≈ -3.73	≈ -0.27	
$x-(-2-\sqrt{3})$	$----$ 0	$++++$ $+$	$++++$
$x-(-2+\sqrt{3})$	$----$ $-$	$----$ 0	$++++$
$[x-(-2-\sqrt{3})][x-(-2+\sqrt{3})]$	$++++$ 0	$----$ 0	$++++$

The inequality is strict, so $-2-\sqrt{3}$ and $-2+\sqrt{3}$ are not part of the solution. Now,

$\left[x-\left(-2-\sqrt{3}\right)\right]\left[x-\left(-2+\sqrt{3}\right)\right]$ is less than zero where the product is negative. The solution is $\left\{x\mid -2-\sqrt{3}<x<-2+\sqrt{3}\right\}$ in set-builder notation; the solution is $\left(-2-\sqrt{3},\ -2+\sqrt{3}\right)$ in interval notation.

29. $-2a^2+7a\geq -4$

$0\geq 2a^2-7a-4$

$2a^2-7a-4\leq 0$

Solve: $2a^2-7a-4=0$

$\left(2a+1\right)\left(a-4\right)=0$

$2a+1=0$ or $a-4=0$

$a=-\dfrac{1}{2}$ or $a=4$

Determine where each factor is positive and negative and where the product of these factors is positive and negative.

Interval	$\left(-\infty,-\frac{1}{2}\right)$	$-\frac{1}{2}$	$\left(-\frac{1}{2}, 4\right)$	4	$(4, \infty)$
$2a+1$	$----$	0	$++++$	$+$	$++++$
$a-4$	$----$	$-$	$----$	0	$++++$
$(2a+1)(a-4)$	$++++$	0	$----$	0	$++++$

The inequality is non-strict, so $-\dfrac{1}{2}$ and 4 are part of the solution. Now, $\left(2a+1\right)\left(a-4\right)$ is less than zero where the product is negative. The solution is $\left\{a\mid -\dfrac{1}{2}\leq a\leq 4\right\}$ in set-builder notation; the solution is $\left[-\dfrac{1}{2},\ 4\right]$ in interval notation.

31. $z^2+2z+3>0$

Solve: $z^2+2z+3=0$

$z=\dfrac{-2\pm\sqrt{2^2-4(1)(3)}}{2(1)}$

$=\dfrac{-2\pm\sqrt{-8}}{2}$

$=\dfrac{-2\pm 2\sqrt{2}\ i}{2}=-1\pm\sqrt{2}\ i$

The solutions to the equation are non-real. This means that z^2+2z+3 will not divide the number line into positive and negative intervals. Instead, z^2+2z+3 will either be positive on the entire number line or be negative on the entire number line. The graph below shows that $f(z)=z^2+2z+3$ is always positive.

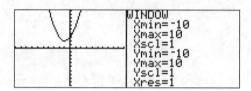

This means that z^2+2z+3 is always greater than zero. The solution is $\{z\mid z$ is any real number$\}$; the solution is $(-\infty,\ \infty)$ in interval notation.

33. $2b^2+5b\leq -6$

$2b^2+5b+6\leq 0$

Solve: $2b^2+5b+6=0$

$b=\dfrac{-5\pm\sqrt{5^2-4(2)(6)}}{2(1)}$

$=\dfrac{-5\pm\sqrt{-23}}{2}$

$=\dfrac{-5\pm\sqrt{23}\ i}{2}$

$=-\dfrac{5}{2}\pm\dfrac{\sqrt{23}}{2}i$

The solutions to the equation are non-real. This means that $2b^2+5b+6$ will not divide the number line into positive and negative intervals. Instead, $2b^2+5b+6$ will either be positive on the entire number line or be negative on the entire number line. The graph below shows that $f(b) = 2b^2 + 5b + 6$ is always positive.

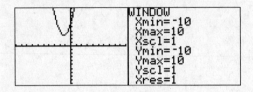

This means that $2b^2 + 5b + 6$ is never less than or equal to zero. The quadratic inequality has no solution: { } or $\varnothing$.

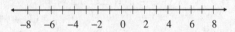

35. $x^2 - 6x + 9 > 0$

Solve: $x^2 - 6x + 9 = 0$

$(x-3)(x-3) = 0$

$x - 3 = 0$ or $x - 3 = 0$

$x = 3$ or $x = 3$

Determine where each factor is positive and negative and where the product of these factors is positive and negative.

Interval	$(-\infty, 3)$	3	$(3, \infty)$
$x - 3$	$----$	0	$++++$
$x - 3$	$----$	0	$++++$
$(x-3)(x-3)$	$++++$	0	$++++$

The inequality is strict, so 3 is not part of the solution. Now, $(x-3)(x-3)$ is greater than zero where the product is positive. Thus, $x^2 - 6x + 9$ is always greater than zero when x is not equal to 3. The solution is $\{x|x \neq 3\}$ in set-builder notation; the solution is $(-\infty, 3) \cup (3, \infty)$ in interval notation.

37. $f(x) < 0$

$x^2 - 5x < 0$

Solve: $x^2 - 5x = 0$

$x(x-5) = 0$

$x = 0$ or $x - 5 = 0$

or $x = 5$

Determine where each factor is positive and

negative and where the product of these factors is positive and negative.

Interval	$(-\infty, 0)$	0	$(0, 5)$	5	$(5, \infty)$
x	$----$	0	$++++$	$+$	$++++$
$x - 5$	$----$	$-$	$----$	0	$++++$
$x(x-5)$	$++++$	0	$----$	0	$++++$

The inequality is strict, so 0 and 5 are not part of the solution. Now, $x(x-5)$ is less than zero where the product is negative. The solution is $\{x|0 < x < 5\}$ in set-builder notation; the solution is $(0, 5)$ in interval notation.

39. $f(x) \geq 0$

$x^2 - 3x - 28 \geq 0$

Solve: $x^2 - 3x - 28 = 0$

$(x+4)(x-7) = 0$

$x + 4 = 0$ or $x - 7 = 0$

$x = -4$ or $x = 7$

Determine where each factor is positive and negative and where the product of these factors is positive and negative.

Interval	$(-\infty, -4)$	-4	$(-4, 7)$	7	$(7, \infty)$
$x + 4$	$----$	0	$++++$	$+$	$++++$
$x - 7$	$----$	$-$	$----$	0	$++++$
$(x+4)(x-7)$	$++++$	0	$----$	0	$++++$

The inequality is non-strict, so –4 and 7 are part of the solution. Now, $(x+4)(x-7)$ is greater than zero where the product is positive. The solution is $\{x\,|\,x \leq -4 \text{ or } x \geq 7\}$ in set-builder notation; the solution is $(-\infty, -4] \cup [7, \infty)$ in interval notation.

41. $g(x) > 0$

$2x^2 + x - 10 > 0$

Solve: $2x^2 + x - 10 = 0$

$(2x+5)(x-2) = 0$

$2x + 5 = 0$ or $x - 2 = 0$

$2x = -5$ or $x = 2$

$x = -\dfrac{5}{2}$

Determine where each factor is positive and negative and where the product of these factors is positive and negative.

Interval

	$\left(-\infty, -\frac{5}{2}\right)$	$-\frac{5}{2}$	$\left(-\frac{5}{2}, 2\right)$	2	$(2, \infty)$
$2x+5$	$----$	0	$++++$	$+$	$++++$
$x-2$	$----$	$-$	$----$	0	$++++$
$(2x+5)(x-2)$	$++++$	0	$----$	0	$++++$

The inequality is strict, so $-\dfrac{5}{2}$ and 2 are not part of the solution. Now, $(2x+5)(x-2)$ is greater than zero where the product is positive. The solution is $\left\{x \mid x < -\dfrac{5}{2} \text{ or } x > 2\right\}$ in set-builder notation; the solution is $\left(-\infty, -\dfrac{5}{2}\right) \cup (2, \infty)$ in interval notation.

43. The domain of $f(x) = \sqrt{x^2 + 8x}$ will be the solution set of $x^2 + 8x \geq 0$.

Solve: $x^2 + 8x = 0$
$x(x+8) = 0$
$x = 0$ or $x + 8 = 0$
or $x = -8$

Determine where each factor is positive and negative and where the product of these factors is positive and negative.

Interval

	$(-\infty, -8)$	-8	$(-8, 0)$	0	$(0, \infty)$
x	$----$	$-$	$----$	0	$++++$
$x+8$	$----$	0	$++++$	$+$	$++++$
$x(x+8)$	$++++$	0	$----$	0	$++++$

The inequality is non-strict, so -8 and 0 are part of the solution. Now, $x(x+8)$ is greater than zero where the product is positive. Thus, the domain of f is $\{x \mid x \leq -8 \text{ or } x \geq 0\}$ in set-builder notation; the domain is $(-\infty, -8] \cup [0, \infty)$ in interval notation.

45. The domain of $g(x) = \sqrt{x^2 - x - 30}$ will be the solution set of $x^2 - x - 30 \geq 0$.

Solve: $x^2 - x - 30 = 0$
$(x-6)(x+5) = 0$
$x - 6 = 0$ or $x + 5 = 0$
$x = 6$ or $x = -5$

Determine where each factor is positive and

negative and where the product of these factors is positive and negative.

Interval

	$(-\infty, -5)$	-5	$(-5, 6)$	6	$(6, \infty)$
$x-6$	$----$	$-$	$----$	0	$++++$
$x+5$	$----$	0	$++++$	$+$	$++++$
$(x-6)(x+5)$	$++++$	0	$----$	0	$++++$

The inequality is non-strict, so -6 and 5 are part of the solution. Now, $(x-6)(x+5)$ is greater than zero where the product is positive. Thus, the domain of g is $\{x \mid x \leq -5 \text{ or } x \geq 6\}$ in set-builder notation; the domain is $(-\infty, -5] \cup [6, \infty)$ in interval notation.

47. The ball will be more than 596 feet above sea level when $s(t) > 596$.

$-16t^2 + 80t + 500 > 596$
$0 > 16t^2 - 80t + 96$
$16t^2 - 80t + 96 < 0$
$\dfrac{16t^2 - 80t + 96}{16} < \dfrac{0}{16}$
$t^2 - 5t + 6 < 0$

Solve: $t^2 - 5t + 6 = 0$
$(t-2)(t-3) = 0$
$t - 2 = 0$ or $t - 3 = 0$
$t = 2$ or $t = 3$

Determine where each factor is positive and negative and where the product of these factors is positive and negative.

Interval

	$(-\infty, 2)$	2	$(2, 3)$	3	$(3, \infty)$
$t-2$	$----$	0	$++++$	$+$	$++++$
$t-3$	$----$	$-$	$----$	0	$++++$
$(t-2)(t-3)$	$++++$	0	$----$	0	$++++$

The inequality is strict, so 2 and 3 are not part of the solution. Now, $(t-2)(t-3)$ is less than zero where the product is negative. The solution is $\{t \mid 2 < t < 3\}$ in set-builder notation; the solution is $(2, 3)$ in interval notation. Thus, the ball will be more than 596 feet above sea level when the time is between 2 and 3 seconds after the ball is thrown.

49. The revenue will exceed \$35,750 when
$R(p) > 35,750$.

$$-2.5p^2 + 600p > 35,750$$
$$0 > 2.5p^2 - 600p + 35,750$$
$$2.5p^2 - 600p + 35,750 < 0$$
$$\frac{2.5p^2 - 600p + 35,750}{2.5} < \frac{0}{2.5}$$
$$p^2 - 240t + 14,300 < 0$$

Solve: $p^2 - 240t + 14,300 = 0$
$$(p - 110)(p - 130) = 0$$
$$p - 110 = 0 \quad \text{or} \quad p - 130 = 0$$
$$p = 110 \quad \text{or} \qquad p = 130$$

Determine where each factor is positive and negative and where the product of these factors is positive and negative.

Interval	$(-\infty, 110)$	110	$(110, 130)$	130	$(130, \infty)$
$p - 110$	$----$	0	$++++$	+	$++++$
$p - 130$	$----$	$-$	$----$	0	$++++$
$(p-110)(p-130)$	$++++$	0	$----$	0	$++++$

The inequality is strict, so 110 and 130 are not part of the solution. Now, $(p-110)(p-130)$ is less than zero where the product is negative. The solution is $\{p \mid 110 < p < 130\}$ in set-builder notation; the solution is $(110, 130)$ in interval notation. Thus, the revenue will exceed \$35,750 when the DVD is sold for a price between \$110 and \$130.

51. By inspection, the only solution is $x = -3$. That is, the solution set is $\{-3\}$.

Explanation: The expression on the left side of the inequality is a perfect square. A perfect square cannot be negative (less than zero). Therefore, the only solution will be where the perfect square expression equals zero, which is -3.

53. By inspection, the solution set is the set of all real numbers.

Explanation: The expression on the left side of the inequality is a perfect square. A perfect square must always be zero or greater. Therefore, it must always be larger than -2. Thus, all values of x will make the inequality true. The solution is the set of all real numbers.

55. Answers may vary. One possibility follows: We want $x \geq -3$ and $x \leq 2$. Now, $x \geq -3$ means $x + 3 \geq 0$ (positive), and $x \leq 2$ means $x - 2 \leq 0$ (negative). If we multiply a positive by a negative, we get a negative result. Thus, $(x+3)(x-2) \leq 0$. Multiplying out the expression on the left, we get $x^2 + x - 6 \leq 0$. The solution set of $x^2 + x - 6 \leq 0$ is $[-3, 2]$.

NOTE: Because the solution contains the endpoints -3 and 2, the inequality must be non-strict.

57. Answer will vary. One possibility follows: The inequalities have the same solution set because they are equivalent:

$$(3x+2)^{-2} > \frac{1}{2}$$
$$\frac{1}{(3x+2)^2} > \frac{1}{2}$$
$$2(3x+2)^2 \cdot \frac{1}{(3x+2)^2} > 2(3x+2)^2 \cdot \frac{1}{2}$$
$$2 > (3x+2)^2$$
$$(3x+2)^2 < 2$$

Note: In the third step, we multiplied both sides by $2(3x+2)^2$ which must be positive, leaving the direction of the inequality symbol unchanged.

59. $(x+3)(x-1)(x-3) < 0$

Solve: $(x+3)(x-1)(x-3) = 0$
$$x + 3 = 0 \quad \text{or} \quad x - 1 = 0 \quad \text{or} \quad x - 3 = 0$$
$$x = -3 \quad \text{or} \qquad x = 1 \quad \text{or} \qquad x = 3$$

Determine where each factor is positive and negative and where the product of these factors is positive and negative.

Interval	$(-\infty, -3)$	-3	$(-3, 1)$	1	$(1, 3)$	3	$(3, \infty)$
$x + 3$	$---$	0	$+++$	+	$+++$	+	$+++$
$x - 1$	$---$	$-$	$---$	0	$+++$	+	$+++$
$x - 3$	$---$	$-$	$---$	$-$	$---$	0	$+++$
$(x+3)(x-1)(x-3)$	$---$	0	$+++$	0	$---$	0	$+++$

The inequality is strict, so -3, 1, and 3 are not part of the solution. Now, $(x+3)(x-1)(x-3)$ is less than zero where the product is negative. The solution is $\{x \mid x < -3 \text{ or } 1 < x < 3\}$ in set-builder notation; the solution is $(-\infty, -3) \cup (1, 3)$ in interval notation.

61. $(3x+4)(x-2)(x-6) \geq 0$

Solve: $(3x+4)(x-2)(x-6) = 0$

$3x+4 = 0$ or $x-2 = 0$ or $x-6 = 0$

$x = -\dfrac{4}{3}$ or $x = 2$ or $x = 6$

Determine where each factor is positive and negative and where the product of these factors is positive and negative.

Interval	$\left(-\infty, -\frac{4}{3}\right)$	$-\frac{4}{3}$	$\left(-\frac{4}{3}, 2\right)$	2	(2, 6)	6	(6, ∞)
$3x+4$	− − −	0	+ + +	+	+ + +	+	+ + +
$x-2$	− − −	−	− − −	0	+ + +	+	+ + +
$x-6$	− − −	−	− − −	−	− − −	0	+ + +
$(3x+4)(x-2)(x-6)$	− − −	0	+ + +	0	− − −	0	+ + +

The inequality is non-strict, so $-\dfrac{4}{3}$, 2, and 6 are part of the solution. Now, $(3x+4)(x-2)(x-6)$ is greater than zero where the product is positive. The solution is $\left\{ x \mid -\dfrac{4}{3} \leq x \leq 2 \text{ or } x \geq 6 \right\}$ in set-builder notation; the solution is $\left[-\dfrac{4}{3}, 2 \right] \cup [6, \infty)$ in interval notation.

63. $(x^2 - 3x - 10)(x+1) < 0$
$(x+2)(x-5)(x+1) < 0$
The rational expression will equal 0 when $x = -2$, $x = 5$, and when $x = -1$.
Determine where each factor is positive and negative and where the product of these factors is positive and negative.

Interval	$(-\infty, -2)$	−2	(−2, −1)	−1	(−1, 5)	5	(5, ∞)
$x+2$	− − −	0	+ + +	+	+ + +	+	+ + +
$x-5$	− − −	−	− − −	−	− − −	0	+ + +
$x+1$	− − −	−	− − −	0	+ + +	+	+ + +
$(x+2)(x-5)(x+1)$	− − −	0	+ + +	∅	− − −	0	+ + +

The inequality is strict, so -2, -1, and 5 are not part of the solution. Now, $(x+2)(x-5)(x+1)$ is less than zero where the quotient is negative. The solution is $\{ x \mid x < -2 \text{ or } -1 < x < 5 \}$ in set-builder notation; the solution is $(-\infty, -2) \cup (-1, 5)$ in interval notation.

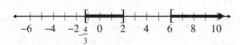

65. Answers may vary. One possibility follows:
$x^2 \geq 0$ for all real values of x, so $x^2 - 1 \geq -1$ for all real values of x. That is, $x^2 - 1$ is always -1 or larger.

67. No. If $x = 0$, we obtain the following:
$x^2 + 1 > 1$
$0^2 + 1 > 1$
$\quad 1 > 1 \leftarrow$ False
Thus, 0 is not a solution to the inequality. The inequality is true for all other real numbers.

Thus, the solution set to the inequality $x^2 + 1 > 1$ is all real numbers except 0. That is, the solution set is $\{ x \mid x \neq 0 \}$ or, using interval notation, $(-\infty, 0) \cup (0, \infty)$.

69. $\left(4mn^{-3} \right)\left(-2m^4 n \right) = -8m^{4+1}n^{-3+1}$

$= -8m^5 n^{-2} = \dfrac{-8m^5}{n^2}$

71. $\left(\dfrac{9a^{\frac{2}{3}}b^{\frac{1}{2}}}{a^{-\frac{1}{9}}b^{\frac{3}{4}}} \right)^{-1} = \left(\dfrac{a^{-\frac{1}{9}}b^{\frac{3}{4}}}{9a^{\frac{2}{3}}b^{\frac{1}{2}}} \right)^{1} = \dfrac{a^{-\frac{1}{9}}b^{\frac{3}{4}}}{9a^{\frac{2}{3}}b^{\frac{1}{2}}}$

$= \dfrac{b^{\frac{3}{4}-\frac{1}{2}}}{9a^{\frac{2}{3}+\frac{1}{9}}} = \dfrac{b^{\frac{3}{4}-\frac{2}{4}}}{9a^{\frac{6}{9}+\frac{1}{9}}} = \dfrac{b^{\frac{1}{4}}}{9a^{\frac{7}{9}}}$

73. $2x^2 + 7x - 49 > 0$

Let $Y_1 = 2x^2 + 7x - 49$. Graph the quadratic function. Use the **ZERO** feature to find the x-intercepts.

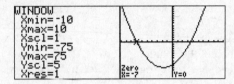

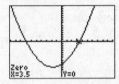

The inequality is strict, so -7 and 3.5 are not part of the solution. From the graph, we can see that $2x^2 + 7x - 49 > 0$ for $x < -7$ or $x > 3.5$. Thus, the solution set is $\{x \mid x < -7 \text{ or } x > 3.5\}$ in set-builder notation; the solution is $(-\infty, -7) \cup (3.5, \infty)$ in interval notation.

75. $6x^2 + x \le 40$

$6x^2 + x - 40 \le 0$

Let $Y_1 = 6x^2 + x - 40$. Graph the quadratic function. Use the **ZERO** feature to find the x-intercepts.

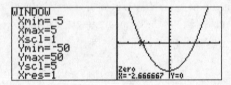

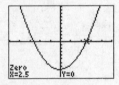

The inequality is non-strict, so $-2.\overline{6} = -\dfrac{8}{3}$ and $2.5 = \dfrac{5}{2}$ are part of the solution. From the graph, we can see that $6x^2 + x - 40 \le 0$ for $-\dfrac{8}{3} \le x \le \dfrac{5}{2}$. Thus, the solution set is

$\left\{x \mid -\dfrac{8}{3} \le x \le \dfrac{5}{2}\right\}$ in set-builder notation; the solution is $\left[-\dfrac{8}{3}, \dfrac{5}{2}\right]$ in interval notation.

Section 10.7

Preparing for Rational Inequalities

P1. The inequality $-1 < x \le 8$ in interval notation is $(-1, 8]$.

P2. $2x + 3 > 4x - 9$

$-4x + 2x + 3 > -4x + 4x - 9$

$-2x + 3 > -9$

$-2x + 3 - 3 > -9 - 3$

$-2x > -12$

$\dfrac{-2x}{-2} < \dfrac{-12}{-2}$

$x < 6$

The solution set is $\{x \mid x < 6\}$ or, using interval notation, $(-\infty, 6)$.

Section 10.7 Quick Checks

1. rational

2. $\dfrac{x-7}{x+3} \ge 0$

The rational expression will equal 0 when $x = 7$. It is undefined when $x = -3$. Thus, we separate the real number line into the intervals $(-\infty, -3)$, $(-3, 7)$, and $(7, \infty)$. Determine where the numerator and denominator are positive and negative and where the quotient is positive and negative.

Interval	$(-\infty, -3)$	-3	$(-3, 7)$	7	$(7, \infty)$
$x - 7$	Neg	Neg	Neg	0	Pos
$x + 3$	Neg	0	Pos	Pos	Pos
$\dfrac{x-7}{x+3}$	Pos	Undef	Neg	0	Pos

The rational function is undefined at $x = -3$, so -3 is not part of the solution. The inequality is non-strict, so 7 is part of the solution. Now, $\dfrac{x-7}{x+3}$ is greater than zero where the quotient is positive. The solution is $\{x \mid x < -3 \text{ or } x \ge 7\}$ in set-builder notation; the solution is

$(-\infty, -3) \cup [7, \infty)$ in interval notation.

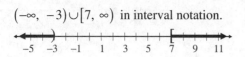

3. $\dfrac{1-x}{x+5} > 0$

The rational expression will equal 0 when $x = 1$. It is undefined when $x = -5$. Thus we separate the real number line into intervals $(-\infty, -5)$, $(-5, 1)$, and $(1, \infty)$. Determine where the numerator and denominator are positive and negative and where the quotient is positive and negative.

Interval	$(-\infty, -5)$	-5	$(-5, 1)$	1	$(1, \infty)$
$1 - x$	Pos	Pos	Pos	0	Neg
$x + 5$	Neg	0	Pos	Pos	Pos
$\dfrac{1-x}{x+5}$	Neg	Undef	Pos	0	Neg

The rational function is undefined at $x = -5$, so -5 is not part of the solution. The inequality is strict, so 1 is not part of the solution. Now, $\dfrac{1-x}{x+5}$ is greater than zero where the quotient is positive. The solution is $(-5, 1)$ in interval notation.

4.

$$\dfrac{4x+5}{x+2} < 3$$

$$\dfrac{4x+5}{x+2} - 3 < 0$$

$$\dfrac{4x+5}{x+2} - \dfrac{3(x+2)}{x+2} < 0$$

$$\dfrac{4x+5-3x-6}{x+2} < 0$$

$$\dfrac{x-1}{x+2} < 0$$

The rational expression will equal 0 when $x = 1$. It is undefined when $x = -2$. Thus, we separate the real number line into the intervals $(-\infty, -2)$, $(-2, 1)$, and $(1, \infty)$. Determine where the numerator and denominator are positive and negative and where the quotient is positive and negative.

Interval	$(-\infty, -2)$	-2	$(-2, 1)$	1	$(1, \infty)$
$x - 1$	Neg	Neg	Neg	0	Pos
$x + 2$	Neg	0	Pos	Pos	Pos
$\dfrac{x-1}{x+2}$	Pos	Undef	Neg	0	Pos

The rational function is undefined at $x = -2$, so -2 is not part of the solution. The inequality is strict, so 1 is not part of the solution. Now, $\dfrac{x-1}{x+2}$ is less than zero where the quotient is negative. The solution is $\{x \mid -2 < x < 1\}$ in set-builder notation; the solution is $(-2, 1)$ in interval notation.

10.7 Exercises

5. $\dfrac{x-4}{x+1} > 0$

The rational expression will equal 0 when $x = 4$. It is undefined when $x = -1$.
Determine where the numerator and denominator are positive and negative and where the quotient is positive and negative.

Interval	$(-\infty, -1)$	-1	$(-1, 4)$	4	$(4, \infty)$
$x - 4$	$----$	$-$	$----$	0	$++++$
$x + 1$	$----$	0	$++++$	$+$	$++++$
$\dfrac{x-4}{x+1}$	$++++$	$\varnothing$	$----$	0	$++++$

The rational function is undefined at $x = -1$, so -1 is not part of the solution. The inequality is strict, so 4 is not part of the solution. Now, $\dfrac{x-4}{x+1}$ is greater than zero where the quotient is positive. The solution is $\{x \mid x < -1 \text{ or } x > 4\}$ in set-builder notation; the solution is $(-\infty, -1) \cup (4, \infty)$ in interval notation.

7. $\dfrac{x+9}{x-3} < 0$

The rational expression will equal 0 when $x = -9$. It is undefined when $x = 3$.
Determine where the numerator and denominator are positive and negative and where the quotient is positive and negative.

Interval

	$(-\infty, -9)$	-9	$(-9, 3)$	3	$(3, \infty)$
$x+9$	$----$	0	$++++$	$+$	$++++$
$x-3$	$----$	$-$	$----$	0	$++++$
$\frac{x+9}{x-3}$	$++++$	0	$----$	$\varnothing$	$++++$

The rational function is undefined at $x = 3$, so 3 is not part of the solution. The inequality is strict, so -9 is not part of the solution. Now, $\frac{x+9}{x-3}$ is less than zero where the quotient is negative. The solution is $\{x \mid -9 < x < 3\}$ in set-builder notation; the solution is $(-9, 3)$ in interval notation.

9. $\frac{x+10}{x-4} \geq 0$

The rational expression will equal 0 when $x = -10$. It is undefined when $x = 4$. Determine where the numerator and denominator are positive and negative and where the quotient is positive and negative.

Interval

	$(-\infty, -10)$	-10	$(-10, 4)$	4	$(4, \infty)$
$x+10$	$----$	0	$++++$	$+$	$++++$
$x-4$	$----$	$-$	$----$	0	$++++$
$\frac{x+10}{x-4}$	$++++$	0	$----$	$\varnothing$	$++++$

The rational function is undefined at $x = 4$, so 4 is not part of the solution. The inequality is non-strict, so -10 is part of the solution. Now, $\frac{x+10}{x-4}$ is greater than zero where the quotient is positive. The solution is $\{x \mid x \leq -10 \text{ or } x > 4\}$ in set-builder notation; the solution is $(-\infty, -10] \cup (4, \infty)$ in interval notation.

11. $\frac{(3x+5)(x+8)}{x-2} \leq 0$

The rational expression will equal 0 when $x = -\frac{5}{3}$ and when $x = -8$. It is undefined when $x = 2$.

Determine where the factors of the numerator and denominator are positive and negative and where the quotient is positive and negative.

Interval

	$(-\infty, -8)$	-8	$\left(-8, -\frac{5}{3}\right)$	$-\frac{5}{3}$	$\left(-\frac{5}{3}, 2\right)$	2	$(2, \infty)$
$3x+5$	$---$	$-$	$---$	0	$+++$	$+$	$+++$
$x+8$	$---$	0	$+++$	$+$	$+++$	$+$	$+++$
$x-2$	$---$	$-$	$---$	$-$	$---$	0	$+++$
$\frac{(3x+5)(x+8)}{x-2}$	$---$	0	$+++$	0	$---$	$\varnothing$	$+++$

The rational function is undefined at $x = 2$, so 2 is not part of the solution. The inequality is non-strict, so -8 and $-\frac{5}{3}$ are part of the solution.

Now, $\frac{(3x+5)(x+8)}{x-2}$ is less than zero where the quotient is negative. The solution is $\left\{x \mid x \leq -8 \text{ or } -\frac{5}{3} \leq x < 2\right\}$ in set-builder notation; the solution is $(-\infty, -8] \cup \left[-\frac{5}{3}, 2\right)$ in interval notation.

13. $\frac{x-5}{x+1} < 1$

$\frac{x-5}{x+1} - 1 < 0$

$\frac{x-5}{x+1} - \frac{x+1}{x+1} < 0$

$\frac{x-5-x-1}{x+1} < 0$

$\frac{-6}{x+1} < 0$

The numerator is always negative, so the rational expression will never equal 0. However, it is undefined when $x = -1$.

Determine where the denominator is positive and negative and where the quotient is positive and negative.

Interval

	$(-\infty, -1)$	-1	$(-1, \infty)$
-6	$----$	$-$	$----$
$x+1$	$----$	0	$++++$
$\frac{-6}{x+1}$	$++++$	$\varnothing$	$----$

The rational function is undefined at $x = -1$, so -1 is not part of the solution. Now, $\frac{-6}{x+1}$ is less than zero where the quotient is negative. The solution is $\{x \mid x > -1\}$ in set-builder notation;

the solution is $(-1, \infty)$ in interval notation.

15.

$$\frac{2x-9}{x-3} > 4$$

$$\frac{2x-9}{x-3} - 4 > 0$$

$$\frac{2x-9}{x-3} - \frac{4(x-3)}{x-3} > 0$$

$$\frac{2x-9-4x+12}{x-3} > 0$$

$$\frac{-2x+3}{x-3} > 0$$

The rational expression will equal 0 when $x = \dfrac{3}{2}$. It is undefined when $x = 3$.

Determine where the numerator and denominator are positive and negative and where the quotient is positive and negative.

Interval	$\left(-\infty, \frac{3}{2}\right)$	$\frac{3}{2}$	$\left(\frac{3}{2}, 3\right)$	3	$(3, \infty)$
$-2x+3$	$++++$	0	$----$	$-$	$----$
$x-3$	$----$	$-$	$----$	0	$++++$
$\frac{-2x+3}{x-3}$	$----$	0	$++++$	$\varnothing$	$----$

The rational function is undefined at $x = 3$, so 3 is not part of the solution. The inequality is strict, so 3 is not part of the solution. Now, $\dfrac{-2x+3}{x-3}$ is greater than zero where the quotient is positive. The solution is $\left\{x \,\middle|\, \dfrac{3}{2} < x < 3\right\}$ in set-builder notation; the solution is $\left(\dfrac{3}{2}, 3\right)$ in interval notation.

17.

$$\frac{3}{x-4} + \frac{1}{x} \geq 0$$

$$\frac{3x}{x(x-4)} + \frac{x-4}{x(x-4)} \geq 0$$

$$\frac{3x+x-4}{x(x-4)} \geq 0$$

$$\frac{4x-4}{x(x-4)} \geq 0$$

The rational expression will equal 0 when $x = 1$. It is undefined when $x = 0$ and when $x = 4$.
Determine where the numerator and the factors

of the denominator are positive and negative and where the quotient is positive and negative.

Interval	$(-\infty, 0)$	0	$(0, 1)$	1	$(1, 4)$	4	$(4, \infty)$
$4x-4$	$---$	$-$	$---$	0	$+++$	$+$	$+++$
x	$---$	0	$+++$	$+$	$+++$	$+$	$+++$
$x-4$	$---$	$-$	$---$	$-$	$---$	0	$+++$
$\frac{4x-4}{x(x-4)}$	$---$	$\varnothing$	$+++$	0	$---$	$\varnothing$	$+++$

The rational function is undefined at $x = 0$ and $x = 4$, so 0 and 4 are not part of the solution. The inequality is non-strict, so 1 is part of the solution. Now, $\dfrac{4x-4}{x(x-4)}$ is greater than zero where the quotient is positive. The solution is $\{x \mid 0 < x \leq 1 \text{ or } x > 4\}$ in set-builder notation; the solution is $(0, 1] \cup (4, \infty)$ in interval notation.

19.

$$\frac{3}{x-2} \leq \frac{4}{x+5}$$

$$\frac{3}{x-2} - \frac{4}{x+5} \leq 0$$

$$\frac{3(x+5)}{(x+5)(x-2)} - \frac{4(x-2)}{(x+5)(x-2)} \leq 0$$

$$\frac{3(x+5)-4(x-2)}{(x+5)(x-2)} \leq 0$$

$$\frac{3x+15-4x+8}{(x+5)(x-2)} \leq 0$$

$$\frac{23-x}{(x+5)(x-2)} \leq 0$$

The rational expression will equal 0 when $x = 23$. It is undefined when $x = 2$ and when $x = -5$.

Determine where the numerator and the factors of the denominator are positive and negative and where the quotient is positive and negative.

Interval	$(-\infty, -5)$	-5	$(-5, 2)$	2	$(2, 23)$	23	$(23, \infty)$
$23-x$	$+++$	$+$	$+++$	$+$	$+++$	0	$---$
$x+5$	$---$	0	$+++$	$+$	$+++$	$+$	$+++$
$x-2$	$---$	$-$	$---$	0	$+++$	$+$	$+++$
$\frac{23-x}{(x+5)(x-2)}$	$+++$	$\varnothing$	$---$	$\varnothing$	$+++$	0	$---$

The rational function is undefined at $x = -5$ and $x = 2$, so -5 and 2 are not part of the solution. The inequality is non-strict, so 23 is part of the solution. Now, $\dfrac{23-x}{(x+5)(x-2)}$ is less than zero

where the quotient is negative. The solution is $\{x \mid -5 < x < 2 \text{ or } x \geq 23\}$ in set-builder notation; the solution is $(-5, 2) \cup [23, \infty)$ in interval notation.

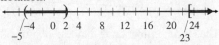

21. $\dfrac{(2x-1)(x+3)}{x-5} > 0$

The rational expression will equal 0 when $x = \dfrac{1}{2}$

and when $x = -3$. It is undefined when $x = 5$. Determine where the factors of the numerator and denominator are positive and negative and where the quotient is positive and negative.

Interval	$(-\infty, -3)$	-3	$\left(-3, \frac{1}{2}\right)$	$\frac{1}{2}$	$\left(\frac{1}{2}, 5\right)$	5	$(5, \infty)$
$2x-1$	$---$	$-$	$---$	0	$+++$	$+$	$+++$
$x+3$	$---$	0	$+++$	$+$	$+++$	$+$	$+++$
$x-5$	$---$	$-$	$---$	$-$	$---$	0	$+++$
$\frac{(2x-1)(x+3)}{x-5}$	$---$	0	$+++$	0	$---$	$\varnothing$	$+++$

The rational function is undefined at $x = 5$, so 5 is not part of the solution. The inequality is strict, so -3 and $\dfrac{1}{2}$ are not part of the solution. Now,

$\dfrac{(2x-1)(x+3)}{x-5}$ is greater than zero where the

quotient is positive. The solution is

$\left\{x \; \middle| \; -3 < x < \dfrac{1}{2} \text{ or } x > 5\right\}$ in set-builder notation;

the solution is $\left(-3, \dfrac{1}{2}\right) \cup (5, \infty)$ in interval

notation.

23. $3 - 4(x+1) < 11$
$$3 - 4x - 4 < 11$$
$$-4x - 1 < 11$$
$$-4x < 12$$
$$x > -3$$

The solution is $\{x \mid x > -3\}$ in set-builder notation; the solution is $(-3, \infty)$ in interval notation.

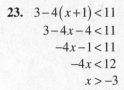

25. $\dfrac{x+7}{x-8} \leq 0$

The rational expression will equal 0 when $x = -7$. It is undefined when $x = 8$. Determine where the numerator and denominator are positive and negative and where the quotient is positive and negative.

Interval	$(-\infty, -7)$	-7	$(-7, 8)$	8	$(8, \infty)$
$x+7$	$----$	0	$++++$	$+$	$++++$
$x-8$	$----$	$-$	$----$	0	$++++$
$\frac{x+7}{x-8}$	$++++$	0	$----$	$\varnothing$	$++++$

The rational function is undefined at $x = 8$, so 8 is not part of the solution. The inequality is non-strict, so -7 is part of the solution. Now, $\dfrac{x+7}{x-8}$ is

less than zero where the quotient is negative. The solution is $\{x \mid -7 \leq x < 8\}$ in set-builder notation; the solution is $[-7, 8)$ in interval notation.

27. $(x-2)(2x+1) \geq 2(x-1)^2$
$$2x^2 - 3x - 2 \geq 2(x^2 - 2x + 1)$$
$$2x^2 - 3x - 2 \geq 2x^2 - 4x + 2$$
$$2x^2 - 3x - 2x^2 + 4x \geq 2 + 2$$
$$x \geq 4$$

The solution is $\{x \mid x \geq 4\}$ in set-builder notation; the solution is $[4, \infty)$ in interval notation.

29.
$$\frac{3x-1}{x+4} \geq 2$$

$$\frac{3x-1}{x+4} - 2 \geq 0$$

$$\frac{3x-1}{x+4} - \frac{2(x+4)}{x+4} \geq 0$$

$$\frac{3x-1-2x-8}{x+4} \geq 0$$

$$\frac{x-9}{x+4} \geq 0$$

The rational expression will equal 0 when $x = 9$. It is undefined when $x = -4$.
Determine where the numerator and denominator are positive and negative and where the quotient is positive and negative.

Interval	$(-\infty, -4)$	-4	$(-4, 9)$	9	$(9, \infty)$
$x - 9$	$----$	$-$	$----$	0	$++++$
$x + 4$	$----$	0	$++++$	$+$	$++++$
$\frac{x-9}{x+4}$	$++++$	$\emptyset$	$----$	0	$++++$

The rational function is undefined at $x = -4$, so -4 is not part of the solution. The inequality is non-strict, so 9 is part of the solution. Now, $\frac{x-9}{x+4}$ is greater than zero where the quotient is positive. The solution is $\{x \mid x < -4 \text{ or } x \geq 9\}$ in set-builder notation; the solution is $(-\infty, -4) \cup [9, \infty)$ in interval notation.

31. $R(x) \leq 0$

$$\frac{x-6}{x+1} \leq 0$$

The rational expression will equal 0 when $x = 6$. It is undefined when $x = -1$.
Determine where the numerator and denominator are positive and negative and where the quotient is positive and negative.

Interval	$(-\infty, -1)$	-1	$(-1, 6)$	6	$(6, \infty)$
$x - 6$	$----$	$-$	$----$	0	$++++$
$x + 1$	$----$	0	$++++$	$+$	$++++$
$\frac{x-6}{x+1}$	$++++$	$\emptyset$	$----$	0	$++++$

The rational function is undefined at $x = -1$, so -1 is not part of the solution. The inequality is non-strict, so 6 is part of the solution. Now, $\frac{x-6}{x+1}$ is less than zero where the quotient is negative. The solution is $\{x \mid -1 < x \leq 6\}$ in set-

builder notation; the solution is $(-1, 6]$ in interval notation.

33. $R(x) < 0$

$$\frac{2x-5}{x+2} < 0$$

The rational expression will equal 0 when $x = \frac{5}{2}$. It is undefined when $x = -2$.

Determine where the numerator and denominator are positive and negative and where the quotient is positive and negative.

Interval	$(-\infty, -2)$	-2	$\left(-2, \frac{5}{2}\right)$	$\frac{5}{2}$	$\left(\frac{5}{2}, \infty\right)$
$2x - 5$	$----$	$-$	$----$	0	$++++$
$x + 2$	$----$	0	$++++$	$+$	$++++$
$\frac{2x-5}{x+2}$	$++++$	$\emptyset$	$----$	0	$++++$

The rational function is undefined at $x = -2$, so -2 is not part of the solution. The inequality is strict, so $\frac{5}{2}$ is not part of the solution. Now, $\frac{2x-5}{x+2}$ is less than zero where the quotient is negative. The solution is $\left\{x \mid -2 < x < \frac{5}{2}\right\}$ in set-builder notation; the solution is $\left(-2, \frac{5}{2}\right)$ in interval notation.

35. The average cost will be no more than $130 when $\overline{C}(x) \leq 130$.

$$\frac{80x + 5000}{x} \leq 130$$

$$\frac{80x + 5000}{x} - 130 \leq 0$$

$$\frac{80x + 5000}{x} - \frac{130x}{x} \leq 0$$

$$\frac{5000 - 50x}{x} \leq 0$$

The rational expression will equal 0 when $x = 100$. It is undefined when $x = 0$.
Determine where the numerator and denominator are positive and negative and where the quotient is positive and negative.

Interval

| | $(-\infty, 0)$ | 0 | $(0, 100)$ | 100 | $(100, \infty)$ |

$5000 - 50x$ $\quad$ $++++$ $\quad+\quad$ $++++$ $\quad 0 \quad$ $----$

x $\qquad ---- \quad 0 \quad ++++ \quad + \quad ++++$

$\dfrac{5000-50x}{x}$ $\quad ---- \quad \varnothing \quad ++++ \quad 0 \quad ----$

The rational function is undefined at $x = 0$, so 0 is not part of the solution. The inequality is non-strict, so 100 is part of the solution. Now,

$\dfrac{5000-50x}{x}$ is less than zero where the quotient

is negative. The solution is

$\{x \mid x < 0 \text{ or } x \geq 100\}$ in set-builder notation;

the solution is $(-\infty, 0) \cup [100, \infty)$ in interval

notation. However, for this problem, $x < 0$ is not meaningful. Thus, the average cost will be no more than \$130 when 100 or more bicycles are produced each day.

37. Answers may vary. One possibility follows:
The left endpoint of the interval is $x = 2$, so we can use $x - 2$ as a factor of the denominator of the rational function. Now, $x - 2$ will be positive for $x > 2$ and negative for $x < 2$. If we use any positive constant for the numerator of the rational function, then the quotient will be positive when $x > 2$ and negative for $x < 2$.

Thus, the rational inequality $\dfrac{10}{x-2} > 0$ will have

$(2, \infty)$ as the solution set.

NOTE: Because the solution set does not contain the endpoint 2, the inequality must be strict.

39. The statement $(-1, 4)$ indicates that neither

endpoint is included in the solution set. The

statement $\{x \mid -1 \leq x \leq 4\}$ indicates that both

endpoints are included in the solution set. In

fact, the solution of the inequality $\dfrac{x-4}{x+1} \leq 0$

should include the endpoint 4, but it should not include the endpoint -1. Therefore, the statement of the solution set should be

$\{x \mid -1 < x \leq 4\}$ using set-builder notation or

$(-1, 4]$ using interval notation.

41. To find the x-intercept(s), we solve $F(x) = 0$:

$$F(x) = 0$$
$$6x - 12 = 0$$
$$6x = 12$$
$$x = 2$$

The x-intercept of F is 2.

43. To find the x-intercept(s), we solve $f(x) = 0$:

$$f(x) = 0$$
$$2x^2 + 3x - 14 = 0$$
$$(2x+7)(x-2) = 0$$
$$2x + 7 = 0 \quad \text{or} \quad x - 2 = 0$$
$$x = -\frac{7}{2} \quad \text{or} \quad x = 2$$

The x-intercepts of f are $-\dfrac{7}{2}$ and 2.

45. To find the x-intercept(s), we solve $R(x) = 0$:

$$R(x) = 0$$
$$\frac{3x-2}{x+4} = 0$$
$$(x+4)\left(\frac{3x-2}{x+4}\right) = (x+4)0$$
$$3x - 2 = 0$$
$$3x = 2$$
$$x = \frac{2}{3}$$

The x-intercept of R is $\dfrac{2}{3}$.

47. $\dfrac{x-5}{x+1} \leq 3$

Let $Y_1 = \dfrac{x-5}{x+1}$ and $Y_2 = 3$. Graph the functions.

Use the **INTERSECT** feature to find the x-coordinates of the point(s) of intersection.

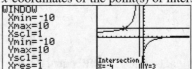

The rational function is undefined at $x = -1$, so -1 is not part of the solution. The inequality is non-strict, so -4 is part of the solution. From

the graph, we can see that $\dfrac{x-5}{x+1} \leq 3$ where

$x \leq -4$ and where $x > -1$. Thus, the solution set

is $\{x \mid x \leq -4 \text{ or } x > -1\}$ in set-builder notation;

the solution is $(-\infty, -4] \cup (-1, \infty)$ in interval

notation.

49. $\dfrac{2x+5}{x-7} > 3$

Let $Y_1 = \dfrac{2x+5}{x-7}$ and $Y_2 = 3$. Graph the

functions. Use the INTERSECT feature to find the x-coordinates of the point(s) of intersection.

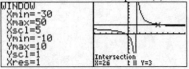

The rational function is undefined at $x = 7$, so 7 is not part of the solution. The inequality is strict, so 26 is not part of the solution. From the

graph, we can see that $\dfrac{2x+5}{x-7} > 3$ for

$7 < x < 26$. Thus, the solution set is $\{x \,|\, 7 < x < 26\}$ in set-builder notation; the

solution is $(7, 26)$ in interval notation.

Chapter 10 Review

1. $m^2 = 169$

$m = \pm\sqrt{169}$

$m = \pm 13$

The solution set is $\{-13,\ 13\}$.

2. $n^2 = 75$

$n = \pm\sqrt{75}$

$n = \pm 5\sqrt{3}$

The solution set is $\left\{-5\sqrt{3},\ 5\sqrt{3}\right\}$.

3. $a^2 = -16$

$a = \pm\sqrt{-16}$

$a = \pm 4i$

The solution set is $\{-4i,\ 4i\}$.

4. $b^2 = \dfrac{8}{9}$

$b = \pm\sqrt{\dfrac{8}{9}}$

$b = \pm\dfrac{\sqrt{8}}{\sqrt{9}}$

$b = \pm\dfrac{2\sqrt{2}}{3}$

The solution set is $\left\{-\dfrac{2\sqrt{2}}{3},\ \dfrac{2\sqrt{2}}{3}\right\}$.

5. $(x-8)^2 = 81$

$x - 8 = \pm\sqrt{81}$

$x - 8 = \pm 9$

$x = 8 \pm 9$

$x = 8 - 9$ or $x = 8 + 9$

$x = -1$ or $x = 17$

The solution set is $\{-1,\ 17\}$.

6. $(y-2)^2 - 62 = 88$

$(y-2)^2 = 150$

$y - 2 = \pm\sqrt{150}$

$y - 2 = \pm 5\sqrt{6}$

$y = 2 \pm 5\sqrt{6}$

The solution set is $\left\{2 - 5\sqrt{6},\ 2 + 5\sqrt{6}\right\}$.

7. $(3z+5)^2 = 100$

$3z + 5 = \pm\sqrt{100}$

$3z + 5 = \pm 10$

$3z = -5 \pm 10$

$3z = -5 - 10$ or $3z = -5 + 10$

$3z = -15$ or $3z = 5$

$z = -5$ or $z = \dfrac{5}{3}$

The solution set is $\left\{-5,\ \dfrac{5}{3}\right\}$.

8. $7p^2 = 18$

$p^2 = \dfrac{18}{7}$

$p = \pm\sqrt{\dfrac{18}{7}}$

$p = \pm\dfrac{\sqrt{18}}{\sqrt{7}} \cdot \dfrac{\sqrt{7}}{\sqrt{7}}$

$p = \pm\dfrac{\sqrt{126}}{\sqrt{49}}$

$p = \pm\dfrac{3\sqrt{14}}{7}$

The solution set is $\left\{-\dfrac{3\sqrt{14}}{7},\ \dfrac{3\sqrt{14}}{7}\right\}$.

9. $3q^2 + 251 = 11$

$$3q^2 = -240$$
$$q^2 = -80$$
$$q = \pm\sqrt{-80}$$
$$q = \pm 4\sqrt{5}\, i$$

The solution set is $\left\{-4\sqrt{5}\, i,\ 4\sqrt{5}\, i\right\}$.

10. $\left(x+\dfrac{3}{4}\right)^2 = \dfrac{13}{16}$

$$x+\frac{3}{4} = \pm\sqrt{\frac{13}{16}}$$
$$x+\frac{3}{4} = \pm\frac{\sqrt{13}}{4}$$
$$x = -\frac{3}{4} \pm \frac{\sqrt{13}}{4}$$

The solution set is $\left\{-\dfrac{3}{4}-\dfrac{\sqrt{13}}{4},\ -\dfrac{3}{4}+\dfrac{\sqrt{13}}{4}\right\}$.

11. Start: $a^2 + 30a$

Add: $\left(\dfrac{1}{2}\cdot 30\right)^2 = 225$

Result: $a^2 + 30a + 225$

Factored Form: $(a+15)^2$

12. Start: $b^2 - 14b$

Add: $\left[\dfrac{1}{2}\cdot(-14)\right]^2 = 49$

Result: $b^2 - 14b + 49$

Factored Form: $(b-7)^2$

13. Start: $c^2 - 11c$

Add: $\left[\dfrac{1}{2}\cdot(-11)\right]^2 = \dfrac{121}{4}$

Result: $c^2 - 11c + \dfrac{121}{4}$

Factored Form: $\left(c-\dfrac{11}{2}\right)^2$

14. Start: $d^2 + 9d$

Add: $\left(\dfrac{1}{2}\cdot 9\right)^2 = \dfrac{81}{4}$

Result: $d^2 + 9d + \dfrac{81}{4}$

Factored Form: $\left(d+\dfrac{9}{2}\right)^2$

15. Start: $m^2 - \dfrac{1}{4}m$

Add: $\left[\dfrac{1}{2}\cdot\left(-\dfrac{1}{4}\right)\right]^2 = \dfrac{1}{64}$

Result: $m^2 - \dfrac{1}{4}m + \dfrac{1}{64}$

Factored Form: $\left(m-\dfrac{1}{8}\right)^2$

16. Start: $n^2 + \dfrac{6}{7}n$

Add: $\left(\dfrac{1}{2}\cdot\dfrac{6}{7}\right)^2 = \dfrac{9}{49}$

Result: $n^2 + \dfrac{6}{7}n + \dfrac{9}{49}$

Factored Form: $\left(n+\dfrac{3}{7}\right)^2$

17.

$$x^2 - 10x + 16 = 0$$
$$x^2 - 10x = -16$$
$$x^2 - 10x + \left(\frac{1}{2}\cdot(-10)\right)^2 = -16 + \left(\frac{1}{2}\cdot(-10)\right)^2$$
$$x^2 - 10x + 25 = -16 + 25$$
$$(x-5)^2 = 9$$
$$x-5 = \pm\sqrt{9}$$
$$x-5 = \pm 3$$
$$x = 5 \pm 3$$
$$x = 2 \text{ or } x = 8$$

The solution set is $\{2,\ 8\}$.

18.

$$y^2 - 3y - 28 = 0$$
$$y^2 - 3y = 28$$
$$y^2 - 3y + \left(\frac{1}{2} \cdot (-3)\right)^2 = 28 + \left(\frac{1}{2} \cdot (-3)\right)^2$$
$$y^2 - 3y + \frac{9}{4} = 28 + \frac{9}{4}$$
$$\left(y - \frac{3}{2}\right)^2 = \frac{121}{4}$$
$$y - \frac{3}{2} = \pm\sqrt{\frac{121}{4}}$$
$$y - \frac{3}{2} = \pm\frac{11}{2}$$
$$y = \frac{3}{2} \pm \frac{11}{2}$$
$$y = -4 \ \text{or} \ y = 7$$

The solution set is $\{-4, 7\}$.

19.

$$z^2 - 6z - 3 = 0$$
$$z^2 - 6z = 3$$
$$z^2 - 6z + \left(\frac{1}{2} \cdot (-6)\right)^2 = 3 + \left(\frac{1}{2} \cdot (-6)\right)^2$$
$$z^2 - 6z + 9 = 3 + 9$$
$$(z - 3)^2 = 12$$
$$z - 3 = \pm\sqrt{12}$$
$$z - 3 = \pm 2\sqrt{3}$$
$$z = 3 \pm 2\sqrt{3}$$

The solution set is $\left\{3 - 2\sqrt{3}, \ 3 + 2\sqrt{3}\right\}$.

20.

$$a^2 - 5a - 7 = 0$$
$$a^2 - 5a = 7$$
$$a^2 - 5a + \left(\frac{1}{2} \cdot (-5)\right)^2 = 7 + \left(\frac{1}{2} \cdot (-5)\right)^2$$
$$a^2 - 5a + \frac{25}{4} = 7 + \frac{25}{4}$$
$$\left(a - \frac{5}{2}\right)^2 = \frac{53}{4}$$
$$a - \frac{5}{2} = \pm\sqrt{\frac{53}{4}}$$
$$a - \frac{5}{2} = \pm\frac{\sqrt{53}}{2}$$
$$a = \frac{5}{2} \pm \frac{\sqrt{53}}{2}$$

The solution set is $\left\{\frac{5}{2} - \frac{\sqrt{53}}{2}, \ \frac{5}{2} + \frac{\sqrt{53}}{2}\right\}$.

21.

$$b^2 + b + 7 = 0$$
$$b^2 + b = -7$$
$$b^2 + b + \left(\frac{1}{2} \cdot 1\right)^2 = -7 + \left(\frac{1}{2} \cdot 1\right)^2$$
$$b^2 + b + \frac{1}{4} = -7 + \frac{1}{4}$$
$$\left(b + \frac{1}{2}\right)^2 = -\frac{27}{4}$$
$$b + \frac{1}{2} = \pm\sqrt{-\frac{27}{4}}$$
$$b + \frac{1}{2} = \pm\frac{3\sqrt{3}}{2}i$$
$$b = -\frac{1}{2} \pm \frac{3\sqrt{3}}{2}i$$

The solution set is $\left\{-\frac{1}{2} - \frac{3\sqrt{3}}{2}i, \ -\frac{1}{2} + \frac{3\sqrt{3}}{2}i\right\}$.

22.
$$c^2 - 6c + 17 = 0$$
$$c^2 - 6c = -17$$
$$c^2 - 6c + \left(\frac{1}{2} \cdot (-6)\right)^2 = -17 + \left(\frac{1}{2} \cdot (-6)\right)^2$$
$$c^2 - 6c + 9 = -17 + 9$$
$$(c - 3)^2 = -8$$
$$c - 3 = \pm\sqrt{-8}$$
$$c - 3 = \pm 2\sqrt{2}\,i$$
$$c = 3 \pm 2\sqrt{2}\,i$$
The solution set is $\left\{3 - 2\sqrt{2}\,i,\ 3 + 2\sqrt{2}\,i\right\}$.

23.
$$2d^2 - 7d + 3 = 0$$
$$\frac{2d^2 - 7d + 3}{2} = \frac{0}{2}$$
$$d^2 - \frac{7}{2}d + \frac{3}{2} = 0$$
$$d^2 - \frac{7}{2}d = -\frac{3}{2}$$
$$d^2 - \frac{7}{2}d + \left[\frac{1}{2} \cdot \left(-\frac{7}{2}\right)\right]^2 = -\frac{3}{2} + \left[\frac{1}{2} \cdot \left(-\frac{7}{2}\right)\right]^2$$
$$d^2 - \frac{7}{2}d + \frac{49}{16} = -\frac{3}{2} + \frac{49}{16}$$
$$\left(d - \frac{7}{4}\right)^2 = \frac{25}{16}$$
$$d - \frac{7}{4} = \pm\sqrt{\frac{25}{16}}$$
$$d - \frac{7}{4} = \pm\frac{5}{4}$$
$$d = \frac{7}{4} \pm \frac{5}{4}$$
$$d = \frac{1}{2} \ \text{ or } \ d = 3$$
The solution set is $\left\{\frac{1}{2},\ 3\right\}$.

24.
$$2w^2 + 2w + 5 = 0$$
$$\frac{2w^2 + 2w + 5}{2} = \frac{0}{2}$$
$$w^2 + w + \frac{5}{2} = 0$$
$$w^2 + w = -\frac{5}{2}$$
$$w^2 + w + \left(\frac{1}{2} \cdot 1\right)^2 = -\frac{5}{2} + \left(\frac{1}{2} \cdot 1\right)^2$$
$$w^2 + w + \frac{1}{4} = -\frac{5}{2} + \frac{1}{4}$$
$$\left(w + \frac{1}{2}\right)^2 = -\frac{9}{4}$$
$$w + \frac{1}{2} = \pm\sqrt{-\frac{9}{4}}$$
$$w + \frac{1}{2} = \pm\frac{3}{2}i$$
$$w = -\frac{1}{2} \pm \frac{3}{2}i$$
The solution set is $\left\{-\frac{1}{2} - \frac{3}{2}i,\ -\frac{1}{2} + \frac{3}{2}i\right\}$.

25.
$$3x^2 - 9x + 8 = 0$$
$$\frac{3x^2 - 9x + 8}{3} = \frac{0}{3}$$
$$x^2 - 3x + \frac{8}{3} = 0$$
$$x^2 - 3x = -\frac{8}{3}$$
$$x^2 - 3x + \left(\frac{1}{2}\cdot(-3)\right)^2 = -\frac{8}{3} + \left(\frac{1}{2}\cdot(-3)\right)^2$$
$$x^2 - 3x + \frac{9}{4} = -\frac{8}{3} + \frac{9}{4}$$
$$\left(x - \frac{3}{2}\right)^2 = -\frac{5}{12}$$
$$x - \frac{3}{2} = \pm\sqrt{-\frac{5}{12}}$$
$$x - \frac{3}{2} = \pm\frac{\sqrt{5}}{\sqrt{12}}i\cdot\frac{\sqrt{3}}{\sqrt{3}}$$
$$x - \frac{3}{2} = \pm\frac{\sqrt{15}}{\sqrt{36}}i$$
$$x - \frac{3}{2} = \pm\frac{\sqrt{15}}{6}i$$
$$x = \frac{3}{2} \pm \frac{\sqrt{15}}{6}i$$

The solution set is $\left\{\frac{3}{2} - \frac{\sqrt{15}}{6}i,\ \frac{3}{2} + \frac{\sqrt{15}}{6}i\right\}$.

26.
$$3x^2 + 4x - 2 = 0$$
$$x^2 + \frac{4}{3}x - \frac{2}{3} = 0$$
$$x^2 + \frac{4}{3}x = \frac{2}{3}$$
$$x^2 + \frac{4}{3}x + \left(\frac{1}{2}\cdot\frac{4}{3}\right)^2 = \frac{2}{3} + \left(\frac{1}{2}\cdot\frac{4}{3}\right)^2$$
$$x^2 + \frac{4}{3}x + \frac{4}{9} = \frac{2}{3} + \frac{4}{9}$$
$$\left(x + \frac{2}{3}\right)^2 = \frac{10}{9}$$
$$x + \frac{2}{3} = \pm\sqrt{\frac{10}{9}}$$
$$x + \frac{2}{3} = \pm\frac{\sqrt{10}}{3}$$
$$x = -\frac{2}{3} \pm \frac{\sqrt{10}}{3}$$

The solution set is $\left\{-\frac{2}{3} - \frac{\sqrt{10}}{3},\ -\frac{2}{3} + \frac{\sqrt{10}}{3}\right\}$.

27.
$$c^2 = 9^2 + 12^2$$
$$= 81 + 144$$
$$= 225$$
$$c = \sqrt{225}$$
$$= 15$$

28.
$$c^2 = 8^2 + 8^2$$
$$= 64 + 64$$
$$= 128$$
$$c = \sqrt{128}$$
$$= 8\sqrt{2}$$

29.
$$c^2 = 3^2 + 6^2$$
$$= 9 + 36$$
$$= 45$$
$$c = \sqrt{45}$$
$$= 3\sqrt{5}$$

30.
$$c^2 = 10^2 + 24^2$$
$$= 100 + 576$$
$$= 676$$
$$c = \sqrt{676}$$
$$= 26$$

31.
$$c^2 = 5^2 + \left(\sqrt{11}\right)^2$$
$$= 25 + 11$$
$$= 36$$
$$c = \sqrt{36}$$
$$= 6$$

32.
$$c^2 = 6^2 + \left(\sqrt{13}\right)^2$$
$$= 36 + 13$$
$$= 49$$
$$c = \sqrt{49}$$
$$= 7$$

33.
$$c^2 = a^2 + b^2$$
$$12^2 = 9^2 + b^2$$
$$144 = 81 + b^2$$
$$63 = b^2$$
$$b = \sqrt{63}$$
$$b = 3\sqrt{7}$$

34. $c^2 = a^2 + b^2$

$10^2 = a^2 + 5^2$

$100 = a^2 + 25$

$75 = a^2$

$a = \sqrt{75}$

$a = 5\sqrt{3}$

35. $c^2 = a^2 + b^2$

$17^2 = a^2 + 6^2$

$289 = a^2 + 36$

$253 = a^2$

$a = \sqrt{253}$

36. For this problem, we are actually looking for the hypotenuse of a right triangle that has legs 90 feet and 90 feet. We use the Pythagorean Theorem to find the desired distance:

$c^2 = 90^2 + 90^2$

$= 8100 + 8100$

$= 16,200$

$c = \sqrt{16,200}$

$= 90\sqrt{2}$

≈ 127.3

The distance from home plate to 2nd base is exactly $90\sqrt{2}$ feet or approximately 127.3 feet.

37. $x^2 - x - 20 = 0$

For this equation, $a = 1$, $b = -1$, and $c = -20$.

$x = \dfrac{-(-1) \pm \sqrt{(-1)^2 - 4(1)(-20)}}{2(1)}$

$= \dfrac{1 \pm \sqrt{1 + 80}}{2}$

$= \dfrac{1 \pm \sqrt{81}}{2}$

$= \dfrac{1 \pm 9}{2}$

$x = \dfrac{1-9}{2} \quad \text{or} \quad x = \dfrac{1+9}{2}$

$= \dfrac{-8}{2} \quad \text{or} \quad = \dfrac{10}{2}$

$= -4 \quad \text{or} \quad = 5$

The solution set is $\{-4,\ 5\}$.

38. $4y^2 = 8y + 21$

$4y^2 - 8y - 21 = 0$

For this equation, $a = 4$, $b = -8$, and $c = -21$.

$y = \dfrac{-(-8) \pm \sqrt{(-8)^2 - 4(4)(-21)}}{2(4)}$

$= \dfrac{8 \pm \sqrt{64 + 336}}{8}$

$= \dfrac{8 \pm \sqrt{400}}{8}$

$= \dfrac{8 \pm 20}{8}$

$y = \dfrac{8-20}{8} \quad \text{or} \quad y = \dfrac{8+20}{8}$

$= \dfrac{-12}{8} \quad \text{or} \quad = \dfrac{28}{8}$

$= -\dfrac{3}{2} \quad \text{or} \quad = \dfrac{7}{2}$

The solution set is $\left\{ -\dfrac{3}{2},\ \dfrac{7}{2} \right\}$.

39. $3p^2 + 8p = -3$

$3p^2 + 8p + 3 = 0$

For this equation, $a = 3$, $b = 8$, and $c = 3$.

$$p = \frac{-8 \pm \sqrt{8^2 - 4(3)(3)}}{2(3)}$$

$$= \frac{-8 \pm \sqrt{64 - 36}}{6}$$

$$= \frac{-8 \pm \sqrt{28}}{6}$$

$$= \frac{-8 \pm 2\sqrt{7}}{6}$$

$$= \frac{-8}{6} \pm \frac{2\sqrt{7}}{6}$$

$$= -\frac{4}{3} \pm \frac{\sqrt{7}}{3}$$

The solution set is $\left\{ -\frac{4}{3} - \frac{\sqrt{7}}{3}, \ -\frac{4}{3} + \frac{\sqrt{7}}{3} \right\}$.

40. $2q^2 - 3 = 4q$

$2q^2 - 4q - 3 = 0$

For this equation, $a = 2$, $b = -4$, and $c = -3$.

$$q = \frac{-(-4) \pm \sqrt{(-4)^2 - 4(2)(-3)}}{2(2)}$$

$$= \frac{4 \pm \sqrt{16 + 24}}{4}$$

$$= \frac{4 \pm \sqrt{40}}{4}$$

$$= \frac{4 \pm 2\sqrt{10}}{4}$$

$$= \frac{4}{4} \pm \frac{2\sqrt{10}}{4}$$

$$= 1 \pm \frac{\sqrt{10}}{2}$$

The solution set is $\left\{ 1 - \frac{\sqrt{10}}{2}, \ 1 + \frac{\sqrt{10}}{2} \right\}$.

41. $3w^2 + w = -3$

$3w^2 + w + 3 = 0$

For this equation, $a = 3$, $b = 1$, and $c = 3$.

$$w = \frac{-1 \pm \sqrt{1^2 - 4(3)(3)}}{2(3)}$$

$$= \frac{-1 \pm \sqrt{1 - 36}}{6}$$

$$= \frac{-1 \pm \sqrt{-35}}{6}$$

$$= \frac{-1 \pm \sqrt{35}\ i}{6}$$

$$= -\frac{1}{6} \pm \frac{\sqrt{35}}{6}i$$

The solution set is $\left\{ -\frac{1}{6} - \frac{\sqrt{35}}{6}i, \ -\frac{1}{6} + \frac{\sqrt{35}}{6}i \right\}$.

42. $9z^2 + 16 = 24z$

$9z^2 - 24z + 16 = 0$

For this equation, $a = 9$, $b = -24$, and $c = 16$.

$$z = \frac{-(-24) \pm \sqrt{(-24)^2 - 4(9)(16)}}{2(9)}$$

$$= \frac{24 \pm \sqrt{576 - 576}}{18}$$

$$= \frac{24 \pm \sqrt{0}}{18}$$

$$= \frac{24}{18}$$

$$= \frac{4}{3}$$

The solution set is $\left\{ \frac{4}{3} \right\}$. It is a double root.

43. $m^2 - 4m + 2 = 0$

For this equation, $a = 1$, $b = -4$, and $c = 2$.

$$m = \frac{-(-4) \pm \sqrt{(-4)^2 - 4(1)(2)}}{2(1)}$$

$$= \frac{4 \pm \sqrt{16 - 8}}{2}$$

$$= \frac{4 \pm \sqrt{8}}{2}$$

$$= \frac{4 \pm 2\sqrt{2}}{2}$$

$$= \frac{4}{2} \pm \frac{2\sqrt{2}}{2}$$

$$= 2 \pm \sqrt{2}$$

The solution set is $\left\{ 2 - \sqrt{2}, \ 2 + \sqrt{2} \right\}$.

44. $5n^2 + 4n + 1 = 0$

For this equation, $a = 5$, $b = 4$, and $c = 1$.

$$n = \frac{-4 \pm \sqrt{4^2 - 4(5)(1)}}{2(5)}$$

$$= \frac{-4 \pm \sqrt{16 - 20}}{10}$$

$$= \frac{-4 \pm \sqrt{-4}}{10}$$

$$= \frac{-4 \pm 2i}{10}$$

$$= -\frac{4}{10} \pm \frac{2i}{10} = -\frac{2}{5} \pm \frac{1}{5}i$$

The solution set is $\left\{ -\frac{2}{5} - \frac{1}{5}i, \ -\frac{2}{5} + \frac{1}{5}i \right\}$.

45. $5x + 13 = -x^2$

$x^2 + 5x + 13 = 0$

For this equation, $a = 1$, $b = 5$, and $c = 13$.

$$x = \frac{-5 \pm \sqrt{5^2 - 4(1)(13)}}{2(1)}$$

$$= \frac{-5 \pm \sqrt{25 - 52}}{2}$$

$$= \frac{-5 \pm \sqrt{-27}}{2}$$

$$= \frac{-5 \pm 3\sqrt{3}\ i}{2} = -\frac{5}{2} \pm \frac{3\sqrt{3}}{2}i$$

The solution set is $\left\{ -\frac{5}{2} - \frac{3\sqrt{3}}{2}i, \ -\frac{5}{2} + \frac{3\sqrt{3}}{2}i \right\}$.

46. $-2y^2 = 6y + 7$

$0 = 2y^2 + 6y + 7$

For this equation, $a = 2$, $b = 6$, and $c = 7$.

$$y = \frac{-6 \pm \sqrt{6^2 - 4(2)(7)}}{2(2)}$$

$$= \frac{-6 \pm \sqrt{36 - 56}}{4}$$

$$= \frac{-6 \pm \sqrt{-20}}{4}$$

$$= \frac{-6 \pm 2\sqrt{5}\ i}{4}$$

$$= -\frac{6}{4} \pm \frac{2\sqrt{5}}{4}i = -\frac{3}{2} \pm \frac{\sqrt{5}}{2}i$$

The solution set is $\left\{ -\frac{3}{2} - \frac{\sqrt{5}}{2}i, \ -\frac{3}{2} + \frac{\sqrt{5}}{2}i \right\}$.

47. $p^2 - 5p - 8 = 0$

For this equation, $a = 1$, $b = -5$, and $c = -8$.

$b^2 - 4ac = (-5)^2 - 4(1)(-8) = 25 + 32 = 57$

Because $b^2 - 4ac = 57$ is positive, but not a perfect square, the quadratic equation will have two irrational solutions.

48. $m^2 + 8m + 16 = 0$

For this equation, $a = 1$, $b = 8$, and $c = 16$.

$b^2 - 4ac = 8^2 - 4(1)(16) = 64 - 64 = 0$

Because $b^2 - 4ac = 0$, the quadratic equation will have one repeated real solution.

49. $3n^2 + n = -4$

$3n^2 + n + 4 = 0$

For this equation, $a = 3$, $b = 1$, and $c = 4$.

$b^2 - 4ac = 1^2 - 4(3)(4) = 1 - 48 = -47$

Because $b^2 - 4ac = -47$ is negative, the quadratic equation will have two complex solutions that are not real. The solutions will be complex conjugates of each other.

50. $7w^2 + 3 = 8w$

$7w^2 - 8w + 3 = 0$

For this equation, $a = 7$, $b = -8$, and $c = 3$.

$b^2 - 4ac = (-8)^2 - 4(7)(3) = 64 - 84 = -20$

Because $b^2 - 4ac = -20$ is negative, the quadratic equation will have two complex solutions that are not real. The solutions will be complex conjugates of each other.

51. $4x^2 + 49 = 28x$

$4x^2 - 28x + 49 = 0$

For this equation, $a = 4$, $b = -28$, and $c = 49$.

$b^2 - 4ac = (-28)^2 - 4(4)(49) = 784 - 784 = 0$

Because $b^2 - 4ac = 0$, the quadratic equation will have one repeated real solution.

52. $11z - 12 = 2z^2$

$0 = 2z^2 - 11z + 12$

For this equation, $a = 2$, $b = -11$, and $c = 12$.

$b^2 - 4ac = (-11)^2 - 4(2)(12) = 121 - 96 = 25$

Because $b^2 - 4ac = 25$ is positive and a perfect square, the quadratic equation will have two rational solutions.

53. $x^2 + 8x - 9 = 0$

Because this equation factors easily, solve by factoring.

$(x+9)(x-1) = 0$

$x+9=0$ or $x-1=0$

$x=-9$ or $x=1$

The solution set is $\{-9, 1\}$.

54. $6p^2 + 13p = 5$

$6p^2 + 13p - 5 = 0$

Because this equation factors easily, solve by factoring.

$(3p-1)(2p+5) = 0$

$3p-1=0$ or $2p+5=0$

$3p=1$ or $2p=-5$

$p=\dfrac{1}{3}$ or $p=-\dfrac{5}{2}$

The solution set is $\left\{-\dfrac{5}{2}, \dfrac{1}{3}\right\}$.

55. $n^2 + 13 = -4n$

$n^2 + 4n + 13 = 0$

Because this equation does not easily factor, solve by using the quadratic formula. For this equation, $a=1$, $b=4$, and $c=13$.

$n = \dfrac{-4 \pm \sqrt{4^2 - 4(1)(13)}}{2(1)}$

$= \dfrac{-4 \pm \sqrt{16 - 52}}{2}$

$= \dfrac{-4 \pm \sqrt{-36}}{2}$

$= \dfrac{-4 \pm 6i}{2}$

$= -\dfrac{4}{2} \pm \dfrac{6i}{2}$

$= -2 \pm 3i$

The solution set is $\{-2-3i,\, -2+3i\}$.

56. $5y^2 - 60 = 0$

Because this equation has no linear term, solve by using the square root method.

$5y^2 = 60$

$y^2 = 12$

$y = \pm\sqrt{12}$

$y = \pm 2\sqrt{3}$

The solution set is $\left\{-2\sqrt{3},\, 2\sqrt{3}\right\}$.

57. $\dfrac{1}{4}q^2 - \dfrac{1}{2}q - \dfrac{3}{8} = 0$

$8\left(\dfrac{1}{4}q^2 - \dfrac{1}{2}q - \dfrac{3}{8}\right) = 8(0)$

$2q^2 - 4q - 3 = 0$

Because this equation does not easily factor, solve by using the quadratic formula. For this equation, $a=2$, $b=-4$, and $c=-3$.

$q = \dfrac{-(-4) \pm \sqrt{(-4)^2 - 4(2)(-3)}}{2(2)}$

$= \dfrac{4 \pm \sqrt{16 + 24}}{4}$

$= \dfrac{4 \pm \sqrt{40}}{4}$

$= \dfrac{4 \pm 2\sqrt{10}}{4}$

$= \dfrac{4}{4} \pm \dfrac{2\sqrt{10}}{4}$

$= 1 \pm \dfrac{\sqrt{10}}{2}$

The solution set is $\left\{1 - \dfrac{\sqrt{10}}{2},\, 1 + \dfrac{\sqrt{10}}{2}\right\}$.

58. $\dfrac{1}{8}m^2 + m + \dfrac{5}{2} = 0$

$8\left(\dfrac{1}{8}m^2 + m + \dfrac{5}{2}\right) = 8(0)$

$m^2 + 8m + 20 = 0$

Because this equation does not easily factor, solve by using the quadratic formula. For this equation, $a=1$, $b=8$, and $c=20$.

$$m = \frac{-8 \pm \sqrt{8^2 - 4(1)(20)}}{2(1)}$$

$$= \frac{-8 \pm \sqrt{64 - 80}}{2}$$

$$= \frac{-8 \pm \sqrt{-16}}{2}$$

$$= \frac{-8 \pm 4i}{2}$$

$$= -\frac{8}{2} \pm \frac{4i}{2}$$

$$= -4 \pm 2i$$

The solution set is $\{-4 - 2i, \ -4 + 2i\}$.

59. $(w-8)(w+6) = -33$

$$w^2 - 2w - 48 = -33$$

$$w^2 - 2w - 15 = 0$$

Because this equation factors easily, solve by factoring.

$$(w-5)(w+3) = 0$$

$$w - 5 = 0 \quad \text{or} \quad w + 3 = 0$$

$$w = 5 \quad \text{or} \qquad w = -3$$

The solution set is $\{-3, \ 5\}$.

60. $(x-3)(x+1) = -2$

$$x^2 - 2x - 3 = -2$$

$$x^2 - 2x - 1 = 0$$

Because this equation does not easily factor, solve by using the quadratic formula. For this equation, $a = 1$, $b = -2$, and $c = -1$.

$$x = \frac{-(-2) \pm \sqrt{(-2)^2 - 4(1)(-1)}}{2(1)}$$

$$= \frac{2 \pm \sqrt{4+4}}{2}$$

$$= \frac{2 \pm \sqrt{8}}{2}$$

$$= \frac{2 \pm 2\sqrt{2}}{2}$$

$$= \frac{2}{2} \pm \frac{2\sqrt{2}}{2}$$

$$= 1 \pm \sqrt{2}$$

The solution set is $\{1 - \sqrt{2}, \ 1 + \sqrt{2}\}$.

61. $9z^2 = 16$

Because this equation has no linear term, solve by using the square root method.

$$z^2 = \frac{16}{9}$$

$$z = \pm\sqrt{\frac{16}{9}}$$

$$z = \pm\frac{4}{3}$$

The solution set is $\left\{-\frac{4}{3}, \ \frac{4}{3}\right\}$.

62.
$$\frac{1-2x}{x^2+5} = 1$$

$$\left(x^2+5\right)\left(\frac{1-2x}{x^2+5}\right) = \left(x^2+5\right)(1)$$

$$1 - 2x = x^2 + 5$$

$$0 = x^2 + 2x + 4$$

Because this equation does not easily factor, solve by using the quadratic formula. For this equation, $a = 1$, $b = 2$, and $c = 4$.

$$x = \frac{-2 \pm \sqrt{2^2 - 4(1)(4)}}{2(1)}$$

$$= \frac{-2 \pm \sqrt{4 - 16}}{2}$$

$$= \frac{-2 \pm \sqrt{-12}}{2}$$

$$= \frac{-2 \pm 2\sqrt{3}\, i}{2}$$

$$= -1 \pm \sqrt{3}\, i$$

The solution set is $\left\{-1 - \sqrt{3}\, i, \ -1 + \sqrt{3}\, i\right\}$.

63.
$$(x+2)^2 + (x-5)^2 = (x+3)^2$$

$$x^2 + 4x + 4 + x^2 - 10x + 25 = x^2 + 6x + 9$$

$$2x^2 - 6x + 29 = x^2 + 6x + 9$$

$$x^2 - 12x + 20 = 0$$

$$(x-2)(x-10) = 0$$

$$x - 2 = 0 \ \text{ or } \ x - 10 = 0$$

$$x = 2 \ \text{ or } \qquad x = 10$$

Disregard $x = 2$ because this value will cause the length of one of the legs to be negative: $x - 5 = 2 - 5 = -3$. Thus, $x = 10$ is the only viable answer. Now, $x - 5 = 10 - 5 = 5$, $x + 2 = 10 + 2 = 12$, and $x + 3 = 10 + 3 = 13$. The three measurements are 5, 12, and 13.

64. Let x represent the length of the rectangle. Then $x-3$ will represent the width.

$$x(x-3)=108$$
$$x^2-3x=108$$
$$x^2-3x-108=0$$
$$(x-12)(x+9)=0$$
$$x-12=0 \quad \text{or} \quad x+9=0$$
$$x=12 \quad \text{or} \quad x=-9$$

Disregard $x=-9$ because x represents the length of the rectangle, which must be positive. Thus, $x=12$ is the only viable answer. Now, $x-3=12-3=9$. Thus, the dimensions of the rectangle are 12 centimeters by 9 centimeters.

65. (a) $-0.2x^2+180x=36,000$
$$0=0.2x^2-180x+36,000$$

For this equation, $a=0.2$, $b=-180$, and $c=36,000$.

$$x=\frac{-(-180)\pm\sqrt{(-180)^2-4(0.2)(36,000)}}{2(0.2)}$$
$$=\frac{180\pm\sqrt{32,400-28,800}}{0.4}$$
$$=\frac{180\pm\sqrt{3600}}{0.4}$$
$$=\frac{180\pm60}{0.4}$$

$$x=\frac{180-60}{0.4} \quad \text{or} \quad x=\frac{180+60}{0.4}$$
$$=300 \qquad \text{or} \qquad =600$$

The revenue will be $36,000 per week if either 300 or 600 cellular phones are sold per week.

(b) $-0.2x^2+180x=40,500$
$$0=0.2x^2-180x+40,500$$

For this equation, $a=0.2$, $b=-180$, and $c=40,500$.

$$x=\frac{-(-180)\pm\sqrt{(-180)^2-4(0.2)(40,500)}}{2(0.2)}$$
$$=\frac{180\pm\sqrt{32,400-32,400}}{0.4}$$
$$=\frac{180\pm\sqrt{0}}{0.4}$$
$$=\frac{180}{0.4}$$
$$=450$$

The revenue will be $40,500 per week if 450 cellular phones are sold per week.

66. (a) $$200=-16t^2+50t+180$$
$$16t^2-50t+20=0$$

For this equation, $a=16$, $b=-50$, and $c=20$.

$$t=\frac{-(-50)\pm\sqrt{(-50)^2-4(16)(20)}}{2(16)}$$
$$=\frac{50\pm\sqrt{2500-1280}}{32}$$
$$=\frac{50\pm\sqrt{1220}}{32}$$
$$=\frac{50\pm2\sqrt{305}}{32}$$
$$=\frac{50}{32}\pm\frac{2\sqrt{305}}{32}$$
$$=\frac{25}{16}\pm\frac{\sqrt{305}}{16}$$

$$t=\frac{25}{16}-\frac{\sqrt{305}}{16} \quad \text{or} \quad t=\frac{25}{16}+\frac{\sqrt{305}}{16}$$
$$\approx 0.471 \qquad \text{or} \qquad \approx 2.654$$

Rounding to the nearest tenth, the height of the ball will be 200 feet after approximately 0.5 seconds and after approximately 2.7 seconds.

(b) $$100=-16t^2+50t+180$$
$$16t^2-50t-80=0$$

For this equation, $a=16$, $b=-50$, and $c=-80$.

$$t=\frac{-(-50)\pm\sqrt{(-50)^2-4(16)(-80)}}{2(16)}$$
$$=\frac{50\pm\sqrt{2500+5120}}{32}$$
$$=\frac{50\pm\sqrt{7620}}{32}$$
$$=\frac{50\pm2\sqrt{1905}}{32}$$
$$=\frac{50}{32}\pm\frac{2\sqrt{1905}}{32}$$
$$=\frac{25}{16}\pm\frac{\sqrt{1905}}{16}$$

$$t=\frac{25}{16}-\frac{\sqrt{1905}}{16} \quad \text{or} \quad t=\frac{25}{16}+\frac{\sqrt{1905}}{16}$$
$$\approx -1.165 \qquad \text{or} \qquad \approx 4.290$$

Because time cannot be negative, we disregard -1.165. Thus, $t \approx 4.290$ is the only viable answer. Rounding to the nearest tenth, the height of the ball will be 100 feet after approximately 4.3 seconds.

(c)
$$300 = -16t^2 + 50t + 180$$
$$16t^2 - 50t + 120 = 0$$

For this equation, $a = 16$, $b = -50$, and $c = 120$.

$$t = \frac{-(-50) \pm \sqrt{(-50)^2 - 4(16)(120)}}{2(16)}$$
$$= \frac{50 \pm \sqrt{2500 - 7680}}{32}$$
$$= \frac{50 \pm \sqrt{-5180}}{32}$$
$$= \frac{50 \pm 2\sqrt{1295}\, i}{32}$$
$$= \frac{50}{32} \pm \frac{2\sqrt{1295}}{32} i$$
$$= \frac{25}{16} \pm \frac{\sqrt{1295}}{16} i$$
$$t = \frac{25}{16} - \frac{\sqrt{1295}}{16} i \text{ or } t = \frac{25}{16} + \frac{\sqrt{1295}}{16} i$$

The ball will never reach a height of 300 feet. This is clear because the solutions to the equation above are complex solutions that are not real.

67. Let x represent the speed the boat would travel in still water.

	Distance	**Rate**	**Time**
Up Stream	10	$x - 3$	$\dfrac{10}{x-3}$
Down Stream	10	$x + 3$	$\dfrac{10}{x+3}$

$$\frac{10}{x-3} + \frac{10}{x+3} = 2$$
$$(x-3)(x+3)\left(\frac{10}{x-3} + \frac{10}{x+3}\right) = (x-3)(x+3)(2)$$
$$10(x+3) + 10(x-3) = \left(x^2 - 9\right)(2)$$
$$10x + 30 + 10x - 30 = 2x^2 - 18$$
$$20x = 2x^2 - 18$$
$$0 = 2x^2 - 20x - 18$$

For this equation, $a = 2$, $b = -20$, and $c = -18$.

$$x = \frac{-(-20) \pm \sqrt{(-20)^2 - 4(2)(-18)}}{2(2)}$$
$$= \frac{20 \pm \sqrt{400 + 144}}{4}$$
$$= \frac{20 \pm \sqrt{544}}{4}$$
$$= \frac{20 \pm 4\sqrt{34}}{4}$$
$$= 5 \pm \sqrt{34}$$
$$x = 5 - \sqrt{34} \text{ or } x = 5 + \sqrt{34}$$
$$\approx -0.831 \text{ or } \approx 10.831$$

Because the speed should be positive, we disregard $x \approx -0.831$. Thus, the only viable answer is $x \approx 10.831$. Rounding to the nearest tenth, the boat would travel approximately 10.8 miles per hour in still water.

68. Let t represent the time required for Beth to wash the car alone. Then $t - 14$ will represent the time required for Tom to wash the car alone.

$$\begin{pmatrix} \text{Part done} \\ \text{by Beth in} \\ \text{1 minute} \end{pmatrix} + \begin{pmatrix} \text{Part done} \\ \text{by Tom in} \\ \text{1 minute} \end{pmatrix} = \begin{pmatrix} \text{Part done} \\ \text{together in} \\ \text{1 minute} \end{pmatrix}$$

$$\frac{1}{t} + \frac{1}{t-14} = \frac{1}{30}$$
$$30t(t-14)\left(\frac{1}{t} + \frac{1}{t-14}\right) = 30t(t-14)\left(\frac{1}{30}\right)$$
$$30(t-14) + 30t = t(t-14)$$
$$30t - 420 + 30t = t^2 - 14t$$
$$60t - 420 = t^2 - 14t$$
$$0 = t^2 - 74t + 420$$

For this equation, $a = 1$, $b = -74$, and $c = 420$.

$$t = \frac{-(-74) \pm \sqrt{(-74)^2 - 4(1)(420)}}{2(1)}$$
$$= \frac{74 \pm \sqrt{5476 - 1680}}{2}$$
$$= \frac{74 \pm \sqrt{3796}}{2}$$
$$= \frac{74 \pm 2\sqrt{949}}{2}$$
$$= 37 \pm \sqrt{949}$$
$$t = 37 - \sqrt{949} \text{ or } t = 37 + \sqrt{949}$$
$$\approx 6.194 \quad \text{or} \quad \approx 67.806$$

Disregard $t \approx 6.194$ because this value makes Tom's time negative:

$t - 14 = 6.194 - 14 = -7.806$. The only viable answer is $t \approx 67.806$ minutes. Working alone, it will take Beth approximately 67.8 minutes to wash the car.

69.
$$x^4 + 7x^2 - 144 = 0$$
$$\left(x^2\right)^2 + 7\left(x^2\right) - 144 = 0$$

Let $u = x^2$.
$$u^2 + 7u - 144 = 0$$
$$(u - 9)(u + 16) = 0$$

$u - 9 = 0$	or	$u + 16 = 0$
$u = 9$	or	$u = -16$
$x^2 = 9$	or	$x^2 = -16$
$x = \pm\sqrt{9}$	or	$x = \pm\sqrt{-16}$
$x = \pm 3$	or	$x = \pm 4i$

Check:
$$x = -3: \quad (-3)^4 + 7(-3)^2 - 144 \stackrel{?}{=} 0$$
$$81 + 7 \cdot 9 - 144 \stackrel{?}{=} 0$$
$$81 + 63 - 144 \stackrel{?}{=} 0$$
$$144 = 144 \checkmark$$

$$x = 3: \quad 3^4 + 7(3)^2 - 144 \stackrel{?}{=} 0$$
$$81 + 7 \cdot 9 - 144 \stackrel{?}{=} 0$$
$$81 + 63 - 144 \stackrel{?}{=} 0$$
$$0 = 0 \checkmark$$

$$x = -4i: \quad (-4i)^4 + 7(-4i)^2 - 144 \stackrel{?}{=} 0$$
$$256i^4 + 7 \cdot 16i^2 - 144 \stackrel{?}{=} 0$$
$$256(1) + 7 \cdot 16(-1) - 144 \stackrel{?}{=} 0$$
$$256 - 112 - 144 \stackrel{?}{=} 0$$
$$0 = 0 \checkmark$$

$$x = 4i: \quad (4i)^4 + 7(4i)^2 - 144 \stackrel{?}{=} 0$$
$$256i^4 + 7 \cdot 16i^2 - 144 \stackrel{?}{=} 0$$
$$256(1) + 7 \cdot 16(-1) - 144 \stackrel{?}{=} 0$$
$$256 - 112 - 144 \stackrel{?}{=} 0$$
$$0 = 0 \checkmark$$

All check; the solution set is $\{-3,\ 3,\ -4i,\ 4i\}$.

70.
$$4w^4 + 5w^2 - 6 = 0$$
$$4\left(w^2\right)^2 + 5\left(w^2\right) - 6 = 0$$

Let $u = w^2$.

$$4u^2 + 5u - 6 = 0$$
$$(4u - 3)(u + 2) = 0$$

$4u - 3 = 0$	or	$u + 2 = 0$
$4u = 3$	or	$u = -2$

$$u = \frac{3}{4}$$

$w^2 = \dfrac{3}{4}$	or	$w^2 = -2$
$w = \pm\sqrt{\dfrac{3}{4}}$	or	$w = \pm\sqrt{-2}$
$w = \pm\dfrac{\sqrt{3}}{2}$	or	$w = \pm\sqrt{2}\,i$

Check:
$$w = -\frac{\sqrt{3}}{2}: \quad 4\left(-\frac{\sqrt{3}}{2}\right)^4 + 5\left(-\frac{\sqrt{3}}{2}\right)^2 - 6 \stackrel{?}{=} 0$$
$$4\left(\frac{9}{16}\right) + 5\left(\frac{3}{4}\right) - 6 \stackrel{?}{=} 0$$
$$\frac{9}{4} + \frac{15}{4} - 6 \stackrel{?}{=} 0$$
$$0 = 0 \checkmark$$

$$w = \frac{\sqrt{3}}{2}: \quad 4\left(\frac{\sqrt{3}}{2}\right)^4 + 5\left(\frac{\sqrt{3}}{2}\right)^2 - 6 \stackrel{?}{=} 0$$
$$4\left(\frac{9}{16}\right) + 5\left(\frac{3}{4}\right) - 6 \stackrel{?}{=} 0$$
$$\frac{9}{4} + \frac{15}{4} - 6 \stackrel{?}{=} 0$$
$$0 = 0 \checkmark$$

$$w = -\sqrt{2}\,i: \quad 4\left(-\sqrt{2}\,i\right)^4 + 5\left(-\sqrt{2}\,i\right)^2 - 6 \stackrel{?}{=} 0$$
$$4 \cdot 4i^4 + 5 \cdot 2i^2 - 6 \stackrel{?}{=} 0$$
$$4 \cdot 4(1) + 5 \cdot 2(-1) - 6 \stackrel{?}{=} 0$$
$$16 - 10 - 6 \stackrel{?}{=} 0$$
$$0 = 0 \checkmark$$

$$w = \sqrt{2}\,i: \quad 4\left(\sqrt{2}\,i\right)^4 + 5\left(\sqrt{2}\,i\right)^2 - 6 \stackrel{?}{=} 0$$
$$4 \cdot 4i^4 + 5 \cdot 2i^2 - 6 \stackrel{?}{=} 0$$
$$4 \cdot 4(1) + 5 \cdot 2(-1) - 6 \stackrel{?}{=} 0$$
$$16 - 10 - 6 \stackrel{?}{=} 0$$
$$0 = 0 \checkmark$$

All check; the solution set is
$$\left\{-\frac{\sqrt{3}}{2},\ \frac{\sqrt{3}}{2},\ -\sqrt{2}\,i,\ \sqrt{2}\,i\right\}.$$

71. $3(a+4)^2 - 11(a+4) + 6 = 0$

Let $u = a + 4$.

$$3u^2 - 11u + 6 = 0$$
$$(3u - 2)(u - 3) = 0$$

$$3u - 2 = 0 \quad \text{or} \quad u - 3 = 0$$
$$3u = 2 \quad \text{or} \quad u = 3$$

$$u = \frac{2}{3}$$

$$a + 4 = \frac{2}{3} \quad \text{or} \quad a + 4 = 3$$

$$a = -\frac{10}{3} \quad \text{or} \quad a = -1$$

Check:

$a = -\dfrac{10}{3}$:

$$3\left(-\frac{10}{3} + 4\right)^2 - 11\left(-\frac{10}{3} + 4\right) + 6 \overset{?}{=} 0$$

$$3\left(\frac{2}{3}\right)^2 - 11\left(\frac{2}{3}\right) + 6 \overset{?}{=} 0$$

$$3\left(\frac{4}{9}\right) - \frac{22}{3} + 6 \overset{?}{=} 0$$

$$\frac{4}{3} - \frac{22}{3} + 6 \overset{?}{=} 0$$

$$0 = 0 \checkmark$$

$a = -1:$ $3(-1+4)^2 - 11(-1+4) + 6 \overset{?}{=} 0$

$$3(3)^2 - 11(3) + 6 \overset{?}{=} 0$$

$$3(9) - 33 + 6 \overset{?}{=} 0$$

$$27 - 33 + 6 \overset{?}{=} 0$$

$$0 = 0 \checkmark$$

Both check; the solution set is $\left\{-\dfrac{10}{3}, \ -1\right\}$.

72. $\left(q^2 - 11\right)^2 - 2\left(q^2 - 11\right) - 15 = 0$

Let $u = q^2 - 11$.

$$u^2 - 2u - 15 = 0$$
$$(u + 3)(u - 5) = 0$$

$$u + 3 = 0 \quad \text{or} \quad u - 5 = 0$$
$$u = -3 \quad \text{or} \quad u = 5$$
$$q^2 - 11 = -3 \quad \text{or} \quad q^2 - 11 = 5$$
$$q^2 = 8 \quad \text{or} \quad q^2 = 16$$
$$q = \pm\sqrt{8} \quad \text{or} \quad q = \pm\sqrt{16}$$
$$q = \pm 2\sqrt{2} \quad \text{or} \quad q = \pm 4$$

Check:

$q = -2\sqrt{2}$:

$$\left(\left(-2\sqrt{2}\right)^2 - 11\right)^2 - 2\left(\left(-2\sqrt{2}\right)^2 - 11\right) - 15 \overset{?}{=} 0$$

$$(8 - 11)^2 - 2(8 - 11) - 15 \overset{?}{=} 0$$

$$(-3)^2 - 2(-3) - 15 \overset{?}{=} 0$$

$$9 + 6 - 15 \overset{?}{=} 0$$

$$0 = 0 \checkmark$$

$q = 2\sqrt{2}$:

$$\left(\left(2\sqrt{2}\right)^2 - 11\right)^2 - 2\left(\left(2\sqrt{2}\right)^2 - 11\right) - 15 \overset{?}{=} 0$$

$$(8 - 11)^2 - 2(8 - 11) - 15 \overset{?}{=} 0$$

$$(-3)^2 - 2(-3) - 15 \overset{?}{=} 0$$

$$9 + 6 - 15 \overset{?}{=} 0$$

$$0 = 0 \checkmark$$

$q = -4$:

$$\left((-4)^2 - 11\right)^2 - 2\left((-4)^2 - 11\right) - 15 \overset{?}{=} 0$$

$$(16 - 11)^2 - 2(16 - 11) - 15 \overset{?}{=} 0$$

$$(5)^2 - 2(5) - 15 \overset{?}{=} 0$$

$$25 - 10 - 15 \overset{?}{=} 0$$

$$0 = 0 \checkmark$$

$q = 4:$ $\left(4^2 - 11\right)^2 - 2\left(4^2 - 11\right) - 15 \overset{?}{=} 0$

$$(16 - 11)^2 - 2(16 - 11) - 15 \overset{?}{=} 0$$

$$(5)^2 - 2(5) - 15 \overset{?}{=} 0$$

$$25 - 10 - 15 \overset{?}{=} 0$$

$$0 = 0 \checkmark$$

All check; the solution set is $\left\{-4, \ -2\sqrt{2}, \ 2\sqrt{2}, \ 4\right\}$.

73.

$$y - 13\sqrt{y} + 36 = 0$$

$$\left(\sqrt{y}\right)^2 - 13\left(\sqrt{y}\right) + 36 = 0$$

Let $u = \sqrt{y}$.

$$u^2 - 13u + 36 = 0$$

$$(u - 4)(u - 9) = 0$$

$u - 4 = 0 \quad$ or $\quad u - 9 = 0$

$u = 4 \quad$ or $\quad u = 9$

$\sqrt{y} = 4 \quad$ or $\quad \sqrt{y} = 9$

$y = 4^2 \quad$ or $\quad y = 9^2$

$y = 16 \quad$ or $\quad y = 81$

Check:

$y = 16:\quad 16 - 13\sqrt{16} + 36 \overset{?}{=} 0$

$16 - 13 \cdot 4 + 36 \overset{?}{=} 0$

$16 - 52 + 36 \overset{?}{=} 0$

$0 = 0 \checkmark$

$y = 16:\quad 81 - 13\sqrt{81} + 36 \overset{?}{=} 0$

$81 - 13 \cdot 9 + 36 \overset{?}{=} 0$

$81 - 117 + 36 \overset{?}{=} 0$

$0 = 0 \checkmark$

Both check, the solution set is $\{16, 81\}$.

74.

$$5z + 2\sqrt{z} - 3 = 0$$

$$5\left(\sqrt{z}\right)^2 + 2\left(\sqrt{z}\right) - 3 = 0$$

Let $u = \sqrt{z}$.

$$5u^2 + 2u - 3 = 0$$

$$(5u - 3)(u + 1) = 0$$

$5u - 3 = 0 \quad$ or $\quad u + 1 = 0$

$5u = 3 \quad$ or $\quad u = -1$

$u = \dfrac{3}{5}$

$\sqrt{z} = \dfrac{3}{5} \quad$ or $\quad \sqrt{z} = -1$

$z = \left(\dfrac{3}{5}\right)^2 \quad$ or $\quad z = (-1)^2$

$z = \dfrac{9}{25} \quad$ or $\quad z = 1$

Check:

$z = \dfrac{9}{25}:\quad 5 \cdot \dfrac{9}{25} + 2\sqrt{\dfrac{9}{25}} - 3 \overset{?}{=} 0$

$\dfrac{9}{5} + 2 \cdot \dfrac{3}{5} - 3 \overset{?}{=} 0$

$\dfrac{9}{5} + \dfrac{6}{5} - 3 \overset{?}{=} 0$

$0 = 0 \checkmark$

$z = 1:\quad 5 \cdot 1 + 2\sqrt{1} - 3 \overset{?}{=} 0$

$5 + 2 \cdot 1 - 3 \overset{?}{=} 0$

$5 + 2 - 3 \overset{?}{=} 0$

$4 \neq 0 \;\text{✗}$

$z = 1$ does not check; the solution set is $\left\{\dfrac{9}{25}\right\}$.

75.

$$p^{-2} - 4p^{-1} - 21 = 0$$

$$\left(p^{-1}\right)^2 - 4\left(p^{-1}\right) - 21 = 0$$

Let $u = p^{-1}$.

$$u^2 - 4u - 21 = 0$$

$$(u + 3)(u - 7) = 0$$

$u + 3 = 0 \quad$ or $\quad u - 7 = 0$

$u = -3 \quad$ or $\quad u = 7$

$p^{-1} = -3 \quad$ or $\quad p^{-1} = 7$

$\dfrac{1}{p} = -3 \quad$ or $\quad \dfrac{1}{p} = 7$

$p = -\dfrac{1}{3} \quad$ or $\quad p = \dfrac{1}{7}$

Check:

$p = -\dfrac{1}{3}:\quad \left(-\dfrac{1}{3}\right)^{-2} - 4\left(-\dfrac{1}{3}\right)^{-1} - 21 \overset{?}{=} 0$

$(-3)^2 - 4(-3) - 21 \overset{?}{=} 0$

$9 + 12 - 21 \overset{?}{=} 0$

$0 = 0 \checkmark$

$p = \dfrac{1}{7}:\quad \left(\dfrac{1}{7}\right)^{-2} - 4\left(\dfrac{1}{7}\right)^{-1} - 21 \overset{?}{=} 0$

$(7)^2 - 4(7) - 21 \overset{?}{=} 0$

$49 - 28 - 21 \overset{?}{=} 0$

$0 = 0 \checkmark$

Both check; the solution set is $\left\{-\dfrac{1}{3},\ \dfrac{1}{7}\right\}$.

76.
$$2b^{2/3} + 13b^{1/3} - 7 = 0$$
$$2\left(b^{1/3}\right)^2 + 13\left(b^{1/3}\right) - 7 = 0$$

Let $u = b^{1/3}$.
$$2u^2 + 13u - 7 = 0$$
$$(2u - 1)(u + 7) = 0$$

$$2u - 1 = 0 \qquad \text{or} \qquad u + 7 = 0$$
$$2u = 1 \qquad \text{or} \qquad u = -7$$
$$u = \frac{1}{2}$$
$$b^{1/3} = \frac{1}{2} \qquad \text{or} \qquad b^{1/3} = -7$$
$$\left(b^{1/3}\right)^3 = \left(\frac{1}{2}\right)^3 \quad \text{or} \quad \left(b^{1/3}\right)^3 = (-7)^3$$
$$b = \frac{1}{8} \qquad \text{or} \qquad b = -343$$

Check:
$$b = \frac{1}{8}:$$

$$2\left(\frac{1}{8}\right)^{2/3} + 13\left(\frac{1}{8}\right)^{1/3} - 7 \overset{?}{=} 0$$

$$2\left(\sqrt[3]{\frac{1}{8}}\right)^2 + 13\left(\sqrt[3]{\frac{1}{8}}\right) - 7 \overset{?}{=} 0$$

$$2\left(\frac{1}{2}\right)^2 + 13\left(\frac{1}{2}\right) - 7 \overset{?}{=} 0$$

$$2\left(\frac{1}{4}\right) + 13\left(\frac{1}{2}\right) - 7 \overset{?}{=} 0$$

$$\frac{1}{2} + \frac{13}{2} - 7 \overset{?}{=} 0$$

$$0 = 0 \ \checkmark$$

$$b = -343: \quad 2(-343)^{2/3} + 13(-343)^{1/3} - 7 \overset{?}{=} 0$$

$$2\left(\sqrt[3]{-343}\right)^2 + 13\left(\sqrt[3]{-343}\right) - 7 \overset{?}{=} 0$$

$$2(-7)^2 + 13(-7) - 7 \overset{?}{=} 0$$

$$2(49) - 91 - 7 \overset{?}{=} 0$$

$$98 - 91 - 7 \overset{?}{=} 0$$

$$0 = 0 \ \checkmark$$

Both check; the solution set is $\left\{ -343, \dfrac{1}{8} \right\}$.

77.
$$m^{1/2} + 2m^{1/4} - 8 = 0$$
$$\left(m^{1/4}\right)^2 + 2\left(m^{1/4}\right) - 8 = 0$$

Let $u = m^{1/4}$.
$$u^2 + 2u - 8 = 0$$
$$(u - 2)(u + 4) = 0$$

$$u - 2 = 0 \qquad \text{or} \qquad u + 4 = 0$$
$$u = 2 \qquad \text{or} \qquad u = -4$$
$$m^{1/4} = 2 \qquad \text{or} \qquad m^{1/4} = -4$$
$$\left(m^{1/4}\right)^4 = 2^4 \quad \text{or} \quad \left(m^{1/4}\right)^4 = (-4)^4$$
$$m = 16 \qquad \text{or} \qquad m = 256$$

Check:
$$m = 16: \quad (16)^{1/2} + 2(16)^{1/4} - 8 \overset{?}{=} 0$$
$$\sqrt{16} + 2\sqrt[4]{16} - 8 \overset{?}{=} 0$$
$$4 + 2 \cdot 2 - 8 \overset{?}{=} 0$$
$$4 + 4 - 8 \overset{?}{=} 0$$
$$0 = 0 \ \checkmark$$

$$m = 256: \quad (256)^{1/2} + 2(256)^{1/4} - 8 \overset{?}{=} 0$$
$$\sqrt{256} + 2\sqrt[4]{256} - 8 \overset{?}{=} 0$$
$$16 + 2 \cdot 4 - 8 \overset{?}{=} 0$$
$$16 + 8 - 8 \overset{?}{=} 0$$
$$16 \neq 0 \ \times$$

$m = 256$ does not check; the solution set is $\{16\}$.

78.
$$\left(\frac{1}{x+5}\right)^2 + \frac{3}{x+5} = 28$$
$$\left(\frac{1}{x+5}\right)^2 + 3\left(\frac{1}{x+5}\right) - 28 = 0$$

Let $u = \dfrac{1}{x+5}$.
$$u^2 + 3u - 28 = 0$$
$$(u - 4)(u + 7) = 0$$

$$u-4=0 \quad \text{or} \quad u+7=0$$
$$u=4 \quad \text{or} \quad u=-7$$
$$\frac{1}{x+5}=4 \quad \text{or} \quad \frac{1}{x+5}=-7$$
$$x+5=\frac{1}{4} \quad \text{or} \quad x+5=-\frac{1}{7}$$
$$x=-\frac{19}{4} \quad \text{or} \quad x=-\frac{36}{7}$$

Check:

$$x=-\frac{19}{4}: \quad \left(\frac{1}{-\frac{19}{4}+5}\right)^2+\frac{3}{-\frac{19}{4}+5} \stackrel{?}{=} 28$$

$$\left(\frac{1}{\frac{1}{4}}\right)^2+\frac{3}{\frac{1}{4}} \stackrel{?}{=} 28$$

$$(4)^2+12 \stackrel{?}{=} 28$$

$$16+12 \stackrel{?}{=} 28$$

$$28=28 \checkmark$$

$$x=-\frac{36}{7}: \quad \left(\frac{1}{-\frac{36}{7}+5}\right)^2+\frac{3}{-\frac{36}{7}+5} \stackrel{?}{=} 28$$

$$\left(\frac{1}{-\frac{1}{7}}\right)^2+\frac{3}{-\frac{1}{7}} \stackrel{?}{=} 28$$

$$(-7)^2-21 \stackrel{?}{=} 28$$

$$49-21 \stackrel{?}{=} 28$$

$$28=28 \checkmark$$

Both check; the solution set is $\left\{-\frac{36}{7}, -\frac{19}{4}\right\}$.

79.
$$4x-20\sqrt{x}+21=0$$
$$4\left(\sqrt{x}\right)^2-20\left(\sqrt{x}\right)+21=0$$

Let $u=\sqrt{x}$.
$$4u^2-20u+21=0$$
$$(2u-3)(2u-7)=0$$

$$2u-3=0 \quad \text{or} \quad 2u-7=0$$
$$2u=3 \quad \text{or} \quad 2u=7$$
$$u=\frac{3}{2} \quad \text{or} \quad u=\frac{7}{2}$$

$$\sqrt{x}=\frac{3}{2} \quad \text{or} \quad \sqrt{x}=\frac{7}{2}$$
$$x=\left(\frac{3}{2}\right)^2 \quad \text{or} \quad x=\left(\frac{7}{2}\right)^2$$
$$x=\frac{9}{4} \quad \text{or} \quad x=\frac{49}{4}$$

Check:
$$f\left(\frac{9}{4}\right)=4\cdot\frac{9}{4}-20\sqrt{\frac{9}{4}}+21$$
$$=4\cdot\frac{9}{4}-20\cdot\frac{3}{2}+21$$
$$=9-30+21$$
$$=0 \checkmark$$

$$f\left(\frac{49}{4}\right)=4\cdot\frac{49}{4}-20\sqrt{\frac{49}{4}}+21$$
$$=4\cdot\frac{49}{4}-20\cdot\frac{7}{2}+21$$
$$=49-70+21$$
$$=0 \checkmark$$

Both check; the zeros of f are $\left\{\frac{9}{4}, \frac{49}{4}\right\}$.

80.
$$x^4-17x^2+60=0$$
$$\left(x^2\right)^2-17\left(x^2\right)+60=0$$

Let $u=x^2$.
$$u^2-17u+60=0$$
$$(u-12)(u-5)=0$$
$$u-12=0 \quad \text{or} \quad u-5=0$$
$$u=12 \quad \text{or} \quad u=5$$
$$x^2=12 \quad \text{or} \quad x^2=5$$
$$x=\pm\sqrt{12} \quad \text{or} \quad x=\pm\sqrt{5}$$
$$x=\pm2\sqrt{3}$$

Check:
$$g\left(-2\sqrt{3}\right)=\left(-2\sqrt{3}\right)^4-17\left(-2\sqrt{3}\right)^2+60$$
$$=144-17\cdot12+60$$
$$=144-204+60$$
$$=0 \checkmark$$

$$g\left(2\sqrt{3}\right)=\left(2\sqrt{3}\right)^4-17\left(2\sqrt{3}\right)^2+60$$
$$=144-17\cdot12+60$$
$$=144-204+60$$
$$=0 \checkmark$$

$$g\left(-\sqrt{5}\right)=\left(-\sqrt{5}\right)^{4}-17\left(-\sqrt{5}\right)^{2}+60$$
$$=25-17\cdot5+60$$
$$=25-85+60$$
$$=0 \checkmark$$

$$g\left(\sqrt{5}\right)=\left(\sqrt{5}\right)^{4}-17\left(\sqrt{5}\right)^{2}+60$$
$$=25-17\cdot5+60$$
$$=25-85+60$$
$$=0 \checkmark$$

All check; the zeros of g are
$\left\{-2\sqrt{3},\ -\sqrt{5},\ \sqrt{5},\ 2\sqrt{3}\right\}$.

81. Begin with the graph of $y=x^{2}$, then shift the graph up 4 units to obtain the graph of $f(x)=x^{2}+4$.

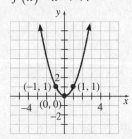 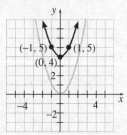

82. Begin with the graph of $y=x^{2}$, then shift the graph down 5 units to obtain the graph of $g(x)=x^{2}-5$.

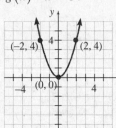 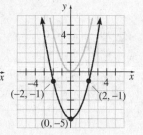

83. Begin with the graph of $y=x^{2}$, then shift the graph to the left 1 unit to obtain the graph of $h(x)=(x+1)^{2}$.

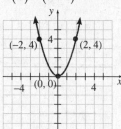 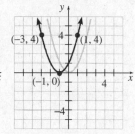

84. Begin with the graph of $y=x^{2}$, then shift the graph to the right 4 units to obtain the graph of $F(x)=(x-4)^{2}$.

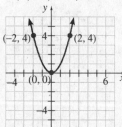 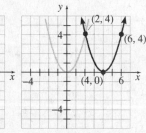

85. Begin with the graph of $y=x^{2}$, then multiply each y-coordinate by -4 to obtain the graph of $G(x)=-4x^{2}$.

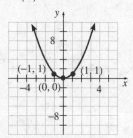 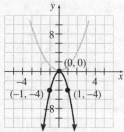

86. Begin with the graph of $y=x^{2}$, then vertically compress the graph by a factor of $\dfrac{1}{5}$ (multiply each y-coordinate by $\dfrac{1}{5}$) to obtain the graph of $H(x)=\dfrac{1}{5}x^{2}$.

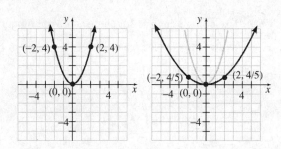

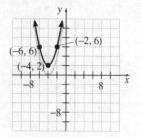

87. Begin with the graph of $y = x^2$, then shift the graph 4 units to the right to obtain the graph of $y = (x-4)^2$. Shift this graph down 3 units to obtain the graph of $p(x) = (x-4)^2 - 3$.

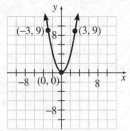

88. Begin with the graph of $y = x^2$, then shift the graph 4 units to the left to obtain the graph of $y = (x+4)^2$. Shift this graph up 2 units to obtain the graph of $P(x) = (x+4)^2 + 2$.

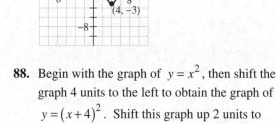

89. Begin with the graph of $y = x^2$, then shift the graph 1 unit to the right to obtain the graph of $y = (x-1)^2$. Multiply the y-coordinates by -1 to obtain the graph of $y = -(x-1)^2$. Shift this graph up 4 units to obtain the graph of $f(x) = -(x-1)^2 + 4$.

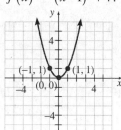

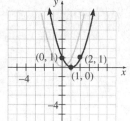

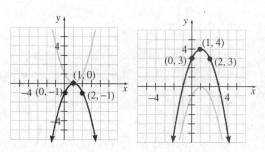

90. Begin with the graph of $y = x^2$, then shift the graph 2 unit to the left to obtain the graph of $y = (x+2)^2$. Multiply the y-coordinates by $\frac{1}{2}$ to obtain the graph of $y = \frac{1}{2}(x+2)^2$. Shift this graph down 1 unit to obtain the graph of $F(x) = \frac{1}{2}(x+2)^2 - 1$

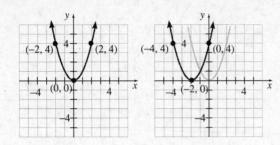

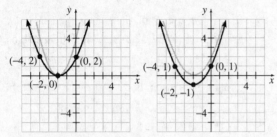

91. Use completing the square to write the function in the form $y = a(x-h)^2 + k$.

$$g(x) = x^2 - 6x + 10$$
$$= \left(x^2 - 6x\right) + 10$$
$$= \left(x^2 - 6x + 9\right) + 10 - 9$$
$$= (x-3)^2 + 1$$

Begin with the graph of $y = x^2$, then shift the graph right 3 units to obtain the graph of $y = (x-3)^2$. Shift this result up 1 unit to obtain the graph of $g(x) = (x-3)^2 + 1$.

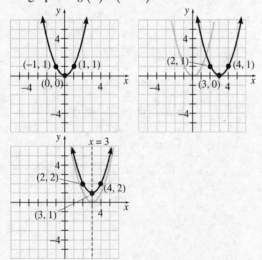

The vertex is $(3, 1)$; The axis of symmetry is $x = 3$.

92. Use completing the square to write the function in the form $y = a(x-h)^2 + k$.

$$G(x) = x^2 + 8x + 11$$
$$= \left(x^2 + 8x\right) + 11$$
$$= \left(x^2 + 8x + 16\right) + 11 - 16$$
$$= (x+4)^2 - 5$$

Begin with the graph of $y = x^2$, then shift the graph left 4 units to obtain the graph of $y = (x+4)^2$. Shift this result down 5 units to obtain the graph of $G(x) = (x+4)^2 - 5$.

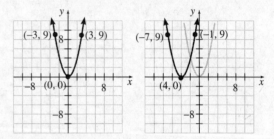

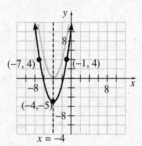

The vertex is $(-4, -5)$. The axis of symmetry is $x = -4$.

93. Use completing the square to write the function in the form $y = a(x-h)^2 + k$.

$$h(x) = 2x^2 - 4x - 3$$
$$= \left(2x^2 - 4x\right) - 3$$
$$= 2\left(x^2 - 2x\right) - 3$$
$$= 2\left(x^2 - 2x + 1\right) - 3 - 2$$
$$= 2(x-1)^2 - 5$$

Begin with the graph of $y = x^2$, then shift the graph right 1 unit to obtain the graph of $y = (x-1)^2$. Vertically stretch this graph by a factor of 2 (multiply the y-coordinates by 2) to obtain the graph of $y = 2(x-1)^2$. Lastly, shift the graph down 5 units to obtain the graph of $H(x) = 2(x-1)^2 - 5$.

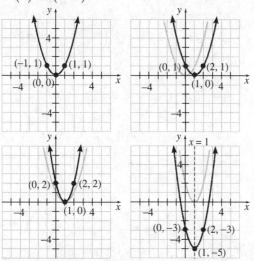

The vertex is $(1, -5)$. The axis of symmetry is $x = 1$.

94. Use completing the square to write the function in the form $y = a(x-h)^2 + k$.

$$H(x) = -x^2 - 6x - 10$$
$$= \left(-x^2 - 6x\right) - 10$$
$$= -\left(x^2 + 6x\right) - 10$$
$$= -\left(x^2 + 6x + 9\right) - 10 + 9$$
$$= -(x+3)^2 - 1$$

Begin with the graph of $y = x^2$, then shift the graph left 3 units to obtain the graph of $y = (x+3)^2$. Multiply the y-coordinates by -1 to obtain the graph of $y = -(x+3)^2$. Lastly, shift the graph down 1 unit to obtain the graph of $H(x) = -(x+3)^2 - 1$.

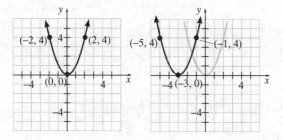

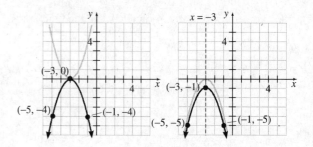

The vertex is $(-3, -1)$. The axis of symmetry is $x = -3$.

95. Use completing the square to write the function in the form $y = a(x-h)^2 + k$.

$$p(x) = -3x^2 + 12x - 8$$
$$= \left(-3x^2 + 12x\right) - 8$$
$$= -3\left(x^2 - 4x\right) - 8$$
$$= -3\left(x^2 - 4x + 4\right) - 8 + 12$$
$$= -3(x-2)^2 + 4$$

Begin with the graph of $y = x^2$, then shift the graph right 2 units to obtain the graph of $y = (x-2)^2$. Multiply the y-coordinates by -3 to obtain the graph of $y = -3(x-2)^2$. Lastly, shift the graph up 4 units to obtain the graph of $p(x) = -3(x-2)^2 + 4$.

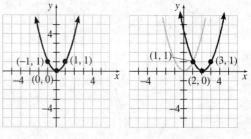

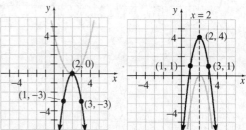

The vertex is $(2, 4)$. The axis of symmetry is $x = 2$.

96. Use completing the square to write the function in the form $y = a(x-h)^2 + k$.

$$P(x) = \frac{1}{2}x^2 - 2x + 5$$
$$= \left(\frac{1}{2}x^2 - 2x\right) + 5$$
$$= \frac{1}{2}\left(x^2 - 4x\right) + 5$$
$$= \frac{1}{2}\left(x^2 - 4x + 4\right) + 5 - 2$$
$$= \frac{1}{2}(x-2)^2 + 3$$

Begin with the graph of $y = x^2$, then shift the graph right 2 units to obtain the graph of $y = (x-2)^2$. Vertically compress this graph by a factor of $\frac{1}{2}$ (multiply the y-coordinates by $\frac{1}{2}$) to obtain the graph of $y = \frac{1}{2}(x-2)^2$. Lastly, shift the graph up 3 units to obtain the graph of $P(x) = \frac{1}{2}(x-2)^2 + 3$.

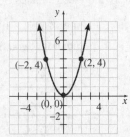
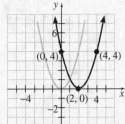

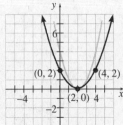
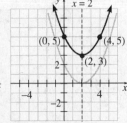

The vertex is $(2,\ 3)$. The axis of symmetry is $x = 2$.

97. Consider the form $y = a(x-h)^2 + k$. From the graph we know that the vertex is $(2,-4)$, so we have $h = 2$ and $k = -4$. The graph also passes through the point $(x,y) = (0,4)$. Substituting values for *x, y, h,* and *k*, we can solve for *a*:

$$4 = a(0-2)^2 - 4$$
$$4 = a(-2)^2 - 4$$
$$8 = 4a$$
$$2 = a$$

The quadratic function is $f(x) = 2(x-2)^2 - 4$ or $f(x) = 2x^2 - 8x + 4$.

98. Consider the form $y = a(x-h)^2 + k$. From the graph we know that the vertex is $(4,3)$, so we have $h = 4$ and $k = 3$. The graph also passes through the point $(x,y) = (3,2)$. Substituting values for *x, y, h,* and *k*, we can solve for *a*:

$$2 = a(3-4)^2 + 3$$
$$2 = a(-1)^2 + 3$$
$$2 = a + 3$$
$$-1 = a$$

The quadratic function is $f(x) = -(x-4)^2 + 3$ or $f(x) = -x^2 + 8x - 13$.

99. Consider the form $y = a(x-h)^2 + k$. From the graph we know that the vertex is $(-2,-1)$, so we have $h = -2$ and $k = -1$. The graph also passes through the point $(x,y) = (0,-3)$. Substituting values for *x, y, h,* and *k*, we can solve for *a*:

$$-3 = a\left(0-(-2)\right)^2 - 1$$
$$-3 = a(0+2)^2 - 1$$
$$-3 = a(2)^2 - 1$$
$$-2 = 4a$$
$$-\frac{1}{2} = a$$

The quadratic function is $f(x) = -\frac{1}{2}(x+2)^2 - 1$ or $f(x) = -\frac{1}{2}x^2 - 2x - 3$.

100. Consider the form $y = a(x - h)^2 + k$. From the graph we know that the vertex is $(-2, 0)$, so we have $h = -2$ and $k = 0$. The graph also passes through the point $(x, y) = (-3, 3)$. Substituting values for x, y, h, and k, we can solve for a:

$$3 = a(-3 - (-2))^2 - 0$$
$$3 = a(-3 + 2)^2$$
$$3 = a(-1)^2$$
$$3 = a$$

The quadratic function is $f(x) = 3(x + 2)^2$ or $f(x) = 3x^2 + 12x + 12$.

101. $f(x) = x^2 + 2x - 8$

$a = 1, b = 2, c = -8$

The graph opens up because the coefficient on x^2 is positive.

vertex:

$$x = -\frac{b}{2a} = -\frac{2}{2(1)} = -1$$

$$f(-1) = (-1)^2 + 2(-1) - 8 = -9$$

The vertex is $(-1, -9)$ and the axis of symmetry is $x = -1$.

y-intercept:

$$f(0) = (0)^2 + 2(0) - 8 = -8$$

x-intercepts:

$$b^2 - 4ac = 2^2 - 4(1)(-8) = 36 > 0$$

There are two distinct x-intercepts. We find these by solving

$$f(x) = 0$$
$$x^2 + 2x - 8 = 0$$
$$(x - 2)(x + 4) = 0$$
$$x - 2 = 0 \text{ or } x + 4 = 0$$
$$x = 2 \text{ or } x = -4$$

Graph:

The y-intercept point, $(0, -8)$, is one unit to the right of the axis of symmetry. Therefore, if we move one unit to the left of the axis of symmetry, we obtain the point $(-2, -8)$ which must also be on the graph.

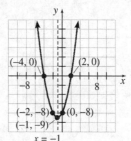

102. $F(x) = 2x^2 - 5x + 3$

$a = 2, b = -5, c = 3$

The graph opens up because the coefficient on x^2 is positive.

vertex:

$$x = -\frac{b}{2a} = -\frac{(-5)}{2(2)} = \frac{5}{4}$$

$$F\left(\frac{5}{4}\right) = 2\left(\frac{5}{4}\right)^2 - 5\left(\frac{5}{4}\right) + 3 = -\frac{1}{8}$$

The vertex is $\left(\frac{5}{4}, -\frac{1}{8}\right)$ and the axis of symmetry is $x = \frac{5}{4}$.

y-intercept:

$$F(0) = 2(0)^2 - 5(0) + 3 = 3$$

x-intercepts:

$$b^2 - 4ac = (-5)^2 - 4(2)(3) = 1 > 0$$

There are two distinct x-intercepts. We find these by solving

$$F(x) = 0$$
$$2x^2 - 5x + 3 = 0$$
$$(2x - 3)(x - 1) = 0$$
$$2x - 3 = 0 \text{ or } x - 1 = 0$$
$$2x = 3 \text{ or } x = 1$$
$$x = \frac{3}{2}$$

Graph:

The y-intercept point, $(0, 3)$, is five-fourths units to the left of the axis of symmetry. Therefore, if we move five-fourths units to the right of the axis of symmetry, we obtain the point $\left(\frac{5}{2}, 3\right)$ which must also be on the graph.

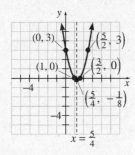

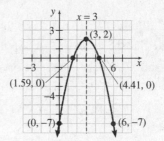

103. $g(x) = -x^2 + 6x - 7$

$a = -1, b = 6, c = -7$

The graph opens down because the coefficient on x^2 is negative.

vertex:

$$x = -\frac{b}{2a} = -\frac{6}{2(-1)} = 3$$

$$g(3) = -(3)^2 + 6(3) - 7 = 2$$

The vertex is $(3, 2)$ and the axis of symmetry is $x = 3$.

y-intercept:

$$g(0) = -(0)^2 + 6(0) - 7 = -7$$

x-intercepts:

$$b^2 - 4ac = 6^2 - 4(-1)(-7) = 8 > 0$$

There are two distinct x-intercepts. We find these by solving

$$g(x) = 0$$

$$-x^2 + 6x - 7 = 0$$

$$x = \frac{-6 \pm \sqrt{8}}{2(-1)}$$

$$= \frac{-6 \pm 2\sqrt{2}}{-2}$$

$$= 3 \pm \sqrt{2}$$

$$x \approx 1.59 \quad \text{or} \quad x \approx 4.41$$

Graph:

The y-intercept point, $(0, -7)$, is three units to the left of the axis of symmetry. Therefore, if we move three units to the right of the axis of symmetry, we obtain the point $(6, -7)$ which must also be on the graph.

104. $G(x) = -2x^2 + 4x + 3$

$a = -2, b = 4, c = 3$

The graph opens down because the coefficient on x^2 is negative.

vertex:

$$x = -\frac{b}{2a} = -\frac{4}{2(-2)} = 1$$

$$G(1) = -2(1)^2 + 4(1) + 3 = 5$$

The vertex is $(1, 5)$ and the axis of symmetry is $x = 1$.

y-intercept:

$$G(0) = -2(0)^2 + 4(0) + 3 = 3$$

x-intercepts:

$$b^2 - 4ac = 4^2 - 4(-2)(3) = 40 > 0$$

There are two distinct x-intercepts. We find these by solving

$$G(x) = 0$$

$$-2x^2 + 4x + 3 = 0$$

$$x = \frac{-4 \pm \sqrt{40}}{2(-2)}$$

$$= \frac{-4 \pm 2\sqrt{10}}{-4}$$

$$= 1 \pm \frac{\sqrt{10}}{2}$$

$$x \approx -0.58 \quad \text{or} \quad x \approx 2.58$$

Graph:

The y-intercept point, $(0, 3)$, is one unit to the left of the axis of symmetry. Therefore, if we move one unit to the right of the axis of symmetry, we obtain the point $(2, 3)$ which must also be on the graph.

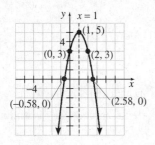

105. $h(x) = 4x^2 - 12x + 9$

$a = 4, b = -12, c = 9$

The graph opens up because the coefficient on x^2 is positive.

<u>vertex:</u>

$$x = -\frac{b}{2a} = -\frac{(-12)}{2(4)} = \frac{3}{2}$$

$$h\left(\frac{3}{2}\right) = 4\left(\frac{3}{2}\right)^2 - 12\left(\frac{3}{2}\right) + 9 = 0$$

The vertex is $\left(\frac{3}{2}, 0\right)$ and the axis of symmetry is

$$x = \frac{3}{2}.$$

<u>y-intercept:</u>

$$h(0) = 4(0)^2 - 12(0) + 9 = 9$$

<u>x-intercepts:</u>

$$b^2 - 4ac = (-12)^2 - 4(4)(9) = 0$$

Since the discriminant is 0, the x-coordinate of the vertex is the only x-intercept, $x = \frac{3}{2}$.

<u>Graph:</u>

The y-intercept point, $(0, 9)$, is three-halves units to the left of the axis of symmetry. Therefore, if we move three-halves units to the right of the axis of symmetry, we obtain the point $(3, 9)$ which must also be on the graph.

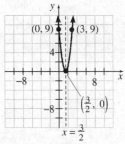

106. $H(x) = \frac{1}{3}x^2 + 2x + 3$

$a = \frac{1}{3}, b = 2, c = 3$

The graph opens up because the coefficient on x^2 is positive.

<u>vertex:</u>

$$x = -\frac{b}{2a} = -\frac{2}{2\left(\frac{1}{3}\right)} = -3$$

$$H(-3) = \frac{1}{3}(-3)^2 + 2(-3) + 3 = 0$$

The vertex is $(-3, 0)$ and the axis of symmetry is $x = -3$.

<u>y-intercept:</u>

$$H(0) = \frac{1}{3}(0)^2 + 2(0) + 3 = 3$$

<u>x-intercepts:</u>

$$b^2 - 4ac = 2^2 - 4\left(\frac{1}{3}\right)(3) = 0$$

Since the discriminant is 0, the x-coordinate of the vertex is the only x-intercept, $x = -3$.

<u>Graph:</u>

The y-intercept point, $(0, 3)$, is three units to the right of the axis of symmetry. Therefore, if we move three units to the left of the axis of symmetry, we obtain the point $(-6, 3)$ which must also be on the graph.

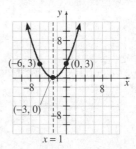

107. $p(x) = \frac{1}{4}x^2 + 3x + 10$

$a = \frac{1}{4}, b = 3, c = 10$

The graph opens up because the coefficient on x^2 is positive.

vertex:

$$x = -\frac{b}{2a} = -\frac{3}{2\left(\frac{1}{4}\right)} = -6$$

$$p(-6) = \frac{1}{4}(-6)^2 + 3(-6) + 10 = 1$$

The vertex is $(-6, 1)$ and the axis of symmetry is $x = -6$.

y-intercept:

$$p(0) = \frac{1}{4}(0)^2 + 3(0) + 10 = 10$$

x-intercepts:

$$b^2 - 4ac = 3^2 - 4\left(\frac{1}{4}\right)(10) = -1 < 0$$

There are no x-intercepts since the discriminant is negative.

Graph:
The y-intercept point, $(0, 10)$, is six units to the right of the axis of symmetry. Therefore, if we move six unit to the left of the axis of symmetry, we obtain the point $(-12, 10)$ which must also be on the graph.

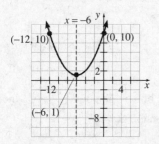

108. $P(x) = -x^2 + 4x - 9$

$a = -1, b = 4, c = -9$
The graph opens down because the coefficient on x^2 is negative.

vertex:

$$x = -\frac{b}{2a} = -\frac{4}{2(-1)} = 2$$

$$P(2) = -(2)^2 + 4(2) - 9 = -5$$

The vertex is $(2, -5)$ and the axis of symmetry is $x = 2$.

y-intercept:

$$P(0) = -(0)^2 + 4(0) - 9 = -9$$

x-intercepts:

$$b^2 - 4ac = 4^2 - 4(-1)(-9) = -20 < 0$$

There are no x-intercepts since the discriminant is negative.

Graph:
The y-intercept point, $(0, -9)$, is 2 units to the left of the axis of symmetry. Therefore, if we move 2 units to the right of the axis of symmetry, we obtain the point $(4, -9)$ which must also be on the graph.

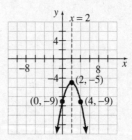

109. If we compare $f(x) = -2x^2 + 16x - 10$ to

$f(x) = ax^2 + bx + c$, we find that $a = -2$,
$b = 16$, and $c = -10$. Because $a < 0$, we know the graph will open down, so the function will have a maximum value.
The maximum value occurs at

$$x = -\frac{b}{2a} = -\frac{16}{2(-2)} = 4.$$

The maximum value is

$$f(4) = -2(4)^2 + 16(4) - 10 = 22.$$

So, the maximum value is 22, and it occurs when $x = 4$.

110. If we compare $g(x) = 6x^2 - 3x - 1$ to

$g(x) = ax^2 + bx + c$, we find that $a = 6$, $b = -3$, and $c = -1$. Because $a > 0$, we know the graph will open up, so the function will have a minimum value.
The minimum value occurs at

$$x = -\frac{b}{2a} = -\frac{(-3)}{2(6)} = \frac{1}{4}.$$

The minimum value is

$$g\left(\frac{1}{4}\right) = 6\left(\frac{1}{4}\right)^2 - 3\left(\frac{1}{4}\right) - 1 = -\frac{11}{8}.$$

So, the minimum value is $-\frac{11}{8}$, and it occurs

when $x = \frac{1}{4}$.

111. If we compare $h(x) = -4x^2 + 8x + 3$ to

$h(x) = ax^2 + bx + c$, we find that $a = -4$, $b = 8$,
and $c = 3$. Because $a < 0$, we know the graph
will open down, so the function will have a
maximum value.
The maximum value occurs at

$$x = -\frac{b}{2a} = -\frac{8}{2(-4)} = 1.$$

The maximum value is

$$h(4) = -4(1)^2 + 8(1) + 3 = 7.$$

So, the maximum value is 7, and it occurs when
$x = 1$.

112. If we compare $F(x) = -\frac{1}{3}x^2 + 4x - 7$ to

$F(x) = ax^2 + bx + c$, we find that $a = -\frac{1}{3}$,

$b = 4$, and $c = -7$. Because $a < 0$, we know the
graph will open down, so the function will have
a maximum value.
The maximum value occurs at

$$x = -\frac{b}{2a} = -\frac{4}{2\left(-\frac{1}{3}\right)} = 6.$$

The maximum value is

$$F(6) = -\frac{1}{3}(6)^2 + 4(6) - 7 = 5.$$

So, the maximum value is 5, and it occurs when
$x = 6$.

113. (a) We first recognize that the quadratic
function has a negative leading coefficient.
This means that the graph will open down
and the function indeed has a maximum
value. The maximum value occurs when

$$p = -\frac{b}{2a} = -\frac{150}{2\left(-\frac{1}{3}\right)} = 225.$$

The revenue will be maximized when the
televisions are sold at a price of $225.

(b) The maximum revenue is obtained by
evaluating the revenue function at the price
found in part (a).

$$R(225) = -\frac{1}{3}(225)^2 + 150(225) = 16,875$$

The maximum revenue is $16,875.

114. (a) We first recognize that the quadratic
function has a negative leading coefficient.
This means that the graph will open down
and the function indeed has a maximum
value. The maximum value occurs when

$$I = -\frac{b}{2a} = -\frac{120}{2(-16)} = 3.75.$$

The power will be maximized when the
current is 3.75 amperes.

(b) The maximum power is obtained by
evaluating the power function for the
current found in part (a).

$$P(3.75) = -16(3.75)^2 + 120(3.75) = 225$$

The maximum power is 225 watts.

115. Let x represent the first number. Then the second
number must be $24 - x$. We can express the
product of the two numbers as the function

$$p(x) = x(24 - x) = -x^2 + 24x$$

This is a quadratic function with $a = -1$,
$b = 24$, and $c = 0$. The function is maximized

when $x = -\frac{b}{2a} = -\frac{24}{2(-1)} = 12$.

The maximum product can be obtained by
evaluating $p(x)$ when $x = 12$.

$$p(x) = 12(24 - 12) = 12(12) = 144$$

Two numbers that sum to 24 have a maximum
product of 144 when both numbers are 12.

116. (a) Let x represent the width of the rectangular
kennel (the side that is not parallel to the
garage). Then $15 - 2x$ is the length of the
kennel (the side that is parallel to the
garage). The area is the product of the
length and width:

$$A = (15 - 2x) \cdot x$$
$$= -2x^2 + 15x$$

The leading coefficient is negative so we
know the graph opens down and there will
be a maximum area. This value occurs when
the width is

$$x = -\frac{b}{2a} = -\frac{15}{2(-2)} = 3.75 .$$

Then the length is $15 - 2(3.75) = 7.5$.

The dimensions that maximize the area of the kennel are 3.75 yards by 7.5 yards.

(b) The maximum area can be found by substituting 3.75 for x into the area function.
$$A = 3.75(15 - 2(3.75)) = 28.125$$

The maximum area of the kennel is 28.125 square yards.

117. (a) First we recognize that the quadratic function has a negative leading coefficient. This means the graph will open down and the function will indeed have a maximum. The maximum height will occur when

$$x = -\frac{b}{2a} = -\frac{1}{2(-0.005)} = 100 .$$

The ball will reach a maximum height when it is 100 feet from Ted.

(b) The maximum height can be found by evaluating $h(x)$ for the value of x found in part (a).

$$h(100) = -0.005(100)^2 + 100 = 50$$

The ball will reach a maximum height of 50 feet.

(c) When the ball is on the ground, it will have a height of 0 feet. Thus, we need to solve $h(x) = 0$.

$$-0.005x^2 + x = 0$$
$$200(-0.005x^2 + x) = 200(0)$$
$$-x^2 + 200x = 0$$
$$-x(x - 200) = 0$$
$$-x = 0 \quad \text{or} \quad x - 200 = 0$$
$$x = 0 \quad \text{or} \qquad x = 200$$

Zero (0) represents the distance from Ted before he kicks the ball. The ball will strike the ground again when it is 200 feet away from Ted.

118. (a) Since $R = x \cdot p$, we have
$$R = x \cdot p$$
$$= (-0.002p + 60) \cdot p$$
$$= -0.002p^2 + 60p$$

(b) The revenue function is quadratic with a negative leading coefficient. This means the graph will open down and the function will have a maximum value. This value occurs when

$$p = -\frac{b}{2a} = -\frac{60}{2(-0.002)} = 15,000 .$$

The maximum revenue can be found by substituting this value for p in the revenue equation.
$$R = -0.002(15,000)^2 + 60(15,000)$$
$$= 450,000$$

The maximum revenue of $450,000 will occur if the price of each automobile is set at $15,000.

(c) To determine how many automobiles will be sold, we substitute the maximizing price into the demand equation.
$$x = -0.002p + 60$$
$$= -0.002(15,000) + 60$$
$$= 30$$

When the price per automobile is $15,000, the dealership will sell 30 automobiles per month.

119. (a) The graph is greater than 0 for $x < -2$ or $x > 3$. The solution is $\{x \mid x < -2 \text{ or } x > 3\}$ in set-builder notation. The solution is $(-\infty, -2) \cup (3, \infty)$ using interval notation.

(b) The graph is less than 0 for $-2 < x < 3$. The solution is $\{x \mid -2 < x < 3\}$ using set-builder notation. The solution is $(-2, 3)$ using interval notation.

120. (a) The graph is 0 or greater for $-\frac{7}{2} \le x \le 1$.

The solution is $\left\{x \mid -\frac{7}{2} \le x \le 1\right\}$ using set-builder notation. The solution is $\left[-\frac{7}{2}, 1\right]$ using interval notation.

(b) The graph is 0 or less for $x \le -\dfrac{7}{2}$ or $x \ge 1$.

The solution is $\left\{ x \,\middle|\, x \le -\dfrac{7}{2} \text{ or } x \ge 1 \right\}$ using set-builder notation. The solution is $\left(-\infty, -\dfrac{7}{2} \right] \cup [1, \infty)$ using interval notation.

121. $x^2 - 2x - 24 \le 0$

Solve: $x^2 - 2x - 24 = 0$

$(x - 6)(x + 4) = 0$

$x - 6 = 0$ or $x + 4 = 0$

$x = 6$ or $x = -4$

Determine where each factor is positive and negative and where the product of these factors is positive and negative.

Interval	$(-\infty, -4)$	-4	$(-4, 6)$	6	$(6, \infty)$
$x - 6$	$----$	$-$	$----$	0	$++++$
$x + 4$	$----$	0	$++++$	$+$	$++++$
$(x-6)(x+4)$	$++++$	0	$----$	0	$++++$

The inequality is non-strict, so -4 and 6 are part of the solution. Now, $(x - 6)(x + 4)$ is less than zero where the product is negative. The solution is $\{ x \mid -4 \le x \le 6 \}$ or, using interval notation, $[-4, 6]$.

122. $y^2 + 7y - 8 \ge 0$

Solve: $y^2 + 7y - 8 = 0$

$(y - 1)(y + 8) = 0$

$y - 1 = 0$ or $y + 8 = 0$

$y = 1$ or $y = -8$

Determine where each factor is positive and negative and where the product of these factors is positive and negative.

Interval	$(-\infty, -8)$	-8	$(-8, 1)$	1	$(1, \infty)$
$y - 1$	$----$	$-$	$----$	0	$++++$
$y + 8$	$----$	0	$++++$	$+$	$++++$
$(y-1)(y+8)$	$++++$	0	$----$	0	$++++$

The inequality is non-strict, so -8 and 1 are part

of the solution. Now, $(y - 1)(y + 8)$ is greater than zero where the product is positive. The solution is $\{ y \mid y \le -8 \text{ or } y \ge 1 \}$ or, using interval notation; $(-\infty, -8] \cup [1, \infty)$.

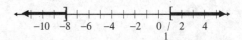

123. $3z^2 - 19z + 20 > 0$

Solve: $3z^2 - 19z + 20 = 0$

$(3z - 4)(z - 5) = 0$

$3z - 4 = 0$ or $z - 5 = 0$

$3z = 4$ or $z = 5$

$z = \dfrac{4}{3}$

Determine where each factor is positive and negative and where the product of these factors is positive and negative.

Interval	$\left(-\infty, \dfrac{4}{3}\right)$	$\dfrac{4}{3}$	$\left(\dfrac{4}{3}, 5\right)$	5	$(5, \infty)$
$3z - 4$	$----$	0	$++++$	$+$	$++++$
$z - 5$	$----$	$-$	$----$	0	$++++$
$(3z-4)(z-5)$	$++++$	0	$----$	0	$++++$

The inequality is strict, so $\dfrac{4}{3}$ and 5 are not part of the solution. Now, $(3z - 4)(z - 5)$ is greater than zero where the product is positive. The solution is $\left\{ z \,\middle|\, z < \dfrac{4}{3} \text{ or } z > 5 \right\}$ or, using interval notation, $\left(-\infty, \dfrac{4}{3} \right) \cup (5, \infty)$.

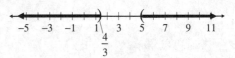

124. $p^2 + 4p - 2 < 0$

Solve: $p^2 + 4p - 2 = 0$

$$p = \frac{-4 \pm \sqrt{4^2 - 4(1)(-2)}}{2(1)}$$

$$= \frac{-4 \pm \sqrt{24}}{2}$$

$$= \frac{-4 \pm 2\sqrt{6}}{2}$$

$$= -2 \pm \sqrt{6}$$

$x = -2 - \sqrt{6}$ or $x = -2 + \sqrt{6}$

≈ -4.45 or ≈ 0.45

Determine where each factor is positive and negative and where the product of these factors is positive and negative.

Interval	$(-\infty, -2-\sqrt{6})$	$-2-\sqrt{6}$ ≈ -4.45	$(-2-\sqrt{6}, -2+\sqrt{6})$	$-2+\sqrt{6}$ ≈ 0.45	$(-2+\sqrt{6}, \infty)$
$p-(-2-\sqrt{6})$	$----$	0	$++++$	$+$	$++++$
$p-(-2+\sqrt{6})$	$----$	$-$	$----$	0	$++++$
$\left[p-(-2-\sqrt{6})\right]\left[p-(-2+\sqrt{6})\right]$	$++++$	0	$----$	0	$++++$

The inequality is strict, so $-2-\sqrt{6}$ and $-2+\sqrt{6}$ are not part of the solution. Now,

$\left[p-\left(-2-\sqrt{6}\right)\right]\left[p-\left(-2+\sqrt{6}\right)\right]$ is less than

zero where the product is negative. The solution

is $\left\{p\mid -2-\sqrt{6} < p < -2+\sqrt{6}\right\}$ or, using

interval notation, $\left(-2-\sqrt{6},\ -2+\sqrt{6}\right)$.

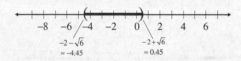

125. $4m^2 - 20m + 25 \geq 0$

Solve: $4m^2 - 20m + 25 = 0$

$(2m-5)(2m-5) = 0$

$2m - 5 = 0$ or $2m - 5 = 0$

$2m = 5$ or $2m = 5$

$m = \dfrac{5}{2}$ or $m = \dfrac{5}{2}$

Determine where each factor is positive and negative and where the product of these factors is positive and negative.

Interval	$\left(-\infty, \frac{5}{2}\right)$	$\frac{5}{2}$	$\left(\frac{5}{2}, \infty\right)$
$2m-5$	$----$	0	$++++$
$2m-5$	$----$	0	$++++$
$(2m-5)(2m-5)$	$++++$	0	$++++$

The inequality is non-strict, so $\dfrac{5}{2}$ is part of the

solution. Now, $(2m-5)(2m-5)$ is greater than

zero where the product is positive. Thus,

$4m^2 - 20m + 25$ is always greater than or equal

to zero. The solution is $\{m \mid m$ is any real

number$\}$ or, using interval notation $(-\infty, \infty)$.

126. $6w^2 - 19w - 7 \leq 0$

Solve: $6w^2 - 19w - 7 = 0$

$(3w+1)(2w-7) = 0$

$3w + 1 = 0$ or $2w - 7 = 0$

$3w = -1$ or $2w = 7$

$w = -\dfrac{1}{3}$ or $w = \dfrac{7}{2}$

Determine where each factor is positive and negative and where the product of these factors is positive and negative.

Interval	$\left(-\infty, -\frac{1}{3}\right)$	$-\frac{1}{3}$	$\left(-\frac{1}{3}, \frac{7}{2}\right)$	$\frac{7}{2}$	$\left(\frac{7}{2}, \infty\right)$
$3w+1$	$----$	0	$++++$	$+$	$++++$
$2w-7$	$----$	$-$	$----$	0	$++++$
$(3w+1)(2w-7)$	$++++$	0	$----$	0	$++++$

The inequality is non-strict, so $-\dfrac{1}{3}$ and $\dfrac{7}{2}$ are

part of the solution. Now, $(3w+1)(2w-7)$ is

less than zero where the product is negative.

The solution is $\left\{w\mid -\dfrac{1}{3} \leq w \leq \dfrac{7}{2}\right\}$ or, using

interval notation, $\left[-\dfrac{1}{3}, \dfrac{7}{2}\right]$.

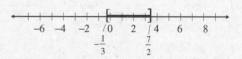

127. $\dfrac{x-4}{x+2} \geq 0$

The rational expression will equal 0 when $x = 4$.
It is undefined when $x = -2$.
Determine where the numerator and denominator are positive and negative and where the quotient is positive and negative.

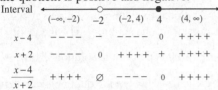

The rational function is undefined at $x = -2$, so -2 is not part of the solution. The inequality is non-strict, so 4 is part of the solution. Now,

$\dfrac{x-4}{x+2}$ is greater than zero where the quotient is

positive. The solution is $\{x \mid x < -2 \text{ or } x \geq 4\}$ in

set-builder notation; the solution is

$(-\infty,\ -2) \cup [4,\ \infty)$ in interval notation.

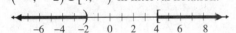

128. $\dfrac{y-5}{y+4} < 0$

The rational expression will equal 0 when
$y = 5$. It is undefined when $y = -4$.
Determine where the numerator and denominator are positive and negative and where the quotient is positive and negative.

Interval	$(-\infty, -4)$	-4	$(-4, 5)$	5	$(5, \infty)$
$y - 5$	$----$	$-$	$----$	0	$++++$
$y + 4$	$----$	0	$++++$	$+$	$++++$
$\dfrac{y-5}{y+4}$	$++++$	$\varnothing$	$----$	0	$++++$

The rational function is undefined at $y = -4$, so -4 is not part of the solution. The inequality is strict, so 5 is not part of the solution. Now,

$\dfrac{y-5}{y+4}$ is less than zero where the quotient is

negative. The solution is $\{y \mid -4 < y < 5\}$ in set-builder notation; the solution is $(-4, 5)$ in interval notation.

129. $\dfrac{4}{z^2 - 9} \leq 0$

$\dfrac{4}{(z-3)(z+3)} \leq 0$

Because the numerator is a constant, this rational expression cannot equal 0. However, it is undefined when $z = 3$ and when $z = -3$.
Determine where the factors of the denominator are positive and negative and where the quotient is positive and negative.

Interval	$(-\infty, -3)$	-3	$(-3, 3)$	3	$(3, \infty)$
4	$++++$	$+$	$++++$	$+$	$++++$
$z - 3$	$----$	$-$	$----$	0	$++++$
$z + 3$	$----$	0	$++++$	$+$	$++++$
$\dfrac{4}{(z-3)(z+3)}$	$++++$	$\varnothing$	$----$	$\varnothing$	$++++$

The rational function is undefined at $z = -3$ and $z = 3$, so -3 and 3 are not part of the solution.

Now, $\dfrac{4}{(z-3)(z+3)}$ is less than zero where the

quotient is negative. The solution is
$\{z \mid -3 < z < 3\}$ in set-builder notation; the

solution is $(-3,\ 3)$ in interval notation.

130. $\dfrac{w^2 + 5w - 14}{w - 4} < 0$

$\dfrac{(w-2)(w+7)}{w-4} < 0$

The rational expression will equal 0 when $w = 2$ and when $w = -7$. It is undefined when $w = 4$.
Determine where the factors of the numerator and the denominator are positive and negative and where the quotient is positive and negative.

Interval	$(-\infty, -7)$	-7	$(-7, 2)$	2	$(2, 4)$	4	$(4, \infty)$
$w - 2$	$---$	$-$	$---$	0	$+++$	$+$	$+++$
$w + 7$	$---$	0	$+++$	$+$	$+++$	$+$	$+++$
$w - 4$	$---$	$-$	$---$	$-$	$---$	0	$+++$
$\dfrac{(w-2)(w+7)}{w-4}$	$---$	0	$+++$	0	$---$	$\varnothing$	$+++$

The rational function is undefined at $w = 4$, so 4 is not part of the solution. The inequality is strict, so -7 and 2 are not part of the solution.

Now, $\dfrac{(w-2)(w+7)}{w-4}$ is less than zero where the

quotient is negative. The solution is

$\{w \mid w < -7 \text{ or } 2 < w < 4\}$ in set-builder notation; the solution is $(-\infty, -7) \cup (2, 4)$ in interval notation.

131.
$$\frac{m-5}{m^2 + 3m - 10} \geq 0$$

$$\frac{m-5}{(m-2)(m+5)} \geq 0$$

The rational expression will equal 0 when $m = 5$. It is undefined when $m = 2$ and when $m = -5$.

Determine where the numerator and the factors of the denominator are positive and negative and where the quotient is positive and negative.

Interval	$(-\infty, -5)$	-5	$(-5, 2)$	2	$(2, 5)$	5	$(5, \infty)$
$m - 5$	$---$	$-$	$---$	$-$	$---$	0	$+++$
$m - 2$	$---$	$-$	$---$	0	$+++$	$+$	$+++$
$m + 5$	$---$	0	$+++$	$+$	$+++$	$+$	$+++$
$\frac{m-5}{(m-2)(m+5)}$	$---$	$\varnothing$	$+++$	$\varnothing$	$---$	0	$+++$

The rational function is undefined at $m = -5$ and when $m = 2$, so -5 and 2 are not part of the solution. The inequality is non-strict, so 5 is part of the solution. Now, $\dfrac{m-5}{(m-2)(m+5)}$ is greater than zero where the quotient is positive. The solution is $\{m \mid -5 < m < 2 \text{ or } m \geq 5\}$ in set-builder notation; the solution is $(-5, 2) \cup [5, \infty)$ in interval notation.

132.
$$\frac{4}{n-2} \leq -2$$

$$\frac{4}{n-2} + 2 \leq 0$$

$$\frac{4}{n-2} + \frac{2(n-2)}{n-2} \leq 0$$

$$\frac{4 + 2n - 4}{n-2} \leq 0$$

$$\frac{2n}{n-2} \leq 0$$

The rational expression will equal 0 when $n = 0$. It is undefined when $n = 2$.

Determine where the numerator and

denominator are positive and negative and where the quotient is positive and negative.

Interval	$(-\infty, 0)$	0	$(0, 2)$	2	$(2, \infty)$
$2n$	$----$	0	$++++$	$+$	$++++$
$n - 2$	$----$	$-$	$----$	0	$++++$
$\frac{2n}{n-2}$	$++++$	0	$----$	$\varnothing$	$++++$

The rational function is undefined at $n = 2$, so 2 is not part of the solution. The inequality is non-strict, so 0 is part of the solution. Now, $\dfrac{2n}{n-2}$ is less than zero where the quotient is negative. The solution is $\{n \mid 0 \leq n < 2\}$ in set-builder notation; the solution is $[0, 2)$ in interval notation.

133.
$$\frac{a+1}{a-2} > 3$$

$$\frac{a+1}{a-2} - 3 > 0$$

$$\frac{a+1}{a-2} - \frac{3(a-2)}{a-2} > 0$$

$$\frac{a+1-3a+6}{a-2} > 0$$

$$\frac{-2a+7}{a-2} > 0$$

The rational expression will equal 0 when $a = \dfrac{7}{2}$. It is undefined when $a = 2$.

Determine where the numerator and denominator are positive and negative and where the quotient is positive and negative.

Interval	$(-\infty, 2)$	2	$\left(2, \frac{7}{2}\right)$	$\frac{7}{2}$	$\left(\frac{7}{2}, \infty\right)$
$-2a + 7$	$++++$	$+$	$++++$	0	$----$
$a - 2$	$----$	0	$++++$	$+$	$++++$
$\frac{-2a+7}{a-2}$	$----$	$\varnothing$	$++++$	0	$----$

The rational function is undefined at $a = 2$, so 2 is not part of the solution. The inequality is strict, so $\dfrac{7}{2}$ is not part of the solution. Now, $\dfrac{-2a+7}{a-2}$ is greater than zero where the quotient is positive. The solution is $\left\{a \mid 2 < a < \dfrac{7}{2}\right\}$ in set-builder notation; the solution is $\left(2, \dfrac{7}{2}\right)$ in

interval notation.

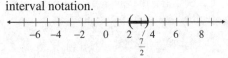

134.
$$\frac{4}{c-2}-\frac{3}{c}<0$$
$$\frac{4c}{c(c-2)}-\frac{3(c-2)}{c(c-2)}<0$$
$$\frac{4c-3c+6}{c(c-2)}<0$$
$$\frac{c+6}{c(c-2)}<0$$

The rational expression will equal 0 when
$c=-6$. It is undefined when $c=0$ and when
$c=2$.
Determine where the numerator and the factors
of the denominator are positive and negative and
where the quotient is positive and negative.

Interval	$(-\infty,-6)$	-6	$(-6,0)$	0	$(0,2)$	2	$(2,\infty)$
$c+6$	$---$	0	$+++$	$+$	$+++$	$+$	$+++$
c	$---$	$-$	$---$	0	$+++$	$+$	$+++$
$c-2$	$---$	$-$	$---$	$-$	$---$	0	$+++$
$\dfrac{c+6}{c(c-2)}$	$---$	0	$+++$	$\varnothing$	$---$	$\varnothing$	$+++$

The rational function is undefined at $c=0$ and
$c=2$, so 0 and 2 are not part of the solution.
The inequality is strict, so -6 is not part of the

solution. Now, $\dfrac{c+6}{c(c-2)}$ is less than zero where

the quotient is negative. The solution is
$\{c\,|\,c<-6 \text{ or } 0<c<2\}$ in set-builder notation;
the solution is $(-\infty,-6)\cup(0,\,2)$ in interval
notation.

135. $Q(x)<0$
$$\frac{2x+3}{x-4}<0$$
The rational expression will equal 0 when

$x=-\dfrac{3}{2}$. It is undefined when $x=4$.

Determine where the numerator and
denominator are positive and negative and where
the quotient is positive and negative.

Interval	$\left(-\infty,-\dfrac{3}{2}\right)$	$-\dfrac{3}{2}$	$\left(-\dfrac{3}{2},4\right)$	4	$(4,\infty)$
$2x+3$	$----$	0	$++++$	$+$	$++++$
$x-4$	$----$	$-$	$----$	0	$++++$
$\dfrac{2x+3}{x-4}$	$++++$	0	$----$	$\varnothing$	$++++$

The rational function is undefined at $x=4$, so 4
is not part of the solution. The inequality is strict,

so $-\dfrac{3}{2}$ is not part of the solution. Now, $\dfrac{2x+3}{x-4}$

is less than zero where the quotient is negative.

The solution is $\left\{x\,\middle|\,-\dfrac{3}{2}<x<4\right\}$ in set-builder

notation; the solution is $\left(-\dfrac{3}{2},\,4\right)$ in interval

notation.

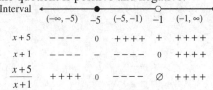

136. $R(x)\ge 0$
$$\frac{x+5}{x+1}\ge 0$$
The rational expression will equal 0 when
$x=-5$. It is undefined when $x=-1$.
Determine where the numerator and the
denominator are positive and negative and where
the quotient is positive and negative.

Interval	$(-\infty,-5)$	-5	$(-5,-1)$	-1	$(-1,\infty)$
$x+5$	$----$	0	$++++$	$+$	$++++$
$x+1$	$----$	$-$	$----$	0	$++++$
$\dfrac{x+5}{x+1}$	$++++$	0	$----$	$\varnothing$	$++++$

The rational function is undefined at $x=-1$, so
-1 is not part of the solution. The inequality is
non-strict, so -5 is part of the solution. Now,

$\dfrac{x+5}{x+1}$ is greater than zero where the quotient is

positive. The solution is $\{x\,|\,x\le -5 \text{ or } x>-1\}$
in set-builder notation; the solution is
$(-\infty,\,-5]\cup(-1,\,\infty)$ in interval notation.

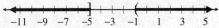

Chapter 10 Test

1. Start: $x^2 - 3x$

Add: $\left[\dfrac{1}{2} \cdot (-3)\right]^2 = \dfrac{9}{4}$

Result: $x^2 - 3x + \dfrac{9}{4}$

Factored Form: $\left(x - \dfrac{3}{2}\right)^2$

2. Start: $m^2 + \dfrac{2}{5}m$

Add: $\left[\dfrac{1}{2}\left(\dfrac{2}{5}\right)\right]^2 = \dfrac{1}{25}$

Result: $m^2 + \dfrac{2}{5}m + \dfrac{1}{25}$

Factored Form: $\left(m + \dfrac{1}{5}\right)^2$

3. $9\left(x + \dfrac{4}{3}\right)^2 = 1$

$\left(x + \dfrac{4}{3}\right)^2 = \dfrac{1}{9}$

$x + \dfrac{4}{3} = \pm\sqrt{\dfrac{1}{9}}$

$x + \dfrac{4}{3} = \pm\dfrac{1}{3}$

$x = -\dfrac{4}{3} \pm \dfrac{1}{3}$

$x = -\dfrac{4}{3} - \dfrac{1}{3}$ or $x = -\dfrac{4}{3} + \dfrac{1}{3}$

$x = -\dfrac{5}{3}$ or $x = -1$

The solution set is $\left\{-\dfrac{5}{3},\ -1\right\}$.

4. $m^2 - 6m + 4 = 0$

Because this equation does not easily factor, solve by using the quadratic formula. For this equation, $a = 1$, $b = -6$, and $c = 4$.

$m = \dfrac{-(-6) \pm \sqrt{(-6)^2 - 4(1)(4)}}{2(1)}$

$= \dfrac{6 \pm \sqrt{36 - 16}}{2}$

$= \dfrac{6 \pm \sqrt{20}}{2}$

$= \dfrac{6 \pm 2\sqrt{5}}{2}$

$= \dfrac{6}{2} \pm \dfrac{2\sqrt{5}}{2}$

$= 3 \pm \sqrt{5}$

The solution set is. $\left\{3 - \sqrt{5},\ 3 + \sqrt{5}\right\}$.

5. $2w^2 - 4w + 3 = 0$

Because this equation does not easily factor, solve by using the quadratic formula. For this equation, $a = 2$, $b = -4$, and $c = 3$.

$w = \dfrac{-(-4) \pm \sqrt{(-4)^2 - 4(2)(3)}}{2(2)}$

$= \dfrac{4 \pm \sqrt{16 - 24}}{4}$

$= \dfrac{4 \pm \sqrt{-8}}{4}$

$= \dfrac{4 \pm 2\sqrt{2}\,i}{4}$

$= \dfrac{4}{4} \pm \dfrac{2\sqrt{2}}{4}i$

$= 1 \pm \dfrac{\sqrt{2}}{2}i$

The solution set is $\left\{1 - \dfrac{\sqrt{2}}{2}i,\ 1 + \dfrac{\sqrt{2}}{2}i\right\}$.

6. $\dfrac{1}{2}z^2 - \dfrac{3}{2}z = -\dfrac{7}{6}$

$6\left(\dfrac{1}{2}z^2 - \dfrac{3}{2}z\right) = 6\left(-\dfrac{7}{6}\right)$

$3z^2 - 9z = -7$

$3z^2 - 9z + 7 = 0$

Because this equation does not easily factor, solve by using the quadratic formula. For this equation, $a = 3$, $b = -9$, and $c = 7$.

$$z = \frac{-(-9) \pm \sqrt{(-9)^2 - 4(3)(7)}}{2(3)}$$

$$= \frac{9 \pm \sqrt{81 - 84}}{6}$$

$$= \frac{9 \pm \sqrt{-3}}{6}$$

$$= \frac{9 \pm \sqrt{3}i}{6}$$

$$= \frac{9}{6} \pm \frac{\sqrt{3}}{6}i$$

$$= \frac{3}{2} \pm \frac{\sqrt{3}}{6}i$$

The solution set is $\left\{ \frac{3}{2} - \frac{\sqrt{3}}{6}i, \ \frac{3}{2} + \frac{\sqrt{3}}{6}i \right\}$.

7. $2x^2 + 5x = 4$

$2x^2 + 5x - 4 = 0$

For this equation, $a = 2$, $b = 5$, and $c = -4$.

$b^2 - 4ac = 5^2 - 4(2)(-4) = 25 + 32 = 57$

Because $b^2 - 4ac = 57$, but not a perfect square, the quadratic equation will have two irrational solutions.

8. $c^2 = a^2 + b^2$

$11^2 = a^2 + 7^2$

$121 = a^2 + 49$

$72 = a^2$

$a = \sqrt{72}$

$a = 6\sqrt{2}$

9. $x^4 - 5x^2 - 36 = 0$

$\left(x^2\right)^2 - 5\left(x^2\right) - 36 = 0$

Let $u = x^2$.

$u^2 - 5u - 36 = 0$

$(u - 9)(u + 4) = 0$

$\begin{array}{lll} u - 9 = 0 & \text{or} & u + 4 = 0 \\ u = 9 & \text{or} & u = -4 \\ x^2 = 9 & \text{or} & x^2 = -4 \\ x = \pm\sqrt{9} & \text{or} & x = \pm\sqrt{-4} \\ x = \pm 3 & \text{or} & x = \pm 2i \end{array}$

Check:

$x = -3: \ (-3)^4 - 5(-3)^2 - 36 \overset{?}{=} 0$

$81 - 5 \cdot 9 - 36 \overset{?}{=} 0$

$81 - 45 - 36 \overset{?}{=} 0$

$0 = 0 \ \checkmark$

$x = 3: \ 3^4 - 5(3)^2 - 36 \overset{?}{=} 0$

$81 - 5 \cdot 9 - 36 \overset{?}{=} 0$

$81 - 45 - 36 \overset{?}{=} 0$

$0 = 0 \ \checkmark$

$x = -2i: \ (-2i)^4 - 5(-2i)^2 - 36 \overset{?}{=} 0$

$16i^4 - 5 \cdot 4i^2 - 36 \overset{?}{=} 0$

$16(1) - 5 \cdot 4(-1) - 36 \overset{?}{=} 0$

$16 + 20 - 36 \overset{?}{=} 0$

$0 = 0 \ \checkmark$

$x = 2i: \ (2i)^4 - 5(2i)^2 - 36 \overset{?}{=} 0$

$16i^4 - 5 \cdot 4i^2 - 36 \overset{?}{=} 0$

$16(1) - 5 \cdot 4(-1) - 36 \overset{?}{=} 0$

$16 + 20 - 36 \overset{?}{=} 0$

$0 = 0 \ \checkmark$

All check; the solution set is $\{-3, \ 3, \ -2i, \ 2i\}$.

10. $6y^{1/2} + 13y^{1/4} - 5 = 0$

$6\left(y^{1/4}\right)^2 + 13\left(y^{1/4}\right) - 5 = 0$

Let $u = y^{1/4}$.

$6u^2 + 13u - 5 = 0$

$(3u - 1)(2u + 5) = 0$

$\begin{array}{lll} 3u - 1 = 0 & \text{or} & 2u + 5 = 0 \\ 3u = 1 & \text{or} & 2u = -5 \\ u = \dfrac{1}{3} & \text{or} & u = -\dfrac{5}{2} \\ y^{1/4} = \dfrac{1}{3} & \text{or} & y^{1/4} = -\dfrac{5}{2} \\ \left(y^{1/4}\right)^4 = \left(\dfrac{1}{3}\right)^4 & \text{or} & \left(y^{1/4}\right)^4 = \left(-\dfrac{5}{2}\right)^4 \\ y = \dfrac{1}{81} & \text{or} & y = \dfrac{625}{16} \end{array}$

Check:

$$y = \frac{1}{81}: \quad 6\left(\frac{1}{81}\right)^{1/2} + 13\left(\frac{1}{81}\right)^{1/4} - 5 \overset{?}{=} 0$$

$$6\sqrt{\frac{1}{81}} + 13\sqrt[4]{\frac{1}{81}} - 5 \overset{?}{=} 0$$

$$6\left(\frac{1}{9}\right) + 13\left(\frac{1}{3}\right) - 5 \overset{?}{=} 0$$

$$\frac{2}{3} + \frac{13}{3} - 5 \overset{?}{=} 0$$

$$0 = 0 \checkmark$$

$$y = \frac{625}{16}: \quad 6\left(\frac{625}{16}\right)^{1/2} + 13\left(\frac{625}{16}\right)^{1/4} - 5 \overset{?}{=} 0$$

$$6\sqrt{\frac{625}{16}} + 13\sqrt[4]{\frac{625}{16}} - 5 \overset{?}{=} 0$$

$$6\left(\frac{25}{4}\right) + 13\left(\frac{5}{2}\right) - 5 \overset{?}{=} 0$$

$$\frac{75}{2} + \frac{65}{2} - 5 \overset{?}{=} 0$$

$$65 \neq 0 \; ✗$$

$y = \dfrac{625}{16}$ does not check; the solution set is

$$\left\{\frac{1}{81}\right\}.$$

11. Begin with the graph of $y = x^2$, then shift the graph 2 units to the left to obtain the graph of $y = (x+2)^2$. Shift this graph down 5 units to obtain the graph of $f(x) = (x+2)^2 - 5$.

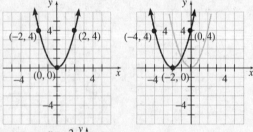

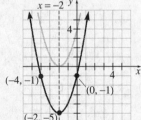

The vertex is $(-2, -5)$. The axis of symmetry is $x = -2$.

12. $g(x) = -2x^2 - 8x - 3$

$a = -2, b = -8, c = -3$

The graph opens down because the coefficient on x^2 is negative.

vertex:

$$x = -\frac{b}{2a} = -\frac{(-8)}{2(-2)} = -2$$

$$g(-2) = -2(-2)^2 - 8(-2) - 3 = 5$$

The vertex is $(-2, 5)$. The axis of symmetry is $x = -2$.

y-intercept:

$$g(0) = -2(0)^2 - 8(0) - 3 = -3$$

x-intercepts:

$$b^2 - 4ac = (-8)^2 - 4(-2)(-3) = 40 > 0$$

There are two distinct x-intercepts. We find these by solving

$$g(x) = 0$$

$$-2x^2 - 8x - 3 = 0$$

$$x = \frac{-(-8) \pm \sqrt{40}}{2(-2)}$$

$$= \frac{8 \pm 2\sqrt{10}}{-4}$$

$$= -2 \pm \frac{\sqrt{10}}{2}$$

$$x \approx -3.58 \quad \text{or} \quad x \approx -0.42$$

Graph:

The y-intercept point, $(0, -3)$, is two units to the right of the axis of symmetry. Therefore, if we move two units to the left of the axis of symmetry, we obtain the point $(-4, -3)$ which must also be on the graph.

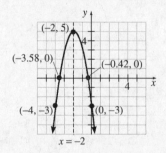

13. Consider the form $y = a(x-h)^2 + k$. From the graph we know that the vertex is $(-3, -5)$, so we have $h = -3$ and $k = -5$. The graph also passes through the point $(x, y) = (0, -2)$. Substituting values for x, y, h, and k, we can solve for a:

$$-2 = a(0 - (-3))^2 - 5$$
$$-2 = a(0+3)^2 - 5$$
$$-2 = a(3)^2 - 5$$
$$-2 = 9a - 5$$
$$3 = 9a$$
$$\frac{3}{9} = a$$
$$\frac{1}{3} = a$$

The quadratic function is $f(x) = \frac{1}{3}(x+3)^2 - 5$

or $f(x) = \frac{1}{3}x^2 + 2x - 2$.

14. If we compare $h(x) = -\frac{1}{4}x^2 + x + 5$ to

$f(x) = ax^2 + bx + c$, we find that $a = -\frac{1}{4}$,

$b = 1$, and $c = 5$. Because $a < 0$, we know the graph will open down, so the function will have a maximum value.
The maximum value occurs at

$$x = -\frac{b}{2a} = -\frac{1}{2\left(-\frac{1}{4}\right)} = -\frac{1}{-\frac{1}{2}} = 2.$$

The maximum value is

$$h(2) = -\frac{1}{4}(2)^2 + 2 + 5 = 6.$$

So, the maximum value is 6, and it occurs when $x = 2$.

15. $2m^2 + m - 15 > 0$

Solve: $2m^2 + m - 15 = 0$
$$(2m - 5)(m + 3) = 0$$
$$2m - 5 = 0 \quad \text{or} \quad m + 3 = 0$$
$$2m = 5 \quad \text{or} \quad m = -3$$
$$m = \frac{5}{2}.$$

Determine where each factor is positive and negative and where the product of these factors is positive and negative.

Interval	$(-\infty, -3)$	-3	$\left(-3, \frac{5}{2}\right)$	$\frac{5}{2}$	$\left(\frac{5}{2}, \infty\right)$
$2m - 5$	$----$	$-$	$----$	0	$++++$
$m + 3$	$----$	0	$++++$	$+$	$++++$
$(2m-5)(m+3)$	$++++$	0	$----$	0	$++++$

The inequality is strict, so -3 and $\frac{5}{2}$ are not part of the solution. Now, $(2m-5)(m+3)$ is greater than zero where the product is positive. The solution is $\left\{ m \mid m < -3 \text{ or } m > \frac{5}{2} \right\}$ or, using interval notation, $(-\infty, -3) \cup \left(\frac{5}{2}, \infty\right)$.

16. $z^2 + 6z - 1 \le 0$

Solve: $z^2 + 6z - 1 = 0$

$$z = \frac{-6 \pm \sqrt{6^2 - 4(1)(-1)}}{2(1)}$$
$$= \frac{-6 \pm \sqrt{40}}{2}$$
$$= \frac{-6 \pm 2\sqrt{10}}{2}$$
$$= -3 \pm \sqrt{10}$$

$z = -3 - \sqrt{10}$ or $z = -3 + \sqrt{10}$
≈ -6.16 or ≈ 0.16

Determine where each factor is positive and negative and where the product of these factors is positive and negative.

Interval	$\left(-\infty, -3-\sqrt{10}\right)$	$-3-\sqrt{10}$ ≈ -6.16	$\left(-3-\sqrt{10}, -3+\sqrt{10}\right)$	$-3+\sqrt{10}$ ≈ 0.16	$\left(-3+\sqrt{10}, \infty\right)$
$z - \left(-3-\sqrt{10}\right)$	$----$	0	$++++$	$+$	$++++$
$z - \left(-3+\sqrt{10}\right)$	$----$	$-$	$----$	0	$++++$
$\left[z-\left(-3-\sqrt{10}\right)\right]\left[z-\left(-3+\sqrt{10}\right)\right]$	$++++$	0	$----$	0	$++++$

The inequality is non-strict, so $-3 - \sqrt{10}$ and $-3 + \sqrt{10}$ are part of the solution. Now, $\left[z - \left(-3 - \sqrt{10}\right)\right]\left[z - \left(-3 + \sqrt{10}\right)\right]$ is less than zero where the product is negative. The solution is $\left\{z \mid -3 - \sqrt{10} \le z \le -3 + \sqrt{10}\right\}$ or, using interval notation, $\left[-3 - \sqrt{10},\ -3 + \sqrt{10}\right]$.

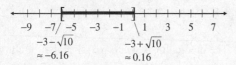

$-3 - \sqrt{10} \approx -6.16$

$-3 + \sqrt{10} \approx 0.16$

17.

$$\frac{x+5}{x-2} \ge 3$$

$$\frac{x+5}{x-2} - 3 \ge 0$$

$$\frac{x+5}{x-2} - \frac{3(x-2)}{x-2} \ge 0$$

$$\frac{x+5-3x+6}{x-2} \ge 0$$

$$\frac{-2x+11}{x-2} \ge 0$$

The rational expression will equal 0 when $x = \frac{11}{2}$. It is undefined when $x = 2$.

Determine where the numerator and denominator are positive and negative and where the quotient is positive and negative.

Interval	$(-\infty, 2)$	2	$\left(2, \frac{11}{2}\right)$	$\frac{11}{2}$	$\left(\frac{11}{2}, \infty\right)$
$-2x+11$	$++++$	$+$	$++++$	0	$----$
$x-2$	$----$	0	$++++$	$+$	$++++$
$\frac{-2x+11}{x-2}$	$----$	$\varnothing$	$++++$	0	$----$

The rational function is undefined at $x = 2$, so 2 is not part of the solution. The inequality is non-strict, so $\frac{11}{2}$ is part of the solution. Now, $\frac{-2x+11}{x-2}$ is greater than zero where the quotient is positive. The solution is $\left\{x \mid 2 < x \le \frac{11}{2}\right\}$ in set-builder notation; the solution is $\left(2, \frac{11}{2}\right]$ in interval notation.

$\frac{11}{2}$

18.

$$50 = -16t^2 + 80t + 20$$

$$16t^2 - 80t + 30 = 0$$

For this equation, $a = 16,\ b = -80,$ and $c = 30$.

$$t = \frac{-(-80) \pm \sqrt{(-80)^2 - 4(16)(30)}}{2(16)}$$

$$= \frac{80 \pm \sqrt{6400 - 1920}}{32}$$

$$= \frac{80 \pm \sqrt{4480}}{32}$$

$$= \frac{80 \pm 8\sqrt{70}}{32}$$

$$= \frac{80}{32} \pm \frac{8\sqrt{70}}{32}$$

$$= \frac{5}{2} \pm \frac{\sqrt{70}}{4}$$

$$t = \frac{5}{2} - \frac{\sqrt{70}}{4} \quad \text{or} \quad t = \frac{5}{2} - \frac{\sqrt{70}}{4}$$

$$\approx 0.408 \quad \text{or} \quad \approx 4.592$$

Rounding to the nearest tenth, the height of the rock will be 50 feet at both 0.4 seconds and 4.6 seconds.

19. Let t represent the time required for Rupert to roof the house alone. Then $t - 4$ will represent the time required for Lex to roof the house alone.

$$\begin{pmatrix} \text{Part done} \\ \text{by Rupert} \\ \text{in 1 hour} \end{pmatrix} + \begin{pmatrix} \text{Part done} \\ \text{by Lex} \\ \text{in 1 hour} \end{pmatrix} = \begin{pmatrix} \text{Part done} \\ \text{together} \\ \text{in 1 hour} \end{pmatrix}$$

$$\frac{1}{t} + \frac{1}{t-4} = \frac{1}{16}$$

$$16t(t-4)\left(\frac{1}{t} + \frac{1}{t-4}\right) = 16t(t-4)\left(\frac{1}{16}\right)$$

$$16(t-4) + 16t = t(t-4)$$

$$16t - 64 + 16t = t^2 - 4t$$

$$32t - 64 = t^2 - 4t$$

$$0 = t^2 - 36t + 64$$

For this equation, $a = 1,\ b = -36,$ and $c = 64$.

$$t = \frac{-(-36) \pm \sqrt{(-36)^2 - 4(1)(64)}}{2(1)}$$

$$= \frac{36 \pm \sqrt{1296 - 256}}{2}$$

$$= \frac{36 \pm \sqrt{1040}}{2}$$

$$= \frac{36 \pm 4\sqrt{65}}{2}$$

$$= 18 \pm 2\sqrt{65}$$

$$t = 18 - 2\sqrt{65} \quad \text{or} \quad t = 18 + 2\sqrt{65}$$

$$\approx 1.875 \quad \text{or} \quad \approx 34.125$$

Disregard $t \approx 1.875$ because this value makes Lex's time negative: $t - 4 = 1.875 - 4 = -2.125$. The only viable answer is $t \approx 34.125$ hours. Rounding to the nearest tenth, Rupert can roof the house in 34.1 hours when working alone.

20. **(a)** We first recognize that the quadratic function has a negative leading coefficient. This means that the graph will open down and the function indeed has a maximum value. The maximum value occurs when

$$p = -\frac{b}{2a} = -\frac{170}{2(-0.25)} = 340.$$

The revenue will be maximized when the product is sold at a price of $340.

(b) The maximum revenue is obtained by evaluating the revenue function at the price found in part (a).

$$R(340) = -0.25(340)^2 + 170(340)$$
$$= 28,900$$

The maximum weekly revenue is $28,900.

21. **(a)** Let x represent the width of the base of the box. Then $\frac{50 - 2x}{2} = 25 - x$ is the length of the base. The volume is the product of the length, width, and height:

$$V = (25 - x) \cdot x \cdot 12$$
$$= -12x^2 + 300x$$

The leading coefficient is negative so we know the graph opens down and there will be a maximum volume. This value occurs when the width is

$$x = -\frac{b}{2a} = -\frac{300}{2(-12)} = 12.5.$$

Then the length is $25 - 12.5 = 12.5$. The dimensions that maximize the volume of the box are 12.5 in. by 12.5 in. by 12 in.

(b) The maximum volume can be found by substituting 12.5 for x into the volume function.

$$V = (25 - 12.5) \cdot 12.5 \cdot 12 = 1875$$

The maximum volume of the box is 1875 cubic inches.

Chapter 11

Preparing for Composite Functions and Inverse Functions

P1. We need to find all values of x that cause the denominator $x^2 + 3x - 28$ to equal 0.

$$x^2 + 3x - 28 = 0$$
$$(x+7)(x-4) = 0$$
$$x + 7 = 0 \quad \text{or} \quad x - 4 = 0$$
$$x = -7 \quad \text{or} \quad x = 4$$

Thus, the domain of $R(x) = \dfrac{x^2 - 9}{x^2 + 3x - 28}$ is

$\{x \mid x \neq -7, x \neq 4\}$.

P2. **(a)** $f(-2) = 2(-2)^2 - (-2) + 1$
$$= 8 + 2 + 1$$
$$= 11$$

(b) $f(a+1) = 2(a+1)^2 - (a+1) + 1$
$$= 2(a^2 + 2a + 1) - (a+1) + 1$$
$$= 2a^2 + 4a + 2 - a - 1 + 1$$
$$= 2a^2 + 3a + 2$$

P3. The graph shown is not a function because it fails the vertical line test.

Section 11.1 Quick Checks

1. composite function

2. $f(x) = 4x - 3$; $g(x) = x^2 + 1$

(a) $g(2) = (2)^2 + 1 = 4 + 1 = 5$
$$f(5) = 4(5) - 3 = 20 - 3 = 17$$
$$(f \circ g)(2) = f(g(2)) = f(5) = 17$$

(b) $f(2) = 4(2) - 3 = 8 - 3 = 5$
$$g(5) = (5)^2 + 1 = 25 + 1 = 26$$
$$(g \circ f)(2) = g(f(2)) = g(5) = 26$$

(c) $f(-3) = 4(-3) - 3 = -12 - 3 = -15$
$$f(-15) = 4(-15) - 3 = -60 - 3 = -63$$
$$(f \circ f)(-3) = f(f(-3)) = f(-15) = -63$$

3. False. $(f \circ g)(x) = f(g(x))$

4. $g(x) = 4x - 3$

5. $f(x) = x^2 - 3x + 1$; $g(x) = 3x + 2$

(a) $(f \circ g)(x) = f(g(x))$
$$= (3x+2)^2 - 3(3x+2) + 1$$
$$= 9x^2 + 12x + 4 - 9x - 6 + 1$$
$$= 9x^2 + 3x - 1$$

(b) $(g \circ f)(x) = g(f(x))$
$$= 3(x^2 - 3x + 1) + 2$$
$$= 3x^2 - 9x + 3 + 2$$
$$= 3x^2 - 9x + 5$$

(c) $(f \circ g)(-2) = 9(-2)^2 + 3(-2) - 1$
$$= 9(4) + 3(-2) - 1$$
$$= 36 - 6 - 1$$
$$= 29$$

6. one-to-one

7. Since the two friends (inputs) Max and Yolanda share the same birthday (output) of November 8, this function is not one-to-one.

8. The function is one-to-one because there are no two distinct inputs that correspond to the same output.

9. **(a)** The graph fails the horizontal line test. For example, the line $y = -1$ will intersect the graph of f in four places. Therefore, the function is not one-to-one.

(b) The graph passes the horizontal line test because every horizontal line will intersect the graph of f exactly once. Thus, the function is one-to-one.

10. The inverse of the one-to-one function is:

Right Tibia		Right Humerus
36.05	⟶	24.80
35.57	⟶	24.59
34.58	⟶	24.29
34.20	⟶	23.81
34.73	⟶	24.87

The domain of the inverse function is $\{36.05, 35.57, 34.58, 34.20, 34.73\}$. The range of the inverse function is $\{24.80, 24.59, 24.29, 23.81, 24.87\}$.

11. To obtain the inverse, we switch the x- and y-coordinates:

$$\{(3,-3),(2,-2),(1,-1),(0,0),(-1,1)\}$$

The domain of the inverse function is $\{3, 2, 1, 0, -1\}$. The range of the inverse function is $\{-3,-2,-1,0,1\}$.

12. To plot the inverse, switch the x- and y-coordinates of each point and connect the corresponding points. The graph of the function (shaded) and the line $y = x$ (dashed) are included for reference.

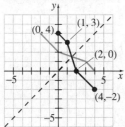

13. True.

14. False. The notation $f^{-1}(x)$ represents the inverse function for the function $f(x)$.

15.
$$g(x) = 5x - 1$$
$$y = 5x - 1$$
$$x = 5y - 1$$
$$x + 1 = 5y$$
$$\frac{x+1}{5} = y$$
$$g^{-1}(x) = \frac{x+1}{5}$$

Check: $g\left(g^{-1}(x)\right) = 5\left(\dfrac{x+1}{5}\right) - 1 = x + 1 - 1 = x$

$$g^{-1}(g(x)) = \frac{(5x-1)+1}{5} = \frac{5x}{5} = x$$

16.
$$f(x) = x^5 + 3$$
$$y = x^5 + 3$$
$$x = y^5 + 3$$
$$x - 3 = y^5$$
$$\sqrt[5]{x-3} = y$$
$$f^{-1}(x) = \sqrt[5]{x-3}$$

Check:

$$f\left(f^{-1}(x)\right) = \left(\sqrt[5]{x-3}\right)^5 + 3 = x - 3 + 3 = x$$

$$f^{-1}(f(x)) = \sqrt[5]{x^5 + 3 - 3} = \sqrt[5]{x^5} = x$$

11.1 Exercises

17. $f(x) = 2x + 5$; $g(x) = x - 4$

(a) $g(3) = 3 - 4 = -1$
$f(-1) = 2(-1) + 5 = 3$
$(f \circ g)(3) = f(g(3)) = f(-1) = 3$

(b) $f(-2) = 2(-2) + 5 = 1$
$g(1) = 1 - 4 = -3$
$(g \circ f)(-2) = g(f(-2)) = g(1) = -3$

(c) $f(1) = 2(1) + 5 = 7$
$f(7) = 2(7) + 5 = 19$
$(f \circ f)(1) = f(f(1)) = f(7) = 19$

(d) $g(-4) = -4 - 4 = -8$
$g(-8) = -8 - 4 = -12$
$(g \circ g)(-4) = g(g(-4)) = g(-8) = -12$

19. $f(x) = x^2 + 4$; $g(x) = 2x + 3$

(a) $g(3) = 2(3) + 3 = 9$
$f(9) = (9)^2 + 4 = 85$
$(f \circ g)(3) = f(g(3)) = f(9) = 85$

(b) $f(-2) = (-2)^2 + 4 = 8$
$g(8) = 2(8) + 3 = 19$
$(g \circ f)(-2) = g(f(-2)) = g(8) = 19$

(c) $f(1) = (1)^2 + 4 = 5$
$f(5) = (5)^2 + 4 = 29$
$(f \circ f)(1) = f(f(1)) = f(5) = 29$

(d) $g(-4) = 2(-4) + 3 = -5$
$g(-5) = 2(-5) + 3 = -7$
$(g \circ g)(-4) = g(g(-4)) = g(-5) = -7$

21. $f(x) = 2x^3$; $g(x) = -2x^2 + 5$

(a) $g(3) = -2(3)^2 + 5 = -2(9) + 5 = -13$
$f(-13) = 2(-13)^3 = -4394$
$(f \circ g)(3) = f(g(3)) = f(-13) = -4394$

(b) $f(-2) = 2(-2)^3 = 2(-8) = -16$
$g(-16) = -2(-16)^2 + 5$
$\qquad = -2(256) + 5$
$\qquad = -507$
$(g \circ f)(-2) = g(f(-2)) = g(-16) = -507$

(c) $f(1) = 2(1)^3 = 2$
$f(2) = 2(2)^3 = 16$
$(f \circ f)(1) = f(f(1)) = f(2) = 16$

(d) $g(-4) = -2(-4)^2 + 5 = -2(16) + 5 = -27$
$g(-27) = -2(-27)^2 + 5$
$\qquad = -2(729) + 5$
$\qquad = -1453$
$(g \circ g)(-4) = g(g(-4)) = g(-27) = -1453$

23. $f(x) = |x - 10|$; $g(x) = \dfrac{12}{x+3}$

(a) $g(3) = \dfrac{12}{3+3} = \dfrac{12}{6} = 2$
$f(2) = |2 - 10| = |-8| = 8$
$(f \circ g)(3) = f(g(3)) = f(2) = 8$

(b) $f(-2) = |-2 - 10| = |-12| = 12$
$g(12) = \dfrac{12}{12+3} = \dfrac{12}{15} = \dfrac{4}{5}$
$(g \circ f)(-2) = g(f(-2)) = g(12) = \dfrac{4}{5}$

(c) $f(1) = |1 - 10| = |-9| = 9$
$f(9) = |9 - 10| = |-1| = 1$
$(f \circ f)(1) = f(f(1)) = f(9) = 1$

(d) $g(-4) = \dfrac{12}{-4+3} = \dfrac{12}{-1} = -12$
$g(-12) = \dfrac{12}{-12+3} = \dfrac{12}{-9} = -\dfrac{4}{3}$
$(g \circ g)(-4) = g(g(-4)) = g(-12) = -\dfrac{4}{3}$

25. $f(x) = x + 1$; $g(x) = 2x$

(a) $(f \circ g)(x) = f(g(x)) = (2x) + 1 = 2x + 1$

(b) $(g \circ f)(x) = g(f(x)) = 2(x+1) = 2x + 2$

(c) $(f \circ f)(x) = f(f(x)) = (x+1) + 1 = x + 2$

(d) $(g \circ g)(x) = g(g(x)) = 2(2x) = 4x$

27. $f(x) = 2x + 7$; $g(x) = -4x + 5$

(a) $(f \circ g)(x) = f(g(x))$
$\qquad = 2(-4x + 5) + 7$
$\qquad = -8x + 10 + 7$
$\qquad = -8x + 17$

(b) $(g \circ f)(x) = g(f(x))$
$\qquad = -4(2x + 7) + 5$
$\qquad = -8x - 28 + 5$
$\qquad = -8x - 23$

(c) $(f \circ f)(x) = f(f(x))$
$\qquad = 2(2x + 7) + 7$
$\qquad = 4x + 14 + 7$
$\qquad = 4x + 21$

(d) $(g \circ g)(x) = g(g(x))$
$$= -4(-4x+5)+5$$
$$= 16x - 20 + 5$$
$$= 16x - 15$$

29. $f(x) = x^2$; $g(x) = x - 3$

(a) $(f \circ g)(x) = f(g(x))$
$$= (x-3)^2 = x^2 - 6x + 9$$

(b) $(g \circ f)(x) = g(f(x)) = (x^2) - 3 = x^2 - 3$

(c) $(f \circ f)(x) = f(f(x)) = (x^2)^2 = x^4$

(d) $(g \circ g)(x) = g(g(x)) = (x-3) - 3 = x - 6$

31. $f(x) = \sqrt{x}$; $g(x) = x + 4$

(a) $(f \circ g)(x) = f(g(x)) = \sqrt{x+4}$

(b) $(g \circ f)(x) = g(f(x)) = \sqrt{x} + 4$

(c) $(f \circ f)(x) = f(f(x)) = \sqrt{\sqrt{x}} = \sqrt[4]{x}$

(d) $(g \circ g)(x) = g(g(x)) = (x+4) + 4 = x + 8$

33. $f(x) = |x+4|$; $g(x) = x^2 - 4$

(a) $(f \circ g)(x) = f(g(x))$
$$= |(x^2 - 4) + 4| = |x^2| = x^2$$

(b) $(g \circ f)(x) = g(f(x))$
$$= (|x+4|)^2 - 4 = (x+4)^2 - 4$$
$$= x^2 + 8x + 16 - 4$$
$$= x^2 + 8x + 12$$

(c) $(f \circ f)(x) = f(f(x)) = ||x+4| + 4|$

(d) $(g \circ g)(x) = g(g(x))$
$$= (x^2 - 4)^2 - 4$$
$$= x^4 - 8x^2 + 16 - 4$$
$$= x^4 - 8x^2 + 12$$

35. $f(x) = \dfrac{2}{x+1}$; $g(x) = \dfrac{1}{x}$

(a) $(f \circ g)(x) = f(g(x))$
$$= \frac{2}{\frac{1}{x}+1} = \frac{2}{\frac{1+x}{x}} = \frac{2x}{x+1}$$
where $x \neq -1, 0$.

(b) $(g \circ f)(x) = g(f(x))$
$$= \frac{1}{\frac{2}{x+1}} = 1 \cdot \frac{x+1}{2} = \frac{x+1}{2}$$
where $x \neq -1$.

(c) $(f \circ f)(x) = f(f(x))$
$$= \frac{2}{\frac{2}{x+1}+1} = \frac{2}{\frac{2+x+1}{x+1}}$$
$$= 2 \cdot \frac{x+1}{x+3} = \frac{2(x+1)}{x+3}$$
where $x \neq -1, -3$.

(d) $(g \circ g)(x) = g(g(x)) = \dfrac{1}{\frac{1}{x}} = 1 \cdot \dfrac{x}{1} = x$

where $x \neq 0$

37. The function is one-to-one. Each element in the range corresponds to exactly one element in the domain.

39. The function is not one-to-one. There is an element in the range ($7.09) that corresponds to more than one element in the domain (30 and 35).

41. The function is one-to-one. Each element in the range corresponds to exactly one element in the domain.

43. The function is not one-to-one. There are elements in the range (2 and 4) that correspond to more than one element in the domain.

45. The function is one-to-one. Each element in the range corresponds to exactly one element in the domain.

47. The graph passes the horizontal line test, so the graph is that of a one-to-one function.

49. The graph fails the horizontal line test. Therefore, the function is not one-to-one.

51. The graph passes the horizontal line test, so the graph is that of a one-to-one function.

53. Inverse:

Weight (g)	U. S. Coin
2.500 $\longrightarrow$	Cent
5.000 $\longrightarrow$	Nickel
2.268 $\longrightarrow$	Dime
5.670 $\longrightarrow$	Quarter
11.340 $\longrightarrow$	Half Dollar
8.100 $\longrightarrow$	Dollar

55. To obtain the inverse, we switch the x- and y-coordinates.

Inverse: $\{(3,0),(4,1),(5,2),(6,3)\}$

57. To obtain the inverse, we switch the x- and y-coordinates.

Inverse: $\{(3,-2),(1,-1),(-3,0),(9,1)\}$

59. To plot the inverse, switch the x- and y-coordinates in each point and connect the corresponding points. The graph of the function (shaded) and the line $y = x$ (dashed) are included for reference.

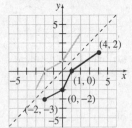

61. To plot the inverse, switch the x- and y-coordinates in each point and connect the corresponding points. The graph of the function (shaded) and the line $y = x$ (dashed) are included for reference.

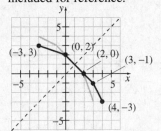

63. To plot the inverse, switch the x- and y-coordinates in each point and connect the corresponding points. The graph of the function (shaded) and the line $y = x$ (dashed) are included for reference.

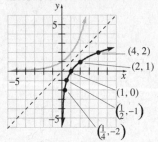

65. $f(g(x)) = (x-5)+5 = x$

$g(f(x)) = (x+5)-5 = x$

Since $f(g(x)) = g(f(x)) = x$, the two functions are inverses of each other.

67. $f(g(x)) = 5\left(\dfrac{x-7}{5}\right)+7 = x-7+7 = x$

$g(f(x)) = \dfrac{(5x+7)-7}{5} = \dfrac{5x}{5} = x$

Since $f(g(x)) = g(f(x)) = x$, the two functions are inverses of each other.

69. $f(g(x)) = \dfrac{3}{\left(\dfrac{3}{x}+1\right)-1} = \dfrac{3}{\dfrac{3}{x}} = 3 \cdot \dfrac{x}{3} = x$

$g(f(x)) = \dfrac{3}{\dfrac{3}{x-1}}+1 = 3 \cdot \dfrac{x-1}{3}+1 = x-1+1 = x$

Since $f(g(x)) = g(f(x)) = x$, the two functions are inverses of each other.

71. $f(g(x)) = \sqrt[3]{(x^3 - 4) + 4} = \sqrt[3]{x^3} = x$

$g(f(x)) = \left(\sqrt[3]{x+4}\right)^3 - 4 = x + 4 - 4 = x$

Since $f(g(x)) = g(f(x)) = x$, the two functions are inverses of each other.

73. $\quad f(x) = 6x$

$y = 6x$

$x = 6y$

$\dfrac{x}{6} = y$

$f^{-1}(x) = \dfrac{x}{6}$

Check:

$f\left(f^{-1}(x)\right) = 6\left(\dfrac{x}{6}\right) = x$

$f^{-1}(f(x)) = \dfrac{6x}{6} = x$

75. $\quad f(x) = x + 4$

$y = x + 4$

$x = y + 4$

$x - 4 = y$

$f^{-1}(x) = x - 4$

Check:

$f\left(f^{-1}(x)\right) = (x - 4) + 4 = x$

$f^{-1}(f(x)) = (x + 4) - 4 = x$

77. $\quad h(x) = 2x - 7$

$y = 2x - 7$

$x = 2y - 7$

$x + 7 = 2y$

$\dfrac{x+7}{2} = y$

$h^{-1}(x) = \dfrac{x+7}{2}$

Check:

$h\left(h^{-1}(x)\right) = 2\left(\dfrac{x+7}{2}\right) - 7 = x + 7 - 7 = x$

$h^{-1}(h(x)) = \dfrac{(2x-7)+7}{2} = \dfrac{2x}{2} = x$

79. $\quad G(x) = 2 - 5x$

$y = 2 - 5x$

$x = 2 - 5y$

$x - 2 = -5y$

$\dfrac{x-2}{-5} = y$

$\dfrac{2-x}{5} = y$

$G^{-1}(x) = \dfrac{2-x}{5}$

Check:

$G\left(G^{-1}(x)\right) = 2 - 5\left(\dfrac{2-x}{5}\right) = 2 - (2 - x) = x$

$G^{-1}(G(x)) = \dfrac{2 - (2 - 5x)}{5} = \dfrac{5x}{5} = x$

81. $\quad g(x) = x^3 + 3$

$y = x^3 + 3$

$x = y^3 + 3$

$x - 3 = y^3$

$\sqrt[3]{x-3} = y$

$g^{-1}(x) = \sqrt[3]{x-3}$

Check:

$g\left(g^{-1}(x)\right) = \left(\sqrt[3]{x-3}\right)^3 + 3 = x - 3 + 3 = x$

$g^{-1}(g(x)) = \sqrt[3]{x^3 + 3 - 3} = \sqrt[3]{x^3} = x$

83. $\quad p(x) = \dfrac{1}{x+3}$

$y = \dfrac{1}{x+3}$

$x = \dfrac{1}{y+3}$

$y + 3 = \dfrac{1}{x}$

$y = \dfrac{1}{x} - 3$

$p^{-1}(x) = \dfrac{1}{x} - 3$

Check:

$$p\left(p^{-1}(x)\right) = \frac{1}{\left(\frac{1}{x}-3\right)+3} = \frac{1}{\frac{1}{x}} = 1 \cdot \frac{x}{1} = x$$

$$p^{-1}\left(p(x)\right) = \frac{1}{\frac{1}{x+3}} - 3 = x+3-3 = x$$

85. $F(x) = \dfrac{5}{2-x}$

$$y = \frac{5}{2-x}$$

$$x = \frac{5}{2-y}$$

$$2-y = \frac{5}{x}$$

$$-y = -2 + \frac{5}{x}$$

$$y = 2 - \frac{5}{x}$$

$$F^{-1}(x) = 2 - \frac{5}{x}$$

Check:

$$F\left(F^{-1}(x)\right) = \frac{5}{2-\left(2-\frac{5}{x}\right)} = \frac{5}{\frac{5}{x}} = 5 \cdot \frac{x}{5} = x$$

$$F^{-1}\left(F(x)\right) = 2 - \frac{5}{\frac{5}{2-x}} = 2-(2-x) = x$$

87. $f(x) = \sqrt[3]{x-2}$

$y = \sqrt[3]{x-2}$

$x = \sqrt[3]{y-2}$

$x^3 = y-2$

$x^3 + 2 = y$

$f^{-1}(x) = x^3 + 2$

Check: $f\left(f^{-1}(x)\right) = \sqrt[3]{x^3+2-2} = \sqrt[3]{x^3} = x$

$$f^{-1}\left(f(x)\right) = \left(\sqrt[3]{x-2}\right)^3 + 2 = x-2+2 = x$$

89. $R(x) = \dfrac{x}{x+2}$

$$y = \frac{x}{x+2}$$

$$x = \frac{y}{y+2}$$

$$x(y+2) = y$$

$$xy + 2x = y$$

$$2x = y - xy$$

$$2x = y(1-x)$$

$$\frac{2x}{1-x} = y$$

$$R^{-1}(x) = \frac{2x}{1-x}$$

Check: $R\left(R^{-1}(x)\right) = \dfrac{\frac{2x}{1-x}}{\frac{2x}{1-x}+2} = \dfrac{\frac{2x}{1-x}}{\frac{2x+2-2x}{1-x}}$

$$= \frac{2x}{1-x} \cdot \frac{1-x}{2} = x$$

$$R^{-1}\left(R(x)\right) = \frac{2\left(\frac{x}{x+2}\right)}{1-\frac{x}{x+2}} = \frac{\frac{2x}{x+2}}{\frac{x+2-x}{x+2}}$$

$$= \frac{2x}{x+2} \cdot \frac{x+2}{2} = x$$

91. $f(x) = \sqrt[3]{x-1}+4$

$y = \sqrt[3]{x-1}+4$

$x = \sqrt[3]{y-1}+4$

$x-4 = \sqrt[3]{y-1}$

$(x-4)^3 = y-1$

$(x-4)^3 + 1 = y$

$f^{-1}(x) = (x-4)^3 + 1$

Check:

$f\left(f^{-1}(x)\right) = \sqrt[3]{(x-4)^3+1-1}+4$

$\qquad = \sqrt[3]{(x-4)^3}+4$

$\qquad = x-4+4$

$\qquad = x$

$$f^{-1}\big(f(x)\big) = \left[\left(\sqrt[3]{x-1}+4\right)-4\right]^3 + 1$$
$$= \left(\sqrt[3]{x-1}\right)^3 + 1$$
$$= x-1+1$$
$$= x$$

93. $A(r) = \pi r^2$; $r(t) = 20t$

$$A(t) = A\big(r(t)\big) = \pi(20t)^2 = 400\pi t^2$$

$$A(3) = 400\pi(3)^2 = 3600\pi \approx 11,309.73 \text{ sq ft}$$

95. $A(x) = \dfrac{x}{9}$; $C(A) = 18A$

 (a) $C(x) = C\big(A(x)\big) = 18\left(\dfrac{x}{9}\right) = 2x$

 (b) The area of the room is $15 \cdot 21 = 315$ sq ft.

  $C(315) = 2(315) = 630$

  It will cost \$630 to install the carpet.

97. $f^{-1}(12) = f^{-1}\big(f(4)\big) = 4$

99. To find the domain and range of the inverse, we simply switch the values for the domain and range of the function.

Domain of f^{-1}: $[-5, \infty)$

Range of f^{-1}: $[0, \infty)$

101. To find the domain and range of the inverse, we simply switch the values for the domain and range of the function.

Domain of g^{-1}: $(-6, 12)$

Range of g^{-1}: $[-4, 10]$

103. $T(x) = 0.15(x - 6000) + 600$

  $T = 0.15(x - 6000) + 600$

  $T = 0.15x - 900 + 600$

  $T = 0.15x - 300$

 $T + 300 = 0.15x$

 $\dfrac{T + 300}{0.15} = x$

 $x(T) = \dfrac{T + 300}{0.15}$

To determine the restrictions, we find

$T(6000) = 0.15(6000 - 6000) + 600 = 600$ and

$T(28,400) = 0.15(28,400 - 6000) + 600 = 3960$.

Thus, $x(T) = \dfrac{T + 300}{0.15}$ for $600 \le T \le 3960$.

105. $f(x) = 2x^2 - x + 5$; $g(x) = x + a$

$$(f \circ g)(x) = f\big(g(x)\big)$$
$$= 2(x+a)^2 - (x+a) + 5$$
$$= 2\big(x^2 + 2ax + a^2\big) - x - a + 5$$
$$= 2x^2 + 4ax + 2a^2 - x - a + 5$$

The *y*-intercept is found by letting $x = 0$ and solving for *y*. We know the *y*-intercept of $(f \circ g)(x)$ is 20. Therefore, we get

$$(f \circ g)(0) = 2(0)^2 + 4a(0) + 2a^2 - 0 - a + 5$$
$$20 = 2a^2 - a + 5$$
$$0 = 2a^2 - a - 15$$
$$0 = (2a + 5)(a - 3)$$

$2a + 5 = 0$ or $a - 3 = 0$

 $2a = -5$ $a = 3$

 $a = -\dfrac{5}{2}$

The solution set for *a* is $\left\{-\dfrac{5}{2}, 3\right\}$.

107. Answers may vary. One possibility follows: A function is one-to-one if each element of the range corresponds to no more than one element in the domain. If a function is not one-to-one, then there is at least one element in the range that corresponds to more than one element in the domain. For the inverse, we would switch the domain and range. Therefore, if a function is not one-to-one, the domain of its inverse will have at least one element that corresponds to more than one element in its range. Hence, the inverse would not be a function.

109. Answers may vary. One possibility follows: By definition, the inverse of a function with ordered pairs of the form (a, b) is the set of ordered pairs of the form (b, a). This means that the set of *y*-coordinates (i.e., the range) of the original function will become the set of *x*-coordinates (i.e., the domain) of the inverse function. Likewise, the set of *x*-coordinates (i.e., the domain) of the original function will become the set of *y*-coordinates (i.e., the range) of the inverse function.

111. $f(x) = 2x + 5$; $g(x) = x - 4$

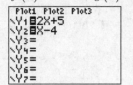

(a) $(f \circ g)(3) = 3$

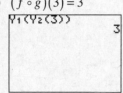

(b) $(g \circ f)(-2) = -3$

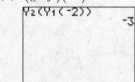

(c) $(f \circ f)(1) = 19$

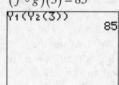

(d) $(g \circ g)(-4) = -12$

113. $f(x) = x^2 + 4$; $g(x) = 2x + 3$

(a) $(f \circ g)(3) = 85$

(b) $(g \circ f)(-2) = 19$

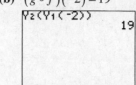

(c) $(f \circ f)(1) = 29$

(d) $(g \circ g)(-4) = -7$

115. $f(x) = 2x^3$; $g(x) = -2x^2 + 5$

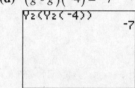

(a) $(f \circ g)(3) = -4394$

(b) $(g \circ f)(-2) = -507$

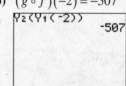

(c) $(f \circ f)(1) = 16$

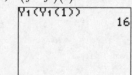

(d) $(g \circ g)(-4) = -1453$

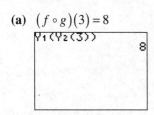

117. $f(x) = |x - 10|$; $g(x) = \dfrac{12}{x + 3}$

```
Plot1 Plot2 Plot3
\Y1■abs(X-10)
\Y2■12/(X+3)
\Y3=
\Y4=
\Y5=
\Y6=
\Y7=
```

(a) $(f \circ g)(3) = 8$

```
Y1(Y2(3))
                 8
```

(b) $(g \circ f)(-2) = \dfrac{4}{5}$

```
Y2(Y1(-2))▶Frac
              4/5
```

(c) $(f \circ f)(1) = 1$

```
Y1(Y1(1))▶Frac
                1
```

(d) $(g \circ g)(-4) = -\dfrac{4}{3}$

```
Y2(Y2(-4))▶Frac
             -4/3
```

119. $f(x) = x + 5$; $g(x) = x - 5$

```
Plot1 Plot2 Plot3
\Y1■X+5
\Y2■X-5
\Y3■X
\Y4=
\Y5=
\Y6=
\Y7=
```

121. $f(x) = 5x + 7$; $g(x) = \dfrac{x - 7}{5}$

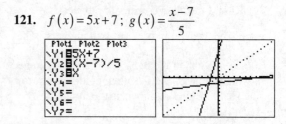

```
Plot1 Plot2 Plot3
\Y1■5X+7
\Y2■(X-7)/5
\Y3■X
\Y4=
\Y5=
\Y6=
\Y7=
```

Section 11.2

Preparing for Exponential Functions

P1. **(a)** $2^3 = 2 \cdot 2 \cdot 2 = 8$

(b) $2^{-1} = \dfrac{1}{2^1} = \dfrac{1}{2}$

(c) $3^4 = 3 \cdot 3 \cdot 3 \cdot 3 = 81$

P2. Locate some points on the graph of $f(x) = x^2$.

x	$f(x) = x^2$	$(x, f(x))$
-3	$f(-3) = (-3)^2 = 9$	$(-3, 9)$
-2	$f(-2) = (-2)^2 = 4$	$(-2, 4)$
-1	$f(-1) = (-1)^2 = 1$	$(-1, 1)$
0	$f(0) = 0^2 = 0$	$(0, 0)$
1	$f(1) = 1^2 = 1$	$(1, 1)$
2	$f(2) = 2^2 = 4$	$(2, 4)$
3	$f(3) = 3^2 = 9$	$(3, 9)$

Plot the points and connect them with a smooth curve.

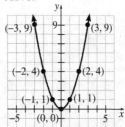

P3. A **rational number** is a number that can be expressed as a quotient $\dfrac{p}{q}$ of two integers. The integer p is called the *numerator*, and the integer q, which cannot be 0, is called the **denominator**. The set of rational numbers is the numbers

$$\mathbb{Q} = \left\{ x \mid x = \frac{p}{q}, \text{ where } p, q \text{ are integers and } q \neq 0 \right\}.$$

P4. An irrational number has a decimal representation that neither repeats nor terminates.

P5. **(a)** Rounding to 4 decimal places, we obtain $3.20349193 \approx 3.2035$.

     (b) Truncating to 4 decimal places, we obtain $3.20349193 \approx 3.2034$.

P6. **(a)** $m^3 \cdot m^5 = m^{3+5} = m^8$

     (b) $\dfrac{a^7}{a^2} = a^{7-2} = a^5$

     (c) $\left(z^3 \right)^4 = z^{3 \cdot 4} = z^{12}$

P7. $x^2 - 5x = 14$
$$x^2 - 5x - 14 = 0$$
$$(x-7)(x+2) = 0$$
$$x - 7 = 0 \quad \text{or} \quad x + 2 = 0$$
$$x = 7 \quad \text{or} \quad \quad x = -2$$
The solution set is $\{-2, 7\}$.

Section 11.2 Quick Checks

1. $>;\ \neq$

2. **(a)** $2^{1.7} \approx 3.249009585$

     (b) $2^{1.73} \approx 3.317278183$

     (c) $2^{1.732} \approx 3.321880096$

     (d) $2^{1.7321} \approx 3.32211036$

     (e) $2^{\sqrt{3}} \approx 3.321997085$

3. Locate some points on the graph of $f(x) = 4^x$.

x	$f(x) = 4^x$	$(x, f(x))$
-2	$f(-2) = 4^{-2} = \dfrac{1}{4^2} = \dfrac{1}{16}$	$\left(-2, \dfrac{1}{16}\right)$
-1	$f(-1) = 4^{-1} = \dfrac{1}{4^1} = \dfrac{1}{4}$	$\left(-1, \dfrac{1}{4}\right)$
0	$f(0) = 4^0 = 1$	$(0, 1)$
1	$f(1) = 4^1 = 4$	$(1, 4)$

Plot the points and connect them with a smooth curve.

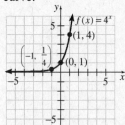

The domain of f is all real numbers or, using interval notation, $(-\infty, \infty)$. The range of f is $\{y \mid y > 0\}$ or, using interval notation, $(0, \infty)$.

4. $\left(-1, \dfrac{1}{a}\right);\ (0,1);\ (1, a)$

5. True.

6. False. The range of the exponential function $f(x) = a^x$, $a > 0$, $a \neq 1$ is $(0, \infty)$.

7. Locate some points on the graph of $f(x) = \left(\dfrac{1}{4}\right)^x$.

x	$f(x) = \left(\dfrac{1}{4}\right)^x$	$(x, f(x))$
-1	$f(-1) = \left(\dfrac{1}{4}\right)^{-1} = 4^1 = 4$	$(-1, 4)$
0	$f(0) = \left(\dfrac{1}{4}\right)^0 = 1$	$(0, 1)$
1	$f(1) = \left(\dfrac{1}{4}\right)^1 = \dfrac{1}{4}$	$\left(1, \dfrac{1}{4}\right)$
2	$f(2) = \left(\dfrac{1}{4}\right)^2 = \dfrac{1}{16}$	$\left(2, \dfrac{1}{16}\right)$

Plot the points and connect them with a smooth curve.

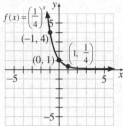

The domain of f is all real numbers or, using interval notation, $(-\infty, \infty)$. The range of f is $\{y \mid y > 0\}$ or , using interval notation, $(0, \infty)$.

8. Locate some points on the graph of $f(x) = 2^{x-1}$.

x	$f(x) = 2^{x-1}$	$(x, f(x))$
-1	$f(-1) = 2^{-1-1} = 2^{-2} = \dfrac{1}{2^2} = \dfrac{1}{4}$	$\left(-1, \dfrac{1}{4}\right)$
0	$f(0) = 2^{0-1} = 2^{-1} = \dfrac{1}{2^1} = \dfrac{1}{2}$	$\left(0, \dfrac{1}{2}\right)$
1	$f(1) = 2^{1-1} = 2^0 = 1$	$(1, 1)$
2	$f(2) = 2^{2-1} = 2^1 = 2$	$(2, 2)$
3	$f(3) = 2^{3-1} = 2^2 = 4$	$(3, 4)$
4	$f(4) = 2^{4-1} = 2^3 = 8$	$(4, 8)$

Plot the points and connect them with a smooth curve.

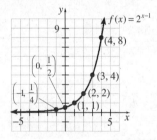

The domain of f is all real numbers or, using interval notation, $(-\infty, \infty)$. The range of f is $\{y \mid y > 0\}$ or , using interval notation, $(0, \infty)$.

9. Locate some points on the graph of $f(x) = 3^x + 1$.

x	$f(x) = 3^x + 1$	$(x, f(x))$
-2	$f(-2) = 3^{-2} + 1 = \dfrac{1}{9} + 1 = \dfrac{10}{9}$	$\left(-2, \dfrac{10}{9}\right)$
-1	$f(-1) = 3^{-1} + 1 = \dfrac{1}{3} + 1 = \dfrac{4}{3}$	$\left(-1, \dfrac{4}{3}\right)$
0	$f(0) = 3^0 + 1 = 1 + 1 = 2$	$(0, 2)$
1	$f(1) = 3^1 + 1 = 3 + 1 = 4$	$(1, 4)$
2	$f(2) = 3^2 + 1 = 9 + 1 = 10$	$(2, 10)$

Plot the points and connect them with a smooth curve.

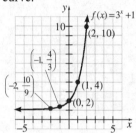

The domain of f is all real numbers or, using interval notation, $(-\infty, \infty)$. The range of f is $\{y \mid y > 1\}$ or , using interval notation, $(1, \infty)$.

10. $e \approx 2.71828$

11. (a) $e^4 \approx 54.598$

    (b) $e^{-4} \approx 0.018$

12. $5^{x-4} = 5^{-1}$
$x - 4 = -1$
$x = 3$
The solution set is $\{3\}$.

13. $3^{x+2} = 81$
$3^{x+2} = 3^4$
$x + 2 = 4$
$x = 2$
The solution set is $\{2\}$.

14.
$$e^{x^2} = e^x \cdot e^{4x}$$
$$e^{x^2} = e^{5x}$$
$$x^2 = 5x$$
$$x^2 - 5x = 0$$
$$x(x-5) = 0$$
$$x = 0 \quad \text{or} \quad x - 5 = 0$$
$$x = 5$$

The solution set is $\{0, 5\}$.

15.
$$\frac{2^{x^2}}{8} = 2^{2x}$$
$$\frac{2^{x^2}}{2^3} = 2^{2x}$$
$$2^{x^2 - 3} = 2^{2x}$$
$$x^2 - 3 = 2x$$
$$x^2 - 2x - 3 = 0$$
$$(x+1)(x-3) = 0$$
$$x + 1 = 0 \quad \text{or} \quad x - 3 = 0$$
$$x = -1 \quad \text{or} \quad x = 3$$

The solution set is $\{-1, 3\}$.

16. (a) $F(10) = 1 - e^{-0.25(10)} \approx 0.918$.

The likelihood that a person will arrive within 10 minutes of 3:00 P.M. is 0.918, or 91.8%.

(b) $F(25) = 1 - e^{-0.25(25)} \approx 0.998$.

The likelihood that a person will arrive within 25 minutes of 3:00 P.M. is 0.998, or 99.8%.

17. (a) $A(10) = 10\left(\frac{1}{2}\right)^{10/18.72} \approx 6.91$.

After 10 days, approximately 6.91 grams of thorium-227 will be left in the sample.

(b) $A(18.72) = 10\left(\frac{1}{2}\right)^{18.72/18.72} = 5$.

After 18.72 days, 5 grams of thorium-227 will be left in the sample.

(c) $A(74.88) = 10\left(\frac{1}{2}\right)^{74.88/18.72} = 0.625$.

After 74.88 days, 0.625 gram of thorium-227 will be left in the sample.

(d) $A(100) = 10\left(\frac{1}{2}\right)^{100/18.72} \approx 0.247$.

After 100 days, approximately 0.247 gram of thorium-227 will be left in the sample.

18. We use the compound interest formula with $P = \$2000$, $r = 0.12$, and $n = 12$, so that
$$A = 2000\left(1 + \frac{0.12}{12}\right)^{12t}$$
$$= 2000(1 + 0.01)^{12t}$$
$$= 2000(1.01)^{12t}$$

(a) The value of the account after $t = 1$ years is
$$A = 2000(1.01)^{12(1)}$$
$$= 2000(1.01)^{12} \approx \$2253.65$$

(b) The value of the account after $t = 15$ years is $A = 2000(1.01)^{12(15)}$
$$= 2000(1.01)^{180} \approx \$11,991.60$$

(c) The value of the account after $t = 30$ years is $A = 2000(1.01)^{12(30)}$
$$= 2000(1.01)^{360} \approx \$71,899.28$$

11.2 Exercises

19. (a) $3^{2.2} \approx 11.212$

(b) $3^{2.23} \approx 11.587$

(c) $3^{2.236} \approx 11.664$

(d) $3^{2.2361} \approx 11.665$

(e) $3^{\sqrt{5}} \approx 11.665$

21. (a) $4^{3.1} \approx 73.517$

(b) $4^{3.14} \approx 77.708$

(c) $4^{3.142} \approx 77.924$

(d) $4^{3.1416} \approx 77.881$

(e) $4^{\pi} \approx 77.880$

23. g. $f(x) = 2^x - 1$ because the following points are on the graph:

x	$f(x) = 2^x - 1$	$(x, f(x))$
-1	$f(-1) = 2^{-1} - 1 = \dfrac{1}{2} - 1 = -\dfrac{1}{2}$	$\left(-1, -\dfrac{1}{2}\right)$
0	$f(0) = 2^0 - 1 = 1 - 1 = 0$	$(0, 0)$
1	$f(1) = 2^1 - 1 = 2 - 1 = 1$	$(1, 1)$

25. e. $f(x) = -2^x$ because the following points are on the graph:

x	$f(x) = -2^x$	$(x, f(x))$
-1	$f(-1) = -2^{-1} = -\dfrac{1}{2}$	$\left(-1, -\dfrac{1}{2}\right)$
0	$f(0) = -2^0 = -1$	$(0, -1)$
1	$f(1) = -2^1 = -2$	$(1, -2)$

27. f. $f(x) = 2^x + 1$ because the following points are on the graph:

x	$f(x) = 2^x + 1$	$(x, f(x))$
-1	$f(-1) = 2^{-1} + 1 = \dfrac{1}{2} + 1 = \dfrac{3}{2}$	$\left(-1, \dfrac{3}{2}\right)$
0	$f(0) = 2^0 + 1 = 1 + 1 = 2$	$(0, 2)$
1	$f(1) = 2^1 + 1 = 2 + 1 = 3$	$(1, 3)$

29. h. $f(x) = -2^{-x}$ because the following points are on the graph:

x	$f(x) = -2^{-x}$	$(x, f(x))$
-1	$f(-1) = -2^{-(-1)} = -2^1 = -2$	$(-1, -2)$
0	$f(0) = -2^{-0} = -2^0 = -1$	$(0, -1)$
1	$f(1) = -2^{-1} = -\dfrac{1}{2}$	$\left(1, -\dfrac{1}{2}\right)$

31. Locate some points on the graph of $f(x) = 5^x$.

x	$f(x) = 5^x$	$(x, f(x))$
-2	$f(-2) = 5^{-2} = \dfrac{1}{5^2} = \dfrac{1}{25}$	$\left(-2, \dfrac{1}{25}\right)$
-1	$f(-1) = 5^{-1} = \dfrac{1}{5^1} = \dfrac{1}{5}$	$\left(-1, \dfrac{1}{5}\right)$
0	$f(0) = 5^0 = 1$	$(0, 1)$
1	$f(1) = 5^1 = 5$	$(1, 5)$

Plot the points and connect them with a smooth curve.

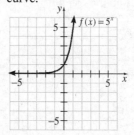

The domain of f is all real numbers or, using interval notation, $(-\infty, \infty)$. The range of f is $\{y \mid y > 0\}$ or, using interval notation, $(0, \infty)$.

33. Locate some points on the graph of $F(x) = \left(\dfrac{1}{5}\right)^x$.

x	$F(x) = \left(\dfrac{1}{5}\right)^x$	$(x, F(x))$
-1	$F(-1) = \left(\dfrac{1}{5}\right)^{-1} = 5^1 = 5$	$(-1, 5)$
0	$F(0) = \left(\dfrac{1}{5}\right)^0 = 1$	$(0, 1)$
1	$F(1) = \left(\dfrac{1}{5}\right)^1 = \dfrac{1}{5}$	$\left(1, \dfrac{1}{5}\right)$
2	$F(2) = \left(\dfrac{1}{5}\right)^2 = \dfrac{1}{25}$	$\left(2, \dfrac{1}{25}\right)$

Plot the points and connect them with a smooth curve.

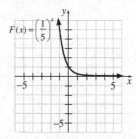

The domain of F is all real numbers or, using interval notation, $(-\infty, \infty)$. The range of F is $\{y \mid y > 0\}$ or, using interval notation, $(0, \infty)$.

35. Locate some points on the graph of $h(x) = 2^{x+2}$.

x	$h(x) = 2^{x+2}$	$(x,\ h(x))$
-5	$h(-5) = 2^{-5+2} = 2^{-3} = \dfrac{1}{2^3} = \dfrac{1}{8}$	$\left(-5,\ \dfrac{1}{8}\right)$
-4	$h(-4) = 2^{-4+2} = 2^{-2} = \dfrac{1}{2^2} = \dfrac{1}{4}$	$\left(-4,\ \dfrac{1}{4}\right)$
-3	$h(-3) = 2^{-3+2} = 2^{-1} = \dfrac{1}{2^1} = \dfrac{1}{2}$	$\left(-3,\ \dfrac{1}{2}\right)$
-2	$h(-2) = 2^{-2+2} = 2^0 = 1$	$(-2,\ 1)$
-1	$h(-1) = 2^{-1+2} = 2^1 = 2$	$(-1,\ 2)$
0	$h(0) = 2^{0+2} = 2^2 = 4$	$(0,\ 4)$
1	$h(1) = 2^{1+2} = 2^3 = 8$	$(1,\ 8)$

Plot the points and connect them with a smooth curve.

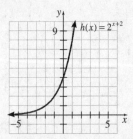

The domain of h is all real numbers or, using interval notation, $(-\infty, \infty)$. The range of h is $\{y \mid y > 0\}$ or, using interval notation, $(0, \infty)$.

37. Locate some points on the graph of $f(x) = 2^x + 3$.

x	$f(x) = 2^x + 3$	$(x,\ f(x))$
-2	$f(-2) = 2^{-2} + 3 = \dfrac{1}{4} + 3 = \dfrac{13}{4}$	$\left(-2,\ \dfrac{13}{4}\right)$
-1	$f(-1) = 2^{-1} + 3 = \dfrac{1}{2} + 3 = \dfrac{7}{2}$	$\left(-1,\ \dfrac{7}{2}\right)$
0	$f(0) = 2^0 + 3 = 1 + 3 = 4$	$(0,\ 4)$
1	$f(1) = 2^1 + 3 = 2 + 3 = 5$	$(1,\ 5)$
2	$f(2) = 2^2 + 3 = 4 + 3 = 7$	$(2,\ 7)$

Plot the points and connect them with a smooth curve.

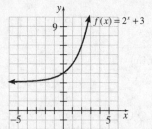

The domain of f is all real numbers or, using interval notation, $(-\infty, \infty)$. The range of f is $\{y \mid y > 3\}$ or, using interval notation, $(3, \infty)$.

39. Locate some points on the graph of
$F(x) = \left(\dfrac{1}{2}\right)^x - 1$.

x	$F(x) = \left(\dfrac{1}{2}\right)^x - 1$	$(x,\ F(x))$
-3	$F(-3) = \left(\dfrac{1}{2}\right)^{-3} - 1 = 8 - 1 = 7$	$(-3,\ 7)$
-2	$F(-2) = \left(\dfrac{1}{2}\right)^{-2} - 1 = 4 - 1 = 3$	$(-2,\ 3)$
-1	$F(-1) = \left(\dfrac{1}{2}\right)^{-1} - 1 = 2 - 1 = 1$	$(-1,\ 1)$
0	$F(0) = \left(\dfrac{1}{2}\right)^0 - 1 = 1 - 1 = 0$	$(0,\ 0)$
1	$F(1) = \left(\dfrac{1}{2}\right)^1 - 1 = \dfrac{1}{2} - 1 = -\dfrac{1}{2}$	$\left(1, -\dfrac{1}{2}\right)$
2	$F(2) = \left(\dfrac{1}{2}\right)^2 - 1 = \dfrac{1}{4} - 1 = -\dfrac{3}{4}$	$\left(2, -\dfrac{3}{4}\right)$
3	$F(3) = \left(\dfrac{1}{2}\right)^3 - 1 = \dfrac{1}{8} - 1 = -\dfrac{7}{8}$	$\left(3, -\dfrac{7}{8}\right)$

Plot the points and connect them with a smooth curve.

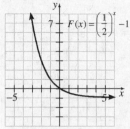

The domain of F is all real numbers or, using interval notation, $(-\infty, \infty)$. The range of F is $\{y \mid y > -1\}$ or, using interval notation, $(-1, \infty)$.

41. Locate some points on the graph of

$$P(x) = \left(\frac{1}{3}\right)^{x-2}.$$

x	$P(x) = \left(\frac{1}{3}\right)^{x-2}$	$(x,\ P(x))$
0	$P(0) = \left(\frac{1}{3}\right)^{0-2} = \left(\frac{1}{3}\right)^{-2} = 3^2 = 9$	$(0,\ 9)$
1	$P(1) = \left(\frac{1}{3}\right)^{1-2} = \left(\frac{1}{3}\right)^{-1} = 3^1 = 3$	$(1,\ 3)$
2	$P(2) = \left(\frac{1}{3}\right)^{2-2} = \left(\frac{1}{3}\right)^{0} = 1$	$(2,\ 1)$
3	$P(3) = \left(\frac{1}{3}\right)^{3-2} = \left(\frac{1}{3}\right)^{1} = \frac{1}{3}$	$\left(3,\ \frac{1}{3}\right)$
4	$P(4) = \left(\frac{1}{3}\right)^{4-2} = \left(\frac{1}{3}\right)^{2} = \frac{1}{9}$	$\left(4,\ \frac{1}{9}\right)$

Plot the points and connect them with a smooth curve.

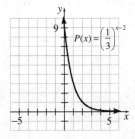

The domain of P is all real numbers or, using interval notation, $(-\infty, \infty)$. The range of P is $\{y \mid y > 0\}$ or, using interval notation, $(0, \infty)$.

43. **(a)** $3.1^{2.7} \approx 21.217$

     (b) $3.14^{2.72} \approx 22.472$

     (c) $3.142^{2.718} \approx 22.460$

     (d) $3.1416^{2.7183} \approx 22.460$

     (e) $\pi^e \approx 22.459$

45. $e^2 \approx 7.389$

47. $e^{-2} \approx 0.135$

49. $e^{2.3} \approx 9.974$

51. Locate some points on the graph of $g(x) = e^{x-1}$.

x	$g(x) = e^{x-1}$	$(x,\ g(x))$
-1	$g(-1) = e^{-1-1} = e^{-2} = \dfrac{1}{e^2} \approx 0.135$	$\left(-1,\ \dfrac{1}{e^2}\right)$
0	$g(0) = e^{0-1} = e^{-1} = \dfrac{1}{e} \approx 0.368$	$\left(0,\ \dfrac{1}{e}\right)$
1	$g(1) = e^{1-1} = e^0 = 1$	$(1,\ 1)$
2	$g(2) = e^{2-1} = e^1 = e \approx 2.718$	$(2,\ e)$
3	$g(3) = e^{3-1} = e^2 \approx 7.389$	$\left(3,\ e^2\right)$

Plot the points and connect them with a smooth curve.

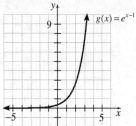

The domain of g is all real numbers or, using interval notation, $(-\infty, \infty)$. The range of g is $\{y \mid y > 0\}$ or, using interval notation, $(0, \infty)$.

53. Locate some points on the graph of $f(x) = -2e^x$.

x	$f(x) = -2e^x$	$(x, f(x))$
-1	$f(-1) = -2e^{-1} = -\dfrac{2}{e} \approx -0.736$	$\left(-1, -\dfrac{2}{e}\right)$
0	$f(0) = -2e^0 = -2$	$(0, -2)$
1	$f(1) = -2e^1 = -2e \approx -5.437$	$(1, -2e)$
2	$f(2) = -2e^2 \approx -14.778$	$\left(2, -2e^2\right)$
3	$f(3) = -2e^3 \approx -40.171$	$\left(3, -2e^3\right)$

Plot the points and connect them with a smooth curve.

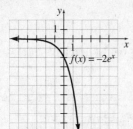

The domain of f is all real numbers, or using interval notation, $(-\infty, \infty)$. The range of f is $\{y \mid y < 0\}$ or, using interval notation, $(-\infty, 0)$.

55. $2^x = 2^5$
$x = 5$
The solution set is $\{5\}$.

57. $3^{-x} = 81$
$3^{-x} = 3^4$
$-x = 4$
$x = -4$
The solution set is $\{-4\}$.

59. $\left(\dfrac{1}{2}\right)^x = \dfrac{1}{32}$
$\left(\dfrac{1}{2}\right)^x = \left(\dfrac{1}{2}\right)^5$
$x = 5$
The solution set is $\{5\}$.

61. $5^{x-2} = 125$
$5^{x-2} = 5^3$
$x - 2 = 3$
$x = 5$
The solution set is $\{5\}$.

63. $4^x = 8$
$\left(2^2\right)^x = 2^3$
$2^{2x} = 2^3$
$2x = 3$
$x = \dfrac{3}{2}$
The solution set is $\left\{\dfrac{3}{2}\right\}$.

65. $2^{-x+5} = 16^x$
$2^{-x+5} = \left(2^4\right)^x$
$2^{-x+5} = 2^{4x}$
$-x + 5 = 4x$
$5 = 5x$
$1 = x$
The solution set is $\{1\}$.

67. $3^{x^2 - 4} = 27^x$
$3^{x^2 - 4} = \left(3^3\right)^x$
$3^{x^2 - 4} = 3^{3x}$
$x^2 - 4 = 3x$
$x^2 - 3x - 4 = 0$
$(x - 4)(x + 1) = 0$
$x - 4 = 0 \quad \text{or} \quad x + 1 = 0$
$x = 4 \quad \text{or} \quad x = -1$
The solution set is $\{-1,\ 4\}$.

69. $4^x \cdot 2^{x^2} = 16^2$
$\left(2^2\right)^x \cdot 2^{x^2} = \left(2^4\right)^2$
$2^{2x} \cdot 2^{x^2} = 2^8$
$2^{x^2 + 2x} = 2^8$
$x^2 + 2x = 8$
$x^2 + 2x - 8 = 0$
$(x + 4)(x - 2) = 0$
$x + 4 = 0 \quad \text{or} \quad x - 2 = 0$
$x = -4 \quad \text{or} \quad x = 2$
The solution set is $\{-4,\ 2\}$.

71. $2^x \cdot 8 = 4^{x-3}$
$2^x \cdot 2^3 = \left(2^2\right)^{x-3}$
$2^{x+3} = 2^{2x-6}$
$x + 3 = 2x - 6$
$-x = -9$
$x = 9$
The solution set is $\{9\}$.

73. $\left(\dfrac{1}{5}\right)^x - 25 = 0$

$\left(\dfrac{1}{5}\right)^x = 25$

$\left(5^{-1}\right)^x = 5^2$

$5^{-x} = 5^2$

$-x = 2$

$x = -2$

The solution set is $\{-2\}$.

75. $\left(2^x\right)^x = 16$

$2^{x^2} = 2^4$

$x^2 = 4$

$x = \pm\sqrt{4}$

$x = \pm 2$

The solution set is $\{-2,\ 2\}$.

77. $e^x = e^{3x+4}$

$x = 3x + 4$

$-2x = 4$

$x = -2$

The solution set is $\{-2\}$.

79. $\left(e^x\right)^2 = e^{3x-2}$

$e^{2x} = e^{3x-2}$

$2x = 3x - 2$

$-x = -2$

$x = 2$

The solution set is $\{2\}$.

81. (a) $f(3) = 2^3 = 8$.

The point $(3, 8)$ is on the graph of f.

(b) $f(x) = \dfrac{1}{8}$

$2^x = \dfrac{1}{8}$

$2^x = \dfrac{1}{2^3} = 2^{-3}$

$x = -3$

The point $\left(-3, \dfrac{1}{8}\right)$ is on the graph of f.

83. (a) $g(-1) = 4^{-1} - 1 = \dfrac{1}{4} - 1 = -\dfrac{3}{4}$.

The point $\left(-1, -\dfrac{3}{4}\right)$ is on the graph of g.

(b) $g(x) = 15$

$4^x - 1 = 15$

$4^x = 16$

$4^x = 4^2$

$x = 2$

The point $(2, 15)$ is on the graph of g.

85. (a) $H(-3) = 3 \cdot \left(\dfrac{1}{2}\right)^{-3} = 3 \cdot 2^3 = 3 \cdot 8 = 24$.

The point $(-3,\ 24)$ is on the graph of H.

(b) $H(x) = \dfrac{3}{4}$

$3 \cdot \left(\dfrac{1}{2}\right)^x = \dfrac{3}{4}$

$\left(\dfrac{1}{2}\right)^x = \dfrac{1}{4}$

$\left(\dfrac{1}{2}\right)^x = \left(\dfrac{1}{2}\right)^2$

$x = 2$

The point $\left(2, \dfrac{3}{4}\right)$ is on the graph of H.

87. (a) $P(2010) = 304(1.011)^{2010-2008} \approx 310.725$

According to the model, the population of the U.S. in 2010 will be approximately 310.7 million people.

(b) $P(2042) = 304(1.011)^{2042-2008} \approx 440.974$

According to the model, the population of the U.S. in 2042 will be approximately 441 million people.

(c) $441 - 400 = 41$. The U.S. Census Bureau's prediction for the population in 2042 is 41 million people fewer than that of the model. Reasons given for the differences may vary. One possibility is that perhaps the U.S. Census Bureau expects the rate of population growth to decline below 1.1% per year over time.

89. We use the compound interest formula with $P = \$5000$, $r = 0.06$, and $n = 12$, so that

$$A = 5000\left(1 + \frac{0.06}{12}\right)^{12t} = 5000(1.005)^{12t} .$$

   (a) The value of the account after $t = 1$ year is
$$A = 5000(1.005)^{12(1)} \approx \$5308.39 .$$

   (b) The value of the account after $t = 3$ years is
$$A = 5000(1.005)^{12(3)} \approx \$5983.40 .$$

   (c) The value of the account after $t = 5$ years is
$$A = 5000(1.005)^{12(5)} \approx \$6744.25 .$$

91. (a) We use the compound interest formula with $P = \$2000$, $r = 0.03$, $t = 5$, and $n = 1$, so that $A = 2000\left(1 + \frac{0.03}{1}\right)^{1(5)} \approx \$2318.55 .$

   (b) We use the compound interest formula with $P = \$2000$, $r = 0.03$, $t = 5$, and $n = 4$, so that $A = 2000\left(1 + \frac{0.03}{4}\right)^{4(5)} \approx \$2322.37 .$

   (c) We use the compound interest formula with $P = \$2000$, $r = 0.03$, $t = 5$, and $n = 12$, so that $A = 2000\left(1 + \frac{0.03}{12}\right)^{12(5)} \approx \$2323.23 .$

   (d) We use the compound interest formula with $P = \$2000$, $r = 0.03$, $t = 5$, and $n = 365$, so that

$$A = 2000\left(1 + \frac{0.03}{365}\right)^{365(5)} \approx \$2323.65 .$$

   (e) Answers may vary. One possibility follows: All other things equal, the number of compounding periods does not have a very significant impact on the future value. In this case, for example, the difference between compounding annually $(n = 1)$ and daily $(n = 365)$ is only $5.10 over 5 years.

93. (a) $V(0) = 14{,}512(0.82)^0 = 14{,}512 .$

   According to the model, the value of a brand-new Aveo is $14,512.

   (b) $V(2) = 14{,}512(0.82)^2 \approx 9{,}757.87 .$

   According to the model, the value of a 2-year-old Aveo is $9,757.87.

   (c) $V(5) = 14{,}512(0.82)^5 \approx 5{,}380.18 .$

   According to the model, the value of a 5-year-old Aveo is $5,380.18.

95. (a) $A(1) = 100\left(\dfrac{1}{2}\right)^{1/13.81} \approx 95.105 .$

   After 1 second, approximately 95.105 grams of beryllium-11 will be left in the sample.

   (b) $A(13.81) = 100\left(\dfrac{1}{2}\right)^{13.81/13.81} = 50 .$

   After 13.81 seconds, 50 grams of beryllium-11 will be left in the sample.

   (c) $A(27.62) = 100\left(\dfrac{1}{2}\right)^{27.62/13.81} = 25 .$

   After 27.62 seconds, 25 grams of beryllium-11 will be left in the sample.

   (d) $A(100) = 100\left(\dfrac{1}{2}\right)^{100/13.81} \approx 0.661 .$

   After 100 seconds, approximately 0.661 gram of beryllium-11 will be left in the sample.

97. (a) $u(5) = 70 + 330e^{-0.072(5)} \approx 300.233 .$

   According to the model, the temperature of the pizza after 5 minutes will be approximately $300.233°F$.

   (b) $u(10) = 70 + 330e^{-0.072(10)} \approx 230.628 .$

   According to the model, the temperature of the pizza after 10 minutes will be approximately $230.628°F$.

   (c) $u(13) = 70 + 330e^{-0.072(13)} \approx 199.424 .$

   According to the model, the temperature of the pizza after 13 minutes will be approximately $199.424°F$. Since this is below $200°F$, the pizza will be ready to eat after cooling for 13 minutes.

99. (a) $L(45) = 200\left(1 - e^{-0.0035(45)}\right) \approx 29.14$.

According to the model, the student will learn approximately 29 words after 45 minutes.

(b) $L(60) = 200\left(1 - e^{-0.0035(60)}\right) \approx 37.88$.

According to the model, the student will learn approximately 38 words after 60 minutes.

101. (a) We use the equation with $E = 120$, $R = 10$, $L = 25$, and $t = 0.05$, so that

$$I = \frac{120}{10}\left[1 - e^{-(10/25)0.05}\right] \approx 0.238 \text{ ampere.}$$

(b) We use the equation with $E = 240$, $R = 10$, $L = 25$, and $t = 0.05$, so that

$$I = \frac{240}{10}\left[1 - e^{-(10/25)0.05}\right] \approx 0.475 \text{ ampere.}$$

103. The exponential function will be of the form $y = a^x$. Now the function contains the points $\left(-1, \frac{1}{3}\right)$, $(0,1)$, and $(1,3)$, so $a = 3$. Thus, the equation of the function is $y = 3^x$.

105. As the base a increases, the steeper the graph of $f(x) = a^x$ $(a > 1)$ is for $x > 0$ and the closer the graph is to the x-axis for $x < 0$.

107. Answers may vary. One possibility follows: For exponential functions, the base is a constant and the exponent is the independent variable. For polynomial functions, the base is the independent variable and the exponents are constants (more specifically, whole numbers).

109. (a) $x = -2$: $x^2 - 5x + 1 = (-2)^2 - 5(-2) + 1$
$= 4 + 10 + 1$
$= 15$

(b) $x = 3$: $x^2 - 5x + 1 = 3^2 - 5(3) + 1$
$= 9 - 15 + 1$
$= -5$

111. (a) $x = -2$: $\frac{4}{x+2} = \frac{4}{-2+2} = \frac{4}{0} = $ undefined

(b) $x = 3$: $\frac{4}{x+2} = \frac{4}{3+2} = \frac{4}{5}$

113. (a) $x = 2$: $\sqrt{2x+5} = \sqrt{2(2)+5}$
$= \sqrt{4+5} = \sqrt{9} = 3$

(b) $x = 11$: $\sqrt{2x+5} = \sqrt{2(11)+5}$
$= \sqrt{22+5}$
$= \sqrt{27} = \sqrt{9 \cdot 3} = 3\sqrt{3}$

115. Let $Y_1 = 1.5^x$.

 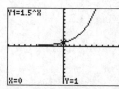

The domain of $f(x) = 1.5^x$ is all real numbers, or using interval notation $(-\infty, \infty)$. The range is $\{y \mid y > 0\}$, or using interval notation $(0, \infty)$.

117. Let $Y_1 = 0.9^x$.

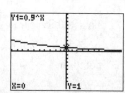

The domain of $H(x) = 0.9^x$ is all real numbers, or using interval notation $(-\infty, \infty)$. The range is $\{y \mid y > 0\}$, or using interval notation $(0, \infty)$.

119. Let $Y_1 = 2.5^x + 3$.

 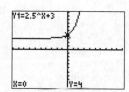

The domain of $g(x) = 2.5^x + 3$ is all real numbers, or using interval notation $(-\infty, \infty)$. The range is $\{y \mid y > 3\}$, or using interval notation $(3, \infty)$.

121. Let $Y_1 = 1.6^{x-3}$.

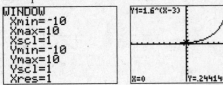

The domain of $F(x) = 1.6^{x-3}$ is all real

numbers, or using interval notation $(-\infty, \infty)$.

The range is $\{y \mid y > 0\}$, or using interval

notation $(0, \infty)$.

Section 11.3

Preparing for Logarithmic Functions

P1. $3x + 2 > 0$

$3x + 2 - 2 > 0 - 2$

$3x > -2$

$\dfrac{3x}{3} > \dfrac{-2}{3}$

$x > -\dfrac{2}{3}$

The solution set is $\left\{ x \mid x > -\dfrac{2}{3} \right\}$ or, using interval

notation, $\left(-\dfrac{2}{3}, \infty \right)$.

P2. $\sqrt{x+2} = x$

$\left(\sqrt{x+2} \right)^2 = (x)^2$

$x + 2 = x^2$

$0 = x^2 - x - 2$

$0 = (x-2)(x+1)$

$x - 2 = 0$ or $x + 1 = 0$

$x = 2$ or $x = -1$

Check:

$x = 2$: $\sqrt{2+2} \overset{?}{=} 2$

$\sqrt{4} \overset{?}{=} 2$

$2 = 2$ ✓

$x = -1$: $\sqrt{-1+2} \overset{?}{=} -1$

$\sqrt{1} \overset{?}{=} -1$

$1 \neq -1$ ✗

The potential solution $x = -1$ does not check;

the solution set is $\{2\}$.

P3. $x^2 = 6x + 7$

$x^2 - 6x - 7 = 0$

$(x-7)(x+1) = 0$

$x - 7 = 0$ or $x + 1 = 0$

$x = 7$ or $x = -1$

The solution set is $\{-1, 7\}$.

Section 11.3 Quick Checks

1. $x = a^y$; $>$; $\neq$

2.   If $4^3 = w$, then $3 = \log_4 w$.

3.   If $p^{-2} = 8$, then $-2 = \log_p 8$.

4.   If $y = \log_2 16$, then $2^y = 16$.

5.   If $5 = \log_a 20$, then $a^5 = 20$.

6.   If $-3 = \log_5 z$, then $5^{-3} = z$.

7.   Let $y = \log_5 25$. Then,

$5^y = 25$

$5^y = 5^2$

$y = 2$

Thus, $\log_5 25 = 2$.

8.   Let $y = \log_2 \dfrac{1}{8}$. Then,

$2^y = \dfrac{1}{8}$

$2^y = \dfrac{1}{2^3} = 2^{-3}$

$y = -3$

Thus, $\log_2 \dfrac{1}{8} = -3$.

9. $g(25)$ means to evaluate $\log_5 x$ at $x = 25$. So,

we want to know the value of $\log_5 25$. Let

$y = \log_5 25$. Then,

$5^y = 25$

$5^y = 5^2$

$y = 2$

Thus, $g(25) = 2$.

10. $g\left(\dfrac{1}{5}\right)$ means to evaluate $\log_5 x$ at $x = \dfrac{1}{5}$. So,

we want to know the value of $\log_5\left(\dfrac{1}{5}\right)$. Let

$y = \log_5\left(\dfrac{1}{5}\right)$. Then,

$5^y = \dfrac{1}{5}$

$5^y = 5^{-1}$

$y = -1$

Thus, $g\left(\dfrac{1}{5}\right) = -1$.

11. The domain of $g(x) = \log_8(x+3)$ is the set of all real numbers x such that

$x + 3 > 0$

$\quad x > -3$

Thus, the domain of $g(x) = \log_8(x+3)$ is

$\{x \,|\, x > -3\}$ or, using interval notation, $(-3, \infty)$.

12. The domain of $F(x) = \log_2(5 - 2x)$ is the set of all real numbers x such that

$5 - 2x > 0$

$\quad -2x > -5$

$\quad\quad x < \dfrac{-5}{-2}$

$\quad\quad x < \dfrac{5}{2}$

Thus, the domain of $F(x) = \log_2(5 - 2x)$ is

$\left\{x \,\middle|\, x < \dfrac{5}{2}\right\}$ or, using interval notation,

$\left(-\infty, \dfrac{5}{2}\right)$.

13. Rewrite $y = f(x) = \log_4 x$ as $x = 4^y$. Locate some points on the graph of $x = 4^y$.

y	$x = 4^y$	(x, y)
-2	$x = 4^{-2} = \dfrac{1}{4^2} = \dfrac{1}{16}$	$\left(\dfrac{1}{16}, -2\right)$
-1	$x = 4^{-1} = \dfrac{1}{4^1} = \dfrac{1}{4}$	$\left(\dfrac{1}{4}, -1\right)$
0	$x = 4^0 = 1$	$(1, 0)$
1	$x = 4^1 = 4$	$(4, 1)$

Plot the points and connect them with a smooth curve.

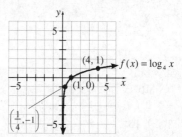

The domain of f is $\{x \,|\, x > 0\}$ or , using interval notation, $(0, \infty)$. The range of f is all real numbers or, using interval notation, $(-\infty, \infty)$.

14. Rewrite $y = f(x) = \log_{1/4} x$ as $x = \left(\dfrac{1}{4}\right)^y$.

Locate some points on the graph of $x = \left(\dfrac{1}{4}\right)^y$.

y	$x = \left(\dfrac{1}{4}\right)^y$	(x, y)
-1	$x = \left(\dfrac{1}{4}\right)^{-1} = 4^1 = 4$	$(4, -1)$
0	$x = \left(\dfrac{1}{4}\right)^0 = 1$	$(1, 0)$
1	$x = \left(\dfrac{1}{4}\right)^1 = \dfrac{1}{4}$	$\left(\dfrac{1}{4}, 1\right)$
2	$x = \left(\dfrac{1}{4}\right)^2 = \dfrac{1}{16}$	$\left(\dfrac{1}{16}, 2\right)$

Plot the points and connect them with a smooth curve.

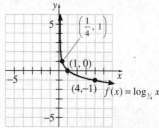

The domain of f is $\{x \,|\, x > 0\}$ or , using interval notation, $(0, \infty)$. The range of f is all real numbers or, using interval notation, $(-\infty, \infty)$.

15. $\log 1400 \approx 3.146$

16. $\ln 4.8 \approx 1.569$

17. $\log 0.3 \approx -0.523$

18. $\log_3(5x+1) = 4$

$5x+1 = 3^4$

$5x+1 = 81$

$5x = 80$

$x = 16$

Check: $\log_3(5 \cdot 16 + 1) \stackrel{?}{=} 4$

$\log_3(80+1) \stackrel{?}{=} 4$

$\log_3(81) \stackrel{?}{=} 4$

$4 = 4 \checkmark$

The solution set is $\{16\}$.

19. $\log_x 16 = 2$

$x^2 = 16$

$x = \pm\sqrt{16}$

$x = \pm 4$

Since the base of a logarithm must always be positive, we know that $a = -4$ is extraneous. We check the potential solution $x = 4$.

Check: $\log_4 16 \stackrel{?}{=} 2$

$2 = 2 \checkmark$

The solution set is $\{4\}$.

20. $\ln x = -2$

$x = e^{-2}$

Check: $\ln e^{-2} \stackrel{?}{=} -2$

$-2 = -2 \checkmark$

The solution set is $\left\{e^{-2}\right\}$.

21. $\log(x-20) = 4$

$x - 20 = 10^4$

$x - 20 = 10,000$

$x = 10,020$

Check: $\log(10,020 - 20) \stackrel{?}{=} 4$

$\log(10,000) \stackrel{?}{=} 4$

$4 = 4 \checkmark$

The solution set is $\{10,020\}$.

22. We evaluate $L(x) = 10\log\dfrac{x}{10^{-12}}$ at $x = 10^{-2}$.

$L\left(10^{-2}\right) = 10\log\dfrac{10^{-2}}{10^{-12}}$

$= 10\log 10^{-2-(-12)}$

$= 10\log 10^{10}$

$= 10(10)$

$= 100$

The loudness of an MP3 player on "full blast" is 100 decibels.

11.3 Exercises

23. If $64 = 4^3$, then $3 = \log_4 64$.

25. If $\dfrac{1}{8} = 2^{-3}$, then $-3 = \log_2\left(\dfrac{1}{8}\right)$.

27. If $a^3 = 19$, then $\log_a 19 = 3$.

29. If $5^{-6} = c$, then $\log_5 c = -6$.

31. If $\log_2 16 = 4$, then $2^4 = 16$.

33. If $\log_3 \dfrac{1}{9} = -2$, then $3^{-2} = \dfrac{1}{9}$.

35. If $\log_5 a = -3$, then $5^{-3} = a$.

37. If $\log_a 4 = 2$, then $a^2 = 4$.

39. If $\log_{1/2} 12 = y$, then $\left(\dfrac{1}{2}\right)^y = 12$.

41. Let $y = \log_3 1$. Then,

$3^y = 1$

$3^y = 3^0$

$y = 0$

Thus, $\log_3 1 = 0$.

43. Let $y = \log_2 8$. Then,

$2^y = 8$

$2^y = 2^3$

$y = 3$

Thus, $\log_2 8 = 3$.

45. Let $y = \log_4\left(\dfrac{1}{16}\right)$. Then,

$$4^y = \dfrac{1}{16}$$
$$4^y = \dfrac{1}{4^2} = 4^{-2}$$
$$y = -2$$

Thus, $\log_4 \dfrac{1}{16} = -2$.

47. Let $y = \log_{\sqrt{2}} 4$. Then,

$$\left(\sqrt{2}\right)^y = 4$$
$$\left(2^{\frac{1}{2}}\right)^y = 2^2$$
$$2^{\frac{1}{2}y} = 2^2$$
$$\dfrac{1}{2}y = 2$$
$$y = 4$$

Thus, $\log_{\sqrt{2}} 4 = 4$.

49. $f(81) = \log_3 81$. To determine the value, let $y = \log_3 81$. Then,

$$3^y = 81$$
$$3^y = 3^4$$
$$y = 4$$

Therefore, $f(81) = 4$.

51. $g\left(\sqrt{5}\right) = \log_5 \sqrt{5}$. To determine the value, let $y = \log_5 \sqrt{5}$. Then,

$$5^y = \sqrt{5}$$
$$5^y = 5^{\frac{1}{2}}$$
$$y = \dfrac{1}{2}$$

Therefore, $g\left(\sqrt{5}\right) = \dfrac{1}{2}$.

53. The domain of $f(x) = \log_2(x-4)$ is the set of all real numbers x such that
$$x - 4 > 0$$
$$x > 4$$
Thus, the domain of $f(x) = \log_2(x-4)$ is $\{x \mid x > 4\}$, or using interval notation, $(4, \infty)$.

55. The domain of $F(x) = \log_3(2x)$ is the set of all real numbers x such that
$$2x > 0$$
$$x > 0$$
Thus, the domain of $F(x) = \log_3(2x)$ is $\{x \mid x > 0\}$, or using interval notation, $(0, \infty)$.

57. The domain of $f(x) = \log_8(3x-2)$ is the set of all real numbers x such that
$$3x - 2 > 0$$
$$3x > 2$$
$$x > \dfrac{2}{3}$$
Thus, the domain of $f(x) = \log_8(3x-2)$ is $\left\{x \mid x > \dfrac{2}{3}\right\}$, or using interval notation, $\left(\dfrac{2}{3}, \infty\right)$.

59. The domain of $H(x) = \log_7(2x+1)$ is the set of all real numbers x such that
$$2x + 1 > 0$$
$$2x > -1$$
$$x > -\dfrac{1}{2}$$
Thus, the domain of $H(x) = \log_7(2x+1)$ is $\left\{x \mid x > -\dfrac{1}{2}\right\}$, or using interval notation, $\left(-\dfrac{1}{2}, \infty\right)$.

61. The domain of $H(x) = \log_2(1-4x)$ is the set of all real numbers x such that
$$1 - 4x > 0$$
$$-4x > -1$$
$$x < \dfrac{-1}{-4}$$
$$x < \dfrac{1}{4}$$
Thus, the domain of $H(x) = \log_2(1-4x)$ is $\left\{x \mid x < \dfrac{1}{4}\right\}$, or using interval notation, $\left(-\infty, \dfrac{1}{4}\right)$.

63. Rewrite $y = f(x) = \log_5 x$ as $x = 5^y$. Locate some points on the graph of $x = 5^y$.

y	$x = 5^y$	(x, y)
-2	$x = 5^{-2} = \dfrac{1}{5^2} = \dfrac{1}{25}$	$\left(\dfrac{1}{25}, -2\right)$
-1	$x = 5^{-1} = \dfrac{1}{5^1} = \dfrac{1}{5}$	$\left(\dfrac{1}{5}, -1\right)$
0	$x = 5^0 = 1$	$(1, 0)$
1	$x = 5^1 = 5$	$(5, 1)$

Plot the points and connect them with a smooth curve.

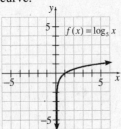

The domain of f is $\{x \mid x > 0\}$ or, using interval notation, $(0, \infty)$. The range of f is all real numbers or, using interval notation, $(-\infty, \infty)$.

65. Rewrite $y = g(x) = \log_6 x$ as $x = 6^y$. Locate some points on the graph of $x = 6^y$.

y	$x = 6^y$	(x, y)
-2	$x = 6^{-2} = \dfrac{1}{6^2} = \dfrac{1}{36}$	$\left(\dfrac{1}{36}, -2\right)$
-1	$x = 6^{-1} = \dfrac{1}{6^1} = \dfrac{1}{6}$	$\left(\dfrac{1}{6}, -1\right)$
0	$x = 6^0 = 1$	$(1, 0)$
1	$x = 6^1 = 6$	$(6, 1)$

Plot the points and connect them with a smooth curve.

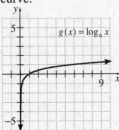

The domain of g is $\{x \mid x > 0\}$ or, using interval notation, $(0, \infty)$. The range of g is all real numbers or, using interval notation, $(-\infty, \infty)$.

67. Rewrite $y = F(x) = \log_{1/5} x$ as $x = \left(\dfrac{1}{5}\right)^y$.

Locate some points on the graph of $x = \left(\dfrac{1}{5}\right)^y$.

y	$x = \left(\dfrac{1}{5}\right)^y$	(x, y)
-1	$x = \left(\dfrac{1}{5}\right)^{-1} = 5^1 = 5$	$(5, -1)$
0	$x = \left(\dfrac{1}{5}\right)^0 = 1$	$(1, 0)$
1	$x = \left(\dfrac{1}{5}\right)^1 = \dfrac{1}{5}$	$\left(\dfrac{1}{5}, 1\right)$
2	$x = \left(\dfrac{1}{5}\right)^2 = \dfrac{1}{25}$	$\left(\dfrac{1}{25}, 2\right)$

Plot the points and connect them with a smooth curve.

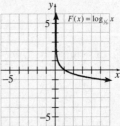

The domain of F is $\{x \mid x > 0\}$ or, using interval notation, $(0, \infty)$. The range of F is all real numbers or, using interval notation, $(-\infty, \infty)$.

69. If $e^x = 12$, then $\ln 12 = x$.

71. If $\ln x = 4$, then $e^4 = x$.

73. $H(0.1) = \log 0.1$. To determine the value, let $y = \log 0.1$. Then,

$10^y = 0.1$

$10^y = \dfrac{1}{10}$

$10^y = 10^{-1}$

$\quad y = -1$

Therefore, $H(0.1) = -1$.

75. $P\!\left(e^3\right) = \ln e^3$. To determine the value, let

$y = \ln e^3$. Then,

$e^y = e^3$

$y = 3$

Therefore, $P\!\left(e^3\right) = 3$.

77. $\log 67 \approx 1.826$

79. $\ln 5.4 \approx 1.686$

81. $\log 0.35 \approx -0.456$

83. $\ln 0.2 \approx -1.609$

85. $\log \dfrac{5}{4} \approx 0.097$

87. $\ln \dfrac{3}{8} \approx -0.981$

89. $\log_3(2x+1) = 2$

$2x+1 = 3^2$

$2x+1 = 9$

$2x = 8$

$x = 4$

The solution set is $\{4\}$.

91. $\log_5(20x-5) = 3$

$20x-5 = 5^3$

$20x-5 = 125$

$20x = 130$

$x = \dfrac{130}{20} = \dfrac{13}{2}$

The solution set is $\left\{\dfrac{13}{2}\right\}$.

93. $\log_a 36 = 2$

$a^2 = 36$

$a = \pm\sqrt{36}$

$a = \pm 6$

Since the base of a logarithm must always be positive, we know that $a = -6$ is extraneous.

The solution set is $\{6\}$.

95. $\log_a 18 = 2$

$a^2 = 18$

$a = \pm\sqrt{18}$

$a = \pm 3\sqrt{2}$

Since the base of a logarithm must always be positive, we know that $a = -3\sqrt{2}$ is extraneous.

The solution set is $\left\{3\sqrt{2}\right\}$.

97. $\log_a 1000 = 3$

$a^3 = 1000$

$a = \sqrt[3]{1000}$

$a = 10$

The solution set is $\{10\}$.

99. $\ln x = 5$

$x = e^5 \approx 148.413$

The solution set is $\left\{e^5\right\}$.

101. $\log(2x-1) = -1$

$2x-1 = 10^{-1}$

$2x-1 = \dfrac{1}{10}$

$2x = \dfrac{11}{10}$

$x = \dfrac{1}{2} \cdot \dfrac{11}{10} = \dfrac{11}{20}$

The solution set is $\left\{\dfrac{11}{20}\right\}$.

103. $\ln e^x = -3$

$e^x = e^{-3}$

$x = -3$

The solution set is $\{-3\}$.

105. $\log_3(81) = x$

$3^x = 81$

$3^x = 3^4$

$x = 4$

The solution set is $\{4\}$.

107. $\log_2\left(x^2-1\right)=3$

$$x^2-1=2^3$$
$$x^2-1=8$$
$$x^2=9$$
$$x=\pm\sqrt{9}$$
$$x=\pm3$$

The solution set is $\{-3,\ 3\}$.

109. (a) $f(16)=\log_2 16$. To determine the value, let $y=\log_2 16$. Then,

$$2^y=16$$
$$2^y=2^4$$
$$y=4$$

Therefore, $f(16)=4$, and the point $(16,4)$ is on the graph of f.

(b) $f(x)=-3$
$\log_2 x=-3$

$$x=2^{-3}$$
$$x=\frac{1}{2^3}=\frac{1}{8}$$

The point $\left(\dfrac{1}{8},-3\right)$ is on the graph of f.

111. (a) $G(7)=\log_4(7+1)=\log_4 8$. To determine the value, let $y=\log_4 8$. Then,

$$4^y=8$$
$$\left(2^2\right)^y=2^3$$
$$2^{2y}=2^3$$
$$2y=3$$
$$y=\frac{3}{2}$$

Therefore, $G(7)=\dfrac{3}{2}$, and the point $\left(7,\dfrac{3}{2}\right)$ is on the graph of G.

(b) $G(x)=2$
$\log_4(x+1)=2$

$$x+1=4^2$$
$$x+1=16$$
$$x=15$$

The point $(15,2)$ is on the graph of G.

113. If the graph of $f(x)=\log_a x$ contains the point $(16,\ 2)$, then

$$\log_a 16=2$$
$$a^2=16$$
$$a=\pm\sqrt{16}=\pm4$$

Since the base of a logarithm must always be positive, we know that $a=-4$ is extraneous. Thus, $a=4$.

115. We evaluate $L(x)=10\log\dfrac{x}{10^{-12}}$ at $x=10^{-10}$.

$$L\left(10^{-10}\right)=10\log\frac{10^{-10}}{10^{-12}}$$
$$=10\log 10^{-10-(-12)}$$
$$=10\log 10^2=10(2)=20$$

The loudness of a whisper is 20 decibels.

117. We evaluate $L(x)=10\log\dfrac{x}{10^{-12}}$ at $x=10^1$.

$$L\left(10^1\right)=10\log\frac{10^1}{10^{-12}}$$
$$=10\log 10^{1-(-12)}$$
$$=10\log 10^{13}=10(13)=130$$

The threshold of pain is 130 decibels.

119. We evaluate $M(x)=\log\left(\dfrac{x}{10^{-3}}\right)$ at $x=63,096$.

$$M(63,096)=\log\left(\frac{63,096}{10^{-3}}\right)$$
$$=\log(63,096,000)\approx7.8$$

The magnitude of the 1906 San Francisco earthquake was approximately 7.8 on the Richter scale.

121. We solve $M(x)=\log\left(\dfrac{x}{10^{-3}}\right)$ for $M(x)=8.8$.

$$M(x)=8.8$$
$$\log\left(\frac{x}{10^{-3}}\right)=8.8$$
$$\frac{x}{10^{-3}}=10^{8.8}$$
$$x=10^{8.8}\cdot 10^{-3}$$
$$=10^{8.8+(-3)}$$
$$=10^{5.8}$$
$$\approx630,957$$

The seismographic reading of the 1906 Ecuador earthquake 100 kilometers from the epicenter was approximately 630,957 millimeters.

123. (a) We evaluate $pH = -\log\left[H^+\right]$ for

$$\left[H^+\right] = 10^{-12}.$$

$$pH = -\log 10^{-12} = -(-12) = 12$$

The pH of household ammonia is 12, so household ammonia is basic.

(b) We evaluate $pH = -\log\left[H^+\right]$ for

$$\left[H^+\right] = 10^{-5}.$$

$$pH = -\log 10^{-5} = -(-5) = 5$$

The pH of black coffee is 5, so black coffee is acidic.

(c) We evaluate $pH = -\log\left[H^+\right]$ for

$$\left[H^+\right] = 10^{-2}.$$

$$pH = -\log 10^{-2} = -(-2) = 2$$

The pH of lemon juice is 2, so lemon juice is acidic.

(d) We solve $pH = -\log\left[H^+\right]$ for $pH = 7.4$.

$$7.4 = -\log\left[H^+\right]$$

$$-7.4 = \log\left[H^+\right]$$

$$\left[H^+\right] = 10^{-7.4}$$

The concentration of hydrogen ions in human blood is $10^{-7.4}$ moles per liter.

125. The domain of $f(x) = \log_2\left(x^2 - 3x - 10\right)$ is the set of all real numbers x such that $x^2 - 3x - 10 > 0$.

Solve: $x^2 - 3x - 10 = 0$
$$(x+2)(x-5) = 0$$
$$x + 2 = 0 \quad \text{or} \quad x - 5 = 0$$
$$x = -2 \quad \text{or} \quad x = 5$$

Determine where each factor is positive and negative and where the product of these factors is positive and negative.

Interval	$(-\infty, -2)$	-2	$(-2, 5)$	5	$(5, \infty)$
$x + 2$	$----$	0	$++++$	$+$	$++++$
$x - 5$	$----$	$-$	$----$	0	$++++$
$(x+2)(x-5)$	$++++$	0	$----$	0	$++++$

The inequality is strict, so -2 and 5 are not part of the solution. Now, $(x+2)(x-5)$ is greater than zero where the product is positive. Thus, the solution of the inequality and the domain of f is $\{x \mid x < -2 \text{ or } x > 5\}$, or using interval notation, $(-\infty, -2) \cup (5, \infty)$.

127. The domain of $f(x) = \ln\left(\dfrac{x-3}{x+1}\right)$ is the set of all real numbers x such that $\dfrac{x-3}{x+1} > 0$.

The rational expression will equal 0 when $x = 3$. It is undefined when $x = -1$. Determine where the numerator and the denominator are positive and negative and where the quotient is positive and negative:

Interval	$(-\infty, -1)$	-1	$(-1, 3)$	3	$(3, \infty)$
$x - 3$	$----$	$-$	$----$	0	$++++$
$x + 1$	$----$	0	$++++$	$+$	$++++$
$\frac{x-3}{x+1}$	$++++$	$\varnothing$	$----$	0	$++++$

The rational function is undefined at $x = -1$, so -1 is not part of the solution. The inequality is strict, so 3 is not part of the solution. Now, $\dfrac{x-3}{x+1}$ is greater than zero where the quotient is positive. Thus, the solution of the inequality and the domain of f is $\{x \mid x < -1 \text{ or } x > 3\}$, or using interval notation, $(-\infty, -1) \cup (3, \infty)$.

129. The base of $f(x) = \log_a x$ cannot equal 1 because $y = \log_a x$ is equivalent to $x = a^y$ and a does not equal 1 in the exponential function because its graph is a vertical line $x = 1$.

131. The domain of $f(x) = \log_a\left(x^2 + 1\right)$ is the set of all real numbers because $x^2 + 1 > 0$ for all x.

133. $\left(2x^2-6x+1\right)-\left(5x^2+x-9\right)$

$=2x^2-6x+1-5x^2-x+9$

$=-3x^2-7x+10$

135. $x^2-1=(x+1)(x-1)$

$x^2+3x+2=(x+1)(x+2)$

LCD $=(x+1)(x-1)(x+2)$

$\dfrac{3x}{x^2-1}+\dfrac{x-3}{x^2+3x+2}$

$=\dfrac{3x}{(x+1)(x-1)}\cdot\dfrac{x+2}{x+2}+\dfrac{x-3}{(x+1)(x+2)}\cdot\dfrac{x-1}{x-1}$

$=\dfrac{3x^2+6x}{(x+1)(x-1)(x+2)}+\dfrac{x^2-x-3x+3}{(x+1)(x-1)(x+2)}$

$=\dfrac{3x^2+6x+x^2-x-3x+3}{(x+1)(x-1)(x+2)}$

$=\dfrac{4x^2+2x+3}{(x+1)(x-1)(x+2)}$

137. $\sqrt{8x^3}+x\sqrt{18x}=\sqrt{4x^2\cdot 2x}+x\sqrt{9\cdot 2x}$

$=2x\sqrt{2x}+3x\sqrt{2x}=5x\sqrt{2x}$

139. Let $Y_1=\log(x+1)$.

 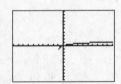

The domain of $f(x)=\log(x+1)$ is $\{x\,|\,x>-1\}$

or, using interval notation, $(-1,\infty)$. The range is all real numbers or, using interval notation, $(-\infty,\infty)$.

Note: The graph shown above is a bit misleading. The curve does not terminate at $x=-1$. Instead, the curve has an asymptote at $x=-1$. More specifically, as x approaches -1 (but stays larger than -1), f goes to $-\infty$.

141. Let $Y_1=\ln(x)+1$.

 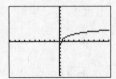

The domain of $G(x)=\ln(x)+1$ is $\{x\,|\,x>0\}$

or, using interval notation, $(0,\infty)$. The range is

all real numbers or, using interval notation, $(-\infty,\infty)$.

Note: The graph shown above is a bit misleading. The curve does not terminate at $x=0$. Instead, the curve has an asymptote at $x=0$. More specifically, as x approaches 0 (but stays larger than 0), G goes to $-\infty$.

143. Let $Y_1=2\log(x-3)+1$.

The domain of $f(x)=2\log(x-3)+1$ is

$\{x\,|\,x>3\}$ or, using interval notation, $(3,\infty)$.

The range is all real numbers or, using interval notation, $(-\infty,\infty)$.

Note: The graph shown above is a bit misleading. The curve does not terminate at $x=3$. Instead, the curve has an asymptote at $x=3$. More specifically, as x approaches 3 (but stays larger than 3), f goes to $-\infty$.

Putting the Concepts Together (Sections 11.1–11.3)

1. $f(x)=2x+3$; $g(x)=2x^2-4x$

(a) $(f\circ g)(x)=f\big(g(x)\big)$

$=2\big(2x^2-4x\big)+3$

$=4x^2-8x+3$

(b) $(g\circ f)(x)=g\big(f(x)\big)$

$=2(2x+3)^2-4(2x+3)$

$=2\big(4x^2+12x+9\big)-4(2x+3)$

$=8x^2+24x+18-8x-12$

$=8x^2+16x+6$

(c) Using the result from part (a):

$(f\circ g)(3)=4(3)^2-8(3)+3$

$=4(9)-8(3)+3$

$=36-24+3$

$=15$

(d) Using the result from part (b):

$$(g \circ f)(-2) = 8(-2)^2 + 16(-2) + 6$$
$$= 8(4) + 16(-2) + 6$$
$$= 32 - 32 + 6$$
$$= 6$$

(e) $f(1) = 2(1) + 3 = 2 + 3 = 5$
$f(5) = 2(5) + 3 = 10 + 3 = 13$
$(f \circ f)(1) = f(f(1)) = f(5) = 13$

2. (a) $\quad f(x) = 3x + 4$

$$y = 3x + 4$$
$$x = 3y + 4$$
$$x - 4 = 3y$$
$$\frac{x-4}{3} = y$$
$$f^{-1}(x) = \frac{x-4}{3}$$

Check:

$$f\left(f^{-1}(x)\right) = 3\left(\frac{x-4}{3}\right) + 4 = x - 4 + 4 = x$$

$$f^{-1}\left(f(x)\right) = \frac{(3x+4)-4}{3} = \frac{3x}{3} = x$$

(b) $\quad g(x) = x^3 - 4$

$$y = x^3 - 4$$
$$x = y^3 - 4$$
$$x + 4 = y^3$$
$$\sqrt[3]{x+4} = y$$
$$g^{-1}(x) = \sqrt[3]{x+4}$$

Check:

$$g\left(g^{-1}(x)\right) = \left(\sqrt[3]{x+4}\right)^3 - 4 = x + 4 - 4 = x$$

$$g^{-1}\left(g(x)\right) = \sqrt[3]{\left(x^3-4\right)+4} = \sqrt[3]{x^3} = x$$

3. To plot the inverse, switch the *x*- and *y*-coordinates in each point and connect the corresponding points. The graphs of the function (shaded) and the line $y = x$ (dashed) are included for reference.

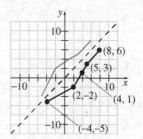

4. (a) $2.7^{2.7} \approx 14.611$

(b) $2.72^{2.72} \approx 15.206$

(c) $2.718^{2.718} \approx 15.146$

(d) $2.7183^{2.7183} \approx 15.155$

(e) $e^e \approx 15.154$

5. (a) If $a^4 = 6.4$, then $\log_a 6.4 = 4$.

(b) If $10^x = 278$, then $\log 278 = x$.

6. (a) If $\log_2 x = 7$, then $2^7 = x$.

(b) If $\ln 16 = M$, then $e^M = 16$.

7. (a) Let $y = \log_5 625$. Then,

$$5^y = 625$$
$$5^y = 5^4$$
$$y = 4$$

Thus, $\log_5 625 = 4$.

(b) Let $y = \log_{\frac{2}{3}}\left(\frac{9}{4}\right)$. Then,

$$\left(\frac{2}{3}\right)^y = \frac{9}{4}$$
$$\left(\frac{2}{3}\right)^y = \left(\frac{3}{2}\right)^2$$
$$\left(\frac{2}{3}\right)^y = \left(\frac{2}{3}\right)^{-2}$$
$$y = -2$$

Thus, $\log_{\frac{2}{3}}\left(\frac{9}{4}\right) = -2$.

8. The domain of $f(x) = \log_{13}(2x+12)$ is the set
of all real numbers x such that
$$2x+12 > 0$$
$$2x > -12$$
$$x > -6$$
Thus, the domain of $f(x) = \log_{13}(2x+12)$ is
$\{x \mid x > -6\}$, or using interval notation, $(-6, \infty)$.

9. Locate some points on the graph of
$$f(x) = \left(\frac{1}{6}\right)^x.$$

x	$f(x) = \left(\frac{1}{6}\right)^x$	$(x, f(x))$
-1	$f(-1) = \left(\frac{1}{6}\right)^{-1} = \left(\frac{6}{1}\right)^1 = 6$	$(-1, 6)$
0	$f(0) = \left(\frac{1}{6}\right)^0 = 1$	$(0, 1)$
1	$f(1) = \left(\frac{1}{6}\right)^1 = \frac{1}{6}$	$\left(1, \frac{1}{6}\right)$
2	$f(2) = \left(\frac{1}{6}\right)^2 = \frac{1}{36}$	$\left(2, \frac{1}{36}\right)$

Plot the points and connect them with a smooth
curve.

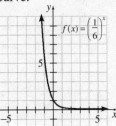

The domain of f is all real numbers or, using
interval notation, $(-\infty, \infty)$. The range of f is
$\{y \mid y > 0\}$ or , using interval notation, $(0, \infty)$.

10. Rewrite $y = g(x) = \log_{\frac{3}{2}} x$ as $x = \left(\frac{3}{2}\right)^y$. Locate
some points on the graph of $x = \left(\frac{3}{2}\right)^y$.

y	$x = \left(\frac{3}{2}\right)^y$	(x, y)
-2	$x = \left(\frac{3}{2}\right)^{-2} = \left(\frac{2}{3}\right)^2 = \frac{4}{9}$	$\left(\frac{4}{9}, -2\right)$
-1	$x = \left(\frac{3}{2}\right)^{-1} = \left(\frac{2}{3}\right)^1 = \frac{2}{3}$	$\left(\frac{2}{3}, -1\right)$
0	$x = \left(\frac{3}{2}\right)^0 = 1$	$(1, 0)$
1	$x = \left(\frac{3}{2}\right)^1 = \frac{3}{2}$	$\left(\frac{3}{2}, 1\right)$
2	$x = \left(\frac{3}{2}\right)^2 = \frac{9}{4}$	$\left(\frac{9}{4}, 2\right)$

Plot the points and connect them with a smooth
curve.

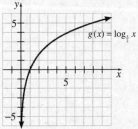

The domain of g is $\{x \mid x > 0\}$ or, using interval
notation, $(0, \infty)$. The range of g is all real
numbers or , using interval notation, $(-\infty, \infty)$.

11. $3^{-x+2} = 27$
$3^{-x+2} = 3^3$
$-x+2 = 3$
$-x = 1$
$x = -1$
The solution set is $\{-1\}$.

12. $e^x = e^{2x+5}$
$x = 2x+5$
$-x = 5$
$x = -5$
The solution set is $\{-5\}$.

13. $\log_2(2x+5)=4$

$$2x+5=2^4$$
$$2x+5=16$$
$$2x=11$$
$$x=\frac{11}{2}$$

The solution set is $\left\{\dfrac{11}{2}\right\}$.

14. $\ln x = 7$

$$x=e^7 \approx 1096.633$$

The solution set is $\left\{e^7\right\}$.

15. $L(90)=150\left(1-e^{-0.0052(90)}\right)\approx 56.06$.

According to the model, the student will learn approximately 56 terms after 90 minutes.

Section 11.4

Preparing for Properties of Logarithms

P1. $3.03468 \approx 3.035$

P2. If $a \neq 0$, then $a^0 = 1$.

Section 11.4 Quick Checks

1. $\log_5 1 = 0$

2. $\ln 1 = 0$

3. $\log_4 4 = 1$

4. $\log 10 = 1$

5. $12^{\log_{12}\sqrt{2}} = \sqrt{2}$

6. $10^{\log 0.2} = 0.2$

7. $\log_8 8^{1.2} = 1.2$

8. $\log 10^{-4} = -4$

9. False. $\log(x \cdot 4) = \log x + \log 4$, but there is no such rule for $\log(x+4)$.

10. $\log_4(9 \cdot 5) = \log_4 9 + \log_4 5$

11. $\log(5w) = \log 5 + \log w$

12. $\log_7\left(\dfrac{9}{5}\right) = \log_7 9 - \log_7 5$

13. $\ln\left(\dfrac{p}{3}\right) = \ln p - \ln 3$

14. $\log_2\left(\dfrac{3m}{n}\right) = \log_2(3m) - \log_2 n$
$$= \log_2 3 + \log_2 m - \log_2 n$$

15. $\ln\left(\dfrac{q}{3p}\right) = \ln q - \ln(3p)$
$$= \ln q - (\ln 3 + \ln p)$$
$$= \ln q - \ln 3 - \ln p$$

16. $\log_2 5^{1.6} = 1.6 \log_2 5$

17. $\log b^5 = 5 \log b$

18. $\log_4(a^2 b) = \log_4 a^2 + \log_4 b$
$$= 2\log_4 a + \log_4 b$$

19. $\log_3\left(\dfrac{9m^4}{\sqrt[3]{n}}\right) = \log_3\left(9m^4\right) - \log_3 \sqrt[3]{n}$
$$= \log_3\left(3^2 m^4\right) - \log_3 n^{\frac{1}{3}}$$
$$= \log_3 3^2 + \log_3 m^4 - \log_3 n^{\frac{1}{3}}$$
$$= 2 + 4\log_3 m - \frac{1}{3}\log_3 n$$

20. $\log_8 4 + \log_8 16 = \log_8(4 \cdot 16)$
$$= \log_8 64$$
$$= 2$$

21. $\log_3(x+4) - \log_3(x-1) = \log_3\left(\dfrac{x+4}{x-1}\right)$

22. $\log_5 x - 3\log_5 2 = \log_5 x - \log_5 2^3$
$$= \log_5 x - \log_5 8$$
$$= \log_5 \frac{x}{8}$$

23. $\log_2(x+1) + \log_2(x+2) - 2\log_2 x$

$= \log_2\left[(x+1)(x+2)\right] - \log_2 x^2$

$= \log_2\left(x^2 + 3x + 2\right) - \log_2 x^2$

$= \log_2\left(\dfrac{x^2 + 3x + 2}{x^2}\right)$

24. $\dfrac{\log 10}{\log 3} = \dfrac{\ln 10}{\ln 3}$

25. Using common logarithms:

$\log_3 32 = \dfrac{\log 32}{\log 3} \approx 3.155$

26. Using natural logarithms:

$\log_{\sqrt{2}} \sqrt{7} = \dfrac{\ln \sqrt{7}}{\ln \sqrt{2}} \approx 2.807$

11.4 Exercises

27. $\log_2 2^3 = 3$

29. $\ln e^{-7} = -7$

31. $3^{\log_3 5} = 5$

33. $e^{\ln 2} = 2$

35. $\log_7 7 = 1$

37. $\log 1 = 0$

39. $\ln 6 = \ln(2 \cdot 3) = \ln 2 + \ln 3 = a + b$

41. $\ln 9 = \ln 3^2 = 2\ln 3 = 2b$

43. $\ln 12 = \ln\left(2^2 \cdot 3\right)$

$= \ln 2^2 + \ln 3$

$= 2\ln 2 + \ln 3$

$= 2a + b$

45. $\ln \sqrt{2} = \ln 2^{1/2} = \dfrac{1}{2}\ln 2 = \dfrac{1}{2}a$

47. $\log(ab) = \log a + \log b$

49. $\log_5 x^4 = 4\log_5 x$

51. $\log_2\left(xy^2\right) = \log_2 x + \log_2 y^2 = \log_2 x + 2\log_2 y$

53. $\log_5(25x) = \log_5 25 + \log_5 x$

$= \log_5 5^2 + \log_5 x$

$= 2 + \log_5 x$

55. $\log_7\left(\dfrac{49}{y}\right) = \log_7 49 - \log_7 y$

$= \log_7 7^2 - \log_7 y$

$= 2 - \log_7 y$

57. $\ln\left(e^2 x\right) = \ln e^2 + \ln x = 2 + \ln x$

59. $\log_3\left(27\sqrt{x}\right) = \log_3 27 + \log_3 \sqrt{x}$

$= \log_3 3^3 + \log_3 x^{1/2}$

$= 3 + \dfrac{1}{2}\log_3 x$

61. $\log_5\left(x^2\sqrt{x^2+1}\right) = \log_5 x^2 + \log_5 \sqrt{x^2+1}$

$= \log_5 x^2 + \log_5\left(x^2+1\right)^{1/2}$

$= 2\log_5 x + \dfrac{1}{2}\log_5\left(x^2+1\right)$

63. $\log\left(\dfrac{x^4}{\sqrt[3]{x-1}}\right) = \log x^4 - \log \sqrt[3]{x-1}$

$= \log x^4 - \log(x-1)^{1/3}$

$= 4\log x - \dfrac{1}{3}\log(x-1)$

65. $\log_7 \sqrt{\dfrac{x+1}{x}} = \log_7\left(\dfrac{x+1}{x}\right)^{\frac{1}{2}}$

$= \dfrac{1}{2}\log_7\left(\dfrac{x+1}{x}\right)$

$= \dfrac{1}{2}\left[\log_7(x+1) - \log_7 x\right]$

$= \dfrac{1}{2}\log_7(x+1) - \dfrac{1}{2}\log_7 x$

67. $\log_2\left[\dfrac{x(x-1)^2}{\sqrt{x+1}}\right]$

$= \log_2\left[x(x-1)^2\right] - \log_2\sqrt{x+1}$

$= \log_2 x + \log_2(x-1)^2 - \log_2\sqrt{x+1}$

$= \log_2 x + \log_2(x-1)^2 - \log_2(x+1)^{\frac{1}{2}}$

$= \log_2 x + 2\log_2(x-1) - \dfrac{1}{2}\log_2(x+1)$

69. $\log 25 + \log 4 = \log(25\cdot 4)$

$\qquad\qquad = \log 100$

$\qquad\qquad = \log 10^2$

$\qquad\qquad = 2$

71. $\log x + \log 3 = \log(3x)$

73. $\log_3 36 - \log_3 4 = \log_3\left(\dfrac{36}{4}\right)$

$\qquad\qquad\qquad = \log_3 9$

$\qquad\qquad\qquad = \log_3 3^2$

$\qquad\qquad\qquad = 2$

75. $10^{\log 8 - \log 2} = 10^{\log\left(\frac{8}{2}\right)} = 10^{\log 4} = 4$

77. $3\log_3 x = \log_3 x^3$

79. $\log_4(x+1) - \log_4 x = \log_4\left(\dfrac{x+1}{x}\right)$

81. $2\ln x + 3\ln y = \ln x^2 + \ln y^3 = \ln\left(x^2 y^3\right)$

83. $\dfrac{1}{2}\log_3 x + 3\log_3(x-1) = \log_3 x^{\frac{1}{2}} + \log_3(x-1)^3$

$\qquad\qquad\qquad\qquad = \log_3\sqrt{x} + \log_3(x-1)^3$

$\qquad\qquad\qquad\qquad = \log_3\left[\sqrt{x}(x-1)^3\right]$

85. $\log x^5 - 3\log x = \log x^5 - \log x^3$

$\qquad\qquad\qquad = \log\left(\dfrac{x^5}{x^3}\right)$

$\qquad\qquad\qquad = \log\left(x^2\right)$

87. $\dfrac{1}{2}[3\log x + \log y] = \dfrac{1}{2}\left[\log x^3 + \log y\right]$

$\qquad\qquad\qquad\qquad = \dfrac{1}{2}\log\left(x^3 y\right)$

$\qquad\qquad\qquad\qquad = \log\left(x^3 y\right)^{\frac{1}{2}}$

$\qquad\qquad\qquad\qquad = \log\sqrt{x^3 y}$

$\qquad\qquad\qquad\qquad = \log\left(x\sqrt{xy}\right)$

89. $\log_8\left(x^2 - 1\right) - \log_8(x+1) = \log_8\left(\dfrac{x^2-1}{x+1}\right)$

$\qquad\qquad\qquad\qquad = \log_8\left[\dfrac{(x-1)(x+1)}{x+1}\right]$

$\qquad\qquad\qquad\qquad = \log_8(x-1)$

91. $18\log\sqrt{x} + 9\log\sqrt[3]{x} - \log 10$

$= 18\log x^{\frac{1}{2}} + 9\log x^{\frac{1}{3}} - \log 10$

$= \log\left(x^{\frac{1}{2}}\right)^{18} + \log\left(x^{\frac{1}{3}}\right)^9 - \log 10$

$= \log x^9 + \log x^3 - \log 10$

$= \log\left(\dfrac{x^9 \cdot x^3}{10}\right)$

$= \log\left(\dfrac{x^{12}}{10}\right)$

93. Using common logarithms:

$\log_2 10 = \dfrac{\log 10}{\log 2} \approx 3.322$

95. Using common logarithms:

$\log_8 3 = \dfrac{\log 3}{\log 8} \approx 0.528$

97. Using natural logarithms:

$\log_{\frac{1}{3}} 19 = \dfrac{\ln 19}{\ln\left(\dfrac{1}{3}\right)} \approx -2.680$

99. Using natural logarithms:

$\log_{\sqrt{2}} 5 = \dfrac{\ln 5}{\ln\sqrt{2}} \approx 4.644$

101. $\log_2 3 \cdot \log_3 4 \cdot \log_4 5 \cdot \log_5 6 \cdot \log_6 7 \cdot \log_7 8$

$= \dfrac{\log 3}{\log 2} \cdot \dfrac{\log 4}{\log 3} \cdot \dfrac{\log 5}{\log 4} \cdot \dfrac{\log 6}{\log 5} \cdot \dfrac{\log 7}{\log 6} \cdot \dfrac{\log 8}{\log 7}$

$= \dfrac{\cancel{\log 3}}{\log 2} \cdot \dfrac{\cancel{\log 4}}{\cancel{\log 3}} \cdot \dfrac{\cancel{\log 5}}{\cancel{\log 4}} \cdot \dfrac{\cancel{\log 6}}{\cancel{\log 5}} \cdot \dfrac{\cancel{\log 7}}{\cancel{\log 6}} \cdot \dfrac{\log 8}{\cancel{\log 7}}$

$= \dfrac{\log 8}{\log 2}$

$= \log_2 8$

$= \log_2 2^3$

$= 3$

103. $\log_2 3 \cdot \log_3 4 \cdot \ldots \cdot \log_n (n+1) \cdot \log_{n+1} 2$

$= \dfrac{\log 3}{\log 2} \cdot \dfrac{\log 4}{\log 3} \cdot \ldots \cdot \dfrac{\log(n+1)}{\log n} \cdot \dfrac{\log 2}{\log(n+1)}$

$= \dfrac{\cancel{\log 3}}{\cancel{\log 2}} \cdot \dfrac{\log 4}{\cancel{\log 3}} \cdot \ldots \cdot \dfrac{\cancel{\log(n+1)}}{\log n} \cdot \dfrac{\cancel{\log 2}}{\cancel{\log(n+1)}}$

$= 1$

Note: The expression "log 4" in the numerator of the second fraction will cancel with the expression "log 4" in the denominator of the third fraction (if the third fraction were written out). Similarly, the expression "log n" in the denominator of the next-to-the-last fraction will cancel with the expression "log n" in the numerator of the third-from-the-last fraction (if it were written out). Likewise, everything in between will cancel leaving an overall value of 1.

105. $\log_a\left(x + \sqrt{x^2 - 1}\right) + \log_a\left(x - \sqrt{x^2 - 1}\right)$

$= \log_a\left[\left(x + \sqrt{x^2-1}\right)\left(x - \sqrt{x^2-1}\right)\right]$

$= \log_a\left[x^2 - x\sqrt{x^2-1} + x\sqrt{x^2-1} - \left(x^2-1\right)\right]$

$= \log_a\left(x^2 - x^2 + 1\right)$

$= \log_a 1$

$= 0$

107. If $f(x) = \log_a x$, then

$f(AB) = \log_a(AB)$

$\qquad = \log_a A + \log_a B$

$\qquad = f(A) + f(B)$

109. Answers may vary. One possibility follows: The log of a product is equal to the sum of the logs.

111. Answers may vary. One example follows: Let $x = 2$ and $y = 1$. Then

$\log_2(x + y) = \log_2(2+1) = \log_2 3$, and

$\log_2 x + \log_2 y = \log_2 2 + \log_2 1 = 1 + 0 = 1$.

However, $\log_2 3 \neq 1$ because $2^1 \neq 3$. Thus,

$\log_2(x+y) \neq \log_2 x + \log_2 y$.

113. $4x + 3 = 13$

$\quad 4x = 10$

$\quad x = \dfrac{10}{4} = \dfrac{5}{2}$

The solution set is $\left\{\dfrac{5}{2}\right\}$.

115. $x^2 + 4x + 2 = 0$

Because this equation does not easily factor, solve by using the quadratic formula. For this equation, $a = 1$, $b = 4$, and $c = 2$.

$x = \dfrac{-4 \pm \sqrt{4^2 - 4(1)(2)}}{2(1)}$

$\quad = \dfrac{-4 \pm \sqrt{16 - 8}}{2}$

$\quad = \dfrac{-4 \pm \sqrt{8}}{2}$

$\quad = \dfrac{-4 \pm 2\sqrt{2}}{2}$

$\quad = -2 \pm \sqrt{2}$

The solution set is $\left\{-2 - \sqrt{2},\ -2 + \sqrt{2}\right\}$.

117. $\sqrt{x+2} - 3 = 4$　　　　　Check:

$\quad \sqrt{x+2} = 7$　　　　　$\sqrt{47+2} - 3 \overset{?}{=} 4$

$\quad \left(\sqrt{x+2}\right)^2 = 7^2$　　　　$\sqrt{49} - 3 \overset{?}{=} 4$

$\quad x + 2 = 49$　　　　　　$7 - 3 \overset{?}{=} 4$

$\quad x = 47$　　　　　　　　$4 = 4\ \checkmark$

The solution set is {47}.

119. Since $\log_3 x = \dfrac{\log x}{\log 3}$, let $Y_1 = \dfrac{\log x}{\log 3}$.

```
WINDOW
Xmin=-10
Xmax=10
Xscl=1
Ymin=-10
Ymax=10
Yscl=1
Xres=1
```

The domain of $f(x) = \log_3 x$ is $\{x \mid x > 0\}$, or using interval notation, $(0, \infty)$; the range is all real numbers or, using interval notation, $(-\infty, \infty)$.

Note: The graph shown above is a bit misleading. The curve does not terminate at $x = 0$. Instead, the curve has an asymptote at $x = 0$. More specifically, as x approaches 0 (but stays larger than 0), f goes to $-\infty$.

121. Since $\log_{1/2} x = \dfrac{\log x}{\log\left(\frac{1}{2}\right)}$, let $Y_1 = \dfrac{\log x}{\log\left(\frac{1}{2}\right)}$.

```
WINDOW
 Xmin=-10
 Xmax=10
 Xscl=1
 Ymin=-10
 Ymax=10
 Yscl=1
 Xres=1
```

The domain of $F(x) = \log_{1/2} x$ is $\{x \mid x > 0\}$, or using interval notation, $(0, \infty)$; the range is all real numbers or, using interval notation, $(-\infty, \infty)$.

Note: The graph shown above is a bit misleading. The curve does not terminate at $x = 0$. Instead, the curve has an asymptote at $x = 0$. More specifically, as x approaches 0 (but stays larger than 0), F goes to ∞.

Section 11.5

Preparing for Exponential and Logarithmic Equations

P1.
$$2x + 5 = 13$$
$$2x + 5 - 5 = 13 - 5$$
$$2x = 8$$
$$\frac{2x}{2} = \frac{8}{2}$$
$$x = 4$$

The solution set is $\{4\}$.

P2.
$$x^2 - 4x = -3$$
$$x^2 - 4x + 3 = 0$$
$$(x-1)(x-3) = 0$$
$$x - 1 = 0 \quad \text{or} \quad x - 3 = 0$$
$$x = 1 \quad \text{or} \quad x = 3$$
The solution set is $\{1, 3\}$.

P3.
$$3a^2 = a + 5$$
$$3a^2 - a - 5 = 0$$
For this equation, $a = 3$, $b = -1$, and $c = -5$.
$$a = \frac{-(-1) \pm \sqrt{(-1)^2 - 4(3)(-5)}}{2(3)}$$
$$= \frac{1 \pm \sqrt{1 + 60}}{6}$$
$$= \frac{1 \pm \sqrt{61}}{6}$$

The solution set is $\left\{ \dfrac{1 - \sqrt{61}}{6}, \dfrac{1 + \sqrt{61}}{6} \right\}$.

P4. $(x+3)^2 + 2(x+3) - 8 = 0$
Let $u = x + 3$.
$$u^2 + 2u - 8 = 0$$
$$(u+4)(u-2) = 0$$
$$u + 4 = 0 \quad \text{or} \quad u - 2 = 0$$
$$u = -4 \quad \text{or} \quad u = 2$$
$$x + 3 = -4 \quad \text{or} \quad x + 3 = 2$$
$$x = -7 \quad \text{or} \quad x = -1$$

Check:
$$x = -7: \ (-7+3)^2 + 2(-7+3) - 8 \stackrel{?}{=} 0$$
$$(-4)^2 + 2(-4) - 8 \stackrel{?}{=} 0$$
$$16 - 8 - 8 \stackrel{?}{=} 0$$
$$0 = 0 \ \checkmark$$

$$x = -1: \ (-1+3)^2 + 2(-1+3) - \stackrel{?}{=} 0$$
$$(2)^2 + 2(2) - 8 \stackrel{?}{=} 0$$
$$4 + 4 - 8 \stackrel{?}{=} 0$$
$$0 = 0 \ \checkmark$$

Both check; the solution set is $\{-7, \ -1\}$.

Section 11.5 Quick Checks

1. $M = N$

2.
$$2\log_4 x = \log_4 9$$
$$\log_4 x^2 = \log_4 9$$
$$x^2 = 9$$
$$x = \pm\sqrt{9}$$
$$x = \pm 3$$

The apparent solution $x = -3$ is extraneous because the argument of a logarithm must be positive. The solution set is $\{3\}$.

3. $\log_4(x-6)+\log_4 x = 2$

$\log_4[x(x-6)]=2$

$\log_4(x^2-6x)=2$

$x^2-6x=4^2$

$x^2-6x=16$

$x^2-6x-16=0$

$(x+2)(x-8)=0$

$x+2=0 \quad$ or $\quad x-8=0$

$x=-2 \quad$ or $\qquad x=8$

The apparent solution $x=-2$ is extraneous because it causes the argument of a logarithm to be negative. The solution set is $\{8\}$.

4. $\log_3(x+3)+\log_3(x+5)=1$

$\log_3[(x+3)(x+5)]=1$

$\log_3(x^2+8x+15)=1$

$x^2+8x+15=3^1$

$x^2+8x+12=0$

$(x+6)(x+2)=0$

$x+6=0 \quad$ or $\quad x+2=0$

$x=-6 \quad$ or $\qquad x=-2$

The apparent solution $x=-6$ is extraneous because it causes the arguments of both logarithms to be negative. The solution set is $\{-2\}$.

5. $2^x=11$

$\log 2^x = \log 11$

$x\log 2 = \log 11$

$x = \dfrac{\log 11}{\log 2} \approx 3.459$

The solution set is $\left\{\dfrac{\log 11}{\log 2}\right\} \approx \{3.459\}$. If we had taken the natural logarithm of both sides, the solution set would be $\left\{\dfrac{\ln 11}{\ln 2}\right\} \approx \{3.459\}$.

6. $5^{2x}=3$

$\log 5^{2x} = \log 3$

$2x\log 5 = \log 3$

$x = \dfrac{\log 3}{2\log 5} \approx 0.341$

The solution set is $\left\{\dfrac{\log 3}{2\log 5}\right\} \approx \{0.341\}$. If we had taken the natural logarithm of both sides, the solution set would be $\left\{\dfrac{\ln 3}{2\ln 5}\right\} \approx \{0.341\}$.

7. $e^{2x}=5$

$\ln e^{2x} = \ln 5$

$2x = \ln 5$

$x = \dfrac{\ln 5}{2} \approx 0.805$

The solution set is $\left\{\dfrac{\ln 5}{2}\right\} \approx \{0.805\}$.

8. $3e^{-4x}=20$

$e^{-4x} = \dfrac{20}{3}$

$\ln e^{-4x} = \ln\left(\dfrac{20}{3}\right)$

$-4x = \ln\left(\dfrac{20}{3}\right)$

$x = \dfrac{\ln\left(\dfrac{20}{3}\right)}{-4} \approx -0.474$

The solution set is $\left\{\dfrac{\ln\left(20/3\right)}{-4}\right\} \approx \{-0.474\}$.

9. (a) We need to determine the time until $A=9$ grams. So we solve the equation

$$9 = 10\left(\dfrac{1}{2}\right)^{t/18.72}$$

$$0.9 = \left(\dfrac{1}{2}\right)^{t/18.72}$$

$$\log 0.9 = \log\left(\dfrac{1}{2}\right)^{t/18.72}$$

$$\log 0.9 = \dfrac{t}{18.72}\log\left(\dfrac{1}{2}\right)$$

$$\dfrac{18.72\cdot\log 0.9}{\log\left(\dfrac{1}{2}\right)} = t$$

So, $t = \dfrac{18.72\cdot\log 0.9}{\log\left(\dfrac{1}{2}\right)} \approx 2.85$ days.

Thus, 9 grams of thorium-227 will be left after approximately 2.85 days.

(b) We need to determine the time until $A = 3$ grams. So we solve the equation

$$3 = 10\left(\frac{1}{2}\right)^{t/18.72}$$

$$0.3 = \left(\frac{1}{2}\right)^{t/18.72}$$

$$\log 0.3 = \log\left(\frac{1}{2}\right)^{t/18.72}$$

$$\log 0.3 = \frac{t}{18.72}\log\left(\frac{1}{2}\right)$$

$$\frac{18.72 \cdot \log 0.3}{\log\left(\frac{1}{2}\right)} = t$$

So, $t = \dfrac{18.72 \cdot \log 0.3}{\log\left(\dfrac{1}{2}\right)} \approx 32.52$ days.

Thus, 3 grams of thorium-227 will be left after approximately 32.52 days.

10. We first write the model with the parameters $P = 2000$, $r = 0.06$, and $n = 12$ to obtain

$$A = 2000\left(1 + \frac{0.06}{12}\right)^{12t} \text{ or } A = 2000(1.005)^{12t}.$$

(a) We need to determine the time until $A = \$3000$, so we solve the equation

$$3000 = 2000(1.005)^{12t}$$

$$1.5 = (1.005)^{12t}$$

$$\log 1.5 = \log(1.005)^{12t}$$

$$\log 1.5 = 12t\log 1.005$$

$$\frac{\log 1.5}{12\log 1.005} = t$$

So, $t = \dfrac{\log 1.5}{12\log 1.005} \approx 6.77$. Thus, after approximately 6.77 years (6 years, 9 months), the account will be worth \$3000.

(b) We need to determine the time until $A = \$4000$, so we solve the equation

$$4000 = 2000(1.005)^{12t}$$

$$2 = (1.005)^{12t}$$

$$\log 2 = \log(1.005)^{12t}$$

$$\log 2 = 12t\log 1.005$$

$$\frac{\log 2}{12\log 1.005} = t$$

So, $t = \dfrac{\log 2}{12\log 1.005} \approx 11.58$. Thus, after approximately 11.58 years (11 years, 7 months), the account will be worth \$4000.

11.5 Exercises

11. $\log_2 x = \log_2 7$

$x = 7$

The solution set is $\{7\}$.

13. $2\log_3 x = \log_3 81$

$\log_3 x^2 = \log_3 81$

$x^2 = 81$

$x = \pm\sqrt{81} = \pm 9$

The apparent solution $x = -9$ is extraneous because the argument of a logarithm must be positive. The solution set is $\{9\}$.

15. $\log_6(3x+1) = \log_6 10$

$3x + 1 = 10$

$3x = 9$

$x = 3$

The solution set is $\{3\}$.

17. $\dfrac{1}{2}\ln x = 2\ln 3$

$\ln x = 4\ln 3$

$\ln x = \ln 3^4$

$\ln x = \ln 81$

$x = 81$

The solution set is $\{81\}$.

19. $\log_2(x+3) + \log_2 x = 2$

$\log_2[x(x+3)] = 2$

$\log_2\left(x^2 + 3x\right) = 2$

$x^2 + 3x = 2^2$

$x^2 + 3x = 4$

$x^2 + 3x - 4 = 0$

$(x+4)(x-1) = 0$

$x + 4 = 0 \quad$ or $\quad x - 1 = 0$

$x = -4 \quad$ or $\quad x = 1$

The apparent solution $x = -4$ is extraneous because it causes the argument of a logarithm to be negative. The solution set is $\{1\}$.

21. $\log_2(x+2)+\log_2(x+5)=\log_2 4$

$\log_2\left[(x+2)(x+5)\right]=\log_2 4$

$x^2+7x+10=4$

$x^2+7x+6=0$

$(x+6)(x+1)=0$

$x+6=0$ or $x+1=0$

$x=-6$ or $x=-1$

The apparent solution $x=-6$ is extraneous because it causes the arguments of both logarithms to be negative. The solution set is $\{-1\}$.

23. $\log(x+3)-\log x=1$

$\log\left(\dfrac{x+3}{x}\right)=1$

$\dfrac{x+3}{x}=10^1$

$x\left(\dfrac{x+3}{x}\right)=x(10)$

$x+3=10x$

$3=9x$

$x=\dfrac{3}{9}=\dfrac{1}{3}$

The solution set is $\left\{\dfrac{1}{3}\right\}$.

25. $\log_4(x+5)-\log_4(x-1)=2$

$\log_4\left(\dfrac{x+5}{x-1}\right)=2$

$\dfrac{x+5}{x-1}=4^2$

$(x-1)\left(\dfrac{x+5}{x-1}\right)=(x-1)16$

$x+5=16x-16$

$-15x=-21$

$x=\dfrac{-21}{-15}=\dfrac{7}{5}$

The solution set is $\left\{\dfrac{7}{5}\right\}$.

27. $\log_4(x+8)+\log_4(x+6)=\log_4 3$

$\log_4\left[(x+8)(x+6)\right]=\log_4 3$

$x^2+14x+48=3$

$x^2+14x+45=0$

$(x+9)(x+5)=0$

$x+9=0$ or $x+5=0$

$x=-9$ or $x=-5$

The apparent solution $x=-9$ is extraneous because it causes the arguments of both logarithms to be negative. The solution set is $\{-5\}$.

29. $2^x=10$

$\log 2^x=\log 10$

$x\log 2=1$

$x=\dfrac{1}{\log 2}\approx 3.322$

The solution set is $\left\{\dfrac{1}{\log 2}\right\}\approx\{3.322\}$. If we had taken the natural logarithm of both sides, the solution set would be $\left\{\dfrac{\ln 10}{\ln 2}\right\}\approx\{3.322\}$.

31. $5^x=20$

$\log 5^x=\log 20$

$x\log 5=\log 20$

$x=\dfrac{\log 20}{\log 5}\approx 1.861$

The solution set is $\left\{\dfrac{\log 20}{\log 5}\right\}\approx\{1.861\}$. If we had taken the natural logarithm of both sides, the solution set would be $\left\{\dfrac{\ln 20}{\ln 5}\right\}\approx\{1.861\}$.

33. $\left(\dfrac{1}{2}\right)^x=7$

$\log\left(\dfrac{1}{2}\right)^x=\log 7$

$x\log\left(\dfrac{1}{2}\right)=\log 7$

$x=\dfrac{\log 7}{\log\left(\dfrac{1}{2}\right)}\approx -2.807$

The solution set is $\left\{ \dfrac{\log 7}{\log\left(\frac{1}{2}\right)} \right\} \approx \{-2.807\}$. If we

had taken the natural logarithm of both sides, the

solution set would be $\left\{ \dfrac{\ln 7}{\ln\left(\frac{1}{2}\right)} \right\} \approx \{-2.807\}$.

35. $e^x = 5$

$\ln e^x = \ln 5$

$x = \ln 5 \approx 1.609$

The solution set is $\{\ln 5\} \approx \{1.609\}$.

37. $10^x = 5$

$\log 10^x = \log 5$

$x = \log 5 \approx 0.699$

The solution set is $\{\log 5\} \approx \{0.699\}$.

39. $3^{2x} = 13$

$\log 3^{2x} = \log 13$

$2x \log 3 = \log 13$

$x = \dfrac{\log 13}{2\log 3} \approx 1.167$

The solution set is $\left\{ \dfrac{\log 13}{2\log 3} \right\} \approx \{1.167\}$. If we

had taken the natural logarithm of both sides, the

solution set would be $\left\{ \dfrac{\ln 13}{2\ln 3} \right\} \approx \{1.167\}$.

41. $\left(\dfrac{1}{2}\right)^{4x} = 3$

$\log\left(\dfrac{1}{2}\right)^{4x} = \log 3$

$4x \log\left(\dfrac{1}{2}\right) = \log 3$

$x = \dfrac{\log 3}{4\log\left(\frac{1}{2}\right)} \approx -0.396$

The solution set is $\left\{ \dfrac{\log 3}{4\log\left(\frac{1}{2}\right)} \right\} \approx \{-0.396\}$. If

we had taken the natural logarithm of both sides,

the solution set would be

$\left\{ \dfrac{\ln 3}{4\ln\left(\frac{1}{2}\right)} \right\} \approx \{-0.396\}$.

$4 \cdot 2^x + 3 = 8$

$4 \cdot 2^x = 5$

$2^x = \dfrac{5}{4}$

$\log 2^x = \log\left(\dfrac{5}{4}\right)$

$x \log 2 = \log\left(\dfrac{5}{4}\right)$

$x = \dfrac{\log\left(\frac{5}{4}\right)}{\log 2} \approx 0.322$

43. The solution set is $\left\{ \dfrac{\log\left(\frac{5}{4}\right)}{\log 2} \right\} \approx \{0.322\}$. If we

had taken the natural logarithm of both sides, the

solution set would be $\left\{ \dfrac{\ln\left(\frac{5}{4}\right)}{\ln 2} \right\} \approx \{0.322\}$.

45. $-3e^x = -18$

$e^x = 6$

$\ln e^x = \ln 6$

$x = \ln 6 \approx 1.792$

The solution set is $\{\ln 6\} \approx \{1.792\}$.

47. $\qquad 0.2^{x+1} = 3^x$

$\log 0.2^{x+1} = \log 3^x$

$(x+1)\log 0.2 = x\log 3$

$x\log 0.2 + \log 0.2 = x\log 3$

$\log 0.2 = x\log 3 - x\log 0.2$

$\log 0.2 = x(\log 3 - \log 0.2)$

$\dfrac{\log 0.2}{\log 3 - \log 0.2} = x$

$-0.594 \approx x$

The solution set is $\left\{ \dfrac{\log 0.2}{\log 3 - \log 0.2} \right\} \approx \{-0.594\}$.

If we had taken the natural logarithm of both

sides, the solution set would be

$\left\{ \dfrac{\ln 0.2}{\ln 3 - \ln 0.2} \right\} \approx \{-0.594\}$.

49. $\log_4 x + \log_4 (x-6) = 2$

$\qquad \log_4 [x(x-6)] = 2$

$\qquad \log_4 (x^2 - 6x) = 2$

$\qquad\qquad x^2 - 6x = 4^2$

$\qquad\qquad x^2 - 6x = 16$

$\qquad x^2 - 6x - 16 = 0$

$\qquad (x-8)(x+2) = 0$

$\qquad x - 8 = 0 \quad$ or $\quad x + 2 = 0$

$\qquad\quad x = 8 \quad$ or $\qquad x = -2$

The apparent solution $x = -2$ is extraneous because it causes the argument of a logarithm to be negative. The solution set is $\{8\}$.

51. $\qquad 5^{3x} = 7$

$\quad \log 5^{3x} = \log 7$

$\quad 3x \log 5 = \log 7$

$\qquad\quad x = \dfrac{\log 7}{3 \log 5} \approx 0.403$

The solution set is $\left\{ \dfrac{\log 7}{3 \log 5} \right\} \approx \{0.403\}$. If we

had taken the natural logarithm of both sides, the

solution set would be $\left\{ \dfrac{\ln 7}{3 \ln 5} \right\} \approx \{0.403\}$.

53. $3 \log_2 x = \log_2 8$

$\quad \log_2 x^3 = \log_2 8$

$\qquad\quad x^3 = 8$

$\qquad\quad x = \sqrt[3]{8}$

$\qquad\quad x = 2$

The solution set is $\{2\}$.

55. $\dfrac{1}{3} e^x = 5$

$\qquad e^x = 15$

$\ln e^x = \ln 15$

$\qquad x = \ln 15 \approx 2.708$

The solution set is $\{\ln 15\} \approx \{2.708\}$.

57. $\left(\dfrac{1}{4} \right)^{x+1} = 8^x$

$\left(\dfrac{1}{2^2} \right)^{x+1} = \left(2^3 \right)^x$

$\left(2^{-2} \right)^{x+1} = \left(2^3 \right)^x$

$\quad 2^{-2(x+1)} = 2^{3x}$

$\quad -2(x+1) = 3x$

$\quad -2x - 2 = 3x$

$\qquad -2 = 5x$

$\qquad -\dfrac{2}{5} = x$

The solution set is $\left\{ -\dfrac{2}{5} \right\}$.

59. $\log_3 x^2 = \log_3 16$

$\qquad x^2 = 16$

$\qquad x = \pm 4$

The solution set is $\{-4, 4\}$.

61. $\log_2 (x+4) + \log_2 (x+6) = \log_2 8$

$\quad \log_2 [(x+4)(x+6)] = \log_2 8$

$\qquad\qquad x^2 + 10x + 24 = 8$

$\qquad\qquad x^2 + 10x + 16 = 0$

$\qquad\qquad (x+8)(x+2) = 0$

$x + 8 = 0 \quad$ or $\quad x + 2 = 0$

$\quad x = -8 \quad$ or $\qquad x = -2$

The apparent solution $x = -8$ is extraneous because it causes the arguments of both logarithms to be negative. The solution set is $\{-2\}$.

63. $\log_4 x + \log_4 (x-4) = \log_4 3$

$\quad \log_4 [x(x-4)] = \log_4 3$

$\quad \log_4 (x^2 - 4x) = \log_4 3$

$\qquad\qquad x^2 - 4x = 3$

$\qquad\qquad x^2 - 4x - 3 = 0$

$a = 1, \ b = -4, \ c = -3$

$x = \dfrac{4 \pm \sqrt{16+12}}{2} = \dfrac{4 \pm \sqrt{28}}{2} = \dfrac{4 \pm 2\sqrt{7}}{2} = 2 \pm \sqrt{7}$

The apparent solution $x = 2 - \sqrt{7}$ is extraneous because it causes the arguments of both logarithms to be negative. The solution set is $\{2 + \sqrt{7}\} = \{4.646\}$.

65. (a) We need to determine the year when $P = 321$ million. So we solve the equation

$$321 = 304(1.011)^{t-2008}$$

$$\frac{321}{304} = (1.011)^{t-2008}$$

$$\log\frac{321}{304} = \log(1.011)^{t-2008}$$

$$\log\frac{321}{304} = (t-2008)\log(1.011)$$

$$\frac{\log\frac{321}{304}}{\log 1.011} = t - 2008$$

$$\frac{\log\frac{321}{304}}{\log 1.011} + 2008 = t$$

$$2012.974 \approx t$$

Thus, according to the model, the population of the United States will reach 321 million people in about the year 2013.

(b) We need to determine the year when $P = 471$ million. So we solve the equation

$$471 = 304(1.011)^{t-2008}$$

$$\frac{471}{304} = (1.011)^{t-2008}$$

$$\log\frac{471}{304} = \log(1.011)^{t-2008}$$

$$\log\frac{471}{304} = (t-2008)\log(1.011)$$

$$\frac{\log\frac{471}{304}}{\log 1.011} = t - 2008$$

$$\frac{\log\frac{471}{304}}{\log 1.011} + 2008 = t$$

$$2048.021 \approx t$$

Thus, according to the model, the population of the United States will reach 471 million people in about the year 2048.

67. We first write the model with the parameters $P = 5000$, $r = 0.06$, and $n = 12$ to obtain

$$A = 5000\left(1 + \frac{0.06}{12}\right)^{12t} \text{ or } A = 5000(1.005)^{12t}.$$

(a) We need to determine the time until $A = \$7000$, so we solve the equation

$$7000 = 5000(1.005)^{12t}$$

$$1.4 = (1.005)^{12t}$$

$$\log 1.4 = \log(1.005)^{12t}$$

$$\log 1.4 = 12t\log 1.005$$

$$\frac{\log 1.4}{12\log 1.005} = t$$

$$5.622 \approx t$$

Thus, after approximately 5.6 years (5 years, 7 months), the account will be worth $7000.

(b) We need to determine the time until $A = \$10,000$, so we solve the equation

$$10,000 = 5000(1.005)^{12t}$$

$$2 = (1.005)^{12t}$$

$$\log 2 = \log(1.005)^{12t}$$

$$\log 2 = 12t\log 1.005$$

$$\frac{\log 2}{12\log 1.005} = t$$

So, $t = \dfrac{\log 2}{12\log 1.005} \approx 11.581$. After about 11.6 years (11 years, 7 months), the account will be worth $7000.

69. (a) We need to determine the time until $V = \$10,000$. So we solve the equation

$$10,000 = 14,512(0.82)^{t}$$

$$\frac{10,000}{14,512} = (0.82)^{t}$$

$$\log\frac{10,000}{14,512} = \log(0.82)^{t}$$

$$\log\frac{10,000}{14,512} = t\log 0.82$$

$$\frac{\log\frac{10,000}{14,512}}{\log 0.82} = t$$

$$1.876 \approx t$$

According to the model, the car will be worth $10,000 after about 1.876 years.

(b) We need to determine the time until $V = \$5000$. So we solve the equation

$$5000 = 14{,}512(0.82)^t$$

$$\frac{5000}{14{,}512} = (0.82)^t$$

$$\log\frac{5000}{14{,}512} = \log(0.82)^t$$

$$\log\frac{5000}{14{,}512} = t\log 0.82$$

$$\frac{\log\dfrac{5000}{14{,}512}}{\log 0.82} = t$$

$$5.369 \approx t$$

According to the model, the car will be worth $5000 in about 5.369 years.

(c) We need to determine the time until $V = \$1000$. So we solve the equation

$$1000 = 14{,}512(0.82)^t$$

$$\frac{1000}{14{,}512} = (0.82)^t$$

$$\log\frac{1000}{14{,}512} = \log(0.82)^t$$

$$\log\frac{1000}{14{,}512} = t\log 0.82$$

$$\frac{\log\dfrac{1000}{14{,}512}}{\log 0.82} = t$$

$$13.479 \approx t$$

According to the model, the car will be worth $1000 in about 13.479 years.

71. (a) We need to determine the time until $A = 90$ grams. So we solve the equation

$$90 = 100\left(\frac{1}{2}\right)^{t/13.81}$$

$$0.9 = \left(\frac{1}{2}\right)^{t/13.81}$$

$$\log 0.9 = \log\left(\frac{1}{2}\right)^{t/13.81}$$

$$\log 0.9 = \frac{t}{13.81}\log\left(\frac{1}{2}\right)$$

$$\frac{13.81\cdot\log 0.9}{\log(1/2)} = t$$

$$2.099 \approx t$$

Thus, 90 grams of beryllium-11 will be left after approximately 2.099 seconds.

(b) We need to determine the time until $A = 25$ grams. So we solve the equation

$$25 = 100\left(\frac{1}{2}\right)^{t/13.81}$$

$$0.25 = \left(\frac{1}{2}\right)^{t/13.81}$$

$$\log 0.25 = \log\left(\frac{1}{2}\right)^{t/13.81}$$

$$\log 0.25 = \frac{t}{13.81}\log\left(\frac{1}{2}\right)$$

$$\frac{13.81\cdot\log 0.25}{\log(1/2)} = t$$

$$27.62 = t$$

Thus, 25 grams of beryllium-11 will be left after 27.62 seconds.

(c) We need to determine the time until $A = 10$ grams. So we solve the equation

$$10 = 100\left(\frac{1}{2}\right)^{t/13.81}$$

$$0.1 = \left(\frac{1}{2}\right)^{t/13.81}$$

$$\log 0.1 = \log\left(\frac{1}{2}\right)^{t/13.81}$$

$$\log 0.1 = \frac{t}{13.81}\log\left(\frac{1}{2}\right)$$

$$\frac{13.81\cdot\log 0.1}{\log(1/2)} = t$$

$$45.876 \approx t$$

Thus, 10 grams of beryllium-11 will be left after approximately 45.876 seconds.

73. (a) $u = 300°F$. So we solve the equation

$$300 = 70 + 330e^{-0.072t}$$

$$230 = 330e^{-0.072t}$$

$$\frac{230}{330} = e^{-0.072t}$$

$$\ln\left(\frac{230}{330}\right) = \ln e^{-0.072t}$$

$$\ln\left(\frac{230}{330}\right) = -0.072t$$

$$\frac{\ln\left(\dfrac{230}{330}\right)}{-0.072} = t$$

$$5.014 \approx t$$

According to the model, the temperature of the pizza will be $300°F$ after about 5 minutes.

(b) We need to determine the time until $u = 220°F$. So we solve the equation

$$220 = 70 + 330e^{-0.072t}$$
$$150 = 330e^{-0.072t}$$
$$\frac{150}{330} = e^{-0.072t}$$
$$\ln\left(\frac{150}{330}\right) = \ln e^{-0.072t}$$
$$\ln\left(\frac{150}{330}\right) = -0.072t$$
$$\frac{\ln\left(\frac{150}{330}\right)}{-0.072} = t$$
$$10.951 \approx t$$

According to the model, the temperature of the pizza will be $220°F$ after about 11 minutes.

75. (a) We need to determine the time at which $L = 50$ words. So we solve the equation

$$50 = 200\left(1 - e^{-0.0035t}\right)$$
$$0.25 = 1 - e^{-0.0035t}$$
$$e^{-0.0035t} = 0.75$$
$$\ln e^{-0.0035t} = \ln 0.75$$
$$-0.0035t = \ln 0.75$$
$$t = \frac{\ln 0.75}{-0.0035}$$
$$t \approx 82.195$$

According to the model, the student must study about 82 minutes in order to learn 50 words.

(b) We need to determine the time at which $L = 150$ words. So we solve the equation

$$150 = 200\left(1 - e^{-0.0035t}\right)$$
$$0.75 = 1 - e^{-0.0035t}$$
$$e^{-0.0035t} = 0.25$$
$$\ln e^{-0.0035t} = \ln 0.25$$
$$-0.0035t = \ln 0.25$$
$$t = \frac{\ln 0.25}{-0.0035}$$
$$t \approx 396.084$$

According to the model, the student must study about 396 minutes (or 6.6 hours) in order to learn 150 words.

77. (a) $t = \frac{72}{8} = 9$

According to the Rule of 72, an investment earning 8% annual interest will take about 9 years to double.

(b) Let $A = 2P$ in the formula $A = P\left(1 + \frac{r}{n}\right)^{nt}$:

$$2P = P\left(1 + \frac{r}{n}\right)^{nt}$$
$$2 = \left(1 + \frac{r}{n}\right)^{nt}$$
$$\log 2 = \log\left(1 + \frac{r}{n}\right)^{nt}$$
$$\log 2 = nt \log\left(1 + \frac{r}{n}\right)$$
$$\frac{\log 2}{n \log\left(1 + \frac{r}{n}\right)} = t$$

(c) Letting $r = 0.08$ and $n = 12$, we obtain

$$t = \frac{\log 2}{12 \cdot \log\left(1 + \frac{0.08}{12}\right)} \approx 8.693 .$$

According to our formula from part (b), an investment earning 8% annual interest will take about 8.693 years to double, which is about the same as the result from the Rule of 72.

79. Since the bacteria doubled every minute, the container would have been half full one minute before it was completely full. Now since it was full after 30 minutes, it would have been half full after $30 - 1 = 29$ minutes.

81. $f(x) = -2x + 7$

(a) $f(3) = -2(3) + 7 = -6 + 7 = 1$

(b) $f(-2) = -2(-2) + 7 = 4 + 7 = 11$

(c) $f(0) = -2(0) + 7 = 0 + 7 = 7$

83. $f(x) = \frac{x}{x-5}$

(a) $f(3) = \frac{3}{3-5} = \frac{3}{-2} = -\frac{3}{2}$

(b) $f(-2) = \dfrac{-2}{-2-5} = \dfrac{-2}{-7} = \dfrac{2}{7}$

(c) $f(0) = \dfrac{0}{0-5} = \dfrac{0}{-5} = 0$

85. $f(x) = 2^x$

(a) $f(3) = 2^3 = 8$

(b) $f(-2) = 2^{-2} = \dfrac{1}{2^2} = \dfrac{1}{4}$

(c) $f(0) = 2^0 = 1$

87. Let $Y_1 = e^x$ and $Y_2 = -2x+5$.

 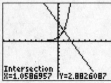

The solution set is approximately $\{1.06\}$.

89. Let $Y_1 = e^x$ and $Y_2 = x^2$.

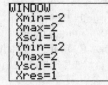 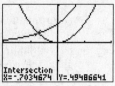

The solution set is approximately $\{-0.70\}$.

91. Let $Y_1 = e^x - \ln x$ and $Y_2 = 4$.

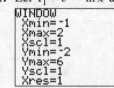 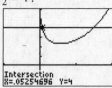

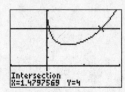

The solution set is approximately $\{0.05, 1.48\}$.

93. Let $Y_1 = \ln x$ and $Y_2 = x^2 - 1$.

 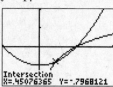

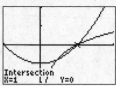

The solution set is approximately $\{0.45, 1\}$.

Chapter 11 Review

1. $f(x) = 3x+5$; $g(x) = 2x-1$

(a) $g(5) = 2(5)-1 = 9$
$f(9) = 3(9)+5 = 32$
$(f \circ g)(5) = f(g(5)) = f(9) = 32$

(b) $f(-3) = 3(-3)+5 = -4$
$g(-4) = 2(-4)-1 = -9$
$(g \circ f)(-3) = g(f(-3)) = g(-4) = -9$

(c) $f(-2) = 3(-2)+5 = -1$
$f(-1) = 3(-1)+5 = 2$
$(f \circ f)(-2) = f(f(-2)) = f(-1) = 2$

(d) $g(4) = 2(4)-1 = 7$
$g(7) = 2(7)-1 = 13$
$(g \circ g)(4) = g(g(4)) = g(7) = 13$

2. $f(x) = x-3$; $g(x) = 5x+2$

(a) $g(5) = 5(5)+2 = 27$
$f(27) = 27-3 = 24$
$(f \circ g)(5) = f(g(5)) = f(27) = 24$

(b) $f(-3) = -3-3 = -6$
$g(-6) = 5(-6)+2 = -28$
$(g \circ f)(-3) = g(f(-3)) = g(-6) = -28$

(c) $f(-2) = -2 - 3 = -5$

$\quad f(-5) = -5 - 3 = -8$

$\quad (f \circ f)(-2) = f(f(-2)) = f(-5) = -8$

(d) $g(4) = 5(4) + 2 = 22$

$\quad g(22) = 5(22) + 2 = 112$

$\quad (g \circ g)(4) = g(g(4)) = g(22) = 112$

3. $f(x) = 2x^2 + 1$; $g(x) = x + 5$

(a) $g(5) = 5 + 5 = 10$

$\quad f(10) = 2(10)^2 + 1 = 201$

$\quad (f \circ g)(5) = f(g(5)) = f(10) = 201$

(b) $f(-3) = 2(-3)^2 + 1 = 19$

$\quad g(19) = 19 + 5 = 24$

$\quad (g \circ f)(-3) = g(f(-3)) = g(19) = 24$

(c) $f(-2) = 2(-2)^2 + 1 = 9$

$\quad f(9) = 2(9)^2 + 1 = 163$

$\quad (f \circ f)(-2) = f(f(-2)) = f(9) = 163$

(d) $g(4) = 4 + 5 = 9$

$\quad g(9) = 9 + 5 = 14$

$\quad (g \circ g)(4) = g(g(4)) = g(9) = 14$

4. $f(x) = x - 3$; $g(x) = x^2 + 1$

(a) $g(5) = (5)^2 + 1 = 26$

$\quad f(26) = 26 - 3 = 23$

$\quad (f \circ g)(5) = f(g(5)) = f(26) = 23$

(b) $f(-3) = -3 - 3 = -6$

$\quad g(-6) = (-6)^2 + 1 = 37$

$\quad (g \circ f)(-3) = g(f(-3)) = g(-6) = 37$

(c) $f(-2) = -2 - 3 = -5$

$\quad f(-5) = -5 - 3 = -8$

$\quad (f \circ f)(-2) = f(f(-2)) = f(-5) = -8$

(d) $g(4) = (4)^2 + 1 = 17$

$\quad g(17) = (17)^2 + 1 = 290$

$\quad (g \circ g)(4) = g(g(4)) = g(17) = 290$

5. $f(x) = x + 1$; $g(x) = 5x$

(a) $(f \circ g)(x) = f(g(x)) = f(5x) = 5x + 1$

(b) $(g \circ f)(x) = g(f(x))$

$\quad = g(x + 1)$

$\quad = 5(x + 1)$

$\quad = 5x + 5$

(c) $(f \circ f)(x) = f(f(x))$

$\quad = f(x + 1)$

$\quad = (x + 1) + 1$

$\quad = x + 2$

(d) $(g \circ g)(x) = g(g(x))$

$\quad = g(5x)$

$\quad = 5(5x)$

$\quad = 25x$

6. $f(x) = 2x - 3$; $g(x) = x + 6$

(a) $(f \circ g)(x) = f(g(x))$

$\quad = f(x + 6)$

$\quad = 2(x + 6) - 3$

$\quad = 2x + 12 - 3$

$\quad = 2x + 9$

(b) $(g \circ f)(x) = g(f(x))$

$\quad = g(2x - 3)$

$\quad = (2x - 3) + 6$

$\quad = 2x + 3$

(c) $(f \circ f)(x) = f(f(x))$

$\quad = f(2x - 3)$

$\quad = 2(2x - 3) - 3$

$\quad = 4x - 6 - 3$

$\quad = 4x - 9$

(d) $(g \circ g)(x) = g(g(x))$
$$= g(x+6)$$
$$= (x+6)+6$$
$$= x+12$$

7. $f(x) = x^2 + 1$; $g(x) = 2x + 1$

(a) $(f \circ g)(x) = f(g(x))$
$$= f(2x+1)$$
$$= (2x+1)^2 + 1$$
$$= 4x^2 + 4x + 1 + 1$$
$$= 4x^2 + 4x + 2$$

(b) $(g \circ f)(x) = g(f(x))$
$$= g(x^2 + 1)$$
$$= 2(x^2 + 1) + 1$$
$$= 2x^2 + 2 + 1$$
$$= 2x^2 + 3$$

(c) $(f \circ f)(x) = f(f(x))$
$$= f(x^2 + 1)$$
$$= (x^2 + 1)^2 + 1$$
$$= x^4 + 2x^2 + 1 + 1$$
$$= x^4 + 2x^2 + 2$$

(d) $(g \circ g)(x) = g(g(x))$
$$= g(2x+1)$$
$$= 2(2x+1) + 1$$
$$= 4x + 2 + 1$$
$$= 4x + 3$$

8. $f(x) = \dfrac{2}{x+1}$; $g(x) = \dfrac{1}{x}$

(a) $(f \circ g)(x) = f(g(x))$
$$= f\left(\frac{1}{x}\right)$$
$$= \frac{2}{\frac{1}{x}+1}$$
$$= \frac{2}{\frac{1+x}{x}}$$
$$= \frac{2x}{x+1} \quad \text{where } x \neq -1, 0.$$

(b) $(g \circ f)(x) = g(f(x))$
$$= g\left(\frac{2}{x+1}\right)$$
$$= \frac{1}{\frac{2}{x+1}}$$
$$= 1 \cdot \frac{x+1}{2}$$
$$= \frac{x+1}{2} \quad \text{where } x \neq -1.$$

(c) $(f \circ f)(x) = f(f(x))$
$$= f\left(\frac{2}{x+1}\right)$$
$$= \frac{2}{\frac{2}{x+1}+1}$$
$$= \frac{2}{\frac{2+x+1}{x+1}}$$
$$= 2 \cdot \frac{x+1}{x+3}$$
$$= \frac{2(x+1)}{x+3} \quad \text{where } x \neq -1, -3.$$

(d) $(g \circ g)(x) = g(g(x))$

$$= g\left(\frac{1}{x}\right)$$

$$= \frac{1}{\frac{1}{x}}$$

$$= 1 \cdot \frac{x}{1}$$

$$= x \text{ where } x \neq 0.$$

9. The function is not one-to-one. There is an element in the range (8) that corresponds to more than one element in the domain (-5 and -1).

10. The function is one-to-one. Each element in the range corresponds to exactly one element in the domain.

11. The graph passes the horizontal line test, so the graph is that of a one-to-one function.

12. The graph fails the horizontal line test. Therefore, the function is not one-to-one.

13. Inverse:

Height (inches)	Age
69 ⟶	24
71 ⟶	59
72 ⟶	29
73 ⟶	81
74 ⟶	37

14. Inverse:

Quantity Demanded	Price ($)
112 ⟶	300
129 ⟶	200
144 ⟶	170
161 ⟶	150
176 ⟶	130

15. To obtain the inverse, we switch the *x*- and *y*-coordinates.

Inverse: $\{(3,-5),(1,-3),(-3,1),(9,2)\}$

16. To obtain the inverse, we switch the *x*- and *y*-coordinates.

Inverse: $\{(1,-20),(4,-15),(3,5),(2,25)\}$

17. To plot the inverse, switch the *x*- and *y*-coordinates in each point and connect the corresponding points. The graphs of the function (shaded) and the line $y = x$ (dashed) are included for reference.

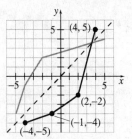

18. To plot the inverse, switch the *x*- and *y*-coordinates in each point and connect the corresponding points. The graphs of the function (shaded) and the line $y = x$ (dashed) are included for reference.

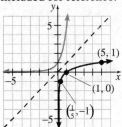

19. $f(x) = 5x$

$$y = 5x$$

$$x = 5y$$

$$\frac{x}{5} = y$$

$$f^{-1}(x) = \frac{x}{5}$$

Check: $f\left(f^{-1}(x)\right) = 5\left(\frac{x}{5}\right) = x$

$$f^{-1}(f(x)) = \frac{5x}{5} = x$$

20. $H(x) = 2x + 7$

$$y = 2x + 7$$

$$x = 2y + 7$$

$$x - 7 = 2y$$

$$\frac{x - 7}{2} = y$$

$$H^{-1}(x) = \frac{x - 7}{2}$$

Check:

$$H\left(H^{-1}(x)\right) = 2\left(\frac{x-7}{2}\right) + 7 = x - 7 + 7 = x$$

$$H^{-1}(H(x)) = \frac{(2x+7)-7}{2} = \frac{2x}{2} = x$$

21.
$$P(x) = \frac{4}{x+2}$$
$$y = \frac{4}{x+2}$$
$$x = \frac{4}{y+2}$$
$$x(y+2) = 4$$
$$y+2 = \frac{4}{x}$$
$$y = \frac{4}{x} - 2$$
$$P^{-1}(x) = \frac{4}{x} - 2$$

Check:
$$P\left(P^{-1}(x)\right) = \frac{4}{\left(\frac{4}{x} - 2\right) + 2} = \frac{4}{\frac{4}{x}} = 4 \cdot \frac{x}{4} = x$$

$$P^{-1}\left(P(x)\right) = \frac{4}{\frac{4}{x+2}} - 2$$
$$= 4 \cdot \frac{x+2}{4} - 2 = (x+2) - 2 = x$$

22.
$$g(x) = 2x^3 - 1$$
$$y = 2x^3 - 1$$
$$x = 2y^3 - 1$$
$$x+1 = 2y^3$$
$$\frac{x+1}{2} = y^3$$
$$\sqrt[3]{\frac{x+1}{2}} = y$$
$$g^{-1}(x) = \sqrt[3]{\frac{x+1}{2}}$$

Check:
$$g\left(g^{-1}(x)\right) = 2\left(\sqrt[3]{\frac{x+1}{2}}\right)^3 - 1$$
$$= 2\left(\frac{x+1}{2}\right) - 1 = (x+1) - 1 = x$$

$$g^{-1}\left(g(x)\right) = \sqrt[3]{\frac{\left(2x^3 - 1\right) + 1}{2}} = \sqrt[3]{\frac{2x^3}{2}} = \sqrt[3]{x^3} = x$$

23. (a) $7^{1.7} \approx 27.332$

(b) $7^{1.73} \approx 28.975$

(c) $7^{1.732} \approx 29.088$

(d) $7^{1.7321} \approx 29.093$

(e) $7^{\sqrt{3}} \approx 29.091$

24. (a) $10^{3.1} \approx 1258.925$

(b) $10^{3.14} \approx 1380.384$

(c) $10^{3.142} \approx 1386.756$

(d) $10^{3.1416} \approx 1385.479$

(e) $10^{\pi} \approx 1385.456$

25. (a) $e^{0.5} \approx 1.649$

(b) $e^{-1} \approx 0.368$

(c) $e^{1.5} \approx 4.482$

(d) $e^{-0.8} \approx 0.449$

(e) $e^{\sqrt{\pi}} \approx 5.885$

26. Locate some points on the graph of $f(x) = 9^x$.

x	$f(x) = 9^x$	$(x, f(x))$
-2	$f(-2) = 9^{-2} = \frac{1}{9^2} = \frac{1}{81}$	$\left(-2, \frac{1}{81}\right)$
-1	$f(-1) = 9^{-1} = \frac{1}{9^1} = \frac{1}{9}$	$\left(-1, \frac{1}{9}\right)$
0	$f(0) = 9^0 = 1$	$(0, 1)$
1	$f(1) = 9^1 = 9$	$(1, 9)$

Plot the points and connect them with a smooth curve.

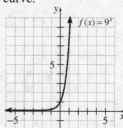

The domain of f is all real numbers or, using interval notation, $(-\infty, \infty)$. The range of f is $\{y \mid y > 0\}$ or , using interval notation, $(0, \infty)$.

27. Locate some points on the graph of $g(x) = \left(\dfrac{1}{9}\right)^x$.

x	$g(x) = \left(\dfrac{1}{9}\right)^x$	$(x,\ g(x))$
-1	$g(-1) = \left(\dfrac{1}{9}\right)^{-1} = 9^1 = 9$	$(-1,\ 9)$
0	$g(0) = \left(\dfrac{1}{9}\right)^0 = 1$	$(0,\ 1)$
1	$g(1) = \left(\dfrac{1}{9}\right)^1 = \dfrac{1}{9}$	$\left(1,\ \dfrac{1}{9}\right)$
2	$g(1) = \left(\dfrac{1}{9}\right)^2 = \dfrac{1}{81}$	$\left(2,\ \dfrac{1}{81}\right)$

Plot the points and connect them with a smooth curve.

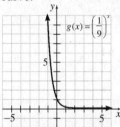

The domain of g is all real numbers or, using interval notation, $(-\infty, \infty)$. The range of g is $\{y \mid y > 0\}$ or , using interval notation, $(0, \infty)$.

28. Locate some points on the graph of $H(x) = 4^{x-2}$.

x	$H(x) = 4^{x-2}$	$(x,\ H(x))$
0	$H(0) = 4^{0-2} = 4^{-2} = \dfrac{1}{4^2} = \dfrac{1}{16}$	$\left(0,\ \dfrac{1}{16}\right)$
1	$H(1) = 4^{1-2} = 4^{-1} = \dfrac{1}{4^1} = \dfrac{1}{4}$	$\left(1,\ \dfrac{1}{4}\right)$
2	$H(2) = 4^{2-2} = 4^0 = 1$	$(2,\ 1)$
3	$H(3) = 4^{3-2} = 4^1 = 4$	$(3,\ 4)$

Plot the points and connect them with a smooth curve.

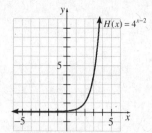

The domain of H is all real numbers or, using interval notation, $(-\infty, \infty)$. The range of H is $\{y \mid y > 0\}$ or , using interval notation, $(0, \infty)$.

29. Locate some points on the graph of $h(x) = 4^x - 2$.

x	$h(x) = 4^x - 2$	$(x,\ h(x))$
-2	$h(-2) = 4^{-2} - 2 = \dfrac{1}{16} - 2 = -\dfrac{31}{16}$	$\left(-2, -\dfrac{31}{16}\right)$
-1	$h(-1) = 4^{-1} - 2 = \dfrac{1}{4} - 2 = -\dfrac{7}{4}$	$\left(-1, -\dfrac{7}{4}\right)$
0	$h(0) = 4^0 - 2 = 1 - 2 = -1$	$(0, -1)$
1	$h(1) = 4^1 - 2 = 4 - 2 = 2$	$(1, 2)$

Plot the points and connect them with a smooth curve.

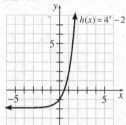

The domain of h is all real numbers or, using interval notation, $(-\infty, \infty)$. The range of h is $\{y \mid y > -2\}$ or , using interval notation, $(-2, \infty)$.

30. The number e is defined as the number that the expression $\left(1 + \dfrac{1}{n}\right)^n$ approaches as n becomes unbounded in the positive direction.

31. $2^x = 64$

$2^x = 2^6$

$x = 6$

The solution set is $\{6\}$.

32. $25^{x-2} = 125$

$\left(5^2\right)^{x-2} = 5^3$

$5^{2(x-2)} = 5^3$

$5^{2x-4} = 5^3$

$2x - 4 = 3$

$2x = 7$

$x = \dfrac{7}{2}$

The solution set is $\left\{\dfrac{7}{2}\right\}$.

33. $27^x \cdot 3^{x^2} = 9^2$

$\left(3^3\right)^x \cdot 3^{x^2} = \left(3^2\right)^2$

$3^{3x} \cdot 3^{x^2} = 3^4$

$3^{x^2+3x} = 3^4$

$x^2 + 3x = 4$

$x^2 + 3x - 4 = 0$

$(x+4)(x-1) = 0$

$x + 4 = 0 \quad \text{or} \quad x - 1 = 0$

$x = -4 \quad \text{or} \quad x = 1$

The solution set is $\{-4,\ 1\}$.

34. $\left(\dfrac{1}{4}\right)^x = 16$

$\left(4^{-1}\right)^x = 4^2$

$4^{-x} = 4^2$

$-x = 2$

$x = -2$

The solution set is $\{-2\}$.

35. $\left(e^2\right)^{x-1} = e^x \cdot e^7$

$e^{2(x-1)} = e^{x+7}$

$e^{2x-2} = e^{x+7}$

$2x - 2 = x + 7$

$x - 2 = 7$

$x = 9$

The solution set is $\{9\}$.

36. $\left(2^x\right)^x = 512$

$2^{x^2} = 2^9$

$x^2 = 9$

$x = \pm\sqrt{9}$

$x = \pm 3$

The solution set is $\{-3,\ 3\}$.

37. **(a)** We use the compound interest formula with $P = \$2500$, $r = 0.045$, $t = 25$, and $n = 1$, so that

$$A = 2500\left(1 + \frac{0.045}{1}\right)^{1(25)} \approx \$7513.59\,.$$

    (b) We use the compound interest formula with $P = \$2500$, $r = 0.045$, $t = 25$, and $n = 4$, so that

$$A = 2500\left(1 + \frac{0.045}{4}\right)^{4(25)} \approx \$7652.33\,.$$

    (c) We use the compound interest formula with $P = \$2500$, $r = 0.045$, $t = 25$, and $n = 12$, so that

$$A = 2500\left(1 + \frac{0.045}{12}\right)^{12(25)} \approx \$7684.36\,.$$

    (d) We use the compound interest formula with $P = \$2500$, $r = 0.045$, $t = 25$, and $n = 365$, so that

$$A = 2500\left(1 + \frac{0.045}{365}\right)^{365(25)} \approx \$7700.01\,.$$

38. **(a)** $A(1) = 100\left(\dfrac{1}{2}\right)^{1/3.5} \approx 82.034\,.$

    After 1 day, approximately 82.034 grams of radon gas will be left in the sample.

    (b) $A(3.5) = 100\left(\dfrac{1}{2}\right)^{3.5/3.5} = 50\,.$

    After 3.5 days, 50 grams of radon gas will be left in the sample.

    (c) $A(7) = 100\left(\dfrac{1}{2}\right)^{7/3.5} = 25\,.$

    After 7 days, 25 grams of radon gas will be left in the sample.

(d) $A(30) = 100\left(\frac{1}{2}\right)^{30/3.5} \approx 0.263$.

After 30 days, approximately 0.263 gram of radon gas will be left in the sample.

39. (a) $P(2010) = 2.496(1.052)^{2010-2006} \approx 3.057$

According to the model, the population of Nevada in 2010 will be approximately 3.057 million people.

(b) $P(2025) = 2.496(1.052)^{2025-2006} \approx 6.539$

According to the model, the population of Nevada in 2025 will be approximately 6.539 million people.

40. (a) $u(15) = 72 + 278e^{-0.0835(15)} \approx 151.449$.

According to the model, the temperature of the cake after 15 minutes will be approximately $151.449°F$.

(b) $u(30) = 72 + 278e^{-0.0835(30)} \approx 94.706$.

According to the model, the temperature of the cake after 30 minutes will be approximately $94.706°F$.

41. If $3^4 = 81$, then $\log_3 81 = 4$.

42. If $4^{-3} = \frac{1}{64}$, then $\log_4\left(\frac{1}{64}\right) = -3$.

43. If $b^3 = 5$, then $\log_b 5 = 3$.

44. If $10^{3.74} = x$, then $\log x = 3.74$.

45. If $\log_8 2 = \frac{1}{3}$, then $8^{1/3} = 2$.

46. If $\log_5 18 = r$, then $5^r = 18$.

47. If $\ln(x+3) = 2$, then $e^2 = x+3$.

48. If $\log x = -4$, then $10^{-4} = x$.

49. Let $y = \log_8 128$. Then,
$$8^y = 128$$
$$\left(2^3\right)^y = 2^7$$
$$2^{3y} = 2^7$$
$$3y = 7$$
$$y = \frac{7}{3}$$
Thus, $\log_8 128 = \frac{7}{3}$.

50. Let $y = \log_6 1$. Then,
$$6^y = 1$$
$$6^y = 6^0$$
$$y = 0$$
Thus, $\log_6 1 = 0$.

51. Let $y = \log\frac{1}{100}$. Then,
$$10^y = \frac{1}{100}$$
$$10^y = \frac{1}{10^2}$$
$$10^y = 10^{-2}$$
$$y = -2$$
Thus, $\log\frac{1}{100} = -2$.

52. Let $y = \log_9 27$. Then
$$9^y = 27$$
$$\left(3^2\right)^y = 3^3$$
$$3^{2y} = 3^3$$
$$2y = 3$$
$$y = \frac{3}{2}$$
Thus, $\log_9 27 = \frac{3}{2}$.

53. The domain of $f(x) = \log_2(x+5)$ is the set of all real numbers x such that
$$x+5 > 0$$
$$x > -5$$
Thus, the domain of $f(x) = \log_2(x+5)$ is $\{x \mid x > -5\}$, or using interval notation, $(-5, \infty)$.

54. The domain of $g(x) = \log_8(7-3x)$ is the set of all real numbers x such that
$$7 - 3x > 0$$
$$-3x > -7$$
$$x < \frac{-7}{-3}$$
$$x < \frac{7}{3}$$
Thus, the domain of $g(x) = \log_8(7-3x)$ is
$\left\{ x \,\middle|\, x < \frac{7}{3} \right\}$, or using interval notation, $\left(-\infty, \frac{7}{3} \right)$.

55. The domain of $h(x) = \ln(3x)$ is the set of all real numbers x such that
$$3x > 0$$
$$x > 0$$
Thus, the domain of $h(x) = \ln(3x)$ is $\{x \,|\, x > 0\}$, or using interval notation, $(0, \infty)$.

56. The domain of $F(x) = \log_{1/3}(4x+10)$ is the set of all real numbers x such that
$$4x + 10 > 0$$
$$4x > -10$$
$$x > \frac{-10}{4}$$
$$x > -\frac{5}{2}$$
Thus, the domain of $F(x) = \log_{1/3}(4x+10)$ is
$\left\{ x \,\middle|\, x > -\frac{5}{2} \right\}$, or using interval notation,
$\left(-\frac{5}{2}, \infty \right)$.

57. Rewrite $y = f(x) = \log_{5/2} x$ as $x = \left(\frac{5}{2}\right)^y$.

Locate some points on the graph of $x = \left(\frac{5}{2}\right)^y$.

y	$x = \left(\dfrac{5}{2}\right)^y$	$(x,\ y)$
-2	$x = \left(\dfrac{5}{2}\right)^{-2} = \left(\dfrac{2}{5}\right)^{2} = \dfrac{4}{25}$	$\left(\dfrac{4}{25}, -2\right)$
-1	$x = \left(\dfrac{5}{2}\right)^{-1} = \left(\dfrac{2}{5}\right)^{1} = \dfrac{2}{5}$	$\left(\dfrac{2}{5}, -1\right)$
0	$x = \left(\dfrac{5}{2}\right)^{0} = 1$	$(1,\ 0)$
1	$x = \left(\dfrac{5}{2}\right)^{1} = \dfrac{5}{2}$	$\left(\dfrac{5}{2}, 1\right)$
2	$x = \left(\dfrac{5}{2}\right)^{2} = \dfrac{25}{4}$	$\left(\dfrac{25}{4}, 2\right)$

Plot the points and connect them with a smooth curve.

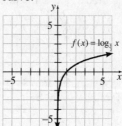

58. Rewrite $y = g(x) = \log_{2/5} x$ as $x = \left(\frac{2}{5}\right)^y$.

Locate some points on the graph of $x = \left(\frac{2}{5}\right)^y$.

y	$x = \left(\dfrac{2}{5}\right)^y$	$(x,\ y)$
-2	$x = \left(\dfrac{2}{5}\right)^{-2} = \left(\dfrac{5}{2}\right)^{2} = \dfrac{25}{4}$	$\left(\dfrac{25}{4}, -2\right)$
-1	$x = \left(\dfrac{2}{5}\right)^{-1} = \left(\dfrac{5}{2}\right)^{1} = \dfrac{5}{2}$	$\left(\dfrac{5}{2}, -1\right)$
0	$x = \left(\dfrac{2}{5}\right)^{0} = 1$	$(1,\ 0)$
1	$x = \left(\dfrac{2}{5}\right)^{1} = \dfrac{2}{5}$	$\left(\dfrac{2}{5}, 1\right)$
2	$x = \left(\dfrac{2}{5}\right)^{2} = \dfrac{4}{25}$	$\left(\dfrac{4}{25}, 2\right)$

Plot the points and connect them with a smooth curve.

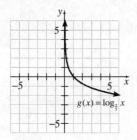

59. $\ln 24 \approx 3.178$

60. $\ln \dfrac{5}{6} \approx -0.182$

61. $\log 257 \approx 2.410$

62. $\log 0.124 \approx -0.907$

63. $\log_7(4x - 19) = 2$

$$4x - 19 = 7^2$$
$$4x - 19 = 49$$
$$4x = 68$$
$$x = 17$$

The solution set is $\{17\}$.

64. $\log_{1/3}\left(x^2 + 8x\right) = -2$

$$x^2 + 8x = \left(\frac{1}{3}\right)^{-2}$$
$$x^2 + 8x = 3^2$$
$$x^2 + 8x = 9$$
$$x^2 + 8x - 9 = 0$$
$$(x + 9)(x - 1) = 0$$
$$x + 9 = 0 \quad \text{or} \quad x - 1 = 0$$
$$x = -9 \quad \text{or} \quad x = 1$$

The solution set is $\{-9, 1\}$.

65. $\log_a \dfrac{4}{9} = -2$

$$a^{-2} = \frac{4}{9}$$
$$a^2 = \frac{9}{4}$$
$$a = \pm\sqrt{\frac{9}{4}} = \pm\frac{3}{2}$$

Since the base of a logarithm must always be positive, we know that $a = -\dfrac{3}{2}$ is extraneous.

The solution set is $\left\{\dfrac{3}{2}\right\}$.

66. $\ln e^{5x} = 30$

$$5x = 30$$
$$x = 6$$

The solution set is $\{6\}$.

67. $\log(6 - 7x) = 3$

$$6 - 7x = 10^3$$
$$6 - 7x = 1000$$
$$-7x = 994$$
$$x = -142$$

The solution set is $\{-142\}$.

68. $\log_b 75 = 2$

$$b^2 = 75$$
$$b = \pm\sqrt{75}$$
$$b = \pm 5\sqrt{3}$$

Since the base of a logarithm must always be positive, we know that $b = -5\sqrt{3}$ is extraneous.

The solution set is $\left\{5\sqrt{3}\right\}$.

69. We evaluate $L(x) = 10\log\dfrac{x}{10^{-12}}$ at $x = 10^{-4}$.

$$L\left(10^{-4}\right) = 10\log\frac{10^{-4}}{10^{-12}}$$
$$= 10\log 10^{-4-(-12)}$$
$$= 10\log 10^8$$
$$= 10(8)$$
$$= 80$$

The loudness of the vacuum cleaner is 80 decibels.

70. We solve $M(x) = \log\left(\dfrac{x}{10^{-3}}\right)$ for $M(x) = 8$.

$$M(x) = 8$$
$$\log\left(\frac{x}{10^{-3}}\right) = 8$$
$$\frac{x}{10^{-3}} = 10^8$$
$$x = 10^8 \cdot 10^{-3}$$
$$= 10^{8+(-3)}$$
$$= 10^5$$
$$= 100{,}000$$

The seismographic reading of the Great New Madrid Earthquake 100 kilometers from the epicenter would have been 100,000 millimeters.

71. $\log_4 4^{21} = 21$

72. $7^{\log_7 9.34} = 9.34$

73. $\log_5 5 = 1$

74. $\log_9 1 = 0$

75. $\log_4 12 - \log_4 3 = \log_4 \dfrac{12}{3} = \log_4 4 = 1$

76. $12^{\log_{12} 2 + \log_{12} 8} = 12^{\log_{12}(2 \cdot 8)} = 12^{\log_{12} 16} = 16$

77. $\log_7 \left(\dfrac{xy}{z} \right) = \log_7 (xy) - \log_7 z$
$\phantom{\log_7 \left(\dfrac{xy}{z} \right)} = \log_7 x + \log_7 y - \log_7 z$

78. $\log_3 \left(\dfrac{81}{x^2} \right) = \log_3 81 - \log_3 x^2$
$\phantom{\log_3 \left(\dfrac{81}{x^2} \right)} = \log_3 3^4 - \log_3 x^2$
$\phantom{\log_3 \left(\dfrac{81}{x^2} \right)} = 4 - 2\log_3 x$

79. $\log 1000 r^4 = \log 1000 + \log r^4$
$ = \log 10^3 + \log r^4$
$ = 3 + 4\log r$

80. $\ln \sqrt{\dfrac{x-1}{x}} = \ln \left(\dfrac{x-1}{x} \right)^{\frac{1}{2}}$
$\phantom{\ln \sqrt{\dfrac{x-1}{x}}} = \dfrac{1}{2} \ln \left(\dfrac{x-1}{x} \right)$
$\phantom{\ln \sqrt{\dfrac{x-1}{x}}} = \dfrac{1}{2} \left[\ln(x-1) - \ln x \right]$
$\phantom{\ln \sqrt{\dfrac{x-1}{x}}} = \dfrac{1}{2} \ln(x-1) - \dfrac{1}{2} \ln x$

81. $4\log_3 x + 2\log_3 y = \log_3 x^4 + \log_3 y^2$
$ = \log_3 \left(x^4 y^2 \right)$

82. $\dfrac{1}{4} \ln x + \ln 7 - 2\ln 3 = \ln x^{\frac{1}{4}} + \ln 7 - \ln 3^2$
$\phantom{\dfrac{1}{4} \ln x + \ln 7 - 2\ln 3} = \ln \sqrt[4]{x} + \ln 7 - \ln 9$
$\phantom{\dfrac{1}{4} \ln x + \ln 7 - 2\ln 3} = \ln \left(7\sqrt[4]{x} \right) - \ln 9$
$\phantom{\dfrac{1}{4} \ln x + \ln 7 - 2\ln 3} = \ln \left(\dfrac{7\sqrt[4]{x}}{9} \right)$

83. $\log_2 3 - \log_2 6 = \log_2 \dfrac{3}{6} = \log_2 \dfrac{1}{2} = \log_2 2^{-1} = -1$

84. $\log_6 \left(x^2 - 7x + 12 \right) - \log_6 (x-3)$
$= \log_6 \left[(x-4)(x-3) \right] - \log_6 (x-3)$
$= \log_6 (x-4) + \log_6 (x-3) - \log_6 (x-3)$
$= \log_6 (x-4)$

85. Using common logarithms:

$\log_6 50 = \dfrac{\log 50}{\log 6} \approx 2.183$

86. Using common logarithms:

$\log_\pi 2 = \dfrac{\log 2}{\log \pi} \approx 0.606$

87. Using natural logarithms:

$\log_{2/3} 6 = \dfrac{\ln 6}{\ln \left(\dfrac{2}{3} \right)} \approx -4.419$

88. Using natural logarithms:

$\log_{\sqrt{5}} 20 = \dfrac{\ln 20}{\ln \sqrt{5}} \approx 3.723$

89. $3\log_4 x = \log_4 1000$
$\log_4 x^3 = \log_4 1000$
$x^3 = 1000$
$x = \sqrt[3]{1000}$
$x = 10$
The solution set is $\{10\}$.

90. $\log_3 (x+7) + \log_3 (x+6) = \log_3 2$
$\log_3 \left[(x+7)(x+6) \right] = \log_3 2$
$x^2 + 13x + 42 = 2$
$x^2 + 13x + 40 = 0$
$(x+8)(x+5) = 0$
$x+8=0 \quad \text{or} \quad x+5=0$
$x=-8 \quad \text{or} \quad x=-5$

The apparent solution $x = -8$ is extraneous because it causes the arguments of both logarithms to be negative. The solution set is $\{-5\}$.

91. $\ln(x+2) - \ln x = \ln(x+1)$

$$\ln\left(\frac{x+2}{x}\right) = \ln(x+1)$$

$$\frac{x+2}{x} = x+1$$

$$x(x+1) = x+2$$

$$x^2 + x = x+2$$

$$x^2 = 2$$

$$x = \pm\sqrt{2} \approx \pm 1.414$$

The apparent solution $x = -\sqrt{2} \approx -1.414$ is extraneous because it causes the argument of a logarithm to be negative. The solution set is $\left\{\sqrt{2}\right\} \approx \{1.414\}$.

92. $\dfrac{1}{3}\log_{12} x = 2\log_{12} 2$

$$\log_{12} x^{\frac{1}{3}} = \log_{12} 2^2$$

$$\log_{12} \sqrt[3]{x} = \log_{12} 4$$

$$\sqrt[3]{x} = 4$$

$$x = 4^3 = 64$$

The solution set is $\{64\}$.

93. $2^x = 15$

$$\log 2^x = \log 15$$

$$x\log 2 = \log 15$$

$$x = \frac{\log 15}{\log 2} \approx 3.907$$

The solution set is $\left\{\dfrac{\log 15}{\log 2}\right\} \approx \{3.907\}$. If we had taken the natural logarithm of both sides, the solution set would be $\left\{\dfrac{\ln 15}{\ln 2}\right\} \approx \{3.907\}$.

94. $10^{3x} = 27$

$$\log 10^{3x} = \log 27$$

$$3x = \log 27$$

$$x = \frac{\log 27}{3} \approx 0.477$$

The solution set is $\left\{\dfrac{\log 27}{3}\right\} \approx \{0.477\}$.

95. $\dfrac{1}{3}e^{7x} = 13$

$$e^{7x} = 39$$

$$\ln e^{7x} = \ln 39$$

$$7x = \ln 39$$

$$x = \frac{\ln 39}{7} \approx 0.523$$

The solution set is $\left\{\dfrac{\ln 39}{7}\right\} \approx \{0.523\}$.

96.
$$3^x = 2^{x+1}$$

$$\log 3^x = \log 2^{x+1}$$

$$x\log 3 = (x+1)\log 2$$

$$x\log 3 = x\log 2 + \log 2$$

$$x\log 3 - x\log 2 = \log 2$$

$$x(\log 3 - \log 2) = \log 2$$

$$x = \frac{\log 2}{\log 3 - \log 2} \approx 1.710$$

The solution set is $\left\{\dfrac{\log 2}{\log 3 - \log 2}\right\} \approx \{1.710\}$. If we had taken the natural logarithm of both sides, the solution set would be

$$\left\{\frac{\ln 2}{\ln 3 - \ln 2}\right\} \approx \{1.710\}.$$

97. (a) We need to determine the time until $A = 75$ grams. So we solve the equation

$$75 = 100\left(\frac{1}{2}\right)^{t/3.5}$$

$$0.75 = \left(\frac{1}{2}\right)^{t/3.5}$$

$$\log 0.75 = \log\left(\frac{1}{2}\right)^{t/3.5}$$

$$\log 0.75 = \frac{t}{3.5}\log\left(\frac{1}{2}\right)$$

$$\frac{3.5\log 0.75}{\log\left(\frac{1}{2}\right)} = t$$

So, $t = \dfrac{3.5\log 0.75}{\log\left(\dfrac{1}{2}\right)} \approx 1.453$ days.

Thus, 75 grams of radon gas will be left after approximately 1.453 days.

(b) We need to determine the time until $A = 1$ gram. So we solve the equation

$$1 = 100\left(\frac{1}{2}\right)^{t/3.5}$$

$$0.01 = \left(\frac{1}{2}\right)^{t/3.5}$$

$$\log 0.01 = \log\left(\frac{1}{2}\right)^{t/3.5}$$

$$\log 0.01 = \frac{t}{3.5}\log\left(\frac{1}{2}\right)$$

$$\frac{3.5\log 0.01}{\log\left(\frac{1}{2}\right)} = t$$

So, $t = \dfrac{3.5\log 0.01}{\log\left(\frac{1}{2}\right)} \approx 23.253$ days.

Thus, 1 gram of radon gas will be left after approximately 23.253 days.

98. (a) We need to determine the year when $P = 3.939$ million. So we solve the equation

$$3.939 = 2.496(1.052)^{t-2006}$$

$$\frac{3.939}{2.496} = (1.052)^{t-2006}$$

$$\log\frac{3.939}{2.496} = \log(1.052)^{t-2006}$$

$$\log\frac{3.939}{2.496} = (t-2006)\log(1.052)$$

$$\frac{\log\dfrac{3.939}{2.496}}{\log 1.052} = t - 2006$$

$$\frac{\log\dfrac{3.939}{2.496}}{\log 1.052} + 2006 = t$$

$$2015.000 \approx t$$

Thus, according to the model, the population of Nevada will reach 3.939 million people in about the year 2015.

(b) We need to determine the year when $P = 8.426$ million. So we solve the equation

$$8.426 = 2.496(1.052)^{t-2006}$$

$$\frac{8.426}{2.496} = (1.052)^{t-2006}$$

$$\log\frac{8.426}{2.496} = \log(1.052)^{t-2006}$$

$$\log\frac{8.426}{2.496} = (t-2006)\log(1.052)$$

$$\frac{\log\dfrac{8.426}{2.496}}{\log 1.052} = t - 2006$$

$$\frac{\log\dfrac{8.426}{2.496}}{\log 1.052} + 2006 = t$$

$$2030.000 \approx t$$

Thus, according to the model, the population of Nevada will reach 8.426 million people in about the year 2030.

Chapter 11 Test

1. The function is not one-to-one. There is an element in the range (4) that corresponds to more than one element in the domain (1 and -1).

2. $\begin{aligned} f(x) &= 4x - 3 \\ y &= 4x - 3 \\ x &= 4y - 3 \\ x + 3 &= 4y \\ \frac{x+3}{4} &= y \\ f^{-1}(x) &= \frac{x+3}{4} \end{aligned}$

Check:

$$f\left(f^{-1}(x)\right) = 4\left(\frac{x+3}{4}\right) - 3 = x + 3 - 3 = x$$

$$f^{-1}\left(f(x)\right) = \frac{(4x-3)+3}{4} = \frac{4x}{4} = x$$

3. **(a)** $3.1^{3.1} \approx 33.360$

   (b) $3.14^{3.14} \approx 36.338$

   (c) $3.142^{3.142} \approx 36.494$

   (d) $3.1416^{3.1416} \approx 36.463$

   (e) $\pi^{\pi} \approx 36.462$

4. If $4^x = 19$, then $\log_4 19 = x$.

5. If $\log_b x = y$, then $b^y = x$.

6. (a) Let $y = \log_3\left(\dfrac{1}{27}\right)$. Then,

$$3^y = \frac{1}{27}$$
$$3^y = \frac{1}{3^3}$$
$$3^y = 3^{-3}$$
$$y = -3$$

Thus, $\log_3\left(\dfrac{1}{27}\right) = -3$.

(b) Let $y = \log 10{,}000$. Then,

$$10^y = 10{,}000$$
$$10^y = 10^4$$
$$y = 4$$

Thus, $\log 10{,}000 = 4$.

7. The domain of $f(x) = \log_5(7 - 4x)$ is the set of all real numbers x such that

$$7 - 4x > 0$$
$$-4x > -7$$
$$x < \frac{-7}{-4}$$
$$x < \frac{7}{4}$$

Thus, the domain of $f(x) = \log_5(7 - 4x)$ is $\left\{x \middle| x < \dfrac{7}{4}\right\}$ or, using interval notation, $\left(-\infty, \dfrac{7}{4}\right)$.

8. Locate some points on the graph of $f(x) = 6^x$.

x	$f(x) = 6^x$	$(x, f(x))$
-2	$f(-2) = 6^{-2} = \dfrac{1}{6^2} = \dfrac{1}{36}$	$\left(-2, \dfrac{1}{36}\right)$
-1	$f(-1) = 6^{-1} = \dfrac{1}{6^1} = \dfrac{1}{6}$	$\left(-1, \dfrac{1}{6}\right)$
0	$f(0) = 6^0 = 1$	$(0, 1)$
1	$f(1) = 6^1 = 6$	$(1, 6)$

Plot the points and connect them with a smooth curve.

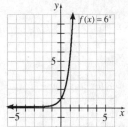

The domain of f is all real numbers or, using interval notation, $(-\infty, \infty)$. The range of f is $\{y \mid y > 0\}$ or, using interval notation, $(0, \infty)$.

9. Rewrite $y = g(x) = \log_{1/9} x$ as $x = \left(\dfrac{1}{9}\right)^y$.

Locate some points on the graph of $x = \left(\dfrac{1}{9}\right)^y$.

y	$x = \left(\dfrac{1}{9}\right)^y$	(x, y)
-1	$x = \left(\dfrac{1}{9}\right)^{-1} = \left(\dfrac{9}{1}\right)^1 = 9$	$(9, -1)$
0	$x = \left(\dfrac{1}{9}\right)^0 = 1$	$(1, 0)$
1	$x = \left(\dfrac{1}{9}\right)^1 = \dfrac{1}{9}$	$\left(\dfrac{1}{9}, 1\right)$
2	$x = \left(\dfrac{1}{9}\right)^2 = \dfrac{1}{81}$	$\left(\dfrac{1}{81}, 2\right)$

Plot the points and connect them with a smooth curve.

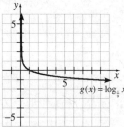

The domain of g is $\{x \mid x > 0\}$ or, using interval notation, $(0, \infty)$. The range of g is all real numbers or, using interval notation, $(-\infty, \infty)$.

10. (a) $\log_7 7^{10} = 10$

(b) $3^{\log_3 15} = 15$

11. $\log_4 \dfrac{\sqrt{x}}{y^3} = \log_4 \sqrt{x} - \log_4 y^3$

$$= \log_4 x^{\frac{1}{2}} - \log_4 y^3$$

$$= \frac{1}{2} \log_4 x - 3 \log_4 y$$

12. $4 \log M + 3 \log N = \log M^4 + \log N^3$

$$= \log \left(M^4 N^3 \right)$$

13. Using common logarithms:

$$\log_{3/4} 10 = \frac{\log 10}{\log \dfrac{3}{4}} \approx -8.004$$

14. $\quad 4^{x+1} = 2^{3x+1}$

$$\left(2^2 \right)^{x+1} = 2^{3x+1}$$

$$2^{2x+2} = 2^{3x+1}$$

$$2x + 2 = 3x + 1$$

$$-x + 2 = 1$$

$$-x = -1$$

$$x = 1$$

The solution set is $\{1\}$.

15. $\quad 5^{x^2} \cdot 125 = 25^{2x}$

$$5^{x^2} \cdot 5^3 = \left(5^2 \right)^{2x}$$

$$5^{x^2+3} = 5^{4x}$$

$$x^2 + 3 = 4x$$

$$x^2 - 4x + 3 = 0$$

$$(x-1)(x-3) = 0$$

$$x - 1 = 0 \quad \text{or} \quad x - 3 = 0$$

$$x = 1 \quad \text{or} \quad x = 3$$

The solution set is $\{1,\ 3\}$.

16. $\log_a 64 = 3$

$$a^3 = 64$$

$$a = \sqrt[3]{64} = 4$$

The solution set is $\{4\}$.

17. $\log_2 \left(x^2 - 33 \right) = 8$

$$2^8 = x^2 - 33$$

$$256 = x^2 - 33$$

$$289 = x^2$$

$$x = \pm\sqrt{289} = \pm 17$$

The solution set is $\{-17,\ 17\}$.

18. $2 \log_7 (x-3) = \log_7 3 + \log_7 12$

$$\log_7 (x-3)^2 = \log_7 (3 \cdot 12)$$

$$\log_7 (x-3)^2 = \log_7 36$$

$$(x-3)^2 = 36$$

$$x - 3 = \pm 6$$

$$x = 3 \pm 6$$

$$x = 9 \quad \text{or} \quad x = -3$$

The apparent solution $x = -3$ is extraneous because it causes the argument of a logarithm to be negative. The solution set is $\{9\}$.

19. $\quad\quad 3^{x-1} = 17$

$$\log 3^{x-1} = \log 17$$

$$(x-1)\log 3 = \log 17$$

$$x \log 3 - \log 3 = \log 17$$

$$x \log 3 = \log 17 + \log 3$$

$$x = \frac{\log 17 + \log 3}{\log 3} \approx 3.579$$

The solution set is $\left\{ \dfrac{\log 17 + \log 3}{\log 3} \right\} \approx \{3.579\}$. If we had taken the natural logarithm of both sides, the solution set would be

$$\left\{ \frac{\ln 17 + \ln 3}{\ln 3} \right\} \approx \{3.579\}.$$

20. $\log (x-2) + \log (x+2) = 2$

$$\log \left[(x-2)(x+2) \right] = 2$$

$$\log \left(x^2 - 4 \right) = 2$$

$$10^2 = x^2 - 4$$

$$x^2 - 4 = 100$$

$$x^2 = 104$$

$$x = \pm\sqrt{104}$$

$$x = \pm 2\sqrt{26} \approx \pm 10.198$$

The apparent solution $x = -2\sqrt{26} \approx -10.198$ is extraneous because it causes the argument of a logarithm to be negative. The solution set is $\left\{ 2\sqrt{26} \right\} \approx \{10.198\}.$

21. **(a)** Evaluate $P(t) = 31.9(1.008)^{t-2002}$ at

$t = 2010$.

$P(2010) = 31.9(1.008)^{2010-2002} \approx 34.000$.

According to the model, the population of Canada in 2010 will be about 34 million people.

(b) We need to determine the year when $P = 50$ million people. So we solve the equation

$$50 = 31.9(1.008)^{t-2002}$$

$$\frac{50}{31.9} = (1.008)^{t-2002}$$

$$\log\frac{50}{31.9} = \log(1.008)^{t-2002}$$

$$\log\frac{50}{31.9} = (t-2002)\log 1.008$$

$$\frac{\log\frac{50}{31.9}}{\log 1.008} = t - 2002$$

$$\frac{\log\frac{50}{31.9}}{\log 1.008} + 2002 = t$$

$$2058.402 \approx t$$

According to the model, the population of Canada will be 50 million people in about 2058.

22. We evaluate $L(x) = 10\log\dfrac{x}{10^{-12}}$ at $x = 10^{-11}$.

$$L(10^{-11}) = 10\log\frac{10^{-11}}{10^{-12}}$$

$$= 10\log 10^{-11-(-12)}$$

$$= 10\log 10^{1}$$

$$= 10(1)$$

$$= 10$$

The loudness of rustling leaves is 10 decibels.

Cumulative Review Chapters 1–11

1. $3(5-2x)+8 = 4(x-7)+1$

$15-6x+8 = 4x-28+1$

$-6x+23 = 4x-27$

$-10x = -50$

$x = 5$

The solution set is {5}.

2. $5-3|x-2| \geq -7$

$-3|x-2| \geq -12$

$|x-2| \leq 4$

$-4 \leq x-2 \leq 4$

$-2 \leq x \leq 6$

The solution set is $\{x \mid -2 \leq x \leq 6\}$ or, using interval notation, $[-2, 6]$.

3. We need to find all values of x that cause the denominator $2x^2 - x - 21$ to equal 0.

$$2x^2 - x - 21 = 0$$

$$(2x-7)(x+3) = 0$$

$$2x-7 = 0 \quad \text{or} \quad x+3 = 0$$

$$x = \frac{7}{2} \qquad\qquad x = -3$$

Thus, the domain of $f(x) = \dfrac{9-x^2}{2x^2-x-21}$ is

$$\left\{ x \,\middle|\, x \neq -3 \text{ and } x \neq \frac{7}{2} \right\}.$$

4. $4x+3y = 6$

$3y = -4x+6$

$y = \dfrac{-4x+6}{3}$

$y = -\dfrac{4}{3}x+2$

The slope is $-\dfrac{4}{3}$ and the y-intercept is 2. Begin at the point $(0, 2)$ and move to the right 3 units and down 4 units to find the point $(3, -2)$.

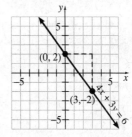

5. The slope of the line we seek is
$$m = \frac{y_2 - y_1}{x_2 - x_1} = \frac{-4 - 17}{5 - (-10)} = \frac{-21}{15} = -\frac{7}{5}.$$
Thus, the equation of the line we seek is:
$$y - y_1 = m(x - x_1)$$
$$y - (-4) = -\frac{7}{5}(x - 5)$$
$$y + 4 = -\frac{7}{5}x + 7$$
$$y = -\frac{7}{5}x + 3 \text{ or } 7x + 5y = 15$$

6. $\begin{cases} x + 2y \geq 8 \\ 2x - y < 1 \end{cases}$

First, graph the inequality $x + 2y \geq 8$. To do so, replace the inequality symbol with an equal sign to obtain $x + 2y = 8$. Because the inequality is not strict, graph $x + 2y = 8$ $\left(y = -\frac{1}{2}x + 4 \right)$ using a solid line.

Test Point: $(0,0)$: $(0) + 2(0) \not\geq 8$

Therefore, the half-plane not containing $(0,0)$ is the solution set of $x + 2y \geq 8$.

Second, graph the inequality $2x - y < 1$. To do so, replace the inequality symbol with an equal sign to obtain $2x - y = 1$. Because the inequality is strict, graph $2x - y = 1$ $(y = 2x - 1)$ using a dashed line.

Test Point: $(0,0)$: $2(0) - (0) < 1$

Therefore, the half-plane containing $(0,0)$ is the solution set of $2x - y < 1$.

The overlapping shaded region (that is, the shaded region in the graph below) is the solution to the system of linear inequalities.

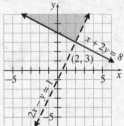

7. $\left(m^2 - 5m + 13 \right) - \left(6 - 2m - 3m^2 \right)$
$$= m^2 - 5m + 13 - 6 + 2m + 3m^2$$
$$= 4m^2 - 3m + 7$$

8. $(2n + 3)\left(n^2 - 4n + 6 \right)$
$$= 2n\left(n^2 \right) + 2n(-4n) + 2n(6) + 3\left(n^2 \right)$$
$$+ 3(-4n) + 3(6)$$
$$= 2n^3 - 8n^2 + 12n + 3n^2 - 12n + 18$$
$$= 2n^3 - 5n^2 + 18$$

9. $16a^2 + 8ab + b^2 = (4a)^2 + 2(4a)(b) + b^2$
$$= (4a + b)^2$$

10. $a \cdot c = 6(7) = 42$.

The factors of 42 that add to -17 (the linear coefficient) are -3 and -14. Thus,
$$6y^2 - 17y + 7 = 6y^2 - 3y - 14y + 7$$
$$= 3y(2y - 1) - 7(2y - 1)$$
$$= (3y - 7)(2y - 1)$$

11. $\dfrac{2x^2 - 9x - 5}{x^2 - 3x - 10} \cdot \dfrac{3x^2 + 2x - 8}{2x^2 - 13x - 7}$
$$= \frac{(2x + 1)(x - 5)}{(x + 2)(x - 5)} \cdot \frac{(3x - 4)(x + 2)}{(2x + 1)(x - 7)}$$
$$= \frac{(2x+1)(x-5)}{(x+2)(x-5)} \cdot \frac{(3x - 4)(x+2)}{(2x+1)(x - 7)}$$
$$= \frac{3x - 4}{x - 7}$$

12. $\dfrac{4}{p^2 - 6p + 5} + \dfrac{2}{p^2 - 3p - 10}$
$$= \frac{4}{(p - 5)(p - 1)} + \frac{2}{(p - 5)(p + 2)}$$
$$= \frac{4(p + 2)}{(p - 5)(p - 1)(p + 2)} + \frac{2(p - 1)}{(p - 5)(p - 1)(p + 2)}$$
$$= \frac{4(p + 2) + 2(p - 1)}{(p - 5)(p - 1)(p + 2)} = \frac{4p + 8 + 2p - 2}{(p - 5)(p - 1)(p + 2)}$$
$$= \frac{6p + 6}{(p - 5)(p - 1)(p + 2)} = \frac{6(p + 1)}{(p - 5)(p - 1)(p + 2)}$$

13. $\dfrac{2}{x-5} = \dfrac{x-2}{x+1} + \dfrac{6x-12}{x^2-4x-5}$

$\dfrac{2}{x-5} = \dfrac{x-2}{x+1} + \dfrac{6x-12}{(x-5)(x+1)}$

$(x-5)(x+1)\left(\dfrac{2}{x-5}\right)$

$\qquad = (x-5)(x+1)\left(\dfrac{x-2}{x+1} + \dfrac{6x-12}{(x-5)(x+1)}\right)$

$2(x+1) = (x-5)(x-2) + (6x-12)$

$2x+2 = x^2 - 2x - 5x + 10 + 6x - 12$

$2x+2 = x^2 - x - 2$

$0 = x^2 - 3x - 4$

$0 = (x+1)(x-4)$

$x+1 = 0 \quad \text{or} \quad x-4 = 0$

$x = -1 \quad \text{or} \qquad x = 4$

Since $x = -1$ is not in the domain of the variable, it is an extraneous solution. Thus, the solution set is {4}.

14. $\sqrt{150} + 4\sqrt{6} - \sqrt{24} = \sqrt{25\cdot 6} + 4\sqrt{6} - \sqrt{4\cdot 6}$

$\qquad\qquad\qquad\qquad = 5\sqrt{6} + 4\sqrt{6} - 2\sqrt{6}$

$\qquad\qquad\qquad\qquad = 7\sqrt{6}$

15. $\dfrac{1+\sqrt{5}}{3-\sqrt{5}} = \dfrac{1+\sqrt{5}}{3-\sqrt{5}} \cdot \dfrac{3+\sqrt{5}}{3+\sqrt{5}}$

$\qquad\qquad = \dfrac{3+\sqrt{5}+3\sqrt{5}+5}{9+3\sqrt{5}-3\sqrt{5}-5} = \dfrac{8+4\sqrt{5}}{4} = 2+\sqrt{5}$

16. $\sqrt{x-8} + \sqrt{x} = 4$

$\sqrt{x-8} = 4 - \sqrt{x}$

$\left(\sqrt{x-8}\right)^2 = \left(4-\sqrt{x}\right)^2$

$x-8 = 16 - 8\sqrt{x} + x$

$8\sqrt{x} = 24$

$\sqrt{x} = 3$

$\left(\sqrt{x}\right)^2 = (3)^2$

$x = 9$

Check:

$\sqrt{9-8} + \sqrt{9} \overset{?}{=} 4$

$\sqrt{1} + \sqrt{9} \overset{?}{=} 4$

$1 + 3 \overset{?}{=} 4$

$4 = 4 \checkmark$

The solution set is {9}.

17. $\qquad 3x^2 = 4x + 6$

$3x^2 - 4x - 6 = 0$

Because this equation does not easily factor, solve by using the quadratic formula. For this equation, $a = 3$, $b = -4$, and $c = -6$.

$x = \dfrac{-(-4) \pm \sqrt{(-4)^2 - 4(3)(-6)}}{2(3)}$

$\quad = \dfrac{4 \pm \sqrt{88}}{6} = \dfrac{4 \pm 2\sqrt{22}}{6} = \dfrac{2 \pm \sqrt{22}}{3}$

The solution set is $\left\{\dfrac{2-\sqrt{22}}{3},\ \dfrac{2+\sqrt{22}}{3}\right\}$.

18. $\qquad 2a - 7\sqrt{a} + 6 = 0$

$2\left(\sqrt{a}\right)^2 - 7\left(\sqrt{a}\right) + 6 = 0$

Let $u = \sqrt{a}$.

$2u^2 - 7u + 6 = 0$

$(2u-3)(u-2) = 0$

$2u - 3 = 0 \quad \text{or} \quad u - 2 = 0$

$u = \dfrac{3}{2} \quad \text{or} \qquad u = 2$

$\sqrt{a} = \dfrac{3}{2} \quad \text{or} \quad \sqrt{a} = 2$

$a = \left(\dfrac{3}{2}\right)^2 \quad \text{or} \qquad a = 2^2$

$a = \dfrac{9}{4} \qquad \text{or} \qquad a = 4$

Check:

$a = \dfrac{9}{4}:\ 2\cdot\dfrac{9}{4} - 7\sqrt{\dfrac{9}{4}} + 6 \overset{?}{=} 0$

$\qquad\qquad \dfrac{9}{2} - 7\cdot\dfrac{3}{2} + 6 \overset{?}{=} 0$

$\qquad\qquad \dfrac{9}{2} - \dfrac{21}{2} + 6 \overset{?}{=} 0$

$\qquad\qquad\qquad\qquad 0 = 0 \checkmark$

$a = 4:\ 2\cdot 4 - 7\sqrt{4} + 6 \overset{?}{=} 0$

$\qquad\qquad 8 - 7\cdot 2 + 6 \overset{?}{=} 0$

$\qquad\qquad 8 - 14 + 6 \overset{?}{=} 0$

$\qquad\qquad\qquad 0 = 0 \checkmark$

Both check; the solution set is $\left\{\dfrac{9}{4},\ 4\right\}$.

19. $f(x) = -x^2 + 6x - 4$

$a = -1, b = 6, c = -4$

The graph opens down because the coefficient on x^2 is negative.

vertex:

$x = -\dfrac{b}{2a} = -\dfrac{6}{2(-1)} = 3$

$y = f(3) = -(3)^2 + 6(3) - 4 = 5$

The vertex is $(3, 5)$. The axis of symmetry is $x = 3$.

y-intercept:

$f(0) = -(0)^2 + 6(0) - 4 = -4$

x-intercepts:

$b^2 - 4ac = 6^2 - 4(-1)(-4) = 20$

There are two distinct x-intercepts. We find these by solving

$$f(x) = 0$$

$-x^2 + 6x - 4 = 0$

$x = \dfrac{-6 \pm \sqrt{20}}{2(-1)} = \dfrac{-6 \pm 2\sqrt{5}}{-2} = 3 \pm \sqrt{5}$

$x \approx 0.76$ or $x \approx 5.24$

Graph:

The y-intercept point, $(0, -4)$, is three units to the left of the axis of symmetry. Therefore, if we move three units to the right of the axis of symmetry, we obtain the point $(6, -4)$ which must also be on the graph.

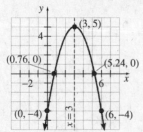

20. $3x^2 + 2x - 8 < 0$

Solve: $3x^2 + 2x - 8 = 0$

$(3x - 4)(x + 2) = 0$

$3x - 4 = 0$ or $x + 2 = 0$

$x = \dfrac{4}{3}$ or $x = -2$

Determine where each factor is positive and negative and where the product of these factors is positive and negative.

Interval	$(-\infty, -2)$	-2	$\left(-2, \frac{4}{3}\right)$	$\frac{4}{3}$	$\left(\frac{4}{3}, \infty\right)$
$3x - 4$	$----$	$-$	$----$	0	$++++$
$x + 2$	$----$	0	$++++$	$+$	$++++$
$(3x-4)(x+2)$	$++++$	0	$----$	0	$++++$

The inequality is strict, so -2 and $\dfrac{4}{3}$ are not part of the solution. Now, $(3x - 4)(x + 2)$ is less than zero where the product is negative. The solution is $\left\{ x \middle| \; -2 < x < \dfrac{4}{3} \right\}$ or, using interval notation, $\left(-2, \dfrac{4}{3}\right)$.

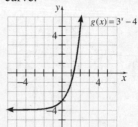

21. Locate some points on the graph of $g(x) = 3^x - 4$.

x	$g(x) = 3^x - 4$	$(x, \; g(x))$
-2	$g(-2) = 3^{-2} - 4 = \dfrac{1}{9} - 4 = -\dfrac{35}{9}$	$\left(-2, \; -\dfrac{35}{9}\right)$
-1	$g(-1) = 3^{-1} - 4 = \dfrac{1}{3} - 4 = -\dfrac{11}{3}$	$\left(-1, \; -\dfrac{11}{3}\right)$
0	$g(0) = 3^0 - 4 = 1 - 4 = -3$	$(0, \; -3)$
1	$g(1) = 3^1 - 4 = 3 - 4 = -1$	$(1, \; -1)$
2	$g(2) = 3^2 - 4 = 9 - 4 = 5$	$(2, \; 5)$

Plot the points and connect them with a smooth curve.

22. Let $y = \log_9\left(\dfrac{1}{27}\right)$. Then,

$$9^y = \frac{1}{27}$$
$$\left(3^2\right)^y = \frac{1}{3^3}$$
$$3^{2y} = 3^{-3}$$
$$2y = -3$$
$$y = -\frac{3}{2}$$

Thus, $\log_9\left(\dfrac{1}{27}\right) = -\dfrac{3}{2}$.

Chapter 12

Section 12.1

Preparing for Distance and Midpoint Formulas

P1. (a) $\sqrt{64} = 8$

(b) $\sqrt{24} = \sqrt{4 \cdot 6} = 2\sqrt{6}$

(c) $\sqrt{(x-2)^2} = |x-2|$

P2. $c^2 = a^2 + b^2$

$c^2 = 6^2 + 8^2$

$c^2 = 36 + 64$

$c^2 = 100$

$c = \sqrt{100} = 10$

The length of the hypotenuse is 10.

Section 12.1 Quick Checks

1. $d = \sqrt{(x_2 - x_1)^2 + (y_2 - y_1)^2}$

2. False. Distance must be positive.

3. Using the distance formula with $P_1 = (3,\ 8)$ and $P_2 = (0,\ 4)$, the length is

$d(P_1, P_2) = \sqrt{(x_2 - x_1)^2 + (y_2 - y_1)^2}$

$= \sqrt{(0-3)^2 + (4-8)^2}$

$= \sqrt{(-3)^2 + (-4)^2}$

$= \sqrt{9 + 16}$

$= \sqrt{25}$

$= 5$

4. Using the distance formula with $P_1 = (-2, -5)$ and $P_2 = (4,\ 7)$, the length is

$d(P_1, P_2) = \sqrt{(x_2 - x_1)^2 + (y_2 - y_1)^2}$

$= \sqrt{(4-(-2))^2 + (7-(-5))^2}$

$= \sqrt{6^2 + 12^2}$

$= \sqrt{36 + 144}$

$= \sqrt{180}$

$= 6\sqrt{5} \approx 13.42$

5. (a)

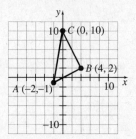

(b) $d(A,B) = \sqrt{(4-(-2))^2 + (2-(-1))^2}$

$= \sqrt{6^2 + 3^2}$

$= \sqrt{36 + 9}$

$= \sqrt{45}$

$= 3\sqrt{5}$

$d(B,C) = \sqrt{(0-4)^2 + (10-2)^2}$

$= \sqrt{(-4)^2 + 8^2}$

$= \sqrt{16 + 64}$

$= \sqrt{80}$

$= 4\sqrt{5}$

$d(A,C) = \sqrt{(0-(-2))^2 + (10-(-1))^2}$

$= \sqrt{2^2 + 11^2}$

$= \sqrt{4 + 121}$

$= \sqrt{125}$

$= 5\sqrt{5}$

(c) To determine if triangle ABC is a right triangle, we check to see if

$[d(A,B)]^2 + [d(B,C)]^2 \overset{?}{=} [d(A,C)]^2$

$\left(3\sqrt{5}\right)^2 + \left(4\sqrt{5}\right)^2 \overset{?}{=} \left(5\sqrt{5}\right)^2$

$9 \cdot 5 + 16 \cdot 5 \overset{?}{=} 25 \cdot 5$

$45 + 80 \overset{?}{=} 125$

$125 = 125 \leftarrow$ True

Therefore, triangle ABC is a right triangle.

(d) The length of the "base" of the triangle is $d(A,B) = 3\sqrt{5}$ and the length of the "height" of the triangle is $d(B,C) = 4\sqrt{5}$. Thus, the area of triangle ABC is

$\text{Area} = \dfrac{1}{2} \cdot \text{base} \cdot \text{height} = \dfrac{1}{2} \cdot 3\sqrt{5} \cdot 4\sqrt{5} = 30$ square units.

6. $M = \left(\dfrac{x_1 + x_2}{2}, \dfrac{y_1 + y_2}{2} \right)$

7. $M = \left(\dfrac{x_1 + x_2}{2}, \dfrac{y_1 + y_2}{2} \right)$

$= \left(\dfrac{3 + 0}{2}, \dfrac{8 + 4}{2} \right)$

$= \left(\dfrac{3}{2}, \dfrac{12}{2} \right)$

$= \left(\dfrac{3}{2}, 6 \right)$

8. $M = \left(\dfrac{x_1 + x_2}{2}, \dfrac{y_1 + y_2}{2} \right)$

$= \left(\dfrac{-2 + 4}{2}, \dfrac{-5 + 10}{2} \right)$

$= \left(\dfrac{2}{2}, \dfrac{5}{2} \right)$

$= \left(1, \dfrac{5}{2} \right)$

12.1 Exercises

9. $d(P_1, P_2) = \sqrt{(x_2 - x_1)^2 + (y_2 - y_1)^2}$

$= \sqrt{(3 - 0)^2 + (4 - 0)^2}$

$= \sqrt{3^2 + 4^2}$

$= \sqrt{9 + 16}$

$= \sqrt{25}$

$= 5$

11. $d(P_1, P_2) = \sqrt{(x_2 - x_1)^2 + (y_2 - y_1)^2}$

$= \sqrt{(4 - (-4))^2 + (1 - 5)^2}$

$= \sqrt{8^2 + (-4)^2}$

$= \sqrt{64 + 16}$

$= \sqrt{80}$

$= 4\sqrt{5} \approx 8.94$

13. $d(P_1, P_2) = \sqrt{(x_2 - x_1)^2 + (y_2 - y_1)^2}$

$= \sqrt{(6 - 2)^2 + (4 - 1)^2}$

$= \sqrt{4^2 + 3^2}$

$= \sqrt{16 + 9}$

$= \sqrt{25}$

$= 5$

15. $d(P_1, P_2) = \sqrt{(x_2 - x_1)^2 + (y_2 - y_1)^2}$

$= \sqrt{(9 - (-3))^2 + (-3 - 2)^2}$

$= \sqrt{12^2 + (-5)^2}$

$= \sqrt{144 + 25}$

$= \sqrt{169}$

$= 13$

17. $d(P_1, P_2) = \sqrt{(x_2 - x_1)^2 + (y_2 - y_1)^2}$

$= \sqrt{(2 - (-4))^2 + (2 - 2)^2}$

$= \sqrt{6^2 + 0^2}$

$= \sqrt{36 + 0}$

$= \sqrt{36}$

$= 6$

19. $d(P_1, P_2) = \sqrt{(x_2 - x_1)^2 + (y_2 - y_1)^2}$

$= \sqrt{(-3 - 0)^2 + (3 - (-3))^2}$

$= \sqrt{(-3)^2 + 6^2}$

$= \sqrt{9 + 36}$

$= \sqrt{45}$

$= 3\sqrt{5} \approx 6.71$

21. $d(P_1, P_2) = \sqrt{(x_2 - x_1)^2 + (y_2 - y_1)^2}$

$= \sqrt{(5\sqrt{2} - 2\sqrt{2})^2 + (4\sqrt{5} - \sqrt{5})^2}$

$= \sqrt{(3\sqrt{2})^2 + (3\sqrt{5})^2}$

$= \sqrt{9(2) + 9(5)}$

$= \sqrt{18 + 45}$

$= \sqrt{63}$

$= 3\sqrt{7} \approx 7.94$

23. $d(P_1, P_2) = \sqrt{(x_2 - x_1)^2 + (y_2 - y_1)^2}$

$= \sqrt{(1.3 - 0.3)^2 + (0.1 - (-3.3))^2}$

$= \sqrt{1^2 + 3.4^2}$

$= \sqrt{1 + 11.56}$

$= \sqrt{12.56} \approx 3.54$

25. $M = \left(\dfrac{x_1 + x_2}{2}, \dfrac{y_1 + y_2}{2} \right)$

$= \left(\dfrac{2 + 6}{2}, \dfrac{2 + 4}{2} \right) = \left(\dfrac{8}{2}, \dfrac{6}{2} \right) = (4, 3)$

27. $M = \left(\dfrac{x_1 + x_2}{2}, \dfrac{y_1 + y_2}{2} \right)$

$= \left(\dfrac{-3 + 9}{2}, \dfrac{2 + (-4)}{2} \right) = \left(\dfrac{6}{2}, \dfrac{-2}{2} \right) = (3, -1)$

29. $M = \left(\dfrac{x_1 + x_2}{2}, \dfrac{y_1 + y_2}{2} \right)$

$= \left(\dfrac{-4 + 2}{2}, \dfrac{3 + 4}{2} \right) = \left(\dfrac{-2}{2}, \dfrac{7}{2} \right) = \left(-1, \dfrac{7}{2} \right)$

31. $M = \left(\dfrac{x_1 + x_2}{2}, \dfrac{y_1 + y_2}{2} \right)$

$= \left(\dfrac{0 + (-3)}{2}, \dfrac{-3 + 3}{2} \right) = \left(\dfrac{-3}{2}, \dfrac{0}{2} \right) = \left(-\dfrac{3}{2}, 0 \right)$

33. $M = \left(\dfrac{x_1 + x_2}{2}, \dfrac{y_1 + y_2}{2} \right)$

$= \left(\dfrac{2\sqrt{2} + 5\sqrt{2}}{2}, \dfrac{\sqrt{5} + 4\sqrt{5}}{2} \right) = \left(\dfrac{7\sqrt{2}}{2}, \dfrac{5\sqrt{5}}{2} \right)$

35. $M = \left(\dfrac{x_1 + x_2}{2}, \dfrac{y_1 + y_2}{2} \right)$

$= \left(\dfrac{0.3 + 1.3}{2}, \dfrac{-3.3 + 0.1}{2} \right)$

$= \left(\dfrac{1.6}{2}, \dfrac{-3.2}{2} \right)$

$= (0.8, -1.6)$

37. (a)

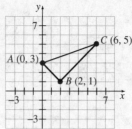

(b) $d(A, B) = \sqrt{(2 - 0)^2 + (1 - 3)^2}$

$= \sqrt{2^2 + (-2)^2}$

$= \sqrt{4 + 4}$

$= \sqrt{8} = 2\sqrt{2} \approx 2.83$

$d(B, C) = \sqrt{(6 - 2)^2 + (5 - 1)^2}$

$= \sqrt{4^2 + 4^2}$

$= \sqrt{16 + 16}$

$= \sqrt{32} = 4\sqrt{2} \approx 5.66$

$d(A, C) = \sqrt{(6 - 0)^2 + (5 - 3)^2}$

$= \sqrt{6^2 + 2^2}$

$= \sqrt{36 + 4}$

$= \sqrt{40} = 2\sqrt{10} \approx 6.32$

(c) To verify that triangle *ABC* is a right triangle, we show that

$[d(A, B)]^2 + [d(B, C)]^2 \overset{?}{=} [d(A, C)]^2$

$\left(2\sqrt{2} \right)^2 + \left(4\sqrt{2} \right)^2 \overset{?}{=} \left(2\sqrt{10} \right)^2$

$4 \cdot 2 + 16 \cdot 2 \overset{?}{=} 4 \cdot 10$

$8 + 32 \overset{?}{=} 40$

$40 = 40 \leftarrow \text{True}$

Therefore, triangle *ABC* is a right triangle.

(d) The length of the "base" of the triangle is $d(B, C) = 4\sqrt{2}$ and the length of the "height" of the triangle is $d(A, B) = 2\sqrt{2}$.

Thus, the area of triangle *ABC* is

$\text{Area} = \dfrac{1}{2} \cdot \text{base} \cdot \text{height}$

$= \dfrac{1}{2} \cdot 4\sqrt{2} \cdot 2\sqrt{2}$

$= 8 \text{ square units.}$

39. (a)

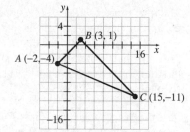

(b) $d(A,B) = \sqrt{(3-(-2))^2 + (1-(-4))^2}$

$= \sqrt{5^2 + 5^2}$

$= \sqrt{25 + 25}$

$= \sqrt{50}$

$= 5\sqrt{2} \approx 7.07$

$d(B,C) = \sqrt{(15-3)^2 + (-11-1)^2}$

$= \sqrt{12^2 + (-12)^2}$

$= \sqrt{144 + 144}$

$= \sqrt{288}$

$= 12\sqrt{2} \approx 16.97$

$d(A,C) = \sqrt{(15-(-2))^2 + (-11-(-4))^2}$

$= \sqrt{17^2 + (-7)^2}$

$= \sqrt{289 + 49}$

$= \sqrt{338}$

$= 13\sqrt{2} \approx 18.38$

(c) To verify that triangle *ABC* is a right triangle, we show that

$[d(A,B)]^2 + [d(B,C)]^2 \overset{?}{=} [d(A,C)]^2$

$(5\sqrt{2})^2 + (12\sqrt{2})^2 \overset{?}{=} (13\sqrt{2})^2$

$25 \cdot 2 + 144 \cdot 2 \overset{?}{=} 169 \cdot 2$

$50 + 288 \overset{?}{=} 338$

$338 = 338 \leftarrow$ True

Therefore, triangle *ABC* is a right triangle.

(d) The length of the "base" of the triangle is $d(B,C) = 12\sqrt{2}$ and the length of the "height" of the triangle is $d(A,B) = 5\sqrt{2}$. Thus, the area of triangle *ABC* is

Area $= \dfrac{1}{2} \cdot$ base $\cdot$ height $= \dfrac{1}{2} \cdot 12\sqrt{2} \cdot 5\sqrt{2} = 60$ square units.

41. We want to find *y* such that the distance between $P_1 = (5,1)$ and $P_2 = (2, y)$ is 5.

$d(P_1, P_2) = \sqrt{(x_2 - x_1)^2 + (y_2 - y_1)^2}$

$5 = \sqrt{(2-5)^2 + (y-1)^2}$

$5 = \sqrt{(-3)^2 + (y-1)^2}$

$5 = \sqrt{9 + (y-1)^2}$

$25 = 9 + (y-1)^2$

$16 = (y-1)^2$

$y - 1 = \pm\sqrt{16}$

$y - 1 = \pm 4$

$y = 1 \pm 4$

$y = -3$ or $y = 5$

Thus, the distance between the points $(5,1)$ and $(2,-3)$ is 5, and the distance between $(5,1)$ and $(2,5)$ is 5.

43. We want to find *x* such that the distance between $P_1 = (2,3)$ and $P_2 = (x,-3)$ is 10.

$d(P_1, P_2) = \sqrt{(x_2 - x_1)^2 + (y_2 - y_1)^2}$

$10 = \sqrt{(x-2)^2 + (-3-3)^2}$

$10 = \sqrt{(x-2)^2 + (-6)^2}$

$10 = \sqrt{(x-2)^2 + 36}$

$100 = (x-2)^2 + 36$

$64 = (x-2)^2$

$x - 2 = \pm\sqrt{64}$

$x - 2 = \pm 8$

$x = 2 \pm 8$

$x = -6$ or $x = 10$

Thus, the distance between the points $(2,3)$ and $(-6,-3)$ is 10, and the distance between $(2,3)$ and $(10,-3)$ is 10.

45. (a) Treating the intersection of Madison and State Streets as the origin, $(0,0)$, of a Cartesian plane, Wrigley Field is located at the coordinates $(-10,36)$.

$$d = \sqrt{(-10-0)^2 + (36-0)^2}$$
$$= \sqrt{(-10)^2 + 36^2}$$
$$= \sqrt{100+1296}$$
$$= \sqrt{1396}$$
$$= 2\sqrt{349} \approx 37.36$$

The distance "as the crow flies" from Madison and State Street to Wrigley Field is approximately 37.36 blocks.

(b) U.S. Cellular Field is located at the coordinates $(-3,-35)$.

$$d = \sqrt{(-3-0)^2 + (-35-0)^2}$$
$$= \sqrt{(-3)^2 + (-35)^2}$$
$$= \sqrt{9+1225}$$
$$= \sqrt{1234} \approx 35.13$$

The distance "as the crow flies" from Madison and State Street to U.S. Cellular Field is approximately 35.13 blocks.

(c) $d = \sqrt{(-10-(-3))^2 + (36-(-35))^2}$
$$= \sqrt{(-7)^2 + 71^2}$$
$$= \sqrt{49+5041}$$
$$= \sqrt{5090} \approx 71.34$$

The distance "as the crow flies" from Wrigley Field to U.S. Cellular Field is approximately 71.34 blocks.

47. (a) The distance from the right fielder to second base is

$$d = \sqrt{(320-90)^2 + (20-90)^2}$$
$$= \sqrt{230^2 + (-70)^2}$$
$$= \sqrt{52,900+4900}$$
$$= \sqrt{57,800}$$
$$= 170\sqrt{2} \approx 240.42 \text{ feet}$$

Now, $\text{time} = \dfrac{\text{distance}}{\text{speed}} \approx \dfrac{240.42}{130} \approx 1.85$.

It will take about 1.85 seconds for the ball to reach the second baseman from the right fielder.

(b) The time for the runner to reach second base will be

$$\text{time} = \frac{\text{distance}}{\text{speed}} = \frac{90}{27} \approx 3.33 \text{ seconds.}$$

The total time for the ball to reach second base from the time the right fielder catches it will be $1.85 + 0.8 = 2.65$ seconds. Since the ball will reach second base before the runner, the first base coach should not send the runner.

49. Answers may vary. One possibility follows: The distance formula is derived from the Pythagorean Theorem. The absolute value of the difference between the x-coordinates, $|x_2 - x_1|$, is the length of one leg of a right triangle. The absolute value of the difference between the y-coordinates, $|y_2 - y_1|$, is the length of the other leg of the right triangle. The length of the hypotenuse is $d(P_1, P_2)$. By the Pythagorean Theorem,

$$[d(P_1, P_2)]^2 = |x_2 - x_1|^2 + |y_2 - y_1|^2$$
$$= (x_2 - x_1)^2 + (y_2 - y_1)^2$$
$$d(P_1, P_2) = \sqrt{(x_2 - x_1)^2 + (y_2 - y_1)^2}$$

51. $3^2 = 3 \cdot 3 = 9$; $\sqrt{9} = 3$

53. $(-3)^4 = (-3)(-3)(-3)(-3) = 81$; $\sqrt[4]{81} = |-3| = 3$

55. If n is a positive integer and $a^n = b$, then
$$\sqrt[n]{b} = \begin{cases} |a|, & \text{if } n \text{ is even} \\ a, & \text{if } n \text{ is odd} \end{cases}$$

Section 12.2

Preparing for Circles

P1. Start: $x^2 - 8x$

Add: $\left(\dfrac{1}{2} \cdot (-8)\right)^2 = 16$

Result: $x^2 - 8x + 16$

Factored Form: $(x-4)^2$

Section 12.2 Quick Checks

1. circle

2. radius

3. We are given that $h = 2$, $k = 4$, and $r = 5$.
Thus, the equation of the circle is

$$(x-h)^2 + (y-k)^2 = r^2$$
$$(x-2)^2 + (y-4)^2 = 5^2$$
$$(x-2)^2 + (y-4)^2 = 25$$

4. We are given that $h = -2$, $k = 0$, and $r = \sqrt{2}$.
Thus, the equation of the circle is

$$(x-h)^2 + (y-k)^2 = r^2$$
$$(x-(-2))^2 + (y-0)^2 = (\sqrt{2})^2$$
$$(x+2)^2 + y^2 = 2$$

5. False. The center is $(-1, 3)$.

6. True.

7. $(x-3)^2 + (y-1)^2 = 4$
$(x-3)^2 + (y-1)^2 = 2^2$
The center is $(h,k) = (3, 1)$; the radius is $r = 2$.

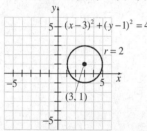

8. $(x+5)^2 + y^2 = 16$
$(x-(-5))^2 + (y-0)^2 = 4^2$
The center is $(h, k) = (-5, 0)$; the radius is $r = 4$.

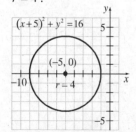

9.
$$x^2 + y^2 - 6x - 4y + 4 = 0$$
$$(x^2 - 6x) + (y^2 - 4y) = -4$$
$$(x^2 - 6x + 9) + (y^2 - 4y + 4) = -4 + 9 + 4$$
$$(x-3)^2 + (y-2)^2 = 9$$
$$(x-3)^2 + (y-2)^2 = 3^2$$

The center is $(h, k) = (3, 2)$; the radius is $r = 3$.

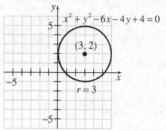

10.
$$2x^2 + 2y^2 - 16x + 4y - 38 = 0$$
$$\frac{2x^2 + 2y^2 - 16x + 4y - 38}{2} = \frac{0}{2}$$
$$x^2 + y^2 - 8x + 2y - 19 = 0$$
$$(x^2 - 8x) + (y^2 + 2y) = 19$$
$$(x^2 - 8x + 16) + (y^2 + 2y + 1) = 19 + 16 + 1$$
$$(x-4)^2 + (y+1)^2 = 36$$
$$(x-4)^2 + (y-(-1))^2 = 6^2$$

The center is $(h, k) = (4, -1)$; the radius is $r = 6$.

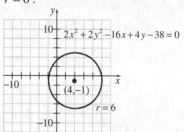

12.2 Exercises

11. The center of the circle is $(1, 2)$. The radius of the circle is the distance from the center point to the point $(-1, 2)$ on the circle. Thus,

$$r = \sqrt{(1-(-1))^2 + (2-2)^2} = 2.$$

The equation of the circle is
$$(x-h)^2 + (y-k)^2 = r^2$$
$$(x-1)^2 + (y-2)^2 = 2^2$$
$$(x-1)^2 + (y-2)^2 = 4$$

13. The center of the circle will be the midpoint of the line segment with endpoints $(-2,-1)$ and $(6,-1)$.

Thus, $(h,k) = \left(\dfrac{-2+6}{2}, \dfrac{-1+(-1)}{2}\right) = (2,-1)$. The radius of the circle will be the distance from the center point $(2,-1)$ to a point on the circle, say $(6,-1)$. Thus, $r = \sqrt{(6-2)^2 + (-1-(-1))^2} = 4$.

The equation of the circle is
$$(x-h)^2 + (y-k)^2 = r^2$$
$$(x-2)^2 + (y-(-1))^2 = 4^2$$
$$(x-2)^2 + (y+1)^2 = 16$$

15.
$$(x-h)^2 + (y-k)^2 = r^2$$
$$(x-0)^2 + (y-0)^2 = 3^2$$
$$x^2 + y^2 = 9$$

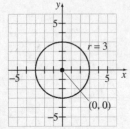

17.
$$(x-h)^2 + (y-k)^2 = r^2$$
$$(x-1)^2 + (y-4)^2 = 2^2$$
$$(x-1)^2 + (y-4)^2 = 4$$

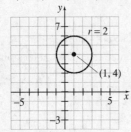

19.
$$(x-h)^2 + (y-k)^2 = r^2$$
$$(x-(-2))^2 + (y-4)^2 = 6^2$$
$$(x+2)^2 + (y-4)^2 = 36$$

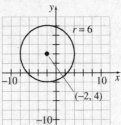

21.
$$(x-h)^2 + (y-k)^2 = r^2$$
$$(x-0)^2 + (y-3)^2 = 4^2$$
$$x^2 + (y-3)^2 = 16$$

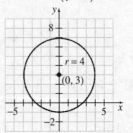

23.
$$(x-h)^2 + (y-k)^2 = r^2$$
$$(x-5)^2 + (y-(-5))^2 = 5^2$$
$$(x-5)^2 + (y+5)^2 = 25$$

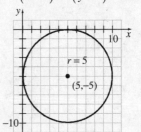

25.
$$(x-h)^2 + (y-k)^2 = r^2$$
$$(x-1)^2 + (y-2)^2 = \left(\sqrt{5}\right)^2$$
$$(x-1)^2 + (y-2)^2 = 5$$

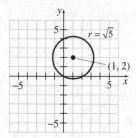

27. $\qquad x^2 + y^2 = 36$

$(x-0)^2 + (y-0)^2 = 6^2$

The center is $(h,k) = (0,0)$, and the radius is $r = 6$.

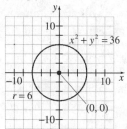

29. $(x-4)^2 + (y-1)^2 = 25$

$(x-4)^2 + (y-1)^2 = 5^2$

The center is $(h,k) = (4,1)$, and the radius is $r = 5$.

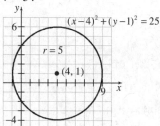

31. $\qquad (x+3)^2 + (y-2)^2 = 81$

$(x-(-3))^2 + (y-2)^2 = 9^2$

The center is $(h,k) = (-3,2)$, and the radius is $r = 9$.

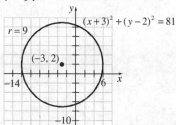

33. $\qquad x^2 + (y-3)^2 = 64$

$(x-0)^2 + (y-3)^2 = 8^2$

The center is $(h,k) = (0,3)$, and the radius is $r = 8$.

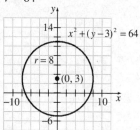

35. $\qquad (x-1)^2 + (y+1)^2 = \dfrac{1}{4}$

$(x-1)^2 + (y-(-1))^2 = \left(\dfrac{1}{2}\right)^2$

The center is $(h,k) = (1,-1)$, and the radius is

$r = \dfrac{1}{2}$.

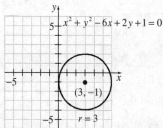

37. $\qquad x^2 + y^2 - 6x + 2y + 1 = 0$

$(x^2 - 6x) + (y^2 + 2y) = -1$

$(x^2 - 6x + 9) + (y^2 + 2y + 1) = -1 + 9 + 1$

$(x-3)^2 + (y+1)^2 = 9$

$(x-3)^2 + (y-(-1))^2 = 3^2$

The center is $(h,k) = (3,-1)$, and the radius is $r = 3$.

39.
$$x^2 + y^2 + 10x + 4y + 4 = 0$$
$$\left(x^2 + 10x\right) + \left(y^2 + 4y\right) = -4$$
$$\left(x^2 + 10x + 25\right) + \left(y^2 + 4y + 4\right) = -4 + 25 + 4$$
$$\left(x+5\right)^2 + \left(y+2\right)^2 = 25$$
$$\left(x-(-5)\right)^2 + \left(y-(-2)\right)^2 = 5^2$$

The center is $(h,k) = (-5,-2)$, and the radius is $r = 5$.

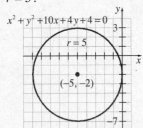

41.
$$2x^2 + 2y^2 - 12x + 24y - 72 = 0$$
$$\frac{2x^2 + 2y^2 - 12x + 24y - 72}{2} = \frac{0}{2}$$
$$x^2 + y^2 - 6x + 12y - 36 = 0$$
$$\left(x^2 - 6x\right) + \left(y^2 + 12y\right) = 36$$
$$\left(x^2 - 6x + 9\right) + \left(y^2 + 12y + 36\right) = 36 + 9 + 36$$
$$\left(x-3\right)^2 + \left(y+6\right)^2 = 81$$
$$\left(x-3\right)^2 + \left(y-(-6)\right)^2 = 9^2$$

The center is $(h,k) = (3,-6)$, and the radius is $r = 9$.

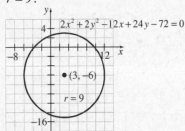

43. The radius of the circle will be the distance from the center point $(0,0)$ to the point on the circle $(4,-2)$. Thus,
$$r = \sqrt{(4-0)^2 + (-2-0)^2} = \sqrt{20} = 2\sqrt{5}.$$
The equation of the circle is
$$\left(x-h\right)^2 + \left(y-k\right)^2 = r^2$$
$$\left(x-0\right)^2 + \left(y-0\right)^2 = \left(2\sqrt{5}\right)^2$$
$$x^2 + y^2 = 20$$

45. Since the center of the circle is $(-3,2)$ and since the circle is tangent to the y-axis, the circle must contain the point $(0,2)$. The radius of the circle will be the distance from the center point $(-3,2)$ to the point on the circle $(0,2)$. Thus,
$$r = \sqrt{(-3-0)^2 + (2-2)^2} = 3.$$
The equation of the circle is
$$\left(x-h\right)^2 + \left(y-k\right)^2 = r^2$$
$$\left(x-(-3)\right)^2 + \left(y-2\right)^2 = 3^2$$
$$\left(x+3\right)^2 + \left(y-2\right)^2 = 9$$

47. The center of the circle will be the midpoint of the diameter with endpoints $(2,3)$ and $(-4,-5)$.

Thus, $(h,k) = \left(\dfrac{2+(-4)}{2}, \dfrac{3+(-5)}{2}\right) = (-1,-1)$.

The radius of the circle will be the distance from the center point $(-1,-1)$ to one of the endpoints of the diameter, say $(2,3)$. Thus,
$$r = \sqrt{(2-(-1))^2 + (3-(-1))^2} = 5.$$
The equation of the circle is
$$\left(x-h\right)^2 + \left(y-k\right)^2 = r^2$$
$$\left(x-(-1)\right)^2 + \left(y-(-1)\right)^2 = 5^2$$
$$\left(x+1\right)^2 + \left(y+1\right)^2 = 25$$

49. The radius of the circle $\left(x-3\right)^2 + \left(y-8\right)^2 = 64$ is $r = \sqrt{64} = 8$. Thus, the area of the circle is $A = \pi r^2 = \pi(8)^2 = 64\pi$ square units.
The circumference of the circle is $C = 2\pi r = 2\pi(8) = 16\pi$ units.

51. To find the area of the square, we must first find the length of the square's sides. To do so, we recognize that the x- and y-coordinates of the corner point in the first quadrant where the square touches the circle must be equal. That is, $x = y$.

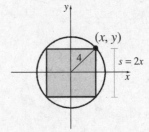

Thus, $x^2 + y^2 = 16$
$$x^2 + x^2 = 16$$
$$2x^2 = 16$$
$$x^2 = 8$$
$$x = \sqrt{8} = 2\sqrt{2}$$
This means the length of each side of the square is $s = 2x = 2\left(2\sqrt{2}\right) = 4\sqrt{2}$, and the area of the square is $A = s^2 = \left(4\sqrt{2}\right)^2 = 32$ square units.

53. The circle shown has its center point on the positive x-axis. Thus, the center point will be of the form $(h, 0)$ with $h > 0$. Also, the entire circle stays to the right of the y-axis, so $h > r$.

(a) $(x-2)^2 + y^2 = 1$ has center $(2, 0)$ and radius $r = 1$, which meets the proper conditions. Thus, this equation <u>could</u> have the graph shown.

(b) $x^2 + (y-2)^2 = 1$ has center $(0, 2)$, which is not on the x-axis. Thus, this equation could <u>not</u> have the graph shown.

(c) $(x+4)^2 + y^2 = 9$ has center $(-4, 0)$, which is on the negative x-axis, not on the positive x-axis. Thus, this equation could <u>not</u> have the graph shown.

(d) $(x-5)^2 + y^2 = 25$ has center $(5, 0)$ and radius $r = 5$. This means the graph of this circle would pass through the origin. Thus, this equation could <u>not</u> have the graph shown.

(e) $x^2 + y^2 - 8x + 7 = 0$
$$\left(x^2 - 8x + 16\right) + y^2 = -7 + 16$$
$$(x-4)^2 + y^2 = 9$$
This circle has center $(4, 0)$ and radius $r = 3$, which meets the proper conditions. Thus, this equation <u>could</u> have the graph shown.

(f) $x^2 + y^2 + 10x + 18 = 0$
$$\left(x^2 + 10x + 25\right) + y^2 = -18 + 25$$
$$(x+5)^2 + y^2 = 7$$
This circle has center $(-5, 0)$, which is on the negative x-axis, not on the positive x-axis. Thus, this equation could <u>not</u> have the graph shown.

55. Answers may vary. One possibility follows: If a circle has center (h, k) and radius r, then the distance from the center to any point (x, y) on that circle must always be r. By the distance formula, we have $r = \sqrt{(x-h)^2 + (y-k)^2}$. Now, if we square both sides of this formula, we obtain the standard equation of the circle: $(x-h)^2 + (y-k)^2 = r^2$.

57. Yes, $x^2 = 36 - y^2$ is the equation of a circle. In standard form, the equation is
$$x^2 + y^2 = 36$$
$$(x-0)^2 + (y-0)^2 = 6^2$$
The center is $(0, 0)$, and the radius is 6.

59. $f(x) = 4x - 3$
To use point plotting, let $x = -1,\ 0,$ and 1.
$$f(-1) = 4(-1) - 3 = -4 - 3 = -7$$
$$f(0) = 4(0) - 3 = 0 - 3 = -3$$
$$f(1) = 4(1) - 3 = 4 - 3 = 1$$
Thus, the points $(-1, -7)$, $(0, -3)$, and $(1, 1)$ are on the graph.

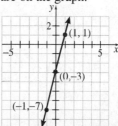

Using properties of linear functions, for $f(x) = 4x - 3$, the slope is 4 and the y-intercept is -3. Begin at $(0, -3)$ and move to the right 1 unit and up 4 units to find point $(1, 1)$.

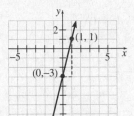

61. $g(x) = x^2 - 4x - 5$

To use point plotting, let $x = 0,\ 1,\ 2,\ 3,$ and 4.

$g(0) = 0^2 - 4(0) - 5 = 0 - 0 - 5 = -5$

$g(1) = 1^2 - 4(1) - 5 = 1 - 4 - 5 = -8$

$g(2) = 2^2 - 4(2) - 5 = 4 - 8 - 5 = -9$

$g(3) = 3^2 - 4(3) - 5 = 9 - 12 - 5 = -8$

$g(4) = 4^2 - 4(4) - 5 = 16 - 16 - 5 = -5$

Thus, the points $(0, -5)$, $(1, -8)$, $(2, -9)$,

$(3, -8)$, and $(4, -5)$ are on the graph.

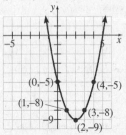

Using properties of quadratic functions, for

$g(x) = x^2 - 4x - 5$, $a = 1$, $b = -4$, and $c = -5$.

The graph opens up because the coefficient on

x^2 is positive.

<u>vertex:</u>

$$x = -\frac{b}{2a} = -\frac{(-4)}{2(1)} = 2$$

$g(2) = (2)^2 - 4(2) - 5 = -9$

The vertex is $(2, -9)$ and the axis of symmetry

is $x = 2$.

<u>y-intercept:</u>

$g(0) = (0)^2 - 4(0) - 5 = -5$

<u>x-intercepts:</u>

$b^2 - 4ac = (-4)^2 - 4(1)(-5) = 36 > 0$

There are two distinct *x*-intercepts. We find these

by solving

$g(x) = 0$

$x^2 - 4x - 5 = 0$

$(x-5)(x+1) = 0$

$x - 5 = 0$ or $x + 1 = 0$

$x = 5$ or $x = -1$

<u>Graph:</u>

The *y*-intercept point, $(0, -5)$, is two units to the

left of the axis of symmetry. Therefore, if we

move two units to the right of the axis of

symmetry, we obtain the point $(4, -5)$ which

must also be on the graph.

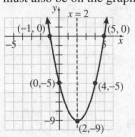

63. $G(x) = -2(x+3)^2 - 5$

To use point plotting, let $x = -4, -3,$ and -2.

$G(-4) = -2(-4+3)^2 - 5$

$\qquad = -2(-1)^2 - 5$

$\qquad = -2(1) - 5$

$\qquad = -2 - 5$

$\qquad = -7$

$G(-3) = -2(-3+3)^2 - 5$

$\qquad = -2(0)^2 - 5$

$\qquad = -2(0) - 5$

$\qquad = 0 - 5$

$\qquad = -5$

$G(-2) = -2(-2+3)^2 - 5$

$\qquad = -2(1)^2 - 5$

$\qquad = -2(1) - 5$

$\qquad = -2 - 5$

$\qquad = -7$

Thus, the points $(-4, -7)$, $(-3, -5)$, and

$(-2, -7)$ are on the graph.

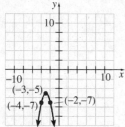

Using properties of quadratic functions, begin with the graph of $y = x^2$, then shift the graph 3 units to the left to obtain the graph of $y = (x+3)^2$. Multiply the y-coordinates by -2 to obtain the graph of $y = -2(x+3)^2$. Lastly, shift the graph down 5 units to obtain the graph of $G(x) = -2(x+3)^2 - 5$.

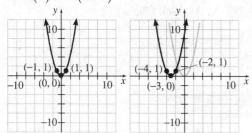

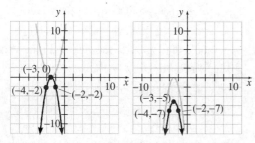

65. $x^2 + y^2 = 36$
$$y^2 = 36 - x^2$$
$$y = \pm\sqrt{36 - x^2}$$

Let $Y_1 = \sqrt{36 - x^2}$ and $Y_2 = -\sqrt{36 - x^2}$.

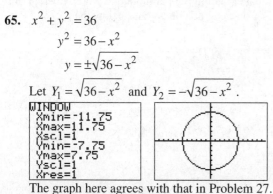

The graph here agrees with that in Problem 27.

67. $(x-4)^2 + (y-1)^2 = 25$
$$(y-1)^2 = 25 - (x-4)^2$$
$$y-1 = \pm\sqrt{25 - (x-4)^2}$$
$$y = 1 \pm \sqrt{25 - (x-4)^2}$$

Let $Y_1 = 1 + \sqrt{25 - (x-4)^2}$ and $Y_2 = 1 - \sqrt{25 - (x-4)^2}$.

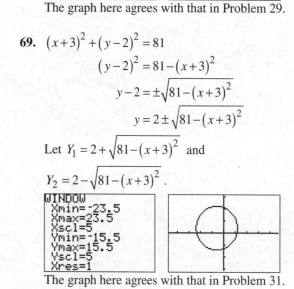

The graph here agrees with that in Problem 29.

69. $(x+3)^2 + (y-2)^2 = 81$
$$(y-2)^2 = 81 - (x+3)^2$$
$$y-2 = \pm\sqrt{81 - (x+3)^2}$$
$$y = 2 \pm \sqrt{81 - (x+3)^2}$$

Let $Y_1 = 2 + \sqrt{81 - (x+3)^2}$ and $Y_2 = 2 - \sqrt{81 - (x+3)^2}$.

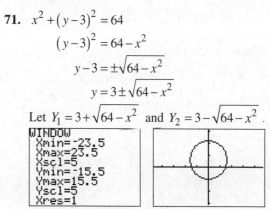

The graph here agrees with that in Problem 31.

71. $x^2 + (y-3)^2 = 64$
$$(y-3)^2 = 64 - x^2$$
$$y-3 = \pm\sqrt{64 - x^2}$$
$$y = 3 \pm \sqrt{64 - x^2}$$

Let $Y_1 = 3 + \sqrt{64 - x^2}$ and $Y_2 = 3 - \sqrt{64 - x^2}$.

The graph here agrees with that in Problem 33.

73. $(x-1)^2 + (y+1)^2 = \dfrac{1}{4}$

$$(y+1)^2 = \dfrac{1}{4} - (x-1)^2$$

$$y+1 = \pm\sqrt{\dfrac{1}{4} - (x-1)^2}$$

$$y = -1 \pm\sqrt{\dfrac{1}{4} - (x-1)^2}$$

Let $Y_1 = -1 + \sqrt{\dfrac{1}{4} - (x-1)^2}$ and

$Y_2 = -1 - \sqrt{\dfrac{1}{4} - (x-1)^2}$.

```
WINDOW
 Xmin=-2.35
 Xmax=2.35
 Xscl=1
 Ymin=-1.55
 Ymax=1.55
 Yscl=1
 Xres=1
```

The graph here agrees with that in Problem 35.

Section 12.3

Preparing for Parabolas

P1. For the function $f(x) = -3(x+4)^2 - 5$, we see that $a = -3$, $h = -4$, and $k = -5$. Thus, the vertex is $(h, k) = (-4, -5)$, and the axis of symmetry is $x = -4$. The parabola opens down since $a = -3 < 0$.

P2. For the function $f(x) = 2x^2 - 8x + 1$, we see that $a = 2$, $b = -8$, and $c = 1$. The x-coordinate of the vertex is $x = -\dfrac{b}{2a} = -\dfrac{(-8)}{2(2)} = 2$. The y-coordinate of the vertex is

$$f\left(-\dfrac{b}{2a}\right) = f(2)$$
$$= 2(2)^2 - 8(2) + 1$$
$$= 8 - 16 + 1$$
$$= -7$$

Thus, the vertex is $(2, -7)$ and the axis of symmetry is the line $x = 2$. The parabola opens up because $a = 2 > 0$.

P3. Start: $x^2 - 12x$

Add: $\left(\dfrac{1}{2} \cdot (-12)\right)^2 = 36$

Result: $x^2 - 12x + 36$

Factored Form: $(x-6)^2$

P4. $(x-3)^2 = 25$

$$x - 3 = \pm\sqrt{25}$$
$$x - 3 = \pm 5$$
$$x = 3 \pm 5$$
$$x = 3 - 5 \quad \text{or} \quad x = 3 + 5$$
$$x = -2 \quad \text{or} \quad x = 8$$

The solution set is $\{-2, 8\}$.

Section 12.3 Quick Checks

1. parabola

2. vertex

3. axis of symmetry

4. Notice that $y^2 = 8x$ is of the form $y^2 = 4ax$, where $4a = 8$, so that $a = 2$. Now, the graph of an equation of the form $y^2 = 4ax$ will be a parabola that opens to the right with the vertex at the origin, focus at $(a, 0)$, and directrix of $x = -a$. Thus, the graph of $y^2 = 8x$ is a parabola that opens to the right with vertex $(0, 0)$, focus $(2, 0)$, and directrix $x = -2$. To help graph the parabola, we plot the two points on the graph above and below the focus. Let $x = 2$:

$$y^2 = 8(2)$$
$$y^2 = 16$$
$$y = \pm 4$$

The points $(2, -4)$ and $(2, 4)$ are on the graph.

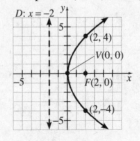

5. Notice that $y^2 = -20x$ is of the form

$y^2 = -4ax$, where $-4a = -20$, so that $a = 5$. Now, the graph of an equation of the form $y^2 = -4ax$ will be a parabola that opens to the left with the vertex at the origin, focus at $(-a, 0)$, and directrix of $x = a$. Thus, the graph of $y^2 = -20x$ is a parabola that opens to the left with vertex $(0,0)$, focus $(-5, 0)$, and directrix $x = 5$. To help graph the parabola, we plot the two points on the graph above and below the focus. Let $x = 5$:

$$y^2 = -20(-5)$$
$$y^2 = 100$$
$$y = \pm 10$$

The points $(-5, -10)$ and $(-5, 10)$ are on the graph.

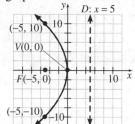

6. Notice that $x^2 = 4y$ is of the form $x^2 = 4ay$, where $4a = 4$, so that $a = 1$. Now, the graph of an equation of the form $x^2 = 4ay$ will be a parabola that opens up with the vertex at the origin, focus at $(0, a)$, and directrix of $y = -a$.

Thus, the graph of $x^2 = 4y$ is a parabola that opens up with vertex $(0,0)$, focus $(0, 1)$, and directrix $y = -1$. To help graph the parabola, we plot the two points on the graph to the left and to the right of the focus. Let $y = 1$:

$$x^2 = 4(1)$$
$$x^2 = 4$$
$$x = \pm 2$$

The points $(-2, 1)$ and $(2, 1)$ are on the graph.

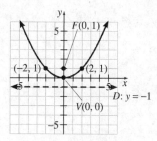

7. Notice that $x^2 = -12y$ is of the form

$x^2 = -4ay$, where $-4a = -12$, so that $a = 3$. Now, the graph of an equation of the form $x^2 = -4ay$ will be a parabola that opens down with the vertex at the origin, focus at $(0, -a)$, and directrix of $y = a$. Thus, the graph of $x^2 = -12y$ is a parabola that opens down with vertex $(0,0)$, focus $(0, -3)$, and directrix $y = 3$. To help graph the parabola, we plot the two points on the graph to the left and to the right of the focus. Let $y = -3$:

$$x^2 = -12(-3)$$
$$x^2 = 36$$
$$x = \pm 6$$

The points $(-6, -3)$ and $(6, -3)$ are on the graph.

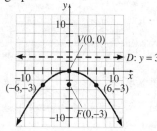

8. The distance from the vertex $(0,0)$ to the focus $(0, -8)$ is $a = 8$. Because the focus lies on the negative y-axis, we know that the parabola will open down and the axis of symmetry is the y-axis. This means the equation of the parabola is of the form $x^2 = -4ay$ with $a = 8$:

$$x^2 = -4(8)y$$
$$x^2 = -32y$$

The directrix is the line $y = 8$. To help graph the parabola, we plot the two points on the graph to the left and right of the focus. Let $y = -8$:

$$x^2 = -32(-8) = 256$$
$$x = \pm 16$$

The points $(-16, -8)$ and $(16, -8)$ are on the graph.

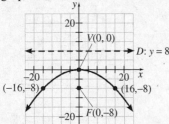

9. The vertex is at the origin and the axis of symmetry is the *x*-axis, so the parabola either opens left or right. Because the graph contains the point $(3, 2)$, which is in quadrant I, the parabola must open right. Therefore, the equation of the parabola is of the form $y^2 = 4ax$. Now $y = 2$ when $x = 3$, so

$$y^2 = 4ax$$
$$2^2 = 4a(3)$$
$$4 = 12a$$
$$a = \frac{4}{12} = \frac{1}{3}$$

The equation of the parabola is

$$y^2 = 4\left(\frac{1}{3}\right)x$$
$$y^2 = \frac{4}{3}x$$

With $a = \frac{1}{3}$, we know that the focus is $\left(\frac{1}{3}, 0\right)$

and the directrix is the line $x = -\frac{1}{3}$. To help graph the parabola, we plot the two points on the graph to the left and right of the focus. Let $x = \frac{1}{3}$:

$$y^2 = \frac{4}{3}\left(\frac{1}{3}\right)$$
$$y^2 = \frac{4}{9}$$
$$y = \pm\frac{2}{3}$$

The points $\left(\frac{1}{3}, -\frac{2}{3}\right)$ and $\left(\frac{1}{3}, \frac{2}{3}\right)$ are on the graph.

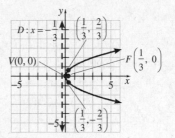

10. $(-3, 2)$

11. We complete the square in *y* to write the equation in standard form:

$$y^2 - 4y - 12x - 32 = 0$$
$$y^2 - 4y = 12x + 32$$
$$y^2 - 4y + 4 = 12x + 32 + 4$$
$$y^2 - 4y + 4 = 12x + 36$$
$$(y-2)^2 = 12(x+3)$$

Notice that the equation is of the form $(y-k)^2 = 4a(x-h)$. The graph is a parabola that opens right with vertex $(h, k) = (-3, 2)$. Since $4a = 12$, we have that $a = 3$. The focus is $(h+a, k) = (-3+3, 2) = (0, 2)$, and the directrix is $x = h - a = -3 - 3 = -6$. To help graph the parabola, we plot the two points on the graph above and below the focus. Let $x = 0$:

$$(y-2)^2 = 12(0+3)$$
$$(y-2)^2 = 12(3)$$
$$(y-2)^2 = 36$$
$$y - 2 = \pm 6$$
$$y = 2 \pm 6$$
$$y = -4 \text{ or } y = 8$$

The points $(0, -4)$ and $(0, 8)$ are on the graph.

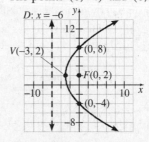

12. The receiver should be located at the focus of the satellite dish, so we need to find where the focus of the satellite dish is. To solve this problem, we draw a parabola on a Cartesian plane so that the vertex is the origin and the focus is on the positive *y*-axis. The width of the parabola is

4 feet, and the depth is 6 inches = 0.5 feet. Therefore, we know two points on the graph of the parabola: $(-2, 0.5)$ and $(2, 0.5)$.

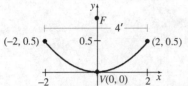

The equation of the parabola has the form $x^2 = 4ay$. Since $(2, 0.5)$ is a point on the graph, we have

$$2^2 = 4a(0.5)$$
$$4 = 2a$$
$$2 = a$$

The receiver should be located 2 feet from the base of the dish along its axis of symmetry.

12.3 Exercises

13. c. The parabola opens upward and has vertex $(0,0)$ and focus $(0,2)$, so the equation is of the form $x^2 = 4ay$ with $a = 2$:

$$x^2 = 4(2)y$$
$$x^2 = 8y$$

15. a. The parabola opens to the right and has vertex $(0,0)$ and focus $(2,0)$, so the equation is of the form $y^2 = 4ax$ with $a = 2$:

$$y^2 = 4(2)x$$
$$y^2 = 8x$$

17. b. The parabola opens to the left and has vertex $(0,0)$ and focus $(-2,0)$, so the equation is of the form $y^2 = -4ax$ with $a = 2$:

$$y^2 = -4(2)x$$
$$y^2 = -8x$$

19. e. The parabola opens to the right and has vertex $(-1,2)$ and focus $(1,2)$, so the equation is of the form $(y-k)^2 = 4a(x-h)$, with $a = 1-(-1) = 2$, $h = -1$, and $k = 2$:

$$(y-2)^2 = 4 \cdot 2(x-(-1))$$
$$(y-2)^2 = 8(x+1)$$

21. Notice that $x^2 = 24y$ is of the form $x^2 = 4ay$, where $4a = 24$, so that $a = 6$. Now, the graph of an equation of the form $x^2 = 4ay$ will be a parabola that opens upward with the vertex at the origin, focus at $(0,a)$, and directrix of $y = -a$.

Thus, the graph of $x^2 = 24y$ is a parabola that opens upward with vertex $(0,0)$, focus $(0, 6)$, and directrix $y = -6$. To help graph the parabola, we plot the two points on the graph to the left and to the right of the focus. Let $y = 6$:

$$x^2 = 24(6)$$
$$x^2 = 144$$
$$x = \pm 12$$

The points $(-12, 6)$ and $(12, 6)$ are on the graph.

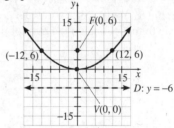

23. Notice that $y^2 = -6x$ is of the form $y^2 = -4ax$, where $-4a = -6$, so that $a = \dfrac{-6}{-4} = \dfrac{3}{2}$. Now, the graph of an equation of the form $y^2 = -4ax$ will be a parabola that opens to the left with the vertex at the origin, focus at $(-a,0)$, and directrix of $x = a$. Thus, the graph of $y^2 = -6x$ is a parabola that opens to the left with vertex $(0,0)$, focus $\left(-\dfrac{3}{2}, 0\right)$, and directrix $x = \dfrac{3}{2}$. To help graph the parabola, we plot the two points on the graph above and below the focus. Let $x = -\dfrac{3}{2}$:

$$y^2 = -6\left(-\dfrac{3}{2}\right)$$
$$y^2 = 9$$
$$y = \pm 3$$

The points $\left(-\dfrac{3}{2}, -3\right)$ and $\left(-\dfrac{3}{2}, 3\right)$ are on the graph.

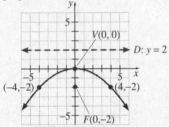

25. Notice that $x^2 = -8y$ is of the form $x^2 = -4ay$, where $-4a = -8$, so that $a = 2$. Now, the graph of an equation of the form $x^2 = -4ay$ will be a parabola that opens downward with the vertex at the origin, focus at $(0, -a)$, and directrix of

$y = a$. Thus, the graph of $x^2 = -8y$ is a parabola that opens downward with vertex $(0, 0)$, focus $(0, -2)$, and directrix $y = 2$. To help graph the parabola, we plot the two points on the graph to the left and to the right of the focus. Let $y = -2$:

$x^2 = -8(-2)$

$x^2 = 16$

$x = \pm 4$

The points $(-4, -2)$ and $(4, -2)$ are on the graph.

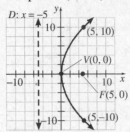

27. The distance from the vertex $(0, 0)$ to the focus $(5, 0)$ is $a = 5$. Because the focus lies on the positive x-axis, we know that the parabola will open to the right and the axis of symmetry is the x-axis. This means the equation of the parabola is of the form $y^2 = 4ax$ with $a = 5$:

$y^2 = 4(5)x$

$y^2 = 20x$

The directrix is the line $x = -5$. To help graph the parabola, we plot the two points on the graph above and below the focus. Let $x = 5$:

$y^2 = 20(5) = 100$

$y = \pm 10$

The points $(5, -10)$ and $(5, 10)$ are on the graph.

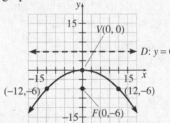

29. The distance from the vertex $(0, 0)$ to the focus $(0, -6)$ is $a = 6$. Because the focus lies on the negative y-axis, we know that the parabola will open downward and the axis of symmetry is the y-axis. This means the equation of the parabola is of the form $x^2 = -4ay$ with $a = 6$:

$x^2 = -4(6)y$

$x^2 = -24y$

The directrix is the line $y = 6$. To help graph the parabola, we plot the two points on the graph to the left and right of the focus. Let $y = -6$:

$x^2 = -24(-6) = 144$

$x = \pm 12$

The points $(-12, -6)$ and $(12, -6)$ are on the graph.

31. The vertex is at the origin and the axis of symmetry is the y-axis, so the parabola either opens upward or downward. Because the graph contains the point $(6, 6)$, which is in quadrant I, the parabola must open upward. Therefore, the equation of the parabola is of the form

$x^2 = 4ay$. Now $y = 6$ when $x = 6$, so

$x^2 = 4ay$

$6^2 = 4a(6)$

$36 = 24a$

$a = \dfrac{36}{24} = \dfrac{3}{2}$

The equation of the parabola is

$$x^2 = 4\left(\frac{3}{2}\right)y$$

$$x^2 = 6y$$

With $a = \frac{3}{2}$, we know that the focus is $\left(0, \frac{3}{2}\right)$

and the directrix is the line $y = -\frac{3}{2}$. To help

graph the parabola, we plot the two points on the graph to the left and right of the focus. Let

$y = \frac{3}{2}$:

$$x^2 = 6\left(\frac{3}{2}\right)$$

$$x^2 = 9$$

$$x = \pm 3$$

The points $\left(-3, \frac{3}{2}\right)$ and $\left(3, \frac{3}{2}\right)$ are on the

graph.

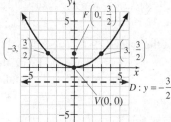

33. The vertex is at the origin and the directrix is the line $y = 3$, so the focus must be $(0, -3)$.

Accordingly, the parabola opens downward with $a = 3$. This means the equation of the parabola is of the form $x^2 = -4ay$ with $a = 3$:

$$x^2 = -4(3)y$$

$$x^2 = -12y$$

To help graph the parabola, we plot the two points on the graph to the left and right of the focus. Let $y = -3$:

$$x^2 = -12(-3) = 36$$

$$x = \pm 6$$

The points $(-6, -3)$ and $(6, -3)$ are on the graph.

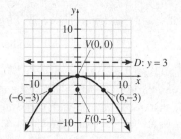

35. Notice that the directrix is the vertical line $x = 3$ and that the focus $(-3, 0)$ is on the x-axis. Thus, the axis of symmetry is the x-axis, the vertex is at the origin, and the parabola opens to the left. Now, the distance from the vertex $(0, 0)$ to the focus $(-3, 0)$ is $a = 3$, so the equation of the parabola is of the form $y^2 = -4ax$ with $a = 3$:

$$y^2 = -4(3)x$$

$$y^2 = -12x$$

To help graph the parabola, we plot the two points on the graph above and below the focus. Let $x = -3$:

$$y^2 = -12(-3) = 36$$

$$y = \pm 6$$

The points $(-3, -6)$ and $(-3, 6)$ are on the graph.

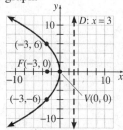

37. The parabola opens to the right and has vertex $(0, 0)$, so the equation must have the form

$y^2 = 4ax$. Because the point $(4, 2)$ is on the parabola, we let $x = 4$ and $y = 2$ to determine a:

$$2^2 = 4a(4)$$

$$4 = 16a$$

$$a = \frac{4}{16} = \frac{1}{4}$$

Thus, the equation of the parabola is

$$y^2 = 4\left(\frac{1}{4}\right)x$$

$$y^2 = x$$

39. Notice that $(x-2)^2 = 4(y-4)$ is of the form

$(x-h)^2 = 4a(y-k)$, where $4a = 4$, so that

$a = 1$ and $(h,k) = (2, 4)$. Now, the graph of an

equation of the form $(x-h)^2 = 4a(y-k)$ will be

a parabola that opens upward with vertex at

(h,k), focus at $(h, k+a)$, and directrix of

$y = k-a$. Note that $k+a = 4+1 = 5$ and

$k-a = 4-1 = 3$. Thus, the graph of

$(x-2)^2 = 4(y-4)$ is a parabola that opens

upward with vertex $(2, 4)$, focus $(2, 5)$, and

directrix $y = 3$. To help graph the parabola, we

plot the two points to the left and to the right of

the focus. Let $y = 5$:

$(x-2)^2 = 4(5-4)$

$(x-2)^2 = 4(1)$

$(x-2)^2 = 4$

$x-2 = \pm 2$

$x = 2 \pm 2$

$x = 0$ or $x = 4$

The points $(0, 5)$ and $(4, 5)$ are on the graph.

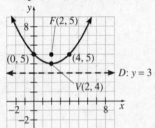

41. Notice that $(y+3)^2 = -8(x+2)$ is of the form

$(y-k)^2 = -4a(x-h)$, where $(h,k) = (-2,-3)$

and $-4a = -8$, so that $a = 2$. Now, the graph of

an equation of the form $(y-k)^2 = -4a(x-h)$

will be a parabola that opens to the left with the

vertex at (h,k), focus at $(h-a, k)$, and directrix

of $x = h+a$. Note that $h-a = -2-2 = -4$ and

$h+a = -2+2 = 0$. Thus, the graph of

$(y+3)^2 = -8(x+2)$ is a parabola that opens to

the left with vertex $(-2,-3)$, focus $(-4,-3)$,

and directrix $x = 0$. To help graph the parabola,

we plot the two points on the graph above and

below the focus. Let $x = -4$:

$(y+3)^2 = -8(-4+2)$

$(y+3)^2 = -8(-2)$

$(y+3)^2 = 16$

$y+3 = \pm 4$

$y = -3 \pm 4$

$y = -7$ or $y = 1$

The points $(-4,-7)$ and $(-4, 1)$ are on the

graph.

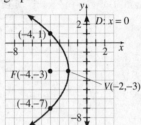

43. Notice that $(x+5)^2 = -20(y-1)$ is of the form

$(x-h)^2 = -4a(y-k)$, where $(h,k) = (-5,1)$ and

$-4a = -20$, so that $a = 5$. Now, the graph of an

equation of the form $(x-h)^2 = -4a(y-k)$ will

be a parabola that opens downward with the

vertex at (h,k), focus at $(h, k-a)$, and directrix

of $y = k+a$. Note that $k-a = 1-5 = -4$ and

$k+a = 1+5 = 6$. Thus, the graph of

$(x+5)^2 = -20(y-1)$ is a parabola that opens

downward with vertex $(-5, 1)$, focus $(-5,-4)$,

and directrix $y = 6$. To help graph the parabola,

we plot the two points on the graph to the left

and right of the focus. Let $y = -4$:

$(x+5)^2 = -20(-4-1)$

$(x+5)^2 = -20(-5)$

$(x+5)^2 = 100$

$x+5 = \pm 10$

$x = -5 \pm 10$

$x = -15$ or $x = 5$

The points $(-15,-4)$ and $(5,-4)$ are on the

graph.

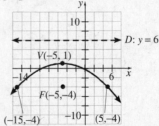

45. We complete the square in x to write the equation in standard form:

$$x^2 + 4x + 12y + 16 = 0$$
$$x^2 + 4x = -12y - 16$$
$$x^2 + 4x + 4 = -12y - 16 + 4$$
$$x^2 + 4x + 4 = -12y - 12$$
$$(x+2)^2 = -12(y+1)$$

Notice that $(x+2)^2 = -12(y+1)$ is of the form $(x-h)^2 = -4a(y-k)$, where $(h,k) = (-2,-1)$ and $-4a = -12$, so that $a = 3$. Now, the graph of an equation of the form $(x-h)^2 = -4a(y-k)$ will be a parabola that opens downward with the vertex at (h,k), focus at $(h,k-a)$, and directrix of $y = k+a$. Note that $k-a = -1-3 = -4$ and $k+a = -1+3 = 2$. Thus, the graph of $(x+2)^2 = -12(y+1)$ is a parabola that opens downward with vertex $(-2,-1)$, focus $(-2,-4)$, and directrix $y = 2$. To help graph the parabola, we plot the two points on the graph to the left and right of the focus. Let $y = -4$:

$$(x+2)^2 = -12(-4+1)$$
$$(x+2)^2 = -12(-3)$$
$$(x+2)^2 = 36$$
$$x+2 = \pm 6$$
$$x = -2 \pm 6$$
$$x = -8 \text{ or } x = 4$$

The points $(-8,-4)$ and $(4,-4)$ are on the graph.

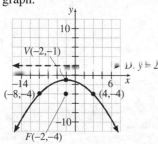

47. We complete the square in y to write the equation in standard form:

$$y^2 - 8y - 4x + 20 = 0$$
$$y^2 - 8y = 4x - 20$$
$$y^2 - 8y + 16 = 4x - 20 + 16$$
$$y^2 - 8y + 16 = 4x - 4$$
$$(y-4)^2 = 4(x-1)$$

Notice that $(y-4)^2 = 4(x-1)$ is of the form $(y-k)^2 = 4a(x-h)$, where $(h,k) = (1,4)$ and $4a = 4$, so that $a = 1$. Now, the graph of an equation of the form $(y-k)^2 = 4a(x-h)$ will be a parabola that opens to the right with the vertex at (h,k), focus at $(h+a,k)$, and directrix of $x = h-a$. Note that $h+a = 1+1 = 2$ and $h-a = 1-1 = 0$. Thus, the graph of $(y-4)^2 = 4(x-1)$ is a parabola that opens to the right with vertex $(1, 4)$, focus $(2, 4)$, and directrix $x = 0$. To help graph the parabola, we plot the two points on the graph above and below the focus. Let $x = 2$:

$$(y-4)^2 = 4(2-1)$$
$$(y-4)^2 = 4$$
$$y-4 = \pm 2$$
$$y = 4 \pm 2$$
$$y = 2 \text{ or } y = 6$$

The points $(2, 2)$ and $(2, 6)$ are on the graph.

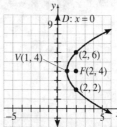

49. We complete the square in x to write the equation in standard form:

$$x^2 + 10x + 6y + 13 = 0$$
$$x^2 + 10x = -6y - 13$$
$$x^2 + 10x + 25 = -6y - 13 + 25$$
$$x^2 + 10x + 25 = -6y + 12$$
$$(x+5)^2 = -6(y-2)$$

Notice that $(x+5)^2 = -6(y-2)$ is of the form $(x-h)^2 = -4a(y-k)$, where $(h,k) = (-5,2)$ and $-4a = -6$, so that $a = \dfrac{-6}{-4} = \dfrac{3}{2}$. Now, the graph of an equation of the form $(x-h)^2 = -4a(y-k)$ will be a parabola that opens downward with the vertex at (h,k), focus at $(h,k-a)$, and directrix of $y = k+a$. Note that $k - a = 2 - \dfrac{3}{2} = \dfrac{1}{2}$ and $k + a = 2 + \dfrac{3}{2} = \dfrac{7}{2}$.

Thus, the graph of $(x+5)^2 = -6(y-2)$ is a parabola that opens downward with vertex $(-5, 2)$, focus $\left(-5, \dfrac{1}{2}\right)$, and directrix $y = \dfrac{7}{2}$.

To help graph the parabola, we plot the two points on the graph to the left and right of the

focus. Let $y = \dfrac{1}{2}$:

$(x+5)^2 = -6\left(\dfrac{1}{2} - 2\right)$

$(x+5)^2 = -6\left(-\dfrac{3}{2}\right)$

$(x+5)^2 = 9$

$\quad x + 5 = \pm 3$

$\qquad x = -5 \pm 3$

$x = -8$ or $x = -2$

The points $\left(-8, \dfrac{1}{2}\right)$ and $\left(-2, \dfrac{1}{2}\right)$ are on the graph.

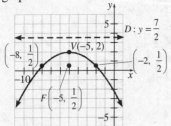

51. The light bulb should be located at the focus of the headlight, so we need to find where the focus of the headlight is. To solve this problem, we draw a parabola on a Cartesian plane so that the vertex is the origin and the focus is on the positive y-axis. The width of the parabola is 4 inches, and the depth is 1 inch. Therefore, we know two points on the graph of the parabola: $(-2, 1)$ and $(2, 1)$.

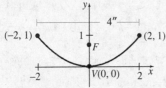

The equation of the parabola has the form $x^2 = 4ay$. Since $(2, 1)$ is a point on the graph, we have

$2^2 = 4a(1)$

$\quad 4 = 4a$

$\quad 1 = a$

The light bulb should be located 1 inch above the vertex, along its axis of symmetry.

53. To solve this problem, we draw a parabola on a Cartesian plane so that the vertex is the origin and the focus is on the positive y-axis. The distance between the towers is 500 feet, and the height of the towers is 60 feet. Therefore, we know two points on the graph of the parabola: $(-250,\ 60)$ and $(250,\ 60)$.

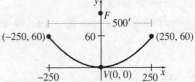

The equation of the parabola has the form $x^2 = 4ay$. Since $(250,\ 60)$ is a point on the graph, we have

$250^2 = 4a(60)$

$62{,}500 = 240a$

$\qquad a = \dfrac{62{,}500}{240} = \dfrac{3125}{12}$

Then the equation of the parabola is

$x^2 = 4\left(\dfrac{3125}{12}\right)y$

$x^2 = \dfrac{3125}{3}y$

To find the height of the cable at a point 150 feet from the center of the bridge, let $x = 150$:

$150^2 = \dfrac{3125}{3}y$

$22{,}500 = \dfrac{3125}{3}y$

$\qquad y = \dfrac{3}{3125} \cdot 22{,}500 = 21.6$

At a point 150 feet from the center of the bridge, the height of the cable is 21.6 feet.

55. To solve this problem, we draw a parabola on a Cartesian plane so that the vertex is on the positive y-axis and the base of the bridge is along the x-axis. The maximum height of the bridge is 30 feet, so we know the vertex of the parabola is $(0,\ 30)$. The span of the bridge is 100 feet, so we know two other points on the graph of the parabola: $(-50,\ 0)$ and $(50,\ 0)$.

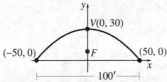

Because the parabola opens downward, its equation has the form $(x-h)^2 = -4a(y-k)$. Substituting the vertex $(h,k) = (0, \ 30)$ into this equation, we obtain

$$(x-0)^2 = -4a(y-30)$$
$$x^2 = -4a(y-30)$$

Because the point $(50, \ 0)$ is on the parabola, we let $x = 50$ and $y = 0$ to determine a:

$$50^2 = -4a(0-30)$$
$$2500 = 120a$$
$$a = \frac{2500}{120} = \frac{125}{6}$$

Thus, the equation of the parabola is

$$x^2 = -4 \cdot \frac{125}{6}(y-30)$$
$$x^2 = -\frac{250}{3}(y-30)$$

To find the height of the bridge at points 10, 30, and 50 feet from the center of the bridge, let $x = 10$, 30, and 50 respectively:

$x = 10:$
$$10^2 = -\frac{250}{3}(y-30)$$
$$100 = -\frac{250}{3}(y-30)$$
$$300 = -250(y-30)$$
$$-1.2 = y-30$$
$$28.8 = y$$

$x = 30:$
$$30^2 = -\frac{250}{3}(y-30)$$
$$900 = -\frac{250}{3}(y-30)$$
$$2700 = -250(y-30)$$
$$-10.8 = y-30$$
$$19.2 = y$$

$x = 50:$
$$50^2 = -\frac{250}{3}(y-30)$$
$$2500 = -\frac{250}{3}(y-30)$$
$$7500 = -250(y-30)$$
$$-30 = y-30$$
$$0 = y$$

The height of the bridge is 28.8 feet at a distance of 10 feet from the center, 19.2 feet at a distance of 30 feet from the center, and 0 feet (i.e. ground level) at a distance of 50 feet from the center.

57. The parabola opens upward, so the equation must have the form $(x-h)^2 = 4a(y-k)$. Substituting the vertex $(h,k) = (3,-2)$ into this equation, we obtain $(x-3)^2 = 4a(y+2)$. Because the point $(5,-1)$ is on the parabola, we let $x = 5$ and $y = -1$ to determine a:

$$(5-3)^2 = 4a(-1+2)$$
$$2^2 = 4a(1)$$
$$4 = 4a$$
$$1 = a$$

Thus, the equation of the parabola is
$$(x-3)^2 = 4 \cdot 1(y+2)$$
$$(x-3)^2 = 4(y+2)$$

59. The parabola opens to the left, so the equation must have the form $(y-k)^2 = -4a(x-h)$. Substituting the vertex $(h,k) = (2, \ 3)$ into this equation, we obtain $(y-3)^2 = -4a(x-2)$. Because the point $(-2,-1)$ is on the parabola, we let $x = -2$ and $y = -1$ to determine a:

$$(-1-3)^2 = -4a(-2-2)$$
$$(-4)^2 = -4a(-4)$$
$$16 = 16a$$
$$1 = a$$

Thus, the equation of the parabola is
$$(y-3)^2 = -4 \cdot 1(x-2)$$
$$(y-3)^2 = -4(x-2)$$

61. $x^2 = 8y$

   (a) Let $x = 4$ and $y = 2$:

   $$4^2 \overset{?}{=} 8 \cdot 2$$
   $$16 = 16 \ \leftarrow \text{True}$$

   Thus, the point $(4, \ 2)$ in on the parabola.

   (b) Notice that $x^2 = 8y$ is of the form

   $x^2 = 4ay$, where $4a = 8$, so that $a = 2$. Thus, the focus of the parabola is $F(0, \ 2)$, and the directrix is $D: y = -2$. Now, the distance from the point $P(4,\ 2)$ to the focus

is $d(F,P)=\sqrt{(0-4)^2+(2-2)^2}=\sqrt{16}=4$,

and the distance from the point $P(4,\ 2)$ to
the directrix $D: y=-2$ is

$d(P,D)=2-(-2)=4$.

Thus, $d(F,P)=d(P,D)=4$.

63. If the distance from the point on the parabola to
the focus is 8 units, then by the definition of a
parabola, the distance from the same point on the
parabola to the directrix must also be 8 units.

65. Answers may vary. One possibility follows:

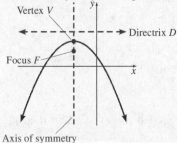

67. Begin with the graph of $y=x^2$, then shift the
graph 3 units to the left to obtain the graph of
$y=(x+3)^2$.

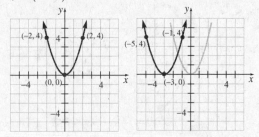

69. Notice that $y=(x+3)^2$ is of the form

$(x-h)^2=4a(y-k)$, where $4a=1$, so that

$a=\dfrac{1}{4}$ and $(h,k)=(-3,\ 0)$. Now, the graph of

an equation of the form $(x-h)^2=4a(y-k)$ will
be a parabola that opens upward with vertex at
(h,k), focus at $(h,k+a)$, and directrix of

$y=k-a$. Note that $k+a=0+\dfrac{1}{4}=\dfrac{1}{4}$ and

$k-a=0-\dfrac{1}{4}=-\dfrac{1}{4}$. Thus, the graph of

$y=(x+3)^2$ is a parabola that opens upward

with vertex $(-3,\ 0)$, focus $\left(-3,\ \dfrac{1}{4}\right)$, and

directrix $y=-\dfrac{1}{4}$. To help graph the parabola,

we plot the two points to the left and to the right

of the focus. Let $y=\dfrac{1}{4}$:

$\dfrac{1}{4}=(x+3)^2$

$\pm\dfrac{1}{2}=x+3$

$-3\pm\dfrac{1}{2}=x$

$x=-\dfrac{7}{2}$ or $x=-\dfrac{5}{2}$

The points $\left(-\dfrac{7}{2},\dfrac{1}{4}\right)$ and $\left(-\dfrac{5}{2},\dfrac{1}{4}\right)$ are on the

graph.

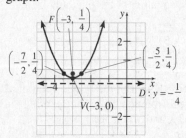

71. Note that $4(y+2)=(x-2)^2$ is of the form

$(x-h)^2=4a(y-k)$, where $4a=4$, so that

$a=1$ and $(h,k)=(2,-2)$. Now, the graph of an

equation of the form $(x-h)^2=4a(y-k)$ will be
a parabola that opens upward with vertex at
(h,k), focus at $(h,k+a)$, and directrix of

$y=k-a$. Note that $k+a=-2+1=-1$ and

$k-a=-2-1=-3$. Thus, the graph of

$4(y+2)=(x-2)^2$ is a parabola that opens

upward with vertex $(2,-2)$, focus $(2,-1)$, and

directrix $y=-3$. To help graph the parabola,

we plot the two points to the left and to the right
of the focus. Let $y=-1$:

$4(-1+2)=(x-2)^2$

$\quad\quad 4=(x-2)^2$

$\quad\quad \pm2=x-2$

$\quad 2\pm2=x$

$\quad x=0$ or $x=4$

The points $(0,-1)$ and $(4,-1)$ are on the graph.

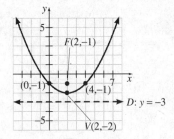

73. $x^2 = 24y$

$\dfrac{1}{24}x^2 = y$

Let $Y_1 = \dfrac{1}{24}x^2$.

 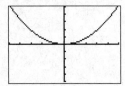

75. $y^2 = -6x$

$y = \pm\sqrt{-6x}$

Let $Y_1 = -\sqrt{-6x}$ and $Y_2 = \sqrt{-6x}$.

 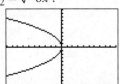

77. $(x-2)^2 = 4(y-4)$

$\dfrac{1}{4}(x-2)^2 = y-4$

$\dfrac{1}{4}(x-2)^2 + 4 = y$

Let $Y_1 = \dfrac{1}{4}(x-2)^2 + 4$.

 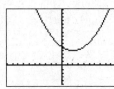

79. $(y+3)^2 = -8(x+2)$

$y+3 = \pm\sqrt{-8(x+2)}$

$y = -3 \pm \sqrt{-8(x+2)}$

Let $Y_1 = -3 - \sqrt{-8(x+2)}$ and

$Y_2 = -3 + \sqrt{-8(x+2)}$.

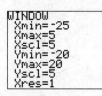 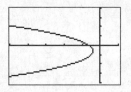

81. $x^2 + 4x + 12y + 16 = 0$

$x^2 + 4x + 16 = -12y$

$-\dfrac{1}{12}\left(x^2 + 4x + 16\right) = y$

Let $Y_1 = -\dfrac{1}{12}\left(x^2 + 4x + 16\right)$.

 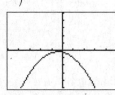

83. $y^2 - 8y - 4x + 20 = 0$

$y^2 - 8y = 4x - 20$

$y^2 - 8y + 16 = 4x - 20 + 16$

$(y-4)^2 = 4x - 4$

$y - 4 = \pm\sqrt{4x-4}$

$y = 4 \pm \sqrt{4x-4}$

Let $Y_1 = 4 - \sqrt{4x-4}$ and $Y_2 = 4 + \sqrt{4x-4}$.

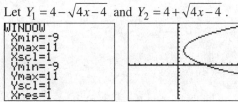

Section 12.4

Preparing for Ellipses

P1. Start: $x^2 + 10x$

Add: $\left(\dfrac{1}{2} \cdot 10\right)^2 = 25$

Result: $x^2 + 10x + 25$

Factored Form: $(x+5)^2$

P2. Begin with the graph of $y = x^2$, then shift the graph 2 units to the left to obtain the graph of $y = (x+2)^2$. Shift this graph down 1 unit to obtain the graph of $f(x) = (x+2)^2 - 1$.

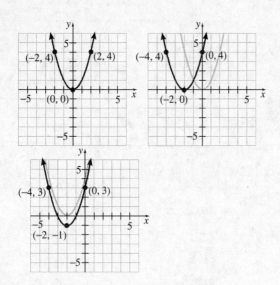

Section 12.4 Quick Checks

1. ellipse; foci

2. major axis

3. vertices

4. False. The equation of an ellipse centered at the origin with vertex $(a,0)$ and focus $(c,0)$ is

   $\dfrac{x^2}{a^2}+\dfrac{y^2}{b^2}=1$, where $b^2=a^2-c^2$.

5. $\dfrac{x^2}{9}+\dfrac{y^2}{4}=1$

   The larger number, 9, is in the denominator of the x^2-term. This means that the major axis is the x-axis and that the equation of the ellipse is of the form $\dfrac{x^2}{a^2}+\dfrac{y^2}{b^2}=1$, so that $a^2=9$ and

   $b^2=4$. The center of the ellipse is $(0,0)$.

   Because $b^2=a^2-c^2$, or $c^2=a^2-b^2$, we have that $c^2=9-4=5$, so that $c=\pm\sqrt{5}$. Since the major axis is the x-axis, the foci are $\left(-\sqrt{5},0\right)$

   and $\left(\sqrt{5},0\right)$. To find the x-intercepts (vertices), let $y=0$; to find the y-intercepts, let $x=0$:

x-intercepts: y-intercepts:

$\dfrac{x^2}{9}+\dfrac{0^2}{4}=1$ $\dfrac{0^2}{9}+\dfrac{y^2}{4}=1$

$\dfrac{x^2}{9}=1$ $\dfrac{y^2}{4}=1$

$x^2=9$ $y^2=4$

$x=\pm3$ $y=\pm2$

The intercepts are $(-3,0)$, $(3,0)$, $(0,-2)$, and $(0,2)$.

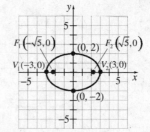

6. $\dfrac{x^2}{16}+\dfrac{y^2}{36}=1$

   The larger number, 36, is in the denominator of the y^2-term. This means that the major axis is the y-axis and that the equation of the ellipse is of the form $\dfrac{x^2}{b^2}+\dfrac{y^2}{a^2}=1$, so that $a^2=36$ and

   $b^2=16$. The center of the ellipse is $(0,0)$.

   Because $b^2=a^2-c^2$, or $c^2=a^2-b^2$, we have that $c^2=36-16=20$, so that

   $c=\pm\sqrt{20}=\pm2\sqrt{5}$. Since the major axis is the y-axis, the foci are $\left(0,-2\sqrt{5}\right)$ and $\left(0,2\sqrt{5}\right)$. To

   find the x-intercepts, let $y=0$; to find the y-intercepts (vertices), let $x=0$:

x-intercepts: y-intercepts:

$\dfrac{x^2}{16}+\dfrac{0^2}{36}=1$ $\dfrac{0^2}{16}+\dfrac{y^2}{36}=1$

$\dfrac{x^2}{16}=1$ $\dfrac{y^2}{36}=1$

$x^2=16$ $y^2=36$

$x=\pm4$ $y=\pm6$

The intercepts are $(-4,0)$, $(4,0)$, $(0,-6)$, and $(0,6)$.

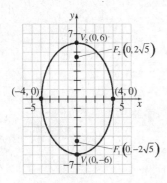

7. The given focus $(0, 3)$ and the given vertex $(0, 7)$ lie on the y-axis. Thus, the major axis is the y-axis, and the equation of the ellipse is of the form $\dfrac{x^2}{b^2} + \dfrac{y^2}{a^2} = 1$. The distance from the center of the ellipse to the vertex is $a = 7$ units. The distance from the center of the ellipse to the focus is $c = 3$ units. Because $b^2 = a^2 - c^2$, we have that $b^2 = 7^2 - 3^2 = 49 - 9 = 40$. So, the equation of the ellipse is $\dfrac{x^2}{40} + \dfrac{y^2}{49} = 1$.

To help graph the ellipse, find the x-intercepts:

Let $y = 0$: $\dfrac{x^2}{40} + \dfrac{0^2}{49} = 1$

$\dfrac{x^2}{40} = 1$

$x^2 = 40$

$x = \pm\sqrt{40} = \pm 2\sqrt{10}$

The x-intercepts are $\left(-2\sqrt{10}, 0\right)$ and $\left(2\sqrt{10}, 0\right)$.

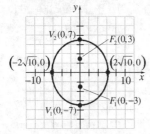

8. $(3, -1)$

9. $9x^2 + y^2 + 54x - 2y + 73 = 0$

$9x^2 + 54x + y^2 - 2y = -73$

$9\left(x^2 + 6x\right) + \left(y^2 - 2y\right) = -73$

$9\left(x^2 + 6x + 9\right) + \left(y^2 - 2y + 1\right) = -73 + 9(9) + 1$

$9(x+3)^2 + (y-1)^2 = 9$

$\dfrac{9(x+3)^2 + (y-1)^2}{9} = \dfrac{9}{9}$

$\dfrac{(x+3)^2}{1} + \dfrac{(y-1)^2}{9} = 1$

The center of the ellipse is $(h, k) = (-3, 1)$. Because the larger number, 9, is the denominator of the y^2-term, the major axis is parallel to the y-axis. Because $a^2 = 9$ and $b^2 = 1$, we have that $c^2 = a^2 - b^2 = 9 - 1 = 8$. The vertices are $a = 3$ units below and above the center at $V_1(-3, -2)$ and $V_2(-3, 4)$. The foci are $c = \sqrt{8} = 2\sqrt{2}$ units below and above the center at $F_1\left(-3,\ 1 - 2\sqrt{2}\right)$ and $F_2\left(-3,\ 1 + 2\sqrt{2}\right)$. We plot the points $b = 1$ unit to the left and right of the center point at $(-4, 1)$ and $(-2, 1)$.

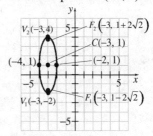

10. To solve the problem, we draw the ellipse on a Cartesian plane so that the center of the ellipse is at the origin and the major axis is along the x-axis. The equation of the ellipse is of the form $\dfrac{x^2}{a^2} + \dfrac{y^2}{b^2} = 1$. Since the length of the hall is 100 feet, the distance from the center of the room to each vertex is $a = \dfrac{100}{2} = 50$ feet. The distance from the center of the room to each focus is $c = 30$ feet.

Now, because $b^2 = a^2 - c^2$, we have that

$b^2 = 50^2 - 30^2 = 2500 - 900 = 1600$. Thus, the equation that describes the room is

$\dfrac{x^2}{2500} + \dfrac{y^2}{1600} = 1$.

The height of the room at its center is

$b = \sqrt{1600} = 40$ feet.

12.4 Exercises

11. c. The center of the ellipse is the origin, the major axis is the x-axis, and the vertices are $(\pm 4, 0)$, and the y-intercepts are $(0, \pm 3)$. Thus,

the equation is of the form $\dfrac{x^2}{a^2} + \dfrac{y^2}{b^2} = 1$, with

$a = 4$ and $b = 3$. Thus, the equation is:

$\dfrac{x^2}{4^2} + \dfrac{y^2}{3^2} = 1$

$\dfrac{x^2}{16} + \dfrac{y^2}{9} = 1$

13. d. The center of the ellipse is the origin, the major axis is the y-axis, and the vertices are $(0, \pm 4)$, and the x-intercepts are $(\pm 3, 0)$. Thus,

the equation is of the form $\dfrac{x^2}{b^2} + \dfrac{y^2}{a^2} = 1$, with

$a = 3$ and $b = 4$. Thus, the equation is:

$\dfrac{x^2}{3^2} + \dfrac{y^2}{4^2} = 1$

$\dfrac{x^2}{9} + \dfrac{y^2}{16} = 1$

15. $\dfrac{x^2}{25} + \dfrac{y^2}{16} = 1$

The larger number, 25, is in the denominator of the x^2-term. This means that the major axis is the x-axis and that the equation of the ellipse is

of the form $\dfrac{x^2}{a^2} + \dfrac{y^2}{b^2} = 1$, so that $a^2 = 25$ and

$b^2 = 16$. The center of the ellipse is $(0,0)$.

Because $b^2 = a^2 - c^2$, or $c^2 = a^2 - b^2$, we have

that $c^2 = 25 - 16 = 9$, so that $c = \pm\sqrt{9} = \pm 3$.

Since the major axis is the x-axis, the foci are $(-3, 0)$ and $(3, 0)$. To find the x-intercepts (vertices), let $y = 0$; to find the y-intercepts, let

$x = 0$:

x-intercepts:　　　　　　y-intercepts:

$\dfrac{x^2}{25} + \dfrac{0^2}{16} = 1$　　　　$\dfrac{0^2}{25} + \dfrac{y^2}{16} = 1$

$\dfrac{x^2}{25} = 1$　　　　　　$\dfrac{y^2}{16} = 1$

$x^2 = 25$　　　　　　$y^2 = 16$

$x = \pm 5$　　　　　　$y = \pm 4$

The intercepts are $(-5, 0)$, $(5, 0)$, $(0, -4)$, and $(0, 4)$.

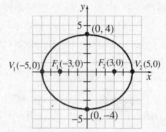

17. $\dfrac{x^2}{36} + \dfrac{y^2}{100} = 1$

The larger number, 100, is in the denominator of the y^2-term. This means that the major axis is the y-axis and that the equation of the ellipse is

of the form $\dfrac{x^2}{b^2} + \dfrac{y^2}{a^2} = 1$, so that $a^2 = 100$ and

$b^2 = 36$. The center of the ellipse is $(0,0)$.

Because $b^2 = a^2 - c^2$, or $c^2 = a^2 - b^2$, we have

that $c^2 = 100 - 36 = 64$, so that $c = \pm\sqrt{64} = \pm 8$.

Since the major axis is the y-axis, the foci are $(0, -8)$ and $(0, 8)$. To find the x-intercepts, let $y = 0$; to find the y-intercepts (vertices), let

$x = 0$:

x-intercepts:　　　　　　y-intercepts:

$\dfrac{x^2}{36} + \dfrac{0^2}{100} = 1$　　　　$\dfrac{0^2}{36} + \dfrac{y^2}{100} = 1$

$\dfrac{x^2}{36} = 1$　　　　　　$\dfrac{y^2}{100} = 1$

$x^2 = 36$　　　　　　$y^2 = 100$

$x = \pm 6$　　　　　　$y = \pm 10$

The intercepts are $(-6, 0)$, $(6, 0)$, $(0, -10)$, and $(0, 10)$.

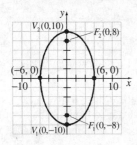

19. $\dfrac{x^2}{49} + \dfrac{y^2}{4} = 1$

The larger number, 49, is in the denominator of the x^2-term. This means that the major axis is the x-axis and that the equation of the ellipse is of the form $\dfrac{x^2}{a^2} + \dfrac{y^2}{b^2} = 1$, so that $a^2 = 49$ and $b^2 = 4$. The center of the ellipse is $(0,0)$.

Because $b^2 = a^2 - c^2$, or $c^2 = a^2 - b^2$, we have that $c^2 = 49 - 4 = 45$, so that $c = \pm\sqrt{45} = \pm 3\sqrt{5}$. Since the major axis is the x-axis, the foci are $\left(-3\sqrt{5}, 0\right)$ and $\left(3\sqrt{5}, 0\right)$. To find the x-intercepts (vertices), let $y = 0$; to find the y-intercepts, let $x = 0$:

x-intercepts: y-intercepts:

$\dfrac{x^2}{49} + \dfrac{0^2}{4} = 1$ $\dfrac{0^2}{49} + \dfrac{y^2}{4} = 1$

$\dfrac{x^2}{49} = 1$ $\dfrac{y^2}{4} = 1$

$x^2 = 49$ $y^2 = 4$

$x = \pm 7$ $y = \pm 2$

The intercepts are $(-7,0)$, $(7,0)$, $(0,-2)$, and $(0,2)$.

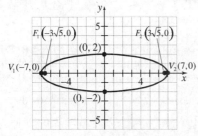

21. $x^2 + \dfrac{y^2}{49} = 1$

$\dfrac{x^2}{1} + \dfrac{y^2}{49} = 1$

The larger number, 49, is in the denominator of

the y^2-term. This means that the major axis is the y-axis and that the equation of the ellipse is of the form $\dfrac{x^2}{b^2} + \dfrac{y^2}{a^2} = 1$, so that $a^2 = 49$ and $b^2 = 1$. The center of the ellipse is $(0,0)$.

Because $b^2 = a^2 - c^2$, or $c^2 = a^2 - b^2$, we have that $c^2 = 49 - 1 = 48$, so that $c = \pm\sqrt{48} = \pm 4\sqrt{3}$. Since the major axis is the y-axis, the foci are $\left(0, -4\sqrt{3}\right)$ and $\left(0, 4\sqrt{3}\right)$. To find the x-intercepts, let $y = 0$; to find the y-intercepts (vertices), let $x = 0$:

x-intercepts: y-intercepts:

$x^2 + \dfrac{0^2}{49} = 1$ $0^2 + \dfrac{y^2}{49} = 1$

$x^2 = 1$ $y^2 = 49$

$x = \pm 1$ $y = \pm 7$

The intercepts are $(-1,0)$, $(1,0)$, $(0,-7)$, and $(0,7)$.

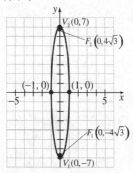

23. $4x^2 + y^2 = 16$

$\dfrac{4x^2 + y^2}{16} = \dfrac{16}{16}$

$\dfrac{x^2}{4} + \dfrac{y^2}{16} = 1$

The larger number, 16, is in the denominator of the y^2-term. This means that the major axis is the y-axis and that the equation of the ellipse is of the form $\dfrac{x^2}{b^2} + \dfrac{y^2}{a^2} = 1$, so that $a^2 = 16$ and $b^2 = 4$. The center of the ellipse is $(0,0)$.

Because $b^2 = a^2 - c^2$, or $c^2 = a^2 - b^2$, we have that $c^2 = 16 - 4 = 12$, so that $c = \pm\sqrt{12} = \pm 2\sqrt{3}$. Since the major axis is the y-axis, the foci are $\left(0, -2\sqrt{3}\right)$ and $\left(0, 2\sqrt{3}\right)$. To find the *x*-intercepts, let $y = 0$; to find the *y*-intercepts (vertices), let $x = 0$:

x-intercepts: *y*-intercepts:

$$4x^2 + 0^2 = 16 \qquad 4(0)^2 + y^2 = 16$$
$$4x^2 = 16 \qquad\qquad y^2 = 16$$
$$x^2 = 4 \qquad\qquad\quad y = \pm 4$$
$$x = \pm 2$$

The intercepts are $(-2, 0)$, $(2, 0)$, $(0, -4)$, and $(0, 4)$.

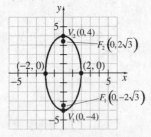

25. The given focus $(4, 0)$ and the given vertex $(6, 0)$ lie on the *x*-axis. Thus, the equation of the ellipse is of the form $\dfrac{x^2}{a^2} + \dfrac{y^2}{b^2} = 1$. The distance from the center of the ellipse to the vertex is $a = 6$ units. The distance from the center of the ellipse to the focus is $c = 4$ units. Because $b^2 = a^2 - c^2$, we have that $b^2 = 6^2 - 4^2 = 36 - 16 = 20$. Thus, the equation of the ellipse is $\dfrac{x^2}{36} + \dfrac{y^2}{20} = 1$.

To help graph the ellipse, find the *y*-intercepts:

Let $x = 0$: $\dfrac{0^2}{36} + \dfrac{y^2}{20} = 1$

$$\dfrac{y^2}{20} = 1$$
$$y^2 = 20$$
$$y = \pm\sqrt{20} = \pm 2\sqrt{5}$$

The *y*-intercepts are $\left(0, -2\sqrt{5}\right)$ and $\left(0, 2\sqrt{5}\right)$.

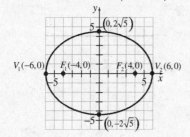

27. The given focus $(0, -4)$ and the given vertex $(0, 7)$ lie on the *y*-axis. Thus, the equation of the ellipse is of the form $\dfrac{x^2}{b^2} + \dfrac{y^2}{a^2} = 1$. The distance from the center of the ellipse to the vertex is $a = 7$ units. The distance from the center of the ellipse to the focus is $c = 4$ units. Because $b^2 = a^2 - c^2$, we have that $b^2 = 7^2 - 4^2 = 49 - 16 = 33$. Thus, the equation of the ellipse is $\dfrac{x^2}{33} + \dfrac{y^2}{49} = 1$.

To help graph the ellipse, find the *x*-intercepts:

Let $y = 0$: $\dfrac{x^2}{33} + \dfrac{0^2}{49} = 1$

$$\dfrac{x^2}{33} = 1$$
$$x^2 = 33$$
$$x = \pm\sqrt{33}$$

The *x*-intercepts are $\left(-\sqrt{33}, 0\right)$ and $\left(\sqrt{33}, 0\right)$.

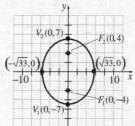

29. The given foci $(\pm 6, 0)$ and the given vertices $(\pm 10, 0)$ lie on the *x*-axis. The center of the ellipse is the midpoint between the two vertices (or foci). Thus, the center of the ellipse is $(0, 0)$, and its equation is of the form $\dfrac{x^2}{a^2} + \dfrac{y^2}{b^2} = 1$. The distance from the center of the ellipse to the vertex is $a = 10$ units. The distance from the center of the ellipse to the focus is $c = 6$ units. Because $b^2 = a^2 - c^2$, we have that $b^2 = 10^2 - 6^2 = 100 - 36 = 64$. Thus, the equation of the ellipse is $\dfrac{x^2}{100} + \dfrac{y^2}{64} = 1$.

To help graph the ellipse, find the *y*-intercepts:

Let $x = 0$: $\dfrac{0^2}{100} + \dfrac{y^2}{64} = 1$

$$\dfrac{y^2}{64} = 1$$

$$y^2 = 64$$

$$y = \pm 8$$

The y-intercepts are $(0, -8)$ and $(0, 8)$.

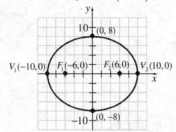

31. The given foci $(0, \pm 5)$ lie on the y-axis. The center of the ellipse is the midpoint between the two foci. Thus, center of the ellipse is $(0, 0)$ and its equation is of the form $\dfrac{x^2}{b^2} + \dfrac{y^2}{a^2} = 1$. Now, the length of the major axis is 16, so the vertices must be $(0, \pm 8)$, and the distance from the center of the ellipse to each vertex is $a = 8$ units. The distance from the center of the ellipse to each focus is $c = 5$ units. Because $b^2 = a^2 - c^2$, we have that $b^2 = 8^2 - 5^2 = 64 - 25 = 39$. Thus, the equation of the ellipse is $\dfrac{x^2}{39} + \dfrac{y^2}{64} = 1$.

To help graph the ellipse, we find the x-intercepts:

Let $y = 0$: $\dfrac{x^2}{39} + \dfrac{0^2}{64} = 1$

$$\dfrac{x^2}{39} = 1$$

$$x^2 = 39$$

$$x = \pm\sqrt{39}$$

The x-intercepts are $\left(-\sqrt{39}, 0\right)$ and $\left(\sqrt{39}, 0\right)$.

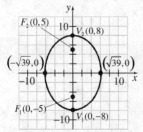

33. $\dfrac{(x-3)^2}{9} + \dfrac{(y+2)^2}{25} = 1$

The center of the ellipse is $(h, k) = (3, -2)$. Because the larger number, 25, is in the denominator of the y^2-term, the major axis is parallel to the y-axis. Because $a^2 = 25$ and $b^2 = 9$, we have that $c^2 = a^2 - b^2 = 25 - 9 = 16$. The vertices are $a = 5$ units below and above the center at $V_1(3, -7)$ and $V_2(3, 3)$. The foci are $c = \sqrt{16} = 4$ units below and above the center at $F_1(3, -6)$ and $F_2(3, 2)$. We plot the points $b = 3$ units to the left and right of the center point at $(0, -2)$ and $(6, -2)$.

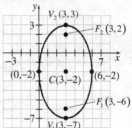

35. $\dfrac{(x+2)^2}{16} + \dfrac{(y-5)^2}{4} = 1$

The center of the ellipse is $(h, k) = (-2, 5)$. Because the larger number, 16, is in the denominator of the x^2-term, the major axis is parallel to the x-axis. Because $a^2 = 16$ and $b^2 = 4$, we have that $c^2 = a^2 - b^2 = 16 - 4 = 12$. The vertices are $a = 4$ units to the left and right of the center at $V_1(-6, 5)$ and $V_2(2, 5)$. The foci are $c = \sqrt{12} = 2\sqrt{3}$ units to the left and right of the center at $F_1\left(-2 - 2\sqrt{3},\ 5\right)$ and $F_2\left(-2 + 2\sqrt{3},\ 5\right)$. We plot the points $b = 2$ units above and below the center point at $(-2, 7)$ and $(-2, 3)$.

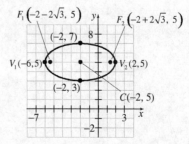

37. $(x-5)^2 + \dfrac{(y+1)^2}{49} = 1$

$\dfrac{(x-5)^2}{1} + \dfrac{(y+1)^2}{49} = 1$

The center of the ellipse is $(h,k) = (5,-1)$. Because the larger number, 49, is in the denominator of the y^2-term, the major axis is parallel to the y-axis. Because $a^2 = 49$ and $b^2 = 1$, we have that $c^2 = a^2 - b^2 = 49 - 1 = 48$. The vertices are $a = 7$ units below and above the center at $V_1(5,-8)$ and $V_2(5,6)$. The foci are $c = \sqrt{48} = 4\sqrt{3}$ units below and above the center at $F_1\left(5, -1 - 4\sqrt{3}\right)$ and $F_2\left(5, -1 + 4\sqrt{3}\right)$. We plot the points $b = 1$ unit to the left and right of the center point at $(4,-1)$ and $(6,-1)$.

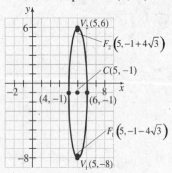

39. $4(x+2)^2 + 16(y-1)^2 = 64$

$\dfrac{4(x+2)^2 + 16(y-1)^2}{64} = \dfrac{64}{64}$

$\dfrac{(x+2)^2}{16} + \dfrac{(y-1)^2}{4} = 1$

The center of the ellipse is $(h,k) = (-2,1)$. Because the larger number, 16, is in the denominator of the x^2-term, the major axis is parallel to the x-axis. Because $a^2 = 16$ and $b^2 = 4$, we have that $c^2 = a^2 - b^2 = 16 - 4 = 12$. The vertices are $a = 4$ units to the left and right of the center at $V_1(-6,1)$ and $V_2(2,1)$. The foci are $c = \sqrt{12} = 2\sqrt{3}$ units to the left and right of the center at $F_1\left(-2 - 2\sqrt{3},\ 1\right)$ and $F_2\left(-2 + 2\sqrt{3},\ 1\right)$. We plot the points $b = 2$ units above and below the center point at $(-2,3)$ and $(-2,-1)$.

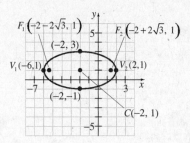

41. $4x^2 + y^2 - 24x + 2y - 63 = 0$

$4x^2 - 24x + y^2 + 2y = 63$

$4\left(x^2 - 6x\right) + \left(y^2 + 2y\right) = 63$

$4\left(x^2 - 6x + 9\right) + \left(y^2 + 2y + 1\right) = 63 + 4(9) + 1$

$4(x-3)^2 + (y+1)^2 = 100$

$\dfrac{4(x-3)^2 + (y+1)^2}{100} = \dfrac{100}{100}$

$\dfrac{(x-3)^2}{25} + \dfrac{(y+1)^2}{100} = 1$

The center of the ellipse is $(h,k) = (3,-1)$. Because the larger number, 100, is in the denominator of the y^2-term, the major axis is parallel to the y-axis. Because $a^2 = 100$ and $b^2 = 25$, we have that $c^2 = a^2 - b^2 = 100 - 25 = 75$. The vertices are $a = 10$ units below and above the center at $V_1(3,-11)$ and $V_2(3,9)$. The foci are $c = \sqrt{75} = 5\sqrt{3}$ units below and above the center at $F_1\left(3, -1 - 5\sqrt{3}\right)$ and $F_2\left(3, -1 + 5\sqrt{3}\right)$. We plot the points $b = 5$ units to the left and right of the center point at $(-2,-1)$ and $(8,-1)$.

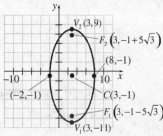

43. (a) To solve this problem, we draw the ellipse on a Cartesian plane so that the x-axis coincides with the water and the y-axis passes through the center of the arch. Thus, the origin is the center of the ellipse. Now, the "center" of the arch is 10 meters above the water, so the point $(0,10)$ is on the ellipse. The river is 30 meters wide, so the two points $(-15,0)$ and $(15,0)$ are the two vertices of the ellipse and the major axis is along the x-axis.

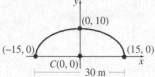

The equation of the ellipse must have the form $\dfrac{x^2}{a^2}+\dfrac{y^2}{b^2}=1$. The distance from the center of the ellipse to each vertex is $a=15$. Also, the height of the semi-ellipse is 10 feet, so $b=10$. Thus, the equation of the ellipse is

$$\frac{x^2}{15^2}+\frac{y^2}{10^2}=1$$

$$\frac{x^2}{225}+\frac{y^2}{100}=1$$

(b) To determine if the barge can fit through the opening of the bridge, we center it beneath the arch so that the points $(-9,7)$ and $(9,7)$ represent the top corners of the barge. Now, we determine the height of the arch above the water at points 9 meters to the left and right of the center by substituting $x=9$ into the equation of the ellipse and solving for y:

$$\frac{9^2}{225}+\frac{y^2}{100}=1$$

$$\frac{9}{25}+\frac{y^2}{100}=1$$

$$\frac{y^2}{100}=\frac{16}{25}$$

$$y^2=64$$

$$y=8$$

At points 9 meters from the center of the river, the arch is 8 meters above the surface of the water. Thus, the barge can fit through the opening with about 1 meter of clearance above the top corners.

(c) No. From part (b), the barge only has 1 meter of clearance when the water level is at it normal stage. Thus, if the water level increases by 1.1 meters, then the barge will not be able to pass under the bridge.

45. To find the perihelion of Earth, we recognize that Perihilion = 2(Mean distance) − Aphelion .
Thus, the perihelion of Earth is
   $2 \cdot 93 - 94.5 = 91.5$ million miles.
To find the equation of the elliptical orbit of Earth, we draw the ellipse on a Cartesian plane so that the center of the ellipse is at the origin and the major axis is along the x-axis. The equation of the ellipse is of the form
$\dfrac{x^2}{a^2}+\dfrac{y^2}{b^2}=1$. Now, the mean distance of Earth from the Sun is 93 million miles, so the distance from the center of the orbit to each vertex is $a=93$ million miles. Since the aphelion of Earth is 94.5 million miles, the distance from the center of the orbit to each focus is $c=94.5-93=1.5$ million miles.

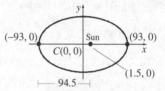

Now, because $b^2=a^2-c^2$, we have that $b^2=93^2-1.5^2=8649-2.25=8646.75$. Thus, the equation that describes the orbit of Earth is:
$$\frac{x^2}{8649}+\frac{y^2}{8646.75}=1.$$

47. The mean distance of Jupiter from the Sun is $507-23.2=483.8$ million miles.
To find the perihelion of Jupiter, we recognize that Perihilion = 2(Mean distance) − Aphelion .
Thus, the perihelion of Jupiter is
   $2 \cdot 483.8 - 507 = 460.6$ million miles.
To find the equation of the elliptical orbit of Jupiter, we draw the ellipse on a Cartesian plane so that the center of the ellipse is at the origin and the major axis is along the x-axis. The equation of the ellipse is of the form
$\dfrac{x^2}{a^2}+\dfrac{y^2}{b^2}=1$. Now, we found above that the mean distance of Jupiter from the Sun is 483.8 million miles, so the distance from the center of the orbit to each vertex is $a=483.8$ million miles. We are given the distance from the center of the orbit to the Sun (that is, to each focus) is $c=23.2$ million miles.

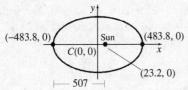

Now, because $b^2 = a^2 - c^2$, we have that

$$b^2 = 483.8^2 - 23.2^2$$
$$= 234,062.44 - 538.24$$
$$= 233,524.2$$

Thus, the equation that describes the orbit of

Jupiter is: $\dfrac{x^2}{234,062.44} + \dfrac{y^2}{233,524.2} = 1$.

49. The major axis of the ellipse is the line $y = 2$, which is parallel to the x-axis, so the equation of

    the ellipse is of the form $\dfrac{(x-h)^2}{a^2} + \dfrac{(y-k)^2}{b^2} = 1$.

    The center is $(h,k) = (1,2)$. The vertices are $(-3, 2)$ and $(5, 2)$, so $a = 5 - 1 = 4$. The points $(1, -1)$ and $(1, 5)$ are the points on the ellipse that are below and above the center, so $b = 5 - 2 = 3$. Thus, the equation is

    $$\frac{(x-1)^2}{4^2} + \frac{(y-2)^2}{3^2} = 1$$
    $$\frac{(x-1)^2}{16} + \frac{(y-2)^2}{9} = 1$$

51. The major axis of the ellipse is the line $x = 2$, which is parallel to the y-axis, so the equation of

    the ellipse is of the form $\dfrac{(x-h)^2}{b^2} + \dfrac{(y-k)^2}{a^2} = 1$.

    The center is $(h,k) = (2,0)$. The vertices are $(2, -4)$ and $(2, 4)$, so $a = 4 - 0 = 4$.

    The points $(0, 0)$ and $(4, 0)$ are the points on the ellipse that are left and right of the center, so $b = 4 - 2 = 2$. Thus, the equation is

    $$\frac{(x-2)^2}{2^2} + \frac{(y-0)^2}{4^2} = 1$$
    $$\frac{(x-2)^2}{4} + \frac{y^2}{16} = 1$$

53. Let $a = b$, then

    $$\frac{x^2}{a^2} + \frac{y^2}{b^2} = 1$$
    $$\frac{x^2}{a^2} + \frac{y^2}{a^2} = 1$$
    $$a^2 \left(\frac{x^2}{a^2} + \frac{y^2}{a^2} \right) = a^2 (1)$$
    $$x^2 + y^2 = a^2$$

    which is the equation of a circle with center $(0,0)$ and radius a.

    Because $c^2 = a^2 - b^2$, we have that $c^2 = a^2 - a^2 = 0$, so that $c = 0$. This means that the foci are a distance of 0 units from the center point. In other words, the foci are located at the center point.

55. Since the center of the ellipse is the origin O, the major axis is along the x-axis, and point B is the y-intercept, then by definition, $F = (c,0)$ and $B = (0,b)$. Thus, $d(O,F) = c$ and $d(O,B) = b$. Now, the sum of the distances from the two foci to a point on the ellipse is defined to be $2a$. Since B is a point on the ellipse that is midway between the two foci, then the distance from F to B must be half this total distance:

    $$d(F,B) = \frac{1}{2}(2a) = a.$$

57. $f(x) = \dfrac{5}{x+2}$

    $f(5) = \dfrac{5}{5+2} = \dfrac{5}{7} \approx 0.71429$

    $f(10) = \dfrac{5}{10+2} = \dfrac{5}{12} \approx 0.41667$

    $f(100) = \dfrac{5}{100+2} = \dfrac{5}{102} \approx 0.04902$

    $f(1000) = \dfrac{5}{1000+2} = \dfrac{5}{1002} \approx 0.00499$

x	5	10	100	100
$f(x)$	0.71429	0.41667	0.04902	0.00499

59. $f(x) = \dfrac{2x+1}{x-3}$

$f(5) = \dfrac{2(5)+1}{5-3} = \dfrac{10+1}{5-3} = \dfrac{11}{2} = 5.5$

$f(10) = \dfrac{2(10)+1}{10-3} = \dfrac{20+1}{10-3} = \dfrac{21}{7} = 3$

$f(100) = \dfrac{2(100)+1}{100-3} = \dfrac{200+1}{100-3} = \dfrac{201}{97} \approx 2.07216$

$f(1000) = \dfrac{2(1000)+1}{1000-3} = \dfrac{2000+1}{1000-3} = \dfrac{2001}{997} \approx 2.00702$

x	5	10	100	100
$f(x)$	5.5	3	2.07216	2.00702

61. $f(x) = \dfrac{x^2+3x+1}{x+1}$; $g(x) = x+2$

$f(5) = \dfrac{5^2+3(5)+1}{5+1} = \dfrac{41}{6} \approx 6.83333$

$f(10) = \dfrac{10^2+3(10)+1}{10+1} = \dfrac{131}{11} \approx 11.90909$

$f(100) = \dfrac{100^2+3(100)+1}{100+1} = \dfrac{10,301}{101} \approx 101.99010$

$f(1000) = \dfrac{1000^2+3(1000)+1}{1000+1}$

$= \dfrac{1,003,001}{1001} \approx 1001.99900$

$g(5) = 5+2 = 7$

$g(10) = 10+2 = 12$

$g(100) = 100+2 = 102$

$g(1000) = 1000+2 = 1002$

x	5	10	100	100
$f(x)$	6.83333	11.90909	101.99010	1001.99900
$g(x)$	7	12	102	1002

63. In Problems 57 and 58, the degrees of the numerators are less than the degrees of the denominators. In Problems 59 and 60, the degrees of the numerators and denominators are the same.
Conjecture 1: If the degree of the numerator of a rational function is less than the degree of the denominator, then as x increases, the value of the function will approach zero (0).
Conjecture 2: If the degree of the numerator of a rational function equals the degree of the

denominator, then as x increases, the value of the function will approach the ratio of the leading coefficients of the numerator and denominator.

65. $\dfrac{x^2}{25} + \dfrac{y^2}{16} = 1$

$\dfrac{y^2}{16} = 1 - \dfrac{x^2}{25}$

$y^2 = 16\left(1 - \dfrac{x^2}{25}\right)$

$y = \pm\sqrt{16\left(1 - \dfrac{x^2}{25}\right)} = \pm 4\sqrt{1 - \dfrac{x^2}{25}}$

Let $Y_1 = 4\sqrt{1 - \dfrac{x^2}{25}}$ and $Y_2 = -4\sqrt{1 - \dfrac{x^2}{25}}$.

```
WINDOW
Xmin=-9.4
Xmax=9.4
Xscl=1
Ymin=-6.2
Ymax=6.2
Yscl=1
Xres=1
```

67. $4x^2 + y^2 = 16$

$y^2 = 16 - 4x^2$

$y = \pm\sqrt{16 - 4x^2}$

Let $Y_1 = \sqrt{16 - 4x^2}$ and $Y_2 = -\sqrt{16 - 4x^2}$.

```
WINDOW
Xmin=-9.4
Xmax=9.4
Xscl=1
Ymin=-6.2
Ymax=6.2
Yscl=1
Xres=1
```

69. $\dfrac{(x-3)^2}{9} + \dfrac{(y+2)^2}{25} = 1$

$\dfrac{(y+2)^2}{25} = 1 - \dfrac{(x-3)^2}{9}$

$(y+2)^2 = 25\left(1 - \dfrac{(x-3)^2}{9}\right)$

$y+2 = \pm\sqrt{25\left(1 - \dfrac{(x-3)^2}{9}\right)}$

$y+2 = \pm 5\sqrt{1 - \dfrac{(x-3)^2}{9}}$

$y = -2 \pm 5\sqrt{1 - \dfrac{(x-3)^2}{9}}$

Let $Y_1 = -2 + 5\sqrt{1 - \dfrac{(x-3)^2}{9}}$ and

$Y_2 = -2 - 5\sqrt{1 - \dfrac{(x-3)^2}{9}}$.

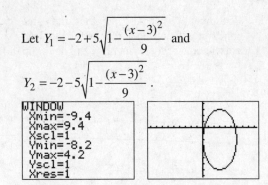

71. $\dfrac{(x+2)^2}{16} + \dfrac{(y-5)^2}{4} = 1$

$$\dfrac{(y-5)^2}{4} = 1 - \dfrac{(x+2)^2}{16}$$

$$(y-5)^2 = 4\left(1 - \dfrac{(x+2)^2}{16}\right)$$

$$y - 5 = \pm\sqrt{4\left(1 - \dfrac{(x+2)^2}{16}\right)}$$

$$y - 5 = \pm 2\sqrt{1 - \dfrac{(x+2)^2}{16}}$$

$$y = 5 \pm 2\sqrt{1 - \dfrac{(x+2)^2}{16}}$$

Let $Y_1 = 5 + 2\sqrt{1 - \dfrac{(x+2)^2}{16}}$ and

$Y_2 = 5 - 2\sqrt{1 - \dfrac{(x+2)^2}{16}}$.

```
WINDOW
 Xmin=-9.4
 Xmax=9.4
 Xscl=1
 Ymin=-4.2
 Ymax=8.2
 Yscl=1
 Xres=1
```

Section 12.5

Preparing for Hyperbolas

P1. Start: $x^2 - 5x$

Add: $\left(\dfrac{1}{2} \cdot (-5)\right)^2 = \dfrac{25}{4}$

Result: $x^2 - 5x + \dfrac{25}{4}$

Factored Form: $\left(x - \dfrac{5}{2}\right)^2$

P2. $y^2 = 64$

$y = \pm\sqrt{64}$

$y = \pm 8$

The solution set is $\{-8,\ 8\}$.

Section 12.5 Quick Checks

1. hyperbola

2. transverse axis

3. conjugate axis

4. $\dfrac{x^2}{36} - \dfrac{y^2}{64} = 1$

Notice the equation is of the form $\dfrac{x^2}{a^2} - \dfrac{y^2}{b^2} = 1$.

Because the x^2-term is first, the transverse axis is the x-axis and the hyperbola opens left and right. The center of the hyperbola is the origin. We have that $a^2 = 36$ and $b^2 = 64$. Because $c^2 = a^2 + b^2$, we have that $c^2 = 36 + 64 = 100$, so that $c = \sqrt{100} = 10$. The vertices are $(\pm a, 0) = (\pm 6, 0)$, and the foci are $(\pm c, 0) = (\pm 10, 0)$. To help graph the hyperbola, we plot the points on the graph above and below the foci. Let $x = \pm 10$:

$$\dfrac{(\pm 10)^2}{36} - \dfrac{y^2}{64} = 1$$

$$\dfrac{100}{36} - \dfrac{y^2}{64} = 1$$

$$\dfrac{25}{9} - \dfrac{y^2}{64} = 1$$

$$-\dfrac{y^2}{64} = -\dfrac{16}{9}$$

$$y^2 = \dfrac{1024}{9}$$

$$y = \pm\dfrac{32}{3}$$

The points above and below the foci are

$\left(-10, -\dfrac{32}{3}\right)$, $\left(-10, \ \dfrac{32}{3}\right)$, $\left(10, -\dfrac{32}{3}\right)$, and

$\left(10, \ \dfrac{32}{3}\right)$.

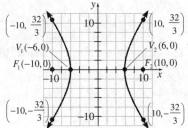

5. $\dfrac{y^2}{9} - \dfrac{x^2}{16} = 1$

Notice the equation is of the form $\dfrac{y^2}{a^2} - \dfrac{x^2}{b^2} = 1$.

Because the y^2-term is first, the transverse axis is the y-axis and the hyperbola opens up and down. The center of the hyperbola is the origin. We have that $a^2 = 9$ and $b^2 = 16$. Because $c^2 = a^2 + b^2$, we have that $c^2 = 9 + 16 = 25$, so that $c = \sqrt{25} = 5$. The vertices are $(0, \pm a) = (0, \pm 3)$, and the foci are $(0, \pm c) = (0, \pm 5)$. To help graph the hyperbola, we plot the points on the graph to the left and right of the foci. Let $y = \pm 5$:

$$\dfrac{(\pm 5)^2}{9} - \dfrac{x^2}{16} = 1$$

$$\dfrac{25}{9} - \dfrac{x^2}{16} = 1$$

$$-\dfrac{x^2}{16} = -\dfrac{16}{9}$$

$$x^2 = \dfrac{256}{9}$$

$$x = \pm \dfrac{16}{3}$$

The points to the left and right of the foci are

$\left(-\dfrac{16}{3}, -5\right)$, $\left(-\dfrac{16}{3}, \ 5\right)$, $\left(\dfrac{16}{3}, -5\right)$, and

$\left(\dfrac{16}{3}, \ 5\right)$.

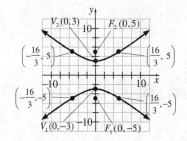

6. False.

7. The given vertices $(\pm 4, 0)$ and focus $(6, 0)$ all lie on the x-axis. Thus, the transverse axis is the x-axis and the hyperbola opens left and right. The center of the hyperbola is the midpoint between the two vertices. Therefore, the center is $(0, 0)$ and the equation of the hyperbola is of

the form $\dfrac{x^2}{a^2} - \dfrac{y^2}{b^2} = 1$. Now, the distance from

the center $(0, 0)$ to each vertex is $a = 4$ units. Likewise, the distance from the center to the given focus $(6, 0)$ is $c = 6$ units. Also, $b^2 = c^2 - a^2$, so $b^2 = 6^2 - 4^2 = 36 - 16 = 20$. Thus, the equation of the hyperbola is

$\dfrac{x^2}{16} - \dfrac{y^2}{20} = 1$.

To help graph the hyperbola, we first find the focus that was not given. Since the center is at $(0, 0)$ and since one focus is at $(6, 0)$, the other focus must be at $(-6, 0)$. Next, plot the points on the graph above and below the foci. Let $x = \pm 6$:

$$\dfrac{(\pm 6)^2}{16} - \dfrac{y^2}{20} = 1$$

$$\dfrac{36}{16} - \dfrac{y^2}{20} = 1$$

$$\dfrac{9}{4} - \dfrac{y^2}{20} = 1$$

$$-\dfrac{y^2}{20} = -\dfrac{5}{4}$$

$$y^2 = 25$$

$$y = \pm 5$$

The points above and below of the foci are $(-6, -5)$, $(-6, \ 5)$, $(6, -5)$, and $(6, \ 5)$.

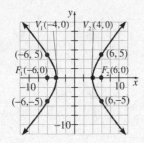

8. $y = -\dfrac{b}{a}x$; $y = \dfrac{b}{a}x$

9. $x^2 - 9y^2 = 9$

$$\frac{x^2 - 9y^2}{9} = \frac{9}{9}$$

$$\frac{x^2}{9} - \frac{y^2}{1} = 1$$

Notice that the equation is of the form
$\dfrac{x^2}{a^2} - \dfrac{y^2}{b^2} = 1$. The center of the hyperbola is
$(0,0)$. Because the x^2-term is first, the
hyperbola opens left and right. The transverse
axis is along the x-axis. We have that $a^2 = 9$
and $b^2 = 1$. Because $c^2 = a^2 + b^2$, we have that
$c^2 = 9 + 1 = 10$, so that $c = \sqrt{10}$. The vertices
are $(\pm a, 0) = (\pm 3, 0)$, and the foci are
$(\pm c, 0) = \left(\pm\sqrt{10}, 0\right)$. The asymptotes are of the
form $y = -\dfrac{b}{a}x$ and $y = \dfrac{b}{a}x$. Since $a = 3$ and
$b = 1$, the equations of the asymptotes are
$y = -\dfrac{1}{3}x$ and $y = \dfrac{1}{3}x$. To help graph the
hyperbola, we form the rectangle using the
points $(\pm a, 0) = (\pm 3, 0)$ and $(0, \pm b) = (0, \pm 1)$.
The diagonals are the asymptotes.

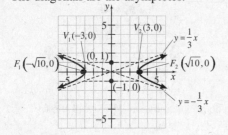

10. $\dfrac{y^2}{16} - \dfrac{x^2}{9} = 1$

Notice the equation is of the form $\dfrac{y^2}{a^2} - \dfrac{x^2}{b^2} = 1$.

Because the y^2-term is first, the transverse is
along the y-axis and the hyperbola opens up and
down. The center of the hyperbola is $(0,0)$. We
have that $a^2 = 16$ and $b^2 = 9$. Because
$c^2 = a^2 + b^2$, we have that $c^2 = 16 + 9 = 25$, so
that $c = \sqrt{25} = 5$. The vertices are
$(0, \pm a) = (0, \pm 4)$, and the foci are
$(0, \pm c) = (0, \pm 5)$. The asymptotes are of the
form $y = -\dfrac{a}{b}x$ and $y = \dfrac{a}{b}x$. Since $a = 4$ and
$b = 3$, the equations of the asymptotes are
$y = -\dfrac{4}{3}x$ and $y = \dfrac{4}{3}x$. To help graph the
hyperbola, we form the rectangle using the
points $(0, \pm a) = (0, \pm 4)$ and $(\pm b, 0) = (\pm 3, 0)$.
The diagonals are the asymptotes.

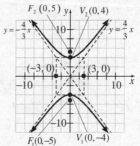

12.5 Exercises

11. b. The transverse axis is the x-axis and the
hyperbola opens left and right. Also, the
distance between the vertices $(\pm 1, 0)$ and the
center $(0,0)$ is $a = 1$ unit. This means that the
equation of the hyperbola is of the form

$$\frac{x^2}{a^2} - \frac{y^2}{b^2} = 1$$

$$\frac{x^2}{1^2} - \frac{y^2}{b^2} = 1$$

$$x^2 - \frac{y^2}{b^2} = 1$$

Of the list of provided equations, only equation

(b): $x^2 - \dfrac{y^2}{4} = 1$ is of this form.

13. a. The transverse axis is the *x*-axis and the hyperbola opens left and right. Also, the distance between the vertices $(\pm 2, 0)$ and the center $(0, 0)$ is $a = 2$ units. This means that the equation of the hyperbola is of the form

$$\frac{x^2}{a^2} - \frac{y^2}{b^2} = 1$$

$$\frac{x^2}{2^2} - \frac{y^2}{b^2} = 1$$

$$\frac{x^2}{4} - \frac{y^2}{b^2} = 1$$

Of the list of provided equations, only equation

(a): $\dfrac{x^2}{4} - y^2 = 1$ is of this form.

15. $\dfrac{x^2}{4} - \dfrac{y^2}{16} = 1$

Notice the equation is of the form $\dfrac{x^2}{a^2} - \dfrac{y^2}{b^2} = 1$.

Because the x^2-term is first, the transverse axis is the *x*-axis and the hyperbola opens left and right. The center of the hyperbola is the origin. We have that $a^2 = 4$ and $b^2 = 16$. Because $c^2 = a^2 + b^2$, we have that $c^2 = 4 + 16 = 20$, so that $c = \sqrt{20} = 2\sqrt{5}$. The vertices are $(\pm a, 0) = (\pm 2, 0)$, and the foci are $(\pm c, 0) = \left(\pm 2\sqrt{5}, 0\right)$. Since $a = 2$ and $b = 4$, the equations of the asymptotes are

$y = \dfrac{4}{2}x = 2x$ and $y = -\dfrac{4}{2}x = -2x$. To help graph the hyperbola, we form the rectangle using the points $(\pm a, 0) = (\pm 2, 0)$ and $(0, \pm b) = (0, \pm 4)$. The diagonals are the asymptotes.

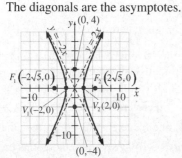

17. $\dfrac{y^2}{25} - \dfrac{x^2}{36} = 1$

Notice the equation is of the form $\dfrac{y^2}{a^2} - \dfrac{x^2}{b^2} = 1$.

Because the y^2-term is first, the transverse axis is the *y*-axis and the hyperbola opens up and down. The center of the hyperbola is the origin. We have that $a^2 = 25$ and $b^2 = 36$. Because $c^2 = a^2 + b^2$, we have that $c^2 = 25 + 36 = 61$, so that $c = \sqrt{61}$. The vertices are $(0, \pm a) = (0, \pm 5)$, and the foci are $(0, \pm c) = \left(0, \pm\sqrt{61}\right)$. To help graph the hyperbola, we plot the points on the graph to the left and right of the foci. Let $y = \pm\sqrt{61}$:

$$\frac{\left(\pm\sqrt{61}\right)^2}{25} - \frac{x^2}{36} = 1$$

$$\frac{61}{25} - \frac{x^2}{36} = 1$$

$$-\frac{x^2}{36} = -\frac{36}{25}$$

$$x^2 = \frac{1296}{25}$$

$$x = \pm\frac{36}{5}$$

The points to the left and right of the foci are

$\left(\dfrac{36}{5}, \sqrt{61}\right)$, $\left(-\dfrac{36}{5}, \sqrt{61}\right)$, $\left(\dfrac{36}{5}, -\sqrt{61}\right)$, and $\left(-\dfrac{36}{5}, -\sqrt{61}\right)$.

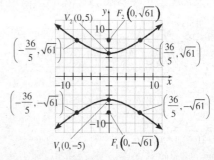

19. $4x^2 - y^2 = 36$

$$\frac{4x^2 - y^2}{36} = \frac{36}{36}$$

$$\frac{x^2}{9} - \frac{y^2}{36} = 1$$

Notice the equation is of the form $\dfrac{x^2}{a^2} - \dfrac{y^2}{b^2} = 1$.

Because the x^2-term is first, the transverse axis is the x-axis and the hyperbola opens left and right. The center of the hyperbola is the origin. We have that $a^2 = 9$ and $b^2 = 36$. Because $c^2 = a^2 + b^2$, we have that $c^2 = 9 + 36 = 45$, so that $c = \sqrt{45} = 3\sqrt{5}$. The vertices are $(\pm a, 0) = (\pm 3, 0)$, and the foci are $(\pm c, 0) = \left(\pm 3\sqrt{5}, 0\right)$. To help graph the hyperbola, we plot the points on the graph above and below the foci. Let $x = \pm 3\sqrt{5}$:

$$\frac{\left(\pm 3\sqrt{5}\right)^2}{9} - \frac{y^2}{36} = 1$$

$$\frac{45}{9} - \frac{y^2}{36} = 1$$

$$5 - \frac{y^2}{36} = 1$$

$$-\frac{y^2}{36} = -4$$

$$y^2 = 144$$

$$y = \pm 12$$

The points above and below the foci are $\left(3\sqrt{5}, 12\right)$, $\left(3\sqrt{5}, -12\right)$, $\left(-3\sqrt{5}, 12\right)$, and $\left(-3\sqrt{5}, -12\right)$.

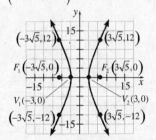

21. $25y^2 - x^2 = 100$

$$\frac{25y^2 - x^2}{100} = \frac{100}{100}$$

$$\frac{y^2}{4} - \frac{x^2}{100} = 1$$

Notice the equation is of the form $\dfrac{y^2}{a^2} - \dfrac{x^2}{b^2} = 1$.

Because the y^2-term is first, the transverse is along the y-axis and the hyperbola opens up and

down. The center of the hyperbola is $(0,0)$. We have that $a^2 = 4$ and $b^2 = 100$. Because $c^2 = a^2 + b^2$, we have that $c^2 = 100 + 4 = 104$, so that $c = \sqrt{104} = 2\sqrt{26}$. The vertices are $(0, \pm a) = (0, \pm 2)$, and the foci are $(0, \pm c) = \left(0, \pm 2\sqrt{26}\right)$. Since $a = 2$ and $b = 10$, the equations of the asymptotes are $y = \dfrac{2}{10}x = \dfrac{1}{5}x$ and $y = -\dfrac{2}{10}x = -\dfrac{1}{5}x$. To help graph the hyperbola, we form the rectangle using the points $(0, \pm a) = (0, \pm 2)$ and $(\pm b, 0) = (\pm 10, 0)$. The diagonals are the asymptotes.

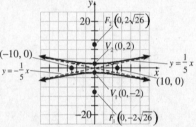

23. The given center $(0,0)$, focus $(3,0)$, and vertex $(2,0)$ all lie on the x-axis. Thus, the transverse axis is the x-axis and the hyperbola opens left and right. This means that the equation of the hyperbola is of the form $\dfrac{x^2}{a^2} - \dfrac{y^2}{b^2} = 1$. Now, the distance between the given vertex $(2,0)$ and the center $(0,0)$ is $a = 2$ units. Likewise, the distance between the given focus $(3,0)$ and the center is $c = 3$ units. Because $b^2 = c^2 - a^2$, we have that $b^2 = 3^2 - 2^2 = 9 - 4 = 5$. Thus, the equation of the hyperbola is $\dfrac{x^2}{4} - \dfrac{y^2}{5} = 1$.

To help graph the hyperbola, we first find the vertex and focus that were not given. Since the center is at $(0,0)$ and since one vertex is at $(2,0)$, the other vertex must be at $(-2,0)$. Similarly, one focus is at $(3,0)$, so the other focus is at $(-3,0)$. Next, plot the points on the graph above and below the foci. Let $x = \pm 3$:

$$\frac{(\pm 3)^2}{4} - \frac{y^2}{5} = 1$$

$$\frac{9}{4} - \frac{y^2}{5} = 1$$

$$-\frac{y^2}{5} = -\frac{5}{4}$$

$$y^2 = \frac{25}{4}$$

$$y = \pm\frac{5}{2}$$

The points above and below the foci are $\left(3, \frac{5}{2}\right)$, $\left(3, -\frac{5}{2}\right)$, $\left(-3, \frac{5}{2}\right)$, and $\left(-3, -\frac{5}{2}\right)$.

$$\frac{(\pm 7)^2}{25} - \frac{x^2}{24} = 1$$

$$\frac{49}{25} - \frac{x^2}{24} = 1$$

$$-\frac{x^2}{24} = -\frac{24}{25}$$

$$x^2 = \frac{576}{25}$$

$$x = \pm\frac{24}{5}$$

The points to the left and right of the foci are $\left(\frac{24}{5}, 7\right)$, $\left(-\frac{24}{5}, 7\right)$, $\left(\frac{24}{5}, -7\right)$, and $\left(-\frac{24}{5}, -7\right)$.

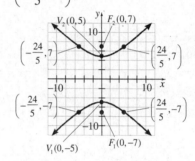

25. The given vertices $(0, \pm 5)$ and focus $(0, 7)$ all lie on the *y*-axis. Thus, the transverse axis is the *y*-axis and the hyperbola opens up and down. The center of the hyperbola is the midpoint between the two vertices. Therefore, the center is $(0, 0)$ and the equation of the hyperbola is of the form $\frac{y^2}{a^2} - \frac{x^2}{b^2} = 1$. Now, the distance from the center $(0, 0)$ to each vertex is $a = 5$ units. Likewise, the distance from the center to the given focus $(0, 7)$ is $c = 7$ units. Also, $b^2 = c^2 - a^2$, so $b^2 = 7^2 - 5^2 = 49 - 25 = 24$. Thus, the equation of the hyperbola is $\frac{y^2}{25} - \frac{x^2}{24} = 1$.

To help graph the hyperbola, we first find the focus that was not given. Since the center is at $(0, 0)$ and since one focus is at $(0, 7)$, the other focus must be at $(0, -7)$. Next, plot the points on the graph to the left and right of the foci. Let $y = \pm 7$:

27. The given foci $(\pm 10, 0)$ and vertex $(-7, 0)$ all lie on the *x*-axis. Thus, the transverse axis is the *x*-axis and the hyperbola opens left and right. The center of the hyperbola is the midpoint between the two vertices. Therefore, the center is $(0, 0)$ and the equation of the hyperbola is of the form $\frac{x^2}{a^2} - \frac{y^2}{b^2} = 1$. Now, the distance between the given vertex $(-7, 0)$ and the center $(0, 0)$ is $a = 7$ units. Likewise, the distance between the given foci and the center is $c = 10$ units. Also, $b^2 = c^2 - a^2$, so $b^2 = 10^2 - 7^2 = 100 - 49 = 51$. Thus, the equation of the hyperbola is $\frac{x^2}{49} - \frac{y^2}{51} = 1$.

To help graph the hyperbola, we first find the vertex that was not given. Since the center is at $(0, 0)$ and since one vertex is at $(-7, 0)$, the other vertex must be at $(7, 0)$. Next, plot the points on the graph above and below the foci. Let $x = \pm 10$:

$$\frac{(\pm 10)^2}{49} - \frac{y^2}{51} = 1$$

$$\frac{100}{49} - \frac{y^2}{51} = 1$$

$$-\frac{y^2}{51} = -\frac{51}{49}$$

$$y^2 = \frac{2601}{49}$$

$$y = \pm\frac{51}{7}$$

The points above and below the foci are

$\left(10, \frac{51}{7}\right), \left(10, -\frac{51}{7}\right), \left(-10, \frac{51}{7}\right),$ and

$\left(-10, -\frac{51}{7}\right).$

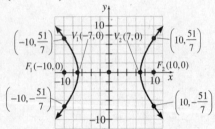

29. $\dfrac{x^2}{25} - \dfrac{y^2}{9} = 1$

Notice the equation is of the form $\dfrac{x^2}{a^2} - \dfrac{y^2}{b^2} = 1$.

Because the x^2-term is first, the transverse axis is the x-axis and the hyperbola opens left and right. The center of the hyperbola is the origin. We have that $a^2 = 25$ and $b^2 = 9$. Because $c^2 = a^2 + b^2$, we have that $c^2 = 25 + 9 = 34$, so that $c = \sqrt{34}$. The vertices are $(\pm a, 0) = (\pm 5, 0)$, and the foci are $(\pm c, 0) = (\pm\sqrt{34}, 0)$. Since $b^2 = 9$, we have that $b = 3$. The equations of the asymptotes are

$y = \dfrac{b}{a}x = \dfrac{3}{5}x$ and $y = -\dfrac{b}{a}x = -\dfrac{3}{5}x$.

To graph the hyperbola, we form a rectangle using the points $(\pm a, 0) = (\pm 5, 0)$ and $(0, \pm b) = (0, \pm 3)$. The diagonals help us draw the asymptotes.

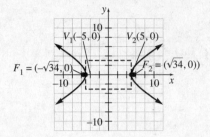

31. $\dfrac{y^2}{4} - \dfrac{x^2}{100} = 1$

Notice the equation is of the form $\dfrac{y^2}{a^2} - \dfrac{x^2}{b^2} = 1$.

Because the y^2-term is first, the transverse axis is the y-axis and the hyperbola opens up and down. The center of the hyperbola is the origin. We have that $a^2 = 4$ and $b^2 = 100$. Because $c^2 = a^2 + b^2$, we have that $c^2 = 4 + 100 = 104$, so that $c = \sqrt{104} = 2\sqrt{26}$. The vertices are $(0, \pm a) = (0, \pm 2)$, and the foci are $(0, \pm c) = \left(0, \pm 2\sqrt{26}\right)$. Since $b^2 = 100$, we have that $b = 10$. The equations of the asymptotes are

$y = \dfrac{a}{b}x = \dfrac{2}{10}x = \dfrac{1}{5}x$ and

$y = -\dfrac{a}{b}x = -\dfrac{2}{10}x = -\dfrac{1}{5}x$.

To graph the hyperbola, we form a rectangle using the points $(0, \pm a) = (0, \pm 2)$ and $(\pm b, 0) = (\pm 10, 0)$. The diagonals help us draw the asymptotes.

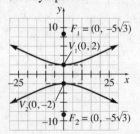

33. $x^2 - y^2 = 4$

$$\frac{x^2}{4} - \frac{y^2}{4} = 1$$

Notice the equation is of the form $\dfrac{x^2}{a^2} - \dfrac{y^2}{b^2} = 1$.

Because the x^2-term is first, the transverse axis is the x-axis and the hyperbola opens left and right. The center of the hyperbola is the origin. We have that $a^2 = 4$ and $b^2 = 4$. Because $c^2 = a^2 + b^2$, we have that $c^2 = 4 + 4 = 8$, so that $c = \sqrt{8} = 2\sqrt{2}$. The vertices are $(\pm a, 0) = (\pm 2, 0)$, and the foci are $(\pm c, 0) = (\pm 2\sqrt{2}, 0)$. Since $b^2 = 4$, we have that $b = 2$. The equations of the asymptotes are

$$y = \frac{b}{a}x = \frac{2}{2}x = x \text{ and } y = -\frac{b}{a}x = -\frac{2}{2}x = -x.$$

To graph the hyperbola, we form a rectangle using the points $(\pm a, 0) = (\pm 2, 0)$ and $(0, \pm b) = (0, \pm 2)$. The diagonals help us draw the asymptotes.

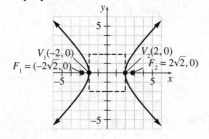

35. The given vertices $(0, \pm 8)$ both lie on the y-axis. Thus, the transverse axis is the y-axis and the hyperbola opens up and down. The center of the hyperbola is the midpoint between the two vertices. Therefore, the center is $(0,0)$ and the equation of the hyperbola is of the form $\frac{y^2}{a^2} - \frac{x^2}{b^2} = 1$. Now, the distance from the center $(0,0)$ to each vertex is $a = 8$ units. We are given that an asymptote of the hyperbola is $y = 2x$. Now the equation of the asymptote is of the form is $y = \frac{a}{b}x$ and $a = 8$. So,

$$\frac{a}{b} = 2$$

$$\frac{8}{b} = 2$$

$$8 = 2b$$

$$4 = b$$

Thus, the equation of the hyperbola is

$$\frac{y^2}{8^2} - \frac{x^2}{4^2} = 1$$

$$\frac{y^2}{64} - \frac{x^2}{16} = 1$$

To help graph the hyperbola, we first find the

foci. Since $c^2 = a^2 + b^2$, we have that $c^2 = 64 + 16 = 80$, so $c = 4\sqrt{5}$. Since the center is $(0,0)$, the foci are $(0, \pm 4\sqrt{5})$. We form the rectangle using the points $(0, \pm a) = (0, \pm 8)$ and $(\pm b, 0) = (\pm 4, 0)$. The diagonals are the two asymptotes $y = 2x$ and $y = -2x$.

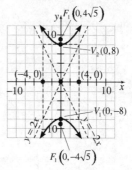

37. The given foci $(\pm 3, 0)$ both lie on the x-axis. Thus, the transverse axis is the x-axis and the hyperbola opens left and right. The center of the hyperbola is the midpoint between the two vertices. Therefore, the center is $(0,0)$ and the equation of the hyperbola is of the form $\frac{x^2}{a^2} - \frac{y^2}{b^2} = 1$. We are given that an asymptote of the hyperbola is $y = x$. Now the equation of the asymptote is of the form $y = \frac{b}{a}x$. So, $\frac{b}{a} = 1$ which means $a = b$. Because each focus is 3 units from the center, we have $c = 3$ and $c^2 = 9$. So,

$$a^2 + b^2 = c^2$$

$$a^2 + a^2 = 9$$

$$2a^2 = 9$$

$$a^2 = \frac{9}{2}$$

$$a = \sqrt{\frac{9}{2}} = \frac{3}{\sqrt{2}} = \frac{3\sqrt{2}}{2}$$

Since $a = b$, we have that $a^2 = b^2 = \frac{9}{2} = 4.5$.

Thus, the equation of the hyperbola is

$$\frac{x^2}{4.5} - \frac{y^2}{4.5} = 1.$$

To help graph the hyperbola, we find the

vertices. Since $a = \dfrac{3\sqrt{2}}{2}$ and the center is

$(0,0)$, the vertices are $\left(\pm\dfrac{3\sqrt{2}}{2},0\right)$. We form

the rectangle using the points

$(\pm a, 0) = \left(\pm\dfrac{3\sqrt{2}}{2}, 0\right)$ and $(0, \pm b) = \left(0, \pm\dfrac{3\sqrt{2}}{2}\right)$.

The diagonals are the two asymptotes $y = x$ and
$y = -x$.

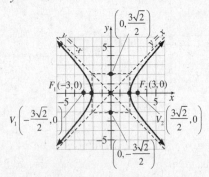

39. The graph of the hyperbola opens left and right
and the center is $(0,0)$, so the equation is of the

form $\dfrac{x^2}{a^2} - \dfrac{y^2}{b^2} = 1$. From the graph, we observe

that the asymptotes are $y = -x$ and $y = x$. Now
the equations of the asymptotes are of the form

$y = -\dfrac{b}{a}x$ and $y = \dfrac{b}{a}x$. So, $\dfrac{b}{a} = 1$ which means

$b = a$. We also observe that the vertices are
$(\pm 1, 0)$, so the distance from the center to each

vertex is $a = 1$. Thus, $a = b = 1$ and

$a^2 = b^2 = 1$. Thus, the equation of the

hyperbola is $\dfrac{x^2}{1} - \dfrac{y^2}{1} = 1$ or $x^2 - y^2 = 1$.

41. The graph of the hyperbola opens up and down
and the center is $(0,0)$, so the equation is of the

form $\dfrac{y^2}{a^2} - \dfrac{x^2}{b^2} = 1$. From the graph, we observe

that the asymptotes are $y = -2x$ and $y = 2x$.
Now the equations of the asymptotes are of the

form $y = -\dfrac{a}{b}x$ and $y = \dfrac{a}{b}x$. So, $\dfrac{a}{b} = 2$ which

means $a = 2b$ or $b = \dfrac{1}{2}a$. We also observe that

the vertices are $(0, \pm 6)$, so the distance from the

center to each vertex is $a = 6$ and $b = \dfrac{1}{2}(6) = 3$.

Thus, $a^2 = 36$ and $b^2 = 9$, and the equation of

the hyperbola is $\dfrac{y^2}{36} - \dfrac{x^2}{9} = 1$.

43. We observe that the equation $\dfrac{x^2}{4} - y^2 = 1$ is of

the form $\dfrac{x^2}{a^2} - \dfrac{y^2}{b^2} = 1$ with $a = 2$ and $b = 1$.

The asymptotes of such a hyperbola are

$y = -\dfrac{b}{a}x$ and $y = \dfrac{b}{a}x$. Thus, the asymptotes

are $y = -\dfrac{1}{2}x$ and $y = \dfrac{1}{2}x$.

Similarly, we observe that the equation

$y^2 - \dfrac{x^2}{4} = 1$ is of the form $\dfrac{y^2}{a^2} - \dfrac{x^2}{b^2} = 1$ with

$a = 1$ and $b = 2$. The asymptotes of such a

hyperbola are $y = -\dfrac{a}{b}x$ and $y = \dfrac{a}{b}x$. Thus, the

asymptotes are $y = -\dfrac{1}{2}x$ and $y = \dfrac{1}{2}x$.

Since the two hyperbolas have the same
asymptotes, they are conjugate.

Now, the vertices of $y^2 - \dfrac{x^2}{4} = 1$ are $(0, \pm 1)$,

each of which is $a = 1$ unit above and below the
center. Utilizing the vertices and the asymptotes,

we draw the graph of $y^2 - \dfrac{x^2}{4} = 1$. Similarly,

the vertices of $\dfrac{x^2}{4} - y^2 = 1$ are $(\pm 2, 0)$, each of

which is $a = 2$ units to the left and right of the

center. We draw the graph of $\dfrac{x^2}{4} - y^2 = 1$ by

utilizing the vertices and the asymptotes.

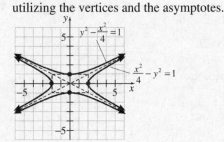

45. Answers may vary. One possibility follows:
The asymptotes provide a means for determining the opening of each branch of the hyperbola without having to find and plot additional points.

47. $\begin{cases} 2x-3y=-9 & (1) \\ -x+5y=8 & (2) \end{cases}$

Because equation (2) is easily solved for x, we use substitution to solve the system.
Equation (2) solved for x is $x=5y-8$.
Substituting $5y-8$ for x in equation (1), we obtain

$$2(5y-8)-3y=-9$$
$$10y-16-3y=-9$$
$$7y-16=-9$$
$$7y=7$$
$$y=1$$

Substituting 1 for y into equation (2), we obtain

$$-x+5(1)=8$$
$$-x+5=8$$
$$-x=3$$
$$x=-3$$

The solution is the ordered pair $(-3,1)$.

49. $\begin{cases} 2x-3y=6 & (1) \\ -6x+9y=-18 & (2) \end{cases}$

Because none of the variables have a coefficient of 1, we use elimination to solve the system.
Multiply both sides of equation (1) by 3, and add the result to equation (2).

$$6x-9y=18$$
$$-6x+9y=-18$$
$$0=0$$

The system is dependent. The solution is $\{(x,y)\mid 2x-3y=6\}$.

51. $\begin{cases} 6x+3y=4 & (1) \\ -2x-y=-\dfrac{4}{3} & (2) \end{cases}$

We use elimination to solve the system.
Multiply both sides of equation (2) by 3, and add the result to equation (1).

$$-6x-3y=-4$$
$$6x+3y=4$$
$$0=0$$

The system is dependent. The solution is $\{(x,y)\mid 6x+3y=4\}$.

53.
$$\frac{x^2}{4}-\frac{y^2}{16}=1$$
$$-\frac{y^2}{16}=1-\frac{x^2}{4}$$
$$-16\left(-\frac{y^2}{16}\right)=-16\left(1-\frac{x^2}{4}\right)$$
$$y^2=16\left(\frac{x^2}{4}-1\right)$$
$$y=\pm\sqrt{16\left(\frac{x^2}{4}-1\right)}$$
$$y=\pm4\sqrt{\frac{x^2}{4}-1}$$

Let $Y_1=4\sqrt{\dfrac{x^2}{4}-1}$ and $Y_2=-4\sqrt{\dfrac{x^2}{4}-1}$.

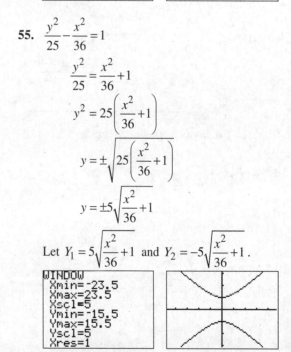

55.
$$\frac{y^2}{25}-\frac{x^2}{36}=1$$
$$\frac{y^2}{25}=\frac{x^2}{36}+1$$
$$y^2=25\left(\frac{x^2}{36}+1\right)$$
$$y=\pm\sqrt{25\left(\frac{x^2}{36}+1\right)}$$
$$y=\pm5\sqrt{\frac{x^2}{36}+1}$$

Let $Y_1=5\sqrt{\dfrac{x^2}{36}+1}$ and $Y_2=-5\sqrt{\dfrac{x^2}{36}+1}$.

```
WINDOW
Xmin=-23.5
Xmax=23.5
Xscl=5
Ymin=-15.5
Ymax=15.5
Yscl=5
Xres=1
```

57. $4x^2 - y^2 = 36$
$$-y^2 = -4x^2 + 36$$
$$y^2 = 4x^2 - 36$$
$$y = \pm\sqrt{4x^2 - 36}$$

Let $Y_1 = \sqrt{4x^2 - 36}$ and $Y_2 = -\sqrt{4x^2 - 36}$.

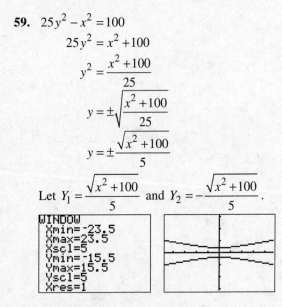

```
WINDOW
  Xmin=-9.4
  Xmax=9.4
  Xscl=1
  Ymin=-6.2
  Ymax=6.2
  Yscl=1
  Xres=1
```

59. $25y^2 - x^2 = 100$
$$25y^2 = x^2 + 100$$
$$y^2 = \frac{x^2 + 100}{25}$$
$$y = \pm\sqrt{\frac{x^2 + 100}{25}}$$
$$y = \pm\frac{\sqrt{x^2 + 100}}{5}$$

Let $Y_1 = \frac{\sqrt{x^2 + 100}}{5}$ and $Y_2 = -\frac{\sqrt{x^2 + 100}}{5}$.

```
WINDOW
  Xmin=-23.5
  Xmax=23.5
  Xscl=5
  Ymin=-15.5
  Ymax=15.5
  Yscl=5
  Xres=1
```

Putting the Concepts Together (Sections 12.1–12.5)

1. $d(P_1, P_2) = \sqrt{(x_2 - x_1)^2 + (y_2 - y_1)^2}$
$$= \sqrt{(3 - (-6))^2 + (-2 - 4)^2}$$
$$= \sqrt{(9)^2 + (-6)^2}$$
$$= \sqrt{81 + 36}$$
$$= \sqrt{117}$$
$$= 3\sqrt{13}$$

2. $M = \left(\frac{x_1 + x_2}{2}, \frac{y_1 + y_2}{2}\right)$
$$= \left(\frac{-3 + 5}{2}, \frac{1 + (-7)}{2}\right) = \left(\frac{2}{2}, \frac{-6}{2}\right) = (1, -3)$$

3. $(x + 2)^2 + (y - 8)^2 = 36$
$$(x - (-2))^2 + (y - 8)^2 = 6^2$$

The center is $(h, k) = (-2, 8)$, and the radius is $r = 6$.

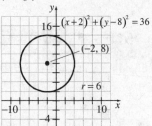

4. $x^2 + y^2 + 6x - 4y - 3 = 0$
$$\left(x^2 + 6x\right) + \left(y^2 - 4y\right) = 3$$
$$\left(x^2 + 6x + 9\right) + \left(y^2 - 4y + 4\right) = 3 + 9 + 4$$
$$(x + 3)^2 + (y - 2)^2 = 16$$
$$(x - (-3))^2 + (y - 2)^2 = 4^2$$

The center is $(h, k) = (-3, 2)$, and the radius is $r = 4$.

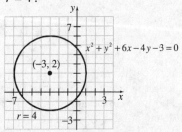

5. The radius of the circle will be the distance from the center point $(0, 0)$ to the point on the circle $(-5, 12)$. Thus, $r = \sqrt{(-5 - 0)^2 + (12 - 0)^2} = 13$. The equation of the circle is
$$(x - h)^2 + (y - k)^2 = r^2$$
$$(x - 0)^2 + (y - 0)^2 = 13^2$$
$$x^2 + y^2 = 169$$

6. The center of the circle will be the midpoint of the diameter with endpoints $(-1, 5)$ and $(5, -3)$.

Thus, $(h, k) = \left(\frac{-1 + 5}{2}, \frac{5 + (-3)}{2}\right) = (2, 1)$. The radius of the circle will be the distance from the center point $(2, 1)$ to one of the endpoints of the diameter, say $(-1, 5)$. Thus,

$r = \sqrt{(-1-2)^2 + (5-1)^2} = \sqrt{25} = 5$.

The equation of the circle is

$(x-h)^2 + (y-k)^2 = r^2$

$(x-2)^2 + (y-1)^2 = 5^2$

$(x-2)^2 + (y-1)^2 = 25$

7. Notice that $(x+2)^2 = -4(y-4)$ is of the form

$(x-h)^2 = -4a(y-k)$, where $(h,k) = (-2,4)$

and $-4a = -4$, so that $a = 1$. Now, the graph of

an equation of the form $(x-h)^2 = -4a(y-k)$

will be a parabola that opens downward with the

vertex at (h,k), focus at $(h, k-a)$, and directrix

of $y = k+a$. Note that $k-a = 4-1 = 3$ and

$k+a = 4+1 = 5$. Thus, the graph of

$(x+2)^2 = -4(y-4)$ is a parabola that opens

downward with vertex $(-2,4)$, focus $(-2,3)$,

and directrix $y = 5$. To help graph the parabola,

we plot the two points on the graph to the left

and right of the focus. Let $y = 3$:

$(x+2)^2 = -4(3-4)$

$(x+2)^2 = 4$

$x+2 = \pm 2$

$x = -2 \pm 2$

$x = -4$ or $x = 0$

The points $(-4,3)$ and $(0,3)$ are on the graph.

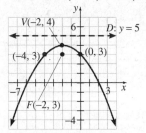

8. We complete the square in y to write the equation in standard form:

$y^2 + 2y - 8x + 25 = 0$

$y^2 + 2y = 8x - 25$

$y^2 + 2y + 1 = 8x - 25 + 1$

$y^2 + 2y + 1 = 8x - 24$

$(y+1)^2 = 8(x-3)$

Notice that $(y+1)^2 = 8(x-3)$ is of the form

$(y-k)^2 = 4a(x-h)$, where $(h,k) = (3,-1)$ and

$4a = 8$, so that $a = 2$. Now, the graph of an

equation of the form $(y-k)^2 = 4a(x-h)$ will be

a parabola that opens to the right with the vertex

at (h,k), focus at $(h+a,k)$, and directrix of

$x = h-a$. Note that $h+a = 3+2 = 5$ and

$h-a = 3-2 = 1$. Thus, the graph of

$(y+1)^2 = 8(x-3)$ is a parabola that opens to the

right with vertex $(3,-1)$, focus $(5,-1)$, and

directrix $x = 1$. To help graph the parabola, we

plot the two points on the graph below and

above the focus. Let $x = 5$:

$(y+1)^2 = 8(5-3)$

$(y+1)^2 = 16$

$y+1 = \pm 4$

$y = -1 \pm 4$

$y = -5$ or $y = 3$

The points $(5,-5)$ and $(5,3)$ are on the graph.

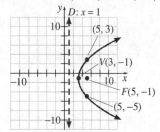

9. The distance from the vertex $(h,k) = (-1,-2)$ to

the focus $(-1,-5)$ is $a = 3$. The vertex and

focus both lie on the vertical line $x = -1$, which

is the axis of symmetry (parallel to the y-axis).

Because the focus is below the vertex, we know

that the parabola will open downward. This

means the equation of the parabola is of the form

$(x-h)^2 = -4a(y-k)$ with $a = 3$, $h = -1$, and

$k = -2$:

$(x-(-1))^2 = -4 \cdot 3(y-(-2))$

$(x+1)^2 = -12(y+2)$

10. The vertex is $(-3,3)$ and the axis of symmetry is

parallel to the x-axis, so the axis of symmetry is

the line $y = 3$ and the parabola either opens to

the left or right. Because the graph contains the

point $(-1,7)$, which is to the right of the vertex,

the parabola must open to the right. Therefore,

the equation of the parabola is of the form

$(y-k)^2 = 4a(x-h)$, with $h = -3$ and $k = 3$.

Now $y = 7$ when $x = -1$, so

$$(7-3)^2 = 4a(-1-(-3))$$
$$(4)^2 = 4a(2)$$
$$16 = 8a$$
$$a = 2$$

The equation of the parabola is

$$(y-3)^2 = 4 \cdot 2(x-(-3))$$
$$(y-3)^2 = 8(x+3)$$

11. $x^2 + 9y^2 = 81$

$$\frac{x^2 + 9y^2}{81} = \frac{81}{81}$$

$$\frac{x^2}{81} + \frac{y^2}{9} = 1$$

The larger number, 81, is in the denominator of the x^2-term. This means that the major axis is the x-axis and that the equation of the ellipse is of the form $\frac{x^2}{a^2} + \frac{y^2}{b^2} = 1$, so that $a^2 = 81$ and $b^2 = 9$. The center of the ellipse is $(0,0)$.

Because $c^2 = a^2 - b^2$, we have that $c^2 = 81 - 9 = 72$, so that $c = \pm\sqrt{72} = \pm 6\sqrt{2}$. Since the major axis is the x-axis, the foci are $\left(-6\sqrt{2},0\right)$ and $\left(6\sqrt{2},0\right)$. To find the x-intercepts (vertices), let $y = 0$; to find the y-intercepts, let $x = 0$:

x-intercepts: y-intercepts:

$x^2 + 9(0)^2 = 81$ $0^2 + 9y^2 = 81$

$\quad\quad x^2 = 81$ $\quad\quad 9y^2 = 81$

$\quad\quad x = \pm 9$ $\quad\quad y^2 = 9$

$\quad\quad\quad\quad\quad\quad\quad\quad y = \pm 3$

The intercepts are $(-9,0)$, $(9,0)$, $(0,-3)$, and $(0,3)$.

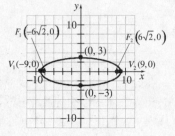

12. $\frac{(x+1)^2}{36} + \frac{(y-2)^2}{49} = 1$

The center of the ellipse is $(h,k) = (-1,2)$. Because the larger number, 49, is the denominator of the y^2-term, the major axis is parallel to the y-axis. Because $a^2 = 49$ and $b^2 = 36$, we have that $c^2 = a^2 - b^2 = 49 - 36 = 13$. The vertices are $a = 7$ units below and above the center at $V_1(-1,-5)$ and $V_2(-1,9)$. The foci are $c = \sqrt{13}$ units below and above the center at $F_1\left(-1,2-\sqrt{13}\right)$ and $F_2\left(-1,2+\sqrt{13}\right)$. We plot the points $b = 6$ units to the left and right of the center point at $(-7,2)$ and $(5,2)$.

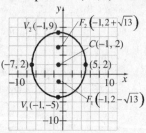

13. The given foci $(0,\pm 6)$ and the given vertices $(0,\pm 9)$ lie on the y-axis. The center of the ellipse is the midpoint between the two vertices (or foci). Thus, the center of the ellipse is $(0,0)$ and its equation is of the form $\frac{x^2}{b^2} + \frac{y^2}{a^2} = 1$. The distance from the center of the ellipse to the vertex is $a = 9$ units. The distance from the center of the ellipse to the focus is $c = 6$ units. Because $b^2 = a^2 - c^2$, we have that $b^2 = 9^2 - 6^2 = 81 - 36 = 45$. Thus, the equation of the ellipse is $\frac{x^2}{45} + \frac{y^2}{81} = 1$.

14. The center is $(h,k) = (3,-4)$. The center, focus, and vertex given all lie on the horizontal line $y = -4$. Therefore, the major axis is parallel to the x-axis, and the equation of the ellipse is of the form $\frac{(x-h)^2}{a^2} + \frac{(y-k)^2}{b^2} = 1$. Now, $a = 4$ is the distance from the center $(3,-4)$ to a vertex $(7,-4)$, and $c = 3$ is the distance from the

center to a focus $(6,-4)$. Also, $b^2 = a^2 - c^2$, so $b^2 = 4^2 - 3^2 = 16 - 9 = 7$. Thus, the equation of the ellipse is

$$\frac{(x-3)^2}{16} + \frac{(y-(-4))^2}{7} = 1$$

$$\frac{(x-3)^2}{16} + \frac{(y+4)^2}{7} = 1$$

15. $\dfrac{y^2}{81} - \dfrac{x^2}{9} = 1$

Notice the equation is of the form $\dfrac{y^2}{a^2} - \dfrac{x^2}{b^2} = 1$.

Because the y^2-term is first, the transverse is along the y-axis and the hyperbola opens up and down. The center of the hyperbola is $(0,0)$. We have that $a^2 = 81$ and $b^2 = 9$. Because $c^2 = a^2 + b^2$, we have that $c^2 = 81 + 9 = 90$, so that $c = \sqrt{90} = 3\sqrt{10}$. The vertices are $(0,\pm a) = (0,\pm 9)$, and the foci are $(0,\pm c) = \left(0,\pm 3\sqrt{10}\right)$. Since $a = 9$ and $b = 3$, the equations of the asymptotes are $y = \dfrac{9}{3}x = 3x$ and $y = -\dfrac{9}{3}x = -3x$. To help graph the hyperbola, we form the rectangle using the points $(0,\pm a) = (0,\pm 9)$ and $(\pm b,0) = (\pm 3,0)$. The diagonals are the asymptotes.

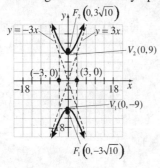

16. $25x^2 - y^2 = 25$

$$\frac{25x^2 - y^2}{25} = \frac{25}{25}$$

$$\frac{x^2}{1} - \frac{y^2}{25} = 1$$

Notice the equation is of the form $\dfrac{x^2}{a^2} - \dfrac{y^2}{b^2} = 1$.

Because the x^2-term is first, the transverse is along the x-axis and the hyperbola opens left and right. The center of the hyperbola is $(0,0)$. We have that $a^2 = 1$ and $b^2 = 25$. Because $c^2 = a^2 + b^2$, we have that $c^2 = 1 + 25 = 26$, so that $c = \sqrt{26}$. The vertices are $(\pm a,0) = (\pm 1,0)$, and the foci are $(\pm c,0) = \left(\pm \sqrt{26},0\right)$. Since $a = 1$ and $b = 5$, the equations of the asymptotes are $y = \dfrac{5}{1}x = 5x$ and $y = -\dfrac{5}{1}x = -5x$. To help graph the hyperbola, we form the rectangle using the points $(\pm a,0) = (\pm 1,0)$ and $(0,\pm b) = (0,\pm 5)$. The diagonals are the asymptotes.

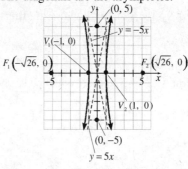

17. The given center $(0,0)$, focus $(0,-5)$, and vertex $(0,-2)$ all lie on the y-axis. Thus, the transverse axis is the y-axis and the hyperbola opens up and down. This means that the equation of the hyperbola is $\dfrac{y^2}{a^2} - \dfrac{x^2}{b^2} = 1$. Now, the distance between the given vertex $(0,-2)$ and the center $(0,0)$ is $a = 2$ units. Likewise, the distance between the given focus $(0,-5)$ and the center is $c = 5$ units. Because $b^2 = c^2 - a^2$, we have that $b^2 = 5^2 - 2^2 = 25 - 4 = 21$. Thus, the equation of the hyperbola is $\dfrac{y^2}{4} - \dfrac{x^2}{21} = 1$.

18. The light bulb should be located at the focus of the flood light, so we need to find where the focus of the flood light is. To solve this problem, we draw a parabola on a Cartesian plane so that the vertex is the origin and the focus is on the positive y-axis. The width of the parabola is 36 inches, and the depth is 12 inches.

Therefore, we know two points on the graph of the parabola: $(-18, 12)$ and $(18, 12)$.

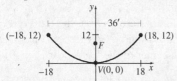

The equation of the parabola has the form $x^2 = 4ay$. Since $(18, 12)$ is a point on the graph, we have: $18^2 = 4a(12)$

$$324 = 48a$$

$$a = \frac{324}{48} = 6.75$$

The light bulb should be located 6.75 inches above the vertex, along its axis of symmetry.

Section 12.6

Preparing for Systems of Nonlinear Equations

P1. $\begin{cases} y = 2x - 5 & (1) \\ 2x - 3y = 7 & (2) \end{cases}$

Substituting $2x - 5$ for y in equation (2), we obtain

$$2x - 3(2x - 5) = 7$$
$$2x - 6x + 15 = 7$$
$$-4x + 15 = 7$$
$$-4x = -8$$
$$x = 2$$

Substituting 2 for x in equation (1), we obtain $y = 2(2) - 5 = 4 - 5 = -1$.

The solution is the ordered pair $(2, -1)$.

P2. $\begin{cases} 2x - 4y = -11 & (1) \\ -x + 5y = 13 & (2) \end{cases}$

Multiply both sides of equation (2) by 2, and add the result to equation (1).

$$2x - 4y = -11$$
$$\underline{-2x + 10y = 26}$$
$$6y = 15$$

$$y = \frac{15}{6} = \frac{5}{2}$$

Substituting $\frac{5}{2}$ for y in equation (1), we obtain

$$2x - 4\left(\frac{5}{2}\right) = -11$$
$$2x - 10 = -11$$
$$2x = -1$$
$$x = -\frac{1}{2}$$

The solution is the ordered pair $\left(-\frac{1}{2}, \frac{5}{2}\right)$.

P3. $\begin{cases} 3x - 5y = 4 & (1) \\ -6x + 10y = -8 & (2) \end{cases}$

Multiply both sides of equation (1) by 2, and add the result to equation (2).

$$6x - 10y = 8$$
$$\underline{-6x + 10y = -8}$$
$$0 = 0$$

The system is dependent. The solution is $\left\{ (x, y) \mid 3x - 5y = 4 \right\}$.

Section 12.6 Quick Checks

1. $\begin{cases} 2x + y = -1 \\ x^2 - y = 4 \end{cases}$

First, graph each equation in the system.

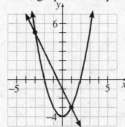

The system apparently has two solutions. Now solve the first equation for y: $y = -2x - 1$.

Substitute the result for y into the second equation:

$$x^2 - (-2x - 1) = 4$$
$$x^2 + 2x + 1 = 4$$
$$x^2 + 2x - 3 = 0$$
$$(x - 1)(x + 3) = 0$$
$$x = 1 \text{ or } x = -3$$

Substitute these x-values into the first equation to find the corresponding y-values:

$$x = 1: \ 2(1) + y = -1$$
$$2 + y = -1$$
$$y = -3$$

$x = -3$: $2(-3) + y = -1$
$$-6 + y = -1$$
$$y = 5$$

Both pairs check, so the solutions are $(-3,\ 5)$ and $(1, -3)$.

2. $\begin{cases} 2x + y = 0 \\ (x-4)^2 + (y+2)^2 = 9 \end{cases}$

First, graph each equation in the system.

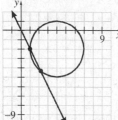

The system apparently has two solutions. Now solve the first equation for y: $y = -2x$.

Substitute the result for y into the second equation:
$$(x-4)^2 + (-2x+2)^2 = 9$$
$$x^2 - 8x + 16 + 4x^2 - 8x + 4 = 9$$
$$5x^2 - 16x + 11 = 0$$
$$(5x - 11)(x - 1) = 0$$
$$x = \frac{11}{5} \quad \text{or} \quad x = 1$$

Substitute these x-values into the first equation to find the corresponding y-values:

$x = \frac{11}{5}$: $2\left(\frac{11}{5}\right) + y = 0$
$$\frac{22}{5} + y = 0$$
$$y = -\frac{22}{5}$$

$x = 1$: $2(1) + y = 0$
$$2 + y = 0$$
$$y = -2$$

Both pairs check, so the solutions are $(1, -2)$ and $\left(\frac{11}{5}, -\frac{22}{5}\right)$.

3. $\begin{cases} x^2 + y^2 = 16 \\ x^2 - 2y = 8 \end{cases}$

First, graph each equation in the system.

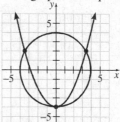

The system apparently has three solutions. Now multiply the second equation by -1 and add the result to the first equation:

$$x^2 + y^2 \qquad = 16$$
$$\underline{-x^2 \qquad + 2y = -8}$$
$$y^2 + 2y = 8$$
$$y^2 + 2y - 8 = 0$$
$$(y + 4)(y - 2) = 0$$
$$y = -4 \quad \text{or} \quad y = 2$$

Substitute these y-values into the first equation to find the corresponding x-values:

$y = -4$: $x^2 + (-4)^2 = 16$
$$x^2 + 16 = 16$$
$$x^2 = 0$$
$$x = 0$$

$y = 2$: $x^2 + 2^2 = 16$
$$x^2 + 4 = 16$$
$$x^2 = 12$$
$$x = \pm\sqrt{12} = \pm 2\sqrt{3}$$

All three pairs check, so the solutions are $(0, -4)$, $\left(-2\sqrt{3},\ 2\right)$ and $\left(2\sqrt{3},\ 2\right)$.

4. $\begin{cases} x^2 - y = -4 \\ x^2 + y^2 = 9 \end{cases}$

First, graph each equation in the system.

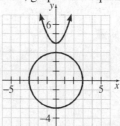

The system apparently has no solution. Now multiply the first equation by -1 and add the result to the second equation:

$$\begin{array}{r} -x^2 \phantom{{}+y} + y = 4 \\ x^2 + y^2 \phantom{{}+y} = 9 \\ \hline y^2 + y = 13 \end{array}$$

$$y^2 + y - 13 = 0$$

$$y = \frac{-1 \pm \sqrt{1^2 - 4(1)(-13)}}{2(1)}$$

$$= \frac{-1 \pm \sqrt{53}}{2}$$

Substitute these y-values into the first equation to find the corresponding x-values:

$$y = \frac{-1 - \sqrt{53}}{2}: \quad x^2 - \left(\frac{-1 - \sqrt{53}}{2}\right) = -4$$

$$x^2 = -4 + \left(\frac{-1 - \sqrt{53}}{2}\right)$$

$$x^2 = \frac{-9 - \sqrt{53}}{2}$$

$$x = \pm\sqrt{\frac{-9 - \sqrt{53}}{2}} \quad \text{(not real)}$$

$$y = \frac{-1 + \sqrt{53}}{2}: \quad x^2 - \left(\frac{-1 + \sqrt{53}}{2}\right) = -4$$

$$x^2 = -4 + \left(\frac{-1 + \sqrt{53}}{2}\right)$$

$$x^2 = \frac{-9 + \sqrt{53}}{2}$$

$$x = \pm\sqrt{\frac{-9 + \sqrt{53}}{2}} \quad \text{(not real)}$$

Since both y-values result in non-real x-values (because the values under each of the radicals are negative), the system of equations is inconsistent. The solution set is $\varnothing$.

12.6 Exercises

5. $\begin{cases} y = x^2 + 4 \\ y = x + 4 \end{cases}$

First, graph each equation in the system.

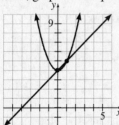

The system apparently has two solutions. Now

substitute $x^2 + 4$ for y into the second equation:

$$x^2 + 4 = x + 4$$

$$x^2 - x = 0$$

$$x(x - 1) = 0$$

$$x = 0 \quad \text{or} \quad x = 1$$

Substitute these x-values into the first equation to find the corresponding y-values:

$$x = 0: \quad y = 0^2 + 4 = 4$$

$$x = 1: \quad y = 1^2 + 4 = 1 + 4 = 5$$

Both pairs check, so the solutions are $(0,\ 4)$ and $(1,\ 5)$.

7. $\begin{cases} y = \sqrt{25 - x^2} \\ x + y = 7 \end{cases}$

First, graph each equation in the system.

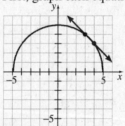

The system apparently has two solutions. Now substitute $\sqrt{25 - x^2}$ for y into the second equation:

$$x + \sqrt{25 - x^2} = 7$$

$$\sqrt{25 - x^2} = 7 - x$$

$$\left(\sqrt{25 - x^2}\right)^2 = (7 - x)^2$$

$$25 - x^2 = 49 - 14x + x^2$$

$$-2x^2 + 14x - 24 = 0$$

$$x^2 - 7x + 12 = 0$$

$$(x - 3)(x - 4) = 0$$

$$x = 3 \quad \text{or} \quad x = 4$$

Substitute these x-values into the first equation to find the corresponding y-values:

$$x = 3: \quad y = \sqrt{25 - 3^2} = \sqrt{25 - 9} = \sqrt{16} = 4$$

$$x = 4: \quad y = \sqrt{25 - 4^2} = \sqrt{25 - 16} = \sqrt{9} = 3$$

Both pairs check, so the solutions are $(3,\ 4)$ and $(4,\ 3)$.

9. $\begin{cases} x^2 + y^2 = 4 \\ y = x^2 - 2 \end{cases}$

First, graph each equation in the system.

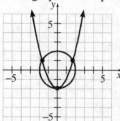

The system apparently has three solutions. Now solve the first equation for x^2: $x^2 = 4 - y^2$.

Substitute the result for x^2 into the second equation:

$$y = \left(4 - y^2\right) - 2$$
$$y = 2 - y^2$$
$$y^2 + y - 2 = 0$$
$$(y + 2)(y - 1) = 0$$
$$y = -2 \quad \text{or} \quad y = 1$$

Substitute these y-values into the first equation to find the corresponding x-values:

$$y = -2: \quad x^2 + (-2)^2 = 4$$
$$x^2 + 4 = 4$$
$$x^2 = 0$$
$$x = 0$$

$$y = 1: \quad x^2 + 1^2 = 4$$
$$x^2 + 1 = 4$$
$$x^2 = 3$$
$$x = \pm\sqrt{3}$$

All three pairs check, so the solutions are $(0, -2)$, $\left(-\sqrt{3}, 1\right)$, and $\left(\sqrt{3}, 1\right)$.

11. $\begin{cases} xy = 4 \\ x^2 + y^2 = 8 \end{cases}$

First, graph each equation in the system.

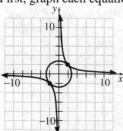

The system apparently has two solutions. Now

solve the first equation for y: $y = \dfrac{4}{x}$.

Substitute the result into the second equation.

$$x^2 + \left(\frac{4}{x}\right)^2 = 8$$
$$x^2 + \frac{16}{x^2} = 8$$
$$x^2\left(x^2 + \frac{16}{x^2}\right) = x^2(8)$$
$$x^4 + 16 = 8x^2$$
$$x^4 - 8x^2 + 16 = 0$$
$$\left(x^2 - 4\right)^2 = 0$$
$$x^2 - 4 = 0$$
$$(x + 2)(x - 2) = 0$$
$$x = -2 \quad \text{or} \quad x = 2$$

Substitute these x-values into the equation $y = \dfrac{4}{x}$ to find the corresponding y-values.

$$x = -2: \quad y = \frac{4}{-2} = -2$$

$$x = 2: \quad y = \frac{4}{2} = 2$$

Both pairs check, so the solutions are $(-2, -2)$ and $(2, 2)$.

13. $\begin{cases} x^2 + y^2 = 4 \\ y^2 - x = 4 \end{cases}$

First, graph each equation in the system.

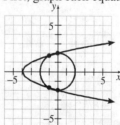

The system apparently has four solutions. Now multiply the second equation by -1 and add the result to the first equation:

$$x^2 + y^2 = 4$$
$$\underline{\quad\quad x - y^2 = -4}$$
$$x^2 + x \quad\quad = 0$$
$$x(x + 1) = 0$$
$$x = 0 \quad \text{or} \quad x = -1$$

Substitute these x-values into the second equation to find the corresponding y-values:

$x = 0: \ y^2 - 0 = 4$

$\qquad\qquad y = \pm\sqrt{4} = \pm 2$

$x = -1: \ y^2 - (-1) = 4$

$\qquad\qquad y^2 + 1 = 4$

$\qquad\qquad y^2 = 3$

$\qquad\qquad\quad y = \pm\sqrt{3}$

All four pairs check, so the solutions are $(0, -2)$, $(0, 2)$, $\left(-1, -\sqrt{3}\right)$ and $\left(-1, \sqrt{3}\right)$.

15. $\begin{cases} x^2 + y^2 = 7 \\ x^2 - y^2 = 25 \end{cases}$

First, graph each equation in the system.

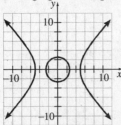

The system apparently has no solution. Now add the two equations:

$x^2 + y^2 = 7$

$\underline{x^2 - y^2 = 25}$

$2x^2 \qquad\ = 32$

$\qquad x^2 = 16$

$\qquad\ x = \pm 4$

Substitute 16 for x^2 into the first equation to find the y values.

$16 + y^2 = 7$

$\qquad y^2 = -9$

$\qquad\ y = \pm\sqrt{-9}$

which yields no real solutions. The system is inconsistent. The solution set is $\varnothing$.

17. $\begin{cases} x^2 + y^2 = 6y \\ x^2 = 3y \end{cases}$

First, graph each equation in the system.

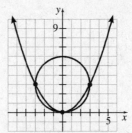

The system apparently has three solutions. Now multiply the second equation by -1 and add the result to the first equation:

$x^2 + y^2 = 6y$

$\underline{-x^2 \qquad\ = -3y}$

$\qquad y^2 = 3y$

$y^2 - 3y = 0$

$y(y - 3) = 0$

$y = 0 \ \text{ or } \ y = 3$

Substitute these y-values into the second equation to find the corresponding x-values:

$y = 0: \ x^2 = 3(0)$

$\qquad\qquad x^2 = 0$

$\qquad\qquad\ x = 0$

$y = 3: \ x^2 = 3(3)$

$\qquad\qquad x^2 = 9$

$\qquad\qquad\ x = \pm 3$

All three pairs check, so the solutions are $(0, 0)$, $(-3, 3)$, and $(3, 3)$.

19. $\begin{cases} x^2 - 2x - y = 8 \\ 6x + 2y = -4 \end{cases}$

First, graph each equation in the system.

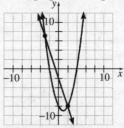

The system apparently has two solutions. Now divide the second equation by 2 and add the result to the first equation:

$x^2 - 2x - y = 8$

$\underline{\qquad 3x + y = -2}$

$x^2 + x \qquad = 6$

$x^2 + x - 6 = 0$

$(x + 3)(x - 2) = 0$

$x = -3$ or $x = 2$

Substitute these results into the second equation to find the corresponding y-values.

$x = -3:$ $6(-3) + 2y = -4$

$$-18 + 2y = -4$$
$$2y = 14$$
$$y = 7$$

$x = 2:$ $6(2) + 2y = -4$

$$12 + 2y = -4$$
$$2y = -16$$
$$y = -8$$

Both pairs check, so the solutions are $(-3, 7)$ and $(2, -8)$.

21. $\begin{cases} y = x^2 - 6x + 4 \\ 5x + y = 6 \end{cases}$

First, graph each equation in the system.

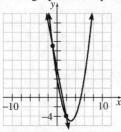

The system apparently has two solutions. Now solve the second equation for y: $y = -5x + 6$.

Substitute the result into the first equation:

$$-5x + 6 = x^2 - 6x + 4$$
$$0 = x^2 - x - 2$$
$$0 = (x + 1)(x - 2)$$
$$x = -1 \text{ or } x = 2$$

Substitute these x-values into the equation $y = -5x + 6$ to find the corresponding y-values:

$x = -1:$ $y = -5(-1) + 6 = 5 + 6 = 11$

$x = 2:$ $y = -5(2) + 6 = -10 + 6 = -4$

Both pairs check, so the solutions are $(-1, 11)$ and $(2, -4)$.

23. $\begin{cases} x^2 + y^2 = 16 \\ x^2 - y^2 = 16 \end{cases}$

First, graph each equation in the system.

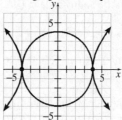

The system apparently has two solutions. Now add the two equations:

$$x^2 + y^2 = 16$$
$$\underline{x^2 - y^2 = 16}$$
$$2x^2 \quad\quad = 32$$
$$x^2 = 16$$
$$x = \pm 4$$

Substitute 16 for x^2 into the first equation to find the y-values.

$$16 + y^2 = 16$$
$$y^2 = 0$$
$$y = 0$$

Both pairs check, so the solutions are $(-4, 0)$ and $(4, 0)$.

25. $\begin{cases} (x - 4)^2 + y^2 = 25 \\ x - y = -3 \end{cases}$

First, graph each equation in the system.

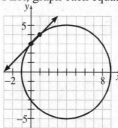

The system apparently has two solutions. Now solve the second equation for y: $y = x + 3$.

Substitute the result into the first equation:

$$(x - 4)^2 + (x + 3)^2 = 25$$
$$x^2 - 8x + 16 + x^2 + 6x + 9 = 25$$
$$2x^2 - 2x = 0$$
$$2x(x - 1) = 0$$
$$x = 0 \text{ or } x = 1$$

Substitute these x-values into the equation

$y = x + 3$ to find the corresponding y-values:

$x = 0$: $y = 0 + 3 = 3$

$x = 1$: $y = 1 + 3 = 4$

Both pairs check, so the solutions are $(0, 3)$ and $(1, 4)$.

27. $\begin{cases} (x-1)^2 + (y+2)^2 = 4 \\ y^2 + 4y - x = -1 \end{cases}$

First, graph each equation in the system.

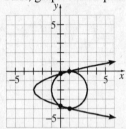

The system apparently has four solutions. Now expand the first equation and simplify:

$$(x-1)^2 + (y+2)^2 = 4$$
$$x^2 - 2x + 1 + y^2 + 4y + 4 = 4$$
$$x^2 - 2x + y^2 + 4y = -1$$

Multiply the second equation by -1 and add the result to the expanded form of the first equation:

$$\begin{array}{l} x^2 - 2x + y^2 + 4y = -1 \\ \underline{\; x - y^2 - 4y = 1} \\ x^2 - x = 0 \\ \; x(x-1) = 0 \\ \; x = 0 \text{ or } x = 1 \end{array}$$

Substitute these x-values into the first equation to find the corresponding y-values.

$x = 0$: $(0-1)^2 + (y+2)^2 = 4$
$$1 + (y+2)^2 = 4$$
$$(y+2)^2 = 3$$
$$y + 2 = \pm\sqrt{3}$$
$$y = -2 \pm \sqrt{3}$$

$x = 1$: $(1-1)^2 + (y+2)^2 = 4$
$$(y+2)^2 = 4$$
$$y + 2 = \pm 2$$
$$y = -2 \pm 2$$
$$y = -4 \text{ or } y = 0$$

All four pairs check, so the solutions are
$\left(0, -2 - \sqrt{3}\right)$, $\left(0, -2 + \sqrt{3}\right)$, $(1, -4)$, and $(1, 0)$.

29. $\begin{cases} (x+3)^2 + 4y^2 = 4 \\ x^2 + 6x - y = 13 \end{cases}$

First, graph each equation in the system.

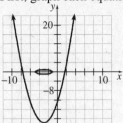

The system apparently has no solution. Now expand the first equation and simplify:

$$(x+3)^2 + 4y^2 = 4$$
$$x^2 + 6x + 9 + 4y^2 = 4$$
$$x^2 + 6x + 4y^2 = -5$$

Multiply the second equation by -1 and add the result to the expanded form of the first equation:

$$\begin{array}{l} x^2 + 6x + 4y^2 = -5 \\ \underline{-x^2 - 6x + y = -13} \\ \; 4y^2 + y = -18 \\ \; 4y^2 + y + 18 = 0 \end{array}$$

$$y = \frac{-1 \pm \sqrt{1^2 - 4(4)(18)}}{2(4)} = \frac{-1 \pm \sqrt{-287}}{8}$$

which yields no real solutions. The system is inconsistent. The solution set is $\varnothing$.

31. $\begin{cases} x^2 - y^2 = 21 \\ x + y = 7 \end{cases}$

First, graph each equation in the system.

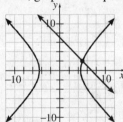

The system apparently has one solution. Now solve the second equation for y: $y = 7 - x$.

Substitute the result into the first equation:

$$x^2 - (7-x)^2 = 21$$
$$x^2 - \left(49 - 14x + x^2\right) = 21$$
$$x^2 - 49 + 14x - x^2 = 21$$
$$14x - 49 = 21$$
$$14x = 70$$
$$x = 5$$

Substitute 5 for x into the equation $y = 7 - x$ to find the corresponding y value: $y = 7 - 5 = 2$.

The pair checks, so the solution is $(5, 2)$.

33. $\begin{cases} x^2 + 2y^2 = 16 \\ 4x^2 - y^2 = 24 \end{cases}$

First, graph each equation in the system.

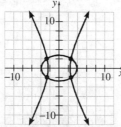

The system apparently has four solutions. Now multiply the second equation by 2 and add the result to the first equation:

$$x^2 + 2y^2 = 16$$
$$\underline{8x^2 - 2y^2 = 48}$$
$$9x^2 \qquad\quad = 64$$
$$x^2 = \frac{64}{9}$$
$$x = \pm\frac{8}{3}$$

Substitute $\frac{64}{9}$ for x^2 into the first equation to find the y-values:

$$\frac{64}{9} + 2y^2 = 16$$
$$9\left(\frac{64}{9} + 2y^2\right) = 9(16)$$
$$64 + 18y^2 = 144$$
$$18y^2 = 80$$
$$y^2 = \frac{80}{18} = \frac{40}{9}$$
$$y = \pm\sqrt{\frac{40}{9}} = \pm\frac{\sqrt{40}}{\sqrt{9}} = \pm\frac{2\sqrt{10}}{3}$$

All four pairs check, so the solutions are

$$\left(-\frac{8}{3}, -\frac{2\sqrt{10}}{3}\right), \left(-\frac{8}{3}, \frac{2\sqrt{10}}{3}\right), \left(\frac{8}{3}, -\frac{2\sqrt{10}}{3}\right),$$

and $\left(\frac{8}{3}, \frac{2\sqrt{10}}{3}\right)$.

35. $\begin{cases} x^2 + y^2 = 25 \\ y = -x^2 + 6x - 5 \end{cases}$

First, graph each equation in the system.

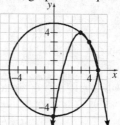

The system apparently has four solutions. Now substitute $-x^2 + 6x - 5$ for y into the first equation:

$$x^2 + \left(-x^2 + 6x - 5\right)^2 = 25$$
$$\left(x^2 - 25\right) + \left(-x^2 + 6x - 5\right)^2 = 0$$
$$(x-5)(x+5) + \left[-1(x-5)(x-1)\right]^2 = 0$$
$$(x-5)(x+5) + (x-5)^2(x-1)^2 = 0$$
$$(x-5)\left[(x+5) + (x-5)(x-1)^2\right] = 0$$
$$(x-5)\left[(x+5) + \left(x^3 - 7x^2 + 11x - 5\right)\right] = 0$$
$$(x-5)\left(x^3 - 7x^2 + 12x\right) = 0$$
$$x(x-5)\left(x^2 - 7x + 12\right) = 0$$
$$x(x-5)(x-3)(x-4) = 0$$
$$x = 0, \ x = 5, \ x = 3, \ \text{or} \ x = 4$$

Substitute these x-values into the second equation to find the corresponding y-values:

$x = 0$: $y = -(0)^2 + 6(0) - 5 = 0 + 0 - 5 = -5$

$x = 3$: $y = -(3)^2 + 6(3) - 5 = -9 + 18 - 5 = 4$

$x = 4$: $y = -(4)^2 + 6(4) - 5 = -16 + 24 - 5 = 3$

$x = 5$: $y = -(5)^2 + 6(5) - 5 = -25 + 30 - 5 = 0$

All four pairs check, so the solutions are $(0, -5)$, $(3, 4)$, $(4, 3)$, and $(5, 0)$.

37. Let x represent the larger of the two numbers and y represent the smaller of the two numbers.
$$\begin{cases} x - y = 2 \\ x^2 + y^2 = 34 \end{cases}$$
Solve the first equation for x: $x = y + 2$.
Substitute the result into the second equation:
$$(y+2)^2 + y^2 = 34$$
$$\left(y^2 + 4y + 4\right) + y^2 = 34$$
$$2y^2 + 4y - 30 = 0$$
$$y^2 + 2y - 15 = 0$$
$$(y+5)(y-3) = 0$$
$$y = -5 \text{ or } y = 3$$
Substitute these y-values into the equation $x = y + 2$ to find the corresponding x-values.
$$y = -5: \ x = -5 + 2 = -3$$
$$y = 3: \ x = 3 + 2 = 5$$
The numbers are either -5 and -3, or 3 and 5.

39. Let x represent the length and y represent the width of the rectangle.
$$\begin{cases} 2x + 2y = 48 \\ xy = 140 \end{cases}$$
Solve the first equation for y: $y = 24 - x$.
Substitute the result into the second equation:
$$x(24 - x) = 140$$
$$24x - x^2 = 140$$
$$x^2 - 24x + 140 = 0$$
$$(x-14)(x-10) = 0$$
$$x = 14 \text{ or } x = 10$$
Substitute these x-values into the second equation to find the corresponding y-values:
$$x = 14: \ 14y = 140$$
$$y = 10$$
$$x = 10: \ 10y = 140$$
$$y = 14$$
Note that the two outcomes result in the same overall dimensions. Assuming the length is the longer of the two sides, the length is 14 feet and the width is 10 feet.

41. Let x represent the length and y represent the width of the rectangular piece of cardboard.

The area of the cardboard is $A = xy = 190 \text{ cm}^2$.

If 2-cm squares are cut from each corner of the rectangle in order to from the box, then the box will be $x-4$ cm long, $y-4$ cm wide, 2 cm high. Thus, the volume of the box will be

$V = 2(x-4)(y-4) = 180 \text{ cm}^3$.

The system of equations follow:
$$\begin{cases} xy = 190 \\ 2(x-4)(y-4) = 180 \end{cases}$$
Solve the first equation for y: $y = \dfrac{190}{x}$.
Substitute the result into the second equation:
$$2(x-4)(y-4) = 180$$
$$(x-4)(y-4) = 90$$
$$xy - 4x - 4y + 16 = 90$$
$$xy - 4x - 4y = 74$$
$$x\left(\frac{190}{x}\right) - 4x - 4\left(\frac{190}{x}\right) = 74$$
$$190 - 4x - \frac{760}{x} = 74$$
$$-4x + 116 - \frac{760}{x} = 0$$
$$x\left(-4x + 116 - \frac{760}{x}\right) = 0$$
$$-4x^2 + 116x - 760 = 0$$
$$-4\left(x^2 - 29x + 190\right) = 0$$
$$x^2 - 29x + 190 = 0$$
$$(x-19)(x-10) = 0$$
$$x = 19 \text{ or } x = 10$$
Substitute these x-values into the equation $y = \dfrac{190}{x}$ to find the corresponding y-values:
$$x = 19: \ y = \frac{190}{19} = 10$$
$$x = 10: \ y = \frac{190}{10} = 19$$
Note that the two outcomes result in the same overall dimensions. The rectangular piece of cardboard must be 19 cm by 10 cm.

43. $\begin{cases} y^2 + y + x^2 - x - 2 = 0 \\ y + 1 + \dfrac{x-2}{y} = 0 \end{cases}$

Multiply the second equation by $-y$:

$$-y\left(y+1+\frac{x-2}{y}\right)=-y(0)$$

$$-y^2-y-x+2=0$$

Add the result to the first equation:

$$y^2+y+x^2-x-2=0$$
$$\underline{-y^2-y\qquad -x+2=0}$$
$$x^2-2x=0$$
$$x(x-2)=0$$
$$x=0 \quad \text{or} \quad x=2$$

Substitute these x-values into the first equation to find the corresponding y-values:

$$x=0: \; y^2+y+0^2-0-2=0$$
$$y^2+y-2=0$$
$$(y+2)(y-1)=0$$
$$y=-2 \quad \text{or} \quad y=1$$

$$x=2: \; y^2+y+2^2-2-2=0$$
$$y^2+y+4-2-2=0$$
$$y^2+y=0$$
$$y(y+1)=0$$
$$y=0 \quad \text{or} \quad y=-1$$

The apparent solution $y=0$ is extraneous because the denominator of a fraction cannot be 0.

The solutions are $(0,-2)$, $(0,1)$, and $(2,-1)$.

45. $\begin{cases} \ln x = 4\ln y \\ \log_3 x = 2+2\log_3 y \end{cases}$

Write the first equation in exponential form:

$$\ln x = 4\ln y$$
$$\ln x = \ln y^4$$
$$x = y^4$$

Write the second equation in exponential form:

$$\log_3 x = 2+2\log_3 y$$
$$\log_3 x = \log_3 9 + \log_3 y^2$$
$$\log_3 x = \log_3 \left(9y^2\right)$$
$$x = 9y^2$$

Thus, we have that

$$y^4 = 9y^2$$
$$y^4 - 9y^2 = 0$$
$$y^2\left(y^2-9\right)=0$$
$$y^2(y+3)(y-3)=0$$

$$y^2=0 \quad \text{or} \quad y+3=0 \quad \text{or} \quad y-3=0$$
$$y=0 \qquad\qquad y=-3 \qquad\quad y=3$$

The apparent y-values 0 and -3 are extraneous because the argument of a logarithm must be positive. Thus, the only possible y-value is 3. Substitute 3 for y into the equation $x=y^4$ to find the corresponding x-value: $x=3^4=81$. The solution is $(81,\,3)$.

47. $\begin{cases} r_1+r_2 = -\dfrac{b}{a} \\ r_1 r_2 = \dfrac{c}{a} \end{cases}$

Solve the second equation for r_1: $r_1 = \dfrac{c}{ar_2}$

Substitute the result into the first equation:

$$\frac{c}{ar_2}+r_2 = -\frac{b}{a}$$
$$ar_2\left(\frac{c}{ar_2}+r_2\right)=ar_2\left(-\frac{b}{a}\right)$$
$$c+a\left(r_2\right)^2 = -br_2$$
$$a\left(r_2\right)^2+br_2+c=0$$
$$r_2 = \frac{-b\pm\sqrt{b^2-4ac}}{2a}$$

Substitute this result into the first equation to find the corresponding value of r_1:

$$r_1+\frac{-b\pm\sqrt{b^2-4ac}}{2a}=-\frac{b}{a}$$
$$r_1-\frac{b}{2a}\pm\frac{\sqrt{b^2-4ac}}{2a}=-\frac{b}{a}$$
$$r_1=-\frac{b}{a}+\frac{b}{2a}\mp\frac{\sqrt{b^2-4ac}}{2a}$$
$$r_1=-\frac{2b}{2a}+\frac{b}{2a}\mp\frac{\sqrt{b^2-4ac}}{2a}$$
$$r_1=-\frac{b}{2a}\pm\frac{\sqrt{b^2-4ac}}{2a}$$
$$r_1=\frac{-b\mp\sqrt{b^2-4ac}}{2a}$$

If $r_1=\dfrac{-b+\sqrt{b^2-4ac}}{2a}$, then

$$r_2 = \frac{-b - \sqrt{b^2 - 4ac}}{2a} \; ; \; \text{if } r_1 = \frac{-b - \sqrt{b^2 - 4ac}}{2a},$$

then $r_2 = \dfrac{-b + \sqrt{b^2 - 4ac}}{2a}$.

49. (a) $f(1) = 3(1) + 4 = 3 + 4 = 7$

(b) $g(1) = 2^1 = 2$

51. (a) $f(3) = 3(3) + 4 = 9 + 4 = 13$

(b) $g(3) = 2^3 = 8$

53. (a) $f(5) = 3(5) + 4 = 15 + 4 = 19$

(b) $g(5) = 2^5 = 32$

55. $\begin{cases} y = x^2 - 6x + 4 \\ 5x + y = 6 \quad (y = -5x + 6) \end{cases}$

Let $Y_1 = x^2 - 6x + 4$ and $Y_2 = -5x + 6$.

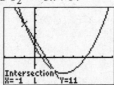

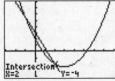

The solutions are $(-1, 11)$ and $(2, -4)$.

57. $\begin{cases} x^2 + y^2 = 16 \quad \left(y = \pm\sqrt{16 - x^2}\right) \\ x^2 - y^2 = 16 \quad \left(y = \pm\sqrt{x^2 - 16}\right) \end{cases}$

Let $Y_1 = \sqrt{16 - x^2}$, $Y_2 = -\sqrt{16 - x^2}$,

$Y_3 = \sqrt{x^2 - 16}$, and $Y_4 = -\sqrt{x^2 - 16}$.

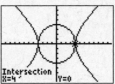

The solutions are $(-4, 0)$ and $(4, 0)$.

59. $\begin{cases} (x-4)^2 + y^2 = 25 \quad \left(y = \pm\sqrt{25 - (x-4)^2}\right) \\ x - y = -3 \qquad (y = x + 3) \end{cases}$

Let $Y_1 = \sqrt{25 - (x-4)^2}$, $Y_2 = -\sqrt{25 - (x-4)^2}$,

and $Y_3 = x + 3$.

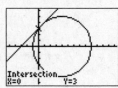

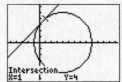

The solutions are $(0, 3)$ and $(1, 4)$.

61. $\begin{cases} (x-1)^2 + (y+2)^2 = 4 \quad \left(y = -2 \pm \sqrt{4 - (x-1)^2}\right) \\ x^2 + 4y - x = -1 \qquad \left(y = \dfrac{-x^2 + x - 1}{4}\right) \end{cases}$

Let $Y_1 = -2 + \sqrt{4 - (x-1)^2}$,

$Y_2 = -2 + \sqrt{4 - (x-1)^2}$, and $Y_3 = \dfrac{-x^2 + x - 1}{4}$.

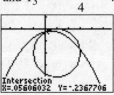

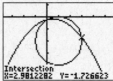

The solutions are approximately $(0.056, -0.237)$ and $(2.981, -1.727)$.

63. $\begin{cases} x^2 + 4y^2 = 4 \\ x^2 + 6x - y = -13 \end{cases}$ $\left(y = \pm\sqrt{1 - \dfrac{x^2}{4}}\right)$ $\left(y = x^2 + 6x + 13\right)$

Let $Y_1 = \sqrt{1 - \dfrac{x^2}{4}}$, $Y_2 = -\sqrt{1 - \dfrac{x^2}{4}}$, and

$Y_3 = x^2 + 6x + 13$.

```
WINDOW
Xmin=-9.4
Xmax=9.4
Xscl=1
Ymin=-6.2
Ymax=6.2
Yscl=1
Xres=1
```

Because the graphs do not intersect, the system
has no solution, $\varnothing$.

Chapter 12 Review

1. $d(P_1, P_2) = \sqrt{(x_2 - x_1)^2 + (y_2 - y_1)^2}$
$= \sqrt{(-4 - 0)^2 + (-3 - 0)^2}$
$= \sqrt{(-4)^2 + (-3)^2}$
$= \sqrt{16 + 9} = \sqrt{25} = 5$

2. $d(P_1, P_2) = \sqrt{(x_2 - x_1)^2 + (y_2 - y_1)^2}$
$= \sqrt{(5 - (-3))^2 + (-4 - 2)^2}$
$= \sqrt{8^2 + (-6)^2}$
$= \sqrt{64 + 36} = \sqrt{100} = 10$

3. $d(P_1, P_2) = \sqrt{(x_2 - x_1)^2 + (y_2 - y_1)^2}$
$= \sqrt{(5 - (-1))^2 + (3 - 1)^2}$
$= \sqrt{6^2 + 2^2}$
$= \sqrt{36 + 4} = \sqrt{40} = 2\sqrt{10} \approx 6.32$

4. $d(P_1, P_2) = \sqrt{(x_2 - x_1)^2 + (y_2 - y_1)^2}$
$= \sqrt{(6 - 6)^2 + (-1 - (-7))^2}$
$= \sqrt{0^2 + 6^2} = \sqrt{0 + 36} = \sqrt{36} = 6$

5. $d(P_1, P_2) = \sqrt{(x_2 - x_1)^2 + (y_2 - y_1)^2}$
$= \sqrt{(4\sqrt{7} - \sqrt{7})^2 + (5\sqrt{3} - (-\sqrt{3}))^2}$
$= \sqrt{(3\sqrt{7})^2 + (6\sqrt{3})^2}$
$= \sqrt{9(7) + 36(3)}$
$= \sqrt{63 + 108} = \sqrt{171} = 3\sqrt{19} \approx 13.08$

6. $d(P_1, P_2) = \sqrt{(x_2 - x_1)^2 + (y_2 - y_1)^2}$
$= \sqrt{(1.3 - (-0.2))^2 + (3.7 - 1.7)^2}$
$= \sqrt{1.5^2 + 2^2}$
$= \sqrt{2.25 + 4} = \sqrt{6.25} = 2.5$

7. $M = \left(\dfrac{x_1 + x_2}{2}, \dfrac{y_1 + y_2}{2}\right)$
$= \left(\dfrac{-1 + (-3)}{2}, \dfrac{6 + 4}{2}\right) = \left(\dfrac{-4}{2}, \dfrac{10}{2}\right) = (-2, 5)$

8. $M = \left(\dfrac{x_1 + x_2}{2}, \dfrac{y_1 + y_2}{2}\right)$
$= \left(\dfrac{7 + 5}{2}, \dfrac{0 + (-4)}{2}\right) = \left(\dfrac{12}{2}, \dfrac{-4}{2}\right) = (6, -2)$

9. $M = \left(\dfrac{x_1 + x_2}{2}, \dfrac{y_1 + y_2}{2}\right)$
$= \left(\dfrac{-\sqrt{3} + (-7\sqrt{3})}{2}, \dfrac{2\sqrt{6} + (-8\sqrt{6})}{2}\right)$
$= \left(\dfrac{-8\sqrt{3}}{2}, \dfrac{-6\sqrt{6}}{2}\right) = (-4\sqrt{3}, -3\sqrt{6})$

10. $M = \left(\dfrac{x_1 + x_2}{2}, \dfrac{y_1 + y_2}{2}\right)$
$= \left(\dfrac{5 + 0}{2}, \dfrac{-2 + 3}{2}\right) = \left(\dfrac{5}{2}, \dfrac{1}{2}\right)$

11. $M = \left(\dfrac{x_1 + x_2}{2}, \dfrac{y_1 + y_2}{2}\right)$
$= \left(\dfrac{\frac{1}{4} + \frac{5}{4}}{2}, \dfrac{\frac{2}{3} + \frac{1}{3}}{2}\right) = \left(\dfrac{\frac{3}{2}}{2}, \dfrac{1}{2}\right) = \left(\dfrac{3}{4}, \dfrac{1}{2}\right)$

12. (a)

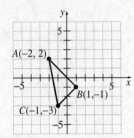

(b) $d(A,B) = \sqrt{(1-(-2))^2 + (-1-2)^2}$

$= \sqrt{3^2 + (-3)^2}$

$= \sqrt{9+9}$

$= \sqrt{18}$

$= 3\sqrt{2} \approx 4.24$

$d(B,C) = \sqrt{(-1-1)^2 + (-3-(-1))^2}$

$= \sqrt{(-2)^2 + (-2)^2}$

$= \sqrt{4+4}$

$= \sqrt{8}$

$= 2\sqrt{2} \approx 2.83$

$d(A,C) = \sqrt{(-1-(-2))^2 + (-3-2)^2}$

$= \sqrt{1^2 + (-5)^2}$

$= \sqrt{1+25}$

$= \sqrt{26} \approx 5.10$

(c) To determine if triangle *ABC* is a right triangle, we check to see if

$[d(A,B)]^2 + [d(B,C)]^2 \overset{?}{=} [d(A,C)]^2$

$(3\sqrt{2})^2 + (2\sqrt{2})^2 \overset{?}{=} (\sqrt{26})^2$

$9 \cdot 2 + 4 \cdot 2 \overset{?}{=} 26$

$18 + 8 \overset{?}{=} 26$

$26 = 26 \leftarrow$ True

Therefore, triangle *ABC* is a right triangle.

(d) The length of the "base" of the triangle is $d(B,C) = 2\sqrt{2}$ and the length of the "height" of the triangle is $d(A,B) = 3\sqrt{2}$. Thus, the area of triangle *ABC* is

Area $= \dfrac{1}{2} \cdot$ base $\cdot$ height $= \dfrac{1}{2} \cdot 2\sqrt{2} \cdot 3\sqrt{2} = 6$ square units.

13. The center of the circle will be the midpoint of the line segment with endpoints $(-6,1)$ and $(2,1)$. Thus, $(h,k) = \left(\dfrac{-6+2}{2}, \dfrac{1+1}{2}\right) = (-2,1)$.

The radius of the circle will be the distance from the center point $(-2,1)$ to a point on the circle, say $(2,1)$. Thus, $r = \sqrt{(2-(-2))^2 + (1-1)^2} = 4$.

The equation of the circle is

$(x-h)^2 + (y-k)^2 = r^2$

$(x-(-2))^2 + (y-1)^2 = 4^2$

$(x+2)^2 + (y-1)^2 = 16$

14. The center of the circle will be the midpoint of the line segment with endpoints $(2,3)$ and $(8,3)$. Thus, $(h,k) = \left(\dfrac{2+8}{2}, \dfrac{3+3}{2}\right) = (5,3)$.

The radius of the circle will be the distance from the center point $(5,3)$ to a point on the circle, say $(8,3)$. Thus, $r = \sqrt{(8-5)^2 + (3-3)^2} = 3$.

The equation of the circle is

$(x-h)^2 + (y-k)^2 = r^2$

$(x-5)^2 + (y-3)^2 = 3^2$

$(x-5)^2 + (y-3)^2 = 9$

15. $(x-h)^2 + (y-k)^2 = r^2$

$(x-0)^2 + (y-0)^2 = 4^2$

$x^2 + y^2 = 16$

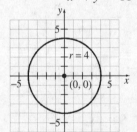

16. $(x-h)^2 + (y-k)^2 = r^2$

$(x-(-3))^2 + (y-1)^2 = 3^2$

$(x+3)^2 + (y-1)^2 = 9$

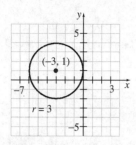

17. $(x-h)^2 + (y-k)^2 = r^2$

$(x-5)^2 + (y-(-2))^2 = 1^2$

$(x-5)^2 + (y+2)^2 = 1$

18. $(x-h)^2 + (y-k)^2 = r^2$

$(x-4)^2 + (y-0)^2 = \left(\sqrt{7}\right)^2$

$(x-4)^2 + y^2 = 7$

19. The radius of the circle will be the distance from the center point $(2,-1)$ to the point on the circle

$(5,3)$. Thus, $r = \sqrt{(5-2)^2 + (3-(-1))^2} = 5$.

The equation of the circle is

$(x-h)^2 + (y-k)^2 = r^2$

$(x-2)^2 + (y-(-1))^2 = 5^2$

$(x-2)^2 + (y+1)^2 = 25$

20. The center of the circle will be the midpoint of the diameter with endpoints $(-3,-1)$ and $(1,7)$.

Thus, $(h,k) = \left(\dfrac{-3+1}{2}, \dfrac{-1+7}{2}\right) = (-1,3)$. The

radius of the circle will be the distance from the

center point $(-1,3)$ to one of the endpoints of the diameter, say $(1,7)$. Thus,

$r = \sqrt{(1-(-1))^2 + (7-3)^2} = \sqrt{20} = 2\sqrt{5}$.

The equation of the circle is

$(x-h)^2 + (y-k)^2 = r^2$

$(x-(-1))^2 + (y-3)^2 = \left(2\sqrt{5}\right)^2$

$(x+1)^2 + (y-3)^2 = 20$

21. $x^2 + y^2 = 25$

$(x-0)^2 + (y-0)^2 = 5^2$

The center is $(h,k) = (0,0)$, and the radius is $r = 5$.

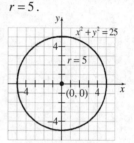

22. $(x-1)^2 + (y-2)^2 = 4$

$(x-1)^2 + (y-2)^2 = 2^2$

The center is $(h,k) = (1,2)$, and the radius is $r = 2$.

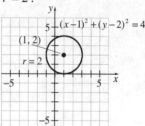

23. $x^2 + (y-4)^2 = 16$

$(x-0)^2 + (y-4)^2 = 4^2$

The center is $(h,k) = (0,4)$, and the radius is $r = 4$.

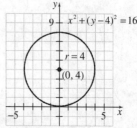

24.
$$(x+1)^2+(y+6)^2=49$$
$$(x-(-1))^2+(y-(-6))^2=7^2$$
The center is $(h,k)=(-1,-6)$, and the radius is $r=7$.

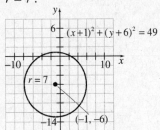

25.
$$(x+2)^2+\left(y-\frac{3}{2}\right)^2=\frac{1}{4}$$
$$(x-(-2))^2+\left(y-\frac{3}{2}\right)^2=\left(\frac{1}{2}\right)^2$$
The center is $(h,k)=\left(-2,\frac{3}{2}\right)$, and the radius is

$r=\frac{1}{2}$.

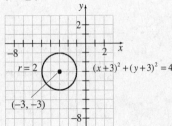

26.
$$(x+3)^2+(y+3)^2=4$$
$$(x-(-3))^2+(y-(-3))^2=2^2$$
The center is $(h,k)=(-3,-3)$, and the radius is $r=2$.

27.
$$x^2+y^2+6x+10y-2=0$$
$$\left(x^2+6x\right)+\left(y^2+10y\right)=2$$
$$\left(x^2+6x+9\right)+\left(y^2+10y+25\right)=2+9+25$$
$$(x+3)^2+(y+5)^2=36$$
$$(x-(-3))^2+(y-(-5))^2=6^2$$
The center is $(h,k)=(-3,-5)$, and the radius is $r=6$.

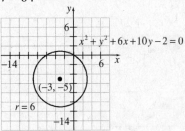

28.
$$x^2+y^2-8x+4y+16=0$$
$$\left(x^2-8x\right)+\left(y^2+4y\right)=-16$$
$$\left(x^2-8x+16\right)+\left(y^2+4y+4\right)=-16+16+4$$
$$(x-4)^2+(y+2)^2=4$$
$$(x-4)^2+(y-(-2))^2=2^2$$
The center is $(h,k)=(4,-2)$, and the radius is $r=2$.

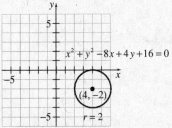

29.
$$x^2+y^2+2x-4y-4=0$$
$$\left(x^2+2x\right)+\left(y^2-4y\right)=4$$
$$\left(x^2+2x+1\right)+\left(y^2-4y+4\right)=4+1+4$$
$$(x+1)^2+(y-2)^2=9$$
$$(x-(-1))^2+(y-2)^2=3^2$$
The center is $(h,k)=(-1,2)$, and the radius is $r=3$.

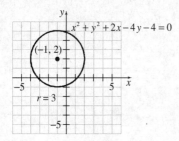

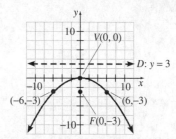

30.

$$x^2 + y^2 - 10x - 2y + 17 = 0$$

$$\left(x^2 - 10x\right) + \left(y^2 - 2y\right) = -17$$

$$\left(x^2 - 10x + 25\right) + \left(y^2 - 2y + 1\right) = -17 + 25 + 1$$

$$\left(x - 5\right)^2 + \left(y - 1\right)^2 = 9$$

$$\left(x - 5\right)^2 + \left(y - 1\right)^2 = 3^2$$

The center is $(h, k) = (5, 1)$, and the radius is $r = 3$.

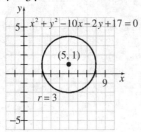

31. The distance from the vertex $(0, 0)$ to the focus $(0, -3)$ is $a = 3$. Because the focus lies on the negative y-axis, we know that the parabola will open downward and the axis of symmetry is the y-axis. This means the equation of the parabola is of the form $x^2 = -4ay$ with $a = 3$:

$$x^2 = -4(3)y$$

$$x^2 = -12y$$

The directrix is the line $y = 3$. To help graph the parabola, we plot the two points on the graph to the left and right of the focus. Let $y = -3$:

$$x^2 = -12(-3) = 36$$

$$x = \pm 6$$

The points $(-6, -3)$ and $(6, -3)$ are on the graph.

32. Notice that the directrix is the vertical line $x = 4$ and that the focus $(-4, 0)$ is on the x-axis. Thus, the axis of symmetry is the x-axis, the vertex is at the origin, and the parabola opens to the left. Now, the distance from the vertex $(0, 0)$ to the focus $(-4, 0)$ is $a = 4$, so the equation of the parabola is of the form $y^2 = -4ax$ with $a = 4$:

$$y^2 = -4(4)x$$

$$y^2 = -16x$$

To help graph the parabola, we plot the two points on the graph above and below the focus. Let $x = -4$:

$$y^2 = -16(-4) = 64$$

$$y = \pm 8$$

The points $(-4, -8)$ and $(-4, 8)$ are on the graph.

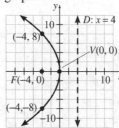

33. The vertex is at the origin and the axis of symmetry is the x-axis, so the parabola either opens to the left or to the right. Because the graph contains the point $(8, -2)$, which is in quadrant IV, the parabola must open to the right. Therefore, the equation of the parabola is of the form $y^2 = 4ax$. Now $y = -2$ when $x = 8$, so

$$y^2 = 4ax$$

$$(-2)^2 = 4a(8)$$

$$4 = 32a$$

$$a = \frac{4}{32} = \frac{1}{8}$$

The equation of the parabola is

$$y^2 = 4\left(\frac{1}{8}\right)x$$

$$y^2 = \frac{1}{2}x$$

With $a = \frac{1}{8}$, we know that the focus is $\left(\frac{1}{8}, 0\right)$

and the directrix is the line $x = -\frac{1}{8}$. To help

graph the parabola, we plot the two points on the

graph above and below the focus. Let $x = \frac{1}{8}$:

$$y^2 = \frac{1}{2}\left(\frac{1}{8}\right) = \frac{1}{16}$$

$$y = \pm\frac{1}{4}$$

The points $\left(\frac{1}{8}, -\frac{1}{4}\right)$ and $\left(\frac{1}{8}, \frac{1}{4}\right)$ are on the

graph.

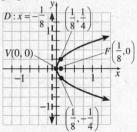

34. The vertex is at the origin and the directrix is the
line $y = -2$, so the focus must be $(0, 2)$.
Accordingly, the parabola opens upward with
$a = 2$. This means the equation of the parabola
is of the form $x^2 = 4ay$ with $a = 2$:

$$x^2 = 4(2)y$$

$$x^2 = 8y$$

To help graph the parabola, we plot the two
points on the graph to the left and right of the
focus. Let $y = 2$:

$$x^2 = 8(2) = 16$$

$$x = \pm 4$$

The points $(-4, 2)$ and $(4, 2)$ are on the graph.

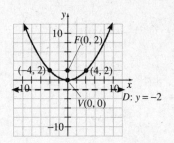

35. Notice that $x^2 = 2y$ is of the form $x^2 = 4ay$,

where $4a = 2$, so that $a = \frac{2}{4} = \frac{1}{2}$. Now, the

graph of an equation of the form $x^2 = 4ay$ will
be a parabola that opens upward with the vertex
at the origin, focus at $(0, a)$, and directrix of

$y = -a$. Thus, the graph of $x^2 = 2y$ is a

parabola that opens upward with vertex $(0, 0)$,

focus $\left(0, \frac{1}{2}\right)$, and directrix $y = -\frac{1}{2}$. To help

graph the parabola, we plot the two points on the
graph to the left and to the right of the focus.

Let $y = \frac{1}{2}$:

$$x^2 = 2\left(\frac{1}{2}\right)$$

$$x^2 = 1$$

$$x = \pm 1$$

The points $\left(-1, \frac{1}{2}\right)$ and $\left(1, \frac{1}{2}\right)$ are on the graph.

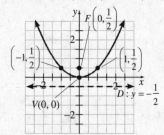

36. Notice that $y^2 = 16x$ is of the form $y^2 = 4ax$,

where $4a = 16$, so that $a = 4$. Now, the graph

of an equation of the form $y^2 = 4ax$ will be a
parabola that opens to the right with the vertex at
the origin, focus at $(a, 0)$, and directrix of

$x = -a$. Thus, the graph of $y^2 = 16x$ is a

parabola that opens to the right with vertex
$(0, 0)$, focus $(4, 0)$, and directrix $x = -4$. To
help graph the parabola, we plot the two points

on the graph above and below the focus. Let $x = 4$:

$$y^2 = 16(4)$$
$$y^2 = 64$$
$$y = \pm 8$$

The points $(4, -8)$ and $(4, 8)$ are on the graph.

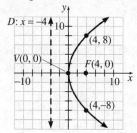

37. Notice that $(x+1)^2 = 8(y-3)$ is of the form $(x-h)^2 = 4a(y-k)$, where $4a = 8$, so that $a = 2$ and $(h,k) = (-1, 3)$. Now, the graph of an equation of the form $(x-h)^2 = 4a(y-k)$ will be a parabola that opens upward with vertex at (h,k), focus at $(h, k+a)$, and directrix of $y = k - a$. Note that $k + a = 3 + 2 = 5$ and $k - a = 3 - 2 = 1$. Thus, the graph of $(x+1)^2 = 8(y-3)$ is a parabola that opens upward with vertex $(-1, 3)$, focus $(-1, 5)$, and directrix $y = 1$. To help graph the parabola, we plot the two points on the left and on the right of the focus. Let $y = 5$:

$$(x+1)^2 = 8(5-3)$$
$$(x+1)^2 = 16$$
$$x + 1 = \pm 4$$
$$x = -1 \pm 4$$
$$x = -5 \text{ or } x = 3$$

The points $(-5, 5)$ and $(3, 5)$ are on the graph.

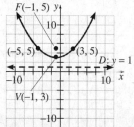

38. Notice that $(y-4)^2 = -2(x+3)$ is of the form $(y-k)^2 = -4a(x-h)$, where $(h,k) = (-3, 4)$ and $-4a = -2$, so that $a = \dfrac{-2}{-4} = \dfrac{1}{2}$. Now, the graph of an equation of the form $(y-k)^2 = -4a(x-h)$ will be a parabola that opens to the left with the vertex at (h,k), focus at $(h-a, k)$, and directrix of $x = h + a$. Note that $h - a = -3 - \dfrac{1}{2} = -\dfrac{7}{2}$ and $h + a = -3 + \dfrac{1}{2} = -\dfrac{5}{2}$. Thus, the graph of $(y-4)^2 = -2(x+3)$ is a parabola that opens to the left with vertex $(-3, 4)$, focus $\left(-\dfrac{7}{2}, 4\right)$, and directrix $x = -\dfrac{5}{2}$. To help graph the parabola, we plot the two points on the graph above and below the focus. Let $x = -\dfrac{7}{2}$:

$$(y-4)^2 = -2\left(-\dfrac{7}{2} + 3\right)$$
$$(y-4)^2 = -2\left(-\dfrac{1}{2}\right)$$
$$(y-4)^2 = 1$$
$$y - 4 = \pm 1$$
$$y = 4 \pm 1$$
$$y = 3 \text{ or } y = 5$$

The points $\left(-\dfrac{7}{2}, 3\right)$ and $\left(-\dfrac{7}{2}, 5\right)$ are on the graph.

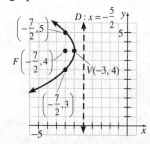

39. We complete the square in x to write the equation in standard form:

$$x^2 - 10x + 3y + 19 = 0$$
$$x^2 - 10x = -3y - 19$$
$$x^2 - 10x + 25 = -3y - 19 + 25$$
$$x^2 - 10x + 25 = -3y + 6$$
$$(x-5)^2 = -3(y-2)$$

Notice that $(x-5)^2 = -3(y-2)$ is of the form $(x-h)^2 = -4a(y-k)$, where $(h,k) = (5,2)$ and $-4a = -3$, so that $a = \dfrac{3}{4}$. Now, the graph of an equation of the form $(x-h)^2 = -4a(y-k)$ will be a parabola that opens downward with the vertex at (h,k), focus at $(h, k-a)$, and directrix of $y = k+a$. Note that $k - a = 2 - \dfrac{3}{4} = \dfrac{5}{4}$ and $k + a = 2 + \dfrac{3}{4} = \dfrac{11}{4}$. Thus, the graph of $(x-5)^2 = -3(y-2)$ is a parabola that opens downward with vertex $(5,2)$, focus $\left(5, \dfrac{5}{4}\right)$, and directrix $y = \dfrac{11}{4}$. To help graph the parabola, we plot the two points on the graph to the left and right of the focus. Let $y = \dfrac{5}{4}$:

$$(x-5)^2 = -3\left(\dfrac{5}{4} - 2\right)$$
$$(x-5)^2 = -3\left(-\dfrac{3}{4}\right)$$
$$(x-5)^2 = \dfrac{9}{4}$$
$$x - 5 = \pm\dfrac{3}{2}$$
$$x = 5 \pm \dfrac{3}{2}$$
$$x = \dfrac{7}{2} \text{ or } x = \dfrac{13}{2}$$

The points $\left(\dfrac{7}{2}, \dfrac{5}{4}\right)$ and $\left(\dfrac{13}{2}, \dfrac{5}{4}\right)$ are on the graph.

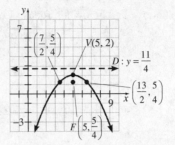

40. The receiver should be located at the focus of the dish, so we need to find where the focus of the dish is. To solve this problem, we draw a parabola on a Cartesian plane so that the vertex is the origin and the focus is on the positive y-axis. The width of the dish is 300 feet, and the depth is 44 feet. Therefore, we know two points on the graph of the parabola: $(-150, 44)$ and $(150, 44)$.

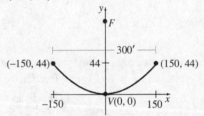

The equation of the parabola has the form $x^2 = 4ay$. Since $(150, 44)$ is a point on the graph, we have

$$150^2 = 4a(44)$$
$$22,500 = 176a$$
$$a = \dfrac{22,500}{176} \approx 127.84$$

The receiver should be located approximately 127.84 feet above the center of the dish, along its axis of symmetry.

41. $\dfrac{x^2}{9} + y^2 = 1$

$$\dfrac{x^2}{9} + \dfrac{y^2}{1} = 1$$

The larger number, 9, is in the denominator of the x^2-term. This means that the major axis is the x-axis and that the equation of the ellipse is of the form $\dfrac{x^2}{a^2} + \dfrac{y^2}{b^2} = 1$, so that $a^2 = 9$ and $b^2 = 1$. The center of the ellipse is $(0,0)$.

Because $b^2 = a^2 - c^2$, or $c^2 = a^2 - b^2$, we have that $c^2 = 9 - 1 = 8$, so that $c = \pm\sqrt{8} = \pm 2\sqrt{2}$.

Since the major axis is the *x*-axis, the foci are $\left(-2\sqrt{2},0\right)$ and $\left(2\sqrt{2},0\right)$. To find the *x*-intercepts (vertices), let $y=0$; to find the *y*-intercepts, let $x=0$:

x-intercepts:

$$\frac{x^2}{9}+0^2=1$$

$$\frac{x^2}{9}=1$$

$$x^2=9$$

$$x=\pm3$$

y-intercepts:

$$\frac{0^2}{9}+y^2=1$$

$$y^2=1$$

$$y=\pm1$$

The intercepts are $(-3,0)$, $(3,0)$, $(0,-1)$, and $(0,1)$.

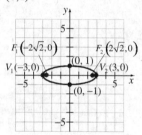

42. $9x^2+4y^2=36$

$$\frac{9x^2+4y^2}{36}=\frac{36}{36}$$

$$\frac{x^2}{4}+\frac{y^2}{9}=1$$

The larger number, 9, is in the denominator of the y^2-term. This means that the major axis is the *y*-axis and that the equation of the ellipse is of the form $\frac{x^2}{b^2}+\frac{y^2}{a^2}=1$, so that $a^2=9$ and $b^2=4$. The center of the ellipse is $(0,0)$.

Because $b^2=a^2-c^2$, or $c^2=a^2-b^2$, we have that $c^2=9-4=5$, so that $c=\pm\sqrt{5}$. Since the major axis is the *y*-axis, the foci are $\left(0,-\sqrt{5}\right)$ and $\left(0,\sqrt{5}\right)$. To find the *x*-intercepts, let $y=0$; to find the *y*-intercepts (vertices), let $x=0$:

x-intercepts:

$$9x^2+4(0)^2=36$$

$$9x^2=36$$

$$x^2=4$$

$$x=\pm2$$

y-intercepts:

$$9(0)^2+4y^2=36$$

$$4y^2=36$$

$$y^2=9$$

$$y=\pm3$$

The intercepts are $(-2,0)$, $(2,0)$, $(0,-3)$, and

$(0,3)$.

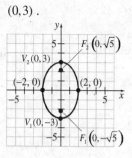

43. The given focus $(0,3)$ and the given vertex $(0,5)$ lie on the *y*-axis. Thus, the equation of the ellipse is of the form $\frac{x^2}{b^2}+\frac{y^2}{a^2}=1$. The distance from the center of the ellipse to the vertex is $a=5$ units. The distance from the center of the ellipse to the focus is $c=3$ units. Because $b^2=a^2-c^2$, we have that $b^2=5^2-3^2=25-9=16$. Thus, the equation of the ellipse is $\frac{x^2}{16}+\frac{y^2}{25}=1$.

To help graph the ellipse, find the *x*-intercepts:

Let $y=0$: $\frac{x^2}{16}+\frac{0^2}{25}=1$

$$\frac{x^2}{16}=1$$

$$x^2=16$$

$$x=\pm4$$

The *x*-intercepts are $(-4,0)$ and $(4,0)$.

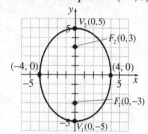

44. The given focus $(-2,0)$ and the given vertex $(-6,0)$ lie on the *x*-axis. Thus, the equation of the ellipse is of the form $\frac{x^2}{a^2}+\frac{y^2}{b^2}=1$. The distance from the center of the ellipse to the vertex is $a=6$ units. The distance from the center of the ellipse to the focus is $c=2$ units. Because $b^2=a^2-c^2$, we have that $b^2=6^2-2^2=36-4=32$. Thus, the equation

of the ellipse is $\dfrac{x^2}{36} + \dfrac{y^2}{32} = 1$.

To help graph the ellipse, find the y-intercepts:

Let $x = 0$: $\dfrac{0^2}{36} + \dfrac{y^2}{32} = 1$

$\dfrac{y^2}{32} = 1$

$y^2 = 32$

$y = \pm\sqrt{32} = \pm 4\sqrt{2}$

The y-intercepts are $\left(0, -4\sqrt{2}\right)$ and $\left(0, 4\sqrt{2}\right)$.

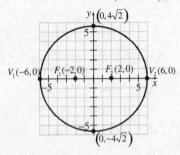

45. The given foci $(\pm 8, 0)$ and the given vertices $(\pm 10, 0)$ lie on the x-axis. The center of the ellipse is the midpoint between the two vertices (or foci). Thus, the center of the ellipse is $(0, 0)$ and its equation is of the form $\dfrac{x^2}{a^2} + \dfrac{y^2}{b^2} = 1$. The distance from the center of the ellipse to the vertex is $a = 10$ units. The distance from the center of the ellipse to the focus is $c = 8$ units. Because $b^2 = a^2 - c^2$, we have that $b^2 = 10^2 - 8^2 = 100 - 64 = 36$. Thus, the equation of the ellipse is $\dfrac{x^2}{100} + \dfrac{y^2}{36} = 1$.

To help graph the ellipse, find the y-intercepts:

Let $x = 0$: $\dfrac{0^2}{100} + \dfrac{y^2}{36} = 1$

$\dfrac{y^2}{36} = 1$

$y^2 = 36$

$y = \pm 6$

The y-intercepts are $(0, -6)$ and $(0, 6)$.

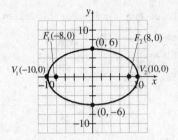

46. $\dfrac{(x-1)^2}{49} + \dfrac{(y+2)^2}{25} = 1$

The center of the ellipse is $(h, k) = (1, -2)$. Because the larger number, 49, is the denominator of the x^2-term, the major axis is parallel to the x-axis. Because $a^2 = 49$ and $b^2 = 25$, we have that $c^2 = a^2 - b^2 = 49 - 25 = 24$. The vertices are $a = 7$ units to the left and right of the center at $V_1(-6, -2)$ and $V_2(8, -2)$. The foci are $c = \sqrt{24} = 2\sqrt{6}$ units to the left and right of the center at $F_1\left(1 - 2\sqrt{6}, -2\right)$ and $F_2\left(1 + 2\sqrt{6}, -2\right)$.

We plot the points $b = 5$ units above and below the center point at $(1, 3)$ and $(1, -7)$.

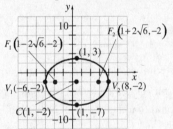

47. $25(x+3)^2 + 9(y-4)^2 = 225$

$\dfrac{25(x+3)^2 + 9(y-4)^2}{225} = \dfrac{225}{225}$

$\dfrac{(x+3)^2}{9} + \dfrac{(y-4)^2}{25} = 1$

The center of the ellipse is $(h, k) = (-3, 4)$. Because the larger number, 25, is the denominator of the y^2-term, the major axis is parallel to the y-axis. Because $a^2 = 25$ and $b^2 = 9$, we have that $c^2 = a^2 - b^2 = 25 - 9 = 16$. The vertices are $a = 5$ units below and above the center at $V_1(-3, -1)$ and $V_2(-3, 9)$. The foci are $c = \sqrt{16} = 4$ units below and above the center at $F_1(-3, 0)$ and $F_2(-3, 8)$. We plot the points

$b = 3$ units to the left and right of the center point at $(-6, 4)$ and $(0, 4)$.

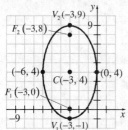

48. **(a)** To solve this problem, we draw the ellipse on a Cartesian plane so that the x-axis coincides with the water and the y-axis passes through the center of the arch. Thus, the origin is the center of the ellipse. Now, the "center" of the arch is 16 feet above the water, so the point $(0, 16)$ is on the ellipse.

The river is 60 feet wide, so the two points $(-30, 0)$ and $(30, 0)$ are the two vertices of the ellipse and the major axis is along the x-axis.

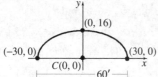

The equation of the ellipse must have the form $\dfrac{x^2}{a^2} + \dfrac{y^2}{b^2} = 1$. The distance from the center of the ellipse to each vertex is $a = 30$. Also, the height of the semi-ellipse is 16 feet, so $b = 16$. Thus, the equation of the ellipse is

$$\frac{x^2}{30^2} + \frac{y^2}{16^2} = 1$$

$$\frac{x^2}{900} + \frac{y^2}{256} = 1$$

(b) To determine if the barge can fit through the opening of the bridge, we center it beneath the arch so that the points $(-12.5, 12)$ and $(12.5, 12)$ represent the top corners of the barge. Now, we determine the height of the arch above the water at points 12.5 feet to the left and right of the center by substituting $x = 12.5$ into the equation of the ellipse and solving for y:

$$\frac{12.5^2}{900} + \frac{y^2}{256} = 1$$

$$\frac{156.25}{900} + \frac{y^2}{256} = 1$$

$$\frac{y^2}{256} = \frac{119}{144}$$

$$y^2 = \frac{1904}{9}$$

$$y = \sqrt{\frac{1904}{9}} \approx 14.54$$

At points 12.5 feet from the center of the river, the arch is approximately 14.54 feet above the surface of the water. Thus, the barge can fit through the opening with about 2.54 feet of clearance above the top corners.

49. $\dfrac{x^2}{4} - \dfrac{y^2}{9} = 1$

Notice the equation is of the form $\dfrac{x^2}{a^2} - \dfrac{y^2}{b^2} = 1$.

Because the x^2-term is first, the transverse axis is the x-axis and the hyperbola opens left and right. The center of the hyperbola is the origin. We have that $a^2 = 4$ and $b^2 = 9$. Because $c^2 = a^2 + b^2$, we have that $c^2 = 4 + 9 = 13$, so that $c = \sqrt{13}$. The vertices are $(\pm a, 0) = (\pm 2, 0)$, and the foci are $(\pm c, 0) = \left(\pm\sqrt{13}, 0\right)$. To help graph the hyperbola, we plot the points on the graph above and below the foci. Let $x = \pm\sqrt{13}$:

$$\frac{\left(\pm\sqrt{13}\right)^2}{4} - \frac{y^2}{9} = 1$$

$$\frac{13}{4} - \frac{y^2}{9} = 1$$

$$-\frac{y^2}{9} = -\frac{9}{4}$$

$$y^2 = \frac{81}{4}$$

$$y = \pm\frac{9}{2}$$

The points above and below the foci are $\left(\sqrt{13}, \dfrac{9}{2}\right)$, $\left(\sqrt{13}, -\dfrac{9}{2}\right)$, $\left(-\sqrt{13}, \dfrac{9}{2}\right)$, and $\left(-\sqrt{13}, -\dfrac{9}{2}\right)$.

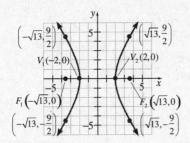

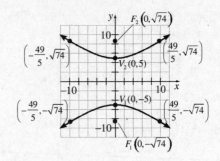

50. $\dfrac{y^2}{25} - \dfrac{x^2}{49} = 1$

Notice the equation is of the form $\dfrac{y^2}{a^2} - \dfrac{x^2}{b^2} = 1$.

Because the y^2-term is first, the transverse axis is the y-axis and the hyperbola opens up and down. The center of the hyperbola is the origin. We have that $a^2 = 25$ and $b^2 = 49$. Because $c^2 = a^2 + b^2$, we have that $c^2 = 25 + 49 = 74$, so that $c = \sqrt{74}$. The vertices are $(0, \pm a) = (0, \pm 5)$, and the foci are $(0, \pm c) = \left(0, \pm \sqrt{74}\right)$. To help graph the hyperbola, we plot the points on the graph to the left and right of the foci. Let $y = \pm\sqrt{74}$:

$$\dfrac{\left(\pm\sqrt{74}\right)^2}{25} - \dfrac{x^2}{49} = 1$$

$$\dfrac{74}{25} - \dfrac{x^2}{49} = 1$$

$$-\dfrac{x^2}{49} = -\dfrac{49}{25}$$

$$x^2 = \dfrac{2401}{25}$$

$$x = \pm\dfrac{49}{5}$$

The points to the left and right of the foci are $\left(\dfrac{49}{5}, \sqrt{74}\right)$, $\left(-\dfrac{49}{5}, \sqrt{74}\right)$, $\left(\dfrac{49}{5}, -\sqrt{74}\right)$, and $\left(-\dfrac{49}{5}, -\sqrt{74}\right)$.

51. $16y^2 - 25x^2 = 400$

$$\dfrac{16y^2 - 25x^2}{400} = \dfrac{400}{400}$$

$$\dfrac{y^2}{25} - \dfrac{x^2}{16} = 1$$

Notice the equation is of the form $\dfrac{y^2}{a^2} - \dfrac{x^2}{b^2} = 1$.

Because the y^2-term is first, the transverse axis is the y-axis and the hyperbola opens up and down. The center of the hyperbola is the origin. We have that $a^2 = 25$ and $b^2 = 16$. Because $c^2 = a^2 + b^2$, we have that $c^2 = 25 + 16 = 41$, so that $c = \sqrt{41}$. The vertices are $(0, \pm a) = (0, \pm 5)$, and the foci are $(0, \pm c) = \left(0, \pm\sqrt{41}\right)$. To help graph the hyperbola, we plot the points on the graph to the left and right of the foci. Let $y = \pm\sqrt{41}$:

$$\dfrac{\left(\pm\sqrt{41}\right)^2}{25} - \dfrac{x^2}{16} = 1$$

$$\dfrac{41}{25} - \dfrac{x^2}{16} = 1$$

$$-\dfrac{x^2}{16} = -\dfrac{16}{25}$$

$$x^2 = \dfrac{256}{25}$$

$$x = \pm\dfrac{16}{5}$$

The points to the left and right of the foci are $\left(\dfrac{16}{5}, \sqrt{41}\right)$, $\left(\dfrac{16}{5}, -\sqrt{41}\right)$, $\left(-\dfrac{16}{5}, \sqrt{41}\right)$, and $\left(-\dfrac{16}{5}, -\sqrt{41}\right)$.

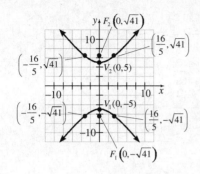

52. $\dfrac{x^2}{36} - \dfrac{y^2}{36} = 1$

Notice the equation is of the form $\dfrac{x^2}{a^2} - \dfrac{y^2}{b^2} = 1$.

Because the x^2-term is first, the transverse is along the x-axis and the hyperbola opens left and right. The center of the hyperbola is $(0,0)$. We have that $a^2 = 36$ and $b^2 = 36$. Because $c^2 = a^2 + b^2$, we have that $c^2 = 36 + 36 = 72$, so that $c = \sqrt{72} = 6\sqrt{2}$. The vertices are $(\pm a, 0) = (\pm 6, 0)$, and the foci are $(\pm c, 0) = \left(\pm 6\sqrt{2}, 0\right)$. Since $a = 6$ and $b = 6$, the equations of the asymptotes are $y = \dfrac{6}{6}x = x$ and $y = -\dfrac{6}{6}x = -x$. To help graph the hyperbola, we form the rectangle using the points $(\pm a, 0) = (\pm 6, 0)$ and $(0, \pm b) = (0, \pm 6)$. The diagonals are the asymptotes.

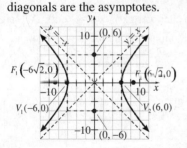

53. $\dfrac{y^2}{25} - \dfrac{x^2}{4} = 1$

Notice the equation is of the form $\dfrac{y^2}{a^2} - \dfrac{x^2}{b^2} = 1$.

Because the y^2-term is first, the transverse is along the y-axis and the hyperbola opens up and down. The center of the hyperbola is $(0,0)$. We

have that $a^2 = 25$ and $b^2 = 4$. Because $c^2 = a^2 + b^2$, we have that $c^2 = 25 + 4 = 29$, so that $c = \sqrt{29}$. The vertices are $(0, \pm a) = (0, \pm 5)$, and the foci are $(0, \pm c) = \left(0, \pm\sqrt{29}\right)$. Since $a = 5$ and $b = 2$, the equations of the asymptotes are $y = \dfrac{5}{2}x$ and $y = -\dfrac{5}{2}x$. To help graph the hyperbola, we form the rectangle using the points $(0, \pm a) = (0, \pm 5)$ and $(\pm b, 0) = (\pm 2, 0)$. The diagonals are the asymptotes.

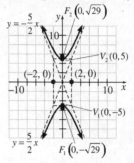

54. The given center $(0,0)$, focus $(-4,0)$, and vertex $(-3,0)$ all lie on the x-axis. Thus, the transverse axis is the x-axis and the hyperbola opens left and right. This means that the equation of the hyperbola is of the form $\dfrac{x^2}{a^2} - \dfrac{y^2}{b^2} = 1$. Now, the distance between the given vertex $(-3,0)$ and the center $(0,0)$ is $a = 3$ units. Likewise, the distance between the given focus $(-4,0)$ and the center is $c = 4$ units. Because $b^2 = c^2 - a^2$, we have that $b^2 = 4^2 - 3^2 = 16 - 9 = 7$. Thus, the equation of the hyperbola is $\dfrac{x^2}{9} - \dfrac{y^2}{7} = 1$.

To help graph the hyperbola, we first find the vertex and focus that were not given. Since the center is at $(0,0)$ and since one vertex is at $(-3,0)$, the other vertex must be at $(3,0)$. Similarly, one the focus is at $(-4,0)$, so the other focus is at $(4,0)$. Next, plot the points on the graph above and below the foci. Let $x = \pm 4$:

$$\frac{(\pm 4)^2}{9} - \frac{y^2}{7} = 1$$

$$\frac{16}{9} - \frac{y^2}{7} = 1$$

$$-\frac{y^2}{7} = -\frac{7}{9}$$

$$y^2 = \frac{49}{9}$$

$$y = \pm \frac{7}{3}$$

The points above and below the foci are $\left(4, \frac{7}{3}\right)$,

$\left(4, -\frac{7}{3}\right)$, $\left(-4, \frac{7}{3}\right)$, and $\left(-4, -\frac{7}{3}\right)$.

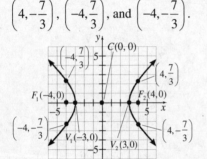

$$\frac{(\pm 5)^2}{9} - \frac{x^2}{16} = 1$$

$$\frac{25}{9} - \frac{x^2}{16} = 1$$

$$-\frac{x^2}{16} = -\frac{16}{9}$$

$$x^2 = \frac{256}{9}$$

$$x = \pm \frac{16}{3}$$

The points to the left and right of the foci are

$\left(\frac{16}{3}, 5\right)$, $\left(-\frac{16}{3}, 5\right)$, $\left(\frac{16}{3}, -5\right)$, and $\left(-\frac{16}{3}, -5\right)$.

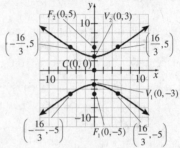

55. The given vertices $(0, \pm 3)$ and focus $(0, 5)$ all lie on the y-axis. Thus, the transverse axis is the y-axis and the hyperbola opens up and down. The center of the hyperbola is the midpoint between the two vertices. Therefore, the center is $(0, 0)$ and the equation of the hyperbola is of the form $\frac{y^2}{a^2} - \frac{x^2}{b^2} = 1$. Now, the distance from the center $(0, 0)$ to each vertex is $a = 3$ units. Likewise, the distance from the center to the given focus $(0, 5)$ is $c = 5$ units. Also, $b^2 = c^2 - a^2$, so $b^2 = 5^2 - 3^2 = 25 - 9 = 16$. Thus, the equation of the hyperbola is $\frac{y^2}{9} - \frac{x^2}{16} = 1$.

To help graph the hyperbola, we first find the focus that was not given. Since the center is at $(0, 0)$ and since one focus is at $(0, 5)$, the other focus must be at $(0, -5)$. Next, plot the points on the graph to the left and right of the foci. Let $y = \pm 5$:

56. The given vertices $(0, \pm 4)$ both lie on the y-axis. Thus, the transverse axis is the y-axis and the hyperbola opens up and down. The center of the hyperbola is the midpoint between the two vertices. Therefore, the center is $(0, 0)$ and the equation of the hyperbola is of the form $\frac{y^2}{a^2} - \frac{x^2}{b^2} = 1$. Now, the distance from the center $(0, 0)$ to each vertex is $a = 4$ units. We are given that an asymptote of the hyperbola is $y = \frac{4}{3} x$. Now the equation of the asymptote is of the form is $y = \frac{a}{b} x$ and $a = 4$. So,

$$\frac{a}{b} = \frac{4}{3}$$

$$\frac{4}{b} = \frac{4}{3}$$

$$4b = 12$$

$$b = 3$$

Thus, the equation of the hyperbola is

$$\frac{y^2}{4^2} - \frac{x^2}{3^2} = 1.$$

$$\frac{y^2}{16} - \frac{x^2}{9} = 1$$

To help graph the hyperbola, we first find the foci. Since $c^2 = a^2 + b^2$, we have that $c^2 = 16 + 9 = 25$, so $c = 5$. Since the center is $(0,0)$, the foci are $(0,\pm5)$. We form the rectangle using the points $(0,\pm a) = (0,\pm4)$ and $(\pm b, 0) = (\pm3, 0)$. The diagonals are the two asymptotes $y = \frac{4}{3}x$ and $y = -\frac{4}{3}x$.

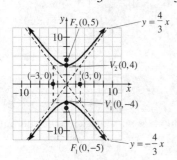

57. $\begin{cases} 4x^2 + y^2 = 10 \\ \qquad y = x \end{cases}$

First, graph each equation in the system.

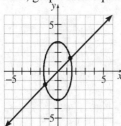

The system apparently has two solutions. Now substitute x for y into the first equation:

$$4x^2 + x^2 = 10$$
$$5x^2 = 10$$
$$x^2 = 2$$
$$x = \pm\sqrt{2}$$

Substitute these x-values into the second equation to find the corresponding y-values:

$x = \sqrt{2}: \quad y = x = \sqrt{2}$
$x = -\sqrt{2}: \quad y = x = -\sqrt{2}$

Both pairs check, so the solutions are $\left(\sqrt{2}, \sqrt{2}\right)$ and $\left(-\sqrt{2}, -\sqrt{2}\right)$.

58. $\begin{cases} y = 2x^2 + 1 \\ y = x + 2 \end{cases}$

First, graph each equation in the system.

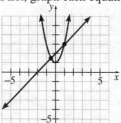

The system apparently has two solutions. Now substitute $2x^2 + 1$ for y into the second equation:

$$2x^2 + 1 = x + 2$$
$$2x^2 - x - 1 = 0$$
$$(2x+1)(x-1) = 0$$
$$x = -\frac{1}{2} \quad \text{or} \quad x = 1$$

Substitute these x-values into the first equation to find the corresponding y-values:

$x = -\frac{1}{2}: \quad y = 2\left(-\frac{1}{2}\right)^2 + 1 = 2\left(\frac{1}{4}\right) + 1 = \frac{1}{2} + 1 = \frac{3}{2}$

$x = 1: \quad y = 2(1)^2 + 1 = 2(1) + 1 = 2 + 1 = 3$

Both pairs check, so the solutions are $\left(-\frac{1}{2}, \frac{3}{2}\right)$ and $(1, 3)$.

59. $\begin{cases} 6x - y = 5 \\ \qquad xy = 1 \end{cases}$

First, graph each equation in the system.

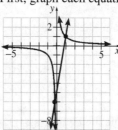

The system apparently has two solutions. Now solve the first equation for y: $y = 6x - 5$.

Substitute the result into the second equation.

$$x(6x - 5) = 1$$
$$6x^2 - 5x - 1 = 0$$
$$(6x+1)(x-1) = 0$$
$$x = -\frac{1}{6} \quad \text{or} \quad x = 1$$

Substitute these x-values into the equation

$y = 6x - 5$ to find the corresponding y-values.

$x = -\dfrac{1}{6}:\ y = 6\left(-\dfrac{1}{6}\right) - 5 = -1 - 5 = -6$

$x = 1:\ y = 6(1) - 5 = 6 - 5 = 1$

Both pairs check, so the solutions are $\left(-\dfrac{1}{6}, -6\right)$

and $(1, 1)$.

60. $\begin{cases} x^2 + y^2 = 26 \\ x^2 - 2y^2 = 23 \end{cases}$

First, graph each equation in the system.

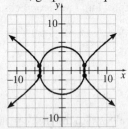

The system apparently has four solutions. Now solve the first equation for x^2: $x^2 = 26 - y^2$.

Substitute the result into the second equation.

$\left(26 - y^2\right) - 2y^2 = 23$

$-3y^2 = -3$

$y^2 = 1$

$y = \pm 1$

Substitute 1 for y^2 into the first equation.

$x^2 + 1 = 26$

$x^2 = 25$

$x = \pm 5$

All four pairs check, so the solutions are $(-5, -1)$, $(-5, 1)$, $(5, -1)$, and $(5, 1)$.

61. $\begin{cases} 4x - y^2 = 0 \\ 2x^2 + y^2 = 16 \end{cases}$

First, graph each equation in the system.

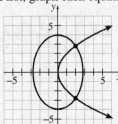

The system apparently has two solutions. Now add the two equations:

$\begin{array}{r} 4x - y^2 = 0 \\ 2x^2 \quad\ + y^2 = 16 \\ \hline 2x^2 \quad\ + 4x = 16 \end{array}$

$2x^2 + 4x - 16 = 0$

$x^2 + 2x - 8 = 0$

$(x + 4)(x - 2) = 0$

$x = -4 \ \text{ or } \ x = 2$

Substitute these x-values into the first equation to find the corresponding y-values:

$x = -4:\ 4(-4) - y^2 = 0$

$-16 - y^2 = 0$

$-16 = y^2$

$y = \pm\sqrt{-16}\ $ (not real)

$x = 2:\ 4(2) - y^2 = 0$

$8 - y^2 = 0$

$8 = y^2$

$y = \pm\sqrt{8} = \pm 2\sqrt{2}$

Both real-number pairs check, so the solutions are $\left(2, -2\sqrt{2}\right)$ and $\left(2, 2\sqrt{2}\right)$.

62. $\begin{cases} x^2 - y = -2 \\ x^2 + y = 4 \end{cases}$

First, graph each equation in the system.

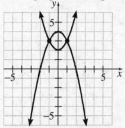

The system apparently has two solutions. Now multiply the first equation by -1 and add the result to the second equation:

$\begin{array}{r} -x^2 + y = 2 \\ x^2 + y = 4 \\ \hline 2y = 6 \\ y = 3 \end{array}$

Substitute this y-value into the second equation to find the corresponding x-values.

$x^2 + 3 = 4$

$x^2 = 1$

$x = \pm 1$

Both pairs check, so the solutions are $(-1, 3)$ and $(1, 3)$.

63. $\begin{cases} 4x^2 - 2y^2 = 2 \\ -x^2 + y^2 = 2 \end{cases}$

First, graph each equation in the system.

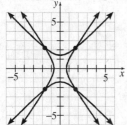

The system apparently has four solutions. Now multiply the second equation by 2 and add the result to the first equation:

$4x^2 - 2y^2 = 2$
$\underline{-2x^2 + 2y^2 = 4}$
$2x^2 \qquad\;\; = 6$
$\qquad x^2 = 3$
$\qquad\; x = \pm\sqrt{3}$

Substitute 3 for x^2 into the second equation:

$-3 + y^2 = 2$
$\quad\; y^2 = 5$
$\qquad y = \pm\sqrt{5}$

All four check, so the solutions are

$\left(-\sqrt{3},\; -\sqrt{5}\right)$, $\left(-\sqrt{3}, \sqrt{5}\right)$, $\left(\sqrt{3},\; -\sqrt{5}\right)$, and

$\left(\sqrt{3}, \sqrt{5}\right)$.

64. $\begin{cases} x^2 + y^2 = 8x \\ y^2 = 3x \end{cases}$

First, graph each equation in the system.

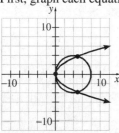

The system apparently has three solutions. Now multiply the second equation by -1 and add the result to the first equation:

$x^2 + y^2 = 8x$
$\underline{\qquad -y^2 = -3x}$
$x^2 \qquad = 5x$
$x^2 - 5x = 0$
$x(x - 5) = 0$
$x = 0$ or $x = 5$

Substitute these x-values into the second equation to find the corresponding y-values:

$x = 0:\; y^2 = 3(0)$
$\qquad\;\; y^2 = 0$
$\qquad\;\;\; y = 0$

$x = 5:\; y^2 = 3(5)$
$\qquad\;\; y^2 = 15$
$\qquad\;\;\; y = \pm\sqrt{15}$

All three pairs check, so the solutions are $(0,0)$,

$\left(5, -\sqrt{15}\right)$, and $\left(5, \sqrt{15}\right)$.

65. $\begin{cases} y = x + 2 \\ y = x^2 \end{cases}$

First, graph each equation in the system.

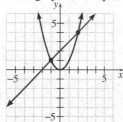

The system apparently has two solutions. Now Substitute x^2 for y into the first equation:

$x^2 = x + 2$
$x^2 - x - 2 = 0$
$(x - 2)(x + 1) = 0$
$x = 2$ or $x = -1$

Substitute these results into the first equation to find the corresponding y values.

$x = 2:\; y - 2 + 2 = 4$
$x = -1: y = -1 + 2 = 1$

Both pairs check, so the solutions are $(2, 4)$ and $(-1, 1)$.

66. $\begin{cases} x^2 + 2y = 9 \\ 5x - 2y = 5 \end{cases}$

First, graph each equation in the system.

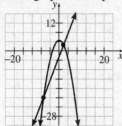

The system apparently has two solutions. Now add the two equations:

$$x^2 \qquad + 2y = 9$$
$$\underline{\quad 5x - 2y = 5 \quad}$$
$$x^2 \qquad + 5x = 14$$
$$x^2 + 5x - 14 = 0$$
$$(x + 7)(x - 2) = 0$$
$$x = -7 \text{ or } x = 2$$

Substitute these results into the first equation to find the corresponding y values.

$$x = -7: \quad (-7)^2 + 2y = 9$$
$$49 + 2y = 9$$
$$2y = -40$$
$$y = -20$$

$$x = 2: \quad (2)^2 + 2y = 9$$
$$4 + 2y = 9$$
$$2y = 5$$
$$y = \frac{5}{2}$$

Both pairs check, so the solutions are $(-7, -20)$ and $\left(2, \dfrac{5}{2}\right)$.

67. $\begin{cases} x^2 + y^2 = 36 \\ x - y = -6 \end{cases}$

First, graph each equation in the system.

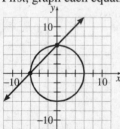

The system apparently has two solutions. Now solve the second equation for y: $y = x + 6$.

Substitute the result into the first equation:

$$x^2 + (x + 6)^2 = 36$$
$$x^2 + (x^2 + 12x + 36) = 36$$
$$2x^2 + 12x = 0$$
$$2x(x + 6) = 0$$
$$2x = 0 \text{ or } x + 6 = 0$$
$$x = 0 \text{ or } \qquad x = -6$$

Substitute these x values into the equation $y = x + 6$ to find the corresponding y values:

$$x = 0: \quad y = 0 + 6 = 6$$
$$x = -6: \quad y = -6 + 6 = 0$$

Both pairs check, so the solutions are $(0, 6)$ and $(-6, 0)$.

68. $\begin{cases} y = 2x - 4 \\ y^2 = 4x \end{cases}$

First, graph each equation in the system.

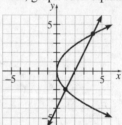

The system apparently has two solutions. Now substitute $2x - 4$ for y into the second equation:

$$(2x - 4)^2 = 4x$$
$$4x^2 - 16x + 16 = 4x$$
$$4x^2 - 20x + 16 = 0$$
$$x^2 - 5x + 4 = 0$$
$$(x - 4)(x - 1) = 0$$
$$x = 4 \text{ or } x = 1$$

Substitute these results into the first equation to find the corresponding y values.

$$x = 4: y = 2(4) - 4 = 4$$
$$x = 1: y = 2(1) - 4 = -2$$

Both pairs check, so the solutions are $(4, 4)$ and $(1, -2)$.

69. $\begin{cases} x^2 + y^2 = 9 \\ x + y = 7 \end{cases}$

First, graph each equation in the system.

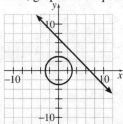

The system apparently has no solution. Now solve the second equation for y: $y = 7 - x$.

Substitute the result into the first equation:

$$x^2 + (7 - x)^2 = 9$$
$$x^2 + (49 - 14x + x^2) = 9$$
$$2x^2 - 14x + 49 = 9$$
$$2x^2 - 14x + 40 = 0$$
$$x = \frac{-(-14) \pm \sqrt{(-14)^2 - 4(2)(40)}}{2(2)}$$
$$= \frac{14 \pm \sqrt{-124}}{4}$$
$$= \frac{14 \pm 2\sqrt{31}\, i}{4}$$
$$= \frac{7}{2} \pm \frac{\sqrt{31}}{2} i$$

Since the solutions are not real, the system is inconsistent. The solution set is $\varnothing$.

70. $\begin{cases} 2x^2 + 3y^2 = 14 \\ x^2 - y^2 = -3 \end{cases}$

First, graph each equation in the system.

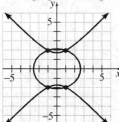

The system apparently has four solutions. Now multiply the second equation by 3 and add the result to the first equation.

$$3x^2 - 3y^2 = -9$$
$$2x^2 + 3y^2 = 14$$

$$5x^2 \qquad = 5$$
$$x^2 = 1$$
$$x = \pm 1$$

Substitute 1 for x^2 into the second equation to find the y values.

$$1 - y^2 = -3$$
$$-y^2 = -4$$
$$y^2 = 4$$
$$y = \pm 2$$

All four pairs check, so the solutions are $(-1, -2)$, $(-1, 2)$, $(1, -2)$, and $(1, 2)$.

71. $\begin{cases} x^2 + y^2 = 16 \\ x^2 + 4y = 16 \end{cases}$

First, graph each equation in the system.

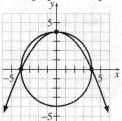

The system apparently has three solutions. Now multiply the second equation by -1 and add the result to the first equation:

$$x^2 + y^2 \qquad = 16$$
$$-x^2 \qquad - 4y = -16$$

$$y^2 - 4y = 0$$
$$y(y - 4) = 0$$
$$y = 0 \quad \text{or} \quad y = 4$$

Substitute these y-values into the first equation to find the corresponding x-values:

$$y = 0: \quad x^2 + 0^2 = 16$$
$$x^2 = 16$$
$$x = \pm 4$$
$$y = 4: \quad x^2 + 4^2 = 16$$
$$x^2 + 16 = 16$$
$$x^2 = 0$$
$$x = 0$$

All three pairs check, so the solutions are $(-4, 0)$, $(4, 0)$, and $(0, 4)$.

72. $\begin{cases} x = 4 - y^2 \\ x = 2y + 4 \end{cases}$

First, graph each equation in the system.

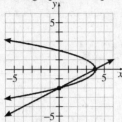

The system apparently has two solutions. Now substitute $2y + 4$ for x into the first equation:

$$2y + 4 = 4 - y^2$$
$$y^2 + 2y = 0$$
$$y(y + 2) = 0$$
$$y = 0 \text{ or } y = -2$$

Substitute these y-values into the second equation to find the corresponding x-values:

$$y = 0 : \ x = 2(0) + 4 = 4$$
$$y = -2 : \ x = 2(-2) + 4 = 0$$

Both pairs check, so the solutions are $(4, \ 0)$ and $(0, -2)$.

73. Let x represent the larger of the two numbers and y represent the smaller of the two numbers.

$$\begin{cases} x + y = 12 \\ x^2 - y^2 = 24 \end{cases}$$

Solve the first equation for y: $y = 12 - x$.

Substitute the result into the second equation:

$$x^2 - (12 - x)^2 = 24$$
$$x^2 - \left(144 - 24x + x^2\right) = 24$$
$$x^2 - 144 + 24x - x^2 = 24$$
$$24x = 168$$
$$x = 7$$

Substitute 7 for x into the first equation to find the corresponding y-value:
$$7 + y = 12$$
$$y = 5$$

The two numbers are 7 and 5.

74. Let x represent the length and y represent the width of the rectangle.

$$\begin{cases} 2x + 2y = 34 \\ xy = 60 \end{cases}$$

Solve the first equation for y: $y = 17 - x$.

Substitute the result into the second equation:

$$x(17 - x) = 60$$
$$17x - x^2 = 60$$
$$x^2 - 17x + 60 = 0$$
$$(x - 12)(x - 5) = 0$$
$$x = 12 \text{ or } x = 5$$

Substitute these x-values into the second equation to find the corresponding y-values:

$$x = 12 : \ 12y = 60$$
$$y = 5$$
$$x = 5 : \ 5y = 60$$
$$y = 12$$

Note that the two outcomes result in the same overall dimensions. Assuming the length is the longer of the two sides, the length is 12 centimeters and the width is 5 centimeters.

75. Let x represent the length and y represent the width of the rectangle.

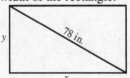

$$\begin{cases} xy = 2160 \\ x^2 + y^2 = 78^2 \end{cases}$$

Solve the first equation for y: $y = \dfrac{2160}{x}$.

Substitute the result into the second equation:

$$x^2 + \left(\frac{2160}{x}\right)^2 = 78^2$$
$$x^2 + \frac{4,665,600}{x^2} = 6084$$
$$x^4 + 4,665,600 = 6084x^2$$
$$x^4 - 6084x^2 + 4,665,600 = 0$$
$$\left(x^2 - 5184\right)\left(x^2 - 900\right) = 0$$
$$(x - 72)(x + 72)(x - 30)(x + 30) = 0$$
$$x = 72 \text{ or } x = -72 \text{ or } x = 30 \text{ or } x = -30$$

Since the length cannot be negative, we discard $x = -72$ and $x = -30$, so we are left with the possible answers $x = 72$ and $x = 30$. Substitute these x-values into the first equation to find the corresponding y-values:

$$x = 72 : \ 72y = 2160$$
$$y = 30$$
$$x = 30 : \ 30y = 2160$$
$$y = 72$$

Note that the two outcomes result in the same overall dimensions. Assuming the length is the longer of the two sides, the length is 72 inches and the width is 30 inches.

76. Let x and y represent the lengths of the two legs of the right triangle.

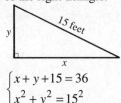

$$\begin{cases} x+y+15=36 \\ x^2+y^2=15^2 \end{cases}$$

Solve the first equation for y: $y=21-x$.

Substitute the result into the second equation:

$$x^2+(21-x)^2=15^2$$
$$x^2+441-42x+x^2=225$$
$$2x^2-42x+216=0$$
$$x^2-21x+108=0$$
$$(x-12)(x-9)=0$$
$$x=12 \text{ or } x=9$$

Substitute these x-values into the first equation to find the corresponding y-values:

$$x=12: \ 12+y+15=36$$
$$y+27=36$$
$$y=9$$
$$x=9: \ 9+y+15=36$$
$$y+24=36$$
$$y=12$$

Note that the two outcomes result in the same overall dimensions. The lengths of the two legs are 12 inches and 9 inches.

Chapter 12 Test

1. $d(P_1,P_2)=\sqrt{(x_2-x_1)^2+(y_2-y_1)^2}$
$$=\sqrt{(3-(-1))^2+(-5-3)^2}$$
$$=\sqrt{4^2+(-8)^2}$$
$$=\sqrt{16+64}$$
$$=\sqrt{80}$$
$$=4\sqrt{5}$$

2. $M=\left(\dfrac{x_1+x_2}{2},\dfrac{y_1+y_2}{2}\right)$
$$=\left(\dfrac{-7+5}{2},\dfrac{6+(-2)}{2}\right)=\left(\dfrac{-2}{2},\dfrac{4}{2}\right)=(-1,2)$$

3. $(x-4)^2+(y+1)^2=9$
$$(x-4)^2+(y-(-1))^2=3^2$$

The center is $(h,k)=(4,-1)$, and the radius is $r=3$.

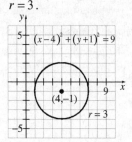

4. $x^2+y^2+10x-4y+13=0$
$$(x^2+10x+25)+(y^2-4y+4)=-13+25+4$$
$$(x+5)^2+(y-2)^2=16$$
$$(x-(-5))^2+(y-2)^2=4^2$$

The center is $(h,k)=(-5,2)$, and the radius is $r=4$.

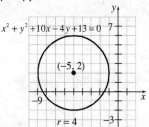

5. $(x-h)^2+(y-k)^2=r^2$
$$(x-(-3))^2+(y-7)^2=6^2$$
$$(x+3)^2+(y-7)^2=36$$

6. The radius of the circle will be the distance from the center point $(-5,8)$ to the point on the circle $(3,2)$. Thus, $r=\sqrt{(3-(-5))^2+(2-8)^2}=10$.
The equation of the circle is
$$(x-h)^2+(y-k)^2=r^2$$
$$(x-(-5))^2+(y-8)^2=10^2$$
$$(x+5)^2+(y-8)^2=100$$

7. Notice that $(y+2)^2=4(x-1)$ is of the form $(y-k)^2=4a(x-h)$, where $(h,k)=(1,-2)$ and $4a=4$, so that $a=1$. Now, the graph of an equation of the form $(y-k)^2=4a(x-h)$ will be a parabola that opens to the right with the vertex at (h,k), focus at $(h+a,k)$, and directrix of $x=h-a$. Note that $h+a=1+1=2$ and $h-a=1-1=0$. Thus, the graph of $(y+2)^2=4(x-1)$ is a parabola that opens to the right with vertex $(1,-2)$, focus $(2,-2)$, and directrix $x=0$. To help graph the parabola, we plot the two points on the graph above and below the focus. Let $x=2$:

$$(y+2)^2=4(2-1)$$
$$(y+2)^2=4$$
$$y+2=\pm 2$$
$$y=-2\pm 2$$
$$y=-4 \text{ or } y=0$$

The points $(2,-4)$ and $(2,0)$ are on the graph.

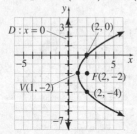

8. We complete the square in x to write the equation in standard form:

$$x^2-4x+3y-8=0$$
$$x^2-4x=-3y+8$$
$$x^2-4x+4=-3y+8+4$$
$$x^2-4x+4=-3y+12$$
$$(x-2)^2=-3(y-4)$$

Notice that $(x-2)^2=-3(y-4)$ is of the form $(x-h)^2=-4a(y-k)$, where $(h,k)=(2,4)$ and $-4a=-3$, so that $a=\dfrac{3}{4}$. Now, the graph of an equation of the form $(x-h)^2=-4a(y-k)$ will be a parabola that opens downward with the vertex at (h,k), focus at $(h,k-a)$, and directrix of $y=k+a$. Note that $k-a=4-\dfrac{3}{4}=\dfrac{13}{4}$ and $k+a=4+\dfrac{3}{4}=\dfrac{19}{4}$. Thus, the graph of $(x-2)^2=-3(y-4)$ is a parabola that opens downward with vertex $(2,4)$, focus $\left(2,\dfrac{13}{4}\right)$, and

directrix $y=\dfrac{19}{4}$. To help graph the parabola, we plot the two points on the graph to the left and right of the focus. Let $y=\dfrac{13}{4}$:

$$(x-2)^2=-3\left(\dfrac{13}{4}-4\right)$$
$$(x-2)^2=-3\left(-\dfrac{3}{4}\right)$$
$$(x-2)^2=\dfrac{9}{4}$$
$$x-2=\pm\dfrac{3}{2}$$
$$x=2\pm\dfrac{3}{2}$$
$$x=\dfrac{1}{2} \text{ or } x=\dfrac{7}{2}$$

The points $\left(\dfrac{1}{2},\dfrac{13}{4}\right)$ and $\left(\dfrac{7}{2},\dfrac{13}{4}\right)$ are on the graph.

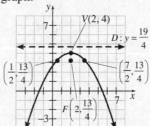

9. The distance from the vertex $(0,0)$ to the focus $(0,-4)$ is $a=4$. Because the focus lies on the negative y-axis, we know that the parabola will open downward and the axis of symmetry is the y-axis. This means the equation of the parabola is of the form $x^2=-4ay$ with $a=4$:

$$x^2=-4(4)y$$
$$x^2=-16y$$

10. Notice that the directrix $x=-1$ is a vertical line and that the focus is $(3,4)$. Therefore, the axis of symmetry must be the line $y=4$. Because the focus is to the right of the directrix, the parabola must open to the right. Since the vertex must be the point on the axis of symmetry that is midway between the focus and the directrix, the vertex is $(1,4)$. Now, the distance from the vertex $(1,4)$ to the focus $(3,4)$ is $a=2$, so the equation of the parabola is of the form

$(y-k)^2 = 4a(x-h)$ with $a=2$, $h=1$, and $k=4$:

$(y-4)^2 = 4 \cdot 2(x-1)$

$(y-4)^2 = 8(x-1)$

11. $9x^2 + 25y^2 = 225$

$\dfrac{9x^2 + 25y^2}{225} = \dfrac{225}{225}$

$\dfrac{x^2}{25} + \dfrac{y^2}{9} = 1$

The larger number, 25, is in the denominator of the x^2-term. This means that the major axis is the x-axis and that the equation of the ellipse is of the form $\dfrac{x^2}{a^2} + \dfrac{y^2}{b^2} = 1$, so that $a^2 = 25$ and $b^2 = 9$. The center of the ellipse is $(0,0)$.

Because $c^2 = a^2 - b^2$, we have that $c^2 = 25 - 9 = 16$, so that $c = \pm\sqrt{16} = \pm 4$. Since the major axis is the x-axis, the foci are $(-4,0)$ and $(4,0)$. To find the x-intercepts (vertices), let $y=0$; to find the y-intercepts, let $x=0$:

x-intercepts:

$9x^2 + 25(0)^2 = 225$

$9x^2 = 225$

$x^2 = 25$

$x = \pm 5$

y-intercepts:

$9(0)^2 + 25y^2 = 225$

$25y^2 = 225$

$y^2 = 9$

$y = \pm 3$

The intercepts are $(-5,0)$, $(5,0)$, $(0,-3)$, and $(0,3)$.

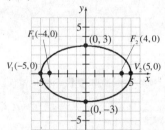

12. $\dfrac{(x-2)^2}{9} + \dfrac{(y+4)^2}{16} = 1$

The center of the ellipse is $(h,k) = (2,-4)$. Because the larger number, 16, is the denominator of the y^2-term, the major axis is parallel to the y-axis. Because $a^2 = 16$ and $b^2 = 9$, we have that $c^2 = a^2 - b^2 = 16 - 9 = 7$.

The vertices are $a = 4$ units below and above the center at $V_1(2,-8)$ and $V_2(2,0)$. The foci are $c = \sqrt{7}$ units below and above the center at $F_1\left(2,-4-\sqrt{7}\right)$ and $F_2\left(2,-4+\sqrt{7}\right)$. We plot the points $b=3$ units to the left and right of the center point at $(-1,-4)$ and $(5,-4)$.

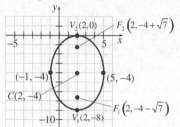

13. The given focus $(0,-4)$ and the given vertex $(0,-5)$ lie on the y-axis. Thus, the equation of the ellipse is of the form $\dfrac{x^2}{b^2} + \dfrac{y^2}{a^2} = 1$. The distance from the center of the ellipse to the vertex is $a = 5$ units. The distance from the center of the ellipse to the focus is $c = 4$ units. Because $b^2 = a^2 - c^2$, we have that $b^2 = 5^2 - 4^2 = 25 - 16 = 9$. Thus, the equation of the ellipse is $\dfrac{x^2}{9} + \dfrac{y^2}{25} = 1$.

14. The given vertices and focus all lie on the vertical line $x = -1$. Therefore, the major axis is parallel to the y-axis, and the equation of the ellipse is of the form $\dfrac{(x-h)^2}{b^2} + \dfrac{(y-k)^2}{a^2} = 1$.

Now, the center will be the midpoint between the two vertices $(-1,7)$ and $(-1,-3)$, which is $(h,k) = (-1,2)$. Now, $a = 5$ is the distance from the center $(-1,2)$ to a vertex $(-1,7)$, and $c = 3$ is the distance from the center to a focus $(-1,-1)$.

Also, $b^2 = a^2 - c^2 = 5^2 - 3^2 = 25 - 9 = 16$. Thus, the equation of the ellipse is

$\dfrac{(x-(-1))^2}{16} + \dfrac{(y-2)^2}{25} = 1$

$\dfrac{(x+1)^2}{16} + \dfrac{(y-2)^2}{25} = 1$

15. $x^2 - \dfrac{y^2}{4} = 1$

$\dfrac{x^2}{1} - \dfrac{y^2}{4} = 1$

Notice the equation is of the form $\dfrac{x^2}{a^2} - \dfrac{y^2}{b^2} = 1$.

Because the x^2-term is first, the transverse is along the x-axis and the hyperbola opens left and right. The center of the hyperbola is $(0,0)$. We have that $a^2 = 1$ and $b^2 = 4$. Because $c^2 = a^2 + b^2$, we have that $c^2 = 1 + 4 = 5$, so that $c = \sqrt{5}$. The vertices are $(\pm a, 0) = (\pm 1, 0)$, and the foci are $(\pm c, 0) = \left(\pm\sqrt{5}, 0\right)$. Since $a = 1$ and $b = 2$, the equations of the asymptotes are

$y = \dfrac{2}{1}x = 2x$ and $y = -\dfrac{2}{1}x = -2x$. To help graph the hyperbola, we form the rectangle using the points $(\pm a, 0) = (\pm 1, 0)$ and $(0, \pm b) = (0, \pm 2)$. The diagonals are the asymptotes.

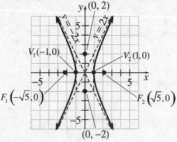

16. $16y^2 - 25x^2 = 1600$

$\dfrac{16y^2 - 25x^2}{1600} = \dfrac{1600}{1600}$

$\dfrac{y^2}{100} - \dfrac{x^2}{64} = 1$

Notice the equation is of the form $\dfrac{y^2}{a^2} - \dfrac{x^2}{b^2} = 1$.

Because the y^2-term is first, the transverse is along the y-axis and the hyperbola opens up and down. The center of the hyperbola is $(0,0)$. We have that $a^2 = 100$ and $b^2 = 64$. Because $c^2 = a^2 + b^2$, we have that $c^2 = 100 + 64 = 164$, so that $c = \sqrt{164} = 2\sqrt{41}$. The vertices are $(0, \pm a) = (0, \pm 10)$, and the foci are $(0, \pm c) = \left(0, \pm 2\sqrt{41}\right)$. Since $a = 10$ and $b = 8$,

the equations of the asymptotes are

$y = \dfrac{10}{8}x = \dfrac{5}{4}x$ and $y = -\dfrac{10}{8}x = -\dfrac{5}{4}x$. To help graph the hyperbola, we form the rectangle using the points $(0, \pm a) = (0, \pm 10)$ and $(\pm b, 0) = (\pm 8, 0)$. The diagonals are the asymptotes.

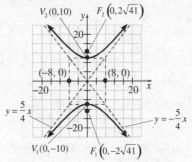

17. The given foci $(\pm 8, 0)$ and vertex $(-3, 0)$ all lie on the x-axis. Thus, the transverse axis is the x-axis and the hyperbola opens left and right. The center of the hyperbola is the midpoint between the two foci. Therefore, the center is $(0,0)$ and the equation of the hyperbola is of the form $\dfrac{x^2}{a^2} - \dfrac{y^2}{b^2} = 1$. Now, the distance between the given vertex $(-3, 0)$ and the center $(0,0)$ is $a = 3$ units. Likewise, the distance between the given foci and the center is $c = 8$ units. Also, $b^2 = c^2 - a^2$, so $b^2 = 8^2 - 3^2 = 64 - 9 = 55$. Thus, the equation of the hyperbola is $\dfrac{x^2}{9} - \dfrac{y^2}{55} = 1$.

18. $\begin{cases} x^2 + y^2 = 17 \\ x + y = -3 \end{cases}$

First, graph each equation in the system.

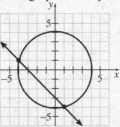

The system apparently has two solutions. Now solve the second equation for y: $y = -x - 3$. Substitute the result into the first equation:

$$x^2 + (-x-3)^2 = 17$$
$$x^2 + (x^2 + 6x + 9) = 17$$
$$2x^2 + 6x - 8 = 0$$
$$x^2 + 3x - 4 = 0$$
$$(x+4)(x-1) = 0$$
$$x = -4 \quad \text{or} \quad x = 1$$

Substitute these results into the first equation to find the corresponding *y*-values.

$$x = -4: \quad -4 + y = -3$$
$$y = 1$$
$$x = 1: \quad 1 + y = -3$$
$$y = -4$$

Both pairs check, so the solutions are $(-4, 1)$ and $(1, -4)$.

19. $\begin{cases} x^2 + y^2 = 9 \\ 4x^2 - y^2 = 16 \end{cases}$

First, graph each equation in the system.

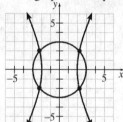

The system apparently has four solutions. Now add the two equations:

$$x^2 + y^2 = 9$$
$$\underline{4x^2 - y^2 = 16}$$
$$5x^2 \quad\quad = 25$$
$$x^2 = 5$$
$$x = \pm\sqrt{5}$$

Substitute 5 for x^2 into the first equation:

$$5 + y^2 = 9$$
$$y^2 = 4$$
$$y = \pm\sqrt{4} = \pm 2$$

All four pairs check, so the solutions are

$\left(-\sqrt{5}, -2\right)$, $\left(-\sqrt{5}, 2\right)$, $\left(\sqrt{5}, -2\right)$, and $\left(\sqrt{5}, 2\right)$.

20. **(a)** To solve this problem, we draw the ellipse on a Cartesian plane so that the *x*-axis coincides with the water and the *y*-axis passes through the center of the arch. Thus, the origin is the center of the ellipse. Now, the "center" of the arch is 10 feet above the water, so the point $(0, 10)$ is on the ellipse.

The creek is 30 feet wide, so the two points $(-15, 0)$ and $(15, 0)$ are the two vertices of the ellipse and the major axis is along the *x*-axis.

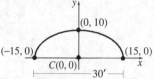

The equation of the ellipse must have the form $\dfrac{x^2}{a^2} + \dfrac{y^2}{b^2} = 1$. The distance from the center of the ellipse to each vertex is $a = 15$. Also, the height of the semi-ellipse is 10 feet, so $b = 10$. Thus, the equation of the ellipse is

$$\frac{x^2}{15^2} + \frac{y^2}{10^2} = 1$$
$$\frac{x^2}{225} + \frac{y^2}{100} = 1$$

(b) Substitute $x = 12$ into the equation, and solve for *y*:

$$\frac{12^2}{225} + \frac{y^2}{100} = 1$$
$$\frac{144}{225} + \frac{y^2}{100} = 1$$
$$\frac{y^2}{100} = \frac{9}{25}$$
$$y^2 = 36$$
$$y = 6$$

Thus, the height of the arch at a distance 12 feet from the center of the creek is 6 feet.

Chapter 13

Section 13.1

Preparing for Sequences

P1. $f(x) = x^2 - 4$

    (a) $f(3) = (3)^2 - 4 = 9 - 4 = 5$

    (b) $f(-7) = (-7)^2 - 4 = 49 - 4 = 45$

P2. $g(x) = 2x - 3$

$$g(1) = 2(1) - 3 = 2 - 3 = -1$$
$$g(2) = 2(2) - 3 = 4 - 3 = 1$$
$$g(3) = 2(3) - 3 = 6 - 3 = 3$$

Therefore,
$$g(1) + g(2) + g(3) = -1 + 1 + 3 = 3$$

P3. In the function $f(n) = n^2 - 4$, the independent variable is n (the input variable).

Section 13.1 Quick Checks

1. sequence

2. infinite; finite

3. True

4. $a_n = 2n - 3$

$$a_1 = 2(1) - 3 = 2 - 3 = -1$$
$$a_2 = 2(2) - 3 = 4 - 3 = 1$$
$$a_3 = 2(3) - 3 = 6 - 3 = 3$$
$$a_4 = 2(4) - 3 = 8 - 3 = 5$$
$$a_5 = 2(5) - 3 = 10 - 3 = 7$$

The first five terms of the sequence are -1, 1, 3, 5, and 7.

5. $b_n = (-1)^n \cdot 4n$

$$b_1 = (-1)^1 \cdot 4(1) = -4$$
$$b_2 = (-1)^2 \cdot 4(2) = 8$$
$$b_3 = (-1)^3 \cdot 4(3) = -12$$
$$b_4 = (-1)^4 \cdot 4(4) = 16$$
$$b_5 = (-1)^5 \cdot 4(5) = -20$$

The first five terms of the sequence are -4, 8, -12, 16, and -20.

6. $5, 7, 9, 11, \ldots$

The terms are consecutive odd numbers with the first term being 5. We can write the terms as follows:

$$5 = 2(1) + 3$$
$$7 = 2(2) + 3$$
$$9 = 2(3) + 3$$
$$11 = 2(4) + 3$$

Notice that each term is 3 more than twice the term number. A formula for the nth term is given by $a_n = 2n + 3$.

7. $\dfrac{1}{2}, -\dfrac{1}{3}, \dfrac{1}{4}, -\dfrac{1}{5}, \ldots$

The terms alternate sign with the first term being positive. So, $(-1)^{n+1}$ must be part of the formula. Each term is a fraction with a numerator of 1 and a denominator that is 1 more than the term number. A formula for the nth term is given by

$$b_n = (-1)^{n+1} \cdot \frac{1}{n+1}.$$

8. partial sum

9. $\displaystyle\sum_{i=1}^{3} (4i - 1) = (4 \cdot 1 - 1) + (4 \cdot 2 - 1) + (4 \cdot 3 - 1)$

$$= 3 + 7 + 11$$
$$= 21$$

10. $\displaystyle\sum_{i=1}^{5} (i^3 + 1) = (1^3 + 1) + (2^3 + 1) + (3^3 + 1)$

$$+ (4^3 + 1) + (5^3 + 1)$$
$$= (1 + 1) + (8 + 1) + (27 + 1)$$
$$+ (64 + 1) + (125 + 1)$$
$$= 2 + 9 + 28 + 65 + 126$$
$$= 230$$

11. $1 + 4 + 9 + \ldots + 144$

Notice that each term is a perfect square. We can rewrite the sum as $1^2 + 2^2 + 3^2 + \ldots + 12^2$. Thus, the sum has 12 terms, each of the form i^2.

$$1 + 4 + 9 + \ldots + 144 = \sum_{i=1}^{12} i^2$$

12. Begin by writing the first term as $\frac{1}{1}$. We notice that each term is a fraction with a numerator of 1 and a denominator that is a power of 2. We can write the sum as

$$\frac{1}{2^0}+\frac{1}{2^1}+\frac{1}{2^2}+...+\frac{1}{2^5}$$

The exponent on 2 in the denominator is always 1 less than the term number. Thus, the sum has 6 terms, each of the form $\frac{1}{2^{i-1}}$.

$$1+\frac{1}{2}+\frac{1}{4}+...+\frac{1}{32}=\sum_{i=1}^{6}\left(\frac{1}{2^{i-1}}\right)$$

13.1 Exercises

13. $\{3n+5\}$

$a_1 = 3\cdot1+5=3+5=8$
$a_2 = 3\cdot2+5=6+5=11$
$a_3 = 3\cdot3+5=9+5=14$
$a_4 = 3\cdot4+5=12+5=17$
$a_5 = 3\cdot5+5=15+5=20$

The first five terms of the sequence are 8, 11, 14, 17, and 20.

15. $\left\{\frac{n}{n+2}\right\}$

$a_1 = \frac{1}{1+2}=\frac{1}{3}$

$a_2 = \frac{2}{2+2}=\frac{2}{4}=\frac{1}{2}$

$a_3 = \frac{3}{3+2}=\frac{3}{5}$

$a_4 = \frac{4}{4+2}=\frac{4}{6}=\frac{2}{3}$

$a_5 = \frac{5}{5+2}=\frac{5}{7}$

The first five terms of the sequence are $\frac{1}{3}$, $\frac{1}{2}$, $\frac{3}{5}$, $\frac{2}{3}$, and $\frac{5}{7}$.

17. $\left\{(-1)^n n\right\}$

$a_1 = (-1)^1 \cdot 1 = -1$

$a_2 = (-1)^2 \cdot 2 = 2$

$a_3 = (-1)^3 \cdot 3 = -3$

$a_4 = (-1)^4 \cdot 4 = 4$

$a_5 = (-1)^5 \cdot 5 = -5$

The first five terms of the sequence are -1, 2, -3, 4, and -5.

19. $\left\{2^n +1\right\}$

$a_1 = 2^1 +1 = 2+1=3$

$a_2 = 2^2 +1 = 4+1=5$

$a_3 = 2^3 +1 = 8+1=9$

$a_4 = 2^4 +1 = 16+1=17$

$a_5 = 2^5 +1 = 32+1=33$

The first five terms of the sequence are 3, 5, 9, 17, and 33.

21. $\left\{\frac{2n}{2^n}\right\}$

$a_1 = \frac{2\cdot1}{2^1}=\frac{2}{2}=1$

$a_2 = \frac{2\cdot2}{2^2}=\frac{4}{4}=1$

$a_3 = \frac{2\cdot3}{2^3}=\frac{6}{8}=\frac{3}{4}$

$a_4 = \frac{2\cdot4}{2^4}=\frac{8}{16}=\frac{1}{2}$

$a_5 = \frac{2\cdot5}{2^5}=\frac{10}{32}=\frac{5}{16}$

The first five terms of the sequence are 1, 1, $\frac{3}{4}$, $\frac{1}{2}$, and $\frac{5}{16}$.

23. $\left\{\frac{n}{e^n}\right\}$

$a_1 = \frac{1}{e^1}=\frac{1}{e}$, $a_2 = \frac{2}{e^2}$, $a_3 = \frac{3}{e^3}$

$a_4 = \frac{4}{e^4}$, $a_5 = \frac{5}{e^5}$

The first five terms of the sequence are $\frac{1}{e}$, $\frac{2}{e^2}$,

$\frac{3}{e^3}$, $\frac{4}{e^4}$, and $\frac{5}{e^5}$.

25. The terms are all multiples of 2 with the first term equaling $2 \cdot 1$, the second term equaling $2 \cdot 2$, and so on. A formula for the nth term is given by $a_n = 2n$.

27. Each term is a fraction with the denominator equaling 1 more than the numerator. When $n = 1$, the numerator equals 1. Each subsequent numerator is one more than previous. A formula for the nth term is given by $a_n = \frac{n}{n+1}$.

29. When $n = 1$, we have that $a_1 = 3 = 1^2 + 2$; when $n = 2$, we have that $a_2 = 6 = 2^2 + 2$; when $n = 3$, we have that $a_3 = 11 = 3^2 + 2$. Notice that each term is equal to 2 more than the square of the term number. Therefore, a formula for the nth term is given by $a_n = n^2 + 2$.

31. Notice that the terms alternate signs with the first term being negative. Ignoring the signs, also notice that the terms are all perfect squares. Therefore, a formula for the nth term is given by $a_n = (-1)^n n^2$.

33. $\sum_{i=1}^{4}(5i+1)$
$= (5 \cdot 1 + 1) + (5 \cdot 2 + 1) + (5 \cdot 3 + 1) + (5 \cdot 4 + 1)$
$= 6 + 11 + 16 + 21$
$= 54$

35. $\sum_{i=1}^{5} \frac{i^2}{2} = \frac{1^2}{2} + \frac{2^2}{2} + \frac{3^2}{2} + \frac{4^2}{2} + \frac{5^2}{2}$
$= \frac{1}{2} + \frac{4}{2} + \frac{9}{2} + \frac{16}{2} + \frac{25}{2}$
$= \frac{55}{2}$

37. $\sum_{k=1}^{3} 2^k = 2^1 + 2^2 + 2^3 = 2 + 4 + 8 = 14$

39. $\sum_{k=1}^{5}\left[(-1)^{k+1} \cdot 2k\right]$
$= (-1)^{1+1} \cdot 2(1) + (-1)^{2+1} \cdot 2(2) + (-1)^{3+1} \cdot 2(3)$
$\quad + (-1)^{4+1} \cdot 2(4) + (-1)^{5+1} \cdot 2(5)$
$= (-1)^2 \cdot 2 + (-1)^3 \cdot 4 + (-1)^4 \cdot 6$
$\quad + (-1)^5 \cdot 8 + (-1)^6 \cdot 10$
$= 2 - 4 + 6 - 8 + 10$
$= 6$

41. $\sum_{j=1}^{10} 5 = 5+5+5+5+5+5+5+5+5+5 = 50$

43. $\sum_{k=3}^{7}(2k-1) = (2 \cdot 3 - 1) + (2 \cdot 4 - 1) + (2 \cdot 5 - 1)$
$\quad + (2 \cdot 6 - 1) + (2 \cdot 7 - 1)$
$= 5 + 7 + 9 + 11 + 13$
$= 45$

45. The sum $1+2+3+...+15$ has 15 terms, each term has the form k, starts at $k = 1$, and ends at $k = 15$.
$$1+2+3+...+15 = \sum_{k=1}^{15} k$$

47. The sum $1 + \frac{1}{2} + \frac{1}{3} + ... + \frac{1}{12}$ has 12 terms, each term has the form $\frac{1}{i}$, starts at $i = 1$, and ends at $i = 12$.
$$1 + \frac{1}{2} + \frac{1}{3} + ... + \frac{1}{12} = \sum_{i=1}^{12} \frac{1}{i}$$

49. To see the pattern, note that the first term can be written as $1 = \frac{1}{1}$. We see that the denominators are all powers of 3 and that the terms alternate signs with the first term being positive. When $n = 1$, we have the denominator $1 = 3^0$; when $n = 2$, we have the denominator $3 = 3^1$; when $n = 3$, we have the denominator $9 = 3^2$; and so on. Thus, the power on 3 in the denominator is always one less than the term number. Looking at the form of the last term, we see that there will be 9 terms and that the power on -1 will be one more than the term number.

$$1 - \frac{1}{3} + \frac{1}{9} - \frac{1}{27} + \ldots + (-1)^{9+1}\left(\frac{1}{3^{9-1}}\right)$$

$$= \sum_{i=1}^{9} (-1)^{i+1}\left(\frac{1}{3^{i-1}}\right)$$

51. To see the pattern, note that the first term can be written as $5 = (5 + 2 \cdot 0)$. Each term is 5 more than a multiple of two. The multiple is one less than the term number. Therefore, there are 11 terms being added together.

$$5 + (5 + 2 \cdot 1) + (5 + 2 \cdot 2) + (5 + 2 \cdot 3) + \ldots + (5 + 2 \cdot 10)$$

$$= \sum_{k=1}^{11}\left[5 + 2(k-1)\right] \quad \text{or} \quad \sum_{k=1}^{11}(2k+3)$$

53. $a_n = 12,000\left(1 + \dfrac{0.06}{4}\right)^n$

(a) $n = 1$

$$a_1 = 12,000\left(1 + \frac{0.06}{4}\right)^1$$

$$= 12,000(1.015)$$

$$= 12,180$$

After 1 quarter, the account will have a balance of $12,180.

(b) $n = 4$ (4 quarters in each year)

$$a_4 = 12,000\left(1 + \frac{0.06}{4}\right)^4$$

$$= 12,000(1.015)^4$$

$$\approx 12,000(1.06136)$$

$$\approx 12,736.36$$

After 1 year (i.e. 4 quarters), the account will have a balance of $12,736.36.

(c) $n = 40$

$$a_{40} = 12,000\left(1 + \frac{0.06}{4}\right)^{40}$$

$$= 12,000(1.015)^{40}$$

$$\approx 12,000(1.8140184)$$

$$\approx 21,768.22$$

After 10 years (40 quarters), the account will have a balance of $21,768.22.

55. $p_n = 304(1.011)^n$

(a) In 2012, we have $n = 4$.

$$p_4 = 304(1.011)^4 \approx 317.60$$

In 2012, the population of the United States will be an estimated 318 million.

(b) In 2050, we have $n = 42$.

$$p_{42} = 304(1.011)^{42} \approx 481.31$$

In 2050, the population of the United States will be an estimated 481 million.

57. $u_n = \dfrac{\left(1+\sqrt{5}\right)^n - \left(1-\sqrt{5}\right)^n}{2^n \cdot \sqrt{5}}$

$$u_1 = \frac{\left(1+\sqrt{5}\right) - \left(1-\sqrt{5}\right)}{2 \cdot \sqrt{5}} = \frac{2\sqrt{5}}{2 \cdot \sqrt{5}} = 1$$

$$u_2 = \frac{\left(1+\sqrt{5}\right)^2 - \left(1-\sqrt{5}\right)^2}{2^2 \cdot \sqrt{5}} = \frac{4\sqrt{5}}{4 \cdot \sqrt{5}} = 1$$

$$u_3 = \frac{\left(1+\sqrt{5}\right)^3 - \left(1-\sqrt{5}\right)^3}{2^3 \cdot \sqrt{5}}$$

$$= \frac{\left(16+8\sqrt{5}\right) - \left(16-8\sqrt{5}\right)}{8 \cdot \sqrt{5}} = \frac{16\sqrt{5}}{8\sqrt{5}} = 2$$

$$u_4 = \frac{\left(1+\sqrt{5}\right)^4 - \left(1-\sqrt{5}\right)^4}{2^4 \cdot \sqrt{5}}$$

$$= \frac{\left(56+24\sqrt{5}\right) - \left(56-24\sqrt{5}\right)}{16 \cdot \sqrt{5}} = \frac{48\sqrt{5}}{16\sqrt{5}} = 3$$

$$u_5 = \frac{\left(1+\sqrt{5}\right)^5 - \left(1-\sqrt{5}\right)^5}{2^5 \cdot \sqrt{5}}$$

$$= \frac{\left(176+80\sqrt{5}\right) - \left(176-80\sqrt{5}\right)}{32 \cdot \sqrt{5}} = \frac{160\sqrt{5}}{32\sqrt{5}} = 5$$

$$u_6 = \frac{\left(1+\sqrt{5}\right)^6 - \left(1-\sqrt{5}\right)^6}{2^6 \cdot \sqrt{5}}$$

$$= \frac{\left(576+256\sqrt{5}\right) - \left(576-256\sqrt{5}\right)}{64 \cdot \sqrt{5}} = \frac{512\sqrt{5}}{64\sqrt{5}}$$

$$= 8$$

$$u_7 = \frac{\left(1+\sqrt{5}\right)^7 - \left(1-\sqrt{5}\right)^7}{2^7 \cdot \sqrt{5}}$$

$$= \frac{\left(1856+832\sqrt{5}\right) - \left(1856-832\sqrt{5}\right)}{128 \cdot \sqrt{5}}$$

$$= \frac{1664\sqrt{5}}{128\sqrt{5}} = 13$$

$$u_8 = \frac{\left(1+\sqrt{5}\right)^8 - \left(1-\sqrt{5}\right)^8}{2^8 \cdot \sqrt{5}}$$

$$= \frac{\left(6016+2688\sqrt{5}\right) - \left(6016-2688\sqrt{5}\right)}{256 \cdot \sqrt{5}}$$

$$= \frac{5376\sqrt{5}}{256\sqrt{5}} = 21$$

$$u_9 = \frac{\left(1+\sqrt{5}\right)^9 - \left(1-\sqrt{5}\right)^9}{2^9 \cdot \sqrt{5}}$$

$$= \frac{\left(19456+8704\sqrt{5}\right) - \left(19456-8704\sqrt{5}\right)}{512 \cdot \sqrt{5}}$$

$$= \frac{17408\sqrt{5}}{512\sqrt{5}} = 34$$

$$u_{10} = \frac{\left(1+\sqrt{5}\right)^{10} - \left(1-\sqrt{5}\right)^{10}}{2^{10} \cdot \sqrt{5}}$$

$$= \frac{\left(62976+28160\sqrt{5}\right) - \left(62976-28160\sqrt{5}\right)}{1024 \cdot \sqrt{5}}$$

$$= \frac{56320\sqrt{5}}{1024\sqrt{5}} = 55$$

The first 10 terms of the sequence are 1, 1, 2, 3, 5, 8, 13, 21, 34, and 55.
Notice that, beginning with the third term, each term in the sequence can be obtained by adding the two previous terms together. For example, $2 = 1+1$, $3 = 1+2$, $5 = 2+3$, etc.

59. $a_1 = 10$
$a_2 = 1.05a_1 = 1.05(10) = 10.5$
$a_3 = 1.05a_2 = 1.05(10.5) = 11.025$
$a_4 = 1.05a_3 = 1.05(11.025) = 11.57625$
$a_5 = 1.05a_4 = 1.05(11.57625) = 12.1550625$
The first five terms of the sequence are 10, 10.5, 11.025, 11.57625, and 12.1550625.

61. $b_1 = 8$
$b_2 = 2 + b_1 = 2 + 8 = 10$
$b_3 = 3 + b_2 = 3 + 10 = 13$
$b_4 = 4 + b_3 = 4 + 13 = 17$
$b_5 = 5 + b_4 = 5 + 17 = 22$
The first five terms of the sequence are 8, 10, 13, 17, and 22.

63. (a) $\dfrac{u_2}{u_1} = \dfrac{1}{1} = 1$; $\dfrac{u_3}{u_2} = \dfrac{2}{1} = 2$; $\dfrac{u_4}{u_3} = \dfrac{3}{2} = 1.5$;

$\dfrac{u_5}{u_4} = \dfrac{5}{3} = 1.\overline{6}$; $\dfrac{u_6}{u_5} = \dfrac{8}{5} = 1.6$;

$\dfrac{u_7}{u_6} = \dfrac{13}{8} = 1.625$; $\dfrac{u_8}{u_7} = \dfrac{21}{13} \approx 1.615385$;

$\dfrac{u_9}{u_8} = \dfrac{34}{21} \approx 1.619048$; $\dfrac{u_{10}}{u_9} = \dfrac{55}{34} \approx 1.617647$;

$\dfrac{u_{11}}{u_{10}} = \dfrac{89}{55} \approx 1.618182$

(b) The ratio approaches a value around 1.618. Note: As n goes off to infinity, the ratio approaches the exact value $\dfrac{1+\sqrt{5}}{2}$.

(c) $\dfrac{u_1}{u_2} = \dfrac{1}{1} = 1$; $\dfrac{u_2}{u_3} = \dfrac{1}{2} = 0.5$; $\dfrac{u_3}{u_4} = \dfrac{2}{3} = 0.\overline{6}$;

$\dfrac{u_4}{u_5} = \dfrac{3}{5} = 0.6$; $\dfrac{u_5}{u_6} = \dfrac{5}{8} = 0.625$;

$\dfrac{u_6}{u_7} = \dfrac{8}{13} \approx 0.615385$; $\dfrac{u_7}{u_8} = \dfrac{13}{21} \approx 0.619048$;

$\dfrac{u_8}{u_9} = \dfrac{21}{34} \approx 0.617647$;

$\dfrac{u_9}{u_{10}} = \dfrac{34}{55} \approx 0.618182$;

$\dfrac{u_{10}}{u_{11}} = \dfrac{55}{89} \approx 0.617978$

(d) The ratio approaches a value around 0.618. Note: As n goes to infinity, the ratio approaches the exact value $\dfrac{\sqrt{5}-1}{2}$.

65. Answers may vary. A function is any relation that assigns each element of the domain to exactly one element of the range. A sequence is a function whose domain is the set of positive integers.

67. Answers may vary. The symbol Σ represents summation and indicates that we will need to add up the values of several terms.

69. (a) The function is in slope-intercept form so the slope is $m = 4$.

    (b) $f(x) = 4x - 6$
$$f(1) = 4(1) - 6 = -2$$
$$f(2) = 4(2) - 6 = 2$$
$$f(3) = 4(3) - 6 = 6$$
$$f(4) = 4(4) - 6 = 10$$

71. (a) The function is in slope-intercept form so the slope is $m = -5$.

    (b) $f(x) = -5x + 8$
$$f(1) = -5(1) + 8 = 3$$
$$f(2) = -5(2) + 8 = -2$$
$$f(3) = -5(3) + 8 = -7$$
$$f(4) = -5(4) + 8 = -12$$

73. $\{3n + 5\}$

```
seq(3X+5,X,1,5,1
)
  {8 11 14 17 20}
```

The first five terms of the sequence are 8, 11, 14, 17, and 20.

75. $\left\{\dfrac{n}{n+2}\right\}$

```
seq(X/(X+2),X,1,   seq(X/(X+2),X,1,
5,1)►Frac          5,1)►Frac
{1/3 1/2 3/5 2/…   …/2 3/5 2/3 5/7}
```

The first five terms of the sequence are $\dfrac{1}{3}$, $\dfrac{1}{2}$, $\dfrac{3}{5}$, $\dfrac{2}{3}$, and $\dfrac{5}{7}$.

77. $\{(-1)^n n\}$

```
seq((-1)^X*X,X,1
,5,1)
  {-1 2 -3 4 -5}
```

The first five terms of the sequence are -1, 2, -3, 4, and -5.

79. $\{2^n + 1\}$

```
seq(2^X+1,X,1,5,
1)
  {3 5 9 17 33}
```

The first five terms of the sequence are 3, 5, 9, 17, and 33.

81. $\displaystyle\sum_{i=1}^{4} (5i + 1)$

```
sum(seq(5X+1,X,1
,4,1)
              54
```

The sum is 54.

83. $\displaystyle\sum_{i=1}^{5} \dfrac{i^2}{2}$

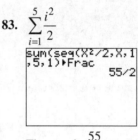

```
sum(seq(X²/2,X,1
,5,1)►Frac
            55/2
```

The sum is $\dfrac{55}{2}$.

85. $\displaystyle\sum_{k=1}^{3} 2^k$

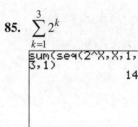

```
sum(seq(2^X,X,1,
3,1)
              14
```

The sum is 14.

87. $\displaystyle\sum_{k=1}^{5} \left[(-1)^{k+1} \cdot 2k\right]$

```
sum(seq((-1)^(X+
1)*2X,X,1,5,1)
               6
```

The sum is 6.

Section 13.2

Preparing for Arithmetic Sequences

P1. $y = -3x + 1$ is in the (slope-intercept) form
$y = mx + b$. Therefore, the slope is $m = -3$.

P2. $g(x) = 5x + 2$
$g(3) = 5(3) + 2$
$\quad = 15 + 2$
$\quad = 17$

P3. $\begin{cases} x - 3y = -17 & (1) \\ 2x + y = 1 & (2) \end{cases}$

Multiply both sides of equation (1) by -2.
$\begin{cases} -2x + 6y = 34 & (1) \\ 2x + y = 1 & (2) \end{cases}$

Add equation (1) and equation (2).
$\begin{array}{r} \begin{cases} -2x + 6y = 34 & (1) \\ 2x + y = 1 & (2) \end{cases} \\ \hline 7y = 35 \\ y = 5 \end{array}$

Substitute this result into the original equation (1) and solve for x.
$x - 3y = -17$
$x - 3(5) = -17$
$x - 15 = -17$
$x = -2$

The solution is the ordered pair $(-2, 5)$.

Section 13.2 Quick Checks

1. arithmetic

2. $-1 - (-3) = -1 + 3 = 2$
$1 - (-1) = 1 + 1 = 2$
$3 - 1 = 2$
$5 - 3 = 2$
The sequence $-3, -1, 1, 3, 5, \ldots$ is arithmetic because the difference between consecutive terms is constant. The first term is $a = -3$ and the common difference is $d = 2$.

3. $9 - 3 = 6$, $27 - 9 = 18$, $81 - 27 = 54$
The sequence is not arithmetic because the difference between consecutive terms is not constant.

4. $a_n - a_{n-1} = [3n - 8] - [3(n-1) - 8]$
$= [3n - 8] - [3n - 11]$
$= 3n - 8 - 3n + 11$
$= 3$
The sequence is arithmetic because the difference between consecutive terms is constant. The first term is $a = a_1 = 3(1) - 8 = -5$ and the common difference is $d = 3$.

5. $b_n - b_{n-1} = [n^2 - 1] - [(n-1)^2 - 1]$
$= [n^2 - 1] - [n^2 - 2n + 1 - 1]$
$= [n^2 - 1] - [n^2 - 2n]$
$= n^2 - 1 - n^2 + 2n$
$= 2n - 1$
The sequence is not arithmetic because the difference between consecutive terms is not constant.

6. $c_n - c_{n-1} = [5 - 2n] - [5 - 2(n-1)]$
$= [5 - 2n] - [5 - 2n + 2]$
$= [5 - 2n] - [7 - 2n]$
$= 5 - 2n - 7 + 2n$
$= -2$
The sequence is arithmetic because the difference between consecutive terms is constant. The first term is $a = c_1 = 5 - 2(1) = 3$ and the common difference is $d = -2$.

7. $a_n = a + (n-1)d$

8. **(a)** We have that $a_5 = 25$ and $d = 6$.
$a_n = a + (n-1)d$
$a_5 = a + (5-1)d$
$25 = a + 4(6)$
$25 = a + 24$
$1 = a$
Therefore, the nth term of the sequence is given by
$a_n = 1 + (n-1)(6) = 1 + 6n - 6$
$a_n = 6n - 5$

(b) $a_{14} = 6(14) - 5 = 84 - 5 = 79$

9. (a) We are given $a_5 = 7$ and $a_{13} = 31$. The nth term of an arithmetic sequence is given by $a_n = a + (n-1)d$. Therefore, we have

$$\begin{cases} a_5 = a + (5-1)d & \text{or} \\ a_{13} = a + (13-1)d \end{cases}$$

$$\begin{cases} 7 = a + 4d & (1) \\ 31 = a + 12d & (2) \end{cases}$$

This is a system of linear equations in a and d. We can solve the system by using elimination. Subtract equation (2) from equation (1) to obtain

$$-24 = -8d$$
$$3 = d$$

Let $d = 3$ in equation (1) and solve for a.

$$7 = a + 4(3)$$
$$7 = a + 12$$
$$-5 = a$$

The first term is $a = -5$ and the common difference is $d = 3$.

(b) A formula for the nth term is

$$a_n = a + (n-1)d = -5 + (n-1)(3)$$
$$= -5 + 3n - 3$$
$$a_n = 3n - 8$$

10. We have $a = 5$ and $d = 2$, and wish to find S_{100}.

$$S_n = \frac{n}{2}\left[2a + (n-1)d\right]$$
$$S_{100} = \frac{100}{2}\left[2(5) + (100-1)(2)\right] = 50[10 + 198]$$
$$= 50(208) = 10,400$$

11. The first term is $a = 1$ and the common difference is $d = 5 - 1 = 4$. We wish to find S_{70}.

$$S_n = \frac{n}{2}\left[2a + (n-1)d\right]$$
$$S_{70} = \frac{70}{2}\left[2(1) + (70-1)(4)\right]$$
$$= 35[2 + 276]$$
$$= 35(278)$$
$$= 9730$$

12. We have $a = 4$ and $a_{50} = 298$, and wish to find the sum of the first 50 terms, S_{50}.

$$S_n = \frac{n}{2}[a + a_n]$$
$$S_{50} = \frac{50}{2}[a + a_{50}]$$
$$= \frac{50}{2}[4 + 298]$$
$$= 25(302)$$
$$= 7550$$

13. $a_n = -3n + 100$

$$a = a_1 = -3(1) + 100 = 97$$
$$a_{75} = -3(75) + 100 = -125$$

$$S_n = \frac{n}{2}[a + a_n]$$
$$S_{75} = \frac{75}{2}[a + a_{75}]$$
$$S_{75} = \frac{75}{2}\left[97 + (-125)\right]$$
$$= \frac{75}{2}(-28)$$
$$= -1050$$

14. If we let a_n = the number of seats in the nth row, then we have an arithmetic sequence with first term $a = 20$ and common difference $d = 2$. Thus, the total number of seats in the 30 rows is given by S_{30}.

$$S_n = \frac{n}{2}\left[2a + (n-1)d\right]$$
$$S_{30} = \frac{30}{2}\left[2a + (30-1)d\right] = 15\left[2(20) + 29(2)\right]$$
$$= 15(98) = 1470$$

There are 1470 seats in the 30 rows.

13.2 Exercises

15. $\{n + 5\}$

$$a_{n-1} = (n-1) + 5 = n + 4$$
$$a_n = n + 5$$
$$a_n - a_{n-1} = (n+5) - (n+4)$$
$$= n + 5 - n - 4$$
$$= 1$$

The difference between *any* consecutive terms is 1, so the sequence is arithmetic with common difference $d = 1$. To find the first term, we evaluate a_1 and find $a_1 = 1 + 5 = 6$.

17. $\{7n+2\}$

$a_{n-1} = 7(n-1)+2 = 7n-5$

$a_n = 7n+2$

$a_n - a_{n-1} = (7n+2)-(7n-5)$

$\qquad\qquad = 7n+2-7n+5$

$\qquad\qquad = 7$

The difference between *any* consecutive terms is 7, so the sequence is arithmetic with common difference $d = 7$. To find the first term, we evaluate a_1 and find $a_1 = 7(1)+2 = 9$.

19. $\{7-3n\}$

$a_{n-1} = 7-3(n-1) = 7-3n+3 = 10-3n$

$a_n = 7-3n$

$a_n - a_{n-1} = (7-3n)-(10-3n)$

$\qquad\qquad = 7-3n-10+3n$

$\qquad\qquad = -3$

The difference between *any* consecutive terms is -3, so the sequence is arithmetic with common difference $d = -3$. To find the first term, we evaluate a_1 and find $a_1 = 7-3(1) = 4$.

21. $\left\{\dfrac{1}{2}n+5\right\}$

$a_{n-1} = \dfrac{1}{2}(n-1)+5 = \dfrac{1}{2}n+\dfrac{9}{2}$

$a_n = \dfrac{1}{2}n+5$

$a_n - a_{n-1} = \left(\dfrac{1}{2}n+5\right)-\left(\dfrac{1}{2}n+\dfrac{9}{2}\right)$

$\qquad\qquad = \dfrac{1}{2}n+5-\dfrac{1}{2}n-\dfrac{9}{2}$

$\qquad\qquad = \dfrac{1}{2}$

The difference between *any* consecutive terms is $\dfrac{1}{2}$, so the sequence is arithmetic with common difference $d = \dfrac{1}{2}$. To find the first term, we evaluate a_1 and find $a_1 = \dfrac{1}{2}(1)+5 = \dfrac{11}{2}$.

23. $a = 4$, $d = 3$

$a_n = a+(n-1)d$

$\quad = 4+(n-1)3$

$\quad = 4+3n-3$

$\quad = 3n+1$

To find the fifth term we let $n = 5$.

$a_5 = 3(5)+1 = 15+1 = 16$

The fifth term of the sequence is 16.

25. $a = 10$, $d = -5$

$a_n = a+(n-1)d$

$\quad = 10+(n-1)(-5)$

$\quad = 10-5n+5$

$\quad = -5n+15$

To find the fifth term we let $n = 5$.

$a_5 = -5(5)+15 = -25+15 = -10$

The fifth term of the sequence is -10.

27. $a = 2$, $d = \dfrac{1}{3}$

$a_n = a+(n-1)d = 2+(n-1)\left(\dfrac{1}{3}\right)$

$\quad = 2+\dfrac{1}{3}n-\dfrac{1}{3}$

$\quad = \dfrac{1}{3}n+\dfrac{5}{3}$

To find the fifth term we let $n = 5$.

$a_5 = \dfrac{1}{3}(5)+\dfrac{5}{3} = \dfrac{5}{3}+\dfrac{5}{3} = \dfrac{10}{3}$

The fifth term of the sequence is $\dfrac{10}{3}$.

29. $a = 5$, $d = -\dfrac{1}{5}$

$a_n = a+(n-1)d = 5+(n-1)\left(-\dfrac{1}{5}\right)$

$\quad = 5-\dfrac{1}{5}n+\dfrac{1}{5}$

$\quad = -\dfrac{1}{5}n+\dfrac{26}{5}$

To find the fifth term we let $n = 5$.

$a_5 = -\dfrac{1}{5}(5)+\dfrac{26}{5} = -1+\dfrac{26}{5} = \dfrac{21}{5}$

The fifth term of the sequence is $\dfrac{21}{5}$.

31. $2, 7, 12, 17, \ldots$

Notice that the difference between consecutive terms is $d = 5$. Since the first term is $a = 2$, the *n*th term can be written as

$$a_n = a + (n-1)d$$
$$= 2 + (n-1)(5)$$
$$= 2 + 5n - 5$$
$$= 5n - 3$$

To find the twentieth term, we let $n = 20$.

$$a_{20} = 5(20) - 3 = 100 - 3 = 97$$

The twentieth term of the sequence is 97.

33. $12, 9, 6, 3, \ldots$

Notice that the difference between consecutive terms is $d = -3$. Since the first term is $a = 12$, the nth term can be written as

$$a_n = a + (n-1)d$$
$$= 12 + (n-1)(-3)$$
$$= 12 - 3n + 3$$
$$= -3n + 15$$

To find the twentieth term, we let $n = 20$.

$$a_{20} = -3(20) + 15 = -60 + 15 = -45$$

The twentieth term of the sequence is -45.

35. $1, \dfrac{5}{4}, \dfrac{3}{2}, \dfrac{7}{4}, \ldots$

Notice that the difference between consecutive terms is $d = \dfrac{1}{4}$. Since the first term is $a = 1$, the nth term can be written as

$$a_n = a + (n-1)d$$
$$= 1 + (n-1)\left(\dfrac{1}{4}\right)$$
$$= 1 + \dfrac{1}{4}n - \dfrac{1}{4}$$
$$= \dfrac{1}{4}n + \dfrac{3}{4}$$

To find the twentieth term, we let $n = 20$.

$$a_{20} = \dfrac{1}{4}(20) + \dfrac{3}{4} = \dfrac{20}{4} + \dfrac{3}{4} = \dfrac{23}{4}$$

The twentieth term of the sequence is $\dfrac{23}{4}$.

37. We know that the nth term of an arithmetic sequence is given by $a_n = a + (n-1)d$ where a is the first term and d is the common difference. Since $a_3 = 17$ and $a_7 = 37$, we have

$$\begin{cases} a_3 = a + (3-1)d \\ a_7 = a + (7-1)d \end{cases} \text{ or } \begin{cases} 17 = a + 2d & (1) \\ 37 = a + 6d & (2) \end{cases}$$

This is a system of linear equations in two variables, a and d. We can solve the system by elimination. If we subtract equation (2) from

equation (1), we obtain

$$-20 = -4d$$
$$5 = d$$

Let $d = 5$ in equation (1) to find a.

$$17 = a + 2(5)$$
$$17 = a + 10$$
$$7 = a$$

The first term is $a = 7$ and the common difference is $d = 5$. Thus, a formula for the nth term is

$$a_n = 7 + (n-1)(5)$$
$$= 7 + 5n - 5$$
$$= 5n + 2$$

39. We know that the nth term of an arithmetic sequence is given by $a_n = a + (n-1)d$ where a is the first term and d is the common difference. Since $a_4 = -2$ and $a_8 = 26$, we have

$$\begin{cases} a_4 = a + (4-1)d \\ a_8 = a + (8-1)d \end{cases} \text{ or } \begin{cases} -2 = a + 3d & (1) \\ 26 = a + 7d & (2) \end{cases}$$

This is a system of linear equations in two variables, a and d. We can solve the system by elimination. If we subtract equation (2) from equation (1), we obtain

$$-28 = -4d$$
$$7 = d$$

Let $d = 7$ in equation (1) to find a.

$$-2 = a + 3(7)$$
$$-2 = a + 21$$
$$-23 = a$$

The first term is $a = -23$ and the common difference is $d = 7$. Thus, a formula for the nth term is

$$a_n = -23 + (n-1)(7)$$
$$= -23 + 7n - 7$$
$$= 7n - 30$$

41. We know that the nth term of an arithmetic sequence is given by $a_n = a + (n-1)d$ where a is the first term and d is the common difference. Since $a_5 = -1$ and $a_{12} = -22$, we have

$$\begin{cases} a_5 = a + (5-1)d \\ a_{12} = a + (12-1)d \end{cases} \text{ or } \begin{cases} -1 = a + 4d & (1) \\ -22 = a + 11d & (2) \end{cases}$$

This is a system of linear equations in two variables, a and d. We can solve the system by elimination. If we subtract equation (2) from equation (1), we obtain

$$21 = -7d$$
$$-3 = d$$

Let $d = -3$ in equation (1) to find a.

$-1 = a + 4(-3)$

$-1 = a - 12$

$11 = a$

The first term is $a = 11$ and the common difference is $d = -3$. Thus, a formula for the nth term is

$a_n = 11 + (n-1)(-3)$

$\quad = 11 - 3n + 3$

$\quad = -3n + 14$

43. We know that the nth term of an arithmetic sequence is given by $a_n = a + (n-1)d$ where a is the first term and d is the common difference. Since $a_3 = 3$ and $a_9 = 0$, we have

$\begin{cases} a_3 = a + (3-1)d \\ a_9 = a + (9-1)d \end{cases}$ or $\begin{cases} 3 = a + 2d \quad (1) \\ 0 = a + 8d \quad (2) \end{cases}$

This is a system of linear equations in two variables, a and d. We can solve the system by elimination. If we subtract equation (2) from equation (1), we obtain

$3 = -6d$

$-\dfrac{1}{2} = d$

Let $d = -\dfrac{1}{2}$ in equation (1) to find a.

$3 = a + 2\left(-\dfrac{1}{2}\right)$

$3 = a - 1$

$4 = a$

The first term is $a = 4$ and the common difference is $d = -\dfrac{1}{2}$. Thus, a formula for the nth term is

$a_n = 4 + (n-1)\left(-\dfrac{1}{2}\right)$

$\quad = 4 - \dfrac{1}{2}n + \dfrac{1}{2} = -\dfrac{1}{2}n + \dfrac{9}{2}$

45. We know that the first term is $a = 2$ and the common difference is $d = 8 - 2 = 6$. The sum of the first $n = 30$ terms of this arithmetic sequence is given by

$S_{30} = \dfrac{30}{2}\left[2a + (30-1)d\right]$

$\quad = 15\left[2(2) + 29(6)\right]$

$\quad = 15(178)$

$\quad = 2670$

47. We know that the first term is $a = -8$ and the common difference is $d = -5 - (-8) = 3$. The sum of the first $n = 25$ terms of this arithmetic sequence is given by

$S_{25} = \dfrac{25}{2}\left[2a + (25-1)d\right]$

$\quad = \dfrac{25}{2}\left[2(-8) + 24(3)\right]$

$\quad = \dfrac{25}{2}(56)$

$\quad = 700$

49. We know that the first term is $a = 10$ and the common difference is $d = 3 - 10 = -7$. The sum of the first $n = 40$ terms of this arithmetic sequence is given by

$S_{40} = \dfrac{40}{2}\left[2a + (40-1)d\right]$

$\quad = 20\left[2(10) + 39(-7)\right]$

$\quad = 20(20 - 273)$

$\quad = 20(-253)$

$\quad = -5060$

51. The first term of the sequence is given by $a_1 = a = 4(1) - 3 = 1$ and the 40th term is given by $a_{40} = 4(40) - 3 = 157$. We can now use the formula $S_n = \dfrac{n}{2}\left[a + a_n\right]$ to find the sum.

$S_{40} = \dfrac{40}{2}\left[1 + 157\right]$

$\quad = 20(158)$

$\quad = 3160$

The sum of the first 40 terms is 3160.

53. The first term of the sequence is given by $a_1 = a = -5(1) + 70 = 65$ and the 75th term is given by $a_{75} = -5(75) + 70 = -305$. We can now use the formula $S_n = \dfrac{n}{2}\left[a + a_n\right]$ to find the sum.

$S_{75} = \dfrac{75}{2}\left[65 + (-305)\right]$

$\quad = \dfrac{75}{2}(-240)$

$\quad = -9000$

The sum of the first 75 terms is -9000.

55. The first term of the sequence is given by

$a_1 = a = 5 + \dfrac{2}{3}(1) = \dfrac{17}{3}$ and the 30th term is given

by $a_{30} = 5 + \dfrac{2}{3}(30) = 25$. We can now use the

formula $S_n = \dfrac{n}{2}\big[a + a_n\big]$ to find the sum.

$S_{30} = \dfrac{30}{2}\left[\dfrac{17}{3} + 25\right]$

$= 15\left(\dfrac{92}{3}\right)$

$= 460$

The sum of the first 30 terms is 460.

57. To be an arithmetic sequence, the difference between successive terms must be constant. Therefore, we start by finding the differences between the terms.

$d_1 = (2x+1) - (x+3)$

$= 2x + 1 - x - 3$

$= x - 2$

$d_2 = (5x+2) - (2x+1)$

$= 5x + 2 - 2x - 1$

$= 3x + 1$

Now we set the two differences equal to each other and solve the resulting equation for x.

$d_1 = d_2$

$x - 2 = 3x + 1$

$-2x = 3$

$x = -\dfrac{3}{2}$

59. To determine the total number of cans, we first need to determine how many rows are in the stack. Letting the bottom row be the first row, we have $a = 35$. Since each row decreases by 1 can, we have $d = -1$.

$a_n = a + (n-1)d$

$1 = 35 + (n-1)(-1)$

$1 = 35 - n + 1$

$-35 = -n$

$35 = n$

The total number of cans in the 35 rows is

$S_{35} = \dfrac{35}{2}(35+1) = \dfrac{35}{2}(36) = 630$

There are 630 cans in the stack.

61. Since the first row has 40 seats, we know that $a = 40$. Each of the $n = 25$ successive rows has 2 more seats, so we also know that $d = 2$. To determine the total number of seats, we first need to know how many seats are in the last row.

$a_{25} = a + (25-1)d$

$= 40 + 24(2)$

$= 40 + 48$

$= 88$

The last row has 88 seats so the total number of seats is given by

$S_{25} = \dfrac{25}{2}[40+88] = \dfrac{25}{2}(128) = 1600$

There are 1600 seats in the auditorium.

63. From the terms listed, we see that the first term is $a = -5$, the last term is $a_n = 244$, and the common difference is $d = -2 - (-5) = 3$. We can determine the number of terms in the sequence by using the formula for the nth term.

$a_n = a + (n-1)d$

$244 = -5 + (n-1)(3)$

$244 = -5 + 3n - 3$

$252 = 3n$

$84 = n$

There are 84 terms in the sequence.

65. From the terms listed, we see that the first term is $a = 108$, the last term is $a_n = -326$, and the common difference is $d = 101 - 108 = -7$. We can determine the number of terms in the sequence by using the formula for the nth term.

$a_n = a + (n-1)d$

$-326 = 108 + (n-1)(-7)$

$-326 = 108 - 7n + 7$

$-441 = -7n$

$63 = n$

There are 63 terms in the sequence.

67. Since your starting salary is $32,000 and you get a $2500 raise each year, your salary each year will form an arithmetic sequence with $a = 32,000$ and $d = 2500$. Your aggregate salary at any point is the sum of what you have earned up to that point. To find how long it would take to have an aggregate of $757,500 we can use the formula

$$S_n = \frac{n}{2}[a + a_n] = \frac{n}{2}[2a + (n-1)d]$$

$$757,500 = \frac{n}{2}[2(32,000) + (n-1)2500]$$

$$757,500 = \frac{n}{2}[64,000 + 2500n - 2500]$$

$$1,515,000 = n(61,500 + 2500n)$$

$$0 = 2500n^2 + 61,500n - 1,515,000$$

$$0 = n^2 + 24.6n - 606$$

We can solve this quadratic equation by using the quadratic formula.

$$n = \frac{-24.6 \pm \sqrt{(24.6)^2 - 4(1)(-606)}}{2(1)}$$

$$= \frac{-24.6 \pm \sqrt{3029.16}}{2}$$

Since n must be positive, we only need to consider the positive solution.

$$n = \frac{-24.6 + \sqrt{3029.16}}{2} \approx 15.22$$

It will take about 15.22 years for you to have an aggregate salary of $757,500.

69. Answers may vary. A sequence is arithmetic only if the difference between consecutive terms is constant. That is, $a_n - a_{n-1} = d$ where d is a constant.

71. **(a)** $f(x) = 3^x$ is in the form $y = a^x$ so the base is 3.

(b) $f(1) = 3^1 = 3$
$f(2) = 3^2 = 9$
$f(3) = 3^3 = 27$
$f(4) = 3^4 = 81$

73. **(a)** $f(x) = 10\left(\frac{1}{2}\right)^x$ is in the form $y = k \cdot a^x$ so the base is $\frac{1}{2}$.

(b) $f(1) = 10\left(\frac{1}{2}\right)^1 = 5$

$f(2) = 10\left(\frac{1}{2}\right)^2 = \frac{5}{2}$

$f(3) = 10\left(\frac{1}{2}\right)^3 = \frac{5}{4}$

$f(4) = 10\left(\frac{1}{2}\right)^4 = \frac{5}{8}$

75. $\{3.45n + 4.12\}$; $n = 20$

```
sum(seq(3.45n+4.
12,n,1,20))
          806.9
■
```

The sum of the first 20 terms of this arithmetic sequence is 806.9.

77. $a = 85.9$; $d = 83.5 - 85.9 = -2.4$
$a_n = 85.9 + (n-1)(-2.4)$
$\quad = 85.9 - 2.4n + 2.4$
$\quad = 88.3 - 2.4n$
$\{88.3 - 2.4n\}$; $n = 25$

```
sum(seq(88.3-2.4
n,n,1,25))
          1427.5
■
```

The sum of the first 25 terms of this arithmetic sequence is 1427.5.

Section 13.3

Preparing for Geometric Sequences and Series

P1. $g(x) = 4^x$

$g(1) = 4^1 = 4$
$g(2) = 4^2 = 16$
$g(3) = 4^3 = 64$

P2. $\dfrac{x^4}{x^3} = x^{4-3} = x$

Section 13.3 Quick Checks

1. geometric

2. $\dfrac{8}{4}=2$, $\dfrac{16}{8}=2$, $\dfrac{32}{16}=2$, $\dfrac{64}{32}=2$

The sequence is geometric because the ratio of consecutive terms is constant. The first term is $a=4$ and the common ratio is $r=2$.

3. $\dfrac{10}{5}=2$, $\dfrac{16}{10}=\dfrac{8}{5}$

The sequence is not geometric because the ratio of consecutive terms is not constant.

4. $\dfrac{3}{9}=\dfrac{1}{3}$, $\dfrac{1}{3}=\dfrac{1}{3}$, $\dfrac{\frac{1}{3}}{1}=\dfrac{1}{3}$, $\dfrac{\frac{1}{9}}{\frac{1}{3}}=\dfrac{1}{9}\cdot 3=\dfrac{1}{3}$

The sequence is geometric because the ratio of consecutive terms is constant. The first term is $a=9$ and the common ratio is $r=\dfrac{3}{9}=\dfrac{1}{3}$.

5. $\dfrac{a_n}{a_{n-1}}=\dfrac{5^n}{5^{n-1}}=\dfrac{5\cdot 5^{n-1}}{5^{n-1}}=5$

The sequence is geometric because the ratio of consecutive terms is constant. The first term is $a=a_1=5^1=5$ and the common ratio is $r=5$.

6. $\dfrac{b_n}{b_{n-1}}=\dfrac{n^2}{(n-1)^2}=\dfrac{n^2}{n^2-2n+1}$

The sequence is not geometric because the ratio of consecutive terms is not constant.

7. $\dfrac{c_n}{c_{n-1}}=\dfrac{5\left(\frac{2}{3}\right)^n}{5\left(\frac{2}{3}\right)^{n-1}}=\dfrac{5\cdot\frac{2}{3}\cdot\left(\frac{2}{3}\right)^{n-1}}{5\left(\frac{2}{3}\right)^{n-1}}=\dfrac{2}{3}$

The sequence is geometric because the ratio of consecutive terms is constant. The first term is $a=a_1=5\left(\dfrac{2}{3}\right)^1=\dfrac{10}{3}$ and the common ratio is $r=\dfrac{2}{3}$.

8. The nth term of a geometric sequence is given by
$a_n=a\cdot r^{n-1}$. We are given $a=5$ and $r=2$. Therefore, a formula for the nth term of the sequence is $a_n=5\cdot 2^{n-1}$.
$a_9=5\cdot 2^{9-1}=5\cdot 2^8=5(256)=1280$

9. The nth term of a geometric sequence is given by $a_n=a\cdot r^{n-1}$. We are given that the first term is $a=50$ and the common ratio is $r=\dfrac{25}{50}=\dfrac{1}{2}$. Therefore, a formula for the nth term of the sequence is $a_n=50\cdot\left(\dfrac{1}{2}\right)^{n-1}$.

$a_9=50\cdot\left(\dfrac{1}{2}\right)^{9-1}=50\cdot\left(\dfrac{1}{2}\right)^8=\dfrac{50}{256}$

$=\dfrac{25}{128}=0.1953125$

10. $S_n=a\cdot\dfrac{1-r^n}{1-r}$

11. We wish to find the sum of the first 13 terms of a geometric sequence with first term $a=3$ and common ratio $r=\dfrac{6}{3}=2$.

$S_n=a\cdot\dfrac{1-r^n}{1-r}$

$S_{13}=3\cdot\dfrac{1-2^{13}}{1-2}=3(8191)=24{,}573$

12. We wish to find the sum of the first 10 terms of a geometric sequence with first term

$a=8\left(\dfrac{1}{2}\right)^1=4$ and common ratio $r=\dfrac{1}{2}$.

$S_n=a\cdot\dfrac{1-r^n}{1-r}$

$S_{10}=4\cdot\dfrac{1-\left(\frac{1}{2}\right)^{10}}{1-\frac{1}{2}}=4\left(\dfrac{\frac{1023}{1024}}{\frac{1}{2}}\right)=4\left(\dfrac{1023}{512}\right)$

$=\dfrac{1023}{128}=7.9921875$

13. $\dfrac{a}{1-r}$

14. This is an infinite geometric series with $a=10$ and $r=\dfrac{\frac{5}{2}}{10}=\dfrac{5}{2}\cdot\dfrac{1}{10}=\dfrac{1}{4}$. Since the common ratio is between -1 and 1, we can use the formula for the sum of an infinite geometric series to find

$$S_\infty = \frac{a}{1-r}$$

$$S_\infty = \frac{10}{1-\frac{1}{4}} = \frac{10}{\frac{3}{4}} = 10 \cdot \frac{4}{3} = \frac{40}{3}$$

15. This is an infinite geometric series with

$a = a_1 = \left(\frac{1}{3}\right)^1 = \frac{1}{3}$ and $r = \frac{1}{3}$. Since the

common ratio is between -1 and 1, we can use the formula for the sum of an infinite geometric series to find

$$S_\infty = \frac{a}{1-r}$$

$$S_\infty = \frac{\frac{1}{3}}{1-\frac{1}{3}} = \frac{\frac{1}{3}}{\frac{2}{3}} = \frac{1}{3} \cdot \frac{3}{2} = \frac{1}{2}$$

16. The line over the 2 indicates that the 2 repeats indefinitely. That is, we can write

$0.\overline{2} = 0.2 + 0.02 + 0.002 + 0.0002 + \ldots$

This is an infinite geometric series with $a = 0.2$ and common ratio $r = 0.1$. Since the common ratio is between -1 and 1, we can use the formula for the sum of a geometric series to find that

$0.\overline{2} = 0.2 + 0.02 + 0.002 + 0.0002 + \cdots$

$$= \frac{0.2}{1-0.1}$$

$$= \frac{0.2}{0.9}$$

$$= \frac{2}{9}$$

17. The total impact of the $500 tax rebate on the U.S. economy is

$\$500 + \$500(0.95) + \$500(0.95)^2 + \$500(0.95)^3 + \ldots$

This is an infinite geometric series with first term $a = 500$ and common ratio $r = 0.95$. The sum of this series is

$\$500 + \$500(0.95) + \$500(0.95)^2 + \cdots$

$$= \frac{\$500}{1-0.95}$$

$$= \frac{\$500}{0.05}$$

$$= \$10,000$$

The U.S. economy will grow by $10,000 because of the child tax credit to Roberta.

18. This is an ordinary annuity with $n = 4 \cdot 30 = 120$ payments with deposits of $P = \$500$. The interest rate per payment period is

$i = \frac{0.08}{4} = 0.02$. This gives

$$A = 500\left[\frac{(1+0.02)^{120}-1}{0.02}\right]$$

$$= 500(488.258152)$$

$$= 244,129.08$$

After 30 years, the IRA will be worth $244,129.08.

13.3 Exercises

19. $\left\{4^n\right\}$

$a_{n-1} = 4^{n-1}$

$a_n = 4^n$

$\frac{a_n}{a_{n-1}} = \frac{4^n}{4^{n-1}} = \frac{4^n}{4^n \cdot 4^{-1}} = \frac{1}{4^{-1}}$

The ratio of *any* two consecutive terms is 4, so the sequence is geometric, with common ratio $r = 4$. To find the first term, we evaluate a_1 and find $a_1 = 4^1 = 4$.

21. $\left\{\left(\frac{2}{3}\right)^n\right\}$

$a_{n-1} = \left(\frac{2}{3}\right)^{n-1}$

$a_n = \left(\frac{2}{3}\right)^n$

$$\frac{a_n}{a_n - 1} = \frac{\left(\frac{2}{3}\right)^n}{\left(\frac{2}{3}\right)^{n-1}}$$

$$= \frac{\left(\frac{2}{3}\right)^n}{\left(\frac{2}{3}\right)^n \cdot \left(\frac{2}{3}\right)^{-1}}$$

$$= \frac{1}{\left(\frac{2}{3}\right)^{-1}}$$

$$= \frac{2}{3}$$

The ratio of *any* two consecutive terms is $\dfrac{2}{3}$, so the sequence is geometric, with common ratio $r = \dfrac{2}{3}$. To find the first term, we evaluate a_1 and find $a_1 = \left(\dfrac{2}{3}\right)^1 = \dfrac{2}{3}$.

23. $\left\{3 \cdot 2^{-n}\right\}$

$a_{n-1} = 3 \cdot 2^{-(n-1)} = 3 \cdot 2^{-n+1}$

$a_n = 3 \cdot 2^{-n}$

$\dfrac{a_n}{a_{n-1}} = \dfrac{3 \cdot 2^{-n}}{3 \cdot 2^{-n+1}}$

$\phantom{\dfrac{a_n}{a_{n-1}}} = \dfrac{3 \cdot 2^{-n}}{3 \cdot 2^{-n} \cdot 2}$

$\phantom{\dfrac{a_n}{a_{n-1}}} = \dfrac{1}{2}$

The ratio of *any* two consecutive terms is $\dfrac{1}{2}$, so the sequence is geometric, with common ratio $r = \dfrac{1}{2}$. To find the first term, we evaluate a_1 and find $a_1 = 3 \cdot 2^{-1} = \dfrac{3}{2}$.

25. $\left\{\dfrac{5^{n-1}}{2^n}\right\}$

$a_{n-1} = \dfrac{5^{(n-1)-1}}{2^{n-1}} = \dfrac{5^{n-2}}{2^{n-1}}$

$a_n = \dfrac{5^{n-1}}{2^n}$

$\dfrac{a_n}{a_{n-1}} = \dfrac{\dfrac{5^{n-1}}{2^n}}{\dfrac{5^{n-2}}{2^{n-1}}} = \dfrac{5^{n-1}}{2^n} \cdot \dfrac{2^{n-1}}{5^{n-2}} = \dfrac{5}{2}$

The ratio of *any* two consecutive terms is $\dfrac{5}{2}$, so the sequence is geometric, with common ratio $r = \dfrac{5}{2}$. To find the first term, we evaluate a_1 and find $a_1 = \dfrac{5^{1-1}}{2^1} = \dfrac{5^0}{2} = \dfrac{1}{2}$.

27. **(a)** $a_n = a \cdot r^{n-1} = 10 \cdot 2^{n-1}$

   (b) $a_8 = 10 \cdot 2^{8-1} = 10 \cdot 2^7 = 10 \cdot 128 = 1280$

29. **(a)** $a_n = a \cdot r^{n-1} = 100 \cdot \left(\dfrac{1}{2}\right)^{n-1}$

   (b) $a_8 = 100 \cdot \left(\dfrac{1}{2}\right)^{8-1} = 100 \cdot \left(\dfrac{1}{2}\right)^7 = \dfrac{100}{128} = \dfrac{25}{32}$

31. **(a)** $a_n = a \cdot r^{n-1} = 1 \cdot (-3)^{n-1} = (-3)^{n-1}$

   (b) $a_8 = (-3)^{8-1} = (-3)^7 = -2187$

33. **(a)** $a_n = a \cdot r^{n-1} = 100 \cdot (1.05)^{n-1}$

   (b) $a_8 = 100 \cdot (1.05)^{8-1} = 100 \cdot (1.05)^7 \approx 140.71$

35. $r = \dfrac{6}{3} = 2$; $a = 3$

$a_n = a \cdot r^{n-1}$

$a_{10} = 3 \cdot 2^{10-1} = 3 \cdot 2^9 = 3 \cdot 512 = 1536$

37. $r = \dfrac{-2}{4} = -\dfrac{1}{2}$; $a = 4$

$a_n = a \cdot r^{n-1}$

$a_{15} = 4 \cdot \left(-\dfrac{1}{2}\right)^{15-1}$

$\phantom{a_{15}} = 4 \cdot \left(-\dfrac{1}{2}\right)^{14}$

$\phantom{a_{15}} = \dfrac{4}{16,384}$

$\phantom{a_{15}} = \dfrac{1}{4096}$

39. $r = \dfrac{0.05}{0.5} = \dfrac{1}{10}$; $a = 0.5$

$a_n = a \cdot r^{n-1}$

$a_9 = 0.5 \cdot (0.1)^{9-1} = 0.5 \cdot (0.1)^8 = 0.000000005$

41. $2+4+8+\cdots+2^{12}$

This is a geometric sequence with $a=2$ and common ratio $r=2$. We wish to find the sum of the first 12 terms, S_{12}.

$$S_n = a \cdot \frac{1-r^n}{1-r}$$

$$S_{12} = 2 \cdot \frac{1-2^{12}}{1-2} = 2 \cdot \frac{-4095}{-1} = 8190$$

$$2+4+8+\cdots+2^{12} = 8190$$

43. $50+20+8+\dfrac{16}{5}+\cdots+50\left(\dfrac{2}{5}\right)^{10-1}$

This is a geometric sequence with $a=50$ and common ratio $r=\dfrac{2}{5}$. We wish to find the sum of the first 10 terms, S_{10}.

$$S_n = a \cdot \frac{1-r^n}{1-r}$$

$$S_{10} = 50 \cdot \frac{1-\left(\dfrac{2}{5}\right)^{10}}{1-\dfrac{2}{5}} = 50 \cdot \frac{\dfrac{9,764,601}{9,765,625}}{\dfrac{3}{5}} \approx 83.3245952$$

$$50+20+8+\frac{16}{5}+\cdots+50\left(\frac{2}{5}\right)^{10-1} \approx 83.3245952$$

45. $\displaystyle\sum_{n=1}^{10}\left[3\cdot2^n\right] = \sum_{n=1}^{10}\left[3\cdot2\cdot2^{n-1}\right] = \sum_{n=1}^{10}\left[6\cdot2^{n-1}\right]$

Here we want to find the sum of the first 10 terms of a geometric sequence with $a=6$ and common ratio $r=2$.

$$S_n = a \cdot \frac{1-r^n}{1-r}$$

$$S_{10} = 6 \cdot \frac{1-2^{10}}{1-2} = 6 \cdot \frac{-1023}{-1} = 6138$$

$$\sum_{n=1}^{10}\left[3\cdot2^n\right] = 6138$$

47. $\displaystyle\sum_{n=1}^{8}\left[\frac{4}{2^{n-1}}\right] = \sum_{n=1}^{8}\left[4\cdot\left(\frac{1}{2}\right)^{n-1}\right]$

Here we want the sum of the first eight terms of a geometric sequence with $a=4$ and common ratio $r=\dfrac{1}{2}$.

$$S_n = a \cdot \frac{1-r^n}{1-r}$$

$$S_8 = 4 \cdot \frac{1-\left(\dfrac{1}{2}\right)^8}{1-\dfrac{1}{2}} = 4 \cdot \frac{\dfrac{255}{256}}{\dfrac{1}{2}} = 8 \cdot \frac{255}{256} = 7.96875$$

$$\sum_{n=1}^{8}\left[\frac{4}{2^{n-1}}\right] = 7.96875$$

49. This is an infinite geometric series with $a=1$ and common ratio $r=\dfrac{\dfrac{1}{2}}{1}=\dfrac{1}{2}$. Since the common ratio, r, is between -1 and 1, we can use the formula for the sum of an infinite geometric series.

$$1+\frac{1}{2}+\frac{1}{4}+\cdots = \frac{a}{1-r} = \frac{1}{1-\dfrac{1}{2}} = \frac{1}{\dfrac{1}{2}} = 2$$

51. This is an infinite geometric series with $a=10$ and common ratio $r=\dfrac{\dfrac{10}{3}}{10}=\dfrac{1}{3}$. Since the common ratio, r, is between -1 and 1, we can use the formula for the sum of an infinite geometric series.

$$10+\frac{10}{3}+\frac{10}{9}+\cdots = \frac{10}{1-\dfrac{1}{3}} = \frac{10}{\dfrac{2}{3}} = 15$$

53. This is an infinite geometric series with $a=6$ and common ratio $r=\dfrac{-2}{6}=-\dfrac{1}{3}$. Since the common ratio, r, is between -1 and 1, we can use the formula for the sum of an infinite geometric series.

$$6-2+\frac{2}{3}-\frac{2}{9}+\cdots = \frac{6}{1-\left(-\dfrac{1}{3}\right)} = \frac{6}{\dfrac{4}{3}} = 6\cdot\frac{3}{4} = \frac{9}{2}$$

55. This is an infinite geometric series with $a=5\cdot\left(\dfrac{1}{5}\right)^1=1$ and common ratio $r=\dfrac{1}{5}$. Since the common ratio, r, is between -1 and 1, we can use the formula for the sum of an infinite geometric series.

$$\sum_{n=1}^{\infty}\left(5\cdot\left(\frac{1}{5}\right)^n\right) = \frac{1}{1-\dfrac{1}{5}} = \frac{1}{\dfrac{4}{5}} = \frac{5}{4}$$

57. This is an infinite geometric series with

$$a = 12 \cdot \left(-\frac{1}{3}\right)^{1-1} = 12 \text{ and common ratio}$$

$r = -\frac{1}{3}$. Since the common ratio, r, is between

-1 and 1, we can use the formula for the sum of an infinite geometric series.

$$\sum_{n=1}^{\infty}\left(12 \cdot \left(-\frac{1}{3}\right)^{n-1}\right) = \frac{12}{1-\left(-\frac{1}{3}\right)} = \frac{12}{\frac{4}{3}}$$

$$= 12 \cdot \frac{3}{4}$$

$$= 9$$

59. $0.\overline{5} = 0.5 + 0.05 + 0.005 + 0.0005 + \cdots$

$$= \sum_{n=1}^{\infty}\left[0.5 \cdot \left(\frac{1}{10}\right)^{n-1}\right]$$

This is an infinite geometric series with $a = 0.5$

and common ratio $r = \frac{1}{10}$. Since the common

ratio, r, is between -1 and 1, we can use the formula for the sum of an infinite geometric series.

$$0.\overline{5} = \frac{0.5}{1-\frac{1}{10}} = \frac{\frac{1}{2}}{\frac{9}{10}} = \frac{1}{2} \cdot \frac{10}{9} = \frac{5}{9}$$

61. $0.\overline{89} = 0.89 + 0.0089 + 0.000089 + \cdots$

$$= \sum_{n=1}^{\infty}\left[0.89 \cdot \left(\frac{1}{100}\right)^{n-1}\right]$$

This is an infinite geometric series with $a = 0.89$

and common ratio $r = \frac{1}{100}$. Since the common

ratio, r, is between -1 and 1, we can use the formula for the sum of an infinite geometric series.

$$0.\overline{89} = \frac{0.89}{1-\frac{1}{100}} = \frac{0.89}{0.99} = \frac{89}{99}$$

63. $\{5n+1\}$

$$a_n - a_{n-1} = [5n+1] - [5(n-1)+1]$$
$$= [5n+1] - [5n-5+1]$$
$$= 5n+1 - 5n+4$$
$$= 5$$

Since the difference between successive terms is constant, this is an arithmetic sequence. The common difference is $d = 5$.

65. $\{2n^2\}$

$$a_n - a_{n-1} = \left[2n^2\right] - \left[2(n-1)^2\right]$$
$$= \left[2n^2\right] - \left[2n^2 - 4n + 2\right]$$
$$= 2n^2 - 2n^2 + 4n - 2$$
$$= 4n - 2$$

Since the difference between consecutive terms is not constant, the sequence is not arithmetic.

$$\frac{a_n}{a_{n-1}} = \frac{2n^2}{2(n-1)^2} = \frac{n^2}{n^2 - 2n + 1}$$

Since the ratio of consecutive terms is not constant, the sequence is not geometric.

67. $\left\{\frac{2^{-n}}{5}\right\}$

$$a_n - a_{n-1} = \frac{2^{-n}}{5} - \frac{2^{-(n-1)}}{5}$$
$$= \frac{2^{-n}}{5} - \frac{2^{-n+1}}{5}$$
$$= \frac{2^{-n}}{5} - \frac{2 \cdot 2^{-n}}{5}$$
$$= -\frac{2^{-n}}{5}$$

Since the difference between consecutive terms is not constant, the sequence is not arithmetic.

$$\frac{a_n}{a_{n-1}} = \frac{\left(\frac{2^{-n}}{5}\right)}{\left(\frac{2^{-(n-1)}}{5}\right)} = \frac{2^{-n}}{5} \cdot \frac{5}{2^{-n+1}}$$
$$= 2^{-n-(-n+1)}$$
$$= 2^{-1}$$
$$= \frac{1}{2}$$

Since the ratio of consecutive terms is constant, the sequence is geometric. The common ratio is

$r = \frac{1}{2}$.

69. $a_2 - a_1 = 36 - 54 = -18$

$a_3 - a_2 = 24 - 36 = -12$

Consecutive differences are not the same so the sequence is not arithmetic.

$\dfrac{a_2}{a_1} = \dfrac{36}{54} = \dfrac{2}{3}; \quad \dfrac{a_3}{a_2} = \dfrac{24}{36} = \dfrac{2}{3}; \quad \dfrac{a_4}{a_3} = \dfrac{16}{24} = \dfrac{2}{3}$

The ratio of consecutive terms is constant so the sequence is geometric. The common ratio is

$r = \dfrac{2}{3}.$

71. $a_2 - a_1 = 6 - 2 = 4$

$a_3 - a_2 = 10 - 6 = 4$

$a_4 - a_3 = 14 - 10 = 4$

The difference between consecutive terms is constant so the sequence is arithmetic. The common difference is $d = 4$.

73. $a_2 - a_1 = 2 - 1 = 1$

$a_3 - a_2 = 3 - 2 = 1$

$a_4 - a_3 = 5 - 3 = 2$

$a_5 - a_4 = 8 - 5 = 3$

The difference between consecutive terms is not constant so the sequence is not arithmetic.

$\dfrac{a_2}{a_1} = \dfrac{2}{1} = 2$

$\dfrac{a_3}{a_2} = \dfrac{3}{2} = 1.5$

$\dfrac{a_4}{a_3} = \dfrac{5}{3} \approx 1.67$

$\dfrac{a_5}{a_4} = \dfrac{8}{5} = 1.6$

The ratio of consecutive terms is not constant so the sequence is not geometric.

75. To be a geometric sequence, the ratio of consecutive terms must be the same. Therefore, we need to solve the equation

$\dfrac{x+2}{x} = \dfrac{x+3}{x+2}$

$x(x+2) \cdot \dfrac{x+2}{x} = x(x+2) \cdot \dfrac{x+3}{x+2}$

$(x+2)^2 = x(x+3)$

$x^2 + 4x + 4 = x^2 + 3x$

$4x + 4 = 3x$

$x + 4 = 0$

$x = -4$

77. Your annual salaries will form a geometric sequence with $a = 40,000$ and common ratio $r = 1.05$.

(a) Your salary at the beginning of the second year is the value of the second term in the sequence.

$a_2 = a \cdot r^{2-1} = 40,000 \cdot (1.05)^1 = 42,000$

Your salary at the beginning of the second year will be \$42,000.

(b) Your salary at the beginning of the tenth year is the value of the tenth term in the sequence.

$a_{10} = a \cdot r^{10-1} = 40,000 \cdot (1.05)^9 \approx 62,053$

Your salary at the beginning of the tenth year will be about \$62,053.

(c) Your cumulative earnings after completing your tenth year is the sum of the first ten terms of the sequence.

$S_{10} = a \cdot \dfrac{1 - r^{10}}{1 - r}$

$= 40,000 \cdot \dfrac{1 - (1.05)^{10}}{1 - 1.05}$

$\approx 40,000 \cdot (12.577893)$

$\approx 503,116$

Your cumulative earnings after finishing your tenth year will be about \$503,116.

79. The value of the car at the beginning of the year forms a geometric sequence with $a = 20,000$ and common ratio $r = 0.92$. The value of the car after you have owned it for five years will be the sixth term of the sequence (the beginning of the sixth year of ownership).

$a_6 = 20,000 \cdot (0.92)^{6-1}$

$= 20,000 \cdot (0.92)^5 \approx 13,182$

After five years of ownership, the car will be worth about \$13,182.

81. The lengths of the arc of the pendulum swings form a geometric sequence with $a = 3$ and common ratio $r = 0.95$.

(a) $a_{10} = a \cdot r^{10-1}$

$= 3 \cdot (0.95)^9$

≈ 1.891

The length of the arc in the 10th swing is about 1.891 feet.

(b) Here we need to determine the term number. We start with the following equation:

$$a_n = 1$$
$$a \cdot r^{n-1} = 1$$
$$3 \cdot (0.95)^{n-1} = 1$$
$$(0.95)^{n-1} = \frac{1}{3}$$
$$\ln\left[(0.95)^{n-1}\right] = \ln\left(\tfrac{1}{3}\right)$$
$$(n-1) \cdot \ln(0.95) = \ln\left(\tfrac{1}{3}\right)$$
$$n-1 = \frac{\ln\left(\tfrac{1}{3}\right)}{\ln(0.95)}$$
$$n = 1 + \frac{\ln\left(\tfrac{1}{3}\right)}{\ln(0.95)}$$
$$n \approx 22.42$$

Since the terms in the sequence are decreasing, the length of the arc will be less than 1 foot on the 23rd swing.

(c)
$$S_{10} = a \cdot \frac{1 - r^{10}}{1 - r}$$
$$= 3 \cdot \frac{1 - (0.95)^{10}}{1 - 0.95} \approx 24.08$$

After 10 swings, the pendulum will have swung a total of about 24.08 feet.

(d) To find the total length the pendulum will swing requires an infinite geometric series.

$$S_\infty = \frac{a}{1 - r}$$
$$= \frac{3}{1 - 0.95}$$
$$= \frac{3}{0.05}$$
$$= 60$$

The pendulum will swing a total of 60 feet.

83. The annual salaries under option A form a geometric sequence with $a = 30,000$ and common ratio $r = 1.05$. The annual salaries under option B form a geometric sequence with $a = 31,000$ and common ratio $r = 1.04$.

To determine which option gives the larger annual salary in the final year of the contract, we need to find the 5th term of each sequence.

Option A: $a_5 = 30,000 \cdot (1.05)^{5-1} \approx 36,465.19$

Option B: $a_5 = 31,000 \cdot (1.04)^{5-1} \approx 36,265.62$

To determine which option gives the largest cumulative salary for five years, we need the sum of the first five terms of each sequence.

Option A: $S_5 = 30,000 \cdot \dfrac{1 - (1.05)^5}{1 - 1.05} \approx 165,768.94$

Option B: $S_5 = 31,000 \cdot \dfrac{1 - (1.04)^5}{1 - 1.04} \approx 167,906.00$

Therefore, option A will yield the larger annual salary in the final year of the contract and option B will yield the larger cumulative salary over the life of the contract.

85. The multiplier in this case is the geometric series with first term $a = 1$ and common ratio $r = 0.98$.

$$1 + 0.98 + (0.98)^2 + \ldots = \frac{1}{1 - 0.98} = 50$$

The multiplier is 50.

87. $\dfrac{1+i}{1+r} = \dfrac{1+.02}{1+.09} = \dfrac{1.02}{1.09} \approx 0.93578$

The price is an infinite geometric series with $a = 2$ and a common ratio of $\dfrac{1.02}{1.09}$. The maximum price is given by

$$\text{Price} = \frac{2}{1 - \dfrac{1.02}{1.09}} \approx 31.14$$

The maximum price you should pay for the stock is $31.14 per share.

89. $A = P \cdot \dfrac{(1+i)^n - 1}{i}$

In this case we have $P = 100$, $n = 30 \cdot 12 = 360$, and $i = \dfrac{0.08}{12}$.

$$A = 100 \cdot \left[\frac{\left(1 + \dfrac{0.08}{12}\right)^{360} - 1}{\dfrac{0.08}{12}}\right]$$
$$= 100 \cdot [1490.359449]$$
$$= \$149,035.94$$

After 30 years, Christina's 401(k) will be worth $149,035.94.

91. $A = P \cdot \dfrac{(1+i)^n - 1}{i}$

In this case we have $P = 500$, $n = 25 \cdot 4 = 100$,

and $i = \dfrac{0.06}{4} = 0.015$.

$A = 500 \cdot \left[\dfrac{(1 + 0.015)^{100} - 1}{0.015} \right]$

$= 500 \cdot [228.8030433]$

$= \$114,401.52$

After 25 years, Jackson's IRA will be worth $114,401.52.

93. $A = P \cdot \dfrac{(1+i)^n - 1}{i}$

In this case we have $A = 1,500,000$,

$n = 35 \cdot 12 = 420$, and $i = \dfrac{0.10}{12}$.

$1,500,000 = P \cdot \left[\dfrac{\left(1 + \dfrac{0.10}{12}\right)^{420} - 1}{\dfrac{0.10}{12}} \right]$

$1,500,000 = P \cdot [3796.638052]$

$P = \$395.09$

Aaliyah will need to contribute $395.09, or about $395, each month in order to achieve her goal.

95. $0.4\overline{9} = 0.4 + 0.09 + 0.009 + 0.0009 + \cdots$

After the 0.4, the rest of the sum forms an infinite geometric series with $a = 0.09$ and common ratio $r = 0.1$. We can find the sum of

this part as $S_\infty = \dfrac{a}{1-r} = \dfrac{0.09}{1 - 0.1} = \dfrac{0.09}{0.9} = 0.1$

Thus, $0.4\overline{9} = 0.4 + 0.1 = 0.5 = \dfrac{1}{2}$.

97. The sum is a geometric series with $a = 2$ and common ratio $r = 2$. We can find the sum by using the summation formula, after we determine how many terms are in the sum.

$a_n = a \cdot r^{n-1}$

$1,073,741,824 = 2 \cdot 2^{n-1}$

$1,073,741,824 = 2^n$

$\ln 1,073,741,824 = \ln 2^n$

$\ln 1,073,741,824 = n \cdot \ln 2$

$\dfrac{\ln 1,073,741,824}{\ln 2} = n$

$30 = n$

$S_{30} = a \cdot \dfrac{1 - r^{30}}{1 - r} = 2 \cdot \dfrac{1 - 2^{30}}{1 - 2} = 2,147,483,646$

99. Answers may vary. The comparison of growth rates really depends on the common difference and the common ratio. It is possible for an arithmetic sequence to grow at a faster rate initially. However, the geometric sequence will always end up growing at a faster rate if given enough time. For example, consider the following sequences:

arithmetic: $\{20 + (n-1) \cdot 2\}$

geometric: $\{20 \cdot 1.05^{n-1}\}$

These two sequences begin at the same value, but then the arithmetic sequence begins growing at a faster rate. After about 27 years, the geometric sequence will take over.

If Malthus' conjecture is correct, it would mean that we would eventually be unable to produce enough food to feed everyone on the planet, even if current supply exceeds demand.

101. A geometric series has a finite sum if $-1 < r < 1$.

103. $\dfrac{1}{3} = 0.333333...$

$\dfrac{2}{3} = 0.666666...$

$\dfrac{1}{3} + \dfrac{2}{3} = (0.333333...) + (0.666666...)$

$= 0.999999...$

Since $\dfrac{1}{3} + \dfrac{2}{3} = \dfrac{3}{3} = 1$, we would conjecture that

$0.999999... = 1$.

105. This is a geometric series with $a = 4$ and common ratio $r = 1.2$. Based on the form of the last term, it appears that there are $n = 15$ terms in the series.

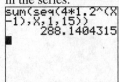

The sum is approximately 288.1404315.

107. $\displaystyle\sum_{n=1}^{20}\left[1.2(1.05)^n\right]$

This is a geometric series with $a = 1.26$ (that is, $1.2(1.05)^1$), common ratio $r = 1.05$, and $n = 20$ terms.

The sum is approximately 41.66310217.

Putting the Concepts Together (Sections 13.1–13.3)

1.

$\dfrac{3/16}{3/4} = \dfrac{3}{16} \cdot \dfrac{4}{3} = \dfrac{1}{4}$

$\dfrac{3/64}{3/16} = \dfrac{3}{64} \cdot \dfrac{16}{3} = \dfrac{1}{4}$

$\dfrac{3/256}{3/64} = \dfrac{3}{256} \cdot \dfrac{64}{3} = \dfrac{1}{4}$

The ratio of consecutive terms is a constant so the sequence is geometric with $a = \dfrac{3}{4}$ and common ratio $r = \dfrac{1}{4}$.

2. $a_n - a_{n-1} = \left[2(n+3)\right] - \left[2(n-1+3)\right]$

$\qquad = 2(n+3) - 2(n+2)$

$\qquad = 2n + 6 - 2n - 4$

$\qquad = 2$

The difference between consecutive terms is a constant so the sequence is arithmetic with $a = 2(1+3) = 8$ and common difference $d = 2$.

3. $a_n - a_{n-1} = \left[\dfrac{7n+2}{9}\right] - \left[\dfrac{7(n-1)+2}{9}\right]$

$\qquad = \left(\dfrac{7n+2}{9}\right) - \left(\dfrac{7n-7+2}{9}\right)$

$\qquad = \dfrac{7n+2-7n+5}{9}$

$\qquad = \dfrac{7}{9}$

The difference between consecutive terms is a constant so the sequence is arithmetic with

$a = \dfrac{7(1)+2}{9} = 1$ and common difference $d = \dfrac{7}{9}$.

4. $-4 - 1 = -5$

$9 - (-4) = 13$

$\dfrac{-4}{1} = -4$

$\dfrac{9}{-4} = -\dfrac{9}{4}$

The sequence is neither arithmetic nor geometric. The difference between consecutive terms is not constant so the sequence is not arithmetic. Likewise, the ratio of consecutive terms is not constant so the sequence is not geometric.

5. $\dfrac{a_n}{a_{n-1}} = \dfrac{3 \cdot 2^{n+1}}{3 \cdot 2^{(n-1)+1}} = \dfrac{3 \cdot 2^n \cdot 2}{3 \cdot 2^n} = 2$

The ratio of consecutive terms is a constant so the sequence is geometric with $a = 3 \cdot 2^{1+1} = 12$ and common ratio $r = 2$.

6. $a_n - a_{n-1} = \left[n^2 - 5\right] - \left[(n-1)^2 - 5\right]$

$\qquad = n^2 - 5 - \left(n^2 - 2n + 1 - 5\right)$

$\qquad = n^2 - 5 - n^2 + 2n + 4$

$\qquad = 2n - 1$

$\dfrac{a_n}{a_{n-1}} = \dfrac{n^2 - 5}{(n-1)^2 - 5} = \dfrac{n^2 - 5}{n^2 - 2n - 4}$

The sequence is neither arithmetic nor geometric. The difference between consecutive terms is not constant so the sequence is not arithmetic. Likewise, the ratio of consecutive terms is not constant so the sequence is not geometric.

7. $\displaystyle\sum_{k=1}^{6}[3k+4]$

$=(3\cdot1+4)+(3\cdot2+4)+(3\cdot3+4)$
$\quad+(3\cdot4+4)+(3\cdot5+4)+(3\cdot6+4)$
$=7+10+13+16+19+22$
$=87$

8. Each term is written in the form $\dfrac{1}{2(6+i)}$, where

i is the term number, and there are $n=12$ terms.

$\dfrac{1}{2(6+1)}+\dfrac{1}{2(6+2)}+\cdots+\dfrac{1}{2(6+12)}$

$=\displaystyle\sum_{i=1}^{12}\dfrac{1}{2(6+i)}$

9. $a_n=a+(n-1)d$
$\quad=25+(n-1)(-2)$
$\quad=25-2n+2$
$\quad=27-2n$
$a_1=27-2(1)=25$
$a_2=27-2(2)=23$
$a_3=27-2(3)=21$
$a_4=27-2(4)=19$
$a_5=27-2(5)=17$
The first five terms of the sequence are 25, 23, 21, 19, and 17.

10. $\quad a_n=a+(n-1)d$
$\quad a_4=a+(4-1)(11)=9$
$a+33=9$
$\quad a=-24$
$a_n=-24+(n-1)(11)$
$\quad=-24+11n-11$
$\quad=11n-35$
$a_1=11(1)-35=-24$
$a_2=11(2)-35=-13$
$a_3=11(3)-35=-2$
$a_4=11(4)-35=9$
$a_5=11(5)-35=20$
The first five terms of the sequence are -24, -13, -2, 9, and 20.

11. $\quad a_n=a\cdot r^{n-1}$

$\quad a_4=a\cdot\left(\dfrac{1}{5}\right)^{4-1}=\dfrac{9}{25}$

$a\cdot\dfrac{1}{125}=\dfrac{9}{25}\qquad\rightarrow\qquad a=125\cdot\dfrac{9}{25}=45$

$a_n=45\cdot\left(\dfrac{1}{5}\right)^{n-1}$

$a_1=45\cdot\left(\dfrac{1}{5}\right)^{1-1}=45$

$a_2=45\cdot\left(\dfrac{1}{5}\right)^{2-1}=9$

$a_3=45\cdot\left(\dfrac{1}{5}\right)^{3-1}=\dfrac{9}{5}$

$a_4=45\cdot\left(\dfrac{1}{5}\right)^{4-1}=\dfrac{9}{25}$

$a_5=45\cdot\left(\dfrac{1}{5}\right)^{5-1}=\dfrac{9}{125}$

The first five terms of the sequence are 45, 9, $\dfrac{9}{5}$,

$\dfrac{9}{25}$, and $\dfrac{9}{125}$.

12. $a_n=a\cdot r^{n-1}$

$\quad=150\cdot(1.04)^{n-1}$

$a_1=150\cdot(1.04)^{1-1}=150$
$a_2=150\cdot(1.04)^{2-1}=156$
$a_3=150\cdot(1.04)^{3-1}=162.24$
$a_4=150\cdot(1.04)^{4-1}=168.7296$
$a_5=150\cdot(1.04)^{5-1}=175.478784$

The first five terms of the sequence are 150, 156, 162.24, 168.7296, and 175.478784.

13. The terms form a geometric sequence with first term $a=2$, common ratio $r=3$, and $n=11$ terms.

$S_{11}=2\cdot\dfrac{1-3^{11}}{1-3}=177,146$

14. The terms form an arithmetic sequence with first term $a = 2$, common difference $d = 5$, and $n = 20$ terms.

$$a_{20} = 2 + (20 - 1) \cdot 5 = 97$$

$$S_{20} = \frac{20}{2}[2 + 97] = 990$$

15. This is an infinite geometric series with $a = 1000$ and $r = \frac{1}{10}$.

$$S_{\infty} = \frac{a}{1-r} = \frac{1000}{1 - \frac{1}{10}} = \frac{1000}{\frac{9}{10}} = \frac{10,000}{9} \text{ or } 1111\frac{1}{9}$$

16. The number of people who can be seated forms an arithmetic sequence with first term $a = 4$ and common difference $d = 2$. The number of tables required is the same as the term number, n.
To find the number of tables required to seat a party of 24 people, we solve the following for n:
$$a_n = a + (n-1)d$$
$$24 = 4 + (n-1)(2)$$
$$24 = 4 + 2n - 2$$
$$24 = 2n + 2$$
$$22 = 2n$$
$$11 = n$$
A party of 24 people would require 11 tables.

Section 13.4

Preparing for the Binomial Theorem

P1. $(x-5)^2 = x^2 - 2 \cdot x \cdot 5 + 5^2$
$$= x^2 - 10x + 25$$

P2. $(2x+3)^2 = (2x+3)(2x+3)$
$$= \left((2x)^2 + 2 \cdot 2x \cdot 3 + 3^2\right)$$
$$= 4x^2 + 12x + 9$$

Section 13.4 Quick Checks

1. $n(n-1)(n-2) \cdot \ldots \cdot 3 \cdot 2 \cdot 1$

2. $1; 1$

3. $5! = 5 \cdot 4 \cdot 3 \cdot 2 \cdot 1 = 120$

4. $\dfrac{7!}{3!} = \dfrac{7 \cdot 6 \cdot 5 \cdot 4 \cdot 3!}{3!} = 7 \cdot 6 \cdot 5 \cdot 4 = 840$

5. False. The numerator should be $n!$ and the denominator should be $j!\,(n-j)!$.

6. True.

7. $\dbinom{7}{1} = \dfrac{7!}{1! \, (7-1)!} = \dfrac{7!}{6!} = \dfrac{7 \cdot 6!}{6!} = 7$

8. $\binom{6}{3} = \dfrac{6!}{3! \cdot (6-3)!} = \dfrac{6 \cdot 5 \cdot 4 \cdot 3!}{3! \cdot 3!} = \dfrac{6 \cdot 5 \cdot 4}{3 \cdot 2 \cdot 1} = 20$

9. $(x+2)^4 = \binom{4}{0}x^4 + \binom{4}{1}2^1 \cdot x^{4-1} + \binom{4}{2}2^2 \cdot x^{4-2} + \binom{4}{3}2^3 \cdot x^{4-3} + \binom{4}{4}2^4$

$\qquad = 1 \cdot x^4 + 4 \cdot 2 \cdot x^3 + 6 \cdot 4 \cdot x^2 + 4 \cdot 8 \cdot x + 1 \cdot 16$

$\qquad = x^4 + 8x^3 + 24x^2 + 32x + 16$

10. $(2p-1)^5$

$\quad = (2p + (-1))^5$

$\quad = \binom{5}{0}(2p)^5 + \binom{5}{1}(-1)^1 \cdot (2p)^{5-1} + \binom{5}{2}(-1)^2 \cdot (2p)^{5-2} + \binom{5}{3}(-1)^3 \cdot (2p)^{5-3} + \binom{5}{4}(-1)^4 \cdot (2p)^{5-4} + \binom{5}{5}(-1)^5$

$\quad = 1 \cdot (2p)^5 + 5 \cdot (-1) \cdot (2p)^4 + 10 \cdot (-1)^2 \cdot (2p)^3 + 10 \cdot (-1)^3 \cdot (2p)^2 + 5 \cdot (-1)^4 \cdot (2p) + 1 \cdot (-1)^5$

$\quad = 32p^5 + 5(-1)\left(16p^4\right) + 10(1)\left(8p^3\right) + 10(-1)\left(4p^2\right) + 5(1)(2p) + 1(-1)$

$\quad = 32p^5 - 80p^4 + 80p^3 - 40p^2 + 10p - 1$

13.4 Exercises

11. $3! = 3 \cdot 2 \cdot 1 = 6$

13. $8! = 8 \cdot 7 \cdot 6 \cdot 5 \cdot 4 \cdot 3 \cdot 2 \cdot 1 = 40{,}320$

15. $\dfrac{10!}{8!} = \dfrac{10 \cdot 9 \cdot 8!}{8!} = 10 \cdot 9 = 90$

17. $\dfrac{8!}{5!} = \dfrac{8 \cdot 7 \cdot 6 \cdot 5!}{5!} = 8 \cdot 7 \cdot 6 = 336$

19. $\binom{7}{2} = \dfrac{7!}{2!(7-2)!} = \dfrac{7!}{2! \, 5!} = \dfrac{7 \cdot 6 \cdot 5!}{2 \cdot 5!} = \dfrac{7 \cdot 6}{2} = 21$

21. $\binom{10}{4} = \dfrac{10!}{4!(10-4)!} = \dfrac{10!}{4! \, 6!}$

$\qquad = \dfrac{10 \cdot 9 \cdot 8 \cdot 7 \cdot 6!}{4 \cdot 3 \cdot 2 \cdot 1 \cdot 6!}$

$\qquad = \dfrac{10 \cdot 9 \cdot 8 \cdot 7}{4 \cdot 3 \cdot 2 \cdot 1}$

$\qquad = 210$

23. $(x+1)^5 = \binom{5}{0}x^5 + \binom{5}{1}1^1 \cdot x^{5-1} + \binom{5}{2}1^2 \cdot x^{5-2} + \binom{5}{3}1^3 \cdot x^{5-3} + \binom{5}{4}1^4 \cdot x^{5-4} + \binom{5}{5}1^5$

$\qquad = x^5 + 5x^4 + 10x^3 + 10x^2 + 5x + 1$

25. $(x-4)^4 = (x+(-4))^4$

$$= \binom{4}{0}x^4 + \binom{4}{1}(-4)^1 \cdot x^{4-1} + \binom{4}{2}(-4)^2 \cdot x^{4-2} + \binom{4}{3}(-4)^3 \cdot x^{4-3} + \binom{4}{4}(-4)^4$$

$$= x^4 + 4(-4)x^3 + 6(16)x^2 + 4(-64)x + 256$$

$$= x^4 - 16x^3 + 96x^2 - 256x + 256$$

27. $(3p+2)^4 = ((3p)+2)^4$

$$= \binom{4}{0}(3p)^4 + \binom{4}{1}2^1 \cdot (3p)^{4-1} + \binom{4}{2}2^2 \cdot (3p)^{4-2} + \binom{4}{3}2^3 \cdot (3p)^{4-3} + \binom{4}{4}2^4$$

$$= 81p^4 + 4 \cdot 2 \cdot 27p^3 + 6 \cdot 4 \cdot 9p^2 + 4 \cdot 8 \cdot 3p + 16$$

$$= 81p^4 + 216p^3 + 216p^2 + 96p + 16$$

29. $(2z-3)^5$

$$= ((2z)+(-3))^5$$

$$= \binom{5}{0}(2z)^5 + \binom{5}{1}(-3)^1 \cdot (2z)^{5-1} + \binom{5}{2}(-3)^2 \cdot (2z)^{5-2} + \binom{5}{3}(-3)^3 \cdot (2z)^{5-3} + \binom{5}{4}(-3)^4 \cdot (2z)^{5-4} + \binom{5}{5}(-3)^5$$

$$= 32z^5 + 5(-3) \cdot 16z^4 + 10(9) \cdot 8z^3 + 10(-27) \cdot 4z^2 + 5(81) \cdot 2z + (-243)$$

$$= 32z^5 - 240z^4 + 720z^3 - 1080z^2 + 810z - 243$$

31. $(x^2+2)^4 = \binom{4}{0}(x^2)^4 + \binom{4}{1}2^1 \cdot (x^2)^{4-1} + \binom{4}{2}2^2 \cdot (x^2)^{4-2} + \binom{4}{3}2^3 \cdot (x^2)^{4-3} + \binom{4}{4}2^4$

$$= x^8 + 4 \cdot 2x^6 + 6 \cdot 4x^4 + 4 \cdot 8x^2 + 16$$

$$= x^8 + 8x^6 + 24x^4 + 32x^2 + 16$$

33. $(2p^3+1)^5 = ((2p^3)+1)^5$

$$= \binom{5}{0}(2p^3)^5 + \binom{5}{1}1^1 \cdot (2p^3)^{5-1} + \binom{5}{2}1^2 \cdot (2p^3)^{5-2} + \binom{5}{3}1^3 \cdot (2p^3)^{5-3} + \binom{5}{4}1^4 \cdot (2p^3)^{5-4} + \binom{5}{5}1^5$$

$$= 32p^{15} + 5 \cdot 16p^{12} + 10 \cdot 8p^9 + 10 \cdot 4p^6 + 5 \cdot 2p^3 + 1$$

$$= 32p^{15} + 80p^{12} + 80p^9 + 40p^6 + 10p^3 + 1$$

35. $(x+2)^6 = \binom{6}{0}x^6 + \binom{6}{1}2^1 \cdot x^{6-1} + \binom{6}{2}2^2 \cdot x^{6-2} + \binom{6}{3}2^3 \cdot x^{6-3} + \binom{6}{4}2^4 \cdot x^{6-4} + \binom{6}{5}2^5 \cdot x^{6-5} + \binom{6}{6}2^6$

$$= x^6 + 6 \cdot 2x^5 + 15 \cdot 4x^4 + 20 \cdot 8x^3 + 15 \cdot 16x^2 + 6 \cdot 32x + 64$$

$$= x^6 + 12x^5 + 60x^4 + 160x^3 + 240x^2 + 192x + 64$$

37. $(2p^2-q^2)^4 = ((2p^2)+(-q^2))^4$

$$= \binom{4}{0}(2p^2)^4 + \binom{4}{1}(2p^2)^{4-1} \cdot (-q^2)^1 + \binom{4}{2}(2p^2)^{4-2} \cdot (-q^2)^2 + \binom{4}{3}(2p^2)^{4-3} \cdot (-q^2)^3 + \binom{4}{4}(-q^2)^4$$

$$= 16p^8 + 4 \cdot 8p^6 \cdot (-q^2) + 6 \cdot 4p^4 \cdot (q^4) + 4 \cdot 2p^2 \cdot (-q^6) + (q^8)$$

$$= 16p^8 - 32p^6q^2 + 24p^4q^4 - 8p^2q^6 + q^8$$

39. $(1.001)^4 = (1+0.001)^4$

$$= \binom{4}{0}1^4 + \binom{4}{1}1^{4-1} \cdot (0.001)^1 + \binom{4}{2}1^{4-2} \cdot (0.001)^2 + \binom{4}{3}1^{4-3} \cdot (0.001)^3 + \binom{4}{4}(0.001)^4$$

$$= 1 + 4(0.001) + 6(0.000001) + 4(0.000000001) + (0.000000000001)$$

$$= 1 + 0.004 + 0.000006 + 0.000000004 + 0.000000000001$$

$$= 1.004006004001$$

$$\approx 1.00401$$

41. $(0.998)^5 = (1-0.002)^5$

$$= (1+(-0.002))^5$$

$$= \binom{5}{0}1^5 + \binom{5}{1}1^{5-1}(-0.002)^1 + \binom{5}{2}1^{5-2}(-0.002)^2 + \binom{5}{3}1^{5-3}(-0.002)^3 + \binom{5}{4}1^{5-4}(-0.002)^4 + \binom{5}{5}(-0.002)^5$$

$$= 1 + 5(-0.002) + 10(0.000004) + 10(-0.000000008) + 5(0.000000000016) + (-0.000000000000032)$$

$$= 1 - 0.01 + 0.00004 - 0.00000008 + 0.00000000008 - 0.000000000000032$$

$$= 1.000040000080000 - 0.010000080000032$$

$$= 0.990039920079968$$

$$\approx 0.99004$$

43. The third term of the expansion of $(x+2)^7$ is

$$\binom{7}{2}2^2 \cdot x^{7-2} = 21 \cdot 4x^5 = 84x^5$$

45. The sixth term of the expansion of $(2p-3)^8 = ((2p)+(-3))^8$ is

$$\binom{8}{5}(-3)^5 \cdot (2p)^{8-5} = 56(-243) \cdot 8p^3$$

$$= -108,864p^3$$

47. $\binom{n}{n-1} = \dfrac{n!}{(n-1)!(n-(n-1))!} = \dfrac{n!}{(n-1)!1!} = \dfrac{n \cdot (n-1)!}{(n-1)!} = n$

$\binom{n}{n} = \dfrac{n!}{n!(n-n)!} = \dfrac{n!}{n!0!} = \dfrac{n!}{n!} = 1$

49.

```
      1
    1   1
  1   2   1
1   3   3   1
```

51. The degree of each monomial (sum of the exponents) in the expansion of $(x+a)^n$ is equal to n.

53. $f(x) = x^4$

$f(a-2) = (a-2)^4$

$= \binom{4}{0}a^4 + \binom{4}{1}(-2)^1 \cdot a^{4-1} + \binom{4}{2}(-2)^2 \cdot a^{4-2} + \binom{4}{3}(-2)^3 \cdot a^{4-3} + \binom{4}{4}(-2)^4$

$= a^4 + 4(-2)a^3 + 6(4)a^2 + 4(-8)a + (16)$

$= a^4 - 8a^3 + 24a^2 - 32a + 16$

55. $H(x) = x^5 - 4x^4$

$H(p+1) = (p+1)^5 - 4(p+1)^4$

$= \left[\binom{5}{0}p^5 + \binom{5}{1}1^1 \cdot p^{5-1} + \binom{5}{2}1^2 \cdot p^{5-2} + \binom{5}{3}1^3 \cdot p^{5-3} + \binom{5}{4}1^4 \cdot p^{5-4} + \binom{5}{5}1^5 \right]$

$\quad -4 \left[\binom{4}{0}p^4 + \binom{4}{1}1^1 \cdot p^{4-1} + \binom{4}{2}1^2 \cdot p^{4-2} + \binom{4}{3}1^3 \cdot p^{4-3} + \binom{4}{4}1^4 \right]$

$= \left[p^5 + 5p^4 + 10p^3 + 10p^2 + 5p + 1 \right]$

$\quad -4 \left[p^4 + 4p^3 + 6p^2 + 4p + 1 \right]$

$= p^5 + 5p^4 + 10p^3 + 10p^2 + 5p + 1 - 4p^4 - 16p^3 - 24p^2 - 16p - 4$

$= p^5 + p^4 - 6p^3 - 14p^2 - 11p - 3$

Chapter 13 Review

1. $a_n = -3n + 2$

$a_1 = -3(1) + 2 = -1$

$a_2 = -3(2) + 2 = -4$

$a_3 = -3(3) + 2 = -7$

$a_4 = -3(4) + 2 = -10$

$a_5 = -3(5) + 2 = -13$

The first five terms of the sequence are -1, -4, -7, -10, and -13.

2. $a_n = \dfrac{n-2}{n+4}$

$a_1 = \dfrac{1-2}{1+4} = \dfrac{-1}{5} = -\dfrac{1}{5}; \quad a_2 = \dfrac{2-2}{2+4} = \dfrac{0}{6} = 0;$

$a_3 = \dfrac{3-2}{3+4} = \dfrac{1}{7}; \quad a_4 = \dfrac{4-2}{4+4} = \dfrac{2}{8} = \dfrac{1}{4};$

$a_5 = \dfrac{5-2}{5+4} = \dfrac{3}{9} = \dfrac{1}{3}$

The first five terms of the sequence are $-\dfrac{1}{5}$, 0, $\dfrac{1}{7}$, $\dfrac{1}{4}$, and $\dfrac{1}{3}$.

3. $a_n = 5^n + 1$

$a_1 = 5^1 + 1 = 5 + 1 = 6$

$a_2 = 5^2 + 1 = 25 + 1 = 26$

$a_3 = 5^3 + 1 = 125 + 1 = 126$

$a_4 = 5^4 + 1 = 625 + 1 = 626$

$a_5 = 5^5 + 1 = 3125 + 1 = 3126$

The first five terms of the sequence are 6, 26, 126, 626, and 3126.

4. $a_n = (-1)^{n-1} \cdot 3n$

$a_1 = (-1)^{1-1} \cdot 3(1) = (-1)^0 \cdot 3 = 3$

$a_2 = (-1)^{2-1} \cdot 3(2) = (-1)^1 \cdot 6 = -6$

$a_3 = (-1)^{3-1} \cdot 3(3) = (-1)^2 \cdot 9 = 9$

$a_4 = (-1)^{4-1} \cdot 3(4) = (-1)^3 \cdot 12 = -12$

$a_5 = (-1)^{5-1} \cdot 3(5) = (-1)^4 \cdot 15 = 15$

The first five terms of the sequence are 3, -6, 9, -12, and 15.

5. $a_n = \dfrac{n^2}{n+1}$

$a_1 = \dfrac{1^2}{1+1} = \dfrac{1}{2}, \qquad a_2 = \dfrac{2^2}{2+1} = \dfrac{4}{3},$

$a_3 = \dfrac{3^2}{3+1} = \dfrac{9}{4}, \qquad a_4 = \dfrac{4^2}{4+1} = \dfrac{16}{5},$

$a_5 = \dfrac{5^2}{5+1} = \dfrac{25}{6}$

The first five terms of the sequence are $\dfrac{1}{2}$, $\dfrac{4}{3}$, $\dfrac{9}{4}$, $\dfrac{16}{5}$, and $\dfrac{25}{6}$.

6. $a_n = \dfrac{\pi^n}{n}$

$a_1 = \dfrac{\pi^1}{1} = \pi; \quad a_2 = \dfrac{\pi^2}{2};$

$a_3 = \dfrac{\pi^3}{3}; \quad a_4 = \dfrac{\pi^4}{4};$

$a_5 = \dfrac{\pi^5}{5}$

The first five terms of the sequence are π, $\dfrac{\pi^2}{2}$, $\dfrac{\pi^3}{3}$, $\dfrac{\pi^4}{4}$, and $\dfrac{\pi^5}{5}$.

7. The terms can be expressed as multiples of -3.

$-3 = -3 \cdot 1$

$-6 = -3 \cdot 2$

$-9 = -3 \cdot 3$

$-12 = -3 \cdot 4$

$-15 = -3 \cdot 5$

The nth term of the sequence is given by $a_n = -3n$.

8. The terms are rational numbers with a denominator of 3. The numerators are consecutive integers beginning with 1. The nth term of the sequence is given by $a_n = \dfrac{n}{3}$.

9. The terms can be expressed as the product of 5 and a power of 2.

$5 = 5 \cdot 2^0$

$10 = 5 \cdot 2^1$

$20 = 5 \cdot 2^2$

$40 = 5 \cdot 2^3$

$80 = 5 \cdot 2^4$

The exponent is one less than the term number. Therefore, the nth term of the sequence is given by $a_n = 5 \cdot 2^{n-1}$.

10. Rewrite as $-\dfrac{1}{2}$, $\dfrac{2}{2}$, $-\dfrac{3}{2}$, $\dfrac{4}{2}$, ...

The terms are rational numbers with alternating signs. The denominator is always 2, and the numerators are consecutive integers beginning with 1. The nth term of the sequence is given by $a_n = (-1)^n \cdot \dfrac{n}{2}$.

11. The terms can be expressed as 5 more than the square of the term number.

$6 = 1^2 + 5$

$9 = 2^2 + 5$

$14 = 3^2 + 5$

$21 = 4^2 + 5$

$30 = 5^2 + 5$

The nth term of the sequence is given by $a_n = n^2 + 5$.

12. Rewrite the terms as $\dfrac{0}{2}, \dfrac{1}{3}, \dfrac{2}{4}, \dfrac{3}{5}, \ldots$

The terms are rational numbers whose numerators are consecutive integers beginning with 0 and whose denominators are consecutive integers beginning with 2. The nth term of the sequence is given by $a_n = \dfrac{n-1}{n+1}$.

13. $\displaystyle\sum_{k=1}^{5}(5k-2)$

$= \big(5(1)-2\big) + \big(5(2)-2\big) + \big(5(3)-2\big)$

$\quad + \big(5(4)-2\big) + \big(5(5)-2\big)$

$= 3 + 8 + 13 + 18 + 23$

$= 65$

14. $\displaystyle\sum_{k=1}^{6}\left(\dfrac{k+2}{2}\right)$

$= \dfrac{1+2}{2} + \dfrac{2+2}{2} + \dfrac{3+2}{2} + \dfrac{4+2}{2} + \dfrac{5+2}{2} + \dfrac{6+2}{2}$

$= \dfrac{3}{2} + \dfrac{4}{2} + \dfrac{5}{2} + \dfrac{6}{2} + \dfrac{7}{2} + \dfrac{8}{2}$

$= \dfrac{33}{2}$

15. $\displaystyle\sum_{i=1}^{5}(-2i)$

$= (-2 \cdot 1) + (-2 \cdot 2) + (-2 \cdot 3) + (-2 \cdot 4) + (-2 \cdot 5)$

$= -2 - 4 - 6 - 8 - 10$

$= -30$

16. $\displaystyle\sum_{i=1}^{4}\dfrac{i^2-1}{3} = \dfrac{1^2-1}{3} + \dfrac{2^2-1}{3} + \dfrac{3^2-1}{3} + \dfrac{4^2-1}{3}$

$= \dfrac{0}{3} + \dfrac{3}{3} + \dfrac{8}{3} + \dfrac{15}{3}$

$= \dfrac{26}{3}$

17. Each term takes on the form $4+3i$ and there are $n = 15$ terms.

$(4+3 \cdot 1) + (4+3 \cdot 2) + \ldots + (4+3 \cdot 15) = \displaystyle\sum_{i=1}^{15}(4+3i)$

18. Each term takes on the form $\dfrac{1}{3^i}$ and there are $n = 8$ terms.

$\dfrac{1}{3^1} + \dfrac{1}{3^2} + \ldots + \dfrac{1}{3^8} = \displaystyle\sum_{i=1}^{8}\dfrac{1}{3^i}$

19. Each term takes on the form $\dfrac{i^3+1}{i+1}$ and there are $n = 10$ terms.

$\dfrac{1^3+1}{1+1} + \dfrac{2^3+1}{2+1} + \ldots + \dfrac{10^3+1}{10+1} = \displaystyle\sum_{i=1}^{10}\dfrac{i^3+1}{i+1}$

20. This is an alternating series. Each term takes on the form $(-1)^{i-1} \cdot i^2$ and there are $n = 7$ terms.

$(-1)^{1-1} \cdot 1^2 + (-1)^{2-1} \cdot 2^2 + \ldots + (-1)^{7-1} \cdot 7^2$

$= \displaystyle\sum_{i=1}^{7}\left[(-1)^{i-1} \cdot i^2\right]$

21. $10 - 4 = 6$

$16 - 10 = 6$

$22 - 16 = 6$

The sequence is arithmetic with a common difference of $d = 6$.

22. $\dfrac{1}{2} - (-1) = \dfrac{3}{2}$

$2 - \dfrac{1}{2} = \dfrac{3}{2}$

$\dfrac{7}{2} - 2 = \dfrac{3}{2}$

The sequence is arithmetic with a common difference of $d = \dfrac{3}{2}$.

23. $-5 - (-2) = -3$

$-9 - (-5) = -4$

$-14 - (-9) = -5$

The differences between consecutive terms is not constant. Therefore, the sequence is not arithmetic.

24. $3 - (-1) = 4$

$-5 - 3 = -8$

$7 - (-5) = 12$

The difference between consecutive terms is not constant. Therefore, the sequence is not arithmetic.

25. $a_n - a_{n-1} = [4n+7] - [4(n-1)+7]$
$$= 4n+7 - (4n-4+7)$$
$$= 4n+7 - 4n+4-7$$
$$= 4$$

The sequence is arithmetic with a common difference of $d = 4$.

26. $a_n - a_{n-1} = \left[\dfrac{n+1}{2n}\right] - \left[\dfrac{(n-1)+1}{2(n-1)}\right]$
$$= \dfrac{n+1}{2n} - \dfrac{n}{2(n-1)}$$
$$= \dfrac{(n+1)(n-1)}{2n(n-1)} - \dfrac{n(n)}{2n(n-1)}$$
$$= \dfrac{n^2 - 1 - n^2}{2n(n-1)}$$
$$= \dfrac{-1}{2n(n-1)}$$

The difference between consecutive terms is not constant. Therefore, the sequence is not arithmetic.

27. $a_n = a + (n-1)d$
$$= 3 + (n-1)(8)$$
$$= 3 + 8n - 8$$
$$= 8n - 5$$

$a_{25} = 8(25) - 5 = 200 - 5 = 195$

28. $a_n = a + (n-1)d$
$$= -4 + (n-1)(-3)$$
$$= -4 - 3n + 3$$
$$= -3n - 1$$

$a_{25} = -3(25) - 1 = -75 - 1 = -76$

29. $d = \dfrac{20}{3} - 7 = -\dfrac{1}{3}$; $a = 7$

$a_n = a + (n-1)d$
$$= 7 + (n-1)\left(-\dfrac{1}{3}\right)$$
$$= 7 - \dfrac{1}{3}n + \dfrac{1}{3}$$
$$= -\dfrac{1}{3}n + \dfrac{22}{3}$$

$a_{25} = -\dfrac{1}{3}(25) + \dfrac{22}{3} = -\dfrac{3}{3} = -1$

30. $d = 17 - 11 = 6$; $a = 11$

$a_n = a + (n-1)d$
$$= 11 + (n-1)(6)$$
$$= 11 + 6n - 6$$
$$= 6n + 5$$

$a_{25} = 6(25) + 5 = 150 + 5 = 155$

31. Since $a_3 = 7$ and $a_8 = 25$, we have

$$\begin{cases} 7 = a + (3-1)d \\ 25 = a + (8-1)d \end{cases} \text{ or } \begin{cases} 7 = a + 2d \\ 25 = a + 7d \end{cases}$$

Subtract the second equation from the first equation to obtain

$-18 = -5d$

$\dfrac{18}{5} = d$

Let $d = \dfrac{18}{5}$ in the first equation to find a.

$$7 = a + 2\left(\dfrac{18}{5}\right)$$
$$7 = a + \dfrac{36}{5}$$
$$-\dfrac{1}{5} = a$$

The nth term of the sequence is given by

$a_n = a + (n-1)d$
$$= -\dfrac{1}{5} + (n-1)\left(\dfrac{18}{5}\right)$$
$$= -\dfrac{1}{5} + \dfrac{18}{5}n - \dfrac{18}{5}$$
$$= \dfrac{18}{5}n - \dfrac{19}{5}$$

$a_{25} = \dfrac{18}{5}(25) - \dfrac{19}{5} = \dfrac{450}{5} - \dfrac{19}{5} = \dfrac{431}{5}$

32. Since $a_4 = -20$ and $a_7 = -32$, we have

$$\begin{cases} -20 = a + (4-1)d \\ -32 = a + (7-1)d \end{cases} \text{ or } \begin{cases} -20 = a + 3d \\ -32 = a + 6d \end{cases}$$

Subtract the second equation from the first equation to obtain

$12 = -3d$

$-4 = d$

Let $d = -4$ in the first equation to find a.

$-20 = a + 3(-4)$

$-20 = a - 12$

$-8 = a$

The nth term of the sequence is given by

$$a_n = a + (n-1)d$$
$$= -8 + (n-1)(-4)$$
$$= -8 - 4n + 4$$
$$= -4n - 4$$

$$a_{25} = -4(25) - 4 = -100 - 4 = -104$$

33. We know the first term is $a = -1$ and the common difference is $d = 9 - (-1) = 10$.

$$S_n = \frac{n}{2}\big[2a + (n-1)d\big]$$
$$S_{30} = \frac{30}{2}\big[2(-1) + (30-1)(10)\big]$$
$$= 15\big[-2 + 29(10)\big]$$
$$= 15(-2 + 290)$$
$$= 15(288)$$
$$= 4320$$

34. We know the first term is $a = 5$ and the common difference is $d = 2 - 5 = -3$.

$$S_n = \frac{n}{2}\big[2a + (n-1)d\big]$$
$$S_{40} = \frac{40}{2}\big[2(5) + (40-1)(-3)\big]$$
$$= 20\big[10 + 39(-3)\big]$$
$$= 20(10 - 117)$$
$$= 20(-107)$$
$$= -2140$$

35. $a_n = -2n - 7$

$$a = a_1 = -2(1) - 7 = -9$$
$$a_{60} = -2(60) - 7 = -127$$

$$S_n = \frac{n}{2}\big[a + a_n\big]$$
$$S_{60} = \frac{60}{2}\big[-9 + (-127)\big]$$
$$= 30(-136)$$
$$= -4080$$

36. $a_n = \dfrac{1}{4}n + 3$

$$a = a_1 = \frac{1}{4}(1) + 3 = \frac{13}{4}$$
$$a_{50} = \frac{1}{4}(50) + 3 = \frac{62}{4}$$
$$S_n = \frac{n}{2}\big[a + a_n\big]$$
$$S_{50} = \frac{50}{2}\left[\frac{13}{4} + \frac{62}{4}\right]$$
$$= 25\left(\frac{75}{4}\right)$$
$$= \frac{1875}{4} \quad \text{or} \quad 468.75$$

37. We could list terms of the sequence formed by the years when the Brood X cicada returns, or we could use the general formula for the nth term. We have $a = 2004$ and $d = 17$. We wish to know the first time when $a_n \geq 2101$.

$$a_n \geq 2101$$
$$a + (n-1)d \geq 2101$$
$$2004 + (n-1)(17) \geq 2101$$
$$2004 + 17n - 17 \geq 2101$$
$$17n + 1987 \geq 2101$$
$$17n \geq 114$$
$$n \geq \frac{114}{17} \approx 6.71$$

The Brood X cicada will first appear in the 22nd century when $n = 7$.

$$a_7 = 2004 + (7-1)(17) = 2106$$

The Brood X cicada will first appear in the 22nd century in the year 2106.

38. Since the distance to each line is the same as the distance back to the goal line, we can consider the sequence of distances run away from the goal line and double our result. The sequence of distances is 10, 20, 30, 40, and 50. This is an arithmetic sequence with first term $a = 10$ and fifth term $a_5 = 50$.

$$S_5 = \frac{5}{2}[10 + 50] = \frac{5}{2}(60) = 150$$

Doubling this result gives a total distance of 300 yards run by each player during wind sprints.

39. $\dfrac{\frac{2}{1}}{\frac{1}{3}} = 2 \cdot \dfrac{3}{1} = 6$

$\dfrac{12}{2} = 6$

$\dfrac{72}{12} = 6$

The ratio of consecutive terms is constant so the sequence is geometric with common ratio $r = 6$.

40. $\dfrac{3}{-1} = -3$

$\dfrac{-9}{3} = -3$

$\dfrac{27}{-9} = -3$

The ratio of consecutive terms is constant so the sequence is geometric with common ratio $r = -3$.

41. $\dfrac{1}{1} = 1$

$\dfrac{2}{1} = 2$

$\dfrac{6}{2} = 3$

The ratio of consecutive terms is not constant so the sequence is not geometric.

42. $\dfrac{4}{6} = \dfrac{2}{3}$

$\dfrac{\frac{8}{3}}{4} = \dfrac{8}{3} \cdot \dfrac{1}{4} = \dfrac{2}{3}$

$\dfrac{\frac{16}{9}}{\frac{8}{3}} = \dfrac{16}{9} \cdot \dfrac{3}{8} = \dfrac{2}{3}$

The ratio of consecutive terms is constant so the sequence is geometric with common ratio $r = \dfrac{2}{3}$.

43. $\dfrac{a_n}{a_{n-1}} = \dfrac{5 \cdot (-2)^n}{5 \cdot (-2)^{n-1}} = \dfrac{(-2)^{n-1} \cdot (-2)}{(-2)^{n-1}} = -2$

The ratio of consecutive terms is constant so the sequence is geometric with common ratio $r = -2$.

44. $\dfrac{a_n}{a_{n-1}} = \dfrac{3n-14}{3(n-1)-14} = \dfrac{3n-14}{3n-17}$

The ratio of consecutive terms is not constant so the sequence is not geometric.

45. $a_n = a \cdot r^{n-1}$

$a_n = 4 \cdot 3^{n-1}$

$a_{10} = 4 \cdot 3^{10-1} = 4 \cdot 3^9 = 4(19,683) = 78,732$

46. $a_n = a \cdot r^{n-1}$

$a_n = 8 \cdot \left(\dfrac{1}{4}\right)^{n-1}$

$a_{10} = 8 \cdot \left(\dfrac{1}{4}\right)^{10-1}$

$= 8 \cdot \left(\dfrac{1}{4}\right)^9$

$= 8 \cdot \left(\dfrac{1}{262,144}\right)$

$= \dfrac{1}{32,768}$

47. $a_n = a \cdot r^{n-1}$

$a_n = 5 \cdot (-2)^{n-1}$

$a_{10} = 5 \cdot (-2)^{10-1} = 5 \cdot (-2)^9 = 5(-512) = -2560$

48. $a_n = a \cdot r^{n-1}$

$a_n = 1000 \cdot (1.08)^{n-1}$

$a_{10} = 1000 \cdot (1.08)^{10-1}$

$= 1000 \cdot (1.08)^9$

$\approx 1000(1.999005)$

$= 1999.005$

49. This is a geometric series with first term $a = 2$, common ratio $r = \dfrac{4}{2} = 2$, and $n = 15$ terms.

$S_n = a \cdot \dfrac{1-r^n}{1-r}$

$S_{15} = 2 \cdot \dfrac{1-2^{15}}{1-2}$

$= 2 \cdot \dfrac{-32,767}{-1}$

$= 65,534$

50. This is a geometric series with first term $a = 40$, common ratio $r = \dfrac{1}{8}$, and $n = 13$ terms.

$$S_n = a \cdot \dfrac{1 - r^n}{1 - r}$$

$$S_{13} = 40 \cdot \dfrac{1 - \left(\dfrac{1}{8}\right)^{13}}{1 - \dfrac{1}{8}}$$

$$\approx 45.71428571$$

51. This is a geometric series with $n = 12$ terms. The first term is $a = \dfrac{3}{4} \cdot 2^{1-1} = \dfrac{3}{4}$ and the common ratio is $r = 2$.

$$S_n = a \cdot \dfrac{1 - r^n}{1 - r}$$

$$S_{12} = \dfrac{3}{4} \cdot \dfrac{1 - 2^{12}}{1 - 2}$$

$$= \dfrac{3}{4} \cdot \dfrac{-4095}{-1}$$

$$= \dfrac{12,285}{4} \quad \text{or} \quad 3071.25$$

52. This is a geometric series with $n = 16$ terms. The first term is $a = -4\left(3^1\right) = -12$ and the common ratio is $r = 3$.

$$S_n = a \cdot \dfrac{1 - r^n}{1 - r}$$

$$S_{16} = -12 \cdot \dfrac{1 - 3^{16}}{1 - 3}$$

$$= -12(21,523,360)$$

$$= -258,280,320$$

53. This is an infinite geometric series with first term $a = 20 \cdot \left(\dfrac{1}{4}\right)^1 = 5$ and common ratio $r = \dfrac{1}{4}$.

Since the ratio is between -1 and 1, we can use the formula for the sum of an infinite geometric series.

$$S_\infty = \dfrac{a}{1 - r} = \dfrac{5}{1 - \dfrac{1}{4}} = \dfrac{5}{\dfrac{3}{4}} = 5 \cdot \dfrac{4}{3} = \dfrac{20}{3}$$

54. This is an infinite geometric series with first term $a = 50 \cdot \left(-\dfrac{1}{2}\right)^{1-1} = 50$ and common ratio $r = -\dfrac{1}{2}$. Since the ratio is between -1 and 1, we can use the formula for the sum of an infinite geometric series.

$$S_\infty = \dfrac{a}{1 - r} = \dfrac{50}{1 - \left(-\dfrac{1}{2}\right)} = \dfrac{50}{\dfrac{3}{2}} = 50 \cdot \dfrac{2}{3} = \dfrac{100}{3}$$

55. This is an infinite geometric series with first term $a = 1$ and common ratio $r = \dfrac{\dfrac{1}{5}}{1} = \dfrac{1}{5}$. The ratio is between -1 and 1, so we can use the formula for the sum of an infinite geometric series.

$$S_\infty = \dfrac{a}{1 - r} = \dfrac{1}{1 - \dfrac{1}{5}} = \dfrac{1}{\dfrac{4}{5}} = \dfrac{5}{4}$$

56. This is an infinite geometric series with first term $a = 0.8$ and common ratio $r = \dfrac{0.08}{0.8} = 0.1$. Since the ratio is between -1 and 1, we can use the formula for the sum of an infinite geometric series.

$$S_\infty = \dfrac{a}{1 - r} = \dfrac{0.8}{1 - 0.1} = \dfrac{0.8}{0.9} = \dfrac{8}{9}$$

57. Here we consider a geometric sequence with first term $a = 200$ and common ratio $r = \dfrac{1}{2}$. Each term in the sequence represents the amount after each half-life, or the amount remaining after every 12 years. Since $\dfrac{72}{12} = 6$, we are interested in the 7th term of the sequence (that is, after 6 common ratios).

$$a_n = a \cdot r^{n-1}$$

$$a_7 = 200 \cdot \left(\dfrac{1}{2}\right)^{7-1}$$

$$= 200 \cdot \left(\dfrac{1}{2}\right)^6$$

$$= \dfrac{200}{64}$$

$$= 3.125$$

After 72 years, there will be 3.125 grams of the Tritium remaining.

58. The number of emails in each cycle forms a geometric sequence with first term $a = 5$ and common ratio $r = 5$. To find the total e-mails sent after 15 minutes, we need to find the sum of the first 15 terms of this sequence.

$$S_n = a \cdot \frac{1 - r^n}{1 - r}$$

$$S_{15} = 5 \cdot \frac{1 - 5^{15}}{1 - 5}$$

$$\approx 3.815 \times 10^{10}$$

After 15 minutes a total of about 38.15 billion e-mails will have been sent.

59. $A = P \cdot \left[\dfrac{(1 + i)^n - 1}{i} \right]$

Note that the total contribution each quarter is the sum of Scott's contribution and the matching contribution from his employer.
In this case we have $P = 900 + 450 = 1350$,

$n = 25 \cdot 4 = 100$, and $i = \dfrac{0.07}{4} = 0.0175$.

$$A = 1350 \cdot \left[\frac{(1 + 0.0175)^{100} - 1}{0.0175} \right]$$

$$= 1350 \cdot [266.7517679]$$

$$= \$360,114.89$$

After 25 years, Scott's 403(b) will be worth $\$360,114.89$.

60. Lump sum option:
For this option, we use the compound interest formula.

$A = P(1 + i)^n$

In this case we have $P = \$28,000,000$,

$n = 26 \cdot 1 = 26$, and $i = \dfrac{0.065}{1} = 0.065$.

$A = 28,000,000 \cdot (1 + 0.065)^{26} = \$143,961,987.40$

Annuity option:
For this option, we use the annuity formula.

$A = P \cdot \left[\dfrac{(1 + i)^n - 1}{i} \right]$

In this case we have $P = 2,000,000$,

$n = 26 \cdot 1 = 26$, and $i = \dfrac{0.065}{1} = 0.065$.

$$A = 2,000,000 \cdot \left[\frac{(1 + 0.065)^{26} - 1}{0.065} \right]$$

$$= 2,000,000 \cdot [63.71537769]$$

$$= \$127,430,755.40$$

The lump sum option would yield more money after 26 years.

61. $A = P \cdot \left[\dfrac{(1 + i)^n - 1}{i} \right]$

In this case we have $A = 2,500,000$,

$n = 40 \cdot 12 = 480$, and $i = \dfrac{0.09}{12} = 0.0075$.

$$2,500,000 = P \cdot \left[\frac{(1 + 0.0075)^{480} - 1}{0.0075} \right]$$

$$2,500,000 = P \cdot [4681.320273]$$

$$P = \$534.04$$

Sheri would need to contribute $\$534.04$, or about $\$534$, each month to reach her goal.

62. $A = P \cdot \left[\dfrac{(1 + i)^n - 1}{i} \right]$

In this case we have $P = 400$, $n = 10 \cdot 12 = 120$,

and $i = \dfrac{0.0525}{12} = 0.004375$.

$$A = 400 \cdot \left[\frac{(1 + 0.004375)^{120} - 1}{0.004375} \right]$$

$$= 400 \cdot [157.3769632]$$

$$= \$62,950.79$$

$$\frac{62,950.79}{340} \approx 185.15$$

When Samantha turns 18, the plan will be worth $\$62,950.79$ and will cover about 185 credit hours.

63. $5! = 5 \cdot 4 \cdot 3 \cdot 2 \cdot 1 = 120$

64. $\dfrac{11!}{7!} = \dfrac{11 \cdot 10 \cdot 9 \cdot 8 \cdot 7!}{7!} = 11 \cdot 10 \cdot 9 \cdot 8 = 7920$

65. $\dfrac{10!}{6!} = \dfrac{10 \cdot 9 \cdot 8 \cdot 7 \cdot 6!}{6!} = 10 \cdot 9 \cdot 8 \cdot 7 = 5040$

66. $\dfrac{13!}{6!7!} = \dfrac{13\cdot12\cdot11\cdot10\cdot9\cdot8\cdot7!}{6\cdot5\cdot4\cdot3\cdot2\cdot1\cdot7!}$

$\qquad = \dfrac{13\cdot12\cdot11\cdot10\cdot9\cdot8}{6\cdot5\cdot4\cdot3\cdot2\cdot1}$

$\qquad = \dfrac{13\cdot\cancel{12}\cdot11\cdot\cancel{10}\cdot\cancel{9}^{3}\cdot\cancel{8}^{4}}{_{12}\cancel{6}\cdot\cancel{5}\cdot\cancel{4}\cancel{3}\cdot\cancel{2}\cdot1}$

$\qquad = 13\cdot11\cdot3\cdot4$

$\qquad = 1716$

67. $\dbinom{7}{3} = \dfrac{7!}{3!4!}$

$\qquad = \dfrac{7\cdot6\cdot5\cdot4!}{3\cdot2\cdot1\cdot4!}$

$\qquad = \dfrac{7\cdot6\cdot5}{3\cdot2\cdot1}$

$\qquad = 35$

68. $\dbinom{10}{5} = \dfrac{10!}{5!5!}$

$\qquad = \dfrac{10\cdot9\cdot8\cdot7\cdot6\cdot5!}{5!\cdot5\cdot4\cdot3\cdot2\cdot1}$

$\qquad = \dfrac{10\cdot9\cdot8\cdot7\cdot6}{5\cdot4\cdot3\cdot2\cdot1}$

$\qquad = 252$

69. $\dbinom{8}{8} = \dfrac{8!}{8!0!} = 1$

70. $\dbinom{6}{0} = \dfrac{6!}{0!6!} = 1$

71. $(z+1)^4 = \dbinom{4}{0}z^4 + \dbinom{4}{1}z^3 + \dbinom{4}{2}z^2 + \dbinom{4}{3}z + \dbinom{4}{4}$

$\qquad = 1\cdot z^4 + 4\cdot z^3 + 6\cdot z^2 + 4\cdot z + 1$

$\qquad = z^4 + 4z^3 + 6z^2 + 4z + 1$

72. $(y-3)^5 = \dbinom{5}{0}y^5 + \dbinom{5}{1}y^4\cdot(-3) + \dbinom{5}{2}y^3\cdot(-3)^2 + \dbinom{5}{3}y^2\cdot(-3)^3 + \dbinom{5}{4}y\cdot(-3)^4 + \dbinom{5}{5}\cdot(-3)^5$

$\qquad = 1\cdot y^5 + 5\cdot y^4\cdot(-3) + 10\cdot y^3\cdot9 + 10\cdot y^2\cdot(-27) + 5\cdot y\cdot81 + (-243)$

$\qquad = y^5 - 15y^4 + 90y^3 - 270y^2 + 405y - 243$

73. $(3y+4)^6$

$= \dbinom{6}{0}(3y)^6 + \dbinom{6}{1}(3y)^5(4) + \dbinom{6}{2}(3y)^4(4)^2 + \dbinom{6}{3}(3y)^3(4)^3 + \dbinom{6}{4}(3y)^2(4)^4 + \dbinom{6}{5}(3y)(4)^5 + \dbinom{6}{6}(4)^6$

$= 1\cdot729y^6 + 6\cdot243y^5\cdot4 + 15\cdot81y^4\cdot16 + 20\cdot27y^3\cdot64 + 15\cdot9y^2\cdot256 + 6\cdot3y\cdot1024 + 4096$

$= 729y^6 + 5832y^5 + 19{,}440y^4 + 34{,}560y^3 + 34{,}560y^2 + 18{,}432y + 4096$

74. $\left(2x^2-3\right)^4 = \binom{4}{0}\left(2x^2\right)^4 + \binom{4}{1}\left(2x^2\right)^3(-3) + \binom{4}{2}\left(2x^2\right)^2(-3)^2 + \binom{4}{3}\left(2x^2\right)(-3)^3 + \binom{4}{4}(-3)^4$

$= 16x^8 + 4\cdot 8x^6\cdot(-3) + 6\cdot 4x^4\cdot 9 + 4\cdot 2x^2\cdot(-27) + 81$

$= 16x^8 - 96x^6 + 216x^4 - 216x^2 + 81$

75. $\left(3p-2q\right)^4 = \binom{4}{0}(3p)^4 + \binom{4}{1}(3p)^3(-2q) + \binom{4}{2}(3p)^2(-2q)^2 + \binom{4}{3}(3p)(-2q)^3 + \binom{4}{4}(-2q)^4$

$= 81p^4 + 4\cdot 27p^3\cdot(-2q) + 6\cdot 9p^2\cdot\left(4q^2\right) + 4\cdot 3p\cdot\left(-8q^3\right) + 16q^4$

$= 81p^4 - 216p^3q + 216p^2q^2 - 96pq^3 + 16q^4$

76. $\left(a^3+3b\right)^5 = \binom{5}{0}\left(a^3\right)^5 + \binom{5}{1}\left(a^3\right)^4(3b) + \binom{5}{2}\left(a^3\right)^3(3b)^2 + \binom{5}{3}\left(a^3\right)^2(3b)^3 + \binom{5}{4}\left(a^3\right)(3b)^4 + \binom{5}{5}(3b)^5$

$= a^{15} + 5a^{12}\cdot 3b + 10a^9\cdot 9b^2 + 10a^6\cdot 27b^3 + 5a^3\cdot 81b^4 + 243b^5$

$= a^{15} + 15a^{12}b + 90a^9b^2 + 270a^6b^3 + 405a^3b^4 + 243b^5$

77. The fourth term of the expansion of $(x-2)^8 = (x+(-2))^8$ is

$\binom{8}{3}(-2)^3 x^5 = 56x^5\cdot(-8) = -448x^5$

78. The seventh term of the expansion of $(2x+1)^{11} = ((2x)+1)^{11}$ is

$\binom{11}{6}1^6(2x)^5 = 462\cdot 32x^5 = 14{,}784x^5$

Chapter 13 Test

1. $-7-(-15) = 8$

$1-(-7) = 8$

$9-1 = 8$

The difference between consecutive terms is a constant so the sequence is arithmetic with $a = -15$ and common difference $d = 8$.

2. $\dfrac{a_n}{a_{n-1}} = \dfrac{(-4)^n}{(-4)^{n-1}} = \dfrac{-4\cdot(-4)^{n-1}}{(-4)^{n-1}} = -4$

The ratio of consecutive terms is a constant so the sequence is geometric with $a = (-4)^1 = -4$ and common ratio $r = -4$.

3. $a_n = \dfrac{4}{n!}$; $a_{n-1} = \dfrac{4}{(n-1)!}$

$a_n - a_{n-1} = \dfrac{4}{n!} - \dfrac{4}{(n-1)!} = \dfrac{4}{n!} - \dfrac{4n}{n(n-1)!} \qquad \dfrac{a_n}{a_{n-1}} = \dfrac{\frac{4}{n!}}{\frac{4}{(n-1)!}} = \dfrac{4}{n!}\cdot\dfrac{(n-1)!}{4} = \dfrac{1}{n}$

$= \dfrac{4-4n}{n!}$

The sequence is neither arithmetic nor geometric. The difference between consecutive terms is not constant so the sequence is not arithmetic. Likewise, the ratio of consecutive terms is not constant so the sequence is not geometric.

4. $a_n - a_{n-1} = \left[\dfrac{2n-3}{5}\right] - \left[\dfrac{2(n-1)-3}{5}\right]$

$= \dfrac{2n-3-(2n-2-3)}{5}$

$= \dfrac{2n-3-2n+5}{5} = \dfrac{2}{5}$

The difference between consecutive terms is constant so the sequence is arithmetic with

$a = \dfrac{2(1)-3}{5} = -\dfrac{1}{5}$ and common difference

$d = \dfrac{2}{5}$.

5. $2-(-3) = 5$

$0-2 = 2$

$\dfrac{2}{-3} = -\dfrac{2}{3}$

$\dfrac{0}{2} = 0$

The sequence is neither arithmetic nor geometric. The difference between consecutive terms is not constant so the sequence is not arithmetic. Likewise, the ratio of consecutive terms is not constant so the sequence is not geometric.

6. $\dfrac{a_n}{a_{n-1}} = \dfrac{7 \cdot 3^n}{7 \cdot 3^{n-1}} = \dfrac{7 \cdot 3 \cdot 3^{n-1}}{7 \cdot 3^{n-1}} = 3$

The ratio of consecutive terms is a constant so the sequence is geometric with $a = 7 \cdot 3^1 = 21$ and common ratio $r = 3$.

7. $\displaystyle\sum_{i=1}^{5}\left[\dfrac{3}{i^2}+2\right]$

$= \left(\dfrac{3}{1^2}+2\right) + \left(\dfrac{3}{2^2}+2\right) + \left(\dfrac{3}{3^2}+2\right) + \left(\dfrac{3}{4^2}+2\right) + \left(\dfrac{3}{5^2}+2\right)$

$= 3+2+\dfrac{3}{4}+2+\dfrac{1}{3}+2+\dfrac{3}{16}+2+\dfrac{3}{25}+2$

$= 13+\dfrac{900+400+225+144}{1200} = 13+\dfrac{1669}{1200}$

$= 14\dfrac{469}{1200}$ or $\dfrac{17269}{1200}$

8. Note that $\dfrac{2}{3} = \dfrac{4}{6}$, $\dfrac{3}{4} = \dfrac{6}{8}$, and $\dfrac{5}{6} = \dfrac{10}{12}$.

Each term can be written as $\dfrac{i+2}{i+4}$, where i is the term number, and there are $n = 8$ terms.

$\dfrac{3}{5}+\dfrac{2}{3}+\dfrac{5}{7}+\dfrac{3}{4}+\ldots+\dfrac{5}{6} = \displaystyle\sum_{i=1}^{8}\dfrac{i+2}{i+4}$

9. $a_n = a+(n-1)d$

$= 6+(n-1)(10)$

$= 6+10n-10$

$= 10n-4$

$a_1 = 10 \cdot 1 - 4 = 6$

$a_2 = 10 \cdot 2 - 4 = 16$

$a_3 = 10 \cdot 3 - 4 = 26$

$a_4 = 10 \cdot 4 - 4 = 36$

$a_5 = 10 \cdot 5 - 4 = 46$

The first five terms of the sequence are 6, 16, 26, 36, and 46.

10. $a_n = a+(n-1)d$

$= 0+(n-1)(-4)$

$= -4n+4$

$= 4-4n$

$a_1 = 4-4 \cdot 1 = 0$

$a_2 = 4-4 \cdot 2 = -4$

$a_3 = 4-4 \cdot 3 = -8$

$a_4 = 4-4 \cdot 4 = -12$

$a_5 = 4-4 \cdot 5 = -16$

The first five terms of the sequence are 0, -4, -8, -12, and -16.

11. $a_n = a \cdot r^{n-1}$

$= 10 \cdot 2^{n-1}$

$a_1 = 10 \cdot 2^{1-1} = 10$

$a_2 = 10 \cdot 2^{2-1} = 20$

$a_3 = 10 \cdot 2^{3-1} = 40$

$a_4 = 10 \cdot 2^{4-1} = 80$

$a_5 = 10 \cdot 2^{5-1} = 160$

The first five terms of the sequence are 10, 20, 40, 80, and 160.

12. $a_n = a \cdot r^{n-1}$

   $a_3 = a \cdot (-3)^{3-1} = 9$

$a(-3)^2 = 9$

   $9a = 9$

   $a = 1$

$a_n = 1 \cdot (-3)^{n-1}$

   $= (-3)^{n-1}$

$a_1 = (-3)^{1-1} = 1$

$a_2 = (-3)^{2-1} = -3$

$a_3 = (-3)^{3-1} = 9$

$a_4 = (-3)^{4-1} = -27$

$a_5 = (-3)^{5-1} = 81$

The first five terms of the sequence are 1, -3, 9, -27, and 81.

13. The terms form an arithmetic sequence with $n = 20$ terms, common difference $d = 4$, and first term $a = -2$.

$$S_n = \frac{n}{2}\left[2a + (n-1)d\right]$$

$$S_{20} = \frac{20}{2}\left[2(-2) + (20-1)(4)\right]$$

$$= 10(72)$$

$$= 720$$

14. The terms form a geometric sequence with $n = 12$ terms, common ratio $r = -3$, and first term $a = \frac{1}{9}$.

$$S_n = a \cdot \frac{1-r^n}{1-r}$$

$$S_{12} = \frac{1}{9} \cdot \frac{1-(-3)^{12}}{1-(-3)}$$

$$= \frac{1}{9} \cdot \frac{1-3^{12}}{1+3}$$

$$= -\frac{132,860}{9}$$

15. The terms form an infinite geometric series with first term $a = 216$ and common ratio $r = \frac{1}{3}$.

$$S_\infty = \frac{a}{1-r} = \frac{216}{1-\left(\frac{1}{3}\right)} = \frac{216}{2/3} = 216 \cdot \frac{3}{2} = 324$$

16. $\dfrac{15!}{8!7!} = \dfrac{15 \cdot 14 \cdot 13 \cdot 12 \cdot 11 \cdot 10 \cdot 9 \cdot 8!}{8! \cdot 7 \cdot 6 \cdot 5 \cdot 4 \cdot 3 \cdot 2 \cdot 1}$

   $= \dfrac{15 \cdot 14 \cdot 13 \cdot 12 \cdot 11 \cdot 10 \cdot 9}{7 \cdot 6 \cdot 5 \cdot 4 \cdot 3 \cdot 2 \cdot 1}$

   $= 6435$

17. $\dbinom{12}{5} = \dfrac{12!}{5!7!} = \dfrac{12 \cdot 11 \cdot 10 \cdot 9 \cdot 8 \cdot 7!}{5 \cdot 4 \cdot 3 \cdot 2 \cdot 1 \cdot 7!}$

$\qquad = \dfrac{12 \cdot 11 \cdot 10 \cdot 9 \cdot 8}{5 \cdot 4 \cdot 3 \cdot 2 \cdot 1}$

$\qquad = 792$

18. $(5m-2)^4 = \dbinom{4}{0}(5m)^4 + \dbinom{4}{1}(5m)^3(-2) + \dbinom{4}{2}(5m)^2(-2)^2 + \dbinom{4}{3}(5m)(-2)^3 + \dbinom{4}{4}(-2)^4$

$\qquad = 625m^4 + 4 \cdot 125m^3 \cdot (-2) + 6 \cdot 25m^2 \cdot 4 + 4 \cdot 5m \cdot (-8) + 16$

$\qquad = 625m^4 - 1000m^3 + 600m^2 - 160m + 16$

19. The average tuition values form a geometric sequence with first term $a = 4694$ and common ratio $r = 1.14$. To determine the average tuition and fees for the 2023-2024 school year, we need the 21st term of the sequence.

$a_{21} = a \cdot r^{21-1}$

$\qquad = 4694 \cdot 1.14^{20}$

$\qquad \approx 64{,}512$

If the percent increase continues, the average tuition and fees for in-state students would be about \$64,512 during the 2023-2024 academic year.

20. $1639 - 1631 = 8$

$1761 - 1639 = 122$

$1769 - 1761 = 8$

$1874 - 1769 = 105$

$1882 - 1874 = 8$

$2004 - 1882 = 122$

Following this pattern, the next three Venus transits should occur after intervals of 8, 105, and 8 years.

$\qquad 2004 + 8 = 2012$

$2012 + 105 = 2117$

$\qquad 2117 + 8 = 2125$

The next three Venus transits should occur in 2012, 2117, and 2125.

Cumulative Review Chapters 1–13

1. $\dfrac{1}{2}(x+2) = \dfrac{5}{4}(x-3y)$

$4 \cdot \dfrac{1}{2}(x+2) = 4 \cdot \dfrac{5}{4}(x-3y)$

$\qquad 2(x+2) = 5(x-3y)$

$\qquad 2x+4 = 5x - 15y$

$\qquad 15y = 3x - 4$

$\qquad y = \dfrac{3x-4}{15} \quad$ or $\quad y = \dfrac{1}{5}x - \dfrac{4}{15}$

2. $f(x) = x^2 - x + 7$

$f(2) = (2)^2 - (2) + 7 = 4 - 2 + 7 = 9$

$f(-3) = (-3)^2 - (-3) + 7 = 9 + 3 + 7 = 19$

3.

$$\frac{1}{2}x - 2 = \frac{1}{3}(x+1) + 3$$

$$6\left(\frac{1}{2}x - 2\right) = 6\left(\frac{1}{3}(x+1) + 3\right)$$

$$3x - 12 = 2(x+1) + 18$$

$$3x - 12 = 2x + 2 + 18$$

$$3x - 2x = 20 + 12$$

$$x = 32$$

The solution set is $\{32\}$.

4.

$$5x^2 - 3x = 2$$

$$5x^2 - 3x - 2 = 0$$

$$(5x+2)(x-1) = 0$$

$$5x + 2 = 0 \quad \text{or} \quad x - 1 = 0$$

$$5x = -2 \qquad\qquad x = 1$$

$$x = -\frac{2}{5}$$

The solution set is $\left\{-\frac{2}{5}, 1\right\}$.

5. $3x^2 + 7x - 2 = 0$

$$a = 3, b = 7, c = -2$$

$$x = \frac{-b \pm \sqrt{b^2 - 4ac}}{2a}$$

$$= \frac{-7 \pm \sqrt{7^2 - 4(3)(-2)}}{2(3)}$$

$$= \frac{-7 \pm \sqrt{49 + 24}}{6}$$

$$= \frac{-7 \pm \sqrt{73}}{6}$$

The solution set is $\left\{\frac{-7 - \sqrt{73}}{6}, \frac{-7 + \sqrt{73}}{6}\right\}$.

6. $\sqrt{2x+1} - 3 = 8$

$$\sqrt{2x+1} = 11$$

$$\left(\sqrt{2x+1}\right)^2 = 11^2$$

$$2x + 1 = 121$$

$$2x = 120$$

$$x = 60$$

The solution set is $\{60\}$.

7.

$$4^{x+1} = 8^{2x-3}$$

$$\left(2^2\right)^{x+1} = \left(2^3\right)^{2x-3}$$

$$2^{2x+2} = 2^{6x-9}$$

Therefore, we get

$$2x + 2 = 6x - 9$$

$$-4x = -11$$

$$x = \frac{11}{4}$$

The solution set is $\left\{\frac{11}{4}\right\}$.

8. $x^2(2x+1) + 40 = \left(x^2 - 8\right)(x-5)$

$$2x^3 + x^2 + 40 = x^3 - 8x - 5x^2 + 40$$

$$x^3 + 6x^2 + 8x = 0$$

$$x\left(x^2 + 6x + 8\right) = 0$$

$$x(x+4)(x+2) = 0$$

$$x = 0 \quad \text{or} \quad x + 4 = 0 \quad \text{or} \quad x + 2 = 0$$

$$x = -4 \qquad\qquad x = -2$$

The solution set is $\{-4, -2, 0\}$.

9.

$$\frac{2}{3}x + 1 > \frac{1}{4}x - \frac{3}{2}$$

$$12\left(\frac{2}{3}x + 1\right) > 12\left(\frac{1}{4}x - \frac{3}{2}\right)$$

$$8x + 12 > 3x - 18$$

$$5x > -30$$

$$\frac{5x}{5} > \frac{-30}{5}$$

$$x > -6$$

Interval: $(-6, \infty)$

10.

$$3x^2 - 2x \le 3 - 10x$$

$$3x^2 + 8x - 3 \le 0$$

$$(3x-1)(x+3) \le 0$$

$$3x - 1 = 0 \qquad x + 3 = 0$$

$$3x = 1 \qquad\qquad x = -3$$

$$x = \frac{1}{3}$$

Interval	$(-\infty,-3)$	-3	$\left(-3,\frac{1}{3}\right)$	$\frac{1}{3}$	$\left(\frac{1}{3},\infty\right)$
$(3x-1)$	$---$	$-$	$---$	0	$+++$
$(x+3)$	$---$	0	$+++$	$+$	$+++$
$(3x-1)(x+3)$	$+++$	0	$---$	0	$+++$

The inequality is non-strict, so $\frac{1}{3}$ and -3 are part of the solution. Now, $(3x-1)(x+3)$ is less than zero where the product is negative. The solution is $\left\{x\middle|\ -3\le x\le\frac{1}{3}\right\}$ or, using interval notation, $\left[-3,\frac{1}{3}\right]$.

11. $2x^2-5x-18$

$ac=2(-18)=-36$

We are looking for two factors of -36 whose sum is -5. Since the product is negative, the factors will have opposite signs. The sum is negative so the factor with the largest absolute value will be negative.

factor 1	factor 2	sum
1	-36	-35
2	-18	-16
3	-12	-9
4	-9	$-5 \leftarrow$ okay

$2x^2-5x-18=2x^2+4x-9x-18$
$=2x(x+2)-9(x+2)$
$=(x+2)(2x-9)$

12. $6x^3-3x^2+4x-2=3x^2(2x-1)+2(2x-1)$
$=(2x-1)(3x^2+2)$

13. $(5x-3)(4x^2-2x+1)$
$=20x^3-10x^2+5x-12x^2+6x-3$
$=20x^3-22x^2+11x-3$

14. $\dfrac{x}{x+4}-\dfrac{3}{x-1}=\dfrac{x(x-1)}{(x+4)(x-1)}-\dfrac{3(x+4)}{(x+4)(x-1)}$

$=\dfrac{x(x-1)-3(x+4)}{(x+4)(x-1)}$

$=\dfrac{x^2-x-3x-12}{(x+4)(x-1)}$

$=\dfrac{x^2-4x-12}{(x+4)(x-1)}$

$=\dfrac{(x-6)(x+2)}{(x+4)(x-1)}$

15. $\dfrac{3-i}{2+i}=\dfrac{(3-i)(2-i)}{(2+i)(2-i)}$

$=\dfrac{6-2i-3i+i^2}{4-i^2}$

$=\dfrac{6-5i-1}{4-(-1)}$

$=\dfrac{5-5i}{5}$

$=1-i$

16. Because of the two radicals, we need
$x-15\ge0 \quad and \quad 2x-5\ge0$
$x\ge15 \qquad\qquad 2x\ge5$
$\qquad\qquad\qquad\qquad x\ge\dfrac{5}{2}$

Therefore, the domain is $\{x\,|\,x\ge15\}$ or $[15,\infty)$.

17. Begin by finding the slope of the line connecting the two points.

$m=\dfrac{y_2-y_1}{x_2-x_1}=\dfrac{4-(-3)}{1-2}=\dfrac{7}{-1}=-7$

$y-y_1=m(x-x_1)$
$y-(-3)=-7(x-2)$
$y+3=-7x+14$
$y=-7x+11$

18. $2x+3y=5$
$x-2y=6$

We can solve this system by using elimination. Multiply the second equation by -2 and add the two equations.

$\begin{array}{r} 2x+3y=5 \\ \underline{-2x+4y=-12} \\ 7y=-7 \\ y=-1 \end{array}$

Let $y = -1$ in the second equation to find x.

$$x - 2(-1) = 6$$
$$x + 2 = 6$$
$$x = 4$$

The ordered pair $(4, -1)$ is the solution to the system.

19. $f(x) = 2x^2 - 8x - 3$

The graph of f will be a parabola. The parabola will open up because $a = 2 > 0$.
The x-coordinate of the vertex is

$$x = -\frac{b}{2a} = -\frac{(-8)}{2(2)} = 2$$

The y-coordinate of the vertex is

$$f(2) = 2(2)^2 - 8(2) - 3 = -11$$

The vertex is $(2, -11)$ and the axis of symmetry is $x = 2$.

$$f(0) = 2(0)^2 - 8(0) - 3 = -3$$

The y-intercept is -3.

To find the x-intercepts we solve the equation $2x^2 - 8x - 3 = 0$.

$$x = \frac{-(-8) \pm \sqrt{(-8)^2 - 4(2)(-3)}}{2(2)}$$

$$= \frac{8 \pm \sqrt{88}}{4}$$

$$= \frac{8 \pm 2\sqrt{22}}{4}$$

$$\frac{4 \pm \sqrt{22}}{2}$$

The graph will have x-intercepts of $x \approx -0.35$ and $x \approx 4.35$.

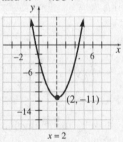

20.
$$(x - h)^2 + (y - k)^2 = r^2$$
$$(x - 4)^2 + (y - (-3))^2 = 6^2$$
$$(x - 4)^2 + (y + 3)^2 = 36$$

To sketch the graph of the circle, we can plot four additional points that are $r = 6$ units above, below, left, and right of the center. These points are $(4, 3), (4, -9), (-2, -3)$, and $(10, -3)$.

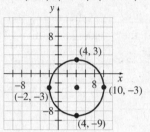

21. Begin by writing the equation in standard form.

$$\frac{4x^2}{64} + \frac{y^2}{64} = \frac{64}{64}$$

$$\frac{x^2}{16} + \frac{y^2}{64} = 1$$

The larger number, 64, is in the denominator of the y^2-term. This means that the major axis is the y-axis and the equation of the ellipse is of the form $\frac{x^2}{b^2} + \frac{y^2}{a^2} = 1$ so that $a^2 = 64$ and $b^2 = 16$.

The center of the ellipse is the origin, $(0, 0)$.

$c^2 = a^2 - b^2 = 64 - 16 = 48$, so that $c = \pm 4\sqrt{3}$.
The foci are $(0, -4\sqrt{3})$ and $(0, 4\sqrt{3})$.

$$\frac{x^2}{16} + \frac{0^2}{64} = 1 \qquad\qquad \frac{0^2}{16} + \frac{y^2}{64} = 1$$

$$\frac{x^2}{16} = 1 \qquad\qquad\qquad \frac{y^2}{64} = 1$$

$$x^2 = 16 \qquad\qquad\qquad y^2 = 64$$

$$x = \pm 4 \qquad\qquad\qquad y = \pm 8$$

The intercepts are $(-4, 0), (4, 0), (0, -8)$, and $(0, 8)$.

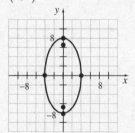

22. $S_n = \dfrac{n}{2}\left[2a + (n-1)d\right]$

$S_{20} = \dfrac{20}{2}\left[2(-47) + (20-1)(12)\right]$

$\phantom{S_{20}} = 10(134)$

$\phantom{S_{20}} = 1340$

23. For this infinite geometric series we have $a = 2$

and $r = \dfrac{3/2}{2} = \dfrac{3}{4}$.

$S_{\infty} = \dfrac{a}{1-r} = \dfrac{2}{1-\frac{3}{4}} = \dfrac{2}{\frac{1}{4}} = 8$

24.

Machine	# of lots	time (hrs)	rate
Robomower	1	5	$\dfrac{1}{5}$
Mowbot	1	6	$\dfrac{1}{6}$
Together	1	t	$\dfrac{1}{t}$

We can't add the times, but we can add the rates.

$\dfrac{1}{5} + \dfrac{1}{6} = \dfrac{1}{t}$

$\dfrac{11}{30} = \dfrac{1}{t}$

$t = \dfrac{30}{11} \approx 2.73$

It would take about 2.73 hours to cut the lot if both machines worked together.

25. Let $x =$ metric tons of pure aluminum .

We can solve the problem by writing an equation for the total amount of manganese.

$$\text{Mn}_{\text{tot}} = \text{Mn}_{\text{init}} + \text{Mn}_{\text{added}}$$

$$(\%)(\text{wt.})_{\text{tot}} = (\%)(\text{wt.})_{\text{init}} + (\%)(\text{wt.})_{\text{added}}$$

$$(1.2)(100 + x) = (2.5)(100) + (0)(x)$$

$$120 + 1.2x = 250$$

$$1.2x = 130$$

$$x = \dfrac{130}{1.2}$$

$$x = \dfrac{325}{3} \quad \text{or} \quad 108\dfrac{1}{3}$$

$108\dfrac{1}{3}$ metric tons of pure aluminum must be added.

Appendix A

Appendix A

Preparing for Synthetic Division

P1. $\dfrac{15x^5}{12x^3} = \dfrac{15}{12} \cdot \dfrac{x^5}{x^3} = \dfrac{5}{4}x^{5-3} = \dfrac{5}{4}x^2$

P2.

$$\begin{array}{r} x^2 + 3x - 10 \\ 2x+1\overline{\smash{\big)}\ 2x^3 + 7x^2 - 17x - 10} \end{array}$$

$$\underline{-(2x^3 +\ x^2)}$$
$$6x^2 - 17x$$
$$\underline{-(6x^2 +\ 3x)}$$
$$-20x - 10$$
$$\underline{-(-20x - 10)}$$
$$0$$

$$\dfrac{2x^3 + 7x^2 - 17x - 10}{2x+1} = x^2 + 3x - 10$$

Appendix A Quick Checks

1. False; to use synthetic division, the divisor must be linear and in the form $x - c$.

2. False; to use synthetic division, the divisor must be linear and in the form $x - c$.

3. The divisor is $x - 2$ so $c = 2$.

$$\begin{array}{r|rrr} 2 & 2 & 1 & -7 & -13 \\ & & 4 & 10 & 6 \\ \hline & 2 & 5 & 3 & -7 \end{array}$$

$$\dfrac{2x^3 + x^2 - 7x - 13}{x - 2} = 2x^2 + 5x + 3 - \dfrac{7}{x - 2}$$

4. The divisor is $x + 3$ so $c = -3$.

$$\begin{array}{r|rrrr} -3 & 1 & 8 & 15 & -2 & -6 \\ & & -3 & -15 & 0 & 6 \\ \hline & 1 & 5 & 0 & -2 & 0 \end{array}$$

$$\dfrac{x^4 + 8x^3 + 15x^2 - 2x - 6}{x + 3} = x^3 + 5x^2 - 2$$

5. $f(x) = 3x^3 + 10x^2 - 9x - 4$

   (a) The divisor is $x - 2$ so the Remainder Theorem says that the remainder is $f(2)$.

   $$\begin{aligned} f(2) &= 3(2)^3 + 10(2)^2 - 9(2) - 4 \\ &= 3(8) + 10(4) - 9(2) - 4 \\ &= 24 + 40 - 18 - 4 \\ &= 42 \end{aligned}$$

   When $f(x) = 3x^3 + 10x^2 - 9x - 4$ is divided by $x - 2$, the remainder is 42.

   (b) The divisor is $x + 4 = x - (-4)$, so the Remainder Theorem says that the remainder is $f(-4)$.

   $$\begin{aligned} f(-4) &= 3(-4)^3 + 10(-4)^2 - 9(-4) - 4 \\ &= 3(-64) + 10(16) - 9(-4) - 4 \\ &= -192 + 160 + 36 - 4 \\ &= 0 \end{aligned}$$

   When $f(x) = 3x^3 + 10x^2 - 9x - 4$ is divided by $x + 4$, the remainder is 0.

6. $f(x) = 2x^3 - 9x^2 - 6x + 5$

   (a)
   $$\begin{aligned} f(-2) &= 2(-2)^3 - 9(-2)^2 - 6(-2) + 5 \\ &= 2(-8) - 9(4) - 6(-2) + 5 \\ &= -16 - 36 + 12 + 5 \\ &= -35 \end{aligned}$$

   Since $f(c) = f(-2) \neq 0$, we know that $x - c = x + 2$ is not a factor of $f(x)$.

   (b)
   $$\begin{aligned} f(5) &= 2(5)^3 - 9(5)^2 - 6(5) + 5 \\ &= 2(125) - 9(25) - 6(5) + 5 \\ &= 250 - 225 - 30 + 5 \\ &= 0 \end{aligned}$$

   Since $f(c) = f(5) = 0$, we know that $x - c = x - 5$ is a factor of $f(x)$.

   $$\begin{array}{r|rrr} 5 & 2 & -9 & -6 & 5 \\ & & 10 & 5 & -5 \\ \hline & 2 & 1 & -1 & 0 \end{array}$$

   $$f(x) = (x - 5)\left(2x^2 + x - 1\right)$$

Appendix A Exercises

7. The divisor is $x-5$ so $c=5$.

$$
\begin{array}{r}
5\,\overline{)\,1 \quad -3 \quad -10} \\
\underline{\quad 5 \quad\; 10} \\
1 \quad\;\; 2 \quad\;\; 0
\end{array}
$$

$$\frac{x^2-3x-10}{x-5}=x+2$$

9. The divisor is $x+4$ so $c=-4$.

$$
\begin{array}{r}
-4\,\overline{)\,2 \quad 11 \quad\; 12} \\
\underline{\;\; -8 \quad -12} \\
2 \quad\;\; 3 \quad\;\;\; 0
\end{array}
$$

$$\frac{2x^2+11x+12}{x+4}=2x+3$$

11. The divisor is $x-6$ so $c=6$.

$$
\begin{array}{r}
6\,\overline{)\,1 \quad -3 \quad -14} \\
\underline{\quad 6 \quad\;\; 18} \\
1 \quad\;\; 3 \quad\;\;\; 4
\end{array}
$$

$$\frac{x^2-3x-14}{x-6}=x+3+\frac{4}{x-6}$$

13. The divisor is $x-5$ so $c=5$.

$$
\begin{array}{r}
5\,\overline{)\,1 \quad 0 \quad -19 \quad\; \cdot-15} \\
\underline{\quad 5 \quad\;\; 25 \quad\;\; 30} \\
1 \quad\; 5 \quad\;\;\; 6 \quad\;\;\; 15
\end{array}
$$

$$\frac{x^3-19x-15}{x-5}=x^2+5x+6+\frac{15}{x-5}$$

15. The divisor is $x-3$ so $c=3$.

$$
\begin{array}{r}
3\,\overline{)\,3 \quad -5 \quad -21 \quad\; 17 \quad\;\; 25} \\
\underline{\quad 9 \quad\;\; 12 \quad -27 \quad -30} \\
3 \quad\;\; 4 \quad -9 \quad -10 \quad -5
\end{array}
$$

$$\frac{3x^4-5x^3-21x^2+17x+25}{x-3}$$
$$=3x^3+4x^2-9x-10-\frac{5}{x-3}$$

17. The divisor is $x+6$ so $c=-6$.

$$
\begin{array}{r}
-6\,\overline{)\,1 \quad 0 \quad -40 \quad\; 0 \quad\;\; 109} \\
\underline{\quad -6 \quad 36 \quad\; 24 \quad -144} \\
1 \quad -6 \quad -4 \quad\; 24 \quad -35
\end{array}
$$

$$\frac{x^4-40x^2+109}{x+6}=x^3-6x^2-4x+24-\frac{35}{x+6}$$

19. The divisor is $x-\dfrac{5}{2}$ so $c=\dfrac{5}{2}$.

$$
\begin{array}{r}
\tfrac{5}{2}\,\overline{)\,2 \quad 3 \quad -14 \quad -15} \\
\underline{\quad 5 \quad\;\; 20 \quad\;\; 15} \\
2 \quad\; 8 \quad\;\;\; 6 \quad\;\;\; 0
\end{array}
$$

$$\frac{2x^3+3x^2-14x-15}{x-\dfrac{5}{2}}=2x^2+8x+6$$

21. The divisor is $x-2$ so $c=2$.

$$
\begin{aligned}
f(2)&=(2)^2-5(2)+1 \\
&=4-10+1 \\
&=-5
\end{aligned}
$$

The remainder is -5.

23. The divisor is $x+4$ so $c=-4$.

$$
\begin{aligned}
f(-4)&=(-4)^3-2(-4)^2+5(-4)-3 \\
&=-64-2(16)-20-3 \\
&=-64-32-20-3 \\
&=-119
\end{aligned}
$$

The remainder is -119.

25. The divisor is $x-5$ so $c=5$.

$$
\begin{aligned}
f(5)&=2(5)^3-4(5)+1 \\
&=2(125)-20+1 \\
&=250-20+1 \\
&=231
\end{aligned}
$$

The remainder is 231.

27. The divisor is $x-1$ so $c=1$.

$$
\begin{aligned}
f(1)&=(1)^4+1 \\
&=1+1 \\
&=2
\end{aligned}
$$

The remainder is 2.

29. $c=2$

$$
\begin{aligned}
f(2)&=(2)^2-3(2)+2 \\
&=4-6+2 \\
&=0
\end{aligned}
$$

Since the remainder is 0, $x-2$ is a factor.

$$
\begin{array}{r}
2\,\overline{)\,1 \quad -3 \quad\; 2} \\
\underline{\quad 2 \quad -2} \\
1 \quad -1 \quad\; 0
\end{array}
$$

$$f(x)=(x-2)(x-1)$$

31. $c = -2$

$$\begin{aligned} f(-2) &= 2(-2)^2 + 5(-2) + 2 \\ &= 2(4) - 10 + 2 \\ &= 8 - 10 + 2 \\ &= 0 \end{aligned}$$

Since the remainder is 0, $x + 2$ is a factor.

$$-2 \overline{)\begin{array}{ccc} 2 & 5 & 2 \\ & -4 & -2 \\ \hline 2 & 1 & 0 \end{array}}$$

$$f(x) = (x+2)(2x+1)$$

33. $c = 3$

$$\begin{aligned} f(3) &= 4(3)^3 - 9(3)^2 - 49(3) - 30 \\ &= 4(27) - 9(9) - 147 - 30 \\ &= 108 - 81 - 147 - 30 \\ &= -150 \end{aligned}$$

Since the remainder is not 0, $x - 3$ is not a factor of $f(x)$.

35. $c = -1$

$$\begin{aligned} f(-1) &= 4(-1)^3 - 7(-1)^2 - 5(-1) + 6 \\ &= -4 - 7 + 5 + 6 \\ &= 0 \end{aligned}$$

Since the remainder is 0, $x + 1$ is a factor.

$$-1 \overline{)\begin{array}{cccc} 4 & -7 & -5 & 6 \\ & -4 & 11 & -6 \\ \hline 4 & -11 & 6 & 0 \end{array}}$$

$$f(x) = (x+1)\left(4x^2 - 11x + 6\right)$$

37. $\dfrac{f(x)}{x-5} = 3x + 5$

$$\begin{aligned} f(x) &= (x-5)(3x+5) \\ &= x \cdot 3x + x \cdot 5 - 5 \cdot 3x - 5 \cdot 5 \\ &= 3x^2 + 5x - 15x - 25 \\ &= 3x^2 - 10x - 25 \end{aligned}$$

39. $\dfrac{f(x)}{x-3} = x + 8 + \dfrac{4}{x-3}$

$$\begin{aligned} f(x) &= (x-3)\left(x + 8 + \frac{4}{x-3}\right) \\ &= x(x-3) + 8(x-3) + \frac{4}{x-3}(x-3) \\ &= x^2 - 3x + 8x - 24 + 4 \\ &= x^2 + 5x - 20 \end{aligned}$$

41. $\dfrac{2x^3 - 3x^2 - 26x - 37}{x+2} = ax^2 + bx + c + \dfrac{d}{x+2}$

Since the divisor is linear, we can use synthetic division.

$$-2 \overline{)\begin{array}{cccc} 2 & -3 & -26 & -37 \\ & -4 & 14 & 24 \\ \hline 2 & -7 & -12 & -13 \end{array}}$$

$$\frac{2x^3 - 3x^2 - 26x - 37}{x+2} = 2x^2 - 7x - 12 + \frac{-13}{x+2}.$$

This gives us $a = 2$, $b = -7$, $c = -12$, and $d = -13$.

Thus,

$$a + b + c + d = 2 + (-7) + (-12) + (-13) = -30$$

43. For polynomial division we have

dividend = divisor $\cdot$ quotient + remainder

The dividend is the polynomial f and has degree n. Since the remainder must be 0 or a polynomial that has a lower degree than f, the degree n must be obtained from the product of the divisor and the quotient. The divisor is $x + 4$ which is of degree 1 so the quotient must be of degree $n - 1$.

45. Answers will vary.

Appendix B

1. If two line segments have the same length, they are said to be <u>congruent</u>.

2. The amount of rotation from one ray to a second ray is called the <u>angle</u> between the rays.

3. An angle that measures 90° is called a <u>right</u> angle.

4. The measure of the angle is between 0° and 90°, so the angle is acute.

5. The measure of the angle is between 90° and 180°, so the angle is obtuse.

6. The measure of the angle is 180°, so the angle is straight.

7. The measure of the angle is 90°, so the angle is right.

8. False; two angles whose measures sum 180° are called supplementary.

9. Two angles are complementary if their sum is 90°. The measure of an angle that is complementary to an angle whose measure is 15° is $90° - 15° = 75°$.
   Two angles are supplementary if their sum is 180°. The measure of an angle that is supplementary to an angle whose measure is 15° is $180° - 15° = 165°$.

10. Two angles are complementary if their sum is 90°. The measure of an angle that is complementary to an angle whose measure is 60° is $90° - 60° = 30°$.
    Two angles are supplementary if their sum is 180°. The measure of an angle that is supplementary to an angle whose measure is 60° is $180° - 60° = 120°$.

11. <u>Parallel</u> lines are lines in the same plane that never meet.

12. False; while two intersecting lines do form four angles, the vertical angles have equal measures, so they are only supplementary when the lines meet at right angles.

13. True

14. $m\angle 1 = 180° - 40° = 140°$ because $\angle 1$ and the 40° angle are supplementary angles.
    $m\angle 2 = 40°$ because $\angle 2$ and the 40° angle are vertical angles.
    $m\angle 3 = 180° - 40° = 140°$ because $\angle 3$ and the 40° angle are supplementary angles.
    $m\angle 4 = 40°$ because $\angle 4$ and the 40° angle are corresponding angles.
    $m\angle 5 = 140°$ because $\angle 5$ and $\angle 3$ are alternate interior angles.
    $m\angle 6 = 40°$ because $\angle 6$ and $\angle 4$ are vertical angles.
    $m\angle 7 = 140°$ because $\angle 7$ and $\angle 5$ are vertical angles.

15. The measure of the angle is between 0° and 90°, so the angle is acute.

17. The measure of the angle is 90°, so the angle is right.

19. The measure of the angle is 180°, so the angle is straight.

21. The measure of the angle is between 90° and 180°, so the angle is obtuse.

23. Two angles are complementary if their sum is 90°. The measure of an angle that is complementary to an angle whose measure is 32° is $90° - 32° = 58°$.

25. Two angles are complementary if their sum is 90°. The measure of an angle that is complementary to an angle whose measure is 73° is $90° - 73° = 17°$.

27. Two angles are supplementary if their sum is 180°. The measure of an angle that is supplementary to an angle whose measure is 67° is $180° - 67° = 113°$.

29. Two angles are supplementary if their sum is 180°. The measure of an angle that is supplementary to an angle whose measure is 8° is $180° - 8° = 172°$.

31. $m\angle 1 = 180° - 50° = 130°$ because $\angle 1$ and the 50° angle are supplementary angles.
$m\angle 2 = 50°$ because $\angle 2$ and the 50° angle are vertical angles.
$m\angle 3 = 180° - 50° = 130°$ because $\angle 3$ and the 50° angle are supplementary angles.
$m\angle 4 = 50°$ because $\angle 4$ and the 50° angle are corresponding angles.
$m\angle 5 = 130°$ because $\angle 5$ and $\angle 3$ are alternate interior angles.
$m\angle 6 = 50°$ because $\angle 6$ and $\angle 4$ are vertical angles.
$m\angle 7 = 130°$ because $\angle 7$ and $\angle 5$ are vertical angles.

Appendix B.2 Quick Checks

1. A triangle in which two sides are congruent is called an <u>isosceles</u> triangle.

2. A <u>right</u> triangle is a triangle that contains a 90° angle.

3. The sum of the measures of the interior angles in a triangle is <u>180</u> degrees.

4. There are 180° in a triangle and the right angle measures 90° so
$m\angle B = 180° - 90° - 20° = 70°$.

5. There are 180° in a triangle so
$m\angle B = 180° - 120° - 18° = 42°$.

6. Two triangles are <u>congruent</u> if the corresponding angles have the same measure and the corresponding sides have the same length.

7. Two triangles are <u>similar</u> if corresponding angles of the triangles are equal and the lengths of the corresponding sides are in proportion.

8. Because the triangles are similar, the corresponding sides are proportional. That is,
$\frac{3}{7} = \frac{6}{x}$.

$$\frac{3}{7} = \frac{6}{x}$$
$$7x \cdot \left(\frac{3}{7}\right) = 7x \cdot \left(\frac{6}{x}\right)$$
$$3x = 42$$
$$x = 14$$

The missing length is 14 units.

9. A <u>radius</u> of a circle is a line segment drawn from the center of the circle to any point on the circle.

10. True

11. The length of the radius is one-half the length of the diameter.
$$r = \frac{1}{2} \cdot d$$
$$r = \frac{1}{2} \cdot 15 \text{ inches}$$
$$r = \frac{15}{2} \text{ or } 7\frac{1}{2} \text{ inches}$$
The radius of the circle is $7\frac{1}{2}$ inches.

12. The length of the radius is one-half the length of the diameter.
$$r = \frac{1}{2} \cdot d$$
$$r = \frac{1}{2} \cdot 24 \text{ feet}$$
$$r = 12 \text{ feet}$$
The radius of the circle is 12 feet.

13. The length of the diameter is twice the length of the radius.
$$d = 2 \cdot r$$
$$d = 2 \cdot 3.6 \text{ yards}$$
$$d = 7.2 \text{ yards}$$
The diameter of the circle is 7.2 yards.

14. The length of the diameter is twice the length of the radius.
$$d = 2 \cdot r$$
$$d = 2 \cdot 9 \text{ centimeters}$$
$$d = 18 \text{ centimeters}$$
The diameter of the circle is 18 centimeters.

Appendix B.2 Exercises

15. There are 180° in a triangle so the measure of the missing angle is $180° - 85° - 40° = 55°$.

17. There are 180° in a triangle and the right angle measures 90° so the measure of the missing angle $180° - 90° - 42° = 48°$.

19. Because the triangles are similar, the corresponding sides are proportional. That is,

$$\frac{4}{8} = \frac{2}{x}.$$

$$\frac{4}{8} = \frac{2}{x}$$

$$8x \cdot \left(\frac{4}{8}\right) = 8x \cdot \left(\frac{2}{x}\right)$$

$$4x = 16$$

$$x = 4$$

The missing length is 4 units.

21. Because the triangles are similar, the corresponding sides are proportional. That is,

$$\frac{20}{30} = \frac{45}{x}.$$

$$\frac{20}{30} = \frac{45}{x}$$

$$30x \cdot \left(\frac{20}{30}\right) = 30x \cdot \left(\frac{45}{x}\right)$$

$$20x = 1350$$

$$x = 67.5$$

The missing length is 67.5 units.

23. The length of the diameter is twice the length of the radius.

$$d = 2 \cdot r$$

$$d = 2 \cdot 5 \text{ inches}$$

$$d = 10 \text{ inches}$$

The diameter of the circle is 10 inches.

25. The length of the diameter is twice the length of the radius.

$$d = 2 \cdot r$$

$$d = 2 \cdot 2.5 \text{ centimeters}$$

$$d = 5 \text{ centimeters}$$

The diameter of the circle is
5 centimeters.

27. The length of the radius is one-half the length of the diameter.

$$r = \frac{1}{2} \cdot d$$

$$r = \frac{1}{2} \cdot 14 \text{ centimeters}$$

$$r = 7 \text{ centimeters}$$

The radius of the circle is 7 centimeters.

29. The length of the radius is one-half the length of the diameter.

$$r = \frac{1}{2} \cdot d$$

$$r = \frac{1}{2} \cdot 11 \text{ yards}$$

$$r = \frac{11}{2} \text{ yards}$$

The radius of the circle is $\frac{11}{2}$ or
5.5 yards.

Appendix B.3 Quick Checks

1. The <u>perimeter</u> of a polygon is the distance around the polygon.

2. The <u>area</u> of a polygon is the amount of surface the polygon covers.

3. $P = 2l + 2w$
$$P = 2 \cdot 8 \text{ ft} + 2 \cdot 3 \text{ ft}$$
$$= 16 \text{ ft} + 6 \text{ ft}$$
$$= 22 \text{ ft}$$
$$A = lw$$
$$A = 8 \text{ ft} \cdot 3 \text{ ft}$$
$$= 24 \text{ square ft}$$
The perimeter of the rectangle is 22 feet, and the area is 24 square feet.

4. $P = 2l + 2w$
$$P = 2 \cdot 10 \text{ m} + 2 \cdot 3 \text{ m}$$
$$= 20 \text{ m} + 6 \text{ m}$$
$$= 26 \text{ m}$$
$$A = lw$$
$$A = 10 \text{ m} \cdot 3 \text{ m}$$
$$= 30 \text{ square m}$$
The perimeter of the rectangle is 26 meters, and the area is 30 square meters.

5. False; the area of a square is the square of the length of a side.

6. $P = 4 \cdot s$
$$P = 4 \cdot 4 \text{ cm}$$
$$= 16 \text{ cm}$$
$$A = s^2$$
$$A = (4 \text{ cm})^2$$
$$= 16 \text{ square cm}$$
The perimeter of the square is 16 centimeters, and the area is 16 square centimeters.

7. $P = 4 \cdot s$
$P = 4 \cdot 1.5$ yd
$\quad = 6$ yd
$A = s^2$
$A = (1.5 \text{ yd})^2$
$\quad = 2.25$ square yd

The perimeter of the square is 6 yards, and the area is 2.25 square yards.

8. The perimeter is the distance around the figure.
Perimeter
$= 45 \text{ yd} + 20 \text{ yd} + 20 \text{ yd} + 10 \text{ yd} + 25 \text{ yd} + 10 \text{ yd}$
$= 130 \text{ yd}$
Area = Area of rectangle + Area of rectangle
$\quad = (45 \text{ yd})(10 \text{ yd}) + (20 \text{ yd})(10 \text{ yd})$
$\quad = 450 \text{ yd}^2 + 200 \text{ yd}^2$
$\quad = 650 \text{ yd}^2$

The perimeter of the figure is 130 yards, and the area is 650 square yards.

9. To find the area of a trapezoid, we use the formula $A = \frac{1}{2}h(b + B)$ where h is the height of the trapezoid and the bases have lengths b and B.

10. $P = 2l + 2w$
$P = 2 \cdot 10 \text{ m} + 2 \cdot 8 \text{ m}$
$\quad = 20 \text{ m} + 16 \text{ m}$
$\quad = 36 \text{ m}$
$A = b \cdot h$
$A = 10 \text{ m} \cdot 7 \text{ m} = 70$ square m

The perimeter of the parallelogram is 36 meters, and the area is 70 square meters.

11. $P = 5 \text{ yd} + 9 \text{ yd} + 12 \text{ yd} + 7 \text{ yd} = 33 \text{ yd}$
$A = \frac{1}{2}h(b + B)$
$A = \frac{1}{2} \cdot 6 \text{ yd} \cdot (12 \text{ yd} + 5 \text{ yd})$
$\quad = \frac{1}{2} \cdot 6 \text{ yd} \cdot 17 \text{ yd}$
$\quad = 51$ square yd

The perimeter of the trapezoid is 33 yards, and the area is 51 square yards.

12. True

13. $P = 6 \text{ mm} + 5 \text{ mm} + 8 \text{ mm} = 19 \text{ mm}$
$A = \frac{1}{2}bh$
$A = \frac{1}{2} \cdot 8 \text{ mm} \cdot 3 \text{ mm} = 12$ square mm

The perimeter of the triangle is 19 millimeters, and the area is 12 square millimeters.

14. $P = 5 \text{ ft} + 12 \text{ ft} + 13 \text{ ft} = 30 \text{ ft}$
$A = \frac{1}{2}bh$
$A = \frac{1}{2} \cdot 12 \text{ ft} \cdot 5 \text{ ft} = 30$ square ft

The perimeter of the triangle is 30 feet, and the area is 30 square feet.

15. The circumference of a circle is the distance around the circle.

16. False: the area of a circle is given by the formula $A = \pi r^2$.

17. We know the length of the radius, so we use the formula $C = 2\pi r$ to find the circumference.
$C = 2\pi r$
$\quad = 2 \cdot \pi \cdot 4 \text{ ft}$
$\quad = 8\pi \text{ ft}$
$\quad \approx 25.13 \text{ ft}$
$A = \pi r^2$
$\quad = \pi \cdot (4 \text{ ft})^2$
$\quad = 16\pi$ square ft
$\quad \approx 50.27$ square ft

The circumference of the circle is exactly 8π feet or approximately 25.13 feet. The area of the circle is exactly 16π square feet or approximately 50.27 square feet.

18. The length of the diameter is given, so we use the formula $C = \pi d$ to find the circumference.
$C = \pi d$
$\quad = \pi \cdot 24 \text{ cm}$
$\quad = 24\pi \text{ cm}$
$\quad \approx 75.40 \text{ cm}$
Since the diameter is 24 centimeters, the radius is 12 centimeters.
$A = \pi r^2$
$\quad = \pi \cdot (12 \text{ cm})^2$
$\quad = 144\pi$ square cm
$\quad \approx 452.39$ square cm

The circumference of the circle is exactly 24π centimeters or approximately

75.40 centimeters. The area of the circle is exactly 144π square centimeters or approximately 452.39 square centimeters.

Appendix B.3 Exercises

19. $P = 2l + 2w$

   $P = 2 \cdot 10 \text{ ft} + 2 \cdot 4 \text{ ft}$

    $= 20 \text{ ft} + 8 \text{ ft}$

    $= 28 \text{ ft}$

   $A = lw$

   $A = 10 \text{ ft} \cdot 4 \text{ ft}$

    $= 40 \text{ square ft}$

The perimeter of the rectangle is 28 feet, and the area is 40 square feet.

21. $P = 2l + 2w$

   $P = 2 \cdot 15 \text{ m} + 2 \cdot 5 \text{ m}$

    $= 30 \text{ m} + 10 \text{ m}$

    $= 40 \text{ m}$

   $A = lw$

   $A = 15 \text{ m} \cdot 5 \text{ m}$

    $= 75 \text{ square m}$

The perimeter of the rectangle is
40 meters, and the area is 75 square meters.

23. $P = 4 \cdot s$

   $P = 4 \cdot 6 \text{ km} = 24 \text{ km}$

   $A = s^2$

   $A = (6 \text{ km})^2 = 36 \text{ square km}$

The perimeter of the square is 24 kilometers, and the area is 36 square kilometers.

25.   The perimeter is the distance around the figure.

   $\text{Perimeter} = 15 \text{ ft} + 6 \text{ ft} + 7 \text{ ft} + 14 \text{ ft} + 22 \text{ ft} + 8 \text{ ft}$

       $= 72 \text{ ft}$

   $\text{Area} = \text{Area of rectangle} + \text{Area of rectangle}$

     $= (7 \text{ ft})(6 \text{ ft}) + (22 \text{ ft})(8 \text{ ft})$

     $= 42 \text{ ft}^2 + 176 \text{ ft}^2$

     $= 218 \text{ ft}^2$

The perimeter of the figure is 72 feet, and the area is 218 square feet.

27.   The perimeter is the distance around the figure.

   $\text{Perimeter} = 13 \text{ m} + 6 \text{ m} + 13 \text{ m} + 2 \text{ m} + 8 \text{ m} + 2 \text{ m} + 8 \text{ m} + 2 \text{ m}$

      $= 54 \text{ m}$

   $\text{Area} = \text{Area of rectangle} + \text{Area of rectangle} + \text{Area of rectangle}$

     $= (13 \text{ m})(2 \text{ m}) + (5 \text{ m})(2 \text{ m}) + (13 \text{ m})(2 \text{ m})$

     $= 26 \text{ m}^2 + 10 \text{ m}^2 + 26 \text{ m}^2$

     $= 62 \text{ m}^2$

The perimeter of the figure is 54 meters, and the area is 62 square meters.

29. $P = 2l + 2w$
$P = 2 \cdot 9 \text{ ft} + 2 \cdot 6 \text{ ft}$
$= 18 \text{ ft} + 12 \text{ ft}$
$= 30 \text{ ft}$
$A = b \cdot h$
$A = 9 \text{ ft} \cdot 5 \text{ ft}$
$= 45 \text{ square ft}$

The perimeter of the parallelogram is 30 feet, and the area is 45 square feet.

31. $P = 2l + 2w$
$P = 2 \cdot 10 \text{ mm} + 2 \cdot 4 \text{ mm}$
$= 20 \text{ mm} + 8 \text{ mm}$
$= 28 \text{ mm}$
$A = b \cdot h$
$A = 4 \text{ mm} \cdot 9 \text{ mm} = 36 \text{ square mm}$

The perimeter of the parallelogram is 28 millimeters, and the area is 36 square millimeters.

33. $P = 8 \text{ in.} + 8 \text{ in.} + 8 \text{ in.} + 16 \text{ in.} = 40 \text{ in.}$
$A = \frac{1}{2} h (b + B)$
$A = \frac{1}{2} \cdot 7 \text{ in.} \cdot (16 \text{ in.} + 8 \text{ in.})$
$= \frac{1}{2} \cdot 7 \text{ in.} \cdot 24 \text{ in.}$
$= 84 \text{ square in.}$

The perimeter of the trapezoid is 40 inches, and the area is 84 square inches.

35. $P = 8 \text{ cm} + 8 \text{ cm} + 10 \text{ cm} + 19 \text{ cm}$
$= 45 \text{ cm}$
$A = \frac{1}{2} h (b + B)$
$A = \frac{1}{2} \cdot 7 \text{ cm} \cdot (19 \text{ m} + 8 \text{ cm})$
$= \frac{1}{2} \cdot 7 \text{ cm} \cdot 27 \text{ cm}$
$= 94.5 \text{ square cm}$

The perimeter of the trapezoid is 45 centimeters, and the area is 94.5 square centimeters.

37. $P = 8 \text{ m} + 12 \text{ m} + 12 \text{ m} = 32 \text{ m}$
$A = \frac{1}{2} bh$
$A = \frac{1}{2} \cdot 12 \text{ m} \cdot 7 \text{ m} = 42 \text{ square m}$

The perimeter of the triangle is 32 meters, and the area is 42 square meters.

39. $P = 11 \text{ ft} + 15 \text{ ft} + 6 \text{ ft} = 32 \text{ ft}$
$A = \frac{1}{2} bh$
$A = \frac{1}{2} \cdot 6 \text{ ft} \cdot 8 \text{ ft} = 24 \text{ square ft}$

The perimeter of the triangle is 32 feet, and the area is 24 square feet.

41. We know the length of the radius, so we use the formula $C = 2\pi r$ to find the circumference.
$C = 2\pi r$
$= 2 \cdot \pi \cdot 16 \text{ in.}$
$= 32\pi \text{ in.}$
$\approx 100.53 \text{ in.}$
$A = \pi r^2$
$= \pi \cdot (16 \text{ in.})^2$
$= 256\pi \text{ square in.}$
$\approx 804.25 \text{ square in.}$

The circumference of the circle is exactly 32π inches or approximately 100.53 inches. The area of the circle is exactly 256π square inches or approximately 804.25 square inches.

43. The length of the diameter is given, so we use the formula $C = \pi d$. Since the diameter is 20 centimeters, the radius is 10 centimeters.
$C = \pi d = \pi \cdot 20 \text{ cm} = 20\pi \text{ cm} \approx 62.83 \text{ cm}$
$A = \pi r^2$
$= \pi \cdot (10 \text{ cm})^2$
$= 100\pi \text{ square cm}$
$\approx 314.16 \text{ square cm}$

The circumference of the circle is exactly 20π centimeters or approximately 62.83 centimeters. The area of the circle is exactly 100π square centimeters or approximately 314.16 square centimeters.

45. Area = Area of Circle = $\pi r^2 = \pi \cdot 1^2 = \pi$
The area of the shaded region is π square units.

47. The distance a wheel travels in one revolution is equal to the circumference of the circle.
$C = \pi d = 20\pi \text{ inches}$
In 5 revolutions, the wheel travels 100π inches, which is about 314.16 inches or 26.18 feet.

Appendix B.4 Quick Checks

1. A <u>polyhedron</u> is a three-dimensional solid formed by connecting polygons.

2. The <u>surface area</u> of a polyhedron is the sum of the areas of the faces of the polyhedron.

3. False; volume is measured in cubic units.

4. True

5. The figure is a cube.

$V = s^3$

$\quad = (5 \text{ m})^3$

$\quad = 125 \text{ cubic m}$

$S = 6s^2$

$\quad = 6(5 \text{ m})^2$

$\quad = 6(25) \text{ square m}$

$\quad = 150 \text{ square m}$

The volume of the cube is 125 cubic meters, and the surface area is 150 square meters.

6. The figure is a sphere.

$V = \dfrac{4}{3}\pi r^3$

$\quad = \dfrac{4}{3} \cdot \pi \cdot (4 \text{ in.})^3$

$\quad = \dfrac{4}{3} \cdot \pi \cdot (64) \text{ cubic in.}$

$\quad = \dfrac{256}{3} \pi \text{ cubic in.}$

$\quad \approx 268.08 \text{ cubic in.}$

$S = 4\pi r^2$

$\quad = 4 \cdot \pi \cdot (4 \text{ in.})^2$

$\quad = 4 \cdot \pi \cdot (16) \text{ square in.}$

$\quad = 64\pi \text{ square in.}$

$\quad \approx 201.06 \text{ square in.}$

The volume of the sphere is exactly $\dfrac{256\pi}{3}$ cubic inches or approximately 268.08 cubic inches. The surface area is exactly 64π square inches or approximately 201.06 square inches.

Appendix B.4 Exercises

7. $V = lwh$

$\quad = 10 \text{ ft} \cdot 5 \text{ ft} \cdot 12 \text{ ft}$

$\quad = 600 \text{ cubic ft}$

$S = 2lw + 2lh + 2wh$

$\quad = 2(10 \text{ ft})(5 \text{ ft}) + 2(10 \text{ ft})(12 \text{ ft}) + 2(5 \text{ ft})(12 \text{ ft})$

$\quad = 460 \text{ square ft}$

The volume is 600 cubic feet, and the surface area is 460 square feet.

9. $V = \dfrac{4}{3}\pi r^3$

$\quad = \dfrac{4}{3} \cdot \pi \cdot (6 \text{ cm})^3$

$\quad = \dfrac{4}{3} \cdot \pi \cdot (216) \text{ cubic cm}$

$\quad = 288\pi \text{ cubic cm}$

$\quad \approx 904.78 \text{ cubic cm}$

$S = 4\pi r^2$

$\quad = 4 \cdot \pi \cdot (6 \text{ cm})^2$

$\quad = 4 \cdot \pi \cdot (36) \text{ square cm}$

$\quad = 144\pi \text{ square cm}$

$\quad \approx 452.39 \text{ square cm}$

The volume is exactly 288π cubic centimeters or approximately 904.78 cubic centimeters. The surface area is exactly 144π square centimeters or approximately 452.39 square centimeters.

11. $V = \pi r^2 h$

$\quad = \pi (2 \text{ in.})^2 (8 \text{ in.})$

$\quad = 32\pi \text{ cubic in.}$

$\quad \approx 100.53 \text{ cubic in.}$

$S = 2\pi r^2 + 2\pi rh$

$\quad = 2\pi (2 \text{ in.})^2 + 2\pi (2 \text{ in.})(8 \text{ in.})$

$\quad = (8\pi + 32\pi) \text{ square in.}$

$\quad = 40\pi \text{ square in.}$

$\quad \approx 125.66 \text{ square in.}$

The volume is exactly 32π cubic inches or approximately 100.53 cubic inches. The surface area is exactly 40π square inches or approximately 125.66 square inches.

13. $V = \dfrac{1}{3}\pi r^2 h$

$\quad = \dfrac{1}{3}\pi (10 \text{ mm})^2 (8 \text{ mm})$

$\quad = \dfrac{800}{3} \pi \text{ cubic mm}$

$\quad \approx 837.76 \text{ cubic mm}$

The volume is exactly $\dfrac{800}{3}\pi$ cubic millimeters or approximately 837.76 cubic millimeters.

15. $V = \dfrac{1}{3}b^2 h$

$\qquad = \dfrac{1}{3}(8 \text{ ft})^2 (10 \text{ ft})$

$\qquad = \dfrac{640}{3} \text{ cubic ft}$

$S = b^2 + 2bs$

$\qquad = (8 \text{ ft})^2 + 2(8 \text{ ft})(12 \text{ ft})$

$\qquad = 64 + 192 \text{ square ft}$

$\qquad = 256 \text{ square ft}$

The volume is $\dfrac{640}{3}$ cubic feet, and the surface area is 256 square feet.

17. We are looking for the volume of the gutter. Note that 12 feet is
$12(12 \text{ inches}) = 144 \text{ inches.}$
$V = lwh$

$\qquad = (144 \text{ in.})(3 \text{ in.})(4 \text{ in.})$

$\qquad = 1728 \text{ cubic in.}$

The gutter will hold 1728 cubic inches of water.

19. $V = \pi r^2 h$

$\qquad = \pi(2 \text{ in.})^2 (6 \text{ in.})$

$\qquad = 24\pi \text{ cubic in.}$

$\qquad \approx 75.40 \text{ cubic in.}$

$S = 2\pi r^2 + 2\pi r h$

$\qquad = 2\pi(2 \text{ in.})^2 + 2\pi(2 \text{ in.})(6 \text{ in.})$

$\qquad = (8\pi + 24\pi) \text{ square in.}$

$\qquad = 32\pi \text{ square in.}$

$\qquad \approx 100.53 \text{ square in.}$

The volume of the can is approximately 75.40 cubic inches. The surface area is approximately 100.53 square inches.

21. We are looking for the volume of the waffle cone.

$V = \dfrac{1}{3}\pi r^2 h$

$\qquad = \dfrac{1}{3}\pi(4 \text{ cm})^2 (16 \text{ cm})$

$\qquad = \dfrac{256}{3}\pi \text{ cubic cm}$

$\qquad \approx 268.08 \text{ cubic cm}$

The cone will hold approximately 268.08 cubic centimeters of ice cream.

Appendix C

Appendix C.1

Preparing for Systems of Linear Equations in Two Variables

P1. Substitute $x = 5$ and $y = 4$, and simplify:

$$2x - 3y = 2(5) - 3(4)$$
$$= 10 - 12$$
$$= -2$$

P2.
$$2(4) - 3(-1) \stackrel{?}{=} 11$$
$$8 + 3 \stackrel{?}{=} 11$$
$$11 = 11 \quad \leftarrow \text{True}$$
Yes, $(4, -1)$ is on the graph of $2x - 3y = 11$.

P3. $y = 3x - 7$

Let $x = 0$, 1, and 2.

$$x = 0: \quad y = 3(0) - 7$$
$$y = 0 - 7$$
$$y = -7$$

$$x = 1: \quad y = 3(1) - 7$$
$$y = 3 - 7$$
$$y = -4$$

$$x = 2: \quad y = 3(2) - 7$$
$$y = 6 - 7$$
$$y = -1$$

Thus, the points $(0, -7)$, $(1, -4)$, and $(2, -1)$ are on the graph.

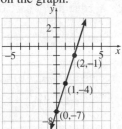

P4. The slope of the line we seek is $m = -3$, the same as the slope of $y = -3x + 1$. The equation of the line is

$$y - y_1 = m(x - x_1)$$
$$y - 3 = -3(x - 2)$$
$$y - 3 = -3x + 6$$
$$y = -3x + 9$$

P5.
$$4x - 3y = 15$$
$$-3y = -4x + 15$$
$$y = \frac{-4x + 15}{-3}$$
$$y = \frac{4}{3}x - 5$$

The slope is $\frac{4}{3}$ and the y-intercept is -5.

P6. The additive inverse of 4 is -4 because $4 + (-4) = 0$.

P7.
$$2x - 3(-3x + 1) = -36$$
$$2x + 9x - 3 = -36$$
$$11x - 3 = -36$$
$$11x = -33$$
$$x = \frac{-33}{11} = -3$$

The solution set is $\{-3\}$.

Appendix C.1 Quick Checks

1. system of linear equations

2. $\begin{cases} 2x + 3y = 7 & (1) \\ 3x + y = -7 & (2) \end{cases}$

(a) Let $x = 2$ and $y = 1$ in both equations (1) and (2).

Equation (1): $2(2) + 3(1) \stackrel{?}{=} 7$
$$4 + 3 \stackrel{?}{=} 7$$
$$7 = 7$$

Equation (2): $3(2) + (1) \stackrel{?}{=} -7$
$$6 + 1 \stackrel{?}{=} -7$$
$$7 \neq -7$$

Although these values satisfy equation (1), they do not satisfy equation (2). Therefore, the ordered pair $(2, 1)$ is not a solution.

(b) Let $x = -4$ and $y = 5$ in both equations (1) and (2).

Equation (1): $2(-4) + 3(5) \stackrel{?}{=} 7$
$$-8 + 15 \stackrel{?}{=} 7$$
$$7 = 7$$

Equation (2): $3(-4) + 5 \stackrel{?}{=} -7$
$$-12 + 5 \stackrel{?}{=} -7$$
$$-7 = -7$$

These values satisfy both equations, so the ordered pair $(-4, 5)$ is a solution.

(c) Let $x = -2$ and $y = -1$ in both equations (1) and (2).

Equation (1): $2(-2) + 3(-1) \stackrel{?}{=} 7$

$$-4 - 3 \stackrel{?}{=} 7$$
$$-7 \neq 7$$

Equation (2): $3(-2) + (-1) \stackrel{?}{=} -7$

$$-6 - 1 \stackrel{?}{=} -7$$
$$-7 = -7$$

Although these values satisfy equation (2), they do not satisfy equation (1). Therefore, the ordered pair $(-2, -1)$ is not a solution.

3. inconsistent

4. consistent; dependent

5. False. If the graphs of the equations are parallel lines, then the system will have no solution.

6. True

7. $\begin{cases} y = -3x + 10 & (1) \\ y = 2x - 5 & (2) \end{cases}$

The two equations are in slope-intercept form. Graph each equation and find the point of intersection.

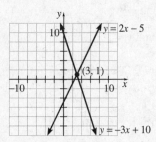

The solution is the ordered pair $(3, 1)$.

8. $\begin{cases} 2x + y = -1 & (1) \\ -2x + 2y = 10 & (2) \end{cases}$

Equation (1) in slope-intercept form is $y = -2x - 1$. Equation (2) in slope-intercept form is $y = x + 5$. Graph each equation and find the point of intersection.

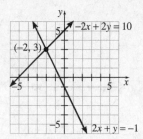

The solution is the ordered pair $(-2, 3)$.

9. $\begin{cases} y = -3x - 5 & (1) \\ 5x + 3y = 1 & (2) \end{cases}$

Substituting $-3x - 5$ for y in equation (2), we obtain

$$5x + 3(-3x - 5) = 1$$
$$5x - 9x - 15 = 1$$
$$-4x - 15 = 1$$
$$-4x = 16$$
$$-x = -4$$

Substituting -4 for x in equation (1), we obtain

$$y = -3(-4) - 5 = 12 - 5 = 7.$$

The solution is the ordered pair $(-4, 7)$.

10. $\begin{cases} 2x + y = -2 & (1) \\ -3x - 2y = -2 & (2) \end{cases}$

Equation (1) solved for y is $y = -2x - 2$.
Substituting $-2x - 2$ for y in equation (2), we obtain

$$-3x - 2(-2x - 2) = -2$$
$$-3x + 4x + 4 = -2$$
$$x + 4 = -2$$
$$x = -6$$

Substituting -6 for y in equation (1), we obtain

$$2(-6) + y = -2$$
$$-12 + y = -2$$
$$y = 10$$

The solution is the ordered pair $(-6, 10)$.

11. $\begin{cases} 2x - 3y = -6 & (1) \\ -8x + 3y = 3 & (2) \end{cases}$

There is no need to multiply either equation by any constant. Add the two equations.

$$\begin{array}{r} 2x - 3y = -6 \\ -8x + 3y = 3 \\ \hline -6x = -3 \end{array}$$

$$x = \frac{1}{2}$$

Substituting $\dfrac{1}{2}$ for x in equation (1), we obtain

$$2\left(\dfrac{1}{2}\right) - 3y = -6$$
$$1 - 3y = -6$$
$$-3y = -7$$
$$y = \dfrac{7}{3}$$

The solution is the ordered pair $\left(\dfrac{1}{2}, \dfrac{7}{3}\right)$.

12. $\begin{cases} -2x + y = 4 & (1) \\ -5x + 3y = 7 & (2) \end{cases}$

Multiply both sides of equation (1) by -3 and add the result to equation (2).

$$\begin{array}{rl} 6x - 3y = -12 \\ -5x + 3y = 7 \\ \hline x = -5 \end{array}$$

Substituting -5 for x in equation (1), we obtain
$$-2(-5) + y = 4$$
$$10 + y = 4$$
$$y = -6$$

The solution is the ordered pair $(-5, -6)$.

13. $\begin{cases} -3x + 2y = 3 & (1) \\ 4x - 3y = -6 & (2) \end{cases}$

Multiply both sides of equation (1) by 4, multiply both sides of equation (2) by 3, and add the results.

$$\begin{array}{rl} -12x + 8y = 12 \\ 12x - 9y = -18 \\ \hline -y = -6 \\ y = 6 \end{array}$$

Substituting 6 for y in equation (1), we obtain
$$-3x + 2(6) = 3$$
$$-3x + 12 = 3$$
$$-3x = -9$$
$$x = 3$$

The solution is the ordered pair $(3, 6)$.

14. $\begin{cases} -3x + y = 2 & (1) \\ 6x - 2y = 1 & (2) \end{cases}$

Multiply both sides of equation (1) by 2 and add the result to equation (2).

$$\begin{array}{rl} -6x + 2y = 4 \\ 6x - 2y = 1 \\ \hline 0 = 5 \end{array}$$

The equation $0 = 5$ is false, so the system has no solution. The solution set is $\varnothing$ or $\{\ \}$. The system is inconsistent.

The graphs of the equations (shown below) are parallel, which supports the statement that the system has no solution.

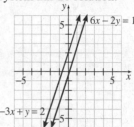

15. $\begin{cases} -3x + 2y = 8 & (1) \\ 6x - 4y = -16 & (2) \end{cases}$

Multiply both sides of equation (1) by 2 and add the results.

$$\begin{array}{rl} -6x + 4y = 16 \\ 6x - 4y = -16 \\ \hline 0 = 0 \end{array}$$

The equation $0 = 0$ is true, so the system is dependent. The solution is
$$\{(x,\ y) \,|\, -3x + 2y = 8\}.$$

The graphs of the equations (shown below) coincide, which supports the statement that the system is dependent.

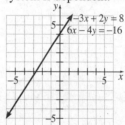

16. $\begin{cases} 2x - 3y = -16 & (1) \\ -3x + 2y = 19 & (2) \end{cases}$

Multiply both sides of equation (1) by 3, multiply both sides of equation (2) by 2, and add the results.

$$\begin{array}{rl} 6x - 9y = -48 \\ -6x + 4y = 38 \\ \hline -5y = -10 \\ y = 2 \end{array}$$

Substituting 2 for y in equation (1), we obtain
$$2x - 3(2) = -16$$
$$2x - 6 = -16$$
$$2x = -10$$
$$x = -5$$

The solution is the ordered pair $(-5, 2)$.
The graphs of the equations (shown below)

intersect at the point (−5, 2), which supports the statement that the solution of the system is (−5, 2).

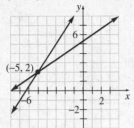

Appendix C.1 Exercises

17. (a) Let $x = 5$, $y = 3$ in both equations (1) and (2).

$$\begin{cases} 2(5)+3=10+3=13 \\ -5(5)+3(3)=-25+9=-16 \neq 6 \end{cases}$$

Although these values satisfy equation (1), they do not satisfy equation (2). Therefore, the ordered pair (5, 3) is not a solution of the system.

(b) Let $x = 3$, $y = 7$ in both equations (1) and (2).

$$\begin{cases} 2(3)+7=6+7=13 \\ -5(3)+3(7)=-15+21=6 \end{cases}$$

Because these values satisfy both equations (1) and (2), the ordered pair (3, 7) is a solution of the system.

19. (a) Let $x = 1$, $y = 2$ in both equations (1) and (2).

$$\begin{cases} 5(1)+2(2)=5+4=9 \\ -10(1)-4(2)=-10-8=-18 \end{cases}$$

Because these values satisfy both equations (1) and (2), the ordered pair (1, 2) is a solution of the system.

(b) Let $x = 2$, $y = -1/2$ in both equations (1) and (2).

$$\begin{cases} 5(2)+2(-1/2)=10+(-1)=9 \\ -10(2)-4(-1/2)=-20+2=-18 \end{cases}$$

Because these values satisfy both equations (1) and (2), the ordered pair $(2, -1/2)$ is a solution of the system.

21. consistent; independent

23. inconsistent

25. $\begin{cases} y = 3x & (1) \\ y = -2x+5 & (2) \end{cases}$

The two equations are in slope-intercept form. Graph each equation and find the point of intersection.

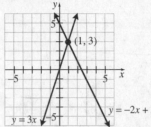

The solution is the ordered pair (1, 3).

27. $\begin{cases} 2x + y = 2 & (1) \\ x + 3y = -9 & (2) \end{cases}$

Equation (1) in slope-intercept form is $y = -2x + 2$. Equation (2) in slope-intercept form is $y = -\dfrac{1}{3}x - 3$. Graph each equation and find the point of intersection.

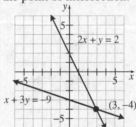

The solution is the ordered pair (3, −4).

29. $\begin{cases} y = -\dfrac{1}{2}x+1 & (1) \\ y+2x = 10 & (2) \end{cases}$

Substituting $-\dfrac{1}{2}x+1$ for y in equation (2), we get

$$-\frac{1}{2}x+1+2x = 10$$
$$2\left(-\frac{1}{2}x+1+2x\right) = 2(10)$$
$$-x+2+4x = 20$$
$$3x = 18$$
$$x = 6$$

Substituting 6 for x in equation (2), we obtain $y = -2(6)+10 = -12+10 = -2$.

The solution is the ordered pair (6, −2).

31. $\begin{cases} x = \dfrac{2}{3}y & (1) \\ 3x - y = -3 & (2) \end{cases}$

Substituting $\dfrac{2}{3}y$ for x in equation (2), we obtain

$3\left(\dfrac{2}{3}y\right) - y = -3$

$\quad 2y - y = -3$

$\qquad\quad y = -3$

Substituting -3 for y in equation (1), we obtain

$x = \dfrac{2}{3}(-3) = -2$.

The solution is the ordered pair $(-2, -3)$.

33. $\begin{cases} 2x - 4y = 2 & (1) \\ x + 2y = 0 & (2) \end{cases}$

Equation (2) solved for x is $x = -2y$.
Substituting $-2y$ for x in equation (1), we obtain

$2(-2y) - 4y = 2$

$\quad -4y - 4y = 2$

$\qquad\quad -8y = 2$

$\qquad\qquad y = \dfrac{2}{-8} = -\dfrac{1}{4}$

Substituting $-\dfrac{1}{4}$ for y in equation (2), we obtain

$x + 2\left(-\dfrac{1}{4}\right) = 0$

$\qquad x - \dfrac{1}{2} = 0$

$\qquad\qquad x = \dfrac{1}{2}$

The solution is the ordered pair $\left(\dfrac{1}{2}, -\dfrac{1}{4}\right)$.

35. $\begin{cases} x + y = 10,000 & (1) \\ 0.05x + 0.07y = 650 & (2) \end{cases}$

Equation (1) solved for x is $x = 10,000 - y$.
Substituting $10,000 - y$ for x in equation (2), we obtain

$0.05(10,000 - y) + 0.07y = 650$

$500 - 0.05y + 0.07y = 650$

$0.02y + 500 = 650$

$0.02y = 150$

$y = 7500$

Substituting 7500 for y in equation (1), we obtain

$x + 7500 = 10,000$

$x = 2500$

The solution is the ordered pair (2500, 7500).

37. $\begin{cases} x + y = -5 & (1) \\ -x + 2y = 14 & (2) \end{cases}$

Add equations (1) and (2).

$\begin{cases} \quad x + y = -5 \\ -x + 2y = 14 \end{cases}$

$\qquad\qquad 3y = 9$

$\qquad\qquad\quad y = 3$

Substituting 3 for y in equation (1), we obtain

$x + 3 = -5$

$x = -8$

The solution is the ordered pair (−8, 3).

39. $\begin{cases} x + 2y = -5 & (1) \\ 3x + 3y = 9 & (2) \end{cases}$

Multiply both sides of equation (1) by -3, and add the result to equation (2).

$\begin{cases} -3x - 6y = 15 \\ \quad 3x + 3y = 9 \end{cases}$

$\qquad\qquad -3y = 24$

$\qquad\qquad\quad y = -8$

Substituting -8 for y in equation (1), we obtain

$x + 2(-8) = -5$

$x - 16 = -5$

$x = 11$

The solution is the ordered pair (11, −8).

41. $\begin{cases} 2x + 5y = -3 & (1) \\ x + \dfrac{5}{4}y = -\dfrac{1}{2} & (2) \end{cases}$

Multiply both sides of equation (2) by -4 and add the result to equation (1).

$\begin{cases} \quad 2x + 5y = -3 \\ -4x - 5y = 2 \end{cases}$

$\quad -2x \qquad\quad = -1$

$\qquad\qquad x = \dfrac{1}{2}$

Substituting $1/2$ for x in equation (1), we obtain

$$2\left(\frac{1}{2}\right)+5y=-3$$
$$1+5y=-3$$
$$5y=-4$$
$$y=-\frac{4}{5}$$

The solution is the ordered pair $\left(\frac{1}{2},-\frac{4}{5}\right)$.

43. $\begin{cases} 0.05x+0.1y=5.25 & (1) \\ 0.08x-0.02y=1.2 & (2) \end{cases}$

Multiply both sides of equation (1) by 0.2, and add the result to equation (2).

$$\begin{cases} 0.01x+0.02y=1.05 \\ \underline{0.08x-0.02y=1.2} \end{cases}$$
$$0.09x\qquad\quad=2.25$$
$$x=25$$

Substituting 25 for x in equation (1), we obtain
$$0.05(25)+0.1y=5.25$$
$$1.25+0.1y=5.25$$
$$0.1y=4$$
$$y=40$$

The solution is the ordered pair (25, 40).

45. $\begin{cases} 3x+y=1 & (1) \\ -6x-2y=-4 & (2) \end{cases}$

Equation (1) solved for y is $y=-3x+1$.
Substituting $-3x+1$ for y in equation (2), we obtain
$$-6x-2(-3x+1)=-4$$
$$-6x+6x-2=-4$$
$$-2=-4$$

The system has no solution. The solution set is $\varnothing$ or $\{\ \}$. The system is inconsistent.

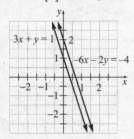

47. $\begin{cases} 5x-2y=2 & (1) \\ -10x+4y=3 & (2) \end{cases}$

Multiply both sides of equation (1) by 2 and add the result to equation (2).

$$\begin{cases} 10x-4y=4 \\ \underline{-10x+4y=3} \end{cases}$$
$$0=7$$

The system has no solution. The solution set is $\varnothing$ or $\{\ \}$. The system is inconsistent.

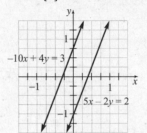

49. $\begin{cases} y=\dfrac{1}{2}x+1 & (1) \\ 2x-4y=-4 & (2) \end{cases}$

Substituting $\dfrac{1}{2}x+1$ for y in equation (2), we obtain
$$2x-4\left(\frac{1}{2}x+1\right)=-4$$
$$2x-2x-4=-4$$
$$-4=-4$$

The system is dependent. The solution is $\{(x,y)\,|\,2x-4y=-4\}$.

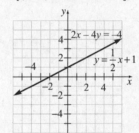

51. $\begin{cases} x+3y=6 & (1) \\ -\dfrac{x}{3}-y=-2 & (2) \end{cases}$

Equation (2) solved for y is $y=-\dfrac{x}{3}+2$.

Substituting $-\dfrac{x}{3}+2$ for y in equation (1), we obtain

$$x+3\left(-\frac{x}{3}+2\right)=6$$
$$x-x+6=6$$
$$6=6$$

The system is dependent. The solution is

$\{(x, y)\mid x+3y=6\}$.

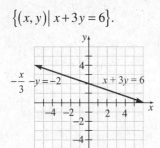

53. $\begin{cases} \dfrac{1}{3}x-2y=6 & (1) \\[2mm] -\dfrac{1}{2}x+3y=-9 & (2) \end{cases}$

Multiply both sides of equation (1) by 3,
multiply both sides of equation (2) by 2, and add
the results.

$x-6y=18$
$-x+6y=-18$
$\qquad 0=0$

The system is dependent. The solution is

$\left\{(x, y)\,\middle|\, \dfrac{1}{3}x-2y=6\right\}$.

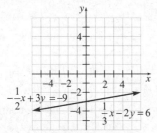

55. $\begin{cases} x+3y=0 & (1) \\ -2x+4y=30 & (2) \end{cases}$

Because the coefficient of x in equation (1) is 1,
we use substitution to solve the system.
Equation (1) solved for x is $x=-3y$.
Substituting $-3y$ for x in equation (2), we
obtain

$-2(-3y)+4y=30$
$\qquad 6y+4y=30$
$\qquad\quad 10y=30$
$\qquad\qquad y=3$

Substituting 3 for y in equation (1), we obtain
$x+3(3)=0$
$\quad x+9=0$
$\qquad x=-9$

The solution is the ordered pair $(-9, 3)$.

57. $\begin{cases} x=5y-3 & (1) \\ -3x+15y=9 & (2) \end{cases}$

Because equation (1) is already solved for x, we
use substitution to solve the system.
Substituting $5y-3$ for x in equation (2), we
obtain

$-3(5y-3)+15y=9$
$\quad -15y+9+15y=9$
$\qquad\qquad\qquad 9=9$

The system is dependent. The solution is
$\{(x, y)\mid x=5y-3\}$.

59. $\begin{cases} 2x-4y=18 & (1) \\ 3x+5y=-3 & (2) \end{cases}$

Because none of variables have a coefficient of
1, we use elimination to solve the system.
Multiply both sides of equation (1) by 3,
multiply both sides of equation (2) by -2, and
add the results.

$\begin{cases} 6x-12y=54 \\ -6x-10y=6 \end{cases}$
$\qquad -22y=60$

$y=\dfrac{60}{-22}=-\dfrac{30}{11}$

Substituting $-\dfrac{30}{11}$ for y in equation (1), we

obtain

$2x-4\left(-\dfrac{30}{11}\right)=18$

$2x+\dfrac{120}{11}=18$

$11\left(2x+\dfrac{120}{11}\right)=11(18)$

$22x+120=198$

$22x=78$

$x=\dfrac{78}{22}=\dfrac{39}{11}$

The solution is the ordered pair $\left(\dfrac{39}{11}, -\dfrac{30}{11}\right)$.

61. $\begin{cases} \dfrac{5}{6}x-\dfrac{1}{3}y=-5 & (1) \\[2mm] -x+\dfrac{2}{5}y=1 & (2) \end{cases}$

Because none of the variables have a coefficient
of 1, we use elimination to solve the system.
Multiply both sides of equation (1) by 6,
multiply both sides of equation (2) by 5, and add
the results.

$$\begin{cases} 5x - 2y = -30 \\ -5x + 2y = 5 \end{cases}$$
$$\overline{0 = -25}$$

The system has no solution. The solution set is $\varnothing$ or $\{\ \}$. The system is inconsistent.

63. $\begin{cases} 2x + y = -5 & (1) \\ 5x + 3y = 1 & (2) \end{cases}$

Write each equation in slope-intercept form.

$$2x + y = -5 \qquad\qquad 5x + 3y = 1$$
$$y = -2x - 5 \qquad\qquad 3y = -5x + 1$$
$$y = \frac{-5x + 1}{3}$$
$$y = -\frac{5}{3}x + \frac{1}{3}$$

Since the equations have different slopes, the lines are neither parallel nor coincident. Thus, the system must have exactly one solution.

65. $\begin{cases} 3x - 2y = -2 & (1) \\ -6x + 4y = 4 & (2) \end{cases}$

Write each equation in slope-intercept form.

$$3x - 2y = -2 \qquad\qquad -6x + 4y = 4$$
$$-2y = -3x - 2 \qquad\qquad 4y = 6x + 4$$
$$y = \frac{-3x - 2}{-2} \qquad\qquad y = \frac{6x + 4}{4}$$
$$y = \frac{3}{2}x + 1 \qquad\qquad y = \frac{3}{2}x + 1$$

Because both equations have the same slope-intercept form, the two lines are coincident. Thus, the system is dependent and has an infinite number of solutions.

67. (a) Equation of line through $(-1, 3)$ and $(3, 1)$:

$$m = \frac{1 - 3}{3 - (-1)} = \frac{-2}{4} = -\frac{1}{2}$$

$$y - 1 = -\frac{1}{2}(x - 3)$$
$$y - 1 = -\frac{1}{2}x + \frac{3}{2}$$
$$y = -\frac{1}{2}x + \frac{5}{2}$$

Equation of line through $(-2, -1)$ and $(4, 5)$:

$$m = \frac{y_2 - y_1}{x_2 - x_1} = \frac{5 - (-1)}{4 - (-2)} = \frac{6}{6} = 1$$

$$y - 5 = 1(x - 4)$$
$$y - 5 = x - 4$$
$$y = x + 1$$

(b) Solve the system formed by the equations found in part a.

$$\begin{cases} y = x + 1 & (1) \\ y = -\dfrac{1}{2}x + \dfrac{5}{2} & (2) \end{cases}$$

Substituting $x + 1$ for y in equation (2), we obtain

$$x + 1 = -\frac{1}{2}x + \frac{5}{2}$$

$$2(x + 1) = 2\left(-\frac{1}{2}x + \frac{5}{2}\right)$$
$$2x + 2 = -x + 5$$
$$3x = 3$$
$$x = 1$$

Substituting 1 for x in equation (1), we obtain $y = 1 + 1 = 2$.
The solution is the ordered pair $(1, 2)$.

69. The two lines intersect in the second quadrant. Of the choices provided, the ordered pairs $(-3, 1)$ and $(-1, 3)$ are the only ones that are in the second quadrant. Therefore, the answers are (c) and (f).

71. $\begin{cases} Ax + 3By = 2 \\ -3Ax + By = -11 \end{cases}$

Substitute 3 for x and 1 for y in the system and solve for A and B.

$$\begin{cases} A(3) + 3B(1) = 2 & \text{or} \quad 3A + 3B = 2 & (1) \\ -3A(3) + B(1) = -11 & \text{or} \quad -9A + B = -11 & (2) \end{cases}$$

Multiply equation (1) by 3 and add the result to equation (2).

$$9A + 9B = 6$$
$$-9A + B = -11$$
$$\overline{10B = -5}$$

$$B = \frac{-5}{10} = -\frac{1}{2}$$

Substituting $-\dfrac{1}{2}$ for B in equation (2), we obtain

$$-9A + \left(-\frac{1}{2}\right) = -11$$

$$-2\left(-9A - \frac{1}{2}\right) = -2(-11)$$
$$18A + 1 = 22$$
$$18A = 21$$

$$A = \frac{21}{18} = \frac{7}{6}$$

Thus, for the system to have $x = 3$, $y = 1$ as a solution, then $A = \dfrac{7}{6}$ and $B = -\dfrac{1}{2}$.

73. Answers will vary. One possibility follows:
$$\begin{cases} x+y=3 \\ x-y=-5 \end{cases}$$

75.
$$\begin{cases} 3x+y=5 & (1) \\ x+y=3 & (2) \\ x+3y=7 & (3) \end{cases}$$

Multiply both sides of equation (2) by -1 and add the result to equation (1).
$$\begin{cases} 3x+y=5 \\ \underline{-x-y=-3} \\ 2x \quad\quad =2 \\ x=1 \end{cases}$$

Substituting 1 for x in equation (1), we obtain
$$3(1)+y=5$$
$$3+y=5$$
$$y=2$$

The point $(1, 2)$ is a solution to the first two equations. Check to see if it is a solution to equation (3).
$$x+3y=7$$
$$(1)+3(2)\overset{?}{=}7$$
$$1+6=7 \checkmark$$

The solution of the system is the ordered pair $(1, 2)$.

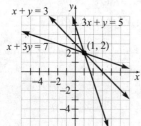

77.
$$\begin{cases} y=\dfrac{2}{3}x-5 & (1) \\ 4x-6y=30 & (2) \\ x-5y=11 & (3) \end{cases}$$

Substituting $\dfrac{2}{3}x-5$ for y in equation (2), we

obtain
$$4x-6\left(\frac{2}{3}x-5\right)=30$$
$$4x-4x+30=30$$
$$30=30$$
These two equations form a dependent system.

Substituting $\dfrac{2}{3}x-5$ for y in equation (3), we

obtain

$$x-5\left(\frac{2}{3}x-5\right)=11$$
$$x-\frac{10}{3}x+25=11$$
$$-\frac{7}{3}x=-14$$
$$x=6$$

Substituting 6 for x in equation (1), we obtain
$$y=\frac{2}{3}(6)-5$$
$$y=4-5$$
$$y=-1$$
The solution of the system is the ordered pair $(6, -1)$.

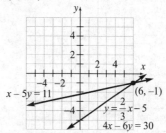

79. Yes. The method of elimination is preferred over the method of substitution if substitution leads to fractions.

81. The lines intersect at the point $(3, -2)$.

83.
$$\begin{cases} y=3x-1 & (1) \\ y=-2x+5 & (2) \end{cases}$$

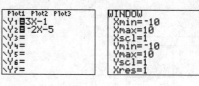

The solution is the ordered pair $(1.2, 2.6)$.

85.
$$\begin{cases} 3x-y=-1 & (1) \\ -4x+y=-3 & (2) \end{cases}$$

Writing each equation in slope-intercept form, we obtain $y=3x+1$ and $y=4x-3$.

The solution is the ordered pair $(4, 13)$.

87. $\begin{cases} 4x - 3y = 1 & (1) \\ -8x + 6y = -2 & (2) \end{cases}$

Writing each equation in slope-intercept form, we obtain the same equation for both,

$$y = \frac{4}{3}x - \frac{1}{3}.$$

The two lines are coincident. The system is dependent. The solution is $\{(x, y) \mid 4x - 3y = 1\}$.

89. $\begin{cases} 2x - 3y = 12 & (1) \\ 5x + y = -2 & (2) \end{cases}$

Writing each equation in slope-intercept form, we obtain $y = \frac{2}{3}x - 4$ and $y = -5x - 2$.

The solution is approximately the ordered pair $(0.35, -3.76)$.

Appendix C.2

Preparing for Systems of Linear Equations in Three Variables

P1. Substitute $x = 1$, $y = -2$, and $z = 3$:
$$3x - 2y + 4z = 3(1) - 2(-2) + 4(3)$$
$$= 3 + 4 + 12$$
$$= 19$$

Appendix C.2 Quick Checks

1. inconsistent; consistent; dependent

2. solution

3. False. For example, if the graphs of the equations are parallel planes, then the system will have no solution.

4. True

5. **(a)** Substitute $x = 3$, $y = 2$, and $z = -2$ into all three equations.
   Equation (1): $3 + 2 + (-2) \overset{?}{=} 3$
   $$3 = 3$$

   Equation (2): $3(3) + 2 - 2(-2) \overset{?}{=} -23$
   $$9 + 2 + 4 \overset{?}{=} -23$$
   $$15 \neq -23$$

   Equation (3): $-2(3) - 3(2) + 2(-2) \overset{?}{=} 17$
   $$-6 - 6 - 4 \overset{?}{=} 17$$
   $$-16 \neq 17$$

   Although these values satisfy equation (1), they do not satisfy equations (2) and (3). Therefore, the ordered triple $(3, 2, -2)$ is not a solution of the system.

   (b) Substitute $x = -4$, $y = 1$, and $z = 6$ into all three equations.
   Equation (1): $-4 + 1 + 6 \overset{?}{=} 3$
   $$3 = 3$$

   Equation (2): $3(-4) + 1 - 2(6) \overset{?}{=} -23$
   $$-12 + 1 - 12 \overset{?}{=} -23$$
   $$-23 = -23$$

   Equation (3): $-2(-4) - 3(1) + 2(6) \overset{?}{=} 17$
   $$8 - 3 + 12 \overset{?}{=} 17$$
   $$17 = 17$$

   Because these values satisfy all three equations, the ordered triple $(-4, 1, 6)$ is a solution of the system.

6. $\begin{cases} x + y + z = -3 & (1) \\ 2x - 2y - z = -7 & (2) \\ -3x + y + 5z = 5 & (3) \end{cases}$

   Multiply both sides of equation (1) by -2 and add the result to equation (2).

$$-2x - 2y - 2z = 6$$
$$2x - 2y - z = -7$$
$$\overline{\qquad -4y - 3z = -1}$$

Multiply both sides of equation (1) by 3 and add the result to equation (3).
$$3x + 3y + 3z = -9$$
$$-3x + y + 5z = 5$$
$$\overline{\qquad 4y + 8z = -4}$$

Rewriting the system, we have
$$\begin{cases} x + y + z = -3 & (1) \\ -4y - 3z = -1 & (2) \\ 4y + 8z = -4 & (3) \end{cases}$$

Add equations (2) and (3).
$$-4y - 3z = -1$$
$$4y + 8z = -4$$
$$\overline{\qquad 5z = -5}$$

Once again rewriting the system, we have
$$\begin{cases} x + y + z = -3 & (1) \\ -4y - 3z = -1 & (2) \\ 5z = -5 & (3) \end{cases}$$

Solving equation (3) for z, we obtain
$$5z = -5$$
$$z = -1$$

Back-substituting -1 for z in equation (2), we obtain
$$-4y - 3(-1) = -1$$
$$-4y + 3 = -1$$
$$-4y = -4$$
$$y = 1$$

Back-substituting 1 for y and -1 for z in equation (1), we obtain
$$x + 1 + (-1) = -3$$
$$x = -3$$
The solution is the ordered triple $(-3, 1, -1)$.

7. $\begin{cases} 2x \qquad - 4z = -7 & (1) \\ x + 6y \qquad = 5 & (2) \\ 2y - z = 2 & (3) \end{cases}$

Multiply both sides of equation (2) by -2 and add the result to equation (1).

$$-2x - 12y \qquad = -10$$
$$2x \qquad - 4z = -7$$
$$\overline{\qquad -12y - 4z = -17}$$

Rewriting the system, we have
$$\begin{cases} 2x \qquad - 4z = -7 & (1) \\ -12y - 4z = -17 & (2) \\ 2y - z = 2 & (3) \end{cases}$$

Multiply both sides of equation (3) by 6 and add the result to equation (2).

$$12y - 6z = 12$$
$$-12y - 4z = -17$$
$$\overline{\qquad -10z = -5}$$

Once again rewriting the system, we have
$$\begin{cases} 2x \qquad - 4z = -7 & (1) \\ -12y - 4z = -17 & (2) \\ -10z = -5 & (3) \end{cases}$$

Solving equation (3) for z, we obtain
$$-10z = -5$$
$$z = \frac{-5}{-10} = \frac{1}{2}$$

Back-substituting $\dfrac{1}{2}$ for z in equation (2), we obtain
$$-12y - 4\left(\frac{1}{2}\right) = -17$$
$$-12y - 2 = -17$$
$$-12y = -15$$
$$y = \frac{-15}{-12} = \frac{5}{4}$$

Back-substituting $\dfrac{1}{2}$ for z in equation (1), we obtain
$$2x - 4\left(\frac{1}{2}\right) = -7$$
$$2x - 2 = -7$$
$$2x = -5$$
$$x = -\frac{5}{2}$$

The solution is the ordered triple $\left(-\dfrac{5}{2}, \dfrac{5}{4}, \dfrac{1}{2}\right)$.

8. $\begin{cases} x - y + 2z = -7 & (1) \\ -2x + y - 3z = 5 & (2) \\ x - 2y + 3z = 2 & (3) \end{cases}$

Multiply both sides of equation (1) by 2 and add the result to equation (2).

$2x - 2y + 4z = -14$
$-2x + y - 3z = 5$
$\overline{\ -y + z = -9}$

Multiply both sides of equation (1) by -1 and add the result to equation (3).

$-x + y - 2z = 7$
$x - 2y + 3z = 2$
$\overline{\ -y + z = 9}$

Rewriting the system, we have

$\begin{cases} x - y + 2z = -7 & (1) \\ -y + z = -9 & (2) \\ -y + z = 9 & (3) \end{cases}$

Multiply both sides of equation (2) by -1 and add the result to equation (3).

$y - z = 9$
$-y + z = 9$
$\overline{\ 0 = 18}$

Once again rewriting the system, we have

$\begin{cases} x - y + 2z = -7 & (1) \\ -y + z = -9 & (2) \\ 0 = 18 & (3) \quad \text{False} \end{cases}$

Equation (3) is a false statement (contradiction). Therefore, the system is inconsistent. The solution set is $\varnothing$ or $\{ \ \}$.

9. $\begin{cases} x - y + 3z = 2 & (1) \\ -x + 2y - 5z = -3 & (2) \\ 2x - y + 4z = 3 & (3) \end{cases}$

Add equations (1) and (2).

$x - y + 3z = 2$
$-x + 2y - 5z = -3$
$\overline{\ y - 2z = -1}$

Multiply both sides of equation (1) by -2 and add the result to equation (3).

$-2x + 2y - 6z = -4$
$2x - y + 4z = 3$
$\overline{\ y - 2z = -1}$

Rewriting the system, we have

$\begin{cases} x - y + 3z = 2 & (1) \\ y - 2z = -1 & (2) \\ y - 2z = -1 & (3) \end{cases}$

Multiply both sides of equation (2) by -1 and add the result to equation (3).

$-y + 2z = 1$
$y - 2z = -1$
$\overline{\ 0 = 0}$

Once again rewriting the system, we have

$\begin{cases} x - y + 3z = 2 & (1) \\ y - 2z = -1 & (2) \\ 0 = 0 & (3) \quad \text{True} \end{cases}$

Thus, the system is dependent and has an infinite number of solutions.

Solve equation (2) for y.
$y = 2z - 1$

Substituting $2z - 1$ for y in equation (1), we obtain

$x - (2z - 1) + 3z = 2$
$x - 2z + 1 + 3z = 2$
$x + z + 1 = 2$
$x + z = 1$
$x = -z + 1$

The solution to the system is

$\{(x, y, z) \,|\, x = -z + 1, y = 2z - 1, z \text{ is any real number}\}$

10. Let x represent the number of 21-inch mowers, let y represent the number of 24-inch mowers, and let z represent the number of 40-inch riding mowers.

$\begin{cases} 2x + 3y + 4z = 81 & (1) \\ 3x + 3y + 4z = 95 & (2) \\ x + y + 2z = 35 & (3) \end{cases}$

Multiply both sides of equation (3) by -2 and add the result to equation (1).

$-2x - 2y - 4z = -70$
$2x + 3y + 4z = 81$
$\overline{\ y = 11}$

Multiply both sides of equation (3) by -3 and add the result to equation (2).

$$-3x - 3y - 6z = -105$$
$$3x + 3y + 4z = 95$$
$$-2z = -10$$

Rewriting the system, we have

$$\begin{cases} 2x + 3y + 4z = 81 & (1) \\ \quad y \quad\quad = 11 & (2) \\ \quad\quad -2z = -10 & (3) \end{cases}$$

From equation (2), we know $y = 11$. Solving equation (3) for z, we obtain
$$-2z = -10$$
$$z = 5$$

Substituting 11 for y and 5 for z and 11 in equation (1), we obtain
$$2x + 3(11) + 4(5) = 81$$
$$2x + 33 + 20 = 81$$
$$2x + 53 = 81$$
$$2x = 28$$
$$x = 14$$

The company can manufacture 14 twenty-one-inch mowers, 11 twenty-four-inch mowers, and 5 forty-inch rider mowers.

Appendix C.2 Exercises

11. (a) Substitute $x = 6$, $y = 2$, $z = -1$ into all three equations.

$$\begin{cases} 6 + 2 + 2(-1) = 6 + 2 - 2 = 6 \\ -2(6) - 3(2) + 5(-1) = -12 - 6 - 5 = -23 \neq 1 \\ 2(6) + 2 + 3(-1) = 12 + 2 - 3 = 11 \neq 5 \end{cases}$$

Although these values satisfy equation (1), they do not satisfy equations (2) and (3). Therefore, the ordered triple $(6, 2, -1)$ is not a solution of the system.

(b) Substitute $x = -3$, $y = 5$, $z = 2$ into all three equations).

$$\begin{cases} -3 + 5 + 2(2) = -3 + 5 + 4 = 6 \\ -2(-3) - 3(5) + 5(2) = 6 - 15 + 10 = 1 \\ 2(-3) + 5 + 3(2) = -6 + 5 + 6 = 5 \end{cases}$$

Because these values satisfy all three equations, the ordered triple $(-3, 5, 2)$ is a solution of the system.

13.
$$\begin{cases} x + y + z = 5 & (1) \\ -2x - 3y + 2z = 8 & (2) \\ 3x - y - 2z = 3 & (3) \end{cases}$$

Multiply both sides of equation (1) by 2 and add

the result to equation (2).
$$2x + 2y + 2z = 10$$
$$-2x - 3y + 2z = 8$$
$$\overline{\quad\quad -y + 4z = 18 \quad (4)}$$

Multiply both sides of equation (1) by -3 and add the result to equation (3).
$$-3x - 3y - 3z = -15$$
$$3x - y - 2z = 3$$
$$\overline{\quad\quad -4y - 5z = -12 \quad (5)}$$

Multiply both sides of equation (4) by -4 and add the result to equation (5).

$$4y - 16z = -72$$
$$-4y - 5z = -12$$
$$\overline{\quad -21z = -84}$$
$$z = 4$$

Substituting 4 for z in equation (4), we obtain
$$-y + 4(4) = 18$$
$$-y + 16 = 18$$
$$-y = 2$$
$$y = -2$$

Substituting -2 for y and 4 for z in equation (1), we obtain
$$x + (-2) + 4 = 5$$
$$x + 2 = 5$$
$$x = 3$$
The solution is the ordered triple $(3, -2, 4)$.

15.
$$\begin{cases} x - 3y + z = 13 & (1) \\ 3x + y - 4z = 13 & (2) \\ -4x - 4y + 2z = 0 & (3) \end{cases}$$

Multiply equation (1) by -3 and add the result to equation (2).
$$-3x + 9y - 3z = -39$$
$$3x + y - 4z = 13$$
$$\overline{\quad 10y - 7z = -26 \quad (4)}$$

Multiply both sides of equation (1) by 4 and add the result to equation (3).
$$4x - 12y + 4z = 52$$
$$-4x - 4y + 2z = 0$$
$$\overline{\quad -16y + 6z = 52 \quad (5)}$$

Multiply equation (4) by 8, multiply equation (5) by 5, and add the results.

$$80y - 56z = -208$$
$$\underline{-80y + 30z = 260}$$
$$-26z = 52$$
$$z = -2$$

Substituting -2 for z in equation (4) and solving for y, we obtain
$$10y - 7(-2) = -26$$
$$10y + 14 = -26$$
$$10y = -40$$
$$y = -4$$

Substituting -2 for z and -4 for y in equation (1), we obtain

$$x - 3(-4) + (-2) = 13$$
$$x + 12 - 2 = 13$$
$$x + 10 = 13$$
$$x = 3$$

The solution is the ordered triple $(3, -4, -2)$.

17. $\begin{cases} x - 4y + z = 5 & (1) \\ 4x + 2y + z = 2 & (2) \\ -4x + y - 3z = -8 & (3) \end{cases}$

Multiply both sides of equation (1) by -4 and add the result to equation (2).
$$-4x + 16y - 4z = -20$$
$$\underline{4x + 2y + \ z = 2}$$
$$18y - 3z = -18 \qquad (4)$$

Add equations (2) and (3).
$$4x + 2y + z = 2$$
$$\underline{-4x + y - 3z = -8}$$
$$3y - 2z = -6 \qquad (5)$$

Multiply both sides of equation (5) by -6 and add the result to equation (4).

$$-18y + 12z = 36$$
$$\underline{18y - 3z = -18}$$
$$9z = 18$$
$$z = 2$$

Substituting 2 for z in equation (5), we obtain
$$3y - 2(2) = -6$$
$$3y - 4 = -6$$
$$3y = -2$$
$$y = -\frac{2}{3}$$

Substituting $-\frac{2}{3}$ for y and 2 for z in equation (1), we obtain

$$x - 4\left(-\frac{2}{3}\right) + 2 = 5$$
$$x + \frac{8}{3} + 2 = 5$$
$$3\left(x + \frac{8}{3} + 2\right) = 3(5)$$
$$3x + 8 + 6 = 15$$
$$3x = 1$$
$$x = \frac{1}{3}$$

The solution is the ordered triple $\left(\frac{1}{3}, -\frac{2}{3}, 2\right)$.

19. $\begin{cases} x - 3y = 12 & (1) \\ 2y - 3z = -9 & (2) \\ 2x + z = 7 & (3) \end{cases}$

Multiply both sides of equation (1) by -2 and add the result to equation (3).
$$-2x + 6y = -24$$
$$\underline{2x + z = 7}$$
$$6y + z = -17 \qquad (4)$$

Multiply both sides of equation (4) by 3 and add the result to equation (2).
$$18y + 3z = -51$$
$$\underline{2y - 3z = -9}$$
$$20y = -60$$
$$y = -3$$

Substituting -3 for y in equation (4), we obtain
$$6(-3) + z = -17$$
$$-18 + z = -17$$
$$z = 1$$
Substituting -3 for y in equation (1), we obtain
$$x - 3(-3) = 12$$
$$x + 9 = 12$$
$$x = 3$$
The solution is the ordered triple $(3, -3, 1)$.

21. $\begin{cases} x + y - 2z = 6 & (1) \\ -2x - 3y + z = 12 & (2) \\ -3x - 4y + 3z = 2 & (3) \end{cases}$

Multiply both sides of equation (1) by 3, and add the result to equation (2).

$$3x+3y-6z=18$$
$$\underline{-2x-3y+z=12}$$
$$x-5z=30 \quad (4)$$

Multiply both sides of equation (1) by 4, and add the result to equation (3).

$$4x+4y-8z=24$$
$$\underline{-3x-4y+3z=2}$$
$$x-5z=26 \quad (5)$$

Multiply both sides of equation (4) by −1, and add the result to equation (5).

$$-x+5z=-30$$
$$\underline{x-5z=26}$$
$$0=-4$$

Because we have arrived at a contradiction, this system of equations has no solution.

23. $\begin{cases} x+y+z=4 & (1) \\ -2x-y+2z=6 & (2) \\ x+2y+5z=18 & (3) \end{cases}$

Add equations (1) and (2).

$$x+y+z=4$$
$$\underline{-2x-y+2z=6}$$
$$-x+3z=10 \quad (4)$$

Multiply both sides of equation (2) by 2, and add the result to equation (3).

$$-4x-2y+4z=12$$
$$\underline{x+2y+5z=18}$$
$$-3x+9z=30 \quad (5)$$

Multiply both sides of equation (4) by −3, and add the result to equation (5).

$$3x-9z=-30$$
$$\underline{-3x+9z=30}$$
$$0=0 \quad \text{True}$$

Thus, the system is dependent and has an infinite number of solutions.
Solve equation (4) for x.

$$-x+3z=10$$
$$x=3z-10$$

Substituting $3z-10$ for x in equation (1), we obtain

$$(3z-10)+y+z=4$$
$$y+4z-10=4$$
$$y=-4z+14$$

The solution to the system is
$$\{(x,y,z)\mid x=3z-10, y=-4z+14,$$
$$z \text{ is any real number}\}$$

25. $\begin{cases} x+3z=5 & (1) \\ -2x+y=1 & (2) \\ y+6z=11 & (3) \end{cases}$

Multiply both sides of equation (1) by 2, and add the result to equation (2).

$$2x+6z=10$$
$$\underline{-2x+y=1}$$
$$y+6z=11 \quad (4)$$

Multiply both sides of equation (3) by −1, and add the result to equation (4).

$$-y-6z=-11$$
$$\underline{y+6z=11}$$
$$0=0 \quad \text{True}$$

Thus, the system is dependent and has an infinite number of solutions.
Solve equation (4) for y.

$$y+6z=11$$
$$y=-6z+11$$

Substituting $-6z+11$ for y in equation (2), we obtain

$$-2x+(-6z+11)=1$$
$$-2x-6z+11=1$$
$$-2x=6z-10$$
$$x=-3z+5$$

The solution to the system is
$$\{(x,y,z)\mid x=-3z+5, y=-6z+11,$$
$$z \text{ is any real number}\}$$

27. $\begin{cases} 2x-y+2z=1 & (1) \\ -2x+3y-2z=3 & (2) \\ 4x-y+6z=7 & (3) \end{cases}$

Add equations (1) and (2).

$$2x-y+2z=1$$
$$\underline{-2x+3y-2z=3}$$
$$2y=4$$
$$y=2$$

Multiply both sides of equation (2) by 2 and add the result to equation (3).

$$-4x+6y-4z=6$$
$$\underline{4x-y+6z=7}$$
$$5y+2z=13 \quad (4)$$

Substituting 2 for y in equation (4), we obtain

$$5(2)+2z=13$$
$$10+2z=13$$
$$2z=3$$
$$z=\frac{3}{2}$$

Substituting 2 for y and $\frac{3}{2}$ for z in equation (1), we obtain

$$2x-2+2\left(\frac{3}{2}\right)=1$$
$$2x-2+3=1$$
$$2x=0$$
$$x=0$$

The solution is the ordered triple $\left(0,\ 2,\ \frac{3}{2}\right)$.

29.
$$\begin{cases} x-y+z=5 & (1) \\ -2x+y-z=2 & (2) \\ x-2y+2z=1 & (3) \end{cases}$$

Multiply both sides of equation (1) by 2 and add the result to equation (2).

$$\begin{array}{r} 2x-2y+2z=10 \\ -2x+y-z=2 \\ \hline -y+z=12 \qquad (4) \end{array}$$

Multiply both sides of equation (3) by 2 and add the result to equation (2).

$$\begin{array}{r} 2x-4y+4z=2 \\ -2x+y-z=2 \\ \hline -3y+3z=4 \qquad (5) \end{array}$$

Multiply both sides of equation (4) by -3 and add the result to equation (5).

$$\begin{array}{r} 3y-3z=-36 \\ -3y+3z=4 \\ \hline 0=-32 \qquad \text{False} \end{array}$$

The system has no solution. The solution set is $\varnothing$ or $\{\ \}$. The system is inconsistent.

31.
$$\begin{cases} 2y-z=-3 & (1) \\ -2x+3y=10 & (2) \\ 4x+3z=-11 & (3) \end{cases}$$

Multiply both sides of equation (2) by 2 and add the result to equation (3).

$$\begin{array}{r} -4x+6y=20 \\ 4x+3z=-11 \\ \hline 6y+3z=9 \qquad (4) \end{array}$$

Multiply both sides of equation (1) by 3 and add the result to equation (4).

$$\begin{array}{r} 6y-3z=-9 \\ 6y+3z=9 \\ \hline 12y=0 \\ y=0 \end{array}$$

Substituting 0 for y in equation (1), we obtain

$$2(0)-z=-3$$
$$-z=-3$$
$$z=3$$

Substituting 0 for y in equation (2), we obtain

$$-2x+3(0)=10$$
$$-2x=10$$
$$x=-5$$

The solution is the ordered triple $(-5,\ 0,\ 3)$.

33.
$$\begin{cases} x-2y+z=5 & (1) \\ -2x+y-z=2 & (2) \\ x-5y-4z=8 & (3) \end{cases}$$

Add equations (1) and (2).

$$\begin{array}{r} x-2y+z=5 \\ -2x+y-z=2 \\ \hline -x-y=7 \quad (4) \end{array}$$

Multiply equation (1) by 4 and add to equation (3).

$$\begin{array}{r} 4x-8y+4z=20 \\ x-5y-4z=8 \\ \hline 5x-13y=28 \quad (5) \end{array}$$

Multiply equation (4) by 5 and add to equation (5).

$$\begin{array}{r} -5x-5y=35 \\ 5x-13y=28 \\ \hline -18y=63 \end{array}$$
$$y=\frac{63}{-18}=-\frac{7}{2}$$

Substituting $-\frac{7}{2}$ for y in equation (4), we obtain

$$-2-\left(-\frac{7}{2}\right)=7$$
$$-x+\frac{7}{2}=7$$
$$-x=\frac{7}{2}$$
$$x=-\frac{7}{2}$$

Substituting $-\frac{7}{2}$ for x and $-\frac{7}{2}$ for y in equation

(1), we obtain

$$-\frac{7}{2}-2\left(-\frac{7}{2}\right)+z=5$$

$$\frac{7}{2}+z=5$$

$$z=\frac{3}{2}$$

The solution is the ordered triple $\left(-\frac{7}{2},-\frac{7}{2},\frac{3}{2}\right)$.

35. $\begin{cases} x+2y-z=1 & (1) \\ 2x+7y+4z=11 & (2) \\ x+3y+z=4 & (3) \end{cases}$

Multiply both sides of equation (1) by –2 and add the result to equation (2).

$$-2x-4y+2z=-2$$
$$\underline{2x+7y+4z=11}$$
$$3y+6z=9 \qquad (4)$$

Multiply both sides of equation (1) by –1 and add the result to equation (3).

$$-x-2y+z=-1$$
$$\underline{x+3y+z=4}$$
$$y+2z=3 \qquad (5)$$

Multiply both sides of equation (5) by –3 and add the result to equation (4).

$$-3y-6z=-9$$
$$\underline{3y+6z=9}$$
$$0=0 \qquad \text{True}$$

Thus, the system is dependent and has an infinite number of solutions.

Solve equation (5) for y.
$$y=-2z+3$$
Substituting $-2z+3$ for y in equation (1), we obtain

$$x+2(-2z+3)-z=1$$
$$x-4z+6-z=1$$
$$x-5z+6=1$$
$$x-5z=-5$$
$$x=5z-5$$

The solution to the system is
$$\{(x,y,z)\,|\,x=5z-5,$$

$$y=-2z+3, z \text{ is any real number}\}.$$

37. $\begin{cases} x+y+z=5 & (1) \\ 3x+4y+z=16 & (2) \\ -x-4y+z=-6 & (3) \end{cases}$

Add equations (1) and (3).

$$x+y+z=5$$
$$\underline{-x-4y+z=-6}$$
$$-3y+2z=-1 \qquad (4)$$

Multiply both sides of equation (3) by 3 and add the result to equation (2).

$$-3x-12y+3z=-18$$
$$\underline{3x+4y+z=16}$$
$$-8y+4z=-2 \qquad (5)$$

Multiply both sides of equation (4) by –2 and add the result to equation (5).

$$6y-4z=2$$
$$\underline{-8y+4z=-2}$$
$$-2y=0$$
$$y=0$$

Substituting 0 for y in equation (4), we obtain
$$-3(0)+2z=-1$$
$$2z=-1$$
$$z=-\frac{1}{2}$$

Substituting 0 for y and $-\frac{1}{2}$ for z in equation (1), we obtain

$$x+0+\left(-\frac{1}{2}\right)=5$$
$$x=\frac{11}{2}$$

The solution is the ordered triple $\left(\frac{11}{2},\,0,-\frac{1}{2}\right)$.

39. $\begin{cases} x+y+z=3 & (1) \\ -x+\frac{1}{2}y+z=\frac{1}{2} & (2) \\ -x+2y+3z=4 & (3) \end{cases}$

Add equations (1) and (3).

$$x+y+z=3$$
$$\underline{-x+2y+3z=4}$$
$$3y+4z=7 \qquad (4)$$

Multiply both sides of equation (1) by 2, multiply both sides of equation (2) by 2, and add the results.

$$2x + 2y + 2z = 6$$
$$\underline{-2x + y + 2z = 1}$$
$$3y + 4z = 7 \qquad (5)$$

Multiply both sides of equation (4) by –1 and add the result to equation (5).

$$-3y - 4z = -7$$
$$\underline{3y + 4z = 7}$$
$$0 = 0 \qquad \text{True}$$

Thus, the system is dependent and has an infinite number of solutions.

Solve equation (5) for y.
$$3y + 4z = 7$$
$$3y = -4z + 7$$
$$y = -\frac{4}{3}z + \frac{7}{3}$$

Substituting $-\frac{4}{3}z + \frac{7}{3}$ for y in equation (1), we obtain

$$x + \left(-\frac{4}{3}z + \frac{7}{3}\right) + z = 3$$
$$x - \frac{1}{3}z + \frac{7}{3} = 3$$
$$x = \frac{1}{3}z + \frac{2}{3}$$

The solution to the system is
$$\left\{ (x, y, z) \middle| x = \frac{1}{3}z + \frac{2}{3}, \right.$$
$$\left. y = -\frac{4}{3}z + \frac{7}{3}, z \text{ is any real number} \right\}.$$

41. Answers will vary. One possibility follows. For the ordered triple $(2, -1, 3)$, evaluate the expressions $x + y + z$, $x - y + z$, and $x + y - z$:
$$x + y + z = 2 + (-1) + 3 = 4$$
$$x - y + z = 2 - (-1) + 3 = 6$$
$$x + y - z = 2 + (-1) - 3 = -2$$
So, the system below has the solution $(2, -1, 3)$.
$$\begin{cases} x + y + z = 4 \\ x - y + z = 6 \\ x + y - z = -2 \end{cases}$$

43. (a) If $f(-1) = -6$, then $a(-1)^2 + b(-1) + c = -6$ or $a - b + c = -6$.
If $f(2) = 3$, then $a(2)^2 + b(2) + c = 3$ or $4a + 2b + c = 3$.

(b) To find a, b, and c, we must solve the following system:
$$\begin{cases} a + b + c = 4 & (1) \\ a - b + c = -6 & (2) \\ 4a + 2b + c = 3 & (3) \end{cases}$$
Add equations (1) and (2).
$$a + b + c = 4$$
$$\underline{a - b + c = -6}$$
$$2a \quad + 2c = -2 \qquad (4)$$

Multiply both sides of equation (2) by 2 and add the result to equation (3).
$$2a - 2b + 2c = -12$$
$$\underline{4a + 2b + c = 3}$$
$$6a \quad + 3c = -9 \qquad (5)$$

Divide both sides of equation (4) by 2, divide equation (5) by –3, and add the results.
$$a + c = -1$$
$$\underline{-2a - c = 3}$$
$$-a = 2$$
$$a = -2$$
Substituting –2 for a in equation (4), we obtain
$$2(-2) + 2c = -2$$
$$-4 + 2c = -2$$
$$2c = 2$$
$$c = 1$$
Substituting –2 for a and 1 for c in equation (1), we obtain

$$-2 + b + 1 = 4$$
$$b = 5$$
Thus, the quadratic equation that contains $(-1, -6)$, $(1, 4)$, and $(2, 3)$ is
$$f(x) = -2x^2 + 5x + 1.$$

45. Rewrite the system with each equation in standard form.
$$\begin{cases} i_1 - i_2 + i_3 = 0 & (1) \\ -3i_1 \quad + 2i_3 = 3 & (2) \\ 4i_2 + 2i_3 = 22 & (3) \end{cases}$$
Multiply both sides of equation (1) by 3 and add

the result to equation (2).

$$3i_1 - 3i_2 + 3i_3 = 0$$
$$\underline{-3i_1 \qquad + 2i_3 = 3}$$
$$-3i_2 + 5i_3 = 3 \qquad (4)$$

Multiply both sides of equation (3) by 3, multiply both sides of equation (4) by 4, and add the results.

$$12i_2 + 6i_3 = 66$$
$$\underline{-12i_2 + 20i_3 = 12}$$
$$26i_3 = 78$$
$$i_3 = 3$$

Substituting 3 for i_3 in equation (2), we obtain

$$-3i_1 + 2(3) = 3$$
$$-3i_1 + 6 = 3$$
$$-3i_1 = -3$$
$$i_1 = 1$$

Substituting 3 for i_3 and 1 for i_1 in equation (1), we obtain

$$1 - i_2 + (3) = 0$$
$$-i_2 + 4 = 0$$
$$-i_2 = -4$$
$$i_2 = 4$$

The currents are $i_1 = 1$, $i_2 = 4$, and $i_3 = 3$.

47. Let b represent the number of box seats, let r represent the number of reserved seats, and let l represent the number of lawn seats.

$$\begin{cases} b + r + l = 4100 & (1) \\ 9b + 7r + 5l = 28,400 & (2) \\ \frac{1}{2}(9b) + \frac{1}{2}(7r) + 5l = 18,300 & (3) \end{cases}$$

Simplify equation (3) and rewrite the system.

$$\frac{1}{2}(9b) + \frac{1}{2}(7r) + 5l = 18,300$$
$$4.5b + 3.5r + 5l = 18,300$$

Rewriting the system, we have

$$\begin{cases} b + r + l = 4100 & (1) \\ 9b + 7r + 5l = 28,400 & (2) \\ 4.5b + 3.5r + 5l = 18,300 & (3) \end{cases}$$

Multiply both sides of equation (1) by -9 and add the result to equation (2).

$$-9b - 9r - 9l = -36,900$$
$$\underline{9b + 7r + 5l = 28,400}$$
$$-2r - 4l = -8500 \qquad (4)$$

Multiply both sides of equation (3) by -2 and add the result to equation (2).

$$-9b - 7r - 10l = -36,600$$
$$\underline{9b + 7r + 5l = 28,400}$$
$$-5l = -8200$$
$$l = 1640$$

Substituting 1640 for l in equation (4), we obtain

$$-2r - 4(1640) = -8500$$
$$-2r - 6460 = -8500$$
$$-2r = -1940$$
$$r = 970$$

Substituting 1640 for l and 970 for r in equation (1), we obtain

$$b + 970 + 1640 = 4100$$
$$b + 2610 = 4100$$
$$b = 1490$$

There are 1490 box seats, 970 reserve seats, and 1640 lawn seats in the stadium.

49. Let c represent the number servings of Chex® cereal, let m represent the number of servings of 2% milk, and let j represent the number of servings of orange juice.

$$\begin{cases} 220c + 125m \qquad\;\; = 470 & \text{(sodium)} & (1) \\ 26c + 12m + 26j = 89 & \text{(carbs)} & (2) \\ c + 8m + 2j = 20 & \text{(protein)} & (3) \end{cases}$$

Multiply both sides of equation (3) by -13 and add the result to equation (2).

$$-13c - 104m - 26j = -260$$
$$\underline{26c + 12m + 26j = 89}$$
$$13c - 92m \qquad\;\; = -171 \qquad (4)$$

Multiply both sides of equation (1) by 13, multiply both sides of equation (4) by -220, and add the results.

$$2860c + 1625m = 6110$$
$$\underline{-2860c + 20,240m = 37,620}$$
$$21,865m = 43,730$$
$$m = 2$$

Substituting 2 for m in equation (1), we obtain
$$220c + 125(2) = 470$$
$$220c + 250 = 470$$
$$220c = 220$$
$$c = 1$$

Substituting 2 for m and 1 for c in equation (3), we obtain
$$1 + 8(2) + 2j = 20$$
$$17 + 2j = 20$$
$$2j = 3$$
$$j = 1.5$$
Nancy needs 1 serving of Chex® cereal, 2 servings of 2% milk, and 1.5 servings of orange juice.

51. Let t represent the amount to be invested in Treasury bills, let m represent the amount to be invested in municipal bonds, and let c represent the amount to be invested in corporate bonds.
$$\begin{cases} t + m + c = 25,000 & (1) \\ 0.03t + 0.05m + 0.09c = 1210 & (2) \\ t - c = 7000 & (3) \end{cases}$$

Multiply both sides of equation (1) by -0.05 and add the result to equation (2).
$$\begin{aligned} -0.05t - 0.05m - 0.05c &= -1250 \\ 0.03t + 0.05m + 0.09c &= 1210 \\ \hline -0.02t \qquad\quad + 0.04c &= -40 \quad (4) \end{aligned}$$

Multiply both sides of equation (3) by 0.02 and add the result to equation (4).
$$\begin{aligned} 0.02t - 0.02c &= 140 \\ -0.02t + 0.04c &= -40 \\ \hline 0.02c &= 100 \\ c &= 5000 \end{aligned}$$

Substituting 5000 for c in equation (3), we obtain
$$t - 5000 = 7000$$
$$t = 12,000$$

Substituting 5000 for c and 12,000 for t in equation (1), we obtain
$$12,000 + m + 5000 = 25,000$$
$$m + 17,000 = 25,000$$
$$m = 8000$$

Sachi should invest \$12,000 in Treasury bills, \$8000 in municipal bonds, and \$5000 in corporate bonds.

53. Let x represent the lengths of $\overline{AO}$ and $\overline{AM}$, let y represent the lengths of $\overline{OC}$ and $\overline{NC}$, and let z represent the lengths of $\overline{BM}$ and $\overline{BN}$.
Because $\overline{AO} + \overline{OC} = \overline{AC}$, $\overline{BN} + \overline{NC} = \overline{BC}$, and $\overline{AM} + \overline{MB} = \overline{AB}$, we obtain the following system.
$$\begin{cases} x + y \quad\;\; = 14 & (1) \\ x \quad\;\; + z = 6 & (2) \\ \quad\;\; y + z = 12 & (3) \end{cases}$$
Multiply both sides of equation (1) by -1 and add the result to equation (2).
$$\begin{aligned} -x - y \quad\;\; &= -14 \\ x \quad\;\; + z &= 6 \\ \hline -y + z &= -8 \quad (4) \end{aligned}$$
Add equations (3) and (4).
$$\begin{aligned} y + z &= 12 \\ -y + z &= -8 \\ \hline 2z &= 4 \\ z &= 2 \end{aligned}$$
Substituting 2 for z in equation (3), we obtain
$$y + 2 = 12$$
$$y = 10$$
Substituting 2 for z in equation (2), we obtain
$$x + 2 = 6$$
$$x = 4$$
Therefore, $\overline{AM} = 4$, $\overline{BN} = 2$, and $\overline{OC} = 10$.

55. $$\begin{cases} \dfrac{2}{5}x + \dfrac{1}{2}y - \dfrac{1}{3}z = 0 & (1) \\ \dfrac{3}{5}x - \dfrac{1}{4}y + \dfrac{1}{2}z = 10 & (2) \\ -\dfrac{1}{5}x + \dfrac{1}{4}y - \dfrac{1}{6}z = -4 & (3) \end{cases}$$
Simplify each equation and rewrite the system.

(1) $\quad 30\left(\dfrac{2}{5}x + \dfrac{1}{2}y - \dfrac{1}{3}z\right) = 30(0)$
$$12x + 15y - 10z = 0$$

(2) $\quad 20\left(\dfrac{3}{5}x - \dfrac{1}{4}y + \dfrac{1}{2}z\right) = 20(10)$
$$12x - 5y + 10z = 200$$

(3) $\quad 60\left(-\dfrac{1}{5}x + \dfrac{1}{4}y - \dfrac{1}{6}z\right) = 60(-4)$
$$-12x + 15y - 10z = -240$$

Rewriting the system, we have

$$\begin{cases} 12x + 15y - 10z = 0 & (1) \\ 12x - 5y + 10z = 200 & (2) \\ -12x + 15y - 10z = -240 & (3) \end{cases}$$

Add equations (1) and (3).

$$\begin{array}{r} 12x + 15y - 10z = 0 \\ -12x + 15y - 10z = -240 \\ \hline 30y - 20z = -240 \quad\quad (4) \end{array}$$

Add equations (2) and (3).

$$\begin{array}{r} 12x - 5y + 10z = 200 \\ -12x + 15y - 10z = -240 \\ \hline 10y \quad\quad = -40 \\ y = -4 \end{array}$$

Substituting -4 for y in equation (4), we obtain

$$\begin{aligned} 30(-4) - 20z &= -240 \\ -120 - 20z &= -240 \\ -20z &= -120 \\ z &= 6 \end{aligned}$$

Substituting -4 for y and 6 for z in equation (1), we obtain

$$\begin{aligned} 12x + 15(-4) - 10(6) &= 0 \\ 12x - 60 - 60 &= 0 \\ 12x &= 120 \\ x &= 10 \end{aligned}$$

The solution is the ordered triple $(10, -4, 6)$.

57.
$$\begin{cases} x + y + z + w = 3 & (1) \\ -2x - y + 3z - w = -1 & (2) \\ 2x + 2y - 2z + w = 2 & (3) \\ -x + 2y - 3z + 2w = 12 & (4) \end{cases}$$

Add equations (1) and (4).

$$\begin{array}{r} x + y + z + w = 3 \\ -x + 2y - 3z + 2w = 12 \\ \hline 3y - 2z + 3w = 15 \quad\quad (5) \end{array}$$

Add equations (2) and (3).

$$\begin{array}{r} -2x - y + 3z - w = -1 \\ 2x + 2y - 2z + w = 2 \\ \hline y + z \quad\quad = 1 \quad\quad (6) \end{array}$$

Multiply both sides of equation (1) by 2 and add the result to equation (2).

$$\begin{array}{r} 2x + 2y + 2z + 2w = 6 \\ -2x - y + 3z - w = -1 \\ \hline y + 5z + w = 5 \quad\quad (7) \end{array}$$

Multiply both sides of equation (7) by -3 and add the result to equation (5).

$$\begin{array}{r} -3y - 15z - 3w = -15 \\ 3y - 2z + 3w = 15 \\ \hline -17z \quad\quad = 0 \\ z = 0 \end{array}$$

Substituting 0 for z in equation (6), we obtain

$$\begin{aligned} y + 0 &= 1 \\ y &= 1 \end{aligned}$$

Substituting 0 for z and 1 for y equation (7), we obtain

$$\begin{aligned} 1 + 5(0) + w &= 5 \\ 1 + w &= 5 \\ w &= 4 \end{aligned}$$

Substituting 0 for z and 1 for y and 4 for w in equation (1), we obtain

$$\begin{aligned} x + 1 + 0 + 4 &= 3 \\ x + 5 &= 3 \\ x &= -2 \end{aligned}$$

The solution is $(-2, 1, 0, 4)$.

59. Answers will vary. One possibility follows: We must eliminate the same variable in the first step so that we will obtain a system of two equations containing two variables, which we can then solve using the methods of Section 3.1. Once we solve for the two variables, we can then back-substitute in order to find the third variable.

Appendix C.3

Preparing for Using Matrices to Solve Systems

P1. For the expression $4x - 2y + z$, the coefficients are 4, -2, and 1.

P2.
$$\begin{aligned} x - 4y &= 3 \\ x - 4y + 4y &= 3 + 4y \\ x &= 4y + 3 \end{aligned}$$

P3. Substitute $x = 1$, $y = -3$, and $z = 2$:
$$\begin{aligned} 3x - 2y + z &= 3(1) - 2(-3) + 2 \\ &= 3 + 6 + 2 \\ &= 11 \end{aligned}$$

Appendix C.3 Quick Checks

1. matrix

2. augmented

3. 4 rows; 3 columns

4. False. The augmented matrix of a system of two equations containing two unknowns has 2 rows and 3 columns.

5. In the augmented matrix, the first column represents the coefficients on the x variable. The second column represents the coefficients on the y variable. The vertical line signifies the equal signs. The third column represents the constants to the right of the equal sign.

$$\begin{bmatrix} 3 & -1 & | & -10 \\ -5 & 2 & | & 0 \end{bmatrix}$$

6. The system
$$\begin{cases} x+2y-2z=11 \\ -x-2z=4 \\ 4x-y+z-3=0 \end{cases}$$
gets rearranged as
$$\begin{cases} x+2y-2z=11 \\ -x+0y-2z=4 \\ 4x-y+z=3 \end{cases}$$
Thus, the augmented matrix is
$$\begin{bmatrix} 1 & 2 & -2 & | & 11 \\ -1 & 0 & -2 & | & 4 \\ 4 & -1 & 1 & | & 3 \end{bmatrix}.$$

7. Since the augmented matrix has two rows, it represents a system of two equations. Because there are two columns to the left of the vertical bar, the system has two variables. If we call the variables x and y, the system of equations is
$$\begin{cases} x-3y=7 \\ -2x+5y=-3 \end{cases}$$

8. Since the augmented matrix has three rows, it represents a system of three equations. Because there are three columns to the left of the vertical bar, the system has three variables. If we call the variables x, y, and z, the system of equations is
$$\begin{cases} x-3y+2z=4 \\ 3x-z=-1 \\ -x+4y=0 \end{cases}$$

9. $\begin{bmatrix} 1 & -2 & | & 5 \\ -4 & 5 & | & -11 \end{bmatrix}$ $(R_2 = 4r_1 + r_2)$

$$= \begin{bmatrix} 1 & -2 & | & 5 \\ 4(1)+(-4) & 4(-2)+5 & | & 4(5)+(-11) \end{bmatrix}$$

$$= \begin{bmatrix} 1 & -2 & | & 5 \\ 0 & -3 & | & 9 \end{bmatrix}$$

10. We want a 0 in row 1, column 2. We accomplish this by multiplying row 2 by -5 and adding the result to row 1. That is, we apply the row operation $R_1 = -5r_2 + r_1$.

$$\begin{bmatrix} 1 & 5 & | & 13 \\ 0 & 1 & | & 2 \end{bmatrix}$$ $(R_1 = -5r_2 + r_1)$

$$= \begin{bmatrix} -5(0)+1 & -5(1)+5 & | & -5(2)+13 \\ 0 & 1 & | & 2 \end{bmatrix}$$

$$= \begin{bmatrix} 1 & 0 & | & 3 \\ 0 & 1 & | & 2 \end{bmatrix}$$

11. True

12. Write the augmented matrix of the system and then put it in row echelon form.

$$\begin{bmatrix} 2 & -4 & | & 20 \\ 3 & 1 & | & 16 \end{bmatrix}$$ $\left(R_1 = \dfrac{1}{2}r_1\right)$

$$= \begin{bmatrix} 1 & -2 & | & 10 \\ 3 & 1 & | & 16 \end{bmatrix}$$ $(R_2 = -3r_1 + r_2)$

$$= \begin{bmatrix} 1 & -2 & | & 10 \\ 0 & 7 & | & -14 \end{bmatrix}$$ $\left(R_2 = \dfrac{1}{7}r_2\right)$

$$= \begin{bmatrix} 1 & -2 & | & 10 \\ 0 & 1 & | & -2 \end{bmatrix}$$

From row 2, we have that $y = -2$. Row 1 represents the equation $x - 2y = 10$. Back-substitute -2 for y and solve for x.
$$x - 2(-2) = 10$$
$$x + 4 = 10$$
$$x = 6$$
The solution is the ordered pair $(6, -2)$.

13. Write the augmented matrix of the system and then put it in row echelon form.

$$\begin{bmatrix} 1 & -1 & 2 & | & 7 \\ 2 & -2 & 1 & | & 11 \\ -3 & 1 & -3 & | & -14 \end{bmatrix} \quad \begin{pmatrix} R_2 = -2r_1 + r_2 \\ R_3 = 3r_1 + r_3 \end{pmatrix}$$

$$= \begin{bmatrix} 1 & -1 & 2 & | & 7 \\ 0 & 0 & -3 & | & -3 \\ 0 & -2 & 3 & | & 7 \end{bmatrix} \quad (\text{Interchange } r_1 \text{ and } r_2)$$

$$= \begin{bmatrix} 1 & -1 & 2 & | & 7 \\ 0 & -2 & 3 & | & 7 \\ 0 & 0 & -3 & | & -3 \end{bmatrix} \quad \begin{pmatrix} R_2 = -\dfrac{1}{2}r_2 \\ R_3 = -\dfrac{1}{3}r_3 \end{pmatrix}$$

$$= \begin{bmatrix} 1 & -1 & 2 & | & 7 \\ 0 & 1 & -\dfrac{3}{2} & | & -\dfrac{7}{2} \\ 0 & 0 & 1 & | & 1 \end{bmatrix}$$

Write the system of equations that corresponds to the row-echelon matrix

$$\begin{cases} x - y + 2z = 7 & (1) \\ y - \dfrac{3}{2}z = -\dfrac{7}{2} & (2) \\ z = 1 & (3) \end{cases}$$

Substituting 1 for z in equation (2), we obtain

$$y - \frac{3}{2}(1) = -\frac{7}{2}$$
$$y - \frac{3}{2} = -\frac{7}{2}$$
$$y = -\frac{4}{2} = -2$$

Substituting 1 for z and -2 for y in equation (1), we obtain

$$x - (-2) + 2(1) = 7$$
$$x + 2 + 2 = 7$$
$$x + 4 = 7$$
$$x = 3$$

The solution is the ordered triple $(3, -2, 1)$.

14. Write the augmented matrix of the system and then put it in row echelon form.

$$\begin{bmatrix} 2 & 3 & -2 & | & 1 \\ 2 & 0 & -4 & | & -9 \\ 4 & 6 & -1 & | & 14 \end{bmatrix} \quad \left(R_1 = \frac{1}{2}r_1 \right)$$

$$= \begin{bmatrix} 1 & \dfrac{3}{2} & -1 & | & \dfrac{1}{2} \\ 2 & 0 & -4 & | & -9 \\ 4 & 6 & -1 & | & 14 \end{bmatrix} \quad \begin{pmatrix} R_2 = -2r_1 + r_2 \\ R_3 = -4r_1 + r_3 \end{pmatrix}$$

$$= \begin{bmatrix} 1 & \dfrac{3}{2} & -1 & | & \dfrac{1}{2} \\ 0 & -3 & -2 & | & -10 \\ 0 & 0 & 3 & | & 12 \end{bmatrix} \quad \begin{pmatrix} R_2 = -\dfrac{1}{3}r_2 \\ R_3 = \dfrac{1}{3}r_3 \end{pmatrix}$$

$$= \begin{bmatrix} 1 & \dfrac{3}{2} & -1 & | & \dfrac{1}{2} \\ 0 & 1 & \dfrac{2}{3} & | & \dfrac{10}{3} \\ 0 & 0 & 1 & | & 4 \end{bmatrix}$$

Write the system of equations that corresponds to the row-echelon matrix

$$\begin{cases} x + \dfrac{3}{2}y - z = \dfrac{1}{2} & (1) \\ y + \dfrac{2}{3}z = \dfrac{10}{3} & (2) \\ z = 4 & (3) \end{cases}$$

Substituting 4 for z in equation (2), we obtain

$$y + \frac{2}{3}(4) = \frac{10}{3}$$
$$y + \frac{8}{3} = \frac{10}{3}$$
$$y = \frac{2}{3}$$

Substituting 4 for z and $\dfrac{2}{3}$ for y in equation (1), we obtain

$$x + \frac{3}{2}\left(\frac{2}{3}\right) - (4) = \frac{1}{2}$$
$$x + 1 - 4 = \frac{1}{2}$$
$$x = \frac{1}{2} + 3 = \frac{7}{2}$$

The solution is the ordered triple $\left(\dfrac{7}{2}, \dfrac{2}{3}, 4 \right)$.

15. Write the augmented matrix of the system and then put it in row echelon form.

$$\begin{bmatrix} 2 & 5 & | & -6 \\ -6 & -15 & | & 18 \end{bmatrix} \quad \left(R_1 = \frac{1}{2}r_1 \right)$$

$$= \begin{bmatrix} 1 & \dfrac{5}{2} & | & -3 \\ -6 & -15 & | & 18 \end{bmatrix} \quad (R_2 = 6r_1 + r_2)$$

$$= \begin{bmatrix} 1 & \dfrac{5}{2} & | & -3 \\ 0 & 0 & | & 0 \end{bmatrix}$$

Write the system of equations that corresponds to the row-echelon matrix.

$$\begin{cases} x + \dfrac{5}{2}y = -3 & (1) \\ 0 = 0 & (2) \end{cases}$$

The statement $0 = 0$ in equation (2) indicates that the system is dependent and has an infinite number of solutions.
The solution to the system is

$$\{(x, y) \mid 2x + 5y = -6\}.$$

16. Write the augmented matrix of the system and then put it in row echelon form.

$$\begin{bmatrix} 1 & 1 & -3 & 8 \\ 2 & 3 & -10 & 19 \\ -1 & -2 & 7 & -11 \end{bmatrix} \quad \begin{pmatrix} R_2 = -2r_1 + r_2 \\ R_3 = r_1 + r_3 \end{pmatrix}$$

$$= \begin{bmatrix} 1 & 1 & -3 & 8 \\ 0 & 1 & -4 & 3 \\ 0 & -1 & 4 & -3 \end{bmatrix} \quad (R_3 = r_2 + r_3)$$

$$= \begin{bmatrix} 1 & 1 & -3 & 8 \\ 0 & 1 & -4 & 3 \\ 0 & 0 & 0 & 0 \end{bmatrix}$$

Write the system of equations that corresponds to the row-echelon matrix.

$$\begin{cases} x + y - 3z = 8 & (1) \\ y - 4z = 3 & (2) \\ 0 = 0 & (3) \end{cases}$$

The statement $0 = 0$ in equation (3) indicates that the system is dependent and has an infinite number of solutions.

Solve equation (2) for y.
$$y - 4z = 3$$
$$y = 4z + 3$$

Substituting $4z + 3$ for y in equation (1), we obtain
$$x + (4z + 3) - 3z = 8$$
$$x + z + 3 = 8$$
$$x = -z + 5$$

The solution to the system is
$$\{(x, y, z) \mid x = -z + 5,\ y + 4z + 3,\ z \text{ is any real number}\}$$

17. Write the augmented matrix of the system and then put it in row echelon form.

$$\begin{bmatrix} -2 & 3 & 4 \\ 10 & -15 & 2 \end{bmatrix} \quad \left(R_1 = -\dfrac{1}{2}r_1 \right)$$

$$= \begin{bmatrix} 1 & -\dfrac{3}{2} & -2 \\ 10 & -15 & 2 \end{bmatrix} \quad (R_2 = -10r_1 + r_2)$$

$$= \begin{bmatrix} 1 & -\dfrac{3}{2} & -2 \\ 0 & 0 & 22 \end{bmatrix}$$

Write the system of equations that corresponds to the row-echelon matrix.

$$\begin{cases} x - \dfrac{3}{2}y = -2 & (1) \\ 0 = 22 & (2) \end{cases}$$

The statement $0 = 22$ in equation (2) indicates that the system is inconsistent. The system has no solution. The solution set is $\varnothing$ or $\{\ \}$.

18. Write the augmented matrix of the system and then put it in row echelon form.

$$\begin{bmatrix} -1 & 2 & -1 & 5 \\ 2 & 1 & 4 & 3 \\ 3 & -1 & 5 & 0 \end{bmatrix} \quad (R_1 = -1 \cdot r_1)$$

$$= \begin{bmatrix} 1 & -2 & 1 & -5 \\ 2 & 1 & 4 & 3 \\ 3 & -1 & 5 & 0 \end{bmatrix} \quad \begin{pmatrix} R_2 = -2r_1 + r_2 \\ R_3 = -3r_1 + r_3 \end{pmatrix}$$

$$= \begin{bmatrix} 1 & -2 & 1 & -5 \\ 0 & 5 & 2 & 13 \\ 0 & 5 & 2 & 15 \end{bmatrix} \quad \left(R_2 = \dfrac{1}{5}r_2 \right)$$

$$= \begin{bmatrix} 1 & -2 & 1 & -5 \\ 0 & 1 & \dfrac{2}{5} & \dfrac{13}{5} \\ 0 & 5 & 2 & 15 \end{bmatrix} \quad (R_3 = -5r_2 + r_3)$$

$$= \begin{bmatrix} 1 & -2 & 1 & -5 \\ 0 & 1 & \dfrac{2}{5} & \dfrac{13}{5} \\ 0 & 0 & 0 & 2 \end{bmatrix}$$

Write the system of equations that corresponds to the row-echelon matrix.

$$\begin{cases} x - 2y + z = -5 & (1) \\ y + \dfrac{2}{5}z = \dfrac{13}{5} & (2) \\ 0 = 2 & (3) \end{cases}$$

The statement $0 = 2$ in equation (3) indicates that the system is inconsistent. The system has no solution. The solution set is $\varnothing$ or $\{\ \}$.

Appendix C.3 Exercises

19. $\begin{bmatrix} 1 & -3 & | & 2 \\ 2 & 5 & | & 1 \end{bmatrix}$

21. $\begin{bmatrix} 1 & 1 & 1 & | & 3 \\ 2 & -1 & 3 & | & 1 \\ -4 & 2 & -5 & | & -3 \end{bmatrix}$

23. Write each equation in the system in standard form.
$$\begin{cases} -x+y=2 \\ 5x+y=-5 \end{cases}$$
Thus, the augmented matrix is
$$\begin{bmatrix} -1 & 1 & | & 2 \\ 5 & 1 & | & -5 \end{bmatrix}$$

25. Write each equation in the system in standard form.
$$\begin{cases} x & +z=2 \\ 2x+y & =13 \\ x-y+4z=-4 \end{cases}$$
Thus, the augmented matrix is
$$\begin{bmatrix} 1 & 0 & 1 & | & 2 \\ 2 & 1 & 0 & | & 13 \\ 1 & -1 & 4 & | & -4 \end{bmatrix}$$

27. $\begin{cases} 2x+5y=3 \\ -4x+y=10 \end{cases}$

29. $\begin{cases} x+5y-3z=2 \\ 3y\ -z=-5 \\ 4x\ +8z=6 \end{cases}$

31. $\begin{cases} x-2y+9z=2 \\ y-5z=8 \\ z=\dfrac{4}{3} \end{cases}$

33. (a) $\begin{bmatrix} 1 & -3 & | & 2 \\ -2 & 5 & | & 1 \end{bmatrix}$ $\quad (R_2=2r_1+r_2)$

$= \begin{bmatrix} 1 & -3 & | & 2 \\ 2(1)+(-2) & 2(-3)+5 & | & 2(2)+1 \end{bmatrix}$

$= \begin{bmatrix} 1 & -3 & | & 2 \\ 0 & -1 & | & 5 \end{bmatrix}$

(b) $\begin{bmatrix} 1 & -3 & | & 2 \\ 0 & -1 & | & 5 \end{bmatrix}$ $\left(R_2 = -1 \cdot r_2 \right)$

$= \begin{bmatrix} 1 & -3 & | & 2 \\ -1(0) & -1(-1) & | & -1(5) \end{bmatrix}$

$= \begin{bmatrix} 1 & -3 & | & 2 \\ 0 & 1 & | & -5 \end{bmatrix}$

35. (a) $\begin{bmatrix} 1 & 1 & -1 & | & 4 \\ 2 & 5 & 3 & | & -3 \\ -1 & -3 & 2 & | & 1 \end{bmatrix}$ $\left(R_2 = -2r_1 + r_2 \right)$

$= \begin{bmatrix} 1 & 1 & -1 & | & 4 \\ -2(1)+2 & -2(1)+5 & -2(-1)+3 & | & -2(4)+(-3) \\ -1 & -3 & 2 & | & 1 \end{bmatrix}$

$= \begin{bmatrix} 1 & 1 & -1 & | & 4 \\ 0 & 3 & 5 & | & -11 \\ -1 & -3 & 2 & | & 1 \end{bmatrix}$

(b) $\begin{bmatrix} 1 & 1 & -1 & | & 4 \\ 0 & 3 & 5 & | & -11 \\ -1 & -3 & 2 & | & 1 \end{bmatrix}$ $\left(R_3 = r_1 + r_3 \right)$

$= \begin{bmatrix} 1 & 1 & -1 & | & 4 \\ 0 & 3 & 5 & | & -11 \\ 1+(-1) & 1+(-3) & -1+2 & | & 4+1 \end{bmatrix}$

$= \begin{bmatrix} 1 & 1 & -1 & | & 4 \\ 0 & 3 & 5 & | & -11 \\ 0 & -2 & 1 & | & 5 \end{bmatrix}$

37. (a) $\begin{bmatrix} 1 & 1 & 1 & | & 4 \\ 0 & 5 & 3 & | & -3 \\ 0 & -4 & 2 & | & 8 \end{bmatrix}$ $\left(R_2 = r_3 + r_2 \right)$

$= \begin{bmatrix} 1 & 1 & 1 & | & 4 \\ 0+0 & -4+5 & 2+3 & | & 8+(-3) \\ 0 & -4 & 2 & | & 8 \end{bmatrix}$

$= \begin{bmatrix} 1 & 1 & 1 & | & 4 \\ 0 & 1 & 5 & | & 5 \\ 0 & -4 & 2 & | & 8 \end{bmatrix}$

(b) $\begin{bmatrix} 1 & 1 & 1 & | & 4 \\ 0 & 1 & 5 & | & 5 \\ 0 & -4 & 2 & | & 8 \end{bmatrix}$ $\left(R_3 = \dfrac{1}{2}r_3\right)$

$= \begin{bmatrix} 1 & 1 & 1 & | & 4 \\ 0 & 1 & 5 & | & 5 \\ \dfrac{1}{2}(0) & \dfrac{1}{2}(-4) & \dfrac{1}{2}(2) & | & \dfrac{1}{2}(8) \end{bmatrix}$

$= \begin{bmatrix} 1 & 1 & 1 & | & 4 \\ 0 & 1 & 5 & | & 5 \\ 0 & -2 & 1 & | & 4 \end{bmatrix}$

39. Write the augmented matrix of the system and then put it in row echelon form.

$\begin{bmatrix} 2 & 3 & | & 1 \\ -1 & 4 & | & -28 \end{bmatrix}$ $\left(R_1 = \dfrac{1}{2}r_1\right)$

$= \begin{bmatrix} 1 & \dfrac{3}{2} & | & \dfrac{1}{2} \\ -1 & 4 & | & -28 \end{bmatrix}$ $\left(R_2 = r_1 + r_2\right)$

$= \begin{bmatrix} 1 & \dfrac{3}{2} & | & \dfrac{1}{2} \\ 0 & \dfrac{11}{2} & | & -\dfrac{55}{2} \end{bmatrix}$ $\left(R_2 = \dfrac{2}{11}r_2\right)$

$= \begin{bmatrix} 1 & \dfrac{3}{2} & | & \dfrac{1}{2} \\ 0 & 1 & | & -5 \end{bmatrix}$

Write the system of equations that corresponds to the row-echelon matrix.

$\begin{cases} x + \dfrac{3}{2}y = \dfrac{1}{2} & (1) \\ y = -5 & (2) \end{cases}$

Substituting -5 for y in equation (1), we obtain

$x + \dfrac{3}{2}(-5) = \dfrac{1}{2}$

$x - \dfrac{15}{2} = \dfrac{1}{2}$

$x = \dfrac{16}{2} = 8$

The solution is the ordered pair $(8, -5)$.

41. Write the augmented matrix of the system and then put it in row echelon form.

$\begin{bmatrix} 1 & 5 & -2 & | & 23 \\ -2 & -3 & 5 & | & -11 \\ 3 & 2 & 1 & | & 4 \end{bmatrix}$ $\begin{pmatrix} R_2 = 2r_1 + r_2 \\ R_3 = -3r_1 + r_3 \end{pmatrix}$

$= \begin{bmatrix} 1 & 5 & -2 & | & 23 \\ 0 & 7 & 1 & | & 35 \\ 0 & -13 & 7 & | & -65 \end{bmatrix}$ $\left(R_2 = \dfrac{1}{7}r_2\right)$

$= \begin{bmatrix} 1 & 5 & -2 & | & 23 \\ 0 & 1 & \dfrac{1}{7} & | & 5 \\ 0 & -13 & 7 & | & -65 \end{bmatrix}$ $\left(R_3 = 13r_2 + r_3\right)$

$= \begin{bmatrix} 1 & 5 & -2 & | & 23 \\ 0 & 1 & \dfrac{1}{7} & | & 5 \\ 0 & 0 & \dfrac{62}{7} & | & 0 \end{bmatrix}$ $\left(R_3 = \dfrac{7}{62}r_3\right)$

$= \begin{bmatrix} 1 & 5 & -2 & | & 23 \\ 0 & 1 & \dfrac{1}{7} & | & 5 \\ 0 & 0 & 1 & | & 0 \end{bmatrix}$

Write the system of equations that corresponds to the row-echelon matrix.

$\begin{cases} x + 5y - 2z = 23 & (1) \\ y + \dfrac{1}{7}z = 5 & (2) \\ z = 0 & (3) \end{cases}$

By equation (3), we know $z = 0$. Substituting 0 for z in equation (2), we obtain

$y + \dfrac{1}{7}(0) = 5$

$y = 5$

Substituting 0 for z and 5 for y in equation (1), we obtain

$x + 5(5) - 2(0) = 23$

$x + 25 = 23$

$x = -2$

The solution is the ordered triple $(-2, 5, 0)$.

43. Write the augmented matrix of the system and then put it in row echelon form.

$\begin{bmatrix} 4 & 5 & 0 & | & 0 \\ -8 & 10 & -2 & | & -13 \\ 0 & 15 & 4 & | & -7 \end{bmatrix}$ $\left(R_2 = 2r_1 + r_2\right)$

$= \begin{bmatrix} 4 & 5 & 0 & | & 0 \\ 0 & 20 & -2 & | & -13 \\ 0 & 15 & 4 & | & -7 \end{bmatrix}$ $\left(R_2 = \dfrac{1}{20}r_2\right)$

$= \begin{bmatrix} 4 & 5 & 0 & | & 0 \\ 0 & 1 & -\dfrac{1}{10} & | & -\dfrac{13}{20} \\ 0 & 15 & 4 & | & -7 \end{bmatrix}$ $\left(R_3 = -15r_2 + r_3\right)$

$= \begin{bmatrix} 4 & 5 & 0 & | & 0 \\ 0 & 1 & -\dfrac{1}{10} & | & -\dfrac{13}{20} \\ 0 & 0 & \dfrac{11}{2} & | & \dfrac{11}{4} \end{bmatrix}$ $\left(R_3 = \dfrac{2}{11}r_3\right)$

$$= \begin{bmatrix} 4 & 5 & 0 & | & 0 \\ 0 & 1 & -\dfrac{1}{10} & | & -\dfrac{13}{20} \\ 0 & 0 & 1 & | & \dfrac{1}{2} \end{bmatrix}$$

Write the system of equations that corresponds to the row-echelon matrix.

$$\begin{cases} 4x + 5y & = 0 & (1) \\ y - \dfrac{1}{10}z = -\dfrac{13}{20} & (2) \\ z = \dfrac{1}{2} & (3) \end{cases}$$

Substituting $\dfrac{1}{2}$ for z in equation (2), we obtain

$$y - \dfrac{1}{10}\left(\dfrac{1}{2}\right) = -\dfrac{13}{20}$$

$$y - \dfrac{1}{20} = -\dfrac{13}{20}$$

$$y = -\dfrac{12}{20} = -\dfrac{3}{5}$$

Substituting $-\dfrac{3}{5}$ for y in equation (1), we obtain

$$4x + 5\left(-\dfrac{3}{5}\right) = 0$$

$$4x - 3 = 0$$

$$4x = 3$$

$$x = \dfrac{3}{4}$$

The solution is the ordered triple $\left(\dfrac{3}{4}, -\dfrac{3}{5}, \dfrac{1}{2}\right)$.

45. Write the augmented matrix of the system and then put it in row echelon form.

$$\begin{bmatrix} 1 & -3 & | & 3 \\ -2 & 6 & | & -6 \end{bmatrix} \qquad (R_2 = 2r_1 + r_2)$$

$$= \begin{bmatrix} 1 & -3 & | & 3 \\ 0 & 0 & | & 0 \end{bmatrix}$$

Write the system of equations that corresponds to the row-echelon matrix.

$$\begin{cases} x - 3y = 3 & (1) \\ 0 = 0 & (2) \end{cases}$$

The statement $0 = 0$ in equation (2) indicates that the system is dependent and has an infinite number of solutions.
The solution to the system is
$$\{(x, y) \mid x - 3y = 3\}.$$

47. Write the augmented matrix of the system and then put it in row echelon form.

$$\begin{bmatrix} 1 & -2 & 1 & | & 3 \\ -2 & 4 & -1 & | & -5 \\ -8 & 16 & 1 & | & -21 \end{bmatrix} \begin{pmatrix} R_2 = 2r_1 + r_2 \\ R_3 = 8r_1 + r_3 \end{pmatrix}$$

$$= \begin{bmatrix} 1 & -2 & 1 & | & 3 \\ 0 & 0 & 1 & | & 1 \\ 0 & 0 & 9 & | & 3 \end{bmatrix} \qquad (R_3 = -9r_2 + r_3)$$

$$= \begin{bmatrix} 1 & -2 & 1 & | & 3 \\ 0 & 0 & 1 & | & 1 \\ 0 & 0 & 0 & | & -6 \end{bmatrix}$$

Write the system of equations that corresponds to the row-echelon matrix.

$$\begin{cases} x - 2y + z = 3 & (1) \\ z = 1 & (2) \\ 0 = -6 & (3) \end{cases}$$

The statement $0 = -6$ in equation (3) indicates that the system is inconsistent. The system has no solution. The solution set is $\varnothing$ or $\{\ \}$.

49. Write the augmented matrix of the system and then put it in row echelon form.

$$\begin{bmatrix} 1 & 1 & -2 & | & 4 \\ -4 & 0 & 3 & | & -4 \\ -2 & 2 & -1 & | & 4 \end{bmatrix} \begin{pmatrix} R_2 = 4r_1 + r_2 \\ R_3 = 2r_1 + r_3 \end{pmatrix}$$

$$= \begin{bmatrix} 1 & 1 & -2 & | & 4 \\ 0 & 4 & -5 & | & 12 \\ 0 & 4 & -5 & | & 12 \end{bmatrix} \qquad \left(R_2 = \dfrac{1}{4}r_2\right)$$

$$= \begin{bmatrix} 1 & 1 & -2 & | & 4 \\ 0 & 1 & -\dfrac{5}{4} & | & 3 \\ 0 & 4 & -5 & | & 12 \end{bmatrix} \qquad (R_3 = -4r_2 + r_3)$$

$$= \begin{bmatrix} 1 & 1 & -2 & | & 4 \\ 0 & 1 & -\dfrac{5}{4} & | & 3 \\ 0 & 0 & 0 & | & 0 \end{bmatrix}$$

Write the system of equations that corresponds to the row-echelon matrix.

$$\begin{cases} x + y - 2z = 4 & (1) \\ y - \dfrac{4}{5}z = 3 & (2) \\ 0 = 0 & (3) \end{cases}$$

The statement $0 = 0$ in equation (3) indicates that the system is dependent and has an infinite number of solutions.
Solve equation (2) for y.

$$y = \dfrac{5}{4}z + 3$$

Substituting $\frac{5}{4}z+3$ for y in equation (1), we obtain

$$x+\left(\frac{5}{4}z+3\right)-2z=4$$

$$x+\frac{5}{4}z-\frac{8}{4}z=1$$

$$x=\frac{3}{4}z+1$$

The solution is the ordered triple

$$\left\{(x,y,z)\,\middle|\,x=\frac{3}{4}z+1,\,y=\frac{5}{4}z+3,\right.$$

$$z \text{ is any real number}\Big\}$$

51. $\begin{cases} x+4y=-5 & (1) \\ \quad\;\; y=-2 & (2) \end{cases}$

This system is consistent and independent.
Substituting -2 for y in equation (1), we obtain

$$x+4(-2)=-5$$

$$x-8=-5$$

$$x=3$$

The solution is the ordered pair $(3,-2)$.

53. $\begin{cases} x+3y-2z=6 & (1) \\ \quad\;\; y+5z=-2 & (2) \\ \qquad\quad 0=4 & (3) \end{cases}$

This system is inconsistent. The solution is $\varnothing$ or $\{\ \}$.

55. $\begin{cases} x-2y-z=3 & (1) \\ \quad\;\; y-2z=-8 & (2) \\ \qquad\quad z=5 & (3) \end{cases}$

This system is consistent and independent.
Substituting 5 for z in equation (2), we obtain

$$y-2(5)=-8$$

$$y-10=-8$$

$$y=2$$

Substituting 5 for z and 2 for y in equation (1), we obtain

$$x-2(2)-5=3$$

$$x-4-5=3$$

$$x-9=3$$

$$x=12$$

The solution is the ordered triple $(12,\,2,\,5)$.

57. Write the augmented matrix of the system and then put it in row echelon form.

$$\begin{bmatrix} 1 & -3 & | & 18 \\ 2 & 1 & | & 1 \end{bmatrix} \qquad (R_2=-2r_1+r_2)$$

$$=\begin{bmatrix} 1 & -3 & | & 18 \\ 0 & 7 & | & -35 \end{bmatrix} \qquad \left(R_2=\frac{1}{7}r_2\right)$$

$$=\begin{bmatrix} 1 & -3 & | & 18 \\ 0 & 1 & | & -5 \end{bmatrix}$$

Write the system of equations that corresponds to the row-echelon matrix

$$\begin{cases} x-3y=18 & (1) \\ \quad\;\; y=-5 & (2) \end{cases}$$

This system is consistent and independent.
Substituting -5 for y in equation (1), we obtain

$$x-3(-5)=18$$

$$x+15=18$$

$$x=3$$

The solution is the ordered pair $(3,-5)$.

59. Write the augmented matrix of the system and then put it in row echelon form.

$$\begin{bmatrix} 2 & 4 & | & 10 \\ 1 & 2 & | & 3 \end{bmatrix} \qquad (\text{Interchange } r_1 \text{ and } r_2)$$

$$=\begin{bmatrix} 1 & 2 & | & 3 \\ 2 & 4 & | & 10 \end{bmatrix} \qquad (R_2=-2r_1+r_2)$$

$$=\begin{bmatrix} 1 & 2 & | & 3 \\ 0 & 0 & | & 4 \end{bmatrix}$$

The system is inconsistent. The system has no solution. The solution set is $\varnothing$ or $\{\ \}$.

61. Write the augmented matrix of the system and then put it in row echelon form.

$$\begin{bmatrix} 1 & -6 & | & 8 \\ 2 & 8 & | & -9 \end{bmatrix} \qquad (R_2=-2r_1+r_2)$$

$$=\begin{bmatrix} 1 & -6 & | & 8 \\ 0 & 20 & | & -25 \end{bmatrix} \qquad \left(R_2=\frac{1}{20}r_2\right)$$

$$=\begin{bmatrix} 1 & -6 & | & 8 \\ 0 & 1 & | & -\frac{5}{4} \end{bmatrix}$$

Write the system of equations that corresponds to the row-echelon matrix

$$\begin{cases} x-6y=8 & (1) \\ \quad\;\; y=-\frac{5}{4} & (2) \end{cases}$$

This system is consistent and independent.

Substituting $-\frac{5}{4}$ for y in equation (1), we obtain

$$x - 6\left(-\frac{5}{4}\right) = 8$$

$$x + \frac{15}{2} = 8$$

$$x = \frac{1}{2}$$

The solution is the ordered pair $\left(\frac{1}{2}, -\frac{5}{4}\right)$.

63. Write the augmented matrix of the system and then put it in row echelon form.

$$\begin{bmatrix} 4 & -1 & | & 8 \\ 2 & -\frac{1}{2} & | & 4 \end{bmatrix} \quad \left(R_1 = \frac{1}{4}r_1\right)$$

$$= \begin{bmatrix} 1 & -\frac{1}{4} & | & 2 \\ 2 & -\frac{1}{2} & | & 4 \end{bmatrix} \quad (R_2 = -2r_1 + r_2)$$

$$= \begin{bmatrix} 1 & -\frac{1}{4} & | & 2 \\ 0 & 0 & | & 0 \end{bmatrix}$$

The system is dependent. The solution to the system is $\{(x, y)\mid 4x - y = 8\}$.

65. Write the augmented matrix of the system and then put it in row echelon form.

$$\begin{bmatrix} 1 & 1 & 1 & | & 0 \\ 2 & -3 & 1 & | & 19 \\ -3 & 1 & -2 & | & -15 \end{bmatrix} \quad \begin{pmatrix} R_2 = -2r_1 + r_2 \\ R_3 = 3r_1 + r_3 \end{pmatrix}$$

$$= \begin{bmatrix} 1 & 1 & 1 & | & 0 \\ 0 & -5 & -1 & | & 19 \\ 0 & 4 & 1 & | & -15 \end{bmatrix} \quad \left(R_2 = -\frac{1}{5}r_2\right)$$

$$= \begin{bmatrix} 1 & 1 & 1 & | & 0 \\ 0 & 1 & \frac{1}{5} & | & -\frac{19}{5} \\ 0 & 4 & 1 & | & -15 \end{bmatrix} \quad (R_3 = -4r_2 + r_3)$$

$$= \begin{bmatrix} 1 & 1 & 1 & | & 0 \\ 0 & 1 & \frac{1}{5} & | & -\frac{19}{5} \\ 0 & 0 & \frac{1}{5} & | & \frac{1}{5} \end{bmatrix} \quad (R_3 = 5r_3)$$

$$= \begin{bmatrix} 1 & 1 & 1 & | & 0 \\ 0 & 1 & \frac{1}{5} & | & -\frac{19}{5} \\ 0 & 0 & 1 & | & 1 \end{bmatrix}$$

Write the system of equations that corresponds to the row-echelon matrix

$$\begin{cases} x + y + z = 0 & (1) \\ y + \frac{1}{5}z = -\frac{19}{5} & (2) \\ z = 1 & (3) \end{cases}$$

This system is consistent and independent. Substituting 1 for z in equation (2), we obtain

$$y + \frac{1}{5}(1) = -\frac{19}{5}$$

$$y + \frac{1}{5} = -\frac{19}{5}$$

$$y = -\frac{20}{4} = -4$$

Substituting 1 for z and -4 for y in equation (1), we obtain

$$x + (-4) + 1 = 0$$

$$x - 3 = 0$$

$$x = 3$$

The solution is the ordered triple $(3, -4, 1)$.

67. Write the augmented matrix of the system and then put it in row echelon form.

$$\begin{bmatrix} 2 & 1 & -1 & | & 13 \\ -1 & -3 & 2 & | & -14 \\ -3 & 2 & -3 & | & 3 \end{bmatrix} \quad (\text{Interchange } r_1 \text{ and } r_2)$$

$$= \begin{bmatrix} -1 & -3 & 2 & | & -14 \\ 2 & 1 & -1 & | & 13 \\ -3 & 2 & -3 & | & 3 \end{bmatrix} \quad (R_1 = -1 \cdot r_1)$$

$$= \begin{bmatrix} 1 & 3 & -2 & | & 14 \\ 2 & 1 & -1 & | & 13 \\ -3 & 2 & -3 & | & 3 \end{bmatrix} \quad \begin{pmatrix} R_2 = -2r_1 + r_2 \\ R_3 = 3r_1 + r_3 \end{pmatrix}$$

$$= \begin{bmatrix} 1 & 3 & -2 & | & 14 \\ 0 & -5 & 3 & | & -15 \\ 0 & 11 & -9 & | & 45 \end{bmatrix} \quad \left(R_2 = -\frac{1}{5}r_2\right)$$

$$= \begin{bmatrix} 1 & 3 & -2 & | & 14 \\ 0 & 1 & -\frac{3}{5} & | & 3 \\ 0 & 11 & -9 & | & 45 \end{bmatrix} \quad (R_3 = -11r_2 + r_3)$$

$$= \begin{bmatrix} 1 & 3 & -2 & | & 14 \\ 0 & 1 & -\frac{3}{5} & | & 3 \\ 0 & 0 & -\frac{12}{5} & | & 12 \end{bmatrix} \quad \left(R_3 = -\frac{5}{12}r_3\right)$$

$$= \begin{bmatrix} 1 & 3 & -2 & | & 14 \\ 0 & 1 & -\frac{3}{5} & | & 3 \\ 0 & 0 & 1 & | & -5 \end{bmatrix}$$

Write the system of equations that corresponds to the row-echelon matrix

$$\begin{cases} x + 3y - 2z = 14 & (1) \\ y - \frac{3}{5}z = 3 & (2) \\ z = -5 & (3) \end{cases}$$

This system is consistent and independent.

Substituting -5 for z in equation (2), we obtain

$$y - \frac{3}{5}(-5) = 3$$
$$y + 3 = 3$$
$$y = 0$$

Substituting -5 for z and 0 for y in equation (1), we obtain

$$x + 3(0) - 2(-5) = 14$$
$$x + 10 = 14$$
$$x = 4$$

The solution is the ordered triple $(4, 0, -5)$.

69. Write the augmented matrix of the system and then put it in row echelon form.

$$\begin{bmatrix} 2 & -1 & 3 & | & 1 \\ -1 & 3 & 1 & | & -4 \\ 3 & 1 & 7 & | & -2 \end{bmatrix} \quad \text{(Interchange } r_1 \text{ and } r_2\text{)}$$

$$= \begin{bmatrix} -1 & 3 & 1 & | & -4 \\ 2 & -1 & 3 & | & 1 \\ 3 & 1 & 7 & | & -2 \end{bmatrix} \quad (R_1 = -1 \cdot r_1)$$

$$= \begin{bmatrix} 1 & -3 & -1 & | & 4 \\ 2 & -1 & 3 & | & 1 \\ 3 & 1 & 7 & | & -2 \end{bmatrix} \quad \begin{pmatrix} R_2 = -2r_1 + r_2 \\ R_3 = -3r_1 + r_3 \end{pmatrix}$$

$$= \begin{bmatrix} 1 & -3 & -1 & | & 4 \\ 0 & 5 & 5 & | & -7 \\ 0 & 10 & 10 & | & -14 \end{bmatrix} \quad (R_3 = -2r_2 + r_3)$$

$$= \begin{bmatrix} 1 & -3 & -1 & | & 4 \\ 0 & 5 & 5 & | & -7 \\ 0 & 0 & 0 & | & 0 \end{bmatrix}$$

The system is dependent and has an infinite number of solutions.

Write the system of equations that corresponds to the row-echelon matrix

$$\begin{cases} x - 3y - z = 4 & (1) \\ 5y + 5z = -7 & (2) \\ 0 = 0 & (3) \end{cases}$$

Solve equation (2) for y.

$$5y + 5z = 7$$
$$5y = -5z - 7$$
$$y = -z - \frac{7}{5} = -z - 1.4$$

Substituting $-z - 1.4$ for y in equation (1), we obtain

$$x - 3(-z - 1.4) - z = 4$$
$$x + 3z + 4.2 - z = 4$$
$$x + 2z + 4.2 = 4$$
$$x = -2z - 0.2$$

The solution to the system is

$$\{(x, y, z) | x = -2z - 0.2,$$

$y = -z - 1.4$, z is any real number$\}$.

71. Write the augmented matrix of the system and then put it in row echelon form.

$$\begin{bmatrix} 3 & 1 & -4 & | & 0 \\ -2 & -3 & 1 & | & 5 \\ -1 & -5 & -2 & | & 3 \end{bmatrix} \quad \text{(Interchange } r_1 \text{ and } r_3\text{)}$$

$$= \begin{bmatrix} -1 & -5 & -2 & | & 3 \\ -2 & -3 & 1 & | & 5 \\ 3 & 1 & -4 & | & 0 \end{bmatrix} \quad (R_1 = -1 \cdot r_1)$$

$$= \begin{bmatrix} 1 & 5 & 2 & | & -3 \\ -2 & -3 & 1 & | & 5 \\ 3 & 1 & -4 & | & 0 \end{bmatrix} \quad \begin{pmatrix} R_2 = 2r_1 + r_2 \\ R_3 = -3r_1 + r_3 \end{pmatrix}$$

$$= \begin{bmatrix} 1 & 5 & 2 & | & -3 \\ 0 & 7 & 5 & | & -1 \\ 0 & -14 & -10 & | & 9 \end{bmatrix} \quad (R_3 = 2r_2 + r_3)$$

$$= \begin{bmatrix} 1 & 5 & 2 & | & -3 \\ 0 & 7 & 5 & | & -1 \\ 0 & 0 & 0 & | & 7 \end{bmatrix}$$

The system is inconsistent. The system has no solution. The solution set is $\varnothing$ or $\{\ \}$.

73. Write the augmented matrix of the system and then put it in row echelon form.

$$\begin{bmatrix} 2 & -1 & 3 & | & -1 \\ 3 & 1 & -4 & | & 3 \\ 1 & 7 & -2 & | & 2 \end{bmatrix} \quad \text{(Interchange } r_1 \text{ and } r_3\text{)}$$

$$= \begin{bmatrix} 1 & 7 & -2 & | & 2 \\ 3 & 1 & -4 & | & 3 \\ 2 & -1 & 3 & | & -1 \end{bmatrix} \quad \begin{pmatrix} R_2 = -3r_1 + r_2 \\ R_3 = -2r_1 + r_3 \end{pmatrix}$$

$$= \begin{bmatrix} 1 & 7 & -2 & | & 2 \\ 0 & -20 & 2 & | & -3 \\ 0 & -15 & 7 & | & -5 \end{bmatrix} \quad \left(R_2 = -\frac{1}{20} r_2\right)$$

$$= \begin{bmatrix} 1 & 7 & -2 & | & 2 \\ 0 & 1 & -\frac{1}{10} & | & \frac{3}{20} \\ 0 & -15 & 7 & | & -5 \end{bmatrix} \quad (R_3 = 15r_2 + r_3)$$

$$= \begin{bmatrix} 1 & 7 & -2 & \bigm| & 2 \\ 0 & 1 & -\frac{1}{10} & \bigm| & \frac{3}{20} \\ 0 & 0 & \frac{11}{2} & \bigm| & -\frac{11}{4} \end{bmatrix} \quad \left(R_3 = \frac{2}{11} r_3 \right)$$

$$= \begin{bmatrix} 1 & 7 & -2 & \bigm| & 2 \\ 0 & 1 & -\frac{1}{10} & \bigm| & \frac{3}{20} \\ 0 & 0 & 1 & \bigm| & -\frac{1}{2} \end{bmatrix}$$

Write the system of equations that corresponds to the row-echelon matrix

$$\begin{cases} x + 7y - 2z = 2 & \quad (1) \\ \quad y - \dfrac{1}{10} z = \dfrac{3}{20} & \quad (2) \\ \qquad\qquad z = -\dfrac{1}{2} & \quad (3) \end{cases}$$

This system is consistent and independent.

Substituting $-\dfrac{1}{2}$ for z in equation (2), we obtain

$$y - \frac{1}{10}\left(-\frac{1}{2}\right) = \frac{3}{20}$$

$$y + \frac{1}{20} = \frac{3}{20}$$

$$y = \frac{2}{20} = \frac{1}{10}$$

Substituting $-\dfrac{1}{2}$ for z and $\dfrac{1}{10}$ for y in equation (1), we obtain

$$x + 7\left(\frac{1}{10}\right) - 2\left(-\frac{1}{2}\right) = 2$$

$$x + \frac{7}{10} + 1 = 2$$

$$x + \frac{17}{10} = 2$$

$$x = \frac{3}{10}$$

The solution is the ordered triple $\left(\dfrac{3}{10},\ \dfrac{1}{10}, -\dfrac{1}{2} \right)$.

75. Write the augmented matrix of the system and then put it in row echelon form.

$$\begin{bmatrix} 3 & 5 & 2 & \bigm| & 6 \\ 0 & 10 & -2 & \bigm| & 5 \\ 6 & 0 & 4 & \bigm| & 8 \end{bmatrix} \quad \left(R_1 = \frac{1}{3} r_1 \right)$$

$$= \begin{bmatrix} 1 & \frac{5}{3} & \frac{2}{3} & \bigm| & 2 \\ 0 & 10 & -2 & \bigm| & 5 \\ 6 & 0 & 4 & \bigm| & 8 \end{bmatrix} \quad \left(R_3 = -6r_1 + r_3 \right)$$

$$= \begin{bmatrix} 1 & \frac{5}{3} & \frac{2}{3} & \bigm| & 2 \\ 0 & 10 & -2 & \bigm| & 5 \\ 0 & -10 & 0 & \bigm| & -4 \end{bmatrix} \quad \left(R_3 = -\frac{1}{10} r_3 \right)$$

$$= \begin{bmatrix} 1 & \frac{5}{3} & \frac{2}{3} & \bigm| & 2 \\ 0 & 10 & -2 & \bigm| & 5 \\ 0 & 1 & 0 & \bigm| & \frac{2}{5} \end{bmatrix} \quad (\text{Interchange } r_2 \text{ and } r_3)$$

$$= \begin{bmatrix} 1 & \frac{5}{3} & \frac{2}{3} & \bigm| & 2 \\ 0 & 1 & 0 & \bigm| & \frac{2}{5} \\ 0 & 10 & -2 & \bigm| & 5 \end{bmatrix} \quad \left(R_3 = -10r_2 + r_3 \right)$$

$$= \begin{bmatrix} 1 & \frac{5}{3} & \frac{2}{3} & \bigm| & 2 \\ 0 & 1 & 0 & \bigm| & \frac{2}{5} \\ 0 & 0 & -2 & \bigm| & 1 \end{bmatrix} \quad \left(R_3 = -\frac{1}{2} r_3 \right)$$

$$= \begin{bmatrix} 1 & \frac{5}{3} & \frac{2}{3} & \bigm| & 2 \\ 0 & 1 & 0 & \bigm| & \frac{2}{5} \\ 0 & 0 & 1 & \bigm| & -\frac{1}{2} \end{bmatrix}$$

Write the system of equations that corresponds to the row-echelon matrix

$$\begin{cases} x + \dfrac{5}{3} y + \dfrac{2}{3} z = 2 & \quad (1) \\ \qquad y \qquad = \dfrac{2}{5} & \quad (2) \\ \qquad\qquad z = -\dfrac{1}{2} & \quad (3) \end{cases}$$

This system is consistent and independent.

We have $y = \dfrac{2}{5}$ and $z = -\dfrac{1}{2}$. Substituting $\dfrac{2}{5}$ for y and $-\dfrac{1}{2}$ for z in equation (1), we obtain

$$x + \frac{5}{3}\left(\frac{2}{5}\right) + \frac{2}{3}\left(-\frac{1}{2}\right) = 2$$

$$x + \frac{2}{3} - \frac{1}{3} = 2$$

$$x + \frac{1}{3} = 2$$

$$x = \frac{5}{3}$$

The solution is the ordered triple $\left(\dfrac{5}{3},\ \dfrac{2}{5}, -\dfrac{1}{2} \right)$.

77. Write the augmented matrix of the system and then put it in row echelon form.

$$\begin{bmatrix} 1 & 0 & -1 & | & 3 \\ 2 & 1 & 0 & | & -3 \\ 0 & 2 & -1 & | & 7 \end{bmatrix} \quad (R_2 = -2r_1 + r_2)$$

$$= \begin{bmatrix} 1 & 0 & -1 & | & 3 \\ 0 & 1 & 2 & | & -9 \\ 0 & 2 & -1 & | & 7 \end{bmatrix} \quad (R_3 = -2r_2 + r_3)$$

$$= \begin{bmatrix} 1 & 0 & -1 & | & 3 \\ 0 & 1 & 2 & | & -9 \\ 0 & 0 & -5 & | & 25 \end{bmatrix} \quad \left(R_3 = -\frac{1}{5}r_3\right)$$

$$= \begin{bmatrix} 1 & 0 & -1 & | & 3 \\ 0 & 1 & 2 & | & -9 \\ 0 & 0 & 1 & | & -5 \end{bmatrix}$$

Write the system of equations that corresponds to the row-echelon matrix

$$\begin{cases} x & -z = 3 & (1) \\ y + 2z = -9 & (2) \\ z = -5 & (3) \end{cases}$$

This system is consistent and independent. Substituting -5 for z in equation (2), we obtain

$$y + 2(-5) = -9$$
$$y - 10 = -9$$
$$y = 1$$

Substituting -5 for z in equation (1), we obtain

$$x - (-5) = 3$$
$$x + 5 = 3$$
$$x = -2$$

The solution is the ordered triple $(-2, 1, -5)$.

79. (a) If $f(1) = 0$, then $a(1)^2 + b(1) + c = 0$ or

$$a + b + c = 0.$$

If $f(2) = 3$, then $a(2)^2 + b(2) + c = 3$ or

$$4a + 2b + c = 3.$$

(b) To find a, b, and c, we must solve the following system:

$$\begin{cases} a - b + c = 6 & (1) \\ a + b + c = 0 & (2) \\ 4a + 2b + c = 3 & (3) \end{cases}$$

Write the augmented matrix of the system and then put it in row echelon form.

$$\begin{bmatrix} 1 & -1 & 1 & | & 6 \\ 1 & 1 & 1 & | & 0 \\ 4 & 2 & 1 & | & 3 \end{bmatrix} \quad \begin{pmatrix} R_2 = -1r_1 + r_2 \\ R_3 = -4r_1 + r_3 \end{pmatrix}$$

$$= \begin{bmatrix} 1 & -1 & 1 & | & 6 \\ 0 & 2 & 0 & | & -6 \\ 0 & 6 & -3 & | & -21 \end{bmatrix} \quad \left(R_2 = \frac{1}{2}r_2\right)$$

$$= \begin{bmatrix} 1 & -1 & 1 & | & 6 \\ 0 & 1 & 0 & | & -3 \\ 0 & 6 & -3 & | & -21 \end{bmatrix} \quad (R_3 = -6r_2 + r_3)$$

$$= \begin{bmatrix} 1 & -1 & 1 & | & 6 \\ 0 & 1 & 0 & | & -3 \\ 0 & 0 & -3 & | & -3 \end{bmatrix} \quad \left(R_3 = -\frac{1}{3}r_3\right)$$

$$= \begin{bmatrix} 1 & -1 & 1 & | & 6 \\ 0 & 1 & 0 & | & -3 \\ 0 & 0 & 1 & | & 1 \end{bmatrix}$$

Write the system of equations that corresponds to the row-echelon matrix

$$\begin{cases} a - b + c = 6 & (1) \\ b = -3 & (2) \\ c = 1 & (3) \end{cases}$$

This system is consistent and independent. We have $b = -3$ and $c = 1$. Substituting -3 for b and 1 for c in equation (1), we obtain

$$a - (-3) + 1 = 6$$
$$a + 4 = 6$$
$$a = 2$$

Thus, the quadratic equation that contains $(-1, 6)$, $(1, 0)$, and $(2, 3)$ is

$$f(x) = 2x^2 - 3x + 1.$$

81. Let t represent the amount to be invested in Treasury bills, let m represent the amount to be invested in municipal bonds, and let c represent the amount to be invested in corporate bonds.

$$\begin{cases} t + m + c = 20,000 & (1) \\ 0.04t + 0.05m + 0.08c = 1,070 & (2) \\ t - c = 3,000 & (3) \end{cases}$$

Write the augmented matrix of the system and then put it in row echelon form.

$$\begin{bmatrix} 1 & 1 & 1 & \Big| & 20{,}000 \\ 0.04 & 0.05 & 0.08 & \Big| & 1{,}070 \\ 1 & 0 & -1 & \Big| & 3{,}000 \end{bmatrix} \quad \begin{pmatrix} R_2 = -0.04r_1 + r_2 \\ R_3 = -1r_1 + r_3 \end{pmatrix}$$

$$= \begin{bmatrix} 1 & 1 & 1 & \Big| & 20{,}000 \\ 0 & 0.01 & 0.04 & \Big| & 270 \\ 0 & -1 & -2 & \Big| & -17{,}000 \end{bmatrix} \quad (R_2 = 100r_2)$$

$$= \begin{bmatrix} 1 & 1 & 1 & \Big| & 20{,}000 \\ 0 & 1 & 4 & \Big| & 27{,}000 \\ 0 & -1 & -2 & \Big| & -17{,}000 \end{bmatrix} \quad (R_3 = r_2 + r_3)$$

$$= \begin{bmatrix} 1 & 1 & 1 & \Big| & 20{,}000 \\ 0 & 1 & 4 & \Big| & 27{,}000 \\ 0 & 0 & 2 & \Big| & 10{,}000 \end{bmatrix} \quad \left(R_3 = \tfrac{1}{2}r_3\right)$$

$$= \begin{bmatrix} 1 & 1 & 1 & \Big| & 20{,}000 \\ 0 & 1 & 4 & \Big| & 27{,}000 \\ 0 & 0 & 1 & \Big| & 5{,}000 \end{bmatrix}$$

Write the system of equations that corresponds to the row-echelon matrix

$$\begin{cases} t + m + c = 20{,}000 & (1) \\ m + 4c = 27{,}000 & (2) \\ c = 5{,}000 & (3) \end{cases}$$

Substituting 5,000 for c in equation (2), we obtain

$$m + 4(5{,}000) = 27{,}000$$
$$m + 20{,}000 = 27{,}000$$
$$m = 7{,}000$$

Substituting 5,000 for c and 7,000 for m in equation (1), we obtain

$$t + 7{,}000 + 5{,}000 = 20{,}000$$
$$t + 12{,}000 = 20{,}000$$
$$t = 8{,}000$$

Therefore, Carissa should invest $8,000 in Treasury bills, $7,000 in municipal bonds, and $5,000 in corporate bonds.

83. Write the augmented matrix of the system and then put it in reduced row echelon form.

$$\begin{bmatrix} 2 & 1 & \Big| & 1 \\ -3 & -2 & \Big| & -5 \end{bmatrix} \quad (R_1 = 2r_1 + r_2)$$

$$= \begin{bmatrix} 1 & 0 & \Big| & -3 \\ -3 & -2 & \Big| & -5 \end{bmatrix} \quad (R_2 = 3r_1 + r_2)$$

$$= \begin{bmatrix} 1 & 0 & \Big| & -3 \\ 0 & -2 & \Big| & -14 \end{bmatrix} \quad \left(R_2 = -\tfrac{1}{2}r_2\right)$$

$$= \begin{bmatrix} 1 & 0 & \Big| & -3 \\ 0 & 1 & \Big| & 7 \end{bmatrix}$$

Write the system of equations that corresponds to the reduced row echelon matrix

$$\begin{cases} x = -3 \\ y = 7 \end{cases}$$

The solution is the ordered pair $(-3,\ 7)$.

85. Write the augmented matrix of the system and then put it in reduced row echelon form.

$$\begin{bmatrix} 1 & 1 & 1 & \Big| & 3 \\ 2 & 1 & -4 & \Big| & 25 \\ -3 & 2 & 1 & \Big| & 0 \end{bmatrix} \quad \begin{pmatrix} R_2 = -2r_1 + r_2 \\ R_3 = 3r_1 + r_3 \end{pmatrix}$$

$$= \begin{bmatrix} 1 & 1 & 1 & \Big| & 3 \\ 0 & -1 & -6 & \Big| & 19 \\ 0 & 5 & 4 & \Big| & 9 \end{bmatrix} \quad \begin{pmatrix} R_1 = r_1 + r_2 \\ R_3 = 5r_2 + r_3 \end{pmatrix}$$

$$= \begin{bmatrix} 1 & 0 & -5 & \Big| & 22 \\ 0 & -1 & -6 & \Big| & 19 \\ 0 & 0 & -26 & \Big| & 104 \end{bmatrix} \quad \left(R_3 = -\tfrac{1}{26}r_3\right)$$

$$= \begin{bmatrix} 1 & 0 & -5 & \Big| & 22 \\ 0 & -1 & -6 & \Big| & 19 \\ 0 & 0 & 1 & \Big| & -4 \end{bmatrix} \quad \begin{pmatrix} R_1 = 5r_3 + r_1 \\ R_2 = 6r_3 + r_2 \end{pmatrix}$$

$$= \begin{bmatrix} 1 & 0 & 0 & \Big| & 2 \\ 0 & -1 & 0 & \Big| & -5 \\ 0 & 0 & 1 & \Big| & -4 \end{bmatrix} \quad (R_2 = -1 \cdot r_2)$$

$$= \begin{bmatrix} 1 & 0 & 0 & \Big| & 2 \\ 0 & 1 & 0 & \Big| & 5 \\ 0 & 0 & 1 & \Big| & -4 \end{bmatrix}$$

Write the system of equations that corresponds to the reduced row echelon matrix

$$\begin{cases} x = 2 \\ y = 5 \\ z = -4 \end{cases}$$

The solution is the ordered triple $(2,\ 5,\ -4)$.

87. Answers will vary. One possibility follows: First, perform row operations so that the entry in row 1, column 1 is 1. Second, perform row operations so that all the entries below the 1 in row 1, column 1 are 0's. Third, perform row operations so that the entry in row 2, column 2 is 1. Make sure that the entries in column 1 remain unchanged. If it is impossible to place a 1 in row 2, column 2, then use row operations to place a 1 in row 2, column 3. (If any row with all 0's occurs, then place it in the last row of the matrix.) Once a 1 is in place, perform row operations to place 0's below it. Continue this process until the augmented matrix is in row echelon form.

89. Multiply each entry in row 2 by $\dfrac{1}{5}$ (or divide each entry of row 2 by 5). That is, use the row operation $R_2 = \dfrac{1}{5}r_2$.

91. Write the augmented matrix of the system.

$$\begin{bmatrix} 2 & 3 & | & 1 \\ -3 & -4 & | & -3 \end{bmatrix}$$

Enter the system into a 2 by 3 matrix, [A]. Then, use the **ref(** command along with the ▶ **frac** command to write the matrix in row echelon form with the entries in fractional form.

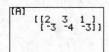 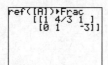

Thus, the row echelon matrix is

$$\begin{bmatrix} 1 & \frac{4}{3} & | & 1 \\ 0 & 1 & | & -3 \end{bmatrix}$$

Write the system of equations that corresponds to the row echelon matrix.

$$\begin{cases} x + \dfrac{4}{3}y = 1 & (1) \\ y = -3 & (2) \end{cases}$$

Substituting –3 for y in equation (1), we obtain

$$x + \frac{4}{3}(-3) = 1$$
$$x - 4 = 1$$
$$x = 5$$

The solution is the ordered pair $(5, -3)$.

93. Write the augmented matrix of the system.

$$\begin{bmatrix} 2 & 3 & -2 & | & -12 \\ -3 & 1 & 2 & | & 0 \\ 4 & 3 & -1 & | & 3 \end{bmatrix}$$

Enter the system into a 3 by 4 matrix, [A]. Then, use the **rcf(** command along with the ▶ **frac** command to write the matrix in row echelon form with the entries in fractional form. Since

the entire matrix does not fit on the screen, we need to scroll right to see the rest of it.

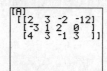

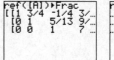

Thus, the row echelon matrix is

$$\begin{bmatrix} 1 & \frac{3}{4} & -\frac{1}{4} & | & \frac{3}{4} \\ 0 & 1 & \frac{5}{13} & | & \frac{9}{13} \\ 0 & 0 & 1 & | & 7 \end{bmatrix}$$

Write the system of equations that corresponds to the row-echelon matrix

$$\begin{cases} x + \dfrac{3}{4}y - \dfrac{1}{4}z = \dfrac{3}{4} & (1) \\ y + \dfrac{5}{13}z = \dfrac{9}{13} & (2) \\ z = 7 & (3) \end{cases}$$

Substituting 7 for z in equation (2), we obtain

$$y + \frac{5}{13}(7) = \frac{9}{13}$$
$$y + \frac{35}{13} = \frac{9}{13}$$
$$y = -\frac{26}{13} = -2$$

Substituting 7 for z and –2 for y in equation (1), we obtain

$$x + \frac{3}{4}(-2) - \frac{1}{4}(7) = \frac{3}{4}$$
$$x - \frac{3}{2} - \frac{7}{4} = \frac{3}{4}$$
$$x - \frac{13}{4} = \frac{3}{4}$$
$$x = \frac{16}{4} = 4$$

The solution is the ordered triple $(4, -2, 7)$.

Appendix C.4

Preparing for Determinants and Cramer's Rule

P1. $4 \cdot 2 - 3 \cdot (-3) = 8 + 9 = 17$

P2. $\dfrac{18}{6} = 3$

Appendix C.4 Quick Checks

1. $ad - bc$

2. square

3. $\begin{vmatrix} 5 & 3 \\ 4 & 6 \end{vmatrix} = 5(6) - 4(3) = 30 - 12 = 18$

4. $\begin{vmatrix} -2 & -5 \\ 1 & 7 \end{vmatrix} = -2(7) - 1(-5) = -14 + 5 = -9$

5. $D = \begin{vmatrix} 3 & 2 \\ -2 & -1 \end{vmatrix} = 3(-1) - (-2)(2) = -3 + 4 = 1$

$D_x = \begin{vmatrix} 1 & 2 \\ 1 & -1 \end{vmatrix} = 1(-1) - 1(2) = -1 - 2 = -3$

$D_y = \begin{vmatrix} 3 & 1 \\ -2 & 1 \end{vmatrix} = 3(1) - (-2)(1) = 3 + 2 = 5$

$x = \dfrac{D_x}{D} = \dfrac{-3}{1} = -3 ; \; y = \dfrac{D_y}{D} = \dfrac{5}{1} = 5$

Thus, the solution is the ordered pair $(-3, \, 5)$.

6. $D = \begin{vmatrix} 4 & -2 \\ -6 & 3 \end{vmatrix} = 4(3) - (-6)(-2) = 12 - 12 = 0$

Since $D = 0$, Cramer's Rule does not apply.

7. $\begin{vmatrix} 2 & -3 & 5 \\ 0 & 4 & -1 \\ 3 & 8 & -7 \end{vmatrix} = 2 \begin{vmatrix} 4 & -1 \\ 8 & -7 \end{vmatrix} - (-3) \begin{vmatrix} 0 & -1 \\ 3 & -7 \end{vmatrix} + 5 \begin{vmatrix} 0 & 4 \\ 3 & 8 \end{vmatrix}$

$= 2 \big[4(-7) - 8(-1) \big] + 3 \big[0(-7) - 3(-1) \big] + 5 \big[0(8) - 3(4) \big]$

$= 2(-28 + 8) + 3(0 + 3) + 5(0 - 12)$

$= 2(-20) + 3(3) + 5(-12)$

$= -40 + 9 - 60$

$= -91$

8. $D = \begin{vmatrix} 1 & -1 & 3 \\ 4 & 3 & 1 \\ -2 & 0 & 5 \end{vmatrix}$

$= 1\begin{vmatrix} 3 & 1 \\ 0 & 5 \end{vmatrix} - (-1)\begin{vmatrix} 4 & 1 \\ -2 & 5 \end{vmatrix} + 3\begin{vmatrix} 4 & 3 \\ -2 & 0 \end{vmatrix}$

$= 1\left[3(5) - 0(1) \right] - (-1)\left[4(5) - (-2)(1) \right] + 3\left[4(0) - (-2)(3) \right]$

$= 1(15 - 0) - (-1)(20 + 2) + 3(0 + 6)$

$= 1(15) - (-1)(22) + 3(6)$

$= 15 + 22 + 18$

$= 55$

$D_x = \begin{vmatrix} -2 & -1 & 3 \\ 9 & 3 & 1 \\ 7 & 0 & 5 \end{vmatrix}$

$= -2\begin{vmatrix} 3 & 1 \\ 0 & 5 \end{vmatrix} - (-1)\begin{vmatrix} 9 & 1 \\ 7 & 5 \end{vmatrix} + 3\begin{vmatrix} 9 & 3 \\ 7 & 0 \end{vmatrix}$

$= -2\left[3(5) - 0(1) \right] - (-1)\left[9(5) - 7(1) \right] + 3\left[9(0) - 7(3) \right]$

$= -2(15 - 0) - (-1)(45 - 7) + 3(0 - 21)$

$= -2(15) - (-1)(38) + 3(-21)$

$= -30 + 38 - 63$

$= -55$

$D_y = \begin{vmatrix} 1 & -2 & 3 \\ 4 & 9 & 1 \\ -2 & 7 & 5 \end{vmatrix}$

$= 1\begin{vmatrix} 9 & 1 \\ 7 & 5 \end{vmatrix} - (-2)\begin{vmatrix} 4 & 1 \\ -2 & 5 \end{vmatrix} + 3\begin{vmatrix} 4 & 9 \\ -2 & 7 \end{vmatrix}$

$= 1\left[9(5) - 7(1) \right] - (-2)\left[4(5) - (-2)(1) \right] + 3\left[4(7) - (-2)(9) \right]$

$= 1(45 - 7) - (-2)(20 + 2) + 3(28 + 18)$

$= 1(38) - (-2)(22) + 3(46)$

$= 38 + 44 + 138$

$= 220$

$D_z = \begin{vmatrix} 1 & -1 & -2 \\ 4 & 3 & 9 \\ -2 & 0 & 7 \end{vmatrix}$

$= 1\begin{vmatrix} 3 & 9 \\ 0 & 7 \end{vmatrix} - (-1)\begin{vmatrix} 4 & 9 \\ -2 & 7 \end{vmatrix} + (-2)\begin{vmatrix} 4 & 3 \\ -2 & 0 \end{vmatrix}$

$= 1\left[3(7) - 0(9) \right] - (-1)\left[4(7) - (-2)(9) \right] + (-2)\left[4(0) - (-2)(3) \right]$

$= 1(21 - 0) - (-1)(28 + 18) + (-2)(0 + 6)$

$= 1(21) - (-1)(46) + (-2)(6)$

$= 21 + 46 - 12$

$= 55$

$x = \dfrac{D_x}{D} = \dfrac{-55}{55} = -1$; $y = \dfrac{D_y}{D} = \dfrac{220}{55} = 4$; $z = \dfrac{D_z}{D} = \dfrac{55}{55} = 1$

Thus, the solution is the ordered triple $(-1,\ 4,\ 1)$.

9. Consistent and dependent. This is because $D = 0$ and all the determinants D_x, D_y, and D_z equal zero.

10. Inconsistent. This is because $D = 0$ and at least one of the determinants D_x, D_y, or D_z is different from zero.

Appendix C.4 Exercises

11. $\begin{vmatrix} 4 & 2 \\ 1 & 3 \end{vmatrix} = 4(3) - 1(2) = 12 - 2 = 10$

13. $\begin{vmatrix} -2 & -4 \\ 1 & 3 \end{vmatrix} = -2(3) - 1(-4) = -6 + 4 = -2$

15. $D = \begin{vmatrix} 1 & 1 \\ 1 & -1 \end{vmatrix} = 1(-1) - 1(1) = -1 - 1 = -2$

$D_x = \begin{vmatrix} -4 & 1 \\ -12 & -1 \end{vmatrix}$
$= -4(-1) - (-12)(1) = 4 + 12 = 16$

$D_y = \begin{vmatrix} 1 & -4 \\ 1 & -12 \end{vmatrix} = 1(-12) - 1(-4) = -12 + 4 = -8$

$x = \dfrac{D_x}{D} = \dfrac{16}{-2} = -8$; $y = \dfrac{D_y}{D} = \dfrac{-8}{-2} = 4$

Thus, the solution is the ordered pair $(-8, 4)$.

17. $D = \begin{vmatrix} 2 & 3 \\ -3 & 1 \end{vmatrix} = 2(1) - (-3)(3) = 2 + 9 = 11$

$D_x = \begin{vmatrix} 3 & 3 \\ -10 & 1 \end{vmatrix} = 3(1) - (-10)(3) = 3 + 30 = 33$

$D_y = \begin{vmatrix} 2 & 3 \\ -3 & -10 \end{vmatrix}$
$= 2(-10) - (-3)(3) = -20 + 9 = -11$

$x = \dfrac{D_x}{D} = \dfrac{33}{11} = 3$; $y = \dfrac{D_y}{D} = \dfrac{-11}{11} = -1$

Thus, the solution is the ordered pair $(3, -1)$.

19. $D = \begin{vmatrix} 3 & 4 \\ -6 & 8 \end{vmatrix} = 3(8) - (-6)(4) = 24 + 24 = 48$

$D_x = \begin{vmatrix} 1 & 4 \\ 4 & 8 \end{vmatrix} = 1(8) - 4(4) = 8 - 16 = -8$

$D_y = \begin{vmatrix} 3 & 1 \\ -6 & 4 \end{vmatrix} = 3(4) - (-6)(1) = 12 + 6 = 18$

$x = \dfrac{D_x}{D} = \dfrac{-8}{48} = -\dfrac{1}{6}$; $y = \dfrac{D_y}{D} = \dfrac{18}{48} = \dfrac{3}{8}$

Thus, the solution is the ordered pair $\left(-\dfrac{1}{6}, \dfrac{3}{8} \right)$.

21. The system in standard form is: $\begin{cases} 2x - 6y = 12 \\ 3x - 5y = 11 \end{cases}$

$D = \begin{vmatrix} 2 & -6 \\ 3 & -5 \end{vmatrix} = 2(-5) - 3(-6) = -10 + 18 = 8$

$D_x = \begin{vmatrix} 12 & -6 \\ 11 & -5 \end{vmatrix}$
$= 12(-5) - 11(-6)$
$= -60 + 66$
$= 6$

$D_y = \begin{vmatrix} 2 & 12 \\ 3 & 11 \end{vmatrix} = 2(11) - 3(12) = 22 - 36 = -14$

$x = \dfrac{D_x}{D} = \dfrac{6}{8} = \dfrac{3}{4}$; $y = \dfrac{D_y}{D} = \dfrac{-14}{8} = -\dfrac{7}{4}$

Thus, the solution is the ordered pair $\left(\dfrac{3}{4}, -\dfrac{7}{4} \right)$.

23. $\begin{vmatrix} 2 & 0 & -1 \\ 3 & 8 & -3 \\ 1 & 5 & -2 \end{vmatrix}$

$= 2 \begin{vmatrix} 8 & -3 \\ 5 & -2 \end{vmatrix} - 0 \begin{vmatrix} 3 & -3 \\ 1 & -2 \end{vmatrix} + (-1) \begin{vmatrix} 3 & 8 \\ 1 & 5 \end{vmatrix}$

$= 2 \big[8(-2) - 5(-3) \big] - 0 \big[3(-2) - 1(-3) \big]$
$\qquad\qquad + (-1) \big[3(5) - 1(8) \big]$

$= 2(-16 + 15) - 0(-6 + 3) + (-1)(15 - 8)$

$= 2(-1) - 0(-3) + (-1)(7)$

$= -2 + 0 - 7$

$= -9$

25. $\begin{vmatrix} -3 & 2 & 3 \\ 0 & 5 & -2 \\ 1 & 4 & 8 \end{vmatrix}$

$= (-3) \begin{vmatrix} 5 & -2 \\ 4 & 8 \end{vmatrix} - 2 \begin{vmatrix} 0 & -2 \\ 1 & 8 \end{vmatrix} + 3 \begin{vmatrix} 0 & 5 \\ 1 & 4 \end{vmatrix}$

$= -3 \big[5(8) - 4(-2) \big] - 2 \big[0(8) - 1(-2) \big]$
$\qquad\qquad + 3 \big[0(4) - 1(5) \big]$

$= -3(40 + 8) - 2(0 + 2) + 3(0 - 5)$

$= -3(48) - 2(2) + 3(-5)$

$= -144 - 4 - 15$

$= -163$

27. $\begin{vmatrix} 0 & 2 & 1 \\ 1 & -6 & -4 \\ -3 & 4 & 5 \end{vmatrix}$

$= 0\begin{vmatrix} -6 & -4 \\ 4 & 5 \end{vmatrix} - 2\begin{vmatrix} 1 & -4 \\ -3 & 5 \end{vmatrix} + 1\begin{vmatrix} 1 & -6 \\ -3 & 4 \end{vmatrix}$

$= 0\left[-6(5) - 4(-4)\right] - 2\left[1(5) - (-3)(-4)\right]$
$\qquad\qquad\qquad\qquad + 1\left[1(4) - (-3)(-6)\right]$

$= 0(-30 + 16) - 2(5 - 12) + 1(4 - 18)$

$= 0(-14) - 2(-7) + 1(-14)$

$= 0 + 14 - 14$

$= 0$

29. $D = \begin{vmatrix} 1 & -1 & 1 \\ 1 & 2 & -1 \\ 2 & 1 & 2 \end{vmatrix}$

$= 1\begin{vmatrix} 2 & -1 \\ 1 & 2 \end{vmatrix} - (-1)\begin{vmatrix} 1 & -1 \\ 2 & 2 \end{vmatrix} + 1\begin{vmatrix} 1 & 2 \\ 2 & 1 \end{vmatrix}$

$= 1\left[2(2) - 1(-1)\right] - (-1)\left[1(2) - 2(-1)\right]$
$\qquad\qquad\qquad\qquad + 1\left[1(1) - 2(2)\right]$

$= 1(4 + 1) - (-1)(2 + 2) + 1(1 - 4)$

$= 1(5) - (-1)(4) + 1(-3)$

$= 5 + 4 - 3$

$= 6$

$D_x = \begin{vmatrix} -4 & -1 & 1 \\ 1 & 2 & -1 \\ -5 & 1 & 2 \end{vmatrix}$

$= -4\begin{vmatrix} 2 & -1 \\ 1 & 2 \end{vmatrix} - (-1)\begin{vmatrix} 1 & -1 \\ -5 & 2 \end{vmatrix} + 1\begin{vmatrix} 1 & 2 \\ -5 & 1 \end{vmatrix}$

$= -4\left[2(2) - 1(-1)\right] - (-1)\left[1(2) - (-5)(-1)\right]$
$\qquad\qquad\qquad\qquad + 1\left[1(1) - (-5)(2)\right]$

$= -4(4 + 1) - (-1)(2 - 5) + 1(1 + 10)$

$= -4(5) - (-1)(-3) + 1(11)$

$= -20 - 3 + 11$

$= -12$

$D_y = \begin{vmatrix} 1 & -4 & 1 \\ 1 & 1 & -1 \\ 2 & -5 & 2 \end{vmatrix}$

$= 1\begin{vmatrix} 1 & -1 \\ -5 & 2 \end{vmatrix} - (-4)\begin{vmatrix} 1 & -1 \\ 2 & 2 \end{vmatrix} + 1\begin{vmatrix} 1 & 1 \\ 2 & -5 \end{vmatrix}$

$= 1\left[1(2) - (-5)(-1)\right] - (-4)\left[1(2) - 2(-1)\right]$
$\qquad\qquad\qquad\qquad + 1\left[1(-5) - 2(1)\right]$

$= 1(2 - 5) - (-4)(2 + 2) + 1(-5 - 2)$

$= 1(-3) - (-4)(4) + 1(-7)$

$= -3 + 16 - 7$

$= 6$

$D_z = \begin{vmatrix} 1 & -1 & -4 \\ 1 & 2 & 1 \\ 2 & 1 & -5 \end{vmatrix}$

$= 1\begin{vmatrix} 2 & 1 \\ 1 & -5 \end{vmatrix} - (-1)\begin{vmatrix} 1 & 1 \\ 2 & -5 \end{vmatrix} + (-4)\begin{vmatrix} 1 & 2 \\ 2 & 1 \end{vmatrix}$

$= 1\left[2(-5) - 1(1)\right] - (-1)\left[1(-5) - 2(1)\right]$
$\qquad\qquad\qquad\qquad + (-4)\left[1(1) - 2(2)\right]$

$= 1(-10 - 1) - (-1)(-5 - 2) + (-4)(1 - 4)$

$= 1(-11) - (-1)(-7) + (-4)(-3)$

$= -11 - 7 + 12$

$= -6$

$x = \dfrac{D_x}{D} = \dfrac{-12}{6} = -2; \quad y = \dfrac{D_y}{D} = \dfrac{6}{6} = 1;$

$z = \dfrac{D_z}{D} = \dfrac{-6}{6} = -1$

Thus, the solution is the ordered triple $(-2, 1, -1)$.

31. $D = \begin{vmatrix} 1 & 1 & 1 \\ 5 & 2 & -3 \\ 2 & -1 & -1 \end{vmatrix}$

$= 1\begin{vmatrix} 2 & -3 \\ -1 & -1 \end{vmatrix} - 1\begin{vmatrix} 5 & -3 \\ 2 & -1 \end{vmatrix} + 1\begin{vmatrix} 5 & 2 \\ 2 & -1 \end{vmatrix}$

$= 1\left[2(-1) - (-1)(-3)\right] - 1\left[5(-1) - 2(-3)\right]$
$\qquad\qquad\qquad\qquad + 1\left[5(-1) - 2(2)\right]$

$= 1(-2 - 3) - 1(-5 + 6) + 1(-5 - 4)$

$= 1(-5) - 1(1) + 1(-9)$

$= -5 - 1 - 9$

$= -15$

$$D_x = \begin{vmatrix} 4 & 1 & 1 \\ 7 & 2 & -3 \\ 5 & -1 & -1 \end{vmatrix}$$

$$= 4\begin{vmatrix} 2 & -3 \\ -1 & -1 \end{vmatrix} - 1\begin{vmatrix} 7 & -3 \\ 5 & -1 \end{vmatrix} + 1\begin{vmatrix} 7 & 2 \\ 5 & -1 \end{vmatrix}$$

$$= 4\left[2(-1)-(-1)(-3)\right] - 1\left[7(-1)-5(-3)\right]$$
$$\qquad +1\left[7(-1)-5(2)\right]$$

$$= 4(-2-3)-1(-7+15)+1(-7-10)$$

$$= 4(-5)-1(8)+1(-17)$$

$$= -20-8-17$$

$$= -45$$

$$D_y = \begin{vmatrix} 1 & 4 & 1 \\ 5 & 7 & -3 \\ 2 & 5 & -1 \end{vmatrix}$$

$$= 1\begin{vmatrix} 7 & -3 \\ 5 & -1 \end{vmatrix} - 4\begin{vmatrix} 5 & -3 \\ 2 & -1 \end{vmatrix} + 1\begin{vmatrix} 5 & 7 \\ 2 & 5 \end{vmatrix}$$

$$= 1\left[7(-1)-5(-3)\right] - 4\left[5(-1)-2(-3)\right]$$
$$\qquad +1\left[5(5)-2(7)\right]$$

$$= 1(-7+15)-4(-5+6)+1(25-14)$$

$$= 1(8)-4(1)+1(11)$$

$$= 8-4+11$$

$$= 15$$

$$D_z = \begin{vmatrix} 1 & 1 & 4 \\ 5 & 2 & 7 \\ 2 & -1 & 5 \end{vmatrix}$$

$$= 1\begin{vmatrix} 2 & 7 \\ -1 & 5 \end{vmatrix} - 1\begin{vmatrix} 5 & 7 \\ 2 & 5 \end{vmatrix} + 4\begin{vmatrix} 5 & 2 \\ 2 & -1 \end{vmatrix}$$

$$= 1\left[2(5)-(-1)(7)\right] - 1\left[5(5)-2(7)\right]$$
$$\qquad +4\left[5(-1)-2(2)\right]$$

$$= 1(10+7)-1(25-14)+4(-5-4)$$

$$= 1(17)-1(11)+4(-9)$$

$$= 17-11-36$$

$$= -30$$

$$x = \frac{D_x}{D} = \frac{-45}{-15} = 3; \quad y = \frac{D_y}{D} = \frac{15}{-15} = -1;$$

$$z = \frac{D_z}{D} = \frac{-30}{-15} = 2$$

Thus, the solution is the ordered triple $(3, -1, 2)$.

33. $D = \begin{vmatrix} 2 & 1 & -1 \\ -1 & 2 & 2 \\ 5 & 5 & -1 \end{vmatrix}$

$$= 2\begin{vmatrix} 2 & 2 \\ 5 & -1 \end{vmatrix} - 1\begin{vmatrix} -1 & 2 \\ 5 & -1 \end{vmatrix} + (-1)\begin{vmatrix} -1 & 2 \\ 5 & 5 \end{vmatrix}$$

$$= 2\left[2(-1)-5(2)\right] - 1\left[(-1)(-1)-5(2)\right]$$
$$\qquad + (-1)\left[(-1)(5)-5(2)\right]$$

$$= 2(-2-10)-1(1-10)+(-1)(-5-10)$$

$$= 2(-12)-1(-9)+(-1)(-15)$$

$$= -24+9+15$$

$$= 0$$

Since $D = 0$, we know that Cramer's Rule does not apply and that this is not a consistent and independent system. Next, we will find D_x, D_y, and D_z in order to investigate the system further.

$$D_x = \begin{vmatrix} 4 & 1 & -1 \\ -6 & 2 & 2 \\ 6 & 5 & -1 \end{vmatrix}$$

$$= 4\begin{vmatrix} 2 & 2 \\ 5 & -1 \end{vmatrix} - 1\begin{vmatrix} -6 & 2 \\ 6 & -1 \end{vmatrix} + (-1)\begin{vmatrix} -6 & 2 \\ 6 & 5 \end{vmatrix}$$

$$= 4\left[2(-1)-5(2)\right] - 1\left[(-6)(-1)-6(2)\right]$$
$$\qquad + (-1)\left[(-6)(5)-6(2)\right]$$

$$= 4(-2-10)-1(6-12)+(-1)(-30-12)$$

$$= 4(-12)-1(-6)+(-1)(-42)$$

$$= -48+6+42$$

$$= 0$$

$$D_y = \begin{vmatrix} 2 & 4 & -1 \\ -1 & -6 & 2 \\ 5 & 6 & -1 \end{vmatrix}$$

$$= 2\begin{vmatrix} -6 & 2 \\ 6 & -1 \end{vmatrix} - 4\begin{vmatrix} -1 & 2 \\ 5 & -1 \end{vmatrix} + (-1)\begin{vmatrix} -1 & -6 \\ 5 & 6 \end{vmatrix}$$

$$= 2\left[(-6)(-1)-6(2)\right] - 4\left[(-1)(-1)-5(2)\right]$$
$$\qquad + (-1)\left[(-1)(6)-5(-6)\right]$$

$$= 2(6-12)-4(1-10)+(-1)(-6+30)$$

$$= 2(-6)-4(-9)+(-1)(24)$$

$$= -12+36-24$$

$$= 0$$

$$D_z = \begin{vmatrix} 2 & 1 & 4 \\ -1 & 2 & -6 \\ 5 & 5 & 6 \end{vmatrix}$$

$$= 2\begin{vmatrix} 2 & -6 \\ 5 & 6 \end{vmatrix} - 1\begin{vmatrix} -1 & -6 \\ 5 & 6 \end{vmatrix} + 4\begin{vmatrix} -1 & 2 \\ 5 & 5 \end{vmatrix}$$

$$= 2\big[2(6)-5(-6)\big] - 1\big[(-1)(6)-5(-6)\big]$$
$$\qquad\qquad + 4\big[(-1)(5)-5(2)\big]$$

$$= 2(12+30) - 1(-6+30) + 4(-5-10)$$

$$= 2(42) - 1(24) + 4(-15)$$

$$= 84 - 24 - 60$$

$$= 0$$

Because $D = 0$ and because all the determinants D_x, D_y, and D_z equal 0, the system is consistent and dependent. We know that there are infinitely many solutions, but we cannot yet tell what the solution set is. Next we use the augmented matrix of the system in order to determine the solution set.

$$\begin{bmatrix} 2 & 1 & -1 & | & 4 \\ -1 & 2 & 2 & | & -6 \\ 5 & 5 & -1 & | & 6 \end{bmatrix} \quad (\text{Interchange } r_1 \text{ and } r_2)$$

$$= \begin{bmatrix} -1 & 2 & 2 & | & -6 \\ 2 & 1 & -1 & | & 4 \\ 5 & 5 & -1 & | & 6 \end{bmatrix} \quad \begin{pmatrix} R_2 = 2r_1 + r_2 \\ R_3 = 5r_1 + r_3 \end{pmatrix}$$

$$= \begin{bmatrix} -1 & 2 & 2 & | & -6 \\ 0 & 5 & 3 & | & -8 \\ 0 & 15 & 9 & | & -24 \end{bmatrix} \quad \begin{pmatrix} R_1 = -1 \cdot r_1 \\ R_3 = -3r_2 + r_3 \end{pmatrix}$$

$$= \begin{bmatrix} 1 & -2 & -2 & | & 6 \\ 0 & 5 & 3 & | & -8 \\ 0 & 0 & 0 & | & 0 \end{bmatrix}$$

We have confirmed that this system is dependent and has an infinite number of solutions.

Write the system of equations that corresponds to the row-echelon matrix

$$\begin{cases} x - 2y - 2z = 6 & \quad (1) \\ \quad 5y + 3z = -8 & \quad (2) \\ \qquad\quad 0 = 0 & \quad (3) \end{cases}$$

Solve equation (2) for y.

$$5y + 3z = -8$$

$$5y = -3z - 8$$

$$y = -\frac{3}{5}z - \frac{8}{5}$$

Substituting $-\dfrac{3}{5}z - \dfrac{8}{5}$ for y in equation (1), we obtain:

$$x - 2\left(-\frac{3}{5}z - \frac{8}{5}\right) - 2z = 6$$

$$x + \frac{6}{5}z + \frac{16}{5} - 2z = 6$$

$$x - \frac{4}{5}z + \frac{16}{5} = 6$$

$$x - \frac{4}{5}z = \frac{14}{5}$$

$$x = \frac{4}{5}z + \frac{14}{5}$$

The solution to the system is

$$\left\{ (x, y, z) \,\middle|\, x = \frac{4}{5}z + \frac{14}{5}, \right.$$

$$\left. y = -\frac{3}{5}z - \frac{8}{5}, z \text{ is any real number} \right\}$$

35. $D = \begin{vmatrix} 3 & 1 & 1 \\ 1 & 1 & -3 \\ -5 & -1 & -5 \end{vmatrix}$

$$= 3\begin{vmatrix} 1 & -3 \\ -1 & -5 \end{vmatrix} - 1\begin{vmatrix} 1 & -3 \\ -5 & -5 \end{vmatrix} + 1\begin{vmatrix} 1 & 1 \\ -5 & -1 \end{vmatrix}$$

$$= 3\big[1(-5)-(-1)(-3)\big] - 1\big[1(-5)-(-5)(-3)\big]$$
$$\qquad\qquad + 1\big[1(-1)-1(-5)\big]$$

$$= 3(-5-3) - 1(-5-15) + 1(-1+5)$$

$$= 3(-8) - 1(-20) + 1(4)$$

$$= -24 + 20 + 4$$

$$= 0$$

$$D_x = \begin{vmatrix} 5 & 1 & 1 \\ 9 & 1 & -3 \\ -10 & -1 & -5 \end{vmatrix}$$

$$= 5\begin{vmatrix} 1 & -3 \\ -1 & -5 \end{vmatrix} - 1\begin{vmatrix} 9 & -3 \\ -10 & -5 \end{vmatrix} + 1\begin{vmatrix} 9 & 1 \\ -10 & -1 \end{vmatrix}$$

$$= 5\big[1(-5)-(-1)(-3)\big]$$
$$\quad - 1\big[9(-5)-(-10)(-3)\big]$$
$$\quad + 1\big[9(-1)-1(-10)\big]$$

$$= 5(-5-3) - 1(-45-30) + 1(-9+10)$$

$$= 5(-8) - 1(-75) + 1(1)$$

$$= -40 + 75 + 1$$

$$= 36$$

Since $D = 0$ and $D_x = 36 \neq 0$, it is not necessary to calculate D_y or D_z. Cramer's Rule tells us that the system is inconsistent and the solution set is $\varnothing$ or $\{\ \}$.

37. $D = \begin{vmatrix} 2 & 0 & 1 \\ -1 & -3 & 0 \\ 1 & -2 & 1 \end{vmatrix}$

$= 2\begin{vmatrix} -3 & 0 \\ -2 & 1 \end{vmatrix} - 0\begin{vmatrix} -1 & 0 \\ 1 & 1 \end{vmatrix} + 1\begin{vmatrix} -1 & -3 \\ 1 & -2 \end{vmatrix}$

$= 2\left[-3(1) - (-2)(0)\right] - 0\left[-1(1) - 1(0)\right]$
$\qquad\qquad + 1\left[-1(-2) - 1(-3)\right]$

$= 2(-3 - 0) - 0(-1 - 0) + 1(2 + 3)$

$= 2(-3) - 0(-1) + 1(5)$

$= -6 - 0 + 5$

$= -1$

$D_x = \begin{vmatrix} 27 & 0 & 1 \\ 6 & -3 & 0 \\ 27 & -2 & 1 \end{vmatrix}$

$= 27\begin{vmatrix} -3 & 0 \\ -2 & 1 \end{vmatrix} - 0\begin{vmatrix} 6 & 0 \\ 27 & 1 \end{vmatrix} + 1\begin{vmatrix} 6 & -3 \\ 27 & -2 \end{vmatrix}$

$= 27\left[-3(1) - (-2)(0)\right] - 0\left[6(1) - 27(0)\right]$
$\qquad\qquad + 1\left[6(-2) - 27(-3)\right]$

$= 27(-3 - 0) - 0(6 - 0) + 1(-12 + 81)$

$= 27(-3) - 0(6) + 1(69)$

$= -81 - 0 + 69$

$= -12$

$D_y = \begin{vmatrix} 2 & 27 & 1 \\ -1 & 6 & 0 \\ 1 & 27 & 1 \end{vmatrix}$

$= 2\begin{vmatrix} 6 & 0 \\ 27 & 1 \end{vmatrix} - 27\begin{vmatrix} -1 & 0 \\ 1 & 1 \end{vmatrix} + 1\begin{vmatrix} -1 & 6 \\ 1 & 27 \end{vmatrix}$

$= 2\left[6(1) - 27(0)\right] - 27\left[-1(1) - 1(0)\right]$
$\qquad\qquad + 1\left[(-1)(27) - 1(6)\right]$

$= 2(6 - 0) - 27(-1 - 0) + 1(-27 - 6)$

$= 2(6) - 27(-1) + 1(-33)$

$= 12 + 27 - 33$

$= 6$

$D_z = \begin{vmatrix} 2 & 0 & 27 \\ -1 & -3 & 6 \\ 1 & -2 & 27 \end{vmatrix}$

$= 2\begin{vmatrix} -3 & 6 \\ -2 & 27 \end{vmatrix} - 0\begin{vmatrix} -1 & 6 \\ 1 & 27 \end{vmatrix} + 27\begin{vmatrix} -1 & -3 \\ 1 & -2 \end{vmatrix}$

$= 2\left[-3(27) - (-2)(6)\right] - 0\left[-1(27) - 1(6)\right]$
$\qquad\qquad + 27\left[-1(-2) - 1(-3)\right]$

$= 2(-81 + 12) - 0(-27 - 6) + 27(2 + 3)$

$= 2(-69) - 0(-33) + 27(5)$

$= -138 - 0 + 135$

$= -3$

$x = \dfrac{D_x}{D} = \dfrac{-12}{-1} = 12; \quad y = \dfrac{D_y}{D} = \dfrac{6}{-1} = -6;$

$z = \dfrac{D_z}{D} = \dfrac{-3}{-1} = 3$

The solution is the ordered triple $(12, -6, 3)$.

39. $D = \begin{vmatrix} 5 & 3 & 0 \\ -10 & 0 & 3 \\ 0 & 1 & -2 \end{vmatrix}$

$= 5\begin{vmatrix} 0 & 3 \\ 1 & -2 \end{vmatrix} - 3\begin{vmatrix} -10 & 3 \\ 0 & -2 \end{vmatrix} + 0\begin{vmatrix} -10 & 0 \\ 0 & 1 \end{vmatrix}$

$= 5\left[0(-2) - 1(3)\right] - 3\left[-10(-2) - 0(3)\right]$
$\qquad\qquad + 0\left[-10(1) - 0(0)\right]$

$= 5(0 - 3) - 3(20 - 0) + 0(-10 - 0)$

$= 5(-3) - 3(20) + 0(-10)$

$= -15 - 60 + 0$

$= -75$

$D_x = \begin{vmatrix} 2 & 3 & 0 \\ -3 & 0 & 3 \\ -9 & 1 & -2 \end{vmatrix}$

$= 2\begin{vmatrix} 0 & 3 \\ 1 & -2 \end{vmatrix} - 3\begin{vmatrix} -3 & 3 \\ -9 & -2 \end{vmatrix} + 0\begin{vmatrix} -3 & 0 \\ -9 & 1 \end{vmatrix}$

$= 2\left[0(-2) - 1(3)\right] - 3\left[-3(-2) - (-9)(3)\right]$
$\qquad\qquad + 0\left[-3(1) - (-9)(0)\right]$

$= 2(0 - 3) - 3(6 + 27) + 0(-3 + 0)$

$= 2(-3) - 3(33) + 0(-3)$

$= -6 - 99 + 0$

$= -105$

$$D_y = \begin{vmatrix} 5 & 2 & 0 \\ -10 & -3 & 3 \\ 0 & -9 & -2 \end{vmatrix}$$

$$= 5\begin{vmatrix} -3 & 3 \\ -9 & -2 \end{vmatrix} - 2\begin{vmatrix} -10 & 3 \\ 0 & -2 \end{vmatrix} + 0\begin{vmatrix} -10 & -3 \\ 0 & -9 \end{vmatrix}$$

$$= 5\left[-3(-2)-(-9)(3)\right] - 2\left[-10(-2)-0(3)\right]$$
$$+ 0\left[-10(-9)-0(-3)\right]$$

$$= 5(6+27) - 2(20-0) + 0(90+0)$$

$$= 5(33) - 2(20) + 0(90)$$

$$= 165 - 40 + 0$$

$$= 125$$

$$D_z = \begin{vmatrix} 5 & 3 & 2 \\ -10 & 0 & -3 \\ 0 & 1 & -9 \end{vmatrix}$$

$$= 5\begin{vmatrix} 0 & -3 \\ 1 & -9 \end{vmatrix} - 3\begin{vmatrix} -10 & -3 \\ 0 & -9 \end{vmatrix} + 2\begin{vmatrix} -10 & 0 \\ 0 & 1 \end{vmatrix}$$

$$= 5\left[0(-9)-1(-3)\right] - 3\left[-10(-9)-0(-3)\right]$$
$$+ 2\left[-10(1)-0(0)\right]$$

$$= 5(0+3) - 3(90+0) + 2(-10-0)$$

$$= 5(3) - 3(90) + 2(-10)$$

$$= 15 - 270 + -20$$

$$= -275$$

$$x = \frac{D_x}{D} = \frac{-105}{-75} = \frac{7}{5}; \quad y = \frac{D_y}{D} = \frac{125}{-75} = -\frac{5}{3};$$

$$z = \frac{D_z}{D} = \frac{-275}{-75} = \frac{11}{3}$$

The solution is the ordered triple $\left(\dfrac{7}{5}, -\dfrac{5}{3}, \dfrac{11}{3}\right)$.

41. $\begin{vmatrix} x & 3 \\ 1 & 2 \end{vmatrix} = 7$

$$x(2) - 1(3) = 7$$
$$2x - 3 = 7$$
$$2x = 10$$
$$x = 5$$

43. $\begin{vmatrix} x & -1 & -2 \\ 1 & 0 & 4 \\ 3 & 2 & 5 \end{vmatrix} = 5$

$$x\begin{vmatrix} 0 & 4 \\ 2 & 5 \end{vmatrix} - (-1)\begin{vmatrix} 1 & 4 \\ 3 & 5 \end{vmatrix} + (-2)\begin{vmatrix} 1 & 0 \\ 3 & 2 \end{vmatrix} = 5$$

$$x\left[0(5)-2(4)\right] - (-1)\left[1(5)-3(4)\right] +$$
$$(-2)\left[1(2)-3(0)\right] = 5$$

$$x(-8) - (-1)(-7) + (-2)(2) = 5$$
$$-8x - 7 - 4 = 5$$
$$-8x - 11 = 5$$
$$-8x = 16$$
$$x = -2$$

45. (a) Triangle *ABC* is graphed below.

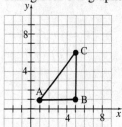

(b) $D = \dfrac{1}{2}\begin{vmatrix} 1 & 5 & 5 \\ 1 & 1 & 6 \\ 1 & 1 & 1 \end{vmatrix}$

$$= \frac{1}{2}\left(1\begin{vmatrix} 1 & 6 \\ 1 & 1 \end{vmatrix} - 5\begin{vmatrix} 1 & 6 \\ 1 & 1 \end{vmatrix} + 5\begin{vmatrix} 1 & 1 \\ 1 & 1 \end{vmatrix}\right)$$

$$= \frac{1}{2}\left[1(1\cdot1-1\cdot6) - 5(1\cdot1-1\cdot6) + 5(1\cdot1-1\cdot1)\right]$$

$$= \frac{1}{2}\left[1(-5) - 5(-5) + 5(0)\right]$$

$$= \frac{1}{2}(-5 + 25 + 0)$$

$$= \frac{1}{2}(20)$$

$$= 10$$

Thus, the area of triangle *ABC* is $|10| = 10$.

47. (a) Parallelogram *ABCD* is graphed below.

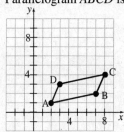

(b) Triangle ABC is formed by the points $(2, 1)$, $(7, 2)$, and $(8, 4)$.

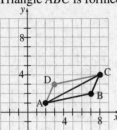

$$D = \frac{1}{2}\begin{vmatrix} 2 & 7 & 8 \\ 1 & 2 & 4 \\ 1 & 1 & 1 \end{vmatrix}$$

$$= \frac{1}{2}\left[2\begin{vmatrix} 2 & 4 \\ 1 & 1 \end{vmatrix} - 7\begin{vmatrix} 1 & 4 \\ 1 & 1 \end{vmatrix} + 8\begin{vmatrix} 1 & 2 \\ 1 & 1 \end{vmatrix}\right]$$

$$= \frac{1}{2}\left[2(2\cdot1-1\cdot4) - 7(1\cdot1-1\cdot4) + 8(1\cdot1-1\cdot2)\right]$$

$$= \frac{1}{2}\left[2(-2) - 7(-3) + 8(-1)\right]$$

$$= \frac{1}{2}(-4+21-8)$$

$$= \frac{1}{2}(9)$$

$$= 4.5$$

Thus, the area of triangle ABC is $|4.5| = 4.5$.

(c) Triangle ADC is formed by the points $(2, 1)$, $(3, 3)$, and $(8, 4)$.

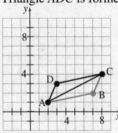

$$D = \frac{1}{2}\begin{vmatrix} 2 & 3 & 8 \\ 1 & 3 & 4 \\ 1 & 1 & 1 \end{vmatrix}$$

$$= \frac{1}{2}\left[2\begin{vmatrix} 3 & 4 \\ 1 & 1 \end{vmatrix} - 3\begin{vmatrix} 1 & 4 \\ 1 & 1 \end{vmatrix} + 8\begin{vmatrix} 1 & 3 \\ 1 & 1 \end{vmatrix}\right]$$

$$= \frac{1}{2}\left[2(3\cdot1-1\cdot4) - 3(1\cdot1-1\cdot4) + 8(1\cdot1-1\cdot3)\right]$$

$$= \frac{1}{2}\left[2(-1) - 3(-3) + 8(-2)\right]$$

$$= \frac{1}{2}(-2+9-16)$$

$$= \frac{1}{2}(-9)$$

$$= -4.5$$

Thus, the area of triangle ADC is $|-4.5| = 4.5$.

(d) The areas of triangle *ABC* and triangle *ADC* are equal. (That is, the diagonal of the parallelogram forms two triangles of equal area.) Thus, the area of parallelogram *ABCD* is $4.5 + 4.5 = 9$.

49. (a)

$$\begin{vmatrix} x & y & 1 \\ 3 & 2 & 1 \\ 5 & 1 & 1 \end{vmatrix} = 0$$

$$x\begin{vmatrix} 2 & 1 \\ 1 & 1 \end{vmatrix} - y\begin{vmatrix} 3 & 1 \\ 5 & 1 \end{vmatrix} + 1\begin{vmatrix} 3 & 2 \\ 5 & 1 \end{vmatrix} = 0$$

$$x\big[2(1) - 1(1)\big] - y\big[3(1) - 5(1)\big] + 1\big[3(1) - 5(2)\big] = 0$$

$$x(1) - y(-2) + 1(-7) = 0$$

$$x + 2y - 7 = 0$$

$$x + 2y = 7$$

(b) $m = \dfrac{y_2 - y_1}{x_2 - x_1} = \dfrac{1 - 2}{5 - 3} = \dfrac{-1}{2} = -\dfrac{1}{2}$

$$y - y_1 = m(x - x_2)$$

$$y - 2 = -\frac{1}{2}(x - 3)$$

$$y - 2 = -\frac{1}{2}x + \frac{3}{2}$$

$$2y - 4 = -x + 3$$

$$x + 2y = 7$$

51. $\begin{vmatrix} 3 & -2 \\ 1 & 4 \end{vmatrix} = 3(4) - 1(-2)$

$$= 12 + 2$$

$$= 14$$

Interchanging rows 1 and 2 and recomputing the determinant, we obtain

$$\begin{vmatrix} 1 & 4 \\ 3 & -2 \end{vmatrix} = 1(-2) - 3(4)$$

$$= -2 - 12$$

$$= -14$$

The two determinants are opposites.

Answers may vary. In general, it is true that the value of a determinant will change signs if any two rows (or any two columns) are interchanged.

53. Using $A = D$, $B = D_x$, and $C = D_y$.

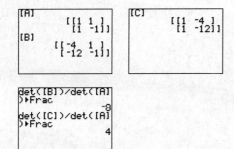

The solution is the ordered pair $(-8, 4)$.

55. Using $A = D$, $B = D_x$, and $C = D_y$.

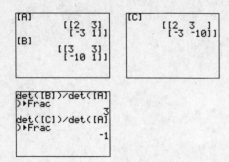

The solution is the ordered pair $(3, -1)$.

57. Using $A = D$, $B = D_x$, $C = D_y$, and $D = D_z$.

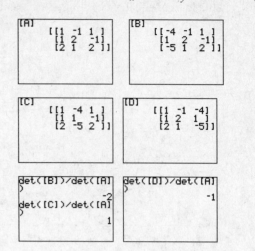

The solution is the ordered triple $(-2, 1, -1)$.